Williams, Charles   Wilfred

W9-CEW-222

Williams, Charles   Wilfred

AAOS

Ninth Edition

Emergency
Care and Transportation of the Sick and Injured

The *Ninth Edition* continues the American Academy of Orthopaedic Surgeons commitment to EMT-Basic education with these exciting enhancements and revisions:

Patient Assessment

* Significant enhancement of patient assessment content.
  - All medical and trauma chapters are organized to follow the patient through the assessment process step-by-step.
  - The organization of these chapters reinforces the patient assessment chapter and flowchart.
* Assessment and Emergency Care summary tables
  - Each medical and trauma chapter concludes with two tables, one summarizing the assessment process, and the other summarizing emergency care for the emergencies discussed in the chapter.

New Chapters

* New chapters on assisting ALS, including:
  - Chapter 40. Assisting With Intravenous Therapy
  - Chapter 41. Assisting With Cardiac Monitoring
* New chapter on Weapons of Mass Destruction:
  - Chapter 38. Response to Terrorism and Weapons of Mass Destruction

New Features

* Progressive case studies:
  Each chapter contains a progressive case study to make students start thinking about what they might do if they encountered a similar case in the field. The case study introduces patients and follows their progress from dispatch to delivery at the emergency department.
* EMT-Basic Practical Skills Review DVD
  Packaged with each copy of the *Ninth Edition*, this DVD provides students with a walk-through of the skills that are required to successfully complete the national EMT-Basic practical examination process. For each skill, students will find helpful information, tips, and pointers designed to facilitate their progression through the practical examination.

AAOS

Ninth Edition

Emergency
Care and Transportation of the Sick and Injured

 American Academy of
Orthopaedic Surgeons

Series Editor:
Andrew N. Pollak, MD, FAAOS

Editors:
Benjamin Gulli, MD
Les Chatelain, BS, MS
Chris Stratford, BS, RN, BCEN, EMT-I

JONES AND BARTLETT PUBLISHERS
Sudbury, Massachusetts
BOSTON   TORONTO   LONDON   SINGAPORE

Jones and Bartlett Publishers

World Headquarters
Jones and Bartlett Publishers
40 Tall Pine Drive, Sudbury, MA 01776
978-443-5000
info@jbpub.com
www.EMSzone.com

Jones and Bartlett Publishers Canada
2406 Nikanna Road
Mississauga, ON L5C 2W6
Canada

Jones and Bartlett Publishers International
Barb House, Barb Mews
London W6 7PA
United Kingdom

Jones and Bartlett's books and products are available through most bookstores and online booksellers. To contact Jones and Bartlett Publishers directly, call 800-832-0034, fax 978-443-8000, or visit our website www.jbpub.com.

Substantial discounts on bulk quantities of Jones and Bartlett's publications are available to corporations, professional associations, and other qualified organizations. For details and specific discount information, contact the special sales department at Jones and Bartlett via the above contact information or send an email to specialsales@jbpub.com.

American Academy of Orthopaedic Surgeons

Editorial Credits
Chief Education Officer: Mark W. Wieting
Director, Department of Publications: Marilyn L. Fox, PhD
Managing Editor: Lynne Roby Shindoll
Managing Editor: Barbara A. Scotese

Board of Directors 2005
Stuart L. Weinstein, MD, President
Richard F. Kyle, MD
James H. Beaty, MD
Edward A. Toriello, MD
Robert W. Bucholz, MD
James H. Herndon, MD
Frank B. Kelly, MD
Dwight W. Burney, III, MD
Matthew S. Shapiro, MD
Mark C. Gebhardt, MD
Andrew N. Pollak, MD
Joseph C. McCarthy, MD
Frances A. Farley, MD
Oheneba Boachie-Adjei, MD
Gordon M. Aamoth, MD
Kristy L. Weber, MD
Leslie L. Altick
William L. Healy, MD (ex-officio)
Karen L. Hackett, FACHE, CAE (ex-officio)

Production Credits

Chief Executive Officer: Clayton E. Jones
Chief Operating Officer: Donald W. Jones, Jr.
President, Higher Education and Professional
    Publishing: Robert W. Holland, Jr.
V.P., Sales and Marketing: William J. Kane
V.P., Production and Design: Anne Spencer
V.P., Manufacturing and Inventory Control: Therese Bräuer
Publisher, Public Safety: Kimberly Brophy
Managing Editor: Carol E. Brewer

Production Editor: Karen Ferreira
Text Design: Anne Spencer, Kristin Ohlin
Composition: Graphic World
Illustrations: Graphic World, Inc., Imagineering, Rolin Graphics
Cover Design: Kristin Ohlin
Photo Research: Kimberly Potvin
Cover Printing: Lehigh Press
Text Printing and Binding: Courier Companies

Copyright © 2005 by the American Academy of Orthopaedic Surgeons

All rights reserved. No part of the material protected by this copyright notice may be reproduced or utilized in any form, electronic or mechanical, including photocopying, recording, or by any information storage and retrieval system, without written permission from the copyright owner.

The procedures and protocols in this book are based on the most current recommendations of responsible medical sources. The American Academy of Orthopaedic Surgeons and the publisher, however, make no guarantee as to, and assume no responsibility for, the correctness, sufficiency, or completeness of such information or recommendations. Other or additional safety measures may be required under particular circumstances.

This textbook is intended solely as a guide to the appropriate procedures to be employed when rendering emergency care to the sick and injured. It is not intended as a statement of the standards of care required in any particular situation, because circumstances and the patient's physical condition can vary widely from one emergency to another. Nor is it intended that this textbook shall in any way advise emergency personnel concerning legal authority to perform the activities or procedures discussed. Such local determinations should be made only with the aid of legal council.

Notice: The patients described in "You are the Provider," "Assessment in Action," and "Points to Ponder" throughout this text are fictitious.

Library of Congress Cataloging-in-Publication Data

Emergency care and transportation of the sick and injured.-9th ed. / American Academy of Orthopaedic Surgeons ; editors, Les Chatelain . . . [et al.].
    p. ; cm.
Includes index.
ISBN 0-7637-4738-6
    1. Medical emergencies. 2. Transport of sick and wounded. I. Chatelain, Les. II. American Academy of Orthopaedic Surgeons.
[DNLM: 1. Emergency Medical Services. 2. Emergency Treatment. 3. Transportation of Patients. WX 215 E487 2005]
RC86.7.A43 2005
616.02'5-dc22
2004056909

Additional illustrations and photo credits appear on page C-2, which constitutes a continuation of the copyright page.

Printed in the United States of America
09 08 07 06 05   10 9 8 7 6 5 4 3 2

Brief Contents

Contents

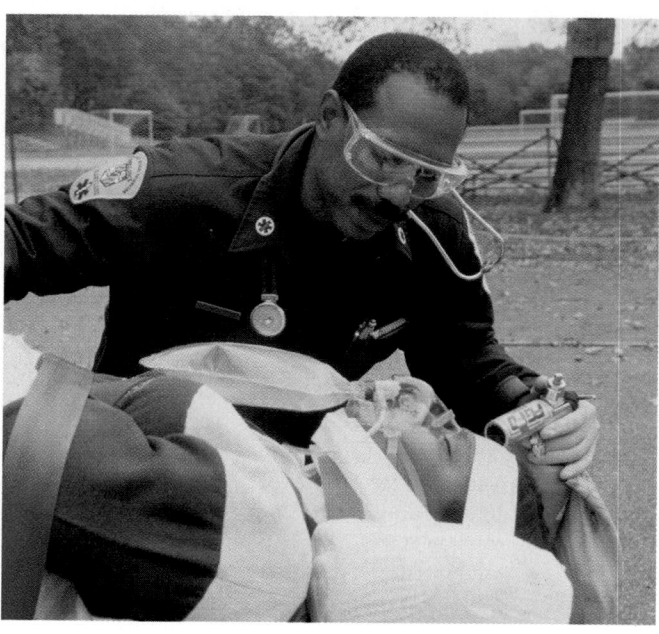

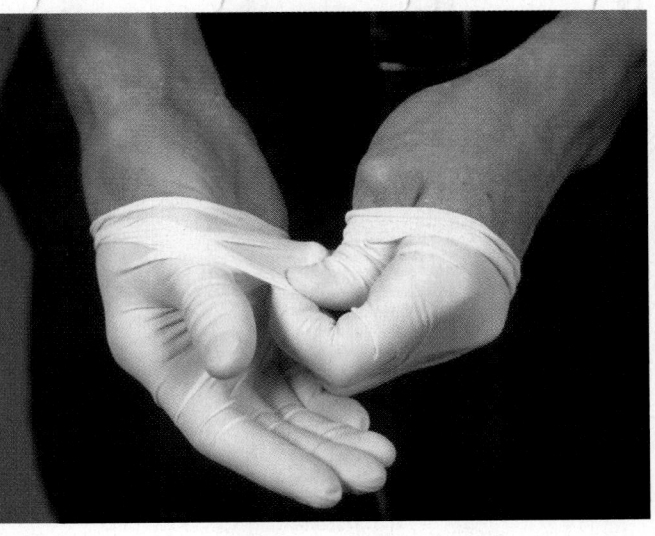

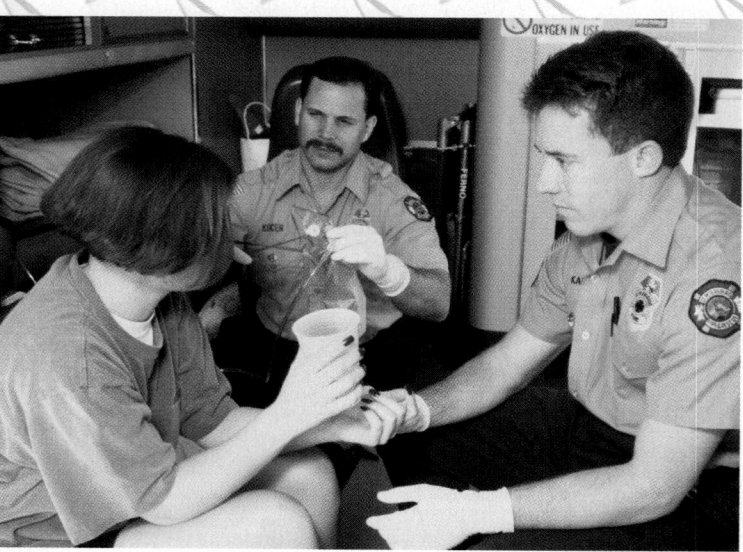

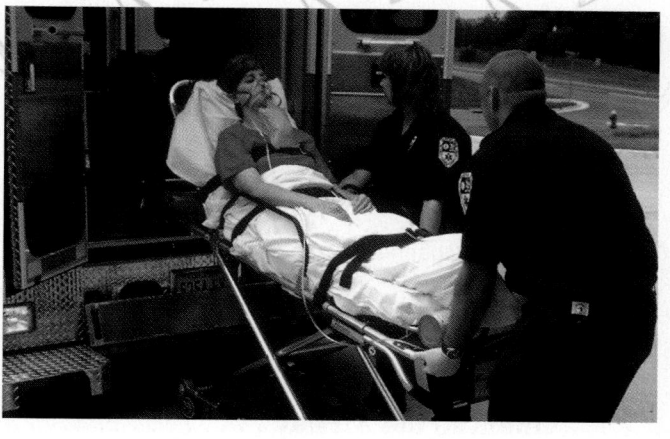

Section 2: Airway   208

Section 3: Patient Assessment   256

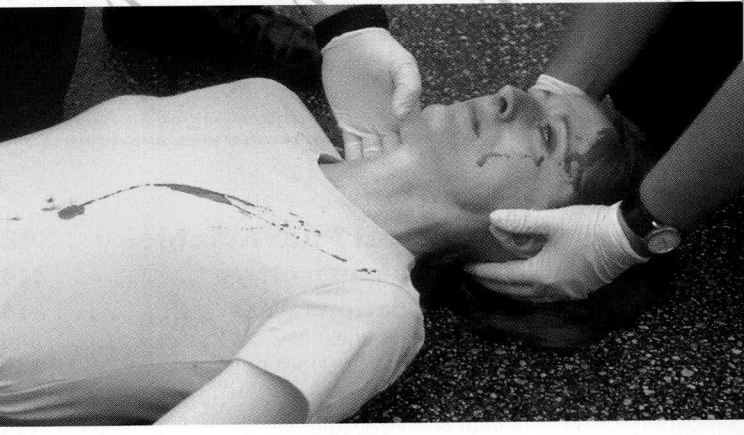

CHAPTER 9 Communications and Documentation. **314**

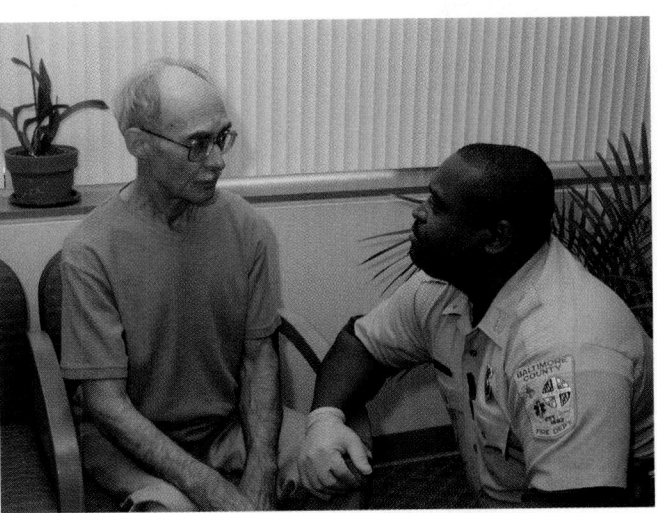

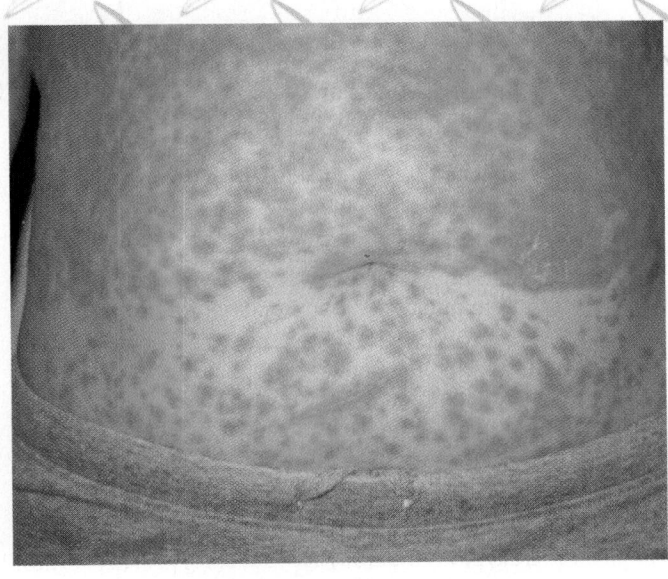

Section 5: Trauma 628

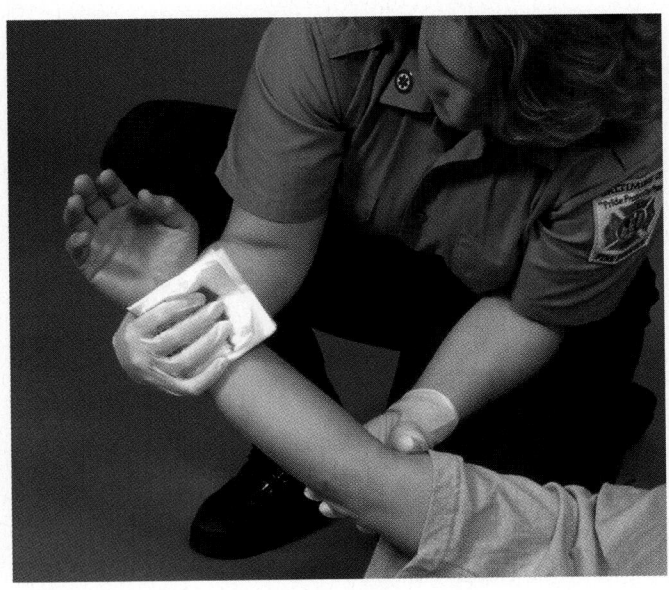

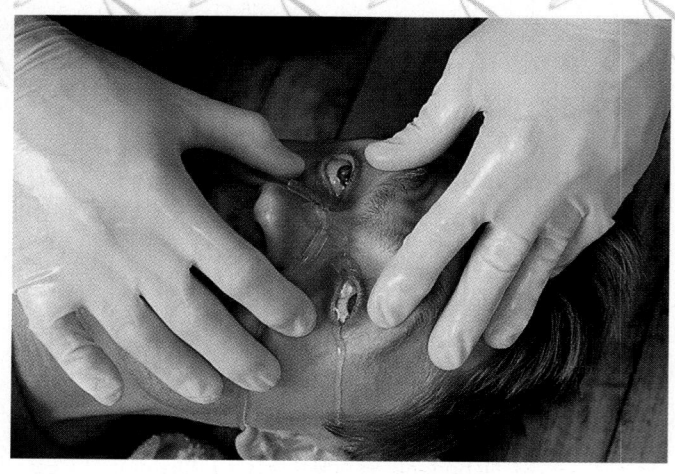

Section 6: Special Populations   912

CHAPTER 31   Pediatric Emergencies.. 914

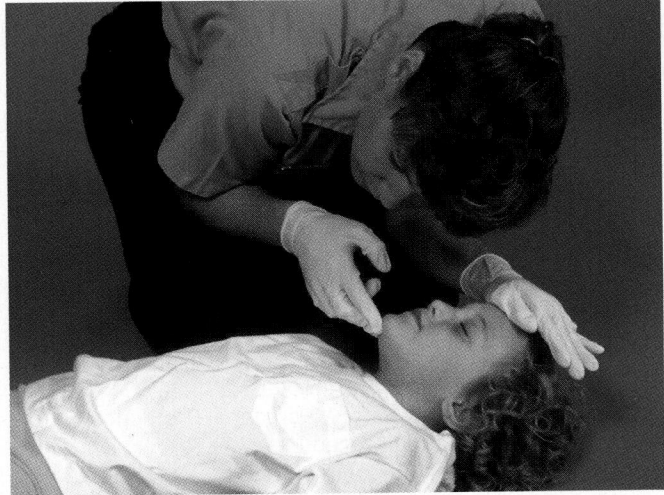

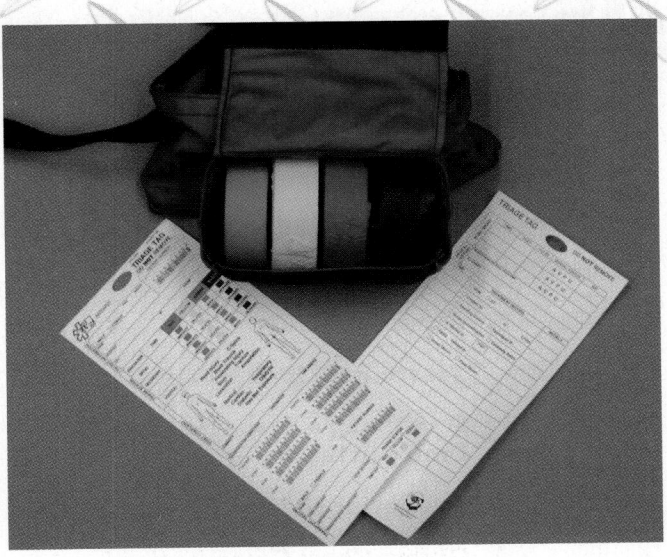

EMT-B Skill Drills

Resource Preview

The American Academy of Orthopaedic Surgeons is pleased to bring you *Emergency Care and Transportation of the Sick and Injured, Ninth Edition,* a modern integrated teaching and learning system. It combines current content with dynamic features, interactive technology, and both instructor and student resources.

Emergency Care and Transportation of the Sick and Injured, Ninth Edition thoroughly addresses the objectives in the DOT EMT-Basic National Standard Curriculum, while also including a wealth of enhancements to enrich EMT-Basic education.

Chapter Resources

The text is the core of the teaching and learning system with features that reinforce and expand on essential information and make information retrieval a snap. These features include:

Chapter Objectives
National Standard Curriculum objectives and additional noncurriculum objectives are provided for each chapter with corresponding page references.

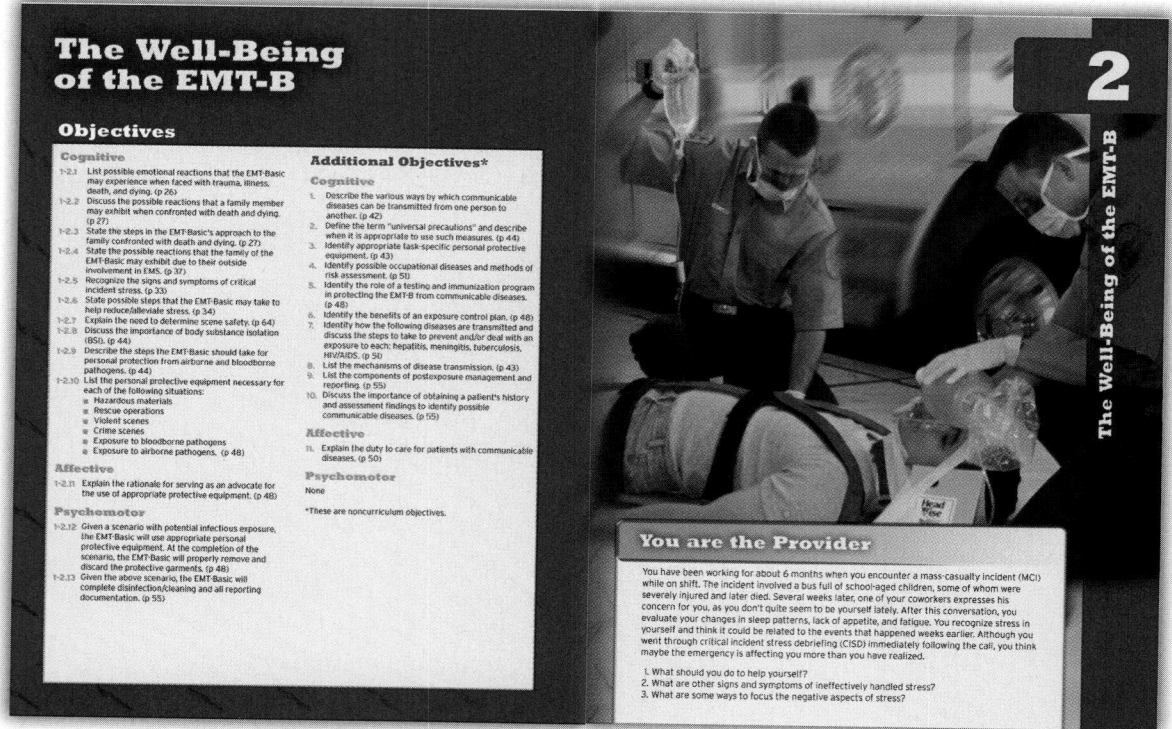

You are the Provider
Each chapter contains a progressive case study to make students start thinking about what they might do if they encountered a similar case in the field.

The case study introduces patients and follows their progress from dispatch to delivery at the emergency department. The case becomes progressively more detailed as new material is presented.

This feature is a valuable learning tool that encourages critical thinking skills. A summary of the case study concludes the chapter.

---

148   Section 1 Preparing to be an EMT-B

EMT-B Tips

Obtaining the respirations or pulse rate

When obtaining a patient's respirations or pulse rate, count the number of breaths or beats in a 30-second period and then multiply by 2. This method produces a significantly more reliable figure than you would get if you counted for only 15 seconds and multiplied by 4. With either method, the result will always be an even number.

consciousness, may be reported by the patient or others at the scene. Because signs and symptoms are essential to understanding the sequence of events and may include signs that are no longer present, they are important parts of the patient history. You should always report how and/or when the signs and symptoms began. This information is important because the reason that signs and symptoms develop often differs, depending on the situation.

Obtaining a SAMPLE History

Once you have stabilized all immediate life threats, provided emergency care, and are ready to further examine the patient, you should try to obtain a key brief history, or SAMPLE history. As part of the assessment of every patient, you should ask the following questions, using the word SAMPLE as a guideline:

- Signs and Symptoms of the episode: What signs and symptoms occurred at onset of the incident? Does the patient report pain?
- Allergies: Is the patient allergic to any medication, food, or other substance? What reactions did the patient have to any of them? If the patient has no known allergies, you should note this on the run report as "no known allergies" or "NKA."
- Medications: What medications was the patient prescribed? What dosage was prescribed? How often is the patient supposed to take the medication? What prescription, over-the-counter medications, and herbal medications has the patient taken in the last 12 hours? How much was taken and when?
- Pertinent past history: Does the patient have any history of medical, surgical, or trauma occurrences? Has the patient had a recent illness or injury, fall, or blow to the head?
- Last oral intake: When did the patient last eat or drink? What did the patient eat or drink, and how much was consumed? Did the patient take any drugs or drink alcohol? Has there been any other oral intake in the last 4 hours?
- Events leading to the injury or illness: What are the key events that led up to this incident? What occurred between the onset of the incident and your arrival? What was the patient doing when this illness started? What was the patient doing when this injury happened?

OPQRST

Another mnemonic device that can be very helpful in remembering questions you should ask in obtaining a patient history is OPQRST. This can be especially helpful when assessing for possible heart attack.

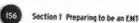

You are the Provider   Part 2

As you are preparing to transport your patient to the hospital, she states that it is becoming increasingly difficult to breathe. Her respirations increase to 34 breaths/min and are shallow and gasping. Even with high-flow oxygen via nonrebreathing mask, her oxygen saturation is only 90% and her capillary refill time is 4 seconds. She is extremely anxious and is having a hard time sitting still.

3. Is there anything in your repeat assessment of vital signs that can help you determine the cause of her distress?
4. What are your treatment priorities?

---

156   Section 1 Preparing to be an EMT-B

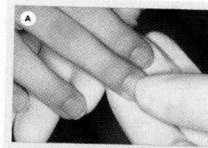

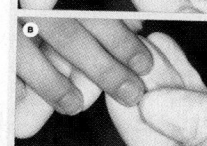

Figure 5-11 A. To test capillary refill, gently compress the fingertip until it blanches. **B.** Release the fingertip, and count until it returns to its normal pink color.

A bluish color are refilling with than with oxygen the test invalid. refill test invalid posed to a cold older. In both may be normal.

To assess cap dren younger th and determine return. In new forehead, chin time. As with a than 2 seconds, dicator of cardi adults and shoe patients.

Blood Pre
Adequate blood circulation and pressure (BP) the walls of th sure may indi
- Loss of
- Loss of striction without
- A cardi

When ar in a drop in c anisms are co arteries cons and by decre tremities, av rected to the perfused. Ha defense me

You should report and document the capillary refill as normal, "CRT < 2." You should suspect poor peripheral circulation when capillary refill takes more than 2 seconds or the nail bed remains blanched. In this instance, you should report and document the capillary refill as delayed or CRT > 2.

You are the Provider

As you are transporting your patient to the hospital, she becom respiratory rate decreases to 16 breaths/min, her pulse goes do pressure is 120/64 mm Hg. She seems less anxious and appears

5. Is your patient's condition improving?
6. What do these changes in vital signs mean? Should you char

---

Chapter 5 Baseline Vital Signs and SAMPLE History   163

You are the Provider   Summary

Frequently, a patient's medications and chief complaint provide valuable clues to the underlying condition. This information will greatly aid you in understanding the source and the severity of the patient's condition(s). You may be unable to obtain information from the patient's medical history either because there is no significant medical history or because the patient is unable to tell you.

Women who are 35 years of age and older who are taking birth control medication and smoke cigarettes are especially high risk of experiencing embolisms. Becoming familiar with commonly prescribed medications and over-the-counter medications can provide important information about a patient's medical history. Sometimes patients will not know why they are taking certain medications, so it would be helpful for you to have a pocket guide on medications for quick and easy reference.

Being able to sort through the clues from the emergency scene itself, from the patient's complaints, and from the patient's signs and symptoms and past medical history will all assist you in understanding the cause of your patient's problem and enable you to make appropriate, timely decisions about your patient's care.

Technology Toolbar
Found at the beginning of each chapter, the technology toolbar guides you through the resources available for that chapter at www.EMTB.com.

Section 1 Preparing to be an EMT-B

Introduction to Emergency Medical Care

This book has been designed to serve as the text and primary resource for the emergency medical technician basic (EMT-Basic) course. This chapter describes the content and objectives of the EMT-Basic course. It also discusses what will be expected of you during the course and what other requirements you will have to meet to be licensed or certified as an EMT-Basic in most states. You will also learn about the differences between first aid training, a Department of Transportation (DOT) First Responder training course, and the training for the EMT-Basic, EMT-Intermediate, and EMT-Paramedic.

Emergency medical services (EMS) is a system. The key components of this system and how they influence and affect the EMT-Basic (EMT-B) and his or her delivery of emergency care are carefully discussed. Next, the administration, medical direction, quality control, and regulation of EMS services are presented. The chapter ends with a detailed discussion of the roles and responsibilities of the EMT-B as a health care professional.

Figure 1-1 As an EMT-B, you will be part of a larger team that responds to a variety of calls and provides a wide range of prehospital emergency care.

emergency medical service is part of a local or regional EMS system that provides the many varied prehospital and hospital components required for the delivery of proper emergency medical care. The standards for prehospital emergency care and the individuals who provide it are governed by the laws in each state and are typically regulated by a state office of EMS.

The individuals who provide the emergency care in the field are trained and, except for licensed physicians, must be state-licensed or certified emergency medical technicians (EMTs). Different states will refer to the authority granted to you to function as an EMT-B as licensure, certification, or credentialing. For the purposes of this text, the term *certification* will be used.

In most states, EMTs are categorized into three training and certification levels: EMT-Basic, EMT-Intermediate, and EMT-Paramedic. An EMT-Basic (EMT-B) has training in basic emergency care skills, including automated external defibrillation, use of airway adjuncts, and assisting patients with certain medications. An EMT-Intermediate (EMT-I) has training in specific aspects of advanced life support, such as intravenous (IV) therapy and cardiac monitoring. An EMT-Paramedic (EMT-P) has extensive training in advanced life support, including IV therapy, pharmacology, cardiac monitoring, and other advanced assessment and treatment skills.

Although the specific training and certification requirements vary from one state to another, the training that is required in almost every state follows or

Course Description

You are about to enter an exciting field. Emergency medical services (EMS) consists of a team of health care professionals who, in each area or jurisdiction, are responsible for and provide emergency care and transportation to the sick and injured (Figure 1-1). Each

Technology
www.EMTB.com

- Interactivities
- Vocabulary Explorer
- Anatomy Review
- Web Links
- Online Review Manual

Skill Drills
Skill Drills provide written step-by-step explanations and visual summaries of important skills and procedures.

Chapter 6 Lifting and Moving Patients 173

adjacent to the plane described by your anterior torso (the anterior torso and imaginary lines extended vertically above and below it). Always keep the weight that you are lifting as close to your body as possible.

Lateral force across the spine and sideways leverage against the lower back must also be avoided. If you lift with only one arm or with the arms extended more to one side than the other, more force will be exerted against one side of the shoulder girdle than the other, causing lateral force to be exerted across the spinal column. To prevent this, keep your arms approximately the same distance apart as when hanging at each side of the body, with the weight distributed equally and properly centered between them. If the weight is not balanced between both arms or properly centered between the shoulders when you are preparing to lift, turn your body and/or move to the left or right until the weight is properly balanced and centered. To lift safely and produce the maximal power lift, you should take the following steps (skill drill 6-1):

1. **Tighten your back in its normal upright position and use your abdominal muscles to lock it in a slight inward curve.**
2. **Spread your legs apart about 15″, and bend your legs to lower your torso and arms.**
3. **With arms extended down each side of the body, grasp the cot or backboard with your hands held palm up and just in front of the plane described by the anterior torso and imaginary lines extending vertically from it to the ground.**
4. **Adjust your orientation and position until the weight is balanced and centered between both arms (Step 1).**
5. **Reposition your feet as necessary so that they are about 15″ apart with one slightly farther forward and rotated so that you and your center of gravity will be properly balanced between them. Be sure to straddle the object, keep your feet flat, and distribute your weight to the balls of the feet or just behind them (Step 2).**
6. **With the arms extended downward, lift by straightening your legs until you are fully standing. Make sure your back is locked in and that your upper body comes up before your hips (Step 3).**

Reverse these steps whenever you are lowering the cot. Always remember to avoid bending at the waist or twisting as you stand.

Your safety, as well as that of the other EMT-Bs and the patient, depends on the use of proper lifting tech-

niques and having and maintaining a proper hold when lifting or carrying a patient. If you do not have proper hold of the cot or of the patient in a body lift, you will not be able to bear a proper share of the weight, and there is an increased chance that you can suddenly lose your grasp with one or both hands. If you temporarily lose your grasp with one or both hands, the position and weight distribution of the cot change suddenly, and the other members of the team must quickly reach beyond a safe distance to avoid dropping the patient. As a result, sudden excessive force may be placed across each one's spine, causing lower back injury.

You should use the power grip to get the maximum force from your hands whenever you are lifting a patient (Figure 6-4). The arm and hand have their greatest lifting strength when facing palm up. Whenever you grasp a cot or backboard, your hands should be at least 10″ apart. Each hand should be inserted under the handle with the palm facing up and the thumb extended upward. You should then advance the hand until the thumb prevents further insertion and the cylindrical handle lies firmly in the crease of the curved palm. Curl your fingers and thumb tightly over the top of the handle. All your fingers should be at the same angle, and have the proper power grip, make sure that the underside of the handle is fully supported on your curved palm with only the fingers and thumb preventing it from being pulled sideways or upward out of the palm.

If you must lift the object higher once you have lifted by extending your legs, you will be able to "curl" the object higher by using your biceps to flex the arms.

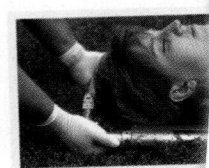

Figure 6-4 To perform the power grip, grasp the handle of the litter with your palms up and your thumbs extended. Make sure your hands are about 10″ apart and that your fingers are all at the same angle. The underside of the handle should be fully supported by the palms of your hands.

174 Section 1 Preparing to be an EMT-B

Performing the Power Lift

Skill Drill 6-1

1. Lock your back into an upright, inward curve. Spread your legs and bend your legs. Grasp the backboard, palms up and just in front of you. Balance and center the weight between your arms.

2. Position your feet, straddle the object, and distribute weight.

3. Straighten your legs and lift, keeping your back locked in.

while maintaining the power grip and weight supported in the palms.

You should never grasp a cot or backboard with the hand placed palm down over the handle unless you are standing at the front end with your back to the cot, as when performing a diamond carry. In lifting with the palm down, the weight is supported by the fingers rather than the palm. This hand orientation places the tips of the fingers and thumb under the handle. If the weight forces them apart, your grasp on the handle will be lost.

When lifting a patient by a sheet or blanket, you should center the patient on the sheet or blanket, and roll up the excess fabric on each side. This produces a cylindrical handle that provides a strong, secure way to grasp the fabric (Figure 6-5). When directly lifting a patient, you should tightly grip the patient in a place and manner that will ensure that you will not lose your grasp on the patient.

Weight and Distribution

Whenever possible, you should use a device that can be rolled to move a patient. However, in case a wheeled device is not available, you must make sure that you understand and follow certain guidelines for carrying a patient on a cot. Table 6-1 shows the guidelines.

If a patient is supine on a backboard or is lying or in a semi-sitting position on the cot, his or her weight is not equally distributed between the two ends of the device. Between 68% and 78% of the body weight of a patient in a horizontal position is in the torso. Therefore, more of the patient's weight rests on the head half of the device than on the foot half.

A patient on a backboard or cot should be lifted and carried by four rescuers in a diamond carry, with one EMT-B at the head end of the device, one at the foot end, and one at each side of the patient's torso

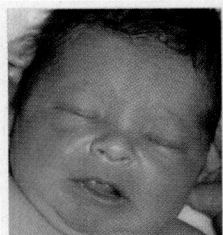

Figure 5-10 Assess skin temperature by feeling the patient's forehead with the back of your hand.

Figure 5-9 Cyanosis occurs when the patient has low levels of oxygen in the blood.

slightly moist. Skin that is only slightly moist but not covered excessively with sweat is described as clammy, damp, or moist. When the skin is bathed in sweat, such as after strenuous exercise or when the patient is in shock, the skin is described as wet or diaphoretic.

Because the skin's color, temperature, and moisture are often related signs, you should consider them together. When recording or reporting your assessment of the skin, you should first describe the color, then the temperature, and last, whether the skin is dry, moist, wet. For example, you could say or write, "Skin: pale, cool, and clammy."

Capillary Refill

Capillary refill is evaluated to assess the ability of the circulatory system to restore blood to the capillary system. When evaluated in an uninjured limb, capillary refill may reflect the patient's perfusion. It should be kept in mind, however, that capillary refill time can be affected by the patient's body temperature, position, preexisting medical conditions, and medications. To test capillary refill, place your thumb on the patient's fingernail with your fingers on the underside of the patient's finger and gently compress Figure 5-11A . The blood will be forced from the capillaries in the nail bed. When you remove the pressure applied against the tip of the patient's finger, the nail bed will remain blanched white for a brief period. As the underlying capillaries refill with blood, the nail bed will be restored to its normal pink color.

Capillary refill should be prompt, and the nail bed color should be pink. With adequate perfusion, the blood in the nail bed should be restored to its normal color within 2 seconds, or about the time it takes to say "capillary refill" at a normal rate of speech Figure 5

is unable to properly dissipate heat will also appear to have red skin.

Changes in skin color may also result from chronic illness. Liver disease or dysfunction may cause jaundice, resulting in the patient's skin and sclera turning yellow. The sclera is the normally white portion of the eye, and may show color changes even before skin color change is visible.

Temperature

Normally, the skin is warm to the touch. When the patient has a significant fever, sunburn, or hyperthermia, the skin feels hot to the touch. The skin will feel cool when the patient is in early shock, has mild hypothermia, or has inadequate perfusion. The skin will feel cold when the patient is in profound shock, has hypothermia, or has frostbite.

Body temperature is normally measured with a thermometer in the hospital. However, in the field, feeling the patient's forehead with the back of your hand is usually adequate to determine whether the patient's temperature is elevated or decreased Figure 5-10 .

Moisture

Dry skin is normal. Skin that is wet, moist (often called diaphoretic), or excessively dry and hot suggests a problem. In the early stages of shock, the skin will become

---

Prep Kit

Ready for Review

- Whenever you are called to the scene of an illness or injury, you should find out the patient's chief complaint.
- Your assessment of the patient should include rapidly evaluating the patient's general condition and identifying any potentially life-threatening injuries or conditions.
- Baseline vital signs are the key signs that you will use to evaluate the patient's general condition.
- You will be assessing the patient's respirations, pulse, skin, blood pressure, level of consciousness, and pupils.
- After you have initially assessed the patient and obtained the baseline vital signs, you should reassess the patient for any changes from your initial findings.
- In addition to determining the chief complaint and assessing the patient's general condition, you should try to obtain a SAMPLE history from the patient or bystanders.
- By asking several important questions, you will be able to determine the patient's signs and symptoms, allergies, medications, pertinent past history, last oral intake, and the events leading up to the incident.

Vital Vocabulary

auscultation A method of listening to sounds within an organ with a stethoscope.

AVPU scale A method of assessing level of consciousness by determining whether the patient is awake and alert, responsive to verbal stimuli or pain, or unresponsive; used principally early in the assessment.

blood pressure (BP) The pressure of circulating blood against the walls of the arteries.

bradycardia Slow heart rate, less than 60 beats/min.

capillary refill The ability of the circulatory system to restore blood to the capillary system; evaluated by using a simple test.

chief complaint The reason a patient called for help. Also, the patient's response to questions such as "What's wrong?" or "What happened?"

conjunctiva The delicate membrane lining the eyelids and covering the exposed surface of the eye.

cyanosis A bluish-gray skin color that is caused by reduced levels of oxygen in the blood.

diaphoretic Characterized by profuse sweating.

diastolic pressure The pressure that remains in the arteries during the relaxing phase of the heart's cycle (diastole) when the left ventricle is at rest.

hypertension Blood pressure that is higher than the normal range.

hypotension Blood pressure that is lower than the normal range.

jaundice A yellow skin or sclera color that is caused by liver disease or dysfunction.

labored breathing Breathing that requires visibly increased effort; characterized by grunting, stridor, and use of accessory muscles.

OPQRST An abbreviation for key terms used in evaluating a patient's signs and symptoms: onset, provocation or palliation, quality, region/radiation, severity, and timing of pain.

www.EMTB.com

Technology

Interactivities
Vocabulary Explorer
Anatomy Review
Web Links
Online Review Manual

Vital Vocabulary
Vocabulary terms are easily identified and defined within the text. A comprehensive list follows each chapter, and Vocabulary Explorer on www.EMTB.com offers interactivities.

Virus has caused some concern recently. This virus' vector is the mosquito, and affects humans and birds. The virus is actually tracked by tests done on birds suspected of being killed by the virus. These diseases are not communicable and do not pose a risk to you during patient care.

Another virus that has caused significant concern of late is best known as SARS, or severe acute respiratory syndrome. SARS is a serious, potentially life-threatening viral infection caused by a recently discovered family of viruses best known as the second most common cause of the common cold. SARS usually starts with flu-like symptoms, which may progress to pneumonia, respiratory failure, and, in some cases, death. The SARS virus strain probably spread from Guangdong province in southern China to Hong Kong, Singapore, and Taiwan. Canada has had a significant outbreak in the Toronto area. SARS is thought to be primarily transmitted by close person-to-person contact. Most cases have involved persons who lived with or cared for a person with SARS or who had exposure to contaminated secretions from a SARS patient.

Multiple antibiotic-resistant organisms have recently been the subject of media scrutiny. These organisms should be viewed as no more or less contagious than other less resistant organisms of the same type. The same precautions apply.

General Postexposure Management

In many instances, you will not know that a patient has an airborne or bloodborne disease, and you could be exposed without knowing it. The Ryan White Law requires that the hospital notify your department's designated officer, the individual in the department who is charged with the responsibility of managing exposures and infection control issues, within 48 hours of the time the hospital identifies the patient's disease. In the event of possible exposure, there should be a protocol in place to obtain information from your local hospital or other medical resource. You should be screened and given information about the necessity of medical follow-up. Treatment depends on the disease. Your designated officer will assist you with the necessary information.

If you experience a needlestick injury or some other unprotected exposure to blood, you must notify your department's designated officer as soon as possible and complete an incident report. The designated officer

Documentation Tips

The ability of your EMS service to support you in case of exposure to a communicable disease depends on your understanding of how exposure can occur and your immediate report of exposure to potentially infectious materials. Make notes right away to ensure that you remember all pertinent information, and report immediately after the response following your service's guidelines.

can contact the hospital for information; the hospital has 48 hours to report back to the designated officer. Depending on your state laws and whether it is possible, patient testing should be done, followed by baseline testing on you.

Because there are many diseases for which there are no outward signs of infection, your protection lies in the use of PPE and/or prompt reporting of exposure. Be familiar with the postexposure protocols outlined in your department's exposure control plan.

Establishing an Infection Control Routine

Infection control, procedures to reduce infection in patients and health care personnel, should be an important part of your daily routine. Take the following steps in dealing with potential exposure situations:

1. **En route to the scene**, make sure that all equipment is out and available.
2. **Upon arrival**, make sure the scene is safe to enter, then do a quick visual assessment of the patient, noting whether any blood is present.
3. **Select the proper PPE** according to the tasks you are likely to perform.
4. **Change gloves and wash hands** between patients; do not unnecessarily delay treatment for use of PPE, thereby potentially putting patients at risk. Remove gloves and other gear after contact with the patient, unless you are in the patient compartment. Remember that good hand washing is always necessary.
5. **Limit the number of people** who are involved in patient care if there are multiple injuries and a substantial amount of blood at the scene.

Documentation Tips
Provide advice on how to document patient care and highlight situations where documentation is especially crucial.

xxxii   **Resource Preview**

Section 1 Preparing to be an EMT-B   `36`

TABLE 2-4 Warning Signs of Stress
Irritability toward coworkers, family, and friends
Inability to concentrate
Difficulty sleeping, increased sleeping, or nightmares
Anxiety
Indecisiveness
Guilt
Loss of appetite (gastrointestinal disturbances)
Loss of interest in sexual activities
Isolation
Loss of interest in work
Increased use of alcohol
Recreational drug use

The following sections provide some suggestions for how to cope better with stress. Some of them may be useful in helping you to prevent problems from developing. Others may help you to solve problems, should they develop.

Lifestyle Changes

Your well-being is of primary importance to effective EMS operations. The effectiveness and efficiency with which you do your job depend on your ability to stay in shape and avoid the risk of personal injury. Burnout is a condition of chronic fatigue and frustration that results from mounting stress over time. To avoid burnout, you need to be in good physical and mental health. Be aware of the potential hazards in rescue and emergency medical care. You must also learn how to avoid or prevent personal injury or illness.

Nutrition

To perform efficiently, you must eat nutritious food. Food is the fuel that makes the body run. The physical exertion and stress that are a part of your job require a high energy output. If you do not have a ready source of fuel, your performance may be less than satisfactory. This can be dangerous for you, your partner, and your patient. Therefore, it is important for you to learn about and follow the rules of good nutrition.

Candy and soft drinks contain sugar. These foods are quickly absorbed and converted to fuel by the body. But simple sugars also stimulate the body's production of insulin, which reduces blood glucose levels. For some people, eating a lot of sugar can actually result in lower energy levels.

EMT-B Tips

Caring for patients can be incredibly rewarding; however, it can also be considerably stressful. Recognizing stress is a key part of successfully coping with it and maintaining a healthy mental attitude.

You will come across situations that you wish you could change. You will encounter abused children, neglected elderly patients, alcoholics, and other upsetting situations. You must find a balance between doing what you can for these patients and protecting yourself from burnout or complete desensitization. However, you will sometimes have the opportunity to help by reporting an abuse case according to local protocol or even by displaying a caring attitude. Even these things can make a profound difference in your patients' lives.

Complex carbohydrates rank next to simple sugars in their ability to produce energy. Complex carbohydrates such as pasta, rice, and vegetables are among the safest, most reliable sources for long-term energy production. However, some carbohydrates take hours to be converted into usable body fuel.

Fats are also easily converted to energy, but eating too much fat can lead to obesity, cardiac disease, and other long-term health problems. The proteins in meat, fish, chicken, beans, and cheese take several hours to convert to energy.

Carry an individual supply of high-energy food to help you maintain your energy levels (Figure 2-7). Try eating several small meals throughout the day to keep your energy resources at constant high levels. Remember, however, that overeating may reduce your physical and mental performance. After a large meal, the blood that is needed for the digestive process is not available for other activities.

You must also make sure that you maintain an adequate fluid intake (Figure 2-8). Hydration is important for proper functioning. Fluids can be easily replenished by drinking any nonalcoholic, non-caffeinated fluid. Water is generally the best fluid available. The body absorbs it faster than any other fluid. Avoid fluids that contain high levels of sugar. These can

---

EMT-B Tips
Provide advice from masters of the trade.

EMT-B Safety
Reinforces safety concerns for both the EMT-B and the patient.

Chapter 2 The Well-Being of the EMT-B   `33`

- Dilated venous vessels near the skin surface (causes cool, clammy skin)
- Dilated pupils
- Tensed muscles
- Increased blood glucose levels
- Perspiration
- Decreased blood flow to the gastrointestinal tract

Stress may also have physical symptoms such as fatigue, changes in appetite, gastrointestinal problems, or headaches. Stress may cause insomnia or hypersomnia, irritability, inability to concentrate, and hyperactivity or underactivity. Additionally, stress may manifest itself in psychological reactions such as fear, dull or nonresponsive behavior, depression, guilt, oversensitivity, anger, irritability, and frustration. Often, today's fast-paced lifestyles compound these effects by not allowing a person to rest and recover after periods of stress. Prolonged or excessive stress has been proven to be a strong contributor to heart disease, hypertension, cancer, alcoholism, and depression.

Many people are subject to cumulative stress, whereby insignificant stressors accumulate to a larger stress-related problem. In the emergency services environment (EMS, police, fire fighters), stressors may also be sudden and more severe. Some events are unusually stressful or emotional, even by emergency services standards. These acute severe stressors result in what is referred to as critical incident stress. Events that can trigger critical incident stress include the following:

- Mass-casualty incidents
- Serious injury or traumatic death of a child
- Crash with injuries, caused by an emergency services provider while responding to or from a call
- Death or serious injury of a coworker in the line of duty

Posttraumatic stress disorder (PTSD) may develop after a person has experienced a psychologically distressing event. It is characterized by re-experiencing the event and overresponding to stimuli that recall the event. PTSD is sometimes referred to as "Vietnam veteran's disease" because of its classification as a mental disorder following the Vietnam conflict. Stressful events in EMS are sometimes psychologically overwhelming. Some of the symptoms include depression, startle reactions, flashback phenomena, and dissociative episodes (eg, amnesia of the event).

A process called critical incident stress management (CISM) was developed to address acute stress situations and potentially decrease the likelihood that PTSD will develop after such an incident (Figure 2-5). The

Figure 2-5 Critical incident stress management is sometimes employed to help providers relieve stress.

process theoretically confronts the responses to critical incidents and defuses them, directing the emergency services personnel toward physical and emotional equilibrium. CISM can occur formally, as a debriefing for those who were on scene. In such situations, trained CISM teams of peers and mental health professionals may facilitate this. Additionally, CISM can occur at an ongoing scene in the following circumstances:

- When personnel are assessed for signs and symptoms of distress while resting
- Before re-entering the scene
- During a scene demobilization in which personnel are educated about the signs of critical incident stress and given a buffer period to collect themselves before leaving

The most common form of CISM is peer defusing, when a group informally discusses events that they experienced together.

EMT-B Safety

Coworkers often notice a change in behavior or attitude before a supervisor does. This is especially true in EMS, where close relationships develop between people who work together and share rooms, meals, and social interaction. This may allow you to help someone before job performance is negatively affected.

breathing normally. However, a patient who can speak only one word at a time or must stop every two to three words to catch his or her breath is having significant difficulty breathing. Patients who are having marked difficulty breathing will instinctively assume a posture in which it is easier for them to breathe. There are two common postures that indicate that the patient is trying to increase airflow. The first is called the tripod position. In this position, a patient sits leaning forward on outstretched arms with the head and chin thrust slightly forward and is having sufficient difficulty breathing that a significant conscious effort is required. The second is most commonly seen in children—the sniffing position. The patient sits upright with the head and chin thrust slightly forward, and the patient appears to be sniffing Figure 5-5 .

Breathing that becomes progressively more difficult requires progressively more effort. When you can see that effort, the patient's breathing is described as labored breathing.

Initially, labored breathing is characterized by the patient's position, concentration on breathing, and the increased effort and depth of each breath. As breathing becomes more labored, accessory muscles in the chest and neck are used, and the patient may make grunting sounds with each breath. In infants and small children, nasal flaring and supraclavicular and intercostal retractions (indentation above the clavicles and in the spaces between the ribs) are commonly associated with labored breathing. Sometimes the patient may be gasping.

Infants and small children will continue to have labored breathing for a sustained period, will then often become exhausted, and finally will no longer have the

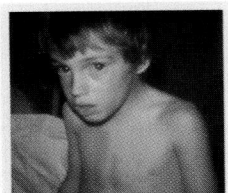

Figure 5-5 A patient in the sniffing position sits upright with the head and chin thrust slightly forward.

strength to maintain the necessary energy to breathe. In infants and small children, cardiac arrest is generally caused by respiratory arrest.

Noisy breathing

Normal breathing is silent or, in a very quiet environment, accompanied only by the sounds of air movement at the mouth and nose. Through a stethoscope, normal breath sounds include only the sound of air movement through the bronchi accompanied by a soft, low-pitched murmur. Breathing accompanied by other sounds indicates a significant respiratory problem. When the airway is partially obstructed by a foreign body or swelling, you may hear stridor, a harsh, high-pitched, crowing sound. If you can hear bubbling or gurgling, the patient probably has fluid in the airway. You may hear other sounds, like wheezes or snoring. The presence of any of these indicates that a serious respiratory problem exists. With a complete airway obstruction, the patient will not be able to move any air and will no longer be able to cough or talk. Sounds are caused by air moving through small spaces or fluid. If you hear nothing, the patient may be moving no air at all.

A patient who coughs up thick, yellowish or greenish sputum (matter from the lungs) most likely has an advanced respiratory infection. A patient with a chest injury may cough up blood or frothy whitish or pinkish foamlike sputum. A patient with congestive heart failure may also cough up frothy sputum. The presence of either substance, regardless of its cause, indicates

⚠ Pediatric Needs

Chest rise in a small child is less marked than that in an adult. However, a small child's abdomen moves more with each breath than an adult's does. Place your hands on the outer margin of the lower anterior chest to feel the chest wall and abdominal movement, and determine whether the depth is normal, shallow, or deep. In a patient of any age, if it is difficult to gauge the depth of breathing from the chest movement, note instead the amount of air that you feel is exhaled with each breath.

Pediatric Needs
Highlight specific concerns and procedures for pediatric patients.

Geriatric Needs
Highlight specific concerns and procedures for geriatric patients.

---

real problem is the droplet nuclei, which are what remains of droplets after excess water has evaporated. These particles are tiny enough to be totally invisible and can remain suspended in the air for a long time. In fact, as long as they are shielded from ultraviolet light, they can remain alive for decades. So you may be at risk by simply entering a closed room that the patient actually left long ago. Particles that are the size of droplet nuclei are not stopped by routine surgical masks. Inhaled, they are carried directly to the alveoli of the lungs, where the bacteria may begin to grow.

Why is tuberculosis not more common than it is? After all, absolute protection from infection with the tubercle bacillus does not exist. Everyone who breathes is at risk. And the vaccine for tuberculosis, called BCG, is only rarely used in the United States. Under normal circumstances, however, the mechanism of

⚠ Geriatric Needs

Everyone has defenses against getting sick, but the aging process can pose a threat to our natural defense mechanisms against invading microorganisms. Our physical defenses weaken or are eliminated as we age. The skin's thinning and loss of supportive collagen, along with a reduction in the number of blood vessels, can allow bacteria or viruses to enter the body with less resistance. The respiratory system cannot trap and eliminate bacteria or viruses in the airways as it once did. Finally, the gastrointestinal system allows an easier entry for bacteria or viruses through the intestines. Not only do our physical barriers to entry weaken, but our immune system deteriorates, and invading organisms are not as easily identified as abnormal. Infectious agents can take hold in the elderly patient much more easily because of reduced defenses.

When transporting an elderly patient, protect the patient from the environment, since extremes in heat or cold can further reduce the body's defenses. If you have a cold or the flu, use respiratory precautions, including a face mask for yourself so that the patient does not get exposed to the viruses. If your patient has a cold or the flu, protect yourself. However, remember that your defense system is probably much stronger than that of the patient.

transmission used by *M tuberculosis* is not very efficient. Infected air is easily diluted with uninfected air. And *M tuberculosis* is one of those germs that typically cause no illness in a new host. In fact, many patients with tuberculosis do not even transmit the infection to family members. In crowded environments with poor ventilation, however, the disease spreads more easily.

If you are exposed to a patient who is found to have pulmonary tuberculosis, you will be given a tuberculin skin test. This simple skin test determines whether a person has been infected with *M tuberculosis*. A positive result means that exposure has occurred; it does not mean that the person has active tuberculosis. It takes at least 6 weeks for the bacteria to show up in the laboratory test. So if you have the test within a few weeks of the exposure and results are positive, this means that you had already acquired the infection from somebody else. You will probably never identify the source. Most transmissions occur silently. This is why health care workers have tuberculin skin tests regularly. If the infection is found before the individual becomes ill, preventive therapy is almost 100% effective. Usually, a daily dose of isoniazid will prevent the development of active infection.

Other Diseases of Concern

Syphilis

Although syphilis is commonly thought of as a sexually transmitted disease, it is also a bloodborne disease. There is a small risk for transmission through a contaminated needlestick injury or direct blood-to-blood contact.

Whooping Cough

Whooping cough, also called pertussis, is an airborne disease caused by bacteria that mostly affects children younger than 6 years. Signs and symptoms include fever and a "whoop" sound that occurs when the patient tries to inhale after a coughing attack.

The best way to prevent exposure is to try to place a mask on the patient and on yourself.

Newly Recognized Diseases

Newly recognized diseases, such as those caused by Hantavirus or enteropathogenic *Escherichia coli*, are being reported. These diseases are not transmitted from person to person directly; rather, they are carried by a vehicle, such as food, or a vector, such as rodents. Although not a newly discovered illness, West Nile

Assessment and Emergency Care Summary Charts

Each medical and trauma chapter concludes with two tables, one summarizing the assessment process, and the other summarizing emergency care for the emergencies discussed in the chapter.

Page 392 (Section 4 Medical Emergencies)

Assessment and Emergency Care

	Respiratory Distress
Scene Size-up	Body substance isolation should include a minimum of gloves and eye protection. Ensure scene safety and determine NOI/MOI. Consider the number of patients, the need for additional help, and c-spine stabilization.
Initial Assessment	
■ General impression	Determine priority of care based on environment and patient's chief complaint. Determine level of consciousness and find/treat any immediate threats to life.
■ Airway	Ensure patent airway.
■ Breathing	Evaluate depth and rate of respirations and provide ventilations as needed. Auscultate and note breath sounds, providing high-flow oxygen.
■ Circulation	Evaluate pulse rate and quality; observe skin color, temperature, and condition. If stable/no life threats, proceed with focused history and physical exam. If unstable/possible life threat, proceed with rapid transportation.
■ Transport decision	If stable/no life threats, proceed with focused history and physical exam. If unstable/possible life threat, proceed with rapid transportation.
Focused History and Physical Exam	NOTE: The order of the steps in the focused history and physical exam differs depending on whether the patient is conscious or unconscious. The order below is for a conscious medical patient. For an unconscious medical patient, perform a rapid physical exam, obtain vital signs, and obtain the history.
■ SAMPLE history	Ask pertinent SAMPLE and OPQRST. Be sure to ask if and what interventions were taken before your arrival, how many, and at what time.
■ Focused physical exam	Perform a focused physical exam, keying in on patient's physical appearance, cyanosis, work of breathing, tripod positioning, pursed lips, use of accessory muscles, adventitious lungs sounds, wheezing, and pedal edema.
■ Baseline vital signs	Take vital signs, noting skin color/temperature as well as patient's level of consciousness. Use pulse oximetry if available.
■ Interventions	Support patient with oxygen, positive pressure ventilations, adjuncts, proper positioning, and assisting with medication(s) as per local protocol. Many of these interventions may need to be performed earlier, in the initial assessment.
Detailed Physical Exam	Consider a detailed physical exam if time and the situation permits.
Ongoing Assessment	Repeat the initial assessment, focused assessment, and reassess interventions performed. Reassess vitals every 5 minutes for the unstable patient, or when an inhaler is used. For the patient who is stable or not using inhalers, reassess vitals every 15 minutes. Reassure and calm the patient.
■ Communications and documentation	Contact medical control with any change in level of consciousness or difficulty breathing. Depending on local protocol, contact medical control prior to assisting with any prescribed medications. Document any changes, the time, and any orders from medical control.

Page 393 (Chapter 11 Respiratory Emergencies)

Assessment and Emergency Care

NOTE: While the steps below are widely accepted, be sure to consult and follow your local protocol.

Respiratory Distress

Administer oxygen by placing a nonrebreathing mask on the patient and supplying oxygen at a rate of 10 to 15 L/min.

For any patient in respiratory distress, use positioning, airway adjuncts (oropharyngeal or nasopharyngeal airway), or positive pressure ventilation as indicated.

Asthma

Administer oxygen. Allow patient to sit in upright position.

Suction large amounts of mucus.

Help patient self-administer a metered-dose inhaler:

1. Obtain order from medical control.
2. Check expiration date and whether patient has taken other doses.
3. Ensure inhaler is at room temperature or warmer.
4. Shake inhaler vigorously several times.
5. Remove oxygen mask. Instruct patient to exhale deeply.
6. Instruct patient to press inhaler and inhale. Instruct patient to hold breath as long as is comfortable.
7. Reapply oxygen.

Infection of Upper or Lower Airway

Administer humidified oxygen if available.

Do not attempt to suction airway or place an oropharyngeal airway.

Transport promptly with patient in position of comfort.

Acute Pulmonary Edema

Administer 100% oxygen and suction any secretions from the airway as necessary.

Place in position of comfort and provide ventilatory support as needed. Transport promptly.

Chronic Obstructive Pulmonary Disease

Provide full-flow oxygen via nonrebreathing mask at 15 L/min.

If patient is prescribed an inhaler, administer it according to local protocol. Document time and effect on patient with each use.

Place in the position of comfort and provide prompt transport.

Spontaneous Pneumothorax

Provide supplemental oxygen and place in position of comfort.

Transport promptly. Support airway, breathing, and circulation as necessary.

Pleural Effusions

Provide high-flow oxygen at 15 L/min and place in position of comfort.

Support airway, breathing, and circulation as necessary.

Transport promptly.

Obstruction of the Upper Airway

For partial or complete foreign body airway obstructions, clear by following BLS guidelines, apply full-flow oxygen at 15 L/min as necessary, and transport promptly.

Pulmonary Embolism

Clear airway and provide full-flow oxygen at 15 L/min. Place in position of comfort and provide prompt transport. Provide ventilatory support as necessary and be prepared for cardiac arrest.

Hyperventilation

Provide full-flow oxygen at 15 L/min and coach respirations slower in a calm manner. Complete an initial assessment and focused history and physical exam. Transport promptly for evaluation.

Prep Kit

End-of-chapter activities reinforce important concepts and improve students' comprehension. **Ready for Review** thoroughly summarizes chapter content. An interactive Ready for Review is provided on www.EMTB.com to help students prepare for exams.

Vital Vocabulary provides key terms and definitions from the chapter. Vocabulary Explorer on www.EMTB.com provides interactivities.

Assessment in Action promotes critical thinking through the use of case studies and provides instructors with discussion points for classroom presentation.

Points to Ponder tackles cultural, social, ethical, and legal issues through case studies.

Chapter 1 Introduction to Emergency Medical Care   19

Prep Kit

Ready for Review

- EMS is the system that provides the emergency medical care that is needed by people who have been injured or have an acute medical emergency. When the dispatcher at the 9-1-1 emergency communications center receives a call for emergency care, he or she dispatches to the scene the designated EMS ambulance squad and any fire, rescue, or police units that may be needed.
- The EMS ambulance is staffed by EMTs who have been trained to the EMT-Basic, EMT-Intermediate, or EMT-Paramedic level according to recommended national standards and have been certified/licensed by the state.
- After the EMTs size up the scene and assess the patient, they provide the emergency care that is indicated based on their findings and ordered by their medical director in the service's standing orders and protocols or by the physician who is providing online medical direction.
- The EMTs then "package" the patient and provide transport to the nearby hospital or designated specialized care facility (eg, trauma center, pediatric hospital) for further evaluation and stabilization in the emergency department and, after admission, definitive surgical or medical care.
- The EMT-B course that you are now taking will present the information and skills that you will need to pass the required examinations for certification and start functioning as an EMT-B in the field. This course will provide you with the training that you need to function as an EMT-B and will serve as the essential foundation upon which you can advance your training and expertise.

- Essential keys to being a good EMT-B include:
  - Compassion and motivation to reduce suffering, pain, and mortality in those who are injured or acutely ill
  - Desire to provide each patient with the best possible care
  - Commitment to obtain the knowledge and skills that this requires
  - The drive to continually increase your knowledge, skills, and ability
- Once you have successfully completed this course and have been certified as an EMT-B, you will enter the next key phase of your training. When you join an EMS service, your first task will be to learn the medical protocols and operating procedures of the squad. You will also have to learn where each piece of equipment is kept on the ambulance and become familiar with how the equipment works.
- From your experience and the guidance provided by your crew chief and the other experienced EMT-Bs you work with, you will gain increased mastery of the skills that you learned in the course and learn how to apply your knowledge and skills in the diverse situations that are encountered in the field.
- Once you have completed the course, you must assume responsibility for directing your own study through continuing education provided by your service's training officer and medical director or through other opportunities available to you. Your commitment to continuing learning is the key to being a good EMT.

Vital Vocabulary

advanced life support (ALS) Advanced lifesaving procedures, some of which are now being provided by the EMT-B.

Americans With Disabilities Act (ADA) Comprehensive legislation that is designed to protect individuals with disabilities against discrimination.

automated external defibrillator (AED) A device that detects treatable life-threatening cardiac arrhythmias (ventricular fibrillation and ventricular tachycardia) and delivers the appropriate electrical shock to the patient.

Technology

Interactivities
Vocabulary Explorer
Anatomy Review
Web Links
Online Review Manual

www.EMTB.com

---

20   Section 1 Preparing to be an EMT-B

Prep Kit continued...

continuous quality improvement (CQI) A system of internal and external reviews and audits of all aspects of an EMS system.

Emergency Medical Dispatch (EMD) A system that assists dispatchers in selecting appropriate units to respond to a particular call for assistance and in providing callers with vital instructions until the arrival of EMS crews.

emergency medical services (EMS) A multidisciplinary system that represents the combined efforts of several professionals and agencies to provide prehospital emergency care to the sick and injured.

emergency medical technician (EMT) A medical professional who is trained and certified/licensed by his or her state to provide emergency life support prior to or with more advanced medical providers.

EMT-Basic (EMT-B) An EMT who has training in basic life support, including automated external defibrillation, use of a definitive airway adjunct, and assisting patients with certain medications.

EMT-Intermediate (EMT-I) An EMT who has training in specific aspects of advanced life support, such as IV (intravenous) therapy, interpretation of cardiac rhythms and defibrillation, and orotracheal intubation.

EMT-Paramedic (EMT-P) An EMT who has extensive training in advanced life support, including IV (intravenous) therapy, pharmacology, cardiac monitoring, and other advanced assessment and treatment skills.

first responder The first trained individual, such as a police officer, fire fighter, lifeguard, or other rescuer, to arrive at the scene of an emergency to provide initial medical assistance.

Health Insurance Portability and Accountability Act (HIPAA) Federal legislation passed in 1996. Its main effect in EMS is in limiting availability of patients' health care information and penalizing violations of patient privacy.

intravenous (IV) therapy The delivery of medication directly into a vein.

medical control Physician instructions that are given directly by radio or cell phone (online/direct) or indirectly by protocol/guidelines (off-line/indirect), as authorized by the medical director of the service program.

medical director The physician who authorizes or delegates to the EMT the authority to provide medical care in the field.

primary service area (PSA) The designated area in which the EMS service is responsible for the provision of prehospital emergency care and transportation to the hospital.

quality control The responsibility of the medical director to ensure that the appropriate medical care standards are met by EMT-Bs on each call.

Points to Ponder

You have been working as an EMT-B in the same area for many years and you know all of the frequent callers. You are dispatched to a man down at 3 o'clock in the morning. You have been on several calls already that night and you just got back into bed. Upon your arrival on scene you recognize the patient immediately as a local drunk. He smells of alcohol and is lying in the front yard of someone's home. Your partner is upset because he transported this man to the hospital two days ago. Your partner calls for police response and starts shaking and yelling at the man to wake up or he's going to have the police arrest him for abusing the 9-1-1 system.

How would you handle this patient? Would you release him to the custody of the police? How do you feel about your partner's behavior towards the patient and how would you handle it?

Issues: Personal Attitudes and Conduct, Professional Interaction, Recognizing Potential Medical Emergencies.

www.EMTB.com

21

Assessment in Action

You are a newly certified EMT-B. Today is your first day on the job at a busy ambulance station. You run several calls and are faced with many new situations. Answer the following questions and discuss them with your instructor.

1. Which of the following is NOT considered an EMT-B skill?
   A. Automated external defibrillation (AED)
   B. Intravenous therapy
   C. Combitube insertion
   D. Assisting patients with the use of their prescribed nitroglycerin

2. You are trained using the Combitube. However, your local protocols state that only a paramedic may insert the device. You are on the scene of a cardiac arrest and the patient needs an airway fast, but you are waiting for a paramedic to arrive in another ambulance. You are anxious to insert the Combitube. What should you do?
   A. Call medical control and ask what you should do.
   B. Wait for the paramedic to arrive, since it is above your scope of practice.
   C. Only do it if you know no one will find out.
   D. Insert the Combitube to practice your skills.

3. Who is medically responsible for establishing written standing orders and protocols?
   A. Department of Transportation
   B. Medical Director
   C. EMS director
   D. Emergency Department physicians

4. Which of the following is the most important skill to an EMT-B?
   A. Knowledge of protocols
   B. Knowledge of location and capabilities of local hospitals
   C. Ability to work with other health care providers
   D. All of the above

5. Which of the following statements is true?
   A. EMT-B training teaches providers everything they need to know.
   B. The main part of being an EMT-B is performing skills; communication should be left to other providers, such as emergency room staff.
   C. Continuing education is vital to successful EMS systems as medicine and technology is ever changing.
   D. An ambulance is only meant to transport people who are having a "real" emergency.

Challenging Questions

6. Why is it important to have continuous quality improvement in an EMS system?

7. What does appropriate patient care mean to an EMT-B?

www.EMTB.com

Instructor Resources

A complete teaching and learning system developed by educators with an intimate knowledge of the obstacles you face each day supports *Emergency Care and Transportation of the Sick and Injured, Ninth Edition.* The supplements provide practical, hands-on, time-saving tools such as PowerPoint presentations, customizable test banks, and web-based distance learning resources to better support you and your students.

Instructor's ToolKit CD-ROM

ISBN: 0-7637-2973-6

Preparing for class is easy with the resources found on this CD-ROM, including:

- **PowerPoint Presentations**, providing you with a powerful way to make presentations that are both educational and engaging. Slides can be modified and edited to meet your needs.
- **Lecture Outlines**, providing you with complete, ready-to-use lesson plans that outline all of the topics covered in the text. Lesson plans can be modified and edited to fit your course.
- **Image and Table Bank**, providing you with many of the images and tables found in the text. You can use them to incorporate more images into the PowerPoint presentations, make handouts, or enlarge a specific image for further discussion.
- **Skill Sheets**, allowing you to track students' skills and conduct skill proficiency exams.

The resources found on the Instructor's ToolKit CD-ROM have been formatted so that you can seamlessly integrate them into the most popular course administration tools. Please contact Jones and Bartlett Publishers technical support at any time with questions.

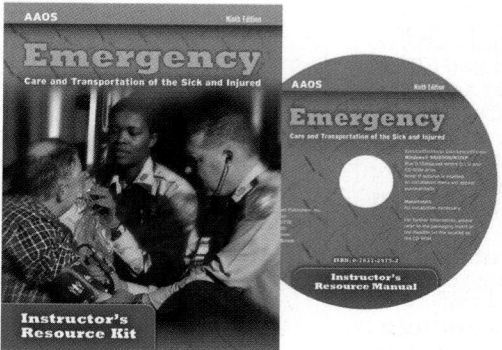

Instructor's Resource Manual

ISBN: 0-7637-2974-4 (8 volumes, printed)
ISBN: 0-7637-2975-2 (CD-ROM)

The Instructor's Resource Manual is your guide to the entire teaching and learning system. It has been designed to assist you with creative ideas and tools to incorporate all of the components of the teaching and learning system. For every chapter, this indispensable manual contains:

- Chapter overview and objectives
- Support materials
- Enhancements
- Teaching tips
- Readings and preparation
- Presentation overview
- Lesson plans and corresponding PowerPoint slide text
- Skill Drill evaluation sheets
- Answers to all end-of-chapter student questions found in the text
- Activities and assignments
- Quizzes
- National Registry Skill Sheets

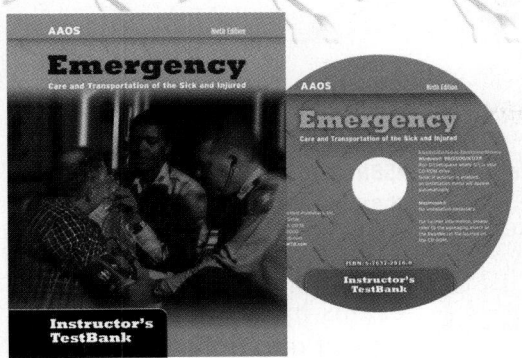

Instructor's Test Bank

ISBN: 0-7637-2977-9 (print)
ISBN: 0-7637-2976-0 (CD-ROM)

Following the DOT EMT-Basic National Standard Curriculum, this TestBank provides you with:

- Multiple-choice, scenario-based questions
- Page references to the *Ninth Edition*

With the TestBank on CD-ROM, you can originate tailor-made tests quickly and easily by selecting, editing, and printing a test along with an answer key.

Instructor's Slide Set

ISBN: 0-7637-2042-9 (CD-ROM)

This easy-to-use CD-ROM will assist you in emphasizing key points and stimulating classroom discussion through realistic examples, chapter-by-chapter. Over 1,200 electronic images are sure to grab students' attention.

Trauma Slide Set

ISBN: 0-7637-2048-8 (CD-ROM)

Bring the "real world" to your classroom discussion with approximately 100 real-life trauma images.

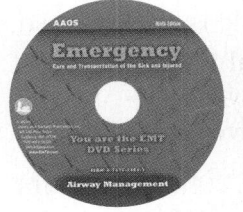
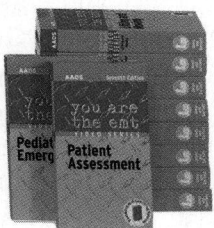

You are the EMT Video Series (10 DVDs or VHS videos)

ISBN: 0-7637-2981-7 (DVD)
ISBN: 0-7637-1025-3 (VHS)

Now available on DVD or VHS, these videos capture real-life scenes from the street and are enriched with thought-provoking questions. Use these videos for initial training to identify correct and incorrect procedures. Each video comes with a booklet that will help you guide class discussions. Videos run approximately 25-30 minutes.

The EMS Call
ISBN: 0-7637-2982-5 (DVD)
ISBN: 0-7637-1014-8 (VHS)

The Well-Being of the EMT
ISBN: 0-7637-2983-3 (DVD)
ISBN: 0-7637-1016-4 (VHS)

Lifting and Moving
ISBN: 0-7637-2984-1 (DVD)
ISBN: 0-7637-1017-2 (VHS)

Airway Management
ISBN: 0-7637-2985-X (DVD)
ISBN: 0-7637-1019-9 (VHS)

Patient Assessment
ISBN: 0-7637-2986-8 (DVD)
ISBN: 0-7637-1018-0 (VHS)

Communication
ISBN: 0-7637-2987-6 (DVD)
ISBN: 0-7637-1071-7 (VHS)

Trauma Management and Spinal Immobilization
ISBN: 0-7637-2989-2 (DVD)
ISBN: 0-7637-1020-2 (VHS)

Pediatric Emergencies I
ISBN: 0-7637-2990-6 (DVD)
ISBN: 0-7637-1021-0 (VHS)

Pediatric Emergencies II
ISBN: 0-7637-2991-4 (DVD)
ISBN: 0-7637-1015-6 (VHS)

Managing the Geriatric Patient
ISBN: 0-7637-2988-4 (DVD)
ISBN: 0-7637-1022-9 (VHS)

Student Resources

To help students retain the most important information and to assist them in preparing for exams, we have developed the following resources:

EMT-Basic Skills DVD

ISBN: 0-7637-2980-9

Packaged with each copy of the *Ninth Edition,* this DVD provides students with a walk-through of the skills that are required to successfully complete the national EMT-Basic practical examination process. For each skill, students can visually learn the steps of each skill and will find helpful information, tips, and pointers designed to facilitate their progression through the practical examination, specifically:

- Objectives
- Equipment
- Key steps to perform to successfully complete the skill
- Critical errors that result in failure

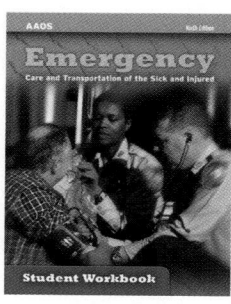

Student Workbook

ISBN: 0-7637-2969-8

This resource is designed to encourage critical thinking and aid comprehension of the course material through:

- Case studies and corresponding questions
- Skill drill activities
- Figure labeling exercises
- Crossword puzzles
- Matching, fill-in-the-blank, short answer, and multiple-choice questions

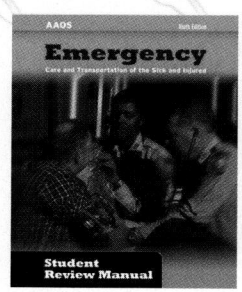

Student Review Manual

ISBN: 0-7637- 2971-X (printed)
ISBN: 0-7637-2970-1 (CD-ROM)
ISBN: 0-7637-2972-8 (online)

This Review Manual has been designed to prepare students for exams by including the same type of questions that they are likely to see on classroom and national examinations. The manual contains multiple-choice question exams with an answer key and page references. It is available in print, on CD-ROM, and online.

EMT-B Field Guide, Second Edition

ISBN: 0-7637-2214-6

This handy reference covers basic information from patient management tips to guidelines on helping a patient use medication. A few special features include a prescription medication reference and documentation tips. The EMT-B Field Guide is pocket-sized, spiral-bound, and water resistant for ready-reference in the field.

EMT-Basic Review Manual for National Certification

ISBN: 0-7637-1829-7

The EMT-Basic Review Manual for National Certification is designed to prepare students to sit for the National Certification Exam by including:

- Practice questions with answers and model exam
- Step-by-step walkthrough of skills, including helpful tips, commonly made errors, and sample scenarios
- Self-scoring guide and winning test-taking tips
- Coverage of the entire DOT EMT-Basic National Standard Curriculum

Technology Resources

A key component to the teaching and learning system are interactivities and simulations to help students become great providers.

www.EMTB.com

Make full use of today's teaching and learning technology with www.EMTB.com. This site has been specifically designed to complement *Emergency Care and Transportation of the Sick and Injured, Ninth Edition* and is regularly updated. Some of the resources available include:

Chapter Pretests prepare students for training. Each chapter has a pretest and provides instant results, feedback on incorrect answers, and page references for further study.

Review Manual allows students to prepare for classroom and national examinations providing instant results, feedback, and page references.

Interactivities allow your students to experiment with the most important skills and procedures in the safety of a virtual environment.

Anatomy Review provides interactive anatomical figure labeling exercises to reinforce students' knowledge of human anatomy.

Vocabulary Explorer is your virtual dictionary. Here, students can review key terms, test their knowledge of key terms through flashcards, and complete crossword puzzles.

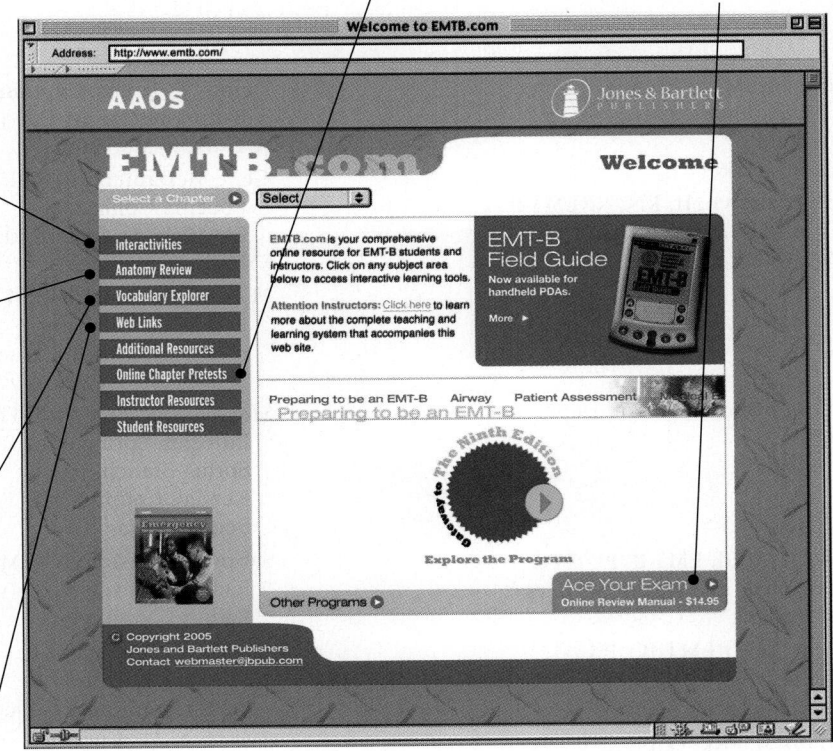

Web Links present current information including trends in health care, the EMS community, and new equipment.

The Web site also provides:

- **National Registry Skill Sheets**—the actual skill sheets that students will be tested with from the National Registry. Use these to test knowledge of skills.
- **Skill Evaluation Sheets**—the steps of each skill from the textbook are delineated here. Use these to test knowledge of skills.
- **Registry Review**—a 75-question test similar to the National Registry Exam. Take this practice test in preparation for exams.

Acknowledgments

The American Academy of Orthopaedic Surgeons would like to acknowledge the contributors and reviewers of *Emergency Care and Transportation of the Sick and Injured, Ninth Edition*.

Editorial Board

Andrew N. Pollak, MD, FAAOS
Medical Director, Baltimore County Fire Department
Associate Professor, University of Maryland School
of Medicine
Baltimore, Maryland

Benjamin Gulli, MD
Northwest Orthopaedic Surgeons
Robbinsdale, Minnesota

Les Chatelain, BS, MS
University of Utah
Health Promotion and Education Dept.
EMS Program
Salt Lake City, Utah

Chris Stratford, BS, RN, BCEN, EMT-I
University of Utah
Health Promotion and Education Dept.
EMS Program
Salt Lake City, Utah

Contributors

Barbara Booton, NREMT-P, EMS-IC
Emergency Education
Brookings Hospital
Brookings, South Dakota

Captain Joseph L. Brown, Jr, RN, NREMT-P
Baltimore County Fire Department
Baltimore, Maryland

Julie F. Chase, BS, BM, NREMT-P
Seattle, Washington

Les Chatelain, BS, MS
University of Utah
Health Promotion and Education Dept.
EMS Program
Salt Lake City, Utah

Ken Fahr
Bedford, Indiana

James M. Floyd, Jr, MEd, EMT-B, PI, CTC
St. Vincent Hospitals and Health Services
Indianapolis, Indiana

Linda J. Gosselin, MS, REMT IC, Ed
MECTA Academy
Millbury, Massachusetts

Donell Harvin, MPA, NREMT-P
NYC*EMS
New York, New York

Guy H. Haskell, PhD, NREMT-P
Director, Emergency Medical and Safety Services
Consultants
Bloomington, Indiana
Paramedic, Bedford Regional Medical Center EMS
Bedford, Indiana

Derek Hunt, NREMT-P, CCEMT-P
Flight Paramedic
AirCare Medevac
Winchester, Virginia

Jay Keefauver, BS, EMT-P, CEMSI
National Emergency Resource Center
Columbus, Nebraska

Dennis R. Krebs, Captain (retired), NREMT-P
Baltimore County Fire Department
Towson, Maryland

Gene McDaniel, BS, NREMT-P
Phoenix College
EMT/Fire Science Department
Phoenix, Arizona

Michael D. Pante, NREMT-P
Paramedic/Educator
Somerset Medical Center
Somerville, New Jersey

Stephen J. Rahm, NREMT-P
Bulverde-Spring Branch EMS Training Center
Spring Branch, TX
Kendall County EMS
Boerne, Texas

Sabra R. Raulston, NREMT-P
Radford University
Radford, Virginia

Angela D. Reed, EMT-B
Program Director
Pima Community College
Tucson, Arizona

Scott Schein, BS, NREMT-P
President/CEO
Florida Medical Training Institute
Melbourne, Florida

Chris Stratford, BS, RN, BCEN, EMT-I
University of Utah
Health Promotion and Education Dept.
EMS Program
Salt Lake City, Utah

David S. Teeter, PharmD
Wishard Hospital Pharmacy
Indianapolis, Indiana

Andrew Turcotte, NREMT-P, CCEMT-P
Flight Paramedic
AirCare Medevac
Winchester, Virginia

Reviewers

Robert Jay Alley EMT-P, CHS-III
School Blue Ridge Community College
Flat Rock, North Carolina

Gary W. Bonewald, MEd, LP
Houston Community College System
Wharton, Texas

James L. Brother, MA, EMSI
Hartford Hospital School of Allied Health
Hartford, Connecticut

Glen Clegg
Leary Technical Center
Tampa, Florida

Sgt John "Chip" Coleman, USAF, EMT-P
United States Air Force Academy
USAF Academy, Colorado

Chris Coughlin, MEd, NREMT-P
Glendale Community College
Glendale, Arizona

Joseph D'Agosto, EMT-B
Rescue Training Institute, Inc.
Fire Department, City of New York
Bayside, New York

Steven K. Frye, BS, NREMT-P
Maryland Fire and Rescue Institute
College Park, Maryland

Phil Giuffre, NREMT-P
Malcolm X College
Chicago, Illinois

John Gosford, EMT-P, AS
State of Florida
Crawfordville, Florida

Linda J. Gosselin, MS, REMT IC, Ed
MECTA Academy
Millbury, Massachusetts

Christopher B. Haber, NREMT-P
MEDPRO EMS Education
Hulmeville, Pennsylvania

Rick Hilinski, BA, EMT-P
Assistant Director Medical Programs
Community College of Allegheny County Public Safety
   Institute
Pittsburgh, Pennsylvania

Dave MacCuish, RN, BSN, EMT-P
Eastern Medical Educators
Quincy, Massachusetts

Lieutenant Craig McElhaney, NREMT
Broward Community College Central
Davie, Florida

Steve Myers, RN, CEN, EMT-P
Indian River Community College
Fort Pierce, Florida

Mary L. Pilling, NREMT-I
Southwest Wisconsin Technical College
Blanchardville, Wisconsin

Monte Posner, MSW, EMT-B, CIC
Training Institute for Medical Emergencies and Rescue
Staten Island, New York

Richard D. Prentiss, MHM, RRT, EMT
Miami Dade College
Miami, Florida

Alice J. Quiroz
349th Medical Group
Travis Air Force Base, California

Angela D. Reed, EMT-B
Program Director
Pima Community College
Tucson, Arizona

Mike Reilley, NREMT-P
Monmouth Ocean Hospital Service Corp
Neptune, New Jersey

Bernadette Royce, BA, FF/EMT-P
Osceola County Fire Rescue
Valencia Community College
Orlando, Florida

Charlene Sayers, EMT-B(I)
Basic Instructor, Salem County EMT;
   NSC, FAI-EH, Instructor; AHA, Instructor
Rutgers University, Department of Biology
Camden, New Jersey

Dr. Raymond Schleif, MMSc, ScD, NREMT
New York State EMS Regional Faculty
Training Institute for Medical Emergencies & Rescue
Staten Island, New York City

Al M. Slarve
Fry Fire District
Cochise Community College
Sierra Vista, Arizona

David H. Sloane AAS, NREMT-P
Georgetown Scott County EMS
Georgetown, Kentucky

**Michael L. Smith, MPA, EMT/SEI (retired), Master
Instructor**
Washington State EMS
Cheney, Washington

Craig S. Spector, EMT-Instructor
President, CPR Heart Starters Inc./ Safety Training
Warrington, Pennsylvania

Nerina J. Stepanovsky, MSN, RN, EMT-P
St. Petersburg College EMS Program
St. Petersburg, Florida

David M. Tauber, NREMT-P, CCEMT-P, I/C
Advanced Life Support Institute
Conway, New Hampshire

Al Tompkins, EMT-D/CIC
SUNY Upstate Medical
East Syracuse, New York

Elliot Velez
Malcolm X College
Chicago, Illinois

Sandy Waggoner, FF/EMTP/EMSI
EHOVE Career Center
Milan, Ohio

Kim R. Weaver, EMT-P
Montgomery County EMS
Conshohocken, Pennsylvania

Preparing to be an EMT-B

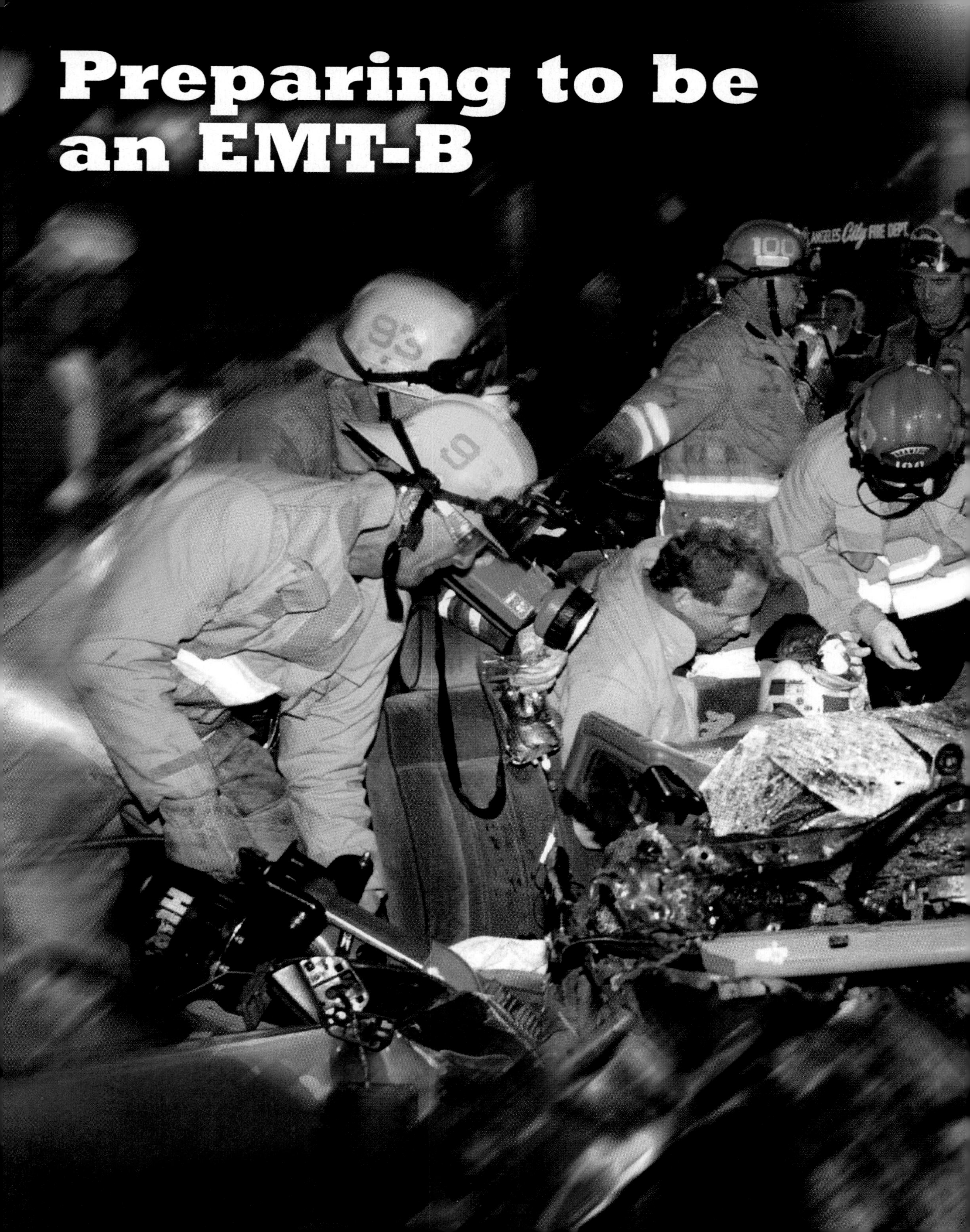

Section

1

Introduction to Emergency Medical Care

Objectives

Cognitive

1-1.1 Define Emergency Medical Services (EMS) systems. (p 4)

1-1.2 Differentiate the roles and responsibilities of the EMT-Basic from other prehospital care providers. (p 4)

1-1.3 Describe the roles and responsibilities related to personal safety. (p 16)

1-1.4 Discuss the roles and responsibilities of the EMT-Basic towards the safety of the crew, the patient, and bystanders. (p 17)

1-1.5 Define quality improvement and discuss the EMT-Basic's role in the process. (p 13)

1-1.6 Define medical direction and discuss the EMT-Basic's role in the process. (p 12)

1-1.7 State the specific statutes and regulations in your state regarding the EMS system. (p 12)

Affective

1-1.8 Assess areas of personal attitude and conduct of the EMT-Basic. (p 16)

1-1.9 Characterize the various methods used to access the EMS system in your community. (p 11)

Psychomotor

None

You are the Provider

You are a recently certified EMT-Basic. You and your partner, a paramedic, are dispatched to a possible heart attack at 200 South 12th Ave. As you get into the ambulance, the dispatcher announces "CPR in progress." You're filled with excitement at the thought of working your first cardiac arrest. En route to the location, you and your partner discuss what roles you will play to assist each other in expediting defibrillation, airway management, intravenous access, and pharmacologic therapies. You feel ready to tackle all the tasks the paramedic has placed in your charge and feel confident that working as a team member you can help save this patient's life.

This chapter will introduce you to the profession of EMS—emergency medical services. It will examine your roles and responsibilities, safety, quality improvement, medical direction, and legal issues in EMS. It will help you answer the following questions:

1. How would you define a "real emergency?"
2. In the overall scheme of prehospital medicine, what are the primary roles and responsibilities of an EMT-B?

Introduction to Emergency Medical Care

This book has been designed to serve as the text and primary resource for the emergency medical technician basic (EMT-Basic) course. This chapter describes the content and objectives of the EMT-Basic course. It also discusses what will be expected of you during the course and what other requirements you will have to meet to be licensed or certified as an EMT-Basic in most states. You will also learn about the differences between first aid training, a Department of Transportation (DOT) First Responder training course, and the training for the EMT-Basic, EMT-Intermediate, and EMT-Paramedic.

Emergency medical services (EMS) is a *system*. The key components of this system and how they influence and affect the EMT-Basic (EMT-B) and his or her delivery of emergency care are carefully discussed. Next, the administration, medical direction, quality control, and regulation of EMS services are presented. The chapter ends with a detailed discussion of the roles and responsibilities of the EMT-B as a health care professional.

Course Description

You are about to enter an exciting field. <u>Emergency medical services (EMS)</u> consists of a team of health care professionals who, in each area or jurisdiction, are responsible for and provide emergency care and transportation to the sick and injured (Figure 1-1 ▶). Each

Figure 1-1   As an EMT-B, you will be part of a larger team that responds to a variety of calls and provides a wide range of prehospital emergency care.

emergency medical service is part of a local or regional EMS system that provides the many varied prehospital and hospital components required for the delivery of proper emergency medical care. The standards for prehospital emergency care and the individuals who provide it are governed by the laws in each state and are typically regulated by a state office of EMS.

The individuals who provide the emergency care in the field are trained and, except for licensed physicians, must be state-licensed or certified <u>emergency medical technicians (EMTs)</u>. Different states will refer to the authority granted to you to function as an EMT-B as licensure, certification, or credentialing. For the purposes of this text, the term *certification* will be used.

In most states, EMTs are categorized into three training and certification levels: EMT-Basic, EMT-Intermediate, and EMT-Paramedic. An <u>EMT-Basic (EMT-B)</u> has training in basic emergency care skills, including automated external defibrillation, use of airway adjuncts, and assisting patients with certain medications. An <u>EMT-Intermediate (EMT-I)</u> has training in specific aspects of advanced life support, such as <u>intravenous (IV) therapy</u> and cardiac monitoring. An <u>EMT-Paramedic (EMT-P)</u> has extensive training in advanced life support, including IV therapy, pharmacology, cardiac monitoring, and other advanced assessment and treatment skills.

Although the specific training and certification requirements vary from one state to another, the training that is required in almost every state follows or

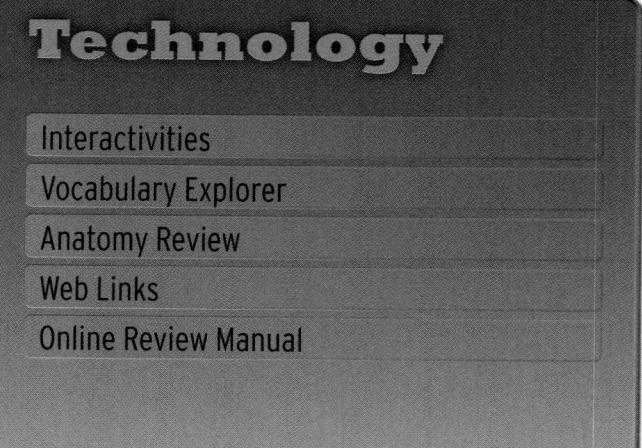

Technology

- Interactivities
- Vocabulary Explorer
- Anatomy Review
- Web Links
- Online Review Manual

www.EMTB.com

exceeds the guidelines that are recommended in the current US DOT (Department of Transportation) National Standard Curriculum for each EMT level.

After you have successfully completed the Basic Life Support/Cardiopulmonary Resuscitation (BLS/CPR) course for health care providers and met the other prerequisites of your training institution, you are ready to take the EMT-B course. Like any introductory course, the EMT-B course covers a great deal of information and introduces many skills. Everything you learn in the course will be important to your ability to provide high-quality emergency care once you are certified and ready to practice. In addition, the knowledge, understanding, and skills that you acquire in the EMT-B course will serve as a foundation for the additional knowledge and training that you will receive in future years.

This textbook covers the material and skills that are identified in the US DOT 1994 EMT-Basic National Standard Curriculum and in the 1994 National EMS Education and Practice Blueprint. In addition to the required core content, it includes additional information that will help you to understand and apply the material and skills that are included in the EMT-B level. Your instructor will furnish you with reading assignments. It is essential that you complete the assigned reading before each class. Your success in this course will depend on it.

In class, the instructor will review the key parts of the reading assignment and clarify and expand on them. He or she will also answer any questions that you have and will clarify any points that you or others find confusing (Figure 1-2 ▼). Unless you have carefully read the assignment and made notes before coming to class,

TABLE 1-1 Study Tips for Using This Textbook
■ Complete each assignment diligently and carefully.
■ Read the textbook like a textbook, not like a newspaper, magazine, or novel.
■ Read each chapter several times and underline key points. Take notes!
■ Ask your instructor to clarify any questions you note in your reading or in class.
■ Take additional notes when the assigned material is expanded upon in class.
■ Remember: The only absurd question is the one that a student has and fails to ask.

you will not fully understand or benefit from the classroom presentation and discussions. You will also need to take additional notes during class (Table 1-1 ▲).

The EMT-B course will include four types of learning activities:

1. **Reading assignments** from the textbook and presentations and discussions held in class will provide you with the necessary knowledge base.
2. **Step-by-step demonstrations** will teach you hands-on skills that you then need to practice repeatedly in supervised small group workshops.
3. **Summary skills sheets** will help you to memorize the sequence of steps in complex skills that contain a large number of steps or variations so that you can perform the skill with no errors or omissions.
4. **Case presentations and scenarios** used in class will help you learn how to apply the knowledge and skills acquired in class situations like those you will find in the field.

EMT-B Training: Focus and Requirements

EMT-B training is divided into three main categories. The first and most important category focuses on the care of life-threatening or potentially life-threatening conditions. To deal with these, you will learn how to do the following:

■ Size up the scene and situation
■ Ensure that the scene is safe
■ Perform an initial assessment of the patient
■ Obtain a history of this episode and a pertinent past medical history

Figure 1-2 In the classroom, you will learn both didactic and practical skills to prepare you for various types of calls.

- Identify life-threatening injuries or conditions
- Establish and maintain an open airway
- Provide adequate ventilation
- Manage conditions that prevent proper ventilation
- Provide high-flow supplemental oxygen
- Perform cardiopulmonary resuscitation (CPR)
- Perform automated or semiautomated external defibrillation (AED)
- Control external bleeding
- Recognize and treat shock
- Care for patients in an acute life-threatening medical emergency
- Assist patients in taking certain medications that they carry and that their physician has prescribed for an acute episode
- Identify and rapidly prepare, or "package," patients (by positioning, covering, and securing them) for rapid initiation of transport when necessary
- Heavy and frequent lifting

The second category of training covers conditions that, although not life-threatening, are key components of emergency care or are necessary to prevent further harm before the patient is moved. You will learn to do the following:

- Identify patients for whom spinal precautions should be taken and immobilize them properly
- Dress and bandage wounds
- Splint injured extremities
- Care for burns
- Care for cases of poisoning
- Deliver a baby
- Assess and care for a newborn
- Manage patients with behavioral or psychological problems
- Cope with the psychological stresses on patients, families, your fellow EMT-Bs, and yourself

The third category covers important issues that are related to your ability to provide emergency care. You will develop the following related skills:

- Understanding the role and responsibilities of the EMT-B
- Understanding your service's protocols and orders from medical direction
- Understanding ethical and medicolegal problems
- Emergency vehicle and defensive driving
- Using equipment carried on the ambulance
- Checking and stocking the ambulance
- Communicating with patients and others at the scene

- Using the radio or cell phone and communicating with the dispatcher or medical control
- Giving a precise patient radio report and obtaining direct medical direction
- Giving a full verbal report when transferring the patient's care at the hospital
- Preparing proper documentation and completing the patient care report
- Working with other responders at a crash scene
- Cooperating with operations at special rescue, mass-casualty, and hazardous materials incidents

Certification Requirements

To be recognized and perform as an EMT-B, you must meet certain requirements. The specific requirements differ from state to state. You should ask your instructor, school, or contact your state EMS office to find out about the requirements in your state. Generally, the criteria to be certified and employed as an EMT will include the following:

- High school diploma or equivalent
- Proof of immunization against certain communicable diseases
- Valid driver's license
- Successful completion of a recognized health care provider BLS/CPR course
- Successful completion of a state-approved EMT-B course
- Successful completion of a state-recognized written certification examination
- Successful completion of a state-recognized practical certification examination
- Demonstrating that you can meet the mental and physical criteria necessary to be able to safely and properly perform all the tasks and functions described in the defined role of an EMT-B
- Compliance with other state and local and employer provisions

The Americans With Disabilities Act (ADA) of 1990 protects individuals who have a disability from being denied access to programs and services that are provided by state or local governments and prohibits employers from failing to provide full and equal employment to the disabled. To obtain further information about the ADA and employment as an EMT-B, you should contact your state EMS office.

In most states, individuals who have been convicted of driving while under the influence of alcohol or other drugs, or have been convicted of certain felonies, may be denied certification as an EMT-B.

States may exclude from certification persons with a history of a health problem that could make their performance of EMT-B tasks dangerous to themselves or others.

Overview of the Emergency Medical Services System

History of EMS

As an EMT-B, you will be joining a long tradition of people who have provided emergency medical care to their fellow human beings. With the early use of motor vehicles in warfare, volunteer ambulance squads were organized and went overseas to provide care for the wounded in World War I. In World War II, the military trained special corpsmen to provide care in the field and bring the casualties to aid stations staffed by nurses and physicians. In the Korean conflict, this evolved to the field medic and rapid helicopter evacuation to nearby Mobile Army Surgical Hospital units, where immediate surgical intervention was provided. Many advances in the immediate care of trauma patients resulted from the casualty experiences in the Korean and Vietnam conflicts.

Unfortunately, emergency care of the injured and ill at home had not progressed to a similar level. As late as the early 1960s, emergency ambulance service and care across the United States varied widely. In some places, it was provided by well-trained advanced first aid squads that had well-equipped modern ambulances. In a few urban areas, it was provided by hospital-based ambulance services that were staffed with interns and early forms of medics. In many places, the only emergency care and ambulance service was provided by the local funeral home using a hearse that could be converted to carry a cot and serve as an ambulance. In other places, the police or fire department used a station wagon that carried a cot and a first aid kit. In most cases, both of these were staffed by a driver and an attendant who had some basic first aid training. In the few areas where a commercial ambulance was available to transport the ill, it was usually similarly staffed and served primarily as a means to transport the patient to the hospital.

Many communities had no formal provision for prehospital emergency care or transportation. Injured persons were given basic first aid by police or fire per-

sonnel at the scene and were transported to the hospital in a police or fire officer's car. Customarily, patients with an acute illness were transported to the hospital by a relative or neighbor and were met by their family physician or an on-call hospital physician, who assessed them and then summoned any specialists and operating room staff that were needed. Except in large urban centers, most hospitals did not have the staffed emergency departments to which we are accustomed today.

EMS as we know it today had its origins in 1966 with the publication of *Accidental Death and Disability: The Neglected Disease of Modern Society*. This report, prepared jointly by the Committees on Trauma and Shock of the National Academy of Sciences/National Research Council, revealed to the public and Congress the serious inadequacy of prehospital emergency care and transportation in many areas. A number of key items were recommended in the report, some of which follow:

- Development of national courses of instruction for prehospital emergency care and transportation by fire, police, rescue, and ambulance personnel
- Development of nationally accepted textbooks and training aids for these courses
- Development of federal guidelines for the design of ambulances and the equipment they carry
- Development and adoption of general policies and regulations pertaining to ambulance services and qualification and supervision of ambulance personnel in each state
- Adoption by each municipality (or district or county) of means to supply the necessary proper prehospital emergency care and transport within its jurisdiction
- Establishment of hospital emergency departments with staffing by physicians, nurses, and other personnel who are trained in resuscitation and the immediate care of the seriously injured and ill

As a result, Congress mandated that two federal agencies address these issues. The National Highway Traffic Safety Administration (NHTSA) of the DOT, through the Highway Safety Act of 1966, and the Department of Health and Human Services, through the Emergency Medical Act of 1973, created funding sources and programs to develop improved systems of prehospital emergency care.

In the early 1970s, the DOT developed and published the first National Standard Curriculum to serve as the guideline for the training of EMTs. To support the EMT course, the American Academy of Orthopaedic Surgeons prepared and published the first EMT textbook—*Emergency Care and Transportation of the Sick and Injured* in 1971, often called the Orange Book. The textbook you are reading is the ninth edition of that publication. Through the 1970s, following the recommended guidelines, each state developed the necessary legislation, and the EMS system was developed throughout the United States. During the same period, emergency medicine became a recognized medical specialty, and the fully staffed emergency departments that we know today became the accepted standard of care.

In the late 1970s and early 1980s, the DOT developed a recommended National Standard Curriculum for the training of paramedics and identified a part of the course to serve as training for basic EMTs.

By 1980, EMS had been established throughout the nation. The system was based on the following two key changes:

- The introduction of legislation that made it the responsibility of each municipality, township, or county to provide proper prehospital emergency care and transportation within its boundaries
- The establishment of recognized and regulated standards for the training of ambulance personnel and equipment required on each ambulance

These changes ensured that, regardless of where an individual became hurt or acutely ill, he or she would receive timely, proper emergency care and transport to the hospital. During the 1980s, many areas enhanced the EMT National Standard Curriculum by adding EMTs with higher levels of training who could provide key components of advanced life support (ALS) care (advanced lifesaving procedures). The availability of paramedics (EMT-Ps) and ALS on calls that require or benefit from advanced care has grown steadily in recent years. In addition, with the evolution in training and technology, the EMT-B and EMT-I can now perform a number of important advanced skills in the field that were formerly reserved for the EMT-P.

The way EMS systems work may differ depending on the geographic area and population served. Regardless of the area, however, the NHTSA is available to evaluate EMS systems, based on the following 10 criteria in their Technical Assistance Program Assessment Standards:

1. Regulation and policy
2. Resource management
3. Human resources and training
4. Transportation equipment and system
5. Medical and support facilities
6. Communications system
7. Public information and education
8. Medical direction
9. Trauma system and development
10. Evaluation

Levels of Training

Certification of EMTs is a state function, subject to the laws and regulations of the state in which the EMT practices. For this reason there is some variation from state to state on the scope of EMT practice, as well as training and recertification requirements.

There are some national guidelines, however. The DOT National Standard Curriculum serves as the basis for the development of state curricula. The EMT-B, EMT-I, and EMT-P curricula can be downloaded from the National Highway Traffic Safety Administration's (NHTSA) website. In addition, the National Registry of Emergency Medical Technicians is a nongovernmental agency that provides a national standard for EMT testing and certification throughout the United States. Many states utilize the National Registry standards in certifying their EMTs and grant licensing reciprocity to nationally registered EMTs. It is important to remember, however, that EMS is regulated entirely by the state in which you are certified.

Public Basic Life Support and Immediate Aid

With the development of EMS and increased awareness of the need for immediate emergency care, millions of laypeople have been trained in BLS/CPR. In addition to CPR, many individuals have taken first aid courses that include control of bleeding and other simple skills that may be required to provide immediate essential care. These courses are designed to train individuals so that those in the workplace, teachers, coaches, and child care providers, among others, can provide the necessary critical care in the minutes before EMTs or other responders arrive at the scene.

In addition, many individuals, such as those who regularly accompany groups on camping trips or are in other situations in which the arrival of EMS may be delayed because of remote location, are trained in advanced first aid. This course includes BLS and the essential additional care and packaging that may be necessary until the help of rescuers and EMTs can be obtained at a remote location.

One of the most dramatic recent developments in prehospital emergency care is the use of an <u>automated external defibrillator (AED)</u>. These remarkable devices, some no larger than a cellular phone, detect treatable life-threatening cardiac arrhythmias (ventricular fibrillation and ventricular tachycardia) and deliver the appropriate electrical shock to the patient. Designed to be used by the untrained layperson, they are now included at every level of prehospital emergency training.

First Responders

Because the presence of a person who is trained to initiate BLS and other urgent care cannot be ensured, the EMS system includes immediate care by <u>first responders</u>, such as law enforcement officers, fire fighters, park rangers, ski patrollers, or other organized rescuers who often arrive at the scene before the ambulance and EMTs (Figure 1-3 ▶). The DOT has established a First Responder Curriculum to provide these individuals with the training necessary to initiate immediate care and then assist the EMTs on their arrival. The course focuses on providing immediate BLS and urgent care with limited equipment. It also familiarizes the student with the additional procedures, equipment, and packaging techniques that EMTs may use and with which the first responder may be called upon to assist.

In addition to professional first responders, EMTs often encounter a variety of people on scene eager to help. You will encounter Good Samaritans trained in first aid and CPR, physicians and nurses, and other well-meaning individuals with or without prior training and experience. Identified and utilized properly,

Figure 1-3 First responders, such as law enforcement officers, are trained to provide immediate basic life support until EMTs arrive on the scene.

these individuals can provide valuable assistance when you are short-handed. At other times, they can interfere with operations and even create problems or danger to themselves or others. It will be your task in your initial scene size-up to identify the various persons on the scene and orchestrate well-meaning attempts to assist.

EMT-Basic

The EMT-B course requires a minimum of 110 hours (more in some states) and includes the essential knowledge and skills required to provide basic emergency care in the field. The course serves as the foundation on which additional knowledge and skills are built in advanced EMT training. On arrival at the scene, you and any other EMTs who have responded should assume

You are the Provider   Part 2

As you arrive on the scene and pull into the driveway, you notice a police officer's vehicle and a pickup truck with a red rotating beacon on the roof used by responders on the volunteer fire department. When you and your partner enter the house, you find a police officer and fire fighter on either side of the patient. The police officer states he was first on the scene and immediately started CPR. After the fire fighter arrived, he applied an AED to the patient and successfully administered three shocks, and now the patient has a carotid pulse.

3. What are your immediate concerns regarding patient care and what are your overall responsibilities as an EMT-B?

responsibility for the assessment and care of the patient, followed by proper packaging and transport of the patient to the emergency department if appropriate. With the continued development of EMS, the definition of what is a basic and what is an advanced skill is ever in flux. With the publication and adoption of the US DOT 1994 EMT-Basic National Standard Curriculum, selected skills formerly considered advanced were added to this level. These skills include automated external defibrillation, use of nonvisualized airways, such as the Combitube or even endotracheal intubation (Figure 1-4 ▼), and assisting patients with the use of their physician-prescribed medications, such as nitroglycerin, epinephrine, and metered-dose inhalers. Additional skills and procedures are continually being evaluated for inclusion in the EMT-Basic scope of practice (discussed in Chapter 3). For example, some states have added aspirin for heart attack, self-injectable epinephrine (EpiPens), and IV therapy to EMT training. For this reason it is essential to keep up with developments in your state and local EMS system. Remember, training does not authorize use—your protocols must give you permission to perform skills and procedures.

EMT-Intermediate

The EMT-Intermediate (EMT-I) course and training are designed to add knowledge and skills in specific aspects of ALS to individuals who have been trained and have experience in providing emergency care as EMT-Bs. These additional skills include IV therapy, interpretation of cardiac rhythms and manual defib-

rillation, orotracheal intubation, and, in many states, the knowledge and skills necessary to administer certain medications. Here again, the scope of practice of this certification is based on the 1985 or 1999 Intermediate National Standard Curriculum and differs from state to state. In the end, it is up to the medical director to determine the appropriate type of care for this training level and to give orders for these actions verbally or in writing (protocols).

EMT-Paramedic

The EMT-Paramedic (EMT-P) has completed an extensive course of training that significantly increases knowledge and mastery of basic skills and covers a wide range of ALS skills based on the 1999 Paramedic National Standard Curriculum (Figure 1-5 ▼). This course ranges from 800 to more than 1,500 hours, usually equally divided between classroom and internship training. Increasingly, this training is offered within the context of an Associate's degree or Bachelor's degree college program.

The skills taught in the EMT-P course include the following:

- Electrocardiogram monitoring and interpretation of cardiac rhythms
- Advanced cardiac life support (ACLS) protocols and skills

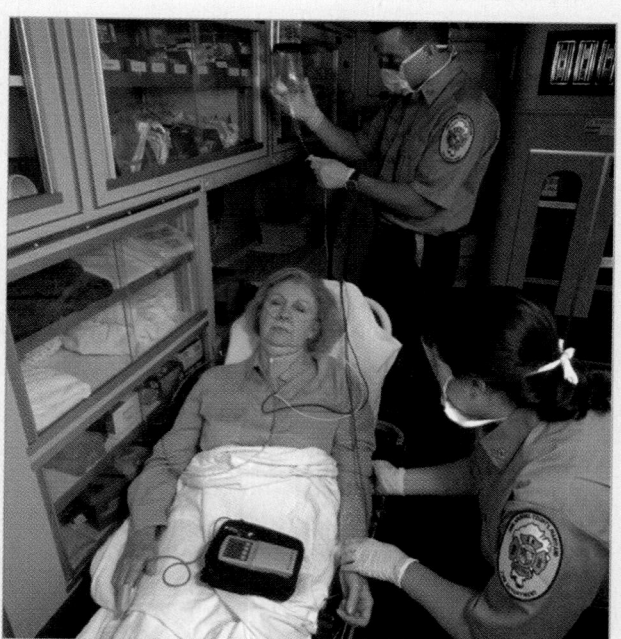

Figure 1-5 Advanced training covers a wide range of ALS skills.

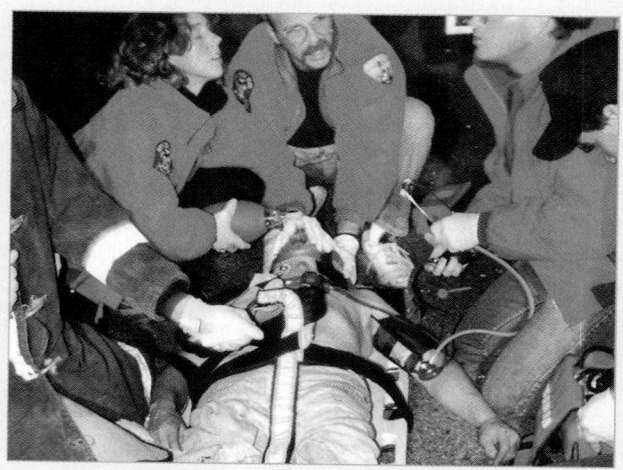

Figure 1-4 EMT-Bs are trained to use airway adjuncts.

- Manual defibrillation and external cardiac pacing
- Orotracheal and nasotracheal intubation
- Needle cricothyroidotomy
- Needle decompression for tension pneumothorax
- IV therapy
- Advanced pharmacology: drug calculations and medication administration

Components of the EMS System

Access

Easy access to help needed in an emergency is essential. In most of the country, an emergency communications center that dispatches fire, police, rescue, and EMS units can be reached by dialing 9-1-1. At the communication center, trained dispatchers obtain the necessary information from the caller and, following dispatch protocols, dispatch the ambulance crew and other equipment and responders that may be needed Figure 1-6 ▼.

In an enhanced 9-1-1 system, the address of the telephone from which the call is made is displayed on a screen. The connection is frozen until the dispatcher releases it so that if the caller is unable to speak, his or her location remains displayed. However, many cellular phones do not yet have this capability. Most emergency communications centers also include special

equipment so that individuals with speech or hearing disabilities can communicate with the dispatcher via a keyboard and printed messages. In some areas, rather than 9-1-1, a different special published emergency number may be used to call for EMS. Training the public in how to summon an EMS unit is an important part of the public education responsibility of each EMS service. Enhanced 9-1-1 systems for cellular phones are now becoming available that identify not only the cellular phone number from which an emergency call is being placed, but also the exact geographic coordinates of the phone at the time the call is made. Such systems use GPS (global positioning system) technology. Because cellular phones capable of transmitting a GPS signal and a system capable of receiving that signal are both required, the technology will require many years to implement.

A system called <u>Emergency Medical Dispatch (EMD)</u> has been developed to assist dispatchers in providing callers with vital instructions to help them deal with a medical emergency until the arrival of EMS crews. Dispatchers are provided with training and scripts to help them relay relevant instructions to the callers. The system also helps dispatchers select appropriately resourced units to respond to a request for assistance. It is the dispatcher's duty to relay all relevant and available information to the responding crews in a timely manner. Keep in mind, however, that current technology does not allow the dispatcher to "see" what is actually going on at the scene, and that it is not uncommon for you to find the reality of the call quite different from the dispatch information.

In most municipalities, EMS is a part of the fire department. In others, it is a part of the police department or is an independent public or private safety service. In some areas, a contractor may provide either BLS or ALS service. In some areas, ALS is provided by paramedics who are based at a hospital or who may cover a number of towns in a region.

New technologies are constantly being developed that can assist responders in locating their patients. As previously described, cellular telephones can be linked to GPS units to display their location. Rescue squads can transmit their position to dispatch and dispatch can transmit the location of a call to a moving digital map in the squad, complete with turn-by-turn directions. Medical databases can be queried and patient information directly downloaded to the EMT's computer, or uploaded from the EMT's laptop to the database. The pace of technological developments in communications makes the latest device soon obsolete, so constant

Figure 1-6 Trained dispatchers obtain information about the call and then send responders to the scene as needed.

training and education are required to keep the EMT's knowledge up to date.

Administration and Policy

Each EMS service operates in a designated primary service area (PSA) in which it is responsible for the provision of prehospital emergency care and the transportation of the sick and injured to the hospital.

EMS services are usually administered by a senior EMS official. Daily operations and overall direction of the service are provided by an appointed chief executive officer and several other officers who serve under him or her. When the EMS service is a part of a fire or police department, the department chief will usually delegate the responsibility for directing EMS to an assistant chief or other officer whose sole responsibility is to manage the EMS activities of the department. To provide clear guidelines, most services have written operating procedures and policies. When you join a service, you will be expected to learn and follow them.

The chief executive of the service is in charge of both the necessary administrative tasks (eg, scheduling, personnel, budgets, purchasing, vehicle maintenance) and the daily operations of the ambulances and crews. Except for medical matters, he or she operates as the chief (similar to a fire chief or police chief) of EMS for the service and the PSA that it covers.

Medical Direction and Control

Each EMS system has a physician medical director who authorizes the EMTs in the service to provide medical care in the field. The appropriate care for each injury, condition, or illness that you will encounter in the field is determined by the medical director and is described in a set of written standing orders and protocols. Protocols are described in a comprehensive guide delineating the EMT's scope of practice. Standing orders are part of protocols and designate what the EMT is required to do for a specific complaint or condition.

The medical director provides the ongoing working liaison between the medical community, hospitals, and the EMTs in the service. If treatment problems arise or different procedures should be considered, these are referred to the medical director for his or her decision and action. To ensure that the proper training standards are met, the medical director determines and approves the continuing education and training that are required of each EMT in the service and approves any that individuals obtain elsewhere.

Medical control is either off-line (indirect) or online (direct), as authorized by the medical director. Online medical control consists of direction given over the phone or radio directly from the medical director or designated physician. The medical direction can be transferred by the physician's designee; it does not have to be transferred by the physician himself or herself. Off-line medical control consists of standing orders, training, and supervision authorized by the medical director. Each EMT must know and follow the protocols developed by his or her medical director.

The service's protocols will also identify an EMS physician, usually at a local hospital, who can be reached by radio or telephone for medical control during a call. This is a type of direct online medical control. On some calls, once the squad has initiated any immediate urgent care and gives its radio report, the online medical control physician may either confirm or modify the proposed treatment plan or may prescribe any additional special orders that the EMT-Bs are to follow for that patient. The

You are the Provider   Part 3

With the teamwork accomplished between you, the police officer, the fire fighter, and the paramedic, a cardiac rhythm has been successfully established, the patient has been intubated, and an IV has been established to administer essential medications. As you are preparing the patient for transport, a family member arrives and appears to be confused and very upset by what she sees. She explains that her father called her complaining of chest pain, so she told him to rest while she called 9-1-1.

4. How can you help her and what should you do or say?

point at which the EMT-Bs should give their radio report or obtain online medical direction will vary.

Quality Control and Improvement

The medical director is responsible for maintaining quality control, ensuring that all staff members who are involved in caring for patients meet appropriate medical care standards on each call. To provide the necessary quality control, the medical director and other involved staff review patient care reports, audit administrative records, and survey patients.

Continuous quality improvement (CQI) is a circular system of continuous internal and external reviews and audits of all aspects of an EMS call. To provide CQI, periodic run review meetings are held in which all those who are involved in patient care review the run reports and then discuss any areas of care that appear to need change or improvement. Positive feedback is also discussed. If a problem appears to be repeated by a single EMT or crew, the medical director will discuss the details with the individuals involved. The CQI process is designed to identify areas of improvement and, if necessary, assign remedial training or develop some other educational activity. The medical director is also responsible for ensuring that appropriate continuing education and training are available.

Information and skills in emergency medical care change constantly. You need refresher training or continuing education as new modalities of care, equipment, and understanding of critical illnesses and trauma develop. Equally, when you have not used a particular procedure for some time, skill decay may occur. Therefore, your medical director might establish a training program to correct the deficit. For example, an emergency department physician noted that despite their assessments, many EMT-Bs were missing a high number of closed long bone fractures, resulting in poor prehospital care. A subsequent audit of calls led to a review and retraining session for assessment and care of fractures. This same process can apply to CPR or any other type of skill that you do not use often. Ensuring that your skills and knowledge are current is one of the ongoing commitments of being an EMT.

Other Physician Input

EMS is an extension of the emergency medical care provided in the emergency department by physicians and the other specialists who provide definitive care in the hospital. Besides the supervision that the medical director and direct online medical control physicians pro-

vide, your training and practices are based on input from many specialty professional associations at the national, state, and local levels.

As an EMT-B, you are part of the professional continuum of care provided to patients who often have life-threatening conditions. Physicians are at the top of the professional continuum pyramid. Many physician experts from the specialties of emergency medicine, traumatology, orthopaedics, cardiology, anesthesiology, radiology, and other medical disciplines participate in the ongoing work of EMS. The efforts of these groups, often through professional associations such as the American Academy of Orthopaedic Surgeons, the American College of Emergency Physicians, the American College of Surgeons, and the National Association of EMS Physicians, include research, the establishment of standards for quality assurance, continuing education, and publications.

Regulation

Although each EMS system, medical director, and training program has latitude, their training, protocols, and practices must conform with the EMS legislation, rules, regulations, and guidelines adopted by each state. The state EMS office is responsible for authorizing, auditing, and regulating all EMS services, training institutions, courses, instructors, and providers within the state. In most states, the state EMS office obtains input from an advisory committee made up of representatives of the services, service medical directors, medical associations, hospitals, training programs, instructors' associations, EMT associations, and the public in that state.

Equipment

As an EMT-B, you will use a wide range of different emergency equipment. During the EMT-B course, you will be introduced to, and learn how to use, a variety of the different appliances and devices that you may need on a call. You will also learn when the use of each is indicated and when it is contraindicated, meaning its use will not be of benefit or may cause harm. Although the use of different models and brands of a given device will follow the same generic principles and methods, some variation and peculiarities exist from one model to another. When you join a service, you should check each key piece of equipment before going on duty to ensure that it is in its assigned place, that it is working properly, and that you are familiar with the specific model carried on your ambulance.

The Ambulance

Each EMT-B may be called upon to drive the ambulance. Therefore, you must familiarize yourself with the roads in your PSA or sector. Before going on duty, you should check all the equipment and supplies and communication equipment that the ambulance carries and make sure that it is fully fueled, that it has sufficient oil and other key fluids, and that the tires are in good condition and properly inflated (Figure 1-7 ▶). You should also test each of the driver's controls and each built-in unit and control in the patient compartment. If you have not driven the specific ambulance before, it is a good idea to take it out and become familiar with it before you respond to a call. Maintenance and safe driving of the ambulance are discussed in detail in Chapter 35.

Transport to Specialty Centers

In addition to hospital emergency departments, many EMS systems include specialty centers that focus on specific types of care, such as trauma, burns, poison, or psychiatric conditions, or specific types of patients, such as children or the elderly. Certain specialty centers maintain in-house staffs of surgeons and other specialists. Other facilities must page operating teams, surgeons, or other specialists from outside the hospital. Typically, only a few hospitals in a region are designated as specialty centers. Transport time to a specialty center may be slightly longer than that to an emergency department, but patients will receive definitive care more quickly at a specialty center. You must know the location of the centers in your area and when, according to your protocol, you must transport the patient directly to one. Sometimes, air medical transport will be preferable. Many factors need to be considered when call-

Figure 1-7 Making sure the ambulance is fueled is part of an EMT-B's responsibility.

ing for air medical transport: weather, time, landing areas, air crew versus ground crew capabilities, among others. Local, regional, and state protocols, department procedures, as well as experience and knowledge of available resources will guide your decision in these instances.

Interfacility Transports

Many EMS services provide interfacility transportation for nonambulatory patients or patients with both acute and chronic medical conditions requiring medical monitoring. This may include transferring patients to and from hospitals, skilled nursing facilities, board and care homes, or even their home residence.

You are the Provider   Part 4

Without delaying patient transport, you briefly explain to the family member that upon your arrival, her father was not awake, was not breathing, and did not have a pulse. You explain that you have helped his heart to start beating again, are breathing for him, but that he is still unconscious. You explain that everything is being done to help him until he arrives at the hospital and physicians take over his care. She seems comforted by your kind words and professional demeanor and asks to travel with you to the hospital. You assist her to the front of the ambulance and help her with her seat belt. The volunteer fire fighter is trained to drive emergency vehicles and offers to drive you and your partner to the hospital. The paramedic requests your help in the patient compartment.

5. What are some considerations for successful patient management during transport?

During ambulance transportation, the health and well-being of the patient is the EMT's responsibility. The EMT should obtain the patient's medical history, chief complaint, and latest vital signs and provide an on-going patient assessment. In certain circumstances, depending on local protocols, a nurse, physician, respiratory therapist, or medical team will accompany the patient. This is especially true when the patient requires care that extends beyond the EMT-B's scope of practice.

Working With Hospital Staff

You should become familiar with the hospital by observing hospital equipment and how it is used, the functions of staff members, and the policies and procedures in all emergency areas of the hospital. You will also learn about advances in emergency care and how to interact with hospital personnel (Figure 1-8 ▶). This experience will help you to understand how your care influences the patient's recovery and will emphasize the importance and benefits of proper prehospital care. It will also show you the consequences of delay, inadequate care, or poor judgment.

Physicians, nurses, and other medical professionals are not likely to be in the field with you routinely to provide personal, on-the-spot instructions. However, you may consult with appropriate medical staff using the radio through established medical control procedures.

In the emergency department, hospital staff may train you by showing you assessment and treatment techniques on patients. A physician or nurse may serve as an instructor for medical subjects in your training program. Through these experiences, you will become more comfortable using medical terms, interpreting patient signs and symptoms, and developing patient management skills.

Hospital staff is usually willing to help you improve your skills and efficiency throughout your career. Some physicians and nurses may have completed the EMT curriculum as part of their formal medical training. The best patient care occurs when all emergency care providers have a close rapport. This allows you and hospital staff the opportunity to discuss mutual problems and to benefit from each other's experiences.

Working With Public Safety Agencies

Some public safety workers have EMS training. As an EMT-B, you must become familiar with all the roles and responsibilities of these agencies. Personnel from certain agencies are better prepared than you to perform

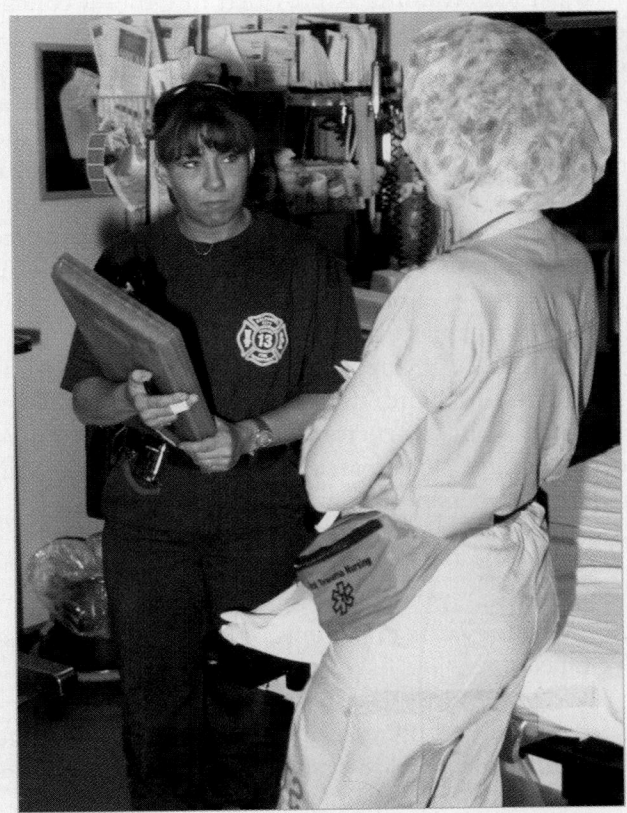

Figure 1-8 As an EMT-B, you will interact with hospital staff.

certain functions. For example, employees of a utility company are better equipped to control downed power lines than you or your partner. Law enforcement personnel are better able to handle violent scenes and traffic control, while you and your partner are better able to provide emergency medical care. Effective scene and patient management will result if you work together and recognize that each person has special talent and a job to do. Remember that the best, most efficient patient care is achieved through cooperation among agencies.

Training

Your training will be conducted by many knowledgeable EMS educators. In most states, the instructors who are responsible for coordinating and teaching the EMT-B course and continuing education courses are approved and certified by the state EMS office or agency. To be certified in some states, an instructor must have extensive medical and educational training and teach for a designated period while being observed and supervised by an experienced instructor.

Most ALS training is provided in either a college adult career center or hospital setting. In most states, educational programs that provide ALS training must be approved by the state and have their own medical director. In these courses, many of the lectures and small group sessions will be presented by the medical director or other physicians, nurses, and EMT instructors. In clinical sessions in which supervised practice is obtained in the emergency department or other in-hospital settings, students are also supervised directly by physicians and nurses.

The quality of care that you will provide depends on your ability and the quality of your training. Therefore, your instructor and the many others who developed and participated in your training program are key members of the emergency care team.

Providing a Coordinated Continuum of Care

The emergency care of patients occurs in four progressive phases:

1. **In the first phase,** the patient or bystanders recognize that there is an emergency and someone initiates the EMS system. Initial patient care is provided by dispatch until EMS arrives.
2. **The second phase** consists of patient assessment, initial prehospital care, proper packaging, and safe transport to the hospital.
3. **In the third phase,** the patient receives continued assessment and stabilization in the hospital emergency department.
4. **In the fourth phase,** the patient receives the necessary definitive specialized care.

These four phases must be provided in a coordinated continuum of care to maximize survival and reduce patient suffering and minimize lasting adverse effects. The EMS system is designed to produce such a coordinated effort among the local EMS services, emergency department staff, and the medical staff who provide definitive care.

Roles and Responsibilities of the EMT-B

As an EMT-B, you will be the first health care professional to assess and treat the patient; as such, you have certain roles and responsibilities (Table 1-2 ▶). Often, patient outcomes are determined by the care that you provide in the field and your identification of patients who need prompt transport.

Professional Attributes

As an EMT-B, whether you are paid or a volunteer, you are a health care professional. Part of your responsibility is to make sure that patient care is given a high priority without endangering your own safety or the safety of others. Another part of the responsibility to yourself, other EMTs, the patient, and other health care professionals is to maintain a professional appearance and manner at all times. Appearance, including uniforms, hair length, and tattoos, are usually regulated by the policies of your department (Figure 1-9 ▶). Your attitude and behavior must reflect that you are knowledgeable and sincerely dedicated to serving anyone who is injured or in an acute medical emergency. A professional appearance and manner help to build confidence and ease the patient's anxiety. You will be expected to perform under pressure with composure and self-confidence. Patients and families who are under stress need to be treated with understanding, respect, and compassion.

Most patients will treat you with respect and appreciation, but some will not. Some patients are uncooperative, demanding, unpleasant, ungrateful, and

You are the Provider

Part 5

Minutes later, you arrive at the hospital. You and your partner have continued stabilizing the patient by reassessing the airway to confirm proper tube placement, establishing continuous infusion of antiarrhythmic medications, and continued monitoring of the cardiac rhythm. One week later, you learn that the patient has made an impressive recovery and has been discharged from the hospital. He and his daughter later visit you and the other responders to express their thanks with homemade treats for everyone to enjoy.

TABLE 1-2 Roles and Responsibilities of the EMT-B

- Ensuring your own safety and the safety of your fellow EMT-Bs, the patient, and others at the scene
- Locating and safely driving to the scene
- Sizing up the scene and situation
- Rapidly assessing the patient's gross neurologic, respiratory, and circulatory status
- Providing any essential immediate intervention
- Performing a thorough, accurate patient assessment
- Obtaining an expanded SAMPLE history
- Reaching a clinical impression and providing prompt, efficient, prioritized patient care based on your assessment
- Communicating effectively with the patient and advising him or her of any procedures you will perform
- Properly interacting and communicating with fire, rescue, and law enforcement responders at the scene
- Identifying patients who require rapid packaging and initiating transport without delay
- Identifying patients who do not need emergency care and will benefit from further detailed assessment and care before they are moved and transported
- Properly packaging the patient
- Safely lifting the patient from his or her initial location into the ambulance for transport
- Providing safe, appropriate transport to the hospital emergency department or other designated facility
- Giving the necessary radio report to the medical control center or receiving hospital emergency department
- Providing any additional assessment or treatment while en route
- Monitoring the patient and checking vital signs while en route
- Documenting all findings and care on the patient care report
- Unloading the patient safely and, after giving a proper verbal report, transferring the patient's care to the emergency department staff
- Safeguarding the patient's rights

Figure 1-9 A. A professional appearance and manner help to build confidence and ease patient anxiety. **B.** An unprofessional appearance may promote distrust.

verbally abusive. You must be nonjudgmental and overcome your instincts to react poorly to such behavior. Remember that when individuals are hurt, ill, under stress, frightened, despondent, under the influence of alcohol or drugs, or feel threatened, they will often react with inappropriate behavior, even toward those who are trying to help and care for them. Every patient, regardless of his or her attitude, is entitled to compassion, respect, and the best care that you can provide.

Most individuals in this country can obtain proper routine medical care when they are ill and are surrounded by relatives and friends who will help to take care of them. However, when you are called to a home for a medical problem that is clearly not an emergency, remember that for some individuals, calling an ambulance and being transported to the emergency department is the only way to obtain medical care.

As a new EMT-B, you will be given a lot of advice and training from the more experienced EMT-Bs with whom you serve. Some may voice a callous disregard for some types of patients. You should not be influenced by the unprofessional attitude of these individuals, regardless of how experienced or skilled they appear.

As a health care professional and an extension of physician care, you are bound by patient confidentiality. You should not discuss your findings or any disclosures made by the patient with anyone but those who are treating the patient or, as required by law, the police or other social agencies. When discussing a call with others, you should be careful to avoid any information that might disclose the name or identity of patients you have treated. Be careful not to gossip about calls and patients with others, even in your own home. Recently, the protection of patient privacy has drawn national attention with the passage of the Health Insurance Portability and Accountability Act (HIPAA). You should become familiar with the requirements of this legislation, especially as it applies to your particular practice situation.

Continuing Education

Once you no longer have the structured learning environment that is provided in your initial training course, you must assume the responsibility for directing your own study and learning. As an EMT-B, you will be required to attend a certain number of hours of continuing education approved for EMT-Bs each year to maintain, update, and expand your knowledge and skills. In many services, the required hours are provided by the training officer and medical director. In addition, most EMS education programs and hospitals offer a number of regular continuing education opportunities in each region. You may also attend state and national EMS conferences to help keep you up-to-date about local, state, and national issues affecting EMS. Since there are many levels of certification, you should ensure that the continuing education you receive is approved for the EMT-B. Whether you take advantage of these opportunities depends on you. Whether you decide to remain an EMT-B or achieve a higher level of training and certification, the key to being a good EMT and providing high-quality care is your commitment to continual learning and ever-increasing knowledge and skills.

EMTs possess special knowledge and skills that are directed to the care of patients in emergency situations. The authority that is delegated to you to care for patients is a very special one. Maintaining your knowledge and skills is a substantial responsibility. Knowledge and skills that are learned in any profession decay and weaken when they are not used on a continual basis. Consider CPR. If you have not used these skills since your original training, it is likely that you will perform CPR in a way that is less than desirable. Continuing education and refresher courses are one way by which you can maintain your skills and knowledge.

You are the Provider   Summary

An emergency can be defined as an event or situation that requires immediate intervention to minimize or prevent serious damage or death. As an EMT-B, your primary roles involve providing basic life support measures, maintaining a state of response readiness, and working as a team member.

This scenario demonstrates the benefits of a tiered-response system and the importance of each link in the chain of survival. Your overall responsibilities as an EMT-B will be to perform the skills within your scope of practice and to assist your ALS partner with advanced measures in accordance with your local protocols.

The patient's family member needs your compassion and support. You must clearly communicate what has happened, avoiding euphemisms or technical language. Achieving a balance between meeting the patient's needs and a family member's needs can be difficult when confronted with all of your responsibilities during a cardiac arrest or similarly urgent situation. Make sure that they receive the support they need, whether that help comes from you, other family members, friends, counselors, or clergy.

Teamwork is essential in saving lives. Each person and agency involved in response to illness or injury contributes to patient care and the overall outcome of the event—the patient or family member who recognizes the emergency, the dispatcher who answers the call and gathers the necessary information and directs you to the appropriate medical resources, the police officers and firefighters who frequently arrive first on the scene, and the EMTs who further stabilize and transport the patient to the hospital. The speed and proficiency of the interventions provided by these persons, in most cases, has a direct impact on patient outcome.

Prep Kit

Ready for Review

- EMS is the system that provides the emergency medical care that is needed by people who have been injured or have an acute medical emergency. When the dispatcher at the 9-1-1 emergency communications center receives a call for emergency care, he or she dispatches to the scene the designated EMS ambulance squad and any fire, rescue, or police units that may be needed.

- The EMS ambulance is staffed by EMTs who have been trained to the EMT-Basic, EMT-Intermediate, or EMT-Paramedic level according to recommended national standards and have been certified/licensed by the state.

- After the EMTs size up the scene and assess the patient, they provide the emergency care that is indicated based on their findings and ordered by their medical director in the service's standing orders and protocols or by the physician who is providing online medical direction.

- The EMTs then "package" the patient and provide transport to the nearby hospital or designated specialized care facility (eg, trauma center, pediatric hospital) for further evaluation and stabilization in the emergency department and, after admission, definitive surgical or medical care.

- The EMT-B course that you are now taking will present the information and skills that you will need to pass the required examinations for certification and start functioning as an EMT-B in the field. This course will provide you with the training that you need to function as an EMT-B and will serve as the essential foundation upon which you can advance your training and expertise.

- Essential keys to being a good EMT-B include:
  - Compassion and motivation to reduce suffering, pain, and mortality in those who are injured or acutely ill
  - Desire to provide each patient with the best possible care
  - Commitment to obtain the knowledge and skills that this requires
  - The drive to continually increase your knowledge, skills, and ability

- Once you have successfully completed this course and have been certified as an EMT-B, you will enter the next key phase of your training. When you join an EMS service, your first task will be to learn the medical protocols and operating procedures of the squad. You will also have to learn where each piece of equipment is kept on the ambulance and become familiar with how the equipment works.

- From your experience and the guidance provided by your crew chief and the other experienced EMT-Bs you work with, you will gain increased mastery of the skills that you learned in the course and learn how to apply your knowledge and skills in the diverse situations that are encountered in the field.

- Once you have completed the course, you must assume responsibility for directing your own study through continuing education provided by your service's training officer and medical director or through other opportunities available to you. Your commitment to continuing learning is the key to being a good EMT.

Vital Vocabulary

advanced life support (ALS) Advanced lifesaving procedures, some of which are now being provided by the EMT-B.

Americans With Disabilities Act (ADA) Comprehensive legislation that is designed to protect individuals with disabilities against discrimination.

automated external defibrillator (AED) A device that detects treatable life-threatening cardiac arrhythmias (ventricular fibrillation and ventricular tachycardia) and delivers the appropriate electrical shock to the patient.

Technology

Interactivities

Vocabulary Explorer

Anatomy Review

Web Links

Online Review Manual

www.EMTB.com

Prep Kit continued...

continuous quality improvement (CQI)   A system of internal and external reviews and audits of all aspects of an EMS system.

Emergency Medical Dispatch (EMD)   A system that assists dispatchers in selecting appropriate units to respond to a particular call for assistance and in providing callers with vital instructions until the arrival of EMS crews.

emergency medical services (EMS)   A multidisciplinary system that represents the combined efforts of several professionals and agencies to provide prehospital emergency care to the sick and injured.

emergency medical technician (EMT)   A medical professional who is trained and certified/licensed by his or her state to provide emergency life support prior to or with more advanced medical providers.

EMT-Basic (EMT-B)   An EMT who has training in basic life support, including automated external defibrillation, use of a definitive airway adjunct, and assisting patients with certain medications.

EMT-Intermediate (EMT-I)   An EMT who has training in specific aspects of advanced life support, such as IV (intravenous) therapy, interpretation of cardiac rhythms and defibrillation, and orotracheal intubation.

EMT-Paramedic (EMT-P)   An EMT who has extensive training in advanced life support, including IV (intravenous) therapy, pharmacology, cardiac monitoring, and other advanced assessment and treatment skills.

first responder   The first trained individual, such as a police officer, fire fighter, lifeguard, or other rescuer, to arrive at the scene of an emergency to provide initial medical assistance.

Health Insurance Portability and Accountability Act (HIPAA)   Federal legislation passed in 1996. Its main effect in EMS is in limiting availability of patients' health care information and penalizing violations of patient privacy.

intravenous (IV) therapy   The delivery of medication directly into a vein.

medical control   Physician instructions that are given directly by radio or cell phone (online/direct) or indirectly by protocol/guidelines (off-line/indirect), as authorized by the medical director of the service program.

medical director   The physician who authorizes or delegates to the EMT the authority to provide medical care in the field.

primary service area (PSA)   The designated area in which the EMS service is responsible for the provision of prehospital emergency care and transportation to the hospital.

quality control   The responsibility of the medical director to ensure that the appropriate medical care standards are met by EMT-Bs on each call.

Points to Ponder

You have been working as an EMT-B in the same area for many years and you know all of the frequent callers. You are dispatched to a man down at 3 o'clock in the morning. You have been on several calls already that night and you just got back into bed. Upon your arrival on scene you recognize the patient immediately as a local drunk. He smells of alcohol and is lying in the front yard of someone's home. Your partner is upset because he transported this man to the hospital two days ago. Your partner calls for police response and starts shaking and yelling at the man to wake up or he's going to have the police arrest him for abusing the 9-1-1 system.

How would you handle this patient? Would you release him to the custody of the police? How do you feel about your partner's behavior towards the patient and how would you handle it?

Issues: Personal Attitudes and Conduct, Professional Interaction, Recognizing Potential Medical Emergencies.

Assessment in Action

You are a newly certified EMT-B. Today is your first day on the job at a busy ambulance station. You run several calls and are faced with many new situations. Answer the following questions and discuss them with your instructor.

1. Which of the following is NOT considered an EMT-B skill?

   A. Automated external defibrillation (AED)
   B. Intravenous therapy
   C. Combitube insertion
   D. Assisting patients with the use of their prescribed nitroglycerin

2. You are trained using the Combitube. However, your local protocols state that only a paramedic may insert the device. You are on the scene of a cardiac arrest and the patient needs an airway fast, but you are waiting for a paramedic to arrive in another ambulance. You are anxious to insert the Combitube. What should you do?

   A. Call medical control and ask what you should do.
   B. Wait for the paramedic to arrive, since it is above your scope of practice.
   C. Only do it if you know no one will find out.
   D. Insert the Combitube to practice your skills.

3. Who is medically responsible for establishing written standing orders and protocols?

   A. Department of Transportation
   B. Medical Director
   C. EMS director
   D. Emergency Department physicians

4. Which of the following is the most important skill to an EMT-B?

   A. Knowledge of protocols
   B. Knowledge of location and capabilities of local hospitals
   C. Ability to work with other health care providers
   D. All of the above

5. Which of the following statements is true?

   A. EMT-B training teaches providers everything they need to know.
   B. The main part of being an EMT-B is performing skills; communication should be left to other providers, such as emergency room staff.
   C. Continuing education is vital to successful EMS systems as medicine and technology is ever changing.
   D. An ambulance is only meant to transport people who are having a "real" emergency.

Challenging Questions

6. Why is it important to have continuous quality improvement in an EMS system?

7. What does appropriate patient care mean to an EMT-B?

The Well-Being of the EMT-B

Objectives

Cognitive

1-2.1 List possible emotional reactions that the EMT-Basic may experience when faced with trauma, illness, death, and dying. (p 26)

1-2.2 Discuss the possible reactions that a family member may exhibit when confronted with death and dying. (p 27)

1-2.3 State the steps in the EMT-Basic's approach to the family confronted with death and dying. (p 27)

1-2.4 State the possible reactions that the family of the EMT-Basic may exhibit due to their outside involvement in EMS. (p 37)

1-2.5 Recognize the signs and symptoms of critical incident stress. (p 33)

1-2.6 State possible steps that the EMT-Basic may take to help reduce/alleviate stress. (p 34)

1-2.7 Explain the need to determine scene safety. (p 64)

1-2.8 Discuss the importance of body substance isolation (BSI). (p 44)

1-2.9 Describe the steps the EMT-Basic should take for personal protection from airborne and bloodborne pathogens. (p 44)

1-2.10 List the personal protective equipment necessary for each of the following situations:
- Hazardous materials
- Rescue operations
- Violent scenes
- Crime scenes
- Exposure to bloodborne pathogens
- Exposure to airborne pathogens. (p 48)

Affective

1-2.11 Explain the rationale for serving as an advocate for the use of appropriate protective equipment. (p 48)

Psychomotor

1-2.12 Given a scenario with potential infectious exposure, the EMT-Basic will use appropriate personal protective equipment. At the completion of the scenario, the EMT-Basic will properly remove and discard the protective garments. (p 48)

1-2.13 Given the above scenario, the EMT-Basic will complete disinfection/cleaning and all reporting documentation. (p 55)

Additional Objectives*

Cognitive

1. Describe the various ways by which communicable diseases can be transmitted from one person to another. (p 42)

2. Define the term "universal precautions" and describe when it is appropriate to use such measures. (p 44)

3. Identify appropriate task-specific personal protective equipment. (p 43)

4. Identify possible occupational diseases and methods of risk assessment. (p 51)

5. Identify the role of a testing and immunization program in protecting the EMT-B from communicable diseases. (p 48)

6. Identify the benefits of an exposure control plan. (p 48)

7. Identify how the following diseases are transmitted and discuss the steps to take to prevent and/or deal with an exposure to each: hepatitis, meningitis, tuberculosis, HIV/AIDS. (p 51)

8. List the mechanisms of disease transmission. (p 43)

9. List the components of postexposure management and reporting. (p 55)

10. Discuss the importance of obtaining a patient's history and assessment findings to identify possible communicable diseases. (p 55)

Affective

11. Explain the duty to care for patients with communicable diseases. (p 50)

Psychomotor

None

*These are noncurriculum objectives.

You are the Provider

You have been working for about 6 months when you encounter a mass-casualty incident (MCI) while on shift. The incident involved a bus full of school-aged children, some of whom were severely injured and later died. Several weeks later, one of your coworkers expresses his concern for you, as you don't quite seem to be yourself lately. After this conversation, you evaluate your changes in sleep patterns, lack of appetite, and fatigue. You recognize stress in yourself and think it could be related to the events that happened weeks earlier. Although you went through critical incident stress debriefing (CISD) immediately following the call, you think maybe the emergency is affecting you more than you have realized.

1. What should you do to help yourself?
2. What are other signs and symptoms of ineffectively handled stress?
3. What are some ways to focus the negative aspects of stress?

The Well-Being of the EMT-B

There is an ancient proverb, "Physician, heal thyself." As providers of health care, doctors need to look after themselves—in all respects—so that they can minister to others. An ill physician is in no position to render care as he or she was trained to do. That dictum applies to all health care providers and goes well beyond just physical issues. In caring for the critically ill and injured, there are many factors and situations that can interfere with the EMT-B's ability to treat the patient.

The personal health, safety, and well-being of all EMT-Bs are vital to an EMS operation. As a part of your training, you will learn how to recognize possible hazards and protect yourself from them. These hazards vary greatly, ranging from personal neglect to environmental and human-made threats to your health and safety. You will also learn about the mental and physical stress that you must cope with as a result of caring for the sick and injured. Death and dying challenge you to deal with the realities of human weaknesses and the emotions of the survivors.

The emotional well-being of the EMT and of the patient are intertwined, especially in high-stress rescues. This chapter covers caring for the patient's well-being and caring for your own.

It is important to remain calm to perform effectively when you are confronted with horrifying events, life-threatening illness, or injury. A special kind of self-control is needed to respond efficiently and effectively to the suffering of others. This self-control is developed through the following:

- Proper training
- Ongoing experience in dealing with all types of physical and mental distress
- Developing healthy strategies to cope with the stresses encountered on the job
- A dedication to serve humanity

Emotional Aspects of Emergency Care

At times, even the most experienced health care providers have difficulty overcoming personal reactions and proceeding without hesitation. Patients need to be removed from life-threatening situations. Life support measures need to be given to patients who are severely injured. You may also be called upon to recover human remains from highway accidents, aircraft disasters, or explosions (Figure 2-1 ▼). In all of these situations, you must be calm and act responsibly as a member of the emergency medical care team. You must also realize that even though your personal emotions must be kept under control, these are normal feelings. Every EMT-B who must deal with such situations has these feelings. The struggle to remain calm in the face of horrible circumstances contributes to the emotional stress of the job.

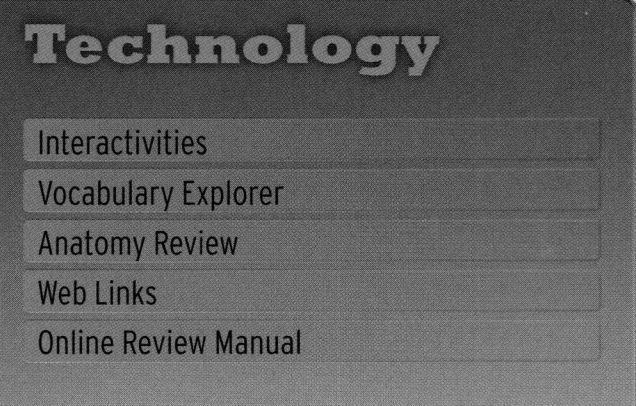

Technology

Interactivities

Vocabulary Explorer

Anatomy Review

Web Links

Online Review Manual

www.EMTB.com

Figure 2-1 As an EMT-B, you will have to deal with the removal of bodies.

Death and Dying

Today, life expectancy has dramatically increased; nearly two-thirds of all deaths occur among those aged 65 years and older. Sixty percent of all deaths today are attributed to heart disease. From the age of 1 year to the age of 34 years, trauma is the leading cause of death. Death today is likely to occur either quite suddenly or after a prolonged terminal illness. The environment of death has changed since our nation's earlier days; it occurs in the home setting less frequently. The setting of death is somewhere else—in the hospital, a hospice, or a convalescent home, at the workplace, or on the highway. For this reason, we are less familiar with death than our ancestors were. We tend to deny death in America. Illness can be much more drawn-out and much more removed from daily life. Life support systems and impersonal care remove the whole experience of death from most people's awareness. The mobility of families also makes it less likely that there will be extended family support when death does occur.

Death in earlier American times was both an expected and accepted part of life. Life expectancy was brief (compared to today's), mortality rates (the ratio of number of deaths to a given population size) were high, and childbirth was hazardous, often resulting in the death of both the mother and the baby (Figure 2-2 ▶). Hardships of the times, both natural and human-made, were great. Children and adults died of disease, injuries, and the traumas of war. Most people had experienced the death of someone close to them. There were no funeral homes; mourning occurred at home in the family setting. The presence of a dead body was a natural event.

You may have significant painful personal experience with death. No matter what the frequency of re-

Figure 2-2 Infant mortality was prevalent in the United States up to the last century. In many countries, it still is.

sponse to emergency calls, death is something that every EMT-B will sometimes face. For some of you, it may be infrequent. Others, in urban settings, may see death many times in responding to motor vehicle crashes, drug overdoses, suicides, or homicides. Some EMTs may have to deal with the mass-casualty incident of an

You are the Provider   Part 2

You decide that further discussion of the MCI is warranted. Your agency offers an employee assistance program free of charge for its employees. This discreet service provides professional counseling for emergency responders. You also decide to seek out a professional fitness trainer to begin a weightlifting and aerobic training program.

4. What are other proven methods for effectively handling stress?
5. How can poorly managed stress affect your job performance and overall career longevity?

airplane crash or a hazardous materials accident. In all these cases, coming to grips with your thoughts, understandings, and adjustment to death is not only important personally, but also a function of delivering emergency medical care.

The Grieving Process

Everyone working as an EMT-B will experience grief. This section discusses how to handle patient grief, as well as how to cope with your own grief that may result from a difficult call.

The death of a human being is one of the most difficult events for another human being to accept. If the survivor is a relative or close friend of the deceased, it is even more difficult. Emotional responses to the loss of a loved one or friend are appropriate and should be expected. In fact, it is expected that you will feel emotional about the death of a patient. Feelings and emotions are part of the grieving process. All of us experience these feelings after a stressful situation that causes us personal pain.

In 1969, Dr Elisabeth Kubler-Ross published research revealing that people go through several stages of grieving. The stages of grieving are as follows:

1. **Denial.** Refusal to accept diagnosis or care, unrealistic demands for miracles, or persistent failure to understand why there is no improvement.
2. **Anger, hostility.** Projection of bad news onto the environment and commonly in all directions, at times almost at random. The person lashes out. Someone must be blamed, and those who are responsible must be punished. This is usually an ugly phase.
3. **Bargaining.** An attempt to secure a prize for good behavior or promise to change lifestyle. "I promise to be a 'perfect patient' if only I can live until 'x' event."
4. **Depression.** Open expression of grief, internalized anger, hopelessness, the desire to die. It rarely involves suicidal threats, complete withdrawal,

or giving up long before the illness seems terminal. The patient is usually silent.

5. **Acceptance.** The simple "yes." Acceptance grows out of a person's conviction that all has been done and the person is ready to die.

Stages may follow one another, occur simultaneously, or a person may jump back and forth between stages. The stages may last different amounts of time.

Even though the event (death) has not yet happened, the patient knows that it will happen. The patient has no control over this process. The patient will die whether or not he or she is ready to die. Furthermore, being ready to die does not mean that the patient will be happy about dying. You may encounter situations in which the patient is close to death, and you may need to provide reassurance and emotional care.

What Can the EMT-B Do?

As patients and bystanders are grieving, you can do helpful things, and make simple suggestions. Ask whether there is anything that you can do that will be of help, such as calling a relative or religious advisor. Provide gentle and caring support. Reinforcing the reality of the situation is important. This can be accomplished by merely saying to a grieving person, "I am so sorry for your loss." It is not important that you have a well-rehearsed script, for it is not likely that your exact words or consolations will be remembered. Being honest and sincere are important.

Some statements tend to be trite, and some suggest a kind of silver lining behind the clouds. Although they may be intended to make the person feel better about a situation, they also can be viewed as an attempt to diminish the person's grief. The grieving person needs to be validated. Statements like these can also indicate our inability to comprehend the profound sadness of grief because we have not experienced that kind of loss. If you have not experienced a death, it is okay to say so; do not pretend that you have.

You are the Provider   Part 3

You have been seeing the counselor for about a month and engaging in a moderate, regular exercise program. You have also slowly changed some eating habits, and are feeling much more energized and seem to have returned to your normal self. Your counselor has given you some tips on how to recognize stress early and how to use other forms of stress-reducing techniques in your everyday life.

6. What are other stress reducers besides exercise and healthy eating habits?

Attempts to take grief away too quickly are not good. If you do not know how the person really feels, you should not say so. People may be offended by responses that give advice or explanations about the death (Table 2-1 ▼). Statements such as "Oh, you shouldn't feel that way" are judgmental. If you judge what the grieving person is feeling, it is likely that he or she will stop talking with you. There is no reason why grieving people should not feel what they are feeling. Remember anger is a stage of grieving. The anger may be directed at you. The anger seems irrational to everyone but the person grieving. A professional attitude is a necessity.

Statements and comments that suggest action on your part are generally helpful. These statements imply a sense of understanding; they focus on the grieving person's feelings. It is not necessary to go into an extensive discussion. All you need to do is be sincere and say, "I am so sorry. I just want you to know that I am thinking about you." What people really appreciate is somebody who will listen to them. Simply ask, "Would you like to talk about how or what you are now feeling?" Then accept the response.

Dealing With the Patient and Family Members

There is no right or wrong way to grieve. Each person will experience grief and respond to it in his or her own way. Family members may express rage, anger, and despair. Many people will be rational and cooperative. Their concerns will usually be relieved by your calm, efficient manner. Your actions and words, even a simple touch, can communicate caring. While you must treat all patients with respect and dignity, use special care with dying patients and their families. Be concerned about their privacy and their wishes, and let them know that you take their concerns seriously. However, it is best to be honest with patients and their families; do not give them false hope.

Initial Care of the Dying, Critically Ill, or Injured Patient

Individuals who are in the process of dying as a result of trauma, an acute medical emergency, or a terminal disease will feel threatened. That threat may be related to their concern about survival. These concerns may involve feelings of helplessness, disability, pain, and separation (Table 2-2 ▼).

Anxiety

Anxiety is a response to the anticipation of danger. The source of the anxiety is often unknown; but in the case of seriously injured or ill patients, the source is usually recognizable. What may increase the anxiety are the unknowns of the current situation. Patients may ask the following:

- What will happen to me?
- What are you doing?
- Will I make it?
- What will my disabilities be?

Patients who are anxious may have the following signs and symptoms:

- Emotional upset
- Sweaty and cool skin (diaphoretic)
- Rapid breathing (hyperventilating)
- Fast pulse (tachycardic)

TABLE 2-1 Responding to Grief

Don't Say...

Give it time. Things will get better.
You should not question God's will.
You have to get on with your life.
You have to keep on going.
You can always have another child.
You're not the only one who suffers.
The living must go on.
I know how you feel.

Try Instead...

I'm sorry.
It is okay to be angry.
It must be hard to accept.
That must be painful for you.
Tell me how you are feeling.
If you want to cry, it's okay.
People really cared for...

TABLE 2-2 Concerns of the Dying, Critically Ill, or Injured Patient

- Anxiety
- Pain and fear
- Anger and hostility
- Depression
- Dependency
- Guilt
- Mental health problems
- Receiving unrelated bad news

- Restlessness
- Tension
- Fear
- Shakiness (tremulous)

For the anxious patient, time seems to be extended; seconds seem like minutes, and minutes seem like hours. Anxiety is never helpful to a patient, and can cause real physiological harm. It is your job to do everything you can to reduce your patient's anxiety and help your patient cope with what may be the most terrifying experience in his or her lifetime.

Pain and Fear

Pain and fear are very closely interrelated. Pain often is associated with illness or trauma. Fear is generally thought of in relation to the oncoming pain and the outcome of the damage. It is often helpful to encourage patients to express their pains and fears, because expression of them begins the process of adjustment to the pain and acceptance of the emergency medical care that may be necessary. Some individuals have difficulty in openly admitting their fear. The fear may be expressed as bad dreams, withdrawal, tension, restlessness, "butterflies" in the stomach, or nervousness. In some cases, it may be expressed as anger.

Often we are tempted to make light of a patient's pain and fear. It is easier to say to the stroke patient, "Oh, you'll be OK," than, "I'm sure you are really scared now not being able to talk, but you should know that we are doing everything we can to help you." Making a connection with your patient through eye contact and the squeeze of a hand can often do more to allay fear than the most eloquent words.

Anger and Hostility

Anger may be expressed by very demanding and complaining behavior. Often, this may be related to the fear and anxiety of the emergency medical care that is being given. Sometimes, the fear is so acute that the patient may want to express anger toward you or others but is unable to do so because of the dependency factor. If you find that you are the target of the patient's anger, make sure that you are safe; do not take the anger or insults personally. Be tolerant, and do not become defensive.

The anger may also be expressed physically, and you may be the target of the displaced aggression. If the patient or a relative becomes so emotionally upset that you are physically assaulted or you believe that this could happen, back out of the situation. Such hostility must be contained. If emergency medical care is not possible under these circumstances, law enforcement intervention is required.

Depression

Depression is a natural physiological and psychological response to illness, especially if the illness is prolonged, debilitating, or terminal. Whether the depression is a temporary sadness or clinical depression that is long-term, there is, of course, little the EMT can do to alleviate the pain of depression during the brief time the patient is being treated and transported. The best one can do in treating and transporting a patient experiencing depression is to be compassionate, supportive, and nonjudgmental.

Dependency

Dependency usually takes longer to develop than during the very brief relationships developed in EMS. When medical care is given to any individual, a sense of dependency may develop. Individuals who are placed in this position may feel helpless and become resentful. The resentfulness may arouse feelings of inferiority, shame, or weakness.

Guilt

Many patients who are dying, their families, or the caregivers of those patients may feel guilty over what has happened to them. Occasionally family members and/or long-term caregivers may feel a degree of relief when an extended illness is finally over. That relief may later turn into guilt. Most of the time, however, no one can explain these feelings. The magnitude of the guilt may be very great. Sometimes, feelings of guilt can result in a delay in seeking emergency medical care. Again, understanding the complex emotions that often come to a head during times of emergency and stress may help you cope with some of the intense and often seemingly bizarre behavior you will encounter in this profession.

Mental Health Problems

As an EMT, you will be called on to treat and transport patients with mental health problems. These problems may be the cause of the patient's distress or may be caused by the stresses of physical illness or injury. Mental health problems such as disorientation, confusion, or delusions may develop in the dying patient. In these instances, the patient may display behavior inconsistent with normal patterns of thinking, feeling, or acting. Common characteristics of such behavior may include the following:

- Loss of contact with reality

- Distortion of perception—patients may have difficulty judging such common factors as time, distance, and relationships.
- Regression—patients may regress to an earlier stage in their development, often infancy or childhood.
- Diminished control of basic impulses and desires—patients may act out on their urges without being able to exercise the normal judgment expected as adults. For example, patients may become violent or inappropriately affectionate.
- Abnormal mental content, including delusions and hallucinations

The normal course of dying can cause a patient to seem disoriented. In some long-term situations, generalized personality deterioration may occur.

Receiving Unrelated Bad News

A patient who is in critical condition or is dying may not want to hear of unrelated bad news, such as the death of someone close to him or her. Such news may depress the patient or cause the patient to give up hope.

Caring for Critically Ill and Injured Patients

Patients need to know who you are and what you are doing. Let the patient know that you are attending to his or her immediate needs and that these are your primary concerns at this moment (Figure 2-3 ▶). As soon as possible, explain to the patient what is going on. Confusion, anxiety, and other feelings of helplessness will be decreased if you keep the patient informed from the start.

Avoid Sad and Grim Comments

EMT-Bs, other safety personnel, family, and bystanders must avoid grim comments about a patient's condition. Remarks such as "This is a bad one" or "The leg is badly damaged, and I think he will lose it" are inappropriate. These remarks may upset or increase anxiety in the patient and compromise possible recovery outcomes. This is especially true for the patient who may be able to hear but not to respond.

Orient the Patient

You should expect a patient to be disoriented in an emergency situation. The aura of the emergency situation—lights, sirens, smells, and strangers—is intense. The impact and effect of injuries or acute illness

Figure 2-3 Let the patient know immediately that you are there to help.

may cause the patient to be confused or unsettled. It is important to orient the patient to his or her surroundings (Figure 2-4 ▼). Use brief, concise statements such as "Mr. Smith, you have had an accident, and I am now splinting your arm. I am John Foxworth of the New Britain EMS; I will be caring for you."

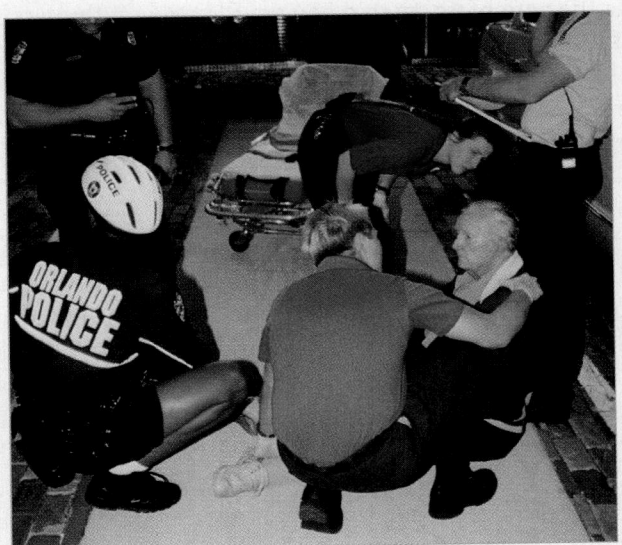

Figure 2-4 The aura of an emergency situation can be confusing and frightening to the patient. Make sure you explain to the patient what has happened.

Be Honest

In approaching any patient, you must decide how much each patient is able to understand and accept. You should be honest without additionally shocking the patient or giving information that is unnecessary or that may not be understood. Simply explain what you are doing, and allow the patient to be part of the care being given; this can relieve feelings of helplessness as well as some of the fear.

Initial Refusal of Care

There may be occasions when a patient may refuse emergency medical care and insist that you do nothing or leave him or her alone. In these cases, it is important to impress on the patient the seriousness of the condition without causing undue alarm. Saying, "Everything will be okay," when it is obvious that it is not, is not being truthful. Generally, seriously ill or injured patients know that they are in trouble.

Allow for Hope

In trauma and acute medical conditions, patients may ask whether they are going to die. You may feel at a loss for words. You may also know, on the basis of past experience or in view of the seriousness of the present situation, that the prognosis is poor. But it is not up to you to tell the patient that he or she is dying. Statements such as "I don't know if you are going to die; let's fight this one out together" or "I am not going to give up on you, so do not give up on yourself" are helpful. These statements transmit a sense of trust and hope, and they let the patient know that you are doing everything possible to save his or her life. If there is the slightest chance of hope remaining, you want that message transmitted in your attitude and in the statements you make to the patient.

Locate and Notify Family Members

Many patients will be concerned and ask you to notify their family or others close to them. The patient may or may not be able to assist you in doing this. You should see to it that an appropriate and responsible person makes an effort to locate the desired persons. Assuring the patient that someone is going to do this may be a significant part of the patient's care.

Injured and Critically Ill Children

Injured and critically ill children who have life-threatening conditions should be cared for as any patient would be, insofar as an assessment of airway, breathing, and circulation (ABC) and immediate life threats are concerned. Due regard should be given to variations in height, weight, and size in providing emergency medical care. Because of the increased excitement and extraordinary nature of the emergency scene for a child, it is important that a relative or responsible adult accompany the child to relieve anxiety and assist in care as appropriate.

Dealing With the Death of a Child

The death of a child is a tragic and dreaded event. It is not unusual to think about the fact that the dead or dying child has a lot more to do and should have many more years to live. In our society, we assume that only old people are supposed to die. Children die less frequently now than they did in earlier times, so most people are unprepared for what they will feel when a child dies. You may think about your own children and those whom you know: nephews, nieces, grandchildren, and children of close friends. And you may think, "Why should this child, who is only 5 years old, die?"

Answering the difficult questions of your own mortality will be of help when dealing with the death of a child. But, still, the death of a child will not be an easy subject to talk about. This will be especially so for the family. And as an EMT-B involved in a call that involves the death of a child, you will also likely experience stress.

One of your responsibilities may be to help the family through the initial period after the death. As an EMT-B, until more definitive and professional help can be available, you may be in the best position to help the family begin to cope with its loss. How a family initially deals with the death of a child will affect its stability and endurance. You can help a family through its initial period of grief and alert the family to the follow-up counseling and support services that are available.

Helping the Family

If the child is dead, acknowledging the fact of the death is important. This should be done in a private place, even if that is inside an ambulance. Often, the parents cannot believe that the death is real, even if they have been preparing for it, as in the case of a terminal illness such as leukemia. Reactions vary, but shock, disbelief, and denial are common. Some parents show little emotion at the initial news.

If it is possible and appropriate, find a place where the mother and father can hold the child. This is important in the parents' grieving process; it helps to lessen the sense of disbelief and makes the death real. Even if the parents do not ask to see the child, you should tell them that they may. Your decision in permitting the parents to see the child may need some discretion. For example, in the case of a traumatic death in which there is significant disfigurement, that decision might have to be delayed. The delay may involve having support services available or contacting the family physician or others who can help the parents through this difficult situation. This may involve preparing the parents for what they will see and the changes brought on by rigor mortis or asphyxiation, for example.

Sometimes, you do not need to say much. In fact, silence can sometimes be more comforting than words. You can express your own sorrow. Do not overload grieving parents with a lot of information; at this point, they cannot handle it. Nonverbal communication, such as holding a hand or touching a shoulder, may also be valuable. Let the family's actions be your guide about what is appropriate. It is important that parents be encouraged to talk about their feelings.

Stressful Situations

Many situations, such as mass-casualty scenes, serious automobile crashes, excavation cave-ins, house fires, infant and child trauma, amputations, abuse of an infant/child/spouse/older person, and death of a coworker or other public safety personnel, will be stressful for everyone involved. During these situations, you must exercise extreme care in both your words and your actions. Be careful to present a professional demeanor in words and actions at the scene. Words that do not seem important, or that are said jokingly, may hurt someone. Conversations at the scene must be professional. You should not say, "Everything will be all right," or "There is nothing to worry about." A person, who is trapped in a wrecked car, hurting from head to foot and worrying about a loved one, knows that all is not well. What will reassure the patient is your calm and caring approach to the emergency situation. Whether you are a brand new EMT or a seasoned veteran, patients expect you to bring some sense of order and stability to the terrifying chaos that has suddenly engulfed them. Briefly explain your plan of action to assist the patient in the crisis. Inform the patient that you need his or her help and the assistance of family members or bystanders to carry out the plan of action.

How a patient reacts to injury or illness may be influenced by certain personality traits. Some patients may become highly emotional over what may seem to be a minor problem. Others may show little or no emotion, even after serious injury or illness. Many other factors influence how a patient reacts to the stress of an EMS incident. Among these factors are the following:

- Socioeconomic background
- Fear of medical personnel
- Alcohol or substance abuse
- History of chronic disease
- Mental disorders
- Reaction to medication
- Age
- Nutritional status
- Feelings of guilt
- Past experience with illness or injury

You are not expected to always know why a patient is having an unusual emotional response. However, you can quickly and calmly assess the actions of the patient, family members, and bystanders. This assessment will help you to gain the confidence and cooperation of everyone at the scene. In addition, you should use a professional tone of voice and show courtesy, along with sincere concern and efficient action. These simple considerations will go far to relieve worry, fear, and insecurity. Calm reassurance will inspire confidence and cooperation. Compassion is important, but you must be careful. Your professional judgment takes priority over compassion. For example, suppose a screaming child with no obvious life-threatening injuries is covered with another patient's blood. This frightened child appeals to your compassion and thus gets your attention. In the meantime, an unconscious, nonbreathing adult nearby could die from lack of care.

Patients must be given the opportunity to express their fears and concerns. You can easily relieve many of these concerns at the scene. Usually, patients are concerned about the safety or well-being of others who are involved in the accident and about the damage or loss of personal property. Your responses must be discreet and diplomatic, giving reassurance when appropriate. If a loved one has been killed or critically injured,

EMT-B Tips

Whether you are a brand new EMT or a seasoned veteran, patients expect you to bring some sense of order and stability to the terrifying chaos that has suddenly engulfed them.

✳ EMT-B Tips

Calm reassurance will inspire confidence and co-operation. Compassion is important, but you must be careful that your compassion does not misdirect you to provide inappropriate care. Your professional judgment takes priority over compassion.

✳ EMT-B Tips

When in doubt, err on the side of caution, acquire the patient's consent, and transport the patient to a medical facility.

you should wait, if possible, until clergy or emergency department staff can give the patient the news. They can then provide the psychological support the patient may need.

Some patients, especially children and the elderly, may be terrified or feel rejected when separated from family members by the uniformed EMS provider team. Other patients may not want family members to share their stress, see their injury, or witness their pain.

It is usually best if parents go with their children and relatives accompany elderly patients.

Religious customs or needs of the patient must also be respected. Some people will cling to religious medals or charms, especially if any attempt is made to remove them. Others will express a strong desire for religious counsel, baptism, or last rites if death is near. You must try to accommodate these requests. Some people have religious convictions that strongly oppose the use of medications, blood, and blood products. If you obtain such information, it is imperative that you report it to the next level of care. ·

In the event of a death, you must handle the body with respect and dignity. It must be exposed as little as possible. Learn your local regulations and protocols about moving the body or changing its position, especially if you are at a possible crime scene. Even in these situations, cardiopulmonary resuscitation (CPR) and appropriate treatment must be given unless there are obvious signs of death.

Uncertain Situations

There will be times when you are unsure whether a true medical emergency exists. If you are unsure, contact medical control about the need to transport. If you cannot reach medical control, it is always best to transport the patient. For both ethical and medicolegal reasons, a physician must examine all patients who are transported and judge the degree of medical need.

You must also realize that the most minor symptoms may be early signs of severe illness or injury.

Symptoms of many illnesses can be similar to those of substance abuse, hysteria, or other conditions. You must accept the patient's complaints and provide appropriate care until you are able to transfer care of the patient to a higher level (eg, paramedic, nurse, or physician). Your local protocols will direct your actions in these uncertain situations. When in doubt, err on the side of caution, acquire the patient's consent, and transport the patient to a medical facility.

Stress Warning Signs and the Work Environment

EMS is a high-stress job. Understanding the causes of stress and knowing how to deal with them are critical to your job performance, health, and interpersonal relationships. To prevent stress from affecting your life negatively, you need to understand what stress is, its physiologic effects, what you can do to minimize these effects, and how to deal with stress on an emotional level.

Stress is the impact of stressors on your physical and mental well-being. Stressors include emotional, physical, and environmental situations or conditions that may cause a variety of physiologic, physical, and psychological responses. The body's response to stress begins with an alarm response, followed by a stage of reaction and resistance, and then recovery or, if the stress is prolonged, exhaustion. This three-stage response is referred to as the general adaptation syndrome.

The physiologic responses involve the interaction of the endocrine and nervous systems, resulting in chemical and physical responses. This is commonly known as the fight-or-flight response. Positive stress, such as exercise, as well as negative forms of stress, such as shift work, long hours, or the frustration of losing a patient, all have the same physiologic manifestations. These include the following:

- Increased respirations and heart rate
- Increased blood pressure

- Dilated venous vessels near the skin surface (causes cool, clammy skin)
- Dilated pupils
- Tensed muscles
- Increased blood glucose levels
- Perspiration
- Decreased blood flow to the gastrointestinal tract

Stress may also have physical symptoms such as fatigue, changes in appetite, gastrointestinal problems, or headaches. Stress may cause insomnia or hypersomnia, irritability, inability to concentrate, and hyperactivity or underactivity. Additionally, stress may manifest itself in psychological reactions such as fear, dull or nonresponsive behavior, depression, guilt, oversensitivity, anger, irritability, and frustration. Often, today's fast-paced lifestyles compound these effects by not allowing a person to rest and recover after periods of stress. Prolonged or excessive stress has been proven to be a strong contributor to heart disease, hypertension, cancer, alcoholism, and depression.

Many people are subject to cumulative stress, whereby insignificant stressors accumulate to a larger stress-related problem. In the emergency services environment (EMS, police, fire fighters), stressors may also be sudden and more severe. Some events are unusually stressful or emotional, even by emergency services standards. These acute severe stressors result in what is referred to as critical incident stress. Events that can trigger critical incident stress include the following:

- Mass-casualty incidents
- Serious injury or traumatic death of a child
- Crash with injuries, caused by an emergency services provider while responding to or from a call
- Death or serious injury of a coworker in the line of duty

Posttraumatic stress disorder (PTSD) may develop after a person has experienced a psychologically distressing event. It is characterized by re-experiencing the event and overresponding to stimuli that recall the event. PTSD is sometimes referred to as "Vietnam veteran's disease" because of its classification as a mental disorder following the Vietnam conflict. Stressful events in EMS are sometimes psychologically overwhelming. Some of the symptoms include depression, startle reactions, flashback phenomena, and dissociative episodes (eg, amnesia of the event).

A process called critical incident stress management (CISM) was developed to address acute stress situations and potentially decrease the likelihood that PTSD will develop after such an incident (Figure 2-5 ▶). The

Figure 2-5 Critical incident stress management is sometimes employed to help providers relieve stress.

process theoretically confronts the responses to critical incidents and defuses them, directing the emergency services personnel toward physical and emotional equilibrium. CISM can occur formally, as a debriefing for those who were on scene. In such situations, trained CISM teams of peers and mental health professionals may facilitate this. Additionally, CISM can occur at an ongoing scene in the following circumstances:

- When personnel are assessed for signs and symptoms of distress while resting
- Before re-entering the scene
- During a scene demobilization in which personnel are educated about the signs of critical incident stress and given a buffer period to collect themselves before leaving

The most common form of CISM is peer defusing, when a group informally discusses events that they experienced together.

EMT-B Safety

Coworkers often notice a change in behavior or attitude before a supervisor does. This is especially true in EMS, where close relationships develop between people who work together and share rooms, meals, and social interaction. This may allow you to help someone before job performance is negatively affected.

Stress and Nutrition

Anyone can respond to a sudden physical stress for a short time. If stress is prolonged, especially if physical action is not a permitted response, the body can quickly be drained of its reserves. This can leave it depleted of key nutrients, weakened, and more susceptible to illness.

Your body's three sources of fuel—carbohydrates, fat, and protein—are consumed in increased quantities during stress, particularly if physical activity is involved. The quickest source of energy is glucose, taken from stored glycogen in the liver. However, this supply will last less than a day. Protein, drawn primarily from muscle, is a long-term source of fuel. Tissues can use fat for energy. The body also conserves water during periods of stress. To do so, it retains sodium by exchanging and losing potassium from the kidneys. Other nutrients that are susceptible to depletion are the vitamins and minerals that are not stored by the body in substantial quan-

tities. These include water-soluble B and C vitamins and most minerals.

As EMS providers, we do not have control of what stressors we will face on any given day. Consequently, stress in one form or another is an unavoidable part of our lives. As one would study for a test, dress properly for a day of snow skiing, or train for a sporting event, we should physically prepare our bodies for stress. Physical conditioning and proper nutrition are the two variables over which we have absolute control. Muscles will grow and retain protein only with sufficient activity. Bones will not passively accumulate calcium. In response to the physical stress of exercise, bones store calcium and become denser and stronger. Regular, well-balanced meals are essential to provide the nutrients that are necessary to keep your body fueled (Figure 2-6 ▼). Vitamin-mineral preparations that provide a balanced mix of all the nutrients may be necessary to supplement a less than perfectly balanced diet.

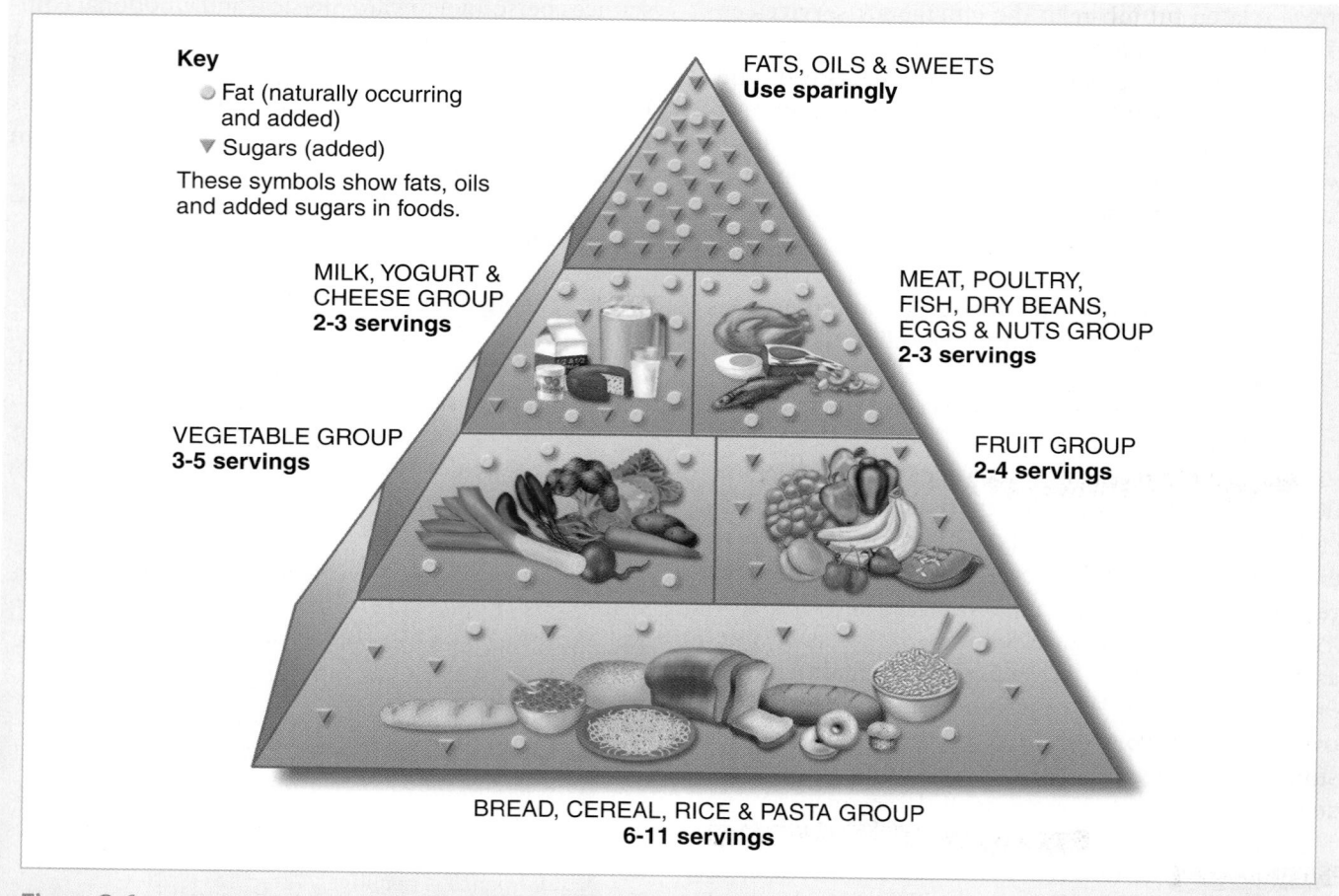

Figure 2-6 A healthy diet is illustrated by the USDA Food Guide Pyramid.

Critical Incident Stress Debriefing

You may be called to a situation so horrible that you find it difficult to respond as you were trained. You may have an immediate or delayed negative response to the incident. Do not be ashamed of such feelings; almost all responders have had the same reaction at one time or another. If you feel overwhelmed, step back and call additional resources. Remember that if you have these feelings from time to time, your partner and other members of the team may have them, too. Keep an eye on other members of your team. Confirm that they are under control and acting appropriately during a major disaster.

After a stressful run or a disaster, there may be an emotional letdown. This letdown is often overlooked. However, it may be more important to deal with than the initial contact response. Critical incident stress debriefing is one way to deal with this emotional letdown phase. A critical incident is any event that causes anxiety and mental stress to emergency workers. Critical incident stress debriefing (CISD) is a program in which severely stressful job-related incidents are discussed. These discussions are conducted in strict confidence with other emergency workers who are trained in CISD. The purpose of CISD is to relieve personal and group anxieties and stress. Although utilized extensively now for many years, CISD has never been demonstrated to be effective. In fact, several recent studies indicate that CISD may actually *increase* rescuer distress and result in worse outcomes.

CISD teams consist of peers and mental health professionals. Usually, CISD meetings are held within 24 to 72 hours of a major incident. CISD meetings may also have to be repeated at a later time.

CISD programs are located throughout the United States. CISD teams usually can be located by calling telephone directory assistance in your area and asking for CISD, or they can be requested through your employer. The International Critical Incident Stress Foundation, Inc has an emergency access number: (410) 313-2473. For general information, call (410) 750-9600.

Stress Management

There are many methods of handling stress. Some are positive and healthy; others are harmful or destructive. Americans consume more than 20 tons of aspirin per day, and doctors prescribe muscle relaxants, tranquilizers, and sedatives more than 90 million times per year to patients in the United States. Although these medications have legitimate uses, they do nothing to com-

TABLE 2-3 Strategies to Manage Stress
Change or eliminate stressors.
Change partners to avoid a negative or hostile personality.
Change work hours.
Cut back on overtime.
Change your attitude about the stressor.
Do not obsess over frustrating situations such as relapsing alcoholics and nursing home transfers. Focus on delivering high-quality care.
Try to adopt a more relaxed, philosophical outlook.
Expand your social support system apart from your coworkers.
Sustain friends and interests outside emergency services.
Minimize the physical response to stress by employing various techniques, including: ■ A deep breath to settle an anger response ■ Periodic stretching ■ Slow, deep breathing ■ Regular physical exercise ■ Progressive muscle relaxation

bat the stress that may cause the medical problems described previously.

The term "stress management" refers to the tactics that have been shown to alleviate or eliminate stress reactions. These may involve changing a few habits, changing your attitude, and perseverance
(Table 2-3 ▲).

A clue to the management of stress comes from the fact that it is not the event itself but the individual's reaction to it that determines how much it will strain the body's resources. Remember that stress is defined as anything that you perceive as a threat to your equilibrium. Stress is an undeniable and unavoidable part of our everyday life. By understanding how it affects you physiologically, physically, and psychologically, you can more successfully manage it.

Supporting patients in emergency situations is difficult. It is stressful for them but also for you. You are vulnerable to all the stresses that go with your profession. It is critical that you recognize the signs of stress so that it does not interfere with your work or life away from work, including your family life. The signs and symptoms of chronic stress may not be obvious at first. Rather, they may be subtle and not present all the time
(Table 2-4 ▶).

TABLE 2-4   Warning Signs of Stress
Irritability toward coworkers, family, and friends
Inability to concentrate
Difficulty sleeping, increased sleeping, or nightmares
Anxiety
Indecisiveness
Guilt
Loss of appetite (gastrointestinal disturbances)
Loss of interest in sexual activities
Isolation
Loss of interest in work
Increased use of alcohol
Recreational drug use

EMT-B Tips

Caring for patients can be incredibly rewarding; however, it can also be considerably stressful. Recognizing stress is a key part of successfully coping with it and maintaining a healthy mental attitude.

You will come across situations that you wish you could change. You will encounter abused children, neglected elderly patients, alcoholics, and other upsetting situations. You must find a balance between doing what you can for these patients and protecting yourself from burnout or complete desensitization. However, you will sometimes have the opportunity to help by reporting an abuse case according to local protocol or even by displaying a caring attitude. Even these things can make a profound difference in your patients' lives.

The following sections provide some suggestions for how to cope better with stress. Some of them may be useful in helping you to prevent problems from developing. Others may help you to solve problems, should they develop.

Lifestyle Changes

Your well-being is of primary importance to effective EMS operations. The effectiveness and efficiency with which you do your job depend on your ability to stay in shape and avoid the risk of personal injury. Burnout is a condition of chronic fatigue and frustration that results from mounting stress over time. To avoid burnout, you need to be in good physical and mental health. Be aware of the potential hazards in rescue and emergency medical care. You must also learn how to avoid or prevent personal injury or illness.

Nutrition

To perform efficiently, you must eat nutritious food. Food is the fuel that makes the body run. The physical exertion and stress that are a part of your job require a high energy output. If you do not have a ready source of fuel, your performance may be less than satisfactory. This can be dangerous for you, your partner, and your patient. Therefore, it is important for you to learn about and follow the rules of good nutrition.

Candy and soft drinks contain sugar. These foods are quickly absorbed and converted to fuel by the body. But simple sugars also stimulate the body's production of insulin, which reduces blood glucose levels. For some people, eating a lot of sugar can actually result in lower energy levels.

Complex carbohydrates rank next to simple sugars in their ability to produce energy. Complex carbohydrates such as pasta, rice, and vegetables are among the safest, most reliable sources for long-term energy production. However, some carbohydrates take hours to be converted into usable body fuel.

Fats are also easily converted to energy, but eating too much fat can lead to obesity, cardiac disease, and other long-term health problems. The proteins in meat, fish, chicken, beans, and cheese take several hours to convert to energy.

Carry an individual supply of high-energy food to help you maintain your energy levels Figure 2-7 ▶ . Try eating several small meals throughout the day to keep your energy resources at constant high levels. Remember, however, that overeating may reduce your physical and mental performance. After a large meal, the blood that is needed for the digestive process is not available for other activities.

You must also make sure that you maintain an adequate fluid intake Figure 2-8 ▶ . Hydration is important for proper functioning. Fluids can be easily replenished by drinking any nonalcoholic, noncaffeinated fluid. Water is generally the best fluid available. The body absorbs it faster than any other fluid. Avoid fluids that contain high levels of sugar. These can

Figure 2-7 Carry a supply of high-energy food with you so that you can maintain your energy levels.

Figure 2-9 A regular program of exercise will increase strength and endurance.

Figure 2-8 Maintain an adequate fluid intake by drinking plenty of water or other nonalcoholic, caffeine-free fluid.

actually slow the rate of fluid absorption by the body. They can also cause abdominal discomfort. One indication of adequate hydration is frequent urination. Infrequent urination or urine that has a deep yellow color indicates dehydration.

Exercise and Relaxation

A regular program of exercise will enhance the benefits of maintaining good nutrition and adequate hydration. When you are in good physical condition, you can handle job stress more easily. A regular program of exercise will increase your strength and endurance (Figure 2-9 ▶). You may wish to practice relaxation techniques, meditation, and visual imagery.

Balancing Work, Family, and Health

As an EMT-B, you will often be called to assist the sick and injured any time of the day or night. Unfortunately, there is no rhyme or reason to the timing of illness, injury, or interfacility transfer. Volunteer EMT-Bs may often be called away from family or friends during social activities. Shift workers may be required to be apart from loved ones for long periods of time. You should never let the job interfere excessively with your own needs. Find a balance between work and family; you owe it to yourself and to them. It is important to make sure that you have the time that you need to relax with family and friends.

It is also important to realize that coworkers, family, and friends often may not understand the stress caused by responding to EMS calls. As a result of a "bad call," you might not feel like going out to a movie or attending a family event that has been planned for some time. In these situations, help from a critical incident stress debriefing team or information sessions conducted by the EMS unit's employee assistance program may assist you in resolving these problems.

When possible, rotate your schedule to give yourself time off. If your EMS system allows you to move from station to station, rotate to reduce or vary your call volume. Take vacations to provide for your good health so that you will be able to respond the next time you are needed. If at any point you feel that the stress of work is more than you can handle, seek help. You may want to discuss your stress informally with your family or coworkers. Help from more experienced team members can be invaluable. You may also wish to get help from peer counselors or other professionals. Seeking this help does not make you weak in the eyes of others. Rather, it shows that you are in control of your life.

Workplace Issues

As our society continues to grow more and more culturally diverse, some groups that may have been satisfied in the past to accept and participate in mainstream American cultural traditions may seek instead to assert, preserve, and nurture their differences. As our society grows more culturally diverse, so do EMS workplaces. There will be challenges as these changes continue to occur. If you have any problem working with any particular group of people, you need to address this before finishing your EMT-B training. EMT-Bs are required to provide an equal standard of care to all patients and also need to be able to work efficiently and effectively with other health care professionals from a variety of different backgrounds.

Cultural Diversity on the Job

Each individual is different, and you should communicate with coworkers and patients in a way that is sensitive to everyone's needs (**Figure 2-10** ▼). Look at cultural diversity as a resource, and make the most of the differences among people in EMS, thus allowing them to provide optimum patient care. As the public safety workplace becomes more culturally diverse, changes may occur that could be considered disruptive. It is possible to build the strength of your workgroup through the use of diversity.

Figure 2-10 Communicate with coworkers in a way that is sensitive and respectful of individual differences.

For many years, EMS and public safety have been dominated by white men, though to a lesser extent than police and fire departments because of the traditional involvement of female nurses in EMS. This trend continues to decline; more women and minorities are working in public safety. The proactive EMT-B understands the benefits of using cultural diversity to improve patient care and expects to work alongside workers with different backgrounds and to accept their differences.

As an arm of public safety, EMS has not been in existence for as long as law enforcement and fire departments. Therefore, there may be less resistance to cultural diversity in EMS than in the other areas of public safety. Depending on your work experience, you may or may not have worked with people of varying backgrounds, attitudes, beliefs, and values.

Compared with traditional workplaces, EMS might seem like chaos. People who work in an office or manufacturing facility can reasonably expect to go to work every day, see the same people, and perform basically the same tasks. In EMS and public safety work, you are exposed to people in crisis. This exposure brings out the traits and qualities that your partners and coworkers use to manage their stress. Coworkers in traditional workplaces may not be willing to show this side of themselves to others. Debriefing after the call will help in this process.

Cultural diversity in EMS allows EMT-Bs to enjoy the benefits of accentuating the skills of a broad range of people. When you accept coworkers as individuals, the need to fit them into rigid roles is eliminated. To be more sensitive to cultural diversity issues, you must first be aware of your own cultural background. Ask yourself, "What are my own issues relative to race, color, religion, and ethnicity?" Because culture is not restricted to different nationalities, you should also consider age, handicap, gender, sexual orientation, marital status, work experience, and education.

In sports, you play to your team's strengths. For example, in football, offensive lines have a fast side and a strong side, and they run plays toward either side depending on the situation. As part of an effective EMS team, you can make it part of your team culture to play to your group's strengths. This may be difficult to do; but once you begin the process, the benefits in terms of improved patient care are immeasurable.

Your Effectiveness as an EMT-B

To be an effective EMT-B, you need to discover the diverse cultural needs of your coworkers, as well as those of your patients and their families. Although it is unrealistic to expect EMT-Bs to become cross-cultural experts with knowledge about all ethnicities, you should learn how to relate effectively.

Teamwork is essential in public safety and EMS. In order to work effectively as a team, you need to communicate to deal with cultural diversity issues.

As a health care professional, you should try to be a role model for new EMT-Bs by showing them the value of diversity. If you are working with a coworker or patient from a particular cultural group, be careful about any opinion you may have formed about that group. Do not assume that there is a language barrier, and do not appear patronizing by saying, "Some of my best friends are. . . . " There are legitimate differences in how various cultures respond to stress. For example, you should be prepared to accept that people of different cultures might respond differently to the death of a loved one.

When you are working with patients or calling the hospital on the radio, other EMT-Bs may be sensitive to how you treat patients from their cultural group. Therefore, when referring to patients, you should use the appropriate terminology. Avoid using terms such as "cripple," "deformed," "deaf," "dumb," "crazy," and "retard" when referring to patients. The word "handicapped" even has a negative connotation. Instead, use the term "disabled," and describe the specific disability.

You might want to consider taking multilingual training classes. This will not only be useful in communicating with your coworkers; it will also help to improve communication with your patients and sensitize you to the cultural richness of the people who are using the language.

Even the perception of discrimination can weaken morale and motivation and negatively affect the goal of EMS. Therefore, to achieve the benefits of cultural diversity in the EMS workplace, EMT-Bs must understand how to communicate effectively with coworkers from various backgrounds.

Avoiding Sexual Harassment

The number of sexual harassment lawsuits skyrocketed in the 1990s because of increased media attention to the problem. Furthermore, guilty verdicts encouraged others to bring suit concerning conduct that once would have gone unchallenged.

Sexual harassment is any unwelcome sexual advance, unwelcome request for sexual favors, or other unwelcome verbal or physical conduct of a sexual nature when submitting is a condition of employment, submitting or rejecting is a basis for an employment decision, or such conduct substantially interferes with performance and/or creates a hostile or offensive work environment.

There are two types of sexual harassment: quid pro quo (the harasser requests sexual favors in exchange for something else, such as a promotion) and hostile work environment (jokes, touching, leering, requests for a date, talking about body parts). Seventy percent of sexual harassment today is considered hostile work environment. Remember, it does not matter what the intent was or who was the harasser. What matters are the other person's perception and what impact that behavior had on that person. For many years, it was not uncommon to walk into a fire station and see sexually suggestive posters, calendars, or cartoons and to hear

You are the Provider   Part 4

Over several months of using stress-reducing tactics like the ones described above, you feel that your personal and professional life have become healthier. Relationships with family and friends are more positive and your work performance has increased. You feel more capable of recognizing stress in yourself, your coworkers, and your patients. Your medical care now addresses the total patient—both physical and emotional needs. You recognize that although you are not a licensed therapist, saying a kind word, acknowledging feelings, and directing patients to a professional counselor when needed can have a positive impact on their quality of life.

7. What are your local resources that could be utilized for total patient care? For example, do you know or have you met with representatives from aging and long-term care agencies in your area?

sexual jokes or comments. This situation is changing because it is not acceptable professional practice.

Because EMT-Bs and other public safety professionals depend on each other for their safety, it is especially important to try to develop nonadversarial relationships with coworkers. Most EMS facilities and fire stations make arrangements for different bunkrooms for men and women. If this is not the case at your facility, you should discuss this with your supervisor and talk openly with coworkers of the opposite gender to allow for their privacy.

If you are concerned about a particular behavior, it may be helpful to ask yourself these questions: "Would I do or say this in front of my spouse, significant other, or parents?" "Would I want my family members to be exposed to this behavior?" "Would I want my behavior videotaped and shown on the evening news?"

If you have been harassed, you should report it to your supervisor immediately and keep notes of what happened and what was said. You should confront the harasser if you feel comfortable doing so; however, this may not be for everyone. If you are asked for a date, say, "I'm not interested." If remarks or touching offend you, say, "Please don't say/do that to me; it offends me."

Substance Abuse

In the past, part of the fire service ritual was to go back to the fire station after the fire, clean and maintain the equipment, and discuss the call. At some locations, having a few beers was not uncommon. EMS today is very different from the ambulance service of 20 years ago.

Drug and alcohol use in the workplace causes an increase in accidents and tension among workers, but most important, it can lead to poor treatment decisions. EMS personnel who use or abuse substances such as alcohol or marijuana are more likely to have problems with their work habits and their drivers' licenses may be revoked as a result. They may be absent from work more often than other workers. If the use or abuse has occurred within hours before the start of their shift, their ability to render emergency medical care may be lessened because of mental or physical impairment. Because of the seriousness of substance abuse or misuse, many EMS systems now require their personnel to undergo periodic random tests for illegal drug use. Since public safety workers depend so much on coworkers for their own safety, it is even more important that ways be found to manage this problem.

As an EMT-B, you will witness firsthand the tremendous effects of violence, trauma, and disease. Beyond CISD, members of the public safety community have a

way of covering for each other. It is important to understand that the problem behavior will usually get worse before it gets better. Unfortunately, the stereotypical image of the alcoholic or addict lying in the gutter in an urban part of town often blinds EMS personnel to the existence of a coworker's drug or alcohol problem. Not all people with a substance abuse problem fit the stereotype.

As a member of the EMS team, you are responsible for responding to the community's emergency medical needs. Hazards in the EMS workplace are many. If you or one of the members of your team has an alcohol or other drug problem, these risks increase. Furthermore, drug use that occurs off the job does not necessarily decrease the risk. While varying state to state, a drug-related or alcohol-related arrest may result in the revocation of some or all driving privileges and even loss of EMT certification. Because of the tremendous risk potential, it is critical that EMT-Bs seek help or find a way to confront their partner or coworker even though there will be great pressure to allow the behavior to continue. Addicts and alcoholics develop great skill at covering their behavior; you might even decide not to bother your coworker because you feel that he or she has caught too many tough calls lately and needs to blow off some steam. Do not let this happen. You have to find a way to confront someone who has a substance abuse problem. Because of the tremendous hazard to patients, the public, and other team members, you have a legitimate right to confront coworkers with drug and alcohol problems.

When confronting a coworker with a potential drug or alcohol problem, make it clear to the worker that if the problem is personal, it is the worker's responsibility to take care of it. You have the power to assist this person. In many workplaces, coworkers are often in a position to notice a change in a coworker's behavior or attitude before a supervisor does. This is even more the case in EMS because of the close relationship that

EMT-B Safety

Substance abuse does not just reduce an EMS responder's ability to provide patient care. It also compromises the safety of that responder and other members of the team. Ignoring a substance abuse problem puts you and those you work with at increased risk.

develops between people who work in the ambulance for so many hours and share rooms, meals, and social interaction while waiting for the next call. This may allow you to help someone before his or her job performance is negatively affected.

To help reduce the potential for drug and alcohol use in the EMS workplace, EMT-Bs can learn about alcohol and other drugs. Beyond following company policy, EMT-Bs can agree among themselves what constitutes unacceptable behavior. The best time to confront these issues is usually after a call. Management sets the tone on these issues, but senior EMT-Bs can also emphasize to new EMT-Bs that drug and alcohol abuse will not be tolerated.

In a manufacturing or office environment, supervisors refer employees with problems to employee assistance programs (EAPs). EMS operations might not lend themselves easily to EAPs. Operations may be geographically spread out with minimal supervision and irregular work hours. Calls may range from a relatively simple 5-minute call to complex mass-casualty incidents that last several hours. Your partners may change regularly. And since you depend so much on each other for your safety, there will be pressure not to rock the boat. You are not "turning someone in." You may be saving his or her life. Your coworker may be a great EMT-B, but if this person has unresolved substance abuse issues, the risk to fellow EMT-Bs and patients is just too great. If a substance abuse-related incident occurs during a call, it may dramatically increase the workload for other emergency responders when they respond to assist you. Early intervention is the best bet to ensure a safe, alcohol- and drug-free workplace.

Scene Safety and Personal Protection

The personal safety of all those involved in an emergency situation is very important. In fact, it is so important that the steps you take to preserve personal safety must become automatic. A second accident at the scene or an injury to you or your partner creates more problems, delays emergency medical care for patients, increases the burden on the other EMT-Bs, and may result in unnecessary injury or death.

You should begin protecting yourself as soon as you are dispatched. Before you leave for the scene, begin preparing yourself both mentally and physically. Make sure you wear seat belts and shoulder harnesses en route to the scene. Wear seat belts and shoulder harnesses at all times unless patient care makes it impossible Figure 2-11 ▼ . Many EMS units have mandatory seat belt policies for the driver at all times, for all EMT-Bs during transit to the scene, and for anyone who is riding with a patient.

Protecting yourself at the scene is also very important. A second accident may damage the ambulance and may result in injury to you, your partner, or additional injury to the patient. The scene must be well marked Figure 2-12 ▼ . If law enforcement has not al-

Figure 2-11 Wear seat belts and shoulder harnesses en route to the scene.

Figure 2-12 Make sure the crash scene is well marked to prevent a second crash that may damage the ambulance or result in injury to you, your partner, or the patient.

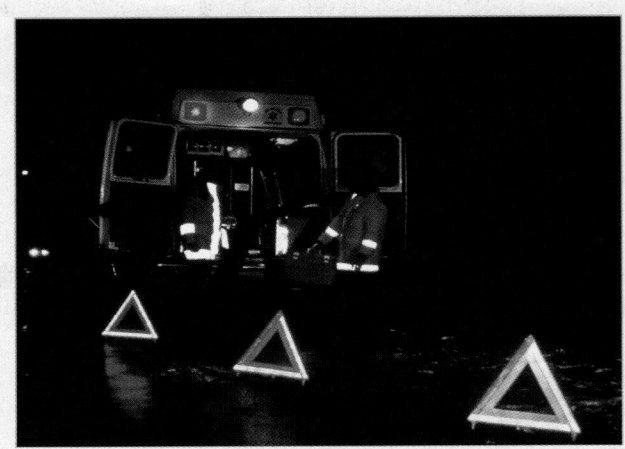

Figure 2-13 Wear reflective emblems or clothing to help make you more visible at night and improve your safety in the dark.

ready done so, you should make sure that proper warning devices are placed at a sufficient distance from the scene. This will alert motorists coming from both directions that a crash has occurred. You should park the ambulance at a safe but convenient distance from the scene. Before attempting to access patients who are trapped in a vehicle, check the vehicle's stability. Then take any necessary measures to secure it. Do not rock or push on a vehicle to find out whether it will move. This can overturn the vehicle or send it crashing into a ditch. If you are uncertain about the safety of a crash scene, wait for appropriately trained individuals to arrive before approaching.

When working at night, you must have plenty of light. Poor lighting increases the risk of injury to both you and the patient. It also results in poor emergency medical care. Reflective emblems or clothing helps to make you more visible at night and decreases your risk of injury (**Figure 2-13 ▲**).

Communicable Diseases

As an EMT-B, you will be called upon to treat and transport patients with a variety of communicable or infectious diseases. Most of these diseases are much harder to catch than is commonly believed. In addition, there are many immunizations, protective techniques, and devices that can minimize the health

EMT-B Tips

Infectious, contagious, or communicable
Many people confuse the terms "infectious" and "contagious." In fact, all contagious diseases are infectious, but only some infectious diseases are contagious. Infectious diseases that are not contagious, such as pneumonia caused by the *pneumococcus* bacteria, will not be transmitted directly from one person to another. However, infectious agents that are contagious, such as the hepatitis B virus, can be transmitted from one person to another.

An <u>infection</u> is an abnormal invasion of a host or host tissue by organisms such as bacteria, viruses, or parasites. A <u>pathogen</u> is a microorganism that is capable of causing disease in a host. A <u>host</u> is simply the organism or individual that is invaded.

An <u>infectious disease</u> is a disease that is caused by an infection. For example, Lyme disease is an infectious disease caused by the *Borrelia burgdorferi* bacterium, which lives in deer ticks. However, Lyme disease is not contagious. Again, a <u>contagious</u> or <u>communicable disease</u> can be transmitted from one person to another. The only way to get Lyme disease is to be bitten by a deer tick.

care provider's risk of infection. When these protective measures are used, the risk of the health care provider contracting a serious communicable disease is negligible.

Routes of Transmission

While all infections result from an abnormal invasion of body spaces and tissues by germs, different germs use different means of attack. We refer to these as the mechanisms of transmission. <u>Transmission</u> is the way an infectious agent is spread. There are three types of transmission: direct, indirect, and airborne. Indirect transmission can be either vehicle- or vector-borne. In this context, "vehicle" means an inanimate object, while "vector" means a living object. Droplets and dust are types of airborne transmission (**Table 2-5 ▶**).

Study This

TABLE 2-5 Mechanisms of Transmission of Infectious Disease

In this table (Data taken from Benenson AS (ed): *Control of Communicable Diseases in Man*, 15th edition, Washington, DC, American Public Health Association, 1990), the routes of transmission and some examples are outlined. Remember, while some germs frequently cause disease after transmission, transmission to a susceptible host is much more likely to cause asymptomatic infection and colonization.

Route	Descriptions	Source	Examples
UNIVERSAL PRECAUTIONS			
Direct	Contact, directly either with the person or with droplets sprayed (eg, by sneezing or coughing)	Ordinary contact	Measles, mumps, chickenpox, bacterial meningitis, influenza, diphtheria, herpes simplex
		Sexual contact	Syphilis, gonorrhea, HIV infection, hepatitis B, herpes simplex
Indirect			
Vehicle-borne	Spread by inanimate objects (eg, food, needles, clothing, transfused blood)	Food or water	Hepatitis A, B, C, salmonella, *Shigella*, poliomyelitis
		Blood	HIV infection
		Other	Measles, tetanus
Vector-borne			
Mechanical	Simple carriage by insects. The vector simply carries the germs.	Houseflies	*Shigella*
Biological	Transmission by insect in which the germ lives and grows	Ticks	Lyme disease, Rocky Mountain spotted fever
		Mosquitoes	Malaria, equine encephalitis
AIRBORNE DISEASE PRECAUTIONS			
Airborne			
Droplet nuclei	Residues after partial evaporation of droplets. Germs may remain viable, and the droplets may remain suspended for long periods		*Mycobacterium tuberculosis*, chickenpox
Dust	Small particles of dust from the soil may carry fungal spores and remain airborne for long periods		*Histoplasma, Coccidioides, Mycobacterium avium-intracellulare*

Risk Reduction and Prevention

Universal Precautions and Body Substance Isolation

The Occupational Safety and Health Administration (OSHA) develops and publishes guidelines concerning reducing risk in the workplace. It is also responsible for enforcing these guidelines. OSHA requires all EMT-Bs to be trained in the handling of bloodborne pathogens and in approaching the patient who may have a communicable or infectious disease. Training must also be provided for issues including blood and body fluid precautions, airborne precautions, and contamination precautions.

Because health care workers are exposed to so many different kinds of infections, the Centers for

Disease Control and Prevention (CDC) developed a set of <u>universal precautions</u> for health care workers to use in treating patients. These protective measures are designed to prevent workers from coming into direct contact with germs carried by patients. <u>Direct contact</u> is the exposure or transmission of a communicable disease from one person to another by physical touching. Gonorrhea is an example of a disease transmitted by direct contact, usually sexual. <u>Exposure</u> is contact with blood, body fluids, tissues, or airborne droplets by direct or indirect contact. <u>Indirect contact</u> is exposure or transmission of a disease from one person to another by contact with a contaminated object. Common colds are probably spread in this way.

The goal of universal precautions is to interrupt the transmission of germs by decreasing the chance that you will come into contact with them. Universal precautions are not universal in the sense that they will help to protect you against all infectious diseases. Instead, the word "universal" is meant to remind you to apply precautions in all situations in which you have direct patient contact. It is impossible to tell whether an individual is free from a communicable disease, even if he or she appears healthy. Therefore, you should always take precautions.

You can also reduce your risk of exposure by following <u>body substance isolation (BSI)</u> precautions. BSI is the preferred infection control concept for fire and EMS personnel. BSI differs from universal precautions in that it is designed to approach all body fluids as being potentially infectious. In observing universal precautions, you assume that only blood and certain body fluids pose a risk for transmission of hepatitis B and human immunodeficiency virus (HIV). In 1988, the CDC removed many body fluids, such as sweat, tears, saliva, urine, feces, vomitus, nasal secretions, and sputum, from the risk category unless these fluids contain visible blood. However, in the dark, you may not be able to see any blood. Therefore, EMS follows the BSI concept rather than relying on universal precautions.

We have learned that infectious diseases are transmitted by direct, indirect, or airborne routes. Common ways disease may spread in the prehospital environment include the following:

- Blood or fluid splash
- Surface contamination
- Needlestick exposure
- Oral contamination due to lack of or improper hand washing

Proper Hand Washing

Proper hand washing is perhaps one of the simplest yet most effective ways to control disease transmission. You should always wash your hands before and after contact with a patient, regardless of whether you wear gloves. The longer the germs remain with you, the greater their chance of getting through your barriers. Although soap and water are not protective in all cases, in certain cases their use provides excellent protection against further transmission from your skin to others.

If no running water is available, you may use waterless hand washing substitutes (Figure 2-14 ▼). If you use a waterless substitute in the field, make sure that you wash your hands once you arrive at the hospital.

The proper procedure for hand washing is as follows:

1. Use soap and warm water.
2. Rub your hands together for at least 10 to 15 seconds to work up a lather. Pay particular attention to your fingernails.
3. Rinse your hands, and dry them with a paper towel.
4. Use the paper towel to turn off the faucet.

Gloves and Eye Protection

Gloves and eye protection are the minimum standard for all patient care if there is any possibility for exposure to blood or body fluids. Both vinyl and latex gloves provide adequate protection. Your department may prefer one type of glove over the other, or you may

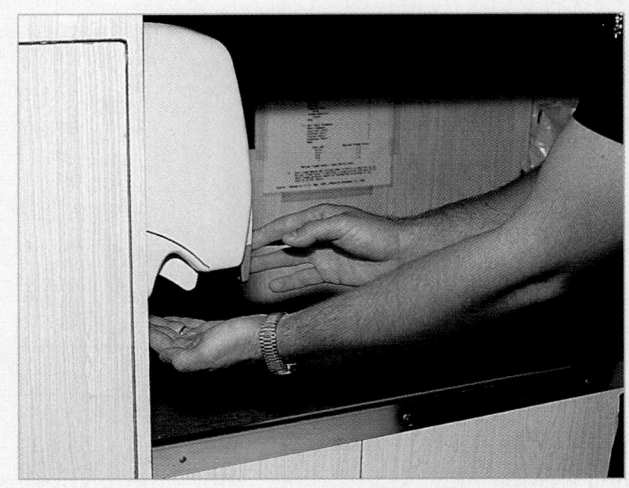

Figure 2-14 Use a waterless handwashing solution if there is no running water available. Be sure to wash your hands with soap once you arrive at the hospital.

choose yourself. You should evaluate each situation and choose the glove that works best. (Some individuals are allergic to latex. If you suspect that you are allergic, consult your supervisor for options.) Vinyl gloves may be best for routine procedures, and latex gloves may be best for invasive procedures. Never use vinyl or latex gloves for cleaning. Change latex gloves if they have been exposed to motor oil, gasoline, or any petroleum-based product. Do not use petroleum jelly with latex gloves. Wear double gloves if there is substantial bleeding. You may also wear double gloves if you will be exposed to large volumes of other body fluids. Be sure to change gloves as you move from patient to patient. For cleaning and disinfecting the unit, you should use heavy-duty utility gloves (Figure 2-15 ▼). *You should never use lightweight latex or vinyl gloves for cleaning.*

Removing used latex or vinyl gloves requires a methodical technique to avoid contaminating yourself with the materials from which the gloves have protected you (Skill Drill 2-1 ▶).

1. **Begin by partially removing one glove.** With the other gloved hand, pinch the first glove at the wrist—being certain to touch only the outside of the first glove—and start to roll it back off the hand, inside out. Leave the exterior of the fingers on that first glove exposed (**Step 1**).
2. **Use the still-gloved fingers** of the first hand to pinch the wrist of the second glove and begin to pull it off, rolling it inside-out toward the fingertips as you did with the first glove (**Step 2**).

3. **Continue pulling the second glove off** until you can pull the second hand free (**Step 3**).
4. **With your now-ungloved second hand**, grasp the exposed inside of the first glove and pull it free of your first hand and over the now-loose second glove. Be sure that you touch only clean, interior surfaces with your ungloved hand (**Step 4**).

Gloves are the most common type of <u>personal protective equipment (PPE)</u>. In many EMS rescue operations, you must also protect your hands and wrists from injury. You may wear puncture-proof leather gloves, with latex gloves underneath. This combination will allow you free use of your hands with added protection from blood and body fluids. Remember that latex or vinyl gloves are considered medical waste and must be disposed of properly. Leather gloves must be treated as contaminated material until they can be properly decontaminated.

Eye protection is important in case blood splatters toward your eyes (Figure 2-16 ▼). If this is a possibility, wearing goggles is your best protection. However, you need not wear goggles if you wear prescription glasses. Prescription glasses are acceptable as eye protection, but you must add removable side shields when on duty. Contact lenses are not considered protective eyewear.

Gowns and Masks ✈

Occasionally, you may need to wear a mask and gown. A mask and gown provide protection from extensive blood splatter. Gowns may be worn in situations such as field delivery of a baby or major trauma. However,

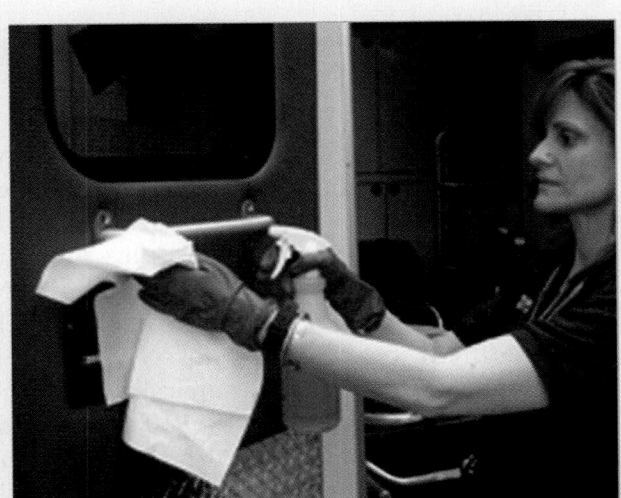

Figure 2-15 Use heavy-duty utility gloves to clean the unit. You should never use lightweight latex or vinyl gloves for cleaning.

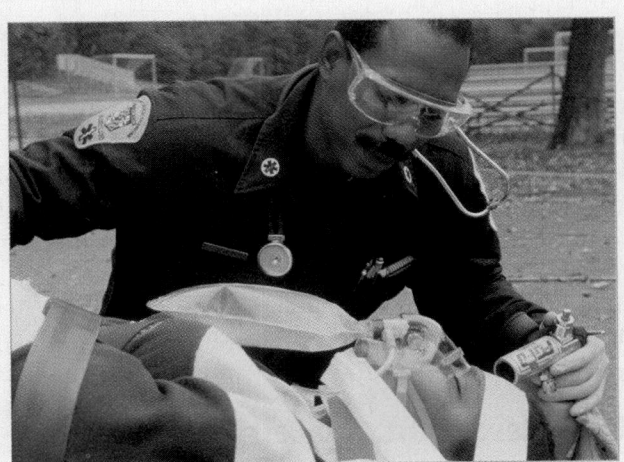

Figure 2-16 Wear eye protection to prevent blood splatter into your eyes.

Skill Drill 2-1

Proper Glove Removal Technique

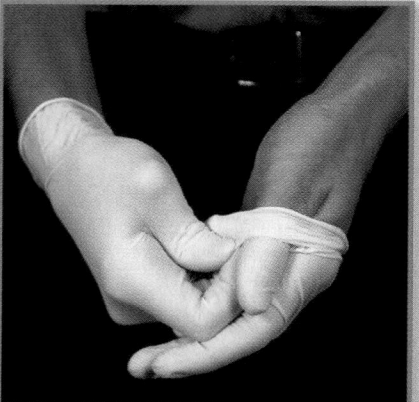

1 Partially remove the first glove by pinching at the wrist. Be careful to touch only the outside of the glove.

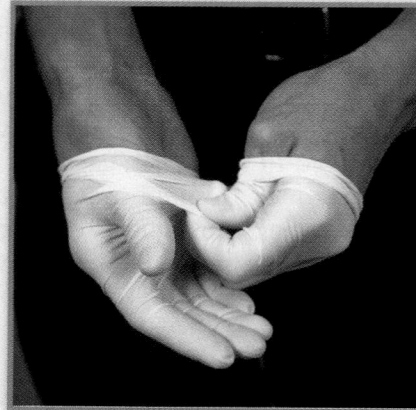

2 Remove the second glove by pinching the exterior with the partially gloved hand.

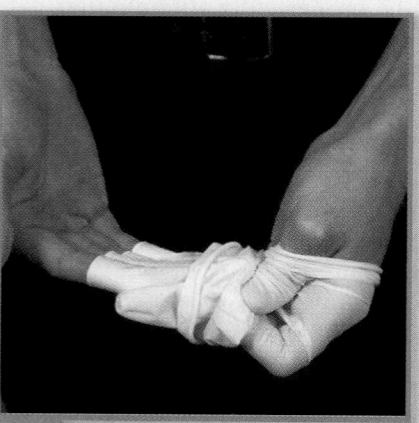

3 Pull the second glove inside-out toward the fingertips.

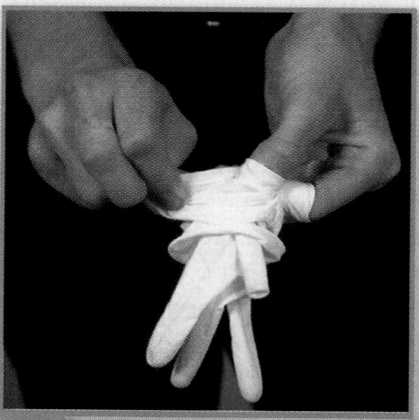

4 Grasp both gloves with your free hand, touching only the clean, interior surfaces.

wearing a gown may not be practical in many situations. In fact, in some instances, a gown may pose a risk for injury. Your department will likely have a policy regarding gowns. Be sure you know your local policy. There are times when a change of uniform is preferred because trying to clean off contaminants is difficult and sometimes impossible without professional cleaning and disinfection or disposing of the uniform entirely.

Masks, Respirators, and Barrier Devices

The use of masks is a complex issue, especially in light of OSHA and CDC requirements regarding protection from tuberculosis. You should wear a standard surgical mask if blood or body fluid spatter is a possibility. If you suspect that a patient has an airborne disease, you should place a surgical mask on the patient. However, if you suspect that the patient has tuberculosis, place a surgical mask on the patient and a High-Efficiency Particulate Air (HEPA) respirator on yourself Figure 2-17 ▶). If the patient needs oxygen, place a nonrebreathing mask instead of a surgical mask on the patient and set the oxygen flow rate at 10 to 15 L/min. Do not place a HEPA respirator on the patient; it is unnecessary and uncomfortable. A simple surgical mask will reduce the risk of transmission of germs from the patient into the air. Use of a HEPA respirator should comply with OSHA standards, which state that facial hair,

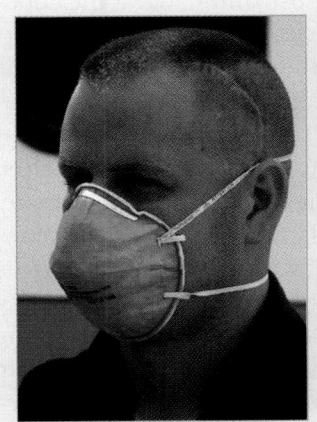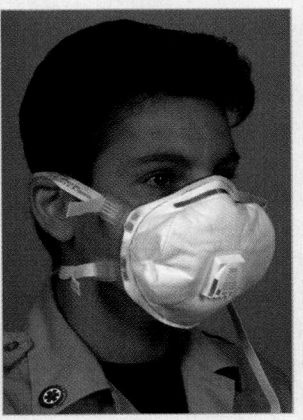

Figure 2-17 Wear a HEPA respirator if you treat a patient whom you suspect has tuberculosis.

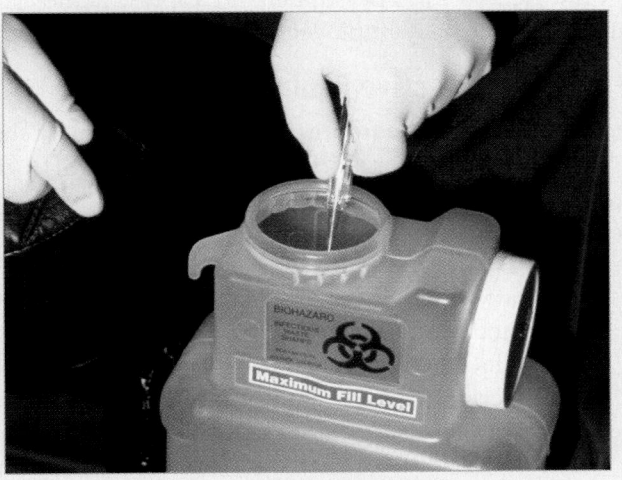

Figure 2-19 Properly dispose of sharps in a closed, rigid, marked container.

such as long sideburns or a mustache, will prevent a proper fit.

Although there are no documented cases of disease transmission to rescuers as a result of performing unprotected mouth-to-mouth resuscitation on a patient with an infection, you should use a pocket mask or bag-valve-mask (BVM) device (Figure 2-18 ▼). Mouth-to-mouth resuscitation is rarely necessary in a work situation.

Remember that the outside surfaces of these items are considered contaminated after they have been exposed to the patient. You must make sure that gloves, masks, gowns, and all other items that have been exposed to infectious processes or blood are properly disposed of according to local guidelines. If you are stuck by a needle, get blood or body fluids in your eye, or have any body fluid contact with the patient, report the incident to your supervisor immediately.

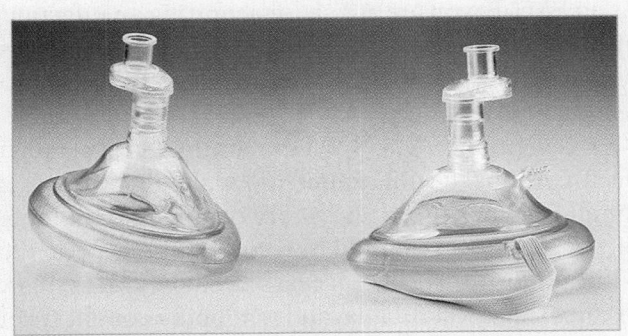

Figure 2-18 Barrier devices such as a pocket mask are necessary in providing artificial ventilations.

Proper Disposal of Sharps

Be careful when handling needles, scalpels, and other sharp items. The spread of HIV and hepatitis in the health care setting can usually be traced to careless handling of sharps.

- Do not recap, break, or bend needles. Even the most careful individuals may stick themselves accidentally.
- Dispose of all sharp items that have been in contact with human secretions in approved closed, rigid containers (Figure 2-19 ▲).

Employer Responsibilities

Your employer cannot guarantee a 100% risk-free environment. Taking the risk of exposure to or acquisition of a communicable disease is a part of your job. You have a right to know about diseases that may pose a risk to you. Remember, though, that your risk for infection is not high; however, OSHA regulations, especially for private and federal agencies, require that all employees be offered a workplace environment that reduces the risk for exposure. Note that in some states that have their own OSHA plans, state and municipal employees must also be covered.

In addition to OSHA guidelines, other national guidelines and standards, including those from the CDC and National Fire Protection Agency (NFPA) Infection Control Standard 1581, address reducing the risk for exposure to bloodborne pathogens (disease-causing organisms) and airborne diseases. These agencies set a standard of care for all fire and EMS personnel and apply whether you are a full-time paid employee or a volunteer.

Personal Protective Equipment

Personal protective equipment (PPE) is equipment that blocks entry of an organism into the body. OSHA requires that the following PPE be made available to you:

- Vinyl and latex gloves
- Heavy-duty gloves for cleaning
- Protective eyewear
- Masks (including a HEPA respirator)
- Cover gowns
- Devices for respiratory assistance

The proper PPE for each task is selected according to the way in which a communicable disease is transmitted. For example, transmission of an airborne disease can be blocked by a mask. Blood spatter into the eye can be prevented by wearing eye protection.

The recommendations for PPE use should be followed; however, OSHA recognizes that there are times when these procedures cannot be performed. There is an "exception" statement in the OSHA regulation that states that when you believe that taking the time to use PPE will delay delivery of care to the patient or will pose a risk to your personal safety, you may choose not to use PPE. Risk to personal safety refers to the likelihood of being attacked by a person or an animal, not to concern over acquiring a communicable disease. If you choose not to use PPE, you may have to justify this action. It is your responsibility to follow the rules in a reasonable, prudent way.

Exposure Control Plan

Another way to reduce risk of exposure is by following your department's exposure control plan, which is a comprehensive plan that incorporates CDC guidelines, OSHA regulations, NFPA Infection Control Standard 1581, and other applicable state and local regulations (Table 2-6 ▶).

Immunity

Even if germs do reach you, they may not infect you. For example, you may be immune, or resistant, to those particular germs. Immunity is a major factor in determining which hosts become ill from which germs (Table 2-7 ▶). One way to gain immunity from many diseases today is to be immunized, or vaccinated, against

them. Vaccinations have almost eliminated some childhood diseases, such as measles and polio.

Another way in which the body becomes immune to a disease is to recover from an infection from that germ. Afterward, the body recognizes and repels that germ when it shows up again. Once exposed, healthy individuals will develop lifelong immunity to many common pathogens. For example, a person who contracts and becomes infected with the hepatitis A virus may be ill for several weeks, but because an immunity will develop, will not have to worry about getting the illness again. Sometimes, however, the immunity is only partial. Partial immunity protects against new infections. But germs that remain in the body from the first illness may still be able to cause the same disease again when the body is stressed or has some impairment in its immune system. For example, tuberculosis can cause a mild, unnoticeable infection before the body builds up a partial immunity. If the infection is never treated, the infection may be reactivated when immunity is weakened; however, such individuals are protected against a new infection from another person.

Humans seem unable to mount an effective immune response to some infections, such as HIV infection, which is infection with the human immunodeficiency virus that can progress to acquired immunodeficiency syndrome (AIDS).

Although hepatitis A immunization is not required by OSHA, you may wish to be vaccinated as a preventive measure. Hepatitis A vaccination is not necessary if you have had hepatitis A in the past. All these vaccines are effective and rarely cause side effects. Many communities require you to show proof that you are up to date with their immunizations.

Remember, germs that cause no symptoms in one person may cause serious illness in another.

Immunizations

As an EMT-B, you are at risk for acquiring an infectious or communicable disease. Using basic protective measures can minimize the risk. You are responsible for protecting yourself.

Prevention begins by maintaining your personal health. Annual health examinations should be required for all EMS personnel. A history of all your childhood infectious diseases should be recorded and kept on file. Childhood infectious diseases include chickenpox, mumps, measles, rubella, and whooping cough. If you have not had one of these diseases, you must be immunized.

TABLE 2-6 Components of an Exposure Control Plan

Determination of Exposure

■ Determines who is at risk for ongoing contact with blood and other body fluids

■ Creates a list of tasks that pose a risk for contact with blood or other body fluids

■ Includes personal protective equipment (PPE) required by OSHA

Education and Training

■ Explains why a qualified individual is required to answer questions about communicable diseases and infection control, rather than relying on packaged training materials

■ Includes availability of an instructor able to train EMTs regarding bloodborne and airborne pathogens, such as hepatitis B and C, HIV, syphilis, and tuberculosis

■ Ensures that the instructor provides appropriate education, which is the best means for correcting many myths surrounding these issues

Hepatitis B Vaccine Program

■ Spells out the vaccine offered, its safety and efficacy, record keeping, and tracking

■ Addresses the need for postvaccine antibody titers to identify individuals who do not respond to the initial three-dose vaccination series

Personal Protective Equipment (PPE)

■ Lists the PPE offered and why it was selected

■ Lists how much equipment is available and where to obtain additional PPE

■ States when each type of PPE is to be used for each risk procedure

Cleaning and Disinfection Practices

■ Describes how to care for and maintain vehicles and equipment

■ Identifies where and when cleaning should be performed, how it is to be done, what PPE is to be used, and what cleaning solution is to be used

■ Addresses medical waste collection, storage, and disposal

Tuberculin Skin Testing/Fit Testing

■ Addresses how often employees should undergo skin testing

■ Addresses how often fit testing should be done to determine the proper size mask to protect the EMT from tuberculosis

■ Addresses all issues dealing with HEPA respirator masks

Postexposure Management

■ Identifies whom to notify when exposure may have occurred, forms to be filled out, where to go for treatment, and what treatment is to be given

Compliance Monitoring

■ Addresses how the service or department evaluates employee compliance with each aspect of the plan

■ Ensures that employees understand what they are to do and why it is important

■ States that noncompliance should be documented

■ Indicates what disciplinary action should be taken in the face of continued noncompliance

Record Keeping

■ Outlines all records that will be kept, how confidentiality will be maintained, and how records can be assessed and by whom

The CDC and OSHA have developed requirements for protection from bloodborne pathogens such as hepatitis B and human immunodeficiency viruses. An immunization program should be in place in your EMS system. Immunizations should be kept up to date and recorded in your file. Recommended immunizations include the following:

■ Tetanus-diphtheria boosters (every 10 years)

■ Measles, mumps, rubella (MMR) vaccine

■ Influenza vaccine (yearly)

■ Hepatitis B vaccine

You should also have a skin test for tuberculosis before you begin working as an EMT-B. The purpose of the test is to identify anyone who has been exposed to tuberculosis in the past. Testing should be repeated every year.

If you know that you will be transporting a patient who has a communicable disease, you have a definite

TABLE 2-7 Immunity to Infectious Diseases

Type of Immunity	Characteristics	Examples	Comments
Lifelong	The illness will not recur.	Measles Mumps Polio Rubella Hepatitis A Hepatitis B	Infection on vaccination provides long-term immunity to new infection. A live vaccine is required for measles only.
Partial	The person who has recovered from a first infection is unlikely to get a new infection from another person but may develop illness from germs that lie dormant from the initial infection.	Chickenpox Tuberculosis	Infection provides lifelong immunity to the patient from acquiring a new infection, but the original illness may recur, or it may recur in a different way. In the case of chickenpox, which is caused by the herpes zoster virus, an infection may recur years later in the form of shingles.
None	Exposure confers no protection from reinfection. The infection may wear down the patient's resistance.	Gonorrhea Syphilis HIV infection	No vaccine is available. Repeated infections are common. For example, there is effective immediate treatment for gonorrhea, and the germs may be eradicated; however, reinfection is likely if the high-risk practices (eg, unprotected sex) continue. For syphilis and HIV infection, the lack of immunity allows the germs to continue to cause damage within the host.

advantage. This is when your health record will be valuable. If you have already had the disease or been vaccinated, you are not at risk. However, you will not always know whether a patient has a communicable disease. Therefore, you should always follow BSI precautions if there is the possibility of exposure to blood or other body fluids.

Duty to Act

You cannot deny care to a patient who you suspect has a communicable disease, even if you believe that the patient poses a risk to your safety. To deny care to such a patient is considered to be abandonment, which is legally and ethically a serious matter that can result in both civil and criminal actions against you. It can

also be considered a breach of duty (a situation in which the EMT-B does not act within an expected and reasonable standard of care). In addition to breach of duty, if the following factors are present, you may be considered negligent in your duties:

- Prudence (acting reasonably in a way that a person with similar training would act)
- Duty (a duty to act existed)
- Damages (physically or psychologically harming a patient)
- Cause (reasonable cause and effect—having a duty and abusing it and harming another individual)

Denying care to a patient who you suspect has a communicable disease can also be considered discrimination according to the Americans With Disabilities Act (ADA), especially when a public department or agency such as EMS is involved.

You must understand the disease process and the factors necessary to put you at risk because your response to them sometimes has legal consequences.

Some Diseases of Special Concern

Herpes Simplex

Herpes simplex is a common virus strain carried by humans. Eighty percent of individuals carrying the virus are asymptomatic, but symptomatic infections can be serious and are on the rise, especially in immunocompromised patients. The primary mode of infection is through close personal contact, so universal precautions are generally sufficient to prevent spread to or from health care workers.

HIV Infection

Exposure to the virus that causes AIDS is the most feared infection risk for EMTs. It is this prospect that led to the development of universal precautions and BSI precautions. There is no vaccine to protect against HIV infection, and despite great progress in drug treatments, AIDS is still fatal. Fortunately, it is not easily transmitted in your work setting. For example, it is far less contagious than hepatitis B. HIV infection is a potential hazard only when deposited on a mucous membrane or directly into the bloodstream. This can occur either via sexual contact or exposure to blood, meaning your risk of infection is limited to exposure to an infected patient's blood and body fluids. Exposure can take place in the following ways:

- The patient's blood is splashed or sprayed into your eyes, nose, or mouth or into an open sore or cut, however tiny; even a microscopic opening in the skin is an invitation for infection with a virus.
- You have blood from the infected patient on your hands and then touch your own eyes, nose, mouth, or an open sore or cut.
- A needle used to inject the patient breaks your skin. The risk to you from a single injection, even with a hollow-bore needle, is small, probably less than 1 in 1,000. However, this is by far the most dangerous form of exposure.
- Broken glass at a motor vehicle crash or other incident may penetrate your glove (and skin), which may have already been covered with blood from an infected patient.

Many patients who are infected with HIV do not show any symptoms. This is why the government requires health care workers to wear certain types of gloves any time they are likely to come into contact with secretions or blood from any patient. You should always put on the proper type of gloves before leaving the ambulance to care for a patient. In addition, you must take great care in handling and disposing of needles and scalpels so that others are not inadvertently exposed to them. Finally, you should cover any open wounds that you have whenever you are on the job.

If you have any reason to think that a patient's blood or secretions may have entered your system, especially through inoculation of a patient's blood, you should seek medical advice as soon as possible. If you know that the patient is infected with HIV, your physician may suggest immediate treatment to try to prevent you from becoming infected. However, if the patient is an unlikely candidate for HIV infection, your physician may recommend that both you and the patient be tested before you undergo therapy. As scientists learn more about HIV infection, testing and treatment recommendations change. So it is important that you immediately see your doctor (or your program's designated doctor) any time you are potentially exposed to a communicable disease. Know the policy for your system and consider now what you would do in the event of exposure.

Hepatitis

The term hepatitis refers to an inflammation (and often infection) of the liver. Hepatitis causes fever, loss of appetite, jaundice, and fatigue. It can be caused by a number of different viruses and toxins. There is no sure way to tell which patients with hepatitis have a contagious form of the disease and which do not. (Table 2-8 ▶) shows the characteristics of different types of hepatitis, from which you can assess your risk of exposure. Hepatitis A can be transmitted only from a patient who has an acute infection, while hepatitis B and hepatitis C can also be transmitted from long-term carriers who have no signs of illness. A carrier is a person (or animal) in whom an infectious organism has taken up permanent residence and who may or may not cause any active disease. Carriers may never know that they harbor the organism; however, they can infect other individuals.

Hepatitis A is transmitted orally through oral or fecal contamination. This means that, generally, you must eat or drink something that is contaminated with the virus. Contamination is the presence of an

TABLE 2-8 Characteristics of Hepatitis

Type	Route of Infection	Incubation Period (time before clinical signs and symptoms appear but infection may still be transmitted)	Acute Disease (when patient usually appears sick)
Viral (infectious)			
Hepatitis A	Fecal-oral, infected food	2 to 6 weeks	Early symptoms of all viral hepatitis include loss of appetite, vomiting, fever, fatigue, sore throat, cough, and muscle and joint pain. Several weeks later, jaundice (yellow eyes and skin) and right upper quadrant abdominal pain develop.
Hepatitis B	Blood, saliva, urine, sexual contact, breast milk	4 to 12 weeks	
Hepatitis C	Blood, sexual contact	2 to 10 weeks	
Hepatitis D	Blood, sexual contact	4 to 12 weeks	
Toxin-Induced			
Medication, Drugs, Alcohol	Inhalation, skin or mucous membrane exposure, oral ingestion, or IV administration	Within hour to days following exposure	Severity of disease depends on amount of agent absorbed and duration of exposure.

infectious organism on or in an object. The organisms that cause hepatitis B and C are transmitted through vehicles other than food or water. For example, these organisms may enter the body through a transfusion or needlestick with infected blood, which puts health care workers at high risk for contracting hepatitis B, the more contagious and virulent form. Virulence is the strength or ability of a pathogen to produce disease. Hepatitis B is far more contagious than HIV. For this reason, vaccination with hepatitis B vaccine is highly recommended for EMTs. Unfortunately, not everyone who is vaccinated develops immediate immunity to the virus. Sometimes, but not always, an additional dose will provide immunity. You should be tested after vaccination (titer) to determine your immune status.

If you are stuck with a needle or injured in some other way while caring for a patient who might have hepatitis, see your doctor immediately.

Meningitis

Meningitis is an inflammation of the meningeal coverings of the brain and spinal cord. Patients with meningitis will have signs and symptoms such as fever, headache, stiff neck, and altered mental status. It is an uncommon but very frightening infectious disease. Meningitis can be caused by viruses or bacteria, most of which are not contagious. However, one form, meningococcal meningitis, is highly contagious. The meningococcus bacterium colonizes the human nose and throat and only rarely causes an acute infection. When it does, it can be lethal. Patients with this kind of infection often have red blotches on their skin; however, many patients with forms of meningitis that are not contagious also have red blotches.

Because only laboratory tests can sort out the different forms of meningitis, you should use universal precautions and follow BSI precautions with any pa-

Chronic Infection (patient may no longer have signs of relevant illness)	Vaccine Available?	Treatment	Comments
Chronic condition does not exist.	Yes	None	Mild illness, approximately 2% of patients die. After acute infection, patient has life-long immunity.
Chronic infection affects up to 10% of patients, up to 90% of newborns who have the disease.	Yes	Yes, but poor	Up to 30% of patients may become chronic carriers. Patients are asymptomatic and without signs of liver disease, but they may infect others. Approximately 1% to 2% of patients die.
Chronic infection affects 90% of patients.	No	Yes, but poor	Cirrhosis of the liver develops in 50% of patients with chronic hepatitis C; chronic infection increases the risk of cancer of the liver.
Chronic infection is very common.	No	None	Occurs only in patients with active hepatitis B infection. Fulminant disease may develop in 20% of patients.
Some chemicals may initiate an inflammatory response that continues to cause liver damage long after the chemical is out of the body.	No	Stop exposure. In patients with overdose of acetaminophen, certain drugs may subdue liver injury if given early enough.	This type of hepatitis is not contagious. Patients with toxin-induced hepatitis may have liver damage, such as jaundice. Not every exposure to a toxin will cause liver damage.

tient who is suspected of having meningitis. Gloves and a mask will go a long way to prevent the patient's secretions from getting into your nose and mouth. Again, the risk of infection is small, even if the organism is transmitted. For this reason, vaccines, which are available for most types of meningococcus, are rarely used. There are no effective treatments for the disease.

After treating a patient with meningitis, you should contact your employer health representative. Many states consider meningitis "reportable" and will notify you that one of your patients was diagnosed with meningitis. Prophylactic treatment is then in order for you.

Tuberculosis

Most patients who are infected with *Mycobacterium tuberculosis* (the tubercle bacillus) are well most of the time. If the disease involves the bone or kidneys, the patient is only slightly contagious. In the United States, however, tuberculosis is a chronic mycobacterial disease that usually strikes the lungs. Disease that occurs shortly after infection is called primary tuberculosis. Except in infants, this infection is not usually serious. After the primary infection, the tubercle bacillus is rendered dormant by the patient's immune system. However, even after decades of lying dormant, this germ can reactivate. Reactive tuberculosis is common and can be much more difficult to treat, especially because an increasing number of tuberculosis strains have grown resistant to most antibiotics.

Although tuberculosis is often hard to distinguish from other diseases, patients who pose the highest risk almost invariably have a cough. Therefore, for your safety, you should consider respiratory tuberculosis to be the only contagious form, as it is the only one that is spread by airborne transmission. The droplets that are produced by coughing are not the real problem. The real problem is the droplet nuclei, which are what re-mains of droplets after excess water has evaporated. These particles are tiny enough to be totally invisible and can remain suspended in the air for a long time. In fact, as long as they are shielded from ultraviolet light, they can remain alive for decades. So you may be at risk by simply entering a closed room that the patient ac-tually left long ago. Particles that are the size of droplet nuclei are not stopped by routine surgical masks. Inhaled, they are carried directly to the alveoli of the lungs, where the bacteria may begin to grow.

Why is tuberculosis not more common than it is? After all, absolute protection from infection with the tubercle bacillus does not exist. Everyone who breathes is at risk. And the vaccine for tuberculosis, called BCG, is only rarely used in the United States. Under nor-mal circumstances, however, the mechanism of

Geriatric Needs

Everyone has defenses against getting sick, but the aging process can pose a threat to our natu-ral defense mechanisms against invading mi-croorganisms. Our physical defenses weaken or are eliminated as we age. The skin's thinning and loss of supportive collagen, along with a reduction in the number of blood vessels, can allow bacte-ria or viruses to enter the body with less resis-tance. The respiratory system cannot trap and eliminate bacteria or viruses in the airways as it once did. Finally, the gastrointestinal system al-lows an easier entry for bacteria or viruses through the intestines. Not only do our physical barriers to entry weaken, but our immune system deteri-orates, and invading organisms are not as easily identified as abnormal. Infectious agents can take hold in the elderly patient much more easily be-cause of reduced defenses.

When transporting an elderly patient, pro-tect the patient from the environment, since ex-tremes in heat or cold can further reduce the body's defenses. If you have a cold or the flu, use respiratory precautions, including a face mask for yourself so that the patient does not get exposed to the viruses. If your patient has a cold or the flu, protect yourself. However, remember that your defense system is probably much stronger than that of the patient.

transmission used by *M tuberculosis* is not very effi-cient. Infected air is easily diluted with uninfected air. And *M tuberculosis* is one of those germs that typ-ically cause no illness in a new host. In fact, many pa-tients with tuberculosis do not even transmit the infection to family members. In crowded environ-ments with poor ventilation, however, the disease spreads more easily.

If you are exposed to a patient who is found to have pulmonary tuberculosis, you will be given a tuberculin skin test. This simple skin test determines whether a per-son has been infected with *M tuberculosis*. A positive result means that exposure has occurred; it does not mean that the person has active tuberculosis. It takes at least 6 weeks for the bacteria to show up in the lab-oratory test. So if you have the test within a few weeks of the exposure and results are positive, this means that you had already acquired the infection from somebody else. You will probably never identify the source. Most transmissions occur silently. This is why health care workers have tuberculin skin tests regularly. If the in-fection is found before the individual becomes ill, pre-ventive therapy is almost 100% effective. Usually, a daily dose of isoniazid will prevent the development of ac-tive infection.

Other Diseases of Concern

Syphilis

Although syphilis is commonly thought of as a sexu-ally transmitted disease, it is also a bloodborne disease. There is a small risk for transmission through a con-taminated needlestick injury or direct blood-to-blood contact.

Whooping Cough

Whooping cough, also called pertussis, is an airborne disease caused by bacteria that mostly affects children younger than 6 years. Signs and symptoms include fever and a "whoop" sound that occurs when the patient tries to inhale after a coughing attack.

The best way to prevent exposure is to try to place a mask on the patient and on yourself.

Newly Recognized Diseases

Newly recognized diseases, such as those caused by Hantavirus or enteropathogenic *Escherichia coli*, are being reported. These diseases are not transmitted from person to person directly; rather, they are carried by a vehicle, such as food, or a vector, such as rodents. Although not a newly discovered illness, West Nile

Virus has caused some concern recently. This virus' vector is the mosquito, and affects humans and birds. The virus is actually tracked by tests done on birds suspected of being killed by the virus. These diseases are not communicable and do not pose a risk to you during patient care.

Another virus that has caused significant concern of late is best known as SARS, or severe acute respiratory syndrome. SARS is a serious, potentially life-threatening viral infection caused by a recently discovered family of viruses best known as the second most common cause of the common cold. SARS usually starts with flu-like symptoms, which may progress to pneumonia, respiratory failure, and, in some cases, death. The SARS virus strain probably spread from Guangdong province in southern China to Hong Kong, Singapore, and Taiwan. Canada has had a significant outbreak in the Toronto area. SARS is thought to be primarily transmitted by close person-to-person contact. Most cases have involved persons who lived with or cared for a person with SARS or who had exposure to contaminated secretions from a SARS patient.

Multiple antibiotic-resistant organisms have recently been the subject of media scrutiny. These organisms should be viewed as no more or less contagious than other less resistant organisms of the same type. The same precautions apply.

General Postexposure Management

In many instances, you will not know that a patient has an airborne or bloodborne disease, and you could be exposed without knowing it. The Ryan White Law requires that the hospital notify your department's designated officer, the individual in the department who is charged with the responsibility of managing exposures and infection control issues, within 48 hours of the time the hospital identifies the patient's disease. In the event of possible exposure, there should be a protocol in place to obtain information from your local hospital or other medical resource. You should be screened and given information about the necessity of medical follow-up. Treatment depends on the disease. Your designated officer will assist you with the necessary information.

If you experience a needlestick injury or some other unprotected exposure to blood, you must notify your department's designated officer as soon as possible and complete an incident report. The designated officer

Documentation Tips

The ability of your EMS service to support you in case of exposure to a communicable disease depends on your understanding of how exposure can occur and your immediate report of exposure to potentially infectious materials. Make notes right away to ensure that you remember all pertinent information, and report immediately after the response following your service's guidelines.

can contact the hospital for information; the hospital has 48 hours to report back to the designated officer. Depending on your state laws and whether it is possible, patient testing should be done, followed by baseline testing on you.

Because there are many diseases for which there are no outward signs of infection, your protection lies in the use of PPE and/or prompt reporting of exposure. Be familiar with the postexposure protocols outlined in your department's exposure control plan.

Establishing an Infection Control Routine

Infection control, procedures to reduce infection in patients and health care personnel, should be an important part of your daily routine. Take the following steps in dealing with potential exposure situations:

1. **En route to the scene**, make sure that all equipment is out and available.
2. **Upon arrival**, make sure the scene is safe to enter, then do a quick visual assessment of the patient, noting whether any blood is present.
3. **Select the proper PPE** according to the tasks you are likely to perform.
4. **Change gloves and wash hands** between patients; do not unnecessarily delay treatment for use of PPE, thereby potentially putting patients at risk. Remove gloves and other gear after contact with the patient, unless you are in the patient compartment. Remember that good hand washing is always necessary.
5. **Limit the number of people** who are involved in patient care if there are multiple injuries and a substantial amount of blood at the scene.

Figure 2-20 Contaminated linen should be bagged appropriately and disposed of according to your local protocols.

6. **If you or your partner is exposed** while providing care, try to relieve one another as soon as possible so that you can seek care. Notify the designated officer and report the incident. This will also help to maintain confidentiality.

Be sure to routinely clean the ambulance after each run and on a daily basis. Cleaning is an essential part of the prevention and control of communicable diseases and will remove surface organisms that may remain in the unit.

You should clean your unit as quickly as possible so that it can be returned to service. Address the high-contact areas, including surfaces that were in direct contact with the patient's blood or body fluids or surfaces that you touched while caring for the patient after having contact with the patient's blood or body fluids.

Whenever possible, cleaning should be done at the hospital. If you clean the unit back at the

station, make sure you have a designated area with good ventilation.

Bag any medical waste and dispose of it in a red bag at the hospital whenever possible. Any contaminated equipment that is left with the patient at the hospital should be cleaned by hospital staff or bagged for transport and cleaning at the station.

Clean the unit with soap and water. After cleaning, disinfect the unit with a bleach and water solution at a 1:10 dilution. Isopropyl alcohol can also be used for disinfecting, as can a hospital-approved disinfectant that is effective against *M tuberculosis*. Use the disinfecting solution in a bucket or pistol-handled spray container. Pay attention to disinfectant directions. Some need to remain on the surface for a few minutes in order to work. Dilute bleach solutions and alcohol should not be used on some soft surfaces since they may corrode or discolor certain fabrics, leathers, vinyl, or other synthetic materials. Note that exam-type gloves are not appropriate for cleaning and disinfecting. These tasks require heavier gloves.

Remove contaminated linen, and place it into an appropriate bag for handling. Each hospital may have a different system for handling contaminated linen; you should learn hospital or department protocols Figure 2-20 ◄.

Learn the regulations defining medical waste in your area. The disposal of infectious waste, such as needles, sharps, and heavily soiled dressings, may vary from hospital to hospital and from state to state.

Scene Hazards

In the course of your career, you will be exposed to many hazards. Some situations will be life-threatening. In these cases, you must be properly protected, or you must avoid the hazard completely.

Hazardous Materials

Your safety is the most important consideration at a hazardous materials incident. Upon your arrival, you should first try to read labels and identification numbers. All hazardous materials should be marked with safety placards, though this is not always done. These placards are marked with colored diamond-shaped labels Figure 2-21 ►. Although it is important for you to obtain information from the placards, you should never approach any object marked with a placard.

It is useful to carry binoculars in the ambulance so that you can read placards from a safe distance. A specially trained and equipped hazardous materials

Figure 2-21 Hazardous materials safety placards are marked with colored diamond-shaped labels.

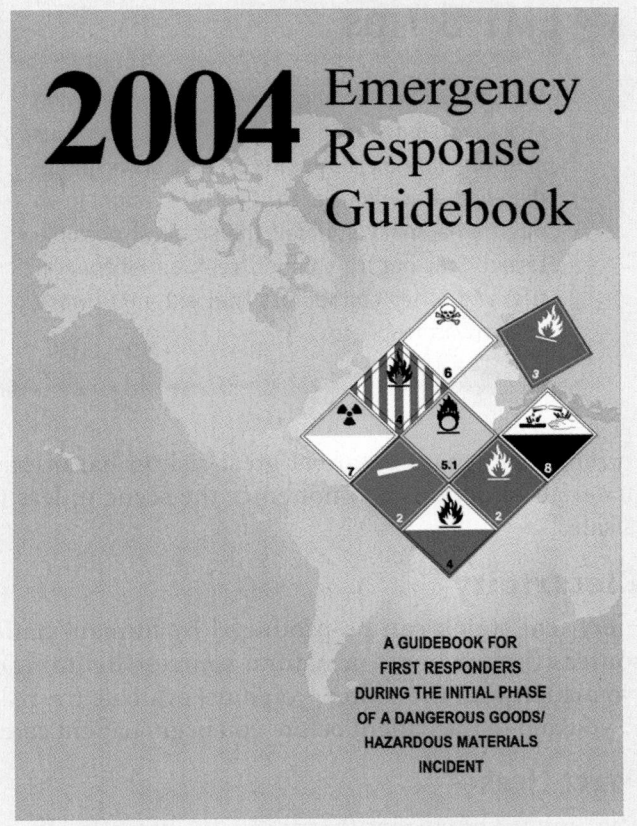

Figure 2-22 The DOT's *2004 Emergency Response Guidebook* lists many hazardous materials and the proper procedures for scene control and emergency care of patients.

team will be called to handle disposal of materials and removal of patients. You should not begin caring for patients until they have been moved away from the scene and are decontaminated or the scene is safe for you to enter.

The DOT's *2004 Emergency Response Guidebook* is an important resource (Figure 2-22 ▶). It lists most hazardous materials and the proper procedures for scene control and emergency care of patients. Several similar resources are available. Some state and local government agencies may also have information about hazardous materials in their areas. A copy of the guidebook and other information relevant to your area should be available in your unit or at the dispatch center. Thus, you should be able to begin proper emergency management as soon as the hazardous material is identified. Again, do not go into an area and risk exposure to yourself. Do not enter the area unless you are absolutely sure that no hazardous spill has occurred.

Hazardous materials are classified according to toxicity levels, which dictate the level of protection required. The toxicity levels—0, 1, 2, 3, 4—measure the risk the substance poses to an individual. The higher the number, the greater is the toxicity and the greater is the protection needed (Table 2-9 ▶). It is important

EMT-B Safety

Experienced EMT-Bs are aware of the potential hazards that may be present when responding to the scene. At times, these situations may be dangerous to you and your crew. The best protection when hazards are present is early recognition that a hazard may exist.

TABLE 2-9 Toxicity Levels of Hazardous Materials*

Level	Hazard	Protection Needed
0	Little to no hazard	None
1	Slightly hazardous	Self-contained breathing apparatus only (level C suit)
2	Slightly hazardous	Self-contained breathing apparatus only (level C suit)
3	Extremely hazardous	Full protection, with no exposed skin (level A or B suit)
4	Minimal exposure causes death	Special HazMat gear (level A suit)

*For more information on classification of hazardous materials, see Chapter 37.

EMT-B Tips

Recognize the warning signs before a lightning strike. You may feel a tingling sensation on your skin, or your hair may stand on end. Move immediately to a low-lying area. If you are caught in an open area, make yourself the smallest possible target. Stay out from under trees, aerial apparatus, and other conductors that extend upward into the atmosphere.

to remember that you are at great risk in hazardous materials situations. Do not enter the scene unless it is safe.

Electricity

Electrical shock can be produced by human-made sources (power lines) or natural sources (lightning). No matter what the source, you must evaluate the risk to you and to the patient before you begin patient care.

Power Lines

The amount of current that is involved greatly affects the level of risk for injury. Your local power company can help you by providing training to evaluate the risks in electrical emergencies. Its staff can also teach you how to deal with power lines once the risks have been established. You should not touch downed power lines. Dealing with power lines is beyond the scope of EMT-B training. However, you should mark off a danger zone around the downed lines.

Energized, or "live" power lines, especially high-voltage lines, behave in unpredictable ways. You need in-depth training to be able to handle the equipment that is used in an electrical emergency. The equipment also has specific storage needs and requires careful cleaning. Dirt or other contaminants can make this equipment useless or dangerous.

At the scene of a motor vehicle crash, above-ground and below-grade power lines may become hazards. Disrupted overhead wires are usually a visible hazard. You must be careful even if you do not see sparks coming from the lines. Visible sparks are not always present in charged wires. The area around downed power lines is always a danger zone. This danger zone extends well beyond the immediate accident scene.

Use the utility poles as landmarks for establishing the perimeter of the danger zone. The danger zone must be a restricted area. Remember, the safety zone is one span of the power pole's distance. Only emergency personnel, equipment, and vehicles are allowed inside this area. Do not approach downed wires or touch anything that downed wires have come in contact with until qualified personnel have concluded that no risk of electrical injury exists. This may mean that you are unable to access a severely injured victim of a motor vehicle crash even though you can see and talk to him or her.

If you must enter this type of situation, be sure to wear the proper protective equipment according to the type of incident. A helmet and turnout gear (Figure 2-23 ▼) are typically called for, though you cannot count on turnout gear for protection from electrical hazards. Other protective equipment may be needed. The Protective Clothing section covers turnout gear and helmets in more depth.

Lightning

Lightning is a complex natural phenomenon. You are unwise to think that "lightning never strikes in the same place twice." If the right conditions remain, a repeat strike in the same area can occur.

Figure 2-23 Wear a helmet made of a certified electrical nonconductor material, making sure that the chin strap is fastened securely.

Lightning is a threat in two ways: through a direct hit and through ground current. After the lightning bolt strikes, the current drains along the earth, following the most conductive pathway. Although you should avoid high ground to avoid a direct strike, to avoid being injured by a ground current, stay away from drainage ditches, moist areas, small depressions, and wet ropes. If you are involved in a rescue operation, you may need to delay it until the storm has passed. Recognize the warning signs just before a lightning strike. As your surroundings become charged, you may feel a slight tingling sensation on your skin, or your hair may even stand on end. In this situation, a strike may be imminent. Move immediately to the lowest possible area.

If you are caught in an open area, try to make yourself the smallest possible target for a direct hit or for ground current. To keep from being hit by the initial strike, stay away from projections from the ground, such as a single tree. Drop all equipment, particularly metal objects that project above your body. Avoid fences and other metal objects. These can transmit current from the initial strike over a long distance. Position yourself in a low crouch. This position exposes only your feet to the ground current. If you sit, both your feet and your buttocks are exposed. Place an object made of nonconductive material, such as a blanket, under your feet. Get inside a car or your unit, if possible, as vehicles will protect you from lightning.

Fire

You will often be called to the scene of a fire. Therefore, you should understand some basic information about fire, if you do not know it already. There are five common hazards in a fire:

1. Smoke
2. Oxygen deficiency
3. High ambient temperatures
4. Toxic gases
5. Building collapse
6. Equipment

Smoke is made up of particles of tar and carbon. These particles irritate the respiratory system on contact. Most smoke particles are trapped in the upper respiratory system, but many smaller particles enter the lungs. Some smoke particles not only irritate the airway, but also may be deadly. You must be trained in the use of proper bunker gear, appropriate airway protection, such as a self-contained breathing apparatus or a disposable short-term device, and following local fire ground protocols before entering a fire scene (Figure 2-24 ▶).

Figure 2-24 You should be trained in the use of self-contained breathing apparatus and have it available if you may be working near fire scenes.

Fire consumes oxygen. Particularly in a closed space, such as a room, fire may consume most of the available oxygen. This will make breathing difficult for anyone in that space. The high ambient temperatures in a fire can result in thermal burns and damage to the respiratory system. Breathing air that is heated above 120°F (49°C) can damage the respiratory system.

A typical building fire emits a number of toxic gases, including carbon monoxide and carbon dioxide. Carbon monoxide is a colorless, odorless gas that is responsible for more fire deaths each year than any other by-product of combustion. Carbon monoxide combines with the hemoglobin in your red blood cells about 200 times more readily than oxygen does. It blocks the ability of the hemoglobin to transport oxygen to your body tissues. Carbon dioxide is also a colorless, odorless gas. Exposure causes increased respirations, dizziness, and sweating. Breathing concentrations of carbon dioxide greater than 10% to 12% will result in death within a few minutes.

During and after a fire, there is always a possibility that all or part of the burned structure will collapse. Often, there are no warning signs. As an EMS provider, you should never enter a burning building without proper breathing apparatus and the approval of the incident commander or safety officer at the scene. Hasty entry into a burning structure may result in serious injury and possibly death. Once inside a burning building, you are subject to an uncontrolled, hostile environment. Fires are not selective about their victims. You must be extremely cautious whenever you are near a burning structure or one in which a fire has just been placed under control. At any fire scene, follow the instructions of the incident commander and safety officer and never undertake any task (ie, enter a burning structure or initiate search and rescue) unless you have been properly trained to do so.

Fuel and fuel systems of vehicles that have been involved in crashes are also a hazard. Although this rarely happens, any leaking car fuel may ignite under the right conditions. If you see or smell a fuel leak, or if people are trapped in the vehicle, you must coordinate appropriate fire protection. Gasoline and other auto fluids are considered hazardous materials.

Make sure that you are properly protected if there is or has been a fire in the vehicle. Wear appropriate respiratory protection and thermal protection, as the smoke from a vehicle fire contains many toxic by-products. The use of appropriate protective gear at a crash scene can reduce your risk of injury. Avoid using oxygen in or near a vehicle that is smoking, smoldering, or leaking fuel.

Protective Clothing: Preventing Injury

Wearing protective clothing and other appropriate gear is critical to your personal safety. Become familiar with the protective equipment that is available to you. Then you will know what clothing and gear are needed for the job. You will also be able to adapt or change items as the situation and environment change. Remember that protective clothing and gear are safe only when they are in good condition. It is your responsibility to inspect your clothing and gear. Learn to recognize how wear and tear can make your equipment unsafe. Be sure to inspect equipment before you use it, even if you must do so at the scene.

Clothing that is worn for rescue must be appropriate for the activity and the environmental conditions in which the activity will take place. For example, bunker gear that is worn for fire fighting may be too restrictive for working in a confined space. In every situation involving blood and/or other body fluids, be sure to follow BSI precautions. You must protect yourself and the patient by wearing gloves and eye protection, as well as any additional protective clothing that may be needed.

EMS coats should provide a body fluids barrier if purchased after 1998.

Cold Weather Clothing

When dressing for cold weather, you should wear several layers of clothing. Multiple layers provide much better protection than a single thick cover. You have more flexibility to control your body temperature by adding or removing a layer. You can also lose a good bit of body heat if you are not wearing a hat. Cold weather protection should consist of at least the following three layers:

1. **A thin inner layer** (sometimes called the transport layer) next to your skin. This layer pulls moisture away from your skin, keeping you dry and warm. Underwear made of polypropylene or polyester material works well. Wool is the best fiber. The goal is to wick moisture away from the skin.
2. **A thermal middle layer** of bulkier material for insulation. Wool has been the material of choice for warmth, but newer materials, such as polyester pile, are also commonly used.
3. **An outer layer** that will resist chilling winds and wet conditions, such as rain, sleet, or snow. The two top layers should have zippers to allow you to vent some body heat if you become too warm.

When choosing clothing to protect yourself from the weather, pay attention to the type of material used. Cotton should be avoided in cold, wet environments. Cotton tends to absorb moisture, causing chilling from wetness. For example, if you wear cotton trousers and walk through wet grass, the cotton soaks up the moisture from the grass. This will chill you in cold weather. However, cotton is appropriate in warm, dry weather because it absorbs moisture and pulls heat away from the body.

As an outer layer in cold weather, you might consider plastic-coated nylon, as it provides good waterproof protection. However, it can also hold in body heat and perspiration, which makes you wet both inside and out. Newer, less airtight materials allow perspiration and some heat to escape while the material retains its water resistance. Avoid flammable or meltable synthetic material anytime there is any possibility of fire.

Turnout Gear

Turnout or bunker gear is a fire service term for protective clothing designed for use in structural fire fighting environments (Figure 2-25 ▶). Turnout gear provides some protection. It uses different layers of fabric or other

Figure 2-25 Turnout or bunker gear is protective clothing designed for use in fire fighting.

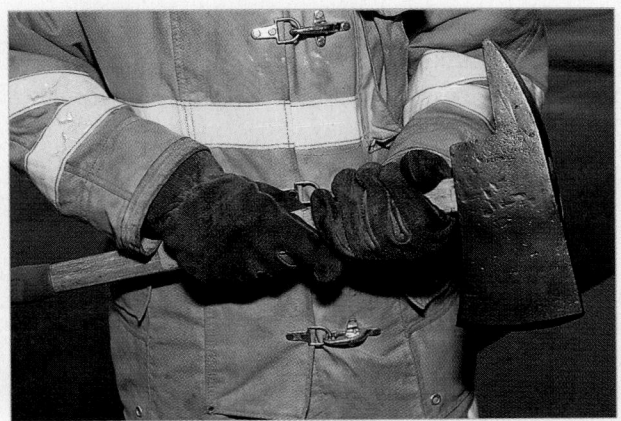

Figure 2-26 Fire fighting gloves protect your hands and wrists from heat, cold, and injury.

material to provide protection from the heat of fire, to reduce trauma from impact or cuts, and to keep water away from the body. Like most protective clothing, turnout gear adds weight and reduces range of motion to some degree.

The exterior fabrics provide increased protection from cuts and abrasions. They also act as a barrier to high external temperatures. In cold weather, an insulated thermal inner layer of material that helps to retain body heat is recommended.

Turnout gear or a bunker jacket provides minimal protection from electrical shock. But it does protect you from heat, fire, possible flashover, and flying sparks. The front opening of the jacket should be fastened, and the jacket should be worn with the collar up and closed in front to protect your neck and upper chest. Proper fit is important so that you can move freely.

Gloves

Fire fighting gloves will provide the best protection from heat, cold, and cuts (Figure 2-26 ▲). Yet these gloves reduce manual dexterity. In addition, fire fighting gloves will not protect you from electrical hazards. In rescue situations, you must be able to use your hands freely to operate rescue tools, provide patient care, and perform other duties. Puncture-proof leather gloves, with latex gloves underneath, will permit free use of your hands with added protection from both injury and body fluids. The Communicable Diseases section in this chapter discusses the use of latex gloves for infectious exposure prevention. Latex gloves do not provide absolute protection.

Helmets

You should wear a helmet any time you are working in a fall zone. A fall zone is an area where you are likely to encounter falling objects. The helmet should provide top and side impact protection. It should also have a secure chin strap (Figure 2-27 ▶). Objects will often fall one after another. If the strap is not secure, the first falling object may knock off your helmet. This leaves your head unprotected as the remaining objects fall.

Construction-type helmets are not well suited for rescue situations. They offer minimal impact protection and have inadequate chin straps. Modern fire helmets offer impact protection. However, the projecting brim at the back of the neck may get in your way in a rescue situation. In cold weather, you can lose a good bit of body heat if you are not wearing a hat.

Figure 2-27 A helmet with side impact protection and a chin strap will not dislodge if struck by an object.

Figure 2-28 Boots should cover and protect your ankles, keeping out stones, debris, and snow. Steel-toed boots are preferred.

An insulated hat made from wool or a synthetic material can be pulled down over the face and the base of the skull to reduce heat loss in extremely cold weather.

In situations that may involve electrical hazard, you should always wear a helmet with chin strap and face shield. The shell of the helmet should be made of a certified electrical nonconductor. The chin strap should not stretch. In fact, it should fasten securely so that the helmet stays in place if you are knocked down or a power line hits your head. You should also be able to lock the face shield on the helmet. This will protect your face and eyes from power lines and flying sparks. A standard fire turnout helmet should meet all of these needs.

Boots

Boots should protect your feet. They should be water resistant, fit well, and be flexible so that you can walk long distances comfortably. If you will be working outdoors, you should choose boots that cover and protect your ankles, keeping out stones, debris, and snow. Steel-toed boots are preferred (Figure 2-28 ▶). In cold weather, your boots must also protect you from the cold. Leather is one of the best materials for boots. However, other materials, such as Gore-Tex water-repellent fabric, are also very good. The soles of your boots must provide traction. Lug-type soles may grip well in snow, but they become very slippery when caked with mud.

The fit of boots and shoes is extremely important, because a minor annoyance can develop into a disabling injury. You may develop painful blisters if your feet slip around inside your boots. However, make sure you have enough room to wiggle your toes.

Boots should be puncture-resistant, protect the toes, and provide foot support. It may be difficult to obtain a good fit with fire fighting boots; shoe inserts or sock layering may be needed for a comfortable fit. Make sure the tops of your boots are sealed off to keep rain, snow, glass, or other materials from getting into your boots. Moisture increases blistering—wool or wicking socks help prevent feet from becoming wet.

Socks will keep your feet warm and provide some cushioning for you as you walk. In cold weather, two pairs of socks are generally preferable to one thick pair. A thin sock next to the foot helps to wick perspiration away to a thicker, outer sock. This tends to keep your feet warmer, drier, and generally more comfortable. When you purchase new shoes or boots, keep these points in mind.

Eye Protection

The human eye is very fragile, and permanent loss of sight can occur from very minor injuries. You need to protect your eyes from blood and other body fluids, foreign objects, plants, insects, and debris from extrication. You may wear eyeglasses with side shields during routine patient care. (The Communicable Diseases section in this chapter covers eye protection against splattering body fluids.)

However, when tools are being used during extrication, you should wear a face shield or goggles. In these instances, prescription eyeglasses do not provide adequate protection. In snow or white sand, particularly at higher altitudes, you must protect your eyes from ultraviolet exposure. Specially designed glasses or goggles can provide this. In addition, your eye protection must be adaptable to the weather and the physical demands of the task. It is critical that you have clear vision at all times.

Ear Protection

Exposure to loud noises for long periods of time can cause permanent hearing loss. Certain equipment, such as helicopters, some extrication tools, and sirens, produces high levels of noise. Wearing soft foam industrial-type earplugs usually provides adequate protection.

Skin Protection

Your skin needs protection against sunburn while you are working outdoors. Long-term exposure to the sun increases the possibility of skin cancer. It might be considered simply an annoyance, but sunburn is a type of burn. In reflective areas such as sand, water, and snow, your risk of sunburn increases. Protect your skin by applying a sunscreen with a minimum rating of SPF 15.

Body Armor

Though policy for most EMT-Bs directs them simply to avoid situations that may involve gunshots, EMS responders in some areas do wear body armor (bulletproof vests) for personal protection. Several types of body armor are available. They range from extremely lightweight and flexible to heavy and bulky. The lighter vests do not stop large-caliber bullets. However, they offer more flexibility and are preferred by most law enforcement personnel. Lighter vests are commonly worn under a uniform shirt or jacket. The larger, heavier vests are worn on the outside of your uniform.

Violent Situations

The safety of you and your team is of primary concern. Civil disturbances, domestic disputes, and crime scenes, especially those involving gangs, can create many haz-

Figure 2-29 Several agencies may respond to large disturbances. It is important for you to know who is in command and will be issuing orders.

ards for EMS personnel. Large gatherings of hostile or potentially hostile people are also dangerous. Several agencies will respond to large civil disturbances. In these instances, it is important for you to know who is in command and will be issuing orders Figure 2-29 ▲). However, you and your partner may be on your own when a group of people seems to grow larger and become increasingly hostile. In these cases, you should call law enforcement immediately if they are not already present. You may need to wait for law enforcement to arrive before you can begin treatment or safely approach a patient.

Remember that you and your partner must be protected from the dangers at the scene before you can provide patient care. Law enforcement must make sure the scene is safe before you and your partner enter. A crime scene often poses potential problems for EMS personnel. If the perpetrator is still somewhere on the scene, this person could reappear and threaten you and your partner or attempt to further injure the patient you are treating. Bystanders who are trying to be helpful may interfere with your emergency medical care. Family members may be very distraught and not understand what you are doing when you attempt to splint an injured extremity and the patient cries out that what you are doing hurts. Be sure that you have adequate assistance from the appropriate public safety agency in these cases.

Sometimes EMT-Bs will be at a scene where a dangerous situation is underway, such as a hostage situation or riot. In these instances, it may be necessary for EMS personnel to be protected from projectiles such as bullets, bottles, and rocks. Law enforcement personnel will ordinarily provide for concealment or cover of personnel who are involved in the response to the incident. Cover and concealment involve the tactical use of an impenetrable barrier for protection. EMT-Bs should not be placed in a position that will endanger their lives or safety during such incidents. Do not depend on someone else for your safety.

Remember that your personal safety is of the utmost importance. You must thoroughly understand the risks of each environment you enter. Whenever you are in doubt about your safety, do not put yourself at risk. Never enter an unstable environment, such as a shooting, a brawl, a hostage situation, or a riot. Therefore, as part of your scene size-up, evaluate the scene for the potential for violence. If further violence is a possibility, call for additional resources. Failure to do so may put you and your partner at serious risk. When appropriate, allow law enforcement personnel to secure the scene before you approach; they have the necessary experience and expertise in handling these situations.

It is also important for you to remember that if you believe that an event is a crime scene, you must attempt to maintain the chain of evidence. Briefly, make sure that you do not disturb the scene unless it is absolutely necessary in caring for the patient.

Behavioral Emergencies

The category of "behavioral emergencies" covers a wide range of situations. This catchall phrase includes emergencies that do not have a clear physical cause and that result in aberrant behavior. Often, the cause turns out to be physical; hypoglycemia, head trauma, hypoxia, and toxic ingestion can all cause altered mental status. Patients with psychiatric diseases, such as certain bipolar disorders or schizophrenia, may have altered sensorium or exhibit abnormal behavior.

Although most behavioral emergencies do not pose a threat to the EMT, the potential of threat to either the patient or the rescuer still exists and caution should be used.

Although Chapter 19 goes into greater depth about behavioral emergencies, consider these questions as you evaluate the patient in terms of a behavioral or psychiatric emergency that may lead to a violent patient reaction:

- How does this patient respond to you? Are your questions answered appropriately? Are the patient's vocabulary and expressions what you would expect under the circumstances?
- Is the patient withdrawn or detached? Is the patient hostile or friendly? Overly friendly?
- Does the patient understand why you are there?
- How is the patient dressed? Is the dress appropriate for the time of the year and occasion? Are the clothes clean? Dirty?
- Does the patient appear relaxed, stiff, or guarded? Are the patient's movements coordinated or jerky and awkward? Is there hyperactivity? Are the patient's movements purposeful, for example, in putting his or her clothes on? Are the actions aimless, such as sitting and rocking back and forth in a chair?
- Has the patient harmed herself or himself? Is there damage to the surroundings?
- What are the patient's facial expressions? Are they bland or flat, or are they expressive? Does the patient show joy, fear, or anger to appropriate stimuli? If so, to what degree?

It might not be possible for you to gather all of the information that these questions suggest. Sometimes, a patient who is experiencing a behavioral emergency will not respond at all. In those cases, the patient's facial expressions, pulse and respirations, tears, sweating, and blushing may be significant indicators of his or her emotional state.

The following principal determinants of violence, though not intended to be all-inclusive, are of value for the EMT-B:

- **Past history.** Has this patient previously exhibited hostile, overly aggressive, or violent behavior? This information should be solicited by EMS personnel at the scene or requested from law enforcement personnel, family, previous EMS records, or hospital information.
- **Posture.** How is this person sitting or standing? Does the patient appear to be tense, rigid, or sitting on the edge of the bed, chair, or wherever he or she is positioned? The observation of

increased tension by physical posture is often a warning signal for hostility.

- **Vocal activity.** What is the nature of the speech the patient is using? Loud, obscene, erratic, and bizarre speech patterns usually indicate emotional distress. The patient who is conversing in quiet, ordered speech is not as likely to strike out against others as is the patient who is yelling and screaming.
- **Physical activity.** Perhaps one of the most demonstrative factors to look for is the motor activity of a person who is undergoing a behavioral crisis. The patient who is pacing, cannot sit still, or is displaying protection of his or her boundaries of personal space needs careful watching. Agitation is a prognostic sign to be observed with great care and scrutiny.

Other factors to take into consideration for potential violence include the following:

- Poor impulse control
- The behavior triad of truancy, fighting, and uncontrollable temper
- Instability of family structure, inability to keep a steady job
- Tattoos, such as those with gang identification or statements like "born to kill" or "born to lose"
- Substance abuse
- Functional disorder (If the patient says that he or she is hearing voices that say to kill, believe it!)
- Depression, which accounts for 20% of violent attacks
- Diagnosed illness such as bipolar disease

You are the Provider   Summary

Your ability to reduce and channel the negative impacts of stress has made you a more healthy person and emergency responder. With increased resilience to stress, you have a stronger immune system and infrequently suffer from illness or injury. You've noticed that the responders who have continued to provide care for 10 years or greater seem to have similar lifestyle habits and you feel encouraged that these choices will assist in your career longevity. There are so many benefits to your new lifestyle you can't imagine rewarding yourself with anything less.

Prep Kit

Ready for Review

- EMT-Bs will encounter death, dying patients, and the families and friends of those who have died.

- When signs of stress such as fatigue, anxiety, anger, feelings of hopelessness, worthlessness, or guilt, and other such indicators manifest themselves, behavioral problems can develop.

- Recognizing the signs of stress is important for all EMT-Bs.

- Every patient encounter should be considered to be potentially dangerous. It is essential that you take all available precautions to minimize exposure and risk to scene hazards and infectious and communicable diseases.

- Infectious diseases can be transmitted in one of four ways: direct transmission, vehicle-borne, vector-borne, and airborne.

- Even if you are exposed to an infectious disease, your risk of becoming ill is small.

- Whether or not an acute infection occurs depends on several factors, including the amount and type of infectious organism and your resistance to that infection.

- You can take several steps to protect yourself against exposure to infectious diseases, including:
  - keeping up to date with recommended vaccinations
  - using universal precautions
  - following BSI precautions at all times
  - handling all needles and other sharp objects with great care

- Because it is often impossible to tell which patients have infectious diseases, you should avoid direct contact with the blood and body fluids of all patients.

- If you think you may have been exposed to an infectious disease, see your physician (or your employer's designated physician) immediately.

- Five infectious diseases of special concern are:
  - HIV infection
  - Hepatitis B
  - Meningitis
  - Tuberculosis
  - SARS

- You should know what to do if you are exposed to an airborne or bloodborne disease. Your department's designated officer will be able to help you follow the protocol set up in your area.

- Infection control should be an important part of your daily routine. Be sure to follow the proper steps when dealing with potential exposure situations.

- Scene hazards include potential exposure to:
  - Hazardous materials
  - Electricity
  - Fire

- At a hazardous materials incident, your safety is the most important consideration. Never approach an object labeled with a hazardous materials placard. Use binoculars to read the placards from a safe distance.

- Do not begin caring for patients until they have been moved away from the scene and decontaminated by the hazardous materials team or the scene has been made safe for you to enter.
  - Electrical shock can be produced by power lines or by lightning.

- There are five common hazards in a fire:
  - Smoke
  - Oxygen deficiency
  - High ambient temperatures
  - Toxic gases
  - Building collapse

- Violent situations such as civil disturbances, domestic disputes, and crime scenes can create many hazards for EMS personnel.

- If you see the potential for violence during a scene size-up, call for additional resources.

www.EMTB.com

Technology

Interactivities

Vocabulary Explorer

Anatomy Review

Web Links

Online Review Manual

Vital Vocabulary

body substance isolation (BSI) An infection control concept and practice that assumes that all body fluids are potentially infectious.

burnout A condition of chronic fatigue and frustration that results from mounting stress over time.

carrier An animal or person who is infected with and may transmit an infectious disease but may not display any symptoms of it; also known as a vector.

communicable disease Any disease that can be spread from person to person, or from animal to person.

contagious disease An infectious disease that is capable of being transmitted from one person to another.

contamination The presence of infectious organisms on or in objects such as dressings, water, food, needles, wounds, or a patient's body.

cover and concealment The tactical use of an impenetrable barrier to conceal EMS personnel and protect them from projectiles (eg, bullets, bottles, rocks).

critical incident stress debriefing (CISD) A confidential peer group discussion of a severely stressful incident that usually occurs within 24 to 72 hours of the incident.

critical incident stress management (CISM) A process that confronts the responses to critical incidents and defuses them, directing the emergency services personnel toward physical and emotional equilibrium.

designated officer The individual in the department who is charged with the responsibility of managing exposures and infection control issues.

direct contact Exposure or transmission of a communicable disease from one person to another by physical contact.

exposure A situation in which a person has had contact with blood, body fluids, tissues, or airborne particles in a manner that suggests that disease transmission may occur.

exposure control plan A comprehensive plan that helps employees to reduce their risk of exposure to or acquisition of communicable diseases.

general adaptation syndrome The body's three-stage response to stress. First, stress causes the body to trigger an alarm response, followed by a stage of reaction and resistance, and then recovery, or if the stress is prolonged, exhaustion.

hepatitis Inflammation of the liver, usually caused by a viral infection, that causes fever, loss of appetite, jaundice, fatigue, and altered liver function.

herpes simplex Infections caused by human herpesviruses 1 and 2, characterized by small blisters whose location depend on the type of virus. Type 2 results in blisters on the genital area, while type 1 results in blisters in nongenital areas.

HIV infection Human immunodeficiency virus (HIV). The virus can cause acquired immunodeficiency syndrome (AIDS).

host The organism or individual that is attacked by the infecting agent.

indirect contact Exposure or transmission of disease from one person to another by contact with a contaminated object (vehicle).

infection The abnormal invasion of a host or host tissues by organisms such as bacteria, viruses, or parasites, with or without signs or symptoms of disease.

infection control Procedures to reduce transmission of infection among patients and health care personnel.

infectious disease A disease that is caused by infection, in contrast to one caused by faulty genes, metabolic or hormonal disturbances, trauma, or something else.

meningitis An inflammation of the meningeal coverings of the brain and spinal cord; it is usually caused by a virus or a bacterium.

Occupational Safety and Health Administration (OSHA) The federal regulatory compliance agency that develops, publishes, and enforces guidelines concerning safety in the workplace.

pathogen A microorganism that is capable of causing disease in a susceptible host.

personal protective equipment (PPE) Protective equipment that OSHA requires to be made available to the EMT. In the case of infection risk, PPE blocks entry of an organism into the body.

posttraumatic stress disorder (PTSD) A delayed stress reaction to a prior incident. This delayed reaction is often the result of one or more unresolved issues concerning the incident.

SARS (severe acute respiratory syndrome) Potentially life-threatening viral infection that usually starts with flu-like symptoms.

www.EMTB.com

transmission The way in which an infectious agent is spread: contact, airborne, by vehicles, or by vectors.

tuberculosis A chronic bacterial disease, caused by *Mycobacterium tuberculosis*, that usually affects the lungs but can also affect other organs such as the brain or kidneys.

universal precautions Protective measures that have traditionally been developed by the Centers for Disease Control and Prevention (CDC) for use in dealing with objects, blood, body fluids, or other potential exposure risks of communicable disease.

virulence The strength or ability of a pathogen to produce disease.

Points to Ponder

You have just been assigned a new partner and it is the beginning of your first shift together. Your new partner is a very experienced EMT-B with more than 15 years on the job. The ambulance has been checked out and you are ready for service. As you respond to the first call of the day, your partner does not put on his safety belt. He tells you to keep alert because he's working an extra shift today and hasn't gotten much sleep the last few days. His wife lost her job a few months ago and he's been taking on as many extra hours as he can. Dispatch informs you that a domestic dispute is in progress and a gunshot was heard. Your partner recognizes the address because he's been to the home twice already this week. As you are putting on your gloves, you give a pair to you partner. He puts them aside and says, "We won't be transporting any patients. I'm so sick of dealing with those two. Nothing we can do will help." You realize the need for patient respect, respect toward your partner, safety procedures, and the need for body substance isolation precautions. What should you do?

Issues: Duty to Act, Infection Control, Burnout, Respect for Patients.

Assessment in Action

This is what you trained for. It is your first shift. After being introduced to your new partner you check out your ambulance and prepare for your first call. The radio goes off and your first call is announced. "Ambulance One, respond to a report of man down."

As you arrive, you put on your gloves and eye protection and walk towards the patient. You find an older man lying on the ground with a group of people surrounding the patient. There are bystanders lingering on the scene. The police officer ensures that the scene is safe. The patient tells you he has been drinking beer all day and wants to go to the hospital. He is very dirty and smells of vomit and old beer. Your partner asks you to move the patient to the ambulance for a more detailed assessment.

1. Personal protection begins:
   - **A.** after seeing the patient.
   - **B.** as soon as you are dispatched.
   - **C.** at the beginning of your shift.
   - **D.** if a large amount of blood is present.

2. By wearing eye protection and gloves to reduce contact with body fluids while treating this patient, you are practicing:
   - **A.** body substance isolation.
   - **B.** common sense.
   - **C.** protocol enforcement practices.
   - **D.** universal precautions.

3. EMS is a high-stress job. In most cases the EMT-B will deal with stress by:
   - **A.** providing absolute confidentiality in patient care.
   - **B.** using general adaptation syndrome.
   - **C.** joking about how stressed out he or she is.
   - **D.** keeping it all inside.

4. The EMT-B must learn how to deal with stressful calls to prevent chronic fatigue and frustration associated with:
   - **A.** burnout.
   - **B.** hopefulness.
   - **C.** critical incident stress debriefing.
   - **D.** multiple changes in assignments.

5. This call requires the EMT-B to communicate with the patient, bystanders, and your partner in a way that is:
   - **A.** forceful and direct.
   - **B.** loud and clear.
   - **C.** soft and slow.
   - **D.** sensitive to everyone's needs.

6. While treating this patient, you get some body fluid on your arms. This contact is called a(n):
   - **A.** disaster.
   - **B.** exposure.
   - **C.** accident.
   - **D.** incident.

7. If you do come in contact with potentially infectious materials, you must follow your:
   - **A.** federal assessment plan.
   - **B.** individual policy and procedures.
   - **C.** supervisor's advice.
   - **D.** agency's exposure control plan.

Challenging Questions

8. Explain the grieving process as defined by Kubler-Ross.

9. How should the EMT-B respond and treat the critically injured child?

10. Define the four major vehicles responsible for the transmission of disease.

11. What is the EMT-B's duty with regard to treating patients with serious infectious diseases?

www.EMTB.com

Medical, Legal, and Ethical Issues

Objectives

Cognitive

1-3.1   Define the EMT-B's scope of practice. (p 72)
1-3.2   Discuss the importance of do not resuscitate (DNR) orders [advance directives] and local or state provisions regarding EMS application. (p 79)
1-3.3   Define consent and discuss the methods of obtaining consent. (p 75)
1-3.4   Differentiate between expressed and implied consent. (p 75)
1-3.5   Explain the role of consent of minors in providing care. (p 76)
1-3.6   Discuss the implications for the EMT-B in patient refusal of transport. (p 77)
1-3.7   Discuss the issues of abandonment, negligence, and battery and their implications for the EMT-B. (p 74)
1-3.8   State the conditions necessary for the EMT-B to have a duty to act. (p 74)
1-3.9   Explain the importance, necessity, and legality of patient confidentiality. (p 80)
1-3.10   Discuss the considerations of the EMT-B in issues of organ retrieval. (p 84)
1-3.11   Differentiate the actions that an EMT-B should take to assist in the preservation of a crime scene. (p 82)
1-3.12   State the conditions that require an EMT-B to notify local law enforcement officials. (p 81)

Affective

1-3.13   Explain the role of EMS and the EMT-B regarding patients with DNR orders. (p 79)
1-3.14   Explain the rationale for the needs, benefits, and usage of advance directives. (p 79)
1-3.15   Explain the rationale for the concept of varying degrees of DNR. (p 79)

Psychomotor

None

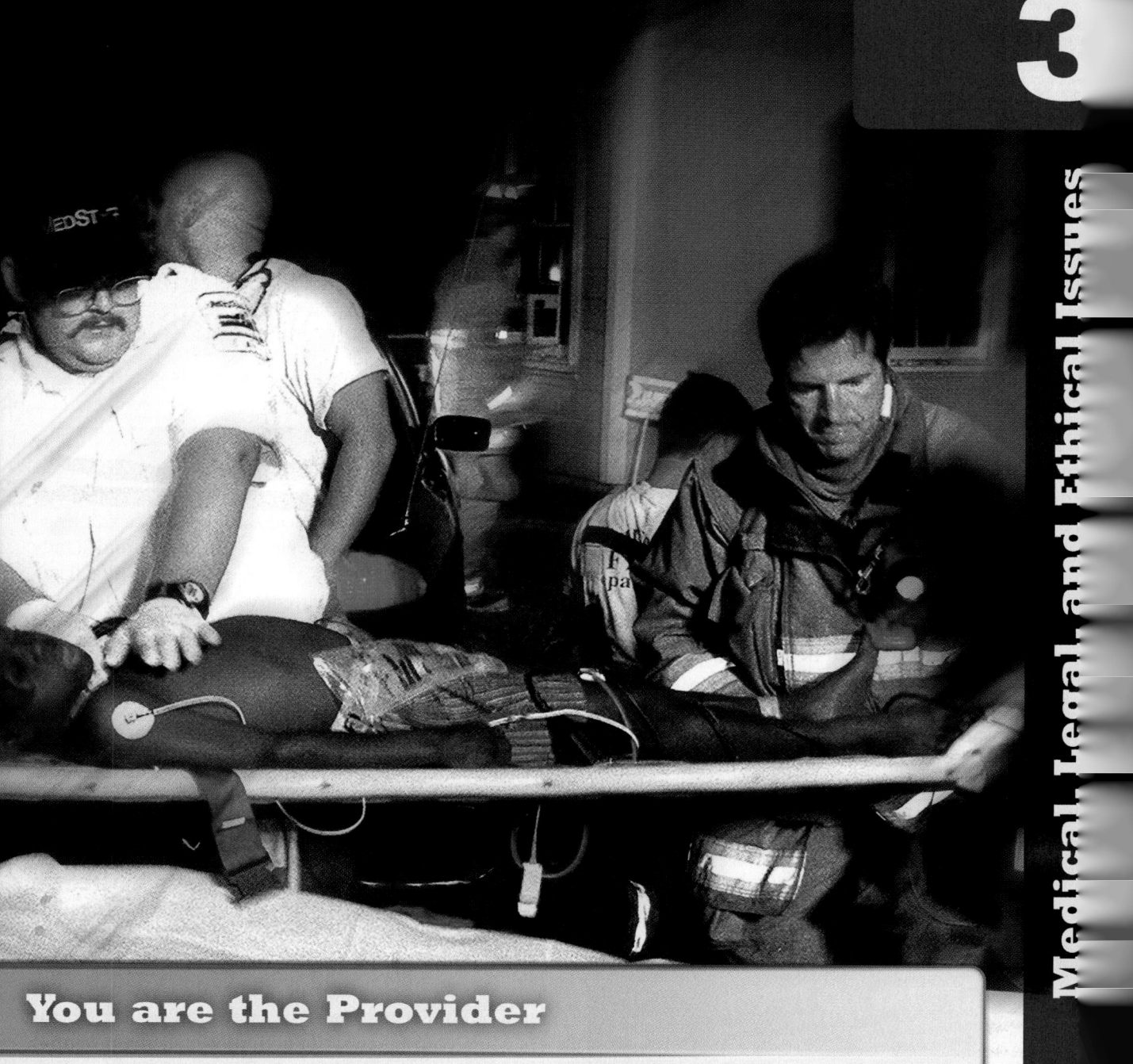

You are the Provider

You are dispatched to 2567 Walnut Lane for a confused male, unknown medical. Upon your arrival, you and your partner find a police officer attempting to calm a 32-year-old man who is somewhat confused and very agitated. When he sees you, he says he doesn't want your help and asks you to leave immediately. You notice a box of syringes in the kitchen. What should you do now?

As an EMT, your job will regularly place you in situations that deal with life and death, ethics and the law, and order and chaos. How you handle these situations, to a large degree, will determine what type of health care professional you will become. This chapter will present the basic legal concepts associated with prehospital medicine along with various ethical considerations.

1. Why is it essential to adhere to the principles of patient confidentiality?
2. What is the difference between informed and implied consent, and why is obtaining patient consent so important?

Medical, Legal, and Ethical Issues

A basic principle of emergency care is to do no further harm. Any health care provider who acts in good faith and according to an appropriate standard of care usually avoids legal exposure. Providing emergency medical care in an organized system is a recent phenomenon. Emergency medical care, or immediate care or treatment, is often provided by an EMT, who may be the first link in the chain of prehospital care. As the scope and nature of emergency medical care become more complex and widely available, litigation involving participants in EMS systems will no doubt increase. Providing competent emergency medical care that conforms with the standard of care taught to you will help you to avoid both civil and criminal actions. Consider the following situations:

- You are transporting a patient, and while the stretcher is being loaded into the ambulance, your partner slips, the stretcher crashes to the ground, and the patient is injured.
- You are about to begin treating a child, and the father commands you to stop.

What should you do? Even when emergency medical care is properly rendered, there are times when you may be sued by a patient who seeks to obtain relief, often in the form of a monetary award, for pain and suffering. Or administrative action, such as suspension of your state license or EMT-B certificate, may be brought against you for failure to abide by the regulations of your state EMS agency. For this reason, you must understand the various legal aspects of emergency medical care.

You must also consider ethical issues. As an EMT-B, should you stop and treat patients who were involved in an automobile crash while you are en route to another emergency call? Should you begin CPR on a patient who, according to the family, has terminal cancer? Should patient information be released to a patient's attorney on the telephone?

Scope of Practice

The scope of practice, which is most commonly defined by state law, outlines the care you are able to provide for the patient. Your medical director further defines the scope of practice by developing protocols and standing orders. The medical director gives you the legal authorization to provide patient care through telephone or radio communication (online) or standing orders and protocols (off-line).

You and other EMS personnel have a responsibility to provide proper, consistent patient care and to report problems, such as possible liability or exposure to infectious disease, to your medical director immediately.

Standards of Care

The law requires you to act or behave toward other individuals in a certain, definable way, regardless of the activity involved. Under given circumstances, you have a duty either to act or not. Generally speaking, you must be concerned about the safety and welfare of others when your behavior or activities have the potential for causing others injury or harm (Figure 3-1 ▶). The manner in which you must act or behave is called a standard of care.

Standard of care is established in many ways, among them local custom, statutes, ordinances, administrative regulations, and case law. In addition, professional or institutional standards have a bearing on determining the adequacy of your conduct.

Standards Imposed by Local Custom

The standard of care is how a reasonably prudent person with similar training and experience would act under similar circumstances, with similar equipment, and

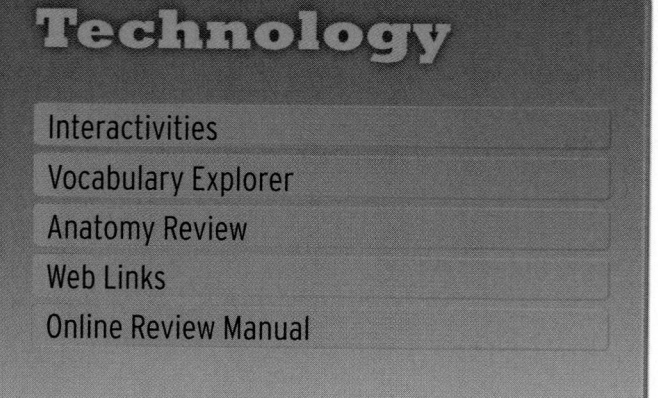

Technology

- Interactivities
- Vocabulary Explorer
- Anatomy Review
- Web Links
- Online Review Manual

www.EMTB.com

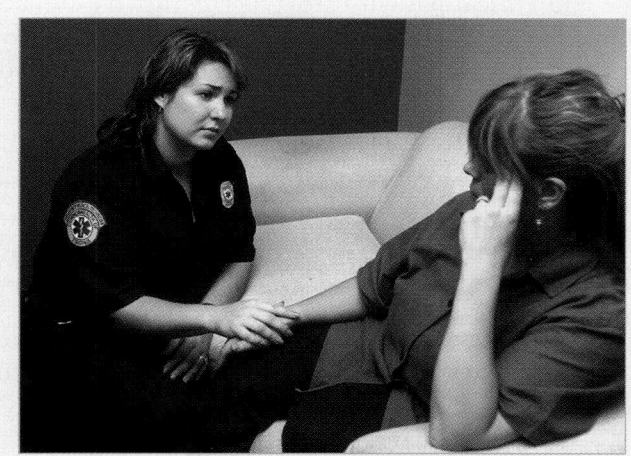

Figure 3-1 Act or behave toward others in a way that shows your concern about their safety and welfare.

Figure 3-2 An emergency is a serious situation that arises suddenly, threatens the life or welfare of one or more individuals, and requires immediate intervention.

in the same place. For example, the conduct of an EMT-B who is employed by an ambulance service is to be judged in comparison with the expected conduct of other EMT-Bs from comparable ambulance services. These standards are often based on locally accepted protocols.

As an EMT-B, you will not be held to the same standard of care as physicians or other more highly trained individuals would. In addition, your conduct must be judged in the light of the given emergency situation, taking into consideration the following factors:

- General confusion at the scene of the emergency
- The needs of other patients
- The type of equipment available

In this context, an <u>emergency</u> is a serious situation, such as injury or illness that arises suddenly, threatens the life or welfare of a person or group of people, and requires immediate intervention (Figure 3-2 ▶).

The prevailing custom of the community is an important element in determining the standard of emergency care required.

Standards Imposed by Law

In addition to local customs, standards of emergency medical care may be imposed by statutes, ordinances, administrative regulation, or case law. In many jurisdictions, violating one of these standards is said to create presumptive negligence. Therefore, you must become familiar with the particular legal standards that may exist in your state. In many states, this may take the form of treatment protocols published by a state agency.

Professional or Institutional Standards

In addition to standards imposed by law, professional or institutional standards may be admitted as evidence in determining the adequacy of an EMT's conduct. Professional standards include recommendations published by organizations and societies that are involved in emergency medical care. Institutional standards include specific rules and procedures of the EMS service, ambulance service, or organization to which you are attached.

Two notes of caution: First, you must be familiar with the standards of your organization. Second, if you are involved in formulating standards for a particular agency, they should be reasonable and realistic so that they do not impose an unreasonable burden on EMTs. Providing the best emergency medical care should be every EMT's goal, but it is not realistic to have institutional standards that demand the best care.

Many standards of care may be imposed on you. State health department regulations usually govern the scope and level of training. Court decisions have resulted in case law defining standards of care. Professional standards are also imposed, such as the American Heart Association's standard for basic life support (BLS) and cardiopulmonary resuscitation (CPR) (Figure 3-3 ▶).

Ordinary care is a minimum standard of care. In general, it is expected that anyone who offers assistance

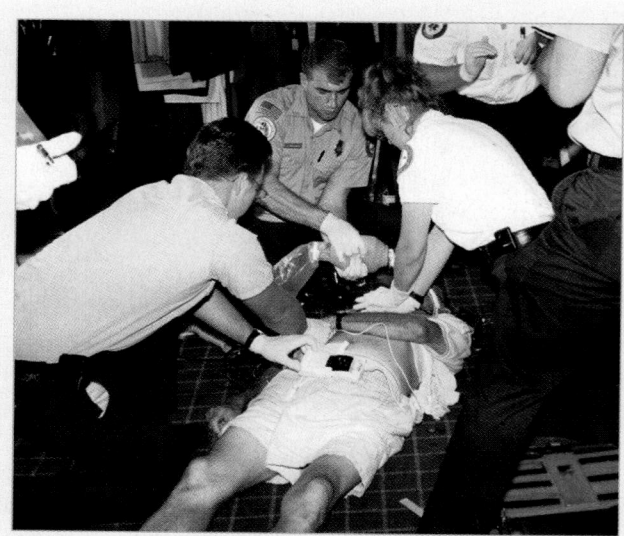

Figure 3-3 Many standards of care are imposed on you, such as those for performing BLS and CPR.

will exercise reasonable care and act prudently. If you act reasonably, according to the accepted standard, the risk of civil suit is small. If you apply the standard practices you have been trained to use, you can likely avoid liability. For example, various organizations have defined standards for performing CPR. If you deviate from these standards, you may be liable for civil and possibly criminal prosecution. In addition, state regulatory agencies that oversee EMS operations can sanction EMS personnel for deviating from the standard of care.

Standards Imposed by States

Medical Practices Act

EMS personnel are exempt from the licensure requirements of the Medical Practices Act in some states because an EMT-B is regarded as a nonmedical professional. The practice of medicine is defined as the diagnosis and treatment of disease or illness. EMT-Bs and others in the prehospital care chain assess the need for life support and begin care. Therefore, the standard of care must be maintained within the scope of your state's provisions and licensing requirements.

Certification

Some states provide certification, licensure, or credentialing of individuals who perform emergency medical care. Certification is the process by which an individual, institution, or program is evaluated and recognized as meeting certain predetermined standards to ensure safe and ethical patient care. Once certified, you are obliged to conform to the standards that are generally recognized nationally by various registry groups and provide an important link in nationwide EMS. You must ensure that your certification or licensure remains current; skill levels must be kept up to date.

Duty to Act

Duty to act is an individual's responsibility to provide patient care. Responsibility comes from either statute or function. A bystander is under no obligation to assist a stranger in distress; there is no duty to act. There may be a duty to act in certain instances, including the following:

- You are charged with emergency medical response.
- Your service or department's policy states that you must assist in any emergency.

Once your ambulance responds to a call or treatment is begun, you have a legal duty to act. In some states, if you are off duty and come upon a crash, you may be legally obligated to stop and assist patients. Check with your local agency to be fully informed about the laws that govern your off duty actions.

Negligence

Negligence is the failure to provide the same care that a person with similar training would provide. It is deviation from the accepted standard of care that may result in further injury to the patient. Determination of negligence is based on the following four factors:

1. **Duty to act.** It is the EMT-B's responsibility to act reasonably within the standards of his or her training.
2. **Breach of duty.** There is a breach of duty when the EMT-B does not act within an expected and reasonable standard of care.
3. **Damages.** There are damages when a patient is physically or psychologically harmed in some noticeable way.
4. **Cause.** There must be a reasonable cause and effect. An example is dropping the patient during lifting, causing a fracture of the patient's leg. If a person has a duty and abuses it, causing harm to another individual, the EMT-B, the agency, and/or the medical director may be sued for negligence.

All four elements must be present for the legal doctrine of negligence to apply.

Abandonment

Abandonment is the unilateral termination of care by the EMT-B without the patient's consent and without making any provisions for continuing care by a medical professional with skills at the same or a higher level. For the EMT-B, once care is started, you have assumed a duty that must not stop until an equally competent person assumes responsibility. Not performing that duty exposes the patient to harm and is a basis for a lawsuit. Abandonment is legally and ethically a serious matter that can result in both civil and criminal actions against an EMT.

For example, suppose you arrive at the scene of a single-car accident and begin care of two injured patients. A passerby tells you of a two-car accident farther down the road in which five people are injured. You turn care of the two injured patients from the first accident over to the passerby and leave to go to the other accident. Abandonment may have occurred because you did not turn care of the patients over to a person with the same or a higher level of skill than yours. Consider the following general questions when you are faced with making a decision such as this one:

- What problems may develop from your actions?
- How might the patient's condition worsen if you leave?
- Does the patient need care?
- Are you neglecting your duty to your patient?
- Is the person assuming care trained to at least the same level as you?
- Are you abandoning the patient if you leave the scene?
- Are you violating a standard of care?
- Are you acting prudently?

Consent

Under most circumstances, consent is required from every conscious, mentally competent adult before care can be started. A person receiving care must give permission, or consent, for treatment. If a person is in control of his or her actions, even though injured, and refuses care, you may not care for the patient. In fact, doing so may be grounds for both criminal and civil

EMT-B Tips

Calls involving refusals of care are among the most frequently litigated. Be sure your patient is competent to refuse care and is fully informed of possible consequences. Always act in the patient's best interest and resist the temptation to obtain a refusal for your convenience.

action such as assault and battery. Consent can be actual or implied and can involve the care of a minor or a mentally incompetent patient.

Expressed Consent

Expressed consent (or actual consent) is the type of consent in which the patient speaks or acknowledges that he or she wants you to provide care or transport. It must be informed consent, which means that the patient has been told of the potential risks, benefits, and alternatives to treatment and has given consent to treatment. The legal basis for this doctrine rests on the assumption that the patient has a right to determine what is to be done with his or her body. The patient must be of legal age and able to make a rational decision.

A patient might agree to certain emergency medical care but not to other care. For example, a patient might agree to be removed from a car but refuse further care. An injured person might agree to emergency care at home but refuse to be transported to a medical facility. Informed consent is valid if given orally; however, it may be difficult to prove. Having the patient sign a consent form does not eliminate your responsibility to fully inform the patient. A witness may be helpful in later proof, and documentation of the consent is always advisable.

Implied Consent

When a person is unconscious and unable to give consent or when a serious threat to life or limb exists, the law assumes that the patient would consent to care and transport to a medical facility Figure 3-4 ▶ . This is called implied consent. Implied consent is limited to life-threatening emergency situations and is appropriate when the patient is unconscious, delusional, unresponsive as a result of drug or alcohol use, or otherwise physically unable to give expressed consent. However,

Figure 3-4   When a serious threat to life or limb exists and the patient is unconscious or otherwise unable to give consent, the law assumes that the patient would give consent to care and transport to the hospital.

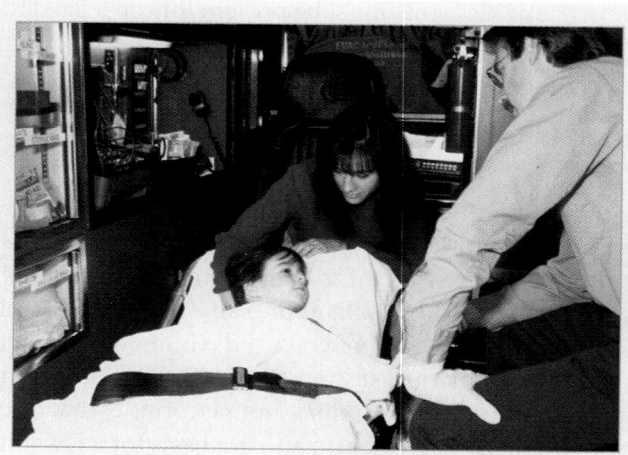

Figure 3-5   The law requires that a parent or legal guardian give consent for treatment or transport of a minor. However, you must never withhold lifesaving care.

many things may be unclear about what represents a "serious threat to life or limb." Legal proceedings would likely revolve around that question. This becomes a medicolegal judgment, which should be supported by the EMT-B's best efforts to obtain consent. Medicolegal is a term that relates to medical jurisprudence (law) or forensic medicine. In most instances, the law allows the spouse, a close relative, or next of kin to give consent for an injured person who is unable to give consent. Refusal of your offer to render emergency care also may be implied. For example, a patient's action in pulling his or her arm from your splint may be an indication of refusal of consent. Again, it is good to document these circumstances and response, and to record witnesses.

Minors and Consent

Because a minor might not have the wisdom, maturity, or judgment to give valid consent, the law requires that a parent or legal guardian give consent for treatment or transport (Figure 3-5 ▶). However, in some states, a minor can give valid consent to receive medical care, depending on the minor's age and maturity. Many states also allow emancipated, married, or pregnant minors to be treated as adults for the purposes of consenting to medical treatment. You should obtain consent from a parent or legal guardian whenever possible; however, if a true emergency exists and the parent or legal guardian is not available, the consent to treat the minor is implied, just as with an adult. You must never withhold lifesaving care.

Mentally Incompetent Adults

Assisting patients who are mentally ill, in behavioral (psychological) crisis, under the influence of drugs or alcohol, or developmentally delayed is complicated. An adult patient who is mentally incompetent is not able to give informed consent. From a legal perspective, this situation is similar to those involving minors. Consent for emergency care should be obtained from someone who is legally responsible, such as a guardian or conservator. In many cases, however, such permission will not be readily obtainable. Many states have protective custody statutes allowing such a person to be taken, under law enforcement authority, to a medical facility. Know the provisions in your area. Remember that when a life-threatening emergency exists, you can assume that implied consent exists.

Forcible Restraint

Forcible restraint is the act of physically preventing an individual from any physical action. Forcible restraint of a mentally disturbed individual may be required before emergency care can be rendered. If you believe that a patient will injure himself, herself, or others, you can legally restrain the patient. However, you must consult medical control, online or off-line depending on local protocol, for authorization to restrain or contact law enforcement personnel who have authority to restrain people. In some states, only a law enforcement officer may forcibly restrain an individual (Figure 3-6 ▶). You

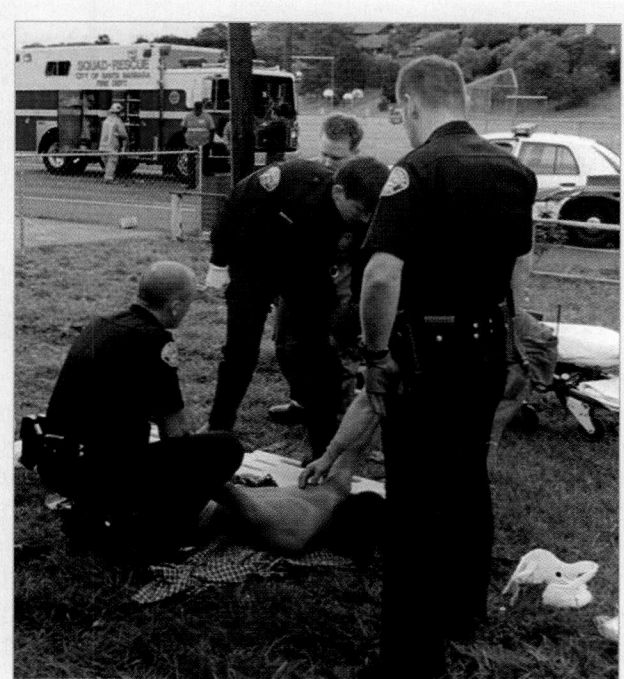

Figure 3-6 Be sure that you know the local laws about forcible restraint of a patient. In some states, only a law enforcement officer has the authority to restrain a patient.

should be clearly informed about local laws. Restraint without authority exposes you to civil and criminal penalties. Restraint may be used only in circumstances of risk to yourself or others.

Your service should have clearly defined protocols to deal with situations involving restraint. After restraints are applied, they should not be removed en route unless they pose a risk to the patient, even if the patient promises to behave.

Remember that if the patient is conscious and the situation is not urgent, consent is required. Adults who appear to be in control of their senses cannot be forced to submit to either care or transportation. Adults are not required by law to make "correct" decisions, or to agree with medical advice.

Assault and Battery

Assault is defined as unlawfully placing a person in fear of immediate bodily harm. Threatening to restrain a patient who does not want to be transported could be considered assault. Battery is unlawfully touching a person; this includes providing emergency care without consent. Serious legal problems may arise in situations in which a patient has not given consent for treatment. Battery could be considered if you apply a splint to a suspected fracture of the lower leg or use an Epi-Pen on a patient without the patient's consent. The patient may have grounds to sue you for assault, battery, or both. To protect yourself from these charges, make sure that you obtain expressed consent or that the situation allows for implied consent. Consult your medical director if you have questions or doubt about a specific situation.

The Right to Refuse Treatment

Mentally competent adults have the right to refuse treatment or withdraw from treatment at any time. However, these patients present you with a dilemma. Should you provide care against their will and risk

You are the Provider   Part 2

You decide to check the house for indications of any medical conditions while the officer and paramedic continue to speak with the man. You look in the refrigerator and see vials of insulin. In the living room, you quietly inform the paramedic of the presence of insulin in the refrigerator. The paramedic asks the patient if he has diabetes and he replies "yes, but it's no one's business" and takes a swing at her. The officer takes the patient to the ground, and you and your partner confirm his blood glucose level is 20 mg/dL. You assist your partner in establishing an IV and administering dextrose.

3. In this case, is restraining the patient considered battery?
4. If you had left the scene as the patient requested, what would you and your partner be accused of?

being accused of battery? Should you leave them alone? If you leave patients alone, you risk being accused of negligence or abandonment if their condition becomes worse.

If a patient refuses treatment or transport, you must make sure that he or she understands, or is informed about, the potential risks, benefits, treatments, and alternatives to treatment. You must also fully inform the patient about the consequences of refusing treatment and encourage the patient to ask questions. Remember that competent adults who refuse specific kinds of treatment for themselves for religious reasons generally have a legal right to do so. Document refusals and obtain a witness' signature, preferably a family member of the patient, to protect yourself later.

When a patient refuses treatment, you must assess whether the patient's mental condition is impaired. If the patient refusing treatment is delusional or confused, you cannot assume that the refusal is an informed refusal. When in doubt, it is always best to proceed with treatment. Providing treatment is a much more defensible position than failing to treat a patient. Failure to treat a patient can be considered negligence. Do not endanger yourself to render care. Utilize law enforcement assistance to ensure your own safety in these cases.

You may also be faced with a situation in which a parent refuses to permit treatment of an ill or injured child. In this situation, you must consider the emotional impact of the emergency on the parent's judgment. In virtually all cases of refusal, you can usually resolve the situation with patience and calm persuasion. You may also need the help of others, such as your supervisor, medical control, or law enforcement officials. In some states, providers may be legally required to report such situations to an appropriate agency as child neglect. EMT-Bs should summon law enforcement officials to further document the need for treatment and transport. This may also be the case with older patients, patients in licensed care facilities, or other custodial care situations. EMT-Bs should be familiar with and adhere to local/state laws regarding mandated reporting of suspected abuse and/or neglect.

There will be times when you are not able to persuade the patient, guardian, conservator, or parent of a minor or mentally incompetent patient to proceed with treatment. In this case, obtain radio contact with the ER physician to further persuade the patient to seek care and further document the refusal. If the patient does refuse, you must obtain the signature of the individual who is refusing treatment on an official release form that acknowledges refusal. Additional documentation may also include a statement, written by the pa-

Documentation Tips

When a patient refuses treatment or transport, protect yourself with a thorough patient care report and an official refusal form. Have the patient sign the form, document in the patient care report what you have done to ensure an informed refusal, and note the involvement of medical control.

tient if possible, from the patient as to why he or she does not want prehospital care and transportation. You must be sure to document any assessment findings and emergency care that you provided. You must also obtain a signature from a witness to the refusal. Keep the refusal with the run report and the patient care report. In addition to the release form itself, write a note about the refusal on the patient care report. If the patient refuses to sign the release form, inform medical control and thoroughly document the situation and the refusal. Report to medical control, and follow your local protocols with regard to this situation.

Good Samaritan Laws and Immunity

Most states have adopted Good Samaritan laws, which are based on the common law principle that when you reasonably help another person, you should not be liable for errors and omissions that are made in giving good faith emergency care. However, Good Samaritan laws do not protect you from a lawsuit. Only a few statutory provisions provide immunity from a lawsuit, and those usually are reserved for governments. Good Samaritan laws provide an affirmative defense if you are sued for rendering care, but they do not protect you from liability or for failure to provide proper care, nor do they pertain to acts outside the scope of care. These laws do not protect anyone from wanton, gross, or willful negligence (eg, the failure to exercise due care).

Another group of laws grants immunity from liability to official emergency medical care providers, such as EMT-Bs. These laws, which vary from state to state, do not provide immunity when injury or damage is caused by gross negligence or willful misconduct.

Most states have also adopted specific laws granting special privileges to EMS personnel, authorizing them to perform certain medical procedures. Many states also grant partial immunity to EMTs and physicians and nurses who give emergency instructions to EMS personnel via radio or other forms of communication. Consult your medical director for more information about the laws in your area.

Advance Directives

Occasionally, you and your partner may respond to a call in which a patient is dying from an illness. When you arrive at the scene, you may find that family members present do not want you to try to resuscitate the patient. Without valid written documentation from a physician, such as an advance directive or a do not resuscitate (DNR) order (also known as a "Do Not Attempt Resuscitation" order), this type of request places you in a very difficult position.

A competent patient is able to make rational decisions about his or her well-being. An advance directive is a written document that specifies medical care that a person would like to have administered should he or she become unable to make medical decisions (incompetent, such as in a coma). An advance directive is also commonly called a living will. There are several types of advance directives. Not all advance directives are directions to withhold care. For example, a comfort care order is an advance directive that specifies care a person should receive in the event that they become incompetent.

DNR orders give you permission not to attempt resuscitation. Although laws can differ from state to state, generally speaking, to be valid, DNR orders must meet the following requirements:

- Clearly state the patient's medical problem(s)
- Be signed by the patient or legal guardian
- Be signed by one or more physicians
- In some states, DNR orders contain an expiration date, while in others, no expiration date is included. DNR orders with expiration dates must be dated in the preceding 12 months in order to be valid.

However, even in the presence of such a DNR order, you are still obligated to provide supportive measures (oxygen, pain relief, and comfort) to a patient who is not in cardiac arrest, whenever possible. Each ambulance service, in consultation with its medical director and legal counsel, must develop a protocol to follow in these circumstances.

Because of terminal nursing home placement and hospice and home health programs, you may be faced with this situation often. Specific guidelines vary from

EMT-B Tips

EMT Oath

"Be it pledged as an Emergency Medical Technician, I will honor the physical and judicial laws of God and man. I will follow that regimen which, according to my ability and judgment, I consider for the benefit of patients and abstain from whatever is deleterious and mischievous, nor shall I suggest any such counsel. Into whatever homes I enter, I will go into them for the benefit of only the sick and injured, never revealing what I see or hear in the lives of men unless required by law.

I shall also share my medical knowledge with those who may benefit from what I have learned. I will serve unselfishly and continuously in order to help make a better world for all mankind.

While I continue to keep this oath unviolated, may it be granted to me to enjoy life, and the practice of the art, respected by all men, in all times. Should I trespass or violate this oath, may the reverse be my lot.

So help me God."

Written by Charles B. Gillespie, MD
Adopted by the National Association of Emergency Medical Technicians, 1978.

state to state, but the following four statements may be considered general guidelines:

1. **Patients have the right to refuse treatment**, including resuscitative efforts, provided that they are able to communicate their wishes.
2. **A written order from a physician is required** for DNR orders to be valid in a health care facility.
3. **You should periodically review** state and local protocols and legislation regarding advance directives.
4. **When you are in doubt** or the written orders are not present, you have an obligation to resuscitate.

Ethical Responsibilities

In addition to legal duties, EMTs have certain ethical responsibilities as health care providers. These responsibilities are to themselves, their coworkers, the public, and the patient. Ethics are related to action, conduct, motive, or character and how they relate to the EMT's responsibilities. From an EMS standpoint, ethics are associated with what the profession of EMS providers deems proper or fitting conduct. Treating a patient ethically means doing so in a manner that conforms to professional standards of conduct. The EMT Code of Ethics, available at the NAEMT website, outlines ethical expectations for EMTs.

How can you make sure that you are acting ethically, especially with all the decisions you have to make in the field (Table 3-1 ▶)?

You must meet your legal and ethical responsibilities while caring for your patients' physical and emotional needs. Patient needs vary depending on the situation.

An unquestionable responsibility is honest reporting. Absolute honesty in reporting is essential. You must provide a complete account of the events and the details of all patient care and professional duties. Accurate records are also important for quality improvement activities.

Confidentiality

Communication between you and the patient is considered confidential and generally cannot be disclosed without permission from the patient or a court order. Confidential information includes the patient history, assessment findings, and treatment provided. You cannot disclose information regarding a patient's diagnosis, treatment, or mental or physical condition without consent; if you do, you may find yourself liable for breach of confidentiality.

In certain situations, you may release confidential information to designated individuals. In most states, records may be released when a legal subpoena is pre-

---

TABLE 3-1 Ethical Decision Making

1. Consider all options available to you and the consequence of each option.
2. What decisions have been made regarding a similar situation? Is this a type of problem that reflects a rule or policy? Can an existing policy or rule be applied? This uses the concept of precedence.
3. How would this action affect you if you were in your patient's or patient's family's place? This is a form of the Golden Rule.
4. Would you feel comfortable having all prehospital care providers apply this action in all similar circumstances?
5. Can you supply a good justification for your action to:
   ■ Your peers?
   ■ The public?
   ■ Your supervisor?
6. How will the consequences of your decision derive the greatest benefit in view of all the alternatives? This is the test of utility.

---

sented or the patient signs a written release. The patient must be mentally competent and fully understand the nature of the release.

Another means for disclosing information is with an automatic release, which does not require a written form. This type of release allows you to share information with other health care providers so that they may continue the patient's care.

You are the Provider   Part 3

A few minutes after the administration of dextrose, the patient is able to answer appropriately when questioned about person, place, time, and event, but he still seems upset. He tells you that he's not going to the hospital and asks that you take the IV out right away. You explain to him that he had a very low blood glucose, or blood sugar, level and that's why the IV was started. You tell him that it would be a good idea to go into the hospital to be evaluated in case adjustments need to be made to his insulin dosages. He again says no. As the paramedic takes out the IV, she reminds him to eat a meal substantial in carbohydrates such as a peanut butter and jelly sandwich to avoid having his blood glucose level drop again. The patient says he doesn't need advice from anyone.

5. Can you continue to restrain this patient and/or transport him to the hospital?
6. What are other considerations to be addressed before you and your partner leave the scene?

In many states, you do not need a written release to report information about cases of rape or abuse to proper authorities. Third-party payment billing forms may also be completed without written consent.

HIPAA

HIPAA is the acronym for the Health Insurance Portability and Accountability Act of 1996. Although this act had many aims, including improving the portability and continuity of health insurance coverage and combating waste and fraud in health insurance and the provision of health care, the section of the act that most affects EMS relates to patient privacy. The aim of this section was to strengthen laws for the protection of the privacy of health care information and to safeguard patient confidentiality. As such, it provides guidance on what types of information is protected, the responsibility of health care providers regarding that protection, and penalties for breaching that protection.

The law has the effect of dramatically limiting the ability of EMS providers to obtain follow-up information about patients they treat including information that would serve to improve their knowledge of medical conditions or help them understand the degree to which they may have been exposed to a communicable disease as a result of a patient encounter.

Most personal health information is protected and should not be released without the patient's permission. If you are not sure, do not give any information to anyone other than those directly involved in the care of the patient. For specific policies, each EMS service is required to have a manual and a compliance officer who can answer questions. You can expect to receive further training on how this act impacts your specific response agency and resource hospital.

Many state privacy laws cover verbal communications not protected under HIPAA. Be familiar with your state's laws regarding patient privacy.

Records and Reports

The government has formulated a policy to protect individuals with health regulations and statutes. Because certain individuals are in a position to observe and gather information about diseases, injuries, and emergency events, an obligation to compile such information and report it to certain agencies may be imposed. Even if there is no such requirement, you should compile a complete and accurate record of all incidents in which you come into contact with sick or injured patients. Most

medical and legal experts believe that a complete and accurate record of an emergency medical incident is an important safeguard against legal complications. The absence of a record or a substantially incomplete record may mean that you have to testify about the events, your findings, and your actions relying on memory alone, which can prove to be wholly inadequate and embarrassing in the face of aggressive cross-examination.

The courts consider the following two rules of thumb regarding reports and records:

- **If an action or procedure** is not recorded on the written report, it was not performed.
- **An incomplete or untidy report** is evidence of incomplete or inexpert emergency medical care.

You can avoid both of these potentially dangerous presumptions by compiling and maintaining accurate reports and records of all events and patients. Patient care reports also help the EMS system evaluate individual and service provider performance. These reports are an integral part of most quality assurance programs.

Special Reporting Requirements

Abuse of Children, Older Persons, and Others

All states and the District of Columbia have enacted laws to protect abused children, and some have added other protected groups such as the older population and "at-risk" adults. Most states have a reporting obligation for certain individuals, ranging from physicians to any person. You must be aware of the requirements of the law in your state. Such statutes frequently grant immunity from liability for libel, slander, or defamation of character to the individual who is obligated to report, even if the reports are subsequently shown to be unfounded, as long as the reports are made in good faith.

✳ EMT-B Tips

When taking a history from a patient you suspect has been abused, you may get more accurate information if your partner interviews the parents or other caregivers separately. Abused patients are usually reluctant to speak openly in front of their abusers.

Injury During the Commission of a Felony

Many states have laws requiring the reporting of any injury that is likely to have occurred during the commission of a crime, such as gunshot wounds, knife wounds, or poisonings. Again, you must be familiar with the legal requirements of your state.

Drug-Related Injuries

In some instances, drug-related injuries must be reported. These requirements may affect the EMT-B. However, it should be stressed that the US Supreme Court has held that drug addiction, in contrast to drug possession or sale, is an illness and not a crime. An injury as a result of a drug overdose, therefore, may not be within the definition of an injury resulting from a crime.

Some states, by statute, specifically establish confidentiality and excuse certain specified individuals from reporting drug cases, either to a government agency or to a minor's parents, if, in the opinion of those individuals, withholding reporting is necessary for the proper treatment of the patient. Once again, you must be familiar with the legal requirements of your state.

Childbirth

Many states require that anyone who attends at a live birth in any place other than a licensed medical facility report the birth. As before, you must be familiar with state requirements.

Other Reporting Requirements

Other reporting requirements may include attempted suicides, dog bites, certain communicable diseases, assaults, and rapes.

Most EMS agencies require that all exposures to infectious diseases be reported. You may be asked to transport certain patients in restraints, which may also need to be reported. Each of these situations can present significant legal problems. You should learn your local protocols regarding these situations.

Scene of a Crime

If there is evidence at an emergency scene that a crime may have been committed, you must notify the dispatcher immediately so that law enforcement authorities can respond. Such circumstances should not stop you from providing life-saving emergency medical care to the patient; however, your safety is a priority, so you must ensure that the scene is safe to enter. At times, you may have to transport the patient to the hospital before law enforcement arrives. While emergency medical care is being provided, you must be careful not to disturb the scene of the crime any more than absolutely necessary. Notes and drawings should be made of the position of the patient and of the presence and position of any weapon or other objects that may be valuable to the investigating officers. If possible, do not cut through holes in clothing that were caused by weapons or gunshot wounds. You should confer periodically with local authorities and be aware of their wishes as to any actions you should take at the scene of the crime. It is best if these guidelines can be established by protocol.

The Deceased

In some states, EMTs do not have the authority to pronounce a patient dead. If there is any chance that life exists or that the patient can be resuscitated, you must make every effort to save the patient at the scene and during transport. However, at times death is obvious. If a victim is clearly dead and the scene of the emergency may be where the crime was committed, do not move the body or disturb the scene.

Physical Signs of Death

Determination of the cause of death is the medical responsibility of a physician. There are both definitive and presumptive signs of death. In many states, death is defined as the absence of circulatory and respiratory function. Many states have also adopted "brain death" provisions; these provisions refer to irreversible cessation of all functions of the brain and brain stem. Questions often arise as to whether to begin basic life support. In the absence of physician orders such as DNR orders, the general rule is: If the body is still warm and intact, initiate emergency medical care. An exception to this rule is cold temperature (hypothermia) emergencies. Hypothermia is a general cooling of the body in which the internal body temperature becomes abnormally low: 95°F (35°C). It is considered a serious condition and is often fatal. At 86°F (30°C), the brain can survive without perfusion for about 10 minutes. When the core temperature drops to 82.4°F (28°C), the patient is in grave danger; however, individuals have survived hypothermic incidents with temperatures as low as 64.4°F (18°C). In cases of hypothermia, the patient should not be considered dead until he or she is warm and dead.

TABLE 3-2 Presumptive Signs of Death

- Unresponsiveness to painful stimuli
- Lack of a carotid pulse or heartbeat
- Absence of breath sounds
- No deep tendon or corneal reflexes
- Absence of eye movement
- No systolic blood pressure
- Profound cyanosis
- Lowered or decreased body temperature

Presumptive Signs of Death

Most medicolegal authorities will consider the presumptive signs of death that are listed in (Table 3-2 ▲) adequate, particularly when they follow a severe trauma or occur at the end stages of long-term illness such as cancer or other prolonged diseases. These signs would not be adequate in cases of sudden death due to hypothermia, acute poisoning, or cardiac arrest. Usually, in these cases, some combination of the signs is needed to declare death, not just one of them alone.

Definitive Signs of Death

Definitive or conclusive signs of death that are obvious and clear to even nonmedical persons include the following:

- Obvious mortal damage, such as a body in parts (decapitation)
- Dependent lividity: blood settling to the lowest point of the body, causing discoloration of the skin (Figure 3-7 ▼). Body surfaces in contact with

---

✚ EMT-B Tips

Definitive Signs of Death

- Obvious mortal damage
- Dependent lividity
- Rigor mortis
- Putrefaction

---

firm surfaces may appear white, as blood cannot pool into capillaries in direct contact with firm surfaces. Do not confuse dependent lividity with cyanosis, mottling, or bruising from trauma.

- Rigor mortis, the stiffening of body muscles caused by chemical changes within muscle tissue. It develops first in the face and jaw, gradually extending downward until the body is in full rigor. The rate of onset is affected by the body's ability to lose heat to its surroundings. A thin body loses heat faster than a fat body. A body on a tile floor loses heat faster than a body wrapped up in a blanket in a bed. Rigor mortis occurs sometime between 2 and 12 hours after death.
- Putrefaction (decomposition of body tissues). Depending on temperature conditions, this occurs sometime between 40 and 96 hours after death.

Medical Examiner Cases

Involvement of the medical examiner, or the coroner in some states, depends on the nature and scene of the death. In most states, when trauma is a factor or the

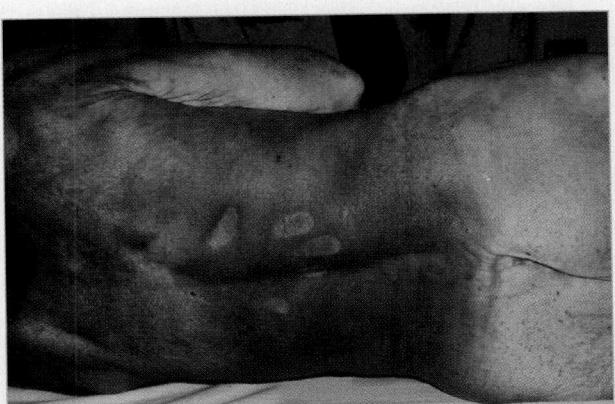

Figure 3-7 Dependent lividity is an obvious sign of death caused by discoloration of the body from pooling of the blood to the lower parts of the body.

Figure 3-8 When trauma is a factor or the death involves a suspected criminal situation, the medical examiner is required.

death involves suspected criminal or unusual situations such as hanging or poisoning, the medical examiner must be notified (Figure 3-8 ▼). When the medical examiner or coroner assumes responsibility of the scene, that responsibility supersedes all others at the scene, including the family's. The following may be considered medical examiner's cases:

- When the person is dead on arrival (DOA) (sometimes referred to as dead on scene [DOS])
- Death without previous medical care or when the physician is unable to state the cause of death
- Suicide (self-destruction)
- Violent death
- Poisoning, known or suspected
- Death resulting from accidents
- Suspicion of a criminal act

If emergency medical care has been initiated, keep thorough notes of what was done or found. These records may be important during a subsequent investigation.

In such instances, there is no urgent reason to move the body. The only immediate action that is required of you is to cover the body and prevent its disturbance. Local protocol will determine your ultimate action in these instances.

Special Situations

Organ Donors

You may be called to a scene involving a potential organ donor. An individual who has expressed a wish to donate organs is a potential organ donor. Consent to organ donation is voluntary and knowing. Consent is evidenced by either a donor card or a driver's license indicating that the individual wishes to be a donor (Figure 3-9 ▶). You may need to consult with medical control when faced with this situation.

You should treat a potential organ donor in the same way that you would any other patient needing treatment. The fact that a patient is a possible donor does not mean that you should not use all means necessary to keep that patient alive. Organs that are often donated, such as a kidney, heart, or liver, need oxygen at all times; you must give the possible donor oxygen, or the organs will be damaged and become useless.

Remember that your priority is to save the patient's life. You may encounter potential organ donor situations at a mass-casualty incident. The potential organ donor should be triaged with other patients and assigned a category; the potential organ donor may have

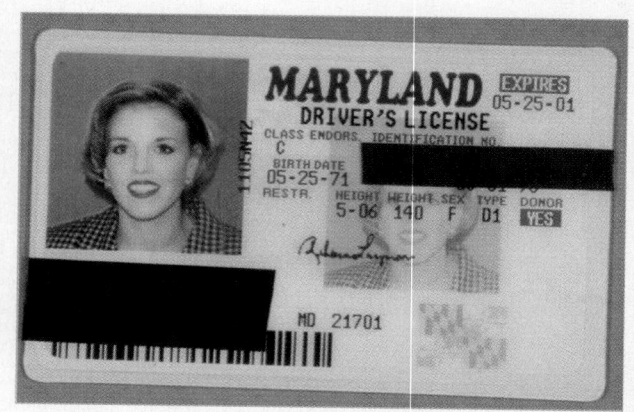

Figure 3-9 The patient may be carrying a donor card or driver's license indicating that he or she wishes to be an organ donor.

to have a lower priority than other less severely injured patients.

Be sure to learn what the specific protocols are in your area regarding these situations. Organ donors may not be able to be maintained properly in a mass-casualty incident. While unfortunate, the compromise of the organs cannot be avoided in this case.

Medical Identification Insignia

Many patients will carry important medical identification and information, often in the form of a bracelet, necklace, or card that will identify whether the patient has allergies, diabetes, epilepsy, or some other serious condition (Figure 3-10 ▼). This information is helpful to you in assessing and treating the patient.

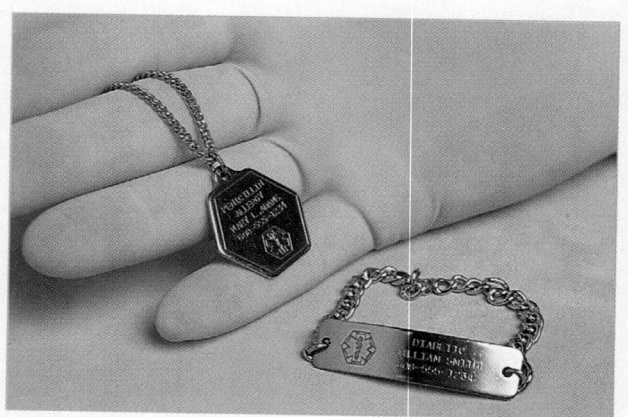

Figure 3-10 The patient may be carrying a medical identification card or wearing a bracelet or necklace that may indicate a serious medical condition.

You are the Provider

Summary

It is important to act in the best interest of a patient. Adults have the right to refuse care if they are fully alert and oriented. If they are impaired due to medical conditions or medications, you are able to take care of them under the rules of implied consent. For those who need help, are alert, and refuse treatment and/or transport, you must explain the ramifications of their decision. Express your genuine concern for their well-being and your concern for the potential effects of not receiving needed medical care. Oftentimes by simply expressing these concerns, patients will change their minds and agree to receiving medical care.

Prep Kit

Ready for Review

- As the scope of emergency medical care becomes more complex and widely available, litigation involving participants in emergency medical services will increase.

- The scope of practice outlines the care you are able to provide to the patient and is most commonly defined by law; the medical director further defines the scope of practice.

- The standard of care is the manner in which you must act or behave when treating sick or injured patients. Some standards are imposed by local custom, the law, and institutions.

- A duty to act is the responsibility of an individual to provide patient care. If you are off duty or out of your jurisdiction, you may not have a legal duty to act.

- Negligence is the failure to provide the same care that a person of similar training would provide. Determination of negligence is based on duty, breach of duty, damages, and cause.

- Abandonment is the termination of care without the patient's consent and without making provisions for the transfer of care to a medical professional with skills at the same or a higher level than yours. Abandonment is legally and ethically a very serious act.

- You must receive consent from a patient before beginning care. A conscious adult patient who can make a rational decision will be able to give you expressed consent. Expressed consent must also be informed consent.

- When a patient is unconscious and unable to give consent, the law assumes implied consent. You should try to obtain consent from a parent or guardian of a minor whenever possible.

- You should never withhold lifesaving care unless a valid DNR order is present.

- Assault is defined as unlawfully placing a person in fear of immediate harm without the person's consent. Battery is unlawfully touching a person; this includes providing emergency care without consent. To protect yourself from these charges, be sure to obtain expressed consent whenever possible.

- Mentally competent patients have the right to refuse treatment. In these instances, be sure to have the patient sign a refusal form, and make sure your department keeps a copy.

- Many states have adopted Good Samaritan laws and other laws that provide immunity to EMS personnel, provided that injury to the patient was not the result of gross negligence or willful misconduct on the part of the EMT-B.

- An advance directive is a written document that specifies medical treatment in case a mentally competent patient becomes unable to make decisions.

- DNR orders give you permission to not attempt resuscitation in the event of cardiac arrest. Your ambulance service should have protocols in place to follow when you are faced with an advance directive or DNR orders.

- Communication between you and the patient is confidential and should not be disclosed without permission from the patient or a court order.

- Records and reports are important; make sure that you compile a complete and accurate record of each incident. The courts consider an action or procedure that was not recorded on the written report as not having been performed, and an incomplete or untidy report is considered evidence of incomplete or inexpert medical care.

- You should know what the special reporting requirements are involving abuse of children, the elderly, and others; injuries related to crimes; drug-related injuries; and childbirth.

- Be sure to note whether patients are carrying some type of medical identification information. If you fail to take this information into account, you may cause harm to the patient.

www.EMTB.com

Technology

- Interactivities
- Vocabulary Explorer
- Anatomy Review
- Web Links
- Online Review Manual

Vital Vocabulary

abandonment Unilateral termination of care by the EMT-B without the patient's consent and without making provisions for transferring care to another medical professional with skills at the same level or higher.

advance directive Written documentation that specifies medical treatment for a competent patient should the patient become unable to make decisions; also called a living will.

assault Unlawfully placing a patient in fear of bodily harm.

battery Touching a patient or providing emergency care without consent.

certification A process in which a person, an institution, or a program is evaluated and recognized as meeting certain predetermined standards to provide safe and ethical care.

competent Able to make rational decisions about personal well-being.

consent Permission to render care.

dependent lividity Blood settling to the lowest point of the body, causing discoloration of the skin.

DNR (Do Not Resuscitate) orders Written documentation by a physician giving permission to medical personnel not to attempt resuscitation in the event of cardiac arrest.

duty to act A medicolegal term relating to certain personnel who either by statute or by function have a responsibility to provide care.

emergency A serious situation, such as injury or illness, that threatens the life or welfare of a person or group of people and requires immediate intervention.

emergency medical care Immediate care or treatment.

expressed consent A type of consent in which a patient gives express authorization for provision of care or transport.

forcible restraint The act of physically preventing an individual from any physical action.

Good Samaritan laws Statutory provisions enacted by many states to protect citizens from liability for errors and omissions in giving good faith emergency medical care, unless there is wanton, gross, or willful negligence.

implied consent Type of consent in which a patient who is unable to give consent is given treatment under the legal assumption that he or she would want treatment.

informed consent Permission for treatment given by a competent patient after the potential risks, benefits, and alternatives to treatment have been explained.

medicolegal A term relating to medical jurisprudence (law) or forensic medicine.

negligence Failure to provide the same care that a person with similar training would provide.

precedence Basing current action on lessons, rules, or guidelines derived from previous similar experiences.

putrefaction Decomposition of body tissues.

rigor mortis Stiffening of the body; a definitive sign of death.

standard of care Written, accepted levels of emergency care expected by reason of training and profession; written by legal or professional organizations so that patients are not exposed to unreasonable risk or harm.

www.EMTB.com

Assessment in Action

You have just arrived at a serious motor vehicle crash (MVC) and find a critically injured patient. As you approach the vehicle you look for any scene hazards. The patient is a young man who appears to be semi-conscious. He is having trouble breathing and is bleeding from his head. Your partner has also surveyed the scene and confirms that you have only one patient.

You move the patient safely to the ambulance for further assessment, treatment, and transportation to the trauma center. Your assessment tells you the patient has a serious head injury. While en route to the hospital the patient becomes agitated and combative. In this situation your agency's protocols allow you to restrain the patient to protect both the patient and yourself. You report to the hospital via the radio, and the trauma center is expecting your arrival in 5 minutes.

1. By accepting the position with the ambulance service and responding to the emergency, your responsibility to treat patients is called:
   A. contractual clause.
   B. duty to act.
   C. right to work.
   D. work ethics

2. When arriving at the trauma center, patient care is transferred to the staff. Leaving a patient without giving a report and ensuring continued care is called:
   A. abandonment.
   B. assault.
   C. late standpoint.
   D. nonsupervision.

3. Before any care can be provided to a patient by you:
   A. approval from medical control must be obtained.
   B. permission or consent must be obtained.
   C. the treatment protocol must be reviewed.
   D. the patient report must be started.

4. Treating patients who are conscious and mentally competent is provided under:
   A. department rules and regulations
   B. expressed or informed consent.
   C. state statutes and ordinances.
   D. the patient's bill of rights.

5. The authority to treat unconscious patients or patients with serious threats to life is considered:
   A. implied consent.
   B. normal policy
   C. routine format.
   D. standard consent.

6. The forcible restraints placed on the above patient served the purpose of:
   A. forcing the patient to behave in an orderly manner.
   B. helping treat the patient.
   C. keeping the ambulance cleaner.
   D. protecting the patient from harming himself.

7. After you complete your written report on the patient, a local reporter asks you for a copy. This would violate:
   A. federal policy.
   B. local standards.
   C. patient confidentiality.
   D. report writing laws.

8. If you treats a patient without obtaining the patient's consent, you could be charged with:
   A. assault and battery.
   B. conspiracy.
   C. fraud.
   D. kidnapping.

Challenging Questions

9. Explain the EMT-B's role in honoring advance DNR orders.

10. What is the EMT-B's ethical responsibility as a health care provider?

11. Why is it important that the EMT-B complete a professional patient care report (PCR)?

12. Explain the special situations that an EMT-B is required to report.

www.EMTB.com

Points to Ponder

You have been dispatched to an incident in which a man was hit by a car. As you arrive at the scene and observe for any hazards, the Sheriff Deputy on scene informs you that the car has left but the pedestrian is present.

You approach the patient and introduce yourself. He tells you right away that his neck and back do not hurt. Blood is moderately flowing from two large cuts (lacerations) on his head. You continue your assessment and find the patient very anxious and also complaining of pain in his legs. On the basis of your assessment findings, you believe the patient needs to be transported to the emergency department for treatment. The patient states, "I'm fine," and refuses to go with you. You request the assistance of the deputy to help convince the patient to go by ambulance.

Issues: Duty to Act, Abandonment, Negligence, Patient Consent, Patient's Right to Refuse Treatment and Transport.

The Human Body

Objectives

You are the Provider

You are dispatched to 340 Tulip Lane for a 45-year-old man complaining of back pain. You arrive to find an alert and oriented man who is unable to sit still and appears to be in severe pain. He states that he thinks he hurt his back, but he's not sure how. He states, "This is the worst pain I've ever had in my life!" When you ask him where it hurts, he points to his right lower back.

1. Why is knowledge of anatomy important in determining potential sources of pain?

The Human Body

A working knowledge of human anatomy is important for you as an EMT-B. Even though you will not make diagnoses, you can help hospital personnel by communicating information using the correct medical terms. All EMT-Bs must be familiar with the language of topographic anatomy. By using the proper medical terms, you will be able to communicate correct information with the least possible confusion.

Using topographic anatomy is actually like using a road map. The terms that are introduced in this chapter will help you to identify the topographic (ie, on the surface) landmarks of the body. These landmarks are used as guides to locate the internal structures that lie under them. These terms also refer to the names of the major regions of the body and the way in which the locations of these regions are described in relation to one another.

Topographic Anatomy

The surface of the body has many definite visible features that serve as guides or landmarks to the structures that lie beneath them. You must be able to identify the superficial landmarks of the body—its topographic anatomy—to perform an accurate assessment. Understanding the terminology is also important so that you can describe patient findings correctly to your team, ALS providers, and hospital personnel.

Learning the terms that are introduced in this chapter will make your job as an EMT-B easier, since you will be able to correctly identify structures as you complete and report your assessment findings. Hospital personnel will use these terms to ask you questions about a patient. Therefore, you must learn what these terms mean and how to use them.

The terms that are used to describe the topographic anatomy are applied to the body when it is in the anatomic position. This is a position of reference in which the patient stands facing you, arms at the side, with the palms of the hands forward.

The Planes of the Body

The anatomic planes of the body are imaginary straight lines that divide the body (Table 4-1 ▼). These planes help you to identify the location of internal structures and understand the relationships between and among the organs.

Anterior and Posterior

Anterior refers to the front surface of the body, the side facing you in the anatomic position. Posterior refers to the back surface of the patient, or the side away from you.

Midline

An imaginary vertical line drawn from the middle of the forehead through the nose and the umbilicus (navel) to the floor is called the midline of the body. This imaginary line divides the body into two halves that are mirror images. The nose, chin, umbilicus (navel), and spine are examples of midline structures.

Midclavicular Line

The midclavicular line is an imaginary line drawn vertically through the middle portion of the clavicle and parallel to the midline. For example, the nipples of the

www.EMTB.com

Technology

Interactivities

Vocabulary Explorer

Anatomy Review

Web Links

Online Review Manual

TABLE 4-1 Directional Landmarks

Term	Definition
Midline	Referring to a line drawn through nose and umbilicus
Midclavicular	Referring to the middle of the clavicle, parallel to the midline
Midaxillary	Referring to the middle of the armpit, parallel to the midline

breasts are in the midclavicular line on either side of the body.

Midaxillary Line

The <u>midaxillary line</u> is an imaginary vertical line drawn through the middle of the axilla (armpit) parallel to the midline.

Directional Terms

In this section, terms that indicate direction are introduced. These terms indicate distance and direction from the midline (Figure 4-1 ▶) (Table 4-2 ▶).

Right and Left

The terms "right" and "left" refer to the patient's right and left sides, not to your right and left sides.

Superior and Inferior

The <u>superior</u> part of the body, or any body part, is the portion nearer to the head. The part nearer to the feet is the <u>inferior</u> portion. These terms are also used to describe the relationship of one structure to another. For example, the knee is superior to the foot and inferior to the pelvis.

Lateral and Medial

Parts of the body that lie farther from the midline are called <u>lateral</u> (outer) structures. The parts that lie closer to the midline are called <u>medial</u> (inner) structures. For example, the knee has medial (inner) and lateral (outer) aspects (surfaces).

Proximal and Distal

The terms "proximal" and "distal" are used to describe the relationship of any two structures on an extremity. <u>Proximal</u> describes structures that are closer to the trunk. <u>Distal</u> describes structures that are farther from the trunk or nearer to the free end of the extremity. For example, the elbow is distal to the shoulder and proximal to the wrist and hand.

Superficial and Deep

<u>Superficial</u> means closer to or on the skin. <u>Deep</u> means farther inside the body and away from the skin.

Ventral and Dorsal

<u>Ventral</u> refers to the belly side of the body, or the anterior surface of the body. <u>Dorsal</u> refers to the spinal side of the body, or the posterior surface of the body, including the back of the hand. These terms are used less frequently than the terms anterior and posterior.

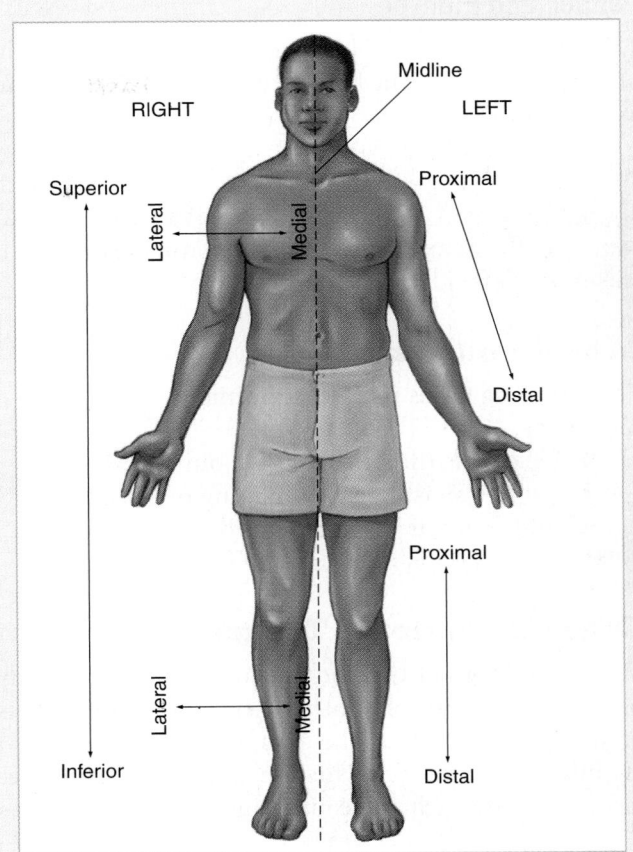

Figure 4-1 Directional terms indicate distance and direction from the midline.

TABLE 4-2 Directional Terms

Term	Definition
Anterior	Front
Posterior	Back
Right	The patient's right
Left	The patient's left
Lateral	Toward the outside
Medial	Toward the inside/middle
Superior	Closer to the head, higher
Inferior	Farther from the head, lower
Proximal	Closer to the midline (in an extremity, closer to the trunk)
Distal	Farther from the midline (in an extremity, further from the trunk)
Dorsal	Toward the spine
Ventral	Toward the superficial abdomen
Palmar	The front region of the hand
Plantar	The bottom of the foot

Palmar and Plantar

The front region of the hand is referred to as the palm or palmar surface. The bottom of the foot is referred to as the plantar surface.

Apex

The apex (plural: apices) is the tip of a structure. For example, the apex of the heart is the top of the ventricles in the left side of the chest.

Movement Terms

The following terms relate to movement Figure 4-2 ▼.

- Flexion is the bending of a joint.
- Extension is the straightening of a joint.
- Adduction is motion toward the midline.
- Abduction is motion away from the midline.

Other Directional Terms

Many structures of the body occur bilaterally. A body part that appears on both sides of the midline is bilateral. For example, the eyes, ears, hands, and feet are bilateral structures. This is also true for structures inside the body, such as the lungs and kidneys. Structures that appear on only one side of the body are said to occur unilaterally. For example, the spleen is on the left side of the body only, and the liver is on the right side. The terms unilateral and bilateral can also refer to something occurring on one side; for example, pain that is occurring on only one side of the body could be called unilateral pain.

As part of the assessment process, you will palpate the abdomen and report your findings. Therefore, it is important that you are able to describe the exact location of areas of the abdomen. The way to describe the sections of the abdominal cavity is by quadrants. Imagine two lines intersecting at the umbilicus, dividing the abdomen into four equal areas Figure 4-3 ►. These are referred to as the right upper quadrant (RUQ), left upper quadrant (LUQ), right lower quadrant (RLQ), and left lower quadrant (LLQ). Remember that here, too, right and left refer to the patient's right and left, not yours.

It is important to learn all of these terms and concepts so that you can describe the location of any injury or assessment findings. When you use these terms properly, any other medical personnel who care for the patient will know immediately where to look and what to expect.

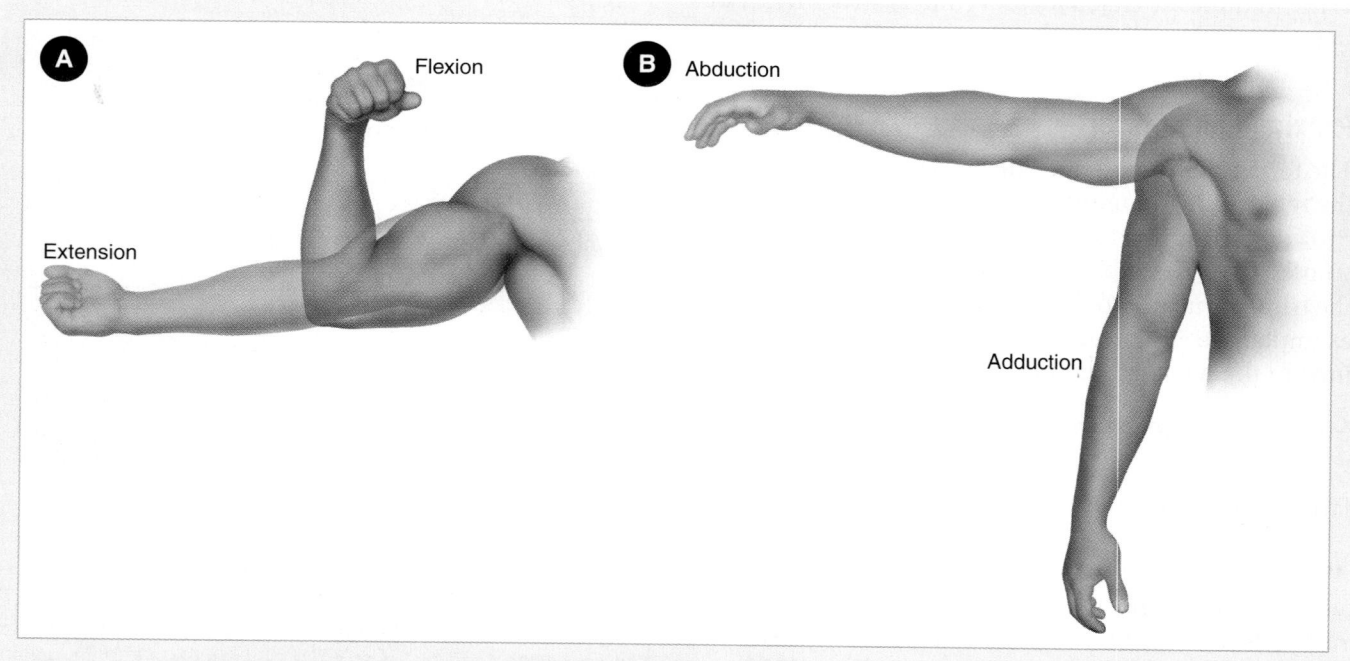

Figure 4-2 **A.** Flexion and extension. **B.** Abduction and adduction.

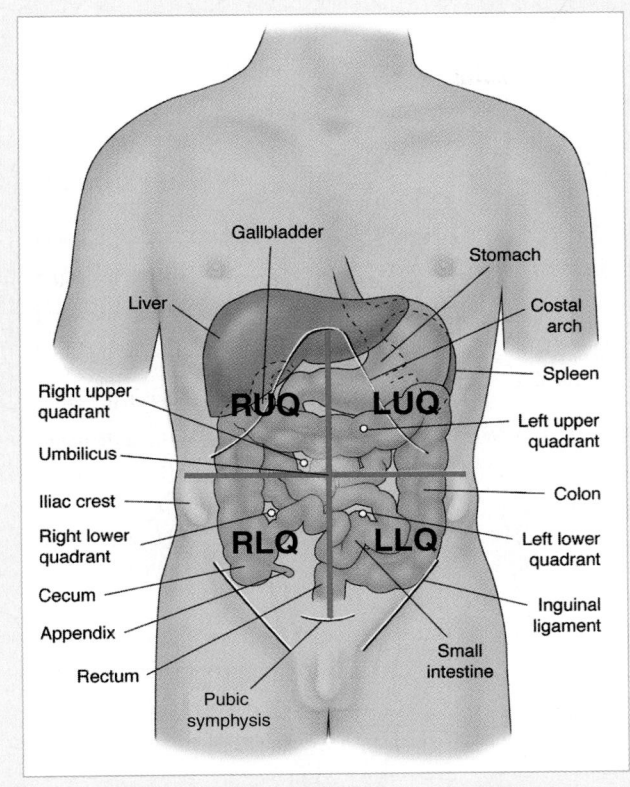

Figure 4-3 The abdomen is divided into four quadrants.

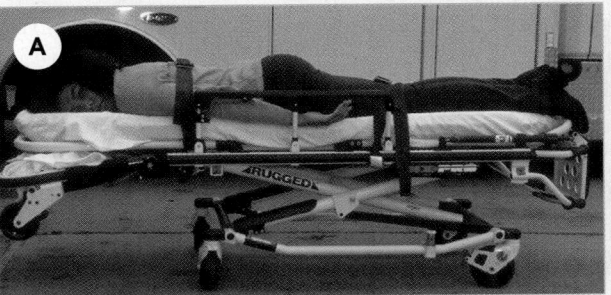

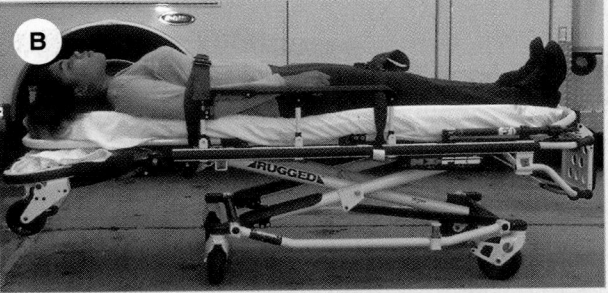

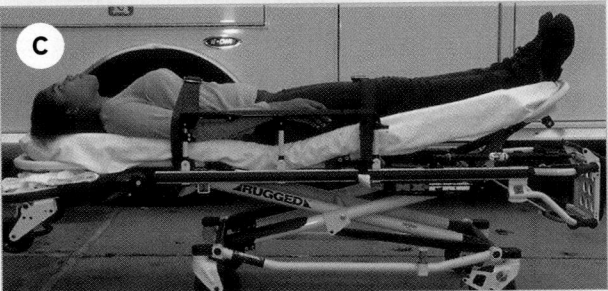

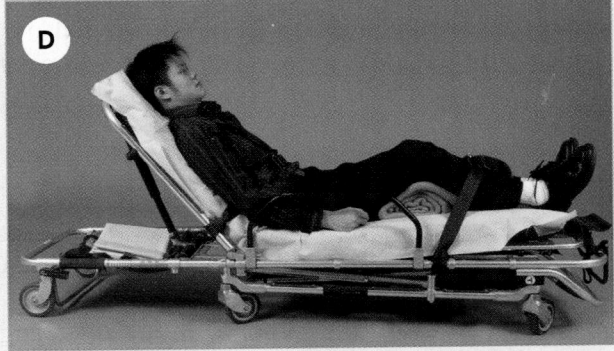

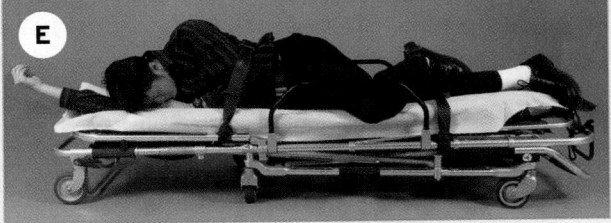

Figure 4-4 A. Prone. **B.** Supine. **C.** Shock position (modified Trendelenburg's position). **D.** Fowler's position. **E.** Recovery position.

Anatomic Positions

You will use these terms to describe the position of the patient as you find him or her or as you are to eventually transport the patient to the emergency department Figure 4-4 ▶ .

Prone and Supine

These terms describe the position of the body. The body is in the prone position when lying face down; the body is in the supine position when lying face up.

Fowler's Position

The Fowler's position was named after a US surgeon, George R. Fowler, MD, at the end of the 19th century. Dr Fowler placed his patients in a semireclining position with their heads elevated to help them breathe easier and to control the airway. A patient who is sitting up with the knees bent is therefore said to be in Fowler's position.

Trendelenburg's Position

Trendelenburg's position was named after a German surgeon, Friedrich Trendelenburg, at the turn of the 20th century. Dr Trendelenburg frequently placed his patients in a supine position on an incline with their feet higher than their head in order to keep blood in the core of the body. Trendelenburg's position is a position in which the body is on a backboard or stretcher with the feet 6″ to 12″ higher than the head.

Shock Position

In the shock position, or modified Trendelenburg's position, the head and torso (the trunk without the head and limbs) are supine, and the lower extremities are elevated 6″ to 12″. This helps to increase blood flow to the brain.

The Skeletal System

The skeleton gives us our recognizable human form and protects our vital internal organs (Figure 4-5 ▶). The brain lies within the skull. The heart, lungs, and great vessels are protected by the thorax, which is part of the torso. Much of the liver and spleen is protected by the lower ribs. The spinal cord is contained within and protected by a bony spinal canal formed by the vertebrae.

The 206 bones of the skeleton provide a framework for the attachment of muscles. The skeleton is also designed to allow motion of the body. Bones come into contact with one another at joints where, with the help of muscles, the body is able to bend and move.

The Skull

The cranium is composed of a number of thick bones that fuse together to form a shell above the eyes and ears that holds and protects the brain (Figure 4-6 ▶). The brain connects to the spinal cord through a large opening at the base of the skull (the foramen magnum). The spinal cord is composed of virtually all the nerves that carry messages between the brain and the rest of the body.

The most posterior portion of the cranium is called the occiput. On each side of the cranium, the lateral portions are called the temples or temporal regions. Between the temporal regions and the occiput lie the parietal regions. The forehead is called the frontal region. Just anterior to the ear, in the temporal region, you can feel the pulse of the superficial temporal ar-

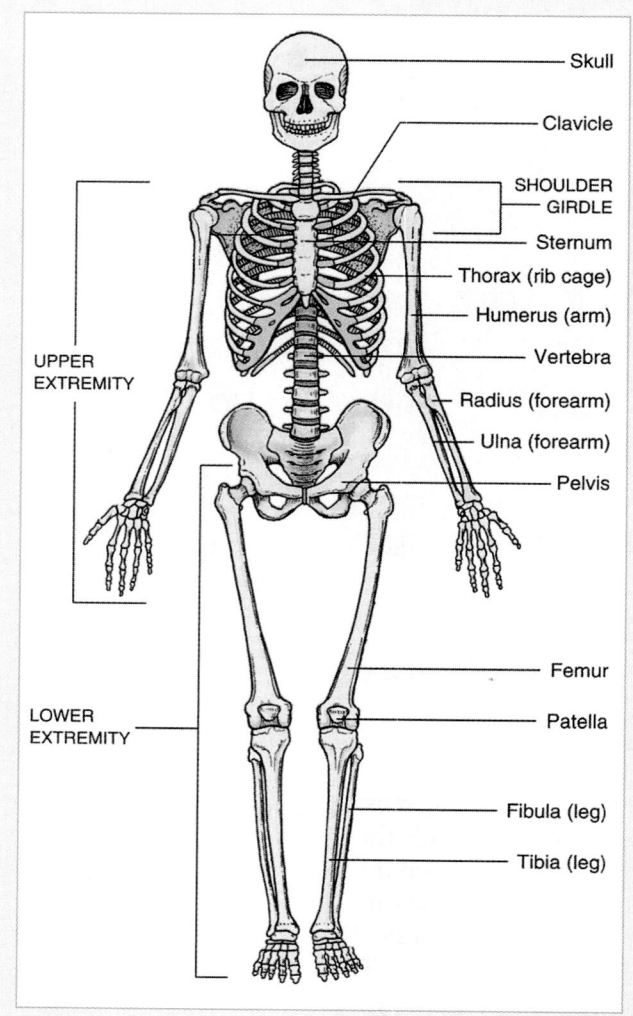

Figure 4-5 The 206 bones of the skeleton give us our form, protect our vital organs, and allow us to move.

tery. The thick skin covering the cranium and usually bearing hair is called the scalp.

The face is composed of the eyes, ears, nose, mouth, and cheeks. Six bones—the nasal bone, the two maxillae (upper jawbones), the two zygomas (cheek bones), and the mandible (lower jawbone)—are the major bones of the face.

The orbit (eye socket) is made up of two facial bones: the maxilla and the zygoma. The orbit also includes the frontal bone of the cranium. Together, these bones form a solid bony rim that protrudes around the eye to protect it. If you look at the face from the side, you can see that the eyeball sits back within the orbit. The nasal bone is

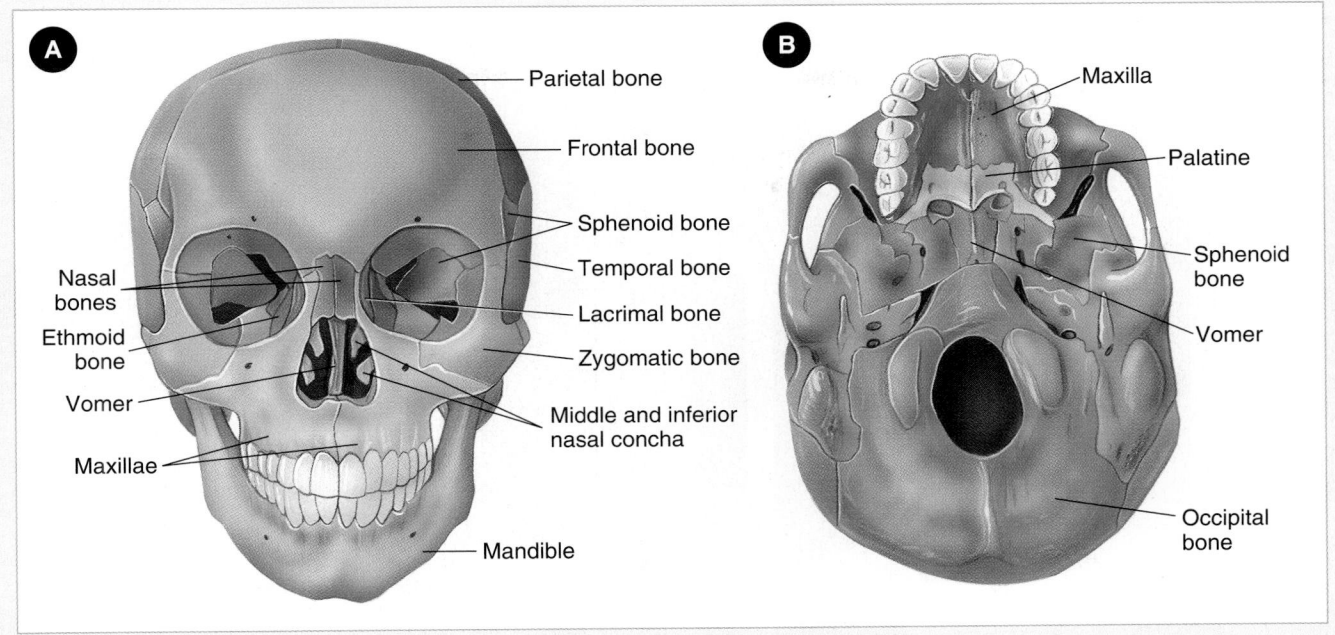

Figure 4-6 The skull. **A.** Front view. **B.** Bottom view.

very short, because most of the nose is made of flexible cartilage. In fact, only the proximal one third of the nose, the bridge, is formed by bone. Unlike the nose, the exposed portion of the ear is made up entirely of cartilage that is covered by skin. The external, visible part of the ear is called the pinna. The ear lobes are the fleshy parts at the bottom of each ear. About 1″ posterior to the external opening of the ear is a prominent bony mass at the base of the skull called the mastoid process.

The maxilla contains the upper teeth and forms the hard palate (roof of the mouth). The mandible is the only movable facial bone that has a joint (temporomandibular joint) where it meets with the temporal bone of the cranium just in front of each ear.

The Neck

The neck contains many important structures. It is supported by the cervical spine, or the first seven vertebrae in the spinal column (C1 through C7). The spinal cord exits from the foramen magnum and lies within the spinal canal formed by the vertebrae. The upper part of the esophagus and the trachea

You are the Provider   Part 2

You determine that this patient's origin of pain is most likely his right kidney. Further questioning of his medical history points to an increased risk for kidney stones. The paramedic agrees with your assessment, starts an IV, and administers morphine per local protocol.

2. Where are the kidneys located in the body? Are there any other organs in the same area?
3. Given the location of the kidneys and other components of the urinary system, what other complaints might this patient express?

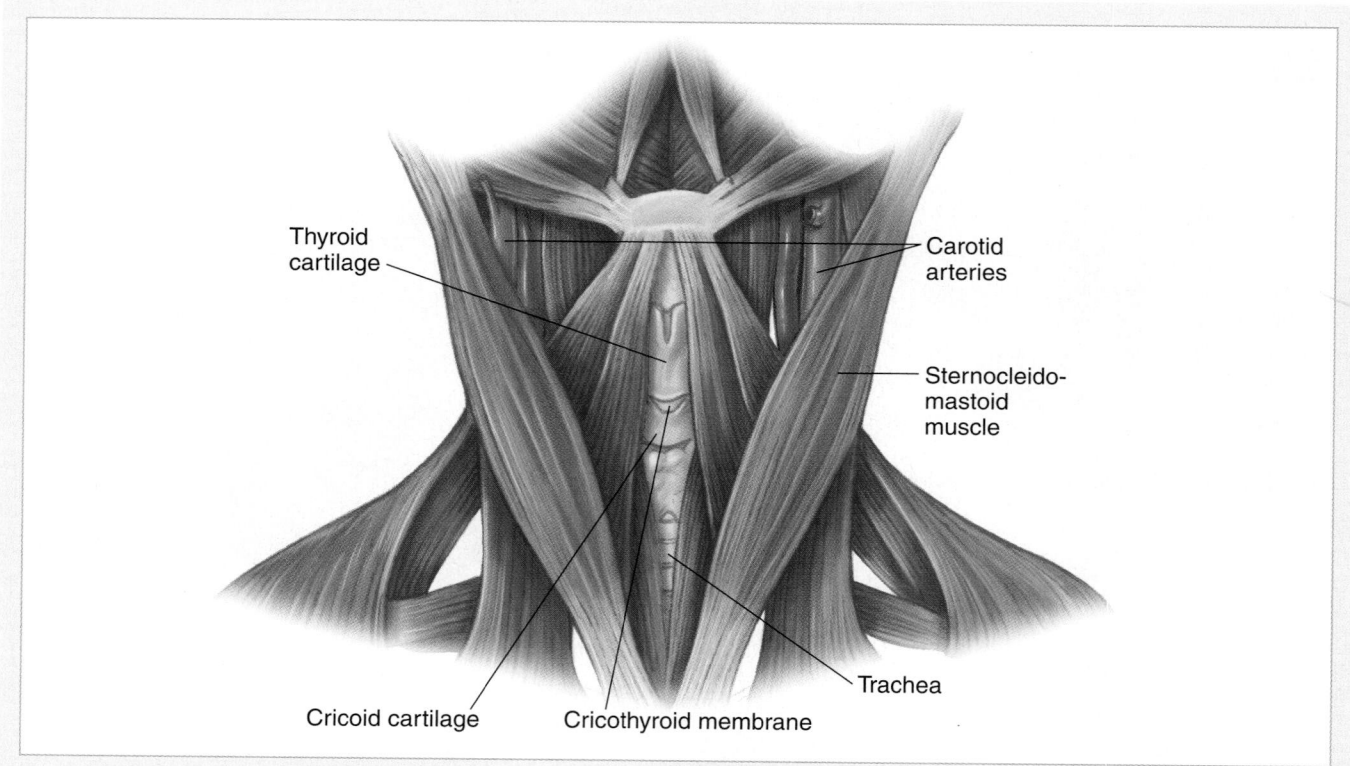

Figure 4-7 The principal structures of the neck include the trachea, along with many blood vessels, muscles, and nerves.

(windpipe) lie in the midline of the neck. The carotid arteries may be found on either side of the trachea, along with the jugular veins and several nerves.

Several useful landmarks can be palpated and seen in the neck (Figure 4-7 ▲). The most obvious is the firm prominence in the center of the anterior surface commonly known as the Adam's apple. Specifically, this prominence is the upper part of the thyroid cartilage. It is more prominent in men than in women. The other, lower portion is the cricoid cartilage, a firm ridge of cartilage inferior to the thyroid cartilage, which is somewhat more difficult to palpate. Between the thyroid cartilage and the cricoid cartilage in the midline of the neck is a soft depression, the cricothyroid membrane. This is a thin sheet of connective tissue (fascia) that joins the two cartilages. The cricothyroid membrane is covered at this point only by skin.

Inferior to the larynx, several additional firm ridges are palpable in the anterior midline. These ridges are the cartilage rings of the trachea. The trachea connects the larynx with the main air passages of the lungs (the

bronchi). On either side of the lower larynx and the upper trachea lies the thyroid gland. Unless it is enlarged, this gland is usually not palpable.

Pulsations of the carotid arteries are palpable in a groove about half an inch lateral to the larynx. Lying immediately adjacent to these arteries, but not palpable, are the internal jugular veins and several important nerves. Lateral to these vessels and nerves lie the sternocleidomastoid muscles, which allow movement of the head. These muscles originate from the mastoid process of the cranium and insert into the medial border of each collarbone and the sternum (breastbone) at the base of the neck. During times of difficulty breathing, use of these muscles may be exaggerated.

A series of bony prominences lie posteriorly, in the midline of the neck. They are the spines of the cervical vertebrae. The lower cervical spines are more prominent than the upper ones, and they are more easily palpable when the neck is flexed or bent forward. At the base of the neck posteriorly, the most prominent spine is the seventh cervical vertebra (Figure 4-8 ▶).

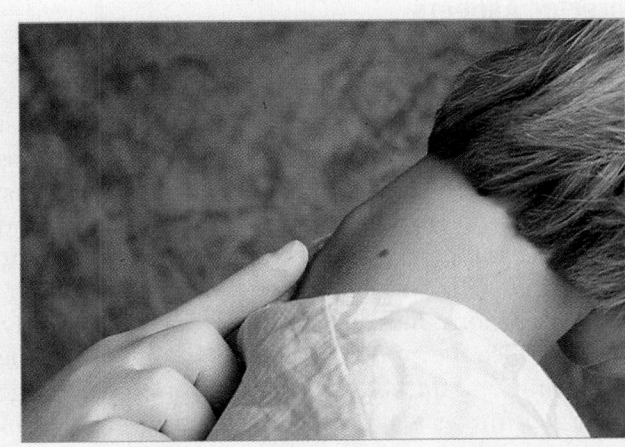

Figure 4-8 The most prominent of the cervical vertebrae is C7.

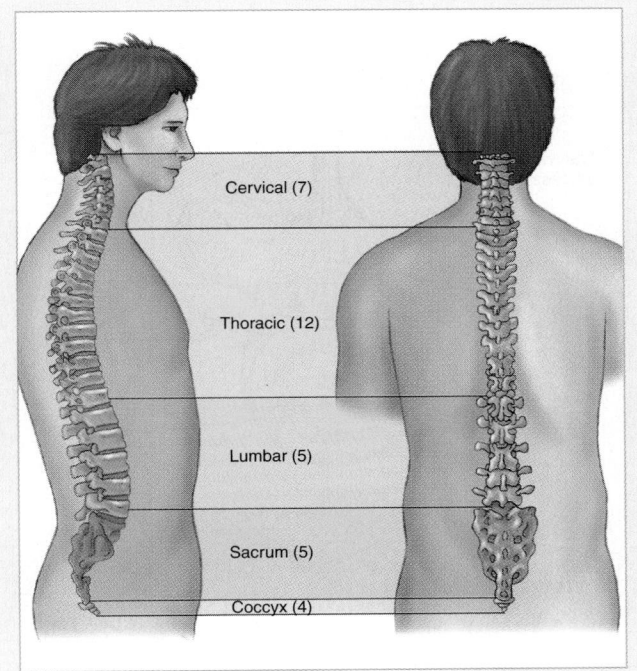

Figure 4-9 The spinal column is composed of 33 bones divided into five sections.

The Spinal Column

The spinal column is the central supporting structure of the body and is composed of 33 bones, each called a vertebra. The  vertebrae are named according to the section of the spine in which they lie and are numbered from top to bottom (Figure 4-9 ▶). From the top down, the spine is divided into five sections:

- Cervical spine. The first seven vertebrae (C1 through C7), which lie in the neck, form the cervical spine. The skull rests on the first cervical vertebra (the atlas) and articulates with it.
- Thoracic spine. The next 12 vertebrae make up the thoracic spine. One pair of ribs is attached to each of the thoracic vertebrae.
- Lumbar spine. The next five vertebrae form the lumbar spine.
- Sacrum. The five sacral vertebrae are fused together to form one bone called the sacrum. The sacrum is joined to the iliac bones of the pelvis with strong ligaments at the sacroiliac joints to form the pelvis.
- Coccyx. The last four vertebrae, also fused together, form the coccyx or tailbone.

The spinal cord is an extension of the brain, composed of virtually all the nerves that carry messages between the brain and the rest of the body. It exits through a large hole in the base of the skull called the foramen magnum and is contained within and protected by the vertebrae of the spinal column. The spinal column is virtually surrounded by muscles. However, the posterior spinous process of each vertebra can be felt as it lies just under the skin in the midline of the back. The most prominent and most easily palpable spinous process is that of the seventh cervical vertebra at the base of the neck.

The anterior part of each vertebra consists of a round, solid block of bone called the body. The posterior part of each vertebra forms a bony arch. This series of rings stacked one vertebra on top of another forms a channel that runs the length of the spine called the spinal canal. The bones of the spinal canal encase and protect the spinal cord (Figure 4-10 ▶). Nerves branch from the spinal cord and exit from the spinal canal between each two vertebrae to form the motor and sensory nerves of the body.

The vertebrae are connected by ligaments, and between each two vertebrae is a cushion called the intervertebral disk. These ligaments and disks allow some motion so that the trunk can bend forward (flex) and back (extend), and also allows for rotation and lateral movement. However, they also limit motion of the vertebrae so that the spinal cord will not be

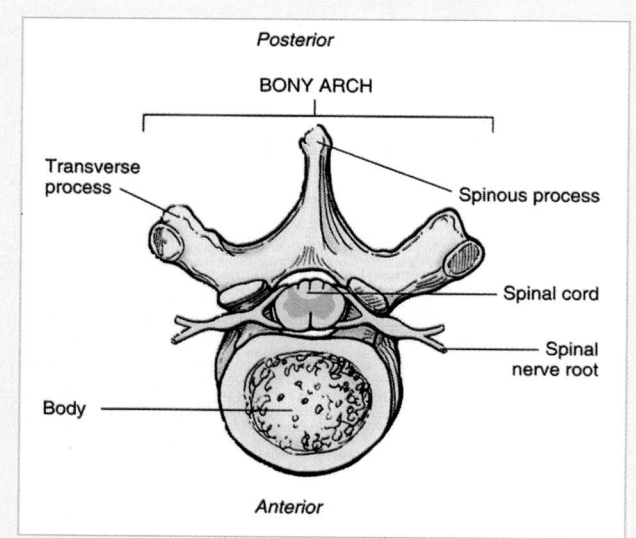

Figure 4-10 The bones of the spinal column encase and protect the spinal cord.

Anterior Aspects

The dimensions of the thorax are defined by the <u>thoracic cage</u> (bony rib cage) and its attachments (Figure 4-11A ▶). Anteriorly, in the midline of the chest is the sternum. The superior border of the sternum forms the easily palpable jugular notch. The sternum has three components: the manubrium, the body, and the xiphoid process. The upper section of the sternum is called the <u>manubrium</u>. The body comprises the rest of the sternum except for a narrow, cartilaginous tip inferiorly, which is called the <u>xiphoid process</u>. The junction of the manubrium and the body forms a very prominent ridge on the sternum, called the angle of Louis. The <u>angle of Louis</u> lies at the level where the second rib is attached to the sternum; it provides a constant and reliable bony landmark on the anterior chest wall.

In the midline of the upper back, the spines of the 12 thoracic vertebrae can be palpated. Twelve ribs on each side form small joints with their respective thoracic vertebrae and extend around to the front to create the walls of the thoracic cage. The upper five ribs connect to the sternum through a short bridge of cartilage. The sixth through tenth ribs insert into the costal arch. The <u>costal arch</u> is a bridge of cartilage that connects the ends of the sixth through tenth ribs with the lower portion of the sternum. The eleventh and twelfth ribs are called <u>floating ribs</u>, because they do not attach to the sternum through the costal arch. The costal arch is easily palpable and represents the boundary between the lower border of the thorax and the upper border of the abdomen.

Posterior Aspects

On the posterior chest wall, the scapulae overlie the thoracic wall and are surrounded by large muscles (Figure 4-11B ▶). When the patient is standing or sitting erect, the two scapulae should lie at approximately the same level, with their inferior tips at about the level of the seventh thoracic vertebra. In the lower part of the

injured. An injury to the spine may damage part of the spinal cord and its nerves that may not be protected by the vertebrae. Therefore, until the injury is stabilized, you must use extreme caution in caring for the patient to prevent injury to the spinal cord.

The Thorax

The <u>thorax</u> (chest) is the cavity that contains the heart, lungs, esophagus, and great vessels (the aorta and two venae cavae). It is formed by the 12 thoracic vertebrae (T1 through T12) and their 12 pairs of ribs. The <u>clavicle</u> (collarbone) overlies the superior boundaries of the thorax in front and articulates (joins) posteriorly with the <u>scapula</u> (shoulder blade), which lies in the muscular tissue of the thoracic wall. The inferior boundary of the thorax is the diaphragm, which separates the thorax from the abdomen.

You are the Provider   Part 3

The patient can now sit relatively still and appears to be more comfortable. You feel that your knowledge of human anatomy helped in determining the source of his pain.

4. If you are unfamiliar with human anatomy, how could your lack of knowledge significantly affect the quality and the timeliness of your patient care?

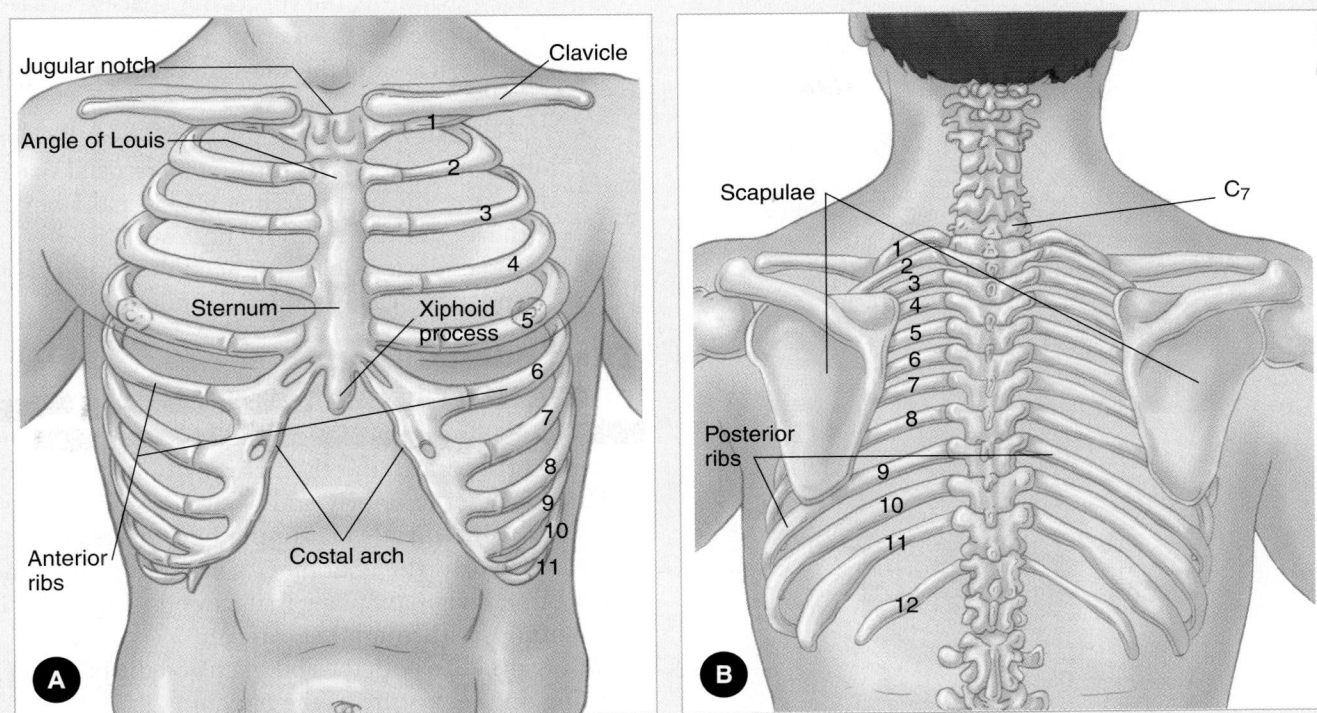

Figure 4-11 **A.** The anterior aspect of the thorax includes the following bony landmarks: the clavicle, the sternum, the xiphoid process, the angle of Louis, and the anterior ribs. **B.** The posterior aspect of the thorax includes the following bony landmarks: the scapulae, the thoracic vertebrae, and the posterior ribs.

thorax on each side, an angle called the <u>costovertebral angle</u> is formed by the junction of the spine and the tenth rib. The kidneys lie deep to (beneath) the back muscles in the costovertebral angle.

Diaphragm

The <u>diaphragm</u> is a muscular dome that forms the inferior boundary of the thorax, separating the chest from the abdominal cavity (Figure 4-12 ▶). Its contraction, along with that of the chest wall muscles, assists with allowing air to be drawn into the lungs. Anteriorly, it attaches to the costal arch; posteriorly, it attaches to the <u>lumbar vertebrae</u>. The diaphragm cannot be seen or palpated.

Organs and Vascular Structures

Within the thoracic cage, the largest structures are the heart and lungs (Figure 4-13 ▶). The heart lies immediately under the sternum. It extends from the second to the sixth ribs anteriorly and from the fifth to the eighth thoracic vertebrae posteriorly. The inferior border of the heart extends into the left side of the chest. Diseased hearts may be larger or smaller. The major blood vessels that travel to and from the heart also lie

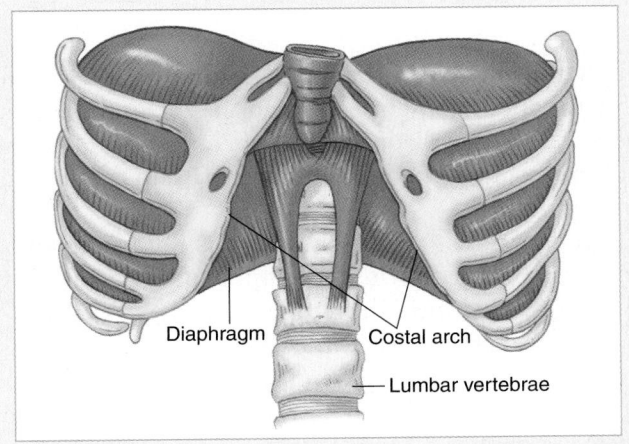

Figure 4-12 The diaphragm forms the undersurface of the thorax, separating the chest from the abdominal cavity.

in the chest cavity. On the right side of the spinal column, the superior and inferior venae cavae carry blood to the heart.

Just beneath the manubrium of the sternum, the arch of the aorta and the pulmonary artery exit the heart. The arch of the aorta passes to the left and lies

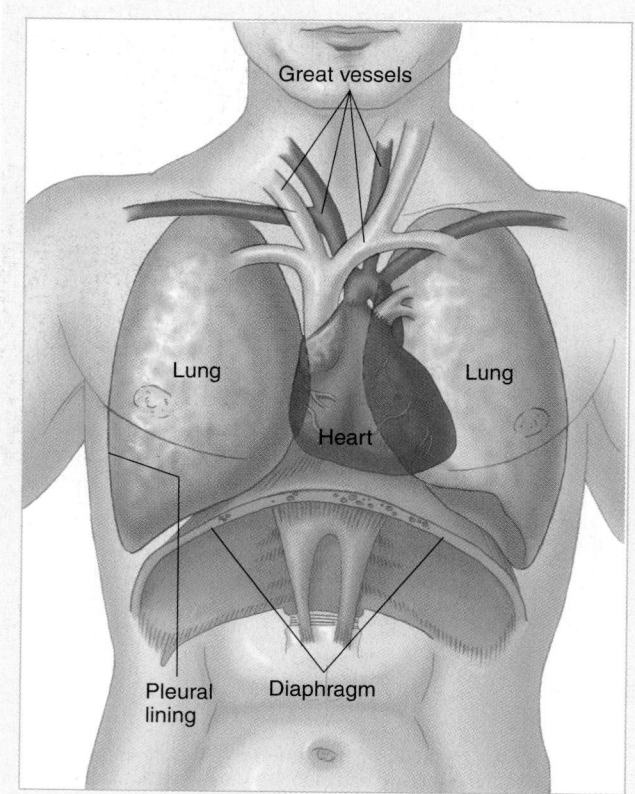

Figure 4-13 The anterior view of the thorax shows the relative positions of the principal organs beneath the surface.

along the left side of the spinal column as it descends into the abdomen. The esophagus lies behind the great vessels and directly on the anterior aspect of the spinal column as it passes through the chest into the abdominal cavity.

All space within the chest that is not occupied by the heart, great vessels, and esophagus is occupied by the lungs. Anteriorly, the lungs extend down to the surface of the diaphragm at the level of the xiphoid process. Posteriorly, the lungs extend farther inferiorly to the surface of the diaphragm at the level of the 12th thoracic vertebra.

Anatomic Landmarks

The major palpable landmarks in the chest are obviously the ribs. Most of them can be easily felt except for the first, which is hidden under and behind the clavicle. The anatomic landmarks are as follows:

- Between each rib is the intercostal space. These spaces can be located by palpating the jugular notch and moving laterally (the first intercostal

space). Counting the successive spaces between the ribs gives us the second, third, and so on.
- Both clavicles and the sternum can be easily palpated.
- The jugular notch is the top portion of the sternum. Lateral to that is the first intercostal space.
- Inferiorly, the costal arch is readily palpable on both sides of the anterior chest wall.
- In the midline, the tip of the xiphoid process is a tender and easily palpated landmark.

The Abdomen

The abdomen is the second major body cavity; it contains the major organs of digestion and excretion. The diaphragm separates the thoracic cavity from the abdominal cavity. Anteriorly and posteriorly, thick muscular abdominal walls create the boundaries of this space. Inferiorly, the abdomen is separated from the pelvis by an imaginary plane that extends from the pubic symphysis through the sacrum (Figure 4-14 ▶). Many organs lie in both the abdomen and the pelvis, depending on the posture of the patient.

The simplest and most common method of describing the portions of the abdomen is by quadrants, the four equal areas formed by two imaginary lines that intersect at right angles at the umbilicus. On the anterior abdominal wall, the quadrants thus formed are the right upper, right lower, left upper, and left lower (Figure 4-15 ▶). The terms "right quadrant" and "left quadrant" refer to the patient's right and left, not to your right and left sides. Pain or injury in a given quadrant usually arises from or involves the organs that lie in that quadrant. This simple means of designation will allow you to identify injured or diseased organs that require emergency attention.

Organs and Vascular Structures

In the right upper quadrant (RUQ), the major organs are the liver, the gallbladder, and a portion of the colon. Most of the liver lies in this quadrant, almost entirely under the protection of the eighth to twelfth ribs. The liver fills the entire anteroposterior depth of the abdomen in this quadrant. Therefore, injuries in this area are frequently associated with injuries of the liver.

In the left upper quadrant (LUQ), the principal organs are the stomach, the spleen, and a portion of the colon. The spleen is almost entirely under the protection of the left rib cage, whereas the stomach may sag well down into the left lower quadrant when full. The spleen lies in the lateral and posterior portion of this

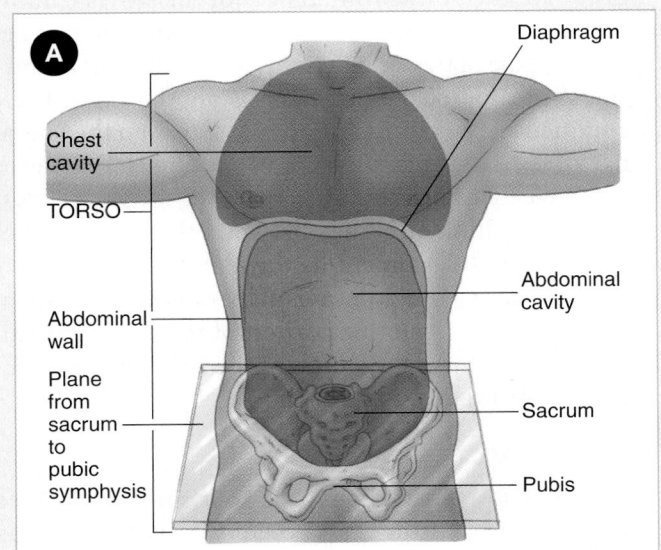
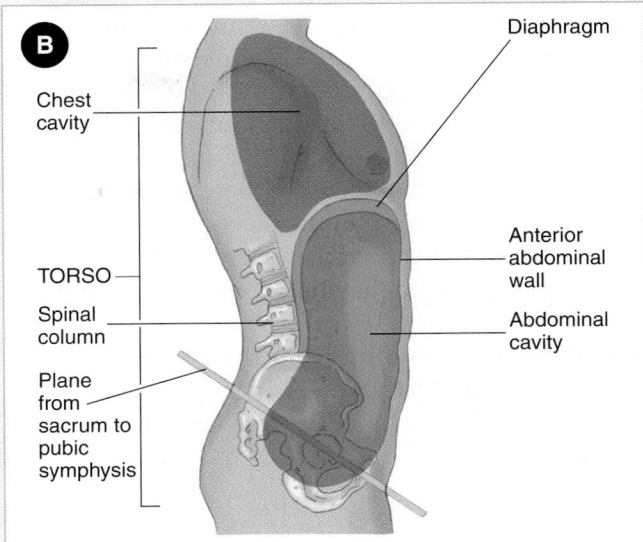

Figure 4-14 The boundaries of the abdomen are the anterior and posterior abdominal cavity walls, the diaphragm, and an imaginary plane from the pubic symphysis to the sacrum. **A.** Anterior view. **B.** Lateral view.

quadrant, under the diaphragm and immediately in front of the ninth to eleventh ribs. The spleen is frequently injured, especially when these ribs are fractured.

The right lower quadrant (RLQ) contains two portions of the large intestine: the cecum, the first portion into which the small intestine (ileum) opens, and the ascending colon. The appendix is a small tubular struc-

ture that is attached to the lower border of the cecum. Appendicitis is the most frequent cause of tenderness and pain in this region. In the left lower quadrant (LLQ) lie the descending and the sigmoid portions of the colon.

Several organs lie in more than one quadrant. The small intestine, for instance, occupies the central part of the abdomen around the umbilicus, and parts of it

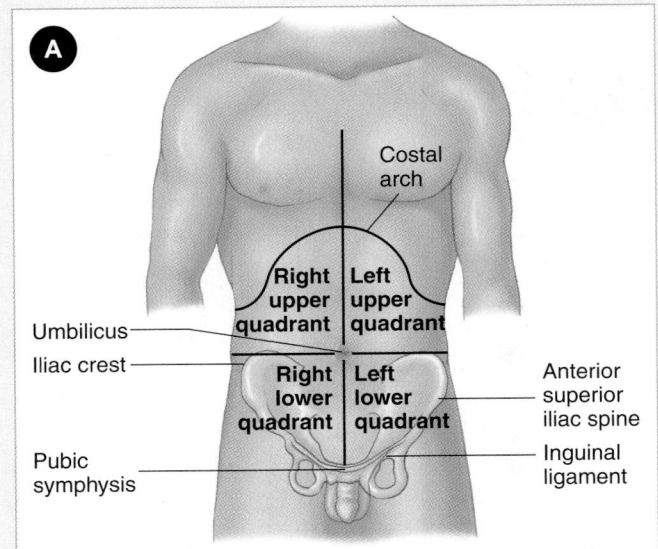
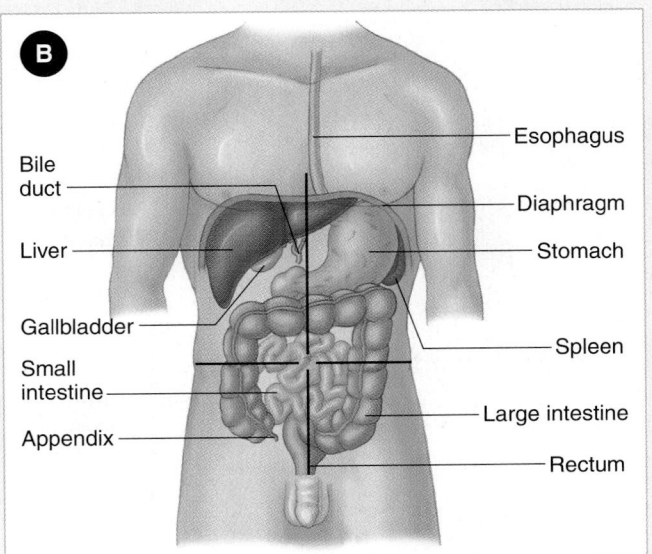

Figure 4-15 A. In the abdomen, quadrants are the easiest system for identifying areas. Major bony landmarks are also shown. **B.** Many of the organs in the abdomen lie in more than one quadrant.

lie in all four quadrants. The pancreas lies just behind the abdominal cavity on the posterior abdominal wall in both upper quadrants. The large intestine also traverses the abdomen, beginning in the RLQ and ending in the LLQ as it passes through all four quadrants. The urinary bladder lies just behind the pubic symphysis in the middle of the abdomen and therefore lies in both lower quadrants and also in the pelvis.

The kidneys and pancreas are called <u>retroperitoneal</u> organs because they lie behind the abdominal cavity (Figure 4-16 ▼). They are above the level of the umbilicus, extending from the eleventh rib to the third lumbar vertebra on each side. They are approximately 5" long and lie just anterior to the costovertebral angle.

Anatomic Landmarks

The chief landmarks in the abdomen are the costal arch, the umbilicus, the anterior superior iliac spines, the iliac crest, and the pubic symphysis. The costal arch, as was noted earlier, is the fused cartilages of the sixth through the tenth ribs. It forms the superior arching boundary of the abdomen. The umbilicus, a constant structure, is in the same horizontal plane as the fourth lumbar vertebra and the superior edge of the iliac crest, the rim of the pelvic bone. The <u>anterior superior iliac</u>

<u>spines</u> are the bony prominences of the pelvis (ilium) at the front on each side of the lower abdomen just below the plane of the umbilicus. In the midline in the lowermost portion of the abdomen is another hard bony prominence, the <u>pubic symphysis</u>. Between the lateral edge of the pubic symphysis and the anterior superior spine on each side, you can palpate the tough <u>inguinal ligament</u>, which stretches between these two structures. Below the ligament lie the femoral vessels.

Posteriorly, you do not usually refer to abdominal quadrants. The posterior portion of the iliac crest can be palpated, as can the spines of the five lumbar vertebrae (L1 through L5) in the midline.

The Pelvis

The pelvis is a closed bony ring that consists of three bones: the sacrum and the two pelvic bones (Figure 4-17 ▼). Much like the skull, each pelvic bone is formed by the fusion of three separate bones. These three bones are called the <u>ilium</u>, the <u>ischium</u>, and the <u>pubis</u>. These bones meet at three joints: the two posterior sacroiliac joints and the anterior midline pubic symphysis. All three joints allow very little motion, as they are firmly held together by strong ligaments. On the lateral side of each pelvic bone—where the three component bones join—is the socket for the hip joint.

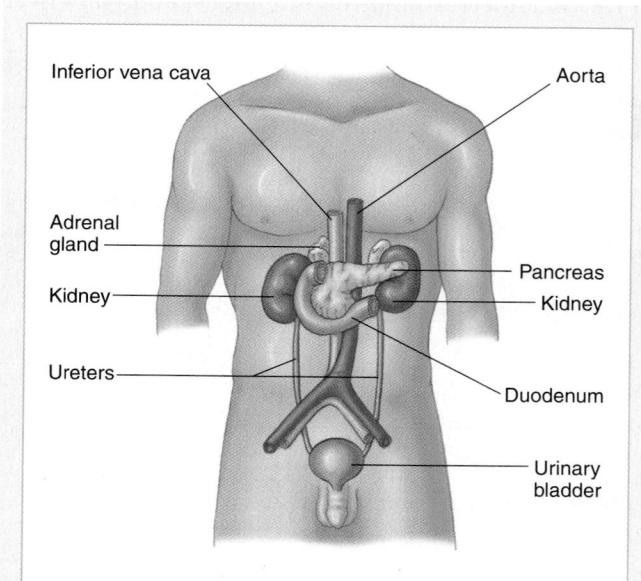

Figure 4-16 The major organs of the retroperitoneal space lie behind the abdominal cavity, above the level of the umbilicus, and extend from the eleventh rib to the third lumbar vertebra. Note that the bladder, inferior vena cava, and aorta also lie in this plane.

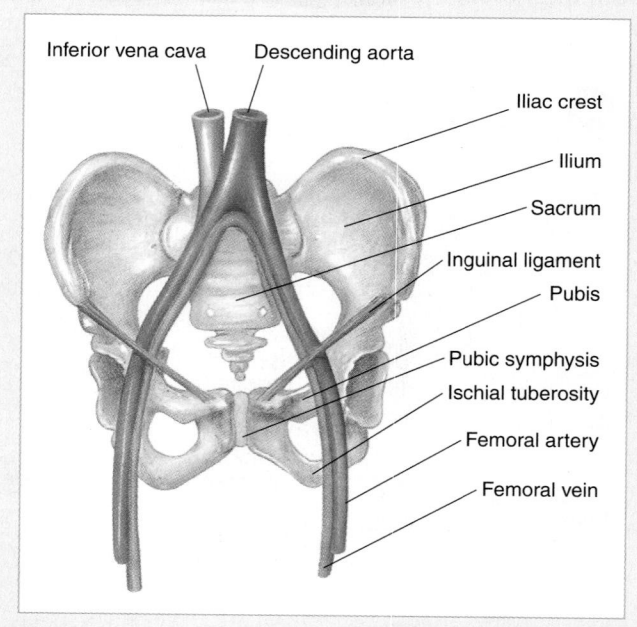

Figure 4-17 The pelvis is a closed bony ring that consists of the sacrum, ilium, ischium, and pubic bones.

This depression, in which the femoral head fits very snugly, is called the acetabulum.✳

The pelvic cavity is bounded superiorly by an imaginary plane that runs from the pubic symphysis to the top of the sacrum. Its lateral walls are formed by the inner borders of the pelvic bone, and its inferior boundary is the pelvic outlet, a layer of muscles with openings for the gastrointestinal tract (the rectum), the female reproductive system (the vagina), and the urinary tract (the urethra). In addition, the pelvis contains the final portions of the gastrointestinal tract (the rectosigmoid colon), the female reproductive organs, and the urinary bladder.

Anterior Aspects

The prominent anterior bony landmarks of the pelvis are the pubic symphysis in the midline and the anterior superior iliac spines. The inguinal ligament attaches to these two bony prominences and can be palpated in a thin person. Just distal to the midpoint of the inguinal ligament, the femoral artery can be palpated as it enters the thigh. From the anterior superior iliac spine, the ilium extends laterally and posteriorly to form the rim of the pelvis. This bony ridge is called the iliac crest, or wing of the pelvis.

Posterior Aspects

Posteriorly, the pelvis appears flat, and in the middle third, the firm bony sacrum can be palpated. Just lateral to the sacrum on either side is a joint with the iliac portion of the pelvic bone (the sacroiliac joint). In the sitting position, a bony prominence is easily felt below the middle of each buttock. These prominences are the ischial tuberosities. The sciatic nerve, which is the major nerve to the lower extremity, lies just lateral to the tuberosity as it enters the thigh.

The Lower Extremity

The main parts of the lower extremity are the thigh, the leg, and the foot (Figure 4-18 ▶). Three joints connect the parts of the lower extremity: the hip, the knee, and the ankle. The joint between the thigh and pelvis is called the hip. The joint between the thigh and the leg is the knee. The joint between the leg and the foot is the ankle.

Thigh

On the proximal lateral side of the thigh, just below the hip joint, is a bony prominence called the greater trochanter. This prominence is sometimes called the "hip bone." During patient assessment, you should always compare the position of the greater trochanter with that on the opposite side as a guide to injury or deformity of the hip.

The femur (thigh bone) is the longest and one of the strongest bones in the body. The femoral head (at the top of the femur) forms the hip joint with the acetabulum of the pelvis. This ball-and-socket joint allows for flexion, extension, and motion toward (adduction) and away (abduction) from the midline. It also allows for internal and external rotation of the entire lower extremity. The shaft of the femur is surrounded by large muscles (the quadriceps in front and the hamstrings in back). Just above the knee, the medial and lateral femoral condyles can be palpated.

You are the Provider   Part 4

You arrive at the hospital and listen to the paramedic provide information to the emergency department physician on the patient's initial presentation. She explains that although the original chief complaint could have led to an incorrect differential diagnosis, your knowledge of anatomy assisted in the appropriate care of this patient. The physician seems to have gained some trust in your knowledge and skills.

5. How can your knowledge of anatomy and correct use of medical terminology when speaking to other professional health care providers affect their opinion of you, your agency, and emergency service professionals in general?

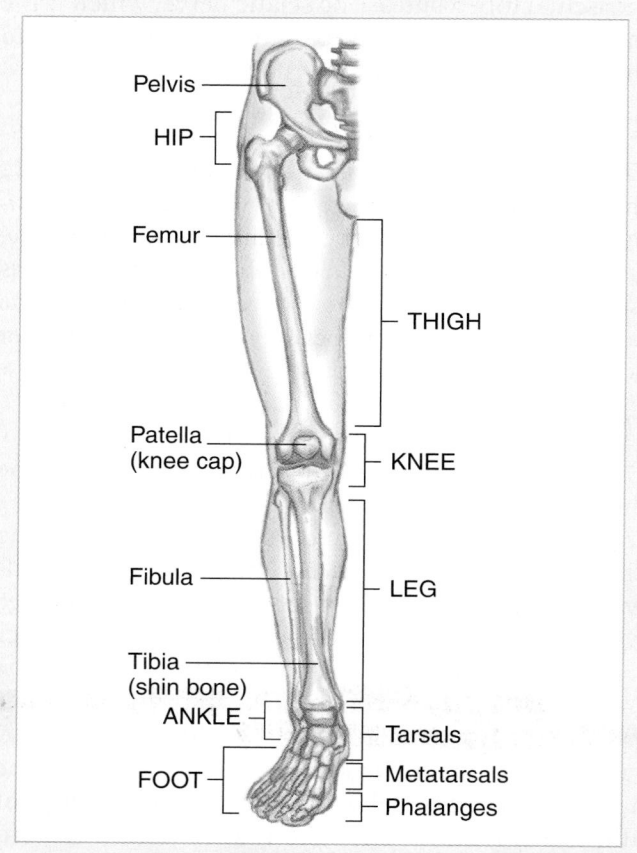

Figure 4-18 The principal parts of the lower extremity include the thigh, leg, and foot. The principal parts of the leg include the tibia and fibula.

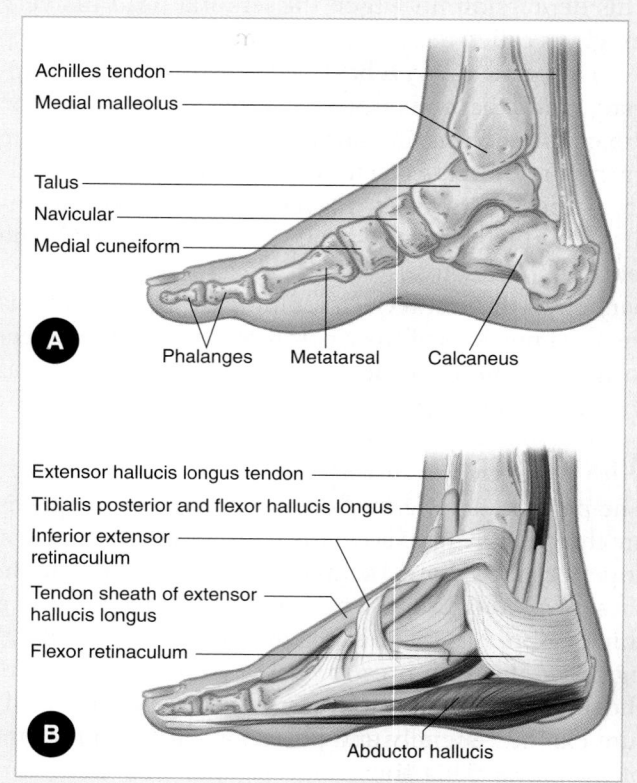

Figure 4-19 **A.** The surface landmarks of the foot and ankle include the medial malleolus, the calcaneus, and the phalanges. **B.** Soft tissue of the ankle.

Knee

Between the thigh and the leg is the largest joint in the body: the knee. The knee is essentially a hinge joint, allowing only flexion and extension between the distal femur and the proximal tibia. Adduction, abduction, and rotation of the knee are resisted by complex ligaments that are quite susceptible to injury. Anterior to the knee is a specialized bone called the patella (kneecap). It lies within the tendon of the quadriceps muscle and protects the front of the knee from injury.

Leg

The leg lies between the knee and the ankle joint and is composed of the tibia and the fibula. The tibia (shin bone) is the larger bone and lies in the front of the leg. You can palpate the entire length of the tibia on the anterior surface of the leg just under the skin. The fibula lies on the lateral side of the leg. You can palpate the head

of the fibula on the lateral aspect of the knee joint. Its distal end forms the lateral malleolus of the ankle joint.

Ankle and Foot

The ankle is a hinge joint that allows flexion and extension of the foot on the leg (Figure 4-19 ▲). The end of the tibia forms the medial malleolus, and the end of the fibula forms the lateral malleolus. These two bony prominences form the socket of the ankle joint. Both are surface landmarks of the ankle joint and are easily palpated. The foot contains seven tarsal bones. The talus is one of the largest; the calcaneus, which forms the prominence of the heel, is the other large tarsal bone. The Achilles tendon inserts into the back of the calcaneus. Five metatarsal bones form the substance of the foot. The five toes are formed by 14 phalanges—two in the great toe and three in each of the smaller toes.

The Upper Extremity

The upper extremity extends from the shoulder girdle to the fingertips and is composed of the arm, forearm, hand, and fingers. The joints are the elbow, wrist, and finger joints. The arm extends from the shoulder to the elbow, the forearm from the elbow to the wrist, the hand from the wrist to the fingertips.

Shoulder Girdle

The proximal portion of the upper extremity is called the shoulder girdle and consists of three bones: the clavicle, the scapula, and the humerus (Figure 4-20 ▼). The shoulder girdle is where the upper extremity attaches to the trunk. The upper extremity can move through a wide range of motion, allowing the arm to be placed in almost any position. This motion occurs at three joints within the shoulder girdle: the sternoclavicular joint, the acromioclavicular (A/C) joint, and the glenohumeral joint. In addition, the scapula can rotate on the thorax, providing additional range of motion. Only slight motion occurs normally at the sternoclavicular and A/C joints. The ball-and-socket arrangement of the glenohumeral joint allows great freedom of motion in almost any direction.

The clavicle is a long, slender bone that lies just under the skin and provides support for the upper extremity. The clavicle is palpable through its entire length from the sternum to its attachment to the scapula. Its medial end is attached by very strong ligaments to the manubrium of the sternum to form the sternoclavicu-lar joint. Its lateral end forms a joint with the acromion process of the scapula to create the A/C joint.

The scapula is a large, flat, triangular bone that overlies the posterior wall of the thorax and is surrounded by large muscles. Because of these muscles, only small parts of this bone are palpable. The scapula has two specially named regions that form joints with the clavicle and the humerus. The acromion process in the front forms part of the A/C joint. The glenoid fossa joins with the humeral head to form the glenohumeral joint. The spine and medial border of the scapula can be seen and palpated posteriorly. The acromion process forms the rounded edge of the shoulder girdle. You can feel this if you slowly move your finger along the clavicle and across the A/C joint.

Arm

The supporting bone of the arm is the humerus. Its long, straight shaft serves as an effective lever for heavy lifting. As in the thigh, there are few bony landmarks in the arm because it is covered by large muscles: the biceps in the front and the triceps in the back. The head of the humerus is covered by muscles that form the rounded prominence of the shoulder girdle laterally. The distal end articulates with both the radius and ulna at the elbow joint (Figure 4-21 ▼).

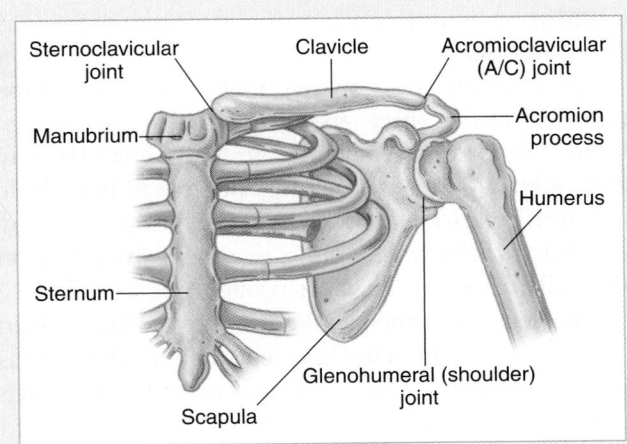

Figure 4-20 The bones of the shoulder girdle include the clavicle, the scapula, and the humerus.

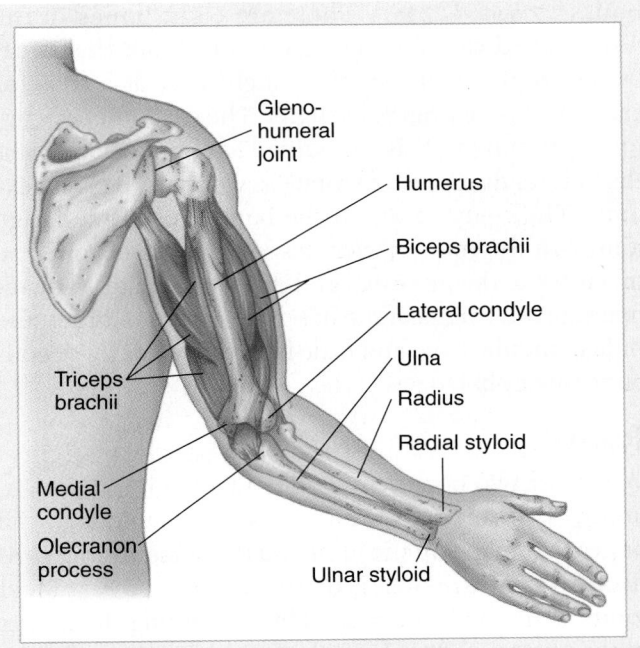

Figure 4-21 The principal bones in the arm and forearm include the humerus, the radius, and the ulna.

The humerus joins with the radius and ulna to form the elbow, which is a relatively simple hinge joint. You can easily see and feel three prominences on the back of the elbow: the medial and lateral condyles of the humerus and the olecranon process of the ulna.

Forearm

The forearm is composed of the radius and the ulna. The ulna is larger in the proximal forearm, and the radius is larger in the distal forearm. The olecranon process of the ulna forms most of the elbow joint. The entire ulnar shaft from the tip of the olecranon process distally can be palpated, because it lies just under the skin on the back of the forearm. The radius is covered by muscles and cannot be palpated except in the lower third of the forearm, where it enlarges to form a major portion of the wrist joint. The radius rotates about the ulna, which allows the palm of the hand to turn up or down. At the wrist, the ends of the radius and ulna (the styloid processes) lie directly under the skin and can be easily palpated. The radial styloid is slightly longer than the ulnar styloid. The radius lies on the lateral, or thumb, side of the forearm, and the ulna is on the medial or little finger side.

Wrist and Hand

The wrist is a modified ball-and-socket joint formed by the ends of the radius and ulna and several small wrist bones (Figure 4-22 ▶). There are eight bones in the wrist, called carpal bones. Extending from the carpal bones are five metacarpals, which serve as a base for each of the five fingers or digits. The carpometacarpal joint (thumb joint) is a modified ball-and-socket joint that allows the thumb to rotate as well as to flex and extend. The other joints in the hand are simple hinge joints. In the thumb, there are two bones beyond the metacarpal: the proximal and distal phalanges. The remaining four digits of the hand are named in order: the index, middle, ring, and little finger. Each of these contains three phalanges.

Joints

Wherever two bones come in contact, a joint (articulation) is formed. A joint consists of the ends of the bones that make up the joint and the surrounding connecting and supporting tissue (Figure 4-23 ▶). Most joints in the body are named by combining the names of the two bones that form that joint. For example, the sternoclavicular joint is the articulation between the sternum and the clavicle. Most joints allow motion—for example, the knee, hip, or elbow—whereas some

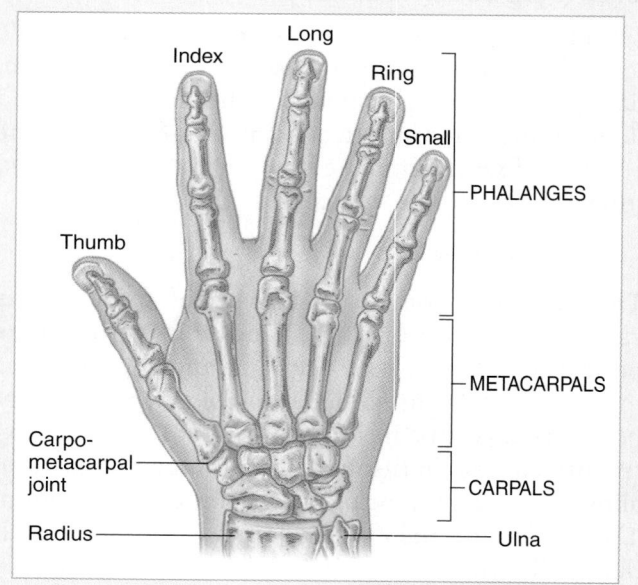

Figure 4-22 The principal bones in the wrist and hand include the carpals, the metacarpals, and the phalanges.

bones fuse with one another at joints to form a solid, immobile, bony structure. For instance, the skull is composed of several bones that fuse as a child grows. An infant, whose skull bones are not yet fused, has fontanels (soft spots) between the bones. The fontanels close as the bones of the infant's skull fuse together. Some joints have slight, limited motion in which the bone ends are held together by fibrous tissue. Such a joint is called a symphysis.

The bone ends of a joint are held together by a fibrous sac called the joint capsule. At certain points around the circumference of the joint, the capsule is lax and thin so that motion can occur. In other areas, it is quite thick and resists stretching or bending. These bands of tough, thick tissue are called ligaments. A joint such as the sacroiliac joint that is virtually surrounded by tough, thick ligaments will have little motion, whereas a joint such as the shoulder, with few ligaments, will be free to move in almost any direction (and will, as a result, be more prone to dislocation).

The degree to which a joint can move is determined by the extent to which the ligaments hold the bone ends together and also by the configuration of the bone ends themselves. The shoulder joint is a ball-and-socket joint, which allows rotation as well as bending (Figure 4-24 ▶). The finger joints, elbow, and knee are hinge joints, with motion restricted to one plane (Figure 4-25 ▶). They can only flex (bend) and extend

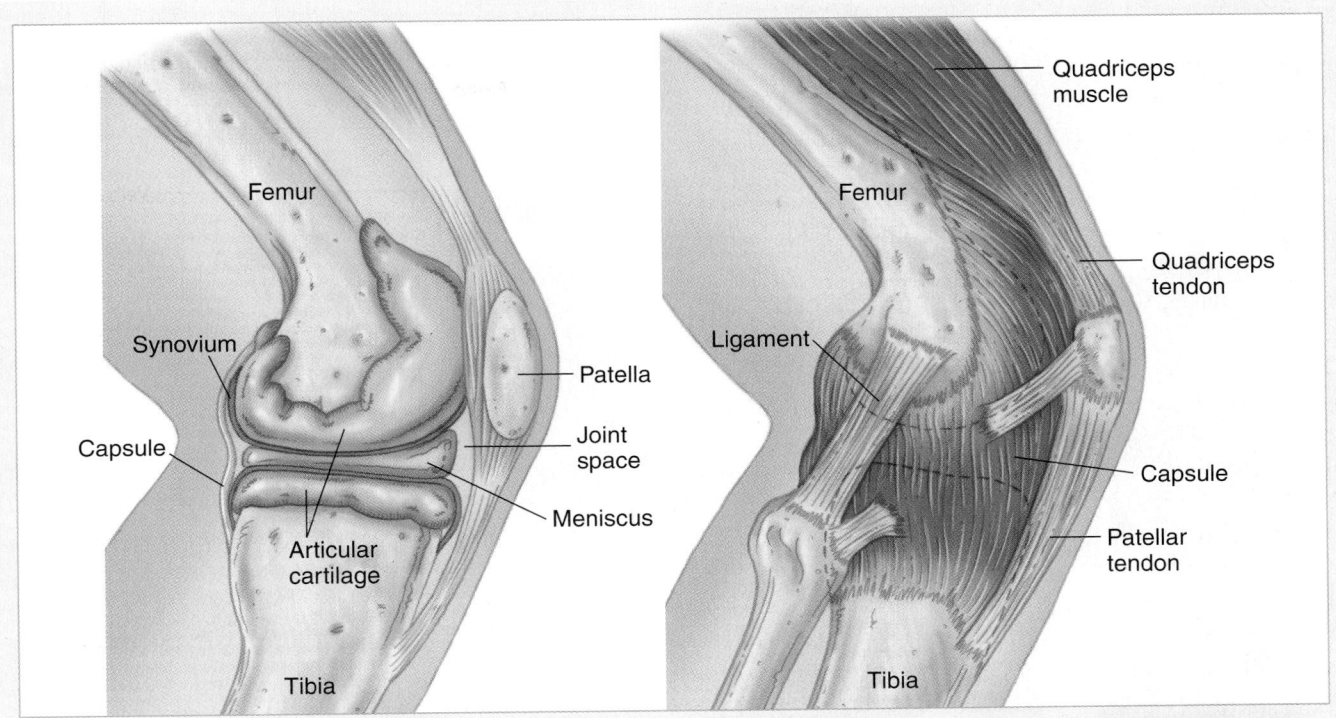

Figure 4-23 A joint consists of bone ends, the fibrous joint capsule, and ligaments. The degree to which a joint can move is determined by how the ligaments hold the bone ends and by the configuration of the bones themselves.

(straighten). Rotation is not possible because of the shape of the joint surfaces and the strong restraining ligaments on both sides of the joint. Thus, although the amount of motion varies from joint to joint, all joints have a definite limit beyond which motion cannot occur. When a joint is forced beyond this limit, damage to some structure must occur. Either the bones that form the joint will break, or the supporting capsule and ligaments will be disrupted.

The Musculoskeletal System

The human body is a well-designed system whose form, upright posture, and movement are provided by the <u>musculoskeletal system</u>. As its combination form suggests, the term musculoskeletal refers to the bones and voluntary muscles of the body. The musculoskeletal system also protects the vital internal organs of the body. Muscles are a form of tissue that allows body movement. There are more than 600 muscles in the musculoskeletal system. The musculoskeletal system contains skeletal muscle. Other types of muscle outside of the musculoskeletal system include smooth muscle and cardiac muscle <u>Figure 4-26 ▶</u>.

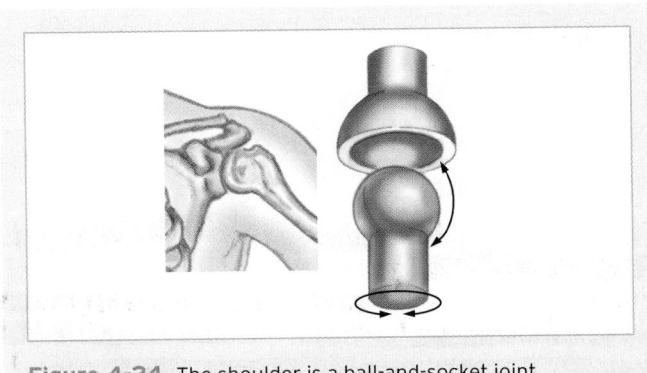

Figure 4-24 The shoulder is a ball-and-socket joint.

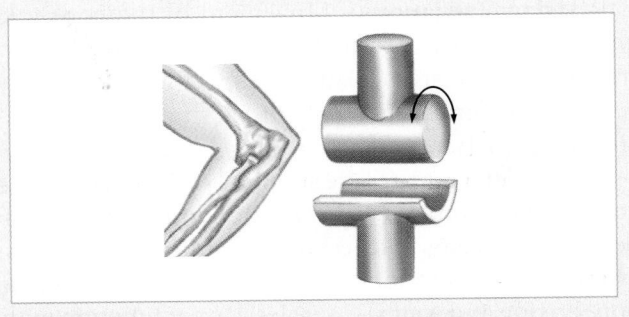

Figure 4-25 The elbow joints are hinge joints, which allow motion in only one plane.

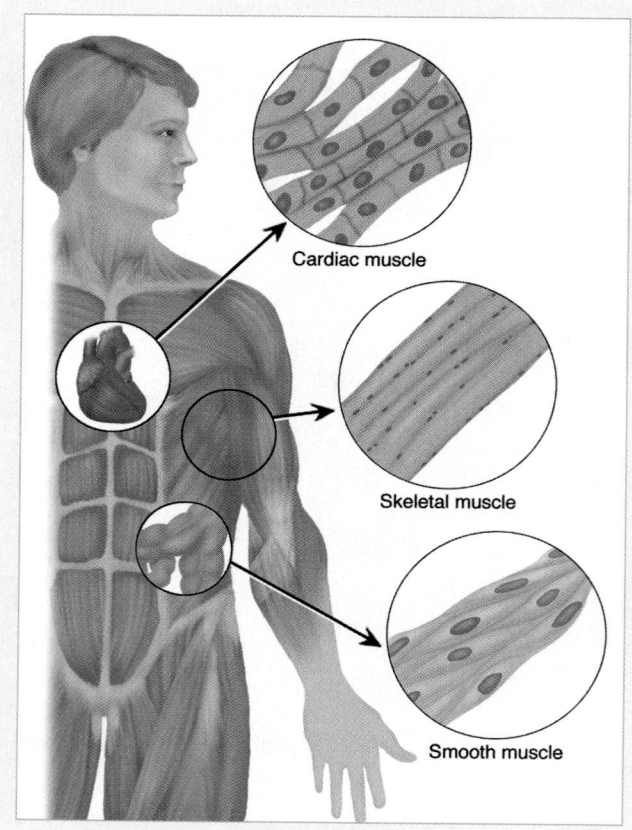

Figure 4-26 The three types of muscle are skeletal, smooth, and cardiac.

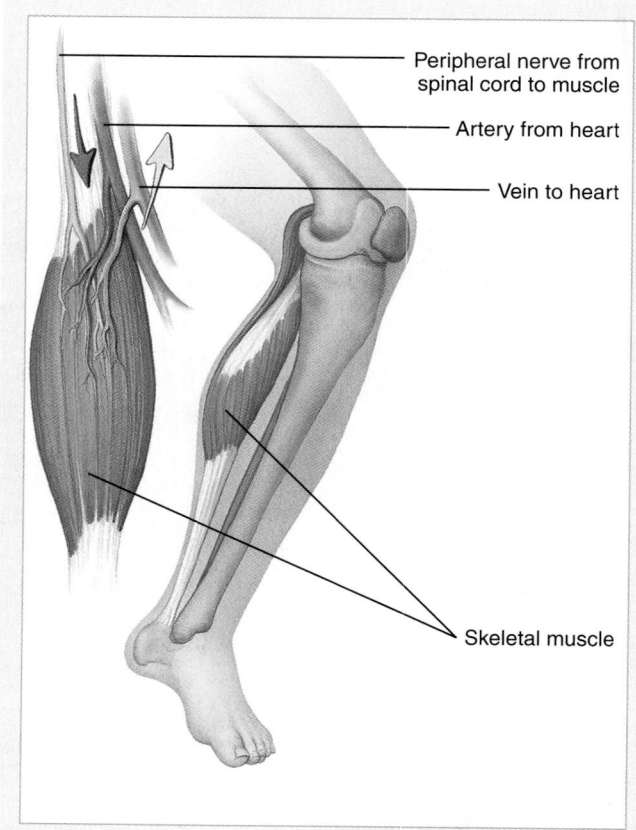

Figure 4-27 All skeletal muscles are supplied with arteries, veins, and nerves.

Skeletal Muscle

Skeletal muscle, so named because it attaches to the bones of the skeleton, forms the major muscle mass of the body. It is also called voluntary muscle, because all skeletal muscle is under direct voluntary control of the brain and can be stimulated to contract or relax at will. Skeletal muscle is also called striated muscle, because when viewed under the microscope, it has characteristic stripes (striations). Movement of the body, like waving or walking, results from skeletal muscle contraction or relaxation. Usually, a specific motion is the result of several muscles contracting and relaxing simultaneously.

All skeletal muscles are supplied with arteries, veins, and nerves (Figure 4-27 ▶). Arterial blood brings oxygen and nutrients to the muscle, and the veins carry away the waste products of muscular contraction (carbon dioxide and water). Muscles cannot function without this ongoing supply of oxygen and nutrients and removal of waste products. Muscle cramps result when insufficient oxygen or food is carried to the muscle or

when acidic waste products accumulate and are not carried away.

Skeletal muscle is under the direct control of the nervous system and responds to a command from the brain to move a specific body part. Specific nerves pass directly from the brain to the spinal cord. There, they connect with other nerves that exit from the spinal cord and pass to each skeletal muscle. Electrical impulses are carried from the cells in the brain and spinal cord along the peripheral nerves to each muscle, signaling it to contract. When this normal nerve supply is lost through injury to the brain, spinal cord, or peripheral nerves, the voluntary control of the muscle is lost, and the muscle becomes paralyzed.

Most skeletal muscles attach directly to bone by tough, ropelike cords of fibrous tissue called tendons, which continue the fascia that covers all skeletal muscles. The fascia is much like the skin of a sausage in that it encases the muscle tissue. At either end of the muscle, the fascia extends beyond the muscle to attach to a bone. This musculotendinous unit crosses a joint

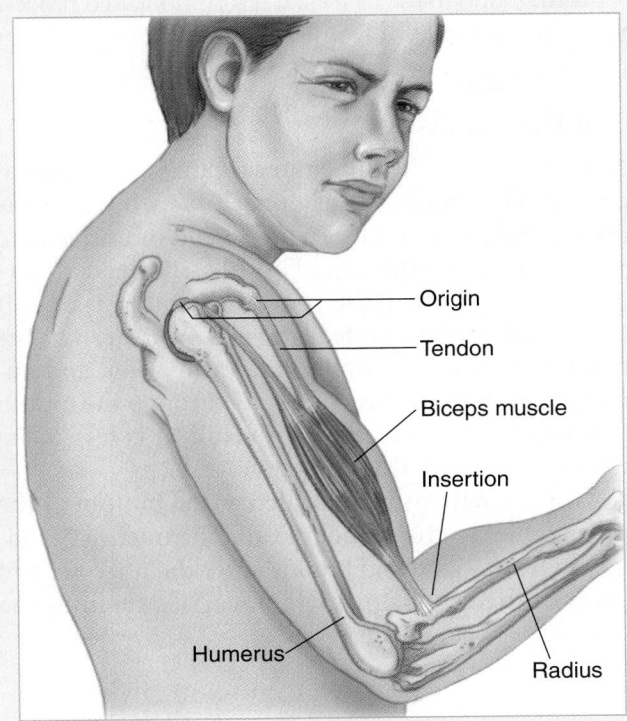

Figure 4-28 The biceps muscle causes the elbow to bend when it contracts. Note the points of tendon origin and insertion. As the muscle contracts and shortens, these points are pulled closer together, with motion occurring at the elbow joint.

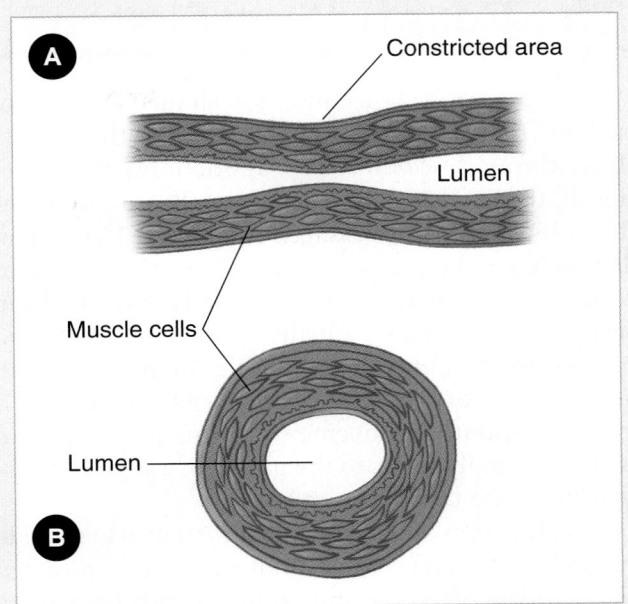

Figure 4-29 A. Smooth muscle lines the walls of the tubular structures of the body. **B.** Contraction of the muscles narrows the diameter of the structure, and relaxation allows the diameter to increase in size.

and is responsible for the motion of that joint. The proximal point of attachment of the musculotendinous unit is its origin, and the distal bony attachment is called the insertion of the muscle. When a muscle contracts, a line of force is created between the origin and the insertion, which pulls the points of origin and insertion closer together (Figure 4-28 ▲). This motion occurs at the joint between the two bones.

Smooth Muscle

Smooth muscle carries out much of the automatic work of the body; therefore, it is also called involuntary muscle. Smooth muscle is found in the walls of most tubular structures of the body, such as the gastrointestinal tract, the urinary system, the blood vessels, and the bronchi of the lungs. Its appearance is smooth. Contraction and relaxation of smooth muscle propel or control the flow of the contents of these structures along their course. For example, the rhythmic contraction and relaxation of the smooth muscles of the wall of the intestine propel ingested food through it,

and smooth muscle in the walls of a blood vessel can alter the diameter of the vessel to control the amount of blood flowing through it (Figure 4-29 ▲).

Smooth muscle responds only to primitive stimuli such as stretching, heat, or the need to relieve waste. An individual cannot exert any voluntary control over this type of muscle.

Cardiac Muscle

The heart is a large muscle composed of a pair of pumps of unequal force: one of lower pressure and one of higher pressure. The heart must function continuously from birth to death. It is a specially adapted involuntary muscle with a very rich blood supply and its own electrical system, which makes it different from both skeletal and smooth muscle. Another difference is that cardiac muscle has the property of "automaticity," which means that the heart muscle can set its own rhythm and rate without influence from the brain. This property is unique to heart muscle. Cardiac muscle can tolerate an interruption of its blood supply for only a few seconds. It requires a continuous supply of oxygen and glucose for normal function. Because of its special structure and function, cardiac muscle is placed in a separate category.

The Respiratory System

The respiratory system consists of all the structures of the body that contribute to respiration, or the process of breathing (Figure 4-30 ▼). It includes the nose, mouth, throat, larynx, trachea, bronchi, and bronchioles, which are all air passages or airways. The system also includes the lungs, where oxygen is passed into the blood and where carbon dioxide is removed from the blood to be exhaled. Finally, the respiratory system includes the diaphragm, the muscles of the chest wall, and the accessory muscles of breathing, which permit normal respiratory movement. In this text, the term "airway" usually refers to the upper airway or the passage above the larynx (voice box).

The function of the respiratory system is to provide the body with oxygen and eliminate carbon dioxide. The exchange of oxygen and carbon dioxide takes place in the lungs and in the tissues. It is a complicated process that occurs automatically unless the airways or the lungs become diseased or damaged.

The Upper Airway

The structures of the upper airway are located anteriorly and at the midline. The upper airway includes the nose, mouth, and throat. The nose and mouth lead to the oropharynx (throat). The nostrils lead to the nasopharynx (above the roof of the mouth, or soft palate), and the mouth leads to the oropharynx. The nasal passages and nasopharynx warm, filter, and humidify air as we breathe. Air enters through the mouth more rapidly and directly. As a result, it is less moist than air that enters through the nose.

Two passageways are located at the bottom of the pharynx: the esophagus behind and the trachea (windpipe) in front. Food and liquids enter the pharynx and pass into the esophagus, which carries them to the

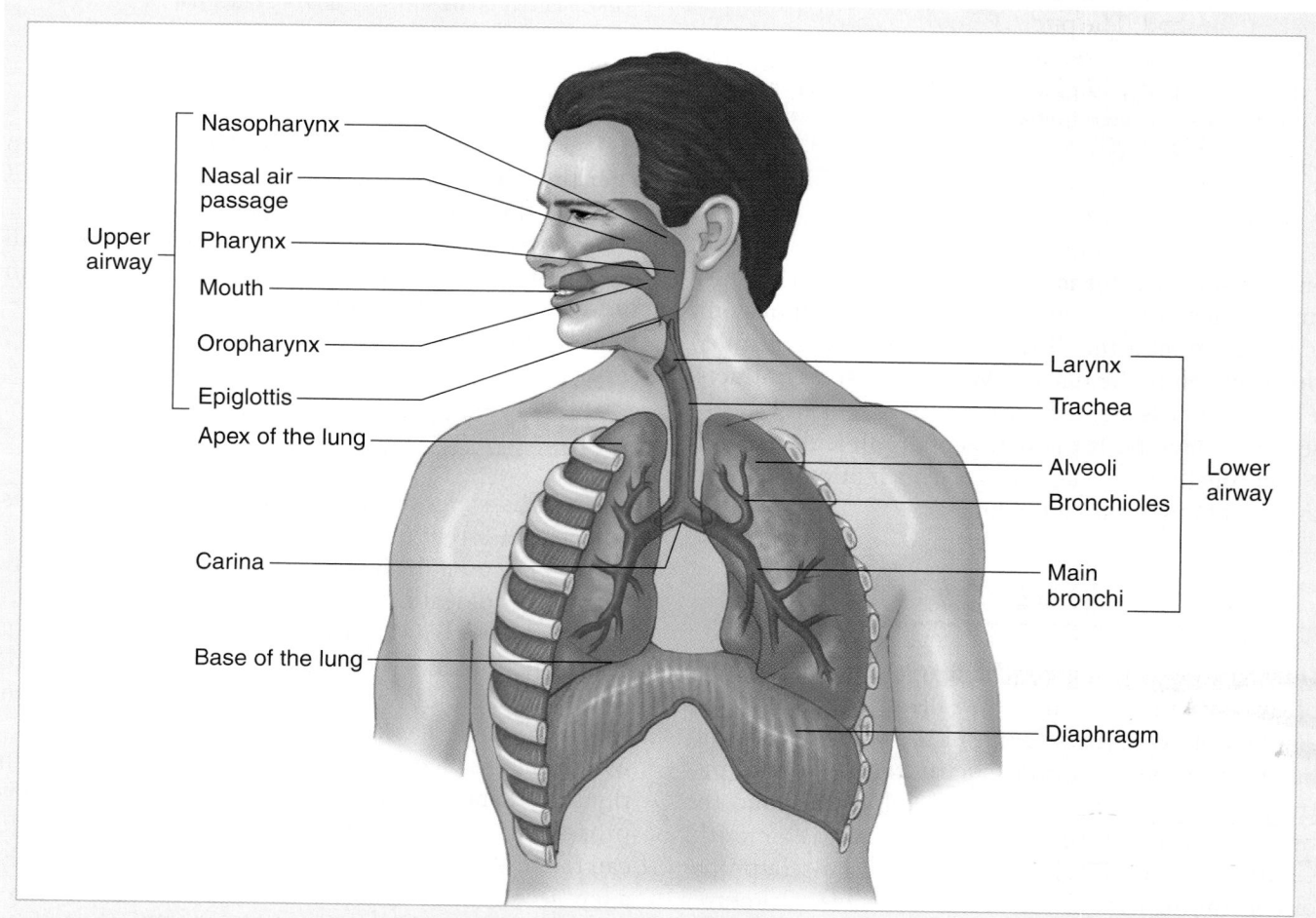

Figure 4-30 The respiratory system consists of all the structures of the body that contribute to the process of breathing.

stomach. Air and other gases enter the trachea and go to the lungs.

Protecting the opening of the trachea is a thin, leaf-shaped valve called the <u>epiglottis</u>. This valve allows air to pass into the trachea but prevents food or liquid from entering the airway under normal circumstances. Air moves past the epiglottis into the larynx and the trachea.

The Lower Airway

The first part of the lower airway is the larynx, a rather complex arrangement of tiny bones, cartilage, muscles, and the two vocal cords. The larynx does not tolerate any foreign solid or liquid material. A violent episode of coughing and spasm of the vocal cords will result from contact with solids or liquids.

The Adam's apple, or thyroid cartilage, is easily seen in the middle of the front of the neck. The thyroid cartilage is actually the anterior part of the larynx. Tiny muscles open and close the vocal cords and control tension on them. Sounds are created as air is forced past the vocal cords, making them vibrate. These vibrations make the sound. The pitch of the sound changes as the cords open and close. You can feel the vibrations if you place your fingers lightly on the larynx as you speak or sing. The vibrations of air are shaped by the tongue and muscles of the mouth to form understandable sounds. Immediately below the thyroid cartilage is the palpable cricoid cartilage.

Between these two prominences lies the cricothyroid membrane, which can be felt as a depression in the midline of the neck just inferior to the thyroid cartilage. Below the cricoid cartilage is the trachea. The trachea is approximately 5″ long and is a semirigid, enclosed air tube made up of rings of cartilage that are open in the back. This enables food to pass through the esophagus, which lies right behind the trachea. The rings of cartilage keep the trachea from collapsing when air moves into and out of the lungs. The trachea ends at the carina and divides into smaller tubes. These tubes are the right and left main bronchi, which enter the lungs. Each main bronchus immediately branches within the lung into smaller and smaller airways. Within the right lung, three major bronchi are formed. Within the left, there are only two. Each bronchus supplies air to one lobe of the lung.

Lungs

The two lungs are held in place within the chest by the trachea, the arteries and veins that run to and from the heart, and the pulmonary ligaments. Each lung is divided into lobes. The right lung has three lobes: the upper, middle, and lower lobes. The left lung has an upper lobe and a lower lobe. Each lobe is divided further into segments. Also within each lung, the main bronchi divide until they end in very fine airways called bronchioles. The bronchioles end in about 700 million tiny grapelike sacs called <u>alveoli</u> (Figure 4-31 ▼). The exchange of oxygen and carbon dioxide occurs within these alveoli. The walls of the alveoli contain a network of tiny blood vessels (pulmonary capillaries) that carry the carbon dioxide from the body to the lungs and the oxygen from the lungs to the body.

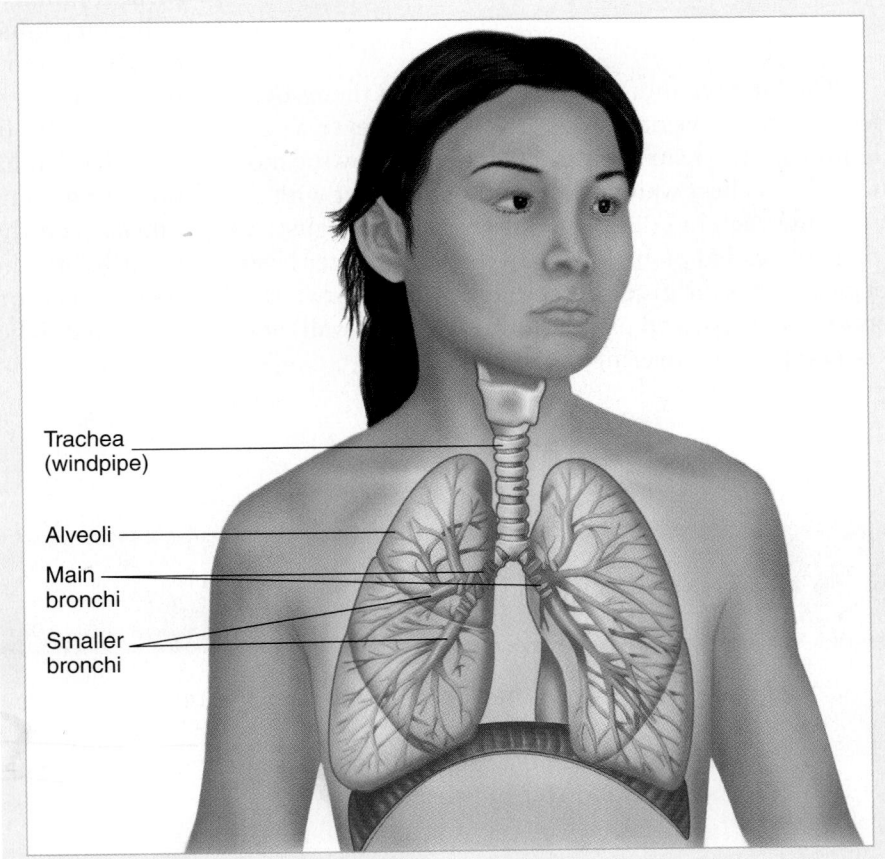

Trachea (windpipe)

Alveoli

Main bronchi

Smaller bronchi

Figure 4-31 The lungs contain millions of air sacs (alveoli), which lie at the ends of air passages.

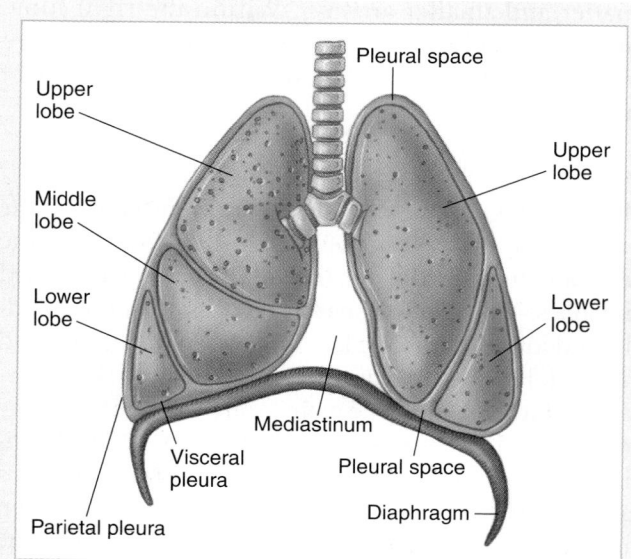

Figure 4-32 The pleura lining the chest wall and covering the lungs is an essential part of the breathing mechanism. The pleural space is not an actual space until blood or air leaks into it, causing the pleural surfaces to separate.

The lungs cannot expand and contract themselves because they have no muscle. There is, however, a very definite mechanism to ensure that they follow the motion of the chest wall and expand or contract with it. Covering each lung is a layer of very smooth, glistening tissue called pleura (Figure 4-32 ▲). Another layer of pleura lines the inside of the chest cavity. The two layers are called parietal pleura (lining the chest wall) and visceral pleura (covering the lungs).

Between the parietal pleura and the visceral pleura is the pleural space, called a "potential" space rather than an actual space in the usual sense because normally these layers are in close contact everywhere. In fact, the layers are sealed tightly against one another by a thin film of fluid. When the chest wall expands, the lung is pulled with it and made to expand by the force exerted through these closely applied pleural surfaces. Normally, the pleural space is quite small and contains only the thin film of pleural fluid as each lung entirely fills its chest cavity.

Diaphragm

The diaphragm is unique because it has characteristics of both voluntary (skeletal) and involuntary (smooth) muscle. It is a dome-shaped muscle that divides the thorax from the abdomen and is pierced by the great vessels and the esophagus (Figure 4-33 ▶). Under the microscope, it has striations like skeletal muscle. Also, it is attached to the costal arch and the lumbar vertebrae like other skeletal muscles. Thus, in many ways, it looks like a voluntary muscle; however, we do not have complete voluntary control over its function. It acts like a voluntary muscle whenever we take a deep breath, cough, or hold our breath. We control these variations in the way we breathe.

However, unlike other skeletal or voluntary muscles, the diaphragm performs an automatic function. Breathing continues while we sleep and at all other times. Even though we can hold our breath or temporarily breathe faster or slower, we cannot continue these variations in breathing pattern indefinitely. Ultimately, when the concentration of carbon dioxide is close to being disturbed, automatic regulation of

You are the Provider   Part 5

Your partner completes the documentation for this response and reviews it with you.

6. How can your documentation skills and use of correct anatomic terminology assist you in reviewing the documentation for this response?

breathing resumes. Therefore, although the diaphragm looks like voluntary skeletal muscle and is attached to the skeleton, it behaves, for the most part, like an involuntary muscle.

During inhalation, the diaphragm and intercostal muscles contract. When the diaphragm contracts, it moves down slightly and enlarges the thoracic cage from top to bottom. When the intercostal muscles contract, they raise the ribs up and out. These actions combine to enlarge the chest cavity in all dimensions. Pressure within the cavity falls, and air rushes into the lungs.

During exhalation, the diaphragm and the intercostal muscles relax. Unlike inhalation, exhalation does not normally require muscular effort. As these muscles relax, all dimensions of the thorax decrease, and the ribs and muscles assume a normal resting position. When the volume of the chest cavity decreases, air in the lungs is compressed into a smaller space. Pressure is increased, and air is pushed out through the trachea.

Respiratory Physiology

Each living cell in the body requires a regular supply of oxygen. Some cells need a constant supply of oxygen to survive. For example, cells in the heart may be damaged if the oxygen supply is interrupted for more than a few seconds. Brain cells and cells in the nervous

system may die after as few as 4 to 6 minutes without oxygen. Dead brain and nerve cells can never be replaced. Permanent changes in the body, such as brain damage, result from the damage caused by a lack of oxygen. Other cells in the body are not as vitally dependent on a constant oxygen supply. They can tolerate short periods without oxygen and still survive. Normally, the air that we breathe contains 21% oxygen and 78% nitrogen. Small amounts of other gases make up the remaining 1%.

The Exchange of Oxygen and Carbon Dioxide

As blood travels through the body, it gives its oxygen and nutrients to various tissues and cells. Oxygen passes from the blood through the capillaries to tissue cells. In the reverse process, carbon dioxide and cell waste pass from tissue cells through capillaries to the blood Figure 4-34 ▶.

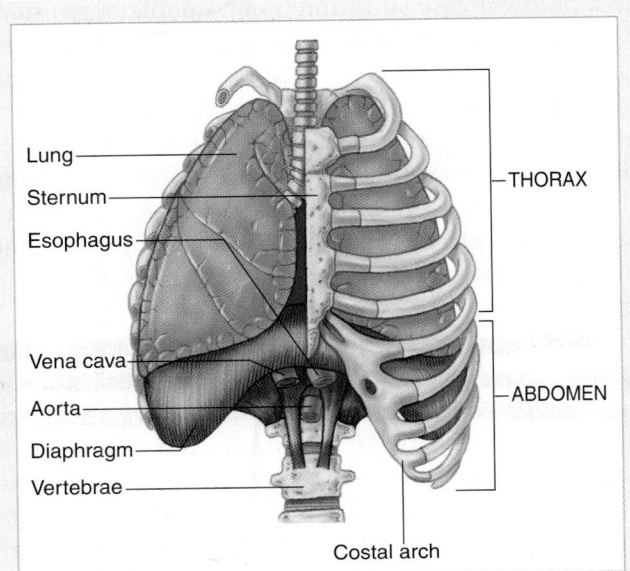

Figure 4-33 The dome-shaped diaphragm divides the thorax from the abdomen. It is pierced by the great vessels and the esophagus.

▲ Pediatric Needs

The anatomy of the respiratory system in children is proportionally smaller and less rigid than that in an adult Figure 4-35 ▶. A child's nose and mouth are much smaller than those of an adult. The larynx, cricoid cartilage, and trachea are smaller, softer, and more flexible as well. This makes the mechanics of breathing much more delicate. A child's pharynx is also smaller and less deeply curved. The tongue takes up proportionally more space in a child's mouth than in an adult's mouth.

These anatomic differences are important for your assessment and treatment. For example, the smaller larynx of a child becomes obstructed more easily. The chest wall in children is softer. Therefore, children depend more heavily on the diaphragm for breathing. You will notice that the abdomen moves in and out considerably with each breath, especially in an infant. Infants younger than age 1 month do not know how to breathe through the mouth. The smaller the child, the larger the head in proportion to the torso. This will affect the way you treat a suspected spinal injury. Therefore, as you assess and treat an infant or a child, you must carefully consider these differences.

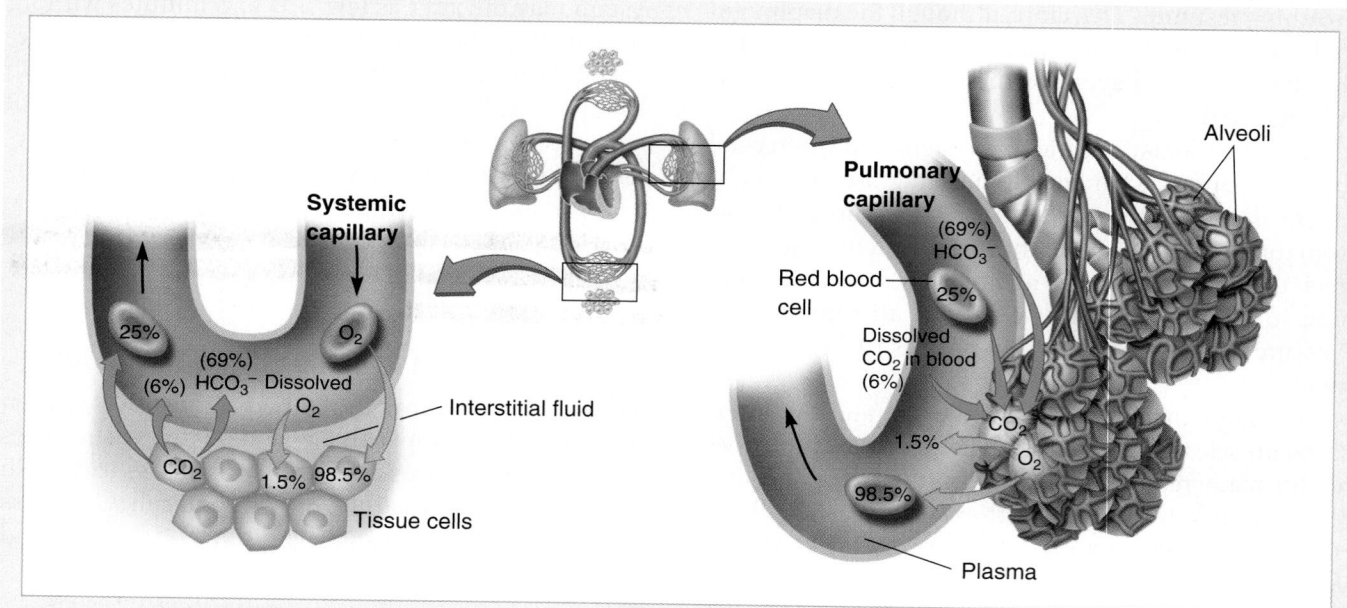

Figure 4-34 In the capillaries of the lungs, oxygen passes from the blood to the tissue cells, and carbon dioxide and waste pass from the tissue cells to the blood.

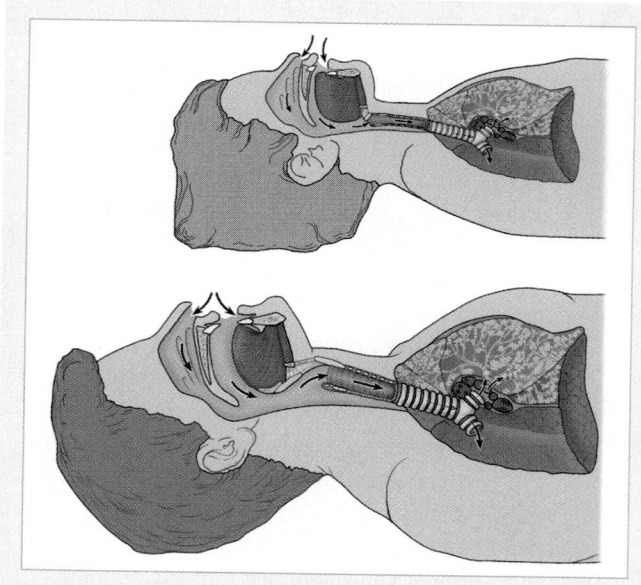

Figure 4-35 The respiratory system of a child is proportionally smaller and less rigid than that of an adult.

Each time we take a breath, the alveoli receive a supply of oxygen-rich air. The oxygen then passes into a fine network of pulmonary capillaries, which are in close contact with the alveoli. In fact, the capillaries in the lungs are located in the walls of the alveoli. The walls of the capillaries and the alveoli are extremely thin. Thus, the air in the alveoli and the blood in the capillaries are separated by two very thin layers of tissue.

Oxygen and carbon dioxide pass rapidly across these thin tissue layers through diffusion. Diffusion is a passive process in which molecules move from an area with higher concentration of molecules to an area of lower concentration. For example, a gas such as hydrogen sulfide moves from an area of high concentration (a rotten egg) by spontaneous movement of the gas molecules until the odor fills the room. There are more oxygen molecules in the alveoli than in the blood. Therefore, the oxygen molecules move from the alveoli into the blood. Because there are more carbon dioxide molecules in the blood than in the inhaled air, carbon dioxide moves from the blood into the alveoli.

The blood does not use all the inhaled oxygen as it passes through the body. Exhaled air contains 16% oxygen and 3% to 5% carbon dioxide; the rest is nitrogen (Figure 4-36 ▶). This 16% concentration of oxygen is adequate to support artificial ventilation. So as you provide artificial ventilations to a patient who is not breathing, that patient is receiving 16% concentration of oxygen with each ventilation.

The Control of Breathing

The brain—or more specifically, an area of the brain stem—controls breathing. This area is in one of the best-protected parts of the nervous system—deep within

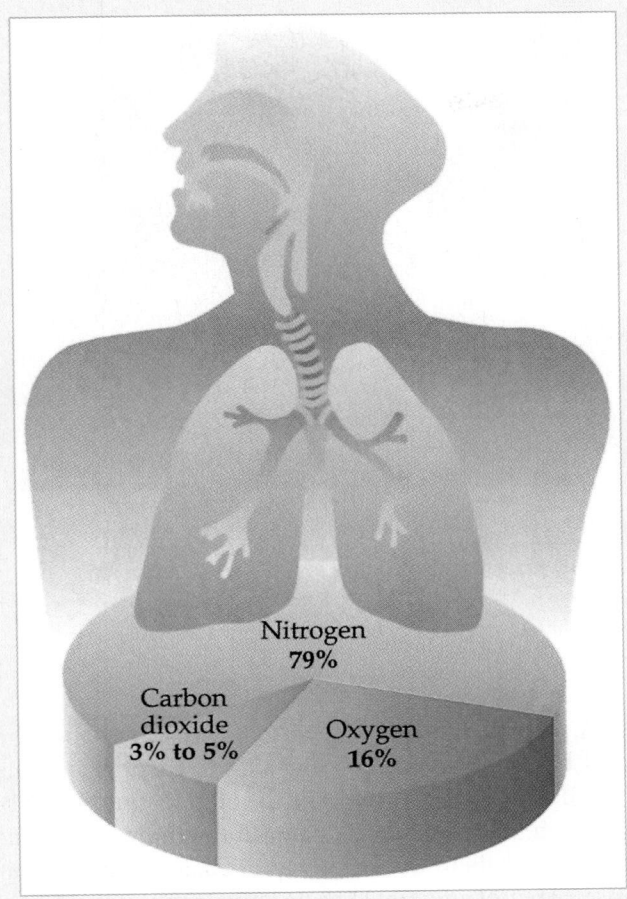

Figure 4-36 The components of exhaled air include oxygen, carbon dioxide, and nitrogen.

Nitrogen 79%
Carbon dioxide 3% to 5%
Oxygen 16%

Pediatric Needs

Normal breathing patterns in infants and children are essentially the same as those in adults. However, infants and children breathe faster than adults. An infant who is breathing normally will have respirations of 25 to 50 breaths/min. A child will have respirations of 15 to 30 breaths/min. Like adults, infants and children who are breathing normally will have smooth, regular inhalation and exhalation, equal breath sounds, and regular rise and fall movement on both sides of the chest.

Breathing problems in infants and children often appear the same as breathing problems in adults. Signs such as increased respirations, an irregular breathing pattern, unequal breath sounds, and unequal chest expansion indicate breathing problems in both adults and children. Other signs that an infant or child is not breathing normally include the following:

- Muscle retractions, in which the muscles of the chest and neck are working extra hard in breathing
- Nasal flaring in children, in which the nostrils flare out as the child breathes
- Seesaw respirations in infants, in which the chest and abdominal muscles alternately contract to look like a seesaw

Exhalation becomes active when infants and children have trouble breathing. Normally, inhalation alone is the active, muscular part of breathing, as described earlier. However, with labored breathing, both inhalation and exhalation are hard work. With labored breathing, exhalation is not passive. Instead, air is forced out of the lungs during exhalation, and the child will often begin to wheeze. This type of labored breathing involves the use of the accessory muscles of breathing.

the skull. The nerves in this area act as sensors of the level of carbon dioxide in the blood. The brain automatically controls breathing if the levels of carbon dioxide or oxygen in the arterial blood are too high or too low. In fact, adjustments can be made in just one breath. For these reasons, you cannot hold your breath indefinitely or breathe rapidly and deeply indefinitely.

When the level of carbon dioxide becomes too high, the brain stem sends nerve impulses down the spinal cord that cause the diaphragm and the intercostal muscles to contract. This increases our breathing, or respirations. The higher the level of carbon dioxide in the blood, the stronger is the impulse to cause breathing. Once the carbon dioxide levels become acceptable, the strength and frequency of respiration decreases.

We also have a "backup system" to control respiration called the hypoxic drive. When oxygen levels fall, this system will also stimulate breathing. There are areas in the brain, the walls of the aorta, and the carotid arteries that act as oxygen sensors. These sensors are easily satisfied by minimal levels of oxygen in the arterial blood. Therefore, our backup system, the hypoxic drive, is much less sensitive and less powerful than the carbon dioxide sensors in the brain stem.

Characteristics of Normal Breathing

You can think of a "normal" breathing pattern as a bellows system. Normal breathing should appear easy, not labored. As with a bellows that is used to move air to start a fire, breathing should be a smooth flow of air moving into and out of the lungs. Normal breathing has the following characteristics:

- A normal rate and depth (tidal volume)
- A regular rhythm or pattern of inhalation and exhalation
- Good audible breath sounds on both sides of the chest
- Regular rise and fall movement on both sides of the chest
- Movement of the abdomen

Inadequate Breathing Patterns in Adults

An adult who is awake, alert, and talking to you has no immediate airway or breathing problems. However, you should keep supplemental oxygen on hand to assist with breathing if it should become necessary. An adult who is not breathing well will appear to be working hard to breathe. This type of breathing pattern is called labored breathing. Labored breathing requires effort and may involve the accessory muscles. The person may also be breathing either much slower (fewer than 8 breaths/min) or much faster (more than 24 breaths/min) than normal. An adult who is breathing normally will have respirations of 12 to 20 breaths/min (Table 4-3 ▼).

With a normal breathing pattern, the accessory muscles are not being used. With inadequate breathing, a person, especially a child, may use the accessory muscles of the chest, neck, and abdomen. Other signs that a person is not breathing normally include the following:

- Muscle retractions above the clavicles, between the ribs, and below the rib cage, especially in children
- Pale or cyanotic (blue) skin
- Cool, damp (clammy) skin
- Tripod position (Figure 4-37 ▶) (a position in which the patient is leaning forward onto two arms stretched forward)

Study This!!!

TABLE 4-3 Normal Respiration Rate Ranges	
Adults	12 to 20 breaths/min
Children	15 to 30 breaths/min
Infants	25 to 50 breaths/min

Note: These ranges are per the US DOT 1994 EMT-Basic National Standard Curriculum. Ranges presented in other courses may vary.

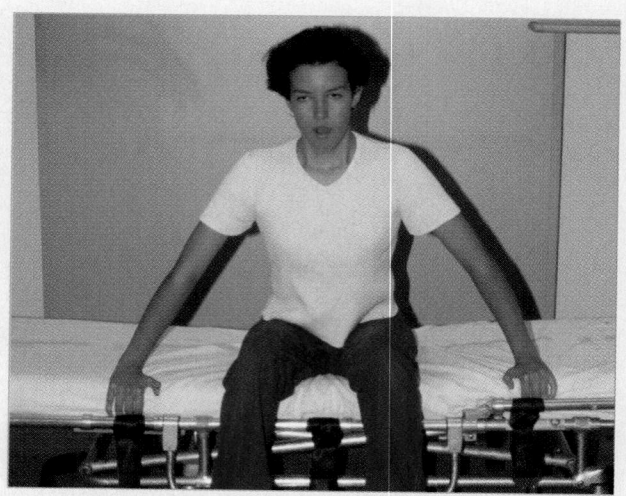

Figure 4-37 A patient in the tripod position will sit leaning forward on outstretched arms with the head and chin thrust slightly forward.

A patient may also appear to be breathing after the heart has stopped. These occasional, gasping breaths are called underlined agonal respirations. Agonal respirations occur when the respiratory center in the brain continues to send signals to the breathing muscles. These respirations are not adequate because they are slow and generally shallow. You should assist ventilations of patients with agonal respirations.

The Circulatory System

The circulatory system is a complex arrangement of connected tubes, including the arteries, arterioles, capillaries, venules, and veins (Figure 4-38 ▶). The circulatory system is entirely closed, with capillaries connecting arterioles and venules. There are two circuits in the body: the systemic circulation in the body and the pulmonary circulation in the lungs. The systemic circulation, the circuit in the body, carries oxygen-rich blood from the left ventricle through the body and back to the right atrium. In the systemic circulation, as blood passes through the tissues and organs, it gives up oxygen and nutrients and absorbs cellular wastes and carbon dioxide. The cellular wastes are eliminated in passages through the liver and the kidneys. The pulmonary circulation, the circuit in the lungs, carries oxygen-poor blood from the right ventricle through the

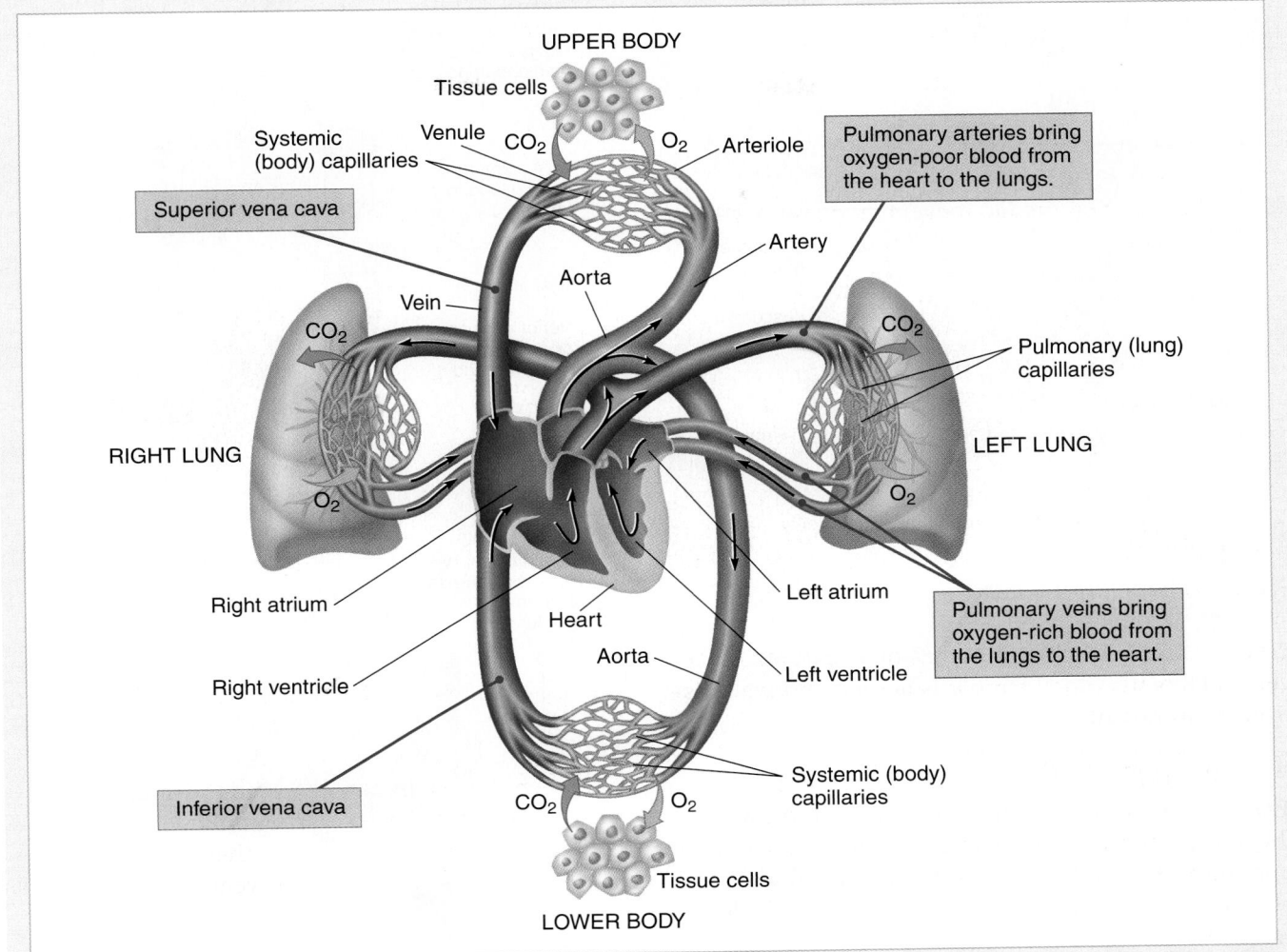

UPPER BODY

Tissue cells

Systemic (body) capillaries

Venule CO_2 O_2 Arteriole

Pulmonary arteries bring oxygen-poor blood from the heart to the lungs.

Superior vena cava

Artery

Aorta

Vein

CO_2

CO_2

Pulmonary (lung) capillaries

RIGHT LUNG

LEFT LUNG

O_2

O_2

Right atrium

Left atrium

Heart

Pulmonary veins bring oxygen-rich blood from the lungs to the heart.

Right ventricle

Aorta

Left ventricle

Inferior vena cava

Systemic (body) capillaries

CO_2 O_2

Tissue cells

LOWER BODY

Figure 4-38 The circulatory system includes the heart, arteries, veins, and interconnecting capillaries. The capillaries are the smallest vessels and connect the venules and arterioles. At the center of the system, and providing its driving force, is the heart. Blood circulates through the body under pressure generated by the two sides of the heart.

lungs and back to the left atrium. In the pulmonary circulation, as blood passes through the lungs, it is refreshed with oxygen and gives up carbon dioxide.

Heart

The heart is a hollow muscular organ approximately the size of an adult's clenched fist. It is made of a unique, adapted tissue called cardiac muscle or myocardium and actually works as two paired pumps, the one on the left side being more muscular. A wall called the septum divides the heart down the middle into right and left sides. Each side of the heart is divided again

into an upper chamber (atrium) and a lower chamber (ventricle).

The heart is an involuntary muscle. As such, it is under the control of the autonomic nervous system. However, it has its own electrical system and continues to function even without its central nervous system control. It is distinct from skeletal or smooth muscle in its requirement for a continuous supply of oxygen and nutrients.

The heart must function continuously from birth to death and has developed special adaptations to meet the needs of this continuous function. It can tolerate a serious interruption of its own blood supply for only a

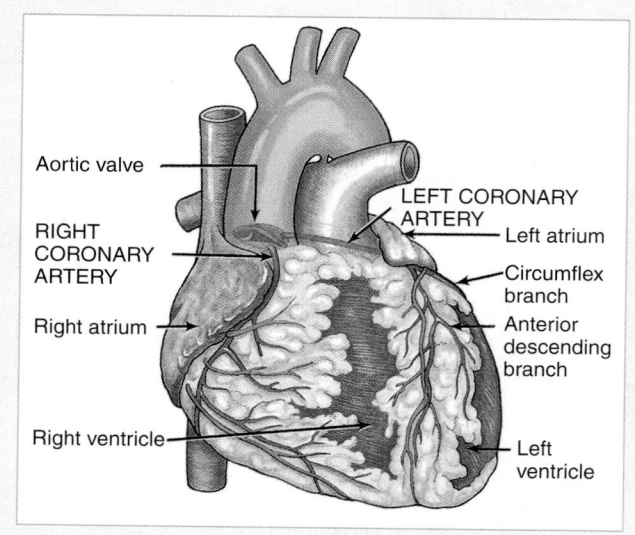

Figure 4-39 The two main coronary arteries supply the heart with blood.

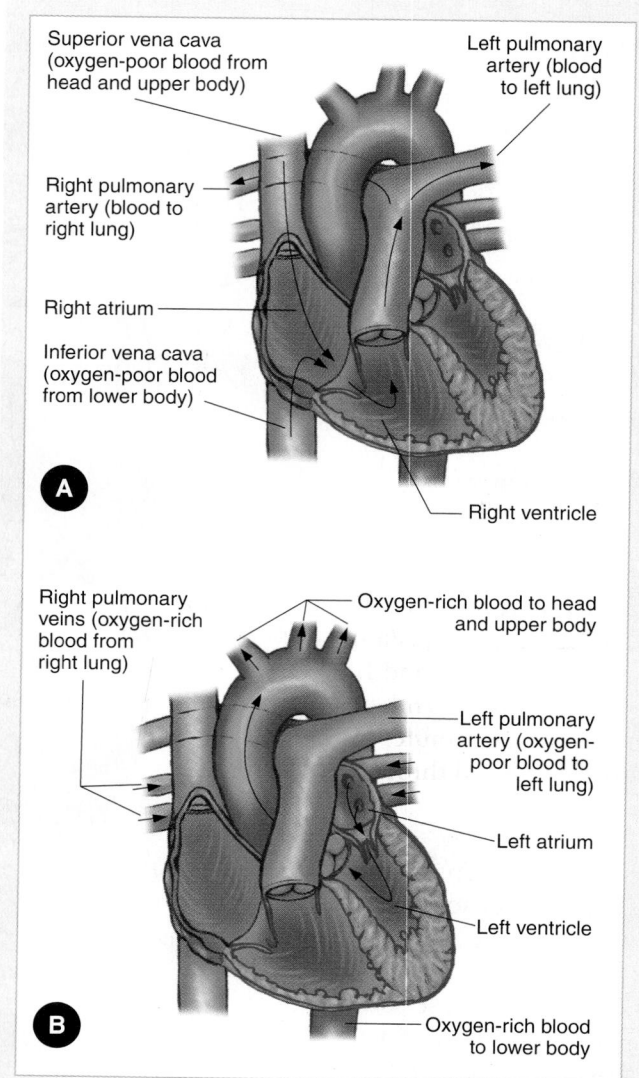

Figure 4-40 A. The right side, or lower pressure side, of the heart pumps blood from the body through the lungs. **B.** The left side, or higher pressure side, of the heart pumps oxygen-rich blood to the rest of the body.

very few seconds before the signs of a heart attack develop. Thus, its blood supply is as rich and well distributed as possible.

How the Heart Works

The heart receives the first blood distribution from the aorta. The two main coronary arteries have their openings immediately above the aortic valve at the beginning of the aorta where the pressures are highest (Figure 4-39 ▲).

The right side of the heart receives blood from the veins of the body (Figure 4-40A ▶). The blood enters from the superior and inferior venae cavae into the right atrium, and then passes through the tricuspid valve to fill the right ventricle. After the right ventricle is filled, the tricuspid valve closes to prevent backflow after the right ventricular muscle contracts. Contraction of the right ventricle causes blood to flow into the pulmonary artery and the pulmonary circulation.

The left side receives oxygenated blood from the lungs through the <u>pulmonary veins</u> into the left atrium, where it passes through the mitral valve into the left ventricle (Figure 4-40B ▶). Contraction of this most muscular of the pumping chambers pumps the blood into the aorta and then to the arteries of the body.

The exit of each of the four heart chambers is governed by a one-way valve. The valves prevent the backflow of blood and keep it moving through the circulatory system in the proper direction. When a valve control-

ling the filling of a heart chamber is open, the other valve allowing it to empty is shut and vice versa. Normally, blood moves in only one direction through the entire system.

When a ventricle contracts, the valve to the artery opens, and the valve between the ventricle and atrium closes. Blood is forced from the ventricle out into the pulmonary artery or aorta. At the end of contraction, the ventricle relaxes. Back pressure causes the valve to the artery to close, and the entry valve to the ventricle opens as the ventricle relaxes. Blood then flows from the

Study This!!!

TABLE 4-4	Normal Heart Rates
Adults	60 to 100 beats/min
Children	70 to 150 beats/min
Infants	100 to 160 beats/min

Study This!!!

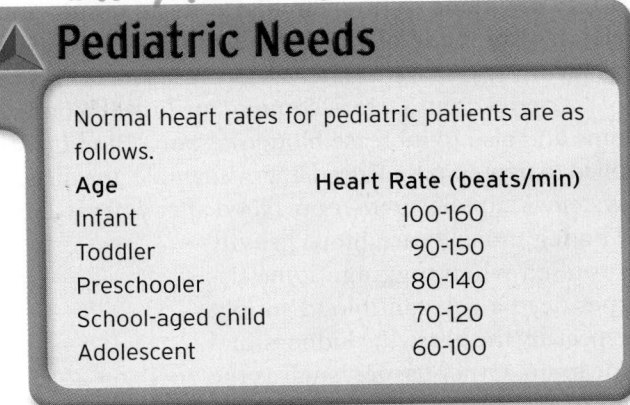

Pediatric Needs

Normal heart rates for pediatric patients are as follows.

Age	Heart Rate (beats/min)
Infant	100-160
Toddler	90-150
Preschooler	80-140
School-aged child	70-120
Adolescent	60-100

atrium into the ventricle. When the ventricle is stimulated to contract, the cycle is repeated.

Normal Heartbeat

In the normal adult, the heartbeat may range from 50 to 180 beats/min, depending on the level of activity. A very well-conditioned athlete may have a normal resting heart rate (pulse) of 50 to 60 beats/min. During vigorous physical activity, the heart rate may rise normally to as fast as 180 beats/min. The usual adult resting heart rate is between 60 and 100 beats/min (Table 4-4 ▲). At each beat, 70 to 80 mL of blood is ejected from the adult heart. In one minute, the entire blood volume of 5 to 6 L is circulated through all the vessels.

Electrical Conduction System

A network of specialized tissue that is capable of conducting electrical current runs throughout the heart (Figure 4-41 ▶). The flow of electrical current through this network causes smooth, coordinated contractions of the heart. These contractions produce the pumping action of the heart. Each mechanical contraction of the heart is associated with two electrical processes. The first is depolarization, during which the electrical charges on the surface of the muscle cell change from positive to negative. The second is repolarization, during which the heart returns to its resting state and the positive charge is restored to the surface.

When the heart is working normally, the electrical impulse begins high in the atria at the sinoatrial (SA) node, then travels to the atrioventricular (AV) node, and moves through the Purkinje fibers to the ventricles. This movement produces a smooth flow of electricity through the heart, which depolarizes the muscle and produces a coordinated pumping contraction. The heart's electrical system becomes disturbed if part of the heart is oxygen deficient, is injured, or dies. As a result, the heart

Study This!!!

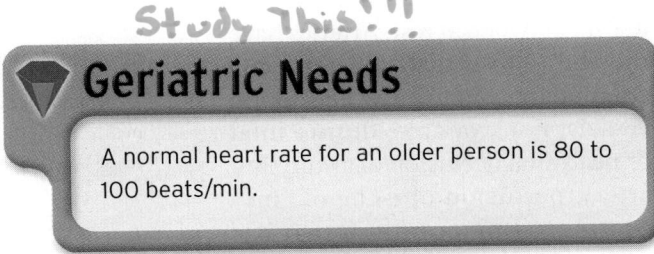

Geriatric Needs

A normal heart rate for an older person is 80 to 100 beats/min.

may not continue to beat properly. Blood pressure decreases, and a patient may lose consciousness.

Arteries

The arteries carry blood from the heart to all body tissues (Figure 4-42 ▶). They branch into smaller arteries and then into arterioles. The arterioles, in turn, branch into the vast network of capillaries. The walls of an

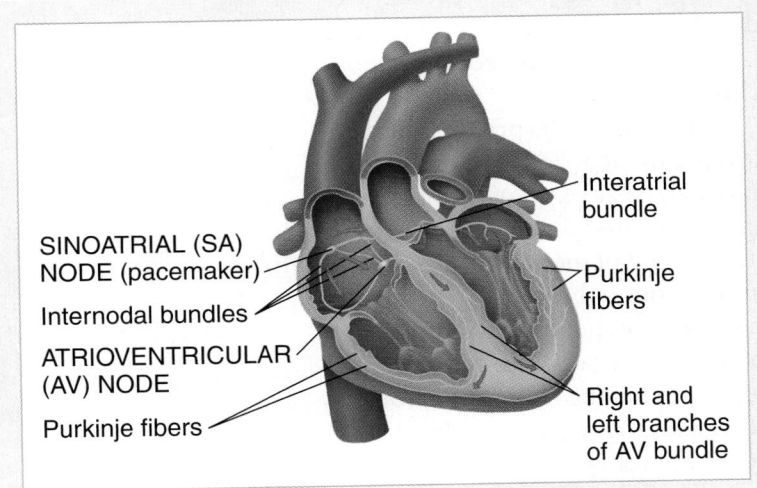

SINOATRIAL (SA) NODE (pacemaker)
Internodal bundles
ATRIOVENTRICULAR (AV) NODE
Purkinje fibers
Interatrial bundle
Purkinje fibers
Right and left branches of AV bundle

Figure 4-41 Electrical current flows through the heart to produce its pumping action.

artery are made of fine, circular muscle tissue. Some arteries are made of fine circular muscle and elastic tissue.

Arteries contract to accommodate loss of blood volume and also to increase blood pressure. Blood is supplied to tissues as they need it. For example, the digestive system is supplied with more blood after you eat a meal. The leg muscles are more heavily supplied when jogging. Some tissues need a constant blood supply, especially the heart, the kidneys, and the brain. Other tissues, such as the muscles in the extremities, the skin, and intestines, can function with less blood when at rest. Arteries have the ability to utilize collateral circulation or even to grow new blood vessels. For example, during total or near-total occlusion of a coronary artery, perfusion of ischemic myocardium—myocardium that has not received enough oxygen to function properly—occurs by way of collateral circulation: vascular channels that interconnect coronary arteries. Preexisting collaterals are thin-walled structures ranging in diameter from 20 to 200 micrometers. Acute coronary occlusion produces no infarction at all in individuals with a well-developed network of collaterals, whereas individuals who lack such a network of collaterals develop rapid and complete infarctions upon acute coronary occlusion.

The aorta is the principal artery leaving the left side of the heart; it carries freshly oxygenated blood to the body. This blood vessel is found just in front of the spine in the chest and abdominal cavities. The aorta has many branches that supply the body's vital organs. The coronary arteries supply the heart, the carotids the head, the hepatic the liver, the renal the kidneys, and the mesenteric the digestive system. The aorta divides at the level of the umbilicus into the two common iliac arteries

that lead to the lower extremities. All of the aorta's branches ultimately become arterioles leading into the body's capillary network.

The pulmonary artery begins at the right side of the heart and carries oxygen-poor blood to the lungs. It divides into finer and finer branches until it meets with the pulmonary capillary system located in the thin

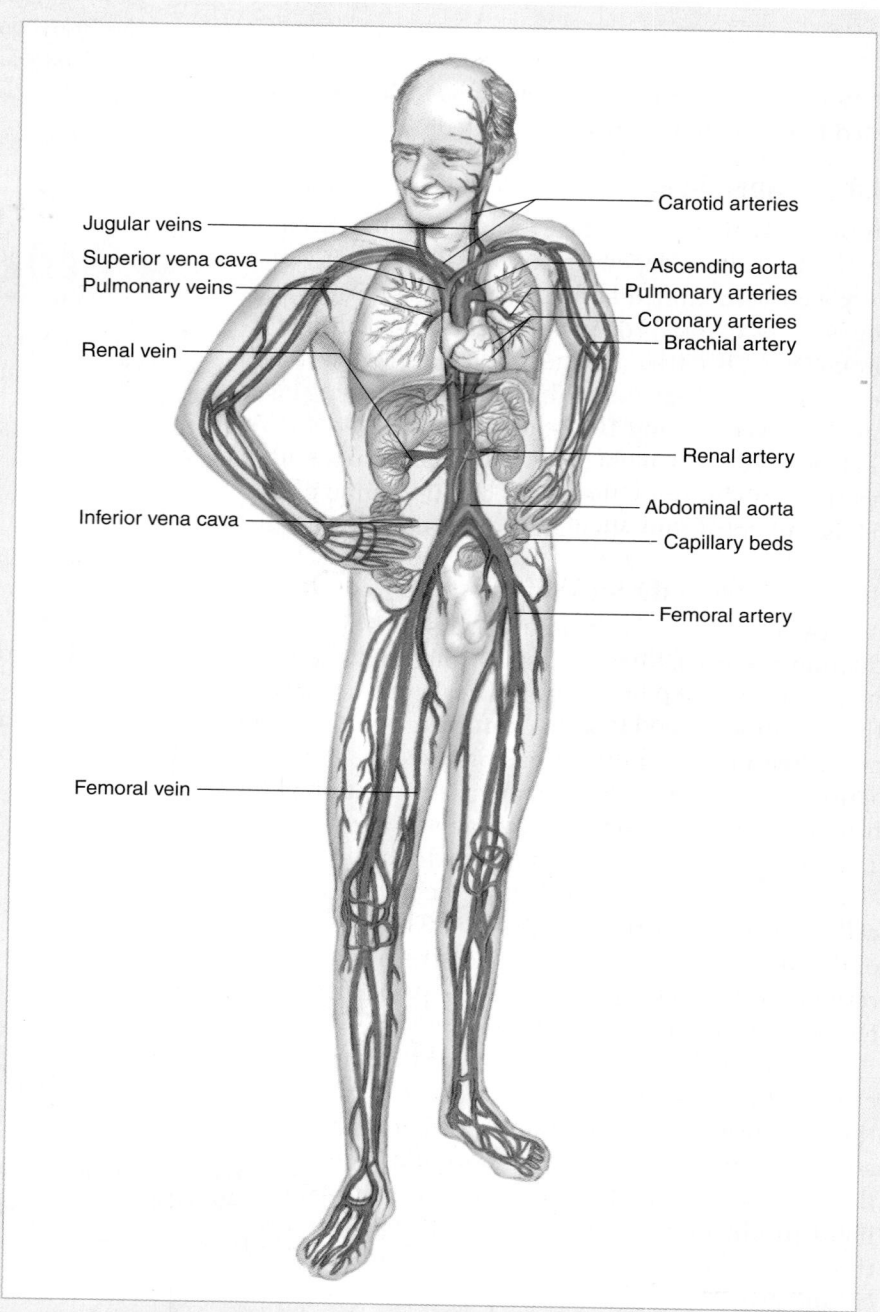

Jugular veins
Superior vena cava
Pulmonary veins
Renal vein
Inferior vena cava
Femoral vein

Carotid arteries
Ascending aorta
Pulmonary arteries
Coronary arteries
Brachial artery
Renal artery
Abdominal aorta
Capillary beds
Femoral artery

Figure 4-42 The principal arteries supply blood to a vast network of smaller arteries and arterioles. Venules deliver oxygen-poor blood to the veins that return blood to the heart.

walls of the alveoli. These arteries are the only ones in the body that carry oxygen-poor blood.

The carotid artery is the major artery that supplies blood to the head and brain. The carotid arteries are located on both sides of the neck. You can easily feel the carotid pulse if you place your fingers at the anterior lateral part of the neck. Since the carotid artery is rather close to the heart, you can feel its pulse even after the pulse in the distal extremities is too weak to feel.

The femoral artery is the major artery that supplies blood to the lower extremities. It is palpable in the groin. It divides at the level of the knee and supplies blood to the leg. At the ankle, two of these branches are palpable. You can feel a pulse at the posterior tibial artery, which is behind the medial prominence of the ankle (medial malleolus). You can also feel a pulse at the dorsalis pedis artery on the anterior surface of the foot (dorsum of the foot).

The brachial artery is the major vessel in the upper extremity that supplies blood to the arm. It divides into two major branches just below the elbow. This is the artery that is used in assessing blood pressure with a blood pressure cuff and stethoscope.

The radial artery is the major artery in the forearm and is palpable at the wrist on the thumb side (radial side). The ulnar artery is also palpable at the wrist on the opposite side (ulnar side, at the base of the fifth finger), although its pulse is not as strong. Both of these arteries supply blood to the hand.

Arteries branch into smaller arteries and then into arterioles. Arterioles are the smallest branches of an artery leading to the vast network of capillaries.

Capillaries

In the body, there are billions of cells and billions of capillaries. Capillary vessels are fine end-divisions of the arterial system that allow contact between the blood and the cells of the tissues. Oxygen and other nutrients pass from blood cells and plasma in the capillaries to the individual tissue cells through the very thin wall of the capillary. Carbon dioxide and other metabolic waste products pass in a reverse direction from the tissue cells to the blood to be carried away. Blood in arteries is characteristically bright red, because its hemoglobin is rich in oxygen. Blood in the veins is dark bluish red, because it has passed through a capillary bed and given up its oxygen to the cells. Capillaries connect directly at one end with the flow-regulating arterioles and at the other with the venules.

Veins

Once oxygen-poor blood passes through the network of capillaries, it moves to the venules, which are the smallest branches of the veins. The blood returns to the heart via a network of larger and larger veins (see Figure 4-42). Veins have much thinner walls than arteries and are generally larger in diameter. The veins become larger and larger and ultimately form two major vessels. These major vessels, part of the great vessels, are located in the midline, just to the left of the spine, and channel blood from the body and collect it just before it enters the heart. Because there is no "flow" from the heart once blood passes the capillaries, venous blood moved by gravity and large muscle contraction and flow is governed by valves within the veins.

The superior vena cava carries blood returning from the head, neck, shoulders, and upper extremities. Blood from the abdomen, pelvis, and lower extremities passes through the inferior vena cava. The superior and inferior venae cavae join at the right atrium of the heart. The right ventricle receives blood from the right atrium and pumps it through the pulmonary arteries into the lungs.

The Spleen

The spleen is a solid organ located under the rib cage in the left upper quadrant. The spleen is particularly susceptible to injury from blunt trauma because it is made of tissue that is delicate and because it is located directly under the flexible lower ribs, with very little soft tissue to cushion it; therefore it is one of the most frequently injured abdominal organs in blunt trauma. Because the spleen is highly vascular, injury can lead to severe internal bleeding. In some cases the spleen begins to hemorrhage one or two days after trauma, which is referred to as "delayed rupture." Delayed rupture of the spleen should be suspected when abdominal pain and signs of internal bleeding develop within a few days of blunt trauma. Virtually all of the blood in the body passes though the spleen, where it is filtered. Worn out blood cells, foreign substances, and bacteria are removed.

Components of Blood

Blood is a complex, thick, red fluid composed of plasma, red blood cells called erythrocytes, white blood cells called leukocytes, and platelets (Figure 4-43 ▶).

■ Plasma is a sticky, yellow fluid that carries the blood cells and nutrients. It also transports cellular waste material to the organs of excretion. It contains most of the compounds needed to produce a blood clot.

- The iron-containing hemoglobin molecules in <u>red blood cells</u> (erythrocytes) give color to the blood and carry oxygen. These make up about 45% of the blood.
- <u>White blood cells</u> (leukocytes) play a role in the body's immune defense mechanisms against infection.
- <u>Platelets</u> are tiny, disk-shaped elements that are much smaller than the cells. They are essential in the initial formation of a blood clot, the mechanism that stops bleeding.

Blood under pressure will gush or spurt intermittently from an artery and is bright red. From a vein, it will flow in a steady stream and is dark bluish-red. From capillaries, it will ooze at many tiny individual points. Clotting normally takes from 6 to 10 minutes.

Physiology of the Circulatory System

The pulse, which is palpated most easily at the neck, wrist, or groin, is created by the forceful pumping of blood out the left ventricle and into the major arteries. It is present throughout the entire arterial system. It can be felt most easily where the larger arteries are near the skin (Figure 4-44 ▶). The central pulses are the carotid artery pulse, which can be felt at the upper portion of the neck, and the femoral artery pulse, which is

felt in the groin. The peripheral pulses are the radial artery pulse, which is felt at the wrist at the base of the thumb; the brachial artery pulse, which is felt on the medial aspect of the arm, midway between the elbow and shoulder; the posterior tibial artery pulse, which is felt posterior to the medial malleolus; and the dorsalis pedis artery pulse, which is felt on the top of the foot.

<u>Blood pressure (BP)</u> is the pressure that the blood exerts against the walls of the arteries as it passes through them. When the cardiac muscle of the left ventricle contracts, it pumps blood from the ventricle into the aorta. This muscular contraction phase is called <u>systole</u>. When the muscle of the ventricle relaxes, the ventricle fills with blood. This phase is called <u>diastole</u>. The pulsed forceful ejection of blood from the left ventricle of the heart into the aorta is transmitted through the arteries as a pulsatile pressure wave. This pressure wave keeps the blood moving through the body. The high and low points of the wave can be measured with a sphygmomanometer (blood pressure cuff) and are expressed numerically in millimeters of mercury (mm Hg). The high

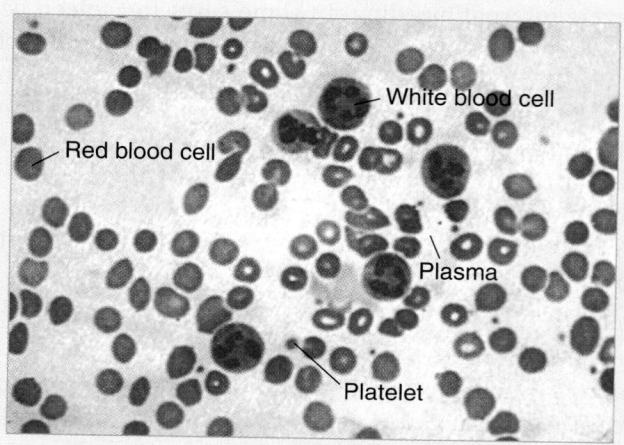

Figure 4-43 The components of blood include red blood cells, white blood cells, platelets, and plasma.

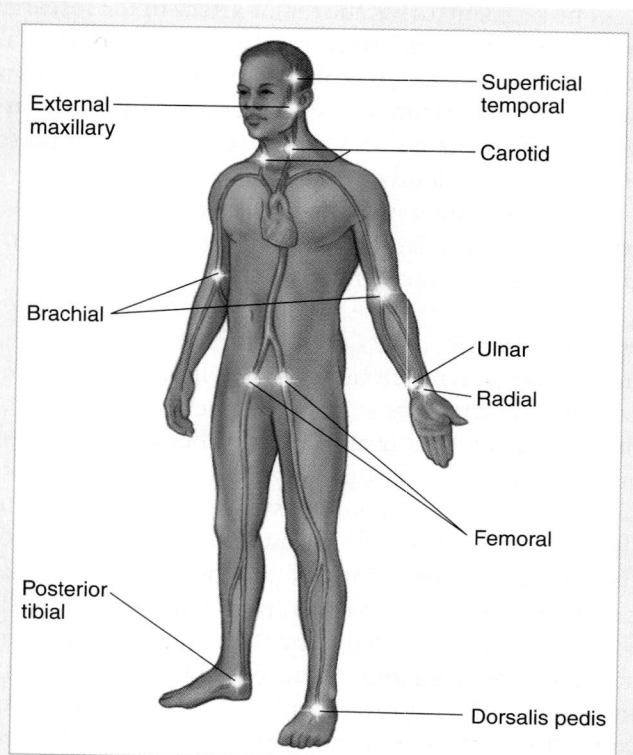

Figure 4-44 The central and peripheral pulses can be felt where the large arteries are near the skin.

point is called the systolic blood pressure (measured as the heart muscle is contracting). The low point is called the diastolic blood pressure (measured when the heart muscle is in its relaxation phase).

The average adult has approximately 6 L of blood in the vascular system. Children have less, 2 to 3 L, depending on their age and size. Infants have only about 300 mL. The loss of an amount of blood that may be negligible for an adult could be fatal for an infant.

Normal Circulation in Adults

In all healthy people, the circulatory system is automatically adjusted and readjusted constantly so that 100% of the capacity of the arteries, veins, and capillaries holds 100% of the blood at that moment. Never are all the vessels fully dilated or constricted. The size of arteries and veins is controlled by the nervous system, according to the amount of blood that is available and many other factors to keep blood pressure normal at all times. Under the condition of normal pressure, with a system that can hold just 100% of the blood available, all parts of the system will have adequate blood supply all of the time.

Perfusion is the circulation of blood within an organ or tissue in adequate amounts to meet the cells' current needs. Blood enters an organ or tissue through the arteries and leaves it through the veins **Figure 4-45** ▶. Loss of normal blood pressure is an indication that the blood is no longer circulating efficiently to every organ in the body. (However, a "good BP" does not indicate that it is reaching all parts of the body.) There are many reasons for loss of blood pressure. The result in each case is the same: Organs, tissues, and cells are no longer adequately perfused or supplied with oxygen and food, and wastes can accumulate. Under these conditions, cells, tissues, and whole organs may die. The state of inadequate circulation, when it involves the entire body, is called shock or hypoperfusion.

Inadequate Circulation in Adults

When a patient loses a small amount of blood, the arteries, veins, and heart automatically adjust to the smaller new volume. The adjustment occurs in an effort to maintain adequate pressure throughout the circulatory system and thereby maintain circulation for every organ. The adjustment occurs very rapidly after the loss, usually within minutes. Specifically, the vessels constrict to provide a smaller bed for the reduced volume of blood to fill. And the heart pumps more rapidly to

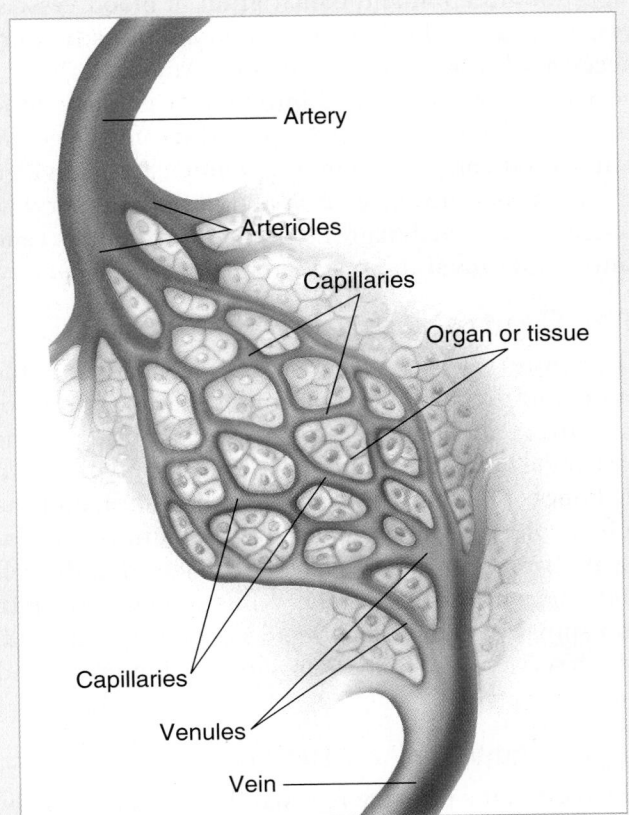

Figure 4-45 Blood enters an organ or tissue through the arteries and leaves through the veins. This process, called perfusion, provides adequate blood flow to the tissue to meet the cells' needs.

circulate the remaining blood more efficiently. As the blood pressure falls, the pulse increases to attempt to keep the cardiac output constant at 5 to 6 L per minute. If the loss of blood is too great, the adjustment fails, and the patient goes into shock.

The Nervous System

The nervous system controls virtually all activities of the body, both voluntary and involuntary. The somatic nervous system is the part of the nervous system that regulates activities over which there is voluntary control. Such activities include walking, talking, and writing. The autonomic nervous system controls the many body functions that occur without voluntary control. These activities include body functions such as digestion, dilation and constriction of blood vessels, sweating, and all other involuntary actions that are necessary for basic body functions. Anatomically, the nervous system is divided into two parts: the central nervous system and the peripheral nervous system. Thus, the nervous system as a whole can be divided anatomically into the central and peripheral nervous systems, and functionally into somatic (voluntary) and autonomic (involuntary) components.

The Central Nervous System

The central nervous system (CNS) is made up of the brain and the spinal cord. From a practical point of view, the central nervous system can be considered the part of the nervous system that is covered and protected by bones. The brain is covered by the skull, and the spinal cord is covered by the spinal column. The major parts of most nerve cells (the nucleus and the cell body) lie within the central nervous system. The brain and spinal column are bathed in cerebrospinal fluid, which serves to cushion them and filter impurities and toxins.

Organs and Vascular Structures

The head is the primary command center of the body, containing the brain, brain stem, and beginning of the spinal cord, all bathed in cerebrospinal fluid. The brain requires a constant supply of oxygen and nutrients, which are supplied by arteries and veins.

Brain

The brain is the controlling organ of the body. It is the center of consciousness. It is responsible for all our voluntary body activities, the perception of our surroundings, and the control of our reactions to the environment. In addition, the brain enables us to experience all the fine shadings of thought and feeling that make us individuals. The brain is subdivided into several areas, all of which have specific functions. Three major subdivisions of the brain are the cerebrum, the cerebellum, and the brain stem (Figure 4-46 ▶).

The cerebrum, which is the largest part of the brain and is sometimes called the "gray matter," makes up about three fourths of the volume of the brain and is itself composed of four lobes: frontal, parietal, temporal, and occipital. The cerebrum on one side of the brain controls activities on the opposite side of the body. Each lobe of the cerebrum is responsible for a specific function. For example, one group of brain cells in the frontal lobe is responsible for the activity of all the voluntary muscles of the body. Brain cells in this area generate impulses that are sent along nerve fibers that extend from each cell into the spinal cord. Another area in the parietal lobe has cells that receive sensory impulses from the peripheral nerves of the body. Other parts of the cerebrum are responsible for other body functions. For instance, the occipital region, on the back of the cerebrum, receives visual impulses for the eyes; other areas control hearing, balance, and speech. Still other parts of the cerebrum are responsible for emotions and other characteristics of an individual's personality.

The cerebellum, which is located underneath the great mass of cerebral tissue, is sometimes called the "little brain." The major function of this area is to coordinate the various activities of the brain, particularly body movements. Without the cerebellum, very specialized muscular activities such as writing or sewing would be impossible.

The brain stem is so called because the brain appears to be sitting on this portion of the central nervous system as a plant sits on its stem. The brain stem

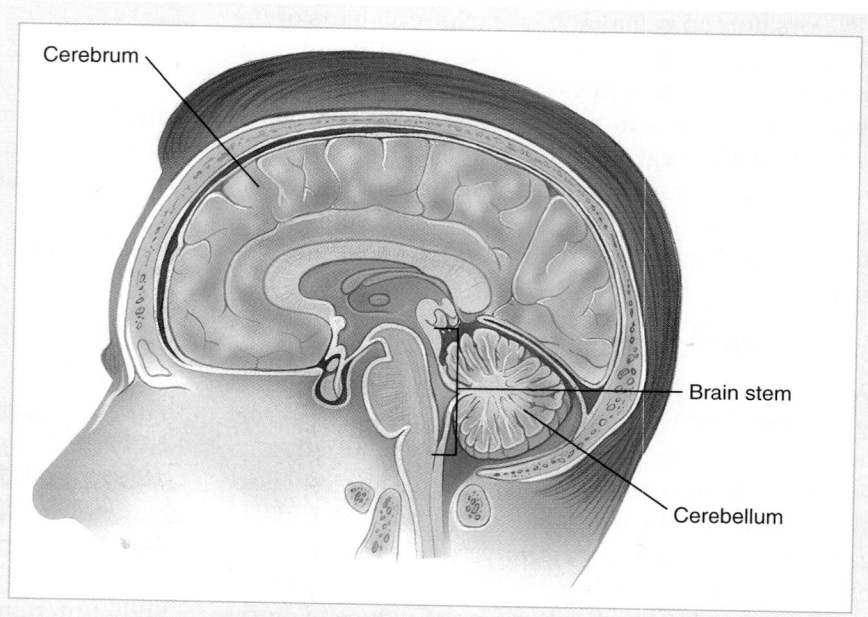

Figure 4-46 The brain lies well protected within the skull. Its principal subdivisions are the cerebrum, the cerebellum, and the brain stem.

Cerebrum

Brain stem

Cerebellum

is the most primitive part of the central nervous system. It lies deep within the cranium and is the best-protected part of the central nervous system. The brain stem is the controlling center for virtually all body functions that are absolutely necessary for life. Cells in this part of the brain control cardiac, respiratory, and other basic body functions.

The brain has many other anatomic areas, all of which have specific and important functions. The brain receives a vast amount of information from the environment, sorts it all out, and directs the body to respond appropriately. Many of the responses involve voluntary muscle action; others are automatic and involuntary.

Cerebrospinal Fluid

Cerebrospinal fluid (CSF) bathes the brain and spinal cord, and serves to cushion these structures and filter out impurities and toxins. For the EMT, a significant finding in trauma indicating skull fracture is cerebrospinal fluid leaking from the ears.

Circulation of the Head

The brain requires a constant flow of oxygenated blood to support brain function. Blood is supplied to the head through the carotid arteries which can be palpated on either side of the neck. Deoxygenated blood drains from the head via the internal and external jugular veins.

Spinal Cord

The spinal cord is the other major portion of the central nervous system (Figure 4-47 ▶). Like the brain, the spinal cord contains nerve cell bodies, but the major portion of the spinal cord is made up of nerve fibers that extend from the cells of the brain. These nerve fibers transmit information to and from the brain. All the fibers join together just below the brain stem to form the spinal cord. The spinal cord exits through a large opening at the base of the skull called the foramen magnum. It is encased within the spinal canal down to the level of the second lumbar vertebra. The spinal canal is created by the vertebrae, stacked one on another. Each vertebra surrounds the cord, and together the vertebrae form the bony spinal canal.

The principal function of the spinal cord is to transmit messages between the brain and the body. These messages are passed along the nerve fibers as electrical impulses, just as messages are passed along a telephone cable. The nerve fibers are arranged in specific bundles within the spinal cord to carry the messages from one specific area of the body to the brain and back.

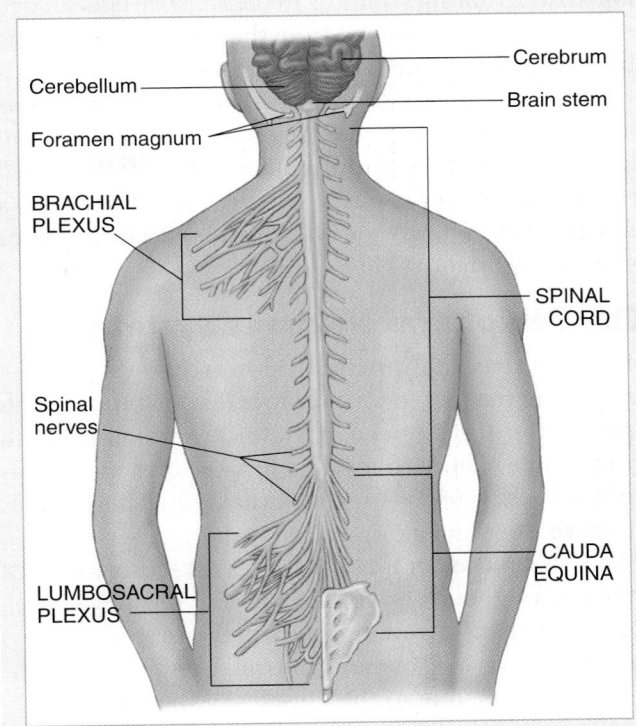

Figure 4-47 The spinal cord is a continuation of the brain stem. It exits the skull at the foramen magnum and extends down to the level of the second lumbar vertebra.

Connecting Nerves

Within the brain and the spinal cord are cells with short fibers that connect the sensory nerves with the motor nerves. In the spinal cord, they connect the sensory and motor nerves directly, bypassing the brain. These connecting nerves allow sensory and motor impulses to be transmitted from one nerve to another within the central nervous system.

The connecting nerves in the spinal cord complete a reflex arc between the sensory and motor nerves of the

Documentation Tips

Using correct anatomic terminology in your patient care report improves patient care by making the report more useful to hospital personnel and enhances your professional image as an EMT-B.

limbs. An irritating stimulus to the sensory nerve, such as heat, will be transmitted from the sensory nerve along the connecting nerve directly to the motor nerve. This will stimulate the sensory nerve. The muscle responds promptly, withdrawing the limb from the irritating stimulus even before this information can be transmitted to the brain. When a physician taps your knee with a rubber hammer, he or she is testing to see whether your reflex arc is intact.

The Peripheral Nervous System

Many of the cells in the central nervous system have long fibers that extend from the cell body out through openings in the bony covering of the spinal canal to form a cable of nerve fibers that link the central nervous system to the various organs of the body. These cables of nerve fibers make up the peripheral nervous system. The three major types of nerves are sensory nerves, motor nerves, and connecting nerves. Sensory nerves carry information from the body to the central nervous system. Motor nerves carry information from the central nervous system to the muscles of the body.

The peripheral nervous system is composed of 31 pairs of peripheral nerves called spinal nerves and 12 pairs called cranial nerves. At each vertebral level from the first cervical to the fifth sacral, on each side of the body, a spinal nerve exits the spinal canal and passes through a bony opening. This spinal nerve is composed of nerve fibers from nerve cells that originate within the spinal cord. The nerve fibers conduct sensory impulses from the skin and other organs to the spinal cord. They also conduct motor impulses from the spinal cord to the muscles that are present in that segment of the body. For example, between the seventh and eighth ribs, the spinal nerve carries sensory fibers from the skin between those two ribs and also has motor nerve fibers to innervate the intercostal muscle between the seventh and eighth ribs. This specific arrangement of nerve fibers becomes more complex and confusing in both the cervical and lumbar regions because of the large number of muscles in the arms and legs that must be supplied with nerve fibers. The spinal nerves combine to form a complex nerve network (called a plexus) in these two areas: the brachial plexus for the upper extremity and the lumbosacral plexus for the lower extremity.

The 12 pairs of peripheral nerves that exit the brain through holes in the skull are called the cranial nerves. For the most part, they are very specialized nerves that provide specific functions. For example, the facial (seventh cranial) nerves send motor impulses to many of the facial muscles.

Sensory Nerves

Sensory nerves of the body are quite complex. There are many different types of sensory cells in the nervous system. One type forms the retina of the eye; others are responsible for the hearing and balancing mechanisms in the ear. Other sensory cells are located within the skin, muscles, joints, lungs, and other organs of the body. When a sensory cell is stimulated, it transmits its own special message to the brain. There are special sensory nerves to detect heat, cold, position, motion, pressure, pain, balance, light, taste, and smell, as well as other sensations. Specialized nerve endings are adapted for each cell so that it perceives only one type of sensation and it transmits only that message.

The sensory impulses constantly provide information to the brain about what the different parts of our body are doing in relation to our surroundings. Thus, the brain is continuously made aware of its surroundings. The cranial nerves supply sensations directly to the brain. Visual sensations (what we see) reach the brain directly by way of the optic nerve (the second cranial nerve) in each eye. The nerve endings for the optic nerve lie in the retina of the eye. The nerve endings are stimulated by light, and the impulses are carried along the nerve that passes through a hole in the back of the eye socket and carries impulses to the occipital portion of the brain.

When sensory nerve endings in the extremities are stimulated, the impulses are transmitted along a peripheral nerve to the spinal cord. The cell body of the peripheral nerve lies in the spinal cord. The impulse is then transmitted from that cell body to another nerve ending in the spinal cord. The impulse is then sent up the spinal cord to the sensory area in the parietal lobe of the brain, where the sensory information can be interpreted and acted on by the brain.

Motor Nerves

Each muscle in the body has its own motor nerve. The cell body for each motor nerve lies in the spinal cord, and a fiber from the cell body extends as part of the peripheral nerve to its specific muscle. Electrical impulses that are produced by the cell body in the spinal cord are transmitted along the motor nerve to the muscle and cause it to contract. The cell body in the spinal cord is stimulated by an impulse produced in the mo-

tor strip of the cerebral cortex. This impulse is transmitted along the spinal cord to the cell body of the motor nerve.

The Skin

The skin, the largest single organ in the body, serves three major functions: to protect the body in the environment, to regulate the temperature of the body, and to transmit information from the environment to the brain.

The protective functions of the skin are numerous. Over 70% of the body is composed of water. The water contains a delicate balance of chemical substances in solution. The skin is watertight and serves to keep this balanced internal solution intact. The skin also protects the body from the invasion of infectious organisms: bacteria, viruses, and fungi. These organisms are everywhere and are routinely found lying on the skin surface and deep in its grooves and glands. However, they never penetrate the skin unless it is broken by injury; thus, the skin provides a constant protection against outside invaders.

The energy of the body is derived from metabolism (chemical reactions) that must take place within a very narrow temperature range. If the body temperature is too low, these reactions cannot proceed, metabolism ceases, and the body dies. If the temperature becomes too high, the rate of metabolism increases. Dangerously high temperatures producing too high a metabolic rate can result in permanent tissue damage and death.

Functions of the Skin

The major organ for regulation of body temperature is the skin. Blood vessels in the skin constrict when the body is in a cold environment and dilate when the body is in a warm environment. In a cold environment, constriction of the blood vessels shunts the blood away from the skin to decrease the amount of heat radiated from the body surface. When the outside environment is hot, the vessels in the skin dilate, the skin becomes flushed or red, and heat radiates from the body surface.

Also, in the hot environment, sweat is secreted to the skin surface from the sweat glands. Evaporation of the sweat requires energy. This energy, as body heat, is taken from the body during the evaporation process, which causes the body temperature to fall. Sweating alone will not reduce body temperature; evaporation of the sweat must also occur.

Information from the environment is carried to the brain through a rich supply of sensory nerves that originate in the skin. Nerve endings that lie in the skin are adapted to perceive and transmit information about heat, cold, external pressure, pain, and the position of the body in space. The skin thus recognizes any changes in the environment. The skin also reacts to pressure, pain, and pleasurable stimuli.

Anatomy of the Skin

The skin is divided into two parts: the superficial epidermis, which is composed of several layers of cells, and the deeper dermis, which contains the specialized skin structures. Below the skin lies the subcutaneous tissue layer (Figure 4-48 ▶). The cells of the epidermis are sealed to form a watertight protective covering for the body.

The epidermis is actually composed of several layers of cells. At the base of the epidermis is the germinal layer, which continuously produces new cells that gradually rise to the surface. On the way to the surface, these cells die and form the watertight covering. The epidermal cells are held together securely by an oily substance called sebum, which is secreted by the sebaceous glands of the dermis. The outermost cells of the epidermis are constantly rubbed away and then replaced by new cells produced by the germinal layer. The deeper cells in the germinal layer also contain pigment granules that (along with the blood vessels lying in the dermis) produce skin color.

The epidermis varies in thickness in different areas of the body. On the soles of the feet, the back, and the scalp, it is quite thick, but in some areas of the body, the epidermis is only two or three cell layers thick. The watertight seal provided by the epidermis prevents the invasion of bacteria and other organisms.

The deeper part of skin, the dermis, is separated from the epidermis by the layer of germinal cells. Within the dermis lie many of the special structures of the skin: sweat glands, sebaceous (oil) glands, hair follicles, blood vessels, and specialized nerve endings.

Sweat glands produce sweat for cooling the body. The sweat is discharged onto the surface of the skin through small pores, or ducts, that pass through the epidermis onto the skin surface. The sebaceous glands produce sebum, the oily material that seals the surface epidermal cells. The sebaceous glands lie next to hair follicles and secrete sebum along the hair follicle to the skin surface. In addition to providing waterproofing for the skin, sebum keeps the skin supple so that it does not crack.

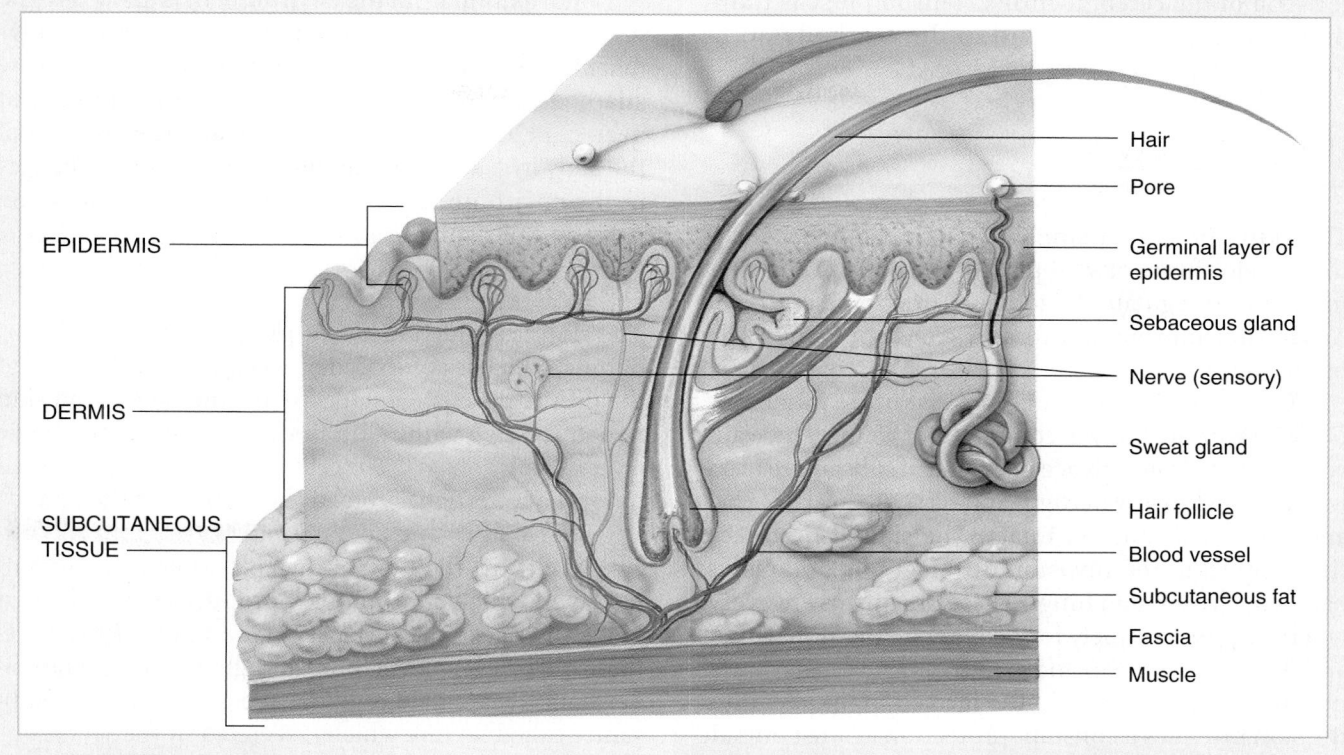

EPIDERMIS

DERMIS

SUBCUTANEOUS
TISSUE

Hair

Pore

Germinal layer of
epidermis

Sebaceous gland

Nerve (sensory)

Sweat gland

Hair follicle

Blood vessel

Subcutaneous fat

Fascia

Muscle

Figure 4-48 The skin has two principal layers: the epidermis and the dermis. Below the skin is a layer of subcutaneous tissue.

Hair follicles are the small organs that produce hair. There is one follicle for each hair connected with a sebaceous gland and also with a tiny muscle. The muscle pulls the hair into an erect position when the individual is cold or frightened. All hair grows continuously and is either cut off or worn away by clothing.

Blood vessels provide nutrients and oxygen to the skin. The blood vessels lie in the dermis. Small branches extend up to the germinal layer. There are no blood vessels in the epidermis. A complex array of nerve endings also lie in the dermis. These specialized nerve endings are sensitive to environmental stimuli; they respond to these stimuli and send impulses along the nerves to the brain.

Beneath the skin, immediately under the dermis and attached to it, lies the subcutaneous tissue. The subcutaneous tissue is composed largely of fat. The fat serves as an insulator for the body and as a reservoir to store energy. The amount of subcutaneous tissue varies greatly from individual to individual. Beneath the subcutaneous tissue lie the muscles and the skeleton.

The skin covers all of the external surface of the body. The various orifices (openings to the body)—including the mouth, nose, anus, and vagina—are not covered by skin. Orifices are lined with mucous membranes. Mucous membranes are quite similar to skin in that they provide a protective barrier against bacterial invasion. Mucous membranes differ from skin in that they secrete mucus, a watery substance that lubricates the openings. Thus, mucous membranes are moist, whereas the skin is dry. A mucous membrane lines the entire gastrointestinal tract from the mouth to the anus.

The Endocrine System

The brain controls the body through both the nervous system and the endocrine system. The endocrine system is a complex message and control system that

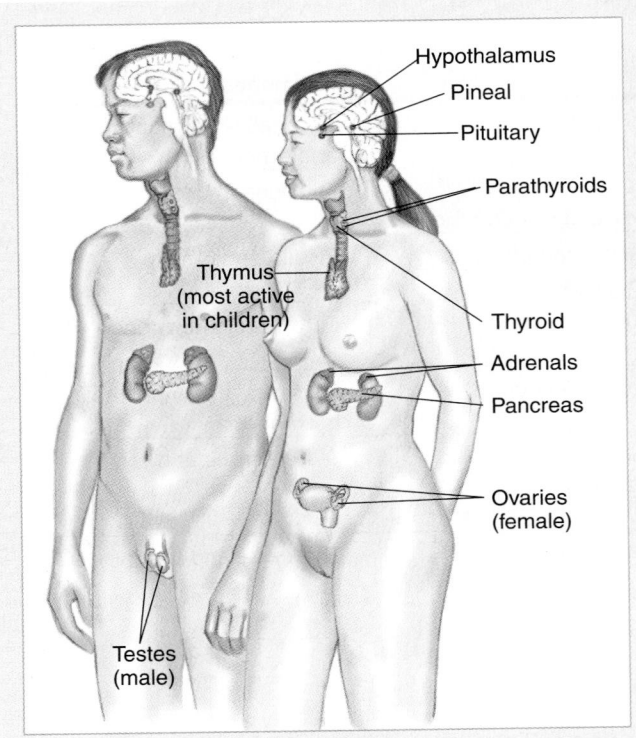

Figure 4-49 The endocrine system controls the release of hormones in the body.

Labels on figure: Hypothalamus, Pineal, Pituitary, Parathyroids, Thymus (most active in children), Thyroid, Adrenals, Pancreas, Ovaries (female), Testes (male)

Excesses or deficiencies in hormones cause various diseases. With endocrine diseases, specific body functions are increased, decreased, or absent. Diabetes mellitus is a common problem. Because production of the hormone insulin is deficient, the body is unable to use glucose normally. This disease also damages the small blood vessels in the body. The tissue damage that results is as much a part of diabetes as is the difficulty in regulating the amount of glucose in the blood.

The Digestive System

The digestive system is composed of the gastrointestinal tract (stomach and intestines), mouth, salivary glands, pharynx, esophagus, liver, gallbladder, pancreas, rectum, and anus. The function of this system is digestion: the processing of food that nourishes the individual cells of the body.

How Digestion Works

Digestion of food, from the time it is taken into the mouth until essential compounds are extracted and delivered by the circulatory system to nourish all of the cells in the body, is a complicated chemical process. In succession, different secretions, primarily enzymes, are added to the food by the salivary glands, the stomach, the liver, the pancreas, and the small intestine to convert the food into basic sugars, fatty acids, and amino acids. These basic products of digestion are carried across the wall of the intestine and transported through the portal vein to the liver. In the liver, the products are processed further and then stored or transported to the heart through veins draining the liver. The heart then pumps the blood with these nutrients throughout the arteries and then to the capillaries, where the nutrients pass through the capillary walls to nourish the body's individual cells.

In normal routine activity, without any food or fluid ingestion at all, between 8 to 10 L of fluid are secreted daily into the gastrointestinal tract. This fluid comes from the salivary glands, stomach, liver, pancreas, and small intestine. In a normal adult, about 7% of the body weight is delivered as fluid daily to the gastrointestinal tract. If significant vomiting or diarrhea occurs for more than 2 or 3 days, the patient will lose a very substantial portion of body composition and become severely ill.

integrates many body functions. It releases substances called hormones, either by target organs or into the bloodstream (Figure 4-49 ▲). Adrenaline and insulin are examples of hormones. Each endocrine gland produces one or more hormones. Each hormone has a specific effect on some organ, tissue, or process (Table 4-5 ▶). The brain controls the release of hormones by the endocrine glands. Hormones can have either a stimulating or inhibiting effect on the body's organs and systems. For example, when we are frightened, the brain stimulates the adrenal gland through a hormone to release adrenaline (epinephrine). Release of adrenaline increases our blood pressure and heart rate. The resulting increase in blood pressure and heart rate decreases the amount of hormone released by the adrenal gland. The brain then reduces the amount of stimulation to the adrenal gland. Thus, a new steady state is achieved at heightened levels of alertness. This cycle is known as a feedback loop, and helps keep the body's systems and functions in balance (Figure 4-50 ▶).

Study This!!!

TABLE 4-5 Endocrine Glands

Gland	Location	Function	Hormones Produced
Adrenal	Kidneys	Regulate salt, sugar, and sexual function	Adrenaline (epinephrine) and others
Ovary	Female pelvis (2 glands)	Regulate sexual function, characteristics, and reproduction	Estrogen and others
Pancreas	Retroperitoneal space	Regulates glucose metabolism and other functions	Insulin and others
Parathyroid	Neck (behind and beside the thyroid) (3-5 glands)	Regulate serum calcium	Parathyroid hormone
Pituitary	Base of skull	Regulates all other endocrine glands	Multiple, very important hormones
Testes	Male scrotum (2 glands)	Regulate sexual function, characteristics, and reproduction	Testosterone and others
Thyroid	Neck (over the larynx)	Regulates metabolism	Thyroxine and other

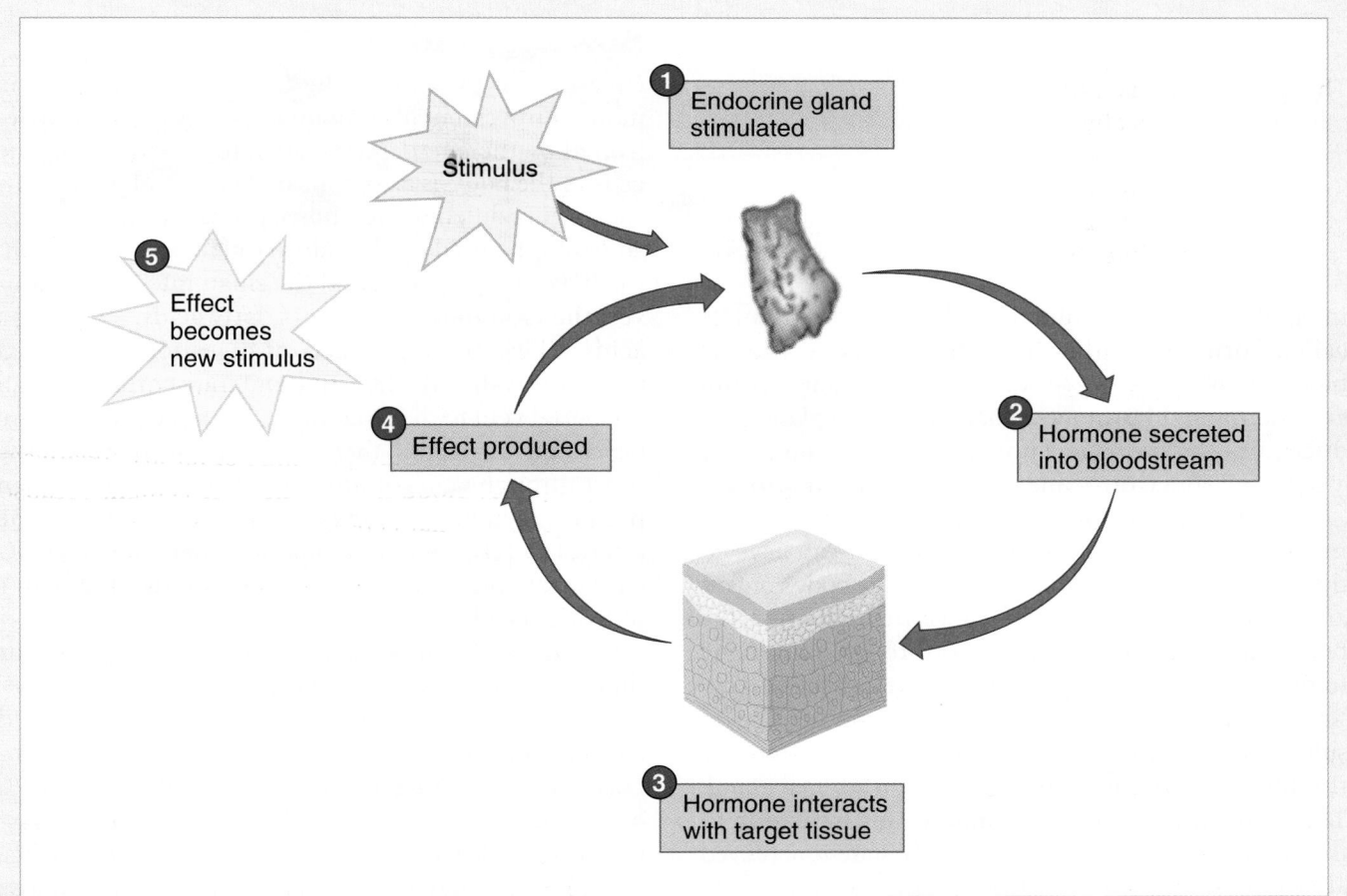

Figure 4-50 The endocrine system is tightly controlled with primary and secondary feedback loops to keep body systems in balance.

Anatomy of the Digestive System

Mouth

The mouth consists of the lips, cheeks, gums, teeth, and tongue. A mucous membrane lines the mouth. The roof of the mouth is formed by the hard and soft palates. The hard palate is a bony plate lying anteriorly; the soft palate is a fold of mucous membrane and muscle that extends posteriorly from the hard palate into the throat. The soft palate is designed to hold food that is being chewed within the mouth and to help initiate swallowing.

Salivary Glands

There are two salivary glands located under the tongue, one on each side of the lower jaw, and one inside each cheek. They produce nearly 1.5 L of saliva daily. Saliva is approximately 98% water. The remaining 2% is composed of mucus, salts, and organic compounds. Saliva serves as a binder for the chewed food that is being swallowed and as a lubricant within the mouth.

Oropharynx

The oropharynx is a tubular structure about 5″ long that extends vertically from the back of the mouth to the esophagus and trachea. An automatic movement of the pharynx during swallowing lifts the larynx to permit the epiglottis to close over it so that liquids and solids are moved into the esophagus and away from the trachea.

Esophagus

The esophagus is a collapsible tube about 10″ long that extends from the end of the pharynx to the stomach and lies just anterior to the spinal column in the chest. Contractions of the muscle in the wall of the esophagus propel food through it to the stomach. Liquids will pass with very little assistance.

Stomach

The stomach is a hollow organ located in the left upper quadrant of the abdominal cavity, largely protected by the lower left ribs. Muscular contractions in the wall of the stomach and gastric juice, which contains much mucus, convert ingested food to a thoroughly mixed semi-solid mass. The stomach produces approximately 1.5 L of gastric juice daily for this process. The principal function of the stomach is to receive food in large quantities intermittently, store it, and provide for its movement into the small bowel in regular, small amounts. In 1 to 3 hours, the semi-solid food mass derived from one meal is propelled by muscular contraction into the duodenum, the first part of the small intestine.

Pancreas

The pancreas, a flat, solid organ, lies below and behind the liver and stomach and behind the peritoneum. It is firmly fixed in position, deep within the abdomen, and is not easily damaged. It contains two kinds of glands. One set of glands secretes nearly 2 L of pancreatic juice daily. This juice contains many enzymes that aid in the digestion of fat, starch, and protein. Pancreatic juice flows directly into the duodenum through the pancreatic ducts. The other gland is the islets of Langerhans, which produces insulin. Insulin regulates the amount of glucose in the blood.

Liver

The liver is a large, solid organ that takes up most of the area immediately beneath the diaphragm in the right upper quadrant and also extends into the upper left quadrant. It is the largest solid organ in the abdomen and has several functions. Poisonous substances produced by digestion are brought to the liver and rendered harmless. Factors that are necessary for blood clotting and for the production of normal plasma are formed here. Between 0.5 and 1 L of bile is made by the liver daily to assist in the normal digestion of fat. The liver is the principal organ for the storage of sugar or starch for immediate use by the body for energy. It also produces many of the factors that aid in the proper regulation of immune responses. Anatomically, the liver is a large mass of blood vessels and cells, packed tightly together. It is fragile and, because of its size, relatively easily injured. Blood flow in the liver is high, because all of the blood that is pumped to the gastrointestinal tract passes into the liver, through the portal vein, before it returns to the heart. In addition, the liver has a generous arterial blood supply of its own. Ordinarily, approximately 25% of the cardiac output of blood (1.5 L) passes through the liver each minute.

Bile Ducts

The liver is connected to the intestine by the bile ducts. The gallbladder is an outpouching from the bile ducts that serves as a reservoir and concentrating organ for bile produced in the liver. Together, the bile ducts and gallbladder form the biliary system. The gallbladder discharges stored and concentrated bile into the

duodenum through the common bile duct. The presence of food in the duodenum triggers a contraction of the gallbladder to empty it. The gallbladder usually contains about 60 to 90 mL of bile.

Small Intestine

The small intestine is the major hollow organ of the abdomen. The cells lining the small intestine produce enzymes and mucus to aid in digestion. Enzymes from the pancreas and the small intestine carry out the final processes of digestion. More than 90% of the products of digestion (amino acids, fatty acids, and simple sugars), together with water, ingested vitamins, and minerals are absorbed across the wall of the lower end of the small intestine into veins to be transported to the liver. The small intestine is composed of the duodenum, the jejunum, and the ileum. The duodenum, which is about 12″ long, is the part of the small intestine that receives food from the stomach. Here, food is mixed with secretions from the pancreas and liver for further digestion. Bile, produced by the liver and stored in the gallbladder, is emptied as needed into the duodenum. It is greenish black, but through changes during digestion, it gives feces its typical brown color. Its major function is in the digestion of fat. The jejunum and ileum together measure more than 20′ on average to make up the rest of the small intestine.

Large Intestine

The large intestine, another major hollow organ, consists of the cecum, the colon, and the rectum. About 5′ long, it encircles the outer border of the abdomen around the small bowel. The major function of the colon, the portion of the large intestine that extends from the cecum to the rectum, is to absorb the final 5% to 10% of digested food and water from the intestine to form solid stool, which is stored in the rectum and passed out of the body through the anus.

Appendix

The appendix is a tube 3″ to 4″ long that opens into the cecum (the first part of the large intestine) in the right lower quadrant of the abdomen. It may easily become obstructed and, as a result, inflamed and infected. Appendicitis, which is the term for this inflammation, is one of the major causes of severe abdominal distress.

Rectum

The lowermost end of the colon is the rectum. It is a large, hollow organ that is adapted to store quantities of feces until it is expelled. At its terminal end is the anus,

a 2″ canal lined with skin. The rectum and anus are supplied with a complex series of circular muscles called sphincters that control, both voluntarily and automatically, the escape of liquids, gases, and solids from the digestive tract.

The Urinary System

The urinary system controls the discharge of certain waste materials filtered from the blood by the kidneys. In the urinary system, the kidneys are solid organs; the ureters, bladder, and urethra are hollow organs Figure 4-51 ▼. Ordinarily, we consider the urinary and genital systems together, because they share many organs.

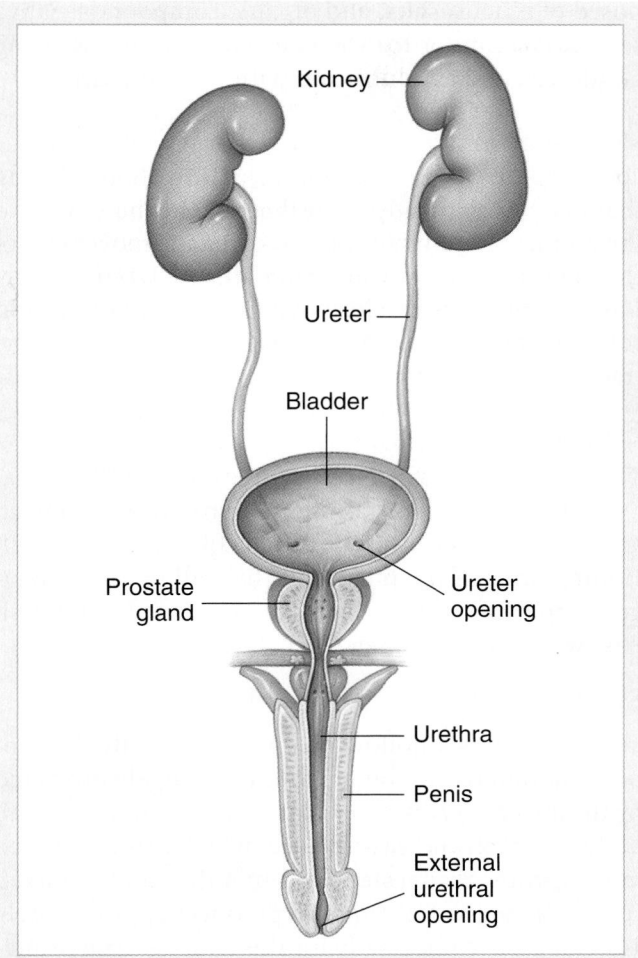

Figure 4-51 The urinary system lies in the retroperitoneal space behind the organs of the digestive system. The urinary system in males and females includes the kidneys, ureters, bladder, and urethra. This diagram shows the male urinary system.

The body has two kidneys that lie on the posterior muscular wall of the abdomen behind the peritoneum in the retroperitoneal space. These organs rid the blood of toxic waste products and control its balance of water and salt. Blood flow in the kidneys is high. Nearly 20% of the output of blood from the heart passes through the kidneys each minute. Large vessels attach the kidneys directly to the aorta and the inferior vena cava. Waste products and water are constantly filtered from the blood to form urine. The kidneys continuously concentrate this filtered urine by reabsorbing the water as it passes through a system of specialized tubes within them. The tubes finally unite to form the renal pelvis, a cone-shaped collecting area that connects the ureter and the kidney. Normally, each kidney drains its urine into one ureter through which the urine passes to the bladder.

A ureter passes from the renal pelvis of each kidney along the surface of the posterior abdominal wall behind the peritoneum to drain into the urinary bladder. The ureters are small (0.2″ in diameter), hollow, muscular tubes. Peristalsis, a wave-like contraction of smooth muscle, occurs in these tubes to move the urine to the bladder.

The urinary bladder is located immediately behind the pubic symphysis in the pelvic cavity and is composed of smooth muscle with a specialized lining membrane. The two ureters enter posteriorly at its base on either side. The bladder empties to the outside of the body through the urethra. In the male, the urethra passes from the anterior base of the bladder through the penis. In the female, the urethra opens at the front of the vagina. The normal adult forms 1.5 to 2 L of urine every day. This waste is extracted and concentrated from the 1,500 L of blood that circulate through the kidneys daily.

The Genital System

The genital system controls the reproductive processes by which life is created. The male genitalia, except for the prostate gland and the seminal vesicles, lie outside the pelvic cavity. The female genitalia, with the exception of the clitoris and labia, are contained entirely within the pelvis. The male and female reproductive organs have certain similarities and, of course, basic differences. They allow the production of sperm and egg cells and appropriate hormones and the act of sexual intercourse and reproduction.

The Male Reproductive System and Organs

The male reproductive system consists of the testicles, epididymis, vasa deferentia, seminal vesicles, prostate gland, urethra, and penis (Figure 4-52 ▶). Each testicle contains specialized cells and ducts; some of these produce male hormones, and others develop sperm. The hormones are absorbed directly into the bloodstream from the testicles. The vasa deferentia (or vas deferens) are ducts that travel from the testicles up beneath the skin of the abdominal wall for a short distance. They then pass through an opening into the abdominal cavity and into the prostate gland to connect with the urethra. The vasa deferentia carry the sperm from the testicles to the urethra. The seminal vesicles are small storage sacs for seminal fluid. The vesicles also empty into the urethra, at the prostate.

Semen, also called seminal fluid, contains sperm cells that are carried up each vas from each testicle to be mixed with fluid from the seminal vesicles and prostate gland. The prostate gland surrounds the urethra, where it emerges from the urinary bladder. Fluids from the prostate gland and from the seminal vesicles mix during sexual intercourse. During intercourse, special mechanisms in the nervous system prevent the passage of urine into the urethra. Only seminal fluid, prostatic fluid, and sperm pass from the penis into the vagina during ejaculation.

The penis contains a special type of tissue called erectile tissue. This specialized tissue is largely vascular and, when filled with blood, causes the penis to distend into a state of erection. As the vessels fill under pressure from the circulatory system, the penis becomes a large, rigid organ that can enter the vagina. Certain spinal injuries and some diseases can cause a painful continuous erection called priapism.

The Female Reproductive System and Organs

The female reproductive organs include the ovaries, fallopian tubes, uterus, cervix, and vagina (Figure 4-53 ▶). The ovaries, like the testicles, produce sex hormones and specialized cells for reproduction. The female sex hormones are absorbed directly into the bloodstream. A specialized ovum, or egg cell, matures and is released regularly during the adult female's reproductive years. The ovaries release a mature egg approximately every 28 days. This egg travels through the fallopian tubes to the uterus.

The fallopian tubes connect with the uterus and carry the ovum into the cavity of this organ. The uterus is pear-shaped and hollow, with muscular walls. The narrow opening from the uterus to the vagina is the cervix. The vagina (birth canal) is a muscular distensible tube that connects the uterus with the vulva (the external female genitalia). The vagina receives the penis during sexual intercourse, when semen is deposited in it. The sperm in the semen may pass into the uterus and fertilize an egg, causing pregnancy. Should the pregnancy come to completion at the end of nine months, the baby will pass through the vagina and be born. The vagina also channels the menstrual flow from the uterus out of the body.

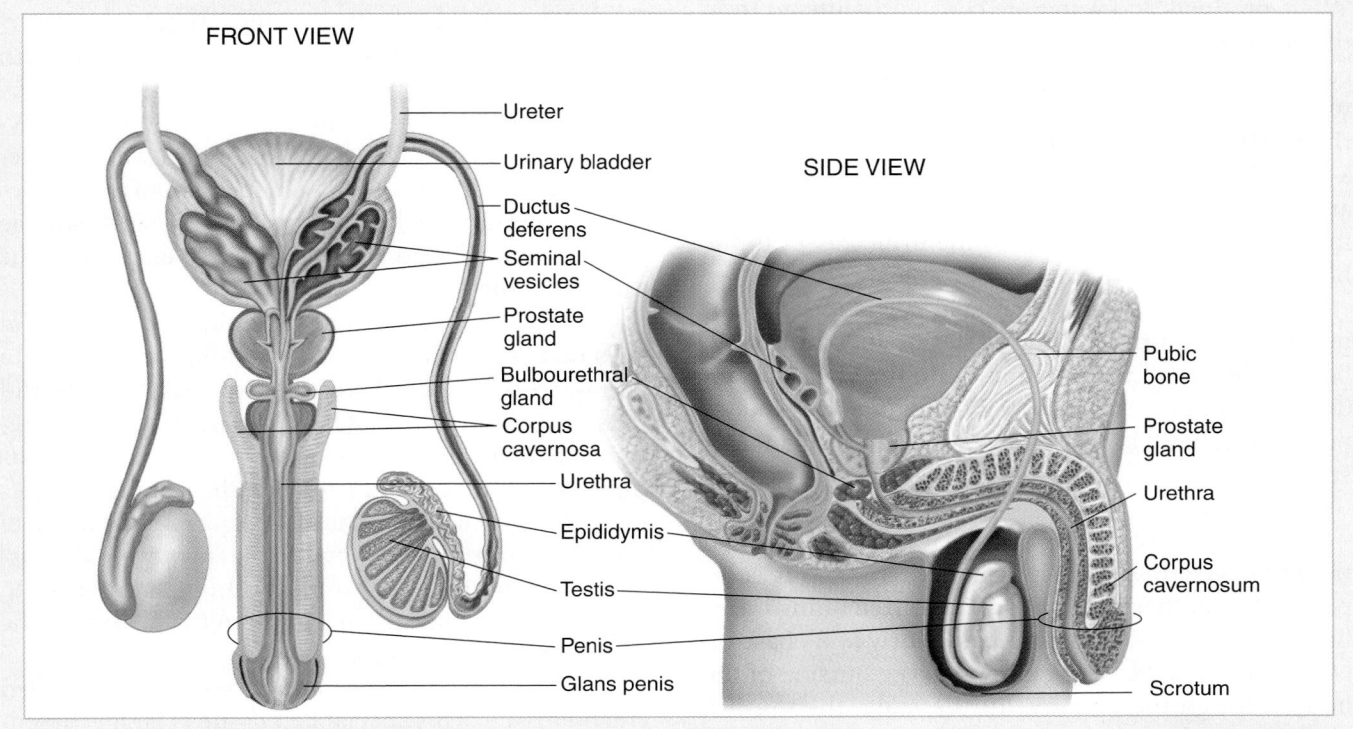

Figure 4-52 The male reproductive system consists of the testicles, vasa deferentia, seminal vesicles, prostate gland, urethra, and penis.

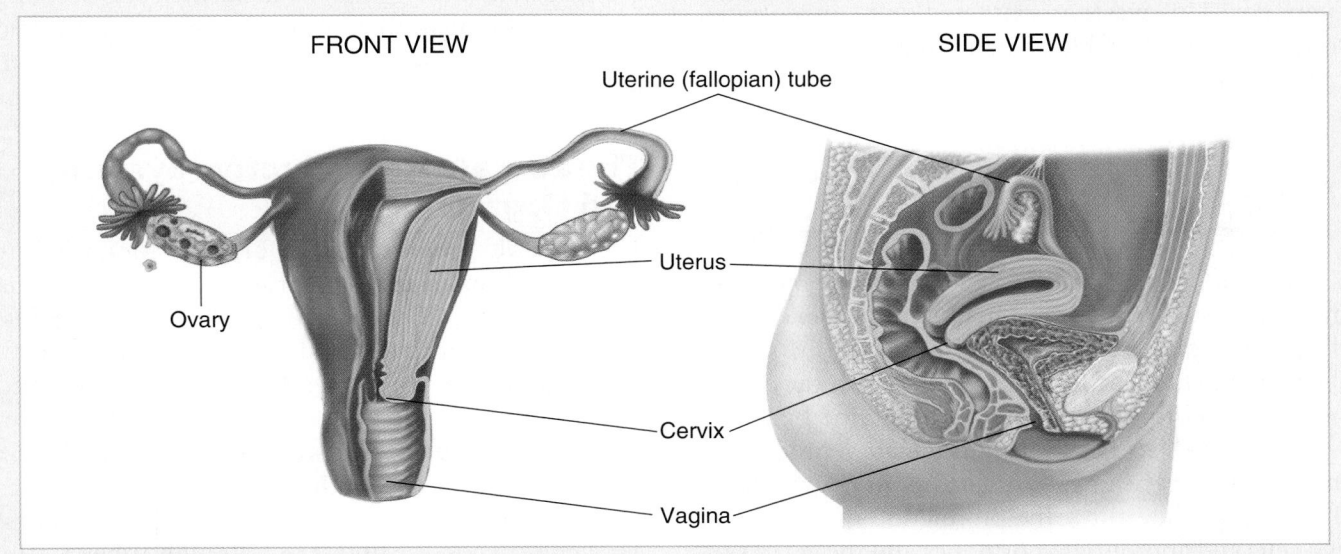

Figure 4-53 The female reproductive system consists of the ovaries, fallopian tubes, uterus, cervix, and vagina.

You are the Provider

Summary

Having an understanding of human anatomy can assist you in determining the source of patient illness and injury and therefore can affect the quality of patient care. Using correct anatomic terminology, your communication with other medical professionals will be clear and also aid in the continuity of care. Extending this level of professionalism to your documentation will contribute to your professional image as a knowledgeable and capable emergency provider.

Prep Kit

Ready for Review

- To do your work as an EMT-B, you must have a working knowledge of human anatomy so that you can communicate with hospital personnel and other health care providers.
- You must be able to identify superficial landmarks of the body and know what lies underneath the skin so that you can perform an accurate assessment.
- Hospital personnel will use medical terms to ask you questions about a patient; therefore, it is critical for the well-being of your patient that you learn them and can use them correctly.

Vital Vocabulary

abdomen The body cavity that contains the major organs of digestion and excretion. It is located below the diaphragm and above the pelvis.

abduction Motion of a limb away from the midline.

acetabulum The depression on the lateral pelvis where its three component bones join, in which the femoral head fits snugly.

Adam's apple The firm prominence in the upper part of the larynx formed by the thyroid cartilage. It is more prominent in men than in women.

adduction Motion of a limb toward the midline.

agonal respirations Slow, gasping respiration, sometimes seen in dying patients.

alveoli The air sacs of the lungs in which the exchange of oxygen and carbon dioxide takes place.

anatomic position The position of reference in which the patient stands facing you, arms at the side, with the palms of the hands forward.

www.EMTB.com

Technology

- Interactivities
- Vocabulary Explorer
- Anatomy Review
- Web Links
- Online Review Manual

angle of Louis A ridge on the sternum that lies at the level where the second rib is attached to the sternum; provides a constant and reliable bony landmark on the anterior chest wall.

anterior The front surface of the body; the side facing you in the standard anatomic position.

anterior superior iliac spines The bony prominences of the pelvis (ilium) at the front on each side of the lower abdomen just below the plane of the umbilicus.

aorta The principal artery leaving the left side of the heart and carrying freshly oxygenated blood to the body.

apex (apices) The pointed extremity of a conical structure.

appendix A small tubular structure that is attached to the lower border of the cecum in the lower right quadrant of the abdomen.

arteriole The smallest branch of an artery leading to the vast network of capillaries.

atrium Upper chamber of the heart.

autonomic nervous system The part of the nervous system that regulates functions, such as digestion and sweating, that are not controlled voluntarily.

ball-and-socket joint A joint that allows internal and external rotation as well as bending.

biceps The large muscle that covers the front of the humerus.

bilateral A body part that appears on both sides of the midline.

bile ducts Ducts that convey bile between the liver and the intestine.

blood pressure (BP) The pressure that the blood exerts against the walls of the arteries as it passes through them.

brachial artery The major vessel in the upper extremity that supplies blood to the arm.

brain The controlling organ of the body and center of consciousness; functions include perception, control of reactions to the environment, emotional responses, and judgment.

brain stem The area of the brain between the spinal cord and cerebrum, surrounded by the cerebellum; controls functions that are necessary for life, such as respirations.

capillary vessels The fine end-divisions of the arterial system that allow contact between cells of the body tissues and the plasma and red blood cells.

carotid artery The major artery that supplies blood to the head and brain.

carpometacarpal joint The joint between the wrist and the metacarpal bones; the thumb joint.

cecum The first part of the large intestine, into which the ileum opens.

central nervous system (CNS) The brain and spinal cord.

cerebellum One of the three major subdivisions of the brain, sometimes called the "little brain"; coordinates the various activities of the brain, particularly fine body movements.

cerebrum The largest part of the three subdivisions of the brain, sometimes called the "gray matter"; made up of several lobes that control movement, hearing, balance, speech, visual perception, emotions, and personality.

cervical spine The portion of the spinal column consisting of the first seven vertebrae that lie in the neck.

circulatory system The complex arrangement of connected tubes, including the arteries, arterioles, capillaries, venules, and veins, that moves blood, oxygen, nutrients, carbon dioxide, and cellular waste throughout the body.

clavicle The collarbone; it is lateral to the sternum and medial to the scapula.

coccyx The last three or four vertebrae of the spine; the tailbone.

connecting nerves Nerves that connect the sensory and motor nerves in the spinal cord.

costal arch A bridge of cartilage that connects the ends of the sixth through tenth ribs with the lower portion of the sternum.

costovertebral angle An angle that is formed by the junction of the spine and the tenth rib.

cranium The area of the head above the ears and eyes; the skull. The cranium contains the brain.

cricoid cartilage A firm ridge of cartilage that forms the lower part of the larynx.

cricothyroid membrane A thin sheet of fascia that connects the thyroid and cricoid cartilages that make up the larynx.

deep Further inside the body and away from the skin.

dermis The inner layer of the skin, containing hair follicles, sweat glands, nerve endings, and blood vessels.

diaphragm A muscular dome that forms the undersurface of the thorax, separating the chest from the abdominal cavity. Contraction of the diaphragm (and the chest wall muscles) brings air into the lungs. Relaxation allows air to be expelled from the lungs.

diastole The relaxation, or period of relaxation, of the heart, especially of the ventricles.

digestion The processing of food that nourishes the individual cells of the body.

distal Structures that are farther from the trunk or nearer to the free end of the extremity.

dorsal The posterior surface of the body, including the back of the hand.

dorsalis pedis artery The artery on the anterior surface of the foot between the first and second metatarsals.

endocrine system The complex message and control system that integrates many body functions, including the release of hormones.

enzymes Protein catalysts designed to speed up the rate of specific biochemical reactions.

epidermis The outer layer of skin, which is made up of cells that are sealed together to form a watertight protective covering for the body.

epiglottis A thin, leaf-shaped valve that allows air to pass into the trachea but prevents food or liquid from entering.

esophagus A collapsible tube that extends from the pharynx to the stomach; contractions of the muscle in the wall of the esophagus propel food and liquids through it to the stomach.

extend To straighten.

extension The straightening of a joint.

fallopian tube Long, slender tube that extends from the uterus to the region of the ovary on the same side, and through which the ovum passes from ovary to uterus.

fascia A sheet or band of tough fibrous connective tissue; lies deep under the skin and forms an outer layer for the muscles.

femoral artery The principal artery of the thigh, a continuation of the external iliac artery. It supplies blood to the lower abdominal wall, external genitalia, and legs. It can be palpated in the groin area.

femoral head The proximal end of the femur, articulating with the acetabulum to form the hip joint.

femur The thighbone; the longest and one of the strongest bones in the body.

flex To bend.

flexion The bending of a joint.

floating ribs The eleventh and twelfth ribs, which do not attach to the sternum through the costal arch.

foramen magnum A large opening at the base of the skull through which the brain connects to the spinal cord.

Fowler's position The position in which the patient is sitting up with the knees bent.

gallbladder A sac on the undersurface of the liver that collects bile from the liver and discharges it into the duodenum through the common bile duct.

genital system The male and female reproductive systems.

greater trochanter A bony prominence on the proximal lateral side of the thigh, just below the hip joint.

hair follicles The small organs in the skin that produce hair.

heart A hollow muscular organ that receives blood from the veins and propels it into the arteries.

heart rate (pulse) The wave of pressure that is created by the heart's contracting and forcing blood out the left ventricle and into the major arteries.

Prep Kit continued...

hinge joints Joints that can bend and straighten but cannot rotate; they restrict motion to one plane.

humerus The supporting bone of the upper arm.

hypoxic drive A "backup system" to control respiration; senses drops in the oxygen level in the blood.

iliac crest The rim, or wing, of the pelvic bone.

ilium One of three bones that fuse to form the pelvic ring.

inferior The part of the body, or any body part, nearer to the feet.

inferior vena cava One of the two largest veins in the body; carries blood from the lower extremities and the pelvic and the abdominal organs into the heart.

inguinal ligament The tough, fibrous ligament that stretches between the lateral edge of the pubic symphysis and the anterior superior iliac spine.

involuntary muscle Muscle over which a person has no conscious control. It is found in many automatic regulating systems of the body.

ischium One of three bones that fuse to form the pelvic ring.

joint (articulation) The place where two bones come into contact.

joint capsule The fibrous sac that encloses a joint.

kidneys Two retroperitoneal organs that excrete the end products of metabolism as urine and regulate the body's salt and water content.

large intestine The portion of the digestive tube that encircles the abdomen around the small bowel, consisting of the cecum, the colon, and the rectum. It helps regulate water and eliminate solid waste.

lateral Parts of the body that lie farther from the midline. Also called outer structures.

ligament A band of the fibrous tissue that connects bones to bones. It supports and strengthens a joint.

liver A large solid organ that lies in the right upper quadrant immediately below the diaphragm; it produces bile, stores sugar for immediate use by the body, and produces many substances that help regulate immune responses.

lumbar spine The lower part of the back, formed by the lowest five nonfused vertebrae; also called the dorsal spine.

lumbar vertebrae Vertebrae of the lumbar spine.

mandible The bone of the lower jaw.

manubrium The upper quarter of the sternum.

mastoid process A prominent bony mass at the base of the skull behind the ear.

maxillae The upper jawbones that assist in the formation of the orbit, the nasal cavity, and the palate, and lodge the upper teeth.

medial Parts of the body that lie closer to the midline; also called inner structures.

metabolism The sum of all the physical and chemical processes of living organisms; the process by which energy is made available for the uses of the organism.

midaxillary line An imaginary vertical line drawn through the middle of the axilla (armpit), parallel to the midline.

midclavicular line An imaginary vertical line drawn through the middle portion of the clavicle and parallel to the midline.

midline An imaginary vertical line drawn from the middle of the forehead through the nose and the umbilicus (navel) to the floor.

motor nerves Nerves that carry information from the central nervous system to the muscles of the body.

mucous membranes The lining of body cavities and passages that communicate directly or indirectly with the environment outside the body.

mucus The opaque, sticky secretion of the mucous membranes that lubricates the body openings.

musculoskeletal system The bones and voluntary muscles of the body.

myocardium The heart muscle.

nasopharynx The part of the pharynx that lies above the level of the roof of the mouth, or soft palate.

nervous system The system that controls virtually all activities of the body, both voluntary and involuntary.

occiput The most posterior portion of the cranium.

orbit The eye socket, made up of the maxilla and zygoma.

oropharynx A tubular structure that extends vertically from the back of the mouth to the esophagus and trachea.

ovary A female gland that produces sex hormones and ova (eggs).

palmar The front region of the hand.

pancreas A flat, solid organ that lies below the liver and the stomach; it is a major source of digestive enzymes and produces the hormone insulin.

parietal regions The areas between the temporal and occiput regions of the cranium.

patella The kneecap; a specialized bone that lies within the tendon of the quadriceps muscle.

perfusion The circulation of oxygenated blood within an organ or tissue in adequate amounts to meet the cells' current needs.

peripheral nervous system The part of the nervous system that consists of 31 pairs of spinal nerves and 12 pairs of cranial nerves. These peripheral nerves may be sensory nerves, motor nerves, or connecting nerves.

peristalsis The wave-like contraction of smooth muscle by which the ureters or other tubular organs propel their contents.

pinna The external, visible part of the ear.

plantar The bottom of the foot.

plasma A sticky, yellow fluid that carries the blood cells and nutrients and transports cellular waste material to the organs of excretion.

platelets Tiny, disk-shaped elements that are much smaller than the cells; they are essential in the initial formation of a blood clot, the mechanism that stops bleeding.

pleura The serous membrane covering the lungs and lining the thoracic cavity, completely enclosing a potential space known as the pleural space.

pleural space The potential space between the parietal pleura and the visceral pleura. It is described as "potential" because under normal conditions, the lungs fill this space.

posterior The back surface of the body; the side away from you in the standard anatomic position.

posterior tibial artery The artery just posterior to the medial malleolus; supplies blood to the foot.

priapism A continuous and painful erection of the penis caused by certain spinal injuries and some diseases.

prone position The position in which the body is lying face down.

prostate gland A small gland that surrounds the male urethra where it emerges from the urinary bladder; it secretes a fluid that is part of the ejaculatory fluid.

proximal Structures that are closer to the trunk.

pubic symphysis A hard bony prominence that is found in the midline in the lowermost portion of the abdomen.

pubis One of three bones that fuse to form the pelvic ring.

pulmonary artery The major artery leading from the right ventricle of the heart to the lungs; it carries oxygen-poor blood.

pulmonary veins The four veins that return oxygenated blood from the lungs to the left atrium of the heart.

pulse The wave of pressure created as the heart contracts and forces blood out the left ventricle and into the major arteries.

quadrants The way to describe the sections of the abdominal cavity. Imagine two lines intersecting at the umbilicus dividing the abdomen into four equal areas.

radial artery The major artery in the forearm; it is palpable at the wrist on the thumb side.

radius The bone on the thumb side of the forearm.

rectum The lowermost end of the colon.

red blood cells Cells that carry oxygen to the body's tissues; also called erythrocytes.

renal pelvis A cone-shaped collecting area that connects the ureter and the kidney.

respiratory system All the structures of the body that contribute to the process of breathing, consisting of the upper and lower airways and their component parts.

retroperitoneal Behind the abdominal cavity.

sacrum One of three bones (sacrum and two pelvic bones) that make up the pelvic ring; consists of five fused sacral vertebrae.

salivary glands The glands that produce saliva to keep the mouth and pharynx moist.

scalp The thick skin covering the cranium, which usually bears hair.

scapula The shoulder blade.

sebaceous glands Glands that produce an oily substance called sebum, which discharges along the shafts of the hairs.

semen Seminal fluid ejaculated from the penis and containing sperm.

seminal vesicles Storage sacs for sperm and seminal fluid, which empty into the urethra at the prostate.

sensory nerves The nerves that carry sensations of touch, taste, heat, cold, pain, or other modalities from the body to the central nervous system.

shock position The position that has the head and torso (trunk) supine and the lower extremities elevated 6" to 12". This helps to increase blood flow to the brain; also referred to as the modified Trendelenburg's position.

shoulder girdle The proximal portion of the upper extremity, made up of the clavicle, the scapula, and the humerus.

skeletal muscle Muscle that is attached to bones and usually crosses at least one joint; striated, or voluntary, muscle.

skeleton The framework that gives us our recognizable form; also designed to allow motion of the body and protection of vital organs.

small intestine The portion of the digestive tube between the stomach and the cecum, consisting of the duodenum, jejunum, and ileum.

smooth muscle Nonstriated, involuntary muscle; it constitutes the bulk of the gastrointestinal tract and is present in nearly every organ to regulate automatic activity.

somatic nervous system The part of the nervous system that regulates activities over which there is voluntary control.

spinal cord An extension of the brain, composed of virtually all the nerves carrying messages between the brain and the rest of the body. It lies inside of, and is protected by, the spinal canal.

sternocleidomastoid muscles The muscles on either side of the neck that allow movement of the head.

sternum The breastbone.

striated muscle Muscle that has characteristic stripes, or striations, under the microscope; voluntary, or skeletal, muscle.

subcutaneous tissue Tissue, largely fat, that lies directly under the dermis and serves as an insulator of the body.

superficial Closer to or on the skin.

www.EMTB.com

Prep Kit continued...

superior The part of the body, or any body part, nearer to the head.

superior vena cava One of the two largest veins in the body; carries blood from the upper extremities, head, neck, and chest into the heart.

supine position The position in which the body is lying face up.

sweat glands The glands that secrete sweat, located in the dermal layer of the skin.

systole The contraction, or period of contraction, of the heart, especially that of the ventricles.

temporal regions The lateral portions on each side of the cranium.

temporomandibular joint The joint where the mandible meets with the temporal bone of the cranium just in front of each ear.

testicle A male genital gland that contains specialized cells that produce hormones and sperm.

thoracic cage The chest or rib cage.

thoracic spine The 12 vertebrae that lie between the cervical vertebrae and the lumbar vertebrae. One pair of ribs is attached to each of the thoracic vertebrae.

thorax The chest cavity that contains the heart, lungs, esophagus, and great vessels (the aorta and the two venae cavae).

thyroid cartilage A firm prominence of cartilage that forms the upper part of the larynx; the Adam's apple.

tibia The shinbone, the larger of the two bones of the lower leg.

topographic anatomy The superficial landmarks of the body that serve as guides to the structures that lie beneath them.

torso The trunk without the head and limbs.

trachea The windpipe; the main trunk for air passing to and from the lungs.

Trendelenburg's position The position in which the body is supine with the head lower than the feet.

triceps The muscle in the back of the upper arm.

ulna The inner bone of the forearm, on the side opposite the thumb.

ulnar artery One of the major arteries of the forearm; it can be palpated at the wrist on the ulnar side (at the base of the fifth finger).

ureter A small, hollow tube that carries urine from the kidneys to the bladder.

urethra The canal that conveys urine from the bladder to outside the body.

urinary bladder A sac behind the pubic symphysis made of smooth muscle that collects and stores urine.

urinary system The organs that control the discharge of certain waste materials filtered from the blood and excreted as urine.

vagina A muscular distensible tube that connects the uterus with the vulva (the external female genitalia); also called the birth canal.

vasa deferentia The spermatic duct of the testicles; also called vas deferens.

ventral The anterior surface of the body.

ventricle Lower chamber of the heart.

vertebrae The 33 bones that make up the spinal column.

voluntary muscle Muscle that is under direct voluntary control of the brain and can be contracted or relaxed at will; skeletal, or striated, muscle.

white blood cells Blood cells that play a role in the body's immune defense mechanisms against infection; also called leukocytes.

xiphoid process The narrow, cartilaginous lower tip of the sternum.

zygomas The quadrangular bones of the cheek, articulating with the frontal bone, the maxillae, the zygomatic processes of the temporal bone, and the great wings of the sphenoid bone.

Points to Ponder

You are responding to a motor vehicle collision involving two persons. You find a 30-year-old woman with her 5-year-old son. They both have airway problems that must be managed effectively. Upon your arrival they are already extricated from the vehicle and you are in charge of the child's airway.

What are the major anatomic differences between the adult and pediatric airway? What is the relationship of the size of the child's head in comparison to the rest of the body? What are good ways to help position an airway in a pediatric patient?

You are now giving respirations to the child. You notice that his stomach is moving a lot more than his chest is.

What should you do? What are the normal breathing rates for a child this age and how fast should you breathe for him? At what age do children learn to breathe through their nose?

Issues: Anatomic Differences in Children vs Adults.

Assessment in Action

You are dispatched to a patient with multiple stab wounds.
Law enforcement has secured the scene. Upon arrival you find a man lying on the ground.

1. The patient is lying on his stomach. What is that position called?

   A. Supine
   B. Prone
   C. Fowler's
   D. Shock

2. The patient has a stab wound on the anterior, right lateral side just distal to the rib cage. Where is this injury located?

   A. Chest
   B. Abdominal LLQ
   C. Abdominal RUQ
   D. Right side of neck

3. The patient has severe bleeding from the wound. What organ is most likely injured?

   A. Liver
   B. Spleen
   C. Pancreas
   D. Small intestine

4. The patient has another stab wound to the anterior side, just medial to the left hip, proximal to the thigh. Where is this injury located?

   A. Knee
   B. RLQ of abdomen
   C. Buttocks
   D. Groin

5. What major artery is located near this injury?

   A. Brachial
   B. Femoral
   C. Dorsalis pedis
   D. Carotid

6. You are instructed to perform chest compressions. You need to find the bony protrusion on the distal end of the sternum. What is that landmark called?

   A. Xiphoid process
   B. Costal arch
   C. Angle of Louis
   D. Jugular notch

7. All arteries carry oxygenated blood and all veins carry deoxygenated blood. What artery and vein combination is the exception to this rule?

   A. Aorta and vena cava
   B. Renal artery and vein
   C. Carotid artery and jugular vein
   D. Pulmonary artery and vein

8. What major organ helps facilitate metabolism and regulate temperature?

   A. Kidneys
   B. Skin
   C. Liver
   D. Pancreas

Challenging Questions

9. Describe the electrical conduction pathway of the heart and how it relates to a person's blood pressure.

10. The brain controls the central nervous system. What kinds of problems would people have if they had an injury, such as a brain attack or CVA, in the cerebrum, cerebellum, and brain stem?

www.EMTB.com

Baseline Vital Signs and SAMPLE History

Objectives

Cognitive

1-5.1 Identify the components of vital signs. (p 149)
1-5.2 Describe the methods to obtain a breathing rate. (p 150)
1-5.3 Identify the attributes that should be obtained when assessing breathing. (p 149)
1-5.4 Differentiate between shallow, labored, and noisy breathing. (p 150)
1-5.5 Describe the methods to obtain a pulse rate. (p 153)
1-5.6 Identify the information obtained when assessing a patient's pulse. (p 153)
1-5.7 Differentiate between a strong, weak, regular, and irregular pulse. (p 154)
1-5.8 Describe the methods to assess skin color, temperature, and condition (capillary refill in infants and children). (p 154)
1-5.9 Identify the normal and abnormal skin colors. (p 154)
1-5.10 Differentiate between pale, blue, red, and yellow skin color. (p 154)
1-5.11 Identify the normal and abnormal skin temperature. (p 155)
1-5.12 Differentiate between hot, cool, and cold skin temperature. (p 155)
1-5.13 Identify normal and abnormal skin conditions. (p 155)
1-5.14 Identify normal and abnormal capillary refill in infants and children. (p 156)
1-5.15 Describe the methods to assess the pupils. (p 161)
1-5.16 Identify normal and abnormal pupil size. (p 161)
1-5.17 Differentiate between dilated (big) and constricted (small) pupil size. (p 161)
1-5.18 Differentiate between reactive and nonreactive pupils and equal and unequal pupils. (p 161)
1-5.19 Describe the methods to assess blood pressure. (p 157)
1-5.20 Define systolic pressure. (p 157)
1-5.21 Define diastolic pressure. (p 157)
1-5.22 Explain the difference between auscultation and palpation for obtaining a blood pressure. (p 157, 158)
1-5.23 Identify the components of the SAMPLE history. (p 148)
1-5.24 Differentiate between a sign and a symptom. (p 147)
1-5.25 State the importance of accurately reporting and recording the baseline vital signs. (p 149)
1-5.26 Discuss the need to search for additional medical identification. (p 149)

Affective

1-5.27 Explain the value of performing the baseline vital signs. (p 149)
1-5.28 Recognize and respond to the feelings patients experience during assessment. (p 150)
1-5.29 Defend the need for obtaining and recording an accurate set of vital signs. (p 149)
1-5.30 Explain the rationale of recording additional sets of vital signs. (p 149, 162)
1-5.31 Explain the importance of obtaining a SAMPLE history. (p 148)

Psychomotor

1-5.32 Demonstrate the skills involved in assessment of breathing. (p 149)
1-5.33 Demonstrate the skills associated with obtaining a pulse. (p 153)
1-5.34 Demonstrate the skills associated with assessing the skin color, temperature, condition, and capillary refill in infants and children. (p 154)
1-5.35 Demonstrate the skills associated with assessing the pupils. (p 160)
1-5.36 Demonstrate the skills associated with obtaining blood pressure. (p 157)
1-5.37 Demonstrate the skills that should be used to obtain information from the patient, family, or bystanders at the scene. (p 149)

Additional Objectives*

Cognitive

None

Affective

1. Explain the rationale for applying pulse oximetry. (p 152)

Psychomotor

None

*These are noncurriculum objectives.

You are the Provider

You are dispatched to a lounge at the airport for a patient with chest pain. Upon arrival you find a 34-year-old woman sitting on a bench complaining of chest pain. She is pale and diaphoretic. Her pulse is 110 beats/min and her blood pressure is 148/84 mm Hg. She is breathing 24 times a minute, with shallow respirations and increased work of breathing. She tells you she is having sharp pain on the left side of her chest and it hurts more every time she breathes in. She denies having any history of heart disease, and the only medication she reports taking is birth control pills.

This chapter will teach you how to obtain and evaluate a patient's baseline vital signs and the SAMPLE history, which are part of the patient assessment process. It will also help you answer the following questions:

1. Why are serial vital signs necessary when caring for seriously ill or injured patients?
2. How can the SAMPLE history you obtain be of value to the emergency department staff at the receiving hospital?

Baseline Vital Signs and SAMPLE History

As an EMT-B, you must perform a quick but thorough assessment to identify a patient's needs and to provide proper emergency medical care. Patient assessment includes many steps and is the most complex skill that you will learn in the EMT-B course. To make the task easier, it is helpful to identify and discuss the key components and skills of patient assessment before you learn the entire process.

As you begin your assessment, you must gather and record some key information about the patient. You will also need to obtain and evaluate the patient's vital signs. The injuries, illnesses, or symptoms that lead to the call to 9-1-1 and the history of what occurred before and since the call was made are key pieces of information that you will have to obtain by asking a series of questions. You must also learn about the patient's medical history and overall health. Bring the necessary equipment for assessment and care to the patient.

This chapter begins by defining the chief complaint and signs and symptoms. It then explains specific information about the patient that you need to obtain at the start of the assessment and why you need it. It also describes each of the vital signs and provides a step-by-step explanation of how to obtain each. Both normal and abnormal vital signs are discussed. The chapter includes a description of the SAMPLE history. Keep in mind that although the DOT National Standard Curriculum separates baseline vital signs and SAMPLE history from patient assessment, they are all a seamless part of the general patient assessment process. They are presented early in your training to allow you plenty of time to practice these skills before learning the complete patient assessment.

Gathering Key Patient Information

During the assessment, you will be using a variety of your senses and a few basic medical instruments to obtain information about your patient. You will need to know which questions to ask and how to ask them (Figure 5-1 ▼). By using your deductive powers, you will be able to interpret the meaning and implications of the information that you have gathered. When assessing the patient, you will have to look, listen, feel, and think.

You will need to obtain the patient's name so that you can properly address the patient. The lead EMT should always introduce himself or herself and his or her partner(s) to the patient. Unless an adult patient is a close friend or relative of yours, you should address him or her as "Mr.," "Ms.," "Miss," or "Mrs," followed by the patient's last name. You may ask the patient how he or she wants to be addressed. Often, relatives or staff members of a nursing home or other extended care facility address geriatric patients by their first names. You should not use such a familiar mode

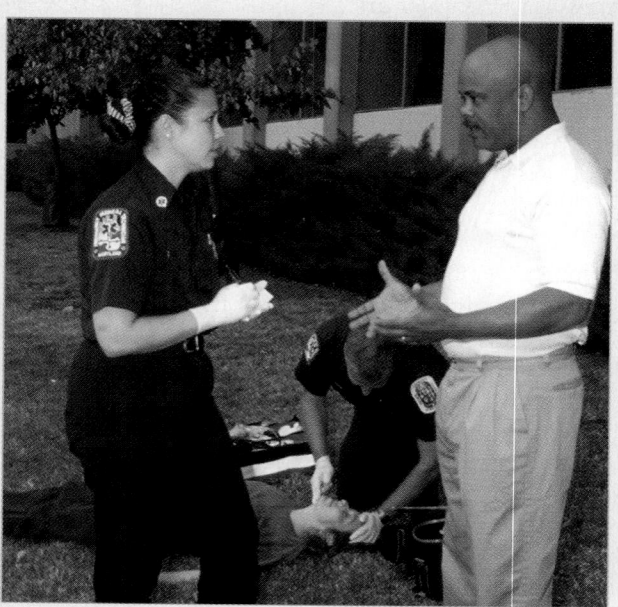

Figure 5-1 You must know how to gather information about the patient by using your senses and by asking relevant questions.

Technology

- Interactivities
- Vocabulary Explorer
- Anatomy Review
- Web Links
- Online Review Manual

www.EMTB.com

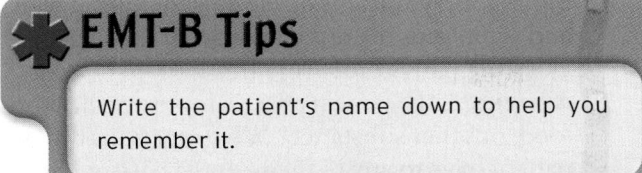

EMT-B Tips

Write the patient's name down to help you remember it.

of address. If the patient's name is difficult to pronounce, you can simply say "Sir" or "Ma'am" instead, to convey a similar respectful and professional manner.

You should try to address children by their first name, especially the name they are customarily called, such as "Johnny," "Betty," or "Joey." Even infants and toddlers who do not yet respond verbally can recognize their name and may be less anxious when it is used.

If an unaccompanied patient is unable to give you his or her name, you should look in his or her wallet or purse for a driver's license or other piece of identification that will tell you the patient's name. At the same time, you should check for any hospital identification or medical alert card. Always look for patient identification in the presence of another EMT or law enforcement officer at the scene. You should also be aware of emerging technologies that enable medical practitioners to access patient information digitally, such as Global MED-NET and OnStar. These technologies will enable EMS providers to obtain histories and medications from a central database while assessing their patient.

Age and gender are also important considerations in assessing a patient. Some conditions and illnesses are found predominantly in younger patients; others are commonly found only in older patients. Some conditions are prevalent in a certain age group in adult men but in a different age group in women. Some are more prevalent in one gender, and some are limited exclusively to either male or female patients. In addition, the normal range of some of the vital signs will be different for different age groups of children, adults, and older patients.

Chief Complaint

The reason for a call to 9-1-1 is vital information. This reason is called the patient's chief complaint. In the most literal definition, chief complaints are the major signs and symptoms that the patient reports when asked, "What seems to be the matter?" or "What's wrong?" A patient who responds "My chest hurts" is stating the chief complaint. What you see must also be considered in deter-

mining the chief complaint. If the patient's response demonstrates that he or she is having significant difficulty breathing, "difficulty breathing" should be included in the chief complaint as if the patient had reported it verbally. In some protocols, a chief complaint also includes any significant gross, apparent injuries.

The problems or feelings the patients report to you, such as, "I feel dizzy," "My leg hurts," or "Ow, that hurts a lot!" are called symptoms (Figure 5-2A ▼). These cannot be felt or observed by others. The severity of a symptom is subjective because it is based on the patient's interpretation and tolerance. Signs are objective conditions that can be seen, heard, felt, smelled, or measured by you or others (Figure 5-2B ▼). Wounds, external bleeding, marked deformities, respirations, and pulse are all signs.

Signs and symptoms that occurred before you arrived, such as dizziness that resulted in a loss of

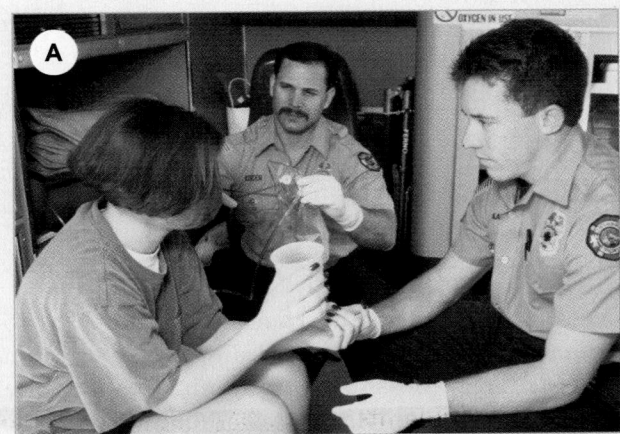

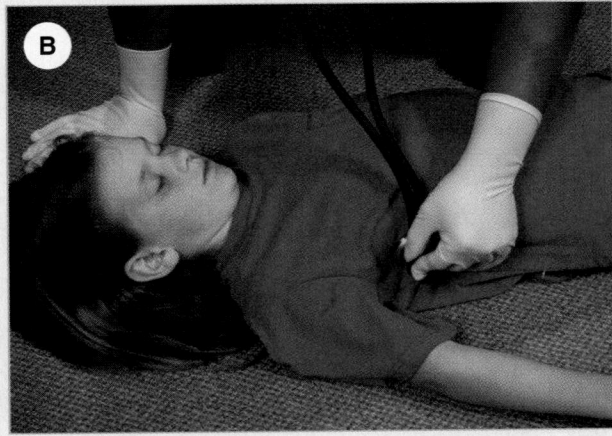

Figure 5-2 A. A symptom is a subjective condition that the patient feels and tells you about. **B.** A sign is an objective condition that you can observe about the patient.

EMT-B Tips

Obtaining the respirations or pulse rate

When obtaining a patient's respirations or pulse rate, count the number of breaths or beats in a 30-second period and then multiply by 2. This method produces a significantly more reliable figure than you would get if you counted for only 15 seconds and multiplied by 4. With either method, the result will always be an even number.

consciousness, may be reported by the patient or others at the scene. Because signs and symptoms are essential to understanding the sequence of events and may include signs that are no longer present, they are important parts of the patient history. You should always report how and/or when the signs and symptoms began. This information is important because the reason that signs and symptoms develop often differs, depending on the situation.

Obtaining a SAMPLE History

Once you have stabilized all immediate life threats, provided emergency care, and are ready to further examine the patient, you should try to obtain a key brief history, or SAMPLE history. As part of the assessment of every patient, you should ask the following questions, using the word SAMPLE as a guideline:

- **S**igns and **S**ymptoms of the episode: What signs and symptoms occurred at onset of the incident? Does the patient report pain?
- **A**llergies: Is the patient allergic to any medication, food, or other substance? What reactions did the patient have to any of them? If the patient has no known allergies, you should note this on the run report as "no known allergies" or "NKA."
- **M**edications: What medications was the patient prescribed? What dosage was prescribed? How often is the patient supposed to take the medication? What prescription, over-the-counter medications, and herbal medications has the patient taken in the last 12 hours? How much was taken and when?
- **P**ertinent past history: Does the patient have any history of medical, surgical, or trauma occurrences? Has the patient had a recent illness or injury, fall, or blow to the head?
- **L**ast oral intake: When did the patient last eat or drink? What did the patient eat or drink, and how much was consumed? Did the patient take any drugs or drink alcohol? Has there been any other oral intake in the last 4 hours?
- **E**vents leading to the injury or illness: What are the key events that led up to this incident? What occurred between the onset of the incident and your arrival? What was the patient doing when this illness started? What was the patient doing when this injury happened?

OPQRST

Another mnemonic device that can be very helpful in remembering questions you should ask in obtaining a patient history is OPQRST. This can be especially helpful when assessing for possible heart attack.

You are the Provider   Part 2

As you are preparing to transport your patient to the hospital, she states that it is becoming increasingly difficult to breathe. Her respirations increase to 34 breaths/min and are shallow and gasping. Even with high-flow oxygen via nonrebreathing mask, her oxygen saturation is only 90% and her capillary refill time is 4 seconds. She is extremely anxious and is having a hard time sitting still.

3. Is there anything in your repeat assessment of vital signs that can help you determine the cause of her distress?
4. What are your treatment priorities?

Any Medical Pain!!

O = Onset, that is, when did the problem begin and what caused it?

P = Provocation or **P**alliation, that is, does anything make it feel better? Worse?

Q = Quality, that is, what is the pain like? Sharp, dull, crushing, tearing?

R = Region/**R**adiation, that is, where does it hurt? Does the pain move anywhere?

S = Severity, that is, on a scale of 1 to 10, how would you rate your pain?

T = Timing of pain, that is, has the pain been constant, or does it come and go? How long have you had the pain (often answered under "O," onset).

With practice, you will be able to obtain, document, and report a meaningful brief history. Be sure to ask the patient and bystanders for information. If the patient is unconscious, look for a medical identification tag or for a medical information card in the patient's wallet or purse. Always look for patient identification in the presence of another EMT or law enforcement officer at the scene.

Baseline Vital Signs

The initial assessment is a rapid evaluation of the patient's general condition to identify any potentially life-threatening conditions. The brain and other vital organs require constant oxygen. Significant problems with breathing or circulation must be considered potentially life-threatening conditions. A critical problem or deficit in any of the body's other vital systems or functions will progressively affect and be reflected by changes in the respiratory, circulatory, and central nervous systems. Therefore, the status of these systems serves as your guideline for evaluating and measuring the patient's general condition.

Vital signs are the key signs that are used to evaluate the patient's condition. The first set of vital signs that you obtain is called the baseline vital signs. By periodically reassessing the vital signs and comparing the findings with the baseline set, you will be able to identify any trends in the patient's condition, particularly whether the patient's condition is becoming worse (Figure 5-3 ▶).

Because key indicators include a quantitative (numeric) objective measurement, you will always include the patient's respirations, pulse, and blood pressure when taking and evaluating the vital signs. Other key indications of the patient's respiratory, cardiovascular, and central nervous system status in-

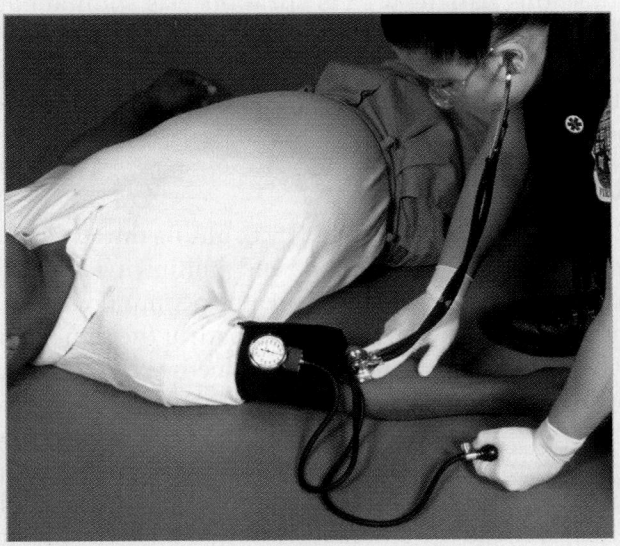

Figure 5-3 Baseline vital signs are key signs that are used to establish the patient's initial condition.

cluded in the baseline vital signs are an evaluation of the following:

- Skin temperature and condition in adults
- Capillary refill in children
- Pupillary reaction
- Level of consciousness

Respirations

A patient who is breathing without assistance is said to have spontaneous respirations or spontaneous breathing. Each complete breath includes two distinct phases: inspiration and expiration. During inspiration (inhalation), the chest rises up and out, drawing oxygenated air into the lungs. During expiration (exhalation), the chest returns to its original position, releasing air with an increased carbon dioxide level out of the lungs. Inhalation and exhalation times occur in a 1:3 ratio; the active inhalation phase lasts one third the amount of time of the passive exhalation phase.

Breathing is a continuous process in which each breath regularly follows the last with no notable interruption. Breathing is normally a spontaneous, automatic process that occurs without conscious thought, visible effort, marked sounds, or pain. You will assess breathing by watching the patient's chest rise and fall, feeling for air through the mouth and nose during exhalation, and listening to breath sounds with a stethoscope over each lung. Chest rise and breath sounds should be equal on both sides of the chest. A conscious patient who is speaking has spontaneous respirations.

When assessing respirations, you must determine the rate, quality (character), and depth of the patient's breathing.

Rate

Respirations are determined by counting the number of breaths in a 30-second period and multiplying by 2. The result equals the number of breaths per minute. For accuracy, you should count each breath at the same point in its cycle. This is most easily done by counting each peak chest rise. Although you can see peak chest rise, it is easier to place your hand on the patient's chest and feel it. However, be aware that a conscious patient who knows that you are evaluating his or her breathing will often override the automatic rate and depth by breathing more slowly and deeply. To prevent this from happening, you should check respirations in a conscious, alert patient without making the patient aware of what you are evaluating. This can be easily done by first taking a radial pulse and then, without releasing the wrist or otherwise suggesting a change, counting the chest rise that you see or feel as the patient's forearm rises and falls with the movement of the chest (Figure 5-4 ▼). If the patient coughs, yawns, sighs, or talks during the 30-second period, you should wait a few seconds and start again. (Table 5-1 ▶) shows the normal range of respiratory rates of patients who are at rest.

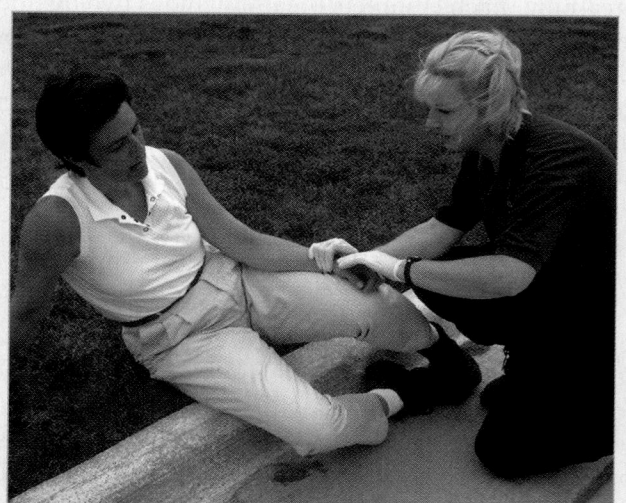

Figure 5-4 Assess respirations in a conscious patient by first taking a radial pulse and then, without releasing the patient's wrist, counting the chest rise and fall for 30 seconds.

Study This [handwritten note]

TABLE 5-1 Normal Ranges for Respirations

Age	Range, breaths/min
Adults	12 to 20
Children	15 to 30
Infants	25 to 50

Note: These ranges are per the US DOT 1994 EMT-Basic National Standard Curriculum. Ranges presented in other courses may vary.

Quality

You can determine the quality or character of respirations as you are counting. (Table 5-2 ▼) shows four ways in which the quality or character can be described.

Rhythm

While counting the patient's respirations, you should also note the rhythm. If the time from one peak chest rise to the next is fairly consistent, respirations are considered regular. If respirations vary or change frequently, they are considered irregular. When you document the vital signs, be sure to note whether the patient's respirations were regular or irregular.

Effort

Normally, breathing is an effortless process that does not affect a patient's speech, posture, or positioning. Speech is a good indicator of whether a conscious patient is having difficulty breathing. A patient who can speak smoothly without unusual extra pauses is

TABLE 5-2 Characteristics of Respirations

Normal	Breathing is neither shallow nor deep
	Equal chest rise and fall
	No use of accessory muscles
Shallow	Decreased chest or abdominal wall motion
Labored	Increased breathing effort
	Use of accessory muscles
	Possible gasping
	Nasal flaring, supraclavicular and intercostal retractions in infants and children
Noisy	Increase in sound of breathing, including snoring, wheezing, gurgling, crowing, grunting, and stridor

breathing normally. However, a patient who can speak only one word at a time or must stop every two to three words to catch his or her breath is having significant difficulty breathing. Patients who are having marked difficulty breathing will instinctively assume a posture in which it is easier for them to breathe. There are two common postures that indicate that the patient is trying to increase airflow. The first is called the <u>tripod position</u>. In this position, a patient sits leaning forward on outstretched arms with the head and chin thrust slightly forward and is having sufficient difficulty breathing that a significant conscious effort is required. The second is most commonly seen in children—the <u>sniffing position</u>. The patient sits upright with the head and chin thrust slightly forward, and the patient appears to be sniffing (Figure 5-5 ▶).

Breathing that becomes progressively more difficult requires progressively more effort. When you can see that effort, the patient's breathing is described as <u>labored breathing</u>.

Initially, labored breathing is characterized by the patient's position, concentration on breathing, and the increased effort and depth of each breath. As breathing becomes more labored, accessory muscles in the chest and neck are used, and the patient may make grunting sounds with each breath. In infants and small children, nasal flaring and supraclavicular and intercostal retractions (indentation above the clavicles and in the spaces between the ribs) are commonly associated with labored breathing. Sometimes the patient may be gasping.

Infants and small children will continue to have labored breathing for a sustained period, will then often become exhausted, and finally will no longer have the

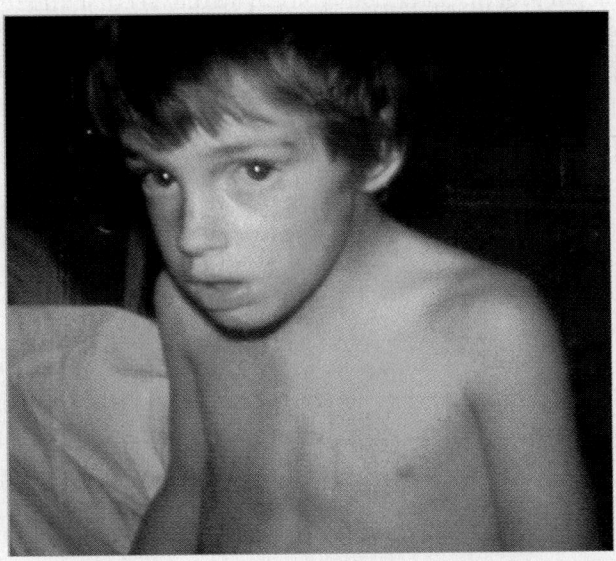

Figure 5-5 A patient in the sniffing position sits upright with the head and chin thrust slightly forward.

strength to maintain the necessary energy to breathe. In infants and small children, cardiac arrest is generally caused by respiratory arrest.

Noisy breathing

Normal breathing is silent or, in a very quiet environment, accompanied only by the sounds of air movement at the mouth and nose. Through a stethoscope, normal breath sounds include only the sound of air movement through the bronchi accompanied by a soft, low-pitched murmur. Breathing accompanied by other sounds indicates a significant respiratory problem. When the airway is partially obstructed by a foreign body or swelling, you may hear <u>stridor</u>, a harsh, high-pitched, crowing sound. If you can hear bubbling or gurgling, the patient probably has fluid in the airway. You may hear other sounds, like wheezes or snoring. The presence of any of these indicates that a serious respiratory problem exists. With a complete airway obstruction, the patient will not be able to move any air and will no longer be able to cough or talk. Sounds are caused by air moving through small spaces or fluid. If you hear nothing, the patient may be moving no air at all.

A patient who coughs up thick, yellowish or greenish sputum (matter from the lungs) most likely has an advanced respiratory infection. A patient with a chest injury may cough up blood or frothy whitish or pinkish foamlike sputum. A patient with congestive heart failure may also cough up frothy sputum. The presence of either substance, regardless of its cause, indicates

▲ Pediatric Needs

Chest rise in a small child is less marked than that in an adult. However, a small child's abdomen moves more with each breath than an adult's does. Place your hands on the outer margin of the lower anterior chest to feel the chest wall and abdominal movement, and determine whether the depth is normal, shallow, or deep. In a patient of any age, if it is difficult to gauge the depth of breathing from the chest movement, note instead the amount of air that you feel is exhaled with each breath.

that an urgent, potentially critical cardiovascular and respiratory problem exists. The patient's condition may deteriorate rapidly to a point at which the patient can no longer breathe.

Depth

The amount of air that the patient is exchanging depends on both the rate and the <u>tidal volume</u>, the amount of air that is exchanged with each breath. The depth of the breath determines whether the tidal volume is normal, less than normal, or more than normal. Respirations are described as shallow when the movement of the chest wall and air that you feel exhaled with each breath is less than normal. Deep respirations occur when chest movement and exhaled air are significantly greater than normal. You should document when the patient's respirations are shallow or deep; however, you do not have to record a normal depth of breathing.

Pulse Oximetry

<u>Pulse oximetry</u> is a recent assessment tool used to evaluate the effectiveness of oxygenation. The pulse oximeter is a photoelectric device that monitors the oxygen saturation of hemoglobin (the iron-containing portion of the red blood cell to which oxygen attaches) in the capillary beds (Figure 5-6 ▼). Parts that make up the pulse oximeter include a monitor and sensing probe. The sensing probe clips onto a finger or ear lobe. The light source must have unobstructed access to a capillary bed, so fingernail polish should be removed. Results appear as a percentage on the display screen. Normally, pulse oximetry values in ambient air will vary depending on the altitude, with the majority being between 95% and 100%.

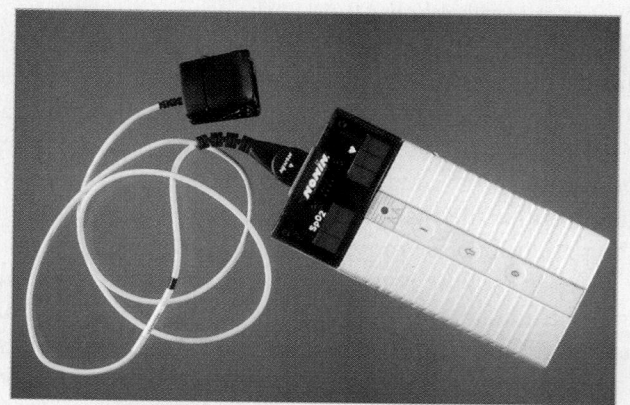

Figure 5-6 The pulse oximeter is a device that measures the saturation of oxygen in the blood as a percentage.

The goal of any oxygen therapy is to increase oxygen saturation to normal levels. This device is a useful assessment tool to determine the effectiveness of oxygen therapy, bronchodilator therapy, and use of the bag-valve-mask (or BVM) device in certain conditions. However, the pulse oximeter does not take the place of good assessment skills and should not prevent the application of oxygen to any patient who complains of difficulty breathing regardless of the pulse oximetry value.

Because the device presumes adequate perfusion and numbers of red blood cells, anything that causes <u>vasoconstriction</u> (narrowing of a blood vessel, such as with hypoperfusion or cold extremities) or loss of red blood cells (such as bleeding or anemia) will result in inaccurate or misleading values. The device also presumes that oxygen is saturating hemoglobin. Therefore, any chemical that displaces oxygen (such as carbon monoxide) will cause misleading values.

The pulse oximeter is a useful tool as long as the EMT-B remembers that the device is only a tool, not a substitute for a good assessment. The device should not be used when hypoperfusion or known anemia is present, carbon monoxide or exposure to other toxic inhalants has occurred, or the patient's extremities are cold.

It is essential to remember that rarely does one sign or symptom reveal to you the patient's status or underlying problem. Rather, it is the combination of many signs and symptoms that reveal the underlying problem or condition of your patient. Therefore, it is essential to have a basic understanding of the causes and presentations of medical emergencies so that you know what to look for.

For example, a patient with chest pain may be having a heart attack. He may also have sustained chest trauma, a lung infection, a pulmonary embolism, or a simple strained muscle in the chest. If he describes his pain as crushing, radiating down his left arm and up into his jaw, he is pale and soaked in sweat, the episode began while he was shoveling snow, and he has a history of coronary bypass surgery and has nitroglycerin in his pocket, your assessment will lean toward myocardial infarction. Therefore, it is essential to collect all pertinent information and be able to interpret how it fits together.

Pulse

With each heartbeat, the ventricles contract, forcefully ejecting blood from the heart and propelling it into the arteries. The <u>pulse</u> is the pressure wave that occurs as each heartbeat causes a surge in the blood circulating through the arteries. The pulse is most easily felt at a pulse point where a major artery lies near the surface and can be pressed gently against a bone or solid organ. To palpate (feel) the pulse, hold together your index and long fingers and place their tips over a pulse point, pressing gently against the artery until you feel intermittent pulsations. Sometimes, you may have to slide your fingertips a little to each side and press again until you feel a pulse. When palpating a pulse, do not allow your thumb to touch the patient. If you do so, you may mistake the strong pulsing circulation in your thumb for the patient's pulse.

In responsive patients who are older than 1 year, you should palpate the radial pulse at the wrist Figure 5-7A ▼ . In unresponsive patients older than 1 year, you should palpate the carotid pulse in the neck Figure 5-7B ▼ . When palpating the carotid pulse, you should place the fingertips of your index and long fingers along the carotid artery in the groove between the trachea and the neck muscle. Use caution when palpating the carotid pulse in a responsive patient, especially an older patient. Only gentle pressure on one side of the neck should be used. Never press on the carotid arteries on both sides of the neck at the same time. Doing so can reduce circulation to the brain.

In infants, both the radial and carotid pulses are difficult to locate. Because of the infant's soft, immature trachea, palpating the carotid pulse is not recommended. Palpate the brachial pulse, located at the medial area (underside) of the upper arm, in children younger than 1 year Figure 5-8 ▼ . With the infant lying supine, you can access the brachial pulse by elevating the arm over the infant's head. Because most infants have chubby arms, you need to press your adjacent fingertips firmly along the brachial artery, which lies parallel to the long axis of the upper arm, to be able to palpate the pulse.

Your first consideration when taking a pulse is to determine whether the patient has a palpable pulse or is pulseless. When taking the pulse, you should assess and report its rate, strength, and regularity.

Rate

To obtain the pulse rate in most patients, you should count the number of pulses felt in a 30-second period and then multiply by 2. A pulse that is weak and difficult to palpate, irregular, or extremely slow should be palpated and counted for a full minute.

A pulse rate is counted as beats per minute; however, in reporting the pulse rate, it is not necessary to state or write "beats per minute" after the number.

The pulse rate in most adults (at rest) averages around 72 beats/min. However, pulse rate can vary

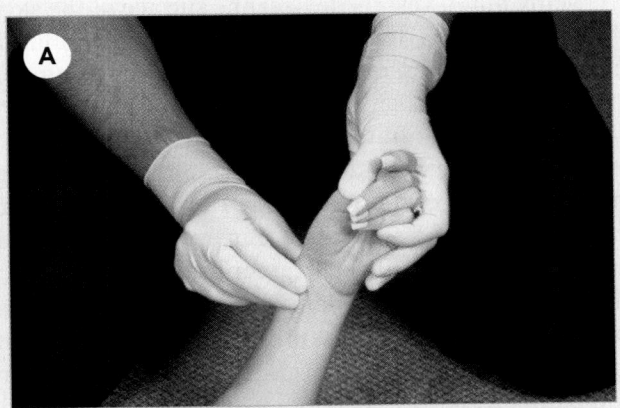

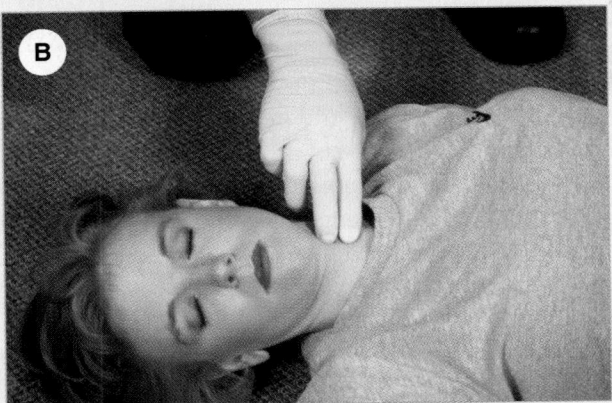

Figure 5-7 A. To palpate the radial pulse, place the tips of your first two fingers over the radial artery, pressing gently until you feel intermittent pulsations. **B.** To palpate the carotid pulse, place the tips of your first two fingers over the carotid artery, pressing gently until you feel intermittent pulsations.

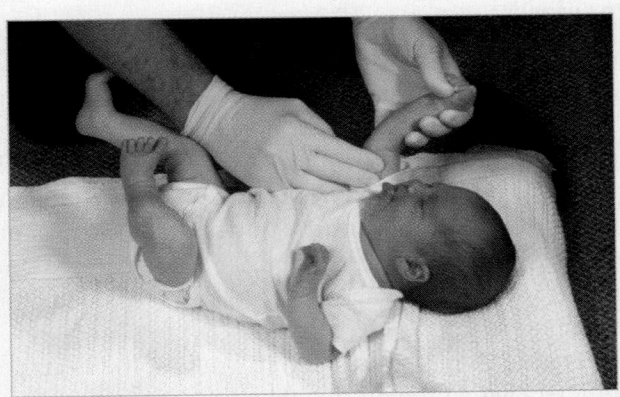

Figure 5-8 To palpate the brachial pulse in an infant, press firmly along the brachial artery at the underside of the upper arm.

Study This !!!

TABLE 5-3 Normal Ranges for Pulse Rate

Age	Range, beats/min
Adults	60 to 100
Children	70 to 150
Infants	100 to 160

significantly from person to person. In the well-conditioned athlete or in individuals taking heart medications such as beta-blockers, the pulse rate may be considerably lower. A pulse rate between 60 and 100 beats/min is considered normal in adults. The average pulse rate in children is generally higher. (Table 5-3 ▲) shows the normal ranges of pulse rates.

In assessing the pulse rate in an adult patient, a rate that is greater than 100 beats/min is described as tachycardia, and a rate of less than 60 beats/min is described as bradycardia.

Strength

You should always report the pulse's strength whenever reporting or recording the pulse. The pulse is generally palpated at the radial or carotid arteries in adults and at the brachial artery in infants, because it is normally strong and easily palpable at these locations. Therefore, if the pulse feels of normal strength, you should describe it as being strong. You should describe a stronger than normal pulse as "bounding" and a pulse that is weak and difficult to feel as "weak" or "thready." With a little experience, you will be able to make the necessary distinctions easily.

Regularity

When assessing the quality of the pulse, you must also determine whether it is regular or irregular. When the interval between each ventricular contraction of the heart is short, the pulse is rapid. When the interval is longer, the pulse is slower. No matter what the rate, the interval between each contraction should be the same, and the pulse that results should occur at a constant, regular rhythm. You should note and document this rhythm as regular.

The rhythm is considered irregular if the heart periodically has a premature or late beat or if a pulse beat is missed. Some individuals have a chronically irregular pulse; however, if an irregular pulse is found in a

patient with signs and symptoms that suggest a cardiovascular problem, the patient likely needs advanced cardiac assessment and life support. Therefore, depending on your protocols, you should call for advanced life support (or ALS) backup, arrange for an intercept by paramedics, or initiate prompt transport to definitive care.

The Skin

The condition of the patient's skin can tell you a lot about the patient's peripheral circulation and perfusion, blood oxygen levels, and body temperature. When assessing the skin condition, you should evaluate its color, temperature, and moisture.

Color

Assessing the skin helps you to determine the adequacy of perfusion. Perfusion is the circulation of blood within an organ or tissue. Adequate perfusion meets the current needs of the cells; inadequate perfusion will cause cells and tissues to die.

Many blood vessels lie near the surface of the skin. The skin's color is determined by the blood circulating through these vessels and the amount and type of pigment that is present in the skin. Blood is red when it is adequately saturated with oxygen. As a result, skin in lightly pigmented individuals is pinkish. The pigmentation in most individuals will not hide changes in the skin's underlying color, regardless of the individual's race. In patients with deeply pigmented skin, changes in color may be apparent only in certain areas, such as the fingernail beds, the mucous membranes in the mouth, the lips, the underside of the arm and palm (which are usually less pigmented), and the conjunctiva of the eyes. The conjunctiva is the delicate membrane lining the eyelids, and it covers the exposed surface of the eye. In addition, the palms of the hands and soles of the feet should be assessed in infants and children.

Poor peripheral circulation will cause the skin to appear pale, white, ashen, or gray, possibly with a waxy translucent appearance like a white candle. Abnormally cold or frozen skin may also appear this way. When the blood is not properly saturated with oxygen, it appears bluish. Therefore, in a patient with insufficient air exchange and low levels of oxygen in the blood, the blood and vessels become bluish, and the lips, mucous membranes, nail beds, and skin over the blood vessels appear blue or gray. This condition is called cyanosis.

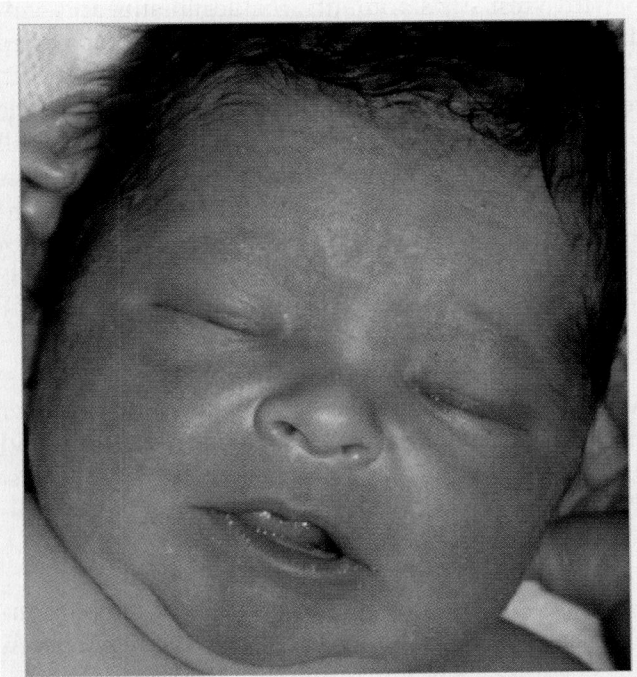

Figure 5-9 High blood pressure may cause the skin to be flushed and red.

High blood pressure may cause the skin to be abnormally flushed and red (Figure 5-9 ▲). In some patients with extremely high blood pressure, all the visible blood vessels will be so full that the skin will appear to be a dark reddish purple. A patient with a significant fever, heatstroke, sunburn, mild thermal burns, or other conditions in which the body is unable to properly dissipate heat will also appear to have red skin.

Changes in skin color may also result from chronic illness. Liver disease or dysfunction may cause jaundice, resulting in the patient's skin and sclera turning yellow. The sclera is the normally white portion of the eye, and may show color changes even before skin color change is visible.

Temperature

Normally, the skin is warm to the touch. When the patient has a significant fever, sunburn, or hyperthermia, the skin feels hot to the touch. The skin will feel cool when the patient is in early shock, has mild hypothermia, or has inadequate perfusion. The skin will feel cold when the patient is in profound shock, has hypother-

mia, or has frostbite. Body temperature is normally measured with a thermometer in the hospital.

Moisture

Dry skin is normal. Skin that is wet, moist (often called diaphoretic), or excessively dry and hot suggests a problem. In the early stages of shock, the skin will become slightly moist. Skin that is only slightly moist but not covered excessively with sweat is described as clammy, damp, or moist. When the skin is bathed in sweat, such as after strenuous exercise or when the patient is in shock, the skin is described as wet or diaphoretic.

Because the skin's color, temperature, and moisture are often related signs, you should consider them together. When recording or reporting your assessment of the skin, you should first describe the color, then the temperature, and last, whether the skin is dry, moist, or wet. For example, you could say or write, "Skin: pale, cool, and clammy."

Capillary Refill

Capillary refill is evaluated to assess the ability of the circulatory system to restore blood to the capillary system. When evaluated in an uninjured limb, capillary refill may reflect the patient's perfusion. It should be kept in mind, however, that capillary refill time can be affected by the patient's body temperature, position, preexisting medical conditions, and medications. To test capillary refill, place your thumb on the patient's fingernail with your fingers on the underside of the patient's finger and gently compress (Figure 5-10A ▶). The blood will be forced from the capillaries in the nail bed. When you remove the pressure applied against the tip of the patient's finger, the nail bed will remain blanched and white for a brief period. As the underlying capillaries refill with blood, the nail bed will be restored to its normal pink color.

Capillary refill should be prompt, and the nail bed color should be pink. With adequate perfusion, the color in the nail bed should be restored to its normal pink within 2 seconds, or about the time it takes to say "capillary refill" at a normal rate of speech (Figure 5-10B ▶). You should report and document the capillary refill as normal, "CRT < 2." You should suspect poor peripheral circulation when capillary refill takes more than 2 seconds or the nail bed remains blanched. In this instance, you should report and document the capillary refill as delayed or CRT > 2.

A bluish color may indicate that the capillaries are refilling with blood drawn from the veins rather

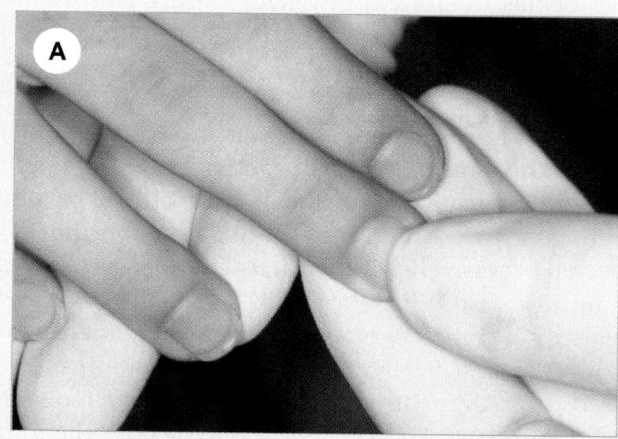

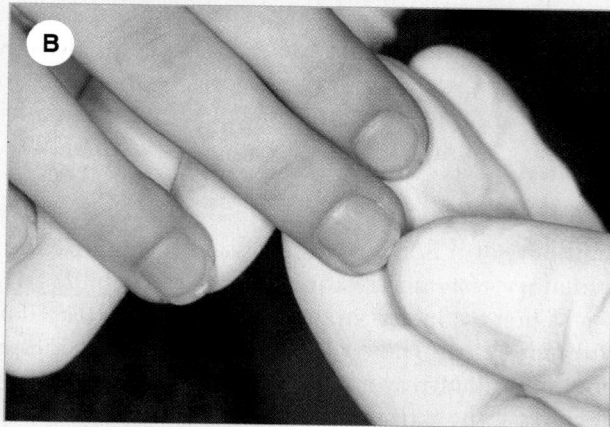

Figure 5-10 A. To test capillary refill, gently compress the fingertip until it blanches. **B.** Release the fingertip, and count until it returns to its normal pink color.

To assess capillary refill in older infants and children younger than 6 years, press on the skin or nail bed and determine how long it takes for the pink color to return. In newborns and young infants, press on the forehead, chin or sternum to determine capillary refill time. As with adults, normal capillary refill takes less than 2 seconds. However, it is a much more reliable indicator of cardiovascular status in children than it is in adults and should be recorded for all of your pediatric patients.

You cannot get a diastolic Pressure from radial you put systolic/p

Blood Pressure (Important Vital sign)

Adequate blood pressure is necessary to maintain proper circulation and perfusion of the vital organ cells. Blood pressure (BP) is the pressure of circulating blood against the walls of the arteries. A decrease in the blood pressure may indicate one of the following:

- Loss of blood or its fluid components
- Loss of vascular tone and sufficient arterial constriction to maintain the necessary pressure even without any actual fluid or blood loss
- A cardiac pumping problem

When any of these conditions occurs and results in a drop in circulation, the body's compensatory mechanisms are activated, the heart rate increases, and the arteries constrict. Normal blood pressure is maintained, and by decreasing the blood flow to the skin and extremities, available blood volume is temporarily redirected to the vital organs so that they remain adequately perfused. However, as shock progresses, and the body's defense mechanisms can no longer keep up, the blood pressure will fall. Decreased blood pressure is a late sign of shock and indicates that the critical decompensated phase has begun. Any patient with a markedly low blood pressure has inadequate pressure to maintain proper perfusion of all the vital organs and needs to have his or her blood pressure and perfusion restored immediately to a normal level.

than with oxygenated blood from the arteries, making the test invalid. You should also consider the capillary refill test invalid if the patient is in or has been exposed to a cold environment or if the patient is older. In both situations, delayed capillary refill may be normal.

You are the Provider   Part 3

As you are transporting your patient to the hospital, she becomes increasingly confused. Her respiratory rate decreases to 16 breaths/min, her pulse goes down to 60 beats/min, and her blood pressure is 120/64 mm Hg. She seems less anxious and appears to be relaxing.

5. Is your patient's condition improving?
6. What do these changes in vital signs mean? Should you change your treatment priorities?

When the blood pressure becomes elevated, the body's defenses act to reduce it. Some individuals have chronically high blood pressure from progressive narrowing of the arteries that occurs with age, and during an acute episode, their blood pressure may increase to even higher levels. Head injury or a number of other conditions may also cause blood pressure to rise to very high levels. Abnormally high blood pressure may result in a rupture or other critical damage in the arterial system.

You should measure blood pressure in all patients older than 3 years.

Blood pressure contains two key separate components: systolic pressure and diastolic pressure. Systolic pressure is the increased pressure that is caused along the artery with each contraction (systole) of the ventricles and the pulse wave that it produces. Diastolic pressure is the residual pressure that remains in the arteries during the relaxing phase of the heart's cycle (diastole), when the left ventricle is at rest. Systolic pressure represents the maximum pressure to which the arteries are subjected, and the diastolic pressure represents the minimum amount of pressure that is always present in the arteries.

Early blood pressure gauges contained a column of mercury and a linear scale that was graduated in millimeters. Even though different gauges are used today, the blood pressure is still measured in millimeters of mercury (mm Hg). Blood pressure is reported as a fraction in the form systolic pressure over diastolic pressure. Therefore, if the patient's systolic pressure is 120 and the diastolic pressure is 78, you would record it as "BP 120/78 mm Hg." You would report the patient's blood pressure verbally as "BP is 120 over 78."

Equipment for Measuring Blood Pressure

You will use a sphygmomanometer (blood pressure cuff) to apply pressure against the artery when measuring the blood pressure. The sphygmomanometer contains the following components:

- A wide outer cuff designed to be fastened snugly around the entire arm or leg
- An inflatable wide bladder sewn into a portion of the cuff
- A ball-pump with a one-way valve that allows air to enter and a turn-valve that can be closed or, when opened, will allow air to be released at a controlled speed from the cuff

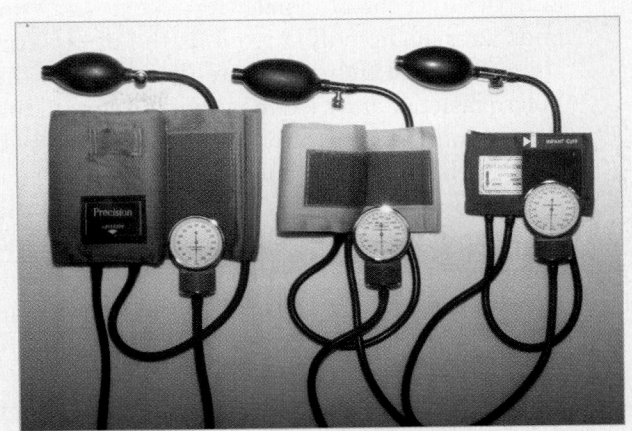

Figure 5-11 Three sizes of blood pressure cuffs: thigh, adult, and pediatric.

- A pressure gauge calibrated in millimeters of mercury, which indicates the pressure that exists in the cuff that is being applied against the underlying artery

Most agencies carry at least three sizes of blood pressure cuffs: adult, thigh, and pediatric (Figure 5-11 ▲). The normal size cuff is designed to wrap around the arm 1 to 1.5 times and take up two thirds the length from the armpit to the crease in the elbow of most adults. Use a thigh cuff with patients who are obese or have exceptionally well-developed arm muscles or to take the blood pressure of the thigh in patients who have injuries in both arms. Use a small pediatric cuff with children and exceptionally small adults.

You must be sure to select the appropriately sized cuff. A cuff that is too small may result in falsely high readings; a cuff that is too large may result in falsely low readings.

Auscultation

Auscultation is the method of listening to sounds within the body with a stethoscope. You will usually measure blood pressure by auscultation (Skill Drill 5-1 ▶). Follow these steps:

1. **With the patient's arm extended** with the palm up, place the cuff so that it lies across the upper arm and is located with its distal edge about 1″ above the crease at the inside of the patient's elbow. Make sure the center of the inflatable bladder, which is usually marked by an arrow on the

cuff, lies over the brachial artery. Next, wrap the ends so that the cuff surrounds the upper arm snugly but not tightly. Secure the cuff with the Velcro fastener attached to it, making sure to rub your hand over the entire area where the two sides of the Velcro fastener are in contact. Once the cuff has been properly secured around the upper arm, the arm should be held at about the same level as the heart (**Step 1**).

2. **Palpate the brachial artery** (in the antecubital fossa, the anterior aspect of the elbow) to determine where to place the stethoscope (**Step 2**).

3. **Place the diaphragm of the stethoscope over the artery**, and hold it firmly against the artery with the fingers of your nondominant hand. Hold the rubber ball-pump in the palm of your other hand and the turn-valve between your thumb and first finger (**Step 3**).

4. **Close the valve tightly, and pump the ball-pump** until you no longer hear pulse sounds. Continue pumping to increase the cuff's pressure by an additional 20 mm Hg. Next, slowly turn the valve, opening it until air is steadily escaping from the cuff and you see the needle of the gauge slowly drop. Watch the gauge, and listen carefully. Note the patient's systolic pressure as the reading on the gauge at which the "taps" or "thumps" of the pulse waves can first be heard clearly. As the pressure in the cuff is progressively reduced, pulse sounds will continue for a time, then suddenly disappear. Note the patient's diastolic pressure as the reading on the gauge at which the sounds stopped (**Step 4**).

5. As soon as the pulse sounds stop, **open the valve, and release the remaining air quickly**. Once you have finished measuring the blood pressure, you should document your findings and the time at which the blood pressure was taken. Blood pressure is most often measured by auscultation with the patient in a sitting or semisitting position. Be sure to note whether a different method or position was used. Occasionally when a patient's blood pressure is very low, you will continue to hear pulse sounds from the reading at which they started all the way until the gauge has reached 0. When this occurs, you should record the diastolic pressure as "0" or "all the way down" to indicate that it was heard until the gauge read 0 (**Step 5**).

Palpation

The auscultation method may be difficult or impossible to use in a very noisy environment, leading to inaccurate findings. The palpation method, which is examination by touch that does not depend on your ability to hear sounds, should be used in these cases (**Skill Drill 5-1, Step 6** ▶).

To measure blood pressure by palpation, secure the appropriately sized cuff around the patient's upper arm in the manner previously described. With your nondominant hand, palpate the patient's radial pulse on the same arm as the cuff, without moving your fingertips once you have located it, until you have completed taking the blood pressure. While holding the ball-pump in your other hand, close the turn-valve and slowly inflate the cuff until the pulse disappears and then continue to inflate another 30 mm Hg. As the cuff inflates, you will no longer feel the pulse under your fingertips. Open the turn-valve so that air slowly escapes from the cuff, and carefully observe the gauge. When you can again feel the radial pulse under your fingertips, you should note the reading on the gauge as the patient's systolic blood pressure. You will not be able to determine the diastolic pressure with this method. Next, open the turn-valve further, and completely deflate the cuff. Document your findings, including the time, and note that the pressure was taken by palpation. On your patient care report, you can record the blood pressure as "120/P" and verbalize it as "120 palpated."

Normal Blood Pressure (Not able to

Blood pressure levels vary with age and gender. (**Table 5-4** ▶) serves as a guideline for normal blood pressure ranges.

A patient has hypotension when the blood pressure is lower than the normal range and hypertension when the blood pressure is higher than the normal range.

✦ **EMT-B Tips**

Record blood pressure readings with the time. Note the arm on which you took the measurement and the patient's position if other than sitting or semisitting. For a blood pressure by palpation, you can abbreviate: "120/P."

Obtaining a Blood Pressure by Auscultation or Palpation

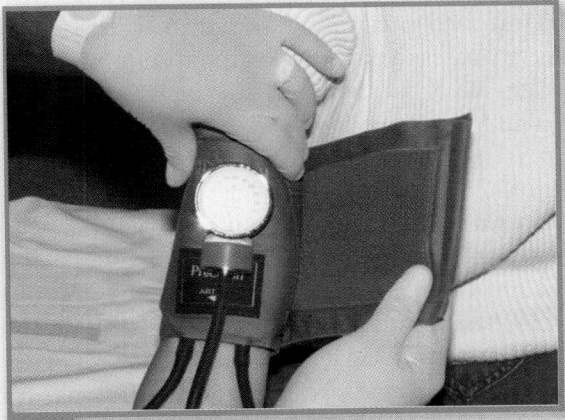

1 Apply the cuff snugly.

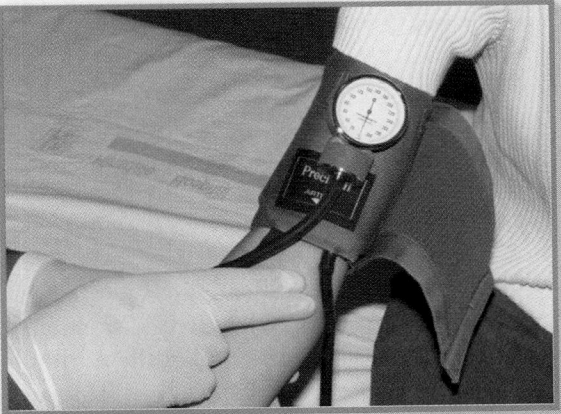

2 Palpate the brachial artery.

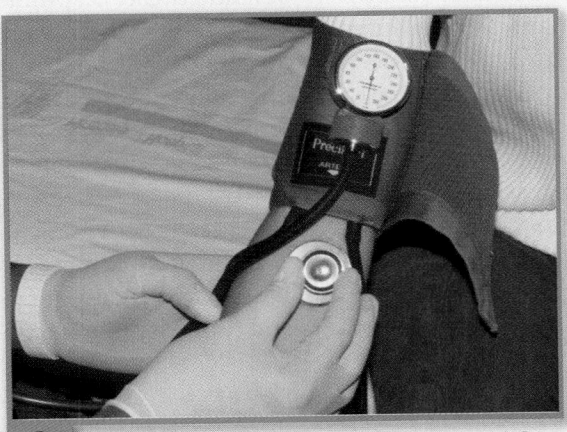

3 Place the stethoscope over the brachial artery, and grasp the ball-pump and turn-valve.

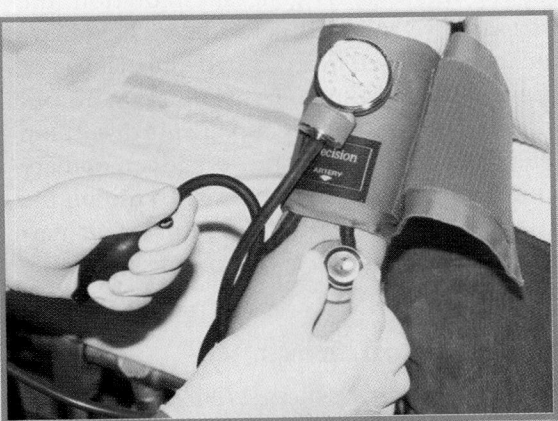

4 Close the valve, and pump to 20 mm Hg above the point at which you stop hearing pulse sounds. Note the systolic and diastolic pressures as you let air escape slowly.

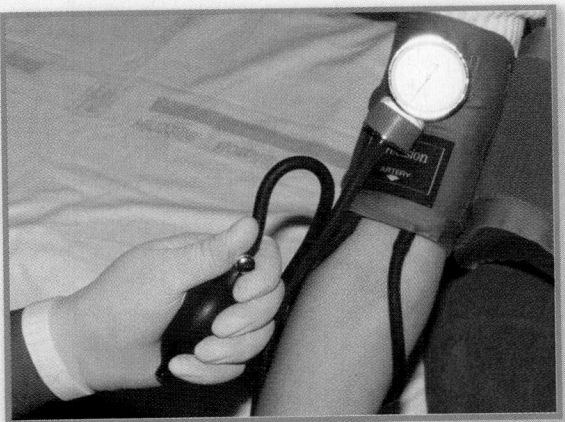

5 Open the valve, and quickly release remaining air.

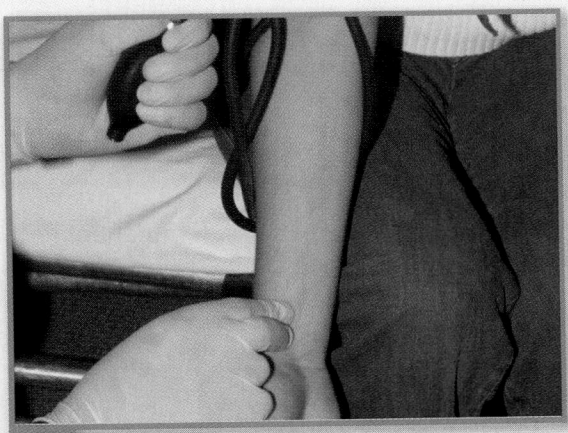

6 When using the palpation method, you should place your fingertips on the radial artery so that you feel the radial pulse.

Study This!!!

TABLE 5-4 Normal Ranges for Blood Pressure

Age	Range, mm Hg
Adults	90 to 140 (systolic)
Children (ages 1 to 8 years)	80 to 110 (systolic)
Infants (newborn to age 1 year)	50 to 95 (systolic)

Typically, you will see children less frequently than adults; therefore, you might not remember the normal ranges for the various age groups. You might want to carry a chart with you that lists normal blood pressure ranges and other vital signs.

In assessing the patient's general circulation, the blood pressure, pulse, skin temperature, and capillary refill should not be assessed in an injured limb. However, once you have obtained these vital signs from an uninjured limb, you might want to compare the distal skin temperature, quality of the distal pulse, and/or capillary refill time in the injured limb with those found on the uninjured side. This information is useful in evaluating whether the injury may have compromised the circulation in the injured limb.

Level of Consciousness

The patient's level of consciousness (LOC) is considered a vital sign because it can tell a great deal about the patient's neurological and physiological status. The brain requires a constant supply of oxygen and sugar to function properly. An alteration in level of consciousness can be one of the first indications of hypoxia. In the early assessment, you need to ascertain only the gross level of consciousness by determining whether the patient is awake and alert with an unaltered LOC, conscious but with an altered LOC, or unconscious.

As you assess a patient, you must determine the appropriateness of a response by how well it demonstrates the patient's understanding and mental activity, not how well it reflects your definition of socially acceptable behavior.

When a patient is conscious with an altered level of consciousness, it may indicate that inadequate perfusion and oxygenation or a chemical or neurologic problem is adversely affecting the brain and its ability to function. An altered level of consciousness in a conscious patient can also be caused by medications, drugs, alcohol, or poisoning.

Your assessment of a patient who is unconscious when you arrive should be focused initially on problems with airway, breathing, and circulation, which are critical life threats, and then on identifying other emergency care that the patient may need. Sustained unconsciousness should warn you that a critical respiratory, circulatory, or central nervous system problem or deficit may exist, and you must assume that the patient has a potentially critical injury or potentially life-threatening condition. Therefore, after rapidly assessing the patient and providing emergency treatment, you should package the patient and provide rapid transport to the hospital.

Pupils

The diameter and reactivity to light of the patient's pupils reflect the status of the brain's perfusion, oxygenation, and condition. The pupil is a circular

You are the Provider

Part 4

As you pull into the hospital bay, your patient becomes unresponsive and apneic. You insert an oropharyngeal airway and begin to ventilate with a bag-valve-mask device. She is intubated in the emergency department and spends 2 weeks in the intensive care unit with a diagnosis of acute pulmonary embolism.

7. What clues were available to you from the scene, history, signs, symptoms, and ongoing assessment and serial vital signs that may have helped you determine what was wrong with your patient?
8. Did you treat her appropriately?
9. Had you realized the cause of her condition, would it have altered your treatment?

Study This !!!

✳ EMT-B Tips

AVPU scale

The AVPU scale is a rapid method of assessing the patient's level of consciousness using one of the following four terms:

A = **A**wake and **A**lert

V = Responsive to **V**erbal Stimuli

P = Responsive to **P**ain

U = **U**nresponsive

You should determine whether a patient is awake and alert. An awake and alert patient is oriented to person, place, time, and mechanism of injury or history of present illness. A patient who is not awake and alert but who is aroused and responds to your voice by opening his or her eyes, moaning, speaking, or moving is responding to verbal stimuli. A patient who does not respond to your normal speaking voice but who responds when you speak loudly is responding to loud verbal stimuli. Be sure to note how the patient responded. Tap a patient who is hearing impaired with your fingers repeatedly. If the patient responds, note that the patient is hearing impaired but responds to being tapped.

To determine whether a patient who does not respond to verbal stimuli will respond to a painful stimulus, you should gently but firmly pinch the patient's skin (Figure 5-12 ▶). A patient who moans or withdraws is responding to painful stimulus. Be sure to

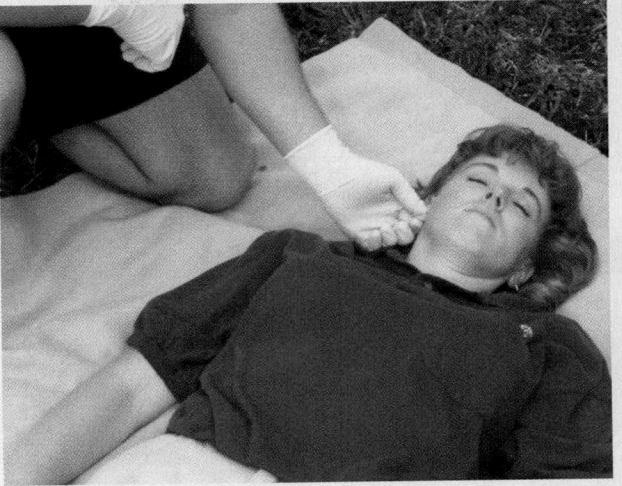

Figure 5-12 To assess whether a patient will respond to a painful stimulus, gently but firmly pinch the patient's skin. This can be done at the neck or on the earlobe.

note the type and location of the stimulus and how the patient responded. Also use caution on the patient with a suspected neck or spinal injury.

If the patient does not respond to a painful stimulus on one side, try to elicit a response on the other side. Note that a patient who remains flaccid without moving or making a sound with no indication of hearing you is unresponsive.

---

opening in the center of the pigmented iris of the eye. The pupils are normally round and of approximately equal size and serve as optical diaphragms, adjusting their size depending on the available light. In normal room light, the pupil appears to be midsize. With less light, the pupils dilate, allowing more light to enter the eye, making it possible to see even in dim light. With high light levels or when a bright light is suddenly introduced, the pupils instantly constrict, allowing less light to enter, protecting the sensitive receptors in the inner eye from damage (Figure 5-13A ▶). When a brighter light is introduced into one eye (or higher levels of light enter one eye only), both pupils should constrict equally to the appropriate size for the pupil receiving the most light.

In the absence of any light, the pupils will become fully relaxed and dilated (Figure 5-13B ▶). When light is introduced, each eye sends sensory signals to

the brain indicating the level of light it is receiving. Pupil size is regulated by a series of continuous motor commands that the brain automatically sends through the oculomotor nerves to each eye, causing both pupils to constrict to the same appropriate size. Normally, pupil size changes instantly to any change in light level.

You must assume the patient has depressed brain function as a result of either central nervous system depression or injury if the pupils react in any of the following ways:

- Become fixed with no reaction to light
- Dilate with introduction of a bright light and constrict when the light is removed
- React sluggishly instead of briskly
- Become unequal in size (Figure 5-13C ▶)
- Become unequal in size when a bright light is introduced into or removed from one eye

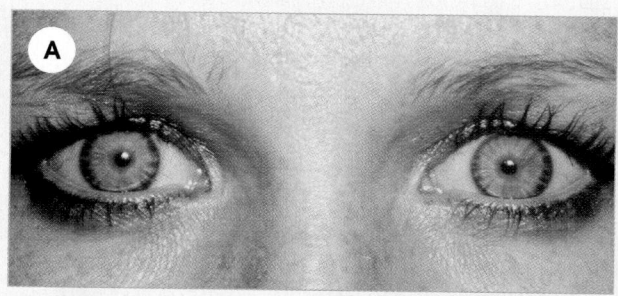

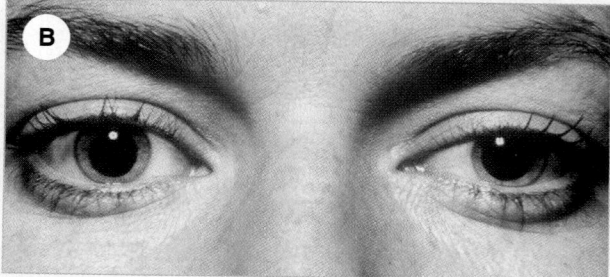

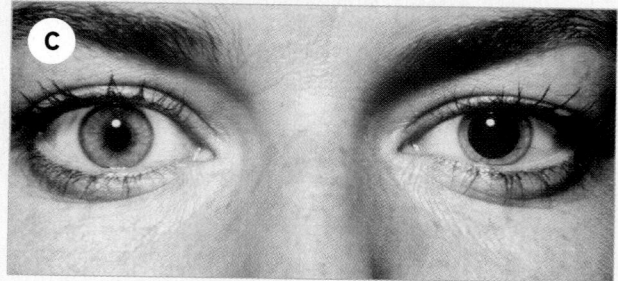

Figure 5-13 **A.** Constricted pupils. **B.** Dilated pupils. **C.** Unequal pupils.

Depressed brain function can be produced by the following situations:

- Injury of the brain or brain stem
- Trauma or stroke
- Brain tumor
- Inadequate oxygenation or perfusion
- Drugs or toxins (central nervous system depressants)

Opiates, which are one category of central nervous system depressants, cause the pupils to constrict so significantly, regardless of light, that they become so small as to be described as pinpoint. Intracranial pressure from intracranial bleeding may cause sufficient pressure against the oculomotor nerve on one side that

the motor commands can no longer pass from the brain to that eye. When this occurs, the eye no longer receives commands to constrict, and its pupil becomes fully dilated and fixed. This is described as a blown pupil.

Pupils may be dilated, unequal as a result of medication placed into one or both eyes or from an injury or condition of the eye, or not reacting appropriately.

The letters PEARRL serve as a useful guide in assessing the pupils. They stand for the following:

P = **Pupils**
E = **Equal**
A = **And**
R = **Round**
R = **Regular in size**
L = react to **Light**

For patients with normal pupils, you can report "Pupils are equal, round, and regular in size, and react properly to light" or "pupils = PEARRL." Describe any abnormal findings using the longer form, such as "Pupils are equal and round, the left pupil is fixed and dilated, the right pupil is regular in size and reacts to light."

Reassessment of the Vital Signs

The vital signs that you obtain serve two important functions. The first set establishes an important initial measurement of the patient's neurological, respiratory, and cardiovascular systems and the quality of perfusion and oxygenation of the brain and other vital organs. The initial vital signs serve as a key baseline.

Throughout your care of the patient, you should monitor the patient's vital signs for any changes from your initial findings. You should reassess and record vital signs at least every 15 minutes in a stable patient and at least every 5 minutes in an unstable patient. You should also reassess and record vital signs following all medical interventions. This ongoing comparative assessment is an important indicator of whether your interventions have restored the patient's vital functions to an acceptable range or are at least preventing further deterioration. Reassessment also indicates whether you should consider more aggressive intervention whenever deterioration continues.

You are the Provider   Summary

Frequently, a patient's medications and chief complaint provide valuable clues to the underlying condition. This information will greatly aid you in understanding the source and the severity of the patient's condition(s). You may be unable to obtain information from the patient's medical history either because there is no significant medical history or because the patient is unable to tell you.

Women who are 35 years of age and older who are taking birth control medication and smoke cigarettes are at especially high risk of experiencing embolisms. Becoming familiar with commonly prescribed medications and over-the-counter medications can provide important information about a patient's medical history. Sometimes patients will not know why they are taking certain medications, so it would be helpful for you to have a pocket guide on medications for quick and easy reference. Being able to sort through the clues from the emergency scene itself, from the patient's complaints, and from the patient's signs and symptoms and past medical history will all assist you in understanding the cause of your patient's problem and enable you to make appropriate, timely decisions about your patient's care.

Prep Kit

Ready for Review

- Whenever you are called to the scene of an illness or injury, you should find out the patient's chief complaint.

- Your assessment of the patient should include rapidly evaluating the patient's general condition and identifying any potentially life-threatening injuries or conditions.

- Baseline vital signs are the key signs that you will use to evaluate the patient's general condition.

- You will be assessing the patient's respirations, pulse, skin, blood pressure, level of consciousness, and pupils.

- After you have initially assessed the patient and obtained the baseline vital signs, you should reassess the patient for any changes from your initial findings.

- In addition to determining the chief complaint and assessing the patient's general condition, you should try to obtain a SAMPLE history from the patient or bystanders.

- By asking several important questions, you will be able to determine the patient's signs and symptoms, allergies, medications, pertinent past history, last oral intake, and the events leading up to the incident.

Technology

Interactivities

Vocabulary Explorer

Anatomy Review

Web Links

Online Review Manual

www.EMTB.com

Vital Vocabulary

auscultation A method of listening to sounds within an organ with a stethoscope.

AVPU scale A method of assessing level of consciousness by determining whether the patient is awake and alert, responsive to verbal stimuli or pain, or unresponsive; used principally early in the assessment.

blood pressure (BP) The pressure of circulating blood against the walls of the arteries.

bradycardia Slow heart rate, less than 60 beats/min.

capillary refill The ability of the circulatory system to restore blood to the capillary system; evaluated by using a simple test.

chief complaint The reason a patient called for help. Also, the patient's response to questions such as "What's wrong?" or "What happened?"

conjunctiva The delicate membrane lining the eyelids and covering the exposed surface of the eye.

cyanosis A bluish-gray skin color that is caused by reduced levels of oxygen in the blood.

diaphoretic Characterized by profuse sweating.

diastolic pressure The pressure that remains in the arteries during the relaxing phase of the heart's cycle (diastole) when the left ventricle is at rest.

hypertension Blood pressure that is higher than the normal range.

hypotension Blood pressure that is lower than the normal range.

jaundice A yellow skin or sclera color that is caused by liver disease or dysfunction.

labored breathing Breathing that requires visibly increased effort; characterized by grunting, stridor, and use of accessory muscles.

OPQRST An abbreviation for key terms used in evaluating a patient's signs and symptoms: onset, provocation or palliation, quality, region/radiation, severity, and timing of pain.

perfusion Circulation of blood within an organ or tissue.

pulse The pressure wave that occurs as each heartbeat causes a surge in the blood circulating through the arteries.

pulse oximetry An assessment tool that measures oxygen saturation of hemoglobin in the capillary beds.

SAMPLE history A brief history of a patient's condition to determine signs and symptoms, allergies, medications, pertinent past history, last oral intake, and events leading to the injury or illness.

sclera The white portion of the eye.

signs Objective findings that can be seen, heard, felt, smelled, or measured.

sniffing position An unusually upright position in which the patient's head and chin are thrust slightly forward.

spontaneous respirations Breathing in a patient that occurs with no assistance.

stridor A harsh, high-pitched, crowing inspiratory sound, such as the sound often heard in acute laryngeal (upper airway) obstruction.

symptoms Subjective findings that the patient feels but that can be identified only by the patient.

systolic pressure The increased pressure along an artery with each contraction (systole) of the ventricles.

tachycardia Rapid heart rhythm, more than 100 beats/min.

tidal volume The amount of air that is exchanged with each breath.

tripod position An upright position in which the patient leans forward onto two arms stretched forward and thrusts the head and chin forward.

vasoconstriction Narrowing of a blood vessel.

vital signs The key signs that are used to evaluate the patient's overall condition, including respirations, pulse, blood pressure, level of consciousness, and skin characteristics.

Points to Ponder

You respond to man who was mowing his lawn around noon on a hot August day. Upon your arrival you find a 54-year-old man who is complaining of chest pain and dizziness. You start a SAMPLE history while your partner obtains baseline vital signs. The patient states he has chest pain that radiates to the left arm and that he feels nauseous. The patient looks pale and diaphoretic. The pain is not reproducible with palpation. The patient has no allergies and takes nitroglycerin for angina. The patient states that he did not eat today and only had a cup of coffee this morning.

Your baseline vitals include a blood pressure of 168/92 mm Hg, a pulse rate of 180 beats/min, and respirations of 20 breaths/min. You move the patient to a cool area and his color returns to normal. Your partner takes another set of vital signs, indicating the patient's blood pressure is 152/88 mm Hg, his pulse is 120 beats/min, and his respirations are 20 breaths/min.

What is the patient's chief complaint? Which of the previous were signs and which were symptoms? What is most concerning about the vital signs? Is your patient's condition improving?

Issues: Recognizing an Emergency, Importance of Obtaining a SAMPLE History.

www.EMTB.com

Assessment in Action

Ambulance 49 is dispatched to an unknown emergency at a local seafood restaurant. It will take you and your EMT-B partner 5 minutes to reach the location. After putting on your personal protective equipment and ensuring scene safety, you enter the restaurant and find an older man lying on the floor.

1. The first set of vital signs you obtain are called the:

   A. chief complaint.
   B. baseline vital signs.
   C. medical history.
   D. SAMPLE history.

2. During the assessment you observe the patient to be unconscious and unresponsive. You open the airway using a head tilt-chin lift and note that the patient is spontaneously breathing. The normal respiratory rate for an adult is:

   A. 15 to 30 breaths/min.
   B. 6 to 12 breaths/min.
   C. 12 to 20 breaths/min.
   D. 25 to 50 breaths/min.

3. As you assess the patient's radial pulse, you note it to be 50 beats/min, strong and regular. A rate that is above 100 beats/min is called:

   A. tachypnea.
   B. bradycardia.
   C. bradypnea.
   D. tachycardia.

4. The patient begins to regain consciousness and states that he was eating lunch when he began "feeling dizzy" and fell to the ground. The patient's statement is known as the:

   A. history of present illness.
   B. severity of illness.
   C. chief complaint.
   D. past medical history.

5. You arrive on the scene of an unknown emergency and find the patient sitting in the tripod position. What does this indicate?

   A. Low blood pressure
   B. Difficulty breathing
   C. Seizure
   D. Low level of blood glucose (sugar)

6. You have a patient who is alert and oriented to all events. His eyes are spontaneous and he is able to obey all commands. What is the patient's GCS score?

   A. 15
   B. 16
   C. 14
   D. 12

7. You have a patient who moans when you apply a sternal rub and speech is garbled. The patient moves his arms away from painful stimuli. What is this patient's GCS score?

   A. 8
   B. 7
   C. 9
   D. 10

Challenging Questions

8. You arrive to find a distressed 2-year-old. The patient's mother states the child isn't eating normally. You take a pulse and find it to be 200 beats/min and irregular. What is your reaction to the pulse rate? Where would you palpate a pulse in this patient?

9. You need to assess the vital signs of a 1-year-old. What is the best way to assess perfusion on an infant?

Lifting and Moving Patients

Objectives

Cognitive

1-6.1 Define body mechanics. (p 171)
1-6.2 Discuss the guidelines and safety precautions that need to be followed when lifting a patient. (p 171, 173)
1-6.3 Describe the safe lifting of cots and stretchers. (p 173)
1-6.4 Describe the guidelines and safety precautions for carrying patients and/or equipment. (p 175, 180)
1-6.5 Discuss one-handed carrying techniques. (p 176)
1-6.6 Describe correct and safe carrying procedures on stairs. (p 176, 180)
1-6.7 State the guidelines for reaching and their application. (p 182)
1-6.8 Describe correct reaching for log rolls. (p 183)
1-6.9 State the guidelines for pushing and pulling. (p 182)
1-6.10 Discuss the general considerations of moving patients. (p 184)
1-6.11 State three situations that may require the use of an emergency move. (p 184)
1-6.12 Identify the following patient-carrying devices:
- Wheeled ambulance stretcher
- Portable ambulance stretcher
- Stair chair
- Scoop stretcher
- Long spine board
- Basket stretcher
- Flexible stretcher (p 197)

Affective

1-6.13 Explain the rationale for properly lifting and moving patients. (p 170)

Psychomotor

1-6.14 Working with a partner, prepare each of the following devices for use, transfer a patient to the device, properly position the patient on the device, move the device to the ambulance, and load the patient into the ambulance:
- Wheeled ambulance stretcher
- Portable ambulance stretcher
- Stair chair
- Scoop stretcher
- Long spine board
- Basket stretcher
- Flexible stretcher (p 197)
1-6.15 Working with a partner, the EMT-B will demonstrate techniques for the transfer of a patient from an ambulance stretcher to a hospital stretcher. (p 182)

You are the Provider

You are dispatched to a motor vehicle crash in which a car ran into a pole. Upon arrival you find a single patient, a 54-year-old woman complaining of neck pain. There is minor damage to the front of the vehicle. The patient was wearing her seat belt, and the air bag deployed. She denies having any other injury, and a rapid trauma assessment reveals no other obvious injuries.

1. What are your treatment priorities?
2. What resources can you call on to help you?

Lifting and Moving Patients

In the course of a call, you will have to move a patient several times to provide emergency medical care in the field and transport the patient to the emergency department. Often, you will have to move the patient into a different position or location. Once you have assessed the patient and provided emergency care, you and your team will have to move the patient onto a long backboard or cot. Then you must move the patient to the waiting ambulance and load the patient into the patient compartment. After you arrive at the hospital, you must unload the patient, move him or her to the correct examining room, and transfer the patient from the cot to the emergency department bed. To avoid injury to the patient, yourself, or your partners, you will have to learn how to lift and carry the patient properly, using proper body mechanics and a power grip. To be able to move a patient safely and properly in the various situations that you will encounter in the field, you will have to learn how to perform emergency body drags and lifts, rapidly move a patient from a car onto the cot, assist a patient from a chair or bed onto the cot, and lift a patient from the floor onto the cot. In addition, you will need to move a patient from the bed onto the cot or carry a patient up or down stairs. You and your team will have to know how to place a patient with a suspected spinal injury onto a long backboard and package patients with and without suspected spinal injury.

At times, you and your team will need to move a patient who is very heavy or carry a patient on a trail or across rugged terrain. You will need to know the special techniques for loading and unloading the cot and transferring the patient from the cot to an examining table or bed in the emergency department.

Lifting and carrying are dynamic processes. To ensure that no individual suddenly bears unexpected, dangerous weight and to reduce the risk of injury to an EMT-B or the patient, you must know where rescuers should be positioned and how to give and receive lifting commands so that all parties act simultaneously. You will also need to know how to prepare patient-moving devices, such as a wheeled ambulance stretcher (also called an ambulance cot, gurney, or simply "the cot"), stair chair, backboard, scoop stretcher, folding ambulance stretcher, basket stretcher, or flexible stretcher, and when and how to use them. This chapter will cover lifting, carrying, and reaching techniques as well as principles of moving patients, including emergency, urgent, and nonurgent moves. In addition, different types of equipment and patient positioning will be discussed in detail.

Moving and Positioning the Patient

Every time you have to move a patient, you must take special care that neither you, your team, nor the patient is injured. Patient packaging and handling are technical skills that you will learn and perfect through practice and training.

Training and practice are required to use all the equipment that is described in this chapter. You must master the skills necessary for their use and understand the advantages and limitations of each device. Practice each technique with your team often so that when you must move a patient, you can perform the move quickly, safely, and efficiently. After each patient transfer, you and your team should evaluate the appropriateness of the technique that you used, as well as your technical skill in completing the transfer. You must also be sure to maintain your equipment according to the manufacturer's instructions. Using clean, well-maintained equipment is but one part of providing high-quality patient care.

After you deliver the patient to the emergency department, you and your team must begin preparing for your next call. Review the positive points about the

www.EMTB.com

Technology

- Interactivities
- Vocabulary Explorer
- Anatomy Review
- Web Links
- Online Review Manual

transport. Discuss changes that would improve the next run. This process of review and evaluation identifies the following:

- Procedures that need more practice
- Equipment that needs to be cleaned or repaired
- Skills that you need to review or acquire

Most important, a critical review helps you and your team to become more confident and better skilled EMT-Bs.

Certain patient conditions, such as head injury, shock, spinal injury, and pregnancy, call for special lifting and moving techniques. Patients with chest pain or who are having difficulty breathing should sit in a position of comfort, as long as they are not hypotensive. Patients with suspected spinal injuries must be immobilized in a supine position on a long backboard. Patients who are in shock should be packaged and moved in a Trendelenburg position or supine with their legs elevated 6″ to 12″. Pregnant patients who are hypotensive should be positioned and transported on their left sides. Place an unresponsive patient with no suspected spinal injury into the recovery position by rolling the patient onto his or her side without twisting the body. Transport a patient who is nauseated or vomiting in a position of comfort, but be sure that you are positioned appropriately to manage the airway.

Body Mechanics

Anatomy Review

The shoulder girdle rests on the rib cage and is supported by the vertebrae that lie inferior to it. The arms are connected to and hang from the shoulder girdle. When the person is standing upright, the individual weight-bearing vertebrae are stacked on top of each other and aligned over the sacrum. The sacrum is both the mechanical weight-bearing base of the spinal column and the fused central posterior section of the pelvic girdle.

When a person is standing upright, the weight of anything being lifted and carried in the hands is reflected onto the shoulder girdle, the spinal column inferior to it, the pelvis, and then the legs Figure 6-1▶. In lifting, if the shoulder girdle is aligned over the pelvis and the hands are held close to the legs, the force that is exerted against the spine occurs in an essentially straight line down the vertebrae in the spinal column. Therefore, with the back properly maintained in an upright position, very little strain occurs against the muscles and ligaments that keep the spinal column in alignment, and significant weight can be lifted and carried without injury to the back Figure 6-2▶. However, you may injure your back if you lift with your back curved or, even if straight, bent significantly forward at the hips. With the back in either of these positions, the shoulder girdle lies significantly anterior to the pelvis, and the force of lifting is exerted primarily across, rather than down, the spinal column. When this occurs, the weight is supported by the muscles of the back and ligaments that run from the base of the skull to the pelvis, keeping the spinal column in alignment, rather than by each vertebral body and disk resting on those aligned below it. In addition, the upper spine and torso serve as a lever so that the force that is exerted against the muscles and ligaments in the lumbar and sacral regions, as a result of the mechanical advantage produced, is many times that of the combined weight of your upper body and the object you are lifting. Therefore, the first key rule of lifting is to always keep the back in a straight, upright (vertical) position and to lift without twisting.

When lifting, you should spread your legs about 15″ apart (shoulder width) and place your feet so that your center of gravity is properly balanced between them. Then, with the back held upright, bring your upper body down by bending the legs. Once you have

You are the Provider   Part 2

Your partner gets into the back seat of the car and holds her head to minimize the possibility of cervical spine injury. Any movement causes your patient to complain of severe pain in her neck and lower back. She complains of tingling in her hands.

3. What is the best way to remove this patient from the vehicle? Are there any devices that can assist you?

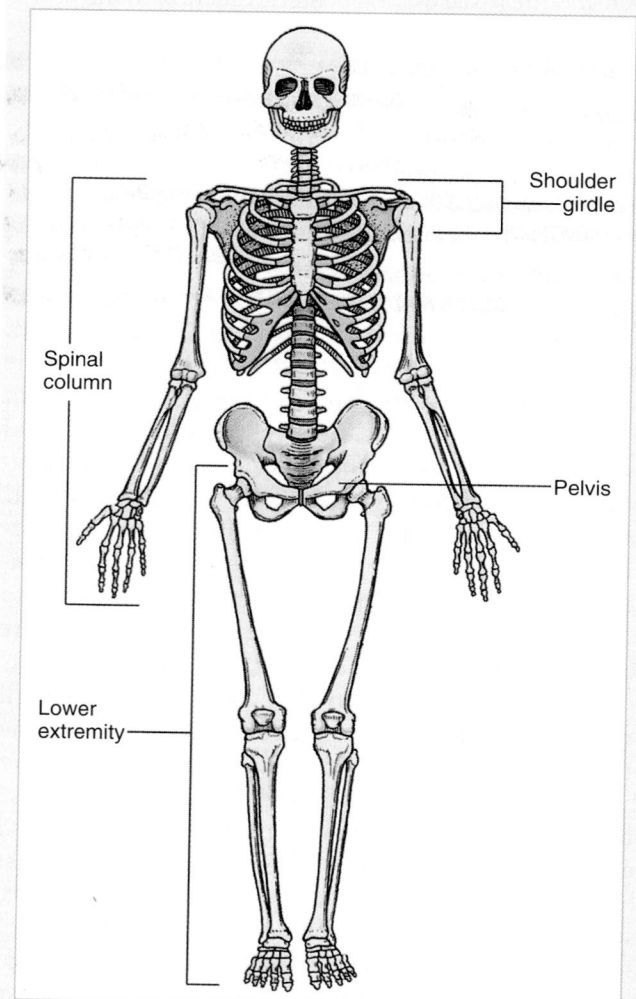

Figure 6-1 When you are standing upright, the weight of anything that you lift and carry in your hands is borne by the shoulder girdle, the spinal column, the pelvis, and the legs.

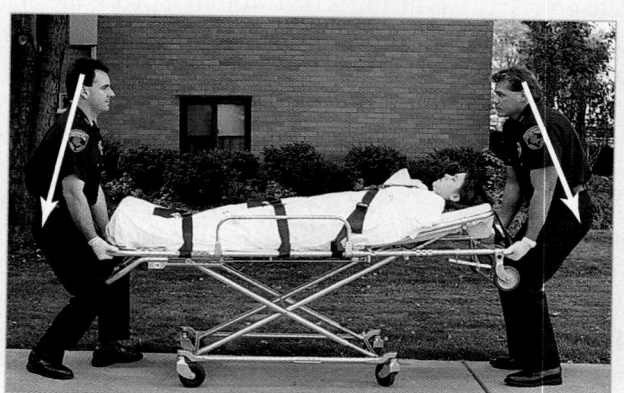

Figure 6-2 If your body is properly aligned when you lift, the line of force exerted against the spine occurs in an essentially straight line down the vertebrae. In this way, the vertebrae support the lift.

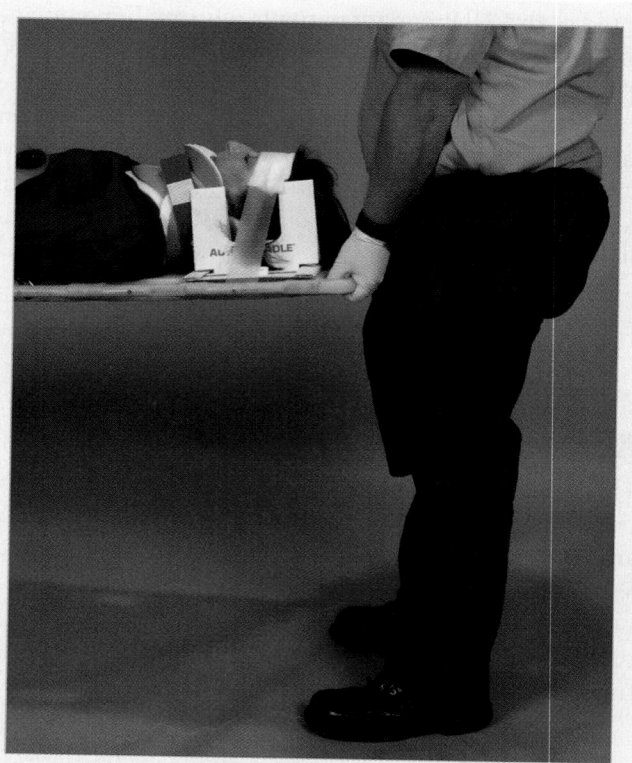

Figure 6-3 Straighten your legs and lift.

properly grasped the patient or cot and made any necessary adjustments in the location of your feet, lift the patient by raising your upper body and arms and by straightening your legs until you are again standing (Figure 6-3 ▶). Because the leg muscles are exercised by walking, climbing stairs, or running, they are well developed and extremely strong. Therefore, as well as being the safest way to lift, lifting by extending the properly placed flexed legs is also the most powerful way to lift. This method is appropriately called a <u>power lift</u>. The power lift position is also useful for individuals who have weak knees or thighs.

Even if the back is held properly upright, the same adverse force across the spinal column and leverage against the lower back will occur if you lift a heavy ob-

ject with your arms outstretched so that your hands are significantly anterior to the plane described by the front of the torso. Therefore, you should never lift a patient or other heavy object while reaching any significant distance in front of your torso or face. Whenever you are lifting or carrying a patient, be sure to hold your arms so that your hands are almost immediately adjacent to the plane described by your anterior torso (the anterior torso and imaginary lines extended vertically above and below it). Always keep the weight that you are lifting as close to your body as possible.

Lateral force across the spine and sideways leverage against the lower back must also be avoided. If you lift with only one arm or with the arms extended more to one side than the other, more force will be exerted against one side of the shoulder girdle than the other, causing lateral force to be exerted across the spinal column. To prevent this, keep your arms approximately the same distance apart as when hanging at each side of the body, with the weight distributed equally and properly centered between them. If the weight is not balanced between both arms or properly centered between the shoulders when you are preparing to lift, turn your body and/or move to the left or right until the weight is properly balanced and centered. To lift safely and produce the maximal power lift, you should take the following steps (Skill Drill 6-1 ▶):

1. **Tighten your back in its normal upright position** and use your abdominal muscles to lock it in a slight inward curve.
2. **Spread your legs apart about 15″**, and bend your legs to lower your torso and arms.
3. **With arms extended down each side of the body**, grasp the cot or backboard with your hands held palm up and just in front of the plane described by the anterior torso and imaginary lines extending vertically from it to the ground.
4. **Adjust your orientation and position** until the weight is balanced and centered between both arms (**Step 1**).
5. **Reposition your feet** as necessary so that they are about 15″ apart with one slightly farther forward and rotated so that you and your center of gravity will be properly balanced between them. Be sure to straddle the object, keep your feet flat, and distribute your weight to the balls of the feet or just behind them (**Step 2**).
6. **With the arms extended downward**, lift by straightening your legs until you are fully standing. Make sure your back is locked in and that your upper body comes up before your hips (**Step 3**).

Reverse these steps whenever you are lowering the cot. Always remember to avoid bending at the waist or twisting as you stand.

Your safety, as well as that of the other EMT-Bs and the patient, depends on the use of proper lifting tech-

niques and having and maintaining a proper hold when lifting or carrying a patient. If you do not have proper hold of the cot or of the patient in a body lift, you will not be able to bear a proper share of the weight, and there is an increased chance that you can suddenly lose your grasp with one or both hands. If you temporarily lose your grasp with one or both hands, the position and weight distribution of the cot change suddenly, and the other members of the team must quickly reach beyond a safe distance to avoid dropping the patient. As a result, sudden excessive force may be placed across each one's spine, causing lower back injury.

You should use the power grip to get the maximum force from your hands whenever you are lifting a patient (Figure 6-4 ▼). The arm and hand have their greatest lifting strength when facing palm up. Whenever you grasp a cot or backboard, your hands should be at least 10″ apart. Each hand should be inserted under the handle with the palm facing up and the thumb extended upward. You should then advance the hand until the thumb prevents further insertion and the cylindrical handle lies firmly in the crease of the curved palm. Curl your fingers and thumb tightly over the top of the handle. All your fingers should be at the same angle. To have the proper power grip, make sure that the underside of the handle is fully supported on your curved palm with only the fingers and thumb preventing it from being pulled sideways or upward out of the palm.

If you must lift the object higher once you have lifted by extending your legs, you will be able to "curl" the object higher by using your biceps to flex the arms

Figure 6-4 To perform the power grip, grasp the handle of the litter with your palms up and your thumbs extending up. Make sure your hands are about 10″ apart and that your fingers are all at the same angle. The underside of the handle should be fully supported by the palms of your hands.

Performing the Power Lift

1 Lock your back into an upright, inward curve.

Spread and bend your legs.

Grasp the backboard, palms up and just in front of you.

Balance and center the weight between your arms.

2 Position your feet, straddle the object, and distribute weight.

3 Straighten your legs and lift, keeping your back locked in.

while maintaining the power grip and weight supported in the palms.

You should never grasp a cot or backboard with the hand placed palm down over the handle unless you are standing at the front end with your back to the cot, as when performing a diamond carry. In lifting with the palm down, the weight is supported by the fingers rather than the palm. This hand orientation places the tips of the fingers and thumb under the handle. If the weight forces them apart, your grasp on the handle will be lost.

When lifting a patient by a sheet or blanket, you should center the patient on the sheet and tightly roll up the excess fabric on each side. This produces a cylindrical handle that provides a strong, secure way to grasp the fabric (Figure 6-5 ▶).

When directly lifting a patient, you should tightly grip the patient in a place and manner that will ensure that you will not lose your grasp on the patient.

Weight and Distribution

Whenever possible, you should use a device that can be rolled to move a patient. However, in case a wheeled device is not available, you must make sure that you understand and follow certain guidelines for carrying a patient on a cot. (Table 6-1 ▶) shows the guidelines.

If a patient is supine on a backboard or is lying or in a semi-sitting position on the cot, his or her weight is not equally distributed between the two ends of the device. Between 68% and 78% of the body weight of a patient in a horizontal position is in the torso. Therefore, more of the patient's weight rests on the head half of the device than on the foot half.

A patient on a backboard or cot should be lifted and carried by four rescuers in a diamond carry, with one EMT-B at the head end of the device, one at the foot end, and one at each side of the patient's torso

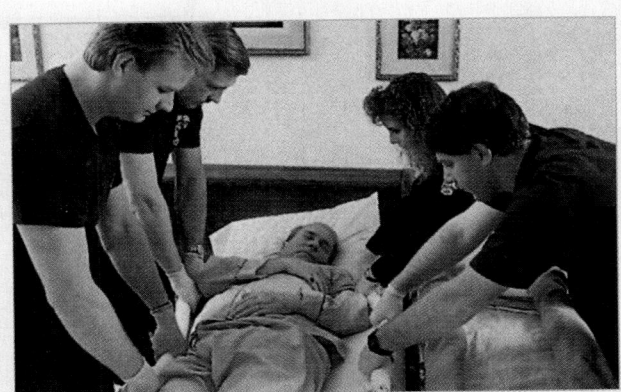

Figure 6-5 When lifting a patient by a bedsheet, you should center the patient on the sheet and tightly roll up the excess fabric on each side. This produces a cylindrical handle that provides a strong way to grasp the fabric.

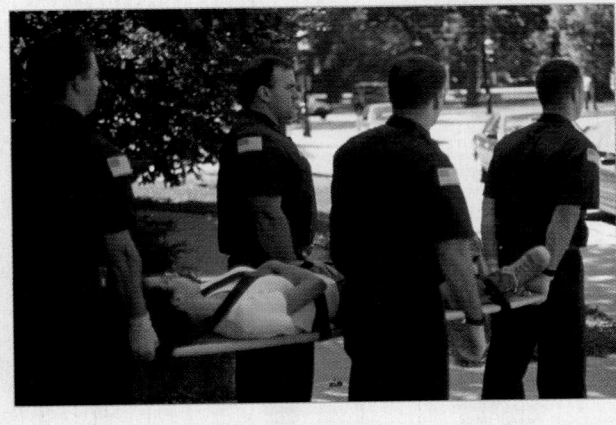

Figure 6-6 The diamond carry requires four rescuers, one each at the head of the backboard, the foot end, and each side of the patient's torso.

(Figure 6-6 ▶). Follow these steps to perform the diamond carry (Skill Drill 6-2 ▶):

1. To best balance the weight, the EMT-Bs at each side should be located so that they are able to grasp the board or stretcher with one hand adjacent to the distal edge of the patient's pelvis and the other midthorax. **All four lift the device while facing in toward the patient (Step 1).**

2. **Once the device has been lifted,** the EMT-B at the foot end turns around to face forward (**Step 2**).

3. **The EMT-B at each side** should grasp the backboard or cot with the head-end hand (**Step 3**).

4. **The EMT-Bs at the sides turn toward the patient's feet.** All four should be facing the same direction and will be walking forward when carrying the patient (**Step 4**).

A patient on a backboard or stretcher should be carried feet first to place the lightest load on the EMT-B at the patient's feet, who, to walk forward, must turn and grasp the handles with his or her back to the device. Carrying the patient feet first will also allow a conscious patient to see in the direction of movement.

It is important that you and your team use the correct lifting techniques to lift the cot. You must also make sure that your team members are of the same approximate height and strength.

TABLE 6-1 Guidelines for Carrying a Patient on a Cot

- Be sure that you know or can find out the weight to be lifted and the limitations of the team's abilities.
- Coordinate your movements with those of the other team members while constantly communicating with them.
- Do not twist your body as you are carrying the patient.
- Keep the weight that you are carrying as close to your body as possible while keeping your back in a locked-in position.
- Be sure to flex at the hips, not at the waist, and bend at the knees, while making sure that you do not hyperextend your back by leaning back from your waist.

You are the Provider   Part 3

You decide that the best way to move this patient is by using a vest-type extrication device. You carefully position the device and apply the straps. Meanwhile, fire department first responders have positioned the cot near the vehicle with the backboard ready to accept the patient.

4. What is the best way to move the patient from the vehicle to the cot? How many rescuers would be ideal for this maneuver?

6-2

Skill Drill

Performing the Diamond Carry

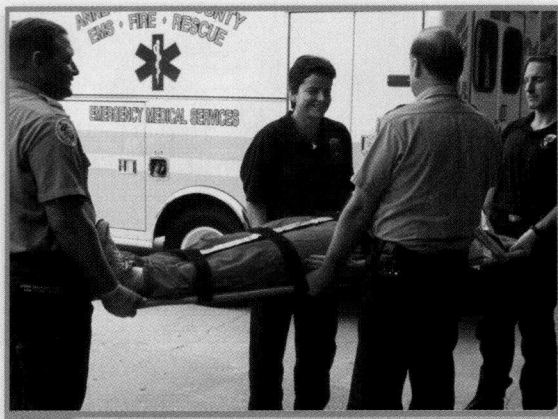

1 Position yourselves facing the patient.

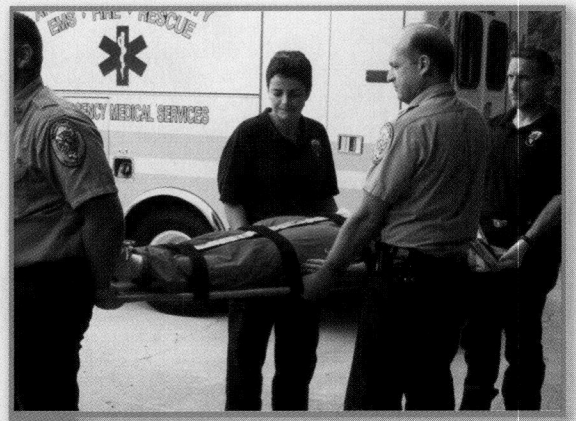

2 After the patient has been lifted, the EMT-B at the foot turns to face forward.

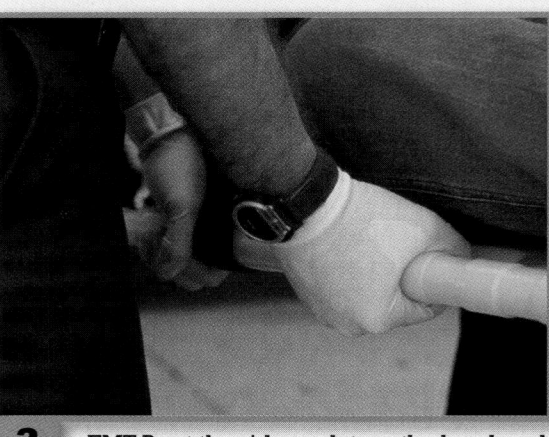

3 EMT-Bs at the side each turn the head-end hand palm down and release the other hand.

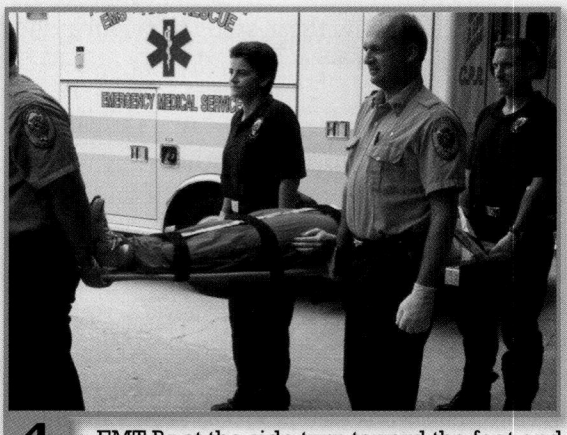

4 EMT-Bs at the side turn toward the foot end.

One method of lifting and carrying a patient on a backboard is the one-handed carrying technique (**Skill Drill 6-3 ▶**). With this method, four or more EMT-Bs each use one hand to support the backboard so that they are able to face forward as they are walking. Here are the steps:

1. **Before lifting the backboard**, be sure that at least two EMT-Bs are on each side of the backboard facing across from each other and using both hands (**Step 1**).
2. **Lift the backboard** to carrying height using correct lifting techniques, including a locked-in back (**Step 2**).
3. Once you have lifted the backboard to carrying height, you and your partners **turn in the direction you will be walking** and switch to using one hand (**Step 3**).

Be sure to pick up and carry the backboard with your back in the locked-in position. If you need to lean to either side to compensate for a weight imbalance, you have probably exceeded your weight limitation. If this occurs, you may need to add helpers or reevaluate the carry, or you might injure yourselves or drop the patient.

When you must carry a patient up or down a flight of stairs or other significant incline, use a stair chair if possible. When you must use a backboard or stretcher, be sure that the patient is anatomically secured to the device in such a way that he or she cannot slide significantly when the stretcher is at an angle (**Skill Drill 6-4 ▶**):

1. **Apply a strap** that passes tightly across the upper torso and through each armpit, but not over the arms, to hold the patient in place while leaving the arms free. The strap is secured to the

Performing the One-Handed Carrying Technique

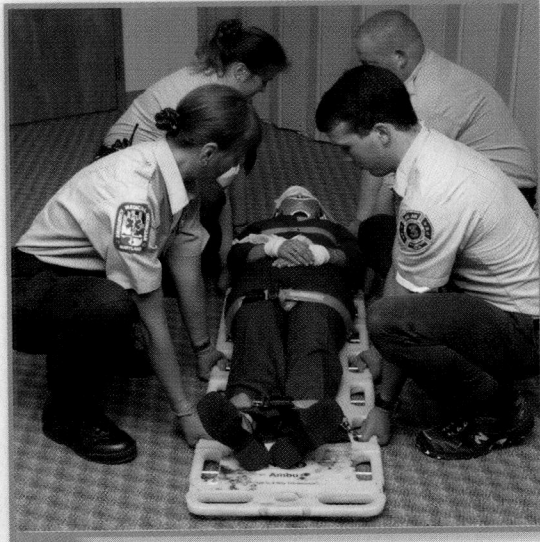

1 Face each other and use both hands.

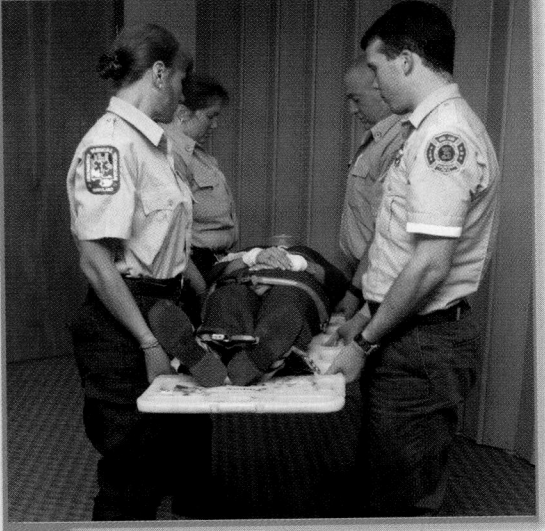

2 Lift the backboard to carrying height.

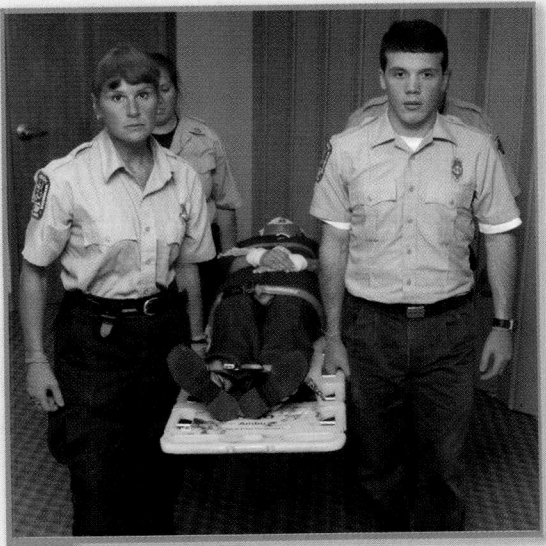

3 Turn in the direction you will walk and switch to using one hand.

handles at both sides of the backboard so that it cannot slide toward the foot end of the board. Strap the patient securely to the backboard (**Step 1**).

2. **When you carry the patient down stairs** or an incline, make sure the backboard or stretcher is carried with the foot end first so that the head end is elevated higher than the foot end. The straps will prevent the patient from sliding down or off the backboard (**Step 2**).

3. **When you carry a patient up stairs** or an incline, the elevated head end of the backboard or stretcher should go first (**Step 3**).

It is helpful to put taller rescuers at the foot of the cot when moving a patient up and down steps. This minimizes bending while lifting and moving the patient.

The <u>wheeled ambulance stretcher</u> or cot, which is a specially designed cot that can be rolled along the

Carrying a Patient on Stairs

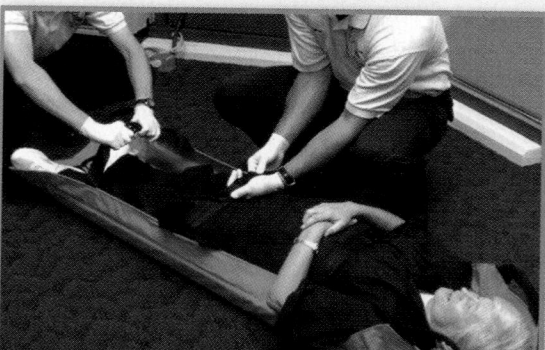

1 Strap the patient securely. Make sure one strap is tight across the upper torso, under the arms, and secured to the handles to prevent the patient from sliding.

2 Carry a patient down stairs with the foot end first, head elevated.

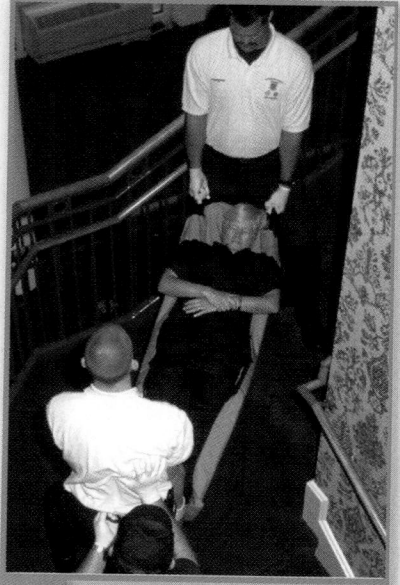

3 Carry the head end first going up stairs, always keeping the head elevated.

✳ EMT-B Tips

Since lifting and moving techniques require a team leader to coordinate and direct the process, it will save time and prevent confusion to establish either informal practices or formal procedures that tell all team members—in advance—who will be in charge of these activities.

ground, weighs between 40 and 70 lb, depending on its design and features (Figure 6-7 ▶). Because its weight must be added to that of the patient, it is generally not taken up or down stairs or to other locations where the patient must be carried for any significant distance. Moving a patient by rolling, using a cot or other rolling device, is preferred when the situation allows and helps prevent injuries from carrying. When the patient is upstairs, you should bring the wheeled ambulance stretcher to the ground floor landing and prepare it for the patient. You should then take either a

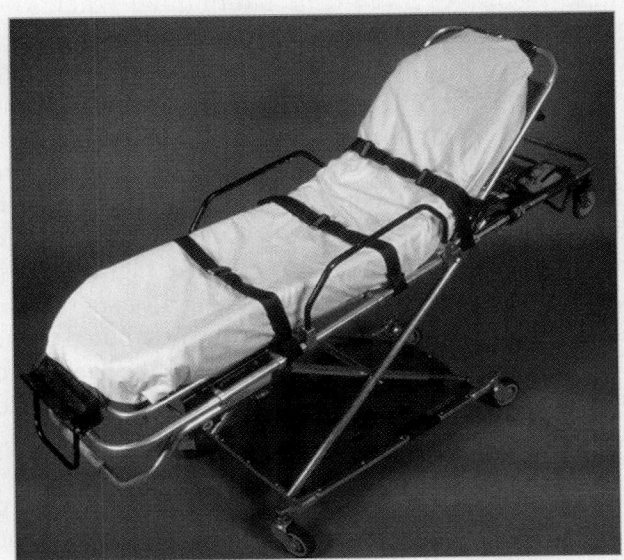

Figure 6-7 The wheeled ambulance stretcher is specially designed to roll along the ground.

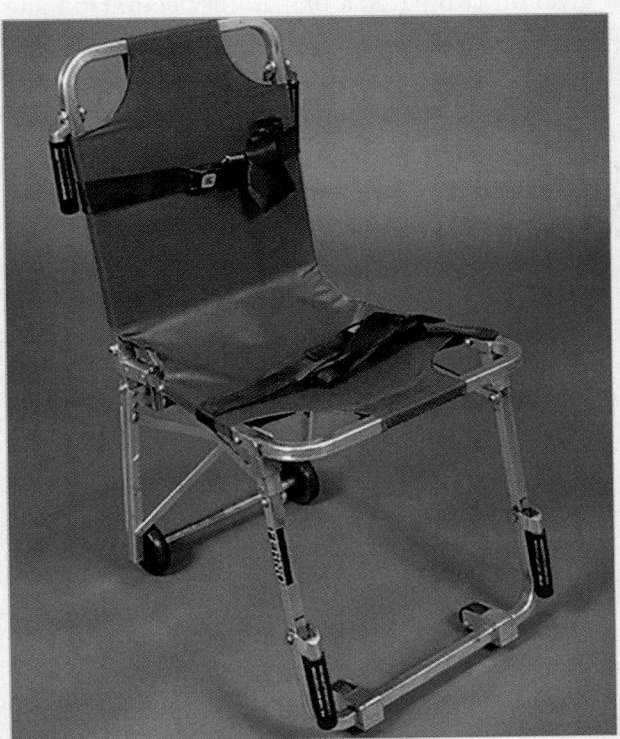

Figure 6-8 A wheeled stair chair can be used to transfer a conscious patient up or down a flight of stairs.

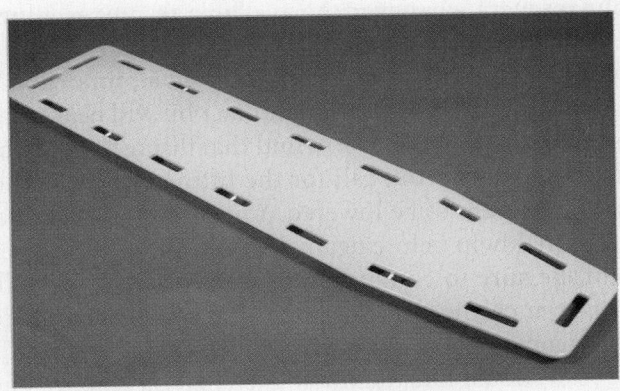

Figure 6-9 A backboard is used to transfer patients who must be moved in a supine or immobilized position.

wheeled stair chair or a backboard upstairs. Both of these devices are considerably lighter than a wheeled cot and may be used to carry the patient down to the waiting cot. Use a wheeled stair chair to bring a conscious patient down to the waiting cot if the patient's condition allows him or her to be placed in a sitting position (Figure 6-8 ▶). Once the cot has been reached, transfer the patient from the stair chair onto the cot. When the patient is in cardiac arrest, must be moved in a lying position, or must be immobilized, secure the patient onto a backboard. A backboard, which is a device that provides support to patients whom you suspect have hip, pelvic, spinal, or lower extremity injuries, is also called a spine board, trauma board, or longboard (Figure 6-9 ▶). You can then carry the patient on the backboard down the stairs to the prepared cot. Once you reach the cot, place both the board and patient on the cot and secure them with additional straps.

Directions and Commands

To safely lift and carry a patient, you and your team must anticipate and understand every move, and each move must be executed in a coordinated manner. The team leader should indicate where each team member is to be located and rapidly describe the sequence of steps that will be performed to ensure that the team knows what is expected of it before any lifting is initiated. If you must lift and move the patient through a number of separate stages, the team leader should first give an abbreviated overview of the stages, followed by a more detailed explanation of each stage just before it will occur.

Orders that will initiate the actual lifting or moving or any significant changes in movement should be given in two parts: a preparatory command and a command of execution. For example, if the team leader says "All ready to stop. STOP!" the "All ready to stop" will get your attention, identify who should act, and prepare them to act; the declarative "STOP!" will indicate the exact moment for execution. Commands of execution should be delivered in a louder voice. Often, a countdown is helpful when you need to lift a patient. To avoid confusion in using a countdown, always clarify whether "three" is to be a part of the preparatory command or whether it is to serve as the order to execute. You can say "We're going to lift on three. One-two-THREE!" or "I'm going to count to three and then we're going to lift. One-two-three-LIFT!"

Additional Lifting and Carrying Guidelines

You should estimate how much the patient weighs before attempting to lift him or her. Remember to add the weight of the carrying device and equipment when calculating weight. Commonly, adult patients weigh between 100 and 210 lb. If you use the correct technique, you and one other EMT-B should be able to safely lift this weight. Depending on your individual strength, you and another EMT-B may be able to safely lift an even heavier patient. However, because it is quite a bit safer to have four rescuers lift, you should try to use four rescuers whenever the available resources allow. You should know how much you can comfortably and safely lift and should not attempt to lift a proportional weight (the share of the weight that you will bear) that exceeds this amount. If you find that lifting the patient places strain on you, call for the lifting to be stopped and the patient to be lowered. You should then obtain additional help before again attempting to lift the patient. Be sure to communicate clearly and frequently with your partner and other rescuers whenever you are lifting a patient.

You should not attempt to lift a patient who weighs more than 250 lb with fewer than four rescuers, regardless of individual strength. Protocols should include a method to rapidly summon additional help to lift and carry such a patient or, as in the case of a cardiac arrest, provide and maintain the necessary care in the field and when moving and transporting the patient. In addition, you must know, or be able to find out, the weight limitations of the equipment you are using and how to handle patients who exceed the weight limitations. Special techniques, equipment, and resources generally are required to move any patient who weighs more than 300 lb to the ambulance.

These resources should be summoned when you arrive.

Because more than half of a patient's weight is distributed to the head end of the backboard or cot, the strongest of the available EMT-Bs should be located at the head end of the device. Even with four or more EMT-Bs carrying the patient, the strain on the EMT-B carrying the head end of the device will be increased when you must negotiate a narrow area or flight of stairs. In carrying a patient up or down a flight of stairs, proportionally greater weight will also be distributed to the EMT-B who is carrying the foot end when the backboard or cot becomes angled because of the incline. You should anticipate this and, in such cases, make sure the two strongest EMT-Bs are positioned at the head and foot ends of the board. Because of the incline of the stairway, if one of the two EMT-Bs is considerably taller than the other, it will be easier if the shorter of the two is at the head end and the taller is at the foot end.

The dynamics that are involved in carrying a patient down a flight of stairs or for any significant distance will not allow you to carry as much proportional weight as you can to safely lift or support the patient during a move onto a nearby backboard or cot. Therefore, if you feel that you are approaching your maximum lifting capacity as you are moving the patient onto a backboard or cot, you should not attempt to lift and carry the patient for any significant distance or down a flight of stairs. You can again attempt to lift and carry the patient after you have decreased the amount of proportional weight you will be carrying by changing your position on the device or that of the others on the team or have obtained additional help.

You should try to use a stair chair instead of a cot, whenever possible, to carry a patient down stairs. Follow these steps (**Skill Drill 6-5** ▶):

1. **Secure the patient to the stair chair with straps.** At a minimum, use a lap belt at the hips and a strap around the chest. You should also use some method to secure the arms and hands so the patient does not reach out to grasp something and throw the carrying team off balance. You can ask the patient to fold his or her hands on the chest (**Step 1**).
2. **Rescuers take their places around the patient** seated on the chair: one at the head and one at the foot. The rescuer at the head will give directions to coordinate the lift and carry (**Step 2**).
3. **A third rescuer precedes the two carrying the chair** to open doors, spot them on stairs, and so

Using a Stair Chair

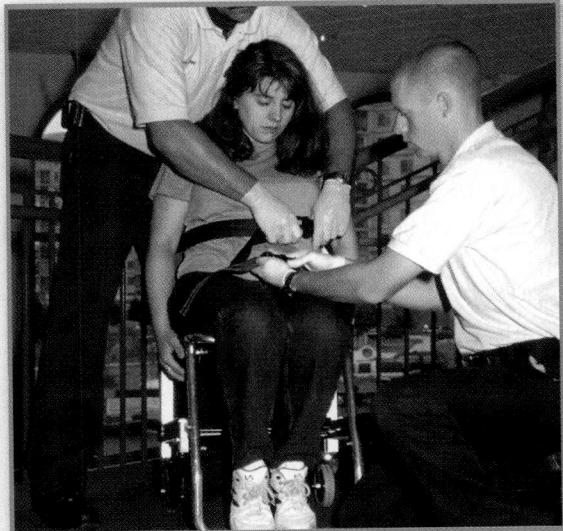

1 Position and secure the patient on the chair with straps.

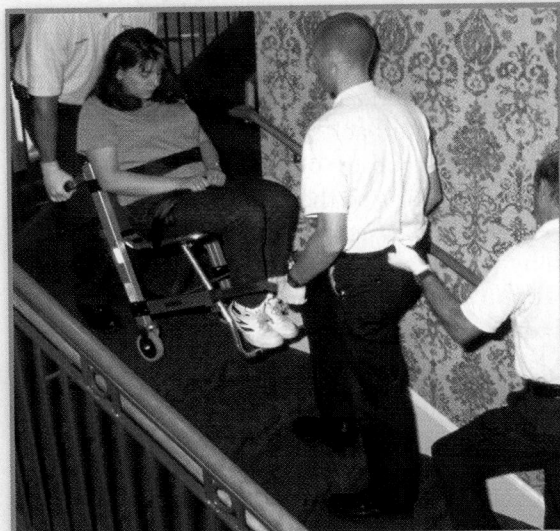

2 Take your places at the head and foot of the chair.

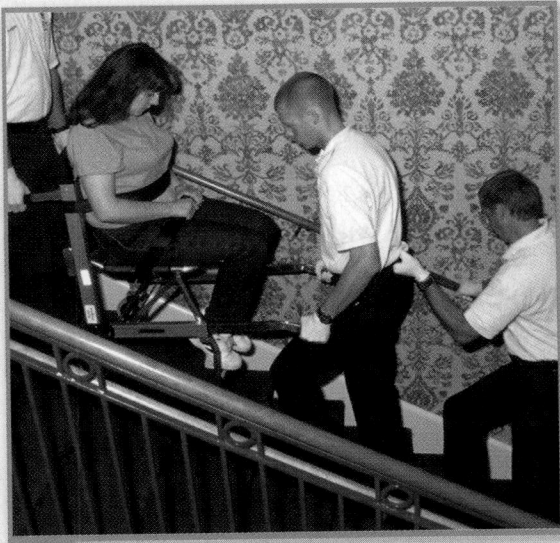

3 A third rescuer "backs up" the rescuer carrying the foot.

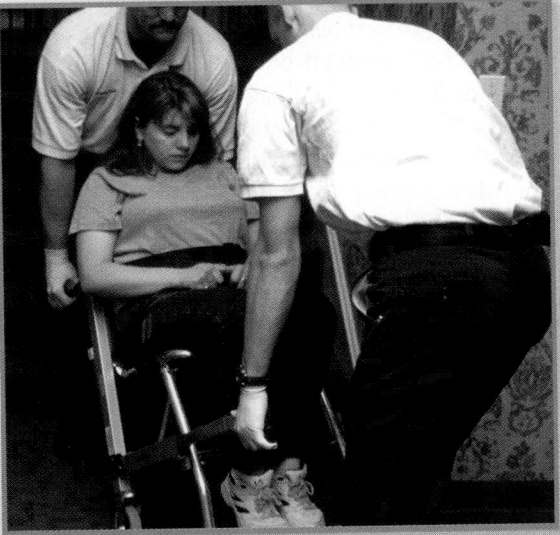

4 Lower the chair to roll on landings, or for transfer to the cot.

on. For lengthy carries, the third responder can also rotate into the carrying team to provide breaks for the other two (**Step 3**).

4. **When reaching landings and other flat intervals** in the carry, lower the chair to the ground and roll it rather than carrying it. When reaching the level where the cot awaits, roll the chair into position next to the cot in preparation for transferring the patient (**Step 4**).

As with other carries, always remember to keep your back in a locked-in position and to flex at the hips, not the waist. You should also bend at the knees and keep the patient's weight and your arms as close to your body as possible. Twisting while carrying or moving a patient will increase the risk of injury. Try to avoid any unnecessary lifting and carrying of the patient. If an assist, log roll, or body drag will not harm or jeopardize the patient, use one to move the patient onto the backboard or cot.

Principles of Safe Reaching and Pulling

When you use a body drag to move a patient, the same basic body mechanics and principles apply as when lifting and carrying. Your back should always be locked and straight, not curved or bent laterally, and you should avoid any twisting so that the vertebrae remain in normal alignment. When you are reaching overhead, avoid hyperextending your back. When you are pulling a patient who is on the ground, you should always kneel to minimize the distance that you will have to lean over (Figure 6-10A ▼). To keep your reach within the recommended distance, reach forward and grasp the patient so that your elbows are just beyond the anterior torso (Figure 6-10B ▼). When you are pulling a patient who is at a different height from you, bend your knees until your hips are just below the height of the plane across which you will be pulling the patient. During pulling, you should extend your arms no more than about 15″ to 20″ in front of your torso. Reposition your feet (or knees, if kneeling) so that the force of pull will be balanced equally between both arms and the line of pull will be centered between them (Figure 6-10C ▼). Pull the patient by slowly flexing your arms. When you can pull no farther because your hands have reached the front of your torso, stop and move back another 15″ to 20″. Then, when properly positioned, repeat the steps. You should alternate between pulling the patient by flexing your arms and then repositioning yourself so that your arms are again extended with your hands about 15″ in front of your torso. By not moving yourself and the patient simultaneously, you will prevent undesirable jostling of the patient and the chance that sudden unscheduled force will occur across your spine. You should also try to prevent injury to yourself by avoiding situations that involve strenuous effort lasting more than 1 minute.

If you must drag a patient across a bed, you will have to kneel on the bed to avoid reaching beyond the recommended distance. Then follow the steps described until the patient is within 15″ to 20″ of the bed's edge (see Figure 6-10). You can then complete the drag while standing at the side of the bed. Rather than dragging the patient by his or her clothing, use the sheet or blanket under the patient for this purpose. You can roll the bedding under the patient until it is about 6 inches wider than the patient. Pull on the rolled bedding smoothly and evenly to glide the patient to the bedside.

Unless the patient is on a backboard, transfer a patient from the cot to a bed in the emergency department or the patient's hospital room with a body drag. With the cot at the same height as the bed and held firmly against its side, you and another EMT-B should kneel on the hospital bed and, in the manner previously described, drag the patient in increments until he or she is properly centered on the bed. When transferring the patient onto a narrow examining table, rather than kneeling on the table, you can usually drag the patient while standing against the opposite side. A third person may need to take both sides of the head to move the patient safely.

Sometimes during a body drag, you and another EMT-B may have to pull the patient with one of you on each side of the patient. You will have to alter the usual pulling technique to prevent pulling sideways and producing adverse lateral leverage against your lower back. You should position yourself by kneeling just beyond the patient's shoulder and facing toward his or her groin (Figure 6-11A ▶). By extending one arm across and in front of your chest, you can grasp the armpit and, with the other arm extended in front and to the side of the

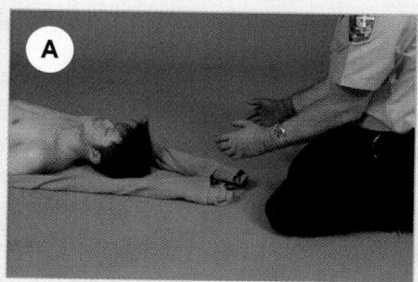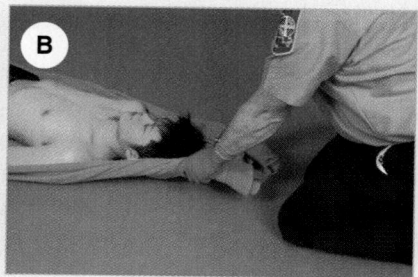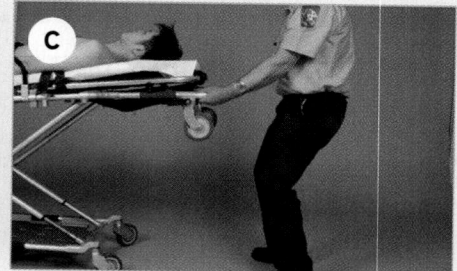

Figure 6-10 Reaching and pulling safely. **A.** Kneel to pull a patient who is on the ground. **B.** When pulling, your elbows should only extend just beyond the anterior torso. **C.** Bend your knees to pull a patient who is at a different height than you are. Position your feet or knees to balance the force of pull.

torso, the patient's belt. Then, by raising your elbows and flexing your arms, you can pull the patient with the line of force at the minimum angle possible Figure 6-11B ▼.

Generally, when log rolling a patient onto his or her side, you will initially have to reach farther than 18″ Figure 6-12 ▶. To minimize this distance, kneel as close to the patient's side as possible, leaving only enough room so that your knees will not prevent the patient from being rolled. When you lean forward, keep your back straight and lean solely from the hips. Be sure to use your shoulder muscles to help with the roll. To minimize the amount of time you are extended like this and to support the patient's weight, roll the patient without stopping until the patient is resting on his or her side. Some EMS experts consider that, during a log roll, you should pull rather than push the patient. Local protocols will guide your training in this area. Pulling toward you allows your legs to prevent the patient from rolling over completely, or from rolling beyond the intended distance.

When you are rolling the wheeled ambulance stretcher, make sure that it is elevated Figure 6-13 ▼. Push the stretcher from the head end. If you are guiding the cot from the foot end, make sure your arms are held close to your body, and be careful to avoid reaching significantly behind you or hyperextending your back. Your back should be locked, straight, and untwisted. While you are walking and guiding the stretcher, bend slightly forward at the hips. As you walk, your

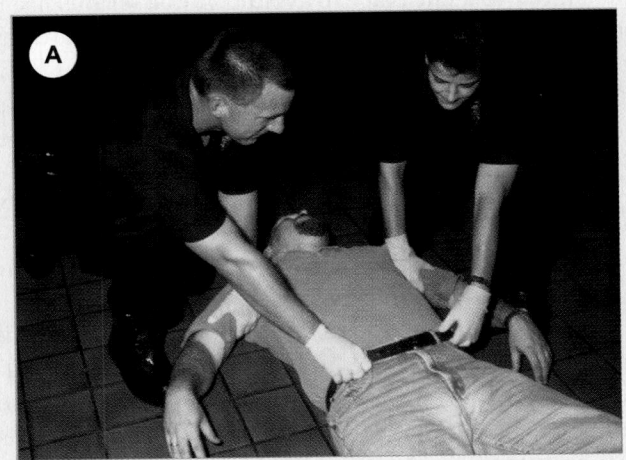
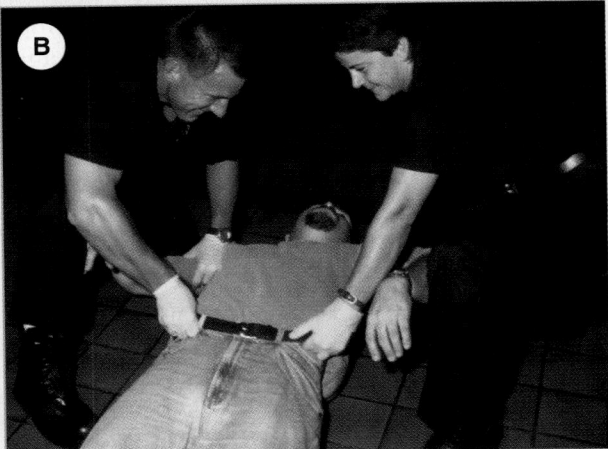

Figure 6-11 A body drag with an EMT-B on each side of the patient. **A.** Kneel just beyond the patient's shoulder facing his or her groin. Extend one arm across and in front of your chest, and grasp the armpit. Extend your other arm in front and to the side of the patient's torso, and grasp the patient's belt. **B.** Raise your elbows and flex your arms to pull the patient.

Figure 6-12 When placing a patient onto a backboard, roll the patient onto his or her side. Kneel as close to the patient's side as possible, leaving only enough room so that your knees will not prevent the patient from being rolled. Lean forward, keeping your back straight and leaning solely from the hips. Use your shoulder muscles to help with the roll.

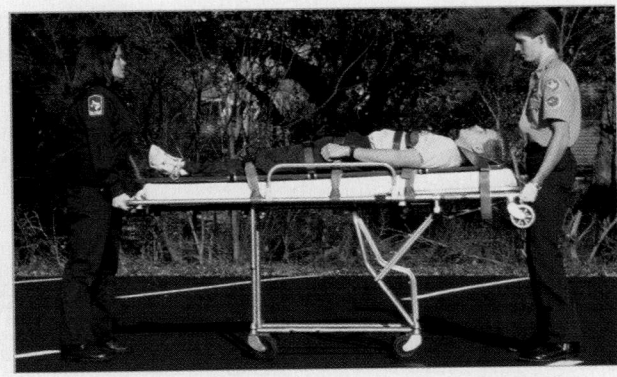

Figure 6-13 Push the stretcher from the head end. If you are guiding the cot from the foot end, make sure your arms are held close to your body, and be careful to avoid reaching significantly behind you or hyperextending your back. Your back should be locked, straight, and untwisted.

legs are pulled back with the feet on the ground, your pelvis is moved forward, and the movement of the pelvis is transferred to the stretcher through your straight torso and firmly held arms. You should try to keep the line of the pull through the center of your body by bending your knees.

A second EMT-B should guide the head end and assist you by pushing with his or her arms held with the elbows bent so that the hands are about 12″ to 15″ in front of the torso. To protect your elbows from injury, you should never push an object with your arms fully extended in a straight line and the elbows locked. When you push with the elbow bent but firmly held from bending further, the strong muscles of the arm serve as a shock absorber if the wheels or foot end of the stretcher strikes an obstacle that causes its progress to be suddenly slowed or stopped. You must be sure that you push from the area of your body that is between the waist and shoulder. If the weight you are pushing is lower than your waist, you should push from a kneeling position. Be careful that you do not push or pull from an overhead position.

General Considerations

Moving a patient should normally be done in an orderly, planned, and unhurried fashion. This approach will protect both you and the patient from further injury and reduce the risk of worsening the patient's condition when he or she is moved. At a minimum, on most calls you will have to lift and carry the patient to the wheeled ambulance stretcher, move the stretcher and patient to the ambulance, and load the stretcher into the patient compartment.

You will often have to include several additional steps to place the patient onto a backboard and/or carry him or her down a flight of stairs. You will also have to add a stop at the top of the stairway so that everyone can reposition for carrying the patient down the stairs. Repositioning usually requires lowering the backboard

to the ground and lifting it again when all EMT-Bs are in their proper places. If you are carrying the patient in a stair chair, the additional step occurs after you have descended the stairs and reached the stretcher. At that point, you will have to assist or lift the patient from the stair chair onto the stretcher.

You should carefully plan ahead and select the methods that will involve the least lifting and carrying. Remember to always consider whether there is an option that will cause less strain to you and the other EMT-Bs.

Emergency Moves

You should use an emergency move to move a patient before initial assessment and care are provided when there is some potential danger, and you and the patient must move to a safe place to avoid possible serious harm or death. The presence of fire, explosives, or hazardous materials and your inability to protect the patient from other hazards or gain access to others in a vehicle that need lifesaving care are all situations in which you should use an emergency move.

The only other time you should use an emergency move is if you cannot properly assess the patient or provide immediate critical emergency care because of the patient's location or position.

If you are alone and danger at the scene makes it necessary for you to use an emergency move, regardless of a patient's injuries, you should use a drag to pull the patient along the long axis of the body. This will help to keep the spinal column in line as much as possible. When performing an emergency move, one of your primary concerns is the danger of aggravating an existing spinal injury. Remember that it is impossible to remove a patient quickly from a vehicle while providing as much protection to the spine as you would give by using an immobilization device. However, if you follow certain guidelines during the move, you can usually move a patient from a life-

You are the Provider   Part 4

As you are loading your patient into the back of the ambulance, she complains that the extrication device is too tight and is making it hard for her to breathe.

5. What should you do?

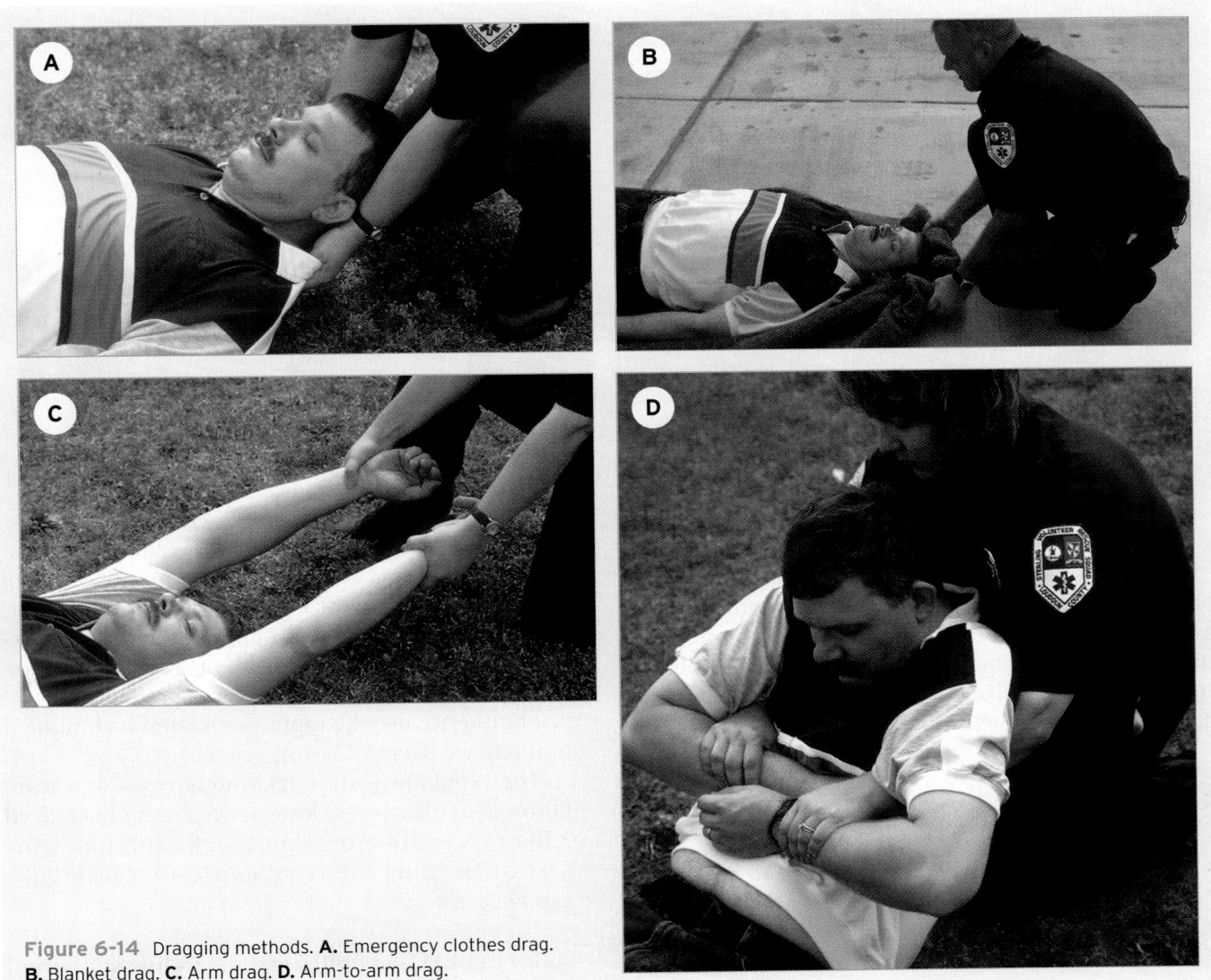

Figure 6-14 Dragging methods. **A.** Emergency clothes drag. **B.** Blanket drag. **C.** Arm drag. **D.** Arm-to-arm drag.

threatening situation without causing further injury to the patient.

You can move a patient on his or her back along the floor or ground by using one of the following methods:

- Pull on the patient's clothing in the neck and shoulder area (Figure 6-14A ▲).
- Place the patient onto a blanket, coat, or other item that can be pulled (Figure 6-14B ▲).
- Rotate the patient's arms so that they are extended straight on the ground beyond his or her head, grasp the wrists, and, with the arms elevated above the ground, drag the patient (Figure 6-14C ▲).
- Place your arms under the patient's shoulders and through the armpits, and, while grasping the patient's arms, drag the patient backward (Figure 6-14D ▲).

If you are alone and must remove an unconscious patient from a car, you should first move the patient's legs so they are clear of the pedals and are against the seat. Then rotate the patient so that his or her back is positioned toward the open car door. Next, place your arms through the armpits and support the patient's head against your body (Figure 6-15A ▶). While supporting the patient's weight, drag the patient from the seat. If the legs and feet clear the car easily, you can rapidly drag the patient to a safe location by continuing this method (Figure 6-15B ▶). If the legs and feet do not clear the car easily, you can slowly lower the patient until he or she is lying on his or her back next to the car, clear the legs from the vehicle, and, as previously described, use a long-axis body drag to move the patient a safe distance from the vehicle.

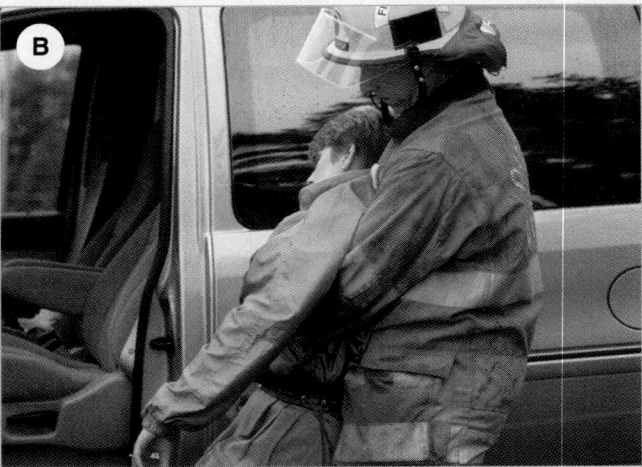

Figure 6-15 One-person technique for moving an unconscious patient from a car. **A.** Grasp the patient under the arms. **B.** Pull the patient down into a supine position.

You should use one-person techniques to move a patient only if an immediately life-threatening danger exists and you are alone or, because of the pressing nature of the danger, your partner is moving a second patient simultaneously. Additional one-rescuer drags, carries, and lifts are shown in (Figure 6-16 ▶).

Urgent Moves

An urgent move may be necessary for moving a patient with an altered level of consciousness, inadequate ventilation, or shock (hypoperfusion). An extreme weather condition may also make an urgent move necessary. In some cases, patients must be urgently moved from the location or position in which they are found. When a patient who is sitting in a car or truck must be urgently moved, you should use the rapid extrication technique.

Rapid Extrication Technique

The long backboard, short backboard, and vest-type devices are known as immobilization devices. Normally, you would use an extrication-type vest or half-backboard device to immobilize a seated patient with a suspected spinal injury before removing the patient from the car. However, using either of these devices usually requires between 6 and 8 minutes, in some cases even longer. By using the rapid extrication technique instead, the patient can be moved from sitting in the car to lying supine on a backboard in 1 minute or less. (Table 6-2 ▶) describes the situations in which you should use the rapid extrication technique.

In such cases, the delay that occurs in applying an extrication-type vest or half-board is contraindicated. However, the manual support and immobilization that you provide when using the rapid extrication technique produce a greater risk of spine movement. You should not use the rapid extrication technique if no urgency exists.

The rapid extrication technique requires a team of three EMT-Bs who are knowledgeable and practiced in the procedure. You should take the following steps when using the rapid extrication technique (Skill Drill 6-6 ▶):

1. **First EMT-B applies manual in-line support** of the patient's head and cervical spine from behind. Support may be applied from the side, if necessary, by reaching through the driver's side doorway (**Step 1**).
2. **Second EMT-B serves as team leader** and, as such, gives the commands until the patient is supine

---

TABLE 6-2 Situations in Which to Use the Rapid Extrication Technique

- The vehicle or scene is unsafe.
- The patient cannot be properly assessed before being removed from the car.
- The patient needs immediate intervention that requires a supine position.
- The patient's condition requires immediate transport to the hospital.
- The patient blocks the EMT-B's access to another seriously injured patient.

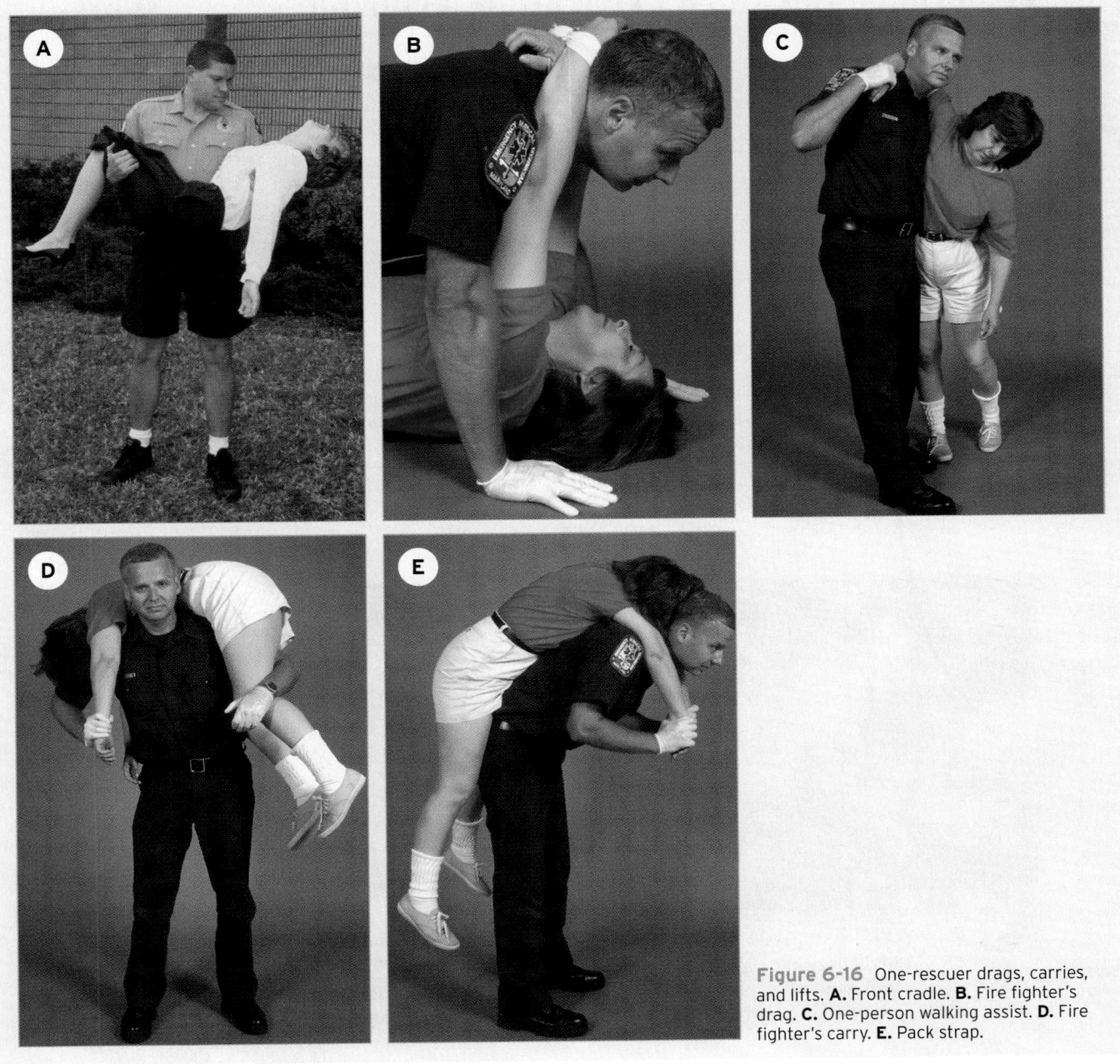

Figure 6-16 One-rescuer drags, carries, and lifts. **A.** Front cradle. **B.** Fire fighter's drag. **C.** One-person walking assist. **D.** Fire fighter's carry. **E.** Pack strap.

on the backboard. Because Second EMT-B lifts and turns the patient's torso, he or she must be physically capable of moving the patient. Second EMT-B works from the driver's side doorway. If First EMT-B is also working from that doorway, Second EMT-B should stand closer to the door hinges toward the front of the vehicle. Second EMT-B applies a cervical immobilization device and may perform the initial assessment (**Step 2**).

3. **Second EMT-B provides continuous support** of the patient's torso until the patient is supine on the backboard. Once Second EMT-B takes con-

trol of the torso, usually in the form of a body hug, he or she should not let go of the patient for any reason. Some type of cross-chest shoulder hug usually works well, but you will have to decide what method works best for you on any given patient. You must remember that you cannot simply reach into the car and grab the patient; this will only twist the patient's torso. You must rotate the patient as a unit.

4. **Third EMT-B works from the front passenger's seat** and is responsible for rotating the patient's legs and feet as the torso is turned, ensuring

Skill Drill 6-6

Performing the Rapid Extrication Technique

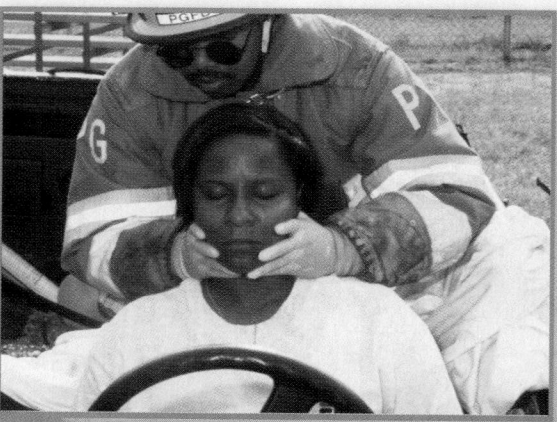

1 First EMT-B provides in-line manual support of the head and cervical spine.

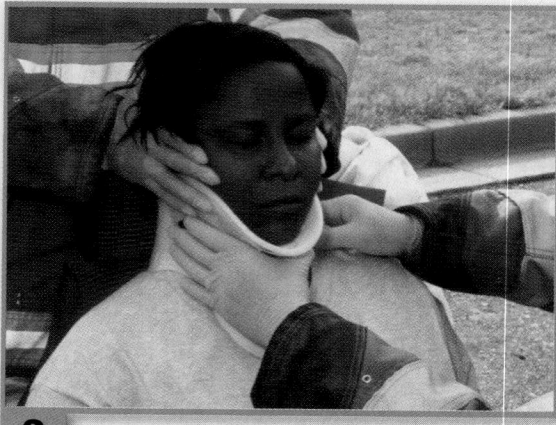

2 Second EMT-B gives commands, applies a cervical collar, and performs the initial assessment.

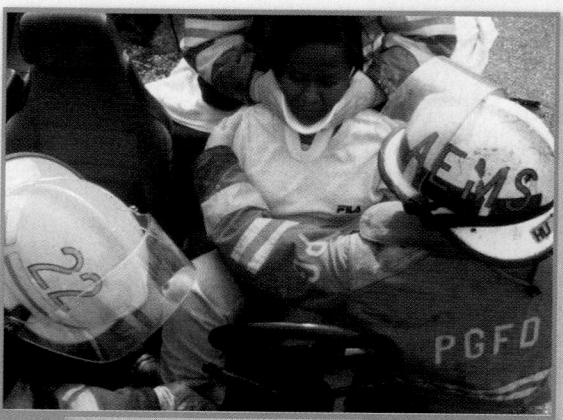

3 Second EMT-B supports the torso.
Third EMT-B frees the patient's legs from the pedals and moves the legs together, without moving the pelvis or spine.

4 Second and Third EMT-Bs rotate the patient as a unit in several short, coordinated moves.
First EMT-B (relieved by Fourth EMT-B or bystander as needed) supports the head and neck during rotation (and later steps).

that they are free of the pedals and any other obstruction. With care, Third EMT-B should first move the patient's nearer leg laterally without rotating the patient's pelvis and lower spine. The pelvis and lower spine rotate only as Third EMT-B moves the second leg during the next step. Moving the nearer leg early makes it much easier to move the second leg in concert with the rest of the body. After Third EMT-B moves the legs together, they should be moved as a unit (**Step 3**).

These first four steps of the rapid extrication technique direct the team to their starting positions and responsibilities. First EMT-B applies in-line support and immobilization of the head and neck. Second EMT-B gives orders and supports the torso. Third EMT-B moves and supports the patient's legs. The team is now ready to move the patient.

5. The patient is rotated 90° so that the back is facing out the driver's door and the feet are on the front passenger's seat. This coordinated move-

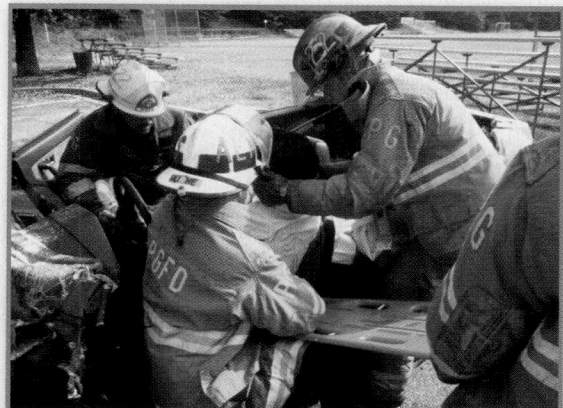

5 First (or Fourth) EMT-B places the backboard on the seat against the patient's buttocks.

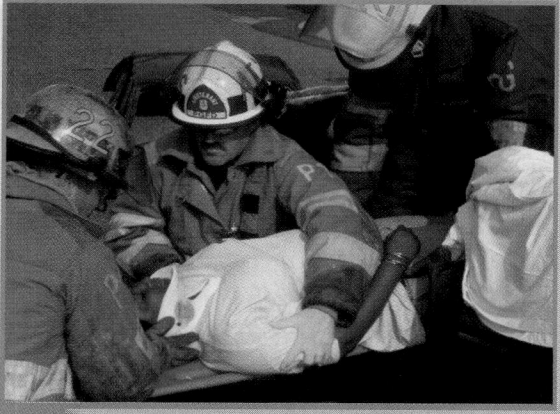

6 Third EMT-B moves to an effective position for sliding the patient.

Second and Third EMT-Bs slide the patient along the backboard in coordinated, 8″ to 12″ moves until the hips rest on the backboard.

7 Third EMT-B exits the vehicle, moves to the backboard opposite Second EMT-B, and they continue to slide the patient until patient is fully on the board.

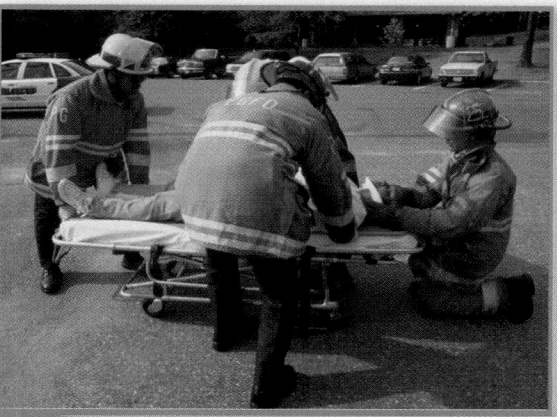

8 First (or Fourth) EMT-B continues to stabilize the head and neck while Second and Third carry the patient away from the vehicle.

ment is done in three or four short, quick "eighth turns." Second EMT directs each quick turn by saying, "Ready, turn" or "Ready, move." Hand position changes should be made between moves.

6. **In most cases, First EMT-B will be working from the back seat.** At some point, either because the doorpost is in the way or because he or she cannot reach farther from the back seat, First EMT-B will be unable to follow the torso rotation. At that time, Third EMT-B should assume temporary in-line support of the head and neck until First EMT-B can regain control of the head from outside the vehicle. If a fourth EMT-B is present, Fourth EMT-B stands next to Second EMT-B. Fourth EMT-B takes control of the head and neck from outside the vehicle without involving Third EMT-B. As soon as the change has been made, the rotation can continue (**Step 4**).

7. **Once the patient has been fully rotated,** the backboard should be placed against the patient's

buttocks on the seat. Do not try to wedge the backboard under the patient. If only three EMT-Bs are present, be sure to place the backboard within arm's reach of the driver's door before the move so that the board can be pulled into place when needed. In such cases, the far end of the board can be left on the ground. When a fourth EMT-B is available, First EMT-B exits the rear seat of the car, places the backboard against the patient's buttocks, and maintains pressure in toward the vehicle from the far end of the board. (Note: When the door opening allows, some EMT-Bs prefer to insert the backboard onto the car seat before the patient is rotated.)

8. **As soon as the patient has been rotated** and the backboard is in place, Second EMT-B and Third EMT-B lower the patient onto the board while supporting the head and torso so that neutral alignment is maintained. First EMT-B holds the backboard until the patient is secured (**Step 5**).

9. **Next, Third EMT-B must move across the front seat** to be in position at the patient's hips. If Third EMT-B stays at the patient's knees or feet, he or she will be ineffective in helping to move the body's weight. The knees and feet follow the hips.

10. **Fourth EMT-B maintains manual in-line support** of the head and now takes over giving the commands. If a fourth EMT-B is not present, you can direct a volunteer to assist you. Second EMT-B maintains direction of the extrication. Second EMT-B stands with his or her back to the door, facing the rear of the vehicle. The backboard should be immediately in front of Third EMT-B. Second EMT-B grasps the patient's shoulders or armpits. Then, on command, Second EMT-B and Third EMT-B slide the patient 8″ to 12″ along the backboard, repeating this slide until the patient's hips are firmly on the backboard (**Step 6**).

11. **At that time, Third EMT-B gets out of the vehicle** and moves to the opposite side of the backboard, across from Second EMT-B. Third EMT-B now takes control at the shoulders, and Second EMT-B moves back to take control of the hips. On command, these two EMT-Bs move the patient along the board in 8″ to 12″ slides until the patient is placed fully on the board (**Step 7**).

12. **First (or Fourth) EMT-B continues to maintain manual in-line support** of the head. Second EMT-B and Third EMT-B now grasp their side of the board, and then carry it and the patient away from the vehicle onto the prepared cot nearby (**Step 8**).

In some cases, you will be able to rest the head end of the backboard on the cot while the patient is moved onto the backboard. In others, you will not. Once the backboard and patient have been placed on the cot, you should begin lifesaving treatment immediately. If you used the rapid extrication technique because the scene was dangerous, you and your team should immediately move the cot a safe distance away from the vehicle before you assess or treat the patient.

The steps of the rapid extrication technique must be considered a general procedure to be adapted as needed. Two-door cars differ from four-door models. Larger cars differ from smaller compact models, pickup trucks, and full-size sedans and four-wheel-drive vehicles. You will handle a large, heavy adult differently from a small adult or child. Every situation will be different—a different car, a different patient, and a different crew. Your resourcefulness and ability to adapt are necessary elements to successfully perform the rapid extrication technique.

Nonurgent Moves

When both the scene and the patient are stable, you should carefully plan how to move the patient. If your patient move is rushed or not well planned, it may result in discomfort or injury to the patient, you, and your team. Before you attempt any move, the team leader must be sure that there are enough personnel, any obstacles have been identified or removed, the proper equipment is available, and the procedure and path to be followed have been clearly identified and discussed.

In nonurgent situations, you and your team may choose one of several methods for lifting and carrying a patient. Three general methods are presented here, which may serve as a basis for your plan. You may adapt these procedures to meet your needs on a case-by-case basis.

Direct Ground Lift

The direct ground lift is used for patients with no suspected spinal injury who are found lying supine on the ground. You should use this lift when you have to lift and carry the patient some distance to be placed on the cot. If you find the patient semi-prone or lying on his or her side, you should first roll the patient onto his or her back. Ideally, the direct ground lift should be performed by three EMT-Bs; however, it can be done with only two. The direct ground lift is performed as follows:

1. **Line up on one side of the patient** with First EMT-B at the patient's head, Second EMT-B at the patient's waist, and Third EMT-B at the patient's knees. All EMT-Bs kneel on one knee, preferably the same knee (Figure 6-17A ▼).
2. **The patient's arms should be placed on his or her chest** if possible.
3. **First EMT-B places one arm under the patient's neck and shoulders** and cradles the patient's head. First EMT-B then places the other arm under the patient's lower back.
4. **Second EMT-B places one hand under the patient's waist,** and the other under the knees.
5. **Third EMT-B places one arm under the patient's knees** and the other under the ankles.
6. **On command, the team lifts the patient** up to knee level as each EMT-B rests an arm on his or her knee (Figure 6-17B ▼).
7. **As a team and on signal,** each EMT-B rolls the patient in toward his or her chest. Again on signal, the team stands and carries the patient to the cot (Figure 6-17C ▼).
8. **The steps are reversed** to lower the patient onto the cot.

Extremity Lift

The extremity lift may also be used for patients with no suspected extremity or spinal injuries that are supine or in a sitting position. The extremity lift may be especially helpful when the patient is in a very narrow space or there is not enough room for the patient and a team of EMTs to stand side by side.

Communication is the key to success with this lift. You and your partner must coordinate your movements through direct verbal commands. You should perform the extremity lift as follows (Skill Drill 6-7 ▶):

1. **First EMT-B kneels behind the patient's head** as Second EMT-B kneels at the patient's feet. The two EMT-Bs are facing each other.
2. **The patient's hands should be crossed** over his or her chest.
3. **First EMT-B places one hand** under each of the patient's armpits. First EMT-B grasps the patient's wrists and pulls the upper torso until the patient is in a sitting position (**Step 1**).
4. **Second EMT-B moves to a position** between the patient's legs, facing in the same direction as the patient, and slips his or her hands under the patient's knees (**Step 2**).
5. **As the EMT-B at the head gives the command,** both stand fully upright and move the patient to the stretcher (**Step 3**).

You will be less likely to injure yourself if you bend at the hips and knees and use your legs for lifting.

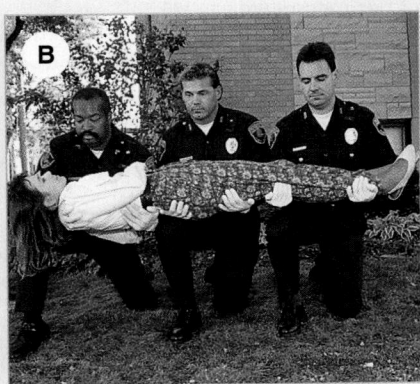

Figure 6-17 The direct ground lift. **A.** Line up on one side of the patient, with one EMT-B at the head, one at the waist, and one at the patient's knees. Place the patient's arms on his or her chest. **B.** On command, lift the patient to knee level. **C.** On command, roll the patient toward your chest, then stand and carry the patient to the cot.

Extremity Lift

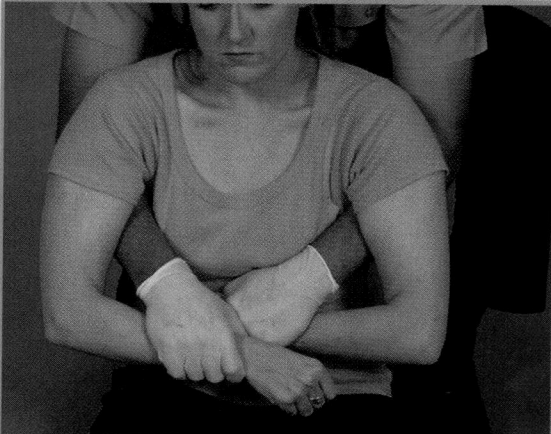

1 Patient's hands are crossed over the chest.

First EMT-B grasps patient's wrists or forearms and pulls patient to a sitting position.

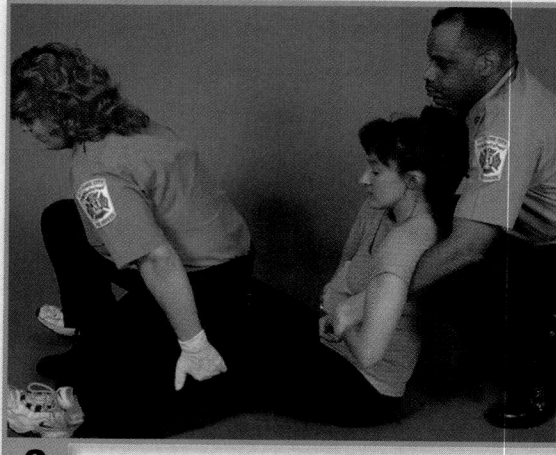

2 When the patient is sitting, First EMT-B passes his or her arms through patient's armpits and grasps the patient's opposite (or his or her own) forearms or wrists.

Second EMT-B kneels between the legs, facing in the same direction as the patient, and places his or her hands under the knees.

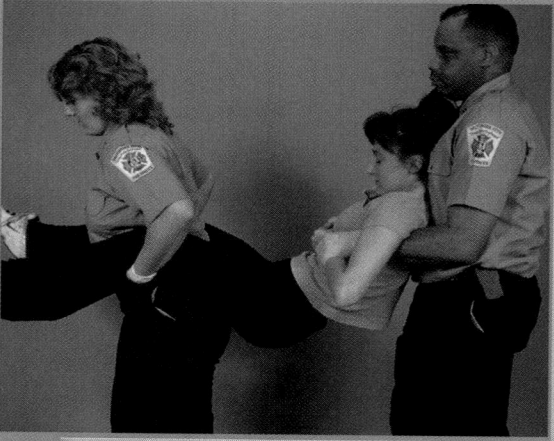

3 Both EMT-Bs rise to crouching.

On command, both lift and begin to move.

However, this lift and carry method increases pressure on the patient's chest, so the patient may be uncomfortable in this position.

Transfer Moves

There are several ways to transfer the patient from a bed onto the cot.

Direct Carry

Transfer a supine patient from a bed to the cot using the direct carry method (Figure 6-18 ▶). Position the cot parallel to the bed, with the head of the cot at the foot of the bed. Be sure that you prepare the cot by unbuck-

ling the straps and removing any other items from it. Both you and your partner should face the patient while standing between the bed and the cot. You should slide one arm under the patient's neck and cup the patient's shoulder. Your partner should slide his or her hand under the patient's hip and lift slightly. You should then slide your other arm under the patient's back, and your partner should place both arms underneath the patient's hips and calves. Slide the patient to the edge of the bed, and lift and curl the patient toward your chests. You and your partner should then rotate to the cot and gently place the patient onto it. This carry can be performed more easily with three providers (as illustrated).

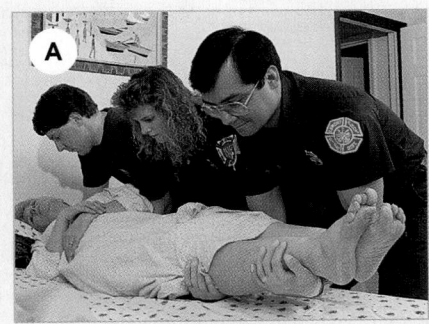
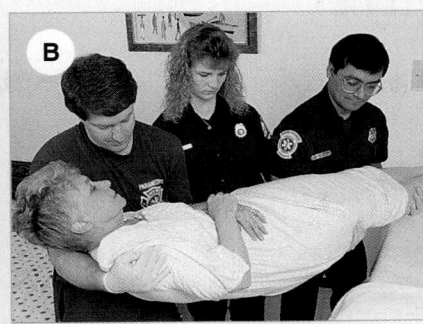
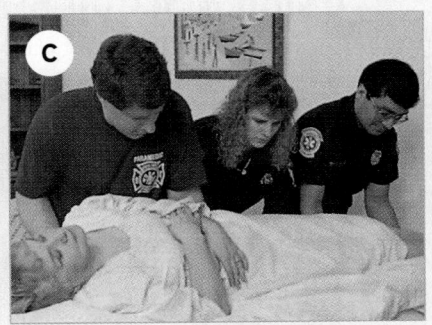

Figure 6-18 The direct carry method. **A.** Bring the cot in parallel to the bed with the patient's feet facing the head of the cot. Secure the cot to prevent it from rolling. **B.** Lift the patient in a smooth, coordinated fashion. Slowly walk the patient around, and position him or her over the cot. **C.** Slowly and gently lower the patient onto the cot.

Draw Sheet Method

To move the patient onto a cot, use the draw sheet method. Place the cot next to the bed, making sure it is at the same height as the bed and that the rails are lowered and straps are unbuckled. Be sure to hold the cot to keep it from moving. Loosen the bottom sheet underneath the patient, or log roll the patient onto a blanket (Figure 6-19A ▼). Reach across the cot, and grasp the sheet or blanket firmly at the patient's head, chest, hips, and knees (Figure 6-19B ▼). Gently slide the patient onto the cot (Figure 6-19C ▼).

Other Carries

Other carries are performed in the following manner:

- Place a backboard next to the patient and, after using a log roll or slide to move the patient onto the backboard, secure the patient and lift and carry the backboard to the nearby prepared cot.
- Insert the halves of a scoop stretcher under each side of the patient, and fasten the two sides together. Lift and carry the patient to the nearby prepared cot. Follow the steps below

(Skill Drill 6-8 ▶). (Note that you can also log roll a patient onto a scoop stretcher that is already locked together.)

1. **With the scoop stretcher separated**, measure the length of the scoop and adjust to the proper length (**Step 1**).
2. **Position the stretcher, one side at a time.** One EMT-B lifts the patient's side slightly by pulling on the far hip and upper arm, while the other EMT-B slides the stretcher into place (**Step 2**).
3. **Lock the stretcher ends together** by engaging their locking mechanisms one at a time and continuing to lift the patient slightly as needed to avoid pinching (**Step 3**).
4. **Apply and tighten straps** to secure the patient to the scoop stretcher before transferring to the cot (**Step 4**).

- Assist an able patient to the edge of the bed, and, placing the patient's legs over the side, help the

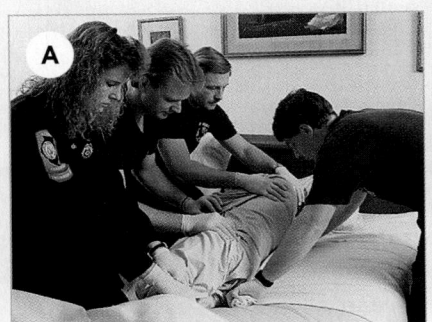
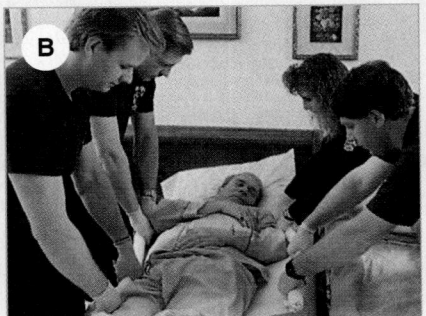
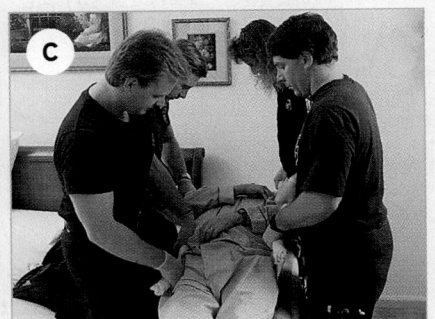

Figure 6-19 The draw sheet method. **A.** Log roll the patient onto a sheet or blanket. **B.** Bring the cot in parallel to the bed. Gently pull the patient to the edge of the bed. **C.** Transfer the patient to the cot.

Using a Scoop Stretcher

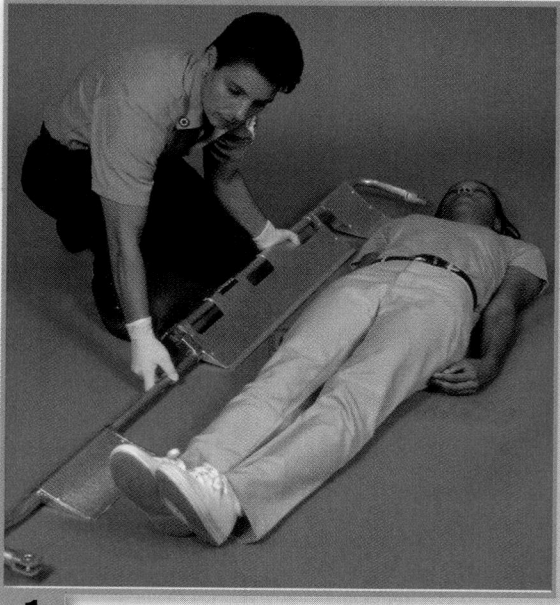

1   Adjust stretcher length.

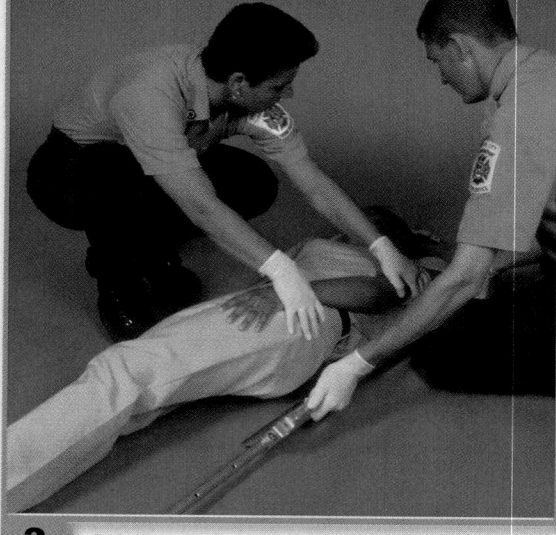

2   Lift the patient slightly and slide stretcher into place, one side at a time.

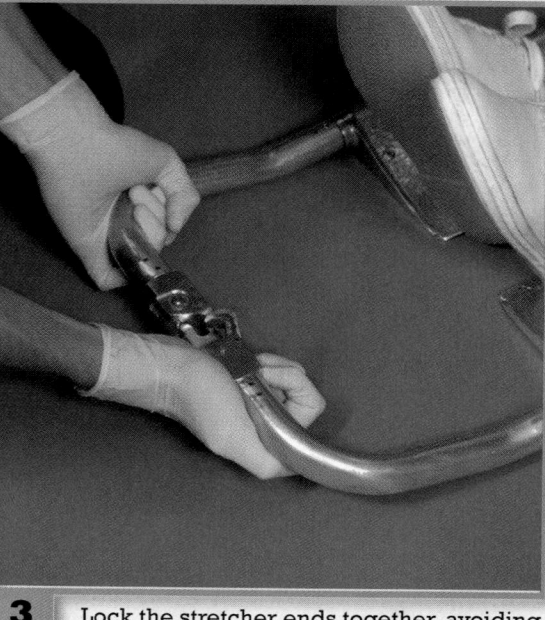

3   Lock the stretcher ends together, avoiding pinching.

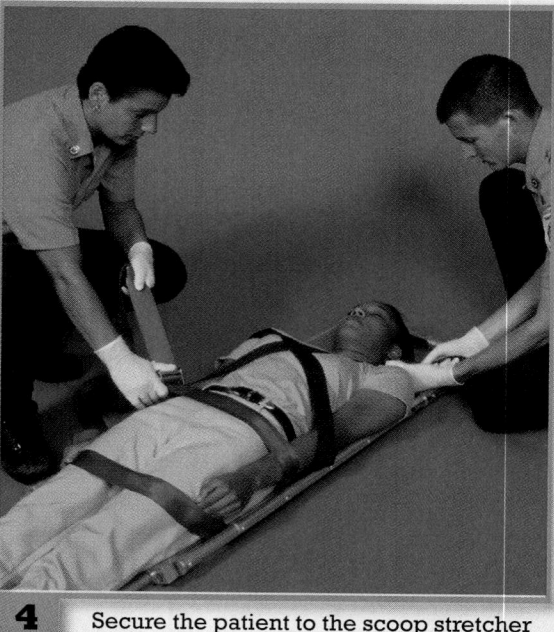

4   Secure the patient to the scoop stretcher and transfer it to the cot.

patient to sit up. Move the cot so that its foot end touches the bed near the patient. Help the patient to stand and rotate so that he or she can sit down on the center of the cot. Lift the patient's legs, and rotate them onto the cot while your partner lowers the torso onto the cot.

To avoid the strain of unnecessary lifting and carrying, you should use the draw sheet method or assist an able patient to the cot whenever possible.

To move a patient from the ground or the floor onto the cot you should use one of the following methods:

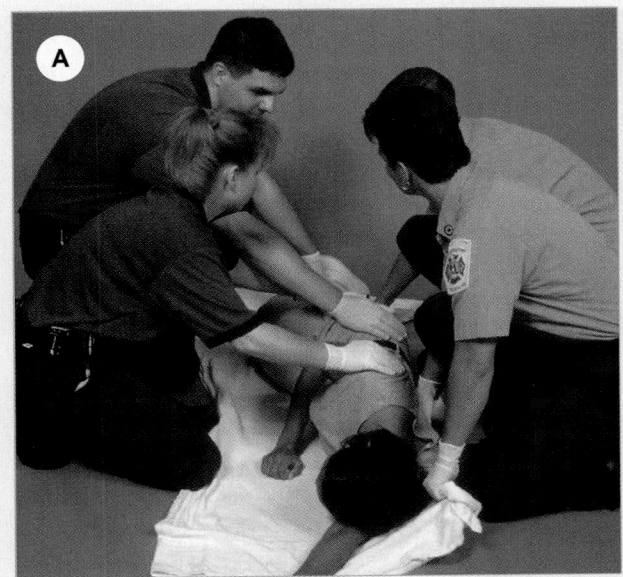

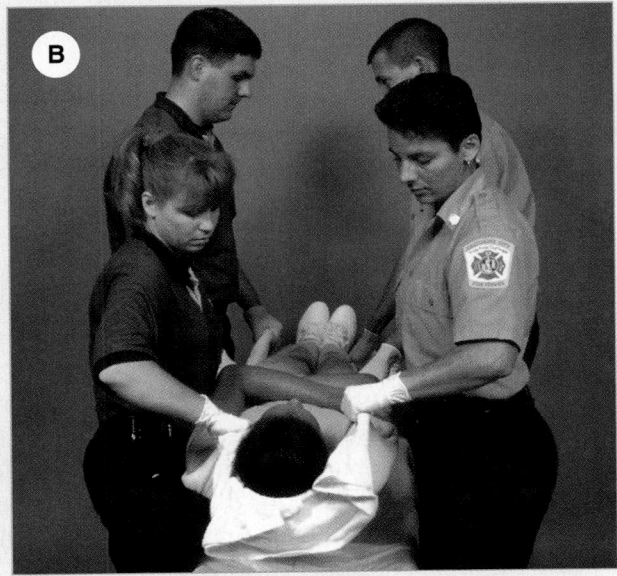

Figure 6-20 Log-rolling a patient on the ground. **A.** Log roll the patient onto a blanket. **B.** Lift the blanket and transfer the patient to the cot.

- Lift and carry the patient to the nearby prepared cot using a direct body carry.
- Use a log roll or long-axis drag to place the patient onto a backboard, and then lift and carry the backboard to the cot. Place both the backboard and the patient onto the cot.
- Use a scoop stretcher.
- Log roll the patient onto a blanket, centering the patient on the blanket and rolling up the excess material on each side (Figure 6-20A ▲). Lift the patient by the blanket, and carry him or her to the nearby cot (Figure 6-20B ▲).

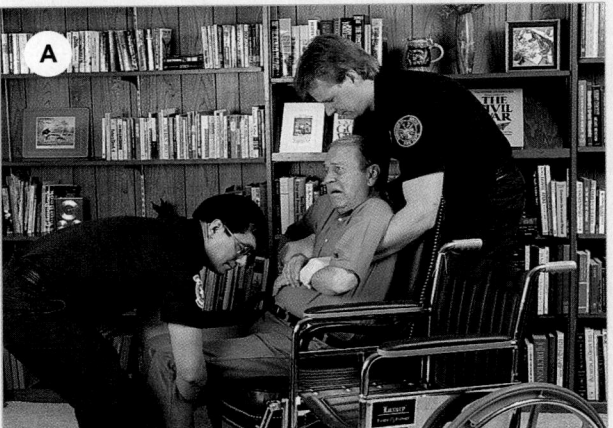

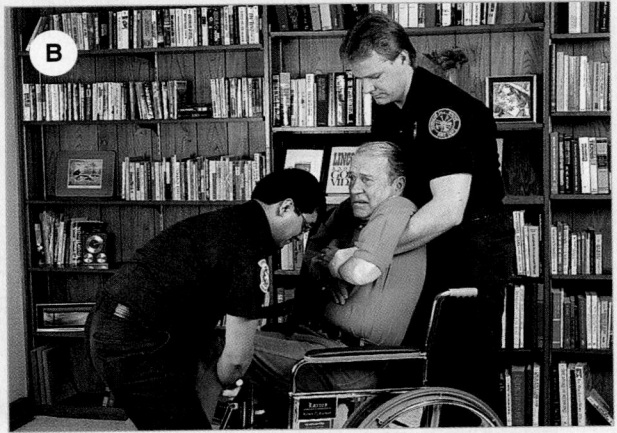

Figure 6-21 Moving a patient from a chair to a wheelchair. **A.** Slide your arms through the patient's armpits, and grasp the patient's crossed forearms. Second EMT-B grasps the patient's legs at the knees. **B.** Gently lift the patient into the locked wheelchair.

If a patient is sitting in a chair and cannot assist you, transfer the patient from the chair to a wheelchair (Figure 6-21 ▲).

Geriatrics

The majority of patients transported by EMS are geriatric patients. For many older patients, the fear of illness and disability is ever present, and an emergency trip to the hospital can be a terrifying and disorienting experience. In addition, there are physiologic changes that occur with aging that require special attention on your part as an EMT.

1. Skeletal changes: brittle bones (osteoporosis), rigidity, and spinal curvatures (kyphosis and spondylosis (Figure 6-22 ▶)) present special challenges in packaging and moving older

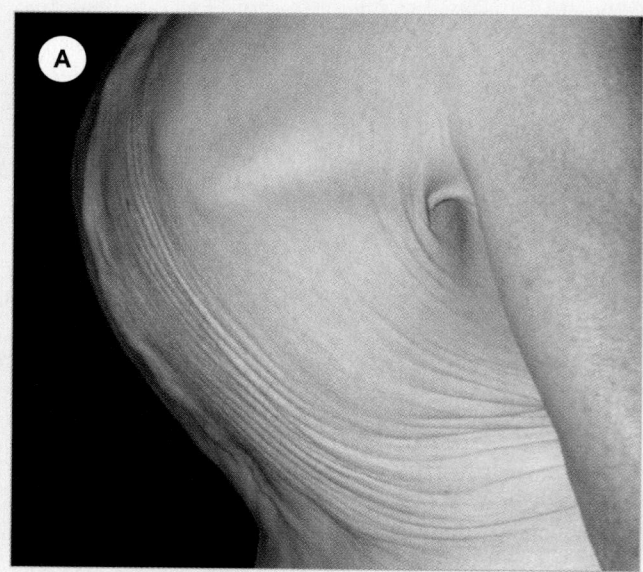
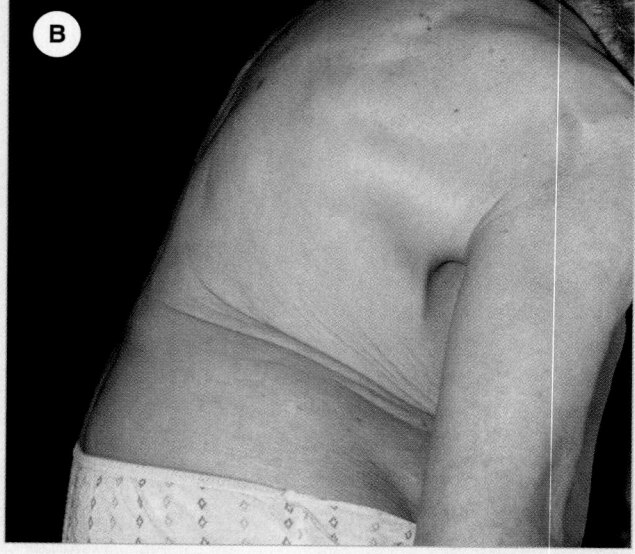

Figure 6-22   **A.** Kyphosis. **B.** Spondylosis.

patients. Many patients cannot lie supine on a backboard without causing additional injury, such as fractures, pressure sores, and skin breakdown. Special care and creativity must be used in immobilizing such patients. For example, a patient with spinal curvature may have to be placed on her side and immobilized in place with towel and blanket rolls to prevent exacerbating her injuries. Be sure to consult your local protocols and medical director about alternative ways of immobilizing such patients.

2. Fear: a sympathetic and compassionate approach can go a long way to allaying the natural fears many older patients experience when interacting with caregivers. Slow down, explain, anticipate: these can go a long way to gaining cooperation and taking some of the anxiety out of the packaging and transportation of your older patient. Imagine how frightening being strapped to a cot and carried down a flight of stairs can be to an individual who lives in constant fear of falls and broken bones.

Bariatrics

Estimates suggest that approximately 100 million adults in the United States are at least overweight or obese. Approximately 35% of women and 31% of men older than 19 years are obese or overweight. The numbers among children are even more imposing. The preva-

lence of obesity in children in the United States has increased markedly between the time of the National Health and Nutrition Examination Survey (NHANES) 2 and 3 trials. Approximately 20% to 25% of children are either overweight or obese, and the prevalence is even greater in some minority groups, including Pima Indians, Mexican Americans, and African Americans. Conservative estimates suggest that the management of obesity consumes approximately $100 billion yearly, without factoring in the costs of various commercial dietary and weight loss programs.

Americans are becoming so large that a new field of medicine has been named for the care of the obese. Bariatrics (bar-e-at′riks) is that branch of medicine concerned with the management (prevention or control) of obesity and allied diseases. It comes from the Greek words *baros*, weight, and *iatreia*, medical treatment. Because there is a direct correlation between degree of obesity and frequency and severity of health problems, the larger the patient the more likely he or she is to need emergency treatment and transportation. This problem is taking an increasing toll on the health and functioning of EMTs, as back injuries account for the largest number of missed days of work and both temporary and permanent disability.

Although ambulance cot and equipment manufacturers are producing equipment with ever higher capacities, this does not address the danger to the users of that equipment. Although European ambulance manufacturers regularly install mechanical lifts on their units, these are not as common in the United States.

Patient-Moving Equipment

The Wheeled Ambulance Stretcher

The wheeled ambulance stretcher, or cot or gurney, is the most commonly used device to move and transport patients. Only when you must transport two patients in the same ambulance should it be necessary to transport one patient on a folding stretcher or backboard placed on the long squad bench.

Most patients are placed directly on the cot. However, you will need to place and secure patients with a possible spinal injury or multiple system trauma onto a backboard. Patients who may need CPR or must be carried down (or up) a flight of stairs while supine should also be placed on a backboard. The backboard and patient are then secured onto the cot.

You can use a stair chair to carry a patient who can tolerate being in a sitting position down a flight of stairs to the prepared cot, which is waiting in the lower hallway. You should then transfer the patient from the chair to the cot.

In most instances, it is best if you push the head of the cot while your partner guides the foot of the cot. When the cot must be carried, it is best if four rescuers are available to carry it. There is more stability with a four-person carry, and the carry requires less strength. One EMT-B should be positioned at each corner of the cot to provide an even lift. A four-person carry is much safer if the cot must be moved over rough ground. If only two EMT-Bs are available, or if limited space will allow room for only two EMT-Bs to carry the cot, there is risk that the cot will become unbalanced. In a two-person carry, the two EMT-Bs should stand facing each other, with one person at the head end of the cot and the other at the foot end. With this type of carry, one EMT-B will have to walk backward.

Features

The modern cot is available in a number of different models, which may include different features (Figure 6-23 ▶). Before going on a call, you should be fully familiar with the specific features of the cot that your ambulance carries. You must know where the controls to adjust and lock each feature are located and how each works.

The cot has a specific head end and foot end. The cot has a strong horizontal rectangular, tubular metal main frame to which all of its other parts are attached. The cot should be pulled, pushed, and lifted only by its main frame or handles, which are attached to the main frame specifically for this purpose.

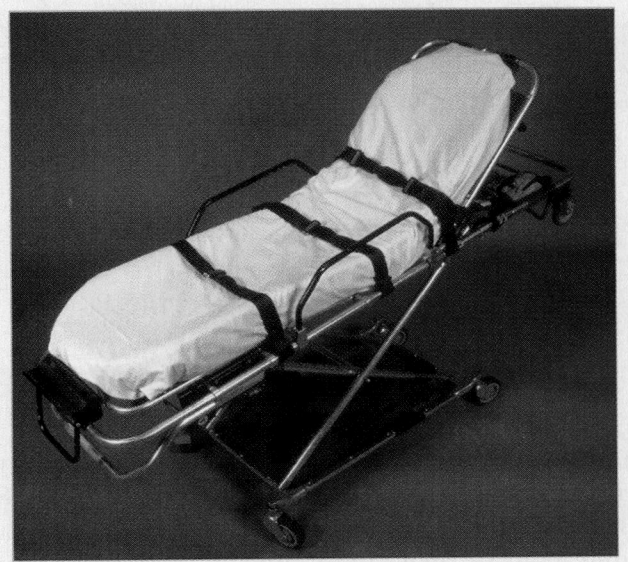

Figure 6-23 An ambulance stretcher (cot).

On most models, a second tubular frame made up of three sections is attached within or above the main frame. A metal plate is fastened to each of the three sections between its sides. This plate serves as the platform on which the cot mattress and patient are supported. The head section runs from the head end of the cot to near the center of the cot, where the patient's hips will be. Hinges at the area where the hips will be allow the head end to be elevated and the patient's back to be positioned at any desired angle from flat to fully upright. The head end of the cot is designed to be elevated or moved down only when a tilt control is purposely released. At all other times, the back will remain locked at the position in which it was placed. The frame and plates that lie from the hips to the foot end of the cot are divided into two hinged sections. These sections may be connected so that the foot end can be drawn in toward the knees, causing the frame and plates to hinge upward under the patient's knees to elevate them as desired. This feature is not found in all models.

A retractable guardrail is attached along the central portion of the main frame of the cot at each side and is lowered out of the way when a patient is being loaded onto the cot. Once the patient has been properly placed on the cot, the handle is drawn up and locked in an elevated position perpendicular to the surface of the cot. The guardrail at each side can be lowered only if its locking handle is released.

The underside of the main frame of the cot is supported on a folding undercarriage that has a smaller

Documentation Tips

Ensure a thorough patient care report by including details of how you moved the patient. For instance: "Moved patient to stretcher with draw-sheet lift."

Be familiar with the terms for anatomic positioning you learned in Chapter 4, such as Fowler's, Trendelenburg, Shock, and use them in your report.

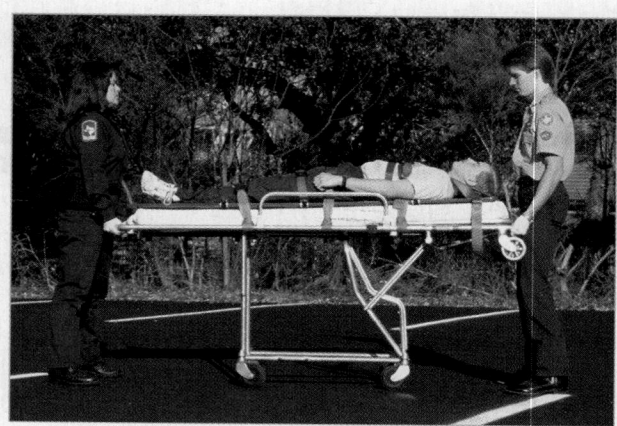

Figure 6-24 Make sure that you hold the main frame of the cot when it is elevated so that even when the patient moves, the cot does not tip.

horizontal rectangular frame and four large rubber casters at its bottom end. The folding undercarriage is designed so that the litter can be adjusted to any height from about 12″ above the ground, which is the desired height when the stretcher is secured in the ambulance, to 32″ to 36″ above the ground, which is the desired height when the stretcher is being rolled. Because you are able to lock the cot at any height between its lowest height and its fully extended height, it can be locked at the same height as any bed or examining table to allow the patient to be slid from one to the other. This permits you to transfer the patient without the need for any additional lifting. The controls for folding the undercarriage are designed so that the cot remains locked at its present height when the controls are not being activated. As an additional safety feature on most cots, the main frame must be slightly lifted so that the undercarriage becomes unweighted before it will fold, even if the control is pulled. Therefore, if the handle is accidentally pulled, the elevated cot will not suddenly drop. Controls for elevating and lowering most cots are located at the foot end and at one or both sides. You and your partner must use the proper lifting mechanics to lift the wheeled ambulance stretcher.

The mattress on a cot must be fluid resistant so that it does not absorb any type of potentially infectious material, including water, blood, or other body fluid.

Moving the Cot

Whenever a patient has been placed onto the cot, one EMT-B must hold the main frame to make sure that it cannot roll. When the cot is elevated, the main frame and the patient extend considerably beyond the wheels at both the head end and foot end of the cot. Therefore, whenever a patient is on an elevated cot, you must ensure that it is held firmly between two hands at all times so that even if the patient moves, the cot cannot tip Figure 6-24 ▶.

If the loaded cot must be carried down a short flight of steps, be sure to first retract the undercarriage; however, this is not necessary when the cot must be lifted over a curb, single step, or obstacle of a similar height Figure 6-25 ▼. Remember, if the patient must be carried up or down a full flight or several flights of stairs, you should prepare the cot and leave it on the ground floor at the bottom (or top) of the stairs. Use a backboard or stair chair to carry the patient up or down the stairs to the waiting cot.

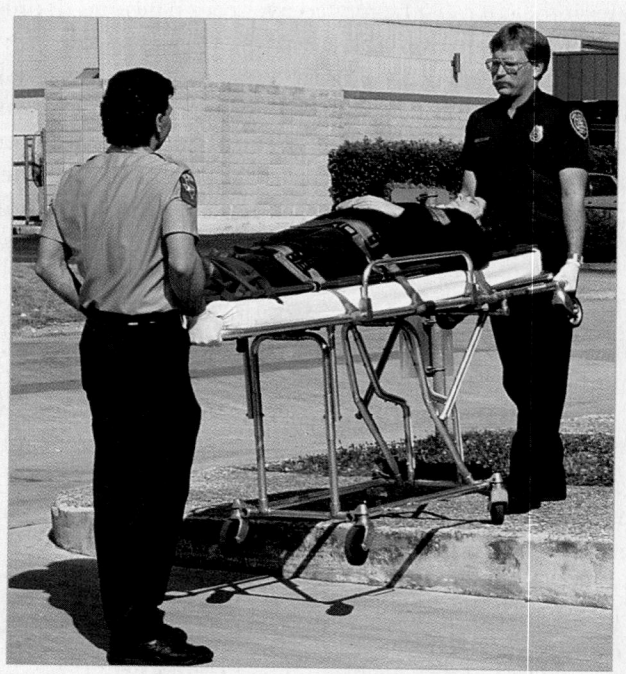

Figure 6-25 You need not retract the undercarriage of the cot when lifting it over a curb, single step, or obstacle of similar height.

Loading a Cot into an Ambulance

6-9 Skill Drill

Skill Drills

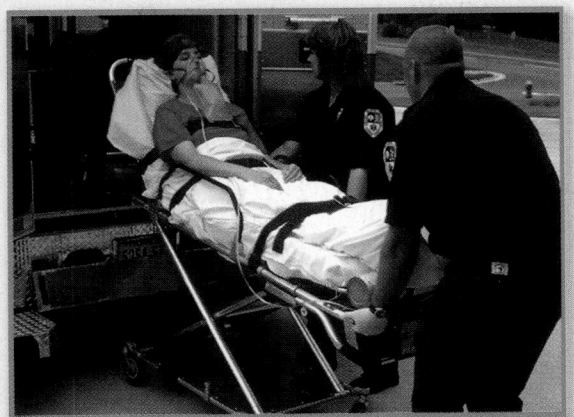

1 Tilt the head of the cot upward, and place it into the patient compartment with the wheels on the floor.

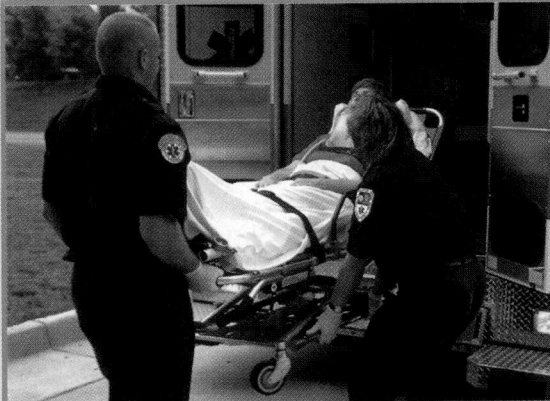

2 Second rescuer on the side of the cot releases the undercarriage lock and lifts the undercarriage.

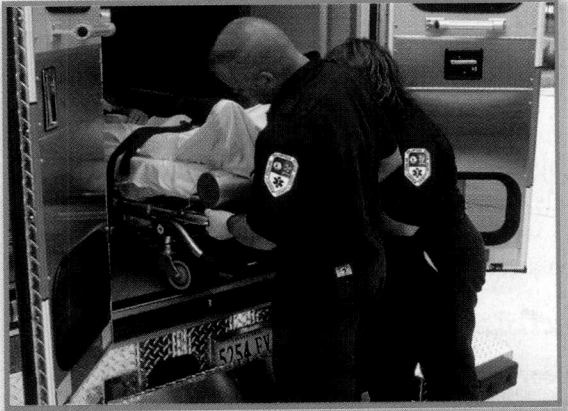

3 Roll the cot into the back of the ambulance.

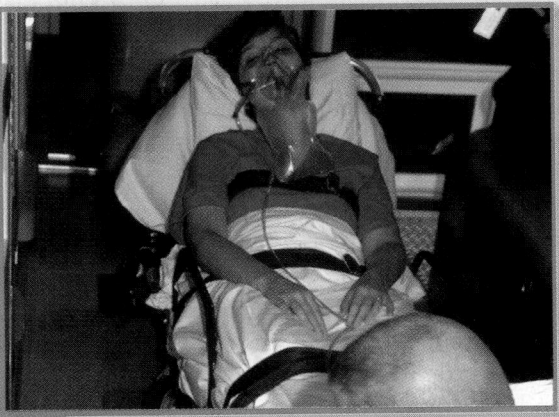

4 Secure the cot to the brackets mounted in the ambulance.

These are the steps to load the cot into an ambulance (Skill Drill 6-9 ▲):

1. **Tilt the head end of the main frame upward** and place it into the patient compartment with the wheels on the floor. The two additional wheels that extend just below the head end are attached to the main frame and will enable this movement (**Step 1**).

2. **With the patient's weight supported** by these two head-end wheels and the EMT-B at the foot end of the cot, move to the side of the main frame and release the undercarriage lock to lift the undercarriage up to its fully retracted position. The wheels of the undercarriage and the two on the head end of the main frame will now be on the same level (**Step 2**).

3. **Simply roll the cot** the rest of the way into the back of the ambulance, where it will rest on all six wheels (**Step 3**).

4. **Secure the cot** in the ambulance with the strong clamps that fasten around the undercarriage when the cot is pushed into them. The clamps are located in a rack on the floor or side of the patient compartment (**Step 4**).

The clamps will hold the cot in place until they are released at the hospital. You can control and release the clamps with a single handle that is positioned so that you can activate it when standing on the ground at the open back doors of the ambulance when the cot is to be unloaded. The cot is designed to be rolled on regular flat surfaces. If the patient must be moved over a lawn or other irregular surface, you must lift and carry the cot over the terrain.

An IV pole is attached to many cots. The IV pole can be unfolded or extended above the main frame to hold an IV bag above the patient while you move the cot to the ambulance. Some wheeled ambulance stretchers even include a carrier to hold an electrocardiogram (ECG) monitor or automated external defibrillator (AED) and portable oxygen unit. If the model you use does not include these features, you will have to secure the portable oxygen unit and ECG monitor or AED to the top surface of the cot mattress at the patient's legs.

The extra wheels below the head end of the main frame of the cot are not featured on some older or less expensive wheeled ambulance stretchers. These cots are not self-loading. When you reach the back of the ambulance with such a cot, you must lower it until the undercarriage is in its lowest retracted position and then, with you and your partner at each side of the cot, lift it to the height of the floor of the ambulance and roll it into the track that locks it into place. (Table 6-3 ▼) shows the guidelines that you must follow to load the cot into the ambulance.

Portable/Folding Stretchers

A portable stretcher is a stretcher with a strong rectangular tubular metal frame and rigid fabric stretched across it (Figure 6-26 ▶). Portable stretchers do not have a second multipositioning frame or adjustable undercarriage. Some models have two wheels that fold down about 4″ underneath the foot end of the frame and legs of a similar length that fold down from the head end at each side. The wheels make it easier to move the loaded stretcher. The legs should not be used as handles.

Some portable stretchers can be folded in half across the center of each side so that the stretcher is only half its usual length during storage. Many ambulances carry a portable stretcher to use if a patient is in an area that is difficult to reach with a wheeled ambulance stretcher

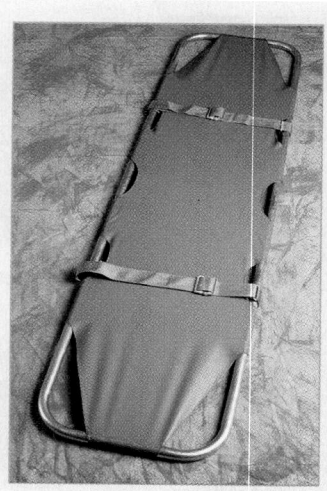

Figure 6-26 A portable stretcher.

or a second patient must be transported on the squad bench of the ambulance.

A portable stretcher weighs much less than a wheeled stretcher and does not have a bulky undercarriage. However, because most models do not have wheels, you and your team must support all of the patient's weight and any equipment along with the weight of the stretcher.

Flexible Stretchers

Several types of flexible stretchers, such as the SKED, Reeves, or Navy stretcher, are available and can be rolled up across either the stretcher's width or, in the case of the SKED, its length, so that the stretcher becomes a smaller tubular package for storage and carrying (Figure 6-27 ▶). When you must carry the equipment a considerable distance from the nearest place that the ambulance can be located, this is an important consideration. A flexible stretcher forms a rigid stretcher that conforms around the patient's sides and does not extend beyond them. When these stretchers are extended, they are particularly useful when you must remove a patient from or through a confined space. The SKED stretcher can also be used if the patient must be belayed or rappelled by ropes.

The flexible stretcher is the most uncomfortable of all the various devices; however, it provides excellent support and immobilization. When the stretcher is wrapped around the patient and the straps are secured, the patient is completely immobilized. The stretcher can then be lowered by rope or slid down a flight of stairs by resting it on the front edge of each step.

TABLE 6-3 Guidelines for Loading the Cot into the Ambulance

- Make sure there is sufficient lifting power.
- Follow the manufacturer's directions for safe and proper use of the cot.
- Make sure that all cots and patients are fully secured before you move the ambulance.

Figure 6-27 A flexible stretcher.

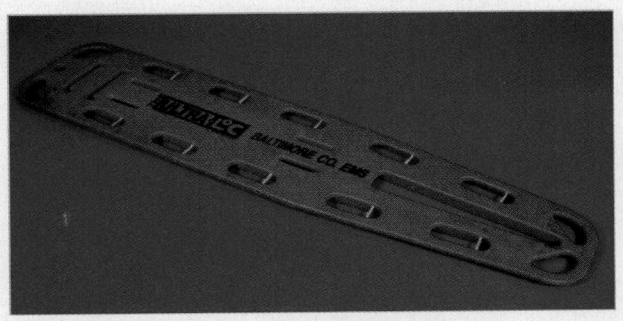

Figure 6-28 A long backboard.

Backboards

Backboards are long, flat boards made of rigid, rectangular material Figure 6-28 ▶. Backboards were originally made of wood but are now made of other materials as well, mostly plastic. They are used to carry patients and to immobilize supine patients with suspected spinal injury or other multiple trauma. Backboards can also be used to move patients out of awkward places. They are 6' to 7' long and are commonly used for patients who are found lying down. Parallel to the sides and ends of the backboard are a number of long holes that are about ½" to 1" from the outer edge. These holes form handles and handholds so that the board can be easily grasped, lifted, and carried. The handles and adjacent holes also allow straps used to secure and immobilize the patient to the backboard to be secured to each side and end of the backboard at any needed location.

For many years, backboards were made of thick marine plywood whose surface was sealed with polyurethane or another marine varnish. Wooden backboards are still used in some places. If wooden backboards are used, you must follow infectious control procedures before you can reuse the backboards. Where wooden backboards are no longer used, they have generally been stored so that they will be available in the event of a mass-casualty situation. Newer backboards are made of plastic materials that will not absorb blood or other infectious substances.

You can use a short backboard, or half-board, to immobilize the torso, head, and neck of a seated patient with a suspected spinal injury until you can immobilize the patient on a long backboard. Short backboards are 3' to 4' long. The original short wooden backboard has generally been replaced with a vest-type device that is specifically designed to immobilize the patient until he or she is moved from a sitting position to supine on a backboard Figure 6-29 ▶. The vest-type devices are easier to use than the wooden backboard.

You are the Provider   Part 5

Once you have secured your patient to the backboard, you may loosen the straps on the extrication device. You pad as best you can under the patient with towels. She complains of lower back pain, so you loosen the leg strap and place a blanket under her knees to achieve a neutral position, then retighten the strap. In the emergency department she is evaluated by the physician and radiographs are ordered. She is eventually released with a diagnosis of muscle strain.

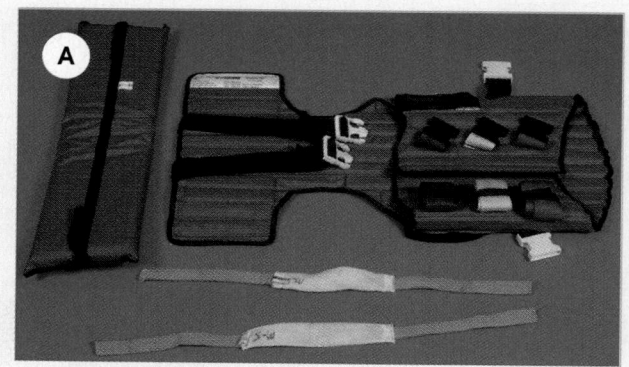

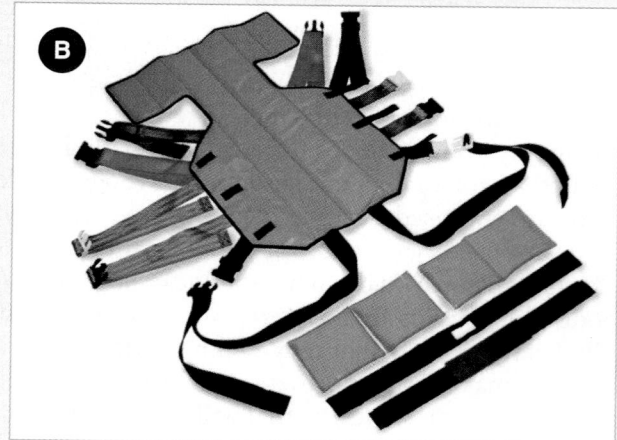

Figure 6-29 Vest-type short immobilization devices. **A.** KED. **B.** Oregon-type backboard.

Figure 6-30 A basket stretcher.

Basket Stretchers

You should use a rigid <u>basket stretcher</u>, often called a Stokes litter, to carry a patient across uneven terrain from a remote location that is inaccessible by ambulance or other vehicle (Figure 6-30 ▶). If you suspect that the patient has a spinal injury, you should first immobilize him or her on a backboard and then place the backboard into the basket stretcher. Once you have reached the ambulance and wheeled ambulance stretcher, you can remove the patient and backboard from the basket stretcher and place them on the cot.

Basket stretchers are made of either plastic with an aluminum frame or have a full steel frame that is connected by a woven wire mesh. The wire basket is very uncomfortable for the patient unless the wire is padded. Either type can be used to carry a patient across fields, rough terrain, or trails or on a toboggan, boat, or all-

terrain vehicle. Basket stretchers surround and support the patient, yet their design allows water to drain through holes in the bottom. Basket stretchers are also used for technical rope rescues and some water rescues. Not all basket stretchers are rated or appropriate for each of these specialized rescue uses. The types of basket stretchers that are acceptable for specialized rescue must be determined by individuals with additional special training.

Scoop Stretcher

The <u>scoop stretcher</u>, or split litter, is designed to be split into two or four pieces (Figure 6-31 ▼). These sections are fitted around a patient who is lying on the ground or another relatively flat surface. The parts are reconnected, and the patient is lifted and placed on a long backboard or stretcher. A scoop stretcher may be

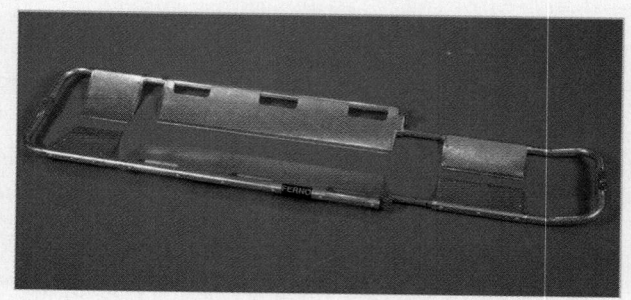

Figure 6-31 A scoop stretcher.

used for patients who have been struck by a motor vehicle. Other uses of the scoop stretcher include patients with hip injuries, patients with multiple injuries, older patients with brittle bones, and carrying patients up or down stairs.

A scoop stretcher is efficient; however, both sides of the patient must be accessible. You must also pay special attention to the closure area beneath the patient so that clothing, skin, or other objects are not trapped. As with the long backboard, you must fully stabilize and secure the patient before moving him or her; however, you cannot slip a scoop stretcher under the long axis of the patient's body. Scoop stretchers are narrow, well constructed, and compact and have excellent body support features. You and your team should practice often with a scoop stretcher to be ready for using it with a patient.

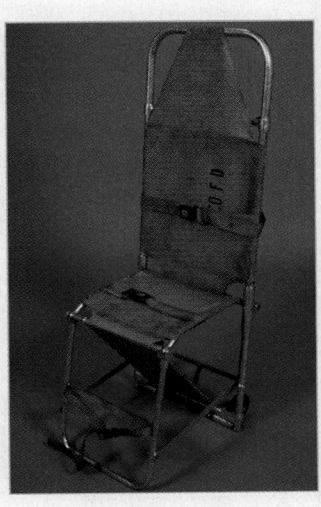

Figure 6-32 A stair chair.

Stair Chairs

Stair chairs are folding aluminum frame chairs with fabric stretched across them to form a seat and seat back (Figure 6-32 ▶). They have fold-out handles to help you carry their head and foot ends up or down a flight of stairs, and most have rubber wheels at their back with casters in front so that they can be rolled along the floor and make turns. Stair chairs serve as an adjunct for moving a patient up or down stairs to the ground floor, where the prepared wheeled ambulance stretcher is waiting. You can roll the stair chair on the floor until you reach the stairwell, then carry it (rather than roll and bump it) up or down the stairs. Once you reach the ground floor, you can roll it to the waiting cot and assist or lift the patient onto the cot.

Maintenance

Be sure to follow manufacturer's directions for maintenance, inspection, repair, and upkeep for any device that you use as patient-handling equipment.

Decontamination

It is essential that you decontaminate your equipment after use, for your own safety, the safety of crew using the equipment after you, and the safety of your patients. Just as we expect a hospital bed to be disinfected after the last patient, so too with our cot and other transport equipment. Know and follow your local standard operating procedures for disinfecting equipment after each call.

You are the Provider   Summary

As with any patient involved in a traumatic injury, cervical spine stabilization should always be considered. It is common for pain, tingling, and discomfort to begin after the initial shock of the incident has subsided. Keep this in mind any time a trauma patient tells you that he or she is feeling no neck or back pain and prior to releasing c-spine stabilization. Your goal is to always provide patient care to the greatest benefit of the patient. Know how your equipment works and maintain good communication with your patient to ensure he or she remains comfortable.

Prep Kit

Ready for Review

- The first key rule of lifting is to always keep your back in an upright position and lift without twisting. You can lift and carry significant weight without injury as long as your back is in the proper upright position.
- The power lift is the safest and most powerful way to lift.
- The safety of you, your team, and the patient depends on the use of proper lifting techniques and maintaining a proper hold when lifting or carrying a patient.
- Pushing is better than pulling.
- If you do not have a proper hold, you will not be able to bear your share of the weight, or you may lose your grasp with one or both hands and possibly cause a lower back injury to one or more EMT-Bs.
- It is always best to move a patient on a device that can be rolled. However, if a wheeled device is not available, you must understand and follow certain guidelines for carrying a patient on a cot.
- You must constantly coordinate your movements with those of the other team members and make sure that you communicate with them.
- When lifting a cot, you must make sure that you and your team use correct lifting techniques.
- You and your team should also be of similar height and strength.
- If you must carry a loaded backboard or cot up or down stairs or other inclines, be sure that the patient is tightly secured to the device to prevent sliding.

- Be sure to carry the backboard or cot foot end first so that the patient's head is elevated higher than the feet.
- Directions and commands are an important part of safe lifting and carrying.
- You and your team must anticipate and understand every move and execute it in a coordinated manner.
- The team leader is responsible for coordinating the moves.
- You should try to use four rescuers whenever resources allow.
- You should also know how much you can comfortably and safely lift and not attempt to lift more than this amount.
- Rapidly summon additional help to lift and carry a weight that is greater than you are able to lift.
- The same basic body mechanics apply for safe reaching and pulling as for lifting and carrying.
- Keep your back locked and straight, and avoid twisting.
- Do not hyperextend your back when reaching overhead.
- You should normally move a patient with nonurgent moves, in an orderly, planned, and unhurried fashion, selecting methods that involve the least amount of lifting and carrying.
- At times, you may have to use an emergency move to maneuver a patient before providing initial assessment and care.
- You should perform an urgent move if a patient has an altered level of consciousness, inadequate ventilation or shock, or in extreme weather conditions.
- The wheeled ambulance stretcher, or cot, is the most commonly used device to move and transport patients.
- Other devices that are used to lift and carry patients include portable stretchers, flexible stretchers, backboards, basket stretchers (Stokes litters), scoop stretchers, and stair chairs.
- Whenever you are moving a patient, you must take special care so that neither you, your team, nor the patient is injured.

www.EMTB.com

Technology

Interactivities

Vocabulary Explorer

Anatomy Review

Web Links

Online Review Manual

- You will learn the technical skills of patient packaging and handling through practice and training.
- Training and practice are also required to use all the equipment that is available to you.
- You must practice each technique with your team often so that you are able to perform the move quickly, safely, and efficiently.

Vital Vocabulary

backboard A device that is used to provide support to a patient who is suspected of having a hip, pelvic, spinal, or lower extremity injury. Also called a spine board, trauma board, or longboard.

basket stretcher A rigid stretcher commonly used in technical and water rescues that surrounds and supports the patient yet allows water to drain through holes in the bottom. Also called a Stokes litter.

diamond carry A carrying technique in which one EMT-B is located at the head end, one at the foot end, and one at each side of the patient; each of the two EMT-Bs at the sides uses one hand to support the stretcher so that all are able to face forward as they walk.

direct ground lift A lifting technique that is used for patients who are found lying supine on the ground with no suspected spinal injury.

emergency move A move in which the patient is dragged or pulled from a dangerous scene before initial assessment and care are provided.

extremity lift A lifting technique that is used for patients who are supine or in a sitting position with no suspected extremity or spinal injuries.

flexible stretcher A stretcher that is a rigid carrying device when secured around a patient but can be folded or rolled when not in use.

portable stretcher A stretcher with a strong rectangular tubular metal frame and rigid fabric stretched across it.

power grip A technique in which the litter or backboard is gripped by inserting each hand under the handle with the palm facing up and the thumb extended, fully supporting the underside of the handle on the curved palm with the fingers and thumb.

power lift A lifting technique in which the EMT-B's back is held upright, with legs bent, and the patient is lifted when the EMT-B straightens the legs to raise the upper body and arms.

rapid extrication technique A technique to move a patient from a sitting position inside a vehicle to supine on a backboard in less than 1 minute when conditions do not allow for standard immobilization.

scoop stretcher A stretcher that is designed to be split into two or four sections that can be fitted around a patient who is lying on the ground or other relatively flat surface; also called a split litter.

stair chair A lightweight folding device that is used to carry a conscious, seated patient up or down stairs.

wheeled ambulance stretcher A specially designed stretcher that can be rolled along the ground. A collapsible undercarriage allows it to be loaded into the ambulance. Also called the cot or an ambulance cot.

Points to Ponder

You are dispatched to a local nursing home. Upon your arrival you find an 88-year-old woman lying supine on the floor. The patient states that she fell out of bed and that her hip hurts. She is lying between the bed and the wall with very little space on either side. The patient's hip is obviously deformed and the patient is in moderate pain.

What would be the best way to move this patient? What would be some good ways to stabilize the patient's hip before moving her?

Issues: Working With Geriatric Patients, Challenging Lifting and Moving Situations.

www.EMTB.com

Assessment in Action

You are dispatched to a medical call at 3 am. The patient is located in the bedroom of an apartment on the third floor. There is no elevator and the patient weighs 150 kg (330 lb).

1. After an assessment you determine there is no immediate life-threatening illness, but transport to a hospital is necessary. You are on scene with your partner. What would be appropriate to consider before moving the patient?
   A. The possibility of needing additional manpower
   B. Discussing a safe means for extricating the patient
   C. Determining what equipment will be needed
   D. All of the above

2. The patient is lying on the floor with hip pain. Moving causes more pain. What would be the best method of lifting the patient?
   A. A backboard
   B. A stair chair
   C. A scoop stretcher
   D. An extremity lift

3. What would be considered important when lifting a patient?
   A. Keeping your back vertical while lifting
   B. Keeping your legs apart about 15"
   C. Keeping your palms up
   D. All of the above

4. The stair chair is often used to carry patient down stairs. What is the job of the third person?
   A. To carry equipment
   B. To guide the person walking backwards down the stairs
   C. To hold onto the patient
   D. To run ahead and ready the stretcher

5. Which of the following carries would be most effective in narrow spaces?
   A. Extremity lift
   B. Fireman's carry
   C. Direct ground lift
   D. No carry; let the patient walk

6. Which of the following are NOT situations in which to use rapid extrication technique?
   A. The vehicle and/or scene is unsafe.
   B. The patient cannot be assessed before being removed from the car.
   C. There is smoke coming from the airbag.
   D. The patient's condition requires immediate transport to the hospital.

7. How many EMT-Bs are necessary to perform a rapid extrication?
   A. One
   B. Two
   C. Four
   D. As many as it takes to be done safely

Challenging Questions

8. Why is it important not to twist your body while lifting and keeping your hands close to your body?

9. You are working a new job and hurt your back lifting on the first day. You do not feel that you are hurt badly and want to keep it to yourself so that you do not cause a problem. What are the possible implications with this decision?

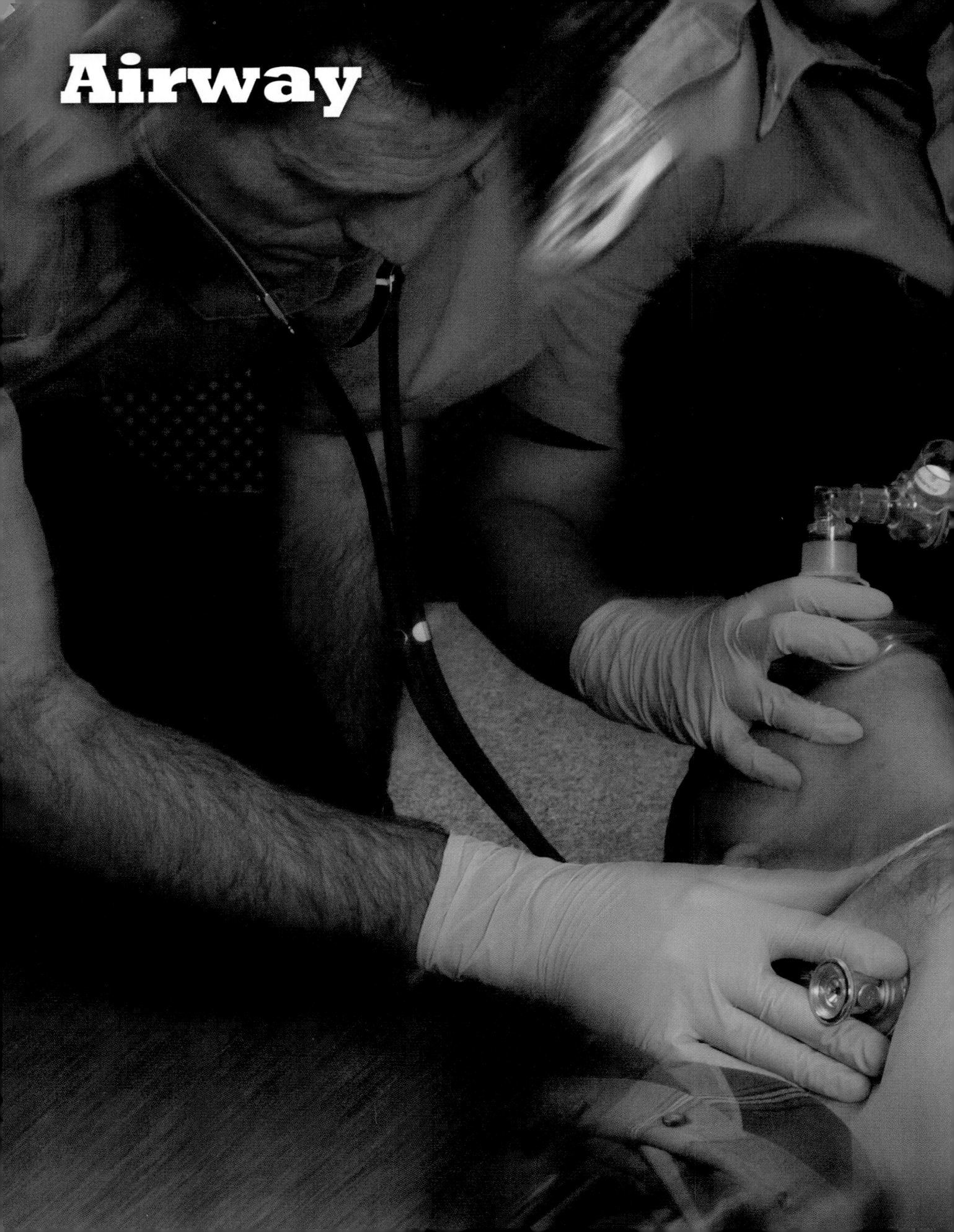

Airway

Airway

Objectives

Cognitive

2-1.1 Name and label the major structures of the respiratory system on a diagram. (p 214)

2-1.2 List the signs of adequate breathing. (p 221)

2-1.3 List the signs of inadequate breathing. (p 221)

2-1.4 Describe the steps in performing the head tilt-chin lift maneuver. (p 224)

2-1.5 Relate mechanism of injury to opening the airway. (p 223)

2-1.6 Describe the steps in performing the jaw-thrust maneuver. (p 225)

2-1.7 State the importance of having a suction unit ready for immediate use when providing emergency care. (p 231)

2-1.8 Describe the techniques of suctioning. (p 232)

2-1.9 Describe how to artificially ventilate a patient with a pocket mask. (p 242)

2-1.10 Describe the steps in performing the skill of artificially ventilating a patient with a bag-valve-mask device while using the jaw-thrust maneuver. (p 242)

2-1.11 List the parts of a bag-valve-mask system. (p 244)

2-1.12 Describe the steps in performing the skill of artificially ventilating a patient with a bag-valve-mask device for one and two rescuers. (p 245)

2-1.13 Describe the signs of adequate artificial ventilation using the bag-valve-mask device. (p 247)

2-1.14 Describe the signs of inadequate artificial ventilation using the bag-valve-mask device. (p 246, 247)

2-1.15 Describe the steps in ventilating a patient with a flow-restricted, oxygen-powered ventilation device. (p 247)

2-1.16 List the steps in performing the actions taken when providing mouth-to-mouth and mouth-to-stoma artificial ventilation. (p 242)

2-1.17 Describe how to measure and insert an oropharyngeal (oral) airway. (p 227)

2-1.18 Describe how to measure and insert a nasopharyngeal (nasal) airway. (p 228)

2-1.19 Define the components of an oxygen delivery system. (p 235)

2-1.20 Identify a nonrebreathing face mask and state the oxygen flow requirements needed for its use. (p 240)

2-1.21 Describe the indications for using a nasal cannula versus a nonrebreathing face mask. (p 241)

2-1.22 Identify a nasal cannula and state the flow requirements needed for its use. (p 241)

Affective

2-1.23 Explain the rationale for basic life support, artificial ventilation, and airway protective skills taking priority over most other basic life support skills. (p 214)

2-1.24 Explain the rationale for providing adequate oxygenation through high inspired oxygen concentrations to patients who, in the past, may have received low concentrations. (p 246)

Psychomotor

2-1.25 Demonstrate the steps in performing the head tilt-chin lift maneuver. (p 224)

2-1.26 Demonstrate the steps in performing the jaw-thrust maneuver. (p 225)

2-1.27 Demonstrate the techniques of suctioning. (p 232)

2-1.28 Demonstrate the steps in providing mouth-to-mouth artificial ventilation with body substance isolation (barrier shields). (p 242)

2-1.29 Demonstrate how to use a pocket mask to artificially ventilate a patient. (p 242)

2-1.30 Demonstrate the assembly of a bag-valve-mask unit. (p 244)

2-1.31 Demonstrate the steps in performing the skill of artificially ventilating a patient with a bag-valve-mask device for one and two rescuers. (p 245)

2-1.32 Demonstrate the steps in performing the skill of artificially ventilating a patient with a bag-valve-mask device while using the jaw-thrust maneuver. (p 245)

2-1.33 Demonstrate artificial ventilation of a patient with a flow-restricted, oxygen-powered ventilation device. (p 247)

2-1.34 Demonstrate how to artificially ventilate a patient with a stoma. (p 248)

2-1.35 Demonstrate how to insert an oropharyngeal (oral) airway. (p 227)

2-1.36 Demonstrate how to insert a nasopharyngeal (nasal) airway. (p 228)

2-1.37 Demonstrate the correct operation of oxygen tanks and regulators. (p 238)

2-1.38 Demonstrate the use of a nonrebreathing face mask and state the oxygen flow requirements needed for its use. (p 240)

2-1.39 Demonstrate the use of a nasal cannula and state the flow requirements needed for its use. (p 241)

2-1.40 Demonstrate how to artificially ventilate the infant and child patient. (p 242)

2-1.41 Demonstrate oxygen administration for the infant and child patient. (p 244)

Additional Objectives*

Cognitive

1. Describe how to perform the Sellick maneuver (cricoid pressure). (p 246)

Affective

2. Explain the rationale for applying cricoid pressure. (p 246)

Psychomotor

3. Demonstrate how to perform the Sellick maneuver (cricoid pressure). (p 247)

*These are noncurriculum objectives.

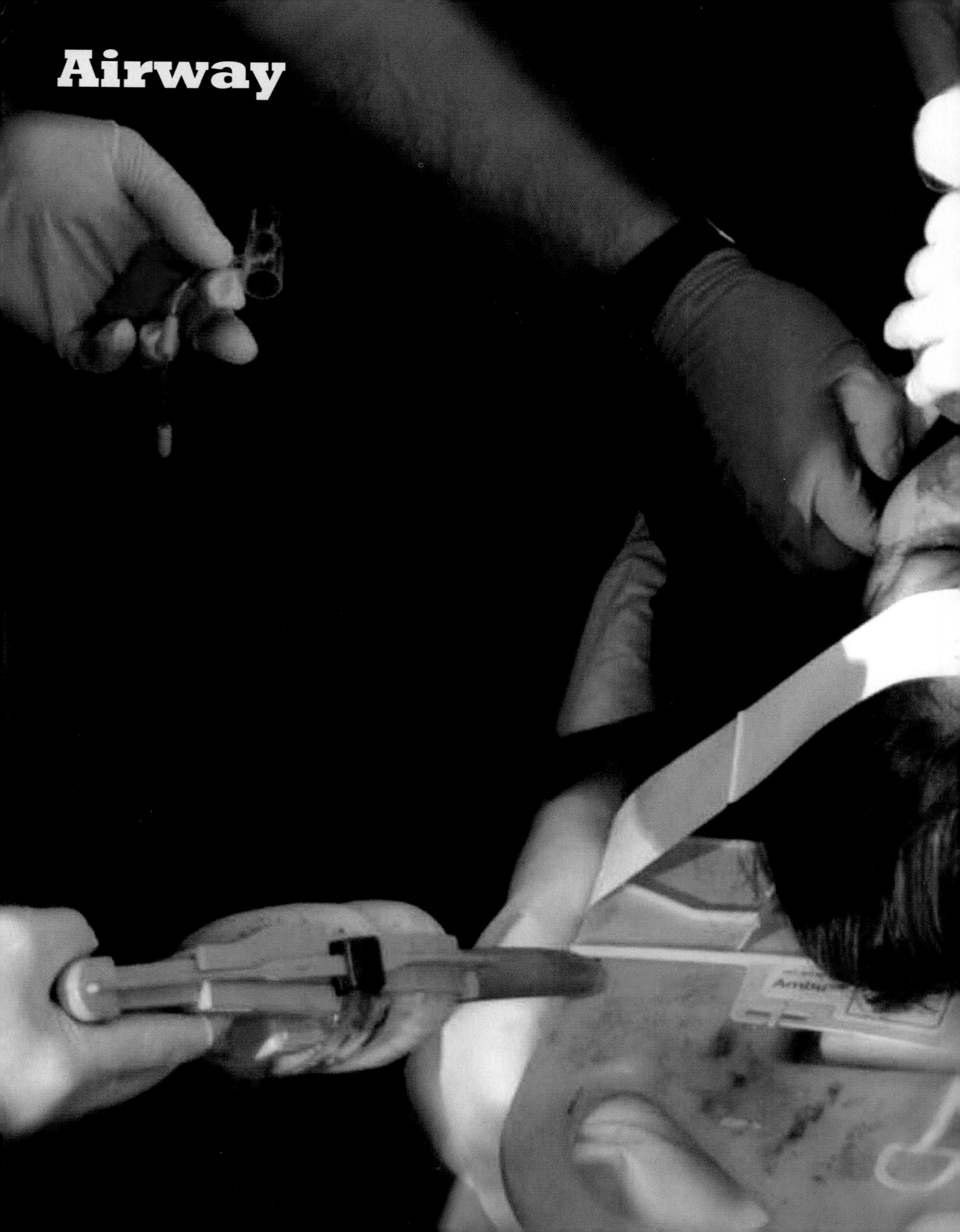

Airway

You are the Provider

It's a warm summer day outside, the temperatures are mild and rain is in the forecast. You are sitting with your partner when the call comes in "University Ambulance 2, respond to the lobby of the Plaza Hotel for a man with respiratory distress."

Approximately one out of every four EMS calls is either airway or respiratory related. This chapter will help prepare you for these frequently encountered calls and help you answer the following questions:

1. Why is it critical to maintain a patient's airway and ensure adequate breathing at all times?
2. How frequently should you assess the condition of a patient's airway and his or her ability to breathe?
3. What impact will inappropriate assessment and management of a patient's airway and breathing have on total patient care?

Airway

The single most important step in caring for any patient is to make sure that he or she can breathe adequately. The patient who cannot breathe effectively is not delivering oxygen to body tissues and cells, which need a constant supply of oxygen to survive. Within seconds of being deprived of oxygen, vital organs such as the heart and brain may not function normally.

Oxygen reaches body tissues and cells through two separate but related processes: breathing and circulation. As we inhale, oxygen moves from the atmosphere into our lungs, then passes from the air sacs in the lungs into the capillaries to oxygenate the blood. At the same time, carbon dioxide, produced by cells in the tissues of the body, moves from the blood into the air sacs. The blood, enriched with oxygen, travels through the body by the pumping action of the heart. The carbon dioxide then leaves our bodies as we exhale.

As an EMT-B, you must be able to locate the parts of the respiratory system, understand how the system works, and be able to recognize which patients are breathing adequately and which ones are breathing inadequately. This will enable you to determine how best to treat your patients.

This chapter will review the anatomy and physiology of the respiratory system, that is, the parts of the system and how they work. It will then describe how to assess patients quickly and to carefully determine

Technology

Interactivities

Vocabulary Explorer

Anatomy Review

Web Links

Online Review Manual

www.EMTB.com

their airway and ventilation status. The equipment, procedures, and guidelines that you will need to manage a patient's airway and breathing are described in detail. You will learn several ways to open a patient's airway and specific techniques for removing foreign objects or fluids that may be blocking the airway. Because airway management equipment can be dangerous if used improperly, the chapter will thoroughly discuss airway adjuncts, oxygen therapy devices, and artificial ventilation methods.

Anatomy of the Respiratory System

The respiratory system consists of all the structures in the body that make up the airway and help us breathe, or ventilate (Figure 7-1 ▶). Structures that help us breathe include the diaphragm, the muscles of the chest wall, accessory muscles of breathing, and the nerves from the brain and spinal cord to those muscles. Ventilation is the exchange of air between the lungs and environment. The diaphragm and muscles of the chest wall are responsible for the regular rise and fall of the chest that accompany normal breathing.

Structures of the Airway

The airway is divided into the upper and lower airways. The upper airway consists of the nose, mouth, throat (pharynx), and a structure called the epiglottis. The epiglottis is a leaf-shaped structure above the larynx that prevents food and liquid from entering the larynx during swallowing. The portion of the throat behind the nose is named the nasopharynx; the portion behind the mouth is the oropharynx.

The lower airway consists of the larynx, trachea, main bronchi, bronchioles (smaller bronchi), and alveoli.

The lower airway begins with the larynx (voice box, vocal cords). The cricoid cartilage is a firm cartilage ring that forms the lower part of the larynx. The trachea is connected to the larynx. The main bronchi and bronchioles branch off from the trachea, extending into each lung. Eventually the bronchioles end in the alveoli. The alveoli are small sacs where the actual exchange of oxygen and carbon dioxide occurs.

The chest (thoracic cage) contains the lungs, one on each side (Figure 7-2 ▶). The lungs hang freely within the chest cavity. Between the lungs is a space called the mediastinum, which is surrounded by tough connective tissue. This space contains the heart, the great vessels, the esophagus, the trachea, the major

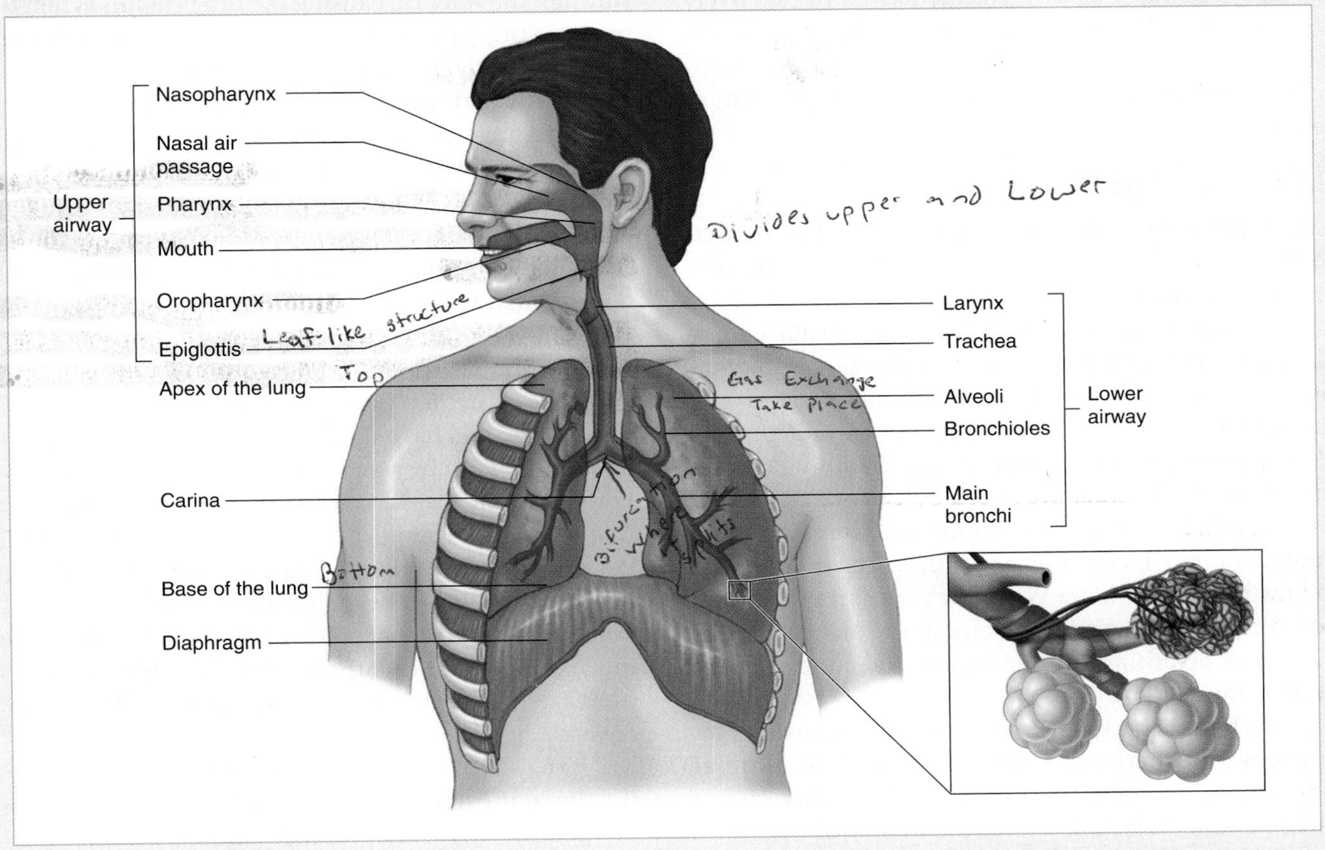

Nasopharynx

Nasal air passage

Pharynx

Mouth

Oropharynx

Epiglottis — *Leaf-like structure*

Upper airway

Apex of the lung — *Top*

Carina

Base of the lung — *Bottom*

Diaphragm

Divides upper and Lower

Larynx

Trachea

Gas Exchange Take Place

Alveoli

Bronchioles

Main bronchi

Lower airway

Bifurcation where it splits

Figure 7-1 The upper and lower airways contain the structures in the body that help us to breathe. The upper airway contains the nose, mouth, throat, and epiglottis.

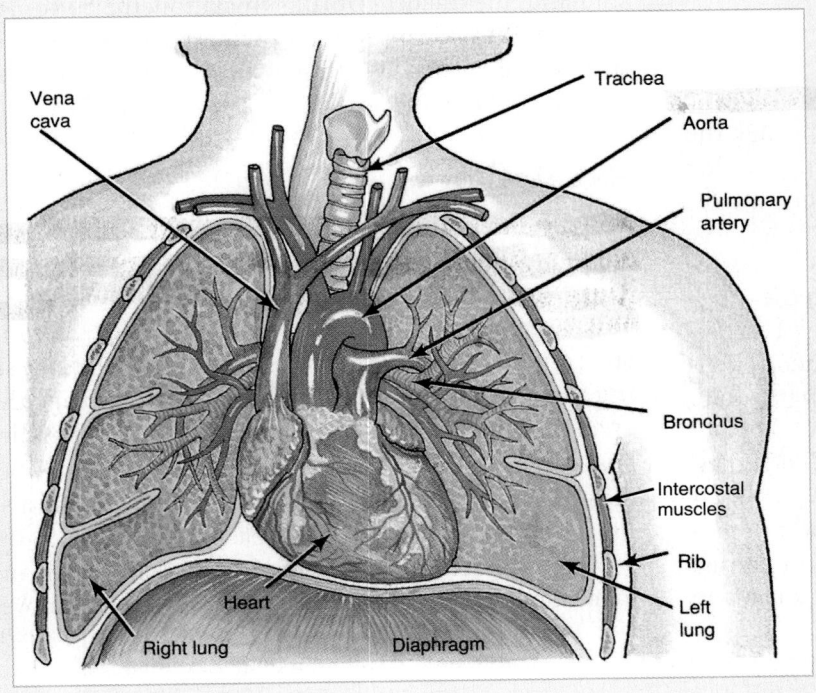

Vena cava

Trachea

Aorta

Pulmonary artery

Bronchus

Intercostal muscles

Rib

Left lung

Heart

Right lung

Diaphragm

Figure 7-2 The thoracic cage contains important anatomic structures for respiration, including the lungs, the heart, the great vessels (the vena cava and aorta), the trachea, and the major bronchi.

bronchi, and many nerves. The mediastinum effectively separates the right lung space from the left lung space. The boundaries of the thorax are the rib cage anteriorly, superiorly, and posteriorly and the diaphragm inferiorly.

Structures of Breathing

The diaphragm is a skeletal muscle because it is attached to the costal arch and the vertebrae. It is considered a specialized muscle because it functions as a voluntary and an involuntary muscle. It acts as a voluntary muscle whenever we take a deep breath, cough, or hold our breath—all actions that we are able to control. However, unlike other skeletal or voluntary muscles, the diaphragm also performs an automatic function. Breathing continues while we sleep and at all other times. Even though we can hold our breath or temporarily breathe more quickly or slowly, we cannot continue these variations in breathing indefinitely. When the concentration of carbon dioxide rises within the blood, the automatic regulation of breathing resumes under the control of the brain stem.

The lungs, because they have no muscle tissue, cannot move on their own. They need the help of other structures to be able to expand and contract as we inhale and exhale. Therefore, the ability of the lungs to function properly is dependent on the movement of the chest and supporting structures. These structures include the thorax, the thoracic cage (chest), the diaphragm, the intercostal muscles, and the accessory muscles of breathing.

Inhalation

The active, muscular part of breathing is called inhalation. As we inhale, air enters the body through the trachea. This air travels to and from the lungs, filling and emptying the alveoli. During inhalation, the diaphragm and intercostal muscles contract. When the diaphragm contracts, it moves down slightly and enlarges the thoracic cage from top to bottom. When the intercostal muscles contract, they lift the ribs up and out. As we inhale, the combined actions of these structures enlarge the thorax in all directions. Take a deep breath to see how your chest expands.

The air pressure outside the body, called the atmospheric pressure, is normally higher than the air pressure within the thorax. As we inhale and the thoracic cage expands, the air pressure within the thorax decreases, creating a slight vacuum. This pulls air in

through the trachea, causing the lungs to fill. When the air pressure outside equals the air pressure inside, air stops moving. Gases, such as oxygen, will move from an area of high pressure to an area of lower pressure until the pressures are equal. At this point, the air stops moving, and we stop inhaling. Tidal volume, a measure of the depth of breathing, is the amount of air in milliliters (mL) that is moved into or out of the lungs during a single breath. The average tidal volume for a man is approximately 500 mL. Minute volume is the amount of air moved through the lungs in 1 minute and is calculated by multiplying tidal volume and respiratory rate. Therefore, if a patient is breathing at a rate of 12 breaths/min and has a tidal volume of 500 mL per breath, his minute volume would be 6,000 mL (6 L). It is important to note that variations in tidal volume, respiratory rate, or both, will affect minute volume. For example, if a patient is breathing at a rate of 12 breaths/min, but his tidal volume is reduced (shallow breathing), minute volume will decrease. Likewise, if a patient is breathing at a rate of 12 breaths/min and his tidal volume is increased (deep breathing), minute volume will increase.

It may help you to understand this if you think of the thoracic cage as a bell jar in which balloons are suspended. In this example, the balloons are the lungs. The base of the jar is the diaphragm, which moves up and down slightly with each breath. The ribs, which are the sides of the jar, maintain the shape of the chest. The only opening into the jar is a small tube at the top, similar to the trachea. During inhalation, the bottom of the jar moves down slightly, causing a decrease in pressure in the jar and creating a slight vacuum. As a result, the balloons fill with air (Figure 7-3 ▶).

Exhalation

Unlike inhalation, exhalation does not normally require muscular effort; therefore, it is a passive process. During exhalation, the diaphragm and the intercostal muscles relax. In response, the thorax decreases in size, and the ribs and muscles assume a normal resting position. When the size of the thoracic cage decreases, air in the lungs is compressed into a smaller space. The air pressure within the thorax then becomes higher than the pressure outside, and air is pushed out through the trachea.

Let's return to the example of the bell jar. During exhalation, the bottom of the jar (the diaphragm) moves

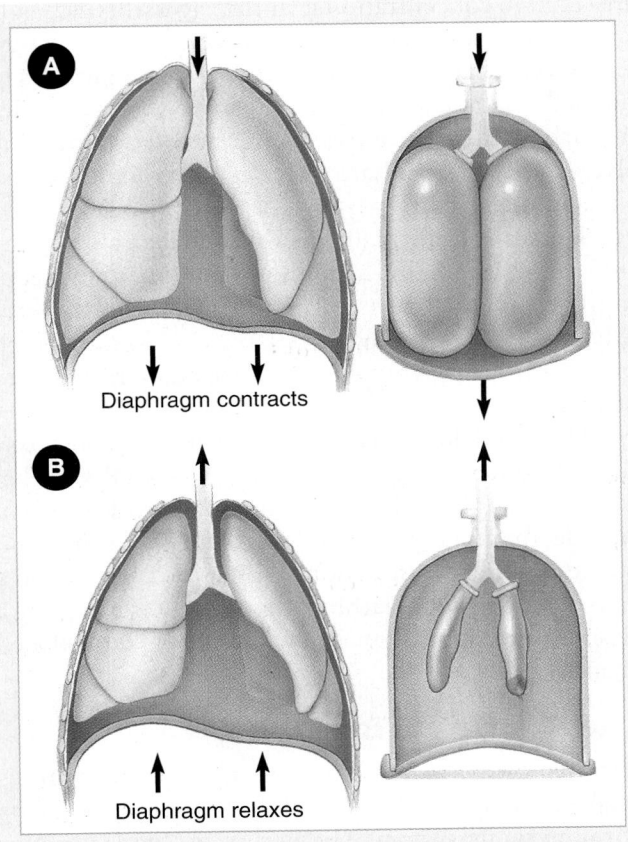

Figure 7-3 The mechanisms of respiration can be illustrated by using a bell jar. **A.** Inhalation and chest expansion, anatomic (left) and bell jar (right). **B.** Exhalation and chest contraction, anatomic (left) and bell jar (right).

Study This

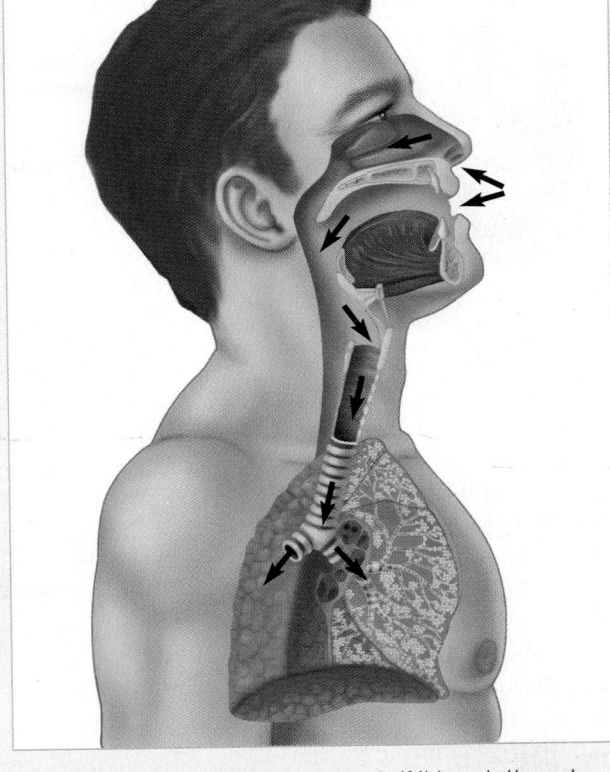

Figure 7-4 Air reaches the lungs only if it travels through the trachea. Maintaining the airway means keeping the airway patent so that air can enter and leave the lungs freely.

up, returning to its normal resting position. This movement increases air pressure within the jar. With this increase in pressure, the sides of the jar contract, and the balloons empty.

Remember that air will reach the lungs only if it travels through the trachea. This is why clearing and maintaining an open airway are so important. Clearing the airway means removing obstructing material, tissue, or fluids from the nose, mouth, or throat. Maintaining the airway means keeping the airway patent (open) so that air can enter and leave the lungs freely (Figure 7-4 ▶).

Air may also pass into the chest cavity through an abnormal opening in the throat or chest wall as a result of trauma, remaining outside of the bronchi and never reaching the alveoli. In Chapter 27, Chest Injuries, you will learn how to recognize and manage these dangerous conditions.

Physiology of the Respiratory System

All living cells need energy to survive. Cells take energy from nutrients through a series of chemical processes. The name given to these processes as a whole is metabolism. During metabolism, each cell combines nutrients and oxygen and produces energy and waste products, primarily water and carbon dioxide.

Each living cell in the body requires a supply of oxygen and a regular means of disposing of waste (carbon dioxide). The body provides these through respiration. Some cells need a constant supply of oxygen to survive. Other cells in the body can tolerate short periods without oxygen and still survive. For example, after 4 to 6 minutes without oxygen, brain cells and

Study This !!!

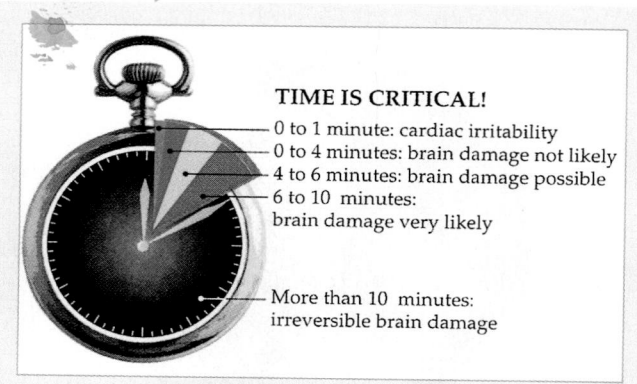

TIME IS CRITICAL!
- 0 to 1 minute: cardiac irritability
- 0 to 4 minutes: brain damage not likely
- 4 to 6 minutes: brain damage possible
- 6 to 10 minutes: brain damage very likely
- More than 10 minutes: irreversible brain damage

Figure 7-5 Cells need a constant supply of oxygen to survive. Some cells may be severely or permanently damaged after 4 to 6 minutes without oxygen.

cells in the nervous system may be severely or permanently damaged and may even die (Figure 7-5 ▲). Dead brain cells can never be replaced. However, cells in the kidney may be without oxygen for 45 minutes or more and still survive. This is why certain organ transplants are possible.

Normally, the air that we breathe contains 21% oxygen and 78% nitrogen. Small amounts of other gases make up the remaining 1%.

The Exchange of Oxygen and Carbon Dioxide

As blood travels through the body, it supplies oxygen and nutrients to various tissues and cells. Oxygen passes from blood in the arteries through the capillaries to tissue cells, while carbon dioxide and cell waste pass in the opposite direction: from tissue cells through capillaries and into the veins (Figure 7-6 ▶).

Each time we inhale, the alveoli receive a supply of oxygen-rich air. The alveoli are surrounded by a network of tiny pulmonary capillaries. These capillaries are, in fact, located in the walls of the alveoli. This means that the air in the alveoli and the blood in the capillaries are separated only by two very thin layers of tissue. Each time we exhale, the carbon dioxide from the bloodstream travels across the same two layers of tissue to the alveoli and is expelled into the atmosphere.

Oxygen and carbon dioxide pass rapidly across the walls of the alveoli and the capillaries through diffusion. Diffusion is a passive process in which molecules move from an area of higher concentration to an area of lower concentration. For example, an entire kitchen can smell like a rotten egg because the molecules of hydrogen sulfide gas have moved spontaneously from an

area of high concentration near the egg to fill the whole space. Molecules of oxygen move from the alveoli into the blood because there are fewer oxygen molecules in the pulmonary capillaries. In the same way, molecules of carbon dioxide move from the blood into the alveoli because there are fewer carbon dioxide molecules in the alveoli (Figure 7-7 ▶).

The alveoli normally produce a chemical, called surfactant, that helps keep the alveoli open. By keeping the alveoli open, diffusion is more efficient. Anything that removes or destroys surfactant (such as water from drowning) will cause acute respiratory distress.

The blood does not distribute all of the inhaled oxygen as it passes through the body. Therefore, the air that we exhale contains 16% oxygen and 3% to 5% carbon dioxide; the rest is nitrogen (Figure 7-8 ▶). When you provide artificial ventilations with a pocket mask to a patient who is not breathing, that patient is receiving a 16% concentration of oxygen with each of your exhaled breaths.

The Control of Breathing

The area of the brain stem that controls breathing is deep within the skull, in one of the best-protected parts of the nervous system. The nerves in this area act as

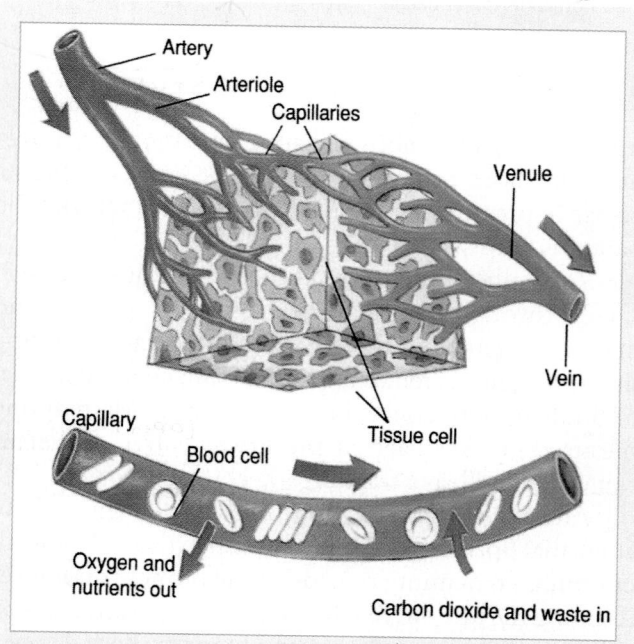

Figure 7-6 Oxygen passes from blood in the arteries through capillaries to tissue cells. Carbon dioxide passes from tissue cells through capillaries and into the veins.

sensors, reacting primarily to the level of carbon dioxide in the arterial blood. If the levels of carbon dioxide become too high or too low, the brain automatically adjusts breathing accordingly. This happens very quickly, after every breath. Again, this is why you cannot hold

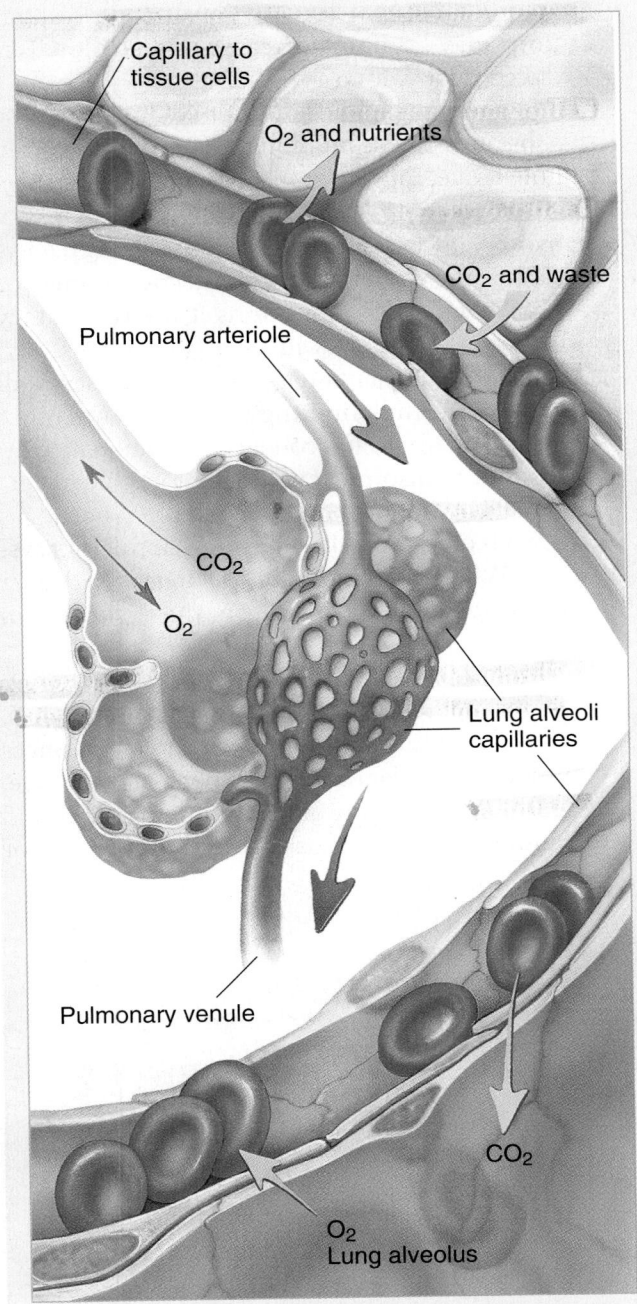

Figure 7-7 With diffusion, molecules of oxygen move from the alveoli into the blood because there are fewer oxygen molecules in the blood. Similarly, molecules of carbon dioxide diffuse from the blood into the alveoli because there are fewer carbon dioxide molecules in the alveoli.

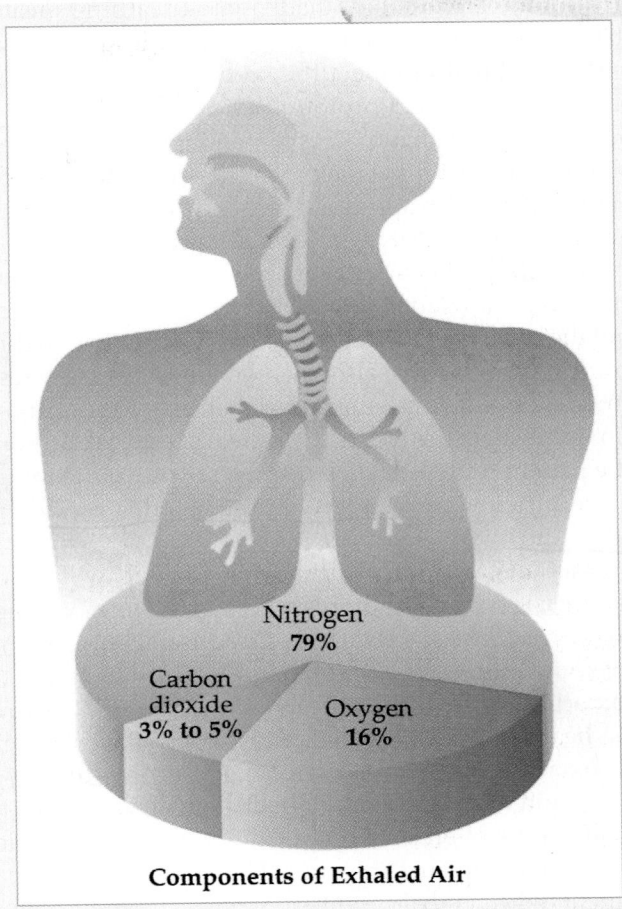

Components of Exhaled Air

Figure 7-8 Exhaled air contains 16% oxygen and 3% to 5% carbon dioxide; 79% is nitrogen.

your breath indefinitely or breathe rapidly and deeply for very long. In a healthy person, this stimulus to breathe is referred to as the primary respiratory drive.

When the level of carbon dioxide becomes too high, the brain stem sends nerve impulses down the spinal cord that cause the diaphragm and the intercostal muscles to contract. This increases our breathing, or respirations. The higher the level of carbon dioxide in the blood, the stronger the impulse is to breathe. Once the carbon dioxide returns to an acceptable level, the strength and frequency of respirations decrease.

Hypoxia (not enough oxygen)

Hypoxia is an extremely dangerous condition in which the body's tissues and cells do not have enough oxygen; unless it is reversed, patients may die in a matter of moments. Hypoxia develops quickly in the vital organs of patients who are not breathing or who are breathing inadequately. Inadequate breathing means that the person cannot move enough air into the lungs with each breath to meet the body's metabolic needs. Hypoxia can have a profound effect on breathing. If the brain senses that there is not enough oxygen in the blood, it will send messages via the spinal cord to the diaphragm and respiratory muscles, thus increasing the patient's respiratory rate and depth.

Patients with chronic respiratory diseases (eg, emphysema) maintain a low oxygen level in their blood, and the sensors in the brain become accustomed to this low level. Unlike a healthy person whose primary respiratory drive is influenced by increasing the carbon dioxide level in the blood, the primary respiratory drive of a patient with a chronic respiratory disease is influenced by a low oxygen level in the blood, a condition called the hypoxic drive.

Patients who are breathing inadequately will show varying signs and symptoms of hypoxia. The onset and the degree of tissue damage caused by hypoxia often depend on the quality of ventilations. Early signs of hypoxia include restlessness, irritability, apprehension, fast heart rate (tachycardia), and anxiety. Late signs of hypoxia include mental status changes, a weak (thready) pulse, and cyanosis. Conscious patients will complain of shortness of breath (dyspnea) and may not be able to talk in complete sentences. The best time to give a patient oxygen is before any signs and symptoms of hypoxia appear.

The following conditions are commonly associated with hypoxia:

- **Heart attack (myocardial infarction).** Ischemia within the heart muscle from myocardial infarction occurs when there is inadequate circulation of oxygen-carrying blood to the tissues of the heart. The weakened heart then pumps oxygenated blood to the remainder of the body less efficiently, resulting in systemic hypoxia.
- **Pulmonary edema.** Fluid accumulates in the lungs, making the exchange of oxygen and carbon dioxide in the alveoli less efficient.
- **Acute narcotic or sedative overdose.** Respirations may become decreased and shallow (reduced tidal volume).
- **Inhalation of smoke and/or toxic fumes.** These substances cause pulmonary edema and destroy lung tissue, causing problems with gas exchange.
- **Stroke (cerebrovascular accident).** The cause of hypoxia in a stroke may be due to facial paralysis leading to potential airway compromise or poor control of respirations if the respiratory center in the brain is affected.
- **Chest injury.** Pain interferes with full chest wall expansion, thus limiting effective ventilation. Lung damage itself secondary to pulmonary contusion can also prevent efficient gas exchange.
- **Shock (hypoperfusion).** Shock often occurs as a result of injuries that affect the circulatory system. When the circulatory system fails to deliver adequate amounts of oxygen, the tissues begin to die.
- **Chronic obstructive pulmonary disease (COPD; for example, chronic bronchitis and emphysema).** Chronic irritation of the lungs and air passages produces alveolar damage and poor gas exchange.
- **Asthma.** Narrowing of respiratory passages and buildup of mucus causes air trapping and poor gas exchange.

You are the Provider

Part 2

En route to the hotel, you consider the potential causes of your patient's respiratory distress. Could it be an asthma attack or a heart attack? Could there be some sort of trauma preventing him from breathing well? You are pleased you thoroughly checked your respiratory equipment and the oxygen cylinders before you left the station that morning. You begin putting on your latex gloves.

4. What are the specific causes of respiratory distress? Which are serious and which are not?
5. What type of equipment would you anticipate needing to treat a patient with difficulty breathing?

■ **Premature birth.** Pulmonary surfactant is decreased in some premature infants, and, therefore, prematurity is often associated with hypoxia. The more premature the infant, the worse the hypoxia.

All hypoxic patients, whatever the cause of their condition, should be treated with high-flow supplemental oxygen. The method of oxygen delivery will vary, depending on the severity of the hypoxia and the adequacy of breathing.

Patient Assessment

Recognizing Adequate Breathing

Earlier, we compared breathing to a bell jar with a movable bottom. You can also think of a normal breathing pattern as a bellows system. Breathing should appear easy, not labored. As with a bellows used to move air to start a fire, breathing should be a smooth flow of air moving into and out of the lungs. As a general rule, unless directly assessing the patient's airway, you should not be able to see or hear a patient breathe. Signs of normal (adequate) breathing for adult patients are as follows:

- A normal rate (between 12 and 20 breaths/min for adults)
- A regular pattern of inhalation and exhalation
- Clear and equal lung sounds on both sides of the chest (bilateral)
- Regular and equal chest rise and fall (chest expansion)
- Adequate depth (tidal volume)

Study This

Recognizing Inadequate Breathing

An adult who is awake, alert, and talking to you generally has no immediate airway or breathing problems. However, you should always have supplemental oxygen and a bag-valve-mask (BVM) device or pocket mask close at hand to assist with breathing if this becomes necessary. An adult who is breathing normally will have respirations of 12 to 20 breaths/min (Table 7-1 ▼). The adult patient who is breathing slower (fewer than 12 breaths/min) or faster (more than 20 breaths/min) than normal should be evaluated for inadequate breathing by assessing the depth of his or her respirations. A patient with a shallow depth of breathing (reduced tidal volume) may require assisted ventilations, even if his or her respiratory rate is within normal limits.

A patient with inadequate breathing may appear to be working hard to breathe. This type of breathing pattern is called labored breathing. It requires effort and, especially among children, may involve the use of accessory muscles. Accessory muscles are secondary

Study This!!!

TABLE 7-1	Normal Respiratory Rate Ranges
Adults	12 to 20 breaths/min
Children	15 to 30 breaths/min
Infants	25 to 50 breaths/min

Note: These ranges are per the US DOT 1994 EMT-Basic National Standard Curriculum. Ranges presented in other courses may vary.

You are the Provider   Part 3

As you arrive at the hotel you are greeted by the hotel security, who report the man was attending a conference when he suddenly began complaining of difficulty breathing and confusion. He then passed out in his chair. The security officer informs you the ushers have carefully moved him into the aisle and are trying to keep his airway open. As you walk into the room you notice the conference session is on break so few people are around and there are no immediate hazards. You see only one large patient who appears to be unresponsive. You ask your partner to call dispatch to send some ALS help.

6. How does the information given by first responders and bystanders help you prepare for your patient?
7. While this may seem to be an obvious situation, what are some potential hazards?

Documentation Tips

The respiratory status of a patient is so important that it should be noted at the beginning of your radio report, after mental status. Any changes during treatment or transport should be immediately reported to the receiving hospital. Respiratory status along with any changes should also be clearly documented in your patient care report.

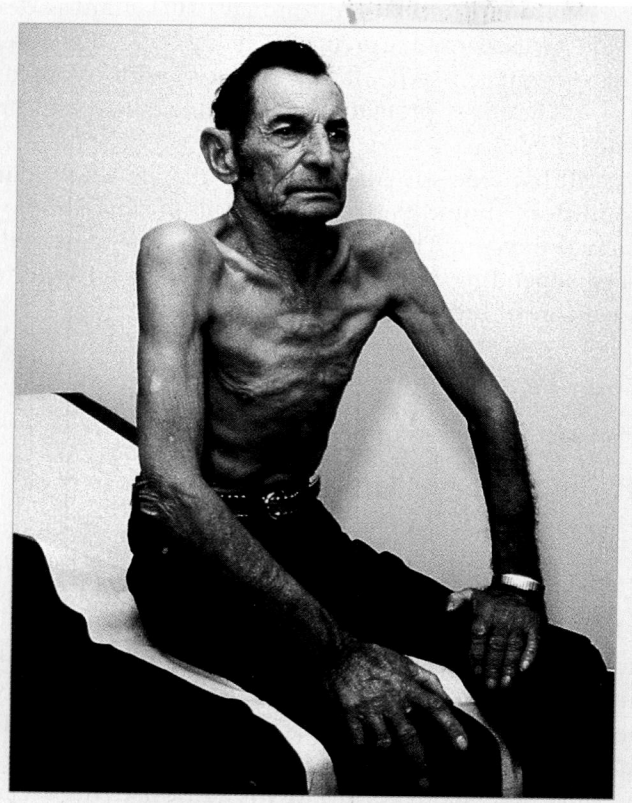

Figure 7-9 The accessory muscles of breathing are used when a patient is having difficulty breathing, but not during normal breathing. The accessory muscles include the sternocleidomastoid, pectoralis major, and abdominal muscles.

muscles of respiration. They include the neck muscles (sternocleidomastoid), the chest pectoralis major muscles, and the abdominal muscles (Figure 7-9 ▶). These muscles are not used during normal breathing. More information about recognizing labored breathing and respiratory distress in children may be found in Chapters 31 and 32. Signs of inadequate breathing in adult patients are as follows:

- Respiratory rate of fewer than 12 breaths/min or more than 20 breaths/min in the presence of dyspnea
- Irregular rhythm, such as a patient taking a series of deep breaths followed by periods of apnea
- Auscultated breath sounds are diminished, absent, or noisy
- Reduced flow of expired air at the nose and mouth
- Unequal or inadequate chest expansion, resulting in reduced tidal volume
- Increased effort of breathing—use of accessory muscles
- Shallow depth (reduced tidal volume)
- Skin that is pale, cyanotic (blue), cool, or moist (clammy)
- Skin pulling in around the ribs or above the clavicles during inspiration (retractions)

You should be aware that a patient may appear to be breathing after his heart has stopped. These occasional, gasping breaths are called agonal respirations. They occur when the respiratory center in the brain continues to send signals to the respiratory muscles. These respirations are not adequate because they are infrequent, gasping respiratory efforts. You will need to provide artificial ventilations to patients with agonal respirations.

Some patients may have irregular respiratory breathing patterns that are related to a specific condition. For example, Cheyne-Stokes respirations are often seen in patients with a stroke and patients with serious head injuries (Figure 7-10 ▶). Cheyne-Stokes respirations are an irregular respiratory pattern in which the patient breathes with an increasing rate and depth of respiration that is followed by a period of apnea or lack of spontaneous breathing, followed again by a pattern of increasing rate and depth of respiration. Serious head injuries may also cause changes in the normal respiratory rate and pattern of breathing. The result may be irregular, ineffective respirations that may or may not have an identifiable pattern (ataxic respirations).

Patients with inadequate breathing have inadequate minute volume and need to be treated immediately. This is most easily recognized in patients who are unable to speak in complete sentences when at rest or who

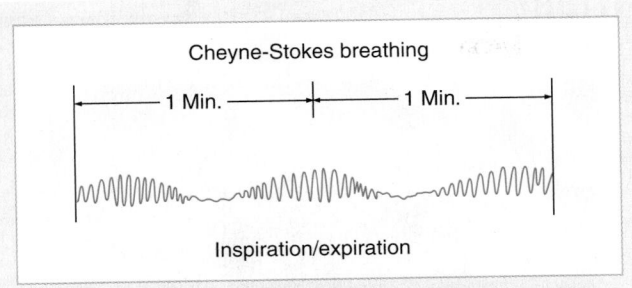

Figure 7-10 Cheyne-Stokes breathing shows irregular respirations followed by a period of apnea.

have a fast or slow respiratory rate, both of which may result in a reduction in tidal volume. Emergency medical care includes airway management, supplemental oxygen, and ventilatory support.

Opening the Airway

Emergency medical care begins with ensuring an open airway. The patient's airway and breathing status are the first steps in your initial assessment for a very good reason: Unless you can immediately open and maintain a patent airway, you cannot provide effective patient care. Regardless of the patient's condition, the airway must remain patent at all times.

When you respond to a call and find an unconscious patient, you need to assess and determine immediately whether the patient has an open airway and breathing is adequate. To most effectively open the airway and assess breathing, the patient should be in the supine position. However, if your patient is in a position that delays placement in a supine position (for example, entrapped in the vehicle), the patient's airway must be opened and assessed in the position in which you find the patient. If your patient is found in the prone position (lying face down), he or she must be repositioned to allow for assessment of airway and breathing and to begin CPR should it become necessary. The patient should be log rolled as a unit so the head, neck, and spine all move together without twisting (Skill Drill 7-1 ▶). Unconscious patients, especially when there are no witnesses who can rule out trauma, should be moved as a unit because of the potential for spinal injury.

1. **Kneel beside the patient.** Have your partner kneel far enough away so that the patient, when rolled toward you, does not come to rest in your lap. Place your hands behind the patient's head and neck to provide in-line stabilization of the cervical spine as your partner straightens the patient's legs (**Step 1**).
2. **Have your partner** place his or her hands on the patient's far shoulder and hip (**Step 2**).
3. **As you call the count** to control movement, have your partner turn the patient toward you by pulling on the far shoulder and hip. Control the head and neck so that they move as a unit with the rest of the torso. In this way, the head and neck stay in the same vertical plane as the back. This single motion will minimize aggravation of any potential spinal injury. At this point, you

You are the Provider
Part 4

The ushers step out of the way as you immediately kneel by the patient. You confirm with the ushers that the patient, in fact, did not fall off his chair but that they lifted him out carefully without hurting him. You then open his airway using a head tilt–chin lift technique and listen for breathing. You hear snoring respirations that persist even after adjusting his head position. You decide to use an oropharyngeal airway to keep his tongue out of the way.

8. What is the most appropriate method to open an unresponsive patient's airway when you are considering a chief complaint of difficulty breathing?
9. How do adventitious (abnormal) sounds help you in evaluating your patient's airway status?

Positioning the Unconscious Patient

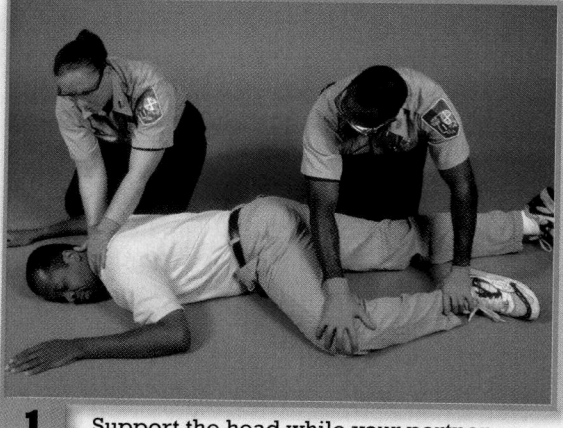

1 Support the head while your partner straightens the patient's legs.

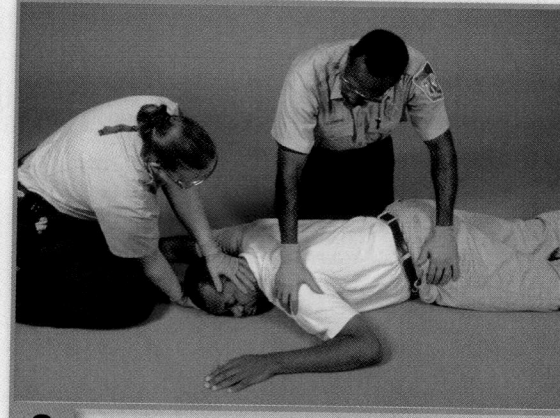

2 Have your partner place his or her hand on the patient's far shoulder and hip.

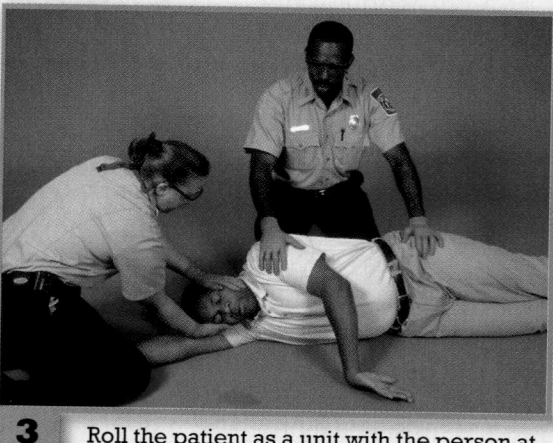

3 Roll the patient as a unit with the person at the head calling the count to begin the move.

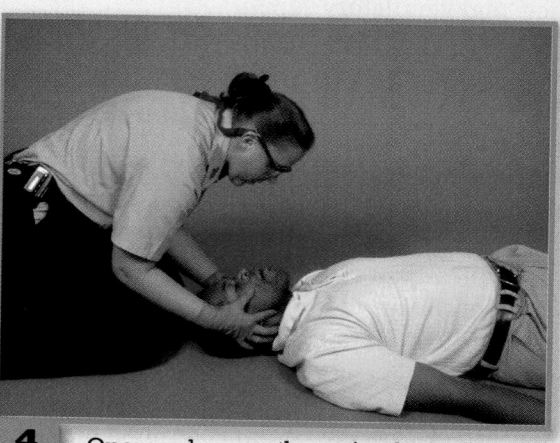

4 Open and assess the patient's airway and breathing status.

should apply a cervical collar. Place the patient's arms at his or her side (**Step 3**).

4. **Once the patient is positioned,** maintain an open airway and check for breathing (**Step 4**).

In an unconscious patient, the most common airway obstruction is the patient's own tongue, which falls back into the throat when the muscles of the throat and tongue relax (Figure 7-11 ▶). Dentures (false teeth), blood, vomitus, mucus, food, and other foreign objects may also create an airway obstruction. Therefore, you should always be prepared to help clear and maintain a patent (open) airway.

Head Tilt–Chin Lift Maneuver

Opening the airway to relieve an obstruction can often be done quickly and easily by simply tilting the patient's head back and lifting the chin in what is known as the head tilt–chin lift maneuver. For patients who have not sustained trauma, this simple maneuver is sometimes all that is needed for the patient to resume breathing.

To perform the head tilt–chin lift maneuver, follow these steps:

1. With the patient in a supine position, position yourself beside the patient's head.
2. Place one hand on the patient's forehead, and apply firm backward pressure with your palm to

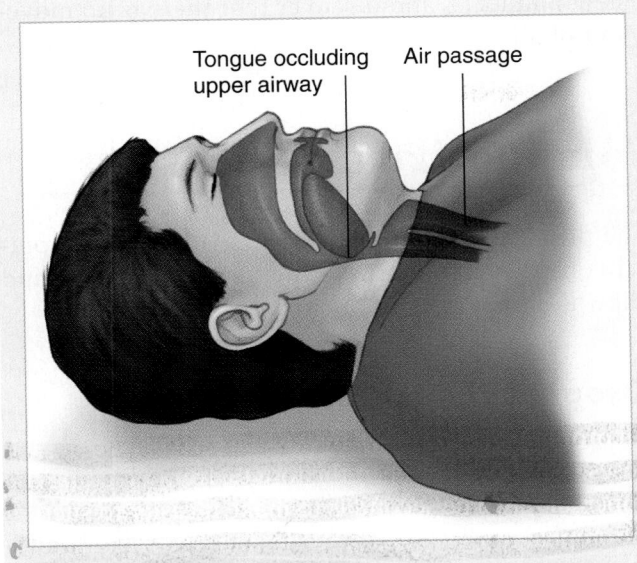

Figure 7-11 The most common airway obstruction is the patient's own tongue, which falls back into the throat when the muscles of the throat and tongue relax.

Tongue occluding upper airway   Air passage

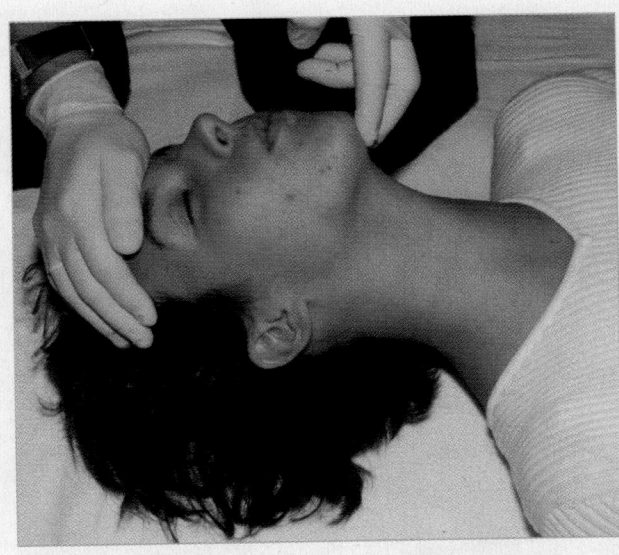

Figure 7-12 The head tilt–chin lift maneuver is a simple technique for opening the airway in a patient without a suspected cervical spine injury.

tilt the patient's head back. This extension of the neck will move the tongue forward, away from the back of the throat, and clear the airway if the tongue is blocking it.

3. Place the tips of the fingers of your other hand under the lower jaw near the bony part of the chin. Do not compress the soft tissue under the chin, as this may block the airway.

4. Lift the chin upward, bringing the entire lower jaw with it, helping to tilt the head back. Do not use your thumb to lift the chin. Lift so that the teeth are nearly brought together, but avoid closing the mouth completely. Continue to hold the forehead to maintain the backward tilt of the head (Figure 7-12 ▶).

Jaw-Thrust Maneuver

The head tilt–chin lift will open the airway in most patients. If you suspect a cervical spine injury, use the jaw-thrust maneuver. The jaw-thrust maneuver is a technique to open the airway by placing the fingers behind the angle of the jaw and lifting the jaw upward. You can easily seal a mask around the mouth while doing the jaw-thrust maneuver. This is the method of choice for patients with suspected cervical spine injury. See Chapter 30, Head and Spine Injuries, for a more detailed discussion of these types of injuries.

Perform the jaw-thrust maneuver in an adult in the following manner (Figure 7-13 ▶):

1. **Kneel above the patient's head**. Place your fingers behind the angles of the lower jaw, and move the jaw upward. Use your thumbs to help position the lower jaw to allow breathing through the mouth and the nose.

2. The completed maneuver should open the airway with the mouth slightly open and the jaw jutting forward.

Once the airway has been opened, the patient may start to breathe on his or her own. Assess whether breathing has returned by using the look, listen, and feel technique (Figure 7-14 ▶).

With complete airway obstruction, there will be no movement of air. However, you may see the chest and abdomen rise and fall considerably with the patient's frantic attempts to breathe. This is why the presence of chest wall movement alone does not indicate breathing is present. Regular chest wall movement indicates a respiratory effort is present. Observing chest and abdominal movement is often difficult with a fully clothed patient. You may see little, if any, chest movement even with normal breathing. This is particularly true in some patients with chronic lung disease. You must begin artificial ventilation immediately if you use the three-part approach—look,

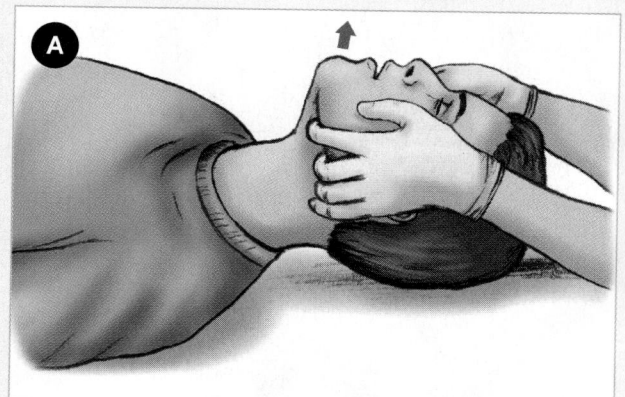

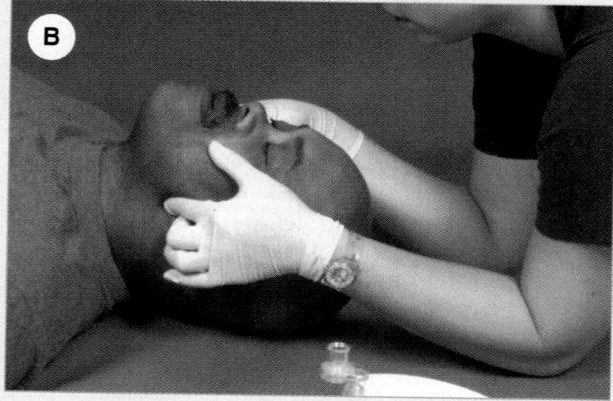

Figure 7-13 Performing the jaw-thrust maneuver. **A.** Kneeling above the patient's head, place your fingers behind the angles of the lower jaw, and move the jaw upward. Use your thumbs to help position the lower jaw. **B.** The completed maneuver should look like this.

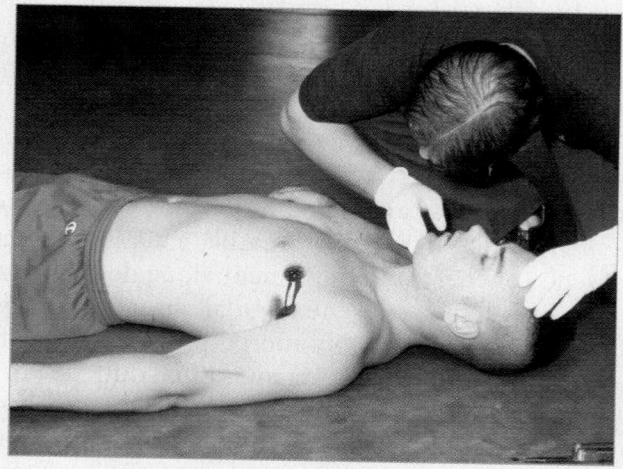

Figure 7-14 The look, listen, and feel technique is used to assess whether breathing has spontaneously returned.

listen, and feel—and discover that there is no movement of air.

Basic Airway Adjuncts

The primary function of an airway adjunct is to prevent obstruction of the upper airway by the tongue and allow the passage of air and oxygen to the lungs.

Oropharyngeal Airways

An oropharyngeal (oral) airway has two principal purposes. The first is to keep the tongue from blocking the upper airway. The second is to make it easier to suction the oropharynx if necessary. Suctioning is possible through an opening down the center or along either side of the oropharyngeal airway (Figure 7-15 ▶).

Indications for the oropharyngeal airway include the following:

- Unconscious patients without a gag reflex (breathing or apneic)
- Any apneic patient being ventilated with a BVM device

Contraindications for the oropharyngeal airway include the following:

- Conscious patients
- Any patient (conscious or unconscious) who has an intact gag reflex

The gag reflex is a protective reflex mechanism that prevents food and other particles from entering the airway. If you try to insert an oral airway in a patient with a gag reflex, the result may be vomiting or a spasm of the vocal cords. If the patient gags while you are attempting to insert an oral airway, immediately remove the oral airway and be prepared to suction the oropharynx, should vomiting occur. An oral airway is also a safe, effective way to help maintain the airway of a patient with a possible spinal injury. The use of an oral airway may make manual airway maneuvers such as the head tilt–chin lift and the jaw-thrust easier to maintain; however, manual maneuvers are often still needed to assure that the airway remains patent.

You must clearly understand when and how this device is used. If the oropharyngeal airway is too large, it could actually push the tongue back into the pharynx, blocking the airway. Conversely, an oral airway that is too small could block the airway directly, just like any other foreign body obstruction. The following

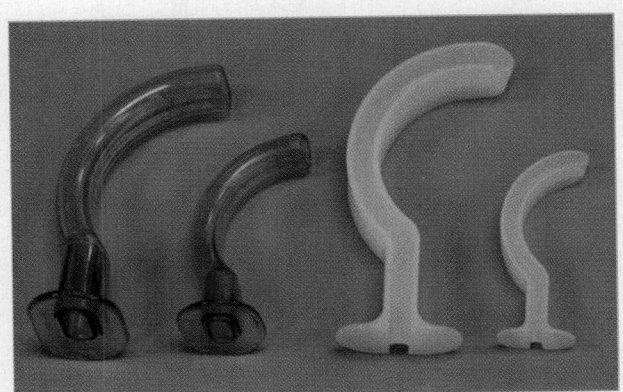

Figure 7-15 An oral airway is used for unconscious patients who have no gag reflex. It keeps the tongue from blocking the airway and makes suctioning the airway easier.

Geriatric Needs

When managing the airway of an older patient, you must be aware of the presence of dentures or other dental appliances. If dentures are tight-fitting and allow for effective airway management, they should be left in place. However, if the dentures are loose, they must be removed to avoid potential airway obstruction.

steps should be used when inserting an oropharyngeal airway (**Skill Drill 7-2** ▶):

1. **To select the proper size**, measure from the patient's earlobe or angle of the jaw to the corner of the mouth on the side of the face (**Step 1**).
2. **Open the patient's mouth** with the cross-finger technique. Hold the airway upside down with your other hand. Insert the airway with the tip facing the roof of the mouth and slide it in until it touches the roof of the mouth (**Step 2**).
3. **Rotate the airway 180°. When inserted properly**, the airway will rest in the mouth with the curvature of the airway following the contour of the airway. The flange should rest against the lips or teeth, with the other end opening into the pharynx (**Step 3**).

Take care to avoid injuring the hard palate (roof of the mouth) as you insert the airway. Roughness can

Pediatric Needs

In children, the alternative method of inserting an oral airway, using a bite stick to hold the tongue down while inserting the airway, is the only acceptable method. Because the airways of children are undeveloped, rotating an oropharyngeal airway in the posterior pharynx may cause damage. For more discussion on pediatric airways, see Chapter 31.

cause bleeding, which may aggravate airway problems or even cause vomiting.

If you encounter difficulty while inserting the oral airway, the following alternative method may be used (**Skill Drill 7-3** ▶):

1. **Use a bite stick** to depress the tongue, ensuring that the tongue remains forward (**Step 1**).
2. **Insert the oral airway sideways** from the corner of the mouth, until the flange reaches the teeth (**Step 2**).
3. **Rotate the oral airway** at a 90° angle, removing the bite stick as you exert gentle backward pressure on the oral airway, until it rests securely in place against the lips and teeth (**Step 3**).

In some cases, a patient may become responsive and regain the gag reflex after you have inserted an oral airway. If this occurs, gently remove the airway by pulling it out, following the normal curvature of the mouth and throat. Be prepared for the patient to vomit. Have suction available, and log roll the patient onto his or her side to allow any fluids to drain out.

Nasopharyngeal Airways

A <u>nasopharyngeal (nasal) airway</u> is usually used with a patient who has an intact gag reflex and is not able to maintain his or her airway spontaneously (**Figure 7-16** ▶). Patients with an altered mental status or those who have just had a seizure may also benefit from this type of airway. If a patient has sustained severe trauma to the head or face, you should consult medical control before inserting a nasopharyngeal airway. Extreme care must be used with such trauma patients. If the airway is accidentally pushed through the hole caused by a fracture of the base of the skull, it may penetrate through the cranium and into the brain.

Inserting an Oral Airway

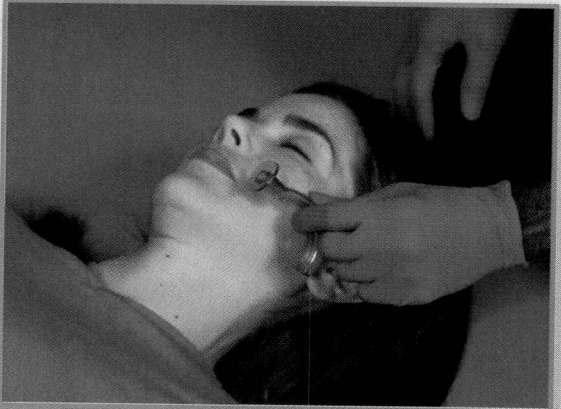

1 Size the airway by measuring from the patient's earlobe to the corner of the mouth.

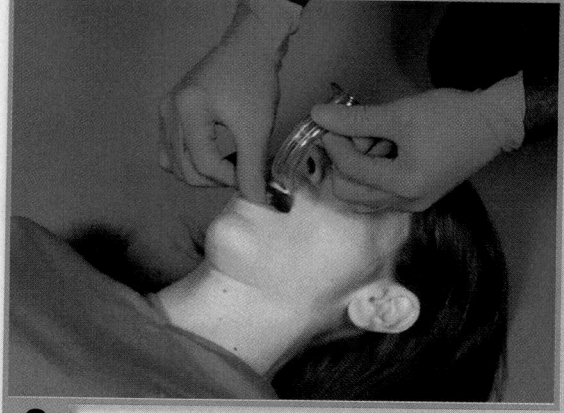

2 Open the patient's mouth with the cross-finger technique. Hold the airway upside down with your other hand. Insert the airway with the tip facing the roof of the mouth and slide it in until it touches the roof of the mouth.

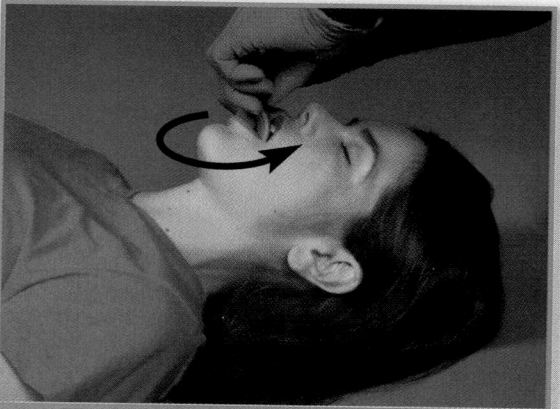

3 Rotate the airway 180°. Insert the airway until the flange rests on the patient's lips and teeth. In this position, the airway will hold the tongue forward.

This type of airway is usually better tolerated by patients who have an intact gag reflex. It is not as likely as the oropharyngeal airway to cause vomiting. You should coat the airway well with a water-soluble lubricant before it is inserted. Be aware that slight bleeding may occur even when the airway is inserted properly. However, you should never force the airway into place.

Indications for the nasopharyngeal airway include the following:

- Semiconscious or unconscious patients with an intact gag reflex

- Patients who otherwise will not tolerate an oropharyngeal airway

Contraindications for the nasopharyngeal airway include the following:

- Severe head injury with blood draining from the nose

- History of fractured nasal bones

Follow these steps to ensure correct placement of the nasopharyngeal airway (**Skill Drill 7-4** ▶):

1. **Before inserting the airway,** be sure you have selected the proper size. Measure from the tip of the

Inserting an Oral Airway With a 90° Rotation

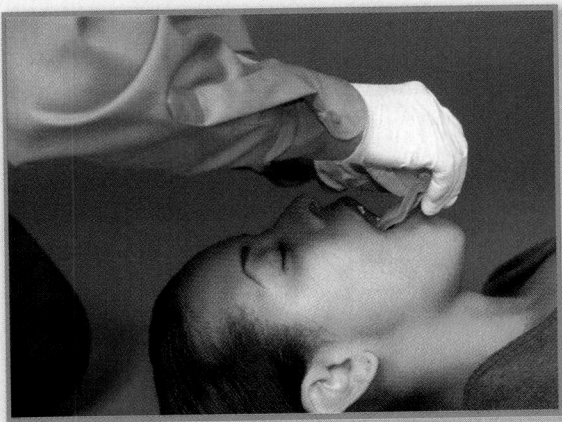

1 Depress the tongue so the tongue remains forward.

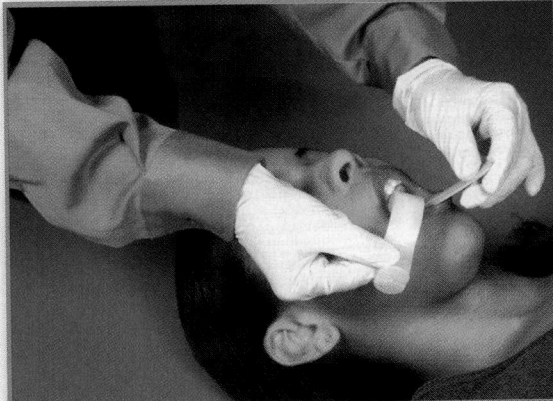

2 Insert the oral airway sideways from the corner of the mouth, until the flange reaches the teeth.

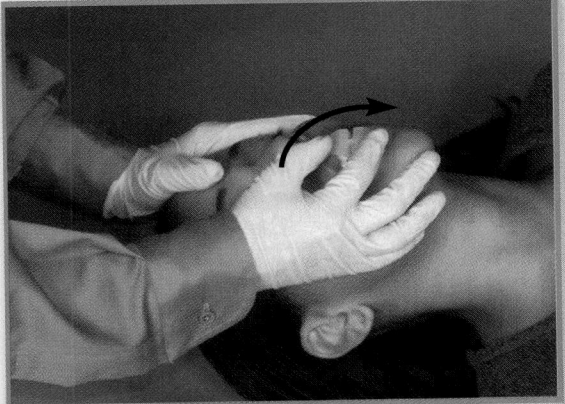

3 Rotate the oral airway at a 90° angle. Remove the bite stick as you exert gentle backward pressure on the oral airway, until it rests securely in place against the lips and teeth.

patient's nose to the earlobe. In almost all individuals, one nostril is larger than the other (**Step 1**).

2. **The airway should be placed** in the larger nostril, with the curvature of the device following the curve of the floor of the nose. If using the right nare, the bevel should face the septum (**Step 2**). If using the left nare, insert the airway with the tip of the airway pointing upward, which will allow the bevel to face the septum.

3. **Advance the airway gently** (**Step 3**). If using the left nare, insert the nasopharyngeal airway until resistance is met. Then rotate the nasopharyn-

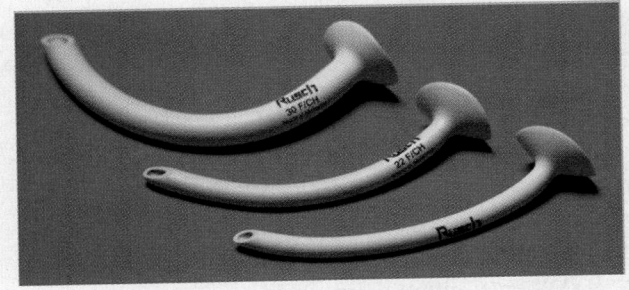

Figure 7-16 A nasal airway is better tolerated than is an oral airway by patients who have an intact gag reflex.

Inserting a Nasal Airway

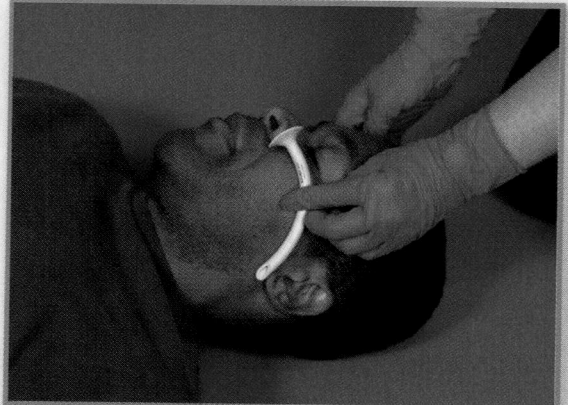

1 Size the airway by measuring from the tip of the nose to the patient's earlobe. Coat the tip with a water-soluble lubricant.

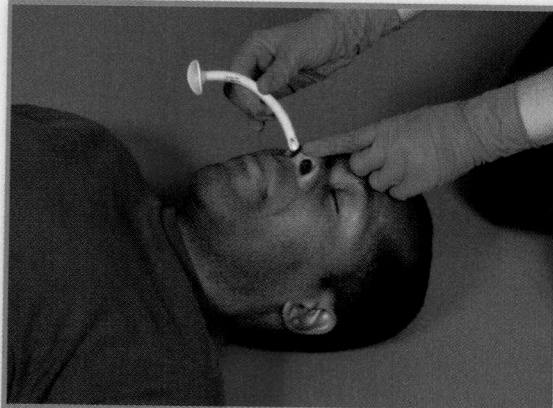

2 Insert the lubricated airway into the larger nostril with the curvature following the floor of the nose. If using the right nare, the bevel should face the septum. If using the left nare, insert the airway with the tip of the airway pointing upward, which will allow the bevel to face the septum.

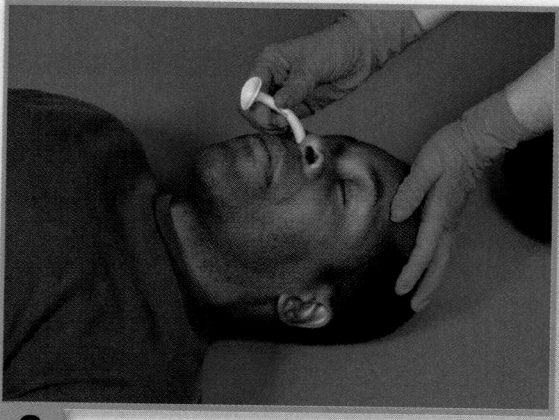

3 Gently advance the airway. If using the left nare, insert the nasopharyngeal airway until resistance is met. Then rotate the nasopharyngeal airway 180° into position. This rotation is not required if using the right nostril.

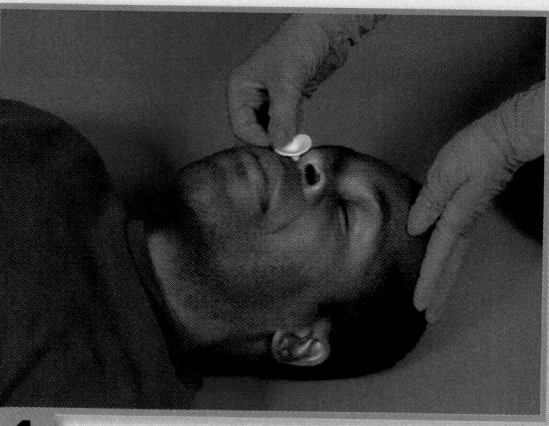

4 Continue until the flange rests against the skin. If you feel any resistance or obstruction, remove the airway and insert it into the other nostril.

geal airway 180° into position. This rotation is not required if using the right nostril.

4. **When completely inserted**, the flange rests against the nostril. The other end of the airway opens into the posterior pharynx (**Step 4**).

If the patient becomes intolerant of the nasal airway, you may have to remove it. Gently withdraw the airway from the nasal passage. Precautions similar to those used when removing an oral airway should be followed.

EMT-B Safety

A mask and protective eyewear should be worn whenever airway management involves suctioning. Body fluids can become aerosolized, and exposure to the mucous membranes of your mouth, nose, and eyes can easily occur.

Suctioning

You must keep the airway clear so that you can ventilate the patient properly. If the airway is not clear, you will force the fluids and secretions into the lungs and possibly cause a complete airway obstruction. Therefore, suctioning is your next priority. If you have any doubt about the situation, remember this rule: If you hear gurgling, the patient needs suctioning!

Suctioning Equipment

Portable, hand-operated, and fixed (mounted) suctioning equipment is essential for resuscitation (Figure 7-17 ▶). A portable suctioning unit must provide enough vacuum pressure and flow to allow you to suction the mouth and nose effectively. Hand-operated suctioning units with disposable chambers are reliable, effective, and relatively inexpensive. A fixed suctioning unit should generate airflow of more than 40 L/min and a vacuum of more than 300 mm Hg when the tubing is clamped.

A portable or fixed suctioning unit should be fitted with the following:

- Wide-bore, thick-walled, nonkinking tubing
- Plastic, rigid pharyngeal suction tips, called tonsil tips or Yankauer tips
- Nonrigid plastic catheters, called French or whistle-tip catheters
- A nonbreakable, disposable collection bottle
- A supply of water for rinsing the tips

A suction catheter is a hollow, cylindrical device that is used to remove fluids from the patient's airway. A tonsil-tip catheter is the best kind of catheter for suctioning the oropharynx in adults and is preferred for infants and children. The plastic tips have a large diameter and are rigid, so they do not collapse (Figure 7-18 ▶). Tips with a curved contour allow for easy, rapid placement in the oropharynx. Soft plastic, nonrigid catheters, sometimes called French or whistle-tip catheters, are used to suction the nose and liquid secretions in the back of the mouth and in situations in which you cannot use a rigid catheter, such as for a patient with a stoma (Figure 7-19 ▶). For example, a rigid catheter could break off a patient's tooth, whereas a flexible catheter may be inserted along the

You are the Provider   Part 5

Several minutes after inserting the oral airway, your patient begins to gag and vomits. You immediately remove the OPA and roll the patient to his side. When he finishes vomiting, you clean the large debris from his face and mouth. Your partner has set up the portable suction for you. As you reevaluate the patient's breathing and airway, you now hear gurgling sounds. You grab a rigid tipped catheter, turn on the suction machine, and open his mouth using a cross-finger technique. After measuring the depth of the catheter against the patient's face, you insert the catheter into the patient's mouth and begin counting the seconds you suction. After about 10 seconds, the mouth appears clear of fluids and the gurgling has stopped.

10. How important is it to reevaluate the interventions you use to treat your patient?
11. If your suction catheter does not remove the large debris from the patient's mouth, how would you remove it?

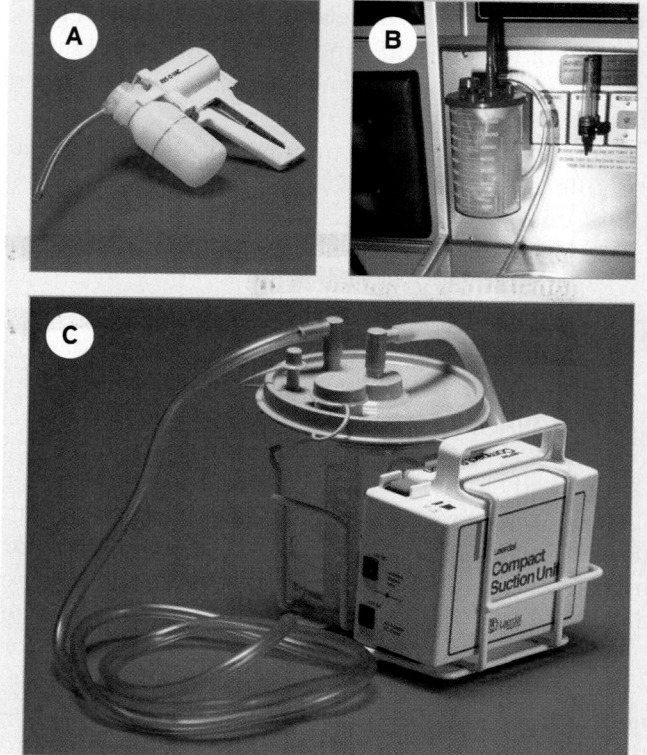

Figure 7-17 Suctioning equipment is essential for resuscitation. **A.** Hand-operated unit. **B.** Fixed unit. **C.** Portable unit.

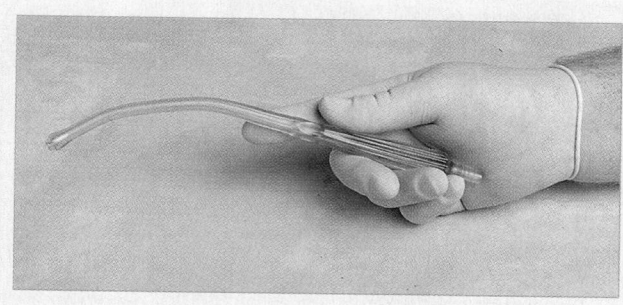

Figure 7-18 Tonsil-tip catheters are the best for suctioning because they have wide-diameter tips and are rigid.

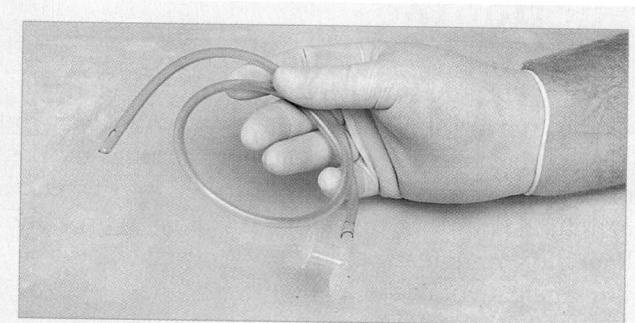

Figure 7-19 French, or whistle-tip, catheters are used in situations in which rigid catheters cannot be used, such as with a patient who has a stoma, patients whose teeth are clenched, or if suctioning the nose is necessary.

cheeks without injury. Before you insert any catheter, make sure to measure for the proper size. Use the same technique as you would use when measuring for an oropharyngeal airway. Be careful not to touch the back of the airway with a suction catheter. This can activate the gag reflex, cause vomiting, and increase the possibility of <u>aspiration</u>.

Techniques of Suctioning ~~(Never for more than 15 sec.)~~

You should inspect your suctioning equipment regularly to make sure it is in proper working condition. Turn on the suction, clamp the tubing, and make sure that the unit generates a vacuum of more than 300 mm Hg. Check that a battery-charged unit has charged batteries. Ensure that your suctioning equipment is at the patient's head and is easily accessible. Follow these general steps to operate the suction unit:

1. Check the unit for proper assembly of all its parts.
2. Turn on the suctioning unit and test it to ensure a vacuum pressure of more than 300 mm Hg.
3. Select and attach the appropriate suction catheter to the tubing.

Never suction the mouth or nose for more than 15 seconds at one time for adult patients, 10 seconds for children, and 5 seconds for infants. Suctioning removes oxygen from the airway along with obstructive material and can result in hypoxia. Rinse the catheter and tubing with water to prevent clogging of the tube with dried vomitus or other secretions. Repeat suctioning only after the patient has been adequately ventilated and reoxygenated.

You should use extreme caution when suctioning a conscious or semiconscious patient. Put the tip of the suction catheter in only as far as you can visualize. Be aware that suctioning may induce vomiting in these patients.

To properly suction a patient (**Skill Drill 7-5** ▶):

1. **Turn on the assembled suction unit** (**Step 1**).
2. **Measure the catheter** to the correct depth by measuring the catheter from the corner of the patient's mouth to the edge of the earlobe or angle of the jaw (**Step 2**).
3. **Open the patient's mouth** using the cross-finger technique or tongue-jaw lift, and insert

Suctioning a Patient's Airway

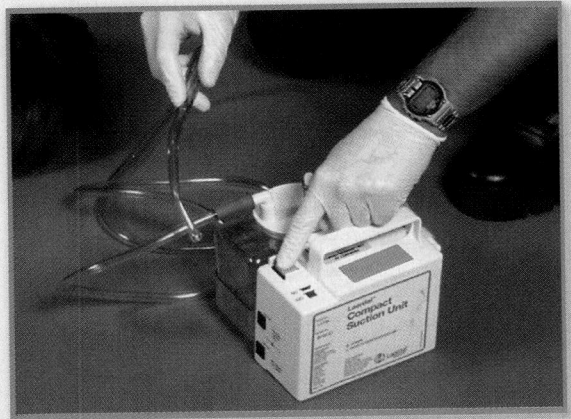

1 Make sure the suctioning unit is properly assembled and turn on the suction unit.

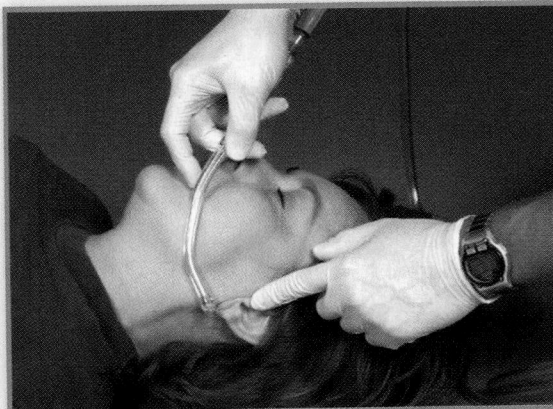

2 Measure the catheter from the corner of the mouth to the earlobe or angle of the jaw.

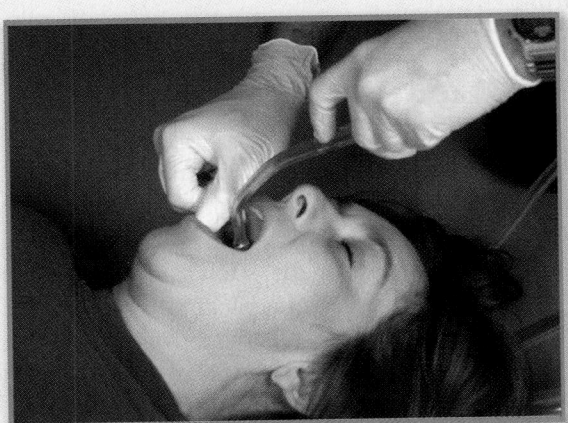

3 Open the patient's mouth and insert the catheter to the depth measured.

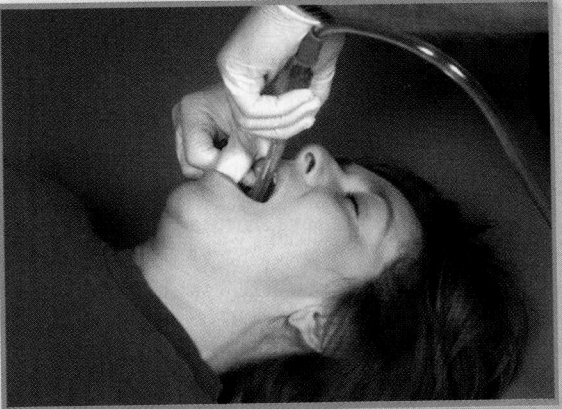

4 Apply suction in a circular motion as you withdraw the catheter. Do not suction an adult for more than 15 seconds.

the tip of the catheter to the depth measured (**Step 3**).

4. **Insert catheter to the premeasured depth and apply suction** in a circular motion as you withdraw the catheter. Do not suction an adult for more than 15 seconds (**Step 4**).

At times, a patient may have secretions or vomitus that cannot be suctioned quickly and easily, and some suction units cannot effectively remove solid objects such as teeth, foreign bodies, and food. In these cases, you should remove the catheter from the patient's mouth, log roll the patient to the side, and then clear the mouth

carefully with your gloved finger. A patient who requires assisted ventilation may also produce frothy secretions as quickly as you can suction them from the airway. In this situation, you should suction the patient's airway for 15 seconds (less time in infants and children), and then ventilate the patient for 2 minutes. This alternating pattern of suctioning and ventilating should continue until all secretions have been cleared from the patient's airway. Continuous ventilation is not appropriate if vomitus or other particles are present in the airway.

You should clean and decontaminate your suctioning equipment after each use according to the

Suctioning Time Limits

Adult	15 seconds
Child	10 seconds
Infant	5 seconds

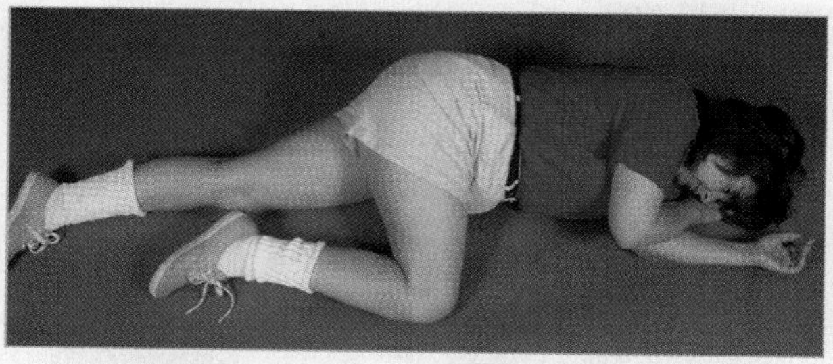

Figure 7-20 In the recovery position, the patient is rolled onto the left side.

manufacturer's guidelines. Place all disposable suctioning equipment (such as catheter, suction tubing) in a biohazard bag.

Maintaining the Airway

The recovery position is used to help maintain a clear airway in a patient who is not injured and is breathing on his or her own with a normal rate and adequate tidal volume (depth of breathing) (Figure 7-20 ▶). Take the following steps to put the patient in the recovery position:

1. Roll the patient onto the left side so that head, shoulders, and torso move at the same time without twisting.
2. Place the patient's extended left arm and right hand under his or her cheek.

Once patients have resumed spontaneous breathing after being resuscitated, the recovery position will prevent the aspiration of vomitus. However, this position is not appropriate for patients with suspected spinal trauma, nor is it adequate for patients who are unconscious and require ventilatory assistance. You must reposition such patients to provide adequate access to the airway while maintaining appropriate spinal immobilization.

Supplemental Oxygen

You should always give supplemental oxygen to patients who are hypoxic because they are not getting enough oxygen to the tissues and cells of the body.

Some tissues and organs, such as the heart, central nervous system, lungs, kidneys, and liver, need a constant supply of oxygen to function normally. **Never withhold oxygen from any patient who might benefit from it, especially if you must assist ventilations.**

When ventilating any patient in cardiac or respiratory arrest, you must always use high-concentration supplemental oxygen.

You are the Provider

Part 6

Now that you have cleared your patient's airway by suctioning, you place the patient in the recovery position and continue with your assessment of the ABCs. You find his breathing to be present and adequate. A pulse is present, and there is no evidence of bleeding. Findings from your focused history and physical exam are normal except for a low pulse oximeter reading of 88%. You place your patient on a nonrebreathing mask at 15 L/min and prepare the patient for transport to the hospital. Dispatch reports that the ALS unit is delayed in traffic due to construction.

12. This patient needs oxygen. What type of patients should not receive oxygen?
13. Earlier in your assessment you called for additional help, but now that help is delayed. How does that change your immediate decisions toward patient care?

Figure 7-21 Oxygen tanks for medical use will have a series of letters and numbers stamped into the metal on the collar of the cylinder.

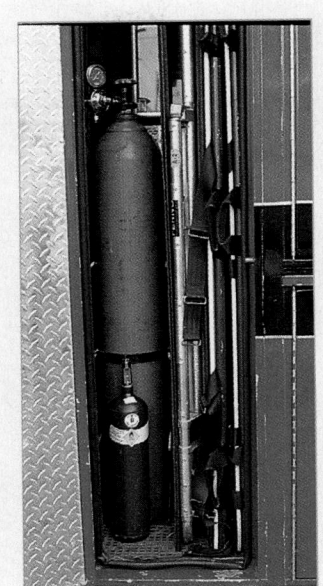

Figure 7-22 The cylinders that are most commonly found on an ambulance are the D (or super D) and M size cylinders.

Supplemental Oxygen Equipment

In addition to knowing when and how to give supplemental oxygen, you must understand how oxygen is stored and the various hazards associated with its use.

Oxygen Cylinders

The oxygen that you will give to patients is usually supplied as a compressed gas in green, seamless, steel or aluminum cylinders. Some cylinders may be silver or chrome with a green area around the valve stem on top. Newer cylinders are often made of lightweight aluminum or spun steel; older cylinders are much heavier.

Check to make sure that the cylinder is labeled for medical oxygen. You should look for letters and numbers stamped into the metal on the collar of the cylinder (Figure 7-21 ▲). Of particular importance are the month and year stamps, which indicate when the cylinder was last tested.

Oxygen cylinders are available in several sizes. The two sizes that you will most often use are the D (or super D) and M cylinders (Figure 7-22 ▶). The D (or super D) cylinder can be carried from your unit to the patient. The M tank remains on board your unit as a main supply tank. Other sizes that you will see are A, E, G, H, and K (Table 7-2 ▶). The length of time you can use an oxygen cylinder depends on the pressure in the cylinder and the flow rate. A method of calculating cylinder duration is shown in (Table 7-3 ▶).

Safety Considerations

Compressed gas cylinders must be handled carefully because their contents are under pressure. Cylinders are fitted with pressure regulators to make sure that patients receive the right amount and type of gas. Make sure that the correct pressure regulator is firmly attached before you transport the cylinders. A puncture or hole in the tank can cause the cylinder to become a deadly missile. Do not handle a cylinder by the neck assembly alone. Cylinders should be secured with mounting brackets when they are stored on the ambulance. Oxygen cylinders that are in use during transport should

TABLE 7-2 Oxygen Cylinder Sizes Carried on the Ambulance

Size	Volume, L
D	350
Super D	500
E	625
M	3,000
G	5,300
H, A, K	6,900

TABLE 7-3 Oxygen Cylinders: Duration of Flow

Formula

$$\frac{(\text{Gauge pressure in psi} - \text{the safe residual pressure}) \times \text{constant}}{\text{Flow rate in L/min}} = \text{duration of flow in minutes}$$

SAFE RESIDUAL PRESSURE = 200 psi

CYLINDER CONSTANT

D = 0.16	G = 2.41
E = 0.28	H = 3.14
M = 1.56	K = 3.14

Determine the life of an M cylinder that has a pressure of 2,000 psi and a flow rate of 10 L/min.

$$\frac{(2,000 - 200) \times 1.56}{10} = \frac{2,808}{10} = 281 \text{ min, or 4 h 41 min}$$

Note: psi indicates pounds per square inch.

be positioned and secured to prevent the tank from falling and to prevent damage to the valve-gauge assembly.

Pin-Indexing System

The compressed gas industry has established a pin-indexing system for portable cylinders to prevent an oxygen regulator from being connected to a carbon dioxide cylinder, a carbon dioxide regulator from being connected to an oxygen cylinder, and so on. In preparing to administer oxygen, always check to be sure that the pinholes on the cylinder exactly match the corresponding pins on the regulator.

The pin-indexing system features a series of pins on a yoke that must be matched with the holes on the valve stem of the gas cylinder. The arrangement of the pins and holes varies for different gases according to accepted national standards (Figure 7-23 ▶). Other gases that are supplied in portable cylinders, such as acetylene, carbon dioxide, and nitrogen, use regulators and flowmeters that are similar to those used with oxygen. Each cylinder of a specific gas type has a given pattern and a given number of pins. These safety measures make it impossible for you to attach a cylinder of nitrous oxide to an oxygen regulator. The oxygen regulator will not fit.

The outlet valves on portable oxygen cylinders are designed to accept yoke-type pressure-reducing gauges, which conform to the pin-indexing system

(Figure 7-24 ▶). The safety system for large cylinders is known as the American Standard System. In this system, oxygen cylinders are equipped with threaded gas outlet valves. The inside and outside thread sizes of these outlets vary depending on the gas in the cylinder. The cylinder will not accept a regulator valve unless it is properly threaded to fit that regulator. The purpose of these safety devices is the same as in the pin-indexing system: to prevent the accidental attachment of a regulator to a wrong cylinder.

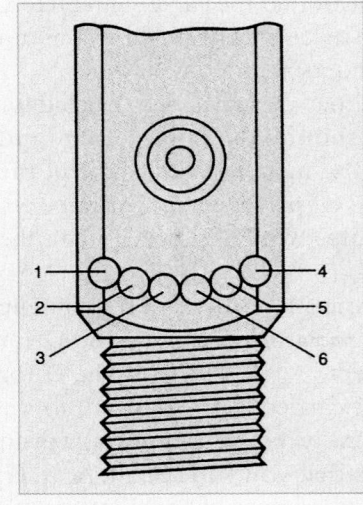

Figure 7-23 The locations of the pin-indexing safety system holes in a cylinder valve face. Each cylinder of a specific gas has a given pattern and a given number of pins.

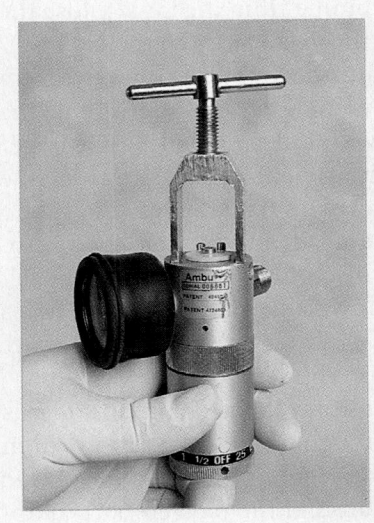

Figure 7-24 A yoke-type pressure-reducing gauge is used with a portable oxygen cylinder.

Figure 7-25 Giving humidified oxygen may be preferred with long transport times. However, the use of this type of oxygen-delivery system is not universal in all EMS systems.

Pressure Regulators

The pressure of gas in a full oxygen cylinder is approximately 2,000 psi. This is far too much pressure to be safe or useful for your purposes. Pressure regulators reduce the pressure to a more useful range, usually 40 to 70 psi. Most pressure regulators in use today reduce the pressure in a single stage, although multistage regulators exist. A two-stage regulator will reduce the pressure first to 700 psi and then to 40 to 70 psi.

After the pressure is reduced to a workable level, the final attachment for delivering the gas to the patient is usually one of the following:

- A quick-connect female fitting that will accept a quick-connect male plug from a pressure hose or ventilator or resuscitator
- A flowmeter that will permit the regulated release of gas measured in liters per minute

Humidification

Some EMS systems provide humidified oxygen to patients during transport (Figure 7-25 ▶). However, humidified oxygen is usually indicated only for long-term oxygen therapies. Dry oxygen is not considered harmful for short-term use. Therefore, many EMS systems do not use humidified oxygen in the prehospital setting. Always refer to medical control or local protocols for guidance involving patient treatment issues.

Flowmeters

Flowmeters are usually permanently attached to pressure regulators on emergency medical equipment. The two types of flowmeters that are commonly used are pressure-compensated flowmeters and Bourdon-gauge flowmeters.

A pressure-compensated flowmeter incorporates a float ball within a tapered calibrated tube. The float rises or falls according to the gas flow within the tube. The flow of gas is controlled by a needle valve located downstream from the float ball. This type of flowmeter is affected by gravity and must always be maintained in an upright position for an accurate flow reading (Figure 7-26 ▼).

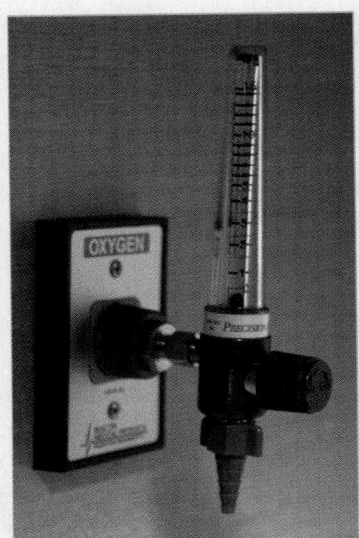

Figure 7-26 A pressure-compensated flowmeter contains a float ball that rises or falls according to the gas flow within the tube. It must be maintained in an upright position for an accurate reading.

Figure 7-27 The Bourdon-gauge flowmeter is not affected by gravity and can be used in any position.

The Bourdon-gauge flowmeter is commonly used because it is not affected by gravity and can be used in any position (Figure 7-27 ▲). It is actually a pressure gauge that is calibrated to record flow rate. The major disadvantage of this flowmeter is that it does not compensate for backpressure. Therefore, it will usually record a higher flow rate when there is any obstruction to gas flow downstream.

Operating Procedures

Before placing an oxygen cylinder into service (Skill Drill 7-6 ▶):

1. **Inspect the cylinder** and its markings. If the cylinder was commercially filled, it will have a plastic seal around the valve stem covering the opening in the stem. Remove the seal, and inspect the opening to make sure that it is free of dirt and other debris. The valve stem should not be sealed or covered with adhesive tape or any petroleum-based substances. These can contaminate the oxygen and can contribute to spontaneous combustion when mixed with the pressurized oxygen.

   "Crack" the cylinder by slowly opening and then reclosing the valve to help make sure that dirt particles and other possible contaminants do not enter the oxygen flow. Never face the tank toward yourself or others when cracking the cylinder. Open the tank by attaching a tank key to the valve and rotating the valve counterclockwise. You should be able to hear clearly the rush of oxygen coming from the tank. Close the tank by rotating the valve clockwise (**Step 1**).

2. **Attach the regulator/flowmeter** to the valve stem after clearing the opening. On one side of the valve stem, you will find three holes. The larger one, on top, is a true opening through which the oxygen flows. The two smaller holes below it do not extend to the inside of the tank. They provide stability to the regulator. Following the design of a pin-indexing system, these two holes are very precisely located in positions that are unique to oxygen cylinders.

   Above the pins on the inside of the collar is the actual port through which oxygen flows from the cylinder to the regulator. A metal or plastic O-ring is placed around the oxygen port to optimize the airtight seal between the collar of the regulator and the valve stem (**Step 2**).

3. **Place the regulator collar** over the cylinder valve, with the oxygen port and pin-indexing pins on the side of the valve stem that has the three holes. Open the screw bolt just enough to allow the collar to fit freely over the valve stem. Move the regulator so that the oxygen port and the pins fit into the correct holes on the valve stem. The screw bolt on the opposite side should be aligned with the dimpled depression. As you hold the regulator securely against the valve stem, tighten the screw bolt until the regulator is firmly attached to the cylinder. At this point, you should not see any open spaces between the sides of the valve stem and the interior walls of the collar (**Step 3**).

4. **With the regulator firmly attached**, open the cylinder, check for air leaking from the regulator-oxygen cylinder connection, and read the pressure level on the regulator gauge. Most portable cylinders have a maximum pressure of approximately 2,000 psi. Most EMS services consider a cylinder with less than 500 to 1,000 psi to be too low to keep in service. Learn your department's policies in this regard and follow them.

   The flowmeter will have a second gauge or a selector dial that indicates the oxygen flow rate. Several popular types of devices are widely used. Attach the selected oxygen device to the flowmeter by connecting the universal oxygen connective tubing to the "Christmas tree" nipple on the flowmeter. Most oxygen-delivery devices come

Placing an Oxygen Cylinder Into Service

1 Using an oxygen wrench, turn the valve counterclockwise to slowly "crack" the cylinder.

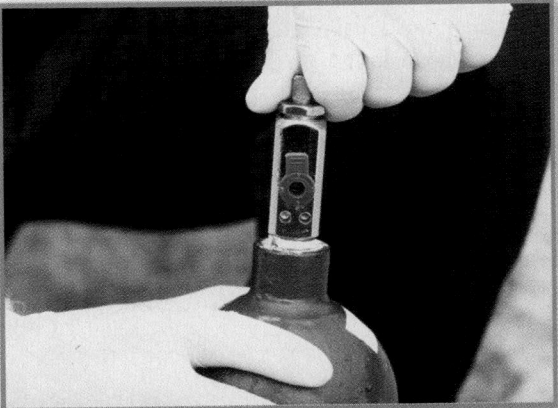

2 Attach the regulator/flowmeter to the valve stem using the two pin-indexing holes and make sure that the washer is in place over the larger hole.

3 Align the regulator so that the pins fit snugly into the correct holes on the valve stem, and hand tighten the regulator.

4 Attach the oxygen connective tubing to the flowmeter.

with this tubing permanently attached. Some oxygen masks do not. You must add this tubing to the oxygen-delivery device if it is not attached (**Step 4**).

Open the flowmeter to the desired flow rate. Flow rates will vary based on the oxygen-delivery device being used. Remember that you must be completely familiar with the equipment before attempting to use it on a patient. Once the oxygen is flowing at the desired rate, apply the oxygen device to the patient and make

any necessary adjustments. Monitor the patient's response to the oxygen and to the oxygen device, and periodically recheck the regulator gauge to make sure there is sufficient oxygen in the cylinder. Disconnect the tubing from the flowmeter nipple and turn off the cylinder valve when oxygen therapy is complete or when the patient has been transferred to the hospital and is using the hospital's oxygen system. In a few seconds, the sound of oxygen flowing from the nipple will cease. This indicates that all the pressurized oxygen has

Study This!!!

✋ EMT-B Safety

Slowly open the oxygen tank after attaching the regulator and check for leaks. Remember that although oxygen itself is not combustible, it supports combustion, and any ignition source may cause fire or an explosion in an oxygen-rich environment—especially if oxygen is being released too quickly from the cylinder at the time or if the seal between the regulator and oxygen cylinder is not secure.

Study This!!

✳ EMT-B Tips

Oxygen-Delivery Devices

Device	Flow Rate	Oxygen Delivered
Nasal cannula	1 to 6 L/min	24% to 44%
Nonrebreathing mask	10 to 15 L/min	Up to 90%
BVM device with reservoir	15 L/min-flush	Nearly 100%

been removed from the flowmeter. Turn off the flowmeter. The gauge on the regulator should read zero with the tank valve closed. This confirms that there is no pressure left above the valve stem. As long as there is a pressure reading on the regulator gauge, it is not safe to remove the regulator from the valve stem.

Hazards of Supplemental Oxygen

Oxygen does not burn or explode. However, it does support combustion. The more oxygen is around, the faster the combustion process. A small spark, even a glowing cigarette, can become a flame in an oxygen-rich atmosphere. Therefore, you must keep any possible source of fire away from the area while oxygen is in use. Make sure the area is adequately ventilated, especially in industrial settings where hazardous materials may be present and where sparks are easily generated. Be extremely cautious in any enclosed environment in which oxygen is being administered, as an oxygen-rich environment increases the chance of fire if a spark or flame is introduced. A bystander who is smoking or sparks generated during vehicle extrication are possible ignition sources. Never leave an oxygen cylinder standing unattended. The cylinder can be knocked over, injuring the patient or damaging the equipment.

Oxygen-Delivery Equipment

In general, the oxygen-delivery equipment that is used in the field should be limited to nonrebreathing masks, BVM devices, and nasal cannulas, depending on local protocol. However, you may encounter other devices during transports between medical facilities.

Nonrebreathing Mask

The nonrebreathing mask is the preferred way of giving oxygen in the prehospital setting to patients who are breathing adequately but are suspected of having or are showing signs of hypoxia. With a good mask-to-face seal, it is capable of providing up to 90% inspired oxygen.

The nonrebreathing mask is a combination mask and reservoir bag system. Oxygen fills a reservoir bag that is attached to the mask by a one-way valve. The system is called a nonrebreathing mask because the exhaled gas escapes through flapper valve ports at the cheek areas of the mask (Figure 7-28 ▼). These valves prevent the patient from rebreathing exhaled gases.

In this system, you must be sure that the reservoir bag is full before the mask is placed on the patient.

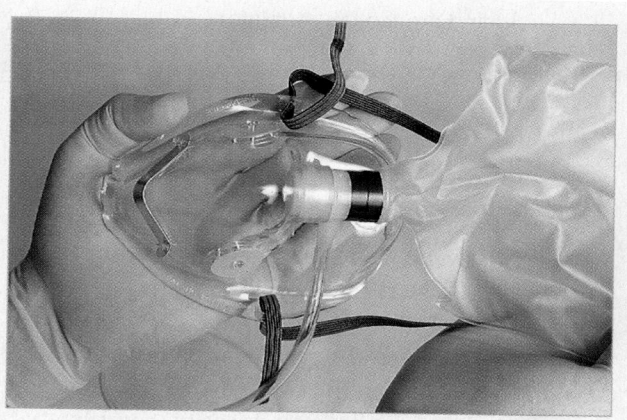

Figure 7-28 The nonrebreathing mask contains flapper valve ports at the cheek areas of the mask to prevent the patient from rebreathing exhaled gases.

Adjust the flow rate so that the bag does not fully collapse when the patient inhales, to about two thirds of the bag volume, or 10 to 15 L/min. Use a pediatric nonrebreathing mask, which has a smaller reservoir bag, with infants and children, as they will inhale a smaller volume.

Nasal Cannula

A nasal cannula delivers oxygen through two small, tubelike prongs that fit into the patient's nostrils (Figure 7-29 ▶). This device can provide 24% to 44% inspired oxygen when the flowmeter is set at 1 to 6 L/min. For the comfort of your patient, flow rates above 6 L/min are not recommended with the nasal cannula.

The nasal cannula delivers dry oxygen directly into the nostrils, which, over prolonged periods, can cause dryness or irritate the mucous membrane lining of the nose. Therefore, when you anticipate a long transport time, you should consider the use of humidification.

A nasal cannula has limited use in the prehospital care setting. For example, a patient who breathes through the mouth or who has a nasal obstruction will get little or no benefit from a nasal cannula. Always try to give high-flow oxygen through a nonrebreathing mask if you suspect that a patient may have hypoxia, coaching him or her as necessary. If the patient will not tolerate a nonrebreathing mask, you will have to use a nasal cannula, which some patients find more comfortable. As always, a good assessment of your patient will guide your decision.

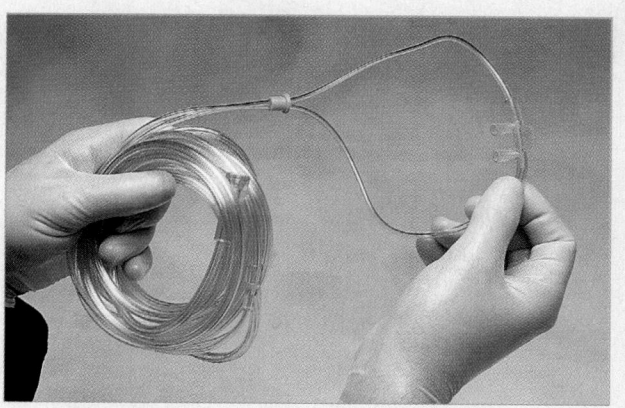

Figure 7-29 The nasal cannula delivers oxygen directly through the nostrils.

Assisted and Artificial Ventilation

Obviously, a patient who is not breathing needs artificial ventilation and 100% supplemental oxygen. Patients who are breathing inadequately, such as those who are breathing too fast or too slow with reduced tidal volume, are unable to speak in complete sentences, or have an irregular pattern of breathing, will also require artificial ventilation to assist them in maintaining adequate minute volume. Keep in mind that fast, shallow breath-

You are the Provider   Part 7

Despite supplemental oxygen therapy, your patient's condition has deteriorated. He is more cyanotic and has shallow, slow respirations. You insert a nasopharyngeal airway and begin assisting his ventilations at 1 breath every 5 seconds with a BVM device attached to 100% supplemental oxygen. He does not resist your attempts to ventilate, and his chest rises and falls with each ventilation. He tolerates the NPA without a problem. Dispatch reports that paramedics will rendezvous in 5 minutes.

14. Is an airway adjunct needed to provide assisted ventilations with a BVM? How does it help?
15. The patient's condition is deteriorating and you have begun ventilations at 1 every 5 seconds. Is this enough? How do you know if your ventilations are effective?

Study This!!!

✳ EMT-B Tips

**Methods of Ventilation
(listed in order of preference)**

- Mouth-to-mask with one-way valve
- Two-person BVM device with reservoir and supplemental oxygen
- Flow-restricted, oxygen-powered ventilation device (manually triggered ventilator)
- One-person BVM device with oxygen reservoir and supplemental oxygen

Note: This order of preference has been stated because research has shown that personnel who infrequently ventilate patients have great difficulty maintaining an adequate seal between the mask and the patient's face.

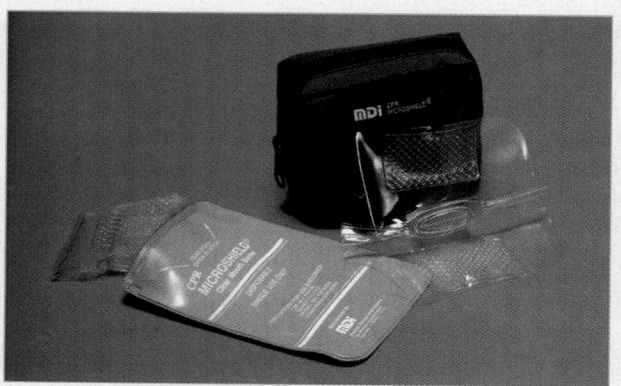

Figure 7-30 Barrier devices such as a plastic shield or a pocket mask with a one-way valve provide adequate BSI.

ing can be as dangerous as very slow breathing. Fast, shallow breathing moves air primarily in the larger airway passages (dead air space) and does not allow for adequate exchange of air and carbon dioxide in the alveoli. Patients with inadequate breathing require assisted ventilations with some form of positive-pressure ventilation. Remember to follow body substance isolation (BSI) precautions as needed when managing the patient's airway.

Once you determine that a patient is not breathing or is breathing inadequately, you should begin artificial ventilation immediately. The methods that an EMT-B may use to provide artificial ventilation include the mouth-to-mask technique, a one- or two-person BVM device, and the flow-restricted, oxygen-powered ventilation device.

Ventilation with a flow-restricted, oxygen-powered ventilation device is not commonly performed because the device is not carried on most ambulances. In addition, it may not be used with all types of patients, especially children.

✳ EMT-B Tips

Ventilation Rates

Adult	1 breath per 5 seconds
Child	1 breath per 3 seconds
Infant	1 breath per 3 seconds

Mouth-to-Mouth and Mouth-to-Mask Ventilation

As you learned in your CPR course, mouth-to-mouth ventilations are now routinely done with a barrier device, such as a mask or face shield. A <u>barrier device</u> is a protective item that features a plastic barrier placed on a patient's face with a one-way valve to prevent the backflow of secretions, vomitus, and gases. Barrier devices provide adequate BSI (Figure 7-30 ▲). Mouth-to-mouth ventilations without a barrier device should be provided only in extreme conditions. Performing mouth-to-mask ventilations with a pocket mask with a one-way valve is a safer method of ventilation to prevent possible disease transmission.

A mask with an oxygen inlet provides oxygen during mouth-to-mask ventilation to supplement the air from your own lungs. Remember that the gas you exhale contains 16% oxygen. With the mouth-to-mask system, however, the patient gets the additional benefit of significant oxygen enrichment with inspired air. This system also frees both your hands to help keep the airway open and helps you to provide a better seal between the mask and the face, thus delivering adequate tidal volume.

The mask may be shaped like a triangle or a doughnut, with the apex (top) placed across the bridge of the nose. The base (bottom) of the mask is placed in the groove between the lower lip and the chin. In the center of the mask is a chimney with a 15-mm connector.

Follow these steps to use mouth-to-mask ventilation (Skill Drill 7-7 ▶):

1. **Kneel at the patient's head.** Open the airway using the head tilt–chin lift maneuver or the jaw-

Performing Mouth-to-Mask Ventilation

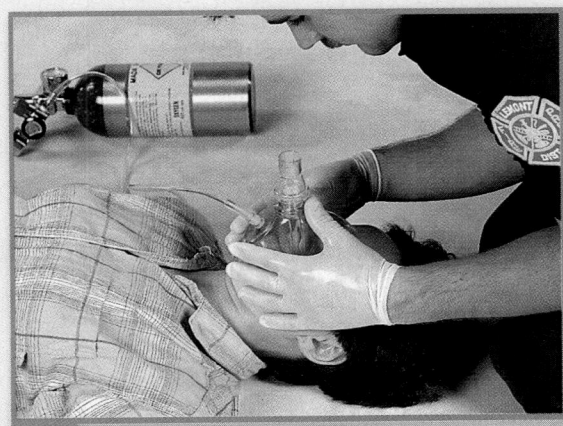

1 Once the patient's head is properly positioned and an airway adjunct is inserted, place the mask on the patient's face. Seal the mask to the face using both hands (EC clamp).

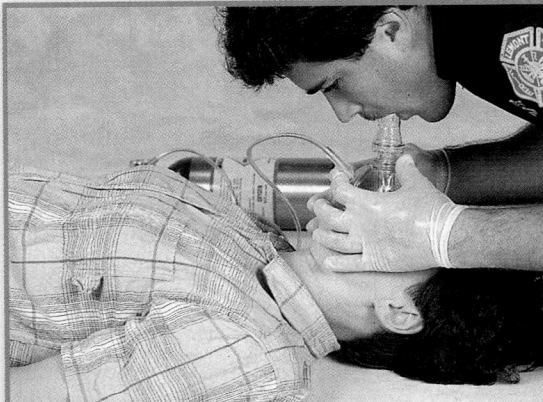

2 Exhale slowly into the open port of the one-way valve for 2 seconds as you observe for adequate chest rise.

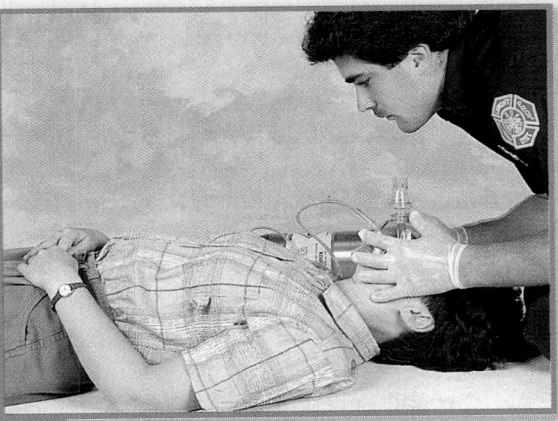

3 Remove your mouth and watch the patient's chest fall during exhalation.

thrust maneuver if indicated. Insert an oral or nasal airway to help maintain airway patency. Connect the one-way valve to the face mask. Place the mask on the patient's face. Make sure the top is over the bridge of the nose and the bottom is in the groove between the lower lip and the chin. Hold the mask in position by placing your thumbs over the top part of the mask and your index fingers over the bottom half. Grasp the lower jaw with the remaining three fingers on each hand. Make an airtight seal by pulling the lower jaw into the mask. Maintain an upward and forward pull on the lower jaw with your fingers to keep the airway open. This method of securing the mask to the patient's face is known as the EC clamp method (**Step 1**).

2. **Take a deep breath and exhale** through the open port of the one-way valve. Breathe slowly into the patient's mask for 2 seconds and observe for adequate chest rise (**Step 2**).

3. **Remove your mouth**, and watch for the patient's chest to fall during passive exhalation (**Step 3**).

You know that you are providing adequate ventilations if you see the patient's chest rise adequately and do not meet resistance when ventilating. You should also hear and feel air escape as the patient exhales. Make sure that you are providing the correct number of breaths per minute for the patient's age.

To increase the oxygen concentration, administer high-flow oxygen at 15 L/min through the oxygen inlet valve of the mask. This, when combined with your exhaled breath, will deliver approximately 55% oxygen to the patient. If supplemental oxygen is available, deliver a tidal volume of approximately 400 to 600 mL (6 to 7 mL/kg) over 1 to 2 seconds. If supplemental oxygen is not available, tidal volumes of approximately 700 to 1,000 mL (10 mL/kg) should be delivered over 2 seconds.

The BVM Device (Positive Pressure Ventilations)

With an oxygen flow rate of 15 L/min and an adequate mask-to-face seal, a <u>bag-valve-mask (BVM) device</u> with an oxygen reservoir can deliver nearly 100% oxygen (Figure 7-31 ▼). Most BVM devices on the market today include modifications or accessories (reservoirs) that permit the delivery of oxygen concentrations approaching 100%. However, the device can deliver only as much volume as you can squeeze out of the bag by hand. The BVM device provides less tidal volume than mouth-to-mask ventilation; however, it delivers a much higher oxygen concentration. The BVM device is the most common method used to ventilate patients in the field. An experienced EMT-B will be able to supply adequate tidal volumes with a BVM device. Be sure

EMT-B Tips

Volume Capabilities of the BVM Device	
Size	Amount, mL
Adult	1,200 to 1,600
Pediatric	500 to 700
Infant	150 to 240

to practice on ventilation manikins several times before using a BVM device on a patient. If you have difficulty adequately ventilating a patient with the BVM, you should immediately switch to an alternate method of ventilation, such as the mouth-to-mask technique.

A BVM device should be used when you need to deliver high concentrations of oxygen to patients who are not ventilating adequately. The device is also used for patients in respiratory arrest, cardiopulmonary arrest, and respiratory failure. The BVM device may be used with or without oxygen. However, to ensure the highest concentration of delivered oxygen, you must attach supplemental oxygen and a reservoir. You should use an oral or nasal airway adjunct in conjunction with the BVM device.

Components

All adult BVM devices should have the following components:

- A disposable self-refilling bag
- No pop-off valve, or if one is present, the capability of disabling the pop-off valve
- An outlet valve that is a true valve for nonrebreathing
- An oxygen reservoir that allows for delivery of high-concentration oxygen
- A one-way, no-jam inlet valve system that provides an oxygen inlet flow at a maximum of 15 L/min with standard 15/22-mm fittings for face mask and endotracheal tube (or other advanced airway adjunct) connection
- A transparent face mask
- Ability to perform under extreme environmental conditions, including extreme heat or cold

The total volume in the bag of an adult BVM device is usually 1,200 to 1,600 mL. The pediatric bag contains 500 to 700 mL, and the infant bag holds 150 to 240 mL.

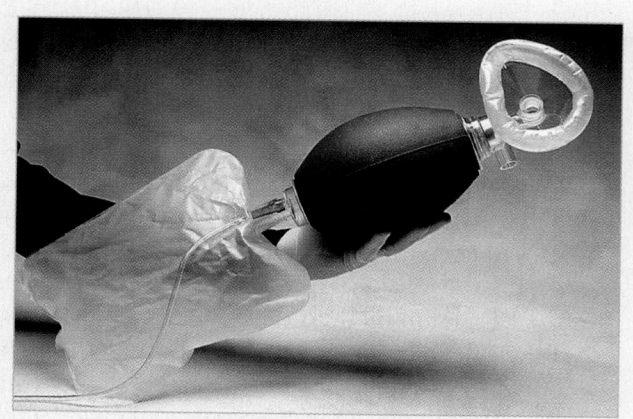

Figure 7-31 A BVM device with an oxygen reservoir can deliver nearly 100% oxygen if a good seal between the mouth and mask is achieved and if supplemental oxygen is used.

The volume of air (oxygen) to deliver to the patient is based on one key observation—chest rise and fall. In most situations, you will be using the BVM device attached to high-flow oxygen (15 L/min). When using the BVM device with high-flow oxygen on an adult patient, you should squeeze the bag enough to cause a noticeable rise of the patient's chest—a volume of 400 to 600 mL (approximately 6 to 7 mL/kg) over 1 to 2 seconds. When oxygen is not available, higher tidal volume amounts are required to cause good rise of the patient's chest—700 to 1,000 mL (approximately 10 mL/kg) over 2 seconds. By delivering smaller tidal volumes when the BVM device is used with oxygen, the risk of gastric distention (and associated complications of vomiting and aspiration) is reduced.

There are two issues to consider with this approach. It is not practical for the EMT-B to accurately measure tidal volumes in milliliters per kilogram for each patient ventilated in the field. There is also a significant risk of hypoxia when ventilating with smaller volumes. For these reasons, the key is to watch for good chest rise and fall—let these observations determine the appropriate amount of volume to deliver.

Technique

Whenever possible, you and your partner should work together to provide BVM device ventilation. One EMT-B can maintain a good mask seal by securing the mask to the patient's face with two hands while the other EMT-B squeezes the bag. Ventilation using a BVM device is a challenging skill: it may be very difficult for one EMT-B to maintain a proper seal between the mask and the face with one hand while squeezing the bag well enough to deliver an adequate volume to the patient. This skill can be difficult to maintain if you do not have many opportunities to practice. Effective one-person BVM device ventilation requires considerable experience. Also, performance of this skill depends on having enough personnel to carry out other actions that need to be done at the same time, such as chest compressions, putting the stretcher in place, or helping to lift the patient onto the stretcher.

Follow these steps to use the two-person BVM device technique:

1. Kneel above the patient's head. If possible, your partner should be at the side of the head to squeeze the bag while you hold a seal between the mask and the patient's face with two hands.
2. Maintain the patient's neck in an extended position unless you suspect a cervical spine injury. In that case, you should immobilize the patient's head and neck and use the jaw-thrust maneuver. Have your partner hold the head, or, if you are alone, use your knees to immobilize the head.
3. Open the patient's mouth, and suction as needed. Insert an oropharyngeal or nasopharyngeal airway to maintain an open airway.
4. Select the proper mask size.
5. Place the mask on the patient's face. Make sure the top is over the bridge of the nose and the bottom is in the groove between the lower lip and the chin. If the mask has a large, round cuff around the ventilation port, center the port over the patient's mouth. Inflate the collar to obtain a better fit and seal to the face if necessary.
6. Hold the mask in position by placing the thumbs over the top part of the mask and the index fingers over the bottom half.
7. Bring the lower jaw up to the mask with the last three fingers of your hand. This will help to maintain an open airway. Make sure you do not grab the fleshy part of the neck, as you may compress structures and create an airway obstruction. If you think the patient may have a spinal injury, make sure your partner immobilizes the cervical spine as you move the lower jaw.
8. Connect the bag to the mask if you have not already done so.
9. Hold the mask in place while your partner squeezes the bag with two hands until the patient's chest rises (Figure 7-32 ▶). If a spinal injury is suspected, immobilize the patient's head and neck with your forearms while maintaining an adequate mask-to-face seal with your hands. Continue squeezing the bag once every 5 seconds for adults and once every 3 seconds for infants and children.
10. If you are alone, hold your index finger over the lower part of the mask, your thumb over the upper part of the mask, and then use your remaining fingers to pull the lower jaw into the mask. This is known as the C-clamp and will maintain an effective face-to-mask seal (Figure 7-33 ▶). Use the head tilt–chin lift maneuver to make sure the neck is extended. If spinal injury is suspected, stabilize the patient's head in a neutral in-line position with your knees as you pull the patient's lower jaw into the mask. Squeeze the bag in a rhythmic manner once every 5 seconds with your other hand. Continue squeezing the bag once every 5 seconds for adults and once every 3 seconds for infants and children.

When using the device to assist ventilations of a patient who is breathing too slowly (hypoventilation) with reduced tidal volume, you should squeeze the bag as the patient tries to breathe in. Then, for the next 5 to 10 breaths, slowly adjust the rate and the delivered tidal volume until an adequate minute volume is achieved.

To assist respirations of a patient who is breathing too fast (hyperventilating) with reduced tidal volume, you must first explain the procedure to the patient if the patient is coherent. Initially assist respirations at the rate at which the patient has been breathing, squeezing the bag each time the patient inhales. Then, for the next 5 to 10 breaths, slowly adjust the rate and the delivered tidal volume until an adequate minute volume is achieved.

Figure 7-32 With two-person BVM device ventilation, you should hold the mask in place while your partner squeezes the bag with two hands until the patient's chest rises.

As you are assisting ventilations with a BVM device, you should evaluate the effectiveness of your delivered ventilations. You will know that artificial ventilation is not adequate if the patient's chest does not rise and fall with each ventilation, the rate at which you are ventilating is too slow or too fast, or the heart rate does not return to normal. If the patient's chest does not rise and fall, you may need to reposition the head, use an airway adjunct, or use the Sellick maneuver, also called cricoid pressure. Note that the Sellick maneuver, however, is contraindicated in a patient who is actively vomiting, as it may cause esophageal rupture.

When using a BVM device or any other ventilation device, be alert for gastric distention, inflation of the stomach with air. To prevent or alleviate distention, you should do the following: (1) ensure that the patient's airway is appropriately positioned, (2) ventilate the patient at the appropriate rate, and (3) ventilate the patient with the appropriate volume. If an additional rescuer is available, use the Sellick maneuver (Figure 7-34 ▼). To perform the Sellick maneuver, have an additional rescuer apply cricoid pressure on the patient by placing the thumb and index finger on either side of the cricoid cartilage (at the inferior border of the larynx) and pressing down. By occluding the esophagus, this will (1) inhibit the flow of air into the stomach, thus reducing gastric distention, and (2) reduce the chance of aspiration by helping block the regurgitation of gastric contents from the esophagus. Cricoid pressure should be performed only on unconscious patients.

If the patient's stomach appears to be distending, you should reposition the head and use cricoid pressure. In a patient with possible spinal injury, you should

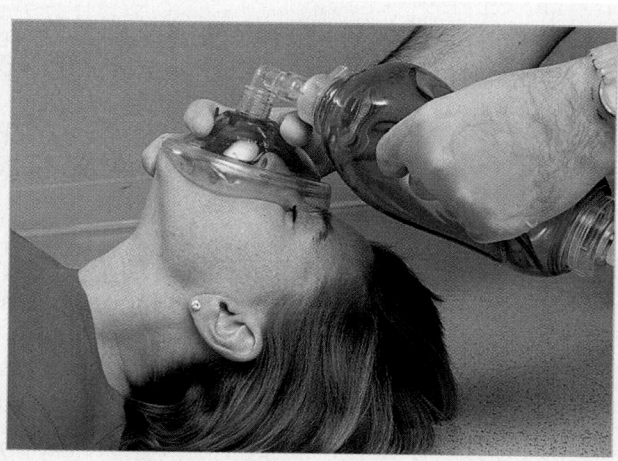

Figure 7-33 Maintain the seal of the mask to the face using the C-clamp if you must ventilate alone.

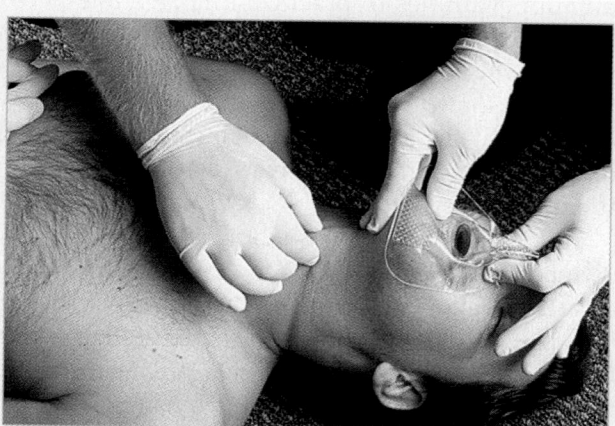

Figure 7-34 The Sellick maneuver, also called cricoid pressure, will help prevent or alleviate gastric distention when artificial ventilations are being performed.

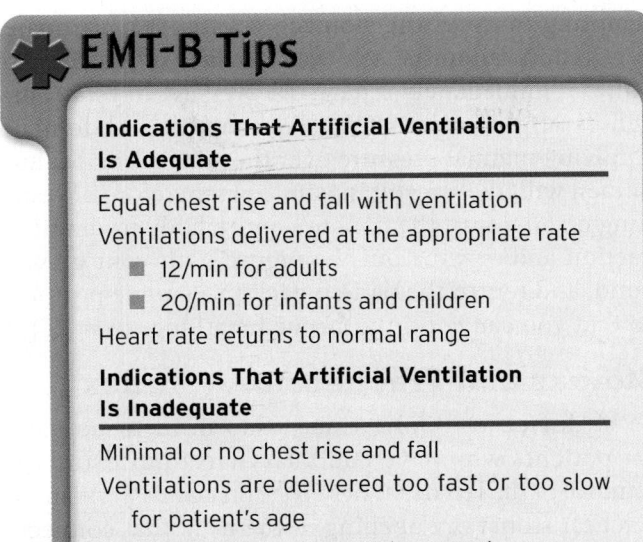

✚ EMT-B Tips

Indications That Artificial Ventilation Is Adequate

Equal chest rise and fall with ventilation

Ventilations delivered at the appropriate rate

- 12/min for adults
- 20/min for infants and children

Heart rate returns to normal range

Indications That Artificial Ventilation Is Inadequate

Minimal or no chest rise and fall

Ventilations are delivered too fast or too slow for patient's age

Heart rate does not return to normal range

reposition the jaw rather than the head (that is, use the jaw-thrust). If too much air is escaping from under the mask, reposition the mask for a better seal. If the patient's chest still does not rise and fall after you have made these corrections, check for an airway obstruction. If an obstruction is not present, you should attempt ventilations using an alternate method, such as the mouth-to-mask technique.

Advanced airway techniques are beneficial when a good seal is difficult to maintain, the patient has a cervical spine injury, or the patient's condition warrants. These techniques are described in Chapter 39, Advanced Airway Management.

The BVM device may also be used in conjunction with an endotracheal tube or with other airway advanced devices such as the esophageal-tracheal Combitube, the pharyngotracheal lumen airway, and the laryngeal mask airway.

Flow-Restricted, Oxygen-Powered Ventilation Devices

Another method of providing artificial ventilation is with flow-restricted, oxygen-powered ventilation devices (Figure 7-35 ▶). These devices are widely available and have been used in EMS for several years. However, recent findings suggest that they should not be used routinely because of the high incidence of gastric distention and possible damage to structures within the chest cavity. Flow-restricted, oxygen-powered devices *should not* be used on infants and children or on

patients with COPD or suspected cervical spine or chest injuries. Cricoid pressure must be maintained whenever flow-restricted, oxygen-powered ventilation devices are used to ventilate a patient. This will help to reduce the amount of gastric distention, the most common and significant complication of the device.

Components

Flow-restricted, oxygen-powered ventilation devices should have the following components:

- A peak flow rate of 100% oxygen at up to 40 L/min
- An inspiratory pressure safety release valve that opens at approximately 60 cm of water and vents any remaining volume to the atmosphere or stops the flow of oxygen
- An audible alarm that sounds whenever you exceed the relief valve pressure
- The ability to operate satisfactorily under normal and varying environmental conditions
- A trigger (or lever) positioned so that both your hands can remain on the mask to provide an airtight seal while supporting and tilting the patient's head and keeping the jaw elevated

Learning how to use these devices correctly requires proper training and considerable practice. As with BVM devices, you must make sure there is an effective seal between the patient's face and mask. The amount of pressure that is necessary to ventilate a patient adequately will vary according to the size of the patient,

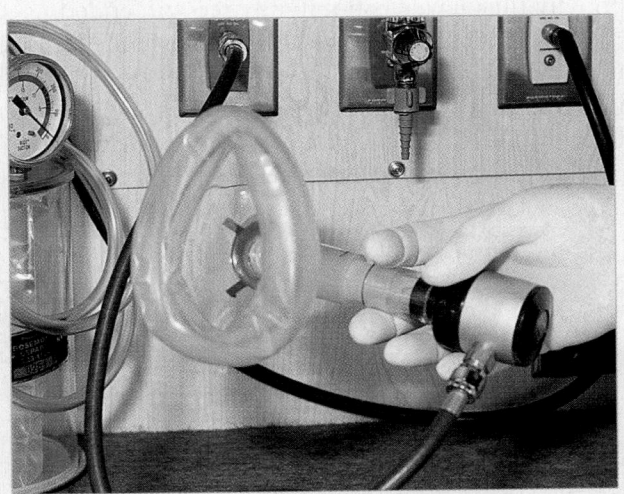

Figure 7-35 A flow-restricted, oxygen-powered ventilation device can provide up to 100% oxygen.

the patient's lung volume, and the condition of the lungs. A patient with COPD will need greater pressure to receive adequate volume than would be necessary for a patient with normal lungs. Pressures that are too great can cause a <u>pneumothorax</u>. Flow-restricted, oxygen-powered ventilation devices are not recommended for use on patients with COPD or suspected cervical spine or chest injuries or on infants and children. Always follow local medical protocols carefully when you use these devices.

Special Considerations

Gastric Distention

Gastric distention occurs when artificial ventilation fills the stomach with air. Although it most commonly affects children, it also affects adults. Gastric distention is most likely to occur when you ventilate the patient too forcefully or too often with a BVM or pocket mask device or when the airway is obstructed as a result of a foreign body or improper head position. For this reason, you should give slow, gentle breaths during artificial ventilation over 2 seconds in the adult patient. Slight gastric distention is not of concern; however, severe inflation of the stomach is dangerous because it may cause vomiting and increase the risk of aspiration during CPR. Gastric distention can also significantly reduce the lung volume by elevating the diaphragm, especially in infants and children. Gastric distention is a common complication associated with the use of flow-restricted, oxygen-powered ventilation devices, a key reason why this device is not highly recommended.

If the patient's stomach becomes distended as a result of rescue breathing, you should recheck and reposition the airway, apply cricoid pressure, and watch for rise and fall of the chest wall as you perform rescue breathing. Continue slow rescue breathing without at-tempting to expel the stomach contents. If adequate ventilation cannot be achieved because of gastric distention, immediately relieve the pressure in the stomach by applying pressure over the upper abdomen. Applying manual pressure over the patient's upper abdomen will likely result in vomiting; therefore, if vomiting occurs, turn the patient's entire body to the side, suction and/or wipe out the mouth with your gloved hand, and return the patient back to a supine position so that you can continue rescue breathing.

Stomas and Tracheostomy Tubes

BVM device ventilation may also need to be used for patients who have had a laryngectomy (surgical removal of the larynx). These patients have a permanent tracheal stoma (an opening in the neck that connects the trachea directly to the skin) Figure 7-36 ▶ . This type of stoma, known as a tracheostomy, is an opening at the center front and base of the neck. Many patients who have had a laryngectomy will have other openings in the neck, according to the type of operation performed. You should ignore any opening other than the midline tracheal stoma. The midline opening is the only one that can be used to put air into the patient's lungs.

Neither the head tilt–chin lift nor the jaw-thrust maneuver is required for ventilating a patient with a stoma. If the patient has a tracheostomy tube, you should ventilate through the tube with a BVM device (the standard 22/15 adapter on the BVM device will fit onto the tube in the tracheal stoma) and 100% oxygen attached directly to the BVM. If the patient has a stoma and no tube is in place, use an infant or child mask with your BVM device to make a seal over the stoma. Seal the patient's mouth and nose with one hand to prevent a leak of air through the upper airway when you ventilate through a stoma. Release the seal of the patient's mouth and nose for exhalation. This allows the air to exhale through the upper airway.

You are the Provider
Part 8

After approximately 2 minutes of assisted ventilation, the patient's cyanosis has resolved and his level of consciousness has improved. You continue BVM ventilations to maintain adequate tidal volume and rendezvous with the paramedics, who intubate the patient and assist you with transporting him to the hospital, where he is diagnosed with a stroke. Following a 2-day stay in the hospital, the patient was discharged to an extended-care facility for continued recovery.

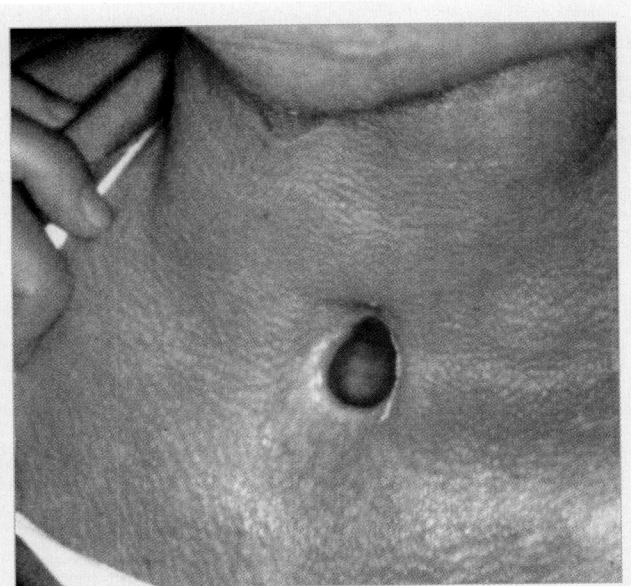

Figure 7-36 A tracheal stoma typically lies in the midline of the neck. The midline opening is the only one that can be used to deliver oxygen to the patient's lungs.

If you are unable to ventilate a patient who has a stoma, try suctioning the stoma and the mouth with a French or soft-tip catheter before giving the patient artificial ventilation through the mouth and nose. If you seal the stoma during mouth-to-mouth ventilation, the ability to ventilate the patient may be improved, or it may help to clear any obstructions.

Foreign Body Airway Obstruction

A foreign body that *completely* blocks the airway in a patient is a true emergency that will result in death if not treated immediately. In an adult, sudden foreign body airway obstruction usually occurs during a meal. In a child, it occurs while eating, playing with small toys, or crawling around the house. An otherwise healthy child who has sudden difficulty breathing has probably aspirated a foreign object.

By far, the most common airway obstruction in an unconscious patient is the tongue, which relaxes and falls back into the throat. There are other causes of airway obstruction that do not involve foreign bodies in the airway. These include swelling (from infection or acute allergic reactions) and trauma (tissue damage from injury). With airway obstruction from medical conditions

such as infection and acute allergic reactions, repeated attempts to clear the airway as if there were a foreign body will be unsuccessful and potentially dangerous. These patients require specific emergency medical care for their condition; therefore, rapid transport to the hospital is critical.

Recognition

Early recognition of airway obstruction is crucial for the EMT-B to be able to provide emergency medical care effectively. Obstruction from a foreign body can result in a <u>partial airway obstruction</u> or a <u>complete airway obstruction</u>.

Patients with a partial airway obstruction are still able to exchange air but will have varying degrees of respiratory distress. Great care must be taken to prevent a partial airway obstruction from becoming a complete airway obstruction. The patient will usually have noisy breathing and may be coughing. You should assess the patient and determine whether the patient has <u>good air exchange</u> or <u>poor air exchange</u>.

With good air exchange, the patient can cough forcefully, although you may hear wheezing between coughs. As long as the patient can breathe, cough forcefully, or talk, you should not interfere with the patient's efforts to expel the foreign object on his or her own. Continue to monitor the patient closely and encourage the patient to continue coughing. Abdominal thrusts are usually not effective for dislodging a partial obstruction. Attempts to remove the object manually could force the object farther down into the airway and cause a complete obstruction. Continually reassess the patient's condition and be prepared to provide treatment if the air exchange becomes poor or a partial obstruction becomes a complete obstruction.

With poor air exchange, the patient has a weak, ineffective (not forceful) cough and may have increased difficulty breathing, stridor (a high-pitched noise heard primarily on inspiration), and cyanosis. You must quickly recognize this situation and provide immediate care.

For patients with partial airway obstruction with poor air exchange, treat immediately as if there is a complete airway obstruction.

Patients with complete airway obstruction cannot breathe, talk, or cough. One sure sign of a complete obstruction is the sudden inability to speak or cough during or immediately after eating. The person may clutch or grasp his or her throat (universal distress signal), begin to turn cyanotic, and have extreme

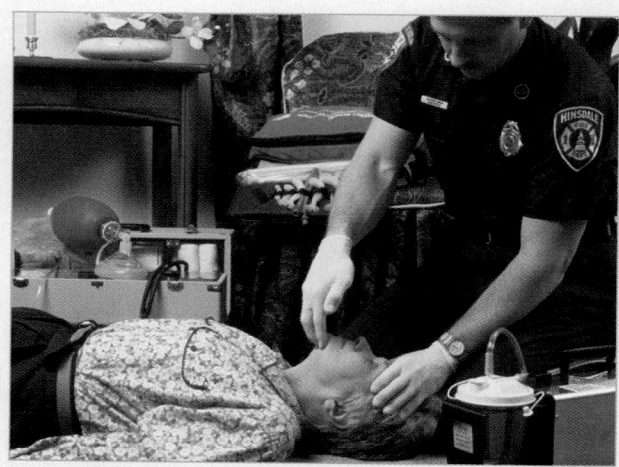

Figure 7-37 The universal sign of choking is a person who grasps his or her throat and has difficulty breathing.

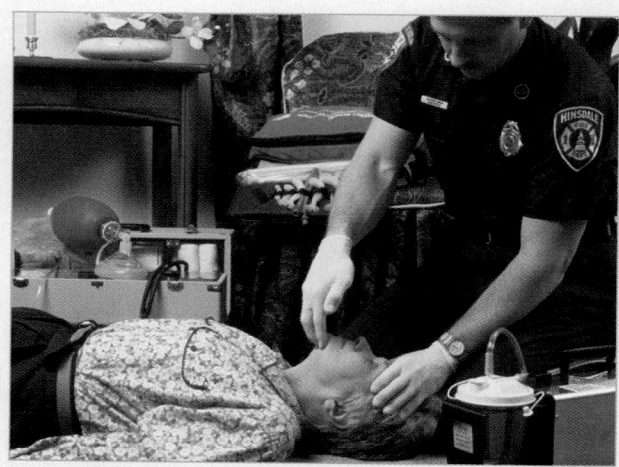
Figure 7-38 Securing and maintaining the airway and ensuring adequate breathing are the first, most important steps in caring for an unconscious patient.

difficulty breathing (Figure 7-37 ▲). There is little or no air movement. Ask the conscious patient, "Are you choking?" If the patient nods "yes," provide immediate treatment. If the obstruction is not cleared quickly, the amount of oxygen in the patient's blood will decrease dramatically. If not treated, the patient will become unconscious and die.

Some patients with a complete airway obstruction will be unconscious during your initial assessment. You may not know that an airway obstruction is the cause of their condition. There are many other causes of unconsciousness and respiratory failure, including stroke, heart attack, trauma, seizures, and drug overdoses. A complete and thorough patient assessment by you, therefore, is key in providing appropriate emergency medical care.

Any person found unconscious must be managed as if he or she has a compromised airway. You must first open the airway and provide artificial ventilation if the patient is not breathing or is breathing inade-

quately (Figure 7-38 ▲). If, after opening the airway, you are unable to ventilate the patient after several attempts (no chest rise and fall) or you feel resistance while ventilating, consider the possibility of an airway obstruction. Resistance to ventilation can also be due to poor lung compliance. Compliance is the ability of the alveoli to expand when air is drawn in during inhalation; poor lung compliance is the inability of the alveoli to fully expand during inhalation.

Emergency Medical Care for Foreign Body Airway Obstruction

Perform the head tilt–chin lift maneuver to clear an obstruction that has been caused by the tongue and throat muscles relaxing back into the airway in any person who is found unconscious, has inadequate breathing or is not breathing, and is not suspected of having spinal trauma. If spinal trauma is suspected, you should open the airway with a jaw-thrust maneuver. Large pieces of vomited food, mucus, loose dentures, or blood clots in the mouth should be swept forward and out of the mouth with your gloved index finger. When available, suctioning should be used to maintain a clear airway.

The Heimlich maneuver (abdominal thrusts) is the most effective method of dislodging and forcing an object out of the airway. Residual air, which is always present in the lungs, is compressed upward and used to expel the object. You should use the Heimlich maneuver followed by finger sweeps and attempts to ventilate in the adult patient with a complete airway obstruction.

If you are unable to clear a complete airway obstruction with your initial attempts, begin rapid trans-

EMT-B Tips

Possible Causes of Airway Obstruction

Relaxation of the tongue in an unconscious patient
Aspirated vomitus (stomach contents)
Foreign objects—food, small toys, dentures
Blood clots, bone fragments, or damaged tissue after an injury
Airway tissue swelling—infection, allergic reaction

port and continue your efforts at relief of the obstruction with abdominal thrusts, finger sweeps, and attempts at ventilation en route to the hospital.

Remember to treat patients with a partial airway obstruction with poor air exchange as if they have a complete obstruction.

Patients with a partial airway obstruction and good air exchange should be monitored closely for deterioration of their condition. If the patient is unable to clear the obstruction and remains conscious, support (or let the patient control) the airway position that is most efficient and comfortable. Provide supplemental oxygen and transport to the hospital.

Dental Appliances

Many dental appliances can cause an airway obstruction. If a dental appliance, such as a crown or bridge, dentures, or even a piece or section of braces, has become loose, you should manually remove it before providing ventilations. Simple manual removal may relieve the obstruction and allow the patient to breathe on his or her own.

Providing BVM device or mouth-to-mask ventilation is usually much easier when dentures can be left in place. Leaving the dentures in place provides more "structure" to the face and will generally assist you in being able to provide a good face-to-mask seal, thus delivering adequate tidal volume. However, loose dentures make it much more difficult to perform artificial ventilation by any method and can easily obstruct the airway. Therefore, dentures and dental appliances that do not stay firmly in place should be removed. Dentures and appliances may become loose or be completely out of place following an accident or as you are providing care. Periodically reassess the patient's airway to make sure the devices are firmly in place.

Facial Bleeding

Airway problems can be especially challenging in patients with serious facial injuries (Figure 7-39 ▼). Because the blood supply in the face is so rich, injuries to the face can result in severe tissue swelling and bleeding into the airway. Control bleeding with direct pressure and suction as necessary. Facial injuries are discussed in detail in Chapter 26.

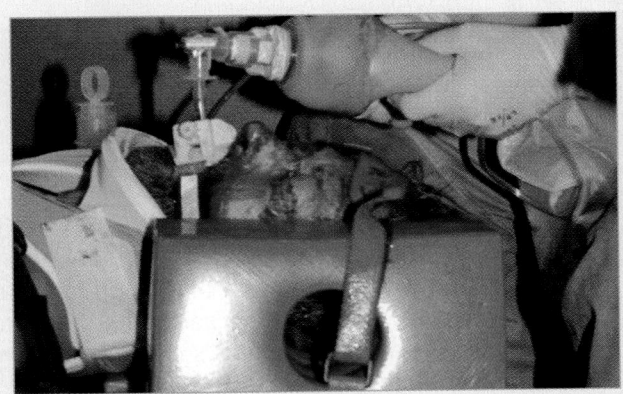

Figure 7-39 Airway problems can be especially challenging in patients with serious facial injuries.

You are the Provider   Summary

Many factors contribute to respiratory problems. Some are as simple as seasonal allergies and allergic reactions. Other factors are more complex, such as trauma, stroke, or an industrial exposure. Oftentimes the situation may be different on scene from what dispatch has reported. Keeping an open mind to all possibilities will help you prepare better. Few situations will require more equipment than a serious problem with an airway or breathing. Inspect your equipment often so you are prepared to use it when needed.

Many hazards may exist on a call. In particular, you should be cautious of unusual odors and the involvement of multiple people. Information from individuals on scene can help you to remain safe and understand the situation better. The assessment and treatment of airway and breathing problems always begin with securing an adequate airway. When treatment is provided, careful reevaluation of the patient is needed to ensure the treatment has been effective. You should be able to decide what is effective and what is not. In most situations, patients should receive oxygen even if it does not seem necessary.

Prep Kit

Ready for Review

- The term "airway" usually means the upper airway, which includes the respiratory structures above the vocal cords. Clearing the airway means removing obstructing material; maintaining the airway means keeping it open.

- Adequate breathing for an adult features a normal rate of 12 to 20 breaths/min, a regular pattern of inhalation and exhalation, adequate depth, bilaterally clear and equal lung sounds, and regular and equal chest rise and fall.

- Inadequate breathing for an adult features a respiratory rate of fewer than 12 breaths/min or more than 20 breaths/min, shallow depth (reduced tidal volume), an irregular pattern of inhalation and exhalation, and breath sounds that are diminished, absent, or noisy.

- Patients who are breathing inadequately show signs of hypoxia, a dangerous condition in which the body's tissues and cells do not have enough oxygen.

- Patients with inadequate breathing need to be treated immediately. Emergency medical care includes airway management, supplemental oxygen, and ventilatory support.

- Basic techniques for opening the airway include the head tilt–chin lift maneuver or, if trauma is suspected, the jaw-thrust maneuver.

- One basic airway adjunct is the oropharyngeal or oral airway, which keeps the tongue from blocking the airway in unconscious patients with no gag reflex. If the oral airway is not the proper size or is inserted incorrectly, it can actually cause an obstruction.

- Another basic airway adjunct is the nasopharyngeal or nasal airway, which is usually used with patients who have a gag reflex and is better tolerated than the oral airway.

- Suctioning is the next priority after opening the airway. Rigid tonsil-tip catheters are the best catheters to use when suctioning the pharynx; soft plastic catheters are used to suction the nose and liquid secretions in the back of the mouth.

- The recovery position is used to help maintain the airway in patients without traumatic injuries who are unconscious and breathing adequately.

- You must provide immediate artificial ventilations with supplemental oxygen to patients who are not breathing on their own. Patients with inadequate breathing may also require artificial ventilations to maintain effective tidal volume.

- Handle compressed gas cylinders carefully; their contents are under pressure. Always make sure the correct pressure regulator is firmly attached before transporting a cylinder. The pin-indexing safety system features a series of pins on a yoke that must be matched with the holes on the valve stem of the gas cylinder. Pressure regulators reduce the pressure of gas in an oxygen cylinder to between 40 and 70 psi. Pressure-compensated flowmeters and Bourdon-gauge flowmeters permit the regulated release of gas measured in liters per minute.

- When oxygen therapy is complete, disconnect the tubing from the flowmeter nipple and turn off the cylinder valve, then turn off the flowmeter. As long as there is a pressure reading on the regulator gauge, it is not safe to remove the regulator from the valve stem. Keep any possible source of fire away from the area while oxygen is in use.

- Nasal cannulas and nonrebreathing masks are used most often to deliver oxygen in the field. The nonrebreathing mask is the delivery device of choice for providing supplemental oxygen to patients who are breathing adequately but are suspected of having or are showing signs of hypoxia. With a flow rate set at 15 L/min and the reservoir bag preinflated, the nonrebreathing mask can provide more than 90% inspired oxygen. If the patient will not tolerate a nonrebreathing mask, apply a nasal cannula.

www.EMTB.com

Technology

- Interactivities
- Vocabulary Explorer
- Anatomy Review
- Web Links
- Online Review Manual

- The methods of providing artificial ventilation include mouth-to-mask ventilation, two-person BVM ventilation, flow-restricted, oxygen-powered ventilation device, and one-person BVM ventilation. The flow-restricted, oxygen-powered ventilation device is not a recommended ventilation device by most standards. Combined with your own exhaled breath, mouth-to-mask ventilation will give your patient up to 55% oxygen; a BVM device with an oxygen reservoir and supplemental oxygen can deliver nearly 100% oxygen.

- When you are providing artificial ventilation, remember that ventilating too forcefully can cause gastric distention. Slow, gentle breaths during artificial ventilation and the use of cricoid pressure can help to prevent gastric distention. Patients who have a tracheal stoma or a tracheostomy tube need to be ventilated through the tube or the stoma.

- Foreign body airway obstruction usually occurs during a meal in an adult or while a child is eating, playing with small objects, or crawling about the house. The earlier you recognize an airway obstruction, the better. You must learn to recognize the difference between airway obstruction caused by a foreign object and that caused by a medical condition.

- A complete airway obstruction can be removed by the Heimlich maneuver, finger sweeps, manual removal of the object, and attempts to ventilate. Treat patients with a partial airway obstruction with poor air exchange as if they had a complete obstruction. Patients with partial airway obstruction and good air exchange should be closely monitored.

- Check for loose dental appliances in a patient before assisting ventilations. Loose appliances should be removed to prevent them from obstructing the airway. Tight-fitting appliances should be left in place.

Vital Vocabulary

agonal respirations Occasional, gasping breaths that occur after the heart has stopped.

airway The upper airway tract or the passage above the larynx, which includes the nose, mouth, and throat.

American Standard System A safety system for oxygen cylinders, designed to prevent the accidental attachment of a regulator to a cylinder containing the wrong type of gas.

apnea A period of not breathing.

aspiration The introduction of vomitus or other foreign material into the lungs.

ataxic respirations Irregular, ineffective respirations that may or may not have an identifiable pattern.

bag-valve-mask (BVM) device A device with a one-way valve and a face mask attached to a ventilation bag; when attached to a reservoir and connected to oxygen, delivers more than 90% supplemental oxygen.

barrier device A protective item, such as a pocket mask with a valve, that limits exposure to a patient's body fluids.

bilateral A body part or condition that appears on both sides of the midline.

complete airway obstruction Occurs when a foreign body completely obstructs the patient's airway. Patients cannot breathe, talk, or cough.

compliance The ability of the alveoli to expand when air is drawn in during inhalation.

cricoid pressure Pressure on the cricoid cartilage; applied to occlude the esophagus in order to inhibit gastric distention and regurgitation of vomitus in the unconscious patient; also called the Sellick maneuver.

diffusion A process in which molecules move from an area of higher concentration to an area of lower concentration.

dyspnea Difficulty breathing.

exhalation The passive part of the breathing process in which the diaphragm and the intercostal muscles relax, forcing air out of the lungs.

gag reflex A normal reflex mechanism that causes retching; activated by touching the soft palate or the back of the throat.

gastric distention A condition in which air fills the stomach, often as a result of high volume and pressure during artificial ventilation.

good air exchange A term used to distinguish the degree of distress in a patient with a partial airway obstruction. With good air exchange, the patient is still conscious and able to cough forcefully, although wheezing may be heard.

head tilt–chin lift maneuver A combination of two movements to open the airway by tilting the forehead back and lifting the chin; not used for trauma patients.

hypoxia A dangerous condition in which the body tissues and cells do not have enough oxygen.

hypoxic drive A condition in which chronically low levels of oxygen in the blood stimulate the respiratory drive; seen in patients with chronic lung diseases.

Prep Kit continued...

inhalation The active, muscular part of breathing that draws air into the airway and lungs.

ischemia A lack of oxygen that deprives tissues of necessary nutrients.

jaw-thrust maneuver Technique to open the airway by placing the fingers behind the angle of the jaw and bringing the jaw forward; used for patients who may have a cervical spine injury.

labored breathing Breathing that requires greater than normal effort; may be slower or faster than normal and usually requires the use of accessory muscles.

metabolism The biochemical processes that result in production of energy from nutrients within the cells.

minute volume The volume of air moved through the lungs in 1 minute; calculated by multiplying tidal volume and respiratory rate.

nasal cannula An oxygen-delivery device in which oxygen flows through two small, tubelike prongs that fit into the patient's nostrils; delivers 24% to 44% supplemental oxygen, depending on the flow rate.

nasopharyngeal (nasal) airway Airway adjunct inserted into the nostril of a conscious patient who is unable to maintain airway patency independently.

nonrebreathing mask A combination mask and reservoir bag system that is the preferred way to give oxygen in the prehospital setting; delivers up to 90% inspired oxygen and prevents inhaling the exhaled gases (carbon dioxide).

oropharyngeal (oral) airway Airway adjunct inserted into the mouth to keep the tongue from blocking the upper airway and to facilitate suctioning the airway.

partial airway obstruction Condition in which an obstruction leaves the patient able to exchange some air, but also causes some degree of respiratory distress.

patent Open, clear of obstruction.

pin-indexing system A system established for portable cylinders to ensure that a regulator is not connected to a cylinder containing the wrong type of gas.

pneumothorax A partial or complete accumulation of air in the pleural space.

poor air exchange A term used to describe the degree of distress in a patient with a partial airway obstruction. With poor air exchange, the patient often has a weak, ineffective cough, increased difficulty breathing, or possible cyanosis and may produce a high-pitched noise during inhalation (stridor).

recovery position A side-lying position used to maintain a clear airway in unconscious patients without injuries who are breathing adequately.

retractions Movements in which the skin pulls in around the ribs during inspiration.

Sellick maneuver A technique that is used to prevent gastric distention in which pressure is applied to the cricoid cartilage; also referred to as cricoid pressure.

stoma An opening through the skin and into an organ or other structure; a stoma in the neck connects the trachea directly to the skin.

suction catheter A hollow, cylindrical device used to remove fluid from the patient's airway.

tidal volume The amount of air moved in or out of the lungs during one breath.

tonsil tips Large, semirigid suction tips recommended for suctioning the pharynx; also called Yankauer tips.

ventilation Exchange of air between the lungs and the environment, spontaneously by the patient or with assistance from another person, such as an EMT-B.

Points to Ponder

You are dispatched to the local nursing home for an older man who "is difficult to wake." You arrive at the nursing home about 5 minutes after the initial call and find the patient to be lying supine in bed with oxygen flowing at 2 L/min via nasal cannula. The nurse states that the patient was fine last evening but they were unable to wake him this morning. They state he has a history of COPD and recent pneumonia. The patient has shallow gurgling respirations at a rate of about 8 breaths/min. You also note cyanosis around the lips. While you are assembling your suction unit, your partner is placing the patient on a pulse oximeter.

Why should this patient be placed on a pulse oximeter? Why is suctioning necessary for this patient? How would you manage this patient's airway and breathing? Would you change the position of the patient?

Issues: Oxygenation, Cricoid Pressure, Potential Elder Abuse.

Assessment in Action

You arrive on the scene of a cardiac arrest and find that bystanders have initiated CPR. You take over the care of the patient, and your assessment reveals a 50-year-old man who is apneic.

1. The bystanders report that the patient was not breathing for about 3 minutes before they started CPR. What does this time frame indicate about the patient's condition?

   A. Cardiac irritability
   B. Brain damage not likely
   C. Brain damage possible
   D. Brain damage very likely

2. Upon your assessment you find the patient has occasional, gasping breaths. This condition is called:

   A. Cheyne-Stokes respirations.
   B. Retractions.
   C. agonal respirations.
   D. Kussmaul's respirations.

3. You decide to open the patient's airway. You have no history of events leading up to the point of cardiac arrest. What is the preferred method of opening the airway?

   A. Head tilt-chin lift
   B. Jaw thrust
   C. Nasal airway
   D. None of the above

4. You start ventilating the patient with a BVM device. What is important to remember during ventilation?

   A. Squeeze the bag slowly; do not force air
   B. Provide a good seal with mask
   C. Be aware of the rate at which you are ventilating the patient
   D. All of the above

5. Approximately how much oxygen was being delivered to the patient using CPR?

   A. 21%
   B. 16%
   C. 40%
   D. 0%

6. The initial CPR has caused some gastric distention and you are worried about possible aspiration. What is the best method to protect the patient's airway?

   A. Cricoid pressure
   B. Sellick maneuver
   C. Recovery position
   D. A and B only

7. You decide to place an oral airway. How do you measure the size of the airway?

   A. From the patient's earlobe to the corner of the mouth
   B. From the patient's nose to the angle of the jaw
   C. From the corner of the mouth to the angle of the jaw
   D. From the corner of the mouth to the tip of the tongue

8. You notice a buildup of fluid in the patient's airway and decide to suction the fluid. Which of the following is NOT correct when providing suctioning on an adult patient?

   A. Suction as you withdraw the catheter
   B. Insert the tip to the base of the tongue
   C. Do not suction for more than 15 seconds
   D. Repeat immediately after initial suctioning if needed

Challenging Questions

9. What are some indications that artificial ventilation is inadequate?

10. You are ventilating a trauma patient with a BVM device and ventilation is becoming more and more ineffective. What are possible causes of this?

11. You have a D cylinder with 1,500 psi. You have a patient who needs 15 L/min. How long will your tank last?

www.EMTB.com

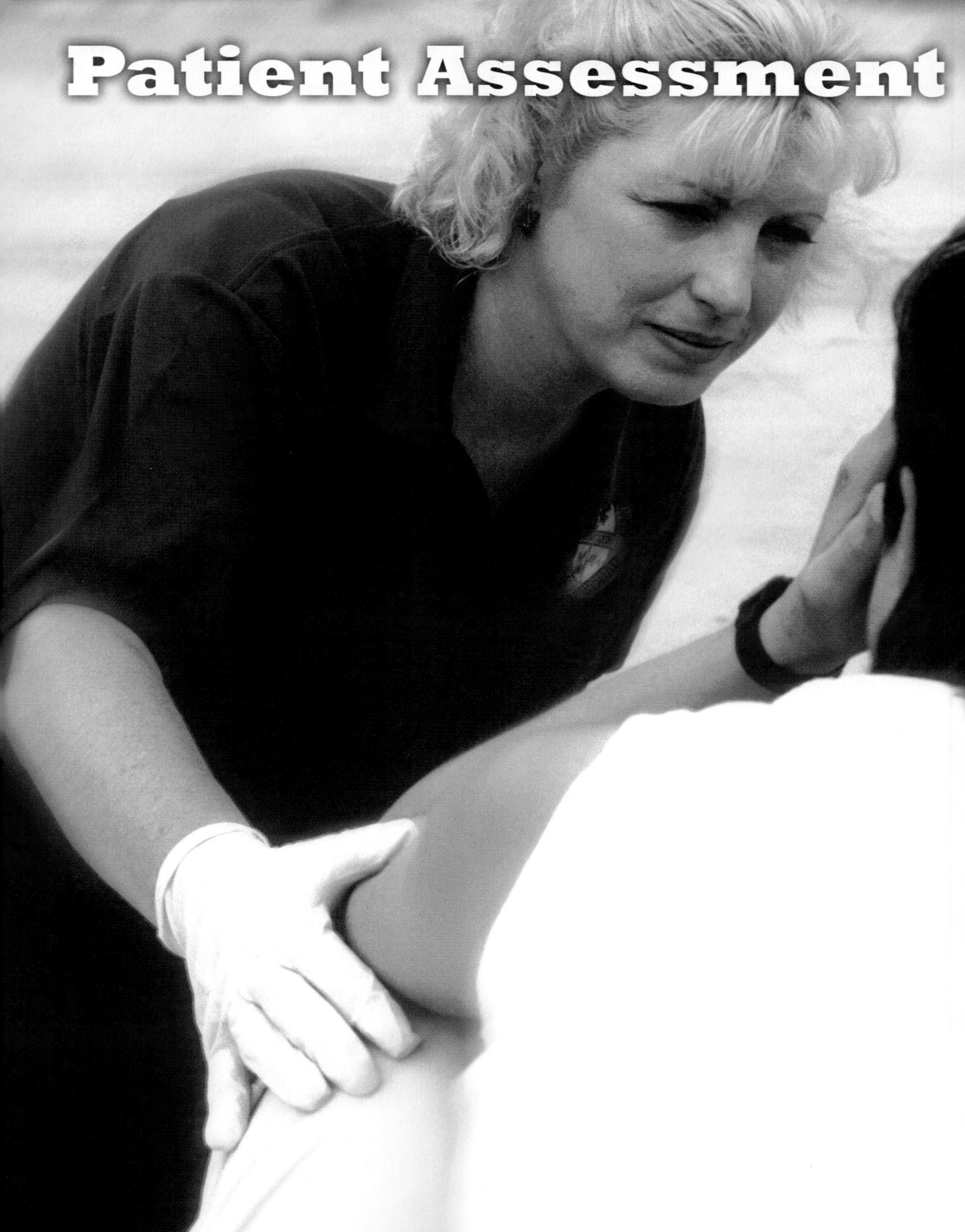

Patient Assessment

Section 3

Patient Assessment

Objectives

Scene Size-up

Cognitive

3-1.1 Recognize hazards/potential hazards. (p 265)

3-1.2 Describe common hazards found at the scene of a trauma and a medical patient. (p 266)

3-1.3 Determine if the scene is safe to enter. (p 265)

3-1.4 Discuss common mechanisms of injury/nature of illness. (p 267)

3-1.5 Discuss the reason for identifying the total number of patients at the scene. (p 268)

3-1.6 Explain the reason for identifying the need for additional help or assistance. (p 268)

Affective

3-1.7 Explain the rationale for crew members to evaluate scene safety prior to entering. (p 266)

3-1.8 Serve as a model for others explaining how patient situations affect your evaluation of mechanism of injury or illness. (p 268)

Psychomotor

3-1.9 Observe various scenarios and identify potential hazards. (p 266)

Initial Assessment

Cognitive

3-2.1 Summarize the reasons for forming a general impression of the patient. (p 271)

3-2.2 Discuss methods of assessing altered mental status. (p 274)

3-2.3 Differentiate between assessing the altered mental status in the adult, child, and infant patient. (p 275)

3-2.4 Discuss methods of assessing the airway in the adult, child, and infant patient. (p 275)

3-2.5 State reasons for management of the cervical spine once the patient has been determined to be a trauma patient. (p 275)

3-2.6 Describe methods used for assessing if a patient is breathing. (p 276)

3-2.7 State what care should be provided to the adult, child, and infant patient with adequate breathing. (p 276)

3-2.8 State what care should be provided to the adult, child, and infant patient without adequate breathing. (p 277)

3-2.9 Differentiate between a patient with adequate and inadequate breathing. (p 276)

3-2.10 Distinguish between methods of assessing breathing in the adult, child, and infant patient. (p 276)

3-2.11 Compare the methods of providing airway care to the adult, child, and infant patient. (p 277)

3-2.12 Describe the methods used to obtain a pulse. (p 277)

3-2.13 Differentiate between obtaining a pulse in an adult, child, and infant patient. (p 277)

3-2.14 Discuss the need for assessing the patient for external bleeding. (p 278)

3-2.15 Describe normal and abnormal findings when assessing skin color. (p 278)

3-2.16 Describe normal and abnormal findings when assessing skin temperature. (p 278)

3-2.17 Describe normal and abnormal findings when assessing skin condition. (p 278)

3-2.18 Describe normal and abnormal findings when assessing skin capillary refill in the infant and child patient. (p 279)

3-2.19 Explain the reason for prioritizing a patient for care and transport. (p 279)

Affective

3-2.20 Explain the importance of forming a general impression of the patient. (p 271)

3-2.21 Explain the value of performing an initial assessment. (p 271)

Psychomotor

3-2.22 Demonstrate the techniques for assessing mental status. (p 274)

3-2.23 Demonstrate the techniques for assessing the airway. (p 275)

3-2.24 Demonstrate the techniques for assessing if the patient is breathing. (p 276)

3-2.25 Demonstrate the techniques for assessing if the patient has a pulse. (p 277)

3-2.26 Demonstrate the techniques for assessing the patient for external bleeding. (p 278)

3-2.27 Demonstrate the techniques for assessing the patient's skin color, temperature, condition, and capillary refill (infants and children only). (p 278)

3-2.28 Demonstrate the ability to prioritize patients. (p 279)

Focused History and Physical Exam: Trauma Patients

Cognitive

3-3.1 Discuss the reasons for reconsideration concerning the mechanism of injury. (p 293)

3-3.2 State the reasons for performing a rapid trauma assessment. (p 293)

3-3.3   Recite examples and explain why patients should receive a rapid trauma assessment. (p 293)

3-3.4   Describe the areas included in the rapid trauma assessment and discuss what should be evaluated. (p 284)

3-3.5   Differentiate when the rapid assessment may be altered in order to provide patient care. (p 293)

3-3.6   Discuss the reason for performing a focused history and physical exam. (p 283)

Affective

3-3.7   Recognize and respect the feelings that patients might experience during assessment. (p 271)

Psychomotor

3-3.8   Demonstrate the rapid trauma assessment that should be used to assess a patient based on mechanism of injury. (p 293)

Focused History and Physical Exam: Medical Patients

Cognitive

3-4.1   Describe the unique needs for assessing an individual with a specific chief complaint with no known prior history. (p 295)

3-4.2   Differentiate between the history and physical exam that are performed for responsive patients with no known prior history and responsive patients with a known prior history. (p 295)

3-4.3   Describe the needs for assessing an individual who is unresponsive. (p 296)

3-4.4   Differentiate between the assessment that is performed for a patient who is unresponsive or has an altered mental status and other medical patients requiring assessment. (p 295)

Affective

3-4.5   Attend to the feelings that these patients might be experiencing. (p 295)

Psychomotor

3-4.6   Demonstrate the patient care skills that should be used to assist a patient who is responsive with no known history. (p 295)

3-4.7   Demonstrate the patient care skills that should be used to assist a patient who is unresponsive or has an altered mental status. (p 296)

Detailed Physical Exam

Cognitive

3-5.1   Discuss the components of the detailed physical exam. (p 299)

3-5.2   State the areas of the body that are evaluated during the detailed physical exam. (p 300)

3-5.3   Explain what additional care should be provided while performing the detailed physical exam. (p 300)

3-5.4   Distinguish between the detailed physical exam that is performed on a trauma patient and that of the medical patient. (p 299)

Affective

3-5.5   Explain the rationale for the feelings that these patients might be experiencing. (p 299)

Psychomotor

3-5.6   Demonstrate the skills involved in performing the detailed physical exam. (p 300)

Ongoing Assessment

Cognitive

3-6.1   Discuss the reason for repeating the initial assessment as part of the ongoing assessment. (p 307)

3-6.2   Describe the components of the ongoing assessment. (p 307)

3-6.3   Describe trending of assessment components. (p 307)

Affective

3-6.4   Explain the value of performing an ongoing assessment. (p 307)

3-6.5   Recognize and respect the feelings that patients might experience during assessment. (p 308)

3-6.6   Explain the value of trending assessment components to other health professionals who assume care of the patient. (p 307)

Psychomotor

3-6.7   Demonstrate the skills involved in performing the ongoing assessment. (p 307)

You are the Provider

You are dispatched to the 4th south viaduct for an unresponsive man. A citizen bystander reports a man acting strangely. The police are already on scene.

You consider the location of the patient and realize a homeless camp is not too far away that has had a recent outbreak of tuberculosis. With this in mind you put on some latex gloves and a HEPA mask.

Good patient care is directly linked to good patient assessment. This chapter will cover this most essential of all EMT-B skills along with helping you answer the following questions:

1. Why is patient assessment considered one of the cornerstones of prehospital care?
2. Is there a difference between assessment of a trauma patient and a patient with a medical condition? If so, what is it?

Patient Assessment

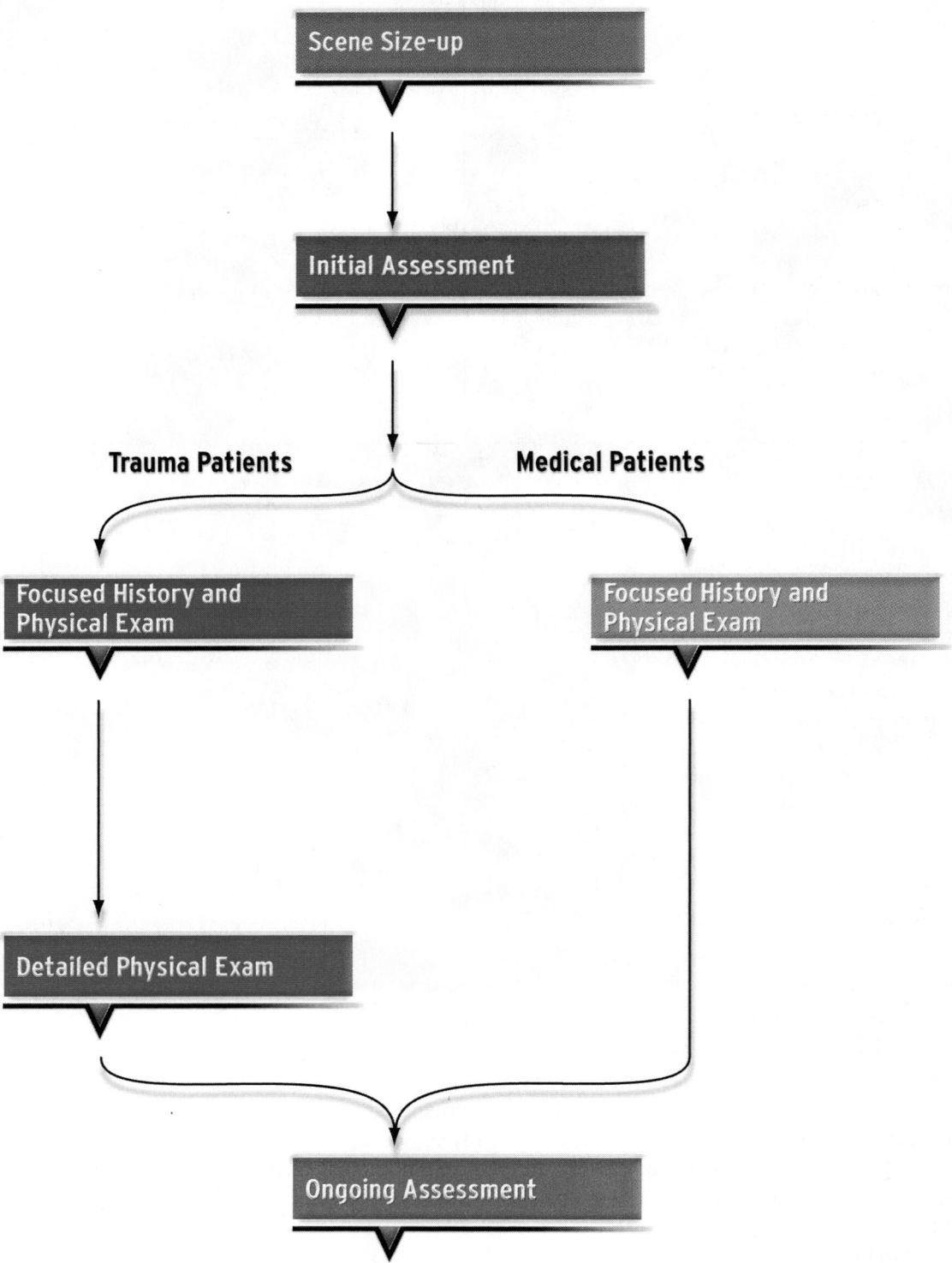

About This Chapter

This chapter will provide a clear and comprehensive approach to patient assessment. A flowchart has been developed to provide a quick, visual reference to guide you through the patient assessment process. The chapter has been divided into six sections. Every section is color coded and numbered for easy reference. The Patient Assessment Flowchart is repeated at every section to show you "at a glance" where you are in the patient assessment process.

Special care has been taken to reflect the current EMT-B National Standard Curriculum (1994), but enhancement information will prepare you for your work in the field. You will also find the special patient assessment needs of pediatric patients in Chapter 32 and of geriatric patients in Chapter 34.

Patient Assessment

From a practical point of view, prehospital emergency care is simply a series of decisions about treatment and transport. The process that guides decision making in EMS is based on your patient assessment findings. For you to make good decisions about how to most effectively care for your patient, you must be able to perform a thorough and accurate patient assessment. The patient assessment process includes the following components:

- Size up the scene to identify safety threats and prepare for the call.

Know This order?

- Identify initial threats to the patient's life and treat them/Identify life-threatening conditions and treat them.
- Perform a physical exam of the patient, looking for signs of illness or injury.
- Obtain vital signs to determine how your patient is tolerating the problem.
- Gather history that may help to explain the physical findings and abnormal vital signs.
- Prepare the patient for transport and continuously assess for changes in his or her condition.

Patient Assessment

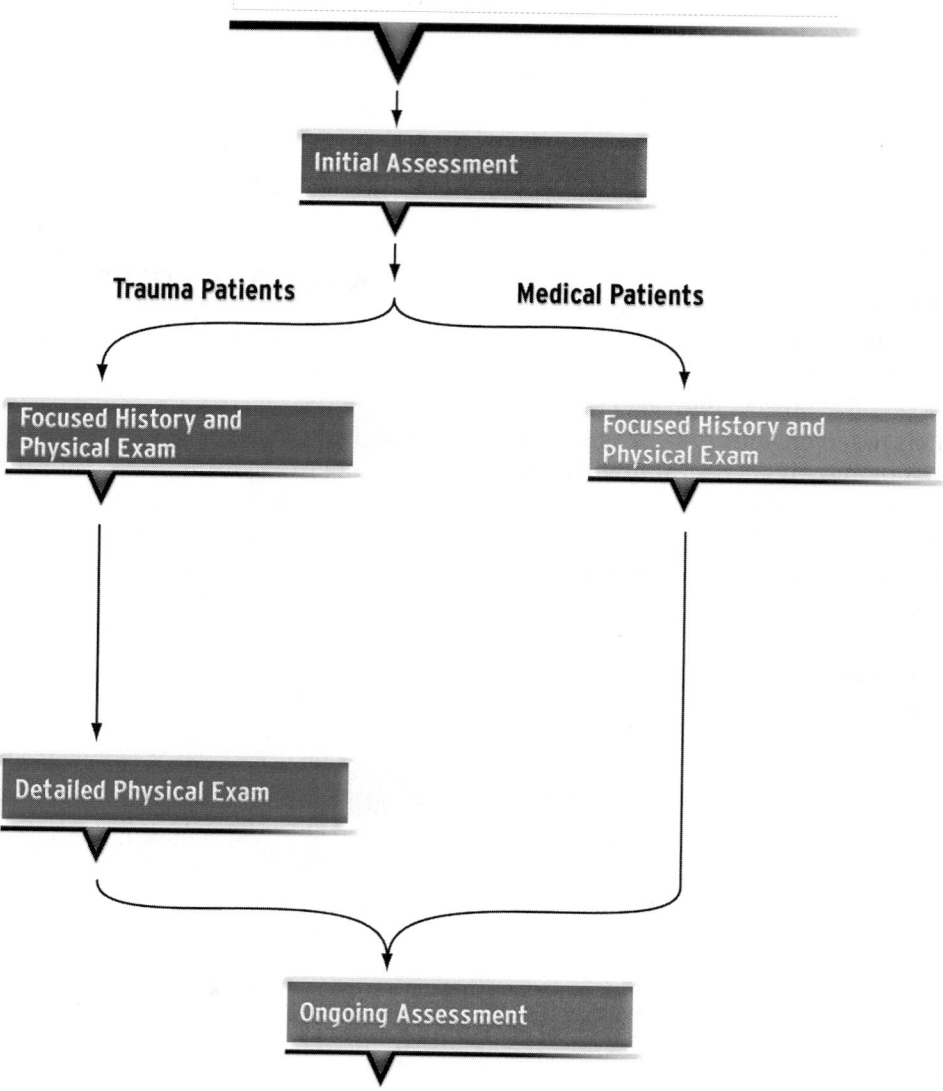

Scene Size-up

Body Substance Isolation
Scene Safety
Consider Mechanism of Injury/Nature of Illness
Determine the Number of Patients
Consider Additional Resources
Consider C-Spine Immobilization

Initial Assessment

Trauma Patients   **Medical Patients**

Focused History and Physical Exam   Focused History and Physical Exam

Detailed Physical Exam

Ongoing Assessment

Scene Size-up

When you are alerted of an emergency call, your dispatcher will provide you with some basic information about the situation that requires your assistance. Your scene size-up begins here. The scene size-up is how you prepare for a specific situation. From the moment you are called into action until you finally reach your patient, you must consider a variety of things that will have an impact on how you begin to care for your patient. The scene size-up includes dispatch information and must be combined with an inspection of the scene to help you identify scene hazards, safety concerns, mechanisms of injury, the natures of illness, and the number of patients you may have, as well as additional resources you might need to safely and effectively care for the patient.

Body Substance Isolation

On every emergency call, you will need to wear personal protective equipment (PPE) because this equipment will reduce your personal risk for injury or illness. The type of PPE you wear will depend a great deal on your specific job responsibilities as an EMT. For example, fire fighters will wear turnout gear to protect them from injury. Hazardous material technicians will wear more sophisticated PPE. As a medical responder, responsible primarily for patient care, you will need to use body substance isolation (BSI) precautions. This is the most effective way to reduce your risk of exposure to potentially infectious substances. The concept of BSI assumes that all body fluids pose a potential risk for infection whether a known infection exists or not.

You should be using BSI precautions when you step out of the vehicle and before you enter the scene. If you have not taken the appropriate BSI precautions when you first approach your patient, the excitement of the call may cause you to begin providing care without the proper protection (Figure 8-1▶). Protective gloves and eye protection are always indicated. Eye protection and masks should be used when blood or fluids may become airborne by coughing or splattering. Masks will protect you from some airborne diseases.

The use of BSI, including gloves, gowns, masks, and eye protection, may be dictated by your local protocol. If a situation requires additional PPE, you must be appropriately trained in the use of that PPE in those specific situations. If you are not appropriately trained, you should not approach the scene and should call for additional help.

Scene Safety

Every scene can potentially cause injury to you, your team, your patients, and bystanders. You will need to evaluate for potential or actual hazards as you approach the scene. Information provided by dispatch may help in determining potential hazards.

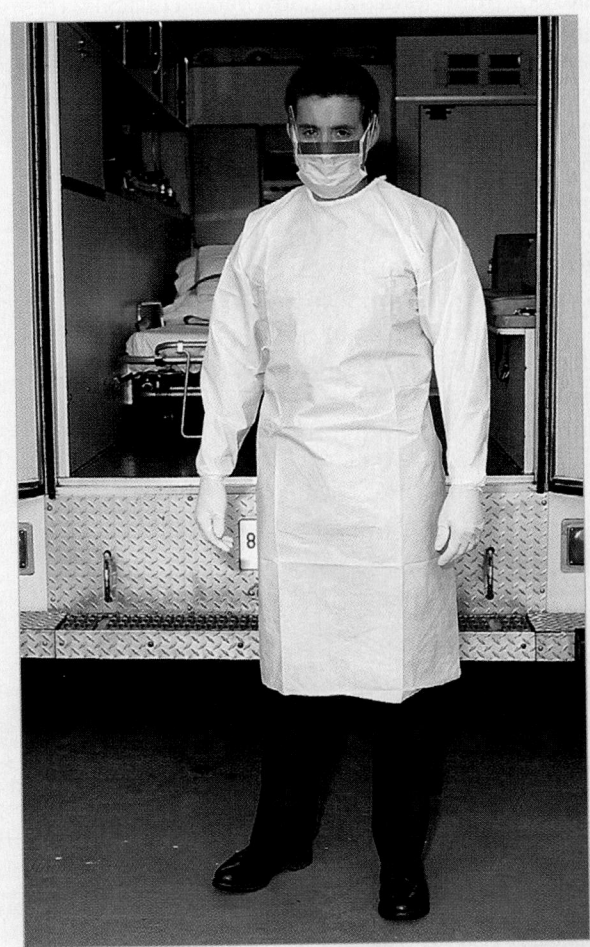

Figure 8-1 Proper protective equipment is vital when you are called to a scene in which you may be exposed to blood or other body fluids.

Making sure the scene safe is not your job, depending on what's going on you call the proper authorities

For example, a call to an industrial site may have chemicals involved or a private residence may have animals that pose a threat. You should be open to many possible risks, ranging from complex hazardous materials spills to slippery grass. If you become injured at the scene, you will not be able to provide appropriate help to your patient. In fact, you may take important resources away from the original patients.

Personal Protection

Ensuring your personal protection begins by looking for possible dangers as you approach the scene and before you step out of the vehicle (Figure 8-2 ▼): *Common Sense*

- Oncoming traffic
- Unstable surfaces (eg, wet or icy patches, loose gravel, slopes)
- Leaking fluids and fumes (eg, gasoline, diesel fuel, battery acid, transmission fluids)
- Broken utility poles and downed electrical wires
- Aggravated or hostile bystanders with a potential for violence
- Smoke or fire
- Possible hazardous or toxic materials (eg, propane, hydrogen chloride)
- Crash or rescue scenes with unstable elements such as unsecured vehicles (paying particular attention to placards)
- Violence and crime scenes

You should consider your ambulance a safe haven for you to care for your patient. Park your unit in a place that provides you and your partner the greatest safety but allows you rapid access to your patient and

Figure 8-2 Before you step out of your unit, be sure to evaluate the scene for any hazards.

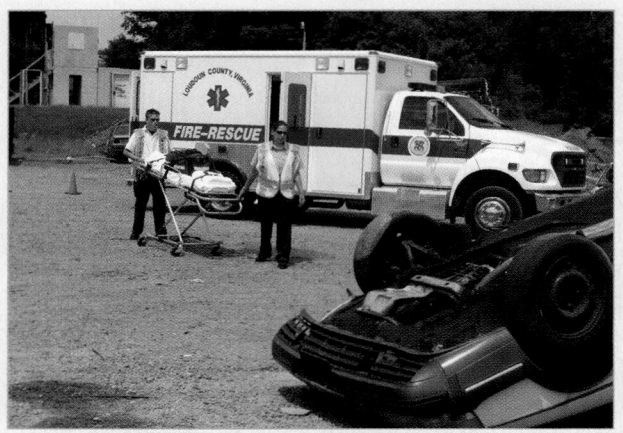

Figure 8-3 Park your unit in a place that is safe, yet allows for rapid access to the patient and your equipment. If law enforcement is already on the scene, make sure to check in with them first.

your equipment (Figure 8-3 ▲). Often, this is in front of the scene. In many instances, law enforcement or fire personnel will be on scene before you arrive. If that is the case, try to talk with them prior to entering the scene to ensure the scene is safe. If the scene is a potential crime scene, follow local protocol before entering. As you enter, stop briefly and take a mental note of where the victim is, where the weapon is if appropriate, and the presence of unusual objects and their location. Careful documentation of these facts may be helpful later if you are asked to testify in a case, but your initial concern must be your own safety. Ask for law enforcement to accompany you if the victim is a suspect in a crime.

Making an Unsafe Scene Safe

Occasionally you and your partner will not be able to enter a scene safely. If the hazard presents a real risk to your health and safety, you should request the appropriate assistance. Do not enter until a professional rescuer (ie, firefighters, utility workers, or hazardous materials crew) has made the scene safe. On other occasions, you may enter a scene that appears safe and then becomes unsafe. If you have the appropriate training and PPE to make the scene safe, you should make it safe. If not, extricate yourself and your patient as quickly as possible, protecting him or her from injury as best you can. These situations can be very difficult to manage when you want to provide medical care as quickly as possible. Remember your safety and the safety of your team come first. Other scenes may have a

EMT-B Safety

Assessing the safety of a scene before entering may be the single most important way in which emergency responders can attend to their own well-being. Subtle signs of danger not recognized and neutralized—or avoided—at this point can grow much more threatening without being noticed once you shift your attention to patient assessment and care. Initial scene assessment often allows you to distinguish between a manageably safe scene and one that could spin dangerously out of control without further warning.

Figure 8-4 With traumatic injuries, the patient has been exposed to some force or energy that results in injury or even possibly death. You can learn a great deal about that force by simply looking at the scene and determining the mechanism of injury.

potential for risk that, as a medical responder, you are able to manage with your training. Remember that hazards do not need to be dramatic situations but could be as simple as a hole in the ground or spilled transmission fluid. Carefully evaluate the scene and request specific help to manage the scene threats (eg, law enforcement, fire personnel, and hazardous materials crews).

Consider Mechanism of Injury/Nature of Illness (Trauma / Medical Patient)

The dispatched complaint will guide you in the direction of the patient's problem. It could be either a traumatic problem that involves a mechanism of injury (MOI) or a medical problem based on the nature of illness (NOI) (Figure 8-4 ▶).

Mechanism of Injury

As an EMT-B, you will be called to motor vehicle crashes or other situations in which patients may have sustained life-threatening traumatic injuries. To care for these patients properly, you must understand how traumatic injuries occur, or the mechanism of injury (MOI). With a traumatic injury, the body has been exposed to some force or energy that has resulted in a temporary injury, permanent damage, or even death.

As you might expect, certain parts of the body are more easily injured than others. The brain and the spinal cord are very fragile and easy to injure. Fortunately, they are protected by the skull, the vertebrae, and several layers of soft tissues. The eyes are also easily injured. Even small forces on the eye may result in serious injury. The bones and certain organs are stronger and can absorb small forces without resulting injury. The

net result of this information is that you can use the mechanism of injury as a kind of guide to predict the potential for a serious injury by evaluating three factors: the amount of force applied to the body, the length of time the force was applied, and the areas of the body that are involved.

You will commonly hear the terms "blunt trauma" and "penetrating trauma." With blunt trauma, the force of the injury occurs over a broad area, and the skin is usually not broken. However, the tissues and organs below the area of impact may be damaged. With penetrating trauma, the force of the injury occurs at a small point of contact between the skin and the object. The object pierces the skin and creates an open wound that carries a high potential for infection. The severity of injury depends on the characteristics of the penetrating object, the amount of force or energy, and the part of the body affected.

Nature of Illness

As an EMT-B, you will care for many medical patients as well as trauma patients. For trauma patients you examine the mechanism of injury as part of your scene size-up. For medical patients, you must examine the nature of the illness (NOI). There are similarities between the mechanism of injury and the nature of illness. Both require you to search for clues regarding how the incident occurred. You must make an effort to determine the general type of illness, which is often best described by the patient's chief complaint: the reason EMS was called. In order to quickly determine the nature of the illness, talk with the patient, family, or bystanders about the problem. But at the same time, use your senses to check the scene for clues as to the possible problem. You may see open or spilled medication containers, poisonous substances, or unsanitary living conditions. You may smell an unusual or strong odor, such as the odor of fresh paint in a closed room. You may hear a hissing sound, such as a leak from a home oxygen system. Keep these observations of the scene in mind as you begin to assess the patient.

The Importance of MOI and NOI

Considering the MOI or NOI early can be of value in preparing to care for your patient. For example, when you begin to gather equipment from the unit to treat your patient, what would you take for an older patient complaining of chest pain? How would that equipment differ from the equipment used for a pedestrian struck by a vehicle? The appearance of the scene may also guide you in your preparation. Other mechanisms of injury may include motor vehicle crashes, assaults, and stabbings or gunshot wounds. Examples of natures of illness include seizures, heart attacks, diabetic problems, and poisonings. Family members, bystanders, or even law enforcement personnel may also provide important trauma or medical information to help you prepare as you approach the patient.

During your prehospital assessment you may be tempted to categorize your patient immediately as a trauma or medical patient. Remember, the fundamentals of a good patient assessment are the same despite the unique aspects of trauma and medical care. If an unconscious patient is found at the bottom of a ladder, did he fall off the ladder, strike his head, and become unconscious or did he climb down the ladder and then lose consciousness? Early in the assessment, it can be difficult to identify with absolute certainty whether the problem is of a traumatic or medical origin. Although further assessment is needed to come to a conclusion, considering the MOI or NOI early will help you prepare for the rest of your assessment.

Determine the Number of Patients

As part of the scene size-up, it is essential that you accurately identify the total number of patients. This evaluation is critical in determining your need for additional resources, such as the fire department, specialized rescue group, or a HazMat team. When there are multiple patients you should establish incident command, call for additional units, and then begin triage (Figure 8-5 ▶). Triage is the process of sorting

Know this

Figure 8-5 With multiple patients, you should establish incident command, call for additional resources, and then begin triage.

patients based on the severity of each patient's condition. Once all the patients have been triaged, you can begin to establish treatment and transport priorities. One EMT-B, usually the most experienced, should be assigned to perform triage. This process will help you allocate your personnel, equipment, and resources to provide the most effective care to everyone. When a large number of patients are present, or if patient needs are greater than your available resources, you should put your mass casualty plan into action based on your local protocols. You should be familiar with the Incident Management System and understand your local protocols regarding responsibilities for establishing and transferring incident command.

Consider Additional Resources

Some trauma or medical situations may simply require more ambulances, while others may have needs for specific additional help. Basic life support (BLS) units may be all that are needed for some patients; however, advanced life support (ALS) should be requested for patients with severe injuries or complex medical problems depending on available resources and local protocols. ALS may be provided by EMT-Is or EMT-Ps, depending on how your EMS system is set up. Air medical support is another good resource for ALS. Follow your local protocols in requesting ALS resources.

Many resources in addition to fire suppression are often available through the fire department, including high-angle rescue, hazardous materials management, complex extrication from motor vehicle crashes, water rescue, or other specific types of rescue, such as swift water rescue. Search and rescue teams can be helpful in finding, packaging, and transporting patients over long distances or through unusual terrain. Law enforcement also may be needed to control traffic or intervene in domestic violence situations.

You should ask yourself the following questions:
- How many patients are there?
- What is the nature of their condition?
- Who contacted EMS?
- Does the scene pose a threat to you, your patient, or others?

Knowing how your EMS system is organized will help you determine what additional resources may be required. The sooner these resources are identified, the sooner they can be requested.

Consider C-Spine Immobilization

If an injury is suspected, consider early spinal immobilization. This is an important step, as moving such a patient without proper spinal immobilization can have serious implications such as lifelong paralysis. When you are uncertain whether spinal immobilization is necessary, err on the side of caution and immobilize the patient.

You are the Provider — Part 2

On arrival you see an unresponsive man lying down on the side of the road. A police officer is standing near the patient. He reports that a utility worker saw the man stumbling along, sit down, and pass out. As you approach the patient, you notice a disheveled-looking man who is approximately 40 years old. An old shopping cart filled with junk is nearby. The patient does not make any attempt to move and is not making any sounds other than some snoring noises as he breathes. He has a radial pulse that is strong. There is no obvious bleeding or signs of trauma. You decide to load the patient in the ambulance and continue your assessment en route to the hospital.

3. What information can be obtained by a quick survey of the scene before greeting your patient?
4. Can you identify life-threatening problems by simply asking the patient what happened?

Patient Assessment ✳

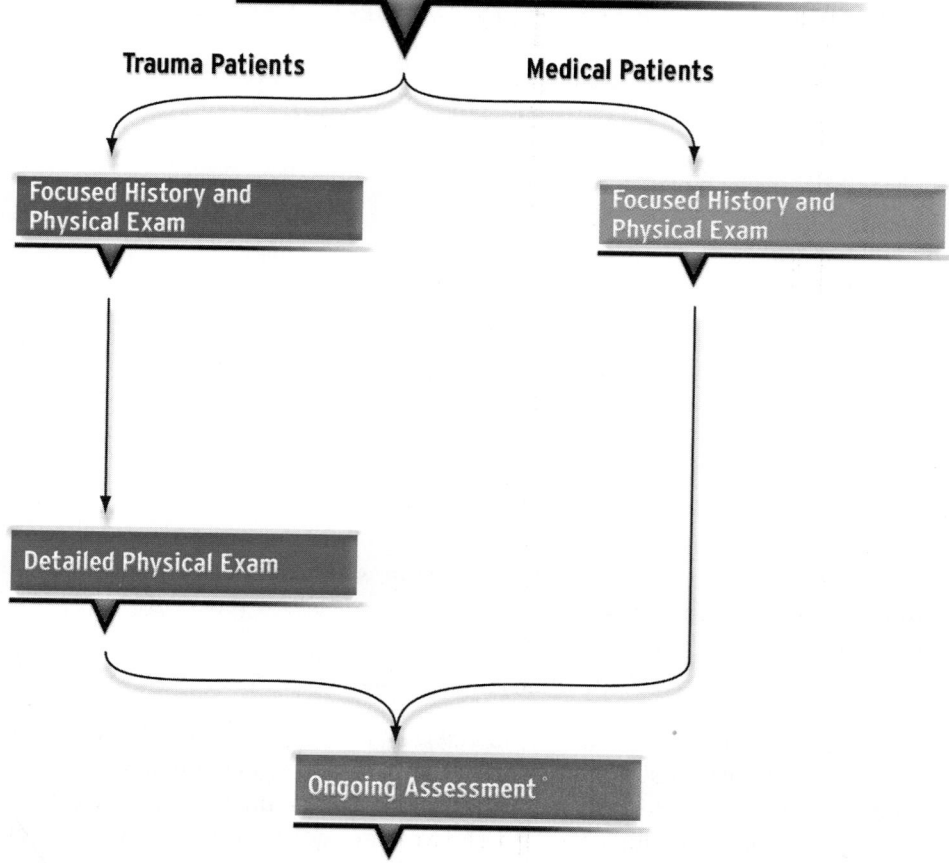

Scene Size-up

Initial Assessment

Approach and Form a General Impression
Assess Mental Status
Assess the Airway   *Know*
Assess Breathing   *This*
Assess Circulation   *study This*
Identify Priority Patients and Make Transport
  Decisions

Trauma Patients   **Medical Patients**

Focused History and   Focused History and
Physical Exam   Physical Exam

Detailed Physical Exam

Ongoing Assessment

Initial Assessment

During the scene size-up, you used dispatch information and your own evaluation of the scene to begin to determine what happened. You also evaluated potential or actual scene threats, how to protect yourself and your team, and whether you need additional resources. These steps are critical in the initiation of patient care. However, your actual patient assessment begins when you greet your patient. The initial assessment has a single, critical, all-important goal: to identify and initiate treatment of immediate or potential life threats. Information concerning life-threatening conditions can be obtained from the visual appearance of the patient, how the patient's complaints relate to the current MOI or NOI, and obvious problems with the patient's airway, breathing, and circulation (ABCs). In all cases, your assessment of the patient's ABCs will determine the extent of your treatment at the scene. Always give priority to the ABCs to ensure life-saving treatment.

Figure 8-6 As you approach the patient, form a general impression of his or her overall condition.

General Impression

Anytime you meet someone new, you form an initial impression about that person. The general impression of your patient is similar but helps to focus your attention towards life-threatening problems
(Figure 8-6 ▶). This general impression includes noting things such as the person's age, gender, race, level of distress, and overall appearance. You may anticipate different problems depending on the patient's age, gender, or race. A woman complaining of abdominal pain, for example, may have more serious implications than a man with the same complaint because of the complexity of the female reproductive system.

You should think of your general impression as a visual assessment, gathering information as you approach the patient (Figure 8-7 ▶). As you approach, make sure that the patient sees you coming to avoid surprising the patient or causing the patient to turn to see you, possibly making any injuries worse. Note the patient's position and whether the patient is moving or still. Make note of odors that suggest chemical hazards or smoke. When you reach your patient, place yourself at a lower position, if possible, to show respect for the patient and help the patient feel comfortable and less threatened as you begin your assessment. The general impression continues during your introduction and questioning of their complaints. For example, they may direct you to a wound on their leg, or demonstrate an airway problem by creating abnormal sounds when they breathe. If a life-threatening problem is found, it should be treated immediately.

Introduce Yourself and Ask Permission to Treat

Introduce yourself to the patient and others by telling them who you are and what you are doing. For example, "Hello, my name is John Smith, I'm an EMT Basic with Anytown EMS. I'm here to help you." After introducing yourself, ask the patient his or her name and address the patient as he or she wishes. Additionally, you should let the patient know how you wish to be addressed. For example, "Hello, my name is Ron Kehl, call me Ron." Using the patient's name frequently will help reduce his or her anxiety about being ill or injured.

Initial Assessment

Obtain Consent to Care for the Patient

Obtain consent to care for the person as discussed in Chapter 3. Conscious patients may allow you to care for them without you needing to ask specifically if you can help them. Other times, you may want to formally ask,

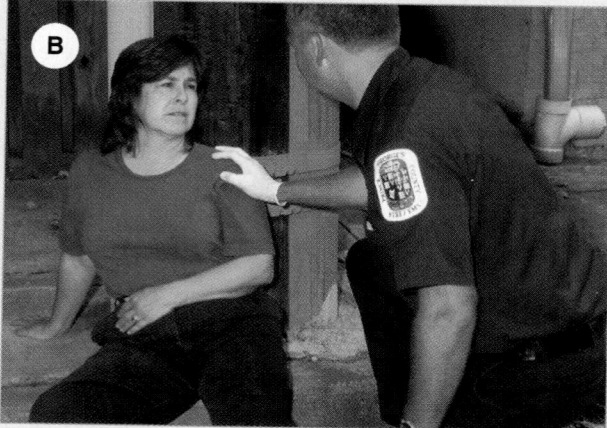

Figure 8-7 A. Observe the patient to form a general impression. **B.** Assess the patient's mental status. **C.** Assess the mental status of a child.

"May I look at your leg? It seems to be injured." This communication establishes a formal relationship between you and the patient and may reduce anxiety and gain cooperation when caring for younger patients. Treatment for unresponsive patients is based on implied consent, assuming that, if they were conscious, they would consent to emergency care. If the patient wakes up, however, you must explain who you are, what you are doing, and why you are doing it.

Determine the Chief Complaint

Now that introductions are over and you have the patient's permission to help, you should determine the patient's chief complaint. This may be done by simply asking, "What happened?" or "How may I help you?" The chief complaint is the most serious thing that the patient is concerned about and is usually expressed in the patient's own words (symptoms); however, it may be something observable by the EMT-B (signs). In unresponsive patients, their chief complaint is often expressed as being "unresponsive." Keep in mind the chief complaint expressed by the patient may not be the most serious thing that is wrong; however, it is a good place to start. For example, a patient with difficulty breathing from a chronic lung condition may complain about leg problems from poor perfusion. Your responsibility is to determine what is more important—the leg problem or the difficulty breathing. The chief complaint gives you a reference point to begin with during your assessment process.

As an EMT-B you will be called on to treat an almost infinite number of different problems. The chief complaint will help you narrow down the MOI or NOI information gathered in the scene size-up. You can try to determine if the patient fell from the ladder or passed out by asking the patient. Evaluating how a person was injured may help predict the types of injuries the person may have. If you suspect a potential for a spinal injury, manually stabilize the patient's head. You will learn that it is not easy to determine whether a patient is a medical or trauma patient until you have completed a more in-depth assessment. This should not prevent you from providing stabilization to a patient with a suspected spinal injury. In many situations, medical emergencies and trauma go hand in hand.

For this reason, the initial assessment does not encourage you to differentiate immediately between medical and trauma patients. Rather, the assessment process begins by assuming that all patients may have both medical and trauma aspects to their condition. This approach is both simpler and safer than an approach that

EMT-B Tips

People call 9-1-1 during some of the most difficult times of their lives. In some cases, they call because of a serious illness or injury. In others, they call because they or their families are frightened, overwhelmed, or unable to cope with a more minor problem any longer. Patients are often fatigued, sick, frightened, angry, or sad, and the families often share some or all of these feelings. Regardless of the exact nature of the call, the patients, families, and bystanders expect you to bring comfort, control, and resolution to these problems—emotional and physical.

In many ways, good communication skills are as or more important than technical proficiency. Each step in the assessment process can be impeded by poor communication, and each can be enhanced by a good connection between you, the patient, and the family. Here are five tips that can vastly improve your communication skills during the assessment process:

1. Do whatever you can, quickly, to make yourself and the patient comfortable. Patients are uncomfortable communicating with someone who is standing over them, pacing, or looking away. When time permits, sit down and/or position yourself near the patient, introduce yourself, and ask the patient's name. This simple action signals to the patient that you are willing to take time to talk; it opens channels of communication. At the same time, be conscious of the patient's personal space. Do not move in too quickly. Ask the patient whether there is something you can do to make him or her more comfortable. Caring gestures and good body language are a visible demonstration of your care and concern.

2. Actively listen to the patient. In many cases, the patient will be able to tell you what is wrong with him if you are paying attention and are truly listening. You can use several skills to actively listen, including leaning in toward the patient, taking selective notes, and periodically repeating important points to the patient to ensure that you understood correctly. Active listening is often more difficult than it might seem, because scenes are often noisy and chaotic, and you will be receiving information from the patient, the family, your partner, and other EMS, fire, or law enforcement individuals on the scene. Try to screen them out for a few minutes so that you can truly listen to the patient; it will pay big dividends.

3. Make eye contact with the person with whom you are speaking. Eye contact signals that you are listening, so the patient is more likely to open up. An added benefit is that you will see facial expressions that, in some cases, communicate more clearly than the patient's words. For instance, you might see a facial grimace of pain or averted eyes indicating embarrassment. Note that some cultures are uncomfortable with or offended by direct eye contact. Be sure to be familiar with the cultural background of citizens in your area.

4. Base your initial questions on the patient's complaints. No one likes to think that he or she is "just another patient." But that is what you communicate if you always ask the same questions of every patient, regardless of the complaint. If you ask questions about a Medicare number while the patient is trying to tell you about his pain, you are communicating that you are not really interested in the problem. Talk about the patient's problem first, then ask paperwork questions.

5. Before you start treatment, stop for a moment and mentally summarize what you have learned and what you are going to do, then tell the patient. By providing necessary information to the patient and family, you help to relieve their anxiety and fear. This will also give them an opportunity to give you additional information if you have missed something.

You should spend your entire EMS career fine-tuning your patient assessment skills, as they are the cornerstone of high-quality prehospital care. A poor assessment almost always results in substandard patient care. Be sure to focus some of your energy on improving the communication process. You will make it easier for patients to feel comfortable around you, which will help them to give honest, direct answers to your questions. As a result, you will get better assessment information in less time.

Initial
Assessment

starts with an unsupported assumption that the patient is either ill or injured. One way to evaluate "the big picture" is to obtain a general impression of the patient before focusing in on specific concerns.

Assess Mental Status

Evaluating a person's mental status is a good way to evaluate brain function. Many conditions, medical or trauma, may alter brain function and therefore the patient's level of consciousness. You will learn more about these conditions as you progress through the EMT-B course. Mental status and level of consciousness can be evaluated in just a few seconds by using two separate tests: responsiveness and orientation. *Know this*

One test for responsiveness uses the AVPU scale to assess how well a patient responds to external stimuli, including verbal stimuli (sound), and painful stimuli (such as pinching the patient's earlobe). The AVPU scale is based on the following criteria:

- Alert. The patient's eyes open spontaneously as you approach, and the patient appears aware of you and responsive to the environment. The patient appears to follow commands, and the eyes visually track people and objects.
- Responsive to Verbal Stimulus. The patient's eyes do not open spontaneously. However, the patient's eyes do open to verbal stimuli, and the patient is able to respond in some meaningful way when spoken to.
- Responsive to Pain. The patient does not respond to your questions but moves or cries out in response to painful stimulus. There are appropri-

ate and inappropriate methods of applying painful stimulus based a great deal on personal preference (Figure 8-8 ▼). Be aware that some methods may not give an accurate result if a spinal cord injury is present.

- Unresponsive. The patient does not respond spontaneously or to verbal or painful stimulus. These patients usually have no cough or gag reflex and lack the ability to protect their airway. If you are in doubt about whether a patient is truly unresponsive, assume the worst and treat appropriately.

For a patient who is alert and responsive to verbal stimuli, you should next evaluate orientation. Orientation tests mental status by checking a patient's memory and thinking ability. The most common test evaluates a patient's ability to remember four things:

- Person. The patient is able to remember his or her name.
- Place. The patient is able to identify his or her current location.
- Time. The patient is able to tell you the current year, month, and approximate date.
- Event. The patient is able to describe what happened (the MOI or NOI).

These questions were not selected at random. They evaluate long-term memory (person and place if the patient is at home), intermediate memory (place and time when asking year or month), and short-term memory (time when asking approximate date and event). If the patient knows these facts, the patient is said to be "alert and fully oriented" or "alert and oriented to person,

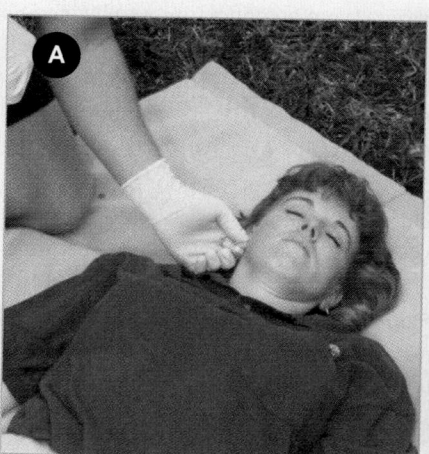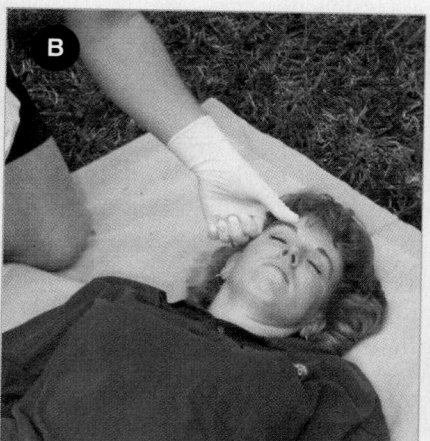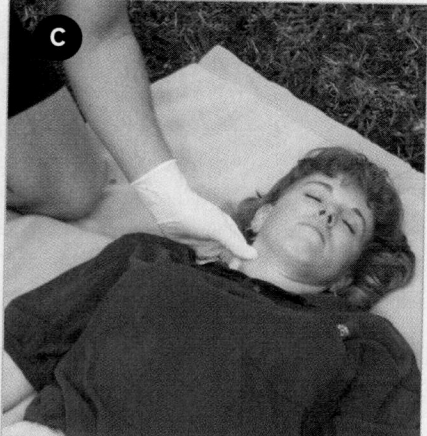

Figure 8-8 **A.** Gently but firmly pinch the patient's earlobe. **B.** Press down on the bone above the eye. **C.** Pinch the muscles of the neck.

Study This !!!

⚠ Pediatric Needs

Mental status may be difficult to evaluate in children. First, determine whether the child is alert. Even infants should be alert to your presence and should follow you with their eyes (a process called "tracking"). Ask the parent whether the child is behaving normally, particularly in regards to alertness. Most children older than 2 years should know their own name and the names of their parents and siblings. Evaluate mental status in school-age children by asking about holidays, recent school activities, or teachers' names.

place, time, and event." If a patient does not know these facts, he or she is considered less than fully oriented.

An altered mental status, anything other than alert, may be caused by a variety of conditions, including head trauma, hypoxemia, hypoglycemia, stroke, cardiac problems, or drug use. If the patient has an altered mental status, you should rapidly complete the initial assessment, provide high-flow supplemental oxygen, consider spinal immobilization if trauma is suspected, and initiate transport. Support the ABCs as required and continually reassess for changes in the patient's condition.

Assess the Airway

As you move through the steps of the initial assessment, you must always be alert for signs of respiratory compromise or airway obstruction. Regardless of the cause, a partial or complete airway obstruction will result in inadequate or absent air flow into and out of the lungs. To prevent permanent damage to the brain, heart, and lungs, or even death, you must determine if the airway is open (patent) and adequate.

Responsive Patients

Patients of any age who are talking or crying have an open airway. However, watching and listening to how patients speak, particularly those with respiratory problems, may provide important clues about the adequacy of their airway and the status of their breathing. For example, sounds of stridor suggest a partially occluded airway caused by swelling. High-pitched crowing sounds may indicate a partial airway obstruction from a for-

eign body. A conscious patient who cannot speak or cry most likely has a complete airway obstruction.

If you identify an airway problem, stop the assessment process and obtain a patent airway. This may be as simple as positioning the patient so the air moves in and out easier or as complex as abdominal thrusts to remove a foreign body from the airway. Although airway and breathing problems are not the same, their signs and symptoms often overlap. If your patient has signs of respiratory difficulty or is not breathing, you should immediately take corrective actions using appropriate airway management techniques.

Unresponsive Patients

With an unresponsive patient or a patient with a decreased level of consciousness, you should immediately assess the patency of the airway. If it is clear, you can continue your assessment. If the airway is not clear, your next priority is to open it using the head tilt–chin lift or jaw-thrust maneuver. An airway obstruction in an unconscious patient is most commonly due to relaxation of the tongue muscles, allowing the tongue to fall to the back of the throat. Dentures, blood clots, vomitus, mucus, food, or other foreign objects may also create an obstruction. Signs of airway obstruction in an unconscious patient include the following:

- Obvious trauma, blood, or other obstruction
- Noisy breathing, such as snoring, bubbling, gurgling, crowing, or other abnormal sounds (Normal breathing is quiet.)
- Extremely shallow or absent breathing (Airway obstructions may impair breathing.)

If the airway is not patent, you should open it using the head tilt–chin lift or jaw-thrust maneuver, suction as necessary, and use an airway adjunct as necessary. The body will not have the necessary oxygen needed to survive if the airway is not managed quickly and efficiently. Remember that airway positioning depends on the age and size of your patient.

Spinal Considerations

Management of a patient's airway can be complicated by the presence of a spinal injury. Trauma patients, those who are conscious or unconscious, should be stabilized to protect their spines. Conscious or unconscious medical patients, however, may have fallen and have a potential for a spinal injury. It is important for you to consider spinal precautions during scene size-up and evaluate the MOI or NOI further when determining the chief complaint. When managing the airway status of

Initial Assessment

a patient, you must decide if you need to protect the spine. If you are uncertain whether a spinal injury exists, even after questioning a conscious patient, assume the worst and stabilize the spine. Airway management and spinal immobilization must be performed simultaneously.

Unconscious or minimally responsive medical patients may not be able to protect their airway. If you have evaluated the scene and have reliable information from witnesses who indicate that a patient does not have a spinal injury, you may consider placing the patient in a recovery position or side-lying position as soon as possible. In this position, secretions will drain out of the patient's mouth rather than into the airway where they could be dangerous. Follow your local protocol in determining who has potential for spinal injury and who may be considered "clear."

Assess Breathing

A patient's breathing status is directly related to the adequacy of his or her airway. Make sure the patient's airway is open, and then make sure the patient's breathing is present and adequate. Oxygen should be administered to patients who are having difficulty breathing and also to patients who are breathing adequately, while positive pressure ventilations should be performed on patients who are apneic or whose breathing is too slow or too shallow.

As you assess a patient's breathing, look, listen, and feel for the presence of breathing and then assess the adequacy of breathing. A normal respiratory rate varies widely in adults, ranging from 12 to 20 breaths/min. Children breathe at even faster rates. However, taking the time to actually count respirations may distract you from assessing more life-threatening problems. With practice, you should be able to estimate the rate and note whether it is too fast or too slow. At times it may be important to actually count the number of respirations in your initial assessment. Remember, the goal of your initial assessment is to identify and treat airway, breathing, and circulation problems as quickly as possible. Measuring vital signs more exactly is accomplished in another part of the assessment once time and life threats are less of an issue.

Observe how much effort is required for the patient to breathe. Normal respirations are not usually shallow or excessively deep. Shallow respirations can be identified by little movement of the chest wall (reduced tidal volume). Deep respirations cause a great deal of chest rise and fall. The presence of retractions or the use of accessory muscles of respiration is also a sign of inad-

EMT-B Tips

Thousands of deaths per year occur from airway obstruction following acute alcohol intoxication or drug overdose. Generally, these patients vomit while lying on their backs and cannot protect their airway because of a severely decreased level of consciousness. Never leave anyone who has passed out unattended. If the person cannot be continually monitored, place the person prone or on their side, not supine.

equate breathing. Nasal flaring and see-saw breathing in pediatric patients indicate inadequate breathing. A patient who can only speak two or three words without pausing to take a breath, a condition known as two-to three-word dyspnea, has a serious breathing problem. As you assess the patient's breathing, you should ask yourself the following questions:

- Does the patient appear to be choking?
- Is the respiratory rate too fast or too slow?
- Are the patient's respirations shallow or deep?
- Is the patient cyanotic (blue)?
- Do you hear abnormal sounds when listening to the lungs?
- Is the patient moving air into and out of the lungs on both sides?

It may be helpful to listen to breath sounds on each side of your patient's chest early in the initial assessment (Figure 8-9 ▼). This can help identify the ade-

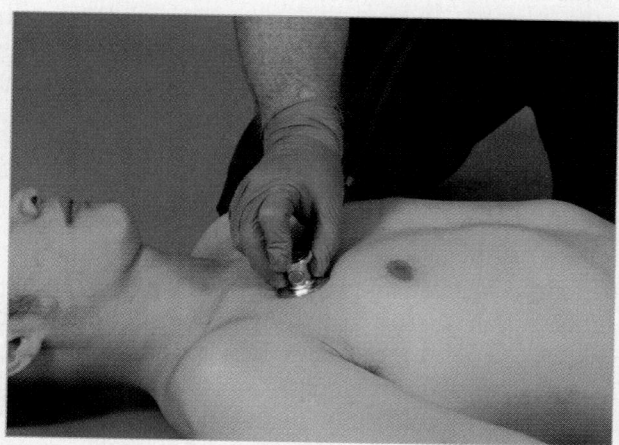

Figure 8-9 Listen for breath sounds on opposite sides of the patient's chest.

quacy of air movement in both lungs. Place the head of your stethoscope on the skin of the upper anterior chest at the midclavicular line and listen to one or two breaths. Repeat on the opposite side. Decreased or absent breath sounds on one side of the chest and decreased movement in the rise and fall on one side indicate inadequate breathing.

If a patient seems to develop difficulty breathing after your initial assessment, you should immediately reevaluate the airway. If the airway is open and breathing is present and adequate, you should consider placing the patient on supplemental oxygen. If breathing is present and inadequate because respirations are too fast (generally more than 20 breaths/min), too shallow, or too slow (generally less than12 breaths/min), you should place the patient on supplemental oxygen and consider providing positive pressure ventilations with an airway adjunct. Remember that what is critical is air exchange, not the number of breaths.

Any patient with a decreased level of consciousness, respiratory distress, or poor skin color should also receive high-flow oxygen. If there is no risk of spinal injury, the patient should remain in a comfortable position that supports breathing; this is usually sitting up with the legs dangling or even a high Fowler's position (sitting up at almost a 90° angle). In any patient with a possible spinal injury, you should immobilize the cervical spine, ensuring that respirations are not compromised.

Oxygen should be delivered to patients using a nonrebreathing mask at 15 L/min. Any patient identified as having potential airway or breathing problems or immediate airway or breathing problems should be given supplemental oxygen and observed for inadequate breathing. Do not withhold oxygen from any patient at the scene!

Assess Circulation

Assessing circulation helps you to evaluate how well blood is circulating to the major organs, including the brain, lungs, heart, kidneys, and the rest of the body. A variety of problems can impair circulation, including blood loss, shock, and conditions that affect the heart and major blood vessels. Circulation is evaluated by assessing the rate and quality of the pulse, identifying external bleeding, and evaluating the skin.

Assess the Pulse

Our first goal in assessing circulation is to determine if the patient's pulse is present and adequate. Assess the pulse by feeling for the radial artery at the distal end of the forearm. If a pulse cannot be felt at the radial artery, check the carotid artery in the neck. If you cannot palpate a pulse in an unresponsive patient, begin CPR. If an AED is available, attach it and follow the voice prompts, following your local protocol. An AED is indicated for use on medical patients who are at least 8 years old and weighing more than 55 lb and who have been assessed to be unresponsive, not breathing, and pulseless. An AED with special pediatric pads is indicated for use on pediatric medical patients between the ages of 1 and 7 years who have been assessed to be unresponsive, not breathing, and pulseless.

If the patient has a pulse but is not breathing, provide ventilations at a rate of at least 12 breaths/min for adults and at least 20 breaths/min for an infant or child. Continue to monitor the pulse to evaluate the effectiveness of your ventilations. If at any time the pulse is lost, start CPR and apply the AED if indicated. The apparent absence of a palpable pulse in a responsive patient is not caused by cardiac arrest. Therefore, never begin CPR or use an AED on a responsive patient.

⚠ Pediatric Needs

You can feel the pulse of a child at the carotid artery, as in an adult. However, palpating the pulse in an infant may present a problem. Because an infant's neck is often very short and fat, and its pulse is often quite fast, you may have a hard time finding the carotid pulse. Therefore, in infants younger than 1 year, you should palpate the brachial artery to assess the pulse. Normal pulse rates for children are shown in (Table 8-1 ▶).

Study This!!!

TABLE 8-1 Normal Pulse Rates in Infants and Children

Age	Range (beats/min)
Infant: 1 month to 1 year	100 to 160
Toddler: 1 to 3 years	90 to 150
Preschool-age: 3 to 6 years	80 to 140
School-age: 6 to 12 years	70 to 120
Adolescent: 12 to 18 years	60 to 100

After determining that a pulse is present, next determine its adequacy. This is done by assessing the rate, rhythm, and strength of the pulse. For an adult, the normal resting pulse rate should be between 60 and 80 beats/min and could be as much as 100 beats/min in geriatric patients. In pediatrics, generally the younger the patient, the faster the pulse rate. The actual number of pulsations per minute is not as important as obtaining a sense of whether the rate is too slow, in the normal range, or too fast. With practice, you can develop a sense for pulse rate without actually counting the pulsations. This will help to speed up your initial assessment and allow you to focus on finding potentially life-threatening problems. A pulse that is too slow or too fast may change decisions related to transporting your patient. The pulse should be easily felt at the radial or carotid artery and have a regular pattern. If it is difficult to feel or irregular, the patient may have problems with his circulatory system that may need further evaluation later in your assessment.

Assess and Control External Bleeding (know This

The next step is to identify any major external bleeding. In some instances blood loss can be very rapid and can quickly result in shock or even death. Therefore, this step demands your immediate attention as soon as the patient's airway is patent and breathing has been stabilized.

Signs of blood loss include active bleeding from wounds and/or evidence of bleeding such as blood on the clothes or near the patient. Serious bleeding from a large vein may be characterized by steady blood flow. Bleeding from an artery is characterized by a spurting flow of blood. When you evaluate an unconscious patient, do a sweep for blood quickly and lightly by running your gloved hands from head to toe, pausing periodically to see if your gloves are bloody.

Controlling external bleeding is often very simple. Initially, direct pressure with your gloved hand and soon thereafter, a sterile bandage over the wound will control bleeding in most instances. This direct pressure stops the bleeding and helps the blood to coagulate, or clot naturally. Most often, bleeding can be adequately controlled by using direct pressure, along with elevating the extremity if bleeding is from the arms or legs. When direct pressure and elevation are not successful, you may apply pressure directly over arterial pressure points.

Assess Perfusion

Assessing the skin is one of the most important and most readily accessible ways of evaluating circulation. A normally functioning circulatory system will perfuse the skin with oxygenated blood. A lack of perfusion or hypoperfusion will result in hypoxia of the brain, lungs, heart, and kidneys. In most situations hypoperfusion is caused by shock. The degree of hypoperfusion and how long it lasts will determine if a patient will suffer permanent damage related to the hypoxia. Perfusion is assessed by evaluating a patient's skin color, temperature, and moisture.

Color Study This!!!

Skin color depends on pigmentation, blood oxygen levels, and the amount of blood circulating through the vessels of the skin. For this reason, skin color is a valuable assessment tool. The normal skin color of lightly pigmented people is pinkish. Deeply pigmented skin may hide skin color changes that result from injury or illness. Therefore, you should look for changes in color in areas of the skin that have less pigment: the fingernail beds, the sclera (white of the eye), the conjunctiva (lining of the eyelid), and the mucous membranes of the mouth. Normal skin color, particularly of the conjunctiva and mucous membranes, is pinkish. Skin colors that should alert you to possible medical problems include cyanosis (blue), flushed (red), pale (white), and jaundice (yellow). Cyanosis and pale skin colors indicate a lack of perfusion.

Temperature Study This !!!

The skin has many functions. It helps maintain the water content of the body, acts as insulation and protection from infection, and also plays a role in regulating body temperature by changing the amount of blood circulating through the surface of skin. With poor perfusion, the body pulls blood away from the surface of the skin and diverts it to the core of the body. The result is cool, pale, clammy skin—a good indication in your initial assessment of hypoperfusion and inadequacy of circulatory system function (shock).

Condition Study This!!!

Assessing the condition of the skin is really assessing the presence of moisture on the skin. The skin is normally warm and dry. Skin that is cool or cold, moist, or clammy suggests shock (hypoperfusion). Again, these characteristics are important findings in your initial assessment because hypoperfusion can lead to serious consequences if treatment is delayed or ignored.

Capillary Refill

Another way to assess perfusion is to check <u>capillary refill</u>. This method is most accurate in children younger than 6 years. Although capillary refill is a quick and very general way to evaluate perfusion, it is important to remember that other conditions, not related to the body's circulation, may also slow capillary refill. These conditions include, but are not limited to, the patient's age as well as exposure to a cold environment (<u>hypothermia</u>), frozen tissue (<u>frostbite</u>), and vasoconstriction. Injuries to bones and muscles of the extremities may cause local circulatory compromise resulting in hypoperfusion of an extremity rather than hypoperfusion of the body in general.

Identify Priority Patients and Make Transport Decisions

As you complete your initial assessment, you have to make some decisions about patient care. You should have already identified and begun treatment of life-threatening injuries and illnesses. Now you should identify the priority status of your patient (Figure 8-10 ▶). Would you consider your patient a high priority or a medium or low priority for transport? Priority designation is used to determine if your patient needs immediate transport or will tolerate a few more minutes on scene. Patients with any of the following conditions are examples of high-priority patients and should be transported immediately:

- Difficulty breathing
- Poor general impression
- Unresponsive with no gag or cough reflexes
- Severe chest pain, especially when the systolic blood pressure is less than 100 mm Hg
- Pale skin, or other signs of poor perfusion
- Complicated childbirth
- Uncontrolled bleeding
- Responsive but unable to follow commands
- Severe pain in any area of the body
- Inability to move any part of the body

Importance (handwritten margin note)

A high-priority patient should be transported as quickly as possible. The decision to transport should be made at this point in the assessment, and preparations for packaging and transport initiated. However, physically loading the patient on the stretcher and leaving the scene may occur shortly after this decision. For example, in a patient with significant trauma, recognizing the need for a surgical consult at the hospital to correct your patient's problems will be important. However, you may need to take 60 to 90 seconds to

Figure 8-10 Identifying priority patients.

identify injuries that must be protected during packaging and loading for transport. Protecting the patient's spine and identifying fractured extremities are an integral part of packaging for transport. These injuries can be made worse if you neglect to assess and treat them before moving the patient.

Recognizing the need to transport serious trauma patients is of such importance that you may hear colleagues refer to the <u>Golden Hour</u>. This refers to the time from injury to definitive care, during which treatment of shock and traumatic injuries should occur because survival potential is best (Figure 8-11 ▶). After the first 60 minutes, the body has increasing difficulty in compensating for shock and traumatic injuries. For this reason you should spend as little time as possible on scene with patients who have sustained significant or severe trauma. Aim to assess, stabilize, package, and begin transport to the appropriate facility within 10 minutes after arrival on scene whenever possible (a difficult or complex extrication may obviously limit possibilities).

Some patients may benefit from remaining on scene and receiving continuing care. For example, an older patient with chest pain may be better served on scene by being administered nitroglycerin and waiting for an ALS vehicle than by immediate transport. ALS should be called for if not already en route to the scene, and depending on the travel distance, can be met while transporting the critical patient. If ALS is delayed or further away, coordinating a rendezvous may be a better decision for your high-priority patient. Your

soon as you get the call (handwritten margin note)

Study This !!!

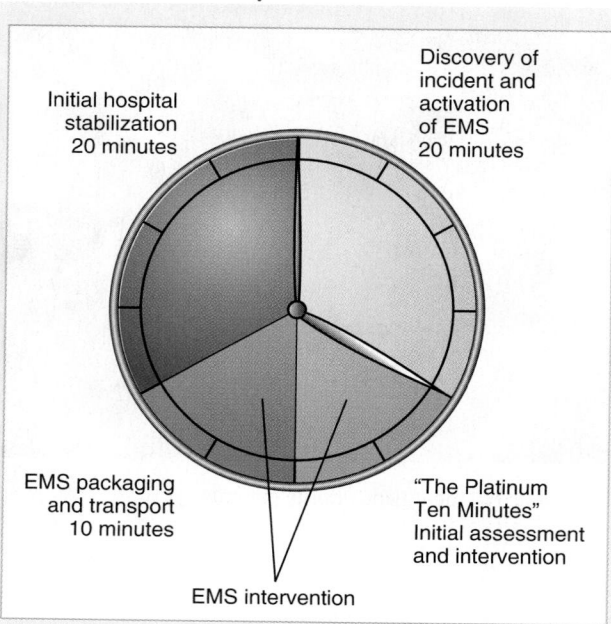

Figure 8-11 The Golden Hour is the time during which treatment of shock or traumatic injuries is most critical and the potential for survival is best.

decision to stay on scene or transport immediately will be based on your patient's condition, the availability of more advanced help, the distance you must transport, and your local protocols.

Correct identification of high-priority patients is an essential aspect of the initial assessment and helps to improve patient outcome. While initial treatment is important, it is essential to remember that immediate transport is one of the keys to the survival of any high-priority patient. Transport should be initiated as soon as practical and possible.

From here, you proceed to the appropriate focused history and physical exam based on your assessment of whether the cause of the patient's problem is a result of trauma or medical emergency, or both.

You are the Provider   Part 3

After evaluating the patient for respiratory difficulty, which improved after placing a nasopharyngeal airway and providing oxygen, the patient is lifted onto the stretcher using an extremity lift and packaged for transport to the local hospital. En route you perform a rapid physical exam looking for an explanation as to the man's unresponsive state and find nothing to explain the situation except that he smells of alcohol. A set of vital signs reveals a blood pressure of 142/96 mm Hg in a supine position, a weak, regular pulse of 88 beats/min, and shallow, labored respirations of 24/min. His skin is pink, warm, and dry, with a normal capillary refill time. History has been difficult to obtain because he is unresponsive. A prescription for Librium, however, written by the homeless clinic, was found in the patient's pocket.

6. How does the patient's chief complaint guide you in your assessment? Would you consider this patient a trauma or a medical patient?
7. What are important clues to focus on when assessing your patient? How do you identify these clues?

Initial Assessment

Patient Assessment

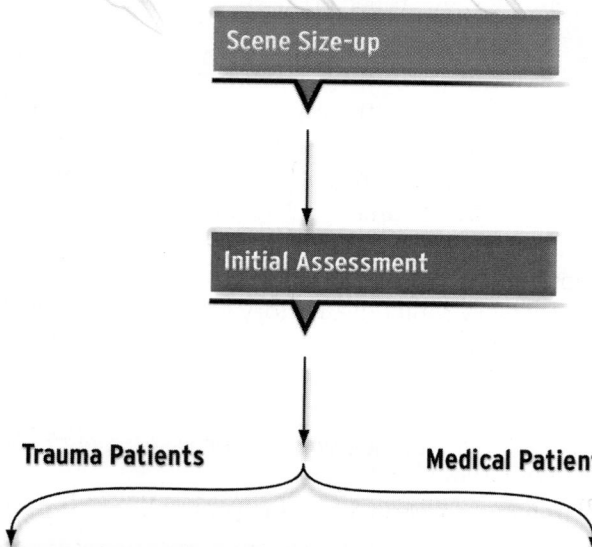

Scene Size-up

Initial Assessment

Trauma Patients

Medical Patients

Focused History and Physical Exam

Reconsider Mechanism of Injury

Significant Mechanism of Injury	No Significant Mechanism of Injury
Rapid Trauma Assessment	Focused Trauma Assessment Based on Chief Complaint
Baseline Vital Signs	Baseline Vital Signs
SAMPLE History	SAMPLE History
Reevaluate Transport Decision	Reevaluate Transport Decision

Focused History and Physical Exam

Evaluate Responsiveness

Responsive	Unresponsive
History of Illness	Rapid Medical Assessment
SAMPLE History	Baseline Vital Signs
Focused Medical Assessment Based on Chief Complaint	SAMPLE History
Baseline Vital Signs	Reevaluate Transport Decision
Reevaluate Transport Decision	

Detailed Physical Exam

Ongoing Assessment

Focused History and Physical Exam

You now have information from the scene size-up and initial assessment. These have provided you with valuable information about the scene, allowing you to prepare to care for your patient. You have stabilized any life-threatening conditions, perhaps provided spinal immobilization, and initiated transport. How do you proceed now? Your patient may have, almost literally, one or more of a million different problems. How do you identify, prioritize, and treat this variety of potential problems?

Study This →

The focused history and physical exam will help you to identify specific problems. It is based on the patient's chief complaint (what happened to this patient) and has the following goals:

Study This

- Understand the specific circumstances surrounding the chief complaint. What key factors were associated with the event? Does the mechanism of injury put the patient at high risk for serious injuries?
- Obtain objective measurements of the patient's condition. Do these measurements validate the seriousness of this patient's condition? How well is the patient dealing with his or her injury or illness?
- Direct further physical examination. What physical clues help us to identify problems?

The focused history and physical exam has three components to meet these goals: an evaluation of the patient's medical history, obtaining baseline vital signs, and performing a physical exam based on the patient's complaint, or, in the case of a critical patient, the MOI or NOI. You have previously learned how to measure baseline vital signs and how to question patients about their medical history.

For many EMT-Bs, taking the patient's history seems to be a bewildering series of questions that seem to bear little or no relationship to the patient's need for help. This becomes worse with patients who have had many medical problems; taking their history is time consuming and may yield little or no information that is useful to you. However, this does not need to be the case. The patient's history can help to tie together your findings from the physical exam and the vital signs. As we explore more of how the focused history and physical exam applies to medical and trauma patients, you will understand how to question patients and obtain a SAMPLE history (general medical history using the mnemonic SAMPLE) and a focused history of specific problems using the OPQRST mnemonic. SAMPLE history and OPQRST are discussed in detail in Chapter 5.

The baseline vital signs provide useful information about the overall functions of the patient's internal organs. They will be an important part of your assessment if your patient appears to have problems related to blood loss, circulation, or breathing. In other patients, you may simply document the vital signs as baseline information. If the patient's condition is stable, you should reassess vital signs every 15 minutes until you reach the emergency department. If the patient is unstable you should reassess at a minimum of every 5 minutes, or as often as the situation permits, looking for trends in the patient's condition, and treat for shock.

Do not be falsely reassured by apparently normal vital signs. The body has amazing abilities to compensate for severe injury or illness, especially in children and young adults. Even patients with severe medical or traumatic conditions may initially present with fairly normal vital signs. However, the body eventually loses its ability to compensate (decompensatory shock), and the vital signs may deteriorate rapidly, especially in children. In fact, this tendency for the vital signs to fall rapidly as the patient decompensates is the reason that it is important to frequently recheck and record vital signs. Treating a patient for shock before obvious signs of shock appear would help to reduce the overall effects of decompensatory shock and therefore potentially increase your patient's survival rate.

There are two types of physical exams performed in this part of the assessment: a rapid physical exam or a focused physical exam. Either one is performed on a medical or a trauma patient depending on the circumstances surrounding his illness or injury.

Rapid Physical Exam

A rapid physical exam is a quick head-to-toe exam to identify any **D**eformities, **C**ontusions, **A**brasions, **P**unctures/penetrations, **B**urns, **T**enderness, **L**acerations, and **S**welling, among other signs, that may indicate a problem. This can be remem-

Know This

bered with the mnemonic <u>DCAP-BTLS</u>. This exam is performed in as quickly as 60 to 90 seconds. The goal of a rapid physical exam is to quickly identify the potential for hidden injuries or identifiable causes that may not have been easily found in the initial assessment. It is usually performed on a trauma patient with a significant MOI or an unresponsive medical patient (Skill Drill 8-1 ▶).

1. Assess the head, looking and feeling for DCAP-BTLS and crepitus (**Step 1**).
2. Assess the neck, looking and feeling for DCAP-BTLS, jugular venous distention, tracheal deviation, and crepitus (**Step 2A**). In trauma patients you should now apply a cervical spinal immobilization device (**Step 2B**).
3. Assess the chest, looking and feeling for DCAP-BTLS, paradoxical motion, and crepitus. You should also listen to breath sounds on both sides of the patient's chest (**Step 3**).
4. Assess the abdomen, looking and feeling for DCAP-BTLS, rigidity (firm or soft), and distention (**Step 4**).
5. Assess the pelvis, looking for DCAP-BTLS. If there is no pain, gently compress the pelvis downward and inward to look for tenderness or instability (**Step 5**).
6. Assess all four extremities, looking and feeling for DCAP-BTLS. Also assess bilaterally for distal pulses, motor, and sensory function (**Step 6**).
7. Assess the back and buttocks, looking and feeling for DCAP-BTLS. In all trauma patients you should maintain in-line stabilization of the spine while rolling the patient on his or her side in one motion (**Step 7**).

Focused Physical Exam

A focused physical exam uses specific assessment techniques to evaluate the patient's chief complaint. The exam generally focuses on the location or body system related to the chief complaint. For example, in a person complaining of a headache, you should carefully assess the head and/or the neurologic system. A person with a laceration to the arm may need only that arm evaluated. The goal of a focused assessment is to focus your attention on the immediate problem. It is usually performed on a trauma patient without a significant mechanism of injury or a responsive medical patient. (Skill Drill 8-2 ▶) summarizes potential steps in a focused physical exam (only the relevant steps for the particular patient will be done):

1. **Head, neck, and cervical spine.** Inspect for abnormalities of the head, neck, and cervical spine. Gently palpate the head and back of the neck for any pain, deformity, tenderness, crepitus, and bleeding (**Step 1**). Ask a responsive patient if he or she feels any pain or tenderness. Check the neck for signs of trauma, swelling, or bleeding. Palpate the neck for subcutaneous emphysema, as well as any abnormal lumps or masses (**Step 2**). In patients where spinal injury is not suspected, you may inspect for pronounced or distended jugular veins with the patient sitting at a 45° angle.
2. **Chest and breath sounds.** Inspect, visualize, and palpate over the chest area for injury or signs of trauma, including bruising, tenderness, or swelling (**Step 3**). Watch for both sides of the chest to rise and fall together with normal breathing. Observe for abnormal breathing signs, including retractions, or paradoxical motion. Feel for crepitus. Palpate the chest for subcutaneous emphysema. Auscultate breath sounds.
3. **Abdomen.** Inspect the abdomen for any obvious injuries, bruising, and bleeding (**Step 4**). Palpate both the front and back of the abdomen, evaluating for tenderness and bleeding.
4. **Pelvis.** Inspect for any obvious signs of injury, bleeding, or deformity (**Step 5**). If the patient reports no pain, gently press downward and inward on the pelvic bones.
5. **Extremities.** Inspect for cuts, bruises, swelling, obvious injuries, and bleeding (**Step 6**). Palpate along each extremity for deformities. Check for pulses and motor and sensory function.
   - **Pulse:** Check the distal pulses on the foot (dorsalis pedis or posterior tibial) and wrist (radial). Also check circulation. Evaluate the skin color in the hands or feet.
   - **Motor function:** Ask the patient to wiggle his or her fingers or toes.
   - **Sensory function:** Evaluate sensory function in the extremity by asking the patient to close his or her eyes. Gently squeeze or pinch a finger or toe, and ask the patient to identify what you are doing.
6. **Posterior body.** Feel the back for tenderness, deformity, and open wounds (**Step 7**). Carefully palpate the spine from the neck to the pelvis for tenderness or deformity, and look under clothing for obvious injuries, including bruising and bleeding.

Performing a Rapid Physical Exam

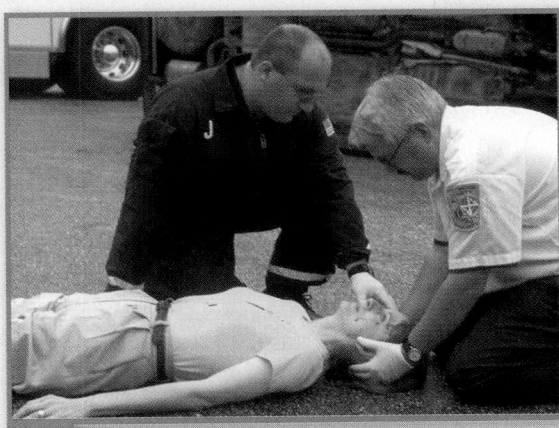

1 Assess the head. Have your partner maintain in-line stabilization.

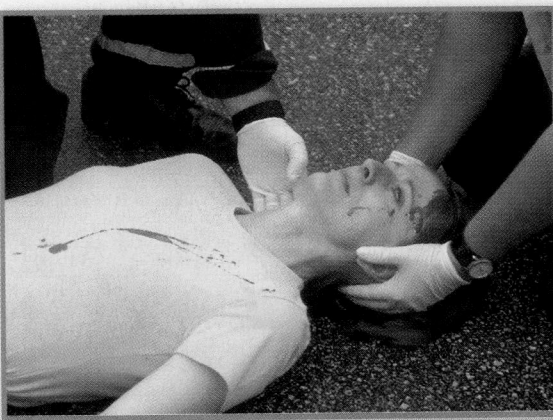

2a Assess the neck.

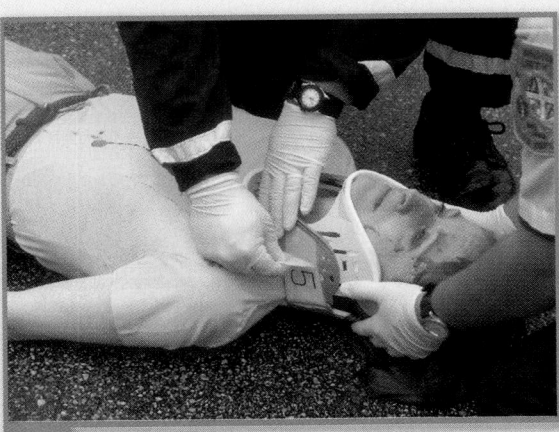

2b Apply a cervical spinal immobilization device on trauma patients.

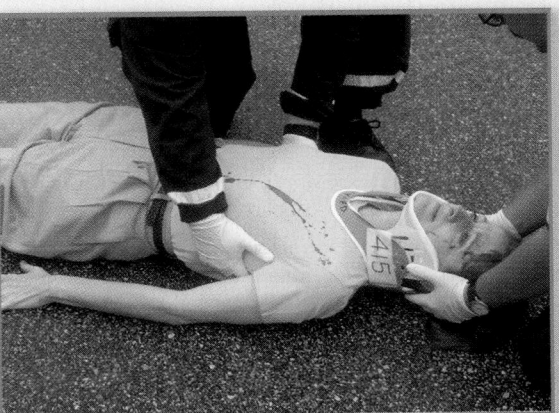

3 Assess the chest. Listen to breath sounds on both sides of the chest.

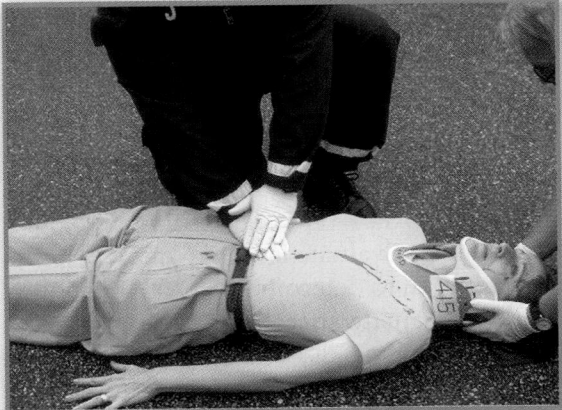

4 Assess the abdomen.

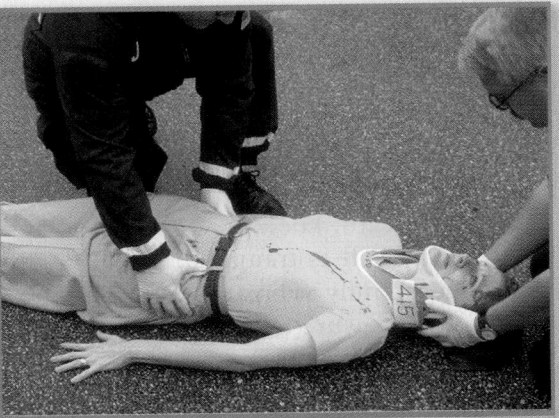

5 Assess the pelvis. If there is no pain, gently compress the pelvis downward and inward to look for tenderness or instability.

Continued.

Skill Drill 8-1

Performing a Rapid Physical Exam continued

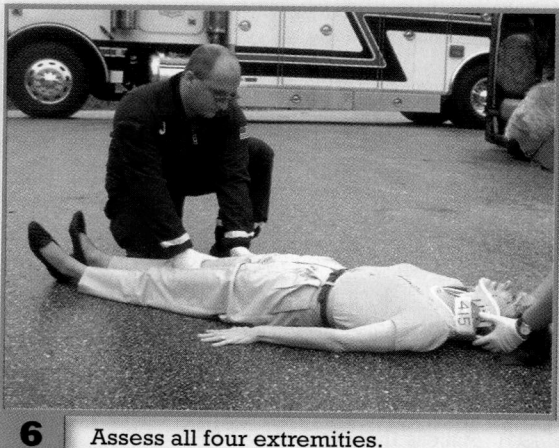

6 Assess all four extremities.
Assess pulse, motor, and sensory function.

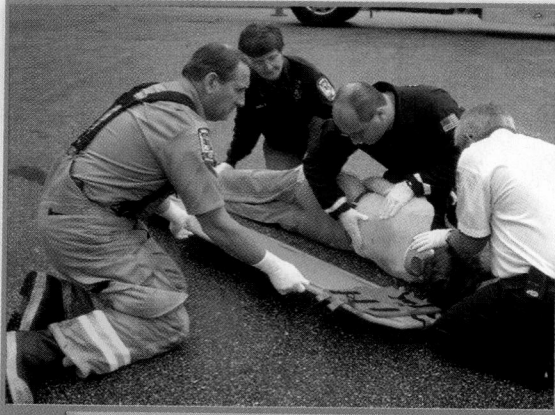

7 Assess the back. In trauma patients, roll the patient in one motion.

Know this page and Exam order

Here are some suggestions for assessing some common chief complaints. Remember that you will also be assessing history and vital signs with each of these.

- **Chest pain:** Look for trauma to the chest and listen for breath sounds. Obtaining a pulse, blood pressure, and respiratory rate and evaluating the skin are good ways to focus on how well the cardiovascular and respiratory systems are functioning.

- **Shortness of breath:** Look for signs of airway obstruction, as well as trauma to the neck or chest. Listen carefully to breath sounds, noting abnormalities. Measure respiratory rate, chest rise and fall (for tidal volume), and effort. Because the location of this complaint is the chest, carefully evaluate pulse, blood pressure, and skin condition.

- **Abdominal pain:** Look for trauma to the abdomen or distention. Palpate the abdomen for tenderness, rigidity, and patient guarding.

- **Any pain associated with bones or joints:** Expose the site and evaluate the pulse, motor, and sensory function adjacent to and below the affected area. Assess range of motion. This should be done by asking the patient how much he or she can move the extremity or joint. Never force a painful joint to move.

- **Dizziness:** Evaluate level of consciousness and orientation to determine the patient's ability to think. Evaluate speech for clarity. Inspect the head for trauma. Pulse, blood pressure, and skin changes may indicate hypoperfusion of the brain.

Focused Physical Exam

1 Gently palpate the head for any pain, deformity, tenderness, crepitus, and bleeding.

Ask a responsive patient if he or she feels any pain or tenderness.

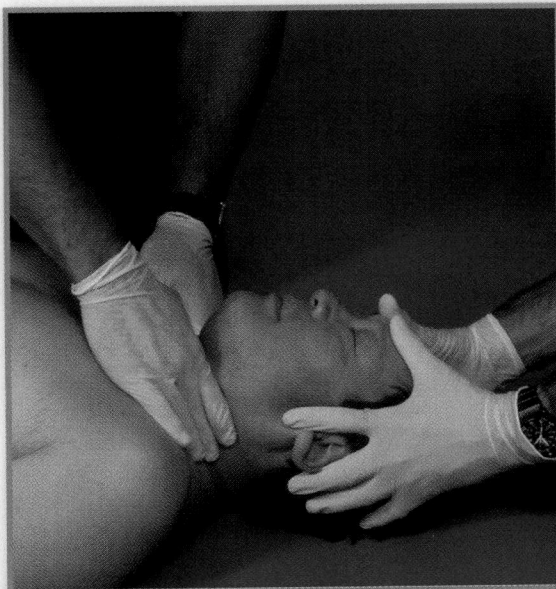

2 Gently palpate the back of the neck.

Ask a responsive patient if he or she feels any pain or tenderness.

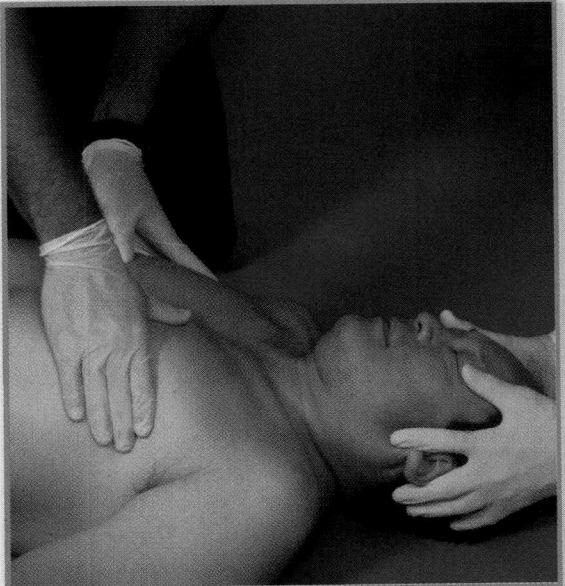

3 Inspect, visualize, and palpate over the chest area for injury or signs of trauma.

Auscultate breath sounds.

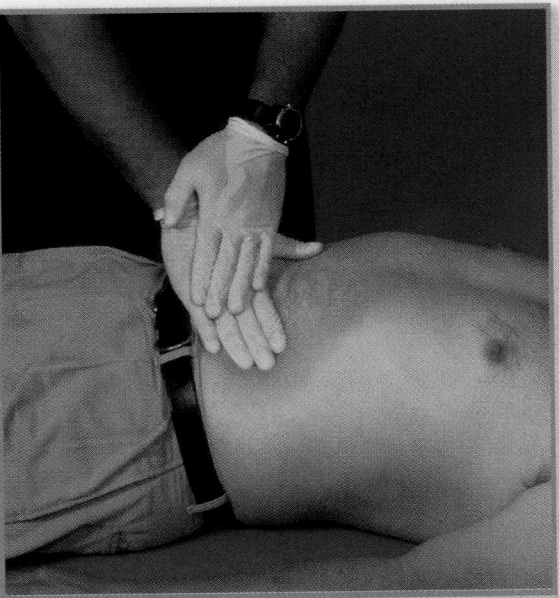

4 Palpate the abdomen, evaluating for tenderness and bleeding.

Continued.

Focused Physical Exam continued

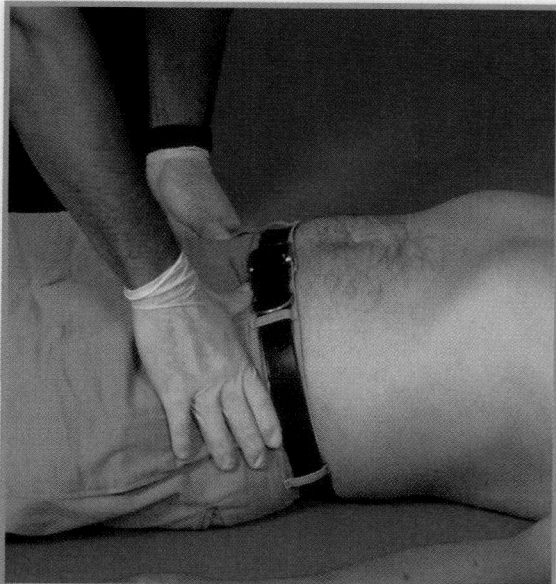

5 Inspect the pelvis for any obvious signs of injury, bleeding, or deformity.

If the patient reports no pain, gently press downward and inward on the pelvic bones.

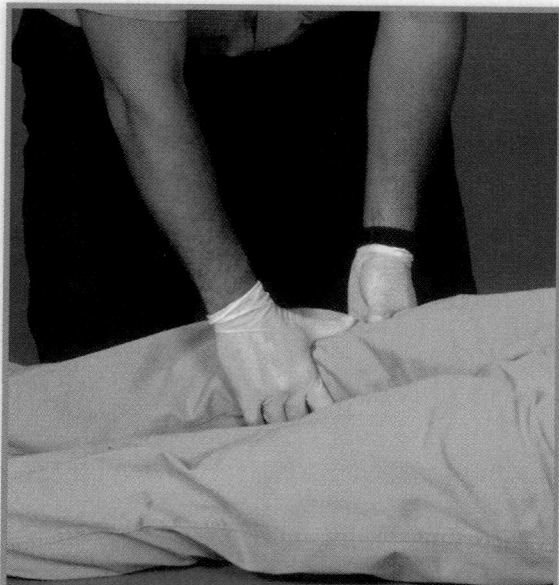

6 Inspect the extremities for cuts, bruises, swelling, obvious injuries, and bleeding.

Palpate along each extremity for deformities.

Check for pulses and motor and sensory function.

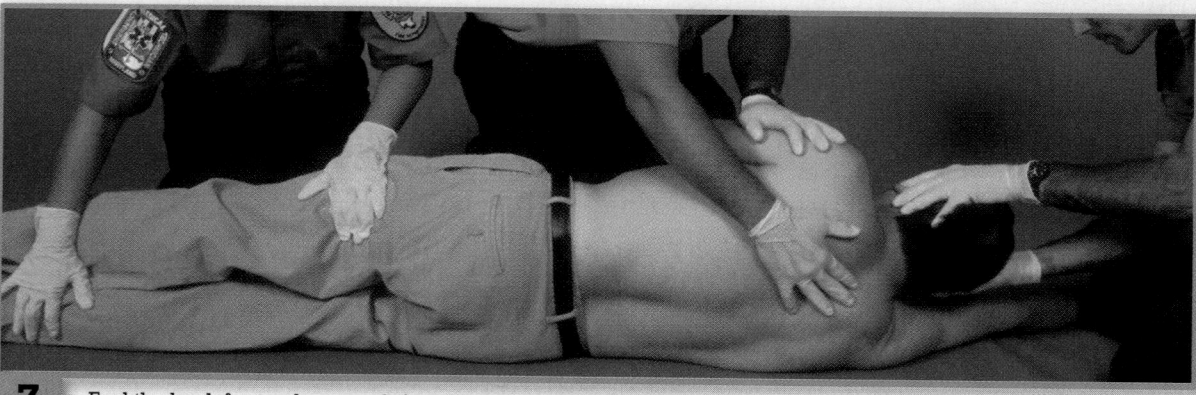

7 Feel the back for tenderness, deformity, and open wounds.

Carefully palpate the spine from the neck to the pelvis for tenderness or deformity.

Look under clothing for obvious injuries, including bruising and bleeding.

Physical Exam Techniques

The type of physical exam you perform is based on the needs of your patient but many of the following assessment techniques may be used:

- **Inspection.** Inspection is simply looking at your patient for abnormalities. This is done by looking for anything that may indicate a problem. For example, swelling in a lower extremity may indicate an acute injury or a chronic illness.
- **Palpation.** Palpation describes the process of touching or feeling the patient for abnormalities. At times palpation is gentle and at other times it is firmer and will help you to identify what hurts.
- **Auscultation.** Auscultation is the process of listening to sounds the body makes by using a stethoscope. For example, when measuring a patient's blood pressure, you listen to flow of blood against the brachial artery with the head of the stethoscope. This is auscultation of a blood pressure.

DCAP-BTLS will help remind you what to look for when inspecting and palpating various body regions.

An integral part of your physical exam is to compare findings on one side of the body to the other side when possible. For example, if one ankle appears swollen, look at the other. If one shoulder feels "out of joint" feel the other one to compare. When listening to breath sounds, listen to both sides of the chest. On some occasions it may be helpful to use your nose in your physical exam. Odors can indicate anything from infections, to certain medical conditions, to scene safety threats.

The following are guidelines on how and what to assess during a physical exam. There may be times when you assess all of these areas quickly (rapid medical or trauma exam). There may be other times when you assess only one or two areas, but in great detail (focused medical or trauma exam).

Head, Neck, and Cervical Spine

Inspect for abnormalities of the head, neck, and cervical spine. Gently palpate the head and back of the neck for any pain, deformity, tenderness, crepitus, and bleeding. Crepitus is the grating or grinding that is often felt or heard when two ends of a broken bone rub together. Ask a responsive patient is he or she feels any pain or tenderness. Next, check the neck for signs of trauma, swelling, or bleeding. Palpate the neck for signs of trauma, such as deformities, bumps, swelling, bruising, or bleeding as well as a crackling sound produced by air bubbles under the skin, also known as subcutaneous emphysema. It is particularly important to assess the neck before covering it with a cervical collar. Also, in patients where spinal injury is not suspected, inspect for pronounced or distended jugular veins with the patient sitting at a 45° angle. This is a normal finding in a person who is lying down; however, jugular venous distention in the patient who is sitting up suggests a problem with blood returning to the heart. Report and record your findings carefully. Do not move on to the next step until you are sure that the airway is secure and you have initiated or continued spinal immobilization.

Chest and Breath Sounds

Next, inspect, visualize, and palpate over the chest area for injury or signs of trauma, including bruising, tenderness, or swelling. Watch for both sides of the chest to rise and fall together with normal breathing. Observe for abnormal breathing signs, including retractions (when the skin pulls in around the ribs during inspiration) or paradoxical motion (when only one section of the chest rises on inspiration while another area of the chest falls).

Retractions indicate the patient has some condition, usually medical, that is impairing the flow of air into and out of the lungs. Paradoxical motion is associated with a fracture of several ribs (flail), causing a section of the chest to move independently from the rest of the chest wall. Feel for grating of the bones as the patient breathes. Crepitus is often associated with rib fractures. Palpate the chest for subcutaneous emphysema, especially in cases of severe blunt chest trauma.

If the patient reports difficulty breathing or has evidence of trauma to the chest, auscultate breath sounds. This helps you to evaluate air movement in and out of the lungs. To auscultate, you need a stethoscope. Make sure you place the ear pieces facing forward in your ears. The position of the patient will determine the way you proceed to check for breathing. Here's how and where to listen.

- First, remember that you can almost always hear a patient's breath sounds better from the patient's back. So if the patient's back is accessible, listen there. If you have immobilized the patient or if the patient is in a supine position, listen from the front (Figure 8-12 ▶).
- Auscultate over the upper lungs (apices), the lower lungs (bases), and over the major airways (midclavicular and midaxillary lines).

Focused History and Physical Exam

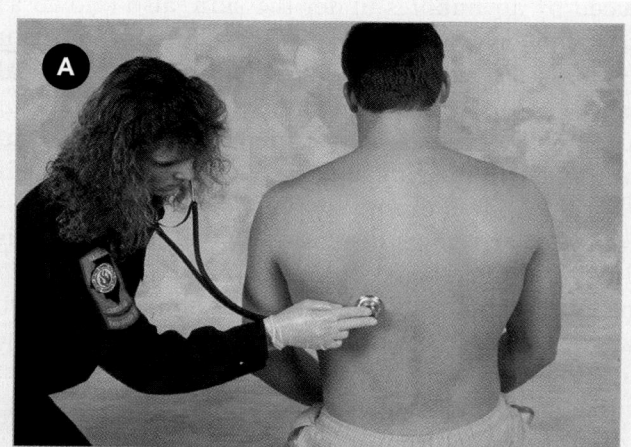

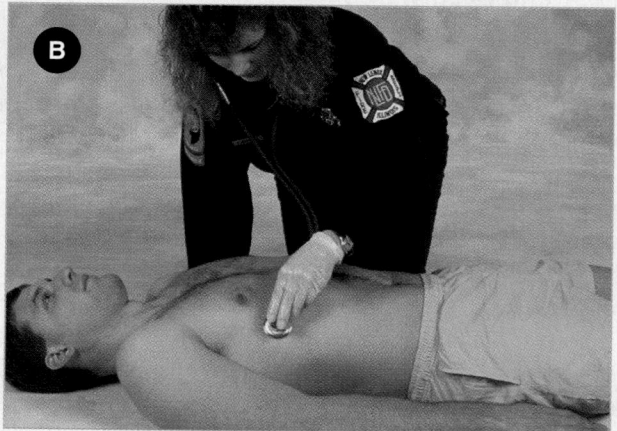

Figure 8-12 A. Listen to breath sounds from the patient's back if possible, over the apices, the bases, and the major airways. **B.** If the patient is immobilized or in a supine position, listen from the front.

- Lift the clothing or slide the stethoscope under the clothing. When you listen over clothing, you will primarily hear the sound of the stethoscope sliding over the fabric because breath sounds are muted by clothing.
- Place the diaphragm of the stethoscope firmly against the skin to hear the breath sounds.

What are you listening for? The goal is to hear and document the presence or absence of breath sounds in the three regions described. It is important to compare one side to the other. If you believe the breathing is abnormal, reassess breathing, and then ensure that the patient is receiving oxygen and, if appropriate, assisted with ventilations.

Abdomen

Inspect the abdomen for any obvious injuries, bruising, and bleeding. Be sure to palpate both the front and back of the abdomen, evaluating for tenderness and

bleeding. As you palpate the abdomen, use the terms "firm," "soft," "tender," or "distended" (swollen) to report your findings. If the patient is awake and alert, ask about pain as you perform the exam. Do not palpate obvious soft-tissue injuries, and be careful not to palpate too firmly.

Pelvis

Inspect for any obvious signs of injury, bleeding, or deformity. If the patient reports no pain, gently press downward and inward on the pelvic bones. Do not rock the pelvis; this motion may result in motion of an unstable spine. If you feel any movement or crepitus, or the patient reports pain or tenderness, severe injury may be present. Injuries to the pelvis and surrounding abdomen may bleed profusely, so continue to monitor the patient's skin color and vital signs, and be sure to give supplemental oxygen to minimize the effects of shock.

Extremities

Inspect for cuts, bruises, swelling, obvious injuries, and bleeding. Next, palpate along each extremity for deformities. Ask the patient about any tenderness or pain. As you evaluate the extremities, check for pulses and motor and sensory function:

- **Pulse:** Check the distal pulses on the foot (dorsalis pedis or posterior tibial) and wrist (radial). Also check circulation. Evaluate the skin color in the hands or feet. Is it normal? How does it compare with the skin color of the other extremities? Pale or cyanotic skin may indicate poor circulation in that extremity.
- **Motor function:** Ask the patient to wiggle his or her fingers or toes. An inability to move a single extremity can be the result of a bone, muscle, or nerve injury. An inability to move several extremities may be a sign of a brain abnormality or spinal cord injury. Verify that spinal precautions have been taken.
- **Sensory function:** Evaluate sensory function in the extremity by asking the patient to close his or her eyes. Gently squeeze or pinch a finger or toe, and ask the patient to identify what you are doing. The inability to feel sensation in the extremity may indicate a local nerve injury. Inability to feel in several extremities may be a sign of a spinal cord injury. Ensure that you have begun and/or are maintaining spinal immobilization.

Posterior Body

Feel the back for tenderness, deformity, and open wounds. If you are placing the patient on a backboard, it is particularly important that you check the back before you log roll the patient, and before you place him or her onto a backboard. Keep the spine in line at all times as you log roll the patient onto his or her side. Carefully palpate the spine from the neck to the pelvis for tenderness or deformity, and look under the patient's clothing for obvious injuries, including bruising and bleeding.

Steps in a Focused History and Physical Exam

At this point in the assessment, the focused history and physical exam guides you to take actions that will stabilize or relieve the patient's problems. It will always have three components: baseline vital signs, SAMPLE history, and a rapid or focused physical exam. The physical exam will tell you what is happening outside the body, the vital signs will tell you what is happening inside the body, and the history will help make sense out of the two by guiding your assessment and treatment. The order in which you will perform these three components will depend on whether your patient is a trauma or a medical patient. The order also depends on the type of trauma or the type of medical patient you encounter.

The next four sections describe how to perform the focused history and physical exam on four types of patients: trauma with significant MOI; trauma without significant MOI; responsive medical patients; and unresponsive medical patients.

Patient Assessment

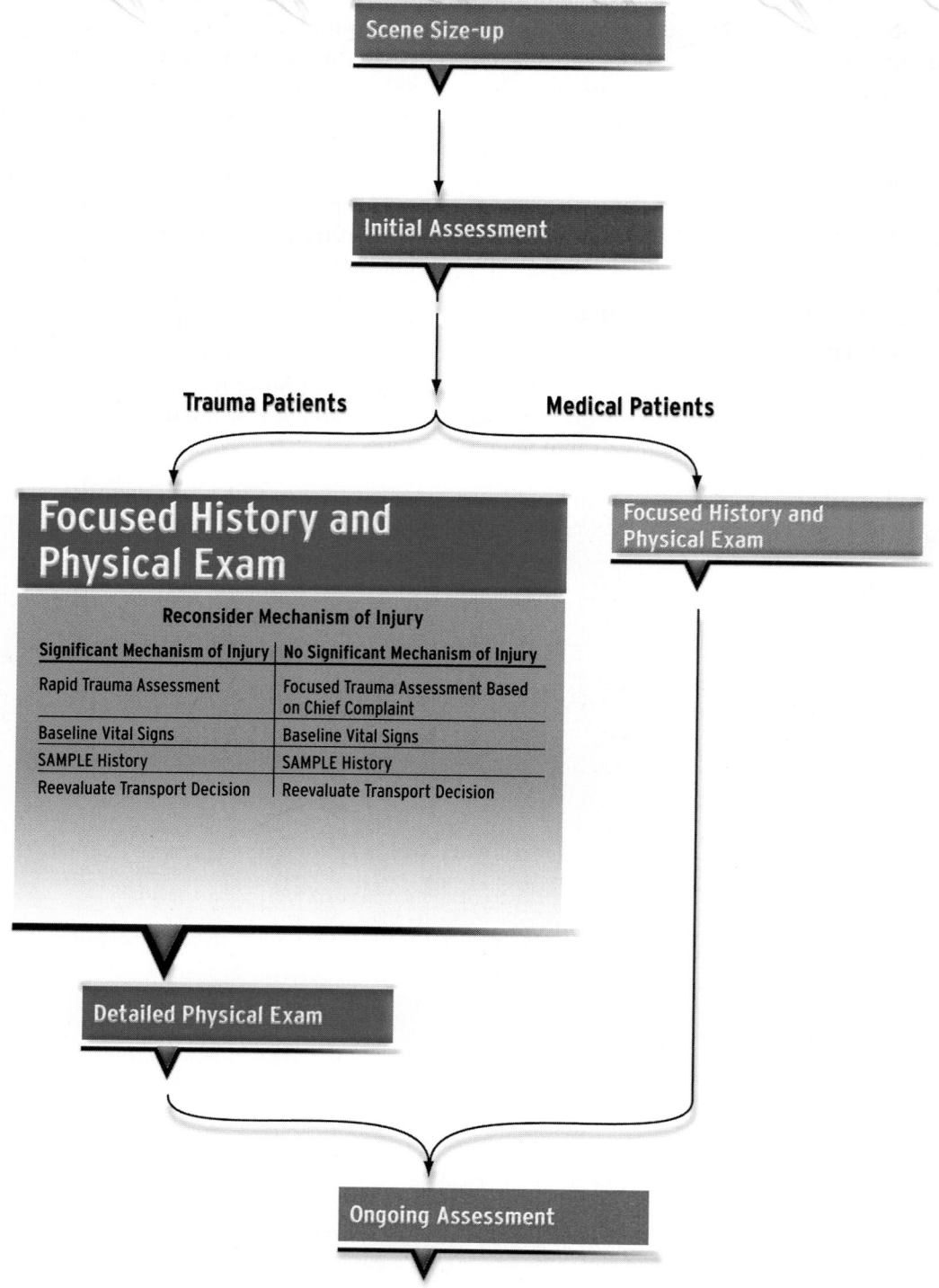

Focused History and Physical Exam

Trauma Patients With a Significant MOI

At this point in the assessment process, you should reconsider the MOI to ensure that you have not missed important information. Understanding the MOI helps you to understand the potential severity of the patient's problem and provide valuable information to hospital staff. Remember that significant mechanisms of injury for adults and children may include the following:

- Ejection from a vehicle
- Death of another occupant of the vehicle
- Fall greater than 15' to 20' or three times the patient's height
- Vehicle rollover
- High-speed vehicle collision
- Vehicle-pedestrian collision
- Motorcycle or bicycle crash
- Unresponsive or altered mental status following trauma
- Penetrating trauma to the head, chest, or abdomen

Rapid Trauma Assessment

In a trauma patient with a significant MOI, you should begin with a rapid trauma assessment. Taking 60 to 90 seconds to identify both hidden and obvious injuries will help you in two ways. First, you can identify and then treat hidden life threats that were not apparent in the initial assessment. Second, you will know better how to prepare your patient for packaging and rapid transport. Remember, the rapid physical exam tells you what is happening outside the body. Review Skill Drill 8-1 for the steps in a rapid physical exam.

Baseline Vital Signs

After the rapid trauma exam is complete, obtain your baseline vital signs. A good baseline set of vital signs will be useful as you continue to monitor changes in the patient's condition. These may be obtained in the ambulance if rapid transport is necessary.

SAMPLE History

In the trauma patient with a significant MOI, the patient's history is not as critical as performing a rapid physical exam or obtaining vital signs; how-

ever, it should not be ignored. Many of these patients are conscious and able to provide some history. A SAMPLE history should be obtained in case the patient becomes unresponsive and is unable to provide the emergency department with this important information. If your patient is unresponsive, continue to gather history from witnesses, bystanders, or from the environment. This information may provide important clues for the emergency department physician.

Reevaluate Transport Decision

If transport is not yet under way, consider transporting the patient at this time.

Trauma Patients With No Significant MOI Study This!!!

Focused Trauma Assessment Based on Chief Complaint

After evaluating the MOI of your trauma patient, you determine the patient has sustained only minor trauma—for example, a twisted ankle or a laceration on the arm. In this case, a focused physical exam guiding you to the specific injury would be appropriate. If your patient has multiple complaints—for example, neck pain, a twisted ankle, and a laceration on the arm—you may want to perform a focused exam on each of these areas. Also suspect other injuries.

Baseline Vital Signs

After evaluating each of the patient's complaints, obtain the patient's pulse, respirations, and blood pressure and assess the patient's pupils and skin, including capillary refill time. These vital signs will serve as a baseline to evaluate changes during transport.

SAMPLE History

A SAMPLE history should be gathered to determine whether a medical problem may have caused the trauma. The mnemonic OPQRST is used to evaluate conditions such as chest pain and headaches; however, it may also be used to evaluate ankle or shoulder pain or pain related to trauma.

Reevaluate Transport Decision

If transport is not yet under way, consider transporting the patient at this time.

Focused History and Physical Exam—Trauma

Patient Assessment

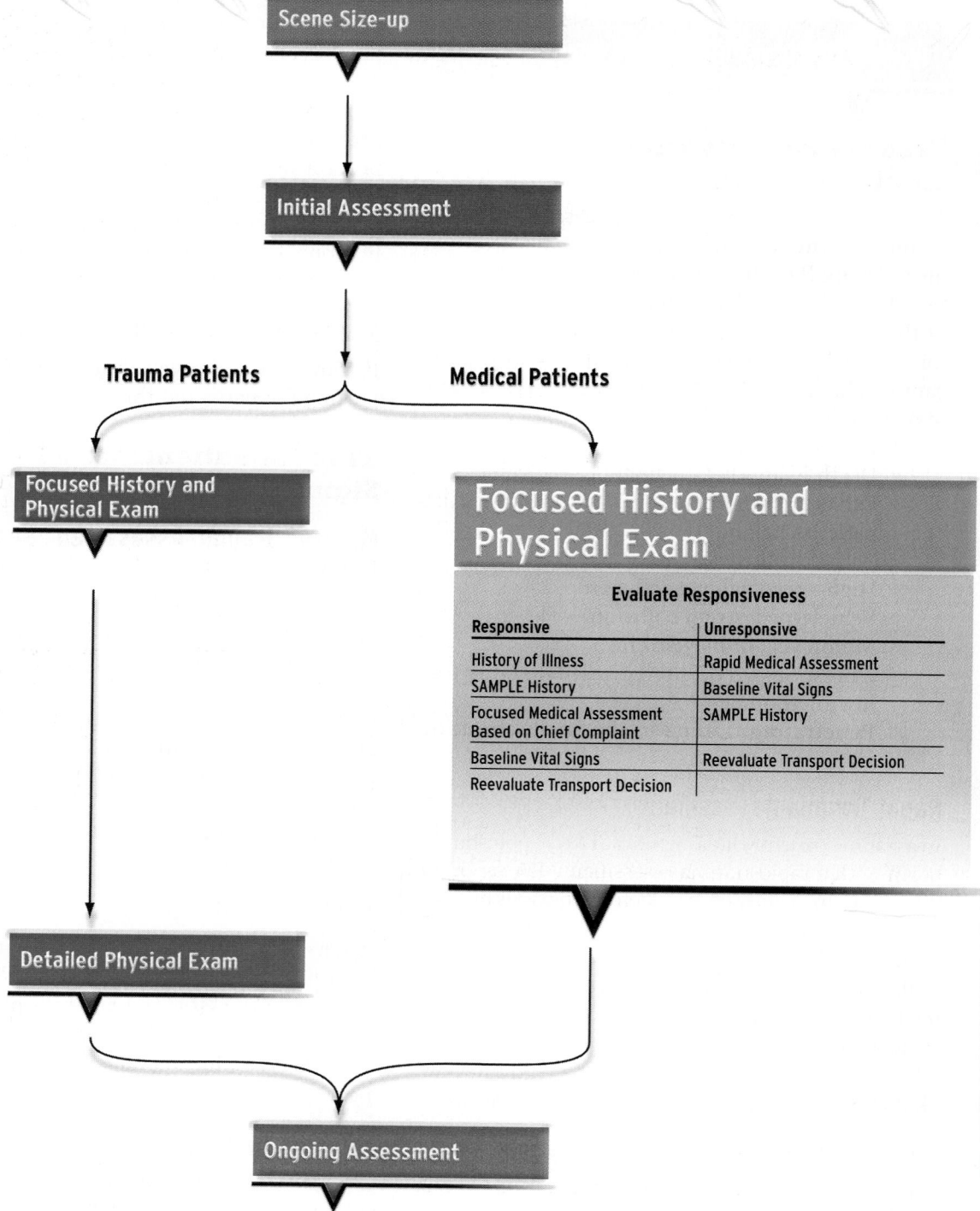

Focused History and Physical Exam

Medical Patients Who Are Responsive

History of Illness

The patient's response to your questions about the chief complaint drives your assessment of the history of the present illness (focused history) and physical exam in the medical patient (Figure 8-13 ▼). If possible, take time to sit down and help the patient get comfortable. Now is the time to listen to develop an increased understanding of the patient's condition. Be careful not to jump to conclusions regarding the chief complaint because of what you have seen or heard about the patient. In many cases, the chief complaint may not be obvious; it may even be different than what the dispatcher reported. When this occurs, stay flexible. Assess and treat the patient's problem rather than simply responding to the dispatch report. Nevertheless, the chief complaint will help you focus your history and physical exam. If the patient cannot tell you what is wrong, perhaps because of a language barrier, altered mental status, or severe respiratory distress, you may ask for the patient's history from a family member or bystander or from your observations of the scene and patient actions. However, remember that information from the patient is far more valuable. You should try, whenever possible, to speak directly to the patient.

SAMPLE History

Evaluate as many signs and symptoms as possible in your SAMPLE history. For example a 50-year-old man with chest pain and dizziness may be having a heart attack. The same person with chest pain and a cough rather than dizziness may be having an asthma attack. The more signs and symptoms you are able to obtain, the better. As you listen to the patient, you might want to make some brief notes to aid your memory and assist with documentation after the call. You should attempt to record the chief complaint in a few of the patient's own words. Be sure to note if your information comes from someone other than the patient.

Focused Medical Assessment Based on Chief Complaint

Now that you have evaluated the chief complaint using OPQRST, and have obtained a thorough SAMPLE history, you should perform a focused

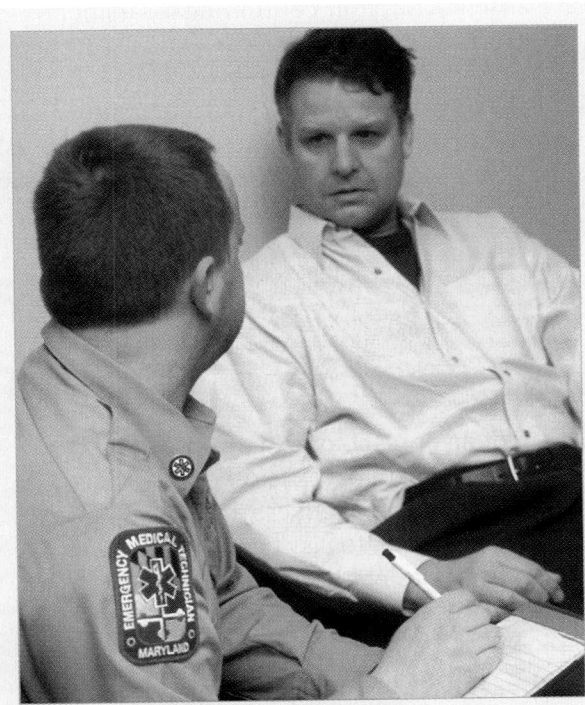

Figure 8-13 The patient's initial response to the question "What's wrong?" is the chief complaint.

Documentation Tips

Your documentation of any pain complaint should include a description in the patient's words, as well as your findings from the other OPQRST questions. Record all pain complaints in detail. Not all pain symptoms are "classic"; the exact description may help hospital personnel make the diagnosis in a case that is not typical.

medical exam. The key to this exam is to emphasize the priorities you have learned during the history. Be logical and investigate problems that you identified during the initial assessment and history. As discussed earlier, you can focus on the region of the problem or the physiological system involved. For example, if the patient's history suggests a heart attack, you may also want to check the patient's chest for indications of trauma and listen to lung sounds.

Baseline Vital Signs

Although vital signs are obtained last in the focused history and physical exam of a responsive medical patient, they are important to establish a baseline for how your patient is compensating with regard to his chief complaint. In problems related to the cardiovascular system and respiratory system, the vital signs will also be a part of your focused physical exam.

Reevaluate Transport Decision

If transport is not yet under way, consider transporting the patient at this time.

Medical Patients Who Are Unresponsive No Difference from RTA

Rapid Medical Assessment

You have just finished the initial assessment on an unresponsive medical patient and have begun treatment for any life-threatening problems. Because an unresponsive patient is unable to provide information, you should perform a rapid medical exam. In 60 to 90 seconds assess

Figure 8-14 If the patient is unresponsive, try to obtain a pertinent history or patient information from family or bystanders.

the patient from head to toe looking for problems and possible life threats that may be hidden.

Baseline Vital Signs

After performing a rapid physical exam, you should evaluate the patient's vital signs to determine if the person is tolerating the unresponsive state well and establish a baseline for your continuing assessment.

SAMPLE History

While packaging the patient for transport, gather what history you can from family, witnesses, and bystanders (Figure 8-14 ▲). Remember, the environment may

You are the Provider   Part 4

En route you contact the hospital and provide a radio report about an unresponsive man suspected of being intoxicated and possibly having tuberculosis. You then perform a detailed physical exam looking for more clues as to the man's current problem. You notice the man has some redness in the conjunctiva of both eyes, but no trauma to the head or scalp. While listening to his chest you hear decreased breath sounds on the right lower part of the chest and crackles on the left lower part of the chest. You notice on his shirt what may be some bloody sputum. His abdomen is soft but appears mildly distended. The pelvis is intact with no evidence of trauma to the extremities or back. Pulses are present in all extremities.

8. Will the patient's injuries and problems always be clear?
9. If the cause of this patient's unconsciousness is unclear, what would you do?

Know This!!!

EMT-B Tips

Common Chief Complaints and Focused Physical Exams

- **Chest pain.** Evaluate skin, pulse, and blood pressure. Look for evidence of trauma to the chest, listen to breath sounds, and palpate the chest.
- **Abdominal pain.** Evaluate skin, pulse, and blood pressure. Look for trauma to the abdomen, and palpate the abdomen to identify any tender spots.
- **Shortness of breath.** Evaluate skin, pulse, blood pressure, and rate and depth of respirations. Assess for airway obstruction. Listen carefully to breath sounds.
- **Dizziness.** Evaluate skin, pulse, blood pressure, and adequacy of respirations. Monitor the level of consciousness and orientation carefully. Check the head for signs of trauma.
- **Any pain associated with bones or joints.** Evaluate skin, pulse, movement, and sensation adjacent and distal to the affected area.

provide important clues as to the patient's condition. For example, drug paraphernalia including syringes may indicate an overdose has occurred. A medical identification device (eg, MedicAlert Tag, Global Med Net Card, Vial of Life Container) may also provide important medical history. Patient medication labels may also be used to help determine the patient's medical condition.

Reevaluate Transport Decision

If transport is not yet under way, consider transporting the patient at this time.

Tying It Together

The focused history and physical exam is the most complex step in the assessment process. Your examination is based on whether your patient has experienced trauma or become ill. It is also based on whether the trauma is significant and whether the patient is responsive or unresponsive. All patients, however, regardless of the type of injury, require the following exam: a focused or rapid physical exam, baseline vital signs, and SAMPLE history.

Focused History and Physical Exam—Medical

Patient Assessment

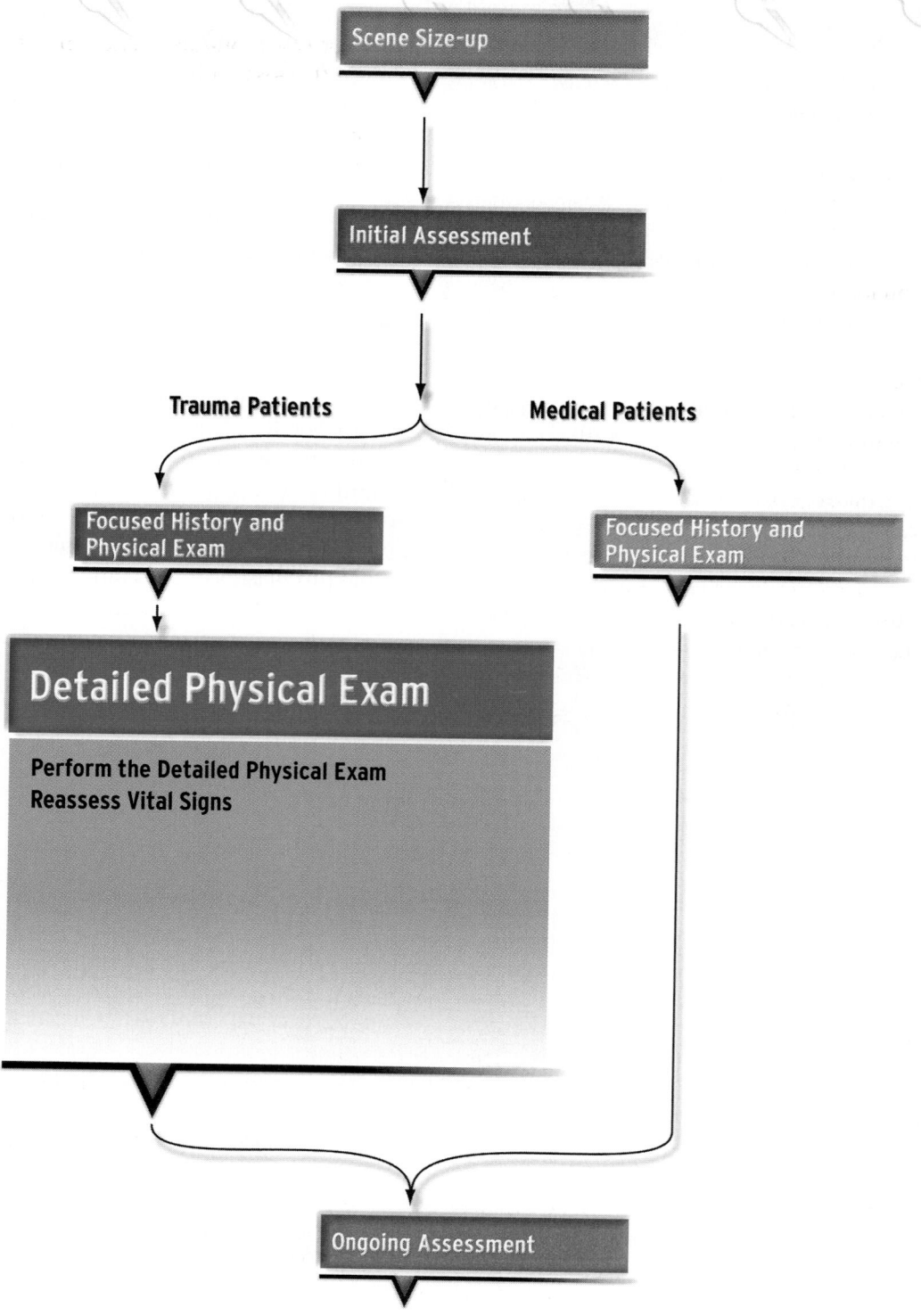

Scene Size-up

Initial Assessment

Trauma Patients

Medical Patients

Focused History and Physical Exam

Focused History and Physical Exam

Detailed Physical Exam

Perform the Detailed Physical Exam
Reassess Vital Signs

Ongoing Assessment

Detailed Physical Exam

Recall that the assessment process began with anticipation and hazard preparation when you received the dispatch information and performed the scene size-up. Then you performed the initial assessment, in which you identified and treated life-threatening conditions. If trauma was a factor, you also initiated spinal immobilization. You also provided transport if your patient had an obvious life-threatening condition. When indicated, you followed up on the initial assessment by gathering history, taking at least one set of vital signs, and performing either a rapid or focused physical exam based on the patient's chief complaint.

At this point, in most cases you are already en route to the hospital. If you are still on the scene, it is because your patient does not have a life-threatening condition and you have not found the cause for the patient's complaints. In the case of a trauma patient with a significant mechanism of injury, you are en route but may still have unanswered questions. In either case, this is the time to perform the detailed physical exam. The goal of this exam is to further define problems that were identified in the focused history and physical exam and to possibly identify the cause of complaints that were not identified during the focused history and physical exam. In most cases, it is the trauma patient with a significant mechanism of injury who receives the detailed exam, because getting the patient to a hospital took priority over performing a detailed exam on scene. The detailed exam can help provide a better understanding of your medical patients as well but should never delay transport, and is also most often performed en route to the hospital.

To achieve the goals of the <u>detailed physical exam</u>, you must simply ask and answer one question: "What additional problems can be identified through a detailed physical exam?" The detailed physical exam will provide you with more information about the nature of the patient's problem. Depending on what is learned, you should be prepared to do the following: (Know the 5 bullets)

- Return to the initial assessment if a potentially life-threatening condition is identified. (This is unlikely this late in the exam, but it

is always possible. Remember, stay focused on the ABCs.)
- Perform spinal immobilization if neck or back pain or abnormality in sensation or movement is identified. (Again, this is unlikely this late in the exam.)
- Modify any treatment that is under way on the basis of any new information.
- Provide treatment for problems that were identified during the detailed exam.
- Provide transport to an appropriate facility, or call for ALS backup.

The detailed physical exam is a more in-depth examination that builds on the focused history and physical exam portion of your assessment. The patient and the particular problem will determine the need for this exam. Many of your patients will not receive a detailed physical exam, either because it will be irrelevant or unnecessary or because it is not possible given the time constraints.

Most patients have isolated problems that can be adequately evaluated earlier in the assessment process. You will identify the problem and treat it, making a more detailed physical exam of the entire body unnecessary. If you do perform a detailed physical exam in these patients, it will be to further explore what you learned during the focused history and physical exam portion of your assessment.

A few patients will have life-threatening conditions that were identified during the initial assessment. You may spend all of your time with these patients, stabilizing ABCs, which means you will never have a chance to perform much of a focused history and physical exam, let alone a detailed assessment.

You will perform a detailed exam only on stable patients with problems that cannot be identified earlier in the patient assessment process. In some cases, this portion identifies only minor, obscure, or isolated problems, which is why you did not identify them earlier. Regardless of the exact situation, the detailed physical exam is usually performed en route to the hospital in order to save time.

Perform the Detailed Physical Exam

Here, organized by body region, are some additional assessments that you might want to perform during the detailed exam. As you evaluate each region, inspect and palpate to find evidence of injury, again using the mnemonic "DCAP-BTLS." Follow the steps in [Skill Drill 8-3 ▶]: *Know these Steps!!! Palpate all the facial structures*

1. Look at the face for obvious lacerations, bruises, or deformities (**Step 1**).
2. Inspect the area around the eyes and eyelids (**Step 2**).
3. Examine the eyes for redness and for contact lenses. Assess the pupils using a penlight (**Step 3**).
4. Look behind the patient's ear to assess for bruising (Battle's sign) (**Step 4**).
5. Use the penlight to look for drainage of spinal fluid or blood in the ears (**Step 5**).
6. Look for bruising and lacerations about the head. Palpate for tenderness, depressions of the skull, and deformities (**Step 6**).
7. Palpate the zygomas for tenderness or instability (**Step 7**).
8. Palpate the maxillae (**Step 8**).
9. Palpate the mandible (**Step 9**).
10. Assess the mouth and nose for cyanosis, foreign bodies (including loose teeth or dentures), bleeding, lacerations, or deformities (**Step 10**).
11. Check for unusual odors on the patient's breath (**Step 11**).
12. Look at the neck for obvious lacerations, bruises, and deformities (**Step 12**).
13. Palpate the front and the back of the neck for tenderness and deformity (**Step 13**).
14. Look for distended jugular veins. Note that distended neck veins are not necessarily significant in a patient who is lying down (**Step 14**).
15. Look at the chest for obvious signs of injury before you begin palpation. Be sure to watch for movement of the chest with respirations (**Step 15**).
16. Gently palpate over the ribs to elicit tenderness. Avoid pressing over obvious bruises or fractures (**Step 16**).
17. Listen for breath sounds over the midaxillary and midclavicular lines (**Step 17**).
18. Listen also at the bases and apices of the lungs (**Step 18**).
19. Look at the abdomen and pelvis for obvious lacerations, bruises, and deformities (**Step 19**).
20. Gently palpate the abdomen for tenderness. If the abdomen is unusually tense, you should describe the abdomen as rigid (**Step 20**).
21. Gently compress the pelvis from the sides to assess for tenderness (**Step 21**).
22. Gently press the iliac crests to elicit instability, tenderness, or crepitus (**Step 22**).
23. Inspect all four extremities for lacerations, bruises, swelling, deformities, and medic alert anklets or bracelets. Also assess distal pulses and motor and sensory function in all extremities (**Step 23**).
24. Assess the back for tenderness or deformities. Remember, if you suspect a spinal cord injury, use spinal precautions as you log roll the patient (**Step 24**).

Head, Neck, and Cervical Spine

A more detailed exam of these areas could include a careful examination of the head, face, scalp, ears, eyes, nose, and mouth for abrasions, lacerations, and contusions. Examine the eyes and eyelids, checking for redness and for contact lenses. Use a penlight to determine whether the pupils are equal and reactive, and look for any fluid drainage or blood, particularly around the ears and nose. Also check for foreign objects and/or blood in the anterior chamber of the eye. Look for bruising or discoloration around the eyes (raccoon eyes) or behind the ears (Battle's sign); these signs may be associated with head trauma.

Next, palpate gently but firmly around the face, scalp, eyes, ears, and nose for tenderness, deformity, or instability. Tenderness or abnormal movement of bones often signals a serious injury and the patient may be at risk for upper airway obstruction. Monitor the airway carefully in these patients. Next, look inside the mouth. Loose or broken teeth or a foreign object may block the airway. It's much safer if a bite block is used. You should also look for lacerations, swelling, bleeding, and any discoloration in the mouth and the tongue. Smell the patient's breath. Any unusual odors, such as a strong alcohol odor or fruity breath odor, should be reported and recorded.

Palpate the front and back of the neck for tenderness and deformity. The sensation of crackling or popping, not unlike palpating the bubbles in bubble-pack packing material, is called subcutaneous emphysema; it indicates that air is leaking into the space under the skin. Usually, this means that the patient has a pneumothorax or has damaged the larynx or trachea. Also look for distended jugular veins. This is normal in a patient who is lying down; however, their presence in the patient who is sitting up may suggest some type of failure.

Chest

Throughout the patient assessment process, you should monitor the patient's breathing. If you have not already done so, you should carefully palpate the patient's chest.

8-3

Skill Drill

Performing the Detailed Physical Exam

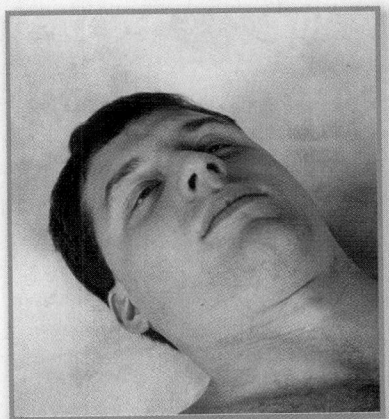

1 Observe the face.

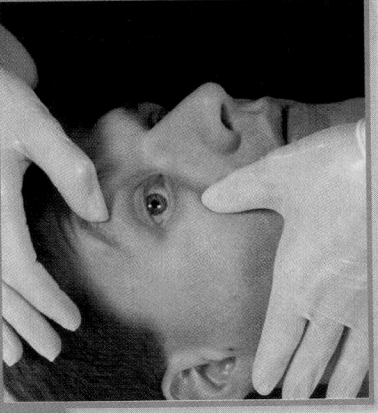

2 Inspect the area around the eyes and eyelids.

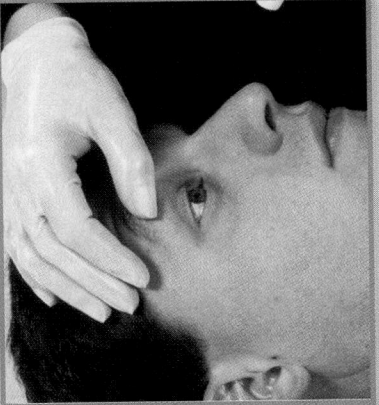

3 Examine the eyes for redness and contact lenses. Check pupil function.

4 Look behind the ears for Battle's sign.

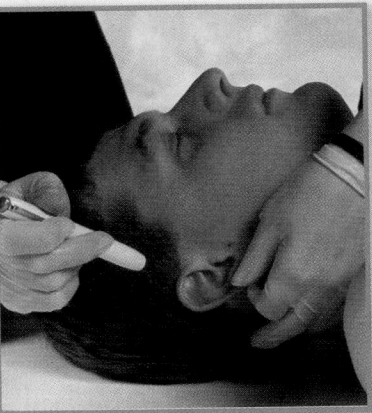

5 Check the ears for drainage or blood.

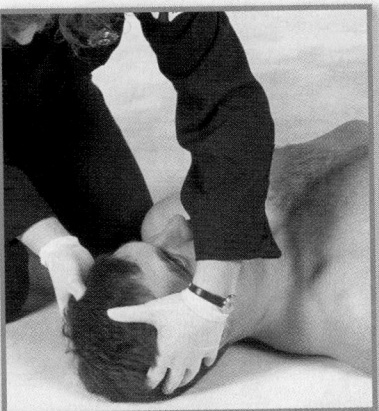

6 Observe and palpate the head.

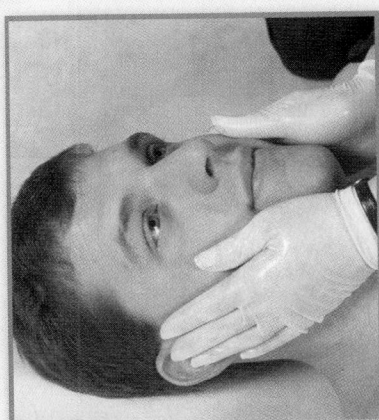

7 Palpate the zygomas.

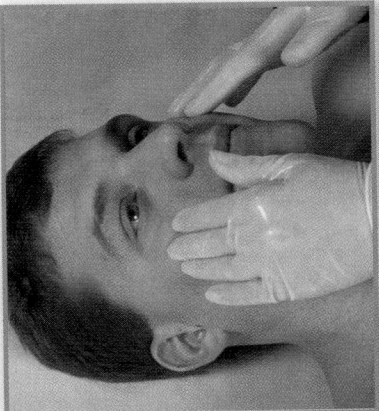

8 Palpate the maxillae.

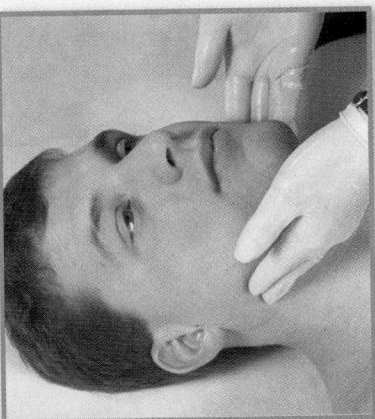

9 Palpate the mandible.

Continued.

Skill Drill 8-3

Performing the Detailed Physical Exam continued

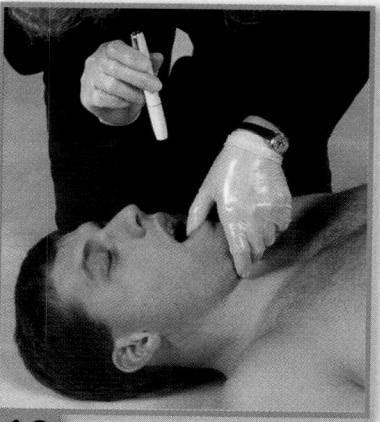

10 Assess the mouth and nose.

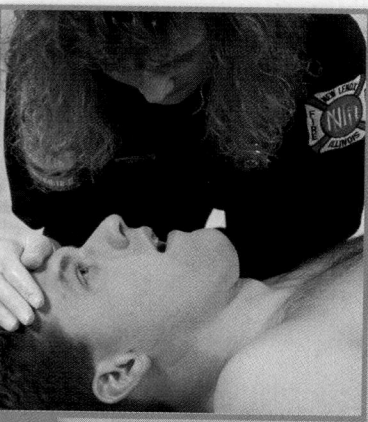

11 Check for unusual breath odors.

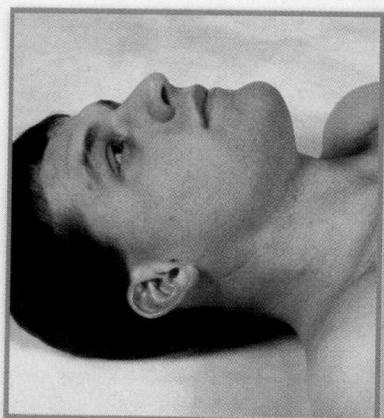

12 Inspect the neck.

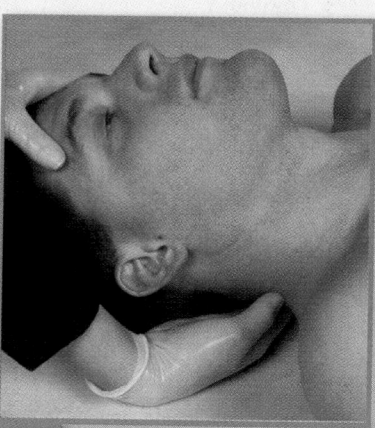

13 Palpate the front and back of the neck.

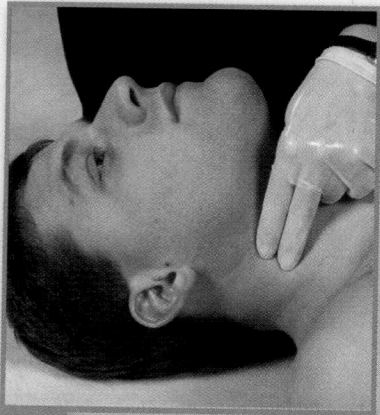

14 Observe for jugular vein distention.

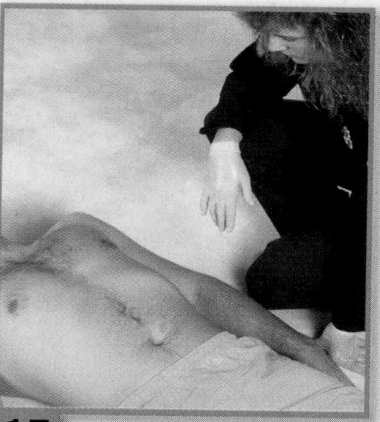

15 Inspect the chest and observe breathing motion.

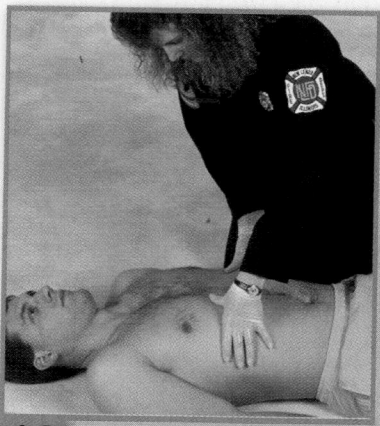

16 Gently palpate over the ribs.

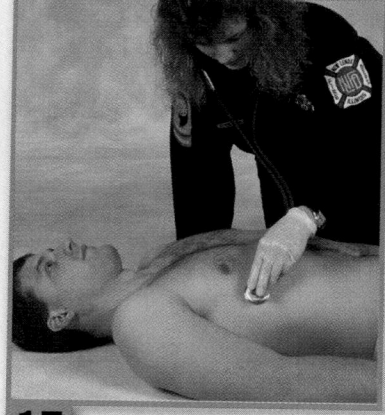

17 Listen to anterior breath sounds (midaxillary, midclavicular).

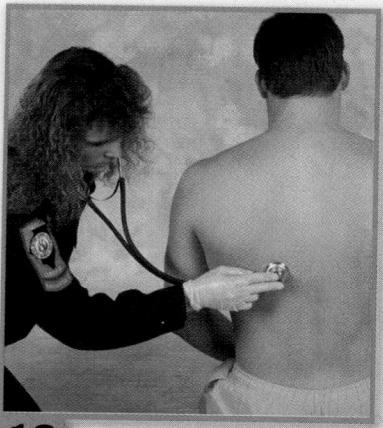

18 Listen to posterior breath sounds (bases, apices).

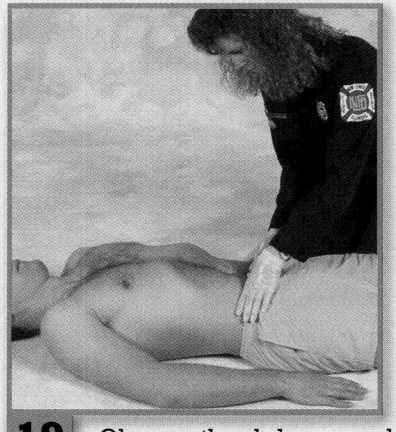

19 Observe the abdomen and pelvis.

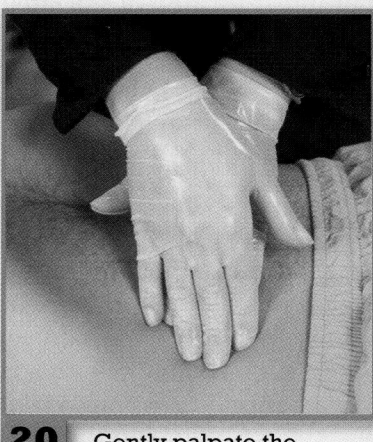

20 Gently palpate the abdomen.

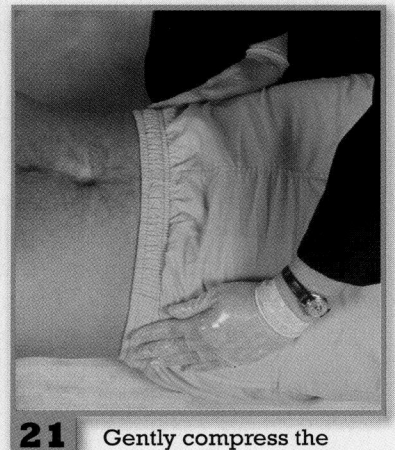

21 Gently compress the pelvis from the sides.

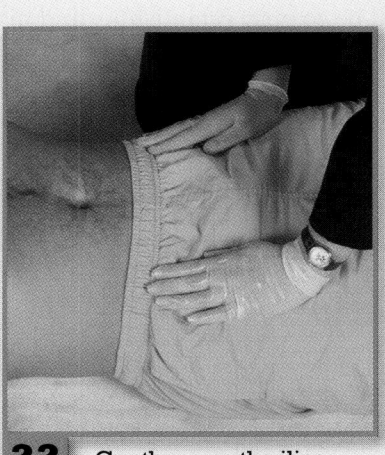

22 Gently press the iliac crests.

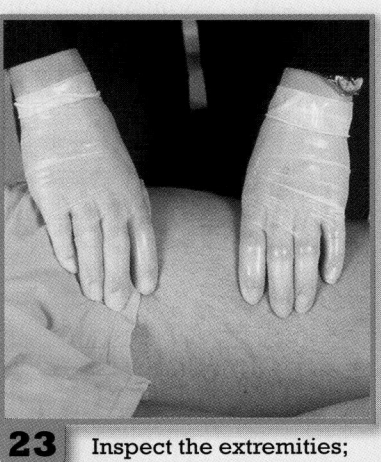

23 Inspect the extremities; assess distal circulation and motor sensory function.

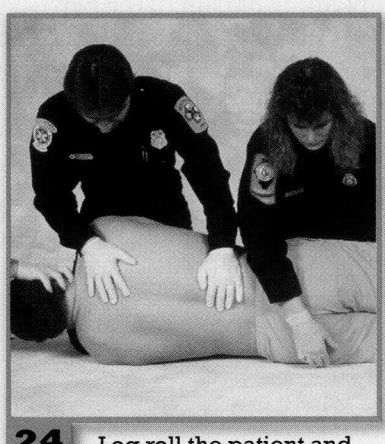

24 Log roll the patient and inspect the back.

Feel for crepitus, as this occurs with a ruptured airway, pneumothorax, or rib fractures. Also evaluate the movement of the chest wall during breathing. Paradoxical motion of the chest wall means that your patient has a flail chest and might need supplemental oxygen and/or assisted ventilation. You might also wish to perform a more detailed evaluation of the patient's breath sounds. Listening to the lungs at the apices, at the midclavicular lines bilaterally, at the bases, and at the midaxillary lines bilaterally, check for the specific sounds of breathing. You may be able to identify one of the following:

- Normal breath sounds. These are clear and quiet on both inspiration and expiration.

- Wheezing breath sounds. These suggest an obstruction of the lower airways. Wheezing is a high-pitched squeal that is most prominent on expiration.

- Wet breath sounds. These may indicate cardiac failure. A moist crackling, usually on both inspiration and expiration, is called <u>rales</u>, or crackles.

- Congested breath sounds. These may suggest the presence of mucus in the lungs. Expect to hear a low-pitched, noisy sound that is most prominent on expiration. This sound may be referred to as <u>rhonchi</u>. The patient often reports a productive cough associated with this sound.

- A crowing sound. This is often heard without a stethoscope and may indicate that the patient has an airway obstruction in the neck or upper part of the chest. Expect to hear a brassy, crowing sound that is most prominent on expiration. This sound may be referred to as stridor. ←

Remember This ⟍

Abdomen

During the detailed physical exam, you may perform a more complete examination of the abdomen. As you palpate the abdomen, use the terms firm, soft, tender, or distended (swollen) to report your findings. Some patients may actively contract their abdominal muscles when you palpate them. This is known as guarding.

Pelvis

If you have not previously identified any pelvic injury, recheck the pelvis to identify problems. If the patient is not complaining of pain in the pelvis, gently press in and down on the pelvis to assess for pain, tenderness, instability, and crepitus; all may indicate a fractured pelvis and the potential for shock.

Extremities

If you have not already done so, you should carefully evaluate the extremities for any signs of trauma, again using the DCAP-BTLS method. You should also evaluate the distal circulation, sensation, and movement. If you have already identified an injury, regular evaluation of the circulation, sensation, and movement below the injury will allow you to be sure that the injury has not compromised neurovascular status.

Back

During the rapid assessment, if performed, you should have visualized and palpated the patient's back for signs of trauma, especially near the spine. You must use spinal precautions when rolling a patient for assessment of back injuries. The presence of spinal deformity or pain suggests that—if you have not already done so—the patient requires spinal immobilization. Look for and document any other conditions that you find on the back.

Reassess Vital Signs

Sometimes, you will be so busy establishing and maintaining the ABCs that you will not have a chance to obtain the patient's vital signs. However, it is important to obtain and record a set of baseline vitals at some time during your patient encounter. If you have not assessed the vital signs, now is the time.

You are the Provider

Part 5

Prior to arrival to the hospital you reevaluate the patient's vital signs and note they appear unchanged significantly from your baseline set. The patient is becoming more responsive by responding to painful stimulus when you apply nail bed pressure with your pen. He continues to tolerate the nasal airway without problem; however, he does have an occasional cough.

10. How would you record this lack of change in the patient's condition and vital signs?
11. How often should you assess vital signs on this patient?
12. What additional assessment would you conduct over the next few minutes until arrival at the hospital?

Detailed
Physical Exam

Patient Assessment

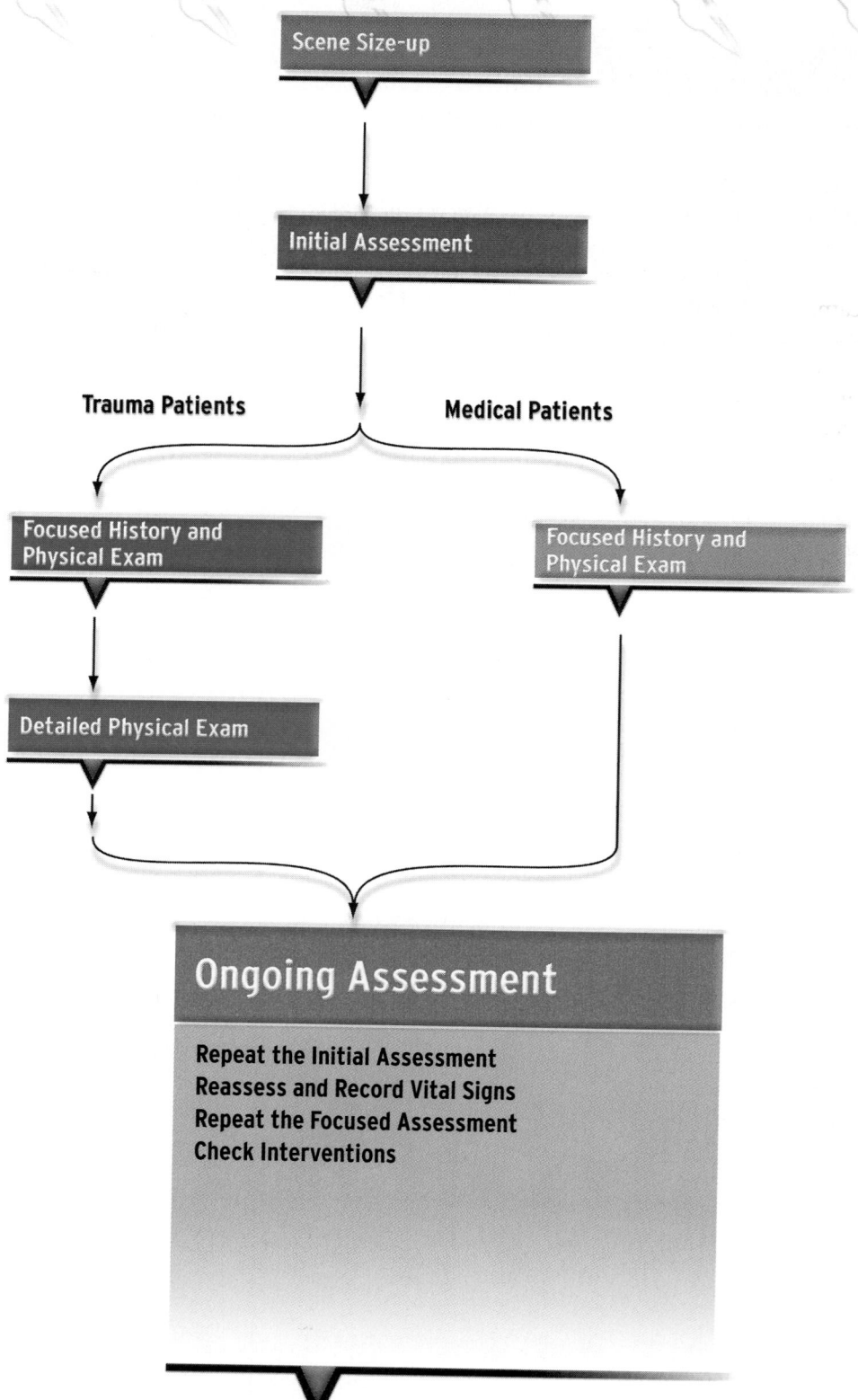

Ongoing Assessment

Ongoing Assessment

Unlike the detailed physical exam, the <u>ongoing assessment</u> is performed on all patients during transport. Its purpose is to ask and answer the following questions: *Know These*

- Is treatment improving the patient's condition?
- Has an already identified problem gotten better? Worse?
- What is the nature of any newly identified problems?

The ongoing assessment helps you to monitor changes in the patient's condition. If the changes are improvements, simply continue whatever treatment you are providing. However, in some instances, the patient's condition will deteriorate. When this happens, you should be prepared to modify treatment as appropriate and then begin new treatment on the basis of the problem identified.

The procedure for the ongoing assessment is simply to repeat the initial assessment and the focused assessment and to check the intervention steps that pertain to the problems you are treating. These steps should be repeated and recorded every 15 minutes for a <u>stable</u> patient and every 5 minutes for an unstable patient (Figure 8-15 ▶). Remember to use your judgment when timing the ongoing assessments. Some patients may require more frequent assessments.

The steps of the ongoing assessment are as follows: *Know these steps !!*

1. Repeat the initial assessment.
   - Reassess mental status.
   - Maintain an open airway.
   - Monitor the patient's breathing.
   - Reassess pulse rate and quality.
   - Monitor skin color and temperature.
   - Reestablish patient priorities.
2. Reassess and record vital signs.
3. Repeat your focused assessment regarding patient complaint or injuries, including questions about the patient's history.
4. Check interventions.
   - Ensure adequacy of oxygen delivery/artificial ventilation.

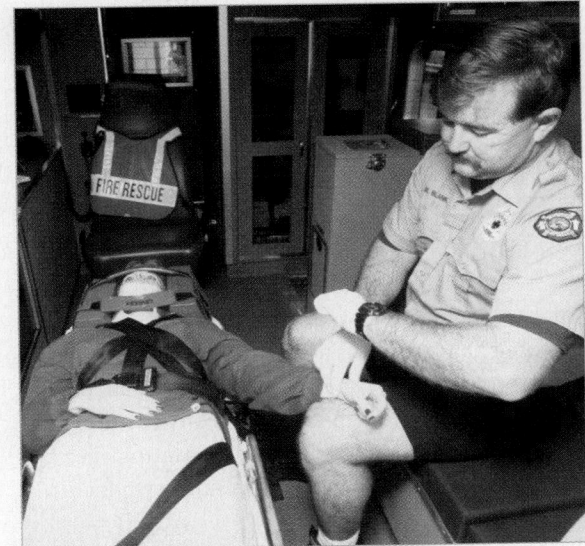

Figure 8-15 During the ongoing assessment, repeat your initial assessment, recheck vital signs, and recheck interventions every 5 minutes if the patient is unstable and every 15 minutes if the patient is stable.

- Ensure management of bleeding.
- Ensure adequacy of other interventions.

Repeat the Initial Assessment

The first step is to repeat the initial assessment. If you have been treating the ABCs, you need to continue monitoring these essential functions. It is particularly important to reassess mental status; changes can be initially subtle and then rapidly decline.

Reevaluate any problems that you have been treating. Reassess the patient's skin color, wound, or anything for which you have begun treatment. If the patient's condition remains stable, great. But, you may discover a need to change a dressing, tighten a strap, or turn up the oxygen. Do it now.

Reassess and Record Vital Signs

Be sure that the patient's vital signs have not changed. Record these so that your documentation is accurate and complete. If the vital signs have changed, evaluate what may have happened and apply the appropriate interventions.

✾ EMT-B Tips

You may notice at times that experienced EMT-Bs and paramedics seem to have a "sixth sense" when it comes to some patients. They seem to be able to recognize severe problems even before they have completed their initial assessment. This clinical intuition is one of the hallmarks of the expert EMT that you can develop as you progress through your career—if you pay attention and remain thorough and diligent in your assessments. The concept of a "sixth sense" is distinct from the concept of "winging it," however. You should be certain that your patient impressions and assessments are reflective of appropriately thorough and structured evaluations as opposed to hastily formed impressions based on limited and disorganized information. The latter will cause you to miss important conditions and form incorrect impressions.

The aspects of clinical intuition include the following:

- The ability to recognize patterns. Intuitive EMTs immediately recognize clinical patterns that they have seen before. For example, you may immediately recognize that a pale, diaphoretic patient looks like another patient you have seen in severe shock.
- Common sense understanding. Experienced, intuitive EMTs who use their knowledge and experience also use common sense in their assessment and treatment. For instance, they refrain from starting CPR on a responsive patient even though they cannot feel a pulse.
- The ability to sense what is important. Good EMTs know how to track problems that are truly important. They avoid the tendency to get lost in unimportant problems and stay focused on the ones that truly matter. For instance, a forehead laceration will not distract them from the serious cardiac problem that caused a fall.
- Deliberate rationally. Good EMTs use their intuition to help them make decisions, but they always temper it by asking, "What if I'm wrong?" For example, a patient who was in a serious motor vehicle crash but has no complaints and appears normal is probably fine. An experienced EMT-B might believe that the patient has no serious injuries and probably does not require spinal immobilization. However, the answer to the question "What if I'm wrong?" could be "Disastrous: permanent spinal cord injury and paralysis." The experienced EMT-B decides to immobilize.

How can you improve your own intuitive powers? First, you need some experience. The truth is that you will not become intuitive until you have been involved in patient care for some time and have evaluated and treated a number of patients.

By following a systematic approach to patient assessment, using your intuition and common sense, and carefully listening to the patient, you will learn to make good decisions about the treatment and transport of the many patients you will care for in your career.

Repeat the Focused Assessment

As you transport your patient, remember to ask the patient about the chief complaint. Is the chest pain getting better or worse? Is leg pain improving with treatment or staying about the same? If you previously asked the patient to rate symptoms on a 1 to 10 scale, ask the patient for an updated rating for comparison.

Check Interventions

Reevaluate any interventions you started. Take a moment to make certain that the airway is still open, the bleeding has been controlled, the oxygen is still flowing, and the backboard straps are still tight. Things often change in the uncontrolled prehospital environment, so this is a good time to be sure that your treatments are still "working" the way you intended.

You are the Provider

Summary

This was a challenging patient because of the lack of information provided by the patient. Although this patient appeared to be a medical patient and not a trauma patient, it is often difficult to tell. He may have been hit by a car, or he may have fallen and hit his head or received a fracture. Also, any history obtained from a patient with an altered level of consciousness should be considered unreliable. Patients do not have to be exclusively trauma or medical.

In cases where the patient cannot tell you what is wrong, you must look for a chief complaint such as obvious bleeding or deformity (DCAP-BTLS). Unconsciousness, although not a medical problem itself, can be a chief complaint. Your focused physical exam, detailed physical exam, and ongoing assessment should help you identify its cause. Your detailed physical exam in this case would look primarily at the central nervous system, but the digestive system and circulatory system should be evaluated also. This means that your detailed physical exam will need to extend to several areas of the body. Remember, do not delay transport of your patient to complete all of the patient assessment. Most of the assessment will be performed during transport.

Ongoing Assessment

Prep Kit

Ready for Review

- The assessment process begins with the scene size-up, which identifies real or potential hazards. The patient should not be approached until these hazards have been dealt with in a way that eliminates or minimizes risk to both the EMTs and the patient(s).

- The initial assessment is performed on all patients. It identifies any life-threatening conditions to the airway, breathing, and circulation (ABCs). Any life threats identified must be treated before moving on to the next step of the assessment.

- The focused history and physical exam includes vital signs, patient history, and a physical exam. The type of physical exam, rapid or focused, and the order of these three components depends on if your patient is a trauma or a medical patient.

- The detailed physical exam is performed on a select group of patients. It helps you to further understand problems that were identified during the focused exam and may also be used to evaluate problems that cannot be identified using the focused exam. The detailed physical exam should be performed en route to the hospital.

- The ongoing assessment is also performed on all patients. It gives you an opportunity to reevaluate problems that are being treated and to recheck treatments to be sure that they are still being delivered correctly. Information from the ongoing assessment may be used to change treatment plans.

- The assessment process is both systematic and dynamic. All patients will be evaluated by using these same steps. However, because the focused history and physical exam will focus your attention on the patient's major problems, each assessment you perform will be slightly different, depending on the needs of the patient. The result will be a process that will enable you to quickly identify and treat the needs of all patients, both medical and trauma related, in a way that meets their unique needs.

Vital Vocabulary

accessory muscles The secondary muscles of respiration.

AVPU A method of assessing a patient's level of consciousness by determining whether a patient is Awake and alert, responsive to Verbal stimulus or Pain, or Unresponsive; used principally in the initial assessment.

body substance isolation (BSI) An infection control concept and practice that assumes that all body fluids are potentially infectious.

breath sounds An indication of air movement in the lungs, usually assessed with a stethoscope.

capillary refill A test that evaluates distal circulatory system function by squeezing (blanching) blood from an area such as a nail bed and watching the speed of its return after releasing the pressure.

chief complaint The reason a patient called for help; also, the patient's response to general questions such as "What's wrong?" or "What happened?"

coagulate To form a clot to plug an opening in an injured blood vessel and stop bleeding.

conjunctiva The delicate membrane that lines the eyelids and covers the exposed surface of the eye.

crepitus A grating or grinding sensation caused by fractured bone ends or joints rubbing together; also air bubbles under the skin that produce a crackling sound or crinkly feeling.

cyanosis Bluish-gray skin color that is caused by reduced oxygen levels in the blood.

DCAP-BTLS A mnemonic for assessment in which each area of the body is evaluated for Deformities, Contusions, Abrasions, Punctures/Penetrations, Burns, Tenderness, Lacerations, and Swelling.

detailed physical exam The part of the assessment process in which a detailed area-by-area exam is performed on patients whose problems cannot be readily identified or when more specific information is needed about problems identified in the focused history and physical exam.

focused history and physical exam The part of the assessment process in which the patient's major complaints or any problems that are immediately evident are further and more specifically evaluated.

www.EMTB.com

Technology

- Interactivities
- Vocabulary Explorer
- Anatomy Review
- Web Links
- Online Review Manual

frostbite Damage to tissues as the result of exposure to cold; frozen or partially frozen body parts.

general impression The overall initial impression that determines the priority for patient care; based on the patient's surroundings, the mechanism of injury, signs and symptoms, and the chief complaint.

Golden Hour The time from injury to definitive care, during which treatment of shock or traumatic injuries should occur because survival potential is best.

guarding Involuntary muscle contractions (spasm) of the abdominal wall in an effort to protect the inflamed abdomen; a sign of peritonitis.

hypothermia A condition in which the internal body temperature falls below 95°F (35°C) after exposure to a cold environment.

initial assessment The part of the assessment process that helps you to identify any immediately or potentially life-threatening conditions so that you can initiate lifesaving care.

jaundice A yellow skin color that is seen in patients with liver disease or dysfunction.

mechanism of injury (MOI) The way in which traumatic injuries occur; the forces that act on the body to cause damage.

nasal flaring Flaring out of the nostrils, indicating that there is an airway obstruction.

nature of illness (NOI) The general type of illness a patient is experiencing.

ongoing assessment The part of the assessment process in which problems are reevaluated and responses to treatment are assessed.

orientation The mental status of a patient as measured by memory of person (name), place (current location), time (current year, month, and approximate date), and event (what happened).

OPQRST The six pain questions: Onset, Provoking factors, Quality, Radiation, Severity, Time.

palpate Examine by touch.

paradoxical motion The motion of the chest wall section that is detached in a flail chest; the motion is exactly the opposite of normal motion during breathing (ie, in during inhalation, out during exhalation).

rales Crackling, rattling, breath sound that signals fluid in the air spaces of the lungs; also called crackles.

responsiveness The way in which a patient responds to external stimuli, including verbal stimuli (sound), tactile stimuli (touch), and painful stimuli.

retractions Movements in which the skin pulls in around the ribs during inspiration.

rhonchi Coarse, low-pitched breath sounds heard in patients with chronic mucus in the upper airways.

SAMPLE history A key brief history of a patient's condition to determine Signs/Symptoms, Allergies, Medications, Pertinent past history, Last oral intake, and Events leading to the illness/injury.

scene size-up A quick assessment of the scene and the surroundings made to provide information about its safety and the mechanism of injury or nature of illness, before you enter and begin patient care.

sclera The white portion of the eye; the tough outer coat that gives protection to the delicate, light-sensitive, inner layer.

stridor A harsh, high-pitched inspiratory sound that is often heard in acute laryngeal (upper airway) obstruction; may sound like crowing and be audible without a stethoscope.

subcutaneous emphysema The presence of air in soft tissues, causing a characteristic crackling sensation on palpation.

triage The process of establishing treatment and transportation priorities according to severity of injury and medical need.

two- to three-word dyspnea A severe breathing problem in which a patient can speak only two to three words at a time without pausing to take a breath.

Points to Ponder

You have been dispatched to a medical call on the east side of the city. The area is known for illegal drug activity, gangs, and the homeless. The patient is in a small house that appears to be rundown. As you walk up the broken wood steps you have to use a flashlight because the house has no power. The house is very dirty and has little furniture. Your patient is an older woman who has been bedridden for several days and is very sick. The patient has been unable to get up to use the bathroom or shower for 3 days. Would the patient's presentation affect how you would complete the patient assessment? What factors may contribute to this patient's condition? Would this patient be treated differently if in an affluent neighborhood?

Issues: Patient Respect, Ethics and Professionalism, Proper Patient Assessment, Possible Neglect or Abuse of Older Patients.

www.EMTB.com

Assessment in Action

You have been dispatched to a report of a man who has fallen from a high building. Due to the type of call you ensure the fire department has been dispatched to assist with any rescue situation. Upon arrival at the scene you meet the fire crew and you are informed that there is only one patient.

The patient fell approximately 24' onto a dirt surface while working construction on a roof. The patient is very pale and is not moving. As you approach, you hear him moaning. You see obvious fractures of his left tibia (lower leg), and blood is flowing from the injured site. You also notice blood flowing from the patient's nose and ears. The fire department first responder reports to you the following patient vital signs: blood pressure, 90/60 mm Hg; pulse rate, 130 beats/min, weak and rapid; and respirations, 24 breaths/min and shallow.

1. The proper body substance isolation precautions for this call would possibly include a gown and:

   A. a face mask and rubber gloves.
   B. gloves and eye protection. ✓
   C. a hard hat and jeans.
   D. steel-toed boots.

2. To better understand what traumatic injuries you would suspect in this patient, you need to determine the patient's:

   A. medical history.
   B. mechanism of injury. ✓
   C. past experiences.
   D. last oral intake.

3. When you first approach the patient, you can quickly determine possible life-threatening injuries by:

   A. asking many detailed questions.
   B. completing an ongoing assessment.
   C. forming a general impression. ✓
   D. performing a scene size-up.

4. The patient does not respond by talking with you and will only respond by moaning every time you touch him. On the basis of the AVPU scale, this patient:

   A. is completely unresponsive.
   B. is alert and orientated.
   C. may be confused.
   D. responds only to painful stimulation. ✓

5. After determining the patient's level of consciousness, you must immediately assess and treat any injuries relating to:

   A. airway, breathing, and circulation. ✓
   B. any past medical problems.
   C. the face and neck.
   D. the arms and legs.

6. On the basis of this patient's general appearance, level of consciousness, pale skin, vital signs, and mechanism of injury, you would identify this patient as:

   A. a challenging priority.
   B. a high priority. ✓
   C. an intermediate priority.
   D. low priority.

7. To meet the goal of the Golden Hour of treatment, you should not spend more than:

   A. 5 minutes on scene.
   B. 10 minutes on scene. ✓
   C. 30 minutes on scene.
   D. 1 hour on scene.

8. After completing the initial assessment on this patient, you should consider a focused history and physical exam consisting of:

   A. a rapid trauma assessment, baseline vital signs, and SAMPLE history. ✓
   B. a quick look at the head and feet.
   C. checking the patient's weight and height.
   D. listening to lung sounds.

Challenging Questions

9. Explain OPQRST as it is applied to the chief complaint of a medical patient.

10. Explain the difference between the rapid trauma assessment and focused trauma assessment.

11. Explain the steps in completing a detailed physical exam.

12. Explain the process of ongoing assessment, including when, why, and how often it would be completed.

Communications and Documentation

Objectives

Cognitive

3-7.1 List the proper methods of initiating and terminating a radio call. (p 325)

3-7.2 State the proper sequence for delivery of patient information. (p 322)

3-7.3 Explain the importance of effective communication of patient information in the verbal report. (p 326)

3-7.4 Identify the essential components of the verbal report. (p 326)

3-7.5 Describe the attributes for increasing effectiveness and efficiency of verbal communications. (p 326)

3-7.6 State legal aspects to consider in verbal communication. (p 326)

3-7.7 Discuss the communication skills that should be used to interact with the patient. (p 327)

3-7.8 Discuss the communication skills that should be used to interact with the family, bystanders, individuals from other agencies while providing patient care and hospital personnel, and the difference between skills used to interact with the patient and those used to interact with others. (p 326, 327)

3-7.9 List the correct radio procedures in the following phases of a typical call:
   ■ To the scene
   ■ At the scene
   ■ To the facility
   ■ At the facility
   ■ To the station
   ■ At the station (p 325)

3-8.1 Explain the components of the written report and list the information that should be included on the written report. (p 334)

3-8.2 Identify the various sections of the written report. (p 332)

3-8.3 Describe what information is required in each section of the prehospital care report and how it should be entered. (p 332, 333)

3-8.4 Define the special considerations concerning patient refusal. (p 336)

3-8.5 Describe the legal implications associated with the written report. (p 333)

3-8.6 Discuss all state and/or local record and reporting requirements. (p 333)

Affective

3-7.10 Explain the rationale for providing efficient and effective radio communications and patient reports. (p 322)

3-8.7 Explain the rationale for patient care documentation. (p 316)

3-8.8 Explain the rationale for the EMS system gathering data. (p 332)

3-8.9 Explain the rationale for using medical terminology correctly. (p 335)

Psychomotor

3-7.11 Perform a simulated, organized, concise radio transmission. (p 325)

3-7.12 Perform an organized, concise patient report that would be given to the staff at a receiving facility. (p 326)

3-7.13 Perform a brief, organized report that would be given to an ALS provider arriving at an incident scene at which the EMT-B was already providing care. (p 332)

3-8.11 Practice completing a prehospital care report. (p 332, 333)

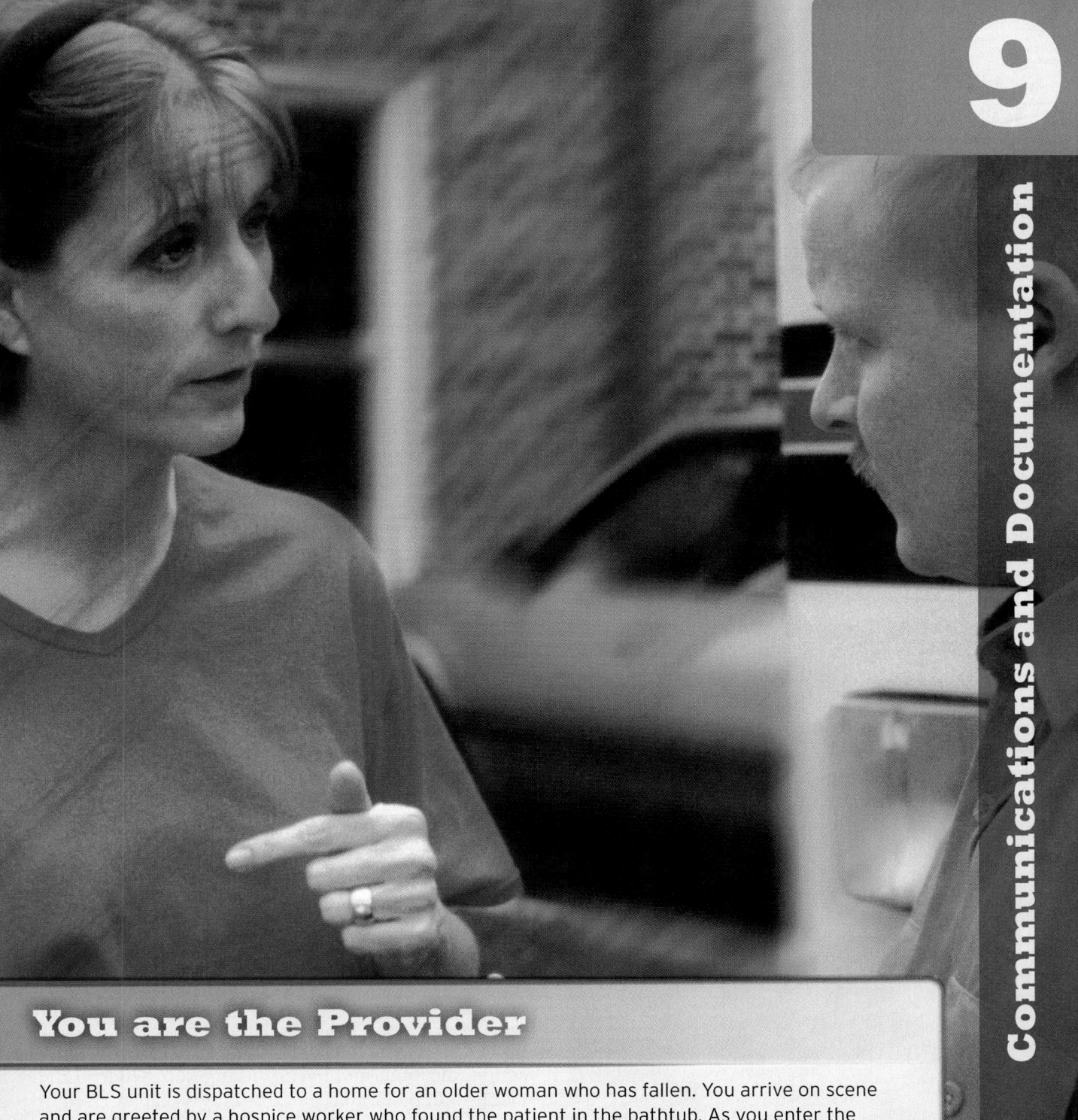

You are the Provider

Your BLS unit is dispatched to a home for an older woman who has fallen. You arrive on scene and are greeted by a hospice worker who found the patient in the bathtub. As you enter the bathroom, the 86-year-old patient is sitting in the bathtub without water; her back is resting on the faucet. She is pale and cold to the touch and complains of having hip pain and back pain and of being very cold. She said she had gotten up around 3 AM to use the restroom, tripped, and fell into the bathtub. She worked herself into the current sitting position and couldn't move after that. She rates her back pain an "8" and her hip pain a "10" of a scale of 1 to 10.

1. What are the primary reasons that a pre-arrival radio report is given to the receiving hospital?
2. Are there advantages/disadvantages to using a narrative-style patient care report rather than a "fill in the bubble"-style format?

Communications and Documentation

Effective communication is an essential component of prehospital care. Radio and telephone communications link you and your team with other members of the EMS, fire, and law enforcement communities. This link helps the entire team to work together more effectively and provides an important layer of safety and protection for each member of the team. You must know what your system can and cannot do, and you must be able to use your system efficiently and effectively. You must be able to send precise, accurate reports about the scene, the patient's condition, and the treatment that you provide.

Verbal communications skills are vitally important for EMT-Bs. Your verbal skills will enable you to gather information from the patient and bystanders. They will also make it possible for you to effectively coordinate the variety of responders who are often present at the scene. Excellent verbal communications are also an integral part of transferring the patient's care to the nurses and physicians at the hospital. You must possess good listening skills to fully understand the nature of the scene and the patient's problem. You must also be able to organize your thoughts to quickly and accurately verbalize instructions to the patient, bystanders, and other responders. Finally, you must be able to organize and summarize the important aspects of the patient's presentation and treatment when reporting to the hospital staff.

A written report is the portion of the EMT's patient care interaction that becomes part of the patient's permanent medical record. It serves many purposes, including demonstrating that the care delivered was appropriate and within the scope and practice of the caregivers involved. Documentation also provides an opportunity to communicate the patient's story to others who may participate in the patient's care in the future. Adequate reporting and accurate records ensure the continuity of patient care. Complete patient records also guarantee proper transfer of responsibility, comply with the requirements of health departments and law enforcement agencies, and fulfill your organization's administrative needs. Reporting and record keeping duties are an essential aspect of patient care, although they are performed only after the patient's condition has been stabilized. Documentation in the field drives both funding and research for EMS. Seat belts are a prime example. Studies gathered from record keeping in the early 1970s showed that patients have a significantly higher survival rate if seatbelts are used during motor vehicle accidents. Armed with this information, laws were passed to enforce seat belt usage, and huge amounts of money were spent on educating the public.

This chapter describes the skills that you need to be an effective communicator. It begins by identifying the kinds of equipment that are used, along with standard radio operating procedures and protocols. Next, the roles of the Federal Communications Commission (FCC) in EMS are described. The chapter concludes with a discussion of a variety of effective methods of verbal communications and guidelines for appropriate written documentation of patient care.

Communications Systems and Equipment

As an EMT-B, you must be familiar with two-way radio communications and have working knowledge of the mobile and hand-held portable radios that are used in your unit. You must also know when to use them and what to say when you are transmitting.

Base Station Radios

The dispatcher usually communicates with field units by transmitting through a fixed radio base station that is controlled from the dispatch center. A <u>base station</u> is any radio hardware containing a transmitter and receiver that is located in a fixed place. The base station may be used in a single place by an operator speaking

www.EMTB.com

Technology

- Interactivities
- Vocabulary Explorer
- Anatomy Review
- Web Links
- Online Review Manual

into a microphone that is connected directly to the equipment. It also works remotely through telephone lines or by radio from a communications center. Base stations may include dispatch centers, fire stations, ambulance bases, or hospitals.

A two-way radio consists of two units: a transmitter and a receiver. Some base stations may have more than one transmitter and/or more than one receiver. They may also be equipped with one multi-channel transmitter and several single channel receivers. A <u>channel</u> is an assigned frequency or frequencies that are used to carry voice and/or data communications. Regardless of the number of transmitters and receivers, they are commonly called base radios or stations. Base stations usually have more power (often 100 watts or more) and higher, more efficient antenna systems than mobile or portable radios. This increased broadcasting range allows the base station operator to communicate with field units and other stations at much greater distances.

The base radio must be physically close to its antenna. Therefore, the actual base station cabinet and hardware are commonly found on the roof of a tall building or at the bottom of an antenna tower. The base station operator may be miles away in a dispatch center or hospital, communicating with the base station radio by dedicated lines or special radio links. A <u>dedicated line</u>, also known as a hot line, is always open or under the control of the individuals at each end. This type of line is immediately "on" as soon as you lift the receiver and cannot be accessed by outside users.

Mobile and Portable Radios

In the ambulance, you will use both mobile and portable radios to communicate with the dispatcher and/or medical control. An ambulance will often have more than one mobile radio, each on a different frequency Figure 9-1 ▶ . One radio may be used to communicate with the dispatcher or other public safety agencies. A second radio is often used for communicating patient information to medical control.

A mobile radio is installed in a vehicle and usually operates at lower power than a base station. Most <u>VHF (very high frequency)</u> mobile radios operate at 100 watts of power. Radios that operate at 800 MHz are increasingly common in EMS systems. <u>UHF (ultra-high frequency)</u> mobile radios usually have only 40 watts of power. Cellular telephones operate on 3 watts of power or less. Mobile antennas are much closer to the ground than base station antennas, so communications from the unit are typically limited to 10 to 15 miles over average terrain.

Figure 9-1 Some ambulances have more than one mobile radio to allow communications with hospitals, mutual aid jurisdictions, and other agencies.

Portable radios are hand-carried or hand-held devices that operate at 1 to 5 watts of power. Because the entire radio can be held in your hand, when in use the antenna is often no higher than the EMT who is using the radio. The transmission range of a portable radio is more limited than that of mobile or base station radios. Portable radios are essential in helping to coordinate EMS activities at the scene of a mass-casualty incident. They are also helpful when you are away from the ambulance and need to communicate with dispatch, another unit, or medical control Figure 9-2 ▼ .

Repeater-Based Systems

A <u>repeater</u> is a special base station radio that receives messages and signals on one frequency and then automatically retransmits them on a second frequency.

Figure 9-2 A portable radio is essential if you need to communicate with the dispatcher or medical control when you are away from the ambulance.

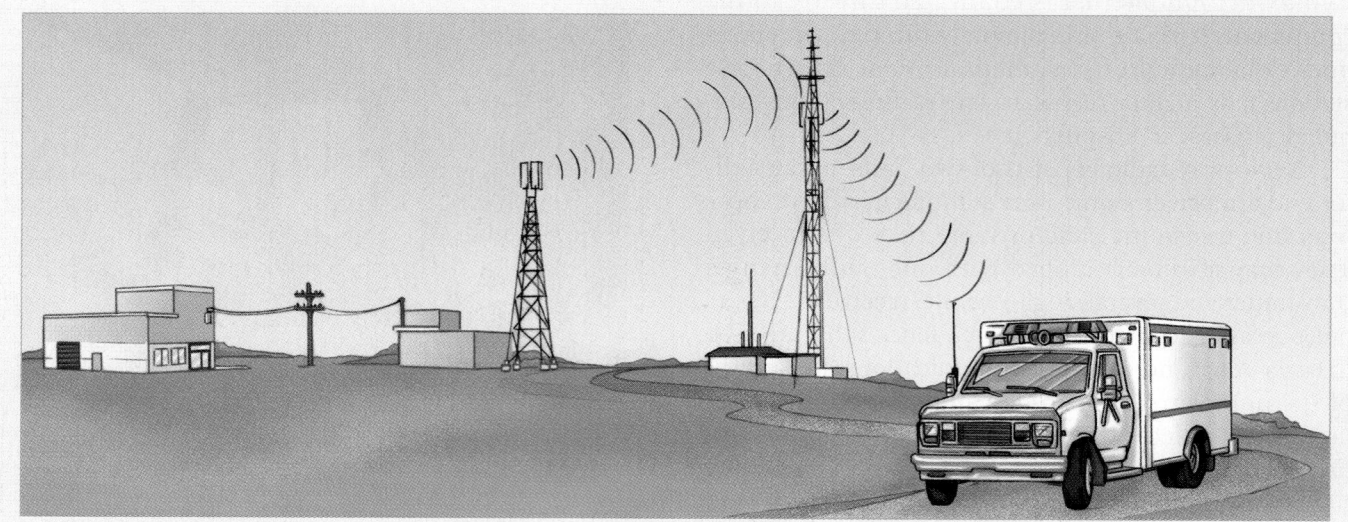

Figure 9-3 A message is sent from the control center by a land line to the transmitter. The radio carrier wave is picked up by the repeater for rebroadcast to outlying units. Return radio traffic is picked up by the repeater and rebroadcast to the control center.

Because a repeater is a base station (with a large antenna), it is able to receive lower power signals, such as those from a portable radio, from a long distance away. The signal is then rebroadcast with all the power of the base station (Figure 9-3 ▲). EMS systems that use repeaters usually have outstanding system-wide communications and are able to get the best signal from portable radios. There are also mobile repeaters that may be found in ambulances or placed in various areas around an EMS system area.

Digital Equipment

Although most people think of voice communications when they think of two-way radios, digital signals are also a part of EMS communications. Some EMS systems use telemetry to send an electrocardiogram from the unit to the hospital. With <u>telemetry</u>, electronic signals are converted into coded, audible signals. These signals can then be transmitted by radio or telephone to a receiver at the hospital with a decoder. The decoder converts the signals back into electronic impulses that can be displayed on a screen or printed. Another example of telemetry is a fax message.

Digital signals are also used in some kinds of paging and tone alerting systems because they transmit faster than spoken words and allow more choices and flexibility.

Cellular Telephones

While dispatchers communicate with field units by transmitting through a fixed radio base station, it is common for EMTs to communicate with receiving facilities by <u>cellular telephone</u>. These telephones are simply low-power portable radios that communicate through a series of interconnected repeater stations called "cells" (hence the name "cellular"). Cells are linked by a sophisticated computer system and connected to the telephone network.

Many cellular systems make equipment and air time available to EMS services at little or no cost as a public service. The public is often able to call 9-1-1 or other emergency numbers on a cellular telephone free of charge. However, this easy access may result in overloading and jamming of cellular systems in mass-casualty and disaster situations.

As with all repeater-based systems, a cellular telephone is useless if the equipment fails, loses power, or is damaged by severe weather or other circumstances. Like all voice radio communications systems, cellular telephones can be easily overheard on scanners. A <u>scanner</u> is a radio receiver that searches or "scans" across several frequencies until the message is completed. Although cellular telephones are more private than most other forms of radio communications, they can still be overheard. Therefore, you must always be careful to appropriately respect patient privacy and to speak in a professional manner every time you use any form of an EMS communications system.

Other Communications Equipment

Ambulances and other field units are usually equipped with an external public address system. This system may be a part of the siren or the mobile radio. The intercom between the cab and the patient compartment may also be a part of the mobile radio. These components do not involve radio wave transmission, but you must understand how they work and practice using them before you really need them.

EMS systems may use a variety of two-way radio hardware. Some systems operate VHF equipment in the simplex (push to talk, release to listen) mode. In this mode, radio transmissions can occur in either direction but not simultaneously in both. When one party transmits, the other can only receive and then wait for the other party to finish before he or she can reply. Other systems conduct duplex (simultaneous talk-listen) communications on UHF frequencies and also use cellular telephones. In the full duplex mode, radios can transmit and receive communications simultaneously on one channel. This is sometimes called "a pair of frequencies." A number of VHF and UHF channels, commonly called MED channels, are reserved exclusively for EMS use. However, hundreds of other commercial, local government, and fire services frequencies are also used for EMS communications.

Some EMS systems rely on dedicated lines (hot lines) as control links for their remotely located base stations and antennas. Other systems are more simply configured and require no off-site control links. No matter what type of equipment is used, all EMS communications systems have some basic limitations. Therefore, you must know what your equipment can and cannot do.

The ability for you to communicate effectively with other units or medical control depends on how well the weaker radio can "talk back." Base and repeater station radios often have much greater power and higher antennas than mobile or portable units do. This increased power affects your communications in two ways. First, their signals are generally heard and understood from a much greater distance than the signal produced from a mobile unit. Second, their signals are received clearly from a much greater distance than is possible with a mobile or portable unit. Remember, when you are at the scene, you may be able

to clearly hear the dispatcher or hospital on your radio, but you may not be heard or understood when you transmit.

Even small changes in your location can significantly affect the quality of your transmission. Also remember that the location of the antenna is critically important for clear transmission. Commercial aircraft flying at 37,000' can transmit and receive signals over hundreds of miles, yet their radios have only a few watts of power. The "power" comes from their 37,000'-high antenna.

At times, you may be able to communicate with a base station radio but you will not be able to hear or transmit to another mobile unit that is also communicating with that base. Repeater base stations eliminate such problems. They allow two mobile or portable units that cannot reach each other directly to communicate through the repeater, using its greater power and antenna.

The success of communications depends on the efficiency of your equipment. A damaged antenna or microphone often prevents high-quality communications. Check the condition and status of your equipment at the start of each shift, and then correct or report any problems.

Radio Communications

All radio operations in the United States, including those used in EMS systems, are regulated by the Federal Communications Commission (FCC). The FCC has jurisdiction over interstate and international telephone and telegraph services and satellite communications—all of which may involve EMS activity.

The FCC has five principal EMS-related responsibilities:

1. **Allocating specific radio frequencies for use by EMS providers.** Modern EMS communications began in 1974. At that time, the FCC assigned 10 MED channels in the 460- to 470-MHz (UHF) band to be used by EMS providers. These UHF channels were added to the several VHF frequencies that were already available for EMS systems. However, these VHF frequencies had to be shared with other "special emergencies" uses, including school buses and veterinarians. In 1993, the FCC created an EMS-only block of frequencies in the 220-MHz portion of the radio spectrum.

2. **Licensing base stations and assigning appropriate radio call signs for those stations.** An FCC license is usually issued for 5 years, after which time it must be renewed. Each FCC license is granted only for a specific operating group. Often, the longitude and latitude (locations) of the antenna and the address of the base station determine the call signs.

3. **Establishing licensing standards and operating specifications for radio equipment used by EMS providers.** Before it can be licensed, each piece of radio equipment must be submitted by its manufacturer to the FCC for type acceptance, based on established operating specification and regulations.

4. **Establishing limitations for transmitter power output.** The FCC regulates broadcasting power to reduce radio interference between neighboring communications systems.

5. **Monitoring radio operations.** This includes making spot field checks to help ensure compliance with FCC rules and regulations.

The FCC's rules and regulations fill many volumes and are written in technical and legal language. Only a very small section (part 90, subpart C) deals with EMS communication issues. You are not responsible for reading these detailed and often confusing documents. For appropriate guidance on technical issues, contact your EMS system supervisor. In fact, many EMS systems look to radio and telephone communications experts for advice on technical issues.

Responding to the Scene

EMS communication systems may operate on several different frequencies and may use different frequency bands. Some EMS systems may even use different radios for different purposes. However, all EMS systems depend on the skill of the dispatcher. The dispatcher receives the first call to 9-1-1. You are part of the team that responds to calls once the dispatcher notifies your unit of an emergency.

The dispatcher has several important responsibilities during the alert and dispatch phase of EMS communications. The dispatcher must do all of the following:

- Properly screen and assign priority to each call (according to predetermined protocols).
- Select and alert the appropriate EMS response unit(s).
- Dispatch and direct EMS response unit(s) to the correct location.
- Coordinate EMS response unit(s) with other public safety services until the incident is over.
- Provide emergency medical instructions to the telephone caller so that essential care (eg, CPR) may begin before the EMTs arrive (according to predetermined protocols).

When the first call to 9-1-1 comes in, the dispatcher must try to judge its relative importance to begin the appropriate EMS response using emergency medical dispatch protocols. First, the dispatcher must find out the exact location of the patient and the nature and severity of the problem. The dispatcher asks for the caller's telephone number, the patient's age and name, and other information, as directed by local protocol. Next, some description of the scene, such as the number of patients or special environmental hazards, is needed.

From this information, the dispatcher will assign the appropriate EMS response unit(s) on the basis of local protocols to determine the level and type of response and the following:

- The dispatcher's determination of the nature and severity of the problem (many emergency

You are the Provider

Part 2

Many different forms of communication must be employed when working in the field. Your responsibility as the "sender" of information must take into consideration that many of those who receive information may have difficulty understanding your instructions and/or requests. The woman you are caring for has rated her pain.

3. In what ways do you anticipate that her pain and anxiety levels will affect her ability to comprehend or comply with your instructions?

4. What can you do to minimize the potential for miscommunication?

medical dispatch systems will determine this automatically based on a caller's answers to a defined series of questions)

- The anticipated response time to the scene
- The level of training (first responder, BLS, ALS) of available EMS response unit(s)
- The need for additional EMS units, fire suppression, rescue, a HazMat team, air medical support, or law enforcement

The dispatcher's next step is to alert the appropriate EMS response unit(s) (Figure 9-4 ▶). Alerting these units may be done in a variety of ways. The dispatch radio system may be used to contact units that are already in service and monitoring the channel. Dedicated lines (hot lines) between the control center and the EMS station may also be used.

The dispatcher may also page EMS personnel. Pagers are commonly used in EMS operations to alert on-duty and off-duty personnel. Paging involves the use of a coded tone or digital radio signal and a voice or display message that is transmitted to pagers (beepers) or desktop monitor radios. Paging signals may be sent to alert only certain personnel or may be blanket signals that will activate all the pagers in the EMS service. Pagers and monitor radios are convenient because they are usually silent until their specific paging code is received. Alerted personnel contact the dispatcher to confirm the message and receive details of their assignments.

Once EMS personnel have been alerted, they must be properly dispatched and sent to the incident. Every EMS system should use a standard dispatching procedure. The dispatcher should give the responding unit(s) the following information:

- The nature and severity of the injury, illness, or incident
- The exact location of the incident
- The number of patients
- Responses by other public safety agencies
- Special directions or advisories, such as adverse road or traffic conditions, severe weather reports, or potential scene hazards
- The time at which the unit or units are dispatched

Your unit must confirm to the dispatcher that you have received the information and that you are en route to the scene. Local protocol will dictate whether it is the job of the dispatcher or your unit to notify other public safety agencies that you are responding to an emergency. In some areas, the emergency department is also notified whenever an ambulance responds to an emergency.

Figure 9-4 You will be assigned to a scene by the dispatcher.

You should report any problems during your response to the dispatcher. You should also inform the dispatcher when you have arrived at the scene. The arrival report to the dispatcher should include any obvious details that you see during scene size-up. For example, you might say, "Dispatcher, BLS Unit Two is on scene at 3010 27th St. Blue house with long driveway." This information is particularly useful if additional units are responding to the same scene.

All radio communications during dispatch, as well as other phases of operations, must be brief and easily understood. Although speaking in plain English is best, many areas find that 10 codes are shorter and simpler for routine communications. The development and use of such codes require strict discipline. When used improperly or not understood, codes create confusion rather than clarity.

Communicating With Medical Control and Hospitals

The principal reason for radio communication is to facilitate communication between you and medical control (and the hospital). Medical control may be located at the receiving hospital, another facility, or sometimes even in another city or state. You must, however, consult with medical control to notify the hospital of an incoming patient, to request advice or orders from medical control, or to advise the hospital of special situations.

It is important to plan and organize your radio communication before you push the transmit button. Remember, a concise, well-organized report is the best method of accurately and thoroughly describing the patient and his medical condition to care providers who will be receiving the patient. It also demonstrates your competence and professionalism in the eyes of all who hear your report. Well-organized radio communications with the hospital will engender confidence in the receiving facility's physicians and nurses, as well as others who are listening. In addition, the patient and family will be comforted by your organization and ability to communicate clearly. A well-delivered radio report puts you in control of the information, which is where you want to be.

Hospital notification is the most common type of communication between you and the hospital. The purpose of these calls is to notify the receiving facility of the patient's complaint and condition (Figure 9-5 ▼). On the basis of this information, the hospital is able to prepare staff and equipment appropriately to receive the patient.

Figure 9-5 Giving the patient report should be done in an objective, accurate, professional manner.

Giving the Patient Report

The patient report should follow a standard format established by your EMS system. The patient report commonly includes the following seven elements:

1. **Your unit identification and level of services.** Example: "Columbus Fire 2-BLS."
2. **The receiving hospital and your estimated time of arrival.** Example: "Columbus Community Hospital, ETA 10 minutes," or "patient transport code" according to local protocols.
3. **The patient's age and gender.** Example: "86-year-old woman." The patient's name should not be given over the radio because it may be overheard. This would be a violation of the patient's privacy.
4. **The patient's chief complaint or your perception of the problem and its severity.** Example: "Patient complains of both severe pelvic and less severe back pain."
5. **A brief history of the patient's current problem.** Example: "Patient fell into bathtub at 3 this morning and hasn't been able to get out." Other important history information that may pertain to the current problem should also be included, such as "The patient has diabetes and takes insulin."
6. **A brief report of physical findings.** This report should include level of consciousness, the patient's general appearance, pertinent abnormalities noted, and vital signs. Example: "The patient is alert and oriented, has pale skin color, and is cold to the touch. We noted crepitus in the pelvic girdle. Her blood pressure is 112 over 84, pulse is 72, and respirations 14."
7. **A brief summary of the care given and any patient response.** Example: "We have immobilized her onto a backboard. She still has pulse, motor, and sensory function distally in all four extremities."

Be sure that you report all patient information in an objective, accurate, and professional manner. People with scanners are listening. You could be successfully sued for slander if you describe a patient in a way that injures his or her reputation.

The Role of Medical Control

The delivery of EMS involves an impressive array of assessments, stabilization, and treatments. In some cases, you may assist patients in taking medications. Intermediate and advanced EMTs go beyond this level by initiating medication therapy based on the patient's presenting signs. For logical, ethical, and legal reasons, the delivery of such sophisticated care must be done in association with physicians. For this reason, every EMS system needs input and involvement from physicians. One or more physicians, including your system or department medical director, will provide medical direction (medical control) for your EMS system. Medical control is either off-line (indirect) or online (direct), as authorized by the medical director. Medical control guides the treatment of patients in the system through protocols, direct orders and advice, and post-call review.

Depending on how the protocols are written, you may need to call medical control for direct orders (permission) to administer certain treatments, to determine the transport destination of patients, or to be allowed to stop treatment and/or not transport a patient. In these cases, the radio or cellular phone provides a vital link between you and the expertise available through the base physician.

To maintain this link 24 hours a day, 7 days a week, medical control must be readily available on the radio at the hospital or on a mobile or portable unit when you call (Figure 9-6 ▶). In most areas, medical control is provided by the physicians who work at the receiving hospital. However, many variations have developed across the country. For example, some EMS units receive medical direction from one hospital even though they are taking the patient to another hospital. In other areas, medical direction may come from a

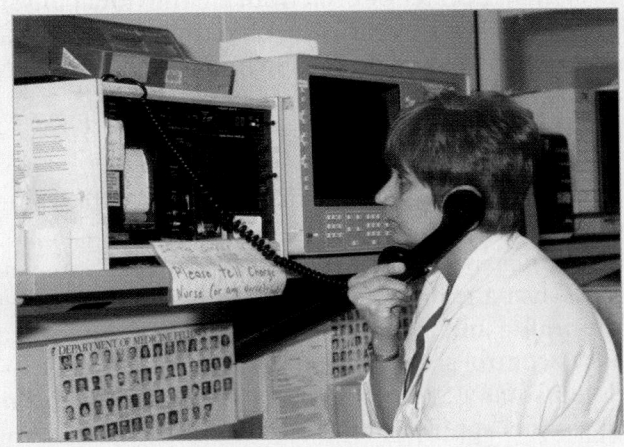

Figure 9-6 Medical control must be readily available on the radio at the hospital.

EMT-B Tips

EMT-Bs in some EMS systems routinely divide duties on a response, between patient care and radio reporting. They must communicate very closely between themselves to make this work. In reality, both EMT-Bs are somewhat involved in each role, but the partial division of responsibilities can be efficient and effective. This approach is most common in systems that employ extensive online medical control.

You are the Provider   Part 3

You provide the following report to the hospital:

"Columbus Fire, Columbus Community Hospital"

"We are en route to your facility with an 86-year-old woman who fell into her bathtub at 3 this morning, where she remained until 10 this morning. Patient is conscious and alert and complains of being severely cold and of having hip pain and mid back pain where she was leaning back on the faucet in the bathtub. The patient is alert and oriented; her skin is pale and cold to the touch. We noted crepitus in the pelvic girdle. Her blood pressure is 112 over 84, pulse is 72, and respirations 14.

We have immobilized her onto a backboard. She still has pulse, motor, and sensory function distally in all four extremities. She has a history of diabetes. Our ETA is 10 minutes. Do you have any questions or orders for us?"

5. What questions do you anticipate emergency department personnel to ask?
6. With the patient's history of diabetes, what else might play a role in your care. How does this disease affect the patient's perception of pain?

free-standing center or even from an individual physician. Regardless of your system's design, your link to medical control is vital to maintain the high quality of care that your patient requires and deserves.

Calling Medical Control

You can use the radio in your unit or a portable radio to call medical control. A cellular telephone can also be used. Regardless of the type of communication, you should use a channel that is relatively free of other radio traffic and interference. There are a number of ways to control access on ambulance-to-hospital channels. In some EMS systems, the dispatcher monitors and assigns appropriate, clear medical control channels. Other EMS systems rely on special communications operations, such as CMEDs (Centralized Medical Emergency Dispatch) or resource coordination centers, to monitor and allocate the medical control channels.

Because of the large number of EMS calls to medical control, your radio report must be well organized and precise and must contain only important information. In addition, because you need specific directions on patient care, the information that you provide to medical control must be accurate. Remember, the physician on the other end bases his or her instructions on the information that you provide.

You should never use codes when communicating with medical control unless you are directed by local protocol to do so. You should use proper medical terminology when giving your report. Never assume that medical control will know what a "10-50" or "Signal 70" means. Most medical control systems handle many different EMS agencies and will most likely not know your unit's special codes or signals.

To ensure complete understanding, once you receive an order from medical control, you must repeat the order back, word for word, and then receive confirmation. Whether the physician gives an order for medication or a specific treatment or denies a request for a particular treatment, you must repeat the order back word for word. This "echo" exchange helps to eliminate confusion and the possibility of poor patient care. Orders that are unclear or seem inappropriate or incorrect should be questioned. Do not blindly follow an order that does not make sense to you. The physician may have misunderstood or may have missed part of your report. In that case, he or she may not be able to respond appropriately to the patient's needs.

EMT-B Tips

Orders that are unclear or seem inappropriate or incorrect should be questioned. Do not blindly follow an order that does not make sense to you.

Information About Special Situations

Depending on your system's procedures, you may initiate communication with one or more hospitals to advise them of an extraordinary call or situation. For instance, a small rural hospital may be better able to respond to multiple victims of a highway crash if it is notified when the ambulance is first responding. At the other extreme, an entire hospital system must be notified of any disaster, such as a plane or train crash, as early as possible to enable activation of its staff call-in system. These special situations might also include HazMat situations, rescues in progress, mass-casualty incidents, or any other situation that might require special preparation on the part of the hospital. In some areas, mutual aid frequencies may be designated in mass-casualty incidents so that responding agencies can communicate with one another on a common frequency.

When notifying the hospital(s) of any special situations, keep the following in mind: The earlier the notification, the better. You should ask to speak to the charge nurse or physician in charge, as he or she is best able to mobilize the resources necessary to respond. Also, whenever possible, provide an estimate of the number of individuals who may be transported to the facility. Be sure to identify any conditions the patient(s) might have that require special needs, such as burns or hazardous materials exposure, to assist the hospital in preparation. In many cases, hospital notification is part of a larger disaster or HazMat plan. Follow the plan for your system.

Standard Procedures and Protocols

You must use your radio communications system effectively from the time you acknowledge a call until you complete your run. Standard radio operating procedures are designed to reduce the number of misunderstood messages, to keep transmissions brief, and to develop effective radio discipline. Standard radio communications protocols help both you and the dispatcher to communicate properly (Table 9-1 ▶). Protocols

TABLE 9-1 Guidelines for Effective Radio Communication

1. **Monitor the channel before transmitting** to avoid interfering with other radio traffic.

2. **Plan your message before pushing the transmit switch.** This will keep your transmissions brief and precise. You should use a standard format for your transmissions.

3. **Press the push-to-talk (PTT) button on the radio**, then wait for 1 second before starting your message. Otherwise, you might cut off the first part of your message before the transmitter is working at full power.

4. **Hold the microphone 2" to 3" from your mouth.** Speak clearly, but never shout into the microphone. Speak at a moderate, understandable rate, preferably in a clear, even voice.

5. **Identify the person or unit you are calling.** Identify both your unit as the sender and the receiving unit as appropriate. You will rarely work alone, so say "we" instead of "I" when describing yourself.

6. **Acknowledge a transmission as soon as you can** by saying, "Go ahead" or whatever is commonly used in your area. You should say, "Over and out," or whatever is commonly used in your area when you are finished. If you cannot take a long message, simply say, "Stand by" until you are ready.

7. **Use plain English.** Avoid meaningless phrases ("Be advised"), slang, or complex codes. Avoid words that are difficult to hear, such as "yes" and "no." Use "affirmative" and "negative."

8. **Keep your message brief.** If your message takes more than 30 seconds to send, pause after 30 seconds and say, "Do you copy?" The other party can then ask for clarification if needed. Also, someone else with emergency traffic can break through if necessary.

9. **Avoid voicing negative emotions**, such as anger or irritation, when transmitting. Courtesy is assumed, making it unnecessary to say "please" or "thank you," which wastes air time. Listen to other communications in your system to get a good idea of the common phrases and their uses.

10. **When transmitting a number with two or more digits, say the entire number first** and then each digit separately. For example, say, "sixty-seven," followed by "six-seven."

11. **Do not use profanity on the radio.** It is a violation of FCC rules and can result in substantial fines and even loss of your organization's radio license.

12. **Use EMS frequencies for EMS communications.** Do not use these frequencies for any other type of communications.

13. **Reduce background noise as much as possible.** Move away from wind, noisy motors, or tools. Close the window if you are in a moving ambulance. If possible, shut off the siren during radio transmissions.

14. **Be sure other radios on the same frequency are turned off or down** to avoid feedback.

---

should include guidelines specifying a preferred format for transmitting messages, definitions of key words and phrases, and procedures for troubleshooting common radio communications problems.

Reporting Requirements (Read This)

(handwritten: Keep This in mind)

Proper use of the EMS communications system will help you to do your job more effectively. From acknowledgment of the call until you are cleared from the medical emergency, you will use radio communications. You must report in to dispatch at least six times during your run:

1. **To acknowledge the dispatch information** and to confirm that you are responding to the scene

2. **To announce your arrival at the scene**

3. **To announce that you are leaving** the scene and are en route to the receiving hospital. (At this point, you typically should also state the number of patients being transported, your estimated arrival time at the hospital, and the run status.)

4. **To announce your arrival at the hospital** or facility

5. **To announce that you are clear of the incident** or hospital and available for another assignment

6. **To announce your arrival back at headquarters** or other off-the-air location

While en route to and from the scene, you should report to the dispatcher any special hazards or road conditions that might affect other responding units. Report any unusual delay, such as roadblocks, traffic, or construction. Once you are at the scene, you may request additional EMS or other public safety assistance and then help to coordinate their response.

During transport, you must periodically reassess the patient's vital signs and response to care provided. You should immediately report any significant changes in the patient's condition, especially if the patient seems worse. Medical control can then give new orders and prepare to receive the patient.

Maintenance of Radio Equipment

Like all other EMS equipment, radio equipment must be serviced by properly trained and equipped personnel. Remember that the radio is your lifeline to other public safety agencies (who function to protect you), as well as medical control, and it must perform under emergency conditions. Radio equipment that is operating properly should be serviced at least once a year. Any equipment that is not working properly should be immediately removed from service and sent for repair.

Sometimes, radio equipment will stop working during a run. Your EMS system must have several backup plans and options. The goal of a backup plan is to make sure that you can maintain contact when the usual procedures do not work. There are quite a few options.

The simplest backup plan relies on written standing orders. Standing orders are written documents that have been signed by the EMS system's medical director. These orders outline specific directions, permissions, and sometimes prohibitions regarding patient care. By their very nature, standing orders do not require direct communication with medical control. When properly followed, standing orders or formal protocols have the same authority and legal status as orders given over the radio. They exist to one extent or another in every EMS system and can be applied to all levels of EMS providers.

Verbal Communications

As an EMT-B, you must master many communication skills, including radio operations and written communications. Verbal communications with the patient, the family, and the rest of the health care team are an essential part of high-quality patient care. And as an EMT-B, you must be able to find out what the patient needs and then tell others. Never forget that you are the vital link between the patient and the remainder of the health care team.

Communicating With Other Health Care Professionals

EMS is the first step in what is often a long and involved series of treatment phases. Effective communication between the EMT-B and health care professionals in the receiving facility is an essential cornerstone of efficient, effective, and appropriate patient care.

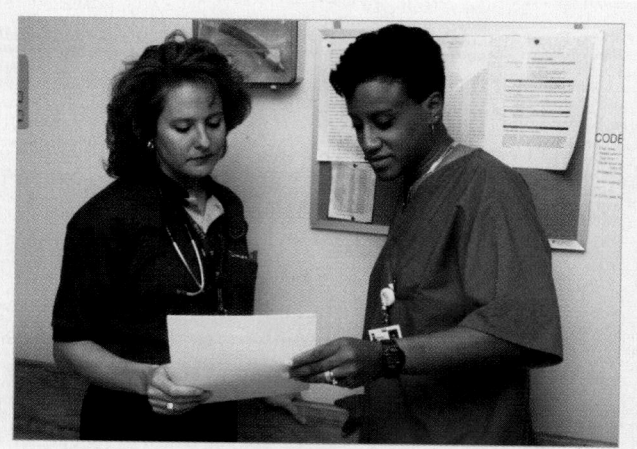

Figure 9-7 Once you arrive at the hospital, a staff member will take responsibility for the patient from you.

Your reporting responsibilities do not end when you arrive at the hospital. In fact, they have just begun. The transfer of care officially occurs during your oral report at the hospital, not as a result of your radio report en route. Once you arrive at the hospital, a hospital staff member will take responsibility for the patient from you (Figure 9-7 ▲). Depending on the hospital and the condition of the patient, the training of the person who takes over the care of the patient varies. However, you may transfer the care of your patient only to someone with at least your level of training. Once a hospital staff member is ready to take responsibility for the patient, you must provide that person with a formal oral report of the patient's condition.

Giving a report is a longstanding and well-documented part of transferring the patient's care from one provider to another. Your oral report is usually given at the same time that the staff member is doing something for the patient. For example, a nurse or physician may be looking at the patient, beginning assessment, or helping you to move the patient from the stretcher to an examination table. Therefore, you must report important information in a complete, precise way. The following six components must be included in the oral report:

1. **The patient's name** (if you know it) and the chief complaint, nature of illness, or mechanism of injury. Example: "This is Mrs McCarty. She woke up around 3 this morning, tripped and fell into the bathtub after using the restroom."
2. **More detailed information** of what you gave in your radio report. Example: "She denies losing

consciousness, no prior history of stroke, TIAs, or cardiac compromise but has been feeling a little light-headed when she stands."

3. **Any important history** that was not given already. Example: "Mrs McCarty lives by herself. She was unable to get out of the tub and was found by the hospice worker at 10 this morning. We suspect hypothermia as she had a core temperature of 94 degrees."

4. **The patient's response to treatment** given en route. It is especially important to report any changes in the patient or the treatment provided since your radio report. Example: "Oxygen was initiated by nonrebreathing face mask at 15 L/min. Though we suspected her mid back pain was due to her leaning against the faucet of the bathtub for 7 hours, we put her in the K.E.D. for both precautionary and extrication reasons. The hot packs wrapped in hand towels were to help warm her up."

5. **The vital signs assessed** during transport and after the radio report. Example: "Her vitals were 112/84, pulse 72, respirations 14, and core body temperature was 94 at time of transport. They are generally unchanged since then except her last temperature was 96."

6. **Any other information** that you may have gathered that was not important enough to report sooner. Information that was gathered during transport, any patient medications you have brought with you, and any other details about the patient that was provided by family members or friends may be included. Example: "Mrs Woods, the home hospice worker, has contacted Mrs McCarty's family and followed us here to answer any questions."

Communicating With Patients

Your communication skills will be put to the test when you communicate with patients and/or families in emergency situations. Remember that someone who is sick or injured is scared and might not understand what you are doing and saying. Therefore, your gestures, body movements, and attitude toward the patient are critically important in gaining the trust of both patient and family. These Ten Golden Rules will help you to calm and reassure your patients:

1. **Make and keep eye contact** with your patient at all times. Give the patient your undivided attention. This will let the patient know that he or she is your top priority. Look the patient straight in the eye to establish <u>rapport</u>. Establishing rapport is building a trusting relationship with your patient. This will make the job of caring for the patient much easier for both you and the patient.

2. **Use the patient's proper name** when you know it. Ask the patient what he or she wishes to be called. Avoid using terms such as "Honey" or "Dear." Use a patient's first name only if the patient is a child or the patient asks you to use his or her first name. Rather, use a courtesy title, such as "Mr Peters," "Mrs Smith," or "Ms Butler." If you do not know the patient's name, refer to him or her as "sir" or "ma'am."

3. **Tell the patient the truth.** Even if you have to say something very unpleasant, telling the truth is better than lying. Lying will destroy the patient's trust in you and decrease your own confidence. You might not always tell the patient everything, but if the patient or a family member asks a specific question, you should answer truthfully. A direct question deserves a direct answer. If you do not know the answer to the patient's question, say so. For example, a patient may ask, "Am I having a heart attack?" "I don't know" is an adequate answer.

4. **Use language that the patient can understand.** Do not talk up or down to the patient in any way. Avoid technical medical terms that the patient might not understand. For example, ask the patient whether he or she has a history of "heart problems." This will usually result in more accurate information than if you ask about "previous episodes of myocardial infarction" or a "history of cardiomyopathy."

5. **Be careful of what you say about the patient to others.** A patient might hear only part of what is said. As a result, the patient might seriously misinterpret (and remember for a long time) what was said. Therefore, assume that the patient can hear every word you say, even if you are speaking to others and even if the patient appears to be unconscious or unresponsive.

6. **Be aware of your body language** (Figure 9-8 ▶). Nonverbal communication is extremely important in dealing with patients. In stressful situations, patients may misinterpret your gestures and movements. Be particularly careful not to appear threatening. Instead, position yourself at a lower level than the patient when practical. Remember that you should always, always conduct yourself in a calm, professional manner.

Figure 9-8 Watch your body language, as patients may misinterpret your gestures, movements, and stance.

7. **Always speak slowly, clearly, and distinctly.** Pay close attention to your tone of voice.

8. **If the patient is hearing impaired, speak clearly,** and face the person so that he or she can read your lips. Do not shout at a person who is hearing impaired. Shouting will not make it any easier for the patient to understand you. Instead, it may frighten the patient and make it even more difficult for the patient to understand you. Never assume that an older patient is hearing impaired or otherwise unable to understand you. Also, never use "baby talk" with older patients or with anyone other than infants.

9. **Allow time for the patient to answer** or respond to your questions. Do not rush a patient unless there is immediate danger. Sick and injured people may not be thinking clearly and may need time to answer even simple questions. This is especially true in treating older patients.

10. **Act and speak in a calm, confident manner** while caring for the patient. Make sure that you attend to the patient's pains and needs. Try to make the patient physically comfortable and relaxed. Find out whether the patient is more comfortable sitting or lying down. Is the patient cold or hot? Does the patient want a friend or relative nearby?

Patients literally place their lives in your hands. They deserve to know that you can provide medical care and that you are concerned about their well-being.

Communicating With Older Patients

According to US Census Data, almost 35 million individuals are older than 65 years. It is projected that by the year 2030, the geriatric population will be greater than 70 million. A person's actual age might not be the most important factor in making him or her geriatric. It is more important to determine a person's functional age. The functional age relates to the person's ability to function in daily activities, the person's mental state, and activity pattern.

As an EMS provider, when you step onto a scene to care for an older patient, you are being asked to take control. You have been called because a person needs help. What you say and how you say it have an impact on the patient's perception of the call. You should present yourself as competent, confident, and concerned. You must take charge of the situation, but do so with compassion. You are there to listen, then act on what you learn. Don't limit your assessment to the obvious problem. Oftentimes, older patients who express that they are not well or who are overly concerned about their health or general condition are at risk for a serious decline in their physical, emotional, or psychological state. (Table 9-2 ▶) provides guidelines for interviewing an older patient.

Most older people think clearly, can give you a clear medical history, and can answer your questions (Figure 9-9 ▶). Do not assume that an older patient is senile or confused. Remember, though, that communicating with some older patients is extremely difficult. Some may be hostile, irritable, and/or confused. Do not assume this is normal behavior for an older patient. These signs may be caused by a simple lack of oxygen (hypoxia), brain injury including a cerebrovascular accident (CVA), unintentional drug overdose, or even

Study This!!!

TABLE 9-2 Interviewing an Older Patient

In general, when interviewing the older patient, the following techniques should be employed:

- Identify yourself. Do not assume an older patient knows who you are.
- Be aware of how you present yourself. Frustration and impatience can be portrayed through body language.
- Look directly at the patient.
- Speak slowly and distinctly.
- Explain what you are going to do before you do it. Use simple terms to explain the use of medical equipment and procedures, avoiding medical jargon or slang.
- Listen to the answer the patient gives you.
- Show the patient respect. Refer to the patient as Mr, Mrs, or Miss.
- Do not talk about the patient in front of him or her; to do so gives the impression that the patient has no choice in his or her medical care. This is easy to forget when the patient has impaired cognitive (thought) processes or has difficulty communicating.
- Be patient!

Figure 9-9 You need a great deal of compassion and patience when caring for older patients. Do not assume that the patient is senile or confused.

hypovolemia. Never attribute altered mental status simply to "old age." Others may have difficulty hearing or seeing you. You need great patience and compassion when you are called upon to care for such a patient. Think of the patient as someone's grandmother or grandfather—or even as yourself when you reach that age.

Approach an older patient slowly and calmly. Allow plenty of time for the patient to respond to your questions. Watch for signs of confusion, anxiety, or impaired hearing or vision. The patient should feel confident that you are in charge and that everything possible is being done for him or her.

Older patients often do not feel much pain. An older person who has fallen or been injured may report no pain. In addition, older patients might not be fully aware of important changes in other body systems. As a result, be especially vigilant for objective changes—no matter how subtle—in their condition. Even minor changes in breathing or mental state may signal major problems.

You are the Provider   Part 4

Although the patient is in extreme pain, you carefully explain all the procedures you perform during her care, which helps lower her anxiety. You also offer reassurance that her anxiety is to be expected. You make direct eye contact when addressing her claustrophobia about being immobilized and fear of falling off the gurney. When you are not performing medical procedures, you take the time to hold her hand, you let her know how far away you are from the hospital, and explain to her what to expect from the doctors and nurses when you arrive.

She tells you that she cannot spend another minute on the board and asks you how long she needs to be tied down. You are unsure.

7. How should you answer her?
8. Is it ever justifiable to misinform a patient to reduce the patient's anxiety temporarily?

When possible (which is more often than you'd think), give the patient some time to pack a few personal items before leaving for the hospital. Be sure to get any hearing aids, glasses, or dentures packed before departure; it will make the patient's hospital stay much more pleasant. You should document on the prehospital care report that these items accompanied the patient to the hospital and were given to a specific staff person in the emergency department.

Communicating With Children

Everyone who is thrust into an emergency situation becomes frightened to some degree. However, fear is probably most severe and most obvious in children. Children may be frightened by your uniform, the ambulance, and the number of people who have suddenly gathered around. Even a child who says little may be very much aware of all that is going on.

Familiar objects and faces will help to reduce this fright. Let a child keep a favorite toy, doll, or security blanket to give the child some sense of control and comfort. Having a family member or friend nearby is also helpful. When not impractical due to the child's condition, it is often helpful to let the parent or an adult friend hold the child during your evaluation and treatment. However, you will have to make sure that this person will not upset the child. Sometimes, adult family members are not helpful because they become too upset by what has happened. An overly anxious parent or relative can make things worse. Be careful about selecting the proper adult for this role.

Children can easily see through lies or deceptions, so you must always be honest with them. Make sure that you explain to the child over and over again what and why certain things are happening. If treatment is going to hurt, such as applying a splint, tell the child ahead of time. Also tell the child that it will not hurt for long and that it will help "make it better."

Respect a child's modesty. Little girls and little boys are often embarrassed if they have to undress or be undressed in front of strangers. This anxiety often intensifies during adolescence. When a wound or site of injury has to be exposed, try to do so out of sight of strangers. Again, it is extremely important to tell the child what you are doing and why you are doing it.

You should speak to a child in a professional yet friendly way. A child should feel reassured that you are there to help in every way possible. Maintain eye contact with a child, as you would with an adult, to let the child know that you are helping and that you can be

Figure 9-10 Maintain eye contact with a child to let the child know that you are there to help and that you can be trusted.

trusted (Figure 9-10 ▲). It is helpful to position yourself at the child's level so that you do not appear to tower above the child.

Communicating With Hearing-Impaired Patients

Patients who are hearing impaired or deaf are usually not ashamed or embarrassed by their disability. Often, it is the people around a deaf or hearing-impaired person who have the problem coping. Remember that you must be able to communicate with hearing-impaired patients so that you can provide necessary or even life-saving care.

The majority of hearing-impaired patients have normal intelligence. Hearing-impaired patients can usually understand what is going on around them, provided that you can successfully communicate with them. Most patients who are hearing impaired can read lips to some extent. Therefore, you should place yourself in a position so that the patient can see your lips. Many hearing-impaired patients have hearing aids to help them communicate. Be careful that hearing aids are not lost during an accident or fall. Not only are they extremely expensive, hearing aids will often make it easier to communicate. Hearing aids may also be forgotten if the patient is confused or ill. Look around, or ask the patient or the family about a hearing aid.

Remember the following five steps to help you efficiently communicate with patients who are hearing impaired:

1. Have paper and a pen available. This way, you can write down questions and the patient can write down answers, if necessary. Be sure to print so that your handwriting is not a communications barrier.
2. If the patient can read lips, you should face the patient and speak slowly and distinctly. Do not cover your mouth or mumble. If it is night or dark, consider shining a light on your face.
3. Never shout.
4. Be sure to listen carefully, ask short questions, and give short answers. Remember that although many hearing-impaired patients can speak distinctly, some cannot.
5. Learn some simple phrases in sign language. For example, knowing the signs for "sick," "hurt," and "help" may be useful if you cannot communicate in any other way (Figure 9-11 ▶).

Communicating With Visually Impaired Patients

Like hearing-impaired patients, visually impaired and blind patients have usually accepted and learned to deal with their disability. Of course, not all visually impaired patients are completely blind. Many can perceive light and dark or can see shadows or movement. Ask the patient whether he or she can see at all. Also remember that, as with other patients who have disabilities, you should expect that visually impaired patients have normal intelligence.

As you begin caring for a visually impaired patient, explain everything that you are doing in detail as you are doing it. Be sure to stay in physical contact with the patient as you begin your care. Hold your hand lightly on the patient's shoulder or arm. Try to avoid sudden movements. If the patient can walk to the ambulance, place his or her hand on your arm, taking care not to rush. Transport any mobility aids, such as a cane, with the patient to the hospital. A visually impaired person may have a guide dog. Guide dogs are easily identified by their special harnesses (Figure 9-12 ▶). They are trained not to leave their masters and not to respond to strangers. A visually impaired patient who is conscious can tell you about the dog and give instructions for its

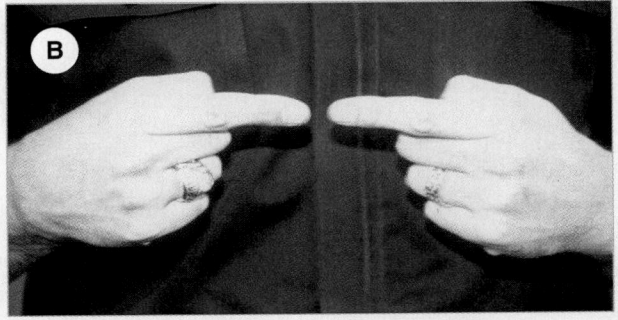
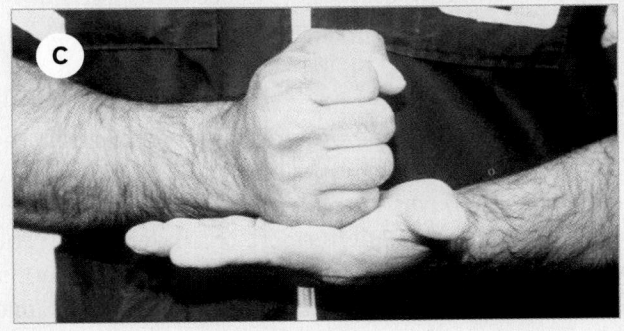

Figure 9-11 Learn simple phrases in sign language. **A.** Sick. **B.** Hurt. **C.** Help.

care. If circumstances permit, bring the guide dog to the hospital with the patient. If the dog has to be left behind, you should arrange for its care.

Communicating With Non-English-Speaking Patients

As part of the focused history and physical exam, you must obtain a medical history from the patient. You cannot skip this step simply because the patient does not speak English. Most patients who do not speak English fluently will still know certain important words or phrases.

Figure 9-12 A guide dog is easily identified by its special harness.

Your first step is to find out how much English the patient can speak. Use short, simple questions and simple words whenever possible. Avoid difficult medical terms. You can help patients to better understand if you point to specific parts of the body as you ask questions.

In many areas, particularly large urban centers, major segments of the population do not speak English. Your job will be much easier if you learn some common words and phrases in their language, especially common medical terms. Pocket cards are available that show the pronunciation of these terms. If the patient does not speak any English, find a family member or friend to act as an interpreter.

Written Communications and Documentation

Along with your radio report and verbal report, you must also complete a formal written report about the patient before you leave the hospital. You might be able to do the written report en route, if the trip is long enough

and the patient needs minimal care. Usually, you will finish the written report after you have transferred the care of the patient to a hospital staff member. Be sure to leave the report at the hospital before you leave.

Data Collection

The information you collect during a call becomes part of the patient's medical record. The National EMS Information System (NEMSIS) has been collecting pre-hospital care information for research purposes since the early 1970s. NEMSIS has identified specific data points needed to enable communication and comparison of EMS runs between agencies, regions, and states. The minimum data set includes both narrative components and check-off boxes (Figure 9-13 ▶). An example of information collected on a Prehospital Care Report (PCR) includes:

- Chief complaint
- Level of consciousness (AVPU) or mental status
- Systolic blood pressure for patients older than 3 years
- Capillary refill for patients younger than 6 years
- Skin color and temperature
- Pulse
- Respirations and effort

Examples of administrative information gathered in a PCR:

- The time that the incident was reported
- The time that the EMS unit was notified
- The time that the EMS unit arrived at the scene
- The time that the EMS unit left the scene
- The time that the EMS unit arrived at the receiving facility
- The time that patient care was transferred

You will begin gathering the patient information as soon as you reach the patient. Continue collecting information as you provide care until you arrive at the hospital.

(Table 9-3 ▶) provides guidelines on how to write the narrative portion of your report. Whether you completed a medical or trauma assessment, the assessment-based approach follows each step of the assessment(s) as a guideline to narrative writing.

Prehospital Care Report

Prehospital care reports help to ensure efficient continuity of patient care. This report describes the nature of the patient's injuries or illness at the scene and the initial treatment you provide. Although this report might not be read immediately at the hospital, it may be referred to later for important information. The prehospital care report serves the following six functions:

1. Continuity of care
2. Legal documentation
3. Education
4. Administrative
5. Essential research record
6. Evaluation and continuous quality improvement

A good prehospital care report documents the care that was provided and the patient's condition on arrival at the scene. It also documents any changes in the patient's condition upon arrival at the hospital. The information in the report also proves that you have provided proper care. In some instances, it also shows that you have properly handled unusual or uncommon situations. Both objective and subjective information is included in this report. It is critical that you document everything in the clearest manner possible. Should you ever be called to give testimony concerning patient care, you and your prehospital care report will be utilized to present evidence. As with your personal appearance, your prehospital care report will reflect a professional or nonprofessional image. A well-written, neat, and concise document—including correct spelling and grammar—will reflect good patient care. Consider using the old adage of "If you didn't write it, it didn't happen," or "If the report looks sloppy, the patient care was also sloppy."

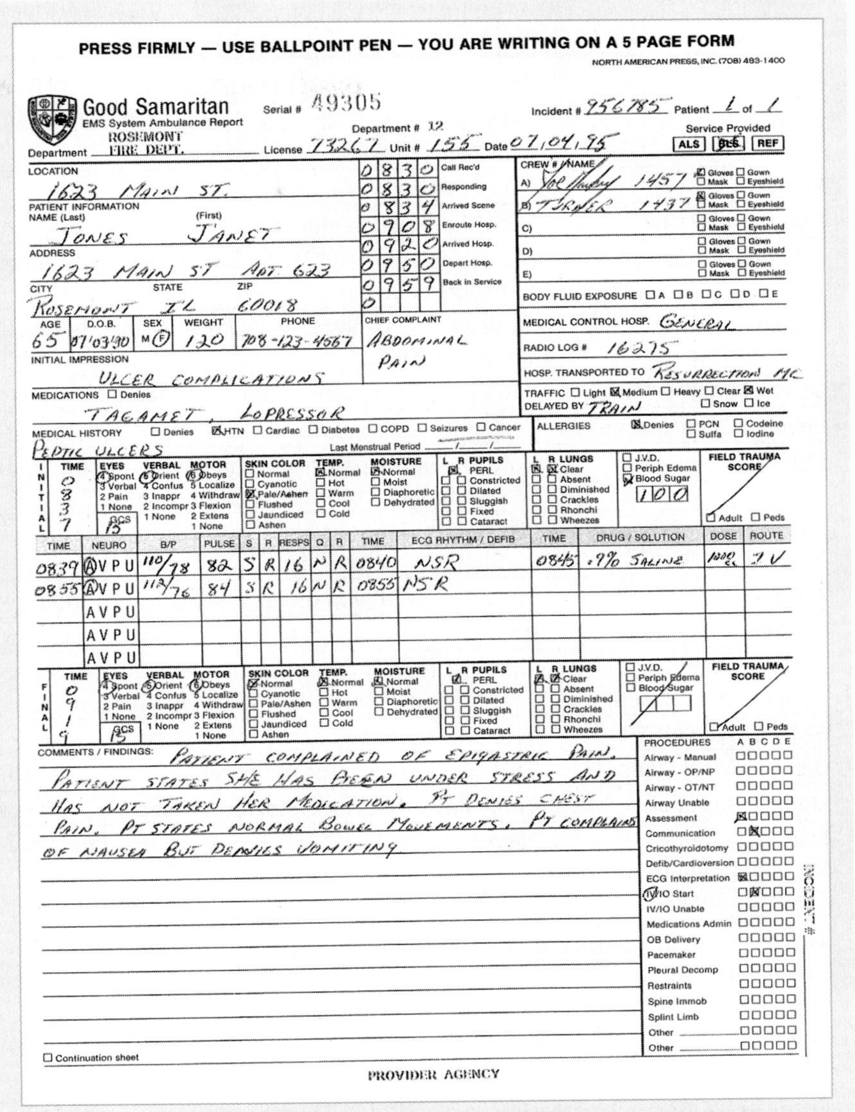
PRESS FIRMLY — USE BALLPOINT PEN — YOU ARE WRITING ON A 5 PAGE FORM

NORTH AMERICAN PRESS, INC. (708) 483-1400

Good Samaritan EMS System Ambulance Report

Serial # 49305 Incident # 956785 Patient 1 of 1

Department ROSEMONT FIRE DEPT. Department # 12 License 73267 Unit # 155 Date 07/04/95 Service Provided ALS BLS REF

LOCATION: 1623 Main St.

PATIENT INFORMATION NAME (Last) Jones (First) Janet

ADDRESS: 1623 Main St Apt 623

CITY Rosemont STATE IL ZIP 60018

AGE 65 D.O.B. 07/03/30 SEX F WEIGHT 120 PHONE 708-123-4567

Call Rec'd 0830 · Responding 0830 · Arrived Scene 0834 · Enroute Hosp. 0908 · Arrived Hosp. 0920 · Depart Hosp. 0950 · Back in Service 0959

CHIEF COMPLAINT: Abdominal Pain

INITIAL IMPRESSION: Ulcer Complications

MEDICATIONS: Tagamet, Lopressor

MEDICAL HISTORY: HTN; Peptic Ulcers

BODY FLUID EXPOSURE A B C D E

MEDICAL CONTROL HOSP. General · RADIO LOG # 16275 · HOSP. TRANSPORTED TO Resurrection MC

TRAFFIC: Medium · Wet · DELAYED BY Train

ALLERGIES: Denies

TIME	NEURO	B/P	PULSE	RESPS	TIME	ECG RHYTHM / DEFIB	TIME	DRUG / SOLUTION	DOSE	ROUTE
0839	A	110/78	82 S R	16 N R	0840	NSR	0845	.9% Saline	1000 cc	IV
0855	A	112/76	84 S R	16 N R	0855	NSR				

COMMENTS / FINDINGS: Patient complained of epigastric pain. Patient states she has been under stress and has not taken her medication. Pt denies chest pain. Pt states normal bowel movements. Pt complains of nausea but denies vomiting.

Continuation sheet

PROVIDER AGENCY

Figure 9-13 The minimum data set includes both patient information and administrative information.

These reports also provide valuable administrative information. For example, the report provides information for patient billing. It can also be used to evaluate response times, equipment usage, and other areas of administrative responsibility.

Data may be obtained from the prehospital care forms to analyze causes, severity, and types of illness or injury requiring emergency medical care. These reports may also be used in an ongoing program for evaluation of the quality of patient care. All records are reviewed periodically by your system. The purpose of this review is to make sure that trauma triage and/or other prehospital care criteria have been met.

There are many requirements of a prehospital care report (Table 9-4 ▶). Often, these requirements vary from jurisdiction to jurisdiction, mainly because so many agencies obtain information from them. While there is no universally accepted form, certain data points (uniform components of a prehospital care report) are common to many reports. The advantages of collecting such information are significant, as national trends can be detected. For example, roughly 5% of the nation's EMS calls involve pediatric patients. Of those 5% of patients, 80% will be in respiratory distress. Such information is invaluable, and, collected, form uniform data points.

TABLE 9-3 How to Write a Report*

BSI	Were extraordinary BSI precautions initiated? If so, state which precautions were used and why.
Scene Safety	Did you have to make your scene safe? If so, what did you do and why did you do it? Was there a delay in patient care?
NOI/MOI	Simply state.
Number of Patients	Record only when more than one patient is present; "This is patient 2 of 3."
Additional Help	Did you call for help? If so, state why, at what time, and what time the help arrived. Was transport delayed?
C-Spine	State what C-spine precautions were initiated. You may want to include why; "Due to the significant MOI..."
General Impression	Simply record, if not already documented on the PCR.
Level of Consciousness	Be sure to report LOC, any changes in LOC, and at what time changes occurred.
Chief Complaint	Note and quote pertinent statements made by patient/and or bystanders. This includes any pertinent denials; "Pt. denies chest pain...."
Life Threats	List all interventions and how the patient responded; "Assisted ventilations with O_2 (15 LPM) at 20 BPM with no change in LOC."
ABCs	Document what you found, and again, any interventions performed.
Oxygen	Record if O_2 was used, how it was applied, and how much was administered.
Focused, Rapid, or Detailed Assessment	State the type of assessment used and any pertinent findings; "Detailed physical exam revealed unequal pupils, crepitus to right ribs, and an apparent closed fracture of left tibia."
SAMPLE/OPQRST-I	Note and quote any pertinent answers.
Baseline Vital Signs	Your service may want you to record vital signs in the narrative as well as other places on the PCR.
Medical Direction	Quote any orders given to you by medical control and who gave them.
Management of Secondary Injuries/ Treat for Shock	Report all patient interventions, at what time they were completed, and how the patient responded.

*Source: Reprinted with permission. Courtesy of Jay C. Keefauver.

TABLE 9-4 Sample Uniform Components of Prehospital Care Report

- Patient's name, gender, date of birth, and address
- Dispatched as
- Chief complaint
- Location of the patient when first seen (including specific details, especially if the incident is a car crash or criminal activity is suspected)
- Rescue and treatment given before your arrival
- Signs and symptoms found during your patient assessment
- Care and treatment given at the site and during transport
- Baseline vital signs
- SAMPLE history changes in vital signs and condition

- Date of the call
- Time of the call
- Location of the call
- Time of dispatch
- Time of arrival at the scene
- Time of leaving the scene
- Time of arrival at the hospital
- Patient's insurance information
- Names and/or certification numbers of the EMT-Bs who responded to the call
- Name of the base hospital involved in the run
- Type of run to the scene: emergency or routine

✚ EMT-B Tips

The Health Insurance Portability and Accountability Act (HIPAA) established mandatory patient privacy rules and regulations to safeguard patient confidentiality. It provides guidance on what types of information is protected, the responsibility of health care providers regarding that protection, and penalties for breaching that protection.

Most personal health information is protected and should not be released without the patient's permission. If you are not sure, do not give any information to anyone other than those directly involved in the care of the patient. Make sure you are aware of all policies and procedures governing your particular agency.

✚ EMT-B Tips

A list of commonly misspelled medical words, along with proper abbreviations, has been provided for you in the companion workbook.

Remember that the report form itself and all the information on it are considered confidential documents. Be sure that you are familiar with state and local laws concerning confidentiality. All prehospital forms must be handled with care and stored in an appropriate manner once you have completed them. After you have completed a report, distribute the copies to the appropriate locations, according to state and local protocol. In most instances, a copy of the report will remain at the hospital and will become a part of the patient's record.

Types of Forms

You will most likely use one of two types of forms. The first type is the traditional written form with check boxes and a narrative section. The second type is a computerized version in which you fill in information using an electronic clipboard or similar device.

If your service uses written forms, be sure to fill in the boxes completely, and avoid making stray marks on the sheet. Make sure that you are familiar with the specific procedures for collecting, recording, and reporting the information in your area.

If you must complete a narrative section, be sure to describe what you see and what you do. Be sure to include significant negative findings and important observations about the scene. Do not record your conclusions about the incident. For example, you may write, "The patient admits to drinking today." This is a clear description that does not make any judgments about the patient's condition. However, a report that says, "The patient was drunk," makes a conclusion about the patient's condition. Also avoid radio codes, and use only standard abbreviations. When information is of a sensitive nature, note the source of the information. Be sure to spell words correctly, especially medical terms.

If you do not know how to spell a particular word, find out how to spell it, or use another word. Also be sure to record the time with all assessment findings.

Reporting Errors

Everyone makes mistakes. If you leave something out of a report or record information incorrectly, do not try to cover it up. Rather, write down what did or did not happen and the steps that were taken to correct the situation. Falsifying information on the prehospital report may result in suspension and/or revocation of your certification/license. More important, falsifying information results in poor patient care, because other health care providers have a false impression of assessment findings or the treatment given. Document only the vital signs that were actually taken. If you did not give the patient oxygen, do not chart that the patient was given oxygen.

If you discover an error as you are writing your report, draw a single horizontal line through the error, initial it, and write the correct information next to it (Figure 9-14 ▶). Do not try to erase or cover the error with correction fluid. This may be interpreted as an attempt to cover up a mistake.

If an error is discovered after you submit your report, draw a single line through the error, preferably in a different color ink, initial it, and date it. Make sure to add a note with the correct information. If you left out information accidentally, add a note with the correct information, the date, and your initials.

When you do not have enough time to complete your report before the next call, you will need to fill it out later.

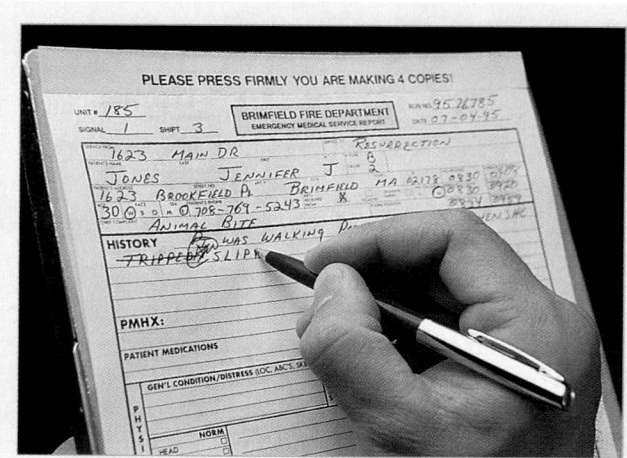

Figure 9-14 If you make a mistake in writing your report, the proper way to correct it is to draw a single horizontal line through the error, initial it, and write the correct information next to it.

Documenting Refusal of Care

Refusal of care is a common source of litigation in EMS. Thorough documentation is crucial. Competent adult patients have the right to refuse treatment and, in fact, must specifically provide permission for treatment to be provided by EMS or any other health care provider. Before you leave the scene, try to persuade the patient to go to the hospital and consult medical direction as directed by local protocol. Also make sure that the patient is able to make a rational, informed decision and is not under the influence of alcohol or other drugs or the effects of an illness or injury. Explain to the patient why it is important to be examined by a physician at the hospital. Also explain what may happen if the patient is not examined by a physician. If the patient still refuses, suggest other means for the patient to obtain proper care. Explain that you are willing to return. If the patient still refuses, document any assessment findings and emergency medical care given, then have the patient sign a refusal form (Figure 9-15 ▶). You should also have a family member, police officer, or bystander sign the form as a witness. If the patient refuses to sign the refusal form, have a family member, police officer, or bystander sign the form verifying that the patient refused to sign.

Be sure to complete the prehospital report, including the patient assessment findings. You'll need to document the advice you gave as to the risks associated with refusal of care. Report clinical information such as level of consciousness (LOC) that suggests competency of the person refusing care. Note pertinent patient comments and any medical advice given to the patient by phone or radio by physician or medical control. Also include a description of the care that you wished to provide for the patient.

Special Reporting Situations

In some instances, you may be required to file special reports with appropriate authorities. These may include incidents involving gunshot wounds, dog bites, certain infectious diseases, or suspected physical, sexual, or substance abuse. Learn your local requirements for reporting these incidents. Failure to report them may have legal consequences. It is important that the report be accurate, objective, and submitted in a timely manner. Also remember to keep a copy for your own records.

Another special reporting situation is a mass-casualty incident (MCI). The local MCI plan should have some means of recording important medical information temporarily (such as a triage tag that can be used later to complete the form). The standard for completing the form in an MCI is not the same as for a typical call. Your local plan should have specific guidelines.

You are the Provider   Summary

You feel that you aided in this patient's care by addressing her pain and fears during the call. Ignoring or failing to validate a patient's feelings can have a negative impact on the overall success of your care. Understanding how pain and fear can cloud a patient's thinking can help you know when to adjust your tone of voice for reassurance and slow down your speaking and simplify your sentences to communicate effectively.

Do not lie to patients or fabricate answers to their questions when you honestly do not know the answer.

RELEASE FROM RESPONSIBILITY WHEN PATIENT REFUSES IV THERAPY

This is to certify that I, _____, am refusing IV treatment. I acknowledge
patient's name
that I have been informed of the risk involved and hereby release the emergency medical services
provider(s), the physician consultant, and the consulting hospital from all responsibility for any
ill effects which may result from this action.

Witness _____   Signed _____
patient name or nearest relative

Witness _____   _____
relationship

RELEASE FROM RESPONSIBILITY WHEN PATIENT REFUSES SERVICE

This is to certify that I, _____, am refusing the services offered by the
patient's name
emergency medical services provider(s). I acknowledge that I have been informed of the risk
involved and hereby release the emergency medical services provider(s), the physician consultant,
and the consulting hospital from all responsibility for any ill effects which may result from this action.

Witness _____   Signed _____
patient name or nearest relative

Witness _____   _____
relationship

RELEASE FROM RESPONSIBILITY WHEN PATIENT REFUSES SERVICES
BUT ACCEPTS TRANSPORT

This is to certify that I, _____, am refusing _____
patient's name
_____. I acknowledge that I have been informed of the risk involved
and hereby release the emergency medical services provider(s), the physician consultant, and the
consulting hospital from all responsibility for any ill effects which may result from this action.
However, I do accept transportation to a medical facility.

Witness _____   Signed _____
patient name or nearest relative

Witness _____   _____
relationship

Figure 9-15 A competent adult patient has the right to refuse medical treatment and must sign a refusal form.

Prep Kit

Ready for Review

- Excellent communication skills are crucial in relaying pertinent information to the hospital before arrival.
- Radio and telephone communication links you and your team to other members of the EMS, fire, and law enforcement communities. This enables your entire team to work together more effectively.
- It is your job to know what your communication system can and cannot handle. You must be able to communicate effectively by sending precise, accurate reports about the scene, the patient's condition, and the treatment that you provide.
- There are many different forms of communication that an EMT-B must understand and be able to use. First, you must be familiar with two-way radio communications and have a working knowledge of mobile and hand-held portable radios. You must know when to use them and what type of information you can transmit.
- Remember, the lines of communication are not always exclusive; therefore, you should speak in a professional manner at all times.
- In addition to radio and oral communications with hospital personnel, EMT-Bs must have excellent person-to-person communication skills. You should be able to interact with the patient and any family members, friends, or bystanders.
- It is important for you to remember that people who are sick or injured may not understand what you are doing or saying. Therefore, your body language and attitude are very important in gaining the trust of both the patient and family. You must also take special care of individuals such as children, geriatric patients, and hearing-impaired, visually impaired, and non-English-speaking patients.
- Along with your radio report and oral report, you must also complete a formal written report about the patient before you leave the hospital. This is a vital part of providing emergency medical care and ensuring the continuity of patient care. This information guarantees the proper transfer of responsibility, complies with the requirements of health departments and law enforcement agencies, and fulfills your administrative needs.
- Reporting and record-keeping duties are essential, but they should never come before the care of a patient.

Vital Vocabulary

base station Any radio hardware containing a transmitter and receiver that is located in a fixed place.

cellular telephone A low-power portable radio that communicates through an interconnected series of repeater stations called "cells."

channel An assigned frequency or frequencies that are used to carry voice and/or data communications.

dedicated line A special telephone line that is used for specific point-to-point communications; also known as a "hot line."

duplex The ability to transmit and receive simultaneously.

Federal Communications Commission (FCC) The federal agency that has jurisdiction over interstate and international telephone and telegraph services and satellite communications, all of which may involve EMS activity.

MED channels VHF and UHF channels that the FCC has designated exclusively for EMS use.

paging The use of a radio signal and a voice or digital message that is transmitted to pagers ("beepers") or desktop monitor radios.

rapport A trusting relationship that you build with your patient.

repeater A special base station radio that receives messages and signals on one frequency and then automatically retransmits them on a second frequency.

scanner A radio receiver that searches or "scans" across several frequencies until the message is completed; the process is then repeated.

simplex Single-frequency radio; transmissions can occur in either direction but not simultaneously in both; when one party transmits, the other can only receive, and the party that is transmitting is unable to receive.

standing orders Written documents, signed by the EMS system's medical director, that outline specific directions, permissions, and sometimes prohibitions regarding patient care; also called protocols.

telemetry A process in which electronic signals are converted into coded, audible signals; these signals can then be transmitted by radio or telephone to a receiver at the hospital with a decoder.

UHF (ultra-high frequency) Radio frequencies between 300 and 3,000 MHz.

VHF (very high frequency) Radio frequencies between 30 and 300 MHz; the VHF spectrum is further divided into "high" and "low" bands.

Technology

- Interactivities
- Vocabulary Explorer
- Anatomy Review
- Web Links
- Online Review Manual

www.EMTB.com

Assessment in Action

You are teaching some new volunteers about your department's communication system. Answer the following questions.

1. How far from your mouth do you want to hold the radio while talking?

   A. 1" to 2"
   B. 2" to 3"
   C. 3" to 4"
   D. More than 4"

2. The Federal Communications Commission (FCC) is responsible for which of the following?

   A. Allocating radio frequencies for EMS
   B. Establishing limitations for transmitter power
   C. Monitoring radio operations
   D. All of the above

3. A special base station radio that receives messages and signals on one frequency and then automatically retransmits them on a second frequency is called a:

   A. repeater.
   B. scanner.
   C. duplex.
   D. flux capacitor.

You are transporting a patient who is a diabetic. The patient took his insulin but did not eat, and his blood glucose level is low. He is alert, but confused. His skin is cool and moist, and his vital signs include a blood pressure of 110/66 mm Hg, a pulse of 82 beats/min, and respirations of 16 breaths/min, nonlabored.

4. You are unable to contact medical control and proceed to treat this patient with oral glucose per protocol. What enables you to provide given care?

   A. Standing orders
   B. Rule of Nines
   C. Standard operating procedures
   D. Your code of ethics

5. You stabilized the patient and are en route to the emergency department. Which of the following reports would be the best to give?

   A. We are en route with a 45 y/o male diabetic. We fixed him and will be there in 5 minutes.
   B. We are en route with a 45 y/o male with a history of IDDM. Pt took his insulin today without eating. We found pt to be confused with a BS of 40 mg/dL. We administered oral glucose with desired effect. Pt is A/O x3. Vital signs BP 120/80, pulse, 80, respirations 12. We have a 5 minute ETA.
   C. We are en route with a 45 y/o male who is a diabetic. Vital signs are stable. Pt is better at this time. ETA 5 minutes.
   D. We are en route with a 45 y/o male with a history of diabetes. Pts blood glucose was low and oral glucose was given and pt is stable. ETA 5 minutes.

Challenging Questions

6. You respond to a scene where several bystanders witnessed a woman take what seemed to be an entire bottle of Tylenol. You find an empty pill bottle. The patient denies taking the medication and refuses treatment. The patient is A/O x 3. How would you handle this patient?

7. You arrive on scene of an unknown problem. The fire department has been on scene for 10 minutes and report the patient is just sleeping. Upon your assessment you find the patient to be apneic with weak pulses. You revive the patient with CPR and transport to the hospital. How do you document this call?

Points to Ponder

You respond to the scene of a homicide. The police have secured the scene and they need you to assess the patient. You find a young man lying face down on the floor. His skin is mottled. Rigor mortis has set in. He has several marks on the back of his neck. Under direction of medical control you call a DOA.

What precautions should you take going into the scene? How should you document the scene in your report?

Issues: Effective Patient Reporting and Documentation, Gathering Data at a Crime Scene.

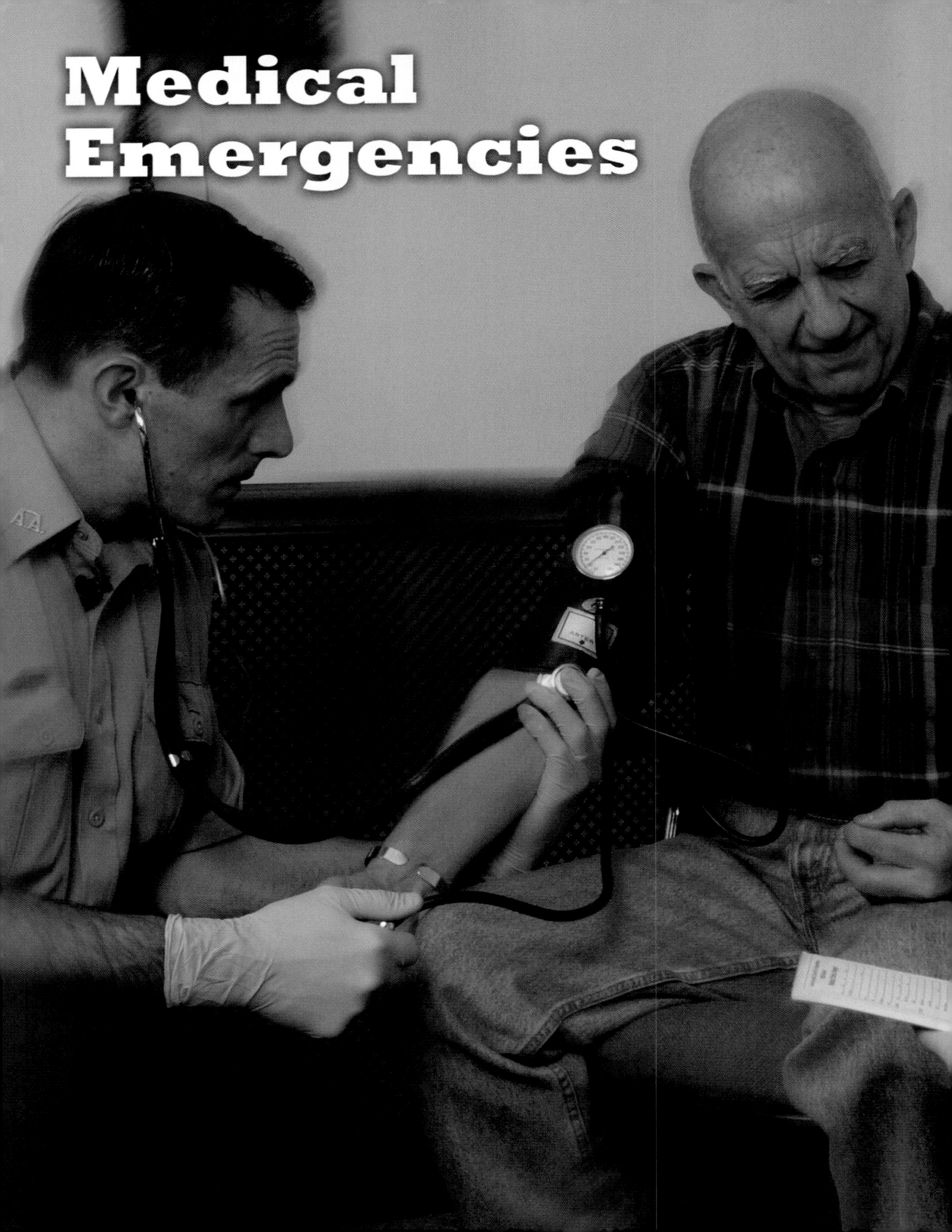

Medical Emergencies

Section

4

General Pharmacology

Objectives

Cognitive

4-1.1 Identify which medications will be carried on the unit. (p 349)

4-1.2 State the medications carried on the unit by the generic name. (p 349)

4-1.3 Identify the medications with which the EMT-B may assist the patient with administering. (p 352)

4-1.4 State the medications the EMT-B can assist the patient with by the generic name. (p 352)

4-1.5 Discuss the forms in which medications may be found. (p 346)

Affective

4-1.6 Explain the rationale for the administration of medications. (p 344)

Psychomotor

4-1.7 Demonstrate general steps for assisting the patient with self-administration of medications. (p 355)

4-1.8 Read the labels and inspect each type of medication. (p 356)

You are the Provider

You and your EMT-B partner are dispatched to a local golf course for a patient complaining of chest pain. Upon arrival you find a 62-year-old man clutching his chest while sitting. He is conscious, alert, and oriented. He is showing no signs of respiratory distress but states that he has a crushing pain in the center of his chest. The pain feels different than any previous episode. As your partner applies high-flow oxygen via a nonrebreathing mask, you continue to ask your patient questions.

1. Why would oxygen be an appropriate treatment?
2. What should your next step be?

General Pharmacology

Administering medications is a serious business. Used appropriately, a medication may alleviate pain and improve a patient's well-being. However, used inappropriately, medication may cause harm and even death. As an EMT-B, you will be responsible for administering certain medications to patients and helping them to self-administer others. You will ask patients about their medications and allergies, and you will report this information to hospital personnel. To act without understanding how medications work is to place patients and yourself in danger.

This chapter describes the various forms of medications, the different ways in which they can be administered, and how they work. It then takes a close look at each of the seven forms of medications you may be asked to administer or help patients to self-administer. It will also explain when it is dangerous to administer these medications.

How Medications Work

Pharmacology is the science of drugs, including their ingredients, preparation, uses, and actions on the body. Although the terms "drugs" and "medications" are often used interchangeably, the term drugs may make some people think of narcotics or illegal substances. For this reason, you should use the word medications,

especially when interviewing patients and families. In general terms, a medication is a chemical substance that is used to treat or prevent disease or relieve pain.

The dose is the amount of the medication that is given. The dose depends on the patient's weight or age; adults and children will get different amounts of the same medication. It also depends on the desired action of the medication. The action is the therapeutic effect that a medication is expected to have on the body. For example, nitroglycerin relaxes the walls of the blood vessels and may dilate the arteries. This increases the blood flow and, thus, the supply of oxygen, to the heart muscle. In this way, nitroglycerin relieves the squeezing or crushing pain that occurs with the cardiac condition called angina. Therefore, nitroglycerin is indicated for chest pain associated with angina. Indications are the reasons or conditions for which a particular medication is given.

There are times when you should not a give a patient medication, even if it usually is indicated for that person's condition. Such situations are called contraindications. A medication is contraindicated when it would harm the patient or have no positive effect on the patient's condition. For example, giving activated charcoal is indicated when a patient has swallowed a poison. Generally, activated charcoal, premixed with water, is used to prevent the body from absorbing a poison. However, activated charcoal would be contraindicated if the patient were unconscious and could not swallow.

Side effects are any actions of a medication other than the desired ones. Side effects may occur even when a medication is administered properly. For example, giving epinephrine to a patient who is having a severe allergic reaction should dilate the bronchioles and decrease wheezing. However, two side effects of epinephrine are cardiac stimulation and constriction of the arteries, which may elevate the patient's heart rate and blood pressure. These side effects are predictable; some others are not.

Medication Names

Medications usually have two types of names. The generic name of a medication (such as ibuprofen) is usually its original given name or the name which it is given by the original manufacturer. The generic name is not capitalized. Sometimes a medication is called by its generic name more often than by any of its trade names. For example, you may hear the term "nitroglycerin" used more often than the trade names Isordil and Nitrostat. All medications that are licensed for use in the United States are listed by their generic names in the United States Pharmacopoeia.

Technology

- Interactivities
- Vocabulary Explorer
- Anatomy Review
- Web Links
- Online Review Manual

www.EMTB.com

✱ *Study this!*

A trade name is the brand name that a manufacturer gives to a medication, such as Tylenol and Lasix. As a proper noun, a trade name begins with a capital letter. Trade names are used in every aspect of our daily lives, not just in medications. Well-known examples include Jell-O gelatin, Band-Aid adhesive bandages, and Hershey chocolate candy. A medication may have many different trade names, depending on how many companies manufacture it. Advil, Nuprin, and Motrin all are trade names for the generic medication, ibuprofen. A trade name sometimes is also designated by a raised registered symbol, that is, Advil®.

Medications may be prescription medications or over-the-counter (OTC) medications. Prescription medications are distributed to patients only by pharmacists according to a physician's order. Medications that are OTC may be purchased directly, such as from a discount store or supermarket, without a prescription. In recent years, the number of prescription medications that have become available OTC has increased dramatically.

You may come into contact with patients who have taken "street" drugs such as heroin or cocaine. Street drugs are pharmacologically active and will cause an effect.

Routes of Administration

Absorption is the process by which medications travel through body tissues until they reach the bloodstream. Often the rate at which a medication is absorbed into the bloodstream depends on its route of administration. (Table 10-1 ▶) lists common routes of medication administration.

- **Intravenous (IV) injection.** Intravenous means into the vein. Medications that need to enter the bloodstream immediately may be injected directly into a vein. This is the fastest way to deliver a chemical substance, but the IV route cannot be used for all chemicals. For example, aspirin, oxygen, and charcoal cannot be given by the IV route.
- **Oral.** Many medications are taken by mouth, or per os (PO), and enter the bloodstream through the digestive system. This process often takes as long as 1 hour.
- **Sublingual (SL).** Sublingual means under the tongue. Medications given by the SL route, such as nitroglycerin tablets, enter through the oral mucosa under the tongue and are absorbed into the bloodstream within minutes. This route is faster than the oral route, and it protects medications from chemicals in the digestive system, such as acids that can weaken or inactivate them.

TABLE 10-1 Routes of Administration and Rates of Absorption

Route	Rate
Intravenous	Immediate
Intraosseous	Immediate
Inhalation	Rapid
Rectal	Rapid
Sublingual	Rapid
Intramuscular	Moderate
Subcutaneous	Slow
Ingestion	Slow
Transcutaneous	Slow

- **Intramuscular (IM) injection.** Intramuscular means into the muscle. Usually, medications that are administered by IM injection are absorbed quickly because muscles have a lot of blood vessels. However, not all medications can be administered by the IM route. Possible problems with IM injections are damage to muscle tissue and uneven, unreliable absorption, especially in people with decreased tissue perfusion or who are in shock.
- **Intraosseous (IO).** Intraosseous means into the bone. Medications that are given by this route reach the bloodstream through the bone marrow. Giving a medication by the IO route, into the marrow, requires drilling a needle into the outer layer of the bone. Because this is painful, the IO route is used most often in patients who are unconscious as a result of cardiac arrest or extreme shock. Most commonly, the IO route is used for children who have fewer available (or difficult to access) IV sites.
- **Subcutaneous (SC) injection.** Subcutaneous means beneath the skin. An SC injection is given into the tissue between the skin and the muscle. Because there is less blood here than in the muscles, medications that are given by this route are generally absorbed more slowly, and their effects last longer. An SC injection is a useful way to give medications that cannot be taken by mouth, as long as they do not irritate or damage the tissue. Daily insulin injections for patients with diabetes are given by the SC route. Some forms of epinephrine can be given by the SC route. (Subcutaneous sometimes is abbreviated as SQ or sub-Q.)

EMT-B Safety

A needle withdrawn from a patient's skin after giving an injection is assumed to be contaminated with potentially infectious fluids. Handle contaminated "sharps" accordingly and dispose of them immediately according to your service's procedures for preventing infectious exposures.

- Transcutaneous. Transcutaneous means through the skin. Some medications can be absorbed transcutaneously, such as the nicotine in patches used by people who are trying to quit smoking. On occasion, a medication that also comes in another form is administered transcutaneously to achieve a longer-lasting effect. An example is an adhesive patch containing nitroglycerin.
- Inhalation. Some medications are inhaled into the lungs so that they can be absorbed into the bloodstream more quickly. Others are inhaled because they work in the lungs. Generally, inhalation helps minimize the effects of the medication in other body tissues. Such medications come in the form of aerosols, fine powders, and sprays.
- Per rectum (PR). Per rectum means by rectum. This route of delivery is frequently used with children because of easier administration and more reliable absorption. (Children often regurgitate some or all of a medication.) For similar reasons, many medications that are used for nausea and vomiting come in a rectal supposi-

tory form. Some medications to control seizures are administered PR when it is impossible to administer them intravenously. The PR route also is used to give some medications when the patient cannot swallow or is unconscious. Table 10-2 ▼ lists the words that are used for routes of medication delivery, along with their meanings.

Medication Forms

The form of a medication usually dictates the route of administration. For example, a tablet or a spray cannot be given through a needle. The manufacturer chooses the form to ensure the proper route of administration, the timing of its release into the bloodstream, and its effects on the target organs or body systems. As an EMT-B, you should be familiar with the following seven medication forms.

Tablets and Capsules

Most medications that are given by mouth to adult patients are in tablet or capsule form (Figure 10-1 ▶). Capsules are gelatin shells filled with powdered or liquid medication. If the capsule contains liquid, the shell is sealed and usually soft. If the capsule contains powder, the shell can usually be pulled apart. In tablets, the medication is compressed under high pressure. Tablets often contain other materials that are mixed with the medication.

Some tablets are designed to dissolve very quickly in small amounts of liquid so that they can be given sublingually and absorbed rapidly. An example is the sublingual nitroglycerin tablet used to treat chest pain

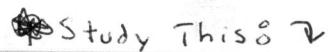 Study This ↓

TABLE 10-2 Routes of Administration: Words and Their Meanings		
This Word...	From These Latin Words...	Means
Inhalation	inhalatio (drawing air into the lungs)	inhaling or breathing in
Intramuscular (IM)	intra (into) and muscularis (of the muscles)	into muscle
Intraosseous (IO)	intra (into) and osse (bone)	into bone
Intravenous (IV)	intra (into) and venosus (of the veins)	into vein
Per os (PO)	per (by) and os (mouth)	by mouth
Per rectum (PR)	per (by) and rectum (rectum)	by rectum
Subcutaneous (SC)	sub (under) and cutis (skin)	under the skin
Sublingual (SL)	sub (under) and lingua (relating to the tongue)	under the tongue
Transcutaneous	trans (through) and cutis (skin)	through the skin

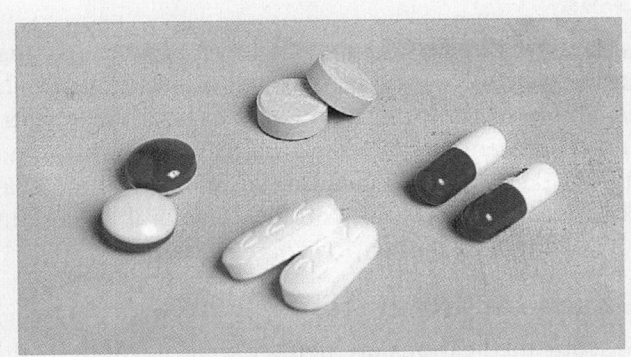

Figure 10-1 Tablets and capsules are typically taken by mouth and enter the bloodstream through the digestive system.

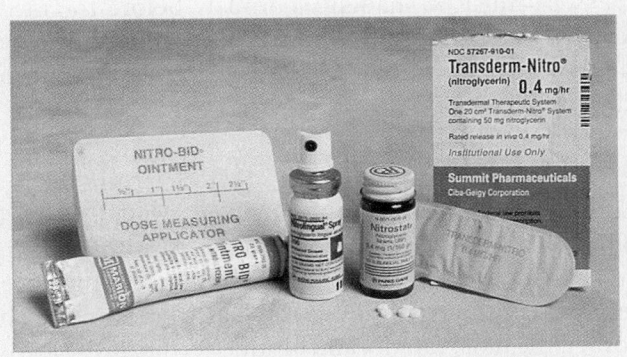

Figure 10-2 Nitroglycerin, which is prescribed for chest pain, is often given sublingually (SL) as a spray or a tablet.

in patients with cardiac conditions. These medications are especially useful in emergency situations. Generally, a medication that must be swallowed is less useful in an emergency because the digestive tract provides a slower route of delivery. For example, an oral pain medication is less useful than an IV pain medication when pain relief is needed within minutes.

Solutions and Suspensions *Study*

A solution is a liquid mixture of one or more substances that cannot be separated by filtering or allowing the mixture to stand. Solutions can be given by almost any route. When given by mouth, solutions may be absorbed from the stomach fairly quickly because the medication is already dissolved. For example, you may need to help in the sublingual delivery of a nitroglycerin spray (Figure 10-2 ▶). Many solutions can be given as an IV, IM, or SC injection. If a patient has a severe allergic reaction, you may help to administer a solution of epinephrine using an autoinjector.

Many substances do not dissolve well in liquids. Some of these can be ground into fine particles and evenly distributed throughout a liquid by shaking or stirring. This type of mixture is called a suspension. An example is activated charcoal, which you may give to patients who have taken overdoses of certain medications or ingested certain poisons.

Suspensions separate if they stand or are filtered. It is very important that you shake or swirl a suspension before administering it to ensure that the patient receives the right amount of medication. For example, if you are a parent, you may have had to shake a suspension of oral antibiotic before giving it to your child.

Suspensions usually are administered by mouth but sometimes are given rectally. Occasionally, suspensions

are applied directly to the skin to treat skin problems. You may have used calamine lotion in this way. Injectable suspensions are given via IM or SC injection only. Certain hormone shots or vaccinations are given this way because of the suspended particles. They cannot be given by IV injection because the suspended particles do not remain dissolved.

Metered-Dose Inhalers

If liquids or solids are broken into small enough droplets or particles, they can be inhaled. A metered-dose inhaler (MDI) is a miniature spray canister used to direct such substances through the mouth and into the lungs (Figure 10-3 ▼). An MDI delivers the same amount of medication each time it is used. Because an inhaled medication usually is suspended in a propellant,

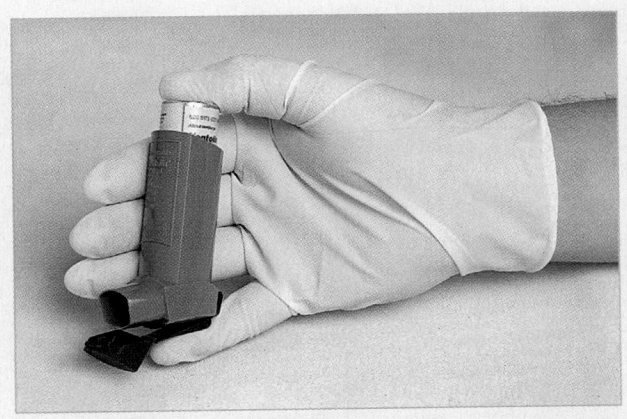

Figure 10-3 Some medications are inhaled into the lungs with a metered-dose inhaler so that they can be absorbed into the bloodstream more quickly.

Nytroglycerine- Vasal dialater

the MDI must be shaken vigorously before the medication is administered. An MDI is often used by a patient with respiratory illnesses such as asthma or emphysema.

Topical Medications

Lotions, creams, and ointments are <u>topical medications</u>, that is, they are applied to the surface of the skin and affect only that area. Lotions contain the most water, and ointments contain the least. Lotions are absorbed the most rapidly and ointments the most slowly. Calamine lotion is an example of a medical lotion. Hydrocortisone cream, to diminish skin itching, is an example of a medical cream that can also be given in ointment form. Neosporin is an example of a first-aid ointment.

Transcutaneous Medications

<u>Transdermal medications</u> are designed to be absorbed through the skin, or transcutaneously. Medications such as nitroglycerin paste usually have properties or delivery systems that help to dilate the blood vessels in the skin and, thus, speed absorption into the bloodstream. In contrast with most topical medicines, which work directly on the application site, transdermal medications are usually intended for systemic (whole-body) effects. A note of caution: If you touch such a medication with your bare skin while administering it, you will absorb it just as readily as the patient will.

One of the newer delivery systems for transcutaneous medications is the adhesive patch. Patches attach to the skin and allow even absorption of a medication for many hours (Figure 10-4 ▼). Prescription and OTC medications come in this form. Common examples are nitroglycerin, nicotine, some pain medications, and some oral contraceptives.

Gels

A <u>gel</u> is a semiliquid substance that is administered orally in capsule form or through plastic tubes. Gels usually have the consistency of pastes or creams but are transparent (clear). "Gelatinous" means thick and sticky, like gelatin. Depending on your local medical directives, as an EMT-B, you may give oral glucose in gel form to a patient with diabetes (Figure 10-5 ▼).

Gases for Inhalation

Gaseous medications are neither solid nor liquid and most often are given in an operating room. The medication that is most commonly used in gas form outside the operating room is oxygen. You might not think of oxygen as a medication because it is all around us and we all use it. However, in its concentrated form, it is a potent medication that has systemic effects, that is, effects throughout the body (Figure 10-6 ▼). You will

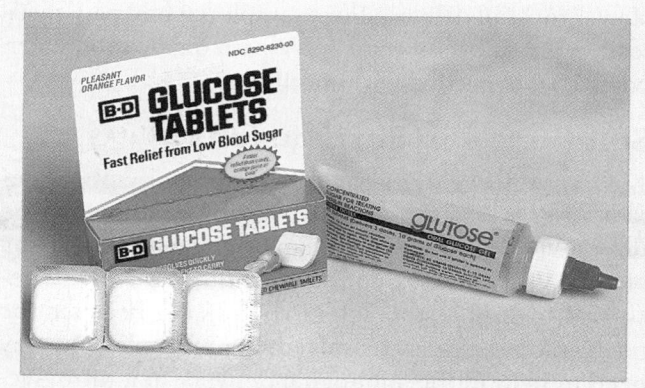

Figure 10-5 Oral glucose, used in diabetic emergencies, is available in gel and tablet forms.

Figure 10-4 Some medications are transcutaneous, or administered through the skin, such as the nitroglycerin patch shown.

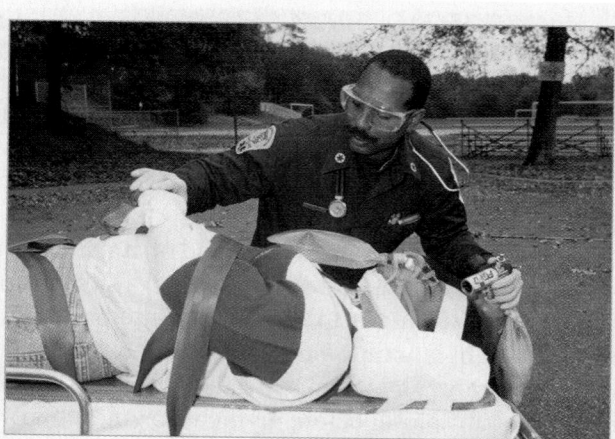

Figure 10-6 Oxygen is a potent medication that you will typically give through a nonrebreathing mask.

Documentation Tips

When documenting the use of oxygen, include the liter flow rate and the type of device used. For example, "Nonrebreathing mask at 15 L/min." Also document the patient's response to oxygen administration.

usually administer oxygen through a nonrebreathing mask or, occasionally, through a nasal cannula.

Medications Carried on the EMS Unit

The five medications that may be carried on the EMS unit are oxygen, oral glucose, activated charcoal, aspirin, and epinephrine. When used wisely, each can be a powerful tool. Keep in mind, however, that you may give these medications only according to standing orders in a protocol (off-line medical control) or a direct order (online medical control).

Oxygen

All cells need oxygen to function properly. The heart and brain, especially, cannot function for long if oxygen levels go down, which is why oxygen was chosen as an on-board medication for EMS units. If a patient is not breathing or is having trouble getting air into the lungs, you should administer supplemental oxygen. In general, you will be giving oxygen via a nonrebreathing mask at 10 to 15 L/min (or via nasal cannula at 2 to 6 L/min if the patient does not tolerate a nonrebreathing mask).

EMT-B Tips

Know These Medications

Under the US Department of Transportation's 1994 EMT-Basic National Standard Curriculum, an EMT-B is allowed to administer or help patients self-administer the following medications. Keep in mind that this list has been expanded by a number of states to include additional medications and other delivery methods.

You may be asked to administer these medications:

- Oxygen
- Activated charcoal
- Oral glucose

You may help patients self-administer these medications:

- Epinephrine autoinjector
- Metered-dose inhaler medications
- Nitroglycerin

However, you may administer or help to administer these medications only under the following conditions:

- A licensed physician gives you a direct order to administer a medication and/or the local medical protocols under which you are working permit you to administer that medication. Some local protocols exclude one or more of the six medications listed above.
- The local medical protocols, developed by a medical physician, under which you are working include standing orders for the use of a medication in defined situations. It is imperative that you do not give or help patients take any other medications under any circumstances.

You are the Provider — Part 2

As you obtain a focused history, you note that the onset of the pain was approximately 10 minutes before you arrived. The pain got worse when he was walking back to the golf cart and has not gotten any better since he has been sitting. The crushing pain is centered below the sternum, but he also has slight discomfort in his left shoulder. On a scale of 1 to 10, the pain is a 9. The patient states he was diagnosed with angina 3 years ago. This is the first time in more than 8 months that he has experienced chest pain. He has no allergies and takes nitroglycerin when needed. His last oral intake was a tuna sandwich about an hour ago, and he has been playing golf for about 45 minutes.

3. How does the information gained in your focused history assist in appropriately treating your patient?
4. What additional information is needed to move forward with your treatment?

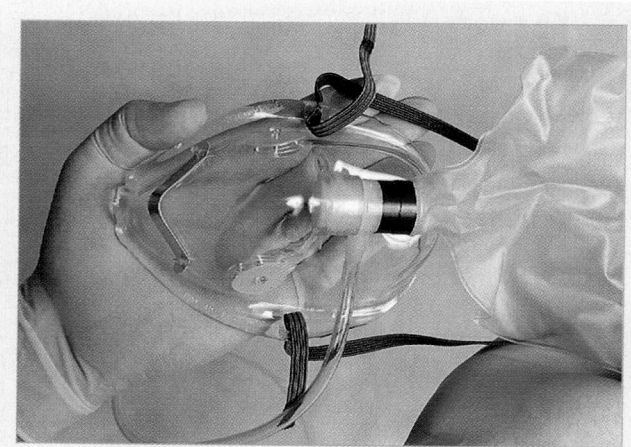

Figure 10-7 A nonrebreathing mask is the preferred method of giving oxygen because it provides up to 90% inspired oxygen.

However, if the patient is not breathing, you must also provide artificial ventilations, so you will need to use a bag-valve-mask device. Oxygen is usually delivered at 15 L/min with this technique.

Outside a hospital, the nonrebreathing mask is the preferred method of giving oxygen to patients who are experiencing significant respiratory difficulties or shock. With a good mask-to-mouth seal, this mask can provide up to 90% inspired oxygen (Figure 10-7 ▲). With a nasal cannula, oxygen flows through two small, tubelike prongs that fit into the patient's nostrils. This device can provide up to 44% inspired oxygen if the flowmeter is set at 6 L/min.

Remember that, although oxygen itself does not burn, it allows other things to burn. If there is extra oxygen in the air, objects will burn more easily. So make sure there are no open flames, lit cigarettes, or sparks in the area in which you are administering oxygen.

Activated Charcoal

Many poisoning emergencies involve overdoses of medications taken by mouth. Many medications bind with activated charcoal, keeping the medications from being absorbed by the body. Adsorption means to bind to or stick to a surface, while absorption is the process by which medications travel through body tissues until they reach the bloodstream. Activated charcoal is ground into a very fine powder to provide the greatest possible surface area for binding. You will probably carry a container with a premixed suspension of activated charcoal powder and water in the EMS unit, if allowed by local protocol (Figure 10-8 ▶).

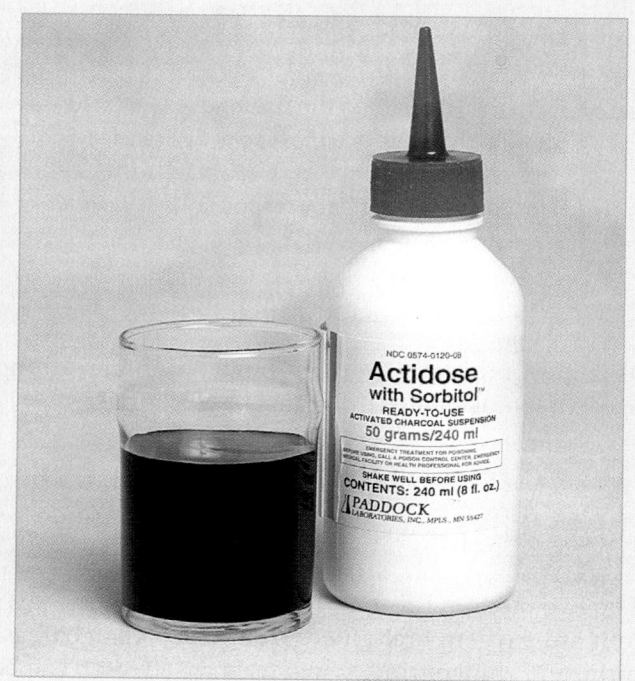

Figure 10-8 Activated charcoal is a suspension that is sometimes used for patients who have taken a medication overdose or swallowed a poison.

The bond between medication and charcoal is not permanent. Because the medication may break free and be absorbed into the bloodstream if activated charcoal remains in the digestive system throughout a normal day, charcoal is frequently suspended with another medication called sorbitol (a complex sugar). This suspension has a laxative effect that causes the entire mixture, including the medication, to move quickly through the digestive system.

Activated charcoal is given by mouth. Although sorbitol sweetens the suspension, the black charcoal makes it look unappealing. For this reason, you should use a covered container and ask the patient to drink the fluid through a straw.

Oral Glucose

Glucose is a sugar that our cells use as fuel. Although some cells can use other sugars, brain cells must have glucose. If the level of glucose in the blood gets too low, a person can lose consciousness, have seizures, and ultimately die.

The medical term for an extremely low blood glucose level is hypoglycemia. Hypoglycemia can be caused by an excess of insulin, which is taken to control blood glucose levels. Patients with diabetes who use insulin regularly understand the effects of this medication on the body. The <u>oral glucose</u> that is carried in the EMS unit can counteract the effects of hypoglycemia (abnormally low blood glucose level) in the same way as a candy bar or sweet drink, but faster. This is because common table sugar (sucrose) and fruit sugars (fructose) are complex sugars and must be broken down before they can be absorbed. Glucose is a simple sugar that is readily absorbed by the bloodstream.

Hospital personnel and paramedics can give glucose through an IV line. As an EMT-B, you can give glucose only by mouth. Glucose is available as a gel designed to be spread on the mucous membranes between the cheek and gum; however, absorption through this route is not as quick as with injection. Because the patient may be conscious one moment and unconscious the next, you must be very careful when administering oral glucose. Never administer oral medications to an unconscious patient or to one who is unable to swallow or protect the airway.

Aspirin

<u>Aspirin (acetylsalicylic acid, or ASA)</u> is an antipyretic (reduces fever), analgesic (reduces pain), and anti-inflammatory (reduces inflammation) and inhibits platelet aggregation (clumping). This last property makes it one of the most used medications today. Because research has shown that platelets aggregating under certain conditions in the coronary arteries is one of the direct causes of heart attack, patients at risk for coronary artery disease are often prescribed one or two "baby" (or children's) aspirins a day. During a potential heart attack, aspirin may be lifesaving.

Contraindications for aspirin include documented hypersensitivity to aspirin, preexistent liver damage, bleeding disorders, and asthma. Because of the association of aspirin with Reye syndrome, it should not be given to children during episodes of fever-causing illnesses.

Epinephrine

<u>Epinephrine</u> is the main hormone used to control the body's fight-or-flight response. It is released inside the body when there is sudden stress, such as during exercise or when the patient is suddenly scared. Because epinephrine is secreted by the adrenal glands, it is also known as adrenaline. Epinephrine has different effects on different body tissues and is used as a medication in several forms. Generally, epinephrine will increase heart rate and blood pressure and dilate passages in the lungs.

Therefore, it can ease breathing problems caused by the bronchial spasms that are common in asthma and allergic reactions. In a person who is close to anaphylactic shock as a result of an allergic reaction, epinephrine may also help to maintain the patient's blood pressure.

Epinephrine has the following characteristics:

- Secreted naturally by adrenal glands
- Dilates passages in lungs
- Constricts blood vessels, causing increased blood pressure
- Increases heart rate and blood pressure

Administering Epinephrine by Injection

Some states and EMS services now authorize the use of epinephrine by EMT-Bs for the treatment of life-threatening anaphylaxis. In certain individuals, insect venom or other allergens cause the body to release histamine, which lowers blood pressure by relaxing the small blood vessels and allowing them to leak. The release of histamine may also cause wheezing from bronchial spasms and swelling of the airway tissues (edema), which make it difficult for the patient to breathe. Epinephrine acts as a specific antidote to histamine, countering both of these harmful effects. It constricts the blood vessels, allowing blood pressure to rise and reducing the swelling. In the lungs, it has the opposite effect; it dilates the air passages, so the flow of air is less restricted. You can also expect the patient's heart rate to increase after administration of epinephrine.

You may be trained to administer SC and IM injections of epinephrine, depending on local protocol. Remember that an SC injection puts the epinephrine into the tissue between the skin and the muscle. Therefore, it is usually helpful to pinch the skin lightly to lift it away from the muscle. A syringe used for SC injections has a short, thin needle, typically between ½" and ⅝" long. The syringe for IM use has a longer, thicker needle that is between 1" and 1½" long so that it can reach into the muscle.

Follow these steps for an SC or IM injection. Before giving an injection, prepare the skin with an appropriate antiseptic (Figure 10-9A ▶). Insert the needle into the skin (or muscle). For an SC injection, pinch the skin while inserting the needle at a 45° angle. Then draw the syringe's plunger back slightly before injecting the medication (Figure 10-9B ▶).

Check to see whether any blood seeps into the syringe. If it does, you have accidentally placed the needle in a small blood vessel and will need to withdraw the needle and start again using the same syringe (assuming that the skin remains sterilized). If no blood

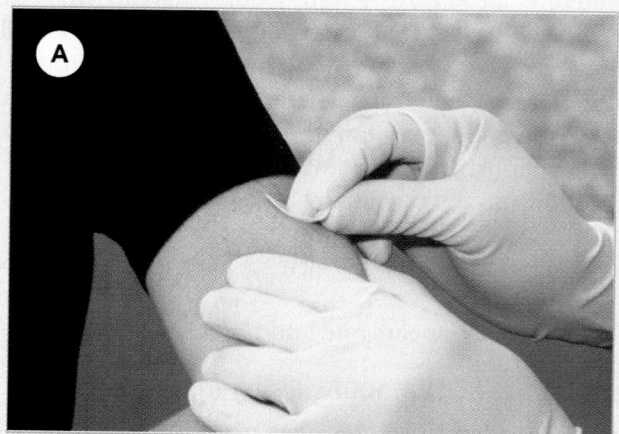

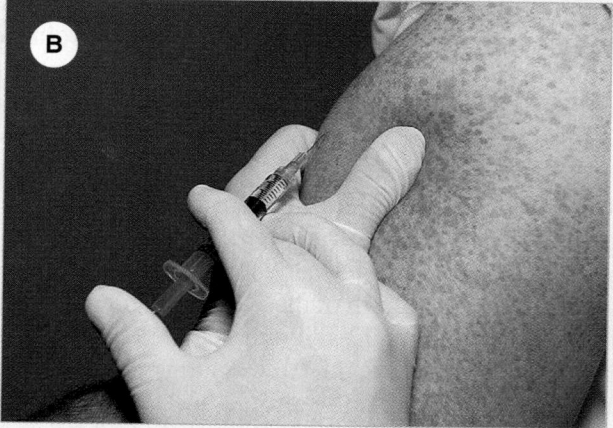

Figure 10-9 Administering an SC or IM injection. **A.** Cleanse the skin. **B.** Insert the needle, then draw the plunger back slightly.

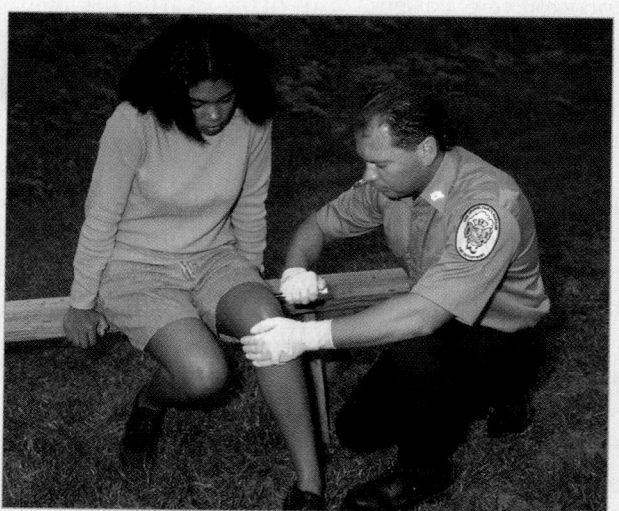

Figure 10-10 An EpiPen autoinjector may be used to administer a preset dose of epinephrine.

Documentation Tips

When documenting a medication, the proper form includes the name of the medication, dose and route, and vital signs before and after administration. For example:

10:30 AM—vital signs, pulse, 88 beats/min; respirations, 18 breaths/min; blood pressure, 128/68 mm Hg; nitroglycerin, 0.4 mg SL

10:35 AM—vital signs, pulse, 80 beats/min; respirations, 18 breaths/min; blood pressure, 124/60 mm Hg

returns when you pull back on the needle, push the plunger on the syringe to inject the medication. Once the needle has been injected into the patient's skin, it becomes contaminated with potential viruses and other infectious agents from the patient. Appropriate disposal precautions must be taken.

Epinephrine may also be dispensed from an autoinjector, which automatically delivers a preset amount of the medication (Figure 10-10 ▶). This is the method that you will most likely use. Be sure to familiarize yourself with the procedures for using the autoinjector on your unit. The general procedure is as follows:

1. Grasp unit with the tip pointing downward.
2. Form a fist around the unit.
3. With the other hand, pull off the activation cap.
4. Hold the tip near the outer part of the patient's thigh.
5. Swing and jab firmly into the outer thigh so that the unit is perpendicular (at a 90° angle) to the thigh. Do not allow it to bounce.

6. Hold firmly in the thigh for several seconds.

Regardless of the method used, epinephrine causes a burning sensation where it is injected, and the patient's heart rate will increase after the injection.

Patient-Assisted Medications

The three prescribed medications that you may help patients self-administer are epinephrine, MDI medications, and nitroglycerin. These medications, prescribed by doctors for self-administration by patients, have risks and benefits, and you must understand them.

Geriatric Needs

Geriatric patients often take many medications. They might also save medications left over from previous medical conditions. Make every effort to identify which medications are current and the conditions they are being used to treat. Ask family members to help distinguish current from outdated medications, or look at the expiration dates on the medication labels. Take a list of all of the patient's current medications or the drugs themselves with you to the emergency department.

Geriatric patients can become confused about their medication regimen. Uncertainty about whether they missed a dose may cause the patients to repeat the medication, possibly leading to an overdose. If you think an overdose has occurred, contact medical control.

Remember, medications can interact with each other, creating potentially harmful conditions. Even though a medication may be indicated for a special condition, it might be contraindicated in the presence of another medication. For example, if the patient is taking the heart medication propranolol (Inderal) and has an acute episode of shortness of breath, any asthma remedy might be made ineffective by the heart medication.

Although medications help people to recover from acute conditions and adjust to chronic diseases, they can pose serious problems for geriatric patients. You should distinguish current from previous medications, suspect accidental or intentional overdoses, and be prepared for potentially lethal medication interactions. Document all findings, and inform medical control.

Pediatric Needs

Children are not small adults, especially when it comes to the administration of medications. The approach to children differs from that for adults. First, doses of medications are different. Sometimes they are smaller; sometimes, however, they can be larger. This has to do with the way that children metabolize medications. Most of the assisted medications will be smaller doses. Let's look at epinephrine, for example: The dose for an adult is 0.3 to 0.5 mg, but for a child, it is 0.2 to 0.3 mg. The approach to the injection is also different; the injection for an adult is given in the lateral part of the thigh and for the child is given in the anterior or lateral part of the thigh. Location of the injection has to do with the development of blood vessels and muscle mass in the thigh. So to avoid hitting a blood vessel, to ensure that you are injecting into the muscle, and to ensure proper distribution of the medication, it is important to use the anterior or lateral part of the thigh in infants and small children. Children may not have the coordination needed to use an MDI. It will be easier if a spacer device is added to the inhaler to ensure the child receives the full benefit of the medicine. The last and most important issue in medication administration in children is affective. Children are not cognitively or emotionally the same as adults. Therefore, a little more time and effort to may be required to explain each procedure. It is also in your best interest to tell the child the truth. It is very important to gain the trust of the child in the short time you have to bond with him or her.

Epinephrine

You have already read the general information on epinephrine and its use by EMT-Bs for anaphylaxis. Some services do not permit EMT-Bs to carry epinephrine but do allow them to assist patients in administering their own epinephrine in life-threatening anaphylactic reactions. In addition, EMT-Bs frequently may assist patients in administering epinephrine through their own MDIs for bronchospasm.

MDI Medications

Sometimes, a respiratory condition such as asthma is not severe enough to require the use of epinephrine. In such cases, patients may use one of the epinephrine "cousins" that are more narrowly focused on the lungs. These medications are delivered with an MDI. An MDI requires a great deal of coordination, something that may be in short supply when an individual is having trouble breathing. Patients must aim properly and spray just as they start to inhale; however, most of the medication tends to end up on the roof of the patient's mouth.

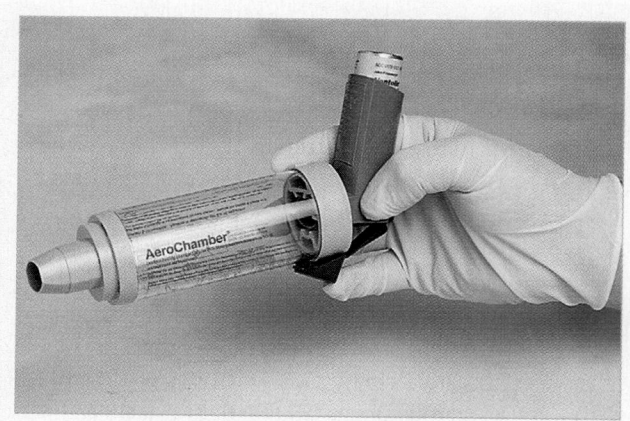

Figure 10-11 Some inhalers have spacer devices to better direct the medication spray.

An adapter, called a "spacer," which fits over the inhaler like a sleeve, can be used to avoid misdirecting the spray (Figure 10-11 ▲). The inhaler fits into an opening on one end of the spacer's chamber, and the mouthpiece fits on the other end. The patient sprays the prescribed dose into the chamber and then breathes in and out of the mouthpiece until the mist is completely inhaled.

You can activate the spray by pressing the canister into the adapter just as the patient starts to inhale. If relief is not achieved, wait about 3 minutes and repeat this sequence according to the patient's prescription for the MDI. Above all, it is important to ensure that the patient inhales all the medication in a single-sprayed dose.

Administering Epinephrine by MDI

Asthma, also known as "reactive airway disease," can be a life-threatening condition. Therefore, some patients use epinephrine inhalers to relieve bronchial spasms quickly. The trade names of some of these inhalers are Primatene Mist, Bronitin Mist, Bronkaid Mist, and Medihaler-Epi. Because epinephrine tends to increase the heart rate and blood pressure, most patients with asthma use certain chemical cousins of epinephrine that produce fewer side effects. Metaproterenol (Alupent or Metaprel) and albuterol (Proventil or Ventolin) work more on the bronchial spasms and less on the cardiovascular system. Give repeated doses per medical control and/or local protocol.

Nitroglycerin

Many patients with cardiac conditions carry some form of fast-acting nitroglycerin to relieve the pain of angina. Nitroglycerin is the same substance as TNT (trinitrotoluene), which is why the term "nitro" is used in medicine and in the explosives industry. The forms of nitroglycerin that you will use have been stabilized so that they are not explosive.

If you have ever run for a prolonged period, you probably remember that your muscles developed a painful, heavy, burning sensation. This is because the demand for oxygen by the muscles exceeded the supply. When heart muscle develops a similar pain, it is called angina pectoris. The cause is the same: not enough oxygen, in this case because of a blockage or narrowing in the blood vessels that supply the heart. Occasionally, the cause is a spasm in these blood vessels. Unlike the runner with sore legs, of course, the

You are the Provider   Part 3

As your partner obtains baseline vital signs, you continue to obtain additional information from your patient. Your patient hands you a small silver pillbox containing what he claims is his nitroglycerin. You find no prescription or patient information on the container. He states that he took one pill as soon as the chest pain began, but he felt no relief. Vital signs show a pulse of 124 beats/min and regular; respirations of 18 breaths/min; a blood pressure of 136/80 mm Hg; and clear and equal breath sounds. His pupils are equal, round, and reactive to light.

5. Of what significance is the blood pressure of 136/80 mm Hg to the use of nitroglycerin?
6. Would an additional dose of nitroglycerin be appropriate in this situation?

heart muscle cannot stop and rest until the pain goes away.

The purpose of nitroglycerin is to increase blood flow by relieving the spasms or causing the arteries to dilate. It does this by relaxing the muscular walls of the coronary arteries and veins. Nitroglycerin also relaxes veins throughout the body, so less blood is returned to the heart and the heart does not have to work as hard each time it contracts. In short, blood pressure is decreased. Because of this, however, it is important that you always take the patient's blood pressure before administering nitroglycerin. If the systolic blood pressure is less than 100 mm Hg, the nitroglycerin may have the harmful effect of lowering the blood flow to the heart's own blood vessels. Even a patient who has adequate blood pressure should sit or lie down with head elevated before taking this medication. If the patient is standing, he or she may faint when blood flow to the brain is reduced as the nitroglycerin starts to work. If a significant drop in the patient's blood pressure (15 to 20 mm Hg) occurs and the patient suddenly feels dizzy or sick, lay the patient down and raise the legs.

During a heart attack (myocardial infarction, or MI), a blood clot forms in a narrowed coronary artery, blocking the blood flow to a section of the heart muscle. If the blockage is not cleared in time, that section of the heart muscle will die. If nitroglycerin no longer brings relief to a person in whom it has previously worked, the person may be experiencing an MI instead of an angina attack. Therefore, it is important to know how much nitroglycerin a patient has needed in the past to relieve chest pain and how much has been taken during the current emergency, including the use of nitroglycerin patches. Always report this information to medical control. Remember, you cannot administer this medication without clearance from medical control or standing orders.

The medication sildenafil (Viagra) can have potentially fatal interactions with nitroglycerin. Ask a patient who has been prescribed nitroglycerin if he has used Viagra within the previous 24 hours. Report this to medical control.

Nitroglycerin has the following effects:

- Relaxes the muscular walls of coronary arteries and veins
- Results in less blood returning to the heart
- Decreases blood pressure
- Relaxes veins throughout the body
- Often causes a mild headache after administration

Administering Nitroglycerin by Tablet

Nitroglycerin is usually taken sublingually. The patient places a tiny tablet under the tongue, where it dissolves. The tablet should create a slight tingling or burning sensation. If the nitroglycerin has lost its usual "bite," it may have lost potency because of aging or improper storage. Be sure to check the expiration date on the bottle.

Sublingual nitroglycerin tablets should be stored in their original glass container with the cap screwed on tightly. Note that what looks like cotton in the container is actually rayon. If real cotton is placed in the container, it can absorb nitroglycerin, thus reducing the potency of the tablets. Other medications placed in the container can likewise rob nitroglycerin of its power. Exposure to light, heat, or air may degrade the strength of the medication as well. If you notice any signs of improper storage, be sure to include that information in the patient's medical history.

Administering Nitroglycerin by Metered-Dose Spray

Some patients who take nitroglycerin use a metered-dose spray, which deposits medication on or under the tongue. Each spray is equivalent to one tablet. To ensure direct, proper dosing on the bottom of the tongue, do not use a spacer with the metered-dose canister when giving nitroglycerin by this method.

Whether using the tablets or the metered-dose spray, you should wait 5 minutes for a response before repeating the dose. Closely monitor the patient's vital signs, particularly the blood pressure. Give repeated doses per medical control and/or local protocol.

General Steps in Administering Medication

As an EMT-B, you must be familiar with the five general steps of administering any medication to a patient:

1. **Obtain an order from medical control.** This order may be given to you directly, through online medical control via telephone or radio. Or it may be indirect, through protocols that contain standing orders for the administration of certain medications. For example, your system may use a protocol that describes how the medical director wants you to deal with a patient who is

having respiratory difficulties. Part of this protocol may direct you to use a nonrebreathing mask to deliver oxygen to such a patient at 15 L/min. You may do this without calling on-line medical control if the patient meets the criteria of the protocol.

2. **Verify the proper medication and prescription.** You have received and confirmed the medication order and determined that the patient is still a candidate for the medication. You must now make sure that the medication you are about to give is the correct medication. Carefully read the label. If it is the patient's own prescription, the bottle may show the trade name or the generic name. If you have any questions, contact online medical control. Make sure that the medication is the patient's own and does not belong to a friend or relative. You should never give a medication to a patient that has been prescribed for someone else.

3. **Verify the form, dose, and route of the medication.** You have confirmed your order and verified that the medication is the correct one to give. Now you must make sure that the form of the medication, the dose, and the route all match the order you received. For example, suppose that you are told to give the patient a sublingual nitroglycerin tablet. The patient's nitroglycerin tablet bottle is empty, but he has another bottle of nitroglycerin capsules. These are to be swallowed four times a day. The medication is the same, but the form, dose, and route of delivery are different from the order given. You may not substitute the capsules for the tablets without specific orders from medical control.

4. **Check the expiration date and condition of the medication.** The last step before administering a medication is to make sure the expiration date has not passed. Prescription and OTC medications alike should have an expiration date on their labels. Check the date. If no date can be found, you should examine the medication with suspicion. In addition, if you find discoloration, cloudiness, or particles in a liquid medication, you should not use it. If a patient with asthma gives you an MDI and the expiration date on it is smudged, you should not use it.

5. **Reassess the vital signs,** especially heart rate and blood pressure, at least every 5 minutes or sooner if the patient's condition changes.

6. **Document.** Remember the EMS rule: The work is not done until the paperwork is done. Once the medication has been given, you must document your actions and the patient's response. This includes the time you gave the medication and the name, dose, and route of administration. Did the patient's condition improve, worsen, or not change? Were there any side effects? A second EMS rule says, "If you did not write it down, it did not happen." If your performance should ever be questioned, documentation is your best defense.

In addition to getting approval to administer a medication, you must verify the online order, which means that you should restate the name of the medication, the dose to be given, and the route of administration to medical control. Verifying the order will help to reduce the chance of making an error in medication administration.

Next, you must reconfirm that the patient can tolerate the medication. For example, suppose that

You are the Provider

Part 4

You are unable to verify the needed information about the nitroglycerin to administer an additional dose. You call for advanced life support backup. While the advanced life support unit is en route, you prepare your patient for transport. You ensure that he continues to receive high-flow oxygen and is in a position of comfort. You reassess his vital signs every 3 to 5 minutes.

7. What information is necessary to administer an additional dose of nitroglycerin?

EMT-B Tips

General Steps in Administering Medication

1. Obtain an order from medical control.
2. Verify the proper medication and prescription.
3. Verify the form, dose, and route of the medication.
4. Check the expiration date and condition of the medication.
5. Reassess the vital signs, especially heart rate and blood pressure, at least every 5 minutes or as the patient's condition changes.
6. Document.

Geriatric Needs

<u>Polypharmacy</u> is a term referring to the use of multiple medications by one person. It is not uncommon today to find patients, especially elderly patients, taking many medications on a regular basis. Often, the prescription regimens can be complex and confusing. The medications may be prescribed by multiple physicians. The person may also be taking nonprescription and herbal medicines. Add to this the possibility of failing memory and confusion, and the potential for overdosing, underdosing, and harmful drug interactions increases exponentially.

you have received and verified the order to give one sublingual nitroglycerin tablet to a patient with a cardiac condition. While you were getting the order, however, the patient began to sweat more and became less responsive. A repeated blood pressure reading is 80/60 mm Hg. Using your knowledge about nitroglycerin, you would decide not to give the medication. Instead, you would notify medical control of the changes in the patient's condition and seek further orders.

Last, if the patient is having chest pain "just like my other heart attack," you may be able to apply oxygen, but you may have to call medical control or follow standing orders before helping to administer the patient's nitroglycerin spray. Knowing and understanding the local protocols under which you will be working are absolutely essential.

Patient Medications

Part of your patient assessment includes finding out what medications your patient is taking. This information may provide vital clues to your patient's condition that may help guide your treatment or be extremely useful to the emergency department physician. Often, knowing what medications a patient takes may be the only way you can determine what chronic or underlying conditions your patient may have, such as when a patient is unable to relate his medical history to you. The patient may be unresponsive, confused, not knowledgeable of his or her medical history, uncooperative, or unable to communicate. Discovering what the patient takes and taking the medications or a list of them with you to the emergency department can be crucial in assessing your patient's needs.

In addition to prescription medications, patients often take nonprescription OTC medications and herbal medications. Many times, they do not consider these substances "medications" and will not report them to you unless you ask about them specifically. Yet, they may be as potent as prescription medications and can have interactions and effects on the patient's health and condition that are just as important. Be sure to ask specifically about these also. (Table 10-3 ▶) lists the top 100 prescribed medications and their uses.

Patients are naturally reluctant to tell you about any illegal drugs they may have taken or overdoses of medication. It is important to ask, and you can assure them that your only interest in asking is to be able to treat them appropriately.

TABLE 10-3 Some Common Prescribed Medications

Category	Drug Name—generic (Trade)	Description
High blood pressure medications	■ atenolol ■ furosemide ■ hydrochlorothiazide ■ triamterene/hydrochlorothiazide (HCTZ) ■ quinapril (Accupril) ■ ramipril (Altace) ■ valsartan (Diovan) ■ amlodipine/benazepril (Lotrel) ■ losartan (Cozaar) ■ valsartan/HCTZ (Diovan HCT) ■ benazepril (Lotensin) ■ clonidine ■ losartan/HCTZ (Hyzaar)	Lower blood pressure by reducing blood volume, affecting the heart, or dilating blood vessels.
Other cardiac medications	■ atorvastatin (Lipitor) ■ amlodipine (Norvasc) ■ lisinopril ■ metoprolol (Lopressor and Toprol XL) ■ simvastatin (Zocor) ■ pravastatin (Pravachol) ■ clopidogrel (Plavix) ■ potassium chloride ■ warfarin (Coumadin) ■ verapamil ■ digoxin ■ diltiazem ■ fenofibrate (Tricor)	Lower blood pressure, decrease work on the heart, strengthen heartbeat, or lower cholesterol (fat) in the blood. Some of these drugs are also used in relieving heart pain.
Respiratory medications	■ albuterol *MDI* ■ cetirizine (Zyrtec) ■ fexofenadine (Allegra) ■ montelukast (Singulair) ■ salmeterol/fluticasone (Advair Diskus) ■ fluticasone (Flonase) ■ mometasone (Nasonex) ■ fexofenadine/pseudoephedrine (Allegra D) ■ fluticasone propionate (Flovent) ■ ipratropium/albuterol (Combivent)	Improve airflow in and out of the lungs or decrease secretions in the respiratory tract. Some of these drugs are used to control allergy symptoms such as sneezing and watery eyes.

Through Inhaler [handwritten annotation]

TABLE 10-3 Some Common Prescribed Medications (continued)

Category	Drug Name—generic (Trade)	Description
Analgesics	▪ hydrocodone/acetaminophen (APAP) ▪ propoxyphene N/APAP ▪ ibuprofen ▪ celecoxib (Celebrex) ▪ acetaminophen/codeine ▪ valdecoxib (Bextra) ▪ naproxen ▪ oxycodone/APAP ▪ oxycodone ▪ rofecoxib (Vioxx; withdrawn from the market, but some patients might have this medication from previous prescriptions)	Decrease pain. A few are narcotics or controlled substances and have potential for abuse. Some also lower fever and fight inflammation.
Behavioral medications	▪ alprazolam ▪ sertraline (Zoloft) ▪ zolpidem (Ambien) ▪ fluoxetine ▪ venlafaxine (Effexor and Effexor SR) ▪ lorazepam ▪ citalopram (Celexa) ▪ bupropion (Wellbutrin, Wellbutrin SR, and Wellbutrin XL) ▪ paroxetine (Paxil) ▪ amitriptyline ▪ escitalopram (Lexapro) ▪ trazodone ▪ risperidone (Risperdal) ▪ olanzapine (Zyprexa) ▪ methylphenidate XR (Concerta)	This group includes sedatives, sleeping medications, and drugs to fight depression or other mental conditions. The sustained release form of bupropion is also used to help people stop smoking tobacco.
Endocrine and hormone medications	▪ levothyroxine ▪ conjugated estrogens (Premarin) ▪ norgestimate/ethinyl estradiol (Ortho Tri-Cyclen) ▪ glipizide (Glucotrol and Glucotrol XL) ▪ norelgestromin/ethinyl estradiol (Ortho Evra) ▪ rosiglitazone (Avandia) ▪ pioglitazone (Actos) ▪ glyburide ▪ metformin (Glucophage and Glucophage XR) ▪ glimepiride (Amaryl) ▪ glyburide/metformin (Glucovance)	Includes hormone replacement (thyroid or estrogen), birth control medications, and drugs used to control blood glucose levels in diabetes.

Continued.

TABLE 10-3 Some Common Prescribed Medications (continued)

Category	Drug Name—generic (Trade)	Description
Antibiotic, antibacterial, and antifungal medications	■ azithromycin (Zithromax) ■ amoxicillin ■ cephalexin ■ amoxicillin/clavulanate ■ levofloxacin (Levaquin) ■ fluconazole ■ penicillin VK ■ ciprofloxacin (Cipro) ■ sulfamethoxazole/trimethoprim	Fight bacterial or fungal infections.
Stomach and intestinal tract medications	■ lansoprazole (Prevacid) ■ esomeprazole (Nexium) ■ pantoprazole (Protonix) ■ ranitidine ■ omeprazole ■ rabeprazole (Aciphex)	Decrease acid production in the stomach and intestinal tract and allow ulcers to heal. May also be used to stop heartburn (burning sensation in throat and upper part of the chest).
Other medications	■ alendronate (Fosamax) ■ prednisone ■ gabapentin (Neurontin) ■ clonazepam ■ sildenafil (Viagra) ■ cyclobenzaprine ■ tamsulosin (Flomax) ■ raloxifene (Evista) ■ risedronate (Actonel) ■ latanoprost (Xalatan)	Alendronate, raloxifene, and risedronate used to prevent osteoporosis (weakened bones) in postmenopausal women. Prednisone used in asthma, allergic reactions, severe arthritis, cancer, and other conditions. Gabapentin used to treat seizures or neurologic (nerve) pain. Clonazepam prevents seizures. Sildenafil aids in erectile dysfunction (lowered male sexual arousal). Cyclobenzaprine used to relax or decrease spasms in muscles. Tamsulosin used to help older men with enlarged prostate glands start and sustain urine flow. Latanoprost used in glaucoma (increased pressure within the eye).

Note: Within each group, the drugs are listed from the most common to the least common. *Data Source:* rx.list.com. Accessed November 29, 2004.

You are the Provider   Summary

When you are administering or assisting in the administration of any medication, it is imperative that you follow certain steps. You must:

1. Obtain an order from medical direction.
2. Verify that you are giving the proper medication and prescription.
3. Verify the form, dose, and route of the medication.
4. Check the expiration date and the condition of the medication.
5. Reassess vital signs, especially heart rate and blood pressure every 3 to 5 minutes or as the patient's condition changes.

Prep Kit

Ready for Review

- Medications come in seven forms: tablets and capsules, solutions and suspensions, metered-dose inhalers, topical medications, transdermal medications, gels, and gases.
- Medications may be administered through nine routes: intravenous, intramuscular, or subcutaneous injection; orally; sublingually; intraosseously; transcutaneously; by inhalation; and by rectum.
- In all but the intravenous injection route, the medication is absorbed into the bloodstream through various body tissues. These routes of administration often determine the speed with which the medication takes effect.
- Three medications are typically carried on the EMS unit: oxygen, oral glucose, and activated charcoal. Two medicines have recently been added to the list by some states and services: aspirin and epinephrine.
- There are three additional medications that you may help the patient self-administer: metered-dose inhaler medications, nitroglycerin, and epinephrine. Remember, though, that the medications may differ depending on local protocol.
- The administration of any medication requires approval by medical control, through direct orders given online or standing orders that are part of the local protocols.
- There are six steps to follow in administering medications, four of which occur before you give the medication: Obtain an order from medical control, verify the proper medication, verify the dose and route, and check the expiration date of the medication. The fifth step is to reassess vital signs, and the sixth is to accurately document the patient's history, assessment, treatment, and response.

Technology

- Interactivities
- Vocabulary Explorer
- Anatomy Review
- Web Links
- Online Review Manual

Vital Vocabulary

absorption The process by which medications travel through body tissues until they reach the bloodstream.

action The therapeutic effect of a medication on the body.

activated charcoal An oral medication that binds and adsorbs ingested toxins in the gastrointestinal tract for treatment of some poisonings and medication overdoses. Charcoal is ground into a very fine powder that provides the greatest possible surface area for binding medications that have been taken by mouth; it is carried on the EMS unit.

adsorption The process of binding or sticking to a surface.

aspirin (acetylsalicylic acid, or ASA) A medication that is an antipyretic (reduces fever), analgesic (reduces pain), anti-inflammatory (reduces inflammation), and potent inhibitor of platelet aggregation (clumping).

contraindications Conditions that make a particular medication or treatment inappropriate, for example, a condition in which a medication should not be given because it would not help or may actually harm a patient.

dose The amount of medication given on the basis of the patient's size and age.

epinephrine A medication that increases heart rate and blood pressure but also eases breathing problems by decreasing muscle tone of the bronchiole tree; you may be allowed to help the patient self-administer the medication.

gel A semiliquid substance that is administered orally in capsule form or through plastic tubes.

generic name The original chemical name of a medication (in contrast with one of its "trade names"); the name is not capitalized.

hypoglycemia An abnormally low blood glucose level.

indications The therapeutic uses for a specific medication.

inhalation Breathing into the lungs; a medication delivery route.

intramuscular (IM) injection An injection into a muscle; a medication delivery route.

intraosseous (IO) Into the bone; a medication delivery route.

Prep Kit continued...

intravenous (IV) injection An injection directly into a vein; a medication delivery route.

metered-dose inhaler (MDI) A miniature spray canister through which droplets or particles of medication may be inhaled.

nitroglycerin A medication that increases cardiac perfusion by causing arteries to dilate; you may be allowed to help the patient self-administer the medication.

oral By mouth; a medication delivery route.

oral glucose A simple sugar that is readily absorbed by the bloodstream; it is carried on the EMS unit.

over-the-counter (OTC) medications Medications that may be purchased directly by a patient without a prescription.

oxygen A gas that all cells need for metabolism; the heart and brain, especially, cannot function without oxygen.

per os (PO) Through the mouth; a medication delivery route; same as oral.

per rectum (PR) Through the rectum; a medication delivery route.

pharmacology The study of the properties and effects of medications.

polypharmacy The use of multiple medications on a regular basis.

prescription medications Medications that are distributed to patients only by pharmacists according to a physician's order.

side effects Any effects of a medication other than the desired ones.

solution A liquid mixture that cannot be separated by filtering or allowing the mixture to stand.

subcutaneous (SC) injection Injection into the tissue between the skin and muscle; a medication delivery route.

sublingual (SL) Under the tongue; a medication delivery route.

suspension A mixture of ground particles that are distributed evenly throughout a liquid but do not dissolve.

topical medications Lotions, creams, and ointments that are applied to the surface of the skin and affect only that area; a medication delivery route.

trade name The brand name that a manufacturer gives a medication; the name is capitalized.

transcutaneous Through the skin; a medication delivery route.

transdermal medications Medications that are designed to be absorbed through the skin (transcutaneously).

Points to Ponder

You have been dispatched to a patient complaining of shortness of breath. Upon arrival you find a 24-year-old man. He states he has had a "bad cold" and cannot stop coughing. He states his chest hurts every time he takes a breath. You obtain a SAMPLE history and learn the following. The patient has no allergies and no medical history except a cold. He has no prescribed medications but has been using a metered-dose inhaler all day, which was given to him by a friend. His blood pressure is 126/80 mm Hg, pulse rate is 78 beats/min, and respiratory rate is 20 breaths/min. The patient refuses transport and asks for another inhaler.

What may be happening with this patient? What could you tell him to convince him to accept emergency care?

Issues: Misuse of Prescription Medication, BSI Precautions, Patient Refusal of Treatment.

Assessment in Action

It has been a quiet shift. As you are completing your station duties, the tones go off. "Ambulance Three respond, 1431 Mariposa Street for a respiratory problem." Your partner locates the call in the map book and you are off. Upon arrival you and your partner put on personal protective equipment and carefully approach the house. The scene appears safe.

You are greeted by the patient, a 28-year-old man, and his mother. The patient appears anxious. You complete an initial assessment. He is pale and obviously short of breath. You also notice raised red spots on his arms. You begin administering high-flow oxygen to the patient and continue your assessment. The patient's mother tells you her son has a history of severe allergic reactions and has been prescribed an EpiPen. Your assessment discloses the following vital signs: a heart rate of 124 beats/min, respirations of 24 breaths/min and labored, a blood pressure of 100/68 mm Hg, and a pulse oximetry reading of 90%. You listen to lung sounds and hear wheezing in both lungs. Additionally the patient tells you it feels like his throat is "closing up."

1. The first step in assisting this patient with his medication is to:
   A. get permission from the family.
   B. place the patient in the ambulance.
   C. obtain an order from medical control.
   D. call the patient's doctor.

2. Epinephrine can be administered by use of an EpiPen or:
   A. orally.
   B. SQ/IM injections.
   C. pill.
   D. intercardiac injection.

3. After administering any patient medication, you must reassess the patient's condition to see whether the medication is working and:
   A. wait and see.
   B. call dispatch to record times.
   C. walk the patient to the ambulance.
   D. document all pertinent information.

4. In some situations medications should not be administered to a patient if there is a:
   A. contraindication.
   B. nonemergent situation.
   C. shortage of the medication.
   D. delay in transporting the patient.

5. When epinephrine is used to treat anaphylaxis (severe allergic reaction), the patient's heart rate increases due to a:
   A. side effect of the medication.
   B. increase in blood pressure.
   C. rise in the patient's body temperature.
   D. defect in the medication.

6. In an effort to protect you and your partner from infectious diseases, the EpiPen must be immediately placed in:
   A. a trash container.
   B. a coffee can.
   C. a sharps container.
   D. the box it came in.

7. The preferred method for administering oxygen to this patient would be:
   A. blow-by method.
   B. nonrebreathing mask.
   C. simple pediatric mask.
   D. nasal cannula.

8. The desired effect of epinephrine will help this patient by:
   A. speeding up the respiratory rate.
   B. changing the lymph tissue.
   C. dilating passages in the lungs.
   D. slowing down the heart rate.

Challenging Questions

9. What are the different routes for administering medications and how do they differ?

10. List the names and uses of the five primary medications that must be carried on the EMS unit.

11. What is "polypharmacy" and where is it most likely found?

12. What is the importance of glucose (sugar) in the body and how can the EMT-B assist with hypoglycemia?

www.EMTB.com

Respiratory Emergencies

Objectives

Cognitive

4-2.1 List the structure and function of the respiratory system. (p 366)

4-2.2 State the signs and symptoms of a patient with breathing difficulty. (p 367)

4-2.3 Describe the emergency medical care of the patient with breathing difficulty. (p 384)

4-2.4 Recognize the need for medical direction to assist in the emergency medical care of the patient with breathing difficulty. (p 385)

4-2.5 Describe the emergency medical care of the patient with breathing distress. (p 384)

4-2.6 Establish the relationship between airway management and the patient with breathing difficulty. (p 384)

4-2.7 List signs of adequate air exchange. (p 367)

4-2.8 State the generic name, medication forms, dose, administration action, indications, and contraindications for the prescribed inhaler. (p 381, 385)

4-2.9 Distinguish between the emergency medical care of the infant, child, and adult patient with breathing difficulty. (p 387)

4-2.10 Differentiate between upper airway obstruction and lower airway disease in the infant and child patient. (p 369)

Affective

4-2.11 Defend EMT-B treatment regimens for various respiratory emergencies. (p 388)

4-2.12 Explain the rationale for administering an inhaler. (p 385)

Psychomotor

4-2.13 Demonstrate the emergency medical care for breathing difficulty. (p 384)

4-2.14 Perform the steps in facilitating the use of an inhaler. (p 386)

You are the Provider

You and your EMT-B partner are dispatched to 1465 Dalles Military Rd for a 33-year-old woman with difficulty breathing. You arrive at the office building and are immediately met by a man who seems very upset and identifies himself as the patient's coworker. As you follow him through a labyrinth of cubicles, he tells you that the patient has had breathing problems before, but he's never seen it this bad. He leads you to a woman who is standing with her arms outstretched on the desk in front of her and a metered-dose inhaler in her right hand. You introduce yourself, and she acknowledges your presence with a nod. When you ask her what is wrong, she is only able to answer you with a two-word response, "can't breathe," and you hear audible wheezes without using your stethoscope.

You are a transporting agency, but because of the nature of the call, ALS was simultaneously dispatched. Although paramedics are en route, they typically have a 20-minute travel time to your location.

1. How significant is the person's response to your question and why?
2. What should you do next? Should you transport this patient or wait for ALS to arrive on the scene?

Respiratory Emergencies

The feeling of being short of breath or having difficulty breathing is a complaint that you will encounter often. It is a symptom of many different conditions, from the common cold and asthma to heart failure and pulmonary embolism. You may or may not be able to determine what is causing dyspnea in a particular patient; this can be difficult even for physicians in a hospital setting. Also, several different problems may contribute to a patient's dyspnea at the same time, including some that are serious or life threatening. Even without a definitive diagnosis, however, you may still be able to save a life.

This chapter begins with a basic explanation of how the lungs function. It then looks at common medical problems that can impede normal functioning and cause dyspnea, including acute pulmonary edema, chronic obstructive pulmonary disease, and asthma.

You will learn the signs and symptoms of each condition. You should keep all these possible medical problems in mind as you take the patient's history and perform a physical assessment, a process that the chapter describes in detail. The information that you collect will help you to decide on the proper treatment, which differs according to the probable cause of the dyspnea.

Technology

Interactivities

Vocabulary Explorer

Anatomy Review

Web Links

Online Review Manual

www.EMTB.com

Remember, the sensation of not getting enough air can be terrifying, regardless of its cause. As an EMT-B, you should be prepared to treat not just the symptom and the underlying problem, but also the anxiety that it produces.

Lung Structure and Function

The respiratory system consists of all the structures of the body that contribute to the breathing process. Important anatomic features include the upper and lower airways, the lungs, and the diaphragm (Figure 11-1 ▶). Air enters the upper airway through the nose and mouth and moves past the epiglottis into the trachea. It then moves along the bronchial tubes to the air spaces, called alveoli, where oxygen and carbon dioxide are exchanged.

The principal function of the lungs is respiration, which is the exchange of oxygen and carbon dioxide. The two processes that occur during respiration are inspiration, the act of breathing in or inhaling, and expiration, the act of breathing out or exhaling. During respiration, oxygen is provided to the blood, and carbon dioxide is removed from it. This exchange of gases takes place rapidly in normal lungs at the level of the alveoli (Figure 11-2 ▶). Alveoli are microscopic, thin-walled air sacs that lie against the pulmonary capillary vessels. Oxygen and carbon dioxide must be able to pass freely between the alveoli and the capillaries. Oxygen entering the alveoli from inhalation passes through tiny passages in the alveolar wall into the capillaries, which carry the oxygen to the heart. The heart pumps the oxygen around the body. Carbon dioxide produced by the body's cells returns to the lungs in the blood that circulates through and around the alveolar air spaces. The carbon dioxide diffuses back into the alveoli and travels back up the bronchial tree and out the upper airways during exhalation (Figure 11-3 ▶). Again, carbon dioxide is "exchanged" for oxygen, which travels in exactly the opposite direction (during inhalation).

The brain stem senses the level of carbon dioxide in the arterial blood. The level of carbon dioxide bathing the brain stem stimulates a healthy person to breathe. If the level drops too low, the person automatically breathes at a slower rate and less deeply. As a result,

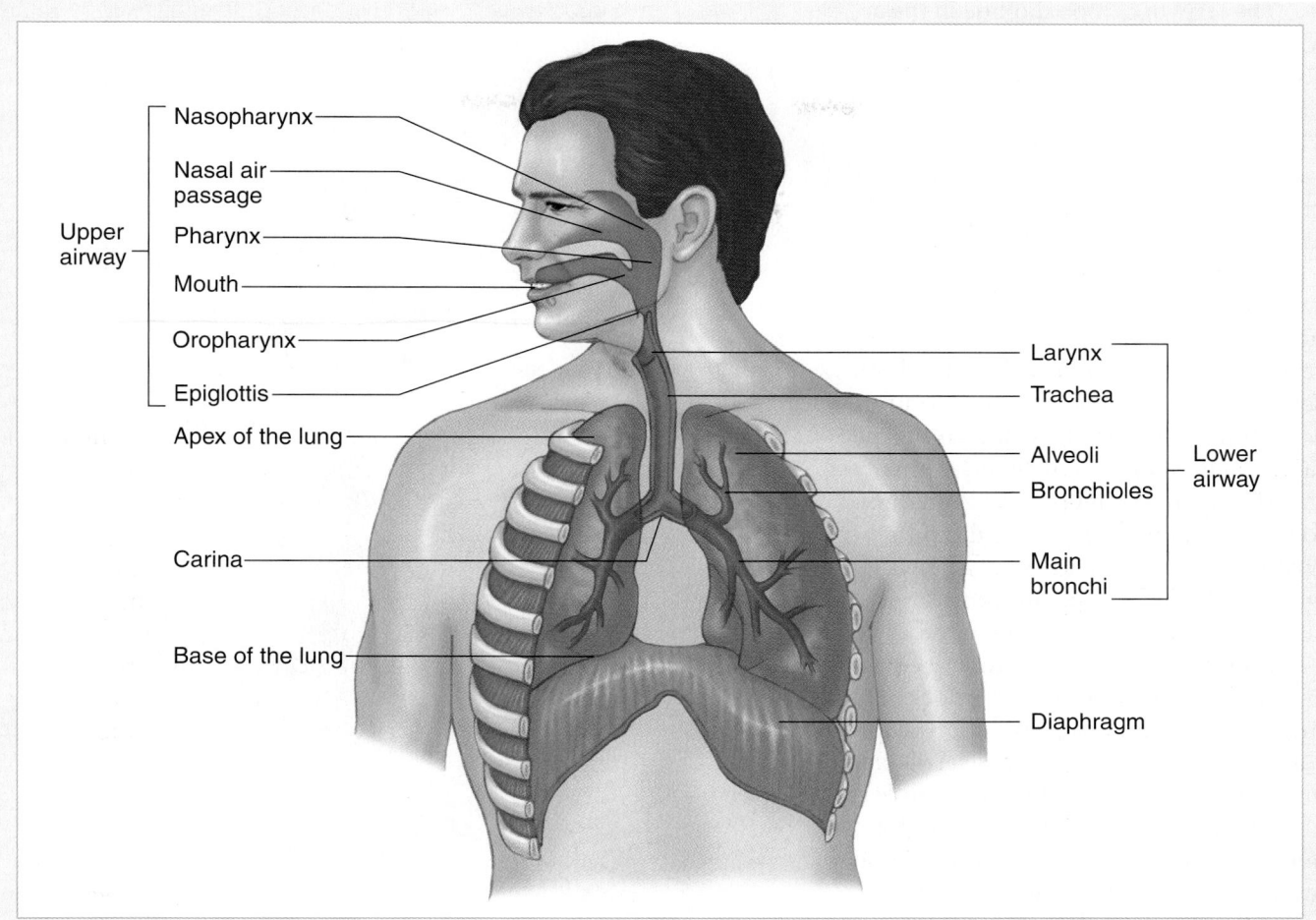

Figure 11-1 The upper airway includes the mouth, nose, pharynx, and larynx. The lower airway includes the trachea, major bronchi, and other air passages within the lungs.

less carbon dioxide is expired, allowing carbon dioxide levels in the blood to return to normal. If the level of carbon dioxide in the arterial blood rises above normal, the patient breathes more rapidly and more deeply. When more fresh air (containing no carbon dioxide) is brought into the alveoli, more carbon dioxide diffuses out of the bloodstream, thereby lowering the level.

The following are the characteristics of adequate breathing:

- A normal rate and depth
- A regular pattern of inhalation and exhalation
- Good audible breath sounds on both sides of the chest
- A regular rise and fall movement on both sides of the chest

- Pink, warm, dry skin

The following are signs of inadequate breathing:

- A rate of breathing that is slower than 12 breaths/min or faster than 20 breaths/min
- Unequal chest expansion
- Decreased breath sounds on one or both sides of the chest
- Muscle retractions above the clavicles, between the ribs, and below the rib cage, especially in children
- Pale or cyanotic skin
- Cool, damp (clammy) skin
- Shallow or irregular respirations
- Pursed lips
- Nasal flaring

The level of carbon dioxide in the arterial blood can rise for a number of reasons. The exhalation process may be impaired by various types of lung disease. The body may also produce too much carbon dioxide, either temporarily or chronically, depending on the disease or abnormality.

If, over a period of years, arterial carbon dioxide levels rise slowly to an abnormally high level and remain there, the respiratory center in the brain, which senses carbon dioxide levels and controls breathing, may work less efficiently. The failure of this center to respond normally to a rise in arterial levels of carbon dioxide is called chronic carbon dioxide retention. If the condition is severe, respiration will stop unless there is a secondary drive, called hypoxic drive, to stimulate the respiratory center. Fortunately, a second stimulus does help in patients with chronically high blood carbon dioxide levels—a low level of oxygen in the blood. Low blood oxygen levels cause the respiratory center to respond and stimulate respiration. If the arterial level of oxygen is then raised, which happens when the patient is given additional oxygen, there is no longer any stimulus to breathe; both the high carbon dioxide and low oxygen drives are lost. Patients with chronic lung diseases frequently have a chronically high level of carbon dioxide in the blood. Therefore, giving too much oxygen to these patients may actually depress, or completely stop, the respirations.

In most disorders of the lung, one or more of the following situations exists:

■ The pulmonary veins and arteries are actually obstructed from absorbing oxygen or releasing carbon dioxide by fluid, infection, or collapsed air spaces.
■ The alveoli are damaged and cannot transport gases properly across their own walls.
■ The air passages are obstructed by muscle spasm, mucus, or weakened floppy airway walls.
■ Blood flow to the lungs is obstructed by blood clots.
■ The pleural space is filled with air or excess fluid, so the lungs cannot properly expand.

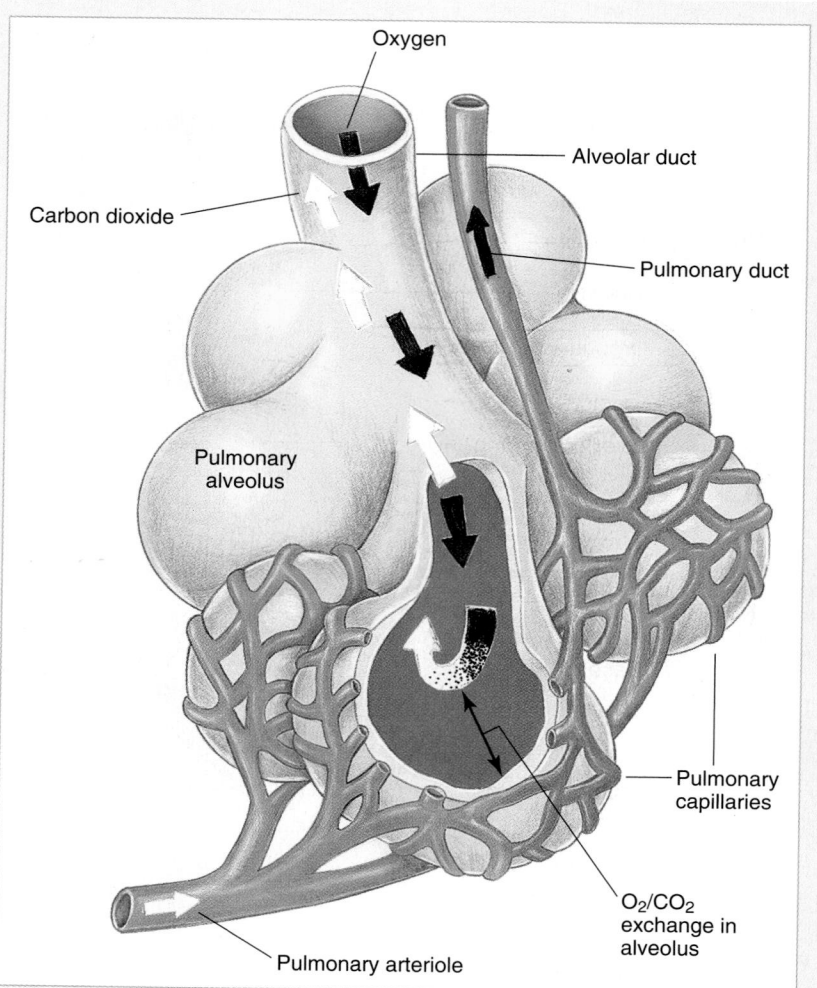

Figure 11-2 An enlarged view of a single alveolus (air sac) showing where the exchange of oxygen and carbon dioxide between air in the sac and blood in the pulmonary capillaries takes place.

All these conditions prevent the proper exchange of oxygen and carbon dioxide. In addition, the pulmonary blood vessels themselves may have abnormalities that interfere with blood flow and thus with the transfer of gases.

Causes of Dyspnea

Dyspnea is shortness of breath or difficulty breathing. Many different medical problems may cause dyspnea. Be aware that if the problem is severe and the brain is deprived of oxygen, the patient may not be alert enough to complain of shortness of breath. More commonly, altered mental status is a sign of hypoxia of the brain.

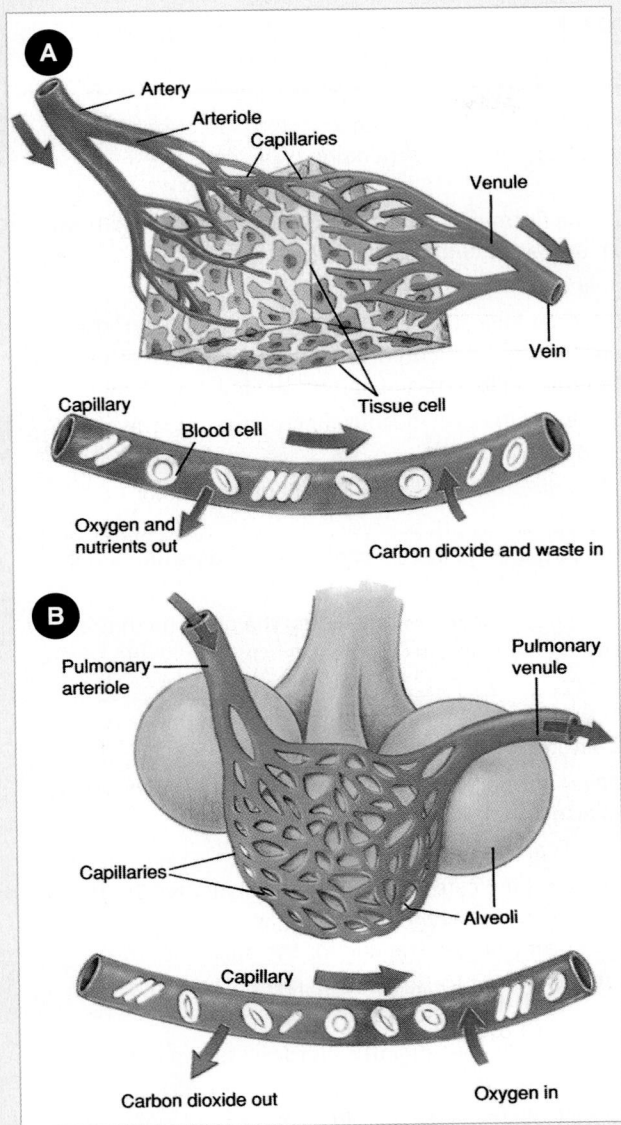

Figure 11-3 The exchange of oxygen and carbon dioxide in respiration. **A.** Oxygen passes from the blood through capillaries to tissue cells. Carbon dioxide passes from tissue cells through capillaries to the blood. **B.** In the lungs, oxygen is picked up by the blood and carbon dioxide is given off.

🖐 EMT-B Safety

If you suspect that a patient has an airborne disease, place a surgical mask (or a nonrebreathing mask if needed) on the patient. When you have specific reason to suspect tuberculosis, do this and also wear a HEPA respirator yourself. See Chapter 2 for a detailed discussion on disease transmission precautions.

Patients often develop breathing difficulty or hypoxia with the following medical conditions:

- Upper or lower airway infection
- Acute pulmonary edema
- Chronic obstructive pulmonary disease (COPD)
- Spontaneous pneumothorax
- Asthma or allergic reactions
- Pleural effusion
- Prolonged seizures
- Obstruction of the airway
- Pulmonary embolism
- Hyperventilation
- Severe pain, particularly chest pain

Upper or Lower Airway Infection

Infectious diseases causing dyspnea may affect all parts of the airway. Some cause mild discomfort. Others obstruct the airway to the point that patients require a full range of respiratory support. In general, the problem is always some form of obstruction, either to the flow of air in the major passages (colds, diphtheria, epiglottitis, and croup) or to the exchange of gases between the alveoli and the capillaries (pneumonia). (Table 11-1 ▶) shows infectious diseases that are associated with some degree of dyspnea. *[handwritten: Right sided Heart failure cannot cause Left-sided heart failure But it can happen the other way around]*

Acute Pulmonary Edema *[handwritten: (Pump problem)]*

Sometimes, the heart muscle is so injured after a heart attack or other illness that it cannot circulate blood properly. In these cases, the left side of the heart cannot remove blood from the lung as fast as the right side delivers it. As a result, fluid builds up within the alveoli as well as in the lung tissue between the alveoli and the pulmonary capillaries. This accumulation of fluid in the space between the alveoli and the pulmonary capillaries, called <u>pulmonary edema</u>, can develop quickly after a major heart attack. By physically separating alveoli from pulmonary capillary vessels, the edema interferes with the exchange of carbon dioxide and oxygen (Figure 11-6 ▶). There is not enough room left in the lung for slow, deep breaths. The patient usually experiences dyspnea with rapid, shallow respirations. In the most severe instances, you will see a frothy pink sputum at the nose and mouth.

In most cases, patients have a longstanding history of chronic congestive heart failure that can be kept under control with medication. However, an acute onset may occur if the patient stops taking the medication, eats food that is too salty, or has a stressful illness, a new heart attack, or an abnormal *[handwritten: pump, container, fluid (3 you need to know)]*

TABLE 11-1 Infectious Diseases Associated With Dyspnea

Disease	Characteristics
<u>Bronchitis</u>	■ An acute or chronic inflammation of the lung that may damage lung tissue, usually associated with cough and production of sputum and, depending on its cause, sometimes fever. ■ Fluid also accumulates in the surrounding normal lung tissue, separating the alveoli from their capillaries. (Sometimes, fluid can also accumulate in the pleural space.) ■ The lung's ability to exchange oxygen and carbon dioxide is impaired. ■ The breathing pattern in bronchitis does not indicate major airway obstruction, but the patient may experience tachypnea, an increase in the breathing rate, which is an attempt to compensate for the reduced amount of normal lung tissue and for the buildup of fluid.
<u>Common cold</u>	■ A viral infection usually associated with swollen nasal mucous membranes and the production of fluid from the sinuses and nose. ■ Dyspnea is not severe; patients complain of "stuffiness" or difficulty breathing through the nose.
<u>Diphtheria</u>	■ Although well controlled in the past decade, it is still highly contagious and serious when it occurs. ■ The disease causes the formation of a diphtheritic membrane lining the pharynx that is composed of debris, inflammatory cells, and mucus. This membrane can rapidly and severely obstruct the passage of air into the larynx.
<u>Pneumonia</u>	■ An acute bacterial or viral infection of the lung that damages lung tissue, usually associated with fever, cough, and production of sputum. ■ Fluid also accumulates in the surrounding normal lung tissue, separating the alveoli from their capillaries. (Sometimes, fluid can also accumulate in the pleural space.) ■ The lung's ability to exchange oxygen and carbon dioxide is impaired. ■ The breathing pattern in pneumonia does not indicate major airway obstruction, but the patient may experience tachypnea, an increase in the breathing rate, which is an attempt to compensate for the reduced amount of normal lung tissue and for the buildup of fluid.
<u>Epiglottitis</u> (Figure 11-4 ▶)	■ A bacterial infection of the epiglottis that can produce severe swelling of the flap over the larynx. ■ In preschool and school-aged children especially, the epiglottis can swell to two to three times its normal size. ■ The airway may become almost completely obstructed, sometimes quite suddenly. ■ <u>Stridor</u> (harsh, high-pitched, continued rough barking inspiratory sounds) may be heard late in the development of airway obstruction. ■ Acute epiglottitis in the adult is characterized by a severe sore throat. ■ The disease is now much less common than it was 20 years ago because of a vaccine that can help to prevent most cases.
<u>Croup</u> (Figure 11-5 ▶)	■ An inflammation and swelling of the whole airway—pharynx, larynx, and trachea—typically seen in children between ages 6 months and 3 years. ■ The common signs of croup are stridor and a seal-bark cough, which signal a significant narrowing of the air passage of the trachea that may progress to significant obstruction. ■ Croup often responds well to the administration of humidified oxygen.
<u>Severe Acute Respiratory Syndrome (SARS)</u>	■ A virus that has caused significant concern. SARS is a serious, potentially life-threatening viral infection caused by a recently discovered family of viruses best known as the second most common cause of the common cold. SARS usually starts with flu-like symptoms, which may progress to pneumonia, respiratory failure, and, in some cases, death. SARS is thought to be transmitted primarily by close person-to-person contact.

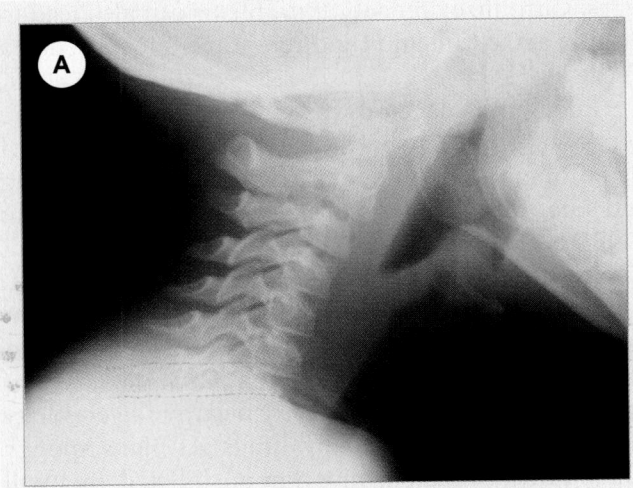

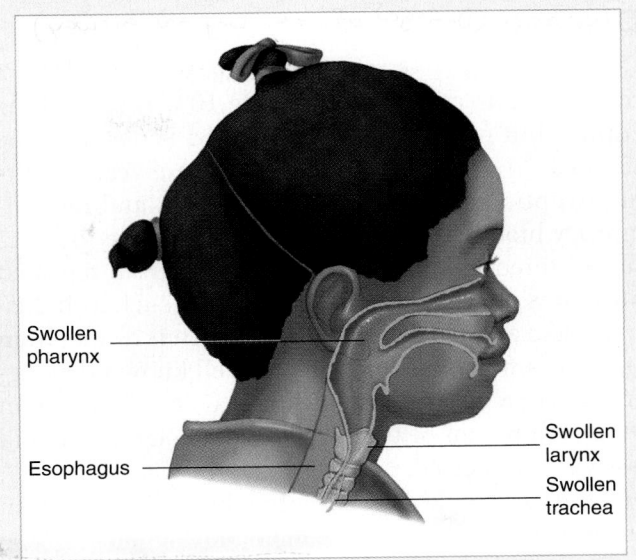

Figure 11-5 Croup results in swelling of the whole airway—pharynx, larynx, and trachea.

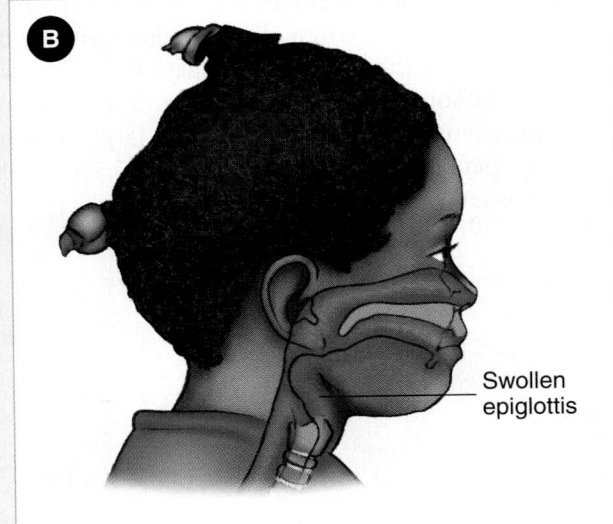

Figure 11-4 Acute epiglottitis. **A.** Epiglottitis is caused by a bacterial infection resulting in severe swelling of the epiglottis. **B.** The epiglottis is massively swollen and almost fully obstructs the airway.

heart rhythm. Pulmonary edema is one of the most common causes of hospital admission in the United States. It is not uncommon for a patient to have repeated bouts.

Some patients who have pulmonary edema do not have heart disease. Poisonings from inhaling large amounts of smoke or toxic chemical fumes can produce pulmonary edema, as can traumatic injuries of the chest. In these cases, fluid collects in alveoli and lung tissue in response to damage to the tissues of the lung or the bronchi.

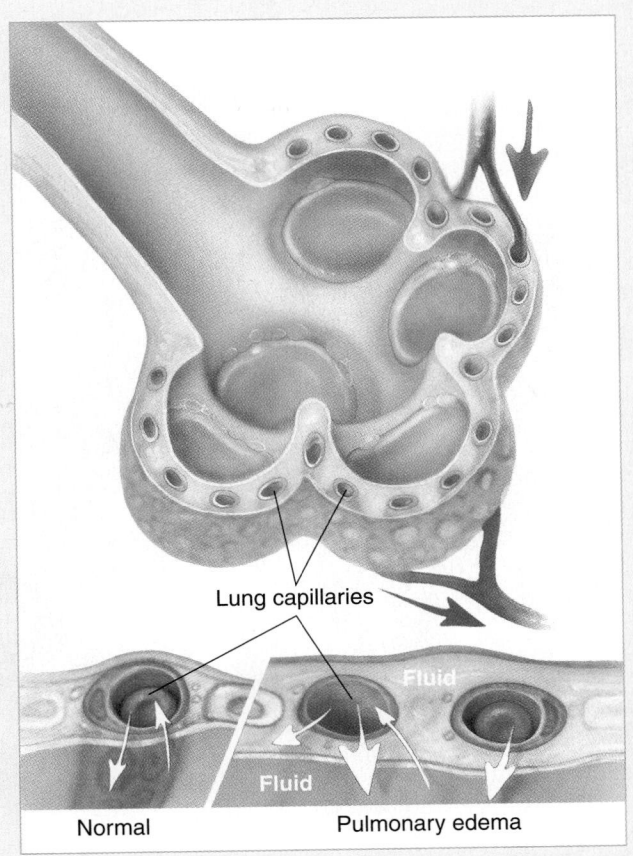

Figure 11-6 In pulmonary edema, fluid fills the alveoli and separates the capillaries from the alveolar wall, interfering with the exchange of oxygen and carbon dioxide.

Chronic Obstructive Pulmonary Disease (Damaged of the lung and airway)

Chronic obstructive pulmonary disease (COPD) is a common lung condition, affecting 10% to 20% of the entire adult population in the United States. It is the end of a slow process, which over several years results in disruption of the airways, the alveoli, and the pulmonary blood vessels. The process itself may be a result of direct lung and airway damage from repeated infections or inhalation of toxic agents such as industrial gases and particles, but most often it results from cigarette smoking. Although it is well known that cigarettes are a direct cause of lung cancer, their role in the development of COPD is far more significant and less well publicized.

Tobacco smoke is itself a bronchial irritant and can create a chronic bronchitis, an ongoing irritation of the trachea and bronchi.

With bronchitis, excess mucus is constantly produced, obstructing small airways and alveoli. Protective cells and lung mechanisms that remove foreign particles are destroyed, further weakening the airways. Chronic oxygenation problems can also lead to right heart failure and fluid retention, such as edema in the leg.

Pneumonia develops easily when the passages are persistently obstructed. Ultimately, repeated episodes of irritation and pneumonia cause scarring in the lung and some dilation of the obstructed alveoli, leading to COPD (Figure 11-7 ▼).

Another type of COPD is called emphysema. Emphysema is a loss of the elastic material around the air spaces as a result of chronic stretching of the alveoli when inflamed airways obstruct easy expulsion of gases. Smoking can also directly destroy the elasticity of the lung tissue. Normally, lungs act like a spongy balloon that is inflated; once they are inflated, they will naturally recoil because of their elastic nature, expelling gas rapidly. However, when they are constantly obstructed or when the "balloon's" elasticity is diminished, air is no longer expelled rapidly, and the walls of the alveoli eventually fall apart, leaving large "holes" in the lung that resemble a large air pocket or cavity. This condition is called emphysema.

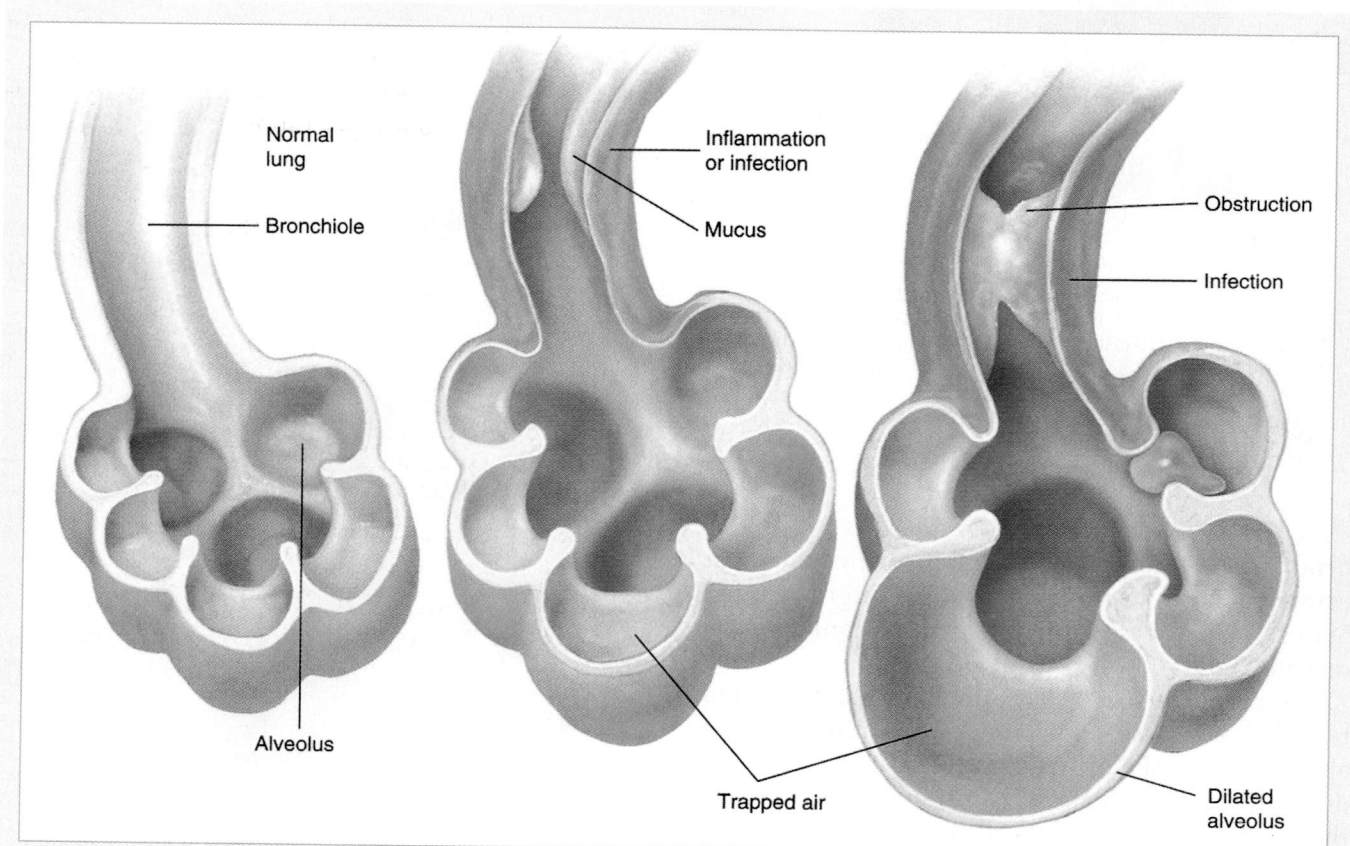

Figure 11-7 Repeated episodes of irritation and inflammation in the alveoli result in the obstruction, scarring, and some dilation of the alveolar sac characteristic of COPD.

Most patients with COPD have elements of both chronic bronchitis and emphysema. Some patients will have more elements of one condition than the other; few patients will have only emphysema or bronchitis. Therefore, most patients with COPD will chronically produce sputum, have a chronic cough, and have difficulty expelling air from their lungs, with long expiration phases and wheezing. These patients present with abnormal breath sounds such as rales, crackles, rhonchi, and wheezes, which are discussed in the section on patient assessment later in this chapter.

Asthma

Asthma is an acute spasm of the smaller air passages called bronchioles, associated with excessive mucus production and with swelling of the mucus lining of the respiratory passages (Figure 11-8 ▼). It is a common but serious disease, affecting about 6 million Americans and killing 4,000 to 5,000 Americans each year. Asthma produces a characteristic wheezing as patients attempt to exhale through partially obstructed air passages. These same air passages open easily during inspiration. In other words, when patients inhale, breathing appears relatively normal; the wheezing is most often heard when they exhale. This wheezing may be so loud that you can hear it without a stethoscope. In other cases, the airways are so blocked that no air movement is heard. In severe cases, the actual work of exhaling is very tiring, and cyanosis and/or respiratory arrest may quickly develop, even within minutes.

Asthma affects patients of all ages and is usually the result of an allergic reaction to an inhaled, ingested, or injected substance. Note that the substance itself is not the cause of the allergic reaction; rather, it is an exaggerated response of the body's immune system to that substance that causes the reaction. In some cases, however, there is no identifiable substance, or allergen, that triggers the body's immune system. Almost anything can be considered an allergen. An allergic response to certain foods or some other allergen may produce an acute asthma attack. Between attacks, patients may breathe normally. In its most severe form, an allergic reaction can produce anaphylaxis and even anaphylactic shock. This, in turn, may cause respiratory distress that is severe enough to result in coma and death. Asthma attacks may also be caused by severe emotional stress, exercise, or respiratory infections.

Most patients with asthma are familiar with their symptoms and know when an attack is imminent. Typically, they will have appropriate medication either with them or at home. You should listen carefully to what these patients tell you; they often know exactly what they need.

Spontaneous Pneumothorax

Normally, the "vacuum" pressure in the pleural space keeps the lung inflated. When the surface of the lung is disrupted, however, air escapes into the pleural cavity, and the negative vacuum pressure is lost; the natural elasticity of the lung tissue causes the lung to collapse. The accumulation of air in the pleural space, which may be mild or severe, is called a pneumothorax (Figure 11-9 ▼). Pneumothorax is most often caused by trauma, but it can also be caused by some medical

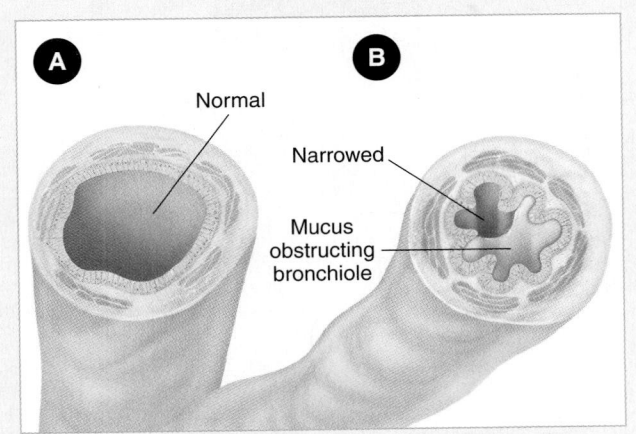

Figure 11-8 Asthma is an acute spasm of the bronchioles. **A.** Cross section of a normal bronchiole. **B.** The bronchiole in spasm; a mucous plug has formed and partially obstructed the bronchiole.

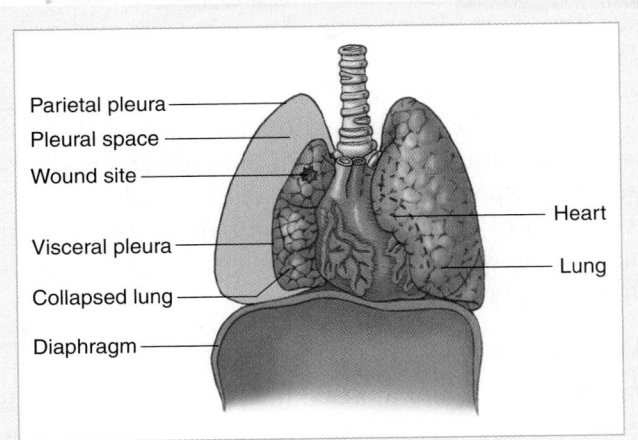

Figure 11-9 A pneumothorax occurs when air leaks into the pleural space from an opening in the chest wall or the surface of the lung. The lung collapses as air fills the pleural space and the two pleural surfaces are no longer in contact.

conditions without any injury. In these patients, the condition is called a "spontaneous" pneumothorax.

Spontaneous pneumothorax may occur in patients with certain chronic lung infections or in young people born with weak areas of the lung. Patients with emphysema and asthma are at high risk for spontaneous pneumothorax when a weakened portion of lung ruptures, often during coughing. A patient with a spontaneous pneumothorax becomes dyspneic (short of breath) and can complain of pleuritic chest pain, a sharp, stabbing pain on one side that is worse during inspiration and expiration, or with certain movement of the chest wall. By listening to the chest with the stethoscope, you can sometimes tell that breath sounds are absent or decreased on the affected side. However, altered breath sounds are very difficult to detect in a patient with severe emphysema. Spontaneous pneumothorax may be the cause of sudden dyspnea in a patient with underlying emphysema.

Anaphylactic Reactions

(Basically, if you are exposed to it for the first time, it may take time for you to gain a reaction to whatever it is.)

Patients who do not have asthma may still have severe allergic reactions. An allergen, a substance that a person is sensitive to, may cause an allergic reaction or may cause anaphylaxis, a reaction characterized by airway swelling and dilation of blood vessels all over the body, which may lower blood pressure significantly. Anaphylaxis may be associated with widespread itching and the same signs and symptoms as asthma. The airway may swell so much that breathing problems can progress from extreme difficulty in breathing to total airway obstruction in a matter of a few minutes. Most anaphylactic reactions occur within 30 minutes of exposure to the allergen, which can be anything from eating certain nuts to receiving a penicillin injection. For some patients, the episode of anaphylaxis may represent the first time they were aware that they had any reaction to the substance. Therefore, they may not know what caused the swelling and allergic reaction. In other cases, the patient may know of the allergen but not be aware of exposure. In severe cases, epinephrine is the treatment of choice. Oxygen and antihistamines are also useful. As always, medical direction should guide appropriate therapy.

Hay Fever

A much milder and more common allergy problem is hay fever. This is caused by an allergic reaction to pollen. In some areas of the country where pollen is present in the air throughout the year, hay fever is almost a universal illness. Generally, it does not produce major emer-

gency problems. It does produce a number of difficulties in the upper respiratory tract, such as a stuffy or runny nose and sneezing.

Pleural Effusions

A pleural effusion is a collection of fluid outside the lung on one or both sides of the chest; in compressing the lung or lungs, it causes dyspnea Figure 11-10 ▼. This fluid may collect in large volumes in response to any irritation, infection, congestive heart failure, or cancer. Though it can build up gradually, over days or even weeks, patients often report that their dyspnea came on suddenly. Pleural effusions should be considered as a contributing diagnosis in any patient with lung cancer and shortness of breath.

When you listen with a stethoscope to the chest of a patient with dyspnea resulting from pleural effusions, you will hear decreased breath sounds over the region of the chest where fluid has moved the lung away from the chest wall. These patients frequently feel better if

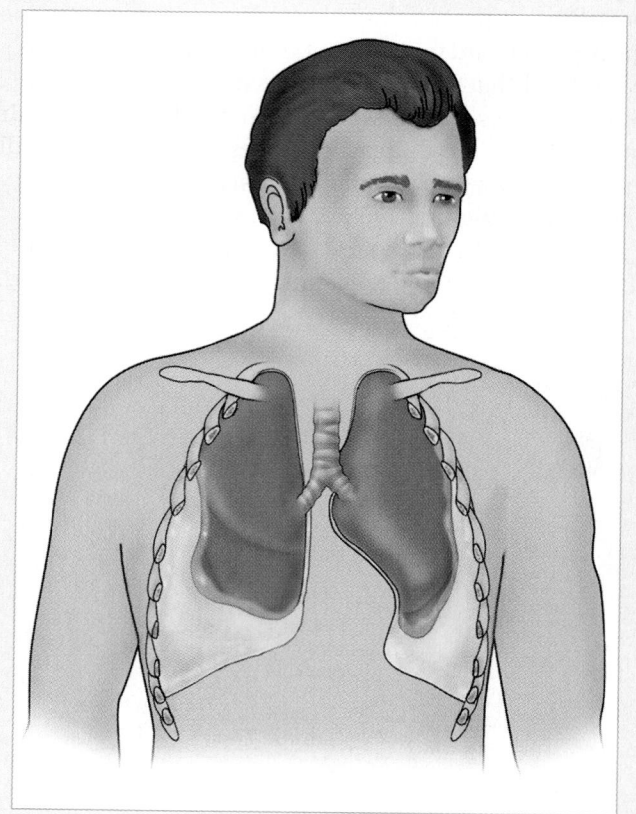

Figure 11-10 With a pleural effusion, fluid may accumulate in large volumes on one or both sides, compressing the lungs and causing dyspnea.

they are sitting upright. Nothing will really relieve their symptoms, however, except removal of the fluid, which must be done by a physician in the hospital.

Mechanical Obstruction of the Airway

As an EMT-B, you should always be aware of the possibility that a patient with dyspnea may have a mechanical obstruction of the airway and be prepared to treat it quickly. In semiconscious and unconscious individuals, the obstruction may be the result of aspiration of vomitus or a foreign object (Figure 11-11A ▼), or of a position of the head that causes obstruction by the tongue (Figure 11-11B ▼). Opening the airway with the head tilt–chin lift maneuver may solve the problem. You should perform this maneuver only after you have ruled out a head or neck injury. If simply opening the airway does not correct the breathing problem, you will have to assess the upper airway for the obstruction.

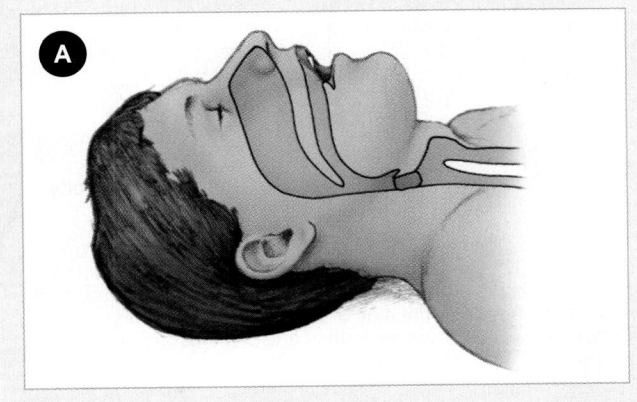

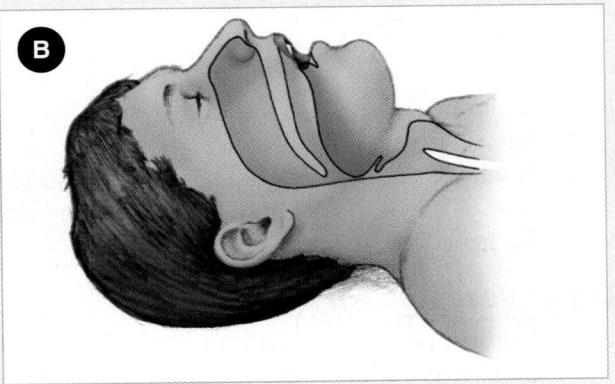

Figure 11-11 A. Foreign body obstruction occurs when an object, such as food, is lodged in the airway. **B.** Mechanical obstruction also occurs when the head is not properly positioned, causing the tongue to fall back into the throat.

Always consider upper airway obstruction from a foreign body first in patients who were eating just before becoming short of breath. The same is true of young children, especially crawling babies, who might have swallowed and choked on a small object.

Pulmonary Embolism

An embolus is anything in the circulatory system that moves from its point of origin to a distant site and lodges there, obstructing subsequent blood flow in that area. Beyond the point of obstruction, circulation can be completely cut off or at least markedly decreased, which can result in a serious, life-threatening condition. Emboli can be fragments of blood clots in an artery or vein that break off and travel through the bloodstream. They also can be foreign bodies that enter the circulation, such as a bullet or a bubble of air.

A pulmonary embolism is the passage of a blood clot formed in a vein, usually in the legs or pelvis, that breaks off and circulates through the venous system. The large clot moves through the right side of the heart and into the pulmonary artery, where it becomes lodged, significantly decreasing or blocking blood flow (Figure 11-12 ▶). Even though the lung is actively involved in inhalation and exhalation of air, no exchange of oxygen or carbon dioxide takes place in the areas of blocked blood flow because there is no effective circulation. In this circumstance, the level of arterial carbon dioxide usually rises, and the oxygen level may drop enough to cause cyanosis. More important, blood clots can inhibit circulation and cause significant dyspnea.

Pulmonary emboli may occur as a result of damage to the lining of vessels, a tendency for blood to clot unusually fast, or, most often, slow blood flow in a lower extremity. Slow blood flow in the legs is usually caused by chronic bed rest, which can lead to the collapse of veins. Patients whose legs are immobilized following a fracture or recent surgery are at risk for pulmonary emboli for days or weeks after the incident. Only rarely do pulmonary emboli occur in active, healthy individuals.

Although they are fairly common, pulmonary emboli are difficult to diagnose. They occur about 650,000 times a year in the United States. Ten percent are immediately fatal, but most often, the patient never notices them. Symptoms and signs, when they do occur, include the following: (Know these)

- Dyspnea
- Acute chest pain
- Hemoptysis (coughing up blood)
- Cyanosis

Study This!!!

◆ Geriatric Needs

As we get older, normal aging processes alter the respiratory system and our ability to exchange oxygen and carbon dioxide. If the patient is a smoker, the disease processes of emphysema or chronic bronchitis can hasten or worsen these changes.

Several changes occur as we age. The chest wall, including the muscles and ribs, becomes less resilient. Additionally, the bronchi and bronchioles lose their muscle mass or tone, and the air sacs (alveoli) become stiffer and less able to recoil (relax and empty) in exhaling. If the chest wall, including muscles and ribs, is weaker or less flexible, the chest cavity cannot expand as easily, and the total amount of air that is allowed into the lungs will be reduced. With decreased recoil of the lungs, alveoli can become distended with air trapped inside. If you are required to ventilate the apneic (nonbreathing) geriatric patient, you will notice that it is more difficult because of increased resistance of the chest and airways as well as reduced compliance of the lungs.

The geriatric patient is at an increased risk of pneumonia or a worsening of asthma or COPD if the airways have lost muscle mass or tone. Secretions might not be expelled from the airways, allowing pneumonia to develop.

The result of normal changes with aging is a reduction of the total amount of air the lungs can hold, air becoming trapped in overstretched alveoli, and increased resistance to airflow into and out of the lungs. Ultimately, all these changes cause a decreased oxygen/carbon dioxide exchange in the respiratory system with reduced oxygen delivery to the cells. Be sure to consider changes in aging that affect the respiratory system, and provide adequate ventilation and oxygenation according to the patient's needs. The geriatric patient may need ventilatory support for conditions that, in the younger adult, are easily accommodated by the respiratory system. Individuals older than 65 years are especially prone to problems with respiration, either from occult (not obvious) stroke, lung disease, cardiovascular disease, liver disease, or certain medications.

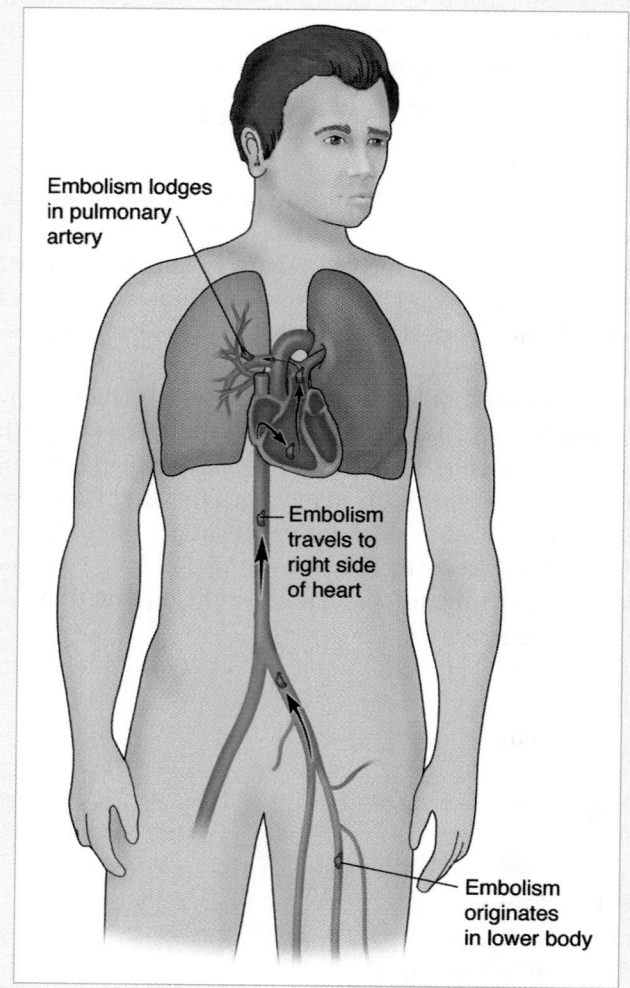

Figure 11-12 A pulmonary embolus is a blood clot from the vein that breaks off, circulates through the venous system, and moves through the right side of the heart into the pulmonary artery. Here, it can become lodged and significantly obstruct blood flow.

- Tachypnea
- Varying degrees of hypoxia

With a large enough embolus, complete, sudden obstruction of the output of blood flow from the right side of the heart can result in sudden death.

Hyperventilation Syndrome

(Emotional or stress)

When dyspnea occurs in a patient with no lung abnormalities, it is called hyperventilation syndrome. Hyperventilation is defined as overbreathing to the point that the level of arterial carbon dioxide falls below normal. This may be an indicator of major, life-threatening illness. For example, a patient with diabetes who has very high blood glucose levels, a patient who has taken an overdose of aspirin, or a patient with a severe infection is likely to hyperventilate. In these patients, rapid, deep breathing is the body's attempt to stay alive. The body is trying to compensate for acidosis, the buildup of excess acid in the blood or body tissues that results from the primary illness. Because carbon dioxide, mixed with water in the bloodstream, can add to the blood's acidity, lowering the level of carbon dioxide helps to compensate for the other acids.

Similarly, in an otherwise healthy person, blood acidity can be diminished by excessive breathing, because it "blows off" too much carbon dioxide. The result is a relative lack of acids. The resulting condition, alkalosis, is the buildup of excess base (lack of acids) in the body fluids.

Alkalosis is the cause of many of the symptoms associated with hyperventilation syndrome, including anxiety, dizziness, numbness, tingling of the hands and feet, and even a sense of dyspnea despite the rapid breathing. Although hyperventilation can be the response to illness and a buildup of acids, hyperventilation syndrome is not the same thing. Instead, this syndrome occurs in the absence of other physical problems. However, it is very common during psychological stress, affecting some 10% of the population at one time or another. The respirations of an individual who is experiencing hyperventilation syndrome may be as high as more than 40 shallow breaths/min or as low as only 20 very deep breaths/min.

The decision whether hyperventilation is being caused by a life-threatening illness or a panic attack should not be made outside the hospital. All patients who are hyperventilating should be given supplemental oxygen and transported to the hospital, where physicians will make that medical decision.

Assessment of the Patient in Respiratory Distress

The assessment of the patient in respiratory distress should be a calm and systematic process. These patients are usually quite anxious and may be some of the most ill and most challenging patients.

Scene Size-up

Your first thought as an EMT-B should be to consider BSI precautions. The minimum for respiratory distress patients is exam gloves. Your consideration should not stop there. The patient could have a respiratory infection that can be passed to you through sputum and/or air droplets. If you suspect the patient has a respiratory disease, then a mask, safety glasses, or face shield should be used.

Scene safety may be as simple as ensuring safe access to the patient and considering safe lifting and moving of the patient. Or you may need to consider that

You are the Provider — Part 2

Given the patient's initial presentation, you immediately choose to rendezvous with ALS. While you apply high-flow oxygen and help her sit on the cot, your partner obtains a quick check of her pulse and blood pressure by palpation. As you roll the cot to the ambulance, you note her respiration rate is 42 breaths/min and apply a pulse oximeter, which reads 90%. When you ask her how many times she's used her inhaler, she holds up two fingers.

3. As an EMT-B, you may assist this patient with her own prescribed MDI. Why is it important to note what medication is in the canister?
4. Why may it be difficult for the patient to correctly use the MDI in this scenario?
5. What methods can be employed to assist with these difficulties and provide a better chance for the medication to work?

YOUR ASS NEEDS TO STUDY!!

the respiratory emergency may have been caused by a toxic substance that was inhaled, absorbed, or ingested.

Once you have determined that the scene is safe, you need to consider the nature of illness or mechanism of injury, and if there is a need to consider taking spinal immobilization precautions. Then determine how many patients there are, and whether you need additional resources. Frequently, in situations where there are multiple people with dyspnea, you should consider the possibility of an airborne hazardous material release.

Initial Assessment

General Impression

As you approach and begin interacting with the patient, you need to gain a general or initial impression of the patient. Does the patient appear calm? Is he or she anxious, restless? Does the patient appear listless and tired? This initial impression will help you decide whether the patient's condition is stable or unstable. A stable condition will not deteriorate during treatment and transport, for example a patient who has had pneumonia for 3 days being transported to the hospital to receive intravenous antibiotics. An unstable condition will deteriorate during treatment and transport, for example a patient who has been stung by a bee and is experiencing increasing difficulty breathing.

At the same time you will be determining the patient's level of consciousness. Using AVPU, you will determine if the patient is alert, responds to verbal stimuli or painful stimuli, or if the patient is unresponsive. If the patient is alert or responding to verbal stimuli, the brain is still receiving oxygen. If the patient is responsive to painful stimuli or unresponsive, the brain may not be oxygenated well and the potential for an airway or breathing problem is more likely. If the patient is alert or responding to verbal stimuli, what is the patient's chief complaint? Within seconds you will be able to determine if there are any immediate threats to life.

Airway and Breathing

Assess the airway. Is it patent? Is it adequate? Air must flow in and out of the chest easily to be considered patent or adequate. If snoring sounds are heard in an unresponsive patient, reposition the airway and insert an oral or nasal airway if necessary to maintain the airway. If stridorous sounds are heard, position the patient so he or she can breathe easily. If gurgling sounds are heard, suction as necessary.

If the airway is adequate or patent, next evaluate your patient's breathing. Is the patient breathing? Is the patient breathing adequately? If the patient is not breathing, give two ventilations immediately. As you ventilate, you need to evaluate if your ventilations are adequate enough to meet the oxygen needs of your patient.

1. Is the air going in?
2. Does the chest expand with each breath?
3. Does the chest fall after each breath?
4. Is the rate adequate for the age of your victim?

If the answer to any of these questions is "no," something is wrong. Try to reposition the patient and insert an oral airway to keep the tongue from blocking the airway. Reposition the patient's head. Reassess your hand position and face mask seal. Slow down or speed up your ventilation rate. Refer to Chapter 7 for a review of positive-pressure ventilation techniques. Remember you will need to continue to monitor the airway for fluid, secretions, or other problems as you move on to assess the adequacy of your patient's breathing.

If the patient is breathing, ensure that the breathing is adequate. Is there adequate rise and fall of the chest? What is the color, temperature, and condition of the patient's skin? Are the patient's respirations labored? If the patient can only speak one or two words at a time before gasping for a breath, ventilations are considered labored. Is the patient using accessory muscles to assist the respiratory effort? If the respiratory effort is inadequate, you must provide the necessary intervention. If the patient is in respiratory distress, place him or her in a position that facilitates breathing easier and begin administering oxygen at 15 L/min via a nonrebreathing mask. If the patient has inadequate depth in the breathing or the rate is too slow, the patient's

Know This!!!

EMT-B Tips

Adventitious breath sounds are sounds heard by auscultation of abnormal lungs. These can include wheezing, rales, rhonchi, gurgling, snoring, crackling, and stridor. Being able to hear and distinguish different kinds of breath sounds can give you important clues as to what is wrong with your patient. The only way to develop your ability to identify breath sounds is through practice. Ask your instructor if you can tag along with a physician, nurse, or respiratory therapist in the hospital to help you develop this experience.

*If your patient looks bad, 9×10 they are bad.

— Know this chart — Might be important to You!!

TABLE 11-2 Signs and Symptoms of Inadequate Breathing

- The patient complains of difficulty breathing.
- The patient has an altered mental status associated with shallow or slow breathing.
- The patient appears anxious or restless. This can happen if the brain is not getting enough oxygen for its needs.
- The patient's respiratory rate is too fast (respirations more than 20 breaths/min).
- The patient's respiratory rate is too slow (respirations are less than 12 breaths/min), you may need to assist ventilations with a BVM device.
- The patient's heart rate is too fast (heart rate more than 100 beats/min).
- The patient's breathing rhythm is irregular. Because the brain controls breathing, an irregular breathing rhythm may indicate a head injury. In this case, the patient will probably be unresponsive.
- The patient's skin is blue (cyanotic). The tongue, nail beds, and inside the lips are good places to look for cyanosis. These all have a large collection of blood vessels and thin skin, making cyanosis more apparent.
- The conjunctivae are pale. Perhaps the patient is short of breath because there are not enough red blood cells to carry oxygen to the tissues.
- The patient is wheezing, gurgling, snoring, stridorous, or crowing. Adventitious sounds can be associated with many types of respiratory problems.
- The patient cannot speak more than few words between breaths. Ask the patient something such as "How are you doing?" If the patient cannot speak at all, he or she probably has a respiratory emergency that will need immediate attention.

- The patient is using accessory muscles to assist breathing. If the patient is using only the diaphragm to breathe, suspect damage to the nerves that carry breathing commands to the chest muscles; the diaphragm may be getting the command to breathe, but because of spinal cord injury, the chest muscles may not.
- The patient is coughing excessively, which might mean that the patient has anything from a mild upper respiratory infection or hay fever to pneumonia, asthma, or heart failure.
- The patient is sitting up, leaning forward with palms flat on the bed or the arms of the chair. This is called the tripod position, because the patient's back and both arms are working together to support the upper body. This position allows the diaphragm the most room to function and helps the patient to use accessory muscles to assist breathing. It is usually a good idea to let the patient stay in the most comfortable position.
- The chest has a barrel shape. In certain chronic lung diseases, because air has been gradually and continuously trapped within the lung in increasing amounts, the distance from front to back gets longer, nearly equaling the side-to-side distance. A barrel chest may indicate a long history of breathing problems.

ventilations may need to be assisted with a BVM device. (Table 11-2 ▲) lists the clues that will help you determine breathing difficulty.

Circulation

If your patient is breathing, he or she will have a pulse; however, evaluating the adequacy of the pulse can give you an indication of the patient's breathing status. If the rate is normal, the patient is most likely receiving enough oxygen to support life. If the pulse rate is too fast or too slow, the patient may not be getting enough oxygen. Assessing a patient's circulation includes an evaluation of shock and bleeding. Respiratory distress in a patient could be from a lack of red blood cells to transport the oxygen. This loss of perfusion may be from chronic anemia, a wound, internal bleeding, or simply from shock overwhelming the body's ability to

compensate for the illness. Recheck everything. Is the oxygen bottle hooked up to the mask? Is the oxygen turned on? Is the flow rate adequate (10 to 15 L/min)? Is there a good face mask seal? Is the chest rising and falling with each breath? Is the airway blocked with vomit or the tongue? Control any bleeding no matter how mild and treat your patient for shock.

Transport Decision

The last step in the initial assessment is to make a transport decision. If the patient's condition is stable and there are no life threats, you may decide to perform a focused history and physical exam on scene. If the patient's condition is unstable and there is a possible life threat, proceed with rapid transport. This means you will keep your scene time short, providing only life-saving interventions on scene. Perform a focused history and physical exam en route to the hospital.

Focused History and Physical Exam

After you have completed your initial assessment, you may or may not be en route to the hospital depending on the seriousness of your patient. Either way, with immediate life threats taken care of, you now have time to focus on why the patient is having dyspnea.

Begin the next step of your assessment, the Focused History and Physical Exam, by asking history questions about the present illness. Use SAMPLE and OPQRST to guide you in your questioning.

SAMPLE History

With patients in respiratory distress, many of the SAMPLE questions can be answered by the family or bystanders. Limit the number of questions to pertinent ones—a patient who is in respiratory distress doesn't need to be using any additional air to answer questions. To help determine the cause of your patient's problem, be a detective. Look for medications, medical alert bracelets, environmental conditions, and other clues to what may be causing the problem. Each part of the SAMPLE history may give you clues. For example, let's say you forget to ask about allergies, only to find out later that your patient has a severe allergy to cat dander and that her 8-year-old son had been playing with a cat shortly before onset of the problem. You would have missed important and possibly life-saving information.

Ask the patient to describe the problem. Begin by asking an open-ended question, "What can you tell me about your breathing?" Pay close attention to OPQRST: when the problem began (onset), what makes the breathing difficulty worse (provocation), how the breathing feels (quality), and whether the discomfort moves (radiation). How much of a problem is the patient having (severity)? Is the problem continuous or intermittent (time)? If it is intermittent, how frequently does it occur and how long does it last?

Find out what the patient has already done for the breathing problem. Does the patient use a prescribed inhaler? If so, when was it used last? How many doses have been taken? Does the patient use more than one inhaler? Be sure to record the name of each inhaler and when it was used.

Different respiratory complaints offer different clues and different challenges. Patients with chronic conditions may have long periods when they are able to live relatively normal lives, but sometimes experience acute worsening of their conditions. That's when you are called, and it is important to be able to determine your patient's baseline status, in other words, their usual condition, and what is different at this time that made them call you. For example, patients with COPD (emphysema and chronic bronchitis) cannot handle pulmonary infections well, because the existing airway damage makes them unable to cough up the mucus or sputum produced by the infection. The chronic lower airway obstruction makes it difficult to breathe deeply enough to clear the lungs. Gradually, the arterial oxygen level falls, and the carbon dioxide level rises. If a new infection of the lung occurs in a patient with COPD, the arterial oxygen level may fall rapidly. In a few patients, the carbon dioxide level may rise high enough to cause sleepiness. These patients require respiratory support and careful administration of oxygen.

You are the Provider   Part 3

You are prepared to coach the patient in using her MDI and can also assist her by placing a spacer on the end of her MDI. A spacer allows the medicine to remain suspended and enables the patient to breathe more normally and take greater advantage of the entire dose of medicine.

You then ask her if she thinks she can self-administer another dosage of her albuterol. She nods. You attach the spacer, then hand her the MDI. You coach and reassure her throughout the next administration of albuterol. You notice continued accessory muscle use as she breathes.

6. What does accessory muscle use indicate?

The patient with COPD usually presents with a long history of dyspnea with a sudden increase in shortness of breath. There is rarely a history of chest pain. More often, the patient will remember having had a recent "chest cold" with fever and either an inability to cough up mucus or a sudden increase in sputum. If the patient is able to cough up sputum, it will be thick and is often green or yellow. The blood pressure of patients with COPD is normal; however, the pulse is rapid and occasionally irregular. Pay particular attention to the respirations. They may be rapid, or they may be very slow.

Patients with asthma may have different "triggers," different causes of acute attacks. These include allergens, cold, exercise, stress, infection and noncompliance with medications. It is important to try to determine what may have triggered the attack so that it can be treated appropriately. For example, an asthma attack that came on while your patient was jogging in the cold will probably not respond to antihistamines, whereas one brought on by a reaction to pollen might.

Patients with congestive heart failure (CHF) often walk a fine line between compensating for their diminished cardiac capacity and decompensating. Many take several medications, most often including diuretics ("water pills") and blood pressure medications. Your history taking should include obtaining a list of all their medications, and paying special attention to the events leading up to the present problem. Your SAMPLE and OPQRST history will be very helpful in helping the emergency department physician plan a course of treatment.

Focused Physical Exam

Patients with COPD usually are older than 50 years. They will always have a history of recurring lung problems and are almost always long-term cigarette smokers. Patients with COPD may complain of tightness in the chest and constant fatigue. Because air has been gradually and continuously trapped in their lungs in increasing amounts, their chests often have a barrel-like appearance (Figure 11-13 ▶). If you listen to the chest with a stethoscope, you will hear abnormal breath sounds. These may include <u>crackles</u>, which are crackling, rattling sounds that are usually associated with fluid in the lungs but here are related to chronic scarring of small airways; <u>rhonchi</u>, which are coarse, gravelly sounds caused by mucus in the upper airways; and <u>wheezing</u>, a high-pitched whistling or crackling sound, most often heard on exhalation, but sometimes heard on both exhalation and inhalation, or inhalation only. Because of large emphysematous air pockets and

✚ EMT-B Tips

Some states allow EMT-Bs to administer inhalers or assist patients in the administration of their own inhalers. With this increased scope of practice comes an increased responsibility to know the names, doses, indications, contraindications, side effects, and precautions of the numerous inhalers available for a variety of conditions. Patients sometimes do not know the difference between their "rescue" inhalers (immediately effective medication, such as albuterol) from their maintenance inhalers (such as corticosteroids, which have no immediate effect). It is essential, then, that you do!

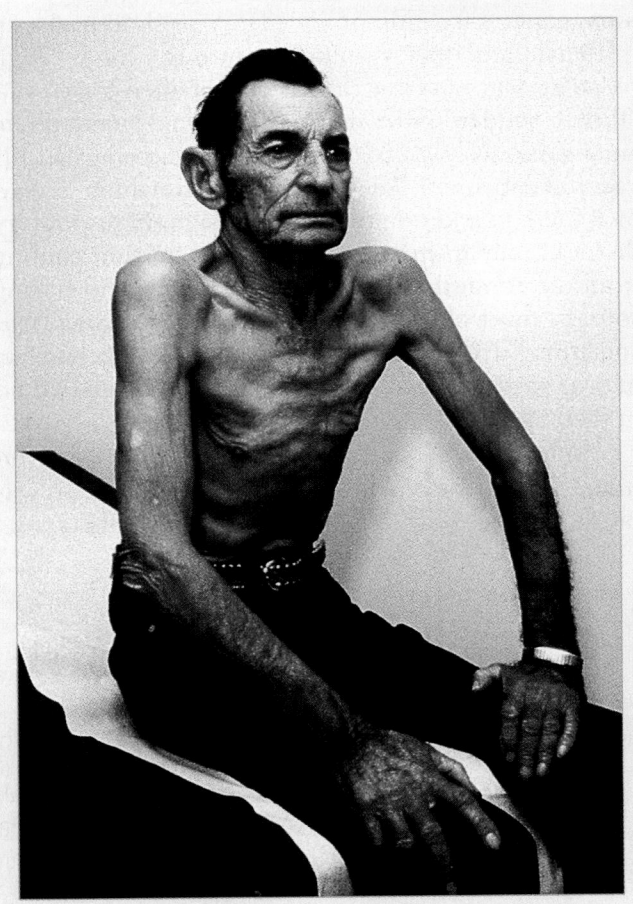

Figure 11-13 Typically, a patient with COPD has a barrel-shaped chest and uses accessory muscles and pursed lips for breathing. Notice, also, that the patient is sitting in the tripod position.

diminished airflow, sounds of breathing are frequently hard to hear and may be detected only high up on the posterior chest. Patients with COPD will often exhale through pursed lips in an unconscious attempt to maintain airway pressures.

In addition to the signs of air hunger present in all patients with respiratory distress, such as tripod positioning, rapid breathing, and use of accessory muscles, restriction of the small lower airways in patients with asthma often causes wheezing. Patients may have a prolonged expiratory phase of breathing as they attempt to exhale trapped air from their lungs. In severe cases, you may actually not hear wheezing because of insufficient airflow. As your patient tires from the effort of breathing and oxygen levels drop, respiratory and heart rates may actually drop, and your patient may seem to relax or go to sleep. These signs indicate impending respiratory arrest, and you must act immediately.

When patients with CHF decompensate, they will often experience pulmonary edema, as fluid backs up in their circulatory system and into the lungs. High blood pressure and low cardiac output often trigger this "flash" (sudden) pulmonary edema. These patients are among the most sick, frightened, and frightening patients you will encounter. They are literally drowning in their own fluid. In addition to the classic signs of respiratory distress, they may have pink, frothy sputum coming from their mouths. They will have adventitious lung sounds, most often wet (rales, rhonchi, crackles) but sometimes dry sounding (wheezes). Their legs and feet may be swollen (pedal edema) from the backup of fluid in their system.

Sometimes it is not possible to quickly and definitively determine what is causing your patient's respiratory distress. The 20-year-old at a picnic who rapidly develops difficulty breathing and hives after being stung by a bee offers a clear-cut diagnostic picture. The older woman receiving 12 medications in a nursing home who has a cough and increasing shortness of breath that developed over a week is more perplexing. Keep an open mind and gather as complete a history as possible and perform a focused physical exam. Remember that in addition to providing you clues to helping your patients, you may be able to obtain information vital to the physician available only at the scene.

Baseline Vital Signs

In addition to pulse, respirations, and blood pressure, other signs such as skin color, capillary refill, level of consciousness, and pain measurement are key in evaluating the respiratory patient. It is essential to look at the whole clinical picture when evaluating the patient in respiratory distress and not fixate on any one vital sign or symptom. This baseline evaluation of vital signs may be used later to determine trends. For example, your patient may present with a rapid respiratory rate to compensate for a failing heart. After you administer oxygen, a decrease in the breathing rate toward normal may indicate that your patient is getting better. On the other hand, it may indicate that your patient is decompensating, no longer able to maintain the effort of rapid breathing, and may quickly deteriorate. Looking at the whole clinical picture, including correlating all the vital signs with your history and findings in the physical exam, will help you make this determination. Patients initially compensate for respiratory distress by increasing their respiratory and heart rates. If they are able to maintain adequate oxygenation, they will be able to maintain their level of consciousness, skin color, and capillary refill time. Blood pressure will vary with the patient's baseline

You are the Provider   Part 4

You have notified the paramedic unit of the patient's situation and vital signs when you begin to notice a change in your patient. Over the next few minutes, she seems very tired and is not as alert as she once was. You notice that her wheezes are less audible, her respiratory rate is decreasing, and her hands and mouth are becoming cyanotic. You ask your partner to notify paramedics of this change, and you begin assisting her respirations with a BVM device connected to high-flow oxygen, taking great care not to force air into the lungs.

7. With cyanosis becoming present in her fingers, what does this tell you about her oxygenation?

Geriatric Needs

Most geriatric patients take medications, sometimes many, to treat various ailments that are part of the aging process. Some of these medications will blunt the body's normal reactions to stress and the mechanisms the body uses to compensate for respiratory compromise and hypoxia. For example, beta blockers, used for a variety of conditions, prevent the heart from speeding up and the veins from constricting to compensate for a loss of blood pressure or oxygenation. Keep this in mind when evaluating vital signs in geriatric patients.

status and condition. It is often elevated in pulmonary edema due to congestive heart failure.

The brain needs a constant, adequate supply of oxygen to function normally. When oxygen levels drop, you will notice an altered level of consciousness. This may manifest itself as confusion, lack of coordination, bizarre behavior, or even combativeness. Change in affect or level of consciousness is one of the early warning signs of respiratory inadequacy.

When there is inadequate oxygen in the blood, the body will attempt to divert blood from the extremities to the core in an attempt to keep the vital organs, including the brain, functioning. This will result in pale skin and delayed capillary refill in the hands and feet. Capillary refill that takes longer than two seconds is considered delayed. Feel for skin temperature and look for color changes both in the extremities and in the core. Cyanosis is a late sign, and can be seen first in the lips and mucous membranes. Cyanosis is an ominous sign that requires immediate, aggressive intervention.

Pulse oximetry is an effective diagnostic tool when used in conjunction with experience, good assessment skills, and clinical judgment. Pulse oximeters measure the percentage of hemoglobin that is saturated by oxygen. In patients with normal levels of hemoglobin, pulse oximetry can be an important tool in evaluating oxygenation. To utilize pulse oximetry properly, it is important for you to be able to evaluate the quality of the reading and correlate it to the patient's condition.

There are various makes and models of pulse oximeters with different features and indicators. Whatever the model, they all have some way of eval-

uating the pulse's waveform, or signal quality, as well as the percentage of hemoglobin saturated with oxygen. Some oximeters have a colored light, others have a bar graph. Regardless of the system your unit uses, you must be sure that you are receiving a clear, strong, regular waveform that corresponds with the patient's pulse and a consistent numerical reading. If the readings jump around, disregard the results. Also, correlate the reading with your patient's clinical condition. It is doubtful a patient with CHF in severe respiratory distress will be able to maintain a pulse oximetry reading of 98%, or that a conscious, alert, active patient with good skin color can be maintained by a reading of 80%.

If you get a good reading consistent with your patient's condition, the pulse oximeter can help you determine the severity of the respiratory component of the patient's problem, and if the reading goes steadily up or down, it can give you an indication of improvement or deterioration of ventilatory status, often even prior to its manifestation in patient appearance or vital signs.

Just remember that the pulse oximeter is a useful tool in the hands of a skilled practitioner. Likewise, it can prove dangerous in the hands of the inexperienced. Pulse oximetry, used in conjunction with sound clinical judgment, can be a useful adjunct to your other assessment skills.

It is important to be aware of conditions that can skew pulse oximeter results. Bright light, dark pigmented skin, and nail polish can cause errors. Remember that it only measures the percentage of hemoglobin that is saturated with oxygen. Therefore, a patient with low hemoglobin, such as an anemic or hypovolemic patient, may have 100% oxygen saturation. This means that the hemoglobin is saturated, but the reading doesn't tell you that the hemoglobin level in the bloodstream is not sufficient enough to sustain organ function. Other conditions that may cause false readings are sickle cell disease and carbon monoxide poisoning.

Interventions

Now that you have completed the focused history and physical exam and have gathered a great deal of information about your patient with difficulty breathing, it is time to provide interventions for those problems found that are not an immediate life threat. Your intervention may be based on standing orders or through contacting the hospital and asking for specific directions. Remember, interventions for immediate life threats should have been completed in the initial assessment and should not require contacting the hospital first. Interventions for respiratory problems may include:

- Oxygen via a nonrebreathing mask at 15 L/min
- Positive-pressure ventilations using a BVM, pocket mask, or a flow-restricted oxygen-powered ventilation device
- Airway management techniques such as use of an oropharyngeal airway, a nasopharyngeal airway, suctioning, or airway positioning
- Positioning the patient in a high Fowler's position or a position of choice to facilitate breathing
- Respiratory medications such as an MDI or other medications

Some of these interventions were performed in the initial assessment as needed to treat immediate life threats. Others are used to support breathing problems until definitive care can be provided at the hospital. Some of your interventions may even correct the problem. Remember to document your assessment, including all medications given.

Detailed Physical Exam

In respiratory emergencies as in all other emergencies, you should only proceed to the detailed physical exam once all life threats have been identified and treated. If you are busy treating airway or breathing problems, you may not have the opportunity to proceed to a detailed physical exam prior to arriving at the emergency department. This is to be expected. Never compromise the assessment and treatment of airway and breathing problems in order to conduct the detailed physical exam.

Keep in mind, though, that there may be additional pieces to the assessment and treatment puzzle that may be revealed in the detailed physical exam. For example, in treating a patient in acute respiratory distress who is breathing 40 times a minute with audible wheezing, you may be unsure as to whether the patient is in CHF

EMT-B Tips

Never compromise the assessment and treatment of airway and breathing problems in order to conduct the detailed physical exam.

or having an asthma attack. The detailed physical exam may provide you with some clues, such as a consistently elevated blood pressure and pedal edema, which would lead you in the direction of CHF.

Ongoing Assessment

You need to carefully watch patients with shortness of breath. Repeat your initial assessment. Have there been any changes in the patient's condition? Obtain vital signs at least every 5 minutes for a patient who is unstable and/or after the patient uses an inhaler. If the patient's condition is stable and no life threat exists, vital signs should be obtained at least every 15 minutes. Perform a focused reassessment of the respiratory system. Ask the patient whether the treatment made any difference. Look at the patient's chest to see whether accessory muscles are still being used to breathe. Listen to the patient's speech pattern. Keep in mind that the patient may get worse instead of better, and be prepared to assist ventilations with a BVM device.

After helping the patient with the inhaler treatment, transport the patient to the emergency department. While en route, continue to assess the patient's breathing. Try talking to calm and reassure the patient and continue to give supplemental oxygen.

Communications and Documentation

Contact medical control with any change in level of consciousness or difficulty breathing. Depending on local protocol, contact medical control prior to assisting with any prescribed medications. Be sure to document any changes (and at what time), and any orders given by medical control.

Emergency Care of Respiratory Emergencies

When taking the initial vital signs of a person with dyspnea, you should pay particular attention to respirations. Always speak with assurance and assume a concerned, professional approach to reassure the patient, who is probably very frightened. You will usually administer oxygen. Take great care in monitoring the patient's respirations as you do so. Reevaluate the respirations and the patient's response to oxygen repeatedly, at least every 5 minutes, until you reach the emergency department. In a person with a chronically high carbon dioxide level (eg, certain patients with COPD), this is critical, because the supplemental oxygen may cause a rapid rise in the arterial oxygen level. This, in turn, may abolish the secondary respiratory oxygen drive and cause respiratory arrest.

Do *not* withhold oxygen for fear of depressing or stopping breathing in a patient with COPD who needs oxygen. Decreased respiratory rate after administration of oxygen does not necessarily mean that the patient no longer needs the oxygen; he or she may need it even more. If respirations slow and the patient becomes unconscious, you should assist breathing with a BVM.

Supplemental Oxygen

If a patient complains of breathing difficulty, you should administer supplemental oxygen during the focused history and physical exam if it was not done during the initial assessment. In general, you do not need to worry about giving too much oxygen. Put a nonrebreathing face mask on the patient and supply oxygen at a rate of 10 to 15 L/min (enough to maintain the reservoir bag) in a patient with severe difficulty breathing.

As was stated previously, there is some concern about suppression of the "hypoxic" drive to breathe in some patients with COPD. Unless these patients are unresponsive, a more conservative approach is suggested. In patients who have longstanding COPD and probable carbon dioxide retention, administration of low-flow oxygen (2 L/min) is a good place to start, with adjustments to 3 L/min, then 4 L/min, and so on until symptoms have improved (for example, the patient has less dyspnea or a better mental status). When in doubt, err on the side of more oxygen, and monitor the patient closely.

Prescribed Inhalers

Patients who call for help because of breathing difficulty are likely to have had the same trouble before. They probably have prescribed medications to use that are delivered by inhaler. If so, you may be able to help them use it. Consult medical control, or go by standing orders if they allow for this. Remember to report what the medication is, when the patient last took a puff, how many puffs were used at that time, and what the label states regarding dosage. If medical control or standing orders permit, you may assist the patient to self-administer the

> ## Documentation Tips
>
> After assisting with the administration of an inhaler treatment, document another set of vital signs as well as the patient's response to the treatment. Be sure to include lung sounds.

medication. Be certain that the inhaler belongs to the patient, it contains the correct medication, the expiration date has not passed, and the correct dose is being administered. Administer repeat doses of the medication if the maximum dose has not been exceeded and the patient is still experiencing shortness of breath.

Some of the most common medications used for shortness of breath are called inhaled beta-agonists, which dilate breathing passages. Typical trade names are Proventil, Ventolin, Alupent, Metaprel, and Brethine. The generic name for Proventil and Ventolin is albuterol; for Alupent and Metaprel, it is metaproterenol; and for Brethine, it is terbutaline. Most of these medications relax the muscles that surround the bronchioles in the lungs, leading to enlargement (dilation) of the airways and easier passage of air. See (Table 11-3 ▶) for a list of medications used for acute symptoms and medication used for chronic symptoms. Those used for acute symptoms are designed to give the patient rapid relief from symptoms if the condition is reversible. Medications used for chronic symptoms are administered for preventative measures or as maintenance doses. The medications for chronic use will provide little relief of acute symptoms. Common side effects of inhalers used for acute shortness of breath include increased pulse rate, nervousness, and muscle tremors.

If the patient has a prescribed MDI, read the label carefully to make sure that the medication is to be used for shortness of breath and that it has, in fact, been prescribed by a physician (Figure 11-14 ▶). When in doubt, consult medical control.

Before helping a patient to self-administer any MDI medication, make sure that the medication is indicated, that is, the patient has signs and symptoms of shortness of breath. Finally, check that there are no contraindications for its use, such as the following:

- The patient is unable to help coordinate inhalation with depression of the trigger, perhaps because the patient is too confused.
- The inhaler is not prescribed for this patient.

TABLE 11-3 Respiratory Inhalation Medications

Medication		Indications			Usage: Acute vs Chronic	
Generic Drug Name	Trade Names	Asthma	Bronchitis	COPD	Acute	Chronic
Albuterol	Proventil, Ventolin, Volmax	Yes	Yes	Yes	Yes	No
Beclomethasone dipropionate	Beclovent	Yes	No	No	No	Yes
Cromolyn sodium	Intal	Yes	No	No	No	Yes
Fluticasone propionate	Flovent	Yes	No	No	No	Yes
Fluticasone propionate, salmeterol xinafoate	Advair Discus	Yes	No	No	No	Yes
Ipratropium bromide	Atrovent	Yes	Yes	Yes	Yes	No
Metaproterenol sulfate	Alupent	Yes	Yes	Yes	Yes	No
Montelukast sodium	Singulair	Yes	No	No	No	Yes
Salmeterol xinafoate	Serevent	Yes	Yes	Yes	No	Yes

- You did not obtain permission from medical control or local protocol.
- The patient had already met the maximum prescribed dose before your arrival.

Administration of a Metered-Dose Inhaler

To help a patient self-administer medication from an inhaler, follow these steps (Skill Drill 11-1 ▶):

1. **Obtain an order** from medical control or local protocol.
2. **Check that you have the right medication**, the right patient, and the right route.
3. **Make sure that the patient is alert** enough to use the inhaler.
4. **Check the expiration date** of the inhaler.
5. **Check to see whether the patient** has already taken any doses.
6. **Make sure the inhaler** is at room temperature or warmer (**Step 1**).
7. **Shake the inhaler** vigorously several times.
8. **Stop administering supplemental oxygen** and remove any mask from the patient's face.
9. **Ask the patient to exhale** deeply and, before inhaling, to put his or her lips around the opening of the inhaler (**Step 2**).

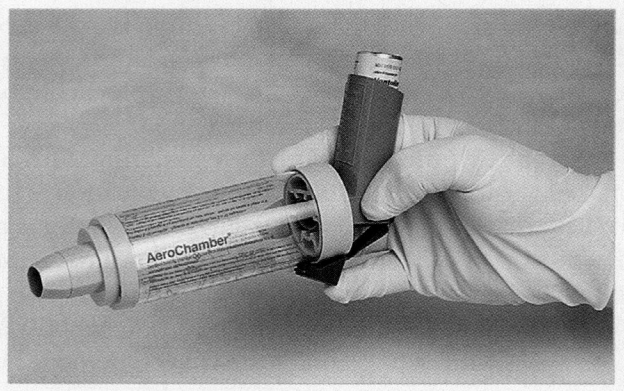

Figure 11-14 Some inhalers have spacer devices to better direct the medication spray.

✱ EMT-B Tips

A spacer device is used to make administering MDIs easier. It is usually a clear, hollow tube that attaches to the MDI. When the MDI is to be used it is attached to one end of the spacer. The inhaler is depressed, releasing the medication into the spacer. The patient then places his or her mouth on the other end of the spacer and inhales the medication. The use of a spacer eliminates the need to coordinate depressing the inhaler at the same time the patient breathes in. Some MDIs have a spacer built into the inhaler.

⚠ Pediatric Needs

Asthma is a common childhood illness. When assessing a pediatric patient, look for retraction of the skin above the sternum and between the ribs. Retractions are typically easier to see in children than in adults. Cyanosis is a late finding in children. Keep in mind that a cough may not be a symptom of a cold; it could signal pneumonia or asthma. Even if you do not hear much wheezing, the presence of a cough can indicate that some degree of reactive airway disease or an acute asthma attack may be taking place.

The emergency care of a child with shortness of breath is the same as it is for an adult, including the use of supplemental oxygen. However, many small children will not tolerate (or may refuse to wear) a face mask. Rather than fighting with the child, provide blow-by oxygen by holding the oxygen mask in front of the child's face or ask the parent to hold the mask (Figure 11-15 ▶). Many children with asthma also will have prescribed hand-held MDIs. Use these inhalers just as you would with an adult. Pediatric patients are more likely to use spacers to assist in inhaler use.

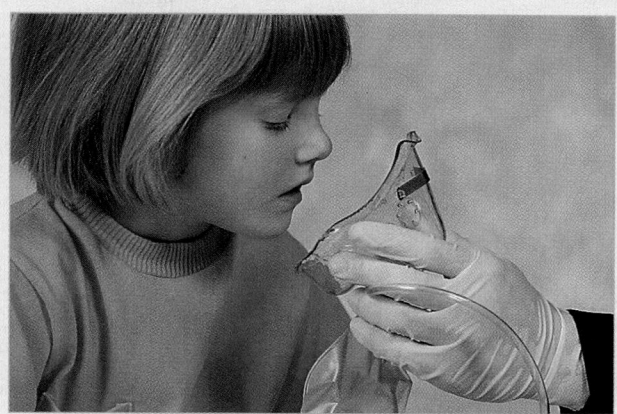

Figure 11-15 Because children may refuse to wear an oxygen mask, you may have to hold the mask in front of the child's face. If the child still refuses, enlist the parents' help.

10. **Have the patient** depress the hand-held inhaler as he or she begins to inhale deeply.
11. **Instruct the patient** to hold his or her breath for as long as is comfortable to help the body absorb the medication (**Step 3**).
12. **Continue to administer** supplemental oxygen.
13. **Allow the patient to breathe** a few times, then repeat second dose per direction from medical control or local protocol (**Step 4**).

You are the Provider   Part 5

You rendezvous with ALS. The paramedic instructs you to continue ventilating. You advise him that the patient's oximetry is now 72% with BVM ventilations. You move the patient into the ALS unit where another paramedic awaits you, prepared for endotracheal intubation.

With the patient semiconscious and a gag reflex intact, one paramedic begins a nebulized treatment of albuterol through your BVM device while the other starts an IV and administers a series of medications. The patient twitches and then becomes flaccid. The paramedic allows you to continue bagging the patient and then asks you to move the BVM device so that she can intubate the patient. The intubation is successful and a few minutes after the procedure, you notice the patient's oximetry reading significantly improves.

8. If intubation had not been successful, what should you have done?
9. What items are important to note in documenting this call?

Assisting a Patient With a Metered-Dose Inhaler

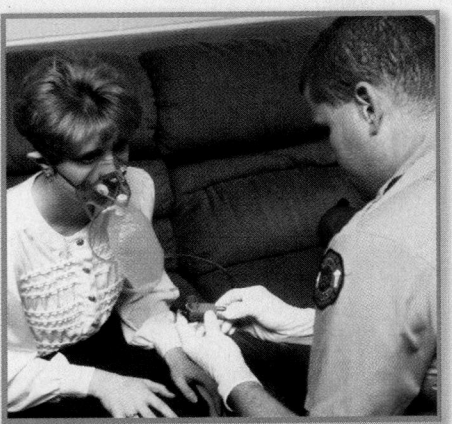

1 Ensure inhaler is at room temperature or warmer.

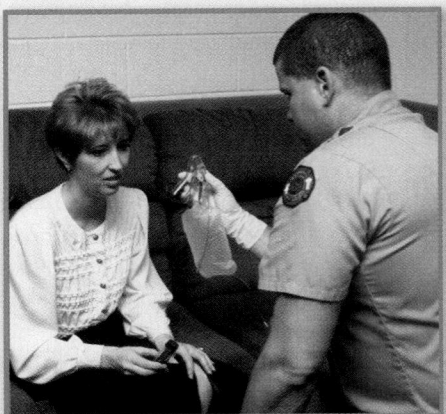

2 Remove oxygen mask.
Hand inhaler to patient. Instruct about breathing and lip seal.

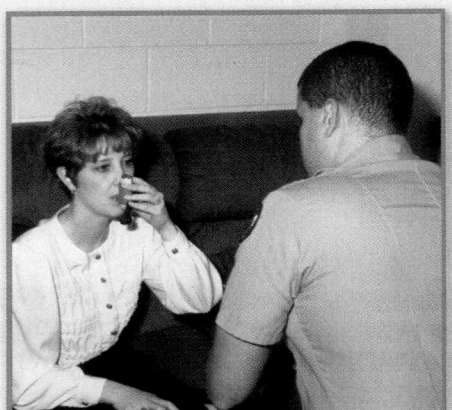

3 Instruct patient to press inhaler and inhale. Instruct about breath holding.

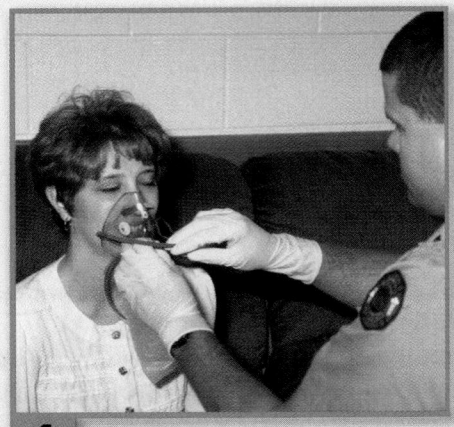

4 Reapply oxygen.
After a few breaths, have patient repeat dose if order/protocol allows.

Treatment of Specific Conditions

Infection of the Upper or Lower Airway

Dyspnea associated with acute infections is quite common. Except for the patient with pneumonia, acute bronchitis, or epiglottitis, it is rarely serious. The acute congestion and stuffiness of a common cold hardly ever require emergency care. Indeed, most people with colds treat themselves with over-the-counter medications. However, individuals with a common cold who have underlying problems such as asthma or heart failure may experience a worsening of their condition as a result of the additional stress of the infection. In addition, cold medications may also have stressful side effects, such as agitation, increased heart rate, and increased blood pressure.

For patients with upper airway infections and dyspnea, administer humidified oxygen (if available). Do not attempt to suction the airway or place an oropharyngeal airway in a patient with suspected epiglottitis. These maneuvers may cause a spasm and complete airway obstruction. Transport the patient promptly to the hospital. Allow the patient to sit in the position that is most comfortable. For someone with epiglottitis, this

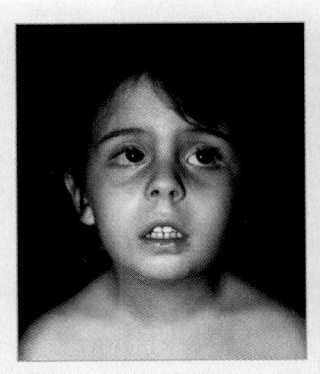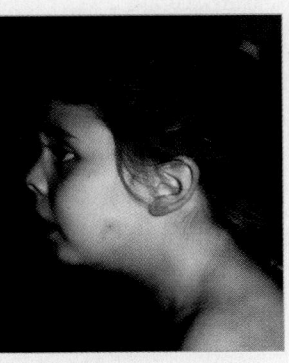

Figure 11-16 A child with epiglottitis may be more comfortable sitting up and leaning forward.

is usually sitting upright and leaning forward in the "sniffing position" (Figure 11-16 ▲). To force a patient with epiglottitis to lie supine may cause upper airway obstruction that could result in death.

The dyspnea of pneumonia is caused not by upper airway obstruction but by the loss of effective lung volume and a need for more rapid air exchange. Here again, the problem will not be helped by the use of artificial airways but may improve with the administration of oxygen.

Acute Pulmonary Edema

Dyspnea caused by acute pulmonary edema may be associated with cardiac disease or direct lung damage. In either case, administer 100% oxygen, and, if necessary, carefully suction any secretions from the airway. Provide prompt transport to the emergency department. The best position for a conscious patient who has a myocardial infarction or direct lung injury is the position in which it is easiest to breathe. Usually, this is sitting up. Rarely will you need to use an artificial airway, because no upper airway obstruction problem exists. However, an unconscious patient with acute pulmonary edema may require full ventilatory support, including airway, positive-pressure ventilation with oxygen, and suctioning.

Chronic Obstructive Pulmonary Disease

Patients with COPD may be semiconscious or unconscious from hypoxia, a condition in which the body's cells and tissues do not get enough oxygen, or from carbon dioxide retention. They may appear to be in respiratory distress and/or be cyanotic. They may have pursed lips and may be using accessory muscles to breathe, including those in the neck and shoulders.

Assist with the patient's prescribed inhaler if there is one. Oftentimes a patient with COPD will overuse an inhaler; watch for side effects. Transport patients with COPD as promptly as possible to the emergency department, allowing them to sit upright if this is most comfortable. Patients with COPD often find breathing difficult when lying down.

Spontaneous Pneumothorax

Patients with spontaneous pneumothorax may have severe respiratory distress, or they may have no distress at all and complain only of pleuritic chest pain. Provide supplemental oxygen, and provide prompt transport to the hospital. Like most dyspneic patients, those with spontaneous pneumothorax are usually more comfortable sitting up. Monitor the patient carefully, watching for any sudden deterioration in the respiratory status. Be ready to support the airway, assist respirations, and give full cardiopulmonary support if it becomes necessary.

Asthma

Many lung problems are incorrectly labeled "asthma"; therefore, your assessment of the patient is critical. A patient who truly has asthma will have a history of repeated episodes of sudden shortness of breath in which he or she had difficulty exhaling. Confirm whether the patient is able to breathe normally at other times. If possible, ask family members to describe the patient's asthma. Even if they only identify wheezing as a problem, be aware that some forms of heart failure, foreign body aspiration, toxic fumes inhalation, or allergic reactions may cause wheezing.

As you assess the patient's vital signs, note that the pulse rate will be normal or elevated, the blood pressure may be slightly elevated, and respirations will be increased. Assist with the patient's prescribed inhaler if there is one. Administer oxygen, and allow the patient to sit in an upright position, which makes breathing easier. Be reassuring; tension and anxiety make asthma attacks worse.

Ask questions about how and when the symptoms began. As you care for the patient, be prepared to suction large amounts of mucus from the mouth and to administer oxygen. If you do suction, do not withhold oxygen for more than 15 seconds for adult patients, 5 seconds for an infant, and 10 seconds for a child. Allow some time for oxygenation between suction attempts. If the patient is unconscious, you may have to provide airway management.

If the patient carries medication, such as an inhaler for an asthma attack, you may help with its administration, as directed by local protocol. Even patients who use their inhaler may continue to get worse. You need to reassess breathing frequently and be prepared to assist ventilations in severe cases. If you must assist ventilations in a patient who is having an asthma attack, use slow, gentle breaths. Remember, the problem in asthma is getting the air out of the lungs, not into them. Resist the temptation to squeeze the bag hard and fast. Always assist with ventilations as a last resort, and then provide only about 10 to 12 shallow breaths/min.

A prolonged asthma attack that is unrelieved may progress into a condition known as *status asthmaticus*. The patient is likely to be frightened, frantically trying to breathe, while using all the accessory muscles. Status asthmaticus is a true emergency, and the patient must be given oxygen and transported immediately to the emergency department.

The effort to breathe during an asthma attack is very tiring, and the patient may be exhausted by the time you arrive. An exhausted patient may have stopped feeling anxious or even struggling to breathe. This patient is not recovering; he or she is at a very critical stage and is likely to stop breathing. Aggressive airway management, oxygen administration, and prompt transport are essential in this situation. ALS support should be considered. Follow local protocol.

Pleural Effusions

Treatment of pleural effusions consists of removal of fluid collected outside the lung, which must be done by a physician in a hospital setting. However, you should provide oxygen and other routine support measures to these patients.

Obstruction of the Upper Airway

If the patient is a small child or someone who was eating just before dyspnea developed, you may assume that the problem is an inhaled or aspirated foreign body. If the patient is old enough to talk but cannot make any noise, upper airway obstruction is the likely cause.

Upper airway obstruction may be either partial or complete. If your patient is able to talk and breathe, the wisest course may be to provide supplemental oxygen and transport carefully in a position of comfort to the hospital. As long as the patient is able to obtain sufficient oxygen, avoid doing anything that might turn a partial airway obstruction into a complete airway obstruction.

There is no condition more immediately life-threatening than a complete airway obstruction. The obstructing body must be removed before any other actions will be effective.

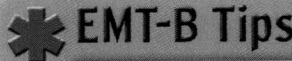

EMT-B Tips

While one EMT-B is getting oxygen ready, the second EMT-B should try to coach the patient with asthma or COPD to use "pursed-lip" breathing. The increase in backpressure will help air flow through narrowed bronchioles.

First you need to clear the patient's upper airway according to BLS guidelines. Then, whether or not you are successful, administer supplemental oxygen and transport the patient promptly to the emergency department.

Pulmonary Embolism

Because a considerable amount of lung tissue may not be functioning, supplemental oxygen is mandatory in a patient with a pulmonary embolism. Place the patient in a comfortable position, usually sitting, and assist breathing as necessary. Hemoptysis, if present, is usually not severe, but any blood that has been coughed up should be cleared from the airway. The patient may have an unusually rapid and possibly irregular heartbeat. Transport the patient to the emergency department promptly. Be aware that pulmonary emboli may cause cardiac arrest.

Hyperventilation

When you respond to a patient who is hyperventilating, complete an initial assessment and history of the event. Is the patient having chest pain? Is there a history of cardiac problems or diabetes? You must always assume a serious underlying problem even if you suspect that the underlying problem is stress. Do not have the patient breathe into a paper bag, even though it is thought to be the traditional technique for managing hyperventilation syndrome. In theory, breathing into a paper bag causes the patient to rebreathe exhaled carbon dioxide, allowing the level of carbon dioxide in the blood to return to normal. In fact, if the patient is hyperventilating because of a serious medical problem, this maneuver could make things worse. A patient with underlying pulmonary disease who breathes into a bag may become severely hypoxic. Treatment should instead consist of reassuring the patient in a calm, professional manner; supplying supplemental oxygen; and providing prompt transport to the emergency department. Patients who hyperventilate need to be evaluated in the hospital setting.

You are the Provider   Summary

All calls involve assessment of the patient's ABCs. Correct evaluation and treatment serve as the foundation of care in EMS. Knowing the capabilities of your scope of practice, being able to anticipate potential changes in patient condition, and requesting the assistance of ALS providers when needed can make the difference between life and death for your patients.

Respiratory Distress

Scene Size-up	Body substance isolation should include a minimum of gloves and eye protection. Ensure scene safety and determine NOI/MOI. Consider the number of patients, the need for additional help, and c-spine stabilization.
Initial Assessment	
■ General impression	Determine priority of care based on environment and patient's chief complaint. Determine level of consciousness and find/treat any immediate threats to life.
■ Airway	Ensure patent airway.
■ Breathing	Evaluate depth and rate of respirations and provide ventilations as needed. Auscultate and note breath sounds, providing high-flow oxygen.
■ Circulation	Evaluate pulse rate and quality; observe skin color, temperature, and condition. If stable/no life threats, proceed with focused history and physical exam. If unstable/possible life threat, proceed with rapid transportation.
■ Transport decision	If stable/no life threats, proceed with focused history and physical exam. If unstable/possible life threat, proceed with rapid transportation.
Focused History and Physical Exam	*NOTE: The order of the steps in the focused history and physical exam differs depending on whether the patient is conscious or unconscious. The order below is for a conscious patient. For an unconscious patient, perform a rapid physical exam, obtain vital signs, and obtain the history.*
■ SAMPLE history	Ask pertinent SAMPLE and OPQRST. Be sure to ask if and what interventions were taken before your arrival, how many, and at what time.
■ Focused physical exam	Perform a focused physical exam, keying in on patient's physical appearance, cyanosis, work of breathing, tripod positioning, pursed lips, use of accessory muscles, adventitious lungs sounds, wheezing, and pedal edema.
■ Baseline vital signs	Take vital signs, noting skin color/temperature as well as patient's level of consciousness. Use pulse oximetry if available.
■ Interventions	Support patient with oxygen, positive pressure ventilations, adjuncts, proper positioning, and assisting with medication(s) as per local protocol. Many of these interventions may need to be performed earlier, in the initial assessment.
Detailed Physical Exam	Consider a detailed physical exam if time and the situation permits.
Ongoing Assessment	Repeat the initial assessment, focused assessment, and reassess interventions performed. Reassess vitals every 5 minutes for the unstable patient, or when an inhaler is used. For the patient who is stable or not using inhalers, reassess vitals every 15 minutes. Reassure and calm the patient.
■ Communications and documentation	Contact medical control with any change in level of consciousness or difficulty breathing. Depending on local protocol, contact medical control prior to assisting with any prescribed medications. Document any changes, the time, and any orders from medical control.

NOTE: While the steps below are widely accepted, be sure to consult and follow your local protocol.

Respiratory Distress

Administer oxygen by placing a nonrebreathing mask on the patient and supplying oxygen at a rate of 10 to 15 L/min.

For any patient in respiratory distress, use positioning, airway adjuncts (oropharyngeal or nasopharyngeal airway), or positive pressure ventilation as indicated.

Asthma

Administer oxygen. Allow patient to sit in upright position.

Suction large amounts of mucus.

Help patient self-administer a metered-dose inhaler:

1. Obtain order from medical control.
2. Check expiration date and whether patient has taken other doses.
3. Ensure inhaler is at room temperature or warmer.
4. Shake inhaler vigorously several times.
5. Remove oxygen mask. Instruct patient to exhale deeply.
6. Instruct patient to press inhaler and inhale. Instruct patient to hold breath as long as is comfortable.
7. Reapply oxygen.

Infection of Upper or Lower Airway

Administer humidified oxygen if available.

Do not attempt to suction airway or place an oropharyngeal airway.

Transport promptly with patient in position of comfort.

Acute Pulmonary Edema

Administer 100% oxygen and suction any secretions from the airway as necessary.

Place in position of comfort and provide ventilatory support as needed. Transport promptly.

Chronic Obstructive Pulmonary Disease

Provide full-flow oxygen via nonrebreathing mask at 15 L/min.

If patient is prescribed an inhaler, administer it according to local protocol. Document time and effect on patient with each use.

Place in the position of comfort and provide prompt transport.

Spontaneous Pneumothorax

Provide supplemental oxygen and place in position of comfort,

Transport promptly. Support airway, breathing, and circulation as necessary.

Pleural Effusions

Provide high-flow oxygen at 15 L/min and place in position of comfort. Support airway, breathing, and circulation as necessary.

Transport promptly.

Obstruction of the Upper Airway

For partial or complete foreign body airway obstructions, clear by following BLS guidelines, apply full-flow oxygen at 15 L/min as necessary, and transport promptly.

Pulmonary Embolism

Clear airway and provide full-flow oxygen at 15 L/min. Place in position of comfort and provide prompt transport. Provide ventilatory support as necessary and be prepared for cardiac arrest.

Hyperventilation

Provide full-flow oxygen at 15 L/min and coach respirations slower in a calm manner. Complete an initial assessment and focused history and physical exam. Transport promptly for evaluation.

Assessment and Emergency Care

Prep Kit

Ready for Review

- Dyspnea is a common complaint that may be caused by numerous medical problems, including infections of the upper or lower airways, acute pulmonary edema, chronic obstructive pulmonary disease, spontaneous pneumothorax, asthma or allergic reactions, pleural effusions, mechanical obstruction of the airway, pulmonary embolism, and hyperventilation.

- Each of these lung disorders interferes in one way or another with the exchange of oxygen and carbon dioxide that takes place during respiration. This interference may be in the form of damage to the alveoli, separation of the alveoli from the pulmonary vessels by fluid or infection, obstruction of the air passages, or air or excess fluid in the pleural space.

- Patients with longstanding lung diseases often have chronically high levels of blood carbon dioxide; in some cases, giving too much oxygen to these patients may depress or stop respirations. However, judicious use of oxygen is always an important priority in patients with dyspnea.

- Signs and symptoms of breathing difficulty include unusual breath sounds, including wheezing, stridor, rales, and rhonchi; nasal flaring; pursed-lip breathing; cyanosis; inability to talk; use of accessory muscles to breathe; and sitting in the tripod position, which allows the diaphragm the most room to function.

- In treating dyspnea, it is important to reassure the patient and provide supplemental oxygen. Remember to maintain the patient in a position that is comfortable for breathing, usually sitting upright.

- If the patient is not breathing, use a BVM device to assist breathing. If the patient is breathing inadequately, apply oxygen through a nonrebreathing face mask with the oxygen flow set at 10 to 15 L/min.

- Next, perform a focused history and physical exam, including vital signs. If the patient has a prescribed inhaler or epinephrine auto injector, consult medical control to assist with its use, or follow standing orders if they allow for this.

- Then transport the patient to the hospital, monitoring his or her condition on the way. Talking with the patient is a good way to monitor a breathing problem.

- Remember, a patient who is breathing rapidly may not be getting enough oxygen as a result of respiratory distress from a variety of problems, including pneumonia or a pulmonary embolism; trying to "blow off" more carbon dioxide to compensate for acidosis caused by a poison, a severe infection, or a high level of blood glucose; or having a stress reaction.

- In every case, prompt recognition of the problem, administration of oxygen, and prompt transport are essential.

Vital Vocabulary

allergen A substance that causes an allergic reaction.

asthma A disease of the lungs in which muscle spasm in the small air passageways and the production of large amounts of mucus with swelling of the mucus lining of the respiratory passages result in airway obstruction.

carbon dioxide retention A condition characterized by a chronically high blood level of carbon dioxide in which the respiratory center no longer responds to high blood levels of carbon dioxide.

chronic bronchitis Irritation of the major lung passageways, from either infectious disease or irritants such as smoke.

chronic obstructive pulmonary disease (COPD) A slow process of dilation and disruption of the airways and alveoli, caused by chronic bronchial obstruction.

Technology

| Interactivities |
| Vocabulary Explorer |
| Anatomy Review |
| Web Links |
| Online Review Manual |

common cold A viral infection usually associated with swollen nasal mucous membranes and the production of fluid from the sinuses and nose.

crackles Crackling, rattling breath sounds signaling fluid in the air spaces of the lungs.

croup An infectious disease of the upper respiratory system that may cause partial airway obstruction and is characterized by a barking cough; usually seen in children.

diphtheria An infectious disease in which a membrane forms, lining the pharynx; this lining can severely obstruct the passage of air into the larynx.

dyspnea Shortness of breath or difficulty breathing.

embolus A blood clot or other substance in the circulatory system that travels to a blood vessel where it causes blockage.

emphysema A disease of the lungs in which there is extreme dilation and eventual destruction of pulmonary alveoli with poor exchange of oxygen and carbon dioxide; it is one form of chronic obstructive pulmonary disease (COPD).

epiglottitis An infectious disease in which the epiglottis becomes inflamed and enlarged and may cause upper airway obstruction.

hyperventilation Rapid or deep breathing that lowers blood carbon dioxide levels below normal.

hypoxia A condition in which the body's cells and tissues do not have enough oxygen.

hypoxic drive Backup system to control respirations when oxygen levels fall.

pleural effusion A collection of fluid between the lung and chest wall that may compress the lung.

pleuritic chest pain Sharp, stabbing pain in the chest that is worsened by a deep breath or other chest wall movement; often caused by inflammation or irritation of the pleura.

pneumonia An infectious disease of the lung that damages lung tissue.

pneumothorax A partial or complete accumulation of air in the pleural space.

pulmonary edema A buildup of fluid in the lungs, usually as a result of congestive heart failure.

pulmonary embolism A blood clot that breaks off from a large vein and travels to the blood vessels of the lung, causing obstruction of blood flow.

rhonchi Coarse breath sounds heard in patients with chronic mucus in the airways.

severe acute respiratory syndrome (SARS) Potentially life-threatening viral infection that usually starts with flu-like symptoms.

stridor A harsh, high-pitched, barking inspiratory sound often heard in acute laryngeal (upper airway) obstruction.

wheezing A high-pitched, whistling breath sound, characteristically heard on expiration in patients with asthma or COPD.

Points to Ponder

It's a cold and damp night, and you have just been dispatched to the home of an older man with breathing difficulties. Upon arrival you find a 78-year-old man with mild to moderate respiratory difficulty. He tells you he has lived with the shortness of breath daily for the past 5 years; however, it has gotten worse in the past few hours. You notice several medications on a table and the house is very messy. You find out the man lives alone and appears to be malnourished. When completing your SAMPLE history, your patient states he does not eat regular meals and does not take his medication as directed. You realize the patient needs to be transported to the emergency department but the patient refuses. He says he wants to stay home and die.

What can you say to encourage this patient to accept emergency care? What is the significance of the condition of the house?

Issues: The Need To Transport Patients for Emergency Treatment and Refusal of Treatment, Patient Non-compliance With Prescription Medication, Depression and Older Persons.

www.EMTB.com

Assessment in Action

You are dispatched to a residence for a man experiencing shortness of breath. Upon arrival you are directed to a bedroom where you find an older man sitting on the end of the bed. From across the room you notice that the patient is leaning forward and obviously having difficulty breathing. He appears pale and is blue (cyanotic) around the lips. Barely able to speak, the patient tells you he had a severe bout of coughing about 20 minutes ago. That is when the severe shortness of breath suddenly began.

Your partner places the patient on oxygen and obtains the following vital signs: pulse, 120 beats/min and weak; blood pressure, 100/70 mm Hg; and respirations, 28 breaths/min and labored. The patient tells you he also feels a stabbing pain in the left side of his chest when he breathes in and out. The patient's wife tells you the patient was diagnosed with COPD and has had asthma for the past 3 years.

1. The term used to describe the patient's difficulty breathing is called:

   A. alveoli shorting.
   B. bradypnea.
   C. dyspnea.
   D. oxygen deprecation.

2. When listening to lung sounds on this patient, you would expect:

   A. absent or decreased lung sounds on one side.
   B. equal lung sounds.
   C. no unusual lung sounds.
   D. wheezing in the top of the lungs.

3. The patient's bout with coughing, his pale, blue coloring, and vital signs are all signs and symptoms of:

   A. a bad cold or flu.
   B. a reaction to drugs.
   C. inadequate breathing.
   D. cardiovascular disease.

4. This patient most likely is suffering from:

   A. an obstructed airway.
   B. asthma.
   C. bronchitis.
   D. a spontaneous pneumothorax.

5. The best method to deliver oxygen to this patient would be:

   A. a nonrebreathing mask at 10 to 15 L/min.
   B. blow-by method at 20 L/min
   C. nasal cannula at 8 to 10 L/min.
   D. through connecting tubing.

6. The best position to transport this patient would be:

   A. face down.
   B. lying flat (supine).
   C. on his side (lateral).
   D. sitting up.

7. During reassessment of the patient you notice a major deterioration of the patient's respiratory status. You should be ready to:

   A. assist the patient's respirations with a BVM device.
   B. call medical control.
   C. move the patient to a backboard.
   D. turn the oxygen off or down.

Challenging Questions

8. Why are geriatric patients at a greater risk for severe respiratory problems?

9. What causes epiglottitis and why is epiglottitis a true emergency?

10. Chronic obstructive pulmonary disease (COPD) affects 10% to 20% of the adult population. Describe two COPD processes and how they differ.

11. How does carbon dioxide retention in the blood affect respirations?

Cardiovascular Emergencies

Objectives

Cognitive

4-3.1 Describe the structure and function of the cardiovascular system. (p 402)

4-3.2 Describe the emergency medical care of the patient experiencing chest pain/discomfort. (p 415)

4-3.3 List the indications for automated external defibrillation (AED). (p 419)

4-3.4 List the contraindications for automated external defibrillation. (p 421)

4-3.5 Define the role of EMT-B in the emergency cardiac care system. (p 402)

4-3.6 Explain the impact of age and weight on defibrillation. (p 420)

4-3.7 Discuss the position of comfort for patients with various cardiac emergencies. (p 411)

4-3.8 Establish the relationship between airway management and the patient with cardiovascular compromise. (p 411)

4-3.9 Predict the relationship between the patient experiencing cardiovascular compromise and basic life support. (p 410)

4-3.10 Discuss the fundamentals of early defibrillation. (p 422)

4-3.11 Explain the rationale for early defibrillation. (p 422)

4-3.12 Explain that not all chest pain patients result in cardiac arrest and do not need to be attached to an automated external defibrillator. (p 421)

4-3.13 Explain the importance of prehospital ACLS intervention if it is available. (p 430)

4-3.14 Explain the importance of urgent transport to a facility with Advanced Cardiac Life Support if it is not available in the prehospital setting. (p 430)

4-3.15 Discuss the various types of automated external defibrillators. (p 420)

4-3.16 Differentiate between the fully automated and the semiautomated defibrillator. (p 421)

4-3.17 Discuss the procedures that must be taken into consideration for standard operations of the various types of automated external defibrillators. (p 421)

4-3.18 State the reasons for assuring that the patient is pulseless and apneic when using the automated external defibrillator. (p 421)

4-3.19 Discuss the circumstances which may result in inappropriate shocks. (p 421)

4-3.20 Explain the considerations for interruption of CPR when using the automated external defibrillator. (p 430)

4-3.21 Discuss the advantages and disadvantages of automated external defibrillators. (p 421)

4-3.22 Summarize the speed of operation of automated external defibrillation. (p 421)

4-3.23 Discuss the use of remote defibrillation through adhesive pads. (p 421)

4-3.24 Discuss the special considerations for rhythm monitoring. (p 421)

4-3.25 List the steps in the operation of the automated external defibrillator. (p 426)

4-3.26 Discuss the standard of care that should be used to provide care to a patient with persistent ventricular fibrillation and no available ACLS. (p 427)

4-3.27 Discuss the standard of care that should be used to provide care to a patient with recurrent ventricular fibrillation and no available ACLS. (p 427)

4-3.28 Differentiate between the single rescuer and multi-rescuer care with an automated external defibrillator. (p 426)

4-3.29 Explain the reason for pulses not being checked between shocks with an automated external defibrillator. (p 423)

4-3.30 Discuss the importance of coordinating ACLS trained providers with personnel using automated external defibrillators. (p 427)

4-3.31 Discuss the importance of postresuscitation care. (p 427)

4-3.32 List the components of postresuscitation care. (p 427)

4-3.33 Explain the importance of frequent practice with the automated external defibrillator. (p 424)

4-3.34 Discuss the need to complete the Automated Defibrillator: Operator's Shift Checklist. (p 425)

4-3.35 Discuss the role of the American Heart Association (AHA) in the use of automated external defibrillation. (p 402)

4-3.36 Explain the role medical direction plays in the use of automated external defibrillation. (p 424)

4-3.37 State the reasons why a case review should be completed following the use of the automated external defibrillator. (p 424)

4-3.38 Discuss the components that should be included in a case review. (p 424)

4-3.39 Discuss the goal of quality improvement in automated external defibrillation. (p 424)

4-3.40 Recognize the need for medical direction of protocols to assist in the emergency medical care of the patient with chest pain. (p 415)

4-3.41 List the indications for the use of nitroglycerin. (p 415)

4-3.42 State the contraindications and side effects for the use of nitroglycerin. (p 415)

4-3.43 Define the function of all controls on an automated external defibrillator, and describe event documentation and battery defibrillator maintenance. (p 420)

Affective

4-3.44 Defend the reasons for obtaining initial training in automated external defibrillation and the importance of continuing education. (p 424)

4-3.45 Defend the reason for maintenance of automated external defibrillators. (p 423)

4-3.46 Explain the rationale for administering nitroglycerin to a patient with chest pain or discomfort. (p 416)

Psychomotor

4-3.47 Demonstrate the assessment and emergency medical care of a patient experiencing chest pain/discomfort. (p 413)

4-3.48 Demonstrate the application and operation of the automated external defibrillator. (p 426)

4-3.49 Demonstrate the maintenance of an automated external defibrillator. (p 423)

4-3.50 Demonstrate the assessment and documentation of patient response to the automated external defibrillator. (p 426)

4-3.51 Demonstrate the skills necessary to complete the Automated Defibrillator: Operator's Shift Checklist. (p 425)

4-3.52 Perform the steps in facilitating the use of nitroglycerin for chest pain or discomfort. (p 416)

4-3.53 Demonstrate the assessment and documentation of patient response to nitroglycerin. (p 416)

4-3.54 Practice completing a prehospital care report for patients with cardiac emergencies. (p 418)

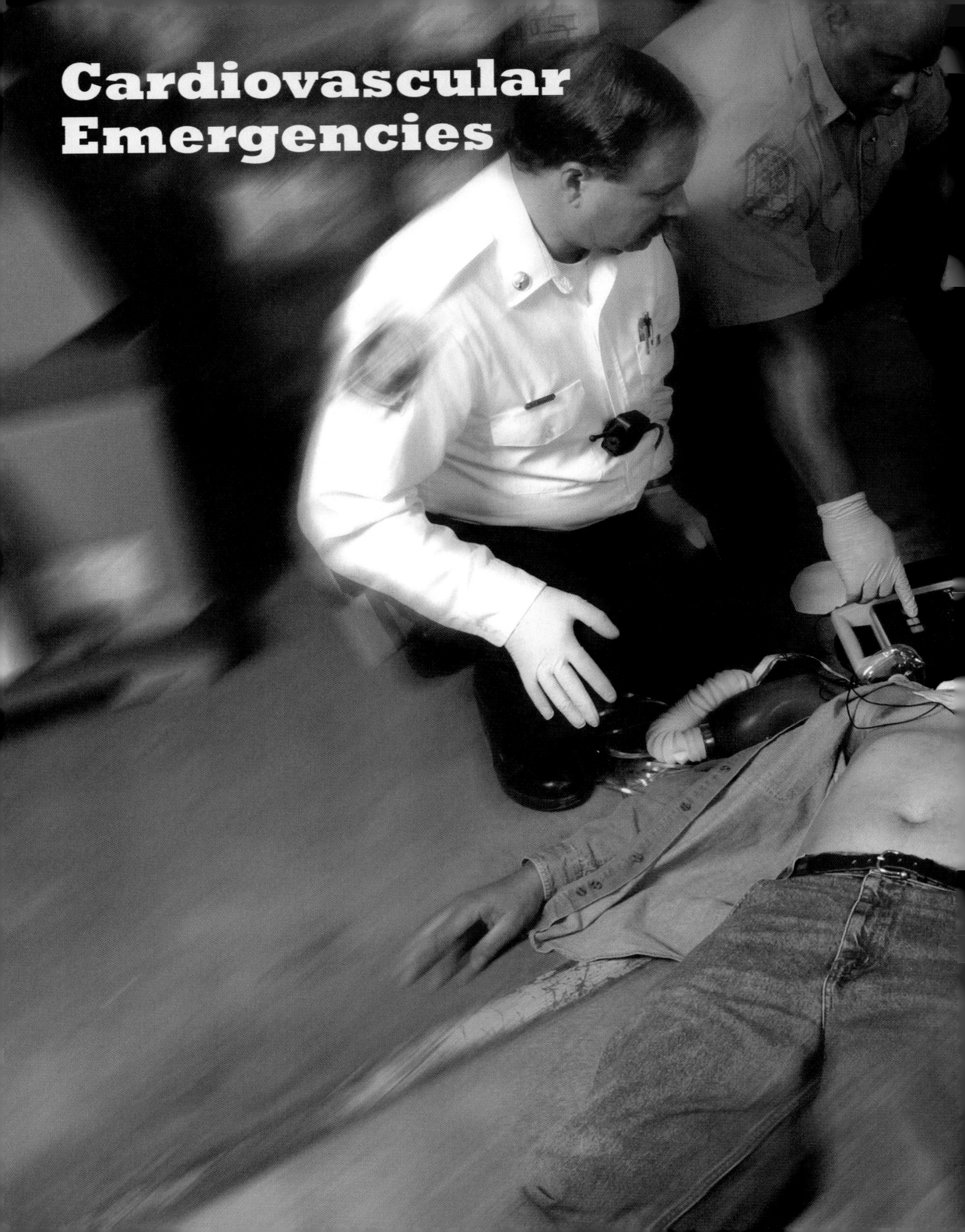

Cardiovascular Emergencies

You are the Provider

You are a volunteer firefighter EMT-B who lives in a rural area. You are at your full-time job when your pager is activated. The dispatcher requests your fire department to respond to 403 McKay Rd for a 65-year-old man complaining of severe chest pain. The address is two blocks from your location, and you respond directly to the scene with your personal jump kit. Advanced life support has been simultaneously dispatched from a location that is 10 to 15 minutes away.

You arrive at the private residence to find an older man in his living room, sitting in a chair clutching his chest. As you introduce yourself the patient says, "This is the worst pain I've ever had in my life!" He tells you that he has had a previous heart attack a couple of years ago, and that he thinks his nitroglycerin is in his bedroom but didn't feel well enough to get it. A fellow EMT-B arrives with your fire department's EMS equipment, including an AED. You ask him to look for the patient's nitroglycerin while you apply high-flow oxygen and take vital signs.

1. What other signs and symptoms might be found in a patient who is having a heart attack?
2. As an EMT-B, you can assist a patient with his or her own prescribed nitroglycerin. What must you know before administering any medications, and what must you specifically know before assisting a patient with nitroglycerin?

Cardiovascular Emergencies

The American Heart Association reports that cardiovascular disease (CVD) claimed 931,108 lives in the United States in 2001. This is 38.5% of all deaths, or 1 of every 2.6 deaths. Heart disease has been the leading killer of Americans since 1900. This statistic is still true today.

It is important for EMS providers to understand that many deaths caused by CVD occur from problems that may have been avoided by people living more prudent lifestyles and by access to improved medical technology. We can help to reduce these numbers of deaths with better public awareness, early access, increased numbers of laypeople trained in CPR, and with public access defibrillation and the recognition of the need for advanced life support services.

This chapter begins with a brief description of the heart and how it works. It then discusses the relationship between chest pain and ischemic heart disease. It explains how to recognize and treat acute myocardial infarction (classic heart attack) and the complications of sudden death, cardiogenic shock, and congestive heart failure. The use of nitroglycerin is described. The last part of the chapter is devoted to the use and maintenance of the automated external defibrillator (AED).

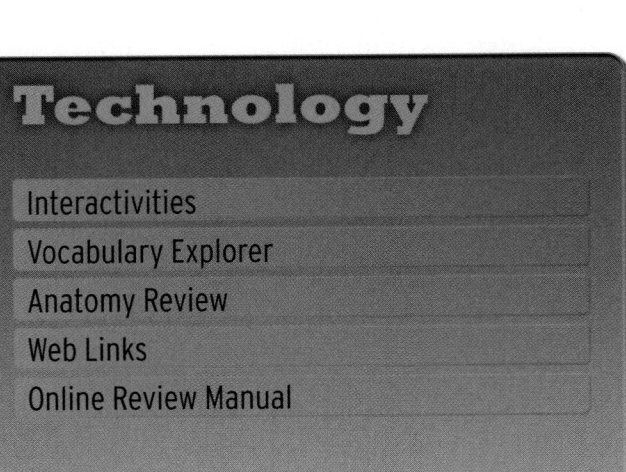

Technology

Interactivities

Vocabulary Explorer

Anatomy Review

Web Links

Online Review Manual

www.EMTB.com

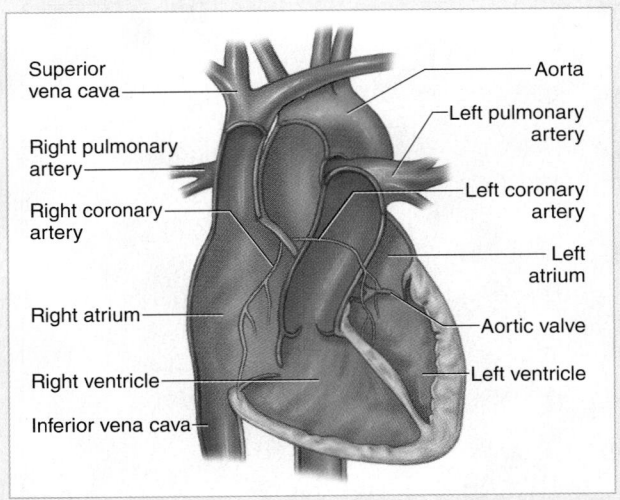

Figure 12-1 The heart is a four-chambered muscle that pumps blood to all parts of the body.

Cardiac Structure and Function

The heart is a relatively simple organ with a simple job. It has to pump blood to supply oxygen-enriched red blood cells to the tissues of the body. The heart is divided down the middle into two sides (left and right) by a wall called the septum. Each side of the heart has an atrium, or upper chamber, to receive incoming blood and a ventricle, or lower chamber, to pump outgoing blood (Figure 12-1 ▲). Blood leaves each of the four chambers of the heart through a one-way valve. These valves keep the blood moving through the circulatory system in the proper direction. The aorta, the body's main artery, receives the blood ejected from the left ventricle and delivers it to all the other arteries so that they can carry blood to the tissues of the body.

The right side of the heart receives oxygen-poor (deoxygenated) blood from the veins of the body (Figure 12-2A ▶). Blood enters into the right atrium from the vena cava, which then fills the right ventricle. After contraction of the right ventricle, blood flows into the pulmonary artery and the pulmonary circulation, where the blood is oxygenated. The left side of the heart receives oxygen-rich (oxygenated) blood from the lungs through the pulmonary veins (Figure 12-2B ▶). Blood enters into the left atrium and then passes into the left ventricle. This side of the heart is more muscular than the other because it must pump blood into the aorta and all the other arteries of the body.

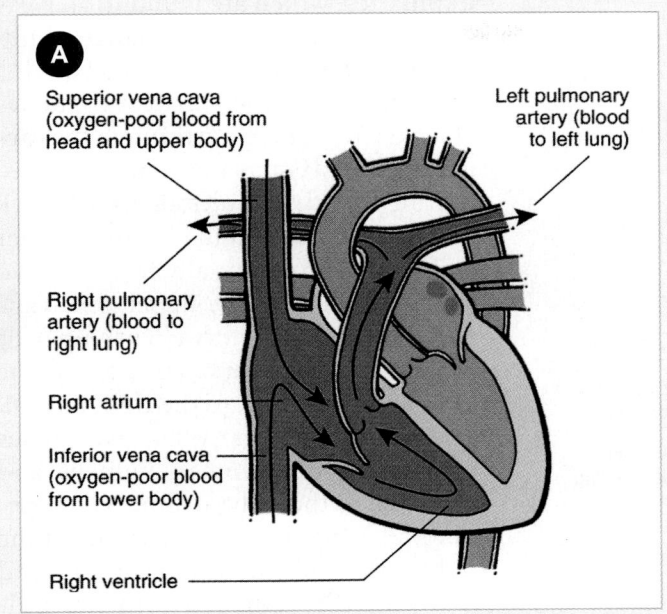
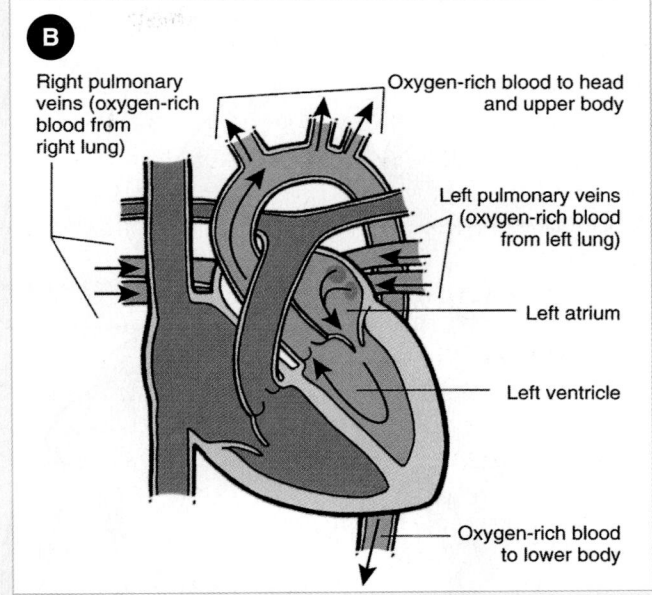

Figure 12-2 **A.** The right side of the heart receives oxygen-poor blood from the veins. **B.** The left side of the heart receives oxygen-rich blood from the lungs through the pulmonary veins.

The heart contains more than muscle tissue. The heart's electrical system, which is distributed throughout the entire heart, controls heart rate and enables the atria and ventricles to work together (Figure 12-3 ▶). Normal electrical impulses begin in the sinus node, just above the atria. The impulses travel across both atria, causing them to contract. Between the atria and the ventricles, the impulses cross over a bridge of special electrical tissue called the atrioventricular (AV) node. Here the signal is slowed down for about one tenth to two tenths of a second to allow blood time to pass from the atria to the ventricles. Then the impulses exit the AV node and spread throughout both ventricles, causing the ventricular muscle cells to contract.

Circulation

To carry out its function of pumping blood, the <u>myocardium</u>, or heart muscle, must have a continuous supply of oxygen and nutrients. During periods of physical exertion or stress, the myocardium requires more oxygen, so the heart must increase its output of blood flow. In the normal heart, the increased need for blood is easily supplied by <u>dilation</u>, or widening, of the coronary arteries, which increases blood flow. The <u>coronary arteries</u> are the blood vessels that supply blood to the heart muscle (Figure 12-4 ▶). They start at the first part of the aorta, just above the <u>aortic valve</u>. The

right coronary artery supplies blood to the right ventricle and, in most people, the bottom part, or inferior wall, of the left ventricle. The left coronary artery divides into two major branches, both of which supply the left ventricle.

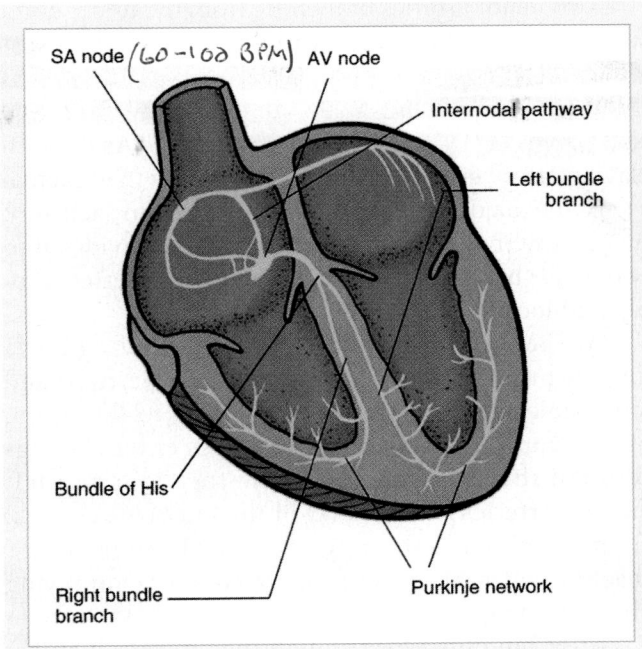

Figure 12-3 The electrical conduction system of the heart controls most aspects of heart rate and enables the four chambers to work together.

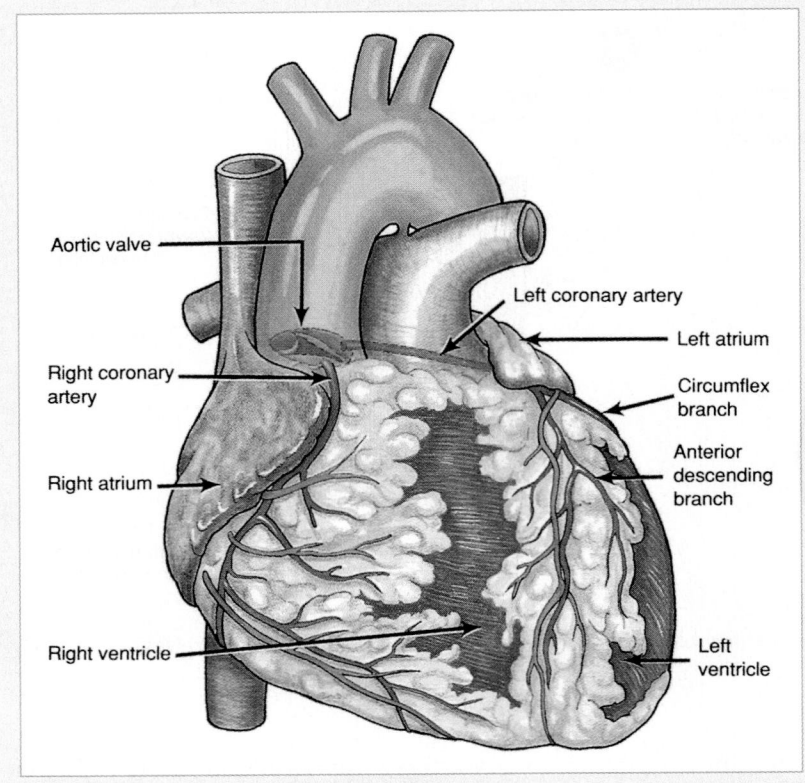

Aortic valve

Right coronary
artery

Right atrium

Right ventricle

Left coronary artery

Left atrium

Circumflex
branch

Anterior
descending
branch

Left
ventricle

Figure 12-4 The coronary arteries carry the blood supply to the heart.

Two major arteries branching from the upper aorta supply blood to the head and arms Figure 12-5 ▶. The right and left carotid arteries supply the head and brain with blood. The subclavian arteries (under the clavicles) supply blood to the upper extremities. As the subclavian artery enters each arm, it becomes the brachial artery, the major vessel that supplies blood to each arm. Just below the elbow, the brachial artery divides into two major branches: the radial and ulnar arteries, supplying blood to the hands.

At the level of the navel, the descending aorta divides into two main branches called the right and left iliac arteries, which supply blood to the groin, pelvis, and legs. As the iliac arteries enter the legs through the groin, they become the right and left femoral arteries. At the level of the knee, the femoral artery divides into the anterior (front) and posterior (back) tibial artery and the peroneal artery, supplying blood to the feet.

After blood travels through the arteries, it enters smaller and smaller vessels called arterioles and capillaries. The capillaries are tiny blood vessels about one cell thick that connect arterioles to venules.

Capillaries, which are found in all parts of the body, allow the exchange of nutrients and waste at the cellular level.

Venules are the smallest branches of veins. After traveling through the capillaries, blood enters the system of veins, starting with the venules, on its way back to the heart. The veins become larger and larger and eventually form the two large venae cavae: the upper vena cava and the lower vena cava. The superior (upper) vena cava carries blood from the head and arms back to the right atrium. The inferior (lower) vena cava carries blood from the abdomen, kidneys, and legs back to the right atrium. The superior and inferior venae cavae join at the right atrium of the heart, where blood is eventually returned into the pulmonary circulation for oxygenation Figure 12-6 ▶.

Blood consists of several types of cells and fluid Figure 12-7 ▶. Red blood cells are the most numerous and give the blood its color. Red blood cells carry oxygen to the body's tissues and then remove carbon dioxide. Larger white blood cells help to fight infection. Platelets, which help the blood to clot, are much smaller than either red or white blood cells. Plasma is the fluid that the cells float in. It is a mixture of water, salts, nutrients, and proteins.

Blood pressure is the pressure of circulating blood against the walls of the arteries. Systolic blood pressure is the maximum pressure exerted by the left ventricle as it contracts. As the left ventricle relaxes, the arterial pressure falls. When the aortic valve closes, blood flow stops. The diastolic blood pressure is the pressure exerted against the walls of the arteries while the left ventricle is at rest. Remember that the top number in a blood pressure reading is the systolic pressure, and the bottom number is the diastolic or resting pressure.

Cardiac Compromise

Chest pain or discomfort that is related to the heart usually stems from a condition called ischemia, or insufficient oxygen. Because of a partial or complete blockage of blood flow through the coronary arteries, heart tissue fails to get enough oxygen and nutrients.

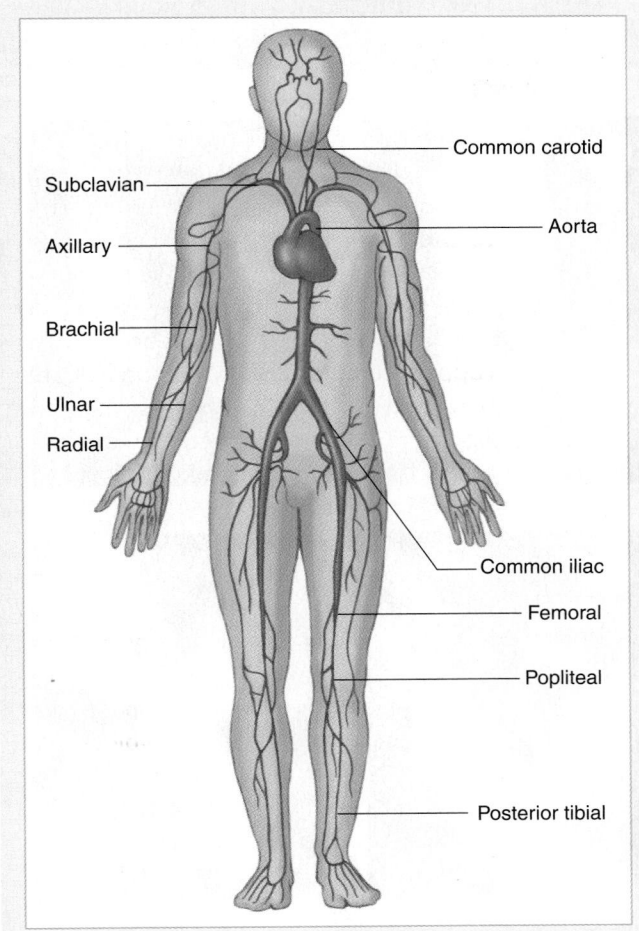

Figure 12-5 The major arteries of the body carry oxygen-rich blood to all parts of the body.

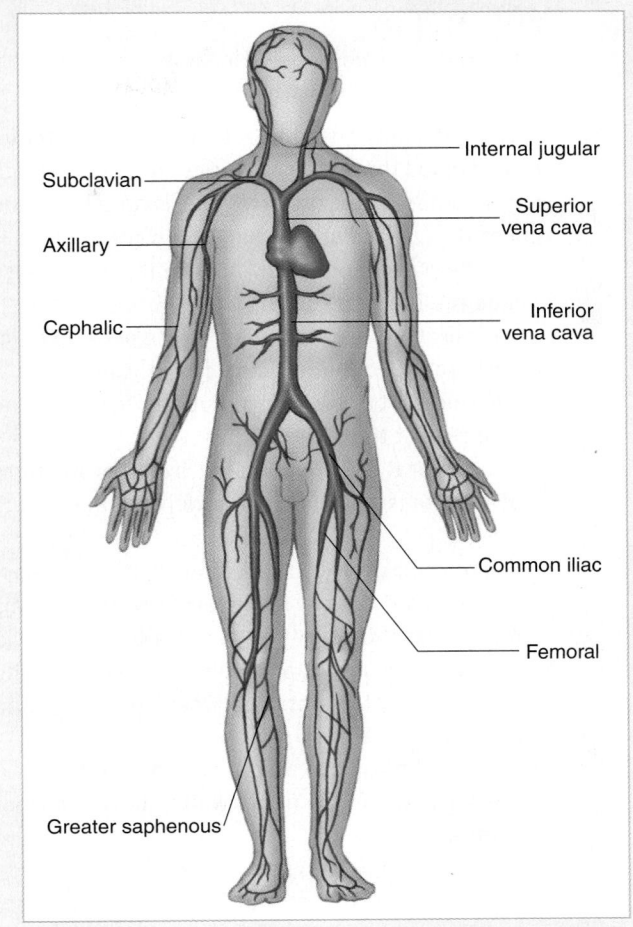

Figure 12-6 The veins carry blood from the body back to the heart, which pumps it through the lungs for oxygenation.

The tissue soon begins to starve and, if blood flow is not restored, eventually dies. Ischemic heart disease, then, is disease involving a decrease in blood flow to one or more portions of the heart muscle.

Atherosclerosis

Most often, the low blood flow to heart tissue is caused by coronary artery atherosclerosis. <u>Atherosclerosis</u> is a disorder in which calcium and a fatty material called cholesterol build up and form a plaque inside the walls of blood vessels, obstructing flow and interfering with their ability to dilate or contract (Figure 12-9 ▶). Eventually, atherosclerosis can even cause complete <u>occlusion</u>, or blockage, of a coronary artery. Atherosclerosis usually involves other arteries of the body as well.

The problem begins when the first deposit of cholesterol is laid down on the inside of an artery. This may happen during the teenage years. As a person ages, more

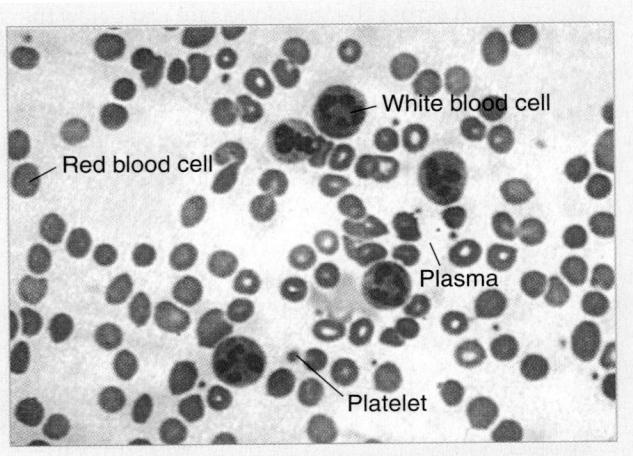

Figure 12-7 Blood consists of several types of cells and fluids, including red blood cells, white blood cells, and platelets.

EMT-B Tips

Pulsation

As the left ventricle contracts, it ejects a forceful wave of blood through the arteries. You can feel that wave in areas where the artery lies over a bone and is near the surface of the skin. This wave of blood is called the pulse. Common places to feel for a pulse include the following (Figure 12-8 ▶):

- The carotid pulse can be felt in the neck, two fingerbreadths on either side of the Adam's apple (thyroid cartilage), and should be taken on the side closest to the EMT-B.
- The femoral pulse can be felt in the groin, right at the crease dividing the lower abdomen from the leg.
- The brachial pulse can be felt on the medial aspect of the elbow, right at the level of the crease. This is the pulse that you listen to when you take blood pressure. Pulsations also can be palpated on the medial side of the arm between the elbow and armpit.
- The radial pulse can be felt on the thumb side of the wrist, about one finger width above the wrist crease.
- The posterior tibial pulse can be felt on the inside of the ankle, just posterior to the medial malleolus. The medial malleolus is the bony bump at the end of the tibia.
- The dorsalis pedis pulse can be felt at the top of the foot. This artery is not in the exact same place in all people. To find its pulse, place your hand across the top of the foot just below the ankle crease. Once you feel something that might be a pulse, use your fingertips to confirm that finding.

Practice feeling for these pulses on yourself and on friends and family members.

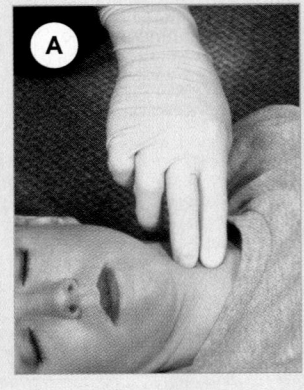
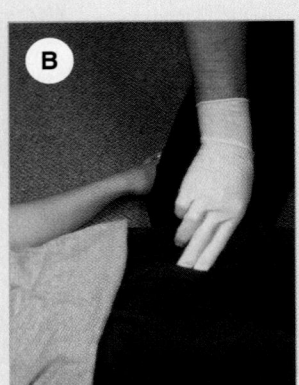
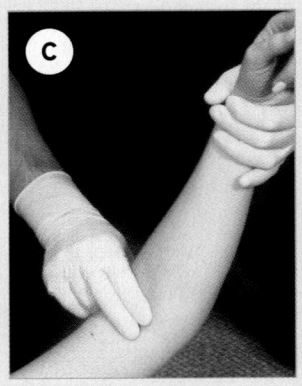
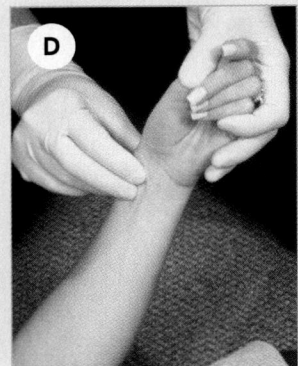
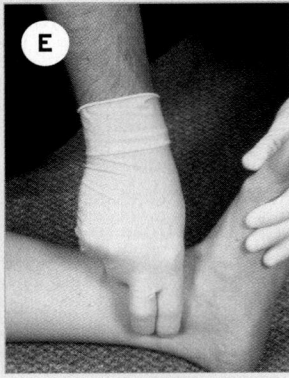
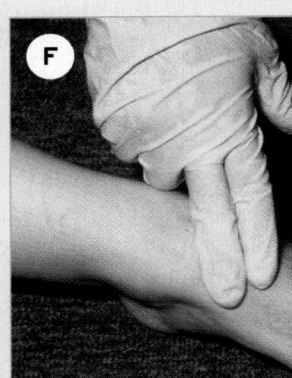

Figure 12-8 Common pulse points.
A. The carotid pulse is taken at the neck.
B. The femoral pulse is felt in the groin area.
C. The brachial pulse can be felt on the inside of the upper arm.
D. The radial pulse can be felt on the thumb side of the wrist.
E. The posterior tibial pulse can be felt on the inside of the ankle.
F. The dorsalis pedis pulse can be felt at the top of the foot.

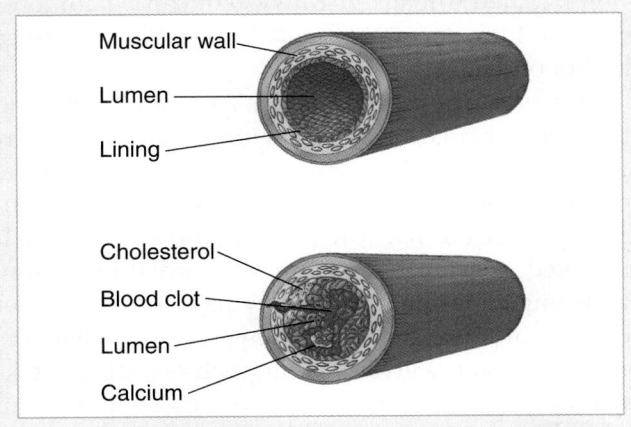

Figure 12-9 In atherosclerosis, calcium and cholesterol build up inside the walls of the blood vessels, causing an obstruction in blood flow to the heart.

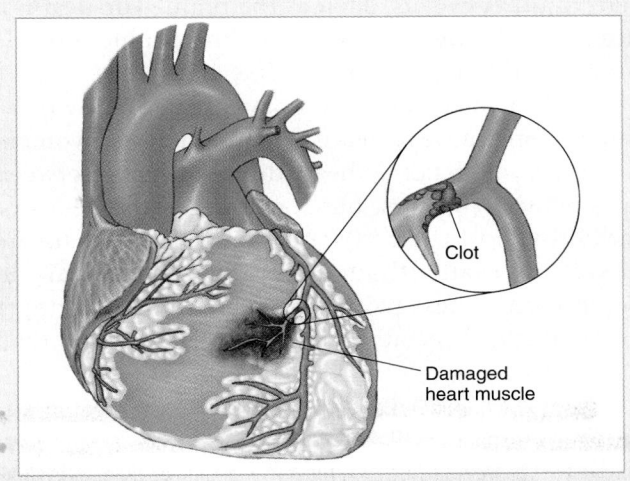

Figure 12-10 An acute myocardial infarction (heart attack) occurs when a blood clot prevents blood flow to an area of the heart muscle. If left untreated, this can result in death of heart tissue.

of this fatty material is deposited; the <u>lumen</u>, or the inside diameter of the artery, narrows. As the cholesterol deposits grow, calcium deposits can form as well. The inner wall of the artery, which is normally smooth and elastic, becomes rough and brittle with these atherosclerotic plaques. Damage to the coronary arteries may become so extensive that they cannot accommodate increased blood flow at times of maximum need.

For reasons that are still not completely understood, a brittle plaque will sometimes develop a crack, exposing the inside of the atherosclerotic wall. Acting like a torn blood vessel, the ragged edge of the crack activates the blood-clotting system, just as it does when

an injury has caused bleeding. In this situation, however, the resulting blood clot will partially or completely block the lumen of the artery. Tissues downstream from the blood clot will suffer from lack of oxygen (ischemia). If blood flow is resumed in a short time, the ischemic tissues will recover. However, if too much time goes by before blood flow is resumed, the tissues will die. This sequence of events is known as an <u>acute myocardial infarction (AMI)</u>, a classic heart attack (Figure 12-10 ▲). <u>Infarction</u> means the death of tissue. The same sequence may also cause the death of

You are the Provider   Part 2

You begin your focused history and physical exam by questioning the patient using the OPQRST mnemonic. You learn that:

- The patient was sitting, watching TV when the chest pain began (Onset)
- Nothing makes the chest pain better or worse, including breathing or body position (Provokes)
- The patient describes the pain as heavy and crushing (Quality)
- The pain radiates from the left side of his chest to his left arm and into his jaw (Radiates)
- 10/10 (Severity)
- The pain started just before he called 9-1-1 and has been constant (Time)

You look for the presence of medication patches or scars indicating previous heart surgeries or the presence of a cardiac pacemaker or defibrillator. None are found. The patient's vital signs are a blood pressure of 160/98 mm Hg, a pulse of 110 beats/min and regular, respirations of 24 breaths/min, and an oximetry reading of 99% (94% originally on room air).

3. What other mnemonic is helpful in obtaining the rest of the information needed and not addressed through OPQRST?

cells in other organs, such as the brain. The death of heart muscle can lead to severe diminishment of the heart's ability to pump, or <u>cardiac arrest</u>.

In the United States, coronary artery disease is the number one cause of death for both men and women. The peak incidence of heart disease occurs between ages 40 and 70 years, but it can also strike teens or individuals in their 90s. You must be alert to the possibility that, although less likely, a 26-year-old person with chest pain could actually be having a heart attack, especially if he or she has a higher than usual risk.

Factors that place a person at higher risk for a myocardial infarction are called risk factors. The major controllable factors are cigarette smoking, high blood pressure, elevated cholesterol levels, elevated blood glucose levels (diabetes), lack of exercise, and stress. The major risk factors that cannot be controlled are older age, family history of atherosclerotic coronary artery disease, and male sex.

Angina Pectoris

Chest pain does not always mean that a person is having an AMI. When, for a brief period of time, heart tissues are not getting enough oxygen, the pain is called <u>angina pectoris</u>, or angina. Although angina can result from a spasm of the artery, it is most often a symptom of atherosclerotic coronary artery disease. Angina occurs when the heart's need for oxygen exceeds its supply, usually during periods of physical or emotional stress when the heart is working hard. A large meal or sudden fear may also trigger an attack. When the increased oxygen demand goes away (eg, the person stops exercising), the pain typically goes away.

Angina pain is typically described as crushing, squeezing, or "like somebody standing on my chest." It is usually felt in the midchest, under the sternum. However, it can radiate to the jaw, the arms (frequently the left arm), the midback, or the epigastrium (the upper-middle region of the abdomen). The pain usually lasts from 3 to 8 minutes, rarely longer than 15 minutes. It may be associated with shortness of breath, nausea, or sweating. It disappears promptly with rest, supplemental oxygen, or nitroglycerin, all of which increase the supply of oxygen to the heart. Although angina pectoris is frightening, it does not mean that heart cells are dying, nor does it usually lead to death or permanent heart damage. It is, however, a warning that you and the patient should both take seriously. Even with angina, because oxygen supply to the heart

is diminished, the electrical system can be compromised and the person is at risk for significant cardiac rhythm problems.

Angina can be further differentiated into "stable" and "unstable" angina. Unstable angina is characterized by pain in the chest of coronary origin that occurs in response to progressively less exercise or fewer other stimuli than those ordinarily required to produce angina. If untreated, it can often lead to myocardial infarction. Stable angina is characterized by pain in the chest of coronary origin that is relieved by the things that normally relieve it in a given patient, such as resting or taking nitroglycerin. EMS usually becomes involved when stable angina becomes unstable, such as when a patient whose pain is normally relieved by sitting down and taking one nitroglycerin tablet has taken three tablets with no relief. Keep in mind that it can be very difficult even for physicians in hospitals to distinguish between the pain of angina and the pain of a myocardial infarction. Patients experiencing chest pain therefore should always be treated as if they are having a myocardial infarction.

Heart Attack

As we have seen, the pain of AMI signals the actual death of cells in the area of the heart where blood flow is obstructed. Once dead, the cells cannot be revived. Instead, they will eventually turn to scar tissue and become a burden to the beating heart. This is why fast action is so critical in treating a heart attack. The sooner the blockage can be cleared, the fewer the cells that may die. About 30 minutes after blood flow is cut off, some heart muscle cells begin to die. After about 2 hours, as many as half of the cells in the area can be dead; in most cases, after 4 to 6 hours, more than 90% will be dead. In many cases, however, opening the coronary artery with either "clot-busting" (thrombolytic) medications or angioplasty (mechanical clearing of the artery) can prevent damage to the heart muscle if done within the first hour after the onset of symptoms. Therefore, immediate treatment and transport to the emergency department are essential.

An AMI is more likely to occur in the larger, thick-walled left ventricle, which needs more blood and oxygen, than in the right ventricle.

Signs and Symptoms of Heart Attack

A patient with a heart attack may show any of the following signs and symptoms:

- Sudden onset of weakness, nausea, and sweating without an obvious cause

- Chest pain/discomfort that is often crushing or squeezing and that does not change with each breath
- Pain in the lower jaw, arms, back, abdomen, or neck
- Sudden arrhythmia with <u>syncope</u> (fainting)
- Shortness of breath or dyspnea
- Pulmonary edema
- Sudden death

The Pain of Heart Attack

The pain of an AMI differs from the pain of angina in three ways:

- **It may or may not be caused by exertion** but can occur at any time, sometimes when a person is sitting quietly or even sleeping.
- **It does not resolve in a few minutes**; rather, it can last between 30 minutes and several hours.
- **It may or may not be relieved** by rest or nitroglycerin.

Note that not all patients who are having an AMI experience pain or recognize it when it does occur. In fact, about a third of patients never seek medical attention. This can be attributed, in part, to the fact that people are afraid of dying and do not wish to face the possibility that their symptoms may be serious (cardiac denial). Middle-aged men, in particular, are likely to minimize their symptoms. However, a few patients, particularly older individuals, women, or those with diabetes do not experience any pain during an AMI but will have other common complaints associated with ischemia discussed earlier. Others may feel only mild discomfort and call it "indigestion." It is not uncommon for the only complaint, especially in older women, to be fatigue.

Therefore, when you are called to a scene where the chief complaint is chest pain, complete a thorough assessment, no matter what the patient says. Any complaint of chest discomfort is a serious matter. In fact, the best thing you can do is to assume the worst.

Physical Findings of AMI and Cardiac Compromise

The physical findings of AMI vary, depending on the extent and severity of heart muscle damage. The following are common:

- **Pulse.** Generally, the pulse rate increases as a normal response to pain, stress, fear, or actual injury to the myocardium. Because arrhythmias are common in AMI, you may feel an irregularity of the pulse.

Documentation Tips

Documenting exactly how a patient describes chest discomfort, in the patient's own words, is a valuable source of information for hospital staff. Remember OPQRST.

- **Blood pressure.** Blood pressure may fall as a result of diminished cardiac output and diminished capability of the left ventricle to pump. However, most patients with AMI will have a normal or, most likely, elevated blood pressure.
- **Respiration.** Respirations are usually normal unless the patient has congestive heart failure. In that case, respirations may become rapid and labored.
- **General appearance.** The patient often appears frightened. There may be nausea, vomiting, and a cold sweat. The skin is often ashen gray because of poor cardiac output and the loss of <u>perfusion</u>, or blood flow through the tissue. Occasionally, the skin will have a bluish tint, called cyanosis; this is the result of poor oxygenation of the circulating blood.
- **Mental status.** Patients with AMI sometimes experience an almost overwhelming feeling of impending doom. If a patient tells you, "I think I am going to die," pay attention.

Consequences of Heart Attack

Heart attack can have three serious consequences:
- Sudden death
- Cardiogenic shock
- Congestive heart failure

Sudden Death

Approximately 40% of all patients with AMI never reach the hospital. Sudden death is usually the result of cardiac arrest, in which the heart fails to generate an effective blood flow. Although you cannot feel a pulse in someone experiencing cardiac arrest, the heart may still be twitching, though erratically. The heart is using up energy without pumping. Such an abnormality of heart rhythm is a ventricular <u>arrhythmia</u>, known as ventricular fibrillation.

A variety of other lethal and nonlethal arrhythmias may follow AMI, usually within the first hour. In most cases, premature ventricular contractions (PVCs), or extra beats in the damaged ventricle, occur. PVCs by themselves may be harmless and are common among healthy, as well as sick, individuals. Other arrhythmias can be much more dangerous. These include the following (Figure 12-11 ▼):

- Tachycardia. Rapid beating of the heart, 100 beats/min or more.
- Bradycardia. Unusually slow beating of the heart, 60 beats/min or less.
- Ventricular tachycardia (VT). Rapid heart rhythm, usually at a rate of 150 to 200 beats/min. The electrical activity starts in the ventricle instead of the atrium. This rhythm usually does not allow adequate time between each beat for the left ventricle to fill with blood. Therefore, the patient's blood pressure may fall, and he or she may lose a pulse altogether. The patient may also feel weak or lightheaded or may even be-

come unresponsive. In some cases, existing chest pain may worsen or chest pain that was not there before onset of the arrhythmia may develop. Most cases of VT will be sustained and may deteriorate into ventricular fibrillation.

- Ventricular fibrillation. Disorganized, ineffective quivering of the ventricles. No blood is pumped through the body, and the patient usually becomes unconscious within seconds. The only way to treat this arrhythmia is to defibrillate the heart. To defibrillate means to shock the heart with a specialized electrical current in an attempt to stop the chaotic, disorganized contraction of the myocardial cells and allow them to start again in a synchronized fashion to restore a normal rhythmic beat. Defibrillation is highly successful in terms of saving a life if delivered within the first few minutes of sudden death. If a defibrillator is not immediately available, CPR must be initiated until the defibrillator arrives. Even if CPR is begun right at the time of collapse, chances of survival diminish 10% each minute until defibrillation is accomplished.

If uncorrected, unstable ventricular tachycardia or ventricular fibrillation will eventually lead to asystole, the absence of all heart electrical activity. Without CPR, this may occur within minutes. Because it reflects a long period of ischemia, nearly all patients you find in asystole will die.

Cardiogenic Shock

Shock is a simple concept but one that few people without medical training really understand. For that reason, Chapter 23 is devoted to a discussion of shock. The discussion of shock in this chapter is limited to that associated with cardiac problems; however, many other medical problems may cause shock as well.

For an EMT-B, shock is also a critical concept. Shock is present when body tissues do not get enough oxygen, causing body organs to malfunction. In cardiogenic shock, often caused by a heart attack, the problem is that the heart lacks enough power to force the proper volume of blood through the circulatory system. Cardiogenic shock can occur immediately or as late as 24 hours after the onset of the AMI. The various signs and symptoms of cardiogenic shock are produced by the improper functioning of the body's organs. The challenge for you is to recognize shock in its early stages, when treatment is much more successful.

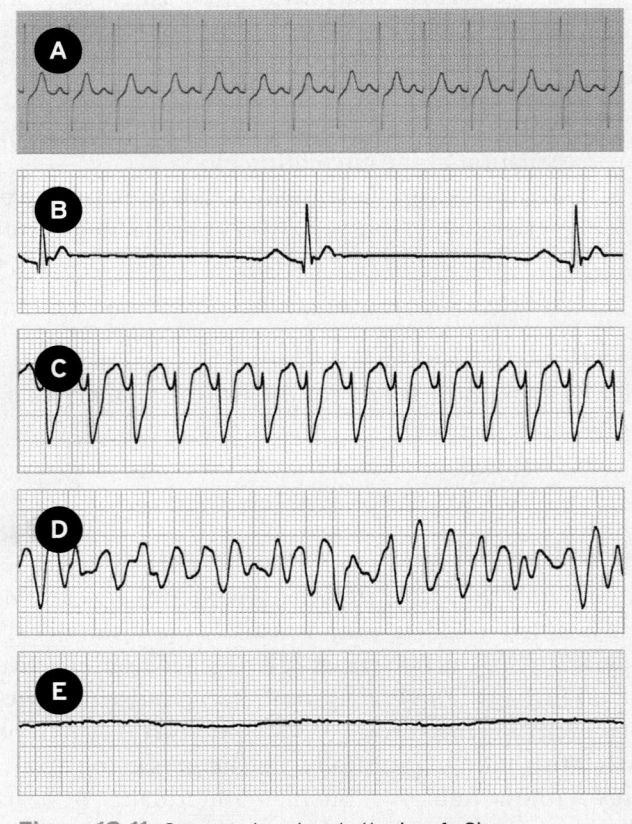

Figure 12-11 Common heart arrhythmias. **A.** Sinus tachycardia. **B.** Bradycardia. **C.** Ventricular tachycardia (VT). **D.** Ventricular fibrillation (VF). **E.** Asystole.

EMT-B Tips

Shock

Signs and symptoms

- One of the first signs of shock is anxiety or restlessness as the brain becomes relatively starved for oxygen. The patient may complain of "air hunger." Think of the possibility of shock when the patient is saying that he or she cannot breathe. Obviously, the patient can breathe, because he or she can talk. However, the patient's brain is sensing that it is not getting enough oxygen.
- As the shock continues, the body tries to send blood to the most important organs, such as the brain and heart, and away from less important organs, such as the skin. Therefore, you may see pale, clammy skin in patients with shock.
- As the shock gets worse, the body will attempt to compensate by increasing the amount of blood pumped through the heart. Therefore, the pulse rate will be higher than normal. In severe shock the heart rate usually, but not always, is greater than 120 beats/min.
- Shock can also be characterized by rapid and shallow breathing, nausea and vomiting, and a decrease in body temperature.
- Finally, as the heart and other organs begin to malfunction, the blood pressure will fall below normal. A systolic blood pressure less than 90 mm Hg is easy to recognize, but it is a late finding that indicates decompensated shock. Do not assume that shock is not present just because the blood pressure is normal (compensated shock).

Treatment of Patients With Cardiogenic Shock

Take the following steps when treating patients with signs and symptoms of shock:

1. Position the patient comfortably. Most patients with heart failure will be more comfortable in semi-Fowler's position; however, those with low blood pressure may not tolerate a semi-upright position. These patients may be more comfortable and be more alert in a supine position.
2. Administer high-flow oxygen.
3. Assist ventilations as necessary.
4. Provide prompt transport to the emergency department.

Congestive Heart Failure

Signs and symptoms

- The patient finds it easier to breathe when sitting up. When the patient is lying down, more blood is returned to the right ventricle and lungs, causing further pulmonary congestion.
- Often, the patient is mildly or severely agitated.
- Chest pain may or may not be present.
- The patient often has distended neck veins that do not collapse even when the patient is sitting.
- The patient may have swollen ankles from dependent edema (back-up of fluid).
- The patient generally will have a high blood pressure, rapid heart rate, and rapid respirations.
- The patient will usually be using accessory breathing muscles of the neck and ribs, reflecting the additional hard work of breathing.
- The fluid surrounding small airways may produce rales (crackles), best heard by listening to either side of the patient's chest, about midway down the back. In severe congestive heart failure, these soft sounds can be heard even at the top of the lung.

Once congestive heart failure develops, it can be treated but not cured. Regular use of medications may alleviate the symptoms. However, these patients often become ill again and are frequently hospitalized. Approximately half will be dead within 5 years of the onset of symptoms.

Treatment of CHF

Treat the patient with congestive heart failure the same way as the patient with chest pain:

1. Take the vital signs, monitor heart rhythm, and give oxygen by nonrebreathing mask with an oxygen flow of 10 to 15 L/min.
2. Allow the patient to remain sitting in an upright position with the legs down.
3. Be reassuring; many patients with CHF are quite anxious because they cannot breathe.
4. Patients who have had problems with CHF before will usually have specific medications for its treatment. Gather these medications and take them along to the hospital.
5. Nitroglycerin may be of value if the patient's blood pressure is above 100 mm Hg systolic. If the patient has been prescribed nitroglycerin, and medical control or standing orders advise you to do so, you can administer it sublingually.
6. Prompt transport to the emergency department is essential.

Congestive Heart Failure

Failure of the heart occurs when the ventricular heart muscle is so damaged that it can no longer keep up with the return flow of blood from the atria. Congestive heart failure (CHF) can occur any time after a myocardial infarction, heart valve damage, or long-standing high blood pressure, but it usually happens between the first few hours and the first few days after a heart attack. (Right-sided Heart failure)

Just as the pumping function of the left ventricle can be damaged by coronary artery disease, it can also be damaged by diseased heart valves or chronic hypertension. In any of these cases, when the muscle can no longer contract effectively, the heart tries other ways to maintain an adequate cardiac output. Two specific changes in heart function occur: The heart rate increases, and the left ventricle enlarges in an effort to increase the amount of blood pumped each minute.

When these adaptations can no longer make up for the decreased heart function, congestive heart failure eventually develops. It is called "congestive" heart failure because the lungs become congested with fluid once the heart fails to pump the blood effectively. Blood tends to back up in the pulmonary veins, increasing the pressure in the capillaries of the lungs. When the pressure in the capillaries exceeds a certain level, fluid (mostly water) passes through the walls of the capillary vessels and into the alveoli. This condition is called pulmonary edema. It may occur suddenly, as in AMI, or slowly over months, as in chronic congestive heart failure. Sometimes, patients with an acute onset of CHF will develop severe pulmonary edema, in which the patient has pink, frothy sputum, and severe dyspnea.

If the right side of the heart is damaged, fluid collects in the body, often showing in the feet and legs. The collection of fluid in the part of the body that is closest to the ground is called dependent edema. The swelling causes relatively few symptoms other than discomfort. However, chronic dependent edema may indicate underlying heart disease even in the absence of pain or other symptoms.

Assessment of the Patient With Chest Pain

While en route, consider the minimum and maximum BSI precautions that will be needed. BSI can be as simple as gloves for the chest pain patient or full BSI precautions for the patient in cardiac arrest. Remember, the patient's condition can change rapidly from the time you are dispatched.

Scene Size-up

Do not let your guard down on medical calls. Always ensure that the scene is safe for you, your partner, your patient, and bystanders. As you approach the scene, determine the nature of illness and how many patients there are. This information can be obtained from

You are the Provider   Part 3

Your partner enters the room with a bottle of nitroglycerin. It is important to know the 5 Rs prior to administering any drug—right drug, right dose, right time, right route, right patient—and you must specifically know that a patient's systolic blood pressure is more than 100 mm Hg prior to administering nitroglycerin. You must also know indications, contraindications, and side effects of medications you assist with or administer.

He examines the nitroglycerin, obtains permission from medical control, and then assists the patient with its administration.

4. What other tactics can you employ to assist this patient?

bystanders, first responders, or the patient. From the nature of the call and first glance at your patient, determine whether you will need additional resources to assist in moving the patient. If you are in a tiered-response system, request that the paramedics be dispatched to your location. You will need to quickly assess the scene to determine if spinal stabilization is needed.

Initial Assessment

General Impression

All patient assessments begin by determining whether or not the patient is responsive. If the patient is not responsive, evaluate the ABCs and assess for use of the automated external defibrillator (AED), which is discussed in the section on cardiac arrest later in this chapter. Generally, the AED should be applied if the patient is pulseless, not breathing (apneic), and unresponsive.

If the patient is responsive, begin by asking the chief complaint. Remember that many patients present differently when experiencing an AMI. A chief complaint of chest pain or discomfort, shortness of breath, or dizziness should be taken seriously. Many patients who suspect that something is wrong appear anxious and perhaps sense an impending doom. Act professionally; be calm. Speak to the patient in a normal voice that is neither too loud nor too soft. Let the patient know that trained individuals, including you, are present to provide care and that he or she will soon be taken to the hospital. Remember, some patients may act carefree, while others may be demanding. Most patients, however, are still frightened. Your professional attitude may be the single most important factor in winning the patient's cooperation and helping the patient through this event. Patients often have a good idea about what is happening, so do not lie and offer false reassurance. If asked, "Am I having a heart attack?" you can say, "I do not know for sure, but in case you are, we are taking care of you. We are going to help you now by giving oxygen, and we will be taking you to the hospital. You are in good hands."

Airway and Breathing

Unless the patient is unresponsive, the airway will most likely be patent. Responsive individuals should be able to maintain their own airway. But some episodes of cardiac compromise may produce dizziness or even fainting spells. If either of these have occurred, be suspicious of spinal injuries from a fall. Assess and treat the patient as appropriate.

Assess the patient's breathing to determine if it is adequate to provide enough oxygen to an ailing heart. Some patients feel short of breath even though there are no obvious signs of respiratory distress. In either situation, apply oxygen with a nonrebreathing mask at 10 to 15 L/min. If the patient is not breathing or has inadequate breathing, ensure adequate ventilations with a BVM device and 100% oxygen.

Circulation

Assess the patient's circulation. Determine the rate and quality of the patient's pulse. Is the pulse rhythm regular or irregular? Is it too fast or too slow? If you find abnormalities in the pulse, you should be more suspicious. Assess the patient's skin condition, color, and temperature as well as capillary refill time. Changes in perfusion may indicate more serious cardiac compromise. Begin treatment for cardiogenic shock early to reduce the workload of the heart. Place the patient in a comfortable position, usually sitting up and well supported. Provide reassurance that appropriate treatment is being given for the condition to reduce the patient's anxiety. Is there any major bleeding that needs to be controlled? If so, utilize direct pressure to control the bleeding and bandage appropriately.

Transport Decision

Make a transport decision. Does the patient need to be transported rapidly? Is the patient's condition life-threatening, or is it stable enough to allow for performing a focused history and physical exam on scene? Generally speaking, most patients with chest pain should be transported immediately. Whether to transport using the lights and siren is determined with each patient and estimated transport time. As a general rule, however, cardiac patients should be transported in the most gentle, stress-relieving manner possible. Very little time is saved by the using the lights and siren, but you can do a lot to calm your patient and reduce the release of heart-damaging adrenaline through your reassurance and by creating a ride to the hospital that is as pleasant as possible. Try not to allow the patient to exert himself or herself, strain, or walk. If necessary, lift the patient, using care.

Your decision of where to transport the patient will depend on your local protocol. Patients are generally transported to the closest appropriate facility. If your service is served by one hospital, the transport decision is an easy one. In larger urban areas, there may be several hospitals within the service areas. Some medical directors have written protocols requiring patients with suspected cardiac emergencies to be transported to medical centers with certain capabilities, such as emergency angioplasty. Others require the patient to be transported to the nearest facility for stabilization prior to transporting to a specialty hospital. Be sure you know your local protocol.

Focused History and Physical Exam

SAMPLE History

For a conscious medical patient, begin with taking a brief history from the patient. Friends or family members who are present often have helpful information. Ask them the following questions:

- Has the patient ever had a heart attack before?
- Has the patient been told about having heart problems?
- Are there any risk factors for coronary artery disease, such as smoking, high blood pressure, or high-stress lifestyle?

The SAMPLE history provides basic information on the patient's overall medical history. You will want to determine as many signs and symptoms as you can. For example, you may determine that the patient has chest pain at rest, or absence of chest pain with respirations or movement. The more signs and symptoms a patient has, the easier it is to identify a particular prob-

lem. In addition, ask whether the patient has had the same pain before. If so, ask "Do you take any medications for the pain?" and "Do you have any of the medication with you?" If the patient has had a heart attack or angina before, ask whether the pain is similar.

Be sure to include the OPQRST questions when you are obtaining the symptoms as part of the SAMPLE history. Using OPQRST helps you to understand the details of specific complaints, such as chest pain (Table 12-1 ▶). Even when a patient may not be able to articulate his or her exact medical condition, knowing the patient's medications may give you important clues. For example, a patient may say he has "heart problems." You see that he is taking furosemide (Lasix), digoxin, and amiodarone. Furosemide is a diuretic, digoxin increases the strength of heart contractions, and amiodarone controls certain types of arrhythmias. These drugs are most often prescribed together for patients with congestive heart failure and may alert you to carefully evaluate lung sounds for pulmonary congestion and increase the amount of oxygen being delivered.

Focused Physical Exam

Pay particular attention to the cardiovascular system, but also check the respiratory system. How well is the heart working? Assess skin color, temperature, and condition. Is it cool, moist? How do the mucous membranes look? Are they pink, ashen, or cyanotic? Are the lung sounds clear? Are the neck veins distended?

Baseline Vital Signs

Measure and record the patient's vital signs. As you obtain the SAMPLE history, have your partner take the patient's baseline vital signs, including pulse, blood

You are the Provider   Part 4

You relay the patient's mental status, age, chief complaint, history of heart attack, vital signs, and your treatment to the incoming ALS unit. Their estimated time of arrival at the hospital is approximately 10 minutes. Because he is not in cardiac arrest, you keep the AED nearby but do not apply it.

You confirm that the patient is not allergic to aspirin and dispense two 81-mg tablets of baby aspirin, noting that the medication has not expired. You instruct the patient to chew and swallow them.

7. What can you do to prepare for the paramedics' arrival?

TABLE 12-1 OPQRST Mnemonic for Assessing Pain

- **Onset.** Determine what time the discomfort that motivated the call for help began.
- **Provocation.** Ask what makes the pain or discomfort worse. Is it positional? Does a deep breath or palpation of the chest make it worse?
- **Quality.** Ask what type of pain it is. Let the patient use his or her own words to describe what is happening. Try to avoid supplying the patient with only one option. Do not ask "Does it feel like an elephant is sitting on your chest?" Instead, say "Tell me what the pain feels like." If the patient cannot answer an open-ended question, then provide a list of alternatives. "There are lots of different kinds of pain. Is your pain more like heaviness, pressure, burning, tearing, dull ache, stabbing, or needlelike?"
- **Radiation.** Ask whether the pain travels to another part of the body.
- **Severity.** Ask the patient to rate the pain on a simple scale. Often, a scale ranging from 0 to 10 is used, in which 0 represents no pain at all and 10 represents the worst pain imaginable. Do not use the patient's answer to determine whether the pain has a serious cause. Instead, use it to check whether the pain is getting better or worse. After a few minutes of oxygen or administration of a nitroglycerin pill, ask the patient to rate the pain again.
- **Time.** Find out how long the pain lasts when it is present and whether it has been intermittent or continuous.

pressure, and respirations. You must obtain readings for both systolic and diastolic blood pressures. If available, use pulse oximetry. Note the time that vital signs are taken.

Communication

Alert the emergency department about the status of your patient's condition and your estimated time of arrival. Report to medical control. Report to the hospital by radio or cellular telephone while en route. Include information about the patient's history, vital signs, repeat vital signs, medications being taken, and any treatment you are giving. Follow the instructions of medical control. Describe the patient's condition to the emergency department staff on arrival.

Interventions

Depending on local protocol, prepare to administer baby aspirin and assist with prescribed nitroglycerin. Check the condition of the medication and its expiration date.

Administer baby aspirin according to local protocol. Baby aspirin comes in 81-mg chewable tablets. Recommended dosage will be either 162 mg (two tablets) or 324 mg (four tablets). Aspirin (acetylsalicylic acid) prevents clots from forming or getting bigger.

After obtaining permission from medical control, help the patient administer prescribed nitroglycerin. Nitroglycerin works in most patients within 5 minutes to relieve the pain of angina. Most patients who have been prescribed nitroglycerin carry a supply with them. Nitrostat is one trade name for nitroglycerin. Patients take one dose of nitroglycerin under the tongue whenever they have an episode of angina that does not immediately go away with rest. If the pain is still present after 5 minutes, patients are typically instructed by their physicians to take a second dose. If the second dose does not work, most patients are told to take a third dose and then call for EMS. If the patient has not taken all three doses, you can help to administer the medication, if you are allowed to do so by local protocol.

Nitroglycerin comes in several forms—as a small white pill, placed sublingually (under the tongue); as a spray, also taken sublingually; or as a skin patch applied to the chest (Figure 12-12 ▶). In any form, the effect is the same. Nitroglycerin relaxes the muscle of blood vessel walls, dilates coronary arteries, increases blood flow and the supply of oxygen to the heart muscle, and decreases the workload of the heart. Nitroglycerin also dilates blood vessels in other parts of the body and can sometimes cause low blood pressure and/or a severe headache. Other side effects include changes in the patient's pulse rate, including tachycardia and bradycardia. For this reason, you should take the patient's blood pressure within 5 minutes after each dose. If the systolic blood pressure is less than 100 mm Hg, do not give more medication. Other contraindications include the presence of a head injury, and the maximum prescribed dose has already been given (usually three doses).

Be aware that nitroglycerin will lose its potency over time, especially if exposed to light (that is why it is supplied in a brown bottle). Patients who take it only rarely may keep a bottle in their pocket for months. It may lose its potency even before its expiration date. When the nitroglycerin tablet loses its potency, patients

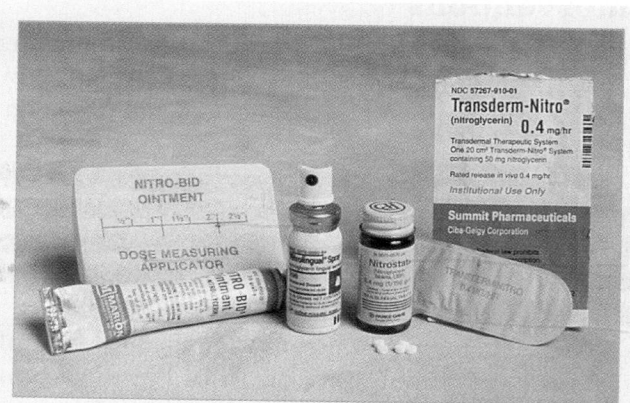

Figure 12-12 Nitroglycerin used to treat AMI comes in many forms, including paste, spray, tablets, and skin patches.

may not feel the fizzing sensation when the tablet is placed under their tongue, and they may not experience the normal burning sensation and headache that often accompany nitroglycerin administration. Note that the fizzing only occurs with a potent tablet, not with the spray form.

To safely assist the patient with nitroglycerin, follow the steps listed below (**Skill Drill 12-1 ▶**):

1. **Obtain an order from medical direction**—either online or offline protocol.
2. **Take the patient's blood pressure.** Continue with administration of nitroglycerin only if the systolic blood pressure is greater than 100 mm Hg (**Step 1**).
3. Check that you have the **right medication, the right patient, and the right delivery route**. Check the expiration date.
4. **Question the patient** about the last dose he or she took and its effects. Make sure that the patient understands the route of administration. **Be prepared to have the patient lie down** to prevent fainting if the nitroglycerin substantially lowers the patient's blood pressure (the patient gets dizzy or feels faint) (**Step 2**).
5. Ask the patient to lift his or her tongue. **Place the tablet or spray the dose underneath the tongue** (while wearing gloves), or have the patient do so. Have the patient keep his or her mouth closed with the tablet under the tongue until it is dissolved and absorbed. Caution the

patient against chewing or swallowing the tablet (**Step 3**).
6. **Recheck blood pressure** within 5 minutes. Record medication and the time of administration. Reevaluate the chest pain and note the response to the medication. If the chest pain persists and the patient still has a systolic blood pressure greater than 100 mm Hg, repeat the dose every 5 minutes as authorized by medical control. In general, a maximum of three doses of nitroglycerin are given for any one episode of chest pain (**Step 4**).

Reevaluate your transport decision. Transport the patient. Early, prompt transport to the emergency department is critical so that treatments such as clot-busting medications or angioplasty can be initiated. To be most effective, these treatments must be started as soon as possible after the onset of the attack. If the patient does not have prescribed nitroglycerin, move ahead with your focused assessment and prepare to transport. Be sure that this process does not consume too much time. Do not delay transport to assist with administration of nitroglycerin. The drug can be given en route.

Detailed Physical Exam

If necessary, perform a detailed physical exam to elicit further information concerning the patient's condition and necessary interventions. If you have time, you can talk with the patient about risk factors for heart disease such as cholesterol level, smoking, activity levels, and family history of heart disease. Do not gather this information unless your patient's condition is stable and everything else is done.

Ongoing Assessment

Repeat your initial assessment by checking to see if the patient's condition has improved or if the patient's condition is deteriorating. Vital signs should be reassessed at least every 5 minutes or as significant changes in the patient's condition occur. It is essential to monitor

Administration of Nitroglycerin

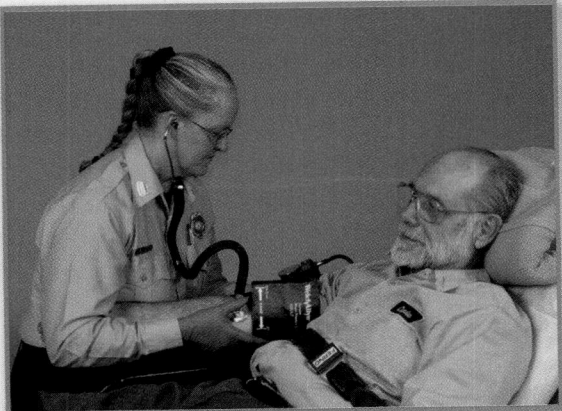

1 Obtain an order from medical direction—either online or offline protocol.

Take the patient's blood pressure. Administer nitroglycerin only if the systolic blood pressure is greater than 100 mm Hg.

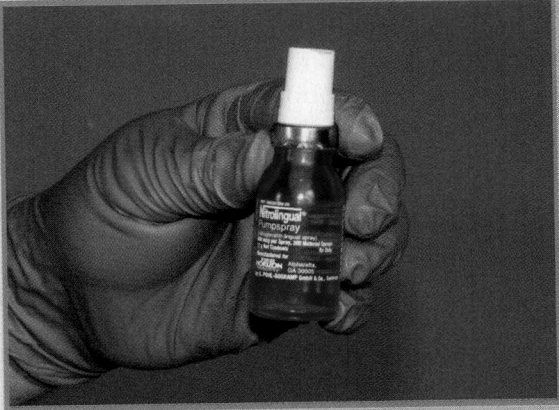

2 Check the medication and expiration date.

Question the patient about the last dose he or she took and its effects. Make sure that the patient understands the route of administration.

Prepare to have the patient lie down to prevent fainting.

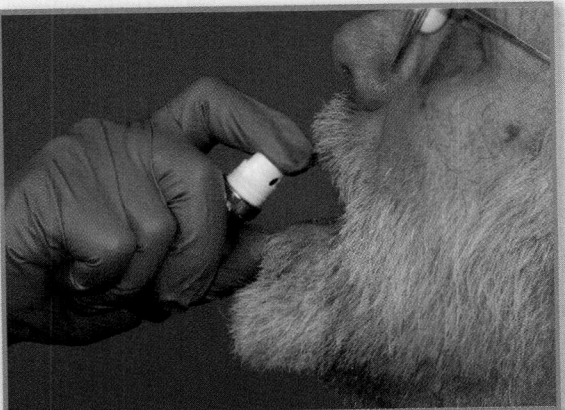

3 Ask the patient to lift his or her tongue.

Place the tablet or spray the dose underneath the tongue (while wearing gloves), or have the patient do so.

Have the patient keep his or her mouth closed with the tablet under the tongue until it is dissolved and absorbed. Caution the patient against chewing or swallowing the tablet.

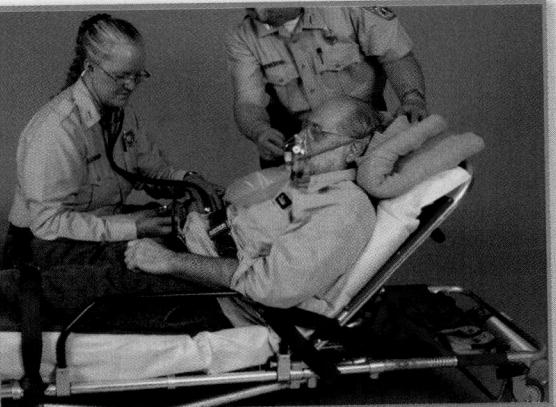

4 Recheck blood pressure within 5 minutes. Record each medication and the time of administration. Reevaluate the chest pain and repeat treatment if necessary.

the patient with a suspected AMI closely because sudden cardiac arrest is always a risk. If cardiac arrest occurs, you must be ready to begin automated defibrillation or CPR immediately. If an AED is immediately available, use it; if not, perform CPR until the AED is available. Reassess your interventions. It is important to continue reassessing in order to see if the interventions are helping and if the patient's condition is improving. Reassess vital signs after administering medications. Reassessment will also determine whether further interventions are indicated or contraindicated.

Communication and Documentation

It is important to document your assessment of the patient. You must record the interventions performed. All interventions should be initiated according to protocol. If the intervention required an order from medical control, document the medication requested and whether approval was granted or not. It must be clear in your documentation that the patient was reassessed appropriately following any intervention. The patient's response to the intervention must also be recorded. Upon completing your documentation, obtain the medical control physician's signature (if required by local protocol) showing approval of medication administration.

Heart Surgeries and Pacemakers

During the last 20 years, hundreds of thousands of open heart surgeries were performed to bypass damaged segments of coronary arteries in the heart. In a coronary artery bypass graft (CABG), a blood vessel from the chest or leg is sewn directly from the aorta to a coronary artery beyond the point of the obstruction. Other patients may have had a procedure called percutaneous transluminal coronary angioplasty (PTCA), which aims to dilate, rather than bypass, the coronary artery. In this procedure, usually called an angioplasty or balloon angioplasty, a tiny balloon is attached to the end of a long, thin tube. The tube is introduced through the skin into a large artery, usually in the groin, and then threaded into the narrowed coronary artery, with radiographs serving as a guide. Once the balloon is in position inside the coronary artery, it is inflated. The balloon is then deflated, and the tube is removed from the body. Sometimes, a metal mesh called a stent is placed inside the artery either instead of or after the balloon. The stent is left in place permanently to help keep the artery from narrowing again.

You are the Provider — Part 5

You hear the paramedics arrive, and you meet them at the door. You give your most recent set of vital signs, explain that a total of two nitroglycerin tablets have been administered along with the aspirin, and now his chest pain is 4/10. You help the paramedic place the patient on the cot, where a 12-lead ECG is performed while his partner prepares an IV in the ambulance.

As the machine prints out the ECG findings, you and the paramedic begin to move the patient to the ambulance. He asks if you would accompany him during transport. Your partner hands the paramedic a grocery bag containing the patient's medicine bottles. As soon as the patient is loaded into the ambulance, the paramedic instructs his partner that he is ready to go. En route, you obtain another set of vital signs, while he starts the IV.

8. Given that the patient's systolic blood pressure remains greater than 100 mm Hg, how many additional nitroglycerin tablets can you expect the paramedic to administer?
9. What other forms of the medication may be given instead of the tablet, and what problems can you encounter when examining/using a patient's nitroglycerin in the field?

A patient who has had an AMI or angina will almost certainly have had one of these procedures. Patients who have had a bypass graft will have a long surgical scar on their chest from the operation. Patients who have had an angioplasty or coronary artery stent usually will not. However, newer "keyhole" surgical techniques may not produce a large scar. You should not assume that a patient who has a small scar has not had bypass surgery. Chest pain in a patient who has had any of these procedures should be treated the same as chest pain in patients who have not had any heart surgery. In any event, chest pain in a patient who has undergone either procedure is treated exactly the same as chest pain in a patient who has not. Carry out all the described tasks, and transport the patient promptly to the emergency department of the hospital. If CPR is required, perform it in the usual way, regardless of the scar on the patient's chest. Likewise, if indicated, an AED should be used as well.

Many people with heart disease in the United States have cardiac pacemakers to maintain a regular cardiac rhythm and rate. Pacemakers are inserted when the electrical control system of the heart is so damaged that it cannot function properly. These battery-powered devices deliver an electrical impulse through wires that are in direct contact with the myocardium. The generating unit is generally placed under a heavy muscle or a fold of skin; it typically resembles a small silver dollar under the skin in the left upper chest (Figure 12-13 ▶).

Normally, you do not need to be concerned about problems with pacemakers. Thanks to modern technology, an implanted unit will not require replacement or a battery charge for years. Wires are well protected and rarely broken. In the past, pacemakers sometimes malfunctioned when a patient got too close to an electrical radiation source, such as a microwave oven, but this is no longer the case. Every patient with a pacemaker should be aware of the precautions, if any, that must be taken to maintain its proper functioning.

If a pacemaker does not function properly, as when the battery wears out, the patient may experience syncope, dizziness, or weakness because of an excessively slow heart rate. The pulse ordinarily will be less than 60 beats/min because the heart is beating without the stimulus of the pacemaker and without the regulation of its own electrical system, which may be damaged. In these circumstances, the heart tends to assume a fixed slow rate that is not fast enough to allow the pa-

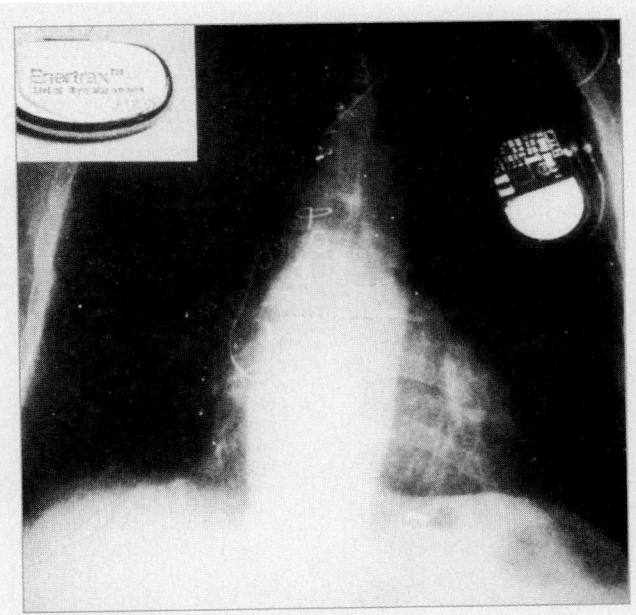

Figure 12-13 A pacemaker, which is typically inserted under the skin in the left upper chest, delivers an electrical impulse to regulate heartbeat.

tient to function normally. A patient with a malfunctioning pacemaker should be promptly transported to the emergency department; repair of the problem may require surgery. When an AED is used, the patches should not be placed directly over the pacemaker. This will ensure a better flow of electricity through the patient's body.

Automatic Implantable Cardiac Defibrillators

More and more patients who survive ventricular fibrillation cardiac arrests have a small automatic implantable cardiac defibrillator (AICD) implanted. Some patients who are at particularly high risk for a cardiac arrest have them as well. These devices are attached directly to the heart and can prolong the lives of certain patients. They continuously monitor the heart rhythm, delivering shocks as needed. Regardless of whether a patient having an AMI has an AICD, he or she should be treated like all other patients having an AMI. Treatment should include performing CPR and using an AED if the patient goes into cardiac arrest. Generally, the electricity from an AICD is so low that it will have no effect on rescuers and therefore should not be of concern to you (Figure 12-14 ▶).

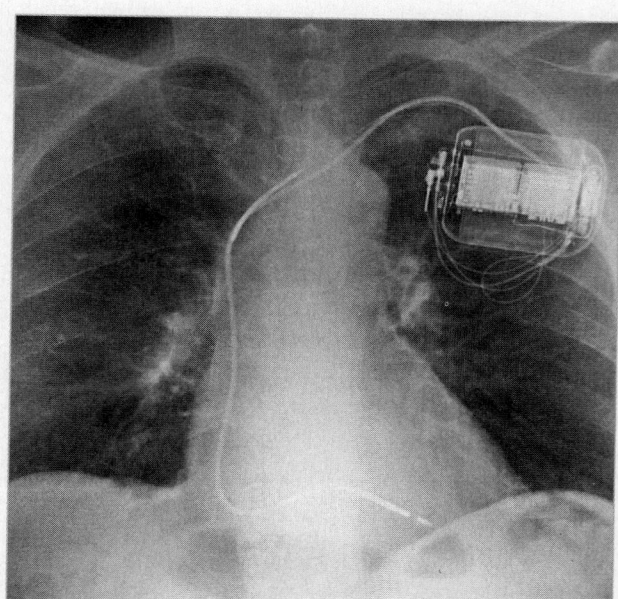

Figure 12-14 An AICD is attached directly to the heart and continuously monitors heart rhythm, delivering shocks as needed. The electricity from the AICD is so low that it has no effect on rescuers.

⚠ Pediatric Needs

Heart problems in childhood are uncommon and usually congenital, meaning that the patient was born with the problem. In general, your approach to these patients should be the same as that for an adult. You should attempt to reassure the patient. If possible, administer oxygen. If the patient will not wear a face mask, have the parent hold the oxygen in front of the child's face.

Cardiac arrest in younger children is less common than in older children and is usually caused by a breathing problem. The energy levels of the electrical shock delivered by most AEDs are too high for children who are younger than 8 years or weigh less than 55 lb (25 kg); therefore, make sure that protocols allow for defibrillation of children 1 to 8 years of age; however, an AED with special pediatric pads is indicated for use on pediatric medical patients between the ages of 1 and 7 years who have been assessed to be unresponsive, not breathing, and pulseless. It is absolutely mandatory that if protocol allows defibrillation of children, you ensure that your equipment is designed to be used on children. Teenagers, who may occasionally have a cardiac arrest related to a heart problem, may benefit from an AED.

Cardiac Arrest

Cardiac arrest is the complete cessation of cardiac activity, either electrical, mechanical, or both. It is indicated in the field by the absence of a carotid pulse. Until the advent of CPR and external defibrillation in the 1960s, cardiac arrest was virtually always a terminal event. Now, although it is still infrequent for a patient to survive a cardiac arrest without neurologic damage, great strides have been made in resuscitation science during the last 40 years.

Automated External Defibrillation

In the late 1970s and early 1980s, scientists developed a small computer that could analyze electrical signals from the heart and determine when ventricular fibrillation was taking place. This development, along with improved battery technology, made the automated portable defibrillator possible—a device that can automatically administer an electrical shock to the heart when needed.

AED machines come in different models with different features (Figure 12-15 ▶). All of them require a certain degree of operator interaction, beginning with applying the pads and turning the machine on. The operator also has to push a button to deliver an electrical shock, regardless of the model. Many AEDs use a computer voice synthesizer to advise the EMT which steps to take on the basis of the AED's analysis. Some have a button that tells the computer to analyze the heart's electrical rhythm; other models start doing this as soon as they are turned on. In the United States, the majority of the AEDs are semiautomated. Even though most defibrillators are now semiautomated, we are using the term automated external defibrillators (AED) as the general term to describe all of these machines.

AEDs also come equipped to give a monophasic shock or a biphasic shock. Monophasic means to send the energy in one direction, from negative to positive and

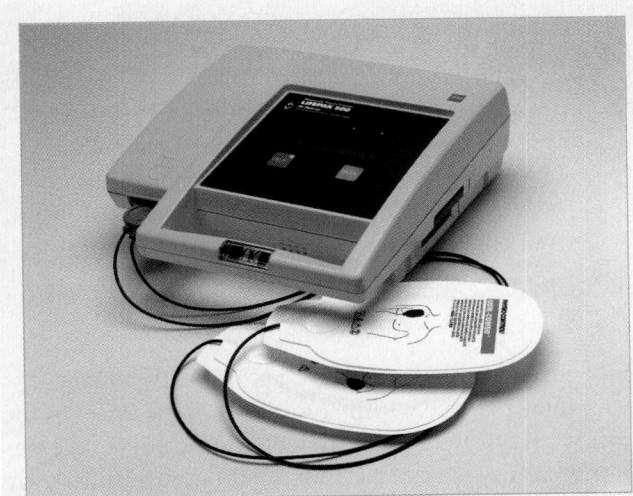

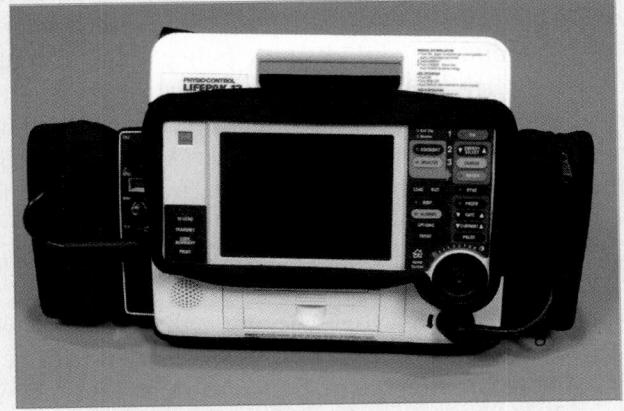

Figure 12-15 AEDs vary in their design, features, and operation.

biphasic means to send the energy in two directions simultaneously. The advantage of biphasic shock is that it produces a more efficient defibrillation and may require a lower energy setting. The energy setting for ventricular fibrillation on a monophasic machine is generally 200 joules to start then moves to 200 or 300 joules, and then to 360 joules. With the biphasic technology the energy can be set at 120 joules for all three shocks and all subsequent shocks after that, or can start at 120 joules and then escalate to 200 joules. The actual setting of the biphasic machines is still being studied and no recommendation for either is currently supported in the literature. The computer inside the AED is specially programmed to recognize rhythms that require defibrillation to correct, most commonly ventricular fibrillation. The current programs are extremely accurate. It would be extremely rare for them to recommend a shock

when a shock would not be called for, and they rarely fail to recommend one when it would be helpful. Therefore, if the AED recommends a shock, you can believe that it is indicated.

When an error does occur, it is usually the operator's fault. The most common error is not having a charged battery. To avoid this problem, many defibrillator companies have built smarter machines that will warn the operator that the battery is unlikely to work. However, some of the older models do not have this feature. You should check the AED daily and exercise the battery as often as the manufacturer recommends.

Another error occurs when the AED is applied to a patient who is moving. The computer may be unable to tell the difference between electrical signals from the heart and electrical signals from the arms and chest muscles that are moving. The way to avoid this error is to apply the AED only to pulseless, unresponsive patients and to stay clear of the patient (do not touch the patient) during analysis and shocking.

A third error can occur when the AED is applied to a responsive patient with a rapid heart rate. Most computers identify a regular rhythm faster than 150 or 180 beats/min as ventricular tachycardia, which should be shocked. Sometimes, though, a patient has another heart rhythm that should not be shocked but that is fast enough to confuse the computer. Again, to avoid this problem, you should apply the AED only to unresponsive patients with no pulse.

Automated external defibrillation offers the EMT-B a number of advantages. First, of course, the machine is fast, and it delivers the most important treatment for the patient in ventricular fibrillation: an electrical shock. It can be delivered within 1 minute of the EMT's arrival at the patient's side. Second, you will find that using an AED is easier than performing CPR. ALS providers do not have to be on the scene to provide this definitive care.

Current AEDs offer two other advantages. The shock can be given through remote, adhesive defibrillator pads, which are safer for you than paddles. Also, the pad area is larger than paddles, which means that the transmission of electricity is more efficient. Usually, there are pictures on the pads to remind you where they go on the patient's chest.

Not all patients in cardiac arrest require an electrical shock. Although all patients in cardiac arrest should be analyzed with an AED, some do not have shockable rhythms (eg, pulseless electrical activity and asystole). Asystole (flatline) indicates that no electrical activity remains. Pulseless electrical activity usually refers to a state of cardiac arrest despite an organized electrical complex. In both cases, CPR should be initiated as soon as possible.

Rationale for Early Defibrillation

Few patients who experience sudden cardiac arrest outside of a hospital survive unless a rapid sequence of events takes place. The chain of survival is a way of describing the ideal sequence of events that can take place when such an arrest occurs.

The four links in the chain of survival are as follows (Figure 12-16 ▼):

- Recognition of early warning signs and immediate activation of EMS
- Immediate bystander CPR
- Early defibrillation
- Early advanced cardiac life support

If any one of the links in the chain is absent, the patient is more likely to die. For example, few patients benefit from defibrillation when more than 10 minutes elapse before administration of the first shock or if CPR is not performed in the first 2 to 3 minutes. If all links in the chain are strong, the patient has the best possible chance of survival. The link with the most determinant for survival is the third link—early defibrillation.

CPR helps patients in cardiac arrest because it prolongs the period of time during which defibrillation can be effective. Rapid defibrillation has successfully resuscitated many patients with cardiac arrest from

EMT-B Safety

When clearing the patient before an AED shock, ensure that no one is touching the patient and that no object is touching the patient, including the stretcher or other furniture if the patient is not on the floor or ground.

ventricular fibrillation. However, defibrillation works best if it takes place within 2 minutes of the onset of the cardiac arrest. To try to achieve better survival rates among cardiac arrest victims, many communities are exploring the idea that nontraditional first responders should be trained to administer early defibrillation. These responders would include police officers, security personnel, lifeguards, maintenance workers, and flight attendants. As an EMT-B, you should support these efforts to shorten the time interval until defibrillation. Remember, seconds really do matter when the patient is in cardiac arrest.

Integrating the AED and CPR

Since most cardiac arrests occur in the home, a bystander at the scene may already have started CPR before you arrive. For this reason, you must know how to work the AED into the CPR sequence. Remember

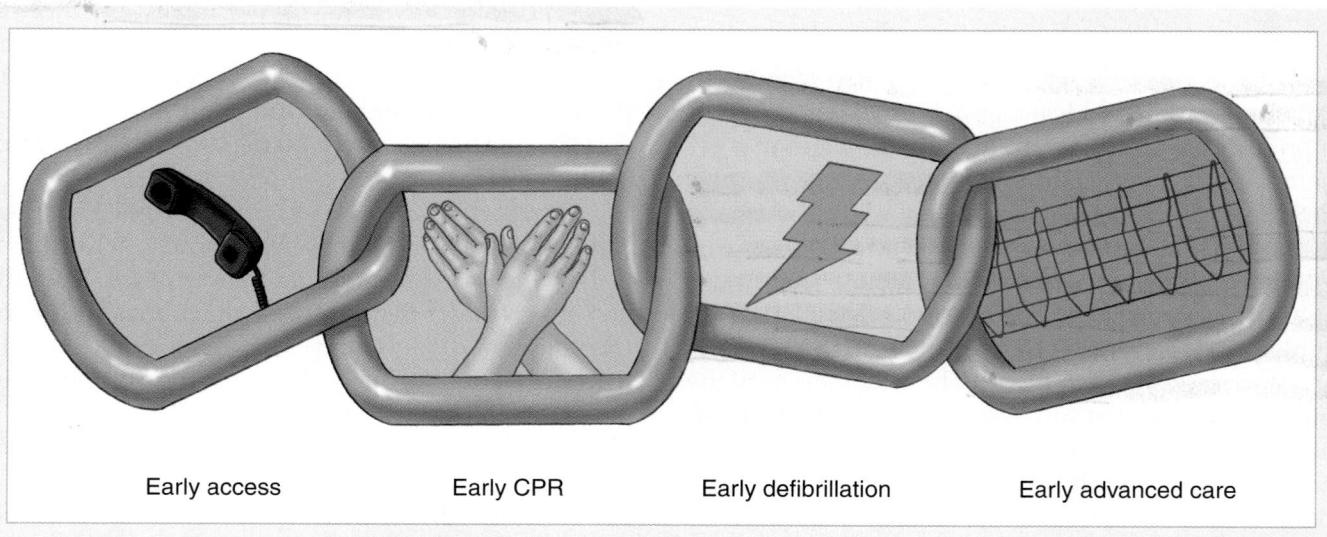

| Early access | Early CPR | Early defibrillation | Early advanced care |

Figure 12-16 The four links of the chain of survival. *Source:* American Heart Association.

that the AED is not very complex; it may not be able to distinguish other movements from ventricular fibrillation. Therefore, do not touch the patient while the AED is analyzing the heart rhythm and delivering shocks. Stop CPR, and let the AED do its job. CPR may be stopped for up to 90 seconds if three shocks are necessary. This is entirely proper; defibrillation is more important than CPR when ventricular fibrillation is present.

AED Maintenance

One of your primary missions as an EMT-B is to deliver an electrical shock to a patient in ventricular fibrillation. To accomplish this mission, you need to have a functioning AED. You must become familiar with the maintenance procedures required for the brand of AED your service uses. Read the operator's manual. If your defibrillator does not work on the scene, someone will want to know what went wrong. That person may be your system's administrator, your medical director, the local newspaper reporter, or the family's attorney. You will be asked to show proof that you maintained the defibrillator properly and attended any mandatory inservices.

The main legal risk in using the AED is failing to deliver a shock when one was needed. The most common reason for this failure is that the battery did not work, usually because it was not properly maintained. Another problem is operator error. This means not pushing the analyze or shock buttons when the machine advises you to do so or failing to apply the AED to a patient in cardiac arrest. Of course, the AED is like any other manufactured item. It can fail, although this is rare. Ideally, you will encounter any such failure while doing routine maintenance, not while caring for a patient in cardiac arrest. Check your equipment, including your AED, at the beginning of each shift. Ask the manufacturer for a checklist of items

Geriatric Needs

Like the other body systems, the cardiovascular system undergoes changes as we get older. The heart, like other major organs, will show the effects of aging. As the heart's muscle mass and tone decrease, the amount of blood pumped out of the heart per beat is decreased. The residual (reserve) capacity of the heart is also reduced; therefore, when the vital organs of the body need additional blood flow, the heart cannot meet the need. When blood flow to the tissues is decreased, the organs suffer. If blood flow to the brain is inadequate, the patient may complain of weakness, fatigue, or dizziness and may develop syncope (fainting).

The power to the heart muscle can fail. The heart runs on electricity and has its own electrical system. Under normal conditions, electrical impulses travel throughout the heart, resulting in the contraction of the heart muscle and the pumping of blood from the heart's chambers. With aging, the electrical system can deteriorate, causing the heart's contraction to weaken or, if blood flow to the heart muscle is affected, extra beats to form. With a decreased strength of contraction, the heartbeat is weaker and blood flow to the tissues is reduced. If extra beats are produced, the patient's heart rhythm will be irregular. While some irregular heart rhythms are acceptable, others can be potentially lethal.

The arteries are also affected by aging. Arteriosclerosis (hardening of the arteries) can develop, affecting perfusion of the tissues. There is an increased chance of heart attack or stroke from decreased blood flow or plaque formation (atherosclerosis) in the narrowed arteries.

Patients with diabetes can experience reduced circulation to the hands and feet; this makes peripheral pulses harder to detect. It also puts the hands and feet at particular risk for developing infection or ulcerations.

In some older patients with angina or AMI, particularly diabetics, chest pain is absent, and the clinical picture can be confused with other, noncardiac conditions.

The cardiovascular system is affected by aging. You should be aware of the changes, seeking to determine what is normal versus what is chronic for the patient as opposed to what is an acute condition. Sometimes, the weakening of the heart muscle, the deterioration of its electrical system, and the hardening of the arteries make the task of assessing and caring for the older patient more difficult.

that should be checked daily, weekly, or less often (Figure 12-17 ▶).

If you do have an AED failure while caring for a patient, you must report that problem to the manufacturer and the US Food and Drug Administration. Be sure to follow the appropriate EMS procedures for notifying these organizations.

Medical Direction

Defibrillation of the heart is a medical procedure. While AEDs have made the process of delivering electricity much simpler, there is still a benefit in having a physician's involvement. The medical director of your service should help to teach you how to use the AED. At the very least, he or she should approve the written protocol that you will follow in caring for patients in cardiac arrest. In most states, successful completion of AED training in an EMT-B course is not permitted without approval by state laws, rules, and local medical direction authority.

There should be a review of each incident in which the AED is used. After returning from the hospital or the scene, sit down with the rest of the team and go over what happened. This discussion will help all members of the team learn from the incident. Review such events by using the written report, any voice-ECG tape recorder, and the device's solid-state memory modules and magnetic tape recordings, if applicable.

There should also be a review of the incident by your service's medical director or quality improvement officer. Quality improvement involves individuals using AEDs and the responsible EMS system managers. This review should focus on speed of defibrillation, that is, the time from call to shock. Few systems will achieve the ultimate goal: shocking 100% of patients within 1 minute of the call. However, all systems continuously work on improving patient care. Mandatory continuing education with skill competency review is generally required for EMS providers, with a continuing competency skill review every 3 to 6 months for the EMT-B.

Emergency Care for Cardiac Arrest

Preparation

En route to the scene, prepare yourself to follow BSI precautions. Upon arrival at the scene, make sure that the scene is safe for you and your partner to enter. If dispatch reports an unresponsive patient with CPR

You are the Provider

Part 6

The patient now rates his pain as 3/10. As you obtain another set of vital signs the paramedic radios the hospital and provides patient updates. You give your blood pressure reading to the paramedic who, during the radio report, administers another dose of sublingual nitroglycerin using a metered-dose spray. After the report, he asks you to complete a head-to-toe physical exam while he faxes the 12-lead ECG to the receiving hospital and prepares morphine for administration.

He checks the blood pressure again and asks the patient to rate his pain, which he says has not changed. The paramedic then confirms the patient is not allergic to morphine and administers a single dose through the IV line. By the time you reach the hospital, the patient's pain is gone.

10. As an EMT-B, why is it extremely important to obtain accurate vital signs?
11. How can an inaccurate blood pressure reading affect patient care, especially during this type of scenario?

AUTOMATED EXTERNAL DEFIBRILLATOR
Daily/Shift Inspection Checklist

Serial # _____   Date _____   Time _____

Model # _____   Inspected by _____

Item	Pass	Fail
Exterior/Cables:		
Nothing stored on top of unit		
Carry case intact and clean		
Exterior/LCD screen clean and undamaged		
Cables/connectors clean and undamaged		
Cables securely attached to unit		
Batteries:		
Unit charger is plugged in and operational (if applicable)		
Fully charged battery in unit		
Fully charged spare battery		
Spare battery charger plugged in and operational (if applicable)		
Valid expiration date on both batteries		
Supplies:		
Two sets of electrodes		
Electrodes in sealed packages with valid expiration dates		
Razor		
Hand towel		
Alcohol wipes		
Memory/voice recording device—module, card, microcassette		
Manual override—module, key (if applicable)		
Printer paper (if applicable)		
Operation:		
Unit self-test per manufacturer's recommendation/instructions		
Display (if applicable)		
Visual indicators		
Verbal prompts		
Printer (if applicable)		
Attach AED to simulator/tester:		
Recognizes shockable rhythm		
Charges to correct energy level within manufacturer's specifications		
Delivers charge		
Recognizes nonshockable rhythm		
Manual override system in working order (if applicable)		

Signature:

Figure 12-17 A sample daily checklist for the AED.

being performed, the AED is probably one of the first pieces of equipment you will gather from the ambulance. As the operator of the AED, you are responsible for making sure that the electricity injures no one, including yourself. Remote defibrillation using pads allows you to distance yourself safely from the patient. As long as you place the pads in the correct position and make sure no one is touching the patient, you should be safe. Do not defibrillate a patient who is in pooled water. While there is some danger to you if you are also in the water, there is another problem. Electricity follows the path of least resistance; instead of traveling between the pads and through the patient's heart, it will diffuse into the water. Therefore, the heart will not receive enough electricity to cause defibrillation. You can defibrillate a soaking wet patient, but try first to dry the patient's chest. Do not defibrillate someone who is touching metal that others are touching, and carefully remove a nitroglycerin patch from a patient's chest and wipe the area with a dry towel before defibrillation to prevent ignition of the patch. It is often helpful to shave a hirsute patient's chest prior to patch placement to increase conductivity. Be sure to consult local protocols for issues such as pad placement and preparation of the pad site.

Determine the nature of illness and/or mechanism of injury. If the incident involves trauma, perform spinal stabilization as you begin the initial assessment. Is there only one patient? If you are in a tiered system and the patient is in cardiac arrest, call for ALS assistance.

Performing Defibrillation

Begin with determining the patient's level of consciousness and chief complaint. If the patient is unresponsive and pulseless, prepare to defibrillate. Ask any bystanders or first responders who are performing CPR to stop so that you can apply the AED and defibrillate the patient. Defibrillation has the best chance of restoring the heart to normal functioning. CPR only buys you time until a defibrillator is available. If an AED is immediately available, take the following steps according to your local protocols Skill Drill 12-2 ▶ : (Learn the steps)

1. **Arrive on scene and perform your initial assessment.** Assess responsiveness. If the patient is responsive, do not apply the AED.
2. **Stop CPR** if it is in progress.
3. **Verify pulselessness and apnea.** Check for breathing and a pulse even if the patient appears to be breathing.

4. If the patient is unresponsive and not breathing or is breathing agonally (slow, gasping breaths), **give two ventilations** using a BVM device or a pocket mask (**Step 1**).
5. If there is a delay in obtaining an AED, **have your partner start or resume CPR.**
6. If an AED is close at hand, **prepare the AED pads.**
7. **Turn on the machine** (**Step 2**).
8. **Remove clothing from the patient's chest area.** Apply the pads to the chest: one just to the right of the breastbone (sternum) just below the collarbone (clavicle), the other on the left chest with the top of the pad 2″ to 3″ below the armpit. Ensure that the pads are attached to the patient cables (and that they are attached to the AED in some models).
9. **Stop CPR** (**Step 3**).
10. State aloud, "**Clear the patient,**" and ensure that no one is touching the patient.
11. **Push the analyze button**, if there is one.
12. **Wait for the computer** in the AED to determine whether a shockable rhythm is present.
13. If a shock is not needed, go to step 18 (CPR only). If a shock is advised, make sure that no one is touching the patient. When the patient and area around them is clear, **push the shock button.**
14. **After the shock is delivered**, most AEDs will automatically reanalyze the rhythm; if not, push the analyze button again.
15. **If the machine advises a shock, deliver a second shock.**
16. **Reanalyze the rhythm.**
17. **If the machine advises a shock, deliver a third shock** (**Step 4**).
18. **Check for a pulse.**
19. **If the patient has a pulse,** check the patient's breathing (**Step 5**).
20. **If the patient is breathing adequately**, give the patient oxygen via nonrebreathing mask and transport. If the patient is not breathing adequately, use necessary airway adjuncts and proper positioning of the head and jaw to ensure an open airway. Provide artificial ventilations with high-concentration oxygen and transport.
21. **If the patient has no pulse,** perform 1 minute of CPR.
22. **Gather additional information** on the arrest event.

23. **After 1 minute of CPR**, make sure no one is touching the patient. Push the analyze button again (as applicable).
24. **If necessary, repeat one cycle of up to three stacked shocks.**
25. **Transport and check with medical control.**
26. **Continue to support the patient** as needed: ventilate until the patient begins to breathe normally, and continue CPR if needed (**Step 6**).

If, after any rhythm analysis, the AED advises no shock, check the patient's pulse. If the patient has a pulse, check the patient's breathing. If the patient is breathing adequately, give high-concentration oxygen via nonrebreathing mask and transport. If the patient is not breathing adequately, provide artificial ventilations with high-concentration oxygen via a BVM device and transport. Ensure that appropriate airway techniques are used at all times.

If the patient has no pulse, resume CPR for 1 minute, then have the AED reanalyze the heart rhythm. If the AED advises shock, deliver up to two sets of three stacked shocks. Separate each set of three shocks with 1 minute of CPR.

If the AED advises no shock and the patient has no pulse, resume CPR for 2 minute, then stop and reanalyze the heart rhythm for a third time. If the AED advises shock, deliver up to two sets of three stacked shocks with 1 minute of CPR between the two sets. If the AED still advises no shock, check with medical control, resume CPR, and transport.

If you are the only rescuer at the scene and you have an AED, take the following steps:

1. Perform an initial assessment. Assess responsiveness. If the patient is responsive, do not apply the AED.
2. Verify that the patient has no pulse and is not breathing (or is breathing with inadequate gasping breaths).
3. If the patient is not breathing or is gasping, give two slow breaths using a BVM device or pocket mask.
4. Expose the patient's chest. Apply one pad just to the right of the breastbone (sternum), just below the collarbone (clavicle), and the other on the left side of the chest with the top of the pad 2″ to 3″ below the armpit.
5. Turn on the AED.
6. Push the analyze button, if there is one.
7. Deliver up to three shocks, if indicated.

8. Follow your local protocol. If the AED indicates no need for shocks, provide CPR.

If another person is available who knows CPR, ask for help. You will perform the steps in the same order. The only difference is that the other person can continue CPR while you are getting the AED out and applied to the patient.

After AED Shocks

The care of the patient after the AED delivers a shock depends on your location and EMS system; therefore, you should follow your local protocols. After the AED protocol is completed, the patient is likely to have had one of the following occur:

- Regained a pulse
- No pulse, and the AED indicates that no shock is advised
- No pulse, and the AED indicates that a shock is advised

Patients who fail to regain a pulse on the scene of the cardiac arrest usually do not survive. What you do with these patients will, again, depend on your EMS system. Whether you should transport the patient or wait for ALS to arrive should be in the local protocols established by medical direction. If paramedics or another advanced life support service is responding to the scene, the best option usually is to stay where you are and continue the sequence of shocks and CPR. Administering CPR while patients are being moved or transported is usually not effective. The best chance for patient survival occurs when the patient is resuscitated where found, unless the location is unsafe.

If an ALS service is not responding to the scene and your local protocols agree, you should begin transport when one of the following occurs:

- The patient regains a pulse.
- Six to nine shocks are delivered.
- The machine gives three consecutive messages (separated by 1 minute of CPR) that no shock is advised.

If you transport a patient while performing CPR, you need a plan for managing the patient in the ambulance. Ideally, you will have two EMT-Bs in the patient compartment while a third drives. You may deliver additional shocks at the scene or en route with the approval of medical control. Keep in mind that AEDs cannot analyze rhythm while the vehicle is in motion. Nor is it as safe to defibrillate in a moving ambulance. Therefore, you should come to a complete stop if more

AED and CPR

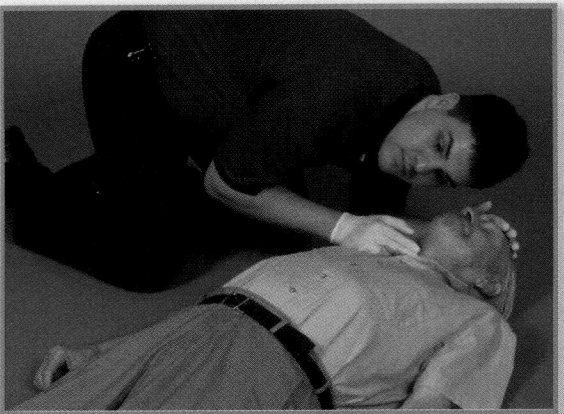

1 Stop CPR if in progress.
Assess responsiveness.
Check breathing and pulse.
If unresponsive and not breathing adequately, give two slow ventilations.

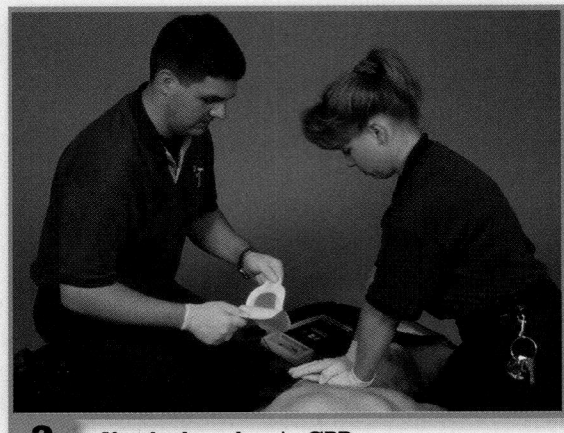

2 If pulseless, begin CPR.
Prepare the AED pads.
Turn on the AED; begin narrative if needed.

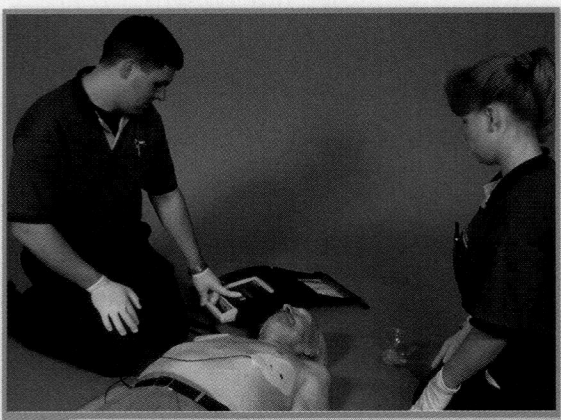

3 Apply AED pads.
Stop CPR.

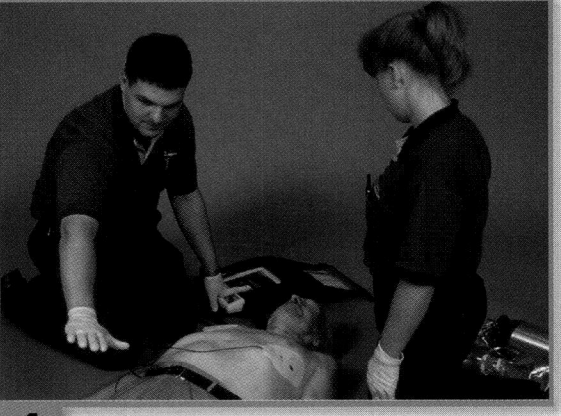

4 Verbally and visually clear the patient.
Push the Analyze button if there is one.
Wait for the AED to analyze rhythm.
If no shock advised, perform CPR for 1 minute.
If shock advised, recheck that all are clear and push the Shock button.
Push the Analyze button, if needed, to analyze rhythm again.
Press Shock if advised (second shock).
Push the Analyze button, if needed, to analyze rhythm again.
Press Shock if advised (third shock).

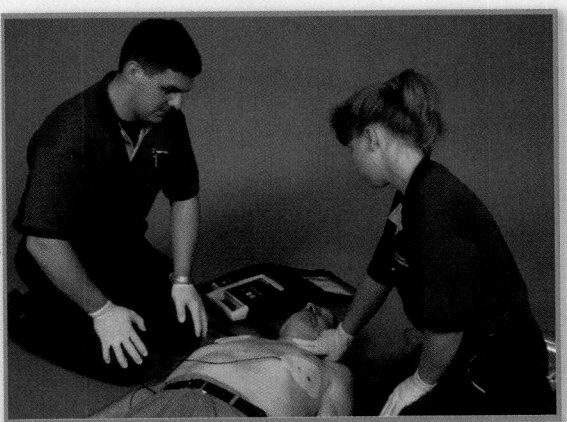

5 Check pulse.

If pulse is present, check breathing.

Gather additional information on the arrest event.

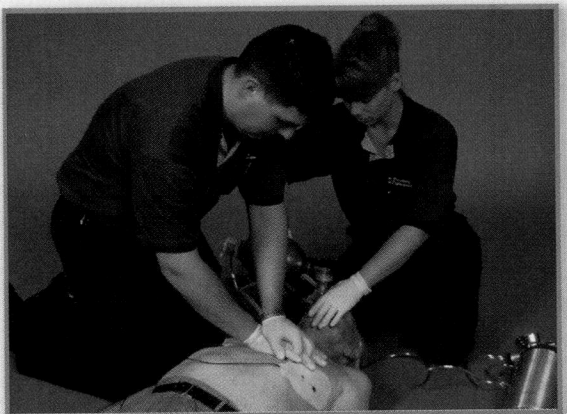

6 If breathing adequately, give oxygen and transport. If not, open airway, ventilate, and transport.

If no pulse, perform CPR for 1 minute.

Clear the patient and analyze again.

If necessary, repeat one cycle of up to three shocks.

Transport and call medical control.

Continue to support breathing or perform CPR, as needed.

shocks are needed. Be sure to memorize the protocol of your EMS service (Figure 12-18 ▶).

Cardiac Arrest During Transport ✄

If you are traveling to the hospital with an unconscious patient, check the pulse at least every 30 seconds. If a pulse is not present, take the following steps:

1. Stop the vehicle.
2. If the AED is not immediately ready, perform CPR until it is available.
3. Analyze the rhythm.
4. Deliver shocks, if indicated.
5. Continue resuscitation according to your local protocol.

If you are en route with a conscious adult patient who is having chest pain and becomes unconscious, take the following steps:

1. Check for a pulse.
2. Stop the vehicle.
3. If the AED is not immediately ready, perform CPR until it is ready.
4. Analyze the rhythm.
5. Deliver up to three shocks, if indicated.
6. Continue resuscitation according to your local protocol. If a "no shock" message is given and no pulse is present, you should start CPR, then transport.

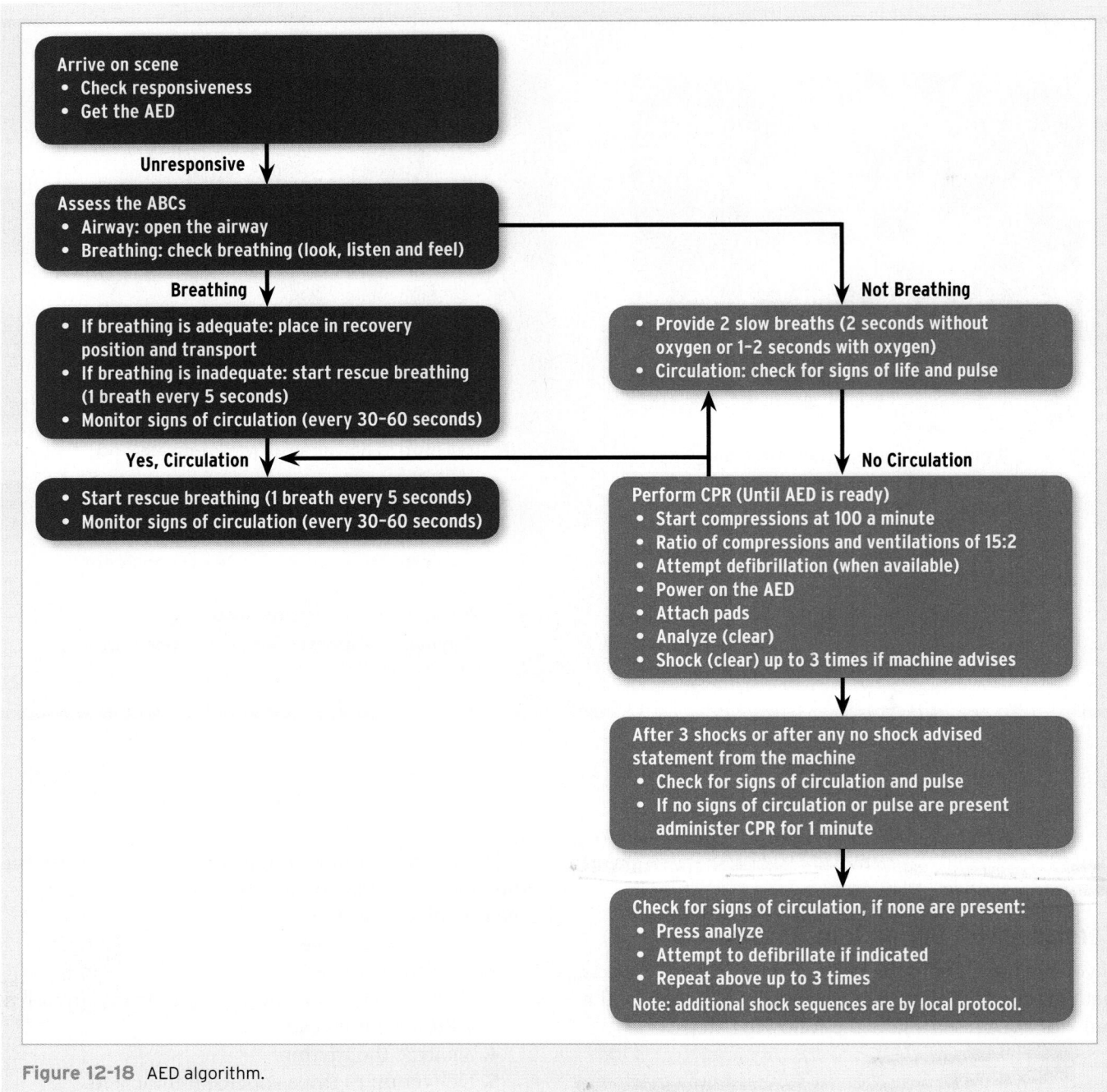

Figure 12-18 AED algorithm.

Coordination With ALS

The time to defibrillation is critical to survival after cardiac arrest. As an EMT-B equipped with an AED, you have the one tool that the dying patient in ventricular fibrillation needs most. Furthermore, it is very hard to hurt someone with an AED. Therefore, if you have an AED available, do not wait for the paramedics to arrive to administer a shock. Waiting might seem like a good idea. It is not. It is throwing away the patient's best chance for survival.

If the patient is unresponsive and does not have a pulse, apply the AED and push the analyze button (if there is one) as quickly as you can. Notify the ALS personnel as soon as possible after you recognize a cardiac arrest, but do not delay defibrillation. After the paramedics arrive at the scene, you should interact with them according to your local protocols.

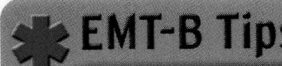
EMT-B Tips

AED Operational Tips

- One EMT-B operates the defibrillator while another does CPR.
- Defibrillation comes first. Do not apply oxygen or do anything else that delays analysis of rhythm or defibrillation.
- Be familiar with the AED device used by your EMS system.
- Avoid all contact with the patient during analysis of the rhythm.
- State, "Clear the patient" before shocking. Another popular phrase is "I'm clear, you're clear, we're all clear" before delivering shocks.
- In applicable models of AEDs, check the batteries at the beginning of your shift; carry an extra charged battery with your AED.
- Only use an AED with special pediatric pads for cardiac arrest in children younger than 8 years or who weigh less than 55 lb (25 kg) and who have been assessed to be unresponsive, not breathing, and pulseless.
- Unless indicated otherwise by local protocol, you do not need to perform pulse checks during rhythm analysis; typically, there will be no pulse check between stacked shocks 1 and 2 and stacked shocks 2 and 3.
- Continued airway maintenance and artificial ventilation are of prime importance.

You are the Provider   Summary

Inaccurate blood pressure measurements can significantly affect treatment options and decisions that must be made regarding patient care. For example, if nitroglycerin is administered because of an inaccurate blood pressure reading, this could have detrimental effects on the patient or exacerbate his condition. Never estimate the reading. Ask your partner for help should it become necessary.

This scenario demonstrates the impact you have as an EMT-B in the first few minutes of a call. Your knowledge and actions can make the difference in a patient avoiding permanent disability or death as a result of a life-threatening situation such as a heart attack. Working as a team with ALS and understanding the needs of these providers can also assist in the overall quality of patient care. The transition from BLS providers to ALS providers to an emergency department physician should be seamless. With knowledge and teamwork, all levels of providers work together to provide the best patient care possible.

Assessment and Emergency Care

	Chest Pain	**Cardiac Arrest**
Scene Size-up	Wear BSI. Ensure scene safety. Determine NOI from patient and/or bystanders. Request additional resources if needed. Determine if spinal stabilization is needed.	Wear BSI. Ensure scene safety. Bring AED. Determine NOI/MOI. Determine if spinal stabilization is needed.
Initial Assessment ■ General impression ■ Airway ■ Breathing ■ Circulation ■ Transport decision	Determine if patient is responsive. If so, ask about chief complaint. If not, evaluate ABCs. If patient has lost consciousness and possibly fallen, consider spinal stabilization. Ensure that the airway is patent. If patient is short of breath or in respiratory distress, provide oxygen via nonrebreathing mask at 10-15 L/min. If patient is not breathing, provide ventilations with a BVM device and 100% oxygen. Assess the pulse and skin. Place patient in position of comfort. Provide reassurance. Transport patients with chest pain immediately in gentle, stress-relieving manner.	Determine patient's LOC and chief complaint. If patient is unresponsive and pulseless, prepare to defibrillate. Check scene safety—do not defibrillate a patient in pooled water.
Focused History and Physical Exam	*NOTE: The order of the steps in the focused history and physical exam differs depending on whether the patient is conscious or unconscious. The order below is for a conscious patient. For an unconscious patient, perform a rapid physical exam, obtain vital signs, and obtain the history.*	
■ SAMPLE history ■ Focused physical exam ■ Baseline vital signs ■ Communication ■ Interventions	If patient is conscious, take brief SAMPLE history and ask OPQRST questions. Specifically, ask if the patient: ■ had heart attack before ■ has heart problems ■ has risk factors: smoking, high blood pressure, high stress ■ takes medications Perform a focused physical exam, focusing on the cardiovascular and respiratory systems. Assess skin color, temperature, and condition. Is cyanosis present? Are neck veins distended? Check mucous membranes. Take vital signs, including systolic and diastolic blood pressures. Use pulse oximetry if available. Report to medical control and the hospital; follow instructions from medical control. Depending on local protocol, administer baby aspirin and assist with prescribed nitroglycerin. Obtain permission from medical control before assisting with prescribed nitroglycerin.	Not applicable to a cardiac arrest patient. See emergency care table on opposite page for summary of AED procedure.
Detailed Physical Exam	If time permits and patient is stable, perform detailed physical exam and ask patient about risk factors for heart disease.	Not applicable to a cardiac arrest patient. See emergency care table on opposite page for summary of AED procedure.
Ongoing Assessment ■ Communication and documentation	Monitor patient very closely. Reassess vital signs every 5 minutes or as patient's condition changes. Reassess interventions. If cardiac arrest occurs, begin defibrillation or CPR immediately. Record all interventions. Obtain medical control physician's signature if required.	After defibrillation, check pulse at least every 30 seconds. If pulse is not present, perform defibrillation and/or CPR again.

NOTE: While the steps below are widely accepted, be sure to consult and follow your local protocol.

Chest Pain	Cardiac Arrest

Chest Pain

Depending on local protocol, prepare to administer baby aspirin and assist with prescribed nitroglycerin. Check condition of medication(s) and expiration date(s).

Aspirin

Administer according to protocols.

Nitroglycerin

1. Obtain permission from medical control.
2. Take patient's blood pressure. Continue only if systolic pressure greater than 100 mm Hg.
3. Check that you have the right medication, right patient, and right delivery route.
4. Question patient about last dose and effects. Ensure patient understands route of administration. Prepare to have the patient lie down to prevent fainting.
5. Ask patient to lift his or her tongue. Place tablet underneath tongue or spray under tongue if medication is in spray form. Have patient keep mouth closed until dissolved/absorbed.
6. Recheck blood pressure within 5 minutes. Record medication and time of administration. If chest pain persists and systolic blood pressure is greater than 100 mm Hg, repeat the dose every 5 minutes as authorized by medical control.

Reevaluate transport decision. Do not delay transport to assist with nitroglycerin.

Cardiac Arrest

Defibrillation

1. Perform initial assessment. If patient is unresponsive and pulseless, prepare to defibrillate. If patient is responsive, do not apply AED.
2. Stop CPR if it is in progress.
3. Verify pulselessness and apnea.
4. Give two ventilations using a BVM device or pocket mask.
5. If the AED is not ready, start CPR.
6. Prepare the AED pads.
7. Turn on the machine.
8. Remove clothing from patient's chest area. Apply the pads: one to the right of the breastbone just below the collarbone, one on the left chest.
9. Stop CPR.
10. State aloud, "Clear the patient." Ensure that no one is touching the patient, including yourself.
11. Push the analyze button.
12. Wait for the computer to determine if a shockable rhythm is present.
13. If a shock is not needed, go to step 18. If a shock is advised, ensure that no one is touching the patient, including yourself. Push shock button.
14. AED will reanalyze rhythm; if not, push analyze button again.
15. If machine advises a shock, deliver second shock.
16. Reanalyze rhythm.
17. If machine advises a shock, deliver third shock.
18. Check pulse.
19. If pulse is present, check breathing.
20. If breathing adequately, give patient oxygen via nonrebreathing mask and transport. If patient is not breathing adequately, open airway and ventilate.
21. If patient has no pulse, perform 1 minute of CPR.
22. Gather additional information on arrest event.
23. After 1 minute of CPR, ensure no one is touching the patient. Push analyze button.
24. If necessary, repeat one cycle of up to three stacked shocks.
25. Transport and check with medical control.
26. Continue to support patient.

Assessment and Emergency Care

Prep Kit

Ready for Review

- The heart is divided down the middle into two sides, right and left, each with an upper chamber called the atrium and a lower chamber called the ventricle.

- The largest of the four heart valves that keep blood moving through the circulatory system in the proper direction is the aortic valve, which lies between the left ventricle and the aorta, the body's main artery.

- The heart's electrical system controls heart rate and helps the atria and ventricles work together.

- During periods of exertion or stress, the myocardium requires more oxygen. This is supplied by dilation of the coronary arteries, which increases blood flow.

- Common places to feel for a pulse include the carotid, femoral, brachial, radial, posterior tibial, and dorsalis pedis arteries.

- Low blood flow to the heart is usually caused by coronary artery atherosclerosis, a disease in which cholesterol plaques build up inside blood vessels, eventually occluding them.

- Occasionally, a brittle plaque will crack, causing a blood clot to form. Heart tissue downstream suffers from a lack of oxygen and, within 30 minutes, will begin to die. This is called an acute myocardial infarction (AMI), or heart attack.

- Heart tissues that are not getting enough oxygen but are not yet dying can cause pain called angina. The pain of AMI is different from the pain of angina in that it can come at any time, not just with exertion; it lasts up to several hours, rather than just a few moments; and it is not relieved by rest or nitroglycerin.

- In addition to crushing chest pain, signs of AMI include sudden onset of weakness, nausea, and sweating; sudden arrhythmia; pulmonary edema; and even sudden death.

- Heart attacks can have three serious consequences. One is sudden death, usually the result of cardiac arrest caused by abnormal heart rhythms called arrhythmias. These include tachycardia, bradycardia, ventricular tachycardia, and, most commonly, ventricular fibrillation.

- The second consequence is cardiogenic shock. Symptoms include restlessness; anxiety; pale, clammy skin; pulse rate higher than normal; and blood pressure lower than normal. Patients with these symptoms should receive oxygen, assisted ventilations as needed, and immediate transport.

- The third consequence of AMI is congestive heart failure, in which damaged heart muscle can no longer contract effectively enough to pump blood through the system. The lungs become congested with fluid, breathing becomes difficult, the heart rate increases, and the left ventricle enlarges.

- Signs include swollen ankles from dependent edema, high blood pressure, rapid heart rate and respirations, rales (crackles), and sometimes the pink sputum and dyspnea of pulmonary edema.

- Treat a patient with CHF as you would a patient with chest pain. Monitor the patient's vital signs. Give the patient oxygen via nonrebreathing face mask. Allow the patient to remain sitting up.

- In treating patients with chest pain, obtain a SAMPLE history, following the OPQRST mnemonic to assess the pain; measure and record vital signs; ensure the patient is in a comfortable position, usually semi-reclining or half sitting up; administer prescribed nitroglycerin and oxygen; and transport the patient, reporting to medical control as you do.

- If a patient is not responsive, you may perform the following, depending on the patient's age, weight, and your local protocol:
  - Unresponsive adult or child older than 8 years and weighing at least 55 lb, perform automated external defibrillation

Technology

Interactivities

Vocabulary Explorer

Anatomy Review

Web Links

Online Review Manual

www.EMTB.com

- Unresponsive child younger than 8 years who weighs less than 55 lb, perform automated external defibrillation with special pediatric pads if protocol allows
- Unresponsive infant, begin CPR.

■ The AED requires the operator to apply the pads, power on the unit, follow the AED prompts, and press the shock button as indicated. The computer inside the AED recognizes rhythms that require shocking and will not mislead you.

■ The three most common errors in using certain AEDs are failure to keep a charged battery in the machine, applying the AED to a patient who is moving, squirming, or being transported, and applying the AED to a responsive patient with a rapid heart rate.

■ Do not touch the patient while the AED is analyzing the heart rhythm or delivering shocks.

■ Application of an AED, analysis of heart rhythm, and defibrillation always are a higher priority than CPR. Stop CPR as soon as an AED is available and treat the patient following the prompts from the AED. Start CPR only if shock is not advised or in 1-minute spurts between rounds of shocks.

■ If advanced life support (ALS) service is responding to the scene, stay where you are and continue the sequence of shocks and CPR. Do not wait for ALS to arrive to begin defibrillation. If ALS is not responding, begin transport after six shocks or after the machine gives three consecutive messages that no shock is advised.

■ If an unconscious patient has a pulse but loses it during transport, you must stop the vehicle, re-analyze the rhythm, and either defibrillate again or begin CPR as appropriate.

■ The chain of survival, which is the sequence of events that must happen for a patient with cardiac arrest to have the best chance of survival, includes recognition of early warning signs and immediate activation of EMS, immediate CPR by bystanders, early defibrillation, and early advanced care. Seconds count at every stage.

Vital Vocabulary

acute myocardial infarction (AMI) Heart attack; death of heart muscle following obstruction of blood flow to it. Acute in this context means "new" or "happening right now."

angina pectoris Transient (short-lived) chest discomfort caused by partial or temporary blockage of blood flow to the heart muscle.

anterior The front surface of the body; the side facing you in the standard anatomic position.

aorta The main artery, which receives blood from the left ventricle and delivers it to all the other arteries that carry blood to the tissues of the body.

aortic valve The one-way valve that lies between the left ventricle and the aorta. It keeps blood from flowing back into the left ventricle after the left ventricle ejects its blood into the aorta. One of four heart valves.

arrhythmia An irregular or abnormal heart rhythm.

atherosclerosis A disorder in which cholesterol and calcium build up inside the walls of blood vessels, eventually leading to partial or complete blockage of blood flow.

asystole Complete absence of heart electrical activity.

atrium One of two (right and left) upper chambers of the heart. The right atrium receives blood from the vena cava and delivers it to the right ventricle. The left atrium receives blood from pulmonary veins and delivers it to the left ventricle.

bradycardia Slow heart rate, less than 60 beats/min.

cardiac arrest A state in which the heart fails to generate an effective and detectable blood flow; pulses are not palpable in cardiac arrest, even if muscular and electrical activity continues in the heart.

cardiogenic shock A state in which not enough oxygen is delivered to the tissues of the body, caused by low output of blood from the heart. It can be a severe complication of a large acute myocardial infarction, as well as other conditions.

Prep Kit continued...

congestive heart failure (CHF) A disorder in which the heart loses part of its ability to effectively pump blood, usually as a result of damage to the heart muscle and usually resulting in a backup of fluid into the lungs.

coronary artery A blood vessel that carries blood and nutrients to the heart muscle.

defibrillate To shock a fibrillating (chaotically beating) heart with specialized electrical current in an attempt to restore a normal rhythmic beat.

dependent edema Swelling in the part of the body closest to the ground, caused by collection of fluid in the tissues; a possible sign of congestive heart failure (CHF).

dilation Widening of a tubular structure such as a coronary artery.

infarction Death of a body tissue, usually caused by interruption of its blood supply.

inferior The part of the body, or any body part, nearer to the feet.

ischemia A lack of oxygen that deprives tissues of necessary nutrients, resulting from partial or complete blockage of blood flow; potentially reversible because permanent injury has not yet occurred.

lumen The inside diameter of an artery or other hollow structure.

myocardium Heart muscle.

occlusion Blockage, usually of a tubular structure such as a blood vessel.

perfusion The flow of blood through body tissues and vessels.

posterior The back surface of the body; the side away from you in the standard anatomical position.

superior The part of the body, or any body part, nearer to the head.

syncope Fainting spell or transient loss of consciousness.

tachycardia Rapid heart rhythm, more than 100 beats/min.

ventricle One of two (right and left) lower chambers of the heart. The left ventricle receives blood from the left atrium (upper chamber) and delivers blood to the aorta. The right ventricle receives blood from the right atrium and pumps it into the pulmonary artery.

ventricular fibrillation Disorganized, ineffective twitching of the ventricles, resulting in no blood flow and a state of cardiac arrest.

ventricular tachycardia (VT) Rapid heart rhythm in which the electrical impulse begins in the ventricle (instead of the atrium), which may result in inadequate blood flow and eventually deteriorate into cardiac arrest.

Points to Ponder

You have been dispatched to a report of a patient having chest pain. Upon arrival you find the patient sitting in a chair. She is pale and slightly short of breath. You complete your SAMPLE history and find the patient has a history of hypertension and CHF and had a heart attack 5 years ago. She takes nitroglycerin and hydrochlorothiazide for her blood pressure. The onset of symptoms has been over the last week. The patient has taken her nitroglycerin as prescribed and her chest pain is relieved.

You complete a detailed physical exam and find crackles in her lungs. Her ankles are very swollen. You recommend transport to the hospital for treatment. The patient refuses and wants to stay at home.

What should you say to this patient? If she refuses treatment, how should you document this?

Issues: Convincing Patients of the Need for Treatment, Right of Refusal, Advocating for the Patient, Assistance from Medical Control.

Assessment in Action

You have been dispatched to City Hall for a report of sick man. En route dispatch gives you additional information. The patient, a 67-year-old man, is at the mayor's office and is possibly having a heart attack.

As you enter the mayor's office you see an older man sitting in a chair. He is pale, sweaty, and appears to be short of breath. You introduce yourself and the patient responds, " I feel like I'm going to die." He tells you he has a "stabbing" pain in his chest that moves to his left arm. The pain started 15 minutes ago while working at his desk. The patient has a pulse of 90 beats/min and irregular, a blood pressure of 180/100 mm Hg, and respirations of 22 breaths/min. He continues to tell you he had open heart surgery two years ago and takes nitroglycerin for angina and Lopressor for high blood pressure.

1. The patient's general appearance, SAMPLE history, and vital signs lead you to believe this patient is having a(n):

   A. acute myocardial infarction (AMI).
   B. asthma attack.
   C. cardiac arrest.
   D. psychological episode.

2. Because this patient's chest pain has lasted longer than 15 minutes, you can rule out:

   A. angina pectoris.
   B. cardiac tamponade.
   C. sudden death.
   D. COPD.

3. The EMT-B can assist this patient with needed heart medication called:

   A. dopamine.
   B. nitroglycerin.
   C. penicillin.
   D. Vasotec.

4. Atherosclerosis can lead to:

   A. pressure sores.
   B. headaches.
   C. lung disease.
   D. a blockage or occlusion of the coronary artery.

5. A severe blockage of a coronary artery that depletes the heart's ability to pump is called:

   A. cardiogenic stoppage.
   B. heart depletion
   C. reduced pump stroke.
   D. cardiac arrest.

6. The poor appearance of this patient is due to lack of cardiac output and:

   A. cold temperatures.
   B. poor skin condition.
   C. loss of perfusion (shock).
   D. past medical history.

7. AMIs (heart attacks) do not always lead to sudden death. The patient can also have cardiogenic shock and:

   A. mini strokes.
   B. congestive heart failure (CHF).
   C. neurogenic myopathy.
   D. lung abscesses.

8. An abnormality in a patient's heart rhythm is called a:

   A. cardiac arrhythmia.
   B. cardiac malfunction.
   C. cardiac seizure.
   D. cardiac standstill.

Challenging Questions

9. Explain the differences in pain between an AMI and angina pectoris.

10. Trace the flow of blood through the cardiovascular system starting from the right atrium.

11. What are signs and symptoms of congestive heart failure (CHF) and the treatment?

12. What are the indications for use and the steps in using an AED?

www.EMTB.com

Neurologic Emergencies

Objectives*

Cognitive

1. Describe the causes of stroke, including the two major types of stroke and the three conditions that cause blockages. (p 442)
2. Describe the sequence of events that occur during a stroke. (p 443)
3. Obtain and interpret the key vital signs in the stroke patient, including the time of onset of the symptoms. (p 448)
4. State the reason stroke must be treated within the first 3 to 6 hours. (p 446)
5. Identify the signs and symptoms of stroke. (p 444)
6. Describe the significance of a transient ischemic attack (TIA). (p 443)
7. Define seizure, including the two major types of seizure. (p 450)
8. Describe the parts of a seizure. (p 451)
9. List possible causes of seizure. (p 451)
10. Explain the importance of recognizing seizures. (p 452)
11. Describe characteristics of the postseizure state. (p 452)
12. Define altered mental status. (p 456)
13. List possible causes of altered mental status. (p 457)

Affective

14. Explain the importance of tolerance and patience when caring for a patient who has had a stroke, seizure, or who has altered mental status. (p 456)

Psychomotor

15. Demonstrate the steps in the emergency medical care for the patient who has had a stroke. (p 445)
16. Demonstrate testing for aphasia, facial weakness, and motor weakness. (p 448)
17. Demonstrate the steps in the emergency medical care for the patient who has had a seizure. (p 453, 455)
18. Demonstrate the steps in the emergency medical care for the patient who has altered mental status. (p 448)

*All of these objectives are noncurriculum objectives.

You are the Provider

You and your paramedic partner are dispatched to 1201 Howard Street, Apartment F, for a 70-year-old man with a severe headache and decreased level of consciousness. You arrive to find the patient seated in his kitchen with his wife standing next to him. When you introduce yourself and ask him what is wrong, he stares at you blankly and you notice that he is drooling from the right side of his mouth. His wife says, "A few minutes ago, he told me that he had a very bad headache. When I came back from the bathroom with some ibuprofen, I tried to hand him a glass of water and he dropped the glass on the floor. I don't know what's wrong with him."

1. What do you suspect is wrong with this patient?
2. What other signs and symptoms would you suspect in this scenario, and what tests could you use to verify your suspicions?

Neurologic Emergencies

Stroke is the third leading cause of death in the United States, after heart disease and cancer. In the past few years, there has been a revolution in the treatment of stroke. Emergency physicians, neurologists, and neurosurgeons can help some patients with acute stroke to avoid the most devastating consequences of this disease, assuming that the patients get to the hospital in time for treatments to be effective.

Seizures and altered mental status may also occur when there is a disorder in the brain. Seizures may occur as a result of a recent or old head injury, a brain tumor, a metabolic problem, or simply a genetic disposition. Your ability to recognize when a seizure has occurred or is occurring is critical for the patient, as it helps to direct appropriate treatment.

Altered mental status (AMS) is a common presentation in patients with a wide variety of medical problems. Although it is tempting, you should not make assumptions about the cause of AMS because there can be many possible causes; some are obvious, some are not: intoxication, head injury, hypoxia, stroke, metabolic disturbances, and many more. Obviously, treatment varies widely as well. Patients with AMS present a particular challenge in that they may be difficult to handle and frustrating to treat at times. Your professionalism is paramount in these situations.

This chapter describes the structure and function of the brain and the most common causes of brain dis-

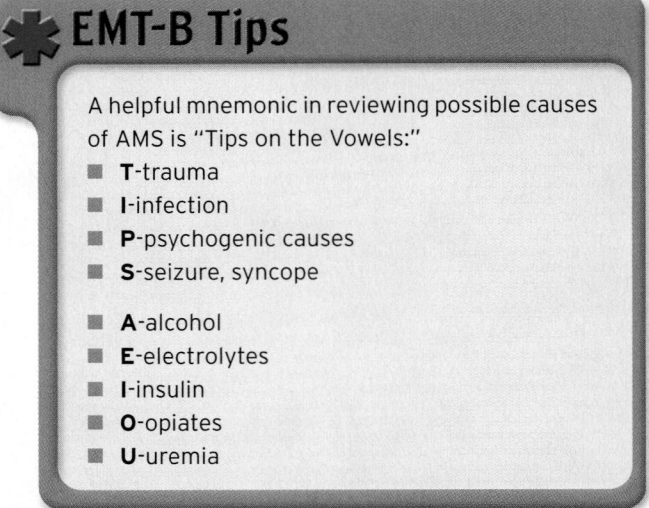

EMT-B Tips

A helpful mnemonic in reviewing possible causes of AMS is "Tips on the Vowels:"

- **T**-trauma
- **I**-infection
- **P**-psychogenic causes
- **S**-seizure, syncope

- **A**-alcohol
- **E**-electrolytes
- **I**-insulin
- **O**-opiates
- **U**-uremia

order, including stroke, seizure, and AMS. It then discusses the signs and symptoms of each condition.

You will learn how to approach and assess a patient with a brain disorder and why prompt transport to an appropriate medical facility is so important. Appropriate management of each is discussed in the context of each emergency.

Brain Structure and Function

The brain is the body's computer. It controls breathing, speech, and all other body functions. All your thoughts, memories, wants, needs, and desires reside in the brain. Different parts of the brain perform different functions. For example, some receive input from the senses, including sight, hearing, taste, smell, and touch; others control the muscles and movement, while others control the formation of speech.

The brain is divided into three major parts: the brain stem, the cerebellum, and the largest part, the cerebrum (Figure 13-1 ▶). The brain stem controls the most basic functions of the body, such as breathing, blood pressure, swallowing, and pupil constriction. Just behind the brain stem, the cerebellum controls muscle and body coordination. It is responsible for coordinating complex tasks that involve many muscles, such as standing on one foot without falling, walking, writing, picking up a coin, or playing the piano.

The cerebrum, located above the cerebellum, is divided down the middle into the right and left cerebral

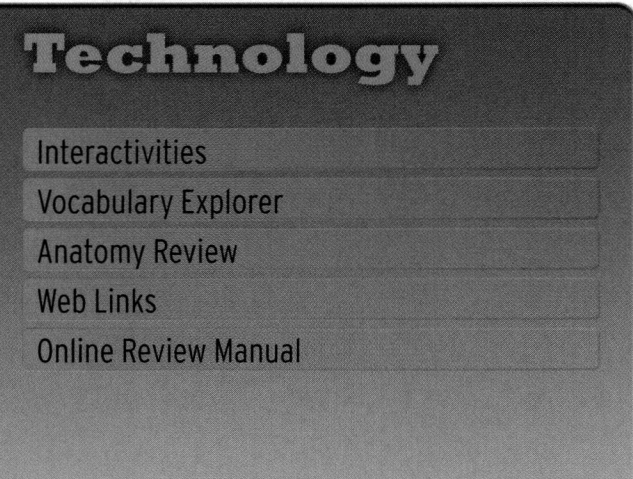

Technology

www.EMTB.com

- Interactivities
- Vocabulary Explorer
- Anatomy Review
- Web Links
- Online Review Manual

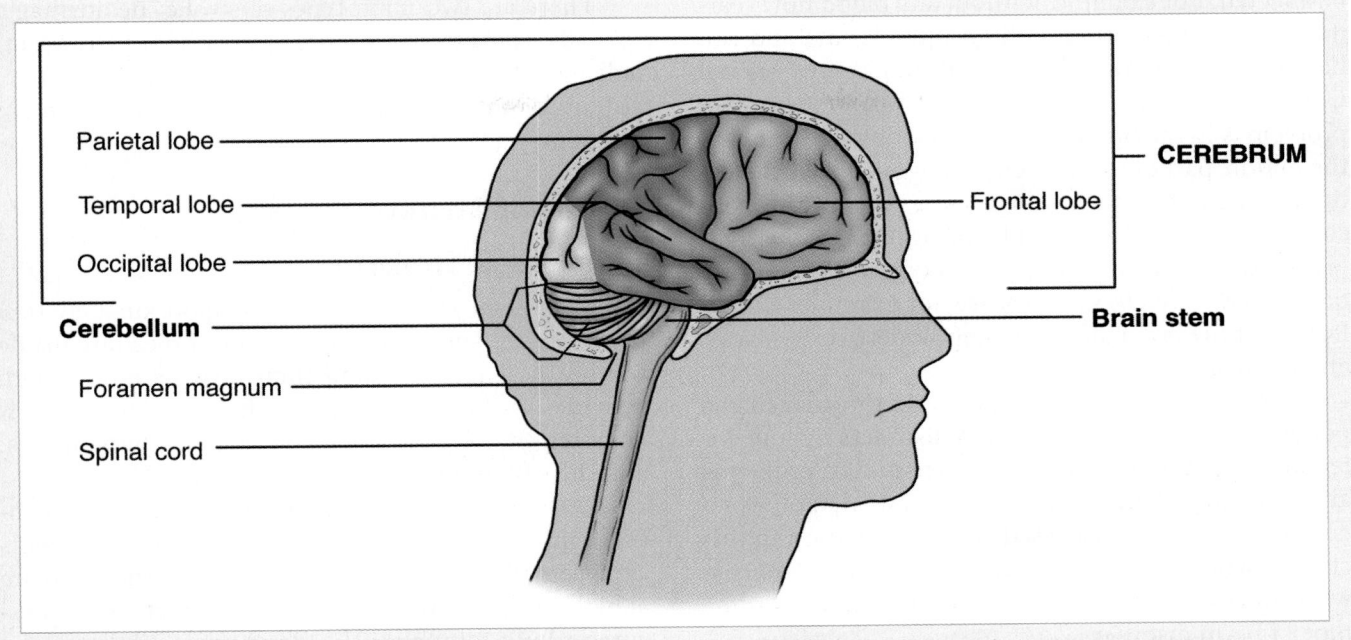

Figure 13-1 The brain lies well protected within the skull. Its major parts are the cerebrum, the cerebellum, and the brain stem.

hemispheres. Each hemisphere controls activities on the opposite side of the body. The front part of the cerebrum controls emotion and thought, and the middle part controls touch and movement. The back part of the cerebrum processes sight. In most people, speech is controlled on the left side of the brain near the middle of the cerebrum.

All the messages traveling to and from the brain travel along nerves. Twelve cranial nerves run directly from the brain to various parts of the head, such as the eyes, ears, nose, and face. All the rest of the nerves join in the spinal cord and exit the brain through a large hole in the base of the skull called the foramen magnum Figure 13-2 ▶ . At each vertebra in the neck and back, two nerves, called spinal nerves, branch out from the spinal cord and carry signals to and from the body.

Common Causes of Brain Disorder

Many different disorders can cause brain dysfunction or other neurologic symptoms and can affect level of consciousness, speech, or voluntary muscle control. As a general rule, if the brain problem is caused primarily by disorders in the heart and lungs, the entire brain will

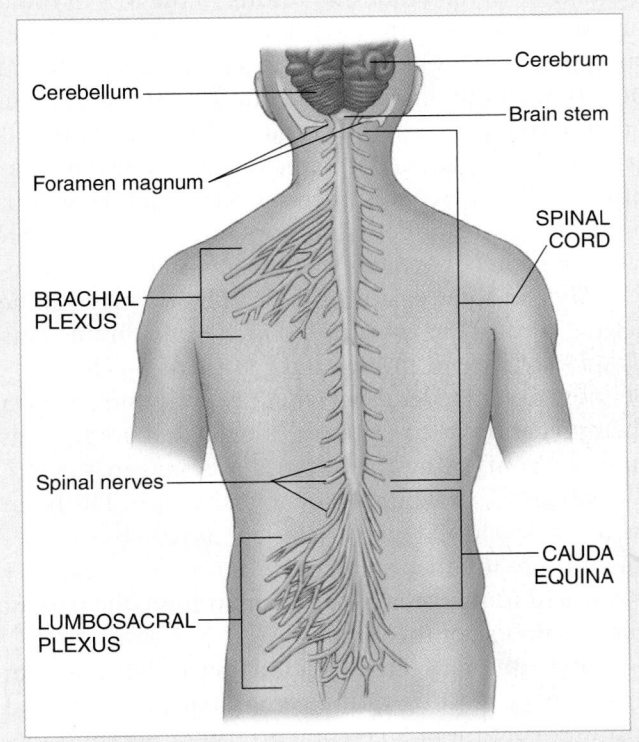

Figure 13-2 The spinal cord is the continuation of the brain stem. It exits the skull at the foramen magnum and extends down to the level of the second lumbar vertebra.

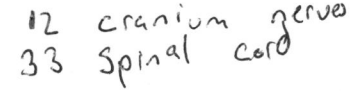

12 cranium nerves
33 spinal cord

be affected. For example, without any blood flow (cardiac arrest), the patient will go into a <u>coma</u> and can have permanent brain damage within minutes, even if CPR is performed immediately. However, if the primary problem is in the brain, such as a poor blood supply to the middle part of the left cerebral hemisphere, the patient may not be able to move some parts of the right side of the body. This might be the right arm, the right leg, or the facial muscles. Low oxygen levels in the bloodstream, due to lung disease, for example, will affect the entire brain, often causing anxiety, restlessness, and confusion.

Stroke is a common cause of brain disorder that is potentially treatable. Other brain disorders include infection and tumor. Although these specific problems are not covered here, the seizures or AMS that often accompany them are discussed. The information in this chapter will help you better understand, communicate with, and care for patients who have experienced some type of brain disorder.

Stroke

A <u>cerebrovascular accident (CVA)</u> is an interruption of blood flow to the brain that results in the loss of brain function. <u>Stroke</u> is the loss of brain function that results from a CVA and occurs when part of the blood flow to the brain is suddenly cut off. Lacking oxygen, brain cells stop working and begin to die; these dead cells are called <u>infarcted cells</u>. Medical science currently has little to offer these cells once they are dead. However, it may take several hours or more for cell death to occur, even when it appears that severe disability will occur. Also, in some cases, a trickle of blood may still be getting through to the affected area of the brain. This blood may supply enough oxygen to keep cells alive, but not enough to let them work properly and perform their given jobs. For example, if cells that are responsible for controlling the left arm are oxygen starved, the patient will not be able to move that arm. The brain cells will develop <u>ischemia</u>, a lack of oxygen that causes the cells not to function properly. If normal blood flow is restored to that area of the brain in time, the patient may regain use of the arm.

Interruption of cerebral blood flow may result from <u>thrombosis</u>, clotting of the cerebral arteries; <u>arterial rupture</u>, rupture of a cerebral artery; or <u>cerebral embolism</u>, obstruction of a cerebral artery caused by a clot that was formed elsewhere and traveled to the brain.

There are two main types of stroke: hemorrhagic (usually from arterial rupture) and ischemic (from embolism or thrombosis). Their symptoms are the same, although the events taking place inside the brain are different.

Types of Stroke

Hemorrhagic Stroke

A <u>hemorrhagic stroke</u> accounts for approximately 10% of all strokes and occurs as a result of bleeding inside the brain. The free blood then forms a clot, which squeezes the brain tissue next to it. When that tissue is compressed, oxygenated blood cannot get into the area, and the surrounding cells begin to die.

Certain types of patients are at higher risk of hemorrhagic stroke, which commonly occurs in patients experiencing stress or exertion. The patients who are at highest risk are those who have very high blood pressure or long-term elevated blood pressure that is not treated. After many years of high pressure, the blood vessels in the brain weaken. Eventually, one of the vessels may rupture, and blood will spurt out of the hole and into the brain, increasing the pressure inside the cranium. Cerebral hemorrhages are often fatal, although proper treatment of high blood pressure can help to prevent this long-term damage to the blood vessels, reducing morbidity and mortality.

Some individuals may have been born with weaknesses, called *aneurysms*, in the walls of the arteries. An aneurysm is a swelling or enlargement of part of an artery, resulting from weakening of the arterial wall. Many of these individuals have a sudden onset of a "bad headache." The headache is from the irritation of blood on the tissue of the brain after the vessel swells and ruptures. When a hemorrhagic stroke occurs in an otherwise healthy young person, the likely cause is often a weakness in a blood vessel called a berry aneurysm. This type of aneurysm resembles a tiny balloon (or berry) that juts out from the artery. When the aneurysm is overstretched and ruptures, blood spurts into an area around the coverings of the brain called the subarachnoid space. Therefore, these types of strokes are called subarachnoid hemorrhages. Again, patients with this type of stroke experience a sudden severe headache, typically described as the worst headache they have ever had. If the patient seeks medical attention immediately, surgeons may be able to repair the aneurysm; yet like other cerebral hemorrhages, these are often fatal.

Ischemic Stroke

When blood flow to a particular part of the brain is cut off by a blockage inside a blood vessel, the result is an ischemic stroke. This can be from a thrombosis or an embolism that blocks blood flow. As with coronary artery disease, atherosclerosis in the blood vessels is often the cause. Atherosclerosis is a disorder in which calcium and cholesterol build up, forming a plaque inside the walls of blood vessels. This plaque may obstruct blood flow and interfere with the vessel's ability to dilate. Eventually, atherosclerosis may cause complete occlusion (blockage) of an artery (Figure 13-3 ▼). In other cases, an atherosclerotic plaque in the carotid artery in the neck will rupture. A blood clot will form over the crack in the plaque, sometimes growing big enough to completely block all blood flow through that artery. Deprived of oxygen, parts of the brain supplied by the artery will stop working. Patients with such ischemic strokes will have dramatic symptoms, including loss of movement on the opposite side of the body.

Even if the blockage in the carotid artery is not complete, smaller pieces of the clot may embolize (break off and be carried by the blood flow) deep into the brain, heart, or lungs. If this piece ends up in the brain, it will lodge in a branch blood vessel. This cerebral embolism then blocks blood flow (Figure 13-4 ▼). Depending on the location of the lodged clot, the patient may experience a range of symptoms from nothing at all to complete paralysis.

Transient Ischemic Attack

In some patients, normal processes in the body will break up a blood clot in the brain. When that happens quickly, blood flow is restored to the affected area, and the patient will regain use of the affected body part; however, this often indicates that the patient has a se-

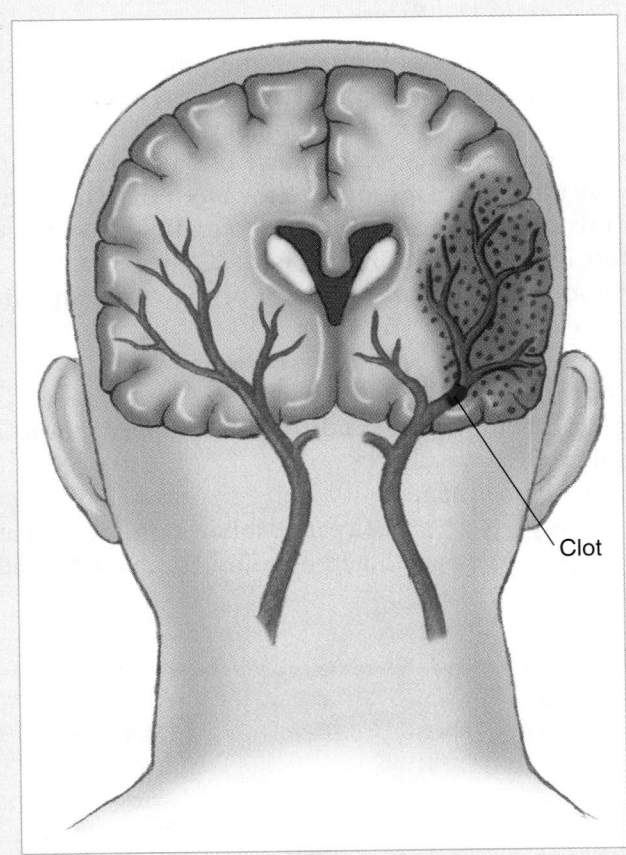

Figure 13-3 Atherosclerosis can damage the wall of a cerebral artery, producing narrowing and/or a clot. When the vessel is narrowed or completely blocked, blood flow to that part of the brain may be blocked, and the cells begin to die due to lack of adequate oxygenation.

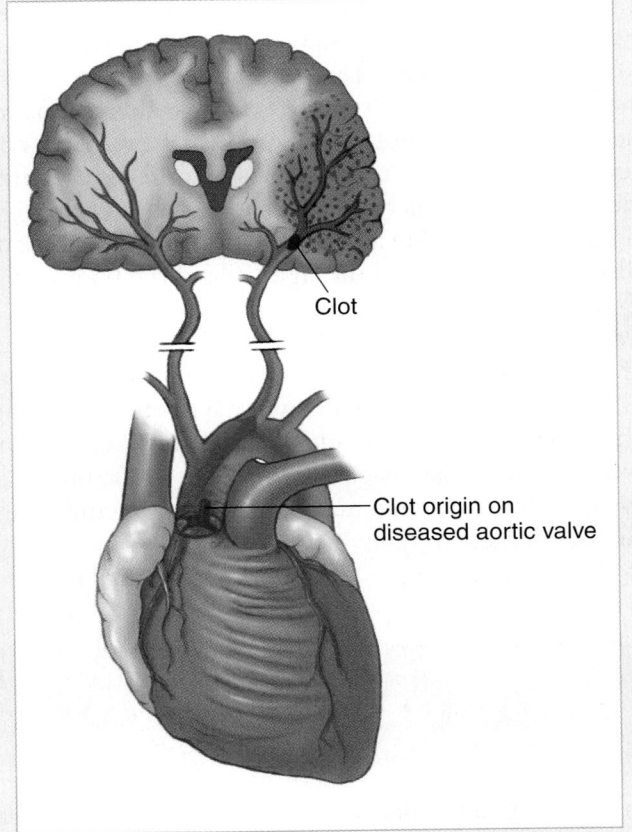

Figure 13-4 An embolus, a blood clot usually formed on a diseased heart valve, can travel through the body's vascular system, lodge in a cerebral artery, and cause a stroke.

rious medial condition that may prove fatal. When stroke symptoms go away on their own in less than 24 hours, the event is called a transient ischemic attack (TIA). Some patients call these mini-strokes.

Although most patients with TIAs do well, every TIA is an emergency. It may be a warning sign that a larger, permanent stroke is about to occur. For this reason, all patients with a new TIA should be evaluated by a physician to determine whether preventive action can be taken.

Signs and Symptoms of Stroke

Left Hemisphere Problems

If the left cerebral hemisphere has been affected, the patient may have a speech disorder called aphasia, an inability to produce or understand speech. Speech problems can vary widely. Some patients will have trouble understanding speech but be able speak clearly. This condition is called receptive aphasia. You can detect this problem by asking the patient a question such as "What day is today?" In response, the patient with aphasia may say, "Green." The speech is clear, but it does not make sense. Other patients will be able to understand the question but cannot produce the right sounds in order to answer. Only grunts or other incomprehensible sounds emerge. These patients have expressive aphasia. Strokes that affect the left side of the brain can also cause paralysis of the right side of the body.

Right Hemisphere Problems

If the right cerebral hemisphere of the brain is not getting enough blood, patients will have trouble moving the muscles on the left side of the body. Usually, they will understand language and be able to speak, but their words may be slurred and hard to understand. This problem is called dysarthria.

Interestingly, patients with right hemisphere strokes may be completely oblivious to their problem. If you ask these patients to lift their left arm and they cannot, they will lift their right arm instead. They seem to have forgotten that the left arm even exists. This symptom is called neglect. Patients with a problem affecting the back part of the cerebrum may neglect certain parts of their vision. Generally, this is hard to detect in the field due to the patient's ability to compensate without conscious effort. Nevertheless, you should be aware of the possibility. Try to sit or stand on the patient's good side because he or she may be unable to see things on the "bad" side.

The problem of neglect causes many patients who have had large strokes to delay seeking help. Strokes may not be painful. Therefore, a patient may be unaware that there is a problem until a family member or friend points out that some part of the patient's body is not working correctly.

Bleeding in the Brain

Patients who have bleeding in their brain, otherwise known as a cerebral hemorrhage, may have very high blood pressure or cerebral aneurysms. Oftentimes, high blood pressure is the cause of the bleeding, but many times it is a response to the bleeding: The brain is raising the blood pressure in an attempt to force more oxygen into its injured parts. Quite often, blood pressure will return to normal or may drop significantly on its own.

Other Conditions

The following three conditions may appear to be a stroke:

- Hypoglycemia
- A postictal state (a period following a seizure that lasts between 5 and 30 minutes, characterized

You are the Provider   Part 2

You utilize a portion of the Cincinnati Stroke Scale by asking him to smile. He seems to understand you and he attempts to smile, but the right side of his face remains flaccid. You know that time is critical in cases of stroke, and you and your partner assist the patient to the cot, taking care to place him upright, slightly on his affected side. As you obtain a quick set of baseline vital signs, your partner applies high-flow oxygen.

3. What other types of disorders or conditions can mimic a stroke?
4. Can all strokes be treated with clot-busting medications?

by labored respirations and some degree of AMS)

- Subdural or epidural bleeding (a collection of blood near the skull that presses on the brain)

Because both oxygen and glucose are needed for brain metabolism, a patient with hypoglycemia may look like a patient who is having a stroke. With good patient assessment, you should find out whether the patient has diabetes and takes insulin or another glucose-lowering medication.

A patient who has experienced a seizure may look like a patient who is having a stroke. This is often referred to as the postictal state. However, in most cases, a patient having a seizure will recover rapidly, within several minutes.

Subdural and epidural bleeding usually occur as a result of trauma. The dura is a leathery covering over the brain, next to the skull. A fracture near the temples may cause an artery to bleed on top of the dura, resulting in pressure on the brain (Figure 13-5A ▼). Onset of this *epidural* bleeding is usually very rapid after injury. In other cases, the veins just below the dura may be torn and bleed, and is referred to as a subdural bleed (Figure 13-5B ▼). Onset occurs more slowly, sometimes over a period of several days.

The onset of stroke-like signs and symptoms may be subtle; the original injury may not even be remembered.

Assessment of the Stroke Patient

The assessment of a patient suspected of having a stroke is similar to that for patients presenting with other complaints. Stay organized in your approach and follow a routine familiar to you. This will help prevent you from forgetting steps and help you organize your information.

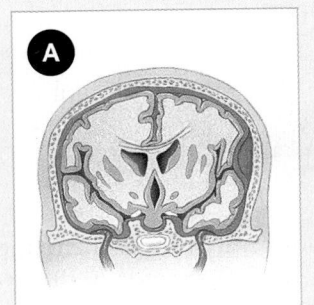
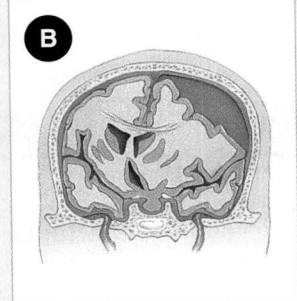

Figure 13-5 Trauma to the head can result in intracranial bleeding. **A.** Bleeding outside the dura and under the skull is epidural. **B.** Bleeding beneath the dura but outside the brain is subdural.

Scene Size-up

Strokes present in different ways. Dispatchers are not trained to diagnose particular problems, but to recognize a set of specific conditions. Because a stroke can present in many different ways, the signs and symptoms may be easily confused with other conditions, particularly over the phone. Be aware of the information that dispatch provides you, but also consider other possibilities such as trauma and other illnesses that might mimic a stroke. For example, both hypoglycemia associated with diabetes and seizures from other causes can present with similar symptoms to a stroke. If this call had described multiple patients, it would be more likely that the incident involves trauma or hazardous materials rather than illness.

Do not be distracted by the seriousness of the situation or by frightened family members who want you to rush. Look first for threats to your safety and follow BSI precautions. Most calls involving AMS require ALS backup or intercept, if available. Call for help early.

Initial Assessment

General Impression

Check the patient's ABCs and care for immediate problems. Problems with the ABCs are initially found by asking the patient, "What is wrong?" or "How may we help you?" The chief complaint will guide you to what the patient is most concerned about. For patients having a stroke, the chief complaint may be highly variable and may include confusion, slurred speech, or unresponsiveness. Determining the patient's level of consciousness should be first in the list of assessment actions for anyone with an AMS. The responsiveness of patients who are not awake and alert needs to be quickly determined by using the AVPU scale.

Airway and Breathing

Strokes affect how the body functions in many ways. Patients may have difficulty swallowing and are at risk for choking on their own saliva. Evaluate the airway of an unresponsive patient to make sure it is patent and will remain that way (Figure 13-6 ▶). If the patient requires assistance maintaining an airway, consider an oropharyngeal or nasopharyngeal airway based on the level of consciousness and the presence of a gag reflex.

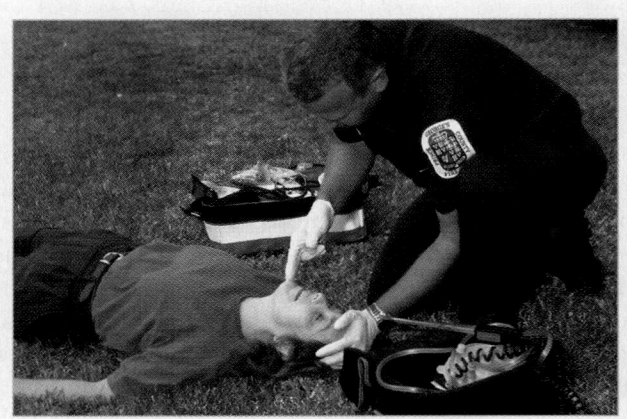

Figure 13-6 Securing and maintaining the airway in a patient who is unconscious is critical; also be sure to have suction readily available in the event that the patient vomits.

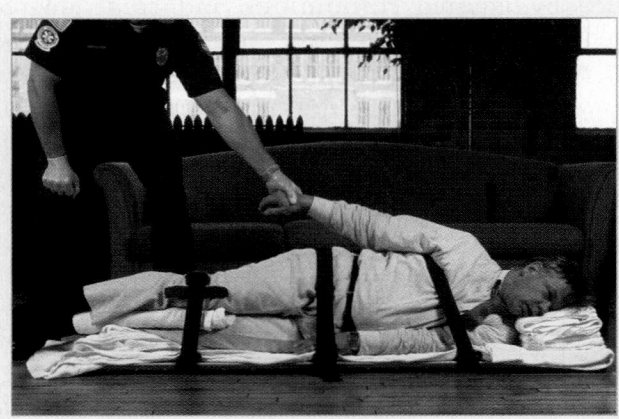

Figure 13-7 A patient who has had a stroke should be positioned with the paralyzed side down and well protected with padding. Elevate the head about 6".

Provide suction and position the patient to prevent aspiration. If you determine that the patient cannot protect his or her own airway, place the patient in the recovery position to help prevent secretions from entering the airway. Suction as necessary.

Although early in your initial assessment you may not have enough information to identify the exact cause of the AMS, you will need to evaluate the patient's breathing. Is the rate within normal parameters? Is the depth adequate? Is the patient using accessory muscles? As you perform the initial assessment, administer supplemental oxygen. If necessary, provide assisted ventilation.

Circulation

Your assessment of the patient's circulation should begin with checking a pulse if the patient is unresponsive. If no pulse is found, immediately begin CPR and attach an AED. If the patient is responsive, determine whether the pulse is fast or slow, weak or strong. Is the patient in shock? Oxygen administration is helpful for limiting the effects of hypoperfusion to the brain. Be careful about aggressively elevating a patient's arms and legs to treat shock. This increases blood in the brain and may aggravate a hemorrhage. Know and follow your local protocol. Evaluate the patient quickly for external bleeding based on the chief complaint. It is unlikely your stroke patient has suffered trauma, but you should consider the possibility and assess appropriately.

Transport Decision

Controversial evidence exists that new therapies, such as thrombolytics (clot dissolvers), may reverse stroke symptoms and even stop the stroke if given within 2 to 3 hours of the onset of symptoms. These therapies may not work for all patients, and they cannot be given to patients with bleeding-type (hemorrhagic) strokes. Because hospital personnel will ultimately make these ongoing treatment decisions, you should proceed under the assumption that an area of the brain can still be saved. The sooner the treatment is begun, the better the prognosis for the patient.

Spend as little time at the scene as possible. Remember, stroke is an emergency. There may be treatment available for the patient at the hospital, and rapid transport is essential to maximize the possibility of recovery. Place the patient on one side, with the paralyzed side down and well protected with padding Figure 13-7 ▲. This will help prevent aspiration of secretions in patients who cannot swallow well and protect their airway. If a recovery position or left supine position is used, the patient's head should be elevated about 6" to maximize drainage of secretions. A patient's paralyzed extremities will require protection from harm. Remember that the patient cannot move if an arm gets pinched or a hand gets caught in a doorway while being transported. Calm and reassure the patient throughout transport.

Focused History and Physical Exam

Once you have concluded your initial assessment and addressed all life threats, begin the next step in the patient assessment process—the focused history and physical exam.

If the patient is unresponsive, begin with a rapid physical exam and then obtain baseline vital signs and a SAMPLE history. The order of steps in the assessment process is different for an unresponsive patient than it is for a responsive patient. A situation in which patients are unresponsive as a result of an AMS is much more serious than when patients are awake but confused. Quickly looking for explanations (trauma, medical tags, track marks) for their AMS may help identify the cause and therefore guide you to an appropriate treatment more quickly. When your rapid physical exam is complete, continue with obtaining the patient's vital signs and history.

In a responsive medical patient, begin with a SAMPLE history, paying special attention to any information that may explain an AMS. Perform a neurologic exam as your focused physical exam, and obtain the patient's baseline vital signs.

SAMPLE History

If the patient is responsive and breathing, obtain a SAMPLE history. Also try to speak with relatives or friends who may be able to explain the events leading to the AMS (Figure 13-8 ▼), remembering that time is critical and making a special effort to determine the exact time that the patient last appeared to be normal. This will help physicians in the emergency department understand whether it is safe to begin certain treatments that must be given within the first hours after onset of symptoms. You may be the only person on the emergency medical team with the opportunity to speak with bystanders to obtain this critical information. Many times, you will be able to find out only that the patient was normal when he or she went to sleep the night before. Note that in such cases, the time the patient was last seen to be normal was at bedtime, not when the patient awoke with symptoms. Collect or list all medications the patient has

EMT-B Tips

Assessing and initiating treatment of a patient with AMS can keep a team of two responders very busy, particularly because history taking often involves interviewing others. Despite the number of demands, at least one responder needs to keep a close eye on the patient for signs of airway compromise, to avert physical danger from falling, or from other risks created by the patient's mental compromise.

taken. When possible, determine allergies and the patient's last oral intake. This information may be helpful if the patient requires surgery for a brain hemorrhage.

Although a patient who has had a stroke may appear to be unconscious and unable to speak, the patient may still be able to hear and understand what is taking place. Therefore avoid all unnecessary or inappropriate remarks. Try to communicate with the patient by looking for indications that the patient can understand you, such as a glance, gaze, motion or pressure of the hand, effort to speak, or head nod. Establishing effective communication can help you to calm the patient and lessen the fear that accompanies an inability to communicate (Figure 13-9 ▼). Try to keep in mind that the patient has just experienced a potentially life-threatening event and that anxiety, frustration, and embarrassment may inhibit communication with you.

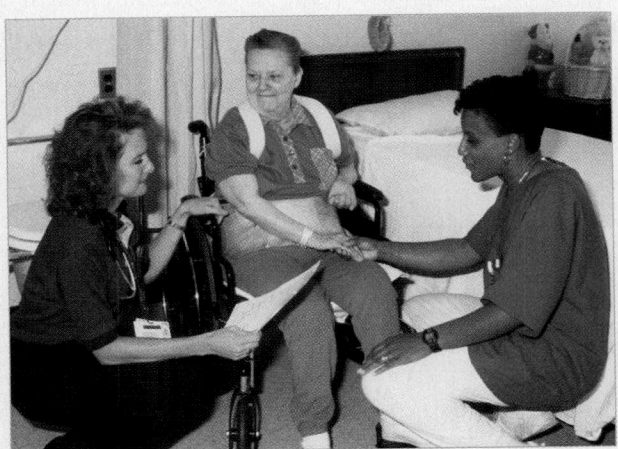

Figure 13-8 Try to speak with family members or bystanders who may have seen what happened. They may also be able to tell you when the patient last appeared normal.

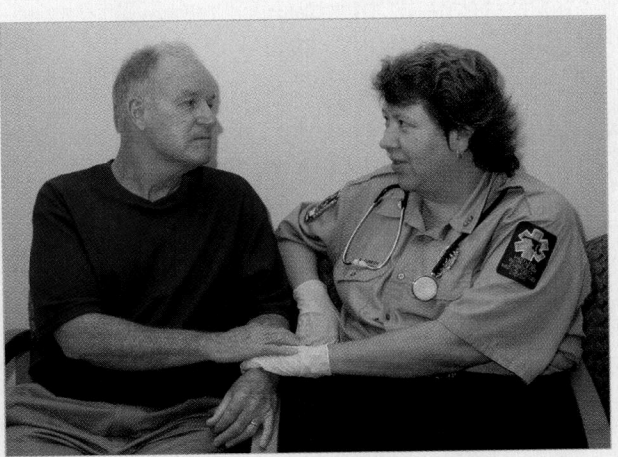

Figure 13-9 Make a special effort to establish communication with a patient who may have had a stroke. Look for indications that the patient understands you, such as a glance, gaze, squeeze of the hand, efforts to speak, or nodding of the head.

Study This ↓

Table 13-1 Cincinnati Stroke Scale

Test	Normal	Abnormal
Facial Droop (Ask patient to show teeth or smile.)	Both sides of face move equally well.	One side of face does not move as well as other.
Arm Drift (Ask patient to close eyes and hold both arms out with palms up.)	Both arms move the same, or both arms do not move.	One arm does not move, or one arm drifts down compared with the other side.
Speech (Ask patient to say, "The sky is blue in Cincinnati.")	Patient uses correct words with no slurring.	Patient slurs words, uses inappropriate words, or is unable to speak.

Focused Physical Exam (Responsive Patients)

You should perform, as your focused physical exam, at least three key physical tests on patients you suspect of having had a stroke: tests of speech, facial movement, and arm movement. If any one of the three is abnormal, the patient may be having a stroke.

Many EMS services utilize the Cincinnati Stroke Scale, which tests speech, facial droop, and arm drift. The entire examination is identified in (Table 13-1 ▲).

To test speech, simply ask the patient to repeat a simple phrase such as "The sky is blue in Cincinnati." If the patient does this correctly, you know that he or she both understands and can produce speech. If the patient cannot repeat the phrase, the problem may be with either function: understanding speech or producing it.

To test facial movement, ask the patient to show his or her teeth (or gums if there are no teeth). Watch to see that both sides of the face around the mouth move equally. If only one side is moving well, you know that something is wrong with the control of the muscles on the other side.

To test arm movement, ask the patient to hold both arms in front of his or her body, palms up toward the sky, with eyes closed and without moving. Over the next 10 seconds, watch the patient's hands. If you see one side drift down toward the ground, you know that side is weak. If both arms stay up and do not move, you know that both sides of the brain are working.

If both arms fall to the ground, you have not really learned anything. Perhaps the patient did not understand your instructions. Try the arm test again, but this time move the patient's arms into position yourself. Another possibility to consider is that the patient is having a problem other than a stroke. This is likely to be the answer if both sides of the brain are not working properly.

All patients who are victims of AMS, including those who have possibly had a stroke, should also have a Glasgow Coma Scale (GCS) calculated (Table 13-2 ▼).

Baseline Vital Signs

The last step of the focused history and physical exam is to obtain a baseline set of vital signs. These will be important to compare with vital signs obtained in the

TABLE 13-2 Glasgow Coma Scale

Eye Opening		Best Verbal Response		Best Motor Response	
Spontaneous	4	Oriented conversation	5	Obeys commands	6
In response to speech	3	Confused conversation	4	Localizes pain	5
In response to pain	2	Inappropriate words	3	Withdraws to pain	4
None	1	Incomprehensible sounds	2	Abnormal flexion	3
		None	1	Abnormal extension	2
				None	1

Score: 14-15 Mild dysfunction
Score: 11-13 Moderate to severe dysfunction
Score: 10 or less is severe dysfunction

ongoing assessment. During severe situations, a great deal of pressure from bleeding in the brain may slow the pulse and cause respirations to be erratic. Blood pressure is usually high to compensate for poor perfusion in the brain. Changes in pupil size and reactivity will be uncommon, but when present indicate severe bleeding and pressure on the brain.

Interventions

The cause of many patients' AMS may be unknown, even after arrival at the hospital. This makes it difficult to provide definitive care in the prehospital environment. Most of your interventions will be based on your assessment findings. For example, if the blood glucose level is low, you may give oral glucose according to protocol, or if a patient is unresponsive you may need to position him or her in the recovery position to protect the airway. Your best treatment in these situations is to perform a thorough assessment and maintain the ABCs.

Detailed Physical Exam

A thorough detailed exam should be performed when time and conditions permit. A detailed physical exam includes inspection, palpation, and auscultation to identify DCAP-BTLS in all areas of the body. Because this exam is thorough and time intensive, it is not often performed when attention is required to continuously treat the ABCs. Every effort should be made to complete this exam, especially when patients are unresponsive and unable to tell you about symptoms. Without a detailed physical exam, subtle or covert problems may go unnoticed. Because of the time sensitivity of treatment options, this is generally performed during transport to the hospital. Do not delay transport to perform this on scene.

Ongoing Assessment

The ongoing assessment should focus on three main goals: reassessing the ABCs, interventions, and vital signs. Patients who have had a stroke can lose an airway or stop breathing without warning. Multiple interventions may be necessary for these patients. The effectiveness of airway adjuncts, positive pressure ventilations, and other treatments can only be determined with both immediate and continuous observation af-

Documentation Tips

Key information to document for a stroke patient:
- Time of onset of the signs and symptoms
- Results of the Glasgow Coma Scale
- Results of the Cincinnati Strokes Scale
- Changes noted upon reassessment

Time of onset is critical because it helps determine whether the stroke patient is a candidate for treatment with clot-dissolving drugs.

ter providing the intervention. If something is not working, try something else.

You have established baseline vital signs already in your assessment as well as a Glasgow Coma Scale score. Now is the time to compare that baseline information with updated information. Any changes may indicate if treatments are effective. Watch carefully for changes in pulse, blood pressure, respirations, and GCS scores.

Communication and Documentation

After you begin transport, you should relay the information you have learned as soon as possible. Notify the receiving facility personnel that it is possible your patient had a stroke, so that staff there can prepare to test and treat the patient without delay. Be sure to include the time that the patient was last seen to be normal, the findings of your neurologic exam, and the time you anticipate arriving at the hospital.

One of the key pieces of information to document is the time of onset of the patient's signs and symptoms. If the physician's diagnosis is an ischemic stroke, time of onset of the signs and symptoms is critical in determining whether the patient is a candidate for treatment with clot-dissolving drugs. It is also important to document the results of the Glasgow Coma Scale as well as your Cincinnati Strokes Scale, along with any changes noted upon reassessment. Document airway management and interventions performed, including the position in which the patient was placed.

Definitive Care for the Patient Who Has Had a Stroke

In most patients with suspected stroke, physicians in the emergency department need to determine whether there is bleeding in the brain. If there is no bleeding, the patient may be a candidate for medication to help break up the blood clot or to help brain cells survive the

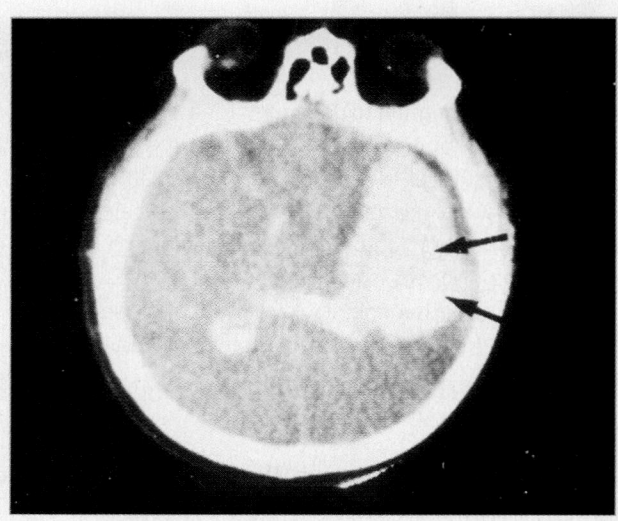

Figure 13-10 A computed tomography scan of a ruptured cerebral aneurysm. The light area represents hemorrhage into the brain tissue (arrows).

TABLE 13-3 Tips on Patient Care

■ Patients who experience a transient ischemic attack (TIA) may exhibit most of the same signs and symptoms as patients who are having a stroke. These signs and symptoms can last from minutes up to 24 hours. Therefore, the signs of stroke that you note on arrival may gradually disappear. Patients who appear to have had a TIA should be transported for further evaluation.

■ Place the patient's affected or paralyzed extremity in a secure and safe position during patient movement and transport.

■ Some patients who have had a stroke may be unable to communicate, but they can often understand what is being said around them. Be aware of this possibility.

■ New therapies for stroke must be used as soon as possible after the start of symptoms. Minimize time on the scene and notify the receiving hospital as soon as possible.

reduced amount of oxygen. The only reliable way to tell whether there is bleeding is with a special type of x-ray test called computed tomography (CT) of the head. Blood is usually easy to see on the CT scan (Figure 13-10 ▲).

Most hospitals have only one CT scanner. The technician who knows how to run the machine may not be in the hospital in the middle of the night. That is why it is important that you recognize the signs and symptoms of stroke. If the emergency department staff knows that you are transporting a patient with a possible stroke, they may be able to call in the technician before you even arrive, or they may decide to delay a CT scan on another patient who has a less critical problem. Keep in mind that most treatments for stroke must be started as soon as possible after the onset of the event

(Table 13-3 ▲). Few, if any, current treatments do any good if they are started more than 3 to 6 hours after the stroke begins. Even if 3 hours have passed, prompt action on your part is essential. Some EMS systems designate specific hospitals for stroke patients. These institutions have CT scanner technicians, radiologists, and neurosurgeons on duty 24 hours a day.

Seizures

Types of Seizures

A seizure, or convulsion, is a temporary alteration in behavior or consciousness and is typically characterized by unconsciousness and a generalized severe twitching

You are the Provider   **Part 3**

Your partner tells you that he will initiate an IV en route. You assist the patient's wife into the ambulance and immediately begin transport using lights and siren. You talk to the patient and his wife about use of the siren and lights, and you do your best to address their concerns about the noise and driving safely. You hear the paramedic performing the rest of the stroke scale.

5. What would this include?

of all of the body's muscles that lasts several minutes or longer. This type of seizure is often called a generalized seizure or *grand mal seizure*. In other cases, the seizure may simply be characterized by a brief lapse of consciousness without loss of composure in which the patient seems to stare and not to respond to anyone. Other characteristics may be lip smacking, eye blinking, or isolated convulsions or jerking of the body. This type of seizure, called an absence (ob-sáhnz) seizure or *petit mal seizure*, typically occurs in children from 4 to 12 years old.

Signs and Symptoms

Some seizures occur on only one side of the body. Others begin on one side and gradually progress to a generalized seizure that affects the entire body. Most individuals with lifelong or chronic seizures tolerate these events reasonably well without complications, but in some situations, seizures may signal life-threatening conditions.

Often, a patient may have experienced a warning prior to the event, which is referred to as an aura. The seizure is characterized by sudden loss of consciousness, chaotic muscle movement and tone, and apnea. The patient may also experience a tonic phase, usually lasting only seconds, in which the there will be a period of extensor muscle tone activity, tongue biting, or bladder or bowel incontinence. During the tonic-clonic phase the patient may exhibit bilateral movement characterized by muscle rigidity and relaxation usually lasting 1 to 3 minutes. Throughout the tonic-clonic phase, the patient exhibits tachycardia, hyperventilation, and intense salivation. Most seizures last 3 to 5 minutes and are followed by a lengthy period (5 to 30 minutes or more) called a postictal state, in which the patient is unresponsive at first and gradually regains consciousness. The postictal state is over when the patient regains a complete level of consciousness. Gradually, in most cases, the patient will begin to recover and awaken, but appear dazed, confused, and fatigued. In contrast, a petit mal seizure can last for just a fraction of minute, after which the patient fully recovers immediately with only a brief lapse of memory of the event.

Seizures that continue every few minutes without regaining consciousness or last longer than 30 minutes are referred to as status epilepticus, also known as status seizures. For obvious reasons, recurring seizures should be considered potentially life-threatening situations in which patients need emergency medical care.

Causes of Seizures

Some seizure disorders, such as epilepsy, are congenital, which means that the patient was born with the condition. Other types of seizures may be due to high fevers, structural problems in the brain, or metabolic or chemical problems in the body (Table 13-4 ▼). Epileptic seizures can usually be controlled with medications such as phenytoin (Dilantin), phenobarbital, or carbamazepine (Tegretol). Patients with epilepsy will often have seizures if they stop taking their medications or if they do not take the prescribed dose on a regular basis.

Seizures may also be caused by an area of abnormality in the brain, such as a benign or cancerous tumor, an infection (brain abscess), or scar tissue from some type of injury. These seizures are said to have a structural cause; in other cases, the seizures are metabolic. Seizures from a metabolic cause can result from abnormal levels of certain blood chemicals (eg, extremely low sodium levels), hypoglycemia (low blood glucose levels), poisons, drug overdoses, or sudden withdrawal from routine heavy alcohol or sedative drug usage or even from prescribed medications. Dilantin, a drug that is used to control seizures, can cause seizures itself if the person takes too much.

Seizures can also result from sudden high fevers, particularly in children. Such convulsions, known as febrile seizures, are usually very unnerving for parents to observe but are generally well tolerated by the child. Nevertheless, you must transport a child who has had a febrile seizure, as this condition needs to be evaluated in the hospital. The fact that a second seizure may

TABLE 13-4 Common Causes of Seizures

Type	Cause
Epileptic	Congenital in origin
Structural	Tumor (benign or cancerous)
	Infection (brain abscess)
	Scar tissue from injury
	Head trauma
	Stroke
Metabolic	Hypoxia
	Abnormal blood chemistry
	Hypoglycemia
	Poisoning
	Drug overdose
	Sudden withdrawal from alcohol, medications
Febrile	Sudden high fever

occur is very worrisome, and if it occurs, the patient requires rapid hospital evaluation to identify possible causes, such as serious infection within the brain or tissues covering the brain.

The Importance of Recognizing Seizures

Regardless of the type of seizure, it is important for you to recognize when a seizure is occurring or whether one has already occurred. You must also determine whether this episode differs from any previous ones. For example, if the previous seizure occurred on only one side of the body and this seizure occurs over the entire body, some additional or new problem may be involved. In addition to recognizing that seizure activity has occurred and/or that something different may now be occurring, you must also recognize the postictal state as well as the complications of seizures.

Because most seizures involve a vigorous twitching of the muscles, the muscles use a lot of oxygen. This excessive demand consumes oxygen being delivered by the circulation to the vital functions of the body. As a result, there is a buildup of acids in the bloodstream, and the patient may turn cyanotic (bluish lips, tongue, and skin) from the lack of oxygen. Often, the seizures themselves prevent the patient from breathing normally, making the problem worse. In the patient with diabetes, blood glucose values may drop because of the excessive muscular contraction of a seizure. Monitor blood glucose levels closely after a patient with diabetes has a seizure.

Recognizing seizure activity also means looking at other problems associated with the seizure. For example, the patient may have fallen during the seizure episode and injured some part of the body; head injury is the most serious possibility. Patients having a generalized seizure may experience incontinence, meaning that they may lose bowel and bladder control. Therefore, one clue that unresponsive or confused patients may have had a seizure is to find that they urinated into their

Documentation Tips

Physician evaluation of a patient who has had a seizure depends heavily on reports of the seizure pattern and changes in that pattern. Record all pertinent information about the seizure in terms of duration, areas of body movement, and possible triggering factors. This requires effective interviewing of available witnesses, family members, or caregivers.

clothing. Although incontinence is possible with other medical conditions, sudden incontinence is very likely a sign that a seizure has occurred. When such patients regain their faculties, they are naturally embarrassed by this temporary loss of control. Do what you can to minimize this discomfort by covering the patient and assuring them that incontinence is part of the loss of control that accompanies a seizure.

The Postictal State

Once a seizure has stopped, the patient's muscles relax, becoming almost flaccid, or floppy, and the breathing becomes labored (fast and deep) in an attempt to compensate for the buildup of acids in the bloodstream. By breathing faster and more deeply, the body can balance the acidity in the bloodstream. With normal circulation and liver function, the acids clear away within minutes, and the patient will begin to breathe more normally. Intuitively, the longer and harder the convulsions are, the longer it will take for this imbalance to correct itself. Likewise, longer and more severe seizures will result in longer postictal unresponsiveness and confusion. Once the patient regains a normal level of consciousness, typically oriented to person, place, and time, the postictal state is over.

In some situations, the postictal state may be characterized by hemiparesis, or weakness on one side of the body, resembling a stroke. Unlike the typical stroke, hypoxic hemiparesis soon resolves itself. Most commonly, the postictal state is characterized by lethargy and confusion to the point that the patient may be combative and appear angry. You must be prepared for these circumstances, both in your approach to scene control and in your treatment of the patient's symptoms. If the patient's condition does not improve, you should consider other possible underlying problems, including hypoglycemia or infection.

EMT-B Safety

Be on the lookout for patients who may behave violently during the postictal phase. Although most patients who have had a seizure pose no threat to EMS responders, signs of alcohol or drug abuse should heighten your awareness of the potential for dangerous behavior.

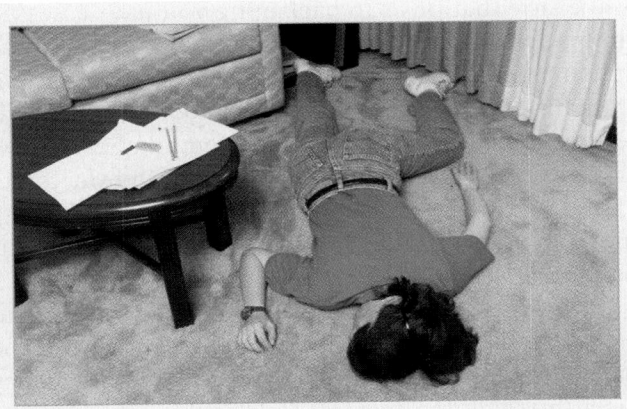

Figure 13-11 A patient who has had a seizure may be found in the postictal state when you arrive. If this is the case, be sure to ask family members or bystanders to verify that a seizure has occurred and how the seizure developed.

Assessing the Seizure Patient

You are typically called to care for a patient who has had a seizure because someone witnessed the seizure. However, you may also be called to see an unresponsive patient when the patient is found in a postictal state (Figure 13-11 ▲). In other situations, you may be called to care for a patient who is having seizures and find that the patient has some other medical problem, such as a cardiac arrest or a psychological problem. Therefore, thorough assessment is key because the information gathered at the scene may be extremely important to the hospital staff who must soon care for the patient.

In most instances, you will arrive sometime after the seizure has occurred, as it only lasts a few minutes. By the time someone recognizes the problem, calls for help, and receives a response, the patient is usually in a postictal state. Thus, you must gather as much information from family or bystanders as possible to verify that a seizure has occurred and to obtain a description of the way the seizure developed.

Scene Size-up

Dispatchers are frequently given information regarding a seizure by the caller. Even if the caller has never seen a seizure before, the description of convulsions or spasms often indicate a seizure is taking place. While this may be an obvious nature of illness problem as reported by bystanders, a mechanism of injury may still be present. Consider the need for spinal precautions

based on dispatch information and your assessment of the scene as you approach the patient. Ensure the scene is safe and wear appropriate BSI protection. Gloves and eye protection, at a minimum, should be worn. ALS is not typically needed for a simple seizure; however, when complications such as severe trauma or prolonged seizures are present, ALS is required. Request ALS earlier rather than later.

Initial Assessment

General Impression

Most seizures last only a few minutes at most. As you approach your patient and observe the level of consciousness you should be able to tell whether the patient is still seizing. Unless you are stationed across the street from where the patient is located and have an arrival time of a minute or less, most seizures should be over by the time you arrive on scene. If the patient is still seizing when you arrive, the potentially life-threatening condition of status epilepticus may be present. If the patient is in the postictal stage of the seizure, he or she may be unresponsive or starting to regain awareness of the surroundings. Use the AVPU scale to determine how well your patient is progressing through the postictal stage.

Airway, Breathing, and Circulation *The whole Thing.*

As with any other situation, you should focus on the patient's ABCs upon arrival. Use of a nasal airway will be well tolerated; an oropharyngeal airway may be quite difficult to insert while the patient is seizing. The patient may have been eating or chewing gum at the time of the seizure, and there may be a foreign body obstruction. Bystanders may have tried to put objects in the patient's mouth "to help them breathe better," even though this practice is ill advised. Assess the patient for adequate ventilation. Even if ventilations are adequate, place the patient on high-flow oxygen at 15 L/min via nonrebreathing mask. Seizures will use up oxygen quickly and cause patients to be hypoxic. Breathing and circulation should be confirmed as normal or treated as necessary. Again, in the immediate postictal state following a major seizure, you should anticipate rapid, deep respirations and an accompanying fast heart rate due to the stress of the severe convulsions. However, both respirations and heart rate should begin to slow to normal rates after several minutes. If not, you might suspect problems beyond the seizure alone.

Transport Decision

It is difficult to package a seizing patient for transport. Because almost all seizures are finished within four or five minutes, you can treat the ABCs while waiting for the seizure to finish prior to attempting packaging.

Geriatric Needs

Over time, the brain will gradually deteriorate and shrink as a part of the normal aging process. This can increase the risk of head injury from minor forces, since the brain can more readily impact the inside of the skull (coup-contrecoup brain injury, discussed in Chapter 21) and since the veins that connect the brain to the dura are stretched. A reduced brain mass can also reduce the patient's mental status and capacity. A smaller brain can impair memory function, and with lapses in short-term memory, the geriatric patient can ask the same or similar questions repeatedly.

When you are called to care for a geriatric patient with an AMS, consider the possibility of a stroke or transient ischemic attack (TIA). At the scene of a motor vehicle crash involving an older driver, consider a stroke or TIA as the precipitating factor in the crash. Be alert for altered mental status or unusual pupil responses (ie, constricted pupils in dim light, unequal pupils).

Beware of headache. Although geriatric patients do get tension headaches, they are far less common in the older population. You should consider any headache as potentially serious.

As with the general population, older people can also experience seizures. Remember that seizures are not necessarily due to epilepsy. You should consider and assess for the possibility of a drug overdose, stroke, head injury, or central nervous system infection. Status epilepticus in a geriatric patient can have harmful effects such as hypoxia, irregular heart rhythm, hypotension, elevated body temperatures, low blood glucose levels, and, if the patient vomits, aspiration.

Remember that the geriatric patient is at higher risk for central nervous system illnesses and injuries, including brain injury, TIA, stroke, and seizures. Do not be surprised to find a serious head injury from what you might consider a simple bump on the head.

Make sure that prior to packaging you have assessed the patient for trauma and have taken appropriate spinal precautions if indicated. Protecting the patient from his or her surroundings prior to, during, and after transport is essential. Never attempt to restrain an actively seizing patient. Injury could result from tonic-clonic movement. Use soft materials for padding and move any objects out of the way that may harm your patient.

Not every patient who has had a seizure wants to be transported. It is usually in the best interest of the patient to be evaluated by a physician in the emergency department after a seizure. Your goal is to encourage the patient to be seen by a physician in the emergency department. Should the patient refuse transport, you should be prepared to discuss the situation with the hospital staff on the radio prior to releasing the patient. Consider the following if a patient in a postictal state refuses transport:

- Is the patient awake and completely oriented after a seizure (GCS of 15)?
- Does your assessment reveal no indication of trauma or complications from the seizure?
- Has the patient ever had a seizure before?
- Was this seizure the "usual" seizure in every way (length, activity, recovery)?
- Is the patient currently being treated with medications and receiving regular evaluations by a physician?

If the answer to all of these questions is "yes," you may consider agreeing to a patient's refusal for transport if the patient can be released to a responsible person and monitored. If any one of the questions has a "no" answer, strongly encourage the patient to be transported and evaluated. Follow your local protocols on releasing patients who refuse care.

Focused History and Physical Exam

If your patient is unresponsive in a postictal state, you will not be able to obtain a history. As with other unresponsive medical patients, perform a rapid physical exam, quickly checking the patient from head to toe, looking for any obvious trauma or explanations as to why they seized. Vital signs and history can then be obtained as the patient regains consciousness.

If patients are already responding to questions and demonstrate a fair level of consciousness, begin with a

SAMPLE history, then perform a focused physical exam looking for injuries, and next obtain vital signs.

SAMPLE History

You should obtain a SAMPLE history, including whether the patient has a history of seizures. If so, it is important to find out how the patient's seizures typically occur and whether this episode differs in some way from previous episodes. You should also ask what medications the patient has been taking. If the patient takes Dilantin and phenobarbital, he or she most likely has chronic problems. You might find that the patient ran out of medication or stopped taking medication for a while. Patients who have a history of both seizures and diabetes may use up all their glucose in the body to fuel the seizure. These patients should have an evaluation by ALS providers as soon as possible to determine their blood glucose levels. If their blood glucose levels are low, intravenous glucose must be given immediately since their level of consciousness will be too low to give oral glucose.

If the patient has no history of convulsions and now has a sudden seizure, a serious condition, such as brain tumor, intracranial bleeding, or serious infection should be suspected. The assessment is also the time to determine whether the patient takes medications that lower blood glucose, such as insulin or oral hypoglycemic agents. In other situations, you may want to inquire about drug use or exposure to poisons.

Focused Physical Exam (Responsive Medical Patients)

A Glasgow Coma Scale score can be obtained as you focus your assessment on the patient's mental status. Other areas to focus on are their speech and thinking ability. Initially they will be cloudy and confused on facts but this will improve over time.

Baseline Vital Signs

During most active seizures it is impossible to evaluate vital signs, nor is this the priority when a patient is actively seizing. Unless the situation is unusual, vital signs in the postictal seizure patient will approximate normal. Obtain pulse rate, rhythm, and quality; respiratory rate, rhythm, and quality; blood pressure; skin color, temperature and condition; and pupil size and reactivity. If they have a history of diabetes and you have the ability to check a finger stick blood glucose level, it should be included in your vital signs. Compare these baseline vital signs to vital signs obtained in the ongoing assessment.

Interventions

Seizures are usually limited in how long they last. Most will not require a lot of intervention because they will be over by the time you arrive. For those who are actively seizing, protect them from harm, maintain a clear airway by suctioning as necessary, and provide oxygen as quickly as possible. Treat any trauma you find as you would any other patient.

For patients who continue to seize, as in status epilepticus, suction the airway according to protocol, provide positive pressure ventilations, and transport quickly to the hospital. If you have the option to rendezvous with ALS, you should do so. ALS providers have medications that can stop a prolonged seizure.

Detailed Physical Exam

Once the initial assessment is complete, any life threats have been treated, and the focused history and physical exam have given you a better picture of the patient's condition through the history, a focused neurologic exam, and vital signs, you can consider performing a detailed physical exam. The patient should be checked for injuries, including head lacerations, shoulder dislocations, tongue lacerations, or extremity fractures. Also assess the patient for weakness or loss of sensation on one side of the body. The detailed physical exam on a patient who has had a seizure and is now awake may not be as important as other parts of the assessment. If time permits, a detailed exam should be performed and may reveal other problems or explanations about your patient's condition.

Ongoing Assessment

If another seizure occurs, note whether the seizure starts at a focal part of the body (eg, one arm or one leg) and then progresses to the rest of the body. Most importantly, evaluate the patient's ABCs, vital signs, and mental status. Monitor the patient's mental status every several minutes to verify progressive improvement. Check to see whether interventions are providing the benefits you want. For example, is the nasal airway still needed?

Communication and Documentation

Report and record your findings of the initial assessment and interventions performed. Give a description of the episode and include bystanders' comments, especially if they witnessed the patient seizing. Document the onset and duration of the seizure. Did the patient notice or express noticing an aura? Record any evidence of trauma and interventions performed. Document whether this is the patient's first seizure or whether the patient has a history of seizures. If the patient has a history of seizure activity, how often does he or she have them, and is there any history of status epilepticus? When you are documenting your interventions, record the time the intervention was performed, the patient's response to the intervention, and continued reassessments.

Definitive Care for the Patient Who Has Had a Seizure

In most situations, patients who have had a seizure require definitive evaluation and treatment in the hospital. Even a patient who has a history of chronic epilepsy that is controlled with medications may have an occasional seizure, commonly referred to as a breakthrough seizure. These patients should also be taken to the hospital for observation. At the hospital, blood levels of seizure medications are checked to ensure that patients are receiving the correct dose. Clearly, patients who have just had their first seizure or those with chronic seizures who have had an episode that is "different" require immediate examination to rule out life-threatening conditions. Unless the patient has a well-established history of seizures and is completely alert and oriented, supplemental oxygen is strongly advised, not only to provide extra oxygen but also to prevent the possibility of a recurrent seizure if there is a hypoxic component to the source of the convulsion.

Depending on local protocols, you should assess and treat the patient for possible hypoglycemia (diabetic with AMS who takes insulin or oral agents that lower blood glucose levels). If trauma is suspected, provide spinal immobilization. With recurrent seizures, protect the patient from further injury and manage the airway once the seizure ceases.

If you are treating a child who you suspect is having a febrile seizure, you should attempt to lower the child's temperature by removing his or her clothing and cooling the child with tepid water, particularly about the head and neck, and then fanning the moistened areas. Be careful not to make the patient shiver, which will increase temperature.

If the patient has been exposed to a toxin or poison, you should safely remove the source if possible. Suction should be readily available in case a patient with a decreased level of consciousness begins to vomit.

In all instances, you should show patience and tolerance with these patients, as many of them are likely to be confused and occasionally frightened. Many patients who experience seizures are frustrated with their condition and may refuse transport. Kindness and professional behavior are required to help convince the patient that transport is necessary for definitive care.

Altered Mental Status

Aside from stroke and seizures, the most common type of neurologic emergency that you will encounter is AMS. Simply put, AMS means that the patient is not thinking clearly or is incapable of being aroused. In some instances, patients will be unconscious (Figure 13-12 ▶); in others, they may be alert but confused. The range of problems is wide, and the causes are many, including problems such as hypoglycemia;

You are the Provider   Part 4

You hear the paramedic ask the patient to hold his hands out in front of him palms up and with his eyes closed. He then asks the patient to repeat a simple declarative statement. The findings in both of these tasks further indicate the presence of stroke. The paramedic assigns this patient a Glasgow Coma Score and obtains another set of vital signs.

6. What would the paramedic have seen if the remaining tests indicated a stroke?
7. If the patient's symptoms began to resolve, would his condition be referred to as a stroke?

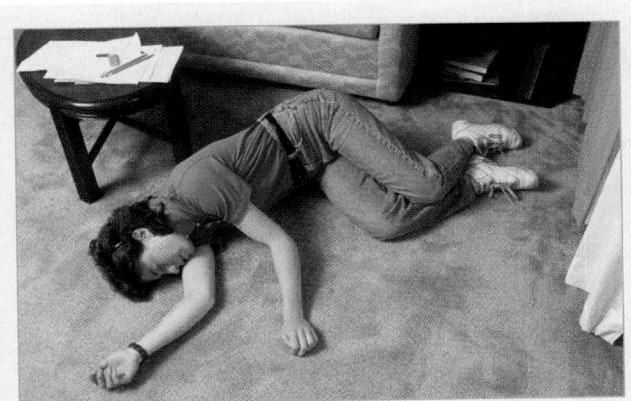

Figure 13-12 A patient with altered mental status can be unconscious in some instances; in others, the patient may be alert by confused.

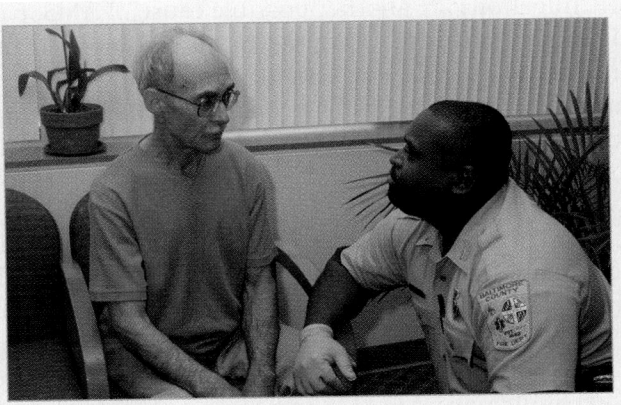

Figure 13-13 During your assessment of a patient with an altered or decreased level of consciousness, consider the possibility of hypoglycemia.

hypoxemia; intoxication; drug overdose; unrecognized head injury; brain infection; body temperature abnormalities; and conditions such as brain tumors, glandular abnormalities, and overdoses/poisonings.

Causes of Altered Mental Status

Hypoglycemia

The clinical picture of patients with AMS due to hypoglycemia is very complex. Patients can have signs and symptoms that mimic stroke and seizures. Because both oxygen and glucose are needed for brain function, hypoglycemia can mimic conditions in the brain such as those associated with stroke. In these instances, the patient may have hemiparesis, similar to what occurs as a result of a stroke. The principal difference, however, is that a patient who has had a stroke may be alert and attempting to communicate normally, whereas a patient with hypoglycemia almost always has an altered or decreased level of consciousness (Figure 13-13 ▶).

Patients with hypoglycemia commonly, but not always, take medications that lower blood glucose levels. Thus, if the patient appears to have signs and symptoms of stroke and an AMS, you should report your findings to medical control and treat the patient accordingly. Check for and report medications, but remember that not all patients who have diabetes take insulin or other medications to lower their blood glucose levels. Remember also that patients with a decreased level of consciousness should not be given anything by mouth. Again, local protocols should guide your actions.

Patients with hypoglycemia can also experience seizures, and you may arrive at the scene to find a patient in a postictal state: confused and disoriented or un-

responsive. The mental status of a patient who has had a typical seizure is likely to improve; however, in a patient with hypoglycemia, the mental status is not likely to improve, even after several minutes. Therefore, you should consider the possibility of hypoglycemia in a patient who has had a seizure, especially if the blood glucose level tested below normal.

Likewise, you should consider hypoglycemia in a patient who has AMS after an injury such as a motor vehicle crash, even when there is the possibility of an accompanying head injury. As with any other patient, you should look for medical identification bracelets or medications that might confirm your suspicions.

Other Causes of Altered Mental Status

AMS can occur as a result of hypoglycemia, but there are many other possibilities as well, including unrecognized head injury or severe alcohol intoxication. Your consideration of other possibilities becomes important because a patient with AMS may be combative and refuse treatment and transport. You should be prepared for difficult patient encounters and follow local protocols for dealing with these situations, recognizing the potential for serious underlying problems.

In most cases, a patient who appears intoxicated most likely is just that; however, you must consider other problems as well. Individuals with chronic alcoholism can have abnormalities in liver function and in their blood-clotting and immune systems, which can predispose them to intracranial bleeding, brain and bloodstream infections, and hypoglycemia.

Psychological problems and complications from medications are also possible causes of AMS. A person who appears to have a psychological problem may also have an underlying medical condition.

Infections are another possible cause of AMS, particularly those involving the brain or bloodstream. Infections in these areas are life threatening and need immediate attention. Patients may not demonstrate typical signs of infection, such as fever, particularly if they are very young or very old or have impaired immune systems.

AMS can also be caused by drug overdose or poisonings; therefore, you should monitor patients closely for accompanying cardiac and breathing problems.

Thus, the presentation of AMS varies widely from simple confusion to coma. No matter what the cause, you should consider AMS to be an emergency that requires immediate attention even when it appears that it may simply be caused by alcohol intoxication or a minor car crash or fall.

Assessment of the Patient With Altered Mental Status

The patient assessment process for patients with AMS is the same as for patients with potential stroke and seizure with a few differences. The most significant difference between AMS and other emergencies is that your patient cannot tell you reliably what is wrong, and there may be more than one cause. Therefore, being vigilant in your ongoing assessment is essential, both to uncover possible causes of your patient's condition and to monitor your patient's condition for changes and deterioration. Prompt transport is necessary, with close monitoring of vital signs en route and careful attention to the airway, positioning the patient to avoid aspiration and maintain comfort.

▲ Pediatric Needs

Children can have AMS caused by strokes, seizures, high or low blood glucose levels, infection (meningitis), poisoning, or tumors. Hemorrhagic strokes are usually caused by congenital defects in blood vessels; these defects are called berry aneurysms. Ischemic strokes can be due to disorders such as sickle cell anemia. However, children who have subarachnoid hemorrhages may not have a berry aneurysm; instead, they may have a congenital problem with the blood vessels in the brain. Children who have sickle cell anemia are at particularly high risk for ischemic stroke. Treat stroke and AMS in children the same way that you do in adults.

As was mentioned earlier in this chapter, seizures can result from sudden high fevers, particularly in children. Remember that although febrile seizures are generally well tolerated by children, you must transport these patients to the hospital. The possibility of a second seizure makes transport mandatory so that if other problems develop, the child is in the hospital and can receive immediate definitive care.

If you suspect that a patient with AMS has hypoglycemia and you have the ability to test for it, you should do so and treat the patient according to local protocols. Also, these patients require close monitoring, particularly of the airway, en route to the hospital.

You are the Provider   Summary

Appropriate assessment and history taking will assist you in ruling out other potential conditions that mimic stroke and will help the patient to receive appropriate and timely care. Preventing further injury of weak or numb limbs is essential, and when aphasia is present, your reassurance and patience will ease some of the patient's anxiety. Utilize other forms of communication by allowing patients to write with the unaffected side, blink their eyelids, or squeeze your hand to respond to questions.

Unfortunately, treatment outside of the hospital for strokes is limited to supportive measures. Time is critical for these patients. Prevention and recognition play a vital role in minimizing death and permanent disability in these cases. Education of the general public on the signs and symptoms of stroke is extremely important in avoiding delays and providing essential medical care. You can begin a program in your community designed to raise awareness of strokes and TIAs that will increase timely recognition and prompt activation of the emergency response system.

	Stroke	**Seizure**	**Altered Mental Status**
Scene Size-up	Body substance isolation precautions should include a minimum of gloves and eye protection. Ensure scene safety and determine NOI/MOI. Consider the number of patients, the need for additional help/ALS, and c-spine stabilization.	Body substance isolation precautions should include a minimum of gloves and eye protection. Ensure scene safety and determine NOI/MOI. Consider the number of patients, the need for additional help, and c-spine stabilization.	Body substance isolation precautions should include a minimum of gloves and eye protection. Ensure scene safety and determine NOI/MOI. Consider the number of patients, the need for additional help, and c-spine stabilization.
Initial Assessment			
■ General impression	Determine level of consciousness and find and treat any immediate threats to life. Determine priority of care based on environment and patient's chief complaint.	Determine level of consciousness and find and treat any immediate threats to life. If the seizure is still taking place on your arrival, the life-threatening status epilepticus may be taking place. Call for ALS.	Determine priority of care based on environment and patient's chief complaint. Determine level of consciousness and find and treat any immediate threats to life.
■ Airway	Ensure patent airway, place in the recovery position, and suction as necessary.	Clear mouth and ensure patent airway. Use airway adjuncts according to local protocol.	Ensure patent airway, place in the recovery position as necessary.
■ Breathing	Provide high-flow oxygen at 15 L/min. Evaluate depth and rate of the respiratory cycle and provide ventilatory support as needed.	Evaluate depth and rate of the respiratory cycle and provide ventilatory support as needed. Provide high-flow oxygen at 15 L/min.	Evaluate depth and rate of the respiratory cycle and provide ventilatory support as needed. Auscultate and note breath sounds, providing high-flow oxygen at 15 L/min.
■ Circulation	Evaluate pulse rate and quality; observe skin color, temperature, and condition and treat accordingly.	Evaluate pulse rate and quality; observe skin color, temperature, and condition.	Evaluate pulse rate and quality; observe skin color, temperature, and condition.
■ Transport decision	Rapid transport to stroke center if available	Transport based on assessment and local guidelines	Rapid transport based on initial assessment
Focused History and Physical Exam	*NOTE: The order of the steps in the focused history and physical exam differs depending on whether the patient is conscious or unconscious. The order below is for a conscious patient. For an unconscious patient, perform a rapid physical exam, obtain vital signs, and obtain the history.*		
■ SAMPLE history	Ask pertinent SAMPLE and OPQRST. Be sure to ask if and what interventions were taken before your arrival, how many interventions, and at what time. Ascertain when the patient last appeared normal.	Ask pertinent SAMPLE and OPQRST. How do seizures typically occur, and did this one differ from the norm? Has patient been compliant with medications?	Ask pertinent SAMPLE and OPQRST. Be sure to ask if and what interventions were taken before your arrival, how many interventions, and at what time.
■ Focused physical exam	Perform a focused neurologic exam using the Cincinnati Stroke Scale, and/or the Glasgow Coma Scale.	Perform focused physical exam for AMS, speech, and thinking ability. Determine blood glucose level and Glasgow Coma Scale score.	Perform a focused neurologic exam for AMS. Determine blood glucose level and Glasgow Coma Scale score.
■ Baseline vital signs	Take vital signs, noting skin color and temperature as well as patient's level of consciousness. Use pulse oximetry if available.	Take vital signs, noting skin color and temperature as well as patient's level of consciousness and pupil size and reactivity. Use pulse oximetry if available.	Take vital signs, noting skin color and temperature as well as patient's level of consciousness. Use pulse oximetry if available.

Assessment and Emergency Care

Assessment and Emergency Care

	Stroke	Seizure	Altered Mental Status
■ Interventions	Support patient as needed. Consider the use of oxygen, positive pressure ventilations, airway adjuncts, and proper positioning of the patient.	Protect patient from harm, maintain a clear airway by suctioning as necessary, and provide oxygen as quickly as possible. Treat any trauma injury you find. For patients who continue to seize, suction airway according to protocol, provide positive pressure ventilations, and transport quickly to the hospital.	Support patient as needed. Consider the use of oxygen, positive pressure ventilations, airway adjuncts, and proper positioning. Treat low glucose levels according to local protocols.
Detailed Physical Exam	Complete a detailed physical exam.	Complete a detailed physical exam.	Complete a detailed physical exam.
Ongoing Assessment	Repeat the initial assessment, focused assessment, and reassess interventions performed. Reassess vital signs and the Glasgow Coma Scale score every 5 minutes for the unstable patient, every 15 minutes for the stable patient. Reassure and calm the patient.	Repeat the initial assessment, focused assessment, and reassess interventions performed. Reassess vital signs every 5 minutes for the unstable patient, every 15 minutes for the stable patient.	Repeat the initial assessment, focused assessment, and reassess interventions performed. Reassess vital signs every 5 minutes for the unstable patient, every 15 minutes for the stable patient. Reassure and calm the patient.
■ Communication and documentation	Contact medical control with a radio report that gives information on patient's condition and the last time patient appeared normal. Relay any change in level of consciousness or difficulty breathing. Be sure to document any changes, at what time, and any Cincinnati Stroke Scale or Glasgow Coma Scale results.	Report and record your findings. Document the onset and duration of the seizure, recording the time of each intervention performed, the patient's response, and continued reassessments.	Report and record your findings. Document each intervention performed, the patient's response, and continued reassessments.

NOTE: While the steps below are widely accepted, be sure to consult and follow your local protocol.

Stroke	Seizure		Altered Mental Status
Cincinnati Stroke Scale	**Glasgow Coma Scale**		Use the Cincinnati Stroke Scale and/or the Glasgow Coma Scale to aid in your assessment.
1. Ask patient to show teeth or smile to determine facial droop.	<u>Eye Opening</u>		
	Spontaneous	4	
	Responsive to speech	3	
	Responsive to pain	2	
2. Ask patient to close eyes and hold both arms out with palms up to measure arm drift.	None	1	
	<u>Best Verbal Response</u>		
	Oriented conversation	5	
	Confused conversation	4	
3. Ask patient to say, "The sky is blue in Cincinnati," to monitor speech.	Inappropriate words	3	
	Incomprehensible sounds	2	
	None	1	
	<u>Best Motor Response</u>		
	Obeys commands	6	
	Localizes pain	5	
	Withdraws to pain	4	
	Abnormal flexion	3	
	Abnormal extension	2	
	None	1	
	Add the total points selected from all three categories to determine the patient's Glasgow Coma Scale score.		

Prep Kit

Ready for Review

- The cerebrum, the largest part of the brain, is divided into right and left hemispheres, each controlling the opposite side of the body.
- Different parts of the brain control different functions. The front part of the cerebrum controls emotion and thought; the middle controls touch and movement; the back part of the cerebrum is involved with vision. In most people, speech is controlled on the left side of the brain, near the middle of the cerebrum.
- Many different disorders can cause brain or other neurologic symptoms. As a general rule, if the problem is primarily in the brain, only part of the brain will be affected. If the problem is in the heart or lungs, the whole brain will be affected.
- Stroke is a significant brain disorder because it is common and potentially treatable.
- Seizures and AMS are also common, and you must learn to recognize the signs and symptoms of each.
- Other causes of neurologic dysfunction include coma, infections, and tumors.
- Strokes occur when part of the blood flow to the brain is suddenly cut off; within minutes, brain cells begin to die.
- Signs and symptoms of stroke include receptive or expressive aphasia, dysarthria, muscle weakness or numbness on one side, facial droop, and sometimes high blood pressure.
- You should always do at least three neurologic tests on patients you suspect of having a stroke: testing speech, facial movement, and arm movement.
- In a transient ischemic attack (TIA), normal body processes break up the blood clot, restoring blood flow and ending symptoms in less than 24 hours. However, patients with TIA are at high risk for a permanent stroke.

- Because current treatments must be administered within 3 to 6 hours (and preferably within 2 hours) of the onset of symptoms to be most effective, you should provide prompt transport.
- Always notify the hospital as soon as possible that you are bringing in a possible stroke patient, so that staff there can prepare to test and treat the patient without delay.
- Seizures are characterized by unconsciousness and generalized twitching of all or part of the body.
- There are types of seizures that you should learn to recognize: generalized, absence, and febrile convulsions.
- Most seizures last between 3 and 5 minutes and are followed by a postictal state in which the patient may be unresponsive, have labored breathing, and have hemiparesis and may have urinated on himself or herself.
- It is important for you to recognize the signs and symptoms of seizures so that you can provide emergency department staff with information as you transport the patient.
- AMS is also a common neurologic problem that you will encounter as an EMT-B. Signs and symptoms vary widely, as do the causes for this condition.
- Among the most common causes are hypoglycemia, intoxication, drug overdose, and poisoning.
- As you assess the patient with AMS, do not always assume intoxication; hypoglycemia is just as likely a cause. Prompt transport with close monitoring of vital signs en route is indicated.

Vital Vocabulary

absence seizure Seizure that may be characterized by a brief lapse of attention in which the patient may stare and does not respond. Also known as petit mal seizure.

aphasia The inability to understand or produce speech.

arterial rupture Rupture of a cerebral artery that may contribute to interruption of cerebral blood flow.

atherosclerosis A disorder in which cholesterol and calcium build up inside the walls of blood vessels, forming plaque, which eventually leads to partial or complete blockage of blood flow; plaque of this type can also become a site where blood clots can form, break off, and embolize elsewhere in the circulation.

aura A sensation experienced prior to a seizure; serves as a warning sign that a seizure is about to occur.

Technology

- Interactivities
- Vocabulary Explorer
- Anatomy Review
- Web Links
- Online Review Manual

www.EMTB.com

Prep Kit continued...

cerebral embolism Obstruction of a cerebral artery caused by a clot that was formed elsewhere in the body and traveled to the brain.

cerebrovascular accident (CVA) An interruption of blood flow to the brain that results in the loss of brain function.

coma A state of profound unconsciousness from which one cannot be roused.

coup-contrecoup brain injury A brain injury that occurs when force is applied to the head and energy transmission through brain tissue causes injury on the opposite side of original impact.

dysarthria The inability to pronounce speech clearly, often due to loss of the nerves or brain cells that control the small muscles in the larynx.

expressive aphasia A speech disorder in which a person can understand what is being said but cannot produce the right sounds in order to speak properly.

febrile seizures Convulsions that result from sudden high fevers, particularly in children.

generalized seizure Seizure characterized by severe twitching of all the body's muscles that may last several minutes or more; also known as a grand mal seizure.

hemiparesis Weakness on one side of the body.

hemorrhagic stroke One of the two main types of stroke; occurs as a result of bleeding inside the brain.

hypoglycemia A condition characterized by low blood glucose levels.

incontinence Loss of bowel and bladder control due to a generalized seizure.

infarcted cells Cells in the brain that die as a result of loss of blood flow to the brain.

ischemia A lack of oxygen in the cells of the brain that cause them to not function properly.

ischemic stroke One of the two main types of stroke; occurs when blood flow to a particular part of the brain is cut off by a blockage (eg, a clot) inside a blood vessel.

postictal state Period following a seizure that lasts between 5 and 30 minutes, characterized by labored respirations and some degree of altered mental status.

receptive aphasia A speech disorder in which a person has trouble understanding speech but is able to speak clearly.

seizure Generalized, uncoordinated muscular activity associated with loss of consciousness; a convulsion.

status epilepticus A condition in which seizures recur every few minutes, or last more than 30 minutes.

stroke A loss of brain function in certain brain cells that do not get enough oxygen during a CVA. Usually caused by obstruction of the blood vessels in the brain that feed oxygen to those brain cells.

thrombosis Clotting of the cerebral arteries that may result in the interruption of cerebral blood flow and subsequent stroke.

tonic-clonic A type of seizure that features rhythmic back-and-forth motion of an extremity and body stiffness.

transient ischemic attack (TIA) A disorder of the brain in which brain cells temporarily stop working because of insufficient oxygen, causing stroke-like symptoms that resolve completely within 24 hours of onset.

Points to Ponder

You are dispatched to a local assisted-living facility. You are greeted by staff members who tell you that the patient seems to have had a stroke sometime during the night. They also tell you that he has a history of TIAs, but has been able to care for himself with a little assistance. You enter the patient's room and find a geriatric man who has obvious right-sided facial droop. You can tell by his facial expression that he is upset. Your partner begins taking vital signs as you ask him what happened. He tries to speak, but only garbled words come out. He then grabs your arm and begins to cry.

What can you do to determine the onset of symptoms? How can you communicate with this patient?

Issues: The Importance of Tolerance and Patience When Caring for Stroke Patients, Using Alternate Methods of Communication, Treating the Whole Patient.

Assessment in Action

You are dispatched 550 Chestnut St, the local senior center, for a woman complaining of a severe headache. You arrive to find an older woman sitting in a chair, moaning and holding her head in her hands. When you ask her what is wrong, she tells you she suddenly developed a terrible headache and she can't seem to see straight.

As you speak with the patient and apply high-flow oxygen, your partner obtains the following set of vital signs: pulse, 100 beats/min and irregular; blood pressure, 198/110 mm Hg; and respirations, 36 breaths/min with adequate tidal volume.

1. What medical emergency is this patient likely experiencing?
   A. Seizure
   B. Hypotension
   C. Hyperglycemia
   D. Stroke

2. When you question her about her past medical history, what condition would you be the **MOST** concerned about?
   A. Seizures
   B. High blood pressure
   C. Emphysema
   D. Anemia

3. If you were to replace the older woman in this scenario with a young, healthy woman, what would likely be the cause of her headache?
   A. Brain tumor
   B. Infection
   C. Hypoglycemia
   D. Berry aneurysm

4. The chief complaint heard in cases of hemorrhagic stroke, "This is the worst headache of my life!" is caused by:
   A. constriction of blood vessels.
   B. dilation of blood vessels.
   C. irritation of brain tissue.
   D. not enough oxygen.

5. What physical tests would you include in your assessment of an older woman patient in this scenario?
   A. Speech
   B. Facial droop
   C. Arm drift
   D. All of the above

6. All patients who have altered mental status, including possible stroke patients, should have what assessment performed?
   A. Cincinnati Stroke Scale
   B. AVPU
   C. Glasgow Coma Scale
   D. B and C

Challenging Questions

7. Stroke patients who have trouble understanding speech but can speak clearly are experiencing what condition?

8. Why is it important to notify the hospital when you are transporting a possible stroke patient?

9. Should you treat TIAs less seriously than CVAs?

10. What common medical emergencies can mimic stroke?

www.EMTB.com

The Acute Abdomen

Objectives*

Cognitive

1. Define the term "acute abdomen." (p 466)
2. Identify the signs and symptoms of the acute abdomen and the necessity for immediate transport of patients with these symptoms. (p 470, 471)
3. Define the concept of "referred pain." (p 466)
4. Describe the areas of pain or referred pain seen with the common causes of the acute abdomen. (p 467, 469)
5. Explain that pain in the abdomen can arise from other body systems. (p 469)

Affective

None

Psychomotor

6. Perform a rapid gentle assessment of the abdomen. (p 471)

*All of these objectives are noncurriculum objectives.

You are the Provider

You and your EMT-I partner are assigned to a football standby at the local high school's homecoming game. As you eat some snacks and enjoy the festivities, you notice that after a tackle, one player remains on the ground with his knees pulled into his chest. You alert your partner and await the player's evaluation by the football team's trainers. After a few moments, the player rises to his feet and walks to the bench with a little help from the assistant coach. You and your partner return to enjoying the game, but a few minutes later you see one of the trainers talking with the same player who now has his helmet off. The trainer then whistles at you and motions you to come over.

1. What sorts of injuries can occur to the abdomen during contact sports such as football?

The Acute Abdomen

Abdominal pain is a common complaint, but the cause is often difficult to identify, even for a physician. As an EMT-B, you do not need to determine the exact cause of acute abdominal pain. You simply need to be able to recognize a life-threatening problem and act swiftly in response. Remember, the patient is in pain and is probably anxious, requiring all your skills of rapid assessment and emotional support.

This chapter begins by explaining the physiology of the abdomen. It then describes the signs and symptoms of the acute abdomen and explains how to examine the abdomen. Next, it discusses the different causes of the acute abdomen and appropriate emergency medical care.

The Physiology of the Abdomen

The abdominal cavity contains the solid and hollow organs that make up the gastrointestinal, genital, and urinary systems (Figure 14-1 ▶). The abdominal cavity is lined by a membrane called the peritoneum. The peritoneum also covers the organs of the abdomen. The *parietal peritoneum* lines the abdominal cavity, and the *visceral peritoneum* covers the organs themselves. The abdominal space normally contains a small amount of *peritoneal fluid* to bathe and lubricate the organs. Any foreign material, such as blood, pus, bile, pancreatic juice, or amniotic fluid, can cause irritation of the peritoneum, called peritonitis. Technically, organs such as kidneys, ovaries, and the pancreas are *retroperitoneal* (behind the peritoneum) (Figure 14-2 ▶). However, because they lie next to the peritoneum, they can cause abdominal pain.

Acute abdomen is a medical term referring to the sudden onset of abdominal pain, generally associated with severe, progressive problems that require medical attention. Peritonitis will usually develop if the acute abdomen is not treated, and can be fatal.

Two different types of nerves supply the peritoneum, and therefore abdominal pain can have different qualities. The parietal peritoneum is supplied by the same nerves from the spinal cord that supply the skin of the abdomen; it can therefore perceive much the same sensations: pain, touch, pressure, heat, and cold. These sensory nerves can easily identify and localize a point of irritation. In contrast, the visceral peritoneum is supplied by the autonomic nervous system. These nerves are far less able to localize sensation. What this means for the EMT is that the patient will not be able to localize it and describe exactly where the pain or injury is. The visceral peritoneum is stimulated when distention or contraction of the hollow abdominal organs activates the stretch receptors. This sensation is often interpreted as colic, a severe, intermittent cramping pain. Other painful sensations that occur because of an irritated visceral peritoneum may be perceived at a distant point on the surface of the body, such as the back or shoulder, or described as a deep pain. This phenomenon is called referred pain.

Referred pain is the result of connections between the body's two separate nervous systems. The spinal cord supplies sensory nerves to the skin and muscles; these nerves are called the somatic nervous system. The autonomic nervous system controls the abdominal organs and the blood vessels. The nerves connecting these two systems cause the stimulation of the autonomic nerves to be perceived as stimulation of the spinal sensory nerves. For example, acute cholecystitis (inflammation of the gallbladder) may cause pain in the right shoulder, because the autonomic nerves serving the gallbladder lie near the spinal cord at the same anatomic level as the spinal sensory nerves that supply the skin of the shoulder (Figure 14-3 ▶).

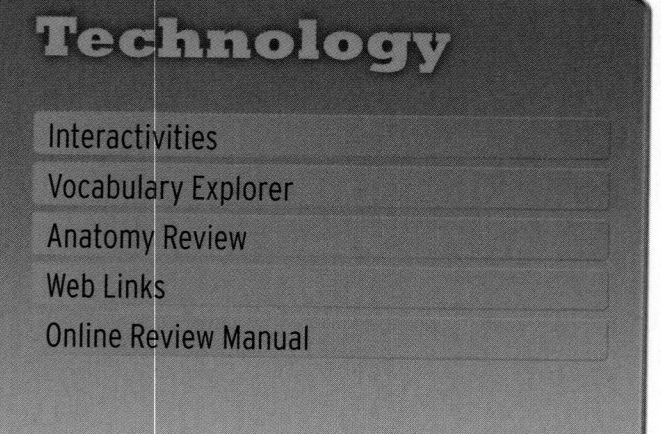

Technology

Interactivities

Vocabulary Explorer

Anatomy Review

Web Links

Online Review Manual

www.EMTB.com

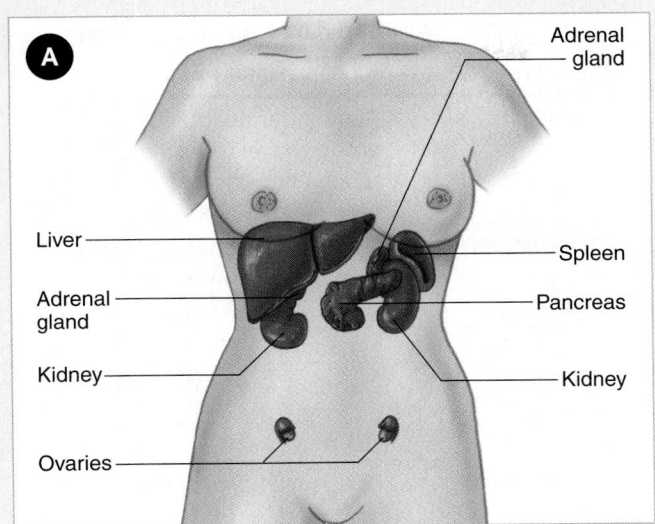

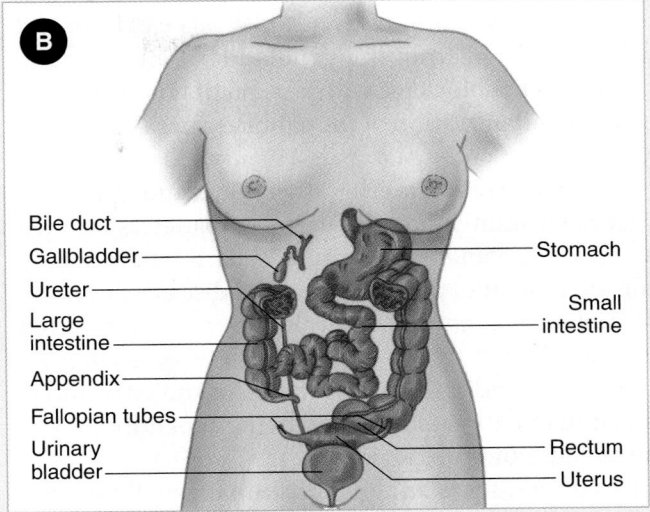

Figure 14-1 The solid and hollow organs of the abdomen. **A.** Solid organs include the liver, spleen, pancreas, kidneys, and ovaries (in women). **B.** Hollow organs include the gallbladder, stomach, small intestine, large intestine, and bladder.

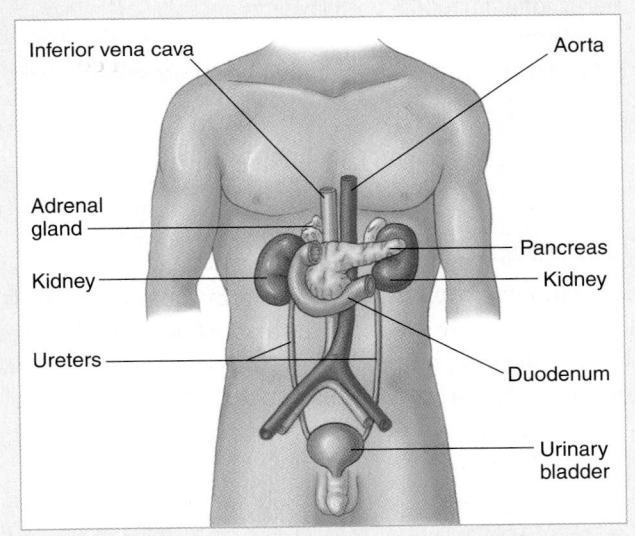

Figure 14-2 The major organs of the retroperitoneal space include the genitourinary structures, pancreas, and great vessels.

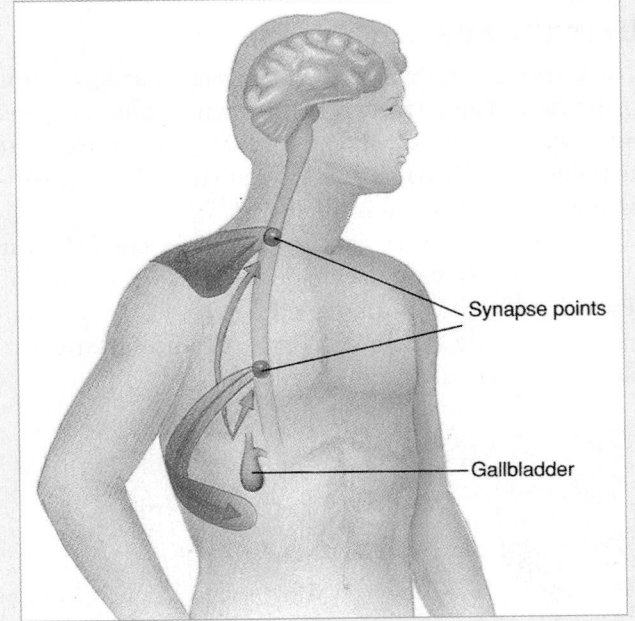

Figure 14-3 Acute cholecystitis causes referred pain in the shoulder as well as the abdomen.

Causes of Abdominal Pain

Almost any problem with an abdominal organ can cause an acute abdomen. Some of the more common causes are discussed here. Because the visceral peritoneum is usually irritated first, early abdominal pain tends to be vague and poorly localized. As the parietal peritoneum becomes irritated, pain becomes more severe and may be more specifically located.

Digestive System

Ulcers are one of the most common abdominal problems. Ulcers are erosions of the stomach or the duodenum, the first part of the small intestine, due to overactivity of digestive juices. Perforation of an ulcer causes severe peritonitis and an acute abdomen.

The gallbladder is a storage pouch for digestive juices and waste from the liver. *Gallstones* can form and block the outlet from the gallbladder, causing pain. Sometimes the blockage will pass, but if not, it can lead to severe inflammation of the gallbladder, called cholecystitis.

The pancreas forms digestive juices and is also the source of insulin. Inflammation of the pancreas is called pancreatitis. Pancreatitis can be caused by a blocking gallstone, alcohol abuse, and other diseases. Because the pancreas is retroperitoneal, pain is often referred to the back.

The appendix is a small recess in the large intestine. Inflammation or infection in the appendix is a frequent cause of acute abdomen.

Diverticulitis is an inflammation of small pouches in the large intestine. These pouches can become blocked and infected, leading to pain, perforation, and severe peritonitis. The more common abdominal emergencies, with most common locations of direct and referred pain, are listed in (Table 14-1 ▶).

Urinary System

The kidneys can be affected by stones that form from materials normally passed in the urine. If a stone passes out of the kidney, it can cause severe pain, referred to as *renal colic*. Passage of a kidney stone is frequently associated with blood in the urine.

Kidney infections can cause severe pain, often in the flank. These patients are often quite ill, with a high fever. Bladder infection, called cystitis, is more common, especially in women. Patients with cystitis usually have lower abdominal pain.

TABLE 14-1 Common Abdominal Conditions

Condition	Localization of Pain
Appendicitis	Right lower quadrant (direct); around navel (referred); rebounding pain (pain felt on the rebound of palpation)
Cholecystitis	Right upper quadrant (direct); right shoulder (referred)
Ulcer	Upper mid-abdomen or upper back
Diverticulitis	Left lower quadrant
Aortic aneurysm (ruptured or dissecting)	Low back and lower quadrants
Cystitis (inflammation of the bladder)	Lower mid-abdomen (retropubic)
Kidney infection	Costovertebral angle
Kidney stone	Right or left flanks, radiating to genitalia (referred)
Pelvic inflammation (in women)	Both lower quadrants
Pancreatitis	Upper abdomen (both quadrants); back

Uterus and Ovaries

Gynecologic problems are a common cause of acute abdominal pain. Always consider that a woman with lower abdominal pain and tenderness may have a problem related to her ovaries, fallopian tubes, or uterus.

You are the Provider   Part 2

When you introduce yourself and ask the patient what happened, he explains that he had the wind knocked out of him during the last tackle when another player's helmet "hit him in the gut." He says he now feels pain in his stomach and is nauseated. Given his description of the events and the trainer's comments regarding the initial assessment, you believe this is an isolated injury to the abdomen.

2. What could be causing his pain and what other signs and symptoms could you expect the patient to have?

Geriatric Needs

Geriatric patients are as susceptible to the acute abdomen as younger adults are. However, the signs and symptoms in geriatric patients might be different. Because of altered pain sensation, geriatric patients with an acute abdomen may not feel any discomfort or may describe the discomfort as mild, even in severe conditions.

Because the older patient has decreased body temperature regulation and response, the patient with an acute abdomen, including peritonitis, may not have a fever. However, if a fever is present, it can be minimal.

Because of the older patient's response to the acute abdomen, a delay in identifying the condition and seeking medical attention is possible, putting the patient at risk for complications. You should ask about the patient's medical history, especially the history of recent illness, to identify a potential illness. Ask about abdominal discomfort, when the patient last had a bowel movement, whether she or he was constipated, or had diarrhea. Inquire as to when the patient last ate, and whether she or he has vomited. Ruling out appendicitis, bowel obstruction, or ruptured bowel can hasten proper treatment and recovery.

Abdominal pains may also be related to the normal menstrual cycle. A common lower abdominal pain, often confused with appendicitis but fairly short lived, is called *mittelschmerz*. It is associated with the release of an egg from the ovary, characteristically occurring in the middle of the menstrual cycle, between menstrual periods. Mittelschmerz may also be associated with lower abdominal tenderness. Some women experience painful cramps at the time of their menstrual periods. In some, the discomfort may be crippling and the menstrual flow severe.

A common cause of an acute abdomen in women is *pelvic inflammatory disease (PID)*, an infection of the fallopian tubes and the surrounding tissues of the pelvis. With PID, acute pain and tenderness in the lower abdomen may be intense and accompanied by a high fever. If you suspect PID, promptly transport the patient to the emergency department for treatment.

Between 1% and 2% of all pregnancies are ectopic. The term *ectopic pregnancy* means that a fertilized egg has come to lie in an area outside the uterus, usually in a fallopian tube. A fallopian tube is not large enough to support the growth of a fetus and placenta for more than about 6 to 8 weeks. When the tube ruptures, it produces massive internal hemorrhage and abrupt abdominal pain. In this situation, the acute abdomen may be associated with the onset of hypovolemic shock. This combination evolves into an emergent situation that requires immediate transport to the hospital.

Other Organ Systems

The aorta lies immediately behind the peritoneum on the spinal column. In older individuals, the wall of the aorta sometimes develops weak areas that swell to form an <u>aneurysm</u>. A pulsating mass may be felt in the abdomen. The development of an aneurysm is rarely associated with symptoms because it occurs slowly, but if the aneurysm ruptures, massive hemorrhage may occur and, with it, the signs of acute peritoneal irritation. The patient may also experience severe back pain, because the peritoneum can, at times, be rapidly stripped away from the wall of the main abdominal cavity by the hemorrhage. Pain can also be associated with the pressure of blood on the back itself. In such instances, bleeding quickly leads to profound shock. Again, the association of acute abdominal signs and symptoms with shock requires prompt transportation. Because this is a fragile situation with a large, leaking artery, avoid unnecessary or vigorous palpation of the abdomen, and remember to handle the patient gently during transport.

Pneumonia, especially in the lower parts of the lung, may cause both ileus and abdominal pain. In this instance, the problem lies in an adjacent body cavity, but the intense inflammatory response can affect the abdomen. Treat and transport this patient as you would any patient with abdominal pain.

A <u>hernia</u> is a protrusion of an organ or tissue through a hole in the body wall covering its normal site. Virtually every organ or tissue in the body will herniate through its covering membranes in certain circumstances. Hernias can occur as a result of the following:

- A congenital defect, as around the umbilicus
- A surgical wound that has failed to heal properly
- Some natural weakness in an area such as in the groin

Hernias do not always produce a mass or lump that the patient will be aware of. At times, the mass will disappear back into the body cavity in which it belongs. In this case, the hernia is said to be *reducible*. If the mass cannot be pushed back within the body, it is said to be *incarcerated*. Note, however, that you should never attempt to push the mass back into the body.

Reducible hernias pose little risk to the patient; some individuals live with them for years. When a hernia is incarcerated, however, its contents may become seriously compressed by the surrounding tissue, eventually compromising the blood supply. This situation, called strangulation, is a serious medical emergency. Immediate surgery is required to remove any dead tissue and repair the hernia.

The following signs and symptoms indicate a serious hernia problem:

- A clear statement that a mass that was reducible can no longer be pushed back inside the body
- Pain at the hernia site
- Tenderness when the hernia is palpated
- Red or blue skin discoloration over the hernia

Any of these signs and symptoms is cause for prompt transport to the emergency department.

Signs and Symptoms of Acute Abdomen

Pain and tenderness are the most common symptom and sign of an acute abdomen. The pain may be sharply localized or diffuse and will vary in its severity. Localized pain gives a clue to the problem organ or area causing it. Tenderness may be minimal or so great that the patient will not allow you to touch the abdomen.

Peritonitis typically causes ileus, or paralysis of the muscular contractions that normally propel material through the intestine. The retained gas and feces, in turn, cause abdominal distention. In the presence of such paralysis, nothing that is eaten can pass normally out of the stomach or through the bowel. The only way the stomach can empty itself, then, is by emesis, or

vomiting. For this reason, peritonitis is almost always associated with nausea and vomiting, usually in that order. These complaints do not point to a particular cause because they can accompany almost every type of gastrointestinal disease or injury.

To gauge the degree of distention, simply look at the patient's abdomen. Distention begins shortly after muscular contractions of the bowel have ceased. Pulse and blood pressure may change significantly or not at all. These findings usually reflect the severity of the process, its duration, and the amount of fluid lost into the abdomen.

Similarly, anorexia, loss of hunger or appetite, is a nonspecific symptom. It, too, is an almost universal complaint in gastrointestinal and abdominal disease or injury. In fact, if a patient does not have anorexia, the situation may not be as serious as it otherwise appears.

Peritonitis is associated with a loss of body fluid into the abdominal cavity. The loss of fluid usually results from abnormal shifts of fluid from the bloodstream into body tissues. This fluid shift decreases the volume of circulating blood and may lead to decreased blood pressure or even shock. Shock is the condition of inadequate perfusion due to the collapse of the cardiovascular system, and is discussed in detail in Chapter 23. The patient may have normal vital signs or, if the peritonitis has progressed farther, may present with tachycardia and hypotension. When peritonitis is accompanied by hemorrhage, the signs of shock are much more apparent.

Fever may or may not be present, depending on the cause of the peritonitis. Patients with diverticulitis (an inflammation of small pockets in the colon) or cholecystitis may have a substantial elevation in temperature. However, patients with acute appendicitis may

You are the Provider

Part 3

You observe the patient's skin, but because he has been involved in recent physical activity, you find it difficult to glean definitive information from his skin signs. You palpate his radial pulse and find it regular and full at a rate of 130 beats/min. You carefully assist him to the cot and suggest that he lay down in a position that is most comfortable.

His complaints are consistent with your initial concern, and you ask him if he is having pain anywhere else in his body. He tells you that as he is speaking with you, his left shoulder is beginning to hurt. You recognize this as a potentially life-threatening emergency and begin transport immediately.

3. What other conditions might have made his injury more likely to occur?

have a temperature within normal limits until the appendix ruptures and an abscess starts to form.

Another sign of the acute abdomen is tenseness of the abdominal muscles over the irritated area. In some instances, the muscles of the abdominal wall become rigid in an involuntary effort to protect the abdomen from further irritation. This board-like muscle spasm, called guarding, can be seen with major problems such as a perforated ulcer or pancreatitis. In some situations, patients are comfortable only when lying in one particular position, which tends to relax muscles adjacent to the inflamed organ and thus lessen the pain. Therefore, the position of the patient may provide an important clue. For example, a patient with appendicitis may draw up the right knee. A patient with pancreatitis may lie curled up on one side.

Remember, the patient with peritonitis usually has abdominal pain, even when lying quietly. The patient can be quiet but have difficulty breathing and may take rapid, shallow breaths because of the pain. Usually, you will find tenderness on palpation of the abdomen or when the patient moves. The degree of pain and tenderness is usually related directly to the severity of peritoneal inflammation.

Assessment of the Patient With Acute Abdomen

Scene Size-up

Ensure that the scene is safe. Acute abdomen can be the result of violence, such as blunt or penetrating trauma. Consider the need for ALS backup. Observe the scene closely and interview bystanders or family members if the cause is not obvious. There may be clues present that only you will be able to report.

Initial Assessment

General Impression

Approach the patient and ask him or her about the chief complaint. A description of the current problem in the patient's own words should help you to identify a place to begin. If the chief complaint indicates a life-threatening problem, assess and treat it immediately. If the chief complaint is a minor problem, it should wait until you have had a chance to assess for and treat any potential life threats. The patient's level of consciousness, using the AVPU scale, should be included in your general impression.

Airway, Breathing, and Circulation

Ensure that the patient's airway is clear and respirations are adequate. Administer oxygen to the patient; often the patient's tidal volume is inadequate due to shallow respirations because deep respirations may cause significant pain.

When you are assessing the patient's circulation remember to assess for major bleeding. The patient's pulse rate and quality, as well as skin condition, may indicate shock. Shock may be hypovolemic or due to a severe infection. If evidence of shock (inadequate perfusion) is present, interventions should include high-flow oxygen, elevating the legs 6″ to 12″ or to a position of comfort, and keeping the patient warm. Ensure that you provide prompt treatment and prompt, gentle transport for the patient; do not delay transport.

Transport Decision

Certain patients should be transported quickly; these include patients who have problems with their airway, breathing, or circulation, including problems with their pulse and perfusion, and patients with suspected internal bleeding. Included in the group to package quickly and transport rapidly are those who have a poor general impression, especially pediatric and geriatric patients.

Focused History and Physical Exam

The signs and symptoms of an acute abdomen signal a serious medical or surgical emergency. The symptoms of an acute abdomen may come on suddenly or may become progressively worse.

The following is a checklist of common signs and symptoms of irritation or inflammation of the peritoneum that you can use to determine whether a patient has an acute abdomen:

- Local or diffuse abdominal pain and/or tenderness
- A quiet patient who is guarding the abdomen (in shock)

- Rapid and shallow breathing
- Referred (distant) pain
- Anorexia, nausea, vomiting
- Tense, often distended, abdomen
- Constipation or bloody diarrhea
- Tachycardia
- Hypotension
- Fever
- Rebound tenderness (may be tender when direct pressure is applied, but very painful when pressure is released)

SAMPLE History

Use OPQRST to ask the patient what makes the pain better or worse:

- **O = O**nset, ie, when did the problem begin and what caused it?
- **P = P**rovocation or **P**alliation, ie, does anything make it feel better? Worse?
- **Q = Q**uality, ie, what is the pain like? Sharp, dull, crushing, tearing?
- **R = R**egion/**R**adiation, ie, where does it hurt? Does the pain move anywhere?
- **S = S**everity, ie, on a scale of 1 to 10, how would you rate your pain?
- **T = T**iming of pain, ie, has the pain been constant or does it come and go? How long have you had the pain? (often answered under "O", onset)

Notice the position that the patient is in. Commonly he or she will have his or her knees drawn up to help alleviate the pain associated with acute abdomen. Ask the patient to locate where the pain is and whether the pain travels or radiates. Utilize a pain scale and ask the patient to rate his or her pain according to that scale (for example, "On a scale of 0 to 10, with 0 meaning no pain at all and 10 being the worst pain you have ever experienced in your life, how would you rate this pain?"). Ask the patient whether the pain has been constant or intermittent. It is important to determine whether this is a medical emergency or trauma related. Therefore, you need to question the patient about any recent trauma. It is also important to determine whether the patient has ingested any substance that could be causing the acute abdomen. This may not affect the interventions you would perform, but it will help the physician in determining the cause.

Do not give the patient anything by mouth. Food or fluid will only aggravate many of the symptoms, because intestinal paralysis will prevent it from passing out of the stomach. The presence of food in the stomach will make any emergency surgery more dangerous.

Focused Physical Exam

Information gathered in the history portion of your assessment may be used to guide you in your focused physical exam of the abdomen. Use the following steps to assess the abdomen:

1. Explain to the patient what you are about to do.
2. Place the patient in a supine position with the legs drawn up and flexed at the knees to relax the abdominal muscles, unless there is any trauma, in which case the patient will remain supine and stabilized.
3. Determine whether the patient is restless or quiet, whether motion causes pain, or whether distention or obvious abnormality is present.
4. Palpate the four quadrants of the abdomen gently to determine whether it is tense (guarded) or soft (Figure 14-4 ▼). The quadrant that you suspect is the source of the pain should be palpated last. If the most painful area is palpated first, the patient may guard against further examination, making evaluation more difficult.
5. Determine whether the patient can relax the abdominal wall on command.
6. Determine whether the abdomen is tender when palpated.

Although such an examination will yield much information, it should not be prolonged. The physician will do a much more detailed examination in the hospital. Remember to be very gentle when palpating the

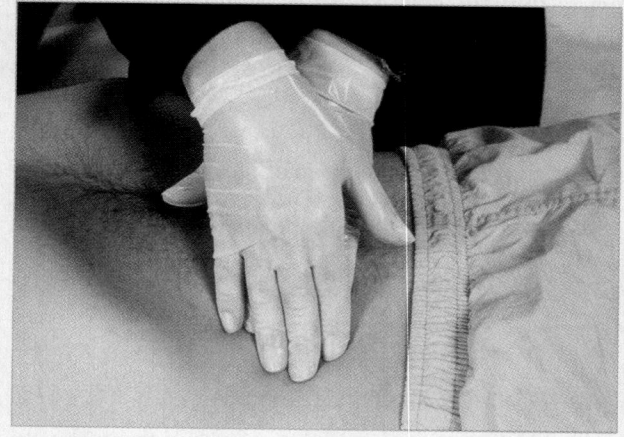

Figure 14-4 Check tenderness or rigidity by gently palpating the abdomen.

Documentation Tips

An acute abdomen usually indicates peritonitis, in which generalized signs can make it challenging to determine exactly where the problem lies, even for physicians. Knowing abdominal assessment steps well, and recording your findings in detail, are important early factors in the process that leads to diagnosis.

abdomen. Occasionally, an organ within the abdomen will be enlarged and very fragile, and rough palpation could cause further damage.

Baseline Vital Signs

Monitor the patient's vital signs for adequate ventilation and shock. A high respiratory rate with a normal pulse and blood pressure may indicate the patient is unable to ventilate properly because of the pain this causes. High respiratory rates and pulse rates with signs of shock, such as pallor and diaphoresis, may indicate septic or hypovolemic shock. Remember, blood pressure is the last vital sign to go. Once your patient becomes hypotensive, shock is severe.

Interventions

Abdominal complaints can vary from very simple problems to very complex problems. Much of your treatment will depend on the signs and symptoms found during your assessment. For example, alterations in breathing may be relieved with oxygen. Nausea is frequently lessened with low-flow oxygen. In addition to symptomatic care, most patients with abdominal pain will benefit from good psychosocial care and proper positioning. If a patient is in shock, you may need to position him or her in a modified shock position or Trendelenburg position. Many will prefer a side-lying position curled up in a ball. Others may need to be more elevated to breathe easier. Being calm and professional will help to reassure the patient he or she is being cared for.

Detailed Physical Exam

The causes of acute abdominal illnesses are often complicated or nonspecific. Identifying the cause of pain may be difficult. Although it is not your responsibility to make a diagnosis, a cause may be identified with a more detailed exam of the patient including a detailed history. Time and experience will guide you in how to perform a detailed physical exam that

You are the Provider   Part 4

You ask the patient whether he has recently had mononucleosis. He seems surprised at your question, and confirms a recent history of 'mono.' He explains that he didn't tell his coach because he was afraid he wouldn't be allowed to play in today's "big game."

An off-duty fire fighter from the local department, who is also watching the football game, notices the commotion on the sideline and approaches you. He offers to drive you and your partner to the hospital. As you begin transport, the teams take a timeout and another ambulance is dispatched to the scene to cover the remainder of the game.

You apply high-flow oxygen and obtain a blood pressure while your partner establishes two large-bore IVs. You note his blood pressure is 96/64 mm Hg, respirations are 36 breaths/min, and his pulse oximetry is 95% receiving 15 L/min via nonrebreathing mask. He complains of feeling dizzy, and you place him in the shock position and cover him with a blanket. He remains awake and alert during transport; he complains of severe abdominal pain throughout the call.

4. What type of a receiving hospital would be most appropriate for this patient?
5. What are the critical elements of the report you will transmit to the receiving hospital and why?

provides useful information. The decision to perform a detailed physical exam should not delay your transport and may even be omitted depending on your transport distance and the seriousness of your patient's condition.

Ongoing Assessment

Because it is often difficult to determine the cause of an acute abdominal emergency, it becomes extremely important to reassess your patient frequently to determine whether the patient's condition has changed. Remember, the condition of a patient with an acute abdomen can change rapidly from stable to unstable.

A patient in shock or with any life-threatening condition should be transported without delay, deferring a thorough history and focused physical exam. Do take the time to determine, to the best of your ability, information on possible causes that might not be available to hospital staff. Do not hesitate to call for ALS backup or intercept if your patient's condition deteriorates during transport.

Usually, patients with acute abdomen will feel most comfortable on their sides with their knees drawn up. This is also a good position to avoid aspiration should they vomit. Shock, from bleeding into the abdomen, will occur more frequently in traumatic situations than in acute illnesses. Anticipate the development of shock. Medical patients can certainly have problems that prevent the circulatory system from meeting the body's oxygen demands. This hypoperfusion can occur from infections, poisonings, chronic vomiting and diarrhea, gastrointestinal bleeding, and numerous other causes. In many of these medical situations, shock progresses more slowly but can be just as dangerous. Treat the patient for shock even though obvious signs of shock are not apparent.

Communication and Documentation

Communicate with your receiving hospital early if the patient's condition is severe to allow hospital staff to recruit the resources necessary to treat your patient on arrival. Relay all relevant information to the receiving physician or nurse, and carefully document your findings in your patient care report.

Emergency Care

Although the EMT-B cannot treat the causes of acute abdomen, you can take steps to provide comfort and lessen the effects of shock by reassuring the patient and making the patient feel at ease. Position patients who are vomiting to maintain a patent airway. Contain vomitus to prevent the spread of infections (by using a red biohazard bag). Airborne bacteria and viruses produced from vomiting can be transmitted to others easily. Ensure you have gloves, eye protection, and a mask to prevent breathing in these infectious organisms. When you have released your patient to the hospital staff, clean the ambulance and any equipment you have used. Do not forget to wash your hands even though you were wearing gloves.

Providing the patient low-flow oxygen often decreases the nausea. If the patient is having problems breathing, high concentrations of oxygen are more appropriate. Elevate the patient's legs to facilitate blood flow to the core of the body and improve circulation. Loosen restrictive clothing and transport gently in a position of comfort. The causes of acute abdominal illnesses are usually not life threatening even though patients feel as if they may die. You should constantly reassess your patient's condition for signs of deterioration.

You are the Provider   Summary

Acute abdominal pain can be caused by many different sources. For example, blunt trauma to the abdomen can cause damage to a variety of solid and hollow organs. Trauma to the spleen can cause an immediate life-threatening situation. If the spleen bursts, significant bleeding will occur within the abdomen and will quickly result in shock. Understanding the signs and symptoms of cases such as a ruptured spleen can give you a better chance of pinpointing the source of a patient's pain.

Understanding the effects of infection on the spleen is useful in this scenario. The spleen produces more white blood cells in response to mononucleosis, causing it to enlarge and making it more susceptible to injury during a significant impact to the abdomen.

Using appropriate questioning techniques, surveying the scene, and performing a thorough physical assessment are essential components to understanding the source and severity of a patient's condition. If you are unable to confidently determine the source of the patient's chief complaint, evaluate airway, breathing, and circulation, provide high-flow oxygen, treat for shock as needed, and promptly transport.

Assessment and Emergency Care

Acute Abdomen	
Scene Size-up	Body substance isolation precautions should include a minimum of gloves and eye protection. Ensure scene safety and determine NOI/MOI. Consider the number of patients, the need for additional help/ALS, and c-spine stabilization.
Initial Assessment	
■ General impression	Determine priority of care based on environment and patient's chief complaint. Determine level of consciousness and find and treat any immediate threats to life.
■ Airway	Ensure patent airway.
■ Breathing	Evaluate depth and rate of the respirations and provide ventilations as needed. Auscultate and note breath sounds, providing high-flow oxygen.
■ Circulation	Evaluate pulse rate and quality; observe skin color, temperature, and condition and treat accordingly. Assess for major bleeding.
■ Transport decision	Rapid transport
Focused History and Physical Exam	*NOTE: The order of the steps in the focused history and physical exam differs depending on whether the patient is conscious or unconscious. The order below is for a conscious patient. For an unconscious patient, perform a rapid physical exam, obtain vital signs, and obtain the history.*
■ SAMPLE history	Note the position in which the patient was found. Ask pertinent SAMPLE and OPQRST questions. Be sure to ask if and what interventions were taken before your arrival, how many interventions, and at what time.
■ Focused physical exam	Perform a focused physical exam on the acute abdomen by observing the patient and palpating all four quadrants, determining whether the patient can relax the abdomen wall on command and whether the abdomen is tender upon palpation. *[See emergency care chart]*
■ Baseline vital signs	Take vital signs, noting skin color and temperature as well as patient's level of consciousness. Use pulse oximetry if available.
■ Interventions	Support patient as needed. Consider the use of oxygen, positive pressure ventilations, airway adjuncts, and proper positioning of the patient.
Detailed Physical Exam	Complete a detailed physical exam.
Ongoing Assessment	Repeat the initial assessment and focused assessment, and reassess interventions performed. Reassess vital signs every 5 minutes for the unstable patient, every 15 minutes for the stable patient. Treat for shock, and reassure and calm the patient.
■ Communication and documentation	Report and record your findings. Document each intervention performed, the patient's response, and continued reassessments.

NOTE: While the steps below are widely accepted, be sure to consult and follow your local protocol.

Acute Abdomen

General Management
1. Explain to the patient what you are about to do.
2. Place the patient in a supine position with the legs drawn up and flexed at the knees unless trauma is suspected.
3. Determine whether the patient is restless or quiet, whether motion causes pain, or whether any characteristic position, distention, or obvious abnormality is present.
4. Palpate the four quadrants of the abdomen.
5. Determine whether the patient can relax the abdominal wall on command.
6. Determine whether the abdomen is tender when palpated.

Prep Kit

Ready for Review

- The acute abdomen is a medical emergency, requiring prompt but gentle transport.

- The pain, tenderness, and abdominal distention associated with an acute abdomen are signs of peritonitis, which may be caused by any condition that allows pus, blood, feces, urine, gastric juice, intestinal contents, bile, pancreatic juice, amniotic fluid, or other foreign material to lie within or adjacent to the peritoneum.

- In addition to abdominal disease or injury, problems in the gastrointestinal, genital, and urinary systems may also cause peritonitis.

- Appendicitis, perforated ulcer, cholecystitis, and diverticulitis are common causes of an acute abdomen. A strangulated hernia is another.

- Signs and symptoms of acute abdomen include pain, nausea, vomiting, and a tense, distended abdomen.

- Pain is common directly over the inflamed area of the peritoneum, or it may be referred to another part of the body. Referred pain occurs because of the connections between the two different nervous systems supplying the parietal peritoneum and the visceral peritoneum.

- Your first priorities are to assess airway, breathing, and circulation and then apply oxygen. Next, obtain a pertinent medical history using OPQRST.

- Take vital signs and gently palpate the abdomen. The presence of abdominal tenderness will confirm the need for rapid transport to the emergency department.

- Do not give the patient with an acute abdomen anything by mouth. In all likelihood, the bowel is paralyzed, making it impossible for food to pass out of the stomach.

- A patient in shock or with any life-threatening condition should be transported without delay. Call for ALS assistance if your patient's condition deteriorates during transport.

Technology

- Interactivities
- Vocabulary Explorer
- Anatomy Review
- Web Links
- Online Review Manual

www.EMTB.com

Prep Kit continued...

Vital Vocabulary

acute abdomen A condition of sudden onset of pain within the abdomen, usually indicating peritonitis; immediate medical or surgical treatment is necessary.

aneurysm A swelling or enlargement of a part of an artery, resulting from weakening of the arterial wall.

anorexia Lack of appetite for food.

appendicitis Inflammation of the appendix.

cholecystitis Inflammation of the gallbladder.

colic Acute, intermittent cramping abdominal pain.

cystitis Inflammation of the bladder.

diverticulitis Bulging out of intestinal rings in small pockets at weak areas in the muscle walls, creating abdominal discomfort.

emesis Vomiting.

guarding Involuntary muscle contractions (spasm) of the abdominal wall, an effort to protect the inflamed abdomen.

hernia The protrusion of a loop of an organ or tissue through an abnormal body opening.

ileus Paralysis of the bowel, arising from any one of several causes; stops contractions that move material through the intestine.

pancreatitis Inflammation of the pancreas.

peritoneum The membrane lining the abdominal cavity (parietal peritoneum) and covering the abdominal organs (visceral peritoneum).

peritonitis Inflammation of the peritoneum.

referred pain Pain felt in an area of the body other than the area where the cause of pain is located.

strangulation Complete obstruction of blood circulation in a given organ as a result of compression or entrapment; an emergency situation causing death of tissue.

ulcer Erosion of the stomach or intestinal lining.

Points to Ponder

You are dispatched to 1313 Military St for a woman with severe abdominal pain. You are greeted at the door by the patient's mother. She is very upset and concerned about her daughter. As you enter the residence, you see a teenage girl lying on the floor in a fetal position with a trashcan next to her. Before you can begin your assessment, the patient's mother grabs you and repeatedly asks, "What's wrong with my daughter?"

As you attempt to calm the mother and ask questions regarding her daughter's medical history, your partner administers high-flow oxygen and assesses the patient's vital signs.

What must you ask of any female patient of childbearing age? How can you ask questions of a sensitive nature so that the patient will likely answer truthfully?

Issues: Obtaining Pertinent Past Medical History in a Discrete Manner, Dealing With Sensitive Issues, Abdominal Pain in Women.

www.EMTB.com

Assessment in Action

You are dispatched to the county jail for man complaining of abdominal pain. As you park your ambulance, a correctional officer informs you that your patient has been complaining of abdominal pain since yesterday. It is now 2:30 AM, and he says the pain is suddenly much worse.

After traveling through the facility, you finally reach the patient. He is lying in his bed, with a knee drawn up to his stomach. He looks pale and is feverish to the touch. He tells you that his stomach hurts. Although he hasn't eaten, he has felt the need to throw up several times. Your partner applies high-flow oxygen and obtains a set of vital signs as you gently palpate the patient's abdomen. You find that it is tender throughout his abdomen, but especially in the right lower quadrant. His vital signs lying down are as follows: pulse, 130 beats/min and weak; blood pressure, 108/60 mm Hg; and respirations, 42 breaths/min and shallow.

1. What condition **MOST ACCURATELY** describes the patient's current condition?

   A. Diverticulitis
   B. Cholecystitis
   C. Appendicitis
   D. Ruptured appendix

2. Pain that is felt in other areas of the body is called:

   A. traveling pain.
   B. visceral pain.
   C. referred pain.
   D. none of the above.

3. Where is someone with this patient's condition likely to feel pain besides the right lower quadrant?

   A. Around the navel
   B. Between the shoulder blades
   C. In the neck
   D. In the jaw

4. What term describes boardlike muscle spasms in the abdomen?

   A. Guarding
   B. Rebound tenderness
   C. Point tenderness
   D. Colicky pain

5. In what position would you expect to find this patient?

   A. Drawing up the right knee
   B. In the supine position
   C. In the prone position
   D. On his left side

6. Extreme pain associated with the release of direct pressure on the abdomen is called:

   A. guarding.
   B. point tenderness.
   C. colicky pain.
   D. rebound tenderness.

Challenging Questions

7. Your patient with an acute abdomen repeatedly asks you for a drink of water. What do you do?

8. How can the ambulance ride have an impact on the pain level of the patient with peritonitis?

9. When assessing the patient with an acute abdomen, how should you examine the abdomen?

10. Why is body position important to note in your assessment?

11. Why is it important to prevent the patient with an acute abdomen from receiving any pain medications in the field?

www.EMTB.com

Diabetic Emergencies

Objectives

Cognitive

4-4.1 Identify the patient taking diabetic medications with altered mental status and the implications of a history of diabetes. (p 488)

4-4.2 State the steps in the emergency medical care of the patient taking diabetic medicine with an altered mental status and a history of diabetes. (p 487)

4-4.3 Establish the relationship between airway management and the patient with altered mental status. (p 487)

4-4.4 State the generic and trade names, medication forms, dose, administration, action, and contraindications for oral glucose. (p 490)

4-4.5 Evaluate the need for medical direction in the emergency medical care of the diabetic patient. (p 488)

Affective

4-4.6 Explain the rationale for administering oral glucose. (p 489)

Psychomotor

4-4.7 Demonstrate the steps in the emergency medical care for the patient taking diabetic medicine with an altered mental status and a history of diabetes. (p 488)

4-4.8 Demonstrate the steps in the administration of oral glucose. (p 491)

4-4.9 Demonstrate the assessment and documentation of patient response to oral glucose. (p 492)

4-4.10 Demonstrate how to complete a prehospital care report for patients with diabetic emergencies. (p 490)

Additional Objective*

1. Demonstrate the steps in the use of a glucometer. (p 484)

*This is a noncurriculum objective.

You are the Provider

You and your partner are dispatched to 1700 Plaza Way for a 43-year-old man who is "very sweaty and acting strangely." Upon your arrival, a police officer directs you to a man who is sitting on the ground, rocking back and forth. He explains to you that he pulled the man over for weaving in and out of traffic. He originally thought the man was drunk, but now he's concerned it could be a medical problem. As you approach the patient, you introduce yourself and ask him what is wrong. He says, "I need to go home." When you ask him further questions such as his name and today's date, he continues to answer, "I need to go home." You note that he is pale, sweaty, and his hands are trembling.

1. What conditions could mimic a decreased level of consciousness that could be mistaken for alcohol intoxication?

Diabetic Emergencies

Diabetes is a very common disease, affecting about 6% of the population. It is a disorder of glucose metabolism or difficulty metabolizing carbohydrates, fats, and proteins. Without treatment, blood glucose levels become too high and can cause coma and death. If properly treated, most people with diabetes can live a relatively normal life. However, diabetes can have many severe complications that affect the length and quality of life, including blindness, cardiovascular disease, and kidney failure. Also, treatment to lower high blood glucose levels can go too far and cause a life-threatening state of hypoglycemia (low blood glucose). Therefore, as an EMT-B, you need to know the signs and symptoms of a blood glucose level that is either too high or too low so that you can administer the proper lifesaving treatment.

This chapter explains two types of diabetes and how they are controlled, including the role of glucose and insulin. You will learn how to distinguish between hyperglycemia and hypoglycemia, which often resemble each other. The chapter discusses how to identify and treat diabetic emergencies in the prehospital setting. Complications, such as seizures, altered mental status, and heart attack, are also briefly discussed.

Greater than 120, called hyperglycemia

Less than 80 hypoglycemia

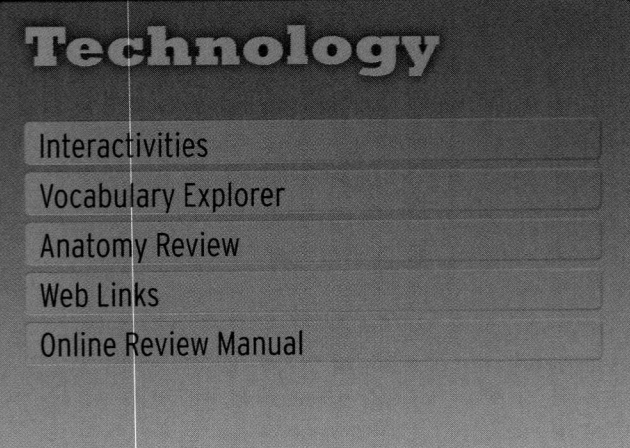

Technology

- Interactivities
- Vocabulary Explorer
- Anatomy Review
- Web Links
- Online Review Manual

www.EMTB.com

Diabetes

Defining Diabetes

Literally, the word "diabetes" means "a passer through; a siphon." Medically, the term refers to a metabolic disorder in which the body's ability to metabolize simple carbohydrates (glucose) is impaired. It is characterized by the passage of large quantities of urine containing glucose, significant thirst, and deterioration of body functions. Glucose, or dextrose, is one of the basic sugars in the body and, along with oxygen, is the primary fuel for cellular metabolism.

The central problem in diabetes is the lack or ineffective action of insulin, a hormone that is normally produced by the endocrine glands on the pancreas that enables glucose to enter the cells. A hormone is a chemical substance produced by a gland that has special regulatory effects on other body organs and tissues. Without insulin, cells begin to "starve" because insulin is needed, like a key, to let glucose into the cells.

The full name of diabetes is diabetes mellitus, which means "sweet diabetes." This refers to the presence of glucose (sugar) in the urine. Diabetes mellitus is considered a metabolic disorder in which the body cannot metabolize glucose, usually because of the lack of insulin; the result is a wasting of glucose in the urine. *Diabetes insipidus*, a rare condition, also involves excessive urination, but here the missing hormone is one that regulates urinary fluid reabsorption. In this book, the term "diabetes" always refers to diabetes mellitus.

Left untreated, diabetes leads to a wasting of body tissues and death. Even with medical care, some patients with particularly aggressive forms of diabetes will die relatively young from one or more complications of the disease. Most patients with diabetes, however, live a normal life span, but they must be willing to adjust their lives to the demands of the disease, especially their eating habits and activities.

Types of Diabetes

Diabetes is a disease with two distinct onset patterns. It may become evident when the patient is a child, or it may develop in later life, usually when the patient is middle-aged.

In type I diabetes, most patients do not produce insulin at all; they are insulin dependent (IDDM). They need daily injections of supplemental, synthetic insulin throughout their lives to control blood glucose. This type generally strikes children as opposed to adults, so in the past it was called "juvenile diabetes." However, it can, in many cases, develop in later life as well. Patients with type I diabetes are more likely to have metabolic problems and organ damage, such as blindness, heart disease, kidney failure, and nerve disorders.

In type II diabetes, which usually appears later in life, patients produce inadequate amounts of insulin or they may produce a normal amount but the insulin does not function effectively. Although some patients with non-insulin-dependent diabetes (NIDD) may require some supplemental insulin, most patients can be treated with diet, exercise, and non-insulin-type oral medications (hypoglycemic agents), such as chlorpropamide (Diabinase), tolbutamide (Orinase), glyburide (Micronase), glipizide (Glucotrol), metformin (Glucophage), and rosiglitazone (Avandia). These medications stimulate the pancreas to produce more insulin and thus lower blood glucose levels. In some cases, these medications can lead to hypoglycemia, particularly when patient activity and exercise levels are too vigorous or excessive. Patients with hypoglycemia have an abnormally low level of blood glucose. Non-insulin-dependent diabetes used to be called adult (maturity)-onset diabetes. Again, some patients with type II diabetes may, in fact, require insulin.

The two types of diabetes are equally serious, although non-insulin-dependent diabetes is easier to regulate. Both can affect many tissues and functions other than the glucose-regulating mechanism. Both require lifelong medical management. Type I diabetes is considered to be an autoimmune problem, in which the body becomes allergic to the insulin-producing cells of the endocrine gland on the pancreas and literally destroys them. The severity of diabetic complications is related to how high the average blood glucose level is and how early in life the diabetes begins.

The Role of Glucose and Insulin

Glucose is the major source of energy for the body, and all cells need it to function properly. Some cells will not function at all without glucose. A constant supply of glucose is as important as oxygen to the brain. Without glucose, or with very low levels, brain cells rapidly suffer permanent damage. With the exception of the brain, insulin is needed to allow glucose to enter individual body cells to fuel their functions. For this reason, insulin is said to be a "cellular key" (Figure 15-1 ▼).

Without insulin, glucose from food remains in the blood and gradually rises to extremely high levels. This condition is called hyperglycemia. Once the blood glucose levels reach 200 mg/dL or more, or twice the usual amount (normal is 80 to 120 mg/dL), excess glucose is excreted by the kidney. This process requires a large amount of water. The loss of water in such large amounts causes the classic symptoms of uncontrolled diabetes, the "3 Ps":

- Polyuria: frequent and plentiful urination
- Polydipsia: frequent drinking of liquid to satisfy continuous thirst (secondary to the loss of so much body water)
- Polyphagia: excessive eating as a result of cellular "hunger"; seen only occasionally

Without glucose to supply energy for cells, the body must turn to other fuel sources. The most abundant is fat. Unfortunately, when fat is used as an immediate energy source, chemicals called *ketones* and *fatty acids* are formed as waste products and are hard for the body to

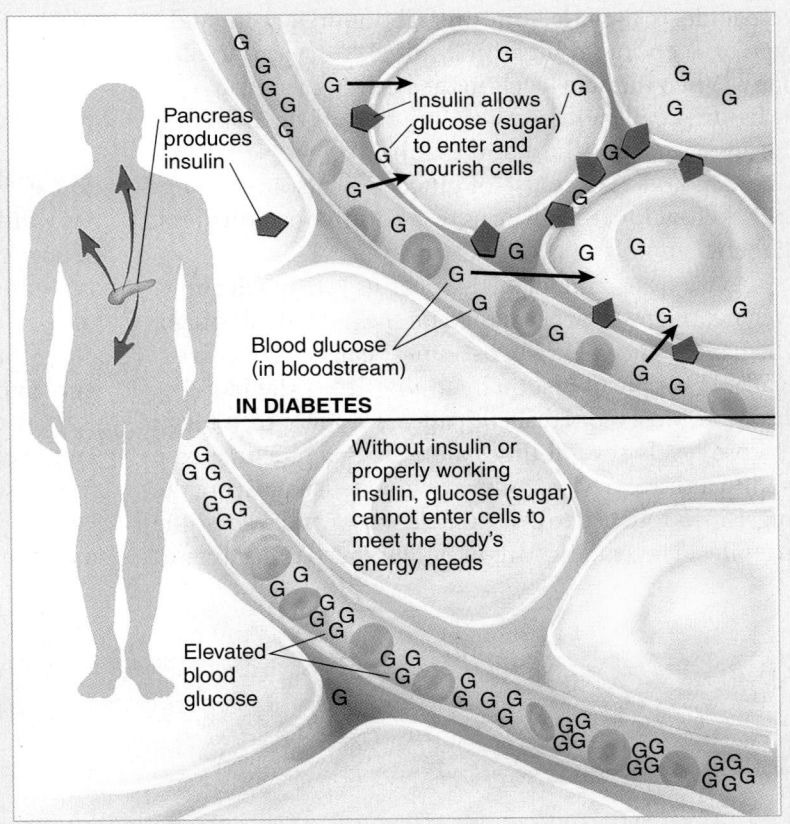

Figure 15-1 Diabetes is defined as a lack of or ineffective action of insulin. Without insulin, cells begin to "starve" because insulin is needed to allow glucose to enter and nourish the cells.

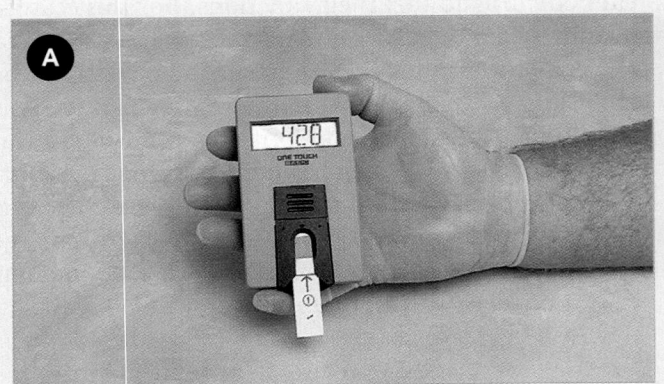
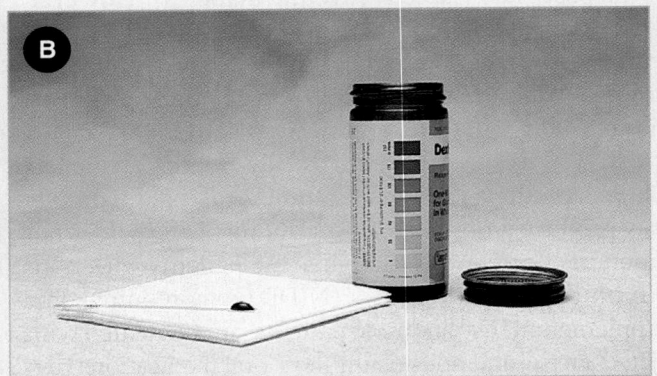

Figure 15-2 A. Blood glucose self-monitoring kit with digital meter is a device used by patients at home or by EMT-Bs in some areas. **B.** Glucose test strips for blood analysis.

excrete. As they accumulate in blood and tissue, certain ketones can produce a dangerous condition called <u>acidosis</u>. The form of acidosis seen in uncontrolled diabetes is called <u>diabetic ketoacidosis (DKA)</u>, in which an accumulation of certain acids occurs when insulin is not available in the body. Signs and symptoms of diabetic ketoacidosis include vomiting, abdominal pain, and a type of deep, rapid breathing called *Kussmaul respirations*. When the acid levels in the body become too high, individual cells will cease to function. If the patient is not given proper fluid and insulin to reverse fat metabolism and restore use of glucose as a source of energy, ketoacidosis will progress to unconsciousness, diabetic coma, and eventually death.

As we have seen, diabetes mellitus is treatable; however, treatment must be tailored for the individual patient. The patient's need for glucose must be balanced with the available supply of insulin by testing either the blood or the urine. Most type I diabetic patients monitor their blood glucose levels several times a day with a glucometer, a credit-card size device. A drop of blood, usually from the fingertip, is touched to a disposable sensor and read by the machine. The readings are in milligrams per deciliter of blood; remember that the normal blood glucose level is between 80 and 120 mg/dL. New measuring devices under development will be worn like a wristwatch or used like a pulse oximeter. EMT-Bs are allowed to use glucometers in some systems across North America (Figure 15-2A ▲). Glucose test strips, in which a drop of blood is placed on a paper strip that changes color, may still be used in some systems (Figure 15-2B ▲). They do not provide the accuracy of glucometers and their readings should be used with caution.

Hyperglycemia and Hypoglycemia

Two different conditions can lead to a diabetic emergency: hyperglycemia and hypoglycemia. <u>Hyperglycemia</u> is a state in which the blood glucose level is above normal. <u>Hypoglycemia</u> is a state in which the blood glucose level is below normal. Extremes of hyperglycemia and hypoglycemia can lead to diabetic emergencies (Figure 15-3 ▶). Ketoacidosis results from prolonged and exceptionally high hyperglycemia. Diabetic coma then results when ketoacidosis is not treated adequately. Hypoglycemia, on the other hand, will progress into unresponsiveness and eventually insulin shock.

You are the Provider   Part 2

As your partner applies oxygen and takes the patient's vital signs, you ready your glucometer.

2. What is one of the simplest ways to discern between hypoglycemia and hyperglycemia?
3. What other indicators can assist you in determining whether this patient is diabetic?

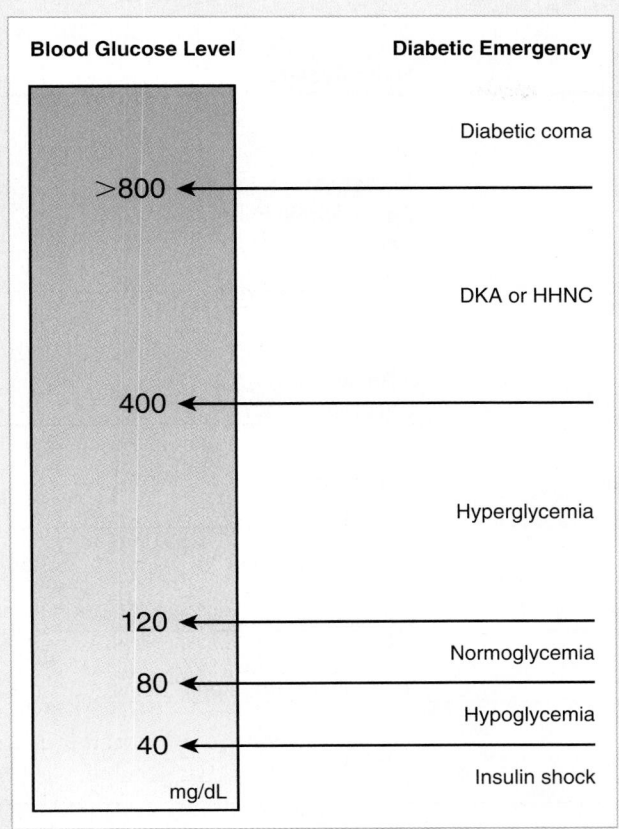

Blood Glucose Level	Diabetic Emergency
>800 ←	Diabetic coma
400 ←	DKA or HHNC
	Hyperglycemia
120 ←	
80 ←	Normoglycemia
40 ←	Hypoglycemia
mg/dL	Insulin shock

Figure 15-3 The two most common diabetic emergencies, ketoacidosis and insulin shock, develop when the patient has either too much or too little glucose in the blood, respectively.

The signs and symptoms of hypoglycemia and hyperglycemia can be quite similar (Table 15-1 ▶). For example, staggering and an intoxicated appearance or complete unresponsiveness are signs and symptoms of both. Note that your assessment of these potential emergencies should not prevent you from providing prompt care and transport as detailed in this chapter. However, in such urgent emergencies, the earlier clues are gathered, the better for the patient. With specific information about the type of emergency, you can help the hospital to prepare prompt, definitive care for the patient.

Diabetic Coma

Diabetic coma is a state of unconsciousness resulting from several problems, including ketoacidosis, dehydration because of excessive urination, and hyperglycemia. Too much blood glucose by itself does not always cause diabetic coma, but on some occasions, it can lead to it.

Diabetic coma may occur in the patient who is not under medical treatment, who takes insufficient insulin, who markedly overeats, or who is undergoing some sort of stress that may involve an infection, illness, overexertion, fatigue, or drinking alcohol. Usually, ketoacidosis develops over a period of time lasting from hours to days. The patient may ultimately be found comatose with the following physical signs:

- Kussmaul respirations
- Dehydration, as indicated by dry, warm skin and sunken eyes
- A sweet or fruity (acetone) odor on the breath, caused by the unusual waste products in the blood (ketones)
- A rapid, weak ("thready") pulse
- A normal or slightly low blood pressure
- Varying degrees of unresponsiveness

Insulin Shock

In insulin shock, the problem is hypoglycemia, insufficient glucose in the blood. When insulin levels remain high, glucose is rapidly taken out of the blood to fuel the cells. If glucose levels get too low, there may be an insufficient amount to supply the brain. If blood glucose remains low, unconsciousness and permanent brain damage can quickly follow.

Insulin shock occurs when the patient has done one of the following:

- Taken too much insulin
- Taken a regular dose of insulin but has not eaten enough food
- Had an unusual amount of activity or vigorous exercise and used up all available glucose

Insulin shock may also occur after the patient vomits a meal after he or she took a regular dose of insulin. At times, insulin shock may occur with no identifiable predisposing factor.

Children who have diabetes may pose a particular management problem. First, their high levels of activity mean that they can use up circulating glucose more quickly than adults do, even after a normal insulin injection. Second, they do not always eat correctly and on schedule. As a result, insulin shock can develop more often and more severely in children than in adults.

Insulin shock develops much more quickly than diabetic coma. In some instances, it can occur in a matter of minutes. Hypoglycemia can be associated with the following signs and symptoms:

- Normal or rapid respirations
- Pale, moist (clammy) skin
- Diaphoresis (sweating)
- Dizziness, headache
- Rapid pulse
- Normal to low blood pressure

Table 15-1 Characteristics of Diabetic Emergencies

	Hyperglycemia	Hypoglycemia
History		
Food intake	Excessive	Insufficient
Insulin dosage	Insufficient	Excessive
Onset	Gradual (hours to days)	Rapid, within minutes
Skin	Warm and dry	Pale and moist
Infection	Common	Uncommon
Gastrointestinal Tract		
Thirst	Intense	Absent
Hunger	Absent	Intense
Vomiting	Common	Uncommon
Respiratory System		
Breathing	Rapid, deep (Kussmaul respirations)	Normal or rapid
Odor of breath	Sweet, fruity	Normal
Cardiovascular System		
Blood pressure	Normal to low	Low
Pulse	Normal or rapid and full	Rapid, weak
Nervous System		
Consciousness	Restless merging to coma	Irritability, confusion, seizure, or coma
Urine		
Sugar	Present	Absent
Acetone	Present	Absent
Treatment		
Response	Gradual, within 6 to 12 hours following medical treatment	Immediately after administration of glucose

- Altered mental status (aggressive, confused, lethargic, or unusual behavior)
- Anxious or combative behavior
- Hunger
- Seizure, fainting, or coma
- Weakness on one side of the body (may mimic stroke)

Both extremes of diabetic coma and insulin shock produce unconsciousness and, in some instances, death. But they call for very different treatment. Diabetic coma is a complex metabolic condition that usually develops over time and involves all the tissues of the body. Correcting this condition may take many hours in a well-controlled hospital setting. Insulin shock, however, is an acute condition that can develop rapidly. A patient with diabetes who has taken his or her standard insulin dose and missed lunch may be in insulin shock before dinner. The condition is just as quickly reversed

You are the Provider   Part 3

As you assemble your equipment, your partner informs you that the patient is wearing a medic alert tag that indicates he is an insulin-dependent diabetic.

4. What are some important considerations to take into account when performing a head-to-toe assessment on a patient with type I diabetes, and how can what you find impact your care?

by giving the patient glucose. Without that glucose, however, the patient will suffer permanent brain damage. Minutes count.

Most individuals with diabetes understand and manage their disease well. Still, emergencies occur. In addition to diabetic coma and insulin shock, patients with diabetes may have "silent," or painless, heart attacks, a possibility that you should always consider. Their only symptom may be "not feeling so well."

Assessment of the Diabetic Patient

Scene Size-up

Although your report from dispatch may be for a patient with an altered mental status, keep open the possibility that trauma may have occurred because of a medical incident. BSI precautions should consist of gloves and an eye shield at a minimum. Remember to evaluate each situation quickly and make sure the necessary BSI equipment is readily available.

Do not let your guard down even on what appears to be a routine call. Evaluate scene safety as you arrive on scene and as you approach the patient. Remember that diabetic patients often use syringes to administer insulin. It is possible you may be stuck by a used needle that was not disposed of properly. Insulin syringes on the bed stand, insulin bottles in the refrigerator, a plate of food, or glass of orange juice are important clues that may help you decide what is possibly wrong with your patient. Question bystanders on events of the situation as you approach. Determine whether this is your only patient, the nature of the illness, and whether there was trauma involved. Decide whether you will need any additional resources. Perform cervical spine stabilization if necessary.

Initial Assessment

General Impression

Form a general impression of the patient. Does the patient appear anxious, restless, or listless? Is the patient apathetic or irritable? Is the patient interacting with his or her environment appropriately? These initial observations may help you to suspect high or low blood glucose values. Determine the patient's level of consciousness. If the patient is conscious, what is the chief complaint? If a suspected diabetic patient is unresponsive, call for ALS immediately.

Airway and Breathing

As you are forming your general impression, assess the patient's airway and breathing. Patients showing signs of inadequate breathing or altered mental status should receive high-flow oxygen at 10 to 15 L/min via nonrebreathing mask. A patient in a diabetic coma who is hyperglycemic may have rapid, deep respirations (Kussmaul respirations) and sweet, fruity breath. A patient in insulin shock who is hypoglycemic will have normal to rapid respirations. If the patient is not breathing or is having difficulty breathing, open the airway, give oxygen, and assist ventilations. Continue to monitor the airway as you provide care.

Circulation

Once you have assessed airway and breathing and have performed the necessary interventions, check the patient's circulatory status. A patient with dry and warm skin indicates diabetic coma, whereas a patient with moist and pale skin indicates insulin shock. The patient in insulin shock will have a rapid, weak pulse.

Transport Decision

Whether you decide to transport will depend on the patient's level of consciousness and ability to swallow. Patients with an altered mental status and impaired ability to swallow should be transported promptly. Patients who have the ability to swallow and are conscious enough to maintain their own airway may be further evaluated on scene and interventions performed.

Focused History and Physical Exam

Assess unresponsive medical patients first from head to toe with a rapid physical exam, looking for clues as to their condition. The patient may have suffered trauma resulting from changes in level of consciousness or dizziness. Next obtain the patient's vital signs and assess history. Remember that the environment, bystanders, and medical identification symbol may provide important clues about your patient's condition.

Responsive medical patients are able to provide their own medical history to help you identify a cause for their altered mental status. Next perform a focused physical exam and obtain your patient's baseline vital signs.

SAMPLE History

Ask the following questions of a known diabetic patient in addition to obtaining a SAMPLE history:

- Do you take insulin or any pills that lower your blood sugar?
- Have you taken your usual dose of insulin (or pills) today?
- Have you eaten normally today?
- Have you had any illness, unusual amount of activity, or stress today?

If the patient has eaten but has not taken insulin, it is more likely that diabetic ketoacidosis is developing. If the patient has taken insulin but has not eaten, the problem is more likely to be insulin shock. A patient with diabetes will often know what is wrong. If the patient is not thinking or speaking clearly (or is unconscious), ask a family member or bystander the same questions.

When you are assessing a patient who might have diabetes, check to see whether he or she has an emergency medical identification symbol—a wallet card, necklace, or bracelet—or ask the patient or a family member. Remember, however, that even though a person has diabetes, the diabetes may not be causing the current problem. A heart attack, stroke, or other medical emergency may be the cause. For this reason, you must always do a thorough, careful assessment, paying attention to the ABCs. Inform medical control that you are at the scene of a diabetic emergency.

Focused Physical Exam

When you suspect a diabetes-related problem, the focused physical exam should focus on the patient's mental status and ability to swallow and protect the airway. Obtain a Glasgow Coma Scale score to track the patient's mental ability. Otherwise, physical signs such as tremors, abdominal cramps, vomiting, a fruity breath odor, or a dry mouth may guide you in determining whether the patient is hypoglycemic or hyperglycemic.

Baseline Vital Signs

Obtain a complete set of vital signs, including a measurement of the patient's blood glucose level using a glucometer or chemical test strips if available. In hypoglycemia respirations are normal to rapid, pulse is weak and rapid, and skin is typically pale and clammy with a low blood pressure. In hyperglycemia respirations are deep and rapid, pulse is full and bounding at a normal or fast rate, and skin is warm and dry with a normal blood pressure. At times the blood pressure may be low. It is easy to identify abnormal vital signs when we know the blood glucose level is too high or too low. Remember, the patient may have abnormal vital signs and a normal blood glucose value. When this is the case, something else may be causing the patient's altered mental status, vomiting, or other complaints.

Interventions

If your patient is conscious and able to swallow without the risk of aspiration, you should encourage him or her to drink juice or milk or other drinks that contain sugar. If you are permitted by local protocol, you may also administer a highly concentrated sugar gel, squirted between the patient's cheek and gums or placed between the cheek and gum on a tongue depressor. The patient will usually become more alert within minutes.

If your patient is unconscious, or if there is any risk of aspiration, the patient will need IV glucose, which you are not authorized to give. Your responsibility is to provide prompt transport to the hospital, where the proper care can be given. If you are working in a tiered system, EMT-Intermediates and paramedics are able to start an IV with glucose.

If no one else is present and you know that the unconscious patient has diabetes, you must use your knowledge of the signs and symptoms to decide whether the problem is diabetic coma or insulin shock. Remember, however, this assessment should not prevent you from providing prompt treatment and transport. The primary visible difference will be the patient's breathing—deep, sighing respirations in diabetic coma and normal or rapid respirations in insulin shock. The patient with diabetes who is unconscious and having convulsions is more likely to be in insulin shock.

✚ EMT-B Tips

Before you give a conscious patient anything to drink or administer instant glucose, you must ensure that there is no danger of aspiration. One rule of thumb: if patients can lift the cup or squirt the glucose into their own mouths, they are probably not in danger of aspiration. Watch them carefully!

Keep in mind that any unconscious patient may have undiagnosed diabetes. In patients with an altered mental status, you may be able to determine this in the field, if you have the proper equipment to test for blood glucose levels. Without this critical knowledge, treat this patient as you would any other unconscious individual. Provide emergency medical care, particularly airway management, and provide prompt transport. At the emergency department, diabetes and its complications can be diagnosed quickly.

A patient in insulin shock (rapid onset of altered mental status, hypoglycemia) needs sugar immediately, and a patient in diabetic coma (acidosis, dehydration, hyperglycemia) needs insulin and IV fluid therapy. These patients need prompt transport to the hospital for appropriate medical care.

For the conscious patient in insulin shock, protocols usually recommend oral glucose. Glucose will usually reverse the reaction within several minutes. Do not be afraid to give too much sugar. The problem often will not be solved with just a sip of juice. An entire candy bar or a full glass of sweetened juice is often needed. Do not give sugar-free drinks that are sweetened with saccharin or other synthetic sweetening compounds, as they will have little or no effect. Remember that even if the patient responds after receiving glucose, he or she may still need additional treatment. Therefore, you must transport the patient to the hospital as soon as possible.

When there is any doubt about whether a conscious patient with diabetes is going into insulin shock or diabetic coma, most protocols will err on the side of giving glucose, even though the patient may have diabetic ketoacidosis. Untreated insulin shock will result in loss of consciousness and can quickly cause significant brain damage or death. The condition of a patient in insulin shock is far more critical and far more likely to cause permanent problems when compared to a patient with diabetic ketoacidosis. Furthermore, the amount of sugar that is typically given to a patient in insulin shock is very unlikely to make a patient in diabetic ketoacidosis significantly worse. When in doubt, consult medical control.

Detailed Physical Exam

As in every call, you should perform a detailed physical exam when time permits. With unconscious patients or patients with an altered mental status, you must play detective and look for problems or injuries that are not obvious because the patient is unable to communicate these to you. Although an altered mental status may be caused by a blood glucose level that is too high or too low, the patient may have sustained trauma or suffer from another metabolic problem. The altered mental status may also be caused by something else, such as intoxication, poisoning, or a head injury. A careful physical exam may provide you with information essential to proper patient care.

EMT-B Safety

Managing problems related to diabetes and altered mental status poses very little risk to you, because exposure to body fluids is generally very limited. However, some patients can become confused and even aggressive at times. Follow BSI precautions, as you would with any other patient. Always use gloves and wash your hands carefully after obtaining and checking a blood sample or performing airway techniques.

Geriatric Needs

You might encounter a geriatric patient who has undiagnosed diabetes. These patients report that they have not been feeling well for a while but have not seen a physician. A patient with undiagnosed diabetes or one who is in denial or ignores the advice of his or her physician may call 9-1-1 when the signs and symptoms become annoying. Nonhealing wounds, blindness, renal failure, and other complications are associated with poorly controlled or uncontrolled diabetes. As an EMT-B, you may be the first to recognize and suggest medical treatment to a geriatric patient who might otherwise ignore his or her condition. It is important that you recognize the signs and symptoms of diabetes.

It is important to reevaluate the diabetic patient frequently to assess changes. Is there an improvement in the patient's mental status? Are the ABCs still intact? How is the patient responding to the interventions performed? How must you adjust or change the interventions? In many patients with diabetes you will note marked improvement with appropriate treatment. Document each assessment, your findings, the time of the interventions, and any changes in the patient's condition. Base your glucose administration on serial readings if you have access to a glucometer. If a glucometer is unavailable, a deteriorating level of consciousness indicates that you need to provide more glucose. Again, the use of glucometers and the administration of glucose will be based on your service's protocols and standing orders.

Communication and Documentation

Determining whether the blood glucose level is too high or too low in a known diabetic patient can be difficult when signs and symptoms are confusing and you have no way to test the actual blood glucose value. In these situations perform a thorough assessment and contact the hospital to help sort out the signs and symptoms. The hospital should be a resource for you to help problem solve situations and provide guidance on how to manage your patient.

Your run report is the only legal document you have to say that appropriate care was provided. Document clearly your assessment findings as the basis for your treatment. Patients who refuse transport because you "cured" them with your oral glucose will require approval from the hospital via radio and even more thorough documentation. Follow your local protocols for patients who refuse treatment or transport.

Emergency Care of Diabetic Emergencies

Giving Oral Glucose

Oral glucose is a commercially available gel that dissolves when placed in the mouth (Figure 15-4 ▼). One toothpaste-type tube of gel equals one dose. Trade names for the gel include Glutose and Insta-Glucose. Glucose gel acts to increase a patient's blood glucose levels. If authorized by your system, you should administer glucose gel to any patient with a decreased level of consciousness who has a history of diabetes. The only contraindications to glucose are an inability to swallow or unconsciousness, because aspiration (inhalation of the substance) can occur. Oral glucose itself has no side ef-

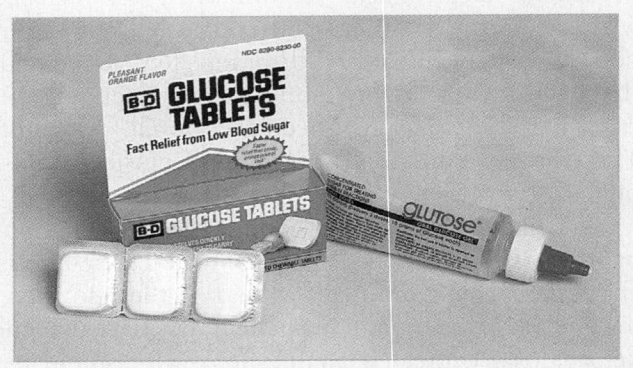

Figure 15-4 Oral glucose is commercially available in gel and tablet form. One tube of gel equals one dose.

You are the Provider   Part 4

You find no insulin pump during your assessment, but the patient's blood glucose level is 45 mg/dL. In anticipation of a low blood glucose reading, you had already prepared the tube of oral glucose. This patient, although confused, is alert enough to swallow and self-administer the entire tube of glucose as per your instructions.

5. If this patient were not alert enough to self-administer the glucose, how would you manage this patient?

Administering Glucose

1 Make sure that the tube of glucose is intact and has not expired.

2 Squeeze a generous amount of oral glucose onto the bottom third of a bite stick or tongue depressor.

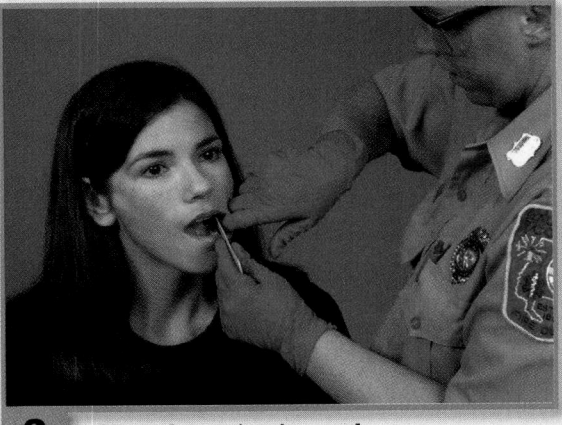

3 Open the patient's mouth.

Place the tongue depressor on the mucous membranes between the cheek and the gum with the gel side next to the cheek.

Repeat until the entire tube has been used.

fects if it is administered properly; however, the risk of aspiration in a patient who does not have a gag reflex can be dangerous. A conscious patient (even if confused) who does not really need glucose will not be harmed by it. Therefore, do not hesitate to give it under these circumstances.

As always, be sure to wear gloves before placing anything into a patient's mouth. After you have confirmed that the patient is conscious and able to swallow, and have obtained an online or off-line order, follow these steps to administer oral glucose (Skill Drill 15-1 ▲):

1. **Examine the tube** to ensure that it is not open or broken. Check the expiration date (**Step 1**).
2. **Squeeze a generous amount** onto the bottom third of a bite stick or tongue depressor (**Step 2**).
3. **Open the patient's mouth**.
4. **Place the tongue depressor** on the mucous membranes between the cheek and gum, with the gel side next to the cheek (**Step 3**). Once the gel is dissolved, or if the patient loses consciousness or has a seizure, remove the tongue depressor. Repeat until the entire tube has been used. Note that the patient should not swallow the glucose; it acts more quickly when dissolved in the mouth.

✖ Reassess the patient's condition regularly after giving glucose, even if you see rapid improvement. Watch for airway problems, sudden loss of consciousness, or seizures. Provide prompt transport to the hospital; do not delay transport just to give additional oral glucose.

Complications of Diabetes

Medical Complications

Diabetes is a systemic disease affecting all tissues of the body, especially the kidneys, eyes, small arteries, and peripheral nerves. Therefore, you are likely to be called to treat patients with a variety of complications of diabetes, such as heart disease, visual disturbances, renal failure, stroke, and ulcers or infections of the feet or toes. With the exception of heart attack and stroke, most of these will not be acute emergencies. Considering that diabetes is a major risk factor for cardiovascular disease, individuals with diabetes should always be suspected of having a potential for heart attack, particularly older patients, even when they do not present with classic symptoms such as chest pain.

Associated Problems

Conditions associated with diabetes include seizures, altered mental status, and airway problems. Remember to consider diabetic emergencies in patients present with these emergencies.

Seizures

Although seizures are rarely life threatening, you should consider them very serious, even in patients with a history of chronic seizures. Seizures, which may be brief or prolonged, are caused by fever, infections, poisoning, hypoglycemia, trauma, or decreased levels of oxygen. They can also be idiopathic (of unknown cause) in children. Although brief seizures are not harmful, they may indicate a more dangerous and potentially life-threatening underlying condition. Because seizures can be caused by a head injury, consider trauma as a cause. In the patient with diabetes, you should also consider hypoglycemia.

Emergency medical care of seizures includes ensuring that the airway is clear and placing the patient on his or her side if there is no possibility of cervical spine trauma. Do not attempt to place anything in the patient's mouth (eg, a bite stick or oral airway). Be sure to have suctioning equipment ready in case the patient vomits. Provide artificial ventilation if the patient is cyanotic or appears to be breathing inadequately, and provide prompt transport.

Altered Mental Status

Although altered mental status is often caused by complications of diabetes, it may also be caused by a variety of conditions, including poisoning, part of the postseizure state, infection, head injury, and decreased levels of oxygen. In diabetes, altered mental status can be caused by hypoglycemia or ketoacidosis.

Begin emergency medical care of altered mental status by ensuring that the airway is clear. Be prepared to provide artificial ventilation and suctioning in case the patient vomits, and provide prompt transport.

Alcoholism

Occasionally, patients in insulin shock or a diabetic coma are thought to be intoxicated, especially if their condition has caused a motor vehicle crash or other incident. Confined by police in a "drunk tank," a patient with diabetes is at risk. In such situations, an emergency medical identification bracelet, necklace, or card

You are the Provider   Part 5

As the patient becomes more alert, he tells you that he was driving home to eat because he realized his blood sugar level was dropping. After a few minutes pass, he is fully alert and refuses to be transported to the hospital. You remind him to eat a meal high in carbohydrates as soon as possible, and a friend who responded to the scene agrees to drive him home and will immediately prepare a substantial meal for him.

6. Is it okay to let this patient return home? What should you do before this patient leaves?

may help to save the patient's life. Often, only a blood glucose test performed at the scene or in the emergency department will identify the real problem. In some EMS systems, you will be trained and allowed to perform the blood glucose test at the scene. Otherwise, you must always suspect hypoglycemia in any patient with altered mental status.

Certainly, diabetes and alcoholism can coexist in a patient. But you must be alert to the similarity in symptoms of acute alcohol intoxication and diabetic emergencies. Likewise, hypoglycemia and a head injury can coexist, and you must appreciate the potential even when the head injury is obvious.

Relationship to Airway Management

Patients with altered mental status, particularly those who are difficult to awaken, are at risk for losing their gag reflex. When the gag reflex is not working, patients cannot reject foreign materials in their mouth (including vomit), and their tongues will often relax and obstruct the airway. Therefore, you must carefully monitor the airway in patients with hypoglycemia, diabetic coma, or a diabetic complication such as stroke or seizure. Place the patient in a lateral recumbent position, and make sure suction is readily available.

You are the Provider   Summary

Calls involving diabetic emergencies, specifically those related to insulin shock, are one of the most common responses you will encounter in the field. Hypoglycemia can cause a myriad of behavioral responses including but not limited to confusion, irritability, anxiousness, and combativeness. One of the simplest factors in determining whether the patient has hypoglycemia or hyperglycemia is onset. Hypoglycemia usually manifests itself very quickly (over minutes), whereas hyperglycemia can take days or weeks to develop. It is common for people who have serious conditions such as diabetes to wear medic alert tags, typically as a bracelet or necklace. They may have insulin pumps designed to provide a continual infusion of insulin, causing blood glucose levels to continue to drop. Some pumps are obvious and are worn on the hip, while others look similar to a pager and can be carried less obviously on the inside of a shoe or boot.

It is also important to note that some responses that involve trauma (such as motor vehicle crashes) can occur as the result of a medical condition. If you cannot find a reasonable explanation for why a crash, fall, or other injury happened, begin to question the presence of a condition such as hypoglycemia, stroke, heart attack, or seizure.

Assessment and Emergency Care

Diabetic Emergencies

Scene Size-up	Body substance isolation precautions should include a minimum of gloves and eye protection. Ensure scene safety and determine NOI/MOI. Consider the number of patients, the need for additional help/ALS, and c-spine stabilization.
Initial Assessment	
■ General impression	Determine level of consciousness and find and treat any immediate threats to life. Determine priority of care based on environment and patient's chief complaint.
■ Airway	Ensure patent airway.
■ Breathing	Provide high-flow oxygen at 15 L/min. Evaluate depth and rate of the respirations and provide ventilations as needed.
■ Circulation	Evaluate pulse rate and quality; observe skin color, temperature, and condition and treat accordingly. Determine if bleeding is present and control if life threatening.
■ Transport decision	Rapid transport
Focused History and Physical Exam	*NOTE: The order of the steps in the focused history and physical exam differs depending on whether the patient is conscious or unconscious. The order below is for a conscious patient. For an unconscious patient, perform a rapid physical exam, obtain vital signs, and obtain the history.*
■ SAMPLE history	Ask pertinent SAMPLE and OPQRST questions. Be sure to ask if and what interventions were taken before your arrival, how many interventions, and at what time.
■ Focused physical exam	Perform focused physical exam and ascertain if patient has taken any insulin or pills for diabetes. Has the patient been compliant with his or her diet and medications? Learn of any recent illness, physical activity, or stress. Determine the blood glucose level and Glasgow Coma Scale score.
■ Baseline vital signs	Take vital signs, noting skin color and temperature as well as patient's level of consciousness. Use pulse oximetry if available.
■ Interventions	A conscious patient who is able to swallow can be given fluids with a high sugar content, or a highly concentrated sugar gel as protocols allow.
Detailed Physical Exam	Complete a detailed physical exam.
Ongoing Assessment	Repeat the initial assessment and focused assessment, and reassess interventions performed. Reassess vital signs and blood glucose levels every 5 minutes for the unstable patient, every 15 minutes for the stable patient. Reassure and calm the patient.
■ Communication and documentation	Contact medical control with a radio report, *informing patient's condition* and blood glucose level(s). Relay any change in level of consciousness or difficulty breathing. Be sure to document any changes, at what time, and blood glucose readings.

NOTE: While the steps below are widely accepted, be sure to consult and follow your local protocol.

Diabetic Emergencies

Administering Glucose

1. Examine the tube to ensure that it is not open or broken. Check the expiration date.
2. Squeeze a generous amount onto the bottom third of a bite stick or tongue depressor.
3. Open the patient's mouth. Place the tongue depressor on the mucous membranes between the cheek and gum, with the gel side next to the cheek.

Prep Kit

Ready for Review

- Diabetes is a disorder of glucose metabolism or difficulty metabolizing carbohydrates, fats, and proteins.
- Diabetes is typically characterized by excessive urination and resulting thirst, along with deterioration of body tissues.
- There are two types of diabetes. Type I diabetes, or insulin-dependent diabetes, usually starts in childhood and requires daily insulin to control blood glucose. Type II diabetes, or non-insulin-dependent diabetes, usually develops in middle age and often can be controlled with diet and oral medications.
- Both types of diabetes are serious systemic diseases, especially affecting the kidneys, eyes, small arteries, and peripheral nerves.
- Patients with diabetes have chronic complications that place them at risk for other diseases such as heart attack, stroke, and infections. Most often, however, you will be called upon to treat the acute complications of blood glucose imbalance. These include hyperglycemia (excess blood glucose) and hypoglycemia (not enough blood glucose).
- Symptoms of hypoglycemia classically include confusion; rapid respirations; pale, moist skin; diaphoresis; dizziness; fainting; and even coma and seizures. This condition, called insulin shock, is rapidly reversible with the administration of glucose or sugar. Without treatment, however, permanent brain damage and death can occur.
- Hyperglycemia is usually associated with dehydration and ketoacidosis. It can result in diabetic coma, marked by rapid (often deep) respirations; warm, dry skin; a weak pulse; and a fruity breath odor. Hyperglycemia must be treated in the hospital with insulin and IV fluids.
- Because a blood glucose level that is either too high or too low can result in altered mental status, you must perform a thorough history and patient assessment. When you cannot determine the nature of the problem, it is best to treat the patient for hypoglycemia.
- Be prepared to give oral glucose to a conscious patient who is confused or has a slightly decreased level of consciousness; however, do not give oral glucose to a patient who is unconscious or otherwise unable to swallow properly or protect his or her own airway.
- Remember, in all cases, providing emergency medical care and prompt transport is your primary responsibility.

Technology

- Interactivities
- Vocabulary Explorer
- Anatomy Review
- Web Links
- Online Review Manual

Vital Vocabulary

acidosis A pathologic condition resulting from the accumulation of acids in the body.

diabetes mellitus A metabolic disorder in which the ability to metabolize carbohydrates (sugars) is impaired, usually because of a lack of insulin.

diabetic coma Unconsciousness caused by dehydration, very high blood glucose levels, and acidosis in diabetes.

diabetic ketoacidosis (DKA) A form of acidosis in uncontrolled diabetes in which certain acids accumulate when insulin is not available.

glucose One of the basic sugars; it is the primary fuel, along with oxygen, for cellular metabolism.

hormone A chemical substance that regulates the activity of body organs and tissues; produced by a gland.

hyperglycemia Abnormally high glucose level in the blood.

hypoglycemia Abnormally low glucose level in the blood.

www.EMTB.com

Prep Kit continued...

insulin A hormone produced by the Islets of Langerhans (an exocrine gland on the pancreas) that enables glucose in the blood to enter the cells of the body; used in synthetic form to treat and control diabetes mellitus.

insulin shock Unconsciousness or altered mental status in a patient with diabetes, caused by significant hypoglycemia; usually the result of excessive exercise and activity or failure to eat after a routine dose of insulin.

Kussmaul respirations Deep, rapid breathing; usually the result of an accumulation of certain acids when insulin is not available in the body.

polydipsia Excessive thirst persisting for long periods of time despite reasonable fluid intake; often the result of excessive urination.

polyphagia Excessive eating; in diabetes, the inability to use glucose properly can cause a sense of hunger.

polyuria The passage of an unusually large volume of urine in a given period; in diabetes, this can result from wasting of glucose in the urine.

type I diabetes The type of diabetic disease that usually starts in childhood and requires insulin for proper treatment and control.

type II diabetes The type of diabetic disease that usually starts in later life and often can be controlled through diet and oral medications.

Points to Ponder

You have been called to the local police station to check out a patient under arrest. Upon your arrival the police officer states that the patient was placed under arrest for drunk driving. The arrest was made over 2 hours ago; however, in the past 10 minutes the patient began to act "strangely."

What potential medical conditions could this patient be experiencing? How can you determine which of these potential conditions is the cause?

Issues: Working With Other Public Safety Responders, Performing a Thorough Assessment on All Patients, Potential Causes of Altered Mental Status, Consent and Refusal.

Assessment in Action

You are in front of the station completing the daily maintenance on your ambulance when a car pulls up. The driver yells out the window, "Please, help my husband—he is really sick." You see a man slumped over in the back seat. You can hear loud snoring respirations. You complete your general impression. The patient is moved to the stretcher and you perform an initial assessment. His airway is opened by inserting an oropharyngeal airway. The respiratory rate is 40 breaths/min and very deep (Kussmaul respirations). As you place the nonrebreathing mask on your patient, you also notice a strong acetone odor with every exhalation. The pulse rate is very rapid and thready at 110 beats/min.

As your partner completes the rapid medical exam, you obtain the SAMPLE history from the patient's wife. The patient has a history of diabetes mellitus and has not been feeling well for the past 5 days. The patient was on his way to a scheduled doctor's appointment when he slumped over. Further information obtained confirms that the patient takes insulin shots two times a day. Because the patient was not feeling well, he might have skipped some injections. You call for ALS assistance and prepare the patient for transport.

1. The assessment leads you to believe the patient is experiencing:
   A. dehydration.
   B. diabetic coma.
   C. sugar compensation.
   D. glucose distress.

2. When checking the blood glucose level on this patient, the glucometer would report:
   A. hyperglycemia.
   B. hypoglycemia.
   C. hypertension.
   D. hypotension.

3. Without glucose in the cells to produce energy, the body uses fatty acids to make energy, thus producing a dangerous condition called:
   A. cellular edema (swelling).
   B. diabetic ketoacidosis (DKA).
   C. metabolic alkalosis.
   D. sugar overload.

4. The body produces a hormone to help move glucose from the bloodstream to the cells. This hormone is called:
   A. adrenalin.
   B. insulin.
   C. the growth hormone.
   D. the antidiuretic hormone.

5. Adult-onset diabetes, often first treated with diet, exercise, and medications, is classified as:
   A. juvenile diabetes.
   B. nonserious.
   C. type I diabetes.
   D. type II diabetes.

6. When reading a glucometer, the normal range is:
   A. 18 to 20 mg/dL.
   B. 550 to 650 mg/dL.
   C. 80 to 120 mg/dL.
   D. 900 to 2,000 mg/dL.

7. A patient with diabetes who has taken too much insulin or has not eaten enough food may experience:
   A. insulin shock.
   B. hyperglycemia
   C. hypotension.
   D. diabetic coma.

8. Insulin shock can quickly be corrected by giving the patient:
   A. epinephrine.
   B. glucose.
   C. oxygen.
   D. water.

Challenging Questions

9. What questions should be asked of the patient with diabetes when completing the SAMPLE history?

10. What are the three steps in administering glucose to the diabetic patient?

11. Diabetes often leads to altered mental status. What are some of the other causes of an altered mental status?

12. What are the medical complications of diabetes?

www.EMTB.com

Allergic Reactions and Envenomations

Objectives

Cognitive

4-5.1   Recognize the patient experiencing an allergic reaction. (p 500)

4-5.2   Describe the emergency medical care of the patient with an allergic reaction. (p 507)

4-5.3   Establish the relationship between the patient with an allergic reaction and airway management. (p 504)

4-5.4   Describe the mechanisms of allergic response and the implications for airway management. (p 504)

4-5.5   State the generic and trade names, medication forms, dose, administration, action, and contraindications for the epinephrine auto-injector. (p 505)

4-5.6   Evaluate the need for medical direction in the emergency medical care of the patient with an allergic reaction. (p 507)

4-5.7   Differentiate between the general category of those patients having an allergic reaction, and those patients having an allergic reaction and requiring immediate medical care, including immediate use of an epinephrine auto-injector. (p 506)

Affective

4-5.8   Explain the rationale for administering epinephrine using an auto-injector. (p 507)

Psychomotor

4-5.9   Demonstrate the emergency medical care of the patient experiencing an allergic reaction. (p 507)

4-5.10   Demonstrate the use of an epinephrine auto-injector. (p 507)

4-5.11   Demonstrate the assessment and documentation of patient response to an epinephrine injection. (p 507)

4-5.12   Demonstrate proper disposal of equipment. (p 508)

4-5.13   Demonstrate completing a prehospital care report for patients with allergic emergencies. (p 507)

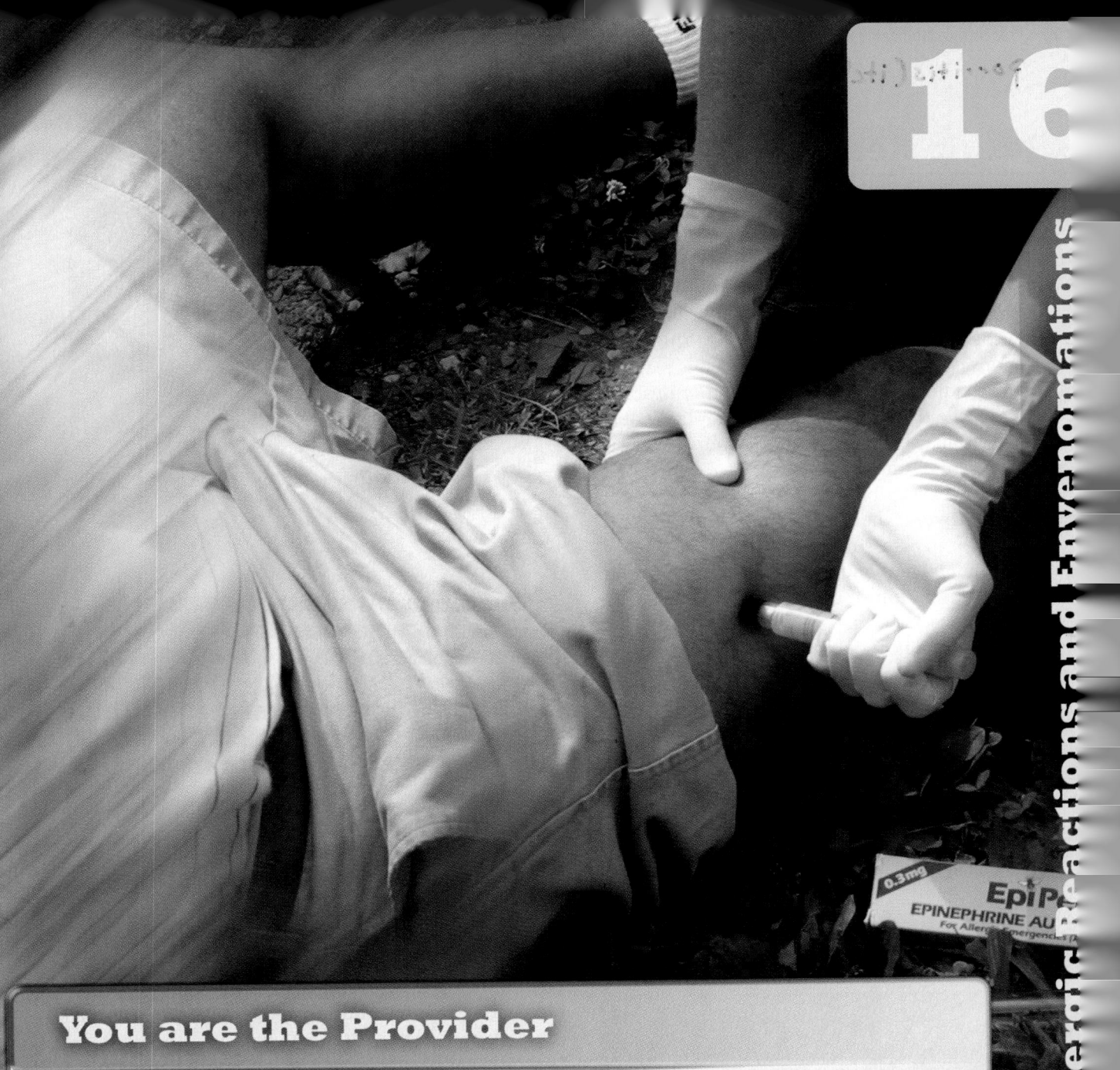

You are the Provider

You and your partner are dispatched to Pioneer Park, a local recreational area where residents frequently picnic and enjoy a variety of outdoor festivities, especially during the spring and summer months. The nature of the call is "possible allergic reaction" for a 25-year-old man.

 You arrive to find a crowd of onlookers who frantically motion for you. As you move through the crowd, you see a man sitting on the ground. He appears to be very anxious and has hives all over his chest and arms. In fragmented sentences, he explains that was playing Frisbee when something stung him on his back. He also says that it's hard to breathe and he feels very dizzy. You see that his breathing is very labored and note the presence of wheezes.

1. How can you discern between a mild or moderate allergic reaction and anaphylaxis?
2. Assuming this person is having a severe, life-threatening allergic reaction, what would you expect his vital signs to indicate? Given his presentation, would you call for other resources?

porritis (itching)

Allergic Reactions

Every year, at least 1,000 Americans die of acute allergic reactions. In dealing with allergy-related emergencies, you must be aware of the possibility of acute airway obstruction and cardiovascular collapse and be prepared to treat these life-threatening complications. You must also be able to distinguish between the body's usual response to a sting or bite and an allergic reaction, which may require epinephrine. Your ability to recognize and manage the many signs and symptoms of allergic reactions may be the only thing standing between a patient's life and imminent death.

This chapter describes the five categories of stimuli that may provoke allergic reactions. You will learn what to look for in assessing patients who may be having an allergic reaction and how to care for them, including administration of epinephrine. The chapter then describes insect bites.

Contrary to what many people think, an <u>allergic reaction</u>, an exaggerated immune response to any substance, is not caused directly by an outside stimulus, such as a bite or sting. Rather, it is a reaction by the body's immune system, which releases chemicals to combat the stimulus. Among these chemicals are <u>histamines</u> and <u>leukotrienes</u>. An allergic reaction may be mild and local, involving hives, itching, or tenderness, or it may be severe and systemic, resulting in shock and respiratory failure.

www.EMTB.com

Technology

Interactivities

Vocabulary Explorer

Anatomy Review

Web Links

Online Review Manual

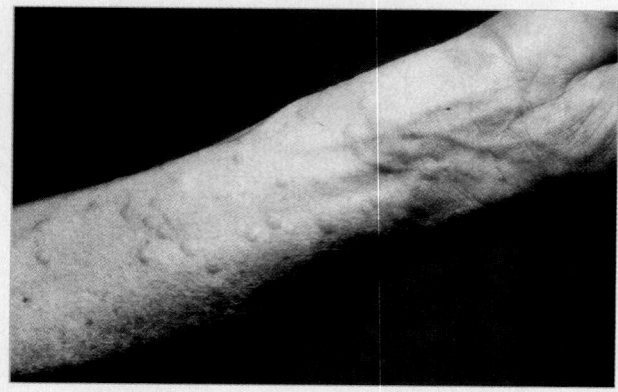

Figure 16-1 Urticaria, or hives, may appear following a sting and are characterized by multiple, small, raised areas on the skin. Urticaria may be one of the warning signs of impending anaphylactic reaction.

<u>Anaphylaxis</u> is an extreme allergic reaction that is usually life threatening and typically involves multiple organ systems. In severe cases, anaphylaxis can rapidly result in death. Two of the most common signs of anaphylaxis are <u>wheezing</u>, a high-pitched, whistling breath sound usually resulting from bronchospasm and typically heard on expiration, and widespread <u>urticaria</u>, or hives. Urticaria consists of small areas of generalized itching or burning that appear as multiple, small, raised areas on the skin (Figure 16-1 ▲).

Given the right person and the right circumstances, almost any substance can trigger the body's immune system and cause an allergic reaction: animal bites, food, latex gloves, and many other substances can be <u>allergens</u>. The most common allergens, however, fall into the following five general categories:

- **Insect bites and stings.** When an insect bites you and injects the bite with its venom, the act is called <u>envenomation</u> or, more commonly, a sting. The sting of a honeybee, wasp, ant, yellow jacket, or hornet may cause a severe reaction with the swiftness of an injected medication. The reaction may be local, causing swelling and itchiness in the surrounding tissue, or it may be systemic, involving the entire body. Such a total body reaction would be considered an anaphylactic reaction.

- **Medications.** Injection of medications such as penicillin may cause an immediate (within 30 minutes) and severe allergic reaction (Figure 16-2 ▶). However, reactions to oral medications, such as oral penicillin, may be slower in onset (more than 30 minutes) but equally

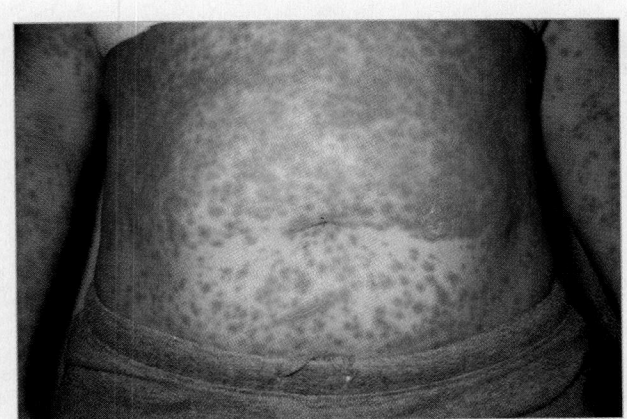

Figure 16-2 A severe allergic reaction to medication.

severe. The fact that a person has taken a medication once without experiencing an allergic reaction is no guarantee that he or she will not have an allergic reaction to it the next time around.

■ **Plants.** Individuals who inhale dusts, pollens, or other plant materials to which they are sensitive may experience a rapid and severe allergic reaction.

■ **Food.** Eating certain foods, such as shellfish or nuts, may result in a relatively slow (more than 30 minutes) reaction that still can be quite severe. The person may be unaware of the exposure or inciting agent.

■ **Chemicals.** Certain chemicals, makeup, soap, latex, and various other substances can cause severe allergic reactions.

Insect Stings

There are more than 100,000 species of bees, wasps, and hornets. Deaths due to anaphylactic reactions to stinging insects far outnumber deaths due to snake bites. The stinging organ of most bees, wasps, yellow jackets, and hornets is a small hollow spine projecting from the abdomen. Venom can be injected through this spine directly into the skin. The stinger of the honeybee is barbed, so the bee cannot withdraw it (Figure 16-3A ▼). Therefore, the bee leaves a part of its abdomen embedded with the stinger and dies shortly after flying away. Wasps and hornets have no such handicap; they can sting repeatedly (Figure 16-3B ▼). Because these insects usually fly away after stinging, it is often impossible to identify which species was responsible for the injury.

Figure 16-3 Most stinging insects inject venom through a small, hollow spine that projects from the abdomen. **A.** The stinger of the honeybee is barbed and cannot be withdrawn once the bee has stung someone. **B.** The wasp's stinger is unbarbed, meaning that it can inflict multiple stings.

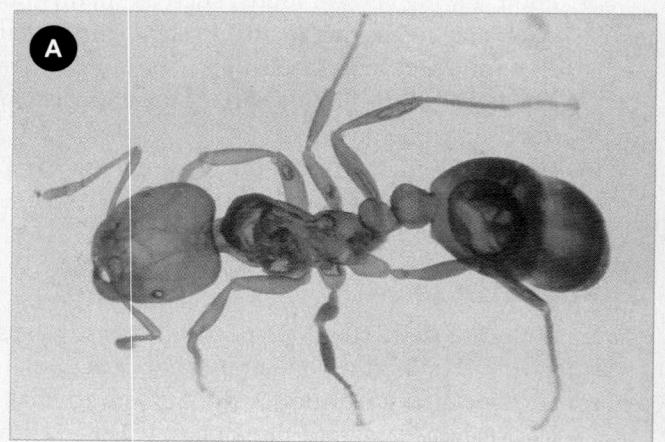
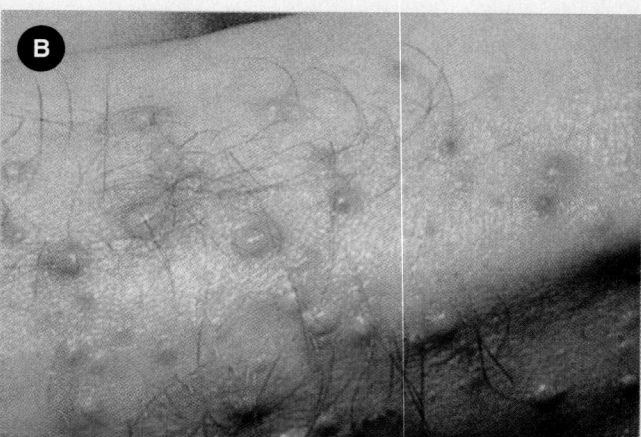

Figure 16-4 A. The fire ant. **B.** Fire ants inject an irritating toxin at multiple sites. Bites are generally found on the feet and the legs and appear as multiple, small, raised pustules.

Some ants, especially the fire ant (*Formicoidea,* Figure 16-4A ▲), also strike repeatedly, often injecting a particularly irritating <u>toxin</u>, or poison, at the bite sites. It is not uncommon for a patient to sustain multiple ant bites, usually on the feet and legs, within a very short time Figure 16-4B ▲ .

Signs and symptoms of insect stings or bites include sudden pain, swelling, localized heat, and redness in light-skinned individuals, usually at the site of injury. There may be itching and sometimes a <u>wheal</u>, which is a raised, swollen, well-defined area on the skin Figure 16-5 ▶ . There is no specific treatment for these injuries, although applying ice sometimes makes them less irritating. The swelling associated with an insect bite may be dramatic and sometimes frightening to patients. However, these local manifestations are usually not serious.

Because the stinger of the honeybee remains in the wound, it can continue to inject venom for up to 20 minutes after the bee has flown away. In caring for

You are the Provider   Part 2

You immediately request advanced life support (ALS) assistance as you ready your EpiPen. You expect to see vital signs that indicate an increased vascular container (hypotension) and the heart's attempts to compensate for that larger container (tachycardia). You also expect to see a subsequent decreased oxygen level secondary to the swelling of the airway and the problems with the circulation.

Your partner takes initial vital signs and reports that the patient's blood pressure is 94/56 mm Hg, pulse is 130 beats/min, respirations are 42 breaths/min, and the pulse oximetry reading is 90%. You explain to the patient that you need to administer epinephrine to reverse the effects of the insect bite. You tell him that the injection may sting but that he should not pull away. You place your knee along the inside of his leg to prevent his moving, and you administer the EpiPen. Your partner gives the patient high-flow oxygen and removes the stinger by scraping it off with a tongue depressor. You then have him lie down, and you prop up his feet on your airway bag.

3. In some states, EMT-Bs have AnaKits that contain multidose epinephrine and antihistamines they can administer by mouth. Is an EpiPen a multidose preparation?

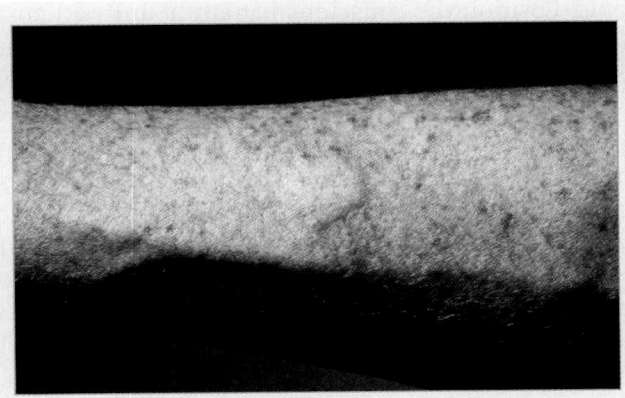

Figure 16-5 A wheal is a whitish, firm elevation of the skin that occurs after an insect sting or bite.

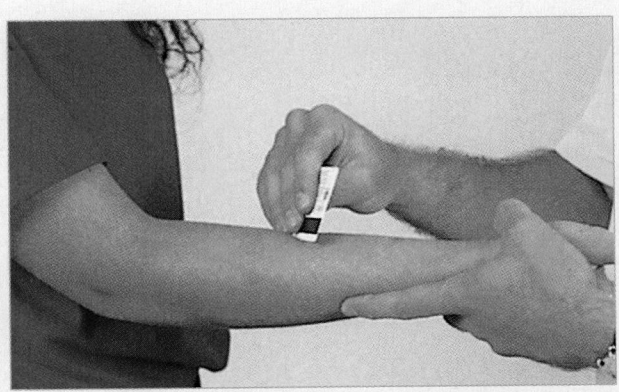

Figure 16-6 To remove the stinger of a honeybee, gently scrape the skin with the edge of a sharp, stiff object such as a credit card.

a patient who has been stung by a honeybee, you should gently attempt to remove the stinger and attached muscle by scraping the skin with the edge of a sharp, stiff object such as a credit card (Figure 16-6 ▶). Generally, you should not use tweezers or forceps because squeezing may cause the stinger to inject still more venom into the wound. Gently wash the area with soap and water or a mild antiseptic. Try to remove any jewelry from the area before swelling begins. Position the injection site slightly below the level of the heart and apply ice or cold packs to the area, but not directly on the skin, to help relieve pain and slow the absorption of the toxin. Be alert for vomiting or any signs of shock or allergic reaction, and do not give the patient anything by mouth. Place the patient in the shock position, and give oxygen if needed. Monitor the patient's vital signs, and be prepared to provide further support as needed.

Anaphylactic Reaction to Stings

Approximately 5% of all people are allergic to the venom of the bee, hornet, yellow jacket, or wasp. This type of allergy, which accounts for about 200 deaths per year, can cause very severe reactions, including anaphylaxis. Patients may experience generalized itching and burning, widespread urticaria, wheals, swelling about the lips and tongue, bronchospasm and wheezing, chest tightness and coughing, dyspnea, anxiety, abdominal cramps, and hypotension. Occasionally, respiratory failure occurs.

If untreated, such an anaphylactic reaction can proceed rapidly to death. In fact, more than two thirds of patients who die of anaphylaxis do so within the first half hour, so speed on your part is essential.

Assessment of a Patient With an Allergic Reaction

Scene Size-up

The environment the patient is in or the activity he or she was performing may indicate the source of the reaction, such as a sting or bite from an insect, a food allergy at a restaurant, or a new medication. A respiratory problem reported by dispatch may be an allergic reaction. If many people are affected, it could be an inhaled poison or terrorist event. As you proceed to the patient, observe for safety threats to yourself and your partner and determine the number of patients at the scene. Gloves and eye protection should be minimum BSI precautions. Call for additional resources earlier rather than later.

Initial Assessment

General Impression

Allergic reactions may present as respiratory distress or as cardiovascular distress in the form of shock. Patients experiencing a severe allergic reaction will often be very

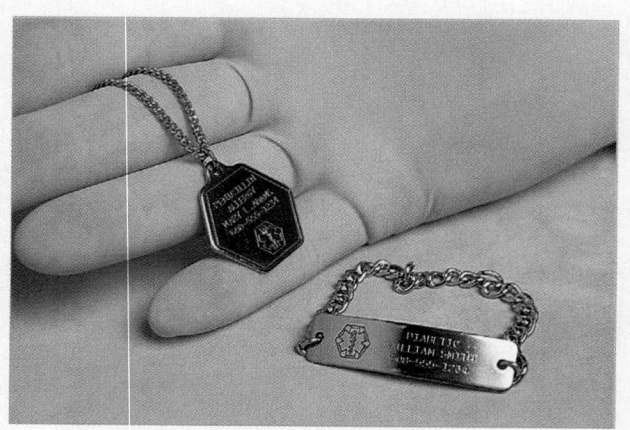

Figure 16-7 Some patients who are known to be severely allergic to bee stings, certain medications, or other substances often wear a medical identification tag.

anxious and feel like they are going to die. If your first impression finds the person anxious and in distress, call for ALS back-up if available. Some patients who are known to be severely allergic to bee stings, certain medications, or other substances wear a medical identification tag (Figure 16-7 ▲). If conscious, they will provide this information and identification as you ask about their chief complaint. Some may have even begun self-treatment with their own medications. If they are unresponsive, immediately evaluate and treat their airway, breathing, and circulation.

Airway and Breathing

The most severe form of allergic reactions, anaphylaxis, can cause rapid swelling of the upper airway. You may have only a few minutes to assess the airway and provide life-saving measures; however, not all allergic reactions are anaphylactic reactions. Work quickly to assess the patient to determine the severity of the symp-

toms. Position the conscious patient in a tripod position leaning forward. This will help to facilitate air entry into the lungs and may help the patient to relax. Quickly listen to the lungs on each side of the chest. If wheezing is heard, the lower airways are also closing, preventing oxygen from entering the circulatory system. Do not hesitate to initiate high-flow oxygen therapy. You may have to assist with ventilations for a patient with a severe allergic reaction. This can be done in a semiresponsive or unresponsive patient. The positive pressure ventilations you provide will force air through the swelling in the throat and into the lungs while you are waiting for more definitive treatment. In severe situations such as these, the definitive care needed is an injection of epinephrine.

If necessary, be prepared to use standard airway procedures and positive pressure ventilation according to the principles identified in Chapter 7.

Circulation

While respiratory complaints are most common, some patients in anaphylaxis may not present with severe respiratory symptoms but primarily with signs and symptoms of circulatory distress. Palpating a radial pulse will help to identify how the circulatory system is responding to the reaction. If the patient is unresponsive and without a pulse, begin basic life support measures or use an automatic external defibrillator if necessary. Assess for rapid heart rate; pale, cool, moist skin; and delayed capillary refill times that indicate hypoperfusion. Your initial treatment for shock should include oxygen, positioning in a shock position, and maintaining normal body temperature. The definitive treatment for anaphylactic shock is epinephrine. Trauma is unlikely with allergic reactions, but if trauma has occurred, bandage all bleeding sites, and take spinal precautions when appropriate.

You are the Provider   Part 3

A few moments after you administer the EpiPen, you notice that the patient is breathing more easily. His blood pressure and pulse oximetry values have risen, and pulse and respiration counts have decreased. The paramedics have now arrived.

4. What treatment would you expect them to provide?

Transport Decision

Always provide prompt transport for any patient who may be having an allergic reaction. Take with you all medications and auto-injectors the patient has at the time. Make your transport decision based on findings in the initial assessment. If the patient has signs of respiratory distress or shock, treat those conditions and transport. If the patient is calm and has no signs of respiratory distress or shock after contact with a substance that causes an allergic reaction, continue with the focused history and physical exam.

Focused History and Physical Exam

Perform a focused history and physical examination. If the allergic reaction has left the patient unresponsive, perform a rapid physical exam to determine hidden trauma or other problems and then obtain the vital signs and history.

If the patient is responsive, begin with obtaining the SAMPLE history and asking questions specific to an allergic reaction. Find out what interventions have been completed. Find out whether the patient has any prescribed, preloaded medications for allergic reactions. After obtaining the necessary history, perform a focused physical exam to look for bee stingers or contact with chemicals and other indications of a reaction. Finish by obtaining a complete set of vital signs.

SAMPLE History

Obtain a SAMPLE history. Things to ask about in relation to the allergic reaction patient may include the following:

- **Symptoms:** Any respiratory symptoms are the most troubling because a patient's condition can rapidly deteriorate from respiratory distress to arrest. Other symptoms may include itching, rash, hives, pallor, bite or sting marks, increased capillary refill time, or altered mental status.
- **Allergies:** Susceptible patients may have a history of the specific allergy involved in this case or of other allergies.
- **Medications:** Patients who have had severe allergic reactions in the past may carry an epinephrine auto-injector or antihistamines, such as chlorophenarimine (found in an AnaKit) or diphenhydramine (Benadryl). Susceptible patients may also carry bronchodilator inhalers, such as albuterol, or allergy medications.
- **Past medical history:** Ask about previous allergic reactions, asthma, and hospitalizations.
- **Last oral intake:** Finding out what and when your patient ate last may help you determine the cause of the reaction. For example, peanuts, chocolate, and shellfish can be potent allergens.
- **Events:** Find out all you can about what the patient was doing and what he or she was exposed to before the onset of symptoms. This information may be key to effective treatment.

Focused Physical Exam

The focused physical exam may help direct treatment. As in all emergencies, your assessment of the patient experiencing an allergic reaction should include evaluations of the respiratory system, circulatory system, mental status, and the skin. Assess for altered mental status, which may be the result of hypoxia or systemic shock. Thoroughly assess breathing, including increased work of breathing, use of accessory muscles, head bobbing, tripod position, nostril flaring, and grunting. Auscultate carefully, both the trachea and the chest.

Wheezing occurs because of narrowing of the air passages, which is mainly due to contraction of muscles around the bronchioles in reaction to the allergen. Exhalation, normally the passive, relaxed part of breathing, becomes harder as the patient tries to cough up the secretions or move air past the constricted airways. The fluid in the air passages and the constricted bronchi together produce the wheezing sound. Breathing rapidly becomes more difficult, and the patient may even stop breathing. Prolonged respiratory difficulty can cause a rapid heartbeat (tachycardia), shock, and even death. Stridor, a harsh, high-pitched inspiratory sound, occurs when swelling in the upper airway (near the vocal cords and throat) closes off the airway and can eventually lead to total obstruction.

Remember, the presence of hypoperfusion (shock) or respiratory distress indicates that the patient is having a severe enough allergic reaction to lead to death. Common signs and symptoms of an allergic reaction are listed in (Table 16-1 ▶).

Carefully assess the skin for swelling, rash, hives, or signs of the source of the reaction: bite, sting, or contact marks. A rapidly spreading rash can be concerning because it may indicate a systemic reaction. Red, hot skin

Auto Injector
.3mg Adult   .15mg Children

Table 16-1 Common Signs and Symptoms of Allergic Reaction

Respiratory System

- Sneezing or an itchy, runny nose (initially)
- Tightness in the chest or throat
- Irritating, persistent dry cough
- Hoarseness
- Respirations that become rapid, labored, or noisy
- Wheezing and/or stridor

Circulatory System

- Decrease in blood pressure as the blood vessels dilate
- Increase in pulse rate (initially)
- Pale skin and dizziness as the vascular system fails
- Loss of consciousness and coma

Skin

- Flushing, itching, or burning skin; especially common over the face and upper part of the chest
- Urticaria over large areas of the body, both internally and externally
- Swelling, especially of the face, neck, hands, feet, and/or tongue
- Swelling and cyanosis or pallor around the lips
- Warm, tingling feeling in the face, mouth, chest, feet, and hands

Other Findings

- Anxiety; a sense of impending doom
- Abdominal cramps
- Headache
- Itchy, watery eyes
- Decreasing mental status

---

hypotension are ominous signs, indicating systemic vascular collapse and shock. Skin signs may be an unreliable indicator of hypoperfusion because of rashes and swelling.

Interventions

In order to treat allergic reactions, you must first identify how much distress the patient is in. Some allergic reactions will produce severe signs and symptoms in a matter of minutes and threaten the patient's life. Others will have a slower onset and cause less severe distress. Epinephrine and ventilatory support are required for severe reactions. Milder reactions, without respiratory or cardiovascular distress, may only require supportive care, such as oxygen. In either situation, the patient should be transported to a medical facility for further evaluation.

Detailed Physical Exam

Consider performing a detailed physical exam if the patient presents with a confusing history or complaint, if there is an extended transport time, or if there is a need to clarify findings from earlier in the assessment. A detailed physical exam may also provide important information in unresponsive patients. In severe reactions, a detailed physical exam may be omitted when time must be spent managing ABCs or when transport distances are short.

Ongoing Assessment

The patient experiencing a suspected allergic reaction should be monitored with vigilance because deterioration of the patient's condition can be rapid and fatal. Special attention should be given to any signs of airway compromise, including increasing work of breathing, stridor, or wheezing. The patient's anxiety level should be monitored because increased anxiety is a good indication that the reaction may be progressing. Also, watch the skin for signs of shock, including pallor and diaphoresis, as well as for flushing because of vascular collapse. Serial vital signs are important in evaluating your patient's status. Any increase in the

may also indicate a systemic reaction as the blood vessels lose their ability to constrict and blood moves to the extremities. If this reaction continues, the body will have difficulty supplying blood and oxygen to the vital organs, and one of the first signs will be altered mental status as the organs are deprived of oxygen and glucose.

Baseline Vital Signs

Vital signs help determine whether the body is compensating for stress. Assess baseline vital signs, including pulse, respirations, blood pressure, skin, and pupils. Rapid, labored breathing indicates airway obstruction. Rapid respiratory and heart rates may indicate respiratory distress or systemic shock. Fast pulses and respiratory or heart rate or decrease in blood pressure should be noted.

If you administered epinephrine, what was the effect? Is the patient's condition improving? Do you need to consider a second dose? You may need to give more than one injection of epinephrine if you note that the patient has decreasing mental status, increased breathing difficulty, or a decreasing blood pressure. Be sure to consult medical control first. Current auto-injectors give only one dose, and a patient who needs more than one dose will need to have more than one injector. Any patient in critical condition should have his or her vital signs taken at least every 5 minutes.

Communication and Documentation

When to contact medical control depends on your assessment findings and the urgency of care required. In some allergic reactions, you may use standing orders to administer epinephrine before ever calling medical control. At other times, the reaction may be less severe and you may question whether the patient needs an injection of epinephrine. Medical control will be most helpful in the latter situation. Follow your local protocols, which may guide you in providing lifesaving care without needing to contact medical control.

Your documentation not only should include the signs and symptoms found during your assessment, but also should clearly show why you chose to provide the care you did. If anyone should question your care, your documentation should show the reasoning for what you did. Be complete in your documentation, including not only assessment findings and treatment, but also the patient's response to your treatment.

Emergency Medical Care

If the patient appears to be having a severe allergic (or anaphylactic) reaction, you should administer BLS at once and provide prompt transport to the hospital. You may want to request ALS backup if you work in a tiered response system. In addition to providing oxygen, you should be prepared to maintain an airway or give CPR. Placing ice over the injury site has been thought to slow absorption of the toxin and diminish swelling, but ice packs placed directly on the skin may freeze it and cause more damage. Like any other attempt to reduce swelling with ice, you should be careful not to overdo the icing. In some areas, you may be allowed to administer epinephrine or assist the patient with epinephrine administration.

The body normally produces epinephrine. Epinephrine works rapidly to raise the pulse rate and blood pressure by constricting the blood vessels. Sometimes the body does not produce enough epinephrine, in which case epinephrine is administered to compensate for the body's slow response in a severe reaction. Epinephrine also inhibits the allergic reaction and dilates the bronchioles. All bee-sting kits should contain a prepared syringe of epinephrine, ready for intramuscular injection, along with instructions for its use. Your EMS service may or may not allow you to help patients self-administer epinephrine to combat allergic reactions or anaphylaxis. In some places, the medical director may authorize you to carry an epinephrine auto-injector (EpiPen) or to assist patients who have

✳ EMT-B Tips

While one EMT-B is getting oxygen ready, the other should be assisting the patient into a comfortable position, generally supine with the head and shoulders elevated. These measures will help perfusion to the brain while easing respiratory effort.

You are the Provider | Part 4

You assist the paramedics in moving the patient to the ambulance. There you see one paramedic has already prepared to initiate intravenous fluids, administer Benadryl, and give a breathing treatment.

5. What can the delay of epinephrine do to a patient who is in anaphylaxis?

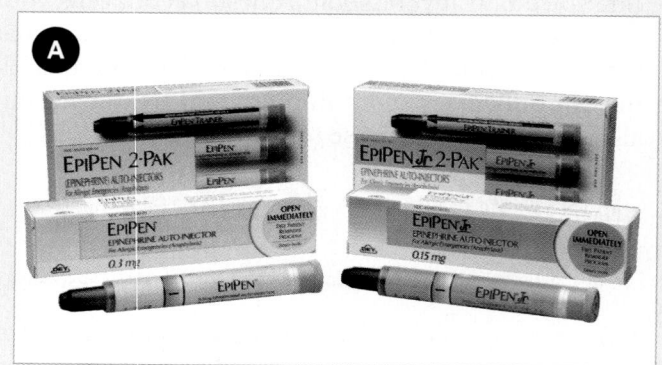

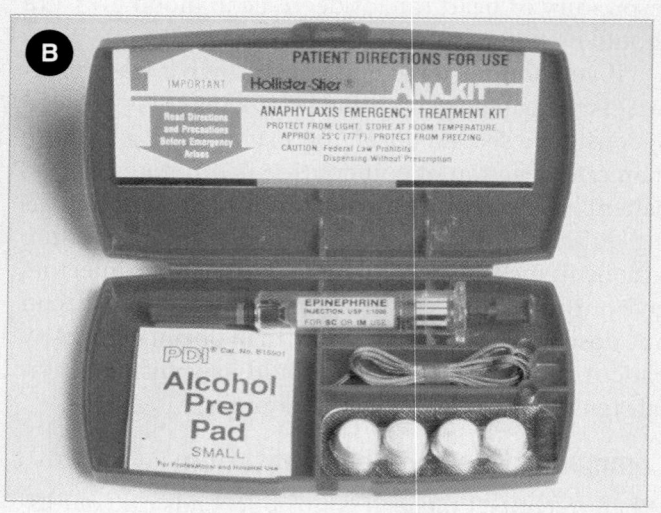

Figure 16-8 Patients who experience severe allergic reactions often carry their own epinephrine, which comes predosed in an auto-injector or a standard syringe. **A.** EpiPen auto-injectors. **B.** AnaKit with epinephrine syringe.

their own EpiPen. The adult system delivers 0.3 mg of epinephrine via an automatic needle and syringe system; the infant-child system delivers 0.15 mg (Figure 16-8A ▲).

If the patient is able to use the auto-injector on his or her own, your role is limited to helping. To use, or help the patient use, the auto-injector, you should first receive a direct order from medical control or follow local protocols or standing orders. Follow BSI precautions, and make sure the medication has been prescribed specifically for that patient. If it has not, do not give the medication: inform medical control, and provide immediate transport. Finally, make sure the medication is not discolored and that the expiration date has not passed.

Once you have done these things, follow the steps in (Skill Drill 16-1 ▶) to use an auto-injector.

1. **Remove the safety cap** from the auto-injector and, if possible, wipe the patient's thigh with alcohol or some other antiseptic. However, do not delay administration of the drug (**Step 1**).
2. **Place the tip of the auto-injector** against the lateral part of the patient's thigh, midway between the waist and the knee (**Step 2**).

3. **Push the injector firmly** against the thigh until the injector activates. Hold steady pressure to prevent kickback from the spring in the syringe, and prevent the needle from being pushed out of the injection site too soon. Hold the injector in place until the medication has been injected (10 seconds) (**Step 3**).
4. **Remove the injector** from the patient's thigh and dispose of it in the proper biohazard container.
5. **Record the time and dose** of the injection on your run sheet.
6. **Reassess and record** the patient's vital signs after using the auto-injector.

If the patient is known to have an allergy, he or she might carry a commercial bee-sting kit (AnaKit) that contains a standard syringe of epinephrine for intramuscular injection (Figure 16-8B ▲). If you will administer this epinephrine, make the same general

EMT-B Tips

If your medical director and protocols allow it and your patient has an inhaler and an epinephrine auto-injector, one EMT-B can help administer the inhaler while the other administers the epinephrine.

Documentation Tips

Allergic reactions and responses to bites and stings can progress quickly to life threats. With good care, severe signs and symptoms may subside just as quickly. Doing a multisystem exam and documenting your findings is important before and after treatment. Give particular attention to skin signs and respiratory, circulatory, and mental functioning.

Using an Auto-injector

1 Remove the auto-injector's safety cap, and quickly wipe the thigh with antiseptic.

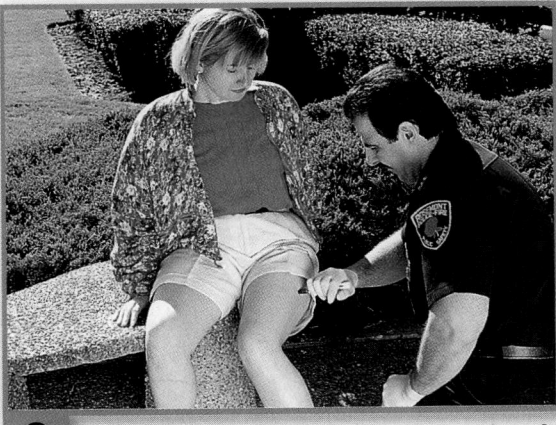

2 Place the tip of the auto-injector against the lateral part of the thigh.

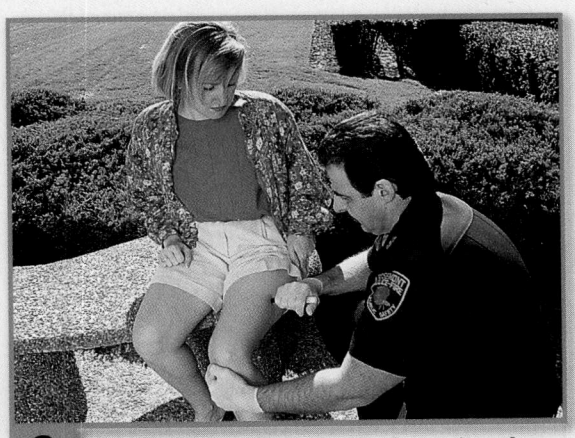

3 Push the auto-injector firmly against the thigh, and hold it in place until all the medication has been injected.

preparations you make for an auto-injector: Get an order, take BSI precautions, and ensure that the medicine is the patient's, is not discolored, and the expiration date has not passed. Follow the steps in (Skill Drill 16-2 ▶) to administer epinephrine from an AnaKit.

1. **Prepare the injection site** with an alcohol wipe or other antiseptic, if there is time. Remove the needle cover (**Step 1**).
2. **Hold the syringe upright** so any air inside rises to the base of the needle. Remove the air by depressing the syringe plunger until it stops (**Step 2**).
3. **Turn the plunger** one-quarter turn (**Step 3**).

4. **Insert the needle quickly**, straight into the injection site, deep enough to place the tip into the muscle beneath the skin and subcutaneous fat (**Step 4**).
5. **Holding the syringe steady**, push the plunger until it stops, to ensure that all medication is injected (**Step 5**). 10 sec.
6. **Have the patient chew and swallow the Chlo-Amine antihistamine tablets** in the kit (**Step 6**).
7. If you have or can **make a cold pack**, apply it to the site of the sting to reduce swelling and minimize the amount of venom that enters the circulation (**Step 7**).

Using an AnaKit

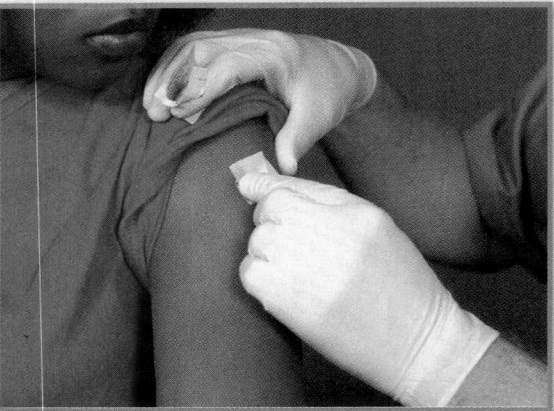

1   Prepare the injection site with antiseptic, and remove the needle cover.

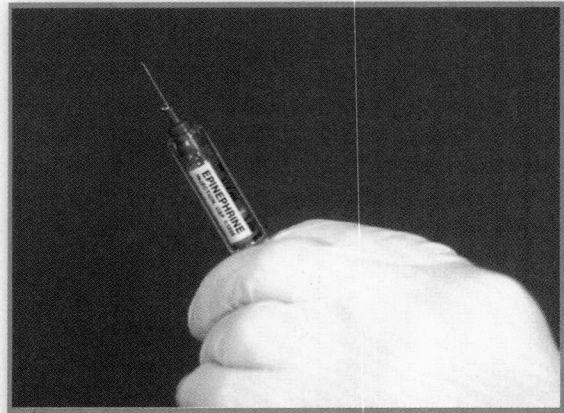

2   Hold the syringe upright, and carefully use the plunger to remove air.

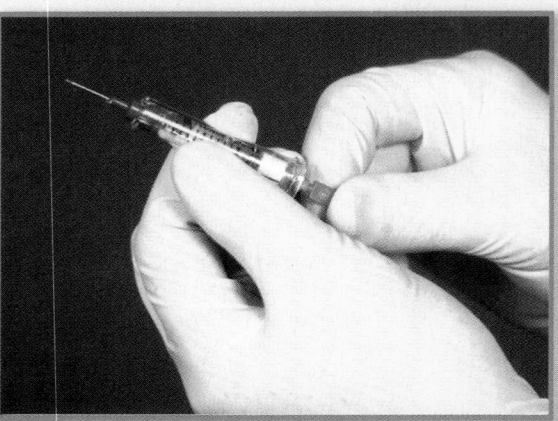

3   Turn the plunger one-quarter turn.

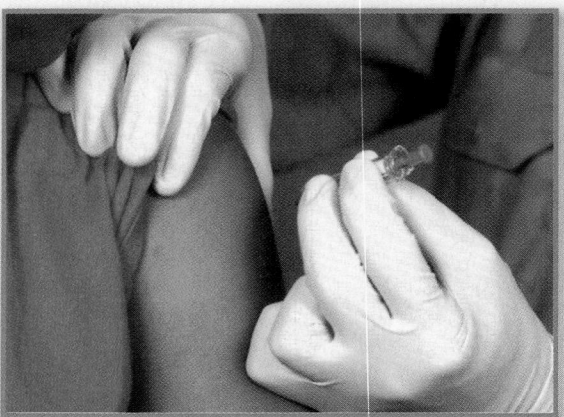

4   Quickly insert the needle into the muscle.

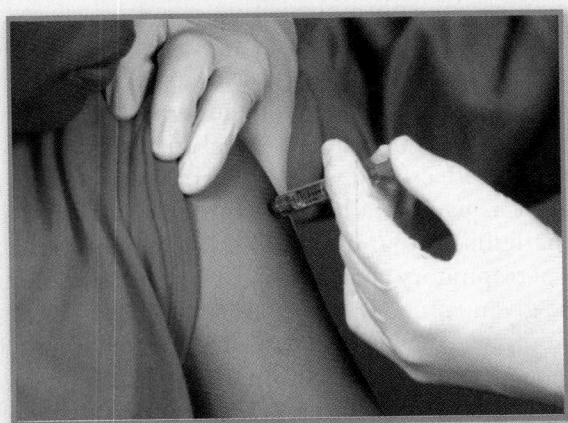

5 Hold the syringe steady, and push the plunger until it stops.

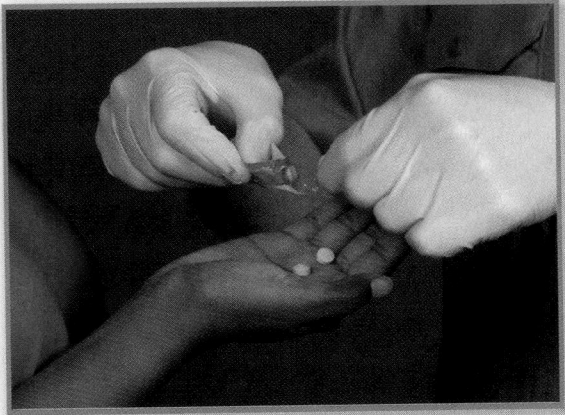

6 Have the patient chew and swallow the Chlo-Amine antihistamine tablets provided in the kit.

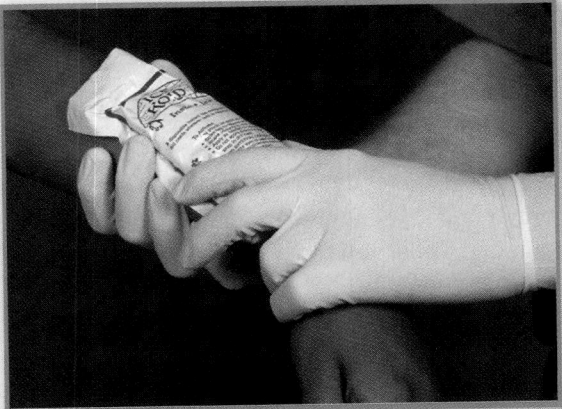

7 If available, apply a cold pack to the sting site.

The syringe from the AnaKit holds a second injection if needed.

Other bee-sting kits contain some oral or intramuscular antihistamines, agents that block the effect of histamine. These work relatively slowly, within several minutes to 1 hour. Because epinephrine can have an effect within 1 minute, it is the primary way to save the life of someone having a severe anaphylactic reaction.

In some areas of the country, EMT-B squads can administer epinephrine for anaphylaxis. Follow your local protocols or medical direction, but generally you will inject 0.3 to 0.5 mL of 1:1,000 epinephrine solution to an adult weighing more than 50 kg intramuscularly or subcutaneously at intervals of 5 to 15 minutes, as needed. Usually not more than two injections are given. Doses for children vary, ranging from 0.1 to 0.3 mL, depending on the patient's weight. Be aware that the med-

Anna Kit

ication may cause significant tachycardia, as well as increased anxiety or nervousness and palpitations.

Because epinephrine constricts blood vessels, it may cause the patient's blood pressure to rise significantly. Other side effects include tachycardia, pallor, dizziness, chest pain, headache, nausea, and vomiting. All these effects may cause the patient to feel anxious or excited. These side effects are worth the trade-off when epinephrine is used in a life-threatening situation. Note that patients who are not wheezing or who have no signs of respiratory compromise or hypotension should not be given epinephrine.

Complete the emergency care outlined above for this patient, and provide prompt transport, closely monitoring vital signs frequently. Remember that all patients with suspected anaphylaxis should be given high-flow, high-concentration oxygen.

You are the Provider   Summary

Anaphylaxis, in simple terms, is the body's attempt to rid itself of an allergen that it perceives as harmful. This normally benign substance (such as pollen or insect sting), causes the body to have leaky capillary beds (the body's attempt to remove the allergen from the bloodstream) and inflammation of the airway (the body's attempt to prevent further exposure of the air passages to this substance). What the body does to help or protect itself, in this case, actually causes great harm.

Mild and moderate allergic reactions are localized. They do not involve entire body systems. In a mild to moderate allergic reaction, you could expect to see itchy, watery eyes, runny nose, sneezing, and, possibly, hives. Although these signs and symptoms are uncomfortable, they do not cause harmful effects to the cardiovascular and respiratory systems.

In cases of anaphylaxis, the delay of epinephrine administration can cost the patient his or her life. You must immediately recognize the situation, know how to correctly administer the epinephrine, and do so in a timely manner. You must be familiar with local protocols because many differ. Some areas have standing orders for EMT-Bs to administer EpiPens when anaphylaxis is expected (history of events with hypotension and breathing difficulty), whereas other localities require EMTs to obtain permission via online medical control.

You can prevent delays by doing the following:

- Be prepared—check your rig every morning for necessary drugs and equipment.
- Be familiar with your apparatus and location of equipment before going on calls.
- Be familiar with local protocols, and train frequently for medical emergencies such as this one to maintain your proficiency.
- Be knowledgeable of your local area to decrease response time. (Always consult your map unless you are 100% sure of the location.)

All of these preparations can significantly impact the quality and timeliness of your response and patient care.

Allergic Reactions	
Scene Size-up	Body substance isolation precautions should include a minimum of gloves and eye protection. Ensure scene safety and determine NOI/MOI. Consider the number of patients, the need for additional help/ALS, and c-spine stabilization.
Initial Assessment	
■ General impression	Determine level of consciousness and find and treat any immediate threats to life. Determine priority of care based on environment and patient's chief complaint. If the patient appears anxious or fears death, call for ALS assistance.
■ Airway	Ensure patent airway.
■ Breathing	Provide high-flow oxygen at 15 L/min. If possible, place in a tripod position and evaluate depth and rate of the respiratory cycle and provide ventilatory support as needed.
■ Circulation	Evaluate pulse rate and quality; observe skin color, temperature, and condition and treat accordingly.
■ Transport decision	Rapid transport
Focused History and Physical Exam	NOTE: The order of the steps in the focused history and physical exam differs depending on whether the patient is conscious or unconscious. The order below is for a conscious patient. For an unconscious patient, perform a rapid physical exam, obtain vital signs, and obtain the history.
■ SAMPLE history	Ask SAMPLE questions and determine if patient has prescribe autoinjector(s)/inhaler. Be sure to ask if and what interventions were taken before your arrival, how many interventions were performed, and at what time.
■ Focused physical exam	Perform focused physical exam keying in on the respiratory drive, adequate ventilation, and the adequacy and effectiveness of the circulatory system, and the patient's mental status.
■ Baseline vital signs	Take vital signs, noting skin color and temperature as well as patient's level of consciousness. Use pulse oximetry if available.
■ Interventions	Support patient as needed. Consider the use of oxygen, positive pressure ventilations, adjuncts, and proper positioning of the patient. Assist with the use of autoinjector(s) or inhaler as defined by local protocol.
Detailed Physical Exam	Consider a detailed physical exam.
Ongoing Assessment	Repeat the initial assessment, focused assessment, and reassess interventions performed. Reassess vital signs every 5 minutes for the unstable patient, every 15 minutes for the stable patient. Note and treat the patient as necessary.
■ Communication and documentation	Contact medical control with a radio report, providing information on patient's condition. Relay any significant changes, including level of consciousness or difficulty in breathing. Be sure to document any changes, at what time, and any interventions performed.

NOTE: While the steps below are widely accepted, be sure to consult and follow your local protocol.

Allergic Reactions

Using an Autoinjector
1. Remove the autoinjector's safety cap, and quickly wipe the thigh with antiseptic.
2. Place the tip of the autoinjector against the lateral part of the thigh.
3. Push the autoinjector firmly against the thigh, and hold it in place until all the medication is injected (about 10 seconds).

Using an AnaKit
1. Prepare the injection site with antiseptic, and remove the needle cover.
2. Hold the syringe upright, and carefully use the plunger to remove air.
3. Turn the plunger one-quarter turn.
4. Quickly insert the needle into the muscle.
5. Hold the syringe steady, and push the plunger until it stops.
6. Have the patient chew and swallow the Chlo-Amine antihistamine tablets provided in the kit.
7. If available, apply a cold pack to the sting site.

Assessment and Emergency Care

Prep Kit

Ready for Review

- An allergic reaction is a response to chemicals the body releases to combat certain stimuli, called allergens.
- Allergic reactions occur most often in response to five categories of stimuli: insect bites and stings, medications, food, plants, and chemicals.
- The reaction may be mild and local, involving itching, redness, and tenderness, or it may be severe and systemic, including shock and respiratory failure.
- Anaphylaxis is a life-threatening allergic reaction mounted by multiple organ systems, which must be treated with epinephrine.
- Wheezing and skin wheals can be signs of anaphylaxis.
- People who know that they are allergic to bee, hornet, yellow jacket, or wasp venom often carry a bee-sting kit that contains epinephrine in an auto-injector. You may help to administer this medication in this form with authorization from medical control.
- All patients with suspected anaphylaxis require oxygen.
- In assessing a person who may be having an allergic reaction, you should check for flushing, itching and swelling skin, hives, wheezing and stridor, a persistent cough, a decrease in blood pressure, a weak pulse, dizziness, abdominal cramps, and headache.
- Always provide prompt transport to the hospital for any patient who is having an allergic reaction. Remember that signs and symptoms can rapidly become more severe. Carefully monitor the patient's vital signs en route, especially for airway compromise.

Vital Vocabulary

allergen A substance that causes an allergic reaction.

allergic reaction The body's exaggerated immune response to an internal or surface agent.

anaphylaxis An extreme, possibly life-threatening systemic allergic reaction that may include shock and respiratory failure.

www.EMTB.com

Technology

Interactivities

Vocabulary Explorer

Anatomy Review

Web Links

Online Review Manual

envenomation The act of injecting venom.

epinephrine A substance produced by the body (commonly called adrenaline), and a drug produced by pharmaceutical companies that increases pulse rate and blood pressure; the drug of choice for an anaphylactic reaction.

histamines Substance released by the immune system in allergic reactions that are responsible for many of the symptoms of anaphylaxis.

leukotrienes Chemical substances that contribute to anaphylaxis; released by the immune system in allergic reactions.

stridor A harsh, high-pitched respiratory sound, generally heard during inspiration, that is caused by partial blockage or narrowing of the upper airway.

toxin A poison or harmful substance.

urticaria Small spots of generalized itching and/or burning that appear as multiple raised areas on the skin; hives.

wheal A raised, swollen, well-defined area on the skin resulting from an insect bite or allergic reaction.

wheezing A high-pitched, whistling breath sound, usually caused by a constriction of the smaller tubes of the lungs and typically heard on expiration.

Points to Ponder

The 9-1-1 communications center receives a call from a baby-sitter at 129 West Hibiscus for a report of a "sick" child. You respond and arrive in just a few minutes. A sheriff deputy arrives just before you and directs you to a porch in the rear of the house.

The patient is a 4-year-old boy who is sitting in a chair crying. The babysitter tells you the children were playing in the backyard when the patient just started crying. She noticed several ants on both legs of the patient. After calling 9-1-1 the sitter brushed off all the ants.

The boy is crying but does not appear to be in any immediate distress. You complete an initial assessment and focused medical assessment and find no immediate concerns. When you complete the detailed physical exam you find several ant bites. Both legs are very red and swollen around the bites. The sitter tells you the patient has no medical problems, takes no medications, and has no known allergies. Vital signs include a pulse rate of 110 beats/min and regular, capillary refill of less than 2 seconds, and a respiratory rate of 30 breaths/min and not labored. The parents are not scheduled to return for several hours and the baby-sitter does not want the child transported.

Can you leave the scene? Is there anything else you need to do before leaving?

Issues: Communicating With Children, Protecting and Treating Pediatric Patients, Implied Consent.

Assessment in Action

Rescue 5 Respond, 5701 East Colorado Ave for a report of a patient having trouble breathing. You head to the scene. Upon arrival a young man meets you at the sidewalk to lead you to the patient. As you move towards the front of the house, the patient's son tells you his mom is having "some type of reaction."

The patient is a 40-year-old woman sitting in a chair with moderate respiratory distress. Her skin is pale. An initial assessment is completed and the following information is obtained. The airway is open but the patient tells you she thinks her throat is closing up. The respiratory rate is 24 breaths/min and dyspneic (labored). The pulse rate is 100 beats/min and regular. The patient is administered oxygen by using a nonrebreathing mask set at 10 to 15 L/min.

You conduct a rapid medical assessment and obtain the patient's vital signs and SAMPLE history. The patient tells you she is allergic to bee stings and may have been stung. The patient has been prescribed an EpiPen but she was scared to use it. When you are assessing her lung sounds, wheezing is heard in all lobes and the patient's arms are covered with urticaria (hives).

1. When treating acute allergic reactions, you must be aware of the possibility cardiovascular collapse and:

   A. acute airway obstruction.
   B. major internal bleeding.
   C. neurologic problems.
   D. skin damage and hair loss.

2. On the basis of your assessment, this patient is experiencing and should be treated for:

   A. addictions.
   B. anaphylaxis.
   C. arteriosclerosis.
   D. ataxia.

3. In addition to oxygen administration and rapid transport, treatment of this patient may include:

   A. obtaining vital signs every 20 minutes.
   B. placing ice around the arms and legs.
   C. elevating the head or feet.
   D. administration of epinephrine.

4. The adult dose of epinephrine for the treatment of anaphylaxis is:

   A. 0.3 mg.
   B. 300 mg.
   C. 30 mg.
   D. 1 mg.

5. When using the EpiPen, remove the safety cap, place the autoinjector against the patient's middle thigh, and:

   A. push firmly against the thigh and hold it in place.
   B. pull the main trigger.
   C. Secure with a bandage.
   D. turn on the switch.

6. The stinger of a honeybee may remain in the skin and continue to:

   A. infect the skin.
   B. paralyze the extremity.
   C. inject venom.
   D. cause discoloration.

7. After a sting, the stinger may still be in the patient. You may remove the stinger by:

   A. sucking it out with a syringe.
   B. making a small incision with a knife.
   C. using tweezers or forceps.
   D. scraping the skin with a stiff object such as a credit card.

8. The irritation of stings is reduced by placing:

   A. the patient flat on his or her back.
   B. a bandage over the sting.
   C. ice over the involved area.
   D. a constricting band below the sting.

Challenging Questions

9. What are the five categories of stimuli that provoke allergic reactions.?

10. What is the difference between an allergic reaction and a response to a bite or sting?

11. What are the common signs and symptoms of an allergic reaction as it relates to the respiratory system, circulatory system, and skin?

12. How does the chemical epinephrine work in the body to treat anaphylaxis?

www.EMTB.com

Substance Abuse and Poisoning

Objectives

Cognitive

4-6.1 List various ways that poisons enter the body. (p 520)

4-6.2 List signs and symptoms associated with poisoning. (p 519, 521)

4-6.3 Discuss the emergency medical care for the patient with possible overdose. (p 530)

4-6.4 Describe the steps in the emergency medical care for the patient with suspected poisoning. (p 522)

4-6.5 Establish the relationship between the patient suffering from poisoning or overdose and airway management. (p 525)

4-6.6 State the generic and trade names, indications, contraindications, medication form, dose, administration, actions, side effects, and reassessment strategies for activated charcoal. (p 527)

4-6.7 Recognize the need for medical direction in caring for the patient with poisoning or overdose. (p 519, 527)

Affective

4-6.8 Explain the rationale for administering activated charcoal. (p 523)

4-6.9 Explain the rationale for contacting medical direction early in the prehospital management of the poisoning or overdose patient. (p 519)

Psychomotor

4-6.10 Demonstrate the steps in the emergency medical care for the patient with possible overdose. (p 530)

4-6.11 Demonstrate the steps in the emergency medical care for the patient with suspected poisoning. (p 527)

4-6.12 Perform the necessary steps required to provide a patient with activated charcoal. (p 527)

4-6.13 Demonstrate the assessment and documentation of patient response. (p 527)

4-6.14 Demonstrate proper disposal of the equipment for the administration of activated charcoal. (p 527)

4-6.15 Demonstrate completing a prehospital care report for patients with a poisoning/overdose emergency. (p 527)

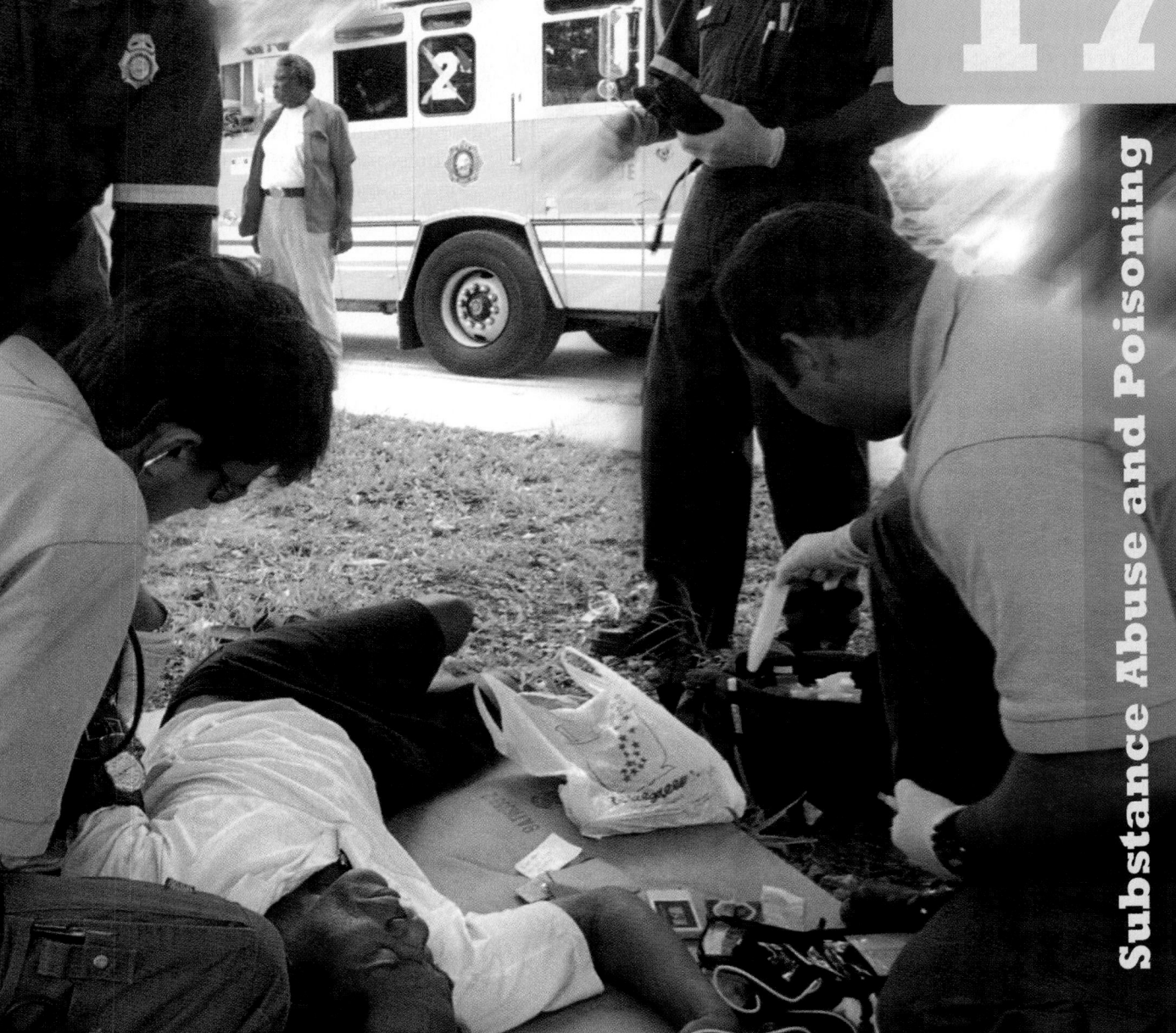

You are the Provider

You and your EMT-B partner are dispatched to the Grandma Jean's Day Care Center for an accidental poisoning. The center director meets you at the front door. She tells you that one of her toddlers ingested a cleaning product. As you enter the building, you hear hysterical crying and screaming. You find a 3-year-old girl in the lap of her teacher in obvious distress.

1. What BSI precautions should you consider using when treating this patient?
2. What type of signs and symptoms are you going to look for?

Substance Abuse and Poisoning

Every day, each of us comes into contact with things that are potentially poisonous. This is not surprising when you consider that almost any substance may be a poison in certain circumstances. Different doses can turn even a remedy into a poison. Consider aspirin. When taken in recommended doses, it is a safe and effective analgesic. Too much aspirin, however, can result in death.

Acute poisoning affects some 5 million children and adults each year. Chronic poisoning, often caused by abuse of medications and other substances, including tobacco and alcohol, is much more common. Fortunately, deaths due to poisoning are fairly rare. Rates of death due to poisoning in children have decreased steadily since the 1960s, when safety caps were introduced for drug bottles and containers. Deaths due to poisoning in adults, though, have been rising, the majority the result of drug abuse.

In this chapter, the term "poisoned" includes acute and chronic poisonings. As an EMT-B, you must recognize that patients with either type of problem may have a variety of injuries. Although you cannot stop a chronic substance abuse problem, you may be able to prevent death due to the acute effects of the poison.

This chapter discusses how to identify a patient who has been poisoned and how to gather clues about the poison. It describes the different ways in which poison is introduced into the body. It then discusses the signs, symptoms, and treatment of specific poisons, including sedatives and opioids (medicines with actions similar to morphine). Food poisoning and plant poisoning are also discussed.

Identifying the Patient and the Poison

A poison is any substance whose chemical action can damage body structures or impair body function. A poison can be introduced into the body through a variety of means. Poisons act by changing the normal metabolism of cells or by actually destroying them. Poisons may act acutely, as in an overdose of heroin, or chronically, as in years of alcohol or other substance abuse. Substance abuse is the misuse of any substance to produce a desired effect (for example, cocaine intoxication).

Your primary responsibility to the patient who has been poisoned is to recognize that a poisoning has occurred. Keep in mind that very small amounts of some poisons can cause considerable damage or death. If you have even the slightest suspicion that a patient has taken a poisonous substance, you should notify medical control and begin emergency treatment at once. Discussion of issues relating to suicide will be covered in the behavioral emergencies chapter.

Symptoms and signs of poisoning vary according to the specific agent, as shown in (Table 17-1 ▶). Some poisons cause the pulse to speed up, while others cause it to slow down; some cause the pupils to dilate, while others cause the pupils to constrict. If respiration is depressed or difficult, cyanosis may occur. Some chemical compounds will irritate or burn the skin or mucous membranes, resulting in burning or blistering. The presence of such injuries at the mouth strongly suggests the ingestion (swallowing) of a poison, such as lye. If possible, consider asking the patient the following questions: *Know This...*

- What substance did you take?
- When did you take it (or become exposed to it)?
- How much did you ingest?
- What actions have been taken?
- How much do you weigh?

Try to determine the nature of the poison. Objects at the scene may provide clues: an overturned bottle, a

www.EMTB.com

Technology

- Interactivities
- Vocabulary Explorer
- Anatomy Review
- Web Links
- Online Review Manual

TABLE 17-1 Toxidromes: Typical Signs and Symptoms of Specific Drug Overdoses

Drug	Signs and Symptoms
Opioid (Examples: heroin, oxycodone)	■ Hypoventilation or respiratory arrest ■ Pinpoint pupils ■ Sedation or coma ■ Hypotension
Sympathomimetics (Examples: epinephrine, albuterol, cocaine, methamphetamine)	■ Hypertension ■ Tachycardia ■ Dilated pupils ■ Agitation or seizures ■ Hyperthermia
Sedative-hypnotics (Examples: diazepam [Valium], secobarbital [Seconal], flunitrazepam [Rohypnol])	■ Slurred speech ■ Sedation or coma ■ Hypoventilation ■ Hypotension
Anticholinergics (Examples: atropine, Jimson weed)	■ Tachycardia ■ Hyperthermia ■ Hypertension ■ Dilated pupils ■ Dry skin and mucous membranes ■ Sedation, agitation, seizures, coma, or delirium ■ Decreased bowel sounds
Cholinergics (Examples: cimetidine, pilocarpine, nerve gas)	■ Excess defecation or urination ■ Muscle fasciculations ■ Pinpoint pupils ■ Excess lacrimation (tearing) or salivation ■ Airway compromise ■ Nausea or vomiting

Know This... ⤵

✚ EMT-B Tips

Poison Control Centers

There are several hundred poison control centers in the United States. The phone number of your local poison control center is typically found on the inside cover of your local phone book. The telephone number for the National Poison Control Center is 1 (800) 222-1222. Staff persons at every center have access to information about virtually all of the commonly used medications, chemicals, and substances that could possibly be poisonous. They know the appropriate emergency treatment for each, including the antidote, if there is one. An antidote is a substance that will counteract the effects of a particular poison.

If you believe that a patient has been poisoned, you should immediately provide medical control with all relevant information: when the poisoning occurred; a description of the suspected poison, including the amount involved; and the patient's size, weight, and age. If necessary, medical control can contact the regional poison control center and relay specific instructions back to you.

A medical toxicologist is a physician who specializes in caring for patients who have been poisoned. About 100 of these specialists work in special hospitals called medical toxicology treatment centers, located throughout the United States. At times, your medical control may divert a patient who meets certain poisoning criteria to one of these centers instead of to the closest hospital.

You and your medical control center should know the telephone number of your regional poison control center and have it available in case you come upon an unexpected case of poisoning.

needle or syringe, scattered pills, chemicals, even an overturned or damaged plant. The remains of any nearby food or drink may also be important. Place any suspicious material in a plastic bag, and take it to the hospital, along with any containers you find.

Containers can provide critical information. In addition to the name and concentration of the drug, a pill bottle label may list specific ingredients, the number of pills that were originally in the bottle, the name of the manufacturer, and the dose that was prescribed. This information can help emergency department physicians to determine how much has been ingested and what specific treatment may be required. For certain food poisonings, a food container that lists the name and location of the maker or the vendor may be of equal importance in saving the life of the patient and possibly other people.

If the patient vomits, examine the contents for pill fragments. Note anything unusual that you see. You may collect the material, called vomitus, in a separate plastic bag so that it can be analyzed at the hospital.

How Poisons Get Into the Body

Emergency care for a patient who has been poisoned may include a range of actions from reassuring an anxious parent to instituting CPR. Most often, it will not include administering a specific antidote, because most poisons do not have one. Therefore, in general, the most important treatment for poisoning is diluting and/or physically removing the poisonous agent. How you do this depends on how the poison gets into the patient's body in the first place. Essentially, the four avenues to consider are as follows:

- Inhalation (Figure 17-1A ▼)
- Absorption (surface contact) (Figure 17-1B ▼)
- Ingestion (Figure 17-1C ▼)
- Injection (Figure 17-1D ▼)

Injection often can be the most worrisome avenue of poisoning. You can administer oxygen to a patient who has inhaled a poison, and you can give activated charcoal to one who has ingested a poison. You can flood the skin with water and wash out the eyes of one who has contacted a poison. However, it is difficult to remove or dilute injected poisons, a fact that makes these cases especially urgent. On the other hand, all routes of poisoning can be deadly, and each should be thought of as being equally serious.

Always consult medical control before you proceed with the treatment of any poisoning victim.

Inhaled Poisons

Patients who have inhaled poison, including natural gas, certain pesticides, carbon monoxide, chlorine, or other gases, should be moved into fresh air immediately (Figure 17-2 ▶). Depending on how long they were

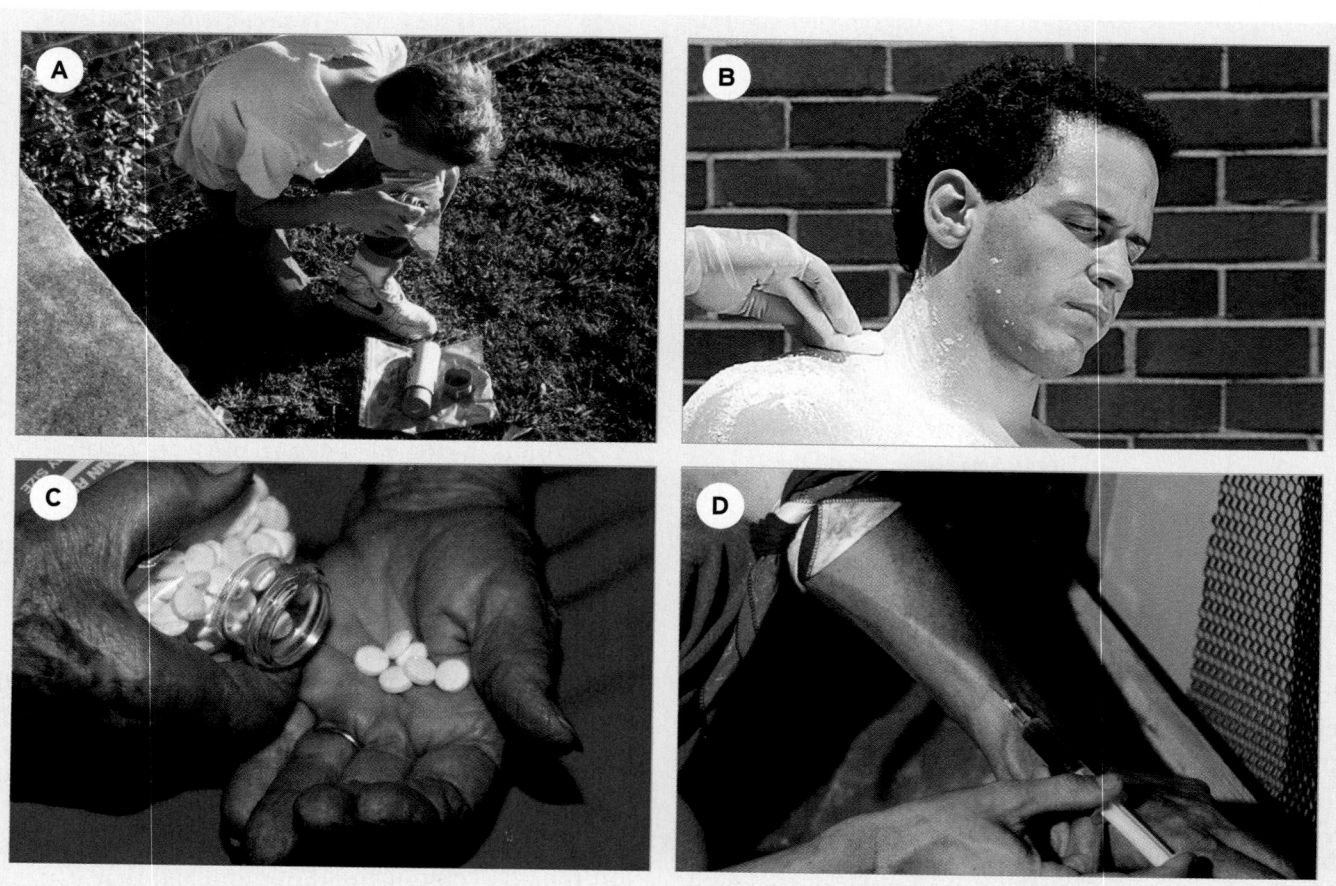

Figure 17-1 There are four routes by which a poison can enter the body. **A.** Inhalation. **B.** Absorption (surface contact). **C.** Ingestion. **D.** Injection.

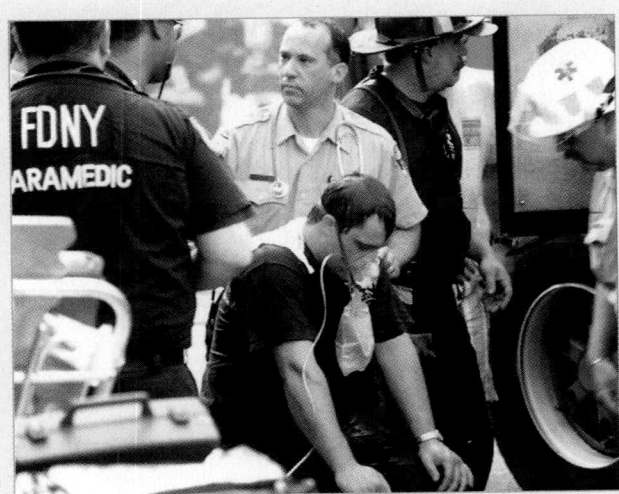

Figure 17-2 Patients who have inhaled poisons need supplemental oxygen and prompt transport to the emergency department.

exposed, they may require supplemental oxygen. Always use self-contained breathing apparatus to protect yourself from poisonous fumes. If you are not specifically trained in the use of this apparatus or do not have appropriately fit-tested equipment available, defer exposure to hazardous environments where inhalational toxins are present to appropriately trained and equipped personnel. Patients may need to be decontaminated by specially trained personnel after they are removed from the toxic environment. You cannot administer emergency care until this step has been completed and the poison cannot contaminate you.

Some inhaled poisons, such as carbon monoxide, are odorless and produce severe hypoxia without damaging or even irritating the lungs. Others, such as chlorine, are very irritating and cause airway obstruction and pulmonary edema. The patient may report the following signs and symptoms: burning eyes, sore throat, cough, chest pain, hoarseness, wheezing, respiratory distress, dizziness, confusion, headache, or stridor in severe cases. The patient may also have seizures or an altered mental status. Some inhaled agents cause progressive lung damage, even after the patient has been removed from direct exposure; the damage may not be evident for a few hours. Meanwhile, it may take 2 or 3 days or more of intensive care to reestablish normal lung function. For this reason, all patients who have inhaled poison require immediate transport to an emergency department. Be prepared to use supplemental oxygen via nonrebreathing mask and/or ventilatory support with a bag-valve-mask (BVM) device, if necessary. Make sure a suctioning unit is available in case the patient vomits. As with other poisonings, it is helpful to take the containers, bottles, and labels when you transport the patient to the hospital.

Absorbed and Surface Contact Poisons

Poisons that come in contact with the surface of the body can affect the patient in many ways. Many corrosive substances will damage the skin, mucous membranes, or eyes, causing chemical burns, telltale rashes, or lesions. Acids, alkalis, and some petroleum (hydrocarbon) products are very destructive. Other substances are absorbed into the bloodstream through the skin and have systemic effects, just like medications or drugs taken via the oral or injectable routes. Other substances such as poison ivy or poison oak may just cause an itchy rash without being dangerous to health. It is important, therefore, to distinguish between contact burns and contact absorption.

EMT-B Safety

Scene safety should be your primary concern when you are called to an inhalation incident. Any time there is more than one patient and no evidence of mechanism of injury, be suspicious. Toxic fumes may be odorless and colorless. If the substance is in the atmosphere, it will affect the rescuers as well as the victims. An incapacitated EMT is no good to anyone. Be suspicious of toxic fumes when encountering patients with changes in level of consciousness, especially at an industrial site or enclosed space.

EMT-B Safety

Absorption of toxic substances through the skin is a common problem in agriculture and manufacturing. Most solvents and "cides"—such as insecticides, herbicides, and pesticides—are toxic and can be readily absorbed through the skin.

Signs and symptoms of absorbed poisoning include a history of exposure, liquid or powder on a patient's skin, burns, itching, irritation, redness of the skin in light-skinned individuals, or typical odors of the substance.

Emergency treatment for a typical contact poisoning includes the following two steps:

1. Avoid contaminating yourself or others.
2. Remove the irritating or corrosive substance from the patient as rapidly as possible.

Remove all clothing that has been contaminated with poisons or irritating substances, thoroughly brush off any dry chemicals, flush the skin with running water, and then wash the skin with soap and water. When a large amount of material has been spilled on a patient, flooding the affected part for at least 20 minutes may be the fastest and most effective treatment. If the patient has a chemical agent in the eyes, you should irrigate them quickly and thoroughly, at least 5 to 10 minutes for acid substances and 15 to 20 minutes for alkalis. To avoid contaminating the other eye as you irrigate the eyes, make sure that the fluid runs from the bridge of the nose outward (Figure 17-3 ▼). This should be started initially on the scene and continued during transport.

Many chemical burns occur in industrial settings, where showers and specific protocols for handling surface burns are available. If you are called to such a scene, trained people usually will be there to assist you. Do not spend time trying to neutralize substances on the skin with additional chemicals. This may actually be more harmful. Instead, wash the substance off immediately with plenty of water. Obtain material safety data

EMT-B Safety

To minimize contamination from a patient who has contacted a hazardous substance, one EMT-B should stay fully protected and assist the patient. This EMT-B is considered "contaminated." The other EMT-B should stay clear and have as little contact as possible, to be able to provide what is needed and drive without contaminating equipment and the front of the vehicle. Do not breathe vapors or powders. Do not get the chemical or runoff on you or others while removing clothes and flushing the patient's skin.

sheets from industrial sites and transport them with the patient, if available.

The only time you should not irrigate the contact area with water is when a poison reacts violently with water, such as contamination with phosphorus or elemental sodium. These substances ignite when they come into contact with water. Instead, brush the chemical off the patient, remove contaminated clothing, and apply a dry dressing to the burn area. Be sure to wear appropriate protective gloves and the proper protective clothing.

Provide prompt transport to the emergency department for definitive care. En route, continue irrigation and provide oxygen if possible.

Ingested Poisons

Approximately 80% of all poisoning is by mouth (ingestion). Ingested poisons include liquids, household cleaners, contaminated food, plants, and, in the majority of cases, drugs. Ingested poisoning is usually accidental in children and, except for contaminated food, deliberate in adults. Plant poisonings are common among children, who like to explore and often bite the leaves of various bushes or shrubs.

Your goal as an EMT-B is to rapidly remove as much of the poison as possible from the gastrointestinal tract. For most poisoning victims, this emergency treatment is sufficient.

In the past, syrup of ipecac was used to cause vomiting, but today it is recommended in only a few situations in which the risk of losing consciousness is clearly low. Because syrup of ipecac induces vomiting, individuals who have ingested substances that may cause diminished alertness over time might vomit and inhale the vomitus into the lungs as they lose consciousness.

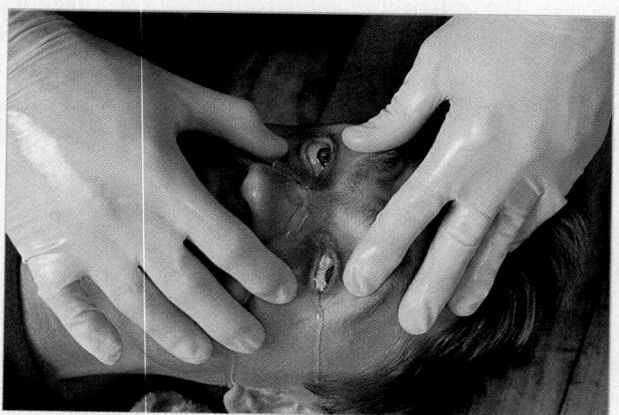

Figure 17-3 If chemical agents are in the patient's eyes, irrigate the eyes quickly and thoroughly, ensuring that the irrigation fluid runs from the bridge of the nose outward. (Use of a nasal cannula is pictured.)

🖐 EMT-B Safety

Be aware that some chemicals react with water. Although small amounts can usually be flushed safely with large quantities of water, larger amounts of such chemicals can give off toxic fumes or explode when wet. Be sure to check the relevant warnings and placards.

✳ EMT-B Tips

While one EMT-B explains the charcoal, the other can prepare a large plastic garbage bag to hang on the patient as a bib. This will help contain the charcoal solution if the patient vomits.

As a result, syrup of ipecac is usually not carried on ambulances. Today, many EMS systems allow you to carry activated charcoal on your unit. Activated charcoal comes as a suspension that binds to the poison in the stomach and carries it out of the system. Therefore, it is more effective and safer than syrup of ipecac. Because activated charcoal is an inky, messy fluid, you may have to do some coaxing to get the patient to drink it; try to give it in a covered cup with a straw (Figure 17-4 ▼). Remember, you should never force this (or any other) liquid into a patient's mouth.

Although every poison will result in a specific set of symptoms and signs, you should always assess the airway, breathing, and circulation (ABCs) of every patient who has been poisoned. Many patients have died as a result of problems with the ABCs that might have been managed easily. Be prepared to provide aggressive ventilatory support and CPR to a patient who has ingested an opiate, sedative, or barbiturate, each of which can cause depression of the central nervous system (CNS) and slow breathing. Whenever poisoning is involved, you should provide prompt transport to the emergency department. The patient may need IV support and other treatments that can be given only in the hospital. If you work in a tiered system, ALS backup also may be appropriate because these providers often carry and can administer additional medications and therapies.

Injected Poisons

Poisoning by injection is usually the result of drug abuse, such as heroin or cocaine (Figure 17-5 ▼). Contrary to the thinking of television detectives, the

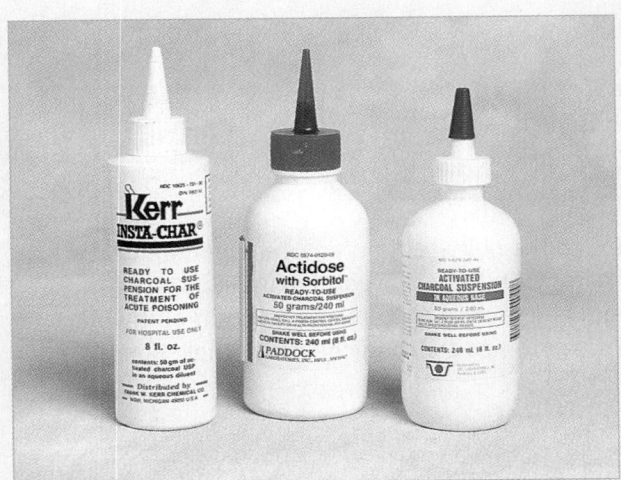

Figure 17-4 Activated charcoal comes as a premixed suspension that you should give, if local protocol allows, in a covered cup with a straw.

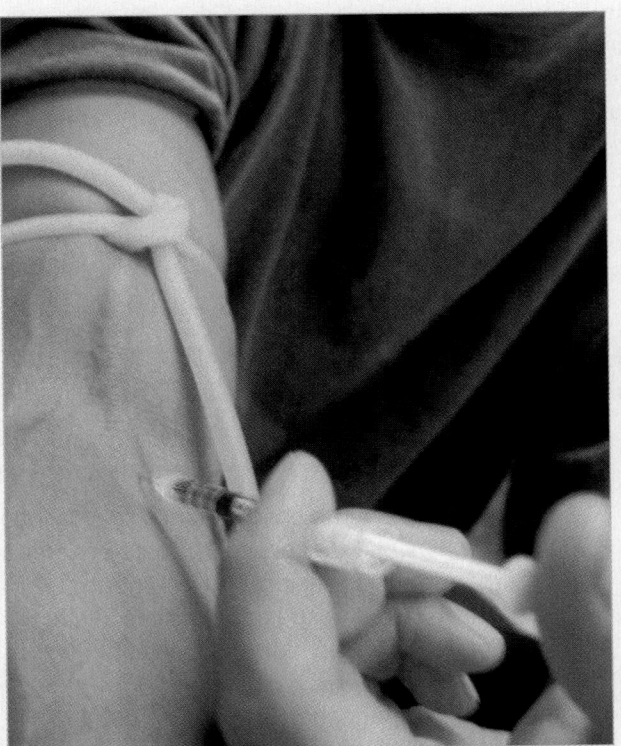

Figure 17-5 Injected poisons are impossible to dilute or remove from the body in the field; therefore, prompt transport to the emergency department is critical.

Documentation Tips

Take time at the scene to make thorough and legible notes about the nature of the poisoning. You can then quickly state the type and amount of substance and the time and route of exposure in your radio, verbal, and written reports. Clear notes that can be handed over on arrival will also be appreciated by busy hospital staff.

Assessment of the Poisoned Patient

Because of the risk of possible cross-contamination by poisons that can be inhaled, ingested, and injected, you must take appropriate BSI precautions. Use the appropriate personal protective equipment necessary to avoid being contaminated.

Scene Size-up

only other parties who are likely to have injected a patient with poison are insects and animals.

Signs and symptoms of poisoning by injection can have a multitude of presentations, including weakness, dizziness, fever, chills, and unresponsiveness, or the patient may be easily excited.

In general, injected poisons are impossible to dilute or remove because they are usually absorbed quickly into the body or cause intense local tissue destruction. If you suspect that rapid absorption has occurred, monitor the patient's airway, provide high-flow oxygen, and be alert for nausea and vomiting. Remove rings, watches, and bracelets from areas around the injection site if swelling occurs. Prompt transport to the emergency department is essential. Take all containers, bottles, and labels with the patient to the hospital.

This is a situation in which a well-trained dispatcher is of great value. Dispatchers with an appropriate set of protocols and excellent interrogation skills can obtain important information pertaining to a poisoning call that will help anticipate the proper protection needed to ensure safety. The dispatcher may be able to obtain information pertaining to the number of patients involved, whether additional resources are needed, and whether trauma is involved. If this information is not obtained before your arrival, you must take the time to assess the scene thoroughly to ensure safety and to determine the nature of the illness, the number of patients involved, the need for additional resources, and whether spine stabilization is required.

You are the Provider   Part 2

As you approach the child, you notice a garbage can next to the child containing what appears to be vomitus. The child is conscious and alert. Her airway is patent, and she is breathing 34 labored breaths/min. The area around her mouth is bright red and has several blisters. Inside her mouth you find red, irritated tissue and multiple blisters. The child's breath has a strong chemical smell. You immediately provide high-flow oxygen.

3. Of what significance is the vomitus?
4. What should you do with the vomitus before leaving the scene?
5. Would you consider the use of activated charcoal in the treatment of this patient?
6. What is the phone number for your local poison control center? What is the National Poison Control Center phone number?

Figure 17-6 A laboratory capable of producing large quantities of methamphetamine.

As you approach the scene, you should look for clues that might indicate the substance and/or poison involved.

- Are there medicine bottles lying around the patient and scene? If so, is there medication missing that might indicate an overdose?
- Are there alcoholic beverage containers present?
- Are there syringes or other drug paraphernalia on the scene?
- Is there an unpleasant or odd odor in the room? If so, is the scene safe? This could be a clue to an inhaled poison too.

A suspicious odor and/or drug paraphernalia may indicate the presence of a drug laboratory. Drug laboratories can be very volatile, so ensure scene safety Figure 17-6 ▲.

Conducting a scene size-up will help ensure safety, help determine the appropriate actions you need to take, and ensure that patient care begins successfully.

Initial Assessment

General Impression

By obtaining a general impression of the patient, chief complaint, apparent life threats, and level of consciousness, you are trying to determine the severity of the patient's condition. With substance abuse and poisonings, do not be fooled into thinking that a conscious, alert, and oriented patient is in stable condition and has no apparent life threats. The patient may have a harmful or even lethal amount of poison in his or her

system that has not had time to produce systemic reactions. An initial assessment that reveals a patient with signs of distress and/or altered mental status gives you early confirmation that the poisonous substance is causing systemic reactions.

Airway and Breathing

Quickly ensure that the patient has an open airway and adequate ventilation. Do not hesitate to begin oxygen therapy for the patient. If the patient is unresponsive to painful stimuli, you need to consider inserting an airway adjunct to ensure an open airway. Have suction available; these patients are susceptible to vomiting. You may also have to assist a patient's ventilations with a BVM device because of substances that act as depressants. As you assess and manage the patient's airway and breathing, you must consider the potential for spinal injury. Spinal precautions in an unresponsive patient must begin when the airway is first opened and continued when positive pressure ventilations are needed.

Circulation

Once the airway and breathing have been assessed and appropriate interventions performed, assess the patient's circulatory status. You will find variations in a patient's circulatory status depending on the substance involved. Assess the pulse and skin condition. Some poisons are stimulants, and others are depressants. Some poisons will cause vasoconstriction and others, vasodilation. Although bleeding may not be obvious, alterations in consciousness may have contributed to trauma and bleeding.

Transport Decision

Patients with obvious alterations in the ABCs or those for whom you have poor general impression should be considered for immediate transport. A delay on the scene to further assess and treat patients is rarely indicated. Some industrial settings may have specific decontamination stations and antidotes available at the site. The majority of time, decontamination and antidote administration will have been initiated by the industrial response team before your arrival and should not delay rapid transport. Consider decontamination before transport depending on the poison your patient was exposed to. This would be necessary if a patient will continue to off-gas or the treating crew has the potential to become exposed in the confined space of the ambulance during transit. This is especially important when transporting exposed patients in a helicopter.

Focused History and Physical Exam

Once initial life threats have been addressed, you may begin the next step of your assessment, the focused history and physical exam. In most situations, this can be performed in the ambulance en route to the hospital. If no trauma is involved and the patient remains unresponsive after your initial assessment, begin with a rapid physical exam to assess for hidden problems or indications of a poisoning or exposure to a chemical. Then follow up with vital signs and gather as much of a SAMPLE history as you can.

If your patient is responsive and can answer questions, begin with an evaluation of the exposure and the SAMPLE history. This will guide you in a focused physical exam of the area exposed or the most concerning problems. Follow up with a complete set of baseline vital signs. In these situations, the history guides you in what to focus on as you continue to assess the patient's complaints; the focused physical exam helps to explain what is happening outside the patient's body, and the vital signs tell you what is happening inside the body. These three assessments give you direction in the interventions your patient might need.

SAMPLE History

As part of the SAMPLE history, you should ask the following questions:

- What is the substance involved? If you know the substance involved, you will be better able to access the appropriate resource, such as the poison control center, to determine lethal doses, time before harmful effects begin, effects of the substance at toxic levels, and appropriate interventions.
- When did the patient ingest or become exposed to the substance? This will let you know if and when the harmful effects will begin. This will also let the emergency physician know what harmful effects can be reversed and which one cannot because of the length of time the patient has been exposed to the substance.
- How much did the patient ingest or what was the level of exposure? With this information, the poison control center will be able to inform you whether the patient has had a harmful or lethal dose.
- Over what period did the patient take the substance? Did the patient take the substance all at once or during minutes or hours?

- Has the patient or a bystander performed any intervention on the patient? Has the intervention helped? The patient's intervention may cause more complications. The emergency physician will also need to know this information to be able to adjust interventions accordingly.
- How much does the patient weigh? If activated charcoal is indicated, you will need to determine the dose based on the patient's weight. The antidote or neutralizing agent given by the emergency physician may be based on the patient's weight as well. For the physical exam, assess the affected body systems, giving particular attention to the respiratory and cardiovascular systems.

Focused Physical Exam

Your focused physical exam should focus on the area of the body or the route of exposure. For example, if a person has ingested a poison, inspect the mouth for indications of poisoning. Are there burns from caustic chemicals? Are there plant or pill fragments? If the person's skin came in contact with a poison, is there a rash or burns? How large an area was involved? If a respiratory exposure occurred, auscultate the lungs. Is there good air movement in and out of the lungs? Do you hear any wheezing or crackles? Much of what you would focus on in your physical exam is based on the route of exposure and the particular drug or chemical the patient was exposed to. Take the time to become familiar with the effects of general classes of drugs and chemicals until you become familiar with specific and common poisons.

Baseline Vital Signs

A complete set of baseline vital signs is an important tool for you to determine how your patient is doing. Many poisons have no outward indications of the seriousness of the exposure. Alterations in the level of consciousness, pulse, respirations, blood pressure, and skin are more sensitive indicators that something serious is wrong.

Interventions Read Whole Thing

The treatment you provide for poisoned patients depends a great deal on what they were exposed to, how they were exposed, and other signs and symptoms found in your assessment. Supporting the ABCs is most important. Some poisons can be easily diluted or decontaminated before transport or en route to the hospital. Dilute airborne exposures with oxygen, remove contact exposures with copious amounts of water unless contraindicated, and consider activated charcoal for ingested poisons. Contact your medical control or a poison control center to discuss treatment options for particular poisonings.

Detailed Physical Exam

Often a detailed exam will provide additional information on the exposure the patient experienced. A general review of all body systems may help to identify systemic problems. This review should be performed, at a minimum, on patients with extensive chemical burns or other significant trauma and on patients who are unresponsive. Management of the ABCs should be the priority assessment and treatment goal. These interventions would take precedence over a detailed physical exam.

Ongoing Assessment

The condition of patients exposed to poisons may change suddenly and without warning. You should continually reassess the adequacy of the patient's ABCs. Repeat the vital signs, and compare them with the baseline set obtained earlier in your assessment. Evaluate the effectiveness of interventions you have provided. If your assessment has provided necessary information about the poisonous substance, you may be able to anticipate changes in the patient's condition. If the patient has consumed a harmful or lethal dose of a poisonous substance, you must repeat the assessment of vital signs every 5 minutes, or constantly if needed. If the patient is in stable condition and there are no life threats, reassess every 15 minutes. If the poison or the level of exposure (for example, the number and type of pills taken) is unknown, careful and frequent reassessment is mandatory.

Communication and Documentation

Once you have completed your focused history and physical exam, including baseline vital signs, contact medical control to request necessary interventions. Report to the hospital as much information as you have about the poison or chemical the patient was exposed to. If a material safety data sheet is immediately available in a work setting, take it with you. If it is not immediately available, ask the company to fax it to the receiving hospital while you are en route. This will help to identify and quickly make available specific interventions and potential antidotes.

Emergency Medical Care

External decontamination is important. Remove tablets or fragments from the patient's mouth, and wash or brush poison from the patient's skin. Treatment focuses on support: assessing and maintaining the patient's ABCs, being sure to monitor the patient's breathing.

In some cases, you will give activated charcoal to patients who have ingested poison, if approved by medical control or local protocol. Charcoal is not indicated for patients who have ingested an acid, an alkali, or a petroleum product; who have a decreased level of consciousness and cannot protect their airway; or who are unable to swallow.

Remember that activated charcoal adsorbs, or sticks to, many commonly ingested poisons, preventing the <u>toxin</u> (poison) from being absorbed into the body by the stomach or intestines. If local protocol permits, you will likely carry plastic bottles of premixed suspension, each containing up to 50 g of activated charcoal. Some common trade names for the suspension form are InstaChar, Actidose, and LiquiChar. The usual dose for an adult or child is 1 g of activated charcoal per kilogram of body weight. The usual adult dose is 25 to 50 g, and the usual pediatric dose is 12.5 to 25 g.

Before you give a patient charcoal, obtain approval from medical control. Next, shake the bottle vigorously to mix the suspension. The medication looks like mud, so it is best to cover the outside of the container so that the fluid is not visible and ask the patient to drink with a straw. You might need to persuade the patient to drink it, particularly if the patient is a child, but never force it. If the patient takes a long time to drink the mixture, you will have to shake the container frequently to keep the medication mixed. Once the patient has finished, discard the container from which the charcoal was administered. Be sure to record the time when you administered the activated charcoal.

The major side effect of ingesting activated charcoal is black stools. If the patient has ingested a poison that causes nausea, he or she may vomit after taking activated charcoal, and the dose will have to be repeated. As you reassess the patient, be prepared for vomiting, nausea, and possible airway problems.

Specific Poisons

Over time, a person who routinely misuses a substance may need increasing amounts of it to achieve the same result. This is called developing a <u>tolerance</u> to the substance. A person with an <u>addiction</u> has an overwhelming desire or need to continue using the agent, at whatever cost, with a tendency to increase the dose. This does not happen only with the classic drugs of abuse, such as cocaine. Almost any substance can be abused, including laxatives, nasal decongestants, vitamins, and food.

The importance of safety awareness and BSI precautions in caring for victims of drug abuse cannot be stressed enough. Known drug abusers have a fairly high incidence of serious and undiagnosed infections, including human immunodeficiency virus and hepatitis. These patients, when intoxicated, may bite, spit, hit, or otherwise injure you, causing you to come into contact with their blood and other body fluids. Always be sure to wear appropriate protective equipment. A calm, professional approach can defuse frightening situations, but keep your safety and that of your team uppermost in mind. Expect the unexpected and remember: the drug user, not the drug, can pose the greatest threat.

Alcohol

The most commonly abused drug in the United States is alcohol (Figure 17-7 ▶). It affects people from all walks of life and kills more than 200,000 people each year. More than 40% of all traffic fatalities or injuries, 67% of murders, and 33% of suicides are related to alcohol, which impairs the capacity to think and function rationally. Alcoholism is one of the greatest national health problems, along with heart disease, cancer, and stroke.

Figure 17-7 Alcohol intoxication causes altered mental status, slowed reflexes, and impaired reaction time.

Alcohol is a powerful CNS depressant. It is a <u>sedative</u>, a substance that decreases activity and excitement, and a <u>hypnotic</u>, meaning that it induces sleep. In general, alcohol dulls the sense of awareness, slows reflexes, and reduces reaction time. It may also cause aggressive and inappropriate behavior and lack of coordination. However, a person who appears intoxicated may have other medical problems as well. Look for signs of head trauma, toxic reactions, or uncontrolled diabetes. Severe acute alcohol ingestion may cause hypoglycemia, which may contribute to the symptoms. At the very least, you should assume that all intoxicated patients are experiencing a drug overdose and require thorough examination by a physician. In most states, such patients cannot legally refuse transport.

You are the Provider   Part 3

As you question the teacher, you discover that your patient ingested CLR cleaner. The incident took place approximately 5 minutes before your arrival. The patient started vomiting almost immediately after the ingestion took place. While you continue to assess your patient and gather additional information, your partner calls the poison control center and medical control for additional assistance. The CLR label states not to induce vomiting but to drink a glass of water followed by a glass of milk. Call a physician immediately.

7. Would you provide the treatment dictated on the bottle?
8. What additional information do you need to know about the incident to assist in providing appropriate treatment?

Alcohol potentiates many other drugs and is commonly not the only drug taken. Over-the-counter drugs including antihistamines and diet medications can cause serious problems when combined with alcohol.

If a patient exhibits signs of serious CNS depression, you must provide respiratory support. This may be difficult, however, because depression of the respiratory system can also cause <u>emesis</u>, or vomiting. The vomiting may be very forceful or even bloody (<u>hematemesis</u>) because large amounts of alcohol irritate the stomach. Internal bleeding should also be considered if the patient appears to be in shock (hypoperfusion) because blood might not clot effectively in a patient who has a prolonged history of alcohol abuse.

A patient in alcohol withdrawal may experience frightening hallucinations, or <u>delirium tremens (DTs)</u>, a syndrome characterized by restlessness, fever, sweating, disorientation, agitation, and even seizures. These conditions may develop if patients no longer have their daily source of alcohol. Alcoholic hallucinations come and go. A patient with an otherwise fairly clear mental state may see fantastic shapes or figures or hear odd voices. Such auditory and visual hallucinations often precede DTs, which are a much more severe complication.

About 1 to 7 days after a person stops drinking or when consumption levels are decreased suddenly, DTs may develop. Patients may experience one or more of the following signs and symptoms:

- Agitation and restlessness
- Fever
- Sweating
- Confusion and/or disorientation
- Delusions and/or hallucinations
- Seizures

Provide prompt transport for these patients after you have completed your assessment and given necessary care. A person who is experiencing hallucinations or DTs is extremely ill. Should seizures develop, treat them as you would any other seizure. The patient should not be restrained, although you must protect him or her from self-injury. Give the patient oxygen, and watch carefully for vomiting. Hypovolemia may develop due to sweating, fluid loss, insufficient fluid intake, or vomiting associated with DTs. If you see signs of hypovolemic shock, elevate the patient's feet slightly, clear the airway, and turn the head to one side to minimize the chance of aspiration during transport. These patients may not respond appropriately to suggestions or conversation; they are often confused and frightened. Therefore, your approach should be calm and relaxed. Reassure the patient, and provide emotional support.

Opioids

The pain relievers called opioid analgesics are named for the opium in poppy seeds, the origin of heroin, codeine, and morphine. On the list of frequently abused drugs, they have been joined by a number of synthetic opioids, with origins in the laboratory. These include meperidine (Demerol), hydromorphone (Dilaudid), propoxyphene (Darvon), oxycodone (Percocet), oxycodone hydrochloride (OxyContin), hydrocodone (Vicodin), and methadone (Table 17-2 ▼). Most of these drugs have legitimate medical uses. With the exception of heroin, which is illegal in the United States, many addicts may have started using any of the opioids with an appropriate medical prescription.

These agents are CNS depressants and can cause severe respiratory depression. When administered intravenously, however, they produce a characteristic "high" or "kick." Tolerance develops rapidly, so some users may require massive doses to experience the same high. In general, emergency medical problems related to opioids are caused by respiratory depression, including a decreased volume of inspired air and decreased respirations. Patients typically appear sedated and cyanotic and have pinpoint pupils.

Treatment includes supporting the airway and breathing. You may try to arouse patients by talking loudly to them or shaking them gently. Always open the airway, give supplemental oxygen, and be prepared for vomiting. Many home remedies are believed to reverse the respiratory depression associated with heroin overdose, including applying ice to the groin or forcing milk into the mouth. None of these work, and they frequently complicate the clinical picture. Nevertheless,

TABLE 17-2 Common Opioid Drugs
Butorphanol (Stadol)
Codeine
Fentanyl derivatives ("China White")
Heroin
Hydrocodone (Hycodan)
Hydromorphone (Dilaudid)
Meperidine (Demerol)
Methadone (Dolophine)
Morphine
Oxycodone (Percodan)
Oxycodone hydrochloride (OxyContin)
Pentazocine (Talwin)
Propoxyphene (Darvon)

you should be aware that a patient's friends may have attempted inappropriate methods of resuscitation. The only effective antidote to reverse the symptoms and signs of opioid overdose are certain narcotic antagonists such as naloxone (Narcan). Patients will respond within 2 minutes to naloxone when it is given intravenously. Naloxone is usually administered by paramedics or by physicians at the emergency department.

Sedative-Hypnotic Drugs

Barbiturates and benzodiazepines have been a part of legitimate medicine for a long time. They are easy to obtain and relatively cheap. People sometimes solicit prescriptions from several physicians for the same hypnotics or a variety of sedative-hypnotics. These drugs are CNS depressants and alter the level of consciousness, with effects similar to those of alcohol so that the patient may appear drowsy, peaceful, or intoxicated (Table 17-3 ▼). By themselves, these drugs do not relieve pain, nor do they produce a specific high, although users often take alcohol or an opioid at the same time to boost their effects.

In general, these agents are taken by mouth. Occasionally, however, contents of capsules are suspended or dissolved in water and injected to produce a rather sudden state of ease and contentment. Use of IV sedative-hypnotic drugs quickly induces tolerance, so an individual requires increasingly larger doses. You are less likely to be called on to treat an acute overdose in someone who chronically abuses these drugs; however, you may be called to a scene of an attempted suicide in which the patient has taken large quantities of these drugs. In these situations, patients will have marked respiratory depression and may be in a coma.

Sedative-hypnotic drugs may also be given to unsuspecting people as a "knock-out" drink, or "Mickey Finn." More recently, drugs such as flunitrazepam (Rohypnol) have been abused as a "date rape drug," causing an unwary individual to become sedated and even unconscious. The individual later awakens, confused and unable to remember what happened.

In general, your treatment of patients who have overdosed with sedative-hypnotics and have respiratory depression is to provide airway clearance, ventilatory assistance, and prompt transport. Give supplemental oxygen, and be ready to assist ventilation. You may attempt to stimulate the person by speaking loudly or gently shaking him or her; remember to watch for vomiting.

A specific antidote is available for acute benzodiazepine overdose. It is called flumazenil and is given intravenously. Although it will reverse the sedation and respiratory depression of the benzodiazepine sedative-hypnotics, it will have no effect on the signs and symptoms of overdose from ethyl alcohol or barbiturates. Almost always, flumazenil is administered in the hospital after a physician's assessment. As multidrug use becomes more common, you may find it increasingly difficult to determine what agents patients have taken. Your best approach is to treat any obvious injuries or illnesses, keeping in mind that drug use may complicate the picture and make full life support necessary. Focus on the ABCs, especially the possibility of airway problems (relaxation of the tongue, causing obstruction), vomiting, respiratory depression, and, in severe cases, cardiac arrest.

Abused Inhalants

Many abused inhalants produce several of the same CNS effects as do other sedative-hypnotics, but these agents are inhaled instead of ingested or injected. Some of the more common agents include acetone, toluene, xylene, and hexane, which are found in glues, cleaning compounds, paint thinners, and lacquers. Similarly, gasoline and various halogenated hydrocarbons, such as Freon,

TABLE 17-3 Examples of Sedative-Hypnotic Drugs

Barbiturates	Benzodiazepines	Others
Amobarbital (Amytal)	Alprazolam (Xanax)	Carisoprodol (Soma)
Butabarbital (Butisol)	Chlordiazepoxide (Librium)	Chloral hydrate ("Mickey Finn")
Pentobarbital (Nembutal)	Diazepam (Valium)	Cyclobenzaprine (Flexeril)
Phenobarbital (Luminal)	Flunitrazepam (Rohypnol)	Ethchlorvynol (Placidyl)
Secobarbital (Seconal)	Lorazepam (Ativan)	Ethyl alcohol (drinking alcohol)
	Oxazepam (Serax)	Glutethimide (Doriden)
	Temazepam (Restoril)	Hydrocarbon inhalants
		Isopropyl alcohol (rubbing alcohol)
		Meprobamate (Equagesic)

used as propellants in aerosol sprays, are also abused as inhalants. None of these inhalants is a medication. Because these are products that can be bought in hardware stores, they are commonly abused by teenagers seeking an alcohol-like high. The effective dose and the lethal dose are very close, making these extremely dangerous drugs. The low cost and relative availability makes them favorites of children and curious experimenters. This is unfortunately an often lethal combination.

Always use special care in dealing with a patient who may have used inhalants. Their effects range from mild drowsiness to coma, but unlike most other sedative-hypnotics, these agents may often cause seizures. Also, halogenated hydrocarbon solvents can make the heart supersensitive to the patient's own adrenaline, putting the patient at high risk for sudden cardiac death due to ventricular fibrillation; even the action of walking may release enough adrenaline to cause a fatal ventricular arrhythmia. You must try to keep such patients from struggling with you or exerting themselves. Give supplemental oxygen, and use a stretcher to move the patient. Prompt transport to the hospital is essential; monitor vital signs en route.

Sympathomimetics

Sympathomimetics are CNS stimulants that frequently cause hypertension, tachycardia, and dilated pupils. A stimulant is an agent that produces an excited state. Amphetamine and methamphetamine ("ice") are commonly taken by mouth. They are also injected by abusers in many cases. They typically are taken to make the user "feel good," improve task performance, suppress appetite, or prevent sleepiness. They may just as easily produce irritability, anxiety, lack of concentration, or seizures. Other common examples include phentermine and amphetamine sulfate (Benzedrine). Caffeine, theophylline, and phenylpropanolamine (a nasal decongestant) are all mild sympathomimetics. So-called designer drugs, such as Ecstasy and Eve, are also frequently abused in certain areas of the United States.

Sympathomimetic drugs are frequently called "uppers" (Table 17-4 ▶). Someone using one of these agents may display disorganized behavior, restlessness, and sometimes anxiety or great fear. Paranoia and delusions are common with sympathomimetic abuse.

Cocaine, also called coke, crack, crystal, snow, freebase, rock, gold dust, blow, and lady, may be taken in a number of different ways. Classically, it is inhaled into the nose and absorbed through the nasal mucosa, damaging tissue, causing nosebleeds, and ultimately destroying the nasal septum. It can also be injected intravenously or subcutaneously (skin-popping).

Cocaine can be absorbed through all mucous membranes and even across the skin. In any form, the immediate effects of a given dose last less than an hour.

Another method of abusing cocaine is by smoking it. Crack is pure cocaine. It melts at 93°F (34°C) and vaporizes at a slightly higher temperature. Therefore, crack is easily smoked. In this form, it reaches the capillary network of the lungs and can be absorbed into the body in seconds. The immediate outflow of blood from the heart speeds the drug to the brain, so its effect is felt at once. Smoked crack produces the most rapid means of absorption and, therefore, the most potent effect.

Cocaine is one of the most addicting substances known. Its immediate effects include excitement and euphoria. Acute cocaine overdose is a genuine emergency because patients are at high risk for seizures and cardiac arrhythmias. Chronic cocaine abuse may cause hallucinations; patients with "cocaine bugs" think that bugs are crawling out of their skin.

In caring for patients who have been poisoned with any of the sympathomimetics, be aware that their severe agitation can lead to tachycardia and hypertension. Patients may also be paranoid, putting you and other health care providers in danger. Law enforcement officers should be at the scene to restrain the patient, if necessary. Do not leave the patient unattended and unmonitored during transport.

TABLE 17-4   Street Names for Amphetamines

Street Name	Drug Name
Adam	3,4-Methylenedioxymethamphetamine (MDMA)
Bennies	Amphetamines
Crank	Crack cocaine, heroin, amphetamine, methamphetamine, methcathinone
DOM	4-Methyl-2,5-dimethoxyamphetamine
Ecstasy	MDMA
Eve	MDMA
Fen-phen	Phentermine
Golden eagle	4-Methylthioamphetamine
Ice	Cocaine, crack cocaine, smokable methamphetamine, methamphetamine, MDMA, phencyclidine (PCP)
MDA	Methaqualone
Meth	Methamphetamine
Speed	Crack cocaine, amphetamine, methamphetamine
STP	PCP
Uppers	Amphetamines

All of these patients need prompt transport to the emergency department because of their risk of seizures, cardiac arrhythmias, and stroke. You may see blood pressures as high as 250/150 mm Hg. Give supplemental oxygen and be ready to provide suctioning. If the patient is already having a seizure, you must protect him or her against self-injury.

Marijuana

The flowering hemp plant, *Cannabis sativa*, called marijuana, is abused throughout the world. It has been estimated that as many as 20 million people use marijuana daily in the United States. Inhaling marijuana smoke from a cigarette or pipe produces euphoria, relaxation, and drowsiness. It also impairs short-term memory and the capacity to do complex thinking and work. In some people, the euphoria progresses to depression and confusion. An altered perception of time is common, and anxiety and panic can occur. With very high doses, patients experience hallucinations.

A person who has been using marijuana rarely needs transport to the hospital. Exceptions may include someone who is hallucinating, very anxious, or paranoid. However, you should be aware that marijuana is often used as a vehicle to get other drugs into the body. For example, it may be covered with crack or PCP, also known as "angel dust."

Hallucinogens

Hallucinogens alter an individual's sensory perceptions (Table 17-5 ▶). The classic hallucinogen is lysergic acid diethylamide (LSD). Abuse of another hallucinogen, PCP, or angel dust, is relatively uncommon among young adults. Phencyclidine is a dissociative anesthetic that is easily synthesized and highly potent. Its effectiveness by oral, nasal, pulmonary, and IV routes makes it easy to add to other street drugs. It is dangerous because it causes severe behavioral changes in which individuals often inflict injury on themselves.

All these agents cause visual hallucinations, intensify vision and hearing, and generally separate the user from reality. The user, of course, expects that the altered sensory state will be pleasurable. Often, however, it can be terrifying. At some point, you are bound to encounter patients who are having a "bad trip." They will usually be hypertensive, tachycardic, anxious, and probably paranoid.

Many hallucinogens have sympathomimetic properties. Indeed, your care for a patient who is having a bad reaction to a hallucinogenic agent is the same as that for a patient who has taken a sympathomimetic. Use a calm, professional manner, and provide emotional support. Do not use restraints unless you or the patient is in danger of injury and then always within the guidelines specified by local authorities. These patients may suddenly experience hallucinations or odd perceptions, so you must watch them carefully throughout transport. Never leave a patient who has taken a hallucinogen unattended and unmonitored. Provide a great deal of reassurance, and request ALS assistance.

Anticholinergic Agents

The classic picture of a person who has taken too much of an anticholinergic medication is "hot as a hare, blind as a bat, dry as a bone, red as a beet, and mad as a hatter." These are medications that have properties that, among other effects, block the parasympathetic nerves. Common drugs with a significant anticholinergic effect include atropine, diphenhydramine (Benadryl), Jimson weed, and certain tricyclic antidepressants. With the exception of Jimson weed, these medications usually are not abused drugs but may be taken as an intentional overdose. You will find that it is often difficult to distinguish between an anticholinergic overdose and a sympathomimetic overdose. Both groups of patients may be agitated and tachycardic and have dilated pupils. Once a pure anticholinergic poisoning has been diagnosed, the patient may be treated with physostigmine intravenously by staff in the emergency department, depending on the severity of the situation.

As newer, safer antidepressants such as fluoxetine (Prozac) and sertraline (Zoloft) crowd the market, you can expect to see fewer overdoses of tricyclic antidepressants such as amitriptyline (Elavil) and imipramine (Tofranil). In addition to its anticholinergic effects, a tricyclic antidepressant overdose may cause more serious, life-threatening, effects. This is because the med-

TABLE 17-5 Commonly Abused Hallucinogens
Bufotenine (toad skin)
Dimethyltryptamine (DMT)
Hashish
Jimson weed
LSD
Marijuana
Mescaline
Morning glory
Mushrooms
Nutmeg
PCP
Psilocybin (mushroom)

Tinitus -

ication may block the electrical conduction system in the heart, leading to lethal cardiac arrhythmias. Patients with acute tricyclic antidepressant overdose must be transported immediately to the emergency department; they may go from appearing "normal" to seizure and death within 30 minutes. The seizures and cardiac arrhythmias caused by a severe tricyclic antidepressant overdose are best treated in the hospital with IV sodium bicarbonate. If you work in a tiered system, you should consider calling for ALS backup when you are en route to the scene.

Cholinergic Agents

The "nerve gases" designed for chemical warfare are cholinergic agents. These agents overstimulate normal body functions that are controlled by parasympathetic nerves, resulting in salivation, mucous secretion, urination, crying, and abnormal heart rate. You are unlikely to encounter nerve gases. However, you may well be called to care for patients who have been exposed to one of the organophosphate insecticides or certain wild mushrooms, which are also cholinergic agents. The signs and symptoms of cholinergic drug poisoning are easy to remember because of the mnemonic DUMBELS: *Study This...*

- Defecation
- Urination
- Miosis (constriction of the pupils)
- Bronchorrhea (discharge of mucus from the lungs)
- Emesis
- Lacrimation (tearing)
- Salivation

Alternatively, you can use the mnemonic SLUDGE:

- Salivation
- Lacrimation
- Urination
- Defecation
- GI (gastrointestinal) irritation
- Eye constriction/emesis

In poisonings, patients will have excessive amounts of these normal functions and body secretions. In addition, patients may have either bradycardia or tachycardia.

The most important consideration in caring for a patient who has been exposed to an organophosphate insecticide or some other cholinergic agent is to avoid exposure yourself. Because such agents may cling to a patient's clothing and skin, decontamination may take priority over immediate transport to the emergency department. Hospital staff or paramedics can use the anticholinergic drug atropine to dry up the patient's

secretions. In the meantime, your priorities after decontamination are to decrease the secretions in the mouth and trachea that threaten to suffocate the patient and provide airway support. Depending on your local EMS protocol, this can be treated as a HazMat (hazardous materials) situation.

Miscellaneous Drugs

Although not as common as it was 30 years ago, aspirin poisoning remains a potentially lethal condition. Ingesting too many aspirin tablets, acutely or chronically, may result in nausea, vomiting, hyperventilation, and ringing in the ears. Patients with this problem are frequently anxious, confused, tachypneic, hyperthermic, and in danger of having seizures. They should be transported quickly to the hospital.

Overdosing with acetaminophen is also very common, probably because acetaminophen is available in so many different preparations, such as Tylenol. The good news is that acetaminophen is generally not very toxic. A healthy patient could ingest 140 mg of acetaminophen for every kilogram of body weight without serious adverse effects. The bad news is that the symptoms of an overdose generally do not appear until it is too late. For example, massive liver failure may not be apparent for a full week. And patients may not provide the information necessary for a correct diagnosis. For this reason, gathering information at the scene is very important. By finding an empty acetaminophen bottle, you may save a patient's life. If given early enough (before liver failure occurs), a specific antidote may prevent liver damage.

Be extremely careful in dealing with a child who has unintentionally ingested a poisonous substance. Although such incidents usually do not lead to death, family members may be distraught, and your professional attitude will help to ease the tension. Remember, however, that a single swallow of some substances can kill a child (Table 17-6 ▶).

Some alcohols, including methyl alcohol and ethylene glycol, are even more toxic than ethyl alcohol (drinking alcohol). Although they may be used as a substitute by a chronic alcoholic who is unable to obtain ethyl alcohol, they are more often taken by someone attempting suicide. In either case, immediate transport to the emergency department is essential. Methyl alcohol is found in dry gas products and Sterno; ethylene glycol is found in some antifreeze products. Both cause a "drunken" feeling. Left untreated, both will also cause severe tachypnea, blindness (methyl alcohol), renal failure (ethylene glycol), and eventually death. Even ethyl alcohol (typical drinking alcohol)

TABLE 17-6 Fatal Ingested Poisons
Benzocaine
Calcium channel blockers (verapamil, nifedipine, diltiazem)
Camphor
Chloroquine
Hydrocarbon solvents
Lomotil
Methanol and ethylene glycol
Methylsalicylate (oil of wintergreen)
Phenothiazines (Thorazine)
Quinine
Theophylline
Tricyclic antidepressants (amitriptyline [Elavil], imipramine [Tofranil], nortriptyline [Pamelor])
Visine

TABLE 17-7 Common Sources of Food Poisoning
Bacillus cereus
Campylobacter
Clostridium botulinum toxin
Clostridium perfringens
Cryptosporidium
Enterococcus
Escherichia coli
Giardia lamblia
Rotavirus
Salmonella
Shigella
Staphylococcus toxin
Vibrio parahaemolyticus
Yersinia enterocolitica

can stop a patient's breathing if taken in too high a dose or too fast, particularly in children.

Food Poisoning

The term "ptomaine poisoning" was coined in 1870 to indicate poisoning by a class of chemicals found in rotting food. It is still used today in many news accounts of food poisoning. Food poisoning is almost always caused by eating food that is contaminated by bacteria. The food may appear perfectly good, with little or no decay or odor to suggest danger.

There are two main types of food poisoning. In one, the organism itself causes disease; in the other, the organism produces toxins that cause disease (Table 17-7 ▲). A toxin is a poison or harmful substance produced by bacteria, animals, or plants.

One organism that produces direct effects of food poisoning is the *Salmonella* bacterium. The condition called salmonellosis is characterized by severe gastrointestinal symptoms within 72 hours of ingestion, including nausea, vomiting, abdominal pain, and diarrhea. In addition, patients with salmonellosis may be systemically ill with fever and generalized weakness. Some people are carriers of certain bacteria; although they may not become ill themselves, they may transmit diseases, particularly if they work in the food services industry. Usually, proper cooking kills bacteria, and proper cleanliness in the kitchen prevents the contamination of uncooked foods.

You are the Provider   Part 4

Your partner has contacted the poison control center. The poison control center recommends giving the patient water to help dilute the chemical. Medical control concurs. Your patient's breathing begins to slow. She is now breathing at a rate of 22 breaths/min. Her pulse is 90 beats/min and weak. Blood pressure is 60 by palpation. Her crying has subsided, and she is becoming lethargic and unresponsive to verbal stimuli. Dispatch confirms that ALS is en route and should be on scene in approximately 2 minutes.

9. Given the change in your patient's mental status, would you attempt to give water as you were directed?
10. What additional treatment would you provide?

The more common cause of food poisoning is the ingestion of powerful toxins produced by bacteria, often in leftovers. The bacterium *Staphylococcus,* a common culprit, is quick to grow and produce toxins in foods that have been prepared in advance and kept too long, even in the refrigerator. Foods prepared with mayonnaise, when left unrefrigerated, are a common vehicle for the development of staphylococcal toxins. Usually, staphylococcal food poisoning results in sudden gastrointestinal symptoms, including nausea, vomiting, and diarrhea. Although timeframes may vary from individual to individual, these symptoms usually may start within 2 to 3 hours after ingestion or as long as 8 to 12 hours after ingestion.

The most severe form of toxin ingestion is botulism. This often-fatal disease usually results from eating improperly canned food, in which the spores of *Clostridium* bacteria have grown and produced a toxin. The symptoms of botulism are neurologic: blurring of vision, weakness, and difficulty in speaking and breathing. Symptoms may develop as long as 4 days after ingestion or as early as the first 24 hours.

In general, you should not try to determine the specific cause of acute gastrointestinal problems. After all, severe vomiting may be a sign of a self-limiting food poisoning, a bowel obstruction requiring surgery, or another poison, such as copper, arsenic, zinc, cadmium, scombrotoxin (fish poison), or *Clitocybe* or *Inocybe* mushrooms. Instead, you should gather as much history as possible from the patient and transport him or her promptly to the hospital. When two or more individuals in one group have the same illness, you should take along some of the suspected food. In advanced cases of botulism, you may have to assist ventilation and give basic life support.

Plant Poisoning

Several thousand cases of poisoning from plants occur each year, some severe. Many household plants are poisonous if ingested, as they may be by children who like to nibble leaves (Table 17-8 ▶). Some poisonous plants cause local irritation of the skin; others can affect the circulatory system, the gastrointestinal tract, or the CNS. It is impossible for you to memorize every plant and poison, let alone their effects (Figure 17-8 ▶). You can and should do the following:

1. Assess the patient's airway and vital signs.
2. Notify the regional poison control center for assistance in identifying the plant.

Geriatric Needs

In an accidental overdose or poisoning, a geriatric patient may have become confused about his or her drug regimen. He or she may have forgotten that the medication had been taken, repeating the dose a number of times. Or the patient could have forgotten the doctor's instructions to discard leftover medication and might have taken the current and the older drug, resulting in an increase in effects or an unwanted drug interaction.

A geriatric patient may also intentionally overdose in an attempt to commit suicide. Geriatric patients have been known to ingest common household chemicals such as insecticides, acetaminophen, aspirin, or caustic substances in an attempt to end their lives. Be alert for any indication of an intentional overdose or poisoning, even though the patient might deny an attempted suicide.

In considering any poisoning, remember the basics. Because of the aging process, the absorption of poisons may change. For example, decreased gastric mobility may delay absorption of ingested poisons, limiting systemic effects, but may result in increased damage to the stomach.

If a senior citizen inhales a poison, even in tiny quantities, lung damage can be severe. Consider the decreased lung capacity and ability to exchange oxygen and carbon dioxide in an older patient's lungs. Pulmonary function could be worsened to potentially fatal levels with the inhalation of minute amounts of poison.

For poisons that are absorbed by or injected into the skin, reduced circulation to the skin can decrease or delay absorption into the body. Watch for an increased reaction or irritation at the skin site.

In a geriatric patient, the liver may not be able to metabolize the poison as effectively or the kidneys may not be able to excrete the poison as quickly. In either case, the drug or poison remains in the body for a longer period, causing additional tissue damage. When a medication is not metabolized or excreted as quickly as before, the drug could accumulate to toxic levels and, ultimately, become fatal in lesser doses than in a younger person.

Figure 17-8 The toxins in these common poisonous plants are often ingested or absorbed through the skin. **A.** Dieffenbachia. **B.** Mistletoe. **C.** Castor bean. **D.** Nightshade. **E.** Foxglove. **F.** Rhododendron. **G.** Jimson weed. **H.** Death camas. **I.** Pokeweed. **J.** Rosary pea. **K.** Poison ivy. **L.** Poison oak. **M.** Poison sumac.

TABLE 17-8 Common Toxic Plants

Scientific Name	Common Name
Abrus precatorius	Jequirity bean/rosary pea
Cicuta species	Water hemlock/wild carrot
Colchicum autumnalel	Autumn crocus
Conium maculatum	Poison hemlock
Convallaria majalis	Lily of the valley
Datura species	Jimson weed/stinkweed
Dieffenbachia	Dumbcane
Digitalis purpurea	Foxglove
Nerium oleander	Oleander or rose laurel
Nicotiana glauca	Tree tobacco
Phoradendron	Mistletoe
Phytolacca americana	Pokeweed
Rhododendron	Rhododendron or azalea
Ricinus communis	Castor bean
Solarium nigrum	Nightshade
Zygadenus species	Death camas

3. Take the plant to the emergency department.
4. Provide prompt transport.

Irritation of the skin and/or mucous membranes is a problem with the common houseplant called dieffenbachia, which resembles "elephant ears." When chewed, a single leaf may irritate the lining of the upper airway enough to cause difficulty swallowing, breathing, and speaking. For this reason, dieffenbachia has been called "dumbcane." In rare circumstances, the airway may be completely obstructed. Emergency medical treatment of dieffenbachia poisoning includes maintaining an open airway, giving oxygen, and transporting the patient promptly to the hospital for respiratory support. You should continue to assess the patient for airway difficulties throughout transport. If necessary, provide positive pressure ventilation.

You are the Provider

Summary

As with any patient, your assessment can mean the difference between life and death. In a situation such as this, it is imperative that you assess the ABCs and vital signs continually, watching for any change. Contact the poison control center and medical control immediately. Remember, only medical control can provide the orders you need to treat a patient.

Substance Abuse and Poisoning

Scene Size-up	Body substance isolation precautions should include a minimum of gloves and eye protection. Ensure scene safety and determine NOI/MOI. Consider the number of patients, the need for additional help/ALS, and c-spine stabilization.
Initial Assessment	
■ General impression	Determine level of consciousness and find and treat any immediate threats to life. Determine priority of care based on environment and patient's chief complaint.
■ Airway	Ensure patent airway.
■ Breathing	Provide high-flow oxygen at 15 L/min. Evaluate depth and rate of the respirations and provide ventilations as needed.
■ Circulation	Evaluate pulse rate and quality; observe skin color, temperature, and condition and treat accordingly. Determine if bleeding is present and control if life threatening.
■ Transport decision	Consider decontamination before rapid transport.
Focused History and Physical Exam	*NOTE: The order of the steps in the focused history and physical exam differs depending on whether the patient is conscious or unconscious. The order below is for a conscious patient. For an unconscious patient, perform a rapid physical exam, obtain vital signs, and obtain the history.*
■ SAMPLE history	Ask SAMPLE and pertinent OPQRST questions. Be sure to ask if and what interventions were taken before your arrival, how many interventions, and at what time.
■ Focused physical exam	Perform focused physical exam on affected body systems and any affected area.
■ Baseline vital signs	Take vital signs, noting skin color and temperature. Use pulse oximetry if available.
■ Interventions	Support patient as needed. Consider the use of oxygen, positive pressure ventilations, adjuncts, and proper positioning.
Detailed Physical Exam	Consider a detailed physical exam.
Ongoing Assessment	Repeat the initial assessment, focused assessment, and reassess interventions performed. Reassess vital signs every 5 minutes for the unstable patient, every 15 minutes for the stable patient. Document and treat the patient as necessary.
■ Communication and documentation	Contact medical control with a radio report giving patient's condition, agent of exposure, and amount ingested or exposed to. Relay any significant changes, including level of consciousness or difficulty in breathing. Contact regional poison control center for information as determined by local protocol. Document any changes, the time they occurred, and any interventions performed.

NOTE: While the steps below are widely accepted, be sure to consult and follow your local protocol.

Substance Abuse and Poisoning

General Management
1. Have trained rescuers remove patient from any poisonous environment.
2. Establish and maintain airway, suctioning as needed. Provide high-flow oxygen.
3. Obtain SAMPLE history and vital signs. Ascertain which drug(s) have been taken.
4. Request ALS when available.
5. Take all containers, bottles, and labels of poisons to the receiving hospital.

For patients who have taken hallucinogens, provide calm, prompt transport.

For cholinergic agents, it is critical to take sufficient body substance isolation precautions.

For patients who have plant poisoning, notify regional poison control center for assistance in identifying the plant.

For patients who have food poisoning, if there are more than two patients with the same illness, transport the food suspected to be responsible for the poisoning.

Administer activated charcoal for poisonous ingestions according to local protocol. Follow these steps:
1. Do not give if the patient exhibits altered mental status, has ingested acids or alkalis, or is unable to swallow.
2. Obtain order from medical direction or follow protocol.
3. Shake container.
4. Place in cup with straw and have patient drink 12.5 to 25 g (for infants and children) or 25 to 50 g (for adults).

Prep Kit

Ready for Review

- Poisons act acutely or chronically to destroy or impair body cells.

- If you believe a patient may have taken a poisonous substance, you should notify medical control and begin emergency treatment at once.
  - This may include administration of an antidote, usually at the hospital, if an antidote exists.
  - It also entails collecting any evidence of the type of poison that was used and taking it to the hospital; diluting and physically removing the poisonous agent; providing respiratory support; and transporting the patient promptly to the hospital.

- A poison can be introduced into the body in one of four ways:
  - Ingestion
  - Inhalation
  - Injection
  - Surface contact (absorption)

- Approximately 80% of all poisoning is by ingestion, including plants, contaminated food, and most drugs. In general, activated charcoal should be used in these patients.

- In the case of surface contact poisons, be sure to avoid contaminating yourself. You should remove all contaminated substances and clothing from the patient, and flood the affected part.

- Move patients who have inhaled poison into the fresh air; be prepared to use supplemental oxygen via nonrebreathing mask and/or ventilatory support via a BVM device.

- BLS may be needed for some patients, especially those who have injected poison, which is almost always a deliberate act.

- People who abuse a substance can develop a tolerance to it or can develop an addiction. Always use BSI precautions when caring for victims of drug abuse. In addition to alcohol and marijuana, commonly abused drugs fall into seven categories:
  - Opioid analgesics
  - Sedative-hypnotics
  - Inhalants
  - Sympathomimetics
  - Hallucinogens
  - Anticholinergics
  - Cholinergics

- Like alcohol, drugs in the first three categories depress the CNS and can cause respiratory depression. You must support the airway in such cases, and be prepared for the patient to vomit.

- Take special care with patients who have used inhalants because the drugs may cause seizures or sudden death.

- Sympathomimetics, including cocaine, stimulate the CNS, causing hypertension, tachycardia, seizures, and dilated pupils. Patients who have taken these drugs may be paranoid, as may patients who have taken hallucinogens.

- Anticholinergic medications, often taken in suicide attempts, can cause a person to become hot, dry, blind, red-faced, and mentally unbalanced. An overdose of tricyclic antidepressants can lead to cardiac arrhythmias.

- The symptoms of cholinergic medications, which include organophosphate insecticides, can be remembered by the mnemonic DUMBELS, for excessive defecation, urination, miosis, bronchorrhea, emesis, lacrimation, and salivation; or SLUDGE, for salivation, lacrimation, urination, defecation, GI (gastrointestinal) irritation, and eye constriction/emesis.

- Two main types of food poisoning cause gastrointestinal symptoms.
  - In one type, bacteria in the food directly cause disease, such as salmonellosis; in the other, bacteria such as *Staphylococcus* produce powerful toxins, often in leftover food.
  - The most severe form of toxin ingestion is botulism, which can produce the first neurologic symptoms as late as 4 days after ingestion.

- Plant poisoning can affect the circulatory system, the gastrointestinal system, and the CNS. Some plants, such as the dieffenbachia, irritate the skin or mucous membranes and even sometimes cause obstruction of the airway.

Technology

Interactivities

Vocabulary Explorer

Anatomy Review

Web Links

Online Review Manual

www.EMTB.com

Prep Kit continued...

Vital Vocabulary

addiction A state of overwhelming obsession or physical need to continue the use of a drug or agent.

antidote A substance that is used to neutralize or counteract a poison.

delirium tremens (DTs) A severe withdrawal syndrome seen in alcoholics who are deprived of ethyl alcohol; characterized by restlessness, fever, sweating, disorientation, agitation, and seizures; can be fatal if untreated.

emesis Vomiting.

hallucinogens Agents that produce false perceptions in any one of the five senses.

hematemesis Vomiting blood.

hypnotic A sleep-inducing effect or agent.

ingestion Swallowing; taking a substance by mouth.

opioids Any drug or agent with actions similar to morphine.

poison A substance whose chemical action could damage structures or impair function when introduced into the body.

sedative A substance that decreases activity and excitement.

stimulant An agent that produces an excited state.

substance abuse The misuse of any substance to produce some desired effect.

tolerance The need for increasing amounts of a drug to obtain the same effect.

toxin A poison or harmful substance produced by bacteria, animals, or plants.

vomitus Vomited material.

Points to Ponder

You are dispatched to 600 Blanchard St, #617, for an unknown medical problem. It is a secure building, and you experience some delays in gaining access. After 10 minutes have passed, you are let into the building by the manager, who also has a key to unit 617. As you enter the apartment, you are met by a young girl. She tells you that her younger brother, a 3-year-old, "got into mom's back pills." As you begin to assess the child, your partner finds the empty pill bottles. One is hydrocodone and the other is cyclobenzaprine. You know the first medication is Vicodin, an opioid analgesic, and you recognize the name of the other drug, but cannot remember what its effects are. Your partner tells you that he thinks it is important to know the effects of both drugs before calling medical control to request administration of activated charcoal. He then tells you he is going to retrieve his field guide, which is in the ambulance.

Should you wait for your partner to get his field guide? Is it necessary to identify both medications before calling the hospital as he suggests?

Issues: Importance of Administering Activated Charcoal, Importance of Contacting Medical Control Early When Managing Overdose and Poisoning Patients.

Assessment in Action

You are dispatched to 1808 Bell St for a suicide attempt. As you respond, the dispatcher informs you that your patient is a 17-year-old adolescent who told a friend over the phone that she wanted to die and took a handful of pills.

As you enter the residence, you find a teenage girl in one of the bedrooms. She is lying in bed, hiding under the covers and crying. As you identify yourself, she tells you to go away. You notice an empty bottle of Tylenol lying on the nightstand. Your partner is initially unable to apply oxygen or assess vital signs because she will not let you near her.

1. What is your first concern regarding this call?
   A. Identifying the substance
   B. Administering charcoal
   C. Inducing vomiting
   D. Scene safety

2. What delayed reaction can occur as the result of an acetaminophen overdose?
   A. Heart failure
   B. Lung failure
   C. Liver failure
   D. Kidney failure

3. Activated charcoal is considered a:
   A. solution.
   B. suspension.
   C. Neither A or B
   D. Both A and B

4. What is the usual adult dose of activated charcoal?
   A. 25-50 mg
   B. 12.5-25 mg
   C. 25-50 g
   D. 12.5-25 g

5. What is the usual pediatric dose of activated charcoal?
   A. 25-50 mg
   B. 12.5-25 mg
   C. 25-50 g
   D. 12.5-25 g

6. Activated charcoal is contraindicated in:
   A. ingested acids, alkalis, or petroleum products.
   B. patients with decreased LOC.
   C. patients who are unable to swallow.
   D. all of the above.

Challenging Questions

7. What problems do you foresee regarding information from the dispatch?

8. How can a patient's emotional state affect your patient care?

9. Can a patient who intentionally overdosed refuse medical care?

10. How should you handle the emotional aspects of a suicidal patient?

www.EMTB.com

Environmental Emergencies

Objectives

Cognitive

4-7.1   Describe the various ways that the body loses heat. (p 545)

4-7.2   List the signs and symptoms of exposure to cold. (p 546)

4-7.3   Explain the steps in providing emergency medical care to a patient exposed to cold. (p 549)

4-7.4   List the signs and symptoms of exposure to heat. (p 555, 557)

4-7.5   Explain the steps in providing emergency care to a patient exposed to heat. (p 555, 557)

4-7.6   Recognize the signs and symptoms of water-related emergencies. (p 560)

4-7.7   Describe the complications of near drowning. (p 562)

4-7.8   Discuss the emergency medical care of bites and stings. (p 572)

Affective

None

Psychomotor

4-7.9   Demonstrate the assessment and emergency medical care of a patient with exposure to cold. (p 548, 549)

4-7.10   Demonstrate the assessment and emergency medical care of a patient with exposure to heat. (p 555, 558)

4-7.11   Demonstrate the assessment and emergency medical care of a near-drowning patient. (p 565)

4-7.12   Demonstrate completing a prehospital care report for patients with environmental emergencies. (p 550, 567)

You are the Provider

At 3:12 PM you and your EMT-B partner are dispatched to the Green Valley Mobile Home Park for a sick person. When you arrive in front of the mobile home you see an older woman standing on the front porch waving at you. As you approach, the woman states that she just got home and found her husband on the couch not responding appropriately. As you enter the residence you note that it is very hot and there appears to be no source of ventilation. You find an older man lying on the couch. He is conscious but disoriented. He has a patent airway and is breathing at a rate of 22 breaths/min. His breathing appears to be very shallow. His skin is red, hot, and dry to the touch.

1. What does the condition of the residence tell you about your patient's condition?
2. What does your index of suspicion tell you about your patient's skin color and condition?

Environmental Emergencies

Heat and cold can both overwhelm the body's mechanisms for regulating temperature, including sweating and radiation of body heat into the atmosphere. A variety of medical emergencies can result from exposure to heat or cold, particularly in children, older people, people with chronic illnesses, and young adults who overexert themselves. There is also a range of medical emergencies that arise from water recreation, and these can sometimes be complicated by the cold. These emergencies include localized injuries and systemic illnesses. As an EMT-B, you can save lives by recognizing and responding properly to these emergencies, most of which require prompt treatment in the hospital.

This chapter describes how the body regulates core temperature, and the ways in which body heat is lost to the environment. It then discusses the various forms of heat-, cold-, and water-related emergencies, including how to diagnose and treat hypothermia, frostbite, hyperthermia, and diving injuries. The chapter concludes with a discussion of bites and stings.

Factors Affecting Exposure

A number of factors will affect how a person deals with a cold or hot environment. These can certainly be used as prevention strategies for those who work or play in extreme environmental temperatures. They can also be useful during the assessment of your patient to determine how prepared they were for a cold or hot environment. A hiker prepared for a warm summer hike in the foothills will present and respond to treatment differently than a traveler stranded in a hot car because the radiator boiled over.

1. **Physical condition**. Patients who are already ill or in poor physical condition will not be able to tolerate extreme temperatures as well as those whose cardiovascular system, metabolic system, and nervous system are functioning well. A well-trained athlete performs much better and is less likely to experience injury or illness than the "weekend warrior" who has not trained well. Increasing your activity will generate more heat when out in the cold but will also produce more heat when it is not needed, as in walking on a hot asphalt road because you ran out of gas.

2. **Age.** Those who are at the extremes of age are more likely to experience illness due to temperatures. Small infants have poor thermoregulation at birth and do not have the ability to shiver and generate heat when needed until about 12 to 18 months of age. Their larger surface area and smaller mass contribute to increased heat loss and heat gain. When you get cold you put on a sweater; a small child may not think to do this or may have difficulty finding and putting one on. On the other end of the age spectrum, older adults loose subcutaneous tissues, reducing the amount of insulation they have. Poor circulation contributes to increased heat loss and gain in either environment. This is why older people often wear extra layers of clothing. Medications taken by older persons can also affect their body's thermostat, putting them at more risk to hot or cold problems.

3. **Nutrition and Hydration.** Your body needs calories for your metabolism to function. Staying well hydrated provides water as a catalyst for much of this metabolism. A decrease in either will aggravate both hot and cold stress. Calories provide fuel to burn, creating heat during the cold, and water provides sweat for evaporation and removing heat. Alcohol use may increase fluid loss and place the patient at greater risk for temperature related problems.

4. **Environmental Conditions.** Conditions such as air temperature, humidity levels, and wind can

www.EMTB.com

Technology

- Interactivities
- Vocabulary Explorer
- Anatomy Review
- Web Links
- Online Review Manual

complicate or improve environmental situations. We all welcome a cool breeze when it is hot outside, but a cold wind when it is cold outside can be uncomfortable. Extremes in temperature and humidity are not needed to produce hot or cold injuries. Many hypothermia cases occur at temperatures between 30° and 50°F. Most heat stroke cases occur when the temperature is 80°F and the humidity is 80%. Be sure to examine the environmental temperature of your patient. Older patients may turn the heat down in the winter or neglect to use air conditioning because of cost concerns. Some people may not open windows in a heat wave for fear of burglars. When evaluating your patient's condition, consider the environment and whether your patient is prepared for that situation. It may help in your treatment decisions and give you an idea on how the patient will respond to your care.

Cold Exposure

Normal body temperature must be maintained within a very narrow range for the body's chemistry to work efficiently. If the body, or any part of it, is exposed to cold environments, these mechanisms may be overwhelmed. Cold exposure may cause injury to individual parts of the body, such as the feet, hands, ears, or nose, or to the body as a whole. When the entire body temperature falls, the condition is called hypothermia.

Because heat always travels from a warmer place to a cooler place, the body will tend to lose heat to the environment. The body can lose heat in the following five ways:

- Conduction is the direct transfer of heat from a part of the body to a colder object by direct contact, as when a warm hand touches cold metal or ice, or is immersed in water with a temperature of less than 98°F (37°C). Heat passes directly from the body to the colder object. Heat can also be gained if the substance being touched is warm. This is why people with chronic medical problems are advised to limit time in hot tubs.
- Convection occurs when heat is transferred to circulating air, as when cool air moves across the body surface. A person standing outside in windy winter weather, wearing lightweight clothing, is losing heat to the environment mostly by convection. A person can gain heat if the air moving across their body is hotter than body temperature such as in deserts or industrial settings like foundries, but it is more common to see rapid heat gain in spas and hot tubs where the water temperature may be well above body temperature.
- Evaporation is the conversion of any liquid to a gas, a process that requires energy, or heat. Evaporation is the natural mechanism by which sweating cools the body. This is why swimmers coming out of the water feel a sensation of cold as the water evaporates from their skin. Individuals who exercise vigorously in a cool environment may sweat and feel warm at first, but later, as their sweat evaporates, they can become exceedingly cool. Measures should be taken to keep a person dry if he or she is too cold.
- Radiation is the transfer of heat by radiant energy. Radiant energy is a type of invisible light that transfers heat. The body can lose heat by radiation, such as when a person stands in a cold room. Heat can also be gained by radiation, for example when a person stands by a fire.
- Respiration causes body heat to be lost as warm air in the lungs is exhaled into the atmosphere and cooler air is inhaled. In warm climates, the air temperature can be well above body temperature, causing an individual to gain heat with each breath.

The rate and amount of heat loss or gain by the body can be modified in three ways:

1. **Increase or decrease heat production**. One way for the body to increase its heat production is to increase the rate of metabolism of its cells, as occurs in shivering. Often people have a natural urge to move around when they are cold. If they are hot, you want to reduce their activity, thus reducing heat production.

2. **Move to an area where heat loss is decreased or increased**. The most obvious way to decrease heat loss from radiation and convection is to move out of a cold environment and seek shelter from wind. Just covering the head will minimize radiation heat loss by up to 70%. The same holds true for a patient who is too hot. Simply moving into the shade can reduce the ambient temperature by 10 degrees or more. If you cannot move the person, create the shade and air movement by fanning.

3. **Wear insulated clothing, which helps to decrease heat loss in several ways.** Insulators, such as specific materials or dry, still air, do not conduct heat. Thus, layers of clothing that trap air provide good insulation, as do wool, down, and synthetic fabrics that have small pockets of trapped air. Protective clothing also traps perspiration and prevents evaporation. Sweating without evaporation will not result in cooling. To encourage heat loss, loosen or remove clothing, particularly around the head and neck.

Hypothermia

Hypothermia literally means "low temperature." It is diagnosed when the core temperature of the body—the temperature of the heart, lungs, and vital organs—falls below 95°F (35°C). The body can usually tolerate a drop in core temperature of a few degrees. However, below this critical point, the body loses the ability to regulate its temperature and to generate body heat. Progressive loss of body heat then begins.

To protect itself against heat loss, the body normally constricts blood vessels in the skin; this results in the characteristic appearance of blue lips and/or fingertips. As a secondary precaution against heat loss, the body tends to create additional heat by shivering, which is the active moving of many muscles to generate heat. As cold exposure worsens and these mechanisms are overwhelmed, many body functions begin to slow down. Eventually, the functioning of key organs such as the heart begins to slow. Untreated, this can lead to death.

Hypothermia can develop either quickly, as when someone is immersed in cold water, or more gradually, as when a lost person is exposed to the cold environment for several hours or more. The temperature does not have to be below freezing for hypothermia to occur. In winter, homeless people and those whose homes lack heating may develop hypothermia at higher temperatures. Even in summer, swimmers who remain in the water for a long time are at risk of hypothermia. Like all heat- and cold-related injuries, hypothermia is more common among geriatric, pediatric, and ill individuals, who are less able to adjust to temperature extremes. Hypothermia is also common among the very young, who are unable to put on clothes to protect themselves against the cold. Infants and children are small, with a relatively large surface area, and have less body fat than do adults. Also, because of their small muscle mass, children may not be able to shiver as effectively as adults, and infants do not shiver at all.

Patients with injuries or illness, such as burns, shock, head injury, stroke, generalized infection, injuries to the spinal cord, diabetes, and hypoglycemia, are more prone to hypothermia, as are patients who have taken certain drugs or poisons.

Signs and Symptoms

Signs and symptoms of hypothermia generally become progressively more severe as the core temperature falls. Hypothermia generally progresses through four general stages, as shown in (Table 18-1 ▼). Although there is no clear distinction among the stages, the different signs and symptoms of each will help you estimate the severity of the problem. When you assess a patient in the field, you should be able to distinguish between mild and severe hypothermia.

To assess the patient's general temperature, pull back on your glove and place the back of your hand on the

Table 18-1 Characteristics of Systemic Hypothermia

Core temperature	90° to 95°F (32° to 35°C)	89° to 92°F (32° to 33°C)	80° to 88°F (32° to 33°C)	<80°F (<27°F)
Signs and symptoms	Shivering, foot stamping	Loss of coordination, muscle stiffness	Coma	Apparent death
Cardiorespiratory response	Constricted blood vessels, rapid breathing	Slowing respirations, slow pulse	Weak pulse, arrhythmias, very slow respirations	Cardiac arrest
Level of consciousness	Withdrawn	Confused, lethargic, sleepy	Unresponsive	Unresponsive

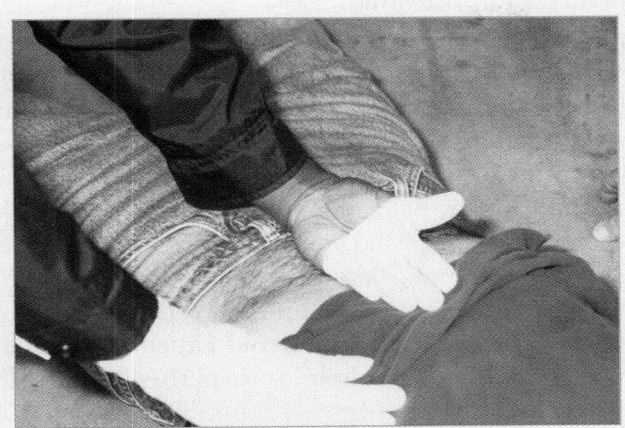

Figure 18-1 To assess a patient's temperature, pull back your glove and place the back of your hand on the patient's skin.

patient's skin at the abdomen (Figure 18-1 ▲). If the skin feels cool, the patient is likely experiencing a generalized cold emergency. If you work in a cold environment, you may carry a hypothermia thermometer, which registers lower core temperatures (Figure 18-2 ▼). It must be inserted in the rectum for an accurate reading. Note that regular thermometers will not register the temperature of a patient who has significant hypothermia. Mild hypothermia occurs when the core temperature is between 90° and 95°F (32° and 35°C). The patient is usually alert and shivering in an attempt to generate more heat through muscular activity. The patient may jump

up and down and stamp his or her feet. Pulse rate and respirations are usually rapid. The skin in light-skinned individuals can be red, but may eventually appear pale, then cyanotic. As was noted previously, individuals in a cold environment may have blue lips or fingertips because of the body's constriction of blood vessels at the skin to retain heat.

More severe hypothermia occurs when the core temperature is less than 90°F (32°C). Shivering stops, and muscular activity decreases. At first, small, fine muscle activity such as coordinated finger motion ceases. Eventually, as the temperature falls further, all muscle activity stops.

As the core temperature drops toward 85°F (29°C), the patient becomes lethargic, usually losing interest in continuing to fight the cold. The level of consciousness decreases, and the patient may try to remove his or her own clothes. Poor coordination and memory loss follow, along with reduced or complete loss of sensation to touch, mood changes, and impaired judgment. The patient becomes less communicative, experiences joint or muscle stiffness, and has trouble speaking. The muscles eventually become rigid, and the patient begins to appear stiff or rigid.

If the temperature continues to fall to 80°F (27°C), vital signs slow; the pulse becomes weaker, and respirations slow to shallow or become absent. Cardiac arrhythmias may occur as the blood pressure decreases or disappears.

At a core temperature less than 80°F (27°C), all cardiorespiratory activity may cease, pupillary reaction is slow, and the patient may appear dead.

Never assume that a cold, pulseless patient is dead. Patients may survive even severe hypothermia, if proper emergency measures are carried out.

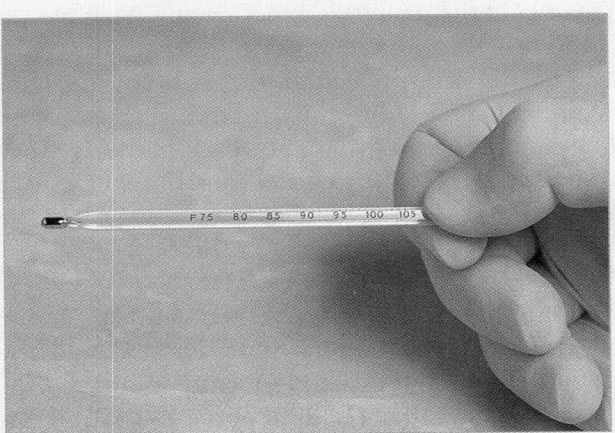

Figure 18-2 A special rectal hypothermia thermometer registers temperatures well below that of a regular thermometer.

Documentation Tips

Recording specific results of your early assessment is particularly valuable in hypothermic patients. If there is a question about beginning CPR, note where you checked the pulse and for how long. Also note the initial body temperature and where it was taken. These points will be important to hospital staff, and will help protect you if medicolegal issues are ever raised.

Assessment of Cold Injuries

Management of hypothermia in the field, regardless of the severity of the exposure, consists of stabilizing the ABCs and preventing further heat loss.

Scene Size-up

Typically, your scene assessment begins with information provided by dispatch. This information helps you consider the MOI or NOI and prepare for the problems your patient may have. Note environmental conditions. Air temperature, wind chill, and whether it is wet or dry are important aspects of scene size-up and will likely affect the patient.

Ensure that the scene is safe for you and other responders. Identify potential safety hazards, such as wet grass, mud, or icy streets. Consider special hazards such as avalanches. Cold environments may present special problems both for you and your patient. Use appropriate BSI precautions and consider the number of patients you may have. Summon additional help, such as a search and rescue team, as quickly as possible.

Initial Assessment

General Impression

In a cold emergency your patient's chief complaint may be only that he or she is cold, or the cold may be an additional complication of an existing medical or trauma problem. Determine whether a life threat exists, and if so, treat it. If the chief complaint is simply being cold, quickly assess how cold the patient is. This is done by feeling the patient's skin on the abdomen. This area of the body is usually well protected and insulated and will give you a general idea of the core body temperature.

Your patient's mental status will usually indicate his or her degree of distress. Evaluate it quickly using the AVPU scale. An altered mental status is related to the intensity of the cold problem.

Airway and Breathing

Your assessment of ABCs should take into account the physiologic changes that occur as a result of hypothermia. Ensure the patient has an adequate airway and is breathing. Consider spinal precautions based on your scene size-up and chief complaint. If your patient's breathing is slow or shallow, BVM ventilations may be necessary. If warmed and humidified oxygen is available, it would be preferred because it helps warm the patient from the inside out. Additional oxygen, even if not warmed and humidified, for cold patients with adequate breathing may help with perfusion of cold tissues.

Circulation

If you cannot feel a radial pulse, gently palpate for a carotid pulse and wait for 30 to 45 seconds before you decide that the patient is pulseless. Physicians disagree about the wisdom of performing BLS (ie, CPR) on a patient with hypothermia who appears to be pulseless. Such a patient actually may be in a kind of "metabolic ice box," having achieved a metabolic balance that BLS may upset. Even a pulse rate of 1 or 2 beats/min indicates cardiac activity, and cardiac activity may spontaneously recover once the core is warmed. However, there is evidence that BLS, when correctly done, will increase blood flow to the critical parts of the body. For this reason, some authorities recommend starting BLS on a patient with hypothermia and no pulse. The American Heart Association recommends that CPR be started if the patient has no detectable pulse or breathing. Again, for a patient with hypothermia, this may require a prolonged pulse check.

Perfusion will be compromised based on the degree of cold the patient is experiencing. Your skin assessment will not be helpful in determining shock. Assume shock is present and treat it appropriately. Bleeding may be difficult to find because of the slow-moving circulation and thick clothing. If the scene size-up, MOI, or chief complaint suggests the potential for bleeding, look for it.

Transport Decision

Even mild degrees of hypothermia can have serious consequences and complications. These include cardiac arrhythmia and blood-clotting abnormalities. Therefore, all patients with hypothermia require immediate transport for evaluation and treatment. Assess the scene for the safest way to quickly move your patient from the cold environment. As you package your patient for transport, work quickly, safely, and gently. Rough handling of a hypothermic patient may cause a cold, slow, weak heart to fibrillate and the patient to lose any pulse that may have existed. If transportation is delayed, protect the patient from further heat loss.

Focused History and Physical Exam

If your patient is a conscious medical patient, as in most cold emergencies, you would begin by asking a SAMPLE history, performing a focused physical exam, and then obtaining baseline vital signs. If your medical patient is unresponsive, begin with a rapid physical exam working from head to toe to look for hidden problems and to determine any body areas that may be frostbitten. Next, obtain baseline vital signs and a SAMPLE history.

SAMPLE History

Obtaining a patient's history in these situations may be difficult but should be attempted. If possible, obtain information as to how long your patient has been exposed to the cold environment, either from the patient or bystanders. Exposures may be acute or chronic. Your SAMPLE history can provide important information affecting both your treatment in the field and the treatment your patient will receive in the hospital. Medications your patient has taken and underlying medical problems may have an impact on the way cold affects his or her metabolism. The patient's last oral intake and what the patient was doing prior to the exposure will help to determine the severity of the cold problem.

Focused Physical Exam

Your focused physical exam should concentrate on the severity of hypothermia, assessing the areas of the body directly affected by cold exposure, and the degree of damage. Is the whole body cold (hypothermia) or just parts (frostbite)? These determinations will have important consequences for your treatment decisions. For example, shivering indicates a protective mechanism to produce more heat because the body is cold. When shivering stops and the patient remains in a cold environment, the cold emergency is more severe.

Baseline Vital Signs

Keep in mind that vital signs may be altered by the effects of hypothermia and can be an indicator of its severity. Respirations may be slow and shallow, resulting in low oxygen levels in the body. Low blood pressure and pulse also indicate moderate to severe hypothermia. Carefully evaluate your patient for changes in mental status.

Determine a core body temperature using a thermometer based on local protocol. A special low-temperature thermometer is required to take a hypothermic patient's temperature, generally done through the rectum. A patient with a moderate form of hypothermia will have a core temperature of less than 90°F (32°C).

Interventions

In most cases, you should move the patient from the cold environment to prevent further heat loss. To prevent further damage to the feet, do not allow the patient to walk. Remove any wet clothing, and place dry blankets over and under the patient (Figure 18-3 ▼). Always make sure to handle the patient gently so that you will not cause any pain or further injury to the skin. Do not massage the extremities. Do not allow the patient to eat, to use any stimulants, such as coffee, tea, or cola, or to smoke or chew tobacco.

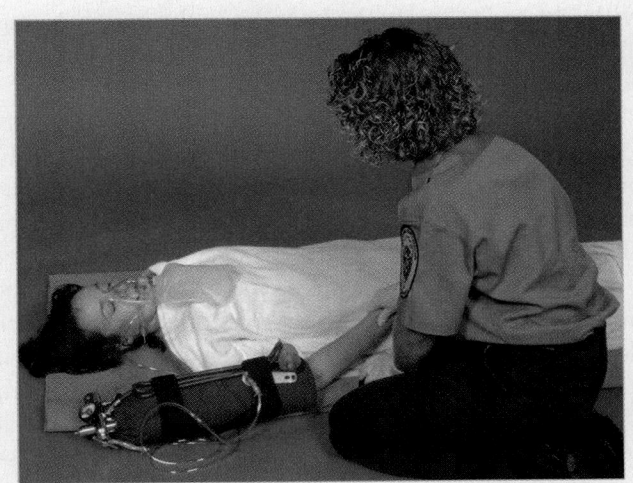

Figure 18-3 Place dry blankets over and under the patient with hypothermia; give warm, humidified oxygen; assess the pulse before considering CPR.

You can give the patient warm, humidified oxygen if you have not already done so as part of the initial assessment. Begin passive rewarming, which includes wrapping the patient in blankets and turning up the heat in the patient compartment of the ambulance.

If the patient is alert and responds appropriately, the hypothermia is mild, and you can begin active rewarming, which includes wrapping the patient in blankets and applying heat packs or hot water bottles to the groin, axillary, and cervical regions. Turn the heat up high in the patient compartment of the ambulance.

You must try to minimize further loss of body heat, especially when you cannot get to a hospital quickly. However, when the patient has moderate or severe hypothermia, you should never try to rewarm the patient actively (placing heat on or into the body). Rewarming too quickly may cause a fatal cardiac arrhythmia that requires defibrillation; for this reason, rewarming should be done in the hospital. Again, your goal is to prevent further heat loss. Remove the patient immediately from the cold environment. Place the patient in the ambulance, and turn up the heat. If you cannot get the patient out of the cold immediately, move him or her out of the wind and away from contact with any object that will conduct heat from the body. Place a protective cover on the patient, and remember that most heat is lost around the head and neck.

If the patient is alert and shivering, you may assume that the hypothermia is relatively mild. If possible, you can give warm fluids by mouth in this case, assuming that the patient can swallow without a problem. Remove all wet clothing, and cover the patient with a blanket. Notify the hospital of the patient's condition so that staff can prepare to start rewarming as soon as you arrive.

When the patient is not shivering and is lethargic, moderate or severe hypothermia is probably present. Remove wet clothing, and protect the patient from the cold and wind with blankets in a warmer environment.

If you are in an area where hypothermia is a common problem, you should have specific protocols for dealing with this situation. In all cases, consult with medical control. Remember that this is complicated in a trauma situation particularly if there is a potential spinal injury. For example, in rural areas where transport time is longer, the patient may have been exposed to cold after being thrown from the vehicle and waiting for a significant time. It is important to keep everything that is wrong with the patient in mind when treating and making transport decisions.

Detailed Physical Exam

Your detailed physical exam should be aimed at determining the degree and extent of cold injury, as well as any other injuries or conditions that may not have been initially detected. The numbing effect of cold, both on the brain and on the body, may affect your patient's ability to tell you about other injuries or illnesses. Therefore, a careful examination of your patient's entire body, with special attention to skin temperatures, textures, and turgor, will help you avoid missing important clues to your patient's condition.

Ongoing Assessment

Keep a very close eye on your patient's level of consciousness and vital signs. As the body rewarms, the sudden redistribution of fluids and the release of built-up chemicals can have harmful effects, including cardiac arrhythmias. Be vigilant and monitor your patient closely, even if the condition appears to be improving.

Communication and Documentation

Communicate all of the information you have gathered to the receiving facility. The conditions you found on scene, what your patient was wearing, and information gathered from bystanders may be essential in evaluating and treating your patient in the hospital. Document not only your patient's physical status, but also the conditions on scene, and carefully document changes in mental status during treatment and transport.

Management of Cold Exposure in a Sick or Injured Person

All patients who are severely injured are at risk for hypothermia. Keep this in mind when you are evaluating a patient with multiple injuries.

A sick or injured person who has been trapped in a cold environment may experience hypothermia or may already have problems related to cold exposure. Such a person is more susceptible than a healthy person to cold injury. Take the following steps promptly to prevent further cold injury:

1. **Remove wet clothing** and keep the patient dry.
2. **Prevent conduction heat loss.** Move the patient away from any wet or cold surfaces, such as a car frame.
3. **Insulate all exposed body parts**, especially the head, by wrapping them in a blanket or any other available dry, bulky material.
4. **Prevent convection heat loss** by erecting a wind barrier around the patient.
5. **Remove the patient** from the cold environment as promptly as possible.

Regardless of the nature or severity of the cold injury, remember that even an unresponsive patient may be able to hear you. Some patients have told of hearing themselves pronounced dead by someone who had forgotten the saying: "No one is dead unless he is warm and dead." If you carry an AED, you should consider defibrillation. Although this heart rhythm is unlikely in patients with hypothermia, it can occur in patients who are rewarmed too rapidly.

Local Cold Injuries

Most injuries from cold are confined to exposed parts of the body. The extremities, particularly the feet, and the exposed ears, nose, and face, are especially vulnerable to cold injury (Figure 18-4 ▼). When exposed parts of the body become very cold but not frozen, the condition is called frostnip, chilblains, or immersion foot

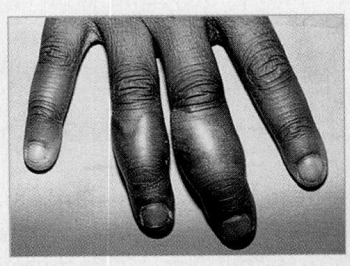
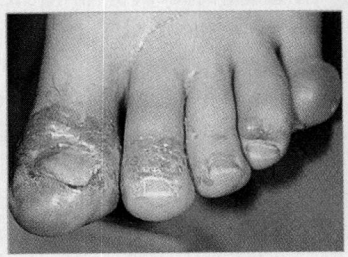
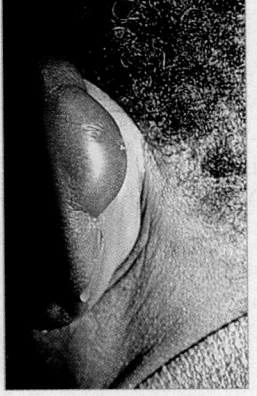

Figure 18-4 The extremities and the ears, nose, and face are particularly susceptible to frostbite.

(trench foot). When the parts become frozen, the injury is called <u>frostbite</u>.

You should try to find out the duration of the exposure, the temperature to which the body part was exposed, and the wind velocity during exposure. These are important factors in determining the severity of a local cold injury. You should also investigate a number of underlying factors:

- Exposure to wet conditions
- Inadequate insulation from cold or wind
- Restricted circulation from tight clothing or shoes or circulatory disease
- Fatigue
- Poor nutrition
- Alcohol or drug abuse
- Hypothermia
- Diabetes
- Cardiovascular disease
- Older age

In hypothermia, blood is shunted away from the extremities in an attempt to maintain the core temperature. This shunting of blood increases the risk of local cold injury to the extremities, ears, nose, and face. Thus, the patient with hypothermia should also be assessed for frostbite or other local cold injury. The reverse is also true. You must remember that both local and systemic cold exposure problems can occur in the same patient.

Frostnip and Immersion Foot

After prolonged exposure to the cold, the skin may be freezing while the deeper tissues are unaffected. This condition, which often affects the ears, nose, and fingers, is called frostnip. Because frostnip is usually not painful, the patient often is unaware that a cold injury has occurred. Immersion foot, also called trench foot, occurs after prolonged exposure to cold water. It is particularly common in hikers or hunters who stand for a long time in a river or lake. With both frostnip and immersion foot, the skin is pale (blanched) and cold to the touch; normal color does not return after palpation of the skin. In some cases, the skin of the foot will be wrinkled, but it can also remain soft. The patient complains of loss of feeling and sensation in the injured area.

As in all other hypothermia cases, the emergency treatment of these less severe local cold injuries consists of removing the patient from the cold, wet environment, but also rewarming the affected part. With frostnip, contact with a warm object may be all that is needed; you can use your hands, your breath, or the patient's own body. During rewarming, the affected part will often tingle and become red in light-skinned individuals. With immersion foot, remove wet shoes, boots, and socks, and rewarm the foot gradually, protecting it from further cold exposure.

Frostbite

Frostbite is the most serious local cold injury, because the tissues are actually frozen. Freezing permanently damages cells, although the exact mechanism by which damage occurs is not known. The presence of ice crystals within the cells may cause physical damage. The change in the water content in the cells may also cause changes in the concentration of critical electrolytes, producing permanent changes in the chemistry of the cell. When the ice thaws, further chemical changes occur in the cell, causing permanent damage or cell death, called gangrene (Figure 18-5 ▼). If gangrene occurs, the dead tissue must be surgically removed, sometimes by amputation. Following less severe damage, the exposed part will become inflamed, tender to touch, and unable to tolerate exposure to cold.

Frostbite can be identified by the hard, frozen feel of the affected tissues. Most frostbitten parts are hard and waxy (Figure 18-6 ▶). The injured part feels firm to frozen as you gently touch it. If the frostbite is only skin deep it will feel leathery or thick, not hard and frozen through. Blisters and swelling may be present. In light-skinned individuals with a deep injury that has thawed or partially thawed, the skin may appear red with purple and white, or it may be mottled and cyanotic.

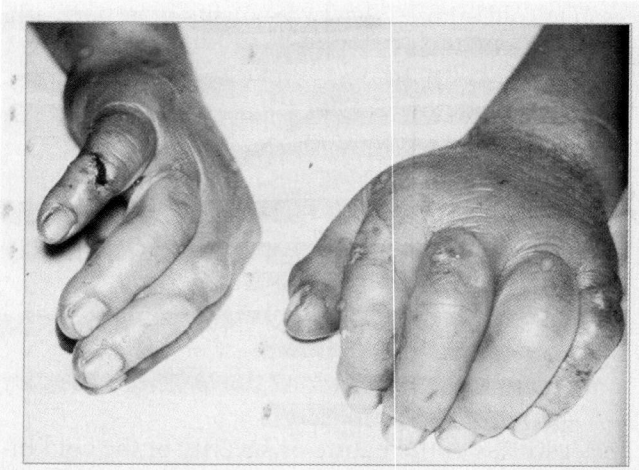

Figure 18-6 Frostbitten parts are hard and usually waxy to touch.

As with a burn, the depth of skin damage will vary. With superficial frostbite, only the skin is frozen; with deep frostbite, the deeper tissues are frozen as well. You may not be able to tell superficial from deep frostbite in the field. Even an experienced surgeon in a hospital setting may not be able to tell until several days have gone by.

Emergency Medical Care and Local Cold Injury

The emergency treatment of local cold injuries in the field should include the following steps:

1. **Remove the patient** from further exposure to the cold.
2. **Handle the injured part gently**, and protect it from further injury.
3. **Administer oxygen**, if this was not already done as part of the initial assessment.
4. **Remove any wet or restricting clothing** over the injured part.

With an early or superficial injury, such as frostnip or immersion foot, splint the extremity and cover it loosely with a dry, sterile dressing. Never rub injured tissues with anything; rubbing causes further damage. Do not reexpose the injury to cold.

With a late or deep cold injury, such as frostbite, be sure to remove any jewelry from the injured part and cover the injury loosely with a dry, sterile dressing. Do not break blisters or rub or massage the area. Do not apply heat or rewarm the part. Unlike frostnip and trenchfoot, rewarming of the frostbitten extremity is best accomplished under controlled circumstances in the emergency department. You can cause a great deal of

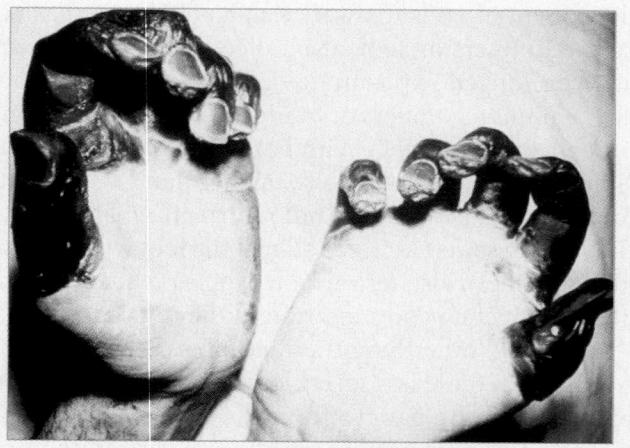

Figure 18-5 Gangrene, or permanent cell death, can occur when tissue is frozen and certain chemical changes occur in the cells.

further injury to fragile tissues by attempting to rewarm a frostbitten part. Never apply something warm or hot, such as the exhaust from the ambulance engine or, even worse, an open flame. Do not allow the patient to stand or walk on a frostbitten foot.

Evaluate the patient's general condition for the signs or symptoms of systemic hypothermia. Support the vital functions as necessary, and transport the patient promptly to the hospital.

If prompt hospital care is not available and medical control instructs you to begin rewarming in the field, use a warm-water bath. Immerse the frostbitten part in water with a temperature between 100° and 112°F (38° and 44.5°C). Check the water temperature with a thermometer before immersing the limb, and recheck it frequently during the rewarming process. The water temperature should never exceed 112°F (44.5°C). Stir the water continuously. Keep the frostbitten part in the water until it feels warm and sensation has returned to the skin. Dress the area with dry, sterile dressings, placing them also between injured fingers or toes. Expect the patient to complain of severe pain.

Never attempt rewarming if there is any chance that the part may freeze again before the patient reaches the hospital. Some of the most severe consequences of frostbite, including gangrene and amputation, have occurred when parts were thawed and then refrozen.

Cover the frostbitten part with soft, padded, sterile cotton dressings. If blisters have formed, do not break them. Remember, you cannot accurately predict the outcome of a case of frostbite early in its course. Even body parts that appear gangrenous may recover following proper emergency and hospital treatment.

Cold Exposure and You

As an EMT-B, you are also at risk for hypothermia if you work in a cold environment. If cold weather search-and-rescue operations are a possibility in your assigned areas, you should receive survival training and precautionary tips. You should be thoroughly familiar with local conditions. Be aware of existing and potential weather conditions, and stay on top of changes that are forecast for the area. Make sure proper clothing is available, and wear it whenever appropriate. Your vehicle, too, must be properly equipped and maintained for a cold environment. You cannot help others if you do not protect yourself. Never allow yourself to become a casualty!

Heat Exposure

Normal body temperature is 98.6°F (37°C). Complicated regulatory mechanisms keep this internal temperature constant, regardless of the ambient temperature, the temperature of the surrounding environment. In a hot environment or during vigorous physical activity, when the body itself produces excess heat, the body will try to rid itself of the excess heat. There are several ways of doing this. The two most efficient are sweating (and evaporation of the sweat) and dilation of skin blood vessels, which brings blood to the skin surface to increase the rate of heat radiation. In addition, of course, the person who becomes overheated can remove clothing and try to find a cooler environment.

Ordinarily, the heat-regulating mechanisms of the body work very well, and individuals are able to

You are the Provider   Part 2

As your partner applies high-flow oxygen via a nonrebreathing mask, you call for ALS backup. Further assessment shows a rapid and thready pulse, low blood pressure, pupils that are sluggish to react, and a temperature of 104°F. You find no signs of traumatic injury. You remove any constricting clothing and jewelry. With your partner's assistance you move your patient out of the hot environment and into your ambulance.

3. Once you and the patient are in the ambulance, what special treatment does your patient need?
4. How serious is this emergency?

EMT-B Safety

Keeping yourself hydrated while on duty is very important, especially during periods of heavy exertion or work in the heat. Drink at least 3 L of water a day and more when exertion or heat is involved. The color of urine (usually darker with dehydration) and frequency of urination correlate directly with the body's fluid level.

tolerate significant temperature changes. When the body is exposed to more heat energy than it loses or generates more heat than it can lose, hyperthermia results. Hyperthermia is a high core temperature, usually 101°F (38.3°C) or higher.

When the body's mechanisms to decrease body heat are overwhelmed and the body is unable to tolerate the excessive heat, illness develops. High air temperature can reduce the body's ability to lose heat by radiation; high humidity reduces the ability to lose heat through evaporation. Another contributing factor is vigorous exercise, during which the body can lose more than 1 L of sweat an hour, causing loss of fluid and electrolytes. Illness from heat exposure can take the following three forms:

- Heat cramps
- Heat exhaustion
- Heatstroke

All three forms of heat illness may be present in the same patient, since untreated heat exhaustion may progress to heatstroke. Heatstroke is a life-threatening emergency.

Persons at greatest risk for heat illnesses are children; geriatric patients; patients with heart disease, COPD, diabetes, dehydration, and obesity; and those with limited mobility. Older people, newborns, and infants exhibit poor thermoregulation. Newborns and infants often wear too much clothing. Alcohol and certain drugs, including medications that dehydrate the body or decrease the ability of the body to sweat, also make a person more susceptible to heat illnesses. When you are treating someone for a heat illness, always obtain a medication history.

Heat Cramps

Heat cramps are painful muscle spasms that occur after vigorous exercise. They do not occur only when it is hot outdoors. They may be seen in factory workers and even well-conditioned athletes. The exact cause of

heat cramps is not well understood. We know that sweat produced during strenuous exercise, particularly in a warm environment, causes a change in the body's electrolyte, or salt, balance. The result may be a loss of essential electrolytes from the cells. Dehydration may also play a role in the development of muscle cramps. Large amounts of water can be lost from the body as a result of excessive sweating. This loss of water may affect muscles that are being stressed and cause them to go into spasm.

Heat cramps usually occur in the leg or abdominal muscles. When the abdominal muscles are involved, the pain and muscle spasm may be so severe that the patient appears to have an acute abdominal problem. If a patient with a sudden onset of abdominal cramps has been exercising vigorously in a hot environment, you should suspect heat cramps.

Take the following steps to treat heat cramps in the field (Figure 18-7 ▼):

1. **Remove the patient** from the hot environment, including sunlight, a source of radiant heat gain. Loosen any tight clothing.
2. **Rest the cramping muscles.** Have the patient sit or lie down until the cramps subside.
3. **Replace fluids by mouth.** Use water or a diluted (half-strength) balanced electrolyte solution, such as Gatorade. In most cases, plain water is the most useful. Do not give salt tablets or solutions that have a high salt concentration. The patient already has an adequate amount of electrolytes circulating; they are just not distributed properly. With adequate rest and fluid replacement,

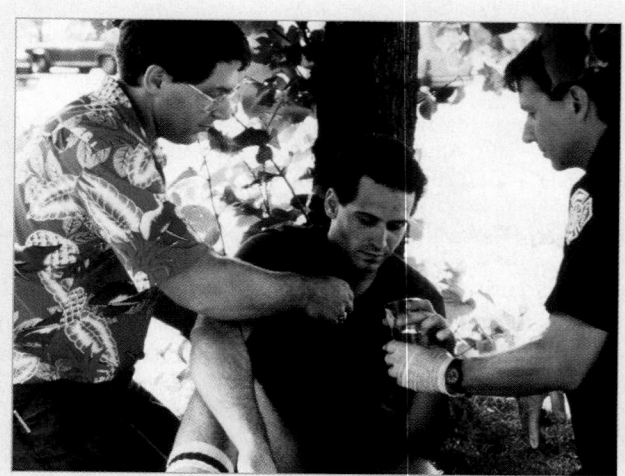

Figure 18-7 A patient with heat cramps should be moved to a cool environment as you begin your assessment and treatment.

the body will adjust the distribution of electrolytes, and the cramps will disappear.

If the cramps do not go away after these measures, transport the patient to the hospital. If you are not sure they are heat cramps or there is anything out of the ordinary, contact medical control or transport to the hospital. Once the cramps are gone, the patient may resume activity. For example, an athlete can return to play once the heat cramps have disappeared. However, heavy sweating may cause the cramps to recur. Hydration by drinking a lot of water is the best preventive and treatment strategy.

Heat Exhaustion

Heat exhaustion, also called heat prostration or heat collapse, is the most common serious illness caused by heat. Heat exposure, stress, and fatigue are causes of heat exhaustion, which is due to hypovolemia as the result of the loss of water and electrolytes from heavy sweating. For sweating to be an effective cooling mechanism, the sweat must be able to evaporate from the body. Otherwise, the body will continue to produce sweat, with further loss of body water. People standing in the hot sun and particularly those wearing several layers of clothing, such as football fans or parade watchers, may sweat profusely but experience little body cooling. High humidity will also decrease the amount of evaporation that can occur. Individuals working or exerting themselves in poorly ventilated areas are unable to release heat through convection. Thus, people who work or exercise vigorously and those who wear heavy clothing in a warm, humid, or poorly ventilated environment are particularly prone to heat exhaustion.

The signs and symptoms of heat exhaustion and those of associated hypovolemia are as follows:

- Dizziness, weakness, or faintness with accompanying nausea or headache
- Onset while working hard or exercising in a hot, humid, or poorly ventilated environment and sweating heavily
- Onset, even at rest, in the older and infant age groups in hot, humid, and poorly ventilated environments or extended time in hot, humid environments. Individuals not acclimatized to the environment may also experience onset at rest.
- Cold, clammy skin with ashen pallor
- Dry tongue and thirst
- Normal vital signs, although the pulse is often rapid and the diastolic blood pressure may be low
- Normal or slightly elevated body temperature; on rare occasions, as high as 104°F (40°C)

To treat the patient, follow the steps in Skill Drill 18-1 ▶ :

1. **Remove any excessive layers of clothing**, particularly around the head and neck (**Step 1**).
2. **Move the patient promptly** from the hot environment, preferably into the back of the air-conditioned ambulance. If outdoors, move out of the sun.

You are the Provider   Part 3

Once in the ambulance, you set the air conditioner on high. You remove the remaining clothing from your patient. You apply cool packs to the patient's neck, groin, and armpits. The patient's wife provides you with additional information about your patient. She states that her husband had been working out in the yard for about 2 hours when he came in complaining that he was hot and felt a bit dizzy. She had told him to sit down at the kitchen table and eat the sandwich that she had made for him. She went to the neighbor's for a few minutes. When she returned she found her husband lying on the couch mumbling inappropriately. She immediately called 9-1-1. She then tells you that he has an allergy to milk and cats. He takes furosemide (Lasix) twice a day and a medication for high blood pressure, but she is unsure of the name. He was diagnosed with high blood pressure 4 years ago. He has been relatively healthy for the last few years.

5. Of what significance is the patient's medication to his current condition?
6. What medical condition do you suspect?

Skill Drill 18-1

Treating for Heat Exhaustion

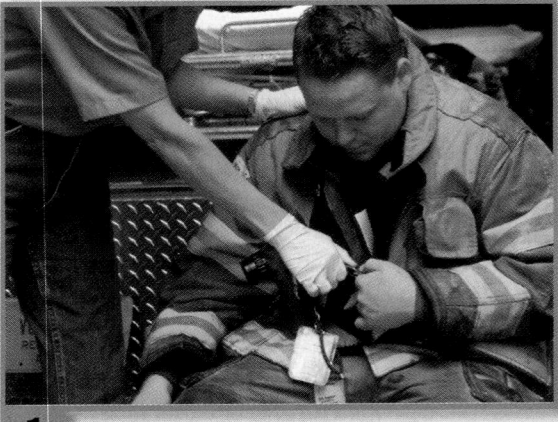

1 Remove extra clothing.

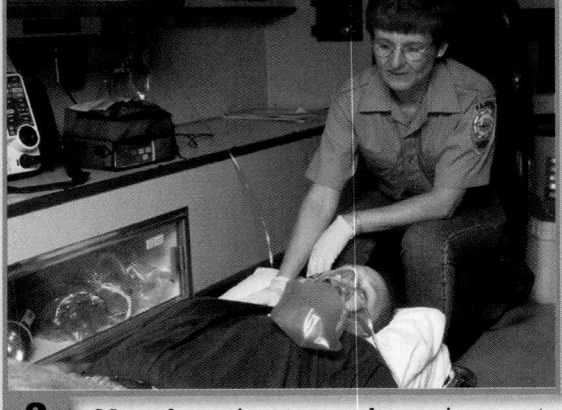

2 Move the patient to a cooler environment. Give oxygen. Place the patient in a supine position, elevate the legs, and fan the patient.

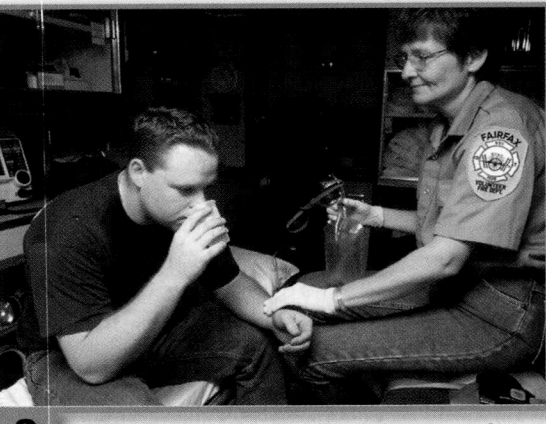

3 If the patient is fully alert, give water by mouth.

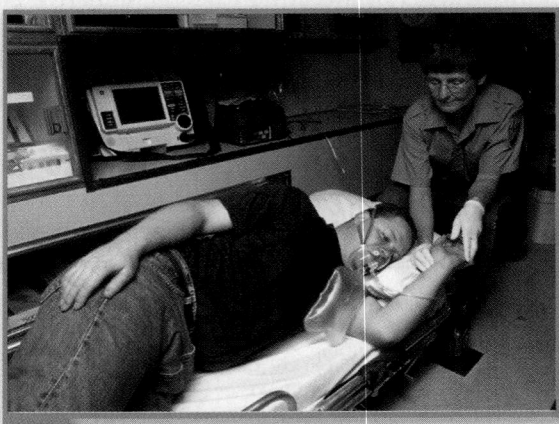

4 If nausea develops, transport on the side.

3. **Give the patient oxygen** if this was not already done as part of the initial assessment.
4. **Encourage the patient to lie down** and elevate the legs (supine position). Loosen any tight clothing and fan the patient for cooling (**Step 2**).
5. **If the patient is fully alert**, encourage him or her to sit up and slowly drink up to a liter of water, as long as nausea does not develop. Never force fluids by mouth on a patient who is not fully alert, or allow drinking while supine, because the patient could aspirate the fluid into the lungs (**Step 3**).

If the patient does become nauseated, transport on the side to prevent aspiration.

In most cases, these measures will reverse the symptoms, causing the patient to feel better within 30 minutes. But you should prepare to transport the patient to the hospital for more aggressive treatment, such as IV fluid therapy and close monitoring, especially in the following circumstances:

- The symptoms do not clear up promptly.
- The level of consciousness decreases.
- The temperature remains elevated.

■ The person is very young, older, or has any underlying medical condition, such as diabetes or cardiovascular disease.

6. **Transport the patient on his or her side** if you think the patient may be nauseated and ready to vomit, but make certain that the patient is secured (**Step 4**).

Heatstroke, the least common but most serious illness caused by heat exposure, occurs when the body is subjected to more heat than it can handle and normal mechanisms for getting rid of the excess heat are overwhelmed. The body temperature then rises rapidly to the level at which tissues are destroyed. Untreated heatstroke always results in death.

Heatstroke can develop in patients during vigorous physical activity or when they are outdoors or in a closed, poorly ventilated, humid space. It also occurs during heat waves among individuals (particularly in geriatric patients) who live in buildings with no air conditioning or with poor ventilation. It may also develop in children who are left unattended in a locked car on a hot day.

Many patients with heatstroke have hot, dry, flushed skin because their sweating mechanism has been overwhelmed. However, early in the course of heatstroke, the skin may be moist or wet. Keep in mind that a patient can have heatstroke even if he or she is still sweating. The body temperature rises rapidly in patients with heatstroke. It may rise to 106°F (41°C) or more. As the body core temperature rises, the patient's level of consciousness falls.

Often, the first sign of heatstroke is a change in behavior. However, the patient then becomes unresponsive very quickly. The pulse is usually rapid and strong at first, but as the patient becomes increasingly unresponsive, the pulse becomes weaker and the blood pressure falls.

Recovery from heatstroke depends on the speed with which treatment is administered, so you must be able to identify this patient quickly. Emergency treatment has one objective: Get the body temperature down by any means available. Take the following steps when treating a patient with heatstroke:

1. **Move the patient** out of the hot environment and into the ambulance.
2. **Set the air conditioning** to maximum cooling.
3. **Remove the patient's clothing.**
4. **Give the patient oxygen** if this was not done as part of the initial assessment.
5. **Apply cool packs** to the patient's neck, groin, and armpits (Figure 18-8 ▶).

Geriatric Needs

As we age, our bodies can lose the ability to respond to the environment. Older adults undergo changes in their ability to compensate for low or high ambient temperatures. For example, if the ambient temperature rises from 85° to 94°F (29.5° to 34.5°C), the older adult may not recognize the change or be able to compensate for it. Therefore, unless the person is accustomed to the heat, heatstroke can develop relatively quickly.

Shivering, a common effect of hypothermia, is the body's attempt to maintain heat. However, because of a decrease in muscle mass or tone, the hypothermic geriatric patient may not shiver. Furthermore, a decrease in muscle mass and body fat means that there is less insulation and protection from the cold. Because of the body's altered response to heat loss and its ability to gain heat, the health care provider might not suspect or report hypothermia. In caring for the geriatric patient in cold climates, be sure to protect the patient against unwanted heat loss. Cover all exposed areas with loose-fitting blankets. Pay particular attention to protecting the patient's head, since heat loss from the head and neck is substantial.

Because of reduced circulation to the skin, heat loss via conduction, convection, and radiation is significantly lower. Additionally, the aging process alters the patient's ability to perspire; therefore, heat loss through evaporation is reduced. Since the elderly patient cannot disperse heat effectively, classic heatstroke can develop rapidly. Typically, the older adult will not go through an initial stage of heat exhaustion. During the summer, you should be acutely aware of the potential for heatstroke and factors that can predispose a patient to heat illness. Factors that increase the possibility of heatstroke include medications, diabetes, alcohol abuse, malnutrition, parkinsonism, hyperthyroidism, and obesity.

Both hypothermia and hyperthermia can appear in older patients in environmental settings that are subtle. These problems appear commonly, for instance, when cost concerns result in keeping heat turned down in the winter or not using air conditioning in hot weather. Thermal emergencies can develop over a period of time for older persons in these indoor, urban environments that may not seem uncomfortable to you.

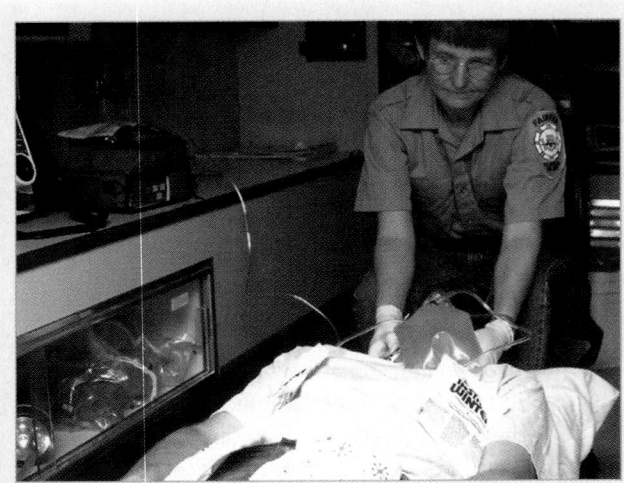

Figure 18-8 As part of treatment of heatstroke, give oxygen and place cool packs about the patient's neck, groin, and armpits.

6. **Cover the patient** with wet towels or sheets, or spray the patient with cool water and fan him or her to quickly evaporate the moisture on the skin.
7. **Aggressively and repeatedly fan** the patient with or without dampening the skin.
8. **Provide immediate transport** to the hospital.
9. **Notify the hospital** as soon as possible so that the staff can prepare to treat the patient immediately on arrival.

Assessment of Heat Injuries

Scene Size-up

As part of your scene size-up, perform an environmental assessment. How hot is it outside? How hot is it in the room where your patient is? How well is the patient tolerating the heat? Dispatch may report the call initially as a medical or trauma problem. The heat illness may only be secondary. Approach the scene looking for hazards as well as clues as to what may have caused your patient's problem. If you anticipate a prolonged scene time, protect yourself from the heat. Use appro-

priate BSI precautions, including gloves and eye protection. Long-sleeved shirts and long pants may not be comfortable in warm weather; however, they can help protect you from being splashed by blood or other fluids. Consider whether you need ALS backup. Intravenous fluids may need to be administered.

Initial Assessment

General Impression

As you approach your patient, observe how the patient interacts with you and the environment. This will help identify the patient's degree of distress. Introduce yourself and ask about the chief complaint. A heat illness may be the primary problem or it may simply be aggravating a medical or trauma condition. Remember, prolonged heat exposure may stress the heart, causing a heart attack. Use this initial interaction to guide you in assessing for immediate life threats and related problems. Avoid tunnel vision.

Assess the patient's mental status using the AVPU scale. Heat stroke is a true life-threatening emergency. The severity of your patient's condition may be identified by gathering clues about his or her mental status. The more altered the patient's mental status is, the more serious the heat problem.

Airway, Breathing, and Circulation

Assess the patient's ABCs and treat any life-threatening problems found. Unless the patient is unresponsive, the airway should be patent. Nausea and vomiting, however, may occur with some heat problems. Position the patient to protect the airway as necessary. If the patient is unresponsive, be cautious how you open the airway; consider spinal precautions. Breathing will be fast depending on the patient's core temperature but should otherwise be adequate. Providing oxygen to the patient will assist with the perfusion of body tissues and may decrease nausea. If your patient is unresponsive, insert an airway and provide BVM ventilations according to protocol.

Circulation is assessed by palpating a pulse. If it is adequate, assess the patient for perfusion and bleeding. Hot, dry, or moist skin that appears red may indicate an elevated core body temperature. Treat the patient aggressively for shock by removing the patient from the heat and positioning the patient to improve circulation. If the patient is bleeding, bandage according to protocol.

Transport Decision

If your patient has any signs of heatstroke (high temperature; red, dry skin; altered mental status; tachycardia; poor perfusion), transport without delay.

Focused History and Physical Exam

If your patient is unresponsive, perform a rapid physical exam from head to toe looking for problems or explanations as to what is wrong. Obtain vital signs to help understand how serious the problem is and gather any available history by asking family or bystanders and look for medical identification. If the patient is conscious, begin with gathering history, then perform a focused physical exam and vital signs.

SAMPLE History

Obtain a SAMPLE history with an eye to any activities, conditions, or medications that may predispose a patient to dehydration or heat-related problems. Patients with inadequate oral intake, or who are taking diuretics, may have difficulty tolerating exposure to heat. Many psychiatric medications used in geriatric patients affect how well they tolerate heat. Be thorough in your questioning. Determine your patient's exposure to heat and humidity and activities prior to the onset of symptoms.

Focused Physical Exam

Exposure to heat has significant effects on the metabolism, muscles, and cardiovascular system. Assess the patient for muscle cramps or confusion. Examine the patient's mental status and skin temperature and wetness. Take the patient's temperature.

Baseline Vital Signs

Patients who are hyperthermic will be tachycardic and tachypneic. As long as they maintain a normal blood pressure, their bodies are compensating for the fluid loss. Once their blood pressure begins to fall, it indicates they are no longer able to compensate for fluid loss and are going into shock. Check the patient's temperature with a thermometer, depending on protocol. Your assessment of the patient's skin will help determine how serious the heat problem is. For example, in heat ex-

haustion, the skin temperature may be normal or may even be cool and clammy; however, in heatstroke, the skin is hot.

Interventions

Remove your patient as quickly as possible from the hot environment. Patients with heat cramps or exhaustion usually respond well to passive cooling and fluids by mouth. Occasionally they will also require intravenous rehydration. Patients with symptoms of heatstroke should be transported immediately and actively cooled. Cover your patient with a sheet and soak it with water. Turn the ambulance air conditioner on high. Place cold packs in the patient's groin and axillae. Use convection to remove heat by fanning.

Detailed Physical Exam

Perform a detailed physical exam if circumstances and time permits. Pay special attention to the patient's skin temperature, turgor, and wetness. Perform a careful neurologic examination.

Skin turgor is the ability of the skin to resist deformation. It is tested by gently pinching skin on the forehead or back of the hand. Normally the skin will quickly flatten out. In dehydration, with poor skin turgor, the skin will remain tented.

Ongoing Assessment

Watch your patient's condition carefully for deterioration. Any decline in level of consciousness is an ominous sign. Monitor the patient's vital signs at least every 5 minutes. Evaluate the effectiveness of your interventions. Be careful not to cause shivering when cooling down a patient with heat problems. Shivering generates more heat and can occur when cooling is not monitored closely.

Communication and Documentation

Let the staff at the receiving facility know early that your patient is experiencing a heatstroke because additional resources may be required. Document the weather conditions and the activities prior to the emergency in your report.

Drowning and Near Drowning

<u>Drowning</u> is death from suffocation after submersion in water; <u>near drowning</u> is defined as survival, at least temporarily (24 hours), after suffocation in water. Drowning is often the last in a cycle of events caused by panic in the water (Figure 18-9 ▶). It can happen to anyone who is submerged in water for even a short period of time. Struggling toward the surface or the shore, the person becomes fatigued or exhausted, which leads him or her to sink even deeper. However, drowning also occurs in mop buckets, puddles, bathtubs, and other places where the individual is not completely submerged. Small children can drown in only a few inches of water if left unattended.

Inhaling very small amounts of either fresh- or saltwater can severely irritate the larynx, sending the muscles of the larynx and the vocal cords into spasm, called <u>laryngospasm</u>. The average person experiences this to a mild degree when a small amount of liquid is inhaled

and the patient coughs and seems to be choking for a few seconds. This is the body's attempt at self-preservation; laryngospasm prevents more water from entering the lungs. In severe cases such as water submersion, however, the patient's lungs cannot be ventilated because significant laryngospasm is present. Instead, progressive hypoxia occurs until the patient becomes unconscious. At this point, the spasm relaxes, making rescue

EMT-B Tips

The patient who experiences near drowning in cold water may require more care than two EMT-Bs can provide by themselves. Airway management and ventilation needs can make it difficult to remove wet clothing, treat hypothermia, or perform further assessment unless additional trained help is available. On this type of response, consider requesting backup before you arrive on scene.

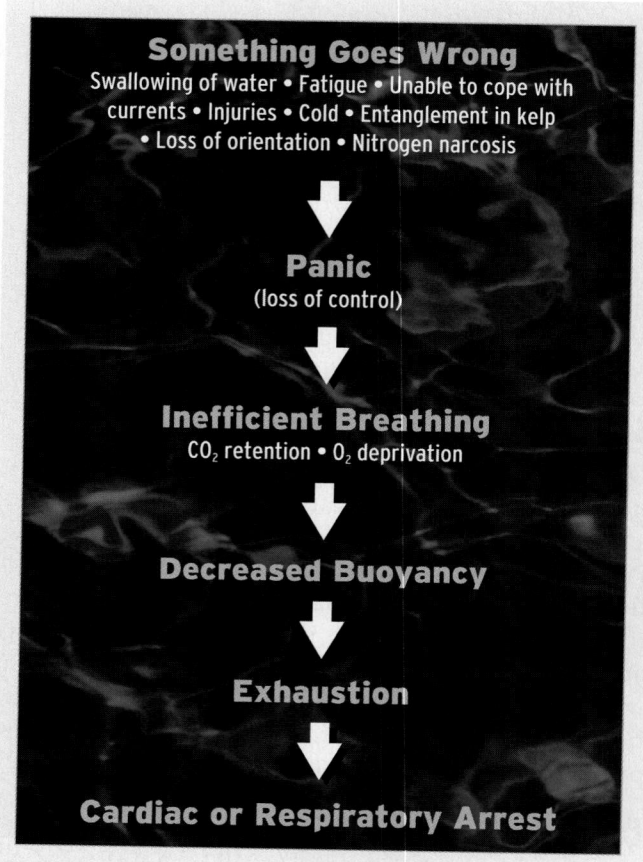

Something Goes Wrong
Swallowing of water • Fatigue • Unable to cope with currents • Injuries • Cold • Entanglement in kelp • Loss of orientation • Nitrogen narcosis

⬇

Panic
(loss of control)

⬇

Inefficient Breathing
CO_2 retention • O_2 deprivation

⬇

Decreased Buoyancy

⬇

Exhaustion

⬇

Cardiac or Respiratory Arrest

Figure 18-9 Panic in the water often precedes drowning.

You are the Provider   Part 4

Your partner tells you that the ALS unit has been delayed and will not be on scene for 25 minutes. You consider the current condition of your patient and you choose to transport immediately and rendezvous with the ALS unit en route. You prepare your patient for transport. While en route, you elevate the patient's legs. You cover the patient with a wet sheet and begin fanning him, constantly reassessing his airway, breathing, circulation, and responsiveness. You notify the hospital personnel of your patient's condition and continue with a rapid transport.

7. What are the possible ramifications of lowering the temperature of a patient with hyperthermia too quickly or lowering the temperature too far?

EMT-B Safety

You must ensure the safety of rescue personnel before a water rescue can begin. If the patient is conscious and still in the water, you should perform a water rescue. The saying: "Reach, throw, and row, and only then go" sums up the basic rule of water rescue. First, try to reach for the patient (Figure 18-10A ▶). If that does not work, throw the patient a rope, a life preserver, or any floatable object that is available (Figure 18-10B ▶). For example, an inflated spare tire, rim and all, will float well enough to support two people in the water. Next, use a boat if one is available (Figure 18-10C ▶). Do not attempt a swimming rescue unless you are trained and experienced in the proper techniques (Figure 18-10D ▶). Even then, you should always wear a helmet and a personal flotation device (Figure 18-11 ▶). Too many well-meaning individuals have themselves become victims while attempting a swimming rescue. In cold climates or cold-water locations, rapid hypothermia is also a concern for rescuers. Be prepared for this potential event.

If you work in a recreation area near lakes, rivers, or the ocean, you should have a prearranged plan for water rescue. This plan should include access to and cooperation with local personnel who are trained and skilled in water rescue; these personnel should help to develop the protocol for water rescue. Because the success of any water rescue depends on how rapidly the patient is removed from the water and ventilated, make sure you always have immediate access to personal flotation devices and other rescue equipment.

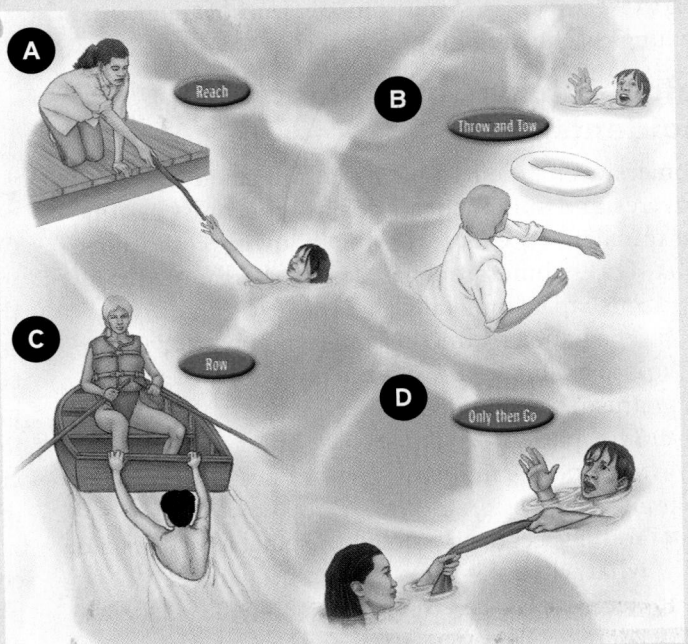

Figure 18-10 Basic rules of water rescue. **A.** Reach for the person from shore. If you cannot reach the person from shore, wade closer. **B.** If an object that floats is available, throw it to the person. **C.** Use a boat if one is available. **D.** If you must swim to the person, use a towel or board for him or her to hold onto. Do not let the person grab you.

Figure 18-11 When performing a water rescue, you must wear proper personal protective equipment, including a personal flotation device.

breathing possible. Of course, if the patient has not already been removed from the water, the patient may now inhale deeply, and more water may enter the lungs. In 85% to 90% of cases, significant amounts of water enter the lungs of the drowning victim.

Spinal Injuries in Submersion Incidents

Submersion incidents may be complicated by spinal fractures and spinal cord injuries. You must assume that spinal injury exists with the following conditions:

- The submersion has resulted from a diving mishap or long fall.
- The patient is unconscious, and no information is available to rule out the possibility of a mechanism causing neck injury.
- The patient is conscious but complains of weakness, paralysis, or numbness in the arms or legs.
- You suspect the possibility of spinal injury despite what witnesses say.

Most spinal injuries in diving incidents affect the cervical spine. When spinal injury is suspected, the neck must be protected from further injury. This means that you will have to stabilize the suspected injury while the patient is still in the water. Follow the steps in (Skill Drill 18-2 ▶):

1. **Turn the patient supine.** Two rescuers are usually required to turn the patient safely, although in some cases one rescuer will suffice. Always rotate the entire upper half of the patient's body as a single unit. Twisting only the head, for example, may aggravate any injury to the cervical spine (**Step 1**).

2. **Restore the airway and begin ventilation.** Immediate ventilation is the primary treatment of all drowning and near-drowning patients as soon as the patient is face up in the water. Use a pocket mask if it is available. Have the other rescuer support the head and trunk as a unit while you open the airway and begin artificial ventilation (**Step 2**).

3. **Float a buoyant backboard under the patient** as you continue ventilation (**Step 3**).

4. **Secure the trunk and head to the backboard** to eliminate motion of the cervical spine. Do not remove the patient from the water until this is done (**Step 4**).

5. **Remove the patient from the water, on the backboard** (**Step 5**).

6. **Cover the patient with a blanket.** Give oxygen if the patient is breathing spontaneously. Begin CPR if there is no pulse. Effective cardiac compression or CPR is extremely difficult to perform when the patient is still in the water (**Step 6**).

Recovery Techniques

On occasion, you may be called to the scene of a drowning and find that the patient is not floating or visible in the water. An organized rescue effort in these circumstances calls for personnel who are experienced with recovery techniques and equipment, including snorkel, mask, and scuba gear. Scuba (self-contained underwater breathing apparatus) gear is a system that delivers air to the mouth and lungs at atmospheric pressures that increase with the depth of the dive.

As a last resort, when standard procedures for recovery are unsuccessful, you may have to use a grappling iron or large hook to drag the bottom for the victim. Although the hook could seriously wound the patient, it may be the only effective way to bring him or her to the surface for resuscitation efforts.

Resuscitation Efforts

You should never give up on resuscitating a cold-water drowning victim. When a person is submerged in water that is colder than body temperature, heat will be conducted from the body to the water. The resulting hypothermia can protect vital organs from the lack of oxygen. In addition, exposure to cold water will occasionally activate certain primitive reflexes, which may preserve basic body functions for prolonged periods.

In one case, a 2½-year-old girl recovered after being submerged in cold water for at least 66 minutes. Continue full resuscitation efforts until the patient recovers or is pronounced dead by a physician.

Also, whenever a person dives or jumps into very cold water, the diving reflex, slowing of the heart rate caused by submersion in cold water, may cause immediate bradycardia, a slow heart rhythm. Loss of consciousness and drowning may follow. However, the person may be able to survive for an extended period of time under water, thanks to a lowering of the metabolic rate associated with hypothermia. For this reason, you should continue full resuscitation efforts no matter how long the patient has been submerged.

Stabilizing a Suspected Spinal Injury in the Water

1 Turn the patient to a supine position by rotating the entire upper half of the body as a single unit.

2 As soon as the patient is turned, begin artificial ventilation using the mouth-to-mouth method or a pocket mask.

3 Float a buoyant backboard under the patient.

4 Secure the patient to the backboard.

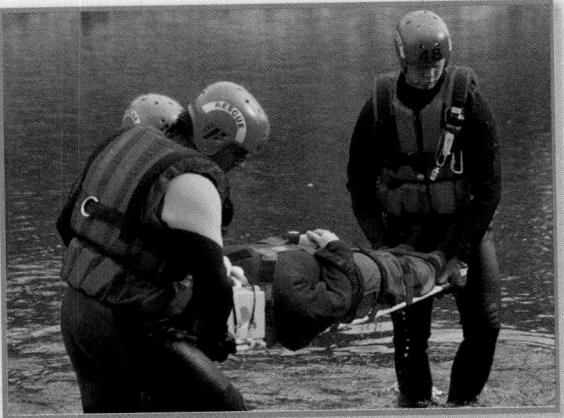

5 Remove the patient from the water.

6 Cover the patient with a blanket and apply oxygen if breathing. Begin CPR if breathing and pulse are absent.

Diving Emergencies

Most serious water-related injuries are associated with dives, with or without scuba gear. Some of these problems are related to the nature of the dive; others result from panic. Panic is not restricted to the person who is frightened by water. It can happen even to the experienced diver or swimmer.

There are more than 3,000,000 scuba sport divers in the United States, and approximately 200,000 new divers being trained annually. Medical problems relating to scuba diving techniques and equipment are becoming increasingly common. These problems are separated into three phases of the dive: descent, bottom, and ascent.

Descent Emergencies

Descent problems are usually due to the sudden increase in pressure on the body as the person dives deeper into the water. Some body cavities cannot adjust to the increased external pressure of the water; the result is severe pain. The usual areas affected are the lungs, the sinus cavities, the middle ear, the teeth, and the area of the face surrounded by the diving mask. Usually, the pain caused by these "squeeze problems" forces the diver to return to the surface to equalize the pressures, and the problem clears up by itself. A diver who continues to complain of pain, particularly in the ear, after returning to the surface should be transported to the hospital.

A person with a perforated tympanic membrane (ruptured eardrum) may develop a special problem while diving. If cold water enters the middle ear through a ruptured eardrum, the diver may lose his or her balance and orientation. The diver may then shoot to the surface and run into ascent problems.

Emergencies at the Bottom

Problems related to the bottom of the dive are rarely seen. They include inadequate mixing of oxygen and carbon dioxide in the air the diver breathes and accidental feeding of poisonous carbon monoxide into the breathing apparatus. Both are the result of faulty connections in the diving gear. These situations can cause drowning or rapid ascent; they require emergency resuscitation and transport of the patient.

Ascent Emergencies

Most of the serious injuries associated with diving are related to ascending from the bottom and are referred to as ascent problems. These emergencies usually require aggressive resuscitation. Two particularly dangerous medical emergencies are air embolism and decompression sickness (also called "the bends").

Air Embolism *Study This?*

The most dangerous, and most common, emergency in scuba diving is air embolism, a condition involving bubbles of air in the blood vessels. Air embolism may occur on a dive as shallow as 6'. The problem starts when the diver holds his or her breath during a rapid ascent. The air pressure in the lungs remains at a high level while the external pressure on the chest decreases. As a result, the air inside the lungs expands rapidly, causing the alveoli in the lungs to rupture. The air released from this rupture can cause the following injuries:

- Air may enter the pleural space and compress the lungs (a pneumothorax).
- Air may enter the mediastinum (the space within the thorax that contains the heart and great vessels), causing a condition called pneumomediastinum.
- Air may enter the bloodstream and create bubbles of air in the vessels called air emboli.

Pneumothorax and pneumomediastinum both result in pain and severe dyspnea. An air embolus will act as a plug and prevent the normal flow of blood and oxygen to a specific part of the body. The brain and spinal cord are the organs most severely affected by air embolism because they require a constant supply of oxygen.

The following are potential signs and symptoms of air embolism:

- Blotching (mottling of the skin)
- Froth (often pink or bloody) at the nose and mouth
- Severe pain in muscles, joints, or abdomen
- Dyspnea and/or chest pain
- Dizziness, nausea, and vomiting
- Dysphasia (difficulty speaking)
- Difficulty with vision
- Paralysis and/or coma
- Irregular pulse and even cardiac arrest

Decompression Sickness

Decompression sickness, commonly called the bends, occurs when bubbles of gas, especially nitrogen, obstruct the blood vessels. This condition results from too rapid an ascent from a dive. During the dive, nitrogen that is being breathed dissolves in the blood and tissues because it is under pressure. When the diver ascends, the external pressure is decreased, and the dissolved nitrogen forms small bubbles within those tissues. These bubbles can lead to problems similar to those that occur in air embolism (blockage of tiny blood vessels, depriving parts of the body of their normal blood supply), but severe pain in certain tissues or spaces in the body is the most common problem.

The most striking symptom is abdominal and/or joint pain so severe that the patient literally doubles up or "bends." Dive tables and computers are available to show the proper rate of ascent from a dive, including the number and length of pauses that a diver should make on the way up. However, even divers who stay within these limits can suffer the bends.

Even after a "safe dive," decompression sickness can occur from driving a car up a mountain or flying in an unpressurized airplane that climbs too rapidly to a great height. However, the risk of this diminishes after 24 to 48 hours. The problem is exactly the same as ascent from a deep dive: a sudden decrease of external pressure on the body and release of dissolved nitrogen from the blood that forms bubbles of nitrogen gas within the blood vessels.

You may find it difficult to distinguish between air embolism and decompression sickness. As a general rule, air embolism occurs immediately on return to the surface, whereas the symptoms of decompression sickness may not occur for several hours. The emergency treatment is the same for both. It consists of BLS followed by recompression in a hyperbaric chamber, a chamber or a small room that is pressurized to more than atmospheric pressure (Figure 18-12 ▼). Recompression treatment allows the bubbles of gas to dissolve into the blood and equalizes the pressures inside and outside the lungs. Once these pressures are equalized, gradual decompression can be accomplished under controlled conditions to prevent the bubbles from reforming.

Assessment of Drowning and Diving Emergencies

Scene Size-up

In managing water emergencies, your BSI precautions should include gloves and eye protection at a minimum. A mask may be necessary if aggressive airway management is necessary. Check for hazards to your crew. Never drive though moving water—a small amount can push the vehicle. Use extreme caution when driving through standing water. Never attempt a water rescue without proper training and equipment. If your patient is still in the water, look for the best, safest means of removal. This may require additional help from search and rescue teams or special extrication equipment. Trauma and spinal stabilization must be considered when the scene is a recreational setting. Check for additional patients based on where and how the problem occurred.

Initial Assessment

General Impression

Use your evaluation of the patient's chief complaint to guide you in your assessment of life threats and determine if spinal precautions are necessary. Pay particular attention to chest pain, dyspnea, and complaints related to sensory changes when a diving emergency is suspected. Determine their LOC using the AVPU scale. Be suspicious of alcohol use and its effects on the patient's LOC.

Airway, Breathing, and Circulation

Usual BLS measures should be employed for any patient found or injured while in water. Begin with opening the airway and assessing breathing in unresponsive patients. Use an airway adjunct to facilitate BVM ventilations as necessary. Suction according to protocol if the patient has vomited or pink, frothy secretions

Figure 18-12 A hyperbaric chamber, usually a small room, is pressurized to more than atmospheric pressure and used in the treatment of decompression sickness and air embolism.

are found in the airway. Provide BVM ventilations for breathing that is inadequate. Check for a pulse. It may be difficult to find a pulse because of constriction of the peripheral blood vessels and low cardiac output. Nevertheless, if the pulse is unmeasurable, start CPR according to BLS guidelines.

If the patient is responsive, provide high-flow oxygen with a nonrebreathing mask and position the patient to protect the airway from aspiration in case of vomiting. Evaluate the patient for adequate perfusion and treat for shock by maintaining normal body temperature and improving circulation through positioning. If the chief complaint suggests trauma, assess for bleeding and treat appropriately.

Transport Decision

Even if resuscitation in the field appears completely successful, you must always transport near-drowning patients to the hospital. Inhalation of any amount of fluid can lead to delayed complications lasting for days or weeks. Patients with decompression sickness and air embolism must be treated in a recompression chamber. If you live in an area with a lot of diving activity, you will have transport protocols in this regard. Usually, the patient will be stabilized in the nearest emergency department. Perform all interventions en route.

Focused History and Physical Exam

If a drowning patient is responsive, assess the history first. On the basis of the chief complaint and the history obtained, perform a focused physical exam. This should include a thorough exam on the patient's lungs, including breath sounds. Last, obtain vital signs. Serious drowning situations typically result in an unresponsive patient. It is important to begin with a rapid physical exam in these situations to look for hidden life threats and potential trauma, even if trauma is not suspected. After you perform a head-to-toe exam, obtain the patient's baseline vital signs and a SAMPLE history.

Rapid Physical Exam

Perform a rapid physical exam on unresponsive medical patients. Look for signs of trauma or complications with the drowning. A diver with problems should be checked from head to toe for indications of the bends or an air embolism. Focus on pain in the joints and the abdomen. Pay attention to whether your patient is get-

ting adequate ventilation and oxygenation, and check for signs of hypothermia. Complete a Glasgow Coma Scale score to assess the patient's neurologic status and thinking.

Baseline Vital Signs

The vital signs are a good indicator of how your patient is tolerating the effects of drowning or diving complications. Check the patient's pulse rate, quality, and rhythm. Pulse and blood pressure may be difficult to palpate in the hypothermic patient. Check carefully for both peripheral and central pulses, and listen over the chest for a heartbeat if pulses are weak. Check the respiratory rate, quality, and rhythm. Assess and document pupil size and reactivity.

SAMPLE History

Obtain a SAMPLE history with special attention to the length of time the drowning victim was under water or the time of onset of symptoms in relation to the last dive. Note any physical activity, alcohol or drug consumption, and other medical conditions. All of these may have an effect on the diving or drowning emergency. In diving emergencies, it is important to determine the dive parameters in your history, including depth, time, and previous diving activity.

Drowning Interventions

Treatment begins with rescue and removal from the water. When necessary, artificial ventilation should begin as soon as possible, even before the victim is removed from the water. At the same time, you must take care to stabilize and protect the patient's spine when a long fall or dive has occurred (or if this is a possibility when no information is provided). Associated cervical spine injuries are possible, especially in diving mishaps. If the patient does not have a possible spinal injury, you can turn the patient quickly to the left side to allow draining from the upper airway. Note that water will not drain from the lungs. If there is evidence of upper airway obstruction by foreign matter, remove the obstruction manually or, if available, by suction. If necessary, use abdominal thrusts, followed by assisted ventilations. Administer oxygen if this was not done as part of the initial assessment, either by mask for patients who are breathing spontaneously or via a BVM device for those requiring assisted ventilation.

Make sure that the patient is kept warm, especially after cold-water immersion. Make sure blankets and protection from the environment are provided as needed.

If ventilation equipment is not available but oxygen is, you can breathe the oxygen in yourself and give mouth-to-mask ventilation until rescue equipment arrives. In this method, your expired air will have a higher percentage of oxygen.

Diving Interventions

In treating patients who are suspected of having air embolism or decompression sickness, you should follow these accepted treatment steps:

1. Remove the patient from the water. Try to keep the patient calm.
2. Begin BLS and administer oxygen.
3. Place the patient in a left lateral recumbent position with the head down.
4. Provide prompt transport to the nearest recompression facility for treatment.

Injury from decompression sickness is usually reversible with proper treatment. However, if the bubbles block critical blood vessels that supply the brain or spinal cord, permanent central nervous system injury may result. Therefore, the key in emergency management of these serious ascent problems is to recognize that an emergency exists and treat as soon as possible. Administer oxygen and provide rapid transport.

Detailed Physical Exam

Time and personnel permitting, complete a detailed physical exam en route to the hospital. A careful head-to-toe exam may reveal additional injuries not initially observable. Examine the patient for respiratory, circulatory, and neurologic compromise. A careful distal circulatory, sensory, and motor function exam will be helpful in assessing the extent of the injury. Examine peripheral pulses, skin color and discoloration, itching, pain, and paresthesias (numbness and tingling).

Ongoing Assessment

The condition of patients who have experienced near drowning may deteriorate rapidly due to pulmonary injury, fluid shifts in the body, cerebral hypoxia, and hypothermia. Patients with air embolism or decompression sickness may decompensate quickly. Assess your patient's mental status constantly, and assess vital signs at least every 5 minutes, paying particular attention to respirations and breath sounds.

Communication and Documentation

Document the circumstances of the drowning and extrication. The receiving facility personnel will need to know how long the patient was submerged, the temperature of the water, the clarity of the water, and whether there was any possibility of cervical spine injury.

The receiving facility personnel will also need a complete dive profile in order to properly treat your patient. This may be available in a dive log or from diving partners. Small diving computers have become standard equipment for most divers, and they record information from the current as well as previous dives. Be sure the computer is brought to the hospital with the patient. If possible, have all of the diver's equipment brought to the hospital. It will be helpful in determining the cause of the accident. Be sure to document the disposition of this equipment.

Other Water Hazards

You must pay close attention to the body temperature of a person who is rescued from cold water. Treat hypothermia caused by immersion in cold water the same way you treat hypothermia caused by cold exposure. Prevent further heat loss from contact with the ground, stretcher, or air, and transport the patient promptly.

A person swimming in shallow water may experience breath-holding syncope, a loss of consciousness caused by a decreased stimulus for breathing. This happens to swimmers who breathe in and out rapidly and deeply before entering the water in an effort to expand their capacity to stay underwater. While increasing the oxygen level, this hyperventilation lowers the carbon dioxide level. Because an elevated level of carbon dioxide in the blood is the strongest stimulus for breathing, the swimmer may not feel the need to breathe even after using up all the oxygen in his or her lungs. The emergency treatment for a breath-holding syncope is the same as that for a drowning or near drowning.

Injuries caused by boat propellers, sharp rocks, water skis, or dangerous marine life may be complicated by immersion in cold water. In these cases, remove the patient from the water, taking care to protect the spine, and administer oxygen. Apply dressings and splints if indicated, and monitor the patient closely for any signs of immersion or cold injury.

You should be aware that a child who is involved in a drowning or near drowning may be the victim of child abuse. Although it may be difficult to prove, such incidents should be handled according to the rules set up for suspected child abuse.

Prevention

Appropriate precautions can prevent most immersion incidents. Each year, many small children drown in residential pools. All pools should be surrounded by a fence that is at least 6′ high, with slats no farther apart than 3″ and self-closing, self-locking gates. The most common problem is lack of adult supervision, even when attention is not given for a few seconds. Half of all teenage and adult drownings are associated with the use of alcohol. As a health care professional, you should be involved in public education efforts to make people aware of the hazards of swimming pools and water recreation.

Lightning

According to the National Weather Service, there are an estimated 25 million cloud-to-ground lightning flashes in the United States each year. On average, lightning kills between 60 and 70 people per year in the United States based on documented cases. While documented lightning injuries in the United States average about 300 per year, undocumented lightning injuries are likely much higher. Lightning is the third most common cause of death from isolated environmental phenomena.

The energy associated with lightning is comprised of direct current (DC) of up to 200,000 amp and a potential of 100 million volts or more. Temperatures generated from lightning vary between 20,000°F and 60,000°F.

Most deaths and injuries caused by lightning occur during the summer months when people are enjoying outdoor activities, despite an approaching thunderstorm. Those most commonly struck by lightning include boaters, swimmers, and golfers; any type of activity that exposes the person to a large, open area increases the risk of being struck by lightning.

Whether or not lightning injures or kills depends on whether a person is in the path of the lightning discharge. The current associated with the lightning discharge travels along the ground. Although some victims are injured or killed by a direct lightning strike, many victims are indirectly struck when standing near an object that has been struck by lightning, such as a tree (splash effect).

The cardiovascular and nervous systems are most commonly injured during a lightning strike; therefore, respiratory or cardiac arrest is the most common cause of lightning-related deaths. The tissue damage caused by lightning is different from that caused by other electrical-related injuries (ie, high-power line injuries). This is because the tissue damage pathway usually occurs over the skin, rather than through it. Additionally, because the duration of a lightning strike is short, skin burns are usually superficial; full-thickness (third-degree) burns are rare. Lightning injuries are categorized as being mild, moderate, or severe:

- **Mild:** loss of consciousness, amnesia, confusion, tingling, and other nonspecific signs and symptoms. Burns, if present, are typically superficial.
- **Moderate:** seizures, respiratory arrest, cardiac standstill (asystole) that spontaneously resolves, and superficial burns.
- **Severe:** cardiopulmonary arrest. Because of the delay in resuscitation, often due to remote locations, many of these patients do not survive.

Emergency Medical Care

As with any scene response, the safety of you and your partner has priority. Take measures to protect yourself from being struck by lightning, especially if the thunderstorm is still in progress. Contrary to popular belief, lightning can, and does, strike in the same place twice. Move the patient to a place of safety, preferably in a sheltered area.

If you are in an open area and adequate shelter is not available, it is important to recognize the signs of an impending lightning strike and take immediate action to protect yourself. If you suddenly feel a tingling sensation or your hair stands on end, the area around you has become charged—a sure sign of an imminent lightning strike. Curl up in a ball and lie on the ground; make yourself as small a target as possible. If you are standing near a tree or other tall object, move away as fast as possible, preferably to a low-lying area. Lightning has an affinity for objects that project from the ground (ie, trees, fences, buildings).

The process of triaging multiple victims of a lightning strike is different than the conventional triage methods used during a mass-casualty incident. When a person is struck by lightning, respiratory or cardiac arrest, if it occurs, usually occurs immediately. Those

✳ EMT-B Tips

Focus your efforts on lightning strike victims who appear "dead." Many of these patients can be successfully resuscitated.

Figure 18-13 Black widow spiders are distinguished by their glossy black color and bright red-orange hourglass marking on the abdomen.

who are conscious following a lightning strike are much less likely to develop delayed respiratory or cardiac arrest; most of these victims will survive. Therefore, you should focus your efforts on those who are in respiratory or cardiac arrest. This, process, called <u>reverse triage</u>, differs from conventional triage, where such patients would ordinarily be classified as deceased.

Emergency care for a lightning injury is the same as it is for other severe electrical injuries. Because the massive DC shock caused by lightning, the patient experiences massive muscle spasms (tetany), which can result in fractures of long bones and spinal vertebrae. Therefore, manually stabilize the patient's head in a neutral in-line position and open the airway with the jaw-thrust maneuver. If the patient is in respiratory arrest with a pulse, begin immediate BVM ventilations with 100% oxygen. If the patient is in cardiac arrest, attach an AED as soon as possible and provide immediate defibrillation if indicated. If severe bleeding is present, control it immediately.

Provide full spinal stabilization and transport the patient to the closest appropriate facility. If CPR or ventilations are not required, address other injuries (ie, splint fractures, dress and bandage burns) and provide continuous monitoring while en route to the hospital.

Bites and Envenomations

This section discusses bites and stings from spiders, snakes, scorpions, ticks, and injuries from marine animals. Insect stings are discussed in Chapter 16 and dog and human bites are discussed in Chapter 24.

Spider Bites

Spiders are numerous and widespread in the United States. Many species of spiders bite. However, only two, the female black widow spider and the brown recluse spider, are able to deliver serious, even life-threatening bites. When you care for a patient who has had some type of bite, be alert to the possibility that the spider may

still be in the area, although it is not likely. Remember that your safety is of paramount importance.

Black Widow Spider

The female black widow spider (*Latrodectus*) is fairly large, measuring approximately 2″ long with its legs extended. It is usually black and has a distinctive, bright red-orange marking in the shape of an hourglass on its abdomen (Figure 18-13 ▲). The female is larger and more toxic than the male. Black widow spiders are found in every state except Alaska. They prefer dry, dim places around buildings, in woodpiles, and among debris.

The bite of the black widow spider is sometimes overlooked. If the site becomes numb right away, the patient may not even recall being bit. However, most black widow spider bites cause localized pain and symptoms, including agonizing muscle spasms. In some cases, a bite on the abdomen causes muscle spasms so severe that the patient may be thought to have an acute abdomen, possibly peritonitis. The main danger with this type of bite, however, is that the black widow's venom is poisonous to nerve tissues (neurotoxic). Other systemic symptoms include dizziness, sweating, nausea, vomiting, and rashes. Tightness in the chest and difficulty breathing develop within 24 hours, as well as severe cramps, with board-like rigidity of the abdominal muscles. Generally, these signs and symptoms subside over 48 hours.

If necessary, a physician can administer a specific <u>antivenin</u>, a serum containing antibodies that counteract the venom, but because of a high incidence of side effects, its use is reserved for very severe bites, for the aged or very feeble, and for children younger than 5 years. The severe muscle spasms are usually treated in the hospital with IV benzodiazepines such as diazepam (Valium) or lorazepam (Ativan). In general, emergency treatment for a black widow spider bite consists of BLS for the patient in respiratory distress. Much more often, the patient will only require relief from pain. Transport the patient to the emergency department as soon as possible for treatment of both pain and muscle rigidity. If possible, bring the spider.

Brown Recluse Spider

The brown recluse spider (*Loxosceles*) is dull brown and, at 1″, smaller than the black widow (Figure 18-14 ▼). The short-haired body has a violin-shaped mark, brown to yellow in color, on its back. Although the brown recluse spider lives mostly in the southern and central parts of the country, it may be found throughout the continental United States. The spider takes its name from the fact that it tends to live in dark areas—in corners of old, unused buildings, under rocks, and in woodpiles. In cooler areas, it moves indoors to closets, drawers, cellars, and old piles of clothing.

In contrast to the venom of the black widow spider, the venom of the brown recluse spider is not neurotoxic but cytotoxic; that is, it causes severe local tissue damage. Typically, the bite is not painful at first but becomes so within hours. The area becomes swollen and tender, developing a pale, mottled, cyanotic center and possibly a small blister (Figure 18-15 ▼). Over the next several days, a scab of dead skin, fat, and debris will form and dig down into the skin, producing a large ulcer that may not heal unless treated promptly. Transport patients with such symptoms as soon as possible.

Brown recluse spider bites rarely cause systemic symptoms and signs. When they do, the initial treatment is BLS and transportation to the emergency department. Again, it is helpful if you can identify the spider and bring it to the hospital with the patient.

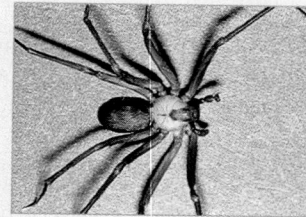

Figure 18-14 Brown recluse spiders are dull brown and have a dark, violin-shaped mark on the back.

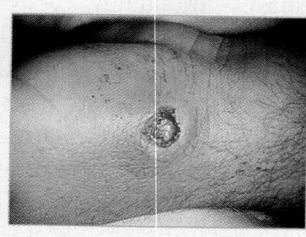

Figure 18-15 The bite of a brown recluse spider is characterized by swelling, tenderness, and a pale, mottled, cyanotic center. There may also be a small blister on the bite.

Snake Bites

Snake bites are a worldwide problem. More than 300,000 injuries from snake bites occur annually, including 30,000 to 40,000 deaths. The greatest number of fatalities occur in Southeast Asia and India (25,000 to 30,000) and in South America (3,000 to 4,000). In the United States, 40,000 to 50,000 snake bites are reported annually, with about 7,000 caused by poisonous snakes. However, snake bite fatalities in the United States are extremely rare, about 15 a year for the entire country.

Of the approximately 115 different species of snakes in the United States, only 19 are venomous. These include the rattlesnake (*Crotalus*), the copperhead (*Agkistrodon contortrix*), the cottonmouth, or water moccasin (*Agkistrodon piscivorus*), and the coral snakes (*Micrurus* and *Micruroides*) (Figure 18-16 ▶). At least one of these poisonous species is found in every state except Alaska, Hawaii, and Maine. As a general rule, these creatures are timid. They usually do not bite unless provoked or accidentally injured, as when they are stepped on. There are a few exceptions to these rules. Cottonmouths are often aggressive, and rattlesnakes are easily provoked. Coral snakes, in contrast, usually bite only when they are being handled.

Most snake bites occur between April and October, when the animals are active, and tend to involve young men who have been drinking alcohol. Texas reports the largest number of bites. Other states with a major concentration of snake bites are Louisiana, Georgia, Oklahoma, North Carolina, Arkansas, West Virginia, and Mississippi. If you work in one of these areas, you should be thoroughly familiar with the emergency handling of snake bites. Remember, almost any time you are caring for a patient with a snake bite, another snake may be in the area and create a second victim—you. Therefore, use extreme caution on these calls and be sure to wear the proper protective equipment for the area.

In general, only a third of snake bites result in significant local or systemic injuries. Often, envenomation does not occur because the snake has recently struck another animal and exhausted its supply of venom for the time being.

With the exception of the coral snake, poisonous snakes native to the United States all have hollow fangs in the roof of the mouth that inject the poison from two sacs at the back of the head. The classic appearance of the poisonous snake bite, therefore, is two small puncture wounds, usually about ½″ apart, with discoloration, swelling, and pain surrounding them (Figure 18-17 ▶). Nonpoisonous snakes can also bite, usually leaving a

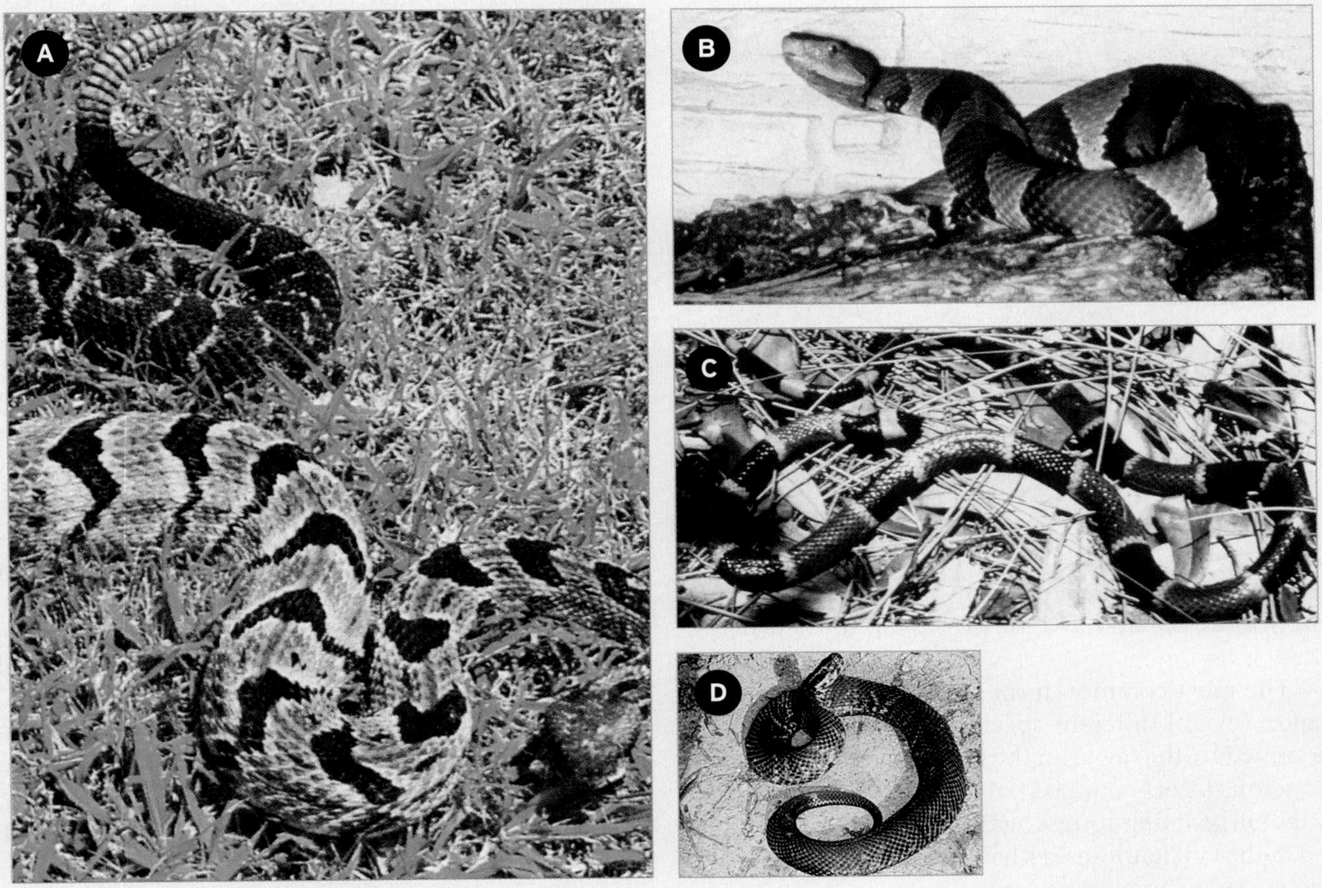

Figure 18-16 **A.** Rattlesnake. **B.** Copperhead. **C.** Coral snake. **D.** Cottonmouth.

horseshoe of tooth marks. However, some poisonous snakes have teeth as well as fangs, making it impossible to say which kind is responsible for a given set of tooth marks. On the other hand, fang marks are a clear indication of a poisonous snake bite.

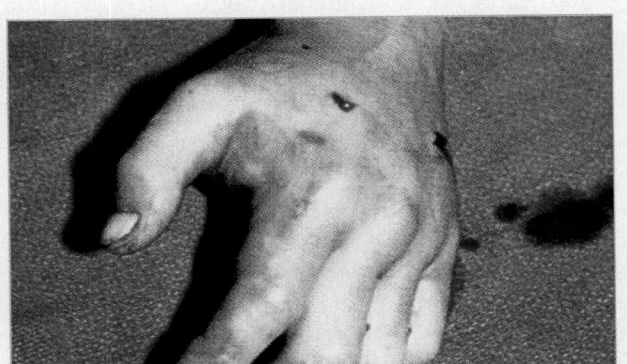

Figure 18-17 A snake bite wound from a poisonous snake has characteristic markings: two small puncture wounds about 1/2″ apart, discoloration, and swelling.

Pit Vipers

Rattlesnakes, copperheads, and cottonmouths are all pit vipers, with triangular-shaped, flat heads (Figure 18-18 ▶). They take their name from the small pits located just behind each nostril and in front of each eye. The pit is a heat-sensing organ that allows the snake to strike accurately at any warm target, especially in the dark, when it cannot see through its vertical, slit-like pupils.

The fangs of the pit viper normally lie flat against the roof of the mouth and are hinged to swing back and forth as the mouth opens. When the snake is striking, the mouth opens wide and the fangs extend; in this way, the fangs penetrate whatever the mouth strikes. The fangs are actually special hollow teeth that act much like hypodermic needles. They are connected to a sac containing a reservoir of venom, which in turn is attached to a poison gland. The gland itself is a specially adapted salivary gland, which produces enzymes that digest and destroy tissue. The primary purpose of the venom is to kill small animals

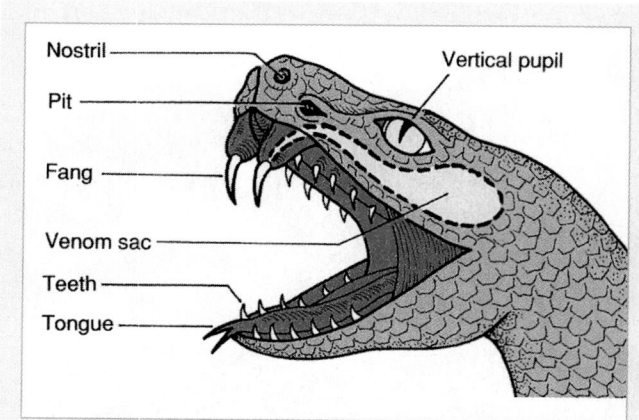

Figure 18-18 Pit vipers have small, heat-sensing organs (pits) located in front of their eyes that allow them to strike at warm targets, even in the dark.

and to start the digestive process prior to their being eaten.

The most common form of pit viper is the rattlesnake. Several different species of rattlesnake can be identified by the rattle on the tail. The rattle is actually numerous layers of dried skin that were shed but failed to fall off, coming to rest against a small knob on the end of the tail. Rattlesnakes have many patterns of color, often with a diamond pattern. They can grow to 6′ or more in length.

Copperheads are smaller than rattlesnakes, usually 2′ to 3′ long, with a reddish coppery color crossed with brown or red bands. These snakes typically inhabit woodpiles and abandoned dwellings, often close to areas of habitation. Although they account for most of the venomous snake bites in the eastern United States, copperhead bites are almost never fatal; however, note that the venom can destroy extremities.

Cottonmouths grow to about 4′ in length. Also called water moccasins, these snakes are olive or brown, with black cross-bands and a yellow undersurface. They are water snakes and have a particularly aggressive pattern of behavior. Although fatalities from these snake bites are rare, tissue destruction from the venom may be severe.

The signs of envenomation by a pit viper are severe burning pain at the site of the injury, followed by swelling and a bluish discoloration (ecchymosis) in light-skinned individuals that signals bleeding under the skin. These signs are evident within 5 to 10 minutes after the bite has occurred and spread over the next 36 hours. In addition to destroying tissues locally, the venom of the

pit viper can also interfere with the body's clotting mechanism and cause bleeding at various distant sites. Other systemic signs, which may or may not occur, include weakness, sweating, fainting, and shock. If the patient has no local signs an hour after being bitten, it is safe to assume that envenomation did not take place. If swelling has occurred, you should mark its edges on the skin. This will allow physicians to assess what has happened and when it happened with greater accuracy.

Occasionally, a patient bitten by a snake will faint from fright. The patient will usually regain consciousness promptly when placed in a supine position. Do not confuse a fainting spell with shock. If shock occurs, it will happen much later.

In treating a snake bite from a pit viper, follow these steps to get the patient to the hospital in a timely manner:

1. Calm the patient; assure him or her that poisonous snake bites are rarely fatal. Place the patient in a supine position and explain that staying quiet will slow the spread of any venom through the system.
2. Locate the bite area; clean it gently with soap and water or a mild antiseptic. **Do not apply ice to the area.**
3. If the bite occurred on an arm or leg, splint the extremity to decrease movement.
4. Be alert for vomiting, which may be a sign of anxiety rather than the toxin itself.
5. Do not give anything by mouth.
6. If, as rarely happens, the patient was bitten on the trunk, keep him or her supine and quiet and transport as quickly as possible.
7. Monitor the patient's vital signs and mark the skin with a pen over the area that is swollen, proximal to the swelling, to note whether swelling is spreading.
8. If there are any signs of shock, place the patient in the shock position and administer oxygen.
9. If the snake has been killed, as is often the case, be sure to bring it with you in a secure container so that physicians can identify it and administer the proper antivenin.
10. Notify the hospital that you are bringing in a patient who has a snake bite; if possible, describe the snake.
11. Transport the patient promptly to the hospital.

If the patient shows no sign of envenomation, provide BLS as needed, place a sterile dressing over the suspected bite area, and immobilize the injury site. All patients with suspected snake bite should be taken to

EMT-B Safety

Evidence of the exact source of an allergic reaction or envenomation may be scarce when you arrive, or bystanders may give you incorrect information. The cause is more likely to pose a risk to responders, and added risk to the patient, if you draw incorrect conclusions about its nature. Keep your eyes and ears open, avoid making unsupported assumptions, and be curious about things that do not quite make sense.

the emergency department, whether they show signs of envenomation or not. Treat the wound as you would any deep puncture wound to prevent infection.

If you work in an area where poisonous snakes are known to live, you should know the local protocol for handling snake bites. You should also know the address of the nearest facility where antivenin is available. This may be a nearby zoo, the local or public state health department, or a local community hospital.

Coral Snakes

The coral snake is a small reptile with a series of bright red, yellow, and black bands completely encircling the body. Many harmless snakes have similar coloring, but only the coral snake has red and yellow bands next to one another, as this helpful rhyme suggests: "Red on yellow will kill a fellow; red on black, venom will lack."

A rare creature that lives in most southern states and in the Southwest, the coral snake is a relative of the cobra. It has tiny fangs and injects the venom with its teeth by a chewing motion, leaving behind one or more puncture or scratch-like wounds. Because of its small mouth and teeth and limited jaw expansion, the coral snake usually bites its victims on a small part of the body, such as a finger or toe.

Coral snake venom is a powerful toxin that causes paralysis of the nervous system. Within a few hours of being bitten, a patient will exhibit bizarre behavior, followed by progressive paralysis of eye movements and respiration. Often, there are limited or no local symptoms.

Successful treatment, either emergency or long term, depends on positive identification of the snake and support of respiration. Antivenin is available, but most hospitals do not stock it. Therefore, you should notify the hospital of the need for it as

soon as possible. The steps for emergency care of a coral snake bite are as follows:

1. Immediately quiet and reassure the patient.
2. Flush the area of the bite with 1 to 2 quarts of warm, soapy water to wash away any poison left on the surface of the skin. **Do not apply ice to the region.**
3. Splint the extremity to minimize movement and the spread of venom at the site.
4. Check the patient's vital signs and continue to monitor them.
5. Keep the patient warm and elevate the lower extremities to help prevent shock.
6. Give supplemental oxygen if needed.
7. Transport the patient promptly to the emergency department, giving advance notice that the patient has been bitten by a coral snake.
8. Give the patient nothing by mouth.

Scorpion Stings

Scorpions are eight-legged arachnids from the biological group Arachnida with a venom gland and a stinger at the end of their tail (Figure 18-19 ▼). Scorpions are rare; they live primarily in the southwestern United States and in deserts. With one exception, a scorpion's sting is usually very painful but not dangerous, causing localized swelling and discoloration. The exception is the *Centruroides sculpturatus*. Although it is found naturally in Arizona and New Mexico, as well as parts of Texas, California, and Nevada, it may be kept as a pet by anyone. The venom of this particular species may produce a severe systemic reaction that brings about circulatory collapse, severe muscle contractions, excessive salivation, hypertension, convulsions, and cardiac failure. Antivenin is available but must be administered by a physician. If you are called to care for a patient with a suspected sting from *C. sculpturatus*, you should notify medical control as soon as possible. Administer BLS and transport the patient to the emergency department as rapidly as possible.

Figure 18-19 The sting of a scorpion is usually more painful than it is dangerous, causing localized swelling and discoloration.

Tick Bites

Found most often on brush, shrubs, trees, sand dunes, or other animals, ticks usually attach themselves directly to the skin (Figure 18-20 ▼). Only a fraction of an inch long, they can easily be mistaken for a freckle, especially since their bite is not painful. Indeed, the danger with a tick bite is not from the bite itself, but from the infecting organisms that the tick carries. Ticks commonly carry two infectious diseases, Rocky Mountain spotted fever and Lyme disease. Both are spread through the tick's saliva, which is injected into the skin when the tick attaches itself.

Rocky Mountain spotted fever, which is not limited to the Rocky Mountains, occurs within 7 to 10 days after a bite by an infected tick. Its symptoms include nausea, vomiting, headache, weakness, paralysis, and possibly cardiorespiratory collapse.

Lyme disease has received extensive publicity. Originally seen only in Connecticut, Lyme disease has now been reported in 35 states. It occurs most commonly in the Northeast, the Great Lake states, and the Pacific Northwest; New York State reports the largest number of cases. The first symptom, a rash that may spread to several parts of the body, begins about 3 days after the bite of an infected tick. The rash may eventually resemble a target bull's-eye pattern in one third of patients (Figure 18-21 ▼). After a few more days or weeks, painful swelling of the joints, particularly the knees, occurs. Lyme disease may be confused with rheumatoid arthritis and, like that disease, may result in permanent disability. However, if it is recognized and treated promptly with antibiotics, the patient may recover completely.

Tick bites occur most commonly during the summer months, when people are out in the woods wearing little protective clothing. Transmission of the infection from tick to person takes at least 12 hours, so if you are called on to remove a tick, you should proceed carefully and slowly. Do not attempt to suffocate the tick with gasoline or Vaseline or burn it with a lighted match; you will only burn the patient. Instead, using fine tweezers, grasp the tick by the body and pull gently but firmly straight up so that the skin is tented. Hold this position until the tick releases. Special tweezers are available for this, but are not necessary. This method will usually remove the whole tick. Even if part of the tick is left embedded in the skin, the part containing the infecting organisms has been removed. Cleanse the area with disinfectant and save the tick in a glass jar or other container so that it can be identified. Do not handle the tick with your fingers. Provide any necessary supportive emergency care, and transport the patient to the hospital.

Injuries From Marine Animals

Coelenterates, including the fire coral, Portuguese man-of-war, sea wasp, sea nettles, true jellyfish, sea anemones, true coral, and soft coral, are responsible for more envenomations than any other marine animals (Figure 18-22 ▶). The stinging cells of the coelenterate are called nematocysts, and large animals may discharge hundreds of thousands of them. Envenomation causes very painful, reddish lesions in light-skinned individuals extending in a line from the site of the sting. Systemic symptoms include headache, dizziness, muscle cramps, and fainting.

To treat a sting from the tentacles of a jellyfish, a Portuguese man-of-war, various anemones, corals, or hydras, remove the patient from the water and pour acetic acid (vinegar) on the affected area. Unlike fresh water, vinegar will inactivate the nematocysts. Do not try to manipulate the remaining tentacles; this will only cause further discharge of the nematocysts. Remove the tentacles by scraping them off with the edge of a sharp, stiff object such as a credit card. Persistent pain may respond to immersion of the area in hot water (110° to 115°F, 43° to 46°C) for 30 minutes. On very rare occa-

Figure 18-20 Ticks typically attach themselves directly to the skin.

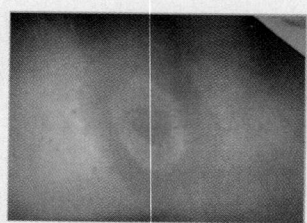

Figure 18-21 The rash associated with Lyme disease has a characteristic "bull's-eye" pattern.

sions, a patient may have a systemic allergic reaction to the sting of one of these animals. Treat such a patient for anaphylactic shock. Provide BLS and immediate transport to the hospital.

Toxins from the spines of urchins, stingrays, and certain spiny fish such as the lionfish, scorpion fish, or stonefish are heat sensitive (Table 18-2 ▶). Therefore, the best treatment for such injuries is to immobilize the affected area and soak it in hot water for 30 minutes. This will often provide dramatic relief from local pain. However, the patient still needs to be transported to the emergency department because an allergic reaction or infection, including tetanus, could develop.

If you work near the ocean, you should be familiar with the marine life in your area. The emergency treatment of common coelenterate envenomations consists of the following steps:

1. **Limit further discharge** of nematocysts by avoiding fresh water, wet sand, showers, or careless manipulation of the tentacles. Keep the patient calm, and reduce motion of the affected extremity.

2. **Inactivate the nematocysts** by applying vinegar. (Isopropyl alcohol may be used if vinegar is not available, but may not be as effective.)

3. **Remove the remaining tentacles** by scraping them off with the edge of a sharp, stiff object such as a credit card. Do not use your ungloved hand to remove the tentacles, because self-envenomation will occur. Persistent pain may respond to immersion in hot water (110° to 115°F, 43° to 46°C) for 30 minutes.

4. **Provide transport** to the emergency department.

TABLE 18-2	Common Marine Envenomations	
Dogfish	Marine snail	Starfish
Dragon fish	Portuguese man-of-war	Stingray
Fire coral	Ratfish	Stonefish
Hydroids	Scorpion fish	Tiger fish
Jellyfish	Sea anemone	Toadfish
Lionfish	Sea urchins	Weever fish

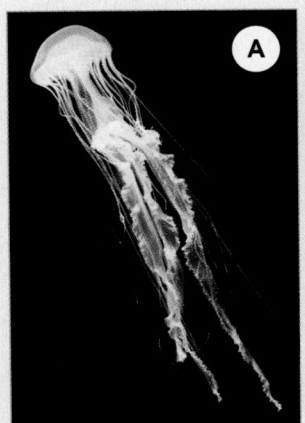
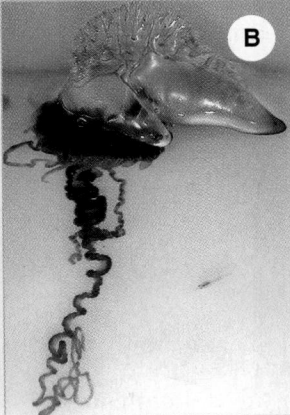

Figure 18-22 Coelenterates are responsible for many marine envenomations. **A.** Jellyfish. **B.** Portuguese man-of-war. **C.** Sea anemone.

You are the Provider

Summary

With a hyperthermic patient, ensure that you rapidly cool your patient but do not cause hypothermia as a result. As with any patient, it is imperative that you survey the scene for clues. Call for backup when needed. Complete an appropriate assessment and treat accordingly. With an environmental emergency such as this, be sure to remove your patient from the hot environment and cool the patient aggressively. Transport immediately and notify the receiving agency of the patient's condition.

Assessment and Emergency Care

	Cold Injuries	Heat Injuries	Drowning and Diving Injuries
Scene Size-up	Body substance isolation precautions should include a minimum of gloves and eye protection. Ensure scene safety and determine NOI/MOI. Consider the number of patients, the need for additional help/ALS, and c-spine stabilization.	Body substance isolation precautions should include a minimum of gloves and eye protection. Ensure scene safety and determine NOI/MOI. Consider the number of patients, the need for additional help/ALS, and c-spine stabilization.	Body substance isolation precautions should include a minimum of gloves and eye protection. Ensure scene safety and determine NOI/MOI. Consider the number of patients, the need for additional help/ALS, and c-spine stabilization.
Initial Assessment ■ General impression	Determine level of consciousness and find and treat any immediate threats to life. Determine priority of care based on environment and patient's chief complaint.	Determine level of consciousness and find and treat any immediate threats to life. Determine priority of care based on environment and patient's chief complaint.	Determine level of consciousness and find and treat any immediate threats to life. Determine priority of care based on environment and patient's chief complaint.
■ Airway	Ensure patent airway.	Ensure patent airway.	Ensure patent airway.
■ Breathing	Provide high-flow oxygen at 15 L/min. Evaluate depth and rate of respirations and provide ventilations as needed. Be prepared to suction.	Provide high-flow oxygen at 15 L/min. Evaluate depth and rate of respirations and provide ventilations as needed. Be prepared to suction.	Provide high-flow oxygen at 15 L/min. Evaluate depth and rate of respirations and provide ventilations as needed. Be prepared to suction.
■ Circulation	Evaluate pulse over a 30- to 45-second period for rate and quality; observe skin color, temperature, and condition and treat for any signs of shock. Determine if bleeding is present and control if life threatening.	Evaluate pulse rate and quality; observe skin color, temperature, and condition and treat aggressively for shock. If bleeding is present, control if life threatening.	Evaluate pulse rate and quality; observe skin color, temperature, and condition and treat for any sign of shock. Determine if bleeding is present and control if life threatening.
■ Transport decision	Rapid transport	Rapid transport	Rapid transport
Focused History and Physical Exam	*NOTE: The order of the steps in the focused history and physical exam differs depending on whether the patient is conscious or unconscious. The order below is for a conscious patient. For an unconscious patient, perform a rapid physical exam, obtain vital signs, and obtain the history.*		
■ SAMPLE history	Obtain SAMPLE history. Be sure to ask if and what interventions were taken before your arrival, how many interventions, and at what time.	Obtain SAMPLE history. Be sure to ask if and what interventions were taken before your arrival, how many interventions, and at what time. Determine patient's activities before onset of symptoms.	Obtain SAMPLE history. Determine if alcohol or recreational drugs have been consumed. Determine how long the patient was submerged and the water temperature and clarity.

	Cold Injuries	Heat Injuries	Drowning and Diving Injuries
■ Focused physical exam	Perform a focused physical exam on the affected area looking for frostbite. Consider the possibility of hypothermia.	Perform focused physical exam on the respiratory drive, adequacy of ventilation, adequacy and effectiveness of the circulatory system, and the patient's mental status. Obtain core body temperature.	Perform focused physical exam on the respiratory drive, adequacy of ventilation, adequacy and effectiveness of the circulatory system, and the patient's mental status. Perform a rapid assessment and obtain a Glasgow Coma Scale score.
■ Baseline vital signs	Take vital signs for 60 seconds, noting skin color and temperature.	Take vital signs, noting skin color and temperature. Use pulse oximetry if available.	Take vital signs, noting skin color and temperature. Assess pupil size and reactivity.
■ Interventions	Handle the patient gently. Remove from cold environment/clothing and begin to rewarm the patient according to local protocol. Support patient as needed. Consider the use of humidified oxygen, positive pressure ventilations, adjuncts, and proper positioning.	Support patient as needed. Consider the use of oxygen, positive pressure ventilations, adjuncts, and proper positioning. Remove clothing, lay a wet sheet over the patient, and turn on air conditioner/fan. Place wrapped ice packs on groin and axillae. Request ALS for fluid therapy for patients suspected of having hypovolemia.	Continue spinal stabilization. Suction as needed. Consider the use of oxygen, positive pressure ventilations, and adjuncts. Make sure the patient is kept warm.
Detailed Physical Exam	Complete a detailed physical exam.	Complete a detailed physical exam.	Complete a detailed physical exam.
Ongoing Assessment	Repeat the initial assessment, focused assessment, and reassess interventions performed. Reassess vital signs every 5 minutes for the unstable patient, every 15 minutes for the stable patient. Note and treat the patient as necessary.	Repeat the initial assessment, focused assessment, and reassess interventions performed. Reassess vital signs every 5 minutes for the unstable patient, every 15 minutes for the stable patient. Note and treat the patient as necessary.	Repeat the initial assessment, focused assessment, and reassess interventions performed. Reassess vital signs every 5 minutes for the unstable patient, every 15 minutes for the stable patient. Note and treat the patient as necessary.
■ Communication and documentation	Contact medical control with a radio report that gives information on patient's condition. Relay any significant changes, including level of consciousness or difficulty in breathing. Note what the patient was wearing and pertinent bystander information. Add any changes in patient condition, at what time, and any interventions performed.	Contact medical control with a radio report that gives information on patient's condition. Relay any significant changes, including level of consciousness or difficulty in breathing. Be sure to document weather conditions and events leading up to the illness. Record any changes, at what time, and any interventions performed.	Contact medical control with a radio report that gives information on patient's condition. Relay any significant changes, including level of consciousness or difficulty in breathing. Be sure to document any changes, at what time, and any interventions performed. If the patient was on a diving expedition, include the patient's diving log and pertinent information from diving partners, as available.

Assessment and Emergency Care

Assessment and Emergency Care

NOTE: While the steps below are widely accepted, be sure to consult and follow your local protocol.

Cold Injuries

1. Remove wet clothing and keep the patient dry.
2. Prevent heat loss. Move the patient away from any wet or cold surface.
3. Insulate all exposed body parts by wrapping them in a blanket or dry, bulky material.
4. Prevent convection heat loss by erecting a wind barrier around the patient.
5. Remove the patient from the cold environment as promptly as possible.

Heat Injuries

1. Remove any excess layers of clothing.
2. Move the patient promptly from the hot environment and out of the sun.
3. Provide oxygen, if not already done during the initial assessment.
4. Encourage the patient to lie supine with legs elevated. Loosen any tight clothing and fan the patient.
5. If the patient is alert, encourage him or her to sit up and slowly drink a liter of water if nausea does not develop.
6. Transport patient in the left-lateral recumbent position.

Drowning and Diving Injuries

1. Turn the patient to a supine position by rotating the entire upper half of the body as a single unit.
2. Begin artificial ventilation using the mouth-to-mouth method or a pocket mask.
3. Float a buoyant backboard under the patient.
4. Secure the patient to the backboard.
5. Remove the patient from the water.
6. Cover the patient with a blanket and apply oxygen if breathing. Begin CPR if breathing and pulse are absent.

Lightning Injuries

1. Move patient to a sheltered area.
2. Those who are conscious following a lightning strike are much less likely to develop delayed respiratory or cardiac arrest; most of these victims will survive. Therefore, you should focus your efforts on those who are in respiratory or cardiac arrest.

Spider Bites

Black Widow Spider

1. Provide BLS for patient in respiratory distress.
2. Transport the patient to the emergency department as soon as possible for treatment of pain and muscle rigidity.
3. If possible, bring the spider.

Brown Recluse Spider
If bite causes systemic symptoms:

1. Provide BLS.
2. Transport to emergency department.
3. If possible, bring the spider.

Snake Bites

Using an AnaKit

1. Prepare the injection site with antiseptic, and remove the needle cover.
2. Hold the syringe upright, and carefully use the plunger to remove air.
3. Turn the plunger one-quarter turn.
4. Quickly insert the needle into the muscle.
5. Holding the syringe steady, push the plunger until it stops.
6. Have the patient chew and swallow the Chlo-Amine antihistamine tablets provided in the kit.
7. If available, apply a cold pack to the sting site.

Injuries From Marine Animals

1. Limit further discharge of nematocysts by avoiding fresh water, wet sand, showers, or careless manipulation of the tentacles. Keep the patient calm and reduce motion of the affected extremity.
2. Inactivate the nematocysts by applying vinegar.
3. Remove the remaining tentacles by scraping them off with the edges of a sharp, stiff object such as a credit card.
4. Provide prompt transport.

Prep Kit

Ready for Review

- Cold illness can be either a local or a systemic problem.
- Local cold injuries include frostbite, frostnip, and immersion foot. Frostbite is the most serious because tissues actually freeze. All patients with a local cold injury should be removed from the cold and protected from further exposure.
- If instructed to do so by medical control, rewarm frostbitten parts by immersing them in water at a temperature between 100° and 112°F (38° and 44.5°C).
- The key to treating hypothermic patients is to stabilize vital functions and prevent further heat loss. Do not attempt to rewarm patients who have moderate to severe hypothermia, because they are prone to developing arrhythmias.
- Do not consider a patient dead until he or she is "warm and dead." Local protocol will dictate whether or not such patients receive CPR or defibrillation in the field.
- The body's regulatory mechanisms normally maintain body temperature within a very narrow range around 98.6°F (37°C). Body temperature is regulated by heat loss to the atmosphere via conduction, convection, evaporation, radiation, and respiration.
- Heat illness can take three forms: heat cramps, heat exhaustion, and heatstroke.
  - Heat cramps are painful muscle spasms that occur with vigorous exercise. Treatment includes removing the patient from the heat, resting the affected muscles, and replacing lost fluids.
  - Heat exhaustion is essentially a form of hypovolemic shock caused by dehydration. Symptoms include cold and clammy skin, weakness, confusion, headache, and rapid pulse. Body temperature can be high, and the patient may or may not still be sweating. Treatment includes removing the patient from the heat and treating for mild hypovolemic shock.
  - Heatstroke is a life-threatening emergency, usually fatal if untreated. Patients with heatstroke are usually dry and will have high body temperatures. Changes in mental status can include coma. Rapid lowering of the body temperature in the field is critical.
- The first rule in caring for drowning or near drowning victims is to be sure not to become a victim yourself. Protect the spine when removing patients from the water because spinal cord injuries often occur in drownings. Be aware of the possibility of hypothermia.
- Injuries associated with scuba diving may immediately apparent or may show up hours later. Patients with air embolism or decompression sickness may have pain, paralysis, or altered mental status. Be prepared to transport such patients to a recompression facility with a hyperbaric chamber.
- Poisonous spiders include the black widow spider and the brown recluse spider.
- Poisonous snakes include pit vipers and coral snakes.
- A person who has been bitten by a pit viper needs prompt transport; clean the bite area and keep the patient quiet to slow the spread of venom.
- Notify the hospital as soon as possible if a patient has been bitten by a coral snake; its venom can cause paralysis of the nervous system, and most hospitals do not have appropriate antivenin on hand.
- Patients who have been bitten by ticks may be infected with Rocky Mountain spotted fever or Lyme disease and should see a doctor within a day or two. Remove the tick using a tweezers, and save it for identification.
- Always provide prompt transport to the hospital for any patient who has been bitten by a poisonous insect or animal. Remember that signs and symptoms can deteriorate rapidly. Carefully monitor the patient's vital signs en route, especially for airway compromise.

Technology

- Interactivities
- Vocabulary Explorer
- Anatomy Review
- Web Links
- Online Review Manual

www.EMTB.com

Prep Kit continued...

Vital Vocabulary

air embolism Air bubbles in the blood vessels.

ambient temperature The temperature of the surrounding environment.

antivenin A serum that counteracts the effect of venom from an animal or insect.

bends Common name for decompression sickness.

bradycardia Slow heart rate, less than 60 beats/min.

breath-holding syncope Loss of consciousness caused by a decreased breathing stimulus.

conduction The loss of heat by direct contact (eg, when a body part comes into contact with a colder object).

convection The loss of body heat caused by air movement (eg, breeze blowing across the body).

core temperature The temperature of the central part of the body (eg, the heart, lungs, and vital organs).

decompression sickness A painful condition seen in divers who ascend too quickly, in which gas, especially nitrogen, forms bubbles in blood vessels and other tissues; also called "the bends."

diving reflex Slowing of the heart rate caused by submersion in cold water.

drowning Death from suffocation by submersion in water.

electrolytes Certain salts and other chemicals that are dissolved in body fluids and cells.

evaporation Conversion of water or another fluid from a liquid to a gas.

frostbite Damage to tissues as the result of exposure to cold; frozen body parts.

heat cramps Painful muscle spasms usually associated with vigorous activity in a hot environment.

heat exhaustion A form of heat injury in which the body loses significant amounts of fluid and electrolytes because of heavy sweating; also called heat prostration or heat collapse.

heatstroke A life-threatening condition of severe hyperthermia caused by exposure to excessive natural or artificial heat, marked by warm, dry skin; severely altered mental status; and often irreversible coma.

hyperbaric chamber A chamber, usually a small room, pressurized to more than atmospheric pressure.

hyperthermia A condition in which core temperature rises to 101°F (38.3°C) or more.

hypothermia A condition in which core temperature falls below 95°F (35°C) after exposure to a cold environment.

laryngospasm A severe constriction of the larynx and vocal cords.

near drowning Survival, at least temporarily, after suffocation in water.

radiation The transfer of heat to colder objects in the environment by radiant energy, for example heat gain from a fire.

respiration The loss of body heat as warm air in the lungs is exhaled into the atmosphere and cooler air is inhaled.

reverse triage A triage process in which efforts are focused on those who are in respiratory and cardiac arrest, and different from conventional triage where such patients would be classified as deceased. Used in triaging multiple victims of a lightning strike.

scuba A system that delivers air to the mouth and lungs at various atmospheric pressures, increasing with the depth of the dive; stands for self-contained underwater breathing apparatus.

turgor The ability of the skin to resist deformation; tested by gently pinching skin on the forehead or back of the hand.

Points to Ponder

You are asked to serve in a "stand-by" mode for a very popular 12-K race. Although athletes compete in the race, it is open to all runners and walkers. Current weather conditions include a temperature of 90°F, humidity at 82%, and a breeze at 1 to 2 mph. A race volunteer motions for you and as you approach you see an overweight woman, who is approximately 40 years old, sitting by the side of the road. She is wearing a long-sleeved shirt and stretch pants. She tells you she feels sick to her stomach, dizzy, and light-headed. Her skin is sweaty and pale and her pulse rate is 128 beats/min.

What is she likely experiencing? What factors have contributed to her condition?

Issues: Contributing Factors in Thermoregulation Problems, Recognizing Heat-Related Emergencies.

www.EMTB.com

Assessment in Action

It is the opening weekend of deer hunting season. As you walk into the station to begin your shift, you hear coworkers talking about reports of a hunter who was possibly lost overnight in the mountains. Moments later, you are dispatched to rendezvous with the county's search and rescue team who has found the lost man. Overnight weather conditions included a temperature low of 32°F and a precipitation mix of snow and rain. En route, the dispatcher informs you that your patient is confused and is slurring his words. The rescue team has been unable to obtain a blood pressure reading in the patient, but reports a weak carotid pulse of 46 beats/min and respirations of 10 breaths/min, shallow and irregular.

1. The patient in the above scenario is most likely experiencing:
   A. mild hypothermia.
   B. moderate hypothermia.
   C. severe hypothermia.
   D. none of the above.

2. Appropriate treatment for the patient in the above scenario includes:
   A. active, aggressive rewarming.
   B. preventing further heat loss.
   C. removing from the environment.
   D. both B and C.

3. A hypothermic patient who is alert and shivering is experiencing:
   A. mild hypothermia.
   B. moderate hypothermia.
   C. severe hypothermia.
   D. heat cramps.

4. Appropriate treatment for the patient in question 3 would include:
   A. giving warm fluids by mouth.
   B. removing all wet clothing.
   C. covering with blankets.
   D. all of the above.

5. Heat loss from wind is an example of:
   A. convection.
   B. radiation.
   C. conduction.
   D. evaporation.

6. The condition that results from exposed body parts becoming very cold is:
   A. frostnip.
   B. chilblains.
   C. immersion foot.
   D. all of the above.

Challenging Questions

7. How can alcohol consumption contribute to the severity of hypothermia?

8. How can age- and health-related conditions affect patients experiencing hypothermia?

9. How can frostnip and frostbite be distinguished from each other, and why is frostbite a serious injury?

10. In regard to hypothermic patients, why is it important to know about the environment in your response area?

www.EMTB.com

Behavioral Emergencies

Objectives

Cognitive

4-8.1   Define behavioral emergencies. (p 584)
4-8.2   Discuss the general factors that may cause an alteration in a patient's behavior. (p 586)
4-8.3   State the various reasons for psychological crises. (p 586)
4-8.4   Discuss the characteristics of an individual's behavior which suggest that the patient is at risk for suicide. (p 590)
4-8.5   Discuss special medical/legal considerations for managing behavioral emergencies. (p 591)
4-8.6   Discuss the special considerations for assessing a patient with behavioral problems. (p 586)
4-8.7   Discuss the general principles of an individual's behavior which suggest that the patient is at risk for violence. (p 594)
4-8.8   Discuss methods to calm behavioral emergency patients. (p 586)

Affective

4-8.9   Explain the rationale for learning how to modify your behavior toward the patient with a behavioral emergency. (p 589)

Psychomotor

4-8.10   Demonstrate the assessment and emergency medical care of the patient experiencing a behavioral emergency. (p 586)
4-8.11   Demonstrate various techniques to safely restrain a patient with a behavioral problem. (p 593)

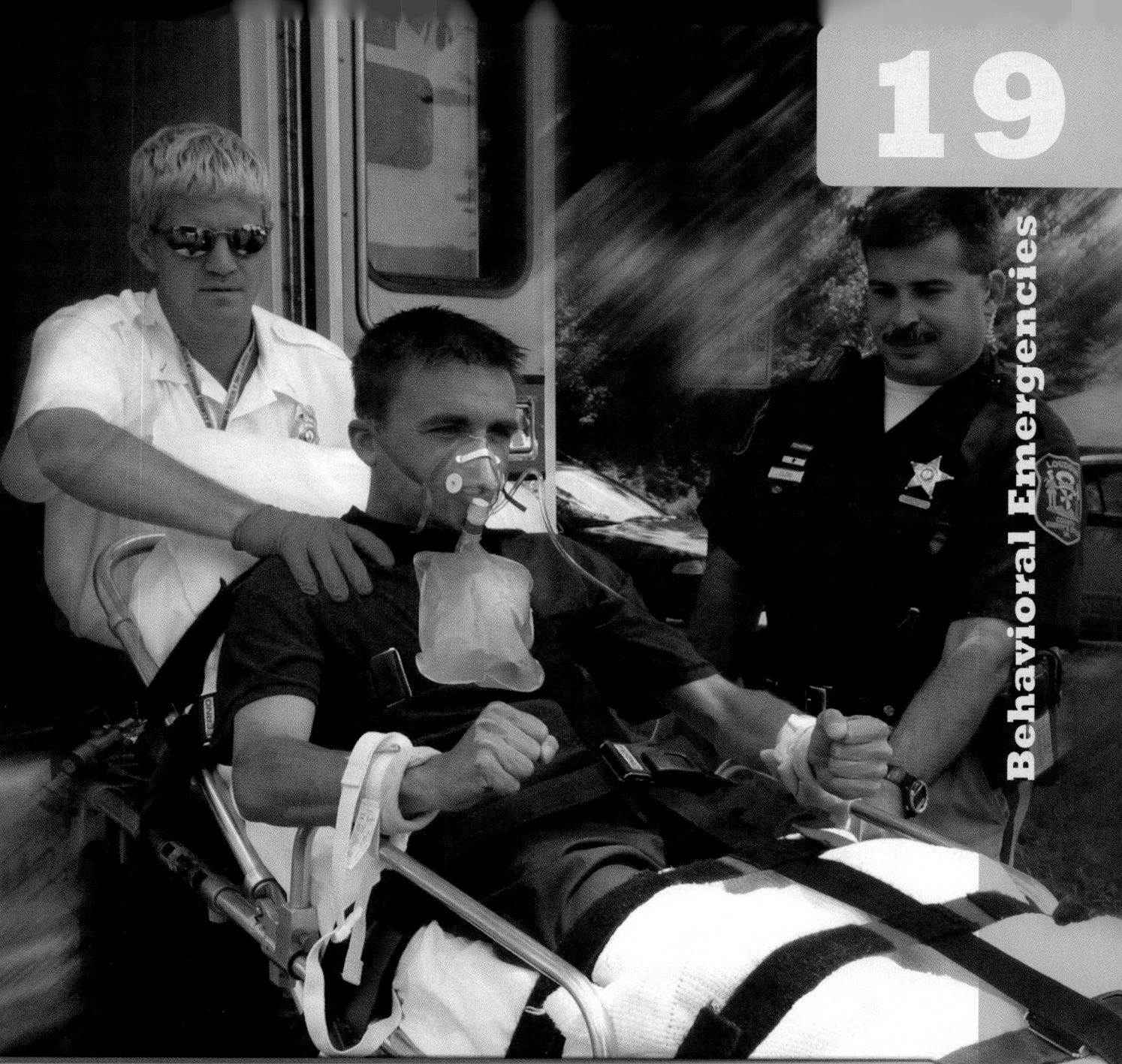

You are the Provider

You and your EMT-B partner are dispatched to 10213 E Desert Crossings for an attempted suicide. As you pull up to the house, a young woman comes running out the front door screaming for you to help her friend. She explains that her friend is in the house crying and threatening to kill herself.

1. What is your first step in treatment?
2. Should you call for law enforcement support?

Behavioral Emergencies

As an EMT-B, you can expect to deal often with patients undergoing a psychological or behavioral crisis. The crisis may be due to the emergency situation, mental illness, mind-altering substances, stress, or many other causes. This chapter discusses various kinds of behavioral emergencies, including those involving overdoses, violent behavior, and mental illness. You will learn how to assess a person who exhibits signs and symptoms of a behavioral emergency and what kind of emergency care may be required in these situations. The chapter also covers legal concerns in dealing with disturbed patients. Finally, it describes how to identify and manage the potentially violent patient, including the use of restraints.

Myth and Reality

Everyone develops some symptoms of mental illness at some point in life, but that does not mean everyone develops mental illness. Perfectly healthy people may have some of the symptoms and signs of mental illness from time to time. Therefore, you should not jump to the conclusion that you are mentally disturbed when you behave in certain ways that are discussed in this chapter. For that matter, you also should not jump to this conclusion about a patient in any given situation.

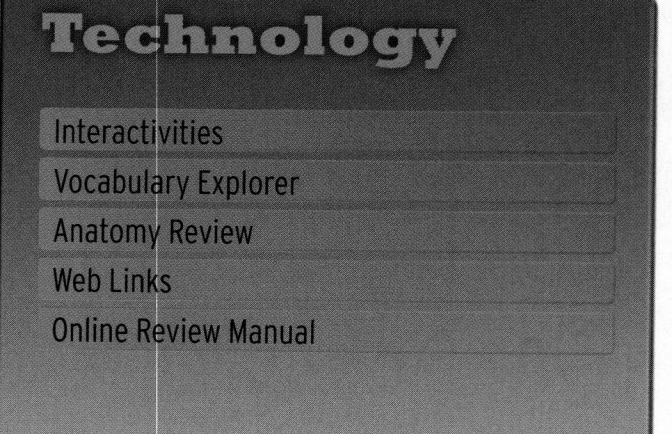

Technology

www.EMTB.com

- Interactivities
- Vocabulary Explorer
- Anatomy Review
- Web Links
- Online Review Manual

The most common misconception about mental illness is that if you are feeling "bad" or "depressed," you must be "sick." That is simply untrue. There are many perfectly justifiable reasons for feeling depressed, including divorce, loss of a job, and the death of a relative or friend. For the teenager who just broke up with his girlfriend of 12 months, it is altogether normal to withdraw from ordinary activities and to feel "blue." This is a normal reaction to a crisis situation. However, when a person finds that Monday morning blues last until Friday, week after week, he or she may indeed have a behavioral problem.

Many people believe that all individuals with mental health disorders are dangerous, violent, or otherwise unmanageable. This is untrue. Only a small percentage of people with mental health problems fall into these categories. As an EMT-B, however, you may be exposed to a higher proportion of violent patients. After all, you are seeing people who are, by definition, considered to be having an emergency; otherwise, you probably would not be seeing them. You are there because family members or friends felt unable to manage the patient by themselves. This may be a result of the use or abuse of drugs or alcohol. It may be that the patient has a long history of mental illness and is reacting to a particularly stressful event.

Although you cannot determine what has caused a person's behavioral problem, you may be able to predict that the person will become violent. The ability to predict violence is an important assessment tool for you.

Defining Behavioral Emergencies

Behavior is what you can see of a person's response to the environment: his or her actions. Sometimes, it is obvious what a person is responding to: A person is punched, and he or she runs away or bursts into tears or hits back. Sometimes, it is less clear, as when someone is depressed for very complex reasons.

Most of the time, individuals respond to the environment in reasonable ways. Over the years, they have learned to adapt to a variety of situations in daily life, including stresses and strains. This is called adjustment. There are times, however, when the stress is so great that the normal ways of adjusting do not work. When this happens, a person's behavior is likely to change, even if only temporarily. The new behavior may not be appropriate, or "normal."

The definition of a behavioral crisis or emergency is any reaction to events that interferes with the activities of daily living (ADL) or has become unacceptable to the patient, family, or community. For example, when someone experiences an interruption of the daily routine, such as bathing, dressing, and eating, chances are his or her behavior has become a problem. For that person, at that time, a behavioral emergency may exist. If the interruption of daily routine tends to recur on a regular basis, the behavior is also considered a mental health problem. It is then a pattern, rather than an isolated incident.

For example, a person who experiences a panic attack after having a heart attack is not necessarily mentally ill. Likewise, you would expect a person who is fired from a job to have some sort of reaction, often sadness and depression. These problems are short-term and isolated events. However, a person who reacts with a fit of rage, attacking people and property or going on a "bender" for a week, has gone beyond what society considers appropriate or normal behavior. That person is clearly undergoing a behavioral emergency. Usually, if an abnormal or disturbing pattern of behavior lasts for at least a month, it is regarded as a matter of concern from a mental health standpoint. For example, chronic depression, a persistent feeling of sadness and despair, may be a symptom of a mental or physical disorder. This type of long-term problem would be labeled a mental health disorder.

When a psychiatric emergency arises, the patient may show agitation or violence or become a threat to himself, herself, or others. This is more serious than a more typical behavioral emergency that causes inappropriate behavior such as interference with ADL or bizarre behavior. An immediate threat to the person involved or to others in the immediate area, including family, friends, bystanders, and EMT-Bs should be considered a psychiatric emergency. For example, a person might respond to the death of a spouse by attempting suicide. On the other hand, although this is a major life disruption, it does not have to involve violence or harm to an individual. Disruption can take many forms; not all involve violence, nor are they all psychiatric emergencies.

The Magnitude of Mental Health Problems

According to the National Institutes of Mental Health, at one time or another, one in five Americans has some type of mental disorder, an illness with psychological or behavioral symptoms that may result in impaired functioning. It can be caused by a social, psychological, genetic, physical, chemical, or biologic disturbance.

Documentation Tips

The medicolegal issues associated with responses to behavioral emergencies put added emphasis on thorough and specific documentation. Record detailed, objective findings that support the conclusion of abnormal behavior (withdrawn, won't talk, crying uncontrollably) and quote the patient's own words when appropriate, for example, "Life isn't worth it any more," or "The voices are telling me to kill people." Avoid judgmental statements; these create the impression that you based your care on personal bias rather than the patient's needs.

You are the Provider   Part 2

The woman tells you that her friend does not have any kind of weapon. Your partner radios in for law enforcement assistance as you are entering the house. Once inside, you find a girl in her early teens sitting on the couch sobbing. She is conscious, alert, and oriented. Her breathing is labored at a rate of 26 breaths/min. You ask if her parents are home and are told that they are both out of town and that she is staying home alone until tomorrow afternoon when her parents will be returning. Your patient tells you that her boyfriend called a short time ago and told her that he no longer wanted to be with her. She again begins crying and saying that she does not want to live if she can't be with him.

3. Do you have consent to treat this patient?
4. What type of information do you need to gain from this patient?

Pathology: Causes of Behavioral Emergencies

As an EMT-B, you are not responsible for diagnosing the underlying cause of a behavioral or psychiatric emergency. However, you should know the two basic categories of diagnosis a physician will use: organic (physical) and functional (psychological).

Organic brain syndrome is a temporary or permanent dysfunction of the brain caused by disturbance in the physical or physiologic functioning of brain tissue. Causes of organic brain syndrome include sudden illness, recent trauma, drug or alcohol intoxication, and diseases of the brain, such as Alzheimer's disease. Altered mental status can arise from low levels of blood glucose, lack of oxygen, inadequate blood flow to the brain, and excessive heat or cold.

A functional disorder is one in which the abnormal operation of an organ cannot be traced to an obvious change in the actual structure, or physiology, of the organ. Something has gone wrong, but the root cause cannot be identified as the working of the organ itself.

Schizophrenia and depression are good examples of functional disorders. There may be a chemical or physical cause for these disorders, but it is not obvious or well understood.

These two types of disorders can look very much alike. An altered mental status, or a change in the way a person thinks or behaves, is one indicator of central nervous system diseases. A patient displaying bizarre behavior may actually have an acute medical illness that is the cause, or a partial cause, of the behavior. Recognizing this possibility may allow you to save a life.

Safe Approach to a Behavioral Emergency

All regular EMT-B skills—assessment, providing care, patient approach, obtaining the history, and patient communication—are used in behavioral emergencies. However, other management techniques also come into play. There is not room in this chapter for a full discussion of these techniques, but you should follow general guidelines to ensure your safety at the scene of a behavioral emergency (Table 19-1 ▼).

TABLE 19-1 Safety Guidelines for Behavioral Emergencies

- **Be prepared to spend extra time.** It may take longer to assess, listen to, and prepare the patient for transport.
- **Have a definite plan of action.** Decide who will do what. If restraint is needed, how will it be accomplished?
- **Identify yourself calmly.** Try to gain the patient's confidence. If you begin shouting, the patient is likely to shout louder or become more excited. A low, calm voice is often a quieting influence.
- **Be direct.** State your intentions and what you expect of the patient.
- **Assess the scene.** If the patient is armed or has potentially harmful objects in his or her possession, have these removed by law enforcement personnel before you provide care.
- **Stay with the patient.** Do not let the patient leave the area, and do not leave the area yourself unless law enforcement personnel can stay with the patient. Otherwise, the patient may go to another room and obtain weapons, lock himself or herself in another room, or take pills.
- **Encourage purposeful movement.** Help the patient to get dressed and gather appropriate belongings to take to the hospital.
- **Express interest in the patient's story.** Let the patient tell you what happened or what is going on now in his or her own words. However, do not play along with auditory or visual disturbances.
- **Do not get too close to the patient.** Everyone needs personal space. Furthermore, you want to be sure you can move quickly if the patient becomes violent or tries to run away. Do not physically talk down to or directly confront the patient. A squatting, 45°-angle approach is usually not confrontational but may hinder your movements. Do not allow the patient to get between you and the exit.
- **Avoid fighting with the patient.** You do not want to get into a power struggle. Remember, the patient is not responding to you in a normal manner; he or she may be wrestling with internal forces over which neither of you has control. You and others may be stimulating these inner forces without knowing it. If you can respond with understanding to the feeling that the patient is expressing, whether this is anger or fear or desperation, you may be able to gain his or her cooperation. If it is necessary to use force, be sure you have adequate training and help and move toward the patient quietly and with assured firmness.
- **Be honest and reassuring.** If the patient asks whether he or she has to go to the hospital, the answer should be, "Yes, that is where you can receive medical help."
- **Do not judge.** You may see behavior that you dislike. Set those feelings aside and concentrate on providing emergency medical care.

Assessment of a Behavioral Emergency

Scene Size-up

Dispatch may provide you with information that suggests a behavioral emergency; however, every situation will have some component of a behavioral emergency. It may be a medical problem made worse by a behavioral issue, or it may be that a behavioral issue has led to trauma. Regardless, the first things to consider are your safety and the patient's response to the environment. Some situations may be more serious than others and, therefore, more threatening to your safety. Is the situation unduly dangerous to you and your partner? Do you need immediate law enforcement backup? Does the patient's behavior seem typical or normal in the circumstances? Are there legal issues involved (crime scene, consent, refusal)? For example, a patient who has just been assaulted has good reason to be fearful of other people, including you. Take appropriate BSI precautions. Request any additional resources you may need (law enforcement, additional personnel) early. You can always send them away if they are not needed. Be vigilant. Avoid tunnel vision.

Initial Assessment

General Impression

Begin your assessment from the doorway. How does the patient appear? Calm? Agitated? Awake or sleepy? Begin with an introduction of who you are and let the patient know that you are there to help. Ask for the chief complaint by asking, "What happened?" or "How can I help?" Allow the patient to tell what happened or how he or she feels. Is the patient alert and oriented? Use the AVPU scale for alertness. To determine orientation, ask the patient, "Where are you?" and "Why are you here?"

Airway, Breathing, and Circulation

If your patient is in physical distress, assess the ABCs as for any other patient. Provide the appropriate interventions based on your assessment findings. Some behavioral situations will involve a compromised airway and inadequate breathing secondary to a suicide attempt from ingesting a handful of sleeping pills with alcohol. A heart attack victim may aggravate cardiac distress because of feeling anxious about possibly dying. A depressed person may slit his or her wrists, causing traumatic bleeding. Almost every situation, medical or trauma, will have some behavioral component. It is

You are the Provider   Part 3

You explain in a calm and reassuring tone that you want to try to understand what she is going through and that you need to ask her some questions and examine her to get her the help she needs. She begins to calm down and allows you to obtain baseline vital signs. Her breathing has slowed to 20 breaths/min. Her skin is pink, warm, and dry, and her pulse is 88 beats/min. She has no obvious bleeding, and her blood pressure is 120/82 mm Hg. Your patient has no known allergies and takes birth control pills daily but no other medication. She attempted suicide last year by taking an entire bottle of her mom's diazepam (Valium).

5. Is the patient's history of attempted suicide pertinent to the current situation?
6. What is your next step in treatment?

just as important to treat the behavioral problem as it is to treat the medical or traumatic problem; however, the focus of the initial assessment is assessing and treating life threats.

Transport Decision

Unless your patient is unstable from a medical problem or trauma, prepare to spend time with your patient. Depending on your local protocol, there may be a specific facility to which such patients are transported.

Focused History and Physical Exam

In an unconscious medical patient, begin with a rapid physical exam to look for a reason for the unresponsiveness. Follow this rapid check for hidden life threats with a complete set of baseline vital signs and then gather what history you can.

When a medical patient is conscious, begin with asking the patient about health history and performing a focused physical exam as necessary; then obtain a full set of baseline vital signs. A focused physical exam for a behavioral problem may be difficult to perform but may provide clues to the patient's state of mind and thinking. Some patients welcome physical contact as reassuring, but others may feel acutely threatened. Avoid touching the patient without permission. In fact, this is a good practice for all of your patients. The majority of your time, at this stage of your assessment, will be asking the patient about medical history.

In trying to determine the reason for the patient's state, your assessment should consider three major areas as possible contributors:

- Is the patient's central nervous system functioning properly? For example, the patient may be experiencing diabetic problems, particularly hypoglycemia. He or she may have been poisoned or may be responding to a physical trauma of some sort. Any of these situations could cause the patient to behave in an unusual or irrational manner.
- Are hallucinogens or other drugs or alcohol a factor? Does the patient see strange things? Is everything distorted? Do you smell alcohol on the patient's breath?
- Are psychogenic circumstances, symptoms, or illness (caused by mental rather than physical factors) involved? These might include the death of a loved one, severe depression, a history of

mental illness, threats of suicide, or some other major interruption of ADL.

SAMPLE History

A complete and careful SAMPLE history will be helpful in treating your patient and passing on information to personnel at the receiving facility. You may be able to elicit information not available to the hospital staff. Ask specifically about previous episodes, treatments, hospitalizations, and medications related to behavioral problems (Table 19-2 ▼).

Is Alzheimer's disease or another type of dementia a possible cause? In geriatric patients, consider Alzheimer's disease and dementia as possible causes of abnormal behavior. In these cases, it is essential to obtain information from relatives, friends, or extended care facility staff. Determining the patient's baseline mental status will be essential in guiding your treatment and transport decisions and will also be extremely helpful to hospital personnel.

Family, friends, and observers may be of great help in answering these questions. Together with your observations and interaction with the patient, they should provide enough data for you to assess the situation. This assessment has two primary goals: recognizing major threats to life and reducing the stress of the situation as much as possible.

Reflective listening is a technique frequently used by mental health professionals to gain insight into a

TABLE 19-2 Questions to Ask in Evaluating a Behavioral Crisis

- Does the patient answer your questions appropriately?
- Does the patient's behavior seem appropriate?
- Does the patient seem to understand you and the surroundings?
- Is the patient withdrawn or detached? Hostile or friendly? Elated or depressed?
- Are the patient's vocabulary and expressions what you would expect under the circumstances?
- Does the patient seem aggressive or dangerous to you or others?
- Is the patient's memory intact? Check orientation to time, place, person, and event: What day, month, and year is it? Who am I?
- Does the patient express disordered thoughts, delusions, or hallucinations?

EMT-B Tips

In assessing a patient in a behavioral emergency, it can be very useful to gather information separately from a relative or caregiver. Splitting up the history-taking process in this way often yields valuable information and can help reduce the potential for violence when there is tension between the people involved. Do not split up if the patient is threatening or uncontrolled unless additional people such as law enforcement personnel are there to help.

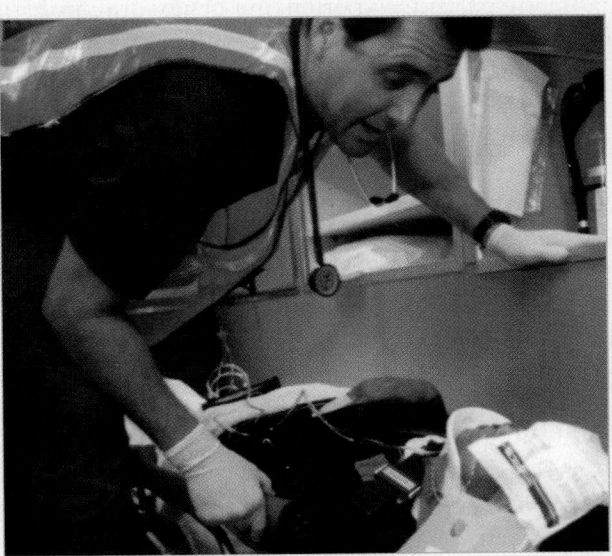

Figure 19-1 Making eye contact with a patient can provide useful clues about a patient's emotional state.

patient's thinking. It involves repeating, in question form, what the patient has said, encouraging the patient to expand on the thoughts. Although it often requires more time to be effective than is available in an EMS setting, it may be a helpful tool for you to use when other techniques are unsuccessful at gathering the patient's history.

Focused Physical Exam

Sometimes even a patient who is conscious in a behavioral or psychiatric emergency will not respond at all to your questions. In those cases, you may be able to tell quite a lot about the patient's emotional state from facial expressions, pulse, and respirations. Tears, sweating, and blushing may be significant indicators of state of mind. Also, make sure that you look at the patient's eyes; a patient who has a blank gaze or rapidly moving eyes may be experiencing central nervous system dysfunction (Figure 19-1 ▶).

A behavioral crisis puts tremendous stress on a person's coping mechanisms, including natural abilities and training. The person is actually incapable of responding reasonably to the demands of the environment. This state may be temporary, as in an acute illness, or longer lived, as in a complex, chronic mental illness. In either case, the patient's perception of reality may be compromised or distorted.

Baseline Vital Signs

Obtain vital signs when doing so will not exacerbate your patient's emotional distress. Make every effort to assess blood pressure, pulse, respirations, skin, and pupils. Remember that behavioral emergencies can be caused or precipitated by physiologic problems, and they can exacerbate preexisting conditions. Don't forget that the physical person and the emotional person are one.

Interventions

As much as your heart may go out to the emotionally distressed patient, there often is little you will be able to do for the patient during the short time you will be treating him or her. Your job is to diffuse and control the situation and safely transport your patient to the hospital. Intervene only as much as it takes to accomplish these tasks. Be caring and careful. If you have determined that it is necessary to restrain your patient, release the restraints only if necessary to provide patient care.

Detailed Physical Exam

Unless there is an accompanying physical complaint, the detailed physical exam is rarely called for in a patient with a behavioral problem and may, in fact, be detrimental to gaining the patient's trust.

Ongoing Assessment

Never let your guard down. Most patients you are called to treat and transport with emotional complaints pose no danger to you or others on your crew, but it is impossible to determine this on the scene. Remember that many patients experiencing behavioral problems will act spontaneously. Be prepared to intervene quickly. If restraints are necessary, reassess and document respirations, as well as pulse and motor and sensory function in all restrained extremities, every 5 minutes. Respiratory and circulatory problems have been known to occur in combative patients who are restrained. When available, have additional personnel such as law enforcement officers or firefighters accompany you in the back of the ambulance during transport. This provides additional assistance should the patient's behavior change rapidly.

Communication and Documentation

Try to give the receiving hospital advance warning when a behavioral emergency patient is coming. Many hospitals require extra preparation to ensure appropriate staff or rooms are available. Report whether restraints will be required when the patient arrives at the hospital. Document thoroughly and carefully. Think about what you are going to write before you write it, so that you can describe what are often confusing scenes as clearly as possible. Because behavioral emergencies rarely have physical signs, yours may be the only documentation about the patient's distress. Because behavioral emergencies are fraught with legal dangers, document everything that occurred on the call, particularly situations that required restraint. When restraints are required to protect you or the patient from harm, include why and what type of restraints were used. This information is essential if the case is reviewed for medicolegal reasons.

Suicide

The single most significant factor that contributes to suicide is depression. Any time you encounter an emotionally depressed patient, you must consider the possibility of suicide. Risk factors for suicide are listed in Table 19-3 ▶.

It is a common misconception that people who threaten suicide never commit it. This is not correct. Suicide is a cry for help. Threatening suicide is an indication that someone is in a crisis he or she cannot handle. Immediate intervention is necessary.

Whether or not the patient has any of these risk factors, you must be alert to the following warning signs:

- Does the patient have an air of tearfulness, sadness, deep despair, or hopelessness that suggests depression?

Study This ↓

TABLE 19-3 Risk Factors for Suicide

- Depression, any age
- Previous suicide attempt (80% of successful suicides were preceded by at least one attempt.)
- Current expression of wanting to commit suicide or sense of hopelessness; specific plan for suicide
- Family history of suicide
- Older than 40 years, particularly for single, widowed, divorced, alcoholic, or depressed people (Men in this category who are older than 55 years have an especially high risk.)
- Recent loss of spouse, significant other, family member, or support system
- Chronic debilitating illness or recent diagnosis of serious illness
- Holidays
- Financial setback, loss of job, police arrest, imprisonment, or some sort of social embarrassment
- Substance abuse, particularly with increasing use
- Children of an alcoholic parent
- Severe mental illness
- Anniversary of death of loved one, job loss, marriage after the death of a spouse, and so forth
- Unusual gathering or new acquisition of things that can cause death, such as purchase of a gun, a large volume of pills, or increased use of alcohol

- Does the patient avoid eye contact, speak slowly or haltingly, and project a sense of vacancy, as if he or she really isn't there?
- Does the patient seem unable to talk about the future? Ask the patient whether he or she has any vacation plans. Suicidal people consider the future so uninteresting that they do not think about it; people who are seriously depressed consider the future so distant that they may not be able to think about it at all.
- Is there any suggestion of suicide? Even vague suggestions should not be taken lightly, even if presented as a joke. If you think that suicide is a possibility, do not hesitate to bring the subject up. You will not "give the patient ideas" if you ask directly, "Are you considering suicide?"
- Does the patient have any specific plans relating to death? Has the patient recently prepared a will? Given away significant possessions or advised close friends what he or she would like done with them? Arranged for a funeral service? These are critical warning signs.

Consider also the following additional risk factors for suicide:

- Are there any unsafe objects in the patient's hands or nearby (for example, a sharp knife, glass, poisons, a gun)?
- Is the environment unsafe (for example, an open window in a high-rise building, a patient standing on a bridge or precipice)?
- Is there evidence of self-destructive behavior (for example, partially cut wrists, large alcohol or drug intake)?
- Is there an imminent threat to the patient or others?
- Is there an underlying medical problem?

Remember, the suicidal patient may be homicidal as well. Do not jeopardize your life or the lives of your fellow EMT-Bs. If you have reason to believe that you are in danger, you must obtain police intervention. In the meantime, try not to frighten the patient or make him or her suspicious.

Medicolegal Considerations

The medical and legal aspects of emergency medical care become more complicated when the patient is undergoing a behavioral or psychiatric emergency. Nevertheless, legal problems are greatly reduced with an emotionally disturbed patient who consents to care. Gaining the patient's confidence is, therefore, a critical task for you.

Mental incapacity can take many forms: unconsciousness (as a result of hypoxia, alcohol, or drugs), temporary but severe stress, or depression. Once you have determined that a patient has impaired mental capacity, you must decide whether he or she requires immediate emergency medical care. A patient in mentally unstable condition may resist your attempts to provide care. Nevertheless, you must not leave this patient alone. Doing so may result in harm to the patient and expose you to civil action for abandonment or negligence. In such situations, you should request that law enforcement personnel handle the patient. Another reason for seeking law enforcement support is for the patient who resists treatment; such a patient often threatens EMT-Bs and others. Violent or dangerous people must be taken into custody by the police before emergency care can be rendered.

Consent

When a patient is not mentally competent to grant consent for emergency medical care, the law assumes that there is implied consent. For example, the consent of an unconscious patient is implied if life or health is at risk. The law refers to this as the emergency doctrine: Consent is implied because of the necessity for immediate emergency treatment. In a situation that is not immediately life threatening, emergency medical care or transportation may be delayed until the proper consent is obtained.

In cases involving psychiatric emergencies, however, the matter is not always clear-cut. Does a life-threatening emergency exist or not? If you are not sure, you should request the assistance of law enforcement personnel.

Limited Legal Authority

As an EMT-B, you have limited legal authority to require or force a patient to undergo emergency medical care when no life-threatening emergency exists. Patients have the right to refuse care. However, most states have

You are the Provider   Part 4

She states that although she has not done anything to injure herself today, she wants to kill herself and will attempt to do so if she gets a chance. She can't see herself living without her boyfriend. As you are caring for your patient, law enforcement personnel arrive. As they enter the house, your patient sees them and becomes hysterical. She jumps up from the couch and runs toward the bathroom, slamming the door behind her. She refuses to open the door. She continues screaming about not wanting to live. Afraid of what she may do to herself, the police officer breaks the door down and restrains the patient.

7. What are your obligations in treating this patient?
8. What are your options in patient treatment now that law enforcement personnel have restrained her?

Geriatric Needs

As the population ages, you will begin to see more patients older than 65 years. In responding to an increasing number of geriatric patients, you will probably notice some behavioral or mental health problems, including depression, dementia, and delirium. These mental status changes can affect your ability to thoroughly assess and treat an ill or injured geriatric patient. Understanding the causes of altered behavior in geriatric patients will help you in patient care.

Depression is one of the more common mental status problems that you will see in older people. As an EMT-B, you can recognize a problem and perhaps prevent a suicide in a depressed older person.

Depression has a number of causes, some organic, some psychological, and some cultural. Organic causes of depression include an emotional response to a major illness such as cancer or dementia. Furthermore, medications can induce a feeling of depression, especially when interacting with other prescription drugs. In addition, changes in the endocrine system, such as menopause, can elicit depression.

With all the possible causes of depression, an older adult can feel helpless and hopeless. A depressed person can be argumentative or placid. He or she might trivialize complaints, not wanting to be a bother to anyone. Someone who sees no way out of his or her situation may turn to suicide. Be alert for a suicidal gesture or ideation, even though it may not be obvious.

Although depression can create behavioral problems in geriatric patients, dementia is another cause of abnormal behavior. The most common cause of dementia is primary progressive dementia, also known as Alzheimer's dementia. It is estimated that 10% of the population older than 65 years and 50% of the population older than 85 years have Alzheimer's dementia. Currently, there is no cure for Alzheimer's.

During the progression of the disease, the patient can develop openly hostile behavior, kicking, yelling, pinching, and hitting you, your partner, or the patient's caregiver. You might need to restrain the violent patient, but do so gently and only to the point at which the violent behavior stops.

Other causes of altered behavior include diabetic emergencies, heat- and cold-related illnesses, poisoning and overdose, strokes and transient ischemic attacks, and infection. It is interesting to note that, although the mechanism is not understood, urinary tract infection and constipation can each alter an older person's behavior.

As the EMT-B responding to a call for help, you should accept the possibility of depression in a geriatric patient. Do not discount the patient's feelings or devalue his or her emotions. Be alert for a suicidal gesture, and pay attention to any statements about death. To get the patient's cooperation, you can elicit his or her help in providing care for the acute illness or injury. A smile and a touch can go a long way in alleviating fear in all of your patients, especially older patients.

legal statutes regarding the emergency care of mentally ill and drug-impaired individuals. These statutory provisions permit law enforcement personnel to place such a person in protective custody so that emergency care can be given. You should be familiar with your local and state laws regarding these situations.

The typical provision states that "Any police officer who has reasonable cause to believe that a person is mentally ill and dangerous to himself, herself, or others or gravely disabled…may take such person into custody and take or cause such person to be taken to a general hospital for emergency examination…." Again, because these provisions vary, you should become familiar with the provisions in your state.

The general rule of law is that a competent adult has the right to refuse treatment, even if lifesaving care is involved. In psychiatric cases, however, a court of law would probably consider your actions in providing life-saving care to be appropriate, particularly if you have a reasonable belief that the patient would harm himself,

herself, or others without your intervention. In addition, a patient who is in any way impaired, whether by mental illness, medical condition, or intoxication, may not be considered competent to refuse treatment or transportation. These situations are among the most perilous you will encounter from a legal standpoint. When in doubt, consult your supervisor, police, or medical control. Always maintain a pessimistic attitude toward your patient's condition—assume the worst and hope for the best. Err on the side of treatment and transport. It is far easier to defend yourself against charges of battery than it is to justify abandonment.

Restraint

Ordinarily, restraint of a person must be ordered by a physician, a court, or a law enforcement officer. If you restrain a person without authority in a nonemergency situation, you expose yourself to a possible lawsuit and to personal danger. Legal actions against you can involve charges of assault, battery, false imprisonment, and violation of civil rights. You may use restraints only to protect yourself or others from bodily harm or to prevent the patient from causing injury to himself or herself (Figure 19-2 ▼). In either case, you may use only reasonable force as necessary to control the patient, something that different courts may define differently. For this reason, you should always consult medical control and contact law enforcement personnel for help before restraining a patient.

In fact, you probably should always involve law enforcement personnel if you are called to assist a patient in a severe behavioral or psychiatric crisis. They will provide physical backup in managing the patient and serve as the necessary witnesses and legal authority to restrain the patient. A patient who is restrained by law enforcement personnel is in their custody.

Always try to transport a disturbed patient without restraints if possible. Once the decision has been made to restrain a patient, however, you should carry it out quickly. Be aware of BSI precautions. If the patient is spitting, place a surgical mask over his or her mouth.

Make sure you have adequate help to restrain a patient safely. At least four people should be present to carry out the restraint, each being responsible for one extremity. Before you begin, discuss the plan of action. As you prepare to restrain the patient, stay outside the patient's range of motion.

In subduing a disturbed patient, use the minimum force necessary. You should avoid acts of physical force that may cause injury to the patient. The level of force will vary, depending on the following factors:

- The degree of force that is necessary to keep the patient from injuring himself, herself, or others
- A patient's gender, size, strength, and mental status

- The type of abnormal behavior the patient is exhibiting. You should use only restraint devices that have been approved by your state's health department for this purpose; soft, wide, leather, or cloth restraints are preferred to police-type handcuffs.

Acting at the same time, the police officers should secure the patient's extremities with approved equipment. Somebody, preferably you or your partner, should continue to talk to the patient throughout the process. Remember to treat the patient with dignity and respect at all times. Also, monitor the patient for vomiting, airway obstruction, and cardiovascular stability because the patient cannot fend for himself or herself. Drug or alcohol intoxication may cause violent behavior but then lead to such physical problems as well. Never place your patient facedown because it is impossible to adequately monitor the patient and this position may inhibit the breathing of an impaired or exhausted patient. Be careful not to place restraints in such a way that respiration is compromised. Reassess airway and breathing continuously. You should make frequent checks of circulation on all restrained extremities, regardless of patient position (Figure 19-3 ▼). Document the reason for the restraint and the technique that was used. Be especially careful if a combative patient suddenly becomes calm and cooperative. This is the time not to relax but to secure the situation. The patient may suddenly become combative again and injure someone. Keep in mind that you may use reasonable force to defend yourself against an attack by an emotionally disturbed patient. It is extremely helpful to have (and document) witnesses in attendance even during transport to protect against false accusations. EMT-Bs have been accused of sexual misconduct and other physical abuse in such circumstances.

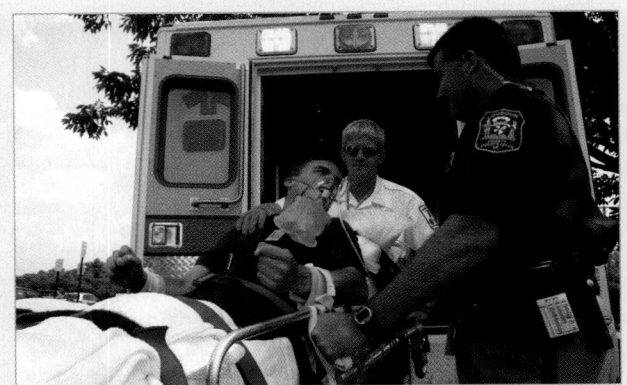

Figure 19-2 You may use restraints only to protect yourself or others or to prevent a patient from causing injury to himself or herself.

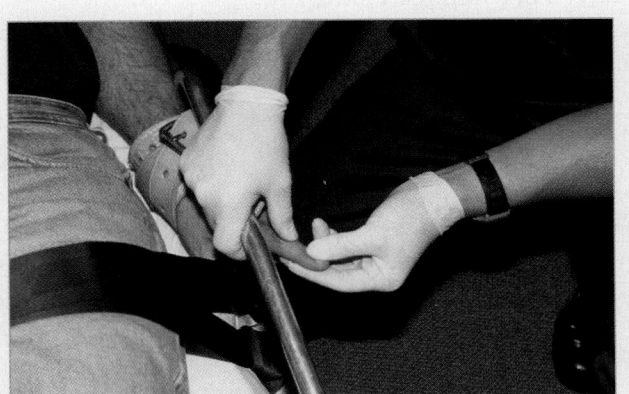

Figure 19-3 Assess circulation frequently while a patient is restrained.

The Potentially Violent Patient

Violent patients make up only a small percentage of those undergoing a behavioral or psychiatric crisis. However, the potential for violence by such a patient is always an important consideration for you (Figure 19-4 ▶).

Use the following list of risk factors to assess the level of danger:

- **History**. Has the patient previously exhibited hostile, overly aggressive, or violent behavior? Ask individuals at the scene, or request this information from law enforcement personnel or family.
- **Posture**. How is the patient sitting or standing? Is the patient tense, rigid, or sitting on the edge of his or her seat? Such physical tension is often a warning signal of impending hostility.
- **The scene**. Is the patient holding or near potentially lethal objects such as a knife, gun, glass, poker, or bat (or near a window or glass door)?
- **Vocal activity**. What kind of speech is the patient using? Loud, obscene, erratic, and bizarre speech patterns usually indicate emotional distress. Someone using quiet, ordered speech is not as likely to strike out as someone who is yelling and screaming.
- **Physical activity**. The motor activity of a person undergoing a psychiatric crisis may be the most telling factor of all. The patient who has tense muscles, clenched fists, or glaring eyes; is pacing; cannot sit still; or is fiercely protecting personal space requires careful watching. Agitation may predict a quick escalation to violence.

Other factors to consider in assessing a patient's potential for violence include the following:

- Poor impulse control
- A history of truancy, fighting, and uncontrollable temper

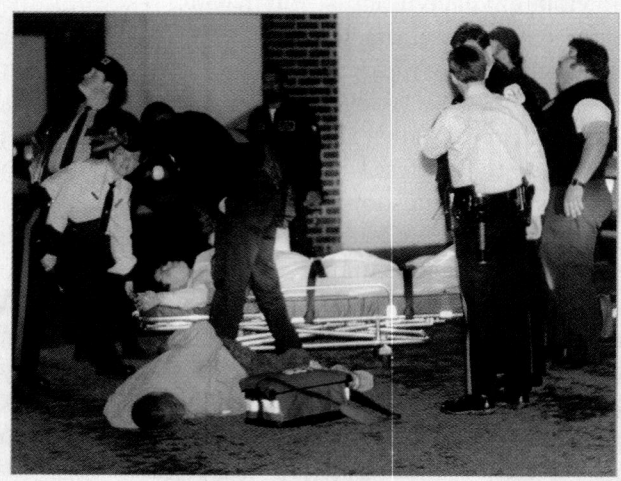

Figure 19-4 The potential for violence is an important consideration for EMT-Bs.

- Low socioeconomic status, unstable family structure, or inability to keep a steady job
- Tattoos, especially those with gang identification or statements such as "Born to Kill" or "Born to Lose"
- Substance abuse
- Depression, which accounts for 20% of violent attacks
- Functional disorder (If the patient says that voices are telling him or her to kill, believe it.)

EMT-B Tips

When working with a potentially hostile or violent patient, remove everyone who is not needed, such as family, friends, or bystanders, from the scene. This will prevent injury or involvement of others.

You are the Provider   Summary

When caring for a patient with a behavioral emergency, it is imperative that your safety and the safety of your partner come first. If there is ever any doubt, call for law enforcement assistance. Be honest and reassuring with your patient. You may be on the scene longer with these types of calls. Take your time; don't be in a hurry to load and go unless there appears to be a life-threatening emergency. Try to relate to your patient and what he or she is going through. Do not be judgmental. Keep the lines of communication open with all parties involved.

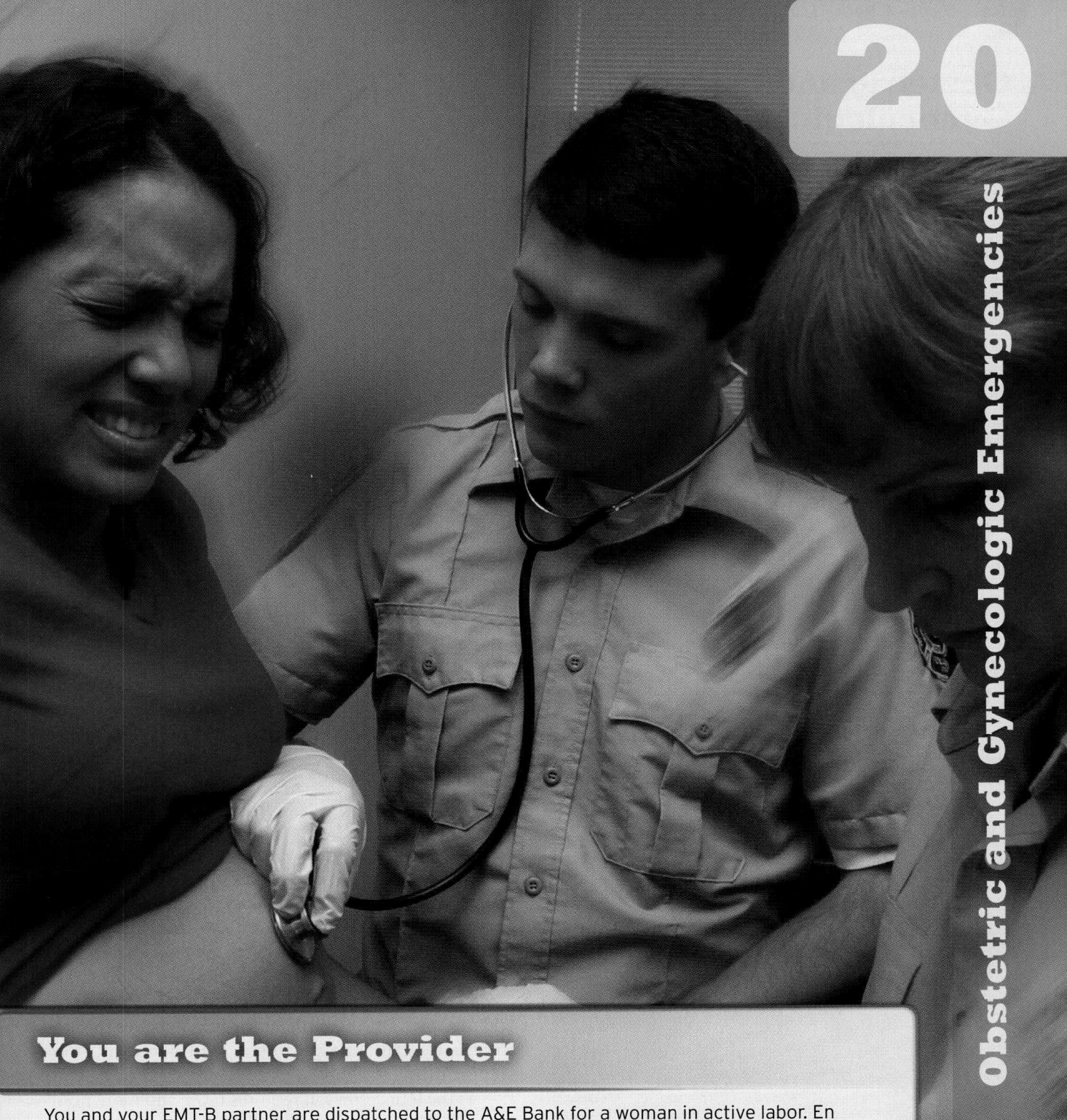

You are the Provider

You and your EMT-B partner are dispatched to the A&E Bank for a woman in active labor. En route, you and your partner discuss your previous experiences with assisting in a delivery and how you can prepare yourselves for your current call.

1. What kind of preparations should you be considering before your arrival?
2. What type of BSI precautions should you take?

Obstetric and Gynecologic Emergencies

Most infants in the United States are delivered in a hospital, with doctors and nurses in attendance to care for not only the mother, but also the newborn infant. Occasionally, the birth process moves faster than the mother expects, and you will find yourself with a decision to make: Should you stay on the scene and deliver the infant or transport the patient to the hospital? Are there other factors that would affect this decision, such as trauma, weather, and distance to the hospital? This chapter will tell you how to make this decision and how to proceed if on-scene delivery is necessary. It describes the normal process of childbirth and discusses common complications so that you will be prepared to handle normal and abnormal deliveries. Next, it describes the evaluation and care of the newborn infant. Finally, the chapter discusses gynecologic emergencies unrelated to childbirth.

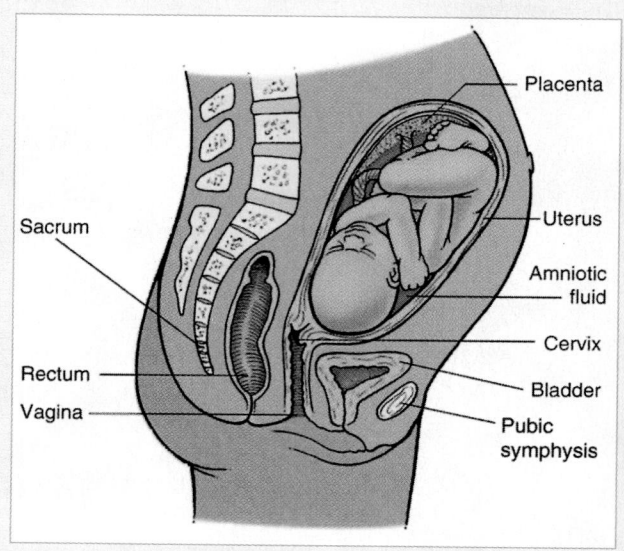

Figure 20-1 Anatomic structures of the pregnant woman.

Anatomy of the Female Reproductive System

The <u>uterus</u>, or womb, is the muscular organ where the fetus grows (Figure 20-1 ▶). It is responsible for contractions during labor and ultimately helps to push the infant through the birth canal. The <u>birth canal</u> is made up of the vagina and the lower third, or neck, of the

uterus, called the <u>cervix</u>. The cervix contains a mucous plug that seals the uterine opening, preventing contamination from the outside world. When the cervix begins to dilate, this plug is discharged as pink-tinged mucus, which may occur with <u>bloody show</u>, a small amount of blood at the vagina that appears at the beginning of labor. This "show" may signal the first stage of labor. The <u>fetus</u> is the developing, unborn infant that grows inside the mother's uterus for approximately 9 months.

The <u>vagina</u> is the outermost cavity of a woman's reproductive system and forms the lower part of the birth canal. It is about 8 to 12 cm in length, begins at the cervix, and ends as an external opening of the body. Essentially, the vagina completes the passageway from the uterus to the outside world for the infant. The <u>perineum</u> is the area of skin between the vagina and the anus. During birth, as the infant moves through the birth canal, the perineum will begin to bulge significantly.

As the fetus grows, it requires more and more nourishment. The <u>placenta</u>, a disk-shaped structure, attaches to the inner lining of the wall of the uterus and is connected to the fetus by the umbilical cord. There is normally no mixing of blood between the fetus and the mother. The placental barrier (Figure 20-2 ▶) consists of two layers of cells, keeping the circulation of the mother and fetus separated but allowing nutrients, oxygen, waste, carbon dioxide, and many toxins and most medications to pass between the fetus and mother.

www.EMTB.com

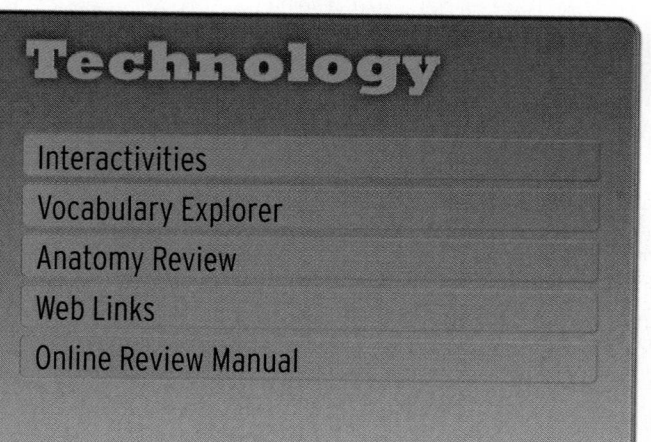

Technology

- Interactivities
- Vocabulary Explorer
- Anatomy Review
- Web Links
- Online Review Manual

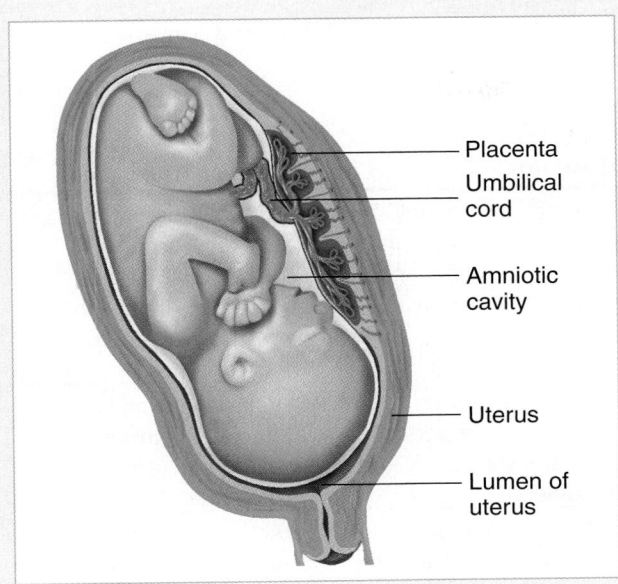

Figure 20-2 The placental barrier allows nutrients, oxygen, waste, carbon dioxide, toxins, and most medications to pass between fetus and mother.

- Placenta
- Umbilical cord
- Amniotic cavity
- Uterus
- Lumen of uterus

✳ EMT-B Tips

Predicting the due date for an infant is not an exact science. Fewer than one half of babies are born on the due date. Many factors influence when a baby is born, and neither the mother nor EMT-Bs have much control over this. One confusing point is the expected date and how "far along" the pregnancy is. Most medical models base the due date on the first day of the last menstrual cycle. That adds approximately 2 weeks to the actual pregnancy because conception occurred sometime after ovulation, which occurred approximately 2 weeks after the beginning of the last menstrual cycle. Most women have a general idea of this date, but young women, women who have very irregular cycles, or women who did not think they were pregnant may not have a very accurate due date. Also, some women talk about weeks from conception instead of menstruation. The important thing to remember is that whether it is 13 weeks or 30 weeks, it is not a firm number. Use this information realizing it may not be a firm date.

Whenever the mother takes anything in, so does the baby. After delivery, the placenta, or afterbirth, separates from the uterus and is delivered. The umbilical cord is the infant's lifeline, connecting mother and infant through the placenta. The umbilical cord contains two arteries and one vein. These vessels supply blood to the fetus: The vein carries blood toward the heart (baby), and the arteries carry blood away from the heart (baby). Oxygen and other nutrients cross from the mother's circulation through the placenta and then along the umbilical cord to support the fetus as it grows. Carbon dioxide and waste products travel the same route in the opposite direction. The remarkable thing about this exchange is that the mother's blood and that of the fetus do not mix during the process.

The fetus develops inside a fluid-filled, baglike membrane called the amniotic sac, or bag of waters. The sac contains about 500 to 1,000 mL of amniotic fluid, which helps insulate and protect the floating fetus as it develops. Released in a gush when the sac ruptures, usually at the onset of labor, this fluid helps to lubricate the birth canal and remove bacteria.

You are the Provider   Part 2

You arrive at the bank and the manager escorts you into his office. You find a woman in her mid 30s lying on the couch, holding her abdomen and moaning. Between labored breaths she tells you that her name is Jane and that she is a teller here at the bank. She is conscious, alert, and oriented. She is breathing in rapid panting breaths. Her pulse is strong and bounding, and her skin is pale and clammy.

3. What should be your first action?

A full-term pregnancy is from 36 to 40 weeks, counting from the first day of the last menstrual cycle. The pregnancy is divided into three trimesters of about 3 months each. Deliveries before 36 weeks are considered premature. Toward the end of the third trimester, the head of the fetus normally descends through the broad upper inlet of the woman's pelvis, positioning itself for the delivery.

Stages of Labor

There are three stages of labor: dilation of the cervix, expulsion of the baby, and delivery of the placenta. The first stage begins with the onset of contractions and ends when the cervix is fully dilated. Because the cervix has to be stretched thin by uterine contractions until the opening is large enough for the infant to pass through into the vagina, the first stage is usually the longest, lasting an average of 16 hours for a first delivery. You will usually have time to transport the mother during the first stage of labor.

The onset of labor starts with contractions of the uterus. Other signs of the beginning of labor are the bloody show and the rupture of the amniotic sac, called breaking of the water. These events may occur before the first contraction or later in the first stage of labor. The uterine contractions may not come at regular intervals at first. The mother may think that she simply has a nagging backache. The frequency and intensity of contractions in true labor increase with time. The uterine contractions become more regular and last about 30 to 60 seconds each. The length of labor varies greatly. As a general rule, it is longer in a primigravida, a woman who is experiencing her first pregnancy, and shorter in a multigravida, a woman who has experienced previous pregnancies. (Two similar terms refer to the outcomes of those pregnancies. A multipara is a woman who has had more than one baby born alive, and a primipara has had one live birth.)

Table 20-1 ▶ discusses how to tell when true labor is occurring. During pregnancy, the mother may experience false labor, or Braxton-Hicks contractions, in which there are contractions but they are not actual labor. In such cases, you do not need to prepare for an emergency delivery, but if real or true labor is occurring, you may need to prepare, depending on the mother's condition and transport time.

The second stage of labor begins when the cervix is fully dilated and ends when the infant is born. During this stage, you will have to make a decision about help-

TABLE 20-1 General Signs and Symptoms of Labor

False Labor and Braxton-Hicks Contractions	Real or True Labor
Contractions are not regular and do not increase in intensity or frequency. Contractions come and go.	Contractions, once started, consistently get stronger and closer together. Change in position does not relieve contractions.
Pain is in the lower abdomen. Contractions start and stay in the lower abdomen.	Pains and contractions start in the lower back and "wrap around" to the lower abdomen.
Activity or changing position will alleviate the pain and contractions.	Activity may intensify the contractions. Pain and contractions are consistent in any position.
If there is any bloody show, it is brownish.	The bloody show will be pink or red and generally accompanied by mucus.
There may be some leakage of fluid, but it is usually urine and will be in small amounts and smell of ammonia.	The bag of waters may have broken just before the contractions started or during contractions and will be of a moderate amount, may smell sweet, and will continue to leak.

ing the mother to deliver on the scene or providing transport to the hospital. Because the infant has to move through the birth canal during this stage, the uterine contractions are usually closer together and last longer. Pressure on the rectum may make the mother feel as if she needs to have a bowel movement. Under no circumstances should you let the mother sit on the toilet. She may also have the uncontrollable urge to push down. The perineum will begin to bulge significantly, and the top of the infant's head should begin to appear at the vaginal opening. This is called crowning.

The third stage begins with the birth of the infant and ends with the delivery of the placenta. This may take up to 30 minutes. Usually, you will not transport the pregnant woman during the third stage—if she is at this stage, you will more likely stay at the scene and perform an emergency delivery. It is important that you always follow BSI precautions, to protect yourself, the baby, and the mother from exposure to body fluids. There is a high potential of exposure because of body fluids released during childbirth.

Emergencies Before Delivery

Most pregnant women are healthy, but some may be ill when they conceive or become ill during pregnancy. You may safely use oxygen to treat any heart or lung disease in the mother without harm to the fetus.

As the time for delivery nears, certain complications can occur. One of these is preeclampsia, or pregnancy-induced hypertension, a condition that can develop after the 20th week of gestation, most commonly in primigravidas. This condition is characterized by the following signs and symptoms:

- Headache
- Seeing spots
- Swelling in the hands and feet
- Anxiety
- High blood pressure

Another condition is eclampsia, or seizures that result from severe hypertension. To treat eclampsia, lay the mother on her side—preferably her left side, maintain an airway, and provide supplemental oxygen; if vomiting occurs, suction the airway. Transport a pregnant patient with seizures promptly. As usual, size up the situation and perform your initial assessment, history, and physical exam, and assess the baseline vital signs. Provide treatment based on signs and symptoms.

If the patient is hypotensive, transport her on her left side. Transporting the mother in this position can prevent supine hypotensive syndrome, a problem due to compression by the pregnant uterus on the inferior vena cava when the mother lies supine, resulting in low blood pressure. In fact, transport the mother in this position and maintain this position whenever she is lying, except during delivery. When she is in the third trimester, she should not be supine.

Internal bleeding may be the sign of an ectopic pregnancy, a pregnancy that develops outside the uterus, most often in a fallopian tube. Ectopic pregnancy occurs about once in every 200 pregnancies. The leading cause of maternal death in the first trimester is internal hemorrhage into the abdomen following rupture of an ectopic pregnancy. For this reason, you should consider the possibility of an ectopic pregnancy in women

who have missed a menstrual cycle and complain of sudden stabbing and usually unilateral pain in the lower abdomen. A history of pelvic inflammatory disease, tubal ligation, or previous ectopic pregnancies should heighten your suspicion of a possible ectopic pregnancy.

Hemorrhage from the vagina that occurs before labor begins may be very serious; call for ALS backup. In early pregnancy, it may be a sign of a spontaneous abortion, or miscarriage. In the later stages of pregnancy, vaginal hemorrhage may indicate problems with the placenta. In placenta abruptio, the placenta separates prematurely from the wall of the uterus (Figure 20-3 ▼). In placenta previa, the placenta develops over and covers the cervix (Figure 20-4 ▼).

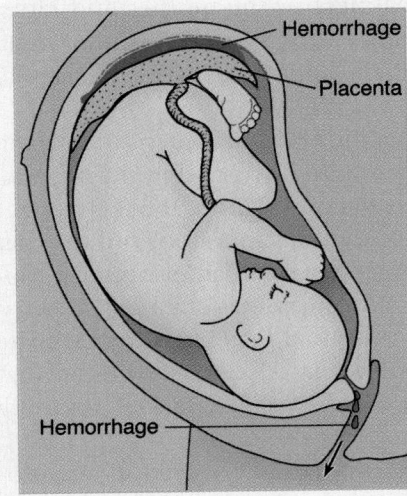

Figure 20-3 In placenta abruptio, the placenta separates prematurely from the wall of the uterus.

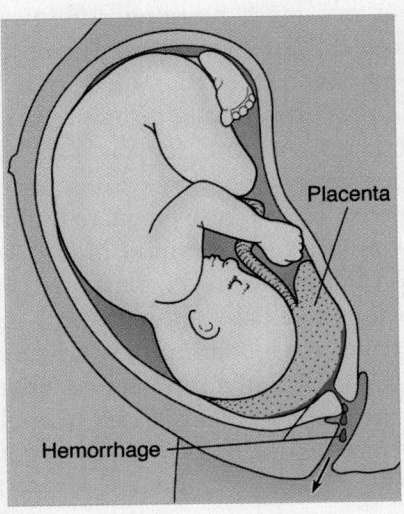

Figure 20-4 In placenta previa, the placenta develops over and covers the cervix.

Many women who did not have diabetes before pregnancy develop diabetes during pregnancy. This is called gestational diabetes and for most women will clear up after delivery of the child. Just like in nongestational diabetes, the mother may control her blood glucose level with diet and exercise or may take medication; in some cases, the mother needs to take insulin. A mother experiencing hyperglycemia or hypoglycemia should be cared for in the same manner as any diabetic, as discussed in Chapter 15. If a pregnant woman is found with an altered level of consciousness, your history should include questions about diabetes, and you should check the blood glucose level if local protocols permit. Remember that labor is hard work. Many mothers experience nausea before labor and may not have eaten. These factors can lead to hypoglycemia and weakness in the mother and fetus. Consult with medical control if delivery is imminent. Rapid transport to the hospital is usually preferred to out-of-hospital delivery.

Any bleeding from the vagina in a pregnant woman is a serious sign and should be treated in the hospital promptly. If the mother shows signs of shock, have her lie on her left side during transportation, and give her high-flow oxygen. Place a sterile pad or sanitary napkin over the vagina, and replace it as often as necessary. Save the pads so that hospital personnel can estimate how much blood she has lost. Also save any tissue that may be passed from the vagina. Do not put anything into the vagina.

When a pregnant woman is involved in an automobile crash, severe hemorrhage may occur from injuries to the pregnant uterus. The resulting oxygen deprivation can cause grave injury to the fetus. Promptly evaluate and transport a pregnant crash victim; support the airway, and if there is any sign of bleeding, administer high-flow oxygen and call for ALS backup. Have the mother lie on her left side rather than on her back; this will relieve the pressure of the uterus on intra-abdominal organs, especially the inferior vena cava and abdominal aorta. Pregnant women have an increased amount of blood volume. Therefore, a pregnant trauma patient may have a significant amount of blood loss before showing signs of shock. However, the infant may be in trouble well before this. Often, if the mother has sustained serious trauma, the blood supply to the fetus is reduced so that the body can supply an adequate amount of blood to the mother. In most cases, the only chance to save the infant is to adequately resuscitate the mother.

Assessment

Scene Size-up

Childbirth is seldom an unexpected event, but there are occasions when childbirth becomes an emergency. Dispatch protocols usually include simple questions to determine whether birth in imminent. Some of this information may get passed on to you to help you prepare for the situation. Contractions may be caused by trauma or medical conditions. It may just be "time" to deliver. Because a pregnant woman's balance is altered by her carrying a fetus and hormones that relax the musculature, falls and spinal stabilization must be considered.

Your first preparation for delivery should be taking appropriate BSI precautions. Gloves and eye protection should be a minimum when delivery has already begun or is complete. If you have time to prepare for the delivery, a mask and gown should also be used.

The usual threats to your safety will be present in this and other medical situations. Do not get lax in your safety observations and precautions because the delivery is in progress or the family is anxious. Rushing around may hurt not only you, but also the child and mother. Remain calm and professional.

Initial Assessment

General Impression

The general impression is a good across-the-room assessment that should tell you whether the mother is in active labor or you have a few minutes to assess for imminent delivery. The chief complaint may be, "The baby is coming!" Take a moment to evaluate the chief complaint and confirm whether the baby is on its way out or, again, whether you have some time to evaluate the situation. When trauma or other medical problems are the presenting complaint, evaluate these first and then assess the impact of these problems on the fetus. Use the AVPU (Awake and alert, responsive to Verbal stimulus or Pain, or Unresponsive) scale to determine the level of consciousness.

Airway, Breathing, and Circulation

During an uncomplicated childbirth, problems with the airway, breathing, and circulation are not usually an issue. However, a motor vehicle crash, an assault, or any number of medical problems in a pregnant woman may initiate a complicated delivery. In these situations, evaluate the **A**irway, **B**reathing, and **C**irculation to ensure they are adequate and treat any airway, breathing, or circulation problem that is identified according to established guidelines and local protocol.

Transport Decision

If delivery is imminent, based on signs and symptoms discussed in this chapter, prepare to deliver on scene. The ideal place to deliver an infant is in the security of your ambulance or the privacy of the mother's home. If these locations are not available, evaluate the scene for the best area to use should delivery be imminent. The area should be warm and private with plenty of room to move around.

If the delivery is not imminent, prepare the patient for transport. Pregnant women in the last two trimesters of pregnancy should be transported lying on the left side when possible. This will keep the weight of the baby off the mother's inferior vena cava, preventing supine hypotensive syndrome. If a spinal injury is suspected and spinal stabilization is indicated, secure the mother as usual and place a towel roll under the right side of the backboard to prevent supine hypotensive syndrome (Figure 20-5 ▼).

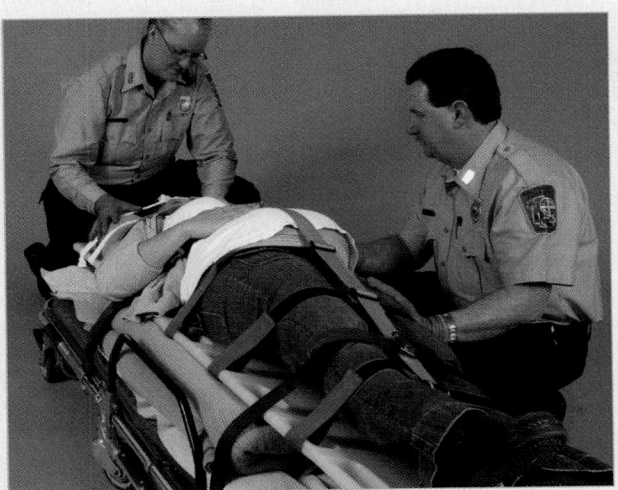

Figure 20-5 Place a towel roll under the right side of the backboard to prevent supine hypotensive syndrome in pregnant patients.

Focused History and Physical Exam

If the mother is unconscious, perform a rapid physical exam to locate injuries and other problems, then obtain vital signs and obtain as much history as is available. Much of this can be performed in the ambulance en route to the hospital. If the mother is conscious, begin with asking history questions, perform a focused physical exam, and then obtain vital signs.

SAMPLE History

Obtain a SAMPLE history. Women with a history of medical problems and who take prescription medications regularly do become pregnant. Some women do not experience medical problems that require medications until they become pregnant. Do not just focus on the pregnancy history but evaluate the SAMPLE history as well. Ask the patient specifically about prenatal care. Identify any complications she may have had during the pregnancy or that her physician has discussed with her about the delivery. These may include Rh factor, size or position of the fetus, or position and health of the placenta. Determine the due date, frequency of contractions, and history of previous pregnancies and deliveries and their complications, if any. Determine whether there is a possibility of twins and whether the mother has taken any drugs or medications. If her "water has broken," ask whether the fluid was green. Green fluid is due to <u>meconium</u> (fetal stool). The presence of meconium can indicate disease in the newborn, and it is possible for the fetus to aspirate meconium during delivery. Any of these are risk factors for fetal distress and indicate the possible need for neonatal resuscitation.

Focused Physical Exam

Your focused physical exam should focus on the abdomen and delivery of the fetus. Assess the length and frequency of contractions by feeling the abdomen. This can be done in a conscious or unconscious woman in labor. Compare what you feel with what the woman experiences in pain with each contraction. When appropriate and according to local protocol, inspect the vaginal opening for rupture of the amniotic sac, bloody show, and crowning if you suspect delivery is imminent. Be sure to protect the woman's privacy during the examination. If the woman has other complaints, such as difficulty breathing, a focused physical exam of her breathing may be important. The focus of this physical exam should be based on her chief complaint and history.

Baseline Vital Signs

Assess baseline vital signs, including heart rate, rhythm, and quality; respiratory rate, rhythm, and quality; skin color, temperature, and condition; capillary refill time; and blood pressure. Pay special attention to tachycardia and hypotension (which could mean hemorrhage or compression of the vena cava) or hypertension (possibly indicating preeclampsia). It is typical for a woman's blood pressure to drop slightly during pregnancy. Compare your findings with previous blood pressures in prenatal visits. Hypertension, even mildly elevated blood pressure, may indicate more serious problems.

Interventions

Remember that in most cases, childbirth is a natural process that does not require your assistance. See the section, Emergency Care. When childbirth is complicated by trauma or other problems, any interventions you provide for the woman will benefit the baby. For example, if a pregnant woman has a low pulse oximetry value, the baby will as well. Giving the woman oxygen will improve oxygen levels in the baby. Be familiar with your local protocols.

Detailed Physical Exam

If delivery is imminent or other assessments or treatments require your attention, defer the detailed physical exam.

Ongoing Assessment

Your ongoing assessment should center on reassessing the woman's ABCs, particularly vaginal bleeding after delivery. Repeat assessment of the vital signs and compare those to the baseline set evaluated earlier in your assessment. Frequent reassessment of vital signs may identify hypoperfusion from excessive blood loss due to delivery. Recheck interventions and treatments to see whether they were effective. Is the vaginal bleeding slowing with uterine massage? (Uterine massage, dis-

cussed later in this chapter, can be used to slow vaginal bleeding after delivery.)

Communication and Documentation

If your assessment determines that delivery is imminent, notify the hospital staff of your preparations for delivery. If delivery has not started within 30 minutes, notify the hospital staff you are going back in transit without delivering. This delay in delivery, when signs and symptoms indicate an immediate birth, usually means a serious problem exists. Request help from hospital staff in assessing and triaging the situation. Be sure to notify staff at the receiving hospital of all relevant information so that they can prepare the proper response. What you tell them may determine whether your patient will be seen in the emergency department or whether you will go directly to the labor and delivery department. Document very carefully, especially on the status of the baby. Obstetrics is among the most litigated specialties in medicine; scrupulous documentation is essential.

Emergency Care

Preparing for Delivery

Consider delivering the baby at the scene in the following circumstances:

- When delivery can be expected within a few minutes
- When a natural disaster, bad weather, or some other type of catastrophe makes it impossible to reach the hospital

How do you determine whether delivery will occur within a few minutes? First, ask the pregnant woman these questions:

- How long have you been pregnant?
- When are you due?
- Is this your first baby?
- Are you having contractions? How far apart are the contractions? How long do the contractions last?
- Do you feel as though you have to strain or move your bowels?
- Have you had any spotting or bleeding?
- Has your water broken?
- Were any of your previous children delivered by cesarean section?

Also consider asking the following questions:

- Have you had a complicated pregnancy in the past?

■ Do you use drugs, drink alcohol, or take any medications?

■ Is there any possibility that this is a multiple birth?

■ Does your doctor expect any complications?

If indicated by the answers to these questions, look for crowning.

If this is not the patient's first child, she may be able to tell you whether she is about to deliver. If she says that she is, make immediate preparations for delivery. Otherwise, does she have an extremely firm abdomen? Does she say that she has to move her bowels or feels the need to push? If so, the infant's head is probably pressing on the rectum, and delivery is about to occur. At this point, you should inspect the vagina to determine whether crowning has occurred; if so, delivery is imminent. Do not touch the vaginal area until you are sure that delivery is, in fact, imminent. In general, do not touch vaginal areas except during delivery (under certain circumstances) and when your partner is present. Spread the pregnant woman's legs apart gently, explaining that you are doing so to decide whether the baby should be delivered immediately or whether she should be transported to the hospital for the delivery.

Once labor has begun, there is no way it can be slowed or stopped. Never attempt to hold the woman's legs together. To do so would only complicate the delivery. Do not let her go to the bathroom. Instead, reassure her that the sensation of needing to move her bowels is normal and that it means she is about to deliver.

If you decide to deliver at the scene, remember that you are only assisting the woman with the delivery. Your part is to help, guide, and support the baby as it is born. Remember to use BSI precautions at all times. Try to limit distractions for yourself and for the patient. You want to appear calm and reassuring while protecting the woman's modesty. Most important, recognize when the situation is beyond your level of training. If delivery is imminent with crowning, contact medical control for a

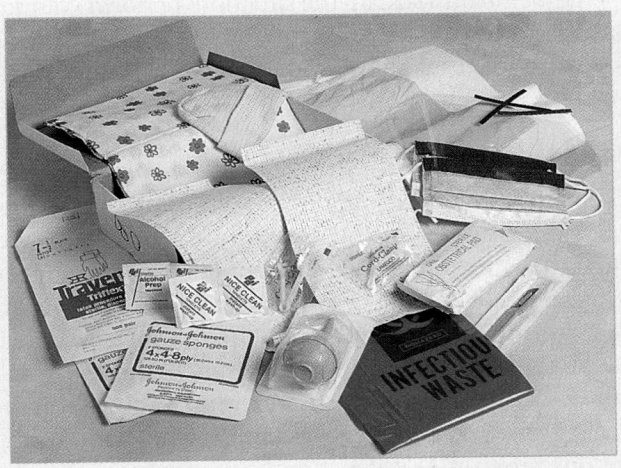

Figure 20-6 Your unit should contain a sterile OB kit. Items usually found in this kit are listed in the text.

decision to deliver on the scene or to transport. When in doubt, contact medical control for further guidance. Always recognize your own limitations, and when you are unsure about what to do, transport the patient even if delivery must occur during transport.

Your emergency vehicle should always be equipped with a sterile emergency obstetric (OB) kit containing the following items (Figure 20-6 ▲):

■ Surgical scissors or a scalpel
■ Umbilical cord clamps
■ Umbilical tape
■ A small rubber bulb syringe
■ Towels
■ 4″ x 4″ x gauze sponges and/or 2″ x 10″ gauze sponges
■ Sterile gloves
■ Infant blanket
■ Sanitary napkins
■ An infant-sized bag-valve-mask (BVM) device
■ Goggles
■ A plastic bag

You are the Provider   Part 3

The woman tells you that she is 1 week past her due date and that she has been having contractions for the past hour. She informs you that her water broke just before your arrival. This is her fourth pregnancy, and she has three children. She tells you she feels like she has to go to the restroom. Your partner applies high-flow oxygen via a nonrebreathing mask and begins timing her contractions.

4. Is your patient's need to use the restroom significant in this situation? If so, why?

5. What additional information should you obtain from your patient?

Remember to size up the situation; perform your initial assessment, focused history, and physical exam; assess baseline vital signs; and provide treatment based on signs and symptoms.

Patient Position

The patient's clothing should be pushed up to her waist or, if she is wearing trousers and undergarments, removed. Remember to preserve her modesty as much as you can while helping her to move into a semi-Fowler's position. Place the patient on a firm surface that is padded with blankets, folded sheets, or towels. Put a pillow or blankets beneath her hips to elevate them about 2″ to 4″. It is sometimes better to put a pillow under one hip to allow the patient to turn to one side. This may also make it easier to suction the baby once it is born. Support the mother's head, neck, and upper back with pillows and blankets. If delivery is occurring in an automobile, the patient should lie on the seat, with one foot on the floor and the other on the seat, with the upper knee and hip bent (Figure 20-7 ▼). This is a

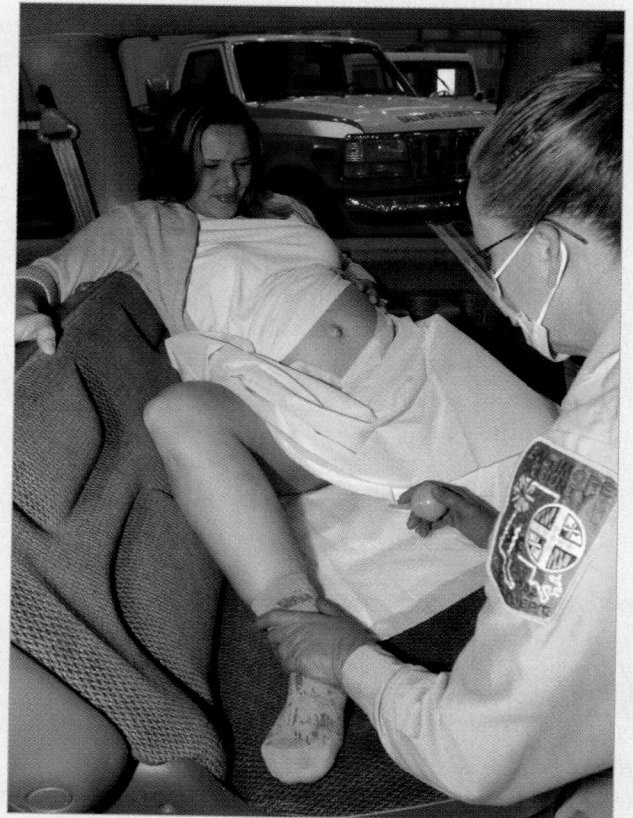

Figure 20-7 For delivery in a car, have the patient lie on the seat with one foot on the floor and the other on the seat. Make sure the upper knee is bent and the upper hip is flexed.

good time to prepare for the baby also. Where will you put the baby? How will you dry it off? How will you keep him or her warm? Planning for these needs makes it much easier when the time comes.

If the emergency delivery is occurring at home, you should move the patient to a sturdy, flat surface. You will find it easier to work with the patient on a firm surface rather than on a bed. Elevate the patient's hips, and support her head with one or two pillows. Have her keep her legs and hips flexed, with her feet flat on the surface beneath her and her knees spread apart. Track the progression of the delivery closely at all times; you do not want an abrupt delivery to occur, when the crowning head pops out uncontrollably.

Preparing the Delivery Field

Take the following steps to prepare the area where the infant will be born:

1. As time allows, place towels or sheets on the floor around the delivery area to help soak up the amniotic fluid that will be released when the amniotic sac ruptures (Figure 20-8A ▶). Note that the amniotic sac may have ruptured before you arrived. Elevate the patient's hips, and support her head and shoulders with folded blankets or pillows.
2. Open the OB kit carefully so that its contents remain sterile.
3. Put on the sterile gloves.
4. Use the sterile sheets and towels from the OB kit to make a sterile delivery field. Place one sheet or towel under the patient's buttocks, and unfold it toward her feet. The other sheet should be draped over her abdomen and upper legs. Alternatively, you can use three sheets: one sheet under the buttocks, one sheet wrapped behind the patient's back and draped over each thigh, and one sheet draped across the abdomen (Figure 20-8B ▶).

Delivering the Baby

Your partner should be at the patient's head to comfort, soothe, and reassure her during the delivery. The patient may want to grip someone's hand. She may yell, cry, or say nothing at all. It is not uncommon for mothers to become nauseated, and some may vomit. If this occurs, have your partner turn the mother's head to the side so that her mouth and airway can be cleared manually or with suction, as needed.

You must continually assess the mother for crowning. Do not allow an abrupt delivery to occur. Position

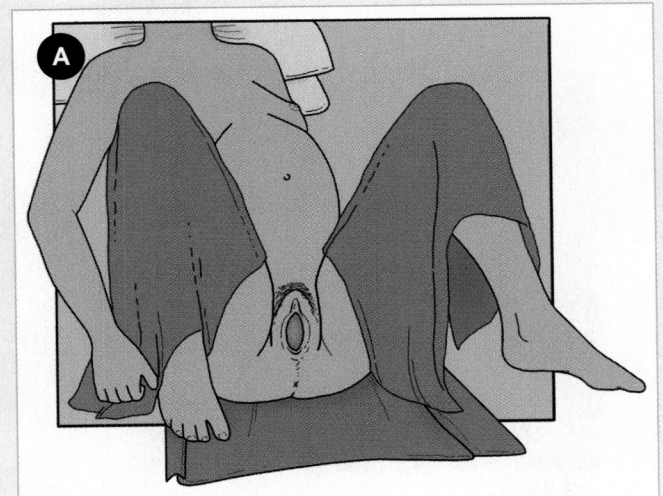
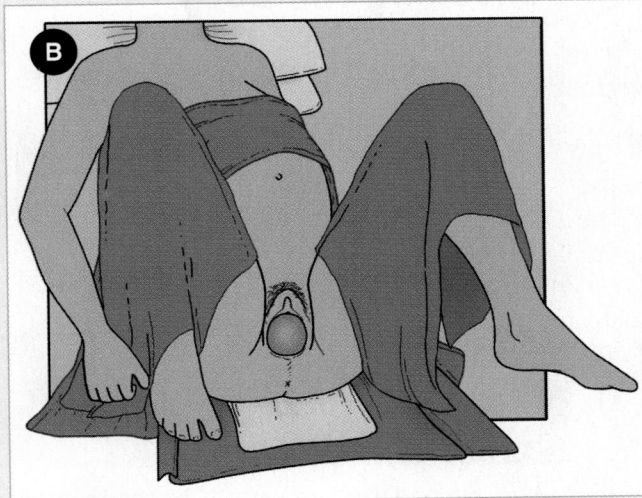

Figure 20-8 Preparing the delivery field. **A.** Place sheets or towels under the mother, elevate the mother's hips, and support her head with one or two pillows. **B.** Use sterile sheets and towels from the OB kit to make a clean delivery field. Place one sheet under her buttocks, drape the other over her abdomen, and wrap a third sheet behind her back with either end draped over the thighs.

yourself so that you can see the vagina at all times. Time the patient's contractions from the beginning of one to the beginning of the next to determine the frequency of the contractions. In addition, time the duration of each contraction. You do this by feeling the patient's abdomen from the moment the contraction begins (uterus and abdomen tightening) to the moment it ends (uterus and abdomen relaxing). Remind the patient to take quick, short breaths during each contraction but not to strain. Between contractions, encourage the mother to rest and breathe deeply through her mouth.

Follow the steps in (Skill Drill 20-1 ▶) to deliver the baby. These steps are described in more detail later.

1. Allow the mother to push the head out. Support it as it emerges, **placing your gloved hand over its bony parts. Suction fluid from the mouth first, then the nostrils (Step 1).**
2. **Feel at the neck to see if the cord is wrapped** around it. If it is, gently lift it over the baby's head without pulling hard on the cord.
3. Once the head is delivered, the upper shoulder will be visible. **Guide the head down slightly, if needed, to help that shoulder deliver (Step 2).**
4. **Support the head and upper body as the shoulders deliver.** You may need to guide the head up slightly to deliver the lower shoulder **(Step 3).**
5. Once the body is delivered, handle the infant firmly but gently. It will be slippery. **Make sure**

the baby's neck is in a neutral position to keep the airway open **(Step 4).**
6. **Place the umbilical cord clamps about 2″ to 4″ apart**, about four fingerbreadths from the infant's body. Depending on local protocol, once they are firmly in place, your protocol may dictate to cut between the clamps **(Step 5).** However, local protocols vary on when to cut—now, after delivery of placenta, or not at all if immediate transport is available. Be sure you know your own local protocol on this.
7. **The placenta delivers itself,** usually within 30 minutes of birth. Never pull on the end of the umbilical cord in an attempt to speed delivery of the placenta **(Step 6).**

Delivering the Head

Watch the head as it begins to exit the vagina because it must be supported as it emerges. It may take two, three, or more contractions for the delivery of the head to occur from the time it begins to crown. Once it is obvious that the head is coming out farther with each contraction, you should place your gloved hand over the emerging bony parts of the head and exert very gentle pressure on it, decreasing the pressure slightly between contractions. This will allow the head to come out smoothly and prevent it and the rest of the infant from suddenly popping out during a strong contraction, possibly causing injury. You may want to move the patient's feet so that you are between the patient's

20-1

Skill Drill

Delivering the Baby

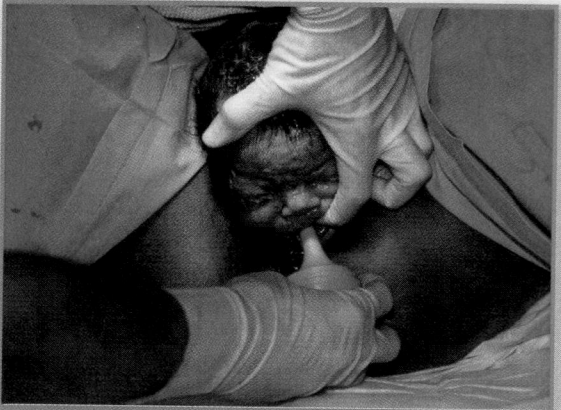

1 Support the bony parts of the head with your hands as it emerges. Suction fluid from the mouth, then nostrils.

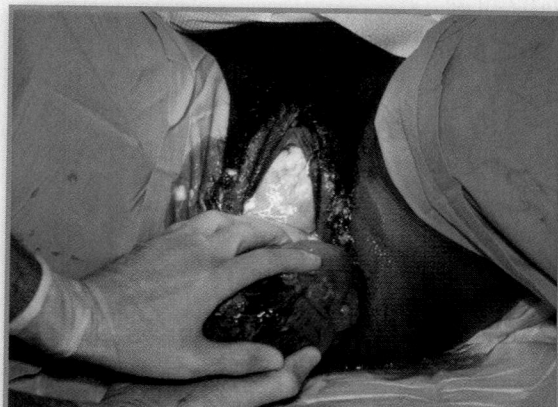

2 As the upper shoulder appears, guide the head down slightly, if needed, to deliver the shoulder.

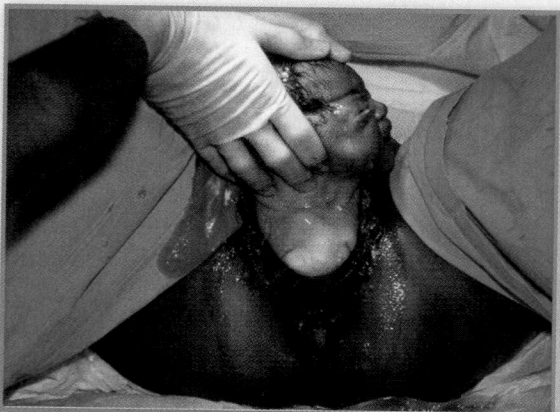

3 Support the head and upper body as the lower shoulder delivers, guiding the head up if needed.

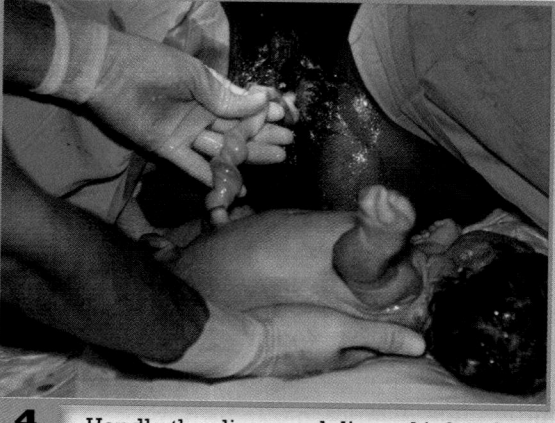

4 Handle the slippery, delivered infant firmly but gently, keeping the neck in neutral position to maintain the airway.

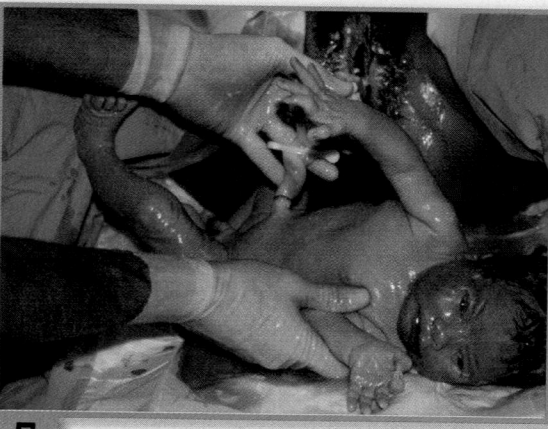

5 Place the umbilical cord clamps 2″ to 4″ apart, and cut between them.

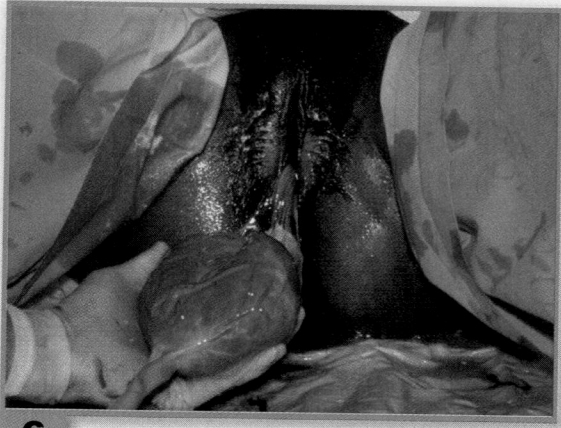

6 Allow the placenta to deliver itself. Do not pull on the cord to speed delivery.

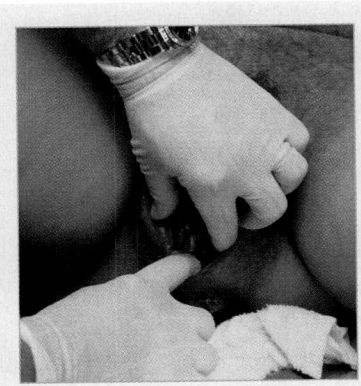

Figure 20-9 One method of reducing the risk of perineal tearing during labor is to apply gentle pressure on the head while gently stretching the perineum.

legs during the delivery. Be careful that you do not poke your fingers into the infant's eyes or into the two soft spots, called fontanels, on the head. One fontanel is located on the top of the head, near the brow, and one is near the back of the head. The brain is covered only by skin and membranes at these spots.

Methods of reducing the risk of perineal tearing during labor include applying gentle pressure horizontally across the perineum with a sterile gauze pad, or applying gentle pressure to the head while gently stretching the perineum (Figure 20-9 ▲). Consult your local protocol regarding the methods used in your area. Also be prepared for the possibility that feces may come out because of the pressure on the rectum.

Unruptured Amniotic Sac

Usually, the amniotic sac will break or rupture at the beginning of labor. The sac may also rupture during contractions. If the amniotic sac has not ruptured by this point, it will appear as a fluid-filled sac (like a water balloon) emerging from the vagina. This situation is serious because the sac will suffocate the baby if it is not removed. If it has not spontaneously ruptured, you may puncture the sac with a clamp, away from the baby's face, only as the head is crowning, not before. As the sac is punctured, amniotic fluid will gush out. Push the ruptured sac away from the infant's face as the head is delivered. Clear the baby's mouth and nose immediately, using the bulb syringe and gauze sponge. If the amniotic fluid is greenish (meconium staining) instead of clear or has a foul odor, make note of this in the information you relay to medical control. Meconium is a sign of two possible problems: a depressed newborn or airway obstruction. Thick meconium can clog the airway of the newborn. Aggressive suctioning of the baby's mouth and oropharynx before delivery of the body may prevent meconium aspira-

tion and respiratory distress. Once the head has been delivered, it usually rotates to one side or the other rather than straight up and down.

Umbilical Cord Around the Neck

As soon as the head is delivered, use the index finger of your other hand to feel whether the umbilical cord is wrapped around the neck. This commonly is called a nuchal cord. A nuchal cord that is wound tightly around the neck could cause the infant to strangle. It must, therefore, be released from the neck immediately. Usually, you can slip the cord gently over the infant's delivered head (or over the shoulder, if necessary). If not, you must cut it by placing two clamps about 2″ apart on the cord and cutting the cord between the clamps. If the cord is wrapped more than once around the neck, a rare event, you have to clamp and cut only once; then you can unwrap the cord from around the neck. Handle the cord very carefully; it is fragile and easily torn. Do not let the clamps come off until the ends of the cord have been tied. Fortunately, the cord is usually not wrapped around the infant's neck and does not have to be cut until after the entire infant has been delivered. However, you must always check for a nuchal cord.

Now that you have delivered the infant's head and verified that no nuchal cord is present, you will need to suction the amniotic fluids from the infant's airway before the delivery proceeds. You must ask the mother not to push while you are doing this, although her desire to do so will be very strong. While supporting the infant's head with one hand, quickly and efficiently suction the fluid from the mouth first and then the nostrils. If you suction the nostrils first, you may stimulate the infant to aspirate the fluid in the mouth or pharynx; because infants are nose breathers, any stimulation of the nose will cause a gasping response. In suctioning the airway, fully compress the bulb syringe before it is inserted 1″ to 1½″ into the infant's mouth, then release the bulb to suction fluids and mucus into the syringe. Make sure the syringe does not touch the back of the mouth. Discard the fluid into a towel, and repeat the procedure, suctioning the mouth and nostrils two or three times each, or until they are clear.

Delivering the Body

By the time you are finished suctioning, the mother will most likely be pushing again, and the upper shoulder will be visible in the vagina. The infant's head is the largest part of the body. Once it is born, the rest of the infant usually delivers easily. Support the head and

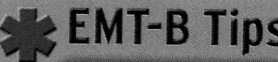

EMT-B Tips

As a baby delivers, you must divide your attention between two patients. This can keep two EMT-Bs busy, even when things go well. To ensure that special care needs do not result in neglecting one patient, designate one member of the crew to pay primary attention to each patient. Call for help early if you suspect that both will need special care or that one will require resuscitation.

Documentation Tips

Recording the time of birth will ensure that the information is available for the birth certificate. It also provides you with a starting point from which to time the intervals for Apgar scores. This is even more important with multiple births. You will be busy; consider asking a family member to act as "timekeeper."

upper body as the shoulders deliver. Do not pull the infant from the birth canal. The abdomen and hips will appear; once these deliver, support them with your other hand. Grasp the infant's feet as they are born. Now the infant is being well supported with both hands. Handle the infant firmly but carefully. It will be slippery with a white, cheesy substance, called vernix caseosa.

Postdelivery Care

As soon as the entire infant is born, dry the baby off and wrap it immediately in a blanket or towel, and place it on one side, with the head slightly lower than the rest of its body. Wrap the baby so that only the face is exposed, making sure that the top of the head is covered. Also make sure that the baby's neck is in a neutral position so the airway remains open. Newborn babies are very sensitive to cold, so if it is at all possible, you should keep the blanket or towel warm before you use it. Use a sterile gauze pad to wipe the infant's mouth, and once again suction the mouth and nose. Suctioning the nose is particularly important because

babies breathe through their noses. If you prefer, you can pick up and cradle the infant in your arm at the level of the mother's vagina while doing this, but always keep the head slightly downward to help prevent aspiration. After suctioning, keep the infant at the same level as the mother's vagina until the umbilical cord is cut. If the infant is higher than the vagina, blood will be siphoned from the infant through the umbilical cord back into the placenta.

A newborn's body temperature can drop very quickly, so dry and wrap the infant as soon as possible. Only then will you clamp and cut the umbilical cord.

Once the infant is born, the umbilical cord is of no further use to mother or infant. Postdelivery care of the umbilical cord is important because infection is easily transmitted through the cord to the baby. Using the two clamps in the OB kit, clamp the cord somewhere between the mother and the infant, preferably four fingerbreadths from the infant. Place the clamps about 2″ to 4″ apart. Once they are firmly in place, cut the cord between them with the sterile scissors or scalpel, using great care. Remember, the cord is fragile; if handled too

You are the Provider

Part 4

You explain to your patient the need to examine her before preparing her for transport to the hospital. While doing so, she tells you that when she went to the doctor yesterday she was dilated to 3 cm and that she lost her mucous plug about 1 hour ago. Your partner tells you that her contractions are 45 seconds long and are 55 seconds apart.

6. What does the length of the contractions and the time between contractions tell you?
7. Of what significance is the loss of her mucous plug?
8. What stage of labor is your patient currently in?

roughly, it could be torn from the infant's abdomen, resulting in a fatal hemorrhage. Once the clamps are in place, there is no need to rush.

After you have cut the cord, tie the end coming from the infant. If it was a nuchal cord and cut during delivery, now is the time to tie it. Do not use ordinary string or twine, which will cut through the soft, fragile tissues of the cord. Place a loop of the special "umbilical tape" around the cord about 1″ nearer to the infant than to the clamp. Tighten the tape slowly so that it does not cut the cord, then tie it firmly with a square knot. Cut the ends of the tape, but do not remove either clamp. The part of the cord that is coming out of the mother's vagina is attached to the placenta and will be delivered when the placenta delivers.

By now, the infant should be pink and breathing on his or her own. Give the infant, wrapped in a warm blanket, to your partner; he or she can monitor the infant and complete its initial care. Alternatively, you can give the infant to the mother if she is alert and in stable condition, if you are allowed to do so by local protocol. The mother may want to begin breastfeeding at this time. You need to return your attention to the mother and the delivery of the placenta.

Delivery of the Placenta

The placenta is attached to the end of the umbilical cord that is coming out of the mother's vagina. Again, you need only assist. Like the infant, the placenta delivers itself, usually within a few minutes of the birth, although it may take as long as 30 minutes. Never pull on the end of the umbilical cord in an attempt to speed delivery of the placenta. You may tear the cord, the placenta, or both and cause serious, perhaps life-threatening, hemorrhage.

The normal placenta is round, about 7″ in diameter, and about 1″ thick. One surface is smooth and covered with a shiny membrane; the other surface is rough and divided into lobes. Wrap the entire placenta and cord in a towel, place them into a plastic bag, and take them to the hospital. Hospital personnel will examine the placenta and the cord to make certain that the entire placenta has been delivered. If a piece of the placenta has been retained inside the mother, it could cause persistent bleeding or infection.

After delivery of the placenta and before transport, place a sterile pad or sanitary napkin over the vagina and straighten the mother's legs. You can help to slow bleeding by gently massaging the mother's abdomen with a firm, circular motion (Figure 20-10 ▶). The ab-

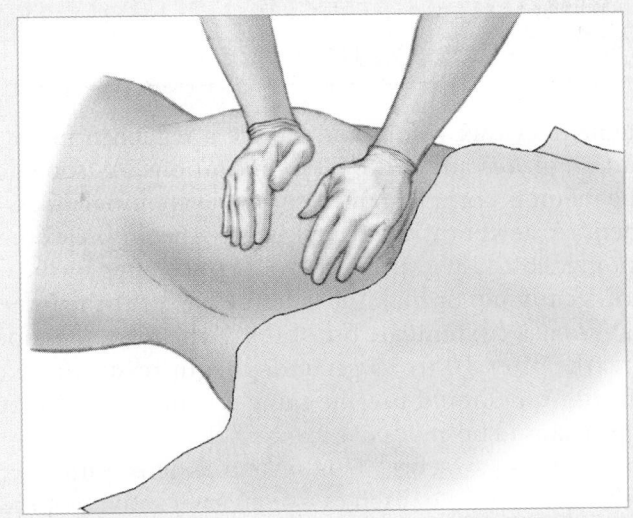

Figure 20-10 After delivery, massage the mother's abdomen in a firm, circular motion.

dominal skin will be wrinkled and very soft. You should be able to feel a firm, grapefruit-sized mass in the lower abdomen. This is called the fundus. As you massage the fundus, the uterus will contract and become firmer. You can also place the infant at the mother's breast to nurse, which stimulates the uterus to contract. Both massaging the uterus and having the baby stimulate the mother's nipples will cause a production of oxytocin, which is a hormone that will help to contract the uterus and slow bleeding. Before taking her, the infant, and the placenta to the hospital, take a minute to congratulate the mother and thank anyone who assisted. In writing your medical report, be sure to record the time of birth for the birth certificate.

Some bleeding, usually less than 500 mL, occurs before the placenta delivers. The following are emergency situations:

- More than 30 minutes elapse, and the placenta has not delivered.
- There is more than 500 mL of bleeding before delivery of the placenta.
- There is significant bleeding after the delivery of the placenta.

If one or more of these events occur, transport mother and infant to the hospital promptly. Place a sterile pad or sanitary napkin over the mother's vagina, place her in shock position, administer oxygen, and monitor her vital signs closely. Never put anything into the vagina.

Neonatal Evaluation and Resuscitation

Remember that before you handle a newborn infant, put on gloves and follow BSI precautions. As soon as the infant is born, you must complete an initial assessment. A newborn infant will usually begin breathing spontaneously within 15 to 20 seconds after birth. If not, gently tap or flick the soles of the feet or rub the baby's back to stimulate breathing. If the baby does not breathe after 10 to 15 seconds, begin resuscitation efforts. You should use the same scoring system that physicians in hospitals use to assess the status of the infant: the Apgar score. This system assigns a number value (0, 1, or 2) to each of five areas of activity of the newborn infant:

- **A**ppearance. Shortly after birth, the skin of a light-skinned newborn infant and the mucous membranes of a dark-skinned infant should turn pink. Newborn infants often have cyanosis of the extremities for a few minutes after birth, but hands and feet should "pink up" quickly. Blue skin all over or blue mucous membranes signal a central cyanosis.
- **P**ulse. If a stethoscope is unavailable, you can measure pulsations with your fingers in the umbilical cord or at the brachial artery. Obviously, the infant with no pulse requires immediate CPR.
- **G**rimace or irritability. Grimacing, crying, or withdrawing in response to stimuli is normal in a newborn and indicates that the newborn infant is doing well. The way to test this is to snap a finger against the sole of the infant's foot.

EMT-B Tips

APGAR scoring system
- **A**ppearance
- **P**ulse
- **G**rimace or irritability
- **A**ctivity or muscle tone
- **R**espirations

- **A**ctivity or muscle tone. The degree of muscle tone indicates the oxygenation of the newborn infant's tissues. Normally, the hips and knees are flexed at birth, and, to some degree, the infant will resist attempts to straighten them out. A newborn should not be floppy or limp.
- **R**espirations. Normally, the newborn's respirations are regular and rapid, with a good strong cry. If the respirations are slow, shallow, or labored, or if the cry is weak, the newborn infant may have respiratory insufficiency and need assistance with ventilation. Complete absence of respirations or crying is obviously a very serious sign; in addition to assisted ventilation, CPR may be necessary.

The total of the five numbers is the Apgar score. A perfect score is 10. The Apgar score should be calculated at 1 minute and 5 minutes after birth. Most newborn infants will have a score of 7 or 8 at one minute and a score of 8 to 10 four minutes later. (Table 20-2 ▼) shows how to calculate an Apgar score.

TABLE 20-2 Apgar Scoring System

Area of Activity	Score		
	2	1	0
Appearance	Entire infant is pink.	Body is pink, but hands and feet remain blue.	Entire infant is blue or pale.
Pulse	More than 100 beats/min	Fewer than 100 beats/min	Absent pulse
Grimace or irritability	Infant cries and tries to move foot away from finger snapped against its sole.	Infant gives a weak cry in response to stimulus.	Infant does not cry or react to stimulus.
Activity or muscle tone	Infant resists attempts to straighten hips and knees.	Infant makes weak attempts to resist straightening.	Infant is completely limp, with no muscle tone.
Respiration	Rapid respirations	Slow respirations	Absent respirations

Consider the following delivery situation. You have assisted a delivery or arrived to find that a delivery has already taken place. You now have two patients who need assessment and care: the mother and the infant. Follow these steps in assessing the newborn infant:

1. **Quickly calculate the Apgar score** to establish a baseline for the newborn infant's vital functions.

2. **Suctioning and stimulation** should result in an immediate increase in respirations. If they do not, you must begin artificial ventilation according to BLS protocols (Table 20-3 ▼). Unlike adults, who may have a sudden cardiac arrest, newborn infants who get into trouble usually have a respiratory arrest first. Therefore, it is essential to keep the infant ventilating and oxygenating well.

3. **If the newborn is breathing well**, you should next check the pulse rate by feeling the brachial pulse or the pulsations in the umbilical cord. The pulse rate should be at least 100 beats/min. If it is not, begin artificial ventilation. This alone may increase the newborn infant's heart rate. Reassess respirations and heart rate at least every 30 seconds to make sure that the pulse rate is increasing and that respirations are becoming spontaneous.

4. **Assess the newborn's skin color.** You are looking for the central cyanosis. If you find it, administer high-flow oxygen (10 to 15 L/min) through oxygen tubing held close to the newborn infant's face.

Pediatric Needs

Current information on neonatal resuscitation varies from what you may have learned in BLS prerequisite courses, which do not differentiate between an infant and a neonate (newborn). Be sure to know your local protocols on neonatal resuscitation.

5. **Remember, you now have two patients.** You should request a second unit as soon as possible if you determine that the newborn infant is having problems.

To assess a newborn's breathing, note whether or not the newborn is crying. Crying is proof that the infant is breathing. The newborn's breathing may be slightly irregular; this is normal. Gasping and grunting are usually signs of increased work of breathing and respiratory distress.

If the baby's breathing is not visible, he or she requires immediate intervention. Sometimes you can stimulate breathing simply by touching the newborn and suctioning. If the baby is still gasping after being dried and suctioned, further stimulation is not likely to improve ventilation. If stimulation is not effective or the baby continues to gasp, assisted ventilation will be required.

In situations in which assisted ventilation is required, you should use an infant BVM device (Figure 20-11 ▼). Cover the newborn infant's mouth and nose with the mask and begin ventilation with high-flow oxygen at a rate of 40 to 60 breaths/min. Make sure you have a good mask-to-face seal. By using gentle pressure, make the chest rise with each breath. Initially, it may be necessary to bypass the pop-off valve to accomplish this.

TABLE 20-3 **Rescue Measures for a Newborn Who Is Not Breathing**	
Assess and support	■ Temperature (warm and dry) ■ Airway (position and suction) ■ Breathing (stimulate to cry) ■ Circulation (heart rate and color)
Basic life support interventions	■ Dry and warm the infant. ■ Clear the airway with a bulb syringe. ■ Stimulate the infant if he or she is unresponsive. ■ Use a BVM device to ventilate the newborn if needed. This is very seldom required. ■ Perform chest compressions if there is no pulse.

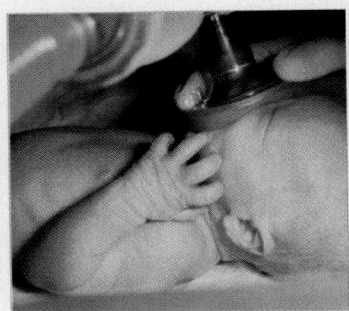

Figure 20-11
Use an infant bag and mask, ensuring that you cover the newborn infant's nose and mouth. Ventilate with high-flow oxygen at a rate of 40 to 60 breaths/min.

Assisted ventilation has been successful if you see both sides of the chest rise and hear breath sounds. After 30 seconds of adequate ventilations, assess the heart rate. If the heart rate is at least 100 beats/min and the newborn infant is breathing spontaneously, you can stop the assisted ventilation. Do not stop suddenly. Instead, gradually decrease the rate and pressure of the assisted ventilation to determine whether the newborn infant will continue to breathe adequately on its own. If not, continue assisted ventilation until it does. You may find that gently stimulating the newborn infant by rubbing it will help it to maintain its respirations.

If the heart rate is less than 60 beats/min and not increasing with ventilation, continue assisted ventilation and start cardiac compressions. Even though this newborn infant has a pulse, the rate and blood output from the heart are not adequate for the needs of a newborn.

There are two ways to give chest compressions to an infant. For the preferred method, follow the steps in (Skill Drill 20-2 ▶) :

1. **Find the proper position:** one fingerbreadth below an imaginary line drawn between the nipples on the middle third of the sternum (**Step 1**).
2. On a normal, full-term-sized infant, **place both hands around the infant so that your thumbs are side by side**, resting on the middle third of the sternum, and the rest of your fingers encircle the thorax. In premature or very small infants, you may have to place one thumb over the other to perform chest compressions (**Step 2**).
3. **Press the two thumbs gently against the sternum.** The newborn's chest is easy to compress. Use only enough force to compress the sternum ½″ to ¾″ (**Step 3**).

If your hands are too small to encircle the chest, you should use the middle and ring fingers of one hand to provide the compressions while your other hand supports the infant's back.

Ventilation with a BVM device is performed during the pause after every third compression. You should deliver a combined total of 120 ventilations and compressions per minute, 100 compressions to 20 ventilations. Keep in mind that ventilation is absolutely crucial to the successful resuscitation of the neonate.

If the infant does not begin breathing on its own or does not have an adequate heart rate, continue CPR on the way to the hospital. Once CPR has been started, do not stop until the infant responds with adequate respirations and heart rates or is pronounced dead by a physician. Do not give up! Many infants have survived without brain damage after prolonged periods of effective CPR. If the infant presents in distress, you should not wait to measure the Apgar score, but begin appropriate care measures immediately.

Abnormal or Complicated Delivery Emergencies

Breech Delivery

The presentation is the position in which an infant is born, the part of the body that comes out first. Most infants are born head first, in what is called a vertex presentation. Occasionally, the buttocks come out first. This is called a breech presentation (Figure 20-12 ▶). With a breech presentation, the infant is at great risk for delivery trauma. In addition, prolapsed cords are more

You are the Provider
Part 5

Upon examination, you find that the baby is crowning. You and your partner prepare for an imminent birth. Your partner notifies dispatch of the situation and requests ALS backup in case of complications. Your partner notifies medical control as well. You quickly assist your patient in moving onto the floor. Using your OB kit, you prepare a sterile delivery field. Your patient tells you that she needs to push. On the next contraction, the baby's head is delivered, facing downward.

9. When suctioning the baby, do you suction the nose or the mouth first?
10. What is your next step?

Giving Chest Compressions to an Infant

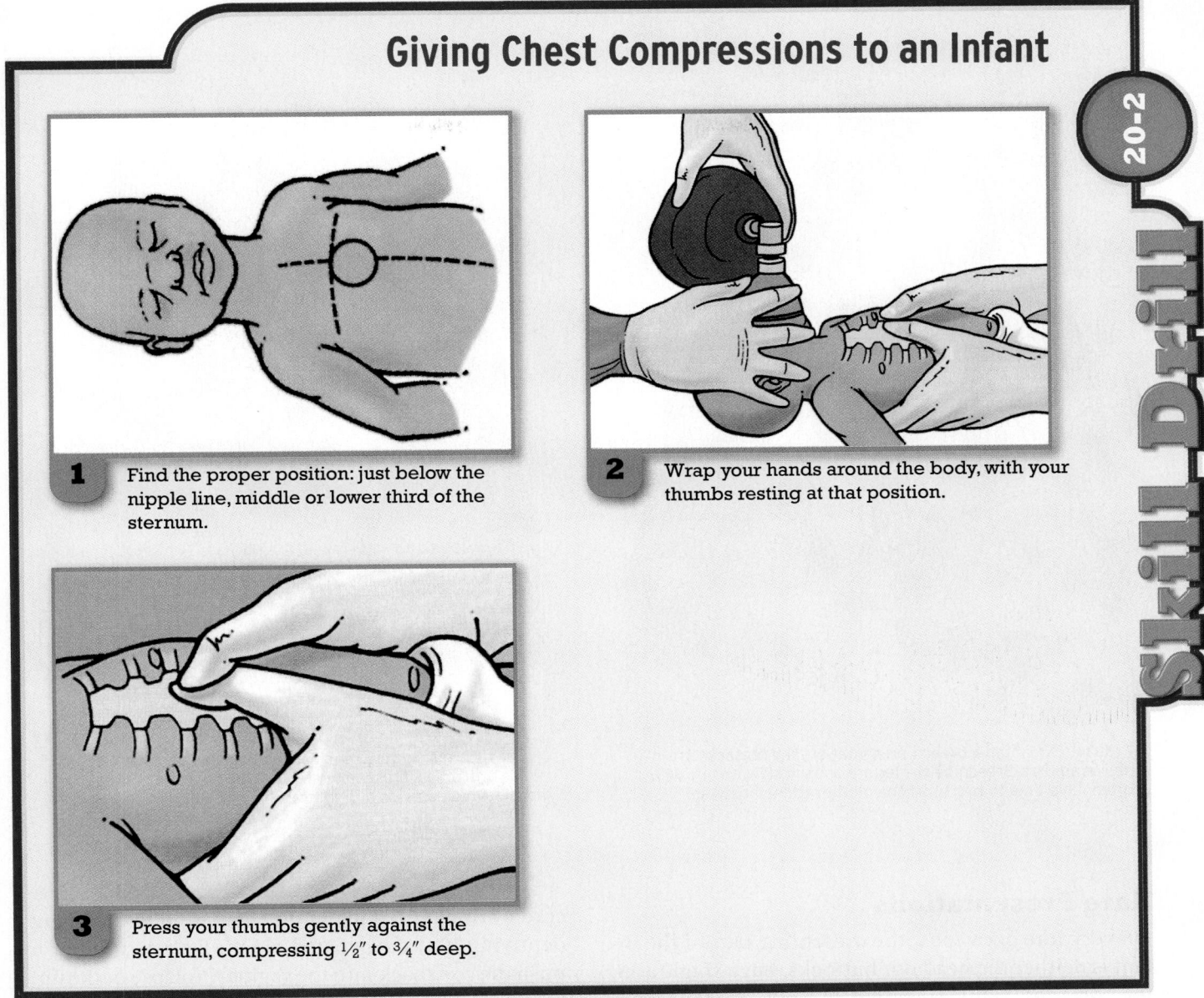

1 Find the proper position: just below the nipple line, middle or lower third of the sternum.

2 Wrap your hands around the body, with your thumbs resting at that position.

3 Press your thumbs gently against the sternum, compressing ½″ to ¾″ deep.

common. Breech deliveries are usually slow, so there is time to get the mother to the hospital. However, if the buttocks have already passed through the vagina, delivery is underway, and you should follow emergency procedures and call for ALS backup. In general, if the mother does not deliver within 10 minutes of the buttocks presentation, provide prompt transport. Have medical control guide you in this difficult situation.

The preparations for a breech delivery are the same as those for a vertex delivery. Position the mother, unwrap the emergency delivery kit, and place yourself and your partner as you would for a normal delivery. Allow the buttocks and legs to deliver spontaneously, supporting them with your hand to prevent rapid expulsion. The buttocks will usually come out easily. Let the legs dangle on either side of your arm while you support the trunk and chest as they are delivered. The head is almost always facedown and should be allowed to deliver spontaneously. As the head is delivering, you should keep the infant's airway open: Make a "V" with your gloved fingers, and then place them into the vagina to keep the walls of the vagina from compressing the airway. This is one of only two circumstances in which you should put your fingers into the vagina.

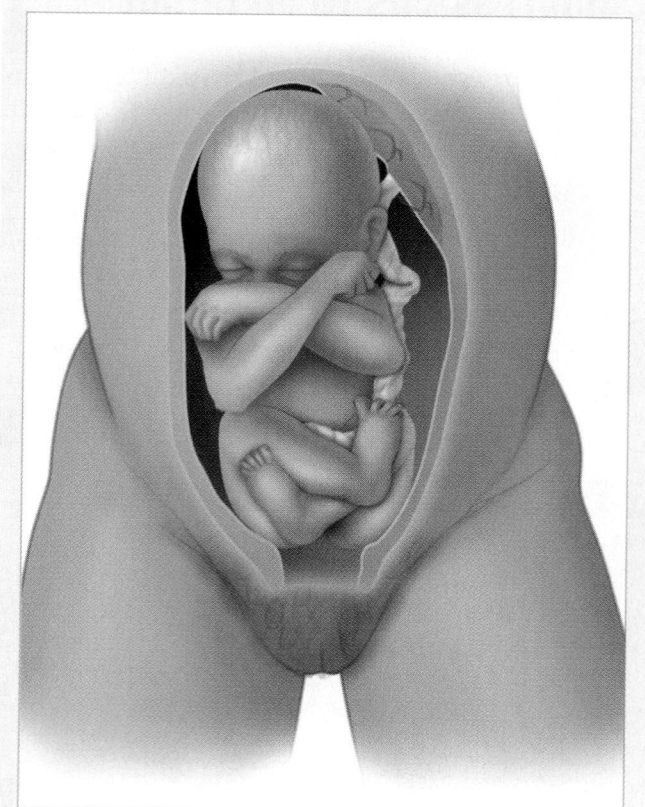

Figure 20-12 In a breech presentation, the buttocks are delivered first. Breech deliveries are usually slow, so you will often have time to transport the mother to the hospital.

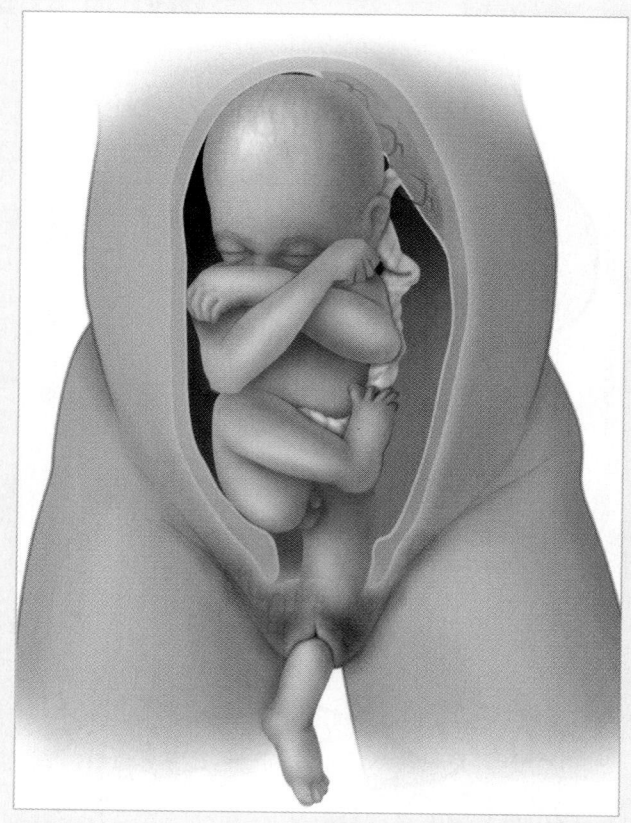

Figure 20-13 In very rare instances, an infant's limb, usually a single arm or leg, presents first. This is a very serious situation, and you must provide prompt transport for hospital delivery.

Rare Presentations

On very rare occasions, the presenting part of the infant is neither the head nor buttocks, but a single arm, leg, or foot. This is called a <u>limb presentation</u> (Figure 20-13 ▶). You cannot successfully deliver such a presentation in the field. These infants usually must be delivered surgically. If you are faced with a limb presentation, you must transport the mother to the hospital immediately. If a limb is protruding, cover it with a sterile towel. Never try to push it back in, and never pull on it. Place the mother on her back, with head down and pelvis elevated. Because both mother and infant are likely to be physically stressed in this situation, remember to give the mother high-flow oxygen.

<u>Prolapse of the umbilical cord</u>, a situation in which the umbilical cord comes out of the vagina before the infant (Figure 20-14 ▶), is another rare presentation that must be handled in the hospital. This situation is very dangerous, because the infant's head will compress the cord during birth and cut off circulation to the infant, depriving it of oxygenated blood. Do not attempt to push the cord back into the vagina. Prolapse of the umbilical cord usually occurs early in labor when the amniotic sac ruptures. As a result, there is time to get the mother to the hospital. Your job is to try to keep the infant's head from compressing the cord.

Place the mother on a backboard in the Trendelenburg position, with her hips elevated on a pillow or folded sheet. Alternatively, the mother may be placed in a knee-chest position: kneeling and bent forward, facedown. Either of these positions is meant to help keep the weight of the infant off the prolapsed cord. Carefully insert your sterile gloved hand into the vagina, and gently push the infant's head away from the umbilical cord. *Note that this is the only other occasion on which you should actually place a hand into the vagina.* Wrap a sterile towel, moistened with saline, around the exposed cord. Give the mother high-flow oxygen, and transport her rapidly.

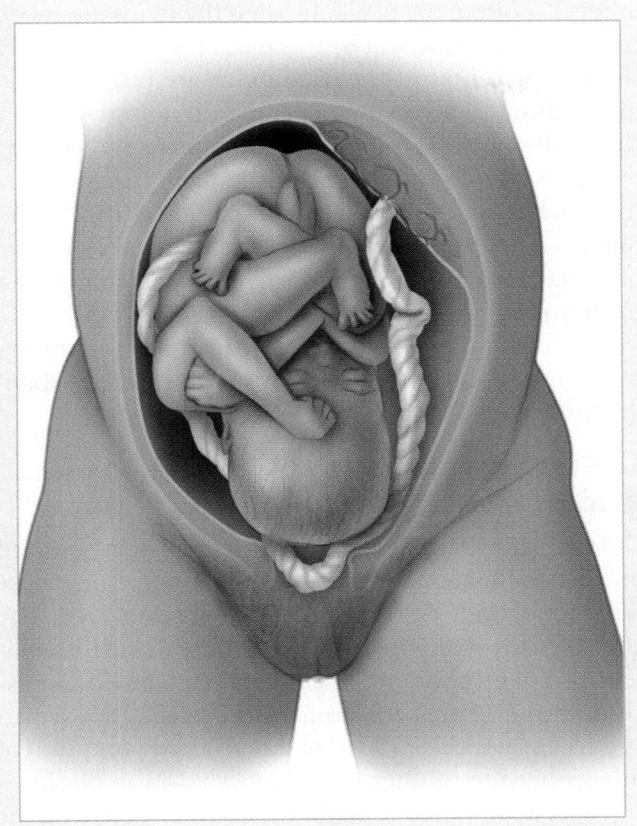

Figure 20-14 A prolapsed umbilical cord, another rare situation, is very dangerous and must be cared for at the hospital.

Excessive Bleeding

Some bleeding always occurs with delivery. However, bleeding that exceeds approximately 500 mL is considered excessive. Although up to 500 mL of blood loss is tolerated, you should continue to massage the uterus after delivery. Be sure to check the massage technique if bleeding continues. If the mother appears to be in shock, treat her accordingly and transport, massaging the uterus en route. There are several other possible causes of excessive bleeding, all of which may be serious and require emergency care. Treat this condition by covering the vagina with a sterile pad, changing the pad as often as necessary. Do not discard these blood-soaked pads; hospital personnel will use them to estimate the amount of blood that the mother has lost. Also save any tissue that may have passed from the vagina.

Place the mother in shock position, administer oxygen, monitor vital signs frequently, and transport her immediately to the hospital. Never hold the mother's legs together in an effort to stop the bleeding, and never pack the vagina with gauze pads in an attempt to control bleeding.

Spina Bifida

Spina bifida is a developmental defect in which a portion of the spinal cord or meninges may protrude outside of the vertebrae and possibly even outside of the body. This is very easily seen on the newborn's back and usually occurs in the lower third of the back in the lumbar area. It is extremely important to cover the open area of the spinal cord with a sterile, moist compress immediately after birth. This prevents infection, which can be fatal for such a newborn. This is something that you can do as an EMT-B and that greatly impacts the newborn's chance of a positive outcome. However, maintenance of body temperature is very important when applying a sterile, moist compress because the moisture can lower the newborn's body temperature. To prevent this, have someone hold the newborn against his or her body. This provides heat to the newborn at the appropriate temperature.

Abortion (Miscarriage)

Delivery of the fetus and placenta before 20 weeks is called abortion, or miscarriage. Abortions may be spontaneous, without any obvious known cause, or deliberate. Deliberate abortions may be self-induced, by the

You are the Provider   Part 6

You successfully deliver a beautiful baby girl. You have suctioned her mouth and then her nose, dried the baby off, and wrapped her in a blanket. The umbilical cord has been cut and the placenta delivered. ALS personnel arrive.

11. What should you do next?

mother herself or by someone else, or planned and performed in a hospital or clinic. Regardless of the reasons for the abortion, it may cause complications that you may be called on to treat.

The most serious complications of abortion are bleeding and infection. Bleeding can result from portions of the fetus or placenta being left in the uterus (incomplete abortion) or from injury to the wall of the uterus (perforation of the uterus and possibly the adjacent bowel or bladder). Infection can result from such perforation and from the use of nonsterile instruments. If the mother is in shock, treat and transport her promptly to the hospital. Collect and bring to the hospital any tissue that passes through the vagina. Never try to pull tissue out of the vagina; instead, cover it with a sterile pad.

Again, as you encounter a patient who is in shock as a result of complications of abortion, be sure to size up the situation; perform your initial assessment, focused history, and physical exam; and assess baseline vital signs.

In rare cases of abortion, massive bleeding may occur and cause severe hypovolemic or hemorrhagic shock. In these cases, provide immediate transport to the emergency department.

Twins

Twins occur about once in every 80 births. Sometimes, there is a family history of twins. The mother may suspect that she is having twins because she has an unusually large abdomen. Usually, however, twins are diagnosed early in pregnancy with modern ultrasound techniques. With twins, always be prepared for more than one resuscitation, and call for assistance.

Twins are smaller than single infants, and delivery is typically not difficult. Consider the possibility that you are dealing with twins any time the first infant is small or the mother's abdomen remains fairly large after the birth. If twins are present, the second one will usually be born within 45 minutes of the first. About 10 minutes after the first birth, contractions will begin again, and the birth process will repeat itself.

The procedure for delivering twins is the same as that for single infants. Clamp and cut the cord of the first infant as soon as it has been born and before the second infant is delivered. The second infant may deliver before or after the first placenta. There may be only one placenta, or there may be two. When the placenta has been delivered, check whether there is one umbilical cord or two. If two cords are coming out of one placenta, the twins are called identical. If only one cord

is coming out of the placenta, then the twins are called fraternal, and there will be two placentas. Occasionally, the two placentas of fraternal twins are fused, so you might think that you are dealing with identical twins. Remember, if you see only one umbilical cord coming out of the first placenta, there is still another placenta to be delivered. However, if both cords are attached to one placenta, the delivery is over. Identical twins are of the same sex; fraternal twins may be of different sexes, or they may be the same.

Record the time of birth of each twin separately. Twins may be so small that they look premature; handle them very carefully, and keep them warm.

Delivering an Infant of an Addicted Mother

Unfortunately, more and more infants are being born to mothers who are addicted to drugs or alcohol. These mothers often have had little or no prenatal care. The effects of the addiction on the infants include prematurity, low birth weight, and severe respiratory depression. Some of these infants will die. Fetal alcohol syndrome is the term used to describe the condition of infants born to mothers who have abused alcohol.

If you are called to handle a delivery of a drug- or alcohol-addicted mother, pay special attention to your own safety. As with all other cases, follow BSI precautions. Wear goggles and sterile gloves at all times. Clues that you are dealing with an addicted mother may include the presence of drug paraphernalia, empty wine or liquor bottles, and statements made by neighbors or by the mother herself. The newborn infant of an addicted mother will probably need immediate hospital care. Carry out the delivery as outlined earlier, but be prepared to support the infant's respirations and administer oxygen during transport. Do not judge or lecture the mother. Your job is to help deliver the infant as best you can and to transport both infant and mother to the hospital.

Premature Infant

The usual gestational period, the period of prenatal development, is 9 calendar months, or 40 weeks. A normal, single infant will weigh approximately 7 lb at birth. Any infant that delivers before 8 months (36 weeks of gestation) or weighs less than 5 lb at birth is considered premature. This determination is not always easy to make. Often, the exact gestation time cannot be determined. Because you probably have no scale to weigh the infant, you will have to use physical guidelines. A premature infant is smaller and thinner than a full-term

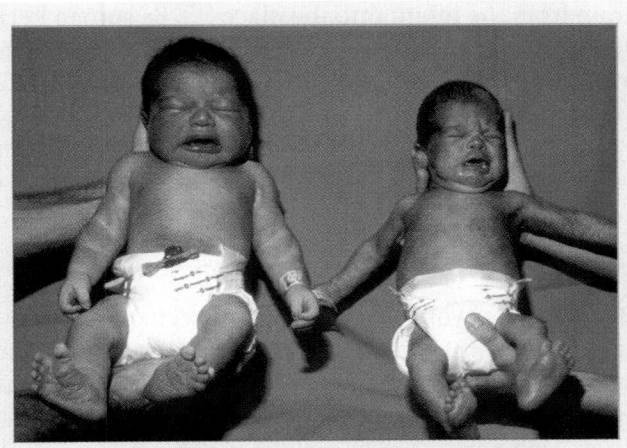

Figure 20-15 Premature infants (right) are smaller and thinner than full-term infants.

infant, and its head is proportionately larger in comparison with the rest of its body (Figure 20-15 ▲). The vernix caseosa, a cheesy white coating on the skin that is found on the full-term infant, will be missing on the premature infant or will be very minimal. There will also be less body hair.

Premature infants need special care to survive. Often they require resuscitation, which should be done unless it is physically impossible. With such care, infants as small as 1 lb have survived and developed normally. Follow these procedures when you are handling a premature infant:

1. **Keep the infant warm.** Dry the infant as soon as it is born, and then remove the wet towels. Wrap it in a warm blanket, exposing the face but covering the head. Keep the infant in a place where the temperature is between 90°F and 95°F (between 32.2°C and 35.0°C).

2. **Keep the mouth and nose clear of mucus.** Like all newborn infants, premature ones are nose breathers, and the small nasal passages can easily be obstructed. Use the bulb syringe to suction the mouth and nostrils frequently. Handle the infant very gently.

3. **Carefully observe** the cut end of the cord attached to the infant, and be sure that it is not bleeding. The loss of even a few drops of blood can be very serious.

4. **Give oxygen.** Open the valve on your oxygen cylinder slowly to give a steady stream of oxygen (about 70 to 100 bubbles per minute through the water bottle that is attached to the oxygen tank).

Direct the stream of oxygen not into the infant's mouth, but into a small tent over the infant's head; you can use a blanket or a piece of aluminum foil to make the tent. Although there is some danger to a premature infant from receiving very high concentrations of oxygen, there is no danger if it is given over a short period in this manner.

5. **Do not infect the infant.** Premature infants are very susceptible to infection. Protect them from contamination. Do not breathe directly into the infant's face. Your mask will help to create a barrier. Keep everyone else as far away from the infant as possible.

6. **Notify the hospital.** Does your system have a neonatal (newborn) transport team with specialized personnel and equipment for the care of premature and sick newborn infants? If so, be sure to contact the hospital before leaving the scene so that medical control can decide whether to call in the team. If not, you should still notify the hospital as soon as possible so that staff can be ready to receive the premature infant and mother. Avoid unnecessary on-scene delays.

You may have access to a specialized premature infant carrier, which can be used for immediate care and for transport. Carrier supplies may include a quilted pad, infant blanket, diaper, thermometer, suction tube and suction bulb, sterile Kelly clamp, and, most important, hot water bottles and an oxygen cylinder with the necessary attachments. Fill the hot water bottles, and pad them well so that they do not come into direct contact with the infant's skin. Place one on the bottom of the carrier and one on each side of the space for the infant. Once you have wrapped the infant in a blanket and placed it inside the carrier, secure the carrier inside the vehicle.

Keep the temperature of the vehicle at 90°F to 95°F (32.2°C to 35.0°C) while the infant and mother are being transported to the hospital. If a special carrier is not available, you must keep the premature infant warm with additional blankets, thermal packets, and warmed patient compartments. Any delays will lower the infant's body temperature.

Fetal Demise

Unfortunately, you may find yourself delivering an infant who died in the mother's uterus before labor. This will be a true test of your medical, emotional, and social abilities. Grieving parents will be emotionally distraught and perhaps even hostile, requiring all your professionalism and support skills.

The onset of labor may be premature, but labor will otherwise progress normally in most cases. If an intrauterine infection has caused the demise, you may note an extremely foul odor. The delivered infant may have skin blisters, skin sloughing, and a dark discoloration, depending on the stage of decomposition. The head will be soft and perhaps grossly deformed.

Do not attempt to resuscitate an obviously dead infant. However, do not confuse such an infant with those who have had a cardiopulmonary arrest as a complication of the birthing process. You must attempt to resuscitate normal-appearing infants.

Delivery Without Sterile Supplies

On rare occasions, you may have to deliver an infant without a sterile OB kit. Even if you do not have a sterile kit, you should always have goggles and sterile gloves with you. These are for your own protection and for that of the mother and infant. Carry out the delivery as if sterile supplies were on hand. If you can, use clean sheets and towels that have not been used since they were laundered. As soon as the infant is born, wipe the inside of its mouth with your finger to clear away blood and mucus. Without the OB kit, you should not cut or tie the umbilical cord. Instead, as soon as the placenta delivers, wrap it in a clean towel or put it in a plastic bag and transport it with the infant and mother to the hospital. Always keep the placenta and the infant at the same level, or elevate the placenta slightly if possible, so that blood does not

drain from the infant into the placenta. Be sure to keep the infant warm. As in the case of other deliveries, note the presence of green-tinged fluid or secretions (meconium staining).

Gynecologic Emergencies

Occasionally, women who are not pregnant will have major gynecologic problems requiring urgent medical care. These include excessive bleeding and soft-tissue injuries to the external genitalia. These genital parts have a rich nerve supply, making injuries very painful.

Treat lacerations, abrasions, and tears with moist, sterile compresses, using local pressure to control bleeding and a diaper-type bandage to hold the dressings in place. Leave any foreign bodies in place after stabilizing them with bandages. Under no circumstances should you ever pack or place dressings in the vagina. Continue to assess these patients while transporting them to the emergency department. Contusions and other blunt trauma will require careful in-hospital evaluation.

Although you might not know the exact cause of a gynecologic emergency, you should treat these individuals as you would any other victim of blood loss: Observe BSI precautions, ensure maintenance of the airway, give oxygen, take and document vital signs, and treat for shock while arranging for prompt transport.

You are the Provider   Summary

When faced with an imminent birth, attempt to make the delivery area as sterile as possible. Maintain good communication with your patient. Always be prepared for the possible need for resuscitation, and provide treatment for the mother and the new baby.

Obstetric and Gynecologic Emergencies	
Scene Size-up	Body substance isolation precautions should include a minimum of gloves, eye protection, mask, and gown. Ensure scene safety and determine NOI/MOI. Consider the number of patients, the need for additional help/ALS, and c-spine stabilization.
Initial Assessment	
■ General impression	Determine level of consciousness and find and treat any immediate threats to life. Determine priority of care based on environment and patient's chief complaint.
■ Airway	Ensure patent airway.
■ Breathing	Provide high-flow oxygen at 15 L/min. Evaluate depth and rate of respirations and provide ventilations as needed.
■ Circulation	Evaluate pulse rate and quality; observe skin color, temperature, and condition and treat accordingly. Treat for any signs of shock.
■ Transport decision	If delivery of a baby is not imminent, provide rapid transport.
Focused History and Physical Exam	NOTE: The order of the steps in the focused history and physical exam differs depending on whether the patient is conscious or unconscious. The order below is for a conscious patient. For an unconscious patient, perform a rapid physical exam, obtain vital signs, and obtain the history.
■ SAMPLE history	Obtain SAMPLE history. If patient is pregnant, determine the due date, frequency of contractions, and history of previous pregnancies and deliveries and complications, if any. Determine whether there is a possibility of twins and whether the patient has taken any drugs or medications. If patient's water has broken, ask whether the fluid was green.
■ Focused physical exam	Your focused physical exam should include the abdomen and assessing the length and frequency of contractions. When appropriate and according to local protocol, inspect the vaginal opening for rupture of the amniotic sac, bloody show, and crowning if you suspect delivery is imminent.
■ Baseline vital signs	Take vital signs, noting skin color and temperature. Use pulse oximetry if available. Watch for increases in blood pressure.
■ Interventions	Support patient as needed. Consider the use of oxygen, positive pressure ventilations, adjuncts, and proper positioning of the patient.
Detailed Physical Exam	If delivery is imminent or other assessments or treatments require your attention, defer the detailed physical exam.
Ongoing Assessment	Repeat the initial assessment and focused assessment, and reassess interventions performed. Reassess vital signs every 5 minutes for the unstable patient and every 15 minutes for the stable patient. Note and treat the patient as necessary.
■ Communication and documentation	Contact medical control with a radio report that gives information on patient's and newborn's condition. Relay any significant changes, including level of consciousness or difficulty in breathing. Record any changes, at what time, and any interventions performed.

Assessment and Emergency Care

NOTE: While the steps below are widely accepted, be sure to consult and follow your local protocol.

Delivering the Baby

1. Support the bony parts of the head with your hands as it emerges. Suction fluid from the baby's mouth, then nostrils.

2. As the upper shoulder appears, guide the head down slightly, if needed, to deliver the shoulder.

3. Support the head and upper body as the lower shoulder delivers, guiding the head up if needed.

4. Handle the slippery, delivered infant firmly but gently, keeping the neck in neutral position to maintain the airway.

5. Place the umbilical cord clamps 2″ to 4″ apart, and cut between them.

6. Allow the placenta to deliver itself. Do not pull on the cord to speed delivery.

Giving Chest Compressions to an Infant

1. Find the proper position—just below the nipple line, middle or lower third of the sternum.

2. Wrap your hands around the body, with your thumbs resting at that position.

3. Press the two thumbs against the sternum, compressing ½″ to ¾″ deep.

Premature Infant

1. Keep the infant warm. Keep the infant in a place where the temperature is between 90°F and 95°F (between 32.2°C and 35°C).

2. Keep the mouth and nose clear of mucus with a bulb syringe.

3. Inspect the cut end of the cord attached to the infant for bleeding.

4. Give oxygen into a small tent over the infant's head.

5. Do not infect the infant. Wear a mask to help prevent you from breathing on the infant.

6. Notify the hospital of a neonatal transport.

Prep Kit

Ready for Review

- Inside the uterus, the developing fetus floats in the amniotic sac. The umbilical cord connects mother and infant through the placenta. Eventually, the uterus will propel the infant through the birth canal.

- The first stage of labor, dilation, begins with the onset of contractions and ends when the cervix is fully dilated. The second stage, expulsion of the baby, begins when the cervix is fully dilated and ends when the infant is born. The third stage, delivery of the placenta, begins with the birth of the infant and ends with the delivery of the placenta.

- Once labor has begun, it cannot be slowed or stopped; however, there is usually time to transport the patient to the hospital during the first stage. During the second stage, you must decide whether to deliver the baby at the scene or transport the mother. During the third stage, once the infant has been born, you will probably not transport until the placenta has delivered.

- Use an infant BVM device to assist ventilation, starting with high-flow oxygen at a rate of 40 to 60 breaths/min. If the infant starts to breathe on its own, attach an oxygen tubing mask and watch for signs of adequate oxygenation. If the heart rate is less than 80 beats/min, start cardiac compressions, using only enough force to compress the sternum $1/2''$ to $3/4''$. Perform a combination of 20 ventilations and 100 compressions per minute.

- Abnormal or complicated deliveries include breech deliveries (buttocks first), limb presentations (arm, leg, or foot first), and prolapse of the umbilical cord (umbilical cord first). Quickly transport the patient with a limb presentation or prolapsed umbilical cord to the hospital.

- The only times you should place a finger or hand into the vagina are to keep the walls of the vagina from compressing the infant's airway during a breech presentation or to push the infant's head away from the cord in a prolapse situation.

- Excessive bleeding is a serious emergency. Cover the vagina with a sterile pad; change the pad as often as necessary, and take all used pads to the hospital for examination.

- Be prepared to support respirations during transport in an infant delivered by a drug- or alcohol-addicted mother. Also use oxygen with premature infants, and keep the temperature of the ambulance at 90°F (32.2°C) or more during transport.

Vital Vocabulary

abortion The delivery of the fetus and placenta before 20 weeks; miscarriage.

amniotic sac The fluid-filled, baglike membrane in which the fetus develops.

Apgar score A scoring system for assessing the status of a newborn that assigns a number value to each of five areas of assessment.

birth canal The vagina and cervix.

bloody show A small amount of blood at the vagina that appears at the beginning of labor and may include a plug of pink-tinged mucus that is discharged when the cervix begins to dilate.

breech presentation A delivery in which the buttocks come out first.

cervix The lower third, or neck, of the uterus.

crowning The appearance of the infant's head at the vaginal opening during labor.

eclampsia Seizures (convulsions) resulting from severe hypertension in a pregnant woman.

ectopic pregnancy A pregnancy that develops outside the uterus, typically in a fallopian tube.

Technology

- Interactivities
- Vocabulary Explorer
- Anatomy Review
- Web Links
- Online Review Manual

www.EMTB.com

Prep Kit continued...

fetal alcohol syndrome A condition of infants who are born to alcoholic mothers; characterized by physical and mental retardation and a variety of congenital abnormalities.

fetus The developing, unborn infant inside the uterus.

gestational diabetes Diabetes that develops during pregnancy in women who did not have diabetes before pregnancy.

limb presentation A delivery in which the presenting part is a single arm, leg, or foot.

meconium A dark green material in the amniotic fluid that can indicate disease in the newborn; the meconium can be aspirated into the infant's lungs during delivery; the baby's first bowel movement.

miscarriage The delivery of the fetus and placenta before 20 weeks; spontaneous abortion.

multigravida A woman who has had previous pregnancies.

multipara A woman who has had more than one live birth.

nuchal cord An umbilical cord that is wrapped around the infant's neck.

pelvic inflammatory disease An infection of the fallopian tubes and the surounding tissues of the pelvis.

perineum The area of skin between the vagina and the anus.

placenta The tissue attached to the uterine wall that nourishes the fetus through the umbilical cord.

placenta abruptio A premature separation of the placenta from the wall of the uterus.

placenta previa A condition in which the placenta develops over and covers the cervix.

preeclampsia A condition of late pregnancy that involves headache, visual changes, and swelling of the hands and feet; also called pregnancy-induced hypertension.

pregnancy-induced hypertension A condition of late pregnancy that involves headache, visual changes, and swelling of the hands and feet; also called preeclampsia.

presentation The position in which an infant is born; the part of the infant that appears first.

primigravida A woman who is experiencing her first pregnancy.

primipara A woman who has had one live birth.

prolapse of the umbilical cord A situation in which the umbilical cord comes out of the vagina before the infant.

spina bifida A developmental defect in which a portion of the spinal cord or meninges may protrude outside of the vertebrae and possibly even outside of the body, usually at the lower third of the spine in the lumbar area.

supine hypotensive syndrome Low blood pressure resulting from compression of the inferior vena cava by the weight of the pregnant uterus when the mother is supine.

umbilical cord The conduit connecting mother to infant via the placenta; contains two arteries and one vein.

uterus The muscular organ where the fetus grows, also called the womb; responsible for contractions during labor.

vagina The outermost cavity of a woman's reproductive system; the lower part of the birth canal.

Points to Ponder

En route to a motor vehicle crash, dispatch reports that the patient is pregnant.

Upon arrival you assess the scene and see two vehicles with moderate damage. You find your patient sitting in the driver's seat of a car that has been struck on the driver's side. Your general impression of the 28-year-old pregnant woman is that she is alert and was wearing her seat belt. The initial assessment shows an open airway, respirations of 22 breaths/min and regular, and a pulse of 110 beats/minute and regular. The patient is complaining of pain in her left pelvic area and her left side. After completing the SAMPLE history and vital signs you ask for consent to transport. The patient is hesitant and does not want to go by ambulance. She states that her husband is on the way and she wants to wait and go by car.

Issues: Transport Protocol for Obstetric Patients, the Need to Protect Both the Patient and Unborn Child.

www.EMTB.com

Assessment in Action

You are called to respond to 5000 East Colorado Ave for a woman in labor. Your partner locates the call in the map book and estimates a 9-minute response time. While en route you request dispatch to obtain any additional information. Dispatch reports back and tells you the patient is 24 years old.

Upon arrival your partner ensures the rescue is safely parked and you scan the scene for any safety issues. After assessing that the scene is safe, you approach the house. You are met at the door by the patient's sister and are escorted to the patient. From across the room you see a female patient sitting on the couch. Your general impression of the patient confirms she is pregnant and appears not to be in any distress.

You complete your initial assessment and find the airway open, normal breathing, and a slightly increased pulse rate. You move on to the focused medical assessment and detailed physical exam. The patient tells you she is 38 weeks' pregnant and it is her first pregnancy. She tells you her contractions are two to three minutes apart. Her water broke just before calling 9-1-1.

1. This patient is in the:
   A. first stage of labor.
   B. second stage of labor.
   C. third stage of labor.
   D. fourth stage of labor.

2. The term used to describe a patient's first pregnancy is:
   A. first trimester.
   B. multigravida.
   C. primigravida.
   D. primapara.

3. The term used to describe the outcome of a pregnancy is:
   A. gravada.
   B. multipara and primipara.
   C. placement.
   D. pg history.

4. During the physical exam of this patient, you examine the patient for:
   A. crowning.
   B. swollen ankles.
   C. hypoglycemia.
   D. trauma.

5. While completing the SAMPLE history, you should determine whether the patient has had any complications during her pregnancy such as:
   A. blurred vision.
   B. chest pain.
   C. extreme thrust.
   D. pre-eclampsia.

6. The patient tells you she had a problem with low blood pressure when she was lying down. This is called:
   A. hypertension sensitivity.
   B. nighttime pressure variance.
   C. orthopenia.
   D. supine hypotension syndrome.

7. A pregnant patient with a history of blood pressure problems should be transported:
   A. in Fowler's position.
   B. face down.
   C. on the left side.
   D. supine.

Challenging Questions

8. What are the five parts of the APGAR score for evaluating a newborn infant?

9. Describe the signs and symptoms and the difference between placenta abruptio and placenta previa.

10. Explain the cause for ectopic pregnancy.

www.EMTB.com

Kinematics of Trauma

Objectives*

Cognitive

1. Describe the "three collisions" associated with motor vehicle crashes. (p 634)
2. Relate how the fundamental principles of physics apply to motor vehicle crashes and other types of accidents. (p 632)
3. State Newton's three laws. (p 634, 635)

Affective

None

Psychomotor

4. Observe various high-energy injuries and identify potential damage to the patient. (p 638)

*All of the objectives in this chapter are noncurriculum objectives.

You are the Provider

You and your EMT partner are dispatched to the scene of a reported rollover-type motor vehicle crash. As you approach the scene, you see skid marks on the roadway continuing off into the dirt. After ensuring the safety of the scene, you approach your patient. You find a 33-year-old man lying supine in the dirt. Bystanders state that they assisted the patient from his vehicle after witnessing the vehicle roll over multiple times. Approximately 30' from the patient you see what appears to be a vehicle on its top in a ravine.

1. How does the basic understanding of kinetic energy contribute to the care of trauma patients?
2. What does your index of suspicion tell you about the potential injuries to your patient?

Kinematics of Trauma

Injuries are the leading cause of death and disability in the United States among children and young adults (ages 1 to 34 years), claiming 140,000 lives annually—more than all other diseases combined. Each year, one person in three sustains an injury that requires medical treatment. Proper prehospital evaluation and care can do much to minimize suffering, long-term disability, and death from trauma.

This chapter introduces the basic physical concepts that dictate how injuries occur and affect the human body. When you understand these concepts, you will be better able to size up a crash scene and use that information as a vital part of patient assessment. This section begins with a basic discussion of energy and trauma. Next, different types of crashes and their impact on the body are explained. By assessing the body of a vehicle that has crashed, you can often determine what happened to the passengers at the time of impact, which may allow you to predict what injuries the passengers sustained at the time of impact. Evaluation of the mechanism of injury for the trauma patient will provide you with an index of suspicion for serious underlying injuries. The index of suspicion is your concern for potentially serious underlying and unseen injuries. Certain injury patterns occur with certain types of injury events. Answers to simple questions will provide you with information on how to identify life-threatening and other serious injuries.

Technology

- Interactivities
- Vocabulary Explorer
- Anatomy Review
- Web Links
- Online Review Manual

www.EMTB.com

Energy and Trauma

Traumatic injury occurs when the body's tissues are exposed to energy levels beyond their tolerance (Figure 21-1 ▶). The mechanism of injury (MOI) is the way in which traumatic injuries occur; it describes the forces (or energy transmission) acting on the body that cause injury. Three concepts of energy are typically associated with injury (not including thermal energy, which causes burns): potential energy, kinetic energy, and work. In considering the effects of energy on the human body, it is important to remember that energy can be neither created nor destroyed, but can only be converted or transformed. It is not the objective of this section to help you to reconstruct the scene of a motor vehicle crash. Rather, you should have a sense of the effects of work on the body and understand, in a broad sense, how that work is related to potential and kinetic energy. For example, when you are assessing a patient who fell, you need not calculate the speed at which the person hit the ground. However, it is important to estimate the height from which he or she fell and to appreciate the injury potential of the fall.

Work is defined as force acting over a distance. For example, the force needed to bend metal multiplied by the distance over which the metal is bent is the work that crushes the front end of an automobile that is involved in a frontal impact. Similarly, forces that bend, pull, or compress tissues beyond their inherent limits result in the work that causes injury.

The energy of a moving object is called kinetic energy and is calculated as follows: kinetic energy = $\frac{1}{2}mv^2$, where m = mass (weight) and v = velocity (speed). Remember that energy cannot be created or destroyed, only converted. In the case of a motor vehicle crash, the kinetic energy of the speeding car is converted into the work of stopping the car, usually by crushing the car's exterior (Figure 21-2 ▶). Similarly, the passengers of the car have kinetic energy because they were traveling at the same speed as the car. Their kinetic energy is converted to the work of bringing them to a stop. It is this work on the passengers that results in injury. Notice that, according to the equation for kinetic energy, the energy that is available to cause injury doubles when an object's weight doubles but quadruples when its speed doubles. Consider the debate over raising the speed limit. Increasing a car's speed from 50 to 70 mph quadruples the energy that is available to cause injury. This point will be even clearer in considering gunshot

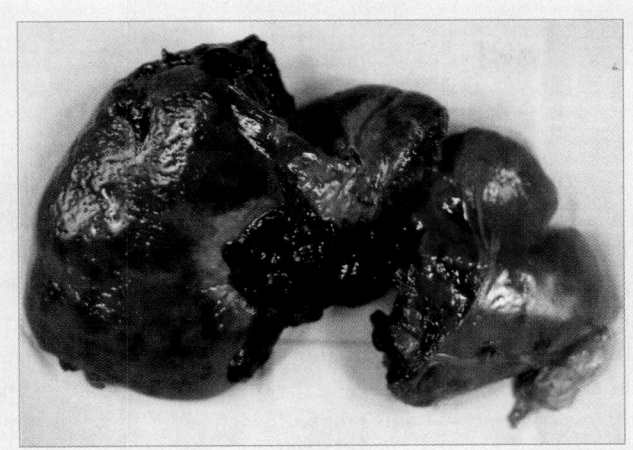

Figure 21-1 Traumatic injury occurs when the body's tissues are exposed to energy levels beyond their tolerance. This photo shows a ruptured spleen.

Figure 21-2 The kinetic energy of a speeding car is converted into the work of stopping the car, usually by crushing the car's exterior.

wounds. The speed of the bullet (high-velocity compared with low-velocity) has a greater impact on producing injury than the mass (size) of the bullet. This is why it is so important to report to the hospital the type of firearm that was used in a shooting. The amount of kinetic energy that is converted to do work on the body dictates the severity of the injury. High-energy injuries often produce such severe damage that patients can be saved only by immediate transport to an appropriate facility.

Potential energy is the product of mass (weight), force of gravity, and height and is mostly associated

EMT-B Tips

Do a "vehicle assessment" if circumstances at the scene allow it. There may be time for one EMT-B to circle the vehicle and assess damage while the other EMT-B begins patient assessment.

with the energy of falling objects. A worker on a scaffold has some potential energy because he or she is some height above the ground. If the worker falls, potential energy is converted into kinetic energy. As the worker hits the ground, the kinetic energy is converted into work, that is, the work of bringing the body to a stop and thereby breaking bones and damaging tissues.

Blunt and Penetrating Trauma

Traumatic injuries can be described in two separate categories: blunt trauma and penetrating trauma. Either type of trauma may occur from a variety of MOIs. It is important for you to consider unseen as well as visible, obvious injuries with either type of trauma. Blunt trauma is the result of force (or energy transmission) to the body that causes injury primarily without penetrating the soft tissues or internal organs and cavities. Penetrating trauma causes injury by objects that primarily pierce and penetrate the surface of the body and cause damage to soft tissues, internal organs, and body cavities.

MOI Profiles

Different types of MOIs will produce many types of injuries. Some will involve an isolated body system; many will result in injury to more than one body system. Whether one body system or more than one system is involved, you should maintain a high index of suspicion for serious unseen injuries. Injuries to trauma patients may be the result of falls, motor vehicle collisions, car versus pedestrian (or bicycle), gunshot wounds, and stabbings. These are a few of the common types of MOI patterns to which you will respond to provide care and treatment to patients.

EMT-B Tips

Newton's Laws

Newton's First Law

Newton's first law states that objects at rest tend to stay at rest and objects in motion tend to stay in motion unless acted on by some force. The first part of the law is fairly clear. An object such as an empty soda can will not move spontaneously unless some force, such as a gust of wind, acts on it. An example will help to illustrate the second part. In a car going 30 mph, the passengers and the car are moving at 30 mph. The passengers do not feel as though they are moving because they are not moving relative to the car. However, when the car strikes a concrete barrier and comes to a sudden stop, the passengers continue to travel at 30 mph. They stay in motion until they are acted on by an external force—most likely the windshield, steering wheel, or dashboard. To appreciate the severity of the impact, think of the driver as sitting motionless while a steering wheel rams into his or her chest at 30 mph. Now consider that the same thing happens to the driver's internal organs. They also are in motion, traveling at 30 mph relative to the ground, until they are acted on by an external force, in this case the sternum, rib cage, or other body structure. This scenario illustrates the three collisions that are associated with blunt trauma.

Newton's Second Law

Newton's second law states that force (F) equals mass (M) times acceleration (A), that is, $F = MA$, in which acceleration is the change in velocity (speed) that occurs over time. Therefore, it is not so much that "speed kills" but that the change in velocity with respect to time generates the forces that cause injury. Simply put, it is not the fall, but the sudden stop at the bottom, that hurts.

In the example of the car traveling at 30 mph, it takes about 3 seconds for the car to decrease its speed from 30 mph to 0 mph when the driver applies the brakes smoothly. If he or she is properly restrained by well-adjusted seat belts, the driver slows, or decelerates, at the same rate as the car. But if the car is stopped not by braking but by hitting a large tree and the driver is not restrained, his or her body will continue to stay in motion at 30 mph until it is stopped by an external force, in this case, the steering wheel. Although the change in the body's velocity is the same as when the car was braking smoothly in 3 seconds (30 to 0 mph), that change now takes place in about 0.01 second. Because the period of deceleration is 300 times less, the average force of impact is 300 times greater. This means that the force is approximately 150 times the force of gravity. Imagine a force 150 times your body weight slamming into your chest.

Now consider the same car striking the same tree, but this time, the driver is restrained with a shoulder and lap belt. The driver is essentially tied to the car and stops during the same period the car stops. It takes some time, although brief,

Blunt Trauma

Blunt trauma results from an object making contact with the body. Motor vehicle crashes and falls are two of the most common MOIs for blunt trauma. Any object, for example a baseball bat, can cause blunt trauma if it is moving fast enough. You should be alert to signs of skin discoloration or complaints of pain because these may be the only signs of blunt trauma. You also should maintain a high index of suspicion during patient assessment for hidden injuries in patients with blunt trauma.

Blunt Trauma: Vehicular Collisions

Motor vehicle crashes are classified traditionally as frontal (head-on), lateral (T-bone), rear-end, rotational (spins), and rollovers. The principal difference among these collision types is the direction of the force of impact; also, with spins and rollovers, there is the possibility of multiple impacts. Motor vehicle crashes typically consist of a series of three collisions. Understanding the events that occur during each collision will help you be alert for certain types of injury patterns. The three collisions in a frontal impact are as follows:

1. **The collision of the car against another car, a tree, or some other object.** Damage to the car

to crush the front of the car and bring it to a halt. The car comes to a stop in approximately 0.05 second. The change in the driver's velocity is the same (30 to 0 mph), but the longer period of deceleration results in a *g* force of only 30 times that of gravity. This is still a substantial force, but it is much less than the force that is experienced by the unrestrained driver. More to the point, it is survivable.

In a final example, the car and driver, as before, are traveling at 30 mph, and the driver is properly restrained with a three-point belt. In this case, however, the car is also equipped with an air bag. When the car hits the tree and suddenly stops, the driver's upper body initially continues forward at 30 mph. The body is partially slowed by the lap and shoulder belts but is finally brought to rest by the air bag. The upper body compresses the air bag, which stops the body's forward motion in about 0.1 second. Thus, the air bag stretches the duration of impact by 0.05 second, buying the body even more time, and the force on the upper body drops to approximately 15 times that of gravity.

The air bag has another advantage. The force of its impact is applied over a much larger area than the area that is affected by the steering wheel or the shoulder belt, shrinking the force per unit area. This point can be illustrated by an analogy. A person standing on one toe on a sheet of ice applies a concentrated load in a very small area, thus breaking the ice and falling through. If the person lies flat on the ice, he or she greatly expands the contact area and reduces the stress on the ice, which, depending on conditions, should not break. The dual action of the air bag (distributing the force of impact over a greater area and increasing the duration of impact) results in less severe injuries.

Newton's Third Law

Newton's third law states that for every action, there is an equal and opposite reaction. Therefore, if you push on a door, the door pushes back (reacts) with an equal force but in the opposite direction. In the case of a dented A-pillar, the force of the driver's head was sufficient to dent the strong metal. But in terms of patient assessment, the more important point is the reaction force of the pillar on the head. Newton's third law states that the two forces are equal but occur in opposite directions. In other words, the head was essentially hit by an A-pillar traveling at 30 mph. Similarly, it takes a substantial force to collapse a steering wheel. When you notice a collapsed steering wheel during scene size-up, you should suspect serious chest injuries even if the driver initially has no visible signs of chest injury. Often, reading the scene and understanding the basic principles of energy transfer will give you as clear a picture of the patient's potential injuries and injury severity as the actual physical patient assessment.

is perhaps the most dramatic part of the collision, but it does not directly affect patient care, except possibly to make extrication difficult (Figure 21-3 ▶). However, it does provide information about the severity of the collision and, therefore, has an indirect effect on patient care. The greater the damage to the car, the greater the energy that was involved and, therefore, the greater the potential to cause injury to the patient. By assessing the body of a vehicle that has crashed, you can often determine the MOI, which may allow you to predict what injuries may have happened to the passengers at the time of impact

Figure 21-3 The first collision in a frontal impact is that of the car against another object (in this case, a utility pole). The appearance of the car can provide you with critical information about the severity of the crash. The greater the damage to the car, the greater the energy that was involved.

according to forces that acted on their bodies. When you arrive at the crash scene and perform your scene size-up, quickly inspect the severity of damage to the vehicle(s). If there is significant damage to a vehicle, your index of suspicion for the presence of life-threatening injuries should automatically increase. A great amount of force is required to crush and deform a vehicle, cause intrusion into the passenger compartment, tear seats from their mountings, and collapse steering wheels. Such damage suggests the presence of high-energy trauma.

2. **The collision of the passenger against the interior of the car.** Just as the kinetic energy produced by the car's mass and velocity is converted into the work of bringing the car to a stop, the kinetic energy produced by the passenger's mass and velocity is converted into the work of stopping his or her body (Figure 21-4 ▼). Just like the obvious damage to the exterior of the car, the injuries that result are often dramatic and usually immediately apparent during your initial assessment. Common injuries include lower extremity fractures (knees into the dashboard), flail chest (rib cage into the steering wheel), and head trauma (head into the windshield). Such injuries occur more frequently if the passenger is not restrained. But even when the passenger is restrained with a properly adjusted seat belt, injuries can occur, especially in lateral and rollover impacts.

Figure 21-4 The second collision in a frontal impact is that of the passenger against the interior of the car. The appearance of the interior of the car can provide you with information about the severity of the patient's injuries.

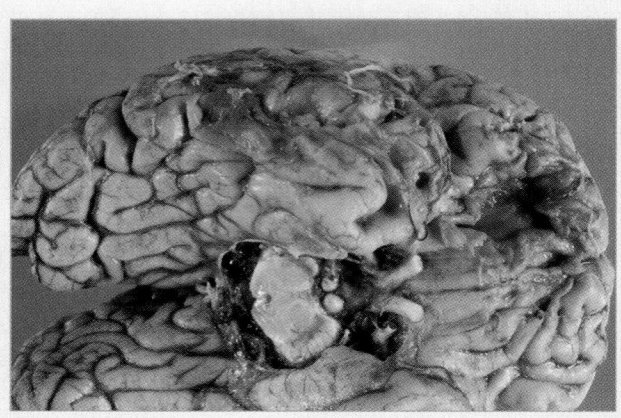

Figure 21-5 A brain with contusions.

3. **The collision of the passenger's internal organs against the solid structures of the body.** The injuries that occur during the third collision may not be as obvious as external injuries, but they are often the most life threatening. For example, as the passenger's head hits the windshield, the brain continues to move forward until it comes to rest by striking the inside of the skull. This results in compression injury (or bruising) to the anterior portion of the brain and stretching (or tearing) of the posterior portion of the brain (Figure 21-5 ▲). This is an example of a coup-contrecoup brain injury (Figure 21-6 ▶). Similarly, in the thoracic cage, the heart may slam into the sternum, occasionally rupturing the aorta and causing fatal bleeding.

Understanding the relationship among the three collisions will help you make the connections between the amount of damage to the exterior of the car and potential injury to the passenger. For example, in a high-speed collision that results in massive damage to the car, you should suspect serious injuries to the passengers, even if the injuries are not readily apparent. A number of potential physical problems may develop as a result of traumatic injuries. Your quick initial assessment of the patient and the evaluation of the MOI can help direct lifesaving care and provide critical information to the hospital staff. Therefore, if you see a contusion on the patient's forehead and the windshield is starred and pushed out, you should strongly suspect an injury to the brain. After you inform medical control about the windshield, hospital staff can prepare the patient by ordering a computed tomography scan of the brain. Without your input, the physician might have

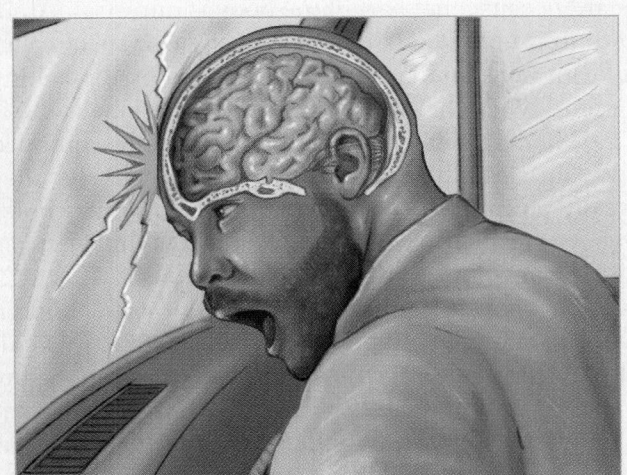

Figure 21-6 The third collision in a frontal impact is that of the passenger's internal organs against the solid structures of the body. In this illustration, the brain continues its forward motion and strikes the inside of the skull, resulting in a compression injury to the anterior portion of the brain and stretching of the posterior portion.

Documentation Tips

In trauma, the MOI is a crucial element of patient history. Be alert to the extent of damage to the interior and exterior of vehicles involved in crashes. Use this observation to paint a picture in written and verbal communication.

- Moderate intrusions from a lateral (T-bone) type of accident
- Severe damage from the rear
- Collisions in which rotation is involved (rollover and spins)

Damage to the vehicle that was involved and information obtained from patient assessment are not the only clues to crash severity. Clearly, if one or more of the passengers are dead, you should suspect that the other passengers have sustained serious injuries, even if the injuries are not obvious. Therefore, you should focus on treating life-threatening injuries and providing transport to a trauma center, because these passengers have likely experienced the same amount of force that caused the death of the others. Polaroid pictures of the crash scene may provide valuable information to the staff and treating physicians at the trauma center.

Frontal Collisions

Understanding the MOI after a frontal collision first involves evaluation of the supplemental restraint system, including seat belts and air bags. You should determine

found the brain injury anyway, but it might have not been detected until the brain had swollen sufficiently to cause clinical signs of the injury.

The amount of damage that is considered significant varies, depending on the type of collision, but any substantial deformity of the vehicle should be enough cause for you to consider transporting the patient to a trauma center. Significant mechanisms of injury include the following:

- Severe deformities of the frontal part of a vehicle, with or without intrusion into the passenger compartment

You are the Provider   Part 2

While you initiate patient care, your partner proceeds to the vehicle to gain a better understanding of the accident and to ensure that there are no additional patients. Your partner returns, stating that the car is in fact on its roof approximately 3' down the side of the ravine. The windshield is shattered, the air bag deployed, and the steering wheel deformed. There is approximately 8" of roof intrusion into the passenger compartment. The driver's side seat belt is intact, and no other patients were found. Bystanders confirm that this is the only patient and that he was unrestrained.

3. How does the mechanism of injury (MOI) influence the treatment of your patient?
4. How does the damage to the interior and exterior of the vehicle relate to potential internal injuries?

whether the passenger was restrained by a full and properly applied three-point restraint. In addition, you should determine whether the air bag was deployed. Identifying the types of restraints used and whether air bags were deployed will help you identify injury patterns related to the supplemental restraint systems.

When properly applied, seat belts are successful in restraining the passengers in a vehicle and preventing a second collision inside the motor vehicle. In addition, they may decrease the severity of the third collision, that of the passenger's organs with the chest or abdominal wall. The very presence of air bags allows seat belts to provide even more "ride down," or the gentle cushioning of the occupant as the body slows, or decelerates. Air bags provide the final capture point of the passengers and again decrease the severity of <u>deceleration</u> injuries by allowing seat belts to be more compliant and by cushioning the occupant as he or she moves forward.

Remember that air bags decrease injury to the chest, face, and head very effectively. However, you should still suspect that other serious injuries to the extremities (resulting from the second collision) and to internal organs (resulting from the third collision) have occurred. Most new motor vehicles are manufactured with air-bag safety systems. These safety devices enhance the safety and survival of forward-facing occupants inside the vehicle during a collision. In an emergency braking event, or collision, the air bag inflates very quickly. Because a rear-facing car seat is in proximity to the dashboard, rapid inflation of the air bag could cause serious injury or death to an infant. All children who are shorter than 4'9" should ride in the rear seat.

When you are providing care to an occupant inside a motor vehicle, it is important to remember that if the air-bag did not inflate, it may deploy during extrication. If this occurs, you may be seriously injured or even killed. Extreme caution must be used when extricating a patient in a vehicle with an air bag that has not deployed.

You should also remember that supplemental restraint systems can cause harm whether they are used properly or improperly. For example, some older models have seat belts that buckle automatically at the shoulder but require the passengers to buckle the lap portion; these can result in the body "submarining" forward underneath the shoulder restraint when the lap portion is not attached. This movement of the body can cause the lower extremities and the pelvis to crash into the dashboard because that part of the body is unrestrained. In addition, individuals of short stature can sustain significant neck and facial injuries caused by the belting systems when their lower torso is unrestrained.

When passengers are riding in vehicles equipped with air bags but are not restrained by seat belts, they are often thrown forward in the act of emergency braking. As a result, they come into contact with the air bag and/or the doors at the time of deployment. This MOI is also responsible for some severe injuries to children who are riding unrestrained in the front seats of vehicles. In addition, some passengers may pass out before impact, and you may find them lying against the air bag when it deploys. You should look for abrasions and/or traction-type injuries on the face, lower part of the neck, and chest ⟨ Figure 21-7 ▼ ⟩.

Contact points are often obvious from a simple quick evaluation of the interior of the vehicle. If there is no intrusion, you might see that an unrestrained front-seat passenger in a frontal collision will come into contact with the dashboard or instrument panel at the knees and transfer loads from the knees through the femur to the pelvis and hip joint ⟨ Figure 21-8A ▶ ⟩. The chest and/or abdomen may also hit the steering wheel ⟨ Figure 21-8B ▶ ⟩. In addition, the passenger's face often hits the steering wheel or may launch forward and up, hitting the windshield and/or the roof header in the area of the visors ⟨ Figure 21-8C ▶ ⟩. Signs of most of these injuries can be found by simply inspecting the interior of the vehicle during extrication of the patient.

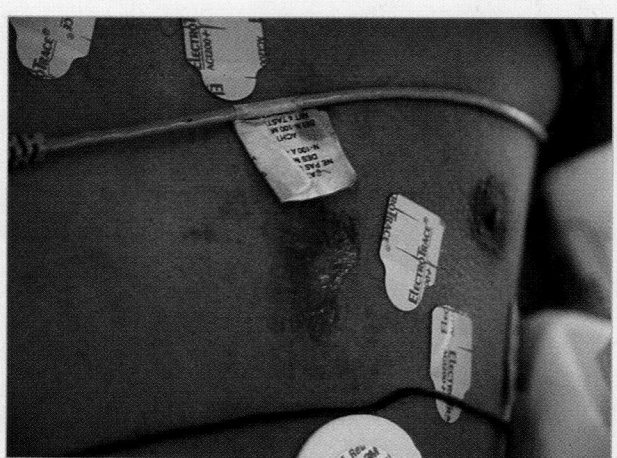

Figure 21-7 Air bags can cause injury in frontal collisions, specifically, abrasions and traction-type injuries to the face, neck, and chest.

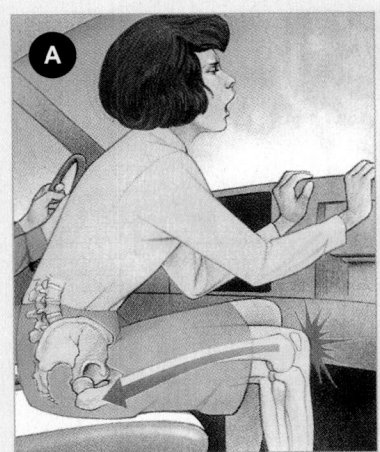
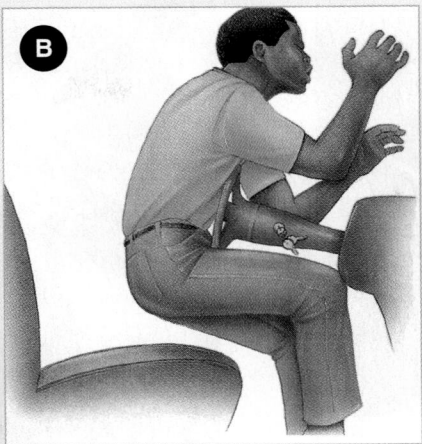
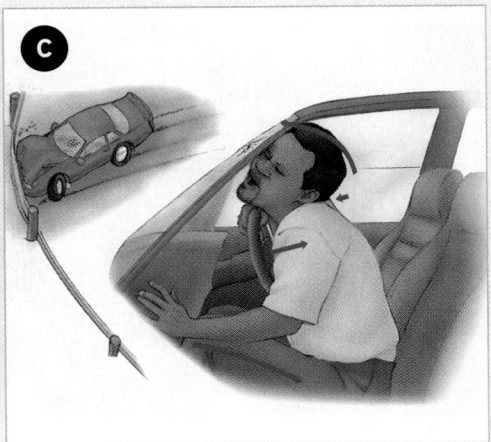

Figure 21-8 Mechanism of injury and condition of the vehicle interior suggest likely areas of injury. **A.** The knee can strike the dashboard, resulting in a hip fracture or dislocation. **B.** Serious chest and abdominal injuries can result from striking the steering wheel. **C.** Head and spinal injuries can result when the face and head strike the windshield.

Rear-End Collisions

Rear-end impacts are known to cause whiplash-type injuries, particularly when the head and/or neck is not restrained by an appropriately placed headrest (Figure 21-9 ▼). On impact, the body and torso move forward. As the body is propelled forward, the head and neck are left behind because the head is relatively heavy, and they appear to be whipped back relative to the torso. As the vehicle comes to rest, the unrestrained passenger moves forward, striking the dashboard. In this type

of collision, the cervical spine and surrounding area may be injured. The cervical spine is less tolerant of damage when it is bent back. Headrests decrease extension during a collision and, therefore, help reduce injury. Other parts of the spine and the pelvis may also be at risk for injury. In addition, the patient may sustain an acceleration-type injury to the brain, that is, the third collision of the brain within the skull. Passengers in the back seat wearing only a lap belt might have a higher incidence of injuries to the thoracic and lumbar spine.

Lateral Collisions

Lateral impacts (commonly called T-bone collisions) are probably now the number one cause of death associated with motor vehicle crashes. When a vehicle is struck from the side, it is typically struck above its center of gravity and begins to rock away from the side of the impact. This results in a lateral whiplash injury (Figure 21-10 ▶). The movement is to the side, and the passenger's shoulders and head whip toward the intruding vehicle. This action may thrust the shoulder, thorax, and upper extremities, and, more important, the skull against the doorpost or the window. The cervical spine has little tolerance for lateral bending.

If there is substantial intrusion into the passenger compartment, you should suspect lateral chest and abdomen injuries on the side of the impact, as well as possible fractures of the lower extremities, pelvis, and ribs. In addition, the organs within the abdomen

Figure 21-9 Rear-end impacts often cause whiplash-type injuries, particularly when the head and/or neck is not restrained by a headrest.

Figure 21-10 In a lateral collision, the car is typically struck above its center of gravity and begins to rock away from the side of impact. This causes a type of lateral whiplash in which the passenger's shoulders and head whip toward the intruding vehicle.

Figure 21-11 Passengers who have been ejected or partially ejected may have struck the interior of the car many times before ejection.

are at risk because of a possible third collision. Approximately 25% of all severe injuries to the aorta that occur in motor vehicle crashes are a result of lateral collisions.

Rollover Crashes

Certain vehicles, such as large trucks and some sport utility vehicles, are more prone to rollover crashes because of their high center of gravity. Injury patterns that are commonly associated with rollover crashes differ, depending on whether the passenger was restrained. The most unpredictable are rollover crashes in which an unrestrained passenger may have sustained multiple strikes within the interior of the vehicle as it rolled

one or more times. The most common life-threatening event in a rollover is ejection or partial ejection of the passenger from the vehicle (Figure 21-11 ▲). Passengers who have been ejected may have struck the interior of the vehicle many times before ejection. The passenger may also have struck several objects, such as trees, a guardrail, or the vehicle's exterior, before landing. Passengers who have been partially ejected may have struck both the interior and exterior of the vehicle and may have been sandwiched between the exterior of the vehicle and the environment as the vehicle rolled. Ejection and partial ejection are significant mechanisms of injury; in these cases, you should prepare to care for life-threatening injuries.

You are the Provider   Part 3

Upon initial assessment, your patient is conscious, alert, and disoriented. You find that he has a patent airway, is breathing at a rate of 12 labored breaths/min, is pale and cool to the touch, and has weak distal pulses. You find no apparent major external bleeding. Pulse oximetry shows 92% oxygen saturation on room air. While you apply high-flow supplemental oxygen, your partner calls for ALS backup.

5. How can potential injuries differ in a rollover collision versus a lateral or frontal collision?
6. What are the three types of collisions that occur during a motor vehicle crash?

Even when restrained, passengers can sustain severe injuries during a rollover crash, although the patterns of injury tend to be more predictable, and when properly used, the restraint system will prevent ejection from the vehicle. A passenger on the outboard side of a vehicle that rolls over is at high risk for injury because of the centrifugal force (the patient is pinned against the door of the vehicle). When the roof hits the ground during a rollover, a passenger who is restrained can still move far enough toward the roof to make contact and sustain a spinal cord injury. Therefore, rollover crashes are particularly dangerous for both restrained and, to a greater degree, unrestrained passengers because these crashes provide multiple opportunities for second and third collisions.

Spins

Spins are conceptually similar to rollovers. The rotation of the vehicle as it spins provides opportunities for the vehicle to strike objects such as utility poles. For example, as a vehicle spins and strikes a pole, the passengers experience not only the rotational motion, but also a lateral impact.

Car Versus Pedestrian

Car-versus-pedestrian collisions often cause serious unseen injuries to underlying body systems. Often they will present with graphic and apparent injuries, such as broken bones. You must maintain a high index of suspicion for unseen injuries. A thorough evaluation of the MOI is critical. The first step is to estimate the speed of the vehicle that struck the patient; next determine whether the patient was thrown through the air and what distance and whether the patient was struck and pulled under the car. You should evaluate the car that struck the patient for structural damage that might indicate contact points with the patient and alert you to potential injuries. Multisystem injuries are common after this type of event.

In a car-versus-bicycle collision, you should evaluate the MOI in much the same manner as car-versus-pedestrian collisions. However, additional evaluation of damage to and the position of the bicycle is warranted. If the patient was wearing a helmet, you should inspect the helmet for damage and suspect potential injury to the head (Figure 21-12 ▶). In both injury profiles, presume that the patient has sustained an injury to the spinal column, or spinal cord, until proven otherwise at the hospital. Spinal immobilization must be initiated and maintained during the encounter.

Figure 21-12 If the patient's bike helmet is damaged, suspect head and spine injuries.

Falls

The injury potential of a fall is related to the height from which the patient fell. The greater the height of the fall, the greater the potential for injury. A fall from more than 15′ or 3 times the patient's height is considered significant. The patient lands on the surface just as an unrestrained passenger smashes into the interior of a vehicle. The internal organs travel at the speed of the patient's body before it hits the ground and stop by smashing into the interior of the body. Again, as in a motor vehicle crash, it is these internal injuries that are the least obvious during assessment but pose the gravest threat to life. Therefore, you should suspect internal injuries in a patient who has fallen from a significant height, just as you would in a patient who has been in a high-speed motor vehicle crash. Always consider syncope or other underlying medical causes of the fall.

⚠ Pediatric Needs

To evaluate the MOI when your patient is a child, remember this:

**A fall greater than
3 times the child's height =
a significant MOI.**

Also note that small children are top-heavy, so they tend to land on their heads even from small falls.

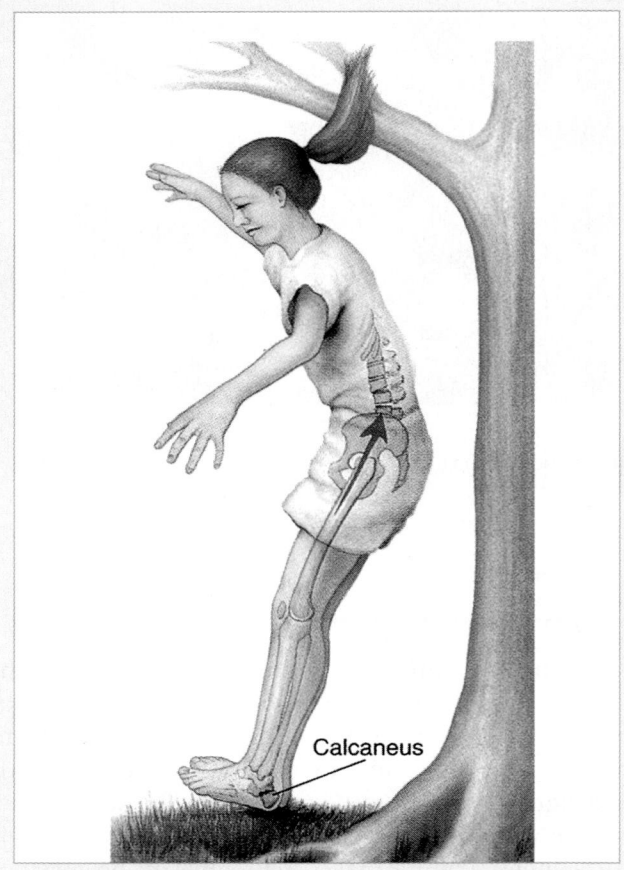

Calcaneus

Figure 21-13 When a patient falls and lands on his or her feet, the energy is transmitted to the spine, sometimes producing a spinal injury in addition to injuries to the legs and pelvis.

Geriatric Needs

Many geriatric patients are seriously injured from falls. Completely assess older patients for all possible injuries, even from low-impact falls.

Patients who fall and land on their feet may have less severe internal injuries because their legs may have absorbed much of the energy of the fall (Figure 21-13 ◀). Of course, as a result, they may have very serious injuries to the lower extremities and pelvic and spinal injuries from energy that the legs do not absorb. Patients who fall onto their heads, as in diving accidents, will likely have serious head and/or spinal injuries. In either case, a fall from a significant height is a serious event with great injury potential, and the patient should be evaluated thoroughly. Take the following factors into account:

- The height of the fall
- The surface struck
- The part of the body that hit first, followed by the path of energy displacement

Some texts consider falls to be the most common form of trauma. Many falls, especially those by older persons, are not considered "true" trauma, even though bones may be broken. Often, these falls occur as a

You are the Provider   Part 4

An ALS unit is en route. Dispatch informed you that the ALS unit is 20 minutes from your location. With this in mind, you finish your rapid trauma assessment. Further assessment shows severe bruising on his chest. His abdomen is rigid and warm to the touch, and his pelvis is unstable. You and your partner quickly prepare your patient for transport. Due to the unstable condition of your patient, you choose to rendezvous with the ALS unit.

7. Of what importance is the MOI and the extent of damage to the vehicle in your written and verbal communication?
8. How does potential patient outcome differ in a motor vehicle crash when a person is properly restrained compared with a person who is not restrained?
9. Should you document the fact that the bystanders assisted the patient away from the scene before you arrived?

result of a fracture. Older patients often have osteoporosis, a condition in which the musculoskeletal system can fail under relatively low stress. Because of this condition, an older patient can sustain a fracture while in a standing position and then fall as a result. Therefore, an older patient may have actually sustained a fracture before the fall. These cases do not constitute true high-energy trauma unless the patient fell from a significant height.

Penetrating Trauma

Penetrating trauma is the second largest cause of death in the United States after blunt trauma. Low-energy penetrating trauma may be caused accidentally by impalement or intentionally by a knife, ice pick, or other weapon (Figure 21-14 ▶). Many times it is difficult to determine entrance and exit wounds from projectiles in a prehospital setting (unless you can determine an obvious exit wound). Determine the number of penetrating injuries and combine that with the important things you already know about the potential pathway of penetrating projectiles to form an index of suspicion about unseen life-threatening injuries. With low-energy penetrations, injuries are caused by the sharp edges of the object moving through the body and are, therefore, close to the object's path. Weapons such as knives, however, may have been deliberately moved around internally, causing more damage than the external wound might suggest.

In medium- and high-velocity penetrating trauma, the path of the object (usually a bullet) may not be as easy to predict. This is because the bullet may flatten out, tumble, or even ricochet within the body before exiting. Also, because of its speed, pressure waves emanate from the bullet, causing damage remote from its path. This phenomenon, called <u>cavitation</u>, can result in serious injury to internal organs distant to the actual path of the bullet. Much like a boat moving through water, the bullet disrupts not only the tissues that are directly in its path but also those in its wake. Therefore, the area that is damaged by medium- and high-velocity projectiles can be many times larger than the diameter of the projectile itself (Figure 21-15 ▶). This is one reason that exit wounds are often many times larger than entrance wounds. As with motor vehicle crashes, the energy available for a bullet to cause damage is more a function of its speed than its mass (weight). If the mass of the bullet is doubled, the energy that is available to

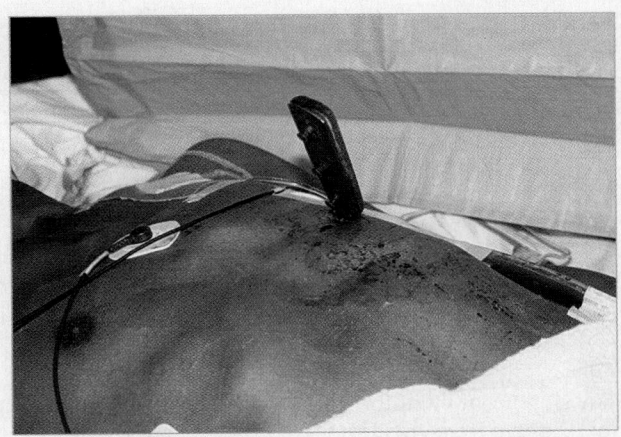

Figure 21-14 Injuries from low-energy penetrations, such as a stab wound, are caused by the sharp edges of the object moving through the body.

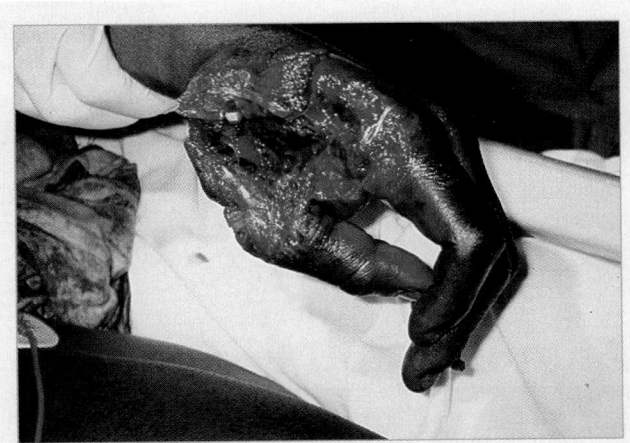

Figure 21-15 The area damaged by high-velocity projectiles, such as bullets, can be many times larger than the diameter of the projectile itself.

cause injury is doubled. If the speed (velocity) of the bullet is doubled, the energy that is available to cause injury is quadrupled. For this reason, it is important for you to try to determine the type of weapon that was used. Although it is not necessary (or always possible) for you to distinguish between medium- and high-velocity injuries, any information regarding the type of weapon that was used should be relayed to medical control. Police at the scene may be a useful source of information regarding the caliber of weapon.

(Table 21-1 ▶) summarizes how to recognize developing problems in trauma patients.

TABLE 21-1 Recognizing Developing Problems in Trauma Patients

Mechanism of Injury	Signs and Symptoms	Index of Suspicion
Blunt or penetrating trauma to the neck	■ Noisy or labored breathing ■ Swelling of the face or neck	■ Significant bleeding or foreign bodies in the upper or lower airway, causing obstruction ■ Be alert for airway compromise.
Significant chest wall blunt trauma from car crashes, car-versus-pedestrian, and other crashes, penetrating trauma to the chest wall	■ Significant chest pain ■ Shortness of breath ■ Asymmetrical chest wall movement	■ Cardiac or pulmonary contusion ■ Pneumothorax or hemothorax ■ Broken ribs, causing breathing compromise
Any significant blunt force trauma, for example from crashes or from penetrating injury	■ Blunt or penetrating trauma to the neck, chest, abdomen, or groin ■ Blows to the head sustained during motor vehicle crashes, falls, or other incidents, producing loss of consciousness, altered mental status, inability to recall events, combativeness, or changes in speech patterns ■ Difficulty moving extremities; headache, especially with nausea and vomiting	■ Injuries in these regions may tear and cause damage to the large blood vessels located in these body areas, resulting in significant internal and external bleeding. ■ Be alert to the possibility of bruising to the brain and bleeding in and around the brain tissue, which may cause serious pressure to accumulate inside the skull around the brain.
Any significant blunt force trauma, falls from a significant height, or penetrating trauma	■ Severe back and/or neck pain, history of difficulty moving extremities, loss of sensation or tingling in the extremities	■ Injury to the bones of the spinal column or to the spinal cord

Anatomy and Physiology

The human body is divided into areas (or systems) based on body function, and its internal organs are subject to unseen injuries when force is applied to the body. For example, the brain may have bruising, the heart and lungs may have bruising or unseen bleeding, and the organs of the abdomen may have life-threatening bleeding.

Injuries to the Head

The brain lies well protected within the skull. However, when the head is injured from trauma, unseen injury to the brain may occur. The brain itself may become bruised or tear, which causes bleeding. The blood vessels around the brain may also tear and produce bleeding. Bleeding or swelling inside the skull from brain injury is often life threatening. Some patients will not have obvious signs or symptoms of unseen brain injury until minutes or hours after the injury has occurred.

Injuries to the Neck and Throat

The neck and throat contain many structures that are susceptible to injuries from trauma that could be serious or deadly to your patients. In this region of the human body, the trachea (or windpipe) may become torn or swell after an injury to the neck. This may result in an airway problem that could quickly become a serious life threat because it would interfere with the patient's ability to breathe.

The neck also contains large blood vessels that supply the brain with oxygen-rich blood. When an injury occurs here, swelling may prevent blood flow to the brain and cause injury to the central nervous system, even though the brain may not have been directly affected by the initial force that caused the injury to the neck. If an open wound is produced from injury, the patient may have significant bleeding, or air may enter the circulatory system and block normal blood flow; either may cause rapid death.

Injuries to the Chest

The chest contains the heart, the lungs, and the large blood vessels of the body. When injury occurs to this area of the body, many life-threatening injuries may occur. For example, when ribs are broken and the chest wall does not expand normally during breathing, this interferes with the body's ability to obtain oxygen for the cells. Bruising may occur to the heart and cause an irregular heartbeat. The large vessels of the heart may be torn inside the chest; this causes massive unseen bleeding that can quickly kill the trauma patient. In some chest injuries the lungs become bruised; this interferes with normal oxygen exchange in the body.

Some chest injuries result in air collecting between the lung tissue and the chest wall. As air accumulates in this space, the lung tissue becomes compressed, again interfering with the body's ability to exchange oxygen. This injury is called a pneumothorax. If left untreated or not recognized, the lung tissue becomes squeezed under pressure until the heart is also squeezed and cannot pump blood. This condition is called a tension pneumothorax and is a life-threatening emergency. Some patients develop bleeding in this portion of the chest. Instead of air collecting, blood collects here and causes interference with breathing. This condition is called hemothorax and also can pose a threat to the patient's life.

Abdominal Injuries

The abdomen is the area of the human body that contains many organs vital to body function. These organs also have a very high amount of blood flow to them so they can perform the functions necessary for life. The organs of the abdomen and retroperitoneum (space immediately behind the true abdomen) can be classified into two simple categories: solid and hollow. The solid organs include the liver, spleen, pancreas, and kidneys. The hollow organs include the stomach, large and small intestines, and urinary bladder.

When injuries from trauma occur in this region of the body, serious and life-threatening problems may occur. The solid organs may tear, lacerate, or fracture. This causes serious bleeding into the abdomen that can quickly cause death. Be alert for a trauma patient who complains of abdominal pain—it may be a symptom of abdominal bleeding. Also be alert to vital signs that begin to worsen; this can be a sign of serious, unseen bleeding inside the abdominal region of the body.

The hollow organs of the body may rupture and leak the acidlike chemicals used for digestion into the abdomen. This not only will cause pain, but also may eventually cause a life-threatening infection.

The abdomen also contains large blood vessels that supply the organs of this region and the lower extremities with oxygen-rich blood. Occasionally these vessels rupture or tear and cause serious unseen bleeding that may cause death.

Trauma patients may have one body area (or system) or several body systems injured. The patient who has more than one body system involved is described as a multisystem trauma patient.

You are the Provider   Summary

As an EMS provider, you will respond to many emergencies resulting from trauma. Understanding the mechanism of injury (MOI) is important in understanding how the injuries occurred and what type of potential injuries there may be. Remember, some of our most serious injuries are those that cannot be seen. It is vital that we not become focused only on the obvious injuries. Use your index of suspicion, recognize the MOI, and always provide the most beneficial and appropriate care.

Prep Kit

Ready for Review

- Determine the MOI as quickly as possible; this will assist you in developing an index of suspicion for the seriousness of unseen injuries.

- Communicate MOI findings in the written patient report and verbally to hospital staff; this will ensure that appropriate treatment continues for the patient at the hospital for potential serious injuries.

- In every crash there are three collisions that occur:
  - The collision of the car against some type of object
  - The collision of the passenger against the interior of the car
  - The collision of the passenger's internal organs against the solid structures of the body

- Maintain a high index of suspicion for serious injury in the patient who has been involved in a car crash with significant damage to the vehicle, has fallen from a significant height, or has sustained penetrating trauma to the body.

Vital Vocabulary

blunt trauma An impact on the body by objects that cause injury without penetrating soft tissues or internal organs and cavities.

cavitation A phenomenon in which speed causes a bullet to generate pressure waves, which cause damage distant from the bullet's path.

coup-contrecoup brain injury A brain injury that occurs when force is applied to the head and energy transmission through brain tissue causes injury on the opposite side of original impact.

deceleration The slowing of an object.

index of suspicion Awareness that unseen life-threatening injuries may exist when determining the mechanism of injury.

kinetic energy The energy of a moving object.

mechanism of injury (MOI) The forces or energy transmission applied to the body that cause injury.

multisystem trauma patient A patient who experienced trauma that affects more than one body system.

penetrating trauma Injury caused by objects, such as knives and bullets, that pierce the surface of the body and damage internal tissues and organs.

potential energy The product of mass, gravity, and height, which is converted into kinetic energy and results in injury, such as from a fall.

Points to Ponder

You are dispatched to the intersection of Reser Road and Rainer Drive for a one-car motor vehicle collision. You arrive to find a four-door sedan that has hit a telephone pole. As you approach the vehicle, you note starring of the windshield. You see a young man behind the wheel who is bleeding from the forehead and is obviously upset. As you begin speaking with him, your partner performs c-spine stabilization from the back seat. The patient is 16 years old, and his parents are on vacation in Hawaii. They specifically told him not to drive to school while they were gone. He is slow to answer your questions, and he keeps repeating, "Please don't tell my parents! They are going to kill me!" As you begin your assessment, he tells you that he is fine and does not need to go to the hospital.

What challenges does this scenario present? How would you handle the patient's concerns?

Issues: Treating the Whole Patient, Dealing With Minors, Implied Consent.

Technology

www.EMTB.com

- Interactivities
- Vocabulary Explorer
- Anatomy Review
- Web Links
- Online Review Manual

Assessment in Action

You are dispatched to a construction site at 3825 Shirrod Lane for a man who has fallen. En route to the location, the dispatcher informs you that your patient fell from the roof of a three-story structure, and he is complaining of severe pain in his legs and back.

Upon arrival, several construction workers excitedly wave to you and yell, "Hurry! Hurry! Over here!" As you approach the crowd of people, you see a man lying on the hard, frost-covered ground. As you begin to introduce yourself, your partner immediately begins c-spine stabilization. The patient tells you that he lost his footing while working on the roof, and the next thing he knew, he was on the ground with agonizing pain in both of his legs and lower back. A coworker who saw the event tells you that the patient landed on his feet. His vital signs are as follows: pulse, 130 beats/min and weak; blood pressure, 108/52 mm Hg; and respirations, 42 breaths/min and shallow.

1. Given patient's landing position, what injuries would you anticipate seeing?

   A. Leg injuries
   B. Spine injuries
   C. Pelvic injuries
   D. All of the above

2. Your concern for potentially serious underlying and unseen injuries is called:

   A. index of belief.
   B. index of suspicion.
   C. index of injury.
   D. index of events.

3. The way in which traumatic injuries occur is referred to as:

   A. mechanism of trauma.
   B. mechanism of event.
   C. mechanism of complaint.
   D. mechanism of injury.

4. What factors should you take into account when evaluating a patient who has fallen from a significant height?

   A. The height of the fall
   B. The surface struck
   C. The part of the body that hit first
   D. All of the above

5. The patient's vital signs indicate that he:

   A. is in pain.
   B. has internal bleeding.
   C. is in shock.
   D. All of the above.

6. What is considered a significant fall?

   A. A fall from >15 feet
   B. A fall from 3 times the patient's height
   C. Both A and B
   D. Neither A nor B

Challenging Questions

7. What clues given to you in the scene description can aid in your understanding of potential injuries?

8. Should you ask about the patient's medical history? Why?

9. Is it important to know the position of his body as he hit the ground?

10. How can a patient's medical condition affect injuries resulting from a traumatic event?

www.EMTB.com

Bleeding

Objectives

Cognitive

5-1.1   List the structure and function of the circulatory system. (p 650)
5-1.2   Differentiate between arterial, venous, and capillary bleeding. (p 655)
5-1.3   State methods of emergency medical care of external bleeding. (p 659)
5-1.4   Establish the relationship between body substance isolation and bleeding. (p 659)
5-1.5   Establish the relationship between airway management and the trauma patient. (p 657)
5-1.6   Establish the relationship between mechanism of injury and internal bleeding. (p 666)
5-1.7   List the signs of internal bleeding. (p 666)
5-1.8   List the steps in the emergency medical care of the patient with signs and symptoms of internal bleeding. (p 670)

Affective

5-1.11 Explain the sense of urgency to transport patients who are bleeding and show signs of shock (hypoperfusion). (p 655)

Psychomotor

5-1.12 Demonstrate direct pressure as a method of emergency medical care of external bleeding. (p 659)
5-1.13 Demonstrate the use of diffuse pressure as a method of emergency medical care of external bleeding. (p 660)
5-1.14 Demonstrate the use of pressure points and tourniquets as a method of emergency medical care of external bleeding. (p 663)
5-1.15 Demonstrate the care of the patient exhibiting signs and symptoms of internal bleeding. (p 670)

You are the Provider

You and your EMT-B partner are dispatched to a cabinet-making shop for a traumatic injury. Once on scene you find a conscious, alert, and oriented 27-year-old man bleeding heavily from his left arm. Your patient explains that he was working with the band saw when he slipped and ran his arm into the blade. He is holding a rag against the wound. The rag is saturated with bright red blood. He states that he cut his arm approximately 10 minutes prior to your arrival and has not been able to make it stop bleeding.

1. What does the color of the blood suggest about the injury?
2. What steps should be taken to stop the bleeding?

Bleeding

After managing the airway, recognizing bleeding and understanding how it affects the body are perhaps the most important skills you will learn as an EMT-B. Bleeding can be external and obvious or internal and hidden. Either way, it is potentially dangerous, first causing weakness and, if left uncontrolled, eventually shock and death. The most common cause of shock after trauma is bleeding. Generally the shock from trauma is caused at least in part from bleeding.

This chapter will help you understand how the cardiovascular system reacts to blood loss. The chapter begins with a brief review of the anatomy and function of the cardiovascular system. It then describes the signs, symptoms, and emergency medical care of both external and internal bleeding. The chapter concludes with a discussion on the relationship between bleeding and hypovolemic shock.

Anatomy and Physiology of the Cardiovascular System

The cardiovascular system circulates blood to all of the body's cells and tissues, delivering oxygen and nutrients and carrying away metabolic waste products

Figure 22-1 ▶). Certain parts of the body, such as the brain, spinal cord, and heart, require a constant flow of blood to live. The cells in these organs cannot tolerate a lack of blood for more than a few minutes. Other organs, such as the lungs and kidneys, can survive for short periods without adequate blood flow. After that, their cells begin to die. This can lead to a permanent loss of function or, if enough cells die, death.

The cardiovascular system, the main system responsible for supplying and maintaining adequate blood flow, consists of three parts:

- The pump (the heart)
- A container (the blood vessels that reach every cell in the body)
- The fluid (blood and body fluids)

The Heart

The heart is a hollow muscular organ about the size of a clenched fist. It is an involuntary muscle that is under the control of the autonomic nervous system, but it has its own regulatory system. Thus, it can function even if the nervous system shuts down.

The heart is always working; all other organs depend on it to provide a rich blood supply. For this reason, it has a number of special features that other muscles do not. First, because the heart cannot tolerate a disruption of its blood supply for more than a few seconds, the heart muscle needs a rich and well-distributed blood supply. Second, the heart works as two paired pumps Figure 22-2 ▶). Each side of the heart has an upper chamber (atrium) and a lower chamber (ventricle), both of which pump blood. Blood leaves each chamber of a normal heart through a one-way valve, which keeps the blood moving in the proper direction by preventing backflow.

The right side of the heart receives oxygen-poor (deoxygenated) blood from the veins of the body. Blood enters the right atrium from the vena cava, then fills the right ventricle. After the right ventricle contracts, blood flows into the pulmonary artery and the pulmonary circulation. The now oxygen-rich (oxygenated) blood returns to the left side of the heart from the lungs through the pulmonary veins. Blood enters the left atrium, then passes into the left ventricle. This side of the heart is more muscular than the other because it must pump blood into the aorta and on to the arteries throughout the body. It is important to remember that the left ventricle is responsible for providing 100% of the body with oxygen-rich blood.

Technology

- Interactivities
- Vocabulary Explorer
- Anatomy Review
- Web Links
- Online Review Manual

www.EMTB.com

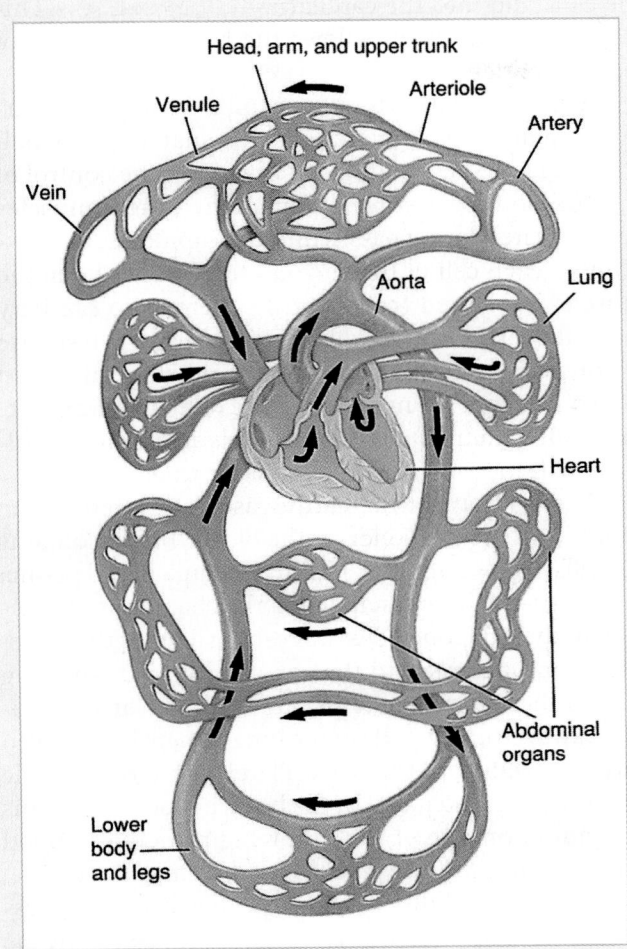

Figure 22-1 The cardiovascular system includes the heart, arteries, veins, and interconnecting capillaries. The exchange of nutrients and waste products that occurs in the capillaries is shown.

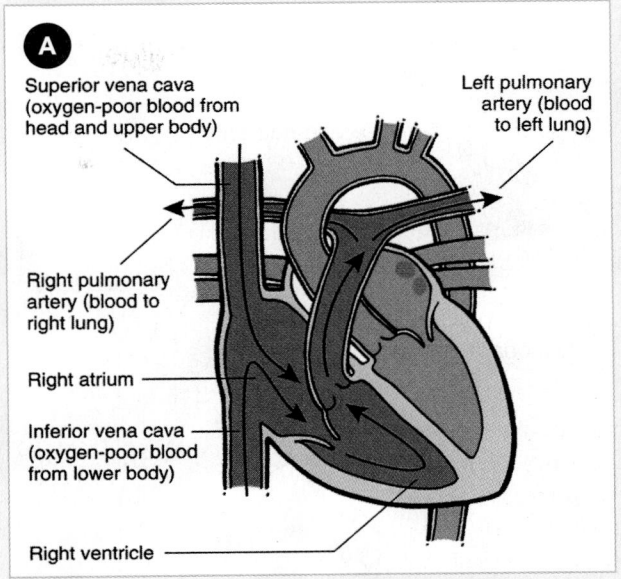

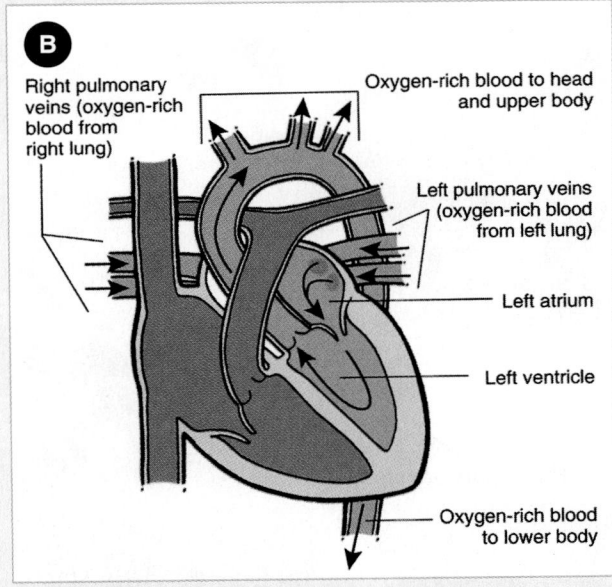

Figure 22-2 A. The right side of the heart circulates blood from the body to the lungs. **B.** The left side of the heart circulates oxygen-rich blood to all parts of the body. It is the more muscular of the two pumps because it must pump blood into the aorta and into the arteries.

Blood Vessels and Blood

There are five types of blood vessels:

- Arteries
- Arterioles
- Capillaries
- Venules
- Veins

As blood flows out of the heart, it passes into the aorta, the largest artery in the body. The arteries become smaller as they move away from the heart. The smaller vessels that connect the arteries and capillaries are called arterioles. Capillaries are small tubes, with the diameter of a single red blood cell, that pass among all the cells in the body, linking the arterioles and the venules. Blood leaving the distal side of the capillaries flows into the venules. These small, thin-walled vessels empty into the veins, and the veins then empty into the vena cava. This is the process that returns blood in the venous side of the circulatory system to the heart. Oxygen and nutrients easily pass from the capillaries into the cells, and waste and carbon dioxide move out of

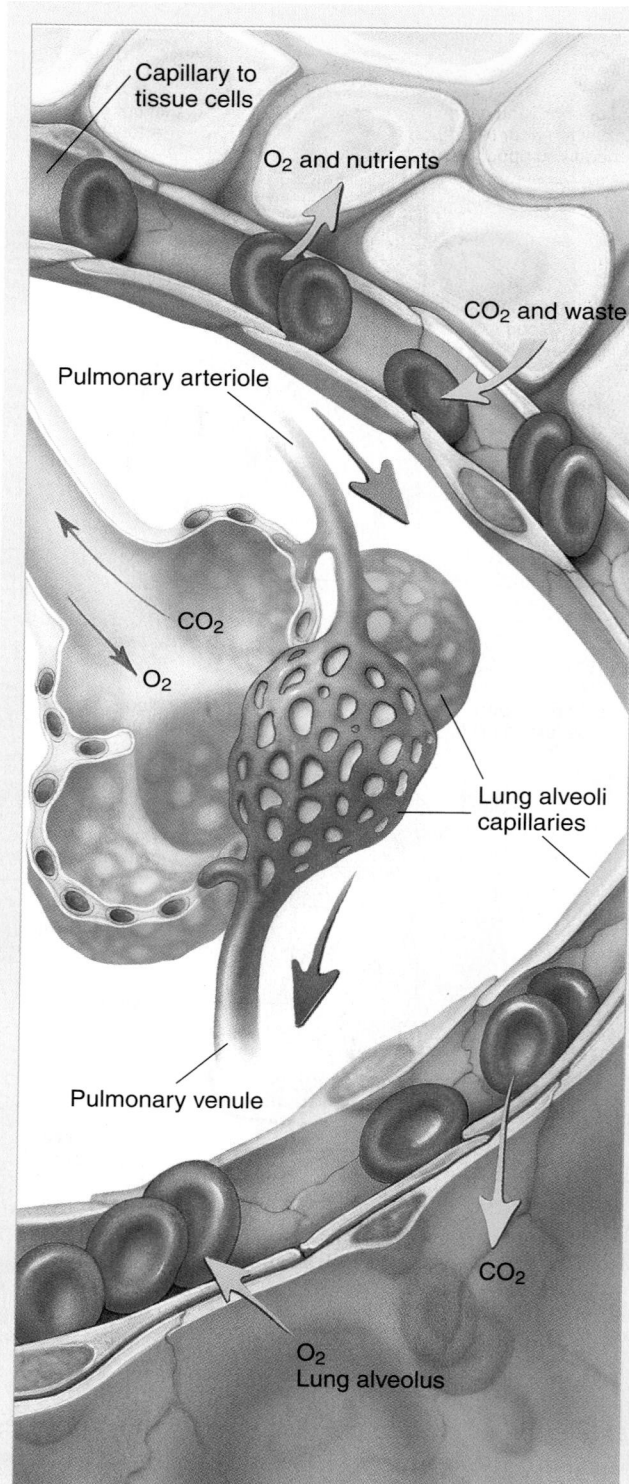

Figure 22-3 Oxygen and nutrients pass easily from the capillaries into the cells, and waste and carbon dioxide move out of the cells into the capillaries (top). Oxygen and carbon dioxide pass freely between the lungs and capillaries (bottom).

the cells and into the capillaries (Figure 22-3 ◀). This transportation system allows the body to rid itself of waste products.

At the arterial ends of the capillaries and in the arteries themselves are circular muscular walls, which constrict and dilate automatically under the control of the autonomic nervous system. When these muscles open (dilate), blood passes into the capillaries in proximity to each cell of the surrounding tissue; when the muscles are closed (constricted), there is no capillary blood flow. The muscles dilate and constrict in response to conditions such as fright, heat, cold, a specific need for oxygen, and the need to dispose of metabolic waste. In a healthy individual, all the vessels are never fully dilated or fully constricted at the same time.

The last part of the cardiovascular system is the contents of the container, or the blood. Blood contains red cells, white cells, platelets, and a liquid called plasma (Figure 22-4 ▼). As discussed in the chapter on the human body, red blood cells are responsible for the transportation of oxygen to the cells and for transporting carbon dioxide (a waste product of cellular metabolism) away from the cells to the lungs where it is exhaled and removed from the body. Platelets are responsible for forming blood clots. In the body, a blood clot forms depending on one of the following principles: blood stasis, changes in the vessel wall (such as a wound), and the blood's ability to clot (due to a disease process or medication). When injury occurs to tissues in the

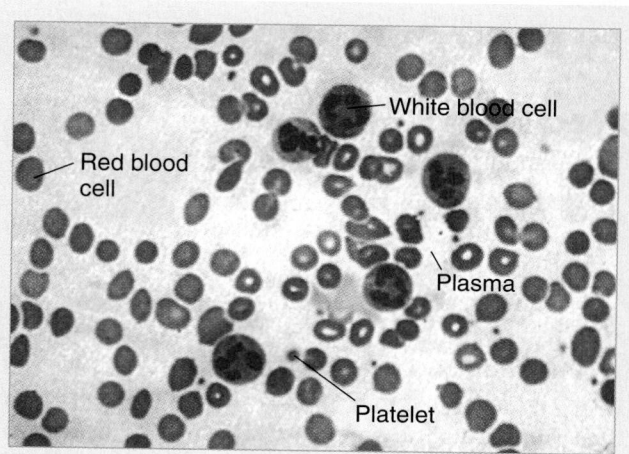

Figure 22-4 The microscopic appearance of the three major elements in blood: red blood cells, white blood cells, and platelets.

body, platelets will begin to collect at the site of injury; this causes red blood cells to become sticky and clump together. As the red blood cells begin to clump, another substance in the body called fibrinogen reinforces the red blood cells. This is the final step in formation of a blood clot. Blood clots are an important response from the body to control blood loss. Certain medical conditions that interfere with the normal clotting process will be discussed later in this chapter.

The autonomic nervous system monitors the body's needs from moment to moment and adjusts the blood flow by adjusting vascular tone as required. During emergencies, the autonomic nervous system automatically redirects blood away from other organs to the heart, brain, lungs, and kidneys. Thus, the cardiovascular system is dynamic and constantly adapting to changing conditions in the body to maintain homeostasis and perfusion. At times, the system fails to provide sufficient circulation for every body part to perform its function. This condition is called hypoperfusion, or shock.

Pathophysiology and Perfusion

Blunt force trauma may cause injury and significant bleeding that is unseen inside a body cavity or region, such as when injury occurs to the liver or the spleen. These injuries cause the patient to lose significant amounts of blood, causing hypoperfusion without visible bleeding. In penetrating trauma, the patient may have only a small amount of bleeding that is visible; however, the patient may have sustained injury to internal organs that will produce significant bleeding that is unseen by you and may cause death quickly. Both of these situations are examples of serious internal bleeding, in which blood volume and supply have been interrupted to the cells of the body; this interruption is the cause of hypoperfusion (or shock) in the trauma patient.

Perfusion is the circulation of blood within an organ or tissue in adequate amounts to meet the cells' current needs for oxygen, nutrients, and waste removal. Blood enters an organ or tissue first through the arteries, then the arterioles, and finally the capillary beds (Figure 22-5 ▶). While passing through the capillaries, the blood delivers nutrients and oxygen to the surrounding cells and picks up the wastes they have gen-

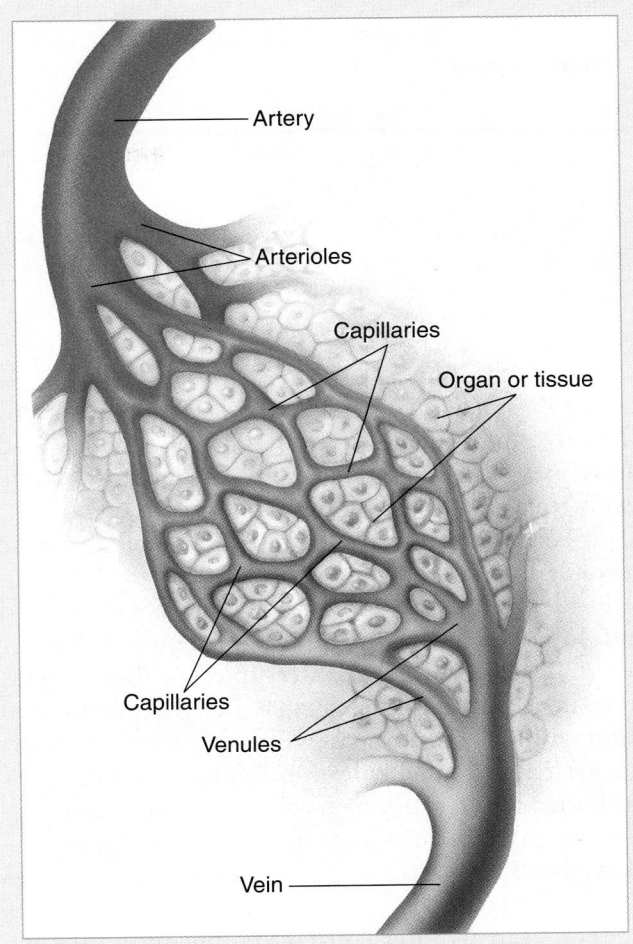

Figure 22-5 Perfusion occurs when blood circulates through tissues or an organ to provide the necessary oxygen and nutrients and remove waste products.

erated. Then the blood leaves the capillary beds through the venules and finally reaches the veins, which take the blood back to the heart. Oxygen and carbon dioxide exchange takes place in the lungs.

Blood must pass through the cardiovascular system at a speed that is fast enough to maintain adequate circulation throughout the body and slow enough to allow each cell time to exchange oxygen and nutrients for carbon dioxide and other waste products. Although some tissues, such as the lungs and kidneys, never rest and require a constant blood supply, most require circulating blood only intermittently, especially when active. Muscles are a good example. When you sleep, they are at rest and require a minimal blood supply. However,

TABLE 22-1 Organs and Corresponding Organ Systems

Organ	Organ System
Heart	Cardiovascular system
Brain	Central nervous system
Lungs	Respiratory system
Kidneys	Renal system

during exercise, they need a very large blood supply. The gastrointestinal tract requires a high flow of blood after a meal. After digestion is completed, it can do quite well with a small fraction of that flow.

All organs and organ systems of the human body are dependent on adequate perfusion to function properly. Some of these organs receive a very rich supply of blood and do not tolerate interruption of blood supply for very long. If perfusion is interrupted to these organs and damage occurs to the organ tissue, dysfunction and failure of that organ system will occur. Death of an organ system can quickly lead to death of the organism, the human. Emergency medical care is designed to support adequate perfusion to these organs and their systems, listed in (Table 22-1 ▲), until the patient arrives at the hospital.

The heart requires constant perfusion to function properly. The brain and spinal cord can be injured after 4 to 6 minutes without perfusion. It is important to remember that cells of the central nervous system do not have the capacity to regenerate. Kidneys can be damaged after 45 minutes of inadequate perfusion. Skeletal muscle demonstrates evidence of injury after 2 hours of inadequate perfusion. The gastrointestinal tract can tolerate slightly longer periods of inadequate perfusion. These times are based on a normal body temperature (98.6°F [37.0°C]). An organ or tissue that is considerably colder may be better able to resist damage from hypoperfusion.

External Bleeding

Hemorrhage means bleeding. External bleeding is visible hemorrhage. Examples include nosebleeds and bleeding from open wounds. As an EMT-B, you must understand how to control external bleeding.

EMT-B Safety

Remember that a bleeding patient may expose you to potentially infectious body fluids; therefore, you must always follow body substance isolation (BSI) precautions when treating patients with external bleeding. Wear gloves and eye protection in all situations, and wear a gown and mask if there is a risk of blood splatter (Figure 22-6 ▼). Avoid direct contact with body fluids if possible. Take special care if you have an open sore, cut, scratch, or ulcer. Also remember that frequent, thorough handwashing between patients and after every run is a simple yet important protective measure. You will be called to respond to emergencies involving more than one patient who needs emergency care. As you complete the assessment and care for each patient, remember to place clean gloves on your hands. Always keep spare gloves with you when responding to these incidents. This approach to patient care will greatly minimize the chance that you could cause cross-contamination of body fluids and blood between patients you may be caring for.

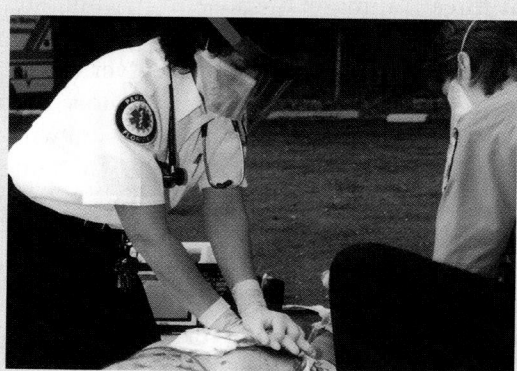

Figure 22-6 Your safety is paramount; therefore, you should always wear proper protective equipment when caring for a patient who is bleeding.

The Significance of Bleeding

When patients have serious external blood loss, it is often difficult to determine the amount of blood that is present. This is a difficult task because blood will look different on different surfaces, such as when it is absorbed

EMT-B Tips

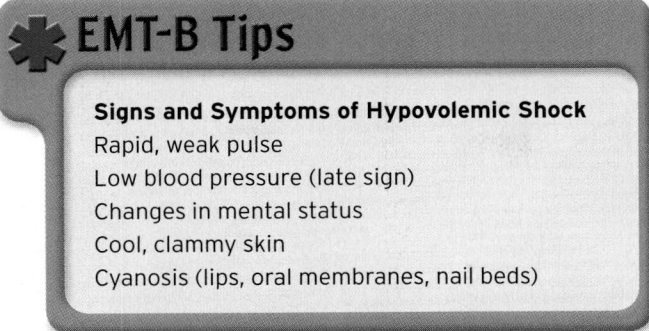

Signs and Symptoms of Hypovolemic Shock
Rapid, weak pulse
Low blood pressure (late sign)
Changes in mental status
Cool, clammy skin
Cyanosis (lips, oral membranes, nail beds)

in clothing or when it has been diluted when mixed in water. Always attempt to determine the amount of external blood loss, but the presentation and assessment of the patient will direct the care and treatment the patient will receive from you as an EMT-B.

The body will not tolerate an acute blood loss of greater than 20% of blood volume (Figure 22-7 ▼). The typical adult has approximately 70 mL of blood per kilogram of body weight, or 6 L (10 to 12 pints) in a body weighing 80 kg (175 lb). If the typical adult loses more than 1 L of blood (about 2 pints), significant changes in vital signs will occur, including increasing heart and respiratory rates and decreasing blood pressure. Because infants and children have less blood volume to begin with, the same effect is seen with smaller amounts of blood loss. For example, a 1-year-old infant has a total blood volume of about 800 mL. Significant symptoms of blood loss will occur after only

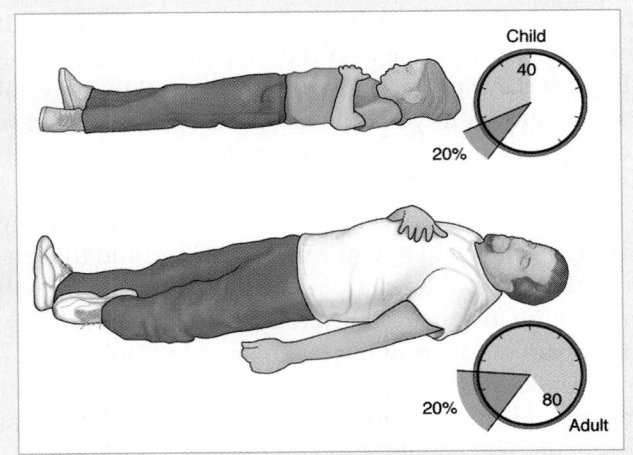

Figure 22-7 Loss of approximately 1 L of blood will cause significant changes in an adult; a much smaller blood loss will result in shock in a child or infant.

100 to 200 mL of blood loss. To put this in perspective, a soft drink can holds roughly 345 mL of liquid.

How well people compensate for blood loss is related to how rapidly they bleed. A normal, healthy adult can comfortably donate 1 unit (500 mL) of blood during a period of 15 to 20 minutes and adapts well to this decrease in blood volume. However, if a similar blood loss occurs in a much shorter period, the person may rapidly develop hypovolemic shock, a condition in which low blood volume results in inadequate perfusion and even death. The body simply cannot compensate for such a rapid blood loss.

You should consider bleeding to be serious if the following conditions are present:

- It is associated with a significant mechanism of injury.
- The patient has a poor general appearance.
- Assessment reveals signs and symptoms of shock (hypoperfusion).
- You note a significant amount of blood loss.
- The blood loss is rapid.
- You cannot control the bleeding.

In any situation, blood loss is an extremely serious problem. It demands your immediate attention as soon as you have cleared the airway and managed the patient's breathing.

Characteristics of Bleeding

Injuries and some illnesses can disrupt blood vessels and cause bleeding. Typically, bleeding from an open artery is brighter red (high in oxygen) and spurts in time with the pulse. The pressure that causes the blood to spurt also makes this type of bleeding difficult to control. As the amount of blood circulating in the body drops, so does the patient's blood pressure and, eventually, the arterial spurting.

Blood from an open vein is darker (low in oxygen) and flows steadily. Because it is under less pressure, most venous blood does not spurt and is easier to manage. Bleeding from damaged capillary vessels is dark red and oozes from a wound steadily but slowly. Venous and capillary blood is more likely to clot spontaneously than arterial blood (Figure 22-8 ▶).

On its own, bleeding tends to stop rather quickly, within about 10 minutes, in response to internal mechanisms and exposure to air. When we are cut, blood flows rapidly from the open vessel. Soon afterward, the cut ends of the vessel begin to narrow, reducing the amount of bleeding. Then a clot forms, plugging

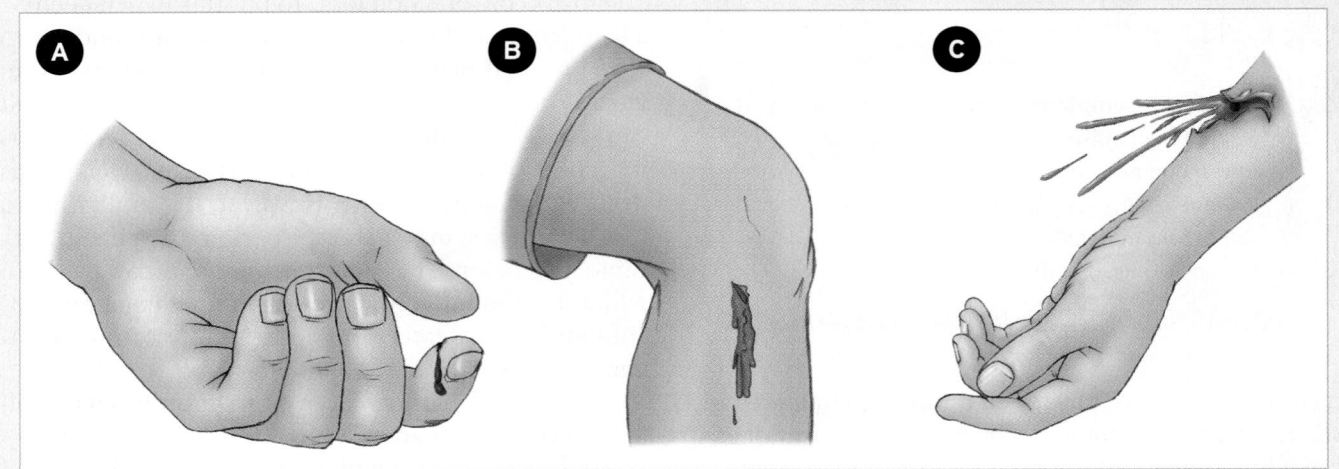

Figure 22-8 **A.** Bleeding from capillary vessels is dark red and oozes from the wound slowly but steadily. **B.** Venous bleeding is darker than arterial bleeding and flows steadily. **C.** Arterial bleeding is characteristically brighter red and spurts in time with the pulse.

the hole and sealing the injured portions of the vessel. This process is called coagulation. Bleeding will never stop if a clot does not form, unless the injured vessel is completely cut off from the main blood supply. Direct contact with body tissues and fluids or the external environment commonly triggers the blood's clotting factors.

Despite the efficiency of this system, it may fail in certain situations. A number of medications, including aspirin, interfere with normal clotting. With a severe injury, the damage to the vessel may be so large that a clot cannot completely block the hole. Sometimes only part of the vessel wall is cut, preventing it from constricting. In these cases, bleeding will continue unless it is stopped by external means. Occasionally, blood loss occurs very rapidly. In these cases, the patient might die before the body's defenses, such as clotting, could help.

A very small portion of the population lacks one or more of the blood's clotting factors. This condition is called hemophilia. There are several forms of hemophilia, most of which are hereditary and some of which are severe. Sometimes bleeding may occur spontaneously in hemophilia. Because the patient's blood does not clot, all injuries, no matter how trivial, are potentially serious. A patient with hemophilia should be transported immediately.

Patient Assessment

Scene Size-up

As you arrive on scene, look for hazards and threats to the safety of the crew, bystanders, and the patient. If this is a trauma scene or bleeding is suspected, put on gloves and eye protection, at a minimum. Stick several pairs of gloves in your pocket for easy access in case your gloves tear or there are multiple patients with bleeding.

At vehicle crashes, ensure that there is no leaking fuel in the area where you will be working and that energized electrical lines are not close to where you will be working. If you are called to a two-car collision, how many patients are possible? Two or eight? Do you have the necessary resources available? Consider early what you may need, and verify as you begin your initial assessment. The sooner you call for help, the sooner it will arrive.

In incidents involving violence, such as assaults or gunshot wounds, make sure that police are on scene. At times you may need to stage several blocks away until law enforcement personnel have secured the area.

✳ EMT-B Tips

If a bandage has already been applied to control bleeding before you arrive on the scene, obtain a description of the wound and the amount of bleeding from the patient or bystanders.

EMT-B Tips

If you put on two pair of gloves before getting bloody, you can take one pair off before getting into bags for equipment or touching other patients. This is easier than trying to put on another pair of gloves on sweaty hands.

Initial Assessment

In trauma patients with suspected significant blood loss, either from a visible wound or from unseen bleeding inside a body cavity, you must not be distracted from the initial assessment. Management of multiple life-threatening concerns in the initial assessment is based on "What is going to kill my patient first?" In some situations, significant bleeding may need management before applying oxygen for a person with adequate breathing. The decision on what to treat first will come with experience. Treating according to the ABCs is always a good choice.

General Impression

As you approach a trauma patient, you must note important indicators that may alert you to the seriousness of the patient's condition. Is the patient interacting with the environment or lying still, making no sounds? Check for responsiveness using the AVPU scale (**A**wake and alert; responsive to **V**erbal stimulus or **P**ain;

Unresponsive). Asking the patient about the chief complaint should help direct you to any apparent life threats such as arterial bleeding. Because of the color of blood and how well it soaks through clothing, you can often identify a patient with bleeding as you approach the patient. How is the patient's skin color? Does he or she "look bad?"

Airway and Breathing

Next, ensure that the patient has a clear airway with adequate breathing, check for breath sounds, and then provide high-flow oxygen with assisted ventilation via a bag-valve-mask device or nonrebreathing mask, depending on the patient's level of consciousness and rate and quality of breathing. Always consider the need for manual spinal stabilization, and provide stabilization at the same time you manage the patient's airway and breathing.

Circulation

You must be able to quickly assess pulse rate and quality; determine the skin condition, color, and temperature; and check the capillary refill time. If visible significant bleeding is seen, you must begin the steps necessary to control bleeding. Significant bleeding, internal or external, is an immediate life threat. If the patient has obvious life-threatening bleeding, it must be controlled quickly and treatment of shock begun as quickly as possible. Non-life-threatening bleeding, such as in abrasions, can be bandaged later in your assessment as necessary.

Transport Decision

If the patient you are treating has an airway or breathing problem or significant bleeding, you must consider quickly transporting the patient to the

You are the Provider — Part 2

During the initial assessment you note that the patient has a patent airway, is showing no signs of respiratory distress, and lung sounds are clear and equal bilaterally. You quickly dress the wound with a sterile bulky dressing. Your partner applies high-flow oxygen via a nonrebreathing mask. While holding direct pressure, you check for a radial pulse. You find the pulse to be rapid and weak. You note a capillary refill of greater than 2 seconds. The patient's skin is pale, cool, and moist to the touch.

3. What is concerning about this patient's pulse and skin condition?
4. How do these findings affect your transport decision?

hospital for treatment. If the patient has signs and symptoms of internal bleeding, you must transport quickly to the appropriate hospital for treatment by a physician. The condition of patients who may have significant bleeding will quickly become unstable; treatment is directed at quickly addressing life threats and at rapid transportation to the closest appropriate hospital. Signs such as tachycardia, tachypnea, low blood pressure, weak pulse, and clammy skin are signs of impending circulatory collapse and imply the need for rapid transport.

Focused History and Physical Exam

Rapid Physical Exam Versus Focused Physical Exam

After the initial assessment is complete, determine which trauma assessment will be performed next. In a responsive patient who has an isolated injury with limited mechanism of injury (MOI), consider a focused physical exam before assessing vital signs and history. Focus your assessment on the isolated injury and complaint. The patient with the large wound to the arm would receive this assessment. Ensure that control of the bleeding is maintained, and note the location of the injury. In an injured extremity, assess pulse and motor and sensory function. If the bleeding is from the chest, assess breathing and circulation. If the bleeding is from the scalp, assess the neurologic system. Assessing the major underlying systems can direct you to other problems that may need your focused attention.

If there is significant trauma, likely affecting multiple systems, start with a rapid trauma assessment looking for DCAP-BTLS (**D**eformities, **C**ontusions, **A**brasions, **P**unctures/**P**enetrations, **B**urns, **T**enderness, **L**acerations, and **S**welling) to be sure that you have found all of the problems and injuries quickly. If any of these injuries are life threatening, treat them immediately. It is important in bleeding cases to avoid focusing just on the bleeding. With significant trauma, you should assess the entire patient, looking for fractures and other problems.

You should not delay transport of a trauma patient, particularly one with significant bleeding, even if controlled, to complete a detailed physical exam. The detailed physical exam can be started during transport. Obtaining vital signs and SAMPLE history should also not delay transportation to the hospital in a significant trauma patient.

Baseline Vital Signs

You must assess baseline vital signs to observe the changes that may occur during treatment. A systolic blood pressure less than 100 mm Hg with a weak, rapid pulse should suggest to you the presence of hypoperfusion in the patient who may have significant bleeding. Cool, moist skin that is pale or gray is an important sign that the patient is experiencing a perfusion problem. Pupillary changes may indicate bleeding inside the skull.

SAMPLE History

Next, obtain a SAMPLE history from your patient. Obtaining a history may be difficult when the patient is unresponsive. Some history may be obtained from medical alert tags or by quickly asking bystanders before transporting the patient.

Interventions

Whenever you suspect significant bleeding, provide high-flow oxygen. If significant bleeding is visible, begin the steps to control bleeding, as shown in Skill Drill 22-1. Using multiple methods to control bleeding

You are the Provider   Part 3

Due to the amount of blood loss and your patient's present condition, you decide on a rapid transport. While gathering additional information about the incident, you find that the dressing is becoming saturated with blood. You elevate the arm and apply additional dressings. With your partner's assistance, you quickly load your patient into the ambulance and begin transporting. Once in the ambulance, your patient tells you that he is feeling nauseous and light-headed.

5. Should saturated dressings be removed prior to applying additional dressings?
6. What type of assessment would be most appropriate for this patient?

usually works best. If the patient has signs of hypoperfusion, treat aggressively for shock and provide rapid transport to the appropriate hospital.

Detailed Physical Exam

Once the initial assessment is complete, all obvious life threats are corrected, and the focused history and physical exam have identified hidden injuries and given a better picture of the patient's condition, consider performing a detailed physical exam as discussed in the chapter on patient assessment. The detailed physical exam of a patient who is bleeding may not be as important as other parts of the assessment. In a patient with significant trauma, performing a detailed physical exam is an important part of identifying and treating all of the patient's injuries. Many times, short transportation times and unstable patient conditions make this assessment impractical.

Ongoing Assessment

The ongoing assessment is an important tool to see how your patient is doing over time. Reassess the patient, especially in areas of abnormal findings during the initial assessment. Are the patient's airway and breathing still adequate? Is your treatment for shock resulting in better perfusion of the vital organs? Vital signs show how well your patient is doing internally over time. In all cases of severe bleeding, take vital signs every 5 minutes. Reassess interventions and treatment you have provided to the patient. Is the bandage controlling the bleeding? Is the oxygen helping the patient to breathe easier?

Communication and Documentation

In cases involving severe external bleeding, it is important to recognize, estimate, and report the amount of blood loss that has occurred and how rapidly or during what time it occurred. This is a challenge, especially if the surface is wet or absorbs fluids or if the environment is dark. For example, you may report that approximately a quart was lost or that the bleeding soaked through three trauma dressings. The examples you use to describe the bleeding are not as important as describing it clearly. With internal bleeding, describe the mechanism of injury and the signs and symptoms

that make you think internal bleeding is occurring. Report this information to hospital personnel during transport to allow the hospital to evaluate needed resources, such as the availability of surgical suites, surgeons, and other specialty providers. Your transfer report at the hospital should update hospital personnel on how your patient has responded to your care. Be sure your paperwork reflects all of the patient's injuries and the care you have provided.

Emergency Medical Care

As you begin to care for a patient with obvious external bleeding, remember to follow BSI precautions. This includes, at a minimum, gloves and eye protection and often a mask and possibly a gown. As with all patient care, make sure that the patient has an open airway and is breathing adequately. Provide high-flow oxygen to the patient. You may then concentrate on controlling the bleeding. In some cases, obvious life-threatening bleeding may be present and should be addressed as an immediate life threat and controlled as quickly as possible.

Several methods are available to control external bleeding. Starting with the most commonly used, these include the following:

- Direct pressure and elevation
- Pressure dressings
- Pressure points (for upper and lower extremities)
- Splints
- Air splints
- Pneumatic antishock garment
- Tourniquets (last resort)

Basic Methods

It will often be useful to combine these methods. Skill Drill 22-1 illustrates the basic techniques that do not require special equipment:

1. Almost all cases of external bleeding can be controlled simply by **applying direct local pressure to the bleeding site**. This method is by far the most effective way to control external bleeding. Pressure stops the flow of blood and permits normal coagulation to occur. You may apply pressure with your gloved fingertip or hand over the top of a sterile dressing if one is immediately available. If there is an object protruding from the wound, apply bulky dressings to stabilize the object in place, and apply pressure as best you can. Never remove an impaled object from a wound. Hold *uninterrupted pressure* for at least 5 minutes.

2. **Elevating a bleeding extremity** by as little as 6″ often stops venous bleeding. Whenever possible, use both techniques: direct pressure and elevation. In most cases, this will stop the bleeding. However, if it does not, you still have several options. Remember to never elevate an open fracture to control bleeding. Fractures can be elevated after splinting, and splinting helps control bleeding (**Step 1**).

3. **Once you have applied a dressing to control the bleeding**, you can create a pressure dressing to maintain the pressure by firmly wrapping a sterile, self-adhering roller bandage around the entire wound. Use 4″ × 4″ sterile gauze pads for small wounds and sterile universal dressings for larger wounds.

   Cover the entire dressing above and below the wound. Stretch the bandage tight enough to control bleeding but not so tight as to decrease blood flow to the extremity. If you were able to palpate a distal pulse before applying the dressing, you should still be able to palpate a distal pulse on the injured extremity after applying the pressure dressing. If bleeding continues, the dressing is probably not tight enough. Do not remove a dressing until a physician has evaluated the patient. Instead, apply additional manual pressure through the dressing. Then add more gauze pads over the first dressing, and secure them both with a second, tighter roller bandage.

   Bleeding will almost always stop when the pressure of the dressing exceeds arterial pressure.

This will assist in controlling bleeding and helping blood to clot (**Step 2**).

4. **If a wound continues to bleed** despite use of direct pressure, elevate the extremity and try placing additional pressure over a proximal pressure point. A pressure point is a spot where a blood vessel lies near a bone. This technique is also useful if you have no material on hand to use for a dressing. Because a wound usually draws blood from more than one major artery, proximal compression of a major artery rarely stops bleeding completely, but it helps to slow the loss of blood. You must be thoroughly familiar with the location of the pressure points for this to work Figure 22-9 ▶ (**Step 3**). If you suspect spinal injury, do not elevate the patient's legs. Instead, elevate the foot end of the backboard and do not cause movement of the spinal column. If the patient has an open fracture of an extremity, use direct pressure to control bleeding. However, do not apply so much pressure as to increase pain or injury.

Special Techniques

Much of the bleeding associated with broken bones occurs because the sharp ends of the bones cut muscles and other tissues. As long as a fracture remains unstable, the bone ends will move and continue to injure partially clotted vessels. Therefore, stabilizing a fracture and decreasing movement is a high priority in the prompt control of bleeding. Often, simple splints will quickly control bleeding associated with a fracture

You are the Provider Part 4

The additional dressings you applied quickly become saturated. You apply more dressings and continue to hold direct pressure. You locate and compress the proximal pressure point. Within a few minutes the bleeding begins to subside. With the bleeding under control, you continue with your assessment. Your vital signs show a rapid, thready pulse of 96 beats/min; shallow rapid respirations of 24 breaths/min; slightly dilated and sluggish pupils; and pale, cool, and clammy skin. Your patient states that he feels like he is going to vomit and is extremely dizzy.

7. What condition does your patient's signs and symptoms suggest?
8. What treatment should you provide?

Controlling External Bleeding

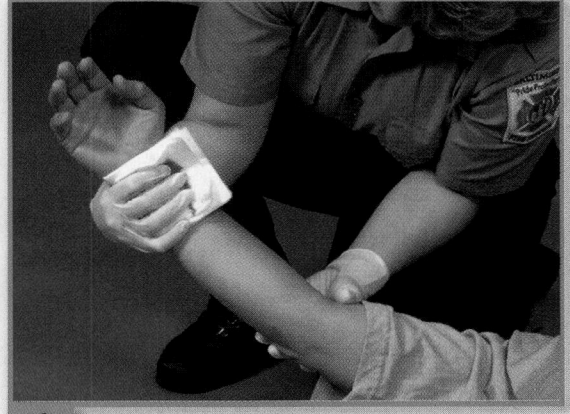

1 Apply direct pressure over the wound. Elevate the injury above the level of the heart if no fracture is suspected.

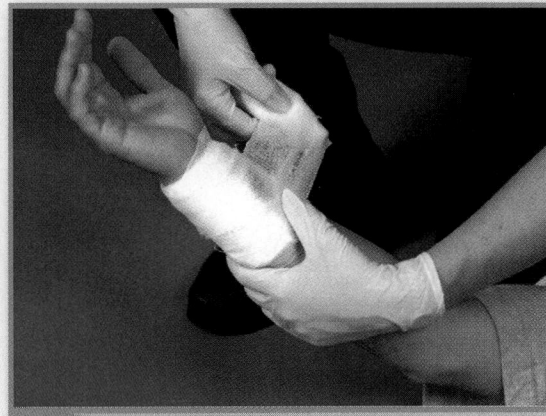

2 Apply a pressure dressing.

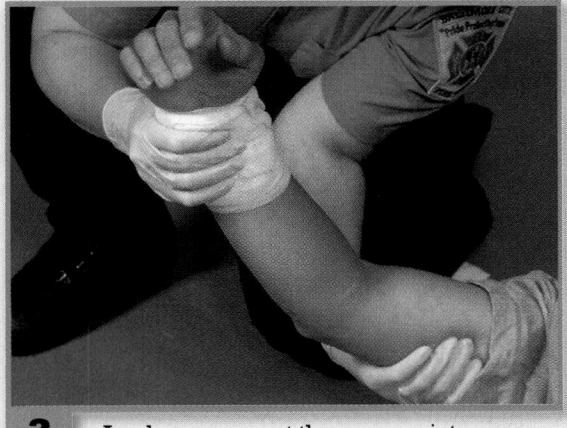

3 Apply pressure at the appropriate pressure point while continuing to hold direct pressure.

Figure 22-10 ▶). If not, you may need to use another splinting device, such as an air splint, pneumatic antishock garment, or a tourniquet, discussed next.

Air Splints
Air splints can control the bleeding associated with severe soft-tissue injuries, such as massive or complex lacerations, or fractures (Figure 22-11 ▶). They also stabilize the fracture itself. An air splint acts like a pressure dressing applied to an entire extremity rather than to a small, local area. Once you have applied an air splint, be sure to monitor circulation in the distal extremity. Use only BSI-approved, clean, or disposable valve stems when orally inflating air splints.

Pneumatic Antishock Garments
If a patient has injuries to the lower extremities or pelvis, you may be able to use a <u>pneumatic antishock garment (PASG)</u> as a splinting device, if local protocol allows. Situations in which use of a PASG is allowed vary widely by locale. Be sure to check with medical control in every case.

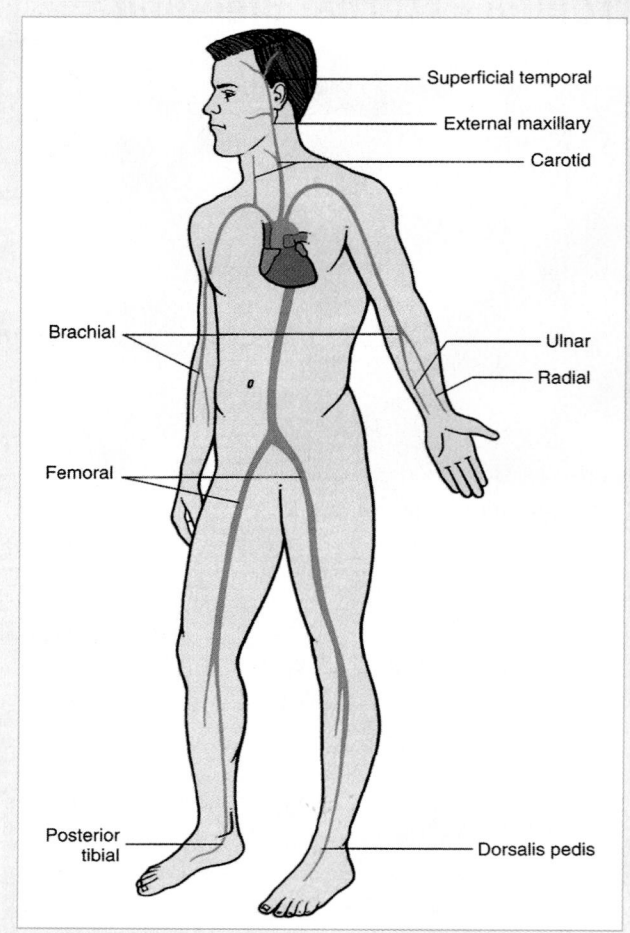

Figure 22-9 You should be familiar with the locations of arterial pressure points.

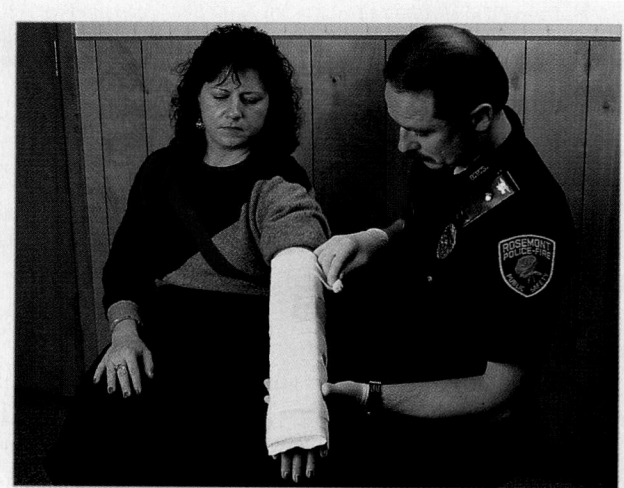

Figure 22-10 Use of a simple splint will often quickly control bleeding associated with a fracture. As long as a fracture is not immobilized, the bone ends are free to move and may continue to injure partially clotted vessels.

Figure 22-11 Air splints can also be used to control bleeding because they act as a pressure bandage for the entire extremity.

The following are the few specific purposes for which a PASG may be effective:

- To stabilize fractures of the pelvis and proximal femurs
- To control significant internal bleeding associated with fractures of the pelvis and proximal femurs
- To control massive soft-tissue bleeding of the lower extremities when direct pressure is not effective

Do not use the PASG if any of the following conditions exist:

- Pregnancy
- Pulmonary edema
- Acute heart failure
- Penetrating chest injuries
- Groin injuries
- Major head injuries
- A transport time of less than 30 minutes

In these situations, the PASG may worsen or complicate the patient's condition. Consult with medical control if you think prolonged use or use in unusual circumstances may be necessary. The PASG works by compressing the abdomen and lower extremities, increasing peripheral resistance in the circulatory system. This increases the amount of blood that is available to perfuse the vital organs. When applying the PASG, you should carefully inflate the device in increments. As a general rule, gradually inflate the legs of the PASG before inflating the abdominal portion. If you are using the device to stabilize a possible pelvic fracture, you must inflate all compartments. Always document all obvious injuries or deformities before application of the PASG.

EMT-B Safety

With a controversial therapy such as PASG, it is particularly important to seek direction from medical control if you have any doubts about a situation. Excellent EMS teams learn how to maintain the flow of care while contacting medical control, often by deciding before arrival on the scene which EMT-B will direct patient care and which one will handle radio communications.

Follow these steps to apply the PASG for bleeding control (Skill Drill 22-2 ▶):

1. **Apply the garment.** If you will immobilize or move the patient on a backboard, lay the PASG out on the board before rolling the patient onto it. Position the top of the abdominal section of the PASG below the lowest rib to ensure that it does not compromise chest expansion (**Step 1**).
2. **Close and fasten both leg compartments** and the abdominal compartment (**Step 2**).
3. **Open the stopcocks** (valves) to the compartments you are preparing to inflate. You will inflate both leg compartments (lower extremity bleeding) or all three compartments together (internal pelvic or abdominal bleeding) (**Step 3**).
4. **Inflate the compartments** with the foot pump. Do not increase the garment's pressure any more than necessary. A PASG is adequately inflated when the Velcro crackles. Higher pressures may cause local tissue damage. Always stop inflating the PASG once the patient's systolic blood pressure exceeds 100 mm Hg (**Step 4**).
5. **Check the patient's blood pressure** during inflation, and continue to monitor vital signs at least every 5 minutes afterward. Remember that the pressure gauges of the PASG measure the air pressure in the device. They *do not* reflect the patient's blood pressure. Be aware of temperature extremes or external pressure changes that can significantly affect the pressure exerted by the PASG, thus requiring frequent monitoring and adjustment (**Step 5**).

You will know that the device has worked if the patient's blood pressure increases.

Do not remove a PASG in the field. It must be deflated gradually in the hospital under careful supervision by a physician and only after appropriate intravenous solutions have been given. Before turning your patient over to hospital personnel, report the patient's blood pressure, the time you applied the PASG, and the results.

Tourniquets

A tourniquet is rarely needed to control bleeding. Applying a tourniquet is considered a last resort because it is rarely necessary and is effective for only a very limited number of injuries. Thus, a tourniquet often creates, rather than solves, problems. Application of a tourniquet can cause permanent damage to nerves, muscles, and blood vessels, resulting in the loss of an extremity. In addition, tourniquets are often improperly applied.

If you cannot control bleeding from the major vessel in an extremity in any other way, a properly applied tourniquet may save a patient's life. Specifically, the tourniquet is useful if a patient is bleeding severely from a partial or complete amputation.

Follow these steps to apply a tourniquet (Skill Drill 22-3 ▶):

1. **Fold a triangular bandage** until it is 4″ wide and six to eight layers thick.
2. **Wrap the bandage** around the extremity twice. Choose an area only slightly proximal to the bleeding, to reduce the amount of tissue damage to the extremity (**Step 1**).
3. **Tie one knot** in the bandage. Then place a stick or rod on top of the knot, and tie the ends of the bandage over the stick in a square knot (**Step 2**).
4. **Use the stick as a handle**, and twist it to tighten the tourniquet until the bleeding has stopped; then stop twisting (**Step 3**).
5. **Secure the stick in place**, and make the wrapping neat and smooth.
6. **Write "TK"** and the exact time (hour and minute) that you applied the tourniquet on a piece of adhesive tape. Use the phrase "time applied." Securely fasten the tape to the patient's forehead. Notify hospital personnel on your arrival that your patient has a tourniquet in place. Record this same information on the ambulance run report form (**Step 4**).
7. **As an alternative**, you can use a blood pressure cuff as an effective tourniquet. Position the cuff proximal to the bleeding point, and inflate it just enough to stop the bleeding. Leave the cuff inflated. If you use a blood pressure cuff, monitor the gauge continuously to make sure that the pressure is not

22-2

Skill Drill

Applying a Pneumatic Antishock Garment (PASG)

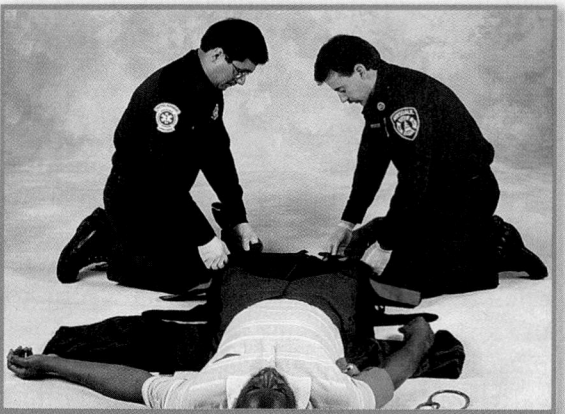

1 Apply the garment so that the top is below the lowest rib.

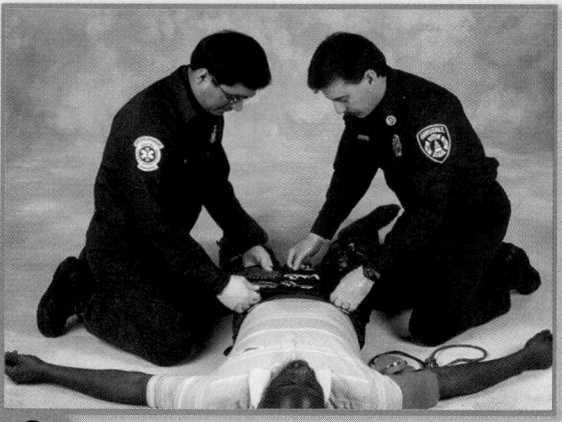

2 Enclose both legs and the abdomen.

3 Open the stopcocks.

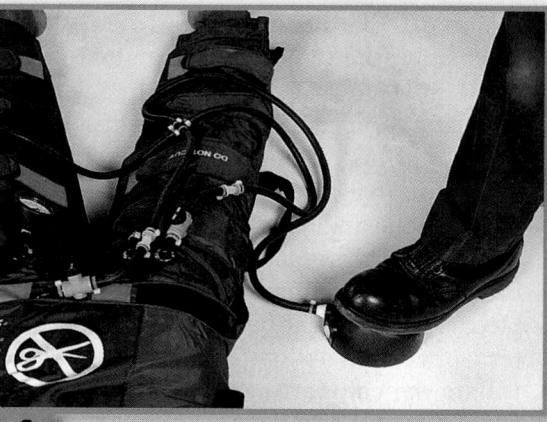

4 Inflate with the foot pump, and close the stopcocks when the patient's systolic blood pressure reaches 100 mm Hg or the Velcro crackles.

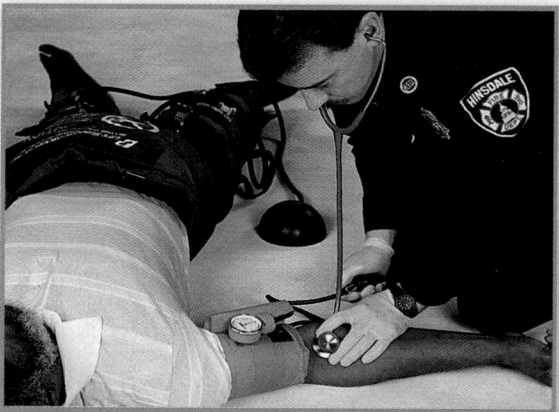

5 Check the patient's blood pressure again. Monitor the vital signs.

Applying a Tourniquet

22-3

Skill Drill

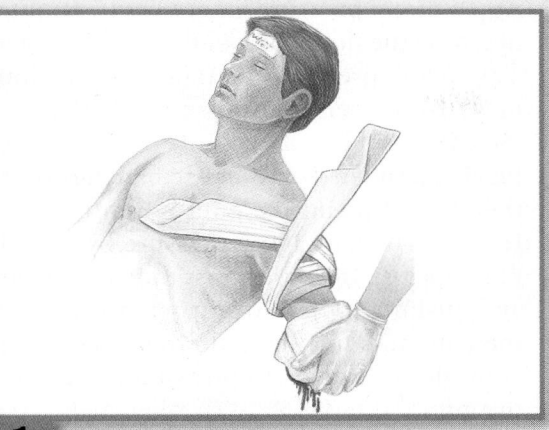

1 Create a 4″-wide, multilayered bandage. Wrap the bandage twice around the extremity, just above the bleeding site.

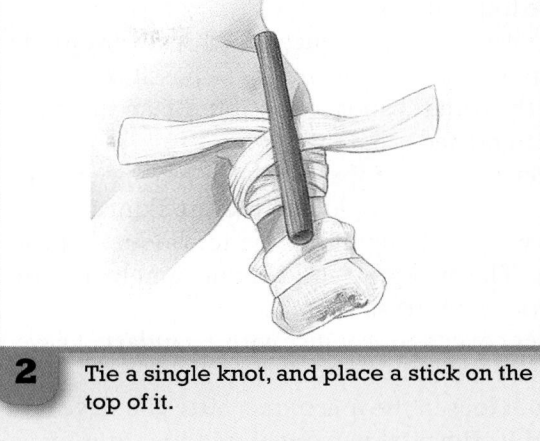

2 Tie a single knot, and place a stick on the top of it.

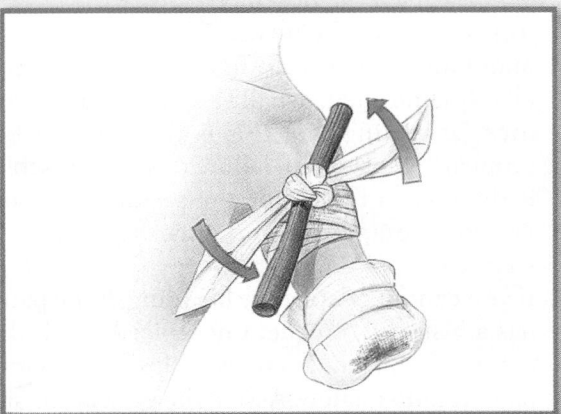

3 Tie a square knot over the stick, and then twist the stick until the bleeding stops.

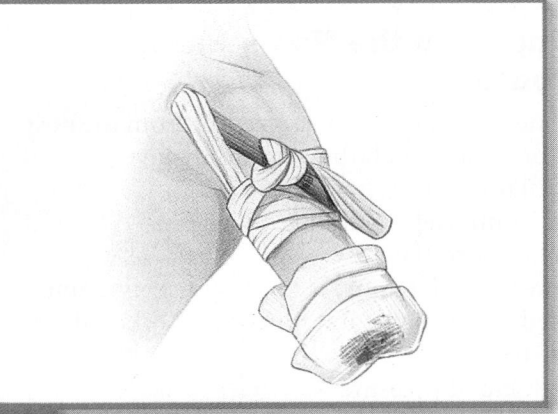

4 Secure the stick so that it will not unwind. Write "TK" and the exact time you applied the tourniquet on a piece of adhesive tape, fasten the tape to the patient's forehead, and notify hospital personnel on arrival.

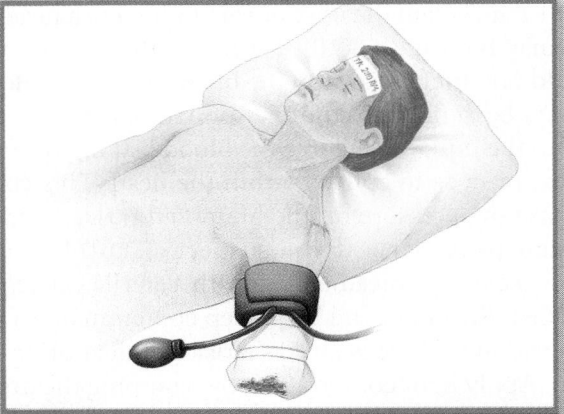

5 You can also use a blood pressure cuff as an effective tourniquet.

gradually dropping. You may have to clamp the tube with a hemostat leading from the cuff to the inflating bulb to prevent loss of pressure (**Step 5**).

Whenever you apply a tourniquet, make sure you observe the following precautions:

- Do not apply a tourniquet directly over any joint. Keep it as close to the injury as possible.
- Use the widest bandage possible. Make sure that it is tightened securely.
- Never use wire, rope, a belt, or any other narrow material. It could cut into the skin.
- Use wide padding under the tourniquet if possible. This will protect the tissues and help with arterial compression.
- Never cover a tourniquet with a bandage. Leave it open and in full view.
- Do not loosen the tourniquet after you have applied it. Hospital personnel will loosen it once they are prepared to manage the bleeding.

Bleeding from the Nose, Ears, and Mouth

Several conditions can result in bleeding from the nose, ears, and/or mouth, including the following:

- Skull fracture
- Facial injuries, including those caused by a direct blow to the nose
- Sinusitis, infections, nose drop use and abuse, dried or cracked nasal mucosa, or other abnormalities
- High blood pressure
- Coagulation disorders
- Digital trauma (nose picking)

Epistaxis, or nosebleed, is a common emergency. Occasionally, it can cause enough of a blood loss to send a patient into shock. Keep in mind that the blood you see may be only a small part of the total blood loss. Much of the blood may pass down the throat into the stomach as the patient swallows. A person who swallows a large amount of blood may become nauseated and start vomiting the blood, which is sometimes confused with internal bleeding. Most nontraumatic nosebleeds occur from sites in the septum, the tissue dividing the nostrils. You can usually handle this type of bleeding effectively by pinching the nostrils together.

Follow these steps to treat a patient with epistaxis (Skill Drill 22-4 ▶):

1. **Follow BSI precautions.**
2. **Help the patient to sit**, leaning forward, with the head tilted forward. This position stops the blood from trickling down the throat or being aspirated into the lungs.
3. **Apply direct pressure** for at least 15 minutes by pinching the fleshy part of the nostrils together. This is the preferred method. This technique may also be self-administered by the patient (**Step 1**).
4. **Placing a rolled 4″ × 4″ gauze bandage** between the upper lip and the gum is another option. Have the patient apply pressure by stretching the upper lip tightly against the rolled bandage and pushing it up into and against the nose. If the patient is unable to do this effectively, use your gloved fingers to press the gauze against the gum.
5. Keep the patient calm and quiet, especially if he or she has high blood pressure or is anxious. Anxiety tends to increase blood pressure, which could worsen the nosebleed (**Step 2**).
6. **Apply ice over the nose.**
7. **Maintain the pressure** until the bleeding is completely controlled, usually no more than 15 minutes (assuming that this is the patient's only problem). Most often, failure to stop a nosebleed is the result of releasing the pressure too soon.
8. **Provide prompt transport** once the bleeding has stopped.
9. **If you cannot control the bleeding**, if the patient has a history of frequent nosebleeds, or if there is a significant amount of blood loss, transport the patient immediately. Assess the patient for signs and symptoms of shock. Treat appropriately for shock, and administer oxygen via mask, if necessary (**Step 3**).

Bleeding from the nose or ears following a head injury may indicate a skull fracture. In these cases, you should not attempt to stop the blood flow. This bleeding may be difficult to control. Applying excessive pressure to the injury may force the blood leaking through the ear or nose to collect within the head. This could increase the pressure on the brain and possibly cause permanent damage. If you suspect a skull fracture, loosely cover the bleeding site with a sterile gauze pad to collect the blood and help keep contaminants away from the site. There is always a risk of infection to the brain. Apply light compression by wrapping the dressing loosely around the head (Figure 22-12 ▶). If blood or drainage contains cerebrospinal fluid, a characteristic staining of the dressing, much like a target or halo, will occur (Figure 22-13 ▶).

Controlling Epistaxis

Skill Drill

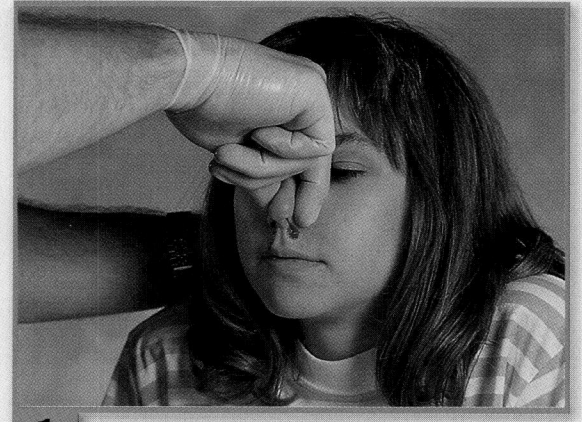

1 Use BSI precautions. Position the patient: sitting, leaning forward. Apply direct pressure, pinching the fleshy part of the nostrils.

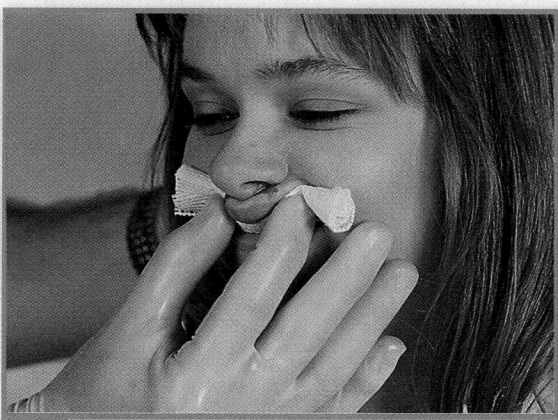

2 Alternative method: Use pressure with a rolled gauze bandage between the upper lip and gum. Calm the patient.

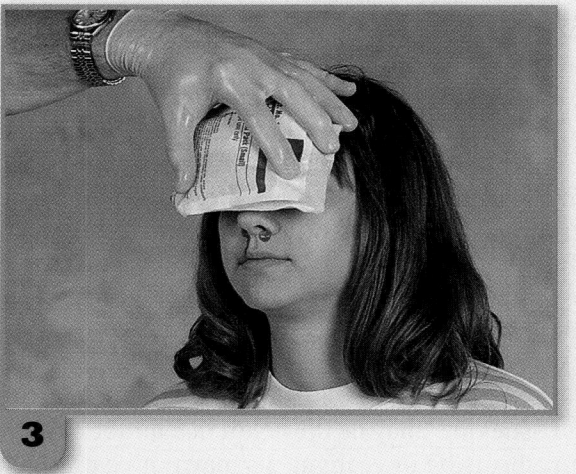

3 Apply ice over the nose. Maintain pressure until bleeding is controlled. Provide prompt transport after bleeding stops.

Transport immediately if indicated. Assess and treat for shock, including oxygen, as needed.

Internal Bleeding

Internal bleeding can be very serious, especially because you might not be aware that it is happening. Injury or damage to internal organs commonly results in extensive internal bleeding, which can cause hypovolemic shock before you realize the extent of blood loss. A person with a bleeding stomach ulcer may lose a large amount of blood very quickly. Similarly, a person who has a lacerated liver or a ruptured spleen may lose a considerable amount of blood within the abdomen. Yet the patient has no outward signs of bleeding.

Broken bones, especially broken ribs, also may cause serious internal blood loss. Sometimes this bleeding ex-

tends into the chest cavity and the soft tissues of the chest wall. A broken femur can easily result in the loss of 1 L or more of blood into the soft tissues of the thigh. Often the only signs of such bleeding are local swelling and bruising due to the accumulation of blood around the ends of the broken bone. Severe pelvic fractures may result in life-threatening hemorrhage.

You must always be alert to the possibility of internal bleeding and assess the patient for related signs and symptoms, particularly if the mechanism of injury is severe. If you suspect that a patient is bleeding internally, you should promptly transport him or her to the hospital.

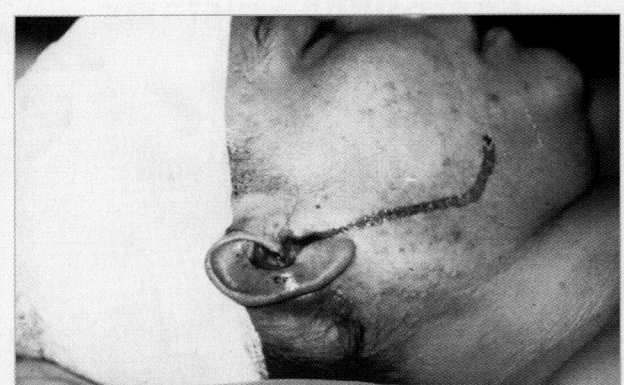

Figure 22-12 Bleeding from the ear after a head injury may indicate a skull fracture. Loosely cover the bleeding site with a sterile gauze pad, and apply light compression by wrapping the dressing loosely around the head.

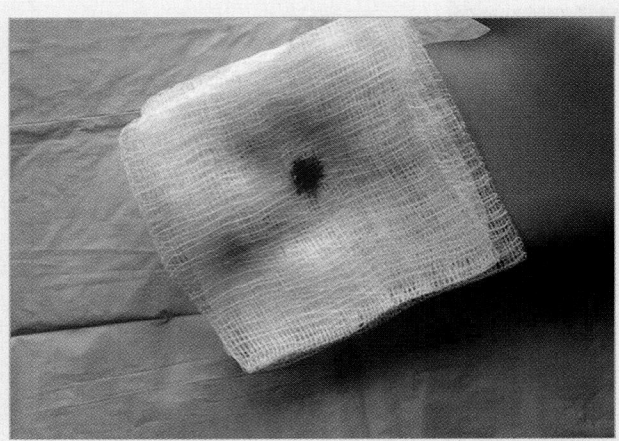

Figure 22-13 When cerebrospinal fluid is present in blood or drainage, a stain in the shape of a target or halo will appear.

Mechanism of Injury

A high-energy mechanism of injury should increase your index of suspicion for the possibility of serious unseen injuries such as internal bleeding in the abdominal cavity. Internal bleeding is possible whenever the mechanism of injury suggests that severe forces affected the body. These forces include blunt and penetrating trauma. Internal bleeding commonly occurs as a result of falls, blast injuries, and automobile or motorcycle crashes. Remember that internal bleeding can result from penetrating trauma as well.

As you assess a patient, look for signs of injury (DCAP-BTLS) over the chest or abdomen, including contusions, abrasions, lacerations, and other signs of injury or deformity. You should always suspect internal bleeding in a patient who has penetrating injury or blunt trauma.

Nature of Illness

Internal bleeding is not always caused by trauma. Many illnesses can cause internal bleeding. Some of the more common causes of nontraumatic internal bleeding include bleeding ulcers, bleeding from the colon, ruptured ectopic pregnancy, and aneurysms.

Abdominal pain and distention are frequent in these situations but are not always present. In older patients, dizziness, faintness, or weakness may be the first sign of nontraumatic internal bleeding. Ulcers or other gastrointestinal problems may cause vomiting of blood or bloody diarrhea.

It is not as important for you to know the specific organ involved as it is to recognize that the patient is in shock and respond appropriately.

Signs and Symptoms

The most common symptom of internal bleeding is pain. Significant internal bleeding will generally cause swelling in the area of bleeding. Intra-abdominal bleeding will often cause pain and distention. Bruising is a sign of internal bleeding. It is most common in head, extremity, and pelvic injuries and can be a sign of significant abdominal trauma. Bleeding into the chest may cause dyspnea in addition to tachycardia and hypotension. A bruise is also called a contusion, or ecchymosis. A hematoma, a mass of blood in the soft tissues beneath the skin, indicates bleeding into soft tissues and may be the result of a minor or a severe injury. Bruising or ecchymosis may not be present initially, and the only sign of severe pelvic or abdominal trauma may be redness, skin abrasions, or pain.

Bleeding, however slight, from any body opening is serious. It usually indicates internal bleeding that is not easy to see or control. Bright red bleeding from the mouth or rectum or blood in the urine (hematuria) may suggest serious internal injury or disease. Nonmenstrual vaginal bleeding is always significant.

Other signs and symptoms of internal bleeding in both trauma and medical patients include the following:

- Hematemesis. This is vomited blood. It may be bright red or dark red, or, if the blood has been partially digested, it may look like coffee-grounds vomitus.
- Melena. This is a black, foul-smelling, tarry stool that contains digested blood.
- Hemoptysis. This is bright red blood that is coughed up by the patient.

- Pain, tenderness, bruising, guarding, or swelling. These signs and symptoms may mean that a closed fracture is bleeding.
- Broken ribs, bruises over the lower part of the chest, or a rigid, distended abdomen. These signs and symptoms may indicate a lacerated spleen or liver. Patients with an injury to either organ may have referred pain in the right shoulder (liver) or left shoulder (spleen). You should suspect internal abdominal bleeding in a patient with referred pain.

The first sign of hypovolemic shock (hypoperfusion) is a change in mental status, such as anxiety, restlessness, or combativeness. In nontrauma patients, weakness, faintness, or dizziness on standing is another early sign. Changes in skin color or pallor (pale skin) are seen often in both trauma and medical patients. Later signs of hypoperfusion suggesting internal bleeding include the following:

- Tachycardia
- Weakness, fainting, or dizziness at rest
- Thirst
- Nausea and vomiting
- Cold, moist (clammy) skin
- Shallow, rapid breathing
- Dull eyes
- Slightly dilated pupils that are slow to respond to light
- Capillary refill in infants and children of more than 2 seconds
- Weak, rapid (thready) pulse
- Decreasing blood pressure
- Altered level of consciousness

Patients with these signs and symptoms are at risk. Some may be in danger. Even if their bleeding stops, it could begin again at any moment. Therefore, prompt transport is necessary.

Patient Assessment

Scene Size-up

As you approach the patient, be alert to potential hazards to yourself and the crew. If you are entering a residence, be alert for anxious bystanders and family members because they may become hostile. Ensure that you are only going to have to provide care for one patient, and be alert to indications of the nature of the illness (such as bloody emesis or bloody stool), or the MOI (such as a turned-over step stool). Consider the need for manual spinal immobilization and the need for additional resources, such as an advanced life support unit.

Initial Assessment

General Impression

As you approach the patient, be aware of obvious signs of injury and distress (such as facial grimace), along with determining gender and age. These indicators will help you determine whether the patient is sick or not so sick; this assists you in developing an index of suspicion for serious illness or injuries. In either case, consider the need for spinal stabilization. Ask the patient what happened to determine the chief complaint. This may point you in the direction of apparent life threats.

Airway and Breathing

Determine the patient's level of consciousness using the AVPU scale, ensure a patent airway, provide the patient with high-flow oxygen, or assist ventilation with a BVM device.

Circulation

Assess the patient's pulse rate and quality, skin color, and temperature to help establish the potential for internal bleeding and shock. Treat the patient for shock if needed by applying oxygen, improving circulation, and maintaining a normal body temperature. If significant external bleeding is found, it should be controlled in the initial assessment.

Transport Decision

The results of your general impression and assessment of ABCs will help you develop a sense of urgency for the patient and guide you in your transport decision to "stay and play" on scene or "play on the way" to the hospital.

Focused History and Physical Exam

Internal bleeding can be found in both medical and trauma patients. If the bleeding is severe, you have identified it in the initial assessment and begun treatment and rapid transport to the hospital. If the signs and symptoms of internal bleeding are not as obvious as described above, you will need to look more carefully in this step of the assessment process.

Rapid Physical Exam Versus Focused Physical Exam

The steps of the focused history and physical exam depend not only on the patient's MOI or nature of illness, but also on the specific type of medical or trauma situation. If the patient is an unresponsive medical patient, perform a rapid physical exam, obtain vital signs, and obtain the patient's medical history if possible. If the patient is a responsive medical patient, obtain the medical history, perform a focused physical exam based on the body region and complaint, use the OPQRST (**O**nset, **P**rovoking factors, **Q**uality, **R**adiation, **S**everity, **T**ime) mnemonic, and obtain vital signs. If the patient is a trauma patient with a significant MOI or multiple complaints or injuries, perform a rapid trauma exam, use the mnemonic DCAP-BTLS, obtain vital signs, and obtain what history is available. If the patient has a limited MOI with an isolated injury, perform a focused physical exam, obtain vital signs, and obtain the patient's history.

Interventions

Determine and initiate the appropriate treatment needed for the patient's condition based on the presentation, complaint, and assessment. If you have not treated for shock yet, you should do so now.

Detailed Physical Exam

If the patient is unstable, problems persist from the initial assessment, and time permits, perform a detailed physical exam of the patient as discussed in Chapter 8.

Ongoing Assessment

Reassess the components of initial assessment and the patient's vital signs. Because of its covert nature, signs and symptoms of internal bleeding are often slow to present. Children especially will compensate well for blood loss and then "crash" quickly. The ongoing assessment is your best opportunity to determine whether your patient's condition is improving or getting worse. Assess the effectiveness of any interventions and treatments provided to the patient.

Communication and Documentation

Communicate with the hospital on your findings and interventions to improve the patient's condition.

Emergency Medical Care

Controlling internal bleeding or bleeding from major organs usually requires surgery or other procedures that must be done in the hospital. It is important for you to calm and reassure the patient. Keeping the patient as still and quiet as possible assists the body's clotting process. Next, if spinal injury is not suspected, place the patient in the shock position. Provide high-flow oxygen; also maintain body temperature. You can usually control internal bleeding into the extremities quite well in the field simply by splinting the extremity, usually most effectively with an air splint, and you should never use a tourniquet to control the bleeding from closed, internal, soft-tissue injuries.

Follow these steps to care for patients with possible internal bleeding:

1. **Follow BSI precautions.**
2. **Maintain the airway** with cervical spine immobilization if the mechanism of injury suggests the possibility of spinal injury.
3. **Administer high-flow oxygen** and provide artificial ventilation as necessary.
4. **Control all obvious external bleeding.**
5. **Treat suspected internal bleeding** in an extremity by applying a splint.
6. **Monitor and record the vital signs** at least every 5 minutes.
7. **Give the patient nothing** (not even small sips of water) by mouth.
8. **Elevate the legs 6″ to 12″ in nontrauma patients** to help the blood return to the vital organs.
9. **Keep the patient warm.**
10. **Provide immediate transport** for all patients with signs and symptoms of shock (hypoperfusion). Report any changes in the patient's condition to emergency department personnel.

You are the Provider

Summary

When you are dealing with a bleeding problem, BSI precautions should always be your first concern. Always ensure that you and your partner are wearing the appropriate personal protective equipment. When assessing your patient, remember that not all major bleeding is external. Consider the mechanism of injury (MOI). Don't get tunnel vision and be distracted by the obvious injury. Listen to your patients— they know better than anyone when something does not feel right. Follow your ABCs and always prepare to treat for shock. Complete a full assessment if patient condition and time allow.

Assessment and Emergency Care

	External Bleeding	Internal Bleeding
Scene Size-up	Wear a minimum of gloves and eye protection to protect from bleeding.	Use appropriate BSI. Consider the need for manual spinal immobilization.
	Consider if additional resources are needed.	Consider if additional resources are needed.
	If incident involved violence, ensure that police are on scene.	
Initial Assessment		
■ General impression	Check for responsiveness.	Ask the patient what happened. Determine level of consciousness.
	Ask the patient about the chief complaint, if responsive.	
■ Airway and breathing	Ensure a patent airway, check for breath sounds, provide high-flow oxygen.	Ensure a patent airway, provide high-flow oxygen, or assist ventilation with a BVM device.
■ Circulation	Control significant bleeding.	Assess pulse rate and quality, skin color, and temperature.
	Consider shock in patients with blood loss. Assess for pale, cool, clammy skin or dizziness. Apply oxygen for significant blood loss.	Treat patient for shock if needed—apply oxygen, improve circulation, maintain normal body temperature.
■ Transport decision	Transport quickly if breathing problem or significant bleeding exists.	Transport quickly if signs of shock are present.
Focused History and Physical Exam	NOTE: The order of the steps in the focused history and physical exam differs depending on whether or not the patient has a significant MOI. The order below is for a patient with a significant MOI. For a patient without a significant MOI, perform a focused trauma assessment, obtain vital signs, and obtain the history.	
■ Focused physical exam or rapid physical exam	Type of exam will depend on the type of patient.	Type of exam will depend on the type of patient.
	Perform the focused physical exam if the patient is responsive and has an isolated injury.	Perform the focused physical exam on a responsive medical patient.
	Perform a rapid physical exam if the patient has significant mechanism of injury. Look for DCAP-BTLS. Treat life-threatening problems immediately.	Perform a rapid physical exam if the patient has significant mechanism of injury. Look for DCAP-BTLS. Treat life-threatening problems immediately.
■ Baseline vital signs	Assess baseline vital signs. Look for signs of hypoperfusion (shock).	Assess baseline vital signs. Look for signs of shock: systolic BP less than 100 mm Hg with weak, rapid pulse and cool, moist skin.
■ SAMPLE history	Obtain via bystanders or medic alert tags if patient is unresponsive.	For a responsive medical patient, use the OPQRST mnemonic.
■ Interventions	Provide high-flow oxygen if significant bleeding is suspected.	Treat for shock if not yet done.
	Control significant bleeding if it is visible.	
	If signs of shock are present, treat aggressively for shock. Rapidly transport to hospital.	
Detailed Physical Exam	Perform detailed physical exam during transport. If time allows, help to identify all injuries.	If patient is unstable, problems persist from initial assessment, and time permits, perform detailed physical exam.
Ongoing Assessment	Reassess patient. In cases of severe bleeding, take vital signs at least every 5 minutes.	Reassess vital signs. Determine whether patient's condition is improving or deteriorating. Assess effectiveness of interventions.
■ Communication and documentation	Report approximate amount of blood lost. Report all injuries and how patient has responded to care.	Communicate with hospital regarding findings, interventions, and patient's response. Include the MOI in your report.

NOTE: While the steps below are widely accepted, be sure to consult and follow your local protocol.

External Bleeding

Follow BSI precautions—minimum of gloves and eye protection.

Ensure that patient has open airway.

Maintain the cervical stabilization if MOI suggests possible spinal injury.

Provide high-flow oxygen.

Control bleeding using one of the following methods:
- Direct pressure and elevation
- Pressure dressings
- Pressure points
- Splints
- Air splints
- PASG
- Tourniquets

Controlling External Bleeding
1. Apply direct local pressure to bleeding site.
2. Elevate the bleeding extremity.
3. Create a pressure dressing.
4. Apply pressure at the appropriate pressure point while continuing to hold direct pressure.
5. If the wound continues to bleed, elevate extremity and place additional pressure over proximal pressure point.

Using PASG for Control of Massive Soft-tissue Bleeding in the Extremities
1. Apply the garment.
2. Close and fasten both leg compartments and the abdominal compartment.
3. Open the stopcocks.
4. Inflate the compartments similar to an air splint.
5. Check the patient's circulation, motor function, and sensation in distal lower extremities.

Applying a Tourniquet
1. Fold a triangular bandage.
2. Wrap the bandage around the extremity twice.
3. Tie one knot in the bandage. Place a stick or rod on top of the knot. Tie the ends of the bandage on the stick in a square knot.
4. Use the stick as a handle and twist it to tighten the tourniquet until bleeding has stopped.
5. Secure the stick in place with another triangular bandage.
6. Write "TK" and the exact time the tourniquet was applied on a piece of adhesive tape. Fasten the tape to the patient's forehead.
7. As an alternative, use a blood pressure cuff. Inflate enough to stop bleeding.

Treating Epistaxis
1. Follow BSI precautions.
2. Help the patient to sit, leaning forward.
3. Apply direct pressure for at least 15 minutes by pinching nostrils together.
4. Keep the patient calm and quiet.
5. Apply ice over the nose.
6. Maintain the pressure until bleeding is completely controlled.
7. Provide prompt transport.
8. If bleeding cannot be controlled, transport patient immediately. Treat for shock and administer oxygen via mask if necessary.

Internal Bleeding

Steps to Caring for Patient With Internal Bleeding
1. Follow BSI precautions.
2. Maintain the airway with cervical immobilization if MOI suggests possible spinal injury.
3. Administer high-flow oxygen.
4. Control all obvious external bleeding.
5. Apply a splint to an extremity where internal bleeding is suspected.
6. Monitor and record vital signs at least every 5 minutes.
7. Give the patient nothing by mouth.
8. Elevate the legs 6" to 12" in non-significant trauma patients.
9. Keep the patient warm.
10. Provide immediate transport for patients with signs and symptoms of shock. Report changes in condition to hospital personnel.

Using PASG for Treatment of Shock
1. Apply the garment.
2. Close and fasten both leg compartments and the abdominal compartment.
3. Open the stopcocks.
4. Contact medical control for specific verbal orders to inflate or use standing orders specific to inflation of PASG.
5. Inflate the compartments based on the patient's blood pressure.
6. Recheck the patient's blood pressure and inflate more based on patient response and blood pressure.

Prep Kit

Ready for Review

- Perfusion is the circulation of blood in adequate amounts to meet each cell's current needs for oxygen, nutrients, and waste removal.
- The three arms of the perfusion triad must be functioning to meet this demand: a working pump (heart), a set of intact pipes (blood vessels), and fluid volume (enough oxygen-carrying blood).
- Hypoperfusion, or shock, occurs when one or more of these three arms is not working properly and the cardiovascular system fails to provide adequate perfusion.
- Both internal and external bleeding can cause shock. You must know how to recognize and control both.
- The six methods to control bleeding, in order, are:
  - Direct local pressure
  - Elevation
  - Pressure dressing
  - Pressure points
  - Splinting device
  - Pneumatic antishock garment (PASG) (as allowed by local protocol)
- Use of a tourniquet is always a last resort and should be avoided if possible.
- Bleeding from the nose, ears, and/or mouth may result from skull fracture. Other causes include high blood pressure and sinus infection. Evaluate the MOI and consider the more serious problem of skull fracture.
- Bleeding around the face always presents a risk for airway obstruction or aspiration. Maintain a clear airway by positioning the patient appropriately and using suction when indicated.
- If bleeding is present at the nose and skull fracture is suspected, place a gauze pad loosely under the nose.
- If bleeding from the nose is present and skull fracture is not suspected, pinch both nostrils together for 15 minutes. If the patient is awake and has a patent airway, place a gauze pad inside the upper lip against the gum.
- Any patient you suspect of having internal bleeding or significant external bleeding should be transported promptly.

- If the mechanism of injury is significant, be alert to signs of unseen bleeding in the chest or abdomen—signs such as serious bruising or symptoms such as complaints of difficulty breathing or abdominal pain.
- Signs of serious internal bleeding include the following:
  - Vomiting blood (hematemesis)
  - Black tarry stools (melena)
  - Coughing up blood (hemoptysis)
  - Distended abdomen
  - Broken ribs

Vital Vocabulary

aorta The main artery, which receives blood from the left ventricle and delivers it to all the other arteries that carry blood to the tissues of the body.

arterioles The smallest branches of arteries leading to the vast network of capillaries.

artery A blood vessel, consisting of three layers of tissue and smooth muscle that carries blood away from the heart.

capillaries The small blood vessels that connect arterioles and venules; various substances pass through capillary walls, into and out of the interstitial fluid, and then on to the cells.

coagulation The formation of clots to plug openings in injured blood vessels and stop blood flow.

contusion A bruise, or ecchymosis.

ecchymosis Discoloration of the skin associated with a closed wound; bruising.

epistaxis A nosebleed.

hematoma A mass of blood in the soft tissues beneath the skin.

hemophilia A congenital condition in which the patient lacks one or more of the blood's normal clotting factors.

hemorrhage Bleeding.

hypovolemic shock A condition in which low blood volume, due to massive internal or external bleeding or extensive loss of body water, results in inadequate perfusion.

perfusion Circulation of blood within an organ or tissue in adequate amounts to meet the current needs of the cells.

pneumatic antishock garment (PASG) An inflatable device that covers the legs and abdomen; used to splint the lower extremities or pelvis or to control bleeding in the lower extremities, pelvis, or abdominal cavity.

pressure point A point where a blood vessel lies near a bone; useful when direct pressure and elevation do not control bleeding.

shock A condition in which the circulatory system fails to provide sufficient circulation so that every body part can perform its function; also called hypoperfusion.

tourniquet The bleeding control method of last resort that occludes arterial flow; used only when all other methods have failed and the patient's life is in danger.

veins The blood vessels that carry blood from the tissues to the heart.

Technology

- Interactivities
- Vocabulary Explorer
- Anatomy Review
- Web Links
- Online Review Manual

Assessment in Action

You are dispatched to 1035 Lenore St for a stabbing. While en route, you are notified by the dispatcher that the scene is secure per law enforcement. You arrive to find a 20-year-old woman who is crying and holding a blood-soaked towel on her left lower arm. A police officer explains that this injury was self-inflicted and shows you the kitchen knife she used to cut her arm. She appears pale, and she tells you that she feels cold and faint. You briefly expose the wound in order to replace the towel with sterile dressings, when you glimpse spurting blood.

You immediately initiate direct pressure. Your partner instructs the patient to lay on the gurney (which he has placed in Trendelenburg position) and places her on high-flow oxygen. As you take steps to control the bleeding, you begin immediate transport to the closest appropriate medical facility. While en route, you are able control the patient's bleeding through the use of elevation, a pressure dressing, and pressure points. Her vital signs en route are the following: blood pressure, 96/52 mm Hg; pulse, 130 beats/min and respirations, 42 breaths/min.

1. What is your first concern upon hearing the nature of the call?

    A. BSI precautions
    B. Scene safety
    C. Location of the injury
    D. Extent of the injury

2. At minimum, what personal protective equipment should you use during this call?

    A. Gloves
    B. Gloves and gown
    C. Eye protection
    D. Gloves and eye protection

3. What clues lead you to the understanding that this patient has lost a significant amount of blood without having to visualize the wound?

    A. Skin color
    B. Patient complaint(s)
    C. Condition of the towel
    D. All of the above

4. What actions should you complete before transport?

    A. Detailed inspection of the scene
    B. Getting specifics from the officers
    C. Obtaining a thorough medical history
    D. None of the above

5. Choose the correct sequence used to control external bleeding.

    1. Pressure points
    2. Direct pressure
    3. Elevation
    4. Pressure dressings
        A. 1, 2, 3, 4
        B. 2, 1, 3, 4
        C. 2, 3, 4, 1
        D. 2, 3, 1, 4

6. Which answer correctly describes the Trendelenburg position?

    A. Lying flat
    B. Head up and feet down
    C. Head down and feet up
    D. None of the above

Challenging Questions

7. What other physical signs can provide clues as to the medical history regarding this patient?

8. What steps should you take to ensure your safety and the safety of your crew during transport?

9. If dressings had been applied prior to your arrival, should you remove them to visualize the wound?

10. Is it important to know what instrument or weapon caused her laceration?

Points to Ponder

You are dispatched to an area ranch for a 26-year-old man who was kicked in the abdomen by a horse. This ranch is approximately 15 miles from your station, and you are told that the patient is located in the main ranch house and has told his family he's feeling fine. You have a tiered-response system and first responders have been simultaneously dispatched from a nearby fire department. Shortly after their arrival, they report the patient has the following vital signs: blood pressure, 96/54 mm Hg; pulse, 136 beats/min and weak and rapid; and respiratory rate, 36 breaths/min and shallow.

What do his vital signs indicate? What may cause a delay in his transport?

Issues: Recognizing the Signs of Shock, Explaining the Urgency to Transport Patients Who Have Suffered Significant Blood Loss, Minimizing Delays in Transport and Treatment.

www.EMTB.com

Shock

Objectives

Cognitive

5-1.9   List signs and symptoms of shock (hypoperfusion). (p 683, 684)

5-1.10   State the steps in the emergency medical care of the patient with signs and symptoms of shock (hypoperfusion). (p 686)

Affective

5-1.11   Explain the sense of urgency to transport patients who are bleeding and show signs of shock (hypoperfusion). (p 688, 690)

Psychomotor

5-1.16   Demonstrate the care of the patient exhibiting signs and symptoms of shock (hypoperfusion). (p 686)

5-1.17   Demonstrate completing a prehospital care report for the patient with bleeding and/or shock (hypoperfusion). (p 686)

You are the Provider

You and your partner respond to a motor vehicle crash involving two cars. En route, you follow BSI precautions in preparation for your arrival at the scene, where you find a 25-year-old man. Law enforcement informs you that the other car left the scene. The patient was restrained and is sitting outside the car. He is pale. You observe the car for damage. The air bag has deployed and the steering wheel has some damage.

1. On the basis of your scene size-up, what may potentially be wrong with the patient?
2. What is your general impression of this patient?

Shock

Shock has a number of meanings. For example, we often say that a person who has been frightened or received bad news is in shock. An electric current passing through the body delivers a shock. In this chapter, shock describes a state of collapse and failure of the cardiovascular system. When the circulation of the blood becomes inadequate, the oxygen and nutrient needs of the cells cannot be met. In the early stages of shock, the body will attempt to maintain homeostasis (a balance of all systems of the body); however, as shock progresses, blood circulation slows and eventually ceases. If not treated promptly, shock (hypoperfusion) can be fatal. Shock is the result of hypoperfusion to the cells of the body that causes organs and then organ systems to fail. Unless treated successfully, this process will ultimately result in the death of the organism.

Shock can occur because of several medical or traumatic events such as a heart attack, severe allergic reaction, an automobile crash, or a gunshot wound. As an EMT-B, you will respond to these different types of emergencies to provide care and transportation for these patients. Therefore, you must be constantly alert to the signs and symptoms of shock. In general, you cannot go wrong assuming that every patient is in shock or may go into shock; treat every patient for shock.

This chapter begins with a close-up look at perfusion, the function that fails in shock. Next it looks at the physiologic causes of shock and describes each of its major forms. Finally, it discusses the emergency treatment of shock in general and of each kind of shock in particular.

Perfusion

Perfusion is the cardiovascular system's circulation of blood and oxygen to all cells in different tissues and organs in the body. Perfusion is also an important part of the process by which waste products made by the cells are removed. Shock, or hypoperfusion, refers to a state of collapse and failure of the cardiovascular system that leads to inadequate circulation. Like internal bleeding, shock is an unseen underlying life threat caused by a medical disorder or traumatic injury. However, you can recognize the signs and symptoms of shock early and initiate treatment for this life threat soon after the onset of shock. Inadequate circulation can lead to cell death. To protect vital organs, the body attempts to compensate by directing blood flow from organs that are more tolerant of low flow (such as the skin and intestines) to organs that cannot tolerate low blood flow (such as heart, brain, and lungs). If the conditions causing shock are not promptly addressed, the patient will soon die.

The cardiovascular system consists of three parts: a pump (the heart), a set of pipes (the blood vessels), and the contents of the container (the blood) (Figure 23-1 ▶). These three parts can be referred to as the "perfusion triangle" (Figure 23-2 ▶). When a patient is in shock, one or more of the three sides is not working properly.

Blood is the vehicle for carrying oxygen and nutrients through the vessels to the capillary beds, where these supplies are exchanged for waste products. The blood keeps moving as a result of pressure that is generated by the contractions of the heart and affected by the dilating and constricting of the vessels. This pressure, which we call blood pressure, is usually carefully controlled by the body so that there is always sufficient circulation, or perfusion, in the various tissues and organs. Blood pressure is, in fact, a rough measure of perfusion.

Remember that blood pressure is really the pressure of blood within the vessels at any one time. The systolic pressure is the peak arterial pressure, or pressure generated every time the heart contracts; the diastolic pressure is the pressure maintained within the arteries while the heart rests between heartbeats.

Blood flow through the capillary beds is regulated by the capillary sphincters, circular muscular walls that constrict and dilate. These sphincters are under the control of the autonomic nervous system, which

www.EMTB.com

Technology

Interactivities

Vocabulary Explorer

Anatomy Review

Web Links

Online Review Manual

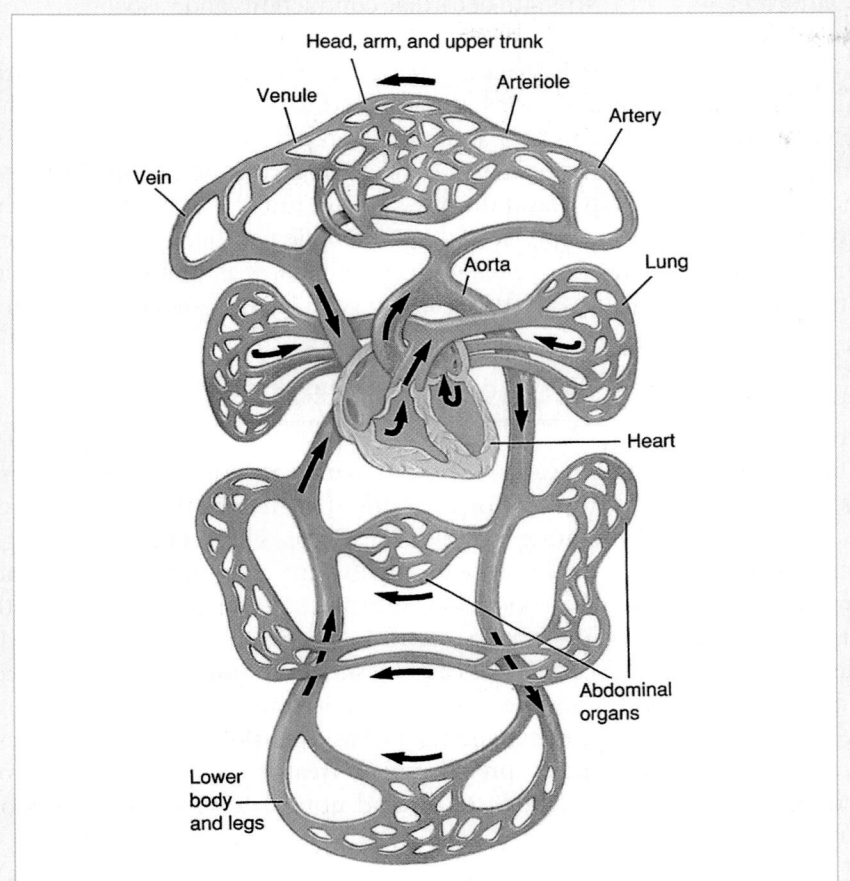

Figure 23-1 The cardiovascular system consists of three parts: the pump (heart), the container (vessels), and the contents (blood). The blood carries oxygen and nutrients through the vessels to the capillary beds, where they are exchanged for waste products.

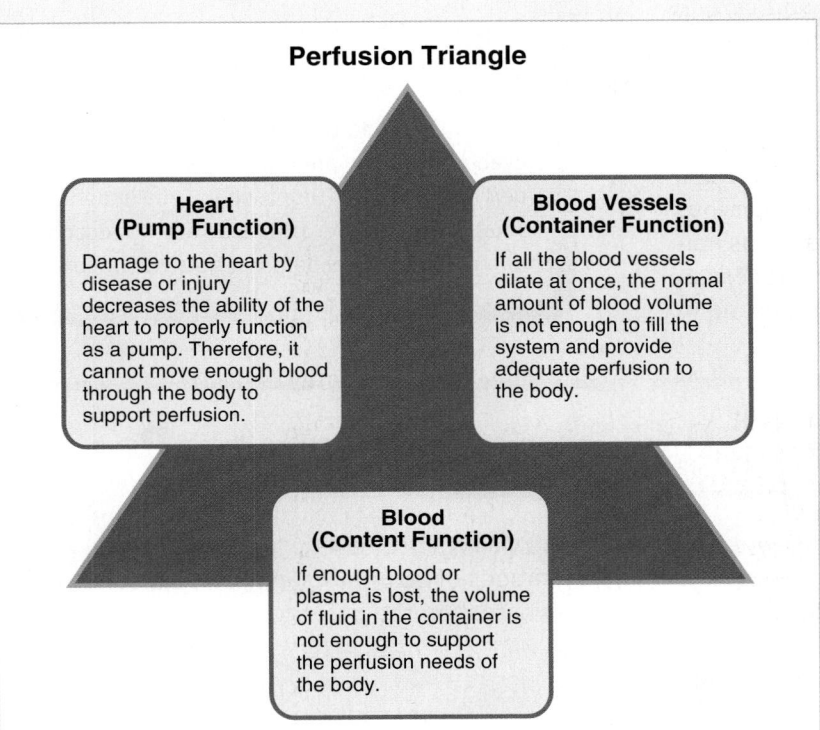

Figure 23-2 The heart, the blood vessels, and the blood represent the three legs of the perfusion triangle.

regulates involuntary functions such as sweating and digestion. Capillary sphincters also respond to other stimuli such as heat, cold, the need for oxygen, and the need for waste removal. Keep in mind that, under normal circumstances, not all cells have the same needs at the same time. For example, the stomach and intestines have a high need for blood flow during and shortly after eating, when digestion is at a peak. Between meals, blood flow is lessened, and blood is diverted to other areas. The brain, by contrast, needs a constant and consistent supply of blood to function.

Thus, regulation of blood flow is determined by cellular need and is accomplished by vessel constriction or dilation, together with sphincter constriction or dilation. Maintenance of blood flow, or perfusion, is accomplished by the heart, blood vessels, and blood working together.

Perfusion requires more than just having a working cardiovascular system, however. It also requires adequate oxygen exchange in the lungs, adequate nutrients in the form of glucose in the blood, and adequate waste removal, primarily through the lungs. Carbon dioxide is one of the primary waste products of cellular work (metabolism) in the body and is removed from the body by the lungs. This is the reason adequate ventilation and oxygenation is one of your primary concerns. The body has mechanisms in place to help support the respiratory and cardiovascular systems when the need for perfusion of vital organs is increased. These mechanisms, including the autonomic nervous system and certain chemicals called hormones, are triggered when the body senses that the pressure in the system is falling. The sympathetic side of the autonomic nervous system, which is responsible for the fight-or-flight response, will assume more control of the body's functions during a state of shock. This response by the autonomic nervous system will cause the release of hormones such as epinephrine. These hormones cause changes in certain body functions such as an increase in heart rate and in

the strength of cardiac contractions and vasoconstriction in nonessential areas, primarily in the skin and gastrointestinal tract (peripheral vasoconstriction). Together, these actions are designed to maintain pressure in the system and, as a result, perfusion of all vital organs.

Eventually, there is also a shifting of body fluids to help maintain pressure within the system. However, the response of the autonomic nervous system and hormones comes within seconds. It is this response that causes all the signs and symptoms of shock in a patient.

Causes of Shock

Shock can result from many conditions, including respiratory failure, acute allergic reactions, and overwhelming infection. In all cases, however, the damage occurs because of insufficient perfusion of organs and tissues. As soon as perfusion stops or becomes impaired, tissues start to die, affecting all local body processes. If the conditions causing shock are not promptly arrested and reversed, death soon follows.

Understanding the basic physiologic causes of shock will better prepare you to treat it (Figure 23-3 ▶). There are cardiovascular and noncardiovascular causes of

✳ EMT-B Tips

Shock is a complex physiologic process that gives subtle signs to its presence before it becomes severe. These early signs relate very closely to the events that lead to more severe shock, so it is even more important than usual to know the underlying processes thoroughly. If you understand what causes shock, you will be able to recognize it in many patients before it gets out of control.

You are the Provider   Part 2

You approach the patient and introduce yourself. He appears visibly upset but he lets you take his vital signs. His vital signs are a pulse of 115 beats/min, respirations of 26 breaths/min, and a blood pressure of 110/75 mm Hg. He has a laceration on his knee where it hit the dashboard.

3. What are the next steps in assessing this patient?
4. Is there a significant mechanism of injury?

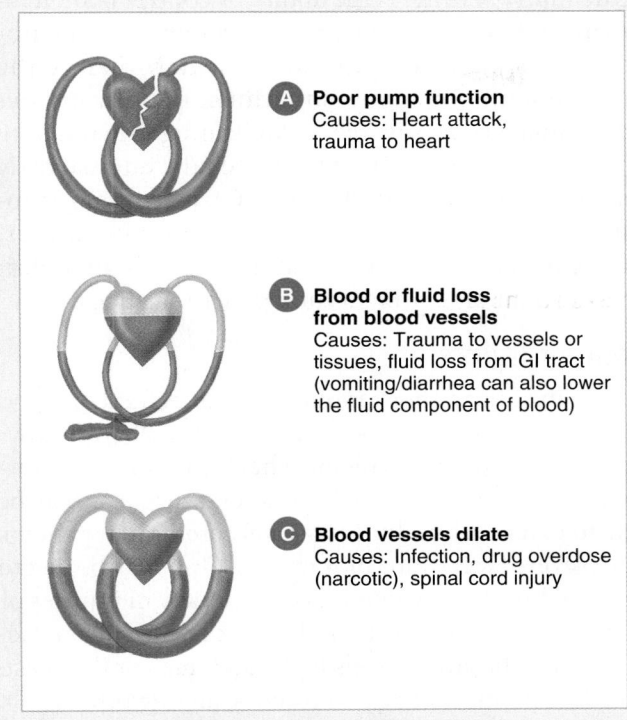

Figure 23-3 There are three basic causes of shock and impaired tissue perfusion. **A.** Pump failure occurs when the heart is damaged by disease or injury. The heart may not generate enough energy to move the blood through the system. **B.** Decreased blood volume, often a result of bleeding, leads to inadequate perfusion. **C.** The blood vessels can dilate excessively so that the blood within them, even though it is of normal volume, is inadequate to fill the system and provide efficient perfusion.

Inside the figure:

A **Poor pump function**
Causes: Heart attack, trauma to heart

B **Blood or fluid loss from blood vessels**
Causes: Trauma to vessels or tissues, fluid loss from GI tract (vomiting/diarrhea can also lower the fluid component of blood)

C **Blood vessels dilate**
Causes: Infection, drug overdose (narcotic), spinal cord injury

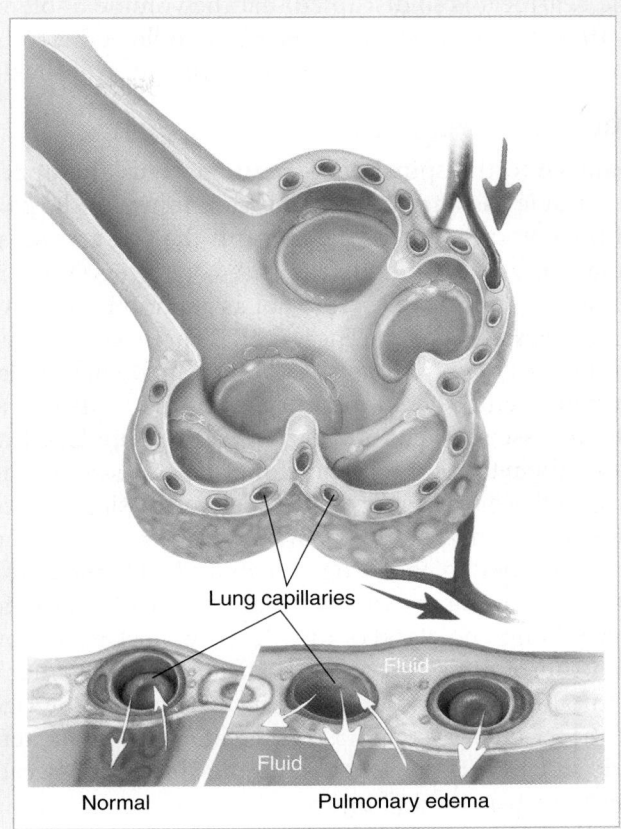

Lung capillaries

Normal   Fluid   Pulmonary edema   Fluid

Figure 23-4 Pulmonary edema develops as a result of fluid buildup within the pulmonary tissue. The edema causes swelling and leads to impaired ventilation.

shock. Cardiovascular causes of shock include heart attack, disease, and injury. Noncardiovascular causes include respiratory insufficiency and anaphylaxis, an unusual or exaggerated allergic reaction to foreign protein or other substances. Noncardiovascular causes of shock will ultimately affect one of the three sides of the perfusion triangle and interrupt the body's normal perfusion status.

Cardiovascular Causes of Shock

Pump Failure

Cardiogenic shock is caused by inadequate function of the heart, or pump failure. Circulation of blood throughout the vascular system requires the constant pumping action of a normal and vigorous heart muscle. Many diseases can cause destruction or inflammation of this muscle. Within certain limits, the heart can adapt to these problems. If too much muscular damage occurs, however, as sometimes happens after a heart attack, the heart no longer functions well. A major ef-

fect is the backup of blood into the lungs. The resulting buildup of fluid within the pulmonary tissue is called pulmonary edema. Edema is the presence of abnormally large amounts of fluid between cells in body tissues, causing swelling of the affected area (Figure 23-4 ▲). Pulmonary edema leads to impaired ventilation, which may be manifested by an increased respiratory rate and abnormal lung sounds.

The muscular contraction of the heart moves blood through the vessels at distinct pressures. For blood to circulate efficiently throughout the entire system, there must be the right amount of pressure and an adequate number of heartbeats. For this reason, the heart has its own electrical system that initiates and regulates its beating. Disease or injury can damage or destroy this system, causing irregular and uncoordinated beats, beats that are too slow (fewer than 60/min), or beats that are too fast (more than 150/min).

Cardiogenic shock develops when the heart muscle can no longer generate enough pressure to circulate the blood to all organs or when the regularity of the heartbeat is so disrupted that the volume of blood within the system can no longer be handled efficiently. In either case, direct pump failure is the cause of shock.

Poor Vessel Function

Damage to the spinal cord, particularly at the upper cervical levels, may cause significant injury to the part of the nervous system that controls the size and muscular tone of the blood vessels. Neurogenic shock is usually the result. Although not as common, there are medical causes as well. These include brain conditions, tumors, pressure on the spinal cord, and spina bifida. In neurogenic shock, the muscles in the walls of the blood vessels are cut off from the nerve impulses that cause them to contract. Therefore, all vessels below the level of the spinal injury dilate widely, increasing the size and capacity of the vascular system (Figure 23-5 ▼) and causing blood to pool. The available 6 L of blood in the body can no longer fill the enlarged vascular system. Even though no blood or fluid has been lost, perfusion of organs and tissues becomes inadequate, and shock occurs. In this condition, a radical change in the size of the vascular system has caused shock. A characteristic sign of this type of shock is the absence of sweating below the level of injury.

With this type of injury, many other functions that are under the control of the same part of the nervous system are also lost. The most important of them, in an

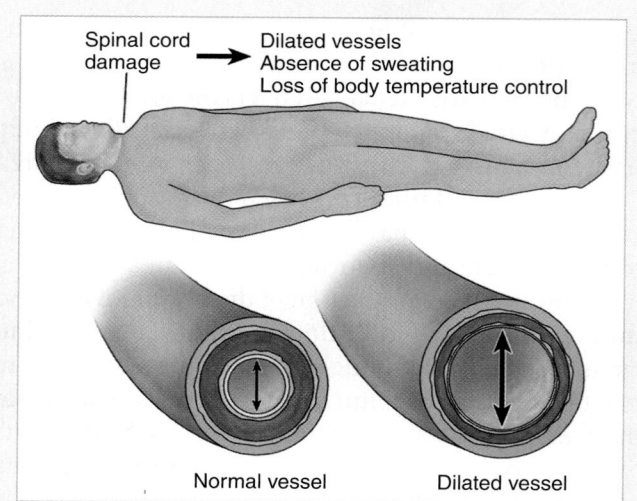

Figure 23-5 Damage to the spinal cord can cause significant injury to the part of the nervous system that controls the size and muscle tone of blood vessels. If the muscles in the blood vessels are cut off from their impulses to contract, the vessels dilate widely, increasing the size and capacity of the vascular system. The blood in the body can no longer fill the enlarged vessels; inadequate perfusion results.

acute injury setting, is the ability to control body temperature. Body temperature in a patient with neurogenic shock can rapidly fall to match that of the environment. In many situations, significant hypothermia occurs, severely complicating the situation. Hypothermia is a condition in which the internal body temperature falls below 95°F (35°C), usually after prolonged exposure to cool or freezing temperatures. Maintenance of body temperature is always an important element of treatment for a patient in shock.

Content Failure

Following injury, shock is often a result of fluid or blood loss. This type of shock is called hypovolemic (low-volume) shock; hypovolemic shock caused by hemorrhage is known as hemorrhagic shock. The loss may be due to external bleeding, which is common in patients with severe lacerations or fractures. Or it may be due to internal bleeding, which follows a variety of injuries or diseases, such as rupture of the liver or the spleen, lacerations of the great vessels in the abdomen or the chest, bleeding peptic ulcers, and tumors, among others.

Hypovolemic shock also occurs with severe thermal burns. In this case, it is intravascular plasma (the colorless part of the blood) that is lost, leaking from the circulatory system into the burned tissues that lie adjacent to the injury. Likewise, crushing injuries may result in the loss of blood and plasma from damaged vessels into injured tissues. Dehydration, the loss of water from body tissues, aggravates shock. In these circumstances, the common factor is an insufficient volume of blood within the vascular system to provide adequate circulation to the organs of the body.

Combined Vessel and Content Failure

In some patients who have severe infections, usually bacterial, toxins (poisons) generated by the bacteria or by infected body tissues produce a condition called septic shock. In this condition, the toxins damage the vessel walls, causing them to become leaky and unable to contract well. Widespread dilation of vessels, in combination with plasma loss through the injured vessel walls, results in shock.

Septic shock is a complex problem. First, there is an insufficient volume of fluid in the container, because much of the plasma has leaked out of the vascular system (hypovolemia). Second, the fluid that has leaked out often collects in the respiratory system, interfering with ventilation. Third, there is a larger-than-normal vascular bed to contain the smaller-than-normal volume of intravascular fluid.

Septic shock is almost always a complication of some very serious illness, injury, or surgery.

Noncardiovascular Causes of Shock

There are two causes of shock that do not result from disturbances of the cardiovascular system: respiratory insufficiency and anaphylaxis.

Respiratory Insufficiency

A patient with a severe chest injury or obstruction of the airway may be unable to breathe in an adequate amount of oxygen.

An insufficient concentration of oxygen in the blood can produce shock as rapidly as vascular causes, even if the volume of blood, the volume of the vessels, and the action of the heart are all normal. Without oxygen, the organs in the body cannot survive, and their cells promptly start to deteriorate.

This is why the first two steps in resuscitation are always securing an airway and restoring respirations. Circulation of nonoxygenated blood will not benefit the patient.

Anaphylactic Shock

Anaphylaxis, or anaphylactic shock, occurs when a person reacts violently to a substance to which he or she has been sensitized. Sensitization means becoming sensitive to a substance that did not initially cause a reaction. Do not be misled by a patient who reports no history of allergic reaction to a substance on first or second exposure. Each subsequent exposure after sensitization tends to produce a more severe reaction.

Instances that cause severe allergic reactions commonly fall into the following four categories of exposure:

- Injections (tetanus antitoxin, penicillin)
- Stings (honeybee, wasp, yellow jacket, hornet)
- Ingestion (shellfish, fruit, medication)
- Inhalation (dusts, pollens)

Anaphylactic reactions can develop in minutes or even seconds after contact with the substance to which the patient is allergic. The signs of such allergic reactions are very distinct and not seen with other forms of shock. (Table 23-1) shows the signs of anaphylactic shock in the order in which they typically occur. Note that cyanosis (bluish color of the skin) is a late sign of anaphylactic shock.

In anaphylactic shock, there is no loss of blood, no mechanical vascular damage, and only a slight possibility of direct cardiac muscular injury. Instead, there is widespread vascular dilation. The combination of poor oxy-

genation and poor perfusion in anaphylactic shock may easily prove fatal.

Psychogenic Shock

A patient in psychogenic shock has had a sudden reaction of the nervous system that produces a temporary, generalized vascular dilation, resulting in fainting, or syncope. Blood pools in the dilated vessels, reducing the blood supply to the brain; as a result, the brain ceases to function normally, and the patient faints. While there are many causes of syncope, it is important to realize that some are of a serious nature but others are not. Causes of syncope that are potentially life threatening result from events such as an irregular heartbeat or a brain aneurysm. Other non–life-threatening events that cause syncope may be the receipt of bad news or experiencing fear or unpleasant sights (like the sight of blood).

TABLE 23-1 Signs of Anaphylactic Shock

Skin
- Flushing, itching, or burning, especially over the face and upper part of the chest
- Urticaria (hives), which may spread over large areas of the body
- Edema, especially of the face, tongue, and lips
- Pallor
- Cyanosis (a bluish cast to the skin resulting from poor oxygenation of circulating blood) about the lips

Circulatory System
- Dilation of peripheral blood vessels
- A drop in blood pressure
- A weak, barely palpable pulse
- Dizziness
- Fainting and coma

Respiratory System
- Sneezing or itching in the nasal passages
- Tightness in the chest, with a persistent dry cough
- Wheezing and dyspnea (difficulty breathing)
- Secretions of fluid and mucus into the bronchial passages, alveoli, and lung tissue, causing coughing
- Constriction of the bronchi; difficulty drawing air into the lungs
- Forced expiration, requiring exertion and accompanied by wheezing
- Cessation of breathing

The Progression of Shock

Although you cannot see shock, you can see its signs and symptoms (Table 23-2 ▶). The early stage of shock, while the body can still compensate for blood loss, is called compensated shock. The late stage, when blood pressure is falling, is called decompensated shock. The last stage, when shock has progressed to a terminal stage, is called irreversible shock. A transfusion during irreversible shock will not save the patient's life.

Remember that blood pressure may be the last measurable factor to change in shock. As we have seen, the body has several automatic mechanisms to compensate for initial blood loss and to help maintain blood pressure. Thus, by the time you detect a drop in blood pressure, shock is well developed. This is particularly true in infants and children, who can maintain their blood pressure until they have lost more than half their blood volume. By the time blood pressure drops in infants and children who are in shock, they are close to death.

You should expect shock in many emergency medical situations. For example, you would expect shock to accompany massive external or internal bleeding.

You should also expect shock if a patient has any one of the following conditions:

- Multiple severe fractures
- Abdominal or chest injury
- Spinal injury
- A severe infection
- A major heart attack
- Anaphylaxis

Study This !!!

✳ EMT-B Tips

Taking and recording frequent vital signs—and observing perfusion indicators such as skin condition and mental status—will give you a window into the progression of shock. Use your documentation to remind you to suspect shock early and treat it aggressively.

TABLE 23-2 Progression of Shock

Compensated Shock

Agitation
Anxiety
Restlessness
Feeling of impending doom
Altered mental status
Weak, rapid (thready), or absent pulse
Clammy (pale, cool, moist) skin
Pallor, with cyanosis about the lips
Shallow, rapid breathing
Air hunger (shortness of breath), especially if there is a chest injury
Nausea or vomiting
Capillary refill of longer than 2 seconds in infants and children
Marked thirst

Decompensated Shock

Falling blood pressure (systolic blood pressure of 90 mm Hg or lower in an adult)
Labored or irregular breathing
Ashen, mottled, or cyanotic skin
Thready or absent peripheral pulses
Dull eyes, dilated pupils
Poor urinary output

You are the Provider   Part 3

You place a bandage on his laceration as he explains what happened. You also ask questions to obtain a SAMPLE history. The patient says that he recently had strep throat. As you are talking, you take another set of vital signs. His pulse is 118 beats/min, respirations are 28 breaths/min, and blood pressure is 108/70 mm Hg. His skin is clammy.

5. What are your concerns for this patient?
6. What is the significance of his recent illness?

Assessment of Shock

Scene Size-up

As you approach the scene, be alert to potential hazards to your safety such as power lines down, fast moving traffic, or anything else that threatens your safety. When you first see the patient, observe the scene and patient for clues to determine the nature of the illness or the mechanism of injury. Medical complaints typically involve only one patient, but always ensure that you only have one patient to care for. It is not uncommon that trauma incidents involve more than one patient; obtain an accurate account of all patients.

Initial Assessment

General Impression

When you first visualize your patient, quickly form a general impression. This includes age, sex, signs of distress, obvious life-threatening injuries, abnormal positioning, and skin color. These observations will help you develop an early sense of urgency for care of a patient who appears "sick."

Once you are close to the patient, determine the need for manual spinal immobilization and assess the patient's level of consciousness using the AVPU (Awake and alert, responsive to Verbal stimuli or Pain, or Unresponsive) scale. A patient who has an altered level of consciousness (LOC) may need emergency airway management. If the patient is awake and alert, determine a chief complaint.

Airway and Breathing

Next, quickly assess the airway to ensure it is patent. If the patient is awake and answering questions, the airway is patent. Be alert to abnormal airway sounds such as gurgling (suction the airway) or stridor, indicating partial airway obstruction. If the patient is awake and answering questions, an airway adjunct is not needed; consider an adjunct such as an oropharyngeal or nasopharyngeal airway for a patient with an altered LOC.

Next, you must quickly assess breathing in the patient. You must inspect and palpate the chest wall to assess for DCAP-BTLS (**D**eformities, **C**ontusions, **A**brasions, **P**unctures/Penetrations, **B**urns, **T**enderness,

Lacerations, and **S**welling). Observe the patient for signs of accessory muscle use such as the muscles of the neck, intercostal retractions, or abnormal use of the abdominal muscles. Increased respiratory rate is often an early sign of impending shock. You must assess the patient's breath sounds with a stethoscope, listening for wheezes or other abnormal breath sounds. Once you have quickly completed this assessment of breathing, give the patient high-flow oxygen, or, if needed, assist respirations with a bag-valve-mask device.

Circulation

Next you must quickly assess the patient's circulatory status. Check for the presence of a distal pulse. If you cannot obtain a distal pulse, assess for a central pulse. Make a rapid determination if the pulse is fast, slow, weak, strong, or altogether absent. A rapid pulse suggests compensated shock. In shock or compensated shock, the skin may be cool, clammy, or ashen. If the patient has no pulse and is not breathing, immediately begin CPR. In trauma patients, ensure you have assessed for and identified any life-threatening bleeding; if serious bleeding is discovered, treat it at once. You must also quickly assess skin temperature, condition, and color; also check for capillary refill time.

Transport Decision

Once you have assessed perfusion, you can determine whether the patient should be treated as high priority, whether ALS is needed, and which facility to transport to. Trauma patients with shock, or a suspicious MOI, generally should go to a trauma center. Sometimes, local protocols dictate that a patient should be transported to the nearest hospital for stabilization prior to transfer to a definitive treatment center.

Focused History and Physical Exam

Once you have determined that the initial assessment is complete and any life threats found during that assessment have been addressed, begin the focused history and physical exam based on the patient's complaint and the extent of trauma incurred. Obtain a history, and perform a continued assessment specific to the patient's complaint or problem and the body region(s) affected.

Rapid or Focused Physical Exam

If your patient is a trauma patient with a significant mechanism of injury or multiple injuries, one who gives you a poor general impression, or you found problems in the initial assessment, perform a rapid physical exam. If your patient has a medical problem but is not responsive or problems were noted in the initial assessment, perform a rapid physical exam. These rapid assessments should be performed quickly but thoroughly to ensure that you will not miss any significant or life-threatening problems or delay needed care.

If your patient has only a simple mechanism of injury, such as a twisted ankle, perform a focused physical exam on the area affected. Whether you perform a rapid or a focused exam, if a life-threatening problem is found, treat it immediately. The focused physical exam and the rapid physical exam will also help you to identify injuries that must be addressed when packaging the patient for transport.

Baseline Vital Signs

Obtain a complete set of baseline vital signs. If the patient's condition is unstable or could become unstable, reassess vital signs every 5 minutes. If the patient is in stable condition, reassess vital signs every 10 to 15 minutes. Baseline vital signs will help you trend changes in your patient.

SAMPLE History

You must now quickly obtain a SAMPLE history from the patient. Remember, if the patient has a significant change in LOC before arrival at the hospital, you will be able to provide the hospital personnel with this important information.

Interventions

You must determine what interventions are needed for your patients at this point based on the findings of your assessment. You should focus on supporting the cardiovascular system. Treating for shock early and aggressively will help to prevent inadequate perfusion from harming your patient. Provide oxygen and put the patient in the shock position.

Detailed Physical Exam

When time permits and the patient's condition is stable, perform a detailed physical exam. Perform a thorough head-to-toe assessment of the patient. This includes a complete neurologic assessment. If the patient is critically ill and problems are found in the initial assessment, you may not have the opportunity to perform a detailed physical exam.

Ongoing Assessment

This portion of patient assessment is very important in patient care. The rule of thumb is assess-intervene-reassess. This portion of the assessment revisits the initial assessment, the vital signs, and any treatment performed on the patient, including oxygen administration. You must assess the patient to determine whether the interventions performed are having any effect on the patient. This step prepares you to present the patient at the hospital with a complete, concise account of the patient encounter and care.

Communication and Documentation

Patients who are in decompensated shock will need rapid interventions to restore adequate perfusion. The hospital may or may not have suggestions on how best to support the ailing cardiovascular system. Most of the interventions used to treat shock do not require a specific physician's order; however, some do. For example, many areas allow EMTs to apply a pneumatic antishock garment (PASG) but require a direct verbal order from medical control before inflating them. Know and follow your own agency's guidelines on PASG use. Determine, based on the signs and symptoms found in your assessment, whether your patient is in compensated or decompensated shock. Document these findings after you have treated for shock.

Emergency Medical Care

You must begin immediate treatment for shock as soon as you realize that the condition may exist. Follow the steps in (**Skill Drill 23-1** ▶):

1. As with any type of patient care, you should **begin by following body substance isolation precautions**, making sure the patient has an open airway, and checking breathing and pulse. In general, keep the patient in a supine position. Patients who have had a severe heart attack or who have lung disease may find it easier to breathe in a sitting or semisitting position (**Step 1**).

Treating Shock

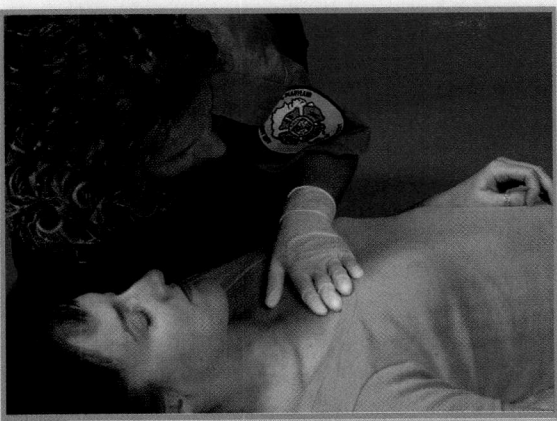

1 Keep the patient supine, open the airway, and check breathing and pulse.

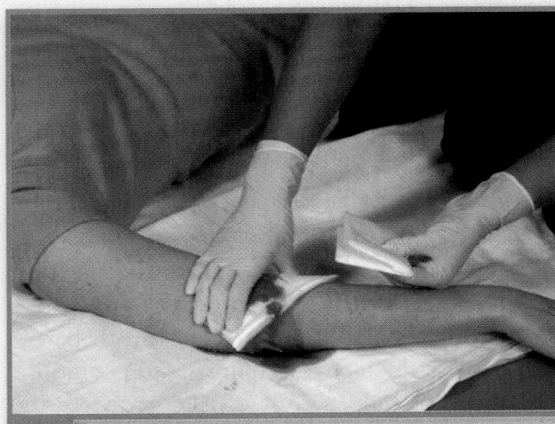

2 Control obvious external bleeding.

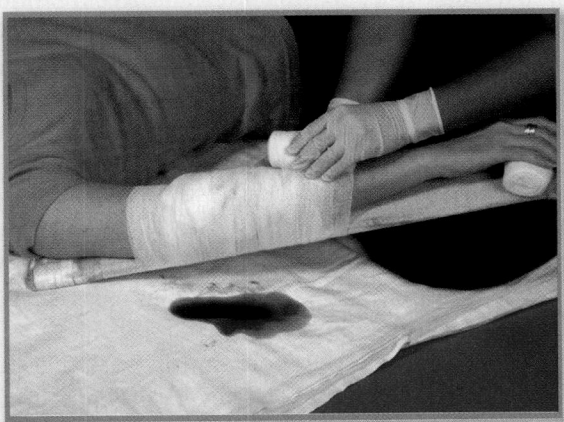

3 Splint any broken bones or joint injuries.

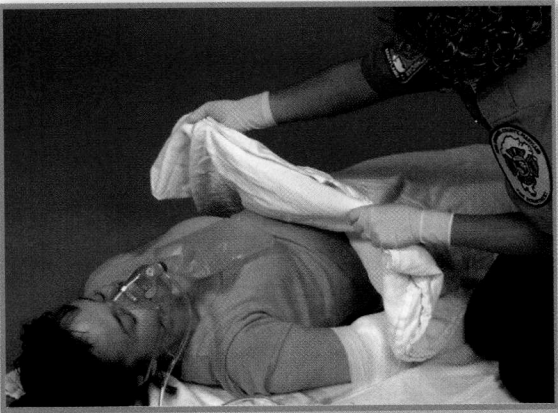

4 Give high-flow oxygen if you have not already done so, and place blankets under and over the patient.

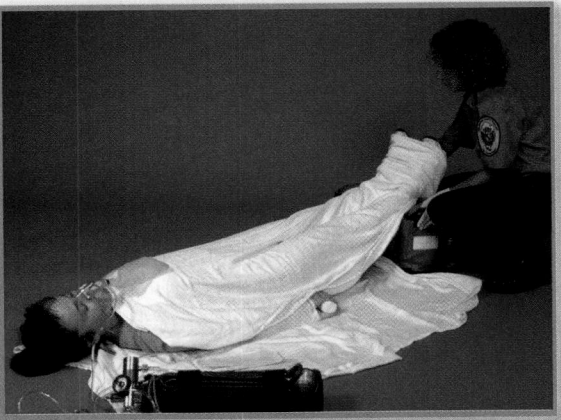

5 If no fractures are suspected, elevate the legs 6″ to 12″.

2. Next, **control all obvious external bleeding.** Place dry, sterile dressings over the bleeding sites, and secure with bandages (**Step 2**).

3. **Splint any bone or joint injuries.** This minimizes pain, bleeding, and discomfort, all of which can aggravate shock. It also prevents the broken bone ends from further damaging adjacent soft tissue. In general, splinting also makes it easier to move the patient. Handle the patient gently and no more than is necessary (**Step 3**).

   There is some controversy surrounding the use of the PASG. When used improperly, the device can increase bleeding from chest injuries, interfere with adequate air exchange, and promote cardiovascular collapse. When used properly, it can effectively control bleeding from fractures and massive soft-tissue wounds. In general, the PASG should not be used without the approval of medical control or established local protocols.

4. **Remember that inadequate ventilation** may be the primary cause of shock or a major factor in its development. Always provide oxygen, assist with ventilations as needed, and continue to monitor the patient's breathing. To prevent the loss of body heat, place blankets under and over the patient. Be careful not to overload the patient with covers or attempt to warm the body too much; it is best for the patient to maintain a normal body temperature. Do not use external heat sources, such as hot water bottles or heating pads. They may harm a patient in shock by causing vasodilation and decreasing blood pressure even more (**Step 4**).

5. Once you have positioned the patient on a backboard or a stretcher, **place the patient in the Trendelenburg position.** This technique is easily accomplished by raising the foot of the backboard or stretcher about 6″ to 12″. If the patient is not on a backboard and no lower extremity fractures are suspected, place the patient in the shock position. This is accomplished by elevating the patient's legs 6″ to 12″ by propping them up on a several blankets or other stable objects. These positions help to return blood from the extremities back to the core of the body where it is needed most. Patients with respiratory distress may benefit from the Trendelenburg position, but the lower extremities should be raised only 6″ to 8″. Raising the lower extremities any higher may aggravate a patient's breathing because the abdominal organs push against the diaphragm (**Step 5**).

EMT-B Tips

You are never wrong to treat for shock, and many patients will experience some degree of shock. Consider treating all patients for shock.

Do not give the patient anything by mouth, no matter how urgently you are asked. To relieve the intense thirst that often accompanies shock, give the patient a moistened piece of gauze to chew or suck. Never give a patient in shock an alcoholic drink or other depressant. A stimulant, such as coffee, also has little value in treating shock.

Accurately record the patient's vital signs approximately every 5 minutes throughout treatment and transport. It is essential to transport trauma patients to the hospital as rapidly as possible for definitive treatment. The Golden Hour refers to the first 60 minutes after injury, which is thought to be a critically important period for the early resuscitation and treatment of severely injured trauma patients. This concept underscores the importance of rapid evaluation, stabilization, and transport. The goal of EMS is to limit on-scene time (time on-scene until transport to hospital is started) to 10 minutes or less. Remember to speak calmly and reassuringly to a conscious patient throughout assessment, care, and transport.

(Table 23-3 ▶) lists the general supportive measures for the major types of shock. Not every measure is used for every type of shock.

Treating Cardiogenic Shock

The patient who is in shock as a result of a heart attack does not require a transfusion of blood, intravenous fluids, elevation of the legs, or a PASG. There is already a greater volume of blood in circulation than the heart can handle. The damaged heart muscle simply cannot generate the necessary power to pump blood throughout the circulatory system.

Keep in mind that chronic lung disease will aggravate cardiogenic shock. If the patient has chronic obstructive pulmonary disease and heart disease, oxygenation of the blood passing through the lungs is impaired. Because fluid is collecting in the lungs, this patient is often able to breathe better in a sitting or semi-sitting position and may tell you so.

Usually, patients with cardiogenic shock do not have any injury, but they may be having chest pain. Such a patient may have taken nitroglycerin before your arrival and

TABLE 23-3 Types of Shock

Types of Shock	Examples of Potential Causes	Signs and Symptoms	Treatment
Anaphylactic	Extreme life-threatening allergic reaction	Can develop within seconds Mild itching or rash Burning skin Vascular dilation Generalized edema Coma Rapid death	Manage the airway Assist ventilations Administer high-flow oxygen Determine cause Assist with administration of epinephrine Transport promptly
Cardiogenic	Inadequate heart function Disease of muscle tissue Impaired electrical system Disease or injury	Chest pain Irregular pulse Weak pulse Low blood pressure Cyanosis (lips, under nails) Cool, clammy skin Anxiety	Position comfortably Administer oxygen Assist ventilations Transport promptly
Hypovolemic	Loss of blood or fluid	Rapid, weak pulse Low blood pressure Change in mental status Cyanosis (lips, under nails) Cool, clammy skin Increased respiratory rate	Secure airway Assist ventilations Administer high-flow oxygen Control external bleeding Elevate legs Keep warm Transport promptly
Respiratory Insufficiency	Severe chest injury, airway obstruction	Rapid, weak pulse Low blood pressure Change in mental status Cyanosis (lips, under nails) Cool, clammy skin Increased respiratory rate	Secure airway Clear air passages Assist ventilations Administer high-flow oxygen Transport promptly
Neurogenic	Damaged cervical spine, which causes widespread blood vessel dilation	Bradycardia (slow pulse) Low blood pressure Signs of neck injury	Secure airway Spinal stabilization Assist ventilations Administer high-flow oxygen Preserve body heat Transport promptly
Psychogenic (fainting)	Temporary, generalized vascular dilation Anxiety, bad news, sight of injury or blood, prospect of medical treatment, severe pain, illness, tiredness	Rapid pulse Normal or low blood pressure	Determine duration of unconsciousness Record initial vital signs and mental status Suspect head injury if patient is confused or slow to regain consciousness Transport promptly
Septic	Severe bacterial infection	Warm skin Tachycardia Low blood pressure	Transport promptly Administer oxygen en route Provide full ventilatory support Elevate legs Keep patient warm

may want to take more. Before helping the patient self-administer nitroglycerin, be sure to consult with medical control for instructions. You will also need to perform an accurate assessment to ensure that the patient's blood pressure meets the criteria for this medication. If the blood pressure is too low, nitroglycerin may increase the problem. Remember that patients in cardiogenic shock usually have a low blood pressure. Other signs include a weak, irregular pulse; cyanosis about the lips and underneath the fingernails; anxiety; and nausea.

Treatment of cardiogenic shock should begin by placing the patient in the position in which breathing is easiest as you give high-flow oxygen. Be ready to assist ventilations as necessary, and have suction nearby in case the patient vomits. Provide prompt transport to the emergency department. Remember also to approach a patient who has had a suspected heart attack with calm reassurance. Frequently checking for a pulse in an unresponsive patient is important to identify early whether an automated external defibrillator is needed.

Treating Neurogenic Shock

Shock that accompanies spinal cord injury is best treated by a combination of all known supportive measures. The patient who has sustained this kind of injury usually will require hospitalization for a long time. Emergency treatment must be directed at obtaining and maintaining a proper airway, providing spinal immobilization, assisting inadequate breathing as needed, conserving body heat, and providing the most effective circulation possible.

This patient usually is not losing blood. However, the capacity of his or her blood vessels has become significantly larger than the volume of blood they contain. Slight elevation of the foot end of the spine board will help bring the blood that is pooling in the vessels of the legs to the vital organs. Placing the patient's arms across his or her chest without moving the spine will also return some pooled blood. Be sure to monitor the patient for breathing problems, and, if they appear,

lower the spine board. Supplemental oxygen will boost the concentration of oxygen in the blood. If respirations are weak or inadequate, provide assisted ventilations. Keep the patient as warm as possible with blankets, because the injury may have disabled the body's normal temperature controls. Transport promptly.

Treating Hypovolemic Shock

The emergency treatment of hypovolemic or hemorrhagic shock includes the control of all obvious external bleeding. To prevent continued bleeding, you must apply sufficient pressure to control obvious external bleeding, splint any bone and joint injuries, and ensure that you use great care to handle the patient gently. If there are no fractured extremities, you should place the patient in the Trendelenburg position, raising the legs 6″ to 12″, keeping the torso in a horizontal position. This will increase blood flow to the heart from the lower body and keep unwanted pressure off the diaphragm. This method combats shock by using the patient's own blood to its best advantage.

Although you cannot control internal bleeding in the field, you must recognize its existence and provide aggressive general support. Secure and maintain an airway, and provide respiratory support, including supplemental oxygen and, if needed, assisted ventilations. Start the oxygen as soon as you suspect shock, and continue it during transport; with too little circulating blood, additional oxygen may be lifesaving. Be sure the patient does not aspirate blood or vomitus. Most important, you must transport the patient as rapidly as possible to the emergency department.

Treating Septic Shock

The proper treatment of septic shock requires complex hospital management, including antibiotics. If you suspect that a patient has septic shock, you must use appropriate body substance isolation precautions and transport as promptly as possible. Use high-flow

You are the Provider   Part 4

You and your partner immobilize the patient on a long backboard. The patient seems anxious; he states that he does not feel well and needs to get to the hospital immediately. He is loaded into the ambulance, and your partner reassesses vital signs. Your partner reports that the vital signs are a pulse of 122 beats/min, respirations of 30 breaths/min, and a blood pressure of 106/68 mm Hg. Your partner places a blanket over the patient and administers high-flow oxygen.

7. What do his vital sign changes indicate?

oxygen during transport. Ventilatory support may be necessary to maintain adequate tidal volume. Use blankets to conserve body heat.

Treating Respiratory Insufficiency

In treating the patient who is in shock as a result of inadequate respiration, you must immediately secure and maintain the airway. Clear the mouth and throat of anything obstructing the air passages, including mucus, vomitus, and foreign material. If necessary, provide ventilations with a bag-valve-mask device. Give supplemental oxygen, and transport the patient promptly.

Treating Anaphylactic Shock

The only really effective treatment for a severe, acute allergic reaction is to administer epinephrine by way of subcutaneous or intramuscular injection. For more information on the emergency care for allergic reactions, see Chapter 16. A patient who is aware of having a specific sensitivity may carry a bee-sting kit containing epinephrine (Figure 23-6 ▶). If he or she is unable to inject the medication, you may have to do so if you are allowed by local protocol. If the patient's signs and symptoms recur or the patient's condition deteriorates, you should repeat the injection after consulting with medical control.

Promptly transport the patient to the emergency department while providing all possible support, primarily supplemental oxygen and ventilatory assistance. You should also try to find out what agent caused the reaction (for example, a drug, an insect bite or sting, a food item) and how it was received (for example, by mouth, by inhalation, or by injection). The severity of allergic reactions can vary greatly, with symptoms ranging from mild itching to profound coma and rapid death.

Keep in mind that a mild reaction may worsen suddenly or over time. Consider requesting advanced life support backup, if available.

Treating Psychogenic Shock

In an uncomplicated case of fainting, once the patient collapses and becomes supine, circulation to the brain is usually restored and with it, a normal state of functioning. Remember that psychogenic shock can significantly worsen other types of shock. If the attack has caused the patient to fall, you must check for injuries, especially in older patients. However, you should also assess the patient thoroughly for any other abnormality. If, after regaining consciousness, the patient is unable to walk without weakness, dizziness, or pain, you should suspect another problem, such as head injury. You should transport this patient promptly.

Be sure to record your initial observations of vital signs and level of consciousness. In addition, try to learn from bystanders whether the patient complained of anything before fainting and how long he or she was unconscious.

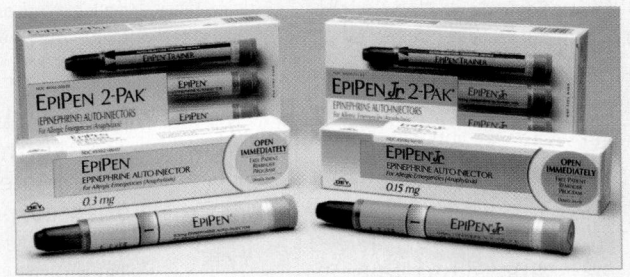

Figure 23-6 Patients who are allergic to bee stings often carry commercial bee-sting kits, such as an intramuscular injector or auto-injector, containing epinephrine.

You are the Provider   Summary

The mechanism of injury in this scenario is significant. Although the patient was restrained, air bag deployment and damage to the steering wheel indicate that there was force involved. Possible injuries include damage to the abdominal organs, such as the spleen, liver, diaphragm, and also damage to the chest, spine, and skeleton.

Shock is even more likely if a patient has had a severe infection, which is why this patient's recent strep throat is significant. Other conditions that make shock more likely to develop are multiple severe fractures, abdominal or chest injury, spinal injury, a heart attack, and anaphylaxis.

As the scenario progresses, the patient's pulse and respirations increase, while his blood pressure decreases. This indicates that he is developing shock. He has additional signs of compensated shock, including anxiety and pale and clammy skin. The patient likely has internal bleeding and must be transported immediately to the emergency department. It is crucial to maintain the airway and provide respiratory support, including supplemental oxygen and, if needed, assisted ventilations. These measures can be lifesaving.

Assessment and Emergency Care

Shock	
Scene Size-up	Body substance isolation precautions should include a minimum of gloves and eye protection. Ensure scene safety and determine NOI/MOI. Consider the number of patients, the need for additional help/ALS, and c-spine stabilization.
Initial Assessment	
■ General impression	Determine level of consciousness and find and treat any immediate threats to life. Determine priority of care based on environment and patient's chief complaint.
■ Airway	Ensure patent airway.
■ Breathing	Listen for abnormal breath sounds and evaluate depth and rate of the respirations. Maintain ventilations as needed. Provide high-flow oxygen at 15 L/min.
■ Circulation	Evaluate distal pulse rate and quality; observe skin color, temperature, and condition and treat accordingly. Consider external bleeding based on chief complaint and MOI.
■ Transport decision	Rapid transport based on a poor general impression or problems with ABCs.
Focused History and Physical Exam	NOTE: The order of the steps in the focused history and physical exam differs depending on whether or not the patient has a significant MOI. The order below is for a patient with a significant MOI. For a patient without a significant MOI, perform a focused trauma assessment, obtain vital signs, and obtain the history.
■ Significant MOI	Reevaluate the mechanism of injury(s). Perform rapid trauma assessment to identify hidden injuries.
■ No significant MOI	Reevaluate the mechanism of injury. If the patient is alert and oriented, perform a focused assessment on the body system or affected area.
■ Baseline vital signs	Take vital signs, noting skin color and temperature and patient's level of consciousness. Use pulse oximetry if available.
■ SAMPLE history	Ask pertinent SAMPLE and OPQRST. Be sure to ask if and what interventions were taken before your arrival, how many interventions, and at what time.
■ Interventions	Support patient and cardiovascular system as needed. Consider the use of oxygen, eliminate the cause of the shock, maintain normal body temperature, and use proper positioning of the patient to improve circulation.
Detailed Physical Exam	Complete a detailed physical exam.
Ongoing Assessment	Repeat the initial assessment, focused assessment, and reassess interventions performed. Reassess vital signs every 5 minutes for the unstable patient, and every 15 minutes for the stable patient. Reassure and calm the patient.
■ Communication and documentation	Contact medical control with a radio report. Relay any change in level of consciousness or difficulty breathing. Be sure to document physician's orders and changes in patient's condition, and at what time they occurred.

NOTE: While the steps below are widely accepted, be sure to consult and follow your local protocol.

General Shock	Anaphylactic Shock

General Shock

Minimum treatment for shock, with the exception of anaphylactic shock, is as follows:

1. Provide spinal stabilization if needed.
2. Keep the patient supine, open the airway, and check breathing and pulse.
3. Control obvious external bleeding.
4. Splint any broken bones or joint injuries.
5. Give high-flow oxygen if you have not already done so, and place blankets under and over the patient.
6. If no extremity fractures are suspected, elevate the legs 6″ to 12″.
7. Provide rapid transport in a position that best supports circulation and breathing.

Treatment for cardiogenic shock also requires the following:
For cardiogenic shock, place the patient in a position of comfort. Assist with ventilation and suction as needed. Monitor the pulse closely and provide rapid, calm transport.

Anaphylactic Shock

Follow the steps for using an autoinjector (discussed in Chapter 16):

1. Remove the autoinjector's safety cap, and quickly wipe the thigh with antiseptic.
2. Place the tip of the autoinjector against the lateral part of the thigh.
3. Push the autoinjector firmly against the thigh, and hold it in place until all the medication is injected.

Assessment and Emergency Care

Prep Kit

Ready to Review

- Perfusion requires an intact cardiovascular system and a functioning respiratory system.
- Remember, most types of shock (hypoperfusion) are caused by dysfunction in one or more sides of the perfusion triangle:
    - The pump (the heart)
    - The pipes, or container (blood vessels)
    - The content, or volume (blood)
- Shock (hypoperfusion) is the collapse and failure of the cardiovascular system, when blood circulation slows and eventually stops.
- Signs of compensated shock include anxiety or agitation; tachycardia; pale, cool, moist skin; increased respiratory rate; nausea and vomiting; and increased thirst. If there is any question, treat for shock. It is never wrong to treat for shock.
- Signs of decompensated shock include labored or irregular respirations, ashen gray or cyanotic skin color, weak or absent distal pulses, dilated pupils, and profound hypotension.
- Remember, by the time a drop in blood pressure is detected, shock is usually in an advanced stage.
- Anticipate shock in patients who may have the following conditions:
    - Severe infection
    - Significant blunt force trauma or penetrating trauma
    - Massive external bleeding or index of suspicion for major internal bleeding
    - Spinal injury
    - Chest or abdominal injury
    - Major heart attack
    - Anaphylaxis
- Treat all patients suspected to be in shock from any cause as follows and in this order:
    - Open and maintain the airway.
    - Provide high-flow oxygen and as needed, provide bag-valve-mask assisted ventilations.
    - Control all obvious external bleeding.
    - Place the patient in the shock position or, if on a backboard or stretcher, in the Trendelenburg position.
    - Maintain normal body temperature with blankets.
    - Provide prompt transport to the appropriate hospital.

Vital Vocabulary

anaphylactic shock Severe shock caused by an allergic reaction.

anaphylaxis An unusual or exaggerated allergic reaction to foreign protein or other substances.

aneurysm A swelling or enlargement of a part of an artery, resulting from weakening of the arterial wall.

autonomic nervous system The part of the nervous system that regulates involuntary functions, such as heart rate, blood pressure, digestion, and sweating.

cardiogenic shock Shock caused by inadequate function of the heart, or pump failure.

compensated shock The early stage of shock, in which the body can still compensate for blood loss.

cyanosis Bluish color of the skin resulting from poor oxygenation of the circulating blood.

decompensated shock The late stage of shock when blood pressure is falling.

dehydration Loss of water from the tissues of the body.

edema The presence of abnormally large amounts of fluid between cells in body tissues, causing swelling of the affected area.

homeostasis A balance of all systems of the body.

hypothermia A condition in which the internal body temperature falls below 95°F (35°C), usually as a result of prolonged exposure to cool or freezing temperatures.

hypovolemic shock Shock caused by fluid or blood loss.

irreversible shock The final stage of shock, resulting in death.

neurogenic shock Circulatory failure caused by paralysis of the nerves that control the size of the blood vessels, leading to widespread dilation; seen in spinal cord injuries.

perfusion Circulation of blood within an organ or tissue in adequate amounts to meet the cells' current needs.

psychogenic shock Shock caused by a sudden, temporary reduction in blood supply to the brain that causes fainting (syncope).

sensitization Developing a sensitivity to a substance that initially caused no allergic reaction.

septic shock Shock caused by severe infection, usually a bacterial infection.

shock A condition in which the circulatory system fails to provide sufficient circulation to enable every body part to perform its function; also called hypoperfusion.

sphincters Circular muscles that encircle and, by contracting, constrict a duct, tube, or opening.

syncope Fainting.

Technology

| Interactivities |
| Vocabulary Explorer |
| Anatomy Review |
| Web Links |
| Online Review Manual |

www.EMTB.com

Assessment in Action

You are working on a beautiful fall weekend for a rural basic life support ambulance service. As you and your EMT-B partner sit down for breakfast, a call comes in for a collision between a motor vehicle and a bicycle. En route, you and your partner are discussing possible causes of shock.

1. Given the above dispatch information, all of the following injuries and insults can ultimately lead to shock except:

   A. closed head injury.
   B. laceration of the radial artery.
   C. closed femur fracture.
   D. contusion of the right ankle.

2. After your partner maintains in-line cervical immobilization of the patient, your next course of action after the initial ABCs would be to:

   A. control any life-threatening bleeding.
   B. splint any fractures that you observe.
   C. place the patient on a backboard.
   D. take the patient's blood pressure.

3. The patient's bleeding is bright red and squirting. This type of bleeding is most likely:

   A. capillary.
   B. arterial.
   C. cardiac.
   D. venous.

4. If bleeding continues through the dressing despite direct pressure, you should then do all of the following to control the bleeding except:

   A. elevate the extremity.
   B. apply additional dressings.
   C. use a pressure point.
   D. lower the extremity.

5. The patient's baseline vital signs include a pulse of 120 beats/min, respirations of 24 breaths/min, and a blood pressure of 90/64 mm Hg. On the basis of these vital signs, you determine that the patient is in shock and you begin to treat the patient by:

   A. giving the patient high-flow oxygen via a nonrebreathing mask.
   B. placing the patient in the shock (Trendelenburg) position.
   C. calling for advanced life support backup.
   D. all of the above.

6. The patient starts complaining that his feet are feeling cold. This process is known as:

   A. pooling.
   B. shunting.
   C. collecting.
   D. cooling.

7. The classic signs of shock include an increased heart rate, increased respiratory rate, decreased blood pressure, and cold extremities. All of the following are additional signs of shock except:

   A. decreased thirst.
   B. dilated pupils.
   C. restlessness.
   D. cyanosis.

Challenging Questions

8. Beta-blockers such as metoprolol (Lopressor) have been used for years to treat hypertension. How could these medications affect the signs and symptoms of shock?

Points to Ponder

A crying mother brings her 2-year-old daughter into your local ambulance station stating that she accidentally ran her over while backing out of the driveway. You observe the child to be crying and you see obvious open femur fractures. No other associated injuries are noted in your observation. The vital signs are as follows: pulse, 140 beats/min, weak and regular; respiratory rate, 30 breaths/min and nonlabored; and blood pressure, 88 by palpation. Is this patient in shock? Are these vital signs normal for a 2-year-old?

Issues: Interacting With Parents, Pediatric Patients, Child Abuse.

www.EMTB.com

Soft-Tissue Injuries

Objectives

Cognitive

5-2.1   State the major functions of the skin. (p 699)

5-2.2   List the layers of the skin. (p 698)

5-2.3   Establish the relationship between body substance isolation (BSI) and soft-tissue injuries. (p 701, 707)

5-2.4   List the types of closed soft-tissue injuries. (p 700)

5-2.5   Describe the emergency medical care of the patient with a closed soft-tissue injury. (p 703)

5-2.6   State the types of open soft-tissue injuries. (p 705)

5-2.7   Describe the emergency medical care of the patient with an open soft-tissue injury. (p 710)

5-2.8   Discuss the emergency medical care considerations for a patient with a penetrating chest injury. (p 707)

5-2.9   State the emergency medical care considerations for a patient with an open wound to the abdomen. (p 712)

5-2.10 Differentiate the care of an open wound to the chest from an open wound to the abdomen. (p 708, 711)

5-2.11 List the classification of burns. (p 716)

5-2.12 Define superficial burn. (p 716)

5-2.13 List the characteristics of a superficial burn. (p 716)

5-2.14 Define partial-thickness burn. (p 716)

5-2.15 List the characteristics of a partial-thickness burn. (p 716)

5-2.16 Define full-thickness burn. (p 716)

5-2.17 List the characteristics of a full-thickness burn. (p 716)

5-2.18 Describe the emergency medical care of the patient with a superficial burn. (p 720)

5-2.19 Describe the emergency medical care of the patient with a partial-thickness burn. (p 720)

5-2.20 Describe the emergency medical care of the patient with a full-thickness burn. (p 720)

5-2.21 List the functions of dressing and bandaging. (p 726)

5-2.22 Describe the purpose of a bandage. (p 727)

5-2.23 Describe the steps in applying a pressure bandage. (p 710)

5-2.24 Establish the relationship between airway management and the patient with chest injury, burns, and blunt and penetrating injuries. (p 702, 708, 719)

5-2.25 Describe the effects of improperly applied dressings, splints, and tourniquets. (p 727)

5-2.26 Describe the emergency medical care of a patient with an impaled object. (p 713)

5-2.27 Describe the emergency medical care of a patient with an amputation. (p 714)

5-2.28 Describe the emergency care for a chemical burn. (p 722)

5-2.29 Describe the emergency care for an electrical burn. (p 703, 725)

Affective

None

Psychomotor

5-2.29 Demonstrate the steps in the emergency medical care of closed soft-tissue injuries. (p 703, 725)

5-2.30 Demonstrate the steps in the emergency medical care of open soft-tissue injuries. (p 710)

5-2.31 Demonstrate the steps in the emergency medical care of a patient with an open chest wound. (p 708)

5-2.32 Demonstrate the steps in the emergency medical care of a patient with open abdominal wounds. (p 712)

5-2.33 Demonstrate the steps in the emergency medical care of a patient with an impaled object. (p 713)

5-2.34 Demonstrate the steps in the emergency medical care of a patient with an amputation. (p 714)

5-2.35 Demonstrate the steps in the emergency medical care of an amputated part. (p 714)

5-2.36 Demonstrate the steps in the emergency medical care of a patient with superficial burns. (p 720)

5-2.37 Demonstrate the steps in the emergency medical care of a patient with partial-thickness burns. (p 720)

5-2.38 Demonstrate the steps in the emergency medical care of a patient with full-thickness burns. (p 720)

5-2.39 Demonstrate the steps in the emergency medical care of a patient with a chemical burn. (p 722)

5-2.40 Demonstrate completing a prehospital care report for patients with soft-tissue injuries. (p 703, 710, 720)

You are the Provider

You are dispatched to a local garage for a 27-year-old man with burns from an accident involving a car, with possible entrapment. On arrival, you are led by the owner into one of the bays where a man is supine on the floor in a pool of antifreeze. He is complaining of pain to his right thigh and ankle.

You are told by the owner that on hearing a crashing noise followed by screams for help, he found his employee with his ankle pinned under a tire, while the other side of the car was still on the jack. After using the jack to lift the car off the employee, the owner pulled the injured mechanic clear of the car. The patient had been trapped while antifreeze drained over his thigh. You immediately take BSI precautions. The patient denies experiencing any loss of consciousness. His respiratory effort is rapid and pursed lipped. A quick check of his radial pulse shows this too is rapid.

1. What are your first impressions of this patient?
2. Is the patient's condition life threatening? What will you do to confirm or rule this out?

Soft-Tissue Injuries

The skin is our first line of defense against external forces and infection. Although it is relatively tough, skin is still quite susceptible to injury. Injuries to soft tissues range from simple bruises and abrasions to serious lacerations and amputations. Soft-tissue injury may result in loss of soft tissue, exposing deep structures such as blood vessels, nerves, and bones. In all instances, you must control bleeding, prevent further contamination to decrease the risk of infection, and protect the wound from further damage. Therefore, you must know how to apply dressings and bandages to various parts of the body.

The Anatomy and Function of the Skin

The skin is the largest organ in the body. It varies in thickness, depending on age and its location. The skin of the very young and very old is thinner than the skin of a young adult. The skin covering your scalp, your back, and the soles of your feet is quite thick, while the skin of your eyelids, lips, and ears is very thin. Thin skin is more easily damaged than thick skin.

Anatomy of the Skin

The skin has two principal layers: the epidermis and the dermis (Figure 24-1 ▶). The epidermis is the tough, external layer that forms a watertight covering for the body. The epidermis is itself composed of several layers. The cells on the surface layer of the epidermis are constantly worn away. They are replaced by cells that are pushed to the surface when new cells form in the germinal layer at the base of the epidermis. Deeper cells in the germinal layer contain pigment granules. Along with blood vessels in the dermis, these granules produce skin color.

The dermis is the inner layer of the skin. It lies below the germinal cells of the epidermis. The dermis contains the structures that give the skin its characteristic appearance: hair follicles, sweat glands, and sebaceous glands. The sweat glands act to cool the body. They discharge sweat onto the surface of the skin through small pores, or ducts, that pass through the epidermis. Sebaceous glands produce sebum, the oily material that waterproofs the skin and keeps it supple. Sebum travels to the skin's surface along the shaft of adjacent hair follicles. Hair follicles are small organs that produce hair. There is one follicle for each hair, each connected with a sebaceous gland and a tiny muscle. This muscle pulls the hair erect whenever you are cold or frightened.

Blood vessels in the dermis provide the skin with nutrients and oxygen. Small branches reach up to the germinal cells, but no blood vessels penetrate farther into the epidermis. There are also specialized nerve endings within the dermis.

Technology

www.EMTB.com

- Interactivities
- Vocabulary Explorer
- Anatomy Review
- Web Links
- Online Review Manual

✳ EMT-B Tips

Although it may be tempting to think of skin injuries as unimportant, the skin plays several crucial protective and regulatory roles. When you are dealing with skin injuries, remember the importance of this organ in protecting against infection and maintaining internal temperature and fluid balance. The skin can also be very important emotionally to the patient; concerns about how bruising and scarring will look later may require your attention and communication skills during a response.

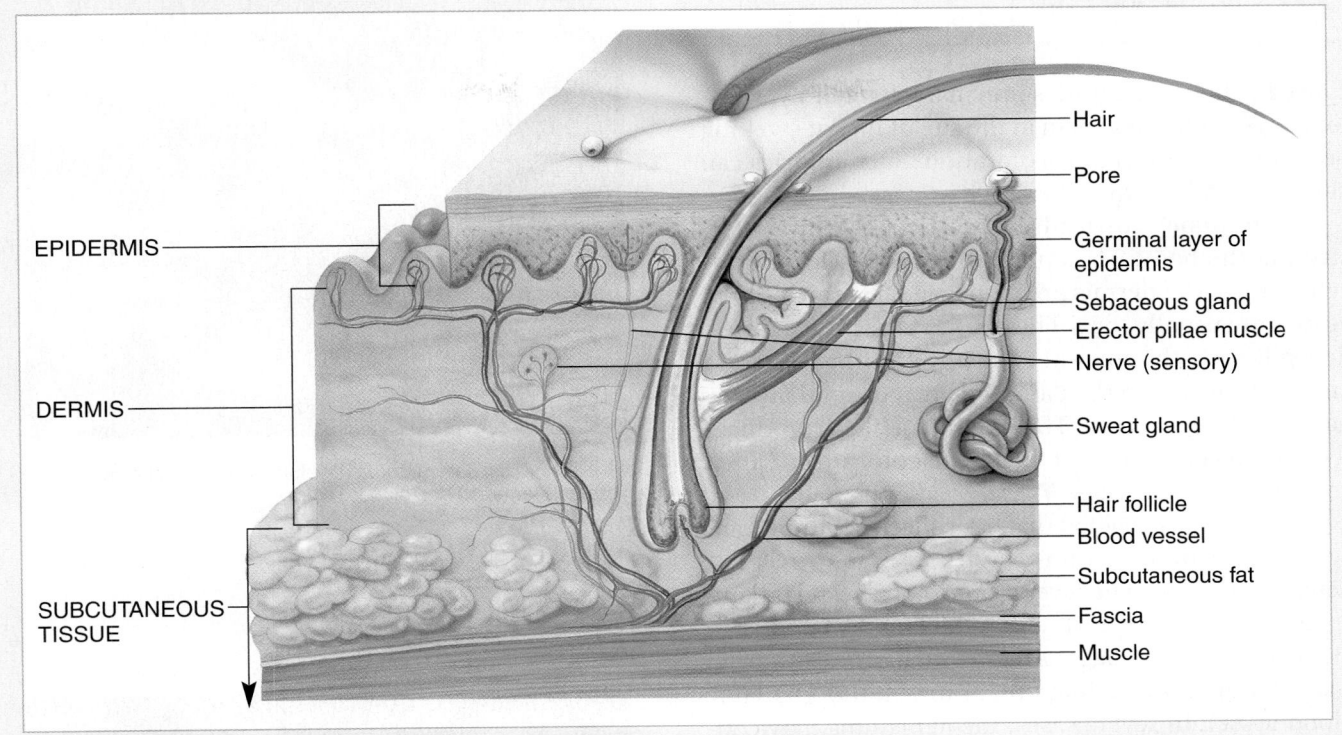

EPIDERMIS

DERMIS

SUBCUTANEOUS TISSUE

Hair
Pore
Germinal layer of epidermis
Sebaceous gland
Erector pillae muscle
Nerve (sensory)
Sweat gland
Hair follicle
Blood vessel
Subcutaneous fat
Fascia
Muscle

Figure 24-1 The skin is composed of a tough external layer called the epidermis and a vascular inner layer called the dermis.

The skin covers all external surfaces of the body. The various openings in our body, including the mouth, nose, anus, and vagina, are not covered by skin. Instead, these openings are lined with <u>mucous membranes</u>. These membranes are similar to skin in that they, too, provide a protective barrier against bacterial invasion. But mucous membranes differ from skin in that they secrete a watery substance that lubricates the openings. Therefore, mucous membranes are moist, while skin is dry.

Functions of the Skin

The skin serves many functions. It protects the body by keeping pathogens out, water in, and assisting in body temperature regulation. The nerves in the skin report to the brain on the environment and on many sensations.

The skin is also the body's major organ for regulating temperature. In a cold environment, the blood vessels in the skin constrict, diverting blood away from the skin and decreasing the amount of heat that is radiated from the body's surface. In hot environments, the vessels in the skin dilate. The skin becomes flushed or red, and heat radiates from the body's surface. Also, sweat glands secrete sweat. As the sweat evaporates from the skin's surface, your body temperature drops, and you begin to cool down.

Any break in the skin allows bacteria to enter and raises the possibilities of infection, fluid loss, and loss of temperature control. Any one of these problems can cause serious illness and even death.

Types of Soft-Tissue Injuries

Soft tissues are often injured because they are exposed to the environment. There are three types of soft-tissue injuries:

- <u>Closed injuries</u>, in which soft-tissue damage occurs beneath the skin or mucous membrane but the surface remains intact.
- <u>Open injuries</u>, in which there is a break in the surface of the skin or the mucous membrane, exposing deeper tissue to potential contamination.
- <u>Burns</u>, in which the soft tissue receives more energy than it can absorb without injury. The source of this energy can be thermal heat, frictional heat, toxic chemicals, electricity, or nuclear radiation.

Closed Injuries

Closed soft-tissue injuries are characterized by a history of blunt trauma, pain at the site of injury, swelling beneath the skin, and discoloration. Such injuries can vary from mild to quite severe.

A <u>contusion</u>, or bruise, results from blunt force striking the body. The epidermis remains intact, but cells within the dermis are damaged, and small blood vessels are usually torn. The depth of the injury varies, depending on the amount of energy absorbed. As fluid and blood leak into the damaged area, the patient may have swelling and pain. The buildup of blood produces a characteristic blue or black discoloration called <u>ecchymosis</u> (Figure 24-2 ▼).

A <u>hematoma</u> is blood that has collected within damaged tissue or in a body cavity (Figure 24-3 ▶). A hematoma occurs whenever a large blood vessel is damaged and bleeds rapidly. It is usually associated with extensive tissue damage. A hematoma can result from a soft-tissue injury, a fracture, or any injury to a large blood vessel. In severe cases, the hematoma may contain more than a liter of blood.

A crushing injury occurs when a great amount of force is applied to the body (Figure 24-4 ▶). The extent of the damage depends on how much force is applied and the amount of time over which it is applied. In addition to causing some direct soft-tissue damage,

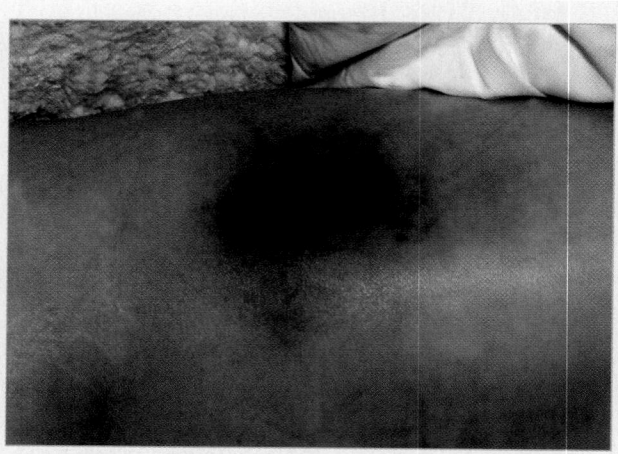

Figure 24-3 A hematoma develops whenever a large blood vessel is damaged and bleeds rapidly.

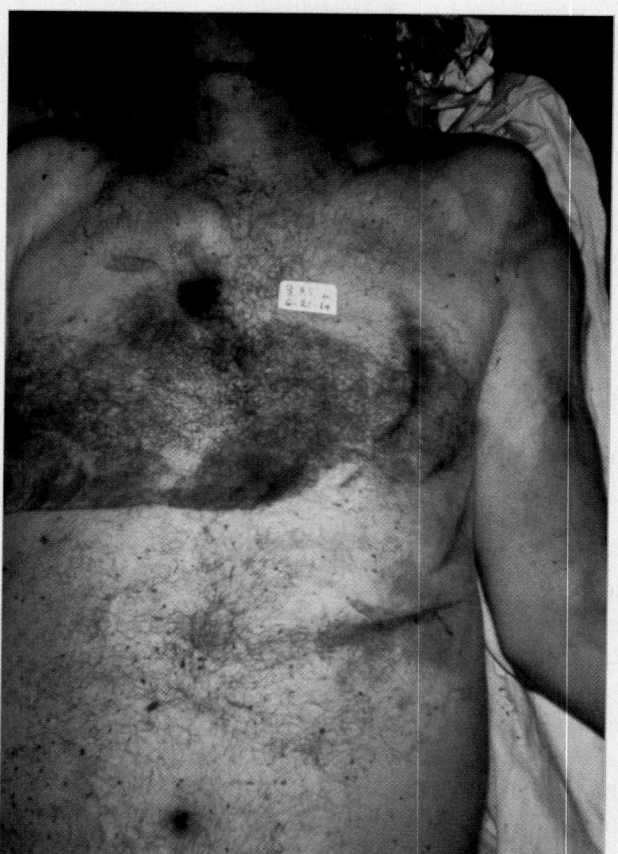

Figure 24-4 The damage associated with a crush or compression injury varies depending on the direct damage to the soft tissues and on how long the tissue was cut off from circulation.

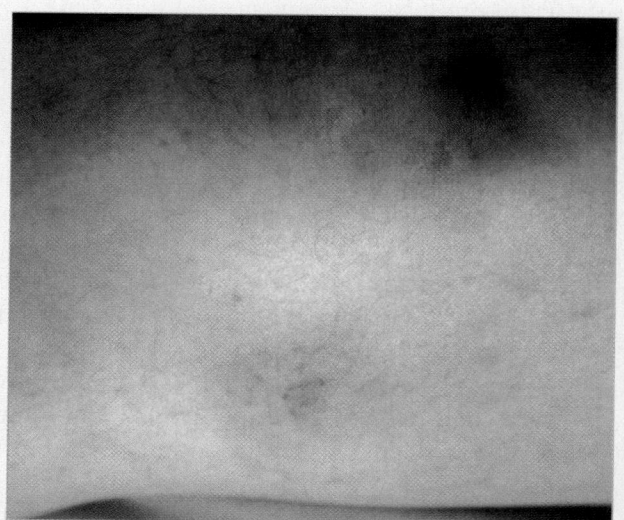

Figure 24-2 Contusions, more commonly known as bruises, occur as a result of a blunt force striking the body. The buildup of blood produces a characteristic blue or black discoloration (ecchymosis).

continued compression of the soft tissues will cut off their circulation, producing further tissue destruction. For example, if a patient's legs are trapped under a collapsed pile of rocks, damage to the leg tissues will continue until the rocks are removed.

Another form of compression can result from the swelling that occurs whenever tissues are injured. The cells that are injured leak watery fluid into the spaces between the cells. The pressure of the fluid may become great enough to compress the tissue and cause further damage. This is especially true if the blood vessels become compressed, cutting off blood flow to the tissue. This condition is called <u>compartment syndrome</u>. Excessive swelling often follows significant injury to the extremities.

Severe closed injuries can also damage internal organs. The greater the amount of energy absorbed from the blunt force, the greater is the risk of injury to deeper structures. Therefore, you must assess all patients with closed injuries for more serious hidden injuries. Remain alert for signs of shock or internal bleeding, and begin treatment of these conditions if necessary.

Assessment of Closed Injuries

Scene Size-up

As you arrive on scene, observe the scene for hazards and threats to the safety of the crew, bystanders, and patient. Ensure you and your crew have taken BSI precautions—a minimum of gloves and eye protection.

Place several pairs of gloves in your pocket for easy access in case your gloves tear or there are multiple patients with bleeding. Because of the color of blood and how well it soaks through clothing, you can often identify bleeding patients as you approach the scene. As you observe the scene, look for indicators of the MOI. This helps you develop an early index of suspicion for underlying injuries in the patient who has sustained a significant MOI.

Initial Assessment

General Impression

As you approach the trauma patient, important indicators will alert you to the seriousness of the patient's condition. Is the patient awake and interacting with his or her surroundings, or laying still, making no sounds? Does the patient have any apparent life threats such as significant bleeding? How is the patient's skin color? Is he or she responding to you appropriately or inappropriately? Your general impression will help you develop an index of suspicion for serious injuries and determine how urgently your patient needs care.

Trauma patients with closed soft-tissue injuries may have what appear to be minor injuries; however, you must not be distracted from looking for more serious hidden injuries during the initial assessment. For example, the general impression of a patient with a hematoma on the head and a decreased level of consciousness may indicate a serious head injury.

You are the Provider   Part 2

You and your partner prepare to move your patient off the floor and out of the dirt and antifreeze before further exposing his burn injuries. However, because he is not in a life-threatening environment, you opt to perform a rapid head-to-toe assessment to determine whether he has other injuries. While your partner maintains cervical spine precautions, you determine that your patient has no obvious life-threatening injuries. Your patient also tells you that he takes no medications, is allergic to penicillin, and has no significant past medical history.

3. At what point should you place oxygen on this patient and how much?
4. Is this the time to take vital signs?
5. What will determine which injury you address first?

Airway and Breathing

Next, ensure that the patient has a clear and patent airway. Because trauma was involved, protect the patient from further spinal injury as you manage the airway by preventing the head and torso from moving. If the patient is unresponsive or has a significant altered level of consciousness, consider inserting an oropharyngeal airway or nasopharyngeal airway. You must also quickly assess for adequate breathing. Palpate the chest wall for DCAP-BTLS. If a soft-tissue injury is discovered on the chest or abdomen, check for clear and symmetrical breath sounds and then provide high-flow oxygen, or provide assisted ventilations using a BVM device as needed, depending on the level of consciousness and if your patient is breathing inadequately.

Circulation

You must quickly assess the patient's pulse rate and quality; determine the skin condition, color, and temperature; and check the capillary refill time. Closed soft-tissue injuries do not have visible signs of bleeding. Because the bleeding is occurring inside the body, shock may be present. Your assessment of the pulse and skin will give you an indication as to how aggressively you need to treat your patient for shock.

Transport Decision

During your initial assessment, determine whether your patient needs immediate transport or stabilization on scene. If the patient you are treating has an airway or breathing problem or signs and symptoms of shock or internal bleeding, you must consider quickly transporting him or her to the hospital for treatment or requesting ALS support. While treatment in the initial assessment is directed at quickly addressing life threats, you should not delay transport of a trauma patient, particularly one where a closed soft-tissue injury may be a sign of a more serious deeper injury. Patients with a significant MOI may require a rapid physical exam to identify these injuries.

Focused History and Physical Exam

After the initial assessment is complete, determine which type of physical exam needs to be performed. A rapid physical exam is based on a significant MOI, whereas a focused physical exam is based on a nonsignificant MOI. For example, an industrial accident during which a worker slips and twists his ankle may not require a complete assessment of the entire body, but just the ankle. Another worker who slips and falls 15′, however, would need a complete and rapid exam to identify all the injuries and to prepare for packaging and rapid transport.

Focused Physical Exam

Focus your assessment on the isolated closed injury, the patient's complaint, and the body region affected. Assess all underlying systems. Assess pulse, motor, and sensory function in the injured extremity. On the trunk, assess respiratory, circulatory, and neurologic systems in the affected area.

Rapid Physical Exam

If significant trauma has likely affected multiple systems, start with a rapid trauma assessment, quickly looking from head to toe for DCAP-BTLS to be sure that you have found all of the problems and injuries.

If a life-threatening problem is found, treat it immediately. If a non–life-threatening injury is found, continue with the rapid physical exam. Begin with the head and neck while manually holding the head in place. When you are done, apply a cervical spine immobilization device if you have not done so already. Quickly assess the chest, abdomen, and extremities for hidden bleeding and injuries. Log roll the patient and assess the posterior torso for injuries as well. Once the back has been assessed, the patient can be log rolled down onto a backboard and complete spinal stabilization finished. Log rolling and securing the patient to a backboard or other full-body stabilization device should take into consideration injuries found during the rapid physical exam. For example, appropriately stabilizing a fractured lower leg during packaging for transport would not be possible if the fracture was not found during the rapid trauma exam. Stabilizing the injury with a splint during packaging increases the delay in initiating transport when a backboard will provide basic immobilization until time permits a more thorough splinting job.

Baseline Vital Signs

Patients who have hidden injuries under a closed soft-tissue injury may have internal bleeding and may rapidly become unstable. Determining a baseline set of vital signs will be important to identifying how quickly the patient's condition is changing. Signs such as tachycardia, tachypnea, low blood pressure, weak pulse, and cool, moist, and pale skin indicate hypoperfusion and imply the need for rapid treatment inside the hospital. Remember that soft-tissue injuries, even without a significant MOI, can cause shock. The patient's vital signs will give you a good understanding of how well your patient is tolerating the injury.

SAMPLE History

Make every attempt to obtain a SAMPLE history from your patient. If the patient is not responsive, attempt to obtain the SAMPLE history from other sources, such as friends or family members. Medical identification jewelry and cards in wallets may also provide information about the patient's medical history. Using OPQRST may provide some background on isolated extremity injuries. You have the opportunity to interview the patient well in advance of the emergency physician. Any information you receive will be very valuable if the patient loses consciousness.

Interventions

Provide complete spinal immobilization early if you suspect that your patient has spinal injuries. Providing high-flow oxygen to patients with closed soft-tissue injuries may help reduce the effects of shock and assist in perfusion of damaged tissues, particularly crush injuries. If the patient has signs of hypoperfusion, treat him or her aggressively for shock and provide rapid transport to the hospital. Request ALS as necessary to assist with more aggressive shock management. Do not delay transport of the seriously injured trauma patient to complete nonlifesaving treatments in the field, such as splinting extremity fractures; instead, complete these types of treatments en route to the hospital.

Detailed Physical Exam

Any time there is a significant mechanism of injury, a detailed physical exam should be done. Many times short transportation times and unstable patient conditions may make this assessment impractical. A detailed physical exam can help identify some injuries that were not quite evident earlier in the assessment.

Ongoing Assessment

Repeat the initial assessment. Is the airway still adequate? Is breathing still adequate? Is the pulse still adequate? Is perfusion adequate? Are the treatments you provided for problems with the ABCs still effective? How is the patient's condition improving with these interventions? Reassess vital signs. Heart rate, respirations, blood pressure, and level of consciousness are good indications of how well the patient is tolerating the stress of the injury. Assess these signs frequently and note trends that indicate whether the patient's condition is getting better or worse.

Communication and Documentation

Verbally describe all the injuries you found and explain their significance in the patient's condition to hospital personnel. Use appropriate anatomic descriptions and terminology. Provide an accurate account of how you treated these injuries. These are all important findings to include in your verbal and written communication. Your ability to communicate clearly and accurately enables the physicians and nurses at the hospital to continue quality care.

Emergency Medical Care

Small contusions require no special emergency medical care. More extensive closed injuries may involve significant swelling and bleeding beneath the skin, which could lead to hypovolemic shock. Before treating a closed injury, make sure to follow BSI precautions.

Soft-tissue injuries may look rather dramatic. However, you must still focus on airway and breathing first. Always maintain the airway and provide oxygen in patients with potentially serious injuries. If the patient has inadequate breathing you may have to assist ventilations with a BVM device. Treat a closed soft-tissue injury by applying the acronym RICES:

- **Rest**—keep the patient as quiet and as comfortable as possible.
- **Ice** (or a cold pack) slows bleeding by causing blood vessels to constrict and also reduces pain.
- **Compression** over the injury site slows bleeding by compressing the blood vessels.
- **Elevation** of the injured part just above the level of the patient's heart decreases swelling.
- **Splinting** decreases bleeding and also reduces pain by immobilizing a soft-tissue injury or an injured extremity.

In addition to using these measures to control bleeding and swelling, you should also be alert for signs of developing shock, including anxiety or agitation, changes in mental status, increased heart rate, increased respiratory rate, diaphoresis, cool or clammy skin, and decreased blood pressure. Any or all of these signs may indicate internal bleeding resulting from injuries to internal organs. If the patient appears to be in shock, you should place the patient in the shock position (supine with legs elevated 6″ to 12″); or if the patient is on a backboard or stretcher, use the Trendelenburg position (on a backboard or stretcher with the feet 6″ to 12″ higher than the head), provide supplemental high-flow oxygen and prompt transport to the hospital.

Open Injuries

Open injuries differ from closed injuries in that the protective layer of skin is damaged. This can produce more extensive bleeding. More important, however, a break in the protective skin layer or mucous membrane means that the wound is contaminated and may become infected. <u>Contamination</u> describes the presence of infectious organisms (pathogens) or foreign bodies, such as dirt, gravel, or metal. You must address these two problems in your treatment of open soft-tissue wounds. There are four types of open soft-tissue wounds that you must be prepared to manage: abrasions, lacerations, avulsions, and penetrating wounds.

You are the Provider   Part 3

With high-flow oxygen in place and no other apparent injuries, you immobilize the victim with a cervical collar, long backboard, and cervical immobilization devices.

You decide to immobilize the injured ankle before rolling the patient. In the process, you notice a bruised area to the outside and top portion of the ankle. Swelling masks any other deformity, and the site is sensitive to gentle palpation. You have an excellent distal pulse at the dorsalis pedis. You suspect a crushing fracture and place the ankle in a firm pillow splint with a cold pack placed over the bruising.

After splinting, the patient expresses relief and you reconfirm the presence of a distal pulse. With the help of other bystanders you carefully log roll the patient onto his left side. Your partner cuts away the patient's coveralls from his back and rapidly checks the area for injury, bruising, or deformity before you and the bystanders roll the patient back down onto the backboard. He presents with no head trauma and his neck is supple and without pain or deformity.

You place the patient into the ambulance, where you remove the remainder of his outer clothing and perform a focused physical exam. Your findings include a respiratory rate of 22 breaths/min and unlabored, a pulse of 96 beats/min and regular, a blood pressure of 134/88 mm Hg, and a pulse oximeter reading of 100%. His pupils are midpoint, equal, and reactive to light. No jugular venous distention (JVD) is present. He has good lung sounds and his abdomen, while firm, is not tender to palpation. His pelvis is stable and not deformed. He has no injury to his groin. He has full range of motion in his uninjured limbs and is able to move them with purpose.

You see that one of his thighs has large, fist-sized blisters. The blisters are obviously fragile due to their size and fullness, and there are three of them surrounded by smaller blisters and a red border, where the skin is not blistered but still thermally burned. The patient states that the area with blisters presents with a lot of pain, but moving towards the outer areas the pain is not as intense.

6. What degree(s) of burns are these? Given the location and area of the burns, are they potentially life threatening?

7. How will you treat his burns?

An <u>abrasion</u> is a wound of the superficial layer of the skin, caused by friction when a body part rubs or scrapes across a rough or hard surface. An abrasion usually does not penetrate completely through the dermis, but blood may ooze from the injured capillaries in the dermis. Known by a variety of names, including road rash, road burn, strawberry, and rug burn, abrasions can be extremely painful (Figure 24-5 ▼).

A <u>laceration</u> is a jagged cut caused by a sharp object or a blunt force that tears the tissue, while an <u>incision</u> is a sharp, smooth cut. The depth of the injury can vary, extending through the skin and subcutaneous tissue even into the underlying muscles and adjacent nerves and blood vessels (Figure 24-6 ▼). Lacerations and incisions may appear linear (regular) or stellate (irregular) and may occur along with other types of soft-tissue injury. Lacerations or incisions that involve arteries or large veins may result in severe bleeding.

An <u>avulsion</u> is an injury that separates various layers of soft tissue (usually between the subcutaneous layer and fascia) so that they become either completely detached or hang as a flap (Figure 24-7 ▶). Often there is significant bleeding. If the avulsed tissue is hanging from a small piece of skin, the circulation through the flap may be at risk. If you can, replace the avulsed flap in its original position. If an avulsion is complete, you

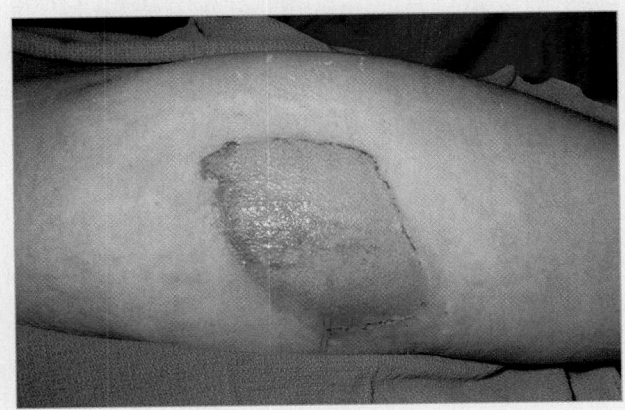

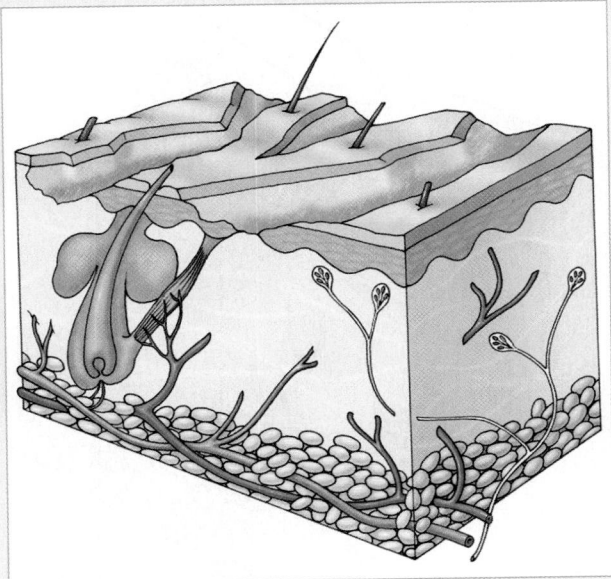

Figure 24-5 Abrasions usually do not penetrate completely through the dermis, but blood may ooze from the capillaries. These wounds are typically superficial and result from rubbing or scraping across a hard, rough surface.

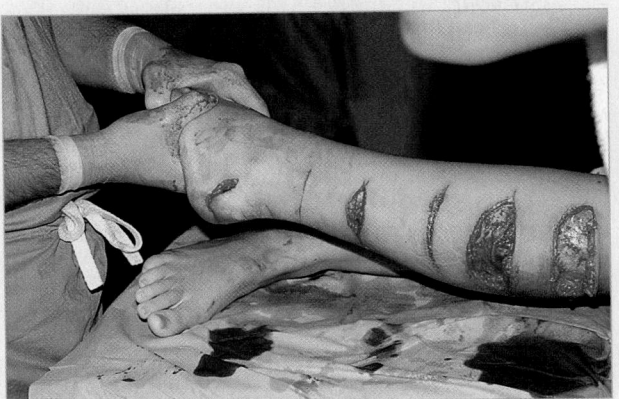

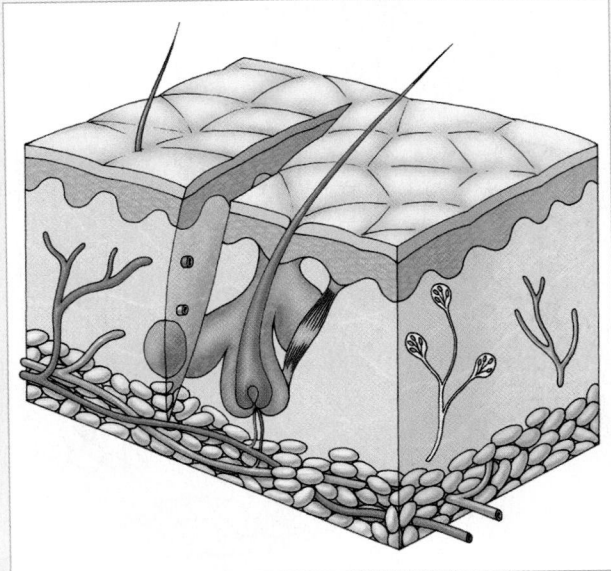

Figure 24-6 Lacerations vary in depth and can extend through the skin and subcutaneous tissue to the underlying muscles, nerves, and blood vessels. These wounds can be smooth or jagged as a result of a cut by a sharp object or a blunt force that tears the tissue.

should wrap the separated tissue in sterile gauze and bring it with you to the emergency department.

We usually think of amputations as involving the upper and lower extremities. But other body parts, such as the scalp, ear, nose, penis, or lips, may also be totally avulsed, or amputated. You can easily control the bleeding from some amputations, such as the fingers, with pressure dressings. But if an avulsion involves a large area of muscle mass, such as a thigh, there may be massive bleeding. In this situation, you need to treat the patient for hypovolemic shock. The use of pressure points may also be necessary to control bleeding not controlled with a standard pressure dressing (see Skill Drill 22-1 in Chapter 22).

A <u>penetrating wound</u> is an injury resulting from a sharp, pointed object, such as a knife, ice pick, splinter, or bullet. Such objects leave relatively small entrance wounds, so there may be little external bleeding (Figure 24-8 ▼). However, these objects can damage structures deep within the body and cause unseen bleeding. If the wound is to the chest or abdomen, the injury can cause rapid, fatal bleeding. Assessing the amount of damage a puncture wound has created is very difficult and is reserved for the physician at the hospital.

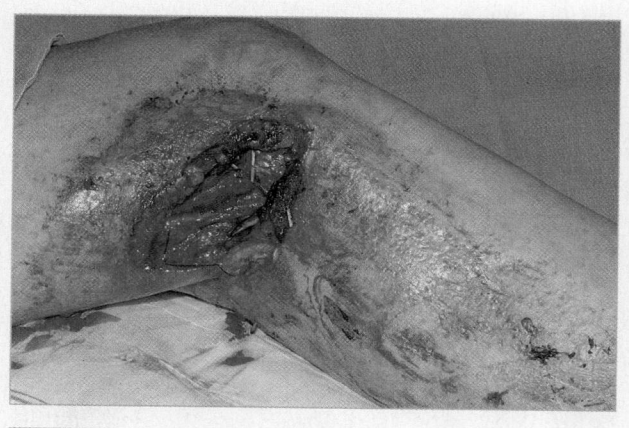
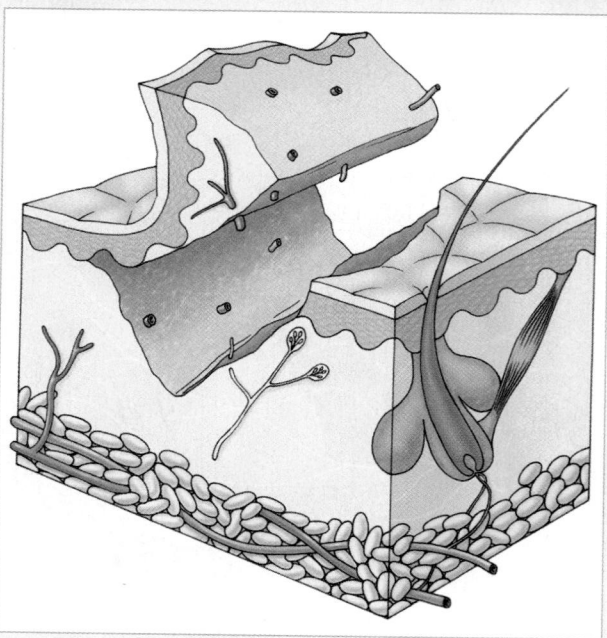

Figure 24-7 Avulsions are injuries characterized by either complete separation of tissue or tissue hanging as a flap. Significant bleeding is common.

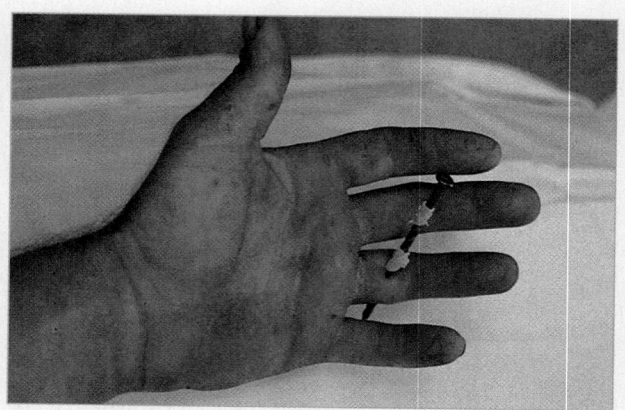
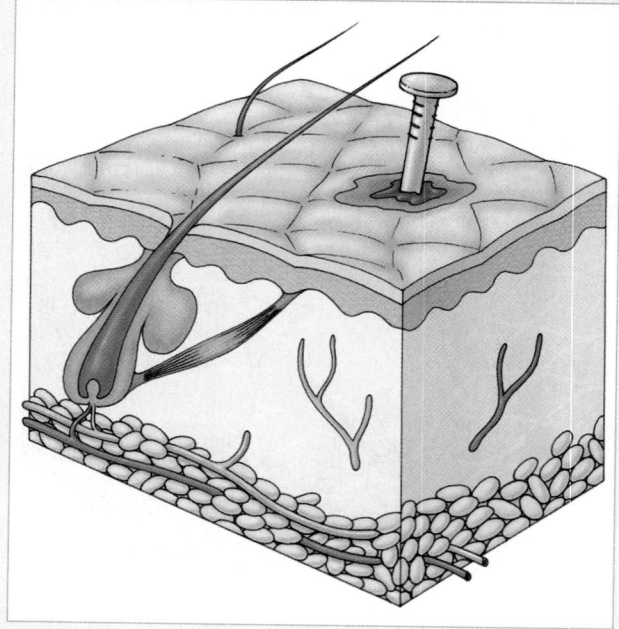

Figure 24-8 Penetrating wounds may cause very little external bleeding but can damage structures deep within the body.

Stabbings and shootings often result in multiple penetrating injuries. You must assess these patients carefully to identify all wounds. Since a penetrating object can pass completely through the body, always count the number of penetrating injuries (or holes), especially with gunshot wounds. Entrance wounds and exit wounds may be difficult to tell apart in a prehospital setting, especially with the different types of ammunition available. While entrance wounds are often smaller than exit wounds (Figure 24-9 ▼), it is better to simply count the number of penetrating injuries, and leave the distinction between entrance and exit to the physician who is working in a more controlled environment. Gunshot wounds have some unique characteristics that require special care. The amount of energy transmitted by a gunshot injury is directly related to the speed of the bullet. Thus, it is important to find out the type of gun that was used in the shooting. Sometimes, the patient or bystanders can tell you how many rounds were fired. This information can help hospital person-

nel to better care for the patient. Shotgun wounds create multiple paths of missiles (shot) and create a larger surface area and volume of tissue damage.

Many shootings end up in court at some point, and you may be called to testify. For this reason, you must carefully document the circumstances surrounding any gunshot injury, the patient's condition, and the treatment you give.

As with closed wounds caused by crushing, open wounds caused by crushing may involve damaged internal organs or broken bones, as well as extensive soft-tissue damage (Figure 24-10 ▼). While external bleeding may be minimal, internal bleeding may be severe, even life-threatening. The crushing force damages soft tissues as well as vessels and nerves. This frequently results in a painful, swollen, deformed area.

Assessment of Open Injuries

Scene Size-up

Open soft-tissue injuries can be very messy. Control of the blood and bloody contaminants can be difficult unless you are careful about what you touch and where. Using BSI precautions can minimize your direct exposure to body fluids. However, reaching into a medical kit for supplies with bloody gloves will extend the area of contamination and increase the risk of exposure to you or other rescuers. Place several pairs of gloves in

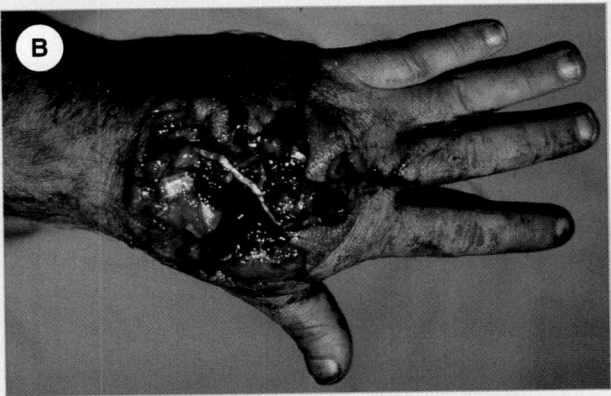

Figure 24-9 **A.** An entrance wound from a gunshot may have burns around the edges. **B.** An exit wound is often larger and associated with greater damage to soft tissues locally.

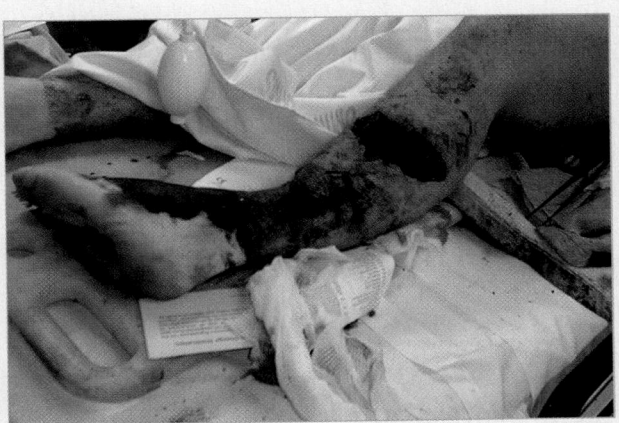

Figure 24-10 A crushing open wound is characterized by extensive tissue damage and deformity that is often accompanied by swelling and extreme pain.

✚ EMT-B Tips

One trick to controlling bloody contaminants is to wear two pairs of gloves. When one gets dirty, peel it off the other layer and you still have immediate protection.

your pocket for easy access in case you need another pair. If your gloves tear or there are multiple patients with bleeding you may need more gloves immediately available. It would be unfortunate for you to contaminate one patient with another patient's blood.

Because of the color of blood and how well it soaks through clothing, you can often identify patients with bleeding as you approach the scene. However, bleeding can be hidden under thick clothing such as denim and leather. Eye exposures may occur from splashes and droplets at a busy scene. Eye protection should be required when managing open wounds.

As you put together information from dispatch and your observations of the scene, consider how the MOI produced the injuries expected. This helps you develop an early index of suspicion for underlying injuries in the patient who has sustained a significant MOI. For example, in a vehicle crash, a patient who has sustained abrasions and lacerations to the face from an impact with the steering wheel or windshield may have experienced enough force to injure the cervical spine as well. In this case and in many trauma situations, spinal precautions should be employed very early in your care of the patient. You may even want to take spinal stabilization equipment with you as you exit the ambulance and approach the patient based on your scene survey. The MOI may also provide indications of safety threats. For example, gunshot wounds may indicate angry individuals. Make sure the scene is safe and consider requesting additional help early.

Initial Assessment

General Impression

Trauma patients may present with obvious significant injuries that indicate a serious condition. However, other injuries may not be as obvious but may still indicate a very serious condition. Your "impression" about

how the patient is doing is based on information as simple as the patient's age, the MOI, and his or her level of consciousness. Observations such as bleeding from open wounds, skin color and condition, and gasping also contribute to your general impression and help to determine your treatment priorities and the urgency of care needed. A good question to ask is, "How sick is my patient based on what I know right now?"

Airway and Breathing

Open soft-tissue injuries of the face and neck have a potential to interfere with the effectiveness of the airway and breathing. Evaluate the patient's voice and speaking ability to identify throat injuries. A patient with a crush injury to the foot or to another area distant to the airway may have other less obvious injuries that may interfere with airway and breathing also. Do not be distracted and assume the patient's condition is stable except for an isolated bleeding injury. An injury to the leg, for example, may have caused the patient to fall and injure the neck or back. If a spinal injury is suspected, stabilize the spine to protect from further injury as you manage airway problems and continue your assessment.

You must also quickly assess for adequate breathing. Observe the rate and depth of respirations. Listen to breath sounds quickly on each side of the chest. If an open wound is found on the chest, evaluate for air movement through the wound in the form of bubbling or sucking sounds that indicate a deep penetrating injury. Quickly place an <u>occlusive dressing</u> over the wound. Provide high-flow oxygen or assisted ventilations with a BVM device as needed, depending on the patient's level of consciousness and on the adequacy of the patient's breathing.

Circulation

Assess the patient's pulse rate and quality, determine the skin condition, color, and temperature, and check the capillary refill time. These assessments will help determine the presence of circulatory problems or shock. If visible significant bleeding is seen, you must begin the steps necessary to control bleeding. Significant bleeding is an immediate life threat and must be controlled quickly using appropriate methods. In dark environments bleeding can be hidden because of its color. Thick clothing may also hide bleeding. Consider the MOI and be suspicious as to where bleeding might occur and expose it. Blood flowing freely from veins in a large gash can be as much of a threat as blood spurting from an artery. Life-threatening bleeding must be controlled in the initial assessment. Should it be controlled before you administer the patient oxygen? You must decide where the priorities lie. Control of oozing blood from damaged capillaries in an abrasion may be controlled later if more important problems are at hand. Covering these lesser wounds is important to prevent infection even when bleeding is minimal.

Transport Decision

If the patient you are treating has an airway or breathing problem or significant bleeding, you must consider quickly transporting this patient to the hospital for treatment. It is easy to become distracted by the gore of significant soft-tissue injuries and large amounts of blood. Frightened, screaming patients can also prevent you from focusing on the problems at hand. The ABCs are simple enough to remember and treat. Follow the protocols you have learned.

Patients who have visible significant bleeding, or signs of significant internal bleeding, may quickly become unstable. Treatment must be directed at quickly addressing life threats and providing rapid transportation to the closest appropriate hospital. Signs such as tachycardia, tachypnea, weak pulse, and cool, moist, and pale skin are signs of hypoperfusion and imply the need for rapid transport. You should be alert to these signs and reassess your priority and transport decision if they develop.

Focused History and Physical Exam

After the initial assessment is complete, determine which type of exam will be performed next—a focused physical exam or a rapid physical exam—based on MOI.

Focused Physical Exam

In the responsive patient who has a simple open injury with limited MOI, consider a focused physical exam. Focus your assessment on the isolated injury, the patient's complaint, and the body region affected. Ensure that control of the bleeding is maintained and note the location of the injury. Assess all underlying systems. In an extremity, assess pulse, motor, and sensory function in the injured extremity.

Rapid Physical Exam

If there is significant trauma, likely affecting multiple systems, start with a rapid physical exam looking for DCAP-BTLS to be sure that you have found all of the problems and injuries. With significant trauma

you should quickly assess the entire patient from head to toe.

You should not delay transport of a trauma patient, particularly a patient with significant bleeding even if controlled. Identifying injuries during a rapid physical exam may help in how you prepare your patient for transport. For example, noting a hip or extremity injury in your patient during this exam would suggest log rolling the patient away from the injured extremity when possible. Spinal stabilization should be completed here, including application of a cervical spine stabilization device and securing the patient to a backboard, if not already done in the initial assessment.

Baseline Vital Signs

You must assess the patient's baseline vital signs to know whether changes in the patient's condition occur during treatment. The findings identified earlier in your assessment, such as tachycardia, tachypnea, weak pulse, and cool, moist, and pale skin, must now be quantified and recorded. When a blood pressure and pupillary response is also assessed, your baseline vital signs are complete. A blood pressure of less than 100 mm Hg with a weak, rapid pulse and cool, moist skin that is pale or gray should alert you to the presence of hypoperfusion in the patient who may have significant bleeding. Remember you must be concerned with both visible bleeding and bleeding that is unseen.

SAMPLE History

Next obtain a SAMPLE history from your patient. Conditions such as anemia (low quantity of hemoglobin in the blood) and hemophilia (a disorder in which blood has a diminished ability to clot) can complicate open soft-tissue injuries. Medications such as aspirin and other blood-thinning medications frequently taken by older patients may make clotting and bleeding control difficult. If the injury was self-inflicted it is also a behavioral problem. If the patient is not responsive, attempt to obtain the SAMPLE history from friends or family members.

Interventions

If bleeding is found, cover the wound and control the bleeding as quickly as possible. If the bleeding is serious, it should be controlled in the initial assessment. If the bleeding is not significant, such as from an abrasion, it can be treated later in the assessment. Spinal stabilization and assistance with breathing or perfusion problems may also need to be provided. Splint a painful, swollen, deformed extremity.

The decision to apply oxygen first or to apply direct pressure and a bandage to a profusely bleeding wound in a patient in shock can be a difficult one. You are the one who will have to decide which treatment takes priority. The protocols discussed here are designed to be flexible and adjust to each situation while still providing some structure. Rely on the experience and judgment of your partner or team until you can confidently make decisions based on your own experiences. Always doing what is best for your patient is a good rule to follow. Do not delay transport of the seriously injured trauma patient to complete nonlifesaving treatments in the field, such as splinting extremity fractures. Instead, complete these types of treatment en route to the hospital.

Detailed Physical Exam

If the patient's condition is stable and problems do not persist after the initial assessment, perform a thorough detailed physical exam of the patient as discussed in the chapter on patient assessment. Many times short transportation times and unstable patient conditions that require continuous monitoring and treatment may make this assessment impractical.

Ongoing Assessment

Reassessing the patient with an open soft-tissue injury is extremely important, especially if you did not put on the bandage. Frequently, other emergency care personnel, such as first responders, may have dressed and bandaged the wound before your arrival. If their bandaging job was ineffective, you may need to consider taking everything off and starting fresh. Remember, starting from scratch may disrupt blood clots that have begun to form and may make bleeding worse.

Assess all bandaging frequently. If blood continues to soak through bandages, use additional methods to control bleeding as discussed later in the chapter. Reassess airway, breathing, and circulation often. Reassess other interventions and treatments you have provided to the patient in addition to reassessing vital signs. Compare your baseline assessment to repeated assessments to see whether your patient's condition is getting better or worse.

Communication and Documentation

You must include a description of the MOI and the position in which you found the patient as you arrived on scene. In cases involving severe external bleeding, it is important to recognize, estimate, and report the amount of blood loss that has occurred and how rapidly or how much time has passed since the bleeding started. This is a challenge, especially if the surface is wet, absorbs fluids, or is dark. You should attempt to report blood loss using terms that you are comfortable with and that will be easily understood by other personnel. For example, you may say "approximately a liter was lost," or "the bleeding has soaked through three trauma dressings." It is not as important how you describe it, but that you describe it accurately. You must include the location and description of any soft-tissue injuries or other wounds you have located and treated. Describe the size and depth of the injury. Provide an accurate account of how you treated these injuries. All of this information is important to include in your verbal and written communication.

Emergency Medical Care

Before you begin caring for a patient with an open wound, you should be sure to protect yourself by following BSI precautions. Wear gloves, eye protection, and, if necessary, a gown and a mask. Remember that you must be sure the patient has an open airway and administer high-flow oxygen as necessary. If life-threatening bleeding is observed, assign a team member to apply direct pressure to control the bleeding. Then assess the severity of the wound. If the wound is in the chest or upper abdomen, place an occlusive dressing on the wound.

Your treatment priority is the initial assessment, including controlling the bleeding, which can be extensive and severe. Then follow the steps in (Skill Drill 24-1 ▶):

1. **Apply a dry, sterile dressing** over the entire wound. Apply pressure to the dressing with your gloved hand (**Step 1**).
2. **Maintain the pressure** and secure the dressing with a roller bandage (**Step 2**).
3. **If bleeding continues or recurs, leave the original dressing in place.** Apply a second dressing on top of the first, and secure it with another roller bandage (**Step 3**).
4. **Splint the extremity** to stabilize the injury even if there is no suspected fracture, to help minimize movement, further control the bleeding, and keep the dressing in place (**Step 4**).

Controlling Bleeding from a Soft-Tissue Injury

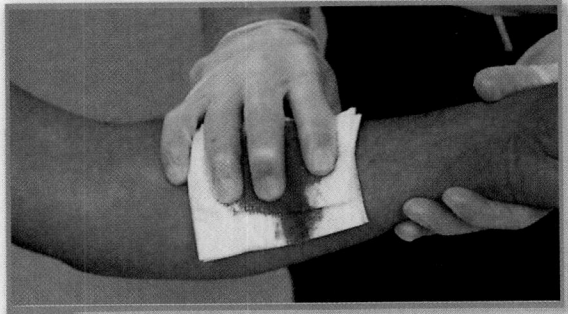

1 Apply direct pressure with a sterile bandage.

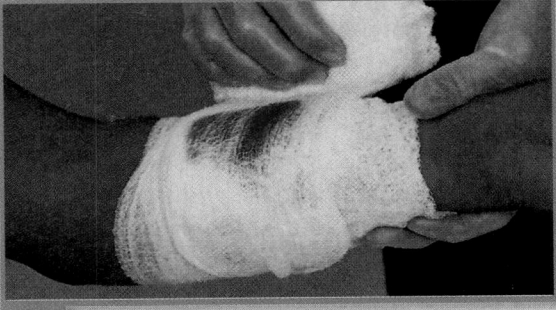

2 Maintain pressure with a roller bandage.

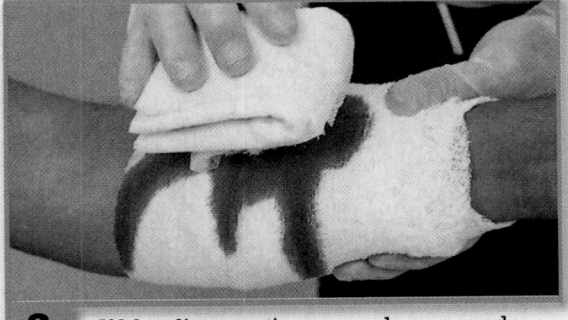

3 If bleeding continues, apply a second dressing and roller bandage over the first.

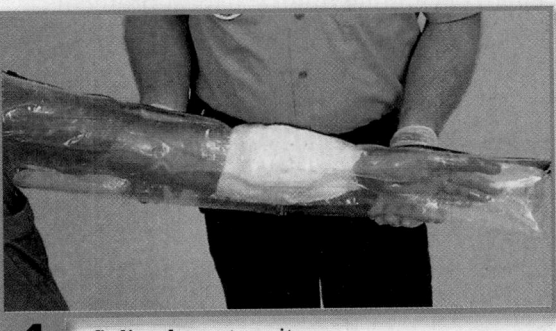

4 Splint the extremity.

Skill Drill 24-1

All open wounds are assumed to be contaminated and present a risk of infection. By applying a sterile dressing, you are reducing the risk of further contamination. This keeps foreign material, such as hair, clothing, and dirt, out of the wound and decreases the risk of infection. In general, you should not try to remove material from an open wound, no matter how dirty the wound is. Rubbing, brushing, or washing an open wound can cause additional bleeding. Chemical burns and contamination should be flushed to remove remaining chemicals. Only hospital personnel should clean out an open wound. To prevent a wound from drying, you may apply sterile dressings moistened with sterile saline solution and then cover the moist dressing with a dry, sterile dressing.

Often, you can better control bleeding from open soft-tissue wounds by splinting the extremity, even if there is no fracture. Splinting can also help you to keep the patient calm and quiet, as it typically reduces pain.

In addition, splinting keeps sterile dressings in place, minimizes damage to an already injured extremity, and makes moving the patient easier.

Severe bleeding often accompanies significant trauma. When significant trauma exists, do not spend time on scene splinting the wound. Apply a pressure bandage and splint during transport if time allows.

Keep in mind that a patient who is bleeding significantly from an open wound is at risk for hypovolemic shock. You must be alert for this possibility and provide treatment, as needed, in all cases of significant trauma and in patients with moderate to severe bleeding.

Abdominal Wounds

An open wound in the abdominal cavity may expose internal organs. In some cases, the organs may even protrude through the wound, an injury called an <u>evisceration</u>

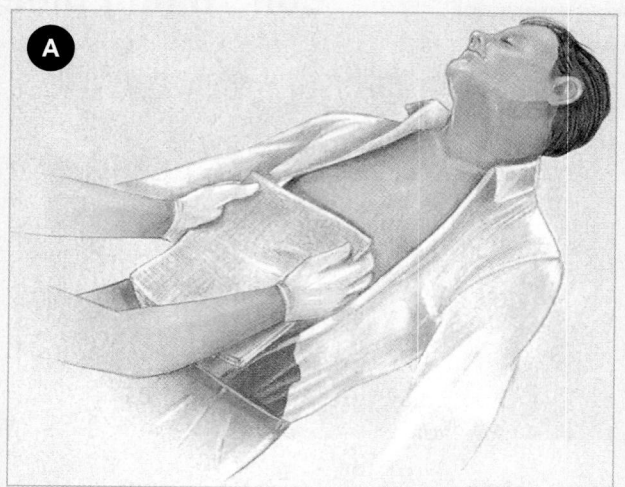

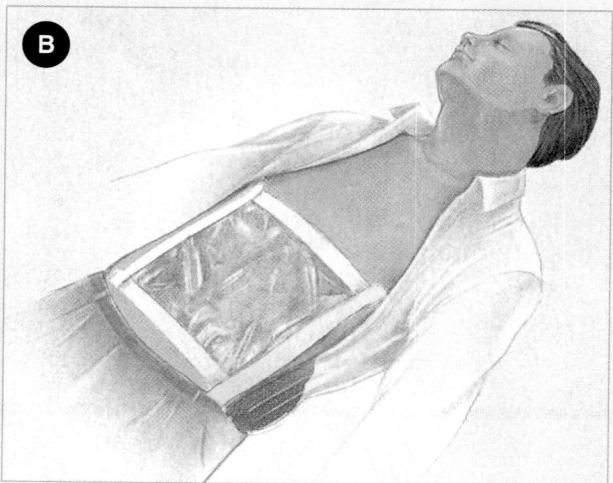

Figure 24-11 ▼). Do not touch or move the exposed organs. Rather, cover the wound with sterile gauze compresses moistened with sterile saline solution and secured with a sterile dressing (Figure 24-12 ▶). Because the open abdomen radiates body heat very effectively, and because exposed organs lose fluid rapidly, you must keep the organs moist and warm. If you do not have gauze compresses, you may use moist sterile dressings, covered and secured in place with a bandage and tape. Do not use any material that is adherent or loses its substance when wet, such as toilet paper, facial tissue, paper towels, or absorbent cotton. If the patient's legs and knees are uninjured, and spinal injury is NOT suspected, flex them to relieve pressure on the abdomen. Most patients with abdominal wounds require immediate transport to a trauma center, depending upon the local protocol.

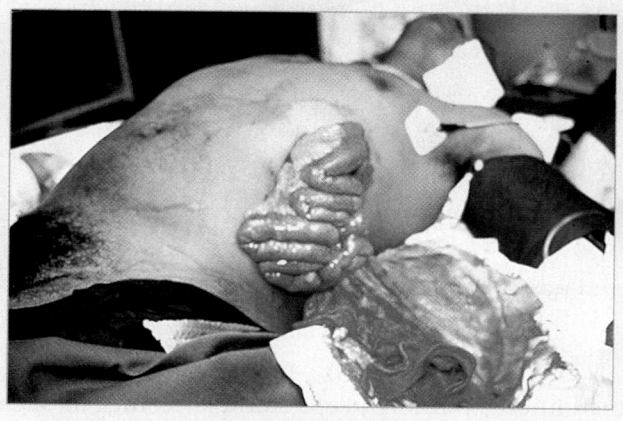

Figure 24-11 An abdominal evisceration is an open wound to the abdomen in which organs protrude through the wound.

Figure 24-12   A. Cover exposed organs with sterile gauze compresses moistened with sterile saline solution. B. Place a dressing over the compresses, and secure it in place by taping all four sides.

You are the Provider   Part 4

You gently pour water over the burn area to flush away any remaining antifreeze. You then place a large, dry, loose dressing over the burned area to protect and keep the injured site clean. On reevaluation of his vital signs, the patient has a respiratory rate of 20 breaths/min and still unlabored, a regular pulse rate of 84 beats/min, a blood pressure of 128/78 mm Hg, and a pulse oximeter reading of 99%. His ankle still presents with a good, if faint, distal pulse, but the patient tells you his foot feels cold and tingly.

8. Does this patient appear to be developing shock?
9. What information will you provide on calling a report in to the receiving facility?

Stabilizing an Impaled Object

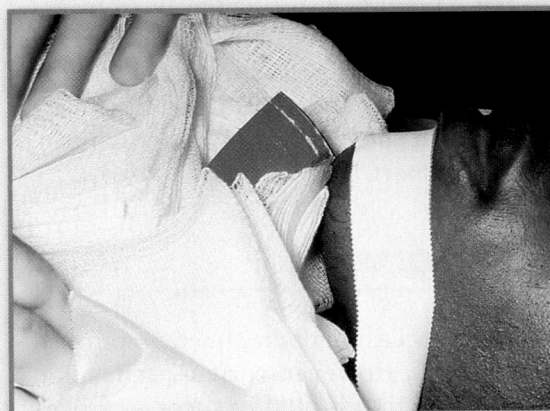

1 Do not attempt to move or remove the object. Stabilize the impaled body part.

2 Control bleeding and stabilize the object in place using soft dressings, gauze, and/or tape.

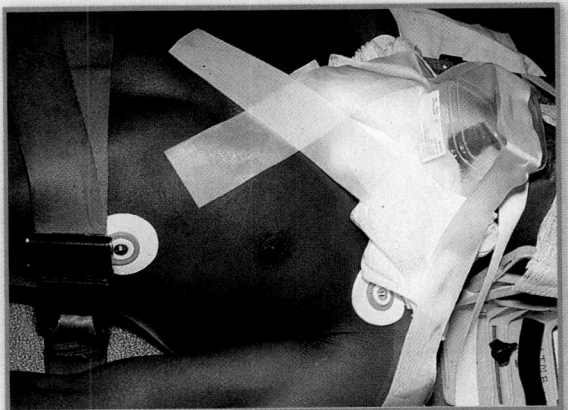

3 Tape a rigid item over the stabilized object to protect it from movement during transport.

Impaled Objects

Occasionally, a patient will have an object, such as a knife, fishhook, wood splinter, or piece of glass, impaled in his or her body. To treat this, follow the steps in (Skill Drill 24-2 ▲):

1. **Do not attempt to move or remove the object** unless it is impaled through the cheek causing airway obstruction, or if the object is in the chest and interferes with CPR. In most cases a surgeon will have to remove the object; removing it in the field may cause more bleeding or damaged nerves, blood vessels, or muscles within the wound (**Step 1**). Stabilize the impaled body part.

2. **Remove any clothing covering the injury. Control bleeding, and apply a bulky dressing** to stabilize the object. Some combination of soft dressings, gauze, and tape may be effective, depending on the location and size of the object. To prevent further injury, manually secure the object by incorporating it into the dressing (**Step 2**).

3. **Protect the impaled object** from being bumped or moved during transport by taping a rigid item such as a plastic cup, a section of a plastic water bottle, or a supply container over the stabilized object and its bandaging (**Step 3**).

The only exception to the rule of not removing an impaled object is an object in the cheek that obstructs breathing. In this situation, restoring the airway takes priority. If the object is very long, cut off (shorten) the exposed portion, first securing it to minimize motion and thus internal damage and pain. Once the object has been secured and the bleeding is under control, provide prompt transport.

Amputations

Surgeons today can occasionally reattach amputated parts (Figure 24-13 ▼). However, correct prehospital care of the amputated part is vital to successful reattachment. With partial amputations, make sure to immobilize the part with bulky compression dressings and a splint to prevent further injury. Do not sever any partial amputations; this may complicate later reattachment.

With a complete amputation, make sure to wrap the part in a sterile dressing and place it in a plastic bag. Follow your local protocols regarding how to preserve amputated parts. In some areas, dry sterile dressings are recommended for wrapping amputated parts; in other areas, dressings moistened with sterile saline are recommended. Put the bag in a cool container filled with ice. Lay the wrapped part on a bed of ice; do not pack it in ice. The goal is to keep the part cool without allowing it to freeze or develop frostbite. The amputated part should be transported with the patient. Remember that the wound on the body side of the amputation needs to be cared for including control of bleeding and appropriate bandage.

Neck Injuries

An open neck injury can be life threatening. If the veins of the neck are open to the environment, they may suck in air (Figure 24-14 ▼). If enough air is sucked into a blood vessel, it can actually block the flow of blood in the lungs, sending the patient into cardiac arrest. This condition is called air embolism. To control bleeding and prevent the possibility of air embolism, cover the wound with an occlusive dressing. Apply manual pressure, but do not compress both carotid vessels at the same time; if you do, this may impair circulation to the brain and cause a stroke. Secure a pressure dressing over the wound by wrapping roller gauze loosely around the neck and then firmly through the opposite axilla (Figure 24-15 ▼).

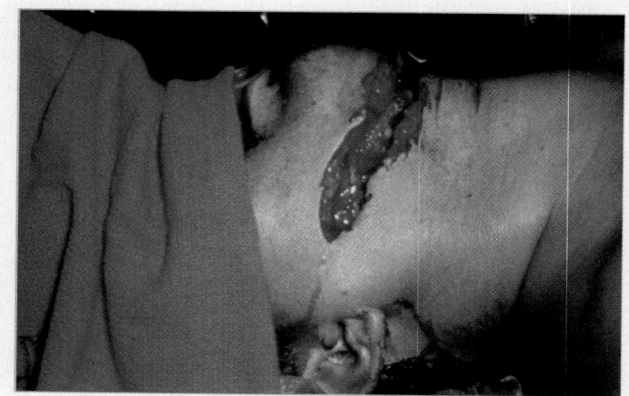

Figure 24-14 Open injuries to the neck can be very dangerous. If veins are open to the environment, they can suck in air, resulting in a potentially fatal condition called air embolism.

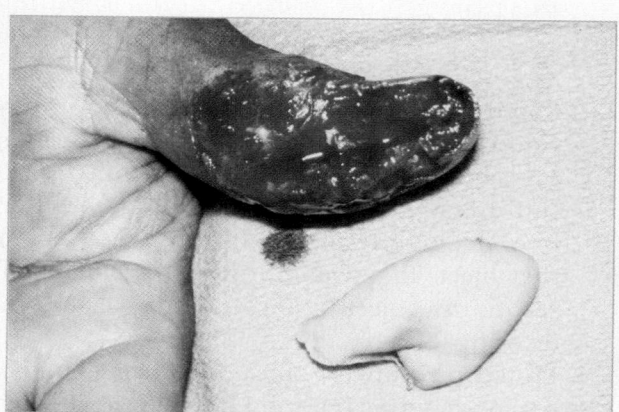

Figure 24-13 Amputated parts can occasionally be reimplanted, so you should make every attempt to find the part and transport it to the emergency department along with the patient.

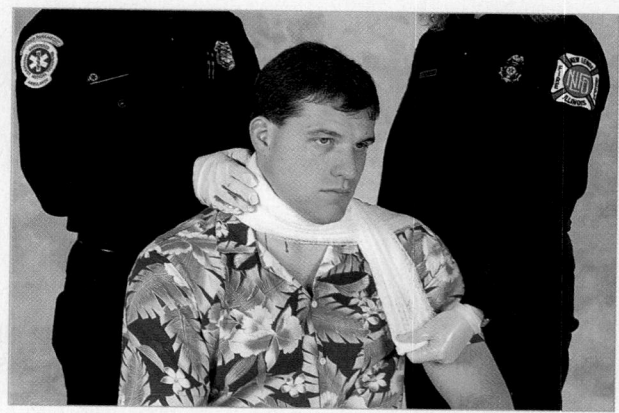

Figure 24-15 Cover neck wounds with an airtight dressing, and apply manual pressure. Be sure that you do not compress both carotid arteries at the same time, as this may impair circulation to the brain.

Burns

As an EMT-B, you will often provide care to patients who have been burned. Burns account for over 10,000 deaths a year. Burns are also among the most serious and painful of all injuries. A burn occurs when the body, or a body part, receives more radiant energy than it can absorb without injury. Potential sources of this energy include heat, toxic chemicals, and electricity. The proper emergency care of a burn may increase a patient's chances of survival and decrease the risk or duration of a long-term disability. Although a burn may be the patient's most obvious injury, you should always perform a complete assessment to determine whether there are other serious injuries.

Burn Severity

The seriousness of a burn may influence the choice of a treatment facility. Five factors will help you to determine the severity of a burn.

1. What is the depth of the burn?
2. What is the extent of the burn?

These first two factors are the most important. After gauging these, ask yourself the remaining questions.

3. Are any critical areas (face, upper airway, hands, feet, genitalia) involved? Also included in critical areas would be any circumferential burns, which are burns that go completely around a body part such as an arm or foot.
4. Are there any preexisting medical conditions or other injuries?
5. Is the patient younger than 5 years or older than 55 years?

If the answer to any of these last three questions is yes, you should upgrade the burn's classification (Table 24-1 ▼).

TABLE 24-1 Classification of Burns in Adults

Critical Burns
- Full-thickness burns involving the hands, feet, face, upper airway, genitalia, or circumferential burns of other areas
- Full-thickness burns covering more than 10% of the body's total surface area
- Partial-thickness burns covering more than 30% of the body's total surface area
- Burns associated with respiratory injury (smoke inhalation or inhalation injury)
- Burns complicated by fractures
- Burns on patients younger than 5 years or older than 55 years that would be classified as "moderate" on young adults

Moderate Burns
- Full-thickness burns involving 2% to 10% of the body's total surface area (excluding hands, feet, face, genitalia, or upper airway)
- Partial-thickness burns covering 15% to 30% of the body's total surface area
- Superficial burns covering more than 50% of the body's total surface area

Minor Burns
- Full-thickness burns covering less than 2% of the body's total surface area
- Partial-thickness burns covering less than 15% of the body's total surface area
- Superficial burns covering less than 50% of the body's total surface area

You are the Provider   Part 5

You slightly loosen the cravats used to apply pressure to the pillow splint and the patient states that the foot is not as tingly. His pulse is still faint but easily palpated. You realize that you will need to evaluate this area often while en route. You deliver your dispatch to the hospital; it includes your patient's chief complaint, the circumstances of his injury, his extrication and immobilization, the degree and depth of his burns along with a description of his blisters, and the care you have provided. You describe the patient's ankle injury, stabilization, and initial and last vital signs. You give your ETA and ask if the facility has any instructions for you.

10. Why is determining the degree and nature of the patient's burn injury and including this information in your verbal report vital to both his short- and long-term outcome?

Geriatric Needs

When treating geriatric patients with burns, it is important to be vigilant for the possibility of geriatric abuse. Geriatric patients who are institutionalized, disoriented, or incapable of clear communication are particularly susceptible to abuse.

Signs of abuse in the geriatric patient include evidence of multiple injuries in various stages of healing (ie, multiple bruises of different colors, new and old fractures involving more than one extremity), injuries that do not seem to correspond to the history provided by caregivers, and burns associated with a suspicious history.

Burns that appear in a "pattern" are suspicious for intentional injuries. Multiple, small circular burns may be indicative of cigarette or cigar injuries. Other patterns may indicate irons, stovetops, or other hot surfaces not easily encountered accidentally. Scalding injuries to hands or feet may also be indicative of abuse. It is important to remember that these injuries are often inflicted in areas not readily seen. If the situation is suspicious for geriatric abuse, be sure to fully examine the patient under his or her clothing for signs of abuse. As always, appropriate support and transport of the patient in timely fashion remain a priority.

Depth

Burns are first classified according to their depth (Figure 24-16 ▶). You must be able to identify the following three types of burns:

- **Superficial (first-degree) burns** involve only the top layer of skin, the epidermis. The skin turns red but does not blister or actually burn through. The burn site is painful. A sunburn is a good example of a superficial burn.
- **Partial-thickness (second-degree) burns** involve the epidermis and some portion of the dermis. These burns do not destroy the entire thickness of the skin, nor is the subcutaneous tissue injured. Typically, the skin is moist, mottled, and white to red. Blisters are common. Partial-thickness burns cause intense pain.

- **Full-thickness (third-degree) burns** extend through all skin layers and may involve subcutaneous layers, muscle, bone, or internal organs. The burned area is dry and leathery and may appear white, dark brown, or even charred. Some full-thickness burns feel hard to the touch. Clotted blood vessels or subcutaneous tissue may be visible under the burned skin. If the nerve endings have been destroyed, a severely burned area may have no feeling. However, the surrounding, less severely burned areas may be extremely painful.

A pure full-thickness burn is unusual. Severe burns are typically a combination of superficial, partial-thickness, and full-thickness burns. Superficial burns heal well without scarring. Small partial-thickness burns also heal without scarring. However, deep partial-thickness burns and all full-thickness burns are best managed surgically.

Significant airway burns are also serious. They may be associated with singeing of the hair within the nostrils, soot around the nose and mouth, hoarseness, or hypoxia.

It may be impossible to accurately estimate the depth of a particular burn. Even experienced burn surgeons sometimes underestimate or overestimate the extent of a particular burn.

Extent

One quick way to estimate the surface area that has been burned is to compare it to the size of the patient's palm, which is roughly equal to 1% of the patient's total body surface area. This technique is called "The Palmer Method." Another useful measurement system is the Rule of Nines, which divides the body into sections, each of which is approximately 9% of the total sur-

Documentation Tips

Burn patterns often require description beyond calculating the amount of body surface involved. If you find written description difficult or too lengthy, try drawing in the affected areas on two outlines of the body, front and back. Your report form may include an area with the outlines provided; if not, don't hesitate to draw them in yourself. One picture can be worth many words.

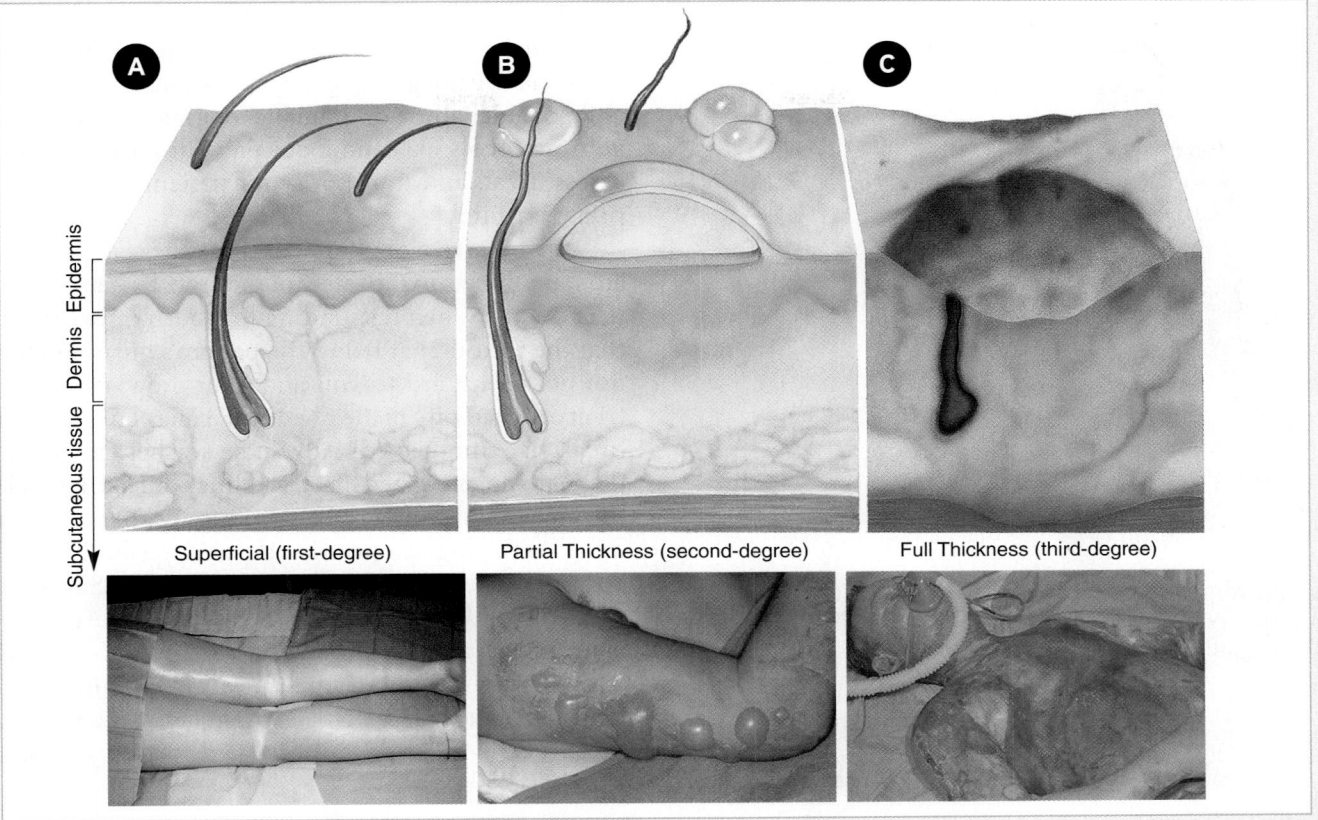

Figure 24-16 Classification of burns. **A.** Superficial or first-degree burns involve only the epidermis. The skin turns red but does not blister or actually burn through. **B.** Partial-thickness or second-degree burns involve some of the dermis, but they do not destroy the entire thickness of the skin. The skin is mottled, white to red, and is often blistered. **C.** Full-thickness or third-degree burns extend through all layers of the skin and may involve subcutaneous tissue and muscle. The skin is dry, leathery, and often either white or charred.

⚠ Pediatric Needs

Burns to children are generally considered more serious than burns to adults (Table 24-2 ▶). This is because infants and children have more surface area relative to total body mass, which means greater fluid and heat loss. In addition, children do not tolerate burns as well as adults do. Children are also more likely to go into shock, develop hypothermia, and experience airway problems because of the unique differences of their ages and anatomy.

Many burns in infants and children result from child abuse. The classic burn resulting from deliberate immersion involves the hands and wrists, as well as the feet, lower legs, and buttocks. Similarly, burns around the genitals and multiple cigarette burns should be viewed as possible abuse. You should report all suspected cases of abuse to the proper authorities (see Chapter 31).

TABLE 24-2 Classification of Burns in Infants and Children

Critical Burns
- Full-thickness or partial-thickness burns covering more than 20% of the body's total surface area
- Burns involving the hands, feet, face, airway, or genitalia

Moderate Burns
- Partial-thickness burns covering 10% to 20% of the body's total surface area

Minor Burns
- Partial-thickness burns covering less than 10% of the body's total surface area

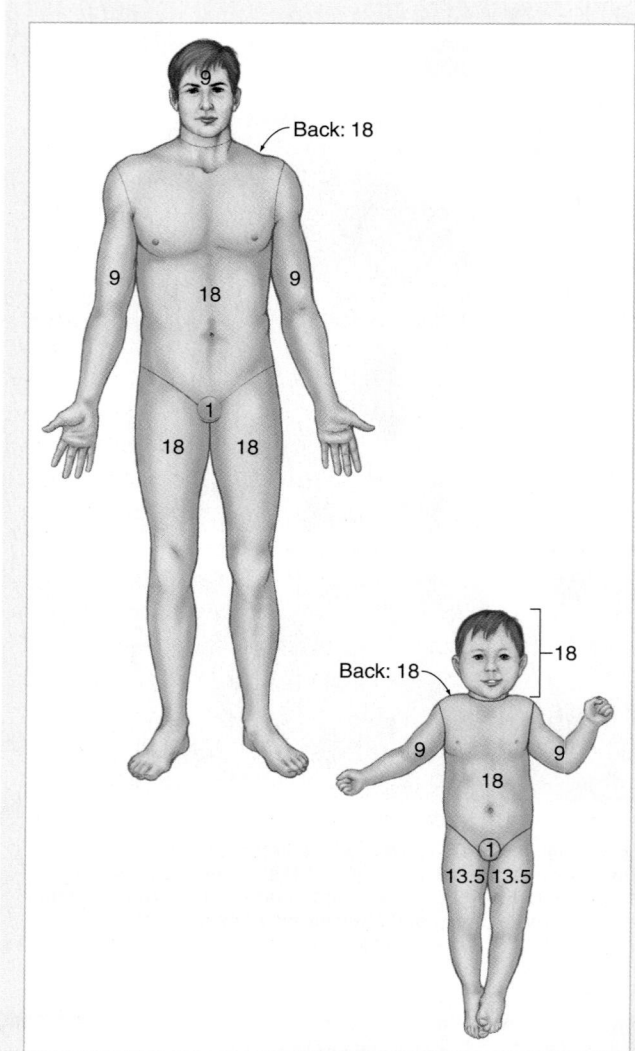

Figure 24-17 The Rule of Nines is a quick way to estimate the amount of surface area that has been burned. It divides the body into sections, each representing approximately 9% of the total body surface area.

face area (Figure 24-17 ▲). Remember that the head of an infant or child is relatively larger than the head of an adult, and the legs are relatively smaller.

Assessment of Burns

Burn patients can be very difficult to manage both physically and emotionally. It is easy to become overwhelmed by the sight, sounds, and smells of burn victims. As you prepare mentally for a burn patient, prepare yourself emotionally as well. Focus on your patient and immediate care. Do not get caught up in the issues surrounding child or elder abuse or dramatic looking wounds. Learn your protocols well and focus on your goal: to quickly assess, treat, and transport your patient to the appropriate hospital.

As you arrive on scene, observe the scene for hazards and threats to the safety of the crew, bystanders, and the patient. When responding to a burn injury, ensure that the factors that led to the patient's burn injury do not pose a hazard to you and your crew. Is the electricity turned off? Is the chemical leak secure? Has the fire been extinguished? At vehicle crashes, ensure that there are no energized electrical lines or leaking fuel in the area where you will be working. Begin with scene safety as the highest priority. If you determine that the power company, the fire department, or ALS units are needed, call for additional resources early. Anticipate using gloves and eye protection with any burn patient and gowns when serious injuries are expected. Remember, the burn patient is a trauma patient. Consider the potential for spinal injuries and other injuries.

Initial Assessment

General Impression

As you approach the burn trauma patient, simple clues can help identify how serious the injuries are and how quickly you need to assess and treat. If your patient greets you in your introduction with a hoarse voice or is reported to have been in an enclosed space with a fire or intense heat source, these should be indications of a significant MOI. Similarly, if the patient has singed facial hair, eyebrows, nasal hair, or moustache, your general impression might be that the patient has a potential airway and/or breathing problem.

Child abuse and elder abuse are unpleasant situations to handle. Unfortunately, they are often situations that involve burns. As you enter a scene where burns are involved, be suspicious of red flags that may indicate abuse.

The burned patient you encounter may have graphic injuries; however, you must not be distracted from the initial assessment. As you begin the initial assessment always consider the need for manual spinal stabilization and determine responsiveness using the AVPU scale.

Airway and Breathing

Ensure that the patient has a clear and patent airway. If the patient is unresponsive or has a significantly altered level of consciousness, consider inserting a properly sized oropharyngeal or nasopharyngeal airway. Be alert to signs that the patient has inhaled hot gases or vapors, such as singed facial hair or soot present in or around the airway. Copious secretions and frequent coughing may also indicate a respiratory burn. You must also quickly assess for adequate breathing. Palpate the chest wall for DCAP-BTLS. Check for clear and symmetrical breath sounds and provide high-flow oxygen or provide assisted ventilations using a BVM device as needed, depending on the level of consciousness and breathing rate/quality of your patient. Burn patients are trauma patients. Evaluate and treat these patients for spinal injuries and airway problems concurrently. How you open the airway depends on whether a neck injury is suspected or not. Could the patient have fallen? Do the circumstances surrounding the MOI suggest a possible spinal injury?

Circulation

You must quickly assess the pulse rate and quality and determine perfusion based on the skin condition, color, temperature, and capillary refill time. If you see significant bleeding, you must take the necessary steps to control it. Significant bleeding is an immediate life threat. If the patient has obvious life-threatening bleeding it must be controlled quickly. Shock frequently develops in burn patients. Support their circulation by elevating the arms and legs as appropriate or placing them in a Trendelenburg position. You should also treat the shock by preventing heat loss. This is very important because the damaged skin has only limited ability to regulate body temperature.

Transport Decision

If the patient you are treating has an airway or breathing problem, significant burn injuries, significant external bleeding, or signs and symptoms of internal bleeding, you must consider quickly transporting this patient to the hospital for treatment. Rendezvous with ALS providers may be appropriate for burn patients with moderate or severe burns and burns of the airway or lungs. ALS providers can treat these patients with endotracheal intubation and intravenous fluids to support airway, breathing, and circulation (shock) problems. These problems can progress so rapidly that immediate ALS help can make the difference between life and death.

Focused History and Physical Exam

Rapid or Focused Physican Exam

After the initial assessment is complete, determine which assessment will be performed next, a rapid physical exam or a focused physical exam. In the responsive patient who has an isolated injury with limited MOI, consider a focused physical exam. Focus your assessment on the isolated injury, the patient's complaint, and the body region affected. If the patient has sustained a small burn area to the body, focus on that injury. Dress the burn with the appropriate bandage, note the location, and estimate size of the injury. Assess all underlying systems. In an extremity, assess pulse, motor, and sensory function in the injured extremity.

If there is a significant amount of body surface burned or significant trauma that may affect multiple systems, start with a rapid trauma assessment, quickly assessing the patient from head to toe looking for DCAP-BTLS to be sure that you have found all of the problems and injuries. Make a rough estimate, using the Rule of Nines, of the extent of the burned area to report to medical control. Package the patient for transport based on your findings. Remember to stabilize your patient for spinal injuries as appropriate. You should not delay transport of a seriously injured patient to complete a detailed physical exam. The detailed physical exam can be started during transport.

Baseline Vital Signs and SAMPLE History

Your physical exam helps you to understand what has happened to the outside of your patient. Vital signs are a good indication of how your patient is doing on the inside. Determining an early set of baseline vital signs will help you to know how your patient is tolerating his or her injuries while en route to the hospital. These can be done in the ambulance on the way to the hospital, decreasing the delay to definitive care in a patient with moderate to severe burns. Because shock is often pronounced in a burn patient, blood pressure, pulse, and skin assessment for perfusion are important signs to obtain.

Interventions

The goal in treating patients with burns is to stop the burning process, assess and treat breathing, support circulation, and provide rapid transport. Because burn patients are also trauma patients, provide complete spinal stabilization if you suspect spinal injuries. Oxygen is mandatory for inhalation burns but is also helpful with smaller burns. If the patient has signs of hypoperfusion, treat aggressively for shock and provide rapid transport to the appropriate hospital. Cover all burns according to your local protocols. The risk of infection is very high and can be reduced if you cover large areas that are burned with sterile burn sheets or clean linen. Do not delay transport of the seriously injured patient to complete nonlifesaving treatments in the field, such as splinting extremity fractures. Instead, complete these types of treatment en route to the hospital.

Detailed Physical Exam

If the patient's condition is stable and problems detected during the initial assessment do not persist, perform a thorough detailed physical examination of the patient as discussed in the chapter on patient assessment. Many times short transportation times and unstable patient conditions may make this assessment impractical.

Ongoing Assessment

Repeat the initial assessment and vital signs. Reevaluate interventions and treatment you have provided to the patient, particularly those used to treat shock.

Communication and Documentation

Provide hospital personnel with a description of how the burn occurred. Many times they can determine the appropriate dilutant for chemical burns or calculate appropriate treatments for other types of burns with enough advanced notice. Your report and documentation should include the extent of the burns. This should include the amount of body surface area involved, the depth of the burn, and the location. For example, you may say 10% full-thickness burns, 15% partial-thickness burns, and 25% superficial burns to the chest, abdomen, and left lower extremity. If special areas are involved (genitalia, feet, hands, face, or circumferential), they should be specifically mentioned and documented.

Emergency Medical Care

Your first responsibility in caring for a patient with a burn is to stop the burning process and prevent additional injury. Follow these steps in caring for a burn patient (**Skill Drill 24-3** ▶):

1. **Follow BSI precautions.** Because a burn destroys the patient's protective skin layer, always wear gloves and eye protection when treating a burn patient.
2. **Move the patient away from the burning area.** If any clothing is on fire, wrap the patient in a blanket or follow specific guidelines outlined by your local fire department protocol to put out the flames, then remove any smoldering clothing and/or jewelry.
3. If allowed by local protocol, **immerse the area in cool, sterile water or saline solution,** or cover with a clean, wet, cool dressing if the skin or clothing is hot. This not only stops the burning, it also relieves pain. Prolonged immersion, however, may increase the risk of infection and hypothermia. For this reason, you should not keep the affected part under water for more than 10 minutes. If the burning has stopped before you arrive, *do not immerse the affected part at all.* As an alternative to immersion, the burned area can be irrigated until the burning stops, followed by the application of a sterile dressing (**Step 1**).
4. **Provide high-flow oxygen.** Also remember that more fire victims die of smoke inhalation than of skin burns. A patient who has facial burns or has inhaled smoke or fumes may experience respiratory distress. Therefore, you should provide high-flow oxygen. Keep in mind that a patient who appears to be breathing well at first may suddenly experience severe respiratory distress. Continually assess the airway therefore for possible problems (**Step 2**).
5. **Rapidly estimate the burn's severity.** Then cover the burned area with a dry, sterile dressing to pre-

✚ EMT-B Tips

General emergency medical care of burns:
1. Follow BSI precautions.
2. Move the patient away from the burning area.
3. Immerse the burned skin in cool, sterile water.
4. Provide high-flow oxygen.
5. Cover the patient with a clean blanket.
6. Rapidly estimate the burn's severity.
7. Check for traumatic injuries.
8. Treat the patient for shock.
9. Provide prompt transport.

Caring for Burns

24-3

Skill Drill

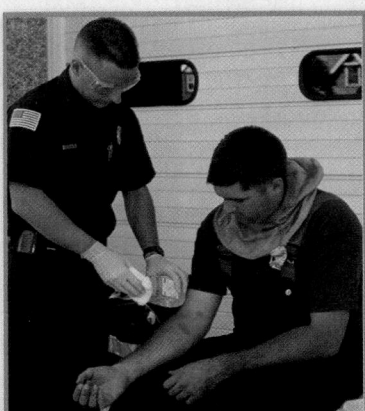

1 Follow BSI precautions to help prevent infection.

If safe to do so, remove the patient from the burning area; extinguish or remove hot clothing and jewelry as necessary.

If the wound(s) is still burning or hot, immerse the hot area in cool, sterile water, or cover with a wet, cool dressing.

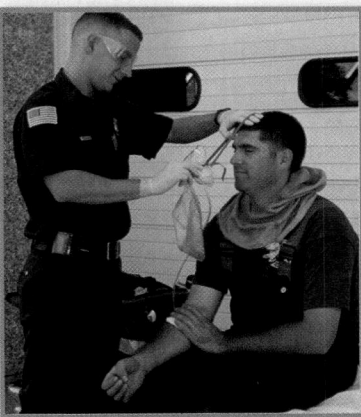

2 Provide high-flow oxygen and continue to assess the airway.

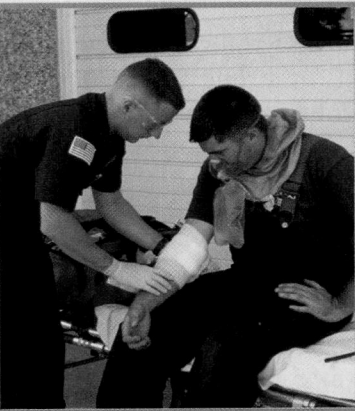

3 Estimate the severity of the burn, then cover the area with a dry, sterile dressing or clean sheet.

Assess and treat the patient for any other injuries.

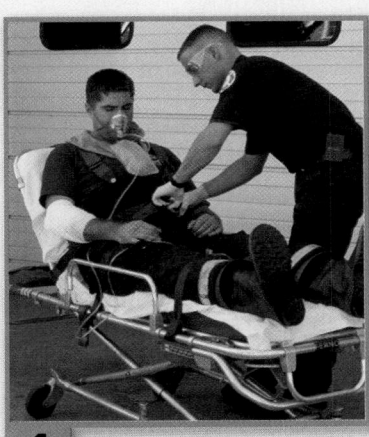

4 Prepare for transport. Treat for shock.

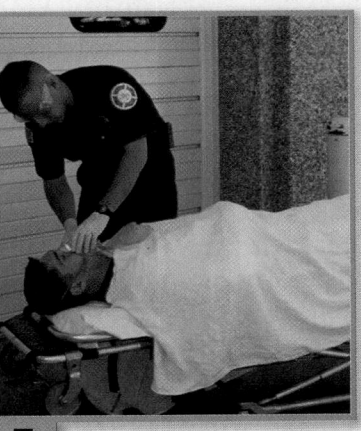

5 Cover the patient with blankets to prevent loss of body heat.

Transport promptly.

vent further contamination. Sterile gauze is best if the area is not too large. You may cover larger areas with a clean, white sheet. Most important, do not put anything else on the burned area. Never use ointments, lotions, or antiseptics of any kind. In addition, do not intentionally break any blisters.

6. **Check for traumatic injuries or other medical conditions** that may be more immediately life threatening. Most patients who have been burned have

normal vital signs and can communicate at first, which will make your assessment easier (**Step 3**).

7. **Treat the patient for shock** (**Step 4**).
8. An extensive burn can produce hypothermia (loss of body heat). **Prevent further heat loss by covering the patient with warm blankets.**
9. **Provide prompt transport by local protocol.** Do not delay transport to do a prolonged assessment or to apply coverings to burns in a critical patient (**Step 5**).

Chemical Burns

A chemical burn can occur whenever a toxic substance contacts the body. Most chemical burns are caused by strong acids or strong alkalis. The eyes are particularly vulnerable to chemical burns (Figure 24-18 ▼). Sometimes the fumes alone from strong chemicals can cause burns, especially to the respiratory tract.

To prevent exposure to hazardous materials, you must wear the appropriate chemical-resistant gloves and eye protection whenever you are caring for a patient with a chemical burn. Be particularly careful not to get any chemical, dry or liquid, on yourself or on your uniform; consider wearing a protective gown when this is a possibility. Remember that exposure risk is also present when you are cleaning up after a call. In cases of severe chemical burns or exposure, consider mobilization of the HazMat team, if appropriate.

The emergency care of a chemical burn is basically the same as that for a thermal burn. To stop the burning process, remove any chemical from the patient. A dry chemical that is activated by contact with water may damage the skin more when it is wet than when it is dry. Therefore, always brush dry chemicals off the skin and clothing before flushing the patient with water (Figure 24-19 ▶). Remove the patient's clothing, including shoes, stockings, and gloves and any jewelry or glasses, because there may be small amounts of chemicals in the creases.

Immediately begin to flush the burned area with large amounts of water (Figure 24-20 ▶), taking care not to contaminate uninjured areas or make the patient hypothermic. Never direct a forceful stream of water

Figure 24-19 Brush dry chemicals off the patient before you flush the burned area with water.

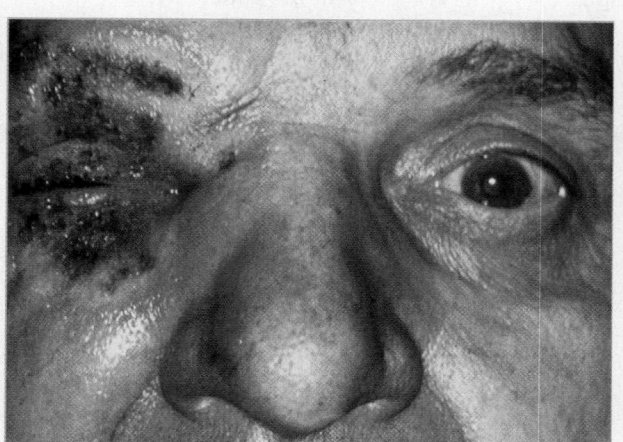

Figure 24-18 The eyes are particularly vulnerable to chemical burns.

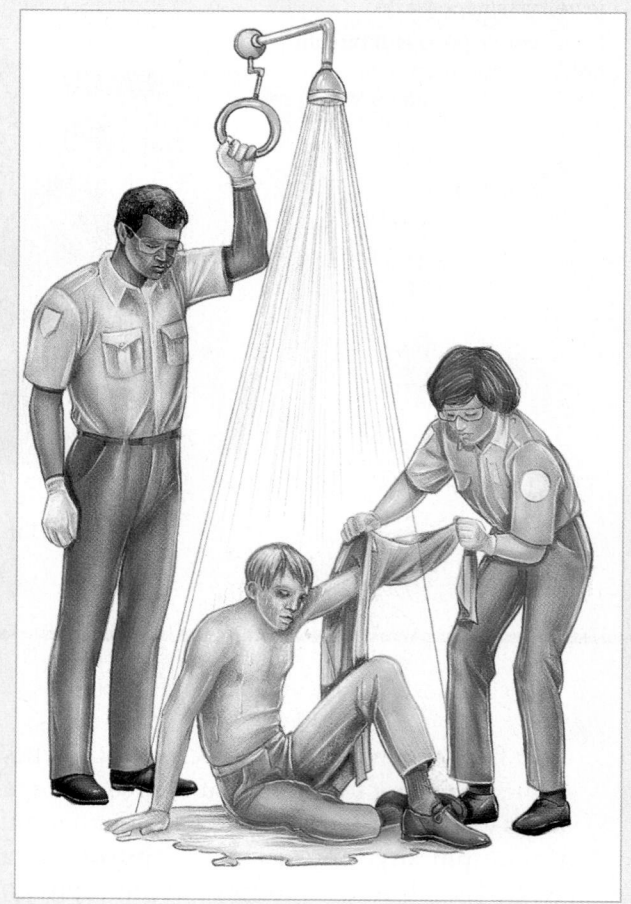

Figure 24-20 Flush the burned area with large amounts of water for 15 to 20 minutes after the patient says that the burning pain has stopped. Be careful to avoid contaminating uninjured areas.

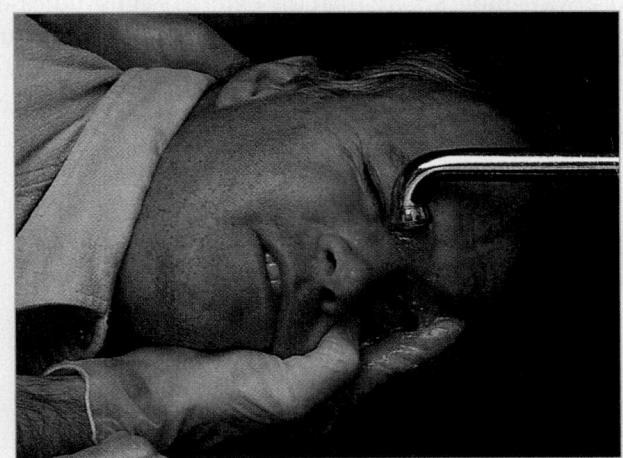

Figure 24-21 Flood the affected eye with a gentle stream of water. Hold the eyelids open, a challenging task because the patient's reflex is to keep the eye shut. Take care to prevent any of the chemical from getting into the other eye during flushing.

Figure 24-22 The human body is a good conductor of electricity. An electrical burn usually occurs when the body, acting as a conductor, completes a circuit.

from a hose at the patient; the extreme water pressure may mechanically injure the burned skin. Continue flooding the area with gallons of water for 15 to 20 minutes after the patient says the burning pain has stopped. If an eye has been burned, hold the eyelid open while flooding the eye with a gentle stream of water (Figure 24-21 ▲). Continue flushing the contaminated area on the way to the hospital.

Electrical Burns

Electrical burns may be the result of contact with high- or low-voltage electricity. High-voltage burns may occur when utility workers make direct contact with power lines. However, ordinary household current is powerful enough to cause severe burns.

For electricity to flow, there must be a complete circuit between the electrical source and the ground. Any substance that prevents this circuit from being completed, such as rubber, is called an insulator. Any substance that allows a current to flow through it is called a conductor. The human body, which is primarily water, is a good conductor. Thus, electrical burns occur when the body, or a part of it, completes a circuit connecting a power source to the ground (Figure 24-22 ▶).

Your safety is of particular importance when you are called to the scene of an emergency involving electricity. Obviously, you can be fatally injured by coming into contact with power lines. But you can also be fatally injured by touching a patient who is still in contact with a live power line or any other electrical source. For this reason, you must never attempt to remove

someone from an electrical source unless you are specially trained to do so. Likewise, you should never move a downed power line unless you have the special training and equipment necessary for the job. Before even approaching someone who may still be in contact with a power line or an electrical appliance, make certain that the power is turned off. Always assume that any downed power line is live.

A burn injury appears where the electricity enters (an entrance wound) and exits (an exit wound) the body. The entrance wound may be quite small

EMT-B Safety

Your safety is the first priority when an electrical hazard may be involved. Do not try to remove someone from an electrical source or move a downed power line unless you are specially trained and equipped to do so. Before approaching someone who may still be in contact with a power line or electrical appliance, ensure that all power is turned off.

Figure 24-23A ▼), but the exit wound can be extensive and deep (Figure 24-23B ▼). Always look for both entrance and exit wounds. There are two dangers specifically associated with electrical burns. First, there may be a large amount of deep tissue injury. Electrical burns are always more severe than the external signs indicate. The patient may have only a small burn to the skin but may have massive damage to the deeper tissues (Figure 24-24 ▼). Second, the patient may go into cardiac or respiratory arrest from the electric shock.

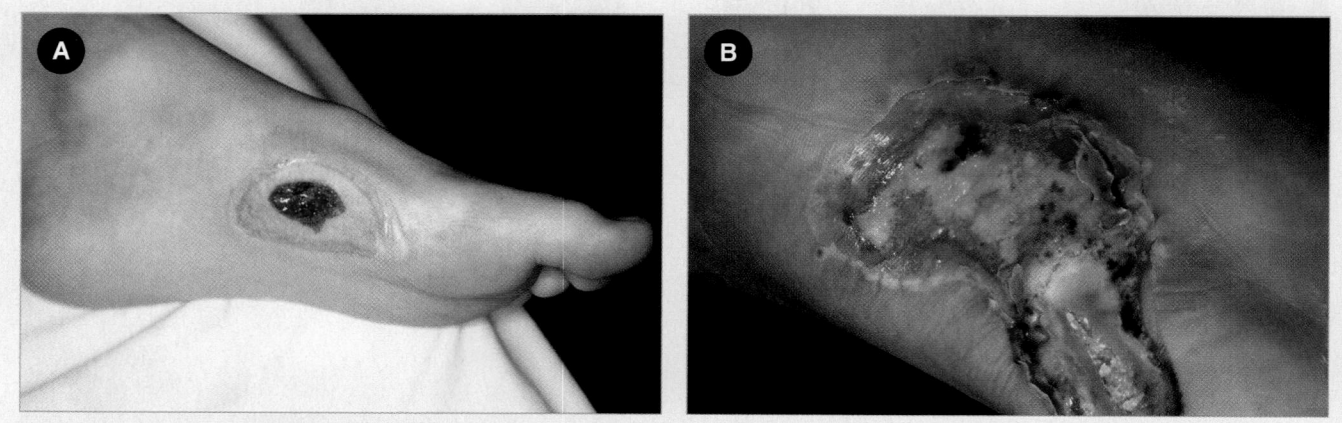

Figure 24-23 Electrical burns, like gunshot wounds, have entrance and exit wounds. **A.** An entrance wound is often quite small. **B.** The exit wound can be extensive and deep.

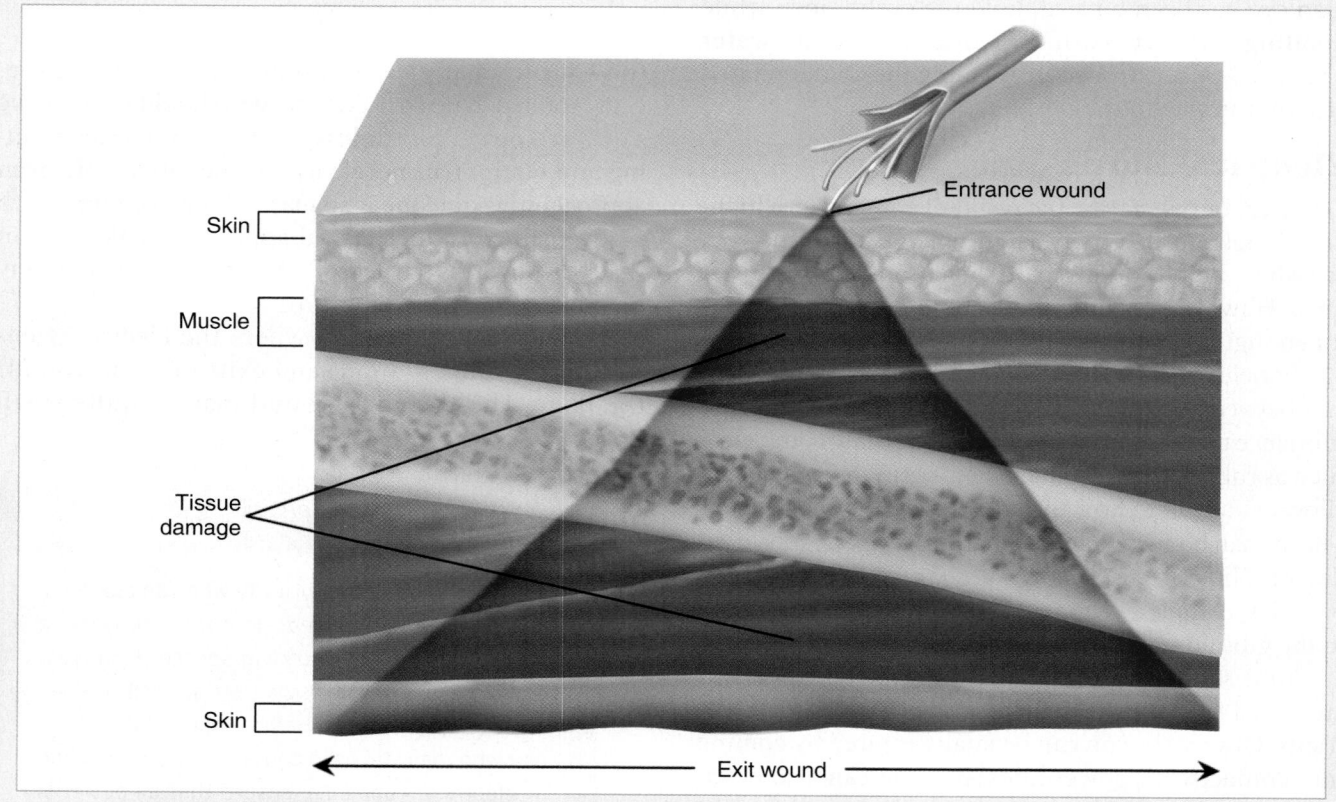

Figure 24-24 External signs of an electrical burn may be deceiving. The entrance wound may be a small burn, while the damage to deeper tissue may be massive.

If indicated, begin CPR and apply the AED. Although CPR may need to be quite prolonged in electrical burn cases, it has a high success rate if started promptly. You should be prepared to defibrillate if necessary. If neither CPR nor defibrillation is indicated, give supplemental oxygen, and monitor the patient closely for respiratory and cardiac arrest. Treat the soft-tissue injuries by placing dry, sterile dressings on all burn wounds and splinting suspected fractures. Provide prompt transport; all electrical burns are potentially severe injuries that require further treatment in the hospital.

Bites

Small Animal Bites and Rabies

At times you may be called to care for a person who has been bitten by a small animal such as a dog, cat, raccoon, squirrel, or other small nonlivestock animal.

Most people who are bitten by small animals do not report the incident to a physician, believing that these bites are not serious. They can be very serious, however. A small animal's mouth is heavily contaminated with virulent bacteria. You should consider all small animal bites as contaminated and potentially infected wounds that may require antibiotics, tetanus prophylaxis, and suturing (Figure 24-25 ▼). Occasionally, small animal bites result in mangled, complex wounds that require surgical repair. For these reasons, all small animal bites should be evaluated by a physician. Place a dry, sterile dressing over the wound, and promptly transport the patient to the emergency department. If an arm or leg was injured, splint that extremity. Often, the patient will be extremely upset and frightened, a situation that calls for reassurance on your part.

A major concern with small animal bites is the spread of rabies, an acute, potentially fatal viral infection of the central nervous system that can affect all warm-blooded animals. Although rabies is extremely rare today, particularly with widespread inoculation of pets, it still exists. Stray dogs that have not been inoculated can be carriers of the disease, as can squirrels, bats, foxes, skunks, and raccoons. The virus is in the saliva of a rabid, or infected, animal and is transmitted through biting or licking an open wound. Infection can be prevented in a person who has been bitten by such an animal only by a series of special vaccine injections, a painful procedure that must be started soon after the bite. Since animals that have rabies do not always demonstrate symptoms immediately, a person's only chance to avoid the vaccine is to find the animal and turn it over to the health department for observation and/or testing. Refer to your local animal control procedures.

Children, particularly young ones, may be seriously injured or even killed by dogs. These dogs are not always vicious or rabid; sometimes a child unknowingly provokes the animal. However, you must assume that the animal may turn and attack you as well. Therefore, you generally should not enter the scene until the animal has been secured by either the police or an animal control officer. Then you may carry out the necessary emergency care and transport the child to the emergency department.

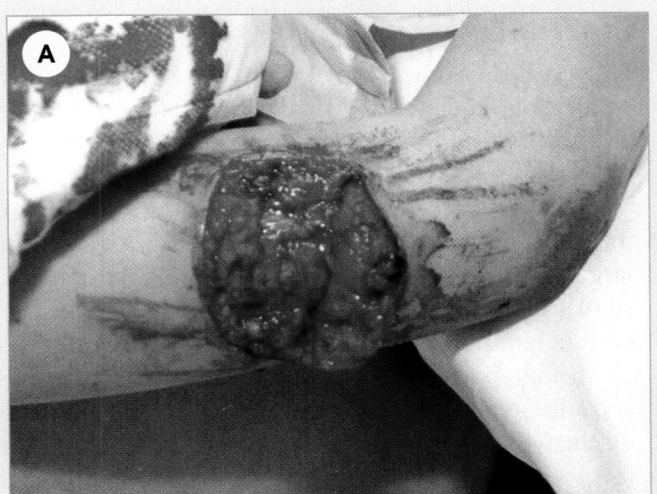
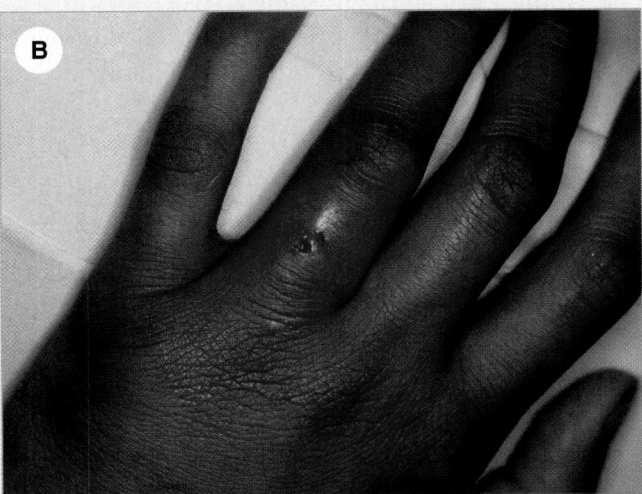

Figure 24-25 Small animal bite wounds should be examined at the hospital, as these wounds are heavily contaminated with virulent bacteria. **A.** Dog bite. **B.** Cat bite.

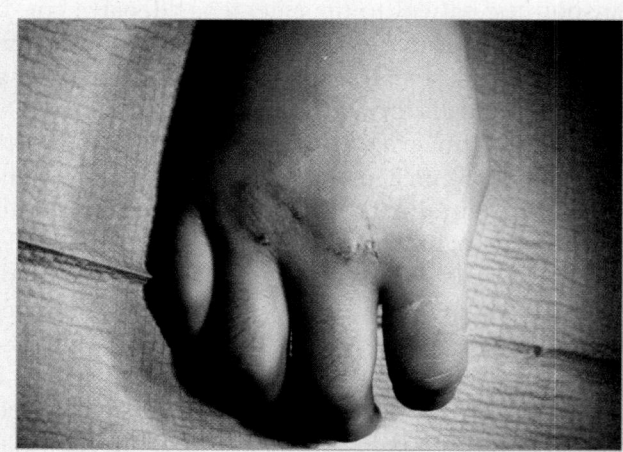

Figure 24-26 Human bites can result in serious, spreading infection. Thus, patients must be evaluated at the hospital.

Human Bites

The human mouth, more so than even the small animal's, contains an exceptionally wide range of virulent bacteria and viruses. For this reason, you should regard any human bite that has penetrated the skin as a very serious injury. Similarly, any laceration caused by a human tooth can result in a serious, spreading infection (Figure 24-26 ▲). Remember this if you have occasion to treat someone who has been punched in the mouth: The person who delivered the punch may also need treatment.

The emergency treatment of human bites consists of the following steps:

1. Promptly immobilize the area with a splint or bandage.
2. Apply a dry, sterile dressing.
3. Provide transport to the emergency department for surgical cleansing of the wound and antibiotic therapy.

Dressing and Bandaging

All wounds require bandaging. In most instances, splints help to control bleeding and provide firm support for the dressing. There are many different types of dressings and bandages (Figure 24-27 ▶). You should be familiar with the function and proper application of each.

In general, dressings and bandages have three primary functions:

- To control bleeding
- To protect the wound from further damage
- To prevent further contamination and infection

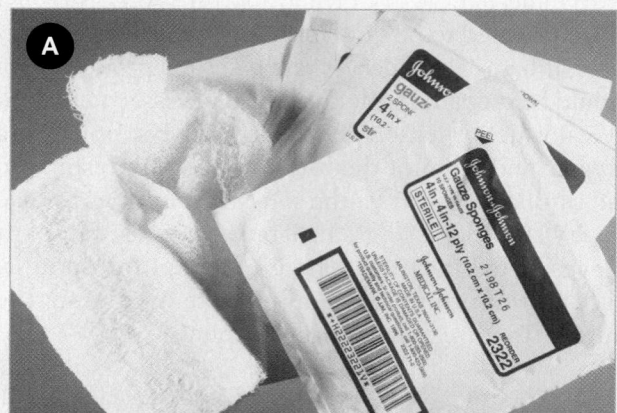

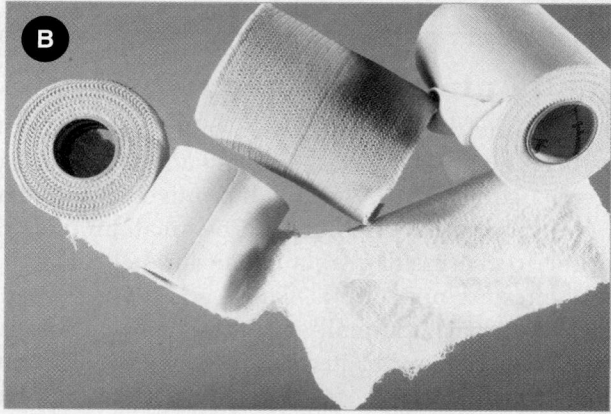

Figure 24-27 A. Many types of sterile dressings are used for covering open wounds, including universal dressings, gauze pads, adhesive dressings, and occlusive dressings. **B.** Bandages keep dressings in place and include soft roller bandages, triangular bandages, and adhesive tape. Splints may also be used to hold dressings in place.

Sterile Dressings

Universal dressings, conventional 4″ × 4″ and 4″ × 8″ gauze pads, and assorted small adhesive-type dressings and soft self-adherent roller dressings will cover most wounds.

Measuring 9″ × 36″ and made of thick, absorbent material, the universal dressing is ideal for covering large open wounds. It also makes an efficient pad for rigid splints. These dressings are available in compact, commercially sterilized packages.

Gauze pads are appropriate for smaller wounds, and adhesive-type dressings are useful for minor wounds. Occlusive dressings, made of Vaseline gauze, aluminum foil, or plastic, prevent air and liquids from entering (or exiting) the wound. They are used to cover sucking chest wounds, abdominal eviscerations, and neck injuries.

Bandages

To keep dressings in place during transport, you can use soft roller bandages, rolls of gauze, triangular bandages, or adhesive tape. The self-adherent, soft roller bandages are probably easiest to use. They are slightly elastic, which makes them easy to apply, and you can tuck the end of the roll into a deeper layer to secure it in place. The layers adhere somewhat but should not be applied too tightly to one another.

Adhesive tape holds small dressings in place and helps to secure larger dressings. Some people, however, are allergic to adhesive tape. If you know that a patient has this problem, use paper or plastic tape instead.

Do not use elastic bandages to secure dressings. If the injury swells, the bandage may become a tourniquet and cause further damage. Any improperly applied bandage that impairs circulation can result in additional tissue damage or even the loss of a limb. For this reason, you should always check a limb distal to a bandage for signs of impaired circulation or loss of sensation. Air splints are useful in stabilizing broken extremities, and they can be used with dressings to help control bleeding from soft-tissue injuries.

You are the Provider · Summary

A patient with multiple injuries is a treatment challenge. The information provided by dispatch, while accurate, was misleading, which can distract you if you have already drawn a "mental image" based on the information. You had to rapidly assess the scene for your safety and the patient's and decide how best to move this patient and at what point to begin treatment. With little warning (as is often the case with young and older patients) the trauma patient can become very sick, very rapidly, for reasons you may be unable to detect. It is vital that you treat trauma patients with a very high and continuous level of suspicion. Reevaluate the patient often regardless of how stable he or she appears.

Assessment and Emergency Care

	Closed Injuries	Open Injuries	Burns
Scene Size-up	Body substance isolation precautions should include a minimum of gloves and eye protection. Ensure scene safety and determine NOI/MOI. Consider the number of patients, the need for additional help/ALS, and c-spine stabilization.	Body substance isolation precautions should include a minimum of gloves and eye protection. Ensure scene safety and determine NOI/MOI. Consider the number of patients, the need for additional help/ALS, and c-spine stabilization.	Body substance isolation precautions should include a minimum of gloves and eye protection. Ensure scene safety and determine NOI/MOI. Consider the number of patients, the need for additional help/ALS, and c-spine stabilization.
Initial Assessment			
■ General impression	Determine level of consciousness and treat any immediate threats to life. Determine priority of care based on environment and patient's chief complaint.	Determine level of consciousness and treat any immediate threats to life. Determine priority of care based on environment and patient's chief complaint.	Determine level of consciousness and treat any immediate threats to life. Determine priority of care based on environment and patient's chief complaint.
■ Airway	Ensure patent airway. Maintain spinal stabilization as necessary.	Ensure patent airway. Maintain spinal stabilization as necessary.	Ensure patent airway. Maintain spinal stabilization as necessary.
■ Breathing	Listen for abnormal breath sounds and evaluate depth and rate of respirations. Maintain ventilations as needed. Provide high-flow oxygen at 15 L/min and inspect the chest wall, assessing for DCAP-BTLS.	Listen for abnormal breath sounds and evaluate depth and rate of respirations. Maintain ventilations as needed. Provide high-flow oxygen at 15 L/min and inspect the chest wall, assessing for DCAP-BTLS.	Listen for abnormal breath sounds and evaluate depth and rate of respirations. Maintain ventilations as needed. Provide high-flow oxygen at 15 L/min and inspect the chest wall, assessing for DCAP-BTLS.
■ Circulation	Evaluate distal pulse rate and quality; observe skin color, temperature, and condition and treat accordingly.	Evaluate distal pulse rate and quality; observe skin color, temperature, and condition. Observe patient for shock and treat bleeding appropriately.	Evaluate distal pulse rate and quality; observe skin color, temperature, and condition and treat accordingly. Prevent heat loss.
■ Transport decision	Prompt transport	Prompt transport	Prompt transport
Focused History and Physical Exam	*NOTE: The order of the steps in the focused history and physical exam differs depending on whether or not the patient has a significant MOI. The order below is for a patient with a significant MOI. For a patient without a significant MOI, perform a focused trauma assessment, obtain vital signs, and obtain the history.*		
■ Rapid trauma assessment	Reevaluate the MOI. Perform a rapid trauma assessment, treating all life-threats immediately. Log roll and secure patient to backboard for all patients with a significant MOI from the clavicles up.	Reevaluate the MOI. Perform a rapid trauma assessment, treating all life-threats immediately. Log roll and secure patient to backboard for all patients with a significant MOI from the clavicles up.	Reevaluate the MOI. Perform a rapid trauma assessment, treating all life-threats immediately. Estimate amount of body surface injured. Stabilize the patient for spinal injuries as appropriate.

	Closed Injuries	Open Injuries	Burns
■ Baseline vital signs	Take vital signs, noting skin color and temperature as well as patient's level of consciousness. Use pulse oximetry if available.	Take vital signs, noting skin color and temperature as well as patient's level of consciousness. Be alert to potential internal bleeding. Use pulse oximetry if available.	Take vital signs, noting skin color and temperature as well as patient's level of consciousness. Use pulse oximetry if available.
■ SAMPLE history	Obtain SAMPLE history. If the patient is not responsive, attempt to get history from family members, friends, or bystanders.	Obtain SAMPLE history. If the patient is not responsive, attempt to get history from family members, friends, or bystanders.	Obtain SAMPLE history. If the patient is not responsive, attempt to get history from family members, friends, or bystanders.
■ Interventions	Provide complete spinal stabilization early if you suspect that your patient has spinal injuries. Treat signs of hypoperfusion (shock) and consider ALS if available. Splint a painful, swollen, deformed extremity.	Provide complete spinal stabilization early if you suspect that your patient has spinal injuries. Treat signs of hypoperfusion (shock) and consider ALS if available. Splint a painful, swollen, deformed extremity.	Stop the burning process. Provide complete spinal stabilization if you suspect spinal injuries. Aggressively treat signs of hypoperfusion (shock) and consider ALS if available. Cover burns with dry, sterile dressings or per protocol.
Detailed Physical Exam	Complete a detailed physical exam.	Complete a detailed physical exam.	Complete a detailed physical exam.
Ongoing Assessment	Repeat the initial assessment, rapid or focused assessment, and reassess interventions performed. Reassess vital signs every 5 minutes for the unstable patient, every 15 minutes for the stable patient.	Repeat the initial assessment, rapid or focused assessment, and reassess interventions performed. Reassess vital signs every 5 minutes for the unstable patient, every 15 minutes for the stable patient.	Repeat the initial assessment, rapid or focused assessment, and reassess interventions performed. Reassess vital signs every 5 minutes for the unstable patient, every 15 minutes for the stable patient.
■ Communication and documentation	Contact medical control with a radio report. Relay any change in level of consciousness or difficulty breathing. Be sure to document physician's orders and changes in patient condition, and at what time they occurred.	Contact medical control with a radio report. Relay any change in level of consciousness or difficulty breathing. Document the MOI and the position in which the patient was found. Report location and description of injury, significant blood losses, and how you treated the injury. Document physician's orders and changes in patient condition, and at what time they occurred.	Contact medical control with a radio report. Describe how the burn occurred, and the extent of the burn(s). Relay any change in level of consciousness or difficulty breathing. Be sure to document physician's orders and changes in patient condition, and at what time they occurred.

Assessment and Emergency Care

Assessment and Emergency Care

NOTE: While the steps below are widely accepted, be sure to consult and follow your local protocol.

Closed Injuries

1. Keep the patient as quiet and comfortable as possible.
2. Apply ice (or cold packs).
3. Apply direct pressure.
4. Elevate the injured part just above the level of the patient's heart.
5. Splint the injured area.

Open Injuries

1. Apply direct pressure with a sterile bandage.
2. Maintain pressure with a roller bandage.
3. If bleeding continues, apply a second dressing and roller bandage over the first.
4. Splint the extremity.

Burns

1. Follow BSI precautions.
2. Move the patient away from the burning area.
3. Immerse the burned skin in cool, sterile water.
4. Provide high-flow oxygen.
5. Cover the patient with a clean blanket.
6. Rapidly estimate the burn's severity.
7. Check for traumatic injuries.
8. Treat the patient for shock.
9. Provide prompt transport.

Abdominal Injuries

1. Do not touch or move exposed organs.
2. Keep organs moist. Use moist sterile dressings, cover and secure in place.
3. If the patient's legs and knees are uninjured, flex them to relieve pressure on the abdomen.

Impaled Objects

1. Do not attempt to move or remove the object.
2. Control bleeding and stabilize the object in place using soft dressings, gauze, and/or tape.
3. Tape a rigid item over the stabilized object to protect it from movement during transport.

Neck Wounds

1. Cover wound with occlusive dressing.
2. Apply manual pressure, but do not compress both carotid vessels at the same time.
3. Secure dressing over the wound.

Chemical Burns

1. Stop the burning process; safely remove any chemical from the patient, always brushing off a dry chemical.
2. Remove all of patient's clothing.
3. Flush the burn area with large amounts of water for 15 to 20 minutes after the patient says the burning has stopped.

Electrical Burns

1. Ensure that the scene is safe.
2. If indicated, begin CPR and apply the AED.
3. Treat soft-tissue areas by placing dry, sterile dressings on all burn wounds and splinting fractures.

Small Animal and Human Bites

1. Promptly stabilize the area with a splint or bandage.
2. Apply a dry, sterile dressing.
3. Provide transport to the emergency department for surgical cleansing of the wound and antibiotic therapy.

Prep Kit

Ready for Review

- The skin has three layers; the epidermis (tough outer layer), the dermis (inner layer containing hair follicles, sweat glands, and sebaceous glands), and the subcutaneous (containing fat and muscle layers).

- The main functions of the skin are to keep pathogens out, fluid in, maintain body temperature, and provide environmental information to the brain.

- Soft-tissue injuries are classified into three groups: closed injuries, open injuries, and burns.

- Closed injuries include hematomas and crushing injuries. Treatment includes RICES (rest, ice, compression, elevation, splinting).

- Open injuries produce more extensive bleeding and increase the risk for infection. There are five types of open injuries: abrasions, lacerations, incisions, avulsions, and penetrating injuries.

- To treat open injuries, control bleeding as indicated and apply sterile dressing; avoid cleaning out an open wound as this may aggravate bleeding.

- Burns are serious and painful soft-tissue injuries caused by heat (thermal), chemicals, electricity, and radiation.

- Burns are classified primarily by the depth and extent of the burn injury and the body area involved. Burns are considered to be superficial, partial thickness, or full thickness based on severity of depth involved.

- When providing emergency care for burns do the following:
  - Use BSI precautions to protect yourself from potentially contaminated body fluid and to protect the patient from potential infection.
  - Ensure you have cooled the burned area to prevent further cellular damage.
  - Remove jewelry or constrictive clothing; never attempt to remove any synthetic material that may have melted into the burned skin.
  - Ensure an open and clear airway, provide high-flow oxygen, and be alert to signs and symptoms of inhalation injury such as difficulty breathing, stridor, or wheezing.
  - Place sterile dressings over the burned area(s); prevent hypothermia by covering the patient with a clean blanket. Provide prompt transport.

- Small animal and human bites can lead to serious infection and must be evaluated by a physician. Small animals can carry rabies.

- Dressing and bandages are designed to control bleeding, provide protection to the wound from further damage, prevent further contamination, and prevent infection.

Technology

- Interactivities
- Vocabulary Explorer
- Anatomy Review
- Web Links
- Online Review Manual

www.EMTB.com

Prep Kit continued...

Vital Vocabulary

abrasion Loss or damage of the superficial layer of skin as a result of a body part rubbing or scraping across a rough or hard surface.

avulsion An injury in which soft tissue either is torn completely loose or is hanging as a flap.

burn An injury in which the soft tissue receives more energy than it can absorb without injury from thermal heat, frictional heat, toxic chemicals, electricity, or nuclear radiation.

closed injury Injury in which damage occurs beneath the skin or mucous membrane but the surface remains intact.

compartment syndrome Swelling in a confined space that produces dangerous pressure; may cut off blood flow or damage sensitive tissue.

contamination The presence of infective organisms or foreign bodies such as dirt, gravel, or metal.

contusion A bruise without a break in the skin.

dermis The inner layer of the skin, containing hair follicles, sweat glands, nerve endings, and blood vessels.

ecchymosis Discoloration associated with a closed wound; signifies bleeding.

epidermis The outer layer of skin that acts as a watertight protective covering.

evisceration The displacement of organs outside the body.

full-thickness burn A burn that affects all skin layers and may affect the subcutaneous layers, muscle, bone, and internal organs, leaving the area dry, leathery, and white, dark brown, or charred; traditionally called a third-degree burn.

hematoma Blood collected within the body's tissues or in a body cavity.

incision A sharp or smooth cut.

laceration A jagged open wound.

mucous membrane The lining of body cavities and passages that are in direct contact with the outside environment.

occlusive dressing Dressing made of Vaseline gauze, aluminum foil, or plastic that prevents air and liquids from entering or exiting a wound.

open injury An injury in which there is a break in the surface of the skin or the mucous membrane, exposing deeper tissue to potential contamination.

partial-thickness burn A burn affecting the epidermis and some portion of the dermis but not the subcutaneous tissue, characterized by blisters and skin that is white to red, moist, and mottled; traditionally called a second-degree burn.

penetrating wound An injury resulting from a sharp, pointed object.

rabid Describes an animal that is infected with rabies.

Rule of Nines A system that assigns percentages to sections of the body, allowing calculation of the amount of skin surface involved in the burn area.

superficial burn A burn affecting only the epidermis, characterized by skin that is red but not blistered or actually burned through; traditionally called a first-degree burn.

Points to Ponder

You have been dispatched to a motor vehicle crash at the intersection of Jefferson Street and State Road 12. En route, the highway patrol reports only minor injuries.

Upon arrival you see two vehicles with only minor damage. Your patient, a 16-year-old male, is complaining of a small laceration to his arm. He is very upset and crying. Your general impression shows nothing significant. The patient is alert and oriented, and your initial assessment indicates no problems with airway, breathing, or circulation. The focused trauma assessment and detailed physical exam finds only a 5-inch laceration to the patient's right arm. His vital signs are normal, and nothing significant is found in the SAMPLE history. The patient refuses treatment. Due to his age, he cannot refuse or legally sign a refusal.

Issues: Legal Age of Consent, Emotional Support for Minors, Notification of Parents or Guardians.

Assessment in Action

A structure fire is in progress at 1400 Mariposa Dr. The fire department is already on scene. Upon arrival you place your ambulance out of the way of the fire apparatus and report to the command post. The incident commander tells you the rescue sector has found a victim and they are moving the patient to the front of the building. You gather all of your needed equipment and stage in front of the building.

A few minutes later four fire fighters carry the victim out front. You start your initial assessment. The general impression tells you the patient is a geriatric man, his skin is pale, and he is burned in several places. You also note a large laceration to his forearm. The patient is alert to your voice, his airway is open, and his breathing is labored (dyspneic) at a rate of 22 breaths/min. His pulse rate is 110 beats/min, weak and regular. Your partner places the patient on high-flow oxygen at a rate of 15 L/min. As you do a rapid trauma assessment, you find heavy bleeding from an 8" laceration to the left forearm. The patient has full-thickness burns to his anterior (front) chest and abdomen. Both legs are severely burned. You wrap the patient in a sterile burn sheet and prepare for rapid transport to the emergency department.

1. The amount of body surface area burned can be estimated using the:
   A. ABA guidelines.
   B. burn rate chart.
   C. Parkland formula.
   D. Rule of Nines.

2. The first step in controlling the bleeding from the laceration is to:
   A. apply direct pressure with a sterile bandage.
   B. place ice on the wound.
   C. put on a tourniquet.
   D. lower the arm below the heart.

3. If the hemorrhaging continues through the bandage, you should:
   A. apply a second dressing.
   B. call for another ambulance.
   C. rush the patient to surgery.
   D. try a different type of dressing.

4. To determine the severity of the burns, consider the depth of the burn and:
   A. color of the burn.
   B. how long ago the burn occurred.
   C. the extent or area of the burn.
   D. what caused the burn.

5. An additional factor that makes this patient's burn a critical situation includes:
   A. the cause of the burn.
   B. his age.
   C. his weight.
   D. the type of burn.

6. Burns that are classified as full-thickness (third-degree) can be identified by:
   A. blisters with fluid discharge.
   B. blue and red skin.
   C. being dry and leathery.
   D. lots of bleeding.

7. A burn that involves the epidermis and part of the dermis, is red, and is accompanied with blisters is classified as a:
   A. circumstantial burn.
   B. complete-thickness burn.
   C. superficial (first-degree) burn.
   D. partial-thickness (second-degree) burn.

8. After treating the burn patient for shock, you must protect the patient from infection and:
   A. hyperglycemia.
   B. hypertension.
   C. hypothermia.
   D. hypouria.

Challenging Questions

9. Using the Rule of Nines, what is the percentage of body surface area burned in the patient?

10. Why should impaled objects not be removed when treating soft-tissue injuries?

11. Describe the treatment for sucking chest wounds.

www.EMTB.com

Eye Injuries

Objectives*

Cognitive

1. List the main anatomic features of the eye. (p 736)
2. Describe the principal functions of the eye. (p 736)
3. Describe the signs and symptoms of eye injuries. (p 738-740)
4. List the steps necessary to assess eye injuries. (p 738)
5. Describe the steps for managing foreign objects in the eye. (p 741-743)
6. Describe the steps for managing puncture wounds to the eye. (p 742, 743)
7. Describe how to manage burns to the eye. (p 744-745)
8. Describe how to remove contact lenses from the eye. (p 742, 750)
9. Recognize abnormalities of the eyes that may indicate an underlying head injury. (p 749)
10. Recognize and manage a patient with an artificial eye. (p 750)

Affective

None

Psychomotor

11. Demonstrate the use of irrigation to flush out foreign bodies lying on the surface of the eye. (p 742)
12. Demonstrate the care of the patient with chemical burns to the eye. (p 744)
13. Demonstrate the steps in the emergency care of the patient with lacerations of the eyelids. (p 747)

*All of the objectives in this chapter are noncurriculum objectives.

You are the Provider

You and your EMT-B partner are dispatched to the scene for an injured person at Frank's Auto Repair. On scene, you discover a patient with a large metal fragment in the right eye. Bystanders tell you he was working with a piece of machinery that exploded in his face.

1. Why is it important to bandage both eyes when only one eye is injured?
2. When is it appropriate to remove contact lenses from your patient's eyes?

Eye Injuries

Injuries to the eye are very common, and you will encounter many in your work as an EMT-B. Proper emergency medical care for these injuries can minimize damage, which can often be severe. Fortunately, most eye injuries are relatively minor, such as foreign objects in the eye, corneal abrasions, and contusions. Some injuries are more serious, such as rupture of the globe, which requires immediate expert management. This chapter first reviews the structure and function of the eye and then looks at the different types of eye injuries, describing the emergency management of each. The handling of contact lenses and artificial eyes is also discussed.

Anatomy and Physiology of the Eye

The eye is globe-shaped, approximately 1″ in diameter, and located within a bony socket in the skull called the <u>orbit</u> (Figure 25-1 ▶). The orbit is composed of the adjacent bones of the face and skull; the orbit forms the base of the floor of the cranial cavity, and directly above it are the frontal lobes of the brain. In the adult, more than 80% of the eyeball is protected within this bony orbit. Between and below the orbits are the nasal bones and the sinuses, respectively. Therefore, any severe injury to the face or head can potentially damage the eye-

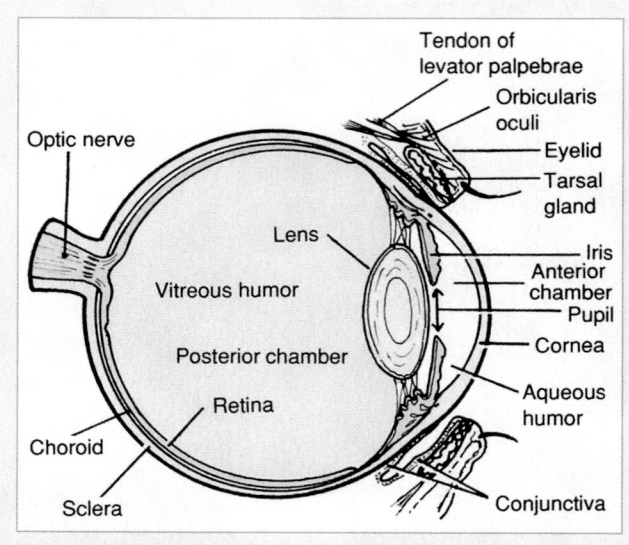

Figure 25-1 The major components of the eye.

ball or the muscles attached to the eyeball that cause the eye to move.

The eyeball, or <u>globe</u>, keeps its global shape as a result of the pressure of the fluid contained within its two chambers. The clear, jellylike fluid near the back of the eye is called the vitreous humor. If the globe is ruptured and this gel leaks out, it cannot be replaced. In front of the lens is a clear fluid called the aqueous humor, named for its watery appearance; in Latin, *aqua* means water. In penetrating injuries of the eye, aqueous humor can also leak out, but with time and appropriate medical treatment, the body can make more.

The inner surface of the eyelids and the exposed surface of the eye itself, which are covered by a delicate membrane, the <u>conjunctiva</u>, are kept moist by fluid produced by the <u>lacrimal glands</u>, often called tear glands (Figure 25-2 ▶). Humans blink unconsciously many times per minute. This action sweeps fluid from the lacrimal glands over the surface of the eye, cleaning it. The tears drain on the inner side of the eye through two lacrimal (tear) ducts into the nasal cavity. This is why, when people cry, they sometimes need to blow their nose.

The white of the eye, called the <u>sclera</u>, extends over the surface of the globe. This is extremely tough, fibrous tissue that helps maintain the eye's globular shape and protect the more delicate inner structures. On the front of the eye, the sclera is replaced by a clear, transparent membrane called the <u>cornea</u>, which allows light to enter the eye. A circular muscle lies behind the cornea

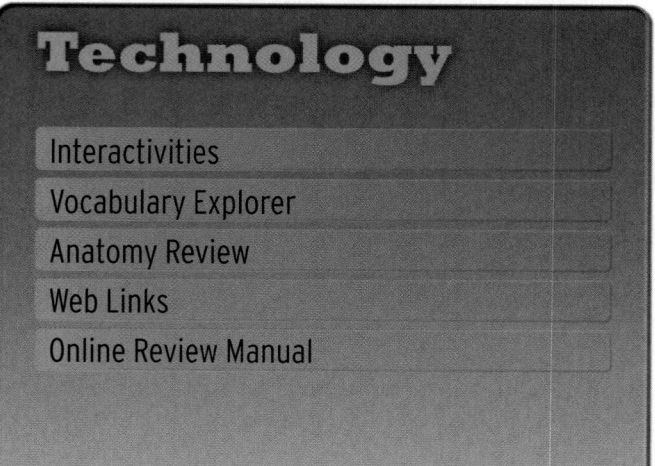

Technology

Interactivities

Vocabulary Explorer

Anatomy Review

Web Links

Online Review Manual

www.EMTB.com

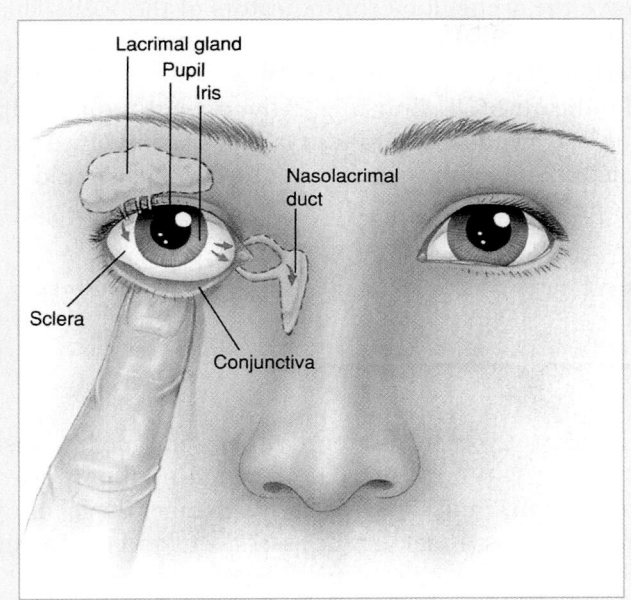

Figure 25-2 The lacrimal system consists of tear glands and ducts. Tears act as lubricants and keep the front of the eye from drying out.

larger when the person is looking at objects near at hand and farther away; these adjustments occur almost instantaneously. Normally, the pupils in both eyes are equal in size. Some people are born with pupils that are not equal; however, particularly in unconscious patients, unequal pupil size may indicate serious injury or illness to the brain or eye.

Behind the iris is the lens. Like the lens of a camera, this lens focuses images on the light-sensitive area at the back of the globe, called the retina. You can think of the retina as the film in the camera. Within the retina are numerous nerve endings, which respond to light by transmitting nerve impulses through the optic nerve to the brain. In the brain, the impulses are interpreted as vision.

The retina is nourished by a layer of blood vessels between it and the sclera at the back of the globe. This layer is called the choroid. If, as sometimes happens, the retina detaches from the underlying choroid and sclera, the nerve endings are not nourished, and the patient experiences blindness. This may be partial blindness, depending on how much of the retina is separated. This condition is called retinal detachment.

Assessment of Eye Injuries

with an opening in its center. Like the shutter in a camera, this muscle adjusts the size of the opening to regulate the amount of light that enters the eye. This circular muscle and surrounding tissue are called the iris. The iris is pigmented, giving the eye its characteristic brown, green, or blue color.

The opening in the center of the iris, which allows light to move to the back of the eye, is called the pupil. Normally, the pupil appears black. Like the opening in a camera, the pupil becomes smaller in bright light and larger in dim light. The pupil also becomes smaller and

Eye injuries are common, particularly in sports. An eye injury can produce severe lifelong complications, including blindness. Proper emergency treatment will minimize pain and may very well help to prevent permanent loss of vision.

In a normal, uninjured eye, the entire circle of the iris is visible. The pupils are round, usually equal in size, and react equally when exposed to light

You are the Provider   Part 2

You take BSI precautions and ask your patient to tell you what happened. He explains that the machine "blew apart" and that he felt something hit his right eye. The patient did not fall, never lost consciousness, and is currently alert and oriented. Your patient has a patent airway, is breathing, and you observe moderate bleeding from his right eye. You perform a rapid assessment and determine the only injury present is the large metal fragment in the sclera of the right eye.

3. Would you put direct pressure on the right eye to control the bleeding?
4. How would you treat the eye injury?

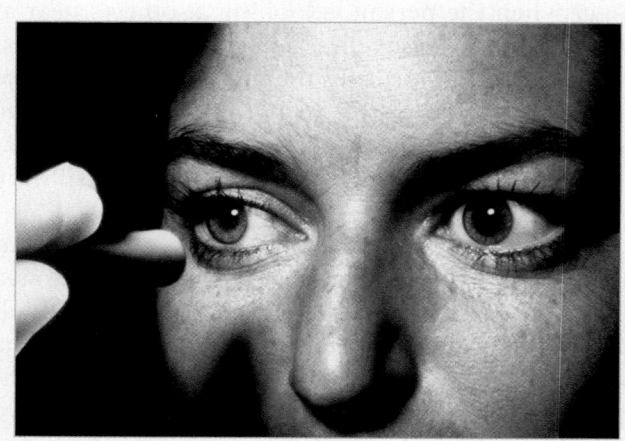

Figure 25-3 Normally, the pupils are round, equal in size, and equally reactive when exposed to light.

Figure 25-3 ▲). Both eyes move together in the same direction when following your moving finger. After an injury, pupil reaction or shape and eye movement are often disturbed. Any of these conditions should cause you to suspect an injury of the globe or its associated tissues. Remember, though, that abnormal pupil reactions sometimes are a sign of brain injury rather than eye injury.

Certain elements of the patient's history are particularly important. Therefore, as you perform your assessment, always note and record the patient's signs and symptoms, including their severity and duration, the details of how the injury occurred, any reported changes in vision, the use of any eye medications, and any history of eye surgery.

Scene Size-up

As you arrive on scene, observe for hazards and threats to the safety of the crew, bystanders, and the patient. Eye injuries often occur in industrial environments or in situations with a potential for other injuries to the patient or even injuries to emergency responders unfamiliar with a specific workplace. Ensure that you and your crew use BSI precautions, including at least gloves and eye protection. Latex gloves provide a good barrier against body fluids; however, they can provide poor protection against many chemicals. Eye injuries are often the result of sudden unexpected changes in energy—mechanical, chemical, or thermal. As you ob-serve the scene, look for indicators of the MOI. This helps you develop an early index of suspicion for underlying injuries in the patient who has sustained a significant MOI. Consider early any additional help you may need. If hazardous materials are involved and uncontained, appropriately trained crews should be requested.

Initial Assessment

The patient you encounter may have graphic injuries; however, you must not be distracted by the injuries found in the initial assessment. Remember the focus of the initial assessment is the assessment and treatment of life-threatening problems.

General Impression

For most people, eye injuries have the potential for permanent disability and, therefore, create a great deal of anxiety. This anxiety will affect how they interact with you as a care provider. Remain calm and professional. As you approach the trauma patient, you must observe for important indicators to alert you to the seriousness of the patient's condition. An eye injury may be an obvious chief complaint as you approach the patient; however, the patient may have other injuries or medical conditions that need assessment and possibly treatment. The general impression will help you develop an index of suspicion for serious injuries and determine your sense of urgency for medical intervention. Is the patient awake or unresponsive? Is speech clear or slurred? Use the AVPU (Awake and alert, responsive to Verbal stimulus or Pain, or Unresponsive) scale to determine the patient's initial level of consciousness. These simple interactions with the patient as you approach and begin care will provide information about how you manage problems in the initial assessment, specifically how you manage airway, breathing, and bleeding problems.

Airway, Breathing, and Circulation

When you begin the initial assessment, consider the need for manual spinal immobilization as you determine whether the patient has a clear and patent airway. If the patient is unresponsive or has a significantly altered level of consciousness, consider inserting a properly sized oropharyngeal or nasopharyngeal airway as necessary to help maintain the airway. Check for clear and symmetric breath sounds, and then provide high-flow oxygen or provide bag-valve-mask device–assisted ventilation as needed, depending on your patient's level of consciousness and breathing rate and quality. Remember that traumatic injuries to the eye, like any other facial injury, have the potential to affect the airway and should be monitored.

Palpate the chest wall for DCAP-BTLS (Deformities, Contusions, Abrasions, Punctures/Penetrations, Burns, Tenderness, Lacerations, and Swelling). You must quickly assess the pulse rate and quality; determine the skin condition, color, and temperature; and check the capillary refill time. If visible significant bleeding is seen, you must begin the steps necessary to control bleeding. Significant bleeding is an immediate life threat. If the patient has obvious life-threatening bleeding, it must be controlled quickly. This can be difficult to do without causing undue pressure on the eye itself when significant facial trauma is present; however, care should be taken not to put pressure on the eye during bandaging because this may worsen the eye injury. Wounds to the soft tissue of the eyelid and tissue surrounding the eye tend to bleed freely but are generally not life threatening and are usually easy to control.

Transport Decision

If the patient you are treating has an airway or breathing problem or significant bleeding, he or she is a high priority patient and you must consider quickly transporting this patient to the hospital for treatment. If the patient has no signs of hypoperfusion or other life-threatening injuries, you must realize that eye injuries are serious and must be transported to the hospital as quickly and as safely as possible. In some situations, surgery and/or restoration of circulation to the eye will need to be accomplished in 30 minutes or permanent blindness may result. You should not delay transport of a seriously injured patient, particularly one with significant bleeding even if controlled, to complete a focused history and physical exam. Further assessment can continue during transport. For serious, isolated eye injuries, consideration should be given to transport to an eye care specialty center depending on local protocol.

Focused History and Physical Exam

After the initial assessment is complete, determine which assessment will be performed next.

Rapid Physical Exam

If there is significant MOI, likely affecting multiple systems, and multiple injuries besides an injury to the eye, a rapid physical exam should be performed using DCAP-BTLS to be sure that you have found all of the problems and injuries. It is important in bleeding cases to not focus just on the bleeding or graphic injuries involving the eye; with significant trauma, you should quickly assess the entire patient from head to toe. This will help you identify hidden injuries and better prepare your patient for packaging for transport.

Focused Physical Exam

In the patient who has an isolated injury with limited MOI, as is common in many eye injuries, begin with a focused physical exam of the eye and face. Focus your assessment on the isolated injury, the patient's complaint, and the body region affected. Is the patient wearing contact lenses? Ensure that control of the bleeding is maintained, and note the location and extent of the injury. Assess the injured eye and associated region of the face involved in potential injury. Assess the eyes for equal gaze—do the eyes look the same, move in the same direction? If not, the patient may experience diplopia, or double vision. Check pupil shape and response, and note the presence of any foreign objects or fluid draining from the eye. The pupils should be round and react to light and without obvious material in the eye or excessive tearing. A teardrop-shaped pupil could indicate a laceration to the eye itself. Assess the globe of the eye for bleeding. If the patient's eye(s) are swollen closed, do not attempt to force them open.

Baseline Vital Signs

Assess baseline vital signs to observe the changes a patient may display during treatment. Blood pressure, pulse, skin assessment, and respirations can indicate how well your patient is tolerating the situation. A blood pressure of less than 100 mm Hg with a weak, rapid pulse and cool, moist skin that is pale or gray should alert you to the presence of shock. Remember, you must be concerned with visible bleeding from the face and unseen bleeding inside a body cavity. Assessment of the pupils will probably have already been completed during assessment of the eye injury.

SAMPLE History

Next, obtain a SAMPLE history from your patient. If the patient is not responsive, look for medical identification in the form of bracelets, necklaces, or wallet

Geriatric Needs

When working with geriatric patients, ask how their vision was before the injury. Older people often have cataracts, poor vision, and drifting eyes, so it is important to find out how their current condition compares with the condition of their eyes before the injury.

cards. You should also attempt to get the SAMPLE history from friends or family members who may be present.

Interventions

Provide complete spinal immobilization to the patient with suspected spinal injuries. Be cautious in bandaging injuries to the eye itself. Pressure on the eye can reduce the circulation to the eye and may force fluids from the eye to leak out. The result of either can be blindness. If significant bleeding is visible to other areas of the body, control the bleeding as discussed in the previous chapter. Whenever you suspect significant bleeding or if the patient has signs and symptoms of shock, provide high-flow oxygen and provide rapid transport to the appropriate hospital. Do not delay transport of a seriously injured trauma patient to complete nonlifesaving treatments in the field, such as splinting extremity fractures; instead, complete these types of treatments en route to the hospital. If the patient is in stable condition, provide the appropriate emergency care for the injured eye(s) as outlined in the section titled Emergency Care.

Detailed Physical Exam

If the patient is in stable condition and problems identified during the initial assessment have been resolved, perform a thorough, detailed physical exam of the patient as discussed in the chapter on patient assessment. Many times, short transportation times and unstable patient conditions may make this assessment impractical. When longer transport times are present and your patient's condition is stable, performing a detailed physical exam will help sharpen your assessment skills for future medical calls.

Ongoing Assessment

Reassess areas examined during the initial assessment, focusing on the patient's airway, breathing, pulse, perfusion, and bleeding. Ask yourself "Are these still adequate?" Reassess vital signs. The changes in vital signs help you to know whether your patient's condition is improving or worsening. Reassess interventions and treatment you have provided. Ask about any changes in the patient's vision since you began the patient encounter. With serious injuries, make sure that your bandage still covers both eyes and that it is not putting pressure on the eyeball.

Communication and Documentation

The depth of information given in a radio call is based a great deal on local protocol and custom. You should provide enough information for the hospital staff to have a "picture" of what your assessment has found. The information you provide the hospital will help the staff prepare better for the patient. If the trauma patient has several injuries and the eye is not the highest priority, still make the hospital staff aware of the eye injury so specialists can be available to deal with the eye. When you arrive at the hospital, your more detailed verbal report to the staff should mirror your radio report and include additional treatments and changes in the patient's condition since your radio report.

Documentation on the run report should include the information given in your verbal report. Be clear and concise but include necessary details. Remember, it is a legal record of the care you provided your patient. Documentation of how much sight or vision the patient had while under your care can be important later. Write down what the patient says, such as "Everything is blurry," "I can see light but not shapes," or "I can't see out of that eye."

Emergency Care

Treatment starts with a thorough examination to determine the extent and nature of any damage. Always perform your examination using BSI precautions, taking great care to avoid aggravating any problems. You are looking for specific abnormalities or conditions that may suggest the nature of the injury (Figure 25-4 ▶). For example, blunt or penetrating injuries can produce swollen or lacerated eyelids. Bleeding soon after irritation or injury can result in a bright red conjunctiva. A damaged cornea quickly loses its smooth, wet appearance.

The following sections discuss common eye injuries and their emergency care.

Foreign Objects

Large objects are prevented from penetrating the eye by the protective orbit that surrounds it. However, moderately sized and smaller foreign objects of many different types can enter the eye and cause significant damage. Even a very small foreign object, such as a grain of sand lying on the surface of the conjunctiva, may produce severe irritation (Figure 25-5 ▼). The conjunctiva becomes inflamed and red—a condition known as conjunctivitis—almost immediately, and the eye begins to produce tears in an attempt to flush out the object. Irritation of the cornea or conjunctiva causes intense pain. The patient may have difficulty keeping the eyelids open, because the irritation is further aggravated by bright light.

If a small foreign object is lying on the surface of the patient's eye, you should use a normal saline solution to gently irrigate the eye. Irrigation with a sterile saline solution will frequently flush away loose, small particles. If a small bulb syringe is at hand, you can use this, or a nasal airway or cannula, to direct the saline into the affected eye (Figure 25-6 ▼). Always flush from the nose

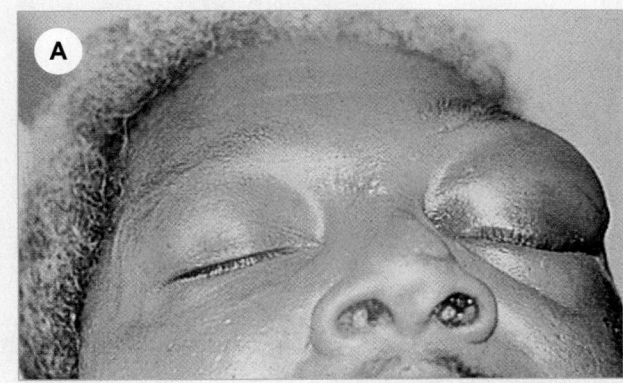
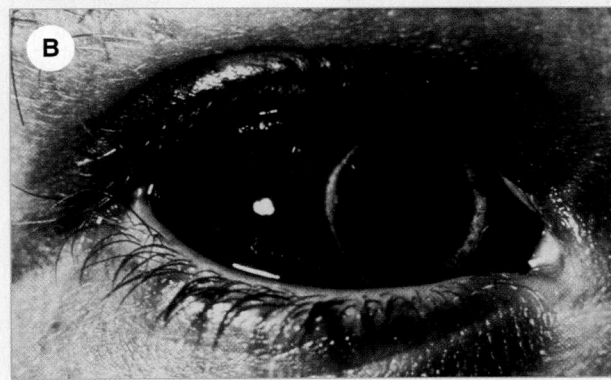
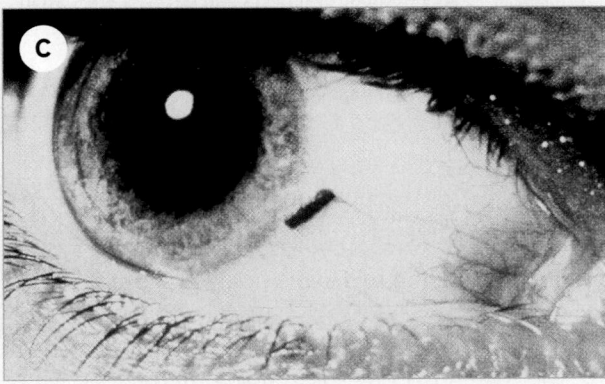

Figure 25-4 Injuries to the eyes are easily detected by **(A)** swelling, **(B)** bleeding, and **(C)** the presence of foreign objects in the eye.

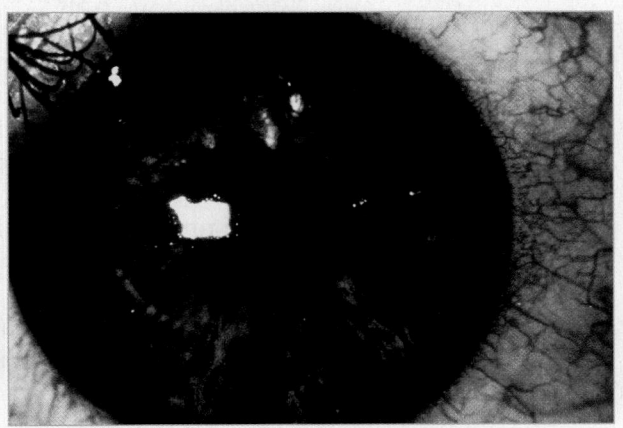

Figure 25-5 Conjunctivitis is often associated with the presence of a foreign object in the eye.

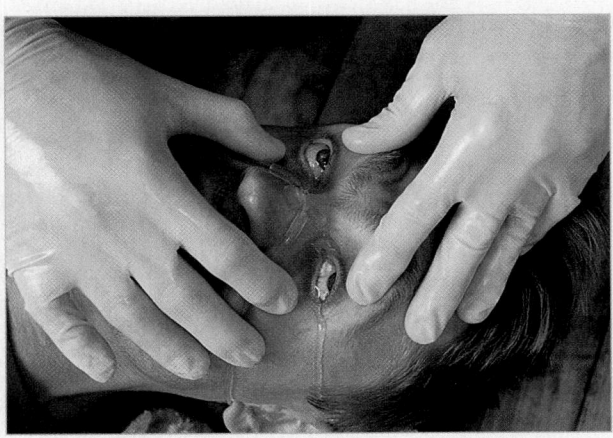

Figure 25-6 One method of irrigation is to direct saline into the injured eye using a round nasal airway or cannula. Always flush from the nose side of the eye toward the outside to avoid flushing material into the other eye.

side of the eye toward the outside to avoid flushing material into the other eye. After it has been flushed away, a foreign body will often leave a small abrasion on the surface of the conjunctiva. For this reason, the patient will complain of irritation even when the particle itself is gone. It is always a good idea to transport the patient to the hospital for further assessment to ensure appropriate medical care to the affected eye.

Gentle irrigation usually will not wash out foreign bodies that are stuck to the cornea or lying under the upper eyelid. To examine the undersurface of the upper eyelid, pull the lid upward and forward. If you spot a foreign object on the surface of the eyelid, you may be able to remove it with a moist, sterile, cotton-tipped applicator (Skill Drill 25-1 ▶). Never attempt to remove a foreign body that is stuck to the cornea.

1. **Tell the patient to look down** while you grasp the lashes of the upper eyelid with your thumb and index finger. Gently pull the eyelid away from the eyeball (**Step 1**).
2. **Gently place a cotton-tipped applicator** horizontally along the center of the outer surface of the upper eyelid (**Step 2**).
3. **Pull the eyelid forward and up**, which causes it to roll or fold back over the applicator, exposing the undersurface of the eyelid (**Step 3**).
4. **If you see a foreign object on the surface of the eyelid, gently remove it** with a moistened, sterile, cotton-tipped applicator (**Step 4**).

Foreign bodies ranging in size from a pencil to a sliver of metal may be impaled in the eye (Figure 25-7 ▶). These objects must be removed by a physician. Your care involves stabilizing the object and preparing the patient for transport to definitive care. The greater the length of the foreign object you can see sticking out of the eye, the more important stabilization becomes in avoiding further damage.

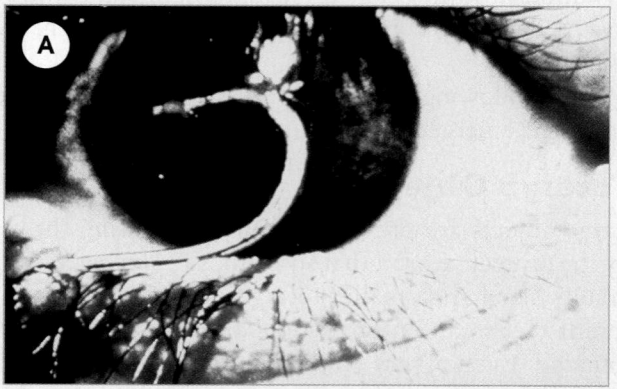

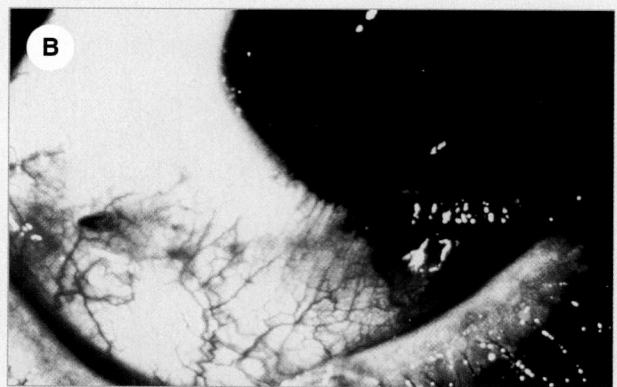

Figure 25-7 Any number of objects can become impaled in the eye. **A.** Fishhook. **B.** Sharp, metal sliver. **C.** Knife blade.

You are the Provider   Part 3

You gently place a moist, sterile dressing on the eye to control the bleeding and avoid applying direct pressure so as to not cause any further damage to the eye. You take a SAMPLE history from your patient and learn that he wears contact lenses. You perform a focused assessment of both eyes and find that the left eye is uninjured, the pupil is round and reactive to light, and the contact lens is in place.

5. Would you remove the contact lens from either eye?
6. Should you bandage one or both eyes?

Removing a Foreign Object From Under the Upper Eyelid

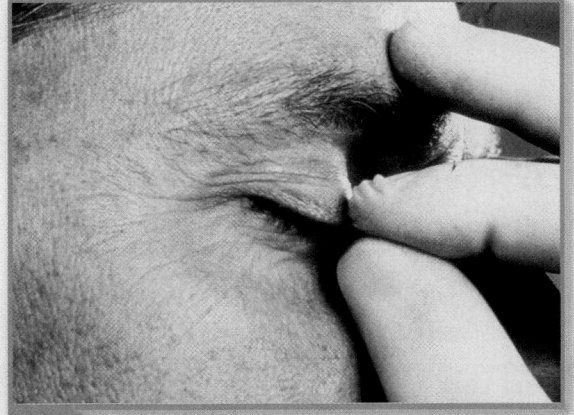

1 Have the patient look down, grasp the upper lashes, and gently pull the lid away from the eye.

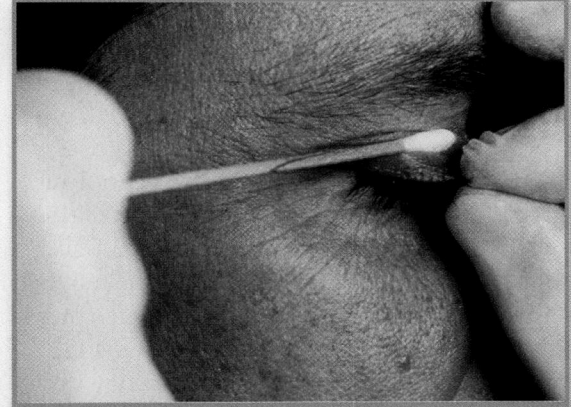

2 Place a cotton-tipped applicator on the outer surface of the upper lid.

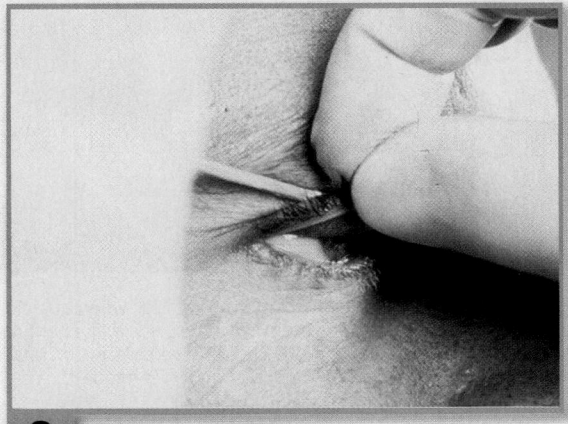

3 Pull the lid forward and up, folding it back over the applicator.

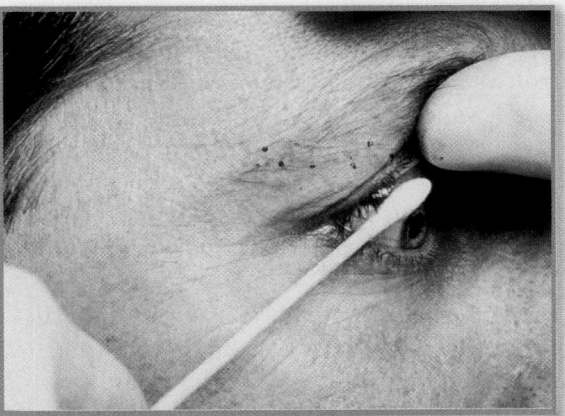

4 Gently remove the foreign object from the eyelid with a moistened, sterile applicator.

Bandage the object in place to support it. Cover the eye with a moist, sterile dressing, and then surround the object with a doughnut-shaped collar made from roller gauze or a small gauze pack. Follow the steps in (Skill Drill 25-2 ▶):

1. **Begin to prepare the doughnut ring** by wrapping a 2″ gauze roll circumferentially around your fingers and thumb enough times to make a thick dressing layer. You can adjust the inner diameter of what will become the ring by spreading your fingers or squeezing them together (**Step 1**).
2. **Remove the gauze** from your hands and start wrapping the remainder of the gauze roll

radially around this ring that you have created (**Step 2**).
3. **Work your way around the ring** until you have wrapped all the way around it and finished the "doughnut" (**Step 3**).
4. **Carefully place the ring over the eye** and impaled object, without bumping the object. You can then stabilize the object and the gauze collar with a roller bandage surrounding the head. Bandage both the injured and uninjured eyes to minimize eye movement and prevent further damage to the globe because when one eye moves, so does the other. Transport to an appropriate medical facility for treatment (**Step 4**).

Sometimes, a variety of types of large and small foreign bodies, particularly small metal fragments, become completely embedded within the eye itself. The patient may not even be aware of the cause of the problem. Suspect such an injury when the history includes metal work (such as hammering, exposure to splinters, grinding, vigorous filing) and when there are other signs of ocular injury. When you see or suspect an impaled object in the eye, bandage both eyes with soft bulky dressings to prevent further injury to the affected eye. Your bandage should be loose enough to hold the eyelid closed but not cause pressure on the eye itself. Using this technique prevents sympathetic motion (the movement of one eye causing both eyes to move), which may cause additional damage to the injured eye. This type of injury must be handled by an ophthalmologist on an urgent basis. X-rays and special equipment may be required to find the foreign body.

Burns of the Eye

Chemicals, heat, and light rays all can burn the delicate tissues of the eye, often causing permanent damage. Your role is to stop the burn and prevent further damage.

Chemical Burns *Study This*

Chemical burns, usually caused by acid or alkaline solutions, require immediate emergency care (Figure 25-8 ▶). This consists of flushing the eye with water or a sterile saline irrigation solution. If sterile saline is not available, you can use any clean water.

The idea is to direct the greatest amount of irrigating solution or water into the eye as gently as possible (Figure 25-9 ▶). Because opening the eye spontaneously may cause the patient pain, you may have to force the lids open to irrigate the eye adequately. Ideally, you will use a bulb or irrigation syringe, a nasal cannula, or some other device that will allow you to control the flow. In some circumstances, you may have to resort to pouring water into the eye by holding the patient's head under a gently running faucet. You can even have the patient immerse his or her face in a large pan or basin of water and rapidly blink the affected eyelid. If only one eye is affected, care must be taken to avoid contaminated water from getting into the unaffected eye.

Irrigate the eye for at least 5 minutes. If the burn was caused by an alkali or a strong acid, you should irrigate the eye continuously for 20 minutes. Follow local protocols on whether to try to irrigate while

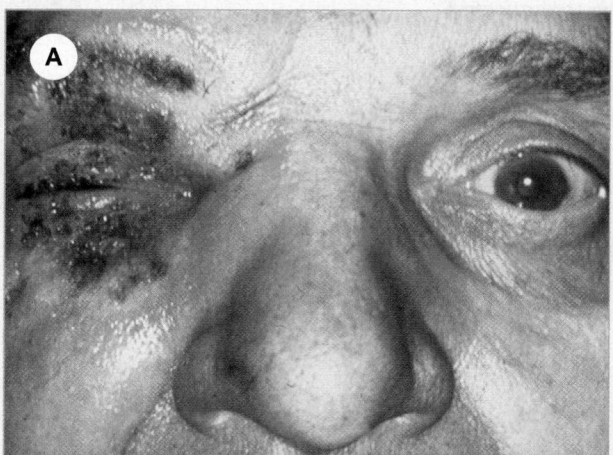

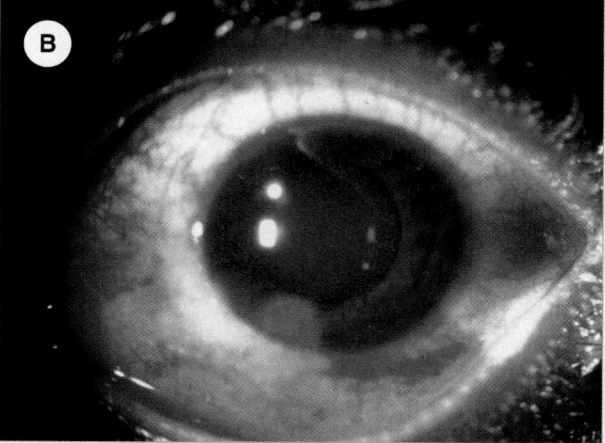

Figure 25-8 **A.** Chemical burns typically occur when an acid or alkali is splashed into the eye. **B.** This figure shows a chemical burn from lye, an alkaline solution. Because lye can continue to damage the eye even when diluted, fast action is needed.

transporting or to stay on scene until flushing is complete. Strong acids and all alkaline solutions can penetrate deeply, requiring a prolonged flush. Again, always take care to protect the uninjured eye and prevent irrigation fluid from running into it.

After you have completed irrigation, apply a clean, dry dressing to cover the eye, and transport the patient promptly to the hospital for further care (Figure 25-10 ▶). If the irrigation can be carried out satisfactorily in the ambulance, it should be done during transport to save time.

Thermal Burns

When a patient is burned in the face during a fire, the eyes usually close rapidly because of the heat. This reaction is a natural reflex to protect the eye from further

Stabilizing a Foreign Object Impaled in the Eye

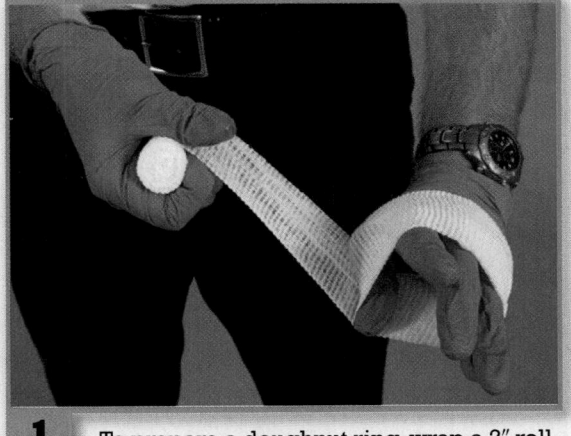

1 To prepare a doughnut ring, wrap a 2″ roll around your fingers and thumb seven or eight times. Adjust the diameter by spreading your fingers.

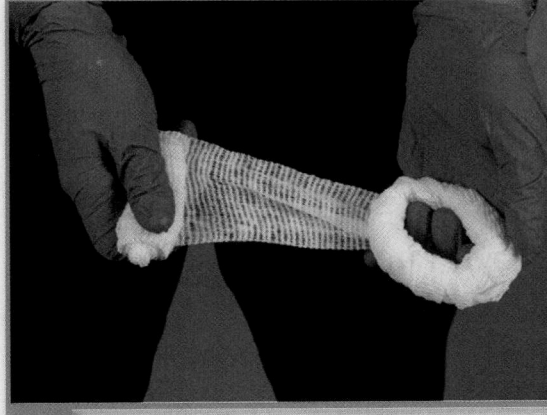

2 Wrap the remainder of the roll, ...

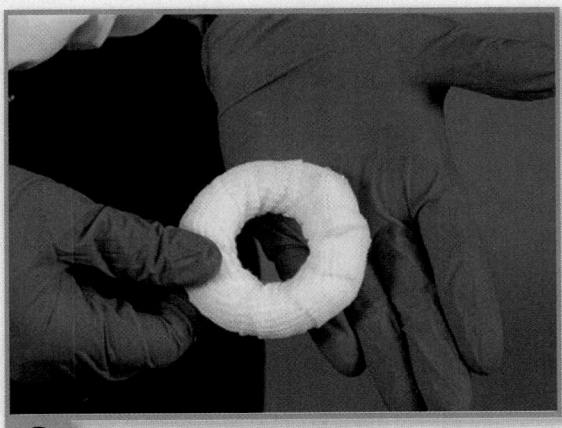

3 ...working around the ring.

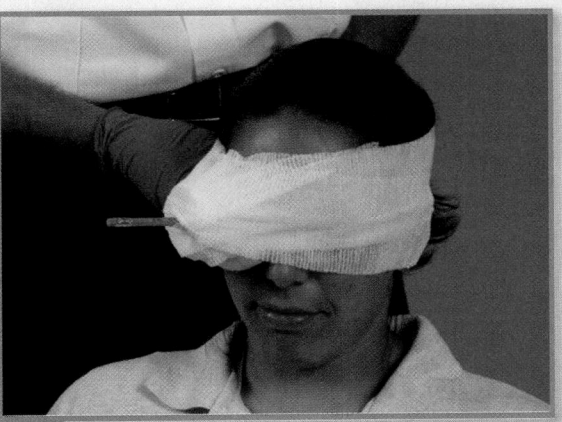

4 Place the dressing over the eye to hold the impaled object in place, then secure it with a gauze dressing.

injury. However, the eyelids remain exposed and are frequently burned (Figure 25-11 ▶). Burns of the eyelids require very specialized care. It is best to provide prompt transport for these patients without further examination. First, however, you should cover both eyes with a sterile dressing moistened with sterile saline. You may apply eye shields over the dressing.

Light Burns

Infrared rays, eclipse light (if the patient has looked directly at the sun), and laser burns all can cause significant damage to the sensory cells of the eye when rays of light become focused on the retina. Retinal injuries that are caused by exposure to extremly bright light are generally not painful but may result in permanent damage to vision.

Superficial burns of the eye can result from ultraviolet rays from an arc welding unit, light from prolonged exposure to a sunlamp, or reflected light from a bright snow-covered area (snow blindness). This kind of burn often is not painful at first but may become so 3 to 5 hours later, as the damaged cornea responds to

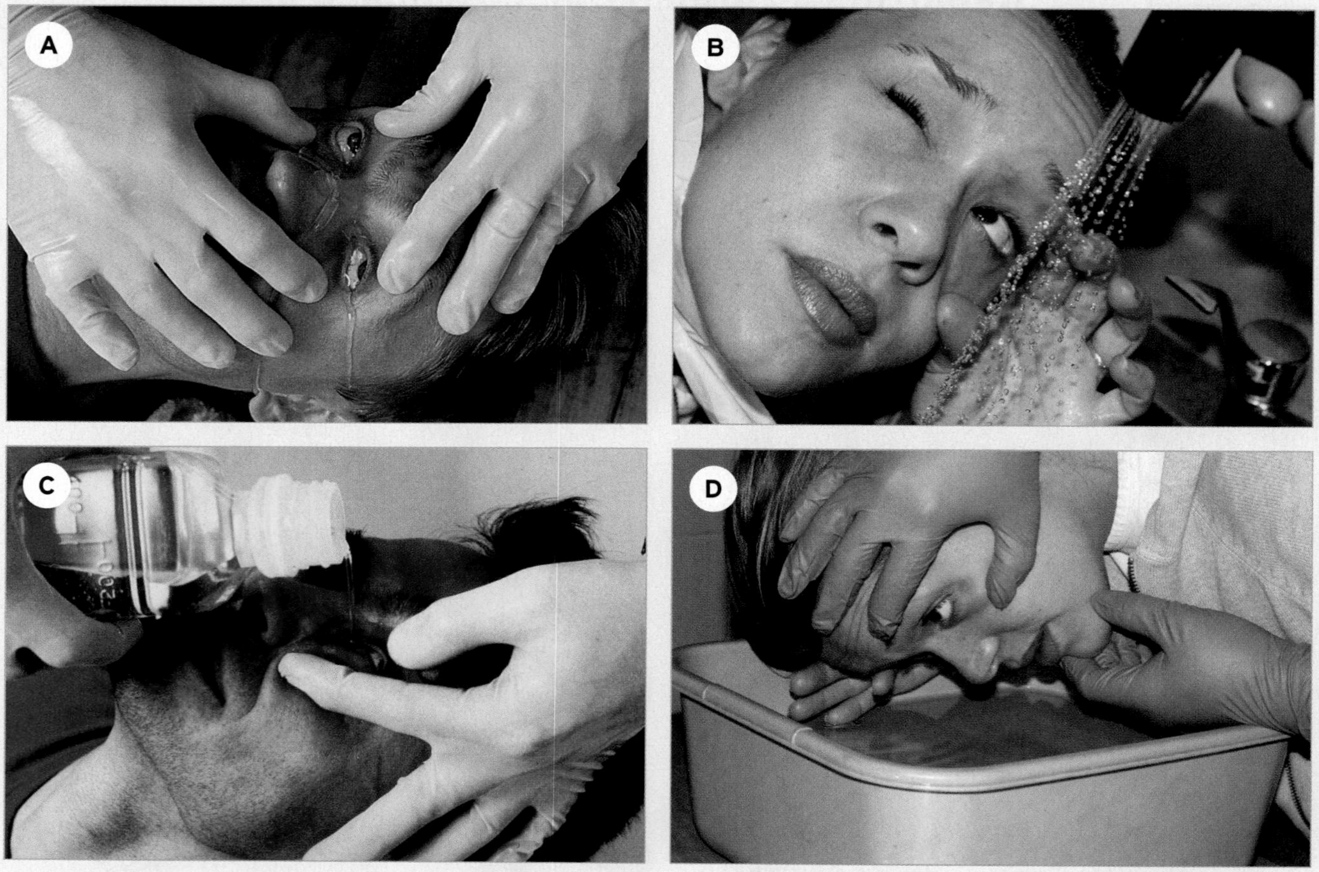

Figure 25-9 The following are four ways to effectively irrigate the eye. **A.** Nasal cannula. **B.** Shower. **C.** Bottle. **D.** Basin. Remember, you must protect the uninjured eye from the irrigating solution to prevent exposure of the unaffected eye to the substance.

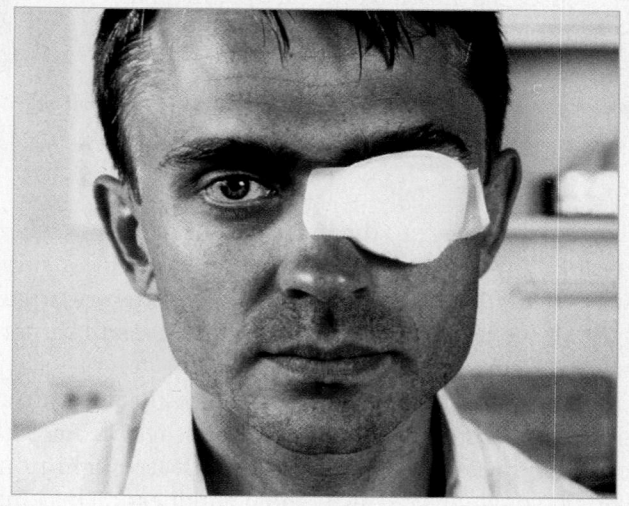

Figure 25-10 Apply a clean, dry dressing to cover the eye after you have finished irrigation.

the injury. Severe conjunctivitis usually develops, with redness, swelling, and excessive tear production. You can ease the pain from these corneal burns by covering each eye with a sterile, moist pad and an eye shield. Have the patient lie down during transport to the hospital, and protect him or her from further exposure to bright light. The patient should be examined by a physician as soon as possible.

Lacerations

Lacerations of the eyelids require very careful repair to restore appearance and function (Figure 25-12 ▶). Bleeding may be heavy, but it usually can be controlled by gentle, manual pressure. If there is a laceration of the globe itself, apply no pressure to the eye; compression can interfere with the blood supply to the back of the eye and result in loss of vision from damage to the retina. Furthermore, pressure may

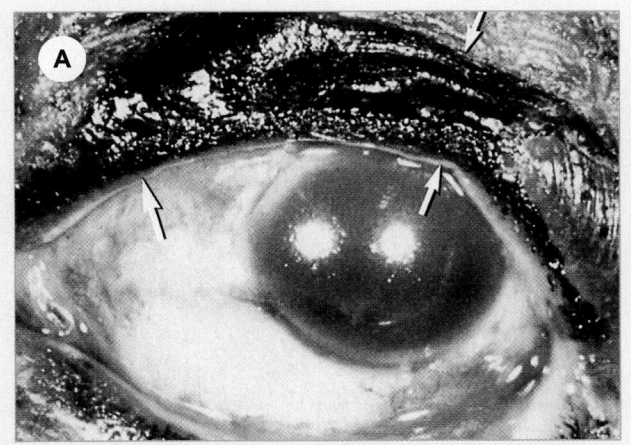

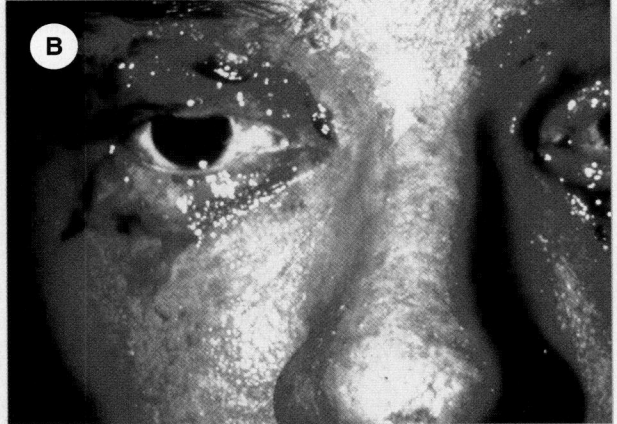

Figure 25-11 Thermal burns occasionally cause significant damage to the eyelids. **A.** Arrows show some full-thickness burns. **B.** Burns of the eyelids require immediate hospital care.

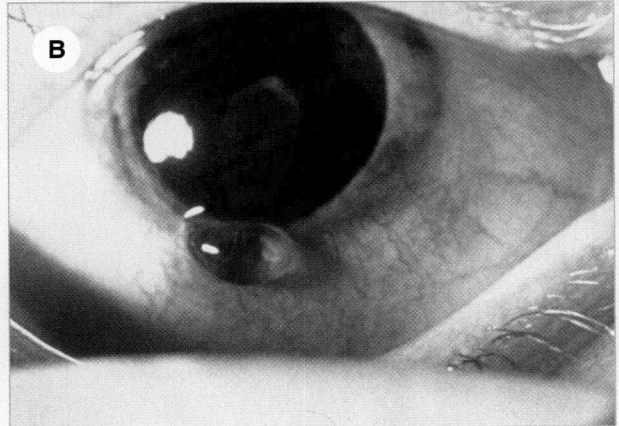

Figure 25-12 Lacerations are serious injuries that require prompt transport. **A.** Although bleeding can be heavy, never exert pressure on the eye. **B.** Pressure may squeeze the vitreous humor, iris, lens, or even the retina out of the eye.

squeeze the vitreous humor, iris, lens, or even the retina out of the eye and cause irreparable damage or blindness.

Follow these three important guidelines in treating penetrating injuries of the eye:

1. **Never exert pressure** on or manipulate the injured eye (globe) in any way.
2. **If part of the eyeball is exposed**, gently apply a moist, sterile dressing to prevent drying.
3. **Cover the injured eye** with a protective metal eye shield, cup, or sterile dressing. Apply soft dressings to both eyes, and provide prompt transport to the hospital.

On rare occasions following a serious injury, the eyeball may be displaced out of its socket. Do not attempt to reposition it. Simply cover the eye and stabi-

lize it with a moist sterile dressing (Figure 25-13 ▶); remember to cover both eyes to prevent further injury due to sympathetic movement. Have the patient lie in a supine position en route to the hospital to prevent further loss of fluid from the eye.

Blunt Trauma

Blunt trauma can cause a number of serious eye injuries. These range from the ordinary "black eye," a result of bleeding into the tissue around the orbit, to a severely damaged globe (Figure 25-14 ▶). You may see an injury called <u>hyphema</u>, or bleeding into the anterior chamber of the eye, that obscures part or all of the iris (Figure 25-15 ▶). This injury is common in blunt trauma and may seriously impair vision. Twenty-five percent of hyphemas are globe injuries, a serious

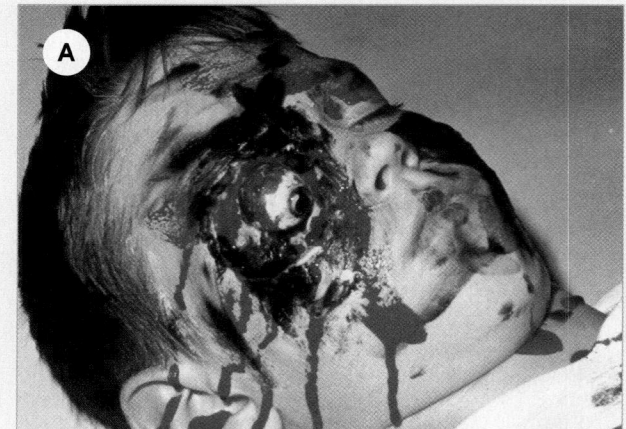

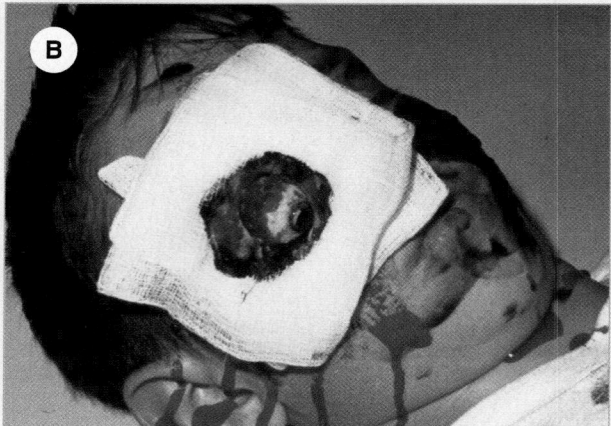

Figure 25-13 An injury that exposes the brain, eye, or other structures **(A)** should be covered with a moist, sterile dressing to prevent further damage **(B)**.

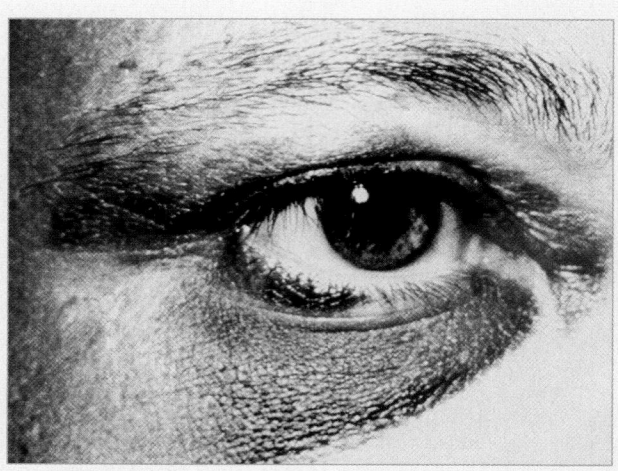

Figure 25-14 The typical "black eye" is caused by bleeding into the tissue around the orbit.

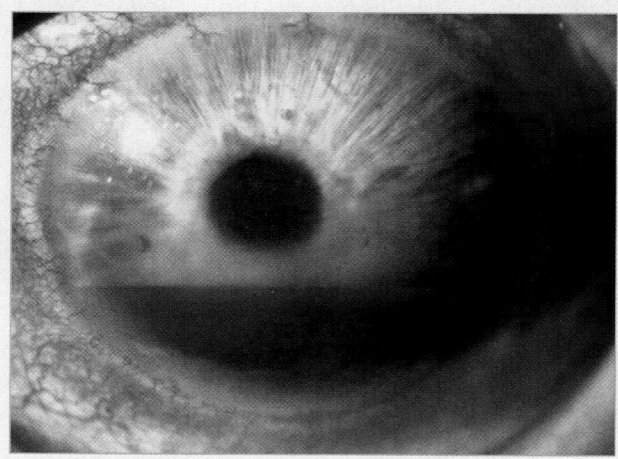

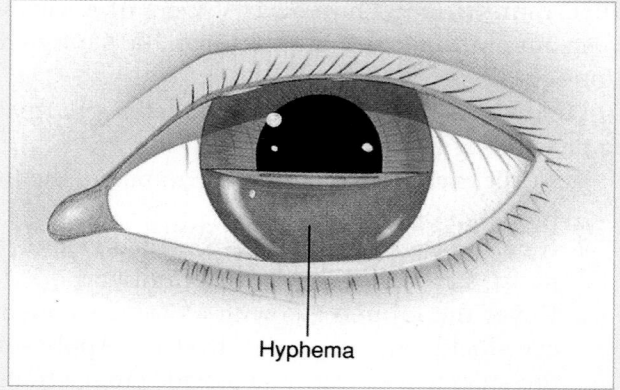

Hyphema

Figure 25-15 A hyphema, characterized by bleeding into the anterior chamber of the eye, is common following blunt trauma to the eye. This condition may seriously impair vision and should be considered a sight-threatening emergency.

injury to the eye. Cover the eye to protect it from further injury and provide transportation to the hospital for further medical evaluation.

Blunt trauma can also cause a fracture of the orbit, particularly of the bones that form its floor and support the globe. This injury is called a <u>blowout fracture</u>. The fragments of fractured bone can entrap some of the muscles that control eye movement, causing double vision (Figure 25-16 ▶). Any patient who reports pain, double vision, or decreased vision following a blunt injury about the eye should be placed on a stretcher and transported promptly to the emergency department. Protect the eye from further injury with a metal shield; cover the other eye to minimize movement on the injured side.

Another possible result of blunt eye injury is retinal detachment. This injury is often seen in sports,

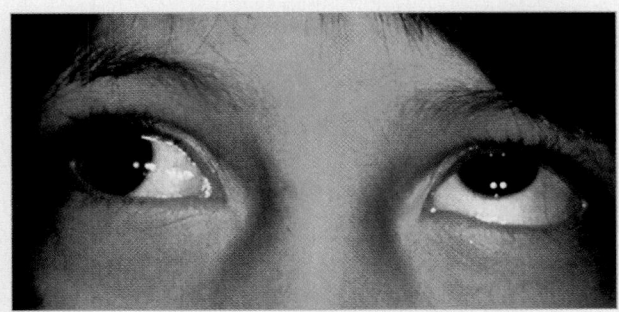

Figure 25-16 A patient with a blowout fracture may not move his or her eyes together because of muscle entrapment. Therefore, the patient sees double images of any object.

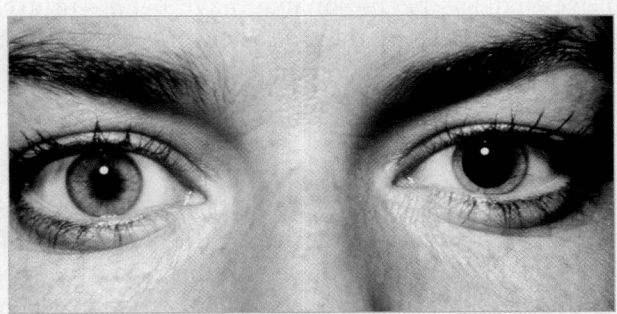

Figure 25-17 Variation of pupil size may indicate a head injury.

especially boxing. It is painless but produces flashing lights, specks, or "floaters" in the field of vision and a cloud or shade over the patient's vision. Because the retina is separated from the nourishing choroid, this injury requires prompt medical attention to preserve vision in the eye.

Eye Injuries Following Head Injury

Abnormalities in the appearance or function of the eyes often occur following a closed head injury. Any of the following eye findings should alert you to the possibility of a head injury:

- One pupil larger than the other (Figure 25-17 ▶)
- The eyes not moving together or pointing in different directions
- Failure of the eyes to follow the movement of your finger as instructed

- Bleeding under the conjunctiva, which obscures the sclera (white portion) of the eye
- Protrusion or bulging of one eye

Record any of these observations, along with the time that you make them. For an unconscious patient, remember to keep the eyelids closed; drying of the ocular tissue can cause permanent injury and may result in blindness. Cover the lids with moist gauze, or hold them closed with clear tape. Normal tears will then keep the tissues moist.

Contact Lenses and Artificial Eyes

Small, hard contact lenses usually are tinted, making them relatively easy to see. Large, soft ones are clear and can be very difficult to see. In general, you should not attempt to remove either kind of lens from a patient. You should never attempt to remove a lens from

You are the Provider   Part 4

After deciding to leave both contact lenses in place, you place a cup over the dressing on the patient's right eye to protect it. You want to minimize the movement of the right eye, so you bandage the dressings in place over both of the patient's eyes.

Your patient becomes anxious because he can no longer see what is happening to him. To calm your patient, you talk to him and explain everything that you are doing as you prepare to place him on the ambulance cot for transportation to the hospital.

7. How will your patient react to having both eyes bandaged?

an eye that has been—or may have been—injured because manipulating the lens can aggravate the problem. The only time that contact lenses should be removed immediately in the field is in the case of a chemical burn of the eye. In this situation, the lens can trap the chemical and make irrigation difficult.

If it is necessary to remove a hard contact lens, use a small suction cup, moistening the end with saline (Figure 25-18A ▶). To remove soft lenses, place one to two drops of saline in the eye (Figure 25-18B ▶), gently pinch it between your gloved thumb and index finger, and lift it off the surface of the eye (Figure 25-18C ▶). Place the contact lens in a container filled with sterile saline solution to prevent damage to the contact lens. Always advise emergency department staff if a patient is wearing contact lenses.

Occasionally, you may find yourself caring for a patient who is wearing an eye prosthesis (an artificial eye). Many people are surprised to find that it can be difficult to distinguish a prosthesis from a natural eye. You should suspect an eye of being artificial when it does not respond to light, move in concert with the opposite eye, or appear quite the same as its mate. If you think that a patient may have an artificial eye but you are not sure, go ahead and ask about it. Although no harm will be done if you care for an artificial eye as you would a normal one, you need to be totally clear about the patient's eye function.

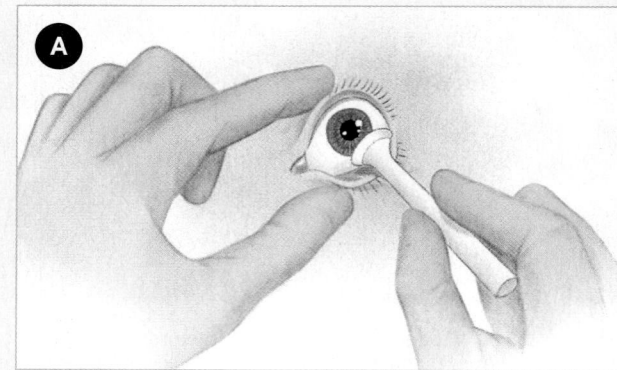
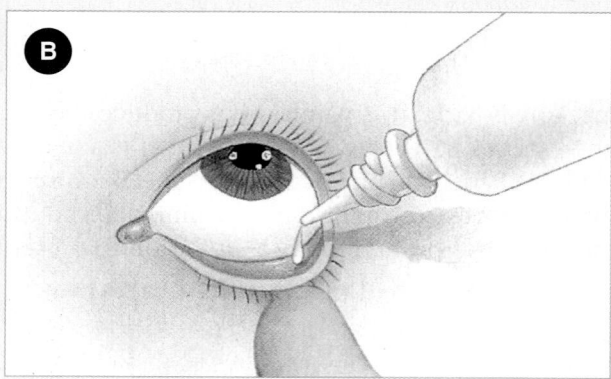
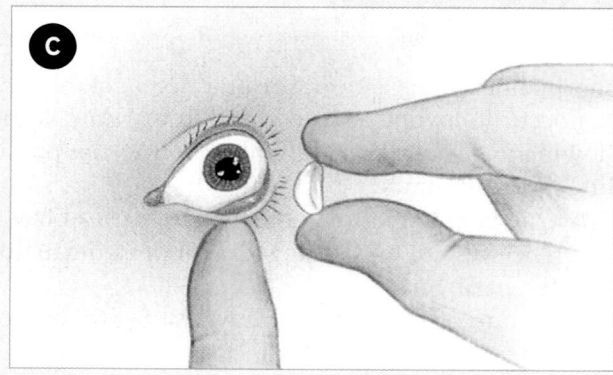

Figure 25-18 Removing contact lenses should be limited to patients with chemical burn injuries. **A.** To remove hard contact lenses, use a specialized suction cup moistened with sterile saline solution. **B.** To remove soft contact lenses, instill one or two drops of saline or irrigating solution. **C.** Next, pinch off the lens with your gloved thumb and index finger.

You are the Provider   Summary

Eye injuries require gentle treatment to prevent further damage. Because allowing an uninjured eye to remain uncovered can cause sympathetic movement in the injured eye, you must cover both of the patient's eyes. This often creates anxiety because the patient no longer has the sense of vision to allow him or her to see what is happening. Professional reassurance and ongoing verbal explanation of your treatment and actions is an important part of your patient care strategy.

	Eye Injuries
Scene Size-up	Body substance isolation precautions should include a minimum of gloves and eye protection. Ensure scene safety and determine NOI/MOI. Consider the number of patients, the need for additional help/ALS, and c-spine stabilization.
Initial Assessment	
■ General impression	Determine level of consciousness and treat any immediate threats to life. Determine priority of care based on environment and patient's chief complaint.
■ Airway	Ensure patent airway. Maintain spinal immobilization as necessary.
■ Breathing	Listen for abnormal breath sounds and evaluate depth and rate of the respiratory cycle. Maintain ventilatory support as needed. Provide high-flow oxygen at 15 L/min and inspect the chest wall, assessing for DCAP-BTLS.
■ Circulation	Evaluate pulse rate and quality; observe skin color, temperature, and condition and treat accordingly.
■ Transport decision	Prompt transport
Focused History and Physical Exam	*NOTE: The order of the steps in the focused history and physical exam differs depending on whether or not the patient has a significant MOI. The order below is for a patient with a significant MOI. For a patient without a significant MOI, perform a focused trauma assessment, obtain vital signs, and obtain the history.*
■ Rapid trauma assessment	Reevaluate the mechanism of injury. Perform a rapid trauma assessment, treating all life threats immediately. Log roll and secure patient to backboard for all patients with a significant MOI from the clavicles on up.
■ Basliine vital signs	Take vital signs, noting skin color and temperature as well as patient's level of consciousness. Use pulse oximetry if available.
■ SAMPLE history	Obtain SAMPLE history. If the patient is not responsive, attempt to get it from family members, friends, or bystanders.
■ Interventions	Provide complete spinal immobilization early if you suspect that your patient has spinal injuries. Control significant bleeding. Without placing pressure on the injury, cover both eyes closed. Treat signs of hypoperfusion.
Detailed Physical Exam	Complete a detailed physical exam.
Ongoing Assessment	Repeat the initial assessment, rapid/focused assessment, and reassess interventions performed. Make sure that your bandage still covers both eyes and that it is not putting pressure on the eyeball. Reassess vital signs every 5 minutes for the unstable patient, every 15 minutes for the stable patient.
■ Communication and documentation	Contact medical control with a radio report. Relay any change in level of consciousness or difficulty breathing. Documentation of how much sight or vision the patient had while under your care will be important. Quote pertinent patient statements, especially those regarding vision. Be sure to document the physician's orders and changes in the patient's condition and at what time they occurred.

Assessment and Emergency Care

Assessment and Emergency Care

NOTE: While the steps below are widely accepted, be sure to consult and follow your local protocol.

Foreign Objects	Burns	Lacerations	Blunt Trauma

Foreign Objects

1. Tell the patient to look down while you grasp the lashes of the upper eyelid with your thumb and index finger. Gently pull the eyelid away from the eyeball.

2. Gently place a cotton-tipped applicator horizontally along the center of the outer surface of the upper eyelid.

3. If you see a foreign object, gently remove it with a moistened, sterile, cotton-tipped applicator.

Stabilizing a Foreign Object Impaled in the Eye

1. To prepare a doughnut ring, wrap a 2″ roll around your fingers and thumb seven or eight times.

2. Wrap the remainder of the roll, working around the ring.

3. Place the dressing over the eye to hold the impaled object in place and then secure it with a gauze dressing.

Burns

Chemical Burns

1. Holding the eyelid open, irrigate the eye for 5 to 20 minutes.

2. Apply a clean, dry sterile dressing to cover the eye.

Thermal Burns

1. Cover both eyes with a sterile dressing moistened with sterile saline.

Light Burns

1. Cover each eye with a sterile, moist pad and eye shield.

2. Transport supine and prevent further exposure to bright light.

Lacerations

1. Never exert pressure on or manipulate the injured eye in any way.

2. If part of the eyeball is exposed, gently apply a moist sterile dressing to prevent drying.

3. Cover the injured eye with a protective metal eye shield or sterile dressing.

Blunt Trauma

1. Protect the eye from further injury with a metal shield.

2. Cover the other eye to minimize movement on the injured side.

Prep Kit

Ready for Review

- The eye is shaped like a globe, about 1" in diameter, and is located inside a bony socket called the orbit. The orbit is made up of the facial bones of the skull.

- Any significant injury to the face or head can potentially cause damage to the eye itself or to the muscles of the eye.

- The fluid in the back chamber of the eye is called vitreous humor and cannot be replaced. The fluid in the front chamber of the eye is called the aqueous humor and can be replaced.

- The pupil functions like a camera. The iris and the pupil make adjustments to light, and the retina of the eye acts like film capturing a picture.

- Nerve endings in the retina send impulses through the optic nerve to the brain, which interprets that message as vision.

- When assessing a patient with a suspected eye injury, assess for swollen or lacerated eyelids, bright red conjunctiva, irregular pupil(s) or eye movements, and changes to the cornea that make it appear rough or dry.

- Foreign bodies on the surface of the eye should be irrigated gently with normal saline solution. Always flush from the region of the eye closest to the nose toward the outside, away from the midline.

- If the foreign body is on the underside of the eyelid, remove it gently with a cotton-tipped applicator. Never remove foreign bodies stuck to the cornea.

- If a foreign body is impaled in the eye, provide the following care in this order:
  - Stabilize the object in place with a bulky dressing using roller gauze or a cup to minimize movement.
  - Transport to the hospital for further medical treatment; small metal fragments embedded in the eye must be treated by an ophthalmologist.

- Chemicals, heat, and light rays can all cause burn injury to the eyes, resulting in permanent damage.

- Irrigate chemical burns with saline solution or clean water for a minimum of 5 minutes; then apply clean, dry dressings to the eyes and transport promptly.

- Immediately transport a patient with heat burns to the eyelid. Cover both eyes with a sterile, moist dressing.

- Burns to the eye from exposure to ultraviolet light rays may become painful several hours after exposure. Ease the patient's pain and discomfort by placing a sterile, moist dressing over the eyes, and then transport.

- For a patient who has sustained a lacerated eyelid that is bleeding, apply gentle manual pressure; do not apply pressure to the globe of the eye itself. Instead apply a moist, sterile dressing to prevent drying, cover the eye(s), and transport.

- You should never attempt to reposition a displaced eyeball. Cover the affected eyeball with a moist, sterile dressing, cover both eyes, and transport to the hospital.

- Blunt trauma can cause a variety of injuries to the eyes and their supporting structures. These injuries include hyphema, retinal detachment, and blowout fractures.

- Any patient with the complaint of painful vision, double vision, or decreased vision after blunt trauma should be transported to the hospital for appropriate treatment.

- Maintain a high index of suspicion for patients with unequal pupils—they may have an illness of or an injury to the brain. Remember, some people are born with one pupil larger than the other. Ask your patient during the assessment whether he or she normally has unequal pupils.

- Never remove contact lenses from an injured eye unless there is a chemical burn.

Technology

- Interactivities
- Vocabulary Explorer
- Anatomy Review
- Web Links
- Online Review Manual

www.EMTB.com

Prep Kit continued...

Vital Vocabulary

blowout fracture A fracture of the orbit or of the bones that support the floor of the orbit.

conjunctiva The delicate membrane that lines the eyelids and covers the exposed surface of the eye.

conjunctivitis Inflammation of the conjunctiva.

cornea The transparent tissue layer in front of the pupil and iris of the eye.

globe The eyeball.

hyphema Bleeding into the anterior chamber of the eye, obscuring the iris.

iris The muscle and surrounding tissue behind the cornea that dilate and constrict the pupil, regulating the amount of light that enters the eye; pigment in this tissue gives the eye its color.

lacrimal glands The glands that produce fluids to keep the eye moist; also called tear glands.

lens The transparent part of the eye through which images are focused on the retina.

optic nerve A cranial nerve that transmits visual information to the brain.

orbit The bony eye socket.

pupil The circular opening in the middle of the iris that admits light to the back of the eye.

retina The light-sensitive area of the eye where images are projected; a layer of cells at the back of the eye that changes the light image into electrical impulses, which are carried by the optic nerve to the brain.

retinal detachment Separation of the retina from its attachments at the back of the eye.

sclera The tough, fibrous, white portion of the eye that protects the more delicate inner structures.

Points to Ponder

You are dispatched to a local industrial complex for an unknown injury. You arrive to find a young man holding a wet towel to his eyes. As you approach him, you notice that there are arc welding tools and materials around him. In a frightened voice he says, "I can't see right." He tells you that he's been welding for a couple of hours without appropriate eye protection.

What patient care issues are associated with a sudden loss of sight? How can you compensate for these issues?

Issues: Anxiety Related to Sudden Loss or Impairment of Sight, Treating the Whole Patient.

Assessment in Action

You are dispatched to 10 East Main St, Kelley's Pub, for an assault victim. Local law enforcement has already arrived on the scene and tells you it is safe to enter. You arrive to find a 22-year-old man who is holding a towel over his right eye. He tells you he was hit in the face with a beer bottle. He says that he can't see out of that eye, and it hurts badly. As your partner begins taking his vital signs, you begin appropriate care for his eye injury.

You then ask the patient to move the towel so you can briefly visualize the injury. You notice even with the eyelid closed that the eyeball has lost its shape. You notice a watery substance and blood the towel and on the patient's face. His vital signs include a blood pressure of 138/88 mm Hg, a pulse of 94 beats/min and regular, and respirations of 36 breaths/min with adequate tidal volume.

1. What injury do you suspect?

   A. Corneal abrasion
   B. Conjunctivitis
   C. Lacerated globe
   D. Both A and B

2. The eye keeps its global shape as a result of:

   A. fluids in the eye.
   B. air in the eye.
   C. the sclera.
   D. both A and C.

3. Vitreous humor is:

   A. the white of the eye.
   B. the jellylike substance in the eye.
   C. the colored part of the eye.
   D. the membrane of the eye.

4. Humans blink because:

   A. air dries the eyes.
   B. this action sweeps fluid from the lacrimal glands.
   C. this action cleans the eye.
   D. both B and C.

5. The white of the eye is called the:

   A. conjunctiva.
   B. iris.
   C. cornea.
   D. sclera.

6. The light-sensitive area at the back of the eye is called the:

   A. lens.
   B. optic nerve.
   C. retina.
   D. choroid.

Challenging Questions

7. Why is it important to cover both eyes even if only one is injured?

8. Why is minimal on-scene time essential for serious eye injuries?

9. For the patient in the above scenario, should you apply direct pressure to the eye to control bleeding?

10. What does the presence of hyphema likely indicate?

www.EMTB.com

Face and Throat Injuries

Objectives*

Cognitive

1. Describe the causes of upper airway obstruction in facial injuries. (p 760)
2. List the steps in the emergency medical care of the patient with soft-tissue wounds of the face and neck. (p 764)
3. List the steps in the emergency medical care of the patient with injuries of the nose and ear. (p 765, 766)
4. List the physical findings of a patient with a facial fracture. (p 767)
5. List the steps in the emergency medical care of the patient with a penetrating injury to the neck. (p 769)
6. List the steps in the emergency medical care of the patient with an upper airway injury. (p 768)
7. List the steps in the emergency medical care of the patient with dental injuries. (p 767)

Affective

None

Psychomotor

8. Demonstrate the care of a patient with soft-tissue wounds of the face and neck. (p 764)
9. Demonstrate the care of a patient with injuries of the nose and ear. (p 765, 766)
10. Demonstrate the care of a patient with a penetrating injury to the neck. (p 769)
11. Demonstrate the care of a patient with an upper airway injury. (p 768)
12. Demonstrate the care of a patient with dental injuries. (p 767)

*All of the objectives in this chapter are noncurriculum objectives.

You are the Provider

You and your EMT-B partner receive a call to the Acme Bar on the report of a person injured in a fight. Dispatch advises you that law enforcement is also en route to the scene.

1. What is your most important concern on arrival at the scene?
2. What are some possible injuries the patient could have sustained in a fight?

Face and Throat Injuries

The face and neck are particularly vulnerable to injury because of their relatively unprotected positions on the body. Soft-tissue injuries and fractures to the bones of the face are common and vary greatly in severity. Some are potentially life threatening, and many leave disfiguring scars if not treated properly. Penetrating trauma to the neck may cause severe bleeding. If a hematoma forms in this area, it may stop or slow blood flow to the brain, causing a stroke. With appropriate prehospital and hospital care, what may at first seem to be a devastating injury can have a surprisingly good outcome.

As an EMT-B, your objectives include prevention of further injury, particularly to the cervical spine, managing any acute airway problems, and controlling bleeding. This chapter first reviews the anatomy of the head and neck, then examines the factors that can produce upper airway obstruction. A discussion of emergency medical care of soft-tissue wounds of the face, nose, and ear; facial fractures; penetrating injuries of the neck; and dental injuries follows.

Technology

www.EMTB.com

- Interactivities
- Vocabulary Explorer
- Anatomy Review
- Web Links
- Online Review Manual

Anatomy of the Head and Neck

The head is divided into two parts: the cranium and the face. The cranium, or skull, contains the brain, which connects to the spinal cord through the foramen magnum, a large opening at the base of the skull. The most posterior portion of the cranium is called the occiput. On each side of the cranium, the lateral portions are called the temples or temporal regions. Between the temporal regions and the occiput lie the parietal regions. The forehead is called the frontal region. Just anterior to the ear, in the temporal region, you can feel the pulse of the superficial temporal artery. The thick skin covering the cranium, which usually bears hair, is called the scalp.

The face is composed of the eyes, ears, nose, mouth, cheeks, and jowls. Six bones—the nasal bone, the two maxillae (upper jawbones), the two zygomas (cheekbones), and the mandible (jawbone)—are the major bones of the face (Figure 26-1 ▶).

The orbit of the eye is composed of the lower edge of the frontal bone of the skull, the zygoma, the maxilla, and the nasal bone. The bony orbit protects the eye from injury. By viewing the face from the side, you can see the eyeball recessed in the orbit. Only the proximal third of the nose—the bridge—is formed by bone. The remaining two thirds are composed of cartilage. Unlike the nose, the exposed portion of the ear is composed entirely of cartilage that is covered by skin. The external, visible part of the ear is called the pinna (Figure 26-2 ▶). The earlobes are the fleshy portions at the bottom of each ear. The tragus is a small, rounded, fleshy bulge immediately anterior to the ear canal. The superficial temporal artery can be palpated just anterior to the tragus. About 1″ posterior to the external opening of the ear is a prominent bony mass at the base of the skull called the mastoid process.

The mandible forms the jaw and chin. Motion of the mandible occurs at the temporomandibular joint (TMJ), which lies just in front of the ear on either side of the face. Below the ear and anterior to the mastoid process, the angle of the mandible is easily palpated.

The neck also contains many important structures. It is supported by the cervical spine, or the first seven vertebrae in the spinal column (C1 through C7). The spinal cord exits from the foramen magnum and lies within the spinal canal formed by the vertebrae. The upper part of the esophagus and the trachea lie in the

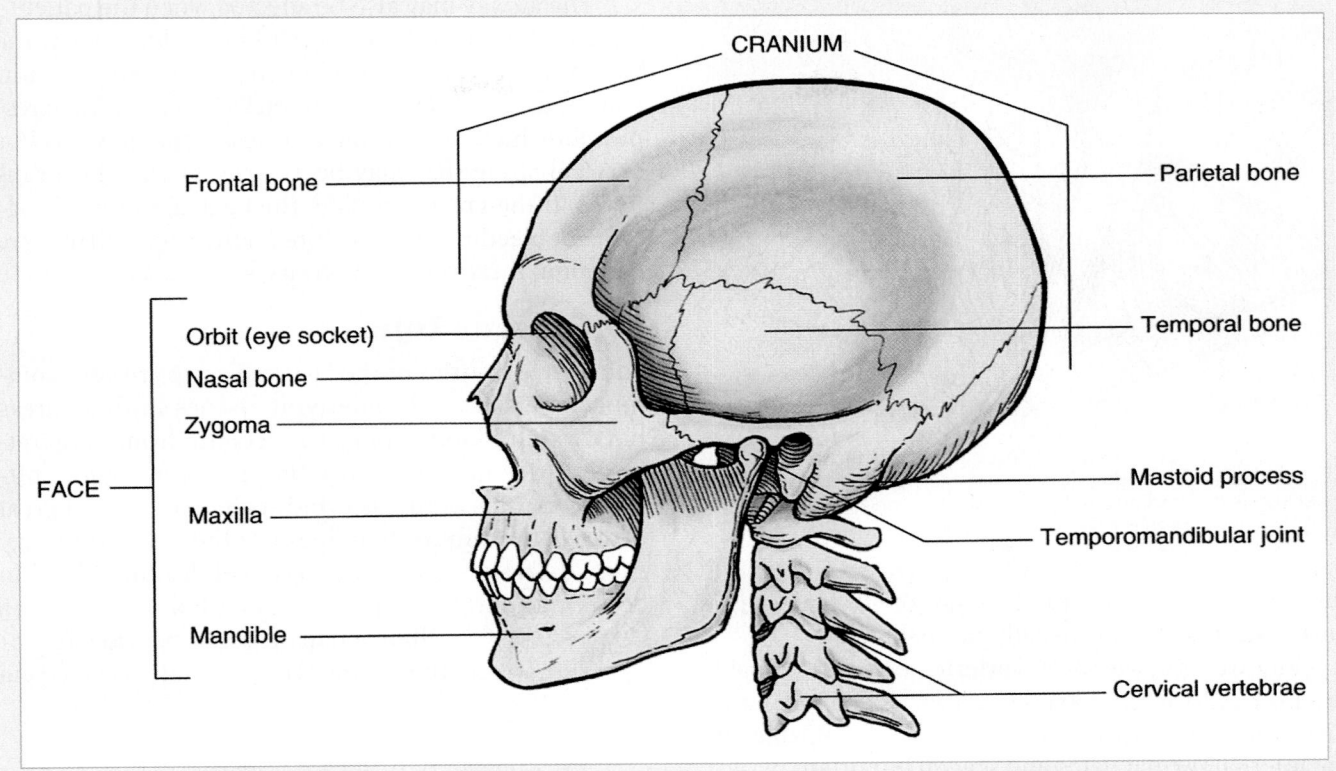

Figure 26-1 The face is composed of six bones: the nasal bone, two maxillae, two zygomas, and the mandible.

midline of the neck. The carotid arteries may be found on either side of the trachea, along with the jugular veins and several nerves.

Several useful landmarks can be palpated and seen in the neck (Figure 26-3 ▶). The most obvious is the firm prominence in the center of the anterior surface commonly known as the Adam's apple. Specifically, this prominence is the upper part of the larynx, formed by the thyroid cartilage. It is more prominent in men than in women. The other portion of the larynx is the cricoid cartilage, a firm ridge of cartilage (the only complete circular cartilage structure of the trachea) below the thyroid cartilage, which is somewhat more difficult to palpate. Between the thyroid cartilage and the cricoid cartilage in the midline of the neck is a soft depression, the cricothyroid membrane. This is a thin sheet of connective tissue (fascia) that joins the two cartilages. The cricothyroid membrane is covered at this point only by skin.

Below the larynx, several additional firm ridges are palpable in the anterior midline. These ridges are the cartilage rings of the trachea. The trachea connects the larynx with the main air passages of the lungs (the bronchi). On either side of the lower larynx and

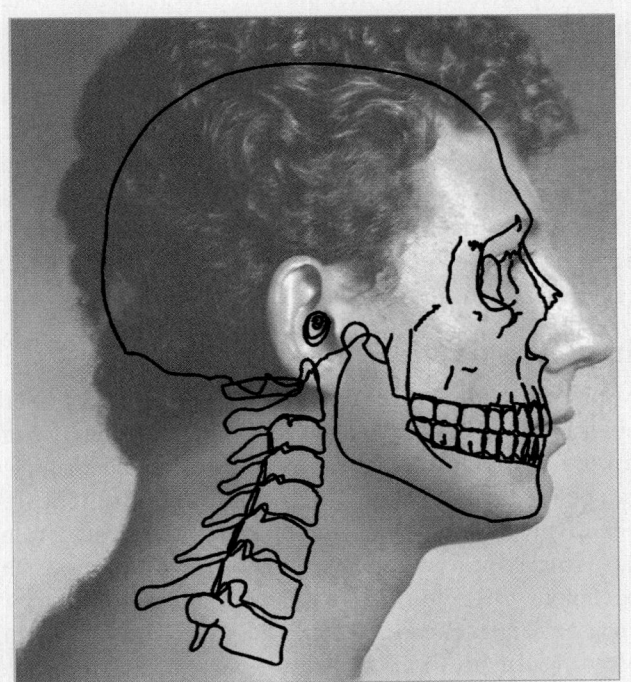

Figure 26-2 Specific landmarks of the head and neck include the pinna, the mandible, the occiput, the seventh cervical vertebra, and the temporomandibular joint.

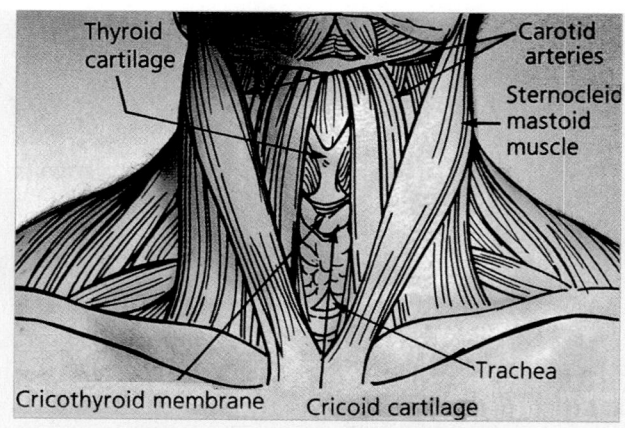

Figure 26-3 Important landmarks in the neck include the cricoid cartilage, the thyroid cartilage, the carotid arteries, the cricothyroid membrane, and the sternocleidomastoid muscles.

the upper trachea lies the thyroid gland. Unless it is enlarged, this gland is usually not palpable.

Pulsations of the carotid arteries are easily palpable in a groove 1 to 2 cm lateral to the larynx. Lying immediately adjacent to these arteries, but not palpable, are the internal jugular veins and several important nerves. Lateral to these vessels and nerves lie the sternocleidomastoid muscles. These muscles originate from the mastoid process of the cranium and insert into the medial border of each collarbone and the sternum at the base of the neck. They allow movement of the head.

A series of bony prominences lie posteriorly, in the midline of the neck. They are the spines of the cervical vertebrae. The lower cervical spines are more prominent than the upper ones. They are more easily palpable when the neck is in flexion. At the base of the neck posteriorly, the most prominent spine is the seventh cervical vertebra.

Injuries of the Face

Injuries about the face often lead to partial or complete obstruction of the upper airway. Several factors may contribute to the obstruction. Bleeding from facial injuries can be very heavy, producing large blood clots in the upper airway. These clots can lead to complete obstruction, particularly in a patient who is not fully conscious. In particular, direct injuries to the nose and mouth, the larynx, or the trachea are often the source of significant bleeding. In addition, as a result of an injury, loosened teeth or dentures may become dislodged into the throat, where they may be swallowed or aspirated. The swelling that often accompanies injury to the soft tissues in these areas can also contribute to the obstruction.

The airway may also be affected when the patient's head is turned to the side, as often is done when the patient has an altered level of consciousness or is unconscious. Other factors that interfere with normal respirations include possible injuries to the brain and/or cervical spine that may be associated with facial injuries. If the great vessels in the neck are injured, significant bleeding and pressure on the upper airway are common; these can result in airway obstruction as well.

Soft-Tissue Injuries

Soft-tissue injuries of the face and scalp are very common. The skin and underlying tissues in these areas have a rich blood supply, so bleeding from penetrating injuries may be heavy. Indeed, even minor soft-tissue wounds of the face and scalp may bleed a great deal. A blunt injury that does not break the skin may cause a break in a blood vessel wall, leading blood to collect under the skin; this is called a hematoma Figure 26-4 ▼ . Often, a flap of skin is peeled back, or avulsed, from the underlying muscle and fascia Figure 26-5 ▶ .

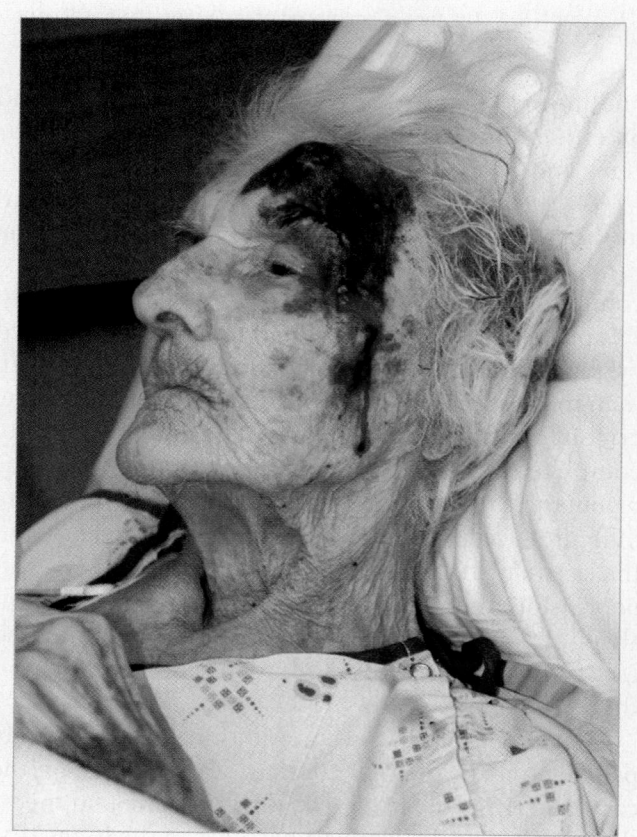

Figure 26-4 Facial hematoma.

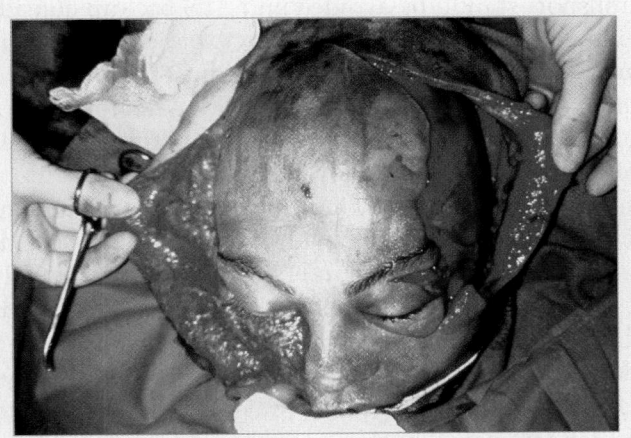

Figure 26-5 A major avulsion injury is characterized by a large flap of skin that is peeled back from the underlying muscle and tissue.

Assessment of Face and Throat Injuries

Scene Size-up

As you arrive on the scene, observe for hazards and threats to the safety of the crew, bystanders, and the patient. Patients who are conscious and supine and have oral or facial bleeding may protect their airway by coughing, projecting the blood at you. Therefore, BSI precautions require eye and oral protection. Also put several pairs of gloves in your pocket for easy access in case your gloves tear or there are multiple patients with bleeding. Because of the color of blood and how well it soaks through clothing, you can often identify patients with bleeding as you approach the scene. As you observe the scene, look for indicators and significance of the mechanism of injury (MOI). This observation helps you develop an early index of suspicion for underlying injuries in the patient who has sustained a significant MOI. Consider spinal immobilization.

Initial Assessment

General Impression

As you approach the trauma patient, you must look for important indicators to alert you to the seriousness of the patient's condition. Is the patient interacting with the environment or lying still, making no sounds? Does the patient have any apparent life threats such as significant bleeding? How is the patient's skin color? Does he or she appear to be "sick" or "not so sick?" The general impression will help you develop an index of suspicion for serious injuries and determine your sense of urgency for medical intervention. The head and face are places we frequently look at when forming our general impression. Injuries to the face and throat may be very obvious, such as bleeding and significant swelling, but may also be hidden under collars and hats. Because

You are the Provider   Part 2

You arrive on the scene and see that law enforcement personnel are present. They advise you that the scene is safe and that they have the suspect in custody. You take BSI precautions and approach your patient who is leaning against the outside wall of the bar. The man, who appears to be approximately 25 years old, is holding his hand to the right side of his neck. You observe blood flowing freely from under his hand. You ask the patient his name and what happened as you pull his hand away from his neck. He replies "Bill... I was stabbed!" Between breaths, you see that he has a puncture wound to the right side of his neck with profuse, bright red bleeding.

3. How would you treat this wound?
4. Is this patient "sick" or "not sick?"

of the likelihood of respiratory distress with these injuries, they should be recognized as early as possible.

As with any injury with life threatening bleeding, assign a team member to control the blood loss with direct pressure. Always consider the need for manual spinal immobilization and check for responsiveness using the AVPU (Awake and alert, responsive to Verbal stimulus or Pain, or Unresponsive) scale.

Airway and Breathing

Next, ensure that the patient has a clear and patent airway. If the patient is unresponsive or has a significant altered level of consciousness, consider inserting a properly sized oropharyngeal or nasopharyngeal airway. If significant face or throat injuries are identified, maintaining a patent airway is very important. You must, however, rule out the chance of basal skull fracture before you insert a nasopharyngeal airway. You must also quickly assess for adequate breathing. Palpate the chest wall for DCAP-BTLS (Deformities, Contusions, Abrasions, Punctures/Penetrations, Burns, Tenderness, Lacerations, and Swelling). If penetrating trauma is discovered, assign a team member to place an occlusive dressing on the wound. If a flail segment is discovered, assign a team member to stabilize the injury with a gloved hand or stabilize the injured chest wall with a bulky dressing. Check for clear and symmetric breath sounds and then provide high-flow oxygen, or provide assisted ventilation using a BVM device as needed, depending on the level of consciousness and your patient's breathing rate and quality. Face and throat injuries increase the need for airway and breathing maintenance, so do not hesitate to place a nonrebreathing mask or rescue breathing mask over facial injuries. The seal may not be as easy to maintain, but airway and breathing take priority over soft-tissue injuries.

Circulation

You must quickly assess the pulse rate and quality; determine the skin condition, color, and temperature; and check the capillary refill time. Significant bleeding is an immediate life threat. If the patient has obvious life-threatening bleeding, you must control it quickly.

Transport Decision

If the patient you are treating has an airway or a breathing problem or significant bleeding, you must consider quickly transporting the patient to the hospital for treatment. Stabilization and maintenance of an airway and breathing and controlling bleeding can be very difficult in patients with facial or throat injuries, so delays in

transport should be avoided and ALS backup considered if the transport time is long. A patient with signs and symptoms of internal bleeding must be transported quickly to the appropriate hospital for treatment by a physician. Internal bleeding in face and throat injuries often involves the brain or major vessels of the throat and can have a serious impact on the patient's airway. The condition of a patient with visible significant bleeding or signs of significant internal bleeding may quickly become unstable. Treatment is directed at quickly addressing life threats and providing rapid transport to the closest appropriate hospital. Signs such as tachycardia, tachypnea, low blood pressure, weak pulse, and cool, moist, pale skin are signs of hypoperfusion and imply the need for rapid transport. The patient who has significant MOI but whose condition appears stable should also be transported promptly to the closest appropriate hospital. Remember that any significant blow to the face or throat should increase your suspicion of spinal or brain injury. You should be alert to these signs and reconsider your priority and transport decision if they develop.

Focused History and Physical Exam

Focused Physical Exam Versus Rapid Trauma Assessment

After the initial assessment is complete, determine which assessment will be performed next. In the responsive patient who has an isolated injury with limited MOI, consider a focused physical exam. Focus your assessment on the isolated injury, the patient's complaint, and the body region affected. In this case, it is the face and throat. Ensure that control of the bleeding is maintained, and note the location of the injury. Assess all underlying systems. This should include neurologic, including brain and major nerves; sensory organs, including the eyes and nose; respiratory system, including mouth, nose, sinuses, and airway; and circulatory system, particularly focusing on the carotid arteries and jugular veins.

If there is significant trauma that likely affects multiple systems, start with a rapid trauma assessment looking for DCAP-BTLS to be sure that you have found all of the problems and injuries. When completed, perform a focused history and physical exam as described above for each injury. It is important in bleeding cases to not focus just on the bleeding. With significant trauma, quickly assess the entire patient from head to toe. Do not delay transport to complete a detailed physical exam.

Baseline Vital Signs

You must assess baseline vital signs to observe the changes a patient may display during treatment. A systolic blood pressure reading of less than 100 mm Hg with a weak, rapid pulse and cool, moist skin that is pale or gray should alert you to the presence of hypoperfusion in a patient who may have significant bleeding. Remember, you must be concerned with visible bleeding and unseen bleeding inside a body cavity. With facial and throat injuries, baseline information about the rate and quality of respirations and pulse is very important, as is monitoring throughout patient care.

SAMPLE History

Next, obtain a SAMPLE history from your patient. If the patient is not responsive, attempt to get the SAMPLE history from friends or family members who may be present.

Interventions

Provide complete spinal immobilization to the patient with suspected spinal injuries. Spinal injuries should be suspected any time there is significant trauma to the face or throat. Maintain an open airway, be prepared to continually suction the patient, and consider an oropharyngeal or nasopharyngeal airway. Whenever you suspect significant bleeding, provide high-flow oxygen. Oxygen and airway maintenance are important for all patients with face and throat injuries. If needed, provide assisted ventilation using a BVM device with high-flow oxygen. When significant bleeding is visible, control the bleeding. If the patient has signs of hypoperfusion, treat the patient aggressively for

shock and provide rapid transport to the appropriate hospital. Do not delay transport of a seriously injured trauma patient to complete nonlifesaving treatments in the field, such as splinting extremity fractures; instead, complete these types of treatment en route to the hospital.

Detailed Physical Exam

If the patient is in stable condition and problems identified in the initial assessment have been resolved, perform a thorough, detailed physical exam of the patient as discussed in Chapter 8. Many times, short transportation times and unstable patient conditions may make this assessment impractical.

Ongoing Assessment

Reassess areas examined during the initial assessment and the vital signs. Reassess interventions and treatment you have provided to the patient. This is particularly important in patients with facial or throat injuries because of the ease in which injuries can affect associated systems, such as the respiratory (airway and breathing), circulatory, and nervous systems. The patient's condition should be reassessed at least every 5 minutes.

Communication and Documentation

You must include a description of the MOI and the position in which you found the patient as you arrived on scene. Document the method used to remove the

You are the Provider   Part 3

You recognize that the patient is "sick," call for ALS backup, and ask your EMT-B partner to immobilize the patient's head for c-spine precautions as you apply direct pressure to the wound using a sterile dressing. However, despite the direct pressure, the bleeding persists. You also notice that the patient's level of consciousness is decreasing.

5. How should you control the bleeding?
6. What is your transport decision?

patient from the vehicle, for example, "prolonged extrication time using tools." In cases involving severe external bleeding, it is important to recognize, estimate, and report the amount of blood loss that has occurred and how rapidly or how much time has passed since the bleeding started. This is a challenge, especially if the surface is wet or absorbs fluids or the environment is dark. It is important that the personnel at the hospital to which you are transporting to know about all of the injuries involving the head and neck. Specialists may need to be called to deal with injuries to the eyes, ears, teeth, mouth, sinuses, larynx, esophagus, or large vessels. These specialists are not always in the hospital, especially during the evening or night, or in smaller hospitals, so informing emergency department personnel of all injuries involving the face and throat can save valuable time.

Emergency Care

The emergency care of soft-tissue injuries to the face and scalp is the same as treatment of soft-tissue injuries elsewhere on the body. You should assess the ABCs and care for any life threats first. Remember also to follow BSI precautions in all cases.

Your first step is to open and clear the airway. Remember that blood draining into the throat can produce vomiting and airway obstruction. Take appropriate precautions if you suspect that the patient has sustained a cervical spine injury; be sure to avoid moving the neck. Use the jaw-thrust maneuver to open the patient's airway, and then suction the mouth. Once the patient is immobilized in a cervical collar and on a backboard, you can turn the backboard to one side to allow any blood or vomitus to drain out of the mouth rather than pool in the pharynx and obstruct the airway.

Control bleeding by applying direct manual pressure with a dry, sterile dressing. Use roller gauze, wrapped around the circumference of the head, to hold a pressure dressing in place (Figure 26-6 ▶). Do not apply excessive pressure if there is a possibility of an underlying skull fracture. When an injury exposes the brain, eye, or other structures, cover the exposed parts with a moist, sterile dressing to protect them from further damage. For injuries in which the skin is not broken, apply ice locally to help control the swelling of bruised tissues.

For soft-tissue injuries around the mouth, you should always check for bleeding inside the mouth. Broken teeth and lacerations to the tongue may cause profuse bleeding and obstruction of the upper airway

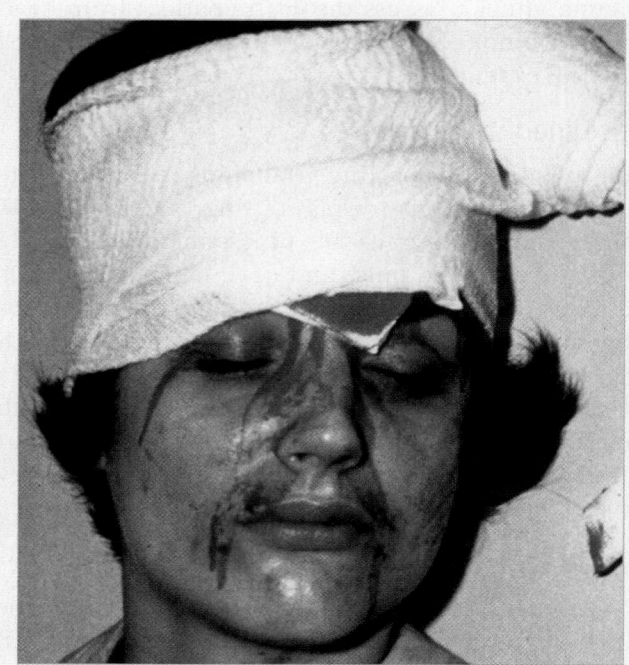

Figure 26-6 Use roller gauze, wrapped around the circumference of the head, to hold a pressure dressing in place.

(Figure 26-7 ▼). Often, the patient will swallow the blood from lacerations inside the mouth, so the hemorrhage may not be apparent. You should also inspect the inside of the mouth for bleeding and hidden injuries in patients who have sustained facial trauma. Remember that patients who swallow blood are prone to vomiting.

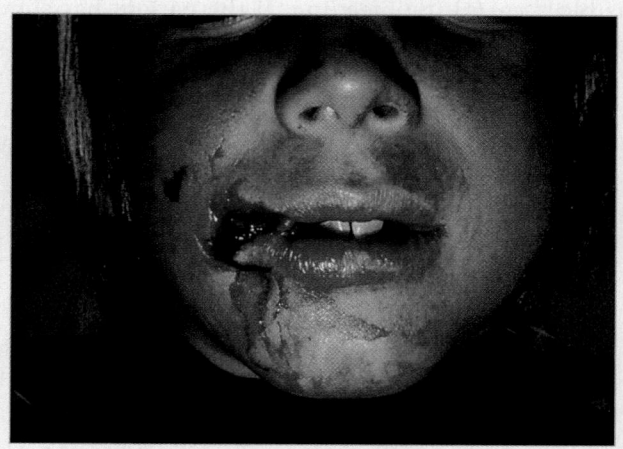

Figure 26-7 Soft-tissue injuries around the mouth can be associated with profuse bleeding inside the mouth and obstruction of the airway.

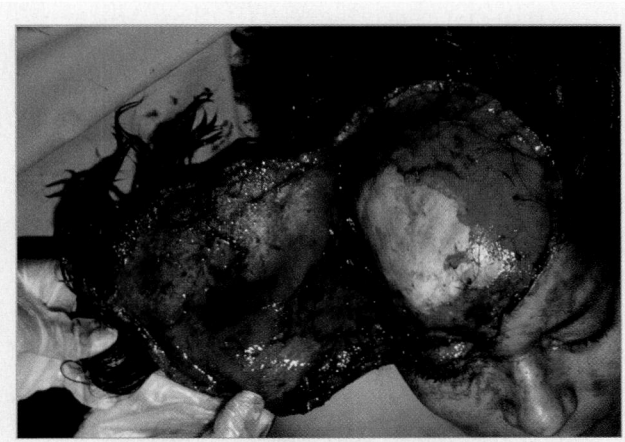

Figure 26-8 If avulsed skin is still attached, place the flap in a position that is as close to normal as possible, and hold it in place with a dry, sterile dressing.

Often, physicians will be able to graft a piece of avulsed skin back into the appropriate position. For this reason, if you find portions of avulsed skin that have become separated, you should wrap them in a sterile dressing, place them in a plastic bag, and keep them cool. Never place tissue directly on ice because freezing will destroy the tissue and make it unusable. Deliver the bag to the emergency department along with the patient. In many avulsion injuries, the skin will still be attached in a loose flap (Figure 26-8 ▲). Place the flap in a position that is as close to normal as possible, and hold it in place with a dry, sterile dressing. These steps will help to increase the patient's chances of having normal appearance restored.

Specific Injuries

Injuries of the Nose

Nosebleeds (epistaxis) are a common problem that can occur spontaneously or from trauma. One of the most common causes of nosebleeds is digital trauma (picking the nose with a finger). Nosebleeds are further classified into anterior and posterior epistaxis. Anterior nosebleeds usually originate from the area of the septum and bleed fairly slowly. These are usually self-limiting and resolve quickly. Posterior nosebleeds are usually more severe and often cause blood to drain into the patient's throat, causing nausea and vomiting. Trauma to the face and skull that results in a basilar skull fracture often will cause the posterior wall of the nasal cavity to become unstable. You should not attempt to

place a nasopharyngeal airway in a patient with a suspected basilar skull fracture or with facial injuries because insertion may permit the airway to enter through the unstable wall of the nasal cavity into the cranial vault.

The nose often takes the brunt of deliberate physical assaults and car crashes. Blunt injuries to the nose caused by a fist or a dashboard may be associated with fractures and soft-tissue injuries of the face, head injuries, and/or injuries to the cervical spine.

In assessing injuries involving the nose, it helps to picture the inside of the nose itself (Figure 26-9 ▼). The nasal cavity is divided into two sections or chambers by the nasal septum, which is made of cartilage. Within each nasal chamber, there are layers of bone called the turbinates, which are covered with a moist lining. Both chambers have a superior turbinate, a middle turbinate, and an inferior turbinate. As we breathe, the air moves through the nasal chambers and is humidified as it passes over the turbinates. Directly above the nose are the frontal sinuses and, on either side, the orbit of the eye.

All of these structures should be assessed for injury. In cases of severe injury, there may also be injury to the cervical spine. Keep in mind that cerebrospinal

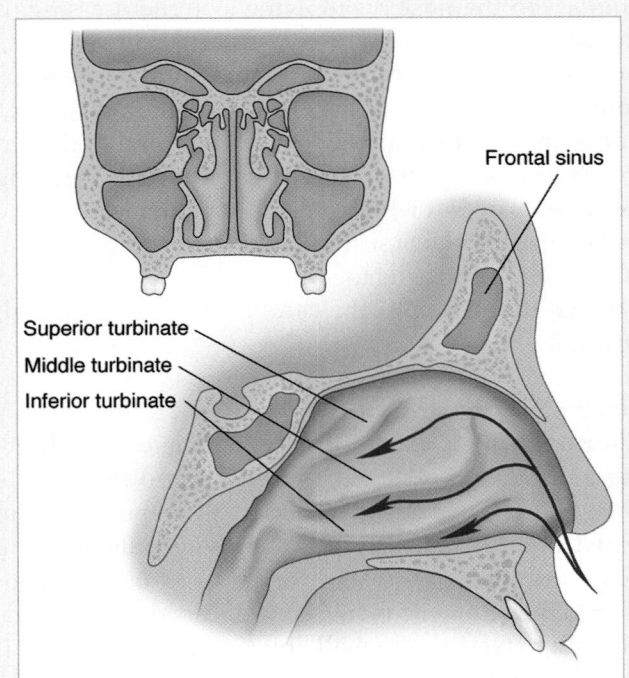

Figure 26-9 The nose has two chambers, divided by the septum. Each chamber is composed of layers of bone called turbinates. Above the nose are the frontal sinuses and, on either side, the orbit of the eye.

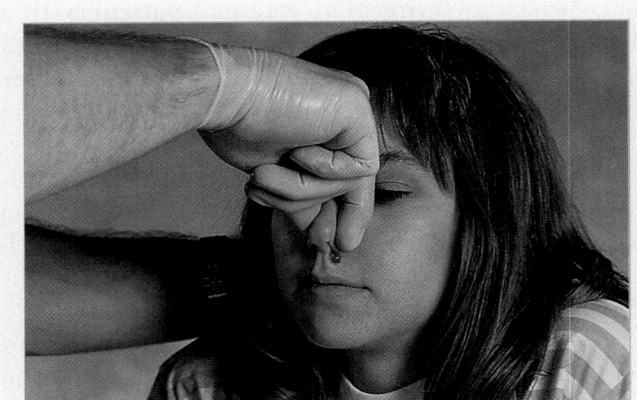

Figure 26-10 Control bleeding from the nose by pinching the nostrils together.

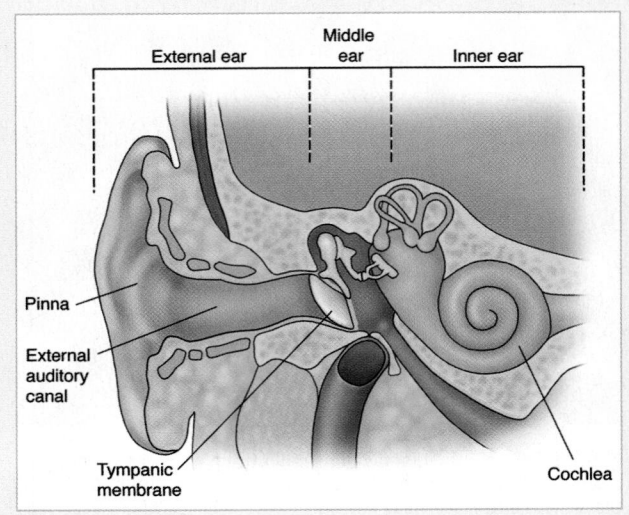

Figure 26-11 The ear has three principal parts: the external ear, composed of the pinna, external auditory canal, and tympanic membrane; the middle ear, including the hammer, anvil, and stirrup; and the inner ear, composed of bony chambers filled with fluid.

fluid (CSF) may escape down through the nose (or ears) following a fracture at the base of the skull. If blood or drainage contains CSF, a characteristic staining of the dressing will occur. This can be seen by using a piece of gauze to absorb blood that is flowing from the nose or ears. If CSF is present, the blood will be surrounded by a lighter ring of fluid. This is often called the halo test.

You can control bleeding from abrasions and lacerations to the nose by applying a sterile dressing. If the patient is bleeding heavily from the nose, this is probably due to significant trauma and you must be concerned with cervical spine injury. The patient should not be moved if the airway can be managed in the patient's present position. For a patient with bleeding from the nose but who have not experienced trauma, place the patient in a sitting position, leaning forward, and pinch the nostrils together (Figure 26-10 ▲). For a detailed discussion of the care for epistaxis, see Skill Drill 22-4.

Injuries of the Ear

The ear is a complex organ that is associated with hearing and balance. The ear is divided into three parts (Figure 26-11 ▶). The external ear is composed of the pinna, or auricle, which is the part lying outside of the head, and the external auditory canal, which leads in toward the tympanic membrane, or eardrum. The middle ear contains three small bones (the hammer, anvil, and stirrup) that move in response to sound waves hitting the tympanic membrane. This is the mechanism by which we hear and differentiate sounds. The middle ear is connected to the nasal cavity by the eustachian tube, which is the internal auditory canal. This connection permits equalization of pressure in the middle ear when external atmospheric pressure changes. The inner ear is composed of bony chambers filled with fluid. As the head moves, so does the fluid. In response, fine nerve endings within the fluid send impulses to the brain indicating the position of the head and the rate of change of position.

Ears are often injured, but they usually do not bleed very much. If local pressure does not control the bleeding, you can apply a roller dressing (Figure 26-12 ▶). First, however, you should place a soft, padded dressing between the ear and the scalp because bandaging the ear against the tender underlying scalp is extremely painful. In the case of an ear avulsion, you should wrap the avulsed part in a moist, sterile dressing and put it in a plastic bag. Often, avulsed tissue from the ear can be reattached.

The external auditory canal is a favorite place for children to place foreign bodies such as peanuts or candy. All such items should be removed by a physician in the emergency department. Never try to manipulate the foreign body because you may press it further into the auditory canal and cause permanent damage to the tympanic membrane.

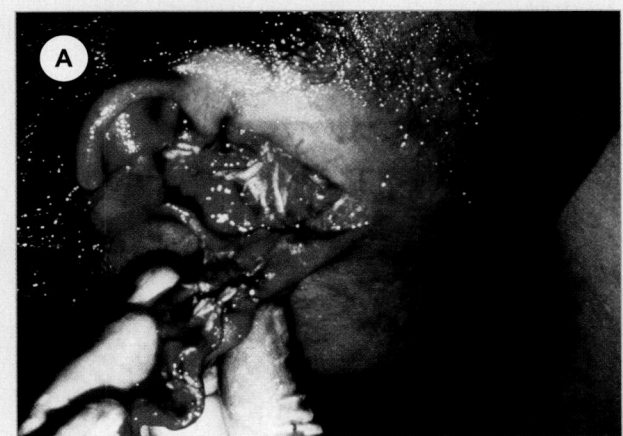

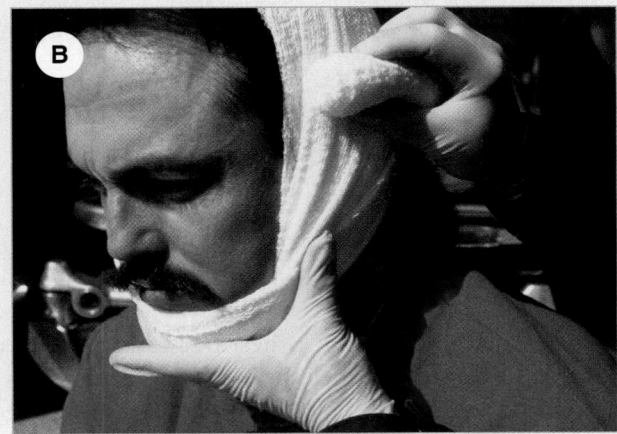

Figure 26-12 **A.** A major laceration of the ear. **B.** Proper treatment includes use of a soft, sterile pad behind the ear, between it and the scalp. Then wrap a roller gauze dressing around the head to include the entire ear.

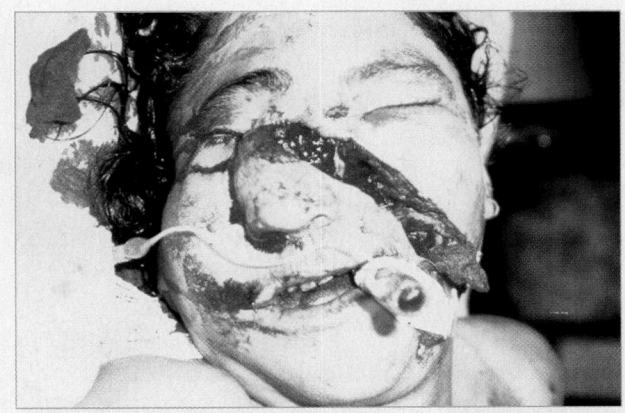

Figure 26-13 Bleeding following a crush injury to the face can be life threatening because, in addition to the external hemorrhage, blood clots in the airway can cause a complete obstruction.

Again, you should note any clear fluid coming from the ear of a severely injured patient because this may indicate a fracture at the base of the skull.

Facial Fractures

Fractures of the facial bones typically result from blunt impact. For example, the patient's head collides with a steering wheel or windshield in an automobile crash or is hit by a baseball bat or pipe in an assault. You should assume that any patient who has sustained a direct blow to the mouth or nose has a facial fracture. Other clues to the possibility of fracture include bleeding in the mouth, inability to swallow or talk, absent or loose teeth, and/or loose or movable bone fragments. Patients may also report that "it doesn't feel right" when they close their jaw, signaling an irregularity of bite.

Facial fractures alone are not acute emergencies unless there is serious bleeding; however, they are an indication of significant blunt force trauma applied to that region of the body (Figure 26-13 ▲). Serious bleeding from a facial fracture can be life threatening. In addition to external hemorrhage, there is the danger of blood clots in the upper airway, leading to obstruction of the upper airway. Fractures around the face and mouth can also produce deformity and loose bone fragments. However, plastic surgeons can repair the damage within 7 to 10 days of the injury. Be sure to remove and save loose teeth or bone fragments from the mouth; it is often possible to reimplant them (Figure 26-14 ▶). Remove dentures and dental bridges if loose to protect against airway obstruction.

Another source of potential airway obstruction is swelling, which can be extreme within the first 24 hours after injury. If you notice swelling during assessment or at any time while the patient is in your care, you should check for airway obstruction.

Dislodged teeth that are not causing airway obstruction should be transported with the patient, in a container with some of the patient's saliva or with milk, if possible.

Injuries of the Neck

The neck contains many structures that are vulnerable to injury by blunt trauma, such as from a steering wheel in a car crash, or by penetrating injury, such as a stab or gunshot wound. These structures include the upper

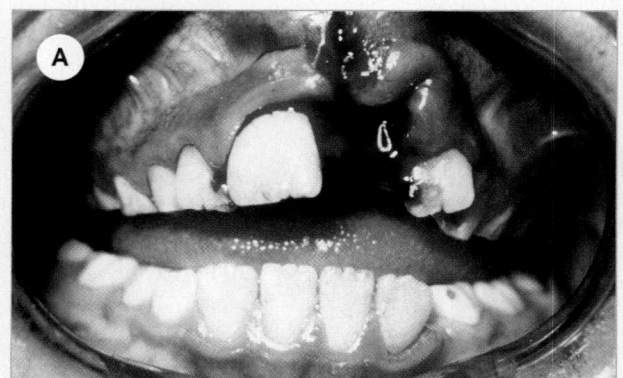

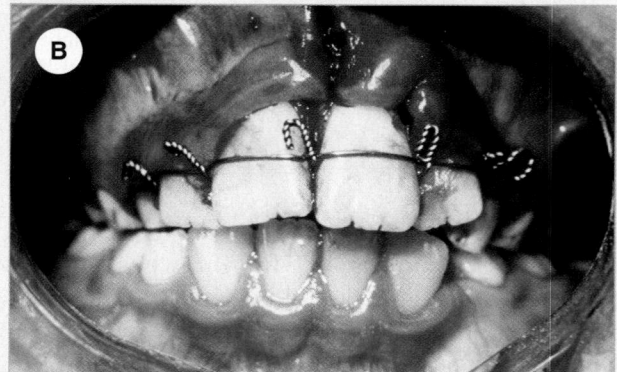

Figure 26-14   **A.** Save any lost teeth or bone fragments following an injury to the mouth. **B.** Even with traumatic loss of a tooth, the possibility of successful reimplantation is very good.

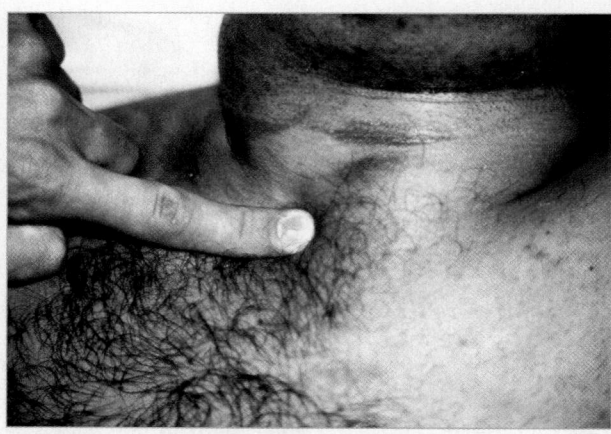

Figure 26-15   Fractures of the larynx or trachea can cause air to leak from the airway into the subcutaneous tissues. The presence of air in the soft tissues produces a crackling sensation called subcutaneous emphysema.

Blunt Injuries

Any crushing injury of the upper part of the neck is likely to involve the larynx or trachea. Examples include a collision with a steering wheel, an attempted suicide by hanging, and a clothesline injury sustained while riding a bicycle. Once the cartilages of the upper airway and larynx are fractured, they do not spring back to their normal position. Such a fracture can lead to loss of voice, severe and sometimes fatal airway obstruction, and leakage of air into the soft tissues of the neck (Figure 26-15 ▲). The presence of air in the soft tissues produces a characteristic crackling sensation called <u>subcutaneous emphysema</u>. If you feel this sensation when you palpate the neck, you should maintain the airway as best you can and provide immediate

airway, the esophagus, the carotid arteries and jugular veins, the thyroid cartilage or Adam's apple, the cricoid cartilage, and the upper trachea. Any injury to the neck is serious and should be considered life threatening until proven otherwise in the emergency department.

You are the Provider   Part 4

You want to immediately transport your patient to the trauma center, but you need to control the bleeding. You apply pressure above and below the wound on the right side of the neck and succeed in controlling the bleeding. You bandage the dressings in place by wrapping them under the patient's arm, and then you place a cervical collar on the patient to apply additional direct pressure and to maintain cervical immobilization. You give the patient high-flow oxygen, place him on a backboard, and move him to the cot. As you begin to load the patient into your ambulance, the ALS unit arrives on the scene.

7. Who should transport the patient to the hospital?
8. What were your treatment priorities with this patient?

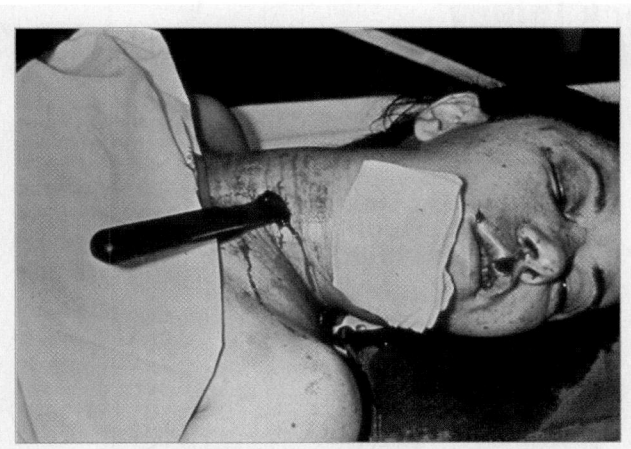

Figure 26-16 Penetrating injuries to the neck can result in profuse bleeding if a carotid artery or jugular vein is damaged.

transport. Be aware that complete airway obstruction can develop very rapidly in these patients as a result of swelling or bleeding into underlying tissues. It may be very difficult to manage the airway in patients with these injuries; some will require a surgical airway at the hospital. You should also keep in mind that an incident involving an injury to the throat may also have caused a cervical spinal injury.

Penetrating Injuries

Penetrating injuries to the neck can cause profuse bleeding from laceration of the great vessels in the neck—the carotid arteries or the jugular veins (Figure 26-16 ▲). Injuries to these large vessels may also allow air to enter the circulatory system and cause a pulmonary embolism. The airway, the esophagus, and even the spinal cord can be damaged by a penetrating injury.

Direct pressure over the bleeding site will control most neck bleeding. Follow the steps in (Skill Drill 26-1 ▶):

1. **Apply direct pressure** to the bleeding site using your gloved fingertips and a sterile occlusive dressing (**Step 1**).
2. **Secure the dressing in place** with roller gauze, adding more dressing if needed (**Step 2**).
3. **Wrap the gauze around and under the patient's shoulder.** Do not wrap the gauze around the neck to avoid possible airway and circulation problems (**Step 3**).

However, the tissues within the neck may still bleed and compress the upper airway, so you should look for signs of airway obstruction. If a vein has been opened, air may be sucked through it to the heart, a clinical situation called air embolism. A large amount of air in the right atrium and right ventricle can lead to cardiac arrest.

You might find it necessary to apply pressure both above and below the penetrating wound to control life-threatening bleeding from the carotid artery (above) and the jugular vein (below). You may also need to treat the patient for shock.

Always maintain cervical spine stabilization, and with the patient fully immobilized to a backboard, provide prompt transport. Ensure that the airway remains open en route, and apply high-flow oxygen.

Skill Drill

26-1

Controlling Bleeding From a Neck Injury

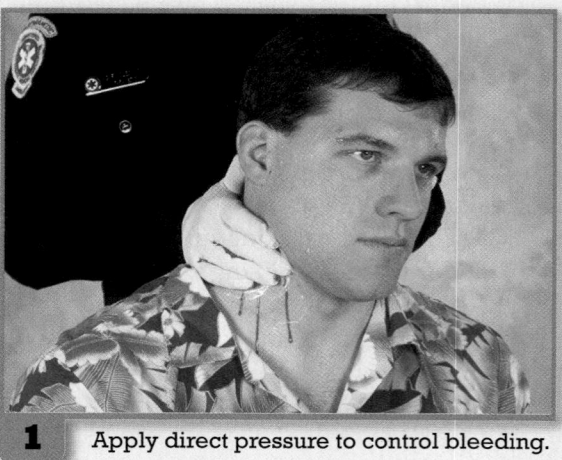

1 Apply direct pressure to control bleeding.

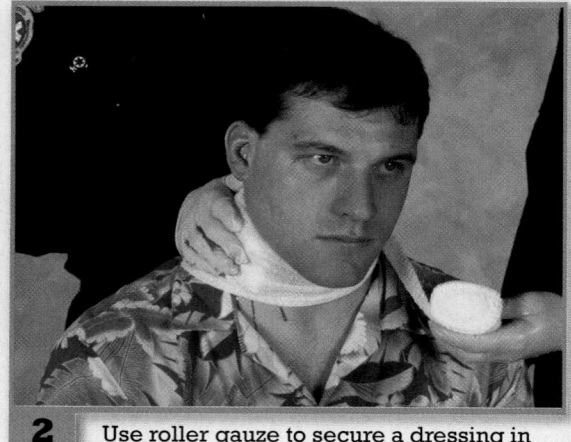

2 Use roller gauze to secure a dressing in place.

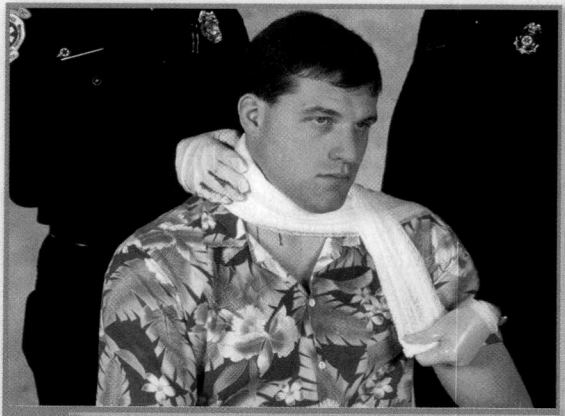

3 Wrap the bandage around and under the patient's shoulder.

You are the Provider   Summary

Penetrating wounds to the neck may require prompt intervention to save the patient's life. In addition to the possibilities of life threatening bleeding and air embolism, damage to the cervical spine, trachea, esophagus, nerves, and other vital structures can result. Attention to the patient's airway, breathing, and circulation are always of the utmost importance!

Face and Throat Injuries	
Scene Size-up	Body substance isolation precautions should include a minimum of gloves and eye protection. Ensure scene safety and determine NOI/MOI. Consider the number of patients, the need for additional help/ALS, and c-spine stabilization.
Initial Assessment	
■ General impression	Determine level of consciousness and treat any immediate threats to life. Determine priority of care based on environment and patient's chief complaint.
■ Airway	Ensure patent airway and provide spinal stabilization as necessary.
■ Breathing	Listen for abnormal breath sounds and evaluate depth and rate of the respirations. Maintain ventilations as needed. Provide high-flow oxygen at 15 L/min and inspect the chest wall, assessing for DCAP-BTLS.
■ Circulation	Evaluate pulse rate and quality; observe skin color, temperature, and condition and treat accordingly. Control life-threatening bleeding.
■ Transport decision	Prompt transport
Focused History and Physical Exam	NOTE: The order of the steps in the focused history and physical exam differs depending on whether or not the patient has a significant MOI. The order below is for a patient with a significant MOI. For a patient without a significant MOI, perform a focused trauma assessment, obtain vital signs, and obtain the history.
■ Rapid trauma assessment	Reevaluate the mechanism of injury. Perform a rapid trauma assessment, treating all life-threats immediately. Log roll and secure patient to backboard for all patients with a significant MOI from the clavicles on up.
■ Baseline vital signs	Take vital signs, noting skin color and temperature as well as patient's level of consciousness. Use pulse oximetry if available.
■ SAMPLE history	Obtain SAMPLE history. If the patient is unresponsive, attempt to obtain information from family members, friends, or bystanders.
■ Interventions	Provide complete spinal immobilization early if you suspect that your patient has spinal injuries. Control significant bleeding. Ensure high-flow oxygen, and apply blankets to conserve body temperature.
Detailed Physical Exam	Complete a detailed physical exam.
Ongoing Assessment	Repeat the initial assessment, rapid/focused assessment, and reassess interventions performed. Reassess vital signs every 5 minutes for the unstable patient, every 15 minutes for the stable patient.
■ Communication and documentation	Contact medical control with a radio report. Relay any change in level of consciousness or difficulty breathing. Describe the MOI and the position in which you found the patient. Provide an estimate of blood volume lost when severe external bleeding has occurred. Be sure to document any physician's orders and changes in the patient's condition and at what time they occurred.

Assessment and Emergency Care

Assessment and Emergency Care

NOTE: While the steps below are widely accepted, be sure to consult and follow your local protocol.

Nose Injuries	Ear Injuries	Facial Fractures	Neck Injuries
Controlling Epistaxis (See Skill Drill 22-4.) 1. Position the patient sitting, leaning forward. 2. Apply direct pressure, pinching the fleshy part of the nostrils for 15 minutes. 3. Keep the patient calm and quiet. 4. Apply ice over the nose.	1. Place a soft, padded dressing between the ear and the scalp. 2. If the ear is avulsed, wrap it in a moist, sterile dressing and place in a plastic bag. 3. Leave any foreign object within the ear for the physician to remove. 4. Note any clear fluid coming from the ear.	1. Remove and save loose teeth and bone fragments from the mouth and transport them with you. 2. Remove dentures and dental bridges to protect them against airway obstruction. 3. Maintain an open airway. 4. When transporting dislodged teeth, place them in a container of the patient's saliva or milk, if possible.	1. Apply direct pressure to control bleeding. 2. Use roller gauze to secure a dressing in place. 3. Wrap the bandage around and under the patient's shoulder.

Prep Kit

Ready for Review

- Soft-tissue injuries and fractures to the bones of the face and neck are common and vary in severity.

- Your priorities are to prevent further injury to the cervical spine and manage the airway and ventilation of the patient.

- Airway compromise may be caused by heavy bleeding into the airway, swelling in and around the structures of the airway located in the face and neck, and injuries to the central nervous system that interfere with normal respiration.

- To control heavy bleeding from soft-tissue injuries to the face and scalp, use direct pressure with a dry, sterile dressing. If brain tissue is exposed, use a moist, sterile dressing.

- Always check for bleeding inside the mouth because this may produce airway obstruction.

- Open the airway using the modified jaw-thrust maneuver (when indicated), and clear the airway in all patients with facial injuries.

- Save avulsed pieces of skin, and transport them with the patient for possible reattachment at the hospital.

- Injuries to the ear usually do not produce significant bleeding. If direct pressure fails to control bleeding, use roller gauze bandages to apply a dressing.

- Often, avulsed portions of the ear can be reattached later at the hospital; save all avulsed parts, and transport them to the hospital with the patient.

- Be alert to clear fluid draining from the ears or nose. This may indicate a basal skull fracture.

- Patients who sustain blunt trauma to the face may sustain a facial fracture. Signs of facial fracture include irregularity of bite, inability to swallow or talk, hoarseness, bleeding in the mouth, and instability of the facial bones.

- Blunt and penetrating trauma to the neck can produce life-threatening injuries. Palpate the neck for signs of subcutaneous emphysema. In patients with this sign, complete airway obstruction may develop in minutes.

- If bleeding is present from a penetrating injury, direct pressure over the site will usually control most forms of bleeding.

- Be alert to the possibility of air embolism from open neck injuries. Place an occlusive dressing over the site, and provide direct pressure.

Technology

Interactivities

Vocabulary Explorer

Anatomy Review

Web Links

Online Review Manual

www.EMTB.com

Prep Kit continued...

Vital Vocabulary

Adam's apple The firm prominence in the upper part of the larynx formed by the thyroid cartilage.

air embolism The presence of air in the veins, which can lead to cardiac arrest if it enters the heart.

avulsed Pulled or torn away.

cranium The skull.

eustachian tube A branch of the internal auditory canal that connects the middle ear to the oropharynx.

external auditory canal The ear canal; leads to the tympanic membrane.

foramen magnum The large opening at the base of the skull through which the brain connects to the spinal cord.

hematoma The collection of blood in a space, tissue, or organ due to a break in the wall of a blood vessel.

mandible The bone of the lower jaw.

mastoid process The prominent bony mass at the base of the skull about 1" posterior to the external opening of the ear.

maxillae The bones that form the upper jaw on either side of the face; they contain the upper teeth and form part of the orbit of the eye, the nasal cavity, and the palate.

occiput The most posterior portion of the skull.

pinna The external, visible part of the ear.

sternocleidomastoid muscles The muscles on either side of the neck that allow movement of the head.

subcutaneous emphysema A characteristic crackling sensation felt on palpation of the skin, caused by the presence of air in soft tissues.

temporomandibular joint (TMJ) The joint formed where the mandible and cranium meet, just in front of the ear.

tragus The small, rounded, fleshy bulge that lies immediately anterior to the ear canal.

turbinates Layers of bone within the nasal cavity.

tympanic membrane The eardrum, which lies between the external and middle ear.

Points to Ponder

You are transporting a patient involved in a motor vehicle crash. The patient was unrestrained and appears to have significant head injuries. You notice bruising over the mastoid process (Battle's sign, indicating basilar skull fracture) but cannot remember the appropriate terminology. You instead tell the emergency department there is bruising to the patient's occiput.

Is the description you provided accurate? How can you decrease confusion if you cannot remember appropriate terminology?

Issues: Importance of Knowing Human Anatomy, Accurate Communication With Hospital Personnel.

Assessment in Action

It is 12:30 AM on a Saturday night. You are dispatched to an area mall parking lot where a person has been assaulted in the head with a baseball bat. Police officers on the scene tell you that your patient is an adolescent male who was beat up by several individuals. Officers further explain that the patient is moaning and making strange grunting sounds.

You arrive to find your patient unconscious, lying on the ground with a pool of blood around his head. When he breathes in, you notice some movement of his facial bones. He is breathing about 6 times per minute and with each breath you hear gurgling sounds.

1. What is your primary concern regarding this patient?

   A. Applying a c-collar
   B. Establishing a patent airway
   C. Checking for other injuries
   D. None of the above

2. The large opening in the skull is called the:

   A. cranium.
   B. occiput.
   C. maxillae.
   D. foramen magnum.

3. The face is comprised of how many facial bones?

   A. 6
   B. 8
   C. 10
   D. 16

4. The bony mass at the base of the skull located just behind the ear is called the:

   A. zygoma.
   B. maxillae.
   C. tragus.
   D. mastoid process.

5. The head is divided into two parts called the:

   A. skull and cranium.
   B. cranium and foramen magnum.
   C. face and cranium.
   D. none of the above

6. The firm prominence in the center of the anterior surface of the neck is called the:

   A. Adam's apple.
   B. upper part of the larynx.
   C. thyroid cartilage.
   D. all of the above

Challenging Questions

7. How can facial injuries affect the patency of the airway?

8. Why is it important to determine the presence of a head injury before choosing between an oral or a nasal airway?

9. How should you care for completely avulsed skin?

10. What are common mistakes most people make when dealing with nosebleeds?

www.EMTB.com

Chest Injuries

Objectives*

Cognitive

1. Differentiate between a pneumothorax, a hemothorax, a tension pneumothorax, and a sucking chest wound. (p 784, 786)
2. Describe the emergency medical care of a patient with a flail chest. (p 787)
3. Describe the emergency medical care of a patient with a sucking chest wound. (p 785)
4. Describe the consequences of blunt injury to the heart. (p 789)
5. List the signs of pericardial tamponade. (p 789)
6. Discuss the complications that can accompany chest injuries. (p 784)

Affective

None

Psychomotor

7. Demonstrate the steps in the emergency medical care of a sucking chest wound. (p 785)

*All of the objectives in this chapter are noncurriculum objectives.

You are the Provider

You and your EMT-B partner are dispatched for an injured person. Dispatch information states you will be responding to a construction site where a worker fell on a piece of metal and has an open chest wound.

1. Why is it important to seal open chest wounds as quickly as possible?
2. What are some of the possible consequences of a blunt injury to the heart? A penetrating injury?

Chest Injuries

Chest injuries are commonly encountered by EMT-Bs. Given the location of the heart, lungs, and great blood vessels within the chest cavity, potentially serious injuries may occur. Any injury that interferes with the body's mechanics of normal breathing must be treated without delay to minimize or prevent permanent damage to tissues that depend on a continuous supply of oxygen. Another major problem with chest injuries may be internal bleeding. Blood from lacerations of the thoracic organs or major blood vessels can collect in the chest cavity, compressing the lungs or heart. This may also occur when air collects in the chest and prevents the lungs from expanding. Your ability to act quickly to care for patients with these injuries can make the difference between a successful outcome and death.

This chapter begins with a review of the anatomy of the chest and the physiology of respiration. It then describes the common signs and symptoms of chest injuries and the proper emergency medical treatment for specific injuries.

Anatomy and Physiology of the Chest

To understand and evaluate chest injuries in the prehospital setting, you must first understand the anatomy of the chest and the mechanism by which gases are exchanged during breathing. A quick review will help you understand the logic in the emergency treatment of chest injuries and the potential complications of that treatment.

A key point to remember is the difference between the concepts of ventilation and respiration. Ventilation is the body's ability to move air in and out of the chest and lung tissue. This is described later in "Mechanics of Ventilation." Any injury that affects the patient's ability to move air in and out of the chest is serious and may be life threatening. Respiration is the exchange of gases in the alveoli of the lung tissue. This is the terminal point of the pulmonary system. Oxygen must be delivered to the cells, and carbon dioxide (a waste product of cell function) must be removed from the body for proper organ system function.

The chest (thoracic cage) extends from the lower end of the neck to the diaphragm (Figure 27-1 ▼). In an individual who is lying down or who has just completed exhalation, the diaphragm may rise as high as the nipple line. Thus, a penetrating injury to the chest, such as

www.EMTB.com

Technology

Interactivities

Vocabulary Explorer

Anatomy Review

Web Links

Online Review Manual

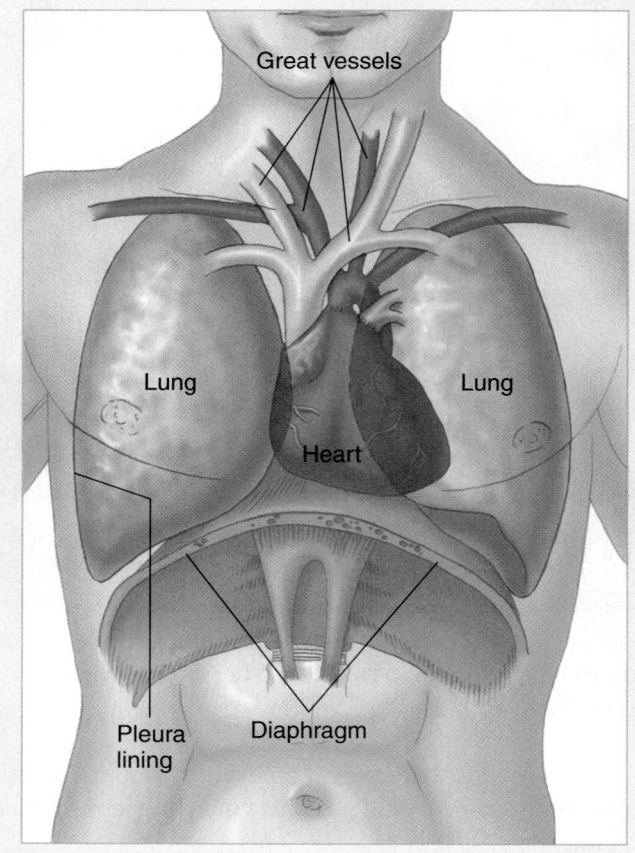

Figure 27-1 A view of the anterior aspect of the chest shows the major organs beneath the surface.

a gunshot or stab wound, may also penetrate the lung and diaphragm and injure the liver or stomach.

Each side of the chest contains lung tissue that is separated into lobes. The right lung has three lobes and the left lung has two lobes. Each of the lungs and lung cavities are covered by a thin membrane called the pleura. The inner chest wall has a lining called the parietal pleura, and the lung is covered by a lining called the visceral pleura. Between these linings is a small amount of fluid that allows the lungs to move freely against the inner chest wall as we breathe.

The contents of the chest are partially protected by the ribs, which are connected in the back to the vertebrae and in the front, through the costal cartilages, to the sternum (Figure 27-2 ▼). The trachea, which is in the middle of the neck, divides into the left and right mainstem bronchi, which supply air to the lungs. Of course, the thoracic cage also contains the heart and the great vessels: the aorta, the right and left subclavian arteries and their branches, and the superior and inferior venae cavae. The esophagus runs through the back of the chest, connecting the pharynx above with the stomach and the abdomen below. At the bottom of the chest, the diaphragm is a muscle that separates the thoracic cavity from the abdominal cavity.

Mechanics of Ventilation

When you inhale, the intercostal muscles between the ribs contract, elevating the rib cage. At the same time, the diaphragm contracts and pushes the contents of the abdomen down. The pressure inside the chest decreases, and air enters the lungs through the nose and mouth. When you exhale, the intercostal muscles and diaphragm relax, and the tissues move back to their normal positions, allowing air to be exhaled. Note that the nerves supplying the diaphragm (the phrenic nerves) exit the spinal cord at C3, C4, and C5. A patient whose spinal cord is injured below the C5 level will lose the power to move his or her intercostal muscles, but the diaphragm will still contract. The patient will still be able to breathe because the phrenic nerves remain intact. Patients with spinal cord injuries at C3 or above can lose their ability to breathe entirely (Figure 27-3 ▼).

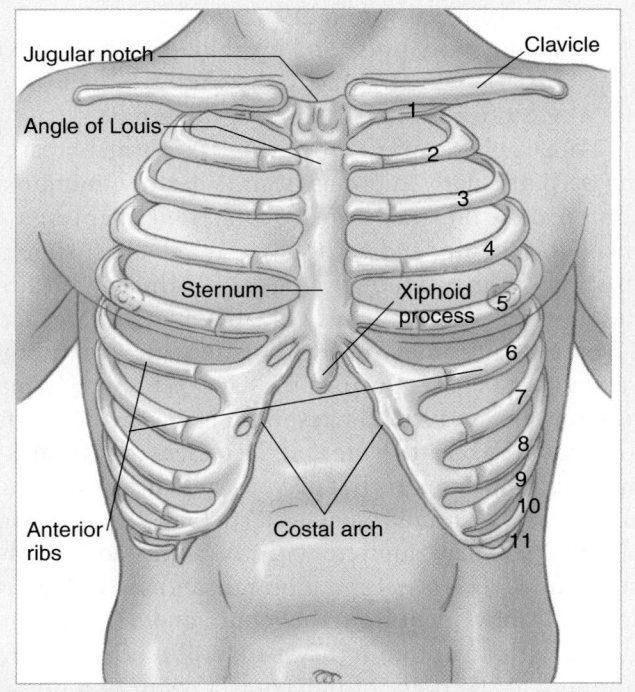

Figure 27-2 The organs within the chest are protected by the ribs, which are connected in back to the vertebrae and in the front, through the costal cartilages, to the sternum.

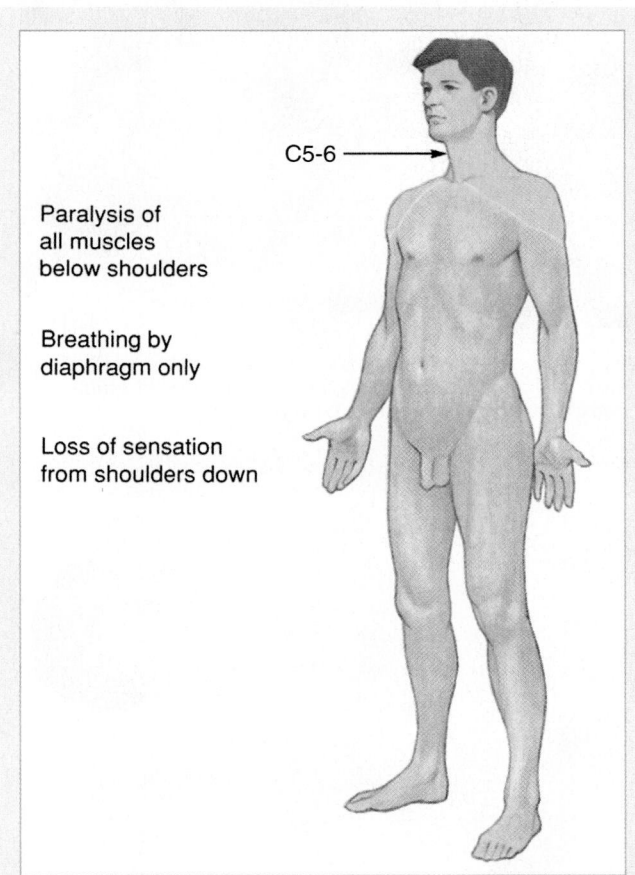

Figure 27-3 A patient who sustains a spinal cord injury below the level of C5 and is paralyzed can still breathe spontaneously because the phrenic nerves originate at the C3, C4, and C5 levels.

Injuries of the Chest

There are two basic types of chest injuries: open and closed. As the name implies, a <u>closed chest injury</u> is one in which the skin is not broken. This type of injury is generally caused by blunt trauma, such as when a driver strikes a steering wheel in a motor vehicle crash, is struck by a falling object, or is struck in the chest by some object during a fight or physical assault (Figure 27-4 ▼). In an <u>open chest injury</u>, the chest wall itself is penetrated by some object such as a knife, a bullet, a piece of metal, or the broken end of a fractured rib (Figure 27-5 ▼).

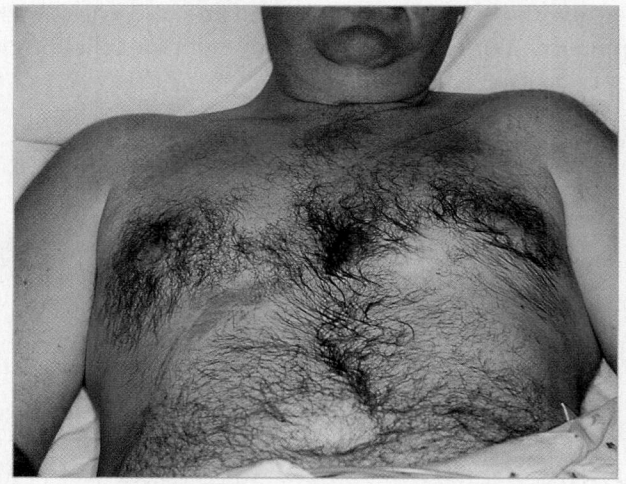

Figure 27-4 Closed injuries usually result from blunt trauma, such as when a patient strikes the steering wheel in a motor vehicle crash or is struck by a falling object.

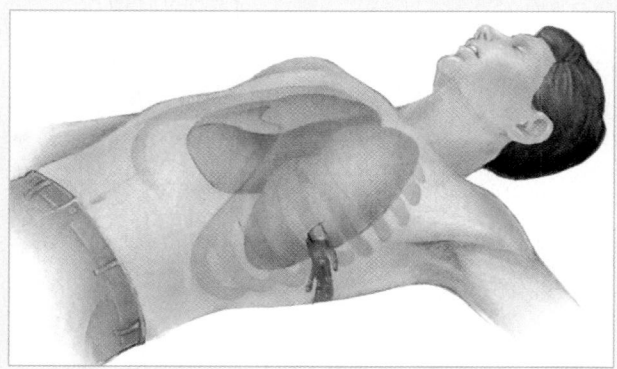

Figure 27-5 Open injuries occur when the chest wall is penetrated by some type of object or the broken end of a fractured rib.

In blunt trauma, a blow to the chest may fracture the ribs, the sternum, or whole areas of the chest wall, bruise the lungs and the heart, and even damage the aorta. Almost one third of people who are killed immediately in car crashes die as a result of traumatic rupture of the aorta. Although the skin and chest wall are not penetrated in a closed injury, the contents of the chest may be lacerated by broken ribs. Damage to the chest wall structures may result in decreased ability of patients to ventilate on their own. Also, vital organs can actually be torn from their attachment in the chest cavity without any break in the skin; this condition can cause serious and life-threatening bleeding that is unseen outside the body.

Signs and Symptoms

Important signs and symptoms of chest injury include the following:

- Pain at the site of injury
- Pain localized at the site of injury that is aggravated by or increased with breathing
- Bruising to the chest wall
- Crepitus with palpation of the chest
- Any penetrating injury to the chest
- Dyspnea (difficulty breathing, shortness of breath)
- Hemoptysis (coughing up blood)
- Failure of one or both sides of the chest to expand normally with inspiration
- Rapid, weak pulse and low blood pressure after experiencing trauma to the chest
- Cyanosis around the lips or fingernails

After a chest injury, any change in normal breathing is a particularly important sign. A healthy, uninjured adult usually breathes from 12 to 20 times per minute without difficulty and without pain. The chest should rise and fall in a symmetric pattern with each breath. Respirations of fewer than 12 breaths/min or more than 20 breaths/min may indicate inadequate breathing. Patients with chest injuries often have <u>tachypnea</u> (rapid respirations) and shallow respirations because it hurts to take a deep breath. Note that the patient may be making breathing attempts but may not actually be moving air. Chest wall trauma such as a sucking chest wound or flail chest may interfere with the ability to actually move air. Check the respiration rate and see if there is actual air movement from the mouth and/or nose.

As with any other injury, pain and tenderness are common at the point of impact as a result of a bruise or fracture. Pain is usually aggravated by the normal process of breathing. Irritation of or damage to the pleural surfaces causes a characteristic sharp or sticking pain with each breath when these normally smooth surfaces slide on one another. This sharp pain is called *pleuritic pain*, or *pleurisy*.

In an injured patient, dyspnea has many causes, including airway obstruction, damage to the chest wall, improper chest expansion due to the loss of normal control of breathing, or lung compression because of accumulated blood or air in the chest cavity. Dyspnea in an injured patient indicates potential compromise of lung function; prompt, vigorous support of oxygenation and ventilation with prompt transport are required.

Hemoptysis, the spitting or coughing up of blood, usually indicates that the lung itself or the air passages have been damaged. With a laceration of the lung tissue, blood can enter the bronchial passages and is coughed up as the patient tries to clear the airway.

A rapid, weak pulse and low blood pressure are the principal signs of hypovolemic shock, which can result from extensive bleeding from lacerated structures within the chest cavity, where the great vessels and heart are located. Shock following a chest injury may also result from insufficient oxygenation of the blood by the poorly functioning lungs.

Cyanosis in a patient with a chest injury is a sign of inadequate respiration. The classic blue or ashen gray appearance around the lips and fingernails indicates that blood is not being oxygenated sufficiently. Patients with cyanosis are unable to provide a sufficient supply of oxygen to the blood through the lungs and require immediate ventilation and oxygenation.

Many of these signs and symptoms occur simultaneously. When any one of them develops as a result of a chest injury, the patient requires prompt hospital care. Remember that the principal reason for concern about a patient who has a chest injury is that his or her body has no means of storing oxygen; it is supplied and used continuously, even during sleep. Any interruption in this supply can be rapidly lethal and must be treated aggressively.

Assessment of Chest Injuries

Scene Size-up

As you arrive on the scene, observe for hazards and threats to the safety of the crew, bystanders, and the patient. Consider the possibility that the area where the patient is located may be a crime scene; disturb potential evidence as little as possible. Ensure that you and your crew use proper body substance isolation precautions, and put on a minimum of gloves and eye protection. Put several pairs of gloves in your pocket for easy access in case your gloves tear or there are multiple patients with bleeding. Because of the color of blood and how well it soaks through clothing, you can often identify patients with bleeding as you approach the scene. As you observe the scene, look for indicators and significance of the MOI. This helps you develop an early index of suspicion for underlying injuries in the patient who has sustained a significant MOI. Chest injuries are common in motor vehicle crashes, falls, and assaults. Knowing the dispatch information and visual inspection of the scene will often increase your suspicion of chest injury. Consider spinal immobilization.

You are the Provider   Part 2

As you arrive on the scene, you and your partner quickly determine that the scene is safe and no other resources are required. Your partner initiates manual spinal immobilization, and you determine that the patient is responsive with a patent airway. Your patient complains, " I can't catch my breath. It hurts." While performing the initial assessment, you find a penetrating injury on the right anterior portion of the chest just below the nipple. You see a small amount of bleeding and notice the blood bubbles as your patient breathes.

3. What is the next step you should take?
4. Does this patient need supplemental oxygen? If so, how much and which delivery device should you use?

Ensure that the police are on scene at incidents involving violence, such as assaults or gunshot wounds. Begin the encounter with scene safety as the highest priority. If you determine that power company, fire department, or ALS units are needed, call for them early.

Initial Assessment

General Impression

During your initial assessment, you must quickly evaluate the patient's ABCs and treat potential life threats. The quickest way to identify these life threats is to begin with the chief complaint. In responsive patients, this may be what they tell you is wrong. Note not only what they say, but also how they say it. Difficulty speaking may indicate several problems, and chest injury is an important one. An unresponsive patient will tell you his or her chief complaint by obvious injuries and the appearance of blood or difficulty breathing. Look for cyanosis, irregular breathing, or chest rise and fall on only one side to indicate chest injuries. If no obvious problems present themselves, begin looking for them by focusing on the ABCs. The general impression will help you develop an index of suspicion for serious injuries and determine your sense of urgency for medical intervention. A good question to ask yourself is "How sick is this patient?" Patients with significant chest injuries will "look" sick and are often frightened or anxious.

Airway and Breathing

Next, ensure that the patient has a clear and patent airway. How you assess and manage the airway depends a great deal on whether you suspect a spinal injury. A significant number of patients with traumatic chest injuries also have spinal injuries, and proper precautions should be taken. Be suspicious, and protect the spine early in your care, even if your assessment later confirms that there is no spinal injury. Once you have determined the patient has a patent airway, determine whether breathing is present and adequate. With chest injuries, begin by inspecting for DCAP-BTLS (Deformities, Contusions, Abrasions, Punctures/Penetrations, Burns, Tenderness, Lacerations, and Swelling), then listen with a stethoscope to each side of the chest. Absent or decreased breath sounds on one side usually indicate significant damage to a lung, preventing it from expanding properly. Be alert to the pattern of symmetric rise and fall of the patient's chest wall. If the chest wall does not expand on each side when the patient inhales, the chest muscles may have lost their ability to work appropriately. Loss of muscle function may be the result of a direct injury to the chest wall, or it may be related to an injury of the nerves that control those muscles. Check also for paradoxical motion, an abnormality associated with multiple fractured ribs, in which one segment of the chest wall moves opposite the remainder of the chest, that is, out with expiration and in with inspiration. If you determine the patient has paradoxical movement of the chest wall, or penetrating trauma, address this life threat at once. These conditions may interfere with the normal mechanics of breathing and can cause the patient's condition to worsen quickly. Apply an occlusive dressing to all penetrating injuries to the chest, and stabilize paradoxical motion with a large bulky dressing and 2″ tape. Apply oxygen with a nonrebreathing mask at 15 L/min. Provide positive-pressure ventilations with 100% oxygen if breathing is inadequate based on the patient's level of consciousness and breathing rate and quality.

Circulation

Assess the patient's pulse. Determine whether it is present and adequate. If it is too fast or too slow or if the skin is pale, cool, or clammy, consider your patient to be in shock and treat aggressively to eliminate the cause and support the patient's circulatory system. External bleeding may not be obvious or may not be significant. However, bleeding inside the chest can be significant and, as illustrated earlier, can be a quick cause of death. Control bleeding with direct pressure and a bulky trauma dressing.

Transport Decision

Priority patients are considered those who have a problem with their airway, breathing, and/or circulation. Sometimes the priority is obvious, and the decision to transport quickly is also easy. At other times, what is happening outside may not provide obvious clues to the seriousness of what is happening inside. Pay attention to subtle clues such as skin signs, level of consciousness, or a sense of impending doom in the patient. These symptoms are not as grand as a large gash across the chest or air being sucked into the chest; however, they can be equally important indicators of a life-threatening condition. When you find signs of poor perfusion or inadequate breathing, transport quickly. A delay on the scene to perform a lengthy assessment will reduce the chances of survival for your patient. With chest injuries, when in doubt, transport rapidly to a hospital.

Rapid Physical Exam Versus Focused Physical Exam

After the initial assessment is complete, determine which physical exam will be performed next in the focused history and physical exam step of your assessment: a rapid physical exam or a focused physical exam. In the patient who has an isolated injury to the chest with limited MOI, such as in a stabbing, consider a focused physical exam. Focus your assessment on the isolated injury, the patient's complaint, and the body region affected. Ensure that wounds are identified and control of the bleeding has been established. Note the location and extent of the injury. Assess all underlying systems. Examine the anterior and posterior aspects of the chest wall, and be alert to changes in the patient's ability to maintain adequate respirations.

If there is significant trauma (such as a blunt trauma or gunshot wound) likely affecting multiple systems, start with a rapid physical assessment looking for DCAP-BTLS to determine the nature and extent of thoracic injury. Inspection for **D**eformities, such as asymmetry of the left and right sides of the chest or shoulder girdle, may reveal the presence of multiple rib fractures, crush injuries, or significant chest wall injury. Identification of discrete areas of **C**ontusion or **A**brasion may pinpoint a specific point of impact. The presence of **P**uncture wounds or other penetrating injuries indicates a possible open chest injury that should be managed accordingly. Be alert for associated **B**urns, which may alter respiratory mechanics. Palpate for **T**enderness to localize the injury and the presence of fractures. Look for **L**acerations and local **S**welling. Application of this systematic approach to patient assessment minimizes the chance of missing significant injury.

It is important in chest injury cases to not focus just on the chest wound. With significant trauma, you should quickly assess the entire patient from head to toe.

Baseline Vital Signs

Once you have stabilized airway, breathing, and circulation problems and have checked the patient from head to toe to identify injuries, obtain a baseline set of vital signs. This should include assessment of pulse, respirations, blood pressure, skin condition, and pupils. Each of these is a sign indicating how your patient is tolerating the injuries. It is a window to the functioning of the vital organs. This baseline set of vital signs will be used to evaluate changes in the patient's condition.

SAMPLE History

Many chest injury patients may be considered high priority and be rapidly transported to a hospital. Obtaining a patient's medical history may not seem very important. Even when faced with managing priority issues like airway, breathing, or circulatory problems, you cannot omit the SAMPLE history. A basic evaluation of allergies, medications, pertinent medical problems, and last oral intake should be completed. Most signs and symptoms have been identified in the initial assessment and the rapid or focused physical exam. The events leading to the emergency may have been identified in the chief complaint of the initial assessment. A SAMPLE history can be obtained quickly in most situations and can certainly be obtained while accomplishing other tasks. However, if the patient loses consciousness, it will no longer be possible to obtain the information.

Interventions

Provide complete spinal immobilization of the patient with suspected spinal injuries. Maintain an open airway, be prepared to suction the patient, and consider an oropharyngeal or nasopharyngeal airway. Whenever you suspect significant bleeding, provide high-flow oxygen. If needed, provide assisted ventilation using a BVM device with high-flow oxygen. If significant bleeding is visible, you must control the bleeding. If you find penetrating trauma to the chest wall, place an occlusive dressing over the wound; if you find a flail segment, stabilize the segment with a bulky dressing. If the patient has signs of hypoperfusion, treat aggressively for shock and provide rapid transport to the appropriate hospital. Do not delay transport of a seriously injured trauma patient to complete nonlifesaving treatments such as splinting extremity fractures; instead, complete these types of treatment en route to the hospital.

Patients with a significant MOI to the chest have a high likelihood of other injuries. During transit, perform a detailed physical exam as discussed in Chapter 8. Your rapid physical exam identified injuries that needed immediate attention and helped you prepare for packaging and transportation. The detailed physical exam can now help to determine all the injuries and the extent of those injuries. If the patient has nonsignificant trauma and no persistent problems are present after the initial assessment, a detailed physical exam may not be necessary. Many times, short transportation times and unstable patient conditions may make this assessment impractical.

Ongoing Assessment

The ongoing assessment identifies how your patient's condition is changing. It should focus on reassessing the patient's airway, breathing, pulse, perfusion, and bleeding. Has breathing improved now that the sucking chest wound is sealed? Or has it become more difficult and associated with the trachea deviating to one side? Will one side of the occlusive dressing need to be released? Is the splint to the flail chest providing stability? Or is it too loose? Other interventions should also be assessed to determine if they are effective. For example, are pulse oximeter values rising now that the patient is receiving oxygen? Vital signs need to be reassessed and compared with the baseline vital signs. Do a drop in blood pressure and tachycardia indicate increasing tension in the chest? Many chest injuries will worsen during transport to the hospital because of the seriousness of the injuries. An astute reassessment will help identify worsening conditions in a timely manner so that they can be addressed.

Communication and Documentation

Communicating with hospital staff early when your patient has a significant MOI to the chest can help them be prepared with appropriate equipment and personnel when you arrive. If a penetrating injury is present, describe it in your report, along with what you have done to care for it. If a flail segment is present, hospital staff may be able to offer assistance on how to manage it. Your documentation should be complete and thorough. Describe all injuries and the treatment given. Remember, your documentation is your legal record of what happened.

Complications of Chest Injuries

Pneumothorax

In any chest injury, damage to the heart, lungs, great vessels, and other organs in the chest can be complicated by the accumulation of air in the pleural space. This is a dangerous condition called a pneumothorax (commonly called a collapsed lung). In this condition, air enters through a hole in the chest wall or the surface of the lung as the patient attempts to breathe, causing the lung on that side to collapse (Figure 27-6 ▶). As a result, any blood that passes through the collapsed portion of the lung is not oxygenated, and hypoxia can develop. Depending on the size of the hole and the rate at which air fills the cavity, the lung may collapse in a few seconds or a few hours. In the uncommon situation when the hole is in the chest wall, you can actually hear a sucking sound as the patient inhales and the sound of rushing air as he or she exhales. For this reason, an open or penetrating wound to the chest wall is often called a sucking chest wound (Figure 27-7 ▶).

This type of open pneumothorax is a true emergency requiring immediate emergency medical care and transport. Initial emergency care, after clearing and maintaining the airway and then providing oxygen, is

You are the Provider   Part 3

You immediately place an occlusive dressing over the wound and begin delivery of high-flow oxygen to the patient. You find no other problems in the initial assessment, and you quickly begin the rapid trauma assessment. No other problems are found, and you and your partner secure the patient to a long backboard and prepare him for transport to the hospital.

5. What would be an important sign to assess in this patient?
6. On the basis of your findings so far, what internal injury do you suspect your patient has?

to rapidly seal the open wound with a sterile <u>occlusive dressing</u> (Figure 27-8 ▶). The purpose of the dressing is to seal the wound and prevent air from being sucked into the chest through the wound. Several sterile materials, including Vaseline gauze or aluminum foil, may be used to seal the wound. Use a dressing that is large enough so that it is not pulled or sucked into the chest cavity. Depending on your local protocol, you may tape the dressing down on all four sides, or you may create a <u>flutter valve</u>, a one-way valve that allows air to leave the chest cavity but not return, by taping only three sides of the dressing.

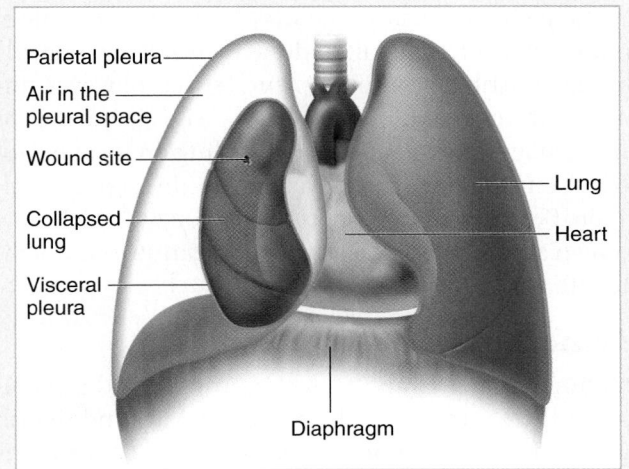

Figure 27-6 Pneumothorax occurs when air leaks into the space between the pleural surfaces from an opening in the chest wall or the surface of the lung. The lung collapses as air fills the pleural space.

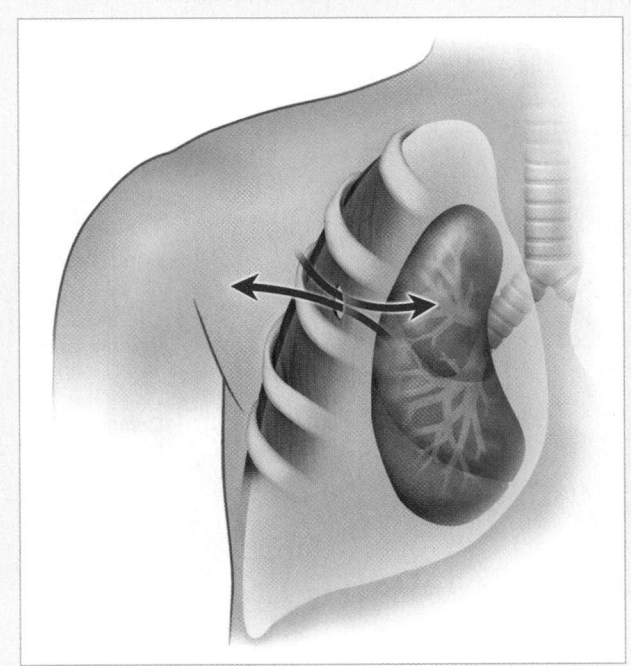

Figure 27-7 With a sucking chest wound, air passes from the outside into the pleural space and back out with each breath, creating a sucking sound.

Documentation Tips

When you use an occlusive dressing to seal an open chest wound, record the type of material used, whether three or four sides were sealed, and any changes noted afterward: skin color, vital signs, breath sounds, and particularly the patient's level of anxiety.

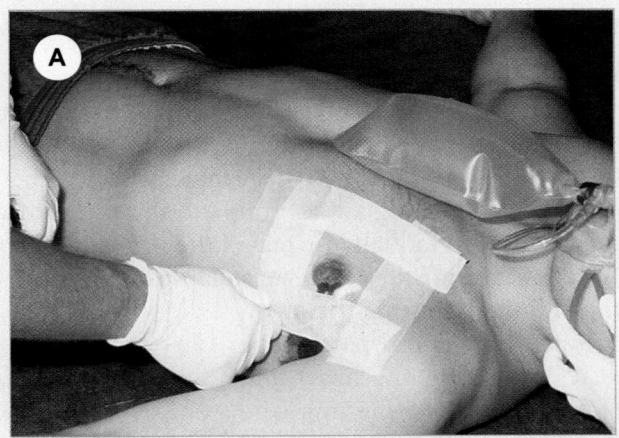

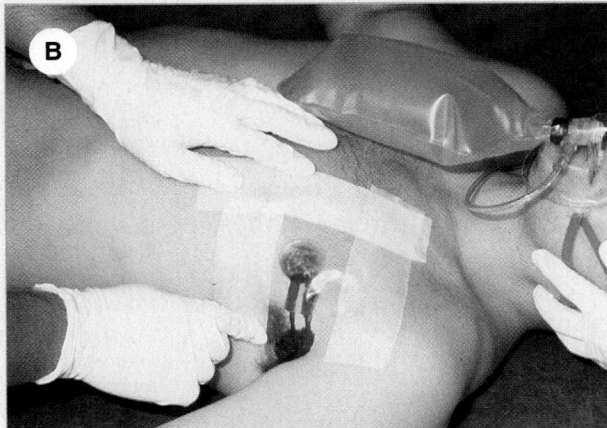

Figure 27-8 A sucking chest wound can be sealed with a large airtight dressing that seals all four sides **(A)** or seals three sides with the fourth left open as a flutter valve **(B)**. Your local protocol will dictate the way you are to care for this injury.

Spontaneous Pneumothorax

Some individuals are born with or develop weak areas on the surface of the lungs. This weakened area of the lung is called a "bleb." Occasionally, such a weak area will rupture spontaneously, allowing air to leak into the pleural space. Usually, this event, called spontaneous pneumothorax, is not related to any major injury but simply happens with normal breathing or may occur during times of strenuous physical activity such as exercise. The patient experiences sudden sharp chest pain and increasing difficulty breathing. A portion of the affected lung collapses, losing its ability to ventilate normally. The amount of pneumothorax that develops varies, as does the amount of respiratory distress the patient experiences.

You should suspect a spontaneous pneumothorax in a patient who experiences sudden chest pain and shortness of breath without a specific known cause. The prehospital treatment that you can provide for this type of pneumothorax is to administer oxygen and transport.

Tension Pneumothorax

A potential complication that may develop following chest injuries with pneumothorax is a tension pneumothorax (Figure 27-9 ▶). This can occur when there is significant ongoing air accumulation in the pleural space. This air gradually increases the pressure in the chest, first causing the complete collapse of the affected lung and then pushing the mediastinum (the central part of the chest containing the heart and great vessels) into the opposite pleural cavity. This prevents blood from returning through the venae cavae to the heart and can cause shock and cardiac arrest.

If signs and symptoms of a tension pneumothorax develop after sealing an open chest wound, you should partly remove the dressing to relieve the tension. As you do so, you may hear a rush of air out of the chest cavity, although this does not occur in all cases.

Tension pneumothorax occurs more commonly as a result of closed, blunt injury to the chest in which a fractured rib lacerates a lung or bronchus. Only very rarely does a tension pneumothorax arise spontaneously.

The common signs and symptoms of tension pneumothorax include increasing respiratory distress, distended neck veins, deviation of the trachea to the side of the chest opposite the tension pneumothorax, tachycardia, low blood pressure, cyanosis, and decreased breath sounds on the side of the pneumothorax.

Relieving a tension pneumothorax due to a blunt trauma in a patient is often done by inserting a needle through the rib cage into the pleural space; however, this procedure must be performed by ALS personnel or emergency department staff depending on local protocols. A tension pneumothorax is a life-threatening condition. Be prepared to support ventilation with high-flow oxygen and request ALS support or transport immediately to the closest hospital.

Hemothorax

In blunt and penetrating chest injuries, blood can collect in the pleural space from bleeding around the rib cage, or from a lung or great vessel. This condition is called a hemothorax (Figure 27-10 ▶). You should suspect a hemothorax if the patient has signs and symptoms of shock or decreased breath sounds on the affected side, an indication that the lung is being compressed by the blood. The presence of air and blood in the pleural space is known as a hemopneumothorax.

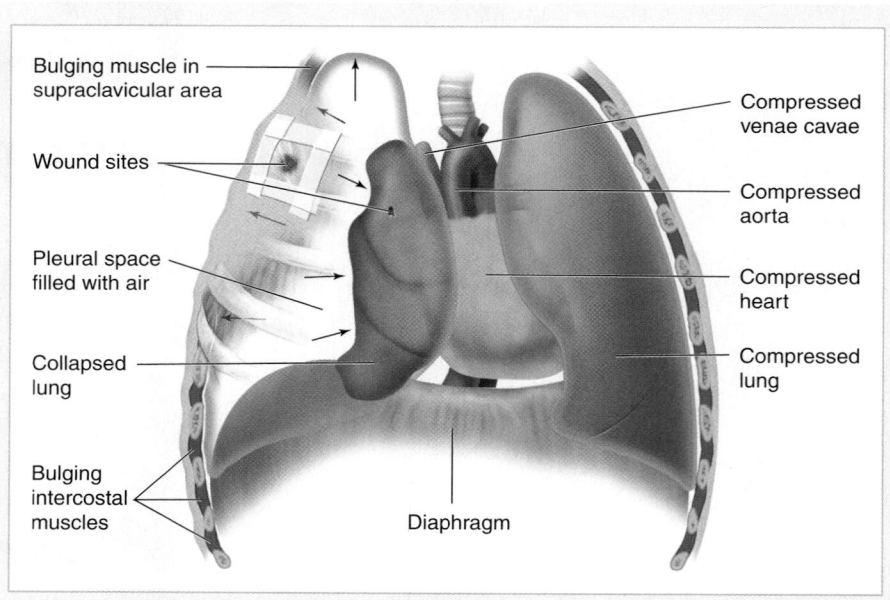

Figure 27-9 A tension pneumothorax can develop if a penetrating chest wound is bandaged tightly and air from a damaged lung cannot escape. The air then accumulates in the pleural space, eventually causing compression of the heart and great vessels.

Rib Fractures

Rib fractures are very common, particularly in older people, whose bones are brittle. Because the upper four ribs are well protected by the bony girdle of the clavicle and scapula, a fracture of one of these upper ribs is a sign of a very substantial MOI.

Be aware that a fractured rib that penetrates into the pleural space may lacerate the surface of the lung, causing a pneumothorax, a tension pneumothorax, a hemothorax, or a hemopneumothorax. One sign of this development can be a crackly feeling to the skin in the area (also called *crepitus or subcutaneous emphysema*), which indicates that air escaping from a lacerated lung is leaking into the chest wall. Be sure to relay this finding to hospital personnel.

Patients with one or more cracked ribs will report localized tenderness and pain when breathing. The pain is the result of broken ends of the fracture rubbing against each other with each inspiration and expiration. Patients will tend to avoid taking deep breaths and their breathing will be rapid and shallow instead. They will often hold the affected portion of the rib cage in an effort to minimize the discomfort. These patients should receive supplemental oxygen during assessment and transport.

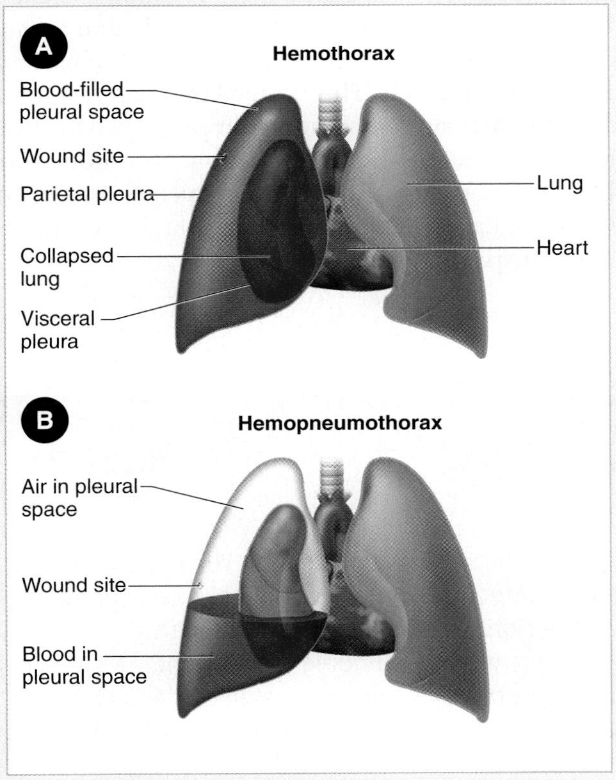

Flail Chest

Ribs may be fractured in more than one place. If two or more ribs are fractured in two or more places or if the sternum is fractured along with several ribs, a segment of chest wall may be detached from the rest of the thoracic cage (Figure 27-11 ▼). This condition is known as <u>flail chest</u>. In what is called <u>paradoxical motion</u>, the detached portion of the chest wall moves opposite of normal: in instead of out during inhalation, out instead of in during exhalation. This occurs because of negative pressure that has built up in the thorax. Breathing with a flail chest can be painful and ineffective, and hypoxemia easily results. A flail segment seriously interferes with the body's normal mechanics of ventilation and must be addressed quickly.

Your treatment of a patient with a flail chest should include maintaining the airway, providing respiratory support if necessary, giving supplemental oxygen, and performing ongoing assessments for possible pneumothorax or other respiratory complications. Treatment may also include positive pressure ventilation with a BVM device.

Figure 27-10 A. A hemothorax is a collection of blood in the pleural space produced by bleeding within the chest. **B.** When both blood and air are present, the condition is a hemopneumothorax.

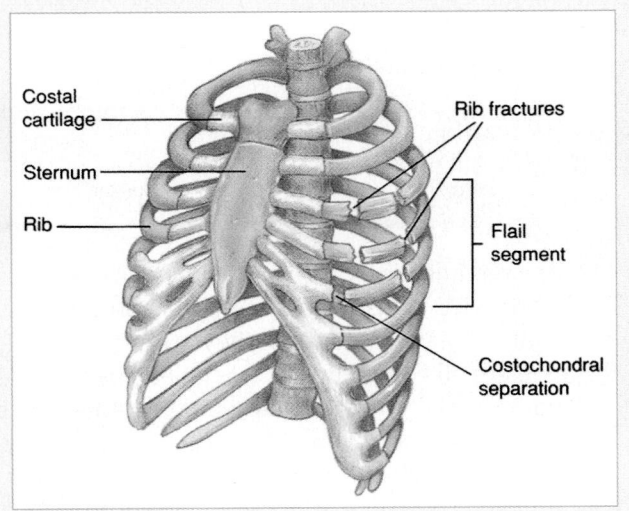

Figure 27-11 When three or more adjacent ribs are fractured in two or more places, a flail chest results. A flail segment will move paradoxically when the patient breathes.

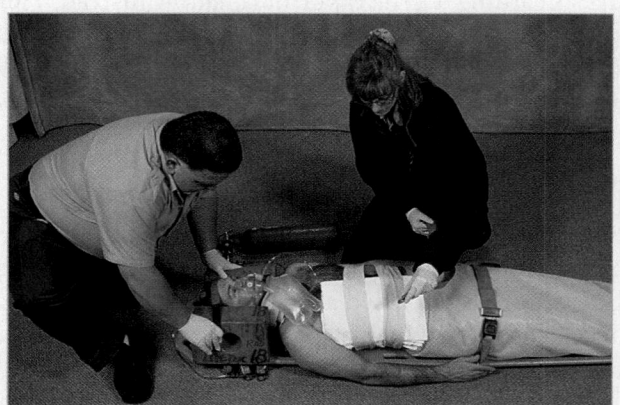

Figure 27-12 A flail anterior chest wall segment can be stabilized by securing (or having the patient hold) a pillow firmly against the chest wall.

The patient may find it easier and less painful to breathe if the flail segment is immobilized. You can tape a bulky pad against that segment of the chest for this purpose, although taping too tightly will also prevent adequate ventilation (Figure 27-12 ▲). You can also immobilize a flail chest by splinting the chest with the patient's arm, placing a sling and swathe on the arm and securing it to the chest wall snugly. Keep in mind that although flail chest itself is a serious condition, it suggests an injury that was forceful enough to cause other serious internal damage and possible spinal injury as well. Often the flail chest contributes less to the patient's ventilation difficulties than does the underlying pulmonary contusion (bruised lung segment).

Other Chest Injuries

Pulmonary Contusion

In addition to fracturing ribs, any severe blunt trauma to the chest can also injure the lung. The pulmonary alveoli become filled with blood, and fluid accumulates in the injured area, leaving the patient hypoxic. Severe pulmonary contusion, bruising of the lung, should always be suspected in patients with a flail chest and usually develops during a period of hours. If you believe that a patient may have a pulmonary contusion, you should provide respiratory support and supplemental oxygen to ensure adequate ventilation.

Traumatic Asphyxia

Sometimes a patient will experience a sudden, severe compression of the chest, which produces a rapid increase in pressure within the chest. This may occur in an unrestrained driver who hits a steering wheel or a pedestrian who is compressed between a vehicle and a wall. The sudden increase in intrathoracic pressure results in a characteristic appearance, including distended neck veins, cyanosis in the face and neck, and hemorrhage into the sclera of the eye, signaling the bursting of small blood vessels. This is called traumatic asphyxia. These findings suggest an underlying injury to the heart and possibly a pulmonary contusion. You should provide ventilatory support with supplemental oxygen and monitor the patient's vital signs, as you provide immediate transport.

You are the Provider   Part 4

The patient is now with you in the back of your transport unit en route to the hospital. The patient has stable vital signs and tells you, "My breathing seems a little better." When auscultating breath sounds bilaterally, you determine they are slightly diminished on the right side and normal on the left side. You are able to complete a detailed physical exam and continue with the ongoing assessment. At the hospital, the physician tells you the patient sustained a pneumothorax from his chest injury, and your prompt recognition and treatment of the injury was a large part of the patient's successful outcome.

7. If your patient was breathing fewer than 12 times per minute or more than 20 times per minute, what should you consider doing?
8. What signs and symptoms would you see if your patient was going into shock from blood loss?

Blunt Myocardial Injury

Blunt trauma to the chest may injure the heart itself, making it unable to maintain adequate blood pressure. There is much debate in the medical literature about how to assess myocardial contusion, or bruising of the heart muscle. Often the pulse rate is irregular, but dangerous rhythms such as ventricular tachycardia and ventricular fibrillation are uncommon. There is no specific diagnostic test at this time, and there is no prehospital treatment for the condition. Still, you should suspect myocardial contusion in all cases of severe blunt injury to the chest. Check the patient's pulse carefully, and note any irregularities. Provide supplemental oxygen, and transport immediately.

Pericardial Tamponade

In pericardial tamponade, blood or other fluid collects in the pericardium, the fibrous sac surrounding the heart (Figure 27-13 ▼). This prevents the heart from filling during the diastolic phase, causing a decrease in the amount of blood pumped to the body and decreased blood pressure. Ultimately, as blood accumulates within the pericardial cavity, it compresses the heart until it can no longer function and the patient goes into cardiac arrest. Signs and symptoms of pericardial tamponade include very soft and faint heart tones, often called muffled heart sounds, a weak pulse, low blood pressure, a decrease in the difference between the systolic and the diastolic blood pressure, and jugular vein distention.

In a trauma situation, even a small amount of fluid in the pericardial sac is enough to cause fatal pericardial tamponade. (Occasionally, fluid in surprisingly large amounts may collect in the pericardial sac as a chronic condition.) Pericardial tamponade is relatively uncommon, seen more often with penetrating injuries to the heart itself than with blunt injuries to the chest. If you suspect this life-threatening condition, provide appropriate respiratory support, supplemental oxygen, and prompt transport. Be sure to notify hospital staff of your suspicions so that preparations can be made for immediate treatment.

Laceration of the Great Vessels

The chest contains several large blood vessels: the superior vena cava, the inferior vena cava, the pulmonary arteries, four main pulmonary veins, and the aorta, with its major branches distributing blood throughout the body. Injury to any of these vessels may be accompanied by massive, rapidly fatal hemorrhage. Any patient with a chest wound who shows signs of shock may have an injury to one or more of these vessels. Frequently,

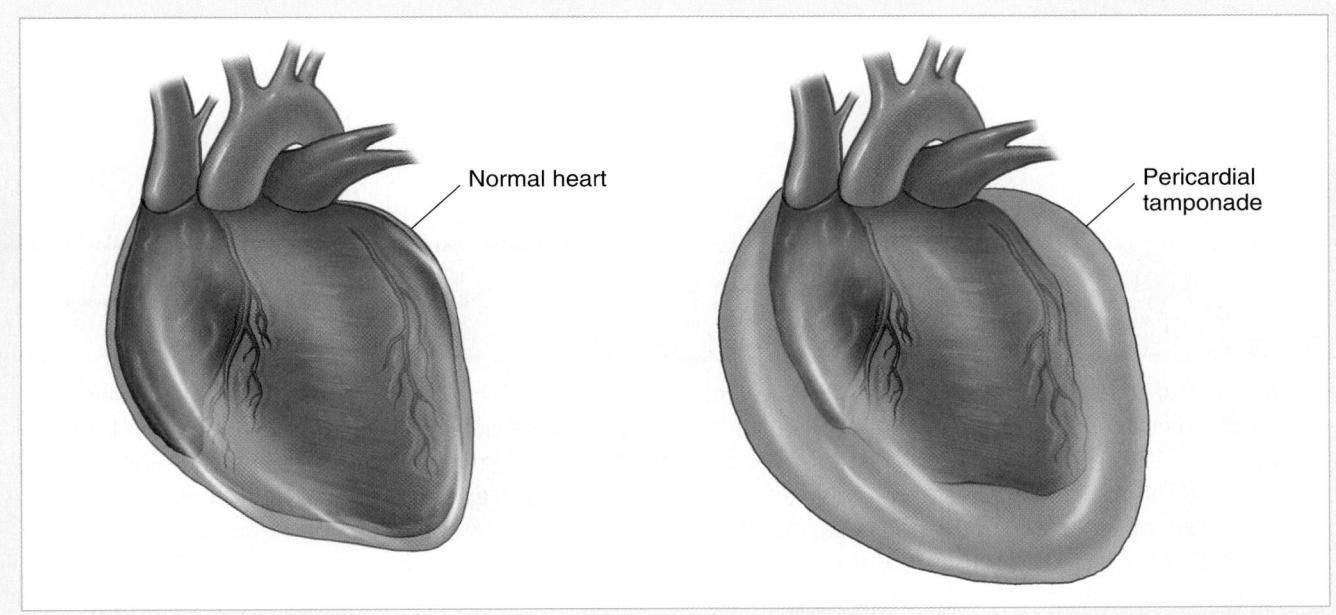

Normal heart

Pericardial tamponade

Figure 27-13 Pericardial tamponade is a potentially fatal condition in which fluid builds up within the pericardial sac, causing compression of the heart's chambers and dramatically impairing its ability to pump blood to the body.

significant blood loss is unseen, because it remains within the chest cavity. You must remain alert to signs and symptoms of shock and to changes in the baseline vital signs, such as tachycardia and hypotension.

Emergency treatment for these patients includes CPR, if appropriate, ventilatory support, and supplemental oxygen. Here, particularly, immediate transport to the hospital may be critical. Occasionally some of these patients can be treated. The overwhelming majority of injuries to the great vessels in the chest are rapidly fatal.

You are the Provider   Summary

Many vital structures are contained within the chest, including the heart, lungs, and the great vessels, all of which are vulnerable to blunt or penetrating injuries. It is critical that you assess your patient for injuries that interfere with airway, breathing, and circulation, and immediately treat these injuries as you find them. Signs of chest injury include tachypnea, bruising, crepitus, dyspnea, hemoptysis, subcutaneous emphysema, paradoxical motion, narrowing pulse pressure, and open wounds. Sometimes a tension pneumothorax can develop if air becomes trapped in the pleural space. If you have taped down all four sides of the occlusive dressing (per your local protocol), you may need to loosen one side to allow the trapped air to escape. Always remember that time is our enemy in EMS, and your rapid recognition and prompt treatment of these injuries can save your patient's life!

Chest Injuries	
Scene Size-up	Body substance isolation precautions should include a minimum of gloves and eye protection. Ensure scene safety and determine NOI/MOI. Consider the number of patients, the need for additional help/ALS, and c-spine stabilization.
Initial Assessment	
■ General impression	Determine level of consciousness and treat any immediate threats to life. Determine priority of care based on environment and patient's chief complaint.
■ Airway	Ensure patent airway. Maintain spinal immobilization as necessary.
■ Breathing	Listen for abnormal breath sounds and evaluate depth and rate of the respirations. Look for symmetric chest rise and fall. Maintain ventilations as needed. Provide high-flow oxygen at 15 L/min and inspect the chest wall assessing for DCAP-BTLS.
■ Circulation	Evaluate pulse rate and quality; observe skin color, temperature, and condition, and treat accordingly. Control bleeding and cover sucking chest wounds with an occlusive dressing.
■ Transport decision	Prompt transport
Focused History and Physical Exam	*NOTE: The order of the steps in the focused history and physical exam differs depending on whether or not the patient has a significant MOI. The order below is for a patient with a significant MOI. For a patient without a significant MOI, perform a focused trauma assessment, obtain vital signs, and obtain the history.*
■ Rapid trauma assessment	Reevaluate the mechanism of injury. Perform a rapid trauma assessment, treating all life-threats immediately. Log roll and secure patient to backboard for all patients with a suspected spinal injury.
■ Baseline vital signs	Take vital signs, noting skin color and temperature as well as patient's level of consciousness. Use pulse oximetry if available.
■ SAMPLE history	Obtain SAMPLE history. If the patient is unresponsive, attempt to obtain information from family members, friends, or bystanders.
■ Interventions	Provide complete spinal immobilization early if you suspect that your patient has spinal injuries. Maintain an open airway and suction as needed. Treat signs of shock and any other life threats.
Detailed Physical Exam	Complete a detailed physical exam.
Ongoing Assessment	Repeat the initial assessment, rapid/focused assessment, and reassess interventions performed. Reassess vital signs every 5 minutes for the unstable patient, every 15 minutes for the stable patient.
■ Communication and documentation	Contact medical control with a radio report. Relay any change in level of consciousness or difficulty breathing. Describe the MOI and any interventions you performed. Be sure to document any physician's orders or changes in the patient's condition and at what time they occurred.

Assessment and Emergency Care

Assessment and Emergency Care

NOTE: While the steps below are widely accepted, be sure to consult and follow your local protocol.

Pneumothorax	Hemothorax	Rib Fractures	Flail Chest
1. Clear and maintain an open airway.	1. Clear and maintain an open airway.	1. Clear and maintain an open airway.	1. Clear and maintain an open airway.
2. Provide high-flow oxygen.	2. Provide high-flow oxygen. Cover the patient with a blanket.	2. Provide high-flow oxygen. Cover the patient with a blanket.	2. Provide respiratory support if necessary.
3. Seal the wound with an occlusive dressing, using a large enough dressing so that it is not pulled or sucked into the chest cavity.	3. Treat the patient for shock.	3. Place in a position of comfort to support breathing unless a spinal injury is suspected.	3. Provide high-flow oxygen.
4. Depending on your local protocol, you may tape the dressing down on all four sides, or create a flutter valve by taping only three sides of the dressing.			4. Stabilize flail segment by securing (or having the patient hold) a pillow firmly against the chest wall

Prep Kit

Ready for Review

- Chest injuries are classified as closed or open. Closed injuries are often the result of blunt force trauma, and open injuries are the result of some object penetrating the skin and/or chest wall.

- Blunt trauma may result in fractures to the ribs and the sternum.

- A flail chest segment is two or more ribs broken in two or more places.

- During the initial assessment, if an injury is encountered that interferes with the ability of the patient to ventilate or oxygenate, the injury must be addressed quickly.

- A flail chest segment should be secured with a large bulky dressing and 2″ tape.

- Any penetrating injury to the chest may result in air entering the pleural space and may cause pneumothorax. An occlusive dressing should be placed on this injury as soon as it is identified.

- A spontaneous pneumothorax may be the result of rupture of a weak spot on the lung allowing air to enter the pleural space and accumulate. This often results from nontraumatic injuries and may occur during times of physical activity such as exercise.

- A pneumothorax may progress to a tension pneumothorax and cause cardiac arrest.

- Hemothorax is the result of blood accumulating in the pleural space after a traumatic injury when the vessels of the lung are lacerated and leak blood.

- The accumulation of blood and air in the pleural space of the chest is called a hemopneumothorax, a potentially fatal condition.

- All patients with chest injuries should receive high-flow oxygen or ventilation with a BVM device.

- Pulmonary contusion, which is bruising of lung tissue after traumatic injury, may interfere with oxygen exchange in the lung tissue.

- Cardiac contusion is bruising of the heart muscle after traumatic injury. This condition may have the same signs and symptoms as a heart attack, including an irregular pulse. Remember that this is an injury to heart muscle from trauma, not from a heart attack.

- Pericardial tamponade is when blood collects in the space between the pericardial sac and the heart. This condition results in pressure building up inside the pericardial sac until the heart cannot pump effectively; cardiac arrest may occur quickly.

- The great vessels of the body are located in the mediastinum. These large vessels may be lacerated or tear after traumatic injury and cause heavy, unseen bleeding inside the patient's chest cavity.

- Any patient who has signs of shock with a chest injury, even with unseen bleeding, should make you suspicious of unseen, life-threatening bleeding inside the chest cavity.

Vital Vocabulary

closed chest injury An injury to the chest in which the skin is not broken, usually due to blunt trauma.

dyspnea Difficulty breathing.

flail chest A condition in which two or more ribs are fractured in two or more places or in association with a fracture of the sternum so that a segment of chest wall is effectively detached from the rest of the thoracic cage.

flutter valve A one-way valve that allows air to leave the chest cavity but not return; formed by taping three sides of an occlusive dressing to the chest wall, leaving the fourth side open as a valve.

hemoptysis The spitting or coughing up of blood.

hemothorax A collection of blood in the pleural cavity.

Technology

- Interactivities
- Vocabulary Explorer
- Anatomy Review
- Web Links
- Online Review Manual

www.EMTB.com

Prep Kit continued...

myocardial contusion A bruise of the heart muscle.

occlusive dressing A dressing made of Vaseline gauze, aluminum foil, or plastic that prevents air and liquids from entering or exiting a wound.

open chest injury An injury to the chest in which the chest wall itself is penetrated, by a fractured rib or, more frequently, by an external object such as a bullet or knife.

paradoxical motion The motion of the portion of the chest wall that is detached in a flail chest; the motion—in during inhalation, out during exhalation—is exactly the opposite of normal chest wall motion during breathing.

pericardial tamponade Compression of the heart due to a buildup of blood or other fluid in the pericardial sac.

pericardium The fibrous sac that surrounds the heart.

pneumothorax An accumulation of air or gas in the pleural cavity.

pulmonary contusion A bruise of the lung.

spontaneous pneumothorax A pneumothorax that occurs when a weak area on the lung ruptures in the absence of major injury, allowing air to leak into the pleural space.

sucking chest wound An open or penetrating chest wall wound through which air passes during inspiration and expiration, creating a sucking sound.

tachypnea Rapid respirations.

tension pneumothorax An accumulation of air or gas in the pleural cavity that progressively increases the pressure in the chest with potentially fatal results.

Points to Ponder

You have been dispatched to assist the Sheriff's Department with a reported shooting. You are told to stage two blocks east of the incident. After the scene has been secured, you are escorted into the building. The sheriff informs you that a suspect was shot by a deputy. The deputy received a superficial gunshot wound to her left arm. She is alert and ambulatory. From across the room you see a man lying on the ground with several gunshot wounds in his chest. You observe minimal respirations. As you approach the patient, the sheriff orders you to treat the deputy first.

Issues: Scene Safety at a Shooting, Prioritizing Patients Based on Injuries, Working With Other Responders.

Assessment in Action

You are called to respond to the report of a vehicle crash at the intersection of University Street and Babcock Avenue. Police and fire fighters are already on scene and report two vehicles and heavy damage. Upon arrival, you park the ambulance a safe distance away from any potential hazards. The lieutenant in command informs you that there are only two patients. You assign your partner to Patient 1 and you take Patient 2.

Your patient was wearing his seat belt while driving when his vehicle was hit on the driver's side. There was heavy damage to the vehicle. During your initial assessment and general impression, you see a young man who is very pale and having difficulty breathing. His airway is open but he is spitting up blood. His respirations are 34 breaths/min and labored. The patient has no palpable radial pulses, and the carotid pulse is very weak and rapid. As the fire department works on extricating the patient, you start assisting ventilations with a BVM device.

1. As you perform a rapid trauma assessment you find that a segment of the chest is moving in the opposite direction of the other ribs. This is called:

   A. chest atrophy.
   B. paradoxical motion.
   C. pulmonary interference.
   D. see-saw breathing.

2. Blunt trauma to the chest most likely leads to a(n):

   A. closed chest injury.
   B. chronic injury.
   C. major disability
   D. open chest injury.

3. One of the signs of a serious chest injury is blood coming from the patient's mouth. This is called:

   A. hemoptysis.
   B. hemoparalysis.
   C. dysphasia.
   D. nucal response.

4. A patient with a serious chest injury may have tachypnea, which is indicated by:

   A. blue skin.
   B. equal chest rise.
   C. rapid respirations.
   D. slow exhalations.

5. If air moves into the pleural space and your patient's lung collapses, his condition is called a(n):

   A. hemothorax.
   B. embolism.
   C. pneumothorax.
   D. throacotomy.

6. As you continue to ventilate the patient, his condition continues to worsen. His level of consciousness is decreasing, he is becoming cyanotic, and you notice his neck veins are sticking out. What does this indicate?

   A. Tension pneumothorax
   B. Pulmonary reflux
   C. Esophageal varicose
   D. Torn diaphragm

7. Your patient has several fractured ribs. You should suspect a:

   A. chest dislocation.
   B. flail chest segment.
   C. lung defacement.
   D. rib forfeiture.

Challenging Questions

8. Explain the procedure for treating an open chest injury.

9. Describe the signs and symptoms of traumatic asphyxia.

Abdomen and Genitalia Injuries

Objectives*

Cognitive

1. State the steps in the emergency medical care of a patient with a blunt or penetrating abdominal injury. (p 801, 802)
2. Describe how solid and hollow organs can be injured. (p 799)
3. State the steps in the emergency medical care of a patient with an object impaled in the abdomen. (p 802)
4. State the steps in the emergency medical care of a patient with an abdominal evisceration wound. (p 802)
5. State the steps in the emergency medical care of the patient with a genitourinary injury. (p 809–812)

Affective

None

Psychomotor

6. Demonstrate proper treatment of a patient who has an object impaled in the abdomen. (p 802)
7. Demonstrate how to apply a dressing to an abdominal evisceration wound. (p 802)

*All of the objectives in this chapter are noncurriculum objectives.

You are the Provider

You have been dispatched to Johnnie's Glass Works for an injured person who was hit with a piece of glass. As you proceed to the scene, dispatch updates you and states that the patient was struck in the abdomen by a piece of broken glass.

1. What are some possible injuries your patient may have?
2. What signs and symptoms of shock might patients with abdominal injuries exhibit?

Abdomen and Genitalia Injuries

The abdomen is the major body cavity extending from the diaphragm to the pelvis. It contains organs that make up the digestive, urinary, and genitourinary systems. Although any of these organs may be injured, some are better protected than others. You must know where these organs are located within the abdominal or pelvic cavities. You must also understand their functions so that when an illness or injury occurs, you can assess its seriousness.

This chapter begins with a brief review of the anatomy of the abdomen, followed by a discussion of common types of abdominal injuries. Next, patient assessment strategies are discussed, followed by a description of specific abdominal injuries that you are likely to encounter and how to treat each. The genitourinary system is then described, and its common injuries and treatment are discussed.

The Anatomy of the Abdomen

The abdomen contains both hollow and solid organs, any of which may be damaged. Hollow organs, including the stomach, intestines, ureters, and bladder, are structures through which materials pass (Figure 28-1 ▶). They usually contain food that is in the process of being digested, urine that is being passed to the

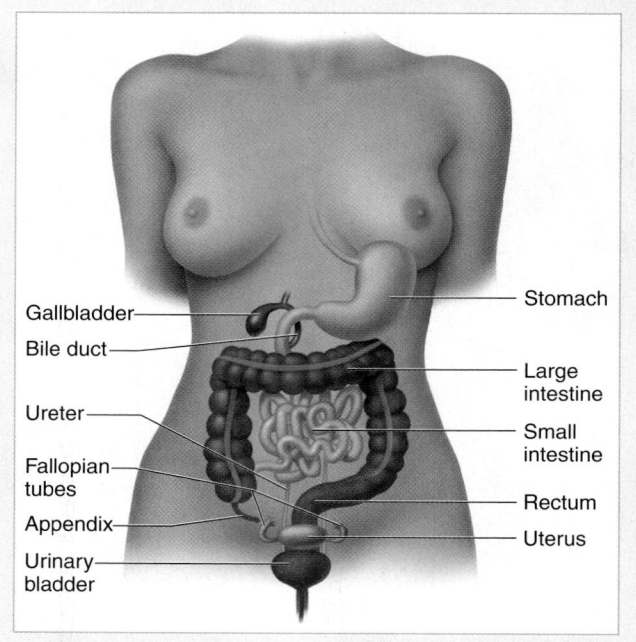

Figure 28-1 The hollow organs in the abdominal cavity are structures through which materials pass.

Labels: Gallbladder, Bile duct, Ureter, Fallopian tubes, Appendix, Urinary bladder, Stomach, Large intestine, Small intestine, Rectum, Uterus

bladder for release, or bile. When ruptured or lacerated, these organs spill their contents into the peritoneal cavity (the abdominal cavity), causing an intense inflammatory reaction and possible infection. Peritonitis is an inflammation of the peritoneum that may be caused by this type of infection. The intestines and stomach contain acid-like substances that aid in the digestive process. When they spill or leak into the peritoneal cavity, pain and irritation to the peritoneum often follow. The first signs of peritonitis are severe abdominal pain, tenderness, and muscular spasm. Later, bowel sounds diminish or disappear as the bowel stops functioning. A patient may feel nauseous and may vomit; the abdomen may become distended and firm to touch, and infection may occur. Peritonitis is serious and may become life threatening to the patient.

The solid organs, as their name suggests, are solid masses of tissue. They include the liver, spleen, pancreas, and kidneys (Figure 28-2 ▶). It is here that much of the chemical work of the body—enzyme production, blood cleansing, and energy production—takes place. Solid organs have a rich blood supply, so injury can cause severe and unseen hemorrhage. The same is true of the aorta or inferior vena cava, whether the injury is open or closed. Blood may irritate the peritoneal cavity and cause the patient to complain of abdominal pain;

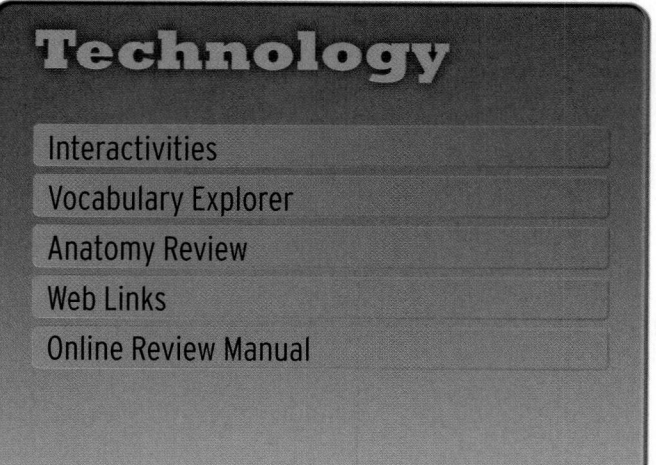

Technology

www.EMTB.com

- Interactivities
- Vocabulary Explorer
- Anatomy Review
- Web Links
- Online Review Manual

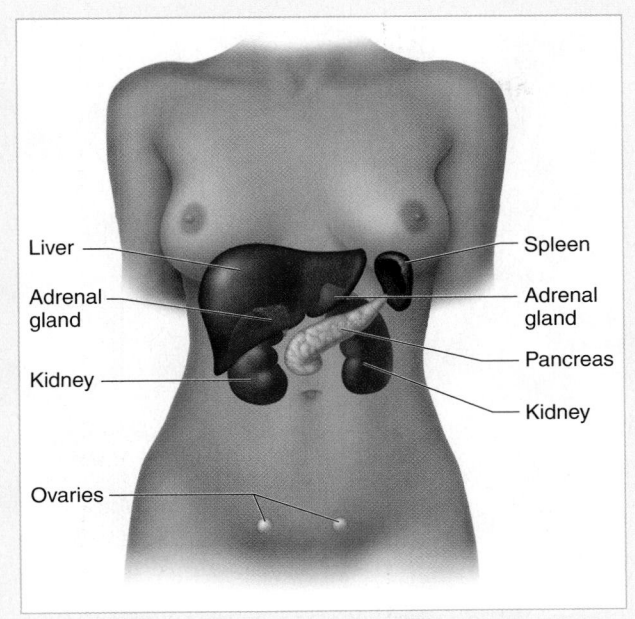

Figure 28-2 The solid organs are solid masses of tissue that do much of the chemical work in the body and receive a large rich supply of blood.

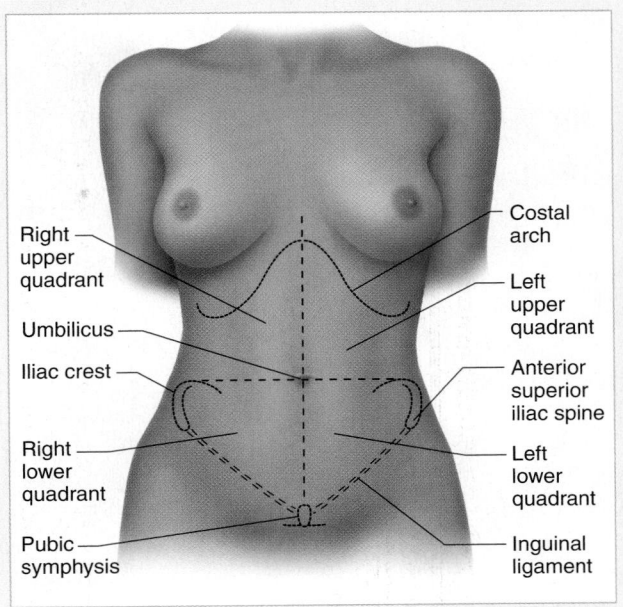

Figure 28-3 The abdominal cavity is divided into four quadrants, which serve as your means of identifying and reporting problems in the abdomen.

however, this may not always occur. Therefore, the absence of pain and tenderness does not necessarily mean the absence of major bleeding in the abdomen.

The bony landmarks in the abdomen include the pubic symphysis, the costal arch, the iliac crests, and the anterior superior iliac spines. The major soft-tissue landmark is the umbilicus, which overlies the fourth lumbar vertebra. The abdomen is divided into four quadrants by two perpendicular lines that intersect at the umbilicus (Figure 28-3 ▶). These quadrants provide a frame of reference for identifying and reporting abdominal signs and symptoms.

Injuries to the Abdomen

Abdominal injuries may be as obvious as loops of intestines protruding from a penetrating injury, or as occult as an unseen injury such as a laceration to the liver or spleen. Traduct injuries to the abdomen are considered as either open or closed, and can involve hollow and/or solid organs. Closed abdominal injuries are those in which blunt force trauma, some type of impact to the body, results in injury to the abdomen without breaking the skin. Such a blow might come from the patient's striking the handlebar of a bicycle or the steering wheel of a car, or when the patient is struck by an item such as a board or baseball bat during a fight or assault (Figure 28-4 ▼). Open abdominal injuries are those in which a foreign object enters the abdomen and opens the peritoneal cavity to the outside; these are also known as penetrating injuries (Figure 28-5 ▶). Open wounds might not be deeper than the muscular wall of the abdomen. This cannot be determined in the pre-

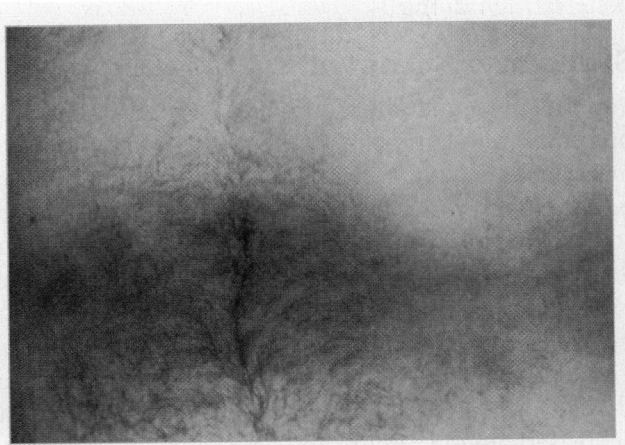

Figure 28-4 Blunt trauma to the abdomen can occur when a patient strikes the steering wheel of an automobile as a result of a crash.

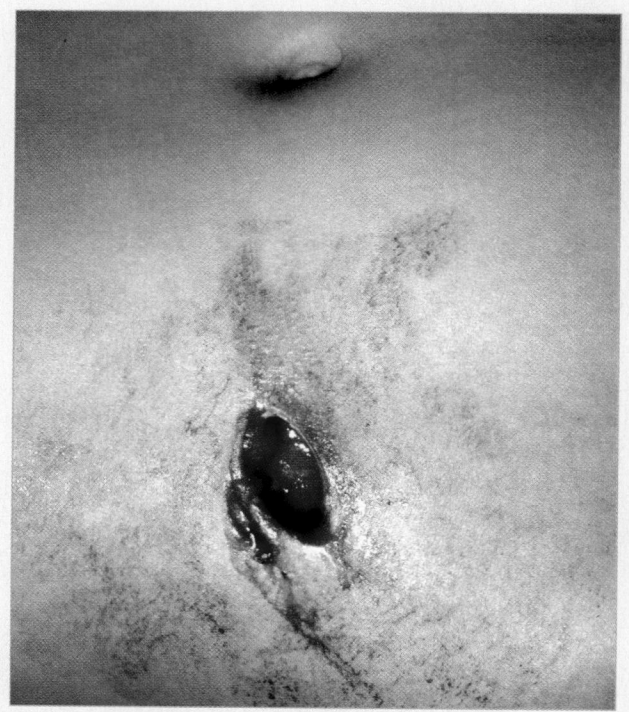

Figure 28-5 Because it is difficult to know how deep a penetrating injury is, assume organ damage and transport promptly.

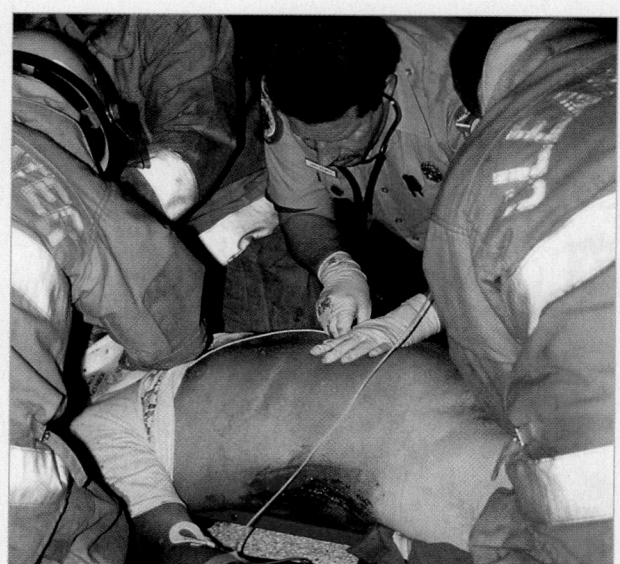

Figure 28-6 Bruising on the abdomen can provide clues to the possible injury of underlying organs.

hospital setting, however, and must be assessed and evaluated at the hospital. Therefore, you should maintain a high index of suspicion for unseen injuries, internal damage to organs, and potential life-threatening injuries and provide prompt transport. Stab wounds and gunshot wounds are examples of open injuries, or penetrating trauma.

Signs and Symptoms

Patients with abdominal injuries generally have one principal complaint: pain. But other significant injuries may mask the pain at first, and some patients may not be able to tell you about pain because they are unconscious or unresponsive, such as after a head injury or drug or alcohol overdose. A very common sign of significant abdominal injury is tachycardia as the heart increases its pumping action to compensate for blood loss, an early indication of compensated blood loss and shock. Later signs include evidence of shock such as decreased blood pressure and pale, cool, moist skin, or changes in the patient's mental status, combined with trauma to the abdomen. In some cases, the abdomen may become distended from the accumulation of blood and fluid. As an EMT-B, you must look for other signs and

symptoms of potential problems and injuries to the abdomen. Blunt injuries include bruises (often indicated by red areas of skin at this early stage) or other visible marks, whose location should guide your attention to underlying structures Figure 28-6 ▲. For example, bruises in the right upper quadrant, left upper quadrant, or flank (the region of the lower rib cage) might suggest an injury to the liver, spleen, or kidney, respectively.

The signs of abdominal injury are usually more definite than the symptoms, including firmness on palpation of the abdomen, obvious penetrating wounds, bruises, and altered vital signs such as increased pulse rate, increased respiratory rate, decreased blood pressure, and shallow respirations (although these might not appear until later). Common symptoms include abdominal tenderness, particularly localized tenderness, and difficulty with movement because of pain.

Types of Abdominal Injuries

Blunt Abdominal Wounds

A patient with a blunt abdominal wound may have one or more of the following:

■ Severe bruising of the abdominal wall

- Laceration of the liver and spleen
- Rupture of the intestine
- Tears in the mesentery, membranous folds that attach the intestines to the walls of the body, and injury to blood vessels within them
- Rupture of the kidneys, or avulsion of the kidneys from their arteries and veins
- Rupture of the bladder, especially in a patient who had a full and distended bladder at the time of the crash
- Severe intra-abdominal hemorrhage
- Peritoneal irritation and inflammation in response to the rupture of hollow organs

A patient who has sustained a blunt abdominal injury should be log rolled to a supine position on a backboard. Ensure that you protect the spine while you roll him. If the patient vomits, turn him or her to one side and clear the mouth and throat of vomitus. Monitor the patient's vital signs for any indication of shock such as pallor; cold sweat; rapid, thready pulse; or low blood pressure. If you see any of these signs, administer high-flow supplemental oxygen via nonrebreathing mask, and take all the appropriate measures to treat for shock. Keep the patient warm with blankets, and provide prompt transport to the emergency department.

Injuries From Seat Belts and Airbags

Seat belts have prevented many thousands of injuries and saved many lives, including those of people who otherwise would have been ejected from a smashed car. However, seat belts occasionally cause blunt injuries of the abdominal organs. When worn properly, a seat belt lies below the anterior superior iliac spines of the pelvis and against the hip joints. If the belt lies too high, it can squeeze abdominal organs or great vessels against the spine when the car suddenly decelerates or stops (Figure 28-7 ▶). Occasionally, fractures of the lumbar spine have been reported. If you are called to the scene of such an accident, keep in mind that the use of seat belts in many cases turns what could have been a fatal injury into a manageable one.

In all current-model automobiles, the lap and diagonal (shoulder) safety belts are combined into one so that they may not be used independently. Of course, people can still place the diagonal portion of the belt behind the back, significantly reducing the effectiveness of this design. In some older cars, only lap belts or two separate belts are provided. Used alone, diagonal shoulder safety belts can cause injuries of the upper part of the trunk, such as a bruised chest, fractured

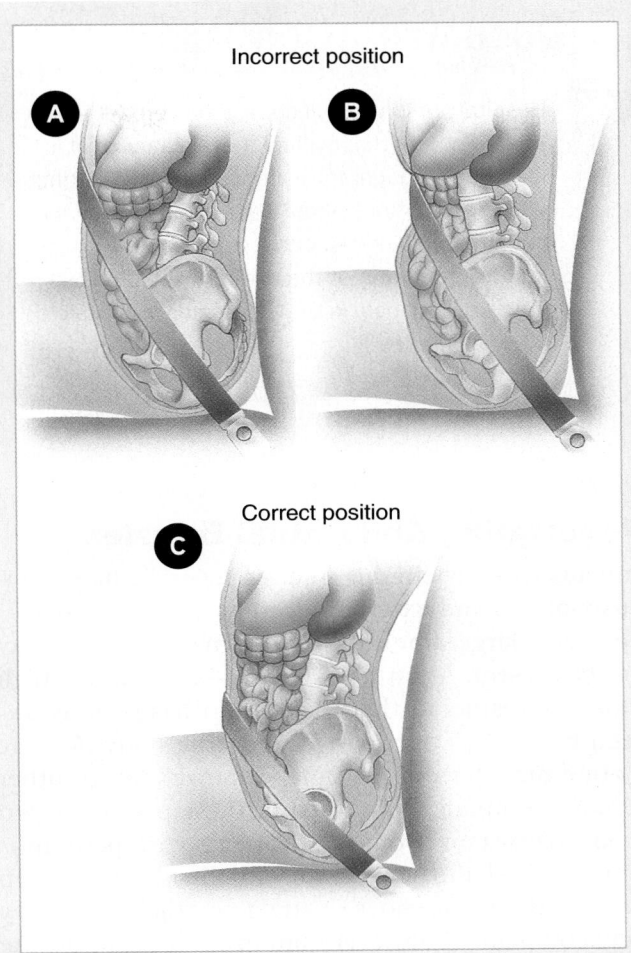

Figure 28-7 The proper position for a seat belt is below the anterior superior iliac spines of the pelvis and against the hip joints, as shown in diagram **C**. Diagrams **A** and **B** show improper positioning of seat belts.

ribs, lacerated liver, or even decapitation. Far fewer head and neck injuries are seen when this belt is used in combination with a lap belt and a headrest.

The airbag, which is standard in today's vehicles, represents a great advance in automotive safety. In head-on collisions, it can be a genuine lifesaver. However, because frontal airbags provide no protection in a side impact or rollover, they must be used in combination with safety belts. Small children and short stature individuals who are in the front seat of the automobile may be at risk of injury if an airbag is deployed. Special attention should be used in evaluating these patients when a deployed airbag is noted. Remember to inspect beneath the airbag for signs of damage to the steering column.

Documentation Tips

Hospital personnel will depend on you to record scene findings that explain the mechanism of injury. Be thorough, for instance, in documenting your observations about the vehicle in which a patient rode. Notes about deployment of air bags and the condition of the exterior and the steering wheel will help in the assessment of possible internal injuries.

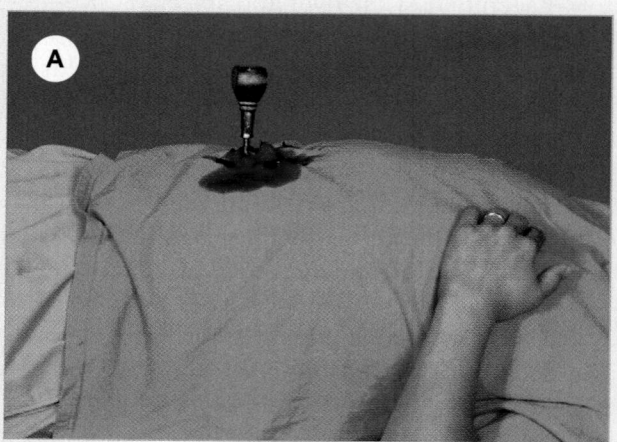

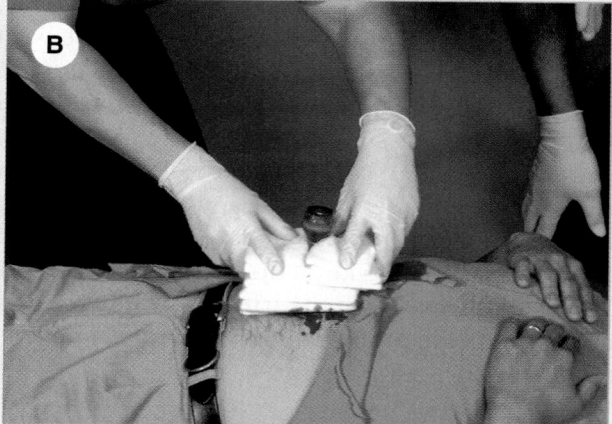

Figure 28-8 A. Penetrating injuries have obvious wounds and may also have external bleeding. **B.** If the penetrating object is still in place, use a roller bandage to stabilize the object and to control bleeding.

Penetrating Abdominal Injuries

Patients with penetrating injuries generally have obvious wounds and external bleeding (Figure 28-8A ▶); however, large amounts of external bleeding may not be present. As an EMT-B, you should have a high index of suspicion that the patient has serious unseen blood loss occurring inside the body. A large wound may have protrusions of bowel, fat, or other structures. In addition to pain, these patients often report nausea and vomiting. Patients with peritonitis generally prefer to lie very still with their legs drawn up because it hurts to move or straighten their legs. They may complain about every bump in the road during transport.

Some penetrating injuries go no deeper than the abdominal wall, but the severity of the injury often cannot be determined in the prehospital setting. Only a surgeon can accurately assess the damage. Therefore, as you care for a patient with this type of wound, you should assume that the object has penetrated the peritoneum, entered the abdominal cavity, and possibly injured one or more organs even if there are no immediate obvious signs.

If major blood vessels are cut or solid organs are lacerated, bleeding may be rapid and severe. Other signs of intra-abdominal injuries may develop slowly, particularly in penetrating wounds to hollow organs. Once such an organ is punctured and its contents are discharged into the abdominal cavity, peritonitis may develop, but this may take several hours.

In caring for a patient with a penetrating wound to the abdomen, follow the general procedures described above for care of a blunt abdominal wound as well as the following specific steps for the penetrating wound. Inspect the patient's back and sides for exit wounds, and apply a dry, sterile dressing to all open wounds. If the penetrating object is still in place, apply a stabilizing bandage around it to control external bleeding and minimize movement of the object (Figure 28-8B ▲).

Abdominal Evisceration

Severe lacerations of the abdominal wall may result in an evisceration, in which internal organs or fat protrude through the wound (Figure 28-9 ▶). Never try to replace an organ that is protruding from an abdominal laceration, whether it is a small fold of peritoneum or nearly all of the intestines. Instead, cover it with sterile gauze compresses moistened with sterile saline solution and secure with a sterile dressing. (Protocols in some EMS systems call for an occlusive dressing over the organs, secured by trauma dressings.) Because the open abdomen radiates body heat very effectively, and because exposed organs lose fluid rapidly, you must keep the organs moist and warm. If you do not have gauze compresses, you may use moist, sterile dressings, covered and secured in place with a bandage and tape (Figure 28-10 ▼). Do not use any material that is adherent or loses its substance when wet, such as toilet paper, facial tissue, paper towels, or absorbent cotton.

Once you have covered the extruding organ, you should provide other emergency care as necessary and provide prompt transport to the emergency department.

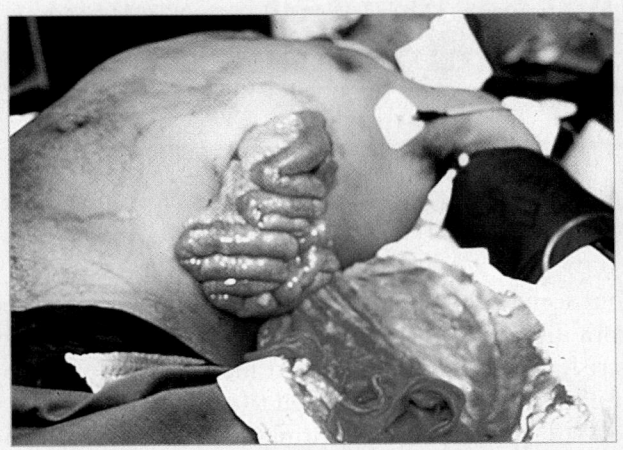

Figure 28-9 An abdominal evisceration is an open abdominal wound from which internal organs or fat protrude.

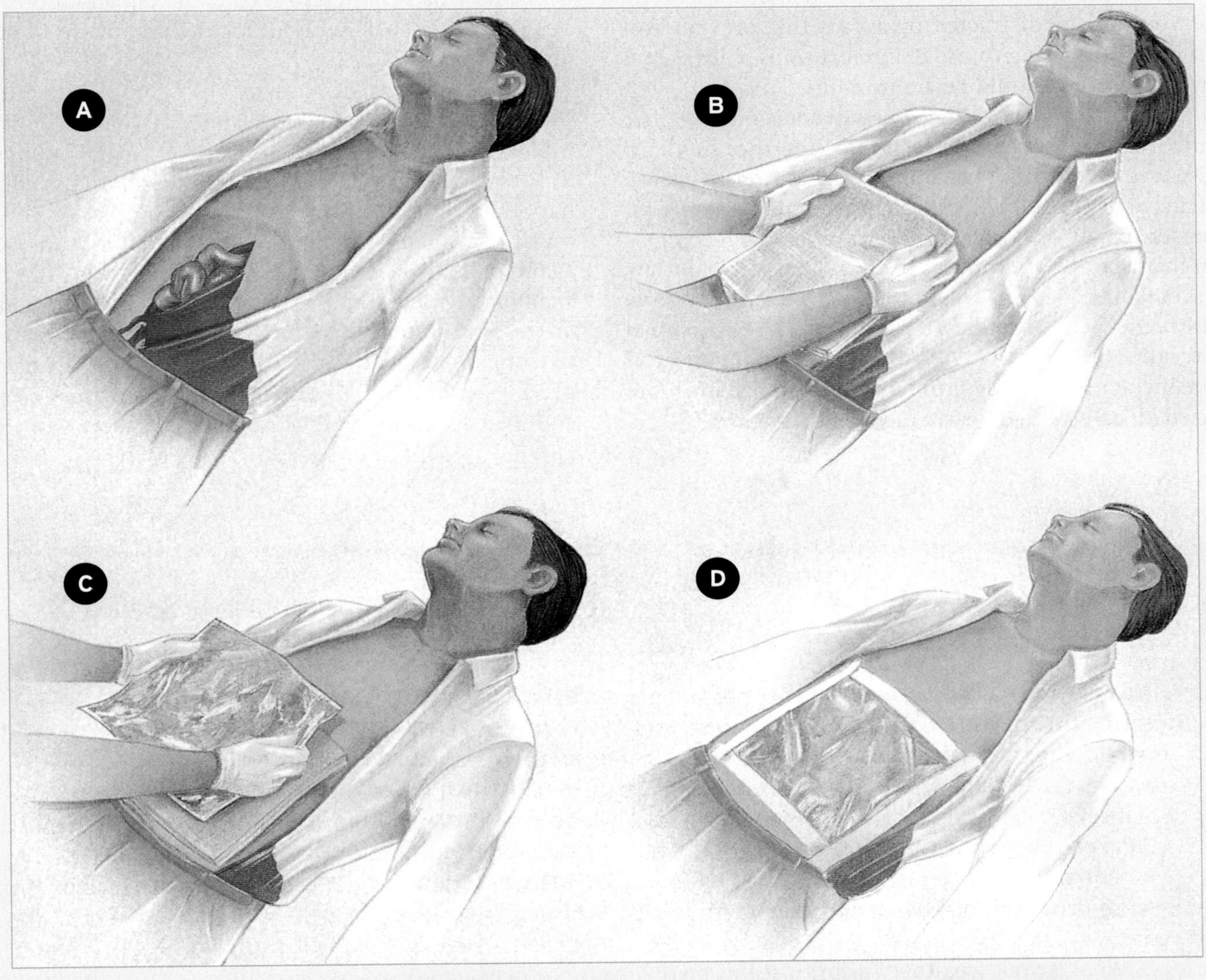

Figure 28-10 A. The open abdomen radiates body heat rapidly and must be covered. **B.** Cover the wound with moistened, sterile gauze, or with an occlusive dressing, depending on local protocol. **C.** Secure the dressing with a bandage. **D.** Secure the bandage with tape.

Assessment of Abdominal Injuries

Scene Size-up

Your scene size-up begins with the information reported from dispatch. This information will help you to prepare for the call. Often the information will be sketchy or even inaccurate as reported to the dispatcher, but it will still provide some information to consider as you respond to the call. For example, is the patient injured or ill? Could one have led to the other? What equipment might you need to assess and treat the patient? BSI precautions should be observed prior to arrival on scene or you may be distracted by events that prevent you from remembering to use BSI precautions. Gloves and eye protection should be a minimum.

As you arrive on scene, you will continue to gather information that will help to manage the incident. Observe the scene for hazards and threats to your safety. If dispatch information indicates a possible assault, domestic dispute, or drive-by shooting, all of which commonly lead to abdominal injuries, be sure that law enforcement has controlled the scene. As you observe the scene, determine the MOI and consider early spinal precautions. How many victims might be involved in the incident? If you determine additional resources are needed, call for them early in your assessment.

Initial Assessment

General Impression

Your goal in the initial assessment is to evaluate the patient's ABCs and then immediately care for any life threats. The general impression, including an evaluation of the patient's age, chief complaint, and level of consciousness, will help you establish the seriousness of the patient's condition. Some abdominal injuries will be obvious and graphic. Most will be very subtle and may go unnoticed. The MOI together with the chief complaint will help you to focus on the immediate problem. Remember, the trauma or blow to the abdomen may have been hours or even days earlier and the pain is just now bad enough to seek help. Ask about previous injuries associated with a chief complaint of abdominal pain.

Quickly assess the patient's chief complaint with a simple inspection, noting the manner in which he or she is lying. If the chief complaint involves sexual or physical assault, the patient may be hesitant to discuss what happened. Bleeding from the reproductive or genitourinary organs is common after sexual assault, but again, patients may be hesitant to discuss this or to be examined to determine severity. Movement of the body or the abdominal organs irritates the inflamed peritoneum, causing additional pain. To minimize this pain, patients will lie still, usually with the knees drawn up, and their breathing will be rapid and shallow. For the

You are the Provider   Part 2

As you arrive on scene, you and your partner quickly determine that the scene is safe and no other resources are required. There is one patient and ALS is en route with an ETA of approximately 10 minutes. No cervical spine involvement is suspected, and your patient is alert and oriented. The patient tells you he was working with a large piece of glass when it broke and a sharp-edged piece struck him in the abdomen. Your general impression is an anxious man, approximately 28 years old, lying on the ground with his hands over his left lower abdominal quadrant. Your assessment reveals a patent airway with equal chest rise and fall with good breath sounds bilaterally. His pulse is rapid, and you see obvious bleeding on and under his hands in the abdominal area.

3. What is your treatment priority at this point?
4. What organs lie in the left lower abdominal quadrant?

same reason, they will contract their abdominal muscles, a sign called <u>guarding</u>.

Airway and Breathing

Next, ensure that the patient has a clear and patent airway. If a spinal injury is suspected, prevent the patient from moving by having a team member hold the patient's head still and verbally reminding the patient not to move. Patients may report that they feel nauseous, and they may vomit. Remember to keep the airway clear of vomitus so that it is not aspirated into the lungs, especially in a patient who is unconscious or has an altered level of consciousness. Turn the patient to one side, using spinal precautions if necessary, and try to clear any material from the throat and mouth. Note the nature of the vomitus: undigested food, blood, mucus, or bile.

You must also quickly assess the patient for adequate breathing. A distended abdomen or pain may prevent adequate inhalation. When these guarded respirations decrease the breathing effectiveness, providing supplemental oxygen with a nonrebreathing mask will help improve oxygenation. If the patient's level of consciousness is decreased and respirations are shallow, consider supplementing respirations with a BVM device. Use airway adjuncts as necessary to make the airway patent and assist with breathing.

Circulation

Superficial abdominal or genitalia injuries usually do not produce significant external bleeding. Internal bleeding from open or closed abdominal injuries, however, can be profound. Trauma to the kidneys, liver, and spleen can cause significant internal bleeding. Evaluate the patient's pulse and skin color, temperature, and condition to determine the stage of shock. If you suspect shock, treat the patient aggressively by providing oxygen, positioning the patient in a modified shock position, and keeping the patient warm. Wounds should be covered and bleeding controlled as quickly as possible.

Transport Decision

Due to the nature of abdominal injuries, a rapid on-scene time and quick transport to the hospital are generally indicated. Abdominal pain together with an MOI that suggests injury to the abdomen or flank is a good indication for rapid transport. In the prehospital environment it is difficult to determine whether the liver, spleen, or kidney has been injured. Hollow organs that have ruptured are also difficult to identify without more advanced diagnostic equipment. A delay in a medical

evaluation may mean shock has a chance to progress unnecessarily. Patients who have visible significant bleeding or signs of significant internal bleeding may quickly become unstable. Treatment is directed at quickly addressing life threats and providing rapid transportation to the closest appropriate hospital.

Focused History and Physical Exam

Normally, you will perform the focused history and physical exam on all patients with abdominal and genitalia injuries in the same manner. Remove or loosen clothes to expose the injured regions of the body for the physical assessment. Provide privacy as needed or wait until you are in the back of the ambulance. The patient without suspected spinal injury should be allowed to stay in the position of comfort—with legs pulled up toward the abdomen. The patient without suspected spinal injury should not be forced to lay flat for the physical exam or transport. Determine which physical assessment process you will utilize—rapid physical exam or focused physical exam.

Use DCAP-BTLS to help identify specific signs and symptoms of injury. Inspect and palpate the abdomen for the presence of **D**eformity, which may be subtle in abdominal injuries. Look for the presence of **C**ontusions and **A**brasions, which can help localize focal points of impact and may indicate significant internal injury. **P**uncture wounds and other **P**enetrating injuries must not be overlooked, as the intra-abdominal extent of these injuries may be life threatening. The presence of **B**urns must be noted and managed appropriately. Palpate for **T**enderness and attempt to localize to a specific quadrant of the abdomen. Identify and treat any **L**acerations with appropriate dressings. **S**welling may involve the abdomen globally and indicate significant

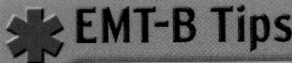
EMT-B Tips

Log rolling onto a backboard always provides a valuable chance to examine the back for signs of injury. Instruct and position helpers to ensure your ability to inspect and palpate the back briefly while the patient is rolled onto his or her side.

intra-abdominal injury. In pediatric patients, the liver and spleen are very large in the abdomen and are more easily injured. The soft, flexible ribs of infants and young children do not protect these two organs very well.

Rapid Physical Exam for Significant MOI

If the patient has been subjected to a significant MOI, a rapid physical exam will help you to quickly identify any injuries your patient may have, not just abdominal injuries. Begin with the head and finish with the lower extremities, moving in a systematic manner. Your goal is not to identify the extent of all the injuries but to determine whether injuries are present. This requires you to work quickly but thoroughly. If you find a life-threatening problem, stop and treat it immediately, otherwise move on. The injuries you find will help you in packaging your patient for transport. Up until now you may have been stabilizing the patient's spine by simply holding the head still and telling him or her not to move. If you have not yet put on a cervical collar, now is the time, before you log roll the patient to inspect the posterior part of the body and place the patient on a backboard.

Focused Physical Exam for Nonsignificant MOI

If the MOI suggests an isolated injury to the abdomen or genitalia, a focused exam of the injured area only may be sufficient. Inspect the skin of the abdomen for wounds through which bullets, knives, or other missile-type foreign bodies may have passed. Keep in mind that the size of the wound does not necessarily indicate the extent of the underlying injuries. If you find an entry wound, you must always check for a corresponding exit wound in the patient's back or sides. If the injury was caused by a very high-velocity missile from a rifle, you may see a small, harmless-looking entrance wound with a large, gaping exit wound. Do not attempt to remove a knife or other object that is impaled in the patient. Instead, stabilize the object with supportive bandaging. Bruises or other visible marks are important clues to the cause and severity of any blunt injury. Steering wheels and seat belts produce characteristic patterns of bruising on the abdomen or chest.

The kidneys are located in the flank region of the back. Inspect and palpate this area for tenderness, bruising, swelling, or other signs of trauma. Genital injuries can be awkward to evaluate and can be even more awkward to treat. Privacy is a genuine concern. Expose only what is needed and cover what has been exposed. Being professional will help reduce anxiety for both you and your patient.

Baseline Vital Signs

Quickly obtain the patient's baseline vital signs. Many abdominal emergencies, in addition to those that cause severe bleeding, can cause a rapid pulse and low blood pressure. Your record of vital signs, made as early as possible and periodically thereafter (every 5 minutes in the patient whom you suspect to have a serious injury), will help you to identify changes in the patient's condition and be alert to signs of decompensation from blood loss.

SAMPLE History

Next, obtain a SAMPLE history from your patient. Using OPQRST to help explain an abdominal injury may provide some helpful information. Some questions can be asked while assessing vital signs, for example while putting on the blood pressure cuff; however, this is the time to confirm that you have all the necessary history to inform the hospital. If the patient is not responsive, attempt to get the SAMPLE history from friends or family members.

Interventions

Manage airway and breathing problems based on signs and symptoms found in your initial assessment. Provide complete spinal stabilization to the patient with suspected spinal injuries. If the patient has signs of hypoperfusion, provide aggressive treatment for shock and rapid transport to the appropriate hospital. If an evisceration is discovered, place a moist, sterile dressing over the wound, apply a bandage, and transport. Never attempt to push eviscerated tissue or organs back into the abdominal cavity. Cover bleeding injuries to the genitalia with a moist dressing. Do not delay transport of the seriously injured trauma patient to complete nonlifesaving treatments such as splinting extremity fractures. Instead, complete these types of treatments en route to the hospital.

Detailed Physical Exam

As time permits, conduct a detailed physical exam. Thoroughly examine the patient from head to toe to identify injuries and determine how serious they are. You may even identify new injuries not previously found in the initial assessment or the focused history and physical exam portions of your assessment. Provide additional treatments as necessary. Short transport times or continuous problems with the ABCs may prevent you from performing a detailed physical exam; however, every effort should be made to thoroughly examine the patient prior to arrival at the hospital.

Ongoing Assessment

Repeat the patient's initial assessment and vital signs. Reassess the interventions and treatment you have provided to the patient. Identifying trends in pain, vital signs, and the progress of treatments will help determine whether the patient's condition is improving or getting worse. Adjustments in care can be based on these objective findings.

Communication and Documentation

Communicate the mechanism of injury and injuries found during your assessment. Use of appropriate medical and anatomic terminology is important; however, when in doubt just describe what you see. The content of your radio report will depend on your local protocols. The information you provide will help the hospital prepare for the patient. Documentation of your assessment and trends in vital signs is a tremendous help to physicians in evaluating the problem when the patient arrives in the emergency department. Continuity of care is maintained when the emergency department has an accurate record of your findings on scene as well as the treatments you have provided. Remember this is also a legal record of your care. If assault is suspected, you may have a legal requirement to inform the hospital staff of your suspicion. This can wait until you have delivered the patient to the hospital and have a chance to discuss it privately with appropriate staff.

Anatomy of the Genitourinary System

The genitourinary system controls both the reproductive functions and the waste discharge system, which are generally considered together.

The urinary system controls the discharge of certain waste materials filtered from the blood by the kidneys. In the urinary system, the kidneys are solid organs; the ureters, bladder, and urethra are hollow organs Figure 28-11 ▼ .

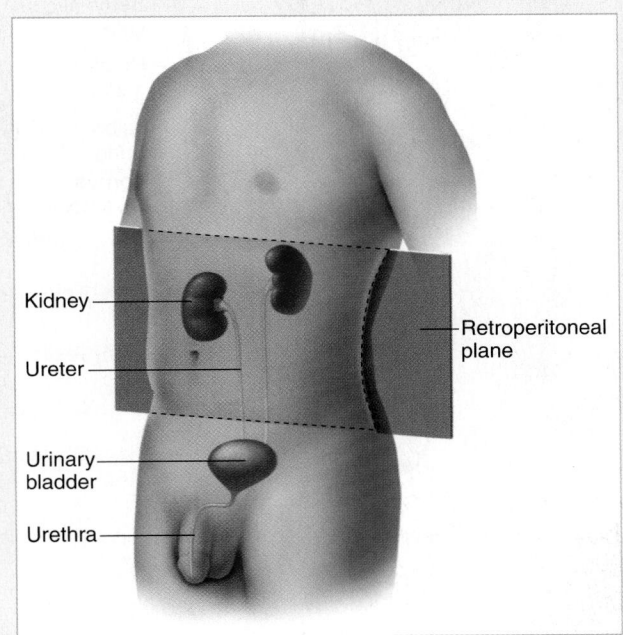

Figure 28-11 The urinary system lies in the retroperitoneal space behind the digestive tract. The kidneys are solid organs; the ureter, bladder, and urethra are hollow organs.

You are the Provider   Part 3

You administer high-flow oxygen using a nonrebreathing mask set to 15 L/min and begin to examine the area under the patient's hands. You carefully remove the patient's shirt and find that he has an approximately 4.5" laceration and subsequent evisceration with a moderately sized section of intestine protruding and visible. While you are examining the wound your partner completes a full-body rapid trauma assessment and finds no other injuries to the patient.

5. How would you treat this wound?
6. If the shirt material were stuck to the internal organs, would you pull it loose? Why or why not?

The genital system is also important to reproductive processes. The male genitalia, except for the prostate gland and the seminal vesicles, lie outside the pelvic cavity (Figure 28-12 ▼). The female genitalia, except for the vulva, clitoris, and labia, are contained entirely within the pelvis (Figure 28-13 ▼). The male and female reproductive organs have certain similarities and, of course, basic differences. They allow for the production of sperm and egg cells and appropriate hormones, the act of intercourse, and, ultimately, reproduction.

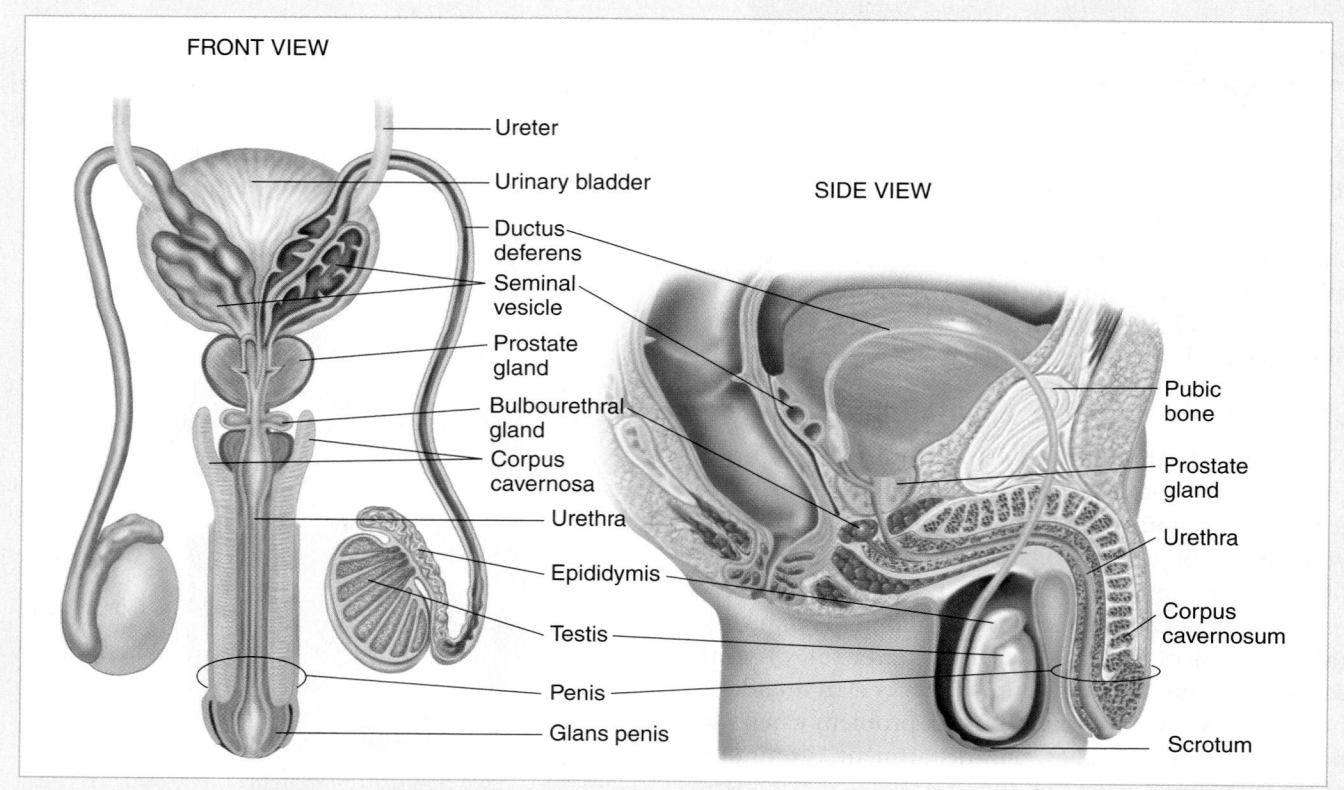

Figure 28-12 The male reproductive system includes the testicles, vasa deferentia, seminal vesicles, prostate gland, urethra, and penis.

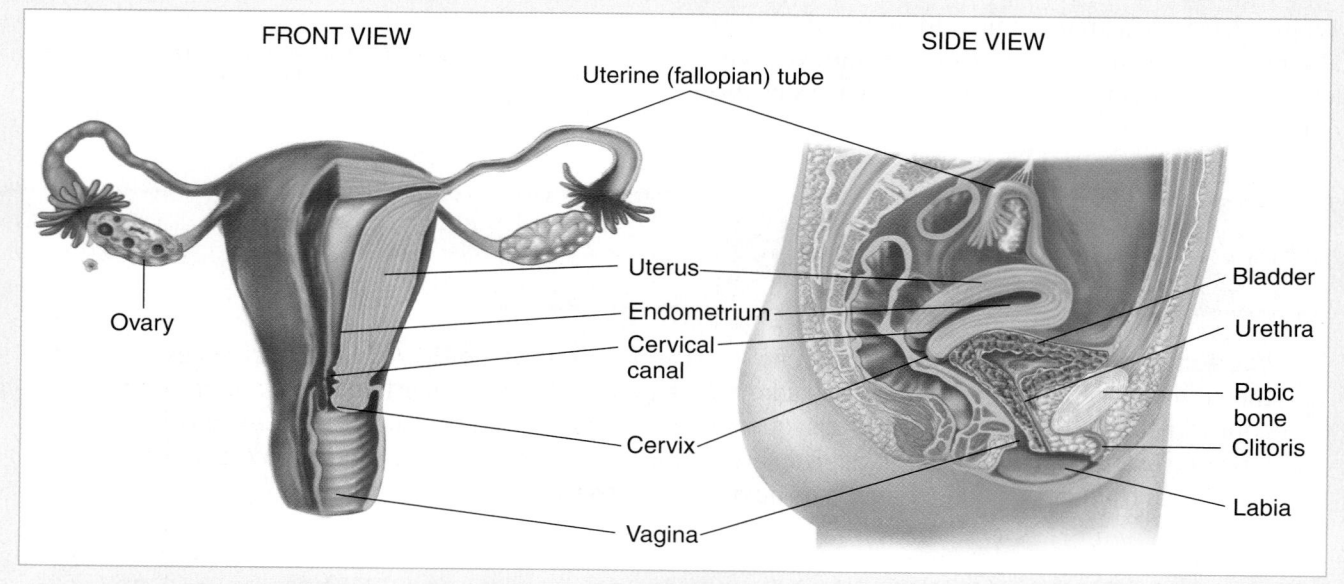

Figure 28-13 The female reproductive system includes the ovaries, fallopian tubes, uterus, cervix, and vagina.

Injuries of the Genitourinary System

Injuries of the Kidney

Injuries of the kidney are not unusual and rarely occur in isolation. This is because the kidneys lie in such a well-protected area of the body. A penetrating wound that reaches the kidneys almost always involves other organs. The same is true with blunt injuries. A blow that is forceful enough to cause significant kidney damage often results in damage to other intra-abdominal organs also. Less significant injuries to the kidneys may result from a direct blow or even from a tackle in football (Figure 28-14 ▼). Suspect kidney damage if the patient has a history or physical evidence of any of the following:

- An abrasion, laceration, or contusion in the flank
- A penetrating wound in the region of the lower rib cage (the flank) or the upper abdomen
- Fractures on either side of the lower rib cage or of the lower thoracic or upper lumbar vertebrae
- A hematoma in the flank region

Damage to the kidneys may not be obvious on inspection of the patient. You may or may not see bruises or lacerations on the overlying skin. However, you will see signs of shock if the injury is associated with significant blood loss. Because one of the functions of the kidney is the formation of urine, another sign of kidney damage is blood in the urine, called hematuria.

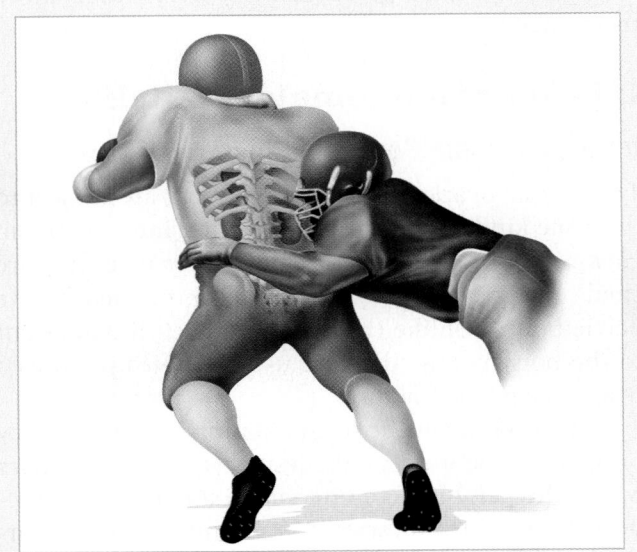

Figure 28-14 A tackle in football that results in blunt trauma to the lower rib cage or the flank can cause kidney injury.

Treat shock and associated injuries in the appropriate manner. Provide prompt transport to the hospital, monitoring the patient's vital signs carefully en route.

Injuries to the Urinary Bladder

Injury to the urinary bladder, either blunt or penetrating, may result in its rupture. When this happens, urine spills into the surrounding tissues, and any urine that passes through the urethra is likely to be bloody. Blunt injuries of the lower abdomen or pelvis often cause rupture of the urinary bladder, particularly when the bladder is full and distended. Sharp, bony fragments from a fracture of the pelvis often perforate the urinary bladder (Figure 28-15 ▶). Penetrating wounds of the lower midabdomen or the perineum (the pelvic floor and associated structures that occupy the pelvic outlet) can directly involve the bladder. In the male, sudden deceleration from a motor vehicle or motorcycle crash can literally shear the bladder from the urethra.

Suspect a possible injury of the urinary bladder if you see blood at the urethral opening or physical signs of trauma on the lower abdomen, pelvis, or perineum. There may be blood at the tip of the penis or a stain on the patient's underwear.

The presence of associated injuries or of shock will dictate the urgency of transport. In most instances, provide prompt transport, and monitor the patient's vital signs en route.

Injuries of the External Male Genitalia

Injuries of the external male genitalia include all types of soft-tissue wounds. Although these injuries are uniformly painful and generally a source of great concern to the patient, they are rarely life threatening. If you encounter a patient with an avulsion (tearing away) of skin of the penis, wrap the penis in a soft, sterile dressing moistened with sterile saline solution, and transport the patient promptly. Use direct pressure to control any bleeding. You should try to save and preserve the avulsed skin, but do not delay treatment or transport for more than a few minutes to do so.

Managing blood loss is your top priority in amputation of the penile shaft, whether partial or complete. You should use local pressure with a sterile dressing on the remaining stump. Never apply a constricting device to the penis to control bleeding. Surgical reconstruction of even a completely amputated penis is possible if you can locate the amputated part. Wrap it in a moist, sterile dressing; place it in a plastic bag; and transport it in a cooled container without allowing it to come in direct contact with ice.

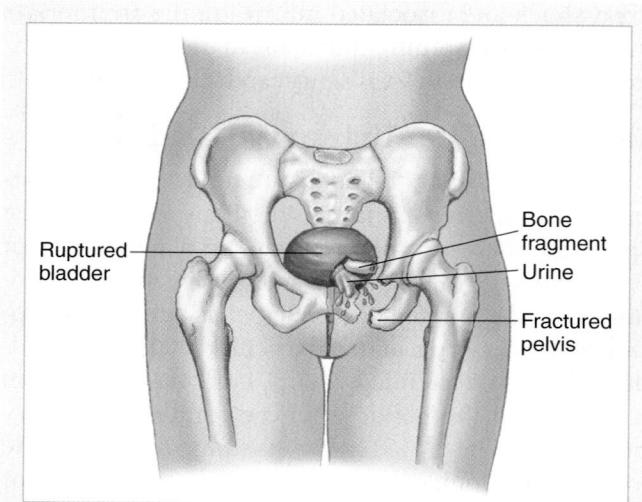

Figure 28-15 Fracture of the pelvis can result in a laceration of the bladder by the bony fragments. Urine then leaks into the pelvis.

Labels in figure: Ruptured bladder, Bone fragment, Urine, Fractured pelvis

If the connective tissue surrounding the erectile tissue in the penis is severely damaged, the shaft of the penis can be fractured or severely angled, sometimes requiring surgical repair. The injury may occur during particularly active sexual intercourse. It is associated with intense pain, bleeding into the tissues, and fear. Provide prompt transport to the emergency department.

Accidental laceration of the skin about the head of the penis usually occurs when the penis is erect and is associated with heavy bleeding. Local pressure with a sterile dressing is usually sufficient to stop the hemorrhage.

It is not uncommon for the skin of the shaft of the penis or the foreskin to get caught in the zipper of pants. If a small segment of the zipper is involved (one or two teeth), you can try to unzip the pants. If a longer segment is involved or the patient is agitated, use heavy scissors to cut the zipper out of the pants to make the patient more comfortable during transport. Be sure to explain how you are going to use the scissors before you begin cutting. Be particularly careful not to cause injury to the scrotum while cutting the zipper away from the penis.

Urethral injuries in the male are not uncommon. Lacerations of the urethra can result from straddle injuries, pelvic fractures, or penetrating wounds of the perineum. These injuries may bleed quite a lot, although bleeding may not be evident externally. Direct pressure with a dry, sterile dressing usually controls

any external hemorrhage. Because the urethra is the channel for urine, it is very important to know whether the patient can urinate and whether hematuria is present. For this reason, you should save any voided urine for later examination at the hospital. Any foreign bodies that may be protruding from the urethra will have to be removed in a surgical setting.

Avulsion of the skin of the scrotum may damage the scrotal contents. If possible, preserve the avulsed skin in a moist, sterile dressing for possible use in reconstruction. Wrap the scrotal contents or the perineal area with a sterile, moist compress, and use a local pressure dressing to control bleeding. Transport this patient promptly to the emergency department.

Direct blows to the scrotum can result in the rupture of a testicle or significant accumulation of blood around the testes. In either case, you should apply an ice pack to the scrotal area while transporting the patient.

A few general rules apply to the treatment of injuries involving the external male genitalia:

- These injuries are very painful. Make the patient as comfortable as possible.
- Use sterile, moist compresses to cover areas that have been stripped of skin.
- Apply direct pressure with dry, sterile gauze dressings to control bleeding.
- Never move or manipulate impaled instruments or foreign bodies in the urethra.
- If possible, always identify and bring avulsed parts to the hospital with the patient.

Remember, these are rarely life-threatening injuries and should not be given priority over other, more severe wounds.

Injuries of the Female Genitalia

Internal Female Genitalia

The uterus, ovaries, and fallopian tubes are subject to the same kinds of injuries as any other internal organ. However, they are rarely damaged because they are small, deep in the pelvis, and well protected by the pelvic bones. Unlike the bladder, which lies adjacent to the bony pelvis, they are usually not injured as a result of a pelvic fracture.

An exception is the pregnant uterus. As pregnancy progresses, the uterus enlarges substantially and rises out of the pelvis, becoming vulnerable to both penetrating and blunt injuries. These injuries can be particularly severe because the uterus has a rich blood supply during pregnancy. You must also keep in mind that the fetus is at risk. You can expect to see the signs and symptoms of shock with these patients; be prepared to provide all necessary support and prompt transport. Note also that contractions may begin. If possible, ask the patient when she is due to deliver, and report this information to the hospital.

In the last trimester of pregnancy, the uterus is large and may obstruct the vena cava, decreasing the amount of blood returning to the heart if the patient is placed in a supine position (supine hypotensive syndrome). As a result, blood pressure may decrease. The patient should be carefully placed on her left side so that the uterus will not lie on the vena cava. If the patient is secured to a backboard, tilt the board to the left.

External Female Genitalia

The external female genitalia include the vulva, the clitoris, and the major and minor labia (lips) at the entrance of the vagina. Injuries of the external female genitalia can include all types of soft-tissue injuries. Because these genital parts have a rich nerve supply, injuries are very painful. Lacerations, abrasions, and avulsions should be treated with moist, sterile compresses. Use local pressure to control bleeding and a diaper-type bandage to hold dressings in place. Under no circumstances should you pack or place dressings into the vagina. Leave any foreign bodies in place after you stabilize them with bandages.

In general, although these injuries are painful, they are not life threatening. Bleeding may be heavy, but it can usually be controlled by local compression. Contusions and other blunt injuries all require careful in-hospital evaluation. However, the urgency of the need for transport will be determined by associated injuries, the amount of hemorrhage, and the presence of shock.

Rectal Bleeding

Rectal bleeding is a common complaint and something that you may hear as a chief complaint or secondary to abdominal or pelvic complaints. Bleeding from the rectum may present as blood in or soaking through undergarments, or patients may complain of blood passed into the toilet associated with a bowel movement or attempted bowel movement. Rectal bleeding can be caused by sexual assault, hemorrhoids, colitis, or ulcers of the digestive track. Significant rectal bleeding can occur after hemorrhoid surgery and can lead to a large amount of blood loss and shock. Acute rectal bleeding should never be passed off as something minor. Pack the crease between the buttocks with compresses and consult with medical control to determine the need for transport.

You are the Provider   Part 4

You moisten a sterile dressing with saline and cover the organs and wound with it. Your partner hands you the "silver swadler" from the OB kit, and you use the aluminum foil-type material to cover and tape over the sterile dressing in order to keep the intestines warm. You and your partner place the patient on a long backboard, cover him with a blanket, and elevate the foot end of the backboard approximately 8" to treat for shock. The ALS unit informs you that they are less than 3 minutes away from your location, so you decide to await their arrival. In the meantime you begin to take a SAMPLE history from the patient and ascertain that he has no allergies, does not take any medications, has no prior medical history, and last ate approximately 3 hours ago. The patient tells you he felt a "cutting sensation" and then saw that he was bleeding. He held the wound and evisceration with his hands until you arrived. The patient's vital signs are a pulse of 120 beats/min, respirations of 22 breaths/min, and a blood pressure of 120/80 mm Hg. You contact the ALS unit and give the paramedic a report on your assessment and treatment. Within a minute the ALS unit arrives on scene and assumes treatment of the patient while complimenting you on a job well done.

7. If the ALS unit were farther away, would you transport your patient or stay at the scene?
8. What would you do with your patient during transport?

Sexual Assault

Sexual assault and rape are all too common. Although most victims are women, men and children are also victims. Often, you can do little beyond providing compassion and transportation to the emergency department. On some occasions, these patients will have suffered multiple-system trauma and will also need treatment for shock.

Do not examine the genitalia of a victim of sexual assault unless obvious bleeding requires you to apply a dressing. Advise the patient not to wash, douche, urinate, or defecate until after a physician has examined him or her; this will help to preserve any evidence of a crime. If oral penetration has occurred, advise the patient not to eat, drink, brush the teeth, or use mouthwash until he or she has been examined.

Treat all other injuries according to appropriate procedures and protocols for your EMS system. Observe BSI precautions. Take the patient's history, perform a limited physical exam, and provide treatment as quickly, quietly, and calmly as possible. Take care to shield the patient from curious onlookers.

The patient may refuse assistance or transport, often because he or she wants to maintain privacy and thus avoid public exposure. For adults, this is the patient's right. In such cases, you should follow your system's refusal of treatment policy or procedure for sexual assault victims without judging or being condescending to the patient. Your compassion is the best tool to gain the patient's confidence to get further help.

Offer to call the local rape crisis center for the patient. The center will have an advocate meet the patient at the hospital and provide support through the rape exam.

In addition to the usual treatment principles that apply to all victims of trauma, you should follow these special steps with patients who have been sexually assaulted:

1. Since you may have to appear in court as much as 2 or 3 years later, you must document the patient's history, assessment, treatment, and response to treatment in detail. Do not speculate. Record only the facts.
2. Make airway maintenance a major priority.
3. Complete the SAMPLE history in an objective and nonjudgmental fashion.
4. Follow any crime scene policy established by your system to protect the scene and any potential evidence for police, particularly that for evidence collection. If the patient will tolerate being wrapped in a sterile burn sheet, this may help investigators to find any hair, fluid, or fiber from the alleged offender.
5. Do not examine the genitalia unless there is major bleeding. If an object has been inserted into the vagina or rectum, do not attempt to remove it.
6. To reduce the patient's anxiety, make sure the EMT-B is the same gender as the patient whenever possible.
7. Discourage the patient from bathing, voiding, or cleaning any wounds until hospital staff has completed an assessment. Handle the patient's clothes as little as possible, placing articles and any other evidence in paper bags. Do not use plastic bags. If the female patient insists on urinating, have her do so in a sterile urine container (if available). Also, have her deposit the toilet paper in a paper bag. Seal and mark the bag for the police. This can be critical evidence.

Remember that victims of sexual assault, whether they are male or female, need medical assistance. In these cases, you must treat the medical injuries but also provide privacy, support, and reassurance.

You are the Provider   Summary

Abdominal injuries may be open or closed and are dangerous to the patient because a life-threatening amount of blood can be lost into the abdominal cavity. When you assess the abdomen, look for bruising and open wounds (including any possible exit wounds) and palpate for rigidity, tenderness, and complaints of pain. Impaled objects are never removed and must be secured in place. Eviscerations are covered with a moist, sterile dressing (or occlusive dressing, depending on your local protocol) and then, if possible, with some type of heat-retaining material. Approach patients with abdominal injuries with a high index of suspicion and constantly reassess your patient's condition and vital signs. Finally, unless ALS is almost on scene, immediately transport your patient and allow ALS to intercept you if they can before you reach the hospital.

Abdomen and Genitalia Injuries	
Scene Size-up	Body substance isolation precautions should include a minimum of gloves and eye protection. Ensure scene safety and determine NOI/MOI. Consider the number of patients, the need for additional help/ALS, and c-spine stabilization.
Initial Assessment	
■ General impression	Determine level of consciousness and treat any immediate threats to life. Determine priority of care based on environment and patient's chief complaint.
■ Airway	Ensure patent airway.
■ Breathing	Listen for abnormal breath sounds and evaluate depth and rate of the respirations. Look for symmetric chest rise and fall. Maintain ventilations as needed. Provide high-flow oxygen at 15 L/min and inspect the chest wall, assessing for DCAP-BTLS.
■ Circulation	Evaluate pulse rate and quality; observe skin color, temperature, and condition, and treat accordingly.
■ Transport decision	Prompt transport
Focused History and Physical Exam	*NOTE: The order of the steps in the focused history and physical exam differs depending on whether or not the patient has a significant MOI. The order below is for a patient with a significant MOI. For a patient without a significant MOI, perform a focused trauma assessment, obtain vital signs, and obtain the history.*
■ Rapid trauma assessment	Reevaluate the mechanism of injury. Perform a rapid trauma assessment, treating all life threats immediately. Log roll and secure patient to backboard if a spinal injury is suspected.
■ Baseline vital signs	Take vital signs, noting skin color and temperature as well as patient's level of consciousness. Use pulse oximetry if available.
■ SAMPLE history	Obtain SAMPLE history. If the patient is unresponsive, attempt to get it from family members, friends, or bystanders.
■ Interventions	Provide complete spinal immobilization early if you suspect that your patient has spinal injuries. Maintain an open airway and suction as needed. If an evisceration is present, place a moist, sterile dressing over the wound and bandage. Cover bleeding injuries to the genitalia with a moist dressing. Treat signs of shock and any other life threats.
Detailed Physical Exam	Complete a detailed physical exam.
Ongoing Assessment	Repeat the initial assessment, rapid/focused assessment, and reassess interventions performed. Reassess vital signs every 5 minutes for the unstable patient, every 15 minutes for the stable patient.
■ Communication and documentation	Contact medical control with a radio report. Relay any change in level of consciousness or difficulty breathing. Describe the MOI and any interventions you performed. Be sure to document any physician's orders and changes in the patient's condition and at what time they occurred.

Assessment and Emergency Care

Assessment and Emergency Care

NOTE: While the steps below are widely accepted, be sure to consult and follow your local protocol.

Kidney and Urinary Bladder Injuries

1. Treat for shock early and aggressively.
2. Place patient in position of comfort unless spinal injury is suspected.
3. Provide prompt transport.
4. Monitor vital signs en route.

External Male Genitalia Injuries

Avulsion

1. Wrap the penis in a soft, sterile dressing moistened with sterile saline solution.
2. Use direct pressure to control any bleeding.
3. Try to save and preserve the avulsed skin, but do not delay transport for more than a few minutes to do so.

Amputation

1. Manage blood loss using local pressure with a sterile dressing on the remaining stump.
2. Wrap the amputated part in a moist, sterile dressing; place it in a plastic bag; transport in a cooled container without allowing it to come in direct contact with ice.

Fractured or Severely Angled

1. Provide prompt transport.

Lacerated Head of the Penis

1. Stop the hemorrhage with a sterile dressing and local pressure.

Penis Caught in a Zipper

1. Attempt to unzip pants.
2. If unable to unzip pants, explain to the patient that you are going to cut the zipper out of the pants to relieve pressure.

Urethral Injuries

1. Save any voided urine for examination.

Avulsion of Scrotum Skin

1. Preserve avulsed skin in moist, sterile dressing.
2. Wrap scrotal contents or the perineal area with a sterile, moist dressing to control bleeding.

Rupture of a Testicle or Significant Accumulation of Blood Around the Testes

1. Apply ice pack to the scrotal area.

External Female Genitalia Injuries

1. Treat with moist, sterile compress.
2. Apply local pressure to control the bleeding and a diaper-type bandage to hold dressings in place.

Internal Female Genitalia Injuries, Pregnancy

1. Carefully place patient on her left side.
2. If the patient is on the backboard, tilt the board to the left.

Rectal Bleeding

1. Pack the crease between the buttocks with compresses and consult with medical control to determine the need for transport.

Sexual Assault

1. Document the patient's history, assessment, treatment, and response to treatment in detail.
2. Make airway maintenance a major priority.
3. Complete the SAMPLE history in an objective, nonjudgmental way.
4. If the patient will tolerate being wrapped in a sterile burn sheet, this may help investigators to find any hair, fluid, or fiber from the alleged offender.
5. Do not examine the genitalia unless there is major bleeding. If an object has been inserted into the vagina or rectum, do not attempt to remove it.
6. Make sure the EMT-B is the same gender as the patient whenever possible.
7. Discourage the patient from bathing, voiding, or cleaning any wounds until hospital staff has completed an assessment. Handle the patient's clothes as little as possible, placing articles and any other evidence in paper bags. Do not use plastic bags. If the female patient insists on urinating, have her do so in a sterile urine container (if available). Also, have her deposit the toilet paper in a paper bag. Seal and mark the bag for the police. This can be critical evidence.

Prep Kit

Ready for Review

- Abdominal injuries are categorized as either open (penetrating trauma) or closed (blunt force trauma).

- Either classification of injury can result in injury to the hollow or solid organs of the abdomen and cause significant life-threatening bleeding.

- Blunt force trauma that causes closed injuries results from an object striking the body without breaking the skin, such as being hit with a baseball bat or when the patient's body strikes the steering wheel during a motor vehicle crash.

- Penetrating trauma is often a result of a gunshot wound or stab wound. Other mechanisms of injury such as a fall on an object can also cause penetrating trauma to the abdomen.

- Injury to the solid internal organs often causes significant unseen bleeding that can be life threatening.

- Injury to the hollow organs of the abdomen may cause irritation and inflammation to the peritoneum as caustic digestive juices leak into the peritoneum. A serious infection may also occur over several hours.

- Always maintain a high index of suspicion for serious intra-abdominal injury in the trauma patient, particularly in the patient who exhibits signs of shock.

- Assess the abdomen for signs of bruising, rigidity, penetrating injuries, and complaints of pain.

- Never remove an impaled object from the abdominal region. Secure it in place with a large bulky dressing and provide prompt transport.

- Be prepared to treat the patient for shock. Place the patient in the shock position, keep the patient warm, and provide high-flow oxygen.

- Never replace an organ that protrudes from an open injury to the abdomen (evisceration). Instead keep the organ moist and warm. Cover the injury site with a large sterile, moist, bulky dressing.

- Injuries to the kidneys may be difficult to detect due to the well-protected region of the body where they are located. Be alert to bruising or hematoma to the flank region.

- Injury to the external genitalia of male and female patients is very painful, but not usually life threatening.

- In the case of sexual assault or rape, treat for shock if necessary, and record all the facts in detail. Follow any crime scene policy established by your system to protect the scene and any potential evidence. Advise the patient not to wash, douche, or void until after a physician has examined him or her.

Vital Vocabulary

closed abdominal injury Any injury of the abdomen caused by a nonpenetrating instrument or force, in which the skin remains intact; also called blunt abdominal injury.

evisceration The displacement of organs outside of the body.

flank The region of the lower rib cage.

guarding Contracting the stomach muscles to minimize the pain of abdominal movement; a sign of peritonitis.

hematuria The presence of blood in the urine.

hollow organs Structures through which materials pass, such as the stomach, small intestines, large intestines, ureters, and bladder.

open abdominal injury An injury of the abdomen caused by a penetrating or piercing instrument or force, in which the skin is lacerated or perforated and the cavity is opened to the atmosphere; also called penetrating injury.

peritoneal cavity The abdominal cavity.

peritonitis Inflammation of the peritoneum.

solid organs Solid masses of tissue where much of the chemical work of the body takes place (eg, the liver, spleen, pancreas, and kidneys).

supine hypotensive syndrome A drop in blood pressure caused when the heavy uterus of a supine patient in the third trimester of pregnancy obstructs the vena cava, decreasing blood return to the heart.

Technology

- Interactivities
- Vocabulary Explorer
- Anatomy Review
- Web Links
- Online Review Manual

www.EMTB.com

Assessment in Action

You are dispatched to the state prison for an unknown medical emergency. Your department frequently responds to this facility, often with little information regarding the nature of the call. On scene, a correctional officer tells you that he isn't sure what happened, but the patient has reportedly been urinating blood for the past few hours.

The patient is in the facility's urgent care area. He is lying down on a gurney and is somewhat confused. He tells you that he fell down a flight of stairs and now his back hurts. You ask him to pull down the top of his jumpsuit to look at his back, where you see multiple large bruises. His vital signs include a blood pressure of 98/60 mm Hg, a radial pulse of 130 beats/min and thready, and respirations of 36 breaths/min.

1. In this scenario, what does hematuria signify?

   A. Liver trauma
   B. Bladder trauma
   C. Kidney trauma
   D. Genitalia trauma

2. The kidneys are located in the:

   A. retroperitoneal cavity.
   B. peritoneal cavity.
   C. abdominal cavity.
   D. none of the above.

3. Abdominal injuries are categorized as either:

   A. open or closed.
   B. penetrating or blunt.
   C. both A and B.
   D. neither A or B.

4. Injury to the solid internal organs often causes:

   A. significant bleeding.
   B. unseen bleeding.
   C. life-threatening bleeding.
   D. all of the above.

5. Injury to hollow abdominal organs may cause irritation and inflammation to the abdominal lining. This condition is referred to as:

   A. sepsis.
   B. cholecystitis.
   C. peritonitis.
   D. eviscerated bowel.

6. The first signs associated with the answer to question 5 include:

   A. abdominal pain.
   B. tenderness.
   C. muscular spasm.
   D. all of the above.

Challenging Questions

7. What parts of the patient's history in the scenario don't seem factual? Is this important?

8. Why is a distended abdomen a significant finding in the presence of abdominal injuries?

9. Do seat belts and airbags ever cause significant injuries?

10. Why is knowing abdominal anatomy and landmarks important to your patient care?

Points to Ponder

You are dispatched to 1212 Pierce Dr for a woman with labor pains. En route to the call, the dispatcher informs you that this patient does not speak English, and there were initially some communication difficulties. Police officers arrived to the scene first and found a young, pregnant woman who appeared to be in some sort of distress. As you arrive, you see a young Hispanic woman holding her abdomen and crying. A translator on-scene tells you that she was kicked in the stomach by her husband, who immediately fled the area. The patient now reports cramping, spotting, and abdominal pain. The patient is 30 weeks' pregnant and the mother of one other child.

How should this patient be transported? What impact can communication barriers have on your patient care?

Issues: Abdominal Trauma in the Pregnant Patient, Communicating With Language Barriers, Transport Considerations for Pregnant Patients.

Musculoskeletal Care

Objectives

Cognitive

5-3.1   Describe the function of the muscular system. (p 820)

5-3.2   Describe the function of the skeletal system. (p 821)

5-3.3   List the major bones or bone groupings of the spinal column, the thorax, the upper extremities, and the lower extremities. (p 822)

5-3.4   Differentiate between an open and closed painful, swollen, deformed extremity (fracture). (p 824)

5-3.5   State the reasons for splinting. (p 836)

5-3.6   List the general rules of splinting. (p 836)

5-3.7   List the complications of splinting. (p 846)

5-3.8   List the emergency medical care for a patient with a swollen, painful, deformed extremity (fracture). (p 836)

Affective

5-3.9   Explain the rationale for splinting at the scene versus load and go. (p 830)

5-3.10 Explain the rationale for immobilization of the painful, swollen, deformed extremity (fracture). (p 836)

Psychomotor

5-3.11 Demonstrate the emergency medical care of a patient with a painful, swollen, deformed extremity (fracture). (p 836)

5-3.12 Demonstrate completing a prehospital care report for patients with musculoskeletal injuries. (p 838)

You are the Provider

You and your EMT-B partner are dispatched to the local skateboarding rink for a fall injury. En route you and your partner discuss how the increase in high-risk adventure sports such as extreme skateboarding and bicycle and rollerblading exhibition or competition has brought an increase in these types of injuries.

1. What is the difference in prehospital care between a fracture and a dislocation?
2. Why is an open fracture more complicated to care for than a closed fracture?

Musculoskeletal Care

The human body is a well-designed system in which form, upright posture, and movement are provided by the musculoskeletal system. This system also protects the vital internal organs of the body. As its combination form suggests, the term "musculoskeletal" refers to the bones and voluntary muscles of the body. However, the bones and muscles themselves are susceptible to external forces that can cause injury. Also at risk are the tendons that attach muscles to bones, the joints that form wherever two bones come into contact, and the ligaments that hold the bone ends of a joint together.

As an EMT-B, you must be familiar with the basic anatomy of the body's musculoskeletal system. Although muscles are technically soft tissue, they are discussed in this chapter because of their close relationship to the skeleton. Therefore, the chapter begins with a review of the musculoskeletal anatomy. Various types and causes of musculoskeletal injuries in general are identified, and the assessment and treatment process for each is explained, followed by a detailed discussion of splinting. The chapter then focuses on specific musculoskeletal injuries, beginning at the clavicle and ending at the feet.

Anatomy and Physiology of the Musculoskeletal System

Muscles

The muscular system includes three types of muscles: skeletal, smooth, and cardiac (Figure 29-1 ▼). Skeletal muscle, also called striated muscle because of its characteristic stripes, attaches to the bones and usually crosses at least one joint, forming the major muscle mass of the body. This type of muscle is also called voluntary muscle, because it is under direct voluntary control of the brain, responding to commands to move specific body parts. Usually, movement is the result of several muscles contracting and relaxing simultaneously. Skeletal muscle is the component of the muscular system that is included in the overall musculoskeletal system. Cardiac muscle contributes to the cardiovascular system and smooth muscle is a component of multiple other body systems, including the digestive system and the cardiovascular system.

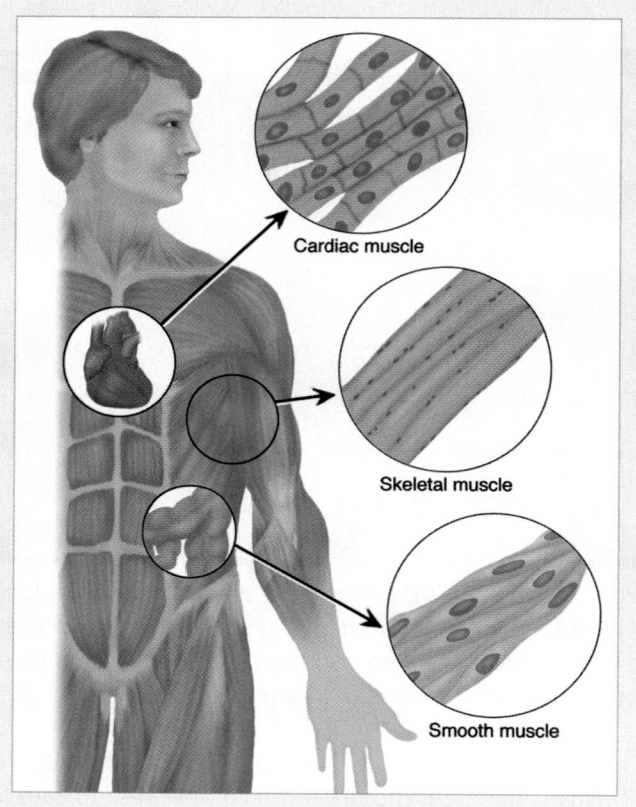

Figure 29-1 The major muscle type of concern for musculoskeletal injuries is skeletal, or voluntary, muscle.

Technology

Interactivities

Vocabulary Explorer

Anatomy Review

Web Links

Online Review Manual

www.EMTB.com

All skeletal muscles are supplied with arteries, veins, and nerves. Blood from the arteries brings oxygen and nutrients to the muscles (Figure 29-2 ▼). Waste products, including carbon dioxide and lactic acid, are carried away in the veins. Either disease or trauma can result in the loss of a muscle's nervous supply; this, in turn, can lead to weakness and eventually atrophy, or a decrease in the size of the muscle and its inherent ability to function. Skeletal muscle tissue is directly attached to the bone by tough, ropelike fibrous structures known as <u>tendons</u>, which are extensions of the fascia that covers all skeletal muscle.

Smooth muscle, also called involuntary muscle because it is not under voluntary control of the brain, performs much of the automatic work of the body. This type of muscle is found in the walls of most tubular structures of the body, such as the gastrointestinal tract and the blood vessels. Smooth muscle contracts and relaxes to control the movement of the contents within these structures (Figure 29-3 ▶).

The heart neither looks nor acts like skeletal or smooth muscle. It is composed largely of cardiac muscle, a specially adapted involuntary muscle with its own regulatory system.

The remainder of this chapter is concerned exclusively with skeletal muscle.

The Skeleton

The skeleton, which gives us our recognizable human form, protects our vital internal organs, and allows us to move, is made up of approximately 206 bones (Figure 29-4 ▶). The bones in the skeleton also produce blood cells (in the bone marrow) and serve as a reservoir for important minerals and electrolytes.

The skull is a solid vault-like structure that surrounds and protects the brain. The thoracic cage protects the heart, lungs, and great vessels; the lower ribs protect the liver and spleen. The bony spinal canal encases and protects the spinal cord. The upper extremity extends from the shoulder to the fingertips and is composed of the arm (humerus), elbow, forearm (radius and ulna), wrist, hand, and fingers. The arm extends from the shoulder to the elbow. The pelvis supports the body weight and protects the structures within the pelvis: the bladder, rectum, and female reproductive organs. The lower extremity consists of the thigh (femur), leg (tibia and fibula), and foot. The joint between the pelvis and the thigh is the hip; the joint between the thigh and lower leg is the knee, and the joint between the lower leg and foot is the ankle.

The bones of the skeleton provide a framework to which the muscles and tendons are attached. Bone is a living tissue that contains nerves and receives oxygen and nutrients from the arterial system. Therefore, when a bone breaks, a patient typically experiences severe pain and bleeding. Bone marrow, located in the center of each bone, is constantly producing red blood cells to

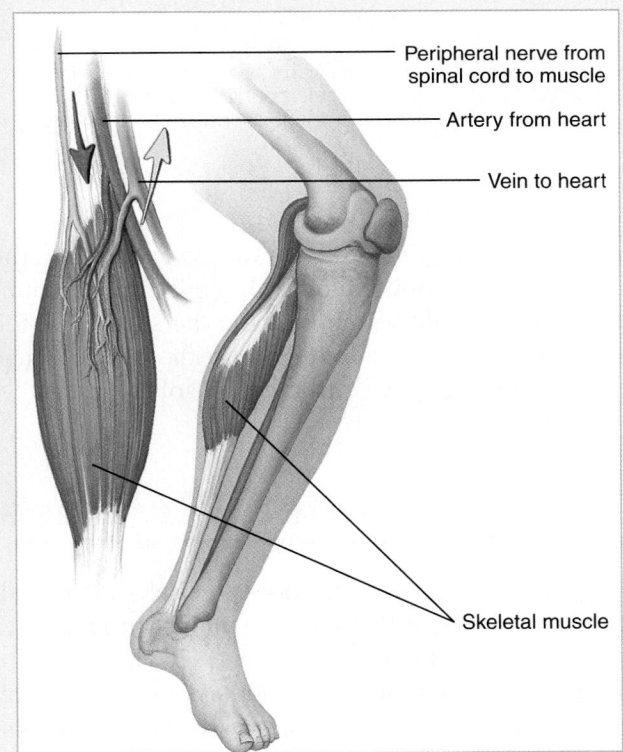

Figure 29-2 Skeletal muscles are supplied with arteries, veins, and nerves that bring oxygen and nutrients, carry away waste products, and supply nervous stimuli.

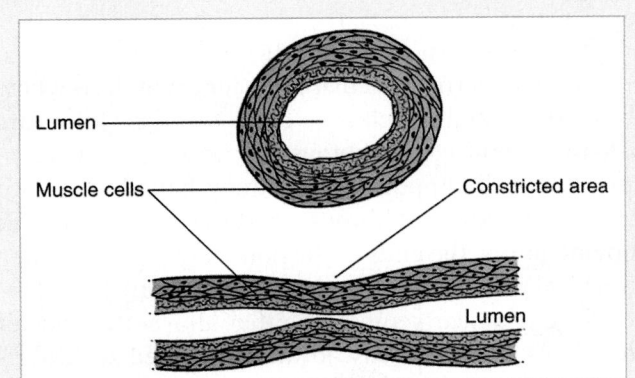

Figure 29-3 Smooth muscle is found in the walls of most tubular structures in the body. These muscles contract and relax to control the movement of the contents within these structures.

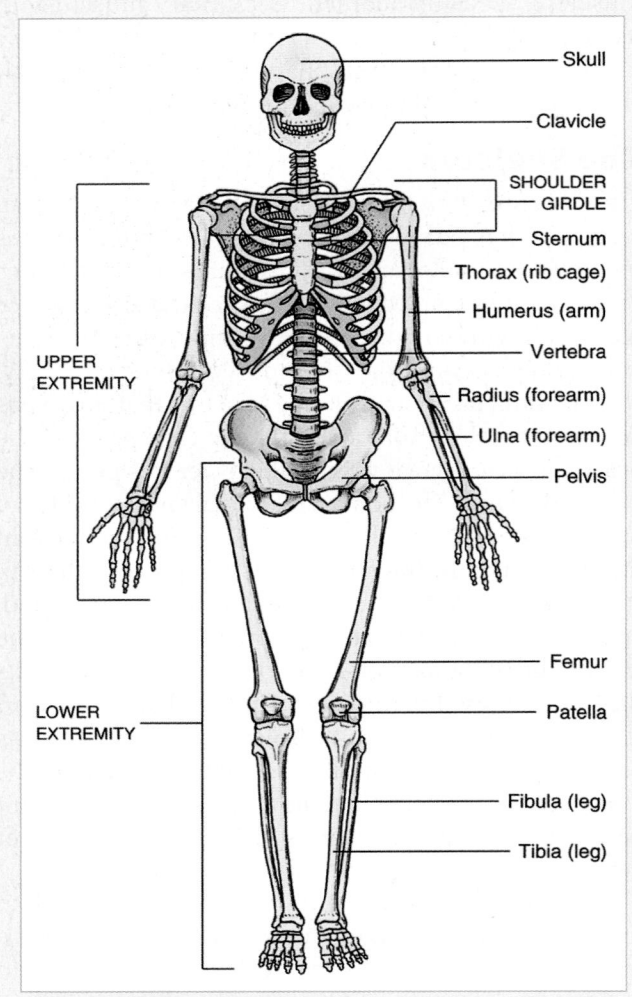

Figure 29-4 The human skeleton, consisting of 206 bones, gives us our form and protects our vital organs.

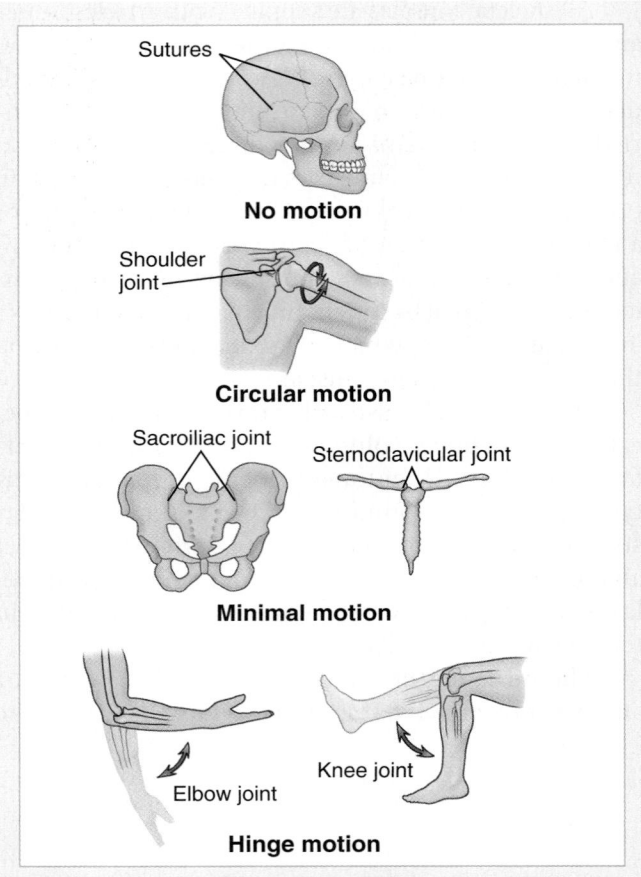

Figure 29-5 Joints have many functions. Some joints allow for motion to occur in a circular fashion; others act as hinges. Still others allow only a minimum amount of motion, or none at all.

provide oxygen and nourishment to the body and remove waste.

A <u>joint</u> is formed wherever two bones come into contact. The sternoclavicular joint, for example, is where the sternum and the clavicle come together. Joints are held together in a tough fibrous structure known as a capsule, which is supported and strengthened in certain key areas by bands of fibrous tissue called <u>ligaments</u>. In moving joints, the ends of the bones are covered with a thin layer of cartilage known as <u>articular cartilage</u>. This cartilage is a pearly substance that allows the ends of the bones to glide easily. Joints are bathed and lubricated by synovial (joint) fluid.

Some joints, such as the shoulder, allow motion to occur in a circular fashion. Other joints, such as the knee and elbow, act as hinges. Still other joints,

including the sacroiliac joint in the lower back and the sternoclavicular joints, allow only a minimum amount of motion. Certain joints, such as the sutures in the skull (present until about 18 months of life), fuse together during growth to create a solid, immobile, bony structure (Figure 29-5 ▲).

Musculoskeletal Injuries

A <u>fracture</u> is a broken bone. More precisely, it is a break in the continuity of the bone, often occurring as a result of an external force (Figure 29-6 ▶). The break can occur anywhere on the surface of the bone and in many different types of patterns. Contrary to a common misconception, there is no difference between a broken bone and a fractured bone.

A <u>dislocation</u> is a disruption of a joint in which the bone ends are no longer in contact. The supporting

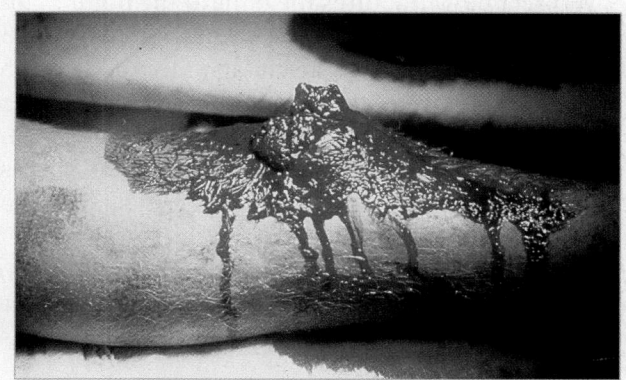

Figure 29-6 A fracture can occur anywhere on the surface of a bone and may or may not break the skin.

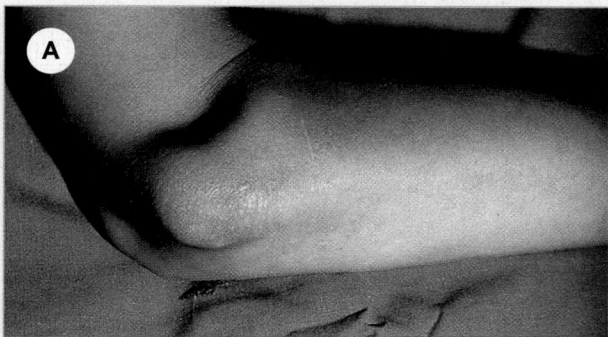

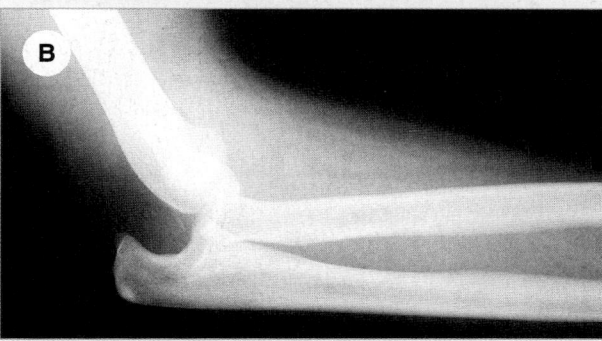

Figure 29-7 A dislocation is a disruption of a joint in which the bone ends are no longer in contact. **A.** The clinical appearance of an elbow dislocation. **B.** X-ray appearance of the same elbow.

ligaments are often torn, usually completely, allowing the bone ends to separate completely from each other (Figure 29-7 ▶). A subluxation is similar to a dislocation except the disruption of the joints is not complete. Therefore, a subluxation is an incomplete dislocation of a joint. A fracture-dislocation is a combination injury at the joint in which the joint is dislocated and there is a fracture of the end of one or more of the bones.

A <u>sprain</u> is a joint injury in which there is both some partial or temporary dislocation of the bone ends and partial stretching or tearing of the supporting ligaments. After the injury, the joint surfaces generally fall back into alignment, so the joint is not significantly displaced. Sprains can range from mild to severe, depending on the amount of damage done to the supporting ligaments. The most severe sprains involve complete dislocation of the joint; mild sprains typically heal rather quickly.

A <u>strain</u>, or muscle pull, is a stretching or tearing of the muscle, causing pain, swelling, and bruising of the soft tissues in the area. Unlike a sprain, no ligament or joint damage typically occurs.

Injury to bones and joints is often associated with injury to the surrounding soft tissues, especially to the adjacent nerves and blood vessels. The entire area is

You are the Provider   Part 2

You arrive on scene. You ensure that the scene is safe, and you approach the patient. You find an 18-year-old man who is holding his left arm close to his chest. He appears to be in a lot of pain. The patient is conscious, alert, and oriented and is talking to a young man standing next to him. You see no major external bleeding.

3. What is your next action in caring for this patient?
4. What information should you determine about the possible injury he has?

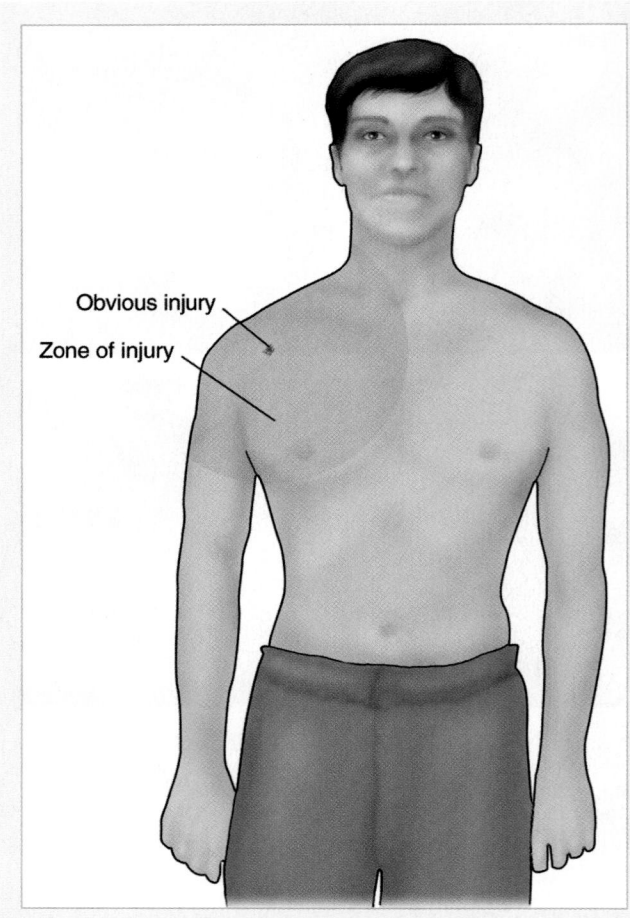

Figure 29-8 The zone of injury is the area of soft tissue, including the adjacent nerves and blood vessels, that surrounds the obvious injury of a bone or joint.

known as the zone of injury (Figure 29-8 ▲). Depending on the amount of kinetic energy the tissues absorb from forces acting on the body, the zone may extend to a distant point. For this reason, you should not focus on a patient's obvious injury without first completing a rapid assessment to check for associated injuries, which may be even more serious. This is especially true in assessing damage from high-energy trauma or gun shots.

Mechanism of Injury

Significant force is generally required to cause fractures or dislocations. This force may be applied to the limb in any of the following ways (Figure 29-9 ▶):

- Direct blows
- Indirect forces
- Twisting forces
- High-energy injury

A direct blow fractures the bone at the point of impact. An example is the patella (kneecap) that fractures when it strikes the dashboard in an automobile crash.

Indirect force may cause a fracture or dislocation at a distant point, as when a person falls and lands on an outstretched hand. The direct impact may cause a wrist fracture, but the indirect force can cause dislocation of the elbow or a fracture of the forearm, humerus, or even clavicle. Therefore, when caring for patients who have fallen, you must identify the point of contact and the mechanism of injury so that you will not overlook associated injuries.

Twisting forces are a common cause of musculoskeletal injury, especially to the anterior cruciate ligament at the knee. Skiing injuries often happen this way. A ski becomes caught, and the skier falls, applying a twisting force to the lower extremity.

High-energy injuries, such as those that occur in automobile crashes, falls from heights, gunshot wounds, and other extreme forces, produce severe damage to the skeleton, surrounding soft tissues, and vital internal organs. A patient may have multiple injuries to many body parts, including more than one fracture or dislocation in a single limb.

A significant MOI is not necessary to fracture a bone. A slight force can easily fracture a bone that is weakened by a tumor or osteoporosis, a generalized bone disease that is common among postmenopausal women. In geriatric patients with osteoporosis, minor falls, simple twisting injuries, or even a muscle contraction can cause a fracture, most often of the wrist, spine, or hip. You should suspect the presence of a fracture in any older patient who has sustained even a mild injury.

Fractures

Fractures are classified as either closed or open. In assessing and treating patients with possible fractures or dislocations, your first priority is to determine whether the overlying skin is damaged. If it is not, the patient has a closed fracture. However, making this determination is not always as easy as it sounds. With an open fracture, there is an external wound, caused either by the same blow that fractured the bone or by the broken bone ends lacerating the skin. The wound may vary in size from a very small puncture to a gaping tear that exposes bone and soft tissue. Regardless of the extent and severity of the damage to the skin, you should treat any injury that breaks the skin as a possible open fracture. Greater blood loss and a higher likelihood of

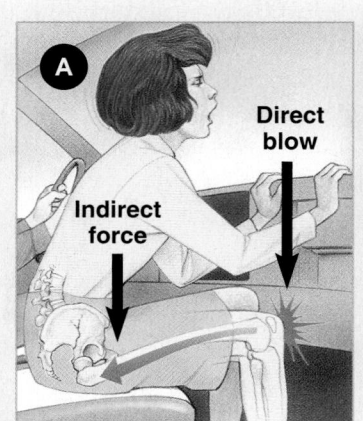

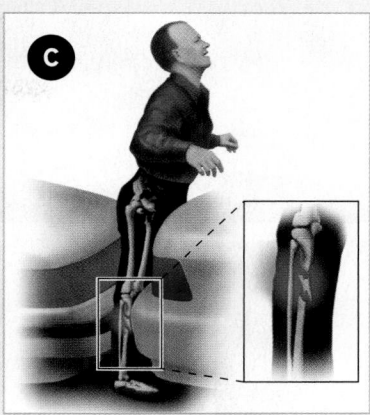

Figure 29-9 Significant force is required to cause fractures or dislocations. Among these are **(A)** Direct blows and indirect forces, **(B)** Twisting forces, and **(C)** High-energy crushing injuries.

infection are complications that you must try to limit; these tend to occur with open fractures.

Fractures are also described by whether the bone is moved from its normal position. A <u>nondisplaced fracture</u> (also known as a hairline fracture) is a simple crack of the bone that may be difficult to distinguish from a sprain or simple contusion. X-rays are required for hospital personnel to diagnose a nondisplaced fracture. A <u>displaced fracture</u> produces actual deformity, or distortion, of the limb by shortening, rotating, or angulating it. Often, the deformity is very obvious and can be associated with crepitus or free movement of a bone that is not normal for that region of the body. However, in some cases the deformity is minimal. Be sure to look for differences between the injured limb and the opposite uninjured limb in any patient with a suspected fracture of an extremity (Figure 29-10 ▶).

Medical personnel often use the following special terms to describe particular types of fractures (Figure 29-11 ▶):

- **Greenstick fracture.** An incomplete fracture that passes only partway through the shaft of a bone but may still cause substantial angulation; occurs in children.
- **Comminuted fracture.** A fracture in which the bone is broken into more than two fragments.
- **Pathologic fracture.** A fracture of weakened or diseased bone, seen in patients with osteoporosis or cancer, generally produced by minimal force.
- **Epiphyseal fracture.** A fracture that occurs in a growth section of a child's bone and may lead to growth abnormalities.

You should suspect a fracture if one or more of the following signs is present in any patient who has a history of injury and reports pain.

Deformity

The limb may appear to be shortened, rotated, or angulated at a point where there is no joint (Figure 29-12 ▶). Always use the opposite limb as a mirror image for comparison.

Tenderness

<u>Point tenderness</u> on palpation in the zone of injury is the most reliable indicator of an underlying fracture, although it does not tell you the type of fracture (Figure 29-13 ▶). Be sure to wear gloves if there are any open wounds.

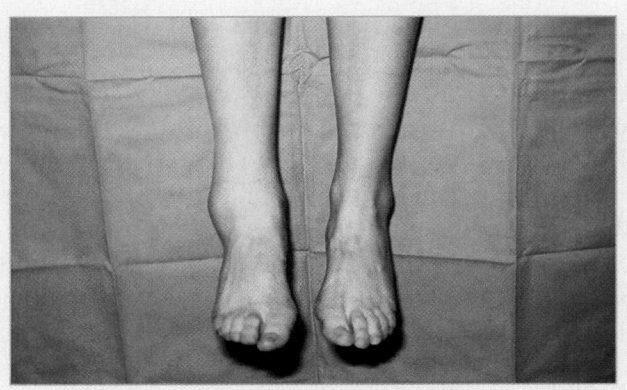

Figure 29-10 You should always compare the injured limb with the uninjured limb when checking for deformity.

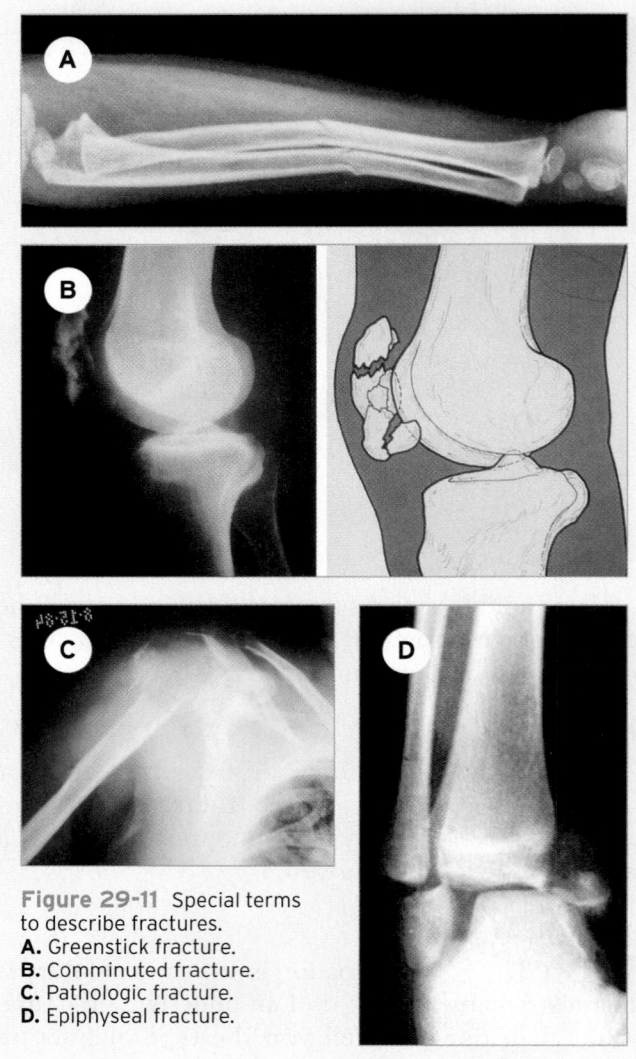

Figure 29-11 Special terms to describe fractures.
A. Greenstick fracture.
B. Comminuted fracture.
C. Pathologic fracture.
D. Epiphyseal fracture.

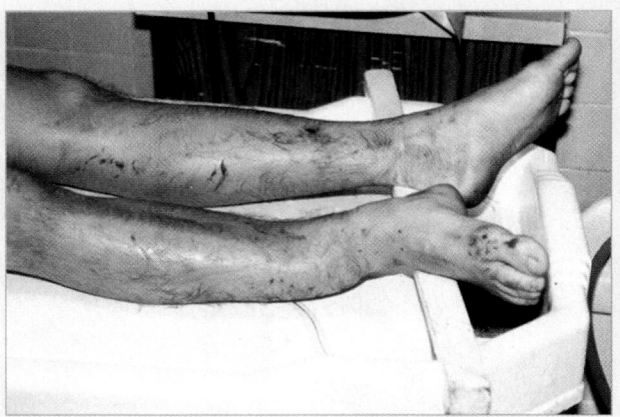

Figure 29-12 Obvious deformity, shortening, rotation, or angulation should increase the EMT-B's index of suspicion for a fracture. Remember to compare the injured limb with the opposite, uninjured limb.

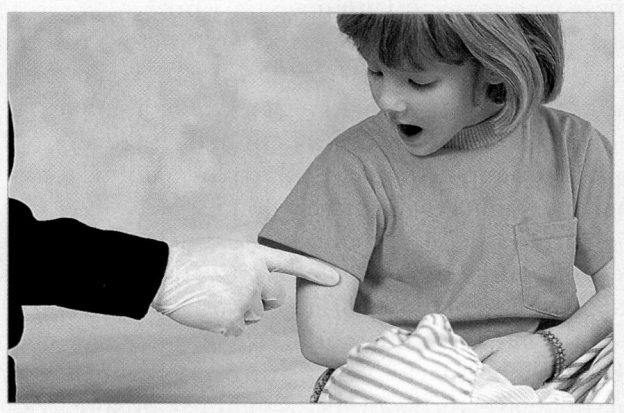

Figure 29-13 Point tenderness is the sensitive spot at the site of injury that can be located by palpation along the bone with the tip of your finger.

Guarding

An inability to use the extremity is the patient's way of immobilizing it to minimize pain. The muscles around the fracture contract in an attempt to prevent any movement of the broken bone. Guarding does not occur with all fractures; some patients may continue to use the injured part for a period of time. Occasionally, nondisplaced fractures are less painful, and there is minimal soft-tissue damage.

Swelling

Rapid swelling usually indicates bleeding from a fracture and is typically followed by substantial pain. Often, if the swelling is severe, it may mask deformity of the limb (Figure 29-14 ▶). Generalized swelling from fluid build up may occur several hours after an injury.

Bruising

Fractures are almost always associated with ecchymosis (discoloration) of the surrounding soft tissues (Figure 29-15 ▶). Bruising may be present after almost any injury and may take hours to develop; it is not specific to bone or joint injuries. The discoloration associated with acute injuries is usually redness as you may have seen with someone who has been punched. In hours to days blue, purple, and black will appear, followed by yellows and greens.

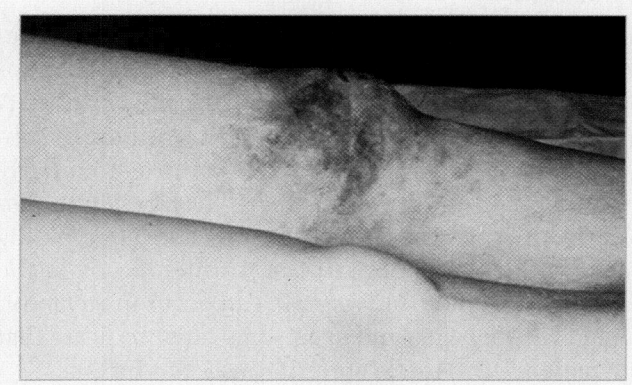

Figure 29-14 Fractures almost always have associated bruising into the surrounding soft tissue.

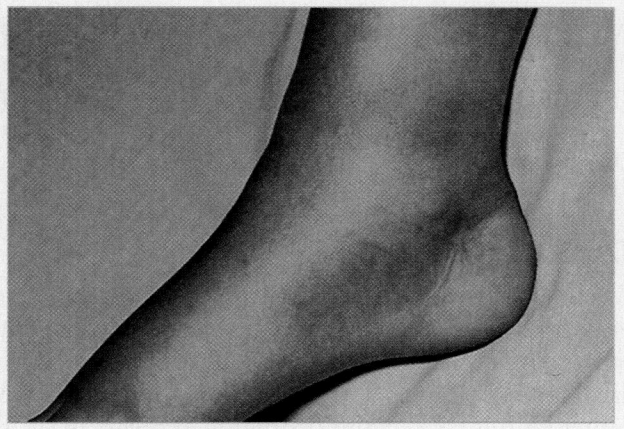

Figure 29-15 Swelling that occurs in association with a fracture can often mask deformity of the limb.

Crepitus

A grating or grinding sensation known as <u>crepitus</u> can be felt and sometimes even heard when fractured bone ends rub together.

False Motion

Also called free movement, this is motion at a point in the limb where there is no joint. It is a positive indication of a fracture.

Exposed Fragments

In open fractures, bone ends may protrude through the skin or be visible within the wound (Figure 29-16 ▶).

Pain

Pain, along with tenderness and bruising, commonly occurs in association with fractures.

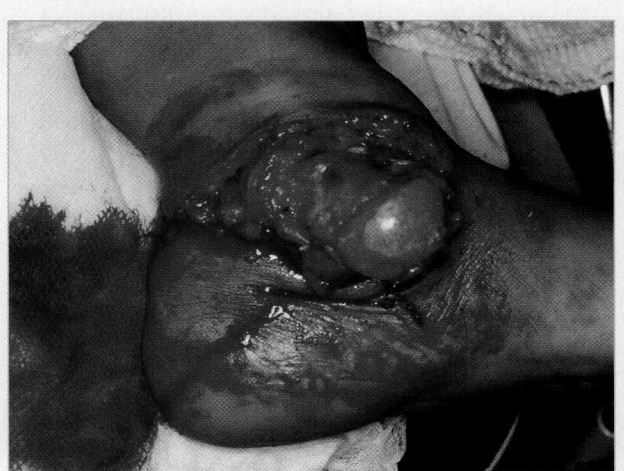

Figure 29-16 Bone ends may protrude through the skin or be visible within the wound of an open fracture.

You are the Provider   Part 3

You introduce yourself to the patient and your partner quickly considers the need for manual spinal stabilization. After you assess his ABCs, you begin providing high-flow oxygen via a nonrebreathing mask. The patient tells you he fell while standing on his skateboard posing for a picture. As he fell he extended his right arm to break the fall. When he made contact with the cement, he heard and felt a "pop." He denies striking his head or experiencing any loss of consciousness. You observe his right forearm to be angulated slightly in the middle, and the patient asks you not to touch it because of pain.

5. Would you continue stabilizing this patient's spine?
6. Is there anything that you can do to ease the patient's pain?

Locked Joint

A joint that is locked into position is difficult and painful to move. Keep in mind that crepitus and false motion appear only when a limb is moved or manipulated and are associated with injuries that are extremely painful. Do not manipulate the limb excessively in an effort to elicit these signs.

Dislocations

A dislocated joint sometimes will spontaneously <u>reduce</u>, or return to its normal position, before your assessment. In this situation, you will be able to confirm the dislocation only by taking a patient history. Often, however, injury to the supporting ligaments and capsule is so severe that the joint surfaces remain completely separated from one another. A dislocation that does not spontaneously reduce is a serious problem. The ends of the bone can be locked in a displaced position, making any attempt at motion of the joint very difficult and very painful. Commonly dislocated joints include the fingers, shoulder, elbow, and knee (patella).

The signs and symptoms of a dislocated joint are similar to those of a fracture (Figure 29-17 ▼):

- Marked deformity
- Swelling
- Pain that is aggravated by any attempt at movement
- Tenderness on palpation
- Virtually complete loss of normal joint motion (locked joint)
- Numbness or impaired circulation to the limb or digit

Sprains

A sprain occurs when a joint is twisted or stretched beyond its normal range of motion. As a result, the supporting capsule and ligaments are stretched or torn. A sprain should be considered a partial dislocation or subluxation. The alignment generally returns to a fairly normal position, although there may be some displacement. Note that severe deformity does not typically occur with a sprain. Sprains most often occur in the knee and the ankle, but a sprain can occur in any joint. The following signs and symptoms often indicate that the patient may have a sprain (Figure 29-18 ▼):

- Point tenderness can be elicited over the injured ligaments.
- Swelling and ecchymosis appear at the point of injury to the ligament as a result of torn blood vessels.
- Pain prevents the patient from moving or using the limb normally.
- Instability of the joint is indicated by increased motion, especially at the knee; however, this may be masked by severe swelling and guarding.

A fracture can look like a sprain, and vice versa. You will frequently not be able to distinguish a nondisplaced fracture from a sprain. Therefore, remember to document the mechanism of injury, as certain sprains and fractures occur more consistently with certain mechanisms. This is especially true at the ankle. In general, your approach should always be to try to rule out the possibility of fracture first. The basic principles of field management for sprains, dislocations, and fractures are essentially the same.

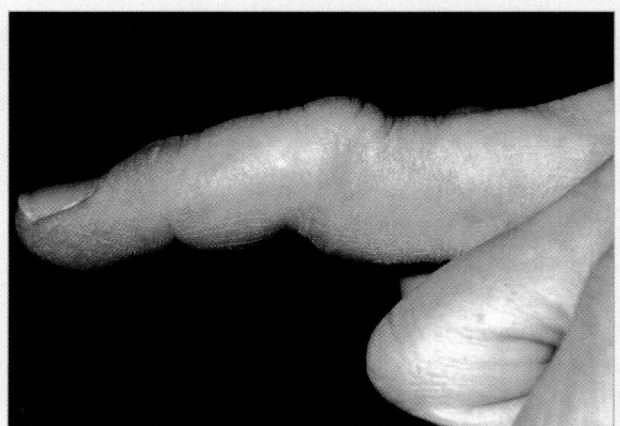

Figure 29-17 Joint dislocations, such as this finger, are characterized by deformity, swelling, pain with any movement, tenderness, locking, and impaired circulation.

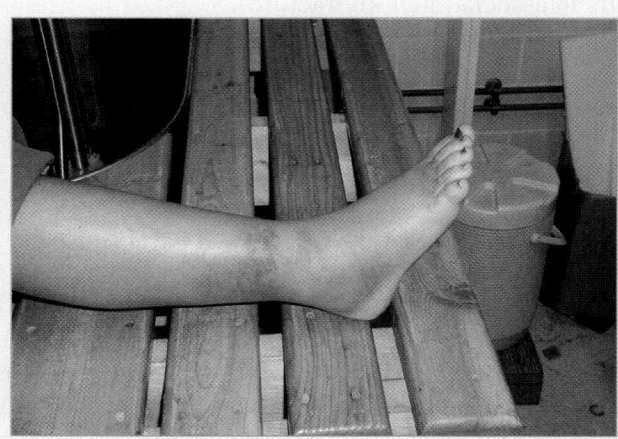

Figure 29-18 Sprains most often occur in the knee or ankle and are characterized by swelling, bruising, point tenderness, pain, and joint instability.

Compartment Syndrome

Be on the alert for compartment syndrome, which most commonly occurs in the fractured tibia or forearm of children and is often overlooked, especially in patients with an altered level of consciousness. The name <u>compartment syndrome</u> refers to elevated pressure within a fascial compartment. Fascia is the fibrous tissue that surrounds and supports the muscles and neurovascular structures.

Compartment syndrome typically develops within 6 to 12 hours after injury, usually as a result of excessive bleeding, a severely crushed extremity, or the rapid return of blood to an ischemic limb. This syndrome is characterized by pain that is out of proportion to the injury, pain on passive stretch of muscles within the compartment, pallor, decreased sensation, and decreased power (ranging from decreased strength and movement of the limb to complete paralysis).

If you suspect that a patient has compartment syndrome, splint the affected limb, keeping it at the level of the heart, and provide immediate transport, reassessing neurovascular status frequently during transport. Compartment syndrome must be managed surgically.

Assessing the Severity of Injury

You must become skilled at quickly and accurately assessing the severity of an injury. The Golden Hour is critical not only for life, but also for preserving limb viability. In an extremity with anything less than complete circulation, prolonged hypoperfusion can cause significant damage. For this reason, any suspected open fracture or vascular injury is considered a medical emergency. In the patient who has multisystem trauma, any additional bleeding can increase problems with underlying injuries or overall perfusion.

Remember that most injuries are not critical; you can identify critical injuries by using the musculoskeletal injury grading system shown in (Table 29-1 ▼).

TABLE 29-1 Musculoskeletal Injury Grading System

Minor Injuries
- Minor sprains
- Fractures or dislocations of digits

Moderate Injuries
- Open fractures of digits
- Nondisplaced long bone fractures
- Nondisplaced pelvic fractures
- Major sprains of a major joint

Serious Injuries
- Displaced long bone fractures
- Multiple hand and foot fractures
- Open long bone fractures
- Displaced pelvic fractures
- Dislocations of major joints
- Multiple digit amputations
- Laceration of major nerves or blood vessels

Severe, Life-Threatening Injuries (survival is probable)
- Multiple closed fractures
- Limb amputations
- Fractures of both long bones of the legs (bilateral femur fractures)

Critical Injuries (survival is uncertain)
- Multiple open fractures of the limbs
- Suspected pelvic fractures with hemodynamic instability

You are the Provider — Part 4

You begin the focused physical exam and history for the injured arm. You note tenderness, swelling, and crepitus with gentle palpation in the right midforearm. The patient can feel you touch his fingers, a distal pulse is found, and the capillary refill time in the injured extremity is normal. Your partner is manually stabilizing the patient's injured extremity as you begin the SAMPLE history and assess baseline vital signs.

7. After you obtain the patient's baseline vital signs, what should you do?

Assessing Musculoskeletal Injuries

As an EMT-B, your assessments, attempts to splint, and work to stabilize the patient are very important. Look at the big picture, evaluating the overall complexity of the situation. Always carefully assess the MOI to try to determine the amount of kinetic energy that an injured limb has absorbed, and maintain a high index of suspicion for associated injuries.

Again, it is not important to distinguish between fractures, dislocations, sprains, and contusions. In most instances, your assessment will be reported as an "extremity injury." However, you must be able to distinguish mild injuries from severe injuries, because some severe injuries may compromise neurovascular function.

Scene Size-up

Information from dispatch may indicate the MOI, the number of patients involved, and any first aid procedures used prior to your arrival. This will be useful information for you to think about as you travel to the scene. Remember, the information given by the dispatcher is only as accurate as what the patient or bystanders report. In addition, the situation may change prior to your arrival at the incident. This information can still be used to help you consider whether spinal stabilization is needed, what equipment you may need, and whether hazards might be present.

As you arrive on scene, observe the scene for hazards and threats to the safety of the crew, bystanders, and the patient. Try to identify the forces associated with the MOI. Could they have produced injuries other than the musculoskeletal injuries reported by dispatch? BSI precautions may be as simple as gloves. With a severe MOI or other risk factors, a mask and gown may be necessary. Consider the possibility that there may be hidden bleeding. Eye protection may also be indicated. Evaluate the need for law enforcement, ALS, or additional ambulances and request them early based on your initial scene assessment.

Initial Assessment

General Impression

Introduce yourself and ask the patient his or her name. This helps you to evaluate the patient's level of consciousness and orientation. Ask the patient his or her chief complaint—the problem that he or she is most concerned about. Ask about the mechanism of injury. Was it a direct blow, indirect force, twisting force, or high-energy injury? In many of these situations, the musculoskeletal complaints will be simple and usually not life threatening; however, some situations, such as those with a significant MOI, will include multiple problems that include musculoskeletal injuries. This initial interaction with your patient will provide you with a starting point and help you to distinguish the simple from the complex injuries. If there was significant trauma and multiple body systems are affected, the musculoskeletal injuries may be a lower priority. Scene time should not be wasted on prolonged musculoskeletal assessment or splinting.

Airway and Breathing

Fractures and sprains usually do not create airway and breathing problems. Other problems, such as injuries to the head, intoxication, or other related illnesses and injuries may cause inadequate breathing. Evaluating the chief complaint and MOI will help you to identify whether the patient has an open airway and whether breathing is present and adequate. In the conscious patient, this is as simple as noting whether the patient can speak normally. In the unconscious patient, it is as simple as opening the airway using the appropriate technique to look, listen, and feel for breathing. If a spinal injury is suspected, take the appropriate precautions and prepare for stabilization. Oxygen may be given to relieve anxiousness and improve perfusion. Even though an injury to the arm or leg may be obvious, take the time to evaluate the adequacy of the airway and breathing. Very little else matters if the patient's airway and breathing are inadequate.

Circulation

Your circulatory assessment should focus on determining whether the patient has a pulse, is perfusing adequately, or is bleeding. If your patient is conscious, as most patients with fractures and dislocations are, he or she will have a pulse. If the patient is unconscious, make sure there is a pulse by palpating the carotid artery. Hypoperfusion (shock) and bleeding problems will most likely be your primary concern. If the skin is pale, cool, or clammy and capillary refill time is slow, treat your patient for shock immediately. Maintain a normal body temperature and improve perfusion with oxygen and by placing the patient in the Trendelenburg position or elevating extremities (shock position). If musculoskeletal injuries in the extremities are suspected they must be at least initially stabilized, if not splinted, prior to moving. Eliminating this cause of shock may need to be done later in your assessment.

Fractures can break through the skin and cause external bleeding. This may occur during the initial injury or during manipulation of the extremity while preparing for splinting or transport. Careful handling of the extremity will minimize this risk. If external bleeding is present, bandage the extremity quickly to control bleeding. The dressings that cover the wound and bone should be kept sterile to reduce the potential for bone infections. The bandage should be secure enough to control bleeding without restricting circulation distal to the injury. Monitor bandage tightness by assessing the circulation, sensation, and movement distal to the bandage. Swelling from fractures and internal bleeding may cause bandages to become too tight.

Transport Decision

If the patient you are treating has an airway or breathing problem, or significant bleeding, you must consider rapid transport to the hospital for treatment. The patient who has a significant MOI but appears otherwise stable should also be transported promptly to the closest appropriate hospital. Patients with bilateral fractures of the long bones (humerus, femur, or tibia) have been subjected to a high amount of kinetic energy. This should dramatically increase your index of suspicion for serious unseen injuries. When a decision for rapid transport is made, a backboard can be used as a splinting device to splint the whole body rather than splinting each extremity individually. Time taken to splint arms and legs individually delays prompt surgical intervention that may be needed for other injuries when a significant MOI has occurred. Individual splints should be applied en route if the ABCs are stable and time permits.

Patients with a simple MOI, such as twisting of an ankle or dislocating a shoulder, may be further assessed and stabilized on scene prior to transport if no other problems exist. In either situation, "load and go" or "stay and play," careful handling of fractures while preparing for transport is necessary to limit pain and prevent sharp bone ends from breaking through the skin or damaging nerves and blood vessels inside the extremity.

Focused History and Physical Exam

The focused history and physical exam is based on the MOI when trauma is present and the NOI when a medical problem exists. The three parts of this step include a history from the patient, vital signs, and physical exam. Patients with musculoskeletal injuries may fall into either a significant trauma or nonsignificant trauma category.

During your assessment of musculoskeletal trauma, use the DCAP-BTLS approach. Identify any extremity **D**eformities that likely represent significant musculoskeletal injury and stabilize appropriately. **C**ontusions and **A**brasions may overlie more subtle injuries and should prompt you to carefully evaluate the stability and neurovascular status of the limb. The presence of **P**uncture wounds or other signs of **P**enetrating injury should alert you to the possibility of an open fracture. Associated **B**urns must be identified and treated appropriately. Palpate for **T**enderness, which, like contusions or abrasions, may be the only significant sign of an underlying musculoskeletal injury. When **L**acerations are present in an extremity, open fracture must be considered, bleeding controlled, and dressings applied. Careful inspection for **S**welling with comparison with the opposite limb may also reveal otherwise occult musculoskeletal injury. You may find a hematoma in the zone of injury during the assessment.

Rapid Physical Exam for Significant Trauma

When significant trauma is involved, you should take a moment to rapidly check your patient from head to toe for any additional injury that may be present. Begin with the head and work systematically toward the feet, checking the head, chest, abdomen, extremities, and back. The goal here is to identify hidden and potentially life-threatening injuries. This rapid exam will also help you to prepare for packaging and rapid transport. Knowing if an arm or leg is broken will be important when log rolling the patient onto a backboard and securing the patient to the board.

If your assessment finds no external signs of injury, ask the patient to move each limb carefully, stopping immediately if a movement causes pain. Skip this step in your evaluation if the patient reports neck or back pain; even slight motion could cause permanent damage to the spinal cord.

Focused Physical Exam for Nonsignificant Trauma

When nonsignificant trauma has occurred and your patient has a simple strain, sprain, dislocation, or fracture, you can take the time to focus your exam on that particular injury. Look for DCAP-BTLS. Evaluate the circulation, motor function, and abnormal sensations distal to the injury. If the patient has two or more extremities injured, treat the patient as a significant trauma patient and provide rapid transport to the hospital. The likelihood of other more severe injuries is greater when two or more bones have been broken. Be sure to assess the entire zone of injury by removing clothing from the area and looking and palpating for injuries. In musculoskeletal injuries this zone generally extends from the joint above (proximal) to the joint below (distal), front and back. Do not forget to check perfusion, motion, and sensation.

Many important blood vessels and nerves lie close to the bone, especially around the major joints. Therefore, any injury or deformity of the bone may be associated with vessel or nerve injury. For this reason, you must assess neurovascular function during the rapid or focused physical exam and repeat it in the detailed exam and every 5 to 10 minutes in the ongoing assessment, depending on the patient's condition, until the patient is at the hospital. Always recheck the neurovascular function before and after you splint or otherwise manipulate the limb. Manipulation can cause a bone fragment to press against or impale a nerve or vessel. Failure to restore circulation in this situation can lead to death of the limb. Always give priority to patients with impaired circulation resulting from bone fragments.

Examination of the injured limb should include assessment of four major signs that are good indicators of circulatory and nervous status distal to the injury: pulse, capillary refill, motor function, and sensory function. Follow the steps in (Skill Drill 29-1 ▶):

1. **Pulse.** Palpate the pulse distal to point of injury. First, palpate the radial pulse in the upper extremity (**Step 1**). Second, in the lower extremity, palpate the posterior tibial and dorsalis pedis pulses (**Step 2**).

2. **Capillary refill.** Note and record the skin color, identifying any pallor or cyanosis. Then apply firm pressure to the tip of the fingernail or toenail, which will cause the skin to blanch (turn white). If normal color does not return within 2 seconds after you release the nail, you can assume that circulation is impaired. This test is typically recommended for use in children, although it can be used in adults also (**Step 3**).

3. **Sensation.** In the hand, check the feeling on the flesh near the tip of the index finger and thumb, as well as the little finger (**Step 4**). In the foot, check the feeling on the flesh of the big toe (**Step 5**) and on the lateral side of the foot (**Step 6**). The patient's ability to sense light touch in the fingers or toes distal to the site of a fracture is a good indication that the nerve supply is intact.

4. **Motor function.** Evaluate muscular activity when the injury is proximal to the patient's hand or foot. Ask the patient to open and close a fist for an upper extremity injury and to wiggle the toes and move the foot up and down for a lower extremity injury (**Steps 7 through 10**). Sometimes, an attempt at motion will produce pain at the injury site. If this happens, do not continue this part of the examination. To avoid causing pain, do not perform this test at all if the injury involves the hand or foot itself.

Because many of the steps require patient cooperation, you will not be able to assess sensory and motor function in an unconscious patient, but you can still evaluate the limb for deformity, swelling, ecchymosis, false motion, and crepitus. If a patient is unconscious, first perform an initial assessment and then examine the extremities.

✱ EMT-B Tips

Extremity injuries that impair circulation or nerve function in distal tissues are urgent conditions. These patients need careful assessment, prompt transport, and frequent reassessment of distal functions. It is also crucial to report this information in your initial radio contact with the hospital to allow personnel to prepare for a condition in which prompt surgery may be necessary to save the limb.

Assessing Neurovascular Status

Skill Drill

29-1

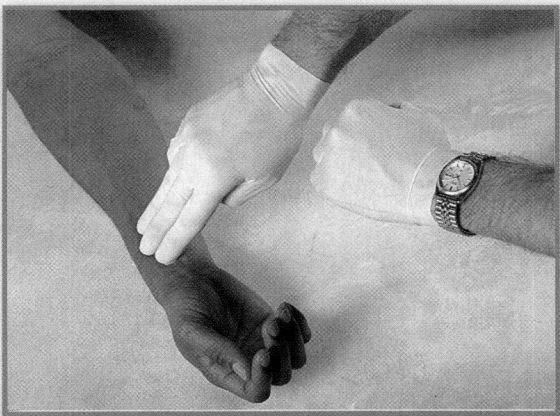

1 Palpate the radial pulse in the upper extremity.

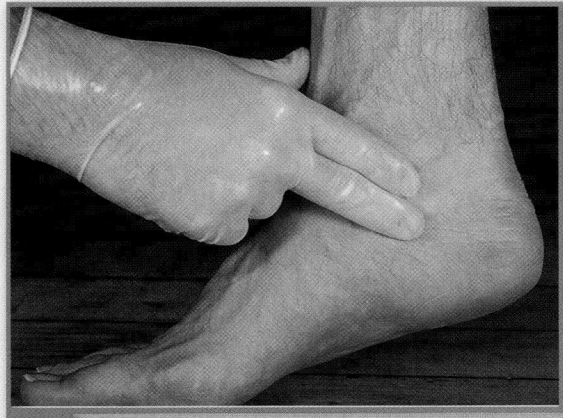

2 Palpate the posterior tibial pulse in the lower extremity.

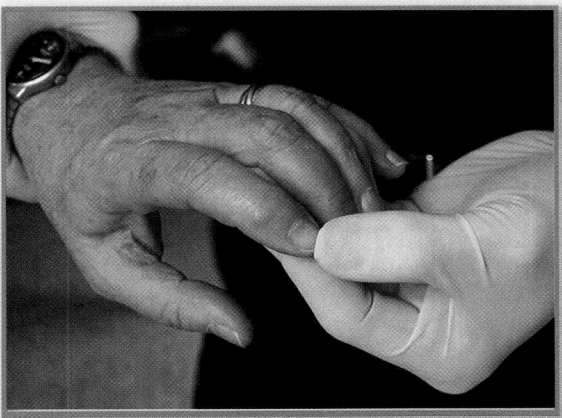

3 Assess capillary refill by blanching a fingernail or toenail.

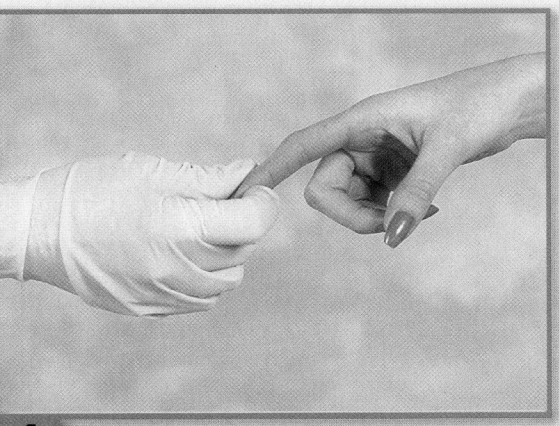

4 Assess sensation on the flesh near the tip of the index finger.

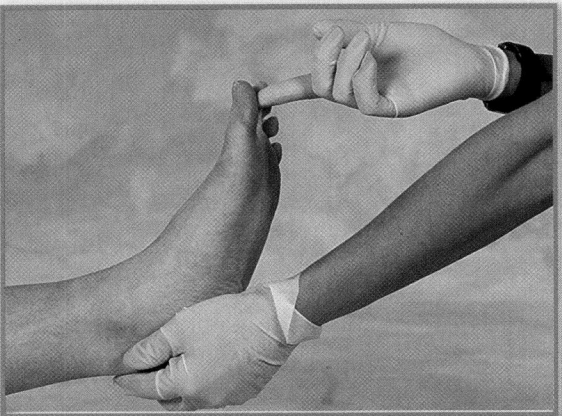

5 On the foot, first check sensation on the flesh near the tip of the great toe.

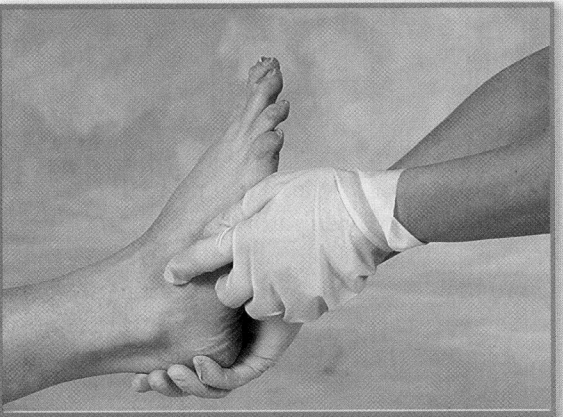

6 Also check foot sensation on the lateral side.

Continued.

Assessing Neurovascular Status continued

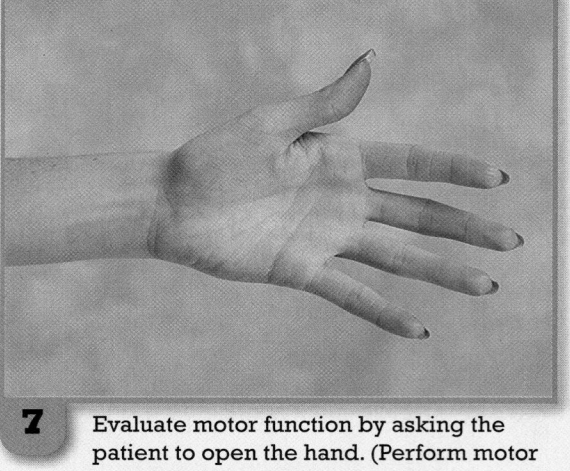

7 Evaluate motor function by asking the patient to open the hand. (Perform motor tests only if the hand or foot is not injured. Stop a test if it causes pain.)

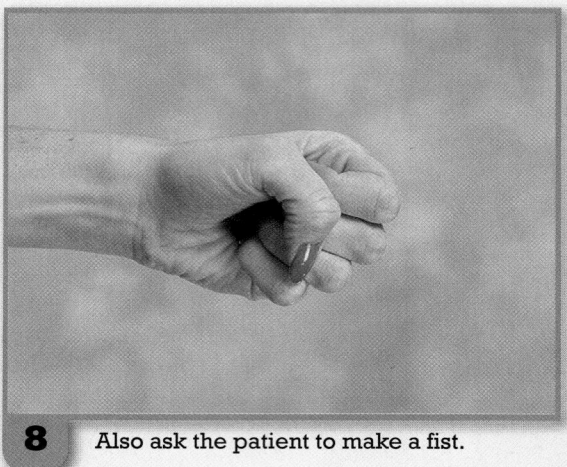

8 Also ask the patient to make a fist.

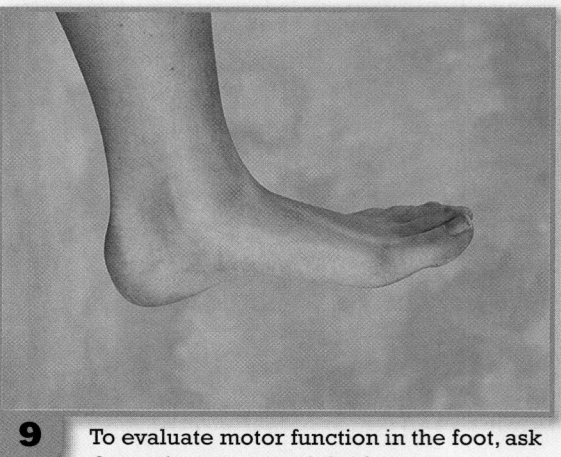

9 To evaluate motor function in the foot, ask the patient to extend the foot.

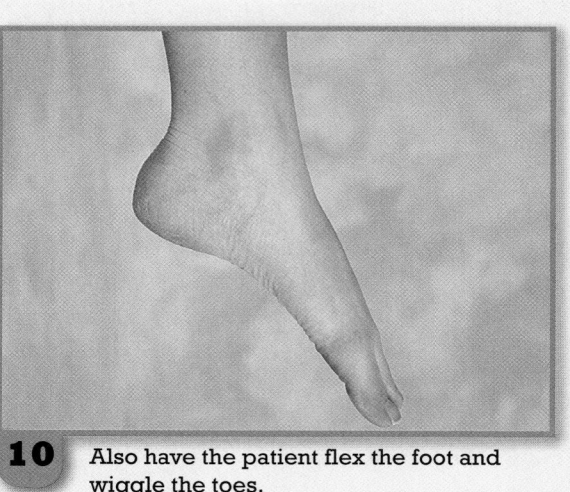

10 Also have the patient flex the foot and wiggle the toes.

Baseline Vital Signs

Determine a baseline set of vital signs including pulse rate, rhythm, and quality; respiratory rate, rhythm, and quality; blood pressure; skin condition; and pupil size and reaction to light. These baseline indicators need to be obtained as soon as possible. Your patient may appear to be tolerating the injury well until you assess these vital signs and they indicate otherwise. Trending these vital signs helps you to understand whether your patient's condition is improving or getting worse over time, particularly during long transports. Shock or hypoperfusion is common in musculoskeletal injuries and this baseline information will be very important in assessing your patient's condition.

SAMPLE History

A SAMPLE history should be obtained for all trauma patients. How much and in what detail you explore this history depends on the seriousness of the patient's condition and how quickly you need to transport the patient to the hospital. Often the hospital is in need of this detailed information for seriously injured patients during times when you do not have enough time to obtain it. For patients with simple fractures or dislocations or sprains, it is easier to obtain a SAMPLE history. Prehospital providers may have, on scene, access to family members and others who have information about the patient's history. Make an attempt to obtain this history without delaying time to definitive care.

OPQRST can be of limited use in cases of severe injury and is usually too lengthy when matters of airway, breathing, circulation, and rapid transport require immediate attention. However, OPQRST may be useful in situations when the MOI is unclear, the patient's condition is stable, or details of the injury are uncertain. This more detailed questioning for simple trauma situations may help you and the hospital to understand the specific injury better.

Interventions

Because trauma patients often have multiple injuries, you must assess their overall condition, stabilize the ABCs, and control any serious bleeding before further treating the injured extremity. In a critically injured patient, you should secure the patient to a long backboard to stabilize the spine, pelvis, and extremities and provide prompt transport to a trauma center. In this situation, extensive evaluation and splinting of limb injuries in the field is a waste of valuable time.

If the patient has no life-threatening injuries, you may take extra time at the scene to stabilize the patient's overall condition and more completely evaluate the injured extremity. If possible, gently and carefully remove the patient's clothing to look for open fractures or dislocations, severe deformity, swelling, and/or ecchymosis. A good rule to follow is to check the patient's circulation, motor function, and sensation prior to and after splinting.

When you have finished assessing the extremity, apply a secure splint, commercial or otherwise, to stabilize the injury prior to transport. The extremity above and below the site of injury should be included in the splint. To minimize the potential for problems, the splint should be well padded. A comfortable and secure splint will reduce pain, improve shock, and minimize compromised circulation.

The main goal in providing care for musculoskeletal injuries is stabilization in the most comfortable position that allows for maintenance of good circulation distal to the injury. This should be done whether you are preparing the patient for rapid transport or when you have as much time as you need to assess and treat the patient.

Detailed Physical Exam

During the detailed physical exam, you can inspect and gently palpate the other extremities and the spine to identify areas of point tenderness that may indicate underlying fractures, dislocations, or sprains. Remember to compare the injured limb with the opposite, uninjured limb.

Ongoing Assessment

Repeat the initial assessment and vital signs. Reassess the interventions and treatment you have provided to the patient. If a splint was applied, reassess the patient's distal neurovascular function and color of the injured extremity distal to the injury site. It is difficult to intervene when problems develop if you do not assess and reassess your patient's condition frequently.

Communication and Documentation

Your radio report to the hospital should include a description of the problems found during your assessment. In particular, you should report problems with the patient's ABCs, if fractures are open, and if circulation is compromised before or after your splinting. Many times the hospital staff can arrange for specialists or consider antibiotics early if they are aware of problems. How much you include in your radio report will depend on your local protocols. Additional details can be given during your verbal report at the hospital when you transfer care to the nursing staff or physician.

Document complete descriptions of injuries and the mechanisms of injury associated with them. Hospital staff may later refer to these notes during confusing situations or when communication problems occur. It is important to assess and document the presence or absence of circulation, motor function, and sensation distal to the injury before you move an extremity, after manipulation or splinting of the injury, and on arrival at the hospital. Careful documentation may prevent you from being included in legal action when patients are unhappy about outcomes from injuries. Do not rely on your memory to remember details from situations; it is unreliable and will not hold up in a court of law.

Emergency Medical Care

Your first steps in providing care for any patient are the initial assessment and stabilizing the patient's ABCs. If needed, perform a rapid trauma assessment, or focus on a specific injury. Remember to always follow BSI precautions.

Follow the steps in (Skill Drill 29-2 ▶) when caring for patients with musculoskeletal injuries:

1. **Completely cover open wounds** with a dry, sterile dressing, and apply local pressure to control bleeding. Once you have applied a sterile dressing, treat an open fracture in the same way as a closed fracture (**Step 1**).

2. **Apply the appropriate splint**, and elevate the extremity. Patients with lower extremity injuries should lie supine with the limb elevated about 6″ to minimize swelling. For any patient, be sure to position the injured limb slightly above the level of the heart. Never allow the injured limb to flop about or dangle from the edge of the backboard. Always assess pulse, motor, and sensory function before and after the application of splints (**Step 2**).

3. **If swelling is present**, apply cold packs to the area; however, avoid placing cold packs directly on the skin or other exposed tissues. Placing a cold pack on top of an air splint or other thick, insulating material will not help to reduce swelling (**Step 3**).

4. **Prepare the patient for transport.** A patient with an isolated upper extremity injury will most likely be more comfortable in a semiseated position rather than lying flat; however, assuming there is no risk of spinal injury, either position is acceptable. Ensure that the extremity is elevated above the level of the heart and secured so that it does not dangle from the edge of the backboard (**Step 4**).

5. **Always inform hospital personnel** about all wounds that have been dressed and splinted.

Splinting

A splint is a flexible or rigid device that is used to protect and maintain the position of an injured extremity (Figure 29-19 ▶). Unless the patient's life is in immediate danger, you should splint all fractures, dislocations, and sprains before moving the patient. By preventing movement of fracture fragments, bone ends, a dislocated joint, or damaged soft tissues, splinting reduces pain and

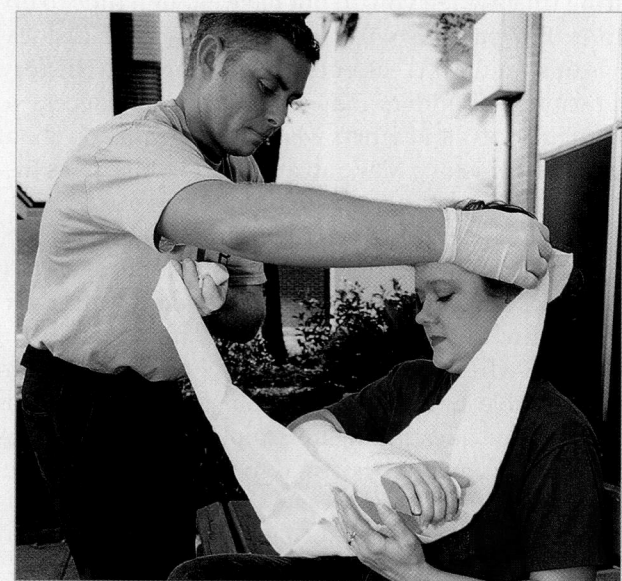

Figure 29-19 Splinting reduces pain and prevents additional damage to the injured extremity.

makes it easier to transfer and transport the patient. In addition, splinting will help to prevent the following:

- Further damage to muscles, the spinal cord, peripheral nerves, and blood vessels from broken bone ends.
- Laceration of the skin by broken bone ends. One of the primary indications for splinting is to prevent a closed fracture from becoming an open fracture (conversion).
- Restriction of distal blood flow resulting from pressure of the bone ends on blood vessels.
- Excessive bleeding of the tissues at the injury site caused by broken bone ends.
- Increased pain from movement of bone ends.
- Paralysis of extremities resulting from a damaged spine.

A splint is simply a device to prevent motion of the injured part. It can be made from any material on occasions when you need to improvise. However, you should have an adequate supply of standard commercial splints on hand.

General Principles of Splinting

The following principles of splinting apply to most situations:

1. **Remove clothing from the area** of any suspected fracture or dislocation so that you can inspect the extremity for DCAP-BTLS.

Caring for Musculoskeletal Injuries

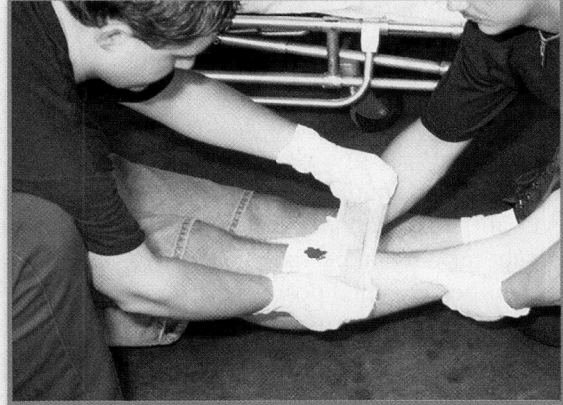

1 Cover open wounds with a dry, sterile dressing, and apply pressure to control bleeding.

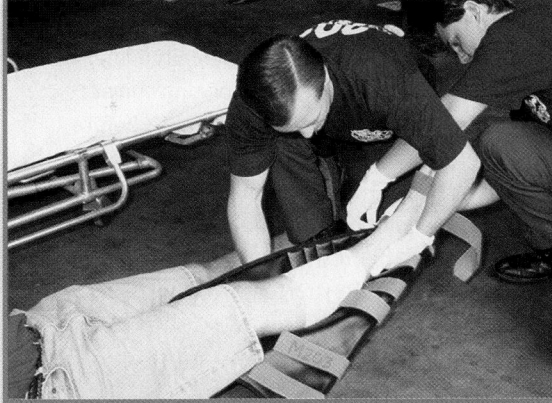

2 Apply a splint and elevate the extremity about 6″ (slightly above the level of the heart).

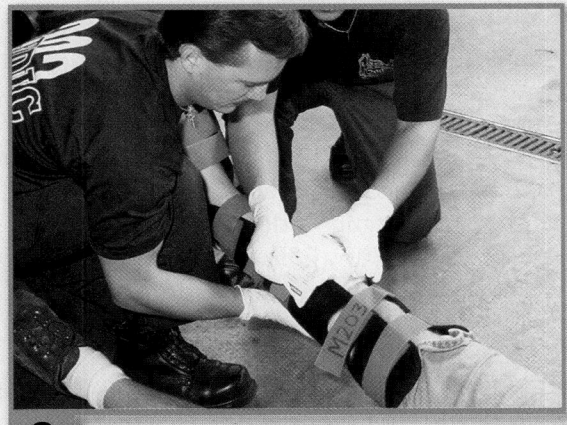

3 Apply cold packs if there is swelling, but do not place them directly on the skin.

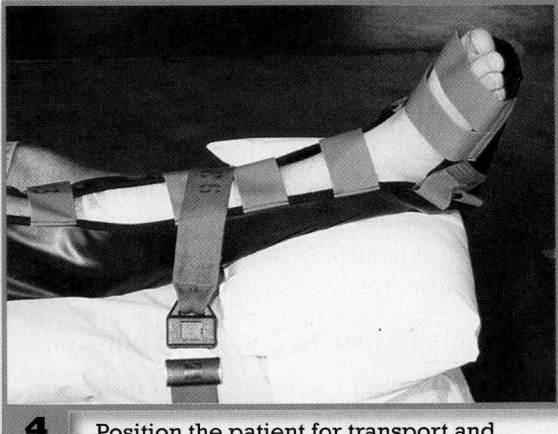

4 Position the patient for transport and secure the injured area.

2. **Note and record the patient's neurovascular status** distal to the site of the injury, including pulse, sensation, and movement. Continue to monitor the neurovascular status until the patient reaches the hospital.
3. **Cover all wounds with a dry, sterile dressing** before splinting. Be sure to follow BSI precautions. Do not intentionally replace protruding bones. Notify the receiving hospital of all open wounds.
4. **Do not move the patient before splinting** an extremity unless there is an immediate hazard to the patient or yourself.

5. In a suspected fracture of the shaft of any bone, be sure to **stabilize the joints** above and below the fracture.
6. With injuries in and around the joint, be sure to **stabilize the bones** above and below the injured joint.
7. **Pad all rigid splints** to prevent local pressure and discomfort to the patient.
8. While applying the splint, **maintain manual stabilization** to minimize movement of the limb and to support the injury site.
9. If fracture of a long bone shaft has resulted in severe deformity, **use constant, gentle manual**

Documentation Tips

Straightening or splinting an injured limb can compromise distal functions, just as the initial injury can. Record the status of distal circulation and nervous function (neurovascular status) both before and after straightening or splinting. At a minimum, your written record should describe these functions before splinting, and confirm that they were normal immediately after splinting and upon hospital arrival. For any but the shortest transports, also indicate the results of reassessments while en route.

traction to align the limb so that it can be splinted. This is especially important if the distal part of the extremity is cyanotic or pulseless.

10. **If you encounter resistance** to limb alignment, splint the limb in its deformed position.
11. **Stabilize all suspected spinal injuries** in a neutral in-line position on a backboard.
12. **If the patient has signs of shock** (hypoperfusion), align the limb in the normal anatomic position and provide transport (total body stabilization).
13. **When in doubt, splint.**

General Principles of In-line Traction Splinting

Application of in-line traction is the act of pulling on a body structure in the direction of its normal alignment. It is the most effective way to realign a fracture of the shaft of a long bone so that the limb can be splinted more effectively. Excessive traction can be harmful to an injured limb. When applied correctly, however, traction stabilizes the bone fragments and improves the overall alignment of the limb. You should not attempt to reduce the fracture or force all the bone fragments back into alignment. This is the physician's responsibility. In the field, the goals of in-line traction are as follows:

1. To **stabilize the fracture** fragments to prevent excessive movement
2. To **align the limb** sufficiently to allow it to be placed in a splint
3. To **avoid** potential neurovascular compromise

The amount of traction that is required to accomplish these objectives varies but often does not exceed 15 lb. You should use the least amount of force necessary. Grasp the foot or hand at the end of the injured limb firmly; once you start pulling, you should not stop until the limb is fully splinted. The direction of traction applied is always along the long axis of the limb. Imagine where the normal, uninjured limb would lie, and pull gently along the line of that imaginary limb until the injured limb is in approximately that position (Figure 29-20 ▼). Grasping the foot or hand and the initial pull of traction usually causes some discomfort as the bone fragments move. It helps if a second person can support the injured limb directly under the site of the fracture. This initial discomfort quickly subsides, and you can then apply further gentle traction. However, if the patient strongly resists the traction or if it causes more pain that persists, you must stop and splint the limb in the deformed position.

Remember that many different materials can be used as splints if necessary. When no splinting materials are available, the arm can be bound to the chest wall, and an injured leg can be bound to the uninjured leg to provide at least temporary stability. The three basic types of splints are rigid, formable, and traction splints.

Rigid Splints

Rigid (nonformable) splints are made from firm material and are applied to the sides, front, and/or back of an injured extremity to prevent motion at the injury site. Common examples of rigid splints include padded board splints, molded plastic and metal splints, padded wire ladder splints, and folded cardboard splints. As always, be sure to follow BSI precautions. It takes two EMT-Bs to apply a rigid splint. Follow the steps in (Skill Drill 29-3 ▶):

1. First EMT-B: **Gently support the limb** at the site of injury as others prepare and begin to position the equipment. Apply steady, in-line traction if necessary. Maintain this support until the splint is completely applied (**Step 1**).

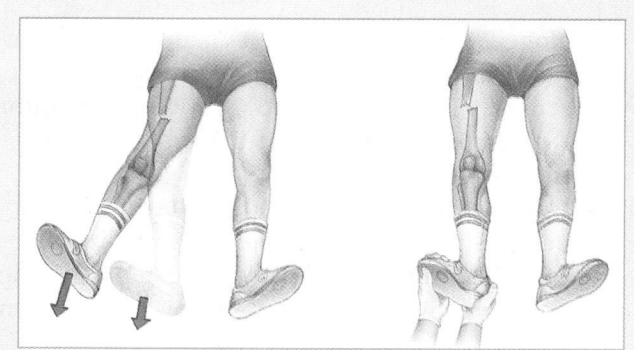

Figure 29-20 To apply traction, imagine the position where the normal uninjured limb would lie, then gently pull along that line until the injured limb is in that position. Do not release traction once you have applied it.

Applying a Rigid Splint

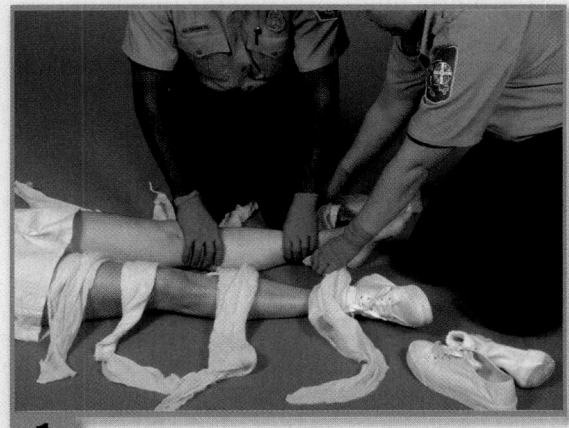

1 Provide gentle support and in-line traction for the limb.

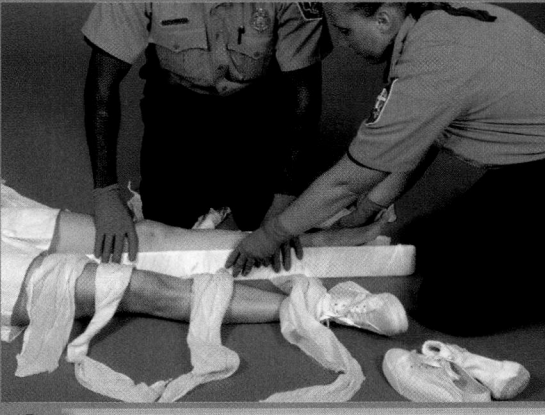

2 Second EMT-B places the splint alongside or under the limb.

Pad between the limb and the splint as needed to ensure even pressure and contact.

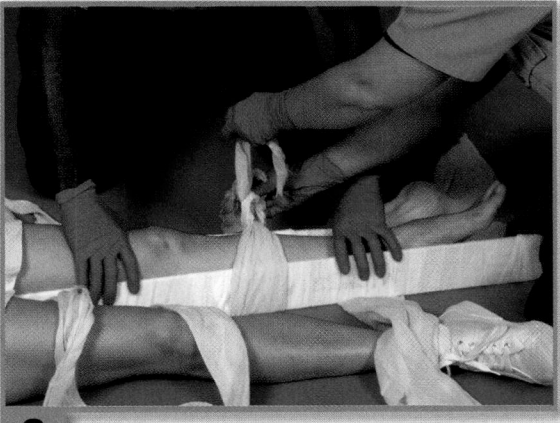

3 Secure the splint to the limb with bindings.

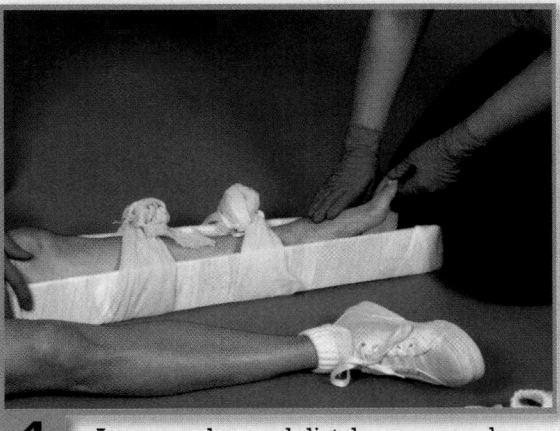

4 Assess and record distal neurovascular function.

2. Second EMT-B: **Place the rigid splint under or alongside the limb** (**Step 2**).
3. Second EMT-B: **Place padding between the limb and the splint** to make sure there is even pressure and even contact. Look for bony prominences, and pad them.
4. Second EMT-B: **Apply bindings** to hold the splint securely to the limb (**Step 3**).
5. Second EMT-B: **Check and record** the distal nervous and circulatory (neurovascular) function (**Step 4**).

There are two situations in which you must splint the limb in the position of deformity—when the deformity is severe, as is the case with many dislocations, or when you encounter resistance or extreme pain when applying gentle traction to the fracture of a shaft of a long bone. In either situation, you should apply padded board splints to each side of the limb and secure them with soft roller bandages (**Figure 29-21** ▶). Most dislocations should be splinted as found, but follow local protocols. Attempts to realign or reduce dislocations can lead to more damage.

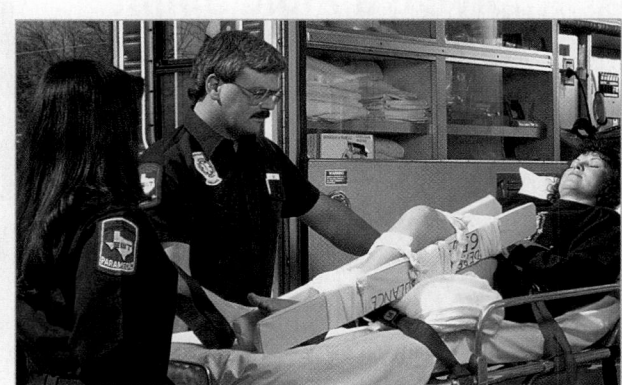

Figure 29-21 If you encounter resistance or extreme pain when applying traction to a long bone, apply padded board splints to each side of the limb, and secure them with soft roller bandages, stabilizing the limb in its deformed position.

Formable Splints

The most commonly used formable or soft splint is the precontoured, inflatable, clear plastic air splint. These are available in a variety of sizes and shapes, with or without a zipper that runs the length of the splint. Always inflate the splint after applying it. The air splint is comfortable, provides uniform contact, and has the added advantage of applying firm pressure to a bleeding wound. Air splints are used to stabilize injuries below the elbow or below the knee.

Air splints have some drawbacks, particularly in cold weather areas. The zipper can stick, clog with dirt, or freeze. Significant changes in the weather affect the pressure of the air in the splint, which decreases as the environment grows colder and increases as the environment grows warmer. The same thing happens when there are changes in altitude, which can be a problem with helicopter transport of patients. Therefore, you should carefully monitor the splint and let air out if the splint becomes overinflated.

The method of applying an air splint depends on whether it has a zipper. With either type, you must first cover all wounds with a dry, sterile dressing, making sure that you use BSI precautions. For a splint that has a zipper, follow the steps in (Skill Drill 29-4 ▶):

1. **Hold the injured limb** slightly off the ground, applying gentle traction and supporting the site of injury. Have your partner place the open, deflated splint around the limb (**Step 1**).
2. **Zip the splint up and inflate it** by pump or by mouth. When this is done, test the pressure in the

splint. With proper inflation, you should just be able to compress the walls of the splint together with a firm pinch between the thumb and index finger near the edge of the splint.
3. **Check and record pulse, motor, and sensory functions**, and monitor them periodically until the patient reaches the hospital (**Step 2**).

If you use an unzipped or partially zippered type of air splint, have another person help you follow the steps in (Skill Drill 29-5 ▶):

1. First EMT-B: **Support the patient's injured limb** until splinting is accomplished.
2. Second EMT-B: **Place your arm through the splint.** Extend your hand beyond the splint, and grasp the hand or foot of the injured limb (**Step 1**).
3. Second EMT-B: **Apply gentle traction** to the hand or foot while sliding the splint onto the injured limb. The hand or foot of the injured limb should always be included in the splint (**Step 2**).
4. First EMT-B: **Inflate the splint** by pump or by mouth (**Step 3**).
5. Second EMT-B: **Test the pressure** in the splint. This is something that you must do with either type of air splint.
6. Second EMT-B: **Check and record pulse, motor, and sensory functions**, and monitor them en route.

Other formable splints include vacuum splints, pillow splints, SAM splints, a sling and swathe, and the pneumatic antishock garment (PASG) for pelvic fractures. Just like an air splint, a vacuum splint can be easily shaped to fit around a deformed limb. Instead of pumping air in, however, you can use a hand pump to pull the air out through a valve. Follow the steps in (Skill Drill 29-6 ▶) to apply a vacuum splint:

1. **Support and stabilize the injured limb**, applying traction if needed, while your partner applies the splint (**Step 1**).
2. **Gently place the injured limb onto the vacuum splint** and wrap the splint around the limb (**Step 2**).
3. **Draw the air out of the splint** through the suction valve, and then seal the valve. Once the valve is sealed, the vacuum splint becomes rigid, conforming to the shape of the deformed limb and stabilizing it (**Step 3**).
4. **Check distal circulation and nervous functions**, and monitor them en route.

Applying a Zippered Air Splint

1 Support the injured limb and apply gentle traction as your partner applies the open, deflated splint.

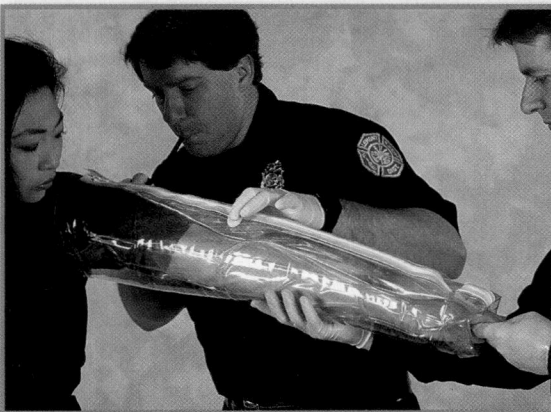

2 Zip up the splint, inflate it by pump or by mouth, and test the pressure.

Check and record distal neurovascular function.

Traction Splints

Traction splints are used primarily to secure fractures of the shaft of the femur, which are characterized by pain, swelling, and deformity of the midthigh. A traction splint should not be used if the patient has an obvious injury of the knee or ankle joint, foot, or lower leg. Several different types of lower extremity traction splints are commercially available, such as the Hare traction splint, the Sager splint, and the Kendrick splint, each with its own unique method of application with which you must be familiar. The use of the Hare and Sager splints is described in this chapter.

The Hare traction splint is not suitable for use on the upper extremity because the major nerves and blood vessels in the patient's axilla cannot tolerate counter-traction forces.

Do not use traction splints for any of the following conditions:

- Injuries of the upper extremity
- Injuries close to or involving the knee

- Injuries of the hip
- Injuries of the pelvis
- Partial amputations or avulsions with bone separation
- Lower leg, foot, or ankle injury

Proper application of a traction splint requires two well-trained EMT-Bs working together. Practice the steps in Skill Drill 29-7 ▶ with your partner until the sequence and necessary teamwork have become routine.

1. **Cut open the patient's pant leg,** or otherwise expose the injured lower extremity. Follow BSI precautions as needed. Be sure to assess and record the pulse, motor function, and sensation distal to the injury.

2. **Place the splint beside the patient's uninjured leg,** and adjust it to the proper length, with the ring at the ischial tuberosity and the splint extending 12″ beyond the foot. Open and adjust the four Velcro support straps, which should be

Applying an Unzippered Air Splint

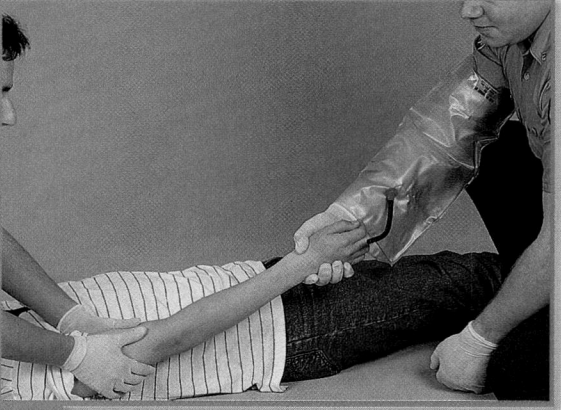

1 Support the injured limb.

Have your partner place his or her arm through the splint to grasp the patient's hand or foot.

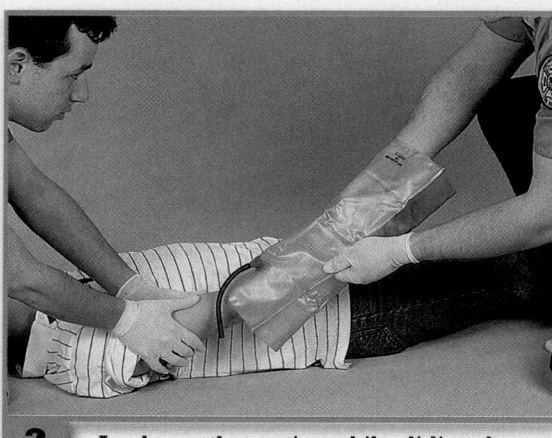

2 Apply gentle traction while sliding the splint onto the injured limb.

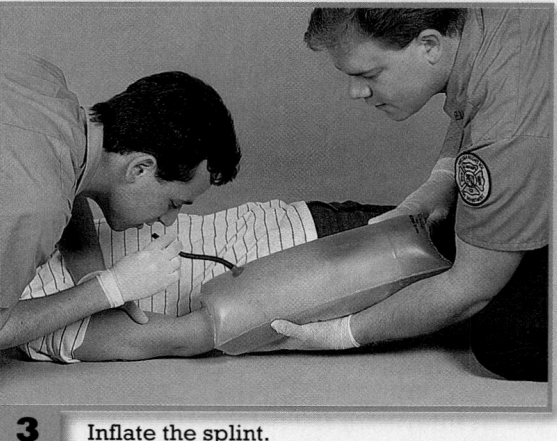

3 Inflate the splint.

positioned at the midthigh, above the knee, below the knee, and above the ankle (**Step 1**).

3. First EMT-B: **Manually support and stabilize the injured limb** so that no motion will occur at the fracture site while the second EMT-B fastens the appropriate-sized ankle hitch about the patient's ankle and foot. Normally, the patient's shoe is removed for this procedure (**Step 2**).

4. First EMT-B: **Support the leg at the site of the suspected injury** while the second EMT-B manually applies gentle longitudinal traction to the ankle hitch and foot. Use only enough force to

align (reposition) the limb so that it will fit into the splint; do not attempt to align the fracture fragments anatomically (**Step 3**).

5. First EMT-B: **Slide the splint into position** under the patient's injured limb, making certain that the ring is seated well on the ischial tuberosity (**Step 4**).

6. **Pad the groin area**, and gently apply the ischial strap (**Step 5**).

7. First EMT-B: While the second EMT-B continues to maintain traction, **connect the loops of the ankle hitch** to the end of the splint. Then

Applying a Vacuum Splint

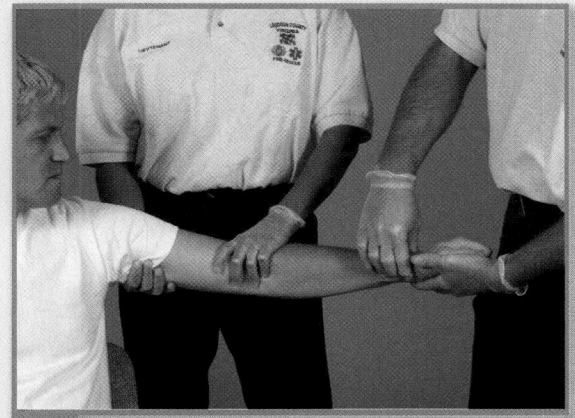

1 Stabilize and support the injury.

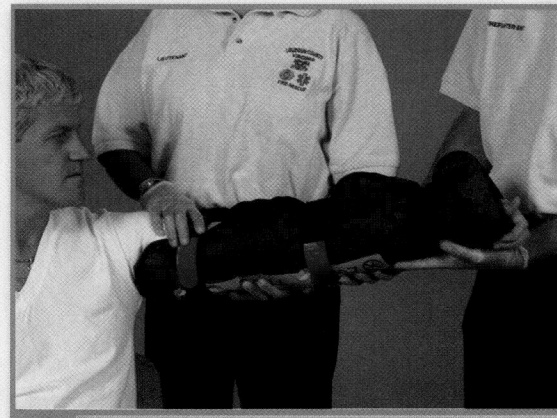

2 Place the splint and wrap it around the limb.

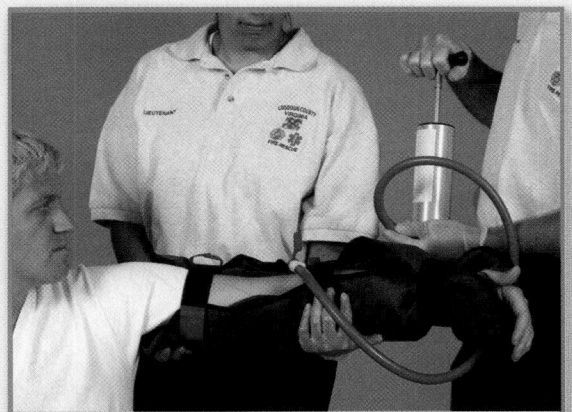

3 Draw the air out of the splint through the suction valve, and then seal the valve.

apply gentle traction to the connecting strap between the ankle hitch and the splint, just strongly enough to maintain limb alignment. Use caution. This splint comes with a ratchet mechanism to tighten the strap, which can overstretch the limb and further injure the patient. Adequate traction has been applied when the leg is the same length as the other leg or the patient feels relief (**Step 6**).

8. Once proper traction has been applied, **fasten the support straps** so that the limb is securely held in the splint. Check all proximal and distal support straps to make sure they are secure (**Step 7**).

9. At this point, **reassess distal pulses**, motor function, and sensation.

10. **Place the patient securely on a long backboard** for transport to the emergency department. You may need to load the patient feet first into the ambulance so that you do not shut the door against the splint (**Step 8**).

Because this traction splint stabilizes the limb by producing countertraction on the ischium and in the groin, use care to pad these areas well. You must avoid excessive pressure on the external genitalia. Always use

Skill Drill

29-7

Applying a Hare Traction Splint

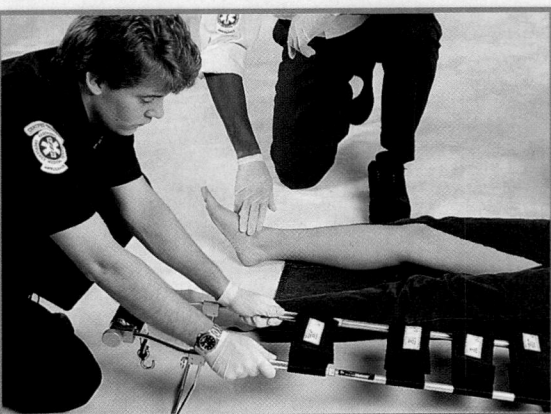

1 Expose the injured limb and check pulse, motor, and sensory function.

Place the splint beside the uninjured limb, adjust the splint to proper length, and prepare the straps.

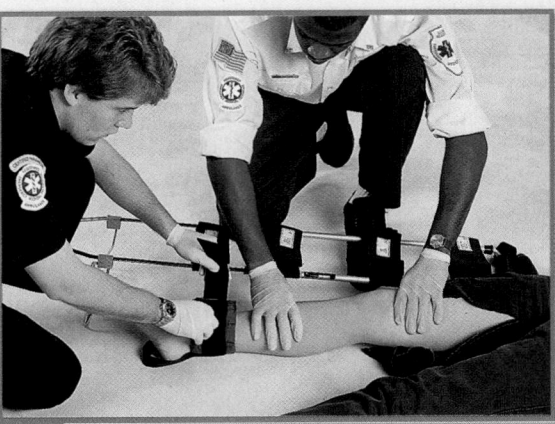

2 Support the injured limb as your partner fastens the ankle hitch about the foot and ankle.

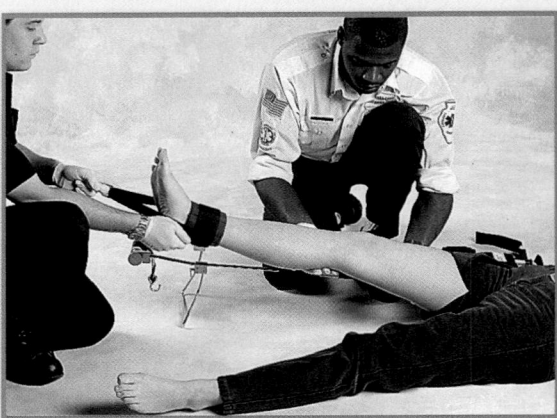

3 Continue to support the limb as your partner applies gentle in-line traction to the ankle hitch and foot.

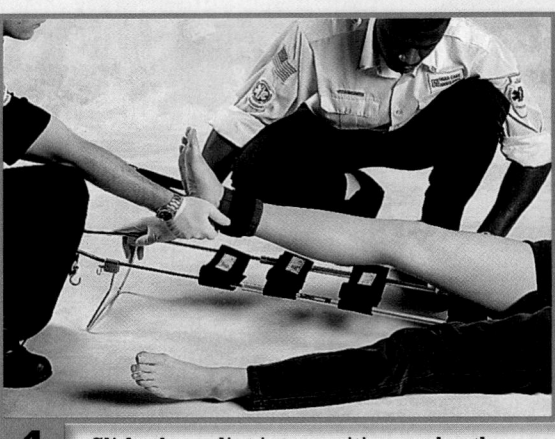

4 Slide the splint into position under the injured limb.

commercially available padded ankle hitches rather than pieces of rope, cord, or tape. Such improvised hitches can sometimes be painful and can potentially obstruct circulation in the foot.

The Sager splint is lightweight, easy to store, applies a measurable amount of traction, and can be used with a PASG. Best of all, you can apply it by yourself when necessary. As with any splint, in addition to knowing the precise sequence of steps to

apply the splint properly, you must practice the splinting technique frequently to maintain the necessary skills. Follow the steps below to apply a Sager splint (Skill Drill 29-8 ▶):

1. **Expose the injured extremity.** Using BSI precautions as needed, assess and record the pulse, motor function, and sensation distal to the injury.

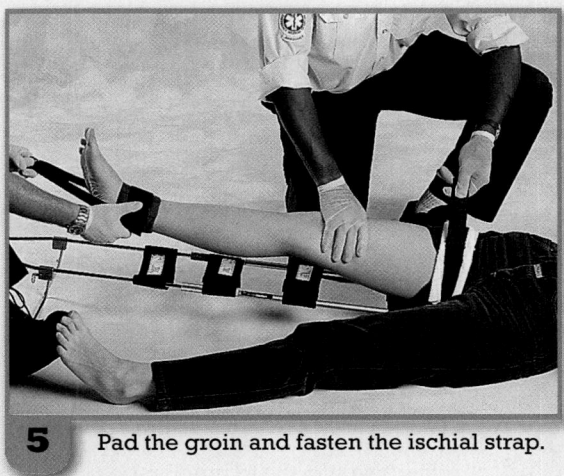

5 Pad the groin and fasten the ischial strap.

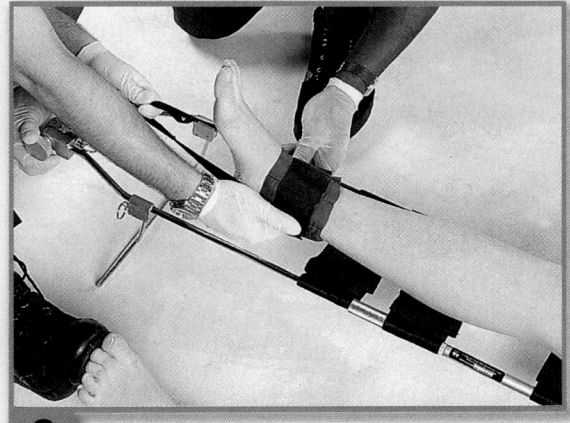

6 Connect the loops of the ankle hitch to the end of the splint as your partner continues to maintain traction. Carefully tighten the ratchet to the point that the splint holds adequate traction.

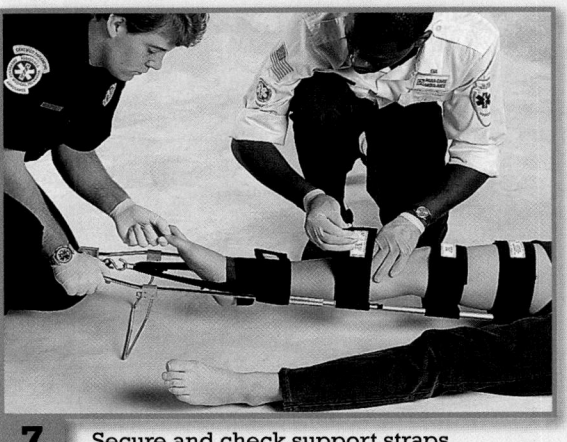

7 Secure and check support straps. Assess pulse, motor, and sensory functions.

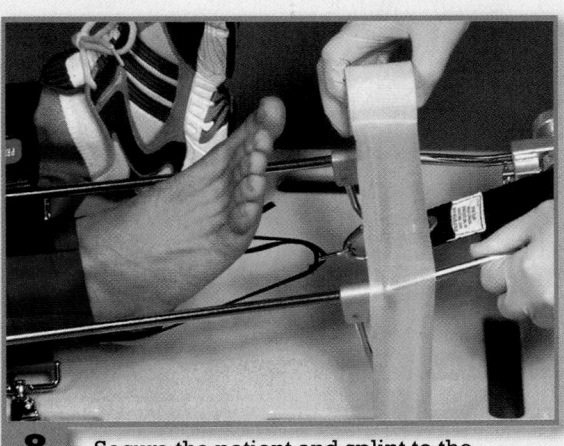

8 Secure the patient and splint to the backboard in a way that will prevent movement of the splint during patient movement and transport.

2. Before applying the splint, **adjust the thigh strap** so that it will lie anteriorly when secured in place (**Step 1**).

3. **Estimate the proper splint length** by placing it alongside the injured limb, so that the wheel is at the level of the heel.

4. **Arrange the ankle pads** to fit the size of the patient's ankle (**Step 2**).

5. **Place the splint along the inner aspect of the limb,** and slide the thigh strap around the upper thigh so that the perineal cushion is snug against the groin and the ischial tuberosity. Tighten the thigh strap snugly (**Step 3**).

6. **Secure the ankle harness** tightly around the patient's ankle just above the malleoli.

7. **Pull the cable ring** snugly up against the bottom of the foot (**Step 4**).

Applying a Sager Traction Splint

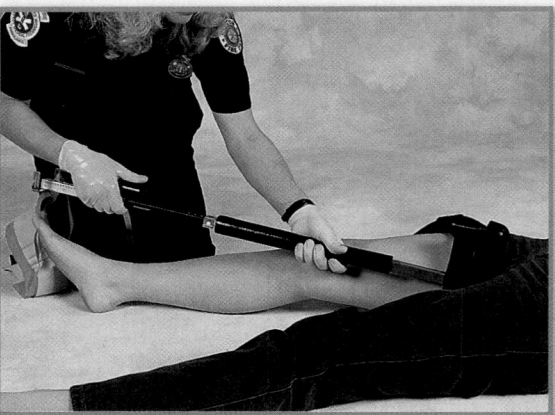

1 After exposing the injured area, check the patient's pulse and motor and sensory function.

Adjust the thigh strap so that it lies anteriorly when secured.

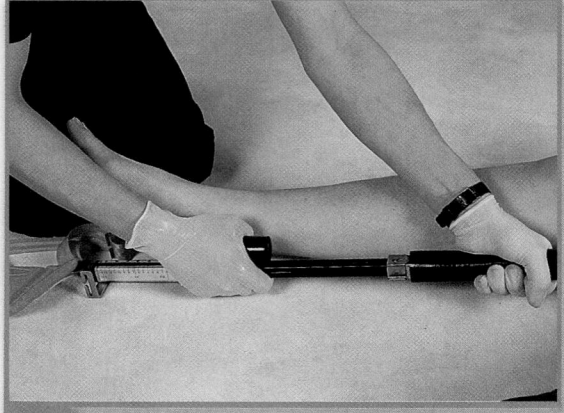

2 Estimate the proper length of the splint by placing it next to the injured limb.

Fit the ankle pads to the ankle.

3 Place the splint at the inner thigh, apply the thigh strap at the upper thigh, and secure snugly.

4 Tighten the ankle harness just above the malleoli.

Snug the cable ring against the bottom of the foot.

8. **Pull out the inner shaft** of the splint to apply traction of approximately 10% of body weight, using a maximum of 15 lb (**Step 5**).
9. **Secure the limb** to the splint using elasticized cravats (**Step 6**).
10. **Secure the patient** to a long backboard.
11. **Check pulse, motor, and sensory function** (**Step 7**).

Hazards of Improper Splinting

You must be aware of the hazards associated with the improper application of splints, including the following:

- Compression of nerves, tissues, and blood vessels
- Delay in transport of a patient with a life-threatening injury
- Reduction of distal circulation
- Aggravation of the injury

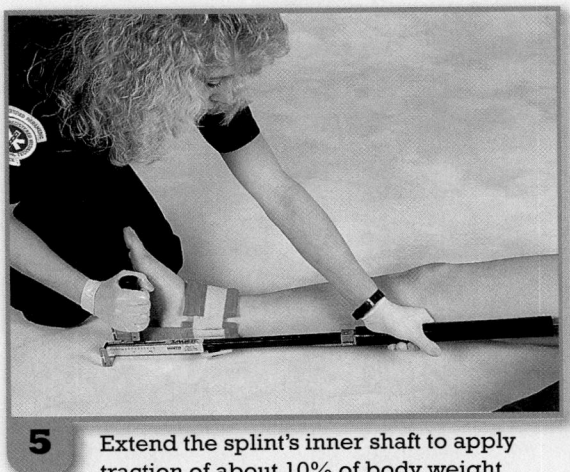

5 Extend the splint's inner shaft to apply traction of about 10% of body weight.

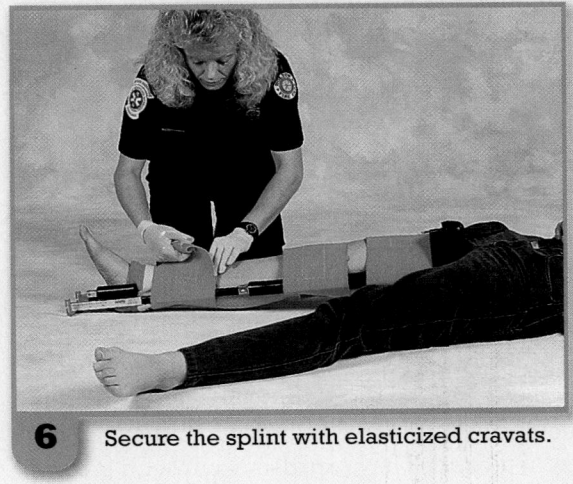

6 Secure the splint with elasticized cravats.

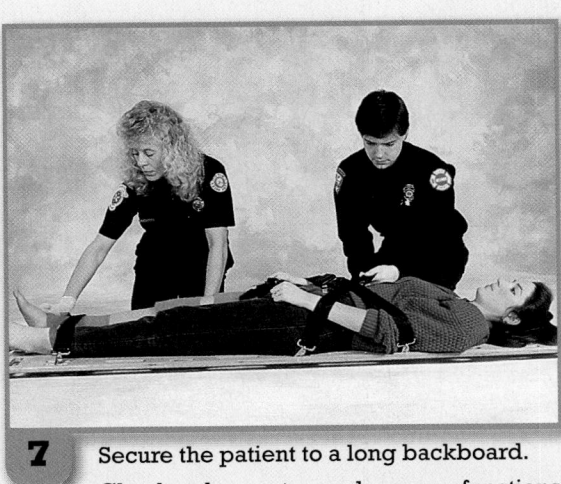

7 Secure the patient to a long backboard. Check pulse, motor, and sensory functions.

■ Injury to tissue, nerves, blood vessels, or muscles as a result of excessive movement of the bone or joint

Transportation

Once an injured limb is adequately splinted, the patient is ready to be transferred to a backboard or stretcher and transported.

Very few, if any, musculoskeletal injuries justify the use of excessive speed during transport. The limb will be stable once a dressing and splint have been applied. However, the patient with a pulseless limb must be given a higher priority. Still, if the hospital is only a few minutes away, speeding to the emergency department will make little or no difference to the patient's eventual outcome. If the treatment facility is an hour or more away, the patient with a pulseless limb should be transported by helicopter or immediate ground transportation. If circulation in the distal limb is impaired, always notify medical control so that proper steps can be taken quickly once the patient arrives in the emergency department.

Specific Musculoskeletal Injuries

Injuries of the Clavicle and Scapula

The clavicle, or collarbone, is one of the most commonly fractured bones in the body. Fractures of the clavicle occur most often in children when they fall on an outstretched hand. They can also occur with crushing injuries of the chest. A patient with a fracture of the clavicle will report pain in the shoulder and will usually hold the arm across the front of his or her body (Figure 29-22 ▶). A young child often reports pain throughout the entire arm and is unwilling to use any part of that limb. These complaints may make it difficult to localize the point of injury, but, generally, swelling and point tenderness occur over the clavicle. Because the clavicle is subcutaneous (just beneath the skin), the skin will occasionally "tent" over the fracture fragment. The clavicle lies directly over major arteries, veins, and nerves; therefore, fracture of the clavicle may lead to neurovascular compromise.

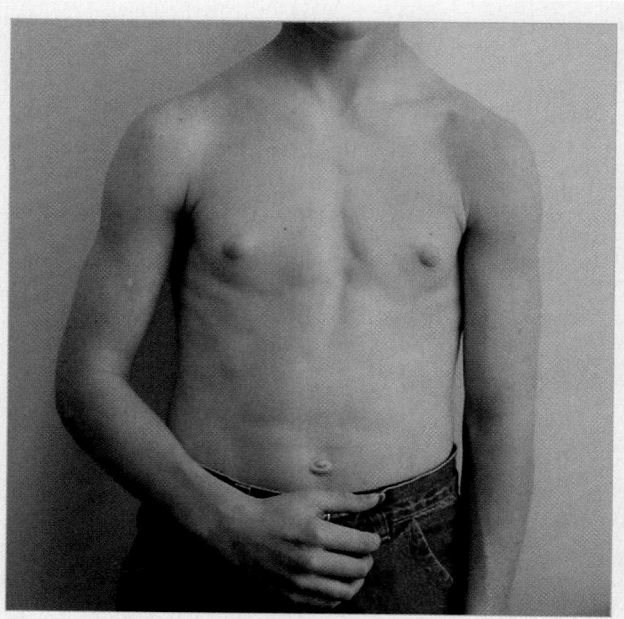

Figure 29-22 A patient with a fracture of the clavicle will usually hold the arm across the front of his or her body.

EMT-B Tips

Point tenderness is the most reliable indicator of an underlying fracture.

You are the Provider   Part 5

The patient's SAMPLE history indicates that he has no allergies to any type of pain medication. He is currently taking no medications and has no significant medical history. His last oral intake was a sandwich about an hour ago. The patient's vital signs are stable, and you and your partner work together to apply a splint to the patient's extremity. You select a splint that stabilizes above and below the area injured. Pulse, motor, and sensory function are present before and after application, and you apply a sling and swathe to stabilize the injured extremity to the patient's body. The patient states he is feeling a little better and you depart for the local hospital.

8. In what position would you transport this patient?
9. How would you document this call?

Fractures of the <u>scapula</u>, or shoulder blade, occur much less frequently because this bone is well protected by many large muscles. Fractures of the scapula are almost always the result of a forceful, direct blow to the back, directly over the scapula, which may also injure the thoracic cage, lungs, and heart. For this reason, you must carefully assess the patient for signs of breathing problems. Provide supplemental oxygen and prompt transport for patients who are having difficulty breathing. Remember, it is the associated chest injuries, not the fractured scapula itself, that pose the greatest threat of long-term disability.

Abrasions, contusions, and significant swelling may also occur, and the patient will often limit use of the arm because of pain at the fracture site. The scapula also has bony projections that may be fractured with a lesser degree of force.

The joint between the outer end of the clavicle and the acromion process of the scapula is called the <u>acromio-clavicular (A/C) joint</u>. This joint is frequently separated during football and hockey play when a player falls and lands on the point of the shoulder, driving the scapula away from the outer end of the clavicle. This dislocation is often called an A/C separation. The distal end of the clavicle will often stick out, and the patient will complain of pain, including point tenderness over the A/C joint (Figure 29-23 ▼).

Fractures of the clavicle and scapula and A/C separations can all be splinted effectively with a sling and swathe. A <u>sling</u> is any bandage or material that helps support the weight of an injured upper extremity, relieving the downward pull of gravity on the injured site. To be effective, a sling must apply gentle upward support to the olecranon process of the ulna. The knot of the sling should be tied to one side of the neck so that it does not press uncomfortably on the cervical spine (Figure 29-24A ▼).

To fully stabilize the shoulder region, a <u>swathe</u>, a bandage that passes completely around the chest, must be used to bind the arm to the chest wall. The swathe should be tight enough to prevent the arm from swinging freely but not so tight as to compress the chest and compromise breathing. Leave the patient's fingers exposed so that you can assess neurovascular function at regular intervals (Figure 29-24B ▼).

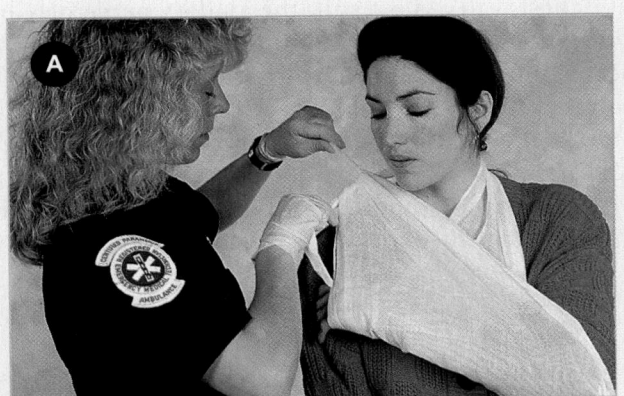
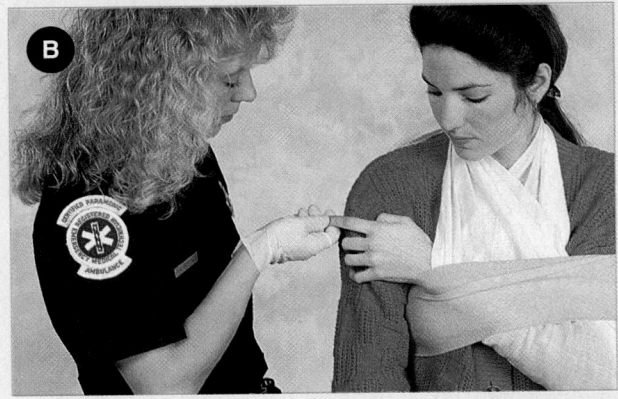

Figure 29-24 **A.** Apply the sling so that the knot is tied to one side of the neck. **B.** Bind the arm to the chest wall with a swathe so that the arm cannot swing freely. Leave the patient's fingers exposed so that you can assess distal circulation.

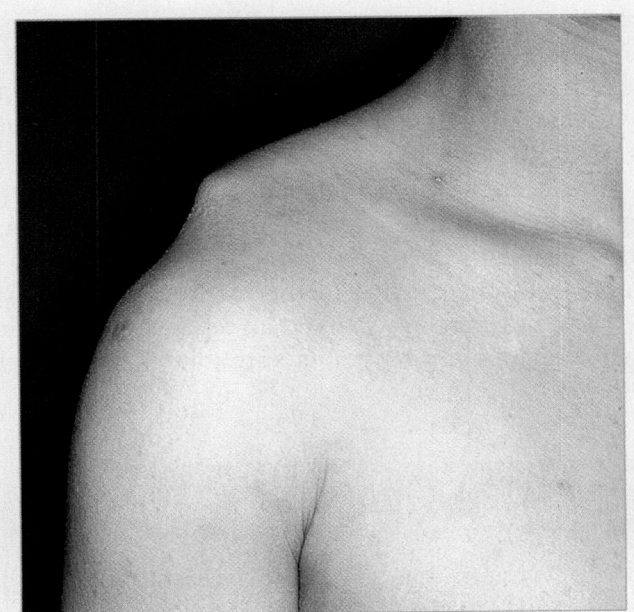

Figure 29-23 With A/C separations, the distal end of the clavicle usually sticks out.

Commercially available shoulder stabilizers or slings will provide adequate splinting for injuries of the shoulder region, as will triangular bandage slings.

Dislocation of the Shoulder

The glenohumeral joint (shoulder joint) is where the head of the <u>humerus</u>, the supporting bone of the upper arm, meets the glenoid fossa of the scapula. The <u>glenoid fossa</u> joins with the humeral head to form the glenohumeral joint. In shoulder dislocations, the humeral head most commonly dislocates anteriorly, coming to lie in front of the scapula as a result of forced abduction (away from the midline) and external rotation of the arm (Figure 29-25 ▼).

Shoulder dislocations are extremely painful. The patient will guard the shoulder and try to protect it by holding the dislocated arm in a fixed position away from the chest wall (Figure 29-26 ▶). The shoulder joint will usually be locked, and the shoulder will appear squared off or flattened. The humeral head will protrude anteriorly underneath the pectoris major on the anterior chest wall. As a result, the axillary nerve may be compressed, causing a numb patch on the outer aspect of the shoulder. Be sure to document this finding. Some patients may also report some numbness in the hand because of either nervous or circulatory compromise.

Stabilizing an anterior shoulder dislocation is difficult, because any attempt to bring the arm in

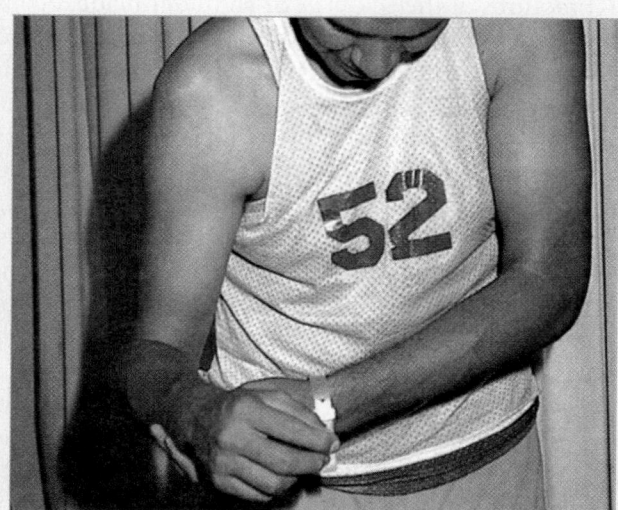

Figure 29-26 A patient with a dislocated shoulder will guard the shoulder, trying to protect it by holding the arm in a fixed position away from the chest wall.

toward the chest will produce pain. You must splint the joint in whatever position is most comfortable for the patient. If necessary, place a pillow or rolled blankets or towels between the arm and chest to fill up the space between them (Figure 29-27 ▶). Once the arm has been stabilized in this way, the elbow can usually be flexed to 90° without causing further pain. At this point, you can apply a sling to the forearm and wrist to support the weight of the arm. Finally, secure the arm in the sling to the pillow and chest with a swathe. Transport the patient in a sitting or semiseated position.

Dislocation of the shoulder disrupts the supporting ligaments of the anterior aspect of the shoulder. Often, these ligaments fail to heal properly, so dislocation recurs, each time causing further neurovascular compromise and joint injury. In certain cases, surgical repair may be required. Some patients are able to reduce (set)

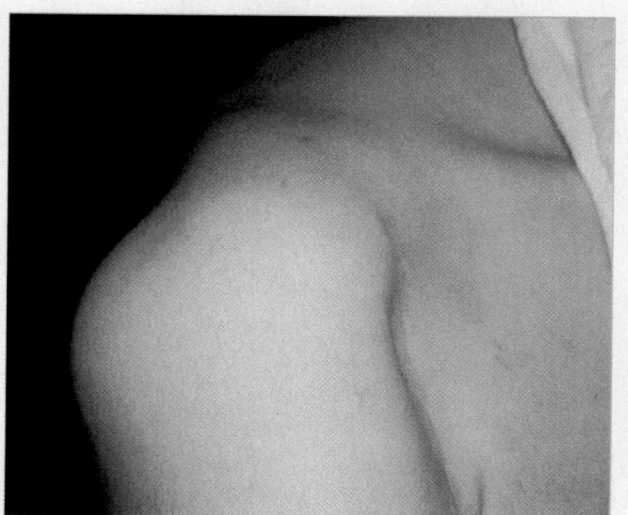

Figure 29-25 Most shoulder dislocations are anterior. Note the absence of the normal rounded appearance of the shoulder.

✚ EMT-B Tips

When assessing a patient with a possible shoulder dislocation, position yourself behind the patient and compare the shoulders. The dislocated side is usually lower than the uninjured side.

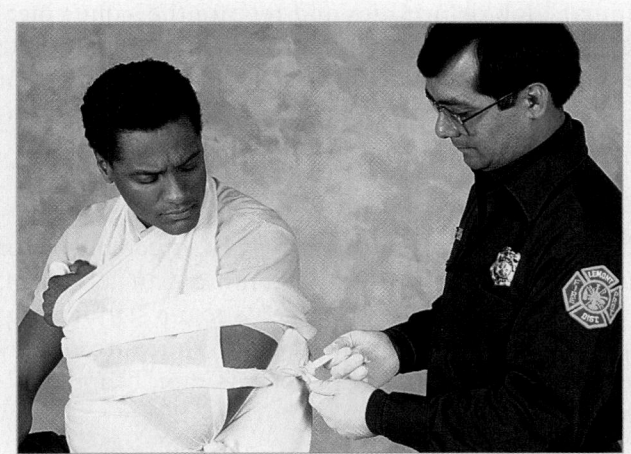

Figure 29-27 Splint the joint in a position of comfort, and place a pillow or towel between the arm and the chest wall to stabilize the arm, after which the elbow can be flexed to 90°. Apply a sling, and secure the arm to the chest with a swathe.

their own dislocated shoulders. Generally, however, this maneuver must be done in a hospital setting and only after X-rays have been obtained.

Posterior dislocation is less common than anterior shoulder dislocation. Football players, especially line-

men, are susceptible to this injury. The arm will often be locked in an adduction (toward the midline), so it cannot be rotated. Reducing the dislocation usually requires medical supervision.

Fracture of the Humerus

Fractures of the humerus occur either proximally, in the midshaft, or distally at the elbow (Table 29-2 ▼). Fractures of the proximal humerus resulting from falls are common among older people. Fractures of the midshaft occur more often in young patients, usually as the result of a violent injury.

With any severely angulated fracture, you should consider applying traction to realign the fracture fragments before splinting them. Check your local protocols for indications and techniques for applying traction to a severely angulated fracture. Support the site of the fracture with one hand, and with the other hand, grasp the two humeral condyles (its lateral and medial protrusions) just above the elbow. Pull gently in line with the normal axis of the limb (Figure 29-28 ▶). Once you achieve gross realignment of the limb, splint the arm with a sling and swathe, supplemented by a padded board splint on the lateral aspect of the arm (Figure 29-29 ▶). If the patient reports significant pain

TABLE 29-2 Characteristics and Treatment of Fractures of the Humerus

Type	Characteristics	Treatment
Proximal Humeral Fractures	■ Significant swelling, but no deformity, of the upper arm ■ Neurovascular compromise ■ Any or all of the brachial plexus affected, depending on the degree of displacement ■ Concurrent soft-tissue injuries ■ Possible rotator cuff injury (if X-rays show no fracture, a tear of the rotator cuff is possible, especially if the patient cannot move the arm toward the medial plane)	■ Stabilize in a sling and swathe or a shoulder stabilizer. ■ Use the chest wall as a splint, and secure the injured arm to the chest wall. ■ Place a short, padded board splint on the lateral side of the arm under the sling and swathe for additional support.
Midshaft Fractures	■ Gross angulation of the arm ■ Marked instability and crepitus of fracture fragments ■ Possible neurovascular compromise ■ Possible entrapment of the radial nerve. (The patient cannot extend or dorsiflex the wrist or fingers and may report numbness on the dorsum of the hand; classic "wrist drop.")	■ Stabilize with a sling and swathe or a shoulder stabilizer. ■ Use the chest wall as a splint, and secure the injured arm to the chest wall. ■ Place a short, padded board splint on the lateral side of the arm under the sling and swathe for additional support.
Distal Humeral Fractures	■ Significant swelling at the elbow ■ Possible neurovascular compromise ■ Possible injury to the ulnar or median nerves (document nerve status before and after any attempt to reduce the fracture)	■ Stabilize in a splint, in addition to a sling and swathe or a shoulder stabilizer.

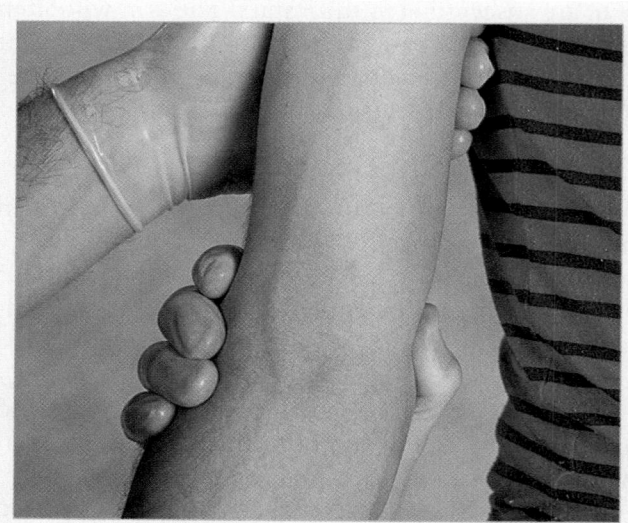

Figure 29-28 To align a severe deformity associated with a humeral shaft fracture, apply gentle pressure to the humeral condyles, as shown in this uninjured arm.

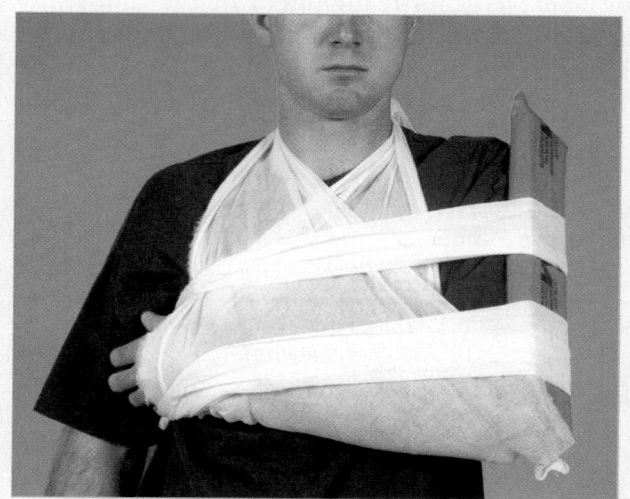

Figure 29-29 Splint a humeral shaft fracture with a sling and swathe supplemented by a padded board splint on the lateral aspect of the arm.

or resists gentle traction, splint the fracture in the deformed position with a padded wire ladder or a padded board splint, using pillows to support the injured limb. Note that compartment syndrome can develop in the forearm in children with these fractures.

Elbow Injuries

Fractures and dislocations often occur around the elbow, and the different types of injuries are difficult to distinguish without X-rays. However, they all produce similar limb deformities and require the same emergency care. Injuries to nerves and blood vessels are quite common in this region. Such injuries can be caused or worsened by inappropriate emergency care, particularly by excessive manipulation of the injured joint.

Fracture of the Distal Humerus

This type of fracture, also known as a supracondylar or intercondylar fracture, is common in children. Frequently, the fracture fragments rotate significantly, producing deformity and causing injuries to nearby vessels and nerves. Swelling occurs rapidly and is often severe.

Dislocation of the Elbow

This type of injury typically occurs in athletes and rarely in young children. It can occur in toddlers when they are lifted or pulled by the arm. The ulna and radius are most often displaced posteriorly. The ulna, the bone on the small finger side of the forearm, and the radius, the bone on the thumb side of the forearm, both join the distal humerus. The posterior displacement makes the olecranon process of the ulna much more prominent (Figure 29-30 ▶). The joint is usually locked, with the forearm moderately flexed on the arm; this position makes any attempt at motion extremely painful. As with a fracture of the distal humerus, there is swelling and significant potential for vessel or nerve injury.

Elbow Joint Sprain

This diagnosis is often mistakenly applied to an occult, nondisplaced fracture.

Fracture of the Olecranon Process of the Ulna

This fracture can result from direct or indirect forces and is often associated with lacerations and abrasions. The patient will be unable to actively extend the elbow.

Fractures of the Radial Head

Often missed during diagnosis, this fracture generally occurs as a result of a fall on an outstretched arm or a direct blow to the lateral aspect of the elbow. Attempts to rotate the elbow or wrist cause discomfort.

Care of Elbow Injuries

All elbow injuries are potentially serious and require careful management. Always assess distal neurovascular functions periodically in patients with elbow injuries. If you find strong pulses and good capillary refill, then splint the elbow injury in the position in which you found it, adding a wrist sling if this seems helpful. Two padded board splints, applied to each side of the limb

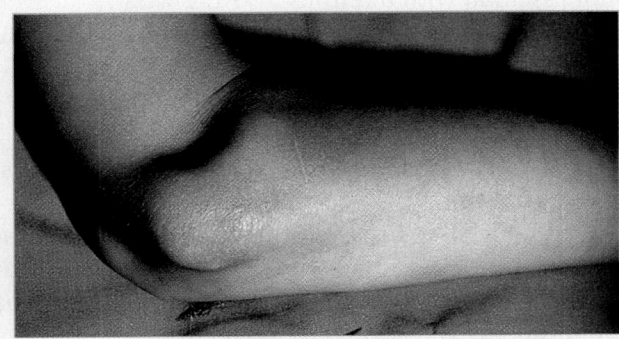

Figure 29-30 Posterior dislocation of the elbow makes the olecranon process of the ulna much more prominent.

Pediatric Needs

Growth plate injuries in children are common, especially around the wrist, elbow, knee, and ankle. Injuries tend to occur through these cartilaginous growth centers because they are inherently weaker than the surrounding bone. Since longitudinal growth of the limb is dependent upon the function of the growth plate, it is extremely important to recognize the possibility of growth plate injuries, stabilize the injured limb, and transport the patient in timely fashion to an appropriate center with pediatric, pediatric orthopaedic, and pediatric surgical coverage. Proper functioning of the injured growth plate throughout the remainder of skeletal growth may depend on timely anatomic reduction of the fracture and close follow-up by an orthopaedist.

Any deformity in close proximity to a joint in children younger than 16 years should be assumed to be a growth plate injury and transported and treated appropriately.

and secured with soft roller bandages, usually are enough to stabilize the arm (Figure 29-31A ▼). Make sure the board extends from the shoulder joint to the wrist joint, stabilizing the entire bone above and below the injured joint. Alternatively, you can mold a padded wire ladder splint or a SAM splint to the shape of the limb (Figure 29-31B ▼). If necessary, you may add further support to the limb with a pillow.

A cold, pale hand or a weak or absent pulse and poor capillary refill indicate that the blood vessels have likely been injured. Further care of this patient must be dictated by a physician. Notify medical control immediately. If you are within 10 to 15 minutes of the hospital, splint the limb in the position in which you found it and provide prompt transport. Otherwise, medical control may direct you to try to realign the limb to improve circulation in the hand.

If the limb is pulseless and significantly deformed at the elbow, apply gentle manual traction in line with the long axis of the limb to decrease the deformity. This maneuver may restore the pulse. Be careful, as excessive manipulation may only worsen the vascular problem. If no pulse returns *after one attempt*, splint the limb in the most comfortable position for the patient. If the pulse is restored by gentle longitudinal traction, splint the limb in whatever position allows the strongest pulse. Provide prompt transport for all patients with impaired distal circulation.

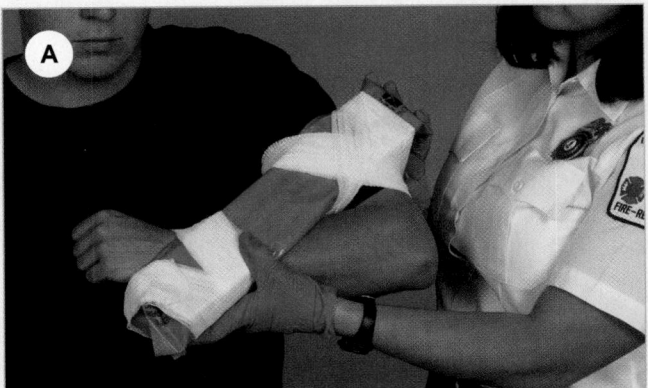
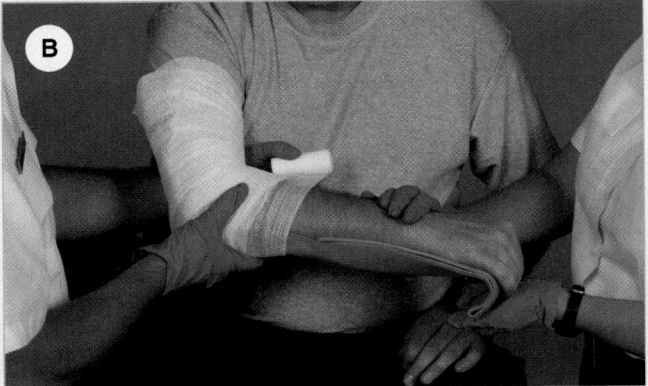

Figure 29-31 A. Two padded board splints provide adequate stabilization for an injured elbow. **B.** A SAM splint can be molded to the shape of the limb so that you can splint it in the position in which it was found.

Fractures of the Forearm

Fractures of the shaft of the radius and ulna are common in people of all age groups but are seen most often in children and older people. Usually, both bones break at the same time when the injury is the result of a fall on an outstretched hand (Figure 29-32 ▼). An isolated fracture of the shaft of the ulna may occur as the result of a direct blow to it; this is known as a nightstick fracture.

Fractures of the distal radius, which are especially common in elderly patients with osteoporosis, are often known as Colles fractures. The term "silver fork deformity" is used to describe the distinctive appearance of the patient's arm (Figure 29-33 ▼). In children, this fracture may occur through the growth plate and can have long-term consequences.

To stabilize fractures of the forearm or wrist, you can use a padded board, air, vacuum, or pillow splint. If the shaft of the bone has been fractured, be sure to include the elbow joint in the splint. Splinting of the elbow joint is not essential with fractures near the wrist; however, the patient will be more comfortable if you add a sling or pillow for more support.

Injuries of the Wrist and Hand

Injuries of the wrist, ranging from dislocations to sprains, must be confirmed by X-ray. Dislocations are usually associated with a fracture, resulting in a fracture dislocation. Another common wrist injury is the isolated, nondisplaced fracture of a carpal bone, especially the scaphoid. Any questionable wrist sprain or fracture should be splinted and evaluated either in the emergency department or an orthopaedic surgeon's office.

Hand injuries vary widely, some with potentially serious consequences. Industrial, recreational, and home accidents often result in dislocations, fractures, lacerations, burns, and amputations. Because the fingers and hands are required to function in such intricate ways, any injury that is not treated properly may result in permanent disability, as well as deformity. For this reason, all injuries to the hand, including simple lacerations, should be evaluated by a physician. For example, you should not attempt to "pop" a dislocated finger joint back in place (Figure 29-34 ▶). Always bring any amputated parts to the hospital with the patient. Be sure to wrap the amputated part in a dry or moist sterile dressing depending on your local protocol and place it in a dry plastic bag. Put the bag in a cooled container; do not soak the part in water or allow it to freeze.

A bulky forearm dressing makes an effective splint for any hand or wrist injury. Follow the steps in (Skill Drill 29-9 ▶):

1. **Follow BSI precautions.**
2. **Cover all wounds** with a dry, sterile dressing.

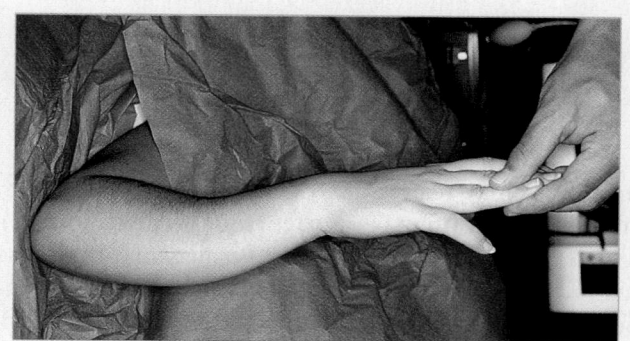

Figure 29-32 Fractures of the forearm often occur in children as a result of a fall on an outstretched hand.

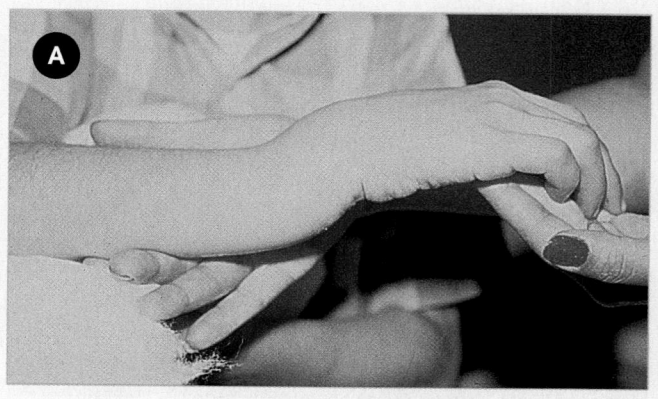
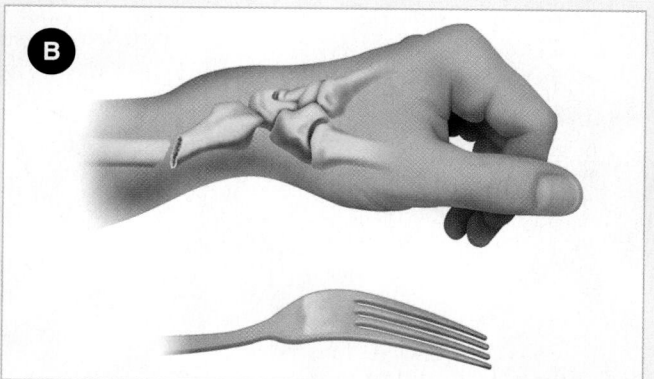

Figure 29-33 A. Fractures of the distal radius produce a characteristic silver fork deformity. **B.** An artist's illustration of same.

Splinting the Hand and Wrist

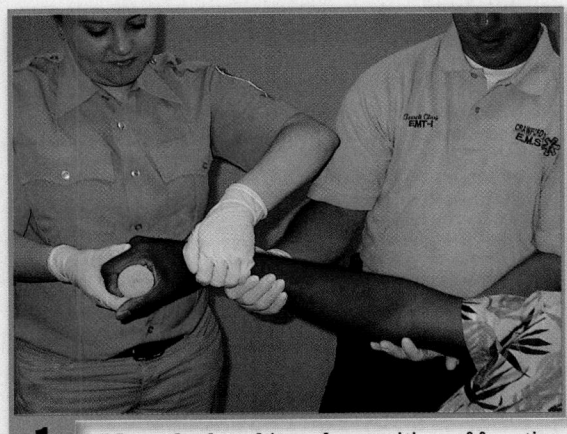

1 Move the hand into the position of function. Place a soft roller bandage in the palm.

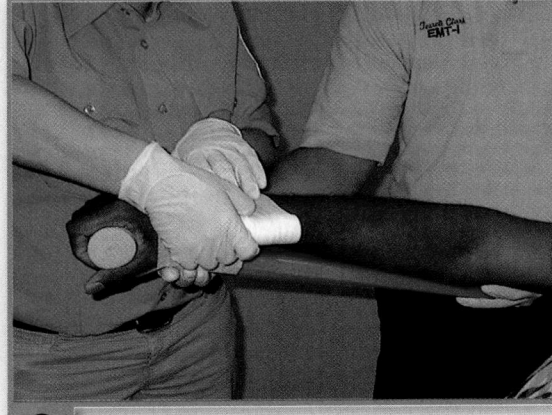

2 Apply a padded board splint on the palmar side with fingers exposed.

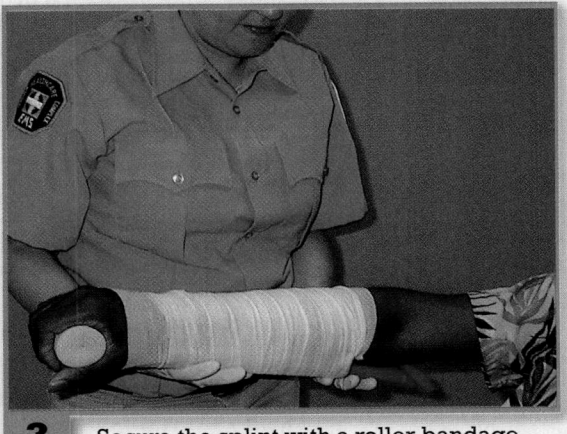

3 Secure the splint with a roller bandage.

3. Supporting the injured limb, **form the injured hand into the <u>position of function</u>**, with the wrist slightly bent down and all finger joints moderately flexed. This is the position that is used to hold a can most comfortably.

4. **Place a soft roller bandage into the palm of the hand** (**Step 1**).

5. **Apply a padded board splint to the palmar side of the wrist**, leaving the fingers exposed (**Step 2**).

6. **Secure the entire length of the splint** with a soft roller bandage (**Step 3**).

7. **Apply a sling and swathe** or prop the splinted hand and wrist on a pillow or on the patient's chest during transport to the hospital.

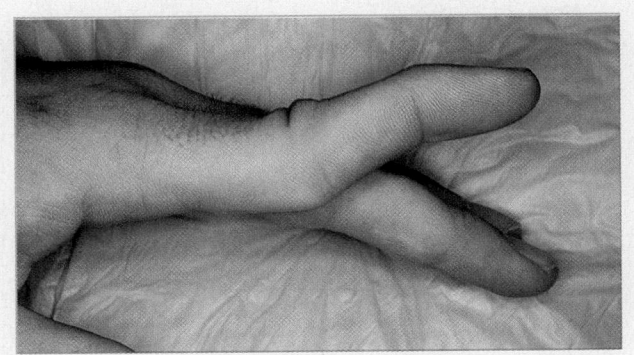

Figure 29-34 Dislocation of the finger joint. Do not be tempted to try to "pop" the joint back into place.

Fractures of the Pelvis

Fracture of the pelvis often results from direct compression in the form of a heavy blow that literally crushes the pelvis. The blow may be from a motor vehicle crash, a weapon used deliberately, a falling object, or a fall from a height. Injuries to the pelvis can also be caused by indirect forces. For example, when the knee strikes the dashboard in an automobile crash, the impact of the force is transmitted along the line of the <u>femur</u>, the thigh bone, which is the longest and largest bone in the body. The head of the femur is driven into the pelvis, causing it to fracture. However, not all pelvic fractures result from violent trauma. Even a simple fall can produce a fracture of the pelvis, especially in older individuals with osteoporosis.

Fractures of the pelvis may be accompanied by life-threatening loss of blood from the laceration of blood vessels affixed to the pelvis at certain key points. Up to several liters of blood may drain into the pelvic space and the <u>retroperitoneal space</u>, which lies between the abdominal cavity and the posterior abdominal wall. The result is significant hypotension, shock, and sometimes death. For this reason, you must take immediate steps to treat shock, even if there is only minimal swelling. Often, there are no visible signs of bleeding until severe blood loss has occurred. You should be prepared to resuscitate the patient rapidly if this becomes necessary.

Because the pelvis is surrounded by heavy muscle, open fractures of the pelvis are quite uncommon. However, pelvis fracture fragments can lacerate the rectum and vagina, creating an open fracture that is often overlooked. Once the protective pelvic ring is broken, the structures it is designed to protect, including the urinary bladder, are more susceptible to injury. The bladder may be lacerated by pelvic bone fragments, but more often, it tears or ruptures as a result of tension on either the bladder or the urethra.

You should suspect a fracture of the pelvis in any patient who has sustained a high-velocity injury and complains of discomfort in the lower back or abdomen. Because the area is covered by heavy muscle and other soft tissue, deformity or swelling may be very difficult to see. The most reliable sign of fracture of the pelvis is simple tenderness or instability on firm compression and palpation. Firm compression on the two iliac crests will produce pain at a fracture site in the pelvic ring. Assess for tenderness by taking the following steps (Figure 29-35 ▼):

1. Place the palms of your hands over the lateral aspect of each iliac crest, and **apply firm but gentle inward pressure** on the pelvic ring.
2. With the patient lying supine, **place a palm over the anterior aspect of each iliac crest, and apply firm downward pressure.**
3. Use the palm of your hand to firmly but gently **palpate the <u>pubic symphysis</u>**, the firm cartilaginous joint between the two pubic bones. This area will be tender if there is injury to the anterior portion of the pelvic ring.

If there has been injury to the bladder or the urethra, the patient will have lower abdominal tenderness and may have evidence of <u>hematuria</u> (blood in the urine) or blood at the urethral opening.

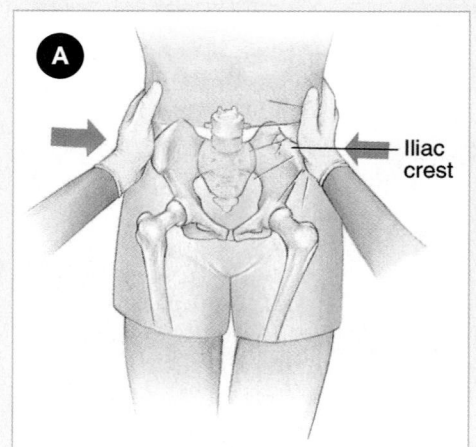
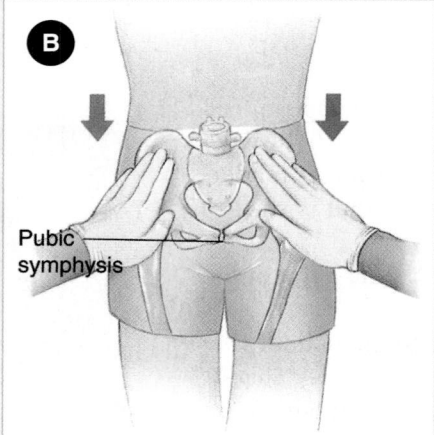
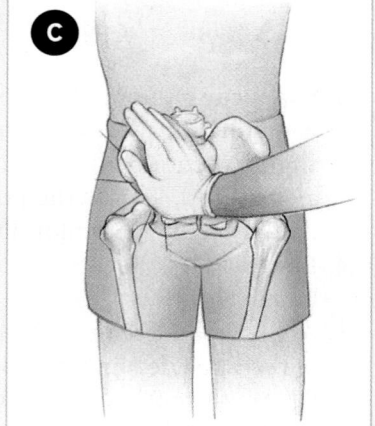

Figure 29-35 **A.** To assess for tenderness or instability in the pelvic region, place your hands over the lateral aspect of each iliac crest, and gently compress the pelvis. **B.** With the patient in a supine position, place your palms over the anterior aspect of each iliac crest, and apply firm but gentle downward pressure. **C.** Palpate the pubic symphysis with the palm of your hand.

Perform an initial assessment, and carefully monitor the general condition of any patient who you suspect has a pelvic fracture, because he or she is at high risk for hypovolemic shock. Stable patients can be secured to a long backboard or a scoop stretcher to stabilize isolated fractures of the pelvis. Place a PASG on the backboard or stretcher before transferring the patient to the backboard (Figure 29-36 ▼). The PASG will be ready to apply and inflate if the patient develops signs of shock. Remember, the PASG is only a temporary stabilization device and must be removed within 24 hours. Such a critically injured patient must be transferred to the hospital immediately.

Dislocation of the Hip

The hip joint is a very stable ball-and-socket joint that dislocates only after significant injury. Most dislocations of the hip are posterior. The femoral head is displaced posteriorly to lie in the muscles of the buttock. Posterior dislocation of the hip most commonly occurs as a result of automobile accidents in which the knee meets with a direct force, such as the dashboard, and the entire femur is driven posteriorly, dislocating the hip joint (Figure 29-37 ▶). Thus, you should suspect a hip dislocation in any patient who has been in an automobile crash and has a contusion, laceration, or obvious fracture in the knee region. Very rarely does the femoral head dislocate anteriorly; in this circumstance, the legs are suddenly and forcibly spread wide apart and locked in this position.

Posterior dislocation of the hip is frequently complicated by injury to the sciatic nerve, which is located directly behind the hip joint. The sciatic nerve is the most important nerve in the lower extremity; it controls the activity of muscles in the posterior thigh and below the knee, as well as sensation in most of the leg and foot. When the head of the femur is forced out of the hip socket, it may compress or stretch the sciatic nerve, leading to partial or complete paralysis of the nerve. The result is decreased sensation in the leg and foot and frequently weakness in the foot muscles. Generally, only the dorsiflexors, the muscles that raise the toes or foot, are involved, causing the "foot drop" that is characteristic of damage to the peroneal portion of the sciatic nerve.

Patients with a posterior dislocation of the hip typically lie with the hip joint flexed (the knee joint drawn up toward the chest) and the thigh rotated inward toward the midline of the body over the top of the opposite thigh (Figure 29-38A ▶). With the less common anterior dislocation, the limb is in the opposite position,

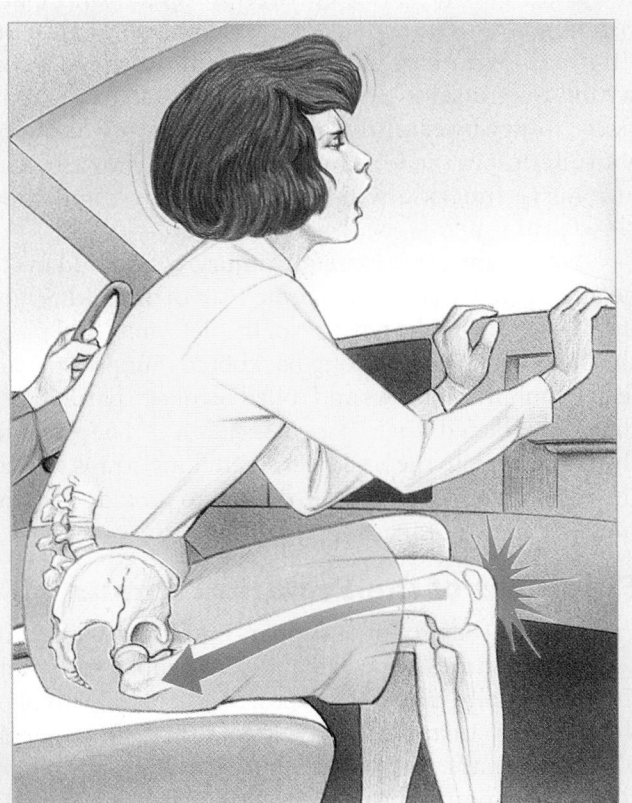

Figure 29-37 Posterior dislocation of the hip can occur as a result of the knee hitting the dashboard in an automobile crash. The impact drives the femur posteriorly (see arrow), dislocating the joint.

Figure 29-36 Place a PASG on the backboard before log-rolling a patient with a suspected pelvic fracture.

Figure 29-38 A. The usual position of a patient with a posterior dislocation of the hip. The hip joint is flexed, and the thigh is rotated inward and adducted across the midline of the body. **B.** Support the affected limb with pillows and blankets, particularly under the flexed knee. Secure the entire limb to a long board with long straps to prevent movement during transport.

extended straight out, externally rotated, and pointing away from the midline of the body.

Dislocation of the hip is associated with very distinctive signs. The patient will have severe pain in the hip and will strongly resist any attempt to move the joint. The lateral and posterior aspects of the hip region will be tender on palpation. With some thin individuals, you can palpate the femoral head deep within the muscles of the buttock. Check for a sciatic nerve injury by carefully assessing sensation and motor function in the lower extremity. Occasionally, sciatic nerve function will be normal at first and then slowly diminish.

As with any other extremity injury, you should make no attempt to reduce the dislocated hip in the field. Splint the dislocation in the position of the deformity and place the patient supine on a long backboard. Support the affected limb with pillows and rolled blankets, particularly under the flexed knee (Figure 29-38B ▲). Then secure the entire limb to the backboard with long straps so that the hip region will not move. Be sure to provide prompt transport.

Fractures of the Proximal Femur

Fractures of the proximal (upper) end of the femur are common fractures, especially in older people. Although they are usually called hip fractures, they rarely involve the hip joint. Instead, the break goes through the neck of the femur, the intertrochanteric (middle) region, or across the proximal shaft of the femur (subtrochanteric fractures). Although these three fracture types occur most often in older patients, particularly patients with osteoporosis, they may also be seen as a result of high-energy injuries in younger patients.

Patients with displaced fractures of the proximal femur display a very characteristic deformity. They lie with the leg externally rotated, and the injured leg is usually shorter than the opposite, uninjured limb (Figure 29-39A ▶). When the fracture is not displaced, this deformity is not present. With any kind of hip fracture, patients typically are unable to walk or move the leg because of pain in the hip region or in the groin or inner aspect of the thigh. The hip region is usually tender on palpation, and gentle rolling of the leg will cause pain but will not do further damage. On occasion, the pain is referred to the knee, and it is not uncommon for a geriatric patient with a hip fracture to complain of knee pain after a fall. You should splint the lower extremity of an older patient who has fallen and complains of pain in either the hip or the knee, even if there is no deformity, and then transport the patient to the emergency department.

The age of the patient and the severity of the injury will dictate how you splint the fracture. With young people, fractures of the hip resulting from violent injury are best stabilized with a traction splint or the combination of a PASG and a backboard. The PASG offers an added advantage: It will help to control bleeding in the region. Apply the traction splint as you would for a femoral shaft fracture, taking special care to protect the injured region from excessive pressure from the ring of a Hare traction splint.

A geriatric patient with an isolated hip fracture does not require a traction splint. You can effectively stabilize such a fracture by placing the patient on a long backboard or scoop stretcher, using pillows or rolled blankets to support the injured limb in the deformed position. Then secure the injured limb carefully to the stretcher with long straps (Figure 29-39B ▶).

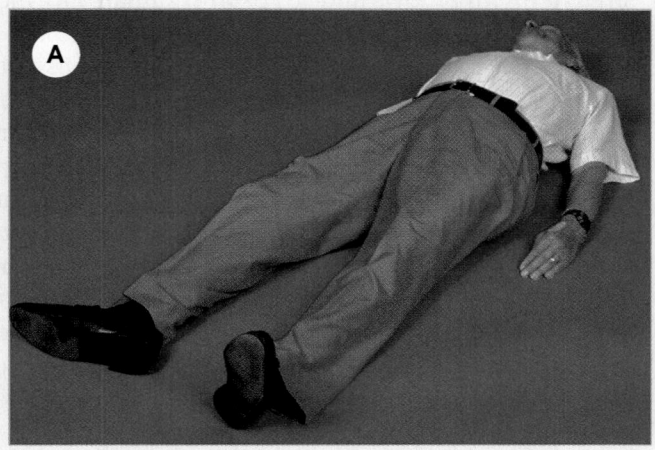 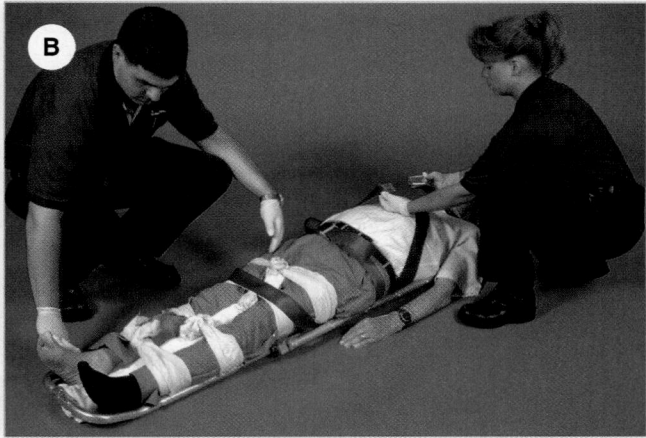

Figure 29-39 **A.** A patient with a fracture of the proximal femur will typically lie still with the extremity externally rotated, making the injured leg appear shorter than the other leg. **B.** Splint the injured leg to the uninjured leg and secure the patient on a scoop stretcher or backboard.

All patients with hip fractures may lose significant amounts of blood. Therefore, you should treat with high-flow oxygen and monitor vital signs frequently, being alert for signs of shock.

Femoral Shaft Fractures

Fractures of the femur can occur in any part of the shaft, from the hip region to the femoral condyles just above the knee joint. Following a fracture, the large muscles of the thigh spasm in an attempt to "splint" the unstable limb. The muscle spasm often produces significant deformity of the limb, with severe angulation or external rotation at the fracture site. Usually, the limb also shortens significantly. Fractures of the femoral shaft may be open, and fragments of bone may protrude through the skin.

There is often a significant amount of blood loss, as much as 500 to 1,000 mL, after a fracture of the shaft of the femur. With open fractures, the amount of blood loss may be even greater. Thus, it is not unusual for hypovolemic shock to develop. Handle these patients with extreme care, because any extra movement or fracture manipulation may increase the amount of blood loss.

Because of the severe deformity that occurs with these fractures, bone fragments may penetrate or press on important nerves and vessels and produce significant damage. For this reason, you must carefully and periodically assess the distal neurovascular function in patients who have sustained a fracture of the femoral shaft. Remove the clothing from the affected limb so that you can adequately inspect the injury site for any open wounds. Remember to follow BSI precautions when any blood or body fluids are present. Monitor the patient's vital signs closely, and continue to watch for the onset of hypovolemic shock. You must provide immediate transport in this situation.

Cover any wound with a dry, sterile dressing. If the foot or leg below the level of the fracture shows signs of impaired circulation (is pale, cold, or pulseless), apply gentle longitudinal traction to the deformed limb in line with the long axis of the limb. Gradually turn the leg from the deformed position to restore the limb's overall alignment. Often, this restores or improves circulation to the foot. If it does not, the patient may have sustained a serious vascular injury and may be in need of prompt medical attention.

A fracture of the femoral shaft is best stabilized with a traction splint, such as a Hare traction splint or a Sager splint (see Skill Drills 29-7 and 29-8).

Injuries of Knee Ligaments

The knee is very vulnerable to injury; therefore, many different types of injuries occur in this region. Ligament injuries, for example, range from mild sprains to complete dislocation of the joint. The patella can also dislocate. In addition, all the bony elements of the knee (distal femur, upper tibia, and patella) can fracture.

The knee is especially susceptible to ligament injuries, which occur when abnormal bending or twisting forces are applied to the joint. Such injuries are often seen in both recreational and competitive athletes. The ligaments on the medial side of the knee are the ones that are most frequently injured, typically when the foot is fixed to the ground and the lateral aspect of the knee is struck by a heavy object, such as when a football player is clipped or tackled from the side.

Usually, the patient with a knee ligament injury will report pain in the joint and be unable to use the extremity normally. When you examine the patient, you will generally find swelling, occasional ecchymosis, point tenderness at the injury site, and a joint effusion (excess fluid in the joint).

You should splint all suspected knee ligament injuries. The splint should extend from the hip joint to the foot, stabilizing the bone above the injured joint (the femur) and the bone below it (the tibia). A variety of splints can be used, including a padded, rigid, long leg splint or two padded board splints securely applied to the medial and lateral aspects of the limb. A long backboard, a pillow splint, or simply binding the injured limb to its uninjured mate are acceptable but less effective splinting techniques. The patient will usually be able to straighten the knee to allow you to apply the splint. However, if you encounter resistance or pain when trying to straighten the knee, splint it in the flexed position. Then continue to monitor the distal neurovascular function until the patient reaches the hospital.

Dislocation of the Knee

Complete disruption of the ligaments supporting the knee may result in dislocation of the joint. When this happens, the proximal end of the tibia completely displaces from its juncture with the lower end of the femur, usually producing a significant deformity. Although substantial ligament damage always occurs with a knee dislocation, the more urgent injury is often to the popliteal artery, which is frequently lacerated or compressed by the displaced tibia. When gross deformity, severe pain, and an inability to move the joint cause you to suspect a dislocation of the knee, always check the distal circulation carefully before taking any other step. If the distal pulses are absent, contact medical control immediately for further stabilization instructions.

If adequate distal pulses are present, splint the knee in the position in which you found it, and transport the patient promptly. Do not attempt to manipulate or straighten any severe knee injury if there are good distal pulses. If the limb is straight, apply standard rigid long leg splints to at least two sides of the limb to stabilize it (Figure 29-40A ▶). If the knee is bent and the foot has a good pulse, splint the joint in the bent position, using parallel padded board splints secured at the hip and ankle joint to provide a stable A-frame (Figure 29-40B ▶). Secure the limb to a backboard or

stretcher with pillows and straps to eliminate any motion during transport.

On rare occasions, medical control may instruct you to realign a deformed, pulseless limb to reduce compression of the popliteal artery and thus restore distal circulation. You should make only one attempt to do this. First, straighten the limb by applying gentle longitudinal traction in the axis of the limb. Once you apply manual traction, maintain it until the limb is fully splinted; otherwise, the limb will return to its deformed position. If traction significantly increases the patient's pain, do not continue. As you apply traction, monitor the posterior tibial pulse to see whether it returns. Splint the limb in the position in which you feel the strongest pulse. If you are unable to restore the distal pulse, splint the limb in the position that is most comfortable for the patient, and then provide prompt transport to the hospital. Notify medical control of the status of the distal pulse so that arrangements to treat the patient can be made in advance.

Fractures About the Knee

Fractures about the knee may occur at the distal end of the femur, at the proximal end of the tibia, or in the patella. Because of local tenderness and swelling, it is easy to confuse a nondisplaced or minimally displaced fracture about the knee with a ligament injury. Likewise, a displaced fracture about the knee may produce significant deformity that makes it look like a dislocation. Management of the two types of injuries is as follows:

- If there is an adequate distal pulse and no significant deformity, splint the limb with the knee straight.
- If there is an adequate pulse and significant deformity, splint the joint in the position of deformity.
- If the pulse is absent below the level of the injury, contact medical control immediately for further instructions.

Dislocation of the Patella

A dislocated patella most commonly occurs in teenagers and young adults who are engaged in athletic activities. Some patients have recurrent dislocations of the patella. As with recurrent dislocation of the shoulder, a minor twisting may be enough to produce the problem. Usually, the dislocated patella displaces to the lateral side, and the knee is held in a partially flexed position. The displacement of the patella produces a significant deformity in which the knee is held in a

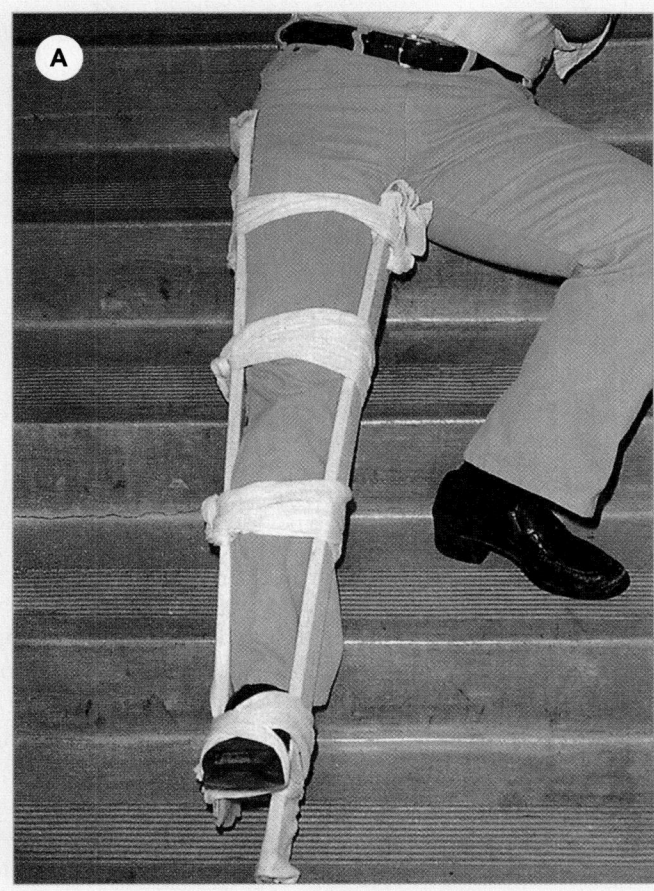

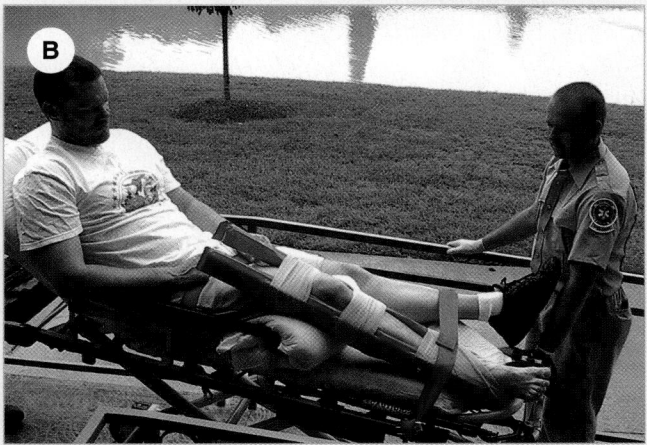

Figure 29-40 **A.** When the injured knee is straight, apply padded board splints extending from the hip to the ankle. **B.** If the knee is flexed and the foot has good pulses, apply padded board splints with the knee in the flexed position.

moderately flexed position and the patella is displaced to the lateral side of the knee (Figure 29-41 ▶).

Splint the knee in the position in which you found it; most often, this is with the knee flexed to a moderate degree. To stabilize the knee, apply padded board splints to the medial and lateral aspects of the joint, extending from the hip to the ankle. Use pillows to support the limb on the stretcher.

Occasionally, as you apply the splint, the patella will return to its normal position spontaneously. When this occurs, stabilize the limb as for a knee ligament injury, in a padded long leg splint. The patient still needs to be transported to the emergency department. Report the spontaneous reduction as soon as you arrive at the hospital so that the medical staff is aware of the severity of the injury.

Injuries of the Tibia and Fibula

The tibia (shinbone) is the larger of the two leg bones that are responsible for supporting the major weight-bearing surface of the knee and ankle; the fibula is the smaller of them. Fracture of the shaft of the tibia or the fibula may occur at any place between the knee joint and the ankle joint. Usually, both bones fracture at the same time. Even a single fracture may result in severe deformity, with significant angulation or rotation. Because the tibia is located just beneath the skin, open fractures of this bone are quite common (Figure 29-42 ▶).

Fractures of the tibia and fibula should be stabilized with a padded, rigid long leg splint or an air splint that extends from the foot to the upper thigh. Traction splints are not indicated for isolated tibial fractures. As with most other fractures of the shaft of long bones, you should correct severe deformity before splinting by applying gentle longitudinal traction. The goal here is to restore a position that will take a standard splint; it is not necessary to replace the fracture fragments in their anatomic position.

Fractures of the tibia and fibula are sometimes associated with vascular injury as a result of the distorted position of the limb following injury. Realigning the

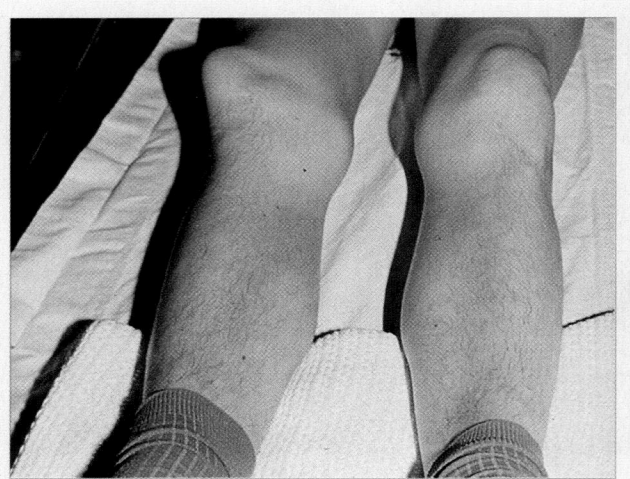

Figure 29-41 Usually, the dislocated patella displaces to the lateral side, and the knee is held in a partially flexed position.

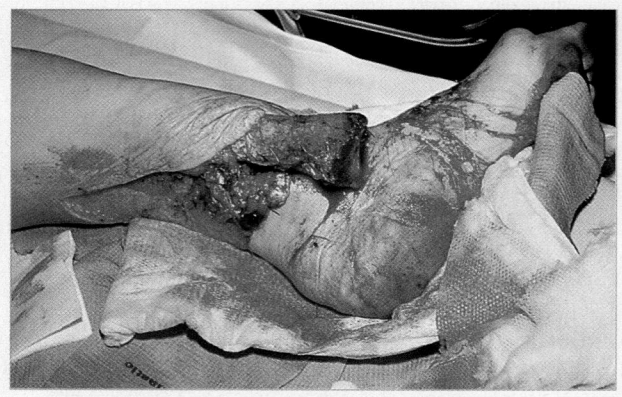

Figure 29-42 Because the tibia is so close to the skin, open fractures are quite common.

limb frequently restores an adequate blood supply to the foot. If it does not, transport the patient promptly and notify medical control while you are en route.

Ankle Injuries

The ankle is a very commonly injured joint. Ankle injuries occur in individuals of all ages and range in severity from a simple sprain, which heals after a few days' rest, to severe fracture-dislocations. As with other joints, it is sometimes difficult to tell a nondisplaced ankle fracture from a simple sprain without X-rays (Figure 29-43 ▶). Therefore, any ankle injury that produces pain, swelling, localized tenderness, or the inability to bear weight must be evaluated by a physician. The most frequent mechanism of ankle injury is twisting, which stretches or tears

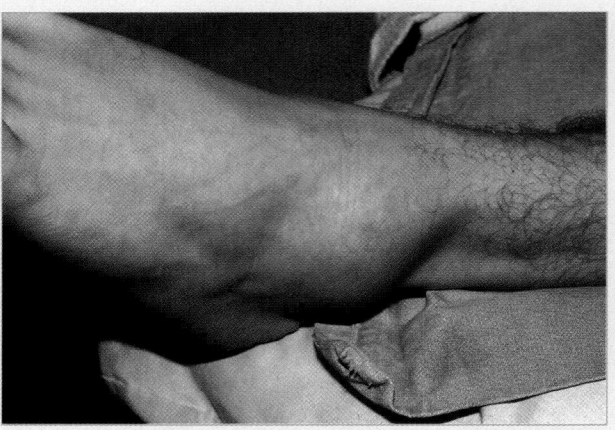

Figure 29-43 Swelling about the ankle is characteristic of both sprains and fractures.

the supporting ligaments. A more extensive twisting force may result in fracture of one or both malleoli. Dislocation of the ankle is usually associated with fractures of one or both malleoli.

You can manage the wide spectrum of injuries to the ankle in the same way, as follows:

1. Dress all open wounds.
2. Assess distal neurovascular function.
3. Correct any gross deformity by applying gentle longitudinal traction to the heel.
4. Before releasing traction, apply a splint.

You can use a padded rigid splint, an air splint, or a pillow splint. Just make sure it includes the entire foot and extends up the leg to the level of the knee joint.

Foot Injuries

Injuries to the foot can result in the fracture of one or more of the tarsals, metatarsals, or phalanges of the toes. Toe fractures are especially common.

Of the tarsal bones, the calcaneus, the heel bone, is the most frequently fractured. Injury often occurs when the patient falls or jumps from a height and lands directly on the heel. The force of injury compresses the calcaneus, producing immediate swelling and ecchymosis. If the force of impact is great enough, as from a fall from a roof or tree, there may also be other fractures.

Frequently, the force of injury is transmitted up the legs to the spine, producing a fracture of the lumbar spine (Figure 29-44 ▶). When a patient who has jumped or fallen from a height complains of heel pain, be sure to question him or her about back pain and carefully check the spine for tenderness or deformity.

Injuries of the foot are associated with significant swelling but rarely with gross deformity. Vascular

Figure 29-44 Frequently after a fall, the force of injury is transmitted up the legs to the spine, sometimes resulting in a fracture of the lumbar spine.

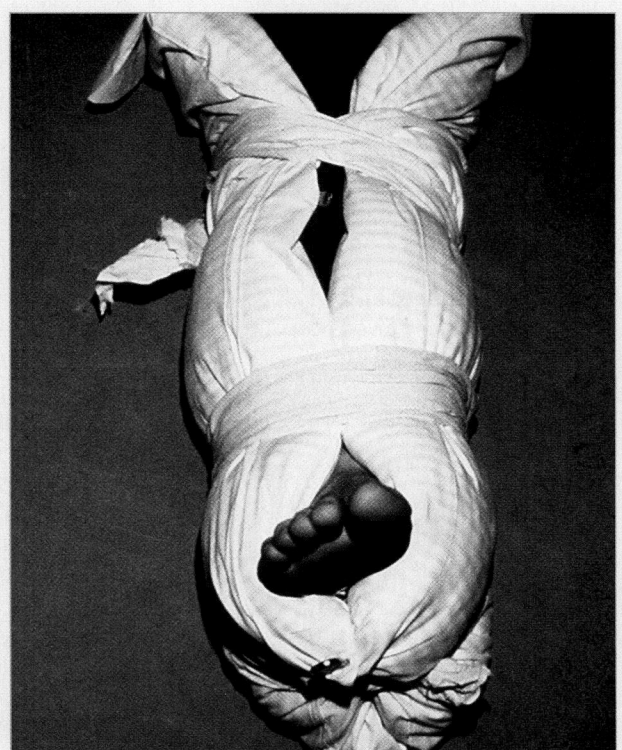

Figure 29-45 A pillow splint can provide excellent stabilization of the foot.

injuries are not common. As in the hand, lacerations about the ankle and foot may damage important underlying nerves and tendons. Puncture wounds of the foot are common and may cause serious infection if not treated early. All of these injuries must be evaluated and treated by a physician.

To splint the foot, apply a rigid padded board splint, an air splint, or a pillow splint, stabilizing the ankle joint as well as the foot (Figure 29-45 ▲). Leave the toes exposed so that you can periodically assess neurovascular function.

When the patient is lying on the stretcher, elevate the foot approximately 6″ to minimize swelling. All patients with lower extremity injuries should be transported in the supine position to allow for elevation of the limb. Never allow the foot and leg to dangle off the stretcher onto the floor or ground.

If a patient has fallen from a height and complains of heel pain, use a long backboard to stabilize any possible spinal injury in addition to splinting the foot.

You are the Provider   Summary

It is important to have a clear understanding of how to treat a joint or bone injury. As an EMT-B, you will not need to distinguish whether your patient has a fracture or a dislocation. You will treat either condition the same way. With any traumatic injury, consider cervical spine stabilization until you are certain the patient's condition does not require it. Complete a full patient assessment and document your findings.

Assessment and Emergency Care

Musculoskeletal Injuries	
Scene Size-up	Body substance isolation precautions should include a minimum of gloves and eye protection. Ensure scene safety and determine NOI/MOI. Consider the number of patients, the need for additional help/ALS, and c-spine stabilization.
Initial Assessment	
■ General impression	Determine level of consciousness and treat any immediate threats to life. Determine priority of care based on environment and patient's chief complaint.
■ Airway	Ensure patent airway.
■ Breathing	Listen for abnormal breath sounds and evaluate depth and rate of the respirations. Look for symmetric chest rise and fall. Maintain ventilations as needed. Provide high-flow oxygen at 15 L/min and inspect the chest wall, assessing for DCAP-BTLS.
■ Circulation	Evaluate pulse rate and quality; observe skin color, temperature, and condition, and treat accordingly.
■ Transport decision	Prompt transport
Focused History and Physical Exam	NOTE: The order of the steps in the focused history and physical exam differs depending on whether or not the patient has a significant MOI. The order below is for a patient with a significant MOI. For a patient without a significant MOI, perform a focused trauma assessment, obtain vital signs, and obtain the history.
■ Rapid trauma assessment	Reevaluate the mechanism of injury. Perform a rapid trauma assessment, treating all life-threats immediately. Log roll and secure patient to backboard for all patients with a suspected spinal injury.
■ Baseline vital signs	Take vital signs, noting skin color and temperature as well as patient's level of consciousness. Use pulse oximetry if available.
■ SAMPLE history	Obtain SAMPLE history. If the patient is unresponsive, attempt to obtain information from family members, friends, or bystanders. Consider obtaining pertinent OPQRST information for patients with minor injuries.
■ Interventions	Provide complete spinal immobilization early if you suspect that your patient has spinal injuries. Maintain an open airway and suction as needed. Immobilize musculoskeletal injuries as per protocol; treat signs of shock and any other life threats.
Detailed Physical Exam	Complete a detailed physical exam.
Ongoing Assessment	Repeat the initial assessment, rapid/focused assessment, and reassess interventions performed. Reassess vital signs every 5 minutes for the unstable patient, every 15 minutes for the stable patient.
■ Communication and documentation	Contact medical control with a radio report. Relay any change in level of consciousness or difficulty breathing. Describe the MOI and any interventions you performed. Be sure to document any physician's orders and changes in the patient's condition and at what time they occurred.

NOTE: While the steps below are widely accepted, be sure to consult and follow your local protocol.

Musculoskeletal Injuries

Caring for Musculoskeletal Injuries
1. Cover open wounds with a dry, sterile dressing, and apply pressure to control bleeding.
2. Assess pulse, motor function, and sensory function prior to splinting.
3. Apply a splint and elevate the extremity about 6" (slightly above the level of the heart).
4. Assess pulse, motor function, and sensory function immediately after splinting and frequently in transit.
5. Apply cold packs if there is swelling, but do not place them directly on the skin.
6. Position the patient for transport and secure the injured area.

Assessing Neurovascular Pulse
1. Palpate the radial pulse in the upper extremity.
2. Palpate the posterior tibial pulse in the lower extremity.
3. Assess capillary refill by blanching a fingernail or toenail.
4. Assess sensation on the flesh near the tip of the index finger.
5. On the foot, first check sensation on the flesh near the tip of the great toe.
6. Also check foot sensation on the lateral side.
7. Evaluate motor function by asking the patient to open the hand. (Perform motor tests only if the hand or foot is not injured. Stop a test if it causes pain.)
8. Also ask the patient to make a fist.
9. To evaluate motor function in the foot, ask the patient to extend the foot.
10. Also have the patient flex the foot and wiggle the toes.

Applying a Rigid Splint
1. Provide gentle support and in-line traction for the limb.
2. Second EMT places the splint alongside or under the limb. Pad between the limb and the splint as needed to ensure even pressure and contact.
3. Secure the splint to the limb with bindings.
4. Assess and record distal neurovascular function.

Applying a Zippered Air Splint
1. Support the injured limb and apply gentle traction as your partner applies the open, deflated splint.
2. Zip up the splint, inflate it by pump or by mouth, and test the pressure. Check and record distal neurovascular function.

Applying an Unzippered Air Splint
1. Support the injured limb. Have your partner place his or her arm through the splint to grasp the patient's hand or foot.
2. Apply gentle traction while sliding the splint onto the injured limb.
3. Inflate the splint.

Applying a Vacuum Splint
1. Stabilize and support the injury.
2. Place the splint and wrap it around the limb.
3. Draw the air out of the splint and seal the valve.

Applying a Hare Traction Splint
1. Expose the injured limb and check pulse, motor, and sensory function. Place the splint beside the uninjured limb, adjust the splint to proper length, and prepare the straps.
2. Support the injured limb as your partner fastens the ankle hitch about the foot and ankle.
3. Continue to support the limb as your partner applies gentle in-line traction to the ankle hitch and foot.
4. Slide the splint into position under the injured limb.
5. Pad the groin and fasten the ischial strap.
6. Connect the loops of the ankle hitch to the end of the splint as your partner continues to maintain traction. Carefully tighten the ratchet to the point that the splint holds adequate traction.
7. Secure and check support straps. Assess pulse, motor, and sensory functions.
8. Secure the patient and splint to the backboard in a way that will prevent movement of the splint during patient movement and transport.

Applying a Sager Traction Splint
1. After exposing the injured area, check the patient's pulse and motor and sensory function. Adjust the thigh strap so that it lies anteriorly when secured.
2. Estimate the proper length of the splint by placing it next to the injured limb. Fit the ankle pads to the ankle.
3. Place the splint at the inner thigh, apply the thigh strap at the upper thigh, and secure snugly.
4. Tighten the ankle harness just above the malleoli. Snug the cable ring against the bottom of the foot.
5. Extend the splint's inner shaft to apply traction of about 10% of body weight.
6. Secure the splint with elasticized cravats.
7. Secure the patient to a long backboard. Check pulse, motor, and sensory functions.

Splinting the Hand and Wrist
1. Move the hand into the position of function. Place a soft roller bandage in the palm.
2. Apply a padded board or similar splint on the palmar side with fingers exposed.
3. Secure the splint with a roller bandage.

Prep Kit

Ready for Review

- Skeletal or involuntary muscle attaches to bone and forms the major muscle mass of the body. This muscle contains veins, arteries, and nerves.

- There are 206 bones in the human body. When this living tissue is fractured, it can produce bleeding and significant pain.

- A joint is a junction where two bones come into contact. Joints are stabilized in key areas by ligaments.

- A fracture is a broken bone, a dislocation is a disruption of a joint, a sprain is a stretching injury to the ligaments around a joint, and strain is stretching of muscle.

- Depending on the amount of kinetic energy absorbed by tissues, the zone of injury may extend beyond the point of contact. Always maintain a high index of suspicion for associated injuries.

- Fractures of the bones are classified as open or closed. Both are splinted in a similar manner, but remember to control bleeding and apply a sterile dressing to the open extremity injury before splinting.

- Fractures and dislocations are often difficult to diagnose without an X-ray. You will treat these injuries similarly. Stabilize the injury with a splint and transport the patient.

- Signs of fractures and dislocations include pain, deformity, point tenderness, false movement, crepitus, swelling, and bruising.

- Signs of sprain include bruising, swelling, and an unstable joint.

- Always assess the trauma patient with the same technique: address the initial assessment and correct problems, determine if you will utilize the rapid trauma exam or focused trauma exam, assess the SAMPLE history and obtain baseline vital signs. When treating musculoskeletal injuries, always assess for pulse, motor, and sensory function before and after applying a splint. Reassess these functions during the ongoing assessment.

- Compare the unaffected extremity to injured extremity for differences whenever possible.

- There are three main types of splints used by the EMT-B: rigid splints, traction splints, and air splints.

- Remember to splint the injured extremity from the joint above to the joint below the injury site for complete stabilization.

- A sling and swathe is used commonly to treat shoulder dislocations and to secure injured upper extremities to the body. Lower extremities can be secured to the unaffected limb or to a long backboard.

- The most common life-threatening musculoskeletal injuries are multiple fractures, open fractures with arterial bleeding, pelvic fractures, bilateral femur fractures, and limb amputations.

www.EMTB.com

Technology

Interactivities

Vocabulary Explorer

Anatomy Review

Web Links

Online Review Manual

Vital Vocabulary

acromioclavicular (A/C) joint A simple joint where the bony projections of the scapula and the clavicle meet at the top of the shoulder.

articular cartilage A pearly layer of specialized cartilage covering the articular surfaces (contact surfaces on the ends) of bones in synovial joints.

calcaneus The heel bone.

clavicle The collarbone.

closed fracture A fracture in which the skin is not broken.

compartment syndrome An elevation of pressure within a closed fascial compartment, characterized by extreme pain, decreased pain sensation, pain on stretching of affected muscles, and decreased power; frequently seen in fractures below the elbow or knee in children.

crepitus A grating or grinding sensation or sound caused by fractured bone ends or joints rubbing together.

dislocation Disruption of a joint in which ligaments are damaged and the bone ends are completely displaced.

displaced fracture A fracture in which bone fragments are separated from one another and not in anatomic alignment.

ecchymosis Bruising or discoloration associated with bleeding within or under the skin.

femur The thigh bone, which extends from the pelvis to the knee and is responsible for formation of the hip and knee; the longest and largest bone in the body.

fibula The outer and smaller bone of the two bones of the lower leg.

fracture A break in the continuity of a bone.

glenoid fossa The part of the scapula that joins with the humeral head to form the glenohumeral joint.

hematuria Blood in the urine.

humerus The supporting bone of the upper arm that joins with the scapula (glenoid) to form the shoulder joint and with the ulna and radius to form the elbow joint.

joint The place where two bones come into contact.

ligament A band of fibrous tissue that connects bones to bones, and supports and strengthens a joint.

nondisplaced fracture A simple crack in the bone that has not caused the bone to move from its normal anatomic position; also called a hairline fracture.

open fracture Any break in a bone in which the overlying skin has been damaged.

patella The kneecap.

point tenderness Tenderness that is sharply localized at the site of the injury, found by gently palpating along the bone with the tip of one finger.

position of function A hand position in which the wrist is slightly dorsiflexed and all finger joints are moderately flexed.

pubic symphysis The firm cartilaginous joint between the two pubic bones.

radius The bone on the thumb side of the forearm; important in both wrist and function.

reduce Return a dislocated joint or fractured bone to its normal position; set.

retroperitoneal space The space between the abdominal cavity and the posterior abdominal wall, containing the kidneys, certain large vessels, and parts of the gastrointestinal tract.

scapula Shoulder blade.

sciatic nerve The major nerve to the lower extremity; controls much of muscle function in the leg, and sensation in most of the leg and foot.

skeletal muscle Striated muscles that are attached to bones and usually cross at least one joint.

sling A bandage or material that helps to support the weight of an injured upper extremity.

splint A flexible or rigid appliance used to protect and maintain the position of an injured extremity.

sprain A joint injury involving damage to supporting ligaments, and sometimes partial or temporary dislocation of bone ends.

strain Stretching or tearing of a muscle; also called a muscle pull.

www.EMTB.com

Prep Kit continued...

swathe A bandage that passes around the chest to secure an injured arm to the chest.

tendon A tough, ropelike cord of fibrous tissue that attaches a skeletal muscle to a bone.

tibia The larger of the two lower leg bones responsible for supporting the major weight-bearing surface of the knee and the ankle; the shinbone.

traction Longitudinal force applied to a structure.

ulna The bone on the small finger side of the forearm; most important for elbow function.

zone of injury The area of potentially damaged soft tissue, adjacent nerves, and blood vessels surrounding an injury to a bone or a joint.

Points to Ponder

You are dispatched to an area nursing home for a man who has fallen. En route to the facility, the dispatcher informs you that your patient is semiconscious. You arrive to find an older man lying on the floor, moaning. You immediately notice his distal portion of left arm is swollen and grossly deformed. The facility's RN tells you that she witnessed the event. The patient tripped, put his arms out to brace his fall, and landed hard on the linoleum floor. Your partner, a new EMT-B, immediately begins to splint the left arm.

What should be your primary focus? Would the priority of splinting change if this man were alert and oriented?

Issues: Splinting at the Scene Versus Load and Go, How Mechanism of Injury Affects Patients Differently.

Assessment in Action

You are dispatched to the soccer field at Gracie High School for a possible leg fracture. You arrive to find a female adolescent surrounded by several people in the middle of the soccer field who hurriedly motion for you. After a quick survey of the field conditions, your partner chooses to drive the ambulance onto the field to gain closer proximity to the patient. A parent tells you that the 15-year-old girl was playing soccer with another student without her normal protective gear. She attempted to kick the ball the same time as another student, and the lower legs of both students collided together.

As you begin talking to the patient, it is easy to see that her right lower leg is angulated. She is crying and trying to protect her leg. She screams when you try to examine it. You explain what you are going to do and why, and you enlist the help of one of the parents. Her vital signs include a blood pressure of 118/66 mm Hg, a pulse of 138 beats/min, and respirations (increasing with pain) of 42 breaths/min.

1. What injury do you suspect?

   A. Femur fracture
   B. Tibia fracture
   C. Ulna fracture
   D. Humerus fracture

2. Exposing and assessing her injury involves:

   A. pulling up her pant leg.
   B. cutting the right pant leg.
   C. cutting both pant legs.
   D. none of the above.

3. When should you check for a distal pulse?

   A. Before splinting the extremity
   B. After splinting the extremity
   C. Any time that is convenient
   D. Both A and B

4. Besides pulses, what should you assess and document regarding perfusion status distal to the injury site?

   A. Motor function
   B. Sensation
   C. Capillary refill
   D. All of the above

5. What type of splint is appropriate to use for this scenario?

   A. Traction splint
   B. Formable splint
   C. Rigid splint
   D. Either B or C

6. Motion at a point in the limb where there is no joint is referred to as:

   A. False motion
   B. True motion
   C. True movement
   D. Both B and C

Challenging Questions

7. What is a sprain and can it mimic other injuries? Does determining the underlying injury have a significant impact on your care?

8. What is guarding, and how does it help the body?

9. Why is it important to splint properly?

10. What is the significance of pelvic and femur fractures?

www.EMTB.com

Head and Spine Injuries

Objectives

Cognitive

5-4.1 State the components of the nervous system. (p 872)

5-4.2 List the functions of the central nervous system. (p 872)

5-4.3 Define the structure of the skeletal system as it relates to the nervous system. (p 876)

5-4.4 Relate mechanism of injury to potential injuries of the head and spine. (p 878)

5-4.5 Describe the implications of not properly caring for potential spine injuries. (p 888)

5-4.6 State the signs and symptoms of a potential spine injury. (p 885)

5-4.7 Describe the method of determining if a responsive patient may have a spine injury. (p 881)

5-4.8 Relate the airway emergency medical care techniques to the patient with a suspected spine injury. (p 882)

5-4.9 Describe how to stabilize the cervical spine. (p 889)

5-4.10 Discuss indications for sizing and using a cervical spine immobilization device. (p 889)

5-4.11 Establish the relationship between airway management and the patient with head and spine injuries. (p 882)

5-4.12 Describe a method for sizing a cervical spine immobilization device. (p 898)

5-4.13 Describe how to log roll a patient with a suspected spine injury. (p 882)

5-4.14 Describe how to secure a patient to a long spine board. (p 891)

5-4.15 List instances when a short spine board should be used. (p 893)

5-4.16 Describe how to immobilize a patient using a short spine board. (p 894)

5-4.17 Describe the indications for the use of rapid extrication. (p 894)

5-4.18 List the steps in performing rapid extrication. (p 186-190)

5-4.19 State the circumstance when a helmet should be left on the patient. (p 900)

5-4.20 Discuss the circumstances when a helmet should be removed. (p 901)

5-4.21 Identify different types of helmets. (p 901)

5-4.22 Describe the unique characteristics of sports helmets. (p 901)

5-4.23 Explain the preferred methods to remove a helmet. (p 901)

5-4.24 Discuss alternative methods for removal of a helmet. (p 902)

5-4.25 Describe how the patient's head is stabilized to remove the helmet. (p 902)

5-4.26 Differentiate how the head is stabilized with a helmet compared to without a helmet. (p 898)

Affective

5-4.27 Explain the rationale for immobilization of the entire spine when a cervical spine injury is suspected. (p 889)

5-4.28 Explain the rationale for utilizing immobilization methods apart from the straps on the cots. (p 889)

5-4.29 Explain the rationale for utilizing a short spine immobilization device when moving a patient from the sitting to the supine position. (p 893)

5-4.30 Explain the rationale for utilizing rapid extrication approaches only when they indeed will make the difference between life and death. (p 894)

5-4.31 Defend the reasons for leaving a helmet in place for transport of a patient. (p 900)

5-4.32 Defend the reasons for removal of a helmet prior to transport of a patient. (p 901)

Psychomotor

5-4.33 Demonstrate opening the airway in a patient with a suspected spinal cord injury. (p 888)

5-4.34 Demonstrate evaluating a responsive patient with a suspected spinal cord injury. (p 888)

5-4.35 Demonstrate stabilization of the cervical spine. (p 889)

5-4.36 Demonstrate the four-person log roll for a patient with a suspected spinal cord injury. (p 892)

5-4.37 Demonstrate how to log roll a patient with a suspected spinal cord injury using two people. (p 891)

5-4.38 Demonstrate securing a patient to a long spine board. (p 891)

5-4.39 Demonstrate using the short board immobilization technique. (p 894)

5-4.40 Demonstrate the procedure for rapid extrication. (p 186-190)

5-4.41 Demonstrate preferred methods for stabilization of a helmet. (p 900)

5-4.42 Demonstrate helmet removal techniques. (p 901)

5-4.43 Demonstrate alternative methods for stabilization of a helmet. (p 902)

5-4.44 Demonstrate completing a prehospital care report for patients with head and spinal injuries. (p 887)

You are the Provider

Your unit is on standby at All American College during a gymnastics tournament. An excited bystander comes to your location and states that a 19-year-old female gymnast has fallen head first from the balance beam. He also tells you the patient is awake "but not moving well or breathing right." You gather your equipment and head to the scene.

1. What is your first reaction to the initial information about this scene?
2. What steps can you take to prepare yourself for treating this patient prior to arriving at the scene?

Head and Spine Injuries

The nervous system is a complex network of nerve cells that enables all parts of the body to function. It includes the brain, the spinal cord, and several billion nerve fibers that carry information to and from all parts of the body. Because the nervous system is so vital, it is well protected. The brain lies within the skull, and the spinal cord is inside the bony spinal canal. Despite this protection, serious injuries can damage the nervous system.

This chapter first briefly reviews the anatomy and function of the central and peripheral nervous systems and of the skeletal system. Discussion of specific head and spinal injuries follows, including signs, symptoms, assessment, and treatment. Extrication of patients with possible spinal injuries and removal of helmets are also described.

Anatomy and Physiology of the Nervous System

The nervous system is divided into two anatomic parts: the central nervous system and the peripheral nervous system (Figure 30-1 ▶). The central nervous system (CNS) includes the brain and the spinal cord, including the nuclei and cell bodies of most nerve cells. Long nerve fibers link these cells to the body's various organs through openings in the spinal column. These ca-

Technology

Interactivities

Vocabulary Explorer

Anatomy Review

Web Links

Online Review Manual

www.EMTB.com

bles of nerve fibers make up the peripheral nervous system.

Central Nervous System

The CNS is composed of the brain and spinal cord. The brain is the organ that controls the body; it is also the center of consciousness. It is divided into three major areas: the cerebrum, the cerebellum, and the brain stem (Figure 30-2 ▶).

The cerebrum, which contains about 75% of the brain's total volume, controls a wide variety of activities, including most voluntary motor function and conscious thought. Underneath the cerebrum lies the cerebellum, which coordinates body movements. The most primitive part of the CNS, the brain stem, controls virtually all the functions that are necessary for life, including the cardiac and respiratory systems. Deep within the cranium, the brain stem is the best-protected part of the CNS.

The spinal cord, the other major portion of the CNS, is mostly made up of fibers that extend from the brain's nerve cells. The spinal cord carries messages between the brain and the body.

Protective Coverings

The cells of the brain and spinal cord are soft and easily injured. Once damaged, they cannot be regenerated or reproduced. Therefore, the entire CNS is contained within a protective framework.

The thick, bony structures of the skull and spinal canal withstand injury very well. The skull is covered by a layer of muscle fascia and above that is the scalp (a thick vascular layer of skin). The spinal canal is also surrounded by a thick layer of skin and muscles.

The CNS is further protected by the meninges, three distinct layers of tissue that suspend the brain and the spinal cord within the skull and the spinal canal (Figure 30-3 ▶). The outer layer, the dura mater, is a tough, fibrous layer that closely resembles leather. This layer forms a sac to contain the CNS, with small openings through which the peripheral nerves exit.

The inner two layers of the meninges, called the arachnoid and the pia mater, are much thinner than the dura mater. They contain the blood vessels that nourish the brain and spinal cord. Cerebral spinal fluid (CSF) is produced in a chamber inside the brain, called the third ventricle. CSF fills the spaces between the meninges and acts as a shock absorber. The brain and spinal cord essentially float in this fluid, buffered from injury. The brain depends on a rich supply of oxygenated blood to

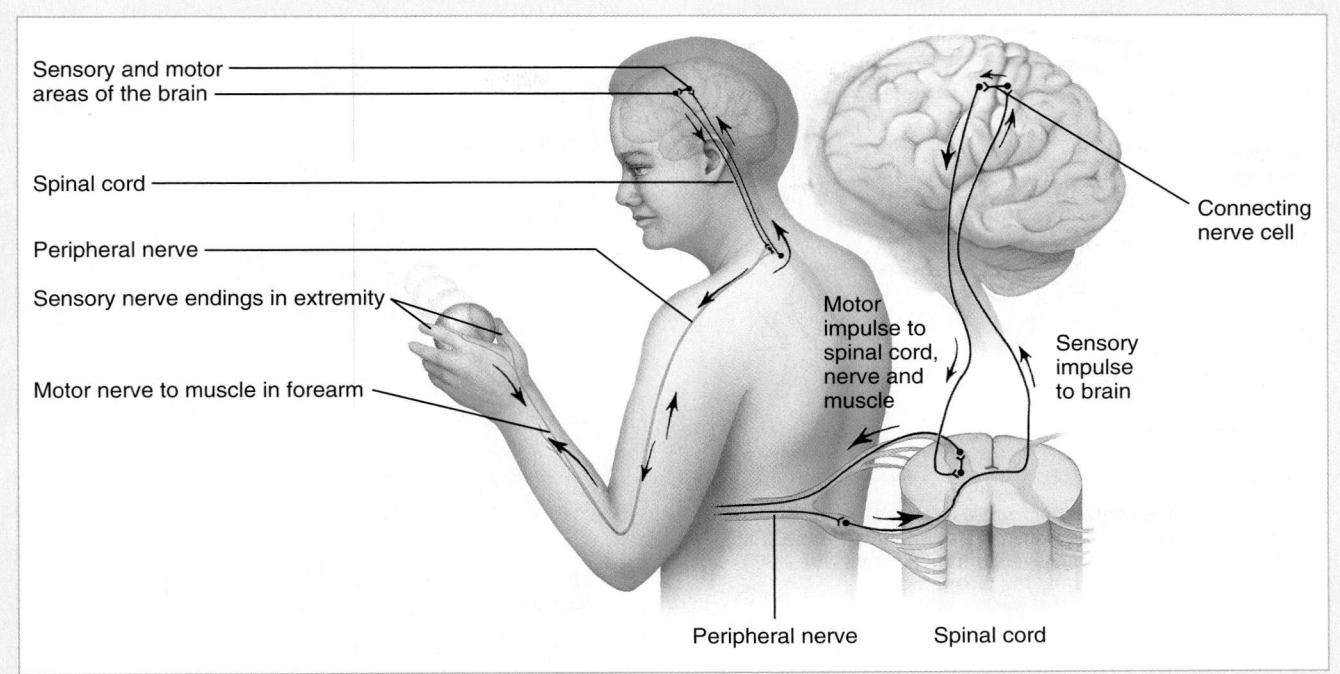

Figure 30-1 The nervous system has two anatomic components: the central nervous system and the peripheral nervous system. The central nervous system is composed of the brain and the spinal cord. The peripheral nervous system conducts sensory and motor impulses from the skin and other organs to the spinal cord.

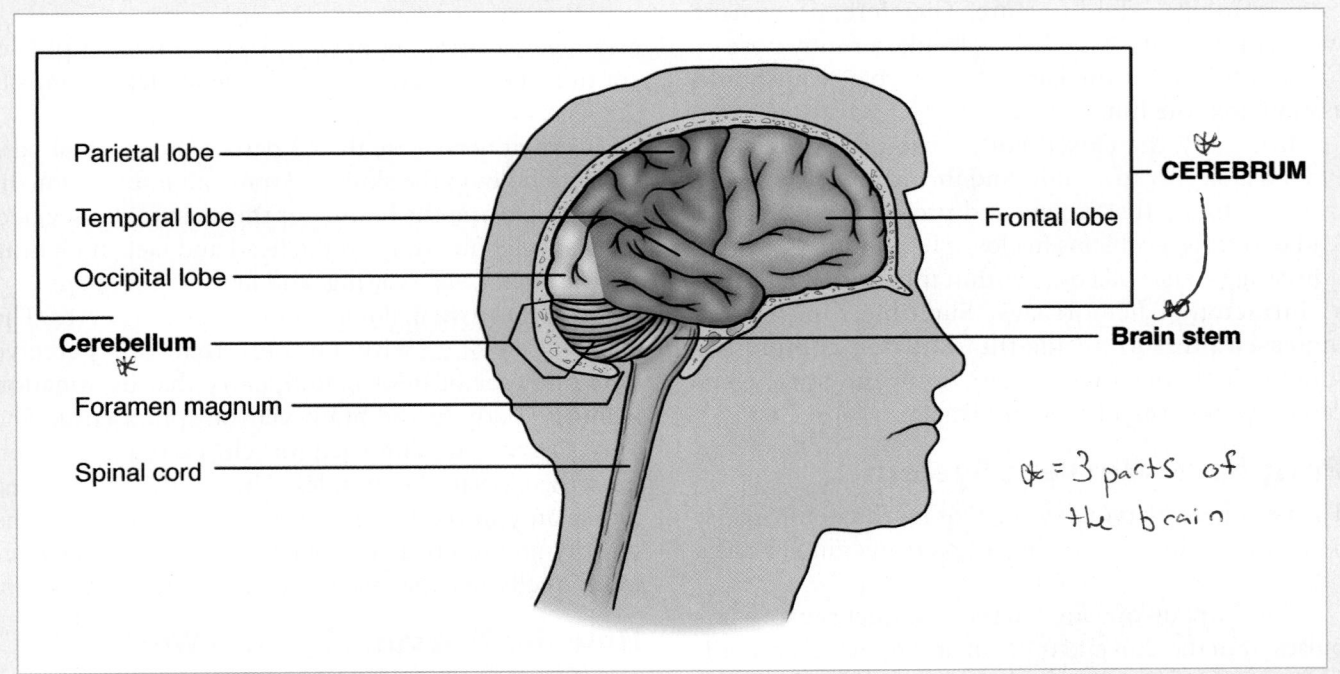

Figure 30-2 The brain is part of the central nervous system and is the organ that controls the body. It is divided into three major areas: the cerebrum, the cerebellum, and the brain stem.

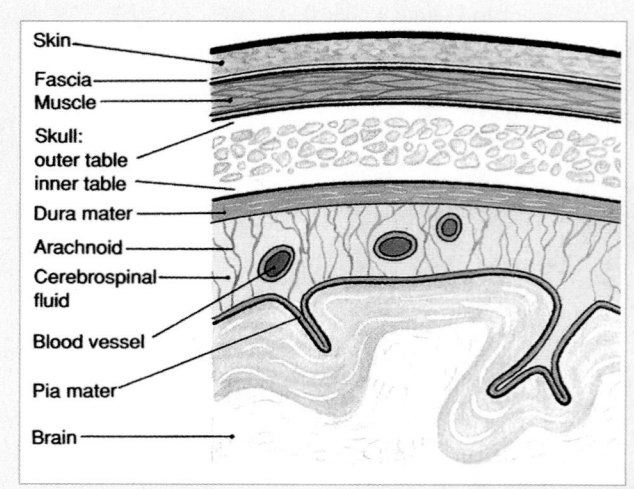

Figure 30-3 The central nervous system has several layers of protective coverings: the skin, muscles and their fascia, bone, and the meninges. The three layers of the meninges are the dura mater, the arachnoid, and the pia mater.

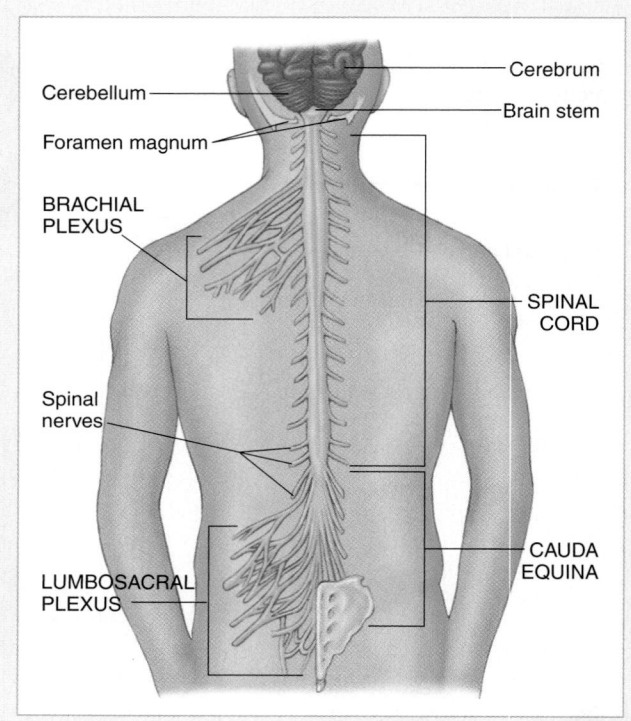

Figure 30-4 The peripheral nervous system is a complex network of motor and sensory nerves. The brachial plexus controls the arms, and the lumbosacral plexus controls the legs.

function properly. When this supply is interrupted, even for short periods of time, serious damage to the brain tissue may occur.

When an injury does penetrate all these protective layers, clear, watery CSF may leak from the nose, the ears, or an open skull fracture. Therefore, if a patient with a head injury has what looks like a runny nose or has a salty taste at the back of the throat, you should assume that the fluid is CSF.

Ironically, the closed bony structure of the skull (which is similar to a vault) and the meninges, the very layers of tissue that isolate and protect the CNS, can lead to serious problems in closed head injuries. Severe injury may cause bleeding within the skull, referred to as intracranial hemorrhage. Such bleeding causes increased pressure inside the skull and compresses softer brain tissue. In many cases, only prompt surgery can prevent permanent brain damage.

Peripheral Nervous System

The peripheral nervous system has two anatomic parts: 31 pairs of spinal nerves and 12 pairs of cranial nerves (Figure 30-4 ▶).

The 31 pairs of spinal nerves conduct sensory impulses from the skin and other organs to the spinal cord. They also conduct motor impulses from the spinal cord to the muscles. Because the arms and legs have so many muscles, the spinal nerves serving the extremities are

arranged in complex networks. The brachial plexus controls the arms, and the lumbosacral plexus controls the legs.

Cranial nerves are the 12 pairs of nerves that pass through holes in the skull and transmit information directly to or from the brain. For the most part, they perform special functions in the head and face, including sight, smell, taste, hearing, and facial expressions.

There are two major types of peripheral nerves. The sensory nerves, with endings that can perceive only one type of information, carry that information from the body to the brain via the spinal cord. The motor nerves, one for each muscle, carry information from the CNS to the muscles. The connecting nerves, found only in the brain and spinal cord, connect the sensory and motor nerves with short fibers, which allow the cells on either end to exchange simple messages.

How the Nervous System Works

The nervous system controls virtually all of our body's activities, including reflex, voluntary, and involuntary activities.

✳ EMT-B Tips

Central nervous system structures, whose bony enclosures protect them quite well, are also very fragile. Protecting them from further damage is vital to the patient's future ability to live a normal life. Lean toward caution and overprotection in assessing and treating possible brain and spinal cord injuries.

In connecting the sensory and motor nerves of the limbs, the connecting nerves in the spinal cord form a reflex arc. If a sensory nerve in this arc detects an irritating stimulus, such as heat, it will bypass the brain and send a message directly to a motor nerve (Figure 30-5 ▼).

Voluntary activities are the actions that we consciously perform, in which sensory input determines the specific muscular activity—for example, reaching across the table for a salt shaker or to pass a dish. Involuntary activities are the actions that are not under the control of our will, such as breathing; in most instances, we inhale and exhale without consciously

thinking about it. Many of our body's functions occur independently of thought, or involuntarily.

The part of the nervous system that regulates or controls our voluntary activities, including almost all coordinated muscular activities, is called the somatic (voluntary) nervous system. The mechanism of the somatic nervous system is simple. The brain interprets the sensory information that it receives from the peripheral and cranial nerves and responds by sending signals to the voluntary muscles.

The body functions that occur without conscious effort are regulated by the much more primitive autonomic (involuntary) nervous system. The autonomic nervous system controls the functions of many of the body's vital organs, over which the brain has no voluntary control.

The autonomic nervous system is divided into two sections: the sympathetic nervous system and the parasympathetic nervous system. Confronted with a threatening situation, the sympathetic nervous system reacts to the stress with the fight-or-flight response. This response causes the pupils to dilate, smooth muscle in the lungs to dilate, heart rate to increase, and blood pressure to rise. This response will also cause the body to shunt blood to vital organs and

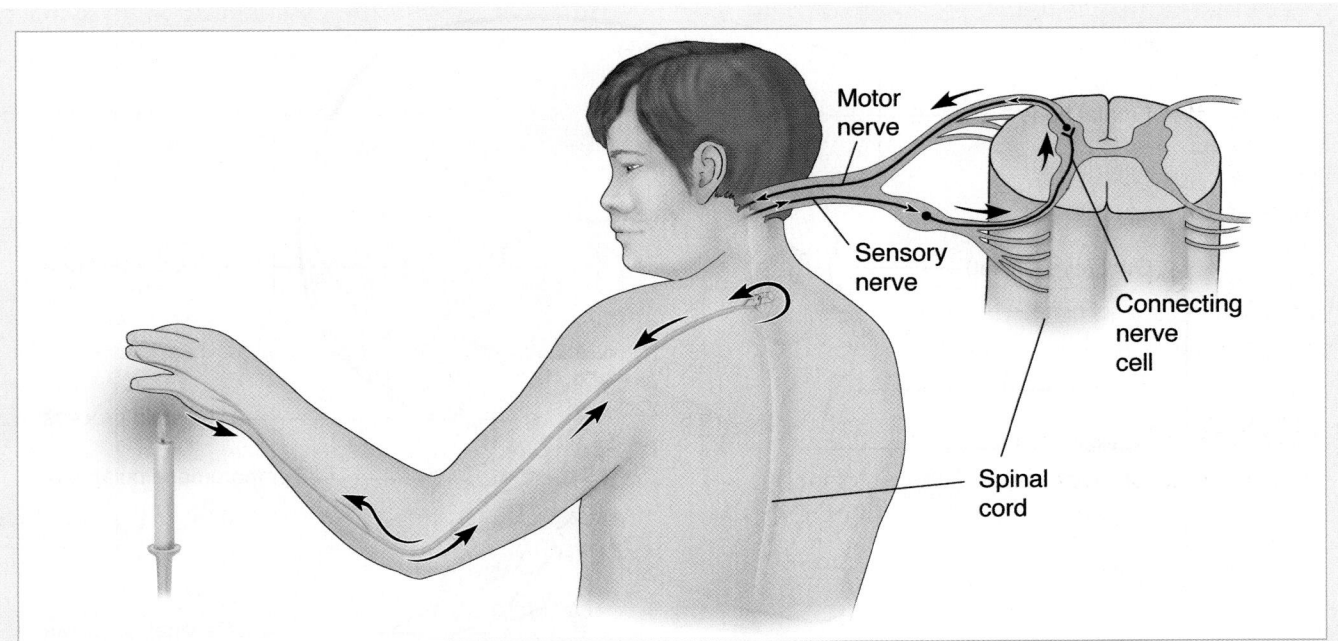

Figure 30-5 The connecting nerves in the spinal cord form a reflex arc. If a sensory nerve in this arc detects an irritating stimulus, it will bypass the brain and send a direct message to a motor nerve.

to skeletal muscle. During this time of stress, a hormone called epinephrine (also known as adrenaline) is released, which is responsible for much of these activities inside the body. The parasympathetic nervous system has the opposite effect on the body, causing blood vessels to dilate, slowing the heart rate, and relaxing the muscle sphincters. When this portion of the autonomic nervous system is activated, the body shunts blood to the organs of digestion. As the body attempts to maintain homeostasis (balance), these two divisions of the autonomic nervous system tend to balance each other so that basic body functions remain stable and effective.

Anatomy and Physiology of the Skeletal System

The skull has two layers of bone, the outer and inner tables, both of which protect the brain. It is divided into two large structures: the cranium and the face (Figure 30-6 ▼). The cranium is occupied by 80% brain tissue, 10% blood supply, and 10% CSF. The mandible

> ## ⚠ Pediatric Needs
>
> The spinal canal is closed by birth and must grow and expand as the child grows. Neural tube deformities are common and serious birth defects. The most discussed is spina bifida, in which the lower portion of the spine does not close prior to birth. As an EMT-B you may be called upon to treat or transport a child with one of these birth defects.

(lower jaw), the only movable facial bone, is connected to the cranium at the temporomandibular joint (TMJ) just in front of each ear.

The spinal column is the body's central supporting structure. It has 33 bones, called vertebrae, and is divided into five sections: cervical, thoracic, lumbar, sacral, and coccygeal (Figure 30-7 ▶). Injury to the vertebrae, depending on the level at which the injury occurs, can result in paralysis if the underlying spinal cord or nervous structures are also damaged.

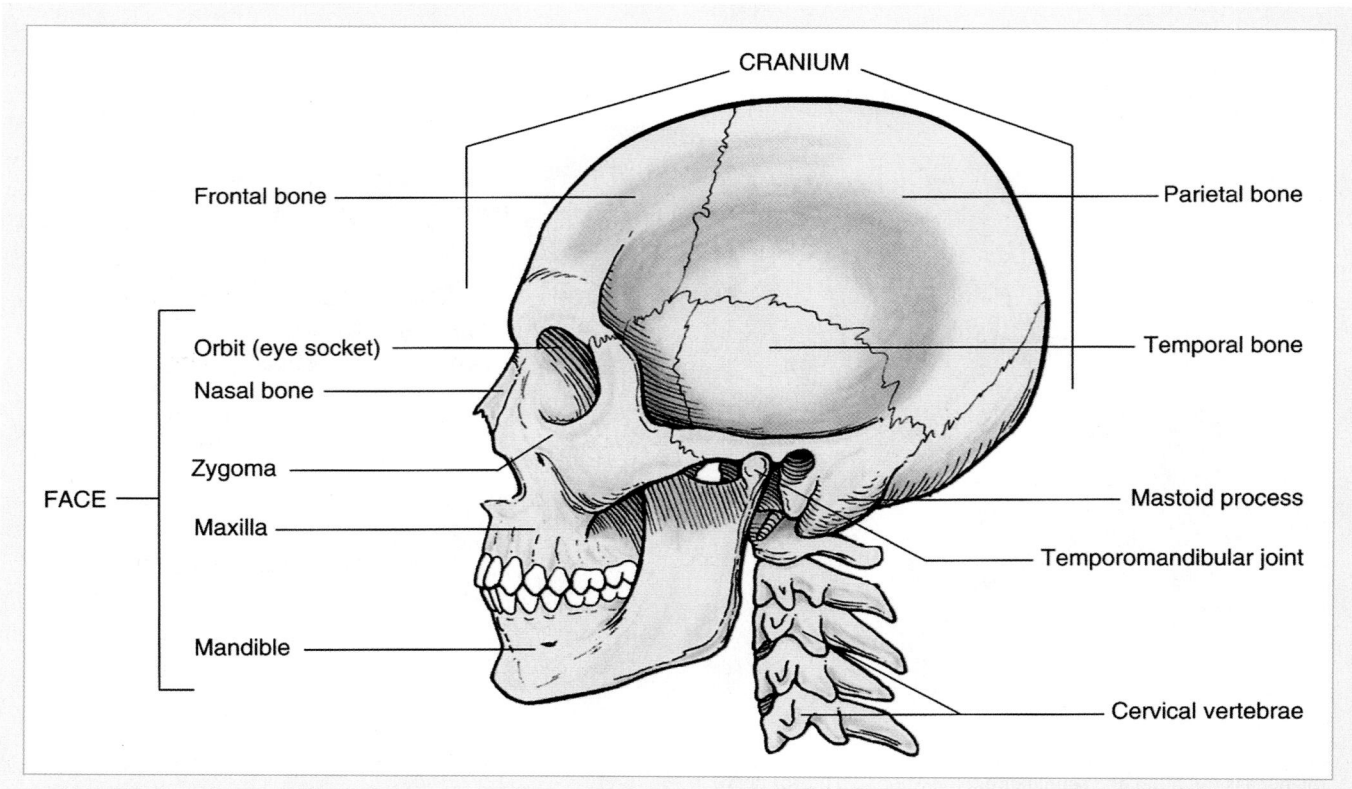

Figure 30-6 The skull includes two large structures: the cranium and the face.

The front part of each vertebra consists of a round, solid block of bone called the vertebral body; the back part forms a bony arch. From one vertebra to the next, the series of arches form a tunnel running the length of the spinal column. This tunnel is the spinal canal, which encases and protects the spinal cord (Figure 30-8 ▼).

The vertebrae are connected by ligaments and separated by cushions, called <u>intervertebral disks</u>. While allowing the trunk to bend forward and back, these ligaments and disks also limit motion so that the spinal cord is not injured. When the spine is injured or fractured, the spinal cord and its nerves are left unprotected. Therefore, until the spine is stabilized, you must keep it aligned as best you can to prevent further injury to the spinal cord.

The spinal column itself is almost entirely surrounded by muscles. However, you can usually palpate the posterior spinous process of each vertebra, which lies just under the skin in the midline of the back. The most prominent and most easily palpable spinous process is at the seventh cervical vertebra at the base of the neck.

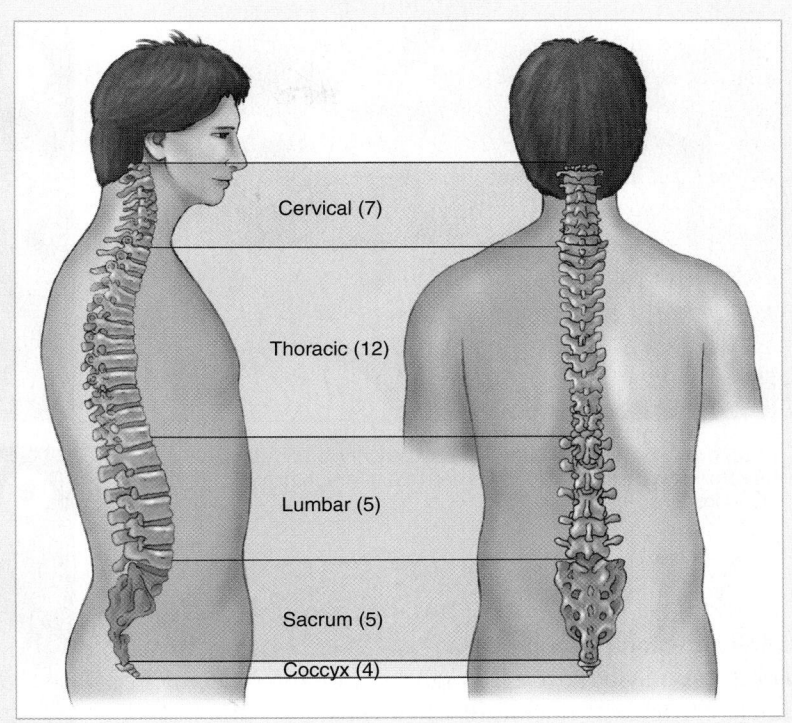

Figure 30-7 The spinal column is the body's central supporting system and consists of 33 bones divided into five sections. Injury to the vertebrae can cause paralysis.

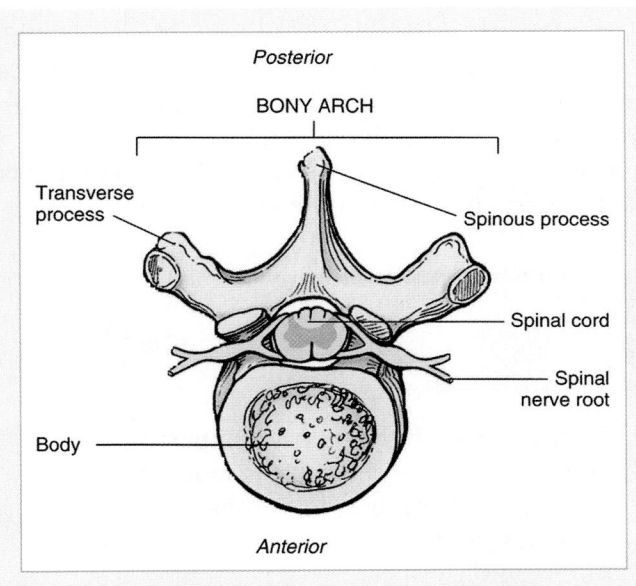

Figure 30-8 The spinal canal is formed by the vertebral body in the front (or anteriorly) and the bony arch in the back (or posteriorly).

Head Injuries

Any head injury is potentially serious. If not properly treated, those that at first seem minor may end up becoming life threatening. On the other hand, severe lacerations of the scalp or fractures of the skull may occur with little or no brain injury and may lead to minimal or no long-term consequences.

There are two general types of head injuries. <u>Closed head injuries</u>, usually associated with blunt trauma, are those in which the brain has been injured but there is no opening into the brain. For example, a severe blow to the head without an open wound would be considered a closed head injury. An <u>open head injury</u> is one in which an opening from the brain to the outside world exists. Obvious skull deformity is a sign of an open head injury, which is often caused by penetrating trauma. There may be bleeding and exposed brain tissue.

Scalp Lacerations

Scalp lacerations can be minor or very serious. Because both the face and the scalp have unusually rich blood supplies, even small lacerations can quickly lead to

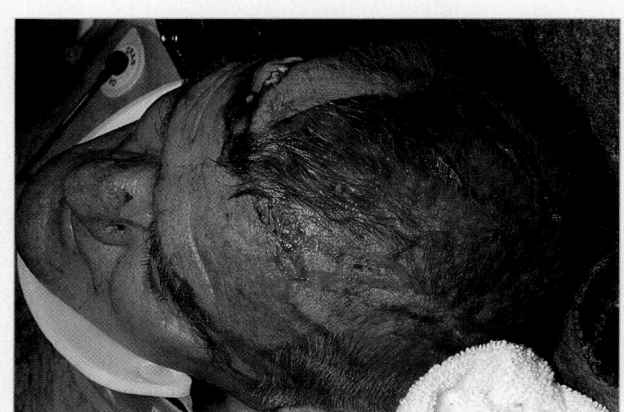

Figure 30-9 The scalp has an unusually rich blood supply; therefore, even small lacerations can result in significant blood loss.

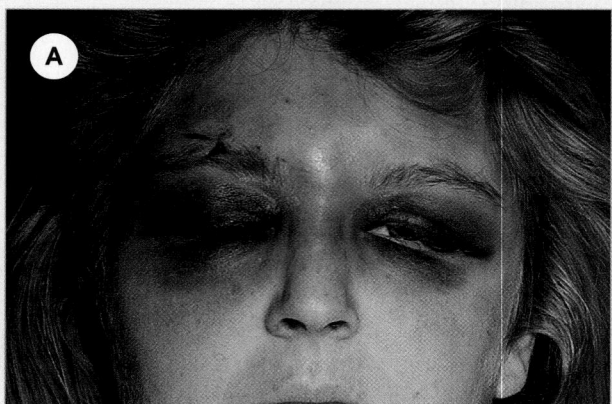

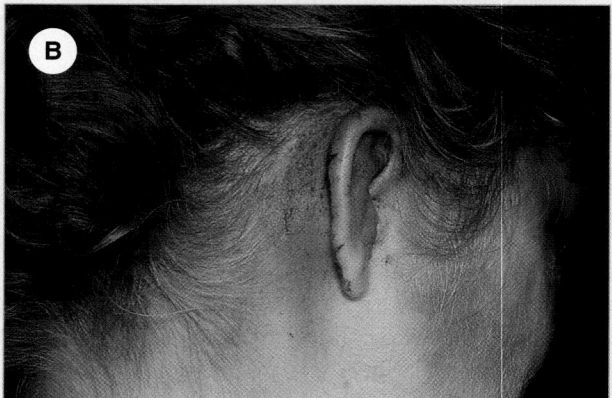

Figure 30-10 Signs of skull fracture include ecchymosis **(A)** under the eyes (raccoon eyes) or **(B)** behind one ear over the mastoid process (Battle's sign).

significant blood loss (Figure 30-9 ▲). Occasionally, this blood loss may be severe enough to cause hypovolemic shock, particularly in children. In any patient with multiple injuries, bleeding from scalp or facial lacerations may contribute to hypovolemia. In addition, since scalp lacerations are usually the result of direct blows to the head, they are often an indicator of deeper, more serious injuries.

Skull Fracture

Significant force applied to the head may cause a skull fracture. As with any fracture, a skull fracture may be open or closed, depending on whether there is an overlying laceration of the scalp. Injuries from bullets or other penetrating weapons frequently result in fracture of the skull. The diagnosis of a skull fracture is usually made in the hospital by CT scan, but you should maintain a high index of suspicion that a fracture is present if the patient's head appears deformed or if there is a visible crack in the skull within a scalp laceration. Additional signs of skull fracture that you may see include ecchymosis (bruising) that develops under the eyes (raccoon eyes) (Figure 30-10A ▲) or behind one ear over the mastoid process (Battle's sign) (Figure 30-10B ▲).

You are the Provider   Part 2

You and your partner arrive at the scene. The scene is safe but many bystanders have gathered around the patient. Several police officers and college security staff arrive and begin to move the bystanders from the area. As you gain access to the patient you find her prone on a rubber mat, awake, and breathing normally. No immediate life threats are observed.

3. What is the next step in providing care to this patient?
4. What is your index of suspicion for unseen injuries?
5. What information must you determine about the patient's MOI?

Brain Injuries

Concussion

A blow to the head or face may cause <u>concussion</u> of the brain. There is no universal agreement on the exact definition of a concussion, but in general, it means a temporary loss or alteration of part or all of the brain's abilities to function without demonstrable physical damage to the brain. For example, a person who "sees stars" after being struck in the head has had a concussion that affects the occipital portion of the brain. A concussion may result in unconsciousness and even the inability to breathe for short periods of time.

A patient with a concussion may be confused or have amnesia (loss of memory). Occasionally, the patient can remember everything but the events leading up to the injury; this is called <u>retrograde amnesia</u>. Inability to remember events after the injury is called <u>anterograde (posttraumatic) amnesia</u>.

Usually, a concussion lasts only a short time. In fact, it has often resolved by the time you arrive. Nevertheless, you should ask about symptoms of concussion in any patient who has sustained an injury to the head; these symptoms include dizziness, weakness, or visual changes.

Patients with symptoms consistent with concussion can also have more serious underlying brain injury. A CT scan is necessary to differentiate between these conditions. You should always assume that a patient with signs or symptoms of concussion has a more serious injury until proven otherwise by CT scan at the hospital or by evaluation by a physician.

Contusion

Like any other soft tissue in the body, the brain can sustain a contusion, or bruise, when the skull is struck. A contusion is far more serious than a concussion because it involves physical injury to the brain tissue, which may suffer long-lasting and even permanent damage. As with contusions elsewhere, there is associated bleeding and swelling from injured blood vessels. Injury of brain tissue or bleeding inside the skull will cause an increase of pressure within the skull. A patient who has sustained a brain contusion may exhibit any or all of the signs of brain injury described later in this chapter.

Intracranial Bleeding

Laceration or rupture of a blood vessel inside the brain or in the meninges that cover the brain will produce intracranial bleeding (hematoma) in one of three areas

Figure 30-11 ▶ :

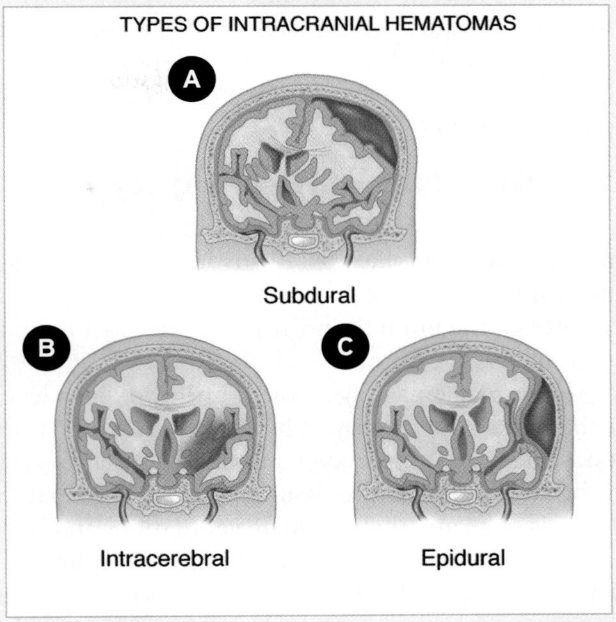

Figure 30-11 Intracranial bleeding can occur in one of three areas. **A.** Beneath the dura but outside the brain (subdural hematoma). **B.** Within the substance of the brain tissue (intracerebral hemorrhage). **C.** Outside the dura and under the skull (epidural hematoma).

- Beneath the dura but outside the brain: a subdural hematoma
- Within the substance of the brain tissue itself: an intracerebral hemorrhage
- Outside the dura and under the skull: an epidural hematoma

A hematoma may develop rapidly such as with an epidural hematoma, which is caused by a tear or laceration of an artery above the dura mater. Or it may develop slowly, as with a subdural hematoma when a vein is lacerated or torn beneath the dura mater. In any case, because the brain occupies nearly the entire space inside the skull, the result is increased pressure inside the skull, leading to compression of the brain tissue. The expanding hematoma will cause progressive loss of brain function and, if not treated properly, death.

Other Brain Injuries

Brain injuries are not always a result of trauma. Certain medical conditions, such as blood clots or hemorrhages, can also cause brain injuries that produce significant bleeding or swelling. Problems with the blood vessels themselves, high blood pressure, or any number of other problems may cause spontaneous bleeding into the brain, affecting the patient's level of consciousness. This is known as altered mental status. The signs and symptoms of nontraumatic injuries are often the same as those of traumatic brain injuries, except that there is no obvious mechanism of injury or any external evidence of trauma.

Complications of Head Injury

Cerebral edema, or swelling of the brain, is one of the most common complications of any head injury. It is also one of the most serious, because swelling in the skull compresses the brain tissue, resulting in a loss of brain function.

Cerebral edema is aggravated by low oxygen levels in the blood and improved by high ones. In fact, the brain consumes more oxygen than any other organ in the body. For this reason, you must make sure that the airway is open and that adequate ventilations and high-flow oxygen are given to any patient with a head injury. This is especially true if the patient is unconscious. Do not wait for cyanosis or other obvious signs of hypoxia to develop.

It is not uncommon for the patient with a head injury to have a convulsion, or seizure. This is the result of excessive excitability of the brain, caused by direct injury or the accumulation of fluid within the brain (edema). You should be prepared to manage seizures in all patients who have had a head injury. Other effects of cerebral edema and increased intracranial pressure may be increased blood pressure, decreased pulse rate, and irregular respirations. This triad of signs is called Cushing's reflex.

Signs and Symptoms of Head Injury

Open and closed head injuries have essentially the same signs and symptoms.

Following an injury, any patient who exhibits one or more of these signs or symptoms has potentially sustained a very serious underlying brain injury:

- Lacerations, contusions, or hematomas to the scalp
- Soft area or depression upon palpation
- Visible fractures or deformities of the skull
- Ecchymosis about the eyes or behind the ear over the mastoid process
- Clear or pink CSF leakage from a scalp wound, the nose, or the ear
- Failure of the pupils to respond to light
- Unequal pupil size
- Loss of sensation and/or motor function
- A period of unconsciousness
- Amnesia

- Seizures
- Numbness or tingling in the extremities
- Irregular respirations
- Dizziness
- Visual complaints
- Combative or other abnormal behavior
- Nausea or vomiting

Spine Injuries

The cervical, thoracic, and lumbar portions of the spine can be injured in a variety of ways. Compression injuries can occur as a result of a fall, regardless of whether the patient landed on his or her feet, coccyx, or on the top of the head. Motor vehicle crashes or other types of trauma can overextend, flex, or rotate the spine. Any one of these unnatural motions, as well as excessive lateral bending, can result in fractures or neurologic deficit.

When the spine is pulled along its length, this is called distraction and can cause injuries. For example, hangings often result in fracture of the vertebrae in the upper portion of the cervical spine.

Assessment of Head and Spine Injuries

You should always suspect a possible spinal injury any time you encounter one of the following mechanisms of injury:

- Motor vehicle crashes
- Pedestrian-motor vehicle collisions
- Falls
- Blunt trauma
- Penetrating trauma to the head, neck, or torso
- Motorcycle crashes
- Hangings
- Diving accidents
- Recreational accidents

Motor vehicle crashes, direct blows, falls from heights, assault, and sports injuries are common causes of spinal injury. A patient who has experienced any of these events may also have sustained a head injury. A deformed windshield or dented helmet may indicate a major blow to the head, which is likely to have caused injury (Figure 30-12 ▶). It is especially important to evaluate and monitor the level of consciousness in patients with suspected head injuries, paying particular attention to any changes that may occur.

Figure 30-12 The classic "star" on the windshield after an automobile crash is a significant indicator of injury. Be alert for the signs and symptoms of head injury.

EMT-B Safety

Many mechanisms that cause head and spine injuries can also entail risk to emergency responders. Before you approach the patient, get the "big picture" of scene safety and take any actions necessary to ensure your own well-being. Do not rely entirely on assistance from fire or police personnel; maintain your own awareness of the scene.

Documentation Tips

Proper care of a patient with a possible spinal injury requires assessment of motor and sensory functions both before and after immobilizing the patient. Likewise, careful observation of level of consciousness at different stages of your care for a head-injured patient can provide crucial information. Document your detailed findings of these repeated neurologic exams to make the information available to hospital personnel and to help establish that your care has been thorough and appropriate.

Scene Size-up

Motor vehicle collisions are a common cause of head and spinal injuries. These situations have the potential to cause injury to rescuers and bystanders as well. Evaluate every scene for hazards to your health and the health of your team or bystanders. Patients with head and spinal injuries can be very sick. When preparing for these patients, expect to encounter airway, breathing, and circulation problems. Have resuscitation and spinal stabilization equipment available. As you evaluate information from dispatch or information you obtain as you arrive on scene, look quickly for indications that the MOI has caused compromised ABCs. Be prepared with appropriate BSI precautions before you approach the patient, before you are tempted by the seriousness of the injuries to intervene without proper protection. You will be spending a great deal of time at the head of the patient. Gloves, a mask, and eye protection should be the minimum BSI precautions that you use. Because these patients can have very complicated injuries, call for advanced life support as soon as possible when a serious MOI or complicated presentation is evident. Law enforcement may be needed to control traffic or unruly people.

In assessing a patient with a possible closed head injury, consider the mechanism of injury. Did the patient fall? Was he or she in an automobile crash or the victim of an assault? Was there deformity of the windshield or deformity of the helmet?

Initial Assessment

General Impression

Begin your initial assessment with a general impression of your patient based on his or her level of consciousness and ask the chief complaint. This will give you a focal point to begin your assessment. Patients with head injuries frequently have spinal problems and vice versa. When assessing a patient for possible head or spinal injury, you should begin by asking the responsive patient these questions to determine his or her chief complaint:

1. What happened?
2. Where does it hurt?

3. Does your neck or back hurt?

4. Can you move your hands and feet?

5. Did you hit your head?

Confused or slurred speech, repetitive questioning, or amnesia in responsive patients are good indicators of a head injury. While other problems may cause similar symptoms, in the setting of trauma, assume a head injury exists until your assessment proves otherwise.

If the patient is found unresponsive, first responders, family members, or bystanders may have helpful information, including when the patient lost consciousness or what his or her previous level of consciousness was. Unresponsive patients with any trauma should be assumed to have a spinal injury. Patients with a decreased level of responsiveness on the AVPU scale (responds to verbal stimulus or responds to painful stimulus) should also be considered to have a spinal injury based on their chief complaint. Unless the patient is absolutely clear in his or her thinking and is lacking other illnesses or injuries that may constitute a distraction, an MOI that suggests a potential spine injury should lead you to provide complete spinal stabilization. A physician is the appropriate person to clear patients with potential spinal injuries. Some jurisdictions allow EMT-Bs to screen patients and to avoid spinal immobilization based on specific criteria in specific patients, although this is not common. Others require that spinal immobilization be provided for every patient whose MOI suggests potential spinal injury regardless of signs or symptoms. Understand and follow your local protocols.

Airway and Breathing

In patients with both head and spinal injuries, airway and breathing problems are common and may result in death if not recognized and treated immediately. When a spinal injury is suspected, how you open and assess the airway is important. Begin by manually holding the head still while you assess the airway. Use a jaw-thrust maneuver to open the airway. When performed correctly, this will prevent movement of the cervical spine. An oral or nasal airway may assist in maintaining an airway; however, the best way to adequately protect the airway is to use advanced airway techniques, usually employed by EMT-Intermediates and paramedics. The decision to use an oropharyngeal or nasopharyngeal airway is based on the patient's ability to maintain his or her own airway, the presence of a gag reflex, and the extent of facial injuries. Review the indications and contraindications for these airway adjuncts in Chapter 7 and the use of advanced airway techniques in Chapter 39.

Vomiting may occur in the patient with a head injury. With large amounts of emesis, the patient may need to be log rolled to the side and the mouth swept of secretions. When necessary to clear the airway, roll the patient as straight as possible to minimize spinal injuries. Suctioning should be performed immediately to remove smaller amounts of secretions.

Apply a cervical spine immobilization device as soon as you have assessed the airway and breathing and provided necessary treatments. A cervical collar may help maintain spinal stabilization as you treat the airway and breathing. The best time to apply the cervical collar depends on the patient's injuries and the seriousness of his or her condition. For some it may be early, while managing the ABCs; in others, manual stabilization may be adequate until the patient is ready to be placed onto a backboard. The key to managing spinal injuries and airway and breathing problems is to move the patient as little as possible and as carefully as possible, maintaining spinal alignment throughout. Place an appropriately sized cervical spine immobilization device on the patient when appropriate. Once it is on, do not remove it unless it causes a problem with maintaining the ABCs. If you must remove the device, you will have to maintain manual stabilization of the cervical spine until it is replaced and the patient has been resecured to the backboard.

Breathing difficulty may result from a high cervical spinal cord injury or from increased pressure on the brain because of bleeding or swelling in the cranium. In either situation, determine if breathing is present and adequate. Oxygen, delivered at a rate of 15 L/min by nonrebreathing mask or via a BVM device, is always indicated for patients with head and spinal injuries. A single episode of hypoxia in a patient with a head injury increases the risk of death or permanent disability significantly. Pulse oximeter values should be maintained above 95%. Positive pressure ventilations are not always necessary; however, if the patient's breathing rate is too slow or too fast and shallow, provide positive pressure ventilations using a BVM or flow-restricted oxygen-powered (FROP) ventilation device (see Chapter 7). The rate of ventilations should be based on the age of the patient and established BLS guidelines. Do not panic and hyperventilate the patient because his or her condition appears severe. Hyperventilation should be reserved for specific conditions and performed under specific guidelines. Be sure to know your local protocols on this.

Circulation

When approaching a patient who is unconscious, the obvious question is, "Is this person alive?" Checking immediately to determine if a pulse is present in this situation is tempting. Remember the ABCs, however, and assess airway and breathing prior to moving on to assessment of circulation by checking a pulse. Patients who are responsive and moving obviously have a pulse; however, you should still check to see if the pulse is weak or strong and if it is generally too fast or too slow. A pulse that is too slow in the setting of a head injury can indicate a serious condition in your patient. If the pulse is present and adequate you can continue to evaluate your patient further.

A single episode of hypoperfusion in a head injury patient can lead to significant brain damage and even death. Assess for signs and symptoms of shock and treat appropriately. Some patients with head and spinal injuries will have a difficult time with blood pressure control. The result can be hypotension and hypoperfusion. Bleeding may also be present from the same injury that caused the spine and/or head injury. That injury may involve blunt or penetrating forces. Consider again the MOI and the effects it has had on your patient. Control bleeding as previously discussed. When bandaging the head, be careful that you do not move the neck if spinal injuries are suspected. Remember that the two injuries often occur together.

Transport Decision

A majority of head injuries are considered mild and result in no or limited permanent disability. A smaller percentage of head injuries are considered moderate, when the patient is left with some permanent disabilities. A still smaller percentage are considered severe. Many of these patients die before ever reaching the hospital or are left in a comatose state despite hospital intervention. There will be a number of patients with head or spine injuries that will not require much intervention other than a thorough assessment and continued observation while being transported to the hospital. In these patients you may choose to take some time on scene to provide careful spine stabilization before transport. In those situations in which your patient has problems with ABCs or has other conditions for which you decide a rapid transport to the hospital is needed, rapid stabilization of the spine and quick loading into the ambulance may be indicated.

Focused History and Physical Exam

Remember that the ability to walk, move the extremities, or feel sensation does not necessarily rule out a spinal cord injury. Similarly, the absence of pain does not always indicate that a spinal injury has not occurred. Do not ask patients with possible spinal injuries to move

You are the Provider   Part 3

You recognize the importance of immediate manual spinal stabilization as you begin patient care. Your patient is awake and able to answer all of your questions appropriately. You quickly begin the initial assessment, use the log roll technique (gaining the assistance of the police who are present), and continue your initial assessment. Nothing significant is abnormal in the ABCs and you place the patient on high-flow oxygen. The patient states she was on the balance beam about 6' off the ground and fell during a tumbling move. She informs you she felt pain right away, and that she has tingling in her arms and legs. You begin a rapid trauma assessment of the patient to check for other injuries she may have sustained. You inform the patient to continue to remain as still as she can.

6. What would be an important finding during the physical exam?
7. During the SAMPLE history, what information is particularly important to obtain?

their necks as a test for pain. Instead, you should instruct them not to move their head or neck.

Rapid Physical Exam for Significant Trauma

Patients with moderate or severe head injuries associated with a significant MOI should receive lifesaving medical or surgical intervention at the hospital without delay. In these situations a spinal injury should be suspected and appropriate measures taken regardless of your physical assessment findings. Perform a rapid physical exam using DCAP-BTLS to identify injuries. This exam should be quick and decisive. If immediate threats to the patient's life are found, they should be treated immediately. Minor injuries can be treated en route to the hospital. Extremities can be stabilized using a long backboard and splinted individually while in the back of the ambulance as time and conditions permit.

The rapid trauma exam should be a quick check of the head, chest, abdomen, extremities, and back. If a spinal injury is suspected, check perfusion, motor function, and sensation in all extremities prior to moving the patient. Make sure that you do not move any body parts excessively. Determine whether the strength in each extremity is equal by asking the patient to squeeze your hands and to gently push each foot against your hands (Figure 30-13 ▶). Also compare the right and left limbs for equality of strength.

Decreased level of consciousness is the most reliable sign of a head injury. Monitor the patient for changes in level of consciousness, including signs of confusion, disorientation, or deteriorating mental status. Is the patient unresponsive or repeating questions? Experiencing seizures? Nauseous or vomiting? Next, determine if there is decreased movement and/or numbness and tingling in the extremities. Assess the vital signs carefully. People with head injuries may have irregular respirations, depending on which region of the brain is affected. Look for blood or CSF leaking from the ears, nose, or mouth and for bruising around the eyes and behind the ears.

You should also evaluate the patient's pupils, especially if he or she has a decreased level of consciousness. Unequal pupil size after a head injury in an unconscious patient often signals a serious problem. Developing blood clots may be compressing the brain and third cranial nerve, causing one pupil to dilate and indicating that the brain is at extreme risk of sustaining catastrophic damage (Figure 30-14 ▶).

Do not probe open scalp lacerations with your gloved finger, as this may push bone fragments into the brain. Do not remove an impaled object from an open head injury.

Focused Physical Exam for Nonsignificant Trauma

A focused physical exam is used when nonsignificant trauma is present. For example, a person falls to the ground while standing or is restrained in a low-speed

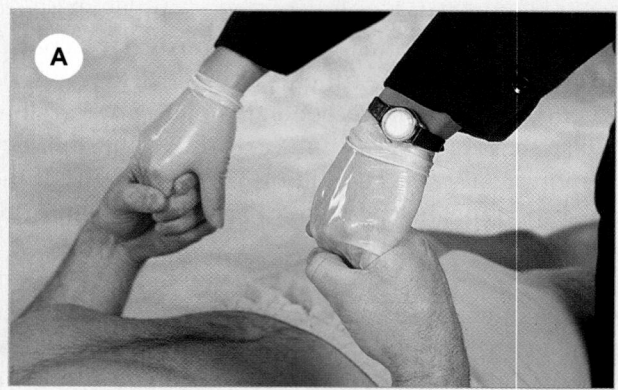
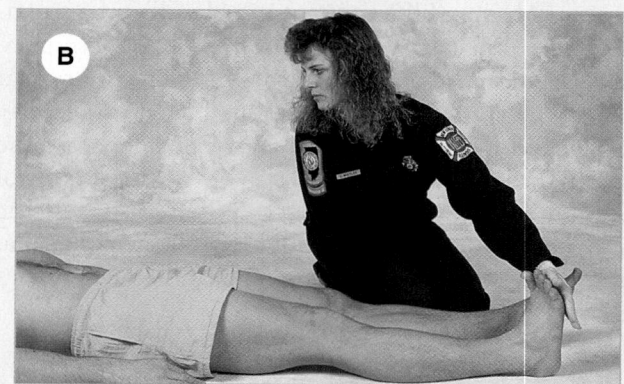

Figure 30-13 **A.** Assess the equality of strength in each extremity by asking the patient to squeeze your hands. **B.** Next, ask the patient to gently push each foot against your hands.

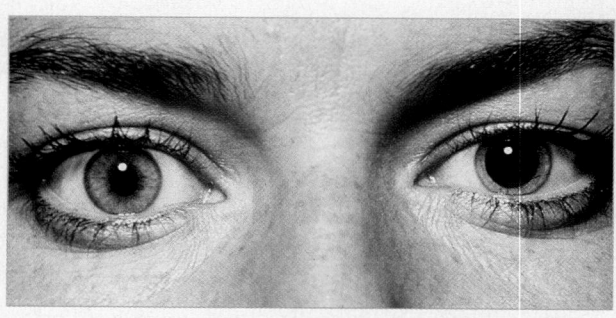

Figure 30-14 Assess pupil size if you suspect a head injury. Unequal pupil size in an unconscious patient may signal a serious problem.

motor vehicle collision in which the air bag has deployed but his or her only complaint is head, neck, or back pain. No other problems or conditions exist. The patient is alert and responds to questions appropriately. In these situations you have the time necessary to focus your assessment on the spine and head.

Change in the level of consciousness is the single-most important observation that you can make in assessing the severity of brain injury. Level of consciousness usually corresponds to the extent of loss of brain function. As soon as you determine that a head injury is present, you should perform a baseline assessment using the AVPU scale and record the time. Reevaluate the patient and record your observations every 15 minutes if the patient's condition is stable and at least every 5 minutes if the patient's condition is unstable, until you reach the hospital.

Frequently, the levels will fluctuate—improving, deteriorating, then improving again over time. On other occasions, there may be a gradual, progressive deterioration in the patient's response to stimuli; this usually indicates serious brain damage that may need aggressive medical and/or surgical treatment. The physicians who treat the patient will need to know when loss of consciousness occurred. They will want to compare their neurologic evaluation with the one you performed in the field.

With a head injury you should use the <u>Glasgow Coma Scale</u> (GCS) instead of the AVPU scale as a focused exam of the neurologic system (Figure 30-15 ▶). The GCS helps you to identify the patient's speech and ability to follow commands. Both are good indicators of how the brain and the spine are functioning. Whether using the GCS or AVPU, always use simple, easily understood terms when reporting the level of consciousness, such as "does not remember events immediately before the injury" or "confused about date and time." Terms such as "obtunded" or "dazed" have different meanings to different people and should not be used in either written or verbal reports.

When a head injury has not occurred but a spine injury is suspected, focus on assessing the spine and back. Inspect for DCAP-BTLS and check the extremities for circulation, motor, or sensory problems.

Pain or tenderness when you palpate the spinal area is certainly a warning sign that a spinal injury may exist. Patients with spinal injuries may complain of constant or intermittent pain along the spinal column or in their extremities. A spinal cord injury may also produce pain independent of movement or palpation.

GLASGOW COMA SCALE

Eye Opening

Spontaneous	4
To Voice	3
To Pain	2
None	1

Verbal Response

Oriented	5
Confused	4
Inappropriate Words	3
Incomprehensible Words	2
None	1

Motor Response

Obeys Command	6
Localizes Pain	5
Withdraws (pain)	4
Flexion (pain)	3
Extension (pain)	2
None	1

Glasgow Coma Score Maximum Total	15
Glasgow Coma Score Minimum Total	3

Figure 30-15 The Glasgow Coma Scale is one method of evaluating level of consciousness. Note that the lower the score, the more severe the extent of brain dysfunction.

Other signs and symptoms of spinal injury include an obvious deformity as you gently palpate the spine; numbness, weakness, or tingling in the extremities; and soft-tissue injuries in the spinal region. Patients with severe spinal injury may lose sensation or experience paralysis below the suspected level of injury or be incontinent (loss of urinary or bowel control) (Figure 30-16 ▶). Obvious injury to the head and neck may indicate injury to the cervical spine. Injury to the shoulders, back, or abdomen (including penetrating trauma) may indicate injury to the thoracic or lumbar spine. Injuries of the lower extremities may indicate a problem with the lumbar spine or sacrum.

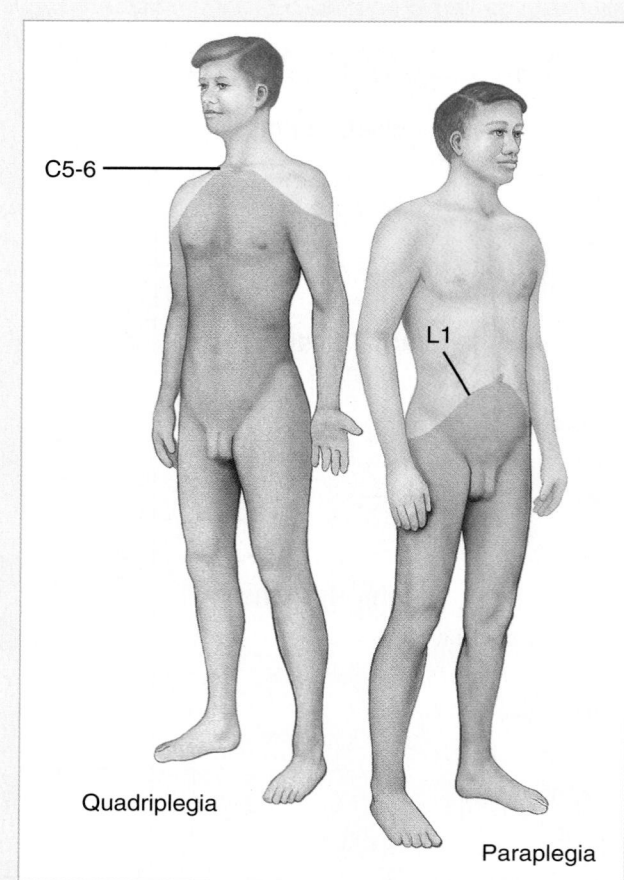

C5-6

L1

Quadriplegia

Paraplegia

Figure 30-16 With severe spinal injuries, patients may lose sensation or experience paralysis below the suspected level of injury.

Baseline Vital Signs

A complete set of baseline vital signs is essential in patients with head and spine injuries. Significant head injuries may cause pulses to slow and blood pressures to rise. With spinal shock, blood pressures may drop and heart rates may increase to compensate. Respirations will become erratic with complications from both head and spine problems.

In head injuries, assess pupil size and reaction to light. The brain controls the diameter of pupils and how quickly they react. If an injury has occurred on one side of the brain, just one pupil will dilate. The pupils are windows to the brain and should be assessed as soon as possible to establish a baseline from which to monitor changes.

As soon as you have assessed the patient's level of consciousness, determine the reaction of each pupil to light. Sketch the size of both pupils on the ambulance report to indicate any difference between the two eyes. Continue to monitor the pupils. Any change in their reactions over time may indicate progressive brain damage.

SAMPLE History

History may be difficult to obtain when a person is confused from a head injury or frightened from a spinal injury. However, the prehospital environment is an excellent place to obtain important history. Do not delay rapid transport for patients who need rapid hospital intervention. Gather as much SAMPLE history as you can while preparing for transport. In less urgent situations, you should have enough time to gather a complete SAMPLE history without compromising patient care.

Interventions

You can almost always control bleeding from a scalp laceration by applying direct pressure over the wound. Remember to follow BSI precautions. Use a dry, sterile dressing, folding any torn skin flaps back down onto the skin bed before applying pressure (Figure 30-17A ▶). In some instances, you will have to apply firm compression for several minutes to control bleeding (Figure 30-17B ▶). If you suspect a skull fracture, do not apply excessive pressure to the open wound. Otherwise, you may increase intracranial pressure or push bone fragments into the brain.

If the dressing becomes soaked, do not remove it. Instead, place a second dressing over the first. Continue applying manual pressure until the bleeding has been controlled, then secure the dressing in place with a soft, self-adhering roller bandage (Figure 30-17C ▶).

Rapid deterioration of neurologic signs following a head injury is a sign of an expanding intracranial hematoma or rapidly progressing brain swelling. You must act quickly to evaluate and treat such patients. The trauma patient with signs and symptoms of head injury who also displays signs of shock has lost blood into another body cavity if hemorrhage is not seen externally. Infants may lose enough blood into the skull region to produce shock, but this is not the case with the older child or the adult patient. Provide oxygen, monitor the airway, treat for shock, and provide immediate transport.

A common response to head injuries, even among children with only very slight head injuries, is vomiting. This is sometimes the result of increased intracranial pressure. In managing such vomiting, you should pay particular attention to protecting the airway.

good time to ask the OPQRST questions about specific injuries. You may find that the pain/injury was there before this episode or that it is getting better or worse. Because this exam is thorough and time-intensive, it is not often performed when attention is required to treat the ABCs. This exam should be performed if time allows, especially when significant trauma is present or when patients are unresponsive and unable to tell you about injuries. Without a detailed physical exam, subtle or covert injuries may go unnoticed.

Ongoing Assessment

The ongoing assessment should focus on three key elements—reassessing the ABCs, interventions, and vital signs. Patients with head or spine injuries can lose an airway or stop breathing without warning. Careful reassessment after moving patients or after providing interventions will help to identify these situations before they have a chance to lead to serious problems.

Multiple interventions may be necessary in these patients. The effectiveness of positive-pressure ventilations, spinal stabilization, and treatments for shock can only be determined with both immediate and continuous observation after providing the intervention. If something is not working, try something else.

You have already established baseline vital signs as part of your assessment. Now is the time to compare those baseline vital signs with repeated vital signs. These changes will often tell you if treatments have been effective. For example, a dilated pupil may constrict with effective positive-pressure ventilations in an apneic head injury patient. Watch carefully for changes in pulse, blood pressure, and respirations. If the pressure in the head increases, the pulse may slow, blood pressure may rise, and respirations may become irregular. Document changes in level of consciousness.

Communication and Documentation

Hospitals may better prepare for seriously injured patients with more advanced warning and a description of the most serious problems found during your assessment. Additional resources can be made available when you arrive. For example, a helicopter may be standing by for transport from a smaller hospital to a Level I trauma center. Larger hospitals may have trauma specialists or neurosurgeons available to meet you on arrival.

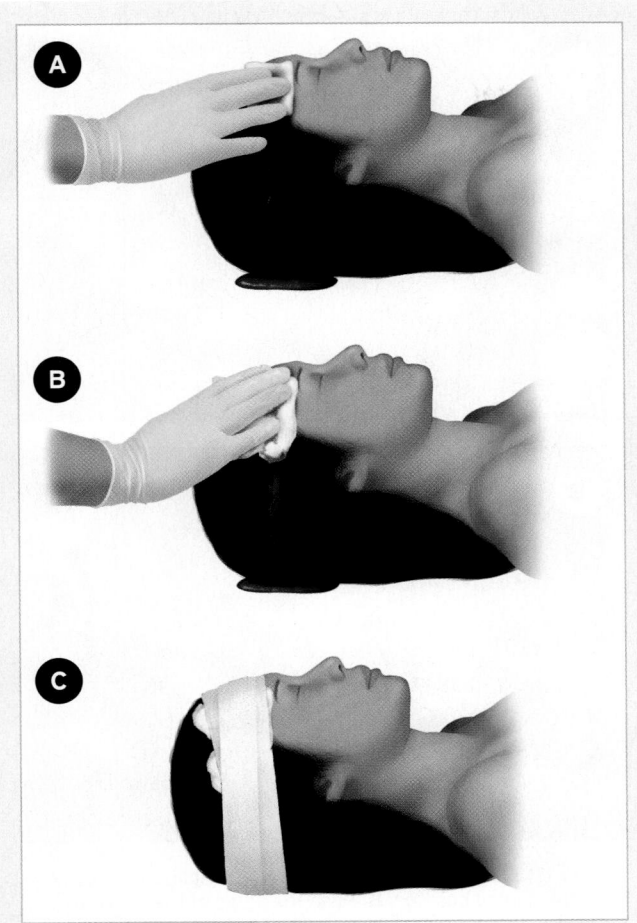

Figure 30-17 A. Use a dry sterile dressing to fold any flaps back down onto the skin bed before applying pressure. **B.** Apply firm compression for several minutes to control the bleeding. **C.** Secure the compression dressing in place with a soft, self-adhering roller bandage.

As discussed earlier, the appearance of clear or pink watery CSF from the nose, the ear, or an open scalp wound indicates that the dura and the skull have both been penetrated. You should make no attempt to pack the wound, ear, or nose in this situation. Cover the scalp wound, if there is one, with sterile gauze to prevent further contamination, but do not bandage it tightly.

Detailed Physical Exam

A thorough detailed exam should be performed if there is time and if the patient is sufficiently stable. A detailed exam uses inspection, palpation, and auscultation to identify DCAP-BTLS in all areas of the body. This is a

Your documentation should include the history you were able to obtain on scene, your findings during your assessment, treatments you provided, and how the patient responded to them. How frequently you document repeat vital signs depends on the condition of your patient. More seriously injured patients should have documented vital signs every 5 minutes, while more stable patients should have documented vital signs every 15 minutes. Take time after your verbal report to hospital staff to sit and make a complete and accurate record of the situation. This will be your only accepted legal memory of the call.

Emergency Medical Care of Spinal Injuries

Emergency medical care of a patient with a possible spinal injury begins, as does all patient care, with your protection; therefore, you must remember to follow BSI precautions. Next, you must maintain the patient's airway while keeping the spine in the proper position, assess respirations, and give supplemental oxygen.

Managing the Airway

Knowing that improper handling of a spinal injury can leave a patient permanently paralyzed must not prevent you from properly addressing an airway obstruction. Remember, all patients without an airway will die. If a patient with a spinal injury has an airway obstruction, you should perform the jaw-thrust maneuver to open the airway (Figure 30-18 ▶). Do not use the head tilt-chin lift maneuver, as it extends the neck and may further damage the cervical spine. If the patient is unconscious, you can lift or pull the tongue forward so that you do not have to move the neck. Once the airway is open, hold the head still in a neutral, in-line position until it can be fully immobilized.

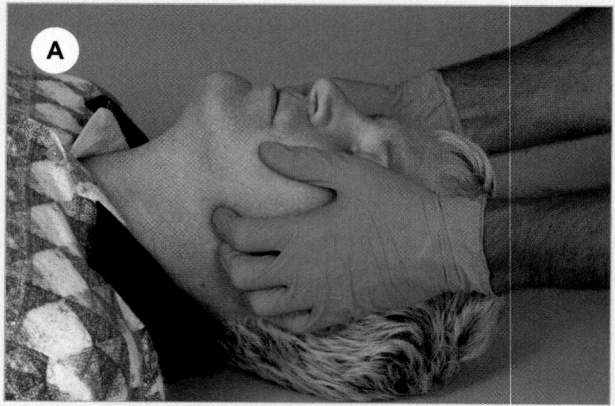

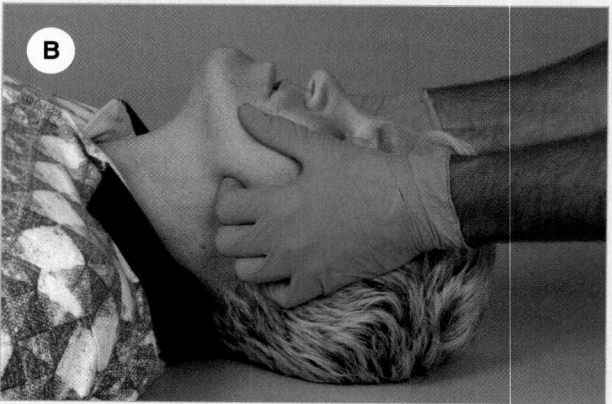

Figure 30-18 Jaw-thrust maneuver **A.** Stabilize the neck in a neutral, in-line position. **B.** Push the angle of the lower jaw forward.

After you open the airway, consider inserting an oropharyngeal airway. If your patient accepts an oropharyngeal airway, be sure to monitor the airway closely; have a suctioning unit available as you will often need to clear away blood, saliva, or vomitus. Provide high-flow oxygen to any patient with suspected head or spine injury, especially those having trouble breathing.

You are the Provider   Part 4

You have completed the rapid trauma assessment and discovered no other injuries. However, you note that the patient has very weak hand grip strength, and weakly pushes against your hands with her feet. During the SAMPLE history your patient denies experiencing any loss of consciousness and is without any significant medical history, is taking no medication, and has no known allergies. Your baseline vital signs are stable and you and your partner prepare to place the patient onto a long backboard for complete spinal immobilization.

8. What should the EMT-B reassess after placing the patient in complete spinal immobilization?
9. What is a concern while caring for the patient who is completely immobilized to a long backboard?

Performing Manual In-Line Stabilization

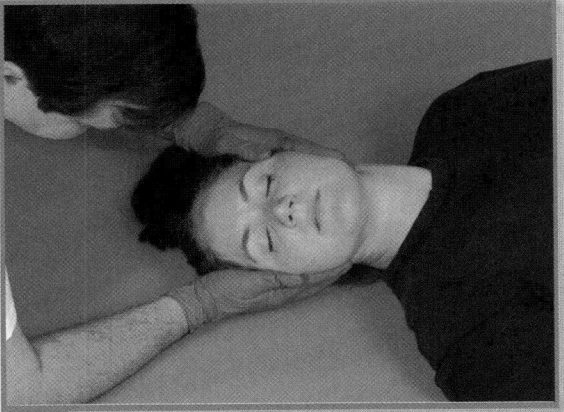

1 Kneel behind the patient and place your hands firmly around the base of the skull on either side.

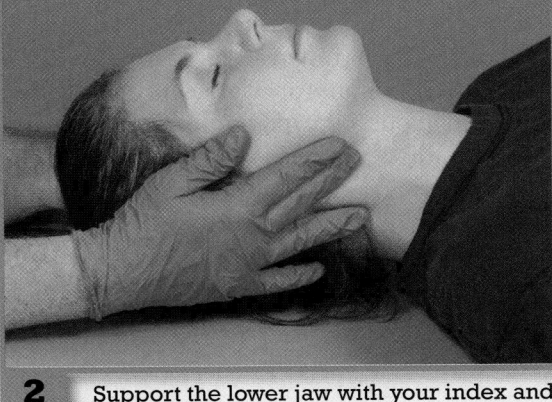

2 Support the lower jaw with your index and long fingers, and the head with your palms.

Gently lift the head into a neutral, eyes-forward position, aligned with the torso. Do not move the head or neck excessively, forcefully, or rapidly.

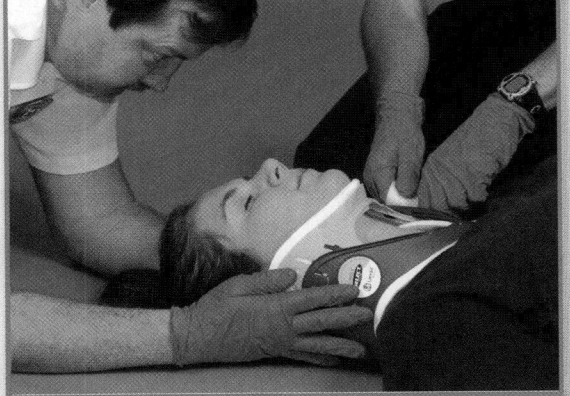

3 Continue to support the head manually while your partner places a rigid cervical collar around the neck.

Maintain manual support until you have completely secured the patient to a backboard.

Stabilization of the Cervical Spine

Stabilizing the airway is your first priority. You must then stabilize the head and trunk so that bone fragments do not cause further damage. Even small movements cause significant injury to the spinal cord. Follow the steps in (Skill Drill 30-1 ▲):

1. **Begin manual in-line stabilization** by holding the head firmly with both hands. Whenever possible, kneel behind the patient, and place your hands around the base of the skull on either side (**Step 1**).
2. **Support the lower jaw** with your index and long fingers, while you are supporting the head with your palms. **Then gently lift the head** until the

patient's eyes are looking straight ahead and the head and torso are in line (**Step 2**).

This neutral <u>eyes-forward position</u> makes stabilization easier. Align the nose with the navel. Never twist, flex, or extend the head or neck excessively.

3. **Manually maintain this position** as you continue to maintain the airway. Have your partner place a rigid cervical collar around the neck to provide more stability. Do not remove your hands from the patient's head until the patient has been completely secured to a backboard and the head has been immobilized. The patient must remain immobilized until he or she has been examined at the hospital (**Step 3**).

Once the patient's head and neck have been manually immobilized, assess the pulse, motor functions, and sensation in all extremities. Then assess the cervical spine area and neck. Keep in mind that the cervical collar is used to provide increased stability to the neck. It is used in addition to, not instead of, manual cervical spine stabilization. An improperly fitting collar will do more harm than good. If you do not have the proper size, place a rolled towel around the head, and tape it to the backboard as you immobilize the patient on the board. In any case, maintain manual support until the patient has been fully secured to a backboard.

You should never force the head into a neutral, in-line position. Do not move the head any further if the patient complains of:

- Muscle spasms in the neck
- Substantial increased pain
- Numbness, tingling, or weakness in the arms or legs
- Compromised airway or ventilations

In these situations, stabilize the patient in his or her current position.

Emergency Medical Care of Head Injuries

Patients with head injuries often have injuries to the cervical spine as well. Therefore, when treating a patient with a head injury, you must keep in mind the need to protect and stabilize the cervical spine at all times. Avoid moving the neck. An initial assessment with spinal stabilization should be done on scene with a complete, detailed physical examination en route.

Beyond this, you should treat the patient with a head injury according to three general principles that are designed to protect and maintain the critical functions of the CNS:

1. **Establish an adequate airway.** If necessary, begin and maintain ventilation and always provide high-flow supplemental oxygen.
2. **Control bleeding**, and provide adequate circulation to maintain cerebral perfusion. Begin CPR, if necessary. Be sure to follow BSI precautions.
3. **Assess the patient's baseline** level of consciousness, and continuously monitor it.

As you continue to treat the patient, do not apply pressure to an open or depressed skull injury. In addition, you must assess and treat other injuries, dress and bandage open wounds as indicated in the treatment of soft-tissue injuries, splint fractures, anticipate and deal

with vomiting to prevent aspiration, be prepared for convulsions and changes in the patient's condition, and transport the patient promptly and with extreme care.

Managing the Airway

The most important step in the treatment of patients with head injury, regardless of the severity, is to establish an adequate airway. If the patient has an airway obstruction, you should perform the jaw-thrust maneuver to open the airway. Once the airway is open, maintain the head and cervical spine in a neutral, in-line position until the patient can be fully immobilized with a cervical collar and backboard (Figure 30-19 ▼). Remove any foreign bodies, secretions, or vomitus from the airway. Make sure a suctioning unit is available, because you will often need to clear blood, saliva, or vomitus from the airway.

Once you have cleared the airway, check ventilation. If the respiratory control center of the brain has been injured, the rate and/or depth of breathing may be ineffective. Ventilation may also be limited by chest injuries or, if the spinal cord is injured, by paralysis of

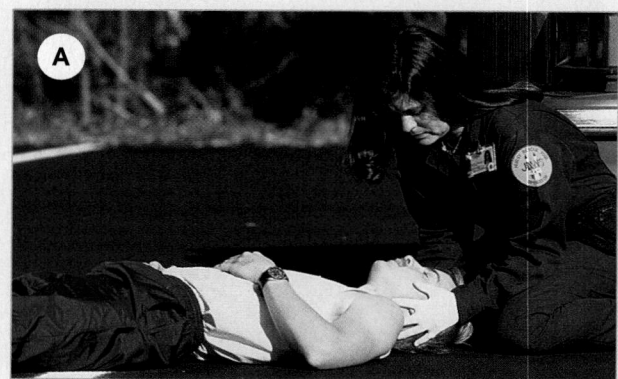

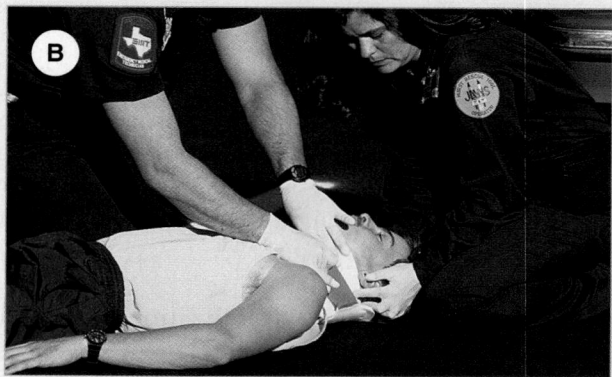

Figure 30-19 A. Maintain the head and cervical spine in a neutral in-line position. **B.** Apply a cervical collar as you finish the initial assessment.

some or all of the muscles of respiration. Give high-flow oxygen to any patient with suspected head injury, particularly anyone who is having trouble breathing. This reduces hypoxia and possible cerebral edema. An injured brain is even less tolerant of hypoxia than a healthy brain, and studies have shown that supplemental oxygen can reduce brain damage; to be effective, however, it must be started as soon as possible. Do not wait until the patient becomes cyanotic. Continue to assist ventilations and administer supplemental oxygen until the patient reaches the hospital.

Circulation

If the heart is not beating, providing airway maintenance, ventilation, and oxygen accomplishes nothing. You must also begin CPR if the patient is in cardiac arrest.

Active blood loss aggravates hypoxia by reducing the available number of oxygen-carrying red blood cells. Although scalp lacerations rarely cause shock except in infants and children, they often cause the loss of large volumes of blood, which must be controlled. Bleeding inside the skull may cause intracranial pressure to rise to life-threatening levels, even though the actual volume of blood lost inside the skull is relatively small.

Shock that develops in a patient with a head injury is usually due to hypovolemia caused by bleeding from other injuries. As with other trauma patients, shock in these cases indicates that the situation is critical. Such patients must be transported immediately to a trauma center. Maintain the airway while you protect the patient's cervical spine, ensure adequate ventilation, administer 100% oxygen, control obvious sites of bleeding with direct pressure, place the patient supine on a spine board, keep the patient warm, and provide immediate transport.

If the patient becomes nauseated or begins to vomit, place him or her on the left side to prevent aspiration. *Be sure to maintain the head in the in-line neutral position, with the cervical collar in place.* You should also have a suctioning unit available.

Preparation for Transport

Supine Patients

A patient who is supine can be effectively immobilized by securing him or her to a long backboard. The ideal procedure for moving a patient from the ground to a backboard is the four-person log roll. This procedure is recommended any time you suspect a spinal injury.

In other cases, you may choose instead to slide the patient onto a backboard or use a scoop stretcher. The patient's condition, the scene, and the available resources will dictate the method you choose.

You should first take the necessary precautions and then direct the team from a kneeling position at the patient's head so that you can maintain manual in-line immobilization. Your job is to ensure that the head, torso, and pelvis move as a unit, with your teammates controlling the movement of the body. If necessary, you may recruit bystanders to the team, but be sure to instruct them fully before moving the patient. To immobilize a patient on a backboard, follow the steps in ⬭ Skill Drill 30-2 ▶ :

1. **Maintain in-line stabilization** from a kneeling position at the patient's head. The EMT-B at the head will direct the log roll.
2. **Assess pulse, motor, and sensory function** in each extremity (**Step 1**).
3. **Apply an appropriately sized cervical collar** (**Step 2**).
4. **The other team members** should position the immobilization device (backboard) and place their hands on the far side of the patient to increase their leverage. Instruct them to use their body weight and their shoulder and back muscles to ensure a smooth, coordinated pull, concentrating their pull on the heavier portions of the patient's body (**Step 3**).
5. **On command** from the EMT-B at the head, the rescuers roll the patient toward themselves. One rescuer quickly examines the back while the patient is rolled on the side, and then slides the backboard behind and under the patient. The team rolls the patient back onto the board, avoiding independent rotation of the head, shoulders, or pelvis (**Step 4**).
6. **Ensure the patient is centered** on the board (**Step 5**).
7. **Secure the upper torso** to the board once the patient is centered on the backboard (**Step 6**). Consider padding voids between the patient and the backboard to make transport more comfortable and protect the patient.
8. **Secure the pelvis and upper legs,** using padding as needed. For the pelvis, use straps over the iliac crests and/or groin loops (**Step 7**).
9. **Begin to immobilize the head to the board** by positioning a commercial immobilization device or towel rolls (**Step 8**).

Skill Drill

30-2

Immobilizing a Patient to a Long Backboard

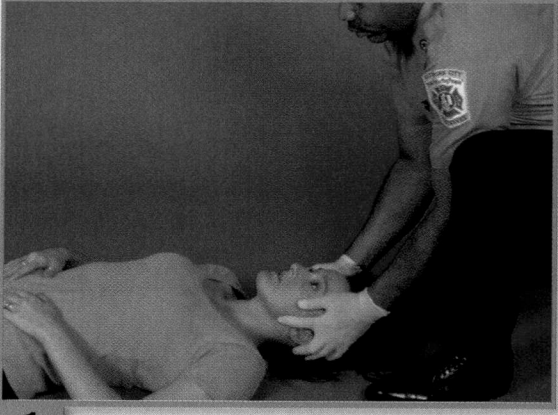

1 Apply and maintain cervical stabilization. Assess distal functions in all extremities.

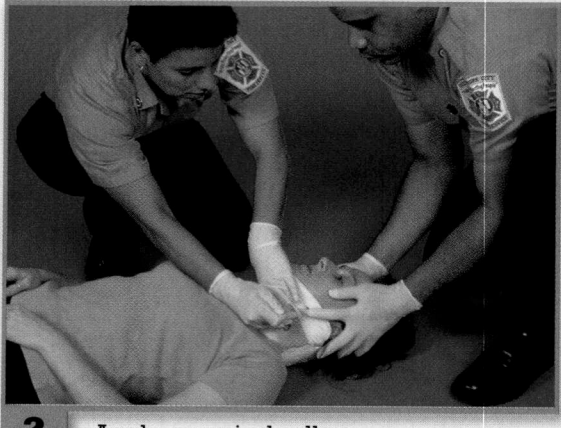

2 Apply a cervical collar.

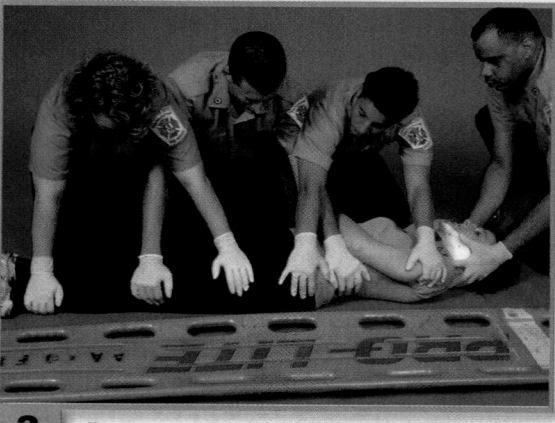

3 Rescuers kneel on one side of the patient and place hands on the far side of the patient.

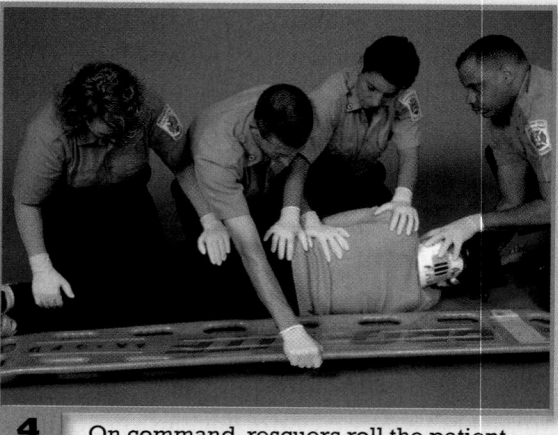

4 On command, rescuers roll the patient toward themselves, quickly examine the back, slide the backboard under the patient, and roll the patient onto the board.

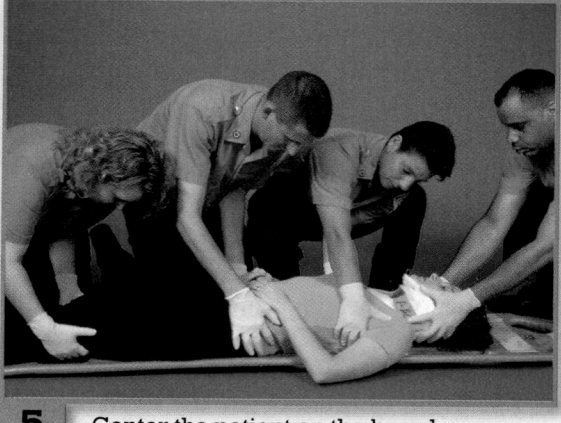

5 Center the patient on the board.

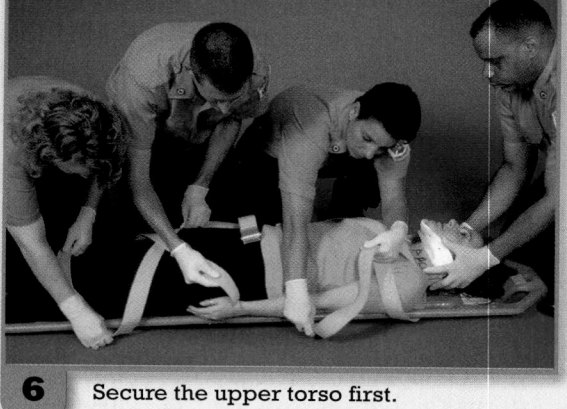

6 Secure the upper torso first.

Immobilizing a Patient Found in a Sitting Position

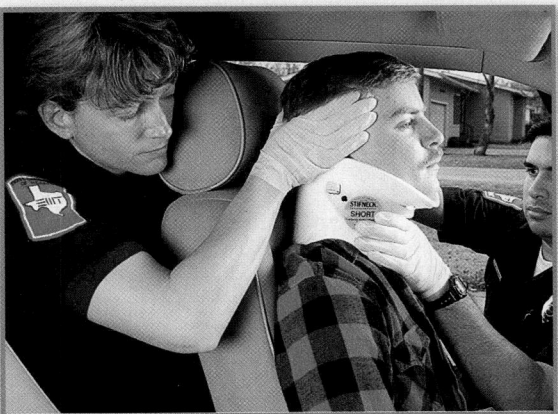

1   Stabilize the head and neck in a neutral, in-line position.

Assess pulse, motor, and sensory function in each extremity.

Apply a cervical collar.

2   Insert a short spine immobilization device between the patient's upper back and the seat.

3   Open the side flaps, and position them around the patient's torso, snug around the armpits.

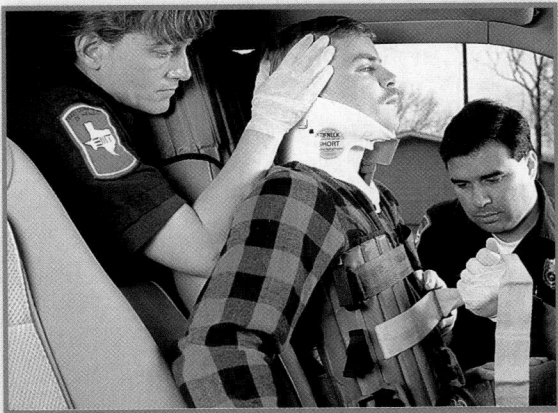

4   Attach the upper torso flaps, leaving the top strap loose.

5   Secure the groin (leg) straps. Check and adjust torso straps.

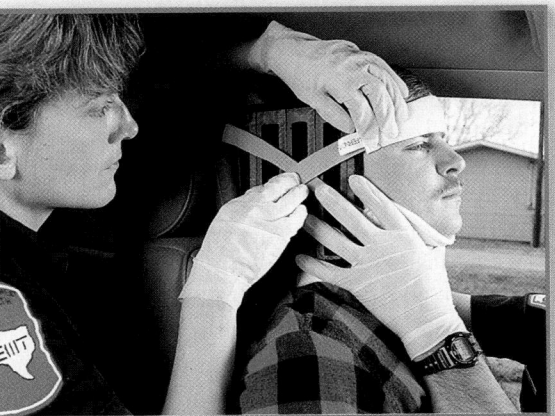

6   Pad between the head and the device as needed.

Secure the forehead strap and fasten the lower head strap around the collar.

Continued.

Skill Drill

30-3

Immobilizing a Patient Found in a Sitting Position

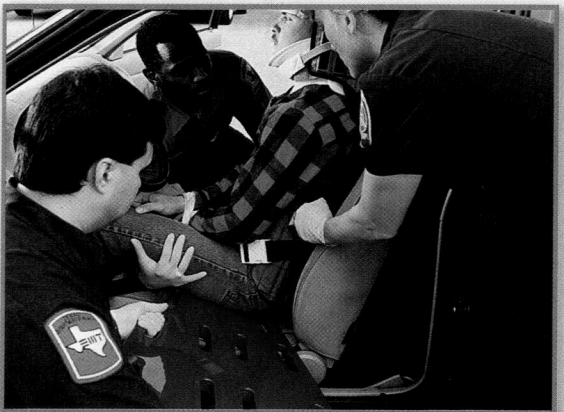

7 Tighten the top torso strap. Wedge a long backboard next to the patient's buttocks.

8 Turn and lower the patient onto the long board.

Lift the patient, and slip the long board under the spine device.

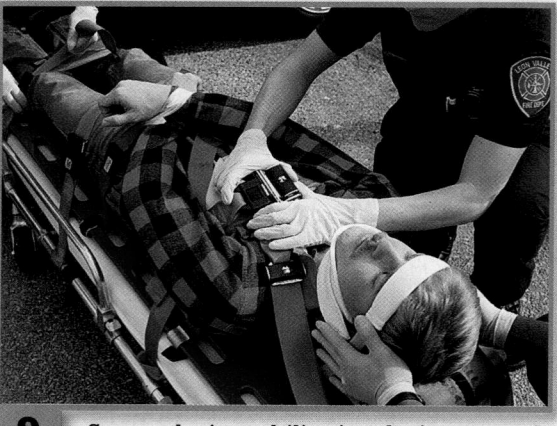

9 Secure the immobilization devices to each other.

Reassess pulse, motor, and sensory functions in each extremity.

2. **Position the board upright** directly behind the patient (**Step 1**).
3. **Two EMT-Bs stand on either side** of the patient and the third is directly behind the patient, maintaining immobilization.
4. **The two EMT-Bs grasp the handholds at shoulder level** or slightly above by reaching under the patient's arms while standing at either side (**Step 2**).
5. **Prepare to lower** the patient to the ground (**Step 3**).
6. **Carefully lower the patient** as a unit under the direction of the EMT-B at the head. The EMT-B

at the head will have to make sure the head stays against the board and carefully rotate his or her hands as the patient is being lowered in order to maintain in-line stabilization (**Step 4**).

Immobilization Devices

An injured spine is often very difficult to evaluate in a patient with a head injury. Sometimes, there is no neurologic loss. Pain in the spine may be missed because of shock or because the patient's attention is directed to more painful injuries. Evaluation is even more difficult if the patient is unconscious. Because any manipulation of the unstable cervical

Immobilizing a Patient Found in a Standing Position

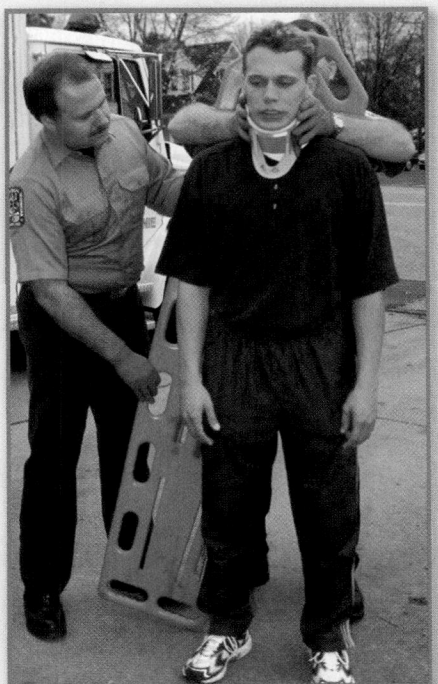

1 While manually stabilizing the head and neck, apply a cervical collar.

Position the board behind the patient.

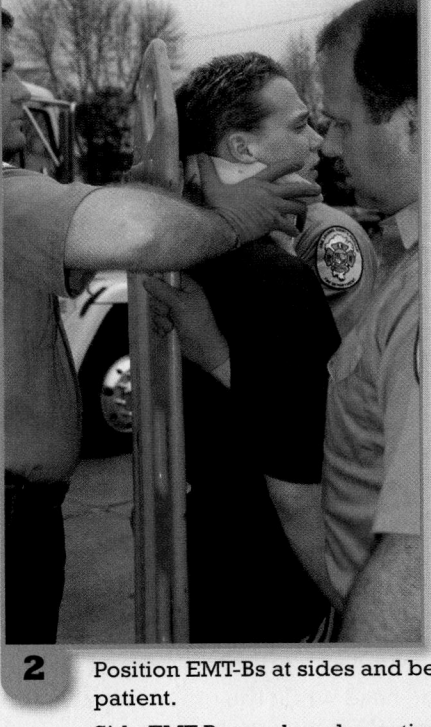

2 Position EMT-Bs at sides and behind the patient.

Side EMT-Bs reach under patient's arms and grasp handholds at or slightly above shoulder level.

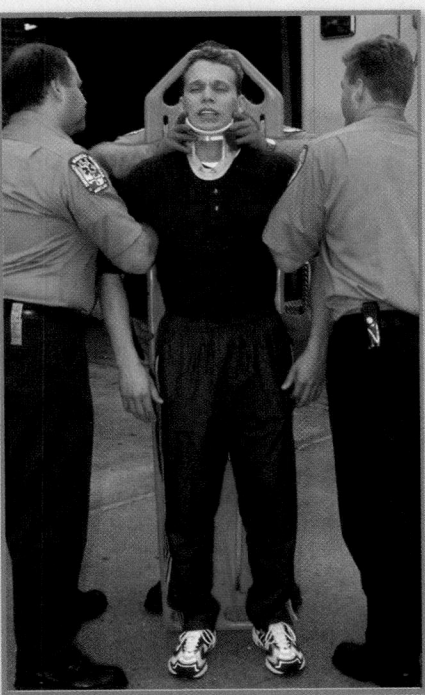

3 Prepare to lower the patient. EMT-Bs on the sides should be facing the EMT-B at the head and wait for his or her direction.

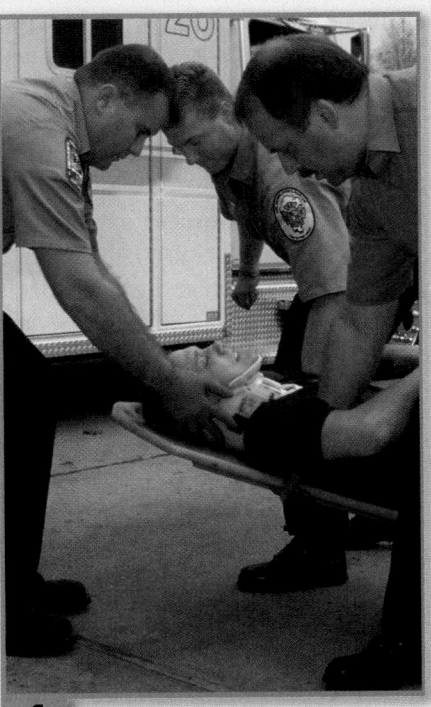

4 On command, lower the backboard to the ground.

spine may cause permanent damage to the spinal cord, you must assume the presence of spinal injury in all patients who have sustained head injuries. Use manual in-line immobilization or a cervical collar and long backboard.

Cervical Collars

Rigid cervical immobilization devices, or cervical collars, provide preliminary, partial support. A cervical collar should be applied to every patient who has a possible spinal injury based on MOI, history, or signs and symptoms. Keep in mind, however, that cervical collars do not fully immobilize the cervical spine. Therefore, you must maintain manual support until the patient has been completely secured to a spinal immobilization device, such as a long or short backboard.

To be effective, a rigid cervical collar must be the correct size for the patient. It should rest on the shoulder girdle and provide firm support under both sides of the mandible, without obstructing the airway or ventilation efforts in any way (Figure 30-20 ▶). Follow the steps in (Skill Drill 30-5 ▶):

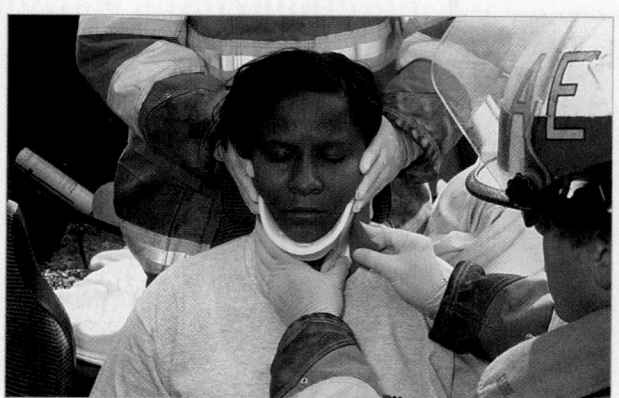

Figure 30-20 Proper fit is essential in applying a cervical collar. The collar should rest on the shoulder girdle and provide firm support under both sides of the mandible without obstructing the airway or any ventilation efforts.

1. **One EMT-B** provides continuous manual in-line support of the head while the other prepares the collar (**Step 1**).
2. **Measure the proper size collar** according to manufacturer's specifications. It is essential that the cervical collar fits properly. An improperly sized immobilization device could allow further injury to occur. If you do not have the correct size, use a rolled towel; tape it to the backboard around the patient's head, and provide continuous manual support (Figure 30-21 ▶) (**Step 2**).
3. **Begin by placing the chin support** snugly underneath the chin (**Step 3**).
4. **Maintaining head stabilization** and neutral neck alignment, wrap the collar around the neck and secure the collar to the far side of the chin support (**Step 4**).
5. **Ensure that the collar fits properly** and recheck that the patient is in a neutral, in-line position. Maintain in-line stabilization until the patient has been completely secured to the board (**Step 5**).

Short Backboards

There are several types of short-board immobilization devices. The most common are the vest-type device and the rigid short board (Figure 30-22 ▶). These devices are designed to stabilize and immobilize the head, neck,

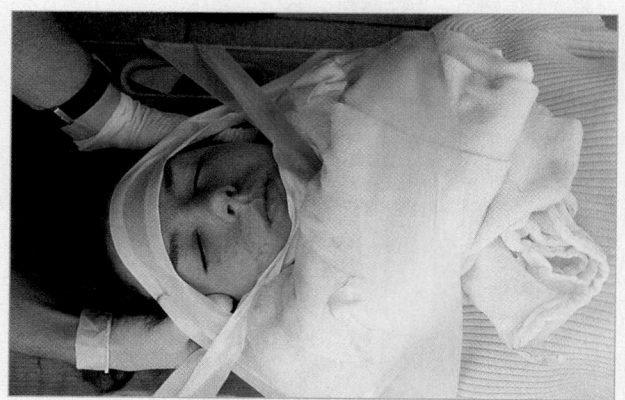

Figure 30-21 If you do not have an appropriately sized cervical collar, you may use a rolled towel. Tape it to the backboard around the patient's head, and provide continuous manual support.

and torso. They are used to immobilize noncritical patients who are found in a sitting position and have possible spinal injuries.

As described earlier in this chapter, the first step in securing a patient to a short board or device is to provide manual, in-line support of the cervical spine. Assess the pulse, motor function, and sensation in all extremities; next assess the cervical area; and then apply an appropriately sized cervical collar.

Position the device behind the patient, and secure it to the torso. Evaluate how well the torso and groin are secured, and make adjustments as necessary. Avoid excessive movement of the patient. Next, evaluate the position of the patient's head. Pad behind the head as needed to maintain neutral, in-line immobilization.

Application of a Cervical Collar

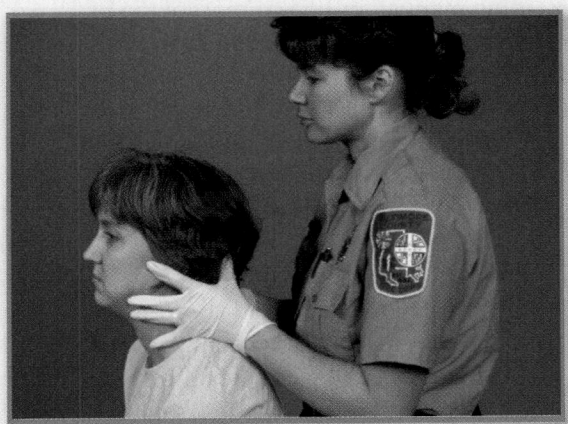

1 Apply in-line stabilization.

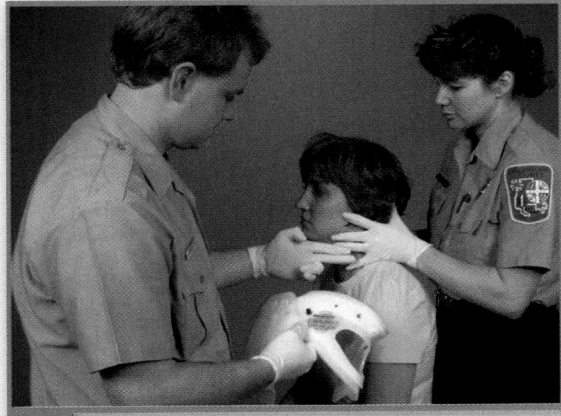

2 Measure the proper collar size.

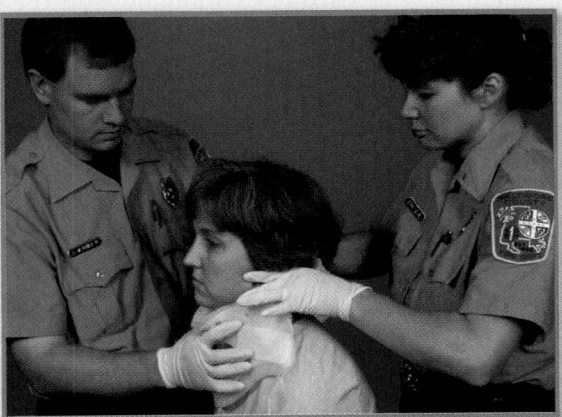

3 Place the chin support first.

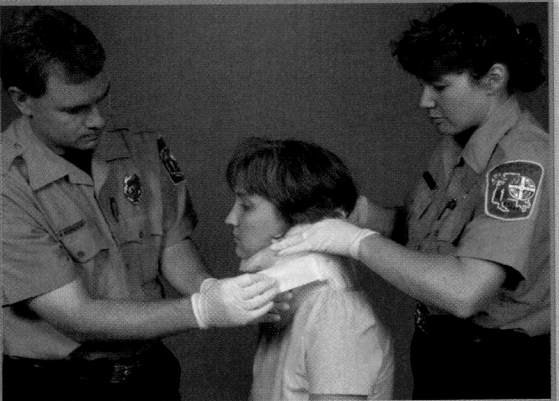

4 Wrap the collar around the neck and secure the collar.

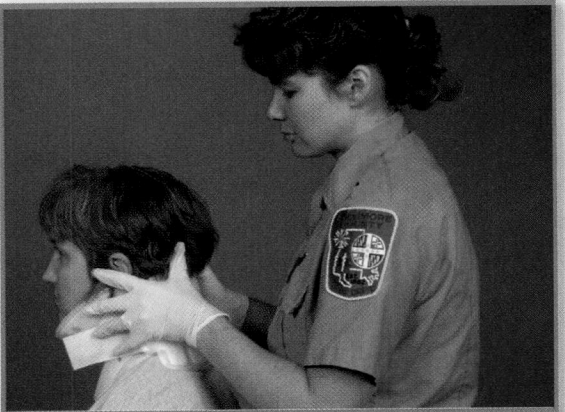

5 Ensure proper fit and maintain neutral, in-line stabilization.

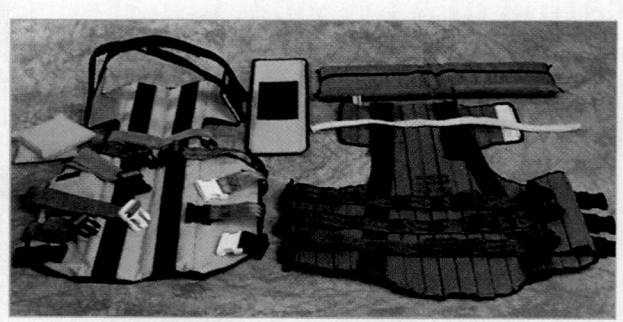

Figure 30-22 The most common types of short-board immobilization devices are vest-type devices.

Now secure the patient's head to the device. Once you have done that, you may release manual support of the head. Rotate or lift the patient to the long backboard. At this point, you must reassess the pulses, motor function, and sensation in all four extremities to determine whether the change in position has affected the patient's vital signs or neurologic status. Finally you should immobilize the patient to the long backboard.

Long Backboards

There are several types of long-board immobilization devices that provide full body spinal immobilization (Figure 30-23 ▼). They also provide stabilization and immobilization to the head, neck and torso, pelvis, and extremities. Long backboards are used to immobilize patients who are found in any position (standing, sitting, supine), sometimes in conjunction with short backboards.

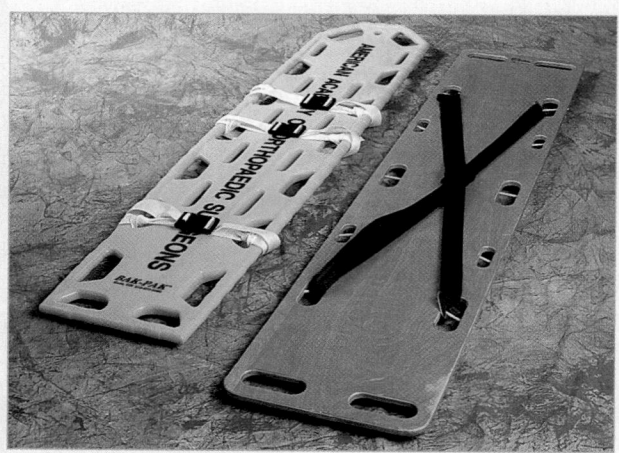

Figure 30-23 Long-board immobilization devices provide full body spinal immobilization, including stabilization of the head, neck and torso, pelvis, and extremities.

Securing a patient to a long board was described in detail earlier in this chapter. Briefly, you should begin by providing manual, in-line support of the head. Assess pulse, motor function, and sensation in all extremities, and assess the cervical area. Then apply an appropriately sized cervical collar, and proceed as follows:

1. **Position the device.**
2. **Log roll the patient onto the device.** You may also move the patient onto the device by suitable lift or slide or by scoop stretcher. As you maintain in-line support, your partner should kneel by the patient's head and direct the other two EMT-Bs as you roll the patient. Your partner's job is to make sure that the head, torso, and pelvis move as a unit. As the patient's back comes into view, quickly assess its condition if you did not do so during initial assessment. One EMT-B should position the device under the patient. Then, at your partner's command, roll the patient onto the board.
3. **If there are spaces** between the patient's head and torso and the board, fill them with padding.
4. **Secure the torso to the device** by applying straps across the chest, pelvis, and legs. Adjust these straps as needed. Then secure the patient's head to the board.
5. **Reassess pulse**, motor function, and sensation in all extremities.
6. **When the patient has been properly secured,** you can safely lift the board or turn it on its side, if necessary.

Helmet Removal

As you plan your care of a patient wearing a helmet, ask yourself the following questions:

- Is the patient's airway clear?
- Is the patient breathing adequately?
- Can you maintain the airway and assist ventilations if the helmet remains in place?
- Can the face guard be easily removed to allow access to the airway without removing the helmet?
- How well does the helmet fit?
- Can the patient move within the helmet?
- Can the spine be immobilized in a neutral position with the helmet on?

A helmet that fits well prevents the patient's head from moving and should be left on, provided (1) there are no impending airway or breathing problems, (2) it

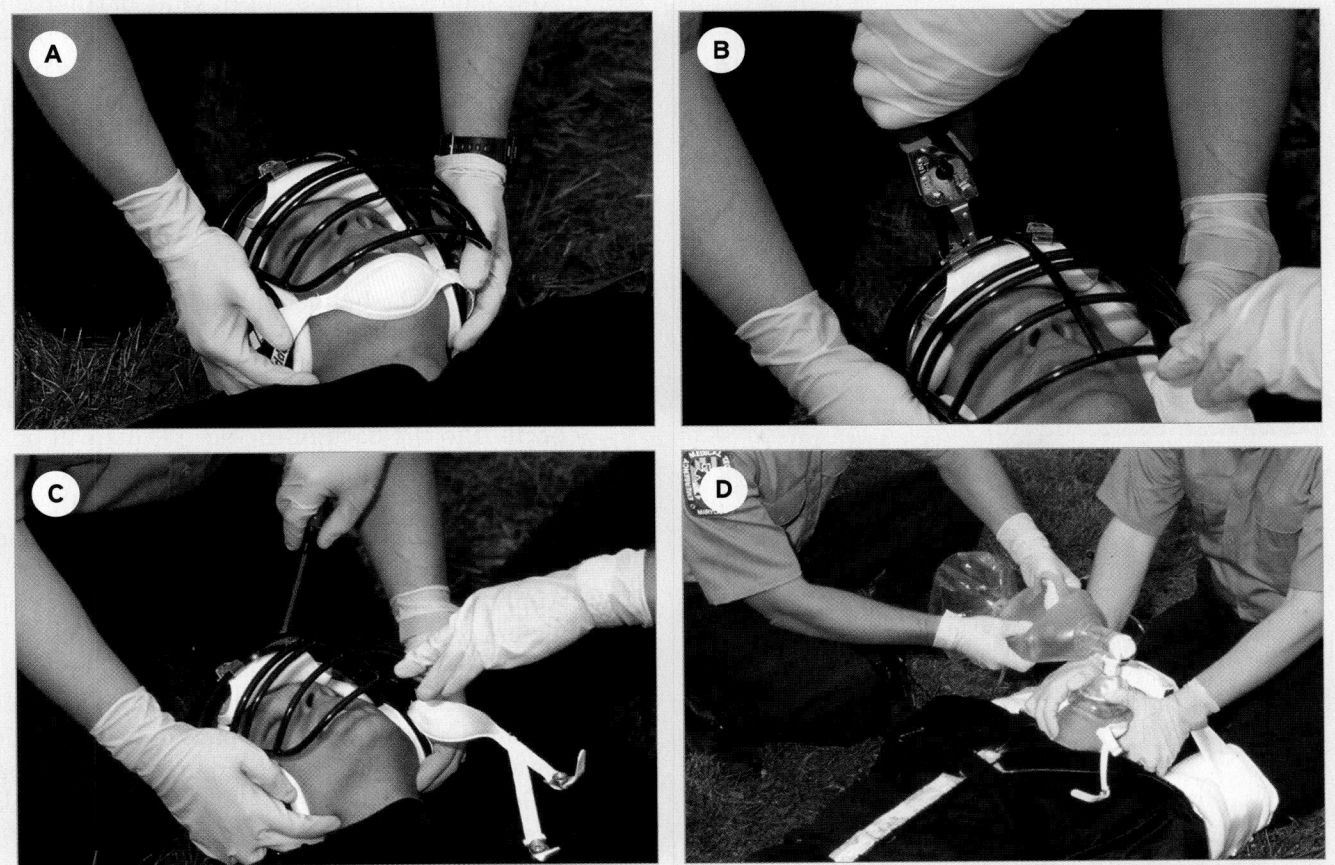

Figure 30-24 The mask on most sports helmets can be removed without affecting helmet position or function. **A.** Stabilize the patient's head and helmet. Then remove the face mask in one of two ways: **B.** Use a trainer's tool designed for cutting retaining clips, or, **C.** Unscrew the retaining clips for the face mask. **D.** Once the face mask has been removed, the helmet can be immobilized against the backboard and a BVM device can be used effectively.

does not interfere with assessment and treatment of airway or ventilation problems, and (3) you can properly immobilize the spine. You should also leave the helmet on if there is any chance that removing it will further injure the patient.

Remove a helmet if (1) it makes assessing or managing airway problems difficult and removal of a face guard to improve airway access is not possible, (2) it prevents you from properly immobilizing the spine, or (3) it allows excessive head movement. Finally, always remove a helmet from a patient who is in cardiac arrest.

Sports helmets are typically open in the front and may or may not include an attached face mask. The mask can be removed without affecting helmet position or function by simply removing or cutting the straps that hold it to the helmet. In this way, sports helmets allow easy access to the airway (Figure 30-24▲).

A patient who is involved in full contact sports may be wearing bulky pads to protect various body regions, such as shoulder pads. Leaving a helmet in place whenever possible is preferred so the body will maintain an inline neutral position. If the helmet must be removed, be sure to provide padding to compensate for shoulder pads and maintain in-line position of the body. Motorcycle helmets often have a shield covering the face. This, too, can be unbuckled to allow access to the airway (Figure 30-25 ▶). If a shield cannot be removed, then the helmet must be removed.

Preferred Method

Removing a helmet is at least a two-person job; however, the technique for helmet removal depends on the actual type of helmet worn by the patient. One EMT-B provides constant in-line support as the other

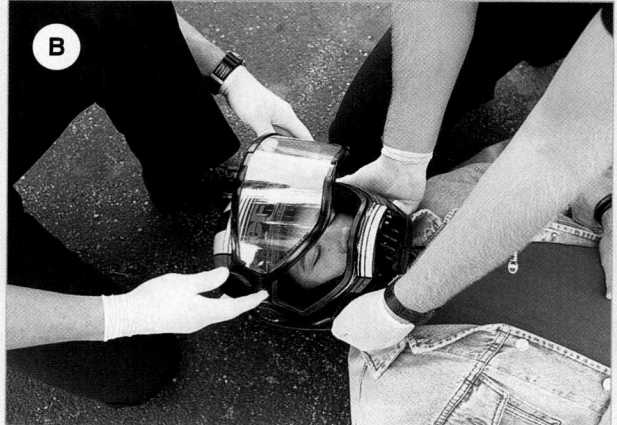

Figure 30-25 Motorcycle helmets often have a shield covering the face that can be removed. **A.** Stabilize the neck in a neutral, in-line position. **B.** Unbuckle or snap off the face shield to access the airway.

moves; you and your partner should not move at the same time. You should first consult with medical control, if possible, about your decision to remove a helmet. When you decide to do so, follow the steps in **Skill Drill 30-6 ▶**):

1. **Begin by kneeling down** at the patient's head. Your partner should kneel on one side of the patient, at the shoulder area.
2. **Open the face shield,** if there is one, and assess the patient's airway and breathing. Remove eyeglasses if the patient is wearing them (**Step 1**).
3. **Stabilize the helmet** by placing your hands on either side of it, with your fingers on the patient's lower jaw to prevent movement of the head. Once

your hands are in position, your partner can loosen the face strap (**Step 2**).

4. **Once the strap has been loosened,** your partner should place one hand on the patient's lower jaw at the angle of the jaw and the other behind the head at the occipital region. Once your partner's hands are in position, you may pull the sides of the helmet away from the patient's head (**Step 3**).
5. **Gently slip the helmet** halfway off the patient's head, stopping when the helmet reaches the halfway point (**Step 4**).
6. **Your partner then slides** his or her hand from the occiput to the back of the head. This will prevent the head from snapping back once the helmet has been completely removed (**Step 5**).
7. With your partner's hand in place, **remove the helmet,** and immobilize the cervical spine.
8. **Apply the cervical collar, and then secure the** patient to the backboard.
9. **With large helmets** or small patients, you may need to pad under the shoulders to prevent flexion of the neck. If shoulder pads or heavy clothing are in place, you may need to pad behind the patient's head to prevent extension of the neck (**Step 6**).

Remember, you do not need to remove a helmet if you can access the patient's airway, the head is snug inside the helmet, and the helmet can be secured to an immobilization device.

Alternate Method

An alternate method for removal of football helmets has also been used. The advantage of this method is that it allows the helmet to be removed with application of less force, therefore reducing the likelihood of motion occurring at the neck. The disadvantage of this method is that it is slightly more time consuming. The first step involves removal of the chinstrap. This can be cut or carefully unsnapped. Be careful during removal of the chinstrap to avoid jarring the neck or head and causing excessive motion. Next, remove the face mask. The face mask is anchored to the helmet by plastic clips (loop straps) secured by screws. These can be removed with a screwdriver or cut with a knife (see Figure 30-24). After the face mask has been removed, the jaw pads can be popped out of place. This can be accomplished with the use of a tongue

Removing a Helmet

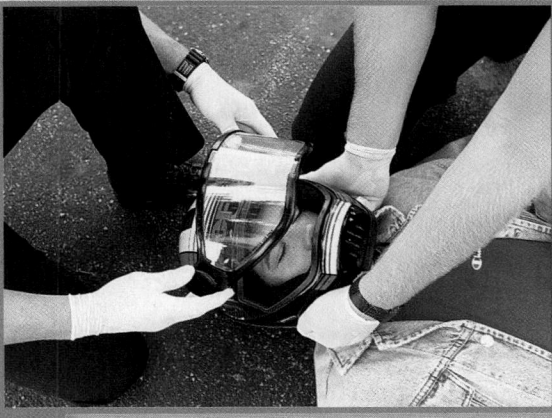

1 Kneel down at the patient's head with your partner at one side.

Open the face shield to assess airway and breathing. Remove eyeglasses if present.

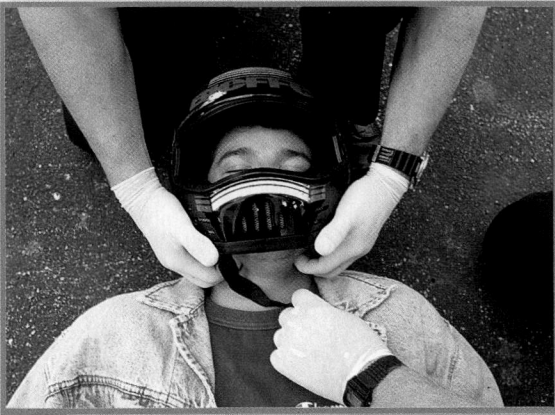

2 Prevent head movement by placing your hands on either side of the helmet and fingers on the lower jaw. Have your partner loosen the strap.

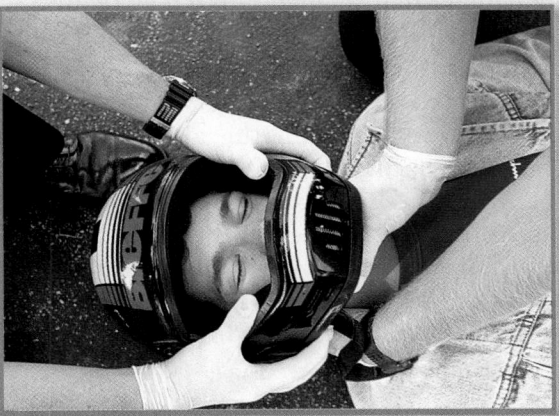

3 Have your partner place one hand at the angle of the lower jaw and the other at the occiput.

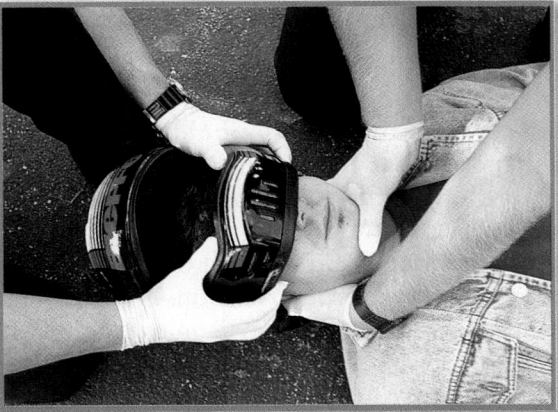

4 Gently slip the helmet about halfway off, then stop.

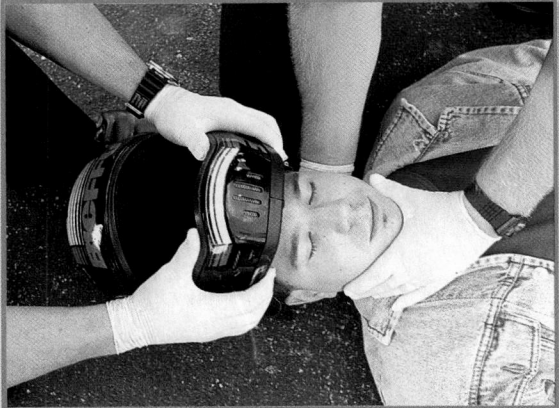

5 Have your partner slide the hand from the occiput to the back of the head to prevent it from snapping back.

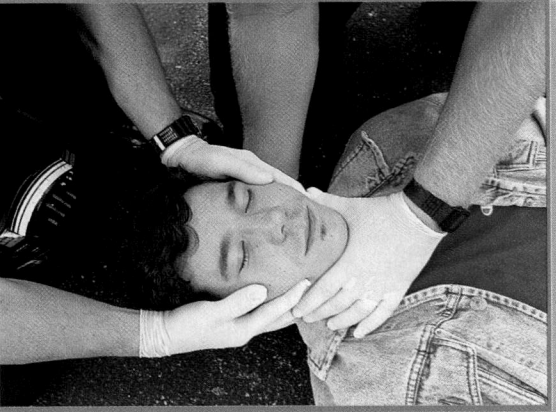

6 Remove the helmet and stabilize the cervical spine.

Apply a cervical collar and secure the patient to a long backboard.

Pad as needed to prevent neck flexion or extension.

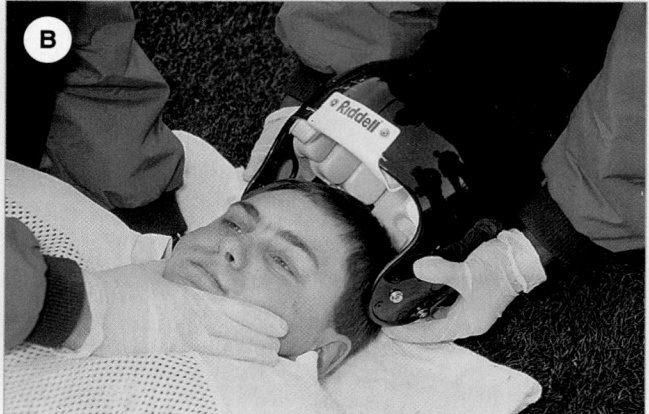

Figure 30-26 **A.** The jaw pads can be removed from the inside of a football helmet with the aid of a tongue depressor. **B.** Place the fingers inside the helmet and gently rock it out of place. The person at the foot controls the lower jaw with one hand and the occiput with the other. Insert padding behind the occiput to prevent neck extension.

depressor (Figure 30-26A ▲). The fingers can then be placed inside the helmet, allowing greater control of the helmet during removal as the helmet is gently rocked back off the top of the head. The person at the foot of the patient controls the head by holding the jaw with one hand and the occiput with the other (Figure 30-26B ▲). Padding is inserted behind the occiput to prevent neck extension. If the shoulder pads are in place, appropriate padding must be applied behind the head to prevent hyperextension. As with the previously described method, the person over the patient's chest is responsible for making sure that the head and neck do not move during removal of the helmet.

Remember that small children may require additional padding to maintain the in-line neutral position. Children are not small adults. They have smaller airways and proportionally larger heads, so padding is important to maintain the airway. Pad under the shoulders to the toes, as needed, to avoid excessive neck flexion (Figure 30-27 ▼). In addition, place blanket rolls between the child and the sides of an adult-sized board to prevent the child from slipping to one side or the other (Figure 30-28 ▼). Appropriately sized backboards are available for children.

Figure 30-27 Children have proportionately larger heads than adults, so you may need to place padding under the shoulders to avoid excessive flexion of the head.

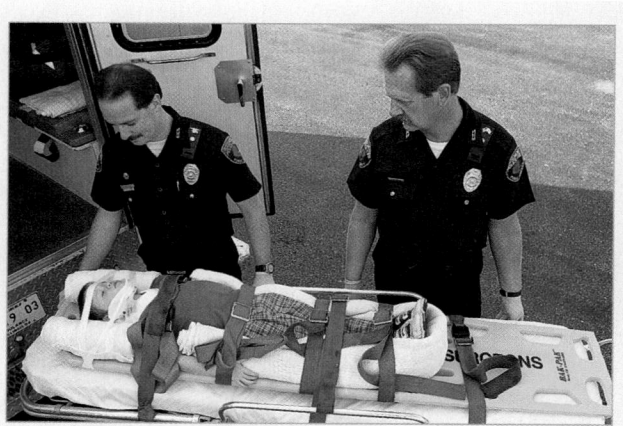

Figure 30-28 Place blanket rolls between the child and the sides of an adult-sized board to prevent the child from slipping to one side or the other.

⚠ Pediatric Needs

You are likely to find infants and children who have been in automobile crashes and are still in their car seats. Your best course of action is to immobilize the child using an appropriately sized pediatric immobilization device, or to consider using a rigid short board device instead. These allow you to completely assess and conduct the ongoing assessment of the injured child while transporting to the hospital. Whenever you apply a cervical collar, make sure it is properly sized. If a properly fitting collar is not available a rolled towel may be used as a substitution (Figure 30-29 ▶). If the child is not in a car seat or was removed before your arrival, use an appropriately sized immobilization device. If the cervical immobilization device does not fit, use a rolled towel, and tape it to the board and manually support the head.

Figure 30-29 If you do not have an appropriately sized cervical collar for a child, you may use a rolled towel and tape it to the car seat. Pad the sides of the car seat, if needed, to prevent lateral movement.

You are the Provider   Summary

Information from bystanders can be crucial to treatment of a patient, but at the same time it can be very misleading. Your patient is usually your best source of information. The MOI is an important piece of information to consider in your treatment. When dealing with a trauma patient, always take spinal precautions until your patient's condition indicates otherwise. Whenever you move a trauma patient you should reassess pulse, motor, and sensory functions to ensure that the move did not cause a change in the patient's presentation.

Talk to your patients. Consider how frightened they may be, and how they may feel about having complete strangers touching them and doing things that they might not understand. Communication is key in good patient care. As with all patients, thorough documentation and a complete patient report are a must.

Head and Spine Injuries

Scene Size-up	Body substance isolation precautions should include a minimum of gloves, mask, and eye protection. Ensure scene safety and determine NOI/MOI. Consider the number of patients, the need for additional help/ALS, and c-spine stabilization.
Initial Assessment	
■ General impression	Determine level of consciousness and treat any immediate threats to life. Determine priority of care based on environment and patient's chief complaint.
■ Airway	Ensure patent airway.
■ Breathing	Listen for abnormal breath sounds and evaluate depth and rate of the respirations. Look for symmetric chest rise and fall. Maintain ventilations as needed. Provide high-flow oxygen at 15 L/min.
■ Circulation	Evaluate pulse rate and quality; observe skin color, temperature, and condition, and treat accordingly.
■ Transport decision	Prompt transport
Focused History and Physical Exam	*NOTE: The order of the steps in the focused history and physical exam differs depending on whether or not the patient has a significant MOI. The order below is for a patient with a significant MOI. For a patient without a significant MOI, perform a focused trauma assessment, obtain vital signs, and obtain the history.*
■ Rapid trauma assessment	Reevaluate the mechanism of injury. Perform a rapid trauma assessment, treating all life threats immediately. Log roll and secure patient to backboard for all patients with a suspected spinal injury.
■ SAMPLE history	Obtain SAMPLE history. If the patient is unresponsive, attempt to obtain information from family members, friends, or bystanders. Consider obtaining pertinent OPQRST information for patients with minor injuries.
■ Baseline vital signs	Take vital signs noting skin color/temperature as well as patient's level of consciousness. Use pulse oximetry if available.
■ Interventions	Provide complete spinal immobilization early if you suspect that your patient has spinal injuries. Maintain an open airway and suction as needed. Treat signs of shock and any other life threats.
Detailed Physical Exam	Complete a detailed physical exam.
Ongoing Assessment	Repeat the initial assessment, rapid/focused assessment, and reassess interventions performed. Reassess vital signs every 5 minutes for the unstable patient, every 15 minutes for the stable patient.
■ Communication and documentation	Contact medical control with a radio report. Relay any change in level of consciousness or difficulty breathing. Describe the MOI and any interventions you performed. Be sure to document any physician's orders and changes in the patient's condition and at what time they occurred.

NOTE: While the steps below are widely accepted, be sure to consult and follow your local protocol.

Head and Spine Injuries

Performing Manual In-line Stabilization
1. Kneel behind the patient and place your hands firmly around the base of the skull on either side.
2. Support the lower jaw with your index and long fingers, and the head with your palms. Gently lift the head into a neutral, eyes-forward position, aligned with the torso. Do not move the head or neck excessively, forcefully, or rapidly.
3. Continue to support the head manually while your partner places a rigid cervical collar around the neck. Maintain manual support until you have secured the patient to a backboard.

Immobilizing a Patient to a Long Backboard
1. Apply and maintain cervical stabilization. Assess distal functions in all extremities.
2. Apply a cervical collar.
3. Rescuers kneel on one side of the patient and place hands on the far side of the patient.
4. On command, rescuers roll the patient toward themselves, quickly examine the back, slide the backboard under the patient, and roll the patient onto the board.
5. Center the patient on the board.
6. Secure the upper torso first.
7. Secure the chest, pelvis, and upper legs.
8. Begin to secure the patient's head using a commercial immobilization device or rolled towels.
9. Place tape across the patient's forehead.
10. Check all straps and readjust as needed. Reassess distal functions in all extremities.

Immobilizing a Patient Found in a Sitting Position
1. Stabilize the head and neck in a neutral, in-line position. Assess pulse, motor, and sensory function in each extremity. Apply a cervical collar.
2. Insert a short spine immobilization device between the patient's upper back and the seat.
3. Open the side flaps, and position them around the patient's torso, snug around the armpits.
4. Secure the upper torso flaps, then the midtorso flaps.
5. Secure the groin (leg) straps. Check and adjust torso straps.
6. Pad between the head and the device as needed. Secure the forehead strap and fasten the lower head strap around the collar.
7. Wedge a long backboard next to the patient's buttocks.
8. Turn and lower the patient onto the long board. Lift the patient, and slip the long board under the spine device.
9. Secure the immobilization devices to each other. Reassess pulse, motor, and sensory functions in each extremity.

Immobilizing a Patient Found in a Standing Position
1. After manually stabilizing the head and neck, apply a cervical collar. Position the board behind the patient.
2. Position EMT-Bs at sides and behind the patient. Side EMT-Bs reach under patient's arms and grasp handholds at or slightly above shoulder level.
3. Prepare to lower the patient. EMT-Bs on the sides should be facing the EMT-B at the head and wait for his or her direction.
4. On command, lower the backboard to the ground.

Application of a Cervical Collar
1. Apply in-line stabilization.
2. Measure the proper collar size.
3. Place the chin support first.
4. Wrap the collar around the neck and secure the collar.
5. Assure proper fit and maintain neutral, in-line stabilization.

Removing a Helmet
1. Kneel down at the patient's head with your partner at one side. Open the face shield to assess airway and breathing. Remove eyeglasses if present.
2. Prevent head movement by placing your hands on either side of the helmet and fingers on the lower jaw. Have your partner loosen the strap.
3. Have your partner place one hand at the angle of the lower jaw and the other at the occiput.
4. Gently slip the helmet about halfway off, then stop.
5. Have your partner slide the hand from the occiput to the back of the head to prevent it from snapping back.
6. Remove the helmet and stabilize the cervical spine. Apply a cervical collar and secure the patient to a long backboard. Pad as needed to prevent neck flexion or extension.

Assessment and Emergency Care

Prep Kit

Ready for Review

- The nervous system of the human can be divided into two parts: the central nervous system (CNS) and the peripheral nervous system.

- The CNS consists of the brain and the spinal cord; the peripheral nervous system consists of a network of nerve fibers, like cables, that transmit information to and from the body's organs to and from the brain.

- The CNS is well protected by bony structures; the brain is protected by the skull and the spinal cord is protected by the bones of the spinal column.

- The CNS is also covered and protected by three layers of tissue called the meninges. The layers are called the dura mater, arachnoid, and the pia mater.

- The peripheral nervous system has two major types of nerves: sensory nerves and motor nerves.

- The nervous system can also be divided into the voluntary nervous system (under our conscious control), and the autonomic nervous system (automatic processes that are not under conscious control).

- The autonomic nervous system is comprised of the sympathetic (fight or flight) system and the parasympathetic (rest and recovery) system. Both systems work together to maintain balance in the body systems and processes.

- The cervical, thoracic, and lumbar portions of the spinal column can be injured through compression such as in a fall, unnatural motions such as overextension from trauma, distraction such as from a hanging, or a combination of mechanisms. Each of these can also cause injury to the spinal cord encased in these regions of bone, causing permanent neurologic injury or death.

- Always begin patient assessment with a high index of suspicion for spinal injury in the trauma patient and the need for manual spinal stabilization.

- Next quickly assess the level of consciousness and perform the initial assessment. Immediately address life threats found during the initial assessment.

- Conduct either a rapid trauma assessment or focused history and physical exam (based on the MOI and potential injuries). During the assessment, be alert to the possibility of neurologic deficits.

www.EMTB.com

Technology

Interactivities

Vocabulary Explorer

Anatomy Review

Web Links

Online Review Manual

- During either the SAMPLE history or the focused history, ask five questions: "Do you have neck or back pain?" "What happened to you?" "Where do you have pain?" "Can you move your hands and feet?" Also, touch the patient's fingers and toes and ask, "Can you feel me touching you? Where?"

- Be alert to complaints of tingling in the upper or lower extremities, numbness, weakness, or paralysis (loss of sensation).

- When applying manual immobilization and immobilizing the patient to a long backboard, keep the head in an in-line neutral position. Be alert to signs of inadequate breathing and vomiting. Use the jaw-thrust maneuver to gain access to the airway. Provide high-flow oxygen.

- Know how to perform spinal immobilization using the long board and rigid short board.

- Common head injuries include scalp and facial lacerations, and fractures of the skull. Brain injuries are also common (concussions, contusions, intracranial bleeding).

- Examples of MOIs that cause these head injuries are falls, motor vehicle crashes, assaults, gunshot wounds, and sport injuries.

- Brain swelling (cerebral edema), seizures, and vomiting are common complications of both closed and open head injuries. CSF may also leak as a result from head injury.

- During the physical assessment, be alert for signs of deformity of the skull, bruising around the eyes or behind the ear (both late signs). Assess for unequal pupil size or reaction, or failure to react to light, loss of sensation and function, and visual complaints.

- Serious head injury patients have an increase in intracranial pressure from brain swelling and may have increased blood pressure (hypertension), decreased heart rate (bradycardia), and irregular respirations. This triad of signs is termed Cushing's reflex and indicates life-threatening pressure within the skull.

- One of the most important signs of head injury is a change in the patient's level of consciousness. Be alert to these changes. Reassess using the AVPU scale or the Glasgow Coma Scale every 5 minutes in the unstable patient and 10 to 15 minutes in the stable patient; monitor pupil size and reaction.

Prep Kit continued...

Vital Vocabulary

anterograde (posttraumatic) amnesia Inability to remember events after an injury.

autonomic (involuntary) nervous system The part of the nervous system that regulates functions that are not controlled by conscious will, such as digestion and sweating.

Battle's sign Bruising behind an ear over the mastoid process that may indicate skull fracture.

brain stem The part of the central nervous system that controls virtually all functions that are necessary for life, including the cardiac and respiratory systems.

central nervous system (CNS) The brain and spinal cord.

cerebellum The part of the brain that coordinates body movements.

cerebral edema Swelling of the brain.

cerebrum The largest part of the brain, containing about 75% of the brain's total volume.

closed head injury Injury in which the brain has been injured but the skin has not been broken and there is no obvious bleeding.

concussion A temporary loss or alteration of part or all of the brain's abilities to function without actual physical damage to the brain.

connecting nerves Nerves in the spinal cord that connect the motor and sensory nerves.

distraction The action of pulling the spine along its length.

eyes-forward position A head position in which the patient's eyes are looking straight ahead and the head and torso are in line.

four-person log roll The recommended procedure for moving a patient with a suspected spinal injury from the ground to a long backboard.

Glasgow Coma Scale A method of evaluating level of consciousness that uses a scoring system for neurologic responses to specific stimuli.

intervertebral disk The cushion that lies between two vertebrae.

involuntary activities The actions that we do not consciously control.

meninges Three distinct layers of tissue that surround and protect the brain and the spinal cord within the skull and the spinal canal.

motor nerves Nerves that carry information from the central nervous system to the muscles.

open head injury Injury to the head often caused by a penetrating object in which there may be bleeding and exposed brain tissue.

peripheral nervous system The 31 pairs of spinal nerves and 12 pairs of cranial nerves that link the body to the central nervous system.

raccoon eyes Bruising under the eyes that may indicate skull fracture.

retrograde amnesia The inability to remember events leading up to a head injury.

sensory nerves Nerves that transmit sensory input, such as touch, taste, heat, cold, and pain, from the periphery to the central nervous system.

somatic (voluntary) nervous system The part of the nervous system that regulates our voluntary activities, such as walking, talking, and writing.

voluntary activities Actions that we consciously perform, in which sensory input or conscious thought determines a specific muscular activity.

Points to Ponder

You are dispatched to the football field at Cloverdale High School for an injured player. You arrive to find the coach and several players crowded around another player who is lying on the ground. The coach tells you that during the football practice, this student tackled another player head-first. The student is alert and oriented and explains that he felt something pop in his neck during the tackle and his arms and legs suddenly went numb. He tells you that he can feel everything right now. An off-duty responder from a neighboring agency tells you that you should immediately remove the player's helmet to begin c-spine stabilization.

When would the other responder's recommendation be appropriate, and what else should you do if you choose to remove the helmet?

Issues: When to Leave a Helmet in Place, When to Remove a Helmet, Working With Other Responders.

www.EMTB.com

Assessment in Action

You are dispatched to the intersection of Ranger Way and Highway 12 for a one-car motor vehicle crash in which a car hit a telephone pole. You arrive to find a 4-door sedan that has crashed head-on into a telephone pole, apparently at great speed. There is significant intrusion of the front end of the car into the driver's compartment, and the windshield is broken out. You look inside the vehicle and see no driver. You then look across the empty field nearby where you locate your patient. He is lying face down in the dirt and does not appear to be moving or breathing.

1. Put the treatment priorities in the most appropriate order.

   1. Circulation
   2. Airway
   3. C-spine
   4. Breathing

   A. 1, 2, 3, 4
   B. 2, 4, 3, 1
   C. 3, 2, 1, 4
   D. 3, 2, 4, 1

2. What is the most appropriate technique to open this patient's airway?

   A. Head tilt-chin lift maneuver
   B. Jaw-thrust maneuver
   C. Tongue-jaw lift
   D. None of the above

3. What is the appropriate order of securing the patient to the long backboard?

   A. Head first
   B. Head last
   C. Body first
   D. Both B and C

4. How many pairs of cranial nerves are there?

   A. 12
   B. 10
   C. 16
   D. 14

5. The autonomic nervous system is divided into the:

   A. sympathetic nervous system.
   B. parasympathetic nervous system.
   C. neither A or B.
   D. both A and B.

6. The spinal column consists of:

   A. 33 bones divided into 4 sections.
   B. 34 bones divided into 3 sections.
   C. 33 bones divided into 5 sections.
   D. 35 bones divided into 3 sections.

Challenging Questions

7. How can significant extremity fractures and other bloody injuries cloud treatment priorities?

8. What is the significance of signs such as raccoon eyes and Battle's sign?

9. How can c-spine stabilization, even when performed correctly, create challenges in patient care and assessment?

10. What should you do if a patient refuses the application of c-spine precautions?

www.EMTB.com

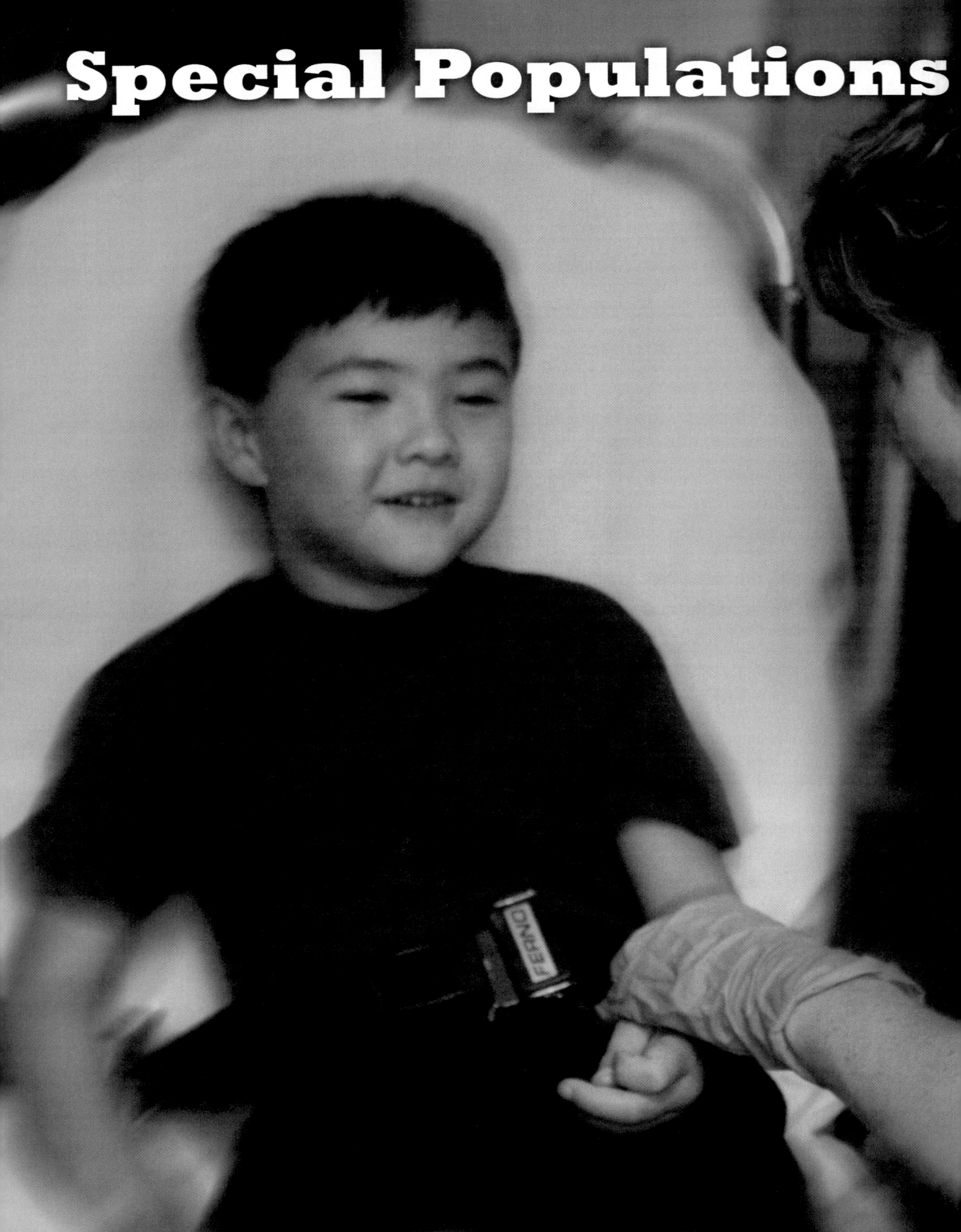

Special Populations

Pediatric Emergencies

Objectives

Cognitive

6-1.1 Identify the developmental considerations for the following age groups:
- infants
- toddlers
- preschool
- school age
- adolescent (p 918)

6-1.2 Describe differences in anatomy and physiology of the infant, child, and adult patient. (p 916)

6-1.3 Differentiate the response of the ill or injured infant or child (age specific) from that of an adult. (p 924)

6-1.8 Identify the signs and symptoms of shock (hypoperfusion) in the infant and child patient. (p 925)

6-1.11 List the common causes of seizures in the infant and child patient. (p 923)

6-1.13 Differentiate between injury patterns in adults, infants, and children. (p 925)

6-1.15 Summarize the indicators of possible child abuse and neglect. (p 927)

6-1.16 Describe the medical/legal responsibilities in suspected child abuse. (p 929)

6-1.17 Recognize the need for EMT-B debriefing following a difficult infant or child transport. (p 931)

Affective

6-1.18 Explain the rationale for having knowledge and skills appropriate for dealing with the infant and child patient. (p 937)

6-1.19 Attend to the feelings of the family when dealing with an ill or injured infant or child. (p 921, 930)

6-1.20 Understand the provider's own response (emotional) to caring for infants or children. (p 933)

Psychomotor

None

You are the Provider

You and your EMT-B partner are dispatched to 818 Nicole Ln for a pedestrian struck. While en route, dispatch provides a patient update stating that your patient is a 6-year-old boy who has been struck by a pick-up truck. He is conscious and breathing. ALS has been dispatched as well but is 15 minutes from the scene.

1. What is your first consideration in the treatment of this patient?
2. Of what significance is it that your patient is a child rather than an adult?

Pediatrics

Children have many unique health problems. Similarly, many problems that are common in adults do not occur in children. Therefore, there is a specialized medical practice devoted to the care of the young, called pediatrics.

Not everyone is comfortable caring for children. In most situations, handling an infant or child means that you must manage the parents as well (Figure 31-1 ▶). Therefore, it is vital that you remain calm, professional, and keep your personal feelings in check as you work with infants, children, and their families throughout the emergency. However, once you learn how to approach children of different ages and what to expect while caring for them, you will find that treating children also offers some very special rewards. Not only are their innocence and openness appealing, they often respond to treatment much more rapidly than adults do.

Anatomy and Physiology

There is no other time in our lives that our bodies are growing and changing as fast as during childhood. Newborns have to quickly change to adapt to the world outside the mother's body. Toddlers learn to walk and talk. School-age children explore the world without thought of consequence. These changes can create difficulties during your assessment of the child if you do not expect them.

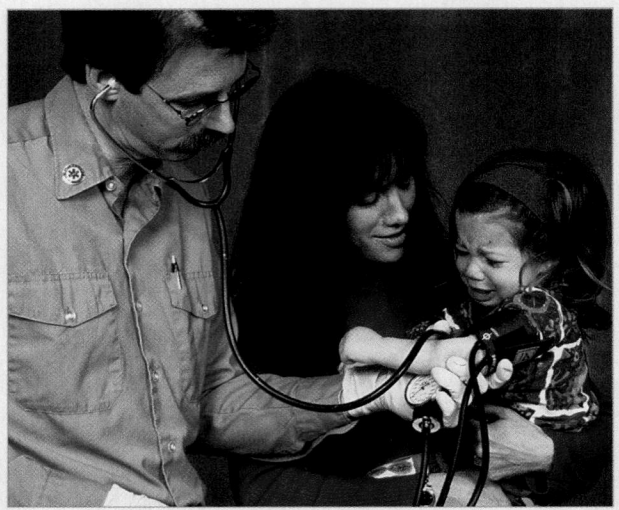

Figure 31-1 Treating a sick or injured child can be extremely challenging. A calm, professional demeanor is of utmost importance as you care for both the child and the parents.

To manage the pediatric airway effectively, you must also understand the anatomic differences between adults and children. To start with, the heart is higher in a child's chest, and the lungs are smaller. The opening to the trachea is higher in the neck, and the neck itself is shorter.

The anatomy of a child's airway differs from that of an adult in five principal ways. These differences will influence the treatment decisions that you make about pediatric patients, including whether or not intervention is needed and, if so, what procedure to use. The anatomy of a child's airway and other important structures differs from that of an adult's in the following ways (Figure 31-2 ▶):

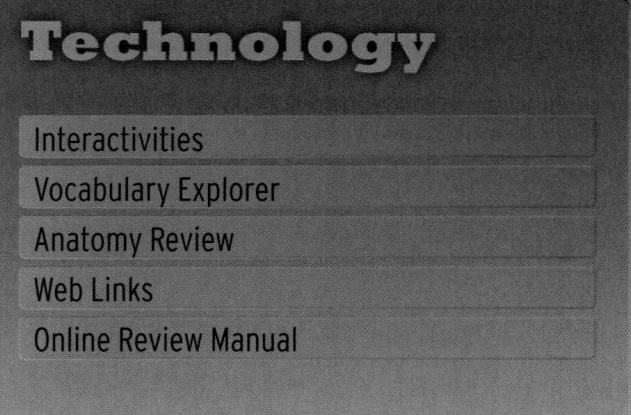

Technology

www.EMTB.com

- Interactivities
- Vocabulary Explorer
- Anatomy Review
- Web Links
- Online Review Manual

Documentation Tips

The unique aspects of caring for children make it wise to carry reference charts or measuring tools to assist you when assessing a child. Many services also carry copies of specialized pediatric protocols in their system. Refer to these resources during your care, and remember to make notes about your specific observations and treatment decisions. This "information-intensive" approach to pediatric care helps ensure both good care and thorough documentation.

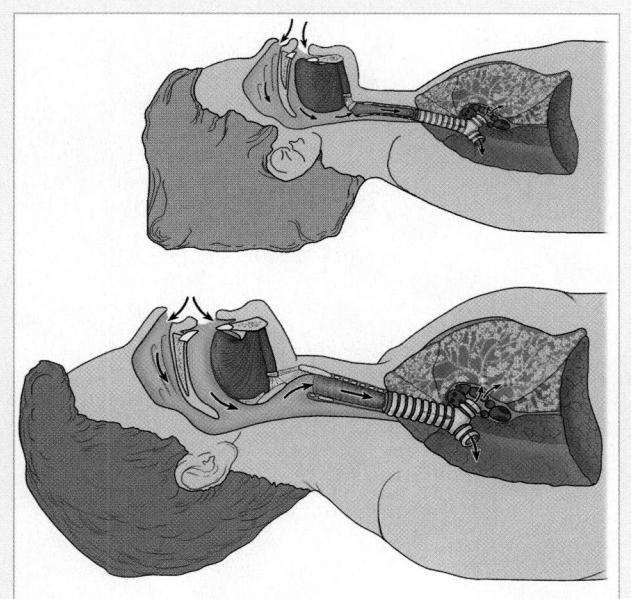

Figure 31-2 The anatomy of a child's airway differs from that of an adult's in several ways. The back of the head is larger in a child, so head positioning requires more care. The tongue is proportionately larger and more anterior in the mouth. The trachea is smaller in diameter and more flexible. The airway itself is lower and narrower.

- A larger, rounder <u>occiput</u>, or back of the head, which requires more careful positioning of the airway.
- A proportionately larger tongue relative to the size of the mouth and a more anterior location in the mouth. The child's tongue is also larger relative to the small mandible and can easily block the airway.
- A floppy, U-shaped epiglottis that is larger than an adult's, relative to the size of the airway.
- Less well-developed rings of cartilage in the trachea that may easily collapse if the neck is flexed or hyperextended.
- A narrower, lower airway.

Because of the smaller diameter of the trachea in infants, which is about the same diameter as a drinking straw, their airway is easily obstructed by secretions, blood, or swelling.

An infant needs to breathe faster than an older child (Table 31-1 ▶). Children's lungs grow and develop increased abilities to handle the exchange of oxygen as they age. A respiratory rate of 40 to 60 breaths/min is normal for the newborn, while the teenager is expected to have rates closer to the adult range (12 to 20 breaths/min). Breathing also requires the use of the

TABLE 31-1 Pediatric Respiratory Rates

Age	Respirations (breaths/min)
Newborn: 0 to 1 month	30 to 60
Infant: 1 month to 1 year	25 to 50
Toddler: 1 to 3 years	20 to 30
Preschool-age: 3 to 6 years	20 to 25
School-age: 6 to 12 years	15 to 20
Adolescent: 12 to 18 years	12 to 16
Older than 18 years	12 to 20

chest muscles and diaphragm. Because intercostal muscles are not well developed in children, movement of the diaphragm, their major muscle of respiration, dictates the amount of air that they inspire. Anything that puts pressure on the abdomen of a young child can block the movement of the diaphragm and cause respiratory compromise. Gastric distention can interfere with movement of the diaphragm and lead to <u>hypoventilation</u>. Young children also experience muscle fatigue much more quickly than older children. This can lead to respiratory failure if a child has to physically fight hard to breathe for long periods of time.

An infant's heart rate can become as high as 200 beats or more per minute if the body needs to compensate for injury or illness. This is the primary method the body uses to compensate for decreased perfusion. It is important to know the normal heart rate ranges when evaluating children (Table 31-2 ▼).

The ability of children to constrict their blood vessels also helps them to compensate for decreased perfusion. Pale skin is an early sign that the child may be compensating for decreased perfusion by constricting the vessels in the skin. Constriction of the vessels can be so profound that blood flow to the extremities can be diminished. Signs of vasoconstriction can include weak distal (eg, radial or pedal) pulses in the extremities, delayed capillary refill, and cool hands or feet.

TABLE 31-2 Pediatric Heart Rates

Age	Heart Rate (beats/min)
Infant	100 to 160
Toddler	90 to 150
Preschooler	80 to 140
School-age child	70 to 120
Adolescent	60 to 100

The skeletal system contains growth plates at the ends of long bones, which enable these bones to grow during childhood. As a result of the active growth plates, children's bones are weaker and more flexible, making them prone to fracture with stress. The bones of the skull also grow during infancy. Infants have two soft openings within the skull called fontanels. These will usually close completely by about 18 months of age; before that time, handle an infant's head with care.

Growth and Development

Adulthood begins at age 21. On this, the medical community has agreed. But when does childhood end? Many EMS systems use 18 years of age, others use 14, and still others use 12 or 16. Between birth and adulthood, many physical and emotional changes occur in children. While each child is unique, the thoughts and behaviors of children as a whole are often grouped into stages: infancy, the toddler years, preschool age, school age, and adolescence. Children in each stage grapple with different developmental issues. Even though there are specific issues that are important to different age groups, there are also some general rules that apply when you care for children of any age.

The Infant

Infancy is usually defined as the first year of life; the first month after birth is called the neonatal or newborn period. At first, infants respond mainly to physical stimuli such as light, warmth, hunger, and sound. Crying is one of their main avenues of expression during this period. After the first few months, however, they learn to coo, smile, roll over, and recognize their parents or caregivers. Infants are usually not afraid of strangers because they become the center of attention in most families. However, by the end of their first year, they may show signs of preferring to be with their caregivers and may cry if they are separated (Figure 31-3 ▶).

Begin your assessment by observing the infant from a distance, preferably in a caregiver's arms. Older infants, from 6 months to a year, may begin to cry when touched or picked up by a stranger, so let the caregiver continue to hold the baby as you start your examination. Provide as much sensory comfort as you can: Warm your hands and the end of the stethoscope, and offer a pacifier if the caregiver allows it. Have a caregiver hold the infant, if possible, during procedures. Plan to complete any painful procedure in an efficient manner. If possible, plan to do any painful procedures at the end

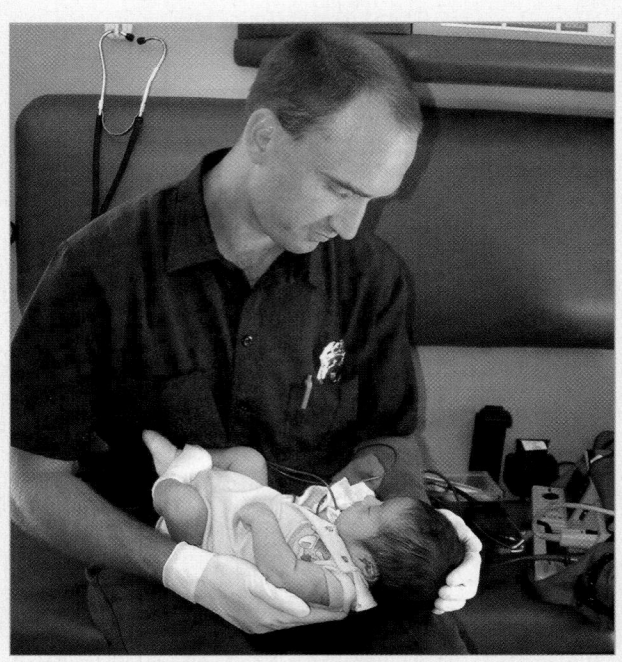

Figure 31-3 Infants are usually not afraid of strangers, but as they reach 6 months to 1 year, they may show signs that they prefer to be with their caregivers.

of the assessment process, so that the child does not become agitated during the assessment. When splinting a suspected fracture, have all of the equipment you will need ready in order to avoid making the procedure take longer than necessary.

Toddlers

After infancy, until about 3 years of age, a child is called a toddler. During this period, children begin to walk and to explore the environment. They are able to open doors, drawers, boxes, and bottles. Because they are explorers by nature and are not afraid, injuries in this age group increase.

Stranger anxiety develops early in this period. Toddlers may resist separation from caregivers and be afraid to let others come near them. Because of their newly found independence, they may also be very unhappy about being restrained or held for procedures (Figure 31-4 ▶). Two-year-olds in particular have a well-deserved reputation for having their own ideas about almost everything, which is why these years are often called the "terrible twos." Toddlers have a hard time describing or localizing pain. Pain in the abdomen may be expressed as, "My tummy hurts," and examination may reveal tenderness throughout the body. This is not because the child is trying to be difficult; the child does not have the verbal ability to be exact.

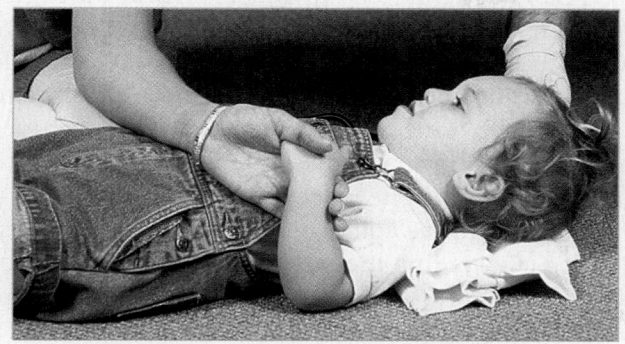

Figure 31-4 Because of their newly found independence, toddlers may be unhappy about being restrained or held for procedures.

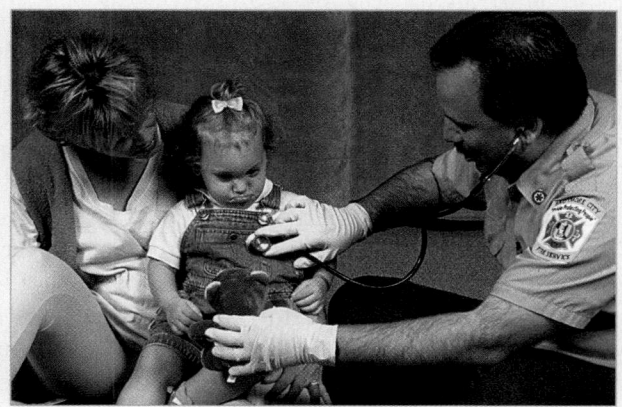

Figure 31-5 Leave a toddler on the caregiver's lap during your assessment, and use a toy to distract him or her.

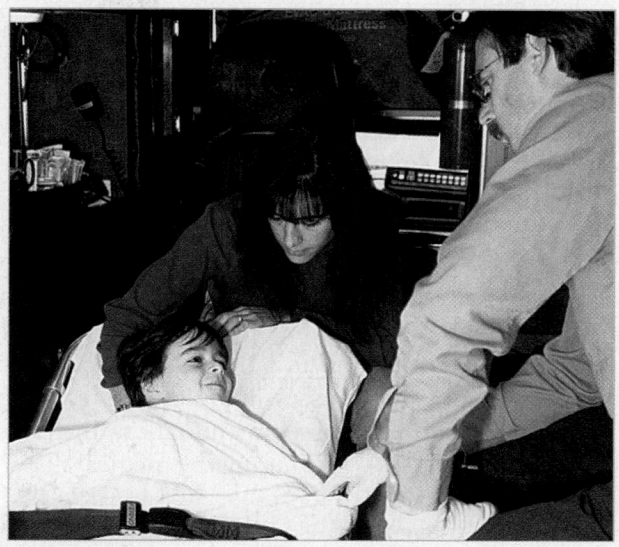

Figure 31-6 Preschool children have a vivid imagination, so much of the history must still be obtained from the caregiver.

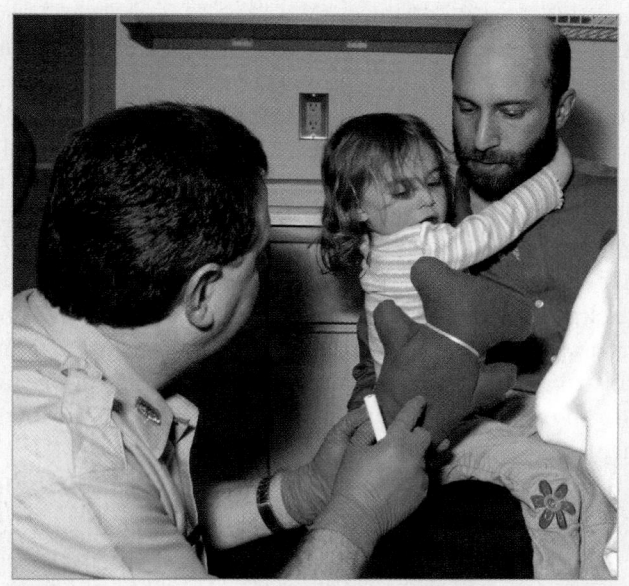

Figure 31-7 A preschool-aged child can be easily distracted by games or conversation.

Toddlers can be curious and adventuresome, so you may able to distract them (Figure 31-5 ▲). For example, you might allow the child to play with a tongue depressor while assessing his or her vital signs. Restrain the child for as short a time as possible, and allow him or her to be comforted by the caregiver immediately after a painful procedure. Begin your assessment at the hands or feet in order to keep from upsetting the child whenever possible.

The Preschool-Age Child

Preschool-age children (ages 3 to 6 years) are able to use simple language quite effectively and have lively imaginations (Figure 31-6 ▶). They can understand directions, be much more specific in describing their sensations, and identify painful areas when questioned. Much of their history must still be obtained from caregivers, however. Preschool-age children have a rich fantasy life, which can make them particularly fearful about pain and change involving their bodies. At this age,

they often believe that their thoughts or wishes can cause injury or harm to themselves or to others. They can believe that an injury was due to a bad deed they did earlier in the day.

Tell the child what you are going to do immediately before you do it; this way, the child has no time to develop frightening fantasies. At this age, children are easily distracted with counting games, small toys, or conversation (Figure 31-7 ▲). Be sure to adjust the level of game to the developmental level of the child; health care providers often assume that preschool children understand more than they actually do. Begin your assessment with the feet and move towards the head, similar to assessing a toddler. Use adhesive bandages to cover the site of an injection or other small wound, because the child might be worried about keeping his or her body together in one piece.

The School-Age Child

School-age children (ages 6 to 12 years) are beginning to act more like adults. They can think in concrete terms, respond sensibly to direct questions, and help take care of themselves. Your assessment, therefore, begins to be more like an adult assessment; talk to the child, not just the caregiver, while taking the medical history (Figure 31-8 ▼).

The school-age child is usually familiar with the process of physical examination. They have been to the doctor for childhood check-ups and immunizations. You may begin at the head and move to the feet similar to assessing adolescent or adult patients. Whenever possible, give the child appropriate choices: Would you like to sit up or lie down? Would you like to take off your clothes yourself? Only ask questions that you can control the answer. If you ask "Can I take your blood pressure?" and the answer is no, you will not be able to take it without upsetting the child. Instead, ask if you may find out their blood pressure on their right or left arm. (Asking if you may "take" the blood pressure may make younger patients think you will not give it back.) Giving them a choice that allows you to still obtain assessment information allows the child some control in a frightening situation. Encourage cooperation by allowing the child to listen to his or her own heartbeat through the stethoscope.

School-age children can understand the difference between emotional and physical pain, and have concerns about what pain means. Give them simple explanations about what is causing their pain and what will be done about it (Figure 31-9 ▼). Games and conversation may distract them. Ask them to describe their favorite place, their pets, or their toys. Ask the caregiver's advice in choosing the right distraction. Rewarding the school-age child after a procedure can be very helpful in his or her future cooperation and recovery. Often, kind words and a smile make a good reward when stuffed toys or books are not available.

Adolescents

Most adolescents (ages 12 to 18 years) are able to think abstractly and can participate in decision making. This is a period where the focus of their strength has moved from parents to peers. They are very concerned about body image and how they appear to their peers and to others their age. They may have very strong feelings against being observed during procedures.

Respect the adolescent's privacy at all times. Remember that adolescents can often understand very complex concepts and treatment options; you should provide them with information when they request it (Figure 31-10 ▶). You will find them more helpful and understanding of necessary procedures than younger patients.

Adolescents have a clear understanding of the purpose and meaning of pain. Whenever possible, explain any necessary procedures well in advance. Assess their pain by facial and body expression as well as by asking questions; adolescents can be very stoic and may not request relief from pain even when

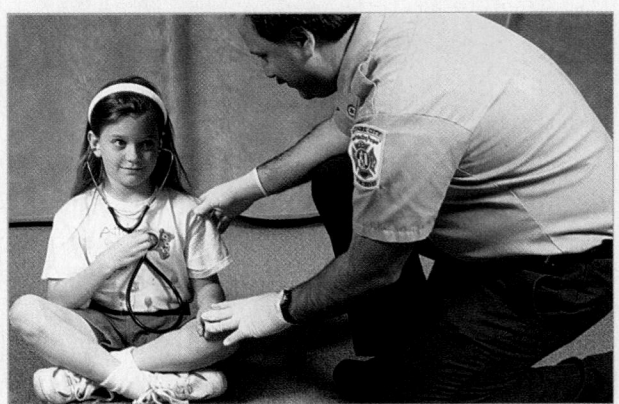

Figure 31-8   School-age children are more like adults in that they can answer your questions and can help to take care of themselves.

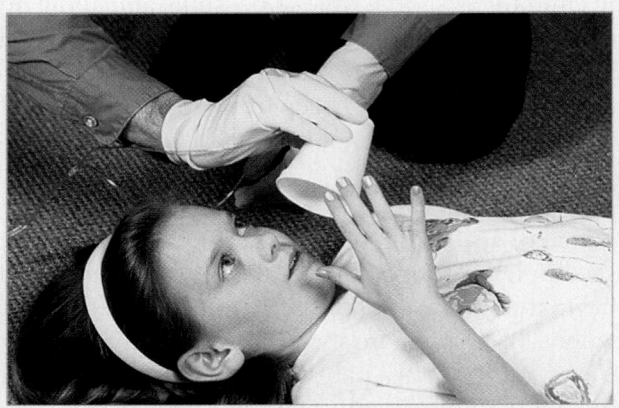

Figure 31-9   School-age children can understand simple explanations about their physical condition and the need for simple procedures.

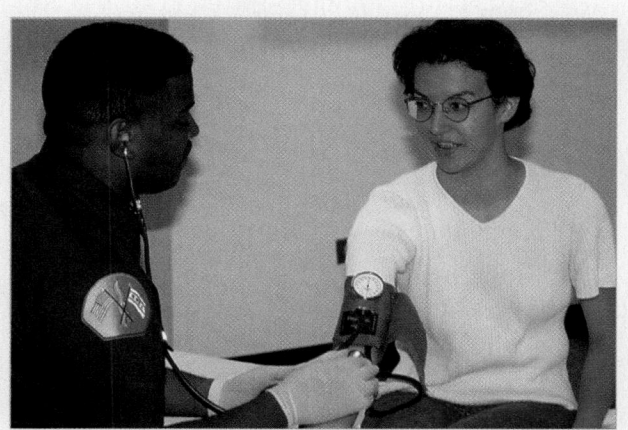

Figure 31-10 Respect the adolescent's privacy at all times; give the patient whatever information he or she requests.

they need it. To distract them, find out what they are interested in, such as sports, books, movies, or friends, and get them talking.

Family Matters

It is important to remember that when children are ill or injured, especially those with chronic illnesses, you may have several patients to treat rather than just one. Family members, especially the parent or primary caregiver, often need help or support when medical emergencies or problems develop. A calm parent usually helps to contribute to a calm child. An agitated parent usually means that the child will act the same way. Make sure that you are calm, efficient, professional, and sensitive as you deal with children and their families.

Pediatric Emergencies

Having explored the developmental and anatomic differences in children, there are many other aspects of emergency care that make dealing with the pediatric patient different. Some of these are dealing with the ability of the young body to compensate for illness or injury and others may be related to how children interact with the world around them.

Dehydration

Dehydration occurs when fluid losses are greater than fluid intake. The most common cause of dehydration in children is vomiting and diarrhea. If left untreated, dehydration can lead to shock and eventually death. Infants and children are at greater risk than adults for dehydration because their fluid reserves are smaller than those in adults. Life-threatening dehydration can overcome an infant in a matter of hours.

Fever

Fever is a common reason why parents call 9-1-1. Simply defined, fever is an increase in body temperature, usually in response to an infection. Body temperatures of 100.4°F (38°C) or higher are considered to be abnormal. Fevers have many causes and are rarely life-threatening events. However, you should not underestimate the potential seriousness of fevers, such as those that occur in conjunction with a rash. You should be suspicious of whether the fever is a sign of serious illness, such as meningitis. Common causes of a high temperature in a child include the following:

- Infection, such as pneumonia, meningitis, or urinary tract infection
- Neoplasm (cancer)
- Drug ingestion

You are the Provider   Part 2

While en route, you and your partner discuss some of the anatomic differences between children and adults. You also discuss some of the differences in the injury patterns of a child struck by a vehicle compared with those of an adult. You begin forming a treatment plan for this patient prior to your arrival on scene.

3. What are some of the anatomic differences between children and adults?
4. Would the age of your patient change the treatment plan?
5. What is your first step in treatment of this patient?

- Arthritis and systemic lupus erythematosus
- High environmental temperature

Fever is due to an internal body mechanism where heat generation is increased and heat loss is decreased. Note that there are other conditions in which the body temperature also increases. Hyperthermia differs from fever in that it is an increase in body temperature caused by an inability of the body to cool itself. Hyperthermia is typically seen in warm environments, such as a closed car on a hot day.

Meningitis

Meningitis is an inflammation of the tissue, called the meninges, that covers the spinal cord and brain. It is caused by an infection by bacteria, viruses, fungi, or parasites. If left untreated, meningitis can lead to permanent brain damage or death. Being able to recognize a patient who may have meningitis is an important skill as an EMT-B.

Meningitis can occur in both children and adults, but some individuals are at greater risk than others, as follows:

- Males
- Newborn infants
- Older people
- People whose immune systems have been weakened by AIDS or cancer
- People who have any history of brain, spinal cord, or back surgery
- Children who have had head trauma
- Children with shunts, pins, or other foreign bodies within their brain or spinal cord

At especially high risk are children with a ventriculoperitoneal (VP) shunt. VP shunts drain excess fluids from around the brain into the abdomen. These special needs children have tubing that can usually be seen and felt just under the scalp.

The signs and symptoms of meningitis vary, depending on the age of the patient. Fever and altered level of consciousness are common symptoms of meningitis in patients of all ages. Changes in level of consciousness can range from a mild or severe headache to confusion, lethargy, and/or an inability to understand commands or interact appropriately. The child may also experience a seizure, which may be a first sign of meningitis. Assess the level of consciousness using the AVPU scale. Infants younger than 2 to 3 months can have apnea, cyanosis, fever, or hypothermia.

In describing children with meningitis, physicians often use the term "meningeal irritation" or "meningeal signs" to describe pain that accompanies movement. Bending the neck forward or back increases the tension

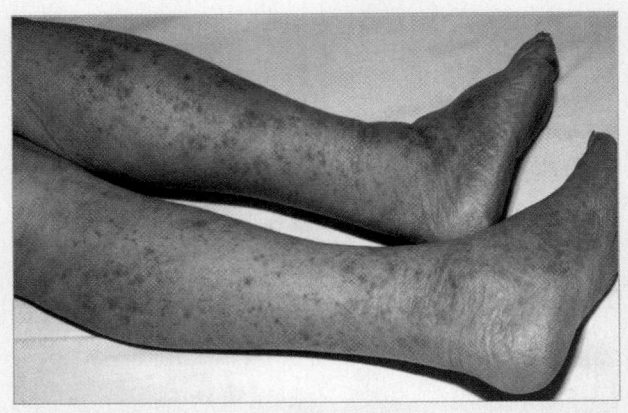

Figure 31-11 Children with *N meningitidis* typically have small, pinpoint, cherry-red spots or a larger purple/black rash.

within the spinal canal and stretches the meninges, causing a great deal of pain. This results in the characteristic stiff neck of children with meningitis, who will often refuse to move their neck, lift their legs, or curl into a "C" position, even if coached to do so. One sign of meningitis in an infant is increasing irritability, especially when being handled. Another sign is a bulging fontanel.

One form of meningitis deserves special attention. *Neisseria meningitidis* is a bacterium that causes a rapid onset of meningitis symptoms, often leading to shock and death. Children with *N meningitidis* typically have small, pinpoint, cherry-red spots or a larger purple/black rash (Figure 31-11 ▲). This rash may be on part of the face or body. These children are at serious risk of sepsis, shock, and death.

All patients with possible meningitis should be considered highly contagious and infectious. Therefore, you should use BSI precautions whenever you suspect meningitis and follow up with the hospital to learn the patient's final diagnosis. If you have been exposed to saliva and respiratory secretions from a child with *N meningitidis,* you should receive antibiotics to protect yourself and others from the bacteria. This is particularly true if you managed the patient's airway. If you were not in close contact with the patient or his or her respiratory secretions, you do not need treatment.

In taking the history of a child with meningitis, pay particular attention to the following details:

- Onset of illness, including any upper respiratory symptoms such as runny nose, cough, or other cold symptoms
- Presence and duration of fever
- Level of activity
- Change in behavior in older children, irritability in younger children

TABLE 31-3 Common Causes of Seizures

- Child abuse
- Electrolyte imbalance
- Fever
- Hypoglycemia (low blood glucose level)
- Idiopathic (no cause can be found)
- Infection
- Ingestion
- Lack of oxygen
- Medications
- Poisoning
- Previous seizure disorder
- Recreational drug use
- Head trauma

TABLE 31-4 Common Sources of Poisoning in Children

- Alcohol
- Aspirin and acetaminophen
- Household cleaning products such as bleach and furniture polish
- Houseplants
- Iron
- Prescription medications of family members
- Street drugs
- Vitamins

Seizures

A seizure is the result of disorganized electrical activity in the brain, causes of which are listed in (Table 31-3 ▲). It can be very frightening to people around the patient. Therefore, it is important to reassure the family and to approach assessment and management in a calm, step-by-step manner.

Febrile Seizures

Febrile seizures are common in children between the ages of 6 months and 6 years. Most pediatric seizures are due to fever alone, which is why they are called febrile seizures. These seizures typically occur on the first day of a febrile illness, are characterized by generalized tonic-clonic seizure activity, and last less than 15 minutes with a short postictal phase or none at all. It may be a sign of a more serious problem, such as meningitis. Obtain a history from the caregivers, as these children may have had a prior febrile seizure.

If you are called to care for a child who has had a febrile seizure, you often will find that the patient is awake, alert, and fully interactive when you arrive. Keep in mind that a persistent fever can lead to another seizure.

Poisoning

Poisoning is common among children, and sources are listed in (Table 31-4 ▶). It can occur by ingesting, inhaling, injecting, or absorbing a toxic substance. The signs and symptoms of poisoning vary widely, depending on the substance and the age and weight of the child. The child may appear normal at first, even in serious cases, or he or she may be confused, sleepy, or unconscious.

Infants may be poisoned as a result of being fed a harmful substance by a sibling or a caregiver or as a result of child abuse. Infants can be exposed to drugs and poisons left on floors and carpeting. They can also be exposed in a room or automobile in which harmful drugs, such as crack, cocaine, or PCP, are being smoked. Toddlers are curious and often ingest poisons when they find them in the home or garage (Figure 31-12 ▼). For example, some people store pe-

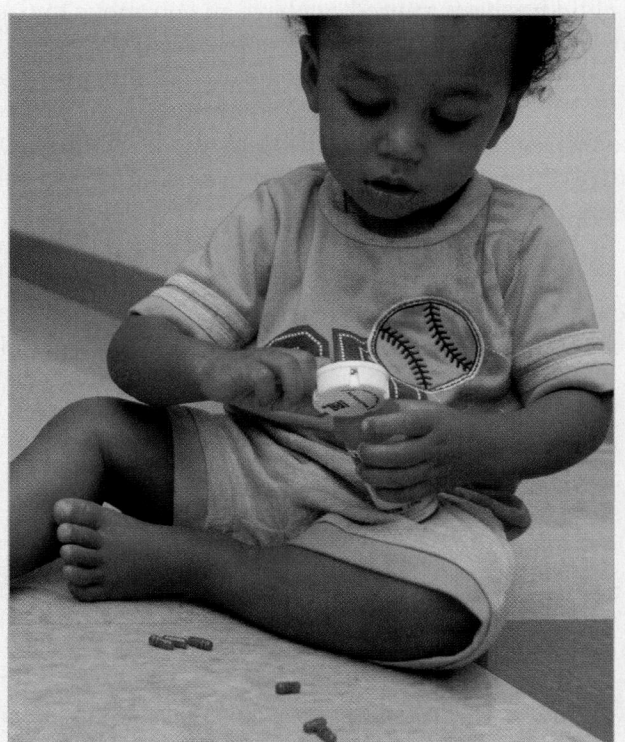

Figure 31-12 A curious child will try to taste or swallow almost any substance. A common victim of accidental ingestion of dangerous compounds is the unwatched toddler.

troleum products in soda bottles. Toddlers may believe the substance to be soda. Adolescents are more likely to have ingested alcohol and street drugs while partying or in a suicide attempt.

After you have completed your initial assessment, you should ask the caregiver the following questions:

- What is the substance(s) involved?
- Approximately how much of the substance was ingested or involved in the exposure (eg, number of pills, amount of liquid)?
- What time did the incident occur?
- Are there any changes in behavior or level of consciousness?
- Was there any choking or coughing after the exposure? (These can be signs of airway involvement.)

Pediatric Trauma

Injuries are the number one killer of children in the United States. More children die of injuries in 1 year than of all other causes combined. As an EMT-B, you will frequently treat injured children; therefore, you must have a thorough understanding of how trauma affects them. The quality of care in the first few minutes after a child has been injured can have an enormous impact on that child's chances for complete recovery.

Infants and toddlers are most commonly hurt as a result of falls or abuse. Older children and adolescents are usually injured as a result of mishaps involving automobiles. According to information collected by the National Pediatric Trauma Registry (NPTR), automobile accidents, including those involving bicycles and pedestrians, are the most significant threat to the well-being of the child. Other common causes of traumatic injury and death include falls, gunshot wounds, blunt injuries, and sports activities. Another extremely serious and troublesome cause of injury is child abuse.

Physical Differences

Children are smaller than adults; therefore, when they are hurt in the same type of accident as an adult, the location of their injuries may differ from those in an adult. For example, the bumper of a car will strike an adult in the lower leg, whereas that same bumper will strike a child in the pelvis. In a crash involving sudden deceleration, an adult might injure a ligament in the knee; in that same accident, a child might injure the bones in the leg.

Children's bones and soft tissues are less well developed than those of adults; therefore, the force of an injury affects these structures somewhat differently than it does in an adult. Because a child's head is proportionately larger than an adult's, it exerts greater stress on the neck structures during a deceleration injury. Because of these anatomic differences, you should always carefully assess children for head and neck injuries.

Psychological Differences

Children are also less mature psychologically than adults; therefore, they are often injured because of their undeveloped judgment and their lack of experience. For example, children are more likely than adults to cross the street without looking for oncoming traffic. As a result, children are more likely than adults to be struck by cars. Children and adolescents are also more likely to sustain injuries from diving into shallow water because they forget to check the depth of the water before they dive. In such situations, you should always assume that the child has serious head and neck injuries.

You are the Provider   Part 3

Upon arrival you find a 6-year-old boy lying in the arms of his crying mother. He is approximately 4' tall and weighs about 65 lb. The vehicle that struck him has no visible damage from the impact. The mother states that she thinks the truck was going about 25 miles an hour when it struck her son. The boy is conscious, alert, and oriented. He is crying uncontrollably and clutching his midsection.

6. Considering the MOI, should this patient be placed in a full spinal immobilization device?
7. Is the height and weight of the patient pertinent information to document?

Injury Patterns

Although you are not responsible for diagnosing injuries in children, your ability to recognize and report serious injuries will provide critical information to hospital staff. For this reason, it is important for you to understand the special physical and psychological characteristics of children and what makes them more likely to have certain kinds of injuries.

Automobile Collisions

Children playing or riding a bicycle can dart out in front of motor vehicles without looking. In such a situation, the driver may have very little time to slow down or stop to prevent hitting the child. The area of greatest injury varies, depending on the size of the child and the height of the bumper at the time of impact. When vehicles slow down at the moment of impact, the bumper dips slightly, causing the point of impact with the child to be lowered. The exact area that is struck depends on the child's height and the final position of the bumper at the time of impact. Children who are injured in these situations often sustain high-energy injuries to the head, spine, abdomen, pelvis, or legs.

Sports Activities

Children, especially those who are older or adolescents, are often injured in organized sports activities. Head and neck injuries can occur after high-speed collisions in contact sports such as football, wrestling, ice hockey, field hockey, soccer, or lacrosse. Remember to stabilize the cervical spine when caring for children with sports-related injuries. You should also be familiar with your local protocols related to helmet removal, and/or follow the guidelines presented in Chapter 30.

Injuries to Specific Body Systems

Head Injuries

Head injuries are common in children. This is because the size of a child's head, in relation to the body, is larger than that of an adult. The signs and symptoms of head injury in a child are similar to those in an adult, but there are some important differences. Nausea and vomiting are common signs and symptoms of head injury in children; however, it is easy to mistake these for an abdominal injury or illness. You should suspect a serious head injury in any child who experiences nausea and vomiting after a traumatic event.

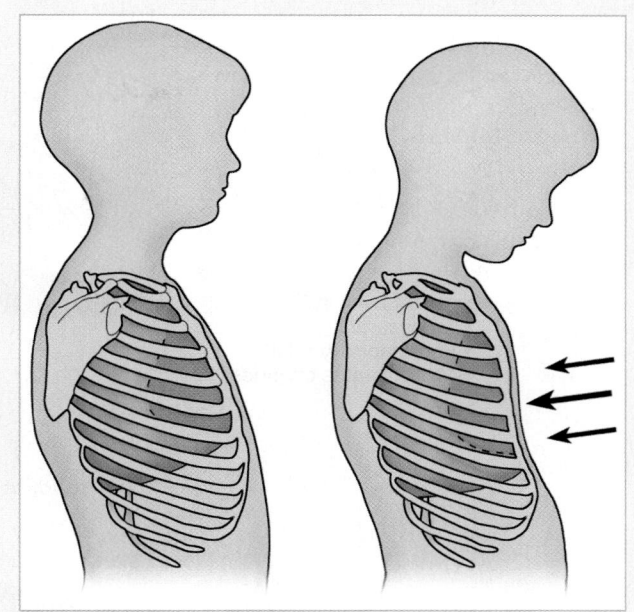

Figure 31-13 A child's ribs are softer and more flexible than an adult's. As a result, they may compress the lungs and heart, causing serious injury with no obvious external damage.

Chest Injuries

Chest injuries in children are usually the result of blunt trauma rather than penetrating objects. Remember that children have very soft, flexible ribs that can be compressed a great deal without breaking. Keep this in mind as you assess a child who has sustained high-energy blunt trauma to the chest. Even though there may be no external sign of injury, such as broken ribs, contusions, or bleeding, there may be significant injuries within the chest (Figure 31-13 ▲).

Abdominal Injuries

Abdominal injuries are very common in children. Remember, though, that children can compensate for significant blood loss better than adults without signs or symptoms of shock developing (Figure 31-14 ▶). They can also have a serious injury without early external evidence of a problem. All children with abdominal injuries should be monitored for signs and symptoms of shock, including a weak, rapid pulse; cold, clammy skin; decreased capillary refill (an early sign); confusion, and decreased systolic blood pressure (a late sign). Even in the absence of signs and symptoms of shock, or with only very few signs and symptoms, you should remain cautious about the possibility of internal injuries.

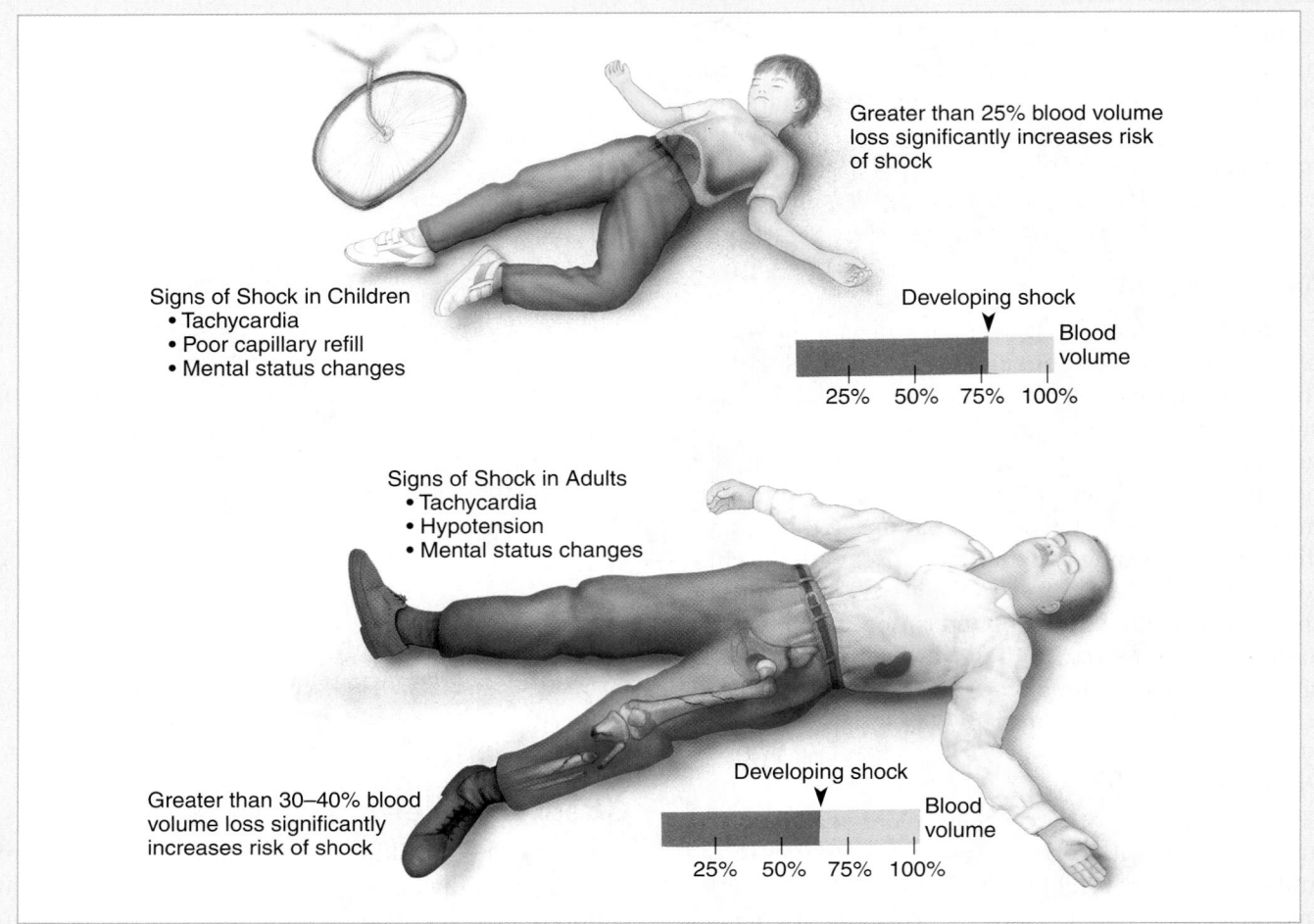

Signs of Shock in Children
- Tachycardia
- Poor capillary refill
- Mental status changes

Greater than 25% blood volume loss significantly increases risk of shock

Developing shock

Blood volume

25% 50% 75% 100%

Signs of Shock in Adults
- Tachycardia
- Hypotension
- Mental status changes

Greater than 30–40% blood volume loss significantly increases risk of shock

Developing shock

Blood volume

25% 50% 75% 100%

Figure 31-14 All children with abdominal injuries should be monitored closely for signs of shock. Although children may compensate for significant blood loss better than adults, they develop shock after proportionally smaller blood losses.

Injuries of the Extremities

Children have immature bones with active growth centers. Growth of long bones occurs from the ends at specialized growth plates. These growth plates are potential weak spots in the bone and are often injured as a result of trauma. In general, children's bones bend more easily than adults' bones. As a result, incomplete or greenstick fractures can occur.

Extremity injuries in children are generally managed in the same manner as those in adults. Painful deformed limbs with evidence of broken bones should be splinted. Specialized splinting equipment, such as a traction splint for fractures of the femur, should be used only if it fits the child. You should not attempt to use adult immobilization devices on a child unless the child is large enough to properly fit in the device.

Other Considerations

Pneumatic Antishock Garments

A pneumatic antishock garment (PASG) is rarely used in treating children. One situation in which you would use a PASG is when the child has obvious lower extremity trauma, particularly to both legs; pelvic instability; and clear signs and symptoms of decompensated shock. However, one problem with the use of this device for children is that it rarely fits. The PASG should be used on children only if it fits properly. Techniques such as placing the child in one leg of the garment are absolutely contraindicated and should never be used. The abdominal compartment of the garment should be inflated with caution on children because excessive pressure on the abdomen will cause pressure on the diaphragm and compromise breathing.

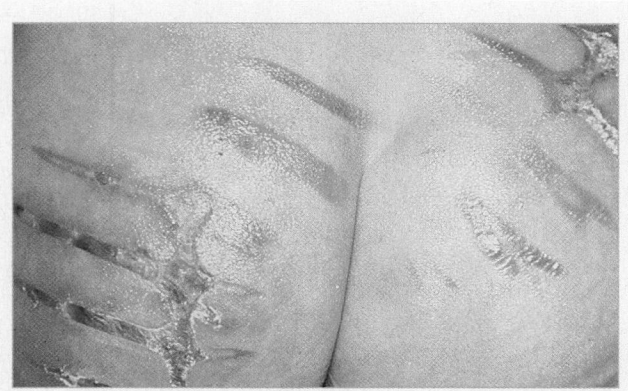

Figure 31-15 The most common burns in children involve exposure to hot surfaces. This child's buttocks were placed against a hot heating grate.

Burns

Children can be burned in a variety of ways. The most common involve exposure to hot substances such as scalding water in a bathtub, hot items on a stove, or exposure to caustic substances such as cleaning solvents or paint thinners (Figure 31-15 ▲). You should suspect possible internal injuries from chemical ingestion when you see a child who has burns, particularly around the face and mouth.

One common problem following burn injuries in children is infection. Burned skin cannot resist infection as effectively as normal skin can. For this reason, sterile techniques should be used in handling the skin of children with burn wounds.

(Table 31-5 ▼) provides some general guidelines to follow in assessing a child who has been burned. These guidelines may help you to determine which children should be treated primarily at specialized burn centers. Also note that you should consider the possibility of child abuse in any burn situation. Make sure you re-

TABLE 31-5 Severity of Burns in Children

Severity of Burn	Body Area Involved
Minor	Partial-thickness burns involving less than 10% of the body surface
Moderate	Partial-thickness burns involving 10% to 20% of the body surface
Critical	Any full-thickness burn Any partial-thickness burn involving more than 20% of the body surface Any burn involving the hands, feet, face, airway, or genitalia

port any information about your suspicions to the appropriate authorities.

Submersion Injury

Submersion injuries include near drowning and drowning. In submersion situations, you must always take steps to ensure your own safety when retrieving the patient from the water.

Drowning is the second most common cause of unintentional death among children in the United States; children younger than 5 years are at particular risk. At this age, children often fall into swimming pools and lakes, but many drown in bathtubs and even buckets. Older adolescents, who account for the most drownings after toddlers, drown when swimming or boating; alcohol is frequently a factor.

The principal injury from submersion is lack of oxygen. Even a few minutes (or less) without oxygen affects the heart, lungs, and brain, causing life-threatening problems such as cardiac arrest, respiratory difficulty, and coma. Submersion in icy water can rob the body of heat, causing hypothermia. While a very few, very cold victims of submersion hypothermia have survived long periods in cardiac arrest in icy water, most people in this situation die. Diving into the water, of course, increases the risk of neck and spinal cord injuries.

Child Abuse

The term child abuse means any improper or excessive action that injures or otherwise harms a child or infant; it includes physical abuse, sexual abuse, neglect, and emotional abuse. The intentional injury of a child, whether physical or emotional, is not rare in our society. More than 2 million cases of child abuse are reported to child protection agencies annually. Many of these children suffer life-threatening injuries, and some die. If suspected child abuse is not reported, the child is likely to be abused again and again, perhaps suffering permanent injuries or even dying. Therefore, you must be aware of the signs of child abuse and neglect, and of your responsibility to report suspected abuse to law enforcement or child protection agencies.

Signs of Abuse

As an EMT-B, you will be called to homes because of a reported injury to a child. If you suspect that physical or sexual abuse is involved, you should ask yourself the following questions:

■ Is the injury typical for the developmental level of the child?

- Is the method of injury reported by the parent or caregiver consistent with the child's injuries?
- Is the caregiver behaving appropriately (concerned about the child's well-being)?
- Is there evidence of drinking or drug use at the scene?
- Was there a delay in seeking care for the child?
- Is there a good relationship between the child and the caregiver?
- Does the child have multiple injuries at different stages of healing?
- Does the child have any unusual marks or bruises that may have been caused by cigarettes, grids, or branding injuries?
- Does the child have several types of injuries, such as burns, fractures, and bruises?
- Does the child have any burns on the hands or feet that involve a glove distribution (marks that encircle a hand or foot in a pattern that looks like a glove)?
- Is there an unexplained decreased level of consciousness?
- Is the child clean and an appropriate weight for his or her age?
- Is there any rectal or vaginal bleeding?
- What does the home look like? Clean or dirty? Is it warm or cold? Is there food?

Your assessment in the field will allow a better assessment by the medical staff later. An easy way to remember these is the mnemonic CHILD ABUSE shown in (Table 31-6 ▼).

TABLE 31-6 Child Abuse

Mnemonic for Assessing Possible Child Abuse

Consistency of the injury with the child's developmental age

History inconsistent with injury

Inappropriate parental concerns

Lack of supervision

Delay in seeking care

Affect

Bruises of varying ages

Unusual injury patterns

Suspicious circumstances

Environmental clues

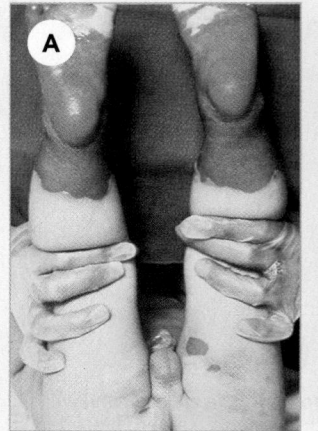
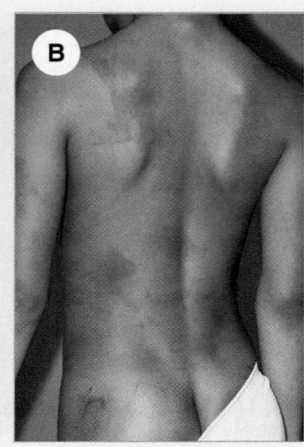

Figure 31-16 Signs of child abuse. **A.** Scald. **B.** Multiple injuries at different stages of healing.

As you assess the child, look for and pay particular attention to the following signs (Figure 31-16 ▲).

Bruises

Observe the color and location of any bruises. New bruises are pink or red. Over time, bruises turn blue, then green, then yellow-brown and faded. Note the location. Bruises to the back, buttocks, or face are suspicious and are usually inflicted by a person.

Burns

Burns to the penis, testicles, vagina, or buttocks are usually inflicted by someone else, as are burns that encircle a hand or foot to look like a glove. You should suspect abuse if the child has cigarette burns or grid pattern burns.

Fractures

Fractures of the humerus or femur do not normally occur without major trauma, such as a fall from a high place or a motor vehicle crash. Falls from bed are not usually associated with fractures.

Shaken Baby Syndrome

Infants may sustain life-threatening head trauma by being shaken or struck on the head, a life-threatening condition called <u>shaken baby syndrome</u>. With this condition, there is bleeding within the head and damage to the cervical spine as a result of intentional, forceful shaking. The infant will be found unconscious, often without evidence of external trauma. The call for help may be for an infant who has stopped breathing or is unresponsive. The infant may appear to be in cardio-pulmonary arrest, but what has likely occurred is that the shaking tore blood vessels in the brain, resulting in bleeding around the brain. The pressure from the blood results in a coma.

Neglect

Children who are neglected are often dirty or too thin or appear developmentally delayed because of lack of stimulation. You may observe such children when you are making calls for unrelated problems. Report all cases of suspicious neglect.

Symptoms and Other Indicators of Abuse

An abused child may appear withdrawn, fearful, or hostile. You should be particularly concerned if the child refuses to discuss how an injury occurred. Occasionally, the parent or caregiver will reveal a history of several "accidents." Be alert for conflicting stories or a marked lack of concern from the parents or caregiver. Remember, the abuser may be a parent, caregiver, relative, or friend of the family. Sometimes the abuser is an acquaintance of a single parent.

EMT-Bs in all states must report all cases of suspected abuse, even if the emergency department fails to do so. Most states have special forms for reporting. Supervisors are generally forbidden to interfere with the reporting of suspected abuse, even if they disagree with the assessment. You do not have to prove that there has been abuse. Law enforcement and child protection agencies are mandated to investigate all reported cases.

Sexual Abuse

Children of any age and either gender can be victims of sexual abuse. Most victims of rape are older than age 10 years, although younger children may be victims as well. This type of sexual abuse is often the result of longstanding abuse by relatives.

Your assessment of a child who has been sexually abused should be limited to determining the type of dressing any injuries require. Sometimes, a sexually abused child is also beaten. Therefore, you should treat any bruises or fractures as well. Do not examine the genitalia of a young child unless there is evidence of bleeding, or there is an injury that must be treated.

In addition, if you suspect that a child is a victim of sexual abuse, do not allow the child to wash, urinate, or defecate before a physician completes an exam. Although this step is difficult, it is important to preserve evidence. If the molested child is a girl, ensure that a female EMT-B or police officer remains with the child unless locating one will delay transport.

You must maintain professional composure the entire time you are assessing and caring for a sexually abused child. Assume a concerned, caring approach, and shield the child from onlookers and curious bystanders. Obtain as much information as possible from the child and any witnesses. The child may be hysterical or unwilling to say anything at all, especially if the abuser is a relative or family friend. You are in the best position to obtain the most accurate firsthand information about the incident. Therefore, you should record any information carefully and completely on the patient care report.

You are the Provider   Part 4

While gaining additional information about the incident, your partner manually stabilizes the boy's head and neck. You assess his airway, breathing, and circulation. He has a patent airway and is breathing at a rate of 22 labored breaths/min. His pulse is 156 beats/min and bounding. He has numerous small abrasions and lacerations on his extremities. There is no apparent major external bleeding. You apply high-flow oxygen via a nonrebreathing mask.

8. What does your index of suspicion tell you about this patient?
9. What is your next step in treatment?

Transport all children who are victims of sexual assault. Sexual abuse of a child is a crime. Cooperate with law enforcement officials in their investigations.

Sudden Infant Death Syndrome

The death of an infant or a young child is called <u>sudden infant death syndrome (SIDS)</u> when, after a complete autopsy, the cause of death remains unexplained. SIDS is the leading cause of death in infants younger than 1 year; most cases occur in infants younger than 6 months.

Although it is impossible to predict SIDS, there are several known risk factors:

- Mother younger than 20 years old
- Mother smoked during pregnancy
- Low birth weight

Deaths due to SIDS can occur at any time of the day; however, these children are often discovered in the morning when the parents go in to check on the infant. If you are the first provider at the scene of suspected SIDS, you will face three tasks: assessment and management of the patient, communication and support of the family, and assessment of the scene.

Assessment and Management

SIDS is a diagnosis of exclusion. All other potential causes must first be ruled out, a process that may take physicians quite a while. An infant who has been a victim of SIDS will be pale or blue, not breathing, and unresponsive. Other causes for such a condition include the following:

- Overwhelming infection
- Child abuse
- Airway obstruction from a foreign object or as a result of infection
- Meningitis
- Accidental or intentional poisoning
- Hypoglycemia (low blood glucose level)
- Congenital metabolic defects

Regardless of the cause, assessment and management of the infant remain the same. Remember that what you find in assessing the infant and the scene may provide important diagnostic information.

Begin with an assessment of the ABCs, and provide interventions as necessary. Depending on how much time has passed since the child was discovered, he or she may show signs of postmortem changes. These include stiffening of the body, called <u>rigor mortis</u>, and

<u>dependent lividity</u>, which is the pooling of blood in the lower parts of body or those that are in contact with the floor or bed.

If the child shows such signs, call medical control. In some EMS systems, a victim of SIDS may be declared dead on the scene. Deciding whether to start CPR on a child who shows clear signs of rigor mortis or dependent lividity can be very difficult. Family members may consider anything less as withholding critical care. In this situation, the best course of action may be to initiate CPR and transport the patient and the family to the nearest emergency department, where the family can receive more extensive support (follow local protocols). If there is no evidence of postmortem changes, begin CPR immediately.

As you assess the infant, pay special attention to any marks or bruises on the child before performing any procedures, including CPR. Also note any intervention such as CPR that was done by the parents before you arrived.

Communication and Support of the Family

The death of a child is a very stressful event for a family; it also tends to evoke strong emotional responses among health care providers, including EMS personnel (Table 31-7 ▶). Part of your job at this point is to allow the family to express their grief in ways that may differ from your own cultural, religious, and personal practices. Provide support in whatever ways you can.

Many times family members will ask specific questions about the event: Why did this happen? How did this happen? Let them know that their concerns will be addressed but that answers are not immediately available (Table 31-8 ▶). Always use the infant's name in speaking to family members. If possible, allow the family to spend time with the infant and to ride in the ambulance to the hospital.

Scene Assessment

Carefully inspect the environment, following local protocols, noting the condition of the scene where the caregivers found the infant. Your assessment of the scene should concentrate on the following:

- Signs of illness, including medications, humidifiers, or thermometers
- The general condition of the house. (Note any signs of poor hygiene.)
- Family interaction. Do not allow yourself to be judgmental about family interactions at this time. Do note and report any behavior that is clearly

TABLE 31-7 How You Can Help the Family of a Deceased Child

When Arriving on Site
- Introduce yourself quickly.
- Obtain a brief history.
- When possible, one provider should stay with the family.

If Resuscitation Is Attempted
- Give brief, frequent updates and explanations.
- Allow family members to stay within viewing distance if they wish.
- Allow family members to accompany child to hospital when possible.

If No Resuscitation Is Performed
- Sit down with the family.
- Inform the family immediately.
- Explain why no resuscitation will be attempted.
- Offer to arrange for religious support, including baptism or last rites.

Beginning the Grieving Process
- Learn and use the child's name.
- Allow family to express emotions; be nonjudgmental.
- Give brief explanations and answers.
- Explain to the family that the cause of death is still unknown.
- Allow time for questions.

DO
- Tell the family how sorry you are.
- Tell the family whom they can call if they have questions later.
- Give written instructions and referrals.

DON'T
- Say, "I know how you feel."
- Say, "You have other children" or "You can have other children."
- Attempt to answer the question "Why did this happen?"
- Try to tell family that they will be feeling better in time.

TABLE 31-8 Common Questions Following Death of a Child

Q Was there pain?
A This often can be answered by a simple "No." If you are uncertain, you may give an indirect answer such as "We really don't know what patients feel in these circumstances."

Q What did he/she die of?
A Do not answer this question; you would probably be guessing at this point.

Q Why did this happen?
A Do not attempt to answer this question either, as the answer depends on one's own individual philosophy, or religion. "I wish I had an answer for you" is usually the most appropriate response.

Q What happens now?
A This question usually concerns the next few minutes or the next hour. If you know, you should give the family a general idea of what will happen. For example, if there is no history of illness, you can say that "a medical examination will be done, and then [child's name] will be taken to the mortuary."

The death of a child is difficult for everyone involved: parents, relatives, friends, and health care professionals. You should arrange for a proper debriefing after your involvement with the case comes to a close. This can be a session with a trained counselor or a group discussion with your colleagues or the entire health care team.

Apparent Life-Threatening Event

Infants who are not breathing and are cyanotic and unresponsive when found by their families sometimes resume breathing and color with stimulation. These children have had what is called an apparent life-threatening event (ALTE), called "near-miss SIDS" in the past. In addition to cyanosis and apnea, a classic ALTE is characterized by a distinct change in muscle tone (limpness) and choking or gagging. After the event, a child may appear healthy and show no signs of illness or distress. Nevertheless, you must complete a careful assessment and provide immediate transport to the emergency department.

Pay strict attention to management of the airway. Assess the infant's history and, if possible, the environment. Allow caregivers to ride in the ambulance. If asked, explain that you cannot say what caused the

not within the acceptable range, such as physical and verbal abuse.
- The site where the infant was discovered. Note all items in the infant's crib or bed, including pillows, stuffed animals, toys, and small objects.

EMT-B Tips

Most parents of children who die suddenly will experience extremely strong emotional responses for a long time after the death. Counseling and support services begin with your care, including immediate referral to longer term services. You can usually make this referral through social services personnel in a hospital you work with, something you should know about in advance. Many communities also have support groups for families who lose children, including deaths caused by SIDS. Make sure the parents are aware of available services, offer to put them in touch while you are there, and leave the contact information in written form for their later reference even if you have helped them make the contact.

event, and that this is something that doctors will have to determine at the hospital.

Death of a Child

As with SIDS, the death of a child from any cause poses special challenges for EMS personnel. In addition to any medical treatment the child may require, you must be prepared to offer the family a high level of support and understanding as they begin the grieving process. First, the family may want you to initiate resuscitation efforts, which may or may not conflict with your EMS protocols. If the child is clearly deceased and, under protocol, can be declared dead in the field, but the family is so distraught that they insist that resuscitation efforts be made, initiate CPR and transport the child.

The extent of your interaction with the family will depend, to some degree, on the number of providers available at the scene. Always introduce yourself to the child's caregivers, and ask about the child's date of birth and medical history. If and when the decision is made to start or stop resuscitation efforts, inform the family immediately. Find a place for family members from where they can watch resuscitation without being in the way. Do not, in any case, speculate on the cause of the child's death. The family will want to see the child and should be asked whether they want to hold the child and say good-bye. Parents may be experiencing strong feelings of denial.

The following interventions are helpful in caring for the family at this time:

- Learn and use the child's name rather than the impersonal "your child."
- Speak to family members at eye level, maintaining good eye contact with them.
- Use the word "dead" or "died" when informing the family of the child's death; euphemisms such as "passed away" or "gone" are not effective.
- Acknowledge the family's feelings ("I know this is devastating for you"), but never say "I know how you feel," even if you have experienced a similar event; the statement will anger many people.
- Offer to call other family members or clergy if the family wishes.
- Keep any instructions short, simple, and basic. Emotional distress may limit their ability to process information.

You are the Provider   Part 5

During your detailed physical exam you find that your patient's abdomen is rigid and tender and also warm to the touch. While your patient is conscious and responsive, he is showing signs of an altered mental status. Given the change in your patient, you re-contact dispatch and get an updated ETA (estimated time of arrival) for the ALS unit. Dispatch notifies you that ALS is approximately 4 minutes out.

10. What do the patient's signs and symptoms indicate?
11. What are your treatment options?

- Ask each adult family member individually whether he or she wants to hold the child.
- Wrap the dead child in a blanket, as you would if he or she were alive, and stay with the family while they hold the child. Ask them not to remove tubes or other equipment that was used in an attempted resuscitation.

Remember that each individual and each culture expresses grief in a different way, some more visibly than others. Some will require intervention; others will not. Most caregivers feel directly or indirectly responsible for the death of a child and may express this immediately; this does not mean that they actually are responsible. Parents often have questions that you should be prepared to answer. Although you should keep the possibility of abuse or neglect in mind, your role is not that of investigator. Any further inquiry is the responsibility of law enforcement.

Some EMS systems arrange for home visits after the death of a child so that EMS providers and family members can come to some sort of closure together. This also gives the family an opportunity to ask any remaining questions about the event. However, you need special training for such visits.

Again, coping with the death of a child can be very stressful for health care professionals. You may find yourself with unexpected feelings of pain and loss. It is helpful to take some time before going back on the job to work through your feelings and to talk about the event with your EMS colleagues. Be alert for signs of posttraumatic stress in yourself and others: nightmares, restlessness, difficulty sleeping, lack of appetite, a constant need for food, and the like. Consider the need for professional help if these signs or symptoms continue. All EMS programs should have critical incident stress management protocols and debriefing teams available for traumatic incidents.

Although you may consider the death of a child to be a failure, your skill at coping with this kind of emotional event can be a great comfort to the family, helping them to accept their loss and begin the long process of grieving.

Infants and Children With Special Needs

The approach to health care in our society continues to focus on decreasing lengths of hospitalization, and technology continues to improve. As a result of these two factors, the number of infants and children with chronic diseases who are living at home or in other environments outside of the hospital continues to grow. You should be familiar with some of the special needs created by these chronic diseases or conditions, particularly as they relate to the potential need for emergency medical care.

Some examples of infants and children with special needs include the following:

- Children who were born prematurely and have associated lung disease problems
- Small children or infants with congenital heart disease
- Children with neurologic disease (occasionally caused by hypoxemia at the time of birth, as with cerebral palsy)
- Children with congenital or acquired diseases resulting in altered body functions such as breathing, eating, urination, and bowel function

The parent or caregiver of a special needs child will be an important part of your assessment. The EMT-B must first determine the child's normal baseline status before an assessment of the current condition can be made. It is often helpful to ask, "What is different today?" These parents or guardians often know their child's history and condition better than some skilled health care workers might.

Occasionally, these children live at home but depend on artificial ventilators or other devices to maintain life. You assess and care for these children the same as for all other patients. Your focus on the ABCs remains the priority.

Tracheostomy Tubes

Children who depend on home artificial ventilators or those who have chronic pulmonary medical conditions may breathe through a tracheostomy tube

---

✳ EMT-B Tips

As EMT-Bs finish initial training and begin their field response work, it is common for them to feel a strong need for additional knowledge about the special needs patients and the special equipment that this section describes. Be aggressive during training in pursuing opportunities to learn the details about these special subjects. Over time, experience will be one of your primary teachers.

Figure 31-17 Some children require a tracheostomy tube to breathe.

Figure 31-18 Children who require frequent IV medications may have a central line in place.

Figure 31-17 ▲). A tracheostomy tube is a tube placed in the neck that passes directly into the major airways. Because this tube bypasses the nose and mouth, the body can build up secretions in or around the tube. These tubes are prone to become obstructed by mucous plugs or foreign bodies. There may be bleeding or air leaking around the tube, which usually happens with new tracheostomies, and the tube can become loose or dislodged. Occasionally, the opening around the tube may become infected. Your care of a patient with a tracheostomy tube includes maintaining an open airway. This can include suctioning the tube if necessary to clear a mucous plug, maintaining the patient in a position of comfort, and providing transport to the hospital. Call for ALS backup.

Artificial Ventilators

Children who are on a ventilator at home cannot breathe without help. If the ventilator malfunctions, remove the child from the ventilator and begin ventilations with a BVM device. To do this, remove the mask from a BVM device and directly attach the bag and valve to the tracheostomy tube; this will allow you to ventilate through the tracheostomy tube.

Children on home artificial ventilators will need artificial ventilation throughout transport. Artificial ventilation is provided through the tracheostomy tube. Remember that the patient's caregivers will know how the ventilator works and will be of great help to you in attaching the BVM device to the tube in preparation for transport.

Central IV Lines

Children with chronic medical conditions such as gastrointestinal disturbances that require prolonged IV feeding or those with infections that require prolonged IV antibiotics will have indwelling IV catheters placed near the heart for long-term use. These catheters can be in the chest or in the arm (Figure 31-18 ▲). Problems associated with these devices may include broken lines, infections around the lines, clotted lines, and bleeding around the line or from the tubing attached to the line. If bleeding occurs, you should apply direct pressure to the tubing and provide transport to the hospital.

Gastrostomy Tubes

Gastrostomy tubes, sometimes referred to as G-tubes, are tubes placed through the wall of the abdomen directly into the stomach for feeding in children who cannot be fed by mouth (Figure 31-19 ▶). Because food is pumped into the stomach, it can back up the esophagus and into the lungs. Breathing problems in these

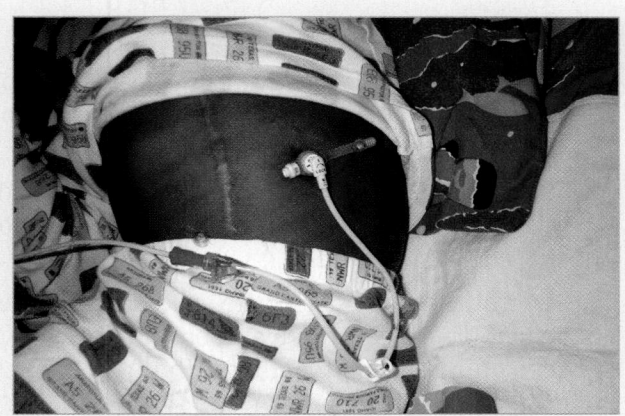

Figure 31-19 Gastrostomy tubes are placed through the skin into the stomach for children who cannot be fed by mouth.

children may be complicated by aspiration of the tube contents into the lungs. You should always have suction readily available to clear any materials from the mouth and to prevent airway problems. Patients with gastrostomy tubes who have difficulty breathing should be transported either sitting or lying on the right side with the head elevated to prevent the contents of the stomach from passing into the lungs. Give supplemental oxygen if the patient has any difficulty breathing. Children with diabetes who receive insulin and tube feedings may become hypoglycemic quickly if tube feedings are discontinued. Be alert for altered mental status.

Shunts

Some children with chronic neurologic conditions may have shunts in place. <u>Shunts</u> are tubes that extend from the brain to the abdomen to drain excess cerebrospinal fluid that may accumulate near the brain. These shunts keep pressures in the head from building up. If a shunt becomes clogged due to infection, changes in mental status and respiratory arrest may occur. Emergency medical care includes airway management and artificial ventilation during transport. During assessment, you will likely feel a device on the side of the head, behind the ear, beneath the skin. This device is a fluid reservoir, and the presence of this device should alert you to the possibility that the child has an underlying shunt. Should the shunt become dysfunctional, the child could be predisposed to respiratory arrest.

You are the Provider   Summary

It is important that you take into consideration the age and size of your patient, the anatomic differences between children and adults, and the possible differences in their injury patterns. Trust your index of suspicion and treat for the worst possible condition. It is always better to do too much than not enough for a patient. When dealing with a pediatric patient, keep in mind that children's bodies tend to compensate longer than an adult's but decompensate much faster, with little to no advance warning. Always consider the possibility of shock and treat accordingly.

Prep Kit

Ready for Review

- Children are not only smaller than adults and more vulnerable, they are also anatomically, physiologically, and psychologically different from adults in some important ways.

- You must understand these differences to provide the best possible care for children who have been injured or are ill.

- Child anatomy contributes to some special challenges.

- The tongue is large relative to other structures, so it poses a higher risk of airway obstruction than in an adult.

- Head size and shape also make airway positioning a specialized task in children. Newborns and infants have high respiratory rates and breathe primarily with their abdomens.

- Vasoconstriction in the skin and increased heart rate are the major means of compensating for decreased perfusion of vital organs.

- The airway in a child has a smaller diameter than the airway in an adult and is therefore more easily obstructed.

- Because the diaphragm is the principal muscle of respiration in children and infants, gastric distention can create breathing difficulties.

- There are five developmental stages in childhood: infancy, the toddler years, preschool age, school age, and adolescence.

- General rules for dealing with children of all ages include appearing confident, being honest, and keeping caregivers together with the patient as much as possible.

- You should remember that children's bones are more flexible and bend more with injury and that the ends of the long bones, where growth occurs, are weaker and may be injured more easily.

- Children's internal organs are not as insulated by fat and may be injured more severely, and children have less circulating blood, so that, although children exhibit the signs of shock more slowly, they go into shock more quickly, with less blood loss.

- Children are not always as cautious as adults and tend to have more accidental poisoning, diving, and bicycle injuries.

- The most common cause of dehydration in children is vomiting and diarrhea. Life-threatening diarrhea can develop in an infant in hours.

Technology

Interactivities

Vocabulary Explorer

Anatomy Review

Web Links

Online Review Manual

- A seizure is the result of disorganized electrical activity in the brain.

- Febrile seizures may be a sign of a more serious problem such as meningitis.

- A victim of sudden infant death syndrome (SIDS) will be pale or blue, not breathing, and unresponsive. He or she may show signs of postmortem changes, including rigor mortis and dependent lividity; if so, call medical control to report the situation.

- If family members insist or protocols mandate, you should initiate CPR and transport infant and family to the emergency department, where the family can receive more extensive support. If the child does not have any evidence of postmortem changes, begin CPR immediately.

- Carefully inspect the environment where a SIDS victim was found, looking for signs of illness, abusive family interactions, and objects in the child's crib.

- Provide support for the family in whatever way you can, but do not make judgmental statements. Allow them to spend time with the child and ride in the ambulance to the hospital.

- Any death of a child is stressful for family members and for health care providers. In dealing with the family, acknowledge their feelings, keep any instructions short and simple, use the child's name, and maintain eye contact.

- Be prepared to respond to philosophical as well as medical questions, in most cases by indicating concern and understanding; do not be specific about the cause of death.

- Be alert for signs of posttraumatic stress in yourself and others after dealing with the death of a child. It can help to talk about the event and your feelings with your EMS colleagues.

- The number of children living at home who have chronic diseases and special needs is increasing.

- These patients will test your knowledge of special equipment and care procedures, your skills in obtaining pertinent history from family and caregivers, and your ability to detect urgent problems when you may not be completely familiar with the technical details involved. Learning everything you can from each special situation will help prepare you for similar responses in the future.

Prep Kit continued...

Vital Vocabulary

adolescents Children between 12 to 18 years of age.

apparent life-threatening event (ALTE) An event that causes unresponsiveness, cyanosis, and apnea in an infant, who then resumes breathing with stimulation.

child abuse Any improper or excessive action that injures or otherwise harms a child or infant; includes neglect and physical, sexual, and emotional abuse.

dehydration A state in which fluid losses are greater than fluid intake into the body, leading to shock and death if untreated.

dependent lividity Pooling of the blood in the lower parts of the body after death.

febrile seizure Seizure relating to a fever.

gastrostomy tube A feeding tube placed directly through the wall of the abdomen; used in patients who cannot ingest liquids or solids.

generalized tonic-clonic seizure A seizure that features rhythmic back-and-forth motion of an extremity and body stiffness.

hypoventilation Reduced minute volume, either from reduced rate and/or depth of breathing.

infancy The first year of life.

meningitis Inflammation of the meninges that cover the spinal cord and the brain.

Neisseria meningitidis A form of bacterial meningitis characterized by rapid onset of symptoms, often leading to shock and death.

neonatal The first month after birth.

occiput The back of the head.

pediatrics A specialized medical practice devoted to the care of the young.

pneumatic antishock garment (PASG) An inflatable device that covers the legs and abdomen; used to splint the lower extremities or pelvis, or to control bleeding in lower extremities, pelvis, or abdominal cavity.

preschool-age Children between 3 to 6 years of age.

rigor mortis Stiffening of the body after death.

school-age Children between 6 to 12 years of age.

shaken baby syndrome Bleeding within the head and damage to the cervical spine of an infant who has been intentionally and forcibly shaken; a form of child abuse.

shunt A tube that diverts excess cerebrospinal fluid from the brain to the abdomen.

sudden infant death syndrome (SIDS) Death of an infant or young child that remains unexplained after a complete autopsy.

toddler The period following infancy until 3 years of age.

tracheostomy tube A tube inserted into the trachea in children who cannot breathe on their own; passes through the neck directly into the major airways.

Points to Ponder

You are dispatched to aid fire fighters at the scene of a fire. As you arrive, three small children have just been rescued from the burning building. One is an infant who is obviously dead. The other two children are badly burned and are screaming. As you begin caring for the children, you overhear that the children were left alone in the home for two days, and the cause seems to have resulted from when the eldest child, who is 8 years old, tried to cook food on the stove for herself and her two brothers.

How does stress play a role in this scenario? How can you manage stress related to calls involving children?

Issues: Calls Involving Child Abuse, Neglect, or Abandonment; Understanding Personal Emotions When Caring for an Ill Child.

Assessment in Action

You are dispatched to 1212 Lionel Ln for an unknown medical problem. En route to the scene, the dispatcher informs you that the 9-1-1 call was placed by the mother of a small child. The mother was so hysterical that the dispatcher could do little more than confirm her location before she hung up the phone. As a precaution, law enforcement has also been dispatched to the scene.

As you arrive, a young woman runs towards the ambulance with what appears to be a bundle of blankets. As you step out of the rig, she shoves the pile of blankets into your arms. You can now see an approximately 1-year-old child cocooned in the layers of blankets. As you take the child to the back of the ambulance, the mother tells you that her baby had been running a fever all day and then she suddenly began to shake all over. The mother also tells you her child has never had any significant medical problems.

1. What is the likely cause of the baby's shaking?
   A. Hypoventilation
   B. Dehydration
   C. Pneumonia
   D. Febrile seizure

2. In this scenario, the presence of a rash could indicate:
   A. an allergic reaction.
   B. a drug ingestion.
   C. dehydration.
   D. meningitis.

3. Children's heads, in proportion to their bodies, are:
   A. smaller than an adult's.
   B. larger than an adult's.
   C. the same as an adult's.
   D. more oblong than an adult's.

4. Compared to an adult's airway, an infant's trachea is:
   A. smaller in diameter and more flexible.
   B. larger in diameter and more flexible.
   C. smaller in diameter and less flexible.
   D. larger in diameter and less flexible.

5. Normal respiratory rates for infants are:
   A. 12 to 20 breaths/min.
   B. 24 to 40 breaths/min.
   C. 30 to 60 breaths/min.
   D. 40 to 60 breaths/min.

6. The normal heart rate range for infants is:
   A. 60 to 100 beats/min.
   B. 70 to 120 beats/min.
   C. 90 to 150 beats/min.
   D. 100 to 160 beats/min.

Challenging Questions

7. Why are children able to compensate for blood loss so effectively?

8. How should you handle parents during your care of their sick or injured children?

9. Why can calls dealing with children be so difficult emotionally for responders?

10. Should you initiate CPR on an infant who shows clear signs of death, such as rigor mortis and dependent lividity?

Pediatric Assessment and Management

Objectives

Cognitive

6-1.4   Indicate various causes of respiratory emergencies. (p 965)

6-1.5   Differentiate between respiratory distress and respiratory failure. (p 975)

6-1.6   List the steps in the management of foreign body airway obstruction. (p 959)

6-1.7   Summarize emergency medical care strategies for respiratory distress and respiratory failure. (p 976)

6-1.8   Identify the signs and symptoms of shock (hypoperfusion) in the infant and child patient. (p 977)

6-1.9   Describe the methods of determining end organ perfusion in the infant and child patient. (p 946)

6-1.10   State the usual cause of cardiac arrest in infants and children versus adults. (p 970)

6-1.12   Describe the management of seizures in the infant and child patient. (p 978)

6-1.14   Discuss the field management of the infant and child trauma patient. (p 971)

Affective

None

Psychomotor

6-1.21   Demonstrate the techniques of foreign body airway obstruction removal in the infant. (p 962)

6-1.22   Demonstrate the techniques of foreign body airway obstruction removal in the child. (p 959)

6-1.23   Demonstrate the assessment of the infant and child. (p 943)

6-1.24   Demonstrate bag-valve-mask artificial ventilations for the infant. (p 956)

6-1.25   Demonstrate bag-valve-mask artificial ventilations for the child. (p 956)

6-1.26   Demonstrate oxygen delivery for the infant and child. (p 954)

Additional Objectives*

Cognitive

1.   Describe the steps in positioning an infant and/or child to maintain an open airway. (p 950)

2.   Summarize neonatal resuscitation procedures. (p 963)

Affective

None

Psychomotor

3.   Demonstrate the techniques necessary in neonatal resuscitation. (p 963)

*These are noncurriculum objectives.

You are the Provider

You and your EMT-B partner are sitting in Casa Molinas having lunch when you hear someone coughing behind you. You turn around to see a young girl standing next to her chair with her hand over her mouth, coughing and turning red in the face. Her mother is yelling, "Jamie, are you ok? Are you ok?" as she pats her briskly on the back.

1. What should you do?
2. Can you provide treatment to this child?

Pediatric Assessment and Management

There are many causes of emergencies in infants and children. Although you might not be able to identify the exact cause, you must be able to intervene appropriately. You will face some special challenges in caring for sick and injured children. First and foremost is the ability to assess the needs of infants and children. Other challenges will include managing the child's airway, ventilations, and the care of injuries.

This chapter examines the importance of assessment and setting priorities when dealing with children. Discussion of procedures for opening and maintaining the airway in infants and children follows, including placement of airway adjuncts and use of oxygen delivery devices, including the bag-valve-mask (BVM) device. The causes and management of airway obstruction from foreign objects are covered next. After the management of children with trauma, seizures, altered level of consciousness, poisoning, meningitis, shock, and dehydration are outlined, a brief review of neonatal resuscitation is provided.

EMT-Bs who are calm when caring for adults often find themselves anxious when dealing with critically ill or injured infants or children. However, treatment of children is the same as that of adults in most emergency situations. Once you understand the differences in anatomy between children and adults and learn to recognize signs of respiratory distress in children, you will find it easier to approach even the youngest patients in a relaxed, professional manner.

Technology

- Interactivities
- Vocabulary Explorer
- Anatomy Review
- Web Links
- Online Review Manual

www.EMTB.com

Because a young child might not be able to speak, your assessment of his or her condition must be based in large part on what you can see and hear yourself. In addition, families may be helpful in providing vital information about an accident or illness. You should include families as part of the caregiving team and, whenever possible, include them in all decisions about care and transportation.

Scene Size-up

As with any EMS call, the scene size-up begins by ensuring that you and your partner have taken the appropriate BSI precautions. As soon as you arrive at the scene, look for any hazards or potential threats to you or your partner. Resist the temptation to hastily access the patient because you know it is a child. Personal safety must always remain your priority.

As you enter the scene, note the position in which the child is found. Observe the area for clues to the mechanism of injury (MOI) or nature of illness (NOI); these observations will help guide your assessment and management priorities.

Note the presence of any pills, medicine bottles, or household chemicals that would suggest possible ingestion by the child. If the child has been injured—a motor vehicle crash, fall, or pedestrian incident—carefully observe the scene or vehicle (if involved) for clues to the potential severity of the child's injuries.

You must not discount the possibility of child abuse. Conflicting information from the parents or caregivers, bruises or other injuries that are not consistent with the MOI described, or injuries that are not consistent with the child's age and developmental abilities should increase your index of suspicion for abuse.

Initial Assessment

Many components of the pediatric initial assessment can be accomplished by simple observation when you first enter the scene or room. As with the adult, the objective of the initial assessment is to identify and treat immediate or potential threats to life.

General Impression

The initial assessment begins as you form a general impression of the child's condition and of the environment in which he or she is found. Determining a chief complaint, often expressed as what the parent is most concerned about, may help to focus your attention toward potential life-threatening problems. Also note the degree of interaction between the parent or caregiver and the child; ask the parent or caregiver if the child is acting normally. Determine whether the child recognizes the parent or caregiver; failure to do so is an ominous sign and indicates a very sick child.

Pediatric Assessment Triangle

The pediatric assessment triangle (PAT) is a structured assessment tool that allows you to rapidly form a general impression of the infant's or child's condition without touching him or her. The intent is to provide a "first glance" assessment to identify the general category of the child's physiologic problem and to establish urgency for treatment and/or transport. The PAT is a visual assessment of the child before performing a hands-on assessment.

The PAT (Figure 32-1 ▼) consists of three elements: appearance (muscle tone and mental status), work of breathing, and circulation to the skin. The only equipment required for the PAT are your own eyes and ears; no stethoscope, blood pressure cuff, cardiac monitor, or pulse oximeter is required.

Appearance

Evaluating the child's appearance involves noting the level of consciousness or interactiveness and muscle tone—signs that will provide you with information about the adequacy of the child's cerebral perfusion and overall function of the central nervous system.

Much of the information regarding the child's level of consciousness can be obtained by using the PAT. In addition, you can evaluate the child's level of consciousness by using the AVPU scale, modified as necessary for the child's age (Table 32-1 ▶).

TABLE 32-1   The AVPU Scale
Alert: Normal interactiveness for age
Verbal
◾ Appropriate: Responds to name
◾ Inappropriate: Nonspecific or confused
Painful
◾ Appropriate: Withdraws from pain
◾ Inappropriate: Sound or movement without purpose or localization of pain
Unresponsive: No response to any stimulus

An infant or child with a normal level of consciousness will act appropriately for his or her age, exhibiting good muscle tone and maintaining good eye contact (Figure 32-2 ▼). An abnormal level of consciousness is characterized by age-inappropriate behavior or interactiveness, poor muscle tone, or poor eye contact with the caregiver or EMT-B (Figure 32-3 ▼).

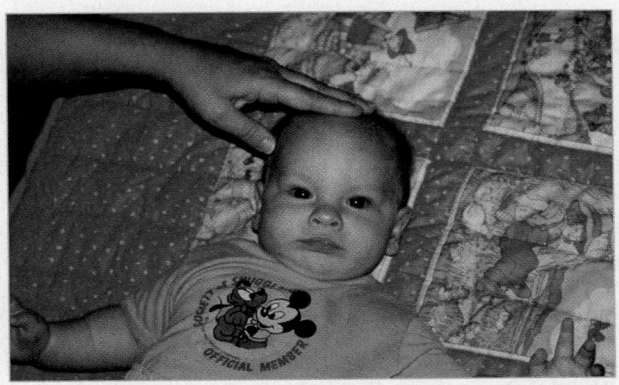

Figure 32-2 An infant or child making good eye contact is not very sick.

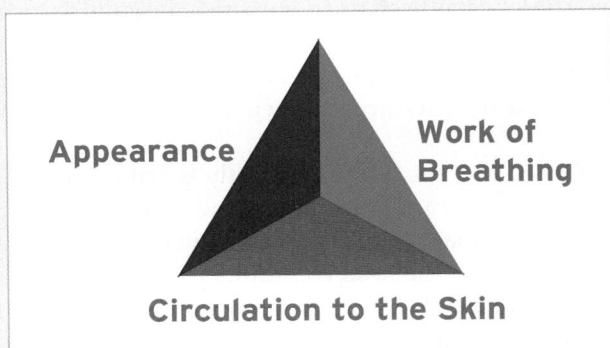

Figure 32-1 The three components of the pediatric assessment triangle (PAT) include appearance, work of breathing, and circulation to the skin.

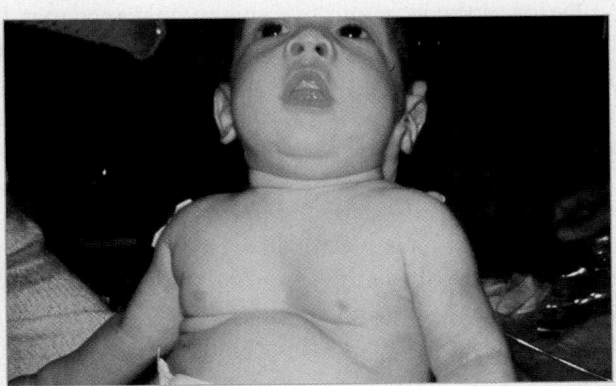

Figure 32-3 A limp child who is unable to maintain eye contact may be critically ill or injured.

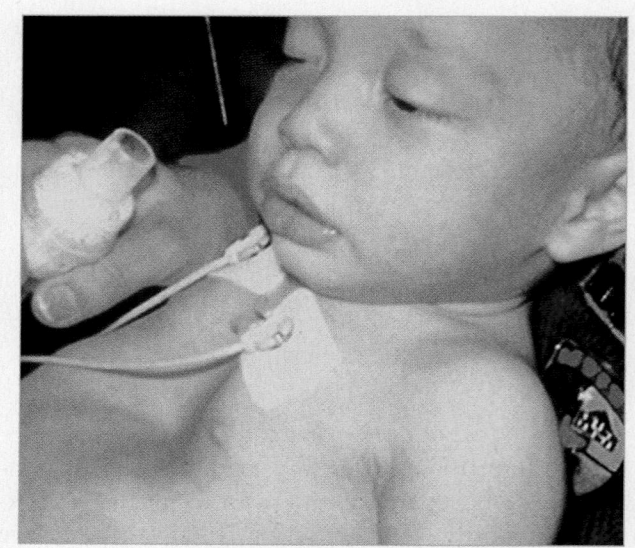

Figure 32-4 Retractions of the intercostal muscles or sternum indicate increased work of breathing.

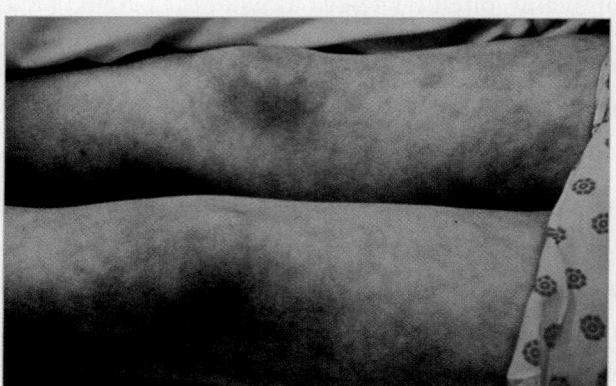

Figure 32-5 Mottling of the skin indicates poor perfusion and is the result of constriction of peripheral blood vessels.

Work of Breathing

A child's work of breathing increases as the body attempts to compensate for abnormalities in oxygenation and ventilation. Increased work of breathing often manifests as <u>tachypnea</u>, <u>retractions</u> of the intercostal muscles or sternum (Figure 32-4 ▲), or the way the child positions himself or herself.

Circulation to the Skin

An important sign of perfusion is circulation to the skin. When cardiac output falls, the body, through vasoconstriction, shunts blood from areas of lesser need (such as the skin) to areas of greater need (such as the brain, heart, and kidneys).

Pallor of the skin and mucous membranes may be seen in compensated shock; it may also be a sign of <u>anemia</u> or hypoxia. Mottling is caused by constriction of peripheral blood vessels and is another sign of poor perfusion (Figure 32-5 ▶).

<u>Cyanosis</u>, a blue discoloration of the skin and mucous membranes, reflects a decreased level of oxygen in the blood. Cyanosis is a late sign of respiratory failure or shock; absence of discoloration, however, does not rule out these conditions. Never wait for the development of cyanosis before administering oxygen!

Airway, Breathing, and Circulation

After forming your general impression of the child's condition using the PAT, perform a hands-on assessment of the child's vital functions—airway, breathing, and circulation—and treat any immediate or potential threats to life. As previously discussed, although your assessment of the child may require some modification based on patient age, the overall assessment flow is essentially the same as for adults.

Airway Assessment

If the infant or child's airway is open and the patient can adequately keep it open (as is often the case in conscious patients), you can proceed with assessment of respiratory adequacy. However, if the child is unresponsive or has difficulty keeping the airway clear, you must ensure that the airway is properly positioned and that it is clear of mucus, vomitus, blood, and foreign bodies.

If trauma has been ruled out, open the child's airway with the head tilt–chin lift maneuver (Figure 32-6 ▶). If the child has been involved in trauma or trauma is suspected, use the jaw-thrust maneuver to open the airway (Figure 32-7 ▶).

Positioning the airway correctly is critical in pediatric emergency care. Position the airway in a <u>sniffing position</u>, which may require the placement of a folded sheet or towel behind the head or shoulders (Figure 32-8 ▶). When the head is bent back (hyperextended) or forward (flexed), the airway may become obstructed because of kinking of the trachea.

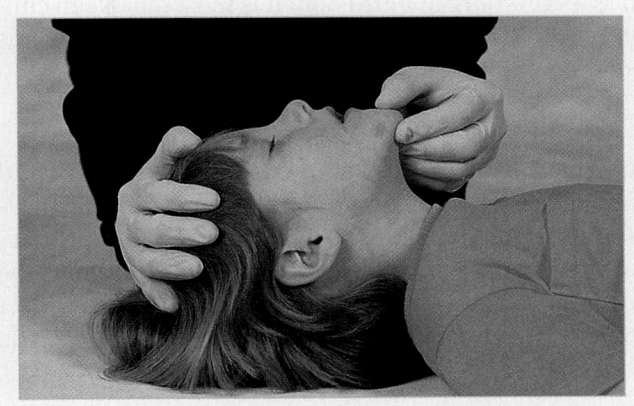

Figure 32-6 Use the head tilt-chin lift maneuver to open the airway of a child without trauma.

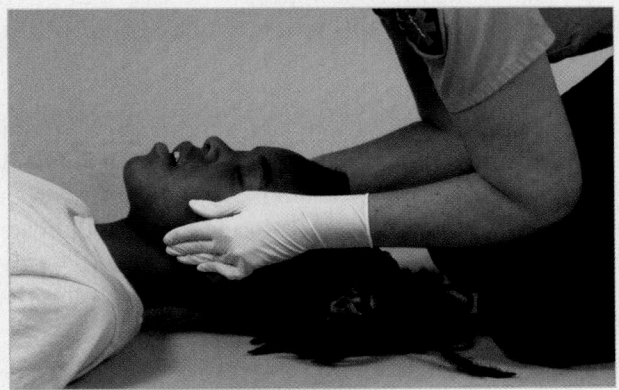

Figure 32-7 Use the jaw-thrust maneuver in a child with possible spinal injury.

After the child's airway has been opened, make sure that it is clear of potential obstructions such as mucus, blood, or foreign bodies. Next, establish whether the child can maintain his or her own airway spontaneously (without the use of airway adjuncts) or whether adjuncts will be necessary to maintain airway patency. Techniques of airway management will be discussed later in this chapter.

Breathing Assessment

Assess the child's breathing by using the look, listen, and feel technique, noting the degree of air movement at the nose and mouth and determining whether the chest is rising adequately. In infants, belly breathing is considered adequate due to the soft pliable bones of the chest and the strong muscular diaphragm.

If the child is conscious and not in need of immediate intervention (such as suctioning or assisted ventilation), assessing respirations is usually easier with the child sitting on the caregiver's lap. Listen for abnormal respiratory sounds (Table 32-2 ▶), and note any signs of increased respiratory effort.

When observing the child's respiratory effort, note any signs of increased work of breathing, including:

- **Accessory muscle use:** Contractions of the muscles above the clavicles (supraclavicular)
- **Retractions:** Drawing in of the muscles between the ribs (intercostal retractions) or of the sternum during inspiration
- Head bobbing: The head lifts and tilts back during inspiration, then moves forward during expiration

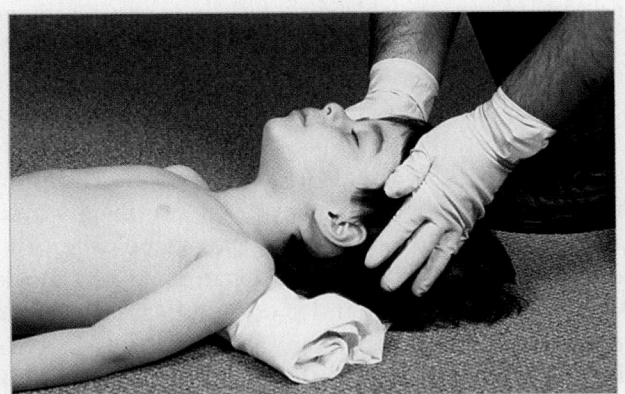

Figure 32-8 The airway should be placed in a neutral position to keep the trachea from kinking when the head is flexed or hyperextended.

TABLE 32-2 Abnormal Respiratory Sounds

- **Stridor:** High-pitched inspiratory sound; indicates a partial upper airway obstruction (such as in croup or from a foreign body)
- **Wheezing:** High- or low-pitched sound heard usually during expiration; indicates a partial lower airway obstruction (such as in asthma or bronchiolitis)
- Grunting: An "uh" sound heard during exhalation; reflects the child's attempt to keep the alveoli open; indicates inadequate oxygenation
- **Absent breath sounds (despite increased work of breathing):** Indicates a complete upper or lower airway obstruction (such as foreign body, severe asthma, or pneumothorax)

- **Nasal flaring**: The nares (the external openings of the nose) widen; usually seen during inspiration
- **Tachypnea**: Increased respiratory rate

As the child begins to tire, retractions often become weak and ineffective and the accessory muscles become less prominent during breathing. Bradypnea, a decrease in the respiratory rate, is an ominous sign and indicates impending respiratory arrest. Do not mistake bradypnea for a sign of improvement; it usually indicates that the child's condition has deteriorated. Therefore, you must be prepared to begin ventilatory assistance.

Circulatory Assessment

When assessing circulation, you must determine if the child has a pulse, is bleeding, or is in shock. Remember, infants and children can tolerate only small amounts of blood loss before circulatory compromise occurs. Assess and control any active bleeding early in your assessment.

Pulses may be difficult to palpate if they are weak, very fast, or very slow. In infants, palpate the brachial pulse or femoral pulse. In children older than 1 year, palpate the carotid pulse (Figure 32-9 ▶). Note the rate and quality of the pulse: Is it weak or strong? Is it normal, slow, or fast? Strong central pulses usually indicate that the child is not hypotensive; however, this does not rule out the possibility of compensated shock. Weak or absent peripheral pulses indicate decreased perfusion. The absence of a central pulse (that is, brachial or femoral in infants, carotid in older children) indicates the need for CPR.

Tachycardia may be an early sign of hypoxia or shock, but it may also reflect less serious conditions such as fever, anxiety, pain, and excitement. Like respiratory rate and effort, heart rate should be interpreted within the context of the overall history, PAT, and entire initial assessment.

A trend of increasing or decreasing heart rate may be quite useful and may suggest worsening hypoxia or shock or improvement after treatment. When hypoxia or shock becomes critical, bradycardia occurs. As with slowing respirations, bradycardia in a child is an ominous sign and often indicates impending cardiopulmonary arrest.

Feel the skin for temperature and moisture at the same time you assess the child's pulse. Is the skin warm and dry, or cold and clammy? Estimate the capillary refill time (CRT) by squeezing the end of a finger or toe for several seconds and then observing the return of blood to the area (Figure 32-10 ▶). Color should return in less than 2 seconds after you let go. The CRT is used to assess end-organ perfusion. It is most reliable in children younger than 6 years; however, factors such as cold temperatures may affect the CRT.

Transport Decision

After you have completed the initial assessment and initiated any treatment, you must make a crucial decision: Is immediate transport to the hospital indicated, or is additional assessment and treatment required at the scene? If the child is in stable condition, you may elect to perform a focused history and physical exam at the scene.

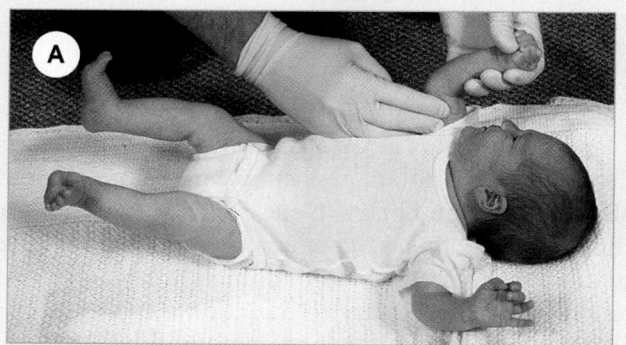
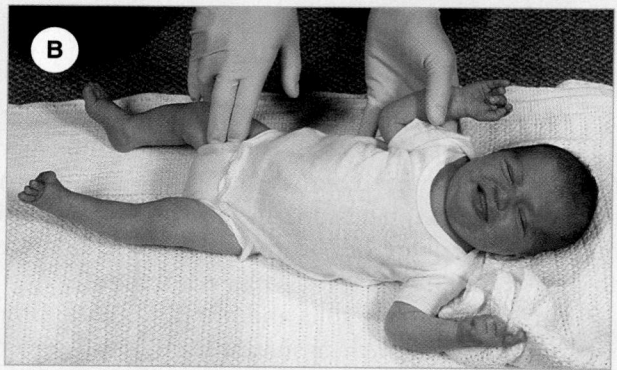
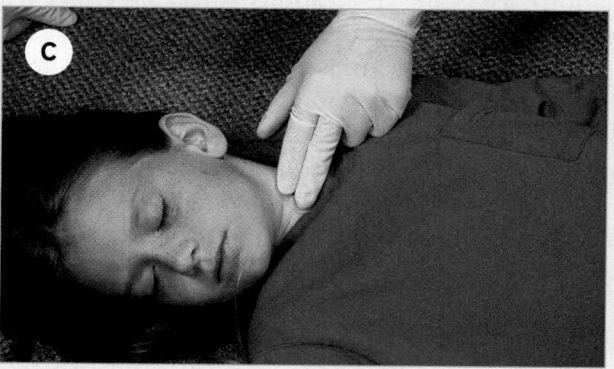

Figure 32-9 A. Palpate the brachial pulse in infants. **B.** Palpate the femoral pulse as a second choice. **C.** In children older than 1 year, palpate the carotid pulse.

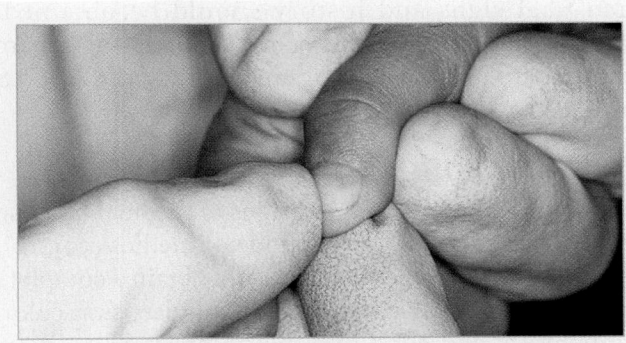
Figure 32-10 Estimate the capillary refill time by squeezing the end of a finger or toe for several seconds until the nailbed blanches. Normal color should return within 2 seconds after you let go.

However, immediate transport is indicated if the scene is unsafe for the child or if any of the following conditions exist:

- A significant MOI—same MOIs as adults (see Chapter 21), with the addition of:
  - Fall from more than 10′ or three times the child's height
  - Bicycle crash
- A history compatible with a serious illness
- A physiologic abnormality noted during the initial assessment
- A potentially serious anatomic abnormality
- Significant pain

In addition to the preceding factors, the EMT-B should also consider the following when making a transport decision:

- The type of clinical problem (injury versus illness)
- The expected benefits of ALS treatment in the field
- Local EMS system treatment and transport protocols
- The EMT-B's comfort level
- Transport time to the hospital

If the child's condition is urgent, perform a rapid assessment, if applicable, and initiate immediate transport. Additional assessment and treatment should occur en route to the hospital.

If the child's condition is nonurgent, perform a focused history and physical exam at the scene, provide additional treatment as needed, and then transport.

Transportation

Children weighing less than 40 lb should be transported in a car seat as long as the situation allows. Many types of seats are available. A seat should be chosen to fit the appropriate weight of the child and should meet the current applicable standards set by your governing agency. There are only a few locations to place a car seat in an ambulance. Seats are designed to be either forward-facing or rear-facing; they cannot be mounted sideways on a bench seat. Seats should not be mounted in the front of an ambulance, especially if the ambulance is equipped with airbags. To mount a car seat to the stretcher, place the head of the cot in an upright position. Place the seat so it is against the back of the cot. Secure one of the cot straps from the upper portion of the cot through the seat belt positions on the seat and strap it tightly to the cot. Repeat on the lower portion of the cot. Push the seat into the cot tightly and retighten the straps.

To secure a seat to the captain's chair, follow the seat manufacturer's instructions. Remember that children younger than 1 year must be transported in a rear-facing position due to the lack of mature neck muscles.

In some situations, it is not appropriate to secure a child in a car seat, for example if the child has to be immobilized on a long board or requires splinting that does not fit in the seat. If the child is unstable and requires airway or ventilatory support, he or she should be positioned to maximize the airway and ventilatory requirements. Children in cardiopulmonary arrest should likewise not be placed in a car seat.

Transition Phase

If the child's condition does not require immediate transport, the transition phase can allow the infant or child to become familiar with you and your equipment. This will help to alleviate the child's anxiety, allowing you to perform a more thorough and accurate assessment.

Remember that sick or injured children are afraid and do not understand why you are there and what you are doing. As a result, they are less likely than an adult to trust you. The transition phase will facilitate the trust-building process between you and the child.

Focused History and Physical Exam

A focused history and physical exam of a child should be performed at the scene, unless his or her condition dictates immediate transport. The purpose of the focused history and physical exam is to obtain additional, specific information about the child's illness or injury. This portion of your assessment includes performing a physical exam (either rapid or focused), obtaining vital signs, and interviewing the patient or guardian about the patient's medical history. The order of these three portions of the assessment will vary according to whether the child is a medical patient (responsive or unresponsive) or a trauma patient (with significant MOI or non-significant MOI). Refer to Chapter 8 to review the appropriate order of assessment steps.

Focused Physical Exam

The focused physical exam should be performed on all children without life-threatening illnesses or injuries who do not require a rapid assessment (for example, responsive children where obtaining a medical history will guide you in your physical exam or trauma patients with a non-significant MOI). Focus your assessment on the area(s) of the body affected by the illness or injury.

Young children should be assessed starting at the feet and ending at the head; older children can be assessed using the head-to-toe approach, as with adults. The extent of the physical exam will depend on the situation and may include the following:

- Pupils
  - Note the size, equality, and reactivity of the pupils to light
- Capillary refill (in children younger than 6 years)
  - Normal CRT should be less than 2 seconds
  - As discussed earlier, assess CRT by blanching the finger or toenail beds; the soles of the feet may also be used
  - Cold temperatures will increase CRT, making it a less reliable sign
- Level of hydration
  - Assess skin turgor, noting the presence of tenting
  - In infants, note whether the fontanels are sunken or flat
  - Ask the parent or caregiver how many diapers the infant has soiled over the last 24 hours
  - Determine whether the child is producing tears when crying; note the condition of the mouth. Is the oral mucosa moist or dry?

Rapid Physical Exam

A rapid physical exam should be used when pediatric patients have potentially life-threatening or hidden injuries, for example, unresponsive medical patients or trauma patients with a significant MOI. This rapid head-to-toe (or toe-to-head) exam may help to identify external bleeding, a distended abdomen, or possible fractures. It should be performed quickly, and

then vital signs and history should be obtained. Identifying these problems early can help to prepare your patient for transport or identify the need for ALS providers.

Pediatric Vital Signs

You should take a child's vital signs in the field because you are the eyes, ears, and hands of medical control. During your assessment, you should obtain a complete set of baseline vital signs, including pulse, skin color, temperature and condition, blood pressure, respirations, and pupils. Guidelines used to assess adult circulatory status—heart rate and blood pressure—have important limitations in children. First, normal heart rates vary with age in children. Second, blood pressure is usually not assessed in children younger than 3 years; it offers little information about the child's circulatory status and is usually difficult to obtain. In these patients, assessment of the skin is a better indication of their circulatory status.

It is important to use appropriately sized equipment when assessing a child's vital signs. To obtain an accurate reading of a child's blood pressure, you must use a cuff that covers two thirds of the patient's upper arm. A blood pressure cuff that is too small may give you a falsely high reading, whereas a cuff that is too large may give you a falsely low reading.

Respiratory rates may be difficult to interpret. Rapid respiratory rates may simply reflect high fever, anxiety, pain, or excitement. Normal rates, on the other hand, may occur in a child who has been breathing rapidly with increased work of breathing for some time and is now becoming tired. Count the respirations for at least 30 seconds and then double that number (if counted for 30 seconds). In infants and children younger than 3 years, evaluate respirations by assessing the rise and fall of the abdomen. Assess the pulse rate by counting for at least 1 minute, noting its quality and regularity. Pulse oximetry can also be used to monitor the patient's status (Figure 32-11 ▶).

Note that normal vital signs in pediatric patients vary with the age of the child (Table 32-3 ▶). Remember that your approach to taking vital signs also varies with the age of the child. Be gentle, talk to the child, assess respirations and then pulse, and assess blood pressure last. Warm your stethoscope on your hands or a cloth before placing it on the skin. You may also want to let the child hold the equipment or stethoscope first; this may help to reduce the child's anxiety.

Evaluate pupils in the child using a small pen light. The response of pupils is a good indication of how well the brain is functioning, particularly when trauma

TABLE 32-3 Vital Signs by Age

Age	Respirations (breaths/min)	Pulse (beats/min)	Systolic Blood Pressure (mm Hg)
Newborn: 0 to 1 month	30 to 60	90 to 180	50 to 70
Infant: 1 month to 1 year	25 to 50	100 to 160	70 to 95
Toddler: 1 to 3 years	20 to 30	90 to 150	80 to 100
Preschool age: 3 to 6 years	20 to 25	80 to 140	80 to 100
School age: 6 to 12 years	15 to 20	70 to 120	80 to 110
Adolescent: 12 to 18 years	12 to 16	60 to 100	90 to 110
Older than 18 years	12 to 20	60 to 100	90 to 140

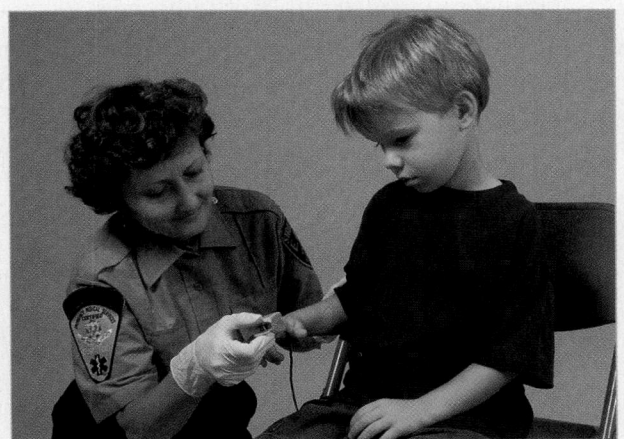

Figure 32-11 Pulse oximetry, which measures the patient's oxygen saturation, can be used to monitor the patient's status.

has occurred. Be sure to compare the size of the pupils against each other.

SAMPLE History

Your approach to the history will depend on the age of the patient. Historical information for an infant, toddler, or preschool-age child will need to be obtained from the parent or caregiver. When dealing with a school-age child or young adolescent, you will usually be able to obtain most of the information from the patient.

Information about sexual activity, the possibility of pregnancy, or the use of illicit drugs or alcohol should be obtained from an older adolescent patient in private. Most of these patients will be reluctant to provide this information in the presence of their parents. When asking such questions, assure the adolescent that this information is important and is needed to provide the most appropriate care.

Questioning of the parent or child about the immediate illness or injury should be based on the child's chief complaint. Together with an evaluation of the child's medical history, this may provide clues to the underlying illness or injury and other conditions that may exist.

When interviewing the parent or older child about the chief complaint, obtain the following pertinent information:

- Nature of the illness or injury
- How long the patient has been sick or injured
- Presence of fever
- Effects of the illness or injury on the child's behavior
- Change in bowel or bladder habits
- Presence of vomiting or diarrhea
- Frequency of urination

When obtaining information about the child's medical history, use SAMPLE to inquire whether the child is currently under the care of a physician, has any chronic illnesses, takes any medications on a regular basis, or has any known drug allergies.

If the caregiver is unable to accompany you to the hospital, get a name and phone number so the staff can call if there are questions. This might be the case when you respond to a daycare facility or babysitter's location. Care may be delayed if this information is not discovered early.

Detailed Physical Exam

Oftentimes pediatric patients will require constant intervention and observation in transit to the hospital. In these situations, or when priority problems require your attention, a detailed exam may not be necessary.

Documentation Tips

Because of the frequency of serious internal injuries in children who show no external signs, it is especially important to investigate and thoroughly document the MOI. Don't let the rush at the scene distract you from determining the mechanism, or at least directing another reliable responder to do so. Hospital care providers need this information.

However, in many situations pediatric patients should have a thorough detailed physical exam, looking over their complete body for signs and symptoms of problems. This is particularly true of patients experiencing trauma from a significant MOI where subtle signs of injury may be present. Use DCAP-BTLS to help remind you what to look for.

Ongoing Assessment

Reassess the child's condition as the situation dictates—every 15 minutes for a child in stable condition and at least every 5 minutes for a child in unstable condition.

The physiologic safeguards in infants and children can decompensate with alarming unpredictability; therefore, continually monitor respiratory effort, skin color and condition, and level of consciousness or interactiveness. Frequently reassess vital signs and temperature. If the child's condition deteriorates, immediately repeat the initial assessment and adjust your treatment accordingly.

The Pediatric Airway

Positioning the Airway

Correct positioning of the airway is critical in pediatric emergency care. Always position the airway in a neutral sniffing position as shown earlier. This accomplishes two goals at once, keeping the trachea from kinking and maintaining the proper alignment should you have to immobilize the spine. If the child has been involved in trauma or trauma is suspected, use the jaw-thrust maneuver to open the airway.

Follow these steps to position the airway in a child or infant (Skill Drill 32-1 ▶):

1. **Place the patient on a firm surface** such as a short backboard or pediatric immobilization device (**Step 1**).
2. **Fold a small towel to a thickness of approximately 1″, and place it under the patient's shoulders and back** (**Step 2**).
3. **Place tape across the child's forehead** to limit rolling of the head during transport (**Step 3**).

Airway Adjuncts

In children with inadequate ventilation, whatever the reason, you should use an airway adjunct to maintain an open airway. Airway adjuncts are devices that help to maintain the airway or assist in providing artificial ventilation, including oral and nasal airways, bite blocks, and BVM devices. Placing the adjuncts correctly starts with choosing the appropriately sized equipment (Table 32-4 ▼).

TABLE 32-4 Pediatric Equipment: Getting the Size Right

The best way to identify the appropriately sized equipment for a pediatric patient is to use the <u>pediatric resuscitation tape measure</u>, which can determine weight as well as height in patients weighing up to 75 lb (34 kg) (Figure 32-12 ▶). The proper sequence for using the tape is the following:

1. Place the patient supine on a flat surface.
2. Lay the tape next to the patient with the multicolored side up.
3. Place the red end of the tape at the top of the patient's head.
4. Place one hand with its side down on top of the patient's head, covering the red box at the end of the tape.
5. Starting from the patient's head, run the side of your free hand down the tape.
6. Stretch the tape out the full length of the child, stopping at the heel. If the child is longer than the tape, stop here and use the appropriate adult technique.
7. Place your free hand, side down, at the bottom of the child's heel.
8. Note the color or letter block and weight range on the edge of the tape where your hand is. Say the color or letter out loud.
9. Select the appropriately sized equipment by matching the color or letter on the tape to the color or letter on the equipment.

Positioning the Airway in a Child

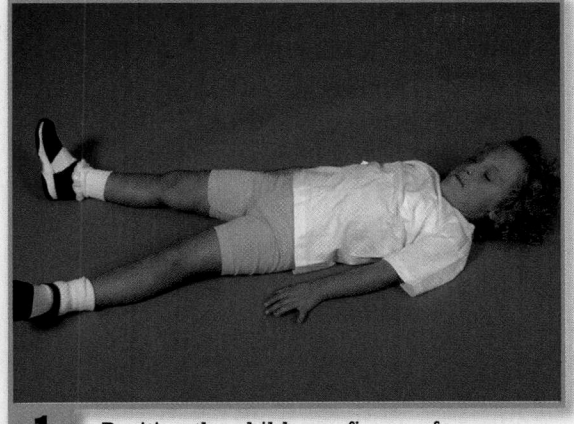

1 Position the child on a firm surface.

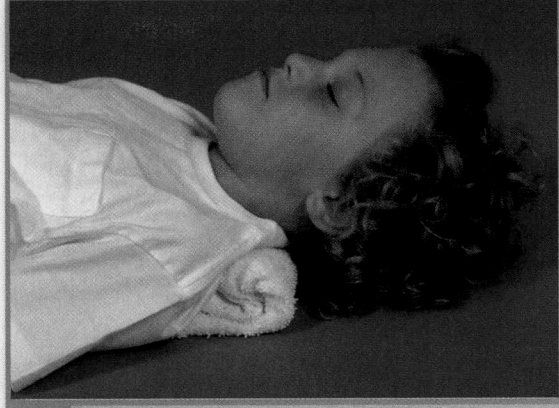

2 Place a folded towel about 1″ thick under the shoulders and back.

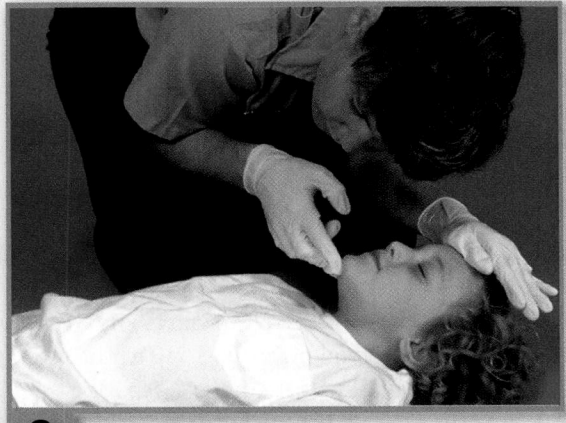

3 Immobilize the forehead to limit movement and use the head tilt–chin lift to open the airway.

Oropharyngeal Airway

An oropharyngeal airway is designed to keep the tongue from blocking the airway, and it makes suctioning the airway, if necessary, easier. An oropharyngeal airway should be used for pediatric patients who are unconscious and in possible respiratory failure. This adjunct should not be used in either conscious patients or those who have a gag reflex. Patients with a gag reflex do not tolerate an oropharyngeal airway. In addition, this adjunct should not be used in children who may have ingested a caustic or petroleum-based product, as it may induce vomiting.

Skill Drill 32-2 ▶ shows the steps for inserting an oropharyngeal airway in a child:

1. **Determine the appropriately sized airway** by measuring from the corner of the patient's mouth to the earlobe, or by using the length-based pediatric resuscitation tape.
2. **Place the airway next to the face** with the flange at the level of the central incisors and the bite block segment parallel to the hard palate. The tip of the airway should reach the angle of the jaw (**Step 1**).

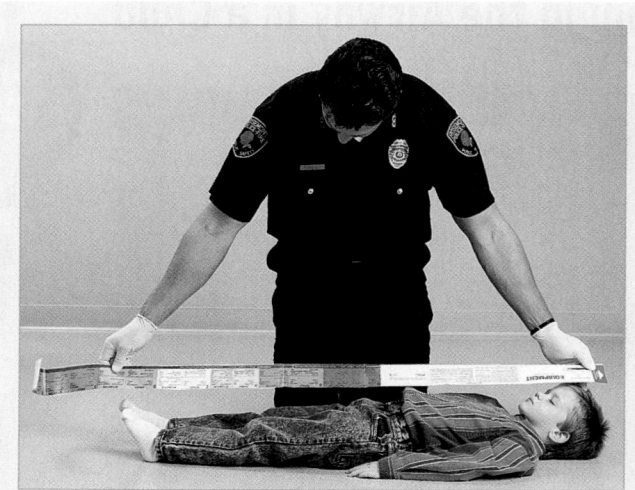

Figure 32-12 Use of a pediatric resuscitation tape measure is one way to identify the correct size for airway adjuncts.

3. **Position the patient's airway.** If the emergency is medical, use the head tilt-chin lift technique, avoiding hyperextension; you may place a towel under the patient's shoulders. If the patient has a traumatic injury, use the jaw-thrust maneuver and provide in-line spinal stabilization (**Step 2**).
4. **Open the mouth** by applying pressure on the chin with your thumb.
5. **Insert the airway** by depressing the tongue with a tongue blade applied to the base of the tongue and inserting the airway directly over the tongue blade. If a tongue blade is not available, point the airway tip toward the roof of the mouth to depress the tongue. Gently rotate the airway into position as it passes through the mouth toward the curve of the tongue. Insert the airway until the flange rests against the lips.
6. **Reassess the airway after insertion** (**Step 3**). Take care to avoid injuring the hard palate as you insert the airway. Rough insertion can cause bleeding, which can aggravate airway problems and may even cause vomiting. Note also that if the patient's airway is too small, the tongue may be pushed back into the pharynx, obstructing the airway. If the airway is too large, it may obstruct the larynx.

Nasopharyngeal Airway

A nasopharyngeal airway is also an airway adjunct. It is usually well tolerated and is not as likely as the oropharyngeal airway to cause vomiting. Unlike the oropharyngeal airway, the nasopharyngeal airway is used for conscious patients or for patients with altered levels of consciousness. In pediatric patients, the nasopharyngeal airway is typically used in association with possible respiratory failure. It is rarely used in infants younger than 1 year.

A nasopharyngeal airway should not be used in patients with nasal obstruction or head trauma (possible basal skull fracture), or in patients with moderate to severe head trauma, as this adjunct could increase intracranial pressure.

Follow the steps in (**Skill Drill 32-3** ▶) to insert a nasopharyngeal airway in a child:

1. **Determine the appropriately sized airway.** The external diameter of the airway should not be larger than the diameter of the nares, and there should be no <u>blanching</u> of the naris after insertion.
2. **Place the airway next to the patient's face** to make sure the length is correct. The airway should extend from the tip of the nose to the tragus of the ear. The <u>tragus</u> is the small cartilaginous projection in front of the opening of the ear.
3. **Position the patient's airway,** using the techniques described above for the oropharyngeal airway (**Step 1**).
4. **Lubricate the airway** with a water-soluble lubricant.
5. **Insert the tip into the right naris** (nostril opening) with the bevel pointing toward the <u>septum</u>, or central divider in the nose (**Step 2**).
6. **Carefully move the tip forward, following the roof of the mouth,** until the flange rests against the outside of the nostril (**Step 3**). If you are inserting the airway on the left side, insert the tip into the left naris upside down, with the bevel pointing toward the septum. Move the airway forward slowly about 1″ until you feel a slight resistance, and then rotate the airway 180°.
7. **Reassess the airway** after insertion.

As with the oropharyngeal airway, there can be problems with the nasopharyngeal airway. An airway with a small diameter may easily become obstructed by mucus, blood, vomitus, or the soft tissues of the pharynx. If the airway is too long, it may stimulate the vagus nerve and slow the heart rate or enter the esophagus, causing gastric distention. Inserting the airway in responsive patients may cause a spasm of the larynx and result in vomiting. Nasopharyngeal airways should not be used when patients have facial trauma, as the airway may tear soft tissues and cause bleeding into the airway.

Inserting an Oropharyngeal Airway in a Child

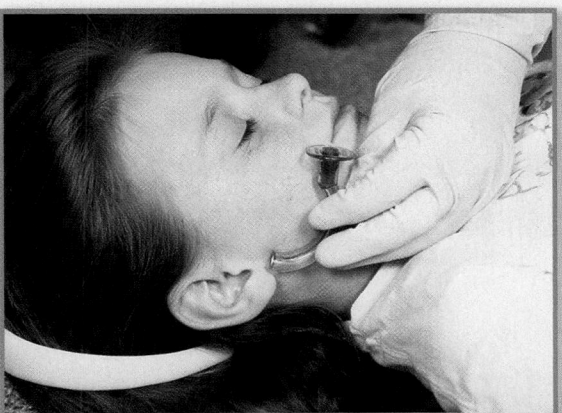

1 Determine the appropriately sized airway. Confirm the correct size visually, by placing it next to the patient's face.

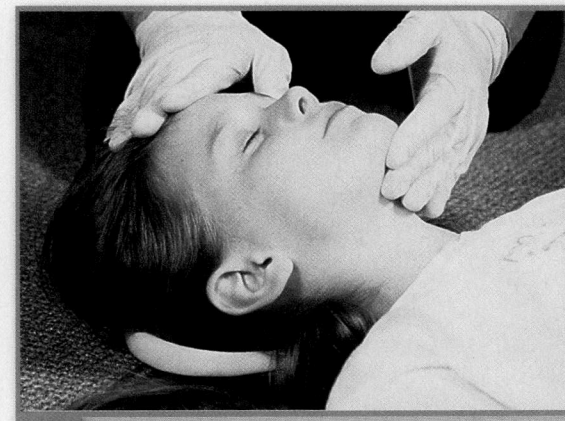

2 Position the patient's airway with the appropriate method.

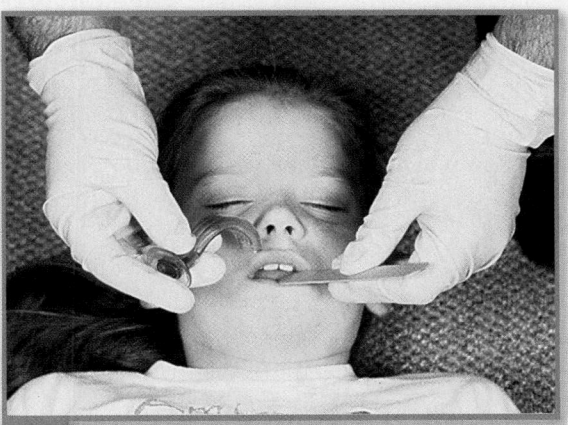

3 Open the mouth. Insert the airway until the flange rests against the lips. Reassess the airway.

You are the Provider   Part 2

Seeing the distressed look in the child's eyes, you quickly get up, introduce yourself, and ask if you can be of assistance. Her mother smacks her on the back again and says, "Please help her! I think she's choking!" Jamie suddenly grasps her throat.

3. What do the child's actions indicate?
4. Are the mother's back blows to the child significant in this situation?

Inserting a Nasopharyngeal Airway in a Child

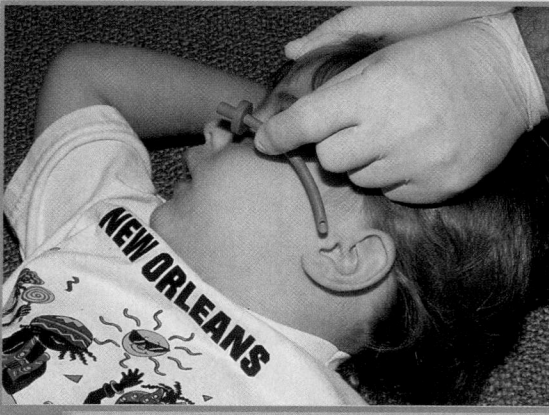

1 Determine the correct airway size by comparing its diameter to the opening of the nostril (naris). Place the airway next to the patient's face to confirm correct length.

Position the airway.

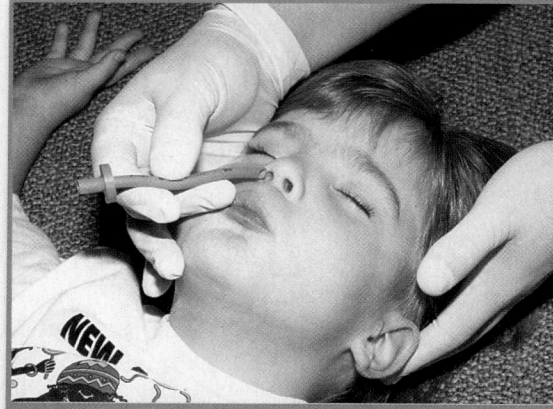

2 Lubricate the airway. Insert the tip into the right naris with the bevel pointing toward the septum.

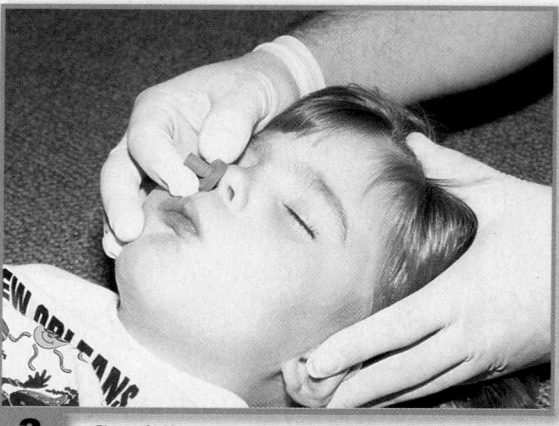

3 Carefully move the tip forward until the flange rests against the outside of the nostril.

Reassess the airway.

Assisting Ventilation and Oxygenation

After opening the airway, you should assess the patient's ventilation status. Look, listen, and feel for breathing. Remember to observe chest rise in older children and abdominal rise in younger children and infants. Skin condition indicates the amount of oxygen getting to the organs of the body. Patients who have pale, mottled, or blue skin may have inadequate levels of oxygen in their blood. All trauma patients should receive oxygen. If the patient has sustained trauma to the face, assisting ventilations may be difficult.

Oxygen Delivery Devices

In treating infants and children who require more than the usual 21% oxygen found in room air, you have several options:

- Nonrebreathing mask at 10 to 12 L/min provides up to 90% oxygen concentration.
- Blow-by technique at 6 L/min provides more than 21% oxygen concentration.
- Nasal cannula at 2 to 6 L/min provides 24% to 44% oxygen concentration.

■ BVM device (with oxygen reservoir) at 10 to 15 L/min provides 90% oxygen concentration. Children need enough air to be delivered for adequate gas exchange in the lungs. Therefore, use of a non-rebreathing mask, a nasal cannula, or a simple face mask is indicated only for patients who have adequate respirations and/or tidal volumes. The tidal volume is the amount of air that is delivered to the lungs and airways in one inhalation. Children with respirations lower than 12 breaths/min or more than 60 breaths/min, an altered level of consciousness, and/or an inadequate tidal volume should receive assisted ventilations with a BVM device.

Blow-by oxygen is not as effective as a face mask or nasal cannula for delivering oxygen. In the blow-by technique, an oxygen tube is held near the infant or child's nose and mouth. It is often used after childbirth to deliver a small amount of oxygen to the newborn. On rare occasions when other adjuncts cannot be used or the child will not tolerate any other adjunct, this technique may be necessary.

Nonrebreathing Mask

A nonrebreathing mask delivers up to 90% oxygen to the patient and allows the patient to exhale all carbon dioxide without rebreathing it (Figure 32-13 ▶). To apply a nonrebreathing mask:

1. **Select the appropriately sized** pediatric nonrebreathing mask. The mask should extend from the bridge of the nose to the cleft of the chin.
2. **Connect the tubing** to an oxygen source set at 10 to 12 L/min.
3. **Adjust oxygen flow** as needed to match the patient's respiratory rate and depth. The reservoir bag should neither deflate completely nor fill to bulging during the respiratory cycle.

Blow-by Technique

As mentioned, the blow-by technique does not provide a high concentration of oxygen but is better than no oxygen. To administer blow-by oxygen:

1. **Place oxygen tubing through a small hole in the bottom of a 6- to 8-oz paper cup** (Figure 32-14 ▶). A cup is a familiar object that is less likely to frighten young children than an oxygen mask. You may be able to use an oxygen mask with an older child if you make it a game. For example, have the child pretend that the mask belongs to a popular action hero or an astronaut.
2. **Connect tubing to an oxygen source** set at 6 L/min.
3. **Hold the cup approximately 1″ to 2″ away** from the child's nose and mouth.

Figure 32-13 A pediatric nonrebreathing mask delivers up to 90% oxygen and allows the patient to exhale carbon dioxide without rebreathing it.

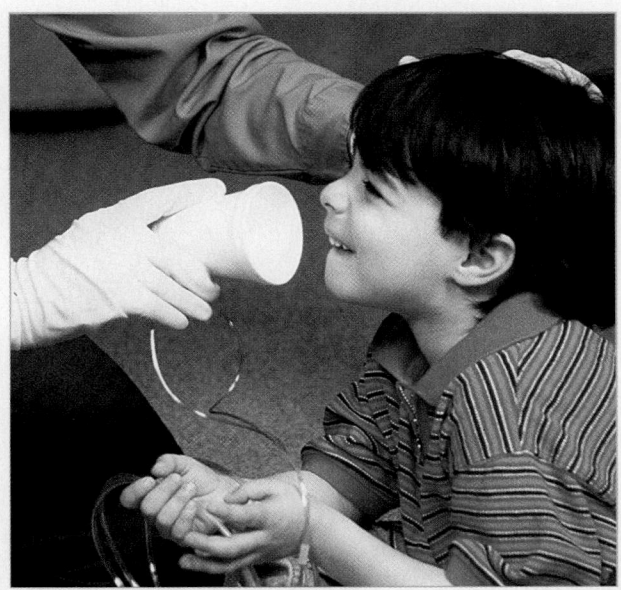

Figure 32-14 Blow-by techniques may be used when oxygen masks frighten children. Make a small hole in a 6- to 8-oz paper cup, or consider using a funnel inserted into the end of the oxygen tubing. Connect tubing to an oxygen source, and hold the cup about 2″ from the child's face.

⚠ Pediatric Needs

One of the problems associated with abdominal injuries in children is the presence of air in the stomach. Children, especially those who have had a traumatic injury, tend to swallow air. Air in the stomach can cause distention and interfere with your assessment. Air can also accumulate in the stomach with artificial ventilation, making it less effective. This is one of the reasons for using the jaw-thrust maneuver to position the airway, as it decreases the amount of air accumulating in the stomach.

Nasal Cannula

Some patients prefer this adjunct while others find it uncomfortable. To apply a nasal cannula:

1. **Choose the appropriately sized pediatric nasal cannula** (Figure 32-15 ▼). The prongs should not fill the nares entirely. If the nares blanch, select a smaller cannula.
2. **Connect the tubing** to an oxygen source set at 2 to 6 L/min.

BVM Device

Assisting ventilations with a BVM device is indicated for patients who have respirations that are either too slow or too fast to provide an adequate volume of inhaled oxygen, who are unresponsive, or who do not respond in a purposeful way to painful stimuli.

Assist ventilation of an infant or child using a BVM device in the following way:

1. **Ensure that you have the appropriate equipment in the right size.** The proper size mask will extend from the bridge of the nose to the cleft of the chin, avoiding compression of the eyes (Figure 32-16 ▼). The mask is transparent, so you can watch for cyanosis and vomiting. In addition, mask volume should be small to decrease dead space and avoid rebreathing; however, the bag should contain at least 450 mL of air. Use an infant bag, not a neonatal bag, for infants younger than 1 year; use a pediatric bag for children older than 1 year. Older children and adolescents may need an adult bag. Make sure that there is no pop-off valve on the bag; if the bag has a pop-off valve, make sure that you can hold it shut as necessary to achieve chest rise.
2. **Maintain a good seal** with the mask on the face.
3. **Ventilate at the appropriate rate and volume** using a slow, gentle squeeze, not a sharp, quick one.

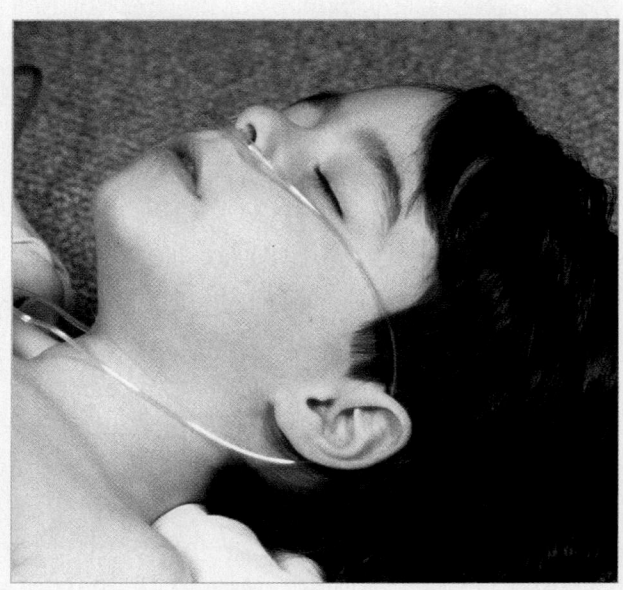

Figure 32-15 The prongs of a pediatric nasal cannula should not fill the nares entirely.

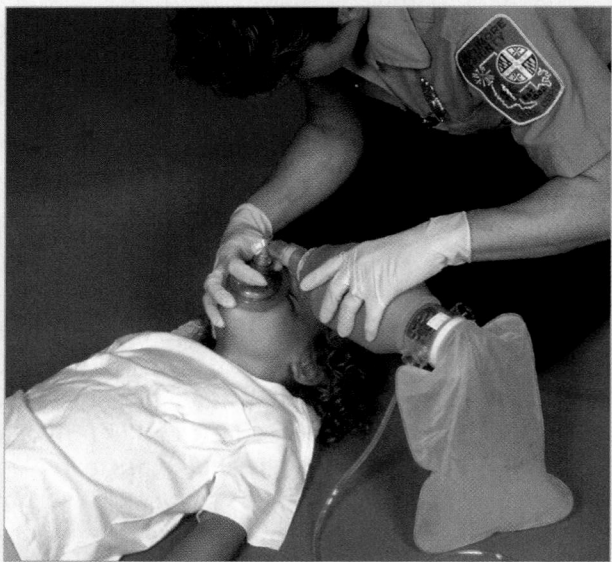

Figure 32-16 Proper mask size for BVM ventilation is critical. The mask should extend from the bridge of the nose to the cleft of the chin, avoiding compression of the eyes.

One-rescuer BVM Ventilation on a Child

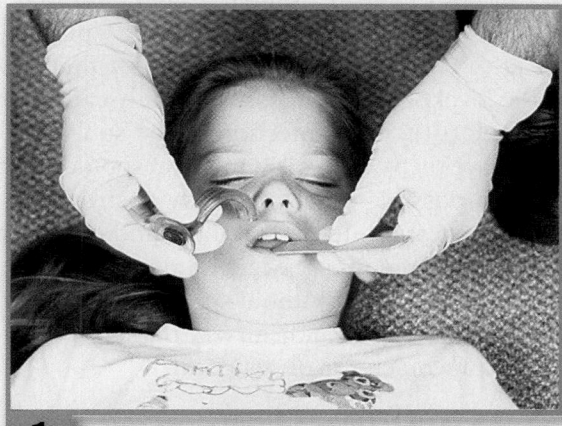

1 Open the airway and insert the appropriate airway adjunct.

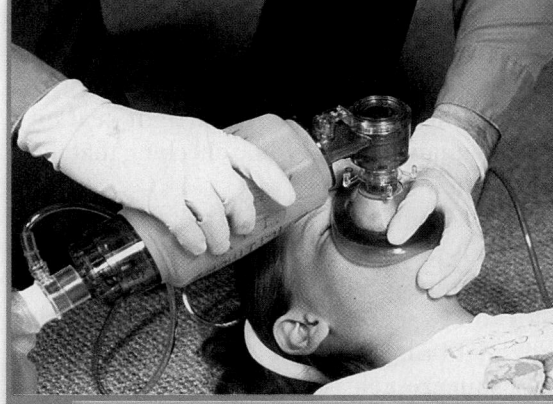

2 Hold the mask on the patient's face with a one-handed head tilt-chin lift technique (E-C grip). Ensure a good mask-face seal while maintaining the airway.

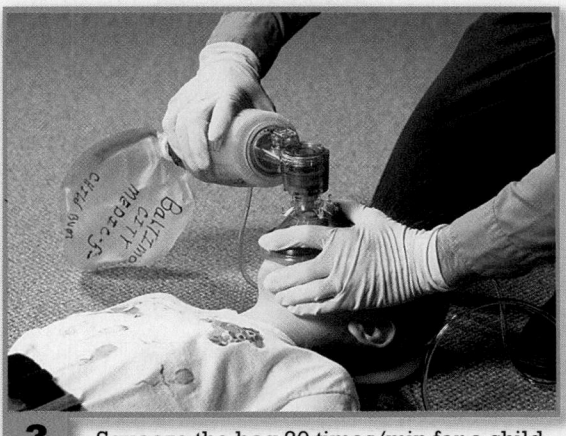

3 Squeeze the bag 20 times/min for a child, or 30 times/min for an infant.

Allow adequate time for exhalation.

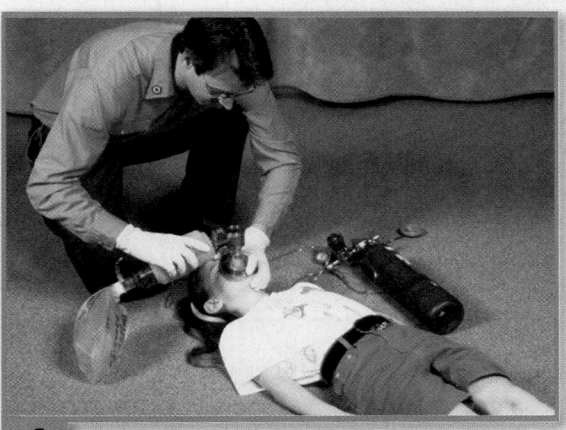

4 Assess effectiveness of ventilation by watching bilateral rise and fall of the chest.

Stop squeezing and begin to release the bag as soon as the chest wall begins to rise, indicating that the lungs are filled to capacity. To keep from ventilating too rapidly, use the phrase "squeeze, release, release." Say "squeeze" as you squeeze the bag; when you see the chest start to rise, release pressure on the bag and slowly say "release, release."

Errors in technique—providing too much volume with each breath, squeezing the bag too forcefully, or ventilating at too fast a rate—can result in gastric distention. An inadequate mask seal or improper head position can lead to hypoventilation or hypoxia. Even with the

best technique in the world, the patient may regurgitate and aspirate the contents of his or her stomach.

One-rescuer BVM Ventilation

Perform one-rescuer BVM ventilation according to these steps (Skill Drill 32-4 ▲):

1. **Open the airway**, and insert the appropriate airway adjunct (**Step 1**).
2. **Hold the mask on the patient's face** by using the E-C grip. Form a C with the thumb and index finger along the mask while the other three fingers

form an E along the mandible. With infants and toddlers, support the jaw with only your third finger. Be careful not to compress the area under the chin, as you may push the tongue into the back of the mouth and block the airway. Keep fingers on the mandible.

3. **Make sure the mask forms an airtight seal** on the face. Maintain the seal while checking that the airway is open (**Step 2**).

4. **Squeeze the bag**, using the correct ventilation rate. This is 20 breaths/min for children and 30 breaths/min for infants.

5. **Allow 1 to 1½ seconds per ventilation**, providing adequate time for exhalation by using the phrase "squeeze, release, release" (**Step 3**).

6. **Assess effectiveness** of ventilation by watching for adequate bilateral rise and fall of the chest (**Step 4**).

Two-rescuer BVM Ventilation

This procedure is similar to one-rescuer ventilation except that it requires two rescuers—one to hold the mask to the patient's face and maintain the patient's head position, the other to ventilate the patient. This technique is usually more effective in maintaining a tight seal.

Airway Obstruction

Children, especially those younger than 5 years, can (and do) obstruct their airway with any object that they can fit into their mouth: hot dogs, balloons, grapes, or coins (Figure 32-17 ▼). In cases of trauma, a child's teeth may have been dislodged into the airway. Blood, vomitus, or other secretions can also cause partial or complete obstruction.

Airway obstructions can also be caused by infections, including pneumonia, croup, and epiglottitis (Figure 32-18 ▼). Croup is an infection of the airway below the level of the vocal cords, usually caused by a virus. Epiglottitis is an infection of the soft tissue in the area above the vocal cords. Infection should be considered as a possible cause of airway obstruction if a child has congestion, fever, drooling, and cold symptoms. Such children must be taken immediately to the emergency department. Without special equipment and training, attempts to clear an airway that is blocked by infection can worsen the obstruction.

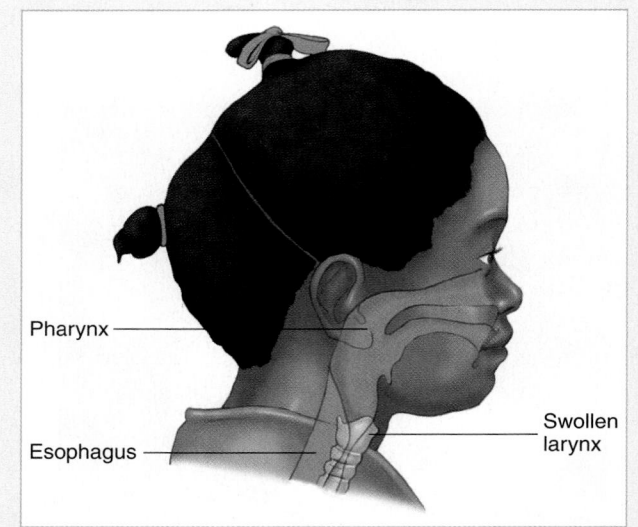

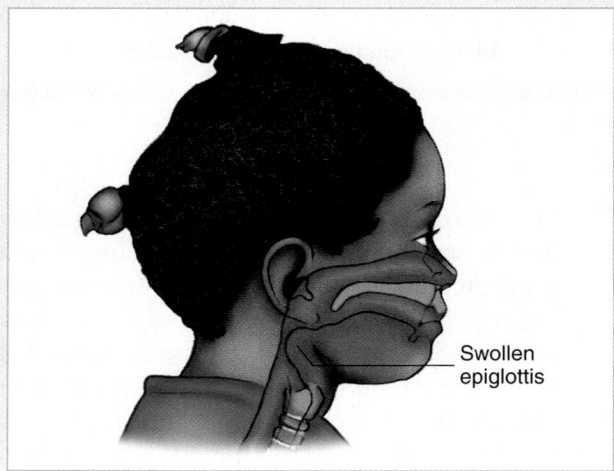

Figure 32-18 Epiglottitis is an infection that can cause airway obstruction in children.

Figure 32-17 Any number of objects can obstruct a child's airway. Some of the more common ones include batteries, coins, toys, buttons, and candy.

Signs and Symptoms

Obstruction by a foreign object may involve the upper or the lower airway. Signs and symptoms that are frequently associated with an upper airway obstruction include decreased or absent breath sounds and stridor. Stridor, a high-pitched noise heard mainly on inspiration, is usually caused by swelling of the area surrounding the vocal cords or upper airway obstruction. In children with croup, it resembles the bark of a seal.

Signs and symptoms of a lower airway obstruction include wheezing, a whistling sound caused by air traveling through narrowed air passages within the bronchioles, and/or crackles. Crackles are caused by the flow of air through liquid, present in the air pouches and smaller airways in the lungs. They produce a crackling sound like that of blowing bubbles through a straw in a glass filled with liquid. The best way to auscultate breath sounds in a child is to listen on both sides of the chest at the level of the armpit Figure 32-19 ▼.

Emergency Medical Care

Treatment of the child with an airway obstruction must begin immediately. If the child is conscious and you know for sure that there is a foreign body in the airway—that is, if someone actually saw the object go into the child's mouth—encourage the child to cough to clear the airway. Abdominal thrusts are also recommended to relieve a complete airway obstruction in a child. If the material in the airway does not completely block the flow of air, the obstruction is not complete and the child may be able to breathe adequately on his or her own without any intervention. If the obstruction is only partial, do not intervene except to provide supplemental oxygen Figure 32-20 ▼. Allow the child to remain in whatever position is most comfortable, and monitor his or her condition.

If you see signs of complete obstruction, however, you must attempt to clear the airway at once. The signs include the following:

- Ineffective cough (no sound)
- Inability to speak or cry
- Increasing respiratory difficulty, with stridor
- Cyanosis
- Loss of consciousness

Management of Airway Obstruction in a Child

If there is reason to believe that an unconscious child has a foreign body obstruction, check the upper airway to see whether the obstructing object is visible. The best way to do this is to grasp the tongue and jaw between your finger and thumb and lift to open the mouth; this is called the tongue-jaw lift Figure 32-21A ▶. If the object is visible, try to remove it using a finger sweep motion Figure 32-21B ▶. Never use finger sweeps in infants or children if you cannot see the object, as you may push it further into the airway.

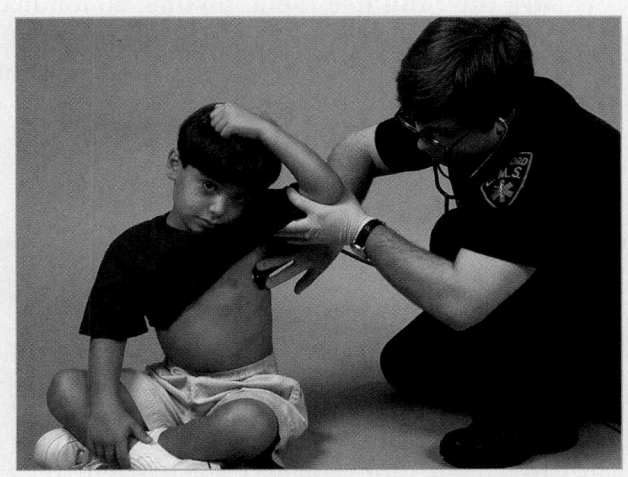

Figure 32-19 The best way to auscultate breath sounds in children is to listen on both sides of the chest at the level of the armpit.

Figure 32-20 If a child has a partial airway obstruction, do not intervene except to give supplemental oxygen and allow the child to remain in whatever position is most comfortable.

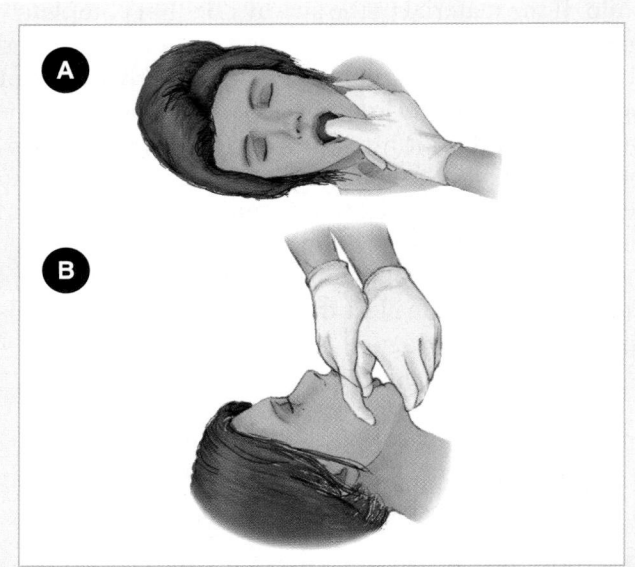

Figure 32-21 A. Use the tongue-jaw lift to open the mouth in an unconscious child. **B.** If the object is visible, try to remove it by using a finger sweep.

3. **Attempt rescue breathing.** If the first try is unsuccessful, reposition the child's head and try again (**Step 3**).
4. **If ventilation is still unsuccessful, kneel beside or straddle the child's hips.** Place the heel of one hand on the front of the child's abdomen just above the navel and well below the rib cage and sternum. Place your other hand on top of the first hand.
5. **Press both hands into the abdomen** in an upward motion, giving five distinct thrusts (**Step 4**).
6. **Open the airway again** using the tongue-jaw lift and visualize the airway.
7. **If you see the foreign body,** remove it. Only attempt to remove objects you can actually see. Blind sweeps may push the object back into the airway.
8. **Attempt rescue breathing.** If the foreign body is not expelled on the first attempt, reposition the child's head and try again.
9. **If the airway remains obstructed,** repeat the abdominal thrusts (**Step 5**).

Abdominal thrusts are recommended to relieve a complete airway obstruction in a child. These thrusts increase the pressure in the chest, creating an artificial cough that may force a foreign body from the airway. (Skill Drill 32-5 ▶) demonstrates the steps for applying abdominal thrusts to an unconscious child who you suspect has a foreign body airway obstruction:

1. **Place the child in a supine position** on a firm, flat surface (**Step 1**).
2. **Inspect the upper airway** using the tongue-jaw lift. If you see the foreign object, try to remove it (**Step 2**).

The following steps are used to remove a foreign body obstruction from a conscious child who is in a standing or sitting position (Figure 32-22 ▶):

1. **Kneel on one knee behind the child,** and circle his or her body with both arms around the patient's chest. Prepare to give abdominal thrusts by placing your fist just above the patient's navel and well below the lower tip of the sternum. Place your other hand over that fist.
2. **Give the child five rapid, distinct abdominal thrusts** in an upward direction. Be careful to avoid applying force to the lower rib cage or sternum.

You are the Provider
Part 3

When you ask the mother to stop patting her on the back, Jamie stops coughing and gets a panicked look on her face. She appears to be gasping for breath with no air movement. You quickly move in to help the little girl. Your partner rushes out to the ambulance to retrieve your jump pack and to radio dispatch to let them know what is happening.

5. What has caused the child to stop coughing?
6. What are your treatment options for this patient?

Removing a Foreign Body Airway Obstruction in an Unconscious Child

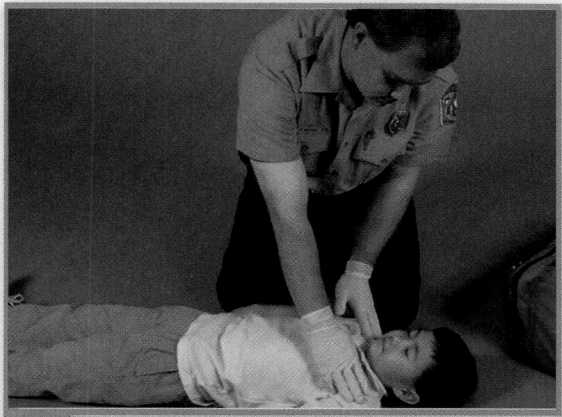

1 Position the child on a firm, flat surface.

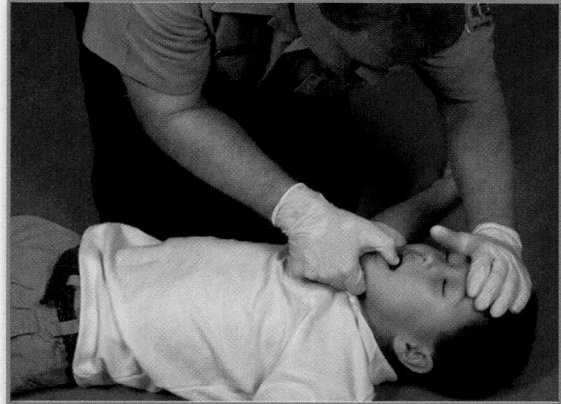

2 Inspect the airway. Remove any foreign object that you can see.

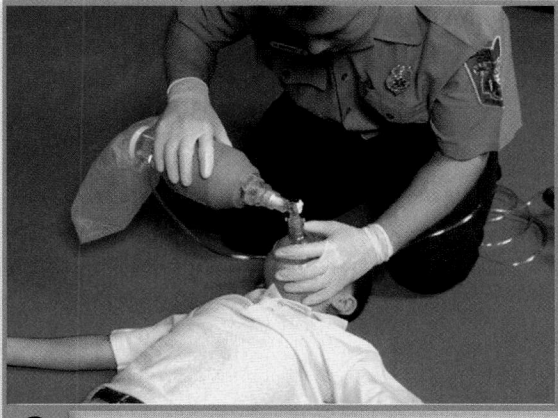

3 Attempt rescue breathing. If unsuccessful, reposition the head and try again.

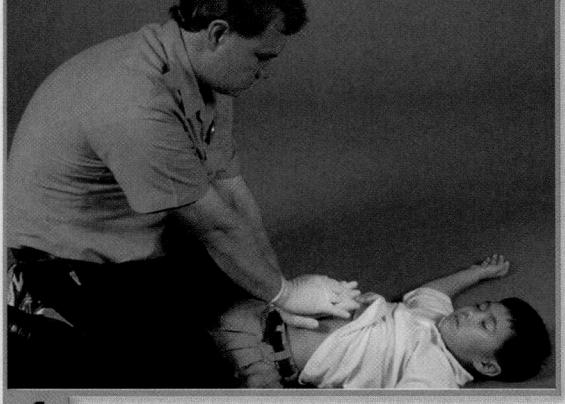

4 If ventilation is unsuccessful, position your hands on the abdomen above the navel and well below the rib cage. Give five abdominal thrusts.

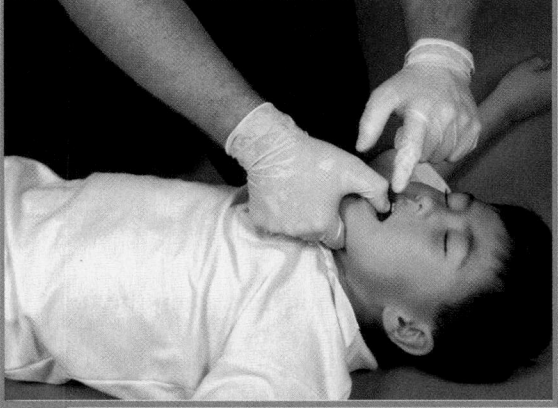

5 Open the airway again and try to see the object. Only try to remove the obstruction if you can see it. Attempt rescue breathing. If unsuccessful, reposition the head and try again. Repeat abdominal thrusts if obstruction persists.

Figure 32-22 Kneel behind the child, wrap your arms around his or her body, and place your fist just above the navel and well below the lower tip of the sternum.

3. **Repeat this standing technique** until the child expels the foreign body or fully loses consciousness.
4. **If the child becomes unconscious**, inspect the airway using the tongue-jaw lift. If you can see the foreign body, try to remove it.
5. **Attempt rescue breathing.** If the first attempt fails, reposition the head and try again.
6. **If the airway remains obstructed**, repeat the abdominal thrusts.

If you manage to clear the airway obstruction in an unconscious child but he or she still has no spontaneous breathing or circulation, perform CPR.

Management of Airway Obstruction in an Infant

Abdominal thrusts are not recommended for infants because of the risk of injury to the immature organs of the abdomen. Instead, use back blows and chest thrusts to try to clear a complete airway obstruction in an infant, as follows (Figure 32-23 ▶):

1. **Hold the infant face down,** with the body resting on your forearm. Support the infant's jaw and

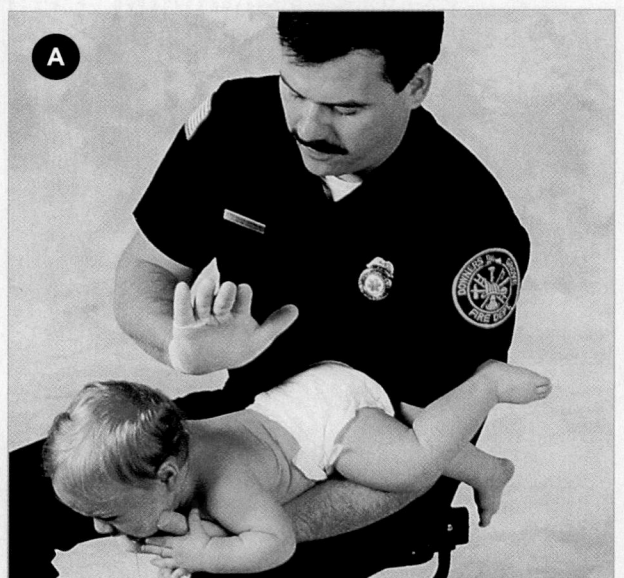

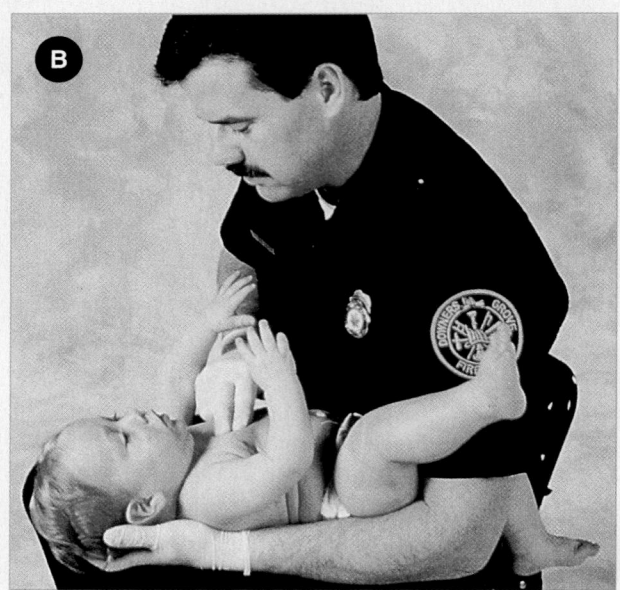

Figure 32-23 A. Hold the infant face down with the body resting on your forearm. Support the jaw and face with your hand, and keep the head lower than the rest of the body. Give the infant five back blows between the shoulder blades, using the heel of your hand. **B.** Give the infant five quick chest thrusts, using two fingers placed on the lower half of the sternum.

face with your hand, and keep the head lower than the rest of the body.
2. **Deliver five back blows** between the shoulder blades, using the heel of your hand.
3. **Place your free hand behind the infant's head and back,** and bring the infant upright on your thigh, sandwiching the infant's body between your two

hands and arms. The infant's head should remain below the level of the body.

4. **Give five quick chest thrusts** in the same location and manner as chest compressions, using two fingers placed on the lower half of the sternum. For larger infants, or if you have small hands, you can perform this step by placing the infant in your lap and turning the infant's whole body as a unit between back blows and chest thrusts.

5. **Check the airway.** If you can see the foreign body now, remove it. If not, repeat the cycle as often as necessary.

6. **If the infant is still unconscious** after removal of the object, check for circulation and perform CPR, if necessary.

If the infant regains consciousness, keep him or her in the recovery position during transport. As you finish the initial assessment, you should have checked the child's level of consciousness, opened the airway, ventilated the child if needed, checked for circulation, and started CPR if required. If you have had to provide any additional treatments to maintain the child's ABCs, the child is a priority patient and transport should be initiated as soon as possible. Otherwise additional assessment is appropriate.

Neonatal Resuscitation

At birth, most infants require resuscitation measures that stimulate the newborn to breathe air and begin circulation of blood through the lungs (Table 32-5 ▶). These measures include positioning of the airway, drying, warming, suctioning, and tactile stimulation. Here are some tips to help you maximize the effects of the measures:

- Position the infant on his or her back with the head down and the neck slightly extended. Place a towel or blanket under the infant's shoulders to help maintain this position.
- Suction the mouth and then nose using a bulb syringe or suction device with an 8- or 10-French catheter. Suction both sides of the back of the mouth, where secretions tend to collect, but avoid deep suctioning of the mouth and throat; this can cause the heart to slow down. Aim blow-by oxygen at the infant's mouth and nose during resuscitation.
- In addition to drying the infant's head, back, and body vigorously with dry towels, you may rub the infant's back and slap the soles of his or her feet.

TABLE 32-5 Rescue Measures for a Newborn Who Is Not Breathing

Assess and support	■ Temperature (warm and dry) ■ Airway (position and suction) ■ Breathing (stimulate to cry) ■ Circulation (heart rate and color)
BLS interventions	■ Dry and warm the infant. ■ Clear the airway with a bulb syringe. ■ Stimulate the infant if he or she is unresponsive. ■ Use a BVM device to ventilate the newborn if needed. This is seldom required. ■ Perform chest compressions if there is no pulse.

In instances when a newborn is in distress, you should be properly equipped for resuscitation measures. All ambulances should have the following equipment and supplies for newborn resuscitation (Figure 32-24 ▼):

- A bulb syringe
- Clean, dry towels
- An infant blanket
- A BVM device with a 450-mL reservoir
- Clear masks in both infant and premature infant sizes
- 2 umbilical clamps
- Sterile 4 × 4 gauze
- A stocking cap
- An oxygen source with tubing

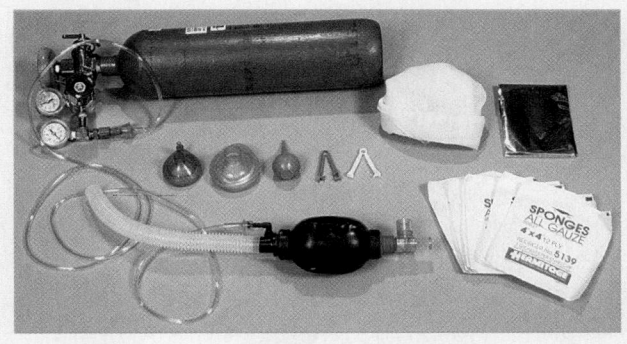

Figure 32-24 The proper equipment for neonatal resuscitation includes a bulb syringe, towels, an infant blanket, a BVM device, clear masks in two sizes, two umbilical clamps, sterile gauze, a stocking cap, and an oxygen source with tubing.

Additional Resuscitation Efforts

Observe the infant for spontaneous respirations, skin color, and movement of the extremities. If the respiratory effort appears appropriate, evaluate the heart rate by palpating the pulse at the base of the umbilical cord or at the brachial artery. The heart rate is the most important measure in determining the need for further resuscitation (Table 32-6 ▼).

If chest compressions are required, give them at a rate of 120 beats/min using either the hand-encircling technique or the two-finger technique (Figure 32-25 ▼). Coordinate chest compressions with ventilations at a ratio of 3:1.

Any newborn who requires more than routine resuscitation requires transport, when possible, to a center with a Level III neonatal intensive care unit. This type of unit is designed for newborns who require specialized care, including mechanical ventilation.

About 12% of deliveries are complicated by the presence of <u>meconium</u>, a dark green material in the amniotic fluid. Meconium can be thick or thin. If the newborn aspirates thick meconium, serious lung disease and sometimes death can occur. Therefore, if you see meconium in the amniotic fluid or meconium staining, you should continue vigorous suctioning of the infant after delivery.

Basic Life Support Review

The reasons for cardiopulmonary arrest differ in children and adults. In adults, cardiac arrest is usually the result of an abnormal cardiac rhythm, which is itself caused by underlying cardiac disease. Because most children have healthy hearts, sudden cardiac arrest is rare. More commonly, children have cardiopulmonary

TABLE 32-6 Additional Neonatal Resuscitation Efforts

If the Heart Rate Is...	More Than 100 Beats/Min	60 to 100 Beats/Min	Fewer Than 60 Beats/Min
Do this:	Keep the newborn warm.	Begin assisted ventilation with a BVM device and 100% oxygen.	Begin assisted ventilation with a BVM device and 100% oxygen.
	Transport the newborn.	Reassess the newborn every 30 seconds until heart rate and respirations are normal.	Reassess the newborn every 30 seconds until heart rate and respirations are normal.
	Assess the newborn continuously.	Continue to reassess the infant. Call for ALS backup. Keep the newborn warm.	Begin chest compressions. Call for ALS backup. If the heart rate does not increase, medication and ALS will be needed.

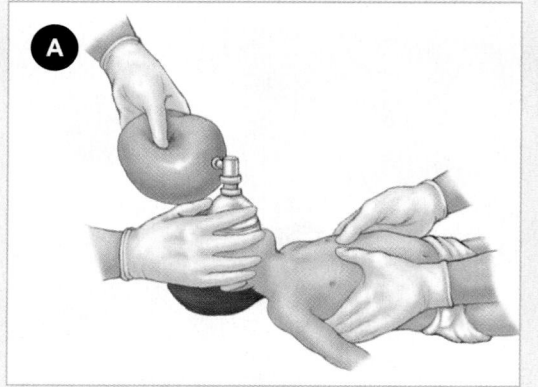
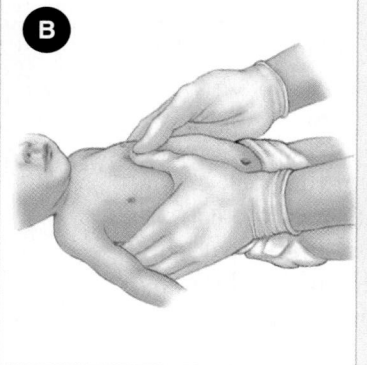
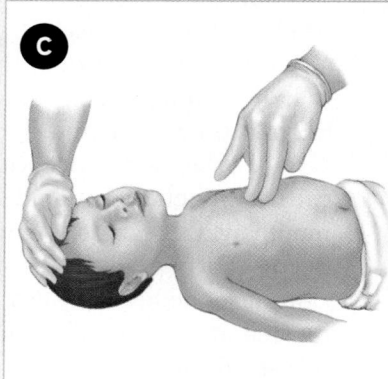

Figure 32-25 A. Chest compressions should be given with the hands encircling the infant and thumbs side by side. **B.** In very small infants, you may need to overlap the thumbs. **C.** In larger infants, you may use the two-finger technique, using the middle and ring fingers.

arrest because of respiratory or circulatory failure from illness or injury. For this reason, the airway and breathing are the focus of pediatric basic life support (BLS) (Table 32-7 ▼).

Respiratory problems leading to cardiopulmonary arrest in children can have a number of different causes, including:

- Injury, both blunt and penetrating
- Infections of the respiratory tract or another organ system
- A foreign body in the airway
- Near drowning
- Electrocution
- Poisoning or drug overdose
- Sudden infant death syndrome (SIDS)

For purposes of pediatric BLS, infancy ends at 1 year of age, and childhood extends to 8 years. Although children older than 8 years are still considered pediatric patients, they are treated with adult BLS methods. The goal, of course, is the same for all patients—to restore breathing and circulation of the blood.

Pediatric BLS can be divided into four steps:

1. Determining responsiveness
2. Airway
3. Breathing
4. Circulation

Determining Responsiveness

Never shake a child to determine whether he or she is responsive, especially if there is a possible neck or back injury. Instead, gently tap the child on the shoulder, and speak loudly (Figure 32-26 ▼). If a child is responsive but struggling to breathe, allow him or her to remain in whatever position is most comfortable.

If you find an unresponsive child while you are alone and not on duty, provide BLS for approximately 1 minute, and then stop to call the EMS system. Why

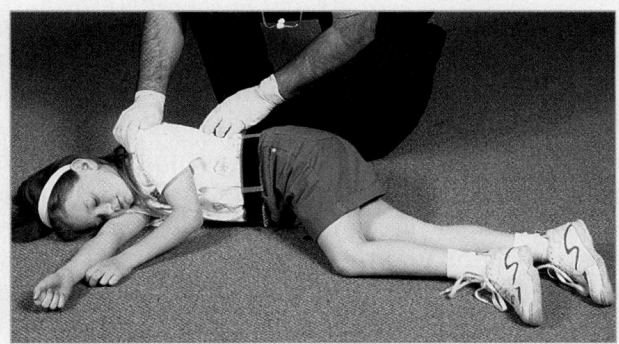

Figure 32-26 Never shake a child to determine responsiveness. Rather, gently tap the child on the shoulder, and speak loudly.

TABLE 32-7 Review of Pediatric BLS

Action	Infants Younger Than 1 Year	Children Between Ages 1 and 8 Years
Airway	Head tilt-chin lift; jaw thrust if spinal injury is suspected	Head tilt-chin lift; jaw thrust if spinal injury is suspected
Breathing		
Initial	2 breaths at a rate of 1 to 1½ seconds/breath (Use a BVM with oxygen connected if available.)	2 breaths at a rate of 1 to 1½ seconds/breath (Use a BVM with oxygen connected if available.)
Subsequent	20 breaths/min	20 breaths/min
Circulation		
Pulse check	Brachial/femoral arteries	Carotid artery
Compression area	Lower half of sternum	Lower half of sternum
Compression width	2 or 3 fingers or 2 thumb encircled hands	Heel of hand
Compression depth	½″ to 1″	1″ to 1½″
Compression rate	At least 100/min	100/min
Ratio of compressions to ventilations	5:1 (pause for ventilation)	5:1 (pause for ventilation)
Foreign body obstruction	Back blows and chest thrusts	Abdominal thrusts

not call right away, as you would with an adult? Because an unconscious child may respond much more quickly to ventilation and oxygenation than an adult. Indeed, these actions may prevent the child from progressing to full cardiopulmonary arrest.

Airway

Because children often put toys and other objects, as well as food, in their mouths, foreign body obstruction of the upper airway is common. The steps for removing a foreign object are reviewed earlier in this chapter. You must make sure that the upper airway is open when dealing with pediatric respiratory emergencies or cardiopulmonary arrest. If the child is unconscious and lying in a supine position, the airway may become obstructed when the tongue and throat muscles relax and the tongue falls backward (Figure 32-27 ▼).

If the child is unconscious but breathing, place him or her on one side or the other in the recovery position, in which the upper leg is flexed and bent forward for stabilization and the head is positioned to allow drainage of saliva or vomitus (Figure 32-28 ▼). Do not

use this position if you suspect a spinal injury unless you can secure the child to a backboard that can be tilted to the side. Do not attempt to open the airway at all if the child is conscious and breathing, but in a labored fashion. Instead, provide immediate transport to the nearest hospital.

There are two common techniques for manually opening the airway in a child who is neither conscious nor breathing: the head tilt-chin lift technique and the jaw-thrust maneuver (Figure 32-29 ▼). The latter is safer if there is a possibility of a neck injury.

Head Tilt-Chin Lift Technique

Perform this technique in a child in the following manner:

1. **Place one hand on the child's forehead**, and tilt the head back gently, with the neck slightly extended.
2. **Place the fingers (not the thumb) of your other hand under the child's chin**, and lift the jaw upward and outward. Do not close the mouth or push under the chin; either move may obstruct rather than open the airway.
3. **Remove any visible foreign body or vomitus.**

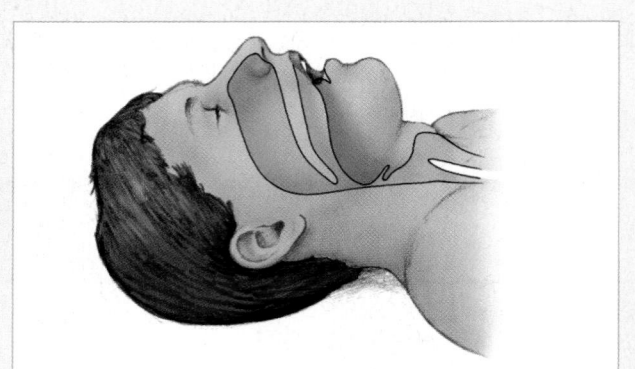

Figure 32-27 The airway may become obstructed when the tongue and throat muscles relax and the tongue falls back into the throat.

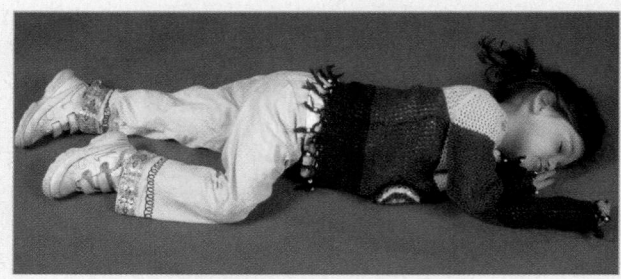

Figure 32-28 A child who is unconscious but breathing should be placed in the recovery position to allow saliva or vomitus to drain from the mouth.

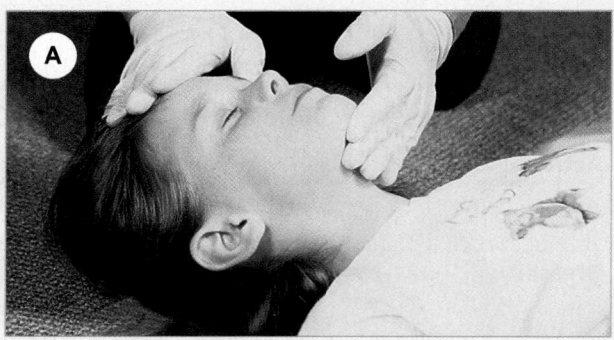

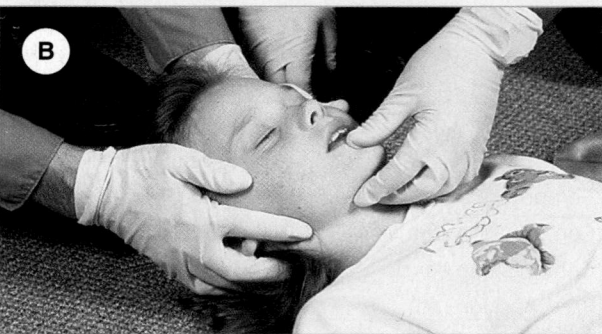

Figure 32-29 **A.** Use the head tilt-chin lift technique to open the airway in a child who has not sustained a traumatic injury. Do not overextend the neck. **B.** Use the jaw-thrust maneuver to open the airway if there is a possibility that a traumatic injury has occurred.

Jaw-Thrust Maneuver

Perform this maneuver in a child in the following manner:

1. **Place two or three fingers** under each side of the angle of the lower jaw; lift the jaw upward and outward.
2. **If the jaw thrust alone** does not open the airway and cervical spine injury is not a consideration, tilt the head slightly. If cervical spine injury is suspected, use a second rescuer to immobilize the cervical spine.

Remember that the head of an infant or young child is disproportionately large in comparison to the chest and shoulders. As a result, when a child is lying flat on his or her back, especially on a backboard, the head will bend forward onto the upper chest. This can partially or completely obstruct the upper airway. To avoid this possibility, place a wedge of padding under the upper chest and shoulders.

Breathing

Once the airway is open, determine if the child is breathing spontaneously, using the look, listen, and feel technique (Figure 32-30 ▶):

- **Look** for rise and fall of the chest or abdomen.
- **Listen** for exhalation of breath.
- **Feel** for exhaled air flow at the mouth.

If an infant or small child is breathing, provide immediate transport. Again, a child who is in respiratory distress should be allowed to stay in whatever position is most comfortable. Larger children who are unconscious and breathing with difficulty should be kept in the recovery position if possible.

If an infant or child is not breathing, provide rescue breathing while keeping the airway open. If you are using mouth-to-mouth resuscitation with an infant, place your mouth over the infant's mouth and nose to create a seal. If you are using a BVM device to assist

ventilations in an infant, use the proper sized mask and the technique described earlier.

When two rescuers are available, use your thumb and index finger to apply pressure over the area just below the Adam's apple (the Sellick maneuver) (Figure 32-31 ▼). This will decrease the risk of gastric

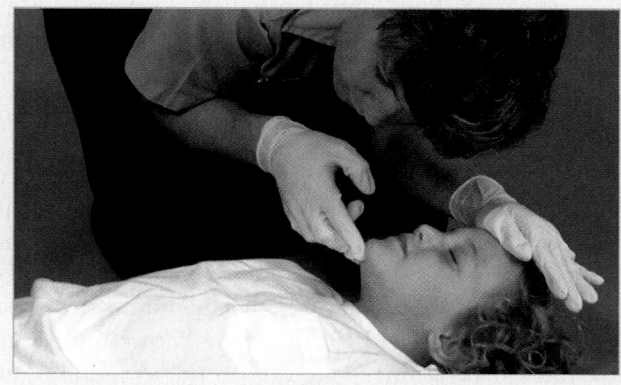

Figure 32-30 After you have opened the airway, use the look, listen, and feel technique to determine if the child is breathing spontaneously.

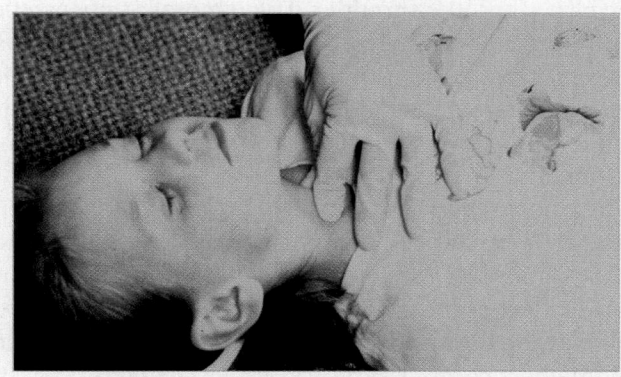

Figure 32-31 Performing the Sellick maneuver decreases the risk of gastric distention and aspiration of vomitus during BVM ventilation.

You are the Provider   Part 4

The partial airway obstruction has become a complete obstruction. You step behind the girl and wrap your arms around her. Finding the appropriate anatomic landmarks, you perform the Heimlich maneuver using forceful inward and upward motions.

7. How many times can you perform the Heimlich maneuver on a child?
8. If you are unsuccessful in alleviating the obstruction by using the Heimlich maneuver, what is your next step in treatment?

distention and aspiration of vomitus by pushing the larynx back to compress and close off the esophagus.

In a child with tracheostomy (breathing) tubes in the neck, remove the mask from the bag-valve device and connect it directly to the tracheostomy tube to ventilate the child. If a BVM device is unavailable, a mask, barrier device, or your mouth over the tracheostomy site can be used. Place your hand firmly over the child's mouth and nose to prevent the artificial breaths from leaking out of the upper airway.

Circulation

Once you have opened the airway and provided two rescue breaths, you must determine the state of the child's circulation. Check for pulses in the carotid artery in older children and the brachial or femoral artery in young children and infants. During your pulse check, also evaluate the child for other signs of circulation; these include breathing, coughing, or movement. Locate the carotid artery by placing one or two fingers over the groove between the Adam's apple and the neck muscles. Locate the brachial artery by placing two or three fingers on the inside of an infant's upper arm, between the elbow and the shoulder. The femoral artery can be felt in the crease between the upper leg and the groin. However, palpating the pulse in an infant or child is difficult; therefore, do not spend more than 10 seconds trying. If an infant or child is not breathing, you can assume that there is no pulse.

For chest compressions to be effective, the patient should be placed on a firm, flat surface with the head at the same level as the body. If you need to carry an infant while providing CPR, your forearm and hand can serve as the flat surface. Follow these steps to perform infant chest compressions (Skill Drill 32-6 ▶):

1. **Place the infant on a firm surface,** using one hand to keep the head in an open airway position. You can also use a pad or wedge under the shoulders and upper body to keep the head from tilting forward.
2. Imagine a line drawn between the nipples. **Place two fingers in the middle of the sternum,** about $1/2''$ below the level of the imaginary line (one fingerwidth) (**Step 1**).
3. **Using two fingers, compress the sternum** about one third to one half the depth of the chest; this is usually about $1/2''$ to $1''$. Compress the chest at a rate of 100 compressions/min. With pauses for ventilation, you will actually compress the chest about 80 compressions/min.

4. **After each compression, allow the sternum to return briefly to its normal position.** Allow equal time for compression and relaxation of the chest. Do not remove your fingers from the sternum, and avoid jerky movements (**Step 2**).

Coordinate rapid compressions and ventilations in a 5:1 ratio, making sure the infant's chest rises with each ventilation. You will find this easier to do if you use your free hand to keep the head in the open airway position. If the chest does not rise, or rises only a little, use a chin lift to open the airway. Reassess the infant for signs of spontaneous breathing or pulses after 1 minute, and again every few minutes.

(Skill Drill 32-7 ▶) shows the steps for performing CPR in children between ages 1 and 8 years:

1. **Place the child on a firm surface,** and use one hand to maintain the head in a tilted-back position (**Step 1**).
2. Using two fingers of the other hand, locate the bottom of the sternum by tracing the lower margin of the rib cage to the notch where the ribs and sternum meet. **Place the heel of your hand over the lower half of the sternum**; this is the area between the meeting place of the ribs and sternum and an imaginary line drawn between the nipples. Avoid compression over the lower tip of the sternum, which is called the <u>xiphoid process</u> (**Step 2**).
3. **Compress the chest** about one third to one half its total depth; this is usually $1''$ to $1^1/_2''$. Compress the chest at a rate of 100 compressions/min. With pauses for ventilation, the actual number of compressions will be about 80/min. Compression and relaxation should be about the same duration. Use smooth movements. Hold your fingers off the child's ribs, and keep the heel of your hand on the sternum.
4. **Coordinate rapid compressions and ventilations** in a 5:1 ratio, making sure the chest rises with each ventilation. At the end of every fifth compression, pause for 1 to $1^1/_2$ seconds for artificial ventilation (**Step 3**).
5. **Reassess the child for signs of spontaneous breathing and pulses** after about a minute and again every few minutes.
6. **If the child resumes effective breathing,** place him or her in the recovery position (**Step 4**).

Performing Infant Chest Compressions

Skill Drill 32-6

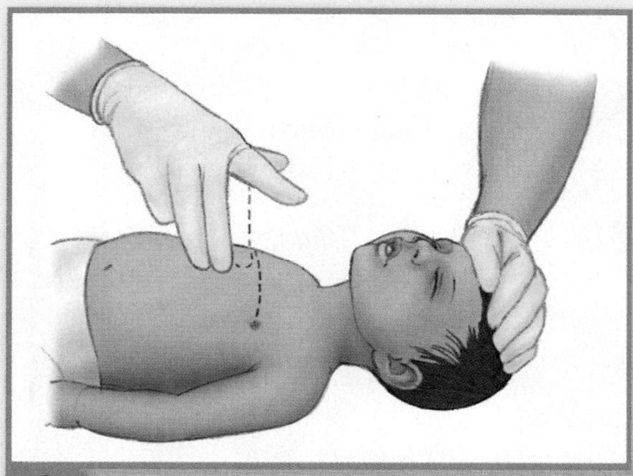

1 Position the infant on a firm surface while maintaining the airway. Place two fingers in the middle of the sternum just below a line between the nipples.

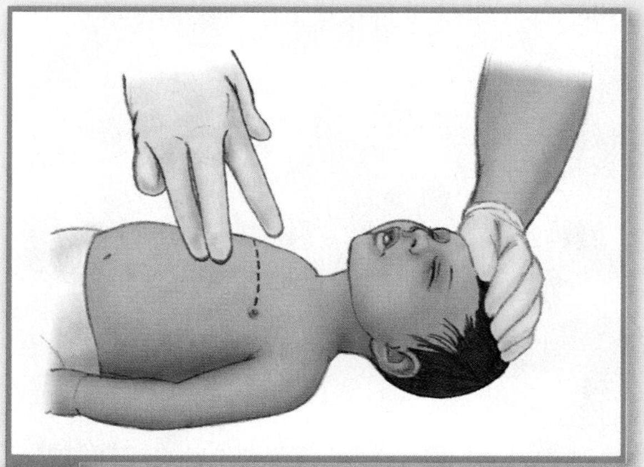

2 Use two fingers to compress the chest about ½″ to 1″ at a rate of 100 times/min.

Allow the sternum to return briefly to its normal position between compressions.

Remember, if the child is older than 8 years or is equivalent to an adult in size, use the adult CPR sequence, including the use of the AED.

AED Usage in Children

Due to the success of the AED programs across the country in adults, pediatric AEDs have been developed and are becoming more accessible in the community. If your service uses a pediatric AED or an AED with a pediatric adaptor, you should be familiar with the local protocols. Children younger than 1 year should not have an AED applied due to their size considerations and the limitations of the AEDs with regard to the settings. During CPR the AED should be applied to children (1 to 8 years old) after the first minute of CPR has been completed. As discussed earlier, cardiac arrest in children is usually due to respiratory causes and oxygenation is vitally important. After the first minute of CPR, the AED should be used to deliver shocks in the same manner as with an adult patient.

EMT-B Tips

An injured child with serious airway or breathing problems is likely to need full-time attention from two EMT-Bs. The need for a driver, and often for added help with patient care, makes it important to start arranging early for backup from another unit—possibly even before you arrive at the scene.

EMT-B Tips

AEDs are becoming more and more accessible in the community. Be familiar with your local protocols on pediatric defibrillation. Your service may use a pediatric AED, or an AED with a pediatric adapter.

Performing CPR on a Child

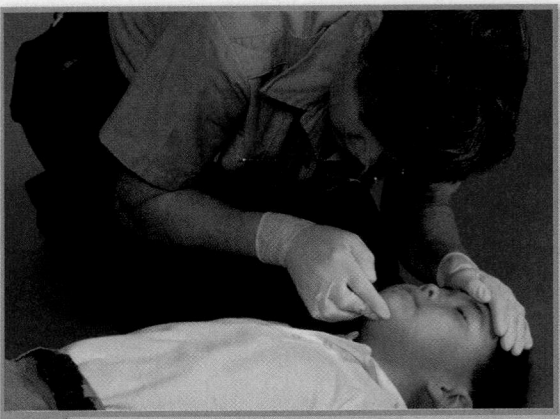

1 Place the child on a firm surface and maintain the airway with one hand.

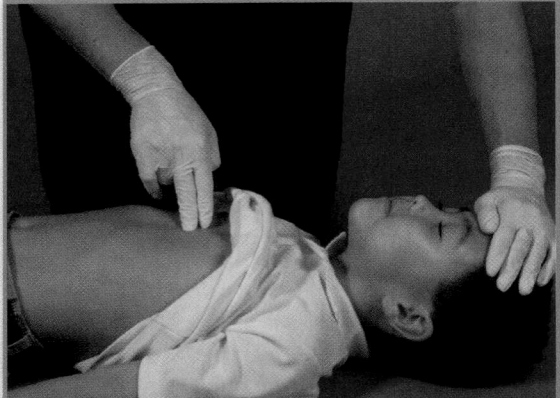

2 Using two fingers of the other hand, locate the bottom of the sternum by tracing the lower margin of the rib cage to the notch where the ribs and sternum meet. Place the heel of your other hand over the lower half of the sternum, avoiding the xiphoid.

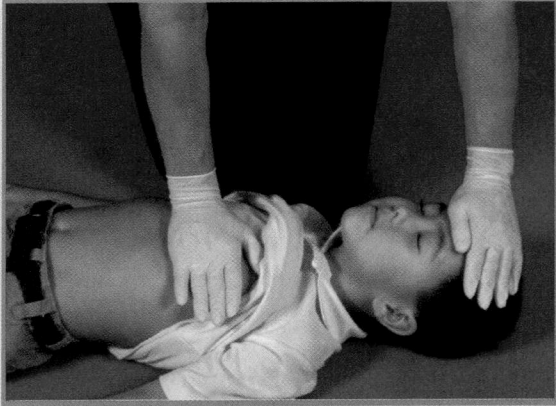

3 Compress the chest about 1″ to 1½″ at a rate of 100/min (about 80/min with pauses for ventilations). Coordinate compressions with ventilations in a 5:1 ratio, pausing for ventilations.

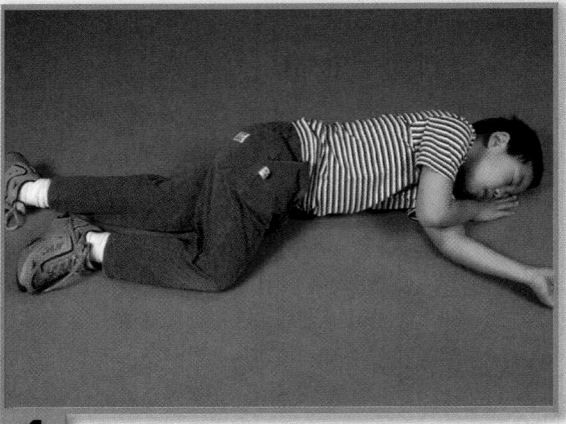

4 Reassess for breathing and pulse after about 1 minute, and then every few minutes.

If the child resumes effective breathing, place him or her in the recovery position.

Cardiopulmonary Arrest

Cardiac arrest in infants and children is most often associated with respiratory failure and respiratory arrest. Children are affected differently than adults when it comes to decreasing oxygen concentrations. An adult becomes hypoxic and the heart gets irritable and sudden cardiac death occurs. This is often in the form of ventricular fibrillation, and is the reason that an AED is the treatment of choice. Children, on the other hand, become hypoxic and their hearts slow down, becoming more and more bradycardic. The heart will beat slower and become weaker with each beat until no pulse is felt. The survival rate from cardiac arrest in the prehospital setting is 3% to 5%. However, the survival rate from respiratory arrest is 75%. Therefore a child who is breathing very poorly with a slowing heart rate must be ventilated with high concentrations of oxygen early to try to oxygenate the heart before cardiac arrest occurs.

Pediatric Trauma

The trauma assessment of a child follows the same format of the adult's; however, several differences make the child more prone to injury. Once the MOI has been determined and the child's level of consciousness assessed, the EMT-B must determine whether to use a focused or head-to-toe exam. Remember that young children cannot be specific about location or severity of pain, thus requiring the EMT-B to do a head-to-toe exam anyway.

When beginning the exam, determine the age of the child. Infants, toddlers, and preschool-age children do not like to be touched. If possible, the exam should start from the toes and move toward the head, leaving any noticeably injured areas for last. Starting at the core may make the child irritable and less likely to assist with your exam.

The head is injured most often and is the most likely injury to cause death. The child's head is large in comparison to the body, and most multisystem trauma will involve the head. During your assessment, concentrate on keeping the airway open as noted earlier. Monitor vital signs often to look for signs of increased intercranial pressure. Hyperventilation should be avoided until normal ventilations have been established and signs of herniation are present.

Cervical spine injury is more prevalent in children than in adults due to the weaker muscles of the neck. This is one of the reasons that children younger than 1 year should always be in a rear-facing car seat when being transported. Careful immobilization should be used to maintain a neutral position as described earlier.

Immobilization

Immobilization is necessary for all children who have possible head or spinal injuries after a traumatic event. Follow these steps (Skill Drill 32-8 ▶):

1. **Maintain the child's head in a neutral position** by placing a towel under the shoulders and torso (**Step 1**).
2. **Place an appropriately sized cervical collar on the patient** (**Step 2**).
3. **Carefully log roll the child** onto the immobilization device (**Step 3**).

4. **Secure the patient's torso** to the immobilization device first (**Step 4**).
5. **Secure the child's head** to the immobilization device (**Step 5**).
6. **Complete immobilization** by ensuring that the child is strapped in properly (**Step 6**).

Immobilization can be difficult to perform due to the child's body proportions. Young children require padding under the torso to maintain a neutral position. At around 8 to 10 years of age, children no longer require padding underneath the torso to create a neutral position. Instead, they can simply lie supine on the board. However, another complication may occur if a child is put onto an adult-sized long board. Because a child's body is narrower than an adult's, padding will be required along the sides in order for the child to be properly secured on an adult-sized long board.

Some infants will be in a car seat when the EMT-B approaches them. There are two methods of transportation that are determined by the patient's severity. If the child has stable vital signs, minimal injury, and the car seat is visibly undamaged, the child can be left in the seat and secured within it for transportation. If the child is unstable, has injuries other than minor ones, or the car seat is visibly damaged, the child must be removed to a board type of device for immobilization and transportation.

Ideally, a cervical collar would be used when immobilizing an infant or toddler in a car seat; however, in most instances an appropriately sized cervical collar will not be available. In this case, place rolled towels on either side of the head to prevent side-to-side movement. Do not place a towel in the shape of an upside-down "U" over the child's head; this may press down on the head and compromise the airway and spinal cord. The steps for immobilizing an infant in a car seat follow (Skill Drill 32-9 ▶):

1. **Carefully stabilize the infant's head in a neutral position.** Leave all car seat straps in place (**Step 1**).
2. **Place an appropriately sized cervical collar on the patient** if available. **Otherwise, place rolled towels or padding alongside the infant** to fill the voids in the car seat (**Step 2**).
3. **Carefully secure the padding,** using tape to keep it in place (**Step 3**).
4. **Secure the car seat to the stretcher** as detailed later in this chapter (**Step 4**).

Immobilizing a Child

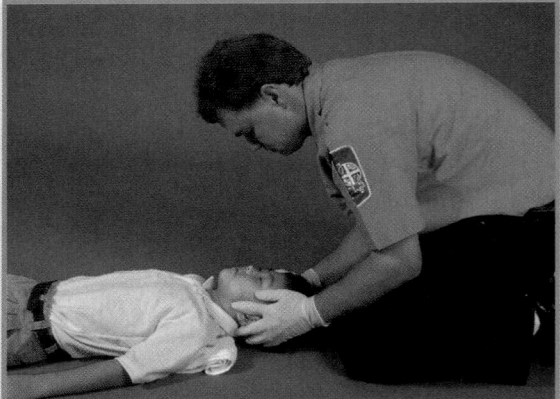

1 Use a towel under the back, from the shoulders to the hips, to maintain the head in a neutral position.

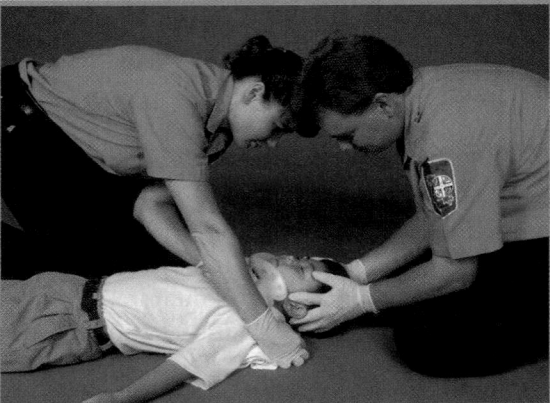

2 Apply an appropriately sized cervical collar.

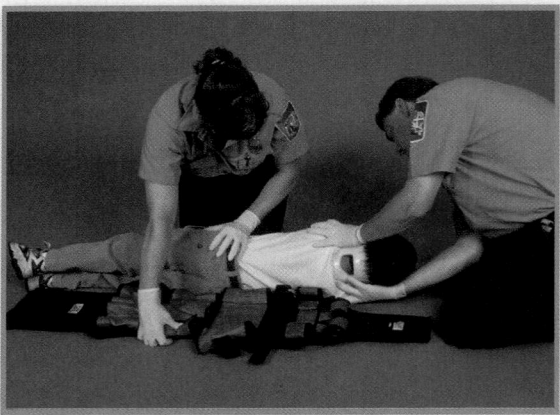

3 Log roll the child onto the immobilization device.

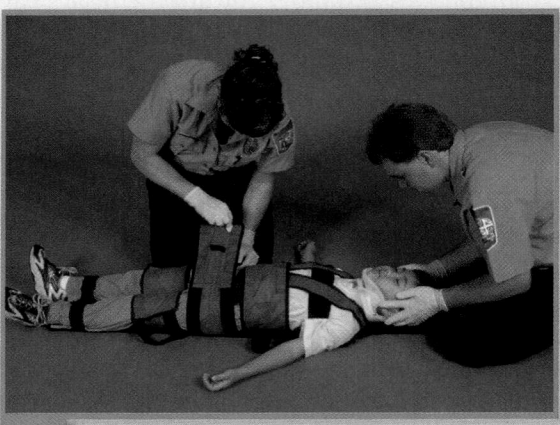

4 Secure the torso first.

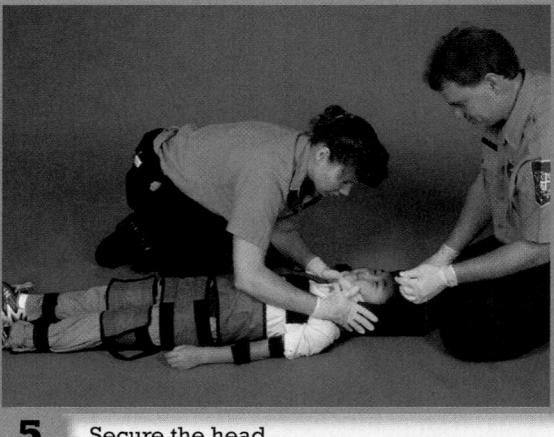

5 Secure the head.

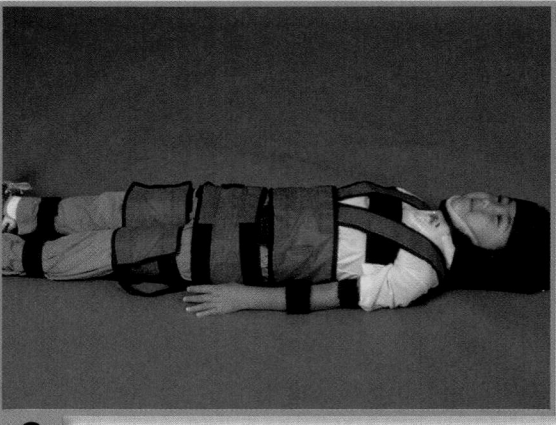

6 Ensure that the child is strapped in properly.

Immobilizing an Infant in a Car Seat

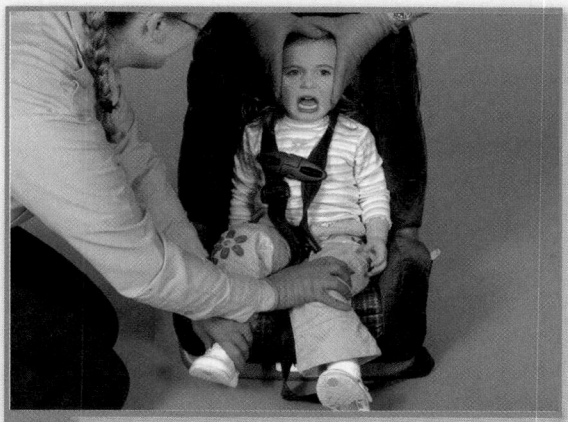

1 Carefully stabilize the infant's head in a neutral position.

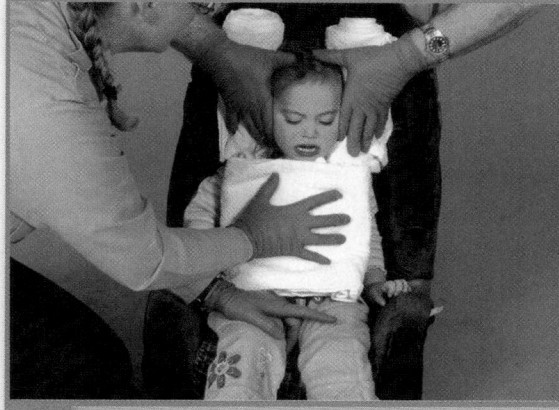

2 Place an appropriately sized cervical collar on the patient if available. Otherwise, place rolled towels or padding alongside the infant.

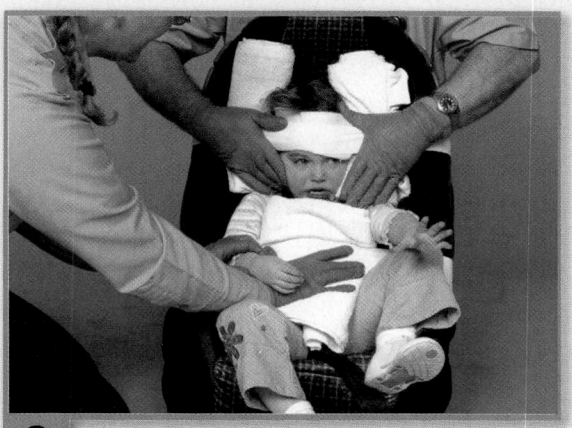

3 Carefully secure the padding, using tape to keep it in place.

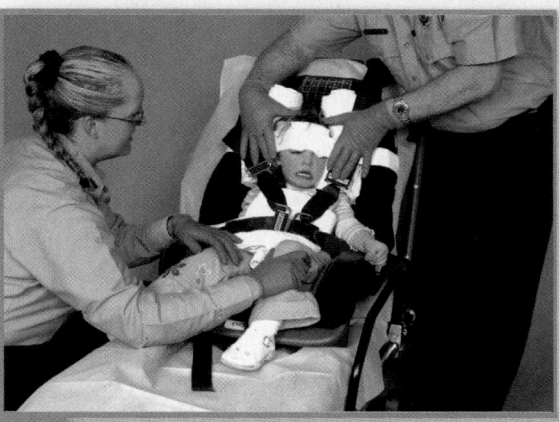

4 Secure the car seat to the stretcher.

Follow these steps to immobilize an infant out of a car seat (Skill Drill 32-10 ▶):

1. **Carefully stabilize the infant's head in a neutral position** and lay the seat down into a reclined position on a hard surface (**Step 1**).
2. **Position a pediatric board or other similar device** between the patient and the surface on which the infant is resting (**Step 2**).
3. **Carefully slide the infant into position** on the board (**Step 3**).
4. **Make sure the infant's head is in a neutral position** by placing a towel under the infant's shoulders (**Step 4**).
5. **Secure the torso first** and place padding to fill any voids (**Step 5**).
6. **Secure the infant's head** to the board (**Step 6**).

Skill Drill

32-10

Immobilizing an Infant Out of a Car Seat

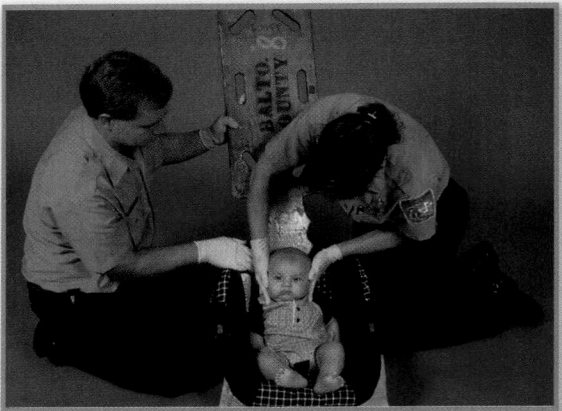

1 Stabilize the head in neutral position.

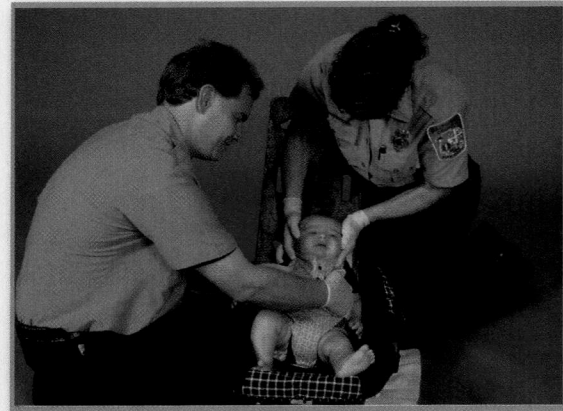

2 Place an immobilization device between the patient and the surface he or she is resting on.

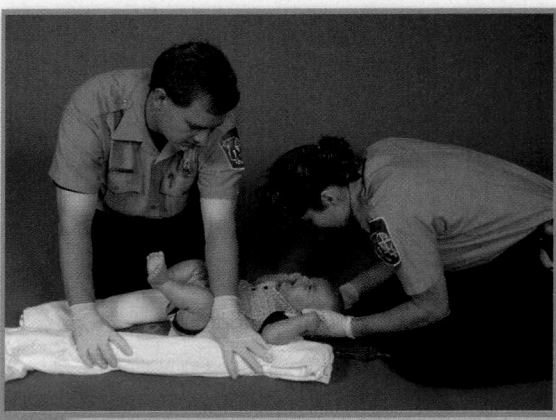

3 Slide the infant onto the board.

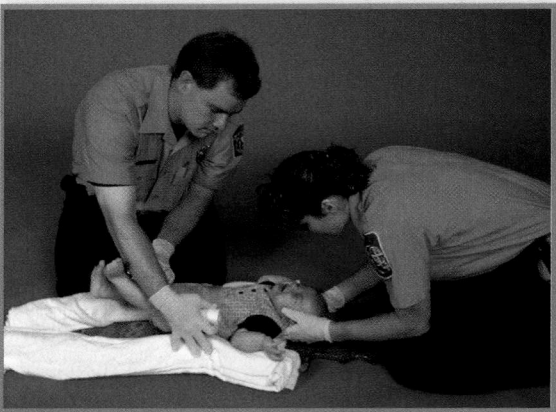

4 Place a towel under the back, from the shoulders to the hips, to ensure neutral head position.

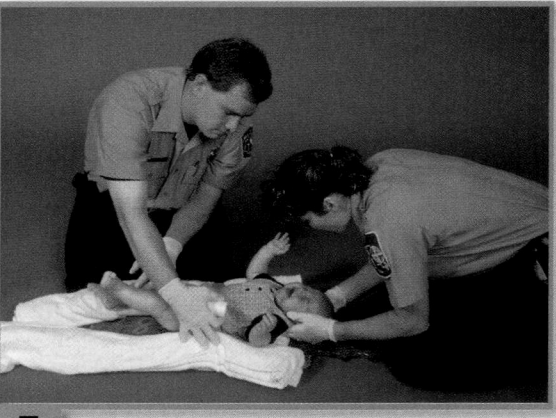

5 Secure the torso first; pad any voids.

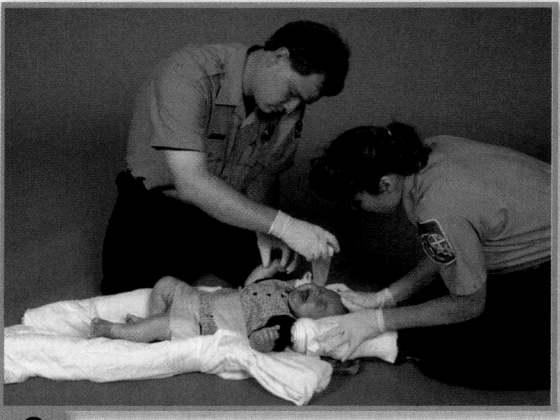

6 Secure the head.

Management of Pediatric Injuries

Extremity injuries in children are generally managed in the same manner as those in adults. Painful deformed limbs with evidence of broken bones should be splinted. Specialized splinting equipment, such as a traction splint for fractures of the femur, should be used only if it fits the child. You should not attempt to use adult immobilization devices on a child unless the child is large enough to properly fit in the device.

Pediatric Medical Emergencies

Like the pediatric trauma assessment, the pediatric medical assessment follows the same pathways as in the adult, with emphasis on the differences in the pediatric patient. Some medical complaints merit additional discussion.

Respiratory Emergencies

In the early stages of respiratory distress or failure, respirations may be too slow or too fast for the patient's age. This suggests that gases are not moving effectively into and out of the lungs. If like most people you find it hard to memorize normal vital sign ranges for infants and children, keep reference charts handy for this purpose. Respirations exceeding 60 breaths/min are a sign of a problem. In most cases, you should begin to assist ventilation immediately, even if the child appears to be breathing adequately. But remember, you are treating the child, not the numbers. A child breathing 60 breaths/min who is playing happily does not need assisted ventilation; a child breathing 60 breaths/min who is lying unconscious on the floor does.

Signs and Symptoms

In the early stages of respiratory distress, you may note changes in the child's behavior, such as combativeness, restlessness, and anxiety. As the body attempts to maximize the amount of air going into the lungs, the work of breathing increases. Signs and symptoms of increased work of breathing include:

- Nasal flaring, as the body tries to increase the size of the airway
- Grunting respirations, as the body attempts to keep the alveoli expanded at the end of expiration
- Wheezing, stridor, or other abnormal airway sounds
- Accessory (intercostal) muscle use; remember that in young children, the diaphragm is the major muscle of ventilation
- Retractions, or movements of the child's flexible rib cage
- The tripod position; in older children, this position will maximize their airway.

You are the Provider   Part 5

You continue performing the Heimlich maneuver with no success. The little girl becomes unconscious. You quickly lay her down on the ground and begin delivering abdominal thrusts. Your partner performs a head tilt-chin lift in an effort to try to visualize the obstruction. While doing so, you continue doing abdominal thrusts. Suddenly a chunk of food shoots out of Jamie's mouth. She gasps for air.

9. Can you perform a blind finger sweep on a child?
10. What is the normal breathing rate for children?

As the child progresses to possible respiratory failure, efforts to breathe decrease; the chest rises less with inspiration. A definitive diagnosis of respiratory failure is made in the hospital. The body has used up its available energy stores and cannot continue to support the extra work of breathing under these conditions. At this point, cyanosis may develop (cyanosis is a late sign). Be aware that not all children become cyanotic. You should be just as concerned about a child with pale skin as one with bluish skin.

Changes in behavior will also occur until the child demonstrates an altered level of consciousness. The patient may experience periods of <u>apnea</u> (absence of breathing). As the lack of oxygen becomes more serious, the heart muscle itself becomes hypoxic and slows down. This leads to bradycardia, a condition in which the heart rate is less than 60 beats/min in children or less than 80 beats/min in newborns. Bradycardia is almost always related to a lack of oxygen and is an ominous sign in pediatric patients. If the heart rate is fast, you need to investigate the cause. However, if the heart rate is slow or absent, you must intervene immediately. Without aggressive airway management, bradycardia may quickly progress to cardiopulmonary arrest.

Of course, respiratory failure does not always indicate airway obstruction. It may indicate trauma, problems with the nervous system, dehydration (often caused by vomiting and diarrhea), or metabolic disturbances. For example, a child with diabetes might have a blood glucose level that is too high or too low; a child might have a pH imbalance, as can happen with some rare childhood diseases. Regardless of the cause, your first step is always to focus on ensuring adequate oxygenation and ventilation.

Never forget that a child can progress from respiratory distress to respiratory failure at any time. For this reason, you must reassess the child frequently.

Emergency Medical Care

A child or infant in respiratory distress or possible respiratory failure needs supplemental oxygen. Remember, anxiety, agitation, or crying may increase the effort or work of breathing, so use whichever method seems least upsetting to the child—mask, blow-by, or nasal cannula (**Figure 32-32 ▶**). You may need to get creative by distracting the child with games, a toy, or talking.

Allow the child to remain in a comfortable position. For a small child, this may mean sitting on the caregiver's lap. Give nothing by mouth, in case the child's condition deteriorates suddenly.

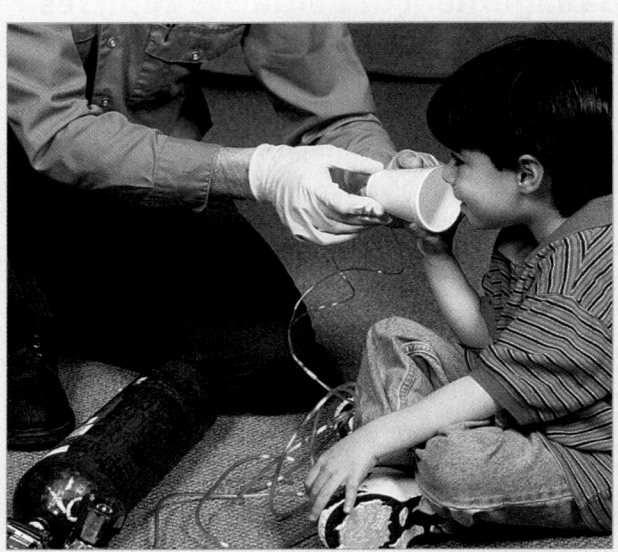

Figure 32-32 A child in respiratory distress needs supplemental oxygen; you should select whichever method seems least upsetting to the child.

If the patient has progressed to respiratory failure, you must begin assisted ventilation immediately and continue to provide supplemental oxygen.

Shock

As discussed in Chapter 23, <u>shock</u> is a condition that develops when the circulatory system is unable to deliver a sufficient amount of blood to the organs of the body. This results in organ failure and eventually cardiopulmonary arrest. In children, shock is rarely due to a primary cardiac event, such as a heart attack. Shock may be due to many things. The most common causes include:

- Traumatic injury with blood loss (especially abdominal)
- Dehydration from diarrhea and vomiting
- Severe infection
- Neurologic injury such as severe head trauma
- A severe allergic reaction to an insect bite or allergy (anaphylaxis)
- Diseases of the heart
- A collapsed lung (pneumothorax)
- Blood or fluid around the heart (cardiac tamponade or pericarditis)

Infants and children have less blood circulating in their bodies than adults do, so the loss of even a small volume of fluid or blood may lead to shock. Pediatric patients also respond differently than adults to fluid loss. They may respond by increasing their heart rate, increasing respirations, and showing signs of pale or blue skin. You must be able to recognize the signs of shock in infants and children.

Loss of more than 25% blood volume significantly increases the risk of shock. Signs of shock in children are:

- Tachycardia
- Poor capillary refill
- Mental status changes

For comparison, signs of shock in adults are:

- Tachycardia
- Hypotension
- Mental status changes

Greater than 30% to 40% blood volume loss significantly increases risk of shock in adults.

Begin by assessing the ABCs, intervening immediately as required; do not wait until you have completed a detailed assessment to take action. Children in shock often have increased respirations but do not demonstrate a fall in blood pressure until shock is severe.

In assessing circulation, you should pay particular attention to the following:

- **Pulse.** Assess both the rate and the quality of the pulse. A weak, "thready" pulse is a sign that there is a problem. The appropriate rate depends on age; anything over 160 beats/min suggests shock.
- **Skin signs.** Assess the temperature and moisture on the hands and feet. How does this compare with the temperature of the skin on the trunk of the body? Is the skin dry and warm, or cold and clammy?
- **CRT.** Squeeze a finger or toe for several seconds until the skin blanches, then release it. Does the fingertip return to its normal color within 2 seconds, or is it delayed?
- **Color.** Assess the patient's skin color. Is it pink, pale, ashen, or blue?

Changes in pulse rate, color, skin signs, and CRT are all important clues suggesting shock.

Blood pressure is the most difficult vital sign to measure in pediatric patients. The cuff must be the proper size—two thirds the length of the upper arm. The value for normal blood pressure is also age-specific. Remember that blood pressure may be normal; this is called compensated shock. Low blood pressure is a sign of decompensated shock, a serious condition that requires care an ALS team can provide.

Part of your assessment should also include talking with the parents or caregivers to determine when the signs and symptoms first appeared and whether any of the following has occurred:

- Decrease in urine output (with infants, are there fewer than 6 to 10 wet diapers?)
- Absence of tears, even when the child is crying
- Changes in level of consciousness and behavior

Limit your management to these simple interventions. Time should not be wasted in field procedures. Ensure that the airway is open, preparing for artificial ventilation; control bleeding; and give supplemental oxygen by mask or nasal cannula as tolerated. Continue to monitor airway and breathing. Position the patient with the head lower than the feet by elevating the feet with blankets. Keep the patient warm with blankets and by turning up the heat in the patient compartment. Provide immediate transport to the nearest appropriate facility and continue monitoring vital signs en route. Contact ALS backup as needed. Allow a caregiver to accompany the child whenever possible.

Seizures

Seizures in children may appear in several different ways, including shaking of the whole body or movement in just a single arm or leg. Seizures can also appear as lip smacking, eye blinking, or staring off into space. In a true seizure, movements cannot be stopped on command or by holding an extremity. The duration of movement varies from patient to patient.

Altered mental status and the inability of others to stop a movement or range of movements in the affected limb are common to all seizures. Some patients may feel pins and needles, hear sounds, and see hallucinations. In all but absence seizures (discussed in Chapter 13), a postictal period of extreme fatigue or unresponsiveness occurs after the seizure for anywhere from a few minutes to several hours. During this time, the patient may appear sleepy and/or confused and is not able to interact appropriately. A short period of seizure activity (under 30 minutes) is not in itself harmful to the patient. After 30 to 45 minutes, however, the brain may run low on energy stores, and continued activity can be harmful. Status epilepticus is a continuous seizure, or multiple seizures without a return to consciousness, for 30 minutes or more.

If you can identify the cause of the seizure, you will be better able to monitor the patient for any potential complications associated with the underlying problem.

In particular, be alert to the presence of medications, possible poisons, and indications of abuse or neglect.

Febrile Seizures

Febrile seizures are common in children between the ages of 6 months and 6 years. Most pediatric seizures are due to fever alone, which is why they are called febrile seizures. These seizures typically occur on the first day of a febrile illness, are characterized by generalized tonic-clonic seizure activity, and last less than 15 minutes with a short postictal phase or none at all. They may be a sign of a more serious problem, such as meningitis. Obtain a history from the caregivers, as these children may have had a prior febrile seizure.

If you are called to care for a child who has had a febrile seizure, you often will find that the patient is awake, alert, and fully interactive when you arrive. Keep in mind that a persistent fever can lead to another seizure. Carefully assess the ABCs, begin cooling measures with tepid (not cold) water, and provide prompt transport; all children with febrile seizures need to be seen in the hospital setting.

Emergency Medical Care

Although medical management of seizures in the hospital setting may vary according to cause, your assessment and management of these patients remain essentially the same. First, ensure that the scene is safe for you and your partner and for the patient. Next, perform an initial assessment, focusing on the ABCs. If possible, obtain a brief history from the caregivers about previous serious illnesses or seizures and current medication or trauma.

Securing and protecting the airway are your priorities. To avoid obstruction from the tongue falling back into the airway, place a child who is having a seizure or who is postictal in the recovery position if you can do so without having to use extreme force against the seizure activity. In the case of trauma, place the head in a neutral in-line position and ensure that the cervical spine is protected. Be ready to use suction to prevent aspiration of stomach contents, blood, or vomitus. Do not place your fingers in the mouth of a patient who is having a seizure.

A patient who is actively seizing or who is postictal may not be breathing adequately. Assessing the rate and depth of respirations in this situation can be difficult but is essential. Patients may have shallow, rapid breathing or may have occasional deep respirations. Signs that a patient is not breathing adequately include:

- Very slow respirations
- Very shallow breaths
- Bluish tint to lips or pale lips
- Snoring respirations caused by the tongue blocking the airway

Deliver oxygen by mask, blow-by, or nasal cannula. If there are no signs of improvement, begin BVM ventilation with appropriately sized equipment.

Patients who are experiencing a seizure usually maintain adequate blood pressure and pulse rate unless the seizure is caused by an underlying circulatory or neurologic problem or trauma, including bleeding, heart problems, or brain injury. Nevertheless, you must evaluate the pulse and blood pressure and re-evaluate them. Once the ABCs have been addressed, assessment and management should proceed. If the patient is actively seizing, note the type of movement and position of the eyes, as this information may be very helpful to hospital staff in making a diagnosis. If there is a fever, begin cooling measures such as removing clothing and placing towels moistened with tepid water on the child. A child with febrile seizures can seize again if the temperature remains high. Do not use alcohol or cold water to cool a patient. Make sure the patient is protected from hitting the sides of the stretcher or nearby equipment. Bring any medications or possible poisons at the scene to the hospital with the patient. If the patient is in status epilepticus, call for ALS backup, as medication is required to stop the seizures.

Dehydration

Dehydration can be described as mild, moderate, or severe. The severity of the dehydration can be gauged by looking at several clues (**Table 32-8 ▶**). For example, an infant with mild dehydration may have dry lips and gums, decreased saliva, and fewer wet diapers throughout the day (**Figure 32-33 ▶**). As the dehydration grows more severe, the lips and gums may become very dry, the eyes may look sunken, and the infant may be sleepy and/or irritable, refusing bottles. The skin may be loose and have no elasticity; this is called poor skin turgor. Also, infants may have sunken fontanels.

Young children can compensate for fluid losses by decreasing blood flow to the extremities and directing it to vital organs such as the brain and heart. Children who are moderately to severely dehydrated

TABLE 32-8   Vital Signs and Symptoms of Dehydration

	Mild Dehydration	Moderate Dehydration	Severe Dehydration
Pulse	Normal	Increased	Increased; 160+ is sign of impending shock
Level of activity	Normal or slowed	Slowed	Variable, weak to unresponsive
Urine output	Decreased	Decreased	No output
Skin	Normal	Cool, mottled; poor turgor	Cool, clammy; poor turgor; delayed CRT
Mouth	Decreased saliva	Dry mucous membranes	Dry mucous membranes
Eyes	Normal	Tears	Sunken eyes
Anterior fontanel	Normal to sunken	Sunken	Very sunken
Level of consciousness	Normal	Altered	Altered; lethargic
Blood pressure	Normal	Normal	Normal to low when shock sets in

may have mottled, cool, clammy skin and delayed CRT. Respirations will usually be increased. Be aware that blood pressure may remain normal while the child is in shock.

Emergency medical care should include careful attention assessing the ABCs and obtaining baseline vital signs. However, if the dehydration is severe, ALS backup may be necessary so that IV access can be obtained and rehydration can begin. All children with signs and symptoms of moderate to severe dehydration should be transported to the emergency department.

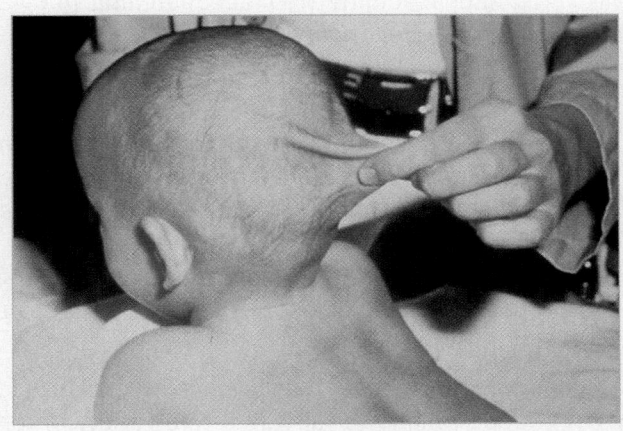

Figure 32-33 An infant with dehydration may exhibit "tenting" or poor skin turgor.

You are the Provider   Summary

You immediately apply high-flow oxygen and begin monitoring her breathing status. Within moments Jamie is breathing normally. She is conscious, alert, and oriented to person, time, place, and event. You offer to transport Jamie to the hospital for observation but her mother refuses, stating that she is ok now. After completing the necessary paperwork and notifying dispatch of your status, you and your partner return to your table to finish your lunch before you receive your next call.

As an EMT-B you will find that you are more aware of your surroundings and those around you. Always ensure that you have consent prior to treating any patient. Make checking the airway, breathing, and circulation your first step in the treatment of every patient. Remember, no airway, no patient.

Prep Kit

Ready for Review

- You will need to carry special sizes of airway equipment for pediatric patients.

- Use a pediatric resuscitation tape measure to determine the appropriately sized equipment for children.

- Use the pediatric assessment triangle (PAT) to obtain a general impression of the infant or child.

- In treating possible respiratory failure in a child, always position the airway in a neutral position.

- Appropriate oxygen delivery devices include the blow-by technique at 6 L/min, a nonrebreathing mask at 10 to 12 L/min, and a BVM device at 10 to 15 L/min.

- Use a BVM device with a child whose breathing and tidal volume are inadequate and who has an altered level of consciousness.

- The three keys to successful use of the BVM device in a child are: (1) have the appropriate equipment in the right size; (2) maintain a good face to mask seal; and (3) ventilate at the appropriate rate and volume—20 breaths/min for an infant or child, 1 to $1\frac{1}{2}$ seconds per ventilation. Squeeze gently, and stop squeezing as the chest wall begins to rise; use the phrase "squeeze, release, release" to maintain a proper rhythm.

- Children younger than 5 years often obstruct their upper and lower airway with a variety of foreign objects.

- If the child is conscious, encourage him or her to cough to clear the airway.

- If the child is unresponsive, you should first use the tongue-jaw lift and finger sweeps to try to remove an object that you can see. Never perform blind finger sweeps in infants or children with an airway obstruction.

- In treating an unresponsive child with complete airway obstruction, use abdominal thrusts (in a series of five), alternating with attempts at artificial breathing; in infants, perform back blows and chest thrusts.

- In a conscious child who is sitting or standing, apply abdominal thrusts from behind. Continue to perform abdominal thrusts until the obstruction is relieved or the child loses consciousness.

- Signs of shock in children are tachycardia, poor CRT, and mental status changes. You must be very alert for signs of shock in a child because children decompensate rapidly.

- Febrile seizures are common in children between the ages of 6 months and 6 years. Most pediatric seizures are due to fever alone, which is why they are called febrile seizures. Treat as you would for an adult patient with a seizure. Carefully assess the ABCs, begin cooling measures, and provide prompt transport.

- The most common cause of dehydration in children is vomiting and diarrhea. Life-threatening diarrhea can develop in an infant in hours. You can determine whether a child's dehydration is mild, moderate, or severe by assessing the child's urine output, level of activity, mental status, skin tone, and pulse.

- BLS for infants and children consists of determining responsiveness and assessing airway, breathing, and circulation.

- If the child is unresponsive but breathing, place him or her in the recovery position unless you suspect a spinal injury. Use the head tilt-chin lift or jaw-thrust maneuver to open the airway in a child who is unresponsive and not breathing.

Technology

Interactivities

Vocabulary Explorer

Anatomy Review

Web Links

Online Review Manual

www.EMTB.com

- If a child is not breathing, provide rescue breathing while keeping the airway open. Breathe for an infant at a rate of 1 to 1½ breaths/second at first, then 20 breaths/min. Breathe for a child between ages 1 and 8 years at a rate of 2 breaths/second at first, then 20 breaths/min.

- To provide CPR in an infant, compress the chest 100 compressions/min, pausing after every five compressions for ventilation; use two or three fingers, and compress the lower half of the sternum to a depth that is one half to one third the diameter of the chest.

- In children, use the same depth and rate of compressions as you did for the infant; however, use the heel of your hand to compress the chest; avoid compressing the xiphoid process.

Vital Vocabulary

anemia A deficiency of red blood cells or hemoglobin.

apnea Absence of breathing.

AVPU scale Used to assess level of consciousness; recorded as being alert, verbally responsive, responsive to pain, or unresponsive.

blanching Turning white.

bradycardia A heart rate of less than 60 beats/min in children or less than 80 beats/min in infants.

bradypnea Slow respiratory rate; ominous sign in a child that indicates impending respiratory arrest.

capillary refill time (CRT) The amount of time that it takes for blood to return to the capillary bed after applying pressure to the skin or nailbed; indicates the status of end-organ perfusion; reliable in children younger than 6 years.

central pulses Pulses that are closest to the core (central) part of the body where the vital organs are located; include the carotid, femoral, and apical pulses.

crackles A crackling breath sound caused by the flow of air through liquid in the lungs; a sign of lower airway obstruction.

croup Infection of the airway below the level of the vocal cords, usually caused by a virus.

cyanosis A blue discoloration of the skin and mucous membranes; indicates decreased levels of oxygen in the blood.

end-organ perfusion The status of perfusion to the vital organs of the body; determined by assessing capillary refill time (CRT).

epiglottitis An infection of the soft tissue in the area above the vocal cords.

grunting An "uh" sound heard during exhalation; reflects the child's attempt to keep the alveoli open; a sign of increased work of breathing.

head bobbing The head lifts and tilts back during inspiration, then moves forward during expiration; a sign of increased work of breathing.

meconium A dark green material in the amniotic fluid that can cause lung disease in the newborn.

nares The external openings of the nostrils. A single nostril opening is called a naris.

nasal flaring Widening of the nares during inspiration; commonly seen in infants; indicates increased work of breathing.

pediatric assessment triangle (PAT) A structured assessment tool that allows you to rapidly form a general impression of the infant or child without touching him or her; consists of assessing appearance, work of breathing, and circulation to the skin.

pediatric resuscitation tape measure A tape used to estimate an infant or child's weight on the basis of length; appropriate drug doses and equipment sizes are listed on the tape.

retractions Drawing in of the intercostal muscles and sternum during inspiration; a sign of increased work of breathing.

septum The central divider in the nose.

sniffing position Optimum neutral head position for the uninjured child who requires airway management.

stridor A high-pitched breath sound heard mainly on inspiration that is a sign of upper airway obstruction.

tachypnea Increased respiratory rate.

tenting A condition in which the skin remains depressed after you remove your finger; indicates dehydration.

Prep Kit continued...

tidal volume The amount of air that is delivered to the lungs and airways in one inhalation.

tragus The small cartilaginous projection in front of the opening of the ear.

transition phase A time period that allows the infant or child to become familiar with you and your equipment; only appropriate if the child's condition is stable.

tripod position An abnormal position to keep the airway open; it involves leaning forward onto two arms stretched forward.

wheezing A whistling breath sound caused by air traveling through narrowed air passages within the bronchioles; a sign of lower airway obstruction.

work of breathing An indicator of oxygenation and ventilation. Work of breathing reflects the child's attempt to compensate for hypoxia.

xiphoid process The lower tip of the sternum.

Points to Ponder

You are dispatched to 722 Chase St for a child with asthma. You arrive to find a 12-year-old girl who is obviously experiencing difficulty breathing. You hear audible wheezing and notice accessory muscle use. Her respiratory rate is 60 breaths/min. She tells you (in 1- to 2-word phrases) that she used her inhaler before you arrived, but it does not seem to be working as well as it usually does. She also tells you that she has been sick recently.

You apply high-flow oxygen and begin transporting her to the nearest hospital. As you obtain your second set of vital signs, you notice her wheezing is gone and her respiratory rate has slowed significantly, but she seems to be staring off into space. You try to get her attention, but she doesn't seem to notice you.

Why is her speaking ability significant? What factors point to the deterioration of her condition?

Issues: Importance of Immediate Transport for Critical Pediatric Patients, Importance of Ongoing Assessments.

Assessment in Action

You are off duty and are at the home of a close friend who has several children. You and your friend are enjoying a cup of coffee in the kitchen while some of the older, teenaged children play a board game in the next room. The family's 7-month-old infant, who can now crawl very easily, is also in the room with the other children. Suddenly you hear one of the teenagers scream, "Mom! Something's wrong with the baby!" You both rush into the living room to find the infant turning blue.

1. What is the first thing should you do?
   A. Ask about the infant's medical history.
   B. Check responsiveness.
   C. Visualize the airway.
   D. Provide back blows.

2. What do you suspect was the cause of the problem?
   A. Seizure
   B. Cardiac arrest
   C. Foreign body airway obstruction
   D. Asthma

3. What would be appropriate care for a 7-year-old in this situation?
   A. Back blows
   B. Chest thrusts
   C. Blind finger sweep
   D. Heimlich maneuver

4. The baby now seems fine. You should:
   A. watch the baby for 1 hour.
   B. put the baby down for a nap.
   C. take the baby to the emergency department.
   D. feed the baby.

5. The pediatric assessment triangle includes:
   A. appearance.
   B. work of breathing.
   C. skin signs.
   D. all of the above.

6. High-pitched inspiratory sounds are called:
   A. wheezing.
   B. grunting.
   C. stridor.
   D. none of the above.

Challenging Questions

7. Why is it important to use appropriate airway adjuncts for the pediatric patient?

8. How is the AVPU scale modified for use in children?

9. What is the significance of bradypnea in a child who has been previously exhibiting signs of increased work of breathing?

10. How do physical exams differ in young children than adults?

www.EMTB.com

Geriatric Emergencies

Objectives*

Cognitive

1. Define the term "geriatric." (p 986)
2. Discuss appropriate ways to communicate with geriatric patients. (p 986)
3. Discuss the GEMS diamond. (p 987)
4. State the leading causes of death of the geriatric population. (p 988)
5. Describe the physiologic changes of aging. (p 988)
6. Define the problem known as polypharmacy. (p 991)
7. State the principles and use of advance directives involving older patients. (p 995)
8. Define elder abuse. (p 996)
9. Discuss the causes of elder abuse. (p 996)
10. Discuss why the extent of elder abuse is not well known. (p 997)

Affective

11. Explain why the special needs of older people and the changes that the aging process brings about in physical structure, body composition, and organ function provide a fundamental knowledge base for maintenance of life support functions. (p 988)

Psychomotor

None

*All of the objectives in this chapter are noncurriculum objectives.

You are the Provider

You and your EMT-B partner are dispatched to 1204 Robbie Ln for a geriatric assist call. While en route, dispatch provides a patient update stating that your patient is an 86-year-old woman complaining of shortness of breath. As you approach the house, your partner tells you that he has responded to the same location a number of times in the past. He states that the patient's name is Ms McCoy. "She is just the sweetest lady."

1. What types of conditions could you expect to find in a patient of this age?
2. Would your treatment for this patient be any different from the treatment for a younger patient?

Geriatrics

Geriatric patients are people who are older than 65 years. It stands to reason that we do not age overnight. A decline in our body systems starts in our late 20s and progresses slowly throughout our lifespan. In this chapter, we use 65 years as the threshold age to be consistent with the definition used by other medical groups and governmental agencies.

No one relishes the thought of growing old, but the reality is that we all will, and the older population will continue to grow into a larger percentage of the population. According to US Census Data, almost 35 million people are older than 65 years. It is projected that by the year 2020, the geriatric population will be greater than 54 million. This is a very significant evolutionary trend for EMT-Bs because older people are major users of the EMS and health care systems in general. These calls may be difficult for you because the "classic" presentation of medical conditions common in younger patients may be altered. An acute myocardial infarction may present as jaw or neck pain, and some older patients may not experience chest pain at all. Their aging body can mask serious medical conditions. Older patients frequently have chronic medical problems and may be taking numerous medications for their illnesses. Providing effective treatment for this growing number of patients will require that all EMT-Bs understand the issues related to aging and that they modify some of their assessment and treatment approaches.

While many EMT-Bs may relish the action of high-profile calls, the reality is that a large percentage of your patient contacts will involve older people. Lifesaving interventions for geriatric patients may include reviewing the home environment to ensure safe and livable conditions exist, providing information on preventing falls, and making referrals to appropriate social services agencies when needed. EMT-Bs who respond to the homes of geriatric patients are in an ideal position to provide not only immediate help, but also key information to others in the health care and social services systems. Often, simple preventive measures can help older people avoid further injury, costly medical treatment, and death. You are on the front line in helping to prevent geriatric emergencies and care for patients when emergencies occur.

Communication and the Older Adult

Good communication is essential to successful assessment and treatment of older patients. Many things make communicating with older patients challenging. The aging process brings about changes in vision, hearing, taste, smell, and touch. Also, there are changes in communication abilities that accompany aging, dementia, and other diseases. These symptoms may be bothersome, but they are considered a normal consequence of aging.

Communication Techniques

Your first words can gain the patient's trust. Speak respectfully when you introduce yourself. If you know the patient's name, use it. Older people may be insulted if you use their first name. Do not take that chance. If they suggest that you call them by their first name, it is fine to do so. If you do not know their first name, use "Sir" or "Ma'am." Do not use "Hon," "Dearie," or "Grandma." Use short words, and ask only one question at a time.

In general, when interviewing an older patient, the following techniques should be used:

- Identify yourself. Do not assume an older patient knows who you are.
- Be aware of how you present yourself. Avoid showing frustration and impatience through body language.
- Look directly at the patient.
- Speak slowly and distinctly.
- Explain what you are going to do before you do it. Use simple terms to explain the use of

www.EMTB.com

Technology

Interactivities

Vocabulary Explorer

Anatomy Review

Web Links

Online Review Manual

medical equipment and procedures, avoiding medical jargon or slang.

- Listen to the answer the patient gives you.
- Show the patient respect. Never use the patient's first name without his or her permission.
- Do not assume that all older patients are hard of hearing. Ask the patient if he or she can hear you, and verify by asking him or her to tell you his or her understanding of what you just said.
- Do not talk about the patient in front of him or her; to do so gives the impression that the patient has no say in any decision making. This is easy to forget when the patient has impaired cognitive (thought) processes or has difficulty communicating.
- Be patient.

As for patients of any age, older patients have more difficulty communicating clearly when they are stressed by an emergency or personal crisis.

The GEMS Diamond

When you are called on to care for older patients, it is important to remember certain key concepts. The GEMS diamond (Table 33-1 ▼) was created to help you remember what is different about the older patient. The GEMS diamond is not intended to be a format for the approach to geriatric patients, nor is it intended to replace the ABCs of care. Instead it serves as an acronym for the issues to be considered when assessing every older patient.

"G" of the GEMS diamond stands for "geriatric." When responding to an emergency involving an older patient, you should consider that older patients are different from younger patients and may present atypically.

"E" of the GEMS diamond stands for an environmental assessment. Assessment of the environment can help give clues to the patient's condition or the cause of the emergency. Is the home too hot or too cold? Is

---

TABLE 33-1 The GEMS Diamond

G Geriatric Patients

- Present atypically.
- Deserve respect.
- Experience normal changes with age.

E Environmental Assessment

- Check the physical condition of the patient's home: Is the exterior of the home in need of repair? Is the home secure?
- Check for hazardous conditions that may be present (for example, poor wiring, rotted floors, unventilated gas heaters, broken window glass, clutter that prevents adequate egress).
- Are smoke detectors present and working?
- Is the home too hot or too cold?
- Is there an odor of feces or urine in the home? Is bedding soiled or urine-soaked?
- Is food present in the home? Is it adequate and unspoiled?
- Are liquor bottles present? If so, are they lying empty?
- If the patient has a disability, are appropriate assistive devices (for example, a wheelchair or walker) present?
- Does the patient have access to a telephone?
- Are medications out of date or unmarked, or are prescriptions for the same or similar medications from many physicians?
- If living with others, is the patient confined to one part of the home?
- If the patient is residing in a nursing facility, does the care appear to be adequate to meet the patient's needs?

M Medical Assessment

- Older patients tend to have a variety of medical problems, making assessment more complex. Keep this in mind in all cases—both trauma and medical. A trauma patient may have an underlying medical condition that could have caused or may be exacerbated by the injury.
- Obtaining a medical history is important in older patients, regardless of the chief complaint.
- Initial assessment
- Ongoing assessment

S Social Assessment

- Assess activities of daily living (eating, dressing, bathing, toileting).
- Are these activities being provided for the patient? If so, by whom?
- Are there delays in obtaining food, medication, or other necessary items? The patient may complain of this, or the environment may suggest this.
- If in an institutional setting, is the patient able to feed himself or herself? If not, is food still sitting on the food tray? Has the patient been lying in his or her own urine or feces for prolonged periods?
- Does the patient have a social network? Does the patient have a mechanism to interact socially with others on a daily basis?

the home well kept and secure? Are there hazardous conditions? Preventive care is also very important for a geriatric patient, who may not carefully study the environment or may not realize where risks exist.

"M" of the GEMS diamond stands for medical assessment. Older patients tend to have a variety of medical problems and may be taking numerous prescription, over-the-counter (OTC), and herbal medications. Obtaining a thorough medical history is very important in older patients.

"S" stands for social assessment. Older people may have less of a social network, because of the death of a spouse, family members, or friends. Older people may also need assistance with activities of daily living, such as dressing and eating. There are numerous social agencies that are readily available to help geriatric patients. Consider obtaining information pamphlets about some of the agencies for older people in your area. If you have these brochures with you and encounter a person in need, you can provide this valuable information. Social agencies that deal with the older population will be more than happy to share a listing of the services they provide.

The GEMS diamond provides a concise way to remember the important issues for older patients. Using this concept will help you make appropriate referrals, and as a result, you will help older patients maintain their quality of life.

Leading Causes of Death

The leading causes of death in the geriatric population include heart disease, cancer, stroke, chronic obstructive pulmonary disease and other respiratory illnesses, diabetes, and trauma. The physiologic aging of older people makes them more vulnerable to the effects of disease and injury than a younger individual. In addition, acute illness and trauma are more likely to involve organ systems beyond those initially involved. For example, in a geriatric patient who has fallen and fractured a hip, pneumonia may develop during recovery.

Physiologic Changes That Accompany Age

As we get older, we experience physiologic changes. In general, a 65-year-old person cannot expect to have the same degree of physical performance as when he or she was 30 years old. By the time a person reaches 65 years, the amount of total body water and the number of body cells has decreased by as much as 30%. Generally, after age 30, organ systems begin to deteriorate at a rate of roughly 1% per year. However, aging does not necessarily mean that a person will experience disease.

Common stereotypes about older people include the presence of mental confusion, illness, a sedentary lifestyle, and immobility. Although these perceptions are common, they are usually far from the norm. Older people can stay fit and be active, even though they will not be able to perform at the same level as they did in their youth. Most older people lead very active lives, participating in the community and in sports, and they are generally healthy despite the aging process (Figure 33-1 ▼).

Skin

As we get older, our skin becomes thinner and wrinkled. Collagen, a protein that is the chief component of connective tissue and bones, and elastin, a protein that helps to make the skin pliable, are lost as we age. The layer of fat under the skin also becomes thinner. As the elasticity of the skin declines, bruising becomes more common because the skin can tear more easily. Without the elasticity of the skin and the cushioning that the fat provides, the skin does not constrict and stop the bleeding as quickly when it is injured. This causes a greater number of bruises and larger ones from minimal trauma. There are also fewer sweat glands, so older skin tends to feel dry. Another problem that affects the skin is pressure ulcers, sometimes referred to as bedsores. Pressure

Figure 33-1 Older people can stay fit and be active.

ulcers form when the patient is lying in the same position for a long time. The pressure from the weight of the body cuts off the blood flow to the area of skin. With no blood flow to the skin, a sore develops. These sores can develop in as little as 45 minutes.

Senses

The pupils of the eyes begin to lose the ability to handle changes in light and require more time to adjust, which can make driving and walking more hazardous (Figure 33-2 ▼). Light changes can cause problems of visual acuity and depth perception. Cataracts, clouding of the lenses or their surrounding membranes, interfere with vision and make it difficult to distinguish colors and see clearly, increasing the likelihood of falls and accidents, and accounting for some mistakes in taking various medications. Hearing is the sensory change that affects the most people. Changes in the inner ear make hearing high-frequency sounds difficult. Changes in the ear can also cause problems with balance and make falls more likely. Changes in appetite may occur because of a decrease in the number of taste buds. The sense of touch lessens from loss of the end nerve fibers. This loss, in conjunction with the slowing of the peripheral nervous system, can create situations in which an older person can be injured. For example, an older person may be slow to react when touching something hot; this delayed response could result in a burn.

Respiratory System

Although the alveoli in an older person's lung tissue become enlarged, their elasticity decreases, making it harder to expel used air. This change in lung tissue

Figure 33-2 Changes in vision, hearing, posture, and motor ability predispose older people to a greater risk of being struck by a vehicle.

quality is comparable to a balloon that has been expanded and then deflated; the balloon loses some of its ability to contract to its original state after inflation. The lack of elasticity results in a decreased ability to exchange oxygen and carbon dioxide. The body's chemoreceptors, which monitor the changes in oxygen and carbon dioxide levels in the blood, slow with age. This can present as lower pulse oximetry readings, even in healthy people. A decrease in the number of cilia that line the bronchial tree lessens the ability to cough and, therefore, increases the chances of infection. Patients lose muscle mass in the chest and may get less help from muscles in the chest wall when they have trouble breathing.

Cardiovascular System

Cardiac output is a measure of the workload of the heart. We normally compensate for an increased demand on the cardiovascular system by increasing our heart rate, increasing the contraction of the heart, and constricting the blood vessels to nonvital organs. Aging decreases a person's ability to speed up contractions, to increase contraction strength, and to constrict, or narrow, blood vessels (called vasoconstriction) because of stiffer vessels. Many geriatric patients are at risk for atherosclerosis, an accumulation of fatty material in the arteries. Major complications of atherosclerosis include myocardial infarction and stroke. The presence of arteriosclerosis, a disease that causes the arteries to thicken, harden, and calcify, makes stroke, heart disease, hypertension, and bowel infarction more likely. Older people are also at an increased risk for aneurysm, an abnormal, blood-filled dilation of the wall of a blood vessel. Severe blood loss can occur when an aneurysm bursts.

Renal System

Kidney function in older people declines from 20% to 50% because of a decrease in the number of nephrons. The kidneys are important in eliminating certain medications from a person's system. With a decrease in renal function, levels of medications may rise, creating the impression of an overdose. Electrolyte disturbances are also more likely to occur with the lowered filtering of the blood, which can often be the cause of altered mental status in older people.

Nervous System

The number of brain cells in some areas may decrease by as much as 45%. By age 85 years, a 10% reduction in brain weight can result in increased risk of head trauma. Short-term memory impairment, a decrease in the ability to perform psychomotor skills, and slower reflex times are normal in the aging process. This decline may make assessment of older patients challenging. Previous injury or illnesses that are not associated with the current problem may also alter the assessment findings. It is important to compare older patients' current status with their normal abilities.

Musculoskeletal System

The disks between the vertebrae begin to narrow, and a decrease in height of between 2″ and 3″ may occur through the lifespan. A decrease in the amount of muscle mass often results in less strength, and fractures are more likely to occur because of a decrease in bone density also known as osteoporosis. Posture also changes as flexion at the neck and a forward curling of the shoulders produce a condition called kyphosis (also called "humpback", "hunchback", or a "Pott curvature"), making immobilization of geriatric patients more challenging (Figure 33-3 ▶).

Gastrointestinal System

A decrease in the volume of saliva and gastric juices causes a dry mouth, making it harder to chew and begin to digest foods. A slowing of the intestinal tract may cause constipation or fecal impaction. Decreased liver function makes it harder to detoxify the blood and eliminate substances such as medications and alcohol. This can make it difficult for patients and their physician to find the appropriate dosage for new medications.

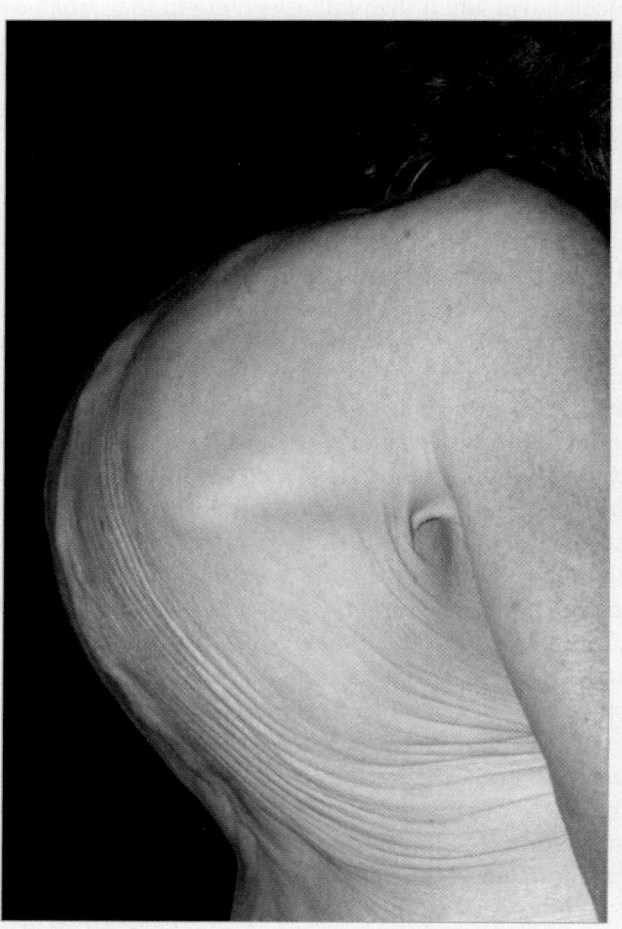

Figure 33-3 Older people often develop kyphosis, in which the back becomes hunched.

You are the Provider   Part 2

A woman meets you at the front door. She appears to be in her mid to late 40s. She states that her mother is not feeling well. "She needs a ride to the hospital but I don't have a car." As you enter the house, you notice a strong smell of urine and feces. There are numerous cats and dogs running around the living room. Your patient is lying on the couch and appears quite lethargic. Your partner approaches her. "Hello Ms McCoy. It's Jim from the fire department. Do you remember me?" She moans in response.

3. What is your first impression of this situation?
4. What is your first step in treatment?

Polypharmacy

Older people account for one eighth of the population but use one quarter of the prescribed medications and one third of the OTC medications sold in the United States. Many medications can have interactions or counter actions when taken together. Polypharmacy refers to the use of multiple prescription drugs by a single patient, causing the potential for negative effects such as overdosing or drug interaction.

Checking a patient's medications should be an easy task; however, physicians face many barriers to this. Many patients have more than one physician, such as a family physician for everyday care, a cardiologist for the heart, and an endocrinologist for care of diabetes, all of whom may prescribe medications for the patient. But what if the patient does not tell each physician about all of the medications that they take? The patient may not remember what medications another doctor prescribed or may not want to tell one doctor about seeing another.

Other sources of medications include OTC medicines such as aspirin, antacids, cough syrups, and decongestants (Figure 33-4 ▶). Herbal remedies can also interact with prescribed or OTC medications. You must be aware of the implications of medication issues when caring for geriatric patients. Complete assessment of a patient includes getting a medication history.

Impact of Aging on Trauma

You must consider the body's decreasing ability to isolate simple trauma when you are assessing and caring for a geriatric patient. An isolated hip fracture in a healthy 25-year-old adult is rarely associated with overall decline. However, the same injury in an 85-year-old patient can produce a wide-ranging, systemic impact that results in deterioration, shock, and life-threatening hypoxia, a dangerous condition in which the body tissues and cells do not have enough oxygen. Although an injury may be considered isolated and not alarming in most adults, an older patient's overall physical condition may lessen the body's ability to compensate for the effects of even simple injuries. Younger patients have the ability to increase their heart rate, constrict their blood vessels, and breathe faster and deeper to

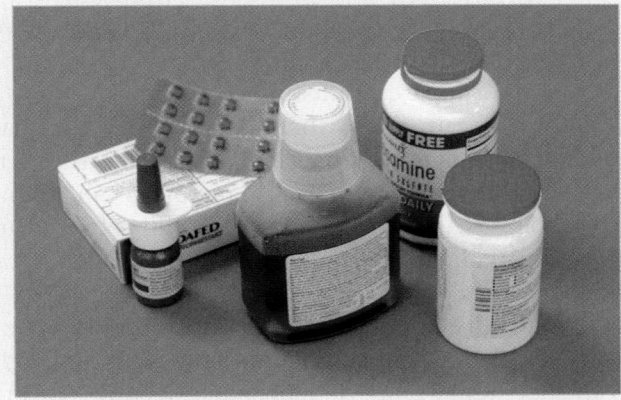

Figure 33-4 Over-the-counter medications such as aspirin, antacids, cough syrups, and decongestants can interact with prescription medications.

compensate for injuries. The aging body has a heart that can no longer beat faster, vessels that cannot constrict due to atherosclerosis, and lungs that do not exchange oxygen as well.

Your assessment of the patient's condition and stability must include past medical conditions, even if they are not currently acute or symptomatic. For example, suppose you respond to a call about a patient with a history of unstable angina who sustains a simple isolated fracture of the ankle. You must consider this patient to be in unstable condition and provide prompt transport before the stress and simple trauma worsen the angina and an unstable condition occurs.

Falls and Trauma

A medical condition such as fainting, a cardiac rhythm disturbance, or a medication interaction may lead to a fall that injures the patient. Whenever you assess a geriatric patient who has fallen, it is important to find out why the fall occurred. Was the patient dizzy before the fall? Does the patient remember the fall? Did a fainting episode cause the fall and injury, or did the patient trip on something or lose balance? Sometimes, a recent history of starting or stopping blood pressure medication is enough to cause a patient to become dizzy and fall. Consider that the fall may have been caused by a medical condition, and look carefully for clues from the patient, bystanders, and the environment. Although the trauma that the patient sustains from the fall can be serious, you should also consider that if a medical condition caused the fall, it may continue to be life threatening.

When you respond to a motor vehicle crash, be alert to the possibility that a medical emergency may have caused the accident, especially in single vehicle collisions with no apparent cause.

Because brain tissue shrinks with age, older patients are more likely to sustain closed head injuries, such as subdural hematomas. These hematomas can go unnoticed because the blood has a void to fill before it can produce pressure in the skull showing the familiar signs of head trauma.

As a result of bone loss from <u>osteoporosis</u>, a generalized bone disease that is commonly associated with postmenopausal women, older patients of both sexes are prone to fractures, especially in areas such as the hip. With age, the spine stiffens as a result of shrinkage of disk spaces, and vertebrae become brittle. Compression fractures of the spine are more likely to occur.

Because of the amount of flexion that occurs in the spinal column, hip, or knee of older patients, use of conventional splints and backboards to immobilize the patient may be difficult or impossible unless a lot of padding is used. What is considered a normal anatomic position for children and adults is often very abnormal for some geriatric trauma patients. You should try to determine the patient's baseline condition and what was normal for the patient before the accident. Trying to force a patient with pronounced joint flexion into "normal" anatomic position can be very painful for the patient and frustrating for you and should never be done. Some devices, such as traction splints, simply do not work on patients with flexed hips and knees. Splinting devices such as vacuum mattresses that conform to body contours may be a good choice for immobilization (Figure 33-5 ▶). Remember that when you treat a geriatric trauma patient, you must assess the injuries and carefully look for the cause of the fall or crash.

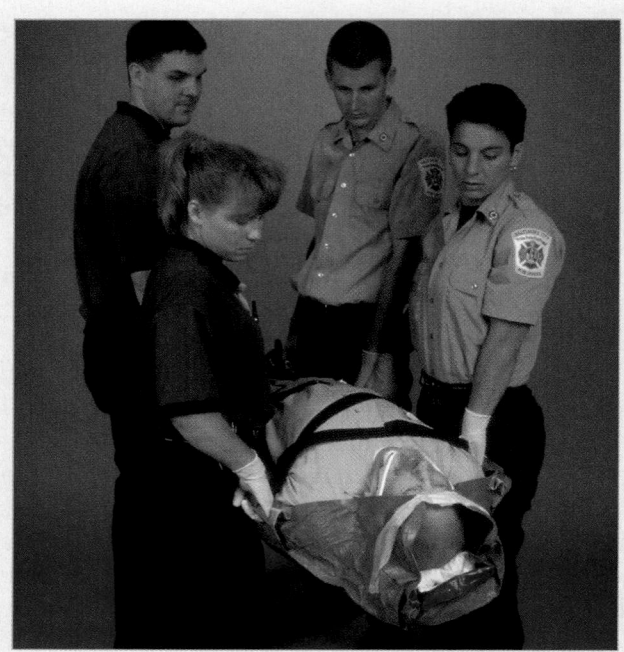

Figure 33-5 Vacuum mattresses that conform to body contours can be a good choice for immobilizing geriatric patients.

Impact of Aging on Medical Emergencies

Syncope

You should always assume that <u>syncope</u>, or fainting, in an older patient is a life-threatening problem until proven otherwise. Syncope is often caused by an interruption of blood flow to the brain. Syncope has many causes; some are serious, and others are not. Regardless, an older person who has a period of unconsciousness should be examined to determine the cause of the syncope. (Table 33-2 ▶) shows some of the causes of syncope in geriatric patients.

Heart Attack

The classic symptoms of a heart attack are often not present in geriatric patients. As many as one third of older patients have "silent" heart attacks in which the usual chest pain is not present. (Table 33-3 ▶) shows signs and symptoms that are commonly noted in geriatric patients who are experiencing a heart attack.

Acute Abdomen

Because of an aging nervous system, abdominal complaints in geriatric patients are extremely difficult to assess. A number of life-threatening abdominal problems are common in older patients. In the field, the one real threat from abdominal complaints is blood loss, which can lead to shock and death. <u>Abdominal aortic aneurysm (AAA)</u> is one of the most rapidly fatal conditions. An AAA tends to develop in people who have a history of hypertension and atherosclerosis. The walls of the aorta weaken and blood begins to leak into the layers of the vessel, causing the aorta to bulge like a bubble on a tire. If enough blood is lost into the vessel wall itself, shock occurs. If the wall bursts, it rapidly

TABLE 33-2 Possible Causes of Syncope in Geriatric Patients	
Arrhythmias and heart attack	The heart is beating too fast or too slowly, the cardiac output drops, and blood flow to the brain is interrupted. A heart attack can also cause syncope.
Vascular and volume changes	Medication interactions can cause venous pooling and vasodilation, widening of a blood vessel, which results in a drop in blood pressure and inadequate blood flow to the brain. Another cause of syncope can be a drop in blood volume because of hidden bleeding from a condition such as an aneurysm.
Neurologic cause	A transient ischemic attack or a "brain attack" can sometimes mimic syncope.

TABLE 33-3 Common Signs and Symptoms of Heart Attack in Geriatric Patients	
Dyspnea	Dyspnea, the feeling of shortness of breath or difficulty breathing, is a common complaint in geriatric patients and is sometimes associated with heart attack. It is often combined with other symptoms, such as nausea, weakness, and sweating. Chest pain associated with angina typically has an onset during periods of stress or exertion. In geriatric patients, chest pain is often not present, but exertional dyspnea is. As the disease progresses, dyspnea may occur without exertion. Dyspnea in older people can be the equivalent of chest pain in younger patients who are having angina or a heart attack. In addition, congestive heart failure and acute pulmonary edema may result from the "silent" heart attack.
A weak feeling	Weakness can have many causes, however, you should suspect a heart attack in a patient with a sudden onset of weakness. Weakness is often associated with sweating.
Syncope, confusion, altered mental status	Syncope can have many causes, and in geriatric patients, none of these causes should be presumed to be minor. Major life-threatening causes of syncope are often cardiac in origin. Altered mental status is usually a signal of poor blood supply to the brain, often from cardiac arrhythmia and heart attack.

leads to fatal blood loss. When the problem is found early, there is a chance to repair the vessel before rupture and fatal blood loss occur.

A patient with an AAA most commonly reports abdominal pain radiating through to the back with occasional flank pain. If the AAA becomes large enough, it can be felt as a pulsating mass just above and slightly to the left of the navel during your physical examination. Occasionally, the AAA causes a decrease in blood flow to one of the legs, and the patient complains of some discomfort in the affected extremity. Assessment may also reveal diminished or absent pulses in the extremity. Compensated shock (early shock) and decompensated shock (late shock) as a result of blood loss are common occurrences. Because of a decrease in blood volume and decreased blood flow to the brain, the patient may experience syncope. You should treat the patient for shock and provide prompt transport to the hospital.

Another cause of abdominal pain and shock is gastrointestinal bleeding, which can occur for a variety of reasons and is usually heralded by the vomiting of blood or material that looks like coffee grounds. Bleeding that travels through the lower digestive tract usually manifests as black or tarry stools, whereas frank red blood usually means a local source of bleeding, such as hemorrhoids. A patient with gastrointestinal bleeding may experience weakness, dizziness, or syncope. Bleeding into the gastrointestinal system can be life threatening because of the potential for blood loss and shock.

Bowel obstructions occur frequently in the geriatric population. The gastrointestinal tract slows with aging, and the patient can experience problems having bowel movements. When these patients go into the bathroom and are straining to have a bowel movement, they can stimulate the vagus nerve and produce a reaction called vasovagal attack, in which the heart rate drops dramatically and the patient becomes dizzy or passes out. This patient will usually be in stable condition on your arrival but requires transport to rule out other conditions.

Altered Mental Status

Because of our stereotypical perceptions about older people, we may expect them to forget names or not be able to remember events or learn new things. However, these types of changes in mental status are not part of the normal aging process. They may be part of a slow deterioration of a condition or a disease of rapid onset, neither of which is normal. To determine the onset of this change in mental status, you must compare the patient's ability to function with that of the recent past. This will help to establish a baseline and give some perspective on the onset of the change. The two terms that are often used to describe a change in mental status are "delirium" and "dementia."

Delirium is a change in mental status that is marked by the inability to focus, think logically, and maintain attention. Acute anxiety may be present in addition to the other symptoms. Usually, memory remains mostly intact. Delirium is commonly marked by acute or recent onset and is a "red flag" for some type of new health problem. Delirium may be caused by tumors, fever, or drug or alcohol intoxication or withdrawal. Delirium can be present from metabolic causes as well. Any time a patient has an acute onset of delirious behavior, you should rapidly assess the patient for the following three conditions:

- Hypoxia
- Hypovolemia
- Hypoglycemia

Any of these three conditions, if left unrecognized or untreated, may be rapidly fatal. Delirium is short in onset and usually curable if identified early.

Dementia is the slow onset of progressive disorientation, shortened attention span, and loss of cognitive function. Dementia develops slowly over a period of years rather than a few days. Alzheimer's disease,

cerebrovascular accidents, and genetic factors may cause dementia. Dementia is usually considered irreversible and is one of the pathophysiologic features of many neurologic diseases. The patient's history and determination of function in the recent past are key factors in determining the baseline. A patient with dementia may be experiencing a delirious event. Delirium is caused by emergency problems; dementia is not.

Impact of Aging on Psychiatric Emergencies

For the majority of older people, the later years are ones of fulfillment and satisfaction with a lifetime of accomplishments. For some older adults, however, later life is characterized by physical pain, psychological distress, doubts about the significance of life's accomplishments, financial concerns, loss of loved ones, dissatisfaction with living conditions, and seemingly unbearable disability. When these factors lead to hopelessness about the possibility for positive change in their lives, depression and, unfortunately, even suicide are possible outcomes. You are often the first health care professional to have contact with older adults with these afflictions.

Depression

Depression is a common, often debilitating psychiatric disorder experienced by approximately 2 million older American adults. Older adults residing in skilled nursing facilities are even more likely to be depressed. Depression is diagnosed three times more commonly in women than in men. In contrast with the normal emotional experiences of sadness, grief, loss, and temporary "bad moods," depression is extreme and persistent

You are the Provider   Part 3

Your patient is conscious and disoriented. She is breathing at a rate of 16 breaths/min. You find no apparent external bleeding. Her skin is pale and cool to the touch. As you check for a radial pulse you notice how thin and frail she is. Her pulse is 82 beats/min and regular. Lung sounds are clear and equal bilaterally. Her daughter states that she refuses to eat or take care of herself. You note a strong smell of body odor, urine, and feces while examining your patient.

5. What is your index of suspicion with this patient?
6. What additional resources could you request in this situation?

and can interfere significantly with an older adult's ability to function. It is impossible to predict which older adults will have depression, but studies indicate that substance abuse, isolation, prescription medication use, and chronic medical conditions all contribute to the onset of significant depression. Treatment of severe depression in older adults usually consists of psychological counseling, medication, or a combination of both. For many older adults, simply reestablishing relationships with the community or with family is enough to lessen the severity of the illness.

Suicide

Older men have the highest suicide rate of any age group in the United States. Equally concerning is the fact that older persons who attempt suicide choose much more lethal means than younger victims and generally have diminished recuperative capacity to survive an attempt. Suicide can happen in any family, regardless of socioeconomic class, culture, race, or religious affiliation. Some common predisposing events and conditions include death of a loved one, physical illness, depression and hopelessness, alcohol abuse and dependence, and loss of meaningful life roles. Frequently, EMT-Bs are the first contact, and for some older patients, the only health care contact. Keep in mind that only a small percentage pursue medical treatment for psychological issues. Not only do many fail to seek care, but they also frequently deny the problem when questioned. It is vital that all members of the health care team be aware of the issues and take appropriate steps to ensure patient safety and initiate effective treatment.

Advance Directives

Many people today are making use of <u>advance directives</u>, specific legal papers that direct relatives and caregivers about what kind of medical treatment may be given to patients who cannot speak for themselves. An advance directive is also commonly called a living will. Mentally competent adults and emancipated minors have the right to consent to or decline treatment, provided they are competent to do so. The definition of competence is often hotly debated, but a person who is older than 18 years, alert, and not intoxicated, and who understands the consequences of his or her decision is generally deemed competent. Unfortunately, patients who are unconscious or in a medical crisis are not able to inform medical personnel about their wishes to consent to or decline treatment. It is dangerous to take

someone else's word for what the patient's wishes are; written advance directives have been developed for this reason.

Advance directives may also take the form of "Do Not Resuscitate" (DNR) orders (Figure 33-6 ▼). A DNR order gives you permission to not attempt resuscitation for a patient in cardiac arrest. However, for a DNR order to be valid, in general, the patient's medical problems must be clearly stated, and the form must be signed by the patient or legal guardian and by one or more physicians. In most states, the form must be dated within the preceding 12 months. Even in the presence of a DNR order, you are still obligated to provide supportive measures that may include oxygen delivery, pain relief, and comfort when you can. Learn and become familiar with your state laws regarding this issue.

A health care power of attorney is an advance directive that is exercised by a person who has been authorized by the patient to make medical decisions for the patient. Be sure to follow your service's protocol when faced with any advance directive.

Dealing with advance directives has become more common for EMS providers because more individuals are electing to use hospice services and spend their final days at home. Although advance directives may be in place, family members or caregivers who are faced with the final moments of life or when the patient's condition worsens often become alarmed and call 9-1-1.

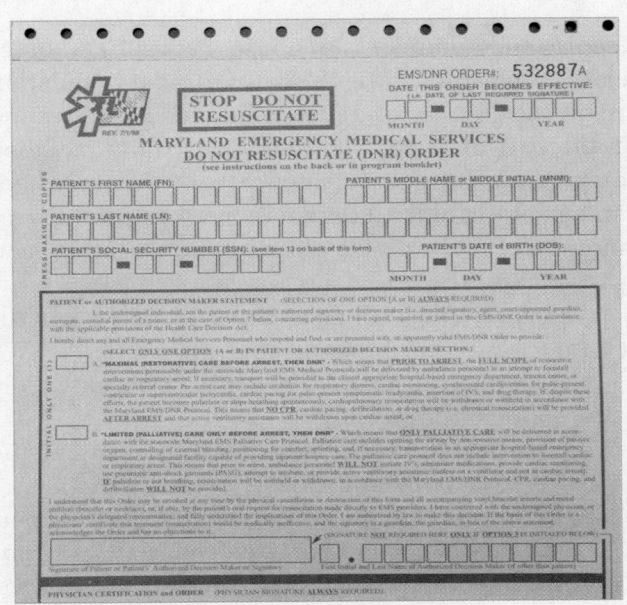

Figure 33-6 A DNR order is commonly used to identify a patient who does not want to be resuscitated.

Family members and caregivers may then become upset when you take resuscitative action and begin transportation to the hospital.

Another common situation is the transportation of patients from nursing facilities. Specific guidelines vary from state to state; however, you should consider the following general guidelines:

- Patients have the right to refuse treatment, including resuscitative efforts, provided that they are able to communicate their wishes.
- A DNR order is valid in a health care facility only if it is in the form of a written order by a physician.
- You should periodically review state and local protocols and legislation regarding advance directives.
- When you are in doubt or when there are no written orders, you should try to resuscitate the patient.

It is absolutely essential that every EMT-B become familiar with his or her state regulations regarding advance directives. Every service should also provide additional training on the actions you should take when presented with advance directives. When in doubt, your best course of action is to take resuscitative action that is appropriate to the situation and to practice sound medical treatment.

Elder Abuse

Reports and complaints of abuse, neglect, and other related problems among the nation's older population are on the rise. Elder abuse is defined as any action on the part of an older individual's family member, caregiver, or other associated person that takes advantage of the older individual's person, property, or emotional state; it is also called "granny beating" and "parent battering."

The exact extent of elder abuse is not known for several reasons, including the following:

- Elder abuse is a problem that has been largely hidden from society.
- The definitions of abuse and neglect among the geriatric population vary.
- Victims of elder abuse are often hesitant to report the problem to law enforcement agencies or human and social welfare personnel.

A parent who feels ashamed or guilty because he or she raised the abuser is a typical victim of elder abuse. The abused person may also feel traumatized by the sit-

uation or be afraid that the abuser will try to get back at him or her. In some areas of the country, there is a lack of formal reporting mechanisms, and some states lack statutory provisions that require that elder abuse be reported.

The physical and emotional signs of abuse, such as rape, spouse beating, or nutritional deprivation, are often overlooked or not accurately identified. Older women in particular are not likely to report incidents of sexual assault to law enforcement agencies. Patients with sensory deficits, senility, and other forms of altered mental status, such as drug-induced depression, may not be able to report abuse.

Elder abuse occurs most often in women older than 75 years. The abused person is often frail with multiple chronic medical conditions, has dementia, and may have an impaired sleep cycle, sleepwalking, and periods of shouting at others. The person may be incontinent and in general is dependent on others for activities of daily living.

Abusers of older people are often products of child abuse themselves, and the abuse that is inflicted on the older person may be retaliatory. Most of these abusers are not trained in the particular care that older people require and have little relief time from the constant care demands of their own family, children, and spouse. Their lives are now complicated by the constant, demanding needs of the older person they have to care for.

The abuser may also have marked fatigue, be unemployed with financial difficulties, and abuse one or more substances. With a careful eye, you can recognize the clues to these stressful situations and help guide the family toward programs in their community that are geared to helping the whole family. Programs such as adult daycare, Meals on Wheels, and many local individualized programs help to decrease the stress put on the family and lower the chances of abuse.

Abuse is not restricted to the home; environments such as nursing, convalescent, and continuing care centers are also sites where older people sustain physical, psychological, or pharmacologic harm. Often, care providers in these environments consider older people to be management problems or categorize them as obstinate and undesirable patients.

Assessment of Elder Abuse

While assessing the patient, you should try to obtain an explanation of what happened. You should suspect abuse when answers to questions about what caused the injury are concealed or avoided.

You must also suspect abuse when you are given unbelievable answers from anyone other than the patient, the possible abuser, or a significant witness. You should be suspicious if you think "Does this make sense?" or "Do I really believe this story?" while reviewing the patient's history. If you see burns, especially cigarette burns or physical marks that indicate that certain parts of the patient's body have been scalded systematically, you must also suspect abuse. As an EMT-B, you may be the first health care provider to observe the signs of possible abuse. Information that may be important in assessing possible abuse includes the following:

- Repeated visits to the emergency department or clinic
- A history of being accident-prone
- Soft-tissue injuries
- Unbelievable or vague explanations of injuries
- Psychosomatic complaints
- Chronic pain
- Self-destructive behavior
- Eating and sleep disorders
- Depression or a lack of energy
- Substance and/or sexual abuse

You should remember that many patients who are being abused are so afraid of retribution that they make false statements. A geriatric patient who is being abused by family members may lie about the origin of abuse for fear of being thrown out of the home. In other cases of elder abuse, sensory deprivation or dementia may hinder adequate explanation.

In addition to the lifesaving care that you can provide the patient, your examination of the patient can help to reduce further trauma from abuse through its very identification. Repeated abuse can lead to a high risk of death. A preventive measure in reducing additional maltreatment of the patient is identification of the abuse by emergency medical providers (Table 33-4 ▲). This may allow for referral and protective services of human, social, and public safety agencies.

TABLE 33-4	Categories of Elder Abuse
Physical	■ Assault
	■ Neglect
	■ Dietary
	■ Poor maintenance of home
	■ Poor personal hygiene
Psychological	■ Benign neglect
	■ Verbal
	■ Treating the person as an infant
	■ Deprivation of sensory stimulation
Financial	■ Theft of valuables
	■ Embezzlement

Signs of Physical Abuse

Signs of abuse may be quite obvious or subtle. Inflicted bruises are usually found on the buttocks and lower back, genitals and inner thighs, cheeks or earlobes, upper lip and inside the mouth, and neck. Pressure bruises caused by the human hand may be identified by oval grab marks, pinch marks, or handprints. Human bites are typically inflicted on the upper

You are the Provider   Part 4

You and your partner continue to attempt to communicate with Ms McCoy but she just moans in response. Her daughter states that her mother has not been feeling very well for the last few weeks. "She just lies on the couch and refuses to get up. That's why I called you." As you continue assessing your patient, you find a number of bedsores on her back and buttocks. Given the obvious signs of neglect, your partner goes out to the ambulance to notify dispatch as well as adult protective services.

7. What treatment can you provide this patient?
8. Should you consider confronting the daughter about her mother's physical condition?

Documentation Tips

As with other legally complex and emotionally charged issues, the possibility of elder abuse demands particularly careful documentation. Be thorough, objective, and factual, avoiding unsupported opinions and personal judgments. You may be called on to explain your report in a legal proceeding.

Figure 33-7 Check for signs of neglect, such as evidence of a lack of hygiene, poor dental hygiene, poor temperature regulation, or lack of reasonable amenities in the home.

extremities and can cause lacerations and infection. You should inspect the patient's ears for indications of twisting, pulling, or pinching or evidence of frequent blows to the outer ears. You should also investigate multiple bruises in various states of healing by asking the patient and reviewing the patient's activities of daily living.

Burns are a common form of abuse. Typical abuse from burns is caused by contact with cigarettes, matches, heated metal, forced immersion in hot liquids, chemicals, and electrical power sources.

It may be difficult to see a failure to thrive in an older patient who has been abused. You should observe the patient's weight and try to determine whether the patient appears undernourished or has been unable to gain weight in the current environment. Does the patient have a ravenous appetite? Has medication been withheld? Is money being withheld, so the patient cannot buy food or medicine? You should also check for signs of neglect, such as evidence of a lack of hygiene, poor dental hygiene, poor temperature regulation, or lack of reasonable amenities in the home (Figure 33-7 ▶).

You must regard injuries to the genitals or rectum with no reported trauma as evidence of sexual abuse in any patient. Geriatric patients with altered mental status may never be able to report sexual abuse. In addition, many women do not report cases of sexual abuse because of shame and the pressure to forget.

You are the Provider   Summary

When dealing with a possible abuse situation, never confront the suspected abuser. Your best and most appropriate course of action is to treat, transport, and report. Removing the patient from the abusive situation is the best treatment you can provide. As with all patients, ensure that you give the receiving facility a complete patient report and document every aspect of the call.

Prep Kit

Ready for Review

- Management of geriatric patients can present you with many challenges that are not encountered with younger patients and with a host of different problems that may be quite difficult and frustrating to solve.

- The GEMS diamond is a tool to help remember key concepts when assessing geriatric patients. G stands for geriatric—older patients present atypically. E is for environment—be aware of the patient's living environment. M reminds you that older patients usually have more medical problems (and medications), so medical history is important. S emphasizes the social aspect of the patient's life, and the importance of the patient's social network.

- The health problems of the older population are multifaceted, and frequent barriers to communication can be expected.

- The leading causes of death in older people include heart disease, cancer, stroke, cardiopulmonary disease, respiratory illnesses, diabetes, and trauma.

- Physiologic changes that accompany aging include the following:
  - Changes to skin quality
  - Weakening of the senses
  - Decrease in respiratory system function
  - Compromised cardiac and renal function
  - Decline in nervous system function
  - Decrease in muscle and bone mass
  - Decrease in gastrointestinal system function

- These changes affect the body's ability to isolate simple trauma and may influence how other medical emergencies present.

Technology

Interactivities

Vocabulary Explorer

Anatomy Review

Web Links

Online Review Manual

- Polypharmacy refers to taking many medications or taking too many medications. Complete assessment of a patient must include getting a medication history.

- To determine the onset of altered mental status, compare the patient's ability to function with that of the recent past. Delirium and dementia are the two terms often used to describe a change in mental status.

- Advance directives, or living wills, specify what kind of medical treatment may be given to patients who cannot speak for themselves. Every EMT-B should be familiar with his or her state regulations regarding advance directives.

- The exact extent of elder abuse is not known because many patients do not report it.

- Abusers are often family members who must care for the older person in addition to caring for their own spouses and children. Elder abuse also occurs in nursing, convalescent, and continuing care centers.

- Elder abuse can be gruesome, vulgar, and barbaric; however, your responsibility is to provide lifesaving care to the patient and try to reduce additional abuse through identification of the problem.

Vital Vocabulary

abdominal aortic aneurysm (AAA) A condition in which the walls of the aorta in the abdomen weaken and blood leaks into the layers of the vessel, causing it to bulge.

advance directives Written documentation that specifies medical treatment for a competent patient should the patient become unable to make decisions; also called living wills.

aneurysm A swelling or enlargement of a part of an artery, resulting from weakening of the arterial wall.

arteriosclerosis A disease that is characterized by hardening, thickening, and calcification of the arterial walls.

atherosclerosis A disorder in which cholesterol and calcium build up inside the walls of blood vessels, forming plaque, which eventually leads to partial or complete blockage of blood flow and the formation of clots that can break off and embolize.

www.EMTB.com

Prep Kit continued...

cataracts Clouding of the lens of the eye or its surrounding transparent membranes.

collagen A protein that is the chief component of connective tissues and bones.

compensated shock The early stage of shock, in which the body can still compensate for blood loss.

decompensated shock The late stage of shock, when blood pressure is falling.

delirium A change in mental status marked by the inability to focus, think logically, and maintain attention.

dementia The slow onset of progressive disorientation, shortened attention span, and loss of cognitive function.

dyspnea Shortness of breath or difficulty breathing.

elastin A protein found in elastic tissues such as skin and artery walls.

elder abuse Any action on the part of an older person's family member, caregiver, or other associated person that takes advantage of the older individual's person, property, or emotional state; also called granny beating and parent battering.

hypoxia A condition in which the body's cells and tissues do not have enough oxygen.

kyphosis A forward curling of the back caused by an abnormal increase in the curvature of the spine.

osteoporosis A generalized bone disease, commonly associated with postmenopausal women, in which there is a reduction in the amount of bone mass leading to fractures after minimal trauma in either sex.

syncope A fainting spell or transient loss of consciousness, often caused by an interruption of blood flow to the brain.

vasoconstriction The narrowing of a blood vessel.

vasodilation The widening of a blood vessel.

Points to Ponder

You are dispatched to 1218 South Main St for an older woman who has fallen. You arrive to find a two-story home overgrown with weeds and in a general state of disrepair. You knock on the door and hear the barking of small dogs. You attempt to open the door, but it is locked. Your partner finds access into the home through an open window. As you walk through the home trying to locate the patient, you notice stacks of papers, unwashed dishes and garbage littered throughout the residence. You finally locate the patient in an upstairs bedroom. She tells you that she is uninjured, but when she fell in her bathroom 3 days ago, she was unable to get up and was unable to reach the phone until today.

Are this patient's living conditions safe? Can she safely live alone?

Issues: Physical Changes in Older Persons that Affect Daily Living, Preventative Measures for Falls and Other Injuries.

Assessment in Action

You are dispatched to 111 Lionel Ln for a person with shortness of breath. You are greeted at the door by a person who tells you she is the live-in caregiver. She says, "She's faking. She always fakes breathing problems. I told her not to call you." The caregiver does not want you to come in the house. She tells you to go away.

You eventually talk the caregiver into letting you see the patient, who is located in a small, dark room in the basement. Her home oxygen tank is empty, and you notice that her clothes are dirty and her bedside commode is full. The caregiver watches from the doorway as you perform your assessment. Each time you ask a question, the patient looks over at the caregiver and does not answer, but she does follow simple commands. Her vital signs disclose the following: a blood pressure of 158/98 mm Hg, an irregular pulse of 110 beats/min, shallow respirations of 60 breaths/min, and a pulse oximetry reading of 78% on room air.

1. Given the nature of the dispatch and the history of home oxygen use, this patient's underlying medical condition is:
   A. diabetes.
   B. stroke.
   C. osteoporosis.
   D. COPD.

2. The "E" in the GEMS diamond stands for:
   A. emergency assessment.
   B. environmental assessment.
   C. elderly assessment.
   D. none of the above.

3. A dangerous condition in which the body's tissues do not have enough oxygen is called:
   A. hypovolemia.
   B. hypoglycemia.
   C. hypoventilation.
   D. hypoxia.

4. Shortness of breath is also referred to as:
   A. dysphagia.
   B. dysuria.
   C. dyspnea.
   D. dysrhythmia.

5. Polypharmacy refers to:
   A. the positive effects of taking multiple medications.
   B. the negative effects of taking multiple mediations.
   C. both A and B.
   D. neither A or B.

6. Written documentation that specifies medical treatment for a competent patient should the patient become unable to make decisions is called:
   A. an advance directive.
   B. an advance measure.
   C. a living will.
   D. both A and C.

Challenging Questions

7. What are the components of the GEMS diamond, and how do they apply to the above scenario?

8. What concerns you about the caregiver's attitude?

9. What do the patient's living conditions suggest?

10. The caregiver tells you not to transport this patient because she has a DNR. How do you respond?

www.EMTB.com

Geriatric Assessment and Management

Objectives*

Cognitive

1. Describe the following basics of patient assessment for the geriatric patient:
   - Scene size-up
   - Initial assessment
   - Focused history and physical exam
   - Detailed physical exam
   - Ongoing assessment (p 1004)
2. Discuss common chief complaints of older patients. (p 1008)
3. Describe trauma assessment in older patients for the following injuries:
   - Injuries to the spine
   - Head injuries
   - Injuries to the pelvis
   - Hip fractures (p 1008)
4. Describe acute illness in older people, including the following conditions:
   - Cardiovascular emergencies
   - Dyspnea
   - Syncope and altered mental status
   - Acute abdomen
   - Septicemia and infectious disease (p 1015)

5. Discuss response to older patients in nursing and skilled care facilities. (p 1016)

Affective

None

Psychomotor

6. Demonstrate the patient assessment skills that should be used to care for an older patient. (p 1004)

*All of the objectives in this chapter are noncurriculum objectives.

You are the Provider

It's 3:17 AM. You and your new EMT-B partner are dispatched to the all-too-familiar address of 7273 Dawn Pl for chest pain. You explain to your partner that you are going to visit Ms Gladys, a 74-year-old woman who lives alone and has no family in the area. She calls an average of once a week for one reason or another, although you have not transported her to the hospital in years. Every time you go to her house, she has freshly baked cookies waiting. Your partner gives you an irritated look and mumbles under his breath about being mad that he got out of bed for nothing.

1. Would this call be considered a medical emergency or system abuse?
2. While en route, how should you and your partner prepare yourselves for this call?

Geriatric Assessment and Management

This chapter covers assessment and management of older patients, building on the concepts introduced in Chapter 33. Assessing an older patient uses the same basic approach as for other patients; the steps are scene size-up, initial assessment, focused history and physical exam, detailed physical exam, and ongoing assessment. However, areas of the exam may require you to modify your approach to become more aware of the conditions around you that may affect geriatric patients. The injury or medical condition may be worse than indicated by the existing signs and symptoms, and the injuries and conditions that are found will have a more profound effect than they would in a younger patient.

In addition to the critical needs that an underlying medical problem may cause, the condition of older patients is more unstable than that of a younger patient and there is an increased possibility for sudden, rapid deterioration.

Assessing an Older Patient

As discussed in Chapter 33, a useful tool to use when assessing the geriatric patient is the GEMS diamond. It is designed to help you remember to look for the little things that can make a big difference in geriatric patients. Older patients require additional time to assess and may have preexisting diseases that affect your current findings.

Scene Size-up

Geriatric calls are more frequent than calls for other age groups, and often the complaints, as reported to dispatch, are more vague. The "G" in the GEMS diamond is for geriatric concerns. Consider carefully the potential nature of illnesses and mechanisms of injury (MOIs) as reported in your dispatch information. Reflect on how this information relates to special situations relative to older patients. For example, if your patient has experienced trauma and spinal immobilization is necessary, what equipment would you need to properly and comfortably stabilize the spine?

As you approach any scene, you must be keenly aware of the environment and the reason you were called. The "E" of the GEMS diamond is for the environment. When you arrive at a patient's residence, you should look for important clues to determine not only your safety, but also that of the occupant. The environment can provide a great deal of important information if you know what to look for. Look for hazards, such as steep stairs, missing handrails, poor lighting, or other things that could cause a fall. Another aspect to consider in the geriatric environment is the overall appearance of the home. Determine whether there is evidence of adequate food, water, heat, lights, and ventilation. Is the home clean? Many older patients may have a hard time physically keeping up with the cleaning or financially keeping food or heat in their home. The general condition of the home will give you some important clues.

Your assessment of the scene will continue even after you begin the assessment of the patient. Activities of daily living such as the ability to move around, talk on the telephone, prepare and eat meals, perform basic cleaning skills, and attend to personal hygiene are essential for continued health in all people. For older people, normal aging or a disease process may make activities of daily living difficult and cause a cascade of problems.

www.EMTB.com

Technology

Interactivities

Vocabulary Explorer

Anatomy Review

Web Links

Online Review Manual

Initial Assessment

General Impression

The sequence of the initial assessment is the same for pediatric, adult, and geriatric patients. Begin with a general impression, including determining the chief complaint and the patient's level of consciousness (LOC). A geriatric patient's chief complaint may be an exacerbation of a chronic problem. Because it is something the patient has always had to deal with, the patient may have waited until the problem is worse than usual. At other times, very subtle and simple complaints may indicate very serious problems. The "S" in the GEMS diamond stands for social situation. We all need a network of people around us for ongoing socialization and to help in times of need. However, as we age and outlive friends and family, our social network may become smaller. When an older patient has fewer people to interact with, he or she may call on EMS for complaints that seem trivial to us but significant to him or her.

Use the AVPU scale to determine a patient's LOC. However, you should not make any assumptions about an older patient's LOC. Never assume that an altered mental status is normal. Altered mental status indicates some level of brain dysfunction and is a serious problem. The best rule of thumb is always to compare the patient's current LOC or ability to function with the level or ability before the problem began. Do not assume that confusion or unresponsiveness is normal behavior for anyone. In many cases, you may have to rely on a family member or caregiver to help establish the patient's baseline LOC before the complaint began (Figure 34-1 ▶).

Airway, Breathing, and Circulation

During the initial assessment, you will assess the patient's ABCs. If a life-threatening condition exists, you will have to perform emergency treatment before continuing your assessment. The treatment is based on treatment protocols discussed previously. If the airway is inadequate, you should make it adequate with positioning and airway adjuncts as necessary. Ensure that breathing is adequate, evaluate for bleeding, and treat for shock.

Transport Decision

The initial assessment sets the tone and helps you decide whether the patient requires a rapid resuscitative approach ("load and go" situation) or a slower, con-

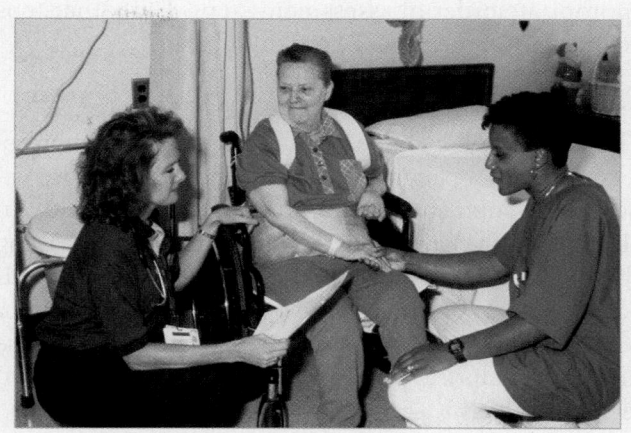

Figure 34-1 Interview family members, friends, and caregivers as part of your assessment of a geriatric patient.

templative one ("stay and play" situation). In most cases, the slower, contemplative approach is all that is needed. Use your assessment of the patient to determine the best way to package and transport. When possible, take the patient to the hospital of his or her choice. This will allow continuity of care if the patient has been there before and can decrease the anxiety of the patient. Allow the patient to assume a position of comfort. This may be very upright for respiratory distress calls or legs raised for a patient with low blood pressure. If transportation on a long board is indicated, use of a folded blanket or inflatable mattress over the board can reduce pressure points that may lead to pressure ulcers.

Focused History and Physical Exam

The focused history and physical exam of a geriatric patient should be performed en route to the hospital to minimize time on scene, when possible. The purpose of the focused history and physical exam is to obtain additional, specific information about the patient's illness or injury. This portion of your assessment includes performing a physical exam (rapid or focused), obtaining vital signs, and interviewing the patient or family members about the patient's medical history. The order of these three portions of the assessment varies if the older patient is a medical patient (responsive or unresponsive) or a trauma patient (with significant or nonsignificant MOI). Refer to Chapter 8 to review the appropriate order of assessment steps. Here, the steps are not presented in any particular order.

SAMPLE History

The history will become one of the best tools when assessing an older patient. To obtain an accurate history, patience and good communication skills are essential. An older patient's diminished sight, hearing, and speaking ability may hamper communication (Figure 34-2 ▼). If possible, take a few moments to have the patient put in dentures or hearing aid and put on glasses. All of these items can help the patient to communicate with you more effectively.

Communicating With an Older Patient

When speaking with an older patient, bend down so that you are eye-to-eye with the patient. Be sure that the patient can see your face. Patients who have difficulty hearing will often look for clues in the speaker's facial expressions to assist in understanding the subject matter. Turn on a light if you are in a dimly lit room. Turn off televisions or radios. Use a normal tone of voice, especially if the patient is wearing a hearing aid. A loud tone may actually cause sound distortion in the hearing aid and make communication worse. Ask as many open-ended questions as possible, and use closed-ended questions to clarify points. It is better to ask an open-ended question such as "Please tell me about the pain you are feeling," rather than a closed-ended one such as "Is the pain sharp or dull?" While taking the history, write down any key points on a notepad so that you do not ask the same question repeatedly because you forgot the answer. After interviewing the patient, ask family members or caregivers to clarify what you just

Figure 34-2 As you assess an older patient, make eye contact, and grasp the patient's hand to feel for temperature, grip, and skin condition.

EMT-B Tips

To minimize distraction and confusion, have only one responder speak to the patient at a time. Another can gather history from relatives or a caregiver or examine the scene for helpful information.

learned from the patient. Be careful not to offend the patient. Taking a few minutes to obtain an accurate history saves time in the long run by providing information on which appropriate decisions can be based.

A poor history-gathering technique can hamper communications. You must be able to gain a patient's confidence, which is best accomplished by treating the patient with respect, taking a slow deliberate approach, and explaining what you are doing. Ask the patient what his or her name is, and then address the patient using courtesy titles, such as "Mr," "Mrs," or "Ms" and his or her last name. Avoid being overly familiar with the patient, and do not use first names or nicknames unless the patient asks you to.

Often when there are multiple responders, everyone asks questions at the same time. This technique may result in obtaining a haphazard history regardless of the patient's age. For geriatric patients who may have communication or perceptual problems, it makes obtaining a thorough SAMPLE history almost impossible. In addition, many people are reluctant to discuss their problems in front of a crowd. Be sure to have one EMT-B obtain the patient's history, one question at a time, providing as much privacy as possible (Figure 34-3 ▶).

Older patients have often had similar episodes and can compare the current condition with them. In geriatric patients, a condition will often present in episodes that will progressively worsen with time. A patient who has cardiac disease and experiences angina may have many episodes of chest pain over time, but the episodes may worsen and the patient might experience a heart attack at some point. There are also conditions that can change the way the patient may present from a condition. A diabetic patient may not have pain from a cardiac event because of the loss of nerve cells over time from the diabetes. The "M" in the GEMS diamond stands for medical history and medications. During your SAMPLE history, you will evaluate the medical history and current medications. These will have an important part in the assessment of the geriatric patient.

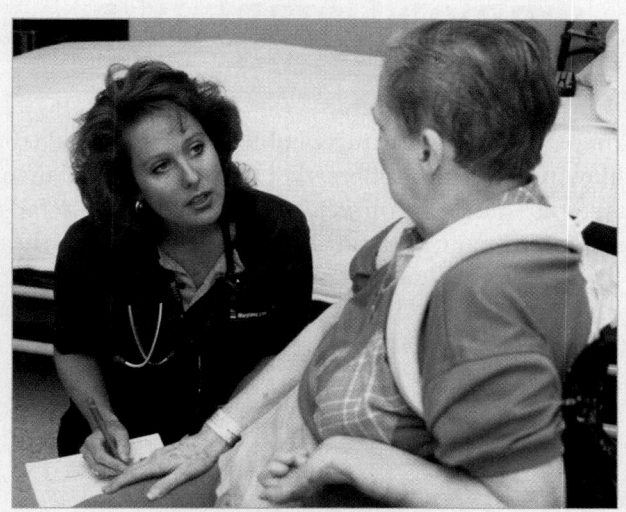

Figure 34-3 A slow, deliberate approach to the patient history, with one EMT-B asking the questions, is generally the best strategy for assessing a geriatric patient.

Medication Use

Medications can become a large problem for older patients. The average patient older than 65 years will be taking four or more medications and be using over-the-counter (OTC) medications as well. <u>Polypharmacy</u> occurs when a patient takes multiple medications that can interact. During the assessment of the patient's medications, you must look at several types of problems that can occur.

List all of the medications that are prescribed for the patient. Then ask the patient whether he or she takes all of the medications on the list. Often the patient may stop taking a medication because of the side effects without talking with the doctor. For example, blood pressure medication may cause the patient to feel dizzy. The patient does not like the feeling and stops

taking the medication. The effect is that the blood pressure rises and causes a medical emergency.

Look for medications that have been started or stopped during the last 2 weeks. Medications that have been used by the patient for a long time are not likely to become a sudden problem; however starting or stopping a medication can affect the body. Ask the patient whether he or she has taken any other medications. Patients may take the a medication prescribed for their spouse or family member if they think they have the same symptoms. Also ask about OTC medications such as aspirin or herbal supplements such as ginkgo biloba. These OTC medications may seem harmless to the patient but may interact with prescribed medications. When possible, the patient's medications should be collected and taken to the hospital for evaluation by the physician.

Focused Physical Exam

The chief complaint and the medical history will help guide you in performing a focused physical exam of the geriatric patient. A focused physical exam is performed on conscious medical patients or trauma patients with a limited MOI. It allows you to focus on specific problems that may need more clarification. Because of the complexity of the geriatric patient and the vagueness of complaints, you should consider broadening your physical exam to include more areas rather than fewer. This may better help to separate chronic problems from more acute problems.

Rapid Physical Exam

A rapid physical exam is typically performed before obtaining the vital signs and history. The goal is to identify hidden threats not picked up in the initial assessment. This rapid head-to-toe assessment is performed on unresponsive medical patients and patients with a more

You are the Provider   Part 2

You arrive on scene and find the front door open. Ms Gladys yells from the front room for you to come in. You find her sitting on the edge of the couch clutching her chest. She is pale and diaphoretic. Her breathing is shallow and rapid. As you approach her, she begins crying and states that she thinks she is having a heart attack. You quickly apply high-flow oxygen, perform a patient assessment, and prepare her for transport.

3. What treatment would you provide this patient?

significant MOI. Remember, though, that the geriatric body is more fragile than a younger body. It will take less of an MOI to produce significant injuries. For this reason, you should consider a rapid head-to-toe exam on most geriatric trauma patients before obtaining vital signs and a SAMPLE history.

Baseline Vital Signs

Normal vital signs will change with age. Irregularity of the pulse will become more prevalent as the patient ages. Blood pressure will increase slightly because the blood vessels have been partially blocked due to atherosclerosis. After a person reaches 65 years old, the increase in systolic pressure is usually only 1 mm Hg per year. Keep in mind that many medications used to treat arrhythmias and hypertension may prevent an increased pulse rate when needed. For example, a patient with low blood pressure and a pulse rate of 84 may be in serious shock. The rate, rhythm, and quality of respirations vary depending on the health of the patient's respiratory system. Pupils will also be slower to react to light. Pupil checks can be complicated by cataracts or surgeries to the eyes.

Detailed Physical Exam

Of all the patients you will care for, the geriatric patient is someone who should have a detailed physical exam performed in most situations. The changes that occur because of aging alter how the patient complains of pain. Chronic changes can mask acute problems, and acute problems may be mistaken for chronic problems. The vagueness of complaints may prevent the identification of serious problems. A thorough detailed exam from head to toe, when correlated with the history and vital signs, can help identify potential conditions.

Ongoing Assessment

Priority patients require frequent reassessment, and geriatric patients are no different. Because older patients cannot get their bodies to mount a good compensatory effort for loss of oxygenation or blood, it is very important to record vital signs every 5 minutes to watch the direction of the trend. Conversing with the patient is a good way to monitor the LOC.

Common Complaints of Older Patients

Common chief complaints of older patients include the following: shortness of breath; chest pain; altered mental status; dizziness or weakness; fever; trauma; falls; generalized pain; and nausea, vomiting, and diarrhea. The complaints that deserve special attention in older patients are covered in the following sections on trauma and medical emergencies.

Geriatric Trauma

Mechanism of Injury

Falls are the leading cause of trauma, death, and disability in older people. Most patients survive; however, a significant number require hospitalization. Motor vehicle trauma is the second leading cause of trauma death in the geriatric population. A geriatric patient is five times more likely than a younger patient to be fatally injured in a car crash, even though excessive speed is rarely a cause in the older age group. Pedestrian accidents and burns are also common MOIs in geriatric patients, resulting in death, serious injury, or disability.

Trauma in older people can be complicated by medical conditions. Often the cause of trauma will be related to a medical condition, such as the patient becoming dizzy and then falling. Try to determine whether the patient had a medical complaint before the trauma.

Trauma Assessment

The priorities in the rapid trauma assessment do not change with older patients. However, there are several confounding factors that have to be reviewed. When assessing the patient's mental status, remember that older patients may have medical conditions such as Alzheimer's disease or previous strokes that may hinder your assessment. Make your judgment based on the patient's baseline when possible. Many older patients have had some type of dental work done over their lifespan. Dentures, bridges, and dental implants may all become loosened following a traumatic event. Check the airway carefully for signs of loose or broken teeth. Older patients have decreased ability to increase the volume or rate of breathing to compensate for hypoperfusion. Assess the ability of the patient to determine whether the breathing is adequate. Assessment of pulses can be complicated by previous injuries or decreased circulation due to medical problems. If you have a hard time finding a radial pulse, ask the patient or family whether this is normal. Many older patients will have changes in their circulation as they age. Finally, if you are in doubt about the patient's priority, believe the patient and ensure that the appropriate care is given.

The physical exam will follow the direction of the rapid trauma or focused trauma exam as needed. (Table 34-1 ▼) lists some changes associated with aging that may affect the exam.

After completing your exam, be sure to obtain vital signs and complete a SAMPLE history because they may show reasons for concern. An example would be a patient who has become unconscious and can no longer tell the emergency department staff that he or she is taking a blood pressure medication. Because an older patient does not compensate well for trauma, small changes in the blood pressure or pulse can indicate a shift in how well the patient continues to compensate for trauma.

Injuries to the Spine

Injuries to the spine are a frequent problem in older people. Spinal injuries may be broadly classified as stable or unstable. A stable spinal injury is one that has a low risk for leading to permanent neurologic deficit or structural deformity; an unstable spinal injury is one that has a high risk of permanent neurologic deficit or structural deformity. The injuries that older patients often incur while performing normal daily activities are most commonly stable ones, whereas unstable injuries tend to be the result of a significant trauma, such as falls from a substantial height or motor vehicle crashes.

Injuries to the upper cervical spine may be potentially lethal because the nerves innervating the diaphragm originate from here. An injury at this level or higher may lead to death secondary to an inability to breathe. Many cervical spine injuries in older patients result from hyperextension of the neck as the result of

TABLE 34-1 Changes Associated With Aging that May Affect the Exam

Part of the Body	Changes Possible With Aging	How to Examine
Pupils	May have cataracts, scars from surgery, or blindness (one or both sides)	Report what you see. It is not always as important to name the condition as it is to note a change.
Neck and back	Kyphosis (the curve in the upper back and neck formed by normal aging)	Note any DCAP-BTLS (**D**eformities, **C**ontusions, **A**brasions, **P**unctures/Penetrations, **B**urns, **T**enderness, **L**acerations, and **S**welling) and position a cervical collar when needed. Extra padding will be required to complete the immobilization. Use of a vacuum mattress or an inflatable pad on a long board can help fill the voids left by the changing shape.
Chest	Examine the chest for symmetry. Changes in lung sounds may be due to previous medical conditions such as congestive heart failure or chronic obstructive pulmonary disease. The chest wall is more brittle in older patients; look for broken ribs.	Listening to lung sounds early and reassessing them periodically may help identify a problem. High-concentration oxygen is always indicated with patients in respiratory distress.
Abdomen	The abdomen senses less pain as we age.	Be careful to look for signs of trauma if the mechanism indicates, even if a patient is not complaining of pain.
Extremities	Examination of the extremities can be complicated by several factors: previous injuries or surgeries that have left the patient deformed, decreased sensation due to lessened pain receptors and a slower nervous system, and decreased circulation to the distal points.	Try to determine whether the signs are new or old. Record your assessment as you find it to provide information for decision making.

a fall or striking the head on the windshield during a motor vehicle crash. Because of the presence of arthritis, relatively small hyperextension injuries can cause the spinal cord to be squeezed, leading to dysfunction known as <u>central cord syndrome</u>. Central cord syndrome results in weak or absent motor function, which is more pronounced in the upper extremities than the lower extremities. Although this type of spinal cord injury is usually not permanent, recovery may take several months or even years.

Osteoporosis in the thoracic and lumbar spine contributes to a high rate of injury in this area in the older population. Three types of fractures are common in the thoracolumbar region: compression fractures, burst fractures, and seat belt–type injuries. <u>Compression fractures</u> are stable injuries in which often only the anterior third of the vertebra is collapsed. This type of fracture often results from minimal trauma, from simply bending over, rising from a chair, or sitting down forcefully. This is by far the most common type of spine fracture seen in the older patient population. <u>Burst fractures</u> typically result from a higher energy mechanism such as a motor vehicle crash or fall from substantial height. These fractures are unstable and may lead to neurologic injury secondary to shifting of the vertebrae with damage to the spinal cord. <u>Seat belt–type fractures</u> involve flexion, and there is a distraction component (energy being dispersed in two opposite directions) that causes a fracture through the entire vertebral body and bony arch. This type of injury typically results from an ejection or occurs in those wearing only a lap belt without a shoulder harness.

Injuries to the spinal cord are usually associated with neurologic deficits. These deficits may be complete or incomplete. An example of an incomplete injury is the previously described central cord syndrome. Patients with incomplete spinal cord injuries may recover function over time, whereas those with a complete spinal cord injury are not likely to regain function.

As with all patients, prompt spinal immobilization is an effective method of reducing further damage to the spinal cord and preserving the older patient's neurologic function. Often, patients may be found in positions in which the neck or body is not in the neutral position, such as the head being rotated to one side. To facilitate the application of a cervical collar, the provider should slowly return the head to the midline. At no time should these attempts continue if the patient develops changes in neurologic status or complains of increasing pain. If these complaints develop, the head should be secured in the position in which it is found

by using blankets and tape to prevent further movement (Figure 34-4 ▼).

Older patients present several unique challenges to EMS providers when treating spinal injuries. To immobilize kyphotic patients, several blankets and pillows or vacuum splints may be required to provide support to the head and upper back (**Skill Drill 34-1 ▶**).

1. **Apply and maintain cervical stabilization.** Assess distal functions in all extremities (**Step 1**).
2. **Apply a cervical collar.** If a collar does not fit, do not attempt to straighten the patient's neck (**Step 2**).
3. **Rescuers kneel on one side of the patient** and place hands on the far side of the patient (**Step 3**).
4. On command, **rescuers roll the patient toward themselves and quickly examine the patient's back** (**Step 4**).
5. Slide the backboard under the patient, and roll the patient onto the board (**Step 5**).
6. **Pad the void space produced by the kyphotic spine** with pillows and blankets. The pillows and blankets should be as wide as the backboard to allow for effective immobilization and support. Place rolled towels or foam padding onto the surface of the backboard next to the patient's head (**Step 6**).
7. Secure the torso to the backboard with straps (**Step 7**).
8. Secure the patient's head and padding to the backboard with 2″ medical tape. Apply the tape across the forehead and cervical collar to prevent the padding from becoming dislodged. Immobilize the remainder of the body as usual (**Step 8**).

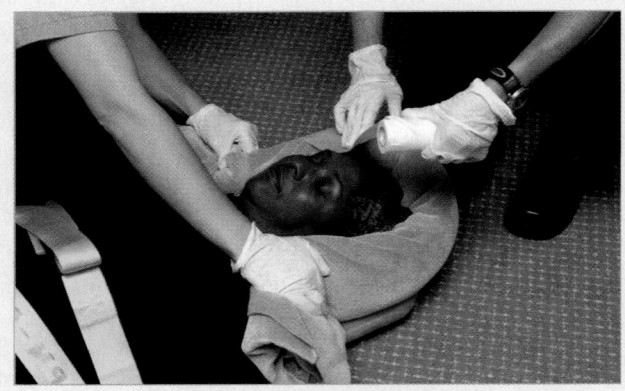

Figure 34-4 If the attempt to move a patient's head to the midline results in changes in neurologic status or complaints of increasing pain, secure the head in the position in which it was found by using blankets and tape.

Immobilizing a Kyphotic Patient on a Long Backboard

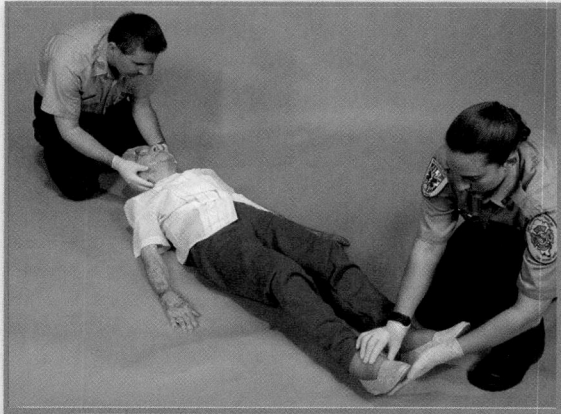

1 Apply and maintain cervical stabilization. Assess distal functions in all extremities.

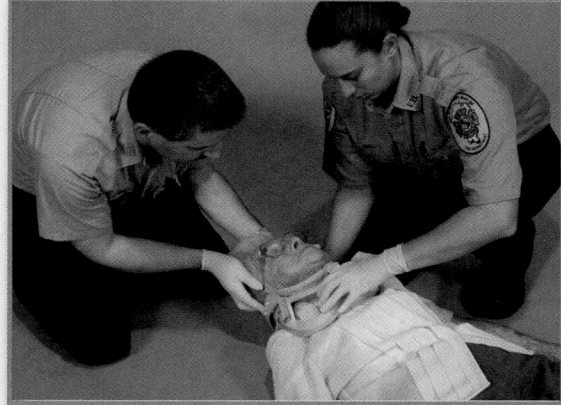

2 Apply a cervical collar. If a collar does not fit, do not attempt to straighten the patient's neck.

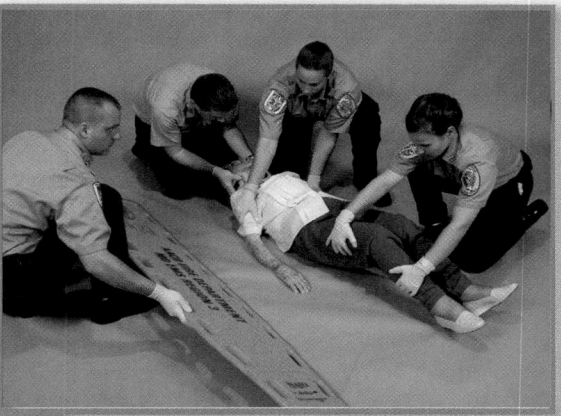

3 Rescuers kneel on one side of the patient and place hands on the far side of the patient.

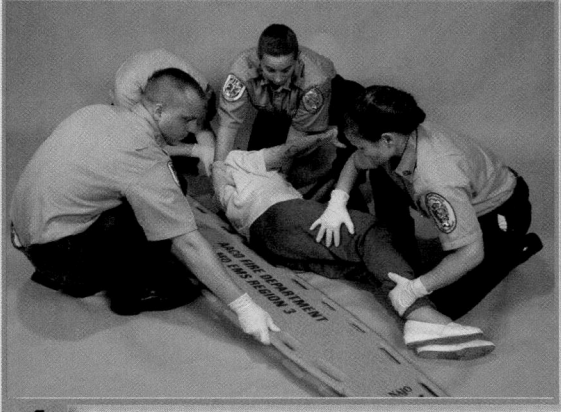

4 On command, rescuers roll the patient toward themselves and quickly examine the patient's back.

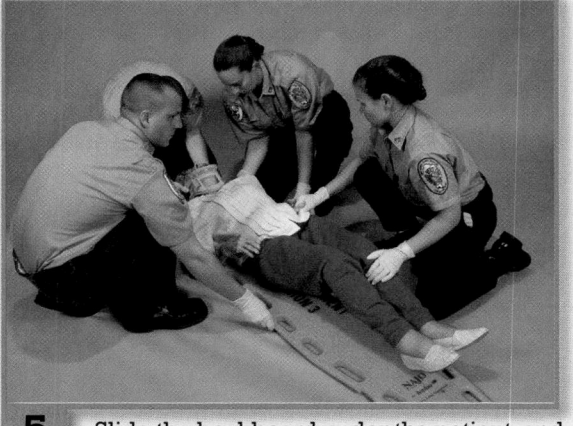

5 Slide the backboard under the patient, and roll the patient onto the board.

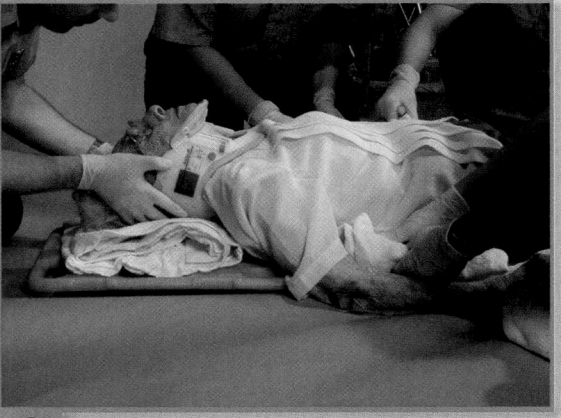

6 Pad the void space below the kyphotic region of the spine with pillows and blankets. Place rolled towels or foam padding onto the surface of the backboard next to the patient's head.

Continued.

Immobilizing a Kyphotic Patient on a Long Backboard continued

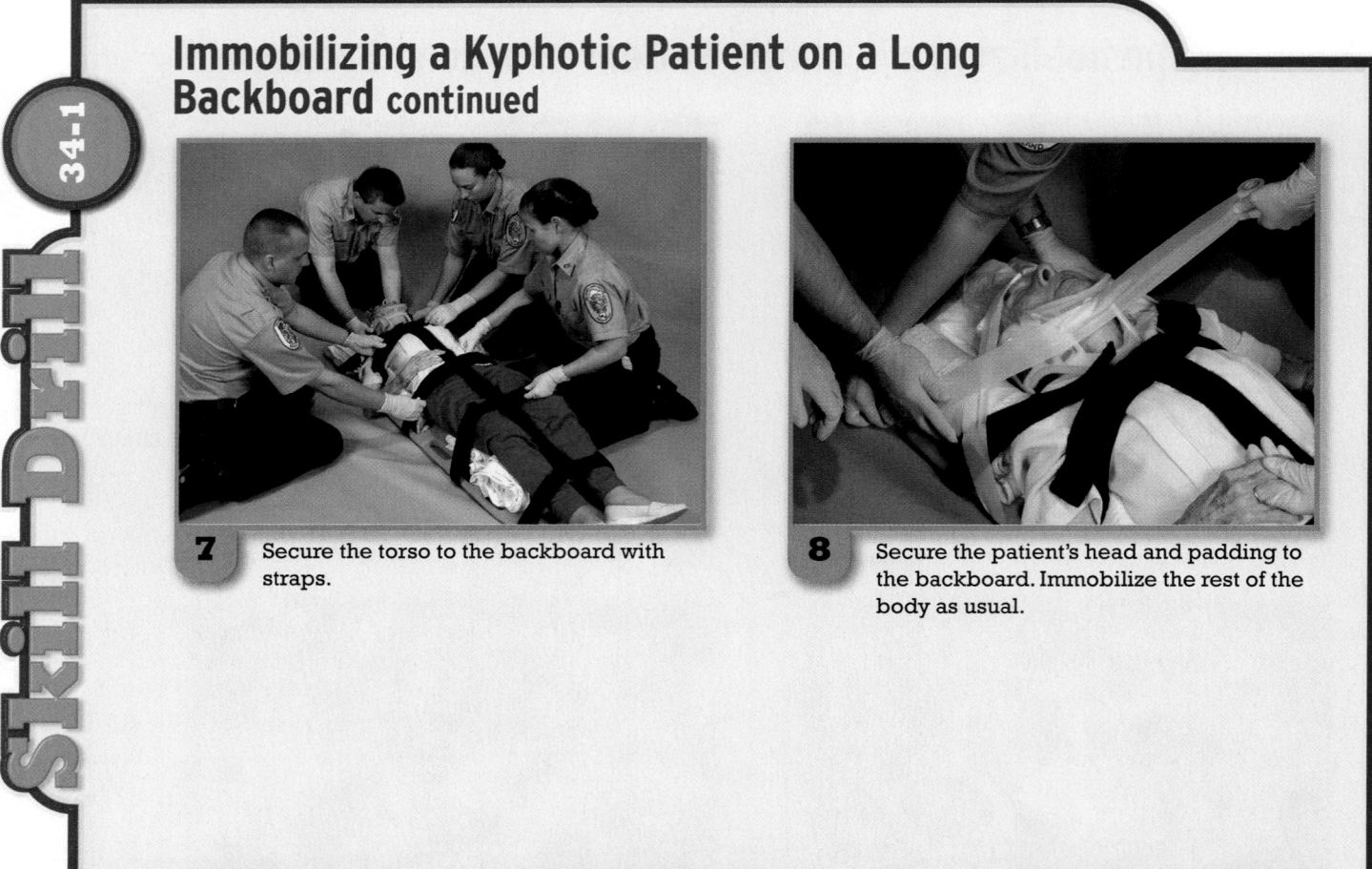

7 Secure the torso to the backboard with straps.

8 Secure the patient's head and padding to the backboard. Immobilize the rest of the body as usual.

Head Injuries

Older patients who have signs or symptoms of a significant head injury, such as loss of consciousness, should be assumed to have sustained a substantial injury even if the patient is neurologically intact at the time of exam. In addition, patients who have sustained even minor-appearing head injuries and who are taking blood thinners should be suspected of having a brain injury and be treated as such. These patients may need persuading to seek medical treatment because they may feel completely normal and may not believe medical treatment is necessary. In situations in which patients refuse care yet have a high risk of brain injury, relatives or neighbors should be instructed to be aware of subtle changes in neurologic status that could indicate deterioration.

Prehospital treatment of older head-injured patients should be aimed at maintaining maximum oxygen delivery to the brain. Patients who have signs of head injury and no evidence of hypotension or shock may also benefit from a slight elevation of the head because this position may help reduce intracranial pressure and increase cerebral perfusion pressure (Figure 34-5 ▶).

Injuries to the Pelvis

Fractures of the pelvis in older patients often occur as the result of a combination of decreased bone strength due to osteoporosis and a low-energy mechanism such as a fall from a standing position. Pelvic fractures often present as hip or buttock pain. More serious high-energy fractures can pose a significant threat to life. Injuries of this type can be encountered in a person who has fallen from a significant height, been involved in a high-speed motor vehicle crash, or been struck by a car and may result in fracture in two places in the pelvis with displacement of a segment of the pelvic ring (pelvic ring disruption). Pelvic ring disruption can lead to hemorrhage from the blood vessels that pass through the pelvis or to injury to the bladder, intestines, or lumbosacral nerve plexus. Older patients are less able than younger patients to tolerate the blood loss or other organ system injuries that are commonly associated with high-energy pelvic ring disruption.

The <u>acetabulum</u> is another site in the pelvis that may be injured as the result of high-energy trauma in older people. Injuries to the acetabulum can occur as the

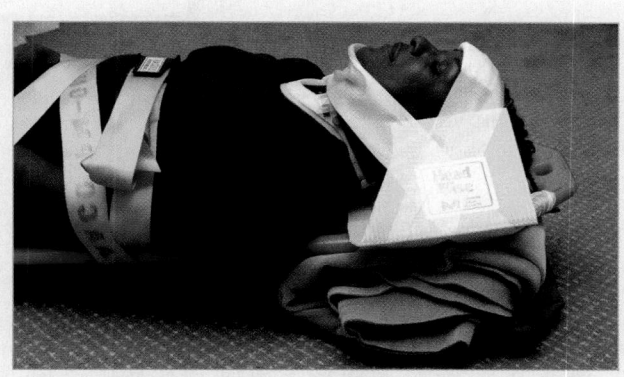

Figure 34-5 Patients who have signs of head injury and no evidence of shock may benefit from a slight elevation of the head because it may help reduce intracranial pressure and increase cerebral perfusion pressure.

result of an injury in which the knee is driven into the dashboard or ground with the head of the femur driven through the acetabulum. Owing to the diminished strength of bone, older patients are at a higher risk for this type of injury from lower energy insults than would be required to cause this type of injury in younger patients.

Hip Fractures

One common debilitating musculoskeletal injury that occurs in older patients is a hip fracture. A hip fracture is a fracture of the head, neck, or proximal portion of the femur. Following a hip fracture, patients often have decreased mobility and independence and can require prolonged rehabilitation. This can be physically and emotionally challenging for the patient and his or her family. Despite the advances made in treatment, a large number of older people will be permanently impaired and nearly 20% will die within the first 12 months following injury.

Fractures of the hip should be treated by splinting the injured extremity with a blanket roll or long board splints. Fractures of the hip do not necessarily require the use of traction splints. The purposes of the blanket roll are to maintain the leg in a static position so that

further injury does not occur and to help control the patient's pain (Skill Drill 34-2 ▶).

1. **Assess pulse, motor, and sensory function of the extremity**. Cover open wounds with dry, sterile dressings, and apply direct pressure, if necessary (**Step 1**).
2. **Place the patient on a scoop stretcher or long backboard** by logrolling the patient onto the uninjured leg while having another provider support the injured extremity (**Step 2**).
3. While continuing to support the injured extremity in its deformed position, your partner should **place a blanket roll between the patient's legs** (**Step 3**).
4. **Place blankets and pillows under the injured extremity** to provide support to the fracture site in the deformed position (**Step 4**).
5. **Secure both legs and the padding to the backboard with at least three cravats or straps** (**Step 5**).
6. **Reassess pulse, motor, and sensory function** (**Step 6**).

Geriatric Medical Emergencies

Determining the chief complaint of an older patient is often challenging. Often, geriatric patients have multiple complaints from multiple conditions. Asking the patient what problem bothers them the most today can make them focus on a single problem. This can be helpful or may require more questions to determine the problem. The patient may complain of foot and ankle pain, but your examination reveals that the legs are swollen from edema from congestive heart failure. This patient may always be somewhat short of breath, and today may be only slightly different.

Be aware that the sensation of pain may be diminished in older patients, leading you to underestimate

You are the Provider   Part 3

Minutes later, you arrive at the hospital. You transfer patient care, including giving a full patient report and completing your run sheet. Before leaving the hospital, you check on Ms Gladys. As you open the door, you see your partner standing at her bedside telling her that the doctors will take good care of her and for her to call 9-1-1 any time she feels the need.

4. Why is it necessary to treat every call with equal importance, even if your patient frequently calls 9-1-1?

Skill Drill

34-2

Splinting a Hip Fracture

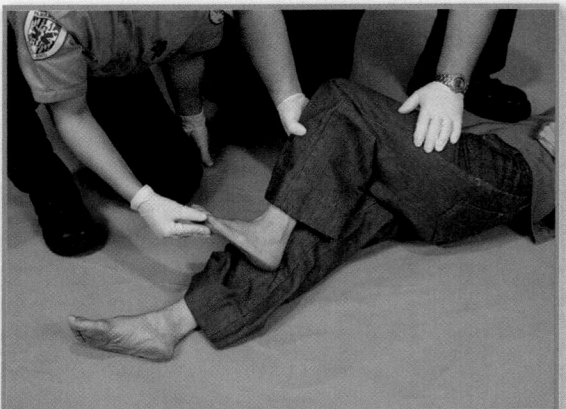

1 Assess pulse, motor, and sensory function. Cover wounds with dry, sterile dressings. Apply direct pressure, if necessary.

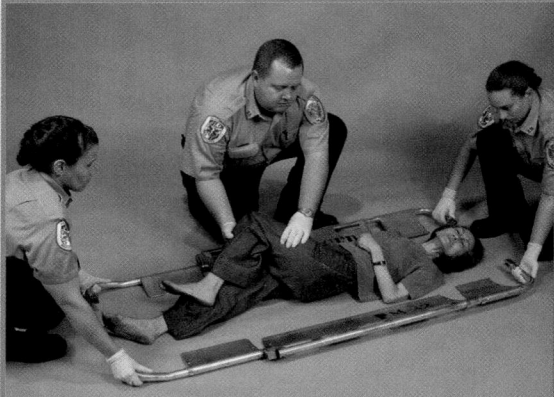

2 Place patient onto a scoop stretcher or long backboard by logrolling onto the uninjured leg while another provider supports injured extremity.

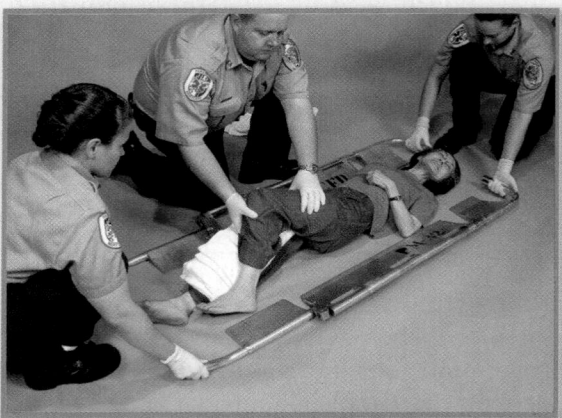

3 Continue to support the injured extremity while another provider places a blanket roll between the patient's legs.

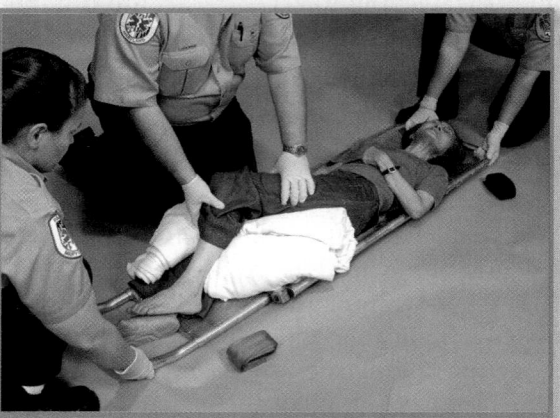

4 Place blankets and pillows under the injured extremity.

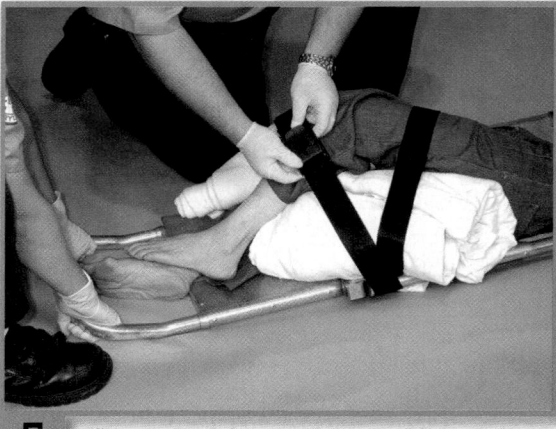

5 Secure both legs and the padding to the backboard with at least three cravats or straps.

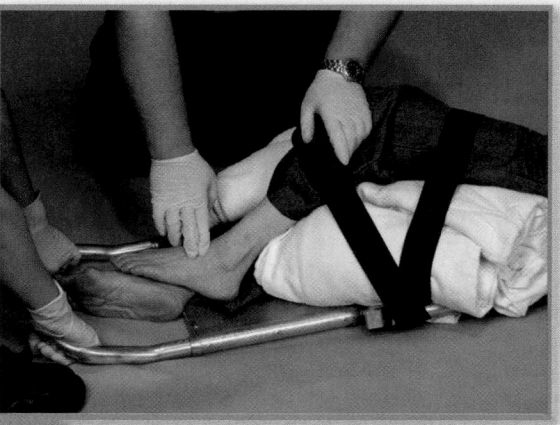

6 Reassess pulse, motor, and sensory function.

the severity of a condition. This is associated with the aging nervous system. For example, 20% to 30% of geriatric patients have "silent" heart attacks, without the typical symptom of chest pain. In addition, fear of hospitalization often causes the patient to understate or minimize symptoms.

Several conditions with which geriatric patients may present differently from a younger patient are discussed in the following paragraphs.

Cardiovascular Emergencies

Because many older patients do not have the crushing chest pain that is experienced by younger patients, a common complaint of an older patient when experiencing a myocardial infarction (MI) is difficulty breathing. Older patients may also complain of a toothache, arm pain, or back pain. This can make cardiac illness difficult to detect. It is often better to ask about chest discomfort rather than pain. Ask the patient whether he or she has experienced this before, and then ask how this is different. If the patient tells you that he or she felt this same way during a heart attack, it is a good bet that this may be another one.

Shortness of Breath

Shortness of breath, or dyspnea, can be related to many causes. Not all of the causes are related to difficulty breathing. The causes for dyspnea include asthma, chronic obstructive pulmonary disease, congestive heart failure, and pneumonia. However, an MI, bleeding, or even hyperglycemia can cause the patient to feel short of breath. To assess these patients, you must look at the entire picture. Complete the SAMPLE history and the physical exam. Assess the patient's work of breathing, including lung sounds, retractions, sitting in a tripod position, and cyanosis. If the patient becomes short of breath with activity, did the distress occur with the same amount of activity or with less activity than before? Does the patient sleep propped up on pillows? This can indicate a problem breathing when the patient lies flat, often fluid buildup in the lungs. All patients experiencing shortness of breath should receive oxygen.

Syncope

Sudden unconsciousness (syncope) or the feeling of almost passing out (near syncope) can occur for many reasons in the older population. Simple causes such as standing up too fast or trying to have a bowel movement while constipated can cause an older patient to faint. More deadly causes of syncope include an MI or diabetic shock (hypoglycemia). The assessment for syncope will include a SAMPLE history and physical exam. During your history taking, ask what was happening before the syncope occurred. Also try taking the blood pressure while the patient is lying flat, seated, and standing. Drops in systolic blood pressure of more than 10 mm Hg can indicate dehydration.

Altered Mental Status

Acute onset of altered mental status is not normal for a patient of any age. Even patients with Alzheimer's disease should not have sudden changes in their mental status. Changes indicate a problem in supplying the brain with the nutrition it needs to work correctly. Infection, hypoglycemia, hypoxia, hypotension, cerebrovascular accident (stroke), trauma, seizures, medication interactions, electrolyte imbalances, and psychotic episodes can all change how the brain functions. Most sudden changes are caused by a reversible condition. Evaluate and treat for hypoxia or hypoglycemia if present. Transport the patient to the most appropriate facility for further evaluation.

Acute Abdomen

In older patients, the nervous system response to pain in the abdomen is lessened. When an older patient complains of abdominal pain, it is usually a more serious event than in a younger patient. Ask whether the patient has had a change in bowel movements. The slowing of the gastrointestinal system in older people can cause constipation or bleeding. When palpating the abdomen, look for a pulsating mass, indicating an abdominal aortic aneurysm. These are most commonly found in patients older than 70 years and can be found slightly above and to the left of the umbilicus.

Septicemia and Infectious Disease

Infections in an older person can be severe and dangerous. Septicemia is the disease state that results from the presence of microorganisms or their toxic products in the bloodstream. Septicemia is a serious problem that every EMS provider should know how to recognize and to treat. Use appropriate BSI precautions when you think a patient may have an infection. Think of septicemia whenever you see a hot, flushed patient who also has tachycardia and tachypnea. Symptoms of infection may be present, such as chills, cough, or burning with urination. Often the infection will cause an altered mental status. Think septic shock when hypotension is also present. The term "sepsis" is used loosely and often interchangeably with septicemia and bacteremia. Bacteremia is the presence of bacteria in the blood, whether or not a disease process is present.

Response to Nursing and Skilled Care Facilities

Responding to a nursing home or skilled care facility is a common patient contact for EMT-Bs. Before you provide transport for the patient, you should obtain the following critical information from the nursing staff:

- What is the patient's chief complaint today?
- What initial problem caused the patient to be admitted to the facility?

To determine the nature of the problem, you will usually have to compare the patient's present condition with his or her condition before onset of the symptoms. Ask the staff about the patient's mobility, activities of daily living, and ability to speak. This will help to paint a picture of the patient's baseline condition and indicate whether today's condition differs.

Many facilities that are transferring patients will include a transfer record that contains the patient's history, medication lists and dosages, previous diagnosis, vital signs, allergies, and more (Figure 34-6 ▼). These records provide members of the medical team with essential information and save time, especially when the patient cannot speak for himself or herself. Be sure to obtain this essential record before leaving for the hospital, and relay it to the hospital staff when giving your report.

DEATON SPECIALTY HOSPITAL & HOME
611 South Charles Street
Baltimore, Maryland 21230
(410) 547-8500

PATIENT INFORMATION AND TRANSFER FORM

Name: ____ Date of Birth: ____ Sex: ____ Race: ____ Religion: ____
Social Security #: ____ Marital Status: M S W D Sep. Deaton Adm. Date: ____
Patient's Home Address: ____
Medicare #: ____ Medicaid #: ____ Other Ins.# ____
Responsible Party: ____ Relationship: ____ Home Phone: ____
Address: ____ Work Phone: ____

Date of Transfer: ____ Time: ____ Transferred To: ____
Unit Transferred From: ____ Nurse Manager: ____ Extension #: ____
Reason for Transfer: ____ Mode of Transport: ____
Physician in Charge at Time of Transfer: ____ Physician's Phone Number: ____

MAJOR DIAGNOSIS: ADVANCE DIRECTIVE: YES NO
CPR STATUS: ____

Vital Signs at Time of Transfer: BP: ____ Pulse: ____ Resp.: ____ Temp.: ____
Pertinent Medical Information: ____

Communicable Disease: YES NO Type: ____
Isolation: YES NO Type: ____
Suctioning: YES NO How Often: ____ Site: Oral ____ Trach ____
Ventilator Dependent: YES NO Vent Orders: ____

DRUGS AND OTHER THERAPY: (at time of discharge)

Name	Frequency	Route

Diet Orders: ____
Method of Feeding: ____

LAB DATA
CXR: Date: ____ Result ____
CBC: Date: ____ Result ____

Form # 612 Rev. 9/94 Original Copy to go with Patient - Yellow Copy to be retained in Patient's Record

ADL'S	Independent	Needs Assistance	Unable to Do
Bed Activity: Turns, Sits			
Personal Hygiene:			
Feeding:			
Transfer: Sitting, Standing, Toilet			
Locomotion: Wheelchair, Walking			

Eyeglasses: YES NO Sent with pt.: YES NO
Dentures: YES NO Sent with pt.: YES NO

Patient Uses: ____ Cane ____ Crutches ____ Prosthesis ____ Walker ____ Chair ____ Stretcher
Bed Type: **Side Rails:** YES NO
____ Regular ____ Clinitron ____ Other ____
Usual Mental Status: (please circle) Alert Minimally Reactive Confused Forgetful Unresponsive
Communication Ability:
Speech ____ Normal ____ Impaired
Hearing ____ Normal ____ Impaired
Appliance: YES NO
Incontinence: Bowel ____ Bladder ____ Colostomy ____ Catheter/Suprapubic ____

ADDITIONAL PERTINENT INFORMATION

ALLERGIES

Skin Condition:

Signature of Nurse Completing Transfer Form ____ Date ____

NOTIFICATION TO EMERGENCY / TRANSPORTING PERSONNEL

This patient is diagnosed as having one of the following communicable diseases: (please circle)
1. HIV 2. Hepatitis B 3. Meningococcal Meningitis 4. Tuberculosis 5. Malaria 6. Rabies 7. Mononucleosis 8. NONE OF THE ABOVE

and is being transported to another facility - emergency or otherwise. Persons handling the transfer must be notified in writing by the transferring nurse of any of the above diseases. Disease notification must be made to:
- Law Enforcement Officer
- Firefighter
- Rescue Squad Personnel
- Emergency Medical Technician(s)

as required by the annotated Code of Maryland Article Health - General - Section 18-213-(D)-(2).

You are hereby notified . . . Signature of Nurse: ____ Date ____
Transport Representative: ____ **Job Classification:** ____ **Date** ____

Form # 612 Rev. 9/94 Original Copy to go with Patient - Yellow Copy to be retained in Patient's Record

Figure 34-6 A transfer record from a long-term care facility contains vital information for members of the health care team.

You are the Provider Summary

At times, older people may call 9-1-1 for complaints that seem trivial to us, yet significant to them. Never assume that any call is a needless call. Approach every scene as a true emergency. Be kind and supportive. Never make patients feel as if they are inconveniencing you.

Prep Kit

Ready for Review

- Although assessment of the geriatric patient involves the same basic approach as that for any other patient, you must take a more wary approach to geriatric patients.

- The injury or medical condition may be worse than indicated by the existing signs and symptoms, and the injuries and conditions that are found will have a more profound effect than they would in a younger patient.

- In addition to the critical needs that an underlying medical problem may cause, the condition of older patients is more unstable than that of a younger patient, and there is an increased possibility for sudden, rapid deterioration.

- To obtain an accurate history for the geriatric patient, patience and good communication skills are essential. A slow, deliberate approach to the patient history, with one EMT-B asking questions, is generally the best strategy.

- Polypharmacy and changes in medications can cause serious problems for the geriatric patient. Conduct a medications history as part of your assessment.

- Common complaints of older patients include shortness of breath; chest pain; altered mental status; dizziness or weakness; fever; trauma; falls; generalized pain; and nausea, vomiting, and diarrhea.

- The priorities in the rapid trauma assessment do not change with older patients, although several confounding factors, such as altered mental status and other physiologic changes associated with aging, must be reviewed.

- Conditions such as cardiovascular emergencies, dyspnea, syncope, altered mental status, sepsis, and the acute abdomen may present differently in older patients.

- When responding to nursing and skilled care facilities, you should determine the patient's chief complaint on that day and what initial problem caused the patient to be admitted to the facility.

Technology

Interactivities

Vocabulary Explorer

Anatomy Review

Web Links

Online Review Manual

Prep Kit continued...

Vital Vocabulary

acetabulum The depression on the lateral pelvis where its three component bones join, in which the femoral head fits snugly.

bacteremia The presence of bacteria in the blood, whether or not a disease process is present.

burst fractures Compression fractures of the vertebrae that typically result from a higher energy mechanism such as a motor vehicle crash or fall from substantial height.

central cord syndrome A form of incomplete spinal cord injury in which some of the signals from the brain to the body are not received; results in weak or absent motor function, which is more pronounced in the upper extremities than the lower extremities.

compression fractures Stable spinal cord injuries in which often only the anterior third of the vertebra is collapsed. This type of fracture often results from minimal trauma, from simply bending over, rising from a chair, or sitting down forcefully.

polypharmacy Simultaneous use of many medications.

seat belt-type fractures Fractures that involve flexion, with a distraction component (energy being dispersed in two opposite directions) that causes a fracture through the entire vertebral body and bony arch; typically results from an ejection or occurs in those wearing only a lap belt without a shoulder harness.

septicemia The disease state that results from the presence of microorganisms or their toxic products in the bloodstream.

stable spinal injury A spinal injury that has a low risk of leading to permanent neurologic deficit or structural deformity.

unstable spinal injury A spinal injury that has a high risk of permanent neurologic deficit or structural deformity.

Points to Ponder

You are dispatched to the intersection of Falcon Way and Hawk Drive for a one-car motor vehicle crash. You arrive to find an older patient at the wheel of a large, older model vehicle. It appears that this was a low-velocity accident; the car has struck a light post, but minimal damage is noted. Your patient was restrained, and although you see no evidence of damage to the interior of the vehicle or any obvious injuries, she is not responding to your questions and is just staring off into the distance.

Is the mechanism related to the motor vehicle crash significant? What could be this patient's problem?

Issues: Mechanism of Injury in the Geriatric Patient, Underlying Medical Conditions in Trauma of the Geriatric Patient.

Assessment in Action

You are dispatched to a local residence for a woman who has fallen. You arrive to find a geriatric woman lying on the tiled kitchen floor of her home. She is alert and oriented to person, place, time, and event. She tells you that she was walking into her living room, when she suddenly tripped on something. You survey the area and see no reason for the fall. The floor is clean and dry and free of any tripping hazards.

She denies having any shortness of breath or dizziness and any significant medical history besides osteoporosis. Her only complaint is right leg pain (especially with movement). Her vital signs include a blood pressure of 104/60 mm Hg, a pulse of 92 beats/min and regular, and respirations of 32 breaths/min.

1. Given her chief complaint, medical history, and the condition of the floor, what most likely caused this patient to trip?

   A. Heart attack
   B. Syncopal episode
   C. Low blood glucose level
   D. None of the above

2. One of the most common debilitating musculoskeletal injuries that occurs in older patients is:

   A. pelvic fractures.
   B. femur fractures.
   C. hip fractures.
   D. both B and C.

3. Hyperextension injuries that result in weak or absent motor function that is more pronounced in the upper extremities are referred to as:

   A. burst fracture.
   B. compression fracture.
   C. central cord syndrome.
   D. both A and B.

4. The most common type of spinal fracture seen in the geriatric population is:

   A. compression fracture.
   B. burst fracture.
   C. seat belt–type fracture.
   D. none of the above.

5. When splinting a possible hip fracture, you should assess pulse, motor, and sensory function:

   A. before immobilizing the extremity.
   B. after immobilizing the extremity.
   C. whenever you have time.
   D. both A and B.

6. When taking a patient's blood pressure to determine the presence of dehydration, the patient should be:

   A. lying.
   B. sitting.
   C. standing.
   D. all of the above.

Challenging Questions

7. What are some common chief complaints from geriatric patients, and what are some potential causes of these complaints?

8. What type of impact can taking a number of prescription medications have on a patient's condition?

9. How do an EMT-B's communication skills influence emergency care of the geriatric population?

10. How can Alzheimer's disease affect your patient assessment and treatment?

www.EMTB.com

Operations

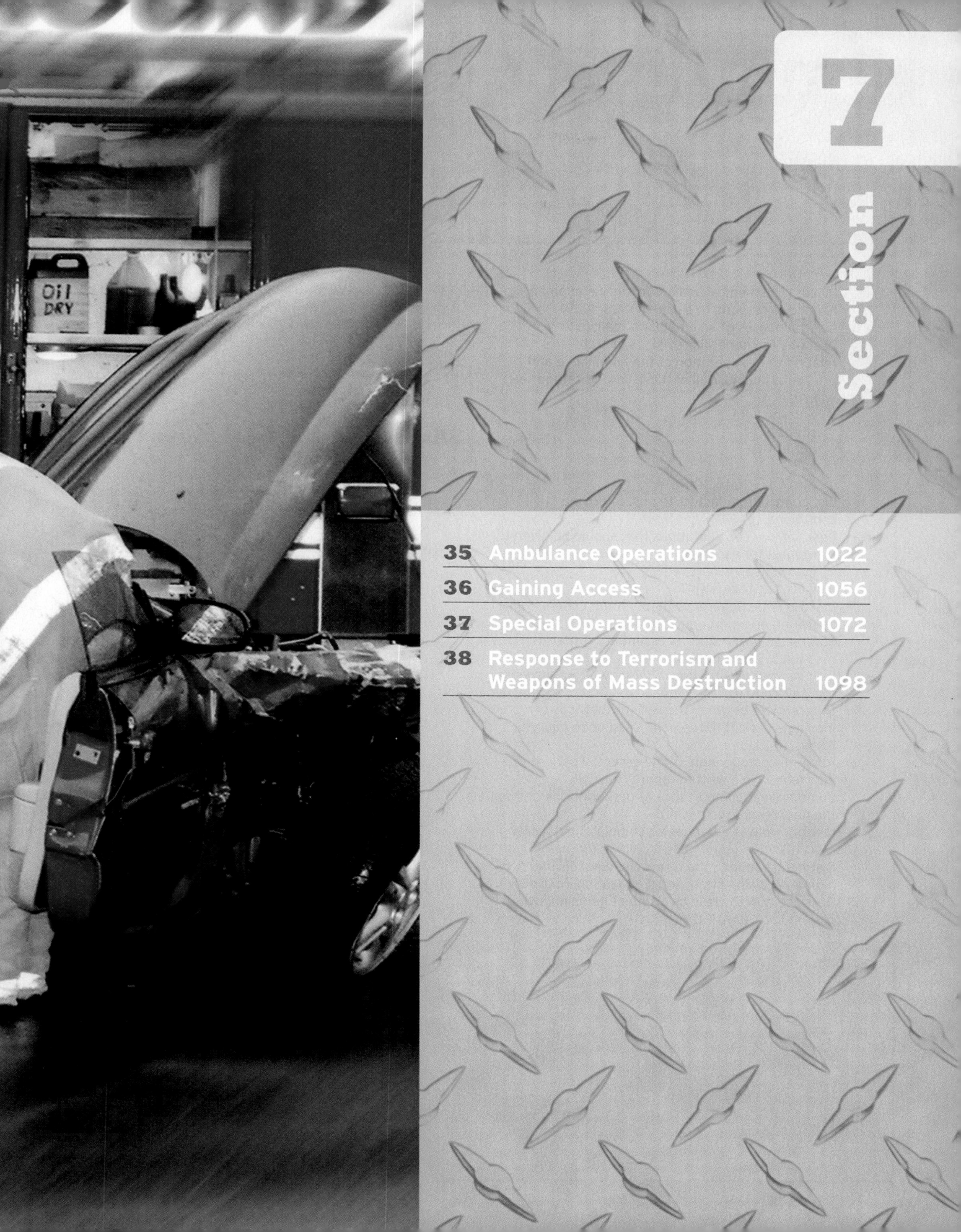

Ambulance Operations

Objectives

Cognitive

7-1.1 Discuss the medical and nonmedical equipment needed to respond to a call. (p 1028, 1029)

7-1.2 List the phases of an ambulance call. (p 1026)

7-1.3 Describe the general provisions of state laws relating to the operation of the ambulance and privileges in any or all of the following categories:
- speed
- warning lights
- sirens
- right-of-way
- parking
- turning (p 1046, 1047)

7-1.4 List contributing factors to unsafe driving conditions. (p 1040)

7-1.5 Describe the considerations that should be given to:
- request for escorts
- following an escort vehicle
- intersections (p 1047)

7-1.6 Discuss "Due Regard for Safety of All Others" while operating an emergency vehicle. (p 1047)

7-1.7 State what information is essential in order to respond to a call. (p 1035)

7-1.8 Discuss various situations that may affect response to a call. (p 1036)

7-1.9 Differentiate between the various methods of moving a patient to the unit based upon injury or illness. (p 1032)

7-1.10 Apply the components of the essential patient information in a written report. (p 1038)

7-1.11 Summarize the importance of preparing the unit for the next response. (p 1040)

7-1.12 Identify what is essential for completion of a call. (p 1040)

7-1.13 Distinguish among the terms cleaning, disinfection, high-level disinfection, and sterilization. (p 1040)

7-1.14 Describe how to clean or disinfect items following patient care. (p 1040)

Affective

7-1.15 Explain the rationale for appropriate reporting of patient information. (p 1038)

7-1.16 Explain the rationale for having the unit prepared to respond. (p 1034)

Psychomotor

None

Additional Objectives*

Cognitive

1. Discuss the elements that dictate the use of lights and siren to the scene and to the hospital. (p 1046)

*This is a noncurriculum objective.

You are the Provider

At 7:30 AM you arrive at the station and begin preparing for your shift. In doing so you carefully inspect your ambulance inside and out. In the driver's compartment you start the engine to ensure that it is running properly. You test all of your interior and exterior lights. This includes all of your scene lights, emergency lights, turn signals, and brake lights. You test the siren and check all vehicle fluid levels. In the patient compartment you must make sure that you have all of the equipment that you could possibly need during your shift. You check not only that the equipment is present but that it is in good working order as well.

1. What possible ramifications could there be if you do not do a complete ambulance check?
2. What is the most important piece of safety equipment on your ambulance?

Gaining Access

As an EMT-B, you will usually not be responsible for rescue and extrication. Rescue involves many different processes and environments. It also requires training beyond the level of the EMT-B. In this chapter, you will learn basic concepts of extrication.

The chapter begins with a discussion of safety at the scene of a rescue incident followed by the 10 phases of extrication. Gaining access is one of the phases examined. This includes how to gain access to patients and how to keep yourself, patients, and bystanders safe in the process. Your main concern is reaching the patient so that you can begin providing care. In most cases, once you have reached the patient, extrication will occur around you and the patient. Communication between the EMT-B caring for the patient and fire personnel performing the extrication is vital.

Safety

You must always be prepared, mentally and physically, for any incident that requires rescue or extrication. The most important part of this preparation is thinking about your safety and the safety of your team. Safety begins with the proper mind-set and the proper protective equipment.

The equipment that you use and the gear that you wear will depend on the situation (Figure 36-1 ▼). However, the importance of wearing blood- and fluid-impermeable gloves at all times during patient contact cannot be emphasized enough. If you will be involved with extrication, you should wear a pair of leather gloves over your disposable gloves to protect you from injury when handling ropes, tools, broken glass, hot or cold objects, or sharp metal. Additional information on protective clothing is given later in this chapter.

Vehicle Safety Systems

A variety of safety systems are used in modern vehicles. Although many of these devices are useful when the automobile is in motion, they can present hazards to you after the car has been involved in a collision.

Shock-absorbing bumpers provide vehicle protection from low-speed impact. Following a front or rear-end collision, the shock absorbers within these bumpers may be compressed or "loaded." You should avoid standing directly in front of such bumpers because they can release and injure your knees and legs.

Manufacturers are now mandated to incorporate supplemental restraint systems or airbags into their vehicles. These airbags fill with a nonharmful gas on impact and quickly deflate after the collision. Airbags are located in the steering wheel and the dash in front of the passenger, and they deploy when the car is struck from the front or rear. Additional bags may be present to protect the driver and passengers from side impacts. These bags may be located in the doors or seats. Airbags

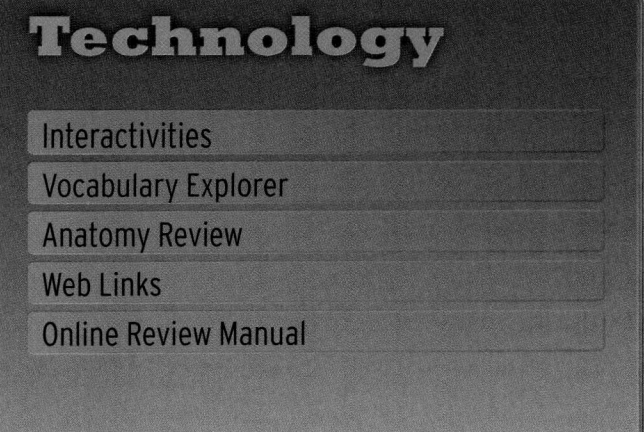

www.EMTB.com

Technology

Interactivities

Vocabulary Explorer

Anatomy Review

Web Links

Online Review Manual

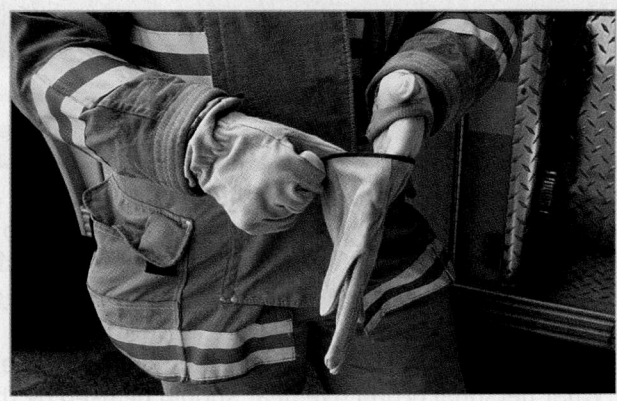

Figure 36-1 Proper protective equipment varies depending on the situation.

EMT-B Safety

A vehicle crash scene can present many hazards to rescuer and patient safety, including fuel spills that pose fire and explosion risks, downed electrical lines, broken glass and torn metal, and exposure to potentially infectious body fluids. Your safety at every type of emergency scene begins with, and depends on, an initial scene size-up that leads to decisions about what kind of personal protective equipment to use and whether to call for additional or specialized assistance.

TABLE 36-1   Ten Phases of Extrication
1. Preparation
2. En route to the scene
3. Arrival and scene size-up
4. Hazard control
5. Support operations
6. Gaining access
7. Emergency care
8. Disentanglement
9. Removal and transfer
10. Termination

should normally deploy and deflate before your arrival. Airbags have, however, inflated while EMT-Bs were providing patient care, causing injury to the medical provider. Caution should be exercised when working in damaged vehicles in which airbags have not inflated. Generally, you should maintain at least a 5″ clearance around side-impact airbags that have not deployed, 10″ around driver airbags that have not deployed, and 20″ around passenger-side airbags that have not deployed. Switching the ignition key to the "off" position and disconnecting the battery, negative side first, should reduce the potential for airbag deployment.

You may notice a haze similar to smoke inside vehicles in which airbags have deployed. Manufacturers use cornstarch or talc on the bags that may cause a minor skin irritation. Appropriate protective gear, including eye protection, will reduce the potential for such irritation.

Fundamentals of Extrication

During all phases of rescue, your primary concern is safety, and your primary roles are to provide emergency medical care and prevent further injury to the patient. You will provide care as extrication goes on around you unless this proves to be too dangerous for you or the patient. Extrication is the removal from entrapment or from a dangerous situation or position. Entrapment means being caught within a closed area with no way out or having a limb or other body part trapped. In the context of this chapter, extrication means removal of a

patient from a wrecked automobile. However, the same principles and concepts apply to other situations.

There are 10 phases to the extrication process (Table 36-1 ▲). Many are similar to the phases of an ambulance call discussed in Chapter 35. Each will be discussed, with emphasis on the phases in which you will participate.

Preparation

Preparing for an incident requiring extrication involves training for the various types of rescue situations your team might face. Some are discussed later in this chapter. Just as you must check the equipment carried on the ambulance, rescue personnel must also routinely check the extrication tools and their response vehicle to ensure its proper operation. Such preparations reduce the possibility of equipment failure at an emergency scene.

En Route to the Scene

Procedures and safety precautions similar to those discussed in the phases of an ambulance call are used when responding to a rescue call.

Arrival and Scene Size-up

When you arrive on the scene, you should position the unit in a safe location that does not add a hazard to the scene. Before proceeding, make sure that the scene is properly marked and that the road is closed or traffic flow is diverted safely around the scene (Figure 36-2 ▶). Size-up is the ongoing process of information gathering and scene evaluation to determine appropriate strategies and tactics to manage an emergency. One of the important

Figure 36-2 The scene of a crash should be marked properly, and traffic should be diverted so that responders have enough room to work.

responsibilities of scene size-up is to determine what, if any, additional resources will be needed. These resources may include additional EMS units and personnel. If you are first on the scene, you may need to initiate a rescue response or call for law enforcement or specialized crews, such as HazMat or utility departments.

You will need to coordinate your efforts with those on the rescue team. If you respect their job, they will respect yours. You should communicate with members of the rescue team throughout the extrication process. Start talking to the rescue team leader as soon as you arrive at the scene. Under the incident command system (described in Chapter 37), rescue operations are integrated as a separate group. You become a member of this group and will enter the vehicle and provide care for the patient(s) when approved by the extrication leader.

The rescue team is responsible for properly securing and stabilizing the vehicle, providing safe entrance and __access__ to patients (the ability to reach the patient), extricating any patients, ensuring that patients are properly protected during extrication or other rescue activities, and providing adequate room so that patients can be removed properly.

EMS personnel are responsible for assessing and providing immediate medical care, triage and assigning priority to patients, packaging patients, providing additional assessment and care as needed once patients are removed, and providing transport to the emergency department.

Hazard Control

A variety of hazards may be present at the extrication scene. Law enforcement personnel are responsible for traffic control and direction, maintaining order at the scene, investigating the crash or crime scene, and establishing and maintaining lines so that bystanders are kept at a safe distance and out of the way of rescuers. Fire fighters are responsible for extinguishing any fire, preventing additional ignition, ensuring that the scene is safe, and removing any spilled fuel (Figure 36-3 ▶).

Downed electrical lines are a common hazard at vehicle crash scenes. You should never attempt to move downed electrical lines. If power lines are touching a vehicle involved in the crash, victims should be instructed to remain in the vehicle until power is removed. You and the ambulance should remain outside the __danger zone (hot zone)__. A danger zone is an area where individuals can be exposed to sharp metal edges, broken glass, toxic substances, lethal rays, or ignition or explosion of hazardous materials.

Bystanders and family members can be hazards themselves. If they are allowed to get too close, they are at risk of injury and may also interfere with the overall management of the incident. For these reasons, the rescue team will set up a danger zone that is off-limits to bystanders (Figure 36-4 ▶). You should help to set

You are the Provider   Part 2

The fire department incident commander advises that the tanker truck is leaking gasoline. Rescue units have been dispatched but are 10 minutes away. The driver of the passenger vehicle appears to be unconscious. The driver's door is jammed, and the rear window is broken out. A blanket of foam is placed beneath the tanker truck and around the vehicle by the fire department.

3. Is the scene safe enough for you to enter?
4. If the area is determined to be safe, should you attempt to gain access to the patient?

Figure 36-3 Every crash requires cooperation, as each responder has a specific role at the scene. Fire fighters, law enforcement personnel, the rescue team, and EMS personnel all have individual roles.

Figure 36-4 A danger zone should be established to prevent bystanders from entering the area around an incident.

Figure 36-5 The exact way to gain access depends on many factors, including the terrain, the way in which the vehicle is situated, and the weather.

up and enforce this zone. If you arrive before the rescue team, you should coordinate crowd control with law enforcement officials.

The vehicle also can be a hazard. An unstable automobile on its side or roof can be a danger to you. Rescue personnel can stabilize the vehicle with a variety of jacks or wooden blocks.

Support Operations

Support operations include lighting the scene, establishing tool and equipment staging areas, and marking helicopter landing zones. Fire and rescue personnel will work together on these functions.

Gaining Access

A critical phase of extrication is gaining access to the patient. Remember, you should not attempt to gain access to the patient or enter the vehicle until you are sure that the vehicle is stable and that any hazards have been identified and properly controlled or eliminated. When there is a rescue leader present, you will be authorized to enter only when these considerations have been met.

The exact way you gain access to or reach the patient(s) depends on the situation. It is up to you to identify the safest, most efficient way to gain access. Darkness, uneven terrain, tall grass, shrubbery, and wreckage may make patients hard to find (Figure 36-5 ▲). Multiple vehicles with multiple patients may be involved. If this is the case, you should locate and rapidly triage each patient to determine who needs urgent care. This step is important before you proceed with any treatment and patient packaging. Be sure to take these factors into account in your scene size-up. Remember that scene size-up is a continuing process, because the situation often changes. As a result, you may need to change your plans for gaining access and providing treatment.

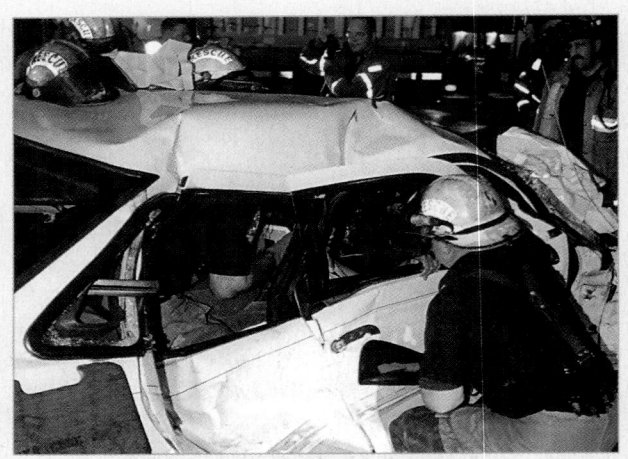

Figure 36-6 Always explain to the patient why you are there and what you are doing.

Figure 36-7 Use a long backboard or blanket to protect the patient and any rescuers who are providing care.

To determine the exact location and position of the patient, you and your team should consider the following questions:

- Is the patient in a vehicle or in some other structure?
- Is the vehicle or structure severely damaged?
- What hazards exist that pose risk to the patient and rescuers?
- In what position is the vehicle? On what type of surface? Is the vehicle stable or is it apt to roll or tip?

You must also take into account the patient's injuries and their severity. You may have to change your course of action as you learn more about the patient's condition. Do not try to access the patient until you are sure that the vehicle is stable and that hazards have been identified and rendered safe. Hazards might include electrical or gas lines.

What should you do if you have to remove a patient quickly because the environment is threatening or you need to perform CPR? CPR is not effective when the patient is in a sitting position or lying on the soft seat of a vehicle. In these cases, you and your team may have to use the rapid extrication technique to move a patient from a sitting position inside a vehicle to a supine position on a long backboard. A team of EMT-Bs who are experienced in using this technique should be able to rapidly remove a patient who is not entrapped, keeping in mind the patient's condition and the group's safety. Use the rapid extrication technique only as a last resort.

While you are gaining access to the patient and during extrication, you must make sure that the patient remains safe. Always talk to the patient and describe what you are going to do before you do it and as you are doing it, even if you think the patient is unconscious (Figure 36-6 ◀). In many cases, you or your partner may be providing cervical spine immobilization or other care during extrication. All EMS personnel should wear proper protective gear while in the working area. The patient and EMS personnel should be covered with a heavy, nonflammable blanket to protect them against flying glass or other objects. A long backboard may also be used as a protective shield (Figure 36-7 ▲). Try to keep heat, noise, and force to a minimum. Use only what is necessary to extricate the patient safely.

Simple Access

Your first step is <u>simple access</u>, trying to get to the patient as quickly and simply as possible without using any tools or breaking any glass. Automobiles are built for easy entry and exit; however, it may be necessary to use tools or other forcible entry methods. Whenever possible, you should first try to unlock the doors (or ask the patient to unlock them) or roll down the windows. Try to open every door using the door handles to gain access before breaking any windows or using other methods of forced entry (Figure 36-8 ▶). Enter through the doors when there is no danger to the patient. The rescue team should provide the entrance you need to gain access to the patient.

Complex Access

<u>Complex access</u> requires the use of special tools and special training and includes breaking windows or other forcible entry. Most of these skills are too advanced for the 110-hour EMT-B course and are not covered in this text.

Figure 36-8 Get to the patient as quickly and simply as possible by opening the door without using tools or breaking any glass.

TABLE 36-2 **Disentanglement Techniques**
■ Brake and gas pedal displacement
■ Dash roll-up
■ Door removal
■ Roof opening and removal
■ Seat displacement
■ Steering column displacement
■ Steering wheel cutting

Emergency Care

Providing medical care to a patient who is trapped in a vehicle is principally the same as for any other patient. Unless there is an immediate threat of fire, explosion, or other danger, once entrance and access to the patient have been provided, you should perform an initial assessment and provide care before further extrication begins, as follows:

1. Provide manual stabilization to protect the cervical spine, as needed.
2. Open the airway.
3. Provide high-flow oxygen.
4. Assist or provide for adequate ventilation.
5. Control any significant external bleeding.
6. Treat all critical injuries.

Good communication among team members and clear leadership are essential to safe, efficient provision of proper emergency care. Although your input at the scene is important, one member of your team must be clearly in charge. The team leader's assessment of the patient and the situation will dictate the way in which medical care, packaging, and transport will proceed. Customarily, the crew chief, who typically is clearly indicated on the shift schedule, is responsible for this role. If not, a team leader must be identified and agreed to before you arrive at the scene.

In some areas, there might not be enough personnel for two or more units. In these areas, you and your team may have two roles. However, one person must still be in charge of the overall rescue operation. A lack of identifiable leadership at the scene hinders the rescue effort and patient care. Leaders should be identified as part of a larger incident command system. They should be medically trained and qualified to judge the priorities of patient care, and they must also be experienced in extrication.

Disentanglement

Disentanglement involves the removal of the motor vehicle from around the patient. Rescue personnel should coordinate with you to determine the best route of removing the patient from the vehicle (Table 36-2 ▲). While one accident may require removal of the patient through the driver's door, a similar accident may require complete removal of the roof.

Disentanglement requires the use of a variety of complex hand and power tools. Specialized education is required for their safe operation.

As a part of your assessment, you should participate in the preparation for removal. Determine how urgently the patient must be extricated, where you should be positioned to best protect the patient during extrication, and, once the patient has been freed, how you will best move the patient from within the vehicle onto the long backboard and onto the stretcher. Carefully examine the exposed area of the limb or other part of the patient that is trapped to determine the extent of injury and whether there is a possibility of hidden bleeding. If possible, you should also evaluate sensation in the trapped area so that you will know whether increased pain indicates that an object is pressing on or impaled in the patient during extrication.

During this time, the rescue team is assessing exactly how the patient is trapped and determining the safest, easiest way to extricate him or her. Your input is essential so that the patient's injuries are considered as the rescue team plans a move that protects the patient from further harm. Once the plan has been devised and everyone understands what will be done, you should determine how best to protect the patient. Often, you or another EMT-B will be placed in the vehicle alongside the patient to monitor his or her condition and well-being as the vehicle is being forcibly cut, bent, or disassembled. Be sure to wear proper protective clothing.

Naturally, your safety and that of the patient are paramount during this process. Both you and the patient should be covered by a thick, fire-resistant canvas or blanket for protection from broken glass, flying particles, tools, or other hazards during any cutting or forceful extrication maneuvers. Extrication is often extremely noisy. You must be sure that you can communicate effectively with the patient and the rescue group so that you can instantly let the rescuers know if it is necessary that they stop.

Removal and Transfer

Once the patient has been freed, rapidly assess any previously inaccessible patient and reassess any patient previously assessed. Make sure that the spine is manually immobilized, and apply a cervical collar if this was not previously done (Figure 36-9 ▶). Reevaluate whether the patient needs to be immediately removed by using manual immobilization and the rapid extrication technique or whether the patient's condition and the scene allow for immobilization using an extrication vest or short backboard before he or she is moved further. In most cases, it is impractical and difficult to properly apply extremity splints within the vehicle. Extremity injuries can generally be rapidly supported and immobilized while the patient is being removed by securing an injured arm to the body and, if a leg is injured, securing one leg to the other. This will be adequate until the patient is secured to the backboard or time permits more detailed assessment and splinting of each injury.

Moving the patient in one fast, continuous step increases the risk of harm and confusion. To ensure that each EMT-B can be positioned so that he or she can lift

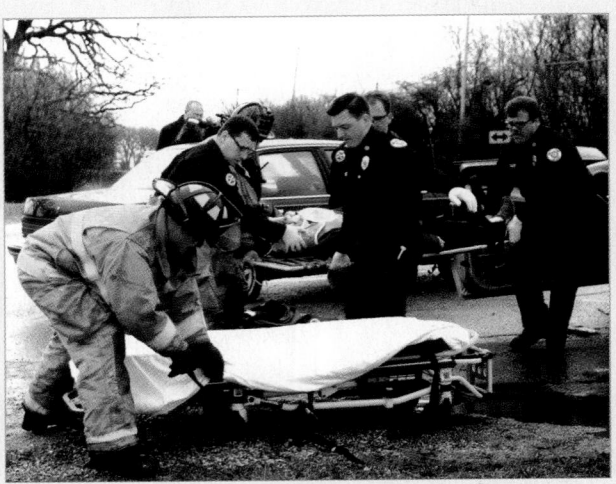

Figure 36-9 Once the patient has been accessed, rapidly assess the patient and make sure that the spine is manually stabilized. Apply a cervical collar if this was not previously done.

and carry properly at all times, move the patient in a series of smooth, slow, controlled steps, with stops designed between them to allow for the repositioning and adjustments that are needed. Plan the exact steps and pathway that you will follow in moving the patient from sitting in the vehicle to lying supine on the backboard and prepared ambulance cot. Choose a path that requires the least manipulation of the patient and equipment. Make sure that sufficient personnel are available. Once you are sure that everyone understands the steps and is ready, you can move the patient safely. Make sure that you move the patient as a unit, resisting the temptation to move the immobilization device instead. While moving the patient, continue to protect him or her from any hazards.

You are the Provider   Part 3

The rescue unit arrives on the scene, assesses the situation, and begins extrication. The driver's door is removed, the dash rolled, and the roof opened. The patient is turned over to you for continued care and transport to a medical facility.

5. What safety concerns remain at this scene?
6. What initial emergency care should be provided to the trapped driver?

Once the patient has been placed on the stretcher, continue with any additional assessment and treatment that was deferred. If it is extremely cold or hot, raining or snowing, you should load the stretcher and patient into the climate-controlled ambulance before continuing assessment and treatment. If the patient's condition requires that transport be initiated without further delay, you should provide only the additional care that is essential or necessary to package the patient. Leave the remaining steps to be performed en route to the hospital.

Termination

Termination involves returning the emergency units to service. For rescue units, this process may be quite involved. All equipment used on the scene, including hydraulic, electrical, and hand tools, must be checked before reloading them on the apparatus. While some tools require only generalized cleaning, others may need to be refueled and various fluid levels checked.

You will also be required to check the ambulance thoroughly, replacing used supplies and conforming to cleaning needs required by bloodborne pathogen standards.

Finally, rescue units and medical units will be required to complete all necessary reports.

Specialized Rescue Situations

On most calls, you can drive the ambulance to within a short distance of the patient's location and, with simple or complex access, you can reach and treat the patient. However, in some situations, the patient can be reached only by teams trained in special technical rescues. Specialized skills of these teams include the following:

- Cave rescue
- Confined space rescue
- Cross-field and trail rescue (park rangers)
- Dive rescue
- Lost person search and rescue
- Mine rescue
- Mountain, rock, and ice-climbing rescue
- Ski slope and cross-country or trail snow rescue (ski patrol)
- Structural collapse rescue
- Tactical response and rescue (SWAT)
- Technical rope rescue (low- and high-angle rescue)
- Trench rescue
- Water and small craft rescue
- White-water rescue

Technical Rescue Situations

Technical rescue situations may contain hidden dangers, and special technical skills are needed for personnel to safely enter and move around. It is not safe to include personnel who do not have the necessary special training and experience in such a rescue. A technical rescue group is made up of individuals from one or more departments in a region who are trained and on call for certain types of technical rescues. Many members of a technical rescue group are also trained as first responders or EMT-Bs so that they can provide the necessary immediate care when only they can safely reach the patient. Even when the technical rescue group includes a paramedic or physician, generally nothing but essential simple care is provided until the rescuers can bring the patient to the nearest point where a safe, stable setting exists.

If a technical rescue group is necessary but is not present when you arrive, you should immediately check with the incident commander to make sure that the group has been summoned and is en route to your location. The incident commander is the individual who has overall command of the scene in the field Figure 36-10 ▼ . If no incident commander is present, follow local guidelines. (Chapter 37 discusses incident command in more detail.)

When you arrive at a scene where a technical rescue is in progress, you will usually be met by a member of the technical rescue group and directed or led to the actual rescue site. If the rescue site is some distance from the road, you may need to leave the ambulance on the road. The use of the ambulance stretcher is impractical in these situations; you should instead take a long backboard and/or basket stretcher

Figure 36-10 The incident commander is the individual who has overall command of the scene.

or similar rescue stretcher to carry the patient back to the waiting ambulance. Be sure that you take all of the carry-in kits and other equipment you may need to treat and immobilize the patient at the rescue site.

When you arrive at the rescue site, identify the stable location to which the technical rescue group will bring the patient, and set up your equipment there. As soon as the technical rescue group has brought the patient to this staging area, you should perform a rapid assessment and, after providing the treatment indicated, package the patient without delay. Although you and the other EMTs who responded with the ambulance will assume the primary responsibility for the patient's care, at this point it usually requires a cooperative effort by the technical rescue group and EMS team to carry the patient to the waiting ambulance. Consider using an air medical unit if the patient will need to be carried or transported an extensive distance.

Lost Person Search and Rescue

When someone is lost in the outdoors and a search effort is initiated, an ambulance is usually summoned to the search base. Each search team will be organized to include a member who is trained at the first responder or EMT-B level, carrying the essential equipment to provide simple immediate care. Your role, and that of the other EMT-Bs who arrive with the ambulance, is to stand by at the search base until the lost person or people have been found.

As soon as you arrive at the scene and have been briefed on the situation, you should isolate and prepare the equipment you will need to carry to the patient's location so that no time is lost once the patient has been found or a member of the search team is injured. The prepared carry-in equipment, including a long backboard and other equipment you will need to immobilize the patient, should be left in the back of the ambulance so that it is protected from the weather. In addition, if the ambulance should need to be relocated, the equipment will not need to be reloaded and will not be left behind. You will usually be given a portable radio that is tuned to the search frequency so that you can monitor the progress of the search and communicate with and be contacted by those in charge of the search operation.

Sometimes, you may be asked to stay with relatives of the lost individual who are at the scene. Find out from relatives whether the lost person has any medical history that may need to be addressed, and pass this information on to those who are in charge of the search.

Unless you have been instructed otherwise, only incident command should communicate any news or progress of the search to the family. For this reason, you must be sure that your radio is set at a discreet volume.

Once the lost person has been found, you will be guided by search personnel to that location or a pre-arranged intersecting point where the patient will be carried to decrease the amount of time you need to reach the patient and begin treatment. You should be sure that the carry-in equipment is evenly distributed among personnel and that the pace is such that all can stay together easily. Sometimes the time and effort that are needed to reach and carry out the patient can be decreased by relocating the ambulance or, if one is available, by using a four-wheel drive or all-terrain vehicle. As with other specialized rescues, although the ambulance crew will assume the responsibility for patient care once they are at the patient's side, a cooperative effort of the EMS and search teams is necessary to safely carry the patient to the base and waiting ambulance.

Trench Rescue

Owing to the physical forces involved, many cave-ins and trench collapses have poor outcomes for victims. Collapses usually involve large areas of falling dirt that weigh approximately 100 pounds per cubic foot. Victims with thousands of pounds of dirt resting on their chests cannot fully expand their lungs and may become hypoxic.

The risk of a secondary collapse during the rescue operation is of concern to rescue personnel and to the EMT-Bs. Safety measures can reduce the potential for injury from this and other hazards. When arriving on the scene of a cave-in or trench collapse, response vehicles should be parked at least 500 feet from the scene. Because vibration is a primary cause of secondary collapse, all vehicles, including on-scene construction equipment, should be turned off. In addition, all road traffic should be diverted from the 500-foot safety area. Other hazards include exposed or downed electric wires and broken gas or water lines. In addition, construction equipment at the collapse may be unstable and could fall into the trench or cave-in site.

Any witnesses to the incident should be identified. They may be valuable in providing information on the number of victims and their location within the collapsed area. Any nontrapped individuals should be assisted from the area. At no time should medical or rescue personnel enter a trench deeper than 4 feet without proper shoring.

During the extrication of any live victims, medical personnel trained in cave-in and trench collapse rescue will provide most medical care. You should be prepared to receive patients once they have been extricated from the site.

Tactical Emergency Medical Support

A steady increase in violence throughout the country has resulted in EMT-Bs taking precautions to ensure personal safety. Normally, when the potential for violence exists—as in shootings, stabbings, and attempted suicides—responding units should stage until the scene is secured by law enforcement personnel. However, some incidents pose an increased risk to EMT-Bs and law enforcement personnel. Hostage incidents, barricaded subjects, and snipers require the use of specialized law enforcement tactical units or the special weapons and tactics team (SWAT).

Owing to the high potential for injuries at these incidents, many communities have incorporated specially trained EMT-Bs, paramedics, nurses, and even physicians into their police SWAT units. These EMS providers provide a special level of care to the sick and injured at such volatile incidents. Their training goes well beyond the practices seen in standard emergency medical care. Thus, the skills used may not seem appropriate or adequate. For example, spinal immobilization is not used within an unsecured area where gunfire may still erupt. The time and manpower necessary to completely secure a victim to a backboard with collar, straps, and head immobilization may expose EMS providers and SWAT officers to injury or death from gunfire

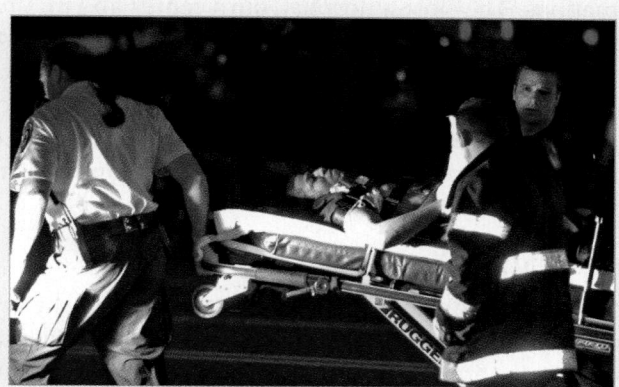

Figure 36-11 EMS providers move a downed officer. In an unsecured area, only the most basic medical care is provided.

(Figure 36-11 ▲). Such altered medical procedures are similar to those used by military EMS providers on the battlefield and are not used in "standard" situations encountered by EMT-Bs.

When called to the scene of a law enforcement tactical situation, you should determine the location of the command post (location of the incident commander) and report to the incident commander for instructions. Lights and siren should be turned off when nearing the scene, and outside radio speakers should not be used. The command post is usually located in an area that cannot be seen by the suspect and is out of range of possible gunfire. You should remain in this area and not wander. Nearby areas may be visible to the suspect, and you could be injured.

A number of planning measures should be started after checking with the incident commander. Such planning will reduce the potential for chaos should a mass-casualty incident occur at the scene. First, have the incident commander identify the specific location of the incident. The information should include the street address and the side of the street on which the house or building is located. The incident commander should determine a safe location where you can meet SWAT team members or tactical EMS providers should an injury occur. Tactical EMS providers or officers will remove the patient to this area for your continued treatment and transport to a medical facility. The incident commander should also determine a safe route to this meeting point.

Designate primary and secondary helicopter landing zones if your region uses aeromedical evacuation. Such preplanning will save valuable time in critical

EMT-B Safety

Physical dangers such as fire, infectious disease, and electricity are not the only risks to your safety during emergency responses. Some calls involve the possibility of deliberate violence against rescuers. Formal tactical situations are obvious examples, but "simple" calls involving assaults, possible alcohol or drug use, and domestic disputes can be just as dangerous. Your training, your attitude when responding to calls, and your routine daily procedures should all take these risks into account. Never become complacent.

situations. The closest hospital, burn center, and trauma center should be identified. The route of travel to these facilities should also be noted. Many of these measures are incorporated into the operational plan used by tactical EMS providers. If tactical EMS providers are used in your jurisdiction, coordinate with them on your arrival at the command post.

Structure Fires

In most areas, an ambulance is dispatched with the fire department apparatus to any structure fire, whether or not injuries are reported. A fire in a house, apartment building, office, school, plant, warehouse, or other building is considered a <u>structure fire</u>. When responding to a major fire scene, you should determine whether, because of the fire, any special route will be necessary. Once you arrive at the scene, you should ask the incident commander where the ambulance should be parked. It is essential that the ambulance be parked far enough from the fire to be safe from the fire itself or a collapsing building. You must also ensure that the ambulance will not block or hinder other arriving equipment or be blocked in by other equipment or hose lines. However, you must also make sure that the ambulance will be close enough to be visible and that patients can be brought to it easily. The fire officer who is the incident commander will determine this location.

Your next step is to determine whether there are any injured patients at the scene or whether you have been called to stand by. A number of ambulances may be dispatched to a major fire to ensure that one or more units will always remain immediately available at the scene if others leave to transport the injured.

As with other specialized rescue situations, search and rescue in a burning building require special training and equipment. Search and rescue are performed by teams of fire fighters wearing full turnout gear and <u>self-contained breathing apparatus (SCBA)</u> and carrying tools and fully charged hose lines. These teams will bring patients out of the burning building to the area where the ambulance is standing by. Therefore, unless otherwise ordered, you should always stay with the ambulance. Do not leave the scene even after the fire is out, in case a fire fighter is injured during salvage and overhaul. The ambulance should leave the scene only if transporting a patient or if the incident commander has released it.

Sometimes the scene at a crash or fire is further complicated by the presence of hazardous materials. A <u>hazardous material</u> is any substance that is toxic, poisonous, radioactive, flammable, or explosive and can cause injury or death with exposure. In addition to posing a threat to you and others at the immediate scene, hazardous materials may pose a threat to a much larger area and population. Whenever there is a possibility that a hazardous material is involved, you will have to follow a number of additional special procedures. Chapter 37 covers the specifics of hazardous material procedures.

You are the Provider   Summary

The incident involving an accident between a passenger vehicle and a tanker truck provides a variety of challenges to emergency responders as a whole and, more specifically, you. A variety of safety issues are of concern at this scene. They include the following:

■ Traffic—oncoming traffic is a concern at all motor vehicle incidents.
■ Electric—the tanker truck's position against the electric pole may have caused power lines to fall.
■ Hazardous material—the tanker truck itself may be leaking an unknown substance. Both the tanker truck and the passenger vehicle may be leaking diesel fuel, gasoline, or another type of hazardous product.

Safety equipment can limit the potential for injury. At this scene, turnout gear should be used—helmet, coat, pants, boots, gloves, and eye protection. Respiratory protection should be used if a hazardous substance is leaking.

The incident commander will advise you when the scene is safe to enter. In this case, the foam from fire department hose lines should allow you to gain access through the broken rear window. There remains a continued concern about leaking gasoline and the potential for electric lines to fall. In addition, the extrication process can cause injury from flying glass, jagged metal, and the rescue tools themselves.

Initial emergency care will involve manual cervical spine immobilization, ensuring an airway, providing adequate ventilation, and providing high-flow oxygen. In addition, you should control significant bleeding and check for critical injuries.

Prep Kit

Ready for Review

- During all phases of rescue, your primary concern is safety, and your primary roles are to provide emergency medical care and prevent further injury to the patient.

- When there are not enough personnel for both an EMS team and a rescue team, you and your team may have to act as rescuers as well.

- Safety during rescue or extrication begins with the proper mind-set and the proper protective equipment.

- During scene size-up, you should identify the safest, most efficient way to access the patient. Try to get to the patient as simply and quickly as possible without using tools or breaking glass.

- Make sure that you and the patient are protected with a fireproof blanket.

- Unless there is immediate danger, perform an initial assessment of a patient while he or she is still in the vehicle. Immobilize the cervical spine before moving the patient from the vehicle.

- If you see that a special rescue team is needed, inform the dispatcher.

- When a scene calls for a search or for specialized rescue, you may have to call for a technical rescue group, or you may find one already at work when you arrive. Your interaction and cooperation with this group, and with an incident commander when one has been designated are important to a smooth rescue.

- You will be involved to some degree in the logistics of vehicle staging and patient movement, in addition to patient care.

- Tactical situations are best directly handled by teams with specialized training. Your role will often largely consist of remaining out of danger, cooperating with the incident commander or with police or other specialized personnel if incident command is not in effect, and remaining ready to care for any patients that are brought to you.

Technology

Interactivities

Vocabulary Explorer

Anatomy Review

Web Links

Online Review Manual

www.EMTB.com

Prep Kit continued...

Vital Vocabulary

access The ability to gain entry to an enclosed area and reach a patient.

command post The location of the incident commander at the scene of an emergency and where command, coordination, control, and communication are centralized.

complex access Complicated entry that requires special tools and training and includes breaking windows or using other force.

danger zone (hot zone) An area where individuals can be exposed to sharp metal edges, broken glass, toxic substances, lethal rays, or ignition or explosion of hazardous materials.

disentanglement The removal of a motor vehicle from around the patient.

entrapment To be caught (trapped) within a vehicle, room, or container with no way out or to have a limb or other body part trapped.

extrication Removal of a patient from entrapment or a dangerous situation or position, such as removal from a wrecked vehicle, industrial accident, or building collapse.

hazardous material Any substance that is toxic, poisonous, radioactive, flammable, or explosive and causes injury or death with exposure.

incident commander The individual who has overall command of the scene in the field.

self-contained breathing apparatus (SCBA) Respirator with independent air supply used by fire fighters to enter toxic and otherwise dangerous atmospheres.

simple access Access that is easily achieved without the use of tools or force.

special weapons and tactics team (SWAT) A specialized law enforcement tactical unit.

structure fire A fire in a house, apartment building, office, school, plant, warehouse, or other building.

tactical situation A hostage, robbery, or other situation in which armed conflict is threatened or shots have been fired and the threat of violence remains.

technical rescue group A team of individuals from one or more departments in a region who are trained and on call for certain types of technical rescue.

technical rescue situation A rescue that requires special technical skills and equipment in one of many specialized rescue areas, such as technical rope rescue, cave rescue, and dive rescue.

Points to Ponder

You are dispatched to a crash between a school bus and a semitractor trailer. Upon your arrival you find a yellow school bus lying on its side with several children still inside and several children walking around the scene. The tractor trailer was transporting an unknown substance that is starting to spill on the ground. As you are assessing the scene, parents of the children start arriving and are upset.

What resources will you need to respond to this scene? How will you gain access to the school bus? How will you handle the leaking fluid and the arrival of parents?

Issues: Interacting With Children, Interacting With Parents, Consent, Scene Safety.

You are the senior EMT-B and have a partner who is working his first day. You are dispatched to a motor vehicle collision.

1. Your truck is the first on scene. After you park your vehicle, what is your first priority?

   A. Performing a scene size-up
   B. Performing a patient assessment
   C. Directing traffic
   D. Stabilizing the vehicle

2. You have determined that you have two vehicles involved with severe structural damage sitting in the middle of a busy intersection. Clear liquid is pouring out of one of the vehicles. What resources are you going to need?

   A. HazMat
   B. Law enforcement
   C. Fire department
   D. All of the above

3. The vehicle has been stabilized and the incident commander states it is safe to enter the vehicle. What is your first priority?

   A. Extricating the patient
   B. Assessing the patient's entrapment
   C. Performing an initial assessment
   D. Breaking all the vehicle's windows

4. During assessment of your patient you find a sucking chest wound. The patient is still maintaining an airway. Treatment should include:

   A. waiting until the patient is extricated.
   B. applying an occlusive dressing and monitoring.
   C. packing the wound with gauze.
   D. all of the above.

5. Your patient goes into sudden cardiac arrest. You should:

   A. start CPR in the vehicle.
   B. pronounce the patient DOA.
   C. place the patient on a KED.
   D. perform a rapid extrication.

6. While you are starting CPR on the patient lying next to the vehicle a bystander runs up and states he is a doctor and wants to help. What is your response?

   A. Let the doctor take over.
   B. Have the police remove him.
   C. Have an EMS provider explain that you must approve his assistance with medical control and that he must follow your protocols and ride in with the patient.
   D. Have the doctor drive the ambulance.

Challenging Questions

You have a tomato truck rollover on the highway with several migrant workers ejected. Upon arrival you have four patients on scene. You are the EMS command. You begin to triage. Patient 1 is in respiratory arrest, but has a pulse rate. Patient 2 has an obvious hip fracture and is screaming in pain. Patient 3 is in cardiac arrest. Patient 4 is walking around confused.

7. Which patient would be the highest priority?

8. Which patient would be transported last?

9. How would you treat Patient 3?

www.EMTB.com

Special Operations

Objectives

Cognitive

7-3.1 Explain the EMT-B's role during a call involving hazardous materials. (p 1088)

7-3.2 Describe what the EMT-B should do if there is reason to believe that there is a hazard at the scene. (p 1087)

7-3.3 Describe the actions that an EMT-B should take to ensure bystander safety. (p 1087)

7-3.4 State the role the EMT-B should perform until appropriately trained personnel arrive at the scene of a hazardous materials situation. (p 1087)

7-3.5 Break down the steps to approaching a hazardous situation. (p 1088)

7-3.6 Discuss the various environmental hazards that affect EMS. (p 1085)

7-3.7 Describe the criteria for a multiple-casualty situation. (p 1079)

7-3.8 Evaluate the role of the EMT-B in the multiple-casualty situation. (p 1079)

7-3.9 Summarize the components of basic triage. (p 1081)

7-3.10 Define the role of the EMT-B in a disaster operation. (p 1084)

7-3.11 Describe basic concepts of incident management. (p 1074)

7-3.12 Explain the methods for preventing contamination of self, equipment, and facilities. (p 1089, 1092)

7-3.13 Review the local mass-casualty incident plan. (p 1074)

Psychomotor

7-3.16 Given a scenario of a mass-casualty incident, perform triage. (p 1082)

Additional Objective*

Affective

1. Discuss the psychological impact of wanting to act but recognizing that a scene is not safe to enter. (p 1085)

*This is a noncurriculum objective.

You are the Provider

You and your EMT-B partner are dispatched to a multi-vehicle collision on Interstate 10 at milepost 86. Dispatch states that there are at least two vehicles involved, one being a semi. Approximately two miles from the scene you note that the eastbound traffic is at a complete standstill. You continue traveling towards the scene by way of the median. As you are approaching the scene you find an 18-wheeler semitrailer truck on its side. Off to the side of the road you see a passenger van that appears to have rolled over and is now sitting upright.

1. What is your first step on this scene?
2. What safety issues should you be considering?

Special Operations

The first section of this chapter will introduce you to incident command systems. The purpose of this section is to give you an idea of what happens during complex incidents. Your role within the system is explained.

The next section describes the several roles of EMT-Bs at mass-casualty incidents. Again under the incident command system, EMS personnel will have one of several roles identified to manage a large number of patients at a single event. The usual EMS response of triaging three or four patients will be difficult when there are 25 or more casualties. To ensure that every patient receives appropriate care and transportation to a hospital consistent with the severity of his or her condition, a more organized operation is required with three major responsibilities assigned: triage, treatment, and transportation.

The final section describes your responsibilities at a hazardous materials incident. When you are responding to this type of incident, you cannot rush in to provide patient care. Rather, you must cooperate with the incident command system, taking time to accurately assess the scene by identifying the size of the hazard area, finding a safe location to which patients can be removed, and taking self-protective measures. Safety is your primary consideration. If a hazardous materials incident is not carefully handled, many people, including rescue personnel, can be injured or die.

Technology

- Interactivities
- Vocabulary Explorer
- Anatomy Review
- Web Links
- Online Review Manual

www.EMTB.com

Incident Command Systems

In recent years, a number of leadership and command systems have been developed to improve the on-scene management of emergency situations. The fire service has taken the lead in developing these programs to help control, direct, and coordinate emergency responders and resources. These programs are called incident command systems. They have been adapted and used by many EMS organizations to better organize their own operations. The incident command system is designed for use in daily operations. However, it is most effective when used to organize large numbers of personnel at complex incidents such as hazardous materials spills and mass-casualty incidents.

Components and Structure of an Incident Command System

At a large fire, a hazardous materials incident, or a mass-casualty incident, fire, rescue, HazMat (hazardous materials), police, and EMS units from many different areas usually will become involved in some way. To ensure clear lines of responsibility and authority, a preestablished system is needed to identify who is in charge of different activities and who reports to whom. Even on a call with only one patient and no need for any other services, the implementation of an incident command system is helpful to identify the roles and responsibilities of each crewmember, particularly if the event begins to escalate.

The incident command system is structured so that there is a single authority with overall responsibility to manage the incident. This person is identified as the incident commander. The incident commander usually remains at a command post, the designated field command center. A field command center is typically a vehicle or building at the scene where the incident commander establishes an "office." From here, the commander oversees and coordinates the activities of the various groups and leaders.

Functions normally centered at the command post include information, safety, and liaison with other agencies and groups who are responding. In a typical incident command system operation, all information to the public and the news media originates at the command post. The incident commander will usually appoint a safety officer who will circulate among responding personnel. It is essential that every EMT-B understand that any order or directive issued by a safety officer has the full authority of the incident commander and must be immediately followed. Many times EMT-Bs cannot see a hazard or problem they are walking into, and the safety officer is responsible for protecting all personnel and any victims of the incident. Finally, an officer may be named by the incident commander to coordinate incoming fire, police, and EMS units.

In the initial response, the incident commander may assume direct control over the groups and task forces being set up. In this circumstance, a medical group supervisor may be named to coordinate all EMS activity, or a rescue group supervisor may be appointed to deal with people entrapped in the wreckage. In extended operations that may go on for hours, days, or longer, the typical incident command structure may have multiple sectors, including operations, planning, logistics, and finance. (Figure 37-1 ▼) shows a sample incident command system organization chart that includes these sectors. Each of these sections will have a single officer acting as the person in charge, the sector commander. Not all positions are used at every incident. The incident commander will select the individual positions and teams and will choose which to use depending on the nature of the incident.

Major incidents often require another level of management, known as unified command. With unified command, the incident commander is joined at the command post by one officer who is in charge of all fire operations, one who is in charge of all rescue or HazMat operations, one who is in charge of all EMS operations, and one who is in charge of all law enforcement. This group, under the direction of the incident commander, directs the overall operations at the scene. Because the different public safety officers are stationed at the command post, they can be easily found and can collectively advise the incident commander of changes and problems that are communicated to them. The incident commander can also involve them in making the necessary decisions and in rapidly conveying orders to those under their command. In addition to unified command, this system ensures that the actions of each different type of responder are properly coordinated.

How these systems work together depends on the nature of the event. For example, with a major airplane crash, the leading agency is typically the fire department. In this situation, EMS is usually one aspect of the overall fire incident command system. Within their own system, EMS personnel establish and carry out

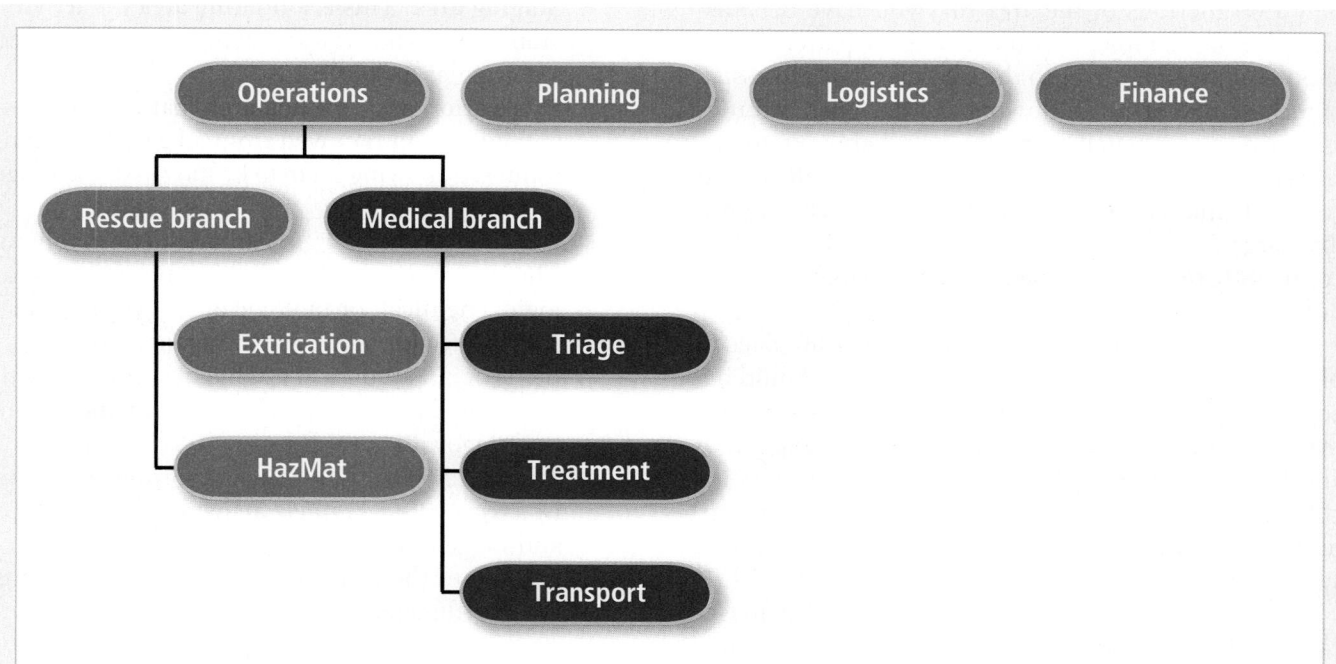

Figure 37-1 Incident command structure. Not all positions will be filled in every incident. However, the incident commander is responsible for all activity. Subordinates may be appointed to assist in managing the incident.

their tasks. However, ultimate control of the incident will rest with the fire commander. Other situations, such as widespread injuries at a rock concert, are primarily civil events. Law enforcement would take the lead, establishing an incident command plan. Fire and EMS would follow the law enforcement commander's decision. EMS is seldom the lead agency.

At one time or another, your unit will probably be the first to respond to an incident that will involve more than one EMS unit or one or more non-EMS agencies. An example might be 10 or 20 teenagers with adverse reactions to pepper spray one student brought to school or another event in which there is a limited role for immediate fire or police intervention. In this situation, the senior EMT-B should establish command and inform the dispatcher by saying, for example, "Dispatch, this is Squad 71, establishing Route 43 command." From that point on, all communications from the dispatcher to the scene will be directed to "Route 43 Command." If you arrive after command has already been established, advise the dispatcher that you are on scene and "reporting to command." Then find the command post, and report for assignment.

When you respond to an incident at which the incident command system has already been established, you will be assigned to a specified area and its designated officer. Report to the area, and perform only the duties that are functions of the area that you have been assigned. For example, if you are triaging patients, you should not wander over to the treatment area to assist. "Freelancing" or not following the designated duties assigned to you (such as not being where you are assigned) places you, your partners, and overall patient care in jeopardy. In many cases, EMT-Bs who do not discharge their duties as assigned within the incident command structure can be held responsible for abandonment of duty.

Incident command systems will vary from place to place, and different terms may be used. You should become familiar with the specific terms and chain of command that are used in your area. When you respond to another community as backup for a mass-casualty incident, you must follow the directions and orders given, regardless of whether the order is given by your direct supervisor.

(Table 37-1 ▶) lists key components at a mass-casualty incident, and (Figure 37-2 ▶) shows the important components of an incident command system in a mass-casualty incident in chart format. As an example

---

TABLE 37-1 Key Components at a Mass-Casualty Incident

- Incident commander, command post, and incident command system
- On-site communications system
- Adequate supply of medical equipment
- Extrication area and retrieval team
- Triage officer and designated triage area
- Staffed patient collection area
- Staffed patient treatment area
- Supply location adjacent to the treatment area
- Transportation officer and transport area
- Staging area to hold resources until they are needed
- Fire and law enforcement personnel
- A secure perimeter

---

of how responsibilities may be assigned at a major EMS incident, consider the following typical assignments:

- **Command center.** This is typically a vehicle or building at the scene where the EMS commander establishes an "office." From here, the commander oversees and coordinates the activities of the various groups and leaders.
- **Staging area.** This is a holding area for arriving ambulances and crews until they can be assigned a particular task.
- **Extrication area.** In this area, patients are disentangled and removed from a hazardous environment, allowing them to be moved to the triage area.
- **Decontamination area.** Any incident involving a hazardous material or use of a nuclear or radiologic, chemical, or biologic agent will require a special area for removing the agent from any patients or responders. If established, this area will be situated after the extrication area and before the triage area.
- **Triage area.** The triage area is a sorting point, run by a triage officer, where all patients are assessed and tagged, using color-coded tags or tape, according to their injuries. These triaged patients are then directed to specific locations in the treatment area(s), according to their assigned priority.
- **Treatment area.** A more thorough assessment is made in this area, and on-scene treatment is

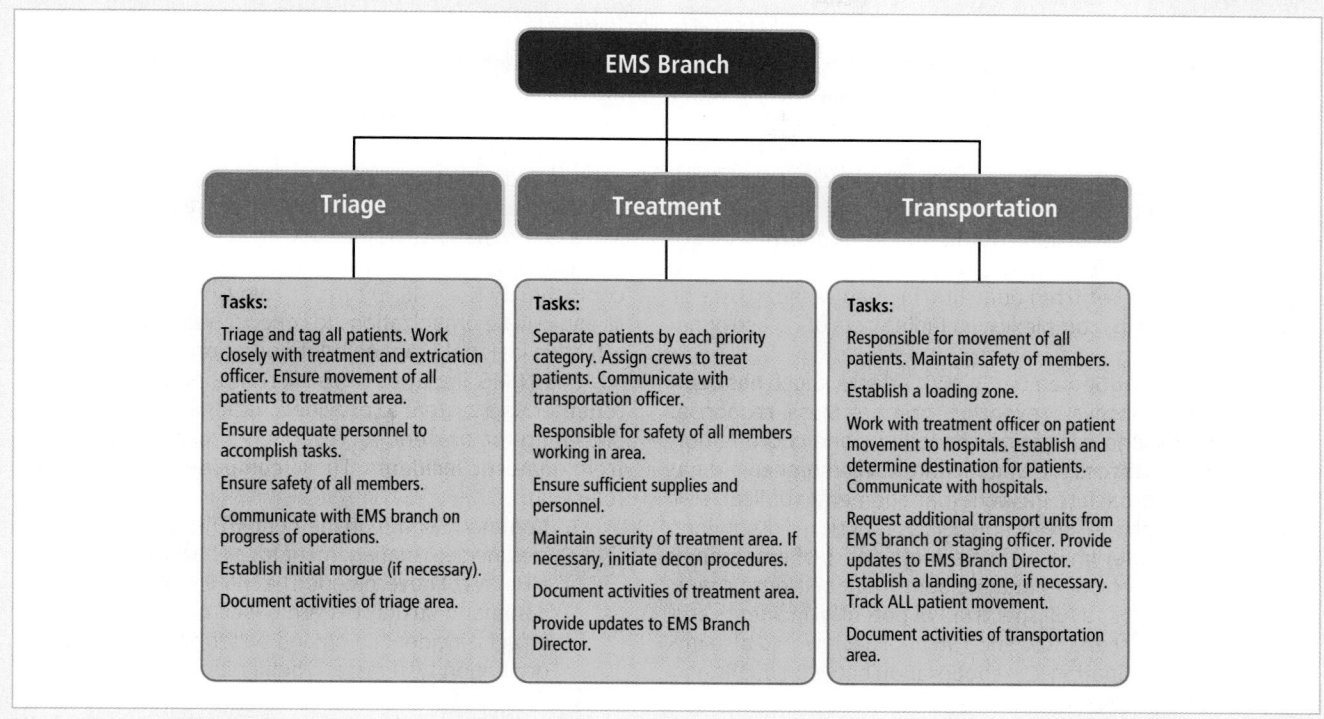

Figure 37-2 Important components of an incident management system at a mass-casualty incident.

begun while transport is being arranged. The treatment area is organized and managed under the authority of the treatment officer. Patients are given care under the standards of the EMS system in the treatment area before being transported. This means that all fractures should be splinted, and all care normally given in a focused assessment should be accomplished before the patient is released for transportation.

- **Supply area**. This is an area in which to assemble extra equipment and supplies, such as blankets, oxygen cylinders, bandages, and backboards, for dispersal to other areas as needed.
- Transportation area. In this area, ambulances and crews are organized to transport patients from the treatment sector to area hospitals. The transportation area is managed by the transportation officer, who will assign patients to waiting ambulances.

You are the Provider   Part 2

You call for immediate backup. You notify dispatch that you will be assuming incident command until further notice. Prior to exiting the ambulance, you and your partner survey the scene for any signs of a hazardous materials spill. You see a red placard on the semitrailer with the numbers 1202 on it. You look up information on this placard in the *Emergency Response Guidebook* and find that it is a classification for diesel fuel. You see no leakage and decide to exit the ambulance and complete a scene size-up.

3. How can you verify what the tanker actually contains?
4. What is your first step in patient care?

✳ EMT-B Tips

National Incident Management System

In 2003, the president directed the secretary of Homeland Security to develop and administer a National Incident Management System (NIMS). This system provides a consistent nationwide template to enable federal, state, and local governments and private-sector and nongovernmental organizations to work together effectively and efficiently to prepare for, prevent, respond to, and recover from domestic incidents, regardless of cause, size, or complexity, including acts of catastrophic terrorism.

Since the September 11, 2001 attacks, much has been done to improve prevention, preparedness, response, recovery, and mitigation capabilities and coordination processes across the United States. A comprehensive national approach to incident management, applicable at all jurisdictional levels and across functional disciplines would further improve the effectiveness of emergency response providers and incident management organizations across a full spectrum of potential incidents and hazardous scenarios. Such an approach would also improve coordination and cooperation between public and private entities in a variety of domestic incident management activities. Incidents can include:

- Acts of terrorism
- Wildland and urban fires
- Floods
- Hazardous materials spills
- Nuclear accidents
- Aircraft accidents
- Earthquakes
- Hurricanes
- Tornadoes
- Typhoons
- War-related disasters

Building on the foundation provided by existing incident management and emergency response systems used by jurisdictions and functional disciplines at all levels, NIMS integrates the practices that have proven most effective over the years into a comprehensive framework for use by incident management organizations in an all-hazards context nationwide. To provide for interoperability and compatibility among federal, state, and local capabilities, the NIMS includes a core set of concepts, principles, terminology, and technologies addressing the following:

- The incident command system
- Multiagency coordination systems
- Unified command
- Training
- Identification and management of resources
- Qualifications and certification
- Collection, tracking, and reporting of incident information and incident resources

While most incidents are generally handled on a daily basis by a single jurisdiction at the local level, there are important cases in which successful domestic incident management operations depend on the involvement of multiple jurisdictions, functional agencies, and emergency responder disciplines. These cases require effective and efficient coordination across this broad spectrum of organizations and activities. The NIMS uses a systems approach to integrate the best of existing processes and methods into a unified national framework for incident management. The framework forms the basis for interoperability and compatibility that will, in turn, enable a diverse set of public and private organizations to conduct well-integrated and effective incident management operations.

The NIMS includes several components that work together as a system to provide a national framework for preparing for, preventing, responding to, and recovering from domestic incidents. These components include the following:

1. Command and management—The NIMS standardizes incident management for all hazards and across all levels of government. The NIMS standard incident command structures are based on three key constructs: incident command system; multiagency coordination systems; and public information systems.
2. Preparedness—The NIMS establishes specific measures and capabilities that jurisdictions and agencies should develop and incorporate into an overall system to enhance operational preparedness for incident management on a steady-state basis in an all-hazards context.
3. Resource management—The NIMS defines standardized mechanisms to describe, inventory, track, and dispatch resources before, during, and after an incident; it also defines standard procedures to recover equipment once it is no longer needed for an incident.
4. Communications and information management—Effective communications, information management, and information and intelligence sharing are critical aspects of domestic incident management. The NIMS communications and information systems enable the essential functions needed to provide a common operating picture and interoperability for incident management at all levels.
5. Supporting technologies—The NIMS promotes national standards and interoperability for supporting technologies to successfully implement the NIMS and standard technologies for specific professional disciplines or incident types. It provides an architecture for science and technology support to incident management.
6. Ongoing management and maintenance—The Department of Homeland Security will establish a multijurisdictional, multidisciplinary NIMS Integration Center. This center will provide strategic direction for and oversight of the NIMS, supporting routine maintenance and continuous improvement of the system in the long term.

- <u>Rehabilitation area</u>. This area provides treatment and rest to emergency responders working at the scene. As workers enter and leave the scene, they are medically monitored and provided any needed care (such as rehydration with fluids or nourished with small snacks). This helps to ensure the safety and health of emergency workers who could become injured or ill while on the job.

Mass-Casualty Incidents

In this text, a <u>mass-casualty incident</u> refers to any call that involves three or more patients, any situation that places such a great demand on available equipment or personnel that the system would require a <u>mutual aid response</u> (an agreement between neighboring EMS systems to respond to mass-casualty incidents or disasters in each other's region when local resources are insufficient to handle the response), or any incident that has the potential to create one of the previously mentioned situations (Figure 37-3 ▶). Bus or train crashes and earthquakes are obvious examples of mass-casualty incidents. However, other causes of these incidents are far more common than such disasters and are usually much smaller in scope. (Figure 37-4 ▶) is a diagrammed example of a residential building fire that is confined to one apartment that may only produce one patient but has the potential to generate dozens of patients from rescuers and residents. Loss of power to a hospital or nursing home with ventilator-dependent and nonambulatory victims is considered a mass-casualty incident, although no one is injured.

All systems have different protocols for when to declare a mass-casualty incident and initiate the incident command system; however, as the EMT-B, ask yourself the following questions when considering whether the call is a mass-casualty incident:

- How many seriously injured or ill patients can you care for effectively and transport in your ambulance? One? Two?
- What happens when you have three patients to deal with?
- How long will it take for additional help to arrive?
- What do you do when a school bus crashes, resulting in eight critically injured patients, and you only have three ambulances available?

Obviously, you and your team cannot treat and transport all injured patients at the same time. At a

Figure 37-3 Mass-casualty incidents can be large, such as the attack on September 11, 2001, or can be much smaller in scope.

mass-casualty incident, you will often experience an increased demand for equipment and personnel. For example, you may realize that there may be 15 or more minutes to wait before the next ambulance will arrive. Should you stay at the scene, placing the patients who are ready to go at some risk? You should never leave the scene with patients who are loaded if there are others who are sick or wounded. This would leave patients at the scene without medical care and can be considered abandonment. If there are multiple patients and not enough resources to handle them without abandoning victims, you should declare a mass-casualty incident

✳ EMT-B Tips

The terminology used to describe an incident with multiple patients varies in different communities. Many communities use the term *multiple-casualty situation* to describe an emergency that involves more than one patient but use the term *mass-casualty incident* to describe larger scale events, such as those with more than 20 patients. In this text, the term *mass-casualty incident* is used to describe any call that involves more than one patient.

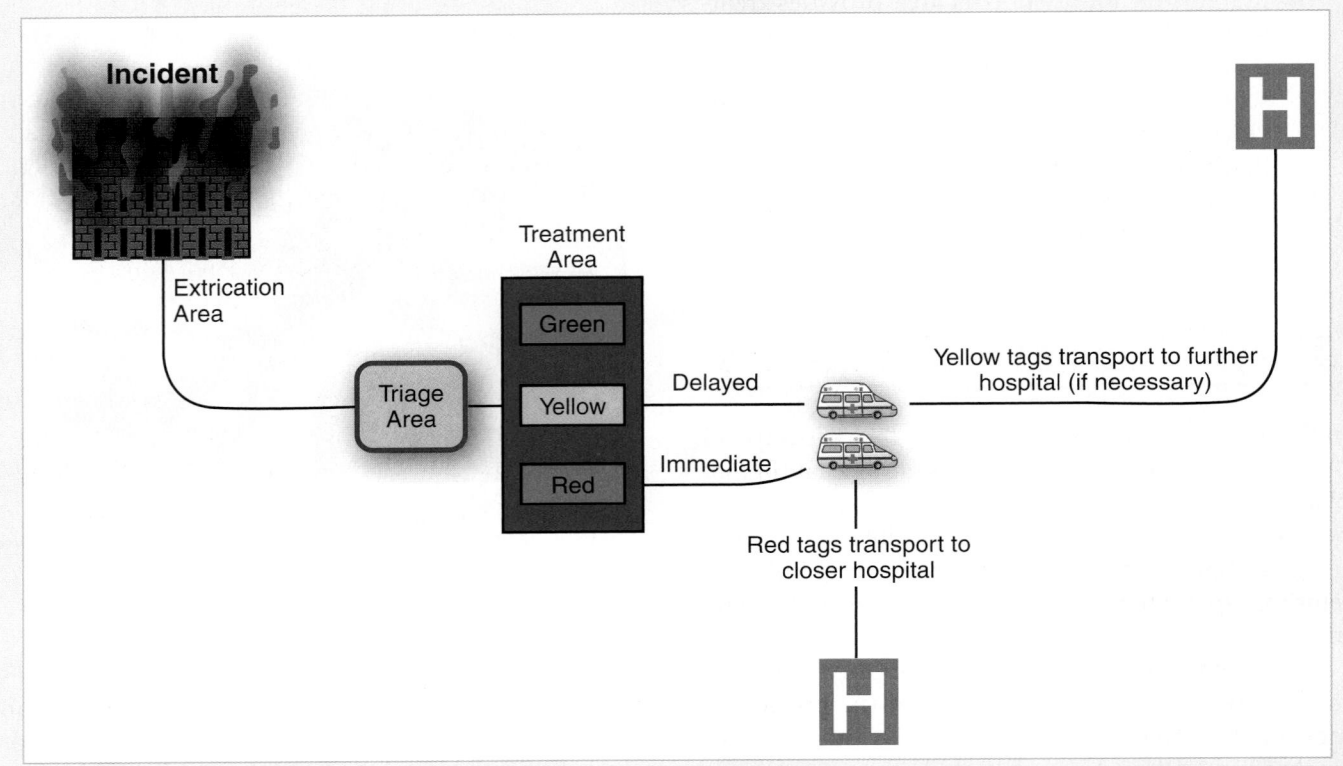

Figure 37-4 Diagram of a mass-casualty incident. The incident command system established at the scene of a building fire may look similar to this diagram.

(at least for the present time), request additional resources, and initiate the incident command system and triage procedures (described below) (Figure 37-5 ▼). Although this may cause some delay in initiating treatment to all patients, it will not adversely affect the pa-

tient care. Always follow your local protocol. Many large EMS systems deploy specialized mass-casualty incident units or mobile emergency room vehicles that are able to treat dozens of patients on the scene (Figure 37-6 ▼).

Figure 37-5 Mass-casualty incidents require additional ambulances and EMS providers from the immediate region.

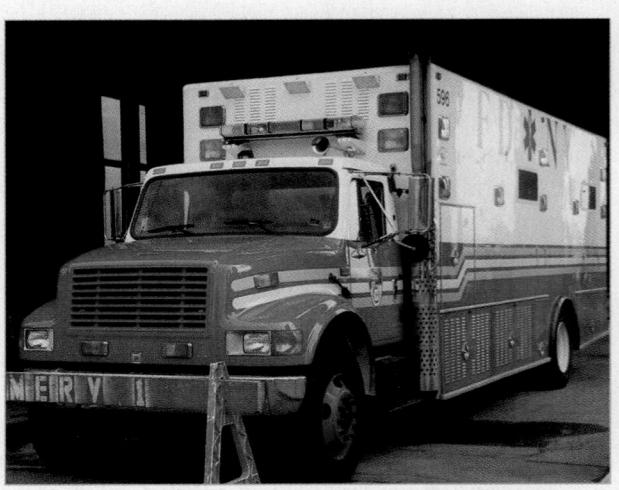

Figure 37-6 This mobile emergency room vehicle is staffed by EMTs and physicians who are able to provide advanced life support to multiple patients simultaneously on the scene of a mass-casualty incident.

Triage

Triage is essential at all mass-casualty incidents. Triage is the sorting of two or more patients based on the severity of their conditions to establish priorities for care based on available resources (Figure 37-7 ▼).

In a smaller scale mass-casualty incident, the first provider on scene with the highest level of training usually begins the triage process. When backup ambulances and crews are readily available, patients are ranked in order of the severity of their conditions. The patient with the most severe injuries (yet viable) is given priority attention. After counting the number of patients and notifying the dispatcher of additional help that is needed, initial assessment of all patients begins. As personnel arrive, you should assign crews and equipment to priority patients first.

Triage at a large-scale mass-casualty incident should be done in several steps. The following triage steps are accepted by the majority of larger scale mass-casualty operations:

- Life saving care rapidly administered to those in need
- Color coding to indicate priority for treatment and transportation at the scene. Red-tagged patients are the first priority, yellow-tagged patients are the second priority, and those tagged green or black are the lowest priority.

- Rapid removal of red-tagged patients for field treatment and transportation as ambulances are available.
- Use of a separate treatment area to care for red-tagged patients if transport is not immediately available. Yellow-tagged patients can also be monitored and cared for in the treatment area while waiting for transportation.
- When there are more patients waiting for transport than there are ambulances, the transportation sector officer decides which patient is the next to be loaded.
- Specialized transportation resources (such as air ambulances, paramedic ambulances) require separate decisions when these resources are available but limited.

Triage Priorities

Patients should be color coded early to visually identify the severity of the condition and to eliminate the need for individual patient assessments to be performed by each EMT-B who comes along later. To accomplish this, triage tags like the ones shown in (Figure 37-8 ▶) are used. Because there are several different manufacturers of triage or "disaster" tags, you should make certain you are familiar with the ones provided and that a sufficient quantity of these tags are available. The tags are perforated, which allows a patient to be triaged upward only and never downgraded. Therefore, for a patient tagged as "yellow" on initial assessment whose condition begins to deteriorate later, the yellow portion of the tag can be ripped off leaving only the red and black sections. Most tags have serial numbers or electronic bar codes, which allow the patient and his or her belongings to be tracked from the scene to the hospital. With recent concerns about the likelihood of an incident involving weapons of mass destruction, some tags are waterproof and can be decontaminated along with the patient.

Patients who are tagged red should later be reassessed in the treatment area to determine who should receive limited resources such as paramedic assessment and care. The sorting of multiple red-tagged patients in the treatment area who need to be seen immediately by paramedics will depend on the number of paramedics available at that time. The order in which the patients will be transported is determined after initial triage and treatment are completed.

If patients are entrapped, extrication is required. If circumstances such as heavy smoke or potential hazardous materials exposure exist, triage will be difficult or impossible. The immediate concern will be the

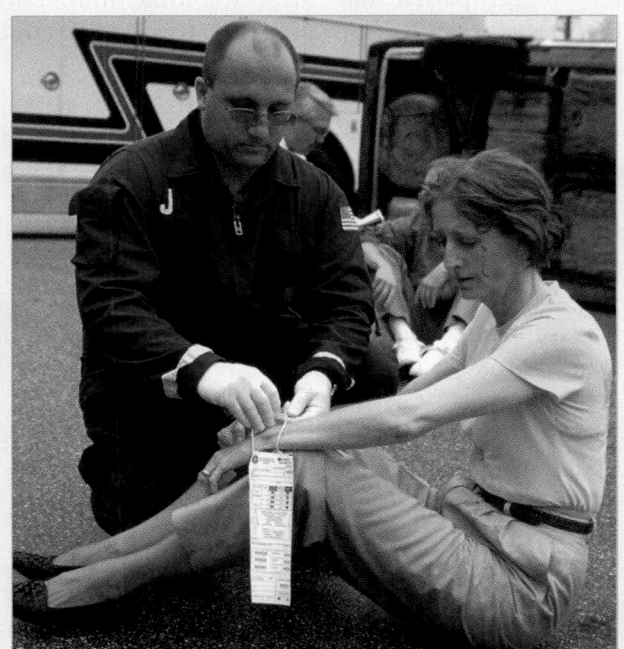

Figure 37-7 Triage is an essential component of operations at a mass-casualty incident.

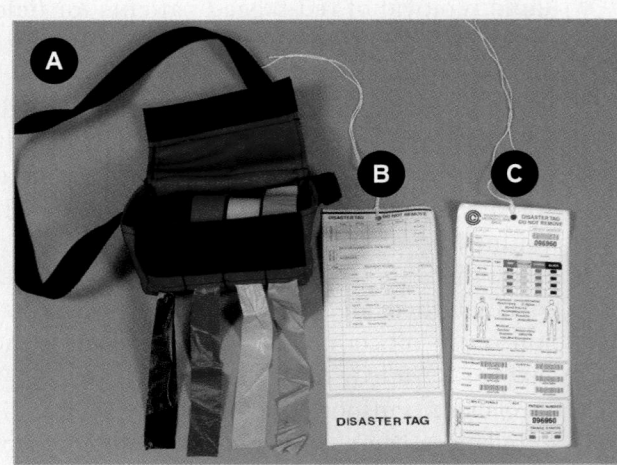

Figure 37-8 Triage tags. **A.** Waterproof weapons of mass destruction tags. **B.** Back. **C.** Front.

removal of the patients to a safe area for later triage. The triage area is the name usually given to such a collecting area for patients to be initially triaged and color-coded tags applied. For patients located in nonhazardous areas, the initial triage can begin right away. Triage priorities are summarized in (Table 37-2 ▶).

Triage Procedures

If medical control is not willing or able to determine the appropriate destination hospital for each patient, a rotation system can be used to distribute patients properly to each hospital on the basis of hospital capacity and capabilities. Of course, any designated trauma cen-

EMT-B Tips

Definitions of mass-casualty incidents vary from one place to another, and training on such specialized topics may not be frequent. As a new EMT-B, you may have to do some research and studying on your own to be sure that you understand and can apply local policy and procedure for large-scale incidents.

ters should be used to receive the most critical patients, following local protocols. In many cases, the closest hospital will be used for red-tagged patients, and the more stable yellow-tagged patients will be transported to further hospitals. This procedure distributes the patient flow throughout the hospitals of the system and, therefore, does not overburden any one hospital. The transport officer is responsible for sending the ambulance to the appropriate hospital or the next hospital in turn. This rotation must occasionally be altered to allow specific patients to be taken to the most appropriate facility, such as a pediatric center, or to another hospital because a hospital has notified the field that it needs to be skipped for one rotation.

Normally, no more than two patients are placed in the same ambulance. However, with severe weather, a green-tagged patient may be seated in an ambulance next to the driver for transportation to a comfortable, safe, indoor holding area at the hospital.

You are the Provider   Part 3

You send your partner to check on the occupants in the passenger van while you see to the driver of the semi. The driver is conscious, alert, and oriented. He states that he is in no major pain but is just a bit shook up. The driver verifies that he is hauling diesel fuel. You consider double-checking the waybill to confirm, once you have performed your initial assessments and any initial lifesaving treatments. Your partner notifies you that there are six passengers in the van and that he needs your assistance. Due to the fact that this is considered a mass-casualty incident and triage must be initiated, you ask the driver to lie on the ground and remain still. You quickly explain why you are leaving him there. You ensure the patient's airway, breathing, and circulation are patent. With the current condition of your patient, you assign him a green tag.

5. What do the colors on a triage tag signify?
6. Is the fact that you left your patient considered abandonment in this situation?

TABLE 37-2 Triage Priorities

Triage Category	Typical Injuries
Red Tag: First Priority (Immediate) Patients who need immediate care and transport. Treat these patients first, and transport as soon as possible.	■ Airway and breathing difficulties ■ Uncontrolled or severe bleeding ■ Decreased level of consciousness ■ Severe medical problems ■ Signs of shock (hypoperfusion) ■ Severe burns
Yellow Tag: Second Priority (Delayed) Patients whose treatment and transportation can be temporarily delayed.	■ Burns without airway problems ■ Major or multiple bone or joint injuries ■ Back injuries with or without spinal cord damage
Green Tag: Third Priority (Walking Wounded) Patients who do not require any treatment or whose treatment and transportation can be delayed until last.	■ Minor fractures ■ Minor soft-tissue injuries
Black Tag: Fourth Priority (DOA) Patients who are already dead or have little chance for survival. If resources are limited, treat salvageable patients before treating these patients.	■ Obvious death ■ Obviously nonsurvivable injury, such as major open brain trauma ■ Respiratory arrest (if limited resources) ■ Full cardiac arrest

As the patients are loaded into the ambulance, the transport officer logs each patient's mass-casualty tag number, each patient's overall condition, and the hospital to which the patient will be taken. As the ambulance leaves the scene, the transport officer radios the receiving hospital and briefly describes the patients, the unit transporting them, and the time they left the scene. To minimize radio traffic during such incidents, personnel on individual ambulances do not usually use their radios except to obtain advice from medical control or to notify the transport officer that they are leaving the hospital and returning to the field.

After giving a verbal report to hospital staff and transferring the patients, the ambulance returns to the staging area without further delay, helping to keep a continuous flow of ambulances moving between the mass-casualty incident site and the hospital. Equipment that is collected at the hospital or additional supplies that are needed in the field are brought to the staging area.

If additional ambulances are needed, the transport officer radios the command center, which then directs the resources from the staging area. If none are available at the staging area, the EMS chief notifies the dispatcher to obtain them elsewhere. To prevent a lack of ambulances, request extra ambulances as early as possible in a triage situation.

After all the first-priority (red-tagged) patients have been transported, the second-priority (yellow-tagged) patients are transported, followed by the third-

✴ **EMT-B Tips**

Special Triage Situations

Patients who have been contaminated by radiation or other hazardous materials are placed in a separate category of triage. This is the highest and most urgent category of all. Contaminated patients must be moved away from all other patients. They must not be allowed to contaminate other patients, EMS personnel, ambulances, or hospitals.

Certain large urban areas that offer regionalized care use another concept of triage. Single patients with specific medical problems, such as burns, trauma, cardiac, or neonatal, are triaged to specialized regional centers for treatment. Making the decision to transport a patient to a special treatment center is difficult. The decision is based on many factors, including (but not limited to) the following:

- The specific illness or injury
- The severity of the illness or injury
- The availability of local resources at the time of the event
- Local rules and protocols

These decisions are often made only after on-line communication with medical control.

If there are special treatment centers in your area, you must know the specific triage protocols that apply. Also note that in the event of a mass-casualty incident, these protocols might not be used. For example, a school bus crash in which all 30 or 40 patients are children can overwhelm a pediatric hospital. Similarly, 10 burn patients from a petroleum plant fire could immobilize a burn center. In these cases, good triage and communication with medical control are essential in providing each patient with the best available treatment.

Most urban areas have regional trauma care centers. Severe, life-threatening injuries should be treated in facilities that are prepared to deal immediately and completely with the problem. Ideally, severely injured patients should be identified in the field and sent to a designated trauma center.

priority (green-tagged) patients. After all patients have been transported, several units usually stay at the incident site to protect the remaining responders in case of injury. Often, ambulances will be needed to transfer patients from the facility to which they were initially brought for stabilization to another, more appropriate facility.

Treatment and triage continue until all patients have been treated and transported. After the mass-casualty incident has ended, all personnel who were involved should be debriefed and evaluated to determine whether they need counseling or medical attention. The success of any incident command system depends on all personnel performing their assigned tasks and working within the system as a member of the team. Therefore, always remember that the cost of working autonomously may include the loss of lives.

Disaster Management

A <u>disaster</u> is a widespread event that disrupts functions and resources of a community and threatens lives and property. Many disasters may not involve personal injuries. Droughts causing widespread crop damage are an example. On the other hand, many disasters such as floods, fires, and hurricanes also result in widespread injuries. Unlike a mass-casualty incident, which generally lasts no longer than a few hours, emergency responders will generally be on the scene of a disaster for days to weeks and sometimes months (as in the events following September 11, 2001). Although you can "declare" a mass-casualty incident, only an elected official can declare a disaster.

Your role in a disaster is to respond when requested and to report to the incident command system for assigned roles. In a disaster with an overwhelming number of casualties, area hospitals may decide that they cannot treat all patients at their facility. In this case, they may mobilize medical and nursing teams with equipment. Using a facility such as a warehouse near the disaster scene, they will set up a <u>casualty collection area</u>. Once at the casualty collection area, triage can be performed, medical care provided, and patients transported to the hospital on a priority basis.

If a casualty collection area is established, it will be coordinated through the incident command system in the same way as all other branches and areas of the operation. This is usually done only in a major disaster

✳ **EMT-B Tips**

Mass-casualty incidents and disasters take a physical and emotional toll on emergency responders. Make certain that you are medically evaluated if you have been injured, come into contact with any hazardous substance, or inhale any dust, fumes, or smoke. Often the health effects of such exposures do not manifest for years and are difficult to link back to a particular event. Also be aware of the signs of stress in yourself and in your coworkers. Take full advantage of the opportunity for stress debriefing after an incident.

requirements, you need to check with your agency for information about additional specific training.

Hazardous materials may be involved in any of the following situations (Figure 37-9 ▼):

■ A truck or train crash in which a substance is leaking from a tank truck or railroad tank car
■ A leak, fire, or other emergency at an industrial plant, refinery, or other complex where chemicals or explosives are produced, used, or stored
■ A leak or rupture of an underground natural gas pipe
■ Deterioration of underground fuel tanks and seepage of oil or gasoline into the surrounding ground

such as an earthquake when transportation to a hospital facility is impossible or involves prolonged delays. It may take several hours to establish a casualty collection area. This delay can limit the number of events in which such an area is an effective method for handling the incident.

Introduction to Hazardous Materials

Your training has taught you that rapid response to the scene of a crash can save lives. However, when you arrive at the scene of a possible hazardous materials incident, you must first step back and assess the situation. This can be very stressful for you, particularly if you can see a patient. However, rushing into such events can have catastrophic results. If you are overcome by a hazardous substance, not only will patients suffer because you will be unable to assist them, but also you will place a strain on your system because you will require emergency care. Because of the unique aspects of responding to and working at a HazMat incident, the Occupational Safety and Health Administration, or OSHA, has set specific additional training requirements in publication "29 CFR 1910.120—Hazardous Waste Operations and Emergency Response Standard," which all individuals, including EMT-Bs, must meet before becoming involved in these situations. Because this text does not include the information to meet these

Figure 37-9 Two examples of hazardous materials incidents.

■ Buildup of methane or other byproducts of waste decomposition in sewers or sewage-processing plants

■ A motor vehicle crash in which a gas tank has ruptured

Often, the presence of hazardous materials is easily recognized from warning signs, placards, or labels found in the following locations (Figure 37-10 ▼):

■ On buildings or areas where hazardous materials are produced, used, or stored

■ On trucks and railroad cars that carry any amount of hazardous material

■ On barrels or boxes that contain hazardous material

Unfortunately, identifying materials can still be difficult. Little consistency is used on labels and placards, and sometimes dishonest transporters will not label containers or vessels appropriately. The laws and regulations that cover labeling of packages and transport vehicles can also be misleading. In most cases, the pack-

Figure 37-10 **A.** Warning placards are found on railroad cars that carry hazardous materials. **B.** Labels are also affixed to boxes that contain hazardous materials.

age or tank must contain a certain amount of a hazardous material before a placard is required. For example, because of the small quantities of hazardous materials that are involved, a truck carrying 99 lb of HazMat #1 and 99 lb of both HazMat #2 and HazMat #3 may not be required by law to display any labels or placards. The truck may show only a "Please drive carefully" placard, implying that it carries no hazardous materials. Therefore, a crash involving this truck is a serious situation, but you would not necessarily know this if you relied on labels and placards. Always maintain a high index of suspicion when approaching the scene of a truck or train tanker accident.

Some substances are not hazardous; however, when mixed with another substance, they may become highly toxic. There may be no regulations against carrying such substances together on one truck or railroad car (or adjacent tank cars). The driver of a commercial truck and the conductor of a train, however, must carry papers that identify what is being transported in their care. These papers may be your first clue that there is a possible HazMat problem, although, depending on the nature of the incident, the papers may not be available to you.

In the event of a leak or spill, a hazardous materials incident is often indicated by the presence of the following:

■ A visible cloud or strange-looking smoke resulting from the escaping substance

■ A leak or spill from a tank, container, truck, or railroad car with or without hazardous material placards or labels

■ An unusual, strong, noxious, acrid odor in the area

To indicate the presence of normally odorless toxic gases or fluids during a leak or spill, manufacturers may add a substance that produces a strong noxious odor. However, a large number of hazardous gases and fluids are essentially odorless (or do not have a distinctive unpleasant smell) even when a substantial leak or spill has occurred. In some incidents, a large number of people are exposed and may be injured or killed before the presence of a hazardous materials incident is identified. If you approach a scene where more than one person has collapsed or is unconscious or in respiratory distress, you should assume that there has been a hazardous materials leak or spill and that it is unsafe to enter the area.

It is important for you to understand the potential danger of hazardous materials and know how to operate safely at a hazardous materials incident. If you do not follow the proper safety measures, you and many

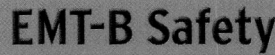

EMT-B Safety

Safety considerations at hazardous materials scenes differ considerably from those involved in emergency response in general. Hazardous materials require an even higher degree of alertness than usual to avoid entering a dangerous environment and to help others avoid it. There is also a need to prevent the spread of contamination to yourself or your ambulance. Understanding these two concepts is a good start toward safe operations in the presence of hazardous materials.

others could end up needlessly injured or dead. The safety of you and your team, the other responders, and the public must be your most important concern.

There will be times when the ambulance is the first to arrive at the scene. If, as you approach, any signs suggest that a hazardous materials incident has occurred, you should stop at a safe distance. After rapidly sizing up the scene, call for a HazMat team. If you do not recognize the danger until you are too close, immediately leave the __danger zone__. Once you have reached a safe place, try to rapidly assess the situation and provide as much information as possible when calling for the HazMat team, including your specific location, the size and shape of the containers of the hazardous material, and what you have observed and been told has occurred. Do not reenter the scene, and do not leave the area until you have been cleared by the HazMat team, or you may contribute to the situation by spreading hazardous materials. Finally, do not allow civilians to enter the scene, if possible.

Identifying Hazardous Materials

Until the HazMat team arrives to determine the hazard zone, you should be aware of the safety perimeters that are necessary for hazardous materials that are toxic (poisonous) and those in which there is danger of fire or explosion. Determining a safe perimeter must involve the assessment of many factors related to the substance, the environment, containment, and training. Some sources suggest staging upwind and at least 100′ from the site. Shifting winds, large volume of material, and properties of the substance may make this minimum too close. Always stay back further than you think is necessary, request the HazMat team in your jurisdiction, and follow local protocols. In a hilly area, you should be uphill and upwind. Remember, wind direction can change quickly. A piece of narrow roller bandage, approximately 2′ long, tied to the top of your antenna will serve as a wind direction guide. Be sure to check the wind direction periodically, and relocate if a change in wind direction dictates.

If you can see and read the placard or other warning sign, note its color, wording, any symbols that it contains, and, if included, the four-digit number that appears on it or on any orange panel near it (Figure 37-11 ▶). This number, which may be preceded by the letters UN or NA, identifies the specific hazardous material. The name of the material may also be displayed along with the number. The same number and name can also be found on the shipping papers and packaging of the material. If you are unable to read the placard or labels, do not move closer and risk exposure. The HazMat team will have binoculars that allow team members to read the placards or labels from a safe distance. If you are able to read the placard with the naked eye, you may be too close and should consider moving farther away. (Figure 37-12 ▶) shows a chart illustrating the hazardous materials warning placards, and (Figure 37-13 ▶) shows

You are the Provider   Part 4

As you are heading toward the van you notify dispatch that you have initiated triage and ask for an estimated time of arrival (ETA) for your backup. The responding units have a 3-minute ETA. In the van you find six patients, all restrained. The driver-side airbag was deployed and there is approximately 10″ of roof intrusion. Your partner tells you that all patients are conscious, alert, and oriented. He has checked the ABCs on all patients. Together you verify the injuries of all patients and assign the appropriate triage tags.

7. What types of injuries or medical conditions would fit into each of the four triage color categories?
8. Who would you provide treatment to first?

a chart illustrating the warning labels. You should study and be familiar with these warning materials.

HazMat Scene Operations

Once you have recognized the incident as one involving hazardous materials and have called for the HazMat team, you should focus your efforts on activities that will

Figure 37-11 The four-digit number that appears on the warning placard identifies the specific hazardous material.

ensure the safety and survival of the greatest number of people. Use the ambulance's public address system to alert individuals who are near the scene and direct them to move to a location where they will be sufficiently far from danger. With the aid of others on your team, try to set up a perimeter to stop traffic and individuals from entering the danger zone.

The HazMat team is equipped to identify the specific substance that is involved and, using a number of complex factors, determine the size, direction or shape, and perimeter of the danger zone. Knowing the type, toxicity or concentration, and quantity of the hazardous material that is involved will help the team to determine the location of and safe distance from the danger zone. Determination of the danger zone will also be affected by the amount of wind and other weather factors, in addition to the potential for fire or explosion. The team will determine which specific hazardous material is involved and will mark the perimeter of the danger or hazard zone with warning tape. Once this area is established, you should not enter it.

Only individuals who are trained in HazMat and wearing the proper level of protective gear should enter the hazard zone. As an EMT-B, your job is to report to a designated area outside of the hazard zone and pro-

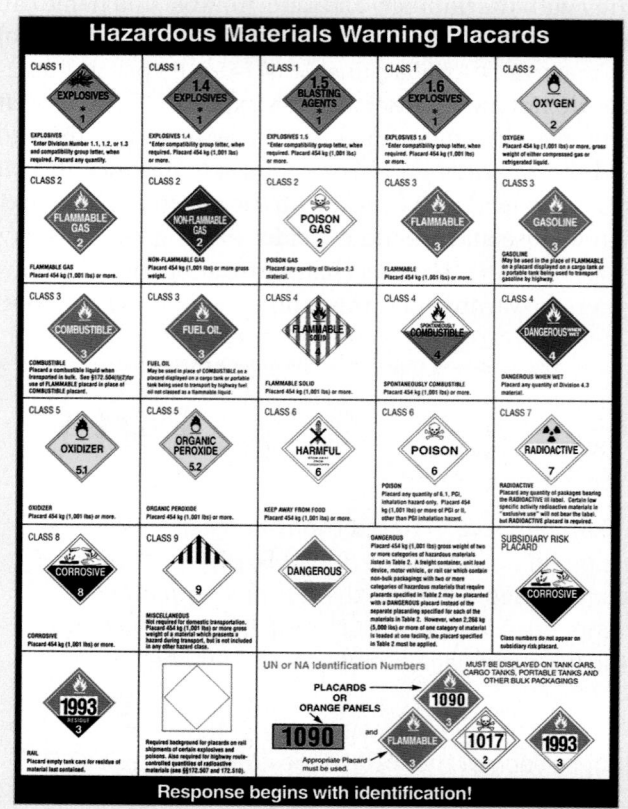

Figure 37-12 Hazardous materials warning placards.

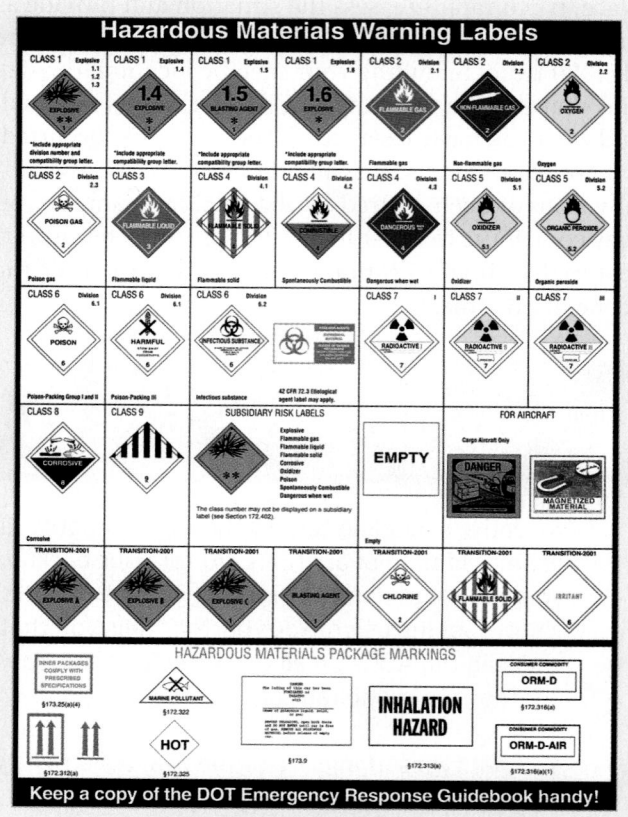

Figure 37-13 Hazardous materials warning labels.

vide triage, treatment, transport, or rehabilitation when HazMat team members bring patients to you.

Patients' skin and clothing may contain hazardous material, so a <u>decontamination area</u> should be set up between the hazard zone and the treatment area. The decontamination area is the designated area where contaminants are removed before an individual can go to another area. <u>Decontamination</u> is the process of removing or neutralizing and properly disposing of hazardous materials from equipment, patients, and rescue personnel (Figure 37-14 ▼). The decontamination area must include special containers for contaminated clothing and special bags to isolate each patient's personal effects safely until they can be decontaminated. The area will also contain a number of special facilities to thoroughly wash and rinse patients and backboards. The water that is used must be captured and delivered into special sealable containers.

Anyone who leaves the hazard zone must pass through the decontamination area. Firefighters' and HazMat team members' outer protective gear is rinsed and washed in the decontamination area before it is removed (Figure 37-15 ▶). To prevent needless contact and communication of splash or residues, different personnel are used in the decontamination and treatment areas. You should not move into the decontamination area unless you are properly trained and equipped. You should wait for the patients to be brought to you.

Classification of Hazardous Materials

The National Fire Protection Association (NFPA) 704 Hazardous Materials Classification standard classifies hazardous materials according to health hazard or toxicity levels, fire hazard, chemical reactive hazard, and special hazards (such as radiation or acids) for fixed facilities that store hazardous materials. Toxicity protection levels are also classified according to the level of personal protection required. For your safety, the type and degree of health, fire, and reactive hazard protection needed to operate safely near these substances must be known before entry to the scene is made. (Figure 37-16 ▶) shows hazardous materials classifications from the NFPA.

Toxicity Level

<u>Toxicity levels</u> are measures of the health risk that a substance poses to someone who comes into contact with it. There are five toxicity levels: 0, 1, 2, 3, and 4. The higher the number, the greater the toxicity, as follows:

- Level 0 includes materials that would cause little, if any, health hazard if you came into contact with them.
- Level 1 includes materials that would cause irritation on contact but only mild residual injury, even without treatment.
- Level 2 includes materials that could cause temporary damage or residual injury unless prompt medical treatment is provided. Both levels 1 and 2 are considered slightly hazardous but require use of self-contained breathing apparatus (SCBA) if you are going to come into contact with them.
- Level 3 includes materials that are extremely hazardous to health. Contact with these materials requires full protective gear so that none of your skin surface is exposed.
- Level 4 includes materials that are so hazardous that minimal contact will cause death. For

Figure 37-14 Patient decontamination prevents contaminants from spreading to others.

Figure 37-15 The decontamination zone is where firefighters' and HazMat team members' outer protective gear is rinsed and washed before removal.

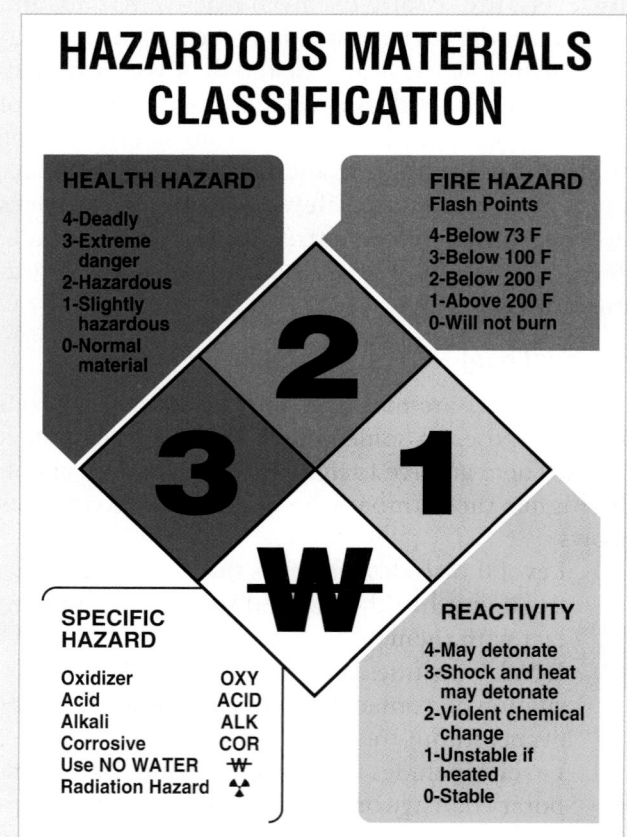

HAZARDOUS MATERIALS CLASSIFICATION

HEALTH HAZARD

4-Deadly
3-Extreme danger
2-Hazardous
1-Slightly hazardous
0-Normal material

FIRE HAZARD
Flash Points

4-Below 73 F
3-Below 100 F
2-Below 200 F
1-Above 200 F
0-Will not burn

SPECIFIC HAZARD

Oxidizer   OXY
Acid   ACID
Alkali   ALK
Corrosive   COR
Use NO WATER   W̶
Radiation Hazard ☢

REACTIVITY

4-May detonate
3-Shock and heat may detonate
2-Violent chemical change
1-Unstable if heated
0-Stable

Figure 37-16 NFPA hazard standards are clearly posted on a facility containing a hazardous material.

level 4 substances, you need specialized gear that is designed for protection against that particular hazard.

You must note that all health hazard levels, with the exception of 0, require respiratory and chemical protective gear that is not standard on most ambulances and specialized training. (Table 37-3 ▼) further describes the four hazard classes.

TABLE 37-3 Toxicity Levels of Hazardous Materials

Level	Health Hazard	Protection Needed
0	Little or no hazard	None
1	Slightly hazardous	SCBA (level C suit) only
2	Slightly hazardous	SCBA (level C suit) only
3	Extremely hazardous	Full protection, with no exposed skin (level A or B suit)
4	Minimal exposure causes death	Special HazMat gear (level A suit)

Caring for Patients at a Hazardous Materials Incident

Generally, HazMat team members who are trained in prehospital emergency care will initiate emergency care for patients who have been exposed to a hazardous material. However, because of the dangers, time constraints, and bulky protective gear that team members wear, it is practical only to provide the simplest assessment and essential care in the hazard zone and the decontamination area. In addition, to avoid entrapment and spread of contaminants, no bandages or splints are applied—except pressure dressings that are needed to control bleeding—until the "clean" (decontaminated) patient has been moved to the treatment area. Therefore, the EMT-Bs providing care in the treatment area should assess and treat the patient in the same way as they would a patient who has not been previously assessed or treated.

Your care of patients at a HazMat incident must address the following two issues:

■ Any trauma that has resulted from other related mechanisms, such as vehicle collision, fire, or explosion

■ The injury and harm that have resulted from exposure to the toxic hazardous substance

Most serious injuries and deaths from hazardous materials result from airway and breathing problems. Therefore, you should be sure to maintain the airway, and, if the patient appears to be in distress, give oxygen at 10 to 15 L/min with a nonrebreathing mask. Monitor the patient's breathing at all times. If you see signs that would indicate that respiratory distress is increasing, you may need to provide assisted ventilation with a bag-valve-mask device and high-flow oxygen.

You should treat the patient's injuries in the same way that you would treat any injury. There are few specific antidotes or treatments for exposure to most hazardous materials. Different people may respond differently to contact with the same hazardous material. Therefore, your treatment for the patient's exposure to the toxic substance should focus mainly on supportive care and initiating transport to the hospital with a minimum of additional delay.

If special antidotes or other special treatments need to be initiated in the field, they will be ordered by medical control and relayed to the officer in charge of EMS operations at the scene. If special treatment includes medications, intravenous fluids, or other advanced care, paramedics or other advanced personnel will be sent to work with you at the treatment area.

Special Care

In some cases, before the decontamination area has been completely set up, the HazMat team will find one or two patients who need immediate treatment and transport without further delay if they are to survive. Even after the decontamination area is set up and functioning, some patients may have such respiratory distress or other urgent critical condition that the time necessary for full decontamination may prove fatal. If additional delay for proper decontamination seems life threatening in nontoxic exposure situations, it may be necessary to simply cut away all of the patient's clothing and do a rapid rinse to remove the majority of the contaminating matter before transport.

If you are treating and transporting a patient who has not been fully and properly decontaminated, you will need to increase the amount of protective clothing you wear, including the use of SCBA. At the least, this should include two pairs of gloves, goggles or a face shield, a protective coat, respiratory protection, and a disposable fluid-impervious apron or similar outfit. Many HazMat teams carry easy-to-use disposable fluid-impervious light protective suits for such a purpose. Remember, however, that transporting a contaminated patient merely increases the size of the event. The decision to transport even a patient with critical injuries rests with the incident commander, who bases his or her decision on recommendations made by the HazMat team.

To make decontaminating the ambulance easier, tape the cabinet doors shut. Any equipment kits, monitors, and other items that will not be used en route should be removed from the patient compartment and placed in the front of the ambulance or in outside compartments. Before loading the patient, you should turn on the power vent ceiling fan and patient compartment air-conditioning unit fan. Unless the weather is too severe, the windows in the driver's area and sliding side windows in the patient compartment should also be partially opened to prevent creating a "closed box" inside the ambulance and to ensure that it is properly ventilated for the safety of the patient and EMT-Bs.

When you leave the scene, inform the hospital that you are transporting a critically injured patient who has not been fully decontaminated at the scene. This will allow the hospital to prepare to receive the patient. Many emergency departments have decontamination facilities and trained personnel for such an event. You may be diverted to a facility with these capabilities if the receiving hospital is not so equipped. Be sure that one EMT-B enters the emergency department and, after giving hospital staff the report and advising them again of the incomplete decontamination, obtains directions before the patient is unloaded and brought in. If there are enough ambulances at a hazardous materials scene, one may be isolated and used only to transport such patients. Remember, the ambulance needs to be decontaminated before transporting another patient.

Resources

Every ambulance and the dispatch center should have a copy of the *Emergency Response Guidebook*, prepared by Transport Canada, the US Department of Transportation, and the Secretariat of Communications and Transportation of Mexico (Figure 37-17 ▼). This publication lists most hazardous materials. For each one, it describes the proper initial emergency action to control the scene and provide emergency medical care. Some state and local government agencies may also have information about hazardous materials that are commonly found in their areas. Be sure to keep these publications up-to-date and close at hand on the unit.

Another valuable resource is the Chemical Transportation Emergency Center (CHEMTREC), located in Washington, DC. CHEMTREC was established by the Chemical Manufacturers Association to assist emergency personnel in identifying and handling hazardous materials transport incidents. The center operates 24 hours a day, 7 days a week. Its toll-free number is 1-800-424-9300 from anywhere in the United States or Canada.

CHEMTREC provides information, warnings, and guidance for proper emergency management and treatment. However, it cannot identify an unknown substance. You must provide CHEMTREC with the correct

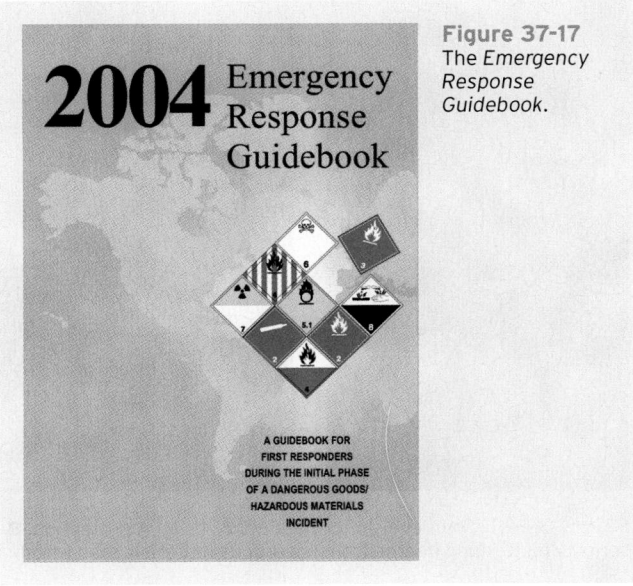

Figure 37-17
The *Emergency Response Guidebook*.

Department of Transportation identification number, the chemical name, or the product name of the material.

If you are interested in learning more about hazardous materials incidents and rescue requirements, you can look to the following:

■ NFPA standard 479
■ OSHA standard 1910.120
■ Federal Emergency Management Agency guidelines for coping with hazardous materials incidents
■ Environmental Protective Agency protective clothing

Personal Protective Equipment Level

Personal protective equipment (PPE) levels indicate the amount and type of protective gear that you need to prevent injury from a particular substance. The four recognized protection levels, A, B, C, and D, are as follows (Figure 37-18 ▼):

■ **Level A**, the most hazardous, requires fully encapsulated, chemical-resistant protective clothing that provides full body protection, as well as SCBA and special, sealed equipment.
■ **Level B** requires nonencapsulated protective clothing or clothing that is designed to protect against a particular hazard (Figure 37-19 ▶). Usually, this clothing is made of material that will let only limited amounts of moisture and vapor pass through (nonpermeable). Level B also

requires breathing devices that contain their own air supply, such as SCBA, and eye protection.
■ **Level C**, like Level B, requires the use of nonpermeable clothing and eye protection. In addition, face masks that filter all inhaled outside air must be used.
■ **Level D** requires a work uniform, such as coveralls, that affords minimal protection.

All levels of protection require the use of gloves. Two pairs of rubber gloves are needed for protection in case one pair must be removed because of heavy contamination.

Figure 37-19 Workers in level B protection.

 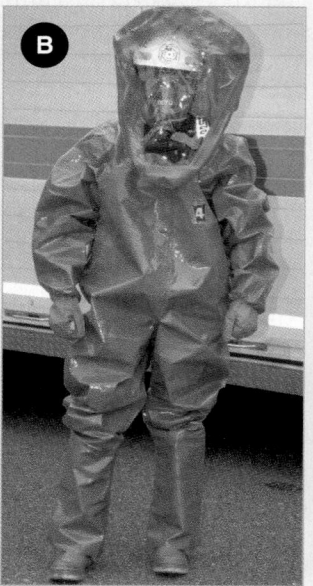 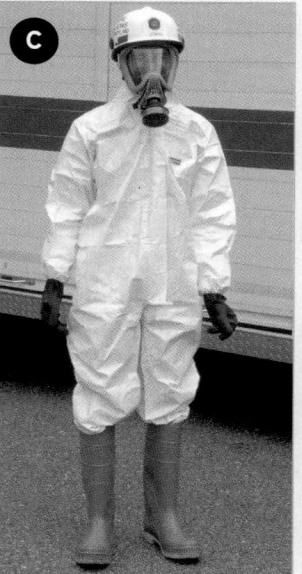 

Figure 37-18 Four levels of protection. **A.** Level A protection. **B.** Level B protection. **C.** Level C protection. **D.** Level D protection. Most serious injuries and deaths from hazardous materials result from airway and breathing problems.

You are the Provider

Summary

The fire engine has arrived. Your partner remains with the patients in the van while you give the report to the captain on the engine. You transfer incident command to the captain and assist with patient care.

As with any scene, safety should always be your number one priority. You are of no help as a rescuer if you yourself become a victim. Call for backup anytime you feel it may be needed. It is better to find out that additional resources are not needed and have them return to the station than to not call them at all or to call them too late. Attempt to identify any hazardous materials before you exit the ambulance. Remember that any time your number of patients outnumber your resources, consider the scene a mass-casualty incident and react accordingly.

Prep Kit

Ready for Review

- Incident command systems allow for coordination of police, fire, and EMS activities in an emergency situation. In major incidents, there is usually a unified command, with a single command post where decisions are made by agency leaders.

- If your unit arrives first at an incident that will involve more than one unit or agency, command should be established. Otherwise, you should report to the command post for assignment.

- You may be assigned to one of the following sectors: staging, extrication, triage, treatment, supply, transportation, or rehabilitation.

- In a mass-casualty incident, the most highly trained medical person on the scene directs triage. This means assigning treatment and transport priorities according to the severity and survivability of patients' injuries.

- There are four triage levels, each with a separate treatment area. Highest priority is given to patients whose injuries are critical but probably survivable with prompt treatment.

- The cardinal rule of triage is to do the greatest good for the greatest number. Treatment and triage continue until all patients have been transported.

- In urban areas with special treatment centers, special protocols are used to triage patients with specific injuries to the appropriate centers.

- Communication with medical control is essential for good triage.

- Disasters are widespread events that can threaten lives and property and sometimes can cause personal injuries. The incident command system is used for disasters just as for other types of mass-casualty incidents. However, if the number of patients overwhelms the area's hospitals, an alternative facility and casualty collection area can be set up for triage and treatment.

- At a hazardous materials incident, safety—of you, your team, the patient, and the public—is your most important concern.

- If you arrive first, assess the situation, taking care to protect yourself, and then call for a trained HazMat team. The most important step in such an incident is to first realize that you are at a hazardous situation and then attempt to identify the substances involved (if possible).

- Do not enter the hazard zone; your job is to provide supportive care once the patient can be safely moved out of the area.

- Hazardous materials are classified according to five toxicity levels. Four protection levels are indicated for the amount and type of protective gear you need. Levels indicate the amount and type of protective gear that you need to prevent injury from a particular substance.

- Most serious injuries and deaths from hazardous materials incidents result from airway and breathing problems. Patients' injuries should be treated in the same way that you would treat any injury.

- Resources for HazMat incidents include the *Emergency Response Guidebook* and the Chemical Transportation Emergency Center (CHEMTREC). CHEMTREC is open 24 hours a day to help you identify and handle hazardous materials transport incidents.

www.EMTB.com

Technology

- Interactivities
- Vocabulary Explorer
- Anatomy Review
- Web Links
- Online Review Manual

Vital Vocabulary

casualty collection area An area set up by physicians, nurses, and other hospital staff near a major disaster scene where patients can receive further triage and medical care.

Chemical Transportation Emergency Center (CHEMTREC) An agency that assists emergency personnel in identifying and handling hazardous materials transport incidents.

command post The designated field command center where the incident commander and support personnel are located.

danger zone An area where individuals can be exposed to toxic substances, lethal rays, or ignition or explosion of hazardous materials.

decontamination The process of removing or neutralizing and properly disposing of hazardous materials from equipment, patients, and rescue personnel.

decontamination area The designated area in a hazardous materials incident where all patients and rescuers must be decontaminated before going to another area.

disaster A widespread event that disrupts community resources and functions, in turn threatening public safety, citizens' lives, and property.

hazardous materials Any substances that are toxic, poisonous, radioactive, flammable, or explosive and cause injury or death with exposure.

hazardous materials incident An incident in which a hazardous material is no longer properly contained and isolated.

incident commander The individual who has overall command of the scene in the field.

incident command system An organizational system to help control, direct, and coordinate emergency responders and resources; also known as an incident management system (IMS).

mass-casualty incident An emergency situation involving three or more patients or that can place great demand on the equipment or personnel of the EMS system or has the potential to produce multiple casualties.

mutual aid response An agreement between neighboring EMS systems to respond to mass-casualty incidents or disasters in each other's region when local resources are insufficient to handle the response.

personal protective equipment (PPE) levels Measures of the amount and type of protective equipment that an individual needs to avoid injury during contact with a hazardous material.

rehabilitation area The area that provides protection and treatment to fire fighters and other personnel working at an emergency. Here, workers are medically monitored and receive any needed care as they enter and leave the scene.

sector commander The individual delegated to oversee and coordinate activity in an incident command sector; works under the incident commander.

toxicity levels Measures of the risk that a hazardous material poses to the health of an individual who comes into contact with it.

transportation area The area in a mass-casualty incident where ambulances and crews are organized to transport patients from the treatment area to receiving hospitals.

transportation officer The individual in charge of the transportation sector in a mass-casualty incident who assigns patients from the treatment area to awaiting ambulances in the transportation area.

treatment area Location in a mass-casualty incident where patients are brought after being triaged and assigned a priority, where they are reassessed, treated, and monitored until transport to the hospital.

treatment officer The individual, usually a physician, who is in charge of and directs EMS personnel at the treatment area in a mass-casualty incident.

triage The process of sorting patients based on the severity of injury and medical need to establish treatment and transportation priorities.

triage area Designated area in a mass-casualty incident where the triage officer is located and patients are initially triaged before being taken to the treatment center.

triage officer The individual in charge of the incident command triage sector who directs the sorting of patients into triage categories in a mass-casualty incident.

www.EMTB.com

Assessment in Action

You have been called to the local YMCA pool for a reported small chlorine spill. When you arrive at the scene, you find that law enforcement officers have already arrived and have asked bystanders to move away from the scene.

A group of parents and children were near the area where the spill occurred. You call for a HazMat team to respond to the scene and await their arrival. Then you tell the parents and children that a special team has been called and is on the way.

1. One of the roles of the EMT-B at the scene of a hazardous materials spill is to assist:

   A. in patient decontamination.
   B. in assessing the amount of spilled material.
   C. in keeping bystanders back.
   D. with patient triage in the hazard zone.

2. Until the HazMat team arrives to determine the hazard zone, you should always be aware of the perimeters that are necessary to keep you safe. The safe area that should be set up when there is an unknown agent should be:

   A. downwind and at least 100'.
   B. upwind and at least 1,000'.
   C. upwind and at least 100'.
   D. downwind and at least 1,000'.

3. Often, the presence of hazardous materials is easily recognized from warning signs, placards, or labels found in which of the following locations?

   A. On buildings or areas where hazardous materials are used
   B. On railroad cars that carry hazardous materials
   C. On boxes that contain hazardous materials
   D. All of the above

4. The process of removing or neutralizing and properly disposing of hazardous materials from equipment, patients, and rescue personnel is known as:

   A. disinfection.
   B. sterilization.
   C. decontamination.
   D. contamination.

5. Protection levels indicate the amount and type of protective gear that you will need to prevent injury from a particular substance. There are how many recognized protection levels?

   A. 3
   B. 4
   C. 5
   D. 6

6. You were just informed that the HazMat technicians are wearing fully encapsulating, chemical-resistant protective clothing that provides full-body protection as well as fully enclosed SCBA. What level of protection is being used?

   A. Level A
   B. Level B
   C. Level C
   D. Level D

7. Most serious injuries and deaths from hazardous materials result from:

   A. cardiac compromise.
   B. airway and breathing problems.
   C. burns.
   D. infection.

Challenging Questions

8. What toxicity level is used to describe a substance that will cause death with minimal exposure?

9. What type of personal protective equipment should be used?

www.EMTB.com

Points to Ponder

You are dispatched to local high school for an unknown medical emergency. As you enter the school you are greeted by the principal, who tells you the chemistry teacher is unresponsive in the lab. The principal informs you that the teacher was found by one of his students and is now being attended to by the school nurse. As you enter the lab you are immediately overcome by a strong chemical odor. You now observe two people on the floor, and the principal begins to complain of a severe headache and nausea.

What are some clues that this is a HazMat incident? Would you attempt to drag out the two patients or perform an immediate evacuation? What additional information do you need to gather? Should you call for additional resources?

Issues: Scene Safety, HazMat Incidents.

Response to Terrorism and Weapons of Mass Destruction

Objectives*

Cognitive

1. Define international and domestic terrorism. (p 1100)
2. List the different terrorist agenda categories. (p 1100)
3. Describe the threat levels (or colors) used by the Department of Homeland Security (DHS) to notify responders of the potential for a terrorist attack. (p 1103)
   - SEVERE (RED)
   - HIGH (ORANGE)
   - ELEVATED (YELLOW)
   - GUARDED (BLUE)
   - LOW (GREEN)
4. On the basis of DHS threat levels, discuss what actions the EMT should take during the course of their work to heighten their ability to respond to and survive a terrorist attack. (p 1102)
5. Recognize the hallmarks of a terrorist event. (p 1103)
6. List potential terrorist targets and their vulnerability. (p 1103)
7. Discuss these key principles to assuring responder safety at the scene of a terrorist event:
   - Establishing scene safety
   - Approaching the scene
   - Protective measures
   - Establishing a safety zone
   - Ongoing reevaluation of scene safety
   - Awareness of secondary devices (p 1104)
8. Discuss the following critical actions that the EMT must perform to operate on the scene following a terrorist attack:
   - Notification
   - Establish command
   - Patient management (p 1105)
9. Describe and list the four weapons of mass destruction (WMD). (p 1101)
10. Describe historical events dealing with WMD. (p 1101)
11. List nuclear/chemical/biological/explosive agents that may be used by a terrorist. (p 1102)
12. Describe the routes of exposure for chemical agents. (p 1106)
13. Describe the routes of exposure for biological agents. (p 1112)
14. Describe the routes of exposure for nuclear/radiological dispersal devices (RDD). (p 1120)
15. Discuss the clinical manifestations of exposure to the various WMD agents. (p 1108, 1117)
16. Describe the treatment to be rendered to a victim of a nuclear/chemical/biological/explosive attack. (p 1108, 1112, 1118, 1122)

Affective

17. Discuss the "new age" terrorist's trend towards apocalyptic violence and indiscriminate death. (p 1101)
18. Explain the rationale behind **NOT** entering the WMD scene or being **UNABLE** to treat contaminated victims, and the possible impact on the EMT-B. (p 1101, 1104)

Psychomotor

19. Demonstrate the patient assessment skills to assist the victim of a nuclear/chemical/biological/explosive agent. (p 1108, 1112, 1118, 1122)
20. Demonstrate the use of the nerve agent antidote (MARK 1) auto-injector kit. (p 1110)
21. Given a scenario of a terrorist event, establish scene safety and begin patient management. (p 1104)

*All of these objectives are noncurriculum objectives.

You are the Provider

You and your EMT-B partner are dispatched to the Lola Johnson Mall for a patient having a seizure. While en route dispatch informs you that they have received numerous 9-1-1 calls from within the mall. Callers are stating that a large number of people are vomiting and having seizures. Many appear to be unconscious. Dispatch is currently trying to make contact with mall security.

1. What is your first priority in this situation?
2. What does the dispatch information indicate as far as the need for additional resources?

Introduction

As a result of the increase in terrorist activity, it is possible that you may witness a terrorist event during your career. International terrorists as well as domestic groups have increased their targeting of civilian populations with acts of terror. The question is not will terrorists strike again, but rather when and where they will strike. The EMT-B must be mentally and physically prepared for the possibility of a terrorist event.

The use of weapons of mass destruction, or weapons of mass casualty, further complicates the management of the terrorist incident and places the EMT-B in greater danger. Although it is difficult to plan and anticipate a response to many terrorist events, there are several key principles that apply to every response. This chapter describes how you can prepare to respond to these events by discussing types of terrorist events, personnel safety, and patient management. You will learn the signs, symptoms, and treatment of patients who have been exposed to nuclear, chemical, or biological agents or an explosive attack. At the end of this chapter, you will be able to answer the following key questions:

- What are your initial actions?
- Who should you notify, and what should you tell them?
- What type of additional resources might you require?

Technology

- Interactivities
- Vocabulary Explorer
- Anatomy Review
- Web Links
- Online Review Manual

www.EMTB.com

- How should you proceed to address the needs of the victims?
- How do you ensure your own and your partner's safety, as well as the safety of the victims?
- What is the clinical presentation of a victim exposed to a WMD?
- How are WMD patients to be assessed and treated?
- How do you avoid becoming contaminated or cross-contaminated with a WMD agent?

What Is Terrorism?

No one is quite sure who the first terrorist was, but terrorist forces have been at work since early civilizations. Today, terrorists pose a threat to nations and cultures everywhere. International terrorism has brought a new fear into the lives of many American citizens.

Modern-day terrorism is common in the Middle East, where terrorist groups have frequently attacked civilian populations. In Ireland terrorist groups have attacked the civilian population for decades under the guise of religious freedom. In Colombia, political terrorist groups target oil resources as a means to instill fear.

In the United States, domestic terrorists have struck multiple times within the last decade. The Centennial Park bombing during the 1996 Summer Olympics and the destruction of the Alfred P. Murrah Federal Building in Oklahoma City in 1995 are examples. Terrorist organizations are generally categorized. Only a small percentage of groups actually turn towards terrorism as a means to achieve their goals, such as the following:

1. **Violent religious groups/doomsday cults**: These include groups such as Aum Shinrikyo, who carried out chemical and biological attacks in Tokyo between 1994 and 1995. Some of these groups may participate in apocalyptic violence (Figure 38-1 ▶).
2. **Extremist political groups**: They may include violent separatist groups and those who seek political, religious, economic, and social freedom, such as many Middle Eastern groups (Figure 38-2 ▶).
3. **Technology terrorists**: Those who would attack a population's technological infrastructure as a means to draw attention to their cause, such as cyber-terrorists.

Figure 38-1 Asahara Shoko, the founder of Aum Shinrikyo.

Figure 38-3 Demonstrators being held back by police near the World Bank in Washington, DC.

Figure 38-2 Palestinian extremist political groups have been associated with terrorism.

4. **Single-issue groups**: These include anti-abortion groups, animal rights groups, anarchists, racists, or even ecoterrorists who threaten or use violence as a means to protect the environment (Figure 38-3 ▶).

Most terrorist attacks require the coordination of multiple terrorists or "actors" working together. However, in a few instances there has been a single terrorist who struck with devastating results. Terrorists who acted alone carried out all of the Atlanta abortion clinic attacks, the 1996 Summer Olympics attack, and the Oklahoma City bombing.

Weapons of Mass Destruction

What Are WMDs?

A <u>weapon of mass destruction (WMD)</u>, or <u>weapon of mass casualty (WMC)</u>, is any agent designed to bring about mass death, casualties, and/or massive damage to property and infrastructure (bridges, tunnels, airports, and seaports). These instruments of death and destruction include nuclear, chemical, biological, and explosive weapons. To date, the preferred WMD for terrorists has been explosive devices. Terrorist groups have favored tactics that use truck bombs or car or pedestrian suicide bombers. Many previous terrorist attempts to use either chemical or biological weapons to their full capacity have been unsuccessful. Nonetheless, as an EMT-B you should understand the destructive potential of these weapons.

As discussed earlier, the motives and tactics of the new-age terrorist groups have begun to change. As with the doomsday cults, many terrorist groups participate in apocalyptic, indiscriminate killing. This doctrine of total carnage would make the use of WMDs highly desirable. WMDs are easy to obtain or create and are specifically geared toward killing large numbers of people. Had the proper techniques been used during the 1995 attack on the Tokyo subway, there may have been tens of thousands of casualties. With the fall of the former Soviet Union, the technology and expertise to produce

EMT-B Tips

Chemical warfare may be in the form of a liquid, powder, or vapor.

WMDs may be available to terrorist groups with sufficient funding. Moreover, the technical recipes for making nuclear, biological, and chemical (NBC) weapons and explosive devices can be found readily on the Internet; in fact, they have even been published on terrorist group websites.

Chemical Terrorism/Warfare

Chemical agents are manmade substances that can have devastating effects on living organisms. They can be produced in liquid, powder, or vapor form depending on the desired route of exposure and dissemination technique. Developed during World War I, these agents have been implicated in thousands of deaths since being introduced on the battlefield, and since then have been used to terrorize civilian populations. These agents consist of:

- Vesicants (blister agents)
- Respiratory agents (choking agents)
- Nerve agents
- Metabolic agents (blood agents)

Biological Terrorism/Warfare

Biological agents are organisms that cause disease. They are generally found in nature; for terrorist use, however, they are cultivated, synthesized, and mutated in a laboratory. The weaponization of biological agents is performed to artificially maximize the target population's exposure to the germ, thereby exposing the greatest number of people and achieving the desired result.

The primary types of biological agents that you may come into contact with during a biological event include:

- Viruses
- Bacteria
- Toxins

Nuclear/Radiological Terrorism

There have been only two publicly known incidents involving the use of a nuclear device. During World War II, Hiroshima and Nagasaki were devastated when they were targeted with nuclear bombs. The awesome destructive power demonstrated by the attack ended World War II and has served as a deterrent to nuclear war.

There are also nations that hold close ties with terrorist groups (known as state-sponsored terrorism) and have obtained some degree of nuclear capability.

It is also possible for a terrorist to secure radioactive materials or waste to perpetrate an act of terror. Such materials are far easier for the determined terrorist to acquire and would require less expertise to use. The difficulties in developing a nuclear weapon are well documented. Radioactive materials, however, such as those in Radiological Dispersal Devices (RDDs), also known as "dirty bombs," can cause widespread panic and civil disturbances. More on these devices will be covered later in this chapter.

EMT-B Response to Terrorism

Recognizing a Terrorist Event (Indicators)

Most acts of terror are covert, which means that the public safety community generally has no prior knowledge of the time, location, or nature of the attack. This element of surprise makes responding to an event more complex. You must constantly be aware of your surroundings and understand the possible risks for terrorism associated with certain locations, at certain times. It is therefore important that you know the current threat level issued by the federal government through the Department of Homeland Security (DHS).

The Homeland Security Advisory System alerts responders to the potential for an attack, although the specifics of the current threat will not be given. On the basis of the current threat level, the EMT-B should take appropriate actions and precautions while continuing to perform daily duties and responding to calls. The system of colors is used to inform the public safety community of the climate of terrorism (derived from intelligence gathering and the amount of terrorist communication) and to heighten the awareness of the potential for a terrorist attack. The system is designed to save lives, including yours.

The DHS has not issued specific recommendations for EMS personnel to follow in response to the alert system. Follow your local protocols, policies, and procedures.

It is your responsibility to make sure you know the advisory level at the start of your workday. Daily

🖐 EMT-B Safety

The Department of Homeland Security Advisory System is posted daily to heighten awareness of the current terrorist threat (Figure 38-4 ▶).
SEVERE (red): Severe risk of terrorist attacks
HIGH (orange): High risk of terrorist attacks
ELEVATED (yellow): Significant risk of terrorist attacks
GUARDED (blue): General risk of terrorist attacks
LOW (green): Low risk of terrorist attacks

newspapers, television news programs, and multiple websites (including the DHS website) all give up-to-date information on the threat level. Many EMS organizations are starting to display the advisory system on boards where they can be seen once staff arrives for a shift.

Understanding and being aware of the current threat is only the beginning of responding safely to calls. Once you are on duty, you must be able to make appropriate decisions regarding the potential for a terrorist event. In determining the potential for a terrorist attack, on every call you should observe the following:

- **Type of location.** Is the location a monument, infrastructure, government building, or a specific type of location such as a temple? Is there a large gathering? Is there a special event taking place?
- **Type of call.** Is there a report of an explosion or suspicious device nearby? Does the call come into dispatch as someone having unexplained coughing and difficulty breathing? Are there reports of people fleeing the scene?
- **Number of patients.** Are there multiple victims with similar signs and symptoms? This is probably the single most important clue that a terrorist attack or an incident involving a WMD has occurred.
- **Victims' statements.** This is probably the second best indication of a terrorist or WMD event. Are the victims fleeing the scene giving statements such as, "Everyone is passing out," "There was a loud explosion," or "There are a lot of people shaking on the ground." If so, something is occurring that you do not want to rush into, even if it is determined not to be a terrorist event.
- **Pre-incident indicators.** Is the terror alert level high (orange) or severe (red)? Has there been a

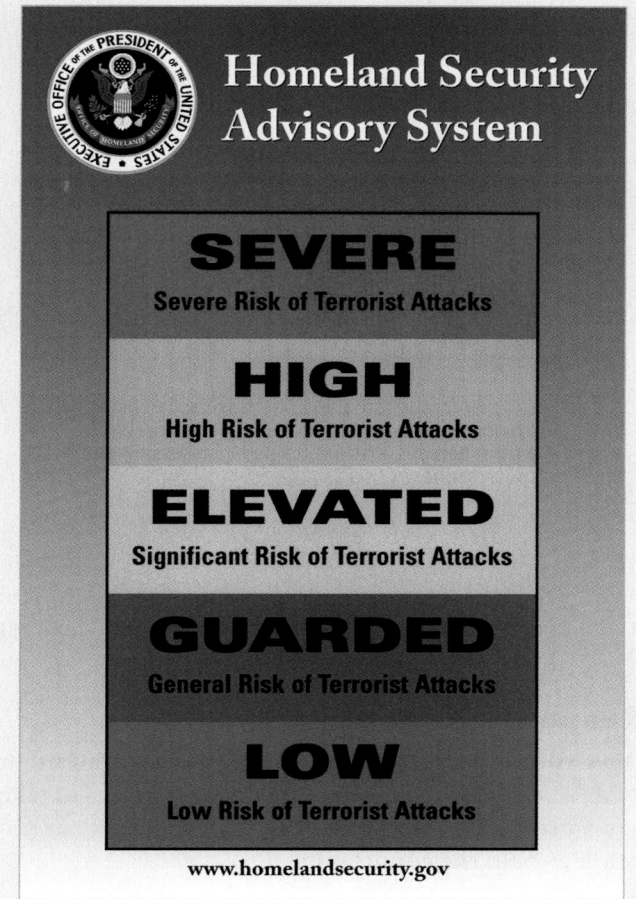

Figure 38-4 Homeland Security Advisory System.

✳ EMT-B Tips

One of the easiest ways to distinguish between a nonterrorist mass-casualty event and a terrorist event is that the intentional use of WMD affects multiple persons. These casualties will generally exhibit the same signs and symptoms. It is highly unlikely for more than one person to experience a seizure at any given time. It is not uncommon to find multiple patients complaining of difficulty breathing at the scene of a fire. However, the same report in the subway at rush hour, when no smell of smoke has been reported, is certainly cause for suspicion. In these situations, you must use good judgment and resist the urge to "rush in and help," especially when there are multiple victims from an unknown cause.

Figure 38-5 Improper staging of a mass-casualty scene could lead to injury or even death of EMS personnel. Wait for assistance from persons who are trained in assessing and managing such scenes.

Figure 38-6 Park your vehicle at a safe location.

recent increase in violent political activism? Are you aware of any credible threats made against the location, gathering, or occasion?

Response Actions

Once you suspect that a terrorist event has occurred or WMD have been used, there are certain actions to take to ensure that you will be safe and be in the proper position to help the community.

Scene Safety

Ensure that the scene is safe. If you have **any** doubt that it may not be safe, **do not enter**. When dealing with a WMD scene, it is safe to assume that you will not be able to enter where the event has occurred—nor do you want to. The best location for staging is upwind and uphill from the incident. Wait for assistance from those who are trained in assessing and managing WMD scenes (Figure 38-5 ▲). Also remember:

- Failure to park your vehicle at a safe location can place you and your partner in danger (Figure 38-6 ▲).
- If your vehicle is blocked in by other emergency vehicles or damaged by a secondary device (or event), you will be unable to provide victims with transportation (Figure 38-7 ▶), or escape yourself.

Responder Safety (Personnel Protection)

The best form of protection from a WMD agent is preventing yourself from coming into contact with the agent. The greatest threats facing an EMT-B in a WMD attack are contamination and <u>cross-contamination.</u> Contamination with an agent occurs when you have direct contact with the WMD or are exposed to it. Cross-contamination occurs when you come into contact with a contaminated person who has not yet been decontaminated.

You are the Provider   Part 2

Prior to your arrival, dispatch states that mall security is reporting a high-pitched whistling sound, much like a gas leak. The HazMat team has been dispatched and has an estimated time of arrival (ETA) of 4 minutes. All units are to hold off until the scene has been secured by HazMat.

3. What precautions can you take to ensure your own safety?
4. What does your index of suspicion tell you about this scene?

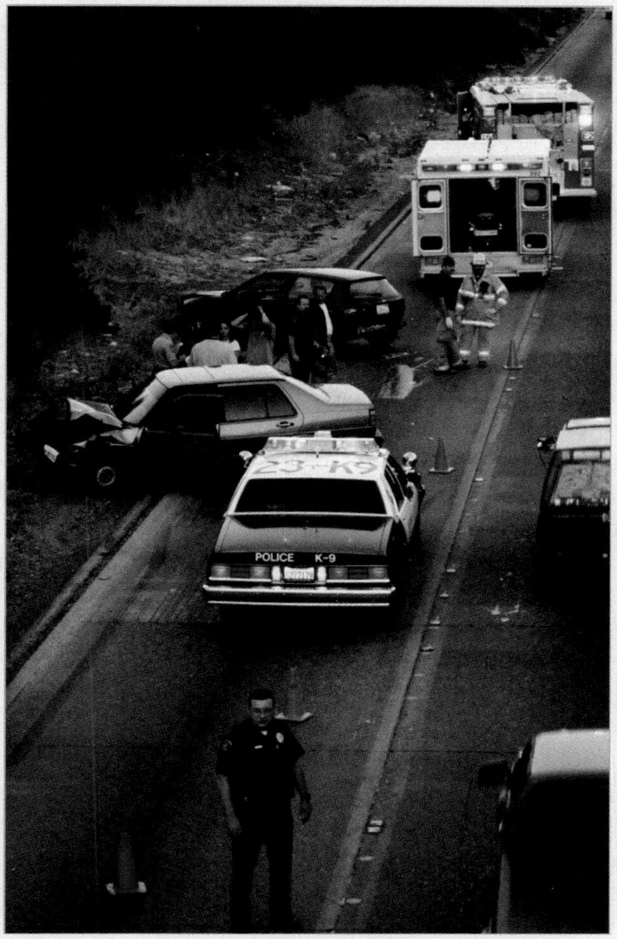

Figure 38-7 Make sure that your vehicle is not blocked in by other emergency vehicles.

Notification Procedures

When you suspect a terrorist or WMD event has taken place, notify the dispatcher, providing that communications function properly. Vital information needs to be communicated effectively if you are to receive the appropriate assistance (see Chapter 9 for information on effective communication). Inform dispatch of the nature of the event, any additional resources that may be required, the estimated number of patients, and the upwind route of approach or optimal route of approach.

It is extremely important to establish a staging area, where other units will converge. Be mindful of access and exit routes when you direct units to respond to a location. It is unwise to have units respond to the front entrance of a hotel or apartment building that has had an explosion (see Chapter 35 on vehicle positioning).

Weapons of Mass Destruction

On September 11th, 2001, communications were severely affected by the collapse of the World Trade Center. The primary repeater was situated atop one of the towers. Additionally, excess radio traffic made transmitting and receiving messages extremely difficult. Not only was radio communications affected, but also most cellular phones and the majority of radio and television stations were disabled. The lesson learned is to have multiple backups to your ability to communicate with your dispatcher. In the event of a terrorist or WMD event, refrain from using the radio unless you have something important to transmit. If you do transmit, gather your thoughts and speak in as calm a tone as possible, avoiding unnecessary chatter. Remember, while you are transmitting, others may be unable to call for help.

Last, trained responders in the proper protective equipment are the only persons equipped to handle the WMD incident. These specialized units, traditionally hazardous materials (HazMat) teams, must be requested as early as possible due to the time required to assemble and dispatch the team and their equipment. Many jurisdictions share HazMat teams, and the team may have to travel a long distance to reach the location of the event. It is always better to be safe than sorry; call the team early and the outcome of the call will be more favorable.

Keep in mind that there may be more than one type of device or agent present.

Establishing Command

The first arriving provider on the scene must begin to sort out the chaos and define his or her responsibilities under the Incident Command System (ICS). As the first person on scene, the EMT-B may need to establish command until additional personnel arrive. Depending on the circumstances, you and other EMT-Bs may function as medical branch officers, triage officers, treatment officers, transportation or logistic officers, or staff. If the ICS is already in place, then you should immediately seek out the medical staging officer to receive your assignment.

EMT-B Tips

Secondary devices may include various types of electronic equipment such as cell phones or pagers that are detonated when "answered."

Secondary Device or Event (Reassessing Scene Safety)

Terrorists have been known to plant additional explosives that are set to explode after the initial bomb. This type of secondary device is intended primarily to injure responders and to secure media coverage, because the media generally arrives on scene just after the initial response. Do not rely on others to secure your safety. It is every EMT-B's responsibility to constantly assess and reassess the scene for safety. It is easy to overlook a suspicious package lying on the floor while you are treating casualties. Stay alert. Something as subtle as a change

EMT-B Tips

You are of no help to the public if you become a patient. More importantly, once you become a victim of the event, you place an additional burden on your fellow responders, who must treat you. Assess the scene and resist the urge to run in and help (do not develop tunnel vision). You may place your life and your partner in danger. Remember... do not become a victim.

in the wind direction during a gas attack or an increase in the number of contaminated patients can place you in danger. Never become so involved with the tasks that you are performing that you do not look around and make sure that the scene remains safe.

Chemical Agents

Chemical agents are liquids or gases that are dispersed to kill or injure. Modern-day chemicals were first developed during WWI and WWII. During the Cold War, many of these agents were perfected and stockpiled. While the United States has long renounced the use of chemical weapons, many nations still develop and stockpile them. These agents are deadly and pose a threat if acquired by terrorists.

Chemical weapons have several classifications. The properties or characteristics of an agent can be described as liquid, gas, or solid material. Persistency and volatility are terms used to describe how long the agent will stay on a surface before it evaporates. Persistent or non-volatile agents can remain on a surface for long periods of time, usually longer than 24 hours. Nonpersistent or volatile agents evaporate relatively fast when left on a surface in the optimal temperature range. An agent that is described as highly persistent (such as VX, a nerve agent) can remain in the environment for weeks to months, whereas an agent that is highly volatile (such as sarin, also a nerve agent) will turn from liquid to gas (evaporate) within minutes to seconds.

Route of exposure is a term used to describe how the agent most effectively enters the body. Chemical agents can have either a vapor or contact hazard. Agents with a vapor hazard enter the body through the respiratory tract in the form of vapors. Agents with a

You are the Provider

Part 3

HazMat arrives on scene and requests that all units report to the staging area until called upon for patient transport. After the HazMat team has decontaminated your patient, you are called to transport her to the nearest facility. Your patient is a 29-year-old woman who was working in the mall when the incident took place. She is conscious, alert, and oriented. She is complaining of nausea and vomiting. She appears diaphoretic and is experiencing excessive tearing (lacrimation).

5. What could your patient's signs and symptoms represent?
6. What are your treatment options for this patient?

contact hazard (or skin hazard) give off very little vapor or no vapors and enter the body through the skin.

Vesicants (Blister Agents)

The primary route of exposure of blister agents, or vesicants, is the skin (contact); however, if vesicants are left on the skin or clothing long enough, they produce vapors that can enter the respiratory tract. Vesicants cause burn-like blisters to form on the victim's skin as well as in the respiratory tract. The vesicant agents consist of sulfur mustard (H), Lewisite (L), and phosgene oxime (CX) (the symbols H, L, and CX are military designations for these chemicals). The vesicants usually cause the most damage to damp or moist areas of the body, such as the armpits, groin, and respiratory tract. Signs of vesicant exposure on the skin include:

- Skin irritation, burning, and reddening
- Immediate intense skin pain (with L and CX)
- Formation of large blisters
- Gray discoloration of skin (a sign of permanent damage seen with L and CX)
- Swollen and closed or irritated eyes
- Permanent eye injury (including blindness)

If vapors were inhaled, the patient may experience the following:

- Hoarseness and stridor
- Severe cough
- Hemoptysis (coughing up of blood)
- Severe dyspnea

Sulfur mustard (agent H) is a brownish, yellowish oily substance that is generally considered very persistent. When released, mustard has the distinct smell of garlic or mustard and is quickly absorbed into the skin and/or mucous membranes. As the agent is absorbed into the skin, it begins an irreversible process of damage to the cells. Absorption through the skin or mucous membranes usually occurs within seconds, and damage to the underlying cells takes place within 1 to 2 minutes.

Mustard is considered a mutagen, which means that it mutates, damages, and changes the structures of cells. Eventually, cellular death will occur. On the surface, the patient will generally not produce any signs or symptoms until 4 to 6 hours after exposure (depending on concentration and amount of exposure) (Figure 38-8 ▶).

The patient will develop a progressive reddening of the affected area, which will gradually develop into large blisters. These blisters are very similar in shape and appearance to those associated with thermal second-degree burns. The fluid within the blisters does not

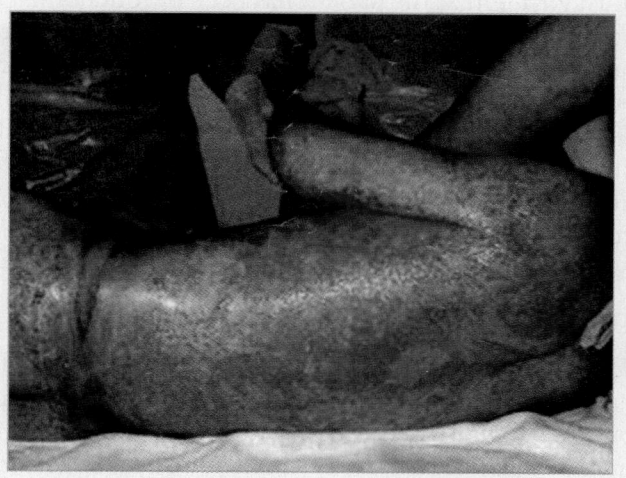

Figure 38-8 Skin damage resulting from exposure to sulfur mustard (agent H).

contain any of the agent; however, the skin covering the area is considered to be contaminated until decontamination by trained personnel has been performed.

Mustard also attacks vulnerable cells within the bone marrow and depletes the body's ability to reproduce white blood cells. As with burns, the primary complication associated with vesicant blisters is secondary infection. If the patient does survive the initial direct injury from the agent, the depletion of the white blood cells leaves the patient with a decreased resistance to infections. Although sulfur mustard is regarded as persistent, it does release enough vapors when dispersed to be inhaled. This creates upper and lower airway compromise. The result is damage and swelling of the airways. The airway compromise makes the patient's condition far more serious.

Lewisite (L) and phosgene oxime (CX) produce blister wounds very similar to mustard. They are highly volatile and have a rapid onset of symptoms, as opposed to the delayed onset seen with mustard. These agents produce immediate intense pain and discomfort when contact is made. The patient may have a grayish discoloration at the contaminated site. While tissue damage also occurs with exposure to these agents, they do not cause the secondary cellular injury that is associated with mustard.

Vesicant Agent Treatment

There are no antidotes for mustard or CX exposure. BAL (British Anti-Lewisite) is the antidote for agent L; however, it is not carried by civilian EMS. The EMT-B must ensure that the patient has been decontaminated before ABCs are initiated. The patient may require prompt airway support if any agent has been inhaled, but this should not occur until after decontamination. Gain IV access and initiate transport as soon as possible. Generally, burn centers are best equipped to handle the wounds and subsequent infections produced by vesicants. Follow your local protocols when deciding what facility to transport the patient to.

Pulmonary Agents (Choking Agents)

The pulmonary agents are gases that cause immediate harm to persons exposed to them. The primary route of exposure for these agents is through the respiratory tract, which makes them an inhalation or vapor hazard. Once inside the lungs, they damage the lung tissue and fluid leaks into the lungs. Pulmonary edema develops in the patient, resulting in difficulty breathing due to the inability for air exchange. These agents produce respiratory-related symptoms such as dyspnea, tachypnea, and pulmonary edema. This class of chemical agents consists of chlorine (CL) and phosgene.

Chlorine (CL) was the first chemical agent ever used in warfare. It has a distinct odor of bleach and creates a green haze when released as a gas. Initially it produces upper airway irritation and a choking sensation. The patient may later experience:

- Shortness of breath
- Chest tightness
- Hoarseness and stridor due to upper airway constriction
- Gasping and coughing

With serious exposures, patients may experience pulmonary edema, complete airway constriction, and death. The fumes from a mixture of household bleach (CL) and ammonia create an acid gas that produces similar effects. Each year, such mixtures overcome hundreds of people when they try to mix household cleaners.

Phosgene should not be confused with phosgene oxime, a blistering agent, or vesicant. Not only has phosgene been produced for chemical warfare, but it is a product of combustion such as might be produced in a fire at a textile factory or house, or from metalwork or burning Freon (a liquid chemical used in refrigeration). Therefore, you may encounter a victim of exposure to this gas during the course of a normal call or at a fire scene. Phosgene is a very potent agent that has a delayed onset of symptoms, usually hours. Unlike CL, when phosgene enters the body, it generally does not produce severe irritation, which would possibly cause the victim to leave the area or hold his or her breath.

In fact, the odor produced by the chemical is similar to that of freshly mown grass or hay. The result is that much more of the gas is allowed to enter the body unnoticed. The initial symptoms of a mild exposure may include:

- Nausea
- Chest tightness
- Severe cough
- Dyspnea upon exertion

The victim of a severe exposure may present with dyspnea at rest, and excessive pulmonary edema (the patient will actually expel large amounts of pulmonary edema from their lungs). The pulmonary edema that is seen with a severe exposure produces such large amounts of fluid from the lungs that the patient may actually become hypovolemic and subsequently hypotensive.

Pulmonary Agent Treatment

The best initial treatment for any pulmonary agent is to remove the patient from the contaminated atmosphere. This should be done by trained personnel in the proper PPE. Aggressive management of the ABCs should be initiated, paying particular attention to oxygenation, ventilation, and suctioning if required. Do not allow the patient to be active, as this will worsen the condition much faster. There are no antidotes to counteract the pulmonary agents. Performing the ABCs, gaining IV access, allowing the patient to rest in a position of comfort with the head elevated, and initiating rapid transport are the primary goals for prehospital emergency care.

Nerve Agents

The nerve agents are among the most deadly chemicals developed. Designed to kill large numbers of people with small quantities, nerve agents can cause cardiac arrest within seconds to minutes of exposure. Nerve agents, discovered while in search of a superior pesticide, are a class of chemical called organophosphates, which are found in household bug sprays, agricultural pesticides, and some industrial chemicals, at far lower strengths than in nerve agents. Organophosphates block an essential enzyme in the nervous system, which cause the body's organs to become over stimulated and burn out.

G agents came from the early nerve agents, the G series, which were developed by German scientists (hence the G) in the period after WWI and into WWII. There are three G series agents, which are all designed with the same basic chemical structure with slight variations to produce different properties. The two variations of these agents are lethality and volatility. The following G agents are listed from high volatility to low volatility:

- **Sarin (GB):** Highly volatile colorless and odorless liquid. Turns from liquid to gas within seconds to minutes at room temperature. Highly lethal, with an LD_{50} of 1,700 mg/70 kg (about 1 drop, depending on the purity). The LD_{50} is the amount that will kill 50% of people who are exposed to this level. Sarin is primarily a vapor hazard, with the respiratory tract as the main route of entry. This agent is especially dangerous in enclosed environments such as office buildings, shopping malls, or subway cars. When this agent comes into contact with skin, it is quickly absorbed and evaporates. When sarin is on clothing, it has the effect of off-gassing, which means that the vapors are continuously released over a period of time (like perfume). This renders the victim as well as the victim's clothing contaminated.

- **Soman (GD):** Twice as persistent as sarin and five times as lethal. It has a fruity odor as a result of the type of alcohol used in the agent and generally has no color. This agent is both a contact and inhalation hazard that can enter the body through skin absorption and through the respiratory tract. A unique additive in GD causes it to bind to the cells that it attacks faster than any other agent. This irreversible binding is called aging, which makes it more difficult to treat patients who have been exposed.

- **Tabun (GA):** Approximately half as lethal as sarin and 36 times more persistent; under the proper conditions it will remain for several days. It also has a fruity smell and an appearance similar to sarin. The components used to manufacture GA are easy to acquire and the agent is easy to manufacture, which make it unique. GA is both a contact and inhalation hazard that can enter the body through skin absorption as well as through the respiratory tract.

- **V agent (VX):** Clear oily agent that has no odor and looks like baby oil. V agent was developed by the British after World War II and has similar chemical properties to the G series agents. The difference is that VX is over 100 times more lethal than sarin and is extremely persistent (Figure 38-9 ▶). In fact, VX is so persistent that given the proper conditions it will remain relatively unchanged for weeks to months. These properties make VX primarily a contact hazard,

Figure 38-9 VX is the most toxic chemical ever produced. The dot on the penny demonstrates the amount needed to achieve the lethal dose.

because it lets off very little vapor. It is easily absorbed into the skin, and the oily residue that remains on the skin's surface is extremely difficult to decontaminate.

Nerve agents all produce similar symptoms, but have varying routes of entry. Nerve agents differ slightly in lethal concentration or dose and also differ in their volatility. Some agents are designed to become a gas quickly (nonpersistent or highly volatile), while others remain liquid for a period of time (persistent or nonvolatile). These agents have been used successfully in warfare and to date represent the only type of chemical agent that has been used successfully in a terrorist act. Once the agent has entered the body through skin contact or through the respiratory system, the patient will begin to exhibit a pattern of predictable symptoms. Like all chemical agents, the severity of the symptoms will depend on the route of exposure and the amount of agent to which the patient was exposed. The resulting symptoms are described below using the military mnemonic SLUDGEM and the medical mnemonic DUMBELS. SLUDGEM/DUMBELS mnemonics are used to describe the symptoms of nerve agent exposure. The medical mnemonic is more useful to you because it lists the more dangerous symptoms associated with exposure to nerve agents.

There are only a handful of medical conditions that are associated with the bilateral pinpoint constricted pupils (miosis) seen with nerve agent exposure. Conditions such as a CVA, direct light to both eyes, and a drug overdose all can cause bilateral constricted pupils. You should therefore assess the patient for all of the SLUDGEM/DUMBELS signs and symptoms to

TABLE 38-1 Symptoms of Persons Exposed to Nerve Agents

Military Mnemonic: SLUDGEM	Medical Mnemonic: DUMBELS
Salivation, **S**weating	**D**iarrhea
Lacrimation (excessive tearing)	**U**rination
Urination	**M**iosis (pinpoint pupils)
Defecation, **D**rooling, **D**iarrhea	**B**radycardia, **B**ronchospasm (spasm of the bronchioles)
Gastric upset and cramps	**E**mesis (vomiting)
Emesis (vomiting)	**L**acrimation (excessive tearing)
Muscle twitching	**S**eizures, **S**alivation, **S**weating

determine whether the patient has been exposed to a nerve agent.

Miosis is the most common symptom of nerve agent exposure and can remain for days to weeks. This symptom, along with the others listed in (Table 38-1 ▲), will help you recognize exposure to a nerve agent early. The seizures that are associated with nerve agent exposure are unlike those found in patients with a history of seizure. The patient will continue to seize until death or until treatment is given with a nerve agent antidote (MARK 1 or NAAK).

Nerve Agent Treatment (MARK 1/NAAK)

Fatalities from severe exposure occur as a result of respiratory complications, which lead to respiratory arrest. Once the patient has been decontaminated, the EMT-B should be prepared to treat these patients aggressively, if they are to be saved. You can greatly increase the patient's chances of survival simply by providing airway and ventilatory support. As with all emergencies, securing the ABCs is the best and most important treatment that you can provide. Often patients exposed to these agents will begin seizing and will not stop. These patients will require administration of nerve agent antidote kits in addition to support of the ABCs.

Fortunately, there is an antidote for nerve agent exposure. MARK 1 kits, also known as Nerve Agent Antidote Kits (NAAK), contain two auto-injector medications: atropine and 2-PAM chloride (pralidoxime chloride). In some regions, the EMT-B may carry MARK 1 kits on the unit and will be called upon to administer one or both of the antidotes. These medications are delivered using the same technique as the EpiPen

EMT-B Safety

On March 20th, 1995, members of a Japanese cult released sarin (GB) in the Tokyo subway. The first arriving medical responders were met with chaos as hundreds and then thousands of people fled the subway system (Figure 38-10 ▶). Many were contaminated and showing signs and symptoms of nerve agent exposure. In the end more than 5,000 people sought medical care for exposure to sarin, and 12 people had died. None of the EMS personnel wore protective clothing and most became cross-contaminated. Remember, you can avoid becoming exposed. Don't become a victim.

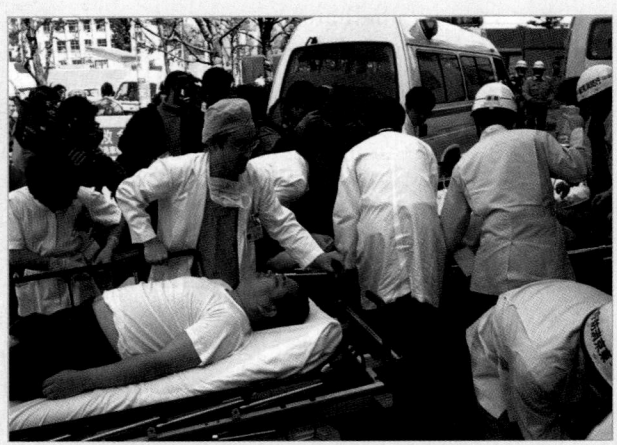

Figure 38-10 Medical professionals responding to an attack in 1995, where cult members released sarin in the Tokyo subway.

TABLE 38-2 The Nerve Agents

Name	Military Designation	Odor	Special Features	Onset of Symptoms	Volatility	Route of Exposure
Tabun	GA	Fruity	Easy to manufacture	Immediate	Low	Both contact and vapor hazard
Sarin	GB	None (if pure) or strong	Will off-gas while on victim's clothing	Immediate	High	Primarily respiratory vapor hazard; extremely lethal if skin contact is made
Soman	GD	Fruity	Ages rapidly, making it difficult to treat	Immediate	Moderate	Contact with skin; minimal vapor hazard
V agent	VX	None	Most lethal chemical agent; difficult to decontaminate	Immediate	Very low	Contact with skin; no vapor hazard (unless aerosolized)

auto-injector; however, multiple doses may need to be administered.

Atropine is used to block the nerve agent from affecting the body. However, because the nerve agent may remain in the body for long periods of time, 2-PAM chloride is used to eliminate the agent from the body. Many of the symptoms described in the DUMBELS mnemonic will be reversed with the use of atropine; however, many doses may need to be administered to see these results. If your service carries a nerve agent antidote, please refer to your local protocols for dose and usage information.

Table 38-2 ▲ has been provided for quick reference and comparison of the nerve agents.

Industrial Chemicals/Insecticides

As previously mentioned, the basic chemical ingredient in nerve agents is organophosphate. This is a common chemical that is used in lesser concentrations for insecticides. While industrial chemicals do not possess sufficient lethality to be effective WMDs, they are easy to acquire, inexpensive, and would have similar effects as the nerve agents. Crop-duster planes could be used to disseminate these chemicals. You should be cautious when responding to calls where insecticide equipment is stored and used, such as a farm or supply store that sells these products. The symptoms and medical management of victims of organophosphate insecticide poisoning are identical to those of the nerve agents.

You are the Provider   Part 4

Your patient has a patent airway and is breathing at a rate of 12 breaths/min. You immediately apply high-flow oxygen via nonrebreathing mask. Her pulse rate is 120 beats/min and her blood pressure is 148/94 mm Hg. Lung sounds are raspy and moist sounding. Her oxygen saturation is at 97% on 15 L/min of oxygen. The patient report that you received from the HazMat team indicates that a nerve agent antidote kit was used on your patient before you received her.

7. What does the nerve agent antidote kit include?
8. Is this kit part of your local protocol in dealing with patients exposed to a nerve agent?

Metabolic Agents (Cyanides)

Hydrogen cyanide (AC) and cyanogen chloride (CK) are both agents that affect the body's ability to use oxygen. Cyanide is a colorless gas that has an odor similar to almonds. The effects of the cyanides begin on the cellular level and are very rapidly seen at the organ system level. Beside the nerve agents, metabolic agents are the only chemical weapons known to kill within seconds to minutes. Unlike nerve agents, however, these deadly gases are commonly found in many industrial settings. Cyanides are produced in massive quantities throughout the United States every year for industrial uses such as gold and silver mining, photography, lethal injections, and plastics processing. They are often present in fires associated with textile or plastic factories. In fact, cyanide is naturally found in the pits of many fruits in very low doses. There is very little difference in the symptoms found between AC and CK. In low doses, these chemicals are associated with dizziness, light-headedness, headache, and vomiting. Higher doses will produce symptoms that include:

- Shortness of breath and gasping respirations
- Tachypnea
- Flushed skin color
- Tachycardia
- Altered mental status
- Seizures
- Coma
- Apnea
- Cardiac arrest

The symptoms associated with the inhalation of a large amount of cyanide will all appear within several minutes. Death is likely unless the patient is treated promptly.

Cyanide Agent Treatment

Cyanide binds with the body's cells, preventing oxygen from being used. Several medications act as antidotes, but most services do not carry them. Once trained personnel in the proper PPE have removed the patient from the source of exposure, even if there is no liquid contamination, all of the patient's clothes must be removed to prevent off-gassing in the ambulance. Trained and protected personnel must decontaminate any patients who may have been exposed to liquid contamination before an EMT-B can initiate treatment. Then you should support the patient's ABCs and gain IV access. Mild effects of cyanide exposure will generally resolve by simply removing the victim from the source of contamination and administering supplemental oxygen. Severe exposure, however, will require aggressive oxy-

EMT-B Safety

Always make sure that your patients have been thoroughly decontaminated by trained personnel before you come into contact with them. Chemical agents are primarily a vapor hazard, and all of the patient's clothing must still be removed to prevent off-gassing. Finally, never perform mouth-to-mouth or mouth-to-mask ventilation on a victim of a chemical agent. Many of the vapors may linger in the patient's airway and cross-contamination may occur.

genation and perhaps ventilation with supplemental oxygen. Always use a BVM device or oxygen-powered ventilator device to ventilate a victim of a metabolic agent. The agent can easily be passed on from the patient to the EMT-B through mouth-to-mouth or mouth-to-mask ventilations. If no antidote is available, initiate transport immediately.

(Table 38-3 ▶) summarizes the chemical agents. The odors of the particular chemicals are provided for informational purposes only. The sense of smell is a poor tool to use to determine whether there is a chemical agent present. Many persons are unable to smell the agents, and the odor could be derived from another source. This information is useful to you if you receive reports from victims claiming to smell of bleach or garlic, for example. You should never enter a potentially hazardous area and "smell" to determine whether a chemical agent is present.

Biological Agents

Biological agents pose many difficult issues when used as a WMD. Biological agents can be almost completely undetectable. Also, most of the diseases caused by these agents will be similar to other minor illnesses commonly seen by EMS providers.

Biological agents are grouped as viruses, bacterium, or neurotoxins and may be spread in various ways. Dissemination is the means by which a terrorist will spread the agent—for example, poisoning the water supply or aerosolizing the agent into the air or ventilation system of a building. A disease vector is an animal that spreads disease, once infected, to another animal. For example, the plague can be spread by infected rats, smallpox by infected persons, and West Nile virus by

TABLE 38-3 Chemical Agents

Name	Military Designations	Odor	Lethality	Onset of Symptoms	Volatility	Primary Route of Exposure
Nerve agents	Tabun (GA) Sarin (GB) Soman (GD) VX	Fruity or none	Most lethal chemical agents; can kill within minutes; effects are reversible with antidotes	Immediate	Moderate (GA, GD) Very high (GB) Low (VX)	GA—both GB—vapor hazard GD—both VX—contact hazard
Vesicants	Mustard (H) Lewisite (L) Phosgene oxime (CX)	Garlic (H) Geranium (L)	Causes large blisters to form on victims; may severely damage upper airway if vapors are inhaled; severe, intense pain and grayish skin discoloration (L and CX)	Delayed (H) Immediate (L, CX)	Very low (H, L) Moderate (CX)	Primarily contact with some vapor hazard
Pulmonary agents	Chlorine (CL) Phosgene (CG)	Bleach (CL) Cut grass (CG)	Causes irritation choking (CL); severe pulmonary edema (CG)	Immediate (CL) Delayed (CG)	Very high	Vapor hazard
Cyanide agents	Hydrogen cyanide (AC) Cyanogen chloride (CK)	Almonds (AC) Irritating (CK)	Highly lethal chemical gases; can kill within minutes; effects are reversible with antidotes	Immediate	Very high	Vapor hazard

infected mosquitoes. How easily the disease is able to spread from one human to another human is called communicability. Some diseases, such as those caused by human immunodeficiency virus, are difficult to spread by routine contact. Therefore communicability is considered low. In other instances when communicability is high, such as with smallpox, the person is considered contagious. Typically, routine BSI precautions are enough to prevent contamination from contagious biological organisms.

Incubation describes the period of time between the person becoming exposed to the agent and when symptoms begin. The incubation period is especially important for the EMT-B to understand. Although your patient may not exhibit signs or symptoms, he or she may be contagious.

EMT-Bs need to be aware of when they should suspect the use of biological agents. If the agent is in the form of a powder, such as in the October 2001 attacks involving anthrax powder mailed in letters, the incident must be handled by HazMat specialists. Patients who have come into direct contact with the agent need to be decontaminated before any EMS contact or treatment is initiated.

Viruses

Viruses are germs that require a living host to multiply and survive. A virus is a simple organism and cannot thrive outside of a host (living body). Once in the body, the virus will invade healthy cells and replicate itself to spread through the host. As the virus spreads, so does the disease that it carries. Viruses survive by moving from one host to another by using its transport system—vectors.

Viral agents that may be used during a biological terrorist release pose an extraordinary problem for health care providers, especially those in EMS. Although some viral agents do have vaccines, there is no treatment for a viral infection other than antivirals for some agents. Because of this characteristic, the following viruses have been used as terrorist agents.

Smallpox

Smallpox is a highly contagious disease. All forms of BSI precautions must be used to prevent cross-contamination to health care providers. Simply by wearing examination gloves, a HEPA-filtered respirator, and eye protection, you will greatly reduce your risk of contamination. The last natural case of smallpox in the world was seen

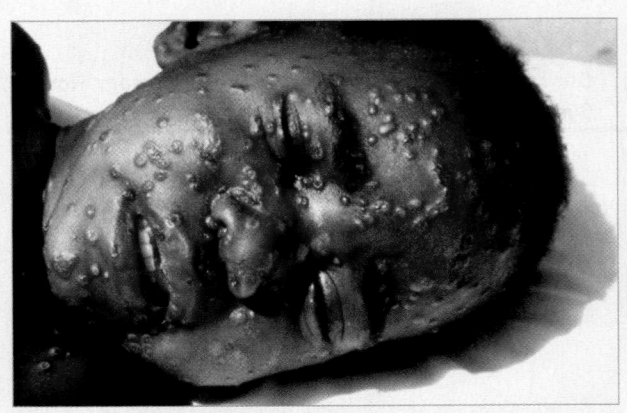

Figure 38-11 In smallpox, all the lesions are identical in their development. In other skin disorders, the lesions will be in various stages of healing and development.

TABLE 38-4	Characteristics of Smallpox
Dissemination	Aerosolized for warfare or terrorist uses.
Communicability	High from infected individuals or items (such as blankets used by infected patients). Person-to-person transmission is possible.
Route of entry	Through inhalation of coughed droplets or direct skin contact with blisters.
Signs and symptoms	Severe fever, malaise, body aches, headaches, small blisters on the skin, bleeding of the skin and mucous membranes. Incubation period is 10 to 12 days and the duration of the illness is approximately 4 weeks.
Medical management	BSI precautions. There is no specific treatment for smallpox victims. Patients should be provided with supportive care (ABCs).

in 1977. Before the rash and blisters show, the illness will start with a high fever and body aches and headaches. The patient's temperature is usually in the range of 101 to 104°F.

An easy, quick way to differentiate the smallpox rash from other skin disorders is to observe the size, shape, and location of the lesions. In smallpox, all the lesions are identical in their development. In other skin disorders, the lesions will be in various stages of healing and development. Smallpox blisters also begin on the face and extremities and eventually move toward the chest and abdomen. The disease is in its most contagious phase when the blisters begin to form (Figure 38-11 ▲). Unprotected contact with these blisters will promote transmission of the disease. There is a vaccine to prevent smallpox; however, it has been linked to medical complications and in very rare cases death (Table 38-4 ▶). Vaccination against the disease is part of a national strategy to respond to a terrorist threat. Because the vaccine does have some risk, only first responders have been offered the vaccine. Should an outbreak occur, vaccine would be offered to people at risk.

Viral Hemorrhagic Fevers

Viral hemorrhagic fevers (VHF) consist of a group of diseases that include the Ebola, Rift Valley, and Yellow Fever viruses, among others. This group of viruses causes the blood in the body to seep out from the tissues and blood vessels (Figure 38-12 ▶). Initially, the patient will have flu-like symptoms, progressing to more serious symptoms such as internal and external

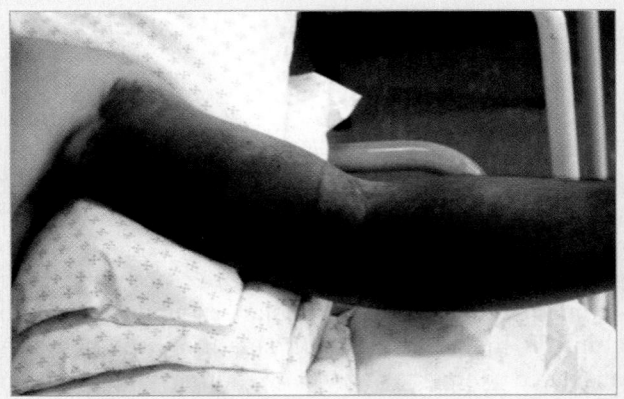

Figure 38-12 Viral hemorrhagic fevers cause the blood vessels and tissues to seep blood. The end result is ecchymosis, hemoptysis, and blood in the patient's stool. Notice the severe discoloration in this patient with Crimean Congo hemorrhagic fever, indicating internal bleeding.

hemorrhaging. Outbreaks are not uncommon in Africa and South America. Outbreaks in the United States, however, are extremely rare. All BSI precautions must be taken when treating these illnesses. Mortality rates can range from 5% to 90%, depending on the strain of virus, the victim's age and health condition, and the availability of a modern health care system (Table 38-5 ▶).

TABLE 38-5 Characteristics of Viral Hemorrhagic Fevers

Dissemination	Direct contact with an infected person's body fluids. It can also be aerosolized for use in an attack.
Communicability	Moderate from person to person, or contaminated items.
Route of entry	Direct contact with an infected person's body fluids.
Signs and symptoms	Sudden onset of fever, weakness, muscle pain, headache, and sore throat. All of these symptoms are followed by vomiting and as the virus runs it course, internal and external bleeding.
Medical management	BSI precautions. There is no specific treatment for viral hemorrhagic fever. Patients should be provided supportive care (ABCs) and treatment for shock and hypotension, if present.

EMT-B Tips

Because humans are acceptable hosts and vectors for many viruses and bacteria, it is important for the EMT-B to use BSI precautions at all times. If you fail to use BSI precautions you may not only become a host for a virus but you may spread it as well. Remember, a virus moves from person to person to survive, and many infectious diseases present like common colds.

Bacteria

Unlike viruses, bacteria do not require a host to multiply and live. Bacteria are much more complex and larger than viruses and can grow up to 100 times larger than the largest virus. Bacteria contain all the cellular structures of a normal cell and are completely self-sufficient. Most importantly, bacterial infections can be fought with antibiotics.

Most bacterial infections will generally begin with flu-like symptoms, which make it quite difficult to identify whether the cause is a biological attack or a natural epidemic. Biological agents have been developed and used for centuries during times of war.

Inhalation and Cutaneous Anthrax (*Bacillus anthracis*)

Anthrax is a deadly bacteria that lays dormant in a spore (protective shell). When exposed to the optimal temperature and moisture, the germ will be released from the spore. The routes of entry for anthrax are inhalation, cutaneous, or gastrointestinal (from consuming food that contain spores) (Figure 38-13 ▶). The inhalational form or pulmonary anthrax is the most deadly and often presents as a severe cold. Pulmonary anthrax infections are associated with a 90% death rate if untreated. Antibiotics can be used to treat anthrax successfully. There is also a vaccine to prevent anthrax infections (Table 38-6 ▶).

Plague (Bubonic/Pneumonic)

Of all the infectious diseases known to humans, none has killed as many as the plague. The 14th century plague that ravaged Asia, the Middle East, and finally Europe (the Black Death) killed an estimated 33 to 42 million people. Later on, in the early 19th century, almost 20 million in India and China perished due to

You are the Provider   Part 5

While en route your patient states that she does not feel right. She is becoming disoriented. She suddenly begins to have a seizure. You quickly place padding around the patient to keep her from injuring herself. She becomes incontinent.

9. What is your next step in treatment?

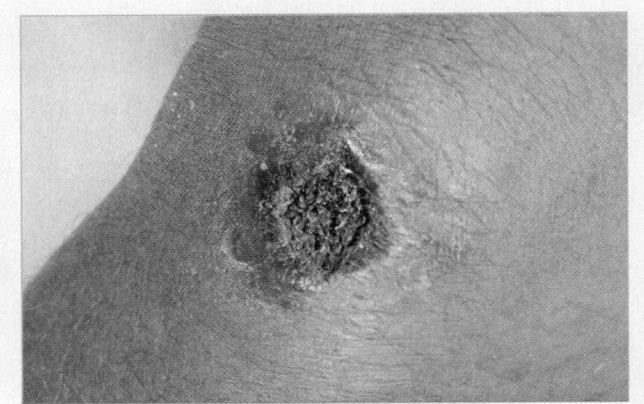

Figure 38-13 Cutaneous anthrax.

TABLE 38-6	Characteristics of Anthrax
Dissemination	Aerosol
Communicability	Only in the cutaneous form (rare)
Route of entry	Through inhalation of spore or skin contact with spore or direct contact with skin wound (cutaneous)
Signs and symptoms	Flu-like symptoms, fever, respiratory distress with tachycardia, shock, pulmonary edema, and respiratory failure after 3 to 5 days of flu-like symptoms
Medical management	Pulmonary/Inhalation: BSI precautions, oxygen, ventilatory support if in pulmonary edema or respiratory failure and transport. Cutaneous: BSI precautions, apply dry sterile dressing to prevent accidental contact with wound and fluids.

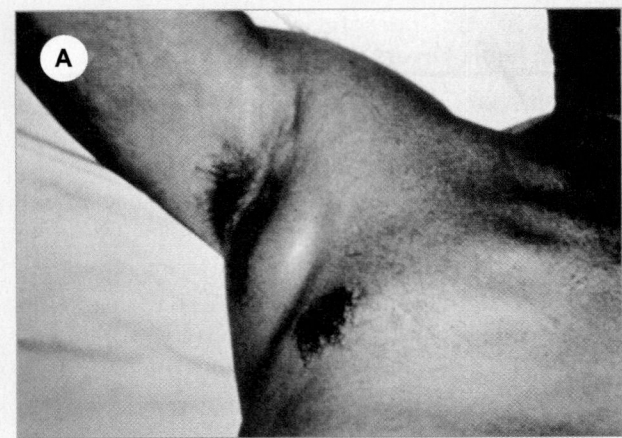

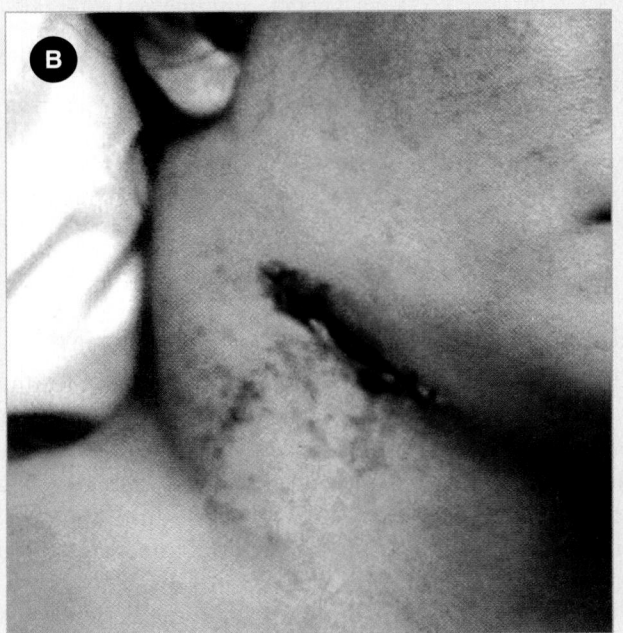

Figure 38-14 A. Plague buboe at lymph node under arm. **B.** Plague buboe at lymph node on neck.

plague. The plague's natural vectors are infected rodents and fleas. When a person is either bit by an infected flea or comes into contact with an infected rodent (or the waste of the rodent), the person can contract bubonic plague.

Bubonic plague infects the lymphatic system (a passive circulatory system in the body that bathes the tissues in lymph and works with the immune system). When this occurs, the patient's lymph nodes (area of the lymphatic system where infection-fighting cells are housed) become infected and grow. The glands of the nodes will grow large (up to the size of a tennis ball) and round, forming buboes (Figure 38-14 ▲). If left untreated, the infection may spread through the body, leading to sepsis and possibly death. This form of plague is not contagious and is not likely to be seen in a bioterrorist incident.

Pneumonic plague is a lung infection, also known as plague pneumonia, that results from inhalation of plague bacteria. This form of the disease is contagious and has a much higher death rate than the bubonic form. This form of plague therefore would be easier to disseminate (aerosolized), has a higher mortality, and is contagious (Table 38-7 ▶).

TABLE 38-7 Characteristics of Plague

Dissemination	Aerosol
Communicability	Bubonic: low, only from contact with fluid in buboe
	Pneumonic: high, from person to person
Route of entry	Ingestion, inhalation, or cutaneous
Signs and symptoms	Fever, headache, muscle pain and tenderness, pneumonia, shortness or breath, extreme lymph node pain and enlargement (bubonic)
Medical management	BSI, ABCs, provide oxygen, and transport

Figure 38-15 These seemingly harmless castor beans contain the key ingredient for ricin, one of the most potent toxins known to humans.

Neurotoxins

Neurotoxins are the most deadly substances known to humans. The strongest neurotoxin is 15,000 times more lethal than VX and 100,000 times more lethal than sarin. These toxins are produced from plants, marine animals, molds, and bacteria. The route of entry for these toxins is through ingestion, inhalation from aerosols, or injection. Unlike viruses and bacteria, neurotoxins are not contagious and have a faster onset of symptoms. Although these biological toxins have immense destructive potential, they have not been used successfully as a WMD.

Botulinum Toxin

The most potent neurotoxin is botulinum, which is produced by bacteria. When introduced into the body, this neurotoxin affects the nervous system's ability to function. Voluntary muscle control will diminish as the toxin spreads. Eventually the toxin will cause muscle paralysis that begins at the head and face and travels downward throughout the body. The patient's accessory muscles and diaphragm will become paralyzed, and the patient will go into respiratory arrest (Table 38-8 ▶).

Ricin

While not as deadly as botulinum, ricin is still five times more lethal than VX. This toxin is derived from mash that is left from the castor bean (Figure 38-15 ▶). When introduced into the body, ricin causes pulmonary edema and respiratory and circulatory failure leading to death (Table 38-9 ▶).

The clinical picture depends on the route of exposure. The toxin is quite stable and extremely toxic by many routes of exposure, including inhalation. Perhaps 1 to 3 mg of ricin can kill an adult, and the ingestion of one seed can probably kill a child.

Although all parts of the castor bean are actually poisonous, it is the seeds that are the most toxic. Castor bean ingestion causes a rapid onset of nausea, vomiting, abdominal cramps, and severe diarrhea, followed by vascular collapse. Death usually occurs on the third day in the absence of appropriate medical intervention.

TABLE 38-8 Characteristics of Botulinum Toxin

Dissemination	Aerosol or food supply sabotage or injection
Communicability	None
Route of entry	Ingestion or gastrointestinal
Signs and symptoms	Dry mouth, intestinal obstruction, urinary retention, constipation, nausea and vomiting, abnormal pupil dilation, blurred vision, double vision, drooping eyelids, difficulty swallowing, difficulty speaking, and respiratory failure due to paralysis
Medical management	ABCs, provide oxygen and transport. Ventilatory support may be needed due to paralysis of the respiratory muscles. A vaccine is available.

TABLE 38-9 Characteristics of Ricin

Dissemination	Aerosol or contamination of a food or water supply by sabotage
Communicability	None
Route of entry	Inhalation, ingestion, injection
Signs and symptoms	Inhaled: cough, difficulty breathing, chest tightness, nausea, muscle aches, pulmonary edema, and hypoxia Ingested: nausea and vomiting, internal bleeding, and death Injection: no signs except swelling at the injection site and death
Medical management	ABCs. No treatment or vaccine exists.

Ricin is least toxic by the oral route. This is probably a result of poor absorption in the gastrointestinal tract, some digestion in the gut, and, possibly, some expulsion of the agent as caused by the rapid onset of vomiting. Ingestion causes local hemorrhage and necrosis of the liver, spleen, kidney, and gastrointestinal tract. Signs and symptoms appear 4 to 8 hours after exposure.

Signs and symptoms of ricin ingestion are as follows:

- Fever
- Chills
- Headache
- Muscle aches
- Nausea
- Vomiting
- Diarrhea
- Severe abdominal cramping
- Dehydration
- Gastrointestinal bleeding
- Necrosis of liver, spleen, kidneys, and gastrointestinal tract

Inhalation of ricin causes nonspecific weakness, cough, fever, hypothermia, and hypotension. Symptoms occur about 4 to 8 hours after inhalation, depending on the inhaled dose. The onset of profuse sweating some hours later signifies the termination of the symptoms.

Signs and symptoms of ricin inhalation are as follows:

- Fever
- Chills
- Nausea
- Local irritation of eyes, nose, and throat
- Profuse sweating
- Headache
- Muscle aches
- Nonproductive cough
- Chest pain
- Dyspnea
- Pulmonary edema
- Severe lung inflammation
- Cyanosis
- Convulsions
- Respiratory failure

Treatment is supportive and includes both respiratory support and cardiovascular support as needed. Early intubation, ventilation, and positive end expiratory pressure, combined with treatment of pulmonary edema, are appropriate. Intravenous fluids and electrolyte replacement are useful for treating the dehydration caused by profound vomiting and diarrhea. Table 38-10 ▶ summarizes the biological agents.

You are the Provider   Part 6

Due to the sudden change in your patient's condition you make contact with dispatch and request ALS backup. The nearest ALS unit is 15 minutes away. You decide to rendezvous with them en route to the hospital. Your patient stops seizing and is in a postictal state. You check for responsiveness. She is unconscious and responsive only to verbal stimuli. You insert an oropharyngeal airway (OPA) to ensure that she maintains a patent airway. She is breathing at a rate of 16 breaths/min. You then place her in the recovery position.

10. Who should you notify about the sudden change in your patient's condition?

TABLE 38-10 Biological Agents

Disease	Transmission Person to Person	Incubation Period	Duration of Illness	Lethality (approximate case fatality rates)
Inhalation anthrax	No	1 to 6 d	3 to 5 d (usually fatal if untreated)	High
Pneumonic plague	High	2 to 3 d	1 to 6 d (usually fatal)	High unless treated within 12 to 24 h
Smallpox	High	7 to 17 d (average 12 d)	4 wk	High to moderate
Viral hemorrhagic fevers	Moderate	4 to 21 d	Death between 7 to 16 d	High to moderate, depending on type of fever
Botulinum	No	1 to 5 d	Death in 24 to 72 h; lasts months if patient does not die	High without respiratory support
Ricin	No	18 to 24 h	Days; death within 10 to 12 d for ingestion	High

Other EMT-B Roles During a Biological Event

Syndromic Surveillance

Syndromic surveillance is the monitoring, usually by local or state health departments, of patients presenting to emergency departments and alternative care facilities, the recording of EMS call volume, and monitoring the use of over-the-counter medications. Patients with signs and symptoms that resemble influenza are particularly important. Local and state health departments monitor for an unusual influx of patients with these symptoms in hopes of catching an outbreak early. The EMS role in syndromic surveillance is a small one, yet valuable in the overall tracking of a biological terrorist event or infectious disease outbreak. Quality assurance and dispatch operations need to be aware of an unusual number of calls from patients with "unexplainable flu" coming from a particular region or community.

Points of Distribution (Strategic National Stockpile)

Points of Distribution (PODs) are strategically placed facilities that have been pre-established for the mass distribution of antibiotics, antidotes, vaccinations, and other medications and supplies. These medications may be delivered in large containers known as "push packs" by the Centers for Disease Control and Prevention National Pharmaceutical Stockpile (Figure 38-16 ▶). These containers have a delivery time of 12 hours anywhere in the country and contain antibiotics, chemical antidotes, antitoxins, life-support medications, IV administration, airway maintenance supplies, and medical/surgical items. In some regions, local and state municipalities have started to stockpile their own supplies to reduce the time delay.

EMT-Bs, EMT-Is, and paramedics may be called on to assist in the delivery of the medications to the public (depending on local emergency management planning). The EMT-B's role may include triage, treatment of seriously ill patients, and patient transport to the hospital. Most plans for PODs include at least one ambulance on standby for the transport of seriously ill patients.

Figure 38-16 The Centers for Disease Control and Prevention Strategic National Stockpile can deliver one of many push packs to any location in the country within 12 hours of an emergency.

Radiological/Nuclear Devices

What Is Radiation?

Ionizing radiation is energy that is emitted in the form of rays, or particles. This energy can be found in radioactive material, such as rocks and metals. Radioactive material is any material that emits radiation. This material is unstable, and attempts to stabilize itself by changing its structure is a natural process called decay. As the substance decays, it gives off radiation, until it stabilizes. The process of radioactive decay can take from as little as minutes to billions of years; meanwhile, the substance remains radioactive.

The energy that is emitted from a strong radiological source is either alpha, beta, gamma (X-rays), or neutron radiation. Alpha is the least harmful penetrating type of radiation and cannot travel fast or through most objects. In fact, a sheet of paper or the body's skin easily stops it. Beta radiation is slightly more penetrating than alpha, and requires a layer of clothing stop it. Gamma or X-rays are far faster and stronger than alpha and beta rays. These rays easily penetrate through the human body and require either several inches of lead or concrete to prevent penetration. Neutron energy is the fastest moving and most powerful form of radiation. Neutrons easily penetrate through lead and require several feet of concrete to stop them (Figure 38-17 ▶).

Sources of Radiological Material

There are thousands of radioactive materials found on the earth. These materials are generally used for purposes that benefit humankind, such as medicine, killing germs in food (irradiating), and construction work. Once radiological material has been used for its purpose, the material remaining is called radiological waste. Radiological waste remains radioactive, but has no more usefulness. These materials can be found at:

- Hospitals
- Colleges and universities
- Chemical and industrial sites

Not all radioactive material is tightly guarded, and the waste is often not guarded. This makes use of radioactive material and substances appealing to terrorists.

Radiological Dispersal Devices (RDD)

A radiological dispersal device (RDD) is any container that is designed to disperse radioactive material. This would generally require the use of a bomb, hence the nickname "dirty bomb". A dirty bomb would carry the potential to injure victims with not only the radioactive material but the explosive material used to deliver it. Just the thought of an RDD creates fear in a population, and so the ultimate goal of the terrorist—fear—is accomplished. In reality, however, the destructive capability of a dirty bomb is limited to the explosives that are attached to it. Therefore, if the explosive is sufficient to kill 10 persons without radioactive material, it will also kill 10 persons with the radioactive material added. There may be long-term injuries and illness associated with the use of an RDD, yet not much more than the bomb by itself would create. In short, the dirty bomb is an ineffective WMD.

Nuclear Energy

Nuclear energy is artificially made by altering (splitting) radioactive atoms. The result is an immense amount of energy that usually takes the form of heat. Nuclear material is used in medicine, weapons, naval vessels, and power plants. Nuclear material gives off all forms of radiation including neutrons (the most deadly type). Like radioactive material, when nuclear material is no longer useful it becomes waste that is still radioactive.

Nuclear Weapons

The destructive energy of a nuclear explosion is unlike any other weapon in the world. That is why nuclear weapons are kept only in secure facilities throughout the world. There are nations that have ties to terrorists and that have actively attempted to build nuclear weapons. Yet the ability of these nations to deliver a nuclear weapon, such as a missile or bomb, is as of yet, incomplete. There is also the deterrent of complete mutual annihilation. Therefore, the likelihood of a nuclear attack is extremely remote.

Unfortunately, however, due to the collapse of the former Soviet Union, the whereabouts of many small nuclear devices is unknown. These small suitcase-sized nuclear weapons are called Special Atomic Demolition Munitions (SADM). The SADM, or "suit-case nuke," was designed to destroy individual targets, such as important buildings, bridges, tunnels, or large ships. The estimate is that perhaps as many as 80 are missing as of 1998. No other information or updates on the whereabouts of these devices have been made public.

Symptomatology

The effects of radiation exposure will vary depending on the amount of radiation that a person receives and the route of entry. Radiation can be introduced into

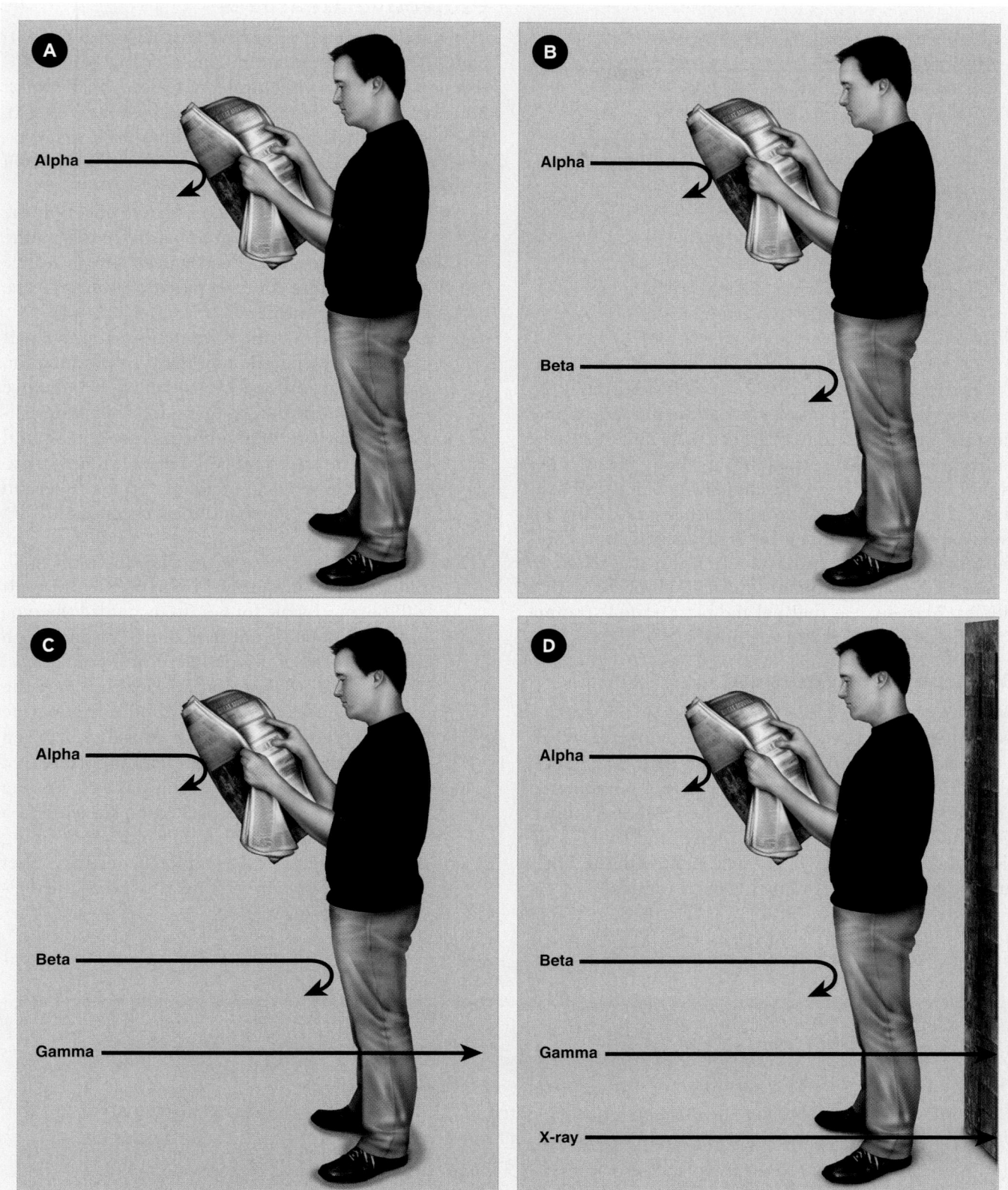

Figure 38-17 The penetrating potential of four different types of radiation. **A.** Alpha. **B.** Beta. **C.** Gamma. **D.** X-ray.

TABLE 38-11 Common Signs of Acute Radiation Sickness

Low exposure	Nausea, vomiting, diarrhea
Moderate exposure	First-degree burns, hair loss, depletion of the immune system (death of white blood cells), and cancer
Severe exposure	Second- and third-degree burns, cancer, and death

the body by all routes of entry as well as through the body (irradiation). The patient can inhale radioactive dust from nuclear fallout or from a dirty bomb, or have radioactive liquid absorbed into the body through the skin. Once in the body, the radiation source will irradiate the person from within rather than from an external source (such as x-ray equipment). Some common signs of acute radiation sickness are listed in (Table 38-11 ▲). Additional injuries will occur with a nuclear blast such as thermal and blast trauma, trauma from flying objects, and eye injuries.

Medical Management

Being exposed to a radiation source does not make a patient contaminated or radioactive. However, when patients have a radioactive source on their body (such as debris from a dirty bomb), they are contaminated and must be initially cared for by a HazMat responder. Once the patient is decontaminated and there is no threat to you, you may begin treatment with the ABCs and treat the patient for any burns or trauma.

Protective Measures

There are no suits or protective gear designed to completely shield from radiation. Those who work in high-risk areas do wear some protection (lead-lined suits); however this equipment is not available to the EMT-B. The best way to protect yourself from the effects of radiation is to use time, distance, and shield yourself in Level C protection from the source.

- **Time.** Radiation has a cumulative effect on the body. The less time that you are exposed to the source, the less the effects will be. If you realize that the patient is near a radiation source, leave the area immediately.
- **Distance.** Radiation is limited as to how far it can travel. Depending on the type of radiation, often moving only a few feet is enough to remove you from immediate danger. Alpha radiation cannot travel more than a few inches. You should take this into account when responding to a nuclear or radiological incident and make certain that responders are stationed far enough from the incident.
- **Shielding.** As discussed earlier, the path of all radiation can be stopped by a specific object. It will be impossible for you to recognize the type of radiation being emitted, or even from which direction it is coming. Therefore, you should always assume that you are dealing with the strongest form of radiation and use concrete shielding (such as buildings or walls) between yourself and the incident. The importance of shielding cannot be overemphasized. In one atomic test, a car was parked on the side of a house, opposite the direction of the oncoming blast. The house was completely destroyed, yet the car that was directly next to it sustained almost no damage.

You are the Provider   Summary

You contact the receiving facility and request a patch to the ALS unit. Once you have met the ALS unit you assist in transferring the patient to the ALS unit and give a full patient report to the paramedic.

Your number one priority while responding to any call is your own personal safety. Given the current threat of terrorism in the world today we as EMS providers must constantly be aware of the various types of terrorist attacks. An attack can affect your day-to-day operation, and in turn affect the way you approach and treat your patients.

Prep Kit

Ready for Review

- As a result of the increase in terrorist activity, it is possible that the EMT-B could witness a terrorist event. You must be mentally and physically prepared for the possibility of a terrorist event.
- The use of weapons of mass destruction or mass casualty further complicates the management of the terrorist incident. Be aware of your surroundings at all times. The best form of protection from a WMD agent is to avoid contact with the agent.
- Types of groups that tend to use terrorism include violent religious groups/doomsday cults, extremist political groups, technology terrorists, and single-issue groups
- A WMD is any agent designed to bring about mass death, casualties, and/or massive damage to property and infrastructure (bridges, tunnels, airports, and seaports). These can be nuclear, chemical, biological, and explosive weapons.
- Chemical agents are manmade substances that can have devastating effects on living organisms. They can be produced in liquid, powder, or vapor form, depending on the desired route of exposure and dissemination technique. These agents consist of vesicants, respiratory, nerve, and metabolic agents.
- Biological agents are organisms that cause disease. They are generally found in nature and can be weaponized to maximize the number of people exposed to the germ. These types of agents include viruses, bacteria, and toxins.
- Nuclear or radiological weapons can create a massive amount of destruction. This type of weapon includes radiological dispersal devices (RDDs), also known as dirty bombs.

- Be aware of the current threat level issued by the federal government through the Department of Homeland Security (DHS). This threat level can be severe, high, elevated, guarded, or low.
- On the basis of the current threat level, take appropriate actions and precautions. Be aware of established policies that your organization may have regarding the current threat level.
- Indicators that may give you clues as to whether the emergency is the result of an attack include the type of location, type of call, number of patients, victims' statements, and preincident indicators.
- If you suspect that a terrorist or WMD event has occurred, ensure that the scene is safe. If you have any doubt that it may not be safe, do not enter. Wait for assistance.
- Notification of the dispatcher is essential. Inform dispatch of the nature of the event, any additional resources that may be required, the estimated number of patients, and the upwind route of approach or optimal route of approach.
- Establish a staging area, where other units will converge. Be mindful of access and exit routes.
- The first arriving provider on the scene must begin to sort out the chaos, and define his responsibilities under the Incident Command System (ICS).
- If the ICS is already in place, the EMT-B should immediately seek out the medical staging officer to receive his or her assignment.
- Terrorists may set secondary devices to explode after the initial bomb, to injure responders and secure media coverage. Constantly assess and reassess the scene for safety.
- Persistent or nonvolatile agents can remain on a surface for long periods of time. A highly persistent agent can remain in the environment for weeks to months.
- Nonpersistent or volatile agents evaporate relatively fast when left on a surface in the optimal temperature range. A highly volatile agent will turn from liquid to gas (evaporate) within minutes to seconds.
- Route of exposure is how the agent most effectively enters the body.
- A vesicant is an agent that enters through the skin and causes burn-like blisters on the victim's skin, as well as in the respiratory tract.
- Vesicant agent treatment includes decontamination first, then the ABCs.

Technology

- Interactivities
- Vocabulary Explorer
- Anatomy Review
- Web Links
- Online Review Manual

www.EMTB.com

Prep Kit continued...

- Pulmonary agents are gases that cause immediate harm by damaging the lung tissue.
- Pulmonary agent treatment is to remove the patient from the contaminated atmosphere. This should be done by trained personnel in the proper PPE. Then begin aggressive management using the ABCs and gaining IV access. Do not allow the patient to be active.
- Nerve agents are among the most deadly chemicals developed and can cause cardiac arrest within seconds to minutes of exposure.
- Securing the ABCs is the best and most important treatment that the EMT-B can render for patients exposed to nerve agents. Patients who will not stop seizing will require administration of nerve agent antidote kits in addition to support of the ABCs.
- Metabolic agents, or cyanides, affect the body's ability to utilize oxygen and are commonly found in many industrial settings.
- Before treatment begins, the patient exposed to a metabolic agent must be removed from the source of exposure by trained personnel in the proper PPE, all of the patient's clothes must be removed, and the patient must be decontaminated. Then support the patient's ABCs and request advance life support immediately.
- Biological agents include viruses such as smallpox and viral hemorrhagic fevers; bacteria such as anthrax and plague; and neurotoxins such as botulinum toxin and ricin.
- Paramedics may be called upon to assist in the delivery of the medications to the public. The EMT-B's role may include triage, treatment of seriously ill patients, and patient transport to the hospital.
- Ionizing radiation is energy that can enter the human body and cause damage.
- Treatment for radiation exposure should begin with making sure that the patient is not contaminated. If the patient is contaminated, he or she must be initially cared for by a HazMat responder.
- There are no suits or protective gear designed to completely shield from radiation. Protect yourself by leaving an area where a radiation source is present, staying as far away as possible, and using concrete shielding when possible.

Vital Vocabulary

alpha Type of energy that is emitted from a strong radiological source; it is the least harmful penetrating type of radiation and cannot travel fast or through most objects.

anthrax A deadly bacteria (*Bacillus anthracis*) that lays dormant in a spore (protective shell); the germ is released from the spore when exposed to the optimal temperature and moisture. The route of entry is inhalation, cutaneous, or gastrointestinal (from consuming food that contains spores).

bacteria Microorganisms that reproduce by binary fission. These single-cell creatures reproduce rapidly. Some can form spores (encysted variants) when environmental conditions are harsh.

beta Type of energy that is emitted from a strong radiological source; is slightly more penetrating than alpha, and requires a layer of clothing stop it.

botulinum Produced by bacteria, this is a very potent neurotoxin. When introduced into the body, this neurotoxin affects the nervous system's ability to function and causes botulism.

buboes Enlarged lymph nodes (up to the size of tennis balls) that were characteristic of people infected with the bubonic plague.

bubonic plague An epidemic that spread throughout Europe in the Middle Ages, causing over 25 million deaths, also called the Black Death, transmitted by infected fleas and characterized by acute malaise, fever, and the formation of tender, enlarged, inflamed lymph nodes that appear as lesions, called buboes.

chlorine (CL) The first chemical agent ever used in warfare. It has a distinct odor of bleach, and creates a green haze when released as a gas. Initially it produces upper airway irritation and a choking sensation.

communicability Describes how easily a disease spreads from one human to another human.

contact hazard A hazardous agent that gives off very little or no vapors; the skin is the primary route for this type of chemical to enter the body; also called a skin hazard.

contagious A person infected with a disease that is highly communicable.

covert Act in which the public safety community generally has no prior knowledge of the time, location, or nature of the attack.

www.EMTB.com

cross-contamination Occurs when a person is contaminated by an agent as a result of coming into contact with another contaminated person.

cyanide Agent that affects the body's ability to use oxygen. It is a colorless gas that has an odor similar to almonds. The effects begin on the cellular level and are very rapidly seen at the organ system level.

decay A natural process in which a material that is unstable attempts to stabilize itself by changing its structure.

dirty bomb Name given to a bomb that is used as a radiological dispersal device (RDD).

disease vector An animal that spreads a disease, once infected, to another animal.

dissemination The means with which a terrorist will spread a disease, for example, by poisoning of the water supply, or aerosolizing the agent into the air or ventilation system of a building.

domestic terrorism Terrorism carried out by native citizens of the country being attacked.

G agents Early nerve agents which were developed by German scientists in the period after WWI and into WWII. There are three such agents: sarin, soman, and tabun.

gamma (X-rays) Type of energy that is emitted from a strong radiological source that is far faster and stronger than alpha and beta rays. These rays easily penetrate through the human body and require either several inches of lead or concrete to prevent penetration.

incubation Describes the period of time from a person being exposed to a disease to the time when symptoms begin.

international terrorism Terrorism that is carried out by those not of the host's country; also known as cross-border terrorism.

ionizing radiation Energy that is emitted in the form of rays, or particles.

LD$_{50}$ The amount of an agent or substance that will kill 50% of people who are exposed to this level.

Lewisite (L) A blistering agent that has a rapid onset of symptoms and produces immediate intense pain and discomfort on contact.

lymph nodes Area of the lymphatic system where infection-fighting cells are housed.

lymphatic system A passive circulatory system that transports a plasma-like liquid called lymph, a thin fluid that bathes the tissues of the body.

MARK 1 A nerve agent antidote kit containing two auto-injector medications, atropine and 2-PAM chloride (pralidoxime chloride); also known as a Nerve Agent Antidote Kit (NAAK).

miosis Bilateral pinpoint constricted pupils.

mutagen Substance that mutates, damages, and changes the structures of DNA in the body's cells.

NAAK A nerve agent antidote kit containing two autoinjector medications, atropine and 2-PAM chloride (pralidoxime chloride); also known as a MARK 1 kit.

nerve agents A class of chemical called organophosphates; they function by blocking an essential enzyme in the nervous system, which causes the body's organs to become overstimulated and burn out.

neurotoxins Biological agents that are the most deadly substances known to humans; they include botulinum toxin and ricin.

neutron radiation Type of energy that is emitted from a strong radiological source; neutron energy is the fastest moving and most powerful form of radiation. Neutrons easily penetrate through lead, and require several feet of concrete to stop them.

off-gassing The emitting of an agent after exposure, for example from a person's clothes that have been exposed to the agent.

persistency Term used to describe how long a chemical agent will stay on a surface before it evaporates.

phosgene A pulmonary agent that is a product of combustion, such as might be produced in a fire at a textile factory or house, or from metalwork or burning Freon. Phosgene is a very potent agent that has a delayed onset of symptoms, usually hours.

phosgene oxime (CX) A blistering agent that has a rapid onset of symptoms and produces immediate intense pain and discomfort on contact.

pneumonic plague A lung infection, also known as plague pneumonia, that is the result of inhalation of plague bacteria.

points of distribution (PODs) Strategically placed facilities that have been pre-established for the mass distribution of antibiotics, antidotes, vaccinations, with other medications and supplies.

radioactive material Any material that emits radiation.

Prep Kit continued...

radiological dispersal device (RDD) Any container that is designed to disperse radioactive material.

ricin Neurotoxin derived from mash that is left from the castor bean; causes pulmonary edema and respiratory and circulatory failure, leading to death.

route of exposure Manner by which a toxic substance enters the body.

sarin (GB) A nerve agent that is one of the G agents; a highly volatile colorless and odorless liquid that turns from liquid to gas within seconds to minutes at room temperature.

secondary device Additional explosives used by terrorists, which are set to explode after the initial bomb.

smallpox A highly contagious disease; it is most contagious when blisters begin to form.

soman (GD) A nerve agent that is one of the G agents; twice as persistent as sarin and five times as lethal; it has a fruity odor, as a result of the type of alcohol used in the agent, and is both a contact and inhalation hazard that can enter the body through skin absorption and through the respiratory tract.

Special Atomic Demolition Munitions (SADM) Small suitcase-sized nuclear weapons that were designed to destroy individual targets, such as important buildings, bridges, tunnels, or large ships.

state-sponsored terrorism Terrorism that is funded and/or supported by nations that hold close ties with terrorist groups.

sulfur mustard (H) A vesicant; it is a brownish-yellowish oily substance that is generally considered very persistent; has the distinct smell of garlic or mustard and, when released, is quickly absorbed into the skin and/or mucous membranes and begins an irreversible process of damaging the cells.

syndromic surveillance The monitoring, usually by local or state health departments, of patients presenting to emergency departments and alternative care facilities, the recording of EMS call volume, and the use of over-the-counter medications.

tabun (GA) A nerve agent that is one of the G agents; is 36 times more persistent than sarin and approximately half as lethal; has a fruity smell and is unique because the components used to manufacture the agent are easy to acquire and the agent is easy to manufacture.

V agent (VX) One of the G agents; it is a clear, oily agent that has no odor and looks like baby oil; over 100 times more lethal than sarin and is extremely persistent.

vapor hazard An agent that enters the body through the respiratory tract.

vesicants Blister agents; the primary route of entry for vesicants is through the skin.

viral hemorrhagic fevers (VHF) A group of diseases that include the Ebola, Rift Valley, and Yellow Fever viruses among others. This group of viruses causes the blood in the body to seep out from the tissues and blood vessels

viruses Germs that require a living host to multiply and survive.

volatility Term used to describe how long a chemical agent will stay on a surface before it evaporates.

weapon of mass casualty (WMC) Any agent designed to bring about mass death, casualties, and/or massive damage to property and infrastructure (bridges, tunnels, airports, and seaports); also known as a weapon of mass destruction (WMD).

weapon of mass destruction (WMD) Any agent designed to bring about mass death, casualties, and/or massive damage to property and infrastructure (bridges, tunnels, airports, and seaports); also known as a weapon of mass casualty (WMC).

weaponization The creation of a weapon from a biological agent generally found in nature and that causes disease; the agent is cultivated, synthesized, and/or mutated to maximize the target population's exposure to the germ.

Points to Ponder

You are responding to a WMD incident where a primary explosion has disseminated chemical agents at the City Bank. You are told by Incident Command (IC) to stage about two blocks from the incident location and await the specialty WMD Hazardous Materials Team. The staging area is near a park and IC wants triage set up in the park. There are about 40 patients confirmed by IC. A total of six ambulances within the city are responding.

What are your concerns with both the location of the triage area and the number of ambulances that are responding? What do you want to know about the chemical agent?

Issues: Scene Safety, Staging Location, Incident Command, Secondary Devices.

www.EMTB.com

Assessment in Action

Events over the past few years have shown that terrorists, foreign and domestic, are willing to attack American interests at home and abroad. Terrorists now have access to a broad array of lethal materials worldwide and can strike a specific target at any given time. Terrorists are no longer limited to conventional weapons.

1. As an EMT-B you must be familiar with the non-conventional agents that may be used in a WMD attack. All of the following are nonconventional weapons except:

   A. chemical.
   B. nuclear.
   C. biological.
   D. explosives.

2. A weapon of mass destruction is any agent that will bring about:

   A. mass casualty.
   B. mass death.
   C. massive damage to infrastructure.
   D. all of the above.

3. Terrorism that is carried out by individuals or groups not of the host country is known as:

   A. domestic terrorism.
   B. doomsday terrorism.
   C. international terrorism.
   D. Al Qaeda.

4. The Department of Homeland Security has posted the threat level to be yellow. What threat level does this color represent?

   A. Low
   B. Elevated
   C. High
   D. Severe

5. Chemical agents are manmade substances that can have devastating effects on living organisms. All of the following are agents that can be used for chemical warfare except:

   A. nerve agents.
   B. choking agents.
   C. bacterial agents.
   D. blood agents.

6. Time, distance, and shielding are the three most important factors in staying safe when dealing with what type of WMD?

   A. Chemical weapon
   B. Radiological weapon
   C. Biological weapon
   D. Bacterial weapon

7. Chemical agents can be produced in which of the following forms?

   A. Liquid
   B. Powder
   C. Vapor
   D. All of the above

Challenging Questions

You have responded to a train station where there was a reported small explosion and now a number of people are complaining of difficulty breathing and nausea per dispatch report. As you walk into the station you observe that two patients are unconscious and seizing, while numerous others are pleading with you to help them.

8. What type of event do you suspect?

9. What concerns do you have for your safety?

www.EMTB.com

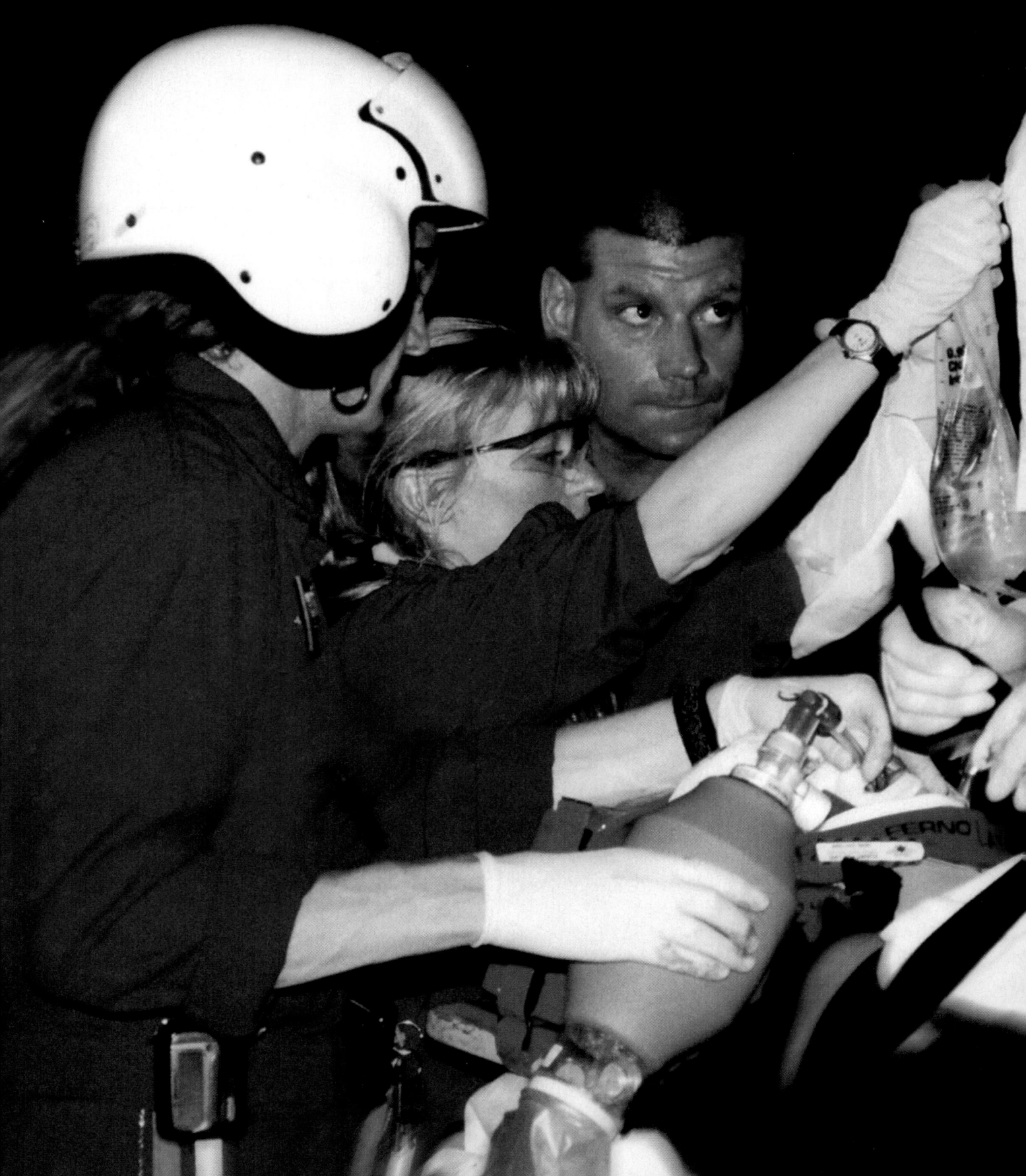

ALS Techniques

Advanced Airway Management

Objectives

Cognitive

8-1.1   Identify and describe the airway anatomy in the infant, child, and the adult. (p 1132)

8-1.3   Explain the pathophysiology of airway compromise. (p 1133)

8-1.4   Describe the proper use of airway adjuncts. (p 1134)

8-1.5   Review the use of oxygen therapy in airway management. (p 1144)

8-1.6   Describe the indications, contraindications, and technique for insertion of nasal gastric tubes. (p 1135)

8-1.7   Describe how to perform the Sellick maneuver (cricoid pressure). (p 1135)

8-1.8   Describe the indications for advanced airway management. (p 1132)

8-1.9   List the equipment required for orotracheal intubation. (p 1138)

8-1.10   Describe the proper use of the curved blade for orotracheal intubation. (p 1139)

8-1.11   Describe the proper use of the straight blade for orotracheal intubation. (p 1139)

8-1.12   State the reasons for and proper use of the stylet in orotracheal intubation. (p 1140, 1142)

8-1.13   Describe the methods of choosing the appropriate size endotracheal tube in an adult patient. (p 1140)

8-1.14   State the formula for sizing an infant or child endotracheal tube. (p 1141)

8-1.15   List complications associated with advanced airway management. (p 1144, 1150)

8-1.17   Describe the skill of orotracheal intubation in the adult patient. (p 1144)

8-1.18   Describe the skill of orotracheal intubation in the infant and child patient. (p 1144)

8-1.19   Describe the skill of confirming endotracheal tube placement in the adult, infant, and child patient. (p 1145)

8-1.20   State the consequences of and the need to recognize unintentional esophageal intubation. (p 1145)

8-1.21   Describe the skill of securing the endotracheal tube in the adult, infant, and child patient. (p 1147)

Affective

8-1.22   Recognize and respect the feelings of the patient and family during advanced airway procedures. (p 1132)

8-1.23   Explain the value of performing advanced airway procedures. (p 1132)

8-1.24   Defend the need for the EMT-B to perform advanced airway procedures. (p 1132)

8-1.25   Explain the rationale for the use of a stylet. (p 1138)

8-1.26   Explain the rationale for having a suction unit immediately available during intubation attempts. (p 1151)

8-1.27   Explain the rationale for confirming breath sounds. (p 1145)

8-1.28   Explain the rationale for securing the endotracheal tube. (p 1147)

Psychomotor

8-1.29   Demonstrate how to perform the Sellick maneuver (cricoid pressure). (p 1136)

8-1.30   Demonstrate the skill of orotracheal intubation in the adult patient. (p 1144)

8-1.31   Demonstrate the skill of orotracheal intubation in the infant and child patient. (p 1144)

8-1.32   Demonstrate the skill of confirming endotracheal tube placement in the adult patient. (p 1145)

8-1.33   Demonstrate the skill of confirming endotracheal tube placement in the infant and child patient. (p 1145)

8-1.34   Demonstrate the skill of securing the endotracheal tube in the adult patient. (p 1147)

You are the Provider

You are called to respond to the Peterson Memorial Auditorium for a patient down. As you arrive at the scene and approach the patient, you see a bystander performing rescue breathing. The patient, a middle-aged man, has a carotid pulse.

   This scenario addresses a common medical situation in which you must obtain and maintain a patent airway in an apneic patient with a pulse in order to prevent full cardiac arrest. This chapter will describe the knowledge and skills needed to manage calls like this and will also help you answer the following questions:

1. Why is it important to secure a patent airway with basic means before performing advanced techniques such as endotracheal intubation?
2. What advantages does endotracheal intubation have over simple bag-valve-mask (BVM) ventilation?

Advanced Airway Management

The single most important manipulative skill you will use as an EMT-B is establishing and maintaining a patient's airway. While the obviously broken leg or amputated finger may be eye-catching, the airway must be secured immediately, or the patient will die. The vast majority of conscious patients with an intact gag reflex can maintain their own airway. Therefore, in managing the conscious patient, you may need only to provide oxygen and monitor the patient closely for any changes. Patients whose consciousness is altered may require an oropharyngeal or nasopharyngeal airway and suctioning. However, patients who are unresponsive and not breathing on their own will fare better with advanced airway techniques. The purpose of advanced airway management is to provide better airway protection and improve ventilation by using a tube to create a direct channel to the trachea. Endotracheal intubation is a difficult skill to master and requires additional training for the EMT-B. Blind intubation using a lighted stylet is another technique for establishing endotracheal (ET) tube placement. Additional options include using a multilumen airway or a laryngeal mask airway. All of these techniques require additional training, appropriate approval for their use, and medical oversight. During advanced procedures, as with basic procedures, you need to recognize and respect the feelings of the patient and family.

Technology

- Interactivities
- Vocabulary Explorer
- Anatomy Review
- Web Links
- Online Review Manual

www.EMTB.com

The chapter begins with a brief review of the anatomy and physiology of the respiratory system, followed by a discussion of gastric tube placement. The chapter concludes with a section on the use of a lighted stylet, multilumen airway devices, and the laryngeal mask airway.

Anatomy and Physiology of the Airway

As you learned earlier, the respiratory system consists of all the structures in the body that are used for breathing (Figure 39-1). The upper airway begins with the nose, mouth, and throat (pharynx). The lower airway includes the larynx (vocal cords), trachea, bronchi, and lungs. The epiglottis is a leaf-shaped structure located at the glottic opening (covering the larynx) that prevents food and liquid from entering the lower airway during swallowing. The bronchi and other air passages branch off from the trachea, extending into each lung, subdividing into bronchioles (smaller passages) down to the alveoli, where the exchange of oxygen and carbon dioxide occurs.

The mechanical process of breathing occurs through the use of the diaphragm and intercostal muscles (muscles between the ribs). The diaphragm is a thin, dome-shaped muscle that separates the thoracic cavity from the abdominal cavity. The diaphragm and intercostal muscles contract during the active phase of breathing (inhalation), increasing the size of the chest cavity. Contraction of the diaphragm pulls the chest cavity down; contraction of the intercostal muscles pulls the rib cage up and out. The increased size of the chest cavity allows air to flow into the lungs. During the passive phase of breathing (exhalation), air flows out of the lungs. The diaphragm and intercostal muscles relax, and the size of the chest cavity decreases. The diaphragm moves up, and the ribs move down.

The respiratory system delivers oxygen to the body and removes carbon dioxide, a process that takes place on two levels: the alveolar-capillary exchange and the capillary-cellular exchange (Figure 39-2).

The alveolar-capillary exchange works in the following way:

1. **Air breathed in during inhalation** travels through the airways to the alveoli.
2. **As this oxygen-rich air** enters the alveoli, oxygen-poor blood is circulated through the capillaries around each alveolus.

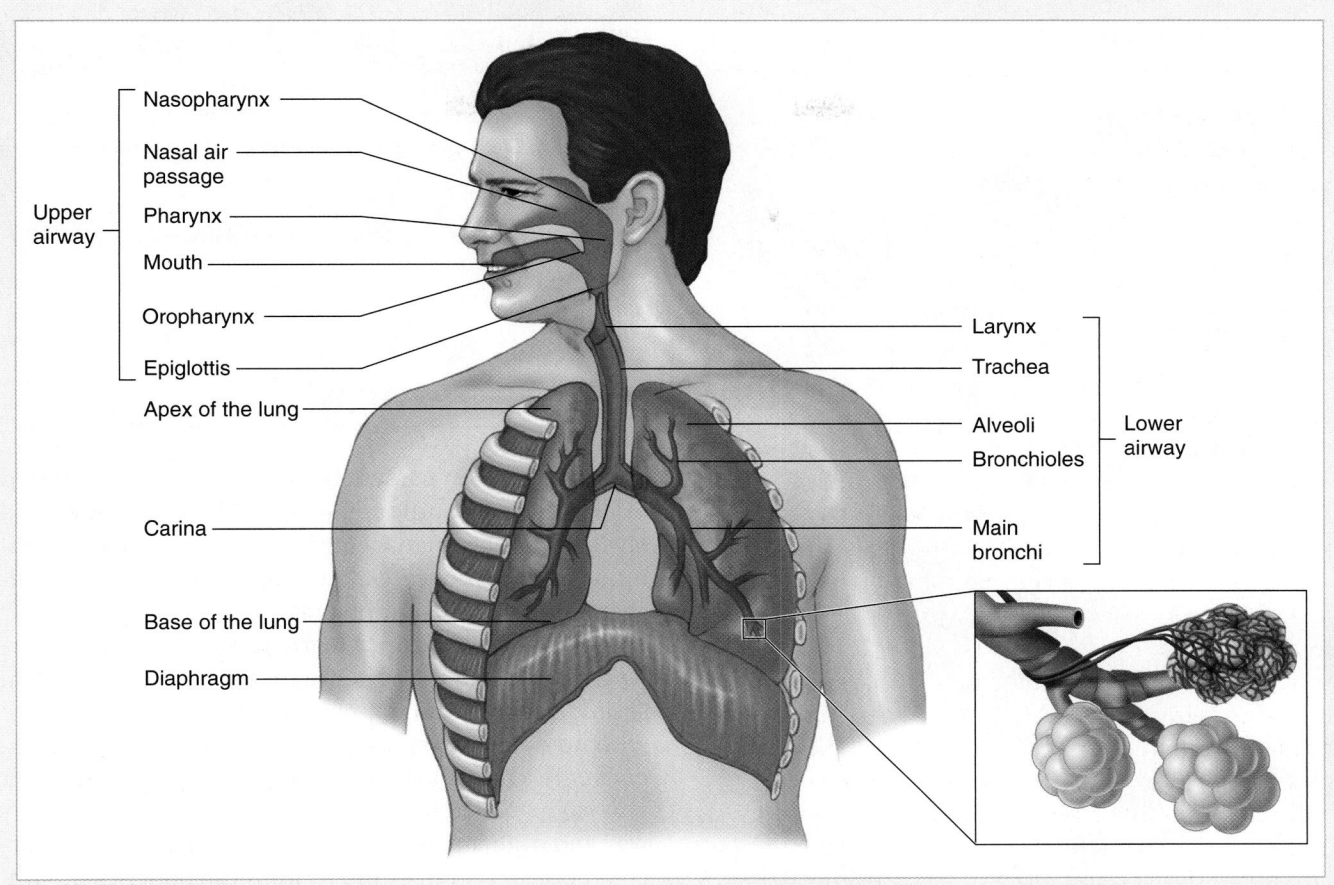

Figure 39-1 The upper and lower airways and other structures in the body that are used for breathing.

3. Oxygen in the alveoli crosses over into the bloodstream. Carbon dioxide in the blood from the capillaries crosses over into the alveoli, creating a shift of oxygen and carbon dioxide.

The capillary-cellular exchange occurs throughout the body's cells. Cells give up carbon dioxide into the capillaries, and capillaries give up oxygen to the cells.

Each living cell in the body requires a regular supply of oxygen; some cells, such as those in the heart, brain, and nervous system, need a constant supply of oxygen to survive. Cells in the heart will be damaged if the oxygen supply is interrupted for more than a few minutes. After 4 to 6 minutes without oxygen, cells in the brain and nervous system begin to die. Dead brain

You are the Provider · Part 2

Your partner inserts an oral airway and continues artificial ventilation with a BVM device attached to 100% oxygen. The patient's wife tells you that her husband, who has a history of congestive heart failure, suddenly began "gasping for air" and then collapsed. You apply a pulse oximeter to the patient. Your partner is having difficulty maintaining a mask-to-face seal with the BVM. You glance at the pulse oximeter, and it reads 85%.

3. What are the complications associated with inadequate artificial ventilation?
4. What should be your next step in managing this patient's airway?

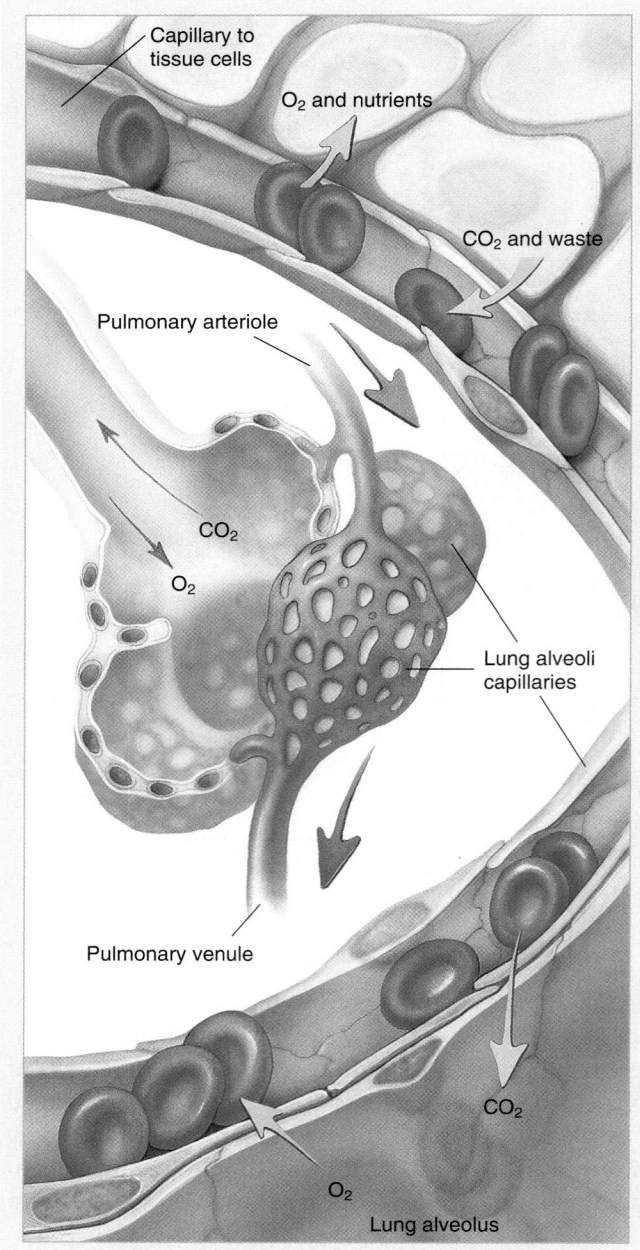

Figure 39-2 The exchange of oxygen and carbon dioxide occurs at the cellular level, where cells pass carbon dioxide into the capillaries and the capillaries transfer oxygen to the cells.

Labels in figure:
- Capillary to tissue cells
- O_2 and nutrients
- CO_2 and waste
- Pulmonary arteriole
- CO_2
- O_2
- Lung alveoli capillaries
- Pulmonary venule
- CO_2
- O_2
- Lung alveolus

✚ **EMT-B Tips**

Advanced airway techniques are begun only after proper basic airway management has been completed.

Basic Airway Management

You should always assess the airway first in an injured or ill patient. This rule applies to the basic and advanced levels of airway management. Advanced airway techniques are begun only after proper basic airway management has been completed.

As you have already learned, the first step in airway management is opening a patient's airway. You should use the head tilt–chin lift maneuver in a patient with no suspected spinal injury and the jaw-thrust maneuver in a patient you suspect has a spinal injury. After you have opened the airway, you should assess the airway and evaluate the need for suctioning to remove foreign bodies, liquid, or blood from the patient's mouth.

After the airway has been cleared, you need to determine whether the patient needs an airway adjunct. The basic airway adjuncts that are already available to you are oropharyngeal and nasopharyngeal airways. The more advanced airway adjuncts that may be available to you, with approval of your medical director, will be discussed in this chapter.

Gastric Tubes

Patients who have gastric distention or are vomiting are especially challenging to manage. You must use basic suctioning techniques to prevent aspiration in a patient who is vomiting. Patients with gastric distention are prone to vomiting; therefore, you should consider using a gastric tube in these patients.

A <u>gastric tube</u> is an advanced airway adjunct that provides a channel directly into a patient's stomach, allowing you to remove gas, blood, and toxins or to instill medications and nutrition. In the field, you will use a gastric tube primarily to <u>decompress</u> the stomach of a patient with gastric distention, a problem that is

cells can never be replaced. Brain damage and other permanent changes in the body result from damage caused by a lack of oxygen.

Other cells in the body that are not as dependent on a constant oxygen supply can tolerate short periods without oxygen and still survive.

most common in children during artificial ventilation but is also seen in adults.

There are two types of gastric tubes: nasogastric tubes, which are inserted through the nose, and orogastric tubes, which are inserted through the mouth. A nasogastric tube is contraindicated in a patient with major facial, head, or spinal trauma. In these patients, an orogastric tube is safer. A nasogastric tube can cause nasal trauma with bleeding, or it can accidentally be passed into the trachea, interfering with the airway and ventilation. In a patient with a basal skull fracture, a nasogastric tube can accidentally be passed into the brain.

Inserting a gastric tube can activate a patient's gag reflex, causing vomiting and aspiration. Clearly, inserting a gastric tube is a delicate task that must be done strictly according to local EMS protocol and medical direction. Special care is called for if the patient has head, spinal, or major facial trauma. Be sure to follow local protocol.

You will need the following equipment for gastric tube insertion (Figure 39-3 ▼):

- Proper-sized tubes
  - newborn and infant, 8 French
  - toddler and preschool, 10 French
  - school-age child, 12 French
  - adolescent, 14 to 16 French
  - adult, 16 to 18 French
- Catheter-tipped 60-mL syringe
- Water-soluble lubricant

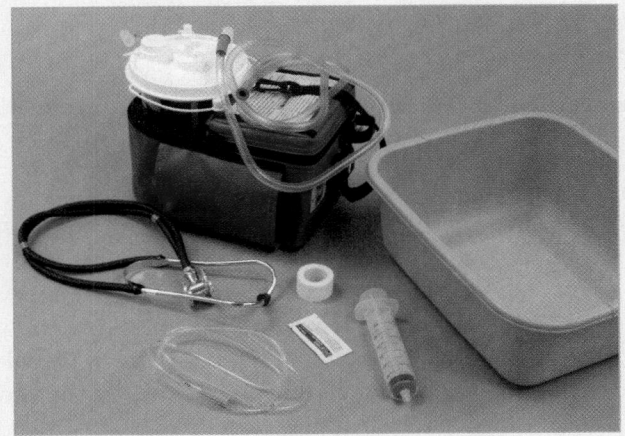

Figure 39-3 Equipment for gastric tube insertion includes gastric tubes, a catheter-tipped 60-mL syringe, water-soluble lubricant, an emesis container, tape, a stethoscope, and a suctioning unit.

- Emesis container
- Tape
- Stethoscope
- Suctioning unit and catheters

Once you have prepared and assembled the proper equipment, use the following procedure to insert a nasogastric tube:

1. **Measure the tube** from the tip of the nose, to the earlobe, to the epigastric area below the xiphoid process. (If you are using an orogastric tube, measure from the teeth to the angle of the jaw and down to the epigastric area.) Mark the measured length on the tube with a piece of tape, or note the number on the tube.
2. **Lubricate the distal end** of the tube with a water-soluble lubricant.
3. **Place the patient** in the proper position. If you do not suspect a spinal injury, place the patient supine, with the head flexed so the chin rests on the chest.
4. **Pass the tube** along the nasal floor (or, for an orogastric tube, over the tongue to the back of the throat) until you reach the tape marker.
5. **Confirm proper tube placement** by aspirating stomach contents with the syringe or injecting 30 to 50 mL of air into the tube and listening for gurgling over the stomach with the stethoscope.
6. **Aspirate air and stomach contents** with the syringe, once placement is confirmed, to decompress the stomach, or attach to on-board suction. Follow local protocols concerning the type of suction to use (intermittent or continuous).
7. **Secure the tube** in place with tape.

The Sellick Maneuver

Intubating an unresponsive patient who has no cough and/or gag reflex may cause vomiting and aspiration, which can ultimately damage airway tissues and block the lower airway passages. A procedure called the Sellick maneuver, or cricoid pressure, was originally developed for intubating patients during surgery, can be helpful in avoiding these complications in the field.

The cricoid cartilage, located just below the thyroid cartilage (Adam's apple), is a rigid, ring-shaped structure that completely encircles the larynx at the

Performing the Sellick Maneuver

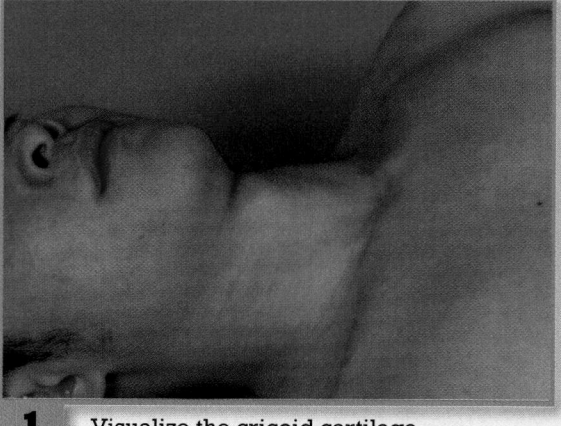

1 Visualize the cricoid cartilage.

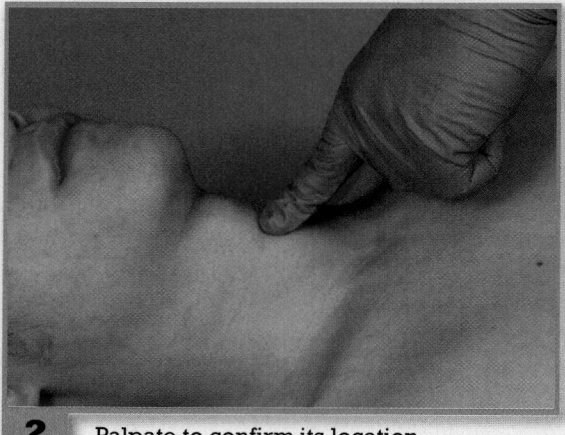

2 Palpate to confirm its location.

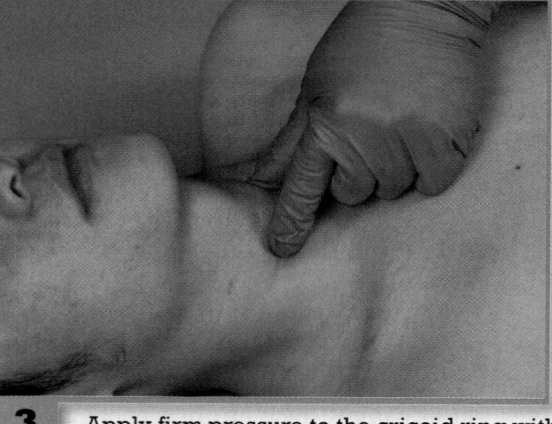

3 Apply firm pressure to the cricoid ring with your thumb and index finger on either side of the midline. Maintain pressure until the patient is intubated.

top of the trachea. It can be difficult to locate in infants, children, and small adults. The depression between the thyroid cartilage and the cricoid cartilage is called the cricothyroid membrane. (In certain cases, advanced life support providers will insert an emergency airway through this membrane.) The esophagus is much softer than the trachea and does not have rings of cartilage to hold it open. It is normally closed, opening only as we eat or drink. By applying pressure on the cricoid cartilage, you can squeeze the esophagus shut and, thus, prevent solids and fluids from leaving the esophagus and eventually being aspirated into the larynx and trachea.

To perform this maneuver, follow the steps in
(Skill Drill 39-1 ▲):

1. **With the patient's neck slightly extended** (only if there is no suspicion of spinal injury), **visualize the cricoid cartilage**, just below the thyroid cartilage (Adam's apple) (**Step 1**).
2. **Confirm location of the cricoid cartilage** by palpating with the tip of your index finger (**Step 2**).
3. **Place a thumb and index finger on either side of the midline of the cricoid cartilage.** Apply firm—but not excessive—posterior pressure on

the cricoid cartilage to compress and shut off the esophagus behind it. Too much pressure could collapse the larynx. Maintain this pressure until the patient is intubated (**Step 3**).

The Sellick maneuver should be performed by a third EMT-B and might not be possible if you and your partner are alone. When performing this maneuver, be sure to correctly identify anatomic landmarks to avoid damaging other structures or inadvertently obstructing the airway.

Endotracheal Intubation

Endotracheal intubation is the insertion of a tube into the trachea to maintain the airway. This can be done through the mouth (called orotracheal intubation) or through the nose (called nasotracheal intubation). In either case, the tube passes directly through the larynx between the vocal cords and then into the trachea.

Endotracheal intubation in patients who are conscious or drifting into and out of consciousness is extremely difficult. As an EMT-B, you will be intubating only patients who are unresponsive with no gag reflex or in cardiac arrest. However, you should not immediately intubate a patient who is unresponsive or in cardiac arrest. First, you must try to open the airway with the appropriate BLS maneuver, clear the airway, and ventilate the patient with a BVM device. If the BLS maneuver fails to open the airway, you should then consider endotracheal intubation based on your local medical protocols. Because of the time that it takes to properly prepare intubation equipment, you must always

secure the patient's airway with basic methods first because they are much quicker to perform and effective in many situations when done properly.

The remainder of this section will focus on orotracheal, rather than nasotracheal, intubation.

Orotracheal Intubation

Orotracheal intubation is the most effective way to control a patient's airway and has many advantages over other airway management techniques (Table 39-1 ▼). It

---

TABLE 39-1 Advantages of Orotracheal Intubation

- Completely controls and protects the airway
- Delivers better minute volume without the difficulty of maintaining an adequate mask seal, as is needed with a BVM device. (The minute volume is the volume of air cycled through the alveoli in 1 minute.)
- May be left in place for several days if prolonged ventilation is required
- Prevents gastric distention, which means less risk of regurgitation of stomach contents and greater opportunity for good tidal volume
- Minimizes the risk of aspiration of stomach contents into the respiratory system because a balloon seals off the trachea
- Allows for direct access to the trachea for suctioning
- Allows for the delivery of high volumes of oxygen at higher than normal pressures
- Provides a route for administration of certain medications

You are the Provider   Part 3

Because of the difficulty that your partner is having with ventilation and the low oxygen saturation, you contact medical control and request to perform endotracheal intubation. Medical control concurs with your decision. After 2 to 3 minutes of preoxygenation while you prepare your equipment, you place a 7.5-mm ET tube without difficulty. On auscultation, you hear breath sounds that are clear and equal bilaterally and no epigastric sounds. The tube is secured, and ventilation is continued. Within a few minutes, you note that the pulse oximeter now reads 95%. After attaching an in-line end-tidal carbon dioxide detector to the ET tube, the patient is loaded into the ambulance and transported to a nearby hospital.

5. What are the indications for endotracheal intubation?
6. What complications are associated with endotracheal intubation?

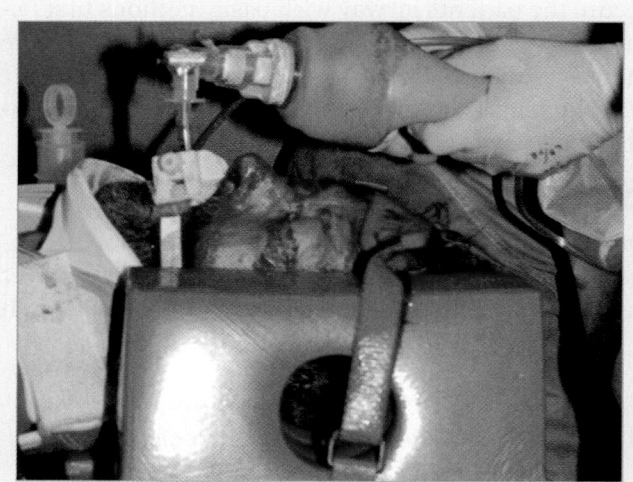

Figure 39-4 Endotracheal intubation is indicated for patients who are unconscious and cannot protect their own airways.

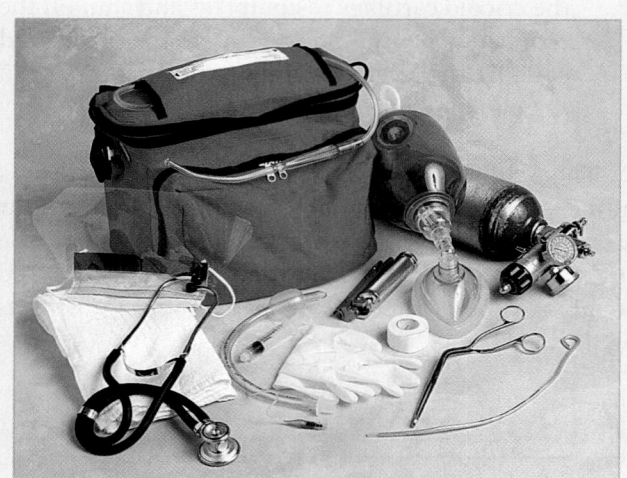

Figure 39-5 Assemble all necessary equipment before you begin intubation.

is indicated for patients who cannot protect their own airways as a result of unconsciousness or cardiac arrest (Figure 39-4 ▲). It is also used for patients who need prolonged artificial ventilation, are unresponsive to painful stimuli, or have no gag reflex or ability to cough. Remember that defibrillation is the priority for patients in cardiac arrest due to ventricular fibrillation. Intubation is performed only after defibrillation and during the 1-minute time lapse for CPR.

Equipment

Endotracheal intubation requires all your attention. You should not be searching for forgotten or misplaced equipment as you work. Therefore, you should assemble all the equipment that you will need before starting the procedure, while the patient is being preoxygenated with a BVM device (Figure 39-5 ▶). As in all patient care situations, be sure to follow BSI precautions. You will be working in proximity to the patient's airway, so the minimum BSI precautions include gloves, eye protection, and a mask.

There are two methods for inserting an ET tube. Both use ET tubes with slight differences in equipment. The visualized technique uses a laryngoscope, which enables visualization of the vocal cords to manually place the ET tube between the vocal cords and into the trachea. The blind technique uses a <u>lighted stylet</u>. The light at the end of the stylet is extremely bright. The light can be visualized on the outside of the body when placed in the trachea. The light cannot be seen if it is placed in the esophagus (there is too much tissue for the light to

penetrate to the outside). Because the vocal cords are normally located in the midline of the neck, a light seen in the sternal notch is below the vocal cords and in the trachea. After inserting the lighted stylet into the ET tube, the tube is then inserted into the airway, with the light as a guide. Because the vocal cords are not actually visualized, the technique is termed "blind."

The equipment for visualized and blind endotracheal intubation is basically the same with the exception of the insertion-visualization device. The following is a list of equipment needed:

- BSI equipment
- Properly sized ET tube
- Laryngoscope handle and blade (visualized technique)
- Stylet (visualized technique) or lighted stylet (blind technique)
- 10-mL syringe
- Oxygen, with BVM device for ventilation before and after intubation
- A suctioning unit with rigid and soft-tip catheters
- Magill forceps
- Towels for raising the patient's head and/or shoulders
- A stethoscope
- Water-soluble lubricant for the ET tube
- A commercial securing device

You must check your equipment daily to ensure that it is all available and to be certain that it is properly assembled and working, especially before you try to intubate a patient.

Laryngoscope

The purpose of a <u>laryngoscope</u> is to sweep the tongue out of the way and align the airway so that you can see the vocal cords and pass the ET tube through them (Figure 39-6 ▼).

The handle of the laryngoscope contains batteries to provide power to the light and has a locking bar to connect the handle to the blade; the blade is detachable from the handle. Blades are curved or straight and range in size from 0 to 4 (Figure 39-7 ▶).

The two blade designs function differently to align the structures so that you can visualize the vocal cords. The curved (Macintosh) blade is inserted just in front of the epiglottis, into the <u>vallecula</u> (the space between the base of the tongue and the epiglottis), allowing you to see the glottic opening and vocal cords (Figure 39-8A ▶). The straight (Miller) blade is inserted past the epiglottis (Figure 39-8B ▶). Because its broader base and flange provide better displacement of the tongue, a curved blade is preferred for use in older

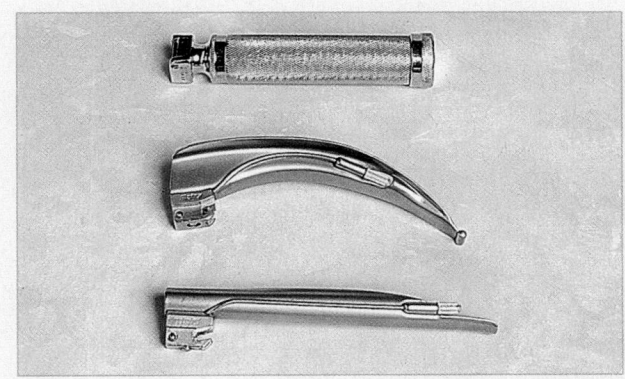

Figure 39-7 Laryngoscope blades can be curved or straight and come in different sizes.

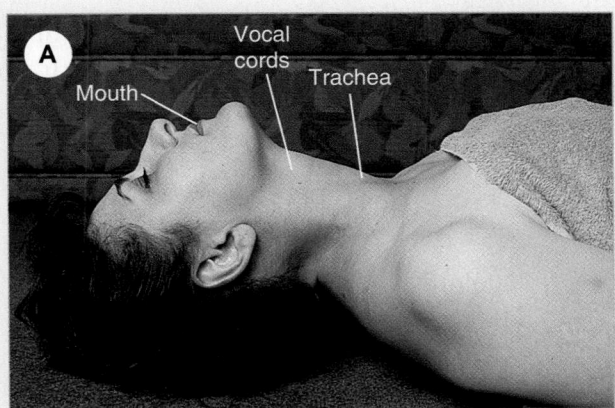

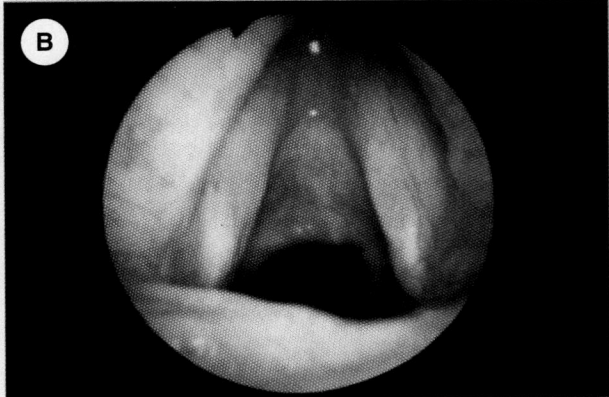

Figure 39-6 A. You must see the vocal cords to pass an ET tube through them. The vocal cords are located in the upper airway at the entrance to the larynx. **B.** A view of the vocal cords.

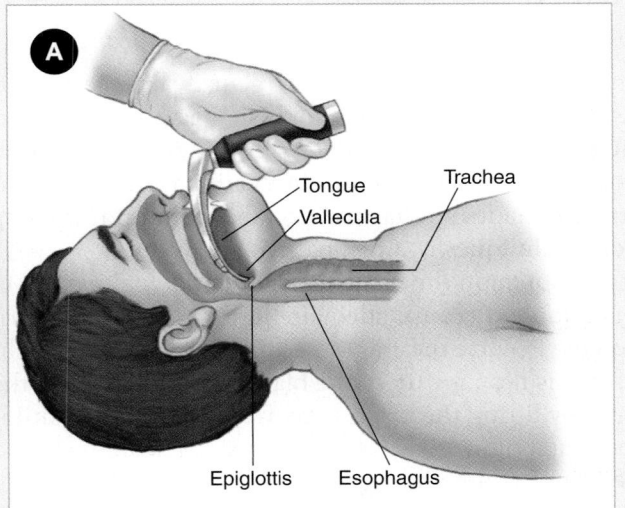

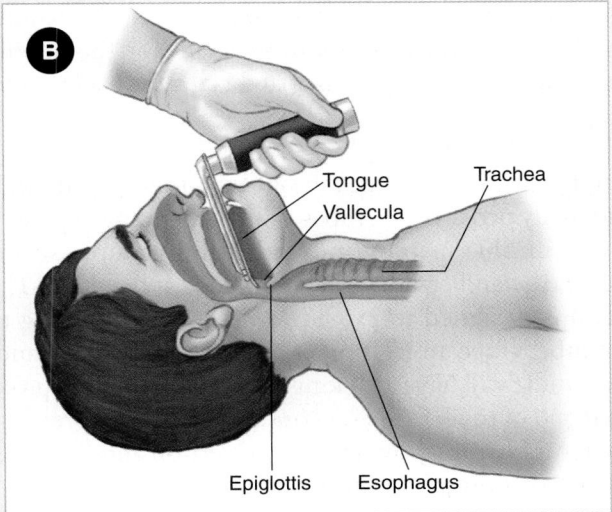

Figure 39-8 A. Insert a curved blade just in front of the epiglottis into the vallecula. **B.** Insert a straight blade past the epiglottis.

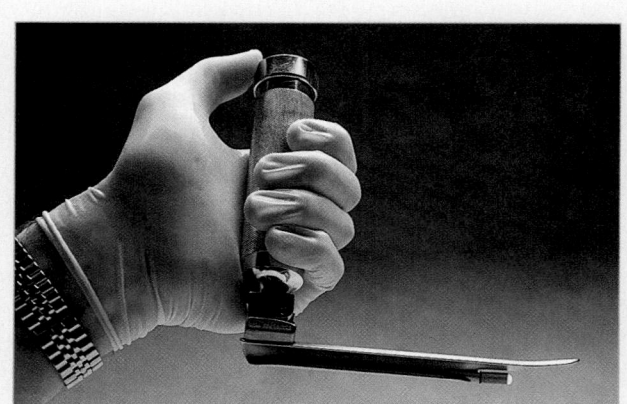

Figure 39-9 Once the blade is locked in place at a 90° angle, the light will turn on.

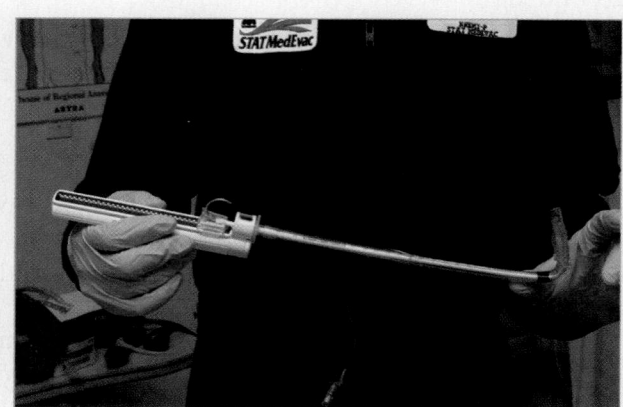

Figure 39-10 A lighted stylet used in conjunction with an ET tube.

children. A straight blade, which actually lifts the epiglottis out of the way to allow for visualization of the glottic opening and vocal cords, is preferred for use in infants. You should practice intubation with curved and straight blades so that you will feel comfortable using both techniques in a real patient situation.

A notch on the blade locks onto the locking bar of the handle. Because adequate lighting is essential for you to visualize the epiglottis and vocal cords, the light source is near the tip of the blade. You can activate the light by lifting the blade away from the handle until it locks at a right angle (Figure 39-9 ▲). The light should be bright white and tight. However, the bulb will not come on if the blade is not attached properly, the bulb is burned out or loose, or the batteries in the handle are dead. Always carry extra batteries for the handle and extra light bulbs in assorted sizes for each blade.

Lighted Stylet

The lighted stylet comes in one size that fits all adult and many pediatric ET tubes. The lighted stylet consists of a malleable metal wand with a light source at one end and a handle with an energy source at the other. The wand is designed to provide stiffness and shaping to the tube. Once an ET tube is threaded onto the wand, the handle will have a locking device to secure the proximal end of the ET tube (Figure 39-10 ▶). This prevents the ET tube from moving before insertion and the stylet from protruding beyond the end of the ET tube.

ET Tubes

Endotracheal tubes come in many sizes; the size is specified by the measurement of the inside diameter of the tube. Sizes range from 2.5 to 9 mm. The length of the

ET tube is marked on the outside of the tube in centimeters. The overall length of a tube for an adult is usually 33 cm. The following are general guidelines to use when you are intubating an average-sized adult patient (Figure 39-11 ▼):

- The centimeter markings on the outside of the tube will usually indicate that it is 15 cm to the

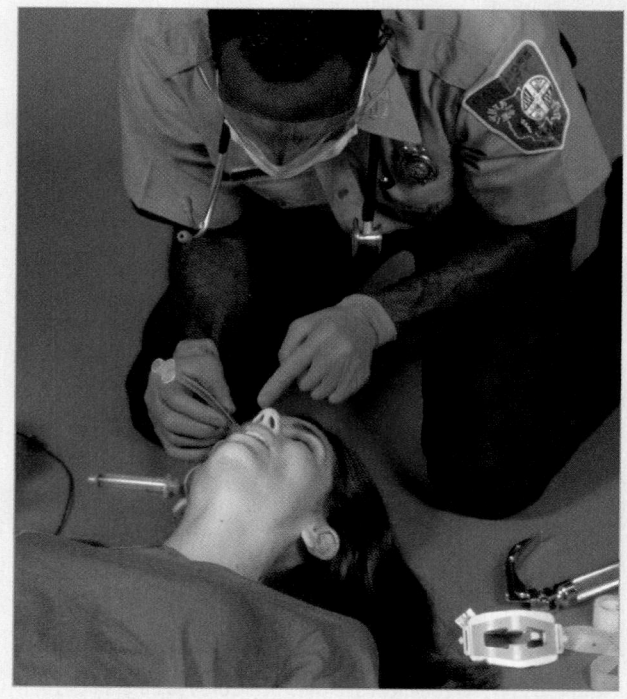

Figure 39-11 Before securing the tube, note the mark on the ET tube. The ET tube typically lines up with the teeth on an intubated patient around the 22-cm mark.

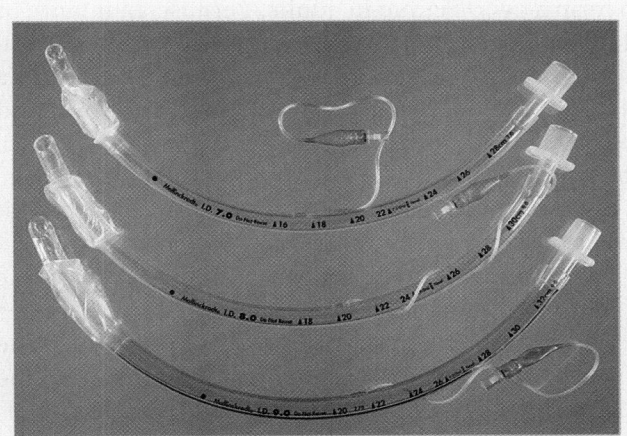

Figure 39-12 Endotracheal tubes that are used for adults generally range from 6.5 to 8.5 mm. Note the centimeter markings.

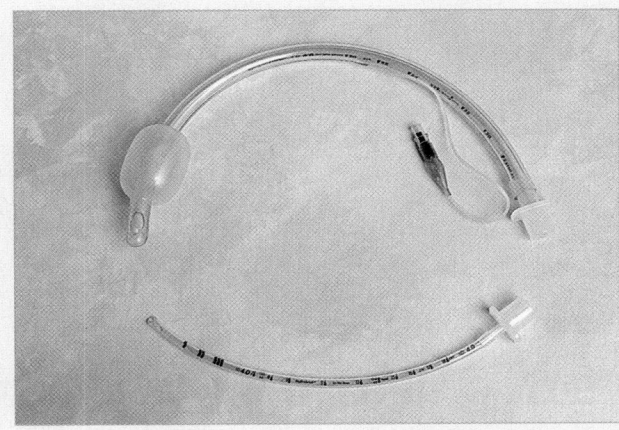

Figure 39-13 The components of the adult ET tube include a 15/22-mm adapter that attaches to a ventilating device, a pilot balloon, the tube, a balloon cuff (shown inflated), and Murphy's eye. The pediatric tube shown at the bottom includes an adapter and Murphy's eye at the uncuffed distal end of the tube.

vocal cords, 20 to 21 cm to the sternal notch, and 25 cm to the carina.

- You can mark the length for placement of the ET tube by looking at where the tube lines up with the teeth on an intubated patient. This tube-to-teeth mark is usually at around 22 cm. Once the patient is intubated, you should monitor the centimeter marking at the patient's teeth because this will determine whether the tube has moved from its original position.

The proper-sized tube for adult men ranges from 7.5 to 8.5 mm; for adult women, it ranges from 6.5 to 8.0 mm (Figure 39-12 ▲). For most efficient use of the tube, use the largest-diameter ET tube that you can. A good rule of thumb is to always have a 7.5-mm ET tube on hand; this size tube will fit most men and women. However, you should carry a complete selection of tube sizes in the unit to ensure that no matter what size you choose, you have one tube smaller and one tube larger, in case you need it.

Tube components include a standard 15/22-mm adapter, which attaches to any ventilation device, such as a BVM device or a mechanical ventilator (Figure 39-13 ▶). Make sure that the adapter is securely pushed into the tube so that it does not pull off when you ventilate the patient. A pilot balloon is attached to the tube to indicate how well the balloon cuff at the distal end of the ET tube is inflated. The cuff at the end of the tube holds about 10 mL of air. The small hole at the distal end of the tube across from the bevel end, called Murphy's eye, helps to prevent tube obstruction by secretions.

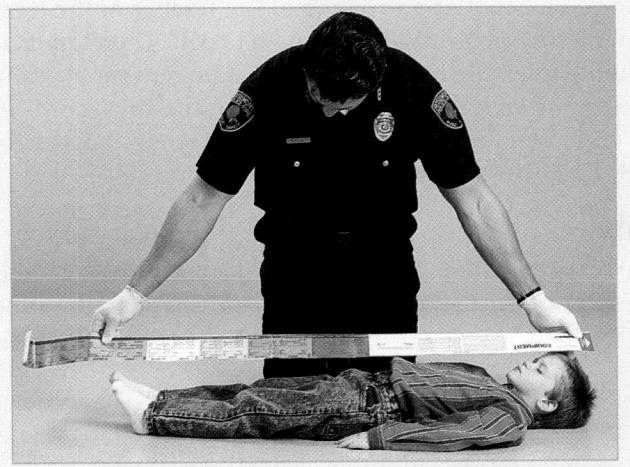

Figure 39-14 A chart or length-based tape device is best for estimating the size of an ET tube for children.

For children, it is best to have a chart or length-based tape device to help you with sizing the ET tube (Figure 39-14 ▲). Generally, for newborns and small infants, the proper tube size ranges from 3.0 to 3.5 mm; for infants up to 1 year, it is 4.0 mm. You may also follow a formula for sizing tubes in children. You can calculate tube size in children by adding 16 to the child's age and then dividing by 4. Another method is to select a tube that roughly equals the size of the diameter of the patient's little finger across the nail bed (Figure 39-15 ▶). No matter what size you decide to use, you should also have one tube larger and one tube smaller available, in case you need it.

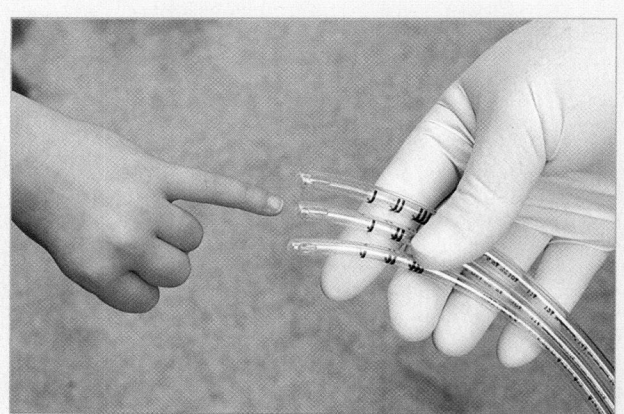

Figure 39-15 Another method of sizing ET tubes in children is to select a tube that roughly equals the size of the diameter of the patient's little finger across the nail bed.

With children older than 8 years and with adults, you will use cuffed tubes. However, in younger children, the circular narrowing of the trachea at the level of the cricoid cartilage functions as a cuff. Therefore, uncuffed tubes are used in children younger than 8 years. Always watch the tube pass through the vocal cords in a child (as well as in an adult) to make sure that the tip of the ET tube is in the proper position.

Stylet

A plastic-coated wire called a <u>stylet</u> may be inserted into the ET tube to add rigidity and shape to the tube (Figure 39-16 ▼). You should bend the tip of the stylet

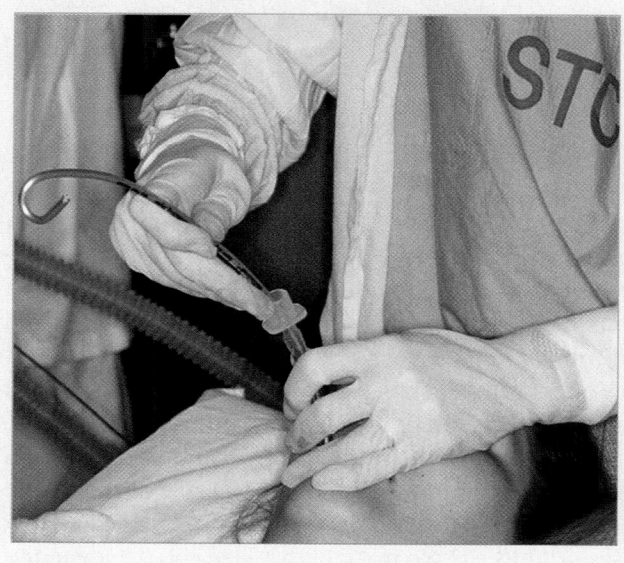

Figure 39-16 A wire stylet, which adds rigidity and shape to the tube, must be removed after tube placement.

to form a gentle curve in adults. Because an infant's or child's airway is more angular and less aligned than an adult's, you should bend the tip of the stylet into a hockey stick shape for use in an infant or child (Figure 39-17 ▼). You should also apply a little water-soluble lubricant to the distal end of the tube to make it easier to insert and to the end of the stylet to make it easier to remove once the tube is in place.

Do not insert the stylet past Murphy's eye because it could puncture or lacerate delicate airway tissues (Figure 39-18 ▼). A good rule of thumb is to keep the stylet 1/4" proximal to the cuff in adults and 1" from the end of the tube in infants and children. Before you attempt intubation, you should always confirm that the stylet is not sticking out past the end of the ET tube.

These principles are also followed for the lighted stylet. The lighted stylet may or may not be disposable, depending on the manufacturer. Because the light source is in the handle, make sure you have replacement handles or batteries, whichever is appropriate for your device.

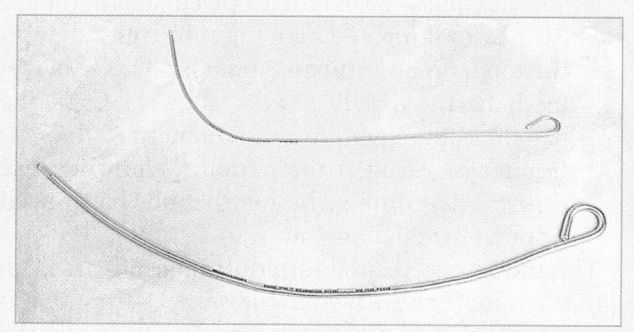

Figure 39-17 Bend the tip of the stylet into a hockey stick shape for a pediatric patient, as shown at the top. Bend the stylet to form a gentle curve for an adult patient, as shown at the bottom.

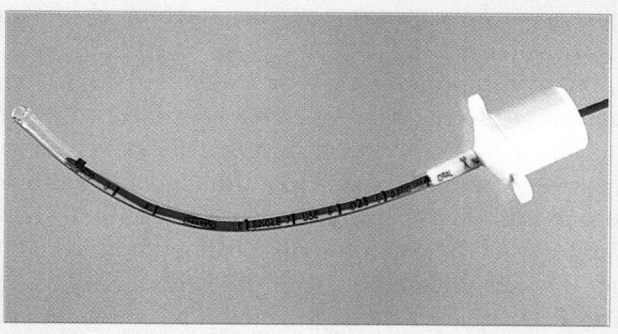

Figure 39-18 Do not insert the stylet past Murphy's eye because it could damage airway tissues. Keep the stylet 1/4" proximal to the cuff in the adult tube and 1" from the end of the pediatric tube.

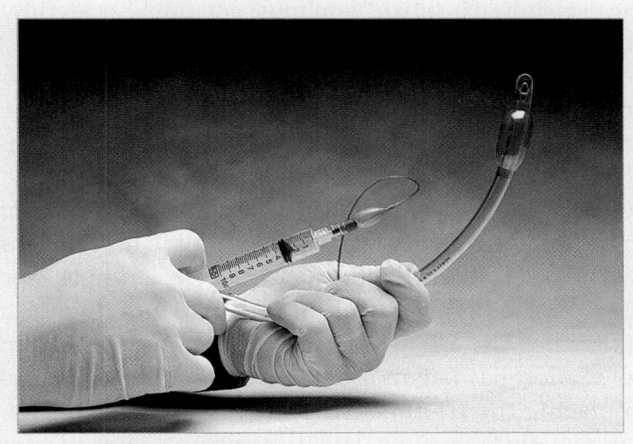

Figure 39-19 Inflate the cuff with 5 to 10 mL of air to check for air leaks.

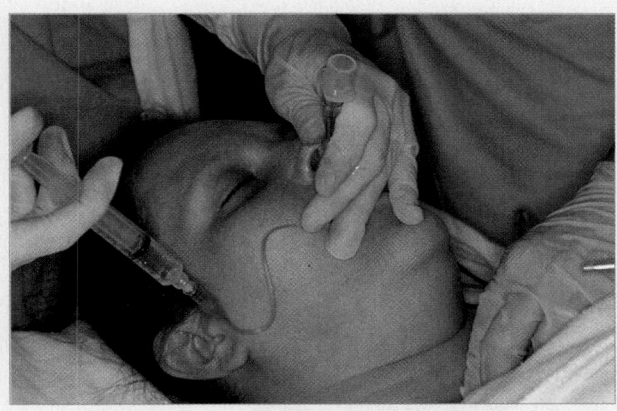

Figure 39-20 Inflate the cuff with 5 to 10 mL of air, and then immediately remove the syringe from the pilot balloon to prevent air from leaking back into the syringe.

Syringe

You will use the 10-mL syringe to test for air leaks in the ET tube before intubation (Figure 39-19 ▲). You will use it again after the ET tube is in place to inflate the cuff and provide a seal inside the trachea to prevent aspiration. Note that this step is done only in ET tubes with cuffs; therefore, you will not perform this step with the uncuffed tubes that are used for infants and children.

Take the following steps to use the syringe properly:

1. **As you are assembling** and checking your equipment before intubation, pull back on the plunger of the syringe to the 10-mL mark to fill the syringe with the amount of air that is needed to inflate the cuff.
2. **Attach the syringe** to the pilot balloon, and test the cuff by inflating it with 5 to 10 mL of air.
3. **Deflate the cuff** after you have confirmed that there are no air leaks.
4. **With the syringe still attached**, remove the air from the cuff by pulling back on the plunger to the 10-mL mark. Be sure that the syringe remains attached to the pilot balloon with the plunger pulled back to the 10-mL mark.

After the ET tube has been properly inserted in the patient, you will inflate the cuff with 5 to 10 mL of air and then immediately remove the syringe from the pilot balloon to prevent air from leaking back into the syringe (Figure 39-20 ▶).

Other Equipment

You must also have a device available to secure the ET tube in place. The use of a commercial securing device may be more effective than using tape. There are many manufacturers of securing devices; you should become familiar with the device used in your system. Some EMS systems use tape to secure the ET tube (Figure 39-21 ▼).

Tape could also be used as a backup system to a commercial device. Medical control may also advise you to use an oral airway or similar device in intubated patients to prevent them from biting down on the tube.

In addition, you will need the following equipment:

- Oxygen
- A suctioning unit
- A BVM device
- Magill forceps (can be used to help guide the tube in the visualized technique) (Figure 39-22 ▶)
- Towels for raising the patient's head or shoulders if necessary
- Secondary confirmation device
- Cervical collar and backboard

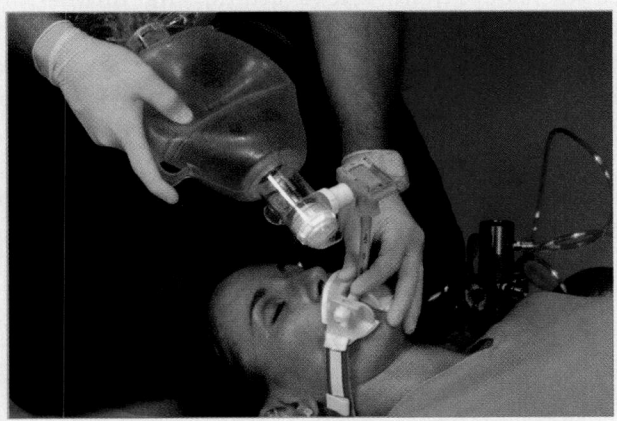

Figure 39-21 Secure the ET tube in place with a commercial device; using tape may not be as effective.

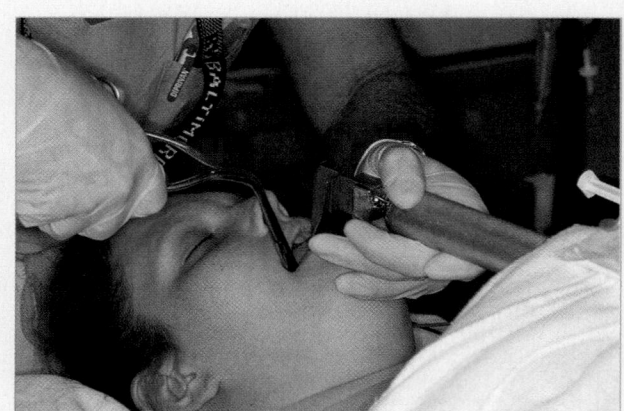

Figure 39-22 Magill forceps can be used to help guide the tube.

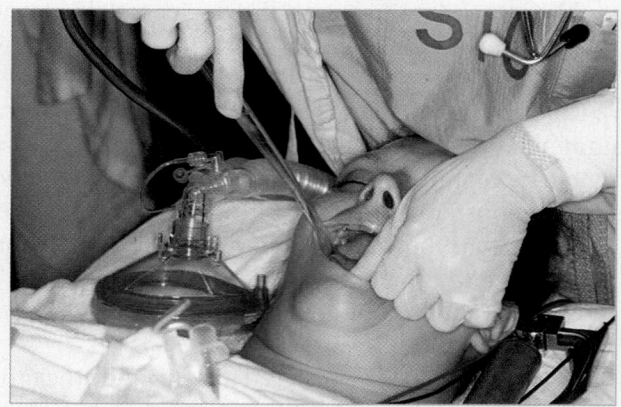

Figure 39-23 Suction any fluid or particles from the mouth before any attempt to intubate.

You will need to use the BVM device to preoxygenate the patient before attempting intubation. Remember, basic ventilation methods must always precede advanced airway management. The suctioning unit should be readily available to clear any fluid or particles from the mouth (Figure 39-23 ▲). Use a rigid, large-bore suction catheter to clear the mouth before intubation. Once the patient is intubated, you might need to use a French catheter to suction any fluids from inside the ET tube.

The Intubation Procedure

Visualized (Oral) Intubation

You may intubate only if authorized to do so by off-line or online medical control, according to your local medical protocols. Once you and medical control have made

the decision to intubate, you must act quickly, carefully, and efficiently. If environmental light is too bright and you are using a lighted stylet, move the patient to your ambulance or another place where ambient light can be controlled. Be sure to follow BSI precautions, including the use of gloves, eye protection, and a mask. You should not use more than 30 seconds in an attempt to intubate a patient. The 30-second time limit begins when you stop ventilation and insert the laryngoscope blade or the lighted stylet into the patient's mouth; it ends when the ET tube has been properly placed and ventilation has begun again. If you are not successful in placing the ET tube, stop, withdraw the tube, hyperventilate the patient, and try again according to your local protocols.

Intubation is a multiple-person task, especially in a situation involving cardiac arrest and use of an automatic external defibrillator (AED). The following tasks should be divided among the EMT-Bs who are present:

- First EMT-B applies and uses the AED.
- Second and third EMT-Bs perform synchronous CPR at a ratio of 15 compressions to 2 ventilations (CPR is performed asynchronously **after** the patient has been intubated).
- Fourth EMT-B prepares and intubates the patient.

You should note that defibrillation with an AED remains the highest priority. Intubating a patient who is in cardiac arrest should occur only after the necessary defibrillations and CPR have been performed for 1 minute. Consult with medical control regarding protocol on the sequence of these events.

Follow the steps in (Skill Drill 39-2 ▶) to perform visualized orotracheal intubation:

1. **Open the patient's airway** with a BLS maneuver, and clear the airway of any foreign material. Be sure to use BSI precautions.
2. **Insert an oropharyngeal airway,** and preoxygenate the patient with a BVM device at the appropriate rate, which will vary depending on the age of the patient. You should preoxygenate the patient at a rate of between 20 and 24 breaths/min for 1 to 2 minutes before attempting intubation (**Step 1**).
3. **As your partner ventilates the patient,** you should quickly assemble and test your equipment. Verify that the bulb on the laryngoscope or lighted stylet is working, select the proper-sized tube, and make sure the ET tube cuff has no leaks. If you are using a nonlighted stylet, in-

sert it in the ET tube. Lubricate the tube and stylet as needed (**Step 2**).

4. **Confirm that the patient has been properly pre-oxygenated.** Stop ventilating the patient and remove the oral airway if it is in place.

5. **If another EMT-B is available,** have him or her perform the Sellick maneuver to improve visualization of the vocal cords or positioning of the lighted stylet. This maneuver will also help prevent vomiting and aspiration. Maintain pressure on the cricoid cartilage until the ET tube cuff is inflated.

6. **Position the patient's head and neck** to allow for the best visualization of the vocal cords. In a patient with no spinal cord injury, use the head tilt–chin lift maneuver to align the structures. When you are using the laryngoscope, place towels under the patient's shoulders, if necessary, to raise the head for a better view of the vocal cords. When you are using the lighted stylet, grasping the tongue and jaw and pulling upward will aid in insertion (**Step 3**).

7. **To intubate a patient** you suspect has a spinal cord injury, you should make sure that your partner maintains manual in-line stabilization of the head and neck in the neutral position with a cervical collar in place while you attempt the intubation. You might need to lie on your stomach or straddle the patient's head while leaning back to visualize the vocal cords adequately (**Step 4**).

8. **Grasp the laryngoscope** handle in your left hand. Make sure the blade is locked into place and the bulb is illuminated. Open the patient's mouth with the gloved fingers of your right hand. Gently place the blade in the right side of the patient's mouth, then move it toward the center of the mouth, gently pushing the tongue to the left. The tongue must be displaced for you to visualize the vocal cords. Visualize the epiglottis. Advance a curved blade along the base of the tongue until its tip rests at the vallecula; advance a straight blade along the base of the tongue until you see it catch the epiglottis. Lift the laryngoscope away from the posterior pharynx so that you can see the vocal cords. The lifting force is directed straight up, parallel to the long axis of the laryngoscope handle, not back toward the patient's head. It should feel as if you are picking up the patient's head by the jaw. To avoid breaking the patient's teeth or lacerating the lips, never use the blade as a lever or fulcrum against the upper teeth. Do not lose sight of the vocal cords at any time after you have visualized them. Proper placement of the ET tube depends on your visualization of the tube as it is placed between the vocal cords.

9. **Insert the ET tube** with your right hand, keeping the vocal cords and the tip of the tube in sight at all times. Do not advance the ET tube down the center of the laryngoscope blade, or your view of the vocal cords will be obstructed.

Advance the tube from the right side of the patient's mouth. Watch the uninflated cuff on the tube as it passes through the vocal cords, then advance the ET tube until the cuff is just past the vocal cords. Note and document the centimeter markings on the outside of the ET tube at the level of the teeth.

Once the tube has been inserted through the vocal cords into the trachea, gently remove the laryngoscope and stylet, if a stylet was used. *Do not let go of the ET tube until it is secured* (**Step 5**).

10. **Inflate the soft balloon cuff** on the end of the tube with 5 to 10 mL of air. This will seal the trachea so that air can be blown directly into the lungs. Gently squeeze the pilot balloon cuff to verify the amount of air you should use. The pilot balloon should be full but easily compressed between your fingers. Immediately detach the syringe so that the air in the cuff will not empty back into it (**Step 6**).

11. **You or your partner** (whoever is not holding the ET tube in place) **should begin ventilating** the patient with a BVM device attached to the ET tube. Confirm placement of the ET tube. Listen with a stethoscope over the stomach, then both lungs as you ventilate the patient through the tube. You should be able to hear equal breath sounds over the right and left lung fields and no sounds over the stomach. Also listen at the sternal notch in children. You should see both sides of the chest rise and fall with each ventilation. This is especially important in children because breath sounds in children may be misleading. You may hear them even if the tube is in the esophagus. You should not be able to hear breath sounds in the stomach.

Proper confirmation of ET tube placement is essential for care of the patient. If you did not actually visualize the ET tube passing through the vocal cords, you may have placed the ET tube in the esophagus rather than in the trachea, which can prove fatal for the patient. The actual

✳ EMT-B Tips

The actual visualization of the ET tube as it passes through the vocal cords is the best way to confirm proper placement.

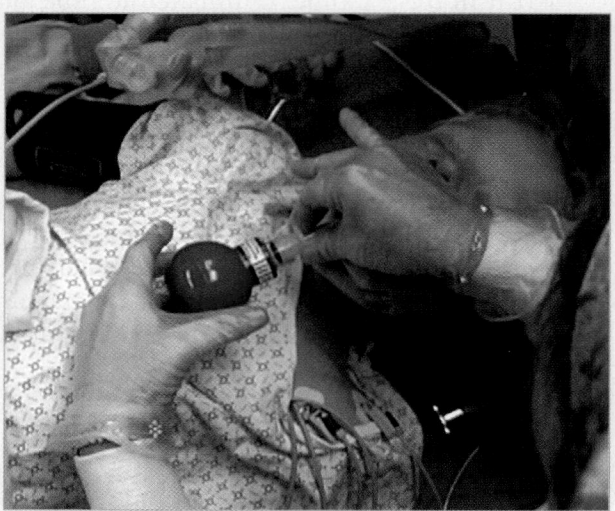

Figure 39-24 If the ET tube is in the trachea, air will freely inflate the bulb of the EDD.

visualization of the ET tube as it passes through the vocal cords is the best way to confirm proper placement.

It is recommended that you use a secondary method of confirming proper tube placement. There are several devices available—esophageal detector devices (EDDs), end-tidal carbon dioxide detectors (sometimes abbreviated as $ETCO_2$, and including colorimetric detectors), and portable capnography monitors.

Esophageal detector devices are designed to connect directly to the ET tube adapter end (where the BVM is attached) once the patient has been intubated. Two common devices use a syringe with plunger or a bulb syringe design. The device works by attempting to withdraw air from the ET tube. If the tube is properly placed, air will freely withdraw through the syringe or will inflate the bulb because the trachea will not collapse around the end of the ET tube (Figure 39-24 ▶). However, if the ET tube is in the esophagus, the soft tissues of the esophagus will collapse around the end of the tube and there will be noticeable resistance to the withdrawal of air through the syringe or plunger or to the inflation of the bulb.

End-tidal carbon dioxide detectors and capnography monitors sense the amount of carbon dioxide during ventilation, specifically during the exhalation phase. Patients in respiratory and/or cardiac arrest will not be producing any carbon dioxide. Therefore, you cannot get an accurate reading until you have restored effective ventilation and circulation to the patient.

One example of an end-tidal carbon dioxide detector is a disposable plastic indicator with chemically treated paper that changes from purple to yellow in the presence of carbon dioxide (usually takes several ventilations to obtain a reading) (Figure 39-25 ▶). The device connects between the ET tube and the BVM. Secondary confirmation is made by verifying the appropriate color change. If carbon dioxide is present, it will change the color of the indicator to yellow. If the ET tube is in the esophagus, ventilation will not produce any carbon dioxide and there will

You are the Provider
Part 4

En route to the hospital, you continue ventilating the patient. His oxygen saturation continues to read 95%, and he still has a carotid pulse. You note that the colorimetric paper in the end-tidal carbon dioxide detector turns yellow when ventilating the patient. You ask your partner to radio ahead to the hospital to notify them of your impending arrival.

7. What methods are used to confirm proper ET tube placement?
8. What should you do if the patient's oxygen saturation suddenly falls and he becomes cyanotic?

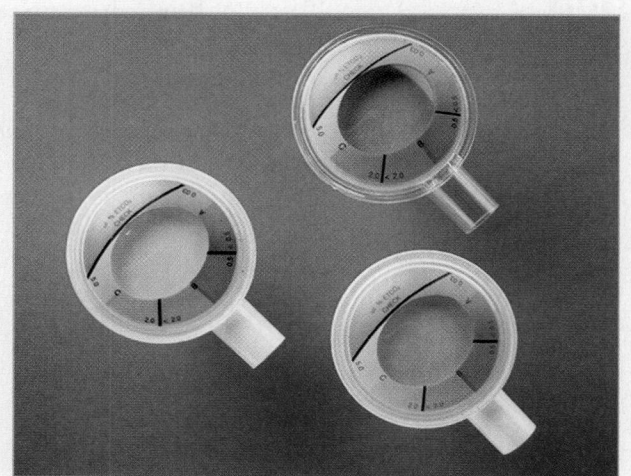

Figure 39-25 Color change on end-tidal carbon dioxide detectors can assist you in determining the location of the ET tube.

EMT-B Tips

The EMT-B who is ventilating the patient is responsible for monitoring tube placement and preventing it from being dislodged. This EMT-B, who should not be physically involved in lifting or moving the patient, has a good viewpoint from which to plan and coordinate team movements.

be no color change on the indicator (may change to tan, but generally will stay purple).

Capnography monitors can also be used for secondary confirmation of ET tube placement. These devices monitor for the presence of carbon dioxide through an adapter that connects between the ET tube and the BVM device. During ventilation, the amount of carbon dioxide present is displayed as a number or as a positive waveform on the monitor. Positive waveforms and/or carbon dioxide readings provide secondary confirmation of proper placement. If the ET tube is in the esophagus, waveforms and readings will not be present.

Remember that these are all devices used as secondary confirmation of ET tube placement. The devices do not give a 100% guarantee that the tube is in the correct location. Your primary confirmation is direct visualization of the tube passing through the vocal cords, auscultating good bilateral breath sounds (no epigastric sounds), and seeing the patient's chest rise and fall with each ventilation (**Step 7**).

12. **Once you have verified** that the tube is properly placed, secure the ET tube in place with the device and technique that have been approved by your medical director.

Keep in mind that, even when placed properly, the ET tube will move if it is not secured. For this reason, you must never let go of the ET tube until it is secured. Even then, you must continuously check the tube to make sure it is secure and in the correct place. Check the centimeter marking on the ET tube at the teeth and assess it frequently for tube movement. Also, frequently reassess the epigastrium and breath sounds and other confirmation methods (that is, capnography, capnometry). It is recommended that the head be immobilized and that cervical collars be placed on patients that are intubated (even if there is no trauma) to avoid unnecessary movement of the head during transport. Excessive head movement is thought to contribute to dislodgment of initially properly placed ET tubes. Continue to artificially ventilate the patient at an age-appropriate rate. Also, again, remember to note the distance the tube has been inserted by frequently reassessing the centimeter marking on the tube at the teeth. Be sure to reassess breath sounds and equal expansion of both sides of the chest each time you move the patient (**Step 8**).

Blind (Nasal) Intubation

Blind nasal intubation is performed without a laryngoscope. Use of a stylet is not recommended because it can damage the nasal mucosa and cause bleeding. Blind nasal intubation can be performed only in patients who are breathing; it should not be performed if there is evidence of a skull fracture, such as blood or cerebrospinal fluid draining from the nose following head or facial trauma. The tube is advanced when the patient inhales because this is when the vocal cords are open at their widest. This facilitates placement of the ET tube into the trachea. When selecting the appropriate size of ET tube, select a tube that is one-half to one full size smaller than you would use for orotracheal intubation (6.0 to 6.5 for a female; 7.0 to 7.5 for a male). This will

Performing Orotracheal Intubation

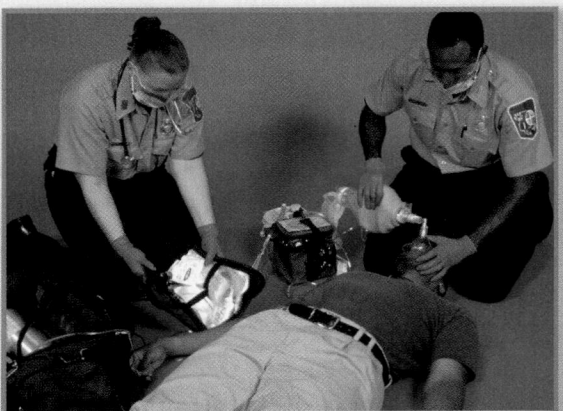

1 Open and clear the airway.

Insert an oropharyngeal airway, and preoxygenate with a BVM device.

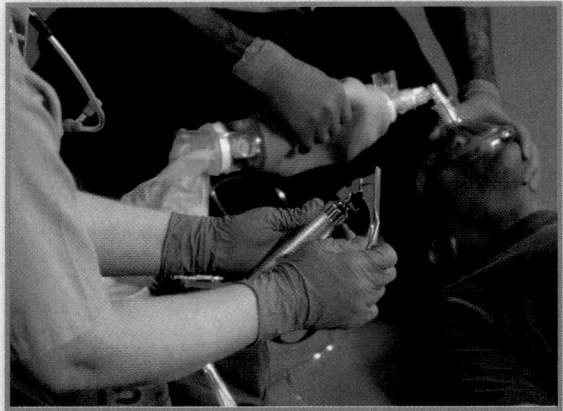

2 Assemble and test intubation equipment as your partner continues to ventilate.

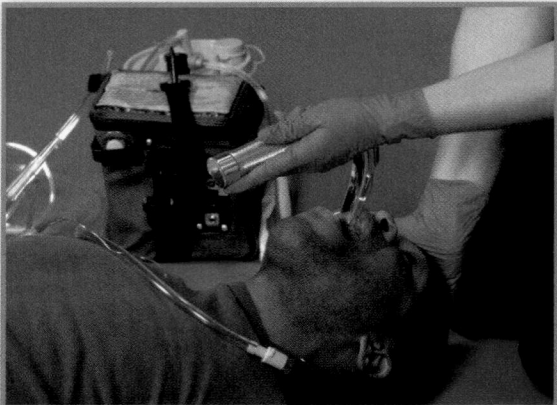

3 Confirm adequate preoxygenation, and remove the oral airway.

If available, have another rescuer perform the Sellick maneuver to improve visualization of the cords.

Use the head tilt–chin lift maneuver to position a nontrauma patient for insertion of the laryngoscope.

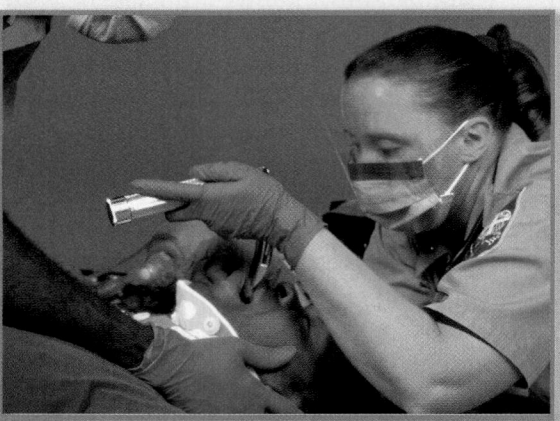

4 In a trauma patient, maintain the cervical spine in-line and neutral as your partner lies down or straddles the patient's head to visualize the vocal cords.

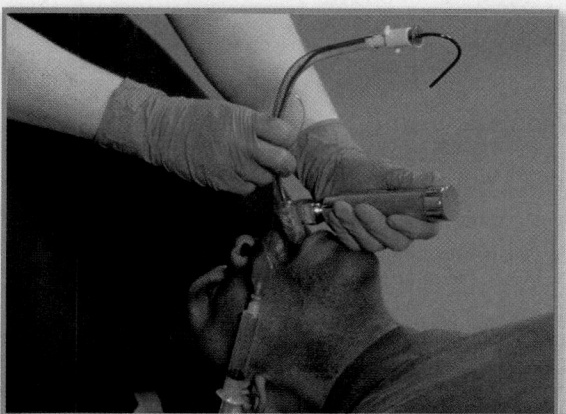

5 Insert the laryngoscope from the right side of the mouth, and move the tongue to the left. Lift the laryngoscope away from the posterior pharynx to visualize the vocal cords. *Do not pry or use the teeth as a fulcrum.*

Insert the ET tube from the right side until the ET tube cuff passes between the vocal cords. Remove the laryngoscope and stylet. Hold the tube carefully until it is secured.

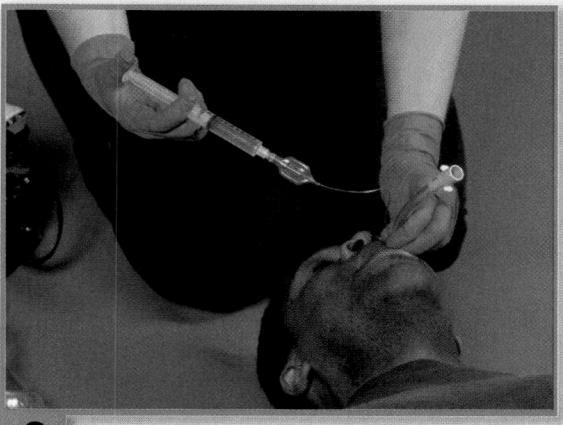

6 Inflate the balloon cuff, and remove the syringe as your partner prepares to ventilate.

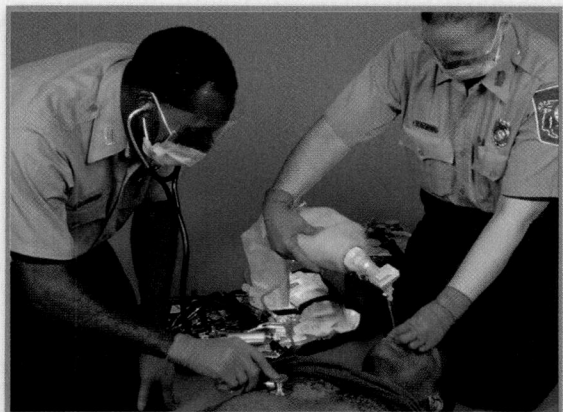

7 Begin ventilating, and confirm placement of the ET tube by listening over the stomach and both lungs. Also confirm placement with an end-tidal carbon dioxide detector or EDD, if available.

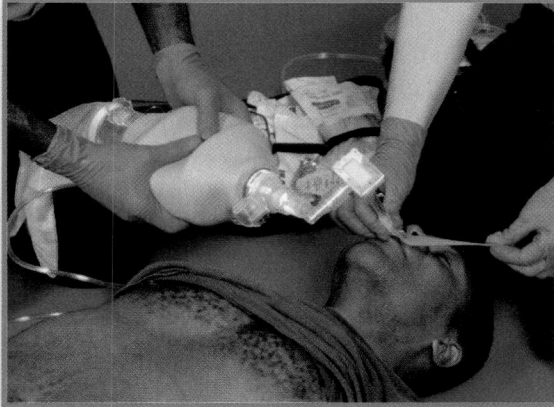

8 Secure the tube, and continue to ventilate.

Note and record depth of insertion (centimeter marking at the teeth), and reconfirm position after each time you move the patient.

✹ EMT-B Tips

Closely monitor the patient's heart rate and oxygen saturation during an intubation attempt. If the patient's heart rate changes significantly or the oxygen saturation falls to less than 90%, abort the intubation attempt and ventilate the patient with a BVM device and 100% oxygen.

minimize the risk of damage to the delicate nasal mucosa. The steps for blind nasal intubation are as follows:

1. Use BSI precautions.
2. Ensure adequate artificial ventilation by using a BVM device and 100% oxygen.
3. Preoxygenate the patient at a rate of 20 to 24 breaths/min before any advanced airway placement. Preoxygenate the patient for 1 to 2 minutes.
4. Assemble and test all equipment, including that for securing the device.
5. Check for a gag reflex by flicking the patient's eyelashes. If the eyelids flutter, the gag reflex is intact.
6. Check for and remove, if possible, any sharp debris in the patient's mouth.
7. Position the patient's head in a neutral position, and move the tongue forward.
8. Grasp the tongue and jaw, and pull upward.
9. Insert the tube into the nostril. If resistance is met, try inserting the tube into the other nostril.
10. Advance the tube when the patient inhales. Note the centimeter marking at the nostril after insertion of the tube.
11. Release the jaw, and hold the tube against the nostril.
12. Inflate the distal cuff with 5 to 10 mL of air, and immediately detach the syringe.
13. Attach the BVM device, and ventilate the patient.
14. Confirm proper tube placement by auscultating over the stomach and both lungs. Also use a secondary confirmation device, such as the EDD or end-tidal carbon dioxide detector.
15. Secure the tube in place.
16. Continue ventilation at the appropriate rate.

Complications

Endotracheal intubation is a difficult skill to master. If intubation takes longer than 30 seconds, the resulting delay in oxygenation may lead to brain damage. Obviously, you will need a great deal of expert instruc-

TABLE 39-2 Benefits and Complications of Endotracheal Intubation

Benefits	Complications
Provides complete protection of airway	Intubation of the right mainstem bronchus
Can be left in for long periods	Intubation of the esophagus, resulting in hypoxia
Delivers better oxygen concentration than a BVM device	Aggravation of a spinal injury
Prevents gastric distention and aspiration	More than 30 seconds to intubate can result in increased hypoxia
Allows for deep suctioning of the trachea	Vomiting and/or tube removal can result in aspiration
Allows for administration of certain medications	Soft-tissue trauma
	Mechanical failure
	Patient intolerance
	Decrease in heart rate

tion and practice to master this skill. Among the possible complications of endotracheal intubation are the following, as summarized in Table 39-2 ▲.

Intubating the Right Mainstem Bronchus

This is the most common error that is made during intubation. If you push or accidentally slip the tube in too far, the ET tube will pass into the right mainstem bronchus because it is shorter and straighter than the left mainstem bronchus. In this position, it will ventilate only the right lung. Therefore, you will hear breath sounds only on the right side. The best way to correct this problem is to deflate the cuff and pull the ET tube back about 1 cm. Listen again for bilateral breath sounds. Make sure you do not completely remove the ET tube.

Intubating the Esophagus

This often occurs when you insert the ET tube without first seeing the vocal cords or without seeing the light in the midline when using a lighted stylet. As a result, the ET tube is inserted into the esophagus rather than the trachea. The result is rapid inflation of the patient's stomach rather than ventilation of the lungs; if the situation is not corrected, the patient will die. To avoid this, carefully watch the ET tube as it passes through the vocal cords or monitor the midline position of the lighted stylet. Next, auscultate over the epigastrium and over the left and right apices and bases of the lungs. Watch for the rise and fall of the chest. If there is any doubt,

EMT-B Tips

If you cannot see the vocal cords, do not insert the ET tube! If you do, the tube will invariably come to rest in the esophagus. If after 30 seconds you are unable to visualize the vocal cords, abort the intubation attempt and immediately ventilate the patient with a BVM device and 100% oxygen.

pull out the ET tube, hyperventilate the patient for 2 to 3 minutes, and try to intubate the patient again. Be sure that you follow the protocol established by your medical director for the number of attempts that you are allowed when intubating a patient. If you are still unsuccessful after two attempts at intubation, consider using a different advanced airway adjunct (laryngeal mask airway [LMA], Combitube, or pharyngeotracheal lumen airway [PtL]) or insert an oral airway and ventilate with a BVM device.

Aggravating a Spinal Injury

Whenever there is concern about a spinal injury, you must intubate without moving the patient's neck from the neutral, in-line position. This makes it difficult to lift the lower jaw and tongue enough to see the vocal cords; indeed, this can only be done by two EMT-Bs working together.

Increased Hypoxia: Taking Too Long to Intubate

Do not take any longer than 30 seconds trying to intubate, or the patient will become more hypoxic. Your partner or another member of the team should actually time the intubation. If you cannot complete the procedure within 30 seconds, you should stop and ventilate the patient with a BVM device and 100% oxygen for 2 to 3 minutes before trying again. Ask a third EMT-B to perform the Sellick maneuver so that you can see the vocal cords more clearly. If after two tries the tube cannot be passed, try another airway technique, or ask another qualified rescuer to try. If the patient is difficult to intubate, you should not waste time in the field. Use another airway adjunct, and provide immediate transport, assisting ventilation as needed.

Patient Vomiting

A patient who is not totally unresponsive may begin to gag or try to remove the tube. Gagging may cause the patient to vomit and aspirate stomach contents. To avoid this, always check for a gag reflex before intubation,

such as by flicking the patient's eyelashes. Patients who tolerate an oral airway do not have an intact gag reflex and may need intubation. You should always have a suctioning unit ready in the event that the patient vomits during the intubation procedure. Trying to insert an ET tube through the vocal cords can cause the cords to spasm, which is called laryngospasm. If this occurs, stop intubating, and ventilate the patient with a BVM device.

Soft-Tissue Trauma

The laryngoscope and the tip of the ET tube can injure the lips, teeth, tongue, gums, and other airway structures. Used as a lever, the laryngoscope blade can easily break teeth, while a tube pushed blindly through the vocal cords can lacerate adjacent structures. Careful attention to your technique will minimize the risk of these complications.

Mechanical Failure

You may hear or feel air coming from the oropharynx when ventilating the patient. In an adult, this means that the cuff does not have enough air in it or that it has been torn and is leaking. If this occurs and ventilation is inadequate, you must get more air into the cuff (check the pilot balloon), or replace the tube. In a child, an air leak means that the uncuffed tube is too small or the child is large enough to need a cuffed tube.

Patient Intolerant of the ET Tube

Because of the reversal of hypoxia from direct oxygenation, a patient may regain a gag reflex or regain consciousness and try to remove the ET tube. Before removing the tube (extubation) in these circumstances, you should determine that the patient can obey commands. If he or she can, ensure that the suctioning unit is nearby and turned on. Then deflate the cuff, and carefully withdraw the ET tube as the patient exhales. Sometimes you may have to ask the patient to cough. Provide immediate suctioning if the patient vomits, then reassess the airway and administer supplemental oxygen. Always consult medical control before removing the ET tube. Generally, the only reason to extubate a patient in the field is if the patient is unreasonably intolerant of the tube. Be aware that conscious patients are at high risk for laryngospasm immediately following extubation.

Decrease in Heart Rate

Be sure to monitor the patient's vital signs carefully and continuously, particularly the heart rate. In children, it is important to check heart rate and skin color. With

✳ EMT-B Tips

You may intubate only if authorized to do so by off-line or online medical direction, according to your local protocols.

endotracheal intubation, the heart rate may decrease when the airway has been stimulated. Use of a straight blade is recommended when intubating small children because it does not stimulate the airway as much as the curved blade.

Also be sure to auscultate over the epigastrium and lungs and evaluate chest wall motion any time you move the patient. It is easy to dislodge the ET tube while moving the patient. Always reconfirm placement after the patient has been moved in any way.

Multilumen Airways

In addition to endotracheal intubation, other advanced airway devices are available to you. These devices are called multilumen airways, and, depending on local protocol and your medical director, you may be trained in the use of these airway devices. The benefits and disadvantages of using multilumen airways are shown in Table 39-3 ▼.

The multilumen airways, which are inserted without direct visualization of the vocal cords, have been designed to provide lung ventilation when placed in the trachea or the esophagus, making them much easier to insert than an ET tube. Two of these adjuncts are described here: the Esophageal Tracheal Combitube (ETC), and the pharyngeotracheal lumen airway (PtL). Remember that you must be trained and authorized to use these devices.

Esophageal Tracheal Combitube

The Esophageal Tracheal Combitube (ETC) consists of a double-lumen tube and two balloon cuffs (Figure 39-26 ▼). The blue lumen (No. 1) is the primary ventilation port when the tube is inserted in the esophagus. The clear lumen (No. 2) is the ventilation port if the tube is placed in the trachea. The clear cuff at the tip of the tube (distal cuff) seals off the esophagus or the trachea; the larger, flesh-colored cuff near the midway point seals off the oropharynx and nasopharynx. A blue pilot balloon and a white pilot balloon correspond to the flesh-colored and distal cuffs. Two syringes are also included in the kit. The ETC is a single-use item and must be discarded after use. It should not be cleaned and reused.

The ETC is inserted blindly. If the tube happens to go into the trachea, ventilation is provided directly into the trachea. If the tube goes into the esophagus, as occurs most often, ventilation can still be provided to the patient (Figure 39-27 ▶). You do not need to maintain a constant face mask seal with either placement. Instead, the ETC forms a seal in the oropharynx by an inflated cuff in the oropharynx, so you can ventilate the trachea via a tube, rather than a mask. This ease of ventilation is an advantage of the ETC.

TABLE 39-3 Benefits and Disadvantages of Multilumen Airways	
Benefits	**Disadvantages**
Ease of proper placement	Loses effectiveness (cuff malfunction)
No mask seal necessary	Requires deeply comatose patient
Requires minimal skill and practice to maintain	Requires constant balloon observation
Easily used in spinal injury patients	Cannot be used on patients shorter than 5' tall
May be inserted blindly	Requires great care in listening for breath sounds
Protects the airway from upper airway secretions	Large balloon is easily broken and tends to push the PtL out of the mouth when inflated

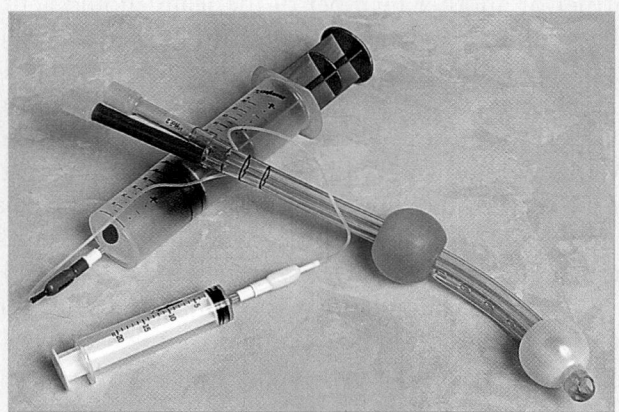

Figure 39-26 The ETC consists of a double-lumen tube and two balloon cuffs. The blue lumen is the primary ventilation port, and the clear lumen is the ventilation port if the tube is placed in the trachea.

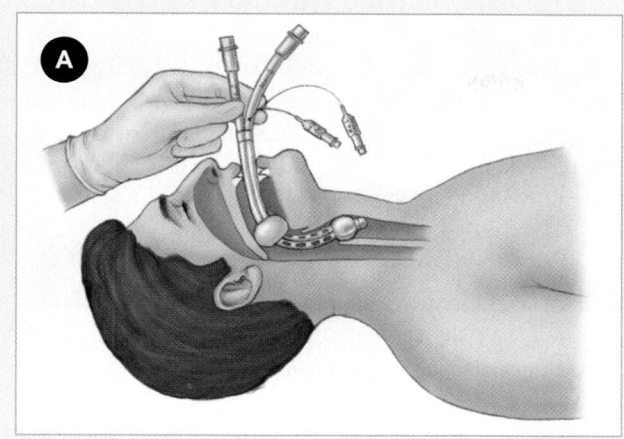

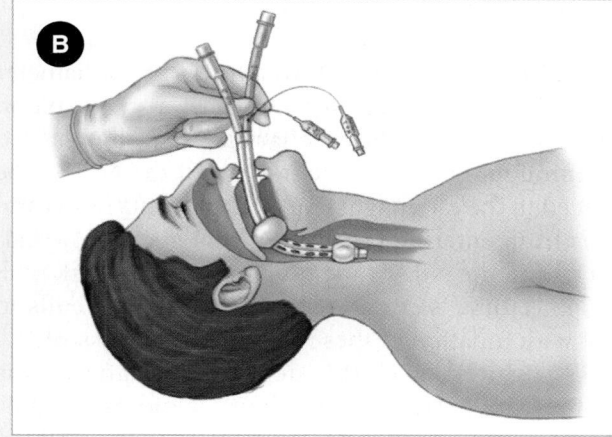

Figure 39-27 A. If the ETC is inserted into the trachea, it functions as an ET tube with ventilation provided directly into the trachea. **B.** If the ETC is inserted into the esophagus, ventilation can still be provided to the patient.

Contraindications

You should not attempt to insert an ETC in the following individuals:

- Conscious or semiconscious patients with a gag reflex
- Children younger than 16 years
- Adults shorter than 5' tall
- Patients who have ingested a caustic substance
- Patients who have a known esophageal disease

Inserting the ETC

As with endotracheal intubation, you must act quickly and carefully once you have received permission from medical control to use the ETC. Remember to follow BSI precautions, including the use of gloves, eye protection, and a mask, any time you may be exposed to blood or other body fluids.

The steps for insertion of the ETC are as follows:

1. **Assemble and check** the proper equipment, including the following:
   - BSI equipment
   - ETC kit with syringes
   - Water-soluble lubricant
   - Suctioning unit with suction catheters
   - BVM device and oxygen
2. **Apply a water-soluble lubricant** to the ETC.
3. **Position the patient.** Open the patient's mouth, and clear it of any foreign objects, including vomitus, dentures, and blood clots. Remove the oral airway if one has been inserted. Open the airway of a patient who has no possible spinal injury by hyperextending the patient's head and neck. Note that with an unconscious patient who may have a spinal injury, you must maintain the neck in a neutral, in-line position during insertion of the ETC.
4. **Preoxygenate the patient** with 100% oxygen using a BVM device.
5. **Lift the lower jaw and tongue** away from the posterior pharynx by inserting your thumb deep into the patient's mouth and grasping the tongue and lower jaw between your thumb and index finger.
6. **Gently guide the ETC** along the base of the tongue and into the airway. Hold the ETC so that it curves in the same direction as the natural curvature of the pharynx. Insert the tip into the mouth, and advance it carefully along the tongue. Do not use force. If you meet resistance, pull back and redirect the ETC. When the ETC is at the proper depth, the teeth will be between the heavy black lines.
7. **Inflate the blue pilot balloon** (and flesh-colored cuff) with the predrawn 100-mL blue-tipped syringe. Once the cuff is inflated and the pilot balloon is tense, immediately inflate the white pilot balloon (and the distal cuff) with the smaller, predrawn 15-mL syringe. The ETC may move forward a bit, but this is normal.
8. **Ventilate the patient** through the blue (No. 1) tube with a BVM device.
9. **Confirm the placement** of the tube. If the chest rises and falls and you hear breath sounds, the ETC is in the esophagus. When this is the case, continue to ventilate through the blue (No. 1) tube. If the chest does not rise and fall, and you do not hear breath sounds, the ETC is in the

Documentation Tips

After placement of any advanced airway device—ET tube, ETC, PtL, or LMA—document the intubation and exactly how you confirmed correct tube position. Also record your reconfirmation of tube position after each patient move. Note tube size, when applicable, and the position of any measurement marks on the tube relative to the patient's teeth.

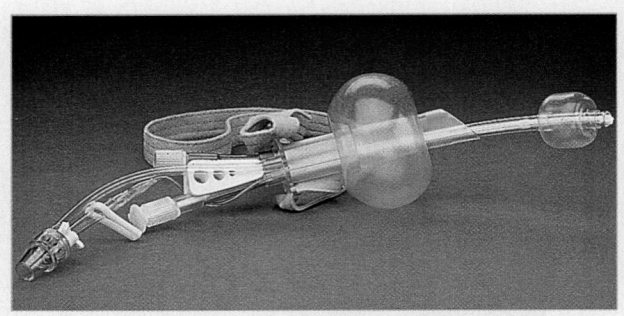

Figure 39-28 The PtL consists of two tubes, two balloon cuffs, a bite block, and a neck-retaining strap.

trachea. In this case, apply the BVM device to the shorter, clear tube, and ventilate the patient through it. Again, listen for breath sounds in all lung fields and in both axillae. Also listen over the stomach to verify that proper placement has occurred.

10. **Continuously monitor the patient.** Watch for balloon cuff leaks by carefully squeezing the pilot balloon. Use the syringes to keep the balloon cuffs properly inflated. Balloon cuffs may be torn by broken teeth, dentures, and bones. Therefore, you must use special care with this device, especially in the event of facial trauma.

Removing the ETC

Removal of the ETC airway is fairly simple. If the patient will no longer tolerate the ETC, you should remove it. Remember that the patient will likely vomit when the ETC is removed, so you must have a suctioning unit readily available. Be sure to turn the patient on his or her side to keep the airway clear of vomitus. When you are ready, simply deflate both balloon cuffs, and gently remove the tube.

Pharyngeotracheal Lumen Airway

The pharyngeotracheal lumen airway (PtL) consists of two tubes, two balloon cuffs, a bite block, and a neck-retaining strap (Figure 39-28 ▶). The long, clear (No. 3) tube contains a stylet and a low-pressure balloon cuff near its tip. The stylet is left in as a plug if the tube is placed in the esophagus but is removed if the tube ends up in the trachea. In either case, the balloon cuff prevents gastric contents from entering the lungs when the cuff is inflated.

The No. 3 tube passes through the larger-diameter green (No. 2) tube. The No. 2 tube has a large balloon cuff designed to seal the oropharynx. This allows air to pass through it and into the trachea (if the No. 3 tube is placed in the esophagus) but also prevents blood and debris from entering the airway from above. The balloon cuff in the No. 2 tube functions as the mask seal. The No. 1 tube is connected to these balloon cuffs to assist with inflation of these important airway seals.

As with the ETC, this device is designed to be inserted blindly into the oropharynx and esophagus, but you must be trained and authorized to use it. If the tube happens to go into the trachea when blindly inserted, it acts like an ET tube. If the long tube goes into the esophagus, you can still provide adequate ventilation to the patient (Figure 39-29 ▶). However, with the PtL, you need not maintain a constant face mask seal. The PtL forms an inflated cuff seal in the oropharynx, and the trachea may be ventilated via a tube rather than a mask.

EMT-B Safety

Placement of advanced airway adjuncts involves placing your own mucous membrane surfaces—eyes, nose, mouth—very near the mouths of patients likely to cough or vomit during the procedure. Without proper use of personal protective equipment, you could have direct exposure to potentially infectious fluids such as blood, vomitus, and mucus. Follow BSI precautions faithfully: Use a mask, eye protection, and gloves.

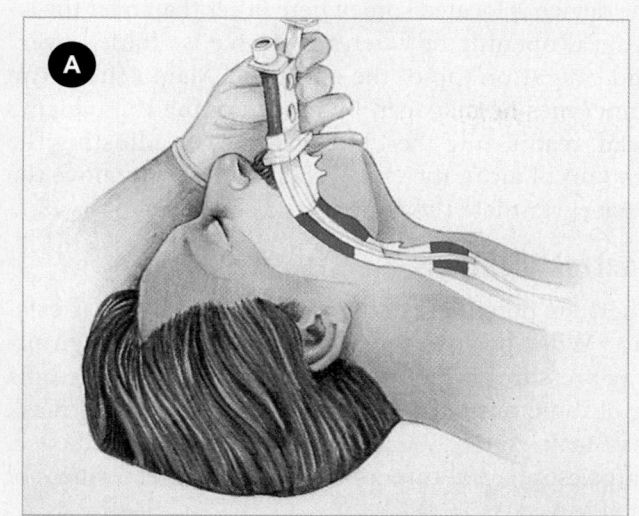

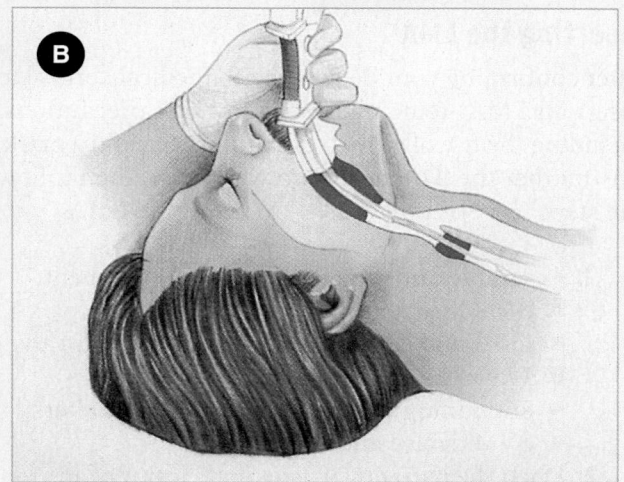

Figure 39-29 The PtL is inserted blindly into the oropharynx. **A.** If the PtL is inserted into the trachea, it functions as an ET tube with ventilation provided directly into the trachea. **B.** If the PtL is inserted into the esophagus, ventilation can still be provided to the patient.

Contraindications

You should not attempt to use the PtL in the following individuals:

- Conscious or semiconscious patients with a gag reflex
- Children younger than 14 years
- Adults shorter than 5′ tall
- Patients who have ingested a caustic substance
- Patients who have a known esophageal disease

Inserting the PtL

Once you have confirmed with medical control your decision to insert a PtL, you must act quickly and carefully. Remember to follow BSI precautions, including gloves, eye protection, and a mask, any time you may be exposed to blood or other body fluids. Take the following steps to insert the PtL:

1. **Assemble and check** the proper equipment, including the following:
   - BSI equipment
   - PtL with syringes
   - Water-soluble lubricant
   - Suctioning unit with suction catheters
   - BVM device and oxygen
2. **Lubricate the tube** on the PtL with a water-soluble lubricant.
3. **Position the patient.** Open the patient's mouth, and clear it of any foreign objects, such as vomitus, dentures, and blood clots. Remove the oral airway if one has been inserted. If the patient has no spinal injury, open the airway by hyperextending the head and neck. With a trauma patient, maintain the neck in a neutral in-line position during insertion of the PtL.
4. **Preoxygenate the patient** with 100% oxygen with the BVM device.
5. **Lift the lower jaw and tongue** away from the posterior pharynx by inserting your thumb deep into the patient's mouth and grasping the tongue and lower jaw between your thumb and index finger.
6. **Hold the PtL** so that it curves in the same direction as the natural curvature of the pharynx. Gently guide the PtL along the base of the tongue and into the airway until the teeth are against the teeth strap. If resistance is met, pull back and redirect the PtL. When the PtL is at the proper depth, place the neck strap over the patient's head and tighten.
7. **Inflate the balloon cuffs** with the No. 1 tube. Be sure to close the white cap.
8. **Ventilate the patient** through the short, green No. 2 tube, and then check the patient's chest for equal breath sounds.
9. **If you see the chest rise** and hear equal breath sounds, the No. 3 tube is in the esophagus, and you should ventilate the patient with the No. 2 tube. If the chest does not rise and fall and you do not hear equal breath sounds, the No. 3 tube is likely in the trachea. In this case, you should remove the stylet from the No. 3 tube and ventilate the patient using the No. 3 tube.
10. **Verify that the patient** is receiving adequate ventilation by listening to the lungs on both sides of the chest anteriorly, in both axillae, and over the stomach as you ventilate the patient with a BVM device.

11. **Continuously monitor the patient.** Watch for balloon cuff leaks, and use inlet tube No. 1 to keep the balloon cuffs properly inflated. Jagged, broken teeth, dentures, and bones can easily tear balloon cuffs, so you must use special care if the patient has facial trauma.

Removing the PtL

Removing the PtL is a simple procedure. You should remove the PtL if the patient will no longer tolerate it. The patient will likely vomit when the PtL is removed, so you must keep a suctioning unit readily available. Turn the patient to one side to help keep the airway clear of vomitus. When you are ready to remove the PtL, simply deflate the balloon cuffs and gently remove the tube.

Laryngeal Mask Airway

The laryngeal mask airway (LMA) was originally developed for use in the operating room. However, since its inception, its use has been expanded to the field, especially as an alternative for BLS providers.

The LMA consists of two parts: the tube and the mask or cuff. The device is made of silicone and is available in reusable (after proper sterilization) and disposable types. After blind insertion, the device molds and seals itself around the laryngeal opening by inflation of the mask. The epiglottis is contained within the mask or cuff. The device comes in seven sizes and can be used in children as well as adults (Table 39-4 ▼).

Use of the device requires training with a manikin. Proponents advocate additional training in an operating room with anesthetized patients to become familiar with the anatomic variations that cannot be duplicated by a manikin. This type of training helps avoid malpositioning. Malpositioning may occur when

the device is located somewhere other than over the laryngeal opening or when the device is "folded over" and caught on top of the epiglottis. Malpositions can sometimes be managed by repositioning the patient's head, readjusting the LMA position, or adjusting the amount of air in the cuff. When in doubt, remove the device, ventilate the patient, and start over.

Contraindications

There are potential contraindications for use of the device. When positive-pressure ventilation with high airway pressures is required (as in a patient with asthma or chronic obstructive pulmonary disease), the mask may leak. Active vomiting may dislodge the device. Large esophageal tumors may prevent effectiveness of the LMA.

Inserting the LMA

After confirming your decision with medical control to insert an LMA, remember to follow BSI precautions, including the use of gloves, eye protection, and a mask. Ensure that the patient has no gag reflex, then follow the steps in (Skill Drill 39-3 ▶) to insert the LMA:

1. **Assemble and check** the proper equipment:
   - BSI equipment
   - LMA and syringe (check the cuff and valve)
   - Water-soluble lubricant
   - Suctioning unit with suctioning catheters
   - BVM device and oxygen
2. **Open the patient's airway**, and clear it of any foreign material. Insert an oropharyngeal airway.
3. **Preoxygenate the patient** with a BVM device at the appropriate rate, depending on the age of the patient (20 to 24 breaths/min is typical for adults) (**Step 1**).
4. **Select the appropriate LMA size** as your partner ventilates the patient, and verify that the mask has no leaks. Fully deflate the mask and lubricate it (**Step 2**).
5. **Position the LMA** for insertion with the mask down, holding it like a pen (**Step 3**).
6. **Remove the oropharyngeal airway**, and, keeping the patient's neck flexed and with your partner holding the mouth open, begin to insert the mask (**Step 4**).
7. **When the widest part of the mask** is past the teeth, use your index finger to maintain a continuous forward pressure, sliding the LMA over the hard palate and soft palate into the hypopharynx until definite resistance is felt (**Step 5**).

TABLE 39-4   LMA Sizes

LMA Size	Patient's Weight or Size	Maximum Air in Mask (mL)
1	<5 kg (11 lb)	4
1.5	5 to 10 kg (11 to 22 lb)	7
2	10 to 20 kg (22 to 44 lb)	10
2.5	20 to 30 kg (44 to 66 lb)	14
3	30 kg (66 lb) to small adult	20
4	Adult	30
5	Large adult	40

Using an LMA

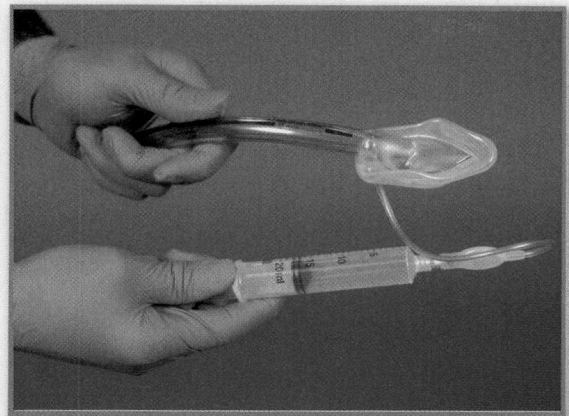

1 Assemble and check equipment.

Open and clear the airway, and insert an oral airway.

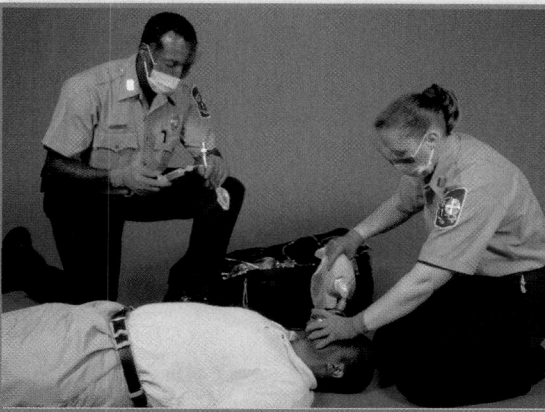

2 Open the airway, insert an oropharyngeal airway, and preoxygenate the patient.

Choose the tube size, check the mask for leaks, and deflate and lubricate it.

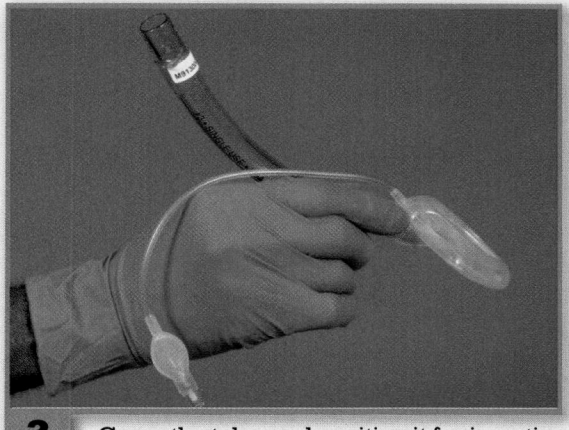

3 Grasp the tube, and position it for insertion, mask down.

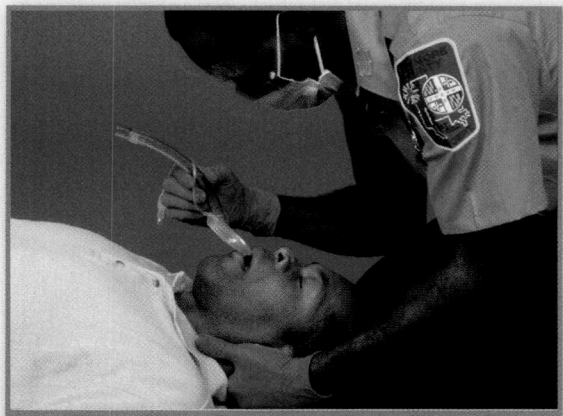

4 Remove the oral airway, and begin to insert the mask.

Continued.

8. **While removing your index finger,** gently press down on the tube with the other hand to prevent the LMA from being pulled out of place (**Step 6**).

9. **After ensuring** that the black line on the LMA is facing the upper lip, inflate the mask without holding the tube. Short outward movement is normal (**Step 7**).

10. **Confirm placement of the LMA** by listening with a stethoscope over the stomach and lungs. You should hear equal breath sounds and see equal chest rise and fall; epigastric sounds should not be heard. Further confirmation with a secondary confirmation device such as an end-tidal carbon dioxide detector may be directed by medical control. Combativeness in a previously unconscious, unresponsive patient indicates that ventilations are successful and the patient has received adequate oxygen. Once this happens, the LMA must be removed immediately to avoid stimulating vomiting.

11. **Once you have verified tube placement,** insert a bite block, such as an oropharyngeal airway, and secure the tube in place with tape or a commercial device. As with other devices, if this device is not properly secured, it will dislodge, preventing oxygen from entering the lungs (**Step 8**).

Using an LMA continued

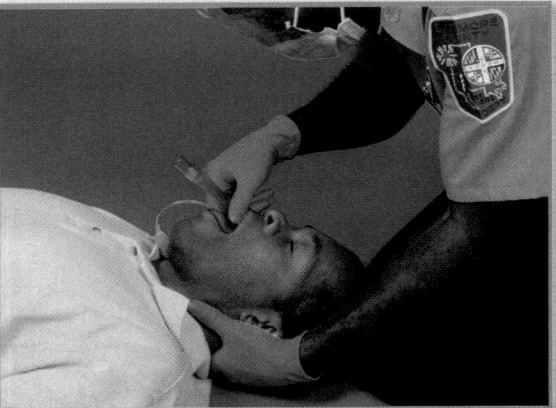

5 By using your index finger, push the mask up onto the hard palate and advance it until you feel resistance.

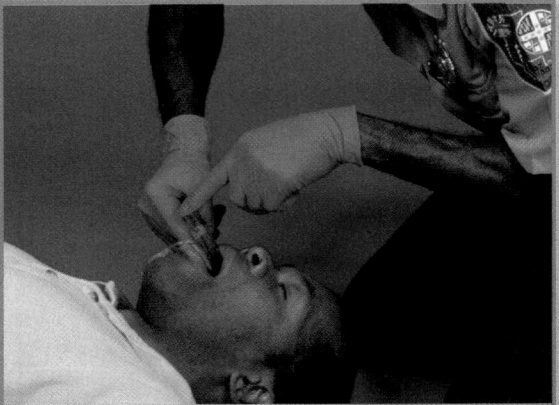

6 Use the other hand to stabilize the tube as you remove your index finger.

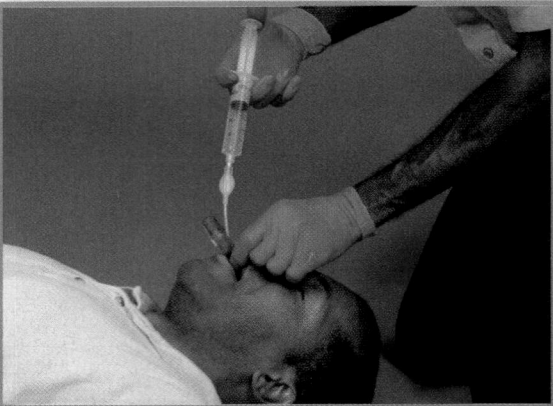

7 Check that the black line on the tube faces the upper lip, then inflate the tube without holding it.

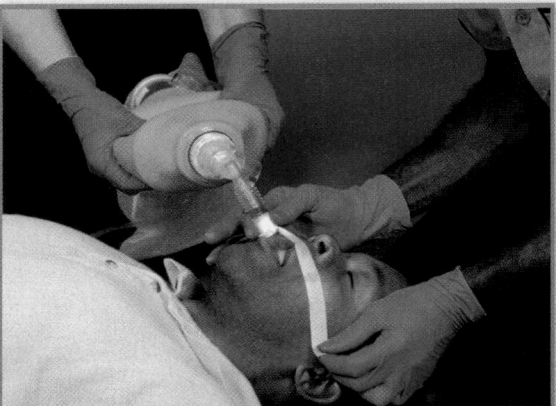

8 Confirm tube placement by listening over the stomach and lungs, and with a secondary confirmation device.

Insert a bite block, and secure the LMA.

You are the Provider

Summary

This scenario involved a patient in respiratory arrest who required advanced airway management. Although the EMT-B in this scenario had difficulty providing adequate ventilations with a BVM device, it is important to note that initial airway management should always begin by using basic techniques (that is, oral airway, BVM device), regardless of your level of certification. Then, if indicated and authorized by medical control, an advanced airway device can be placed.

Relative to simple BVM ventilations and an oral or nasal airway, endotracheal intubation provides definitive control of the airway—preventing gastric distention and aspiration. However, the use of advanced airway devices is not without risk; complications can occur if proper technique is not used. Such complications include improper placement, worsened hypoxia, and damage to the soft tissues in the mouth. Advanced airway placement requires authorization from medical control, considerable practice, and experience; practice with live patients in the operating room is optimal but not always possible.

After inserting an advanced airway device, it is paramount to ensure proper placement—immediately after placement and after any move of the patient. Initial confirmation is performed by auscultating over the epigastrium and lungs; secondary confirmation is performed by using an end-tidal carbon dioxide detector or similar device.

Proper securing of the airway device is also crucial; a commercial securing device is recommended for this purpose. Frequently reassess the patient and observe for signs of inadvertent misplacement of the device (that is, falling oxygen saturation, cyanosis). Should this occur, immediately remove the airway device and ventilate the patient with a BVM device and 100% oxygen.

Because of the risk of exposure to blood or other body fluids, BSI precautions, including gloves, a face shield, and mask, should always be used when managing a patient's airway.

Prep Kit

Ready for Review

- Gastric tubes provide a channel directly into a patient's stomach, allowing you to remove gas, blood, and toxins or to administer medications and nutrition. In the field, gastric tubes are most commonly used to decompress the stomach of a patient with gastric distention.

- There are two types of gastric tubes: nasogastric tubes and orogastric tubes. An orogastric tube, which is inserted through the mouth, is safer and easier to use. A nasogastric tube, which is inserted through the nose, can cause nasal trauma with bleeding and, in patients with a skull fracture, can be accidentally passed into the brain. Either type of tube can be accidentally passed into the trachea.

- Insertion of a gastric tube is a delicate task that must be done strictly according to local EMS protocol, with special care being taken for a patient who has head, spinal, or major facial trauma.

- Endotracheal intubation, the insertion of a tube into the trachea to maintain and protect a patient's airway, can be done through the mouth (orotracheal intubation) or through the nose (nasotracheal intubation).

- Nasal intubation can be performed only in patients who are breathing; it should not be performed in patients with a possible skull fracture. Oral intubation is needed for patients who are unconscious, unresponsive, or in cardiac arrest.

- Orotracheal intubation controls and protects the airway and may be used long-term if necessary. It also permits direct access to the trachea for suctioning, the delivery of high volumes of oxygen at higher than normal pressures, and the administration of certain medications.

- There are two techniques for insertion of an endotracheal tube: using a laryngoscope and using a lighted stylet.

- Some of the equipment needed for endotracheal intubation includes the correctly sized ET tube; a laryngoscope or lighted stylet, stylet, and syringe; oxygen with a BVM device; a suctioning unit; and Magill forceps (if a laryngoscope is used).

- Complications of endotracheal intubation include intubation of the right mainstem bronchus or the esophagus, aggravation of a spinal injury, increased hypoxia because of taking too long to intubate, causing soft-tissue trauma, mechanical failure, and a decrease in heart rate. A patient who regains consciousness may also vomit and/or try to remove the tube.

- A patient who has been intubated must be monitored continuously to evaluate the heart rate (especially in children) and lung sounds. In addition, movement of the patient can dislodge the ET tube. Always reassess breath sounds after any patient movement.

- Multilumen airways, such as the Esophageal Tracheal Combitube and the pharyngeotracheal lumen airway, and the laryngeal mask airway are inserted blindly and are easier to insert than an ET tube. However, you must be trained and authorized to use these devices.

Technology

www.EMTB.com

| Interactivities |
| Vocabulary Explorer |
| Anatomy Review |
| Web Links |
| Online Review Manual |

Vital Vocabulary

cricoid cartilage A rigid, ring-shaped structure that completely encircles the larynx at the top of the trachea.

cricoid pressure A technique that is used with intubation in which pressure is applied on either side of the cricoid cartilage to prevent gastric distention and aspiration and allow better visualization of vocal cords; also called the Sellick maneuver.

decompress To release from pressure or compression.

end-tidal carbon dioxide detector Plastic, disposable indicator that signals by color change when an endotracheal tube is in the proper place.

endotracheal intubation Insertion of an endotracheal (ET) tube directly through the larynx between the vocal cords and into the trachea to maintain and protect an airway.

Esophageal Tracheal Combitube (ETC) A multilumen airway that consists of a single, dual-lumen tube with two cuffs.

extubation Removal of a tube after it has been placed.

gastric tube An advanced airway adjunct that provides a channel directly into a patient's stomach, allowing you to remove gas, blood, and toxins or to instill medications and nutrition.

laryngeal mask airway (LMA) An advanced airway device that is blindly inserted into the mouth to isolate the larynx for direct ventilation; consists of a tube and a mask or cuff that inflates to seal around the laryngeal opening.

laryngoscope An instrument used to give a direct view of the patient's vocal cords during endotracheal intubation.

laryngospasm Spasm of the larynx and surrounding structures.

lighted stylet An instrument used to aid in blind insertion of an endotracheal tube.

multilumen airways Advanced airway devices, such as the Esophageal Tracheal Combitube and the pharyngeotracheal lumen airway, that have multiple tubes to aid in ventilation and will work whether placed in the trachea or esophagus.

nasotracheal intubation Endotracheal intubation through the nose.

orotracheal intubation Endotracheal intubation through the mouth.

pharyngeotracheal lumen airway (PtL) A multilumen airway that consists of two tubes, two masks, and a bite block.

Sellick maneuver A technique that is used with intubation in which pressure is applied on either side of the cricoid cartilage to prevent gastric distention and aspiration and allow better visualization of vocal cords; also called cricoid pressure.

stylet A plastic-coated wire that gives added rigidity and shape to the endotracheal tube.

vallecula The space between the base of the tongue and the epiglottis; receives the tip of a curved laryngoscope blade during endotracheal intubation.

Prep Kit continued...

Points to Ponder

You are providing positive-pressure ventilation to a 55-year-old apneic man when he begins to vomit. You and your partner roll the patient onto his side and suction his airway. After returning him to a supine position, your partner continues positive-pressure ventilation. However, he is having difficulty maintaining an adequate mask-to-face seal, resulting in ineffective ventilation. You look in the airway bag for a pocket mask, but are unable to find one. You and your partner agree that the patient needs to be intubated. You are both trained to intubate; however, you are unable to contact medical control for authorization. The only available paramedic unit is approximately 15 minutes from the scene; the closest hospital is 30 miles away. Should you intubate the patient? Should you wait for the paramedics to arrive to perform intubation?

Issues: Decision Making and Critical Thinking, Ensuring the Presence of Proper Equipment, Best Interest of the Patient.

Assessment in Action

You and your partner have just begun eating dinner during an extremely busy shift when you hear, "Rescue 5, PD requests your assistance at the Pleasant Hills RV Park for an unconscious male." Your response time to the scene is less than 5 minutes.

Upon arriving at the scene and exiting the ambulance, a police officer tells you that he believes the patient to be "extremely intoxicated." Your partner immediately assesses the patient and determines that he is unconscious and not breathing. After inserting an oral airway and initiating ventilation with a BVM device and 100% oxygen, you assess for a carotid pulse, which is regular, weak, and rapid. You contact medical control and receive authorization to intubate the patient. As you are preparing the ET tube and laryngoscope, the patient begins to vomit.

1. Your initial priority is to:

   A. perform immediate intubation.
   B. remove the oral airway.
   C. suction and then remove the oral airway.
   D. provide continuous suctioning until the patient stops vomiting.

2. How would aspirated vomitus be MOST detrimental to this patient?

   A. It causes severe tachycardia.
   B. It causes severe bradycardia.
   C. It will cause pneumonia and fever.
   D. It impairs gas exchange in the lungs.

3. A single endotracheal intubation attempt should not exceed:

   A. 10 seconds.
   B. 15 seconds.
   C. 30 seconds.
   D. 60 seconds.

4. Why is it important to preoxygenate a patient before you attempt to intubate?

   A. It minimizes gastric distention.
   B. It prevents accidental intubation of the esophagus.
   C. It provides additional oxygen for the brain and heart.
   D. It makes passing the tube through the vocal cords easier.

5. Which of the following airway devices features a dual lumen?

   A. Laryngeal mask airway
   B. Oropharyngeal airway
   C. Endotracheal tube
   D. Esophageal Tracheal Combitube

6. What procedure helps prevent vomiting while at the same time improving visualization of the vocal cords?

   A. Heimlich maneuver
   B. Gallop procedure
   C. Sellick maneuver
   D. Marriott procedure

7. Once you have inserted the ET tube, you need to confirm placement by listening with the stethoscope over:

   A. either the right or left lung.
   B. one lung and then the stomach.
   C. both lungs only.
   D. the epigastrium and both lungs.

Challenging Questions

8. What should you suspect if this intubated patient became cyanotic and his oxygen saturation fell to 84%? How would you manage the situation?

9. Why does an adult patient's heart rate improve when he or she is adequately ventilated with 100% oxygen?

10. What is the purpose of the cuff on the distal end of the endotracheal tube? How do you check the integrity of the distal cuff?

www.EMTB.com

Assisting With Intravenous Therapy

Objectives*

Cognitive

1. Know the types of IV fluid used in the prehospital setting. (p 1167)
2. Analyze and differentiate between the various intended applications for each of the IV solutions. (p 1167)
3. Analyze and differentiate between administration sets and their appropriate applications. (p 1167)
4. Analyze and differentiate between the various types of catheters used in IV therapy and their appropriate use. (p 1168)
5. Analyze and discuss the need for properly securing the IV tubing to the patient following IV insertion. (p 1170)
6. Analyze the need for alternative IV insertion sites and equipment, and differentiate between them:
   - Saline locks (buff caps)
   - Intraosseous needles
   - External jugular IVs (p 1170)
7. Analyze and differentiate between the various types of local and systemic complications in IV therapy:
   - Infiltration
   - Phlebitis
   - Occlusion
   - Vein irritation
   - Hematoma
   - Allergic reactions
   - Air embolus
   - Catheter shear
   - Circulatory overload
   - Vasovagal reaction (p 1172-1175)
8. Correctly define the following terms:
   - Access port
   - Crystalloid
   - Piercing spike
   - Drip set
   - Macrodrip
   - Microdrip
   - Drip chamber
   - Keep-vein-open (KVO)
   - Butterfly catheter
   - Over-the-needle catheter (p 1167, 1168)
9. Analyze and appreciate the differences in treatment required for pediatric IV therapy. (p 1175)
10. Analyze and appreciate the differences in treatment required for geriatric IV therapy. (p 1175)

Affective

11. Apply and maintain proper body substance isolation throughout the entire IV therapy process. (p 1166)
12. Explain the concept of IV equipment assembly before any catheter insertion. (p 1166)
13. Explain and appreciate the special requirements and training needed for alternative IV sites:
    - Saline lock (buff cap)
    - Intraosseous needles
    - External jugular IVs (p 1170)
14. Understand possible complications associated with IV therapy. (p 1172, 1175)
15. Explain how to troubleshoot and correct complications associated with IV therapy. (p 1175)
16. Appreciate the limits of fluid administration for both geriatric and pediatric patients. (p 1175)

Psychomotor

17. Demonstrate the proper sterile technique for assembly of the IV equipment, including:
    - Gloves
    - 4" x 4" gauze sponges
    - Proper IV tape (p 1168)
18. Spike the IV bag with the proper IV administration set. Correctly fill the administration set, including the drip chamber. (p 1168)
19. Demonstrate the proper technique for securing the IV tubing to the patient. (p 1170)
20. Demonstrate the proper technique for choosing age-appropriate catheter sizes for pediatric and geriatric patients. (p 1175)

*All of the objectives in this chapter are noncurriculum objectives.

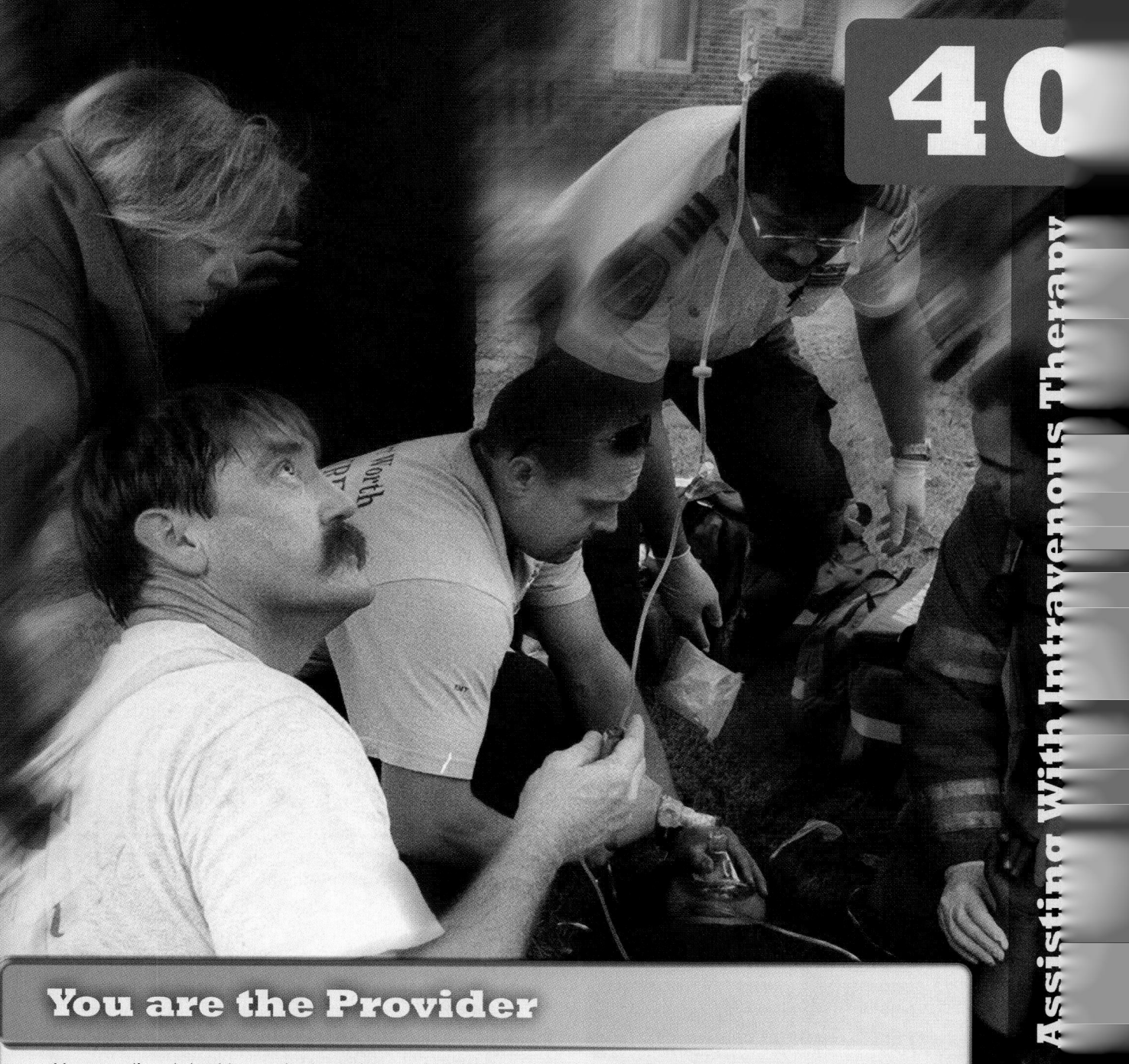

You are the Provider

You are dispatched to a private residence for a call involving respiratory distress. On arrival, you find a 74-year-old woman sitting at her kitchen table in the tripod position complaining of difficulty breathing. The patient's skin is pale and diaphoretic. Her pulse is between 120 and 130 beats/min and irregular; her blood pressure is 90/54 mm Hg; and her respirations are 32 breaths/min and shallow. She is coughing foamy sputum into a tissue. She denies having chest pain, but places her hand on her chest and whispers "pressure." You learn that she became very dyspneic after walking several blocks home after her car had a flat tire. Her list of medications, found in a "Vial of Life" on her refrigerator, includes Lasix, Lanoxin, Coumadin, potassium, and an albuterol inhaler.

1. What other signs and symptoms would you expect to find?
2. What immediate treatment could you initiate to help relieve her distress?

Introduction

This chapter is designed to familiarize you in assisting an advanced life support (ALS) partner in setting up the equipment necessary to gain intravenous (IV) access. You will learn about the equipment used and understand the importance of early access and how to recognize complications when they happen.

IV Techniques and Administration

The most important thing to remember about IV techniques and fluid administration is to keep the IV equipment sterile. Forethought will help prevent mental and procedural errors while inserting an IV needle or catheter.

Teamwork is critical to good patient care. As an EMT-B, you are a most valuable member of the team. Learning how to assemble the equipment is often done on the job. You may hear your ALS partner ask you to "Spike 1,000 LR with a macro and get me a 16-gauge catheter." "Spike" means attaching the tubing to an IV bag; "1,000" means the number of milliliters, or size, of the bag; "LR" means lactated Ringer's solution; "macro" means the size of drip chamber to choose for the tubing; and "16-gauge" refers to the size of catheter.

One way to ensure proper technique is to develop a routine to follow as you assemble the appropriate equipment. A routine will help you keep track of your equipment and the steps necessary to complete successful IV administration.

Assembling Your Equipment

To avoid delays or the possibility of IV site contamination, gather and prepare all your equipment before the attempt to start IV administration. Sometimes the condition and presentation of the patient make full preparation difficult. In this situation, working as a team becomes critical. By anticipating the needs of your ALS partner, you can help make the IV equipment assembly possible. (Table 40-1 ▼) shows a logical sequence of steps for assembling your equipment. Depending on local protocols, your ALS partner will take care of inserting the catheter, but your assistance will be very helpful in other areas.

Choosing an IV Solution

While ALS providers are likely to select the solution to use, this section discusses likely possibilities so that you may become familiar with them. In the

www.EMTB.com

Technology

- Interactivities
- Vocabulary Explorer
- Anatomy Review
- Web Links
- Online Review Manual

TABLE 40-1 EMT-B Steps in Assembling IV Equipment

1. Get your gloves on! BSI precautions cannot be emphasized strongly enough.
2. Obtain the solution requested by your ALS partner—check the bag for clarity, expiration date, and correct solution.
3. Choose an appropriate administration set for the patient.
4. Obtain the catheter requested by your ALS partner. Have a couple of catheters ready for insertion.
5. Spike the bag by inserting the administration set into the port in the fluid bag.
6. Allow fluid to pass through the administration set to completely displace all of the air in the tubing.
7. Tear tape for securing the IV site.
8. Open an alcohol wipe.
9. Have 4″ x 4″ pieces of gauze ready for catching blood.
10. After your ALS partner has inserted the catheter, adequately dispose of sharps.
11. Hook up the IV tubing, and adjust the flow.

prehospital setting, the choice of IV solution is limited to the isotonic crystalloids, normal saline and lactated Ringer's solution. D_5W (5% dextrose in water) is often reserved for administering medication.

Each IV solution bag is wrapped in a protective sterile plastic bag and is guaranteed to remain sterile until the posted expiration date. Once the protective wrap is torn and removed, the IV solution has a shelf life of 24 hours. The bottom of each IV bag has two ports: an injection port for medication, and an access port for connecting the administration set (Figure 40-1 ▼). The sterile access port is protected by a removable pigtail. Once this is removed, the bag must be used immediately or discarded.

Bags of IV solution come in different fluid volumes (Figure 40-2 ▼). The more common prehospital volumes are 1,000 mL and 500 mL. The smaller volumes (250 mL and 100 mL) more commonly contain D_5W and are used for mixing and administering medication.

Choosing an Administration Set

An administration set moves fluid from the IV bag into the patient's vascular system. As with IV solution bags, IV administration sets are sterile as long as they remain in their protective packaging. Once they are removed from the packaging, their sterility cannot be guaranteed. Each IV administration set has a piercing spike

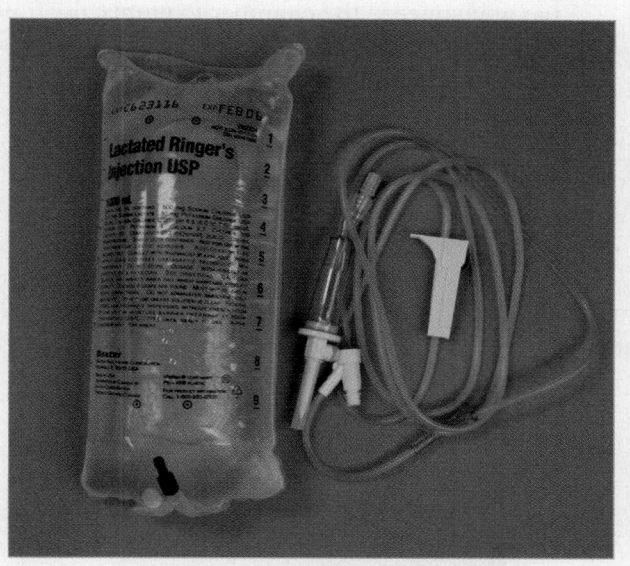

Figure 40-1 An IV bag with an administration set.

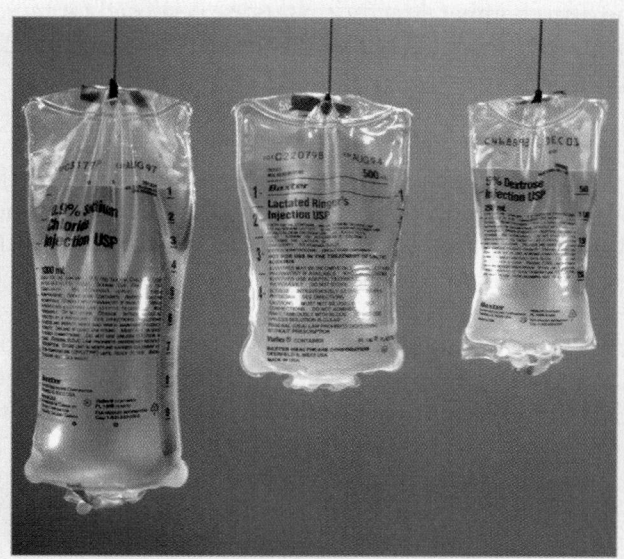

Figure 40-2 Examples of different IV bag sizes.

You are the Provider   Part 2

You notice that the patient's feet and ankles are swollen. She tells you she saw her doctor 3 days ago and was told she has atrial fibrillation, but she refused to be hospitalized. She is taking Lasix, which you recognize as a diuretic, and Lanoxin to control a cardiac condition. She says that her albuterol inhaler is not helping her breathe better. Her pulse oximetry reading is 87%. She appears to be getting quite agitated during your questioning.

3. Is the edema in her lower extremities a significant finding?

protected by a plastic cover. Again, once the piercing spike is exposed and the seal surrounding the cap is broken, the set must be used immediately or discarded.

There are different sizes of administration sets for different situations and patients. Most <u>drip sets</u> have a number visible on the package (Figure 40-3 ▼), which indicates the number of drops it takes for a milliliter of fluid to pass through the orifice and into the <u>drip chamber</u>. Drip sets commonly used in the prehospital environment come in two primary sizes: microdrip and macrodrip. <u>Microdrip sets</u> allow 60 gtt (drops)/mL through the small, needlelike orifice inside the drip chamber. Microdrips are ideal for medication administration or pediatric fluid delivery because it is easy to control their fluid flow. <u>Macrodrip sets</u> allow 10 to 15 gtt/mL through a large opening between the piercing spike and the drip chamber. Macrodrip sets are best used for rapid fluid replacement but can also be used for maintenance and <u>keep-the-vein-open (KVO) IV setups</u>.

Preparing an Administration Set

After choosing the IV administration set and the IV solution bag, verify the expiration date of the solution and check for solution clarity. Prepare to spike the bag with the administration set as follows (Skill Drill 40-1 ▶).

1. **Remove the rubber pigtail** found on the end of the IV bag by pulling on it. The bag is still sealed and will not leak until the piercing spike of the IV administration set punctures this port.
2. **Remove the protective cover** from the piercing spike (remember, this spike is sterile!) (**Step 1**).

3. **Slide the spike into the IV bag port** until you see fluid enter the drip chamber (**Step 2**).
4. **Allow the solution to run freely** through the drip chamber and into the tubing to prime the line and flush the air out of the tubing (**Step 3**).
5. **Twist the protective cover** on the opposite end of the IV tubing to allow air to escape. Carefully remove the cover without breaching sterility. Let the fluid flow until air bubbles are removed from the line before turning the roller clamp wheel to stop the flow (**Step 4**). Replace the cover.
6. Next, go back and **check the drip chamber**; it should be only half filled. The fluid level must be visible to calculate drip rates. If the fluid level is too low; squeeze the chamber until it fills; if the chamber is too full, invert the bag and the chamber and squeeze the chamber to empty the fluid back into the bag (**Step 5**).
7. **Hang the bag** in the appropriate location with the end of the IV tubing easily accessible.

Catheters

A <u>catheter</u> is a hollow, laser-sharpened needle inside a hollow plastic tube inserted into a vein to keep the vein open (Figure 40-4 ▼). The most common types of catheters found in the prehospital setting are <u>butterfly catheters</u> and <u>over-the-needle catheters</u> (Figure 40-5 ▶). Advanced

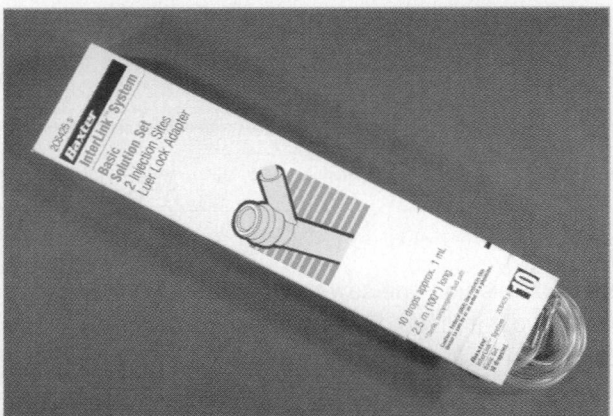

Figure 40-3 The number visible on the drip set refers to the number of drops it takes for a milliliter of fluid to pass through the orifice and into the drip chamber.

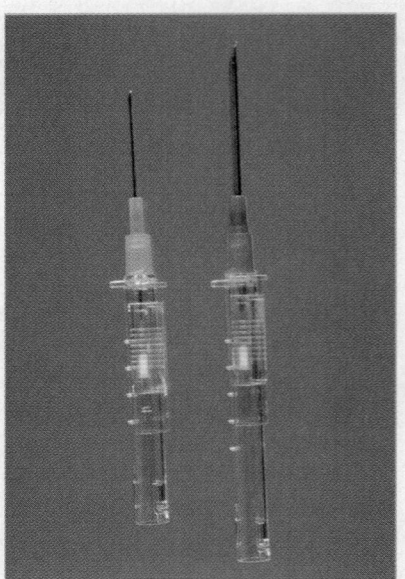

Figure 40-4 Over-the-needle catheters are commonly used in the prehospital environment.

Spiking the Bag

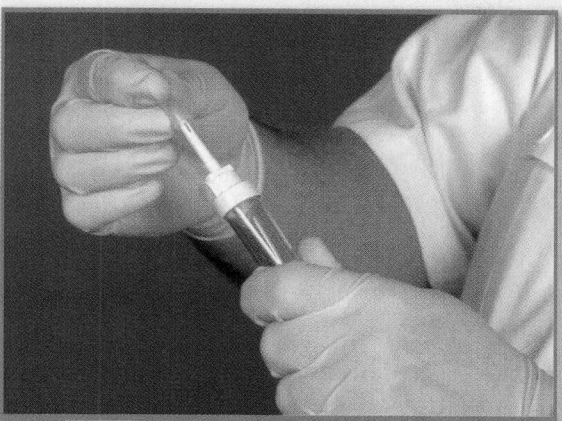

1 Remove the rubber pigtail found on the end of the IV bag by pulling on it.

Remove the protective cover from the piercing spike.

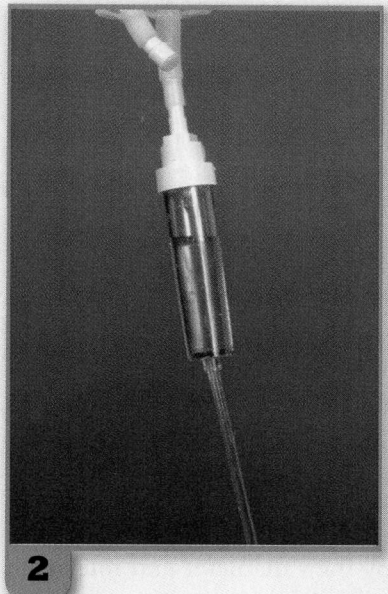

2 Slide the spike into the IV bag port until you see fluid enter the drip chamber.

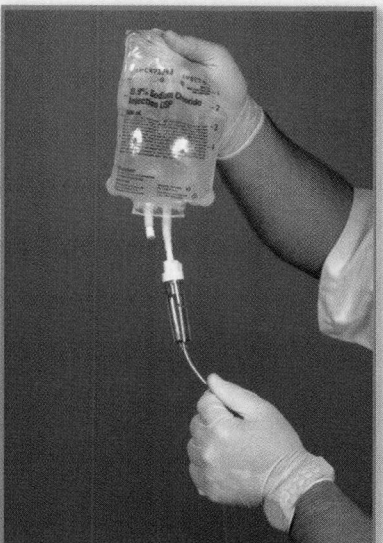

3 Allow the solution to run freely through the drip chamber and into the tubing to prime the line and flush the air out of the tubing.

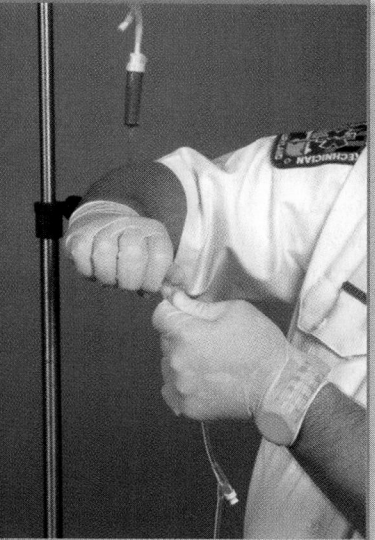

4 Twist the protective cover on the opposite end of the IV tubing to allow air to escape. Remove the cover. Let the fluid flow until air bubbles are removed from the line before turning the roller clamp wheel to stop the flow. Replace the cover.

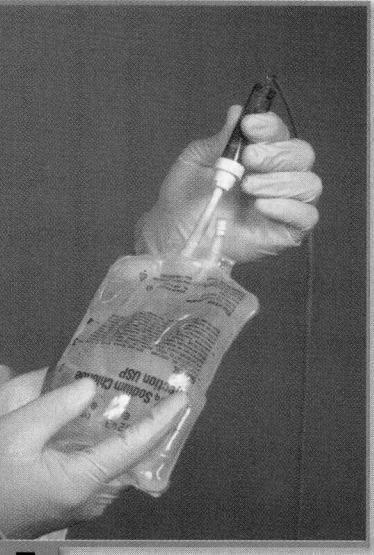

5 Check the drip chamber; it should be only half filled. If the fluid level is too low, squeeze the chamber until it fills; if the chamber is too full, invert the bag and the chamber and squeeze the chamber to empty the fluid back into the bag.

Hang the bag in the appropriate location with the end of the IV tubing easily accessible.

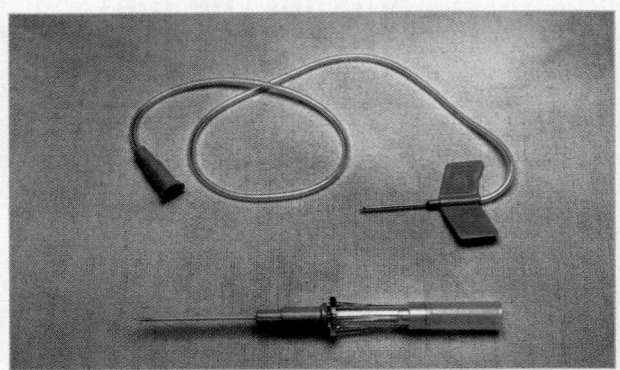

Figure 40-5 A butterfly catheter (top) and an over-the-needle catheter (bottom).

✚ **EMT-B Tips**

To differentiate between macrodrip and micro-drip sets, remember that the prefixes refer to the size of the drops, not the size of the tubing.

Macro means large. A 10-gtt set, which is a macrodrip set, has 10 drops that equal 1 mL of fluid. *Micro* means small. A 60-gtt set, which is a microdrip set, has 60 drops that equal 1 mL of fluid.

life support providers select the catheter based on the need for the IV, the age of the patient, and the location for the IV.

Catheters are sized by their diameter and referred to by the gauge of the catheter. A larger-diameter catheter corresponds to a smaller gauge. Thus, a 14-gauge catheter has a greater diameter than a 22-gauge catheter. With larger-diameter catheters, more fluid can be delivered into the vein faster.

Securing the Line

Once the catheter is in position and the contents of the IV bag are flowing properly, the site must be secured. Tape the area so that the catheter and tubing are securely anchored in case of a sudden pull on the line (Figure 40-6 ▼). You should tear the tape before the IV catheter is inserted, because you will need one hand to stabilize the site while you tape the IV catheter and tubing. Double back the tubing to create a loop that will act as a shock absorber if the line is pulled accidentally. Avoid circumferential taping around any extremity because circumferential taping can act like a constricting band and stop circulation.

Alternative IV Sites and Techniques

Saline locks (buff caps) are a way to maintain an active IV site without having to run fluids through the vein. These access devices are used primarily for patients who do not need additional fluids but may need rapid medication delivery. A saline lock is attached to the end of an IV catheter and filled with approximately 2 mL of normal saline to keep blood from clotting at the end of the catheter (Figure 40-7 ▼). Saline remains in the port without entering the vein.

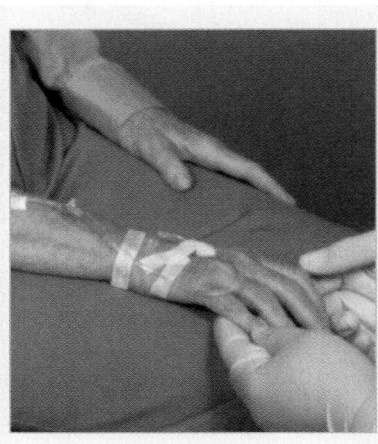

Figure 40-6
Tape the area so that the catheter and tubing are securely anchored.

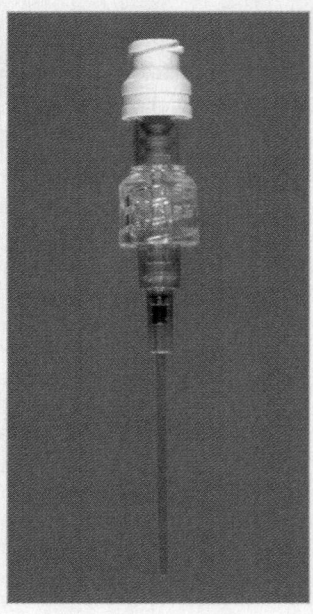

Figure 40-7 A saline lock is attached to the end of an IV catheter and filled with approximately 2 mL of normal saline.

✳ EMT-B Tips

When starting an IV on a patient who is frightened of needles, make sure he or she is lying down before the paramedic begins catheter insertion. Advise the patient of each step—even when you are only cleansing the site.

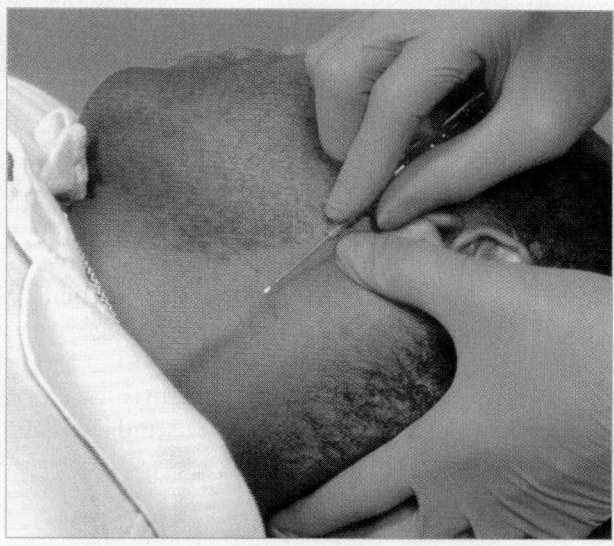

Figure 40-8 The external jugular IV requires a very specific insertion site midway between the angle of the jaw and the midclavicular line with the catheter pointed toward the shoulder of the same side as the puncture.

Intraosseous (IO) needles are used for emergency venous access in pediatric patients as defined by protocol when immediate IV access is difficult or impossible. Often these children are experiencing a life-threatening situation such as cardiac arrest, status epilepticus, or progressive shock. The IO needles are inserted in the proximal tibia with a rigid, boring IV catheter, commonly known as a Jamshedi needle. This double needle, consisting of a solid boring needle inside a sharpened hollow needle, is pushed into the bone with a screwing, twisting action.

External jugular IVs provide venous access through the external jugular veins of the neck. These are the same veins used to assess jugular vein distention (JVD). The vein is compressed by placing a finger on the vein above the clavicle, causing the vein to fill.

The catheter is inserted into the vein in the same manner as any other IV catheter, except the insertion point is very specific. The catheter is inserted midway between the angle of the jaw and the midclavicular line, with the catheter pointed toward the shoulder on the same side as the puncture site (Figure 40-8 ▲). These

You are the Provider   Part 3

You set up a nonrebreathing mask and attempt to apply it to your patient. She slumps in her chair and appears lethargic. She holds the mask to her face as she gasps for more breath. You can hear wet lung sounds without a stethoscope, and your partner listens carefully to determine diminished or absent lung sounds in specific lung fields. You set up the portable suction machine and, with the patient's permission, begin to suction. Your partner has laid out the airway equipment in anticipation of the need to intubate.

After clearing some secretions, the decision is made to provide positive pressure ventilation with 100% oxygen. After explaining to the patient what he is about to do, your partner asks you to set up the IV equipment while he sets up for possible intubation and looks for an IV site.

4. Why will this patient benefit from IV therapy?
5. Why did she become lethargic?

punctures are difficult because these veins are surrounded by a very tough, fibrous sheath that makes access difficult. Understanding the procedure is important because you may need to assist.

Possible Complications of IV Therapy

Peripheral IV insertion carries risks. The problems associated with IV administration can be categorized as local or systemic reactions. Local reactions include problems like infiltration and phlebitis. Systemic complications include allergic reactions and circulatory overload.

Local IV Site Reactions

Most local reactions require that the IV catheter be removed and reinserted at an alternative site. Some examples of common local reactions include the following:

- Infiltration
- Phlebitis
- Occlusion
- Vein irritation
- Hematoma

Infiltration

Infiltration is the escape of fluid into the surrounding tissue. This escape of fluid can cause a localized area of edema or simply swelling. Some of the more common reasons for infiltration include the following:

- The IV catheter has passed completely through the vein and out the other side.
- The patient is moving excessively.
- The tape used to secure the area has become loose or dislodged.
- The catheter was inserted at too shallow an angle and has only entered the tissue surrounding the vein (this is more common with IV catheters in larger veins, such as those in the upper arm and neck).

The following are some of the signs and symptoms of infiltration:

- Edema at the catheter site
- Extremely slow IV flow despite use of a large catheter
- Patient complaint of tightness and pain around the IV site

To correct the infiltration, an ALS provider must remove the IV catheter and reinsert it at an alternative

site. After this is done, you may apply direct pressure over the swollen area to reduce further swelling or bleeding into the tissue. Avoid wrapping tape around the extremity to apply direct pressure.

Phlebitis

Phlebitis is inflammation of the vein. Phlebitis is not usually seen with emergency prehospital patients, although you may encounter it in patients with IV drug abuse and in patients receiving IV therapy in a hospital outpatient treatment or home health care program. Often phlebitis is associated with fever, tenderness, and red streaking along the course of the associated vein. Some of the more common causes for phlebitis include localized irritation and infection from nonsterile equipment, prolonged IV therapy, and irritating IV solutions. If the phlebitis is associated with the IV administration you assisted with, the ALS provider must discontinue and reestablish the IV therapy at another location, using new equipment.

Occlusion

In IV therapy, occlusion is the physical blockage of a vein or catheter. If the flow rate is not sufficient to keep fluid moving out of the catheter tip and if blood enters the catheter, a clot may form and occlude the flow. The first sign of a possible occlusion is a decreasing drip rate or the presence of blood in the IV tubing. Proximity to a valve is often the reason for this problem. Other causes can be related to patient movement that allows the line to become physically blocked, such as resting on the IV line or crossing the arms. Occlusion may also develop if the IV bag is nearly empty and the blood pressure overcomes the flow and causes blood to back up into the line.

Vein Irritation

Occasionally, a patient will experience vein irritation in reaction to the IV fluid. This is more common with IV medication administration and very uncommon with administration of pure IV fluids. Patients who have this problem often complain immediately that the IV is bothering them. It may tingle, sting, or itch. Note these complaints, and observe the patient closely in case a more serious allergic reaction develops.

The cause of venous irritation is usually excessively rapid infusion of an irritating solution. If redness develops at the IV site with early phlebitis, the IV fluid should be discontinued and the administration set and fluid saved for later analysis. The ALS provider should reestablish a new IV site distant to the site of the initial reaction using all new equipment.

Hematoma

A hematoma is an accumulation of blood in the tissues surrounding an IV site. Hematomas result from vein perforation or improper catheter removal, which allows blood to accumulate in the surrounding tissues. Blood can be seen rapidly pooling around the IV site, leading to tenderness and pain (Figure 40-9 ▶). Patients with a history of vascular diseases (such as in diabetes) or patients receiving certain drug therapies (such as corticosteroids) can have a predisposition to vein rupture or have a tendency to develop hematomas rapidly with IV insertion.

If a hematoma develops when IV catheter insertion is attempted, the procedure should stop. Direct pressure should be applied to help minimize bleeding. Application of ice may help. If a hematoma develops after a successful catheter insertion, evaluate the IV flow and the hematoma. This can be done by lowering the IV bag and watching for blood backup into the line. If the hematoma appears to be controlled and the flow is not affected, monitor the IV site and leave the line in place. If the hematoma develops as a result of discontinuing the IV, apply direct pressure to the site with a piece of 4″ × 4″ gauze.

Systemic Complications

Systemic complications can evolve from reactions or complications associated with IV insertion. Systemic complications usually involve other body systems and

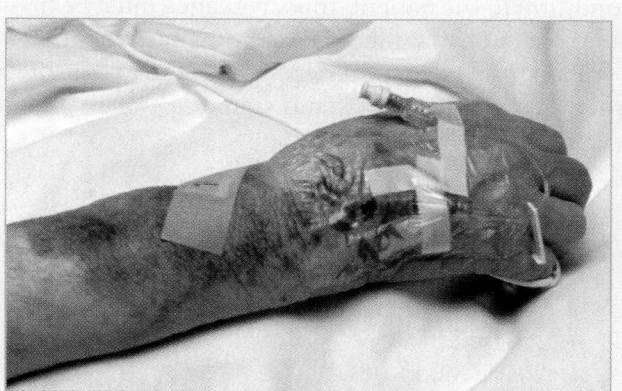

Figure 40-9 Hematomas can be caused by the improper removal of a catheter, resulting in the pooling of blood around the IV site, leading to tenderness and pain.

can be life threatening. Common systemic complications are as follows:

- Allergic reactions
- Air embolus
- Catheter shear
- Circulatory overload
- Vasovagal reactions

Allergic Reactions

Often allergic reactions are minor, but true anaphylaxis is possible and must be treated aggressively. Allergic reactions can be related to an individual's unexpected sensitivity to an IV fluid or (much more commonly) medication. Such a sensitivity could be an unknown

You are the Provider   Part 4

You place your patient in a seated position on the cot and prepare for transport. While en route, you notice the patient's skin color has improved, she appears more alert, and the oxygen saturation is now at 94%. You discontinue the BVM device and reattach the nonrebreathing mask to the oxygen, applying it to the patient. IV access is achieved with a 20-gauge needle and a microdrip set, and your partner administers IV medication, listens to lung sounds again, and takes another set of vital signs. The patient is transported without further incident.

6. Will the nonrebreathing mask keep the patient's oxygen saturation at an acceptable level?
7. Why not let the patient lie down during transport?

condition to the patient; thus, vigilance must be maintained with any IV therapy for a possible reaction.

Patient presentation depends on the extent of the reaction. Common signs and symptoms of an allergic reaction include the following:

- Itching
- Edema of face and hands
- Bronchospasm
- Wheezing
- Shortness of breath
- Urticaria
- Anaphylaxis

If an allergic reaction occurs, the ALS provider must discontinue the IV fluid and remove the solution. The catheter will be left in place as an emergency medication route. Medical control should be notified. Maintain an open airway, and monitor ABCs and vital signs. Keep the solution or medication for evaluation by the hospital.

Air Embolus

Healthy adults can tolerate as much as 200 mL of air introduced into the circulatory system, but patients who are already ill or injured can be affected if any air is introduced. Properly flushing an IV line will help eliminate any potential of introducing air into a patient. IV bags are designed to collapse as they empty to help prevent this problem, but collapse does not always occur. Be sure to replace empty IV bags with full ones.

If your patient begins developing respiratory distress, consider the possibility of an air embolus. Other associated signs and symptoms include the following:

- Cyanosis (even in the presence of high-flow oxygen)
- Signs and symptoms of shock
- Loss of consciousness
- Respiratory arrest

Treat a patient with a suspected air embolus by placing the patient on his or her left side with the head down. Be prepared to ventilate the patient if he or she experiences increasing shortness of breath. Symptomatic air embolus is an extremely rare event and should be considered only after more common explanations for the patient's presenting symptoms have been excluded.

Catheter Shear

Catheter shear occurs when part of the catheter is pinched against the needle, and the needle slices through the catheter, creating a free-floating segment. This allows the catheter segment to travel through the circulatory system and possibly end up in the pulmonary circulation, causing a pulmonary embolus.

Treatment involves surgical removal of the sheared tip. Catheter hubs are radiopaque (that is, they will appear white in an X-ray) to aid in diagnosing this type of problem. This problem is caused by rethreading needles through catheters after they have been removed. To avoid this problem, a catheter should never be rethreaded back into a needle.

Patients who have experienced catheter shear present with sudden shortness of breath and possibly diminished breath sounds. They will mimic the presentations of a patient with an air embolus and can be treated the same way. Such patients will need continued IV access. Another extremity should be used if possible.

Circulatory Overload

An unmonitored IV bag can lead to circulatory overload. Healthy adults can handle as much as 2 to 3 extra liters of fluid without compromise. Problems occur when the patient has cardiac, pulmonary, or renal dysfunction. These types of conditions do not allow the patient to tolerate the additional demands associated with increased circulatory volume. The most common cause for circulatory overload in the prehospital setting is failure to readjust the drip rate after flushing an IV line immediately after insertion. Always monitor IV bags to ensure the proper drip rate.

Patient presentation includes shortness of breath, JVD, and increased blood pressure. Crackles are often heard when evaluating breath sounds. Acute peripheral edema can also indicate circulatory overload.

To treat a patient with circulatory overload, slow the IV rate to keep the vein open and raise the patient's head to ease respiratory distress. Administer high-flow oxygen and monitor vital signs and shortness of breath. Medical control should be contacted immediately and informed of the developing problem, because there are drugs that can be given to reduce the circulatory volume.

Vasovagal Reactions

Some patients have anxiety concerning needles or in response to the sight of blood. Such anxiety may lead to a drop in blood pressure, and the patient may collapse. Patients can present with anxiety, diaphoresis, nausea, and syncopal episodes.

Treatment for patients with vasovagal reactions centers on treating them for shock:

1. Place patient in shock position.
2. Apply high-flow oxygen.

3. Monitor vital signs.
4. The ALS provider should insert an IV catheter in case fluid resuscitation is needed.

Troubleshooting

Several factors can influence the IV flow rate. For example, if the IV bag is not hung high enough, the flow rate will not be sufficient. It is always helpful to perform the following checks after completing IV administration. Also, if there is a flow problem, rechecking these items will help you determine the problem.

- **Check your IV fluid.** Thick, viscous fluids infuse slowly and may be diluted to help speed delivery. Cold fluids run slower than warm fluids. If possible, warm IV fluids should be administered during cold months.
- **Check your administration set.** Macrodrips are used for rapid fluid delivery, whereas microdrips are designed to deliver a more controlled flow.
- **Check the height of the IV bag.** The IV bag must be hung high enough to overcome the patient's own blood pressure. Hang the bag as high as possible.
- **Check the type of catheter used.** The wider the catheter (the smaller the gauge), the more fluid can be delivered; 14 gauge is the widest, 27 gauge the narrowest.
- **Check the constricting band (tourniquet).** The ALS provider applies a constricting band during the IV insertion process. Leaving the constricting band on the patient's arm after establishing IV access can prevent the IV fluid from flowing at the proper rate.

Age-Specific Considerations

Pediatric and geriatric populations warrant specific attention. What sets these populations apart are physical differences specific to these populations and communication barriers that might prevent these patients from expressing themselves.

Accordingly, pediatric and geriatric patients have different medical needs from the general medical population, making it sometimes necessary to use other methods of assessment and treatment.

IV Therapy for Pediatric Patients

The same IV solutions and equipment used for adults can be used for pediatric patients with a few exceptions.

If an over-the-needle IV catheter is used, the 20-, 22-, 24-, or 26-gauge catheters are best for insertions Figure 40-10 ▼. Butterfly catheters are ideal for pediatric patients and can be placed in the same locations as over-the-needle catheters, as well as in visible scalp veins. Scalp veins are best used in infants. Intraosseous needles can be used for difficult and emergency fluid infusions. Intraosseous administration equipment contains special needles that puncture the bone tissue of the proximal tibia, leaving the rigid catheter in place. The IV tubing is attached to the rigid catheter just as it is with flexible catheters. Stabilization is critical for these lines to maintain adequate flow. Once established, these lines work as well as peripheral lines.

Fluid control for pediatric patients is important. Using a special type of microdrip set called a Volutrol IV allows you to fill the large drip chamber with a specific amount of fluid and administer only this amount to avoid fluid overload. The 100-mL calibrated drip chamber can be shut off from the IV bag.

IV Therapy for Geriatric Patients

Smaller catheters may be preferable for geriatric patients unless rapid fluid replacement is needed. Some medications commonly used by older patients have

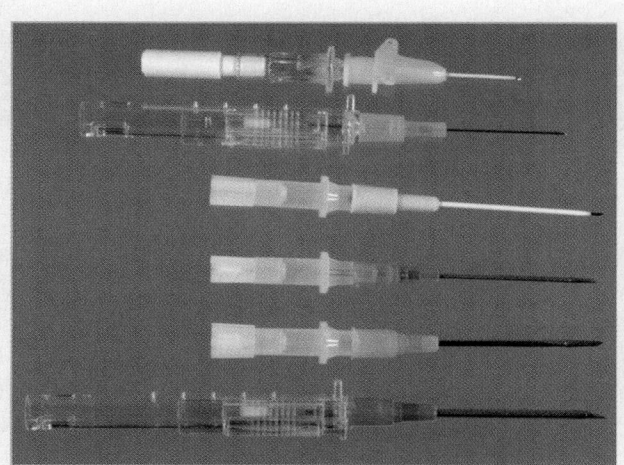

Figure 40-10 Note the difference in sizes of catheters.

the tendency to create fragile skin and veins. Often, simply puncturing the vein will cause a massive hematoma. The use of tape can lead to skin damage, so be careful when taping IV catheters and tubing on older patients.

The use of the smaller catheters (such as 20-, 22-, or 24-gauge catheters) may be more comfortable for the patient and can reduce the risk of extravasation. If fluid resuscitation is necessary, an appropriately sized catheter must be used.

Be careful when using macrodrips, because they can allow rapid infusion of fluids, which may lead to edema if they are not monitored closely. With both geriatric and pediatric patients, fluid overloading is a real possibility. If necessary, use the Volutrol IV set to prevent fluid overloads (Figure 40-11 ▶).

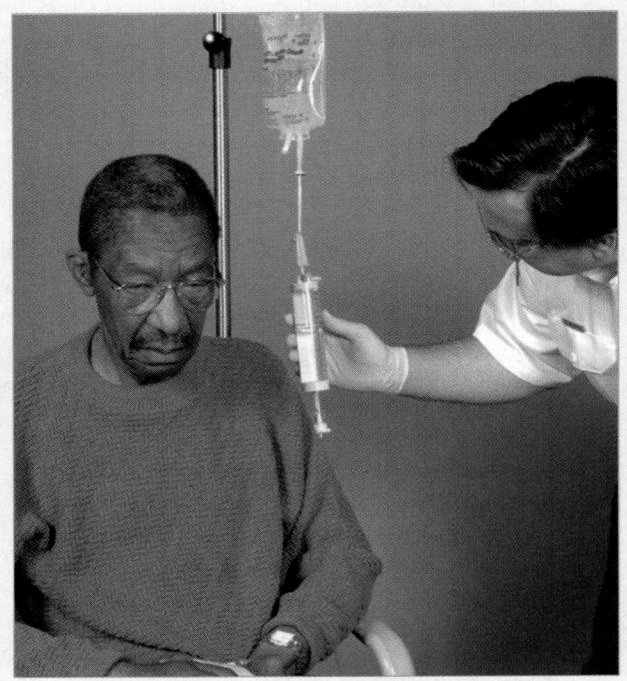

Figure 40-11 It may be necessary to use the Volutrol IV set to prevent fluid overload.

You are the Provider   Summary

Many older people do not seek medical attention, unless it becomes an emergency, out of fear of losing their independence. Discuss your concerns with the patient, explaining why evaluation at the hospital is necessary. IV access is necessary to deliver medications that will help alleviate her symptoms. She does not need additional fluid, despite a low blood pressure. Her decreased level of consciousness was due to the developing hypoxia.

After treatment with 100% oxygen by BVM device, which forced some fluid out of her lungs and improved her oxygenation, her mentation also improved. If the nonrebreathing mask failed to maintain her improved status, BVM ventilation should resume. Patients with difficulty breathing should always be transported in a position of comfort or one that is best for them to breathe easily. If conscious, they will rarely, if ever, want to lie flat.

Prep Kit

Ready for Review

- Understanding the equipment used in IV access and knowing how to prepare it will help you to be a critical member of an ALS team.

- Anticipating your partner's needs and following a logical sequence of steps in assembling the equipment will help to facilitate IV access.

- It is absolutely essential to take proper BSI precautions on any call involving IV therapy.

- In the prehospital setting, the choice of IV solution is limited to normal saline and lactated Ringer's solution.

- Areas in which EMT-Bs can assist with IV therapy include gathering and preparing all equipment; checking the bag for clarity, expiration date, and correct solution; choosing the appropriate administration set; obtaining the catheter requested by the ALS provider; and spiking the bag.

- In addition, once the catheter is in position and the contents of the IV bag are flowing properly, you should secure the site by taping the area.

- An administration set moves fluid from the IV bag into the patient's vascular system. Each IV administration set has a piercing spike protected by a plastic cover.

- Drip sets commonly used in the prehospital environment come in two primary sizes: microdrip and macrodrip. Microdrip sets allow 60 gtt/mL into the drip chamber. Macrodrip sets allow 10 to 15 gtt/mL into the drip chamber.

- A catheter is a hollow, laser-sharpened needle inside a hollow plastic tube inserted into a vein to keep the vein open. The most common types in the prehospital setting are butterfly catheters and over-the-needle catheters.

- Alternative IV sites and techniques include saline locks, IO access, and external jugular IVs. A saline lock is attached to the end of an IV catheter and filled with approximately 2 mL of normal saline to keep blood from clotting at the end of the catheter. Access by the IO route is established in the proximal tibia with a rigid, boring IV catheter. External jugular IVs provide venous access through the external jugular veins of the neck.

- Understanding the possible complications and knowing how to troubleshoot a problem is knowledge that will help make calls run more smoothly.

- Possible complications of IV therapy include local reactions and systemic complications. Local reactions include infiltration, phlebitis, occlusion, vein irritation, and hematoma. Systemic complications include allergic reactions, air embolus, catheter shear, circulatory overload, and vasovagal reactions.

- If you encounter problems with the IV flowing effectively, potential causes are the fluid used, the administration set, the height of the bag, the type of catheter used, and the presence of a constricting band, or tourniquet.

- Pediatric and geriatric populations warrant specific attention. Different equipment and sites may be used for these patients. Older skin is more delicate and must be handled with care.

Technology

- Interactivities
- Vocabulary Explorer
- Anatomy Review
- Web Links
- Online Review Manual

www.EMTB.com

Prep Kit continued...

Vital Vocabulary

access port A sealed hub on an administration set designed for sterile access to the IV fluid.

administration set Tubing that connects to the IV bag access port and the catheter to deliver the IV fluid.

butterfly catheter Rigid, hollow, venous cannulation device identified by plastic "wings" that act as anchoring points for securing the catheter.

catheter A flexible, hollow structure that drains or delivers fluids.

catheter shear The cutting of the catheter by the needle during improper rethreading of the catheter with the needle; the severed piece can then enter the circulatory system.

drip chamber The area of the administration set where fluid accumulates so that the tubing remains filled with fluid.

drip sets Another name for administration sets.

external jugular IV IV access established in the external jugular vein of the neck.

gauge A measure of the interior diameter of the catheter. It is inversely proportional to the true diameter of the catheter.

infiltration The escape of fluid into the surrounding tissue.

intraosseous (IO) needle Rigid, boring catheter placed into a bone to provide IV fluids.

isotonic crystalloids The main type of fluids used in the prehospital setting for fluid replacement because of the ability to support blood pressure by remaining within the vascular compartment.

Jamshedi needle A type of intraosseous double needle consisting of a solid, boring needle inside a sharpened hollow needle.

keep-the-vein-open (KVO) IV setup A phrase that refers to the flow rate of a maintenance IV line established for prophylactic access.

local reaction Mild to moderate reaction to an irritant without systemic consequence.

macrodrip set Administration set named for the large orifice between the piercing spike and the drip chamber; allows for rapid fluid flow into the vascular system.

microdrip set Administration set named for the small orifice between the piercing spike and the drip chamber; allows for carefully controlled fluid flow and is ideally suited for medication administration.

occlusion Blockage, usually of a tubular structure such as a blood vessel.

over-the-needle catheter The prehospital standard for IV cannulation; consists of a hollow tube over a laser-sharpened, steel needle.

phlebitis Inflammation of a vein. Often associated with a clot in the vein.

piercing spike The hard, sharpened plastic spike on the end of the administration set designed to pierce the sterile membrane of the IV bag.

proximal tibia Anatomic location for intraosseous catheter insertion; the wide portion of the tibia located directly below the knee.

saline lock A special type of IV apparatus, also called buff cap, heparin cap, and heparin lock.

systemic complication Moderate to severe complication affecting the systems of the body; after administration of medications, the reaction might be systemic.

vasovagal reaction Sudden hypotension and fainting associated with traumatic or medical events.

Points to Ponder

You are assigned to work with a young, new paramedic who is very aggressive and doesn't always show compassion for patients. In several instances, the paramedic inserted 16- and 14-gauge IV catheters into the hands of medical patients. The patients showed significant pain during insertion and even after the IV catheter was placed.

Is the paramedic doing anything wrong? How would you handle this situation?

Issues: Overly Aggressive Treatment, Compassion for Patients, Addressing Issues With Coworkers.

www.EMTB.com

Assessment in Action

You respond with your paramedic partner to a cardiac patient. Upon arrival, your partner asks you to set up an IV.

1. All of the following has to be checked when selecting an IV solution EXCEPT:

   A. clarity
   B. expiration date
   C. type of solution
   D. blood pressure

2. Your partner wants you to use a microdrip set. Your IV tubing is not marked in such a way. To which set is your partner referring?

   A. 60-gtt set
   B. 15-gtt set
   C. 10-gtt set
   D. Use any set available.

3. Your partner obtains IV access and asks you to secure the line. Which one of the following supplies are NOT necessary?

   A. Bioclusive dressing
   B. Clear tape
   C. Constricting band
   D. Kling bandage

4. After you attach the IV tubing to the IV catheter, you notice that the IV fluid is not running. Which of the following is NOT a possible cause?

   A. Kinked tubing
   B. Low blood pressure
   C. Closed valves on the IV line
   D. The IV bag hanging too low

5. Your patient states that he is scared of needles and does not want an IV. How do you approach this patient?

   A. Explain the benefits and risks of IV therapy. If the patient is alert and oriented to person, time, and place and still refuses, have the patient sign a refusal.
   B. Hold the patient down and insert the IV catheter.
   C. Tell the patient he has to have IV fluid if he wants to go to the hospital.
   D. Do not insert the IV catheter.

Challenging Questions

6. What solution is usually used for administering medication?

7. You have assisted a paramedic in establishing IV access, but the patient begins to have shortness of breath. What are the possible causes?

Assisting With Cardiac Monitoring

Objectives*

1. To gain an understanding of basic terminology and techniques of cardiac monitoring. (p 1182)
2. To give you the knowledge and tools you need to assist the advanced provider with the use and implementation of an ECG. (p 1191)
3. To better understand the basic anatomy and physiology of the heart. (p 1182)
4. Identify the components of basic cardiac arrhythmias. (p 1187)

5. Evaluate the rate and rhythm of a patient's cardiovascular system, and become familiar with the normal ECG. (p 1185)
6. Familiarize yourself with and apply 4-lead electrodes and identify placements for the 12-lead systems. (p 1191)

*All of the objectives in this chapter are noncurriculum objectives.

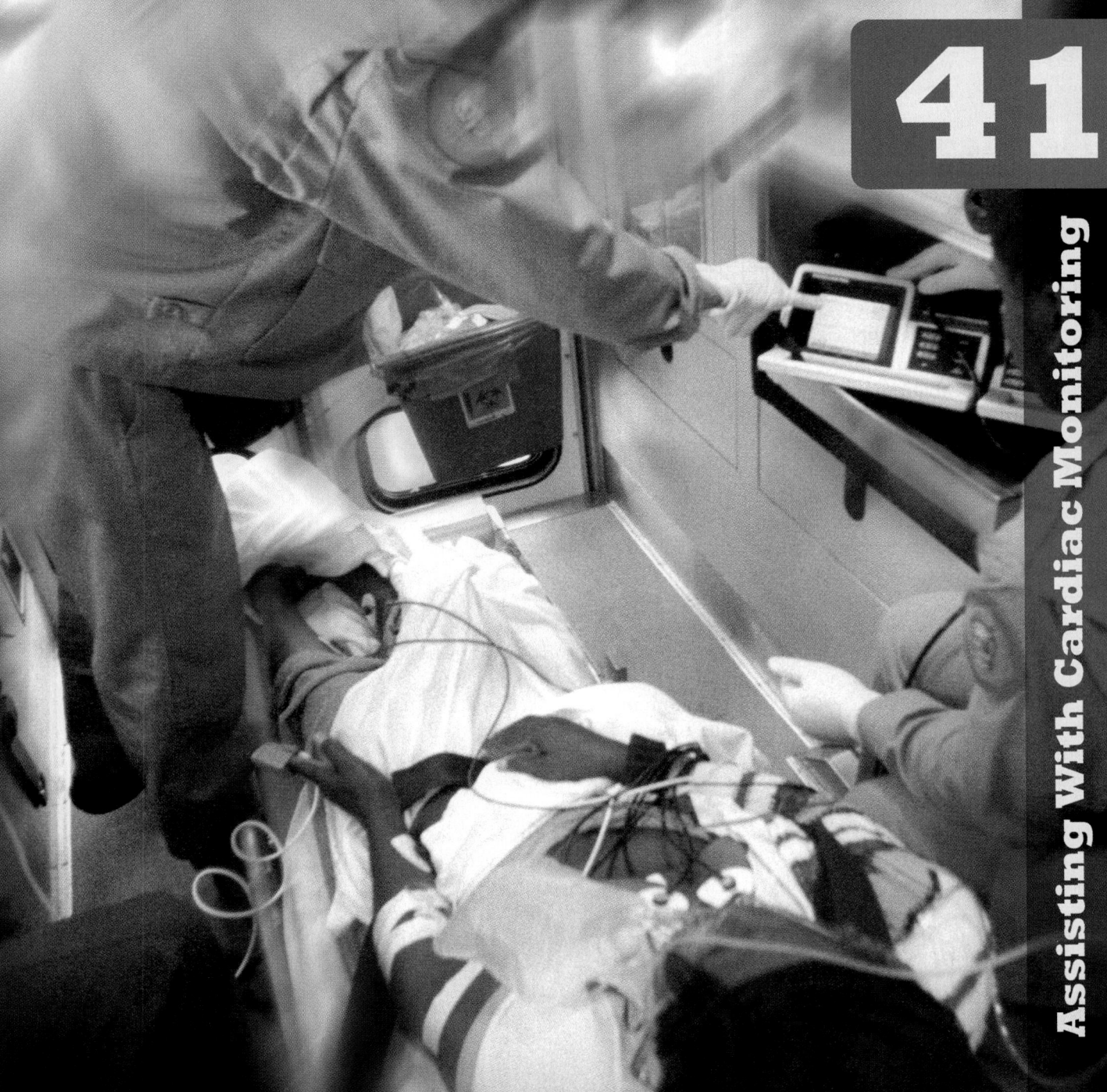

You are the Provider

You and your EMT-B partner are dispatched to a local hotel for a cardiac call. When you arrive at the hotel, the desk clerk meets you outside. She states that a guest in room 13 called the front desk approximately 8 minutes ago saying that he thought he was having a heart attack and for someone to please call 9-1-1. After calling 9-1-1, the desk clerk called the room to let him know help was on the way, but she got no answer.

1. What is your next step?
2. What does your index of suspicion tell you about this call?

Cardiac Monitoring

Within the past few years, a number of studies have examined the various aspects of the use of the 12-lead ECG, or electrocardiograms, in the prehospital setting. In 1993, the National Institutes of Health recommended that "EMS systems should consider providing prehospital 12-lead ECGs to facilitate early identification of AMI [acute myocardial infarction]." In fact, studies have indicated a 95% or better accuracy rate in the diagnosis of myocardial infarction when a 12-lead ECG is used. Why is this so important? The evaluation and treatment of a patient experiencing the signs and symptoms of AMI has progressed to the point that use of a 12-lead ECG is rapidly becoming the norm when one considers the overall patient outcome. Early identification of an impending AMI allows the provider to alert the receiving hospital, which allows the hospital to prepare before receiving the patient, drastically reducing the time to definitive care and overall cell death. Reduction in cell death means a more positive patient outcome.

Although the interpretation of cardiac rhythm may not be an EMT-B skill, you will find it quite helpful to be able to place electrodes and leads on a patient in preparation for cardiac monitoring by an advanced life support (ALS) provider. Whether it be a 3-, 4-, or 12-lead system, it may also be helpful to be able to recognize a normal ECG tracing and become familiar with basic rhythm disturbances, so that you can help identify many of the acute and chronic problems affecting the heart while learning the practice of ECG analyses.

Technology

- Interactivities
- Vocabulary Explorer
- Anatomy Review
- Web Links
- Online Review Manual

www.EMTB.com

In addition, 3- and 4-lead ECG systems are still commonly used and are often all the provider has time to use.

The identification of cardiac conditions and correct management in the field requires that you have a good understanding of the basic anatomy and physiology of the heart and that you understand the basic terminology and techniques of cardiac monitoring so that you will be able to recognize changes in the ECG and call them to an ALS provider's attention.

Electrical Conduction System of the Heart

Before discussing how to assist with cardiac monitoring, basic information on the electrical conduction system of the heart is required to understand what is happening on an ECG.

As discussed in Chapter 4, the heart contains a network of specialized tissue that is capable of conducting electrical current throughout the heart. This is called the electrical conduction system. The flow of electrical current through this network causes contractions of the heart that produce pumping of blood.

When the heart is working normally, the electrical impulse moves through the electrical conduction system and produces a coordinated pumping contraction. If the heart is deprived of oxygen or is injured, its electrical system may not function and the heart may not continue to beat properly. Blood pressure decreases, and the patient may lose consciousness. The next section discusses the pathway of an electrical impulse through the electrical conduction system.

The Process of Electrical Conduction

The main function of the electrical conduction system is to create an electrical impulse and transmit it through the heart in an organized manner. Electrical conduction in the heart occurs through a pathway of special cells (Figure 41-1 ▶). This pathway contains the following components:

- The sinoatrial (SA) node, the heart's main pacemaker, which is located in the wall of the right atrium where it meets the superior vena cava
- Three internodal pathways: anterior, middle, and posterior, which transmit the pacing impulse from the SA node to the atrioventricular (AV) node
- The AV node, which transmits the impulse from the atria to the ventricles
- The bundle of His, which starts at the AV node and then splits into the right and left bundle

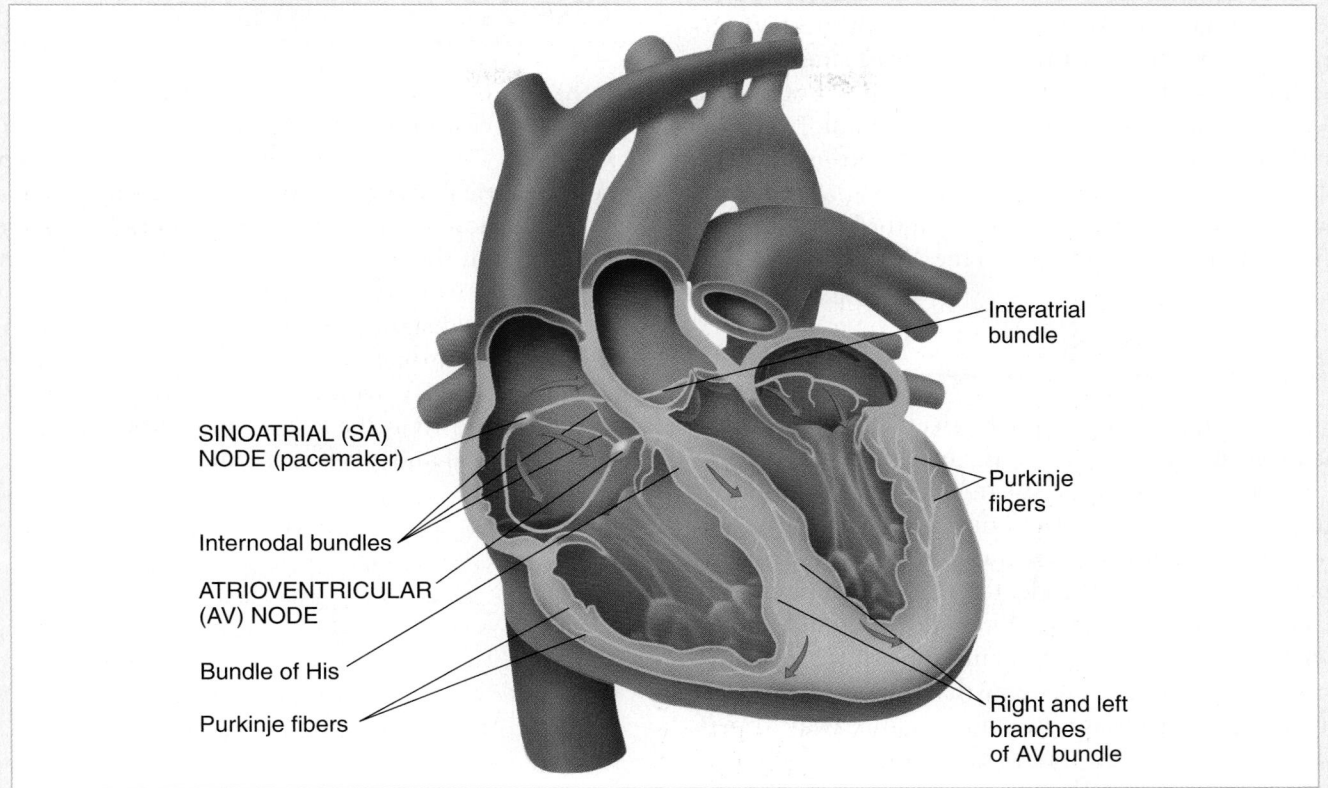

Figure 41-1 The electrical conduction system of the heart.

branches; the only route of communication between the atria and the ventricles
- The left bundle, which travels through the interventricular septum and leads to the left face of the interventricular septum
- The right bundle, which leads to the right side of the interventricular septum

- The left anterior superior fascicle, which travels through the left ventricle to the Purkinje cells
- The left posterior fascicle, which is a fanlike structure leading to parts of the left ventricle
- The Purkinje system, which contains fibers that extend from the right bundle

You are the Provider   Part 2

You knock on the door but get no answer. Out of concern, the desk clerk lets you into the room. As you enter, you see an older man sitting on the end of the bed in the tripod position clutching his chest. As you approach your patient, you notice that his skin is pale and diaphoretic. You see that he is breathing rapidly but taking shallow, labored breaths. You introduce yourself and begin asking your patient questions about his current condition. The patient is conscious, alert, and oriented. He can speak in only 2- to 3-word sentences because of his respiratory difficulties.

3. What does your patient's skin condition indicate?
4. What immediate interventions should you provide for this patient?

For the heart to pump, one of the parts of the electrical conduction system needs to act as the heart's pacemaker. In a normally functioning heart, the SA node performs this function. It paces at a rate of 60 to 100 beats/min, with an average of 70 beats/min. Every cell in the conduction system is capable of setting the pace. However, the rate of each type of cell is slower than the cells that precede it. This means that the fastest pacer is the SA node, the next fastest is the AV node, and so on.

Electrodes and Waves

The ECG electrodes pick up the electrical activity of the heart occurring beneath them, and the ECG machine converts them to waves. When an electrical impulse is moving away from the electrode, the ECG machine converts it into a negative (downward) wave. When a wave moves toward an electrode, the ECG machine records a positive (upward) wave. When the electrode is somewhere in the middle, the ECG machine shows a positive wave for the amount of energy that is coming toward it and a negative wave for the amount going away from it. Therefore, if you see a change in the way the ECG waves look on the printout, this means something is changing in the way the impulse is being conducted in the heart.

The way an ECG looks depends on where the lead is placed (Figure 41-2 ▼). For example, if an electrical impulse is moving toward the patient's left side, a lead on the right arm will create negative a wave on the ECG, and a lead on the left arm will create a positive wave.

The ECG Complex

On the ECG, one complex represents one beat in the heart. The complex consists of several waves: the P, QRS, and T waves (Figure 41-3 ▼). These waves represent electrical activity in the heart. A segment is a specific portion of the complex. For example, the segment between the end of the P wave and the beginning of the Q wave is known as the P-R segment. An interval is the distance, measured as time, occurring between two cardiac events. The time between the beginning of the P wave and the beginning of the QRS complex is known as the P-R interval. Note that there is a P-R interval and a P-R segment.

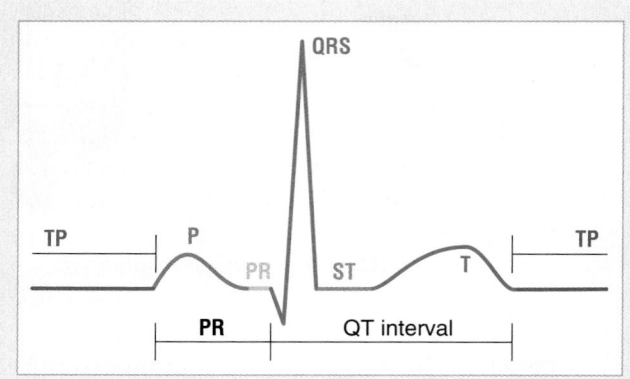

Figure 41-3 Basic components of the ECG complex.

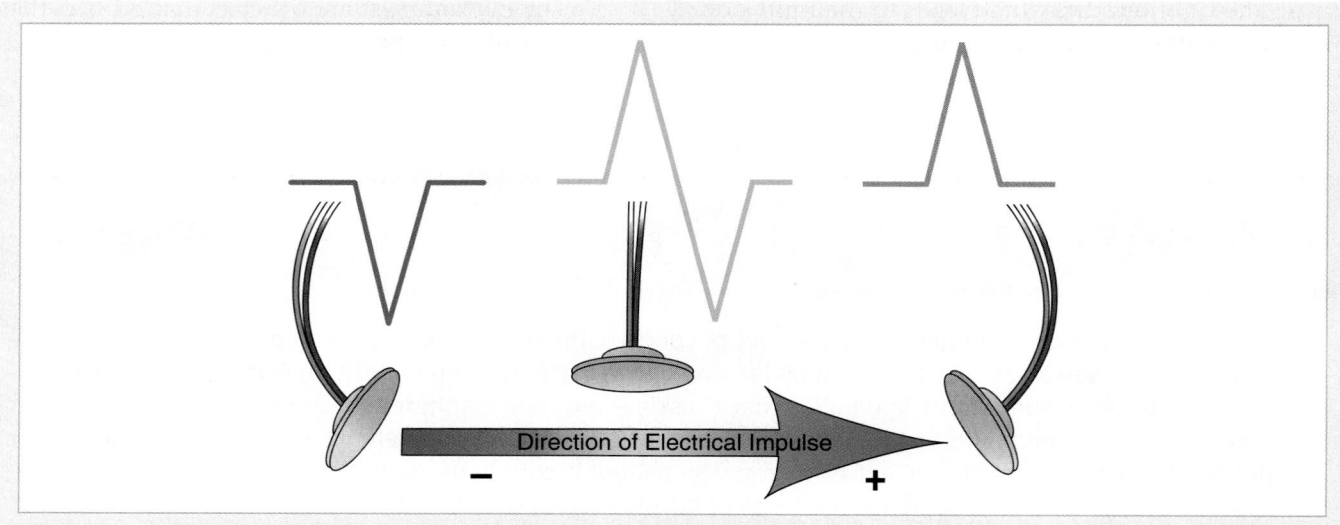

Figure 41-2 Three different ECGs resulting from the same wave owing to different lead placements.

ECG Paper

The paper on which an ECG is recorded contains a grid. As the ECG is recorded, the paper passes underneath the pen at a rate of 25 mm/sec. Therefore, there is a relationship between the ECG grid and the ECG itself. Each little box on the ECG paper represents 1/25 of a second, or 0.04 seconds. Each bigger box on the paper is composed of five smaller boxes, making each big box 5×0.04 seconds, or 0.20 seconds. Finally, five big boxes equal 1 second. By knowing how much time each box on the grid represents, ALS providers can look for problematic waves or intervals that are slower or faster than normal.

Normal Sinus Rhythm

Sinus rhythm is a rhythm in which the SA node acts as the pacemaker. All of the P waves on the patient's ECG should be the same. A normal rate for most people is from 60 to 100 beats/min. A rhythm strip with consistent P waves, consistent P-R intervals, and a regular heart rate between 60 and 100 beats/min is showing normal sinus rhythm (Figure 41-4 ▶).

The Formation of the ECG

This section shows how normal sinus rhythm looks when recorded by an ECG machine. While reviewing this section, keep in mind that production of the heart's rhythm is a continuous process. There is no actual

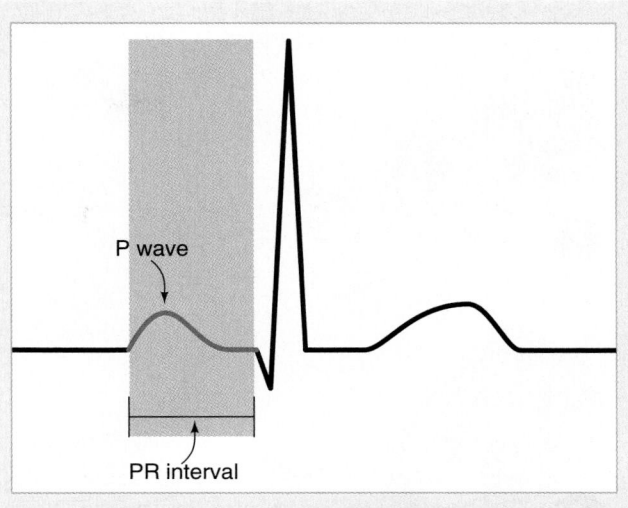

Figure 41-4 In all sinus rhythms, the P waves are identical.

period of rest or inactivity, and the cycle repeats over and over again.

- The baseline is a period when the majority of the cardiac muscle is at rest (Figure 41-5A ▶).
- At this point, the SA node is firing and the electrical impulse is being transmitted through the internodal pathways to the AV node. Note that only a straight line exists so far on the ECG tracing (Figure 41-5B ▶).
- The electrical activity in the right atria begins to create a P wave (Figure 41-5C ▶).
- The formation of the P wave is complete (Figure 41-5D ▶).
- The impulse has traveled through the bundle of His, right and left bundles, the fascicles, and the Purkinje system. As electrical activity occurs in

You are the Provider   Part 3

Considering your patient's condition, you immediately apply high-flow oxygen via a nonrebreathing mask and ask your partner to call for ALS backup. The patient tells you that he is 59 years old. He was taking a nap when the pain woke him up. He has never had chest pain before, but this is definitely the worst pain he has ever experienced. He describes the pain as a crushing feeling in the center of his chest that radiates down his left arm.

5. What do these signs and symptoms indicate?
6. Why would ALS backup be a consideration in the scenario?

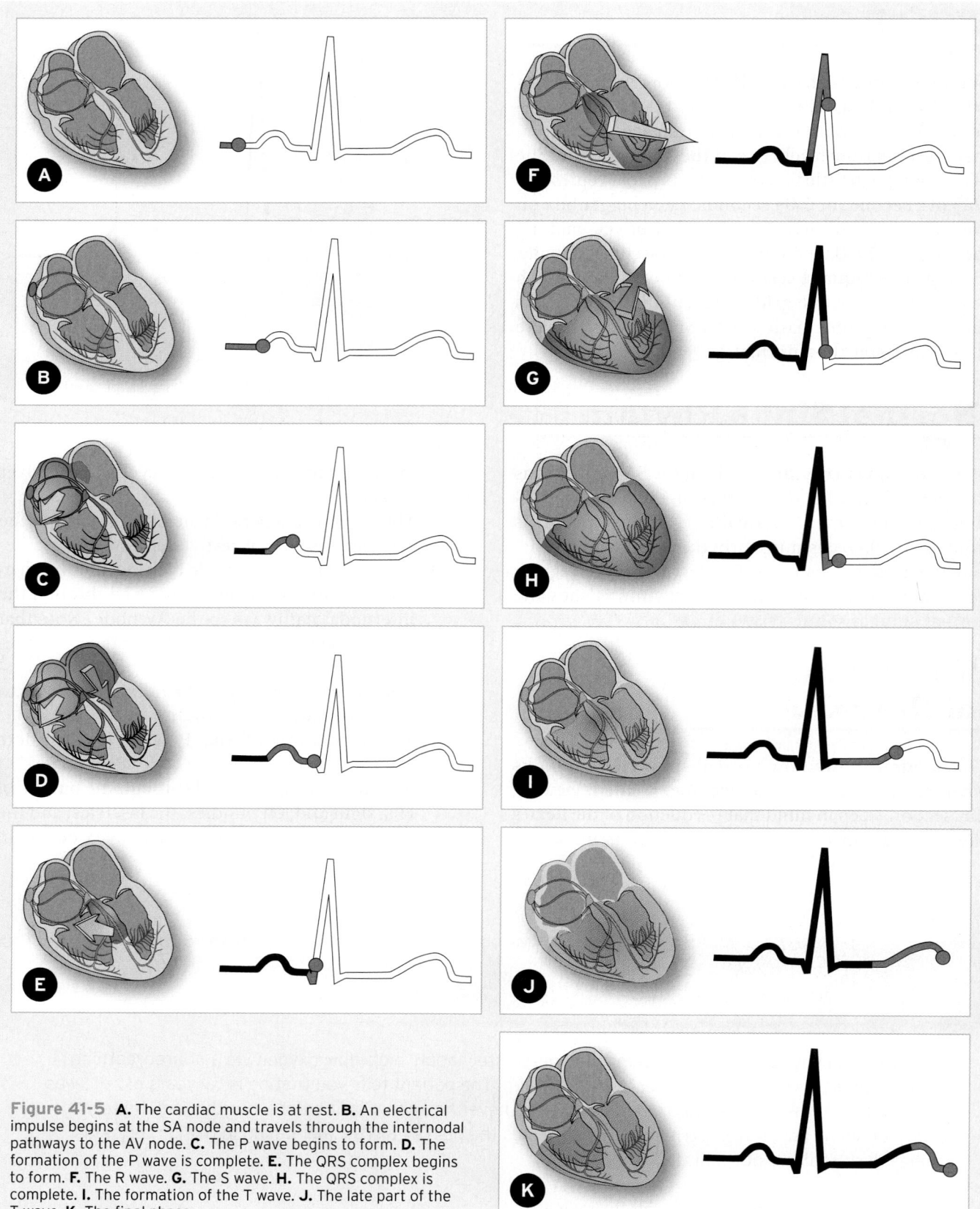

Figure 41-5 **A.** The cardiac muscle is at rest. **B.** An electrical impulse begins at the SA node and travels through the internodal pathways to the AV node. **C.** The P wave begins to form. **D.** The formation of the P wave is complete. **E.** The QRS complex begins to form. **F.** The R wave. **G.** The S wave. **H.** The QRS complex is complete. **I.** The formation of the T wave. **J.** The late part of the T wave. **K.** The final phase.

the ventricles, the QRS complex begins to form (Figure 41-5E ◀).

- Electrical activity continues through the main portion of the left ventricle and gives rise to the large R wave depicted here (Figure 41-5F ◀).
- Electrical activity in the last part of the left ventricle is represented as the S wave or end portion of the QRS complex (Figure 41-5G ◀).
- At this point, the QRS complex is complete. Since the depolarization of the ventricles occurred using the normal electrical conduction system, the QRS complex should be normal in duration. The normal QRS interval is between 0.06 and 0.11 seconds (Figure 41-5H ◀).
- The formation of the T wave (Figure 41-5I ◀).
- The late part of the T wave (Figure 41-5J ◀).
- This is the final phase (Figure 41-5K ◀). The heart is relaxing. If the heart is functioning normally, the above process will repeat over and over continuously.

Arrhythmias

An arrhythmia is an abnormal rhythm of the heart and is sometimes called a dysrhythmia. The following section discusses some basic arrhythmias with which you should be familiar.

Sinus Bradycardia

Bradycardia refers to a slow heart rate, usually less than 60 beats/min. Therefore, sinus bradycardia is a rhythm that has consistent P waves, consistent P-R intervals, and a regular heart rate that is less than 60 beats/min (Figure 41-6 ▼).

Most patients can tolerate heart rates between 50 and 60 beats/min without much difficulty. Sinus bradycardia typically becomes a problem when the heart rate drops to less than 50 beats/min. Under normal circumstances, however, heart rates as slow as the low 40s may be normal for very well-conditioned athletes and for some patients during sleep.

Sinus Tachycardia

Tachycardia refers to a fast heart rate, more than 100 beats/min, to be exact. Sinus tachycardia is a rhythm that has consistent P waves, consistent P-R intervals, and a regular heart rate that is more than 100 beats/min (Figure 41-7 ▶).

Because tachycardia is the body's response to stress, it should not really be considered a pathologic rhythm. Cardiac output is maintained by the heart rate and stroke volume (amount of blood ejected by the heart during each mechanical contraction). Stroke volume depends on the mechanical filling of the heart, which is a passive process and an active process. The heart passively fills when the AV valves open and the blood from the atria floods into the ventricular chambers. The atria actively overfill the ventricles when the atria mechanically contract. This pushes the remaining atrial contents into the already filled ventricles.

Tachycardia causes a decrease in cardiac output when the rate becomes so high that the stroke volume is affected. Tachycardia decreases the amount of passive filling time. In other words, as the amount of time needed to fill the ventricles decreases with increasing heart rates, the less the ventricles are filled, thereby decreasing cardiac output.

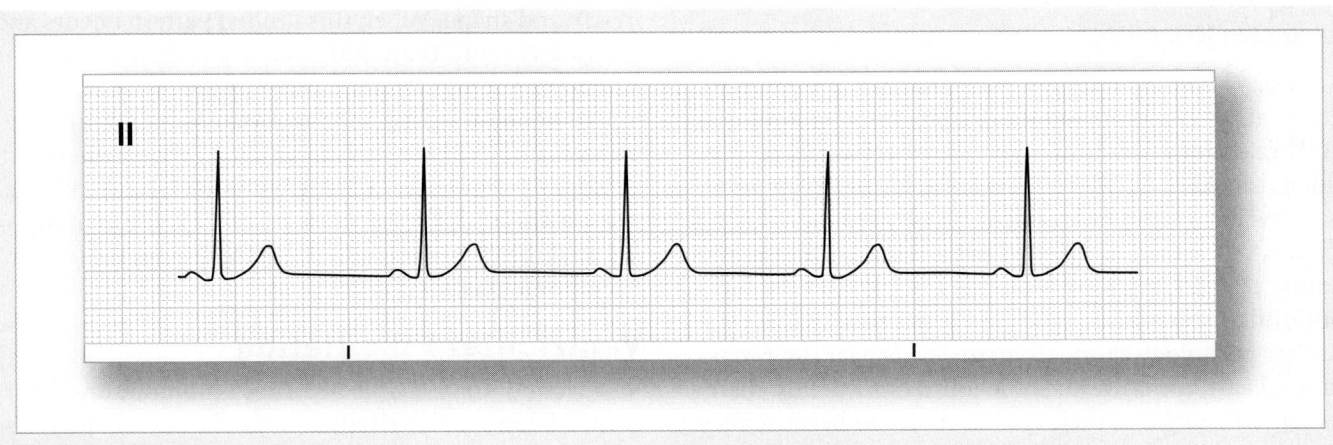

Figure 41-6 Sinus bradycardia.

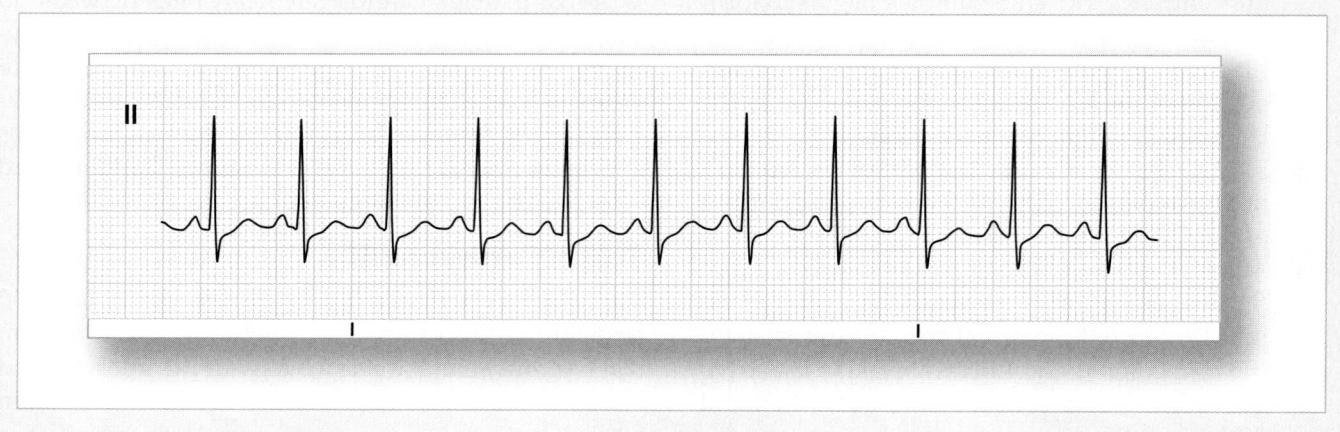

Figure 41-7 Sinus tachycardia.

For any given patient, the heart rate is determined by a constant tug-of-war between the sympathetic and the parasympathetic divisions of the autonomic nervous system. If the parasympathetic dominates, the rhythm is slowed. If the sympathetic dominates, the rhythm is sped up. The effects of the sympathetic stimulation can go beyond the rate; it can sometimes cause minor physiologic differences in the appearance of the QRS complex.

As mentioned, the heart rate in sinus tachycardia is more than 100 beats/min. Generally, it is between 101 and 160 beats/min in most patients. At this rate, the tachycardia itself does not pose significant problems. The heart rate in sinus tachycardia, however, can go up to 200 or even 220 beats/min in rare circumstances. At these rates, the rhythm can pose clinical and diagnostic challenges. In general, the maximum heart rate that can be considered normal for any individual patient is derived by using the following formula:

Maximum Heart Rate = 220 beats/min − Age (in years)

For example, a 20-year-old man has a maximum heart rate of 200 beats/min (220 beats/min − 20 years = 200 beats/min). Anything above that level would be considered abnormal and would require further evaluation.

The maximum heart rate is usually reached during exercise or forced activity of some kind. Athletes and young people are able to tolerate the high levels without difficulty. Geriatric patients or patients with some cardiac pathology cannot tolerate levels near their maximum heart rate without some difficulty. Luckily, many patients never get to their maximum levels because they have some disease in their electrical conduction system that limits the rate they can reach.

Ventricular Tachycardia

The most deadly arrhythmias include ventricular tachycardia, ventricular fibrillation, and asystole. Together, ventricular tachycardia and ventricular fibrillation account for more than 300,000 sudden cardiac deaths in the United States each year. Because of their life-ending potential, these rhythms deserve our full attention.

A basic definition of ventricular tachycardia is simply the presence of three or more abnormal ventricular complexes in a row with a rate of more than 100 beats/min (Figure 41-8 ▶). The rate for ventricular tachycardia is between 100 and 200 beats/min, but the rate most commonly is between 140 and 200 beats/min. Rates of more than 200 beats/min can occur, and when they do, the physical size and shape of the complexes slowly begin to blur with no discernible QRS, ST, or T waves. In fact, the complexes actually become unified in size and shape. When this unified pattern occurs and the rate is more than 200 beats/min, the condition is called ventricular flutter (Figure 41-9 ▶).

In general, ventricular tachycardia is a very regular rhythm. In about 10% of cases, however, there can be slight irregularity. This irregularity causes a variation in the cadence of the complexes and usually is found if the rate is at the lower end of the range for ventricular tachycardia.

Ventricular Fibrillation

Ventricular fibrillation is a rapid, completely disorganized ventricular rhythm with chaotic characteristics. The electrocardiographic characteristics of this arrhythmia are undulations of varying shapes and sizes

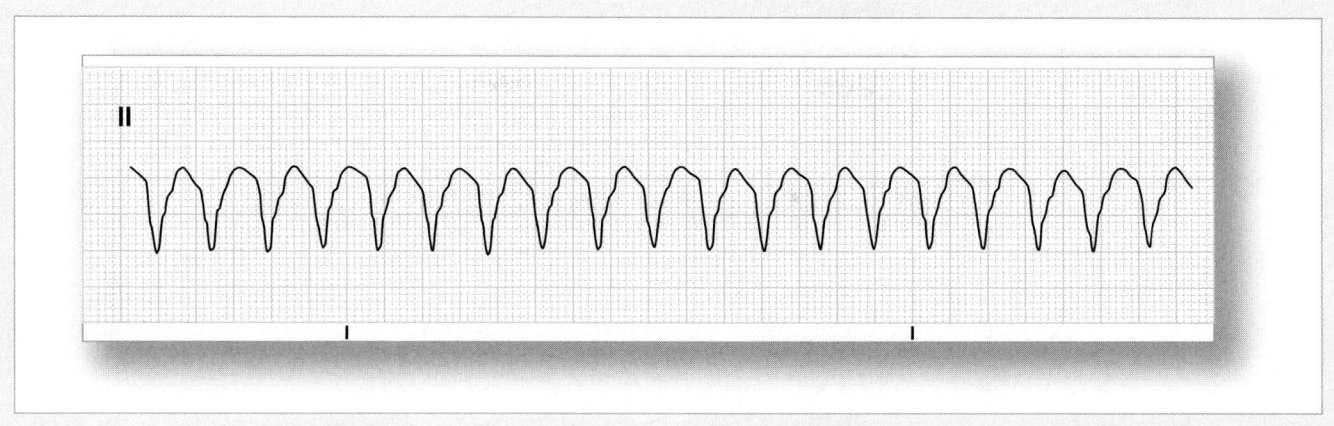

Figure 41-8 Ventricular tachycardia.

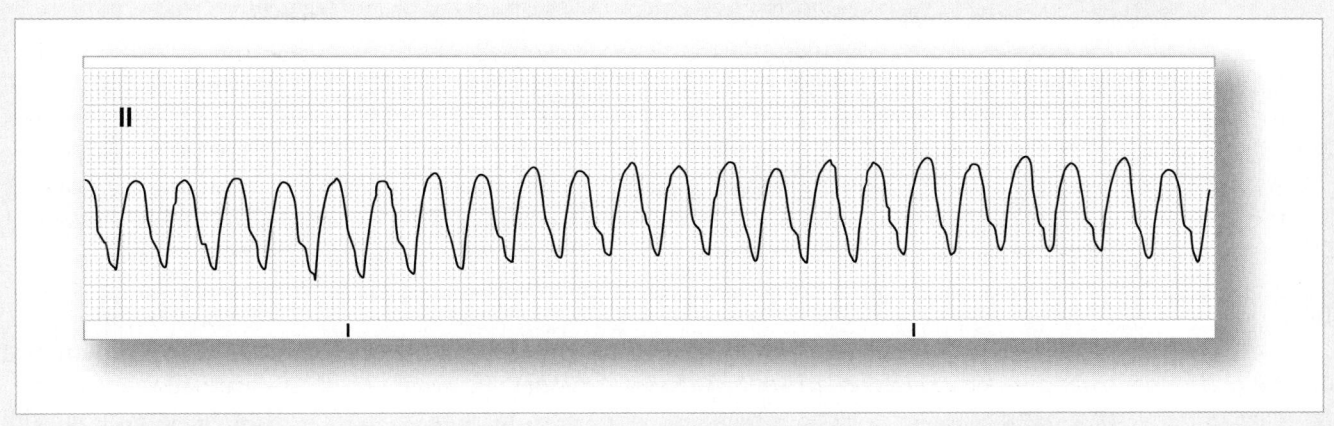

Figure 41-9 Ventricular flutter.

with no specific pattern and no discernable P, QRS, or T waves (Figure 41-10 ▶).

The undulations occur anywhere from 150 to 500 times in a minute. Notice that we did not use the words "beats" to describe the undulations. This is because in ventricular fibrillation (also called V-fib), there is no organized beating of the heart.

Ventricular fibrillation can occur spontaneously in a patient with a normal heart. It is also common for ventricular tachycardia to deteriorate into ventricular fibrillation. However, the most common cause of the arrhythmia is an AMI. For many patients, this lethal arrhythmia is the first sign of an early AMI. Together with ventricular tachycardia, these two arrhythmias account for more than 50% of deaths associated with coronary artery disease.

Ventricular fibrillation is a deadly arrhythmia and does not cease on its own. The only chance of survival is immediate treatment. Within only 10 to 20 seconds of a cessation of cardiac output, life-threatening complications can begin. Immediate defibrillation is the most effective treatment for ventricular fibrillation.

Asystole

Asystole refers to the complete absence of any electrical cardiac activity (Figure 41-11 ▶). It looks like a straight or almost straight line on an ECG strip. There is a complete absence of any P, QRS, and T waves anywhere along the strip because no electrical activity is occurring.

As you can imagine, the patient is clinically dead at this point. There is no electrical or mechanical activity whatsoever. However, the decision about when to call

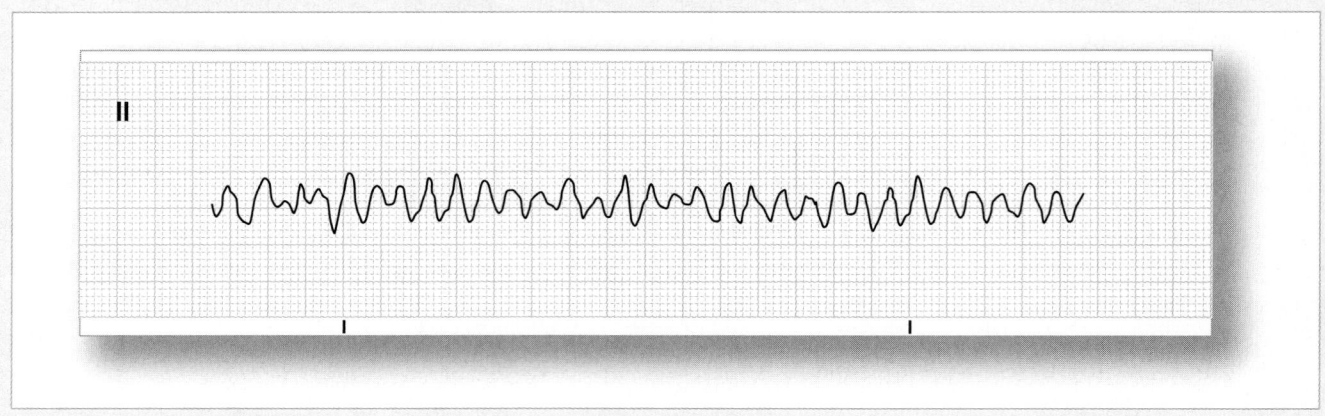

Figure 41-10 Ventricular fibrillation.

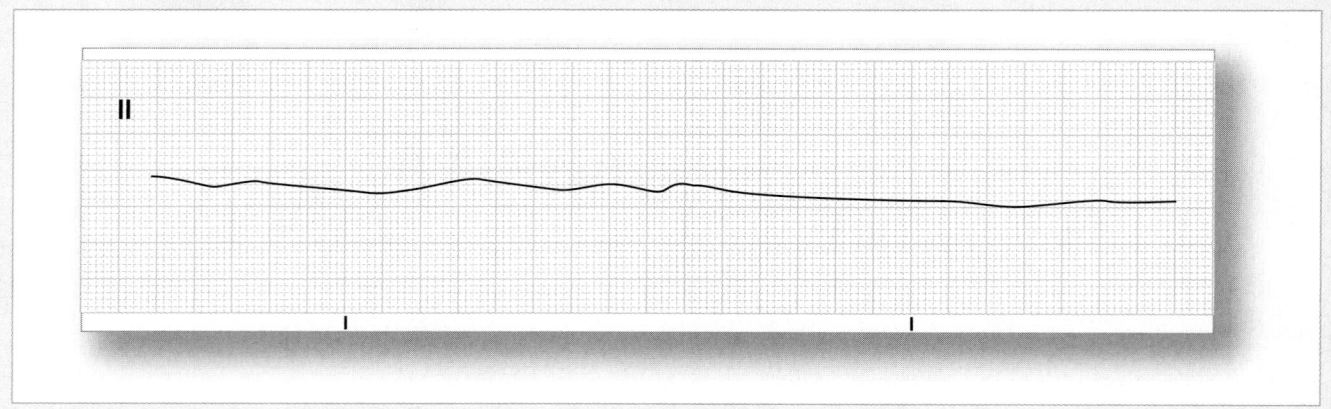

Figure 41-11 Asystole.

You are the Provider   Part 4

Your partner confirms that the estimated time of arrival for the ALS unit is 6 to 7 minutes. She then begins taking a set of baseline vital signs while you continue to assess the patient. Your patient tells you that his only allergy is to penicillin. He takes daily vitamins but no other medications. He has no major medical history. He states that he has always been very healthy and that he just does not understand why this is happening. His last oral intake was about 2 hours ago when he had a hamburger and fries at the hotel restaurant. His baseline vital signs are a blood pressure of 178/96 mm Hg, a pulse of 128 beats/min, and respirations of 28 breaths/min and labored; his lung sounds are clear and equal bilaterally but diminished in the lower lobes.

7. Why is it important to obtain baseline vital signs?
8. What is significant about this patient's vital signs?

or terminate resuscitation efforts generally depends on local protocol. Be sure to know your protocols.

Assisting With Cardiac Monitoring

Now that you have an understanding of how an ECG is formed, this section discusses the cardiac monitor and how to apply the leads. Depending on the patient and your local protocols, the number of leads on your cardiac monitor may vary. You may have a 3-lead, 4-lead, or 12-lead system.

Cardiac Monitors

In recent years, the industry has developed a vast array of cardiac monitoring devices. These cardiac monitors include several new features using modern technology. They are compact, light, and portable and combine non-invasive defibrillation and monitoring capabilities. Many offer the capabilities of pulse oximetry including pulse rate, blood pressure monitoring capabilities, and manual and semiautomatic defibrillation functions.

The 12-Lead System

Although most patients will need only three or four leads attached for standard ECGs, sometimes a 12-lead ECG is needed. The importance of obtaining a 12-lead ECG is for early identification of potential myocardial ischemia, or lack of oxygen to the heart's tissue, so that the cause can be appropriately treated and possibly reversed.

Even though obtaining a 12-lead ECG has distinct benefits, it is important to remember to always treat your patient first. If your patient is in severe distress, do not withhold treatment to do a 12-lead ECG.

Many believe that the time used for taking a 12-lead ECG in the field could be better spent by rapidly transporting the patient to the hospital. This is one of the hardest decisions to make. Studies have shown that obtaining a 12-lead ECG in the field takes very little extra time, and once you are comfortable with proper electrode placement, your confidence will pick up and this will minimize the extra time spent on preparation of the 12-lead systems.

There are some immediate advantages of 12-lead monitoring. An example is early identification of acute ischemia and the accurate identification of arrhythmias. Early identification and early treatment can lead to reperfusion of valuable cardiac muscle cells, preventing tissue death and the potential for a life-threatening arrhythmia.

Lead Placement

One of the areas where you can help in efficient cardiac monitoring is electrode placement. Electrodes can be placed while the ALS provider prepares for other parts of the call.

A 4-lead ECG contains four leads, which are electrodes attached to wires, that are attached to the cardiac monitor. These four leads are called the limb leads because they are placed on the patient's limbs, or close to them. For a 4-lead ECG, the electrodes should be placed as shown in (Figure 41-12 ▶), with the white lead on the patient's right shoulder, the black lead on the patient's left shoulder, the green lead on the right side of the patient's abdomen, and the red lead on the left side of the patient's abdomen. It does not matter if you place the arm leads on the shoulders or arms, as long as they are at least 10 cm from the heart. Likewise, for the abdominal or leg leads, it does not matter whether these

You are the Provider   Part 5

The ALS unit arrives. Your partner begins giving the patient report while you assist the paramedic in hooking the patient up to the cardiac monitor. You are assisting with a 12-lead ECG and begin by placing the extremity electrodes, remembering that they must be at least 10 cm from the heart. You then find the angle of Louis to help you locate the proper placement of the chest leads. With the leads in place, the paramedic is able to get a clear view of what is happening with the patient's heart.

9. Why is the placement of the leads so important?
10. What information can be gained from the use of a cardiac monitor?

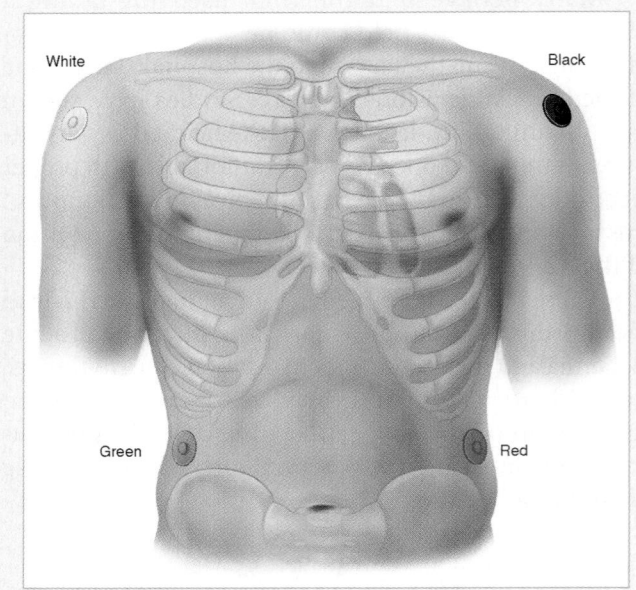

Figure 41-12   4-lead electrode placement.

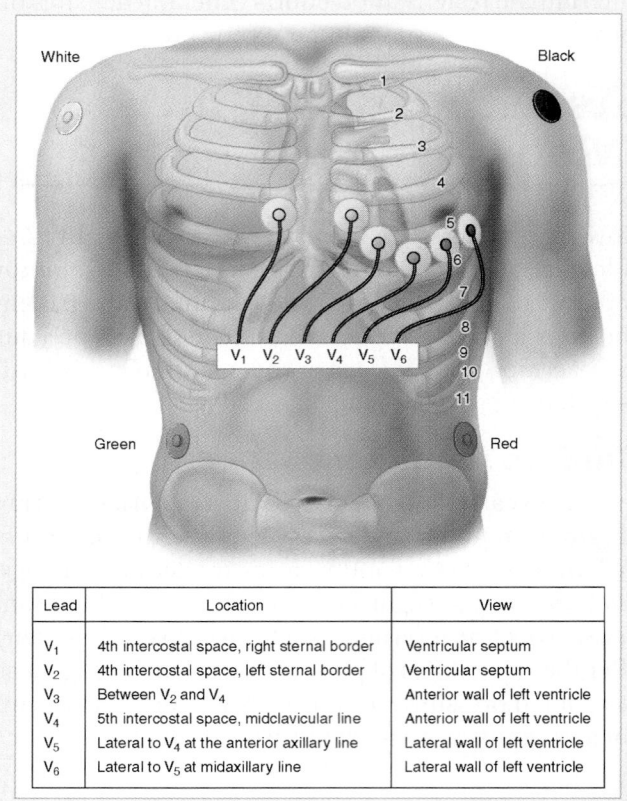

Lead	Location	View
V_1	4th intercostal space, right sternal border	Ventricular septum
V_2	4th intercostal space, left sternal border	Ventricular septum
V_3	Between V_2 and V_4	Anterior wall of left ventricle
V_4	5th intercostal space, midclavicular line	Anterior wall of left ventricle
V_5	Lateral to V_4 at the anterior axillary line	Lateral wall of left ventricle
V_6	Lateral to V_5 at midaxillary line	Lateral wall of left ventricle

Figure 41-13   12-lead electrode placement.

are on the abdomen or legs, as long as they are at least 10 cm from the heart.

When using a 12-lead ECG, electrodes are placed over the areas shown in (Figure 41-13 ▶). As with a 4-lead ECG system, the limb leads are placed at least 10 cm from the heart. The underlined chest leads, however, have to be placed exactly. Position the V_1 and V_2 leads on each side of the sternum at the fourth intercostal space. To find the space, first find the angle of Louis. This is a hump located near the top third of the sternum. Start feeling down your sternum from the top, and you'll feel it. It is located next to the second rib. The space directly beneath it is the second intercostal space. Count down two more spaces and you are there. V_4 is at the fifth intercostal space in the midclavicular line. Follow the diagram in Figure 41-13 for the remaining positions.

Troubleshooting Lead Placement

It is very important to have direct skin contact when using a 12-lead ECG. It may be difficult to place leads on patients in certain situations, for example if the patient's skin is diaphoretic (sweaty), oily, dirty, or hairy. Patients with cardiac emergencies may be sweating as a result of the situation. Wipe and clean the patient's skin thoroughly with a towel. Using benzoin can help keep leads attached. If the patient's hair prevents attachment of the leads, use a razor to remove excess hair and apply the leads as appropriate.

You are the Provider   Summary

Although your scope of practice varies depending on the region in which you are working, it is important to understand how the heart works. A basic knowledge of the heart's electrical system can enable you to have a clearer understanding of what may be happening with a cardiac patient and, in turn, assist you in providing appropriate treatment. The time that can be saved by your ability to assist the ALS provider can often mean the difference between life and death.

Prep Kit

Ready for Review

- Cardiac monitoring is a skill that helps identify patients who are experiencing an AMI (a heart attack) or other dangerous abnormal heart rhythms. Time is of the essence in treating these patients.

- Cardiac monitoring is not an EMT-B skill, but your assistance can make the call run much more efficiently and smoothly, improving the patient's chances of survival.

- The flow of electrical current through the heart's electrical conduction system causes contractions of the heart. If the heart is deprived of oxygen or is injured, its electrical system may not function properly and the heart may not pump.

- Electrical conduction occurs through a pathway that includes the SA node, three internodal pathways, the AV node, the bundle of His, the left bundle, the right bundle, the left anterior superior fascicle, the left posterior fascicle, and the Purkinje system.

- An ECG machine converts electrical activity in the heart to waves.

- The ECG's appearance depends on where the lead is placed.

- The ECG is made of complexes. One complex represents one beat in the heart. The complex consists of the P, QRS, and T waves.

- The paper on which an ECG is recorded contains a grid that represents time.

- Normal sinus rhythm is a rhythm with a regular rate of 60 to 100 beats/min, consistent P waves, and consistent P-R intervals.

- An arrhythmia is an abnormal rhythm of the heart.

- Sinus bradycardia is a rhythm that has consistent P waves, consistent P-R intervals, and a regular heart rate that is less than 60 beats/min.

- Sinus tachycardia is a rhythm that has consistent P waves, consistent P-R intervals, and a regular heart rate that is more than 100 beats/min.

- The most deadly arrhythmias include ventricular tachycardia, ventricular fibrillation, and asystole.

- Ventricular tachycardia is the presence of three or more abnormal ventricular complexes in a row with a rate of more than 100 beats/min.

- Ventricular fibrillation is a rapid, completely disorganized ventricular rhythm with chaotic characteristics. It is deadly and requires immediate defibrillation.

- Asystole is the complete absence of any electrical cardiac activity and looks like a straight or almost straight line on an ECG strip.

- Depending on the patient and your local protocols, the number of leads on your cardiac monitor may vary. You may have a 3-lead, 4-lead, or 12-lead system.

- When using a 4-lead ECG, the white lead should be placed on the patient's right shoulder, the black lead on the patient's left shoulder, the green lead on the right side of the patient's abdomen, and the red lead on the left side of the patient's abdomen.

- When using a 12-lead ECG, the limb leads are placed in the same way as with a 4-lead ECG. The chest leads have to be placed exactly. Position the V_1 and V_2 leads on each side of the sternum at the fourth intercostal space.

- If anything prevents the electrodes from attaching to the patient's skin, this must be resolved. Wipe and clean the patient's skin thoroughly, and shave hair if necessary.

Technology

- Interactivities
- Vocabulary Explorer
- Anatomy Review
- Web Links
- Online Review Manual

www.EMTB.com

Prep Kit continued...

Vital Vocabulary

4-lead ECG An ECG that uses 4 leads attached to the patient's skin; these include the limb leads.

12-lead ECG An ECG that uses 12 leads attached to the patient's skin; these include the limb leads and chest leads.

arrhythmia An abnormal rhythm of the heart, sometimes called a dysrhythmia.

asystole The complete absence of any electrical cardiac activity, appearing as a straight or almost straight line on an ECG strip.

cardiac monitoring The act of viewing the electrical activity of the heart through the use of an ECG machine or cardiac monitor.

chest leads The leads that are used only with a 12-lead ECG and must be placed exactly; includes leads V_1, V_2, V_3, V_4, V_5, and V_6.

ECG Electrocardiogram; an electronic tracing of the heart's electrical activity through leads, which originate in the electrocardiograph machine and contain electrodes that attach to the patient's chest and/or limbs.

electrical conduction system A network of special cells in the heart through which an electrical current flows, causing contractions of the heart that produce pumping of blood.

limb leads The four leads used with a 4-lead ECG; placed on or close to the right arm, left arm, right leg, and left leg.

normal sinus rhythm A rhythm that has consistent P waves, consistent P-R intervals, and a regular heart rate between 60 and 100 beats/min.

sinus bradycardia A rhythm that has consistent P waves, consistent P-R intervals, and a regular heart rate that is less than 60 beats/min.

sinus rhythm A rhythm in which the SA node acts as the pacemaker.

sinus tachycardia A rhythm that has consistent P waves, consistent P-R intervals, and a regular heart rate that is more than 100 beats/min.

ventricular fibrillation A rapid, completely disorganized ventricular rhythm with chaotic characteristics, no specific pattern, and no discernable P, QRS, or T waves.

ventricular tachycardia The presence of three or more abnormal ventricular complexes in a row with a rate of more than 100 beats/min.

Points to Ponder

You are with your paramedic partner on the scene of a patient with chest pain. The patient states the pain started 45 minutes ago without exertion. The pain is an 8 on a scale of 0 to 10 and radiates to his left arm. The patient is nauseous and has had several episodes of emesis. The ECG shows a sinus rhythm, and your partner wants to perform a 12-lead ECG. The patient is diaphoretic, and the leads keep falling off.

Where do you place the leads? How do you attach leads that keep falling off?

Issues: Lead Locations, Difficulty With Lead Placement.

Assessment in Action

You are dispatched to a patient who is having general weakness. Upon your arrival, you find a 65-year-old man who is complaining of weakness and shortness of breath. His skin is pale and diaphoretic.

1. As your partner applies oxygen, you take vital signs, which are as follows: blood pressure, 80/60 mm Hg; pulse, 46 beats/min; and respirations, 32 breaths/min. You apply the ECG monitor, and it shows a rate of 46 beats/min. What is the patient's rhythm?

   A. Sinus tachycardia
   B. Sinus bradycardia
   C. Normal sinus rhythm
   D. Ventricular fibrillation

2. Which of the following is a rhythm with all P waves and P-R intervals identical and a rate of 60 to 100 beats/min?

   A. Normal sinus rhythm
   B. Sinus bradycardia
   C. Sinus tachycardia
   D. Ventricular tachycardia

3. Which of the following is a rhythm with all P waves and P-R intervals identical and a rate more than 100 beats/min?

   A. Normal sinus rhythm
   B. Sinus bradycardia
   C. Sinus tachycardia
   D. Asystole

4. You are in the process of applying the leads of your 4-lead ECG. You are about to place the red lead. Where should this lead be placed?

   A. On the left side of the patient's abdomen, at least 10 cm from the heart
   B. On the patient's left shoulder, at least 12 cm from the heart
   C. In a location that is 10 cm from the heart
   D. On the patient's right leg, at least 12 cm from the heart

Challenging Question

5. Describe the electrical conduction pathway of the heart and its components.

Appendix A: BLS Review

Objectives*

Cognitive

1. Identify the need for basic life support, including the urgency surrounding its rapid application. (p A-3)
2. List the EMT-B's responsibilities in beginning and terminating CPR. (p A-6, A-7)
3. Describe the proper way to position an adult patient to receive basic life support. (p A-8)
4. Describe the proper way to position an infant and child to receive basic life support. (p A-8)
5. Describe the three techniques for opening the airway in infants, children, and adults. (p A-8 to A-10)
6. List the steps in providing artificial ventilation in infants, children, and adults. (p A-17, A-19)
7. Describe how gastric distention occurs. (p A-18)
8. Define the recovery position. (p A-20)
9. Describe infectious disease issues related to rescue breathing. (p A-3)
10. List the steps in providing chest compressions in an adult. (p A-23 to A-25)
11. List the steps in providing chest compressions in an infant and child. (p A-28 to A-30)
12. List the steps in providing one-rescuer CPR in an infant, child, and adult. (p A-23, A-28, A-29)
13. List the steps in providing two-rescuer CPR in an infant, child, and adult. (p A-25 to A-27)
14. Distinguish foreign body airway obstruction from other conditions that cause respiratory failure. (p A-11)
15. Distinguish a complete airway obstruction from a partial airway obstruction. (p A-11)
16. Describe the steps in removing a foreign body obstruction in an infant, child, and adult. (p A-12, A-15)

Affective

17. Recognize and respect the feelings of the patient and family during basic life support. (p A-6)
18. Explain the urgency surrounding the rapid initiation of basic life support measures. (p A-3)
19. Explain the EMT-B's responsibilities in starting and terminating CPR. (p A-6 to A-7)
20. Explain the rationale for removing a foreign body obstruction. (p A-11)

Psychomotor

21. Demonstrate how to position the patient to open the airway. (p A-9)
22. Demonstrate how to perform the head tilt–chin lift maneuver in infants, children, and adults. (p A-10)
23. Demonstrate how to perform the jaw-thrust and modified jaw-thrust maneuvers in infants, children, and adults. (p A-11)
24. Demonstrate how to place a patient in the recovery position. (p A-20)
25. Demonstrate how to perform chest compressions in an adult. (p A-22)
26. Demonstrate how to perform chest compressions in an infant and child. (p A-29)
27. Demonstrate how to perform one-rescuer CPR in an infant, child, and adult. (p A-24 to A-25)
28. Demonstrate how to perform two-rescuer CPR in an infant, child, and adult. (p A-26 to A-27)
29. Demonstrate how to remove a foreign body obstruction in an infant, child, and adult. (p A-12, A-13, A-15)

*All of the objectives in this chapter are noncurriculum objectives.

BLS Review

The principles of BLS were introduced in 1960. Since then, the specific techniques have been reviewed and revised every 5 to 6 years. The updated guidelines are published in the *Journal of the American Medical Association*. The most recent revision occurred as a result of the 2000 Conference on Cardiopulmonary Resuscitation and Emergency Cardiac Care. The guidelines in this appendix follow those proposed at the 2000 conference. Note that the 1994 EMT-Basic National Standard Curriculum requires BLS as a prerequisite to the EMT-B course; a review of BLS is presented here.

This appendix begins with a definition and general discussion of BLS. The next sections describe methods for opening and maintaining an airway, providing artificial ventilation to a person who is not breathing, providing artificial circulation to a person with no pulse, and removing a foreign body airway obstruction. Each of these sections is followed by a review of the changes in technique that are necessary to treat infants and children. A discussion of the methods of preventing the transmission of infectious diseases during CPR is provided in Chapter 2.

Technology

Interactivities

Vocabulary Explorer

Anatomy Review

Web Links

Online Review Manual

www.EMTB.com

EMT-B Safety

Although your chances of contracting a disease during CPR training or actual CPR on a patient are very low, common sense and OSHA (Occupational Safety and Health Administration) guidelines both demand that you take reasonable precautions to prevent unnecessary exposure to infectious disease. Using standard precautions makes the risk extremely low; then you do not have to be anxious about practicing and performing the skill.

Elements of BLS

Basic life support (BLS) is noninvasive (not involving penetration of the body, such as with surgery or a hypodermic needle) emergency lifesaving care that is used to treat medical conditions, including airway obstruction, respiratory arrest, and cardiac arrest. This care focuses on what is often termed the ABCs: airway (obstruction), breathing (respiratory arrest), and circulation (cardiac arrest or severe bleeding) (Figure A-1 ▶). BLS follows a specific sequence for adults and for infants and children (Table A-1 ▶). Ideally, only seconds should pass between the time you recognize that a patient needs BLS and the start of treatment. Remember, brain cells die every second that they are deprived of oxygen. Permanent brain damage may occur if the brain is without oxygen for 4 to 6 minutes. After 6 minutes without oxygen, some brain damage is almost certain (Figure A-2 ▶).

If a patient is not breathing well or at all, you may simply need to open the airway. Very often, this will help the patient to breathe normally again. However, if the patient has no pulse, you must combine artificial ventilation with artificial circulation. If breathing stops before the heart stops, the patient will have enough oxygen in the lungs to stay alive for several minutes. But when cardiac arrest occurs first, the heart and brain stop receiving oxygen immediately.

Cardiopulmonary resuscitation (CPR) is used to establish artificial ventilation and circulation in a patient

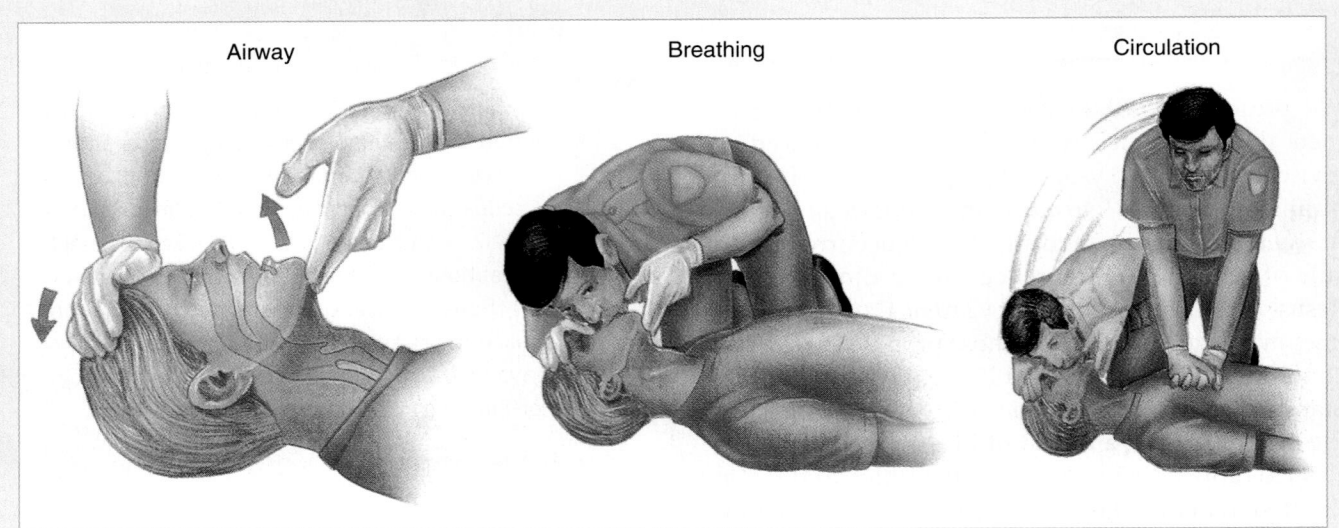

Figure A-1 The ABCs of BLS are airway, breathing, and circulation.

TABLE A-1 Review of Pediatric BLS Procedures

Procedure	Infants (younger than 1 y)	Children (1 to 8 y)
Airway	Head tilt–chin lift; jaw thrust if spinal injury is suspected	Head tilt–chin lift; jaw thrust if spinal injury is suspected
Breathing		
Initial breaths	2 breaths with duration of 1 to 1½ seconds each	2 breaths with duration of 1 to 1½ seconds each
Subsequent breaths	1 breath every 3 seconds; 20 breaths/min	1 breath every 3 seconds; 20 breaths/min
Circulation		
Pulse check	Brachial and femoral arteries	Carotid artery
Compression area	Lower half of sternum	Lower half of sternum
Compression width	2 or 3 fingers	Heel of one hand
Compression depth	½" to 1"	1" to 1½"
Compression rate	At least 100/min	100/min
Ratio of compressions to ventilations	5:1 (pause for ventilation)	5:1 (pause for ventilation)
Foreign body obstruction	Back blows and chest thrusts	Abdominal thrusts

who is not breathing and has no pulse. The steps for CPR include the following:

1. Opening the airway
2. Restoring breathing by means of rescue breathing (mouth-to-mouth ventilation, mouth-to-nose ventilation, or the use of mechanical devices)

3. Restoring circulation by means of chest compressions to circulate blood through the body

For CPR to be effective, you must be able to easily identify a patient who is in respiratory and/or cardiac arrest and begin treatment with BLS measures immediately Figure A-3 ▶.

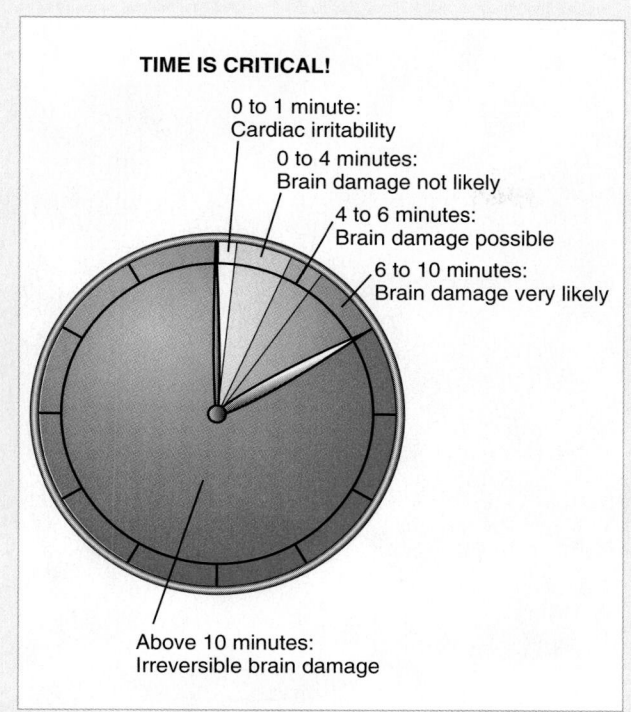

TIME IS CRITICAL!

0 to 1 minute:
Cardiac irritability

0 to 4 minutes:
Brain damage not likely

4 to 6 minutes:
Brain damage possible

6 to 10 minutes:
Brain damage very likely

Above 10 minutes:
Irreversible brain damage

Figure A-2 Time is critical for patients who are not breathing. If the brain is deprived of oxygen for 4 to 6 minutes, brain damage is likely to occur.

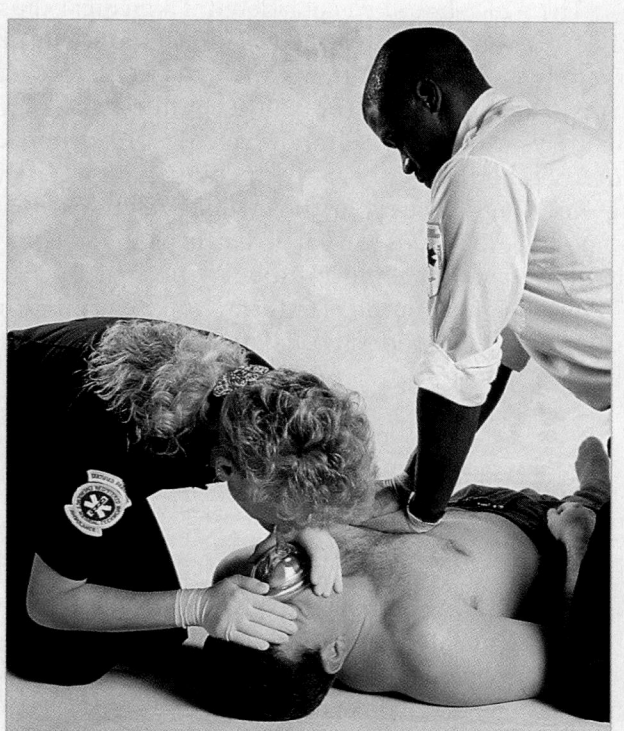

Figure A-3 You must quickly identify patients in respiratory and/or cardiac arrest so that BLS measures can begin immediately.

Rescue breathing can be given by one or two EMT-Bs, by first responders, or by trained bystanders. It does not require any equipment; however, you should use a barrier device when performing rescue breathing. Rescue breathing delivers exhaled gas from you to the patient. This gas contains 16% oxygen, which is sufficient to maintain the patient's life. Once you determine that the patient needs BLS, you should begin rescue breathing immediately, along with efforts to support the circulation and correct cardiac problems.

BLS differs from <u>advanced life support (ALS)</u>, which involves advanced lifesaving procedures, such as cardiac monitoring, administration of intravenous fluids and medications, and use of advanced airway adjuncts. However, when done correctly, BLS can maintain life for a short time until ALS measures can be started. In some cases, such as choking, near drowning, or lightning injuries, early BLS measures may be all that is needed to restore a patient's pulse and breathing. Of course, these patients also require transport to the hospital for evaluation.

The BLS measures are only as effective as the person who is performing them. Your skills will be very good immediately after training. However, as time goes on, your skills will deteriorate unless you practice them regularly.

Automated External Defibrillation

Most out-of-hospital cardiac arrests occur as the result of a sudden cardiac rhythm disturbance (dysrhythmia), such as ventricular fibrillation (V-Fib) or pulseless ventricular tachycardia (V-Tach). According to the American Heart Association, early defibrillation is the link in the chain of survival that is most likely to improve survival rates. For each minute the patient remains in V-Fib or pulseless V-Tach, there is a 7% to 10% less chance of survival.

The automated external defibrillator (AED) should be attached to any adult in nontraumatic cardiac arrest as soon as possible; defibrillation, if indicated, must be performed without delay.

The AED's simple design makes it easy for EMT-Bs, first responders, and laypersons to use; very little training is required.

The AED should be considered if a medical cause of cardiac arrest is suspected; defibrillation will not likely be helpful in patients with traumatic cardiac arrest. Regardless of the cause, children younger than 1 year of age should not have the AED applied. An AED should only be used for children between the ages of 1 and 7 years or those who weigh less than 55 pounds (25 kg) if special pediatric pads and equipment are available and protocols allow it. Refer to Chapter 12 for complete information regarding the AED, including proper use, safety considerations, and the AED algorithm.

Assessing the Need for BLS

As always, begin by surveying the scene. Is it safe? How many patients are there? What is your initial impression of the patients? Are there bystanders who may have information? Maintain open communication with family members during BLS. Do you suspect trauma? If you were dispatched to the scene, does the dispatch information match what you are seeing?

Because of the urgent need to start CPR in a pulseless, nonbreathing patient, you must complete an initial assessment as soon as possible, evaluating the patient's ABCs. The first step is determining unresponsiveness (Figure A-4 ▶). Clearly, a patient who is conscious does not need CPR; a person who is unresponsive may need CPR, based on further assessment.

You may also suspect the presence of a cervical spine injury. If so, you must protect the spinal cord from further injury as you perform CPR. If there is even a remote possibility of this type of injury, you should begin taking appropriate precautions during the initial assessment.

The basic principles of BLS are the same for infants, children, and adults. For the purposes of BLS, anyone younger than 1 year is considered an infant. A child is between ages 1 and 8 years. For children older than 8 years, you can usually use the same techniques that you use for adults. However, these are guidelines, not rules. Children vary in size. Some small children may best be treated as infants, some larger children as adults. There are two basic differences in providing CPR for infants, children, and adults. The first is that the emergencies in which infants and children require CPR have different underlying causes. The second is that

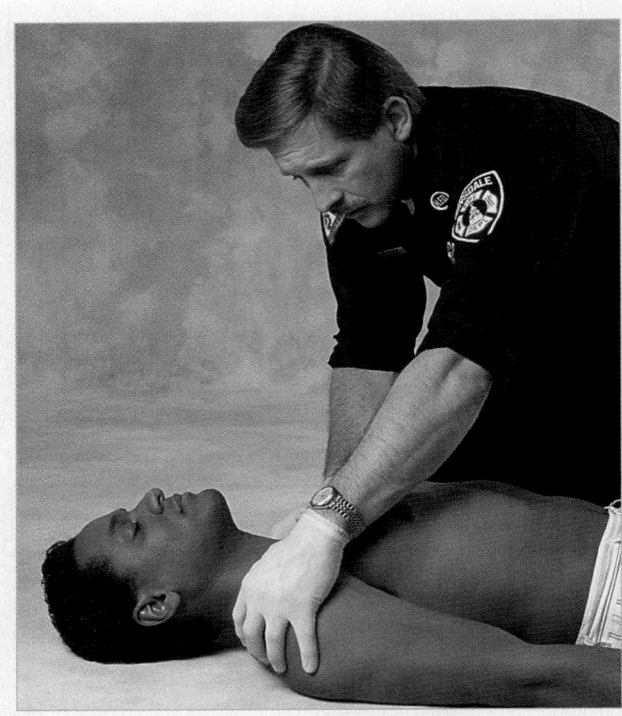

Figure A-4 Assess airway, breathing, and circulation in an unconscious patient by first attempting to rouse the patient.

there are anatomic differences in adults, children, and infants. These differences include smaller airways in infants and children than in adults.

Although cardiac arrest in adults usually occurs before respiratory arrest, the reverse is true in infants and children. In most cases, cardiac arrest in children results from respiratory arrest. If untreated, respiratory arrest will quickly lead to cardiac arrest and death. Respiratory arrest in infants and children has a variety of causes, including aspiration of foreign bodies into the airway, such as parts of hot dogs, peanuts, candy, or small toys; airway infections, such as croup and epiglottitis; near-drowning incidents or electrocution; and sudden infant death syndrome (also known as SIDS).

When to Start and Stop BLS

As an EMT-B, it is your responsibility to start CPR in virtually all patients who are in cardiac arrest. There are only two general exceptions to the rule.

First, you should not start CPR if the patient has obvious signs of death. Obvious signs of death include an

Documentation Tips

Correct handling of situations when you choose not to start CPR on a patient in cardiac arrest begins with compliance with protocols and ends with detailed documentation. In particular, record physical exam signs that led to your decision and make reference to the protocol that states these signs as a reason not to start. If extenuating circumstances such as entrapment physically prevent resuscitation attempts, record the conditions thoroughly. These decisions occasionally give rise to questions that can often be put to rest immediately by reference to a well-written report.

absence of a pulse and breathing, along with any one of the following:

- Rigor mortis, or stiffening of the body after death
- Dependent lividity (livor mortis), a discoloration of the skin due to pooling of blood Figure A-5 ▼
- Putrefaction or decomposition of the body
- Evidence of nonsurvivable injury, such as decapitation, dismemberment, or burned beyond recognition.

Rigor mortis and dependent lividity develop after a patient has been dead for a long period.

Second, you should not start CPR if the patient and his or her physician have previously agreed on DNR (do not resuscitate) or no-CPR orders Figure A-6 ▶. This

may apply only to situations in which the patient is known to be in the terminal stage of an incurable disease. In this situation, CPR serves only to prolong the patient's death. However, this can be a complicated issue. Advance directives, such as living wills, may express the patient's wishes; however, these documents may not be readily producible by the patient's family or caregiver. In such cases, the safest course is to assume that an emergency exists and begin CPR under the rule of implied consent and contact medical control for further guidance. Conversely, if a valid DNR document or living will is produced, resuscitative efforts may be withheld. Learn your local protocols and the standards in your system for treating terminally ill patients. Some EMS systems have computer notes on patients who are preregistered with the system. These notes usually specify the amount and extent of treatment that are desired. Other states have specific EMS DNR forms that allow EMS providers to withhold care when the patient, family, and physician have agreed in advance that such a course is most appropriate. It is critical that you understand your local protocols and are aware of the specific restrictions these advance directives imply.

In all other cases, you should begin CPR on anyone who is in cardiac arrest. It is usually impossible to know how long the patient has been without oxygen to the brain and vital organs. Factors such as air temperature and the basic health of the patient's tissues and organs can affect their ability to survive. Therefore, most legal advisers recommend that, when in doubt, always give too much care rather than too little. You should always start CPR if any doubt exists.

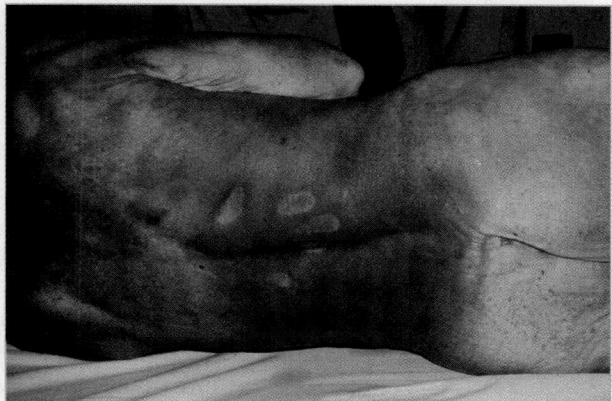

Figure A-5 Dependent lividity is an obvious sign of death, caused by blood settling to the areas of the body not in firm contact with the ground. The lividity in this figure is seen as purple discoloration of the back, except in areas that are in firm contact with the ground (scapula and buttock).

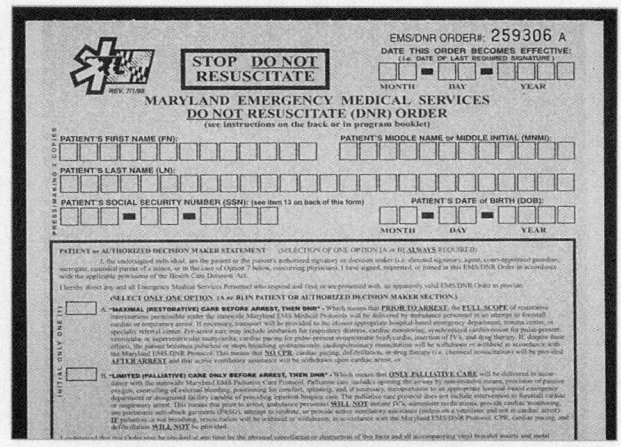

Figure A-6 You should not start CPR if the patient and his or her physician have previously agreed on DNR or no-CPR orders. Learn your local protocols for treating terminally ill patients.

You are not responsible for making the decision to stop CPR. Once you begin CPR in the field, you must continue until one of the following events occurs:

- ■ S The patient **Starts** breathing and has a pulse.
- ■ T The patient is **Transferred** to another person who is trained in BLS, to ALS-trained personnel, or to another emergency medical responder.
- ■ O You are **Out** of strength or too tired to continue.
- ■ P A **Physician** who is present or providing online medical direction assumes responsibility for the patient and gives direction to discontinue CPR.

Pediatric Needs

Opening the airway in an infant or child is done by using the same techniques as used for an adult. However, because a child's neck is so flexible, the head tilt–chin lift maneuver should be modified so that as you tilt the head back, you are moving it only into the neutral position or a slightly extended position (**Figure A-7 ▼**). You may also use the jaw-thrust maneuver without a head tilt. In fact, this is the best method to use if you suspect a spinal injury in a child. If a second rescuer is present, he or she should immobilize the child's cervical spine.

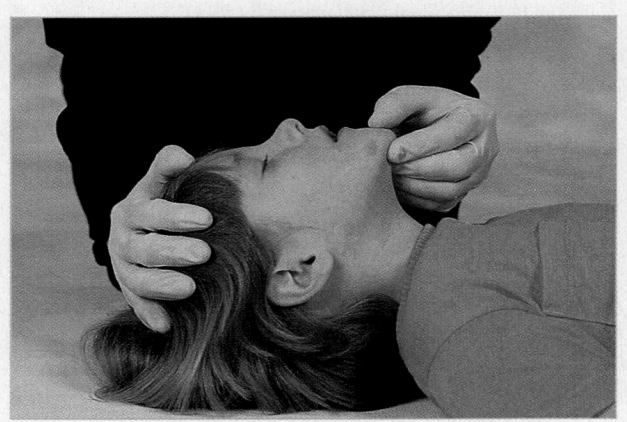

Figure A-7 The head tilt–chin lift maneuver on a child is slightly modified: as you tilt the head back, you move it only into the neutral position or a slightly extended position.

"Out of strength" does not mean merely weary; rather, it means no longer physically able to perform CPR. In short, CPR should always be continued until the patient's care is transferred to a physician or higher medical authority in the field. In some cases, your medical director or a designated medical control physician may order you to stop CPR on the basis of the patient's condition.

Every EMS system should have clear standing orders or protocols that provide guidelines for starting and stopping CPR. Your medical director and your system's legal adviser should agree on these protocols, which should be closely administered and reviewed by your medical director.

Positioning the Patient

The next step in providing CPR is to position the patient to ensure that the airway is open. For CPR to be effective, the patient must be lying supine on a firm surface, with enough clear space around the patient for two rescuers to perform CPR. If the patient is crumpled up or lying face down, you will need to reposition him or her. The few seconds that you spend to position the patient properly will greatly improve the delivery and effectiveness of CPR.

Follow the steps in (**Skill Drill A-1 ▶**) to reposition an unconscious adult for airway management:

1. **Kneel beside the patient.** You and your partner must be far enough away so that the patient, when rolled toward you, does not come to rest in your lap (**Step 1**).
2. First EMT-B: **Place your hands** behind the patient's back, head, and neck to protect the cervical spine if you suspect spinal injury. Second EMT-B: Place your hands on the distant shoulder and the hip (**Step 2**).
3. Second EMT-B: **Turn the patient toward you** by pulling on the distant shoulder and the hip. First EMT-B: Control the head and neck so that they move as a unit with the rest of the torso. This single motion will allow the head, neck, and back to stay in the same vertical plane and will minimize aggravation of any spinal injury (**Step 3**).
4. First EMT-B: **Place the patient in a supine position**, with the legs straight and both arms at the sides (**Step 4**).

If possible, log roll the patient onto a long backboard as you are positioning him or her for CPR. This

Positioning the Patient

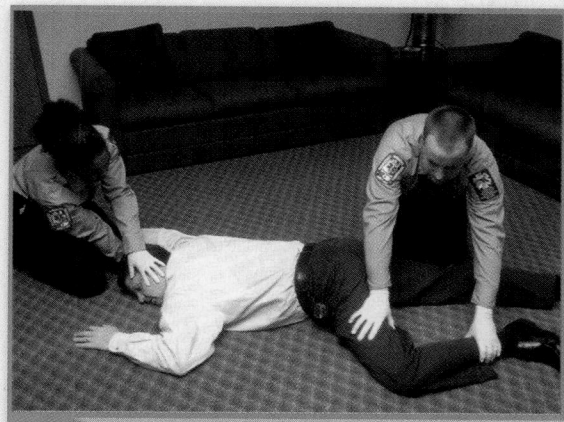

1 Kneel beside the patient, leaving room to roll the patient toward you.

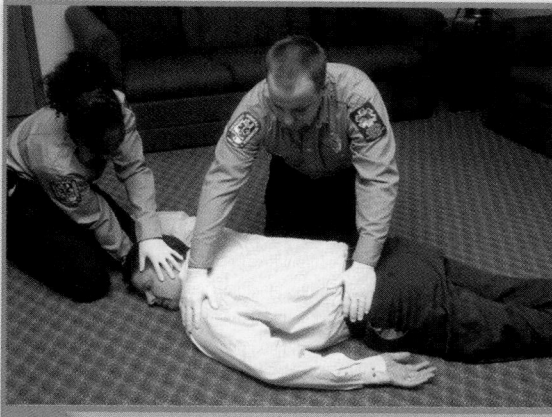

2 Grasp the patient, stabilizing the cervical spine if needed.

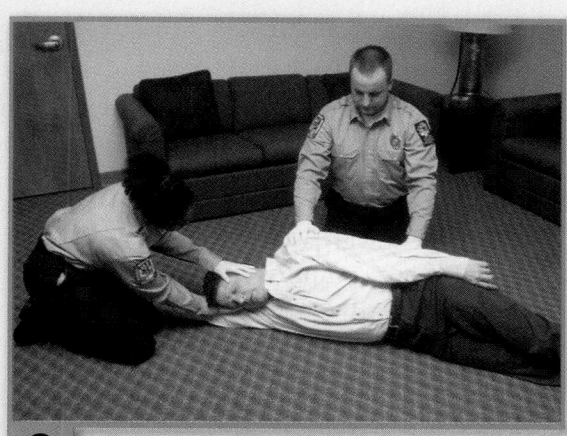

3 Move the head and neck as a unit with the torso as your partner pulls on the distant shoulder and hip.

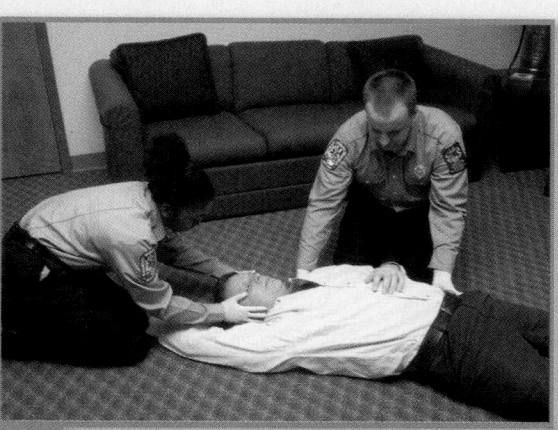

4 Move the patient to a supine position with legs straight and arms at the sides.

device will provide support during transport and emergency department care. Once the patient is properly positioned, you can easily assess airway, breathing, circulation, and the need for defibrillation and start CPR if necessary.

Opening the Airway in Adults

Without an open airway, rescue breathing will not be effective. There are two techniques for opening the airway in adults: the head tilt–chin lift maneuver and the jaw-thrust maneuver.

Head Tilt–Chin Lift Maneuver

Opening the airway to relieve an obstruction caused by relaxation of the tongue can often be accomplished quickly and easily with the head tilt–chin lift maneuver Figure A-8 ▶. In patients who have not sustained trauma, this simple maneuver is sometimes all that is required for the patient to resume breathing. If the patient has any foreign material or vomitus in the mouth, you should quickly remove it. Wipe out any liquid materials from the mouth with a piece of cloth held by your index and middle fingers; use your hooked index finger to remove any solid material. You should perform the head tilt–chin lift maneuver in an adult in the following way Figure A-9 ▶:

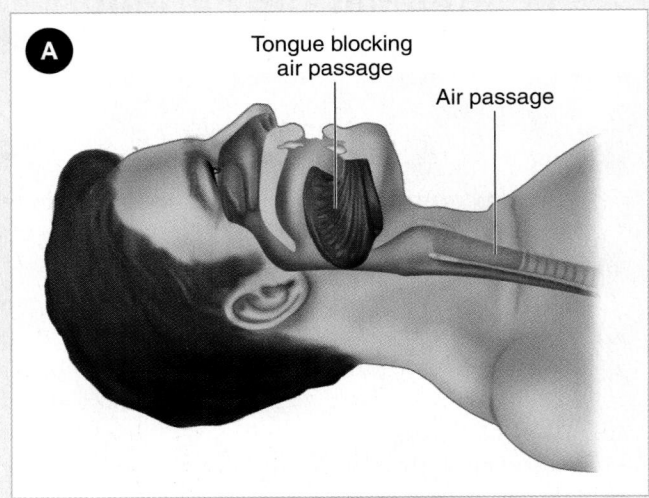
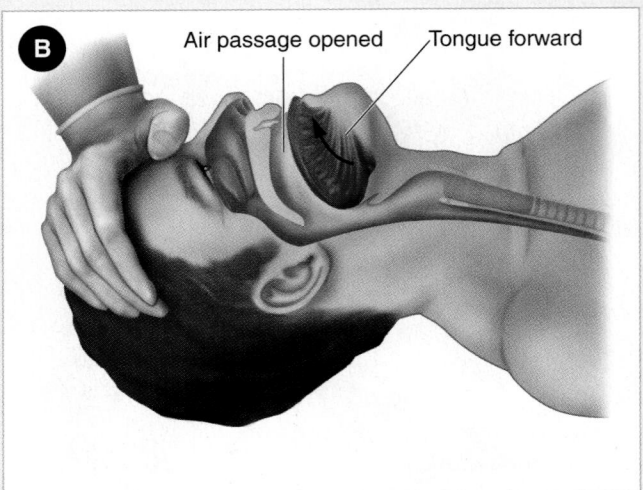

Figure A-8   A. Relaxation of the tongue back into the throat causes airway obstruction. **B.** The head tilt–chin lift maneuver combines two movements of opening the airway.

1. **Make sure the patient is supine.** Kneel close beside the patient.
2. **Place one hand on the patient's forehead,** and apply firm backward pressure with your palm to

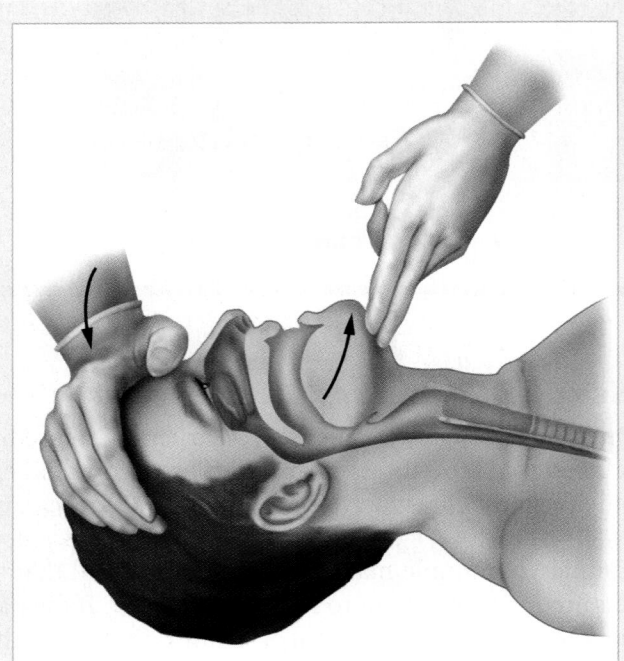

Figure A-9   To perform the head tilt–chin lift maneuver, place one hand on the patient's forehead and apply firm backward pressure with your palm to tilt the head back. Next, place the tips of the fingers of your other hand under the lower jaw near the bony part of the chin. Lift the chin forward, bringing the entire lower jaw with it, helping to tilt the head back.

tilt the patient's head back. This extension of the neck will move the tongue forward, away from the back of the throat, and will clear the airway if the tongue is blocking it.

3. **Place the tips of the fingers** of your other hand under the lower jaw near the bony part of the chin. Do not compress the soft tissue under the chin because this would block the airway.
4. **Lift the chin forward,** bringing the entire lower jaw with it, helping to tilt the head back. Do not use your thumb to lift the chin. Lift so that the teeth are nearly brought together, but avoid closing the mouth completely.

The chin lift has the added advantage of holding loose dentures in place, making obstruction by the lips less likely. Performing ventilation is much easier when dentures are in place. However, dentures that do not stay in place should be removed. Partial dentures (plates) may come loose as a result of an accident or as you are providing care, so check these periodically.

Jaw-Thrust Maneuver

The head tilt–chin lift maneuver is effective for opening the airway in most patients. In cases of suspected spinal injury, you want to minimize movement of the patient's neck. In this case, perform a jaw-thrust maneuver. To perform a jaw-thrust maneuver, place your fingers behind the angles of the patient's lower jaw and then move the jaw forward. Keep the head in a neutral position as you move the jaw forward and open the mouth. If the patient's mouth remains closed, you can

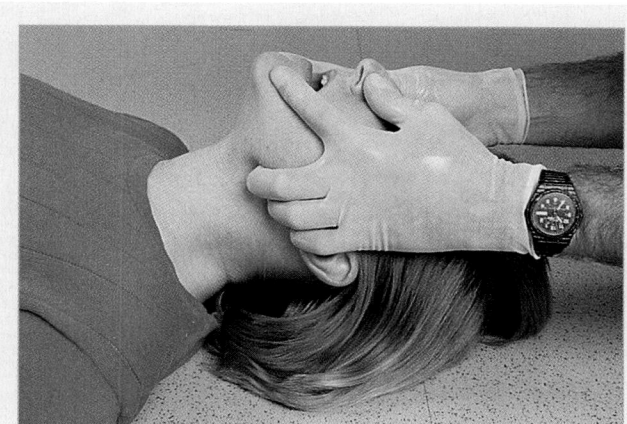

Figure A-10 To perform the jaw-thrust maneuver, maintain the head in neutral alignment and use your index and long fingers to thrust the jaw forward.

use your thumbs to pull the patient's lower lip down, to allow breathing. You can also easily apply a face mask or other barrier device with both hands while performing the jaw thrust.

Perform the jaw-thrust maneuver as follows Figure A-10 ▲ :

1. **Kneel above the patient's head.** Place your index or middle finger behind the angle of the patient's lower jaw on both sides, and forcefully move the jaw forward without manipulating the patient's neck.
2. **Use your thumbs to open the mouth** to allow breathing.
3. **The nose can be sealed with your cheek** when providing rescue breathing using the jaw-thrust maneuver.

Foreign Body Airway Obstruction in Adults

Airway obstruction may be caused by many things, including relaxation of the throat muscles in an unconscious patient, vomited or regurgitated stomach contents, a blood clot, damaged tissue after an injury, dentures, or foreign bodies. Occasionally, a large foreign body will be aspirated and block the upper airway.

Large objects that cannot be removed from the airway with suction, such as loose dentures, large pieces of vomited food, or blood clots, should be swept forward and out with your gloved index finger.

Suctioning can then be used as needed to keep the airway clear of thinner secretions such as blood, vomitus, and mucus.

Recognizing Foreign Body Obstruction

Sudden airway obstruction by a foreign body in an adult usually occurs during a meal. In a child, it usually occurs during mealtime or at play. Children commonly choke on peanuts, large bits of a hot dog, or small toys. If the foreign body is not removed quickly, the lungs will use up their oxygen supply; unconsciousness and death will follow. Your treatment will be based on the cause of the obstruction. Therefore, you must learn to tell the difference between obstructions caused by a foreign body and those due to respiratory failure or arrest, as might occur with fainting, stroke, or heart problems.

Conscious Patients

Sudden airway obstruction is usually easy to recognize in someone who is eating or has just finished eating. The person is suddenly unable to speak or cough, grasps his or her throat, turns cyanotic, and makes exaggerated efforts to breathe. Air is not moving into and out of the airway or the air movement is so slight that it is not detectable. At first, the patient will be conscious and able to clearly indicate the nature of the problem. Ask the patient, "Are you choking?" The patient will usually answer by nodding yes. Alternatively, he or she may use the universal sign to indicate airway blockage Figure A-11 ▼ .

Figure A-11 Hands at the throat are the universal sign to indicate choking.

Unconscious Patients

When you discover an unconscious patient, your first step is to determine whether he or she is breathing and has a pulse. The unconsciousness may be due to airway obstruction, cardiac arrest, or a number of other problems. Remember that you must first clear the patient's airway before addressing other problems, such as cardiac arrest. You must first ensure an open and unobstructed airway before checking for a pulse.

You should suspect an airway obstruction if the standard maneuvers to open the airway and ventilate the lungs are not effective. If you feel resistance to blowing into the patient's lungs or pressure builds up in your mouth, the patient probably has some type of obstruction.

Removing a Foreign Body Obstruction

The manual maneuvers recommended for removing a foreign body airway obstruction in the adult are the abdominal-thrust maneuver (the Heimlich maneuver) and chest thrusts. In either case (abdominal thrusts and chest thrusts), a finger sweep is used to remove the object from the airway.

Abdominal-Thrust Maneuver

The abdominal-thrust maneuver, also called the Heimlich maneuver, is the preferred way to dislodge and force food or other material from the throat of a choking victim. Residual air, which is always present in the lungs, is compressed upward and used to expel the object. In conscious patients with a complete airway obstruction, you should repeat abdominal thrusts until the foreign body is expelled or the patient becomes unconscious. Each thrust should be deliberate, with the intent of relieving the obstruction.

In an unconscious patient, you should perform abdominal thrusts in sets of five, then perform a finger sweep followed by attempts to ventilate the patient. Repeat the sequence of five abdominal thrusts, finger sweep, and attempts to ventilate until the obstruction is relieved.

To perform abdominal thrusts on a conscious adult (Figure A-12 ▶), use the following technique:

1. **Stand behind the patient**, and wrap your arms around his or her waist.
2. **Make a fist with one hand**; grasp the fist with the other hand. Place the thumb side of the fist against the patient's abdomen, just above the umbilicus and well below the xiphoid.

Figure A-12 The abdominal thrust maneuver in a conscious adult. Stand behind the patient, and wrap your arms around the patient's waist. Press your fists into the patient's abdomen, and deliver quick inward and upward thrusts.

3. **Press your fist into the patient's abdomen** with a quick inward and upward thrust.
4. **Continue abdominal thrusts** until the object is expelled from the airway or the patient becomes unconscious.

If the patient becomes unconscious, use the following technique (Figure A-13 ▶):

1. **Place the patient in a supine position.**
2. **Perform a tongue-jaw lift, then a finger sweep** to try to remove the object.
3. **Open the airway and attempt to ventilate.** If unsuccessful, reposition the head, reopen the airway, and try to ventilate again.
4. If you are still unable to ventilate, **straddle the patient's legs.**
5. **Place the heel of one hand against the patient's abdomen** above the umbilicus and well below the xiphoid process. Then place your other hand on top of the first.
6. **Press the hand into the patient's abdomen** with quick inward and upward thrusts, and repeat five times.
7. **Repeat the sequence** of tongue-jaw lift, finger sweep, attempts to ventilate, and abdominal thrusts until the obstruction is cleared or advanced procedures are available to establish a patent airway.

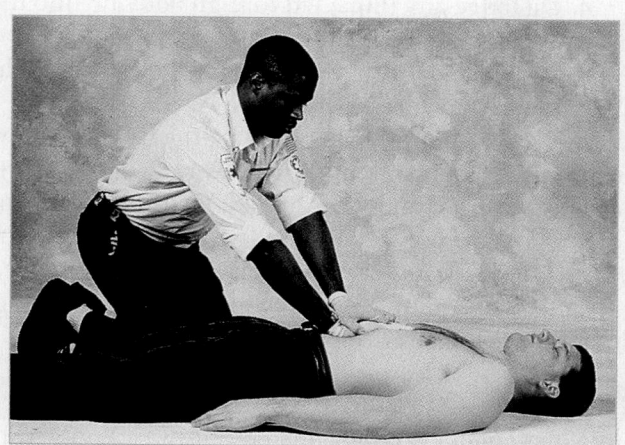

Figure A-13 The abdominal thrust maneuver in an unconscious adult. Straddle the hips or legs. Place the heel of one hand against the patient's abdomen and the other hand on top of the first. Press your hands into the patient's abdomen in a series of five quick inward and upward thrusts.

Chest Thrusts

You can perform the abdominal-thrust maneuver safely on all adults and children. However, you should preferentially use chest thrusts for women in advanced stages of pregnancy, patients who are very obese, and children younger than 1 year.

To perform chest thrusts on the conscious adult, use the following technique (Figure A-14 ▶):

1. **Stand behind the patient** with your arms directly under the patient's armpits, and wrap your arms around the patient's chest.
2. **Make a fist with one hand**; grasp the fist with the other hand. Place the thumb side of the fist against the patient's sternum, avoiding the xiphoid process and the edges of the rib cage.
3. **Press your fist into the patient's chest** with backward thrusts until the object is expelled or the patient becomes unconscious.

If the patient becomes unconscious, use the following technique (Figure A-15 ▶):

1. **Place the patient in a supine position.**
2. **Kneel next to the patient.**
3. **Perform a tongue-jaw lift, then a finger sweep** to try to remove the object.
4. **Open the airway and attempt to ventilate.** If unsuccessful, reposition the head, reopen the airway, and try to ventilate again.

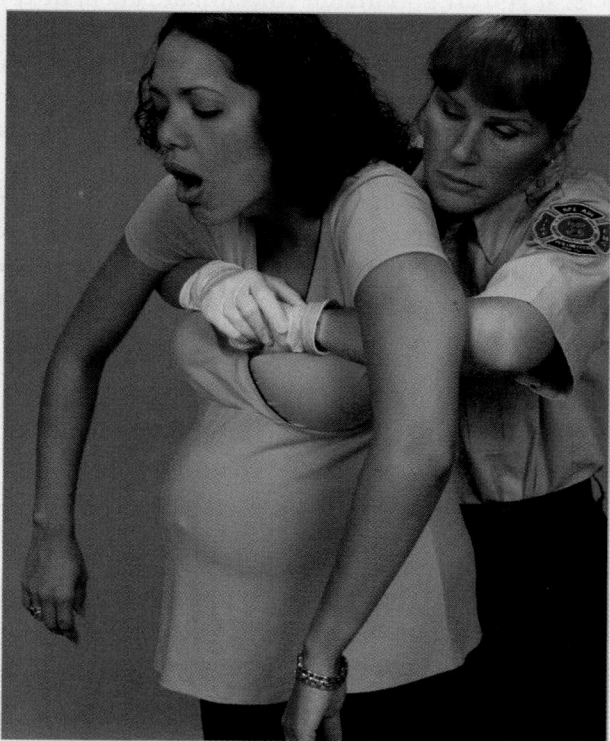

Figure A-14 Removal of foreign body obstruction in a conscious adult using chest thrusts. Stand behind the patient, and wrap your arms around the patient's chest. Place the thumb side of one fist against the chest while holding your fist with the other hand. Press your fists into the patient's chest with backward thrusts.

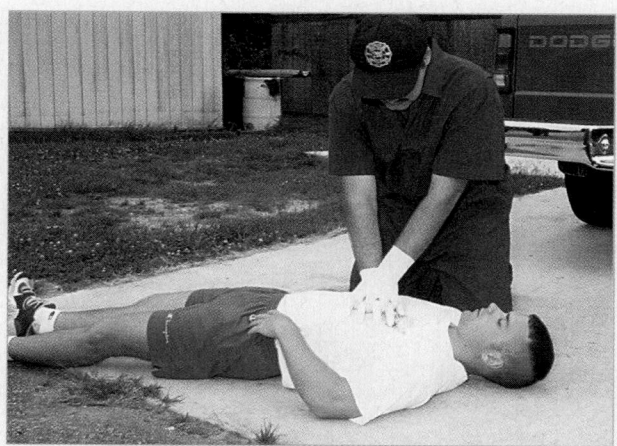

Figure A-15 Chest thrusts in an unconscious adult. Kneel next to the patient. Place your hands as you would to deliver chest compressions for CPR. Deliver slow chest thrusts until the object has been expelled.

5. If you are still unable to ventilate, **kneel next to the patient.**
6. **Use the same hand position** as for chest compressions during CPR.
7. **Deliver slow, deliberate chest thrusts** to expel the object.
8. **Repeat the sequence** of tongue-jaw lift, finger sweep, attempts to ventilate, and chest thrusts until the obstruction is cleared or advanced procedures are available to establish a patent airway.

Manual Removal of Foreign Objects

Use of finger sweeps should be limited to unconscious patients. If you can see a foreign object in the patient's mouth, you should remove it carefully with your gloved fingers. This may be necessary if the abdominal-thrust maneuver dislodges but does not expel the foreign body.

Use the following technique to remove the foreign material manually (Figure A-16 ▼):

1. **Place the patient in a supine position.**
2. **Open the patient's mouth** by grasping the tongue and the lower jaw between your thumb and fingers and lifting them forward (tongue-jaw lift). This pulls the tongue away from the back of the throat and from the foreign body that may be lodged there.
3. **Use the index finger of your opposite hand as a hook to sweep down inside the patient's cheek** to the base of the tongue.

4. **Dislodge any impacted foreign body** up into the mouth.
5. **When the foreign body comes within reach, grasp and carefully remove it.**

Make sure that you do not push the dislodged foreign body farther back into the airway. Because this is very easy to do in infants and small children, blind finger sweeps are not recommended for these patients. Instead, look first, and then perform a finger sweep only after you see the object. If you do not see anything, do not perform a finger sweep.

Partial Airway Obstruction

Some patients may have only a partial airway obstruction. They are able to exchange some air but still have signs of respiratory distress. Breathing is noisy, and the patient may be coughing. Your main concern is to prevent a partial airway obstruction from becoming a complete airway obstruction. Neither the abdominal-thrust maneuver nor chest thrusts are indicated in these situations. Manual removal is dangerous because you could force the object farther down the airway, causing a complete obstruction.

Therefore, for a patient with a partial airway obstruction, you should first encourage the patient to cough. Do not interfere with the patient's attempts to expel the foreign body; rather, give 100% oxygen to the patient using a nonrebreathing mask and provide prompt transport. A partial airway obstruction with poor exchange (shown by ineffective cough, high-pitched

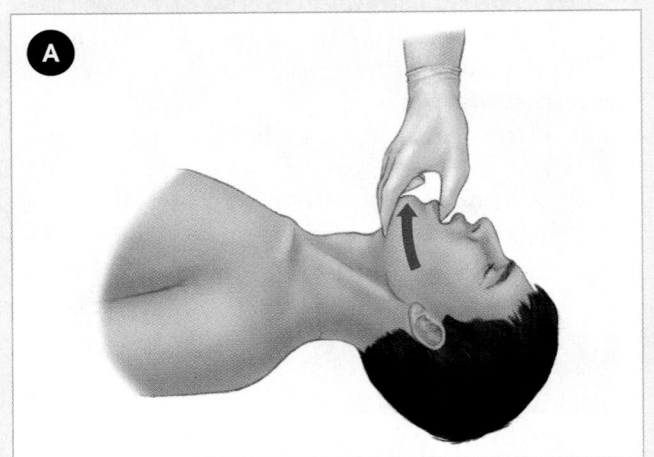
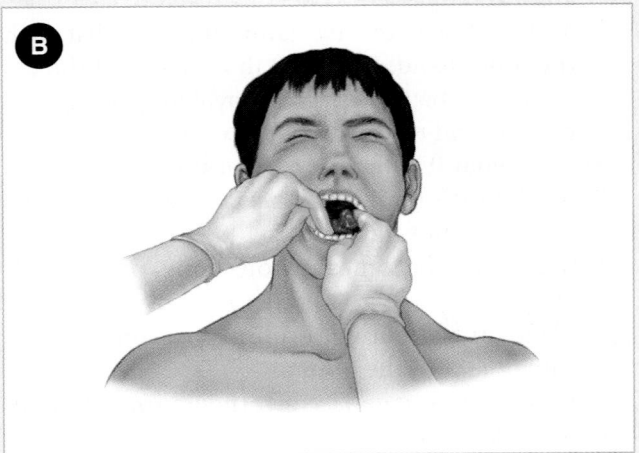

Figure A-16   A. To manually remove a foreign object in an unconscious patient, use the tongue-jaw lift to open the mouth and to help see the object. **B.** Hook the index finger of your opposite hand to sweep down inside the cheek to the base of the tongue.

inspiratory sound, increased respiratory difficulty, and cyanosis) should be treated as a complete airway obstruction.

Foreign Body Obstruction in Infants and Children

Airway obstruction, a common problem in infants and children, usually is caused by a foreign body or an infection, such as croup or epiglottitis, resulting in swelling and narrowing of the airway. You should try to identify the cause of the obstruction as soon as possible. In patients who have signs and symptoms of an airway infection, you should not waste time trying to dislodge a foreign body. The child needs 100% oxygen with a non-rebreathing mask and immediate transport to the emergency department.

A previously healthy child who is eating or playing with small toys or an infant who is crawling about the house and who suddenly has difficulty breathing has probably aspirated a foreign body. As in adults, foreign bodies may cause a partial or complete airway obstruction. With a partial airway obstruction, air exchange can be good or poor.

With good air exchange, the patient can cough forcefully, although there may be wheezing between coughs. As long as the patient can breathe, cough, or talk, you should not interfere with his or her attempts to expel the foreign body. In fact, you should encourage the child to continue coughing and breathing. You should attempt to remove the obstruction only if the cough becomes ineffective; if the patient has stridor, increased respiratory difficulty, or cyanosis; or if the patient loses consciousness.

Give patients with a partial airway obstruction 100% oxygen and be prepared to treat them for a complete airway obstruction if the need arises. Promptly transport the child to the hospital with continuous monitoring en route.

Removing a Foreign Body Airway Obstruction

To perform abdominal thrusts in a conscious child, use the following technique (Figure A-17 ▶):

1. **Stand behind the patient.** Place your arms under the patient's armpits, and wrap your arms around the patient's abdomen and chest.

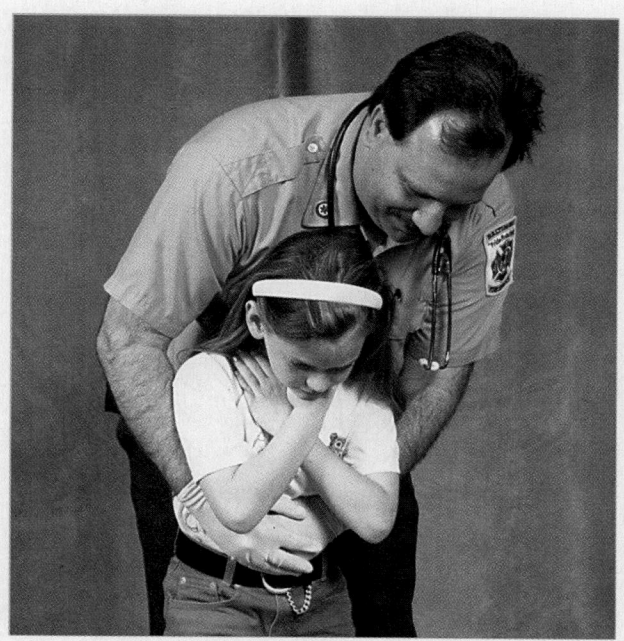

Figure A-17 Stand behind the child, place your arms under the armpits, and wrap your arms around the patient's abdomen and chest. Press your fists into the patient's abdomen, and provide quick, upward thrusts.

2. **Make a fist with one hand**; grasp the fist with the other hand. Place the thumb side of your fist against the patient's abdomen, just above the umbilicus and well below the xiphoid process.
3. **Press your fist into the patient's abdomen** with quick upward thrusts.
4. **Repeat thrusts** until the object is expelled or the patient loses consciousness.

If the child becomes unconscious, use the following technique (Figure A-18 ▶):

1. **Place the patient in a supine position.**
2. **Perform a tongue-jaw lift and look for an object in the pharynx.** If an object is visible, remove it. Do not perform a blind finger sweep.
3. **Open the airway, and attempt to ventilate.** If unsuccessful, reposition the head, reopen the airway, and try to ventilate again.
4. **If you are still unable to ventilate,** kneel beside the patient, or straddle the patient's legs.
5. **Place the heel of one hand** against the patient's abdomen above the umbilicus and well below the xiphoid process. Then place your other hand on top of the first.

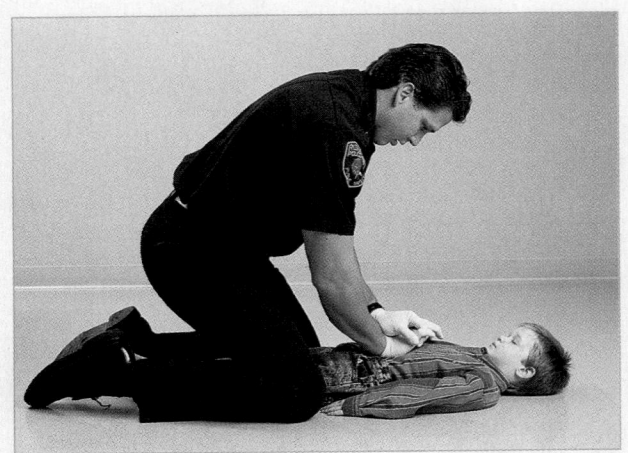

Figure A-18 Place the child in a supine position, and kneel beside the patient or straddle the patient's hips. Place the heel of one hand against the patient's abdomen, with the other hand on top of the first. Press your hands into the patient's abdomen in a series of five quick upward thrusts.

6. **Press the hand into the patient's abdomen** with quick inward and upward thrusts, and repeat five times.

7. **Repeat the sequence** of tongue-jaw lift, airway visualization, attempts to ventilate, and abdominal thrusts until the obstruction is cleared or advanced procedures are available to establish a patent airway.

If you see the foreign body, perform the tongue-jaw lift, and then use a finger sweep to remove it. Do not use blind finger sweeps on infants and children because you may force foreign objects farther into the airway. In rare cases when you cannot remove the foreign body, perform mouth-to-mask ventilation en route to the hospital.

The abdominal-thrust maneuver might injure the liver or other abdominal organs in an infant. Therefore, use the following technique to remove a foreign body in an infant (**Figure A-19** ▶):

1. **Place one hand on the infant's back and neck** and the other on his or her chest, jaws, and face, holding the jaw firmly to support the head at a level lower than the trunk. This sandwiches the infant between your hands and arms. Your forearm should rest on your thigh to support the infant.

2. **Deliver five quick back blows** between the shoulder blades, using the heel of your hand.

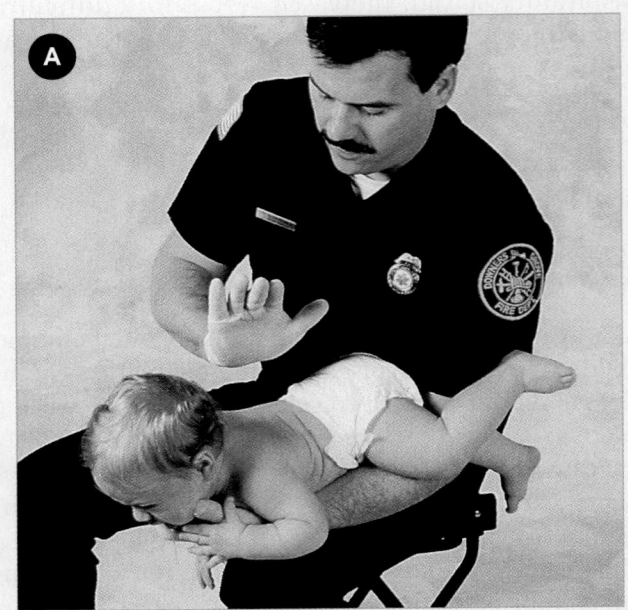

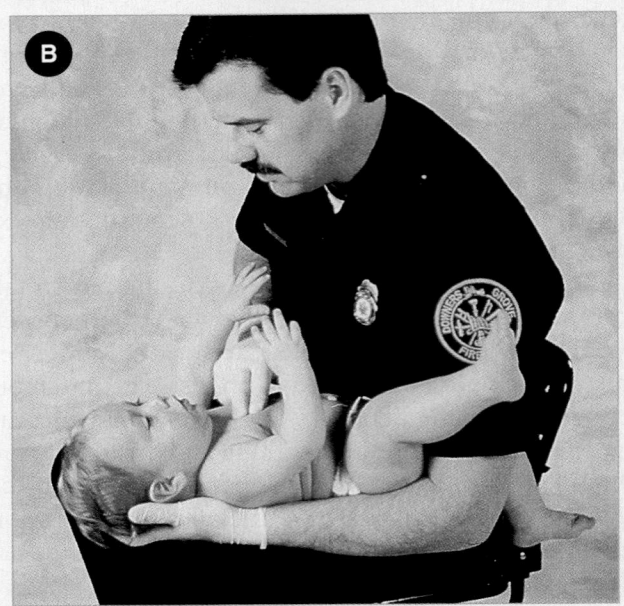

Figure A-19 **A.** Deliver five quick back blows between the shoulder blades, using the heel of your hand. **B.** Give five quick chest thrusts on the sternum at a slightly slower rate than you would for CPR.

3. **Next, turn the infant face up**, making sure that you support the head and neck. Hold the infant in a supine position on your thigh, with the head slightly lower than the trunk.

4. **Give five quick chest thrusts on the sternum** in the same manner as for CPR, except at a slightly slower rate. If the infant is large or your hands

are small, you might need to place the infant on your lap to deliver the chest thrusts.

5. If the infant is unconscious, you should perform the tongue-jaw lift to open the mouth. Remove the object manually if you can see it.

If the infant is unconscious and does not start breathing after these maneuvers, try to open the airway again and give artificial ventilation. If the chest does not rise, reposition the head and attempt ventilation again. If the chest still does not rise, continue giving back blows followed by chest thrusts and attempts to ventilate until the obstruction is cleared or you reach the hospital.

Rescue Breathing in Adults

Once you open the airway, check for breathing by placing your ear about 1″ above the patient's nose and mouth; listen carefully for sounds of breathing. Turn your head so that you can watch for movement of the patient's chest and abdomen (Figure A-20 ▼). This is called the look, listen, and feel technique. You know that the patient is breathing if you see the chest and abdomen rise and fall and, more important, if you feel and hear air move during exhalation. With airway obstruction, there may be no movement of air, even though the chest and abdomen rise and fall as the patient tries to breathe. You may also have difficulty seeing movement of the chest and abdomen if

the patient is fully clothed. Finally, you may see very little or no chest movement in some patients, particularly those with chronic lung disease. Therefore, if you do not feel any air movement as you look, listen, and feel, you must begin artificial ventilation. This evaluation should take no more than 10 seconds.

A lack of oxygen, combined with too much carbon dioxide in the blood, is lethal. To correct this condition, you must provide slow, deliberate ventilations that last 2 seconds. This gentle, slow method of ventilating the patient prevents air from being forced into the stomach.

Ventilation

Ventilations are now done routinely with barrier devices, such as masks. These devices feature a plastic barrier that covers the patient's mouth and nose and a one-way valve to prevent exposure to secretions and exhaled contaminants (Figure A-21 ▼). Such devices also provide good infection control. Providing ventilations without a barrier device should be performed only in extreme conditions. When using a pocket face mask or bag-valve-mask (BVM) device without supplemental oxygen, you should deliver a tidal volume of about 700 to 1000 mL (10 mL/kg) over 2 seconds. If supplemental oxygen is attached to the pocket face mask or BVM device, deliver a tidal volume of about 400 to 600 mL (4 to 6 mL/kg) over 1 to 2 seconds. Regardless of whether you are ventilating the patient with or without supplemental oxygen, you should observe the chest for good rise to assess the effectiveness of your ventilations.

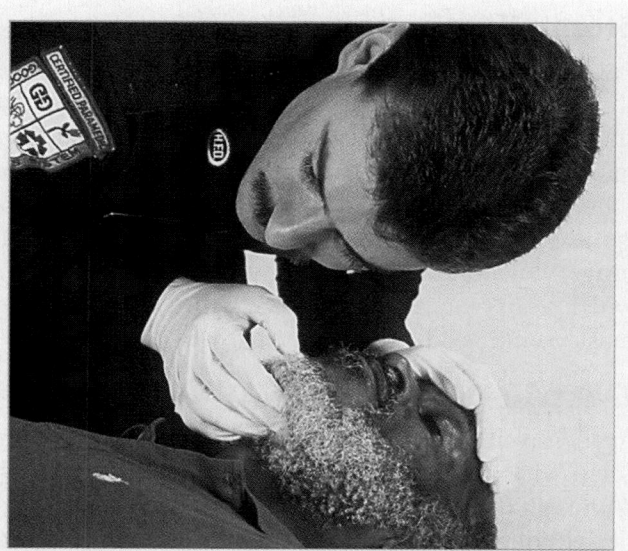

Figure A-20 Look, listen, and feel for signs of breathing.

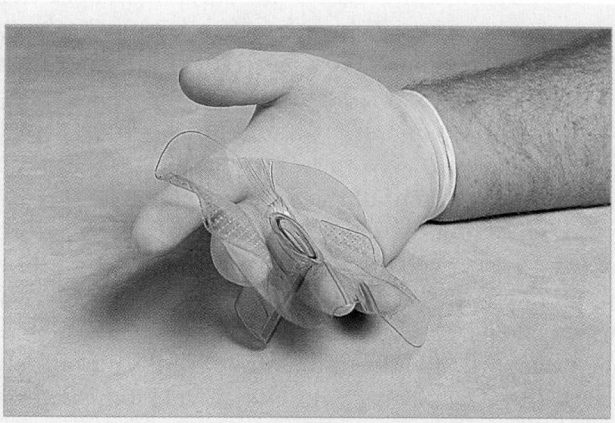

Figure A-21 A barrier device is used in performing ventilation because it prevents exposure to saliva, blood, and vomitus.

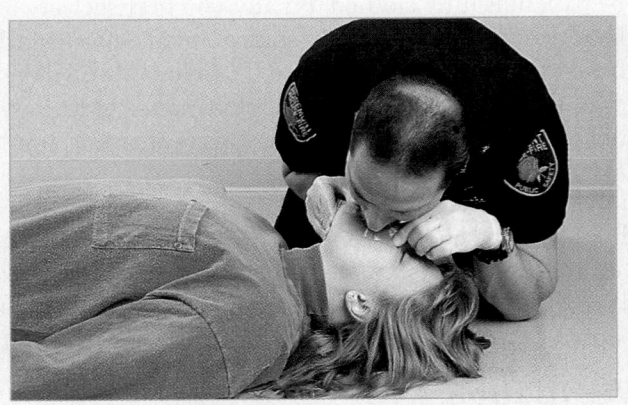

Figure A-22 To perform ventilations, ensure that you make a tight seal with your mouth around the barrier device, and then give two slow gentle breaths, each lasting 2 seconds (1 to 2 seconds if oxygen is attached).

You should perform rescue breathing in an adult with a simple barrier device in the following way (Figure A-22 ▲):

1. **Open the airway** with the head tilt–chin lift maneuver (nontrauma patient).
2. **Press on the forehead** to maintain the backward tilt of the head. Pinch the patient's nostrils together with your thumb and index finger.
3. **Depress the lower lip** with the thumb of the hand that is lifting the chin. This will help to keep the patient's mouth open.
4. **Open the patient's mouth widely**, and place the barrier device over the patient's mouth and nose.
5. **Take a deep breath**, then make a tight seal with your mouth around the barrier device. Give two slow rescue breaths, each lasting 2 seconds (1 to 2 seconds if oxygen is attached), followed by 10 to 12 breaths/min.
6. **Remove your mouth**, and allow the patient to exhale passively. Turn your head slightly to watch for movement of the patient's chest.

When using the jaw-thrust maneuver to open the airway (in suspected neck or spine injury), positioning yourself at the patient's head will facilitate simultaneous c-spine stabilization and adequate ventilation. Keep the patient's mouth open with both thumbs, and seal the nose by placing your cheek against the patient's nostrils (Figure A-23 ▶). Note that this maneuver is somewhat difficult; practicing with a manikin will help you gain familiarity with this technique.

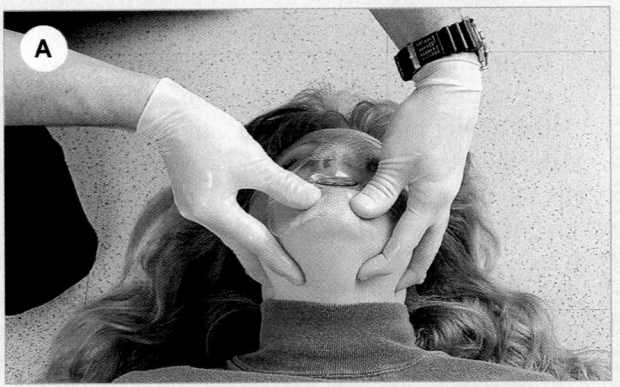

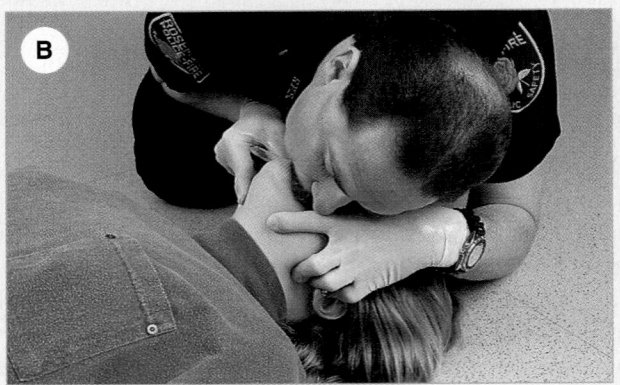

Figure A-23 A. If you use the jaw-thrust maneuver to open the airway, keep the patient's mouth open with both thumbs as you move from above the patient's head to the side. **B.** Seal the nose by placing your cheek against the patient's nostrils.

Stoma Ventilation

Patients who have undergone surgical removal of the larynx often have a permanent tracheal stoma at the midline in the neck. In this case, a stoma is an opening that connects the trachea directly to the skin (Figure A-24 ▶). Because it is at the midline, the stoma is the only opening that will move air into the patient's lungs; you should ignore any other openings. Patients with a stoma should be ventilated with a BVM or pocket mask device, as described in Chapter 7.

Gastric Distention

Artificial ventilation may result in the stomach becoming filled with air, a condition called <u>gastric distention</u>. Although it occurs more easily in children, it also happens frequently in adults. Gastric distention is likely to occur if you blow too fast as you ventilate, if you give too much air, or if the patient's airway is not opened

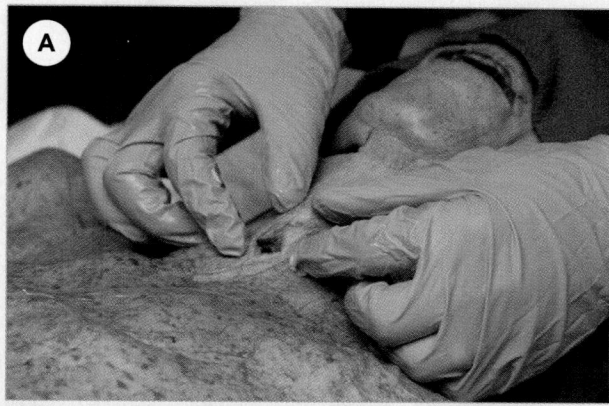

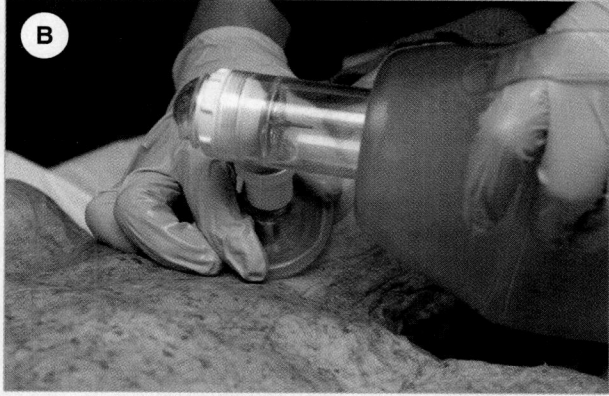

Figure A-24   A. This stoma connects the trachea directly to the skin. **B.** Use a BVM or pocket mask device to ventilate a patient with a stoma.

adequately. Therefore, it is important for you to give slow, gentle breaths. Such breaths are also more effective in ventilating the lungs. Serious inflation of the stomach is dangerous because it can cause the patient to vomit during CPR. It can also reduce lung volume by elevating the diaphragm.

If massive gastric distention interferes with adequate ventilation, you should contact medical control. Check the airway again and reposition the patient, watch for rise and fall of the chest, and avoid giving forceful breaths. Medical control may order you to turn the patient on his or her side and provide gentle manual pressure to the abdomen to expel air from the stomach. Have suction readily available, and be prepared for copious amounts of vomitus. If gastric distention interferes with your ability to perform adequate artificial ventilations, it must be managed.

⚠ Pediatric Needs

Children in respiratory distress are often struggling to breathe. As a result, they usually position themselves in a way that keeps the airway open enough for air to move. Let them stay in that position as long as breathing remains adequate. If you and your partner arrive at the scene and find that the infant or child is not breathing or has cyanosis, immediate management (that is, rescue breathing, supplemental oxygen) is essential. Consider requesting additional assistance, if available.

For infants, the preferred technique of rescue breathing is mouth-to-nose-and-mouth ventilation. With this technique, a seal must be made over the mouth and nose. Various masks and other barrier devices are recommended for this technique. If the patient is a large child (1 to 8 years old) for whom a tight seal cannot be made over both mouth and nose, you should provide mouth-to-mouth ventilation as you would for an adult.

Once you have made an airtight seal over the mouth, give two gentle breaths, each lasting 1 to $1\frac{1}{2}$ seconds. These initial breaths will help you assess for airway obstruction and expand the lungs. Because the lungs of infants and children are much smaller than those of adults, you do not need to blow in a great amount of air. Limit the amount of air to that needed to cause the chest to rise.

Remember, too, that a child's airway is smaller than that of an adult. Therefore, there is greater resistance to airflow. As a result, you will need to use a bit more ventilatory pressure to inflate the lungs. You know you are giving the correct amount of air volume as soon as you see the chest rise. Infants and children should be ventilated once every 3 seconds, or 20 times per minute.

If air enters freely with your initial breaths and the chest rises, the airway is clear. You should then check the pulse. If air does not enter freely, you should check the airway for obstruction. Reposition the patient to open the airway, and attempt to give another breath. If air still does not enter freely, you must take steps to relieve the obstruction.

Figure A-25 The recovery position is used to maintain an open airway in an adequately breathing patient with a decreased level of consciousness who has had no traumatic injuries. It allows vomitus, blood, and any other secretions to drain from the mouth.

Figure A-26 Feel for the carotid artery by locating the larynx, then sliding two fingers toward one side. You can feel the pulse in the groove between the larynx and sternocleidomastoid muscle.

Recovery Position

The recovery position helps to maintain a clear airway in a patient with a decreased level of consciousness who has not had traumatic injuries and is breathing adequately on his or her own (Figure A-25 ▲). It also allows vomitus to drain from the mouth. Roll the patient onto his or her side so that the head, shoulders, and torso move as a unit, without twisting. Then place the patient's hands under his or her cheek. Never place a patient who has a suspected head or spinal injury in the recovery position because maintenance of spinal alignment in this position is not possible and further spinal cord injury could result.

Adult CPR

Once you have arrived at the scene and determined that the patient is unresponsive and not breathing, you must position the patient and begin rescue breathing. After you begin rescue breathing, you must assess the patient's circulation.

Cardiac arrest is determined by the absence of a palpable pulse at the carotid artery. Feel for the carotid artery by locating the larynx at the front of the neck and then sliding two fingers toward one side. The pulse is felt in the groove between the larynx and the sternocleidomastoid muscle, with the pulp of the index and long fingers held side by side (Figure A-26 ▶). Light pressure is sufficient to palpate the pulse. Excessive pressure must not be applied because it can obstruct the carotid circulation, dislodge blood clots, or produce marked reflex slowing of heart rate. Look for other signs of circulation also, such as skin color.

External Chest Compression

You can provide CPR by applying rhythmic pressure and relaxation to the lower half of the sternum. The heart is located slightly to the left of the middle of the chest between the sternum and the spine (Figure A-27 ▶). The blood that circulates through the lungs by chest compressions is likely to receive adequate oxygen to maintain life when accompanied by artificial ventilation. However, keep in mind that, at its best, external chest compression provides only one third of the blood that is normally pumped by the heart, so it is very important to do it properly.

The patient must be placed on a firm, flat surface, in a supine position. The head should not be elevated at a level above the heart because this will further reduce blood flow to the brain. The surface can be the ground, the floor, or a backboard on a stretcher. You cannot perform chest compressions adequately on a bed; therefore, a patient who is in bed should be moved to the floor or have a board placed under the back. Remember, too, that external chest compressions must always be accompanied by artificial ventilation.

Proper Hand Position

Correct hand position is established by sliding the index and middle fingers of the hand that is nearest the patient's feet along the edge of the rib cage until they

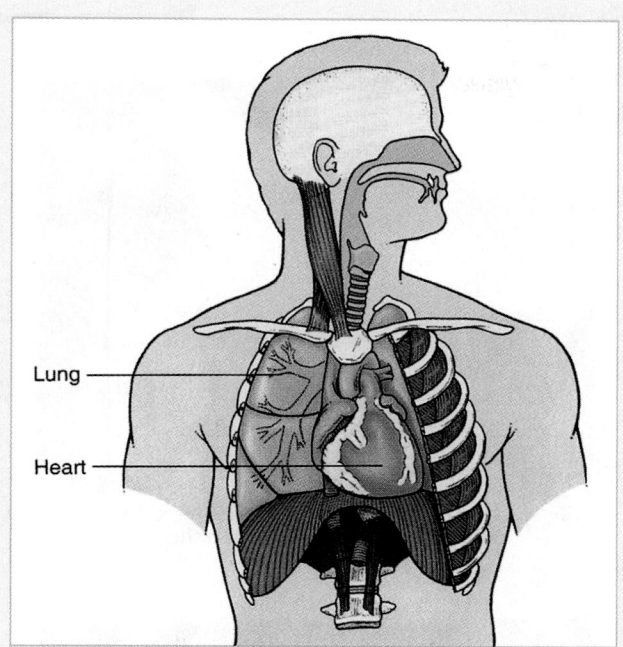

Figure A-27 The heart lies slightly to the left of the middle of the chest between the sternum and spine.

✴ EMT-B Tips

Performing CPR in the field is a different experience than practicing in the classroom and requires special preparations. You and your partner should drill in advance how you will make the best use of your skills, equipment, and personnel available to assist. Besides improving patient care, practicing how to deploy equipment, assign roles, and move patients with fire crews who may respond to help you is also an excellent way to develop good working relations.

◆ Geriatric Needs

Proper hand position and depth of compression, which are always important, take on added priority in geriatric patients who are likely to have brittle bones and chest cartilage. There is no guarantee against causing injury to these tissues, and you must compress adequately to provide adequate perfusion of vital organs. Paying particular attention to your compression technique, however, will help reduce avoidable injuries.

reach the xiphoid notch in the center of the chest. Follow the steps in (Skill Drill A-2 ▶):

1. **Slide your index and middle fingers** of the hand nearest the patient's feet along the center of the patient's rib cage to the notch in the center of the chest (**Step 1**).
2. **Push your middle finger** as high as possible into the notch, and then lay your index finger on the lower portion of the sternum with the two fingers touching (**Step 2**).
3. **Place the heel of your other hand** on the lower half of the sternum so that it touches the index finger of the first hand (**Step 3**).
4. **Remove your first hand** from the notch in the center of the rib cage, and place it over the hand that is resting on the patient's lower sternum (**Step 4**).
5. **With your arms straight**, lock your elbows, and position your shoulders directly over your hands. Depress the sternum 1½″ to 2″, using direct downward movement and then rising gently upward. Avoid using a rocking motion because this will not provide effective compression of the chest. Only the heel of one hand should be in contact with the lower half of the sternum. Take great

care not to place your hand on the xiphoid process, which extends down over the upper abdomen and liver, or beside the sternum onto the ribs or costal cartilage. Your technique may be improved or made more comfortable if you interlock the fingers of your lower hand with the fingers of your upper hand; either way, your fingers should be kept off the patient's chest (**Step 5**).

Proper Compression Technique

Complications from chest compressions are rare but can include fractured ribs, a lacerated liver, and a fractured sternum. Although these injuries cannot be entirely avoided, you can minimize the chance that they will occur if you use good, smooth technique and proper hand placement.

Proper compressions begin by locking your elbows, with your arms straight, and positioning your shoulders

Skill Drill A-2

Performing Chest Compressions

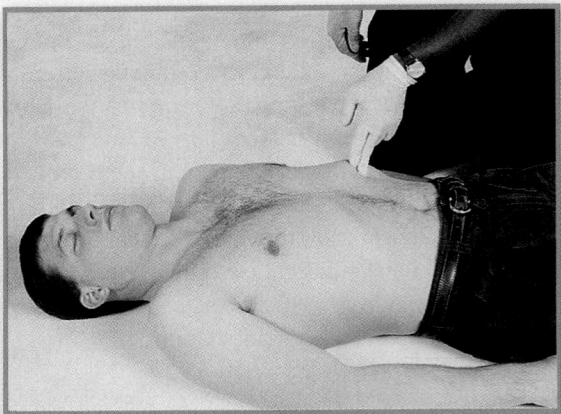

1 Slide your index and middle fingers along the rib cage to the notch in the center of the chest.

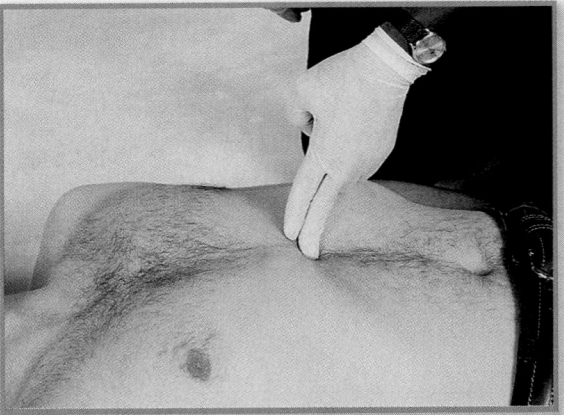

2 Push the middle finger high into the notch, and lay the index finger on the lower portion of the sternum.

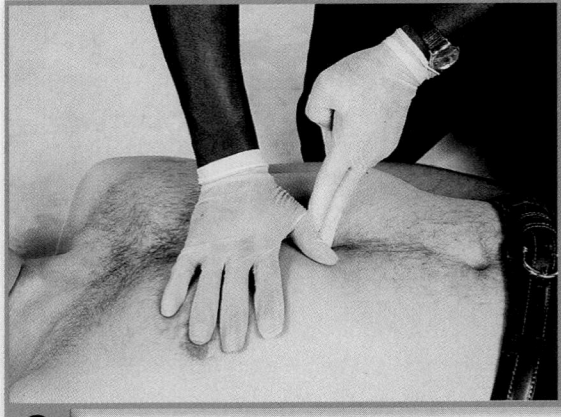

3 Place the heel of the second hand on the lower half of the sternum, touching the index finger of your first hand.

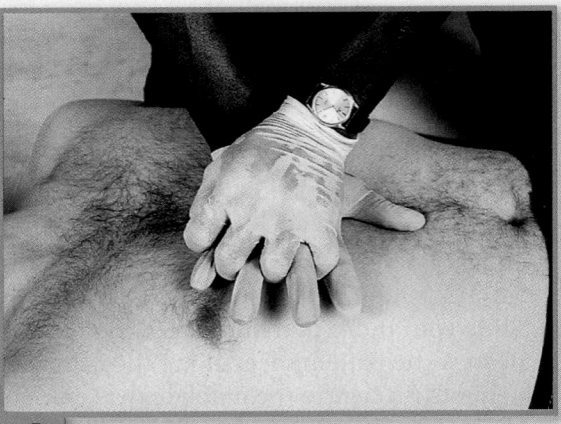

4 Remove your first hand from the notch, and place it over the hand on the sternum.

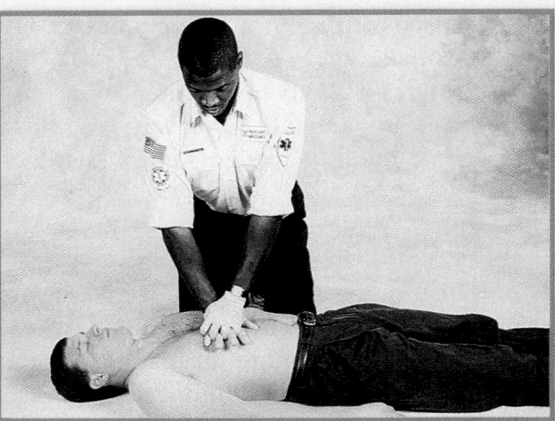

5 With your arms straight, lock your elbows, and position your shoulders directly over your hands. Depress the sternum $1\frac{1}{2}$″ to 2″ using a rhythmic motion.

directly over your hand so that the thrust of each compression is straight down on the sternum. Depress the sternum 1½″ to 2″ in an adult, avoiding a rocking motion and rising gently upward. This motion allows pressure to be delivered vertically down from your shoulders. Vertical downward pressure produces a compression that must be followed immediately by an equal period of relaxation. The ratio of time devoted to compression versus relaxation should be 1:1.

The actual motions must be smooth, rhythmic, and uninterrupted (Figure A-28A ▼). Short, jabbing compressions are not effective in producing artificial blood flow. Do not remove the heel of your hand from the patient's chest during relaxation, but make sure that you completely release pressure on the sternum so that it can return to its normal resting position between compressions (Figure A-28B ▼).

One-rescuer Adult CPR

When you are providing CPR alone, you must give both artificial ventilations and chest compressions in a ratio of compressions to ventilations of 15:2. To perform one-rescuer adult CPR, follow the steps in (Skill Drill A-3 ▶):

1. **Determine unresponsiveness**, and then call for additional help (**Step 1**).
2. Position the patient properly (supine) and **open the airway** according to suspicion of spinal injury (**Step 2**).
3. **Determine breathlessness** by using the look, listen, and feel technique. If the patient is unconscious but breathing adequately, place him or her in the recovery position, and maintain the open airway (**Step 3**).

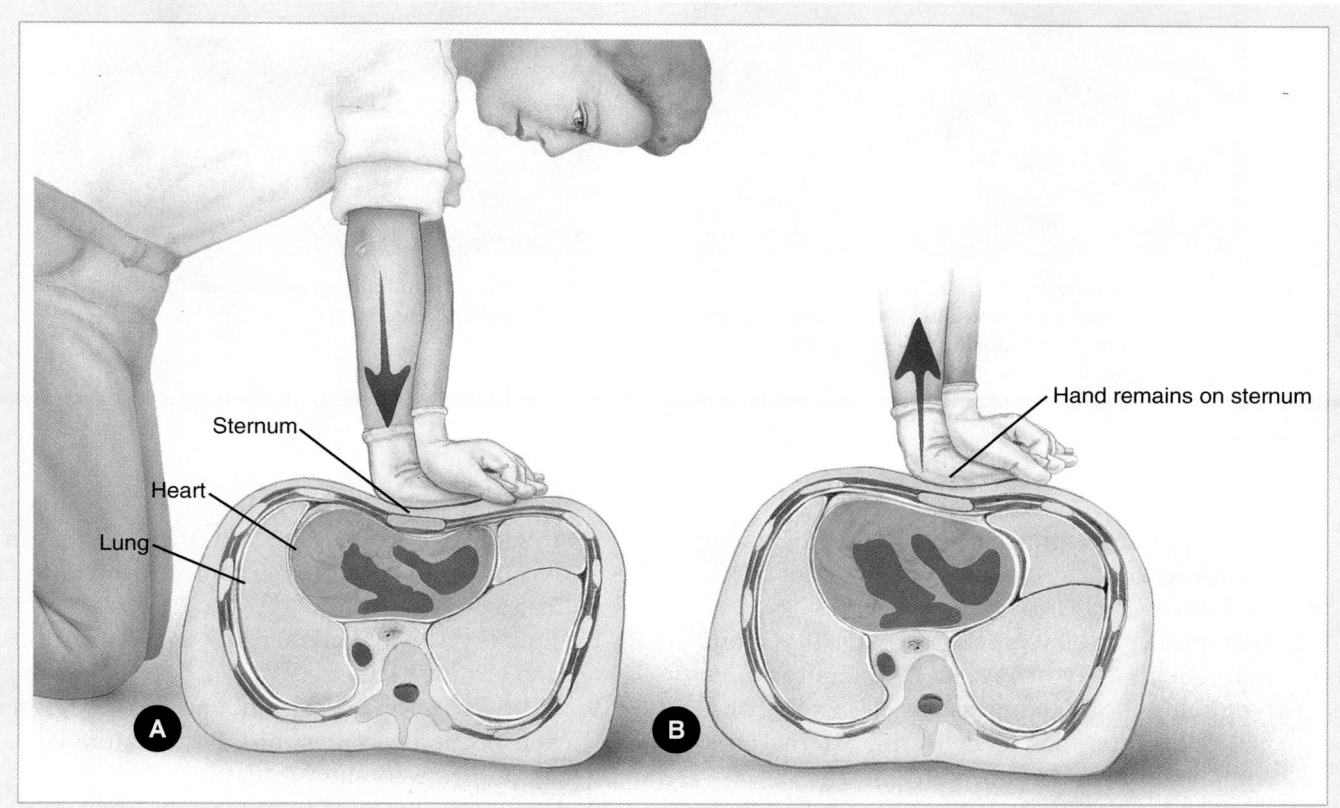

Sternum

Heart

Lung

Hand remains on sternum

A

B

Figure A-28   A. Compression and relaxation should be rhythmic and of equal duration. **B.** Pressure on the sternum must be released so that the sternum can return to its normal resting position between compressions. However, do not remove the heel of the hand from the sternum.

Performing One-rescuer Adult CPR

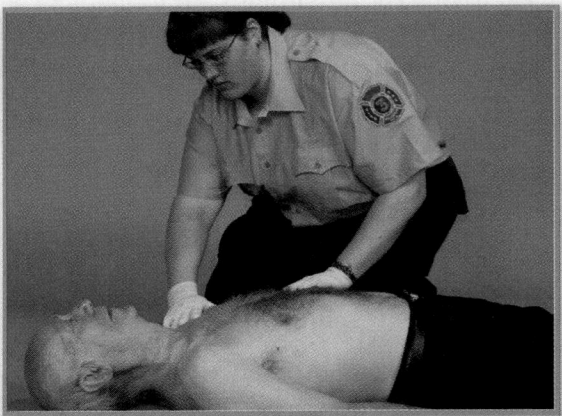

1 Establish unresponsiveness, and call for help.

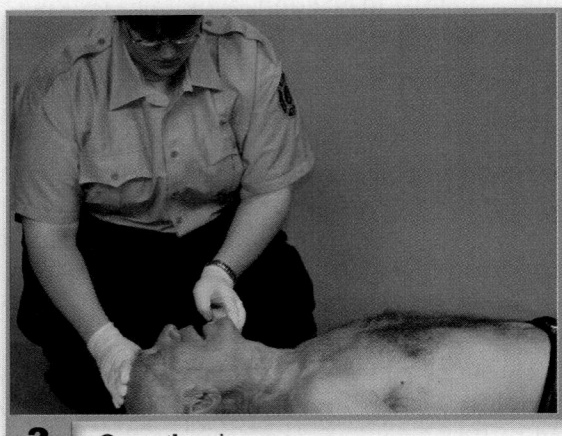

2 Open the airway.

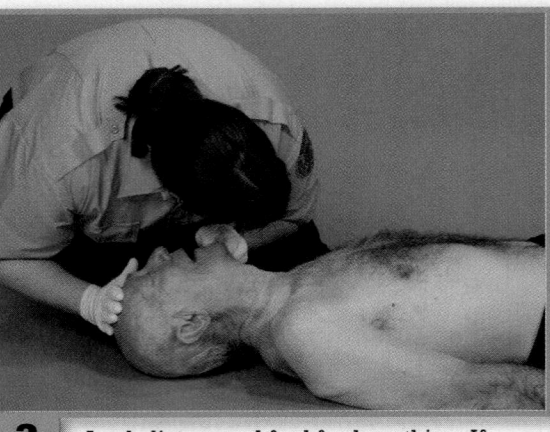

3 Look, listen, and feel for breathing. If breathing is adequate, place the patient in the recovery position and monitor.

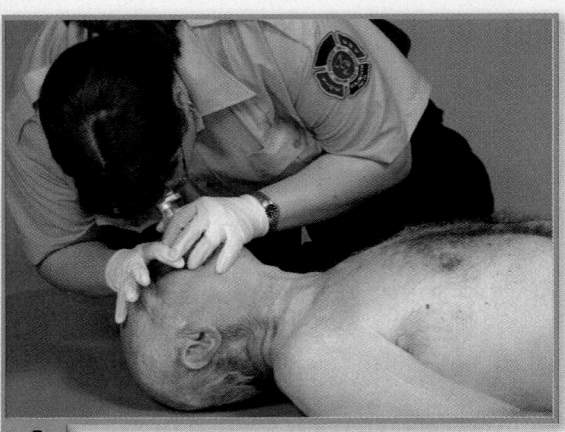

4 If not breathing, give two breaths of 2 seconds each.

4. **If the patient is not breathing, begin rescue breathing** by delivering two breaths, for 2 seconds each (**Step 4**).

5. **Determine pulselessness** by checking the carotid pulse (**Step 5**). If you have an AED, apply it now.

6. **If pulseless, begin compressions.** Place your hands in the proper position for delivering external chest compressions, as described previously.

7. **Give 15 compressions** at a rate of about 100/min for an adult. Each set of 15 compressions should take about 10 seconds. By using a rhythmic motion, apply pressure vertically from your shoulders down through both arms to depress the sternum $1\frac{1}{2}''$ to $2''$ in an adult, then rise up gently and fully. Count the compressions aloud (**Step 6**).

8. **Open the airway**, and then give two ventilations, each lasting 2 seconds.

9. Locate the proper position, and **begin another cycle** of chest compressions. Perform four cycles of compressions and ventilations.

10. **After four cycles** of compressions and ventilations, stop CPR and check for the return of a carotid pulse. If there is no change, resume CPR. If the patient has a pulse, check for breathing. If the patient is not breathing or breathing inadequately,

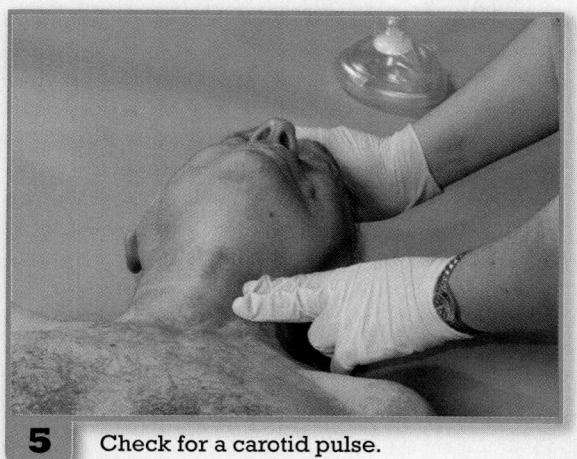

5 Check for a carotid pulse.

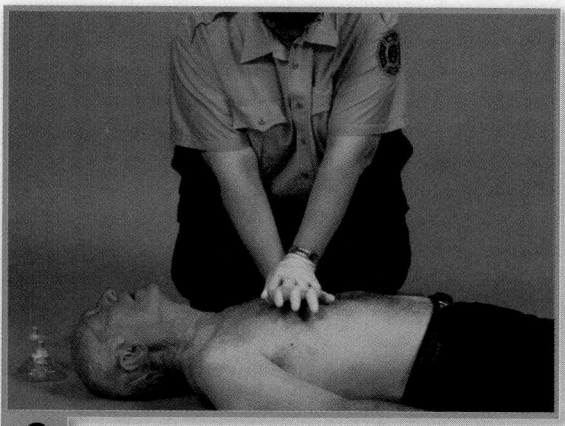

6 If no pulse is found, apply your AED. If there is no AED, place your hands in the proper position for chest compressions.

Give 15 compressions at a rate of about 100/min.

Open the airway, and give two ventilations of 2 seconds each.

Perform four cycles of compressions.

Stop CPR, and check for return of the carotid pulse.

Depending on patient condition, continue CPR, continue rescue breathing only, or place the patient in the recovery position and monitor breathing and pulse.

provide rescue breathing. If the patient has a pulse, is breathing adequately, but remains unresponsive, place the patient in the recovery position and closely monitor the patient's condition.

Two-rescuer Adult CPR

You and your team should be able to perform one-rescuer and two-rescuer CPR with ease. Two-rescuer CPR is always the first choice because it is less tiring for the rescuers. Once one-rescuer CPR is in progress, a second rescuer can be added very easily. He or she should enter the procedure after a cycle of 15 compressions and two ventilations. You should use airway adjuncts, such as mouth-to-mask ventilation, whenever possible. To perform two-rescuer adult CPR, follow the steps in Skill Drill A-4 ▶:

1. While moving to the patient's head, **establish unresponsiveness** as your partner moves to the patient's side to be ready to deliver chest compressions (**Step 1**).
2. Position the patient to **open the airway (Step 2)**.
3. **Check for breathing** by using the look, listen, and feel technique. If the patient is unconscious

Skill Drill A-4

Performing Two-rescuer Adult CPR

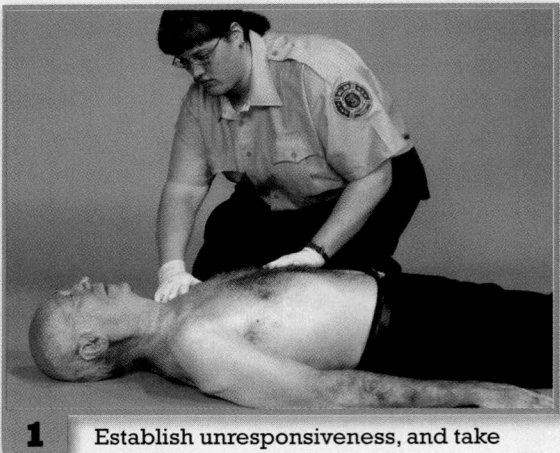

1 Establish unresponsiveness, and take positions.

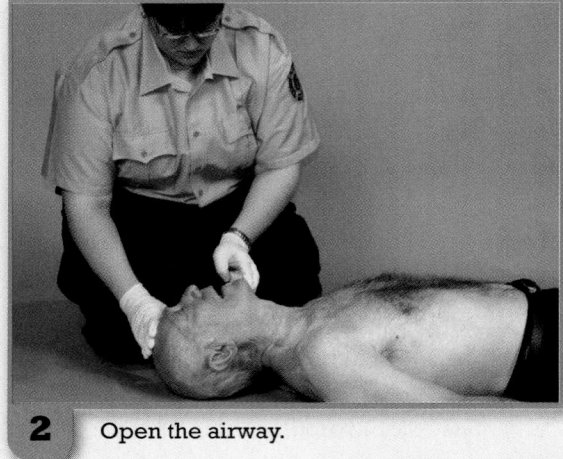

2 Open the airway.

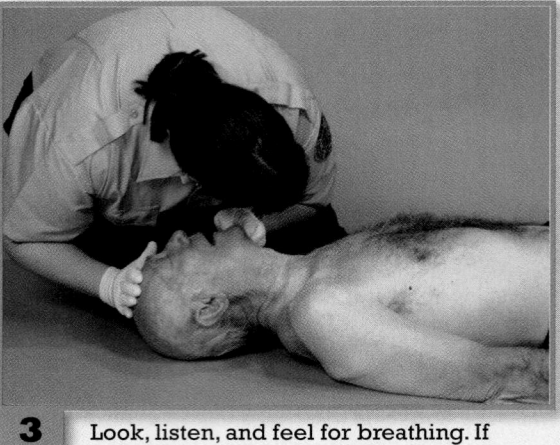

3 Look, listen, and feel for breathing. If breathing is adequate, place the patient in the recovery position and monitor.

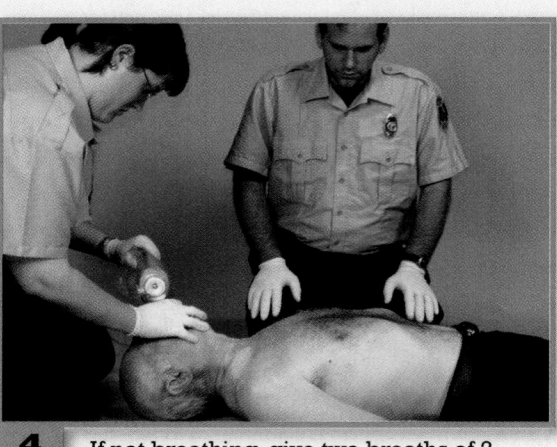

4 If not breathing, give two breaths of 2 seconds each.

but breathing adequately, place him or her in the recovery position and maintain the open airway (**Step 3**).

4. **If the patient is not breathing, begin rescue breathing** by delivering two breaths for 2 seconds each (**Step 4**).

5. **Determine pulselessness** by checking the carotid pulse (**Step 5**). If the patient has no pulse and an AED is available, apply it now.

6. **Begin chest compressions,** as described above, at a compression to ventilation ratio of 15:2 (**Step 6**). Once the airway is secured (intubated) the ratio of

chest compressions to breaths should change to 5:1, without a pause in compressions to perform a ventilation (<u>asynchronous CPR</u>). As in one-rescuer CPR, compressions should be given at a rate of about 100/min.

7. **After 1 minute of CPR,** the rescuer at the head checks the patient's pulse to determine whether the carotid pulse has returned. Because chest compressions produce a pulse, you must check for the pulse after a ventilation but before starting compressions again. Continue to check for a pulse every few minutes.

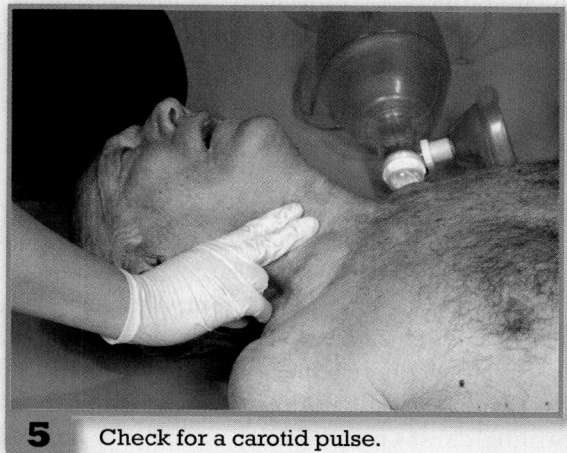

5 Check for a carotid pulse.

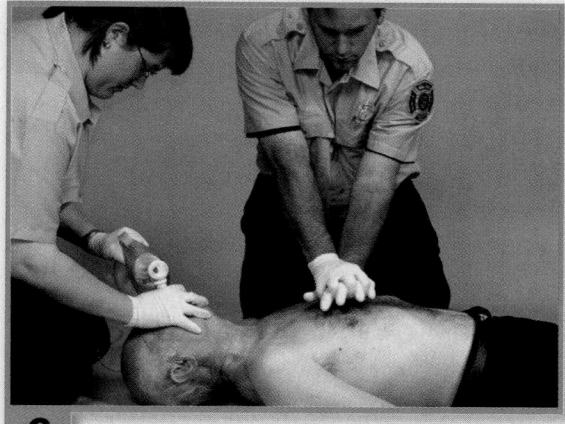

6 If there is no pulse but an AED is available, apply it now. If no AED is available, begin chest compressions at about 100/min (15 compressions to two ventilations).

After 1 minute, check for a carotid pulse. Check every few minutes thereafter.

Depending on patient condition, continue CPR, continue rescue breathing only, or place in the recovery position and monitor.

8. **If there is no change, resume CPR.** If the patient has a pulse, check for breathing. If the patient is not breathing, provide rescue breathing. If the patient has a pulse, is breathing, but remains unresponsive, place the patient in the recovery position and closely monitor the patient's condition.

Switching Positions

The best time to switch positions is during the pulse checks. However, you can switch positions any time one of you needs to change by following the same sequence of steps: The EMT-B who has been performing chest compressions checks for a carotid pulse following the final ventilation by the other EMT-B, who then moves to the chest and establishes his or her hand position on the sternum. Compressions start with the command, "No pulse. Continue CPR." You and your partner should be on opposite sides of the patient so that you can easily switch positions when necessary, as follows:

1. **First EMT-B:** Move into position to begin chest compressions after giving a breath.

2. **Second EMT-B:** Give the 15th compression, then move to the patient's head.
3. **Second EMT-B:** Check the carotid pulse for 5 to 10 seconds. If the patient has no pulse, say "No pulse. Continue CPR."

Infant and Child CPR

In most cases, cardiac arrest in infants and children follows respiratory arrest, which triggers hypoxia and ischemia of the heart. Children consume oxygen two to three times as rapidly as adults. Therefore, you must first focus on opening an airway and providing artificial ventilation. Often, this will be enough to allow the child to resume spontaneous breathing and, thus, prevent cardiac arrest.

Once the airway is open and you have delivered two breaths, you need to assess circulation. As with an adult, you should first check for a palpable pulse in a large central artery. Absence of a palpable pulse in a major central artery means that you must begin external chest compressions. You can usually palpate the carotid pulse in children older than 1 year, but it is difficult in infants, who have short and often fat necks. Therefore, in infants, palpate the brachial artery, which is located on the inner side of the arm, midway between the elbow and shoulder. Place your thumb on the outer surface of the arm between the elbow and shoulder. Then place the tips of your index and long fingers on the inside of the biceps, and press lightly toward the bone (Figure A-29 ▼).

External Chest Compression

Most BLS techniques are the same for infants, small children, larger children, and adults. As with an adult, an infant or child must be lying on a hard surface for

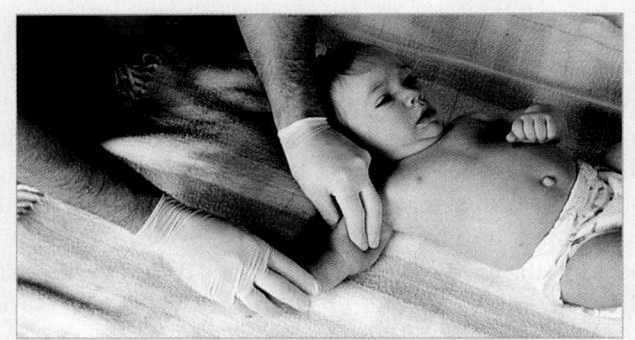

Figure A-29 To assess circulation in an infant, palpate the brachial artery on the inner side of the arm, midway between the elbow and shoulder.

the best results. If you are holding an infant, the hard surface can be your forearm, with your palm supporting the infant's head. In this way, the infant's shoulders are elevated, and the head is slightly tilted back in a position that will keep the airway open. However, you must ensure that the infant's head is not higher than the rest of the body. The technique for chest compressions in infants and children differs because of a number of anatomic differences, including the position of the heart, the size of the chest, and the fragile organs of a child. The liver is relatively large, immediately under the right side of the diaphragm, and very fragile, especially in infants. The spleen, on the left, is much smaller and much more fragile in children than in adults. These organs are easily injured if you are not careful in performing chest compressions, so be sure that your hand position is correct before you begin.

Proper Hand Position and Compression Technique

The chest of an infant or child is smaller and more pliable than that of an adult. Therefore, you should not use both hands to compress the chest.

Infant

In an infant, there are two methods for performing chest compressions: the two-finger technique and the two thumb-encircling hands technique.

The two-finger technique is the preferred method for performing chest compressions if you are alone. Place two fingers of one hand over the lower half of the sternum, approximately 1 fingerbreadth below an imaginary line located between the nipples (Figure A-30A ▶). Compress the sternum approximately one third to one half the depth of the infant's chest, which will correspond to a depth of about $\frac{1}{2}''$ to $1''$. With your middle and ring fingers, perform compressions at a rate of at least 100/min (Figure A-30B ▶). Finger position is important because you must avoid compressing the xiphoid process.

The two thumb-encircling hands technique is the preferred method for performing two-rescuer infant CPR, when physically feasible (Figure A-31 ▶). Place both thumbs side by side over the lower half of the infant's sternum, approximately 1 fingerbreadth below an imaginary line located between the nipples. Ensure that the thumbs do not compress on or near the xiphoid process. In very small infants, you may need to overlap your thumbs. Encircle the infant's chest and support the infant's back with the fingers of both hands. With your hands encircling the chest, use both thumbs

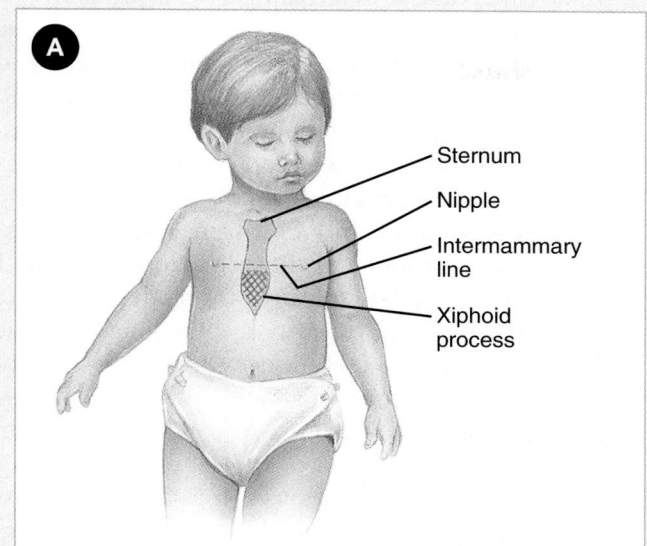

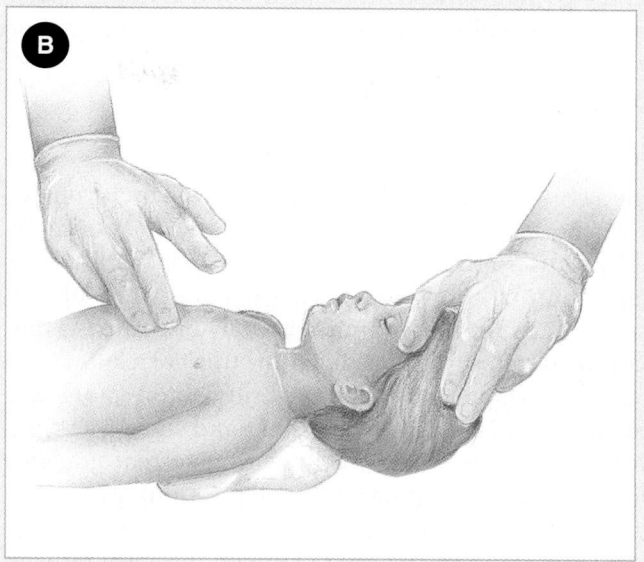

Figure A-30 A. The proper location for chest compressions in an infant is in the midline, one fingerbreadth below an imaginary line drawn between the nipples at the sternum. **B.** With your middle and ring fingers, compress the sternum approximately one third to one half the depth of the infant's chest, which corresponds to a depth of about ¹⁄₂″ to 1″. Perform compressions at a rate of 100/min.

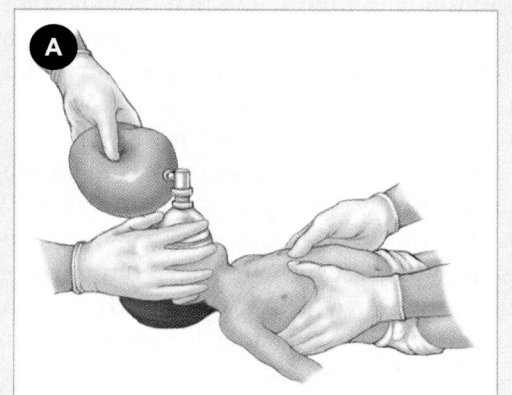

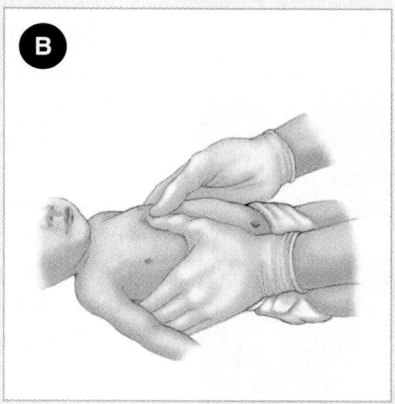

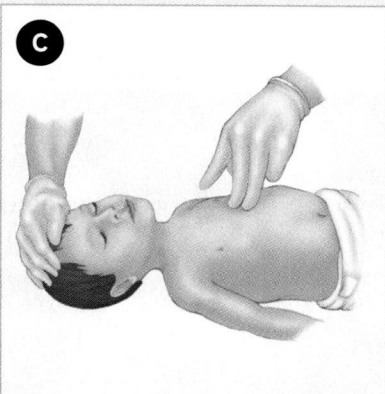

Figure A-31 A. Place both thumbs side by side over the lower half of the infant's sternum, approximately 1 fingerbreadth below an imaginary line located between the nipples. **B.** In very small infants, you may need to overlap your thumbs. **C.** In larger infants, you may use the two-finger technique, using the middle and ring fingers.

to depress the sternum approximately one third to one half the depth of the infant's chest, which will correspond to a depth of about ¹⁄₂″ to 1″. Perform compressions at a rate of at least 100/min. After 5 compressions, pause briefly for the second rescuer to open the airway and deliver a ventilation. Compressions and ventilations should be coordinated to avoid simultaneous delivery and ensure adequate ventilation and chest expansion, especially when the infant's airway has not been definitively protected (intubated).

After each compression, release pressure from the sternum without removing your fingers (or thumbs) from the patient's chest. Use smooth, rhythmic motions to deliver compressions, as with adult CPR.

When performing one-rescuer CPR on an infant on a hard surface (for example, the ground, a table), make sure that the hand closest to the head remains on the infant's forehead during chest compressions. Your other hand may remain on the chest as you give rescue breathing. If the chest does not rise, remove the hand that is on the patient's chest, and reposition the airway using the head tilt–chin lift maneuver. Then relocate the proper anatomic landmark and return the compression hand to the chest.

Child

For a child older than 1 year, the way in which you deliver chest compressions differs somewhat. You might need to use more force with a child than with an infant, compressing the sternum with the heel of one hand to a depth that is approximately one third to one half the depth of the child's chest; this corresponds to a depth of about 1″ to 1½″. Perform compressions at a rate of about 100/min. Your other hand should be used to maintain the child's head position so that you can provide rescue breathing without repositioning the head (Figure A-32 ▼). Compressions should be delivered in a smooth, rhythmic manner in which the chest returns to its resting position after each compression, but you should leave your hand on the patient's chest. Note that large children or those older than 8 years should be given chest compressions as you would in an adult.

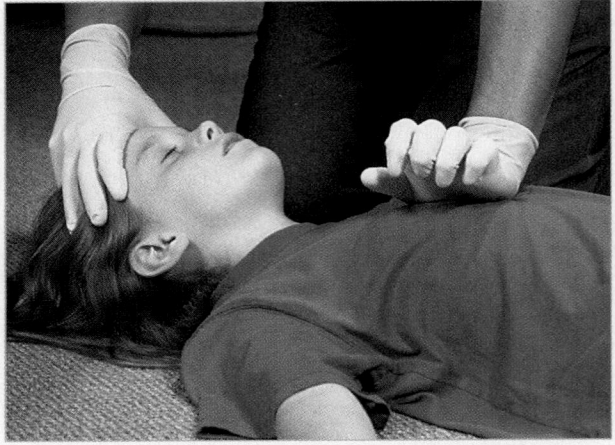

Figure A-32 When performing chest compressions on a child, use the heel of one hand to compress the sternum to a depth that is approximately one third to one half the depth of the child's chest; this corresponds to a depth of about 1″ to 1½″. Perform compressions at a rate of about 100/min. The other hand should remain on the child's head to maintain an open airway.

As with an adult, external chest compressions on a child or infant must be coordinated with ventilations. The rate of compression to ventilation for infants and children is 5:1 for one-rescuer and two-rescuer CPR. This means that you should open the airway and ventilate the patient once after each set of five compressions. One ventilation should take 1 to 1½ seconds.

Reassess infants and children after the first 20 cycles of compressions and ventilations (about 1 minute) and every few minutes thereafter. Evaluate for the return of spontaneous respirations with the look, listen, and feel technique. Assess for signs of spontaneous circulation by palpating the brachial pulse in infants and the carotid pulse in children older than 1 year of age.

Switching rescuer positions is the same for infants and children as for adults. The best time to switch positions is when you reassess the infant or child for breathing and circulation.

Interrupting CPR

CPR is an important holding action that provides minimal circulation and ventilation until the patient can receive definitive care in the form of defibrillation or further care at the hospital. No matter how well it is performed, however, CPR is rarely enough to save a patient's life. If ALS is not available at the scene, you must provide immediate transport, continuing one-rescuer CPR on the way. En route to the hospital, you should consider requesting a rendezvous with an ALS ambulance, if available; this will provide ALS care to the patient earlier, improving his or her chance for survival. Note however, that not all EMS systems have ALS support available to them, especially in rural settings.

Try not to interrupt CPR for more than a few seconds, except when it is absolutely necessary. For example, if you have to move a patient up or down stairs, you should continue CPR until you arrive at the head or foot of the stairs, interrupt CPR at an agreed-on signal, and move quickly to the next level where you can resume CPR. Do not move the patient until all transport arrangements are made so that your interruptions of CPR can be kept to a minimum.

Vital Vocabulary

abdominal-thrust maneuver The preferred method to dislodge and force food or other material from the throat of a choking victim; also called the Heimlich maneuver.

asynchronous CPR The performance of CPR in which chest compressions are not paused for the delivery of a ventilation; performed after the airway has been definitively secured (that is, after intubation).

advanced life support (ALS) Advanced lifesaving procedures, such as cardiac monitoring, administration of intravenous fluids and medications, and use of advanced airway adjuncts.

basic life support (BLS) Noninvasive emergency lifesaving care that is used to treat airway obstruction, respiratory arrest, and cardiac arrest. Although this term represents a wide variety of procedures performed by EMT-Bs, in this chapter, it is used synonymously with CPR.

cardiopulmonary resuscitation (CPR) The combination of rescue breathing and chest compressions used to establish adequate ventilation and circulation in a patient who is not breathing and has no pulse.

gastric distention A condition in which air fills the stomach as a result of high volume and pressure or airway obstruction during artificial ventilation.

head tilt–chin lift maneuver A technique to open the airway that combines tilting back the forehead and lifting the chin.

jaw-thrust maneuver A technique to open the airway by placing the fingers behind the angles of the patient's lower jaw and forcefully moving the jaw forward; can be performed with or without head tilt.

recovery position A position that helps to maintain a clear airway in a patient with a decreased level of consciousness who has had no traumatic injuries and is breathing on his or her own.

Index

Vapor hazard, of chemical weapons, 1106–7
Vasa deferentia, 135
Vasoconstriction, 152, 989
Vasodilation, syncope and, 993*t*
Vasovagal reactions, to IV therapy, 1174–75
Vector-borne disease transmission, 43*t*, 1112–13
Vehicle assessment, 633
Vehicle-borne disease transmission, 43*t*
Vehicle safety systems, 1058–59
Vehicular collisions. *see* Motor vehicle crashes
Vein irritation, IV therapy and, 1172–73
Veins, 123, 405*f*, 651
   bleeding from, 655, 656*f*
   in muscle, 188, 188*f*
Ventilation, 214. *see also* Airway management
   artificial, in children, 934
   assisted and artificial, 241–48
   barrier devices and, 242*f*, A-18–A-19, A-18*f*
   BVM devices for, 244–47, 244*f*, 246*f*
   cricoid pressure for, 246
   dental appliances and, 251
   equipment, 1029*t*
   facial bleeding and, 251, 251*f*
   flow-restricted, oxygen-powered devices, 247–48, 247*f*
   gastric distention and, 248
   mechanics, 779
   methods (in order of preference), 242
   mouth-to-mask, 242–44, 243*f*
   mouth-to-mouth, 242
   one-person BVM, 246*f*
   rates, 242
   respiration *versus*, 778
   two-person BVM, 246*f*
Ventolin, 385, 386*t*
Ventral (body) position or location, 93, 93*t*
Ventricles, 119, 402

Ventricular fibrillation (V-fib), 409, 410, 410*f*, 1188–89, 1190*f*
Ventricular flutter, 1189*f*
Ventricular tachycardia (V-tach), 410, 410*f*, 1188, 1189*f*
Verbal communications, 326–32
Vertebrae, 99, 99*f*, 876–77, 876*f*, 877*f*
   lumbar, 101, 101*f*
Vesicants (blister agents), terrorism and, 1107–8, 1113*t*
VHF (very high frequency), 317
Viagra, nitroglycerin precautions for, 355
Violent behaviors, 594, 594*f*
Violent religious groups, 1100
Violent situations, 63–64, 63*f*
Viral hemorrhagic fevers (VHFs), 1114, 1114*f*, 1115*t*, 1119*t*
Virulence, infectious disease, 52
Viruses, as biological weapons, 1102, 1113
Visceral peritoneum, 466
Visibility, driving during decreased, 1045
Visually-impaired patients, 331, 332*f*
Vital signs
   baseline, 146, 149–62, 149*f*
   cardiovascular emergencies, 415
   detailed physical exam, 291
   focused assessment, 293, 296
   geriatric, 1008
   pediatric, 948, 949*t*
   progression of shock, 684
   reassessment, 162, 307
   respiratory distress, 382–83
Vocal cords, 1139, 1139*f*, 1146, 1151
Volatility, of chemical weapons, 1106
Volmax, 386*t*
Voluntary activities, 875
Voluntary muscles, 110*f*, 156
Volutrol IV set, 1176, 1176*f*
Vomer, 97*f*
Vomiting (emesis), 470, 529, 886
   endotracheal intubation and, 1151
   gastric tubes for management of, 1134–35
Vomitus, bagging of, 519, 523

W

Warning lights, use of, 1046
Water
   chemical reactions with, 523
   panic in, 560, 560*f*
   on the roadway, 1045
Water moccasins, 570
Water rescue safety, 561, 561*f*
Weaponization, 1102
Weapons of mass casualty (WMCs), 1100, 1101–2
Weapons of mass destruction (WMDs), 1100, 1101–2
Weather conditions, 1044–46, 1045*f*
Weight, patient's, 174–79
West Nile virus, 54–55
Wheals, 502, 503*f*
Wheeled ambulance stretchers, 177–79, 179*f*, 197–99*f*, 197–200
Wheezing, 378, 381, 500, 945, 959
Whistle-tip catheters, 231, 232*f*
White blood cells, 124, 124*f*, 652–53, 652*f*
Whooping cough, 54
Work of breathing, 943, 944, 944*f*
Wound care supplies, 1029*t*, 1031
Wrists, 108, 108*f*
   injuries, 854–55, 855*f*
Written communications, 332–33, 333*f*, 334*t*, 335–36, 336–37*f*

X

X-rays (gamma radiation), 1120, 1121*f*
Xiphoid process, 100, 101*f*, 102, 968

Y

Yellow Fever virus, 1114

Z

Zone of injury, 824, 824*f*
Zygomas, 96
Zygomatic bone, 97*f*

Credits

Section 1
Opener © Chris Jensen

Chapter 1
Opener © Corbis; 1-1 Courtesy of District Chief Chris E. Mickal/New Orleans Fire Department, Photo Unit; 1-3 © Craig Jackson/In the Dark Photography; 1-4 © Jeff Hawk/911 Pictures

Chapter 2
2-1 © James Schaffer/PhotoEdit; 2-2 Courtesy of Richard E. Ahlborn (#NV8-RA10-13)/The American Folklife Center, Library of Congress; 2-3 © Robert Brenner/PhotoEdit; 2-4 © Craig Jackson/In the Dark Photography; 2-6 Reproduced with permission of the USDA and DHHS; 2-12 © Craig Jackson/In the Dark Photography; 2-22 Courtesy of U.S. Department of Transportation; 2-29 Courtesy of George Roarty/Virginia Department of Emergency Management

Chapter 3
Opener © Glen E. Ellman; 3-3 © Eddie M. Sperling; 3-4 © Kenneth Murray/Photo Researchers, Inc.; 3-6 © Keith D. Cullom; 3-8 © Mike Alexander/AP Photo

Chapter 4
Opener © Maria Taglienti-Molinari/Brand X Pictures/Alamy Images

Chapter 5
Opener © Michael Heller/911 Pictures; 5-5 Used with permission of the American Academy of Pediatrics, *Pediatric Education for Prehospital Professionals, First Edition,* © American Academy of Pediatrics, 2000; 5-9 © St. Bartholomew's Hospital, London/Photo Researchers, Inc.

Chapter 6
6-22 a and b: © Dr. P. Marazzi/Photo Researchers, Inc.

Chapter 7
Opener © Keith D. Cullom; 7-39 © Eddie M. Sperling

Chapter 8
Opener and 8-2 © Craig Jackson/In the Dark Photography; 8-5 © Peter Willott, The St. Augustine Record/AP Photo; 8-10 © Eddie M. Sperling

Chapter 9
9-9 © Lawrence Migdale/Photo Researchers, Inc.; 9-12 Courtesy of Guide Dog Foundation for the Blind, Inc. (www.guidedog.org)

Chapter 12
12-15 a: Courtesy of Medtronic; 12-16 Source: American Heart Association

Chapter 14
Opener © Craig Jackson/In the Dark Photography

Chapter 15
Opener © Craig Jackson/In the Dark Photography

Chapter 16
16-8 a: Courtesy of Dey, L.P.

Chapter 17
Opener © Michael Heller/911 Pictures; 17-2 © Neil Schneider/911 Pictures; 17-5 © Oscar Knott/FogStock/Alamy Images; 17-7 © Michael Heller/911 Pictures

Chapter 18
Opener © Dennis Wetherhold, Jr.; 18-22 a: © Creatas/Alamy Images, b: Courtesy of NOAA, c: © Photos.com

Chapter 19
19-1 © Craig Jackson/In the Dark Photography; 19-4 © Tom Carter/911 Pictures

Chapter 20
20-11 Courtesy of David J. Burchfield, MD

Section 5
Opener © Craig Jackson/In the Dark Photography

Chapter 21
Opener © Dennis Wetherhold, Jr.; 21-1 © Shout Pictures/Custom Medical Stock Photo; 21-2 © Terry Dickson, Florida Times-Union/AP Photo; 21-4 Courtesy of Captain David Jackson, Saginaw Township Fire Department; 21-5 © Dr. E. Walker/Photo Researchers, Inc.

Chapter 22
Opener © Dave Olsen, The Columbian/AP Photo; 22-6 © Craig Jackson/In the Dark Photography

Chapter 23
Opener © Steve L. Smith; 23-6 Courtesy of Dey, L.P.

Chapter 24
Opener © Craig Jackson/In the Dark Photography; 24-2 © Jones and Bartlett Publishers. Photographed by Kimberly Potvin; 24-3 © Medscan/Visuals Unlimited; 24-6 © English/Custom Medical Stock Photo; 24-10 Courtesy of Andrew N. Pollak, MD/University of Maryland School of Medicine; 24-23 a and b: © Charles Stewart & Associates; 24-25 a: Courtesy of Moose Jaw Police Service, b: © Charles Stewart & Associates

Chapter 25
Opener © Pete Fisher/911 Pictures

Chapter 26
Opener © Eddie M. Sperling; 26-4 Courtesy of Rhonda Beck

Chapter 27
Opener © Craig Jackson/In the Dark Photography

Chapter 28
Opener Courtesy of AAOS; 28-5 © Shout/Custom Medical Stock Photo; 28-6 © Eddie M. Sperling

Chapter 29
Opener © Keith Srakocic/AP Photo; 29-12 and 29-16 © Charles Stewart & Associates; 29-17 © Dr. P. Marazzi/Photo Researchers, Inc.; 29-18 Courtesy of Jeff Oliphant, University of Wisconsin-Eau Claire; 29-19 © Craig Jackson/In the Dark Photography; 29-22 © K. Shea/Custom Medical Stock Photo; 29-43 © Science Photo Library/Photo Researchers, Inc.

Chapter 30
Opener © Dan Myers; 30-12 © Tony Freeman/PhotoEdit; 30-27 © American Academy of Pediatrics

Chapter 31
31-3 © Eddie M. Sperling; 31-11 © Mediscan/Visuals Unlimited; 31-19 Courtesy of Cindy Bissell

Chapter 32
32-2 Courtesy of National EMSC Slide Set; 32-3 Courtesy of Dena Brownstein, MD; 32-4 and 32-5 Courtesy of National EMSC Slide Set; 32-11 © American Academy of Pediatrics; 32-33 © Ron Dieckmann

Chapter 33
33-1 Courtesy of National Cancer Institute; 33-3 © Dr. P. Marazzi/Photo Researchers, Inc.; 33-7 © Jeff Greenberg/PhotoEdit

Chapter 34
Opener © Glen E. Ellman

Section 7
Opener © Dan Myers

Chapter 35
Opener © Comstock Images/Alamy Images; 35-1 © National Library of Medicine; 35-3 c: © Dan Myers; 35-20 © Steve Spak/911 Pictures; 35-22 © Craig Jackson/In the Dark Photography; 35-26 © Stephane Brunet/911 Pictures; 35-28 Courtesy of Bryan Dahlberg/FEMA; 35-29 b: © Pete Fischer/911 Pictures

Chapter 36
Opener © Chris Abraham, Wilkes-Barre Time Leader/AP Photo; 36-2 © Tony Freeman/PhotoEdit; 36-3 © Spencer Grant/PhotoEdit; 36-4 and 36-6 © Craig Jackson/In the Dark Photography; 36-9 © Dan Myers; 36-11 © Kathy Easthagen, The Minnesota Daily/AP Photo

Chapter 37
Opener © John A. Bone, Cumberland Times-News/AP Photo; 37-1 © National Wildfire Coordinating Group; 37-3 Courtesy of Michael Rieger/FEMA; 37-5 © Linda Gheen; 37-9 a: Courtesy of Rob L. Jackson/U.S. Marines; 37-12 and 37-13 © U.S. Department of Transportation; 37-14 Courtesy of Sgt Lee, II. John A./U.S. Marines; 37-15 © Michael Heller/911 Pictures

Chapter 38
Opener © Sean Adair-Files/Reuters/Landov; 38-1 © NHK/AP Photo; 38-2 © Oded Bality/AP Photo; 38-3 © Rick Bowmer/AP Photo; 38-4 Courtesy of U.S. Department of Homeland Security; 38-5 © Gary Stelzer, Middletown Journal/AP Photo; 38-6 © Dennis MacDonald/PhotoEdit; 38-7 © Spencer Grant/PhotoEdit; 38-8 Courtesy of Dr. Saeed Keshavarz/RCCI, Research Center of Chemical Injuries/IRAN; 38-9 © Jones and Bartlett Publishers. Photographed by Kimberly Potvin; 38-10 © Chiaki Tsukumo/AP Photo; 38-11 Courtesy of CDC; 38-12 Courtesy of Professor Robert Swanepoel/National Institute for Communicable Disease, South Africa; 38-13 Courtesy of James H. Steele/CDC; 38-14 a and b courtesy of the CDC; 38-15 Courtesy of Brian Prechtel/USDA; 38-16 Courtesy of the Strategic National Stockpile/CDC

Section 8
Opener © Dan Myers

Chapter 39
Opener © Keith D. Cullom; 39-4 © Eddie M. Sperling

Chapter 40
Opener © Glen E. Ellman

Chapter 41
Opener © Michael Heller/911 Pictures; 41-2, 41-3, 41-4, 41-5, 41-6, 41-7, 41-8, 41-9, 41-10, and 41-11 from *Arrhythmia Recognition: The Art of Interpretation,* Courtesy of Tomas B. Garcia, MD